2024 IEEE 17th International Conference on Solid-State & Integrated Circuit Technology (ICSICT 2024)

Zhuhai, China
22-25 October 2024

Pages 1-443

IEEE Catalog Number: CFP24829-POD
ISBN: 979-8-3503-6184-1

**Copyright © 2024 by the Institute of Electrical and Electronics Engineers, Inc.
All Rights Reserved**

Copyright and Reprint Permissions: Abstracting is permitted with credit to the source. Libraries are permitted to photocopy beyond the limit of U.S. copyright law for private use of patrons those articles in this volume that carry a code at the bottom of the first page, provided the per-copy fee indicated in the code is paid through Copyright Clearance Center, 222 Rosewood Drive, Danvers, MA 01923.

For other copying, reprint or republication permission, write to IEEE Copyrights Manager, IEEE Service Center, 445 Hoes Lane, Piscataway, NJ 08854. All rights reserved.

*** *This is a print representation of what appears in the IEEE Digital Library. Some format issues inherent in the e-media version may also appear in this print version.*

IEEE Catalog Number:	CFP24829-POD
ISBN (Print-On-Demand):	979-8-3503-6184-1
ISBN (Online):	979-8-3503-6183-4
ISSN:	2835-7612

Additional Copies of This Publication Are Available From:

Curran Associates, Inc
57 Morehouse Lane
Red Hook, NY 12571 USA
Phone: (845) 758-0400
Fax: (845) 758-2633
E-mail: curran@proceedings.com
Web: www.proceedings.com

TABLE OF CONTENTS

A High-Performance Multicore Testing Platform for Multi-Scenario Applications 1
Zipeng Ling, Tianshu Zhuo, Zhuoyuan Yang, Jinhong Ye, Jun Han, Jingtao Zhang

S-SIFT: A Simple SIFT Algorithm with High Efficiency 4
Yixue Wang, Yujie Huang, Liyuan Peng, Mingyu Wang, Wenhong Li, Minge Jing, Xiaoyang
Zeng

Design of a High-Speed SAR Processor Based on the Chirp Scaling Algorithm 7
Xianghe Cui, Yukun Song, Yurun Zhang, Jingyi Hu, Zhenmin Li, Duoli Zhang

Accelerating Matrix-Vector Multiplications of Large Language Models Via Efficient Encoding 10
Yongjin Tao, Wendi Sun, Song Chen, Yi Kang

Flexible Yet Efficient Transformer Acceleration with Unified Sparse Attention Support on FPGA 13
Linfeng Zhong, Qingyu Guo, Runsheng Wang, Yuan Wang, Meng Li

Backward-Edge Control Flow Integrity Based on Return Address Encryption 16
Fengshuo Tian, Kaixuan Wang, Jun Han

Stability Enhancement Technique for Monostable PUF Based on Hysteresis Effect of Schmitt
Trigger 19
Ruize Xu, Gang Li, Pengjun Wang, Hui Li, Xudong Wu

A Reliable Current Starved Inverter PUF Based on SRAM Memory Structure 22
Mingze Ren, Pengjun Wang, Yuejun Zhang, Shutong Zhang, Zhikang Chen, Tengfei Yuan

An Efficient Convolutional Neural Network Hardware IP for Epilepsy Detection 25
Yi Gong, Yuejun Zhang, Jiangtao Tu, Rongxin Zou, Liang Wen

TLBshield: A Low-Cost Secure Reinforce on Translation Lookaside Buffer to Mitigate the
Speculative Attacks 28
Yuyang Liu, Runye Ding, Yujie Chen, Pujin Xie, Yao Liu, Zhiyi Yu

One-Step Circuit Analysis Based on LCA for Sparse Coding 31
Hanxi Xu, Zirui Chen, Chenqi, Xiangshui Miao, Yuhui He

A Hybrid-Logic Scheme for High-Performance and Low-Power Decoders in 7nm Process 35
Donghao Xia, Yuejun Zhang, Mengfan Xu, Liang Wen, Yiting Guo

Ternary Logic Units Design Based on the TDDFETs 38
Hua Qiang, Haoran Lu, Xiaotao Liu, Linlin Xing, Bin Lu

Enhancing SRAM Cell Stability Through Single-Carrier CMOS Latch Integration 41
Yuan-Yu Chuang, Pei-Zhang Xie, Jyi-Tsong Lin

A RRAM Based 9T1R NVSRAM for Low-Power Computing in Memory 44
Huimeng Guo, Yujia Li, Tingrui Ren, Chenge Dong, Liang Wang, Yuanfu Zhao, Yanlong
Zhang

A High-Resistance SOT Device Based Computing-In-Memory Macro with High Sensing Margin
and Multi-Bit MAC Operations for AI Edge Inference 47
Junzhan Liu, Jinyao Mi, Haiyan Qin, He Zhang, Wang Kang

RISC-V Domain-Specific Processor for Accelerating SPHINCS+ on Multi-Core Architecture 50
Shengnan Zhang, Yifan Zhao, Xinglong Yu, Jun Han

Design of an Out-Of-Order Superscalar Processor with Improved Register Alias Table Recovery
Method 53
Wu Yang, Jun Zhang

An SDPF RISC-V Processor with Two-Stage Pseudo-Pipelined Architecture for IoT Applications 56
Wenji Mo, Yuchen Wang, Haoning Sun, Jingjing Liu

A Unified Verification Scheme for the Acceleration of RISC-V Processor Design 59
Zixiang Chen, Jiyuan Bai, Yueru Yu, Gengsheng Chen, Xiaofang Zhou

Asynchronous Arbitration Circuit Optimization for Multicore Neuromorphic Processors 62
Jiajie Guo, Guangyao Lin, Bohan Wang, Zhiyi Yu, Shanlin Xiao

A Run-Time Temperature Monitoring with Adaptive Duty Cycle Control for FPGA Applications 65
Weizhi Li, Wangyong Chen, Haifeng Chen, Haoyu Zhang, Linlin Cai

An FPGA-Based Top-K Gradient Compression Accelerator for Distributed Deep Learning Training 68
Ziyao Wang, Jiayu Zhang, Kunyue Li, Jialei Sun, Feng Dong, Ke Chen, Yong Qiao, Jianfei Jiang

Dynamic-Matrix-Encryption Based Secure Strong PUF for Device Authentication Protocols 71
Liangxiao Zhao, Gang Li, Pengjun Wang, Xuejiao Ma, Ziyu Zhou

A Low Latency and High Throughout Hardware Design of Random Matrix Number Generator for
FrodoKEM 74
Shengfei Gu, Jiahao Lu, Tianze Huang, Jiaming Zhang, Kai Li, Cheng Wu, Mingbo Wang, Xianqi Mei, Ang Hu, Dongsheng Liu

A 4K60fps Ultra-Low-Latency Light Compression Encoder for Bandwidth-Constrained Scenarios 77
Yanzhong Li, Leilei Huang, Yibo Fan

Layer Pipelined Neural Network Accelerator Design on 2.5D FPGAs 80
Mengxuan Wang, Chang Wu

Fast and Accurate Partial-Zoom Depth Estimation for SPAD LiDAR Readout on FPGA 83
Lichen Feng, Hongwei Shan, Rundong Cai, Zhangming Zhu

A Self-Adaptive Gamma Voltage Regulation Circuit for AMOLED Displays 86
Zhifeng Mao, Fei Gou, Bin Sheng, Jing Xie, Wenwei Xu, Wei Liu, Jun Xu

A Reconfigurable Thermoelectric Energy Harvesting Interface Based on OPDC and DSCT 89
Peiyuan Fu, Jiabin Wang, Xufeng Liao, Lianxi Liu

A Fixed-Peak-Current Single-Inductor-Multiple-Output DC-DC Converter Achieving 92.6% Peak
Efficiency 92
Fei Liu, Langyuan Wang, Shuyu Zhang, Hanlu Zhang, Na Yan

Buck-Boost Converter with Stable Transition Mode for Low Average Inductor Current 95
Ningning Li, Yushen Zhang, Yibo Zhang, Yizhe Yang, Wenhao Yang, Yimeng Zhang, Yuming Zhang

A SET Sensitive Model of LC and Ring Voltage-Controlled Oscillator in FinFET Technology 98
Liu Heyuan, Yuan Hengzhou, Lu Jianzhuang, Chen Xiaowen, Sang Hao, Liu Jingtian, Guo Yang

A Low Spur Wideband PLL in 65nm CMOS 101
Zijun Wang, Biao Li, Teng Wang, Hong Li, Ruiting Niu, Jinpeng Lin

A Low Power PLL Circuit with Signal 50% Duty Cycle Corrected in 180nm CMOS 104
Bangtian Li, Xueke Li, Liying Chen, Chuantong Cheng, Jian Mei

MTJ Based Compensation for Charge Pump Temperature Drift 107
Yongliang Zhou, Jingxue Zhong, Chengxing Dai, Yingxue Sun, Xin Li, Chunyu Peng

A 112-Gb/s Coherent Receiver with a Novel Modulation Format 110
Tianyuan Zhong, Boyang Zhang, Weixin Gai

Beyond Bandwidth Trade-Off: Simultaneous Wireless Power and Data Transfer System Design for
Biomedical Implants 113
Quanrong Zhuang, Junyi Sun, Xusheng Zhang, Bo Li, Yi Shi, Hao Qiu

A High Precision Operational Amplifier with Improved Bias Current Cancellation Circuit 117
Zhili Zhang, Siyuan Yao, Puyang Liu, Cheng Li, Lu You, Hailong Wei

A 0.11-PJ/Bit True Random Number Generator Based on a Clocked Current-Starved Inverter 122
Kai Cheng, Chaowei Yang, Rui P. Martins, Pui-In Mak, Yong Chen

A Super-Mixed Current Decay Mode for Reducing the Angular Position Error in Stepper Motor 126
Jian Fang, Xurui Chen, Huajie Liu, Yuhan Jin

A 109 dB 44-PArms Current Readout Circuit with Automatic Current Control for Multimodality
Electrochemical Sensing 129
Lina Wang, Jianzheng Li, Weiming Hu, Yajie Qin

A Low Temperature Coefficient Bandgap Reference for Temperature Sensor System 132
Longjiang Jia, Yuanhong Ding, Jian Mei, Lei Deng, Rui Yin

Toward Unification of Digital Error Correction Algorithms for ADCs with Redundancy 135
Haruo Kobayashi, Tomohiko Ogawa, Yutaro Kobayashi, Kentaroh Katoh, Jiangling Wei

A 1.2-V 2-GS/s Trimming-Free Input Buffer with Robust Output Common-Mode Voltage 139
Wei Zhang, Xizhu Peng, He Tang

A 12-Bit 1-MS/s SAR ADC Using Vcm-Based Split MSB Switching and Segmented CDAC 143
Zheng-Han Chen, Ya-Cong Zhang, Wen-Gao Lu, Zhong-Jian Chen

A Simplified and Accelerated Opportunistic Bit-Weight Calibration for High-Resolution ADCs 146
Bingbing Ma, Wei Li, Hongtao Xu

Background Calibration for Bit Weights in Pipelined SAR ADCs Using Split ADC Architecture 149
Zecheng Zhou, Longsheng Wang, Dongxian Ye, Yexin Zhu, Dengquan Li, Zhangming Zhu

When Time Interleaving Encounters Oversampling in ADC 152
Mingqiang Guo, Dongyang Jiang, Shulin Zhao, Sai-Weng Sin, Rui P. Martins

A 0.000355mm^2 4.6μm-Pitch 5.75fJ/Conv 6-Bit SAR ADC for High Throughput Parallel Readout
of Analog SRAM Computing-In-Memory 156
Lin Wu, Lichen Feng, Hongwei Shan, Zhangming Zhu

A 250MS/s, 12 Bit Pipeline-SAR ADC Using Coarse-Fine Ring Amplifier 159
Linghao Liu, Junyan Ren, Fan Ye

A 0.71pJ/b 16Gb/s Equalizer with Inverter_Based CTLE and 4-Tap Speculative DFE 162
Huihong Zhang, Chuangao Yan, Peng Luo, Maliang Liu

A Digital Foreground Calibration Method for Pipeline SAR ADCs Using Extended Kalman Filter 165
Dayan Zhou, Yuguo Xiang, Junyan Ren, Fan Ye

A Second-Order Charge Pump Noise-Shaping SAR ADC 168
Haoning Sun, Yuchen Wang, Wenji Mo, Kangkang Sun, Jingjing Liu

Computing in Memory for Accelerating Light-Weighted On-Chip Learning in IoT Devices 171
Zhiwang Guo, Xiaoyong Xue, Qiang Zhou, Xiaoyang Zeng

A Novel Beamforing Receiver Architecture Combining MASH SDM and BSP 175
Tao Zhong, Yuekang Guo, Jing Jin, Jianjun Zhou

In Situ Localization Techniques of Defects in Advanced Semiconductor Devices from Macro-Scale
to Atomistic-Scale 178
Jialu Huang, Jingming Zhou, Zuoyuan Dong, Runsheng Wang, Junhao Chu, Xing Wu

Wafer-Level Characterization of Ring-Oscillators Frequency Degradation in FinFET Technology 183
Hao Chang, Dan Gao, Yongsheng Sun, Junlin Huang

Exhaustive Application-Dependent Testing for FPGA Interconnect Resources 186
Wenwei Chen, Xinyu He, Tongshu Ding, Jian Wang, Jinmei Lai

A Comprehensive and Efficient Instruction-Level Testing Method for Processor 189
Zixin Yang, Zhichao Wei, Huanlin Luo, Jian Wang, Jinmei Lai

Thermal Effect and Calibration for High Precision On-Wafer Analog IC Probe Testing 192
*Daisuke IImori, Takayuki Nakatani, Shogo Katayama, Gaku Ogihara, Shuhei Yamamoto,
Misaki Takagi, Yujie Zhao, Jianglin Wei, Anna Kuwana, Keno Sato, Takashi Ishida, Toshiyuki
Okamoto, Tamotsu Ichikawa, Kentaroh Katoh, Kazumi Hatayama, Haruo Kobayashi*

Deep Learning Design-Flow with Static and Dynamic Optimizations 195
Zhiqiang Que, Jose G. F. Coutinho, Wayne Luk

A QEMU-Based Virtual Platform of MPSoC 199
Liangquan Qiao, Lei Li, Xingyu Gao, Jinxiang Wang, Fangfa Fu, Keli Long, Jinghan Zhou

A Parallel Harmonic Balance Method Based on GPU for Efficient Periodic Steady-State Analysis 202
Zhengzhuo Wang, Yanliang Sha, Lingyun Ouyang, Quan Chen, Jianguo Hu

Efficient Dynamic Memory Management for High Bandwidth Memory on FPGA 205
Yuwei Qu, Yiqing Mao, Wenbo Yin, Lingli Wang

An Improved Clock-Aware Global Placement Algorithm 208
Ziang Ge, Pingqiang Zhou

Analyzing Timing in Shorter Time: A Journey Through Heterogeneous Parallelism for Static
Timing Analysis 212
Zizheng Guo, Yibo Lin, Runsheng Wang, Ru Huang

TBPart-B: An Effective Hypergraph Partitioning Algorithm Considering Topological Order
Balance for Processor-Based Hardware Emulation 216
Jing Tang, Shunyang Bi, Hailong You

FCE: A Fast CGRA Architecture Exploration Framework 219
    Sichao Chen, Yiqing Mao, Yuan Dai, Xuchen Gao, Wai-Shing Luk, Wenbo Yin, Lingli Wang

Research on Parametric Subthreshold Cell Delay Modeling Based on ANN 222
    Xuelian Zhang, Yuping Wu, Zhiqiang Li, Donglin Liu, Shushan Qiao

A High-Performance Routing Architecture with 16 LUTs Per CLB for Nanoscale FPGAs 225
    Sijing Yang, Jide Zhang, Hao Zhou, Lingli Wang

High-Efficiency Power Amplifier Design for Bluetooth Low Energy Applications 228
    Bharatha Kumar Thangarasu, Li Shuai, Yu Hongshi, Ge Wansi, Liu Yuqing, Nagarajan Mahalingam, Meng Fanyi, Kaixue Ma, Juin J. Liou, Bo Wang, Younan Hua, Xiaomin Li, Lu Zhenghao, Kiat Seng Yeo

A 0.15-6.5GHz Stacked CMOS Power Amplifier with Low-Frequency Bandwidth Extension 232
    Shijiao Dong, Wei Li, Xingyu Ma, Fan Chen, Hongtao Xu

A 2-To-2.7GHz Class-G Switched-Capacitor PA with Cascode Switch-Reused Structure Achieving 25.92dBm Peak Power and 42% Efficiency 235
    Jie Deng, Gengzhen Qi

A X-Band High Linearity Tunable Bandpass Filter in 130nm CMOS 238
    Tianrui Wang, Ziyu Wang, Huiquan Xie, Yifei Chen, Haokun Lan, Maliang Liu, Yintang Yang

Analysis of Polar and Quadrature Digital Transmitters for Wi-Fi7 Applications 241
    Lixuan Cao, Yun Yin

Atomically Thin Graphene Nanopore Based MEMS Iontronic Devices for Sensing, Separation, and Energy Applications 244
    Luda Wang, Ruiyang Song, Ningran Wu

Systems-On-Chips for Invasive Brain-Computer Interfaces: Challenges and Opportunities 248
    Jie Yang, Mohamad Sawan

Multi-Physics Simulation and Application of Ion-Gel Based Triboelectric Nanogenerators 252
    Chen Liu, Ruibo Wang, Ruiyi Gao, Yuming Zhang

A Flexible Graphene Acoustic Sensor for Sound Signal Acquisition and Spiking Neural Network Recognition 255
    Lu-Yu Zhao, Hao-Yuan Shen, Yi-Wen Wu, Lu-Lu Zhang, Yu-Tao Li, Tian-Ling Ren

0.15μm BCD Platform with High Sensitivity Hall Device and Low Noise CMOS for Sensor IC Applications 258
    Guiqiang Zheng, Qingyin Zhong, Jie Ma, Nannan Cheng, Yichen Li, Yongjia Li, Xiaofeng Sun, Dejin Wang, Sen Zhang, Long Zhang, Siyang Liu, Weifeng Sun

A High Dynamic Range Pixel with Inverse Proportional Response 261
    Yuchen Wang, Wenji Mo, Haoning Sun, Jingjing Liu

Enhancement of Image Sensor Pixel Performance Through Ring-Shaped Vertical Transfer Gate Structure 264
    Shuang Yan, Shuai Yuan, Haoping Zheng, Yudi Zhao, Gang Du, Junchen Dong, Kai Zhao

A 10-GHz Low Power Class-C VCO with Long-Term Reliability and Tunable Performance in 28 nm FD-SOI for Satellite Communications 267
    Yann Deval, Henrique Iha Taguti, Ayoub Ait Ihda, Hervé Lapuyade, Stephane Rochette, François Rivet

A 191-GHz Harmonic Oscillator with Self-Feeding Line and Return-Path Gap Coupler Structure in 65nm CMOS.. 270
Xiaohan Shen, Chen Jiang

A Blocker-Tolerant High-Linear Receiver Employing Baseband Noise-Cancelling and Bottom-Plate Switched-Capacitor Techniques 273
Chenxiang Cai, Gengzhen Qi

A High Sensitivity Series-Parallel Rectifier with Pre-Bias for RF Energy Harvesting Systems 276
Haiqin Wu, Dejian Li, Xin Jin, Xufeng Liao, Lianxi Liu

A 24.3-43.7 GHz Variable-Gain Low-Noise Amplifier with Phase Self-Compensation 279
Yue Wu, Wei Li, Mohan Zhou, Hongtao Xu

A Broadband Active Variable Attenuator with Phase Compensation Technique 282
Zhiying Xia, Zhiqun Li, Bofan Chen, Xiaowei Wang

Waterproof and Wearable Power Sources 285
Sixing Xiong, Kenjiro Fukuda, Takao Someya

A CMOS Pixel Sensor for Precise Track and Charge Measurement of Cosmic Ray Nuclei 289
Ruikai Zhang, Wen He, Shanqiang Yang, Min Luo, Chenxu Wang, Cunfeng Feng, Meng Wang, Liang Zhang, Anqing Wang, Jianing Dong, Dong Liu, Yan Niu, Yang Zhou, Yuehong Gong, Xiaoli Wang, Shucheng Shi

$Sc_{0.096}Al_{0.904}N$-Based Bimorph Piezoelectric Micro Machined Ultrasonic Transducers 292
Ziye Zhai, Wenjuan Liu, Chengliang Sun

TeO_2 as Amorphous P-Type Transistor for Back-End-Of-Line Applications 296
John Robertson, Xuewei Zhang, Qingzhong Gui, Yuzheng Guo

Miniaturization of High-Speed GaN Based Laser Diodes 299
Junfei Wang, Chaowen Guan, Leihao Sun, Zhichong Wang, Chao Shen

Impact of Interfacial Layer on the Optoelectronic Performance of $MoTe_2$/Ge Heterojunction 301
Wenyu Lei, Xiaokun Wen, Boyuan Di, Xinyue Xu, Haixin Chang, Wenfeng Zhang

MoS_2-WS_2 Heterostructure-Enabled Optoelectronic Synaptic Diode 304
Mingjie Li, Yingtao Ding, Jianzhi Hu, Hankun Zhao, Yilin Sun

Pseudo-Parallel Symmetrical and Crossed Perovskite Solar Cells for Bifacial Applications 307
Guang-Wei Huang, Hsing-Mao Cheng, Jyi-Tsong Lin

Cryogenic and RF Modeling of On-Chip Passive Devices for Quantum Computer 310
Akira Tsuchiya

Comparison of Nanosheet and Fin Integration in Stacked Induced Tunnel Field-Effect Transistors 313
Ruei-Cheng Tu, Chia-Yo Kuo, Jyi-Tsong Lin

Nonlinear Contact Behavior in MoS_2 Field Effect Transistors at Cryogenic Temperature 316
Shihab Ahammed, Mansun Chan

Experimental Verification of 1D Transport Model by Quantized Current Spectrum of Si JNT Device 319
Zi-Meng Shang, Bo-Wei Wang, Wei-Hua Han

Impact of Gate Overlap Length Modulation on Electrical Characteristics and Subthreshold Swing in Nanosheet TFETs with Varying Tunneling Mechanisms 322
Zheng-Hong Zhong, Wei-Heng Tai, Jyi-Tsong Lin

Recent Progress in the Development of Complementary Field-Effect Transistors 325
Mansun Chan, Shengdong Zhang

Metal-Oxide Thin-Film Transistors for Artificial Neural Networks 329
Yushen Hu, Tengteng Lei, Man Wong

Cryogenic Threshold Voltage and on-Current Variability Analysis of GAA Nanosheet FETs at 4K 332
Zihao Liu, Tomoko Mizutani, Kiyoshi Takeuchi, Takuya Saraya, Hiroshi Oka, Takahiro Mori, Masaharu Kobayashi, Toshiro Hiramoto

Reverse-Biased PN Junction Isolation for Leakage Suppression and Strain Enhancement in Gate-All-Around Nanosheet FETs 335
Boqian Shen, Chunlei Wu, Yumin Xu, Fei Zhao, Hanzhi Gu, Jian Ma, Yueyuan Yu, Yiming Xia, Qingqing Sun, David Wei Zhang

Studies on Selective Deposition of SiO$_2$ by Rapid Atomic Layer Deposition 338
Sicong Shao, Jin Yan, Wang Li, Kun Cao, Rong Chen

Resistance Dependence of Cobalt on Line Width in Advanced Interconnects: First-Principles Modelling 342
Kang Wang, Menglin Huang, Shiyou Chen

Reduction of Specific Contact Resistivity by Employing Pre-Amorphization Implantation and In-Situ Steam Generation Oxidation 345
Chang Liu, Xu Chen, Jinbiao Liu, Yanping He, Wenjuan Xiong, Weibing Liu, Mingshan Liu, Zhe Liu, Yaoqi Dong, Jeffrey Xu, Jing Xu, Jun Luo

High-Performance Ultrathin ITO Thin-Film Transistor with Ultralow Subthreshold Swing 348
Yanheng Liu, Tiaoyang Li

Defect-Centric Insights into Flicker Noise in Ultra-Scaled FETs: From Physics to Compact Model for Circuit Level Simulation (Invited) 351
Chenyang Zhang, Yu Xiao, Pengpeng Ren, Shiyu Xia, Sheng Yang, Lining Zhang, Runsheng Wang, Zhigang Ji

A Modified Virtual-Source Model for Ballistic Transport Characterization of FinFETs at Cryogenic Temperature 355
Hongbo Wang, Zirui Wang, Zixuan Sun, Runsheng Wang, Ru Huang

Investigation on Asymmetric HfO$_2$-ZrO$_2$-HfO$_2$ Superlattice Gate Stacks with Ultra-Low EOT for Advanced Transistors 358
Haiyuan Lyu, Kun Zhong, Zhaohao Zhang, Huaxiang Yin

Evaluation of Contact Resistance with the 'L' Kelvin Test Structure and the Modified Kelvin Test Structure 361
Gui Chen, Yun-Hao Shao, Xin-Ping Qu

Flip 3D (F3D): A Novel 3D Integration Technology with Dual-Side Integration Capabilities 364
Heng Wu, Haoran Lu, Runsheng Wang, Ming Li, Yibo Lin, Weihai Bu, Jin Kang, Ru Huang

Modeling and Simulation of a Conical 3D Monopole Antenna Embedded in Substrate for WNoC 368
Junhao Wang, Ziyu Liu, Zhiyuan Zhu, Lin Chen, Qingqing Sun, Wei David Zhang

Gate Driver ICs for Wide Bandgap Power Transistors 372

Wai Tung Ng, Rophina Li, Wentao Cui, Jingyuan Liang

Suppression of Back-Gating Effect by Integrated Substrate Termination Network for 200V
Monolithic GaN Half-Bridge Power IC 376

*Mengyao Zhao, Yifei Zheng, Yanfeng Ma, Yuan Sun, Denggui Wang, Chuanqi Pan, Jianjun
Zhou, Sheng Li, Long Zhang, Siyang Liu, Weifeng Sun*

High Short-Circuit Capability and Low-Loss SOI-LIGBT with Double-Integrated NMOS 379

*Jialei Tan, Jie Wei, Jinlong Lu, Xindi Liu, Gaoqiang Deng, Wei Song, Pei Guo, Bo Zhang,
Xiaorong Luo*

Body Diode Degradation Mechanism of 1200V SiC Power MOSFETs Under Gamma Rays Total
Ionizing Dose Irradiation 382

Yu Tian, Zhaoxu Song, Hao Fu, Jiaxing Wei, Siyang Liu, Weifeng Sun

Novel Heterojunction Field Plate β-Ga_2O_3 MOSFET with High Breakdown Voltage 385

*Xiangnan Li, Jie Wei, Kai Zhao, Linyao Hao, Xiaosong Peng, Yuxi Wei, Renkuan Liu, Wei
Song, Pei Guo, Xiaorong Luo*

Investigation of SiON Passivation for High Performance AlGaN/GaN HEMTs 388

Difei Fan, Chenkai Deng, Jiming Zhang, Peiran Wang, Nick Tao, Qing Wang, Hongyu Yu

180nm BCD Technology Platform with 8V to 65V Isolated LDMOS 392

*Qi Ding, Renxiong Li, Ning Ning, Jun Huang, Yutuo Guo, Yu Wang, Kunqin He, Yaxin Liu,
Huaishan Wang, Juan Tang, Qiuyue Huo, Minghong Yuan, Pan Peng, Ming Qiao, Lulu Peng,
Bo Zhang*

A Novel Insulated Gate-Triggered Thyristor with Integrated Super-Clamp Gate Transient Voltage
Suppressor for Ultrahigh Di/Dt Pulse Switching 395

*Shiyu Deng, Yuxiao Yang, Xinqi Sun, Pengwei Zhou, Ruize Sun, Chao Liu, Pengcheng Xing,
Xiaoming Wang, Wanjun Chen, Bo Zhang*

Device Instability in the Third Quadrant of Schottky-Type p-GaN Gate HEMTs: The Hole
Defficiency & Trapping Effect 398

*Kuangli Chen, Shuting Huang, Jinggui Zhou, Ning Yang, Jianggen Zhu, Enchuan Duan, Bo
Zhang, Qi Zhou*

Static Characteristic Recovery of SiC MOSFETs Induced by Dynamic Gate Stress After Total
Ionizing Dose Irradiation 401

Jiahao Hu, Xiaochuan Deng, Xing Zeng, Tao Xu, Haibo Wu, Xuan Li, Bo Zhang

The Status of WBG Devices Towards Net-Zero Solutions 404

Mikael Ostling

Impact of the Resistive Silicon Base Wafer on Substrate Coupling in Power Integrated Circuits in
GaN-On-Si Technology 408

Zijin Jiang, Rui Ray Yao, Miao Cui, Zhao Wang, Sang Lam, Stephen Taylor

A Novel Snapback-Free Double-RESURF Reverse-Conducting LIGBT with Dual Conduction
Paths 411

*Yun Xia, Yuxi Wan, Wei Zeng, Yu Shi, Xiaoping Wang, Wei Liu, Haizhao Zhi, Ziwei Zhou, Xixi
Luo, Ruize Sun, Xiaoming Wang, Yan Wang, Wanjun Chen*

Comparsion of SiC Planar and Trench Junction Barrier Schottky Diode with Surge Current Capability 414
  Zi-Ming Zhao, Yan-Cong Liu, Hao Yuan, Feng-Yu Du, Yu Zhou, Ke-Yu Liu, Xiao-Yan Tang, Chao Han, Qing-Wen Song, Yu-Ming Zhang

Comparative Study on Reliability of Conventional SiC MOSFET and JBS Integrated SiC MOSFET 417
  Moufu Kong, Shurui Li, Hongfei Deng, Bo Yi, Hongqiang Yang, Sen Gong

Study on Single Event Effect of SiC MOSFET by Proton Irradiation 420
  Wende Huang, Chengwen Fu, Yao Ma, Mingmin Huang, Xiaoping Dong, Qiang Yu

Investigating Single-Event Burnout in 4H-SiC Inverters: Experiments and Simulations 423
  Yong Gu, Yurui Yang, Yawen Xv, Hongyang Wen, Xiangyu Hou, Runhua Huang, Ao Liu, Song Bai, Jie Ma, Long Zhang, Siyang Liu, Weifeng Sun

Challenges of Design for Reliability in Advanced CMOS Technology: From Single-Mode to Mixed-Mode Mechanisms 426
  Zixuan Sun, Lining Zhang, Ru Huang, Runsheng Wang

Frequency-Dependent Time-Dependent Dielectric Breakdown (TDDB) Behavior and Physical Study in Gate Oxides 427
  Wei Liu, Chu Yan, Xinwei Yu, Yiming Qu, Wenchao Yan, Yi Zhao

Lightning Protection Stacked TVS Structure Based on a Novel Total-Ionizing-Dose Radiation-Hardened Technology 430
  Zhao Qi, Hongquan Chen, Yirui Jia, Nailong He, Zhili Zhang, Sen Zhang, Ming Qiao, Bo Zhang

Characterization and Modeling of Non-Conducting RF Hot Carrier Stress in FinFETs 433
  G. Niu, X. Ding, H. Zhang, W. Wang, K. Imura, F. Dai

Predictive Modelling of Hot Carrier Degradation 437
  James Brown, Kean Hong Tok, Rui Gao, Zhigang Ji, Weidong Zhang, Jian Fu Zhang

New Insights into the Saturation Behavior of the Hot Carrier Degradation in STI-Based N-Type LDMOS 441
  Zhuoqing Yu, Dan Gao, Yongsheng Sun, Junlin Huang

The TID Response and HCI Degradation for Multi-Vt nFinFETs 444
  Ruxue Yao, Hongliang Lu, Yuming Zhang, Yutao Zhang

Modeling and Parameter Extraction of Semiconductor Devices for Simulation and Design Optimization of ESD Protection Circuits on BCD Technologies for Automobile and Industry Applications 447
  Yuhua Cheng, Wei-Wei Yu, Eugene Worley

A Novel Double-Zener Process and Multiplex Design for High-Power Surge and High-Speed ESD Devices Development 451
  Zhao Qi, Yirui Jia, Hongquan Chen, Ming Qiao, Zhaoji Li, Bo Zhang

The Non-Monotonic Instability of V_{TH} and $R_{ds,on}$ in P-GaN Gate HEMTs Under Repetitive Short-Circuit Stress: The Role of Electric-Field & Self-Heating Effect 454
  Long Wang, Ning Yang, Shuting Huang, Jianggen Zhu, Kuangli Chen, Chao Feng, Haolin Hu, Wei Zeng, David Zhou, Yuxi Wan, Bo Zhang, Qi Zhou

Optimizing Flash Memory Towards Storage-Class Memory (SCM) Applications 457
  Xinyi Guo, Yang Feng, Jing Liu, Junyu Zhang, Xuepeng Zhan, Jixuan Wu, Jiezhi Chen

Investigation of Reliability Characteristics of $Hf_xZr_{1-x}O_2$-Based FeFET and AFeFET Non-Volatile Memory 460
Min Liao, Xianzhou Shao, Junshuai Chai, Xiaoqing Sun, Xiaoyu Ke, Hao Xu, Jinjuan Xiang, Xiaolei Wang, Wenwu Wang

Deep Understanding of Charge Trapping Phenomenon in n-FeFET and Endurance Improvement by Interlayer Engineering 463
Saifei Dai, Hao Xu, Fengbin Tian, Xianzhou Shao, Xiaoqing Sun, Junshuai Chai, Xiaolei Wang, Wenwu Wang

An FPGA-Based Dual-Mode SSD for Device-Side Performance Optimization 466
Xingyu Chen, Sirui Peng, Hankun Lv, Zhangbin Yang, Daixiao Peng, Xi Cai, Xueguang Lian, Yong Ding, Xiaoyong Xue

A Simulation Comparison of Channel-All-Around and Gate-All-Around 3D Vertical Structure FeFET with IGZO Channel 469
Xuebin Wang, Zhijian Guo, Yutao Li, Chengji Jin, Jixuan Wu, Guanhua Yang, Yuanxiao Ma, Masaharu Kobayashi, Fei Mo, Yeliang Wang

Low Operating Voltage in HfO_2/ZrO_2 Superlattice Ferroelectric Capacitor Achieved by Thickness Scaling 472
Dongya Li, Huan Liu, Peiyuan Du, Fei Yu, Chengji Jin, Xiao Yu, Yan Liu, Genquan Han, Yue Hao

Co-Optimization of Oxide Semiconductor-Based Ferroelectric Transistors Between Electrical Performance and Ambient Stability by Using TiO_2-IGZO Dual-Channel Layers 475
Shangze Li, Xujin Song, Dijiang Sun, Xiaoyan Liu, Jinfeng Kang

Enhancing Computational Precision in PLRAM-Based In-Memory Computing with High-Low Bit Concatenation 478
Saike Zhu, Xiang Qiu, Yong Gong, Cimang Lu, Yi Zhao

FeFET Based Logic-In-Memory Pipeline-Style Circuits 481
Yang Li, Zhaohui Yang, Yinshui Xia

Study of V_{th} Degradation Mechanism in FeFET with $TiN/Al_2O_3/HfO_2/Al_2O_3/Hf_{0.5}Zr_{0.5}O_2/SiO_x/Si$ Structure 484
Runhao Han, Jia Yang, Tao Hu, Mingkai Bai, Yajing Ding, Xianzhou Shao, Saifei Dai, Xiaoqing Sun, Junshuai Chai, Hao Xu, Kai Han, Xiaolei Wang, Wenwu Wang, Tianchun Ye

Random Number Generation from 3D-NAND Flash Memory Using Shallow Charge Trap Related Short-Term Retention Errors 487
Ruibin Zhou, Jian Huang, Xianping Liu, Yuhan Wang, Xinrui Zhang, Yungen Peng, Zhiyi Yu

Simulation of Program/Erase Cycling and Retention Loss in 3-D CTF NAND Flash 490
Anuj Kumar, Ravi Tiwari, Souvik Mahapatra

Switch-Off Mechanisms in GeAsTe Ovonic Threshold Switching Selector Device 494
Z. Hu, Z. Chai, W. Zhang, Jian Zhang

Orthorhombic-I (Pbca) Phase: Origin of Anti-Ferroelectricity in HfZrO Films 498
Wei Liu, Zeping Weng, Jianguo Li, Wenchao Yan, Yiming Qu, Yi Zhao

The Maximum Storage Capacity of Open-Loop Written RRAM is Around 4 Bits 501
Yongxiang Li, Shiqing Wang, Zhong Sun

High-Density and High-Reliability RRAM for Memory and Computing Applications 504
Yimao Cai, Xiahong Zhou, Zongwei Wang, Lin Bao, Ling Liang, Cuimei Wang, Ru Huang

Impact of Different MAC Schemes on Computing in Memory Based on 1T1R Array 508
*Ruiqing Xie, Gaoqi Yang, Zongwei Wang, Linbo Shan, Jinshan Li, Chaoyi Ban, Lin Bao, Ling
Liang, Cuimei Wang, Yimao Cai, Ru Huang*

Investigation of Gate Injection Charges Behavior on FeFETs with $TiN/Al_2O_3/Hf_{0.5}Zr_{0.5}O_2/SiON/Si$
Structure by Analyzing ISPP/ISPE 511
*Jia Yang, Runhao Han, Saifei Dai, Tao Hu, Xianzhou Shao, Kanyi Li, Wenbo Fan, Xiaoqing
Sun, Junshuai Chai, Hao Xu, Kai Han, Xiaolei Wang, Wenwu Wang, Tianchun Ye*

Electric-Thermal Characteristics of Bottom P-I-N Isolated Nanosheet Gate-All-Around FETs 514
*Chunlei Wu, Hanzhi Gu, Jian Ma, Boqian Shen, Fei Zhao, Yueyuan Yu, Yiming Xia, Qingqing
Sun, David Wei Zhang*

Surface Potential-Based Compact Model for ITO Thin-Film Transistors with Ultra-Thin Channel 517
Wenting Xu, Xinxin Shen, Zuoxu Yu, Tingrui Huang, Yuzhen Zhang, Weifeng Sun, Wangran Wu

A Continuous Full Channel Potential Model for Accurate Synthetic Electricfield Calculating in
Gate-All-Around Devices 520
*Fei Zhao, Chunlei Wu, Yumin Xu, Boqian Shen, Jian Ma, Hanzhi Gu, Yueyuan Yu, Yiming Xia,
Qingqing Sun, David Wei Zhang*

Deep Learning and Adaptive Pattern Search Based BSIM-CMG Parameter Extraction Applicable to
Process Migration 523
Xingyu Li, Wangyong Chen, Linlin Cai

Vertical Channel Transistor (VCT) for Advanced Logic and Memory Applications 526
Mingmin Shi, Ran Bi, Ming Li

High Precision I-V Characteristics SPICE Model for Silicon Carbide MOSFET 530
Jinhong Shi, Yongxi Li, Jincheng Shi, Ruguan Li, Haimeng Huang, Hongqiang Yang

An Analytical Model for Characterizing Density of States of Oxide Transistors 533
Siyuan Hu, Chuanlin Sun, Junchen Dong, Zhensong Li, Kai Zhao, Dedong Han, Xing Zhang

A Physics-Informed Neural Network Model for Body Potential Distribution in MOSFETs Down to
50 K 536
*Honglin Wu, Fangxing Zhang, Xinyue Zhang, Baokang Peng, Wu Dai, Lining Zhang,
Runsheng Wang, Ru Huang*

Modeling the Transient Characteristics with Trap Behaviors in LTPS-TFTs 539
Haolin Li, Zheng Zhou, Xiaoyan Liu

A Novel β-Ga_2O_3-Based Enhancement-Mode Transistor Combining Heterojunction Gate and Fin
Shaped Gate 542
Yu Shao, Yunlong He, Xiaoli Lu, Songyao Wang, Xuefeng Zheng, Xiaohua Ma, Yue Hao

Electrical Characteristics and Thermal Reliability Investigation of TreeFET, FishboneFET,
CombLikeFET and NSFET 545
Mingyu Ma, Wenbin Wang, Haokun Li, Shujun Gao, Hailong You, Cong Li

A Fast-Response Current Source with High-Impedance for Zero-Crossing-Based Circuits 548
Ruoyu Li, Xianglong Wang, Jianqiang Xu, Yintang Yang

A PPG Analog Front-End with PVT-Insensitive High-Pass Frequency 551
Zhaofeng Huang, Zepeng Huang, Hengchang Bi, Xing Wu, Liangjian Lyu

A Fully Integrated FVF Based Low-Noise Voltage Buffer for ADC Reference 554
Ikhwan Kim, Yajie Qin

A Resistor-Free Grounded High-Frequency Memristor Emulator 557
Xinying Su, Bingjun Xiong, Junjie Yu, Jingjing Liu

An Ultra-Low-Leakage Current Sensing Interface for Wide Temperature Range 560
Jinsheng Tang, Chun Zhao, Lin He

A Global Threshold Voltage Finder Technology for the Readout Circuit of Event-Based Vision
Sensor 563
Yanwen Su, Hao Li, Dongsheng Liu, Ang Hu, Kaiyue Li

A Residue Amplifier with 72.27 dB Loop-Gain and 4.64 GHz Closed-Loop Bandwidth Consuming
6.4 mW for 12-Bit 1-Gsps Pipelined ADC 566
Jiangbo Wei, Jin Liu, Wei Tian, Chao Wang, Maliang Liu

A 180 mV–1.6 V Thermoelectric Energy Harvesting Converter with Low-Voltage Cold Start and
Less than 1 µW Power Loss 569
Chunlin Wang, Anzhi Yan, Tianyu Guo, Peng Wan, Houfang Liu, Yi Yang, Tianling Ren

A Signal Conditioning ASIC with High Precision and Low Noise for MEMS Accelerometers 572
Quan Sun, Rui Liu, Zhe Zheng, Lei Dong, Ji-Jiang Wang

Design of a High-Precision Self-Calibration Readout Circuit for CMOS Microbolometer 575
Qianhao Zhang, Jie Liu, Sheng Xu, Yiming Liao, Feng Yan, Xiaoli Ji

A Smooth Two-Stage Soft Start Method for Current Mode Boost Converter 578
Yue Shi, Shi-Dong Wang, Zekun Zhou, Bo Zhang, Zhigang Qin

Design of 12-Bit Low-Power Single-Slope ADC with 2048 Columns for Infrared Focal Plane Array 581
Lixiang Han, Hao Li, Yihui Zhang, Liang Gao, Dongsheng Liu

A High-Voltage Smooth Self-Starting Reference Current Source Circuit 584
Dongyan Zhao, Jie Pan, Chenghao Zhang, Yidong Yuan, Yi Hu, Hongwei Shen, Zekun Zhou

A Temporal and Spatial Reuse Interpolation Hardware for VVC Motion Compensation 587
Huanxiang He, Shushi Chen, Leilei Huang, Yibo Fan

A Broadband Digital Beamforming Method Based on FPGA 590
Yiwen Tang, Guowen Jia, Zhen Zhang, Yue Zhang

Conditional Cycle Termination RANSAC 593
*Tong Jiang, Yujie Huang, Liyuan Peng, Mingyu Wang, Wenhong Li, Minge Jing, Xiaoyang
Zeng*

A Multi-Resolution Propagation Algorithm and Pixel Grouping Storage Strategy for PatchMatch
Stereo 596
Kai Liu, Zhenyu Zhang, Haiwei Wang, Leilei Huang, Chunqi Shi, Long Xu, Runxi Zhang

An XOR Arbiter PUF Based on the IGZO TFT Devices 599
Xiang Chen, Yongliang Chen, Xiaole Cui

Design and Implementation of Hierarchical Storage Structure for MCCSIP-RAA 602
Longmei Nan, Yu Jin, Yiran Du, Tao Chen, Yanjiang Liu, Wei Li

A Highly Scalable Hardware HEVC Encoder Based on FPGA 605
Guohao Xu, Chenlong He, Shiyan Yi, Leilei Huang, Xiaoyang Zeng, Yibo Fan

A Hardware-Friendly Fast Block Partition Decision Algorithm Based on Histogram of Oriented
Gradient for AV1 608
Guohao Xu, Shiyan Yi, Zhijian Hao, Leilei Huang, Hao Zhang, Xiaoyang Zeng, Yibo Fan

A Low Power Narrow-Band Complex-Bandpass Filter Based on Feedforward Compensation
Amplifiers for NB-IoT Applications 611
Xu Zhao, Ziqiang Wang, Jie Gan

A High-Performance MTJ-LUT Circuit Using 4T1M Architecture 614
*Yu Pan, Yuejun Zhang, Shuaicheng Guo, Yuanxin Tian, Bo Hong, Rui Fang, Liang Wen, Yong
Ding*

Optimizing Communication Efficiency of GNN Inference in Distributed Systems...................... 617
Wenqian Zhou, Qiaosha Zou

SST: Simplified Space-Time Transformer Based on Time-Assisted Spatial MSA for 3D Human
Pose Estimation 620
Sheng Lu, Qiyun Dong, Zhenyin Zhang, Gengsheng Chen, Yinna Zhu, Wei Xu

SALTS: An Efficient and Flexible Self-Attention Accelerator with Long Token Support on FPGA 624
Kaiqi Chen, Xinhua Shi, Jun Han

RISC-V Neural Network Instruction Design and Simulation with Cache Scheduling Via ROCC
Interface... 627
Siyao Dai, Zikang Zhou, Jun Han

Impact of External Magnetic Interference on the Performance of MRAM-Based Neuromorphic
Computing.. 630
*Yingtong He, Suihuan An, Yu Chen, Xue Zhou, Xihui Yuan, Weidong Zhang, Zheng Chai, Tai
Min*

A Hardware Accelerator for Image Super-Resolution with Algorithm Lightweighting and Custom
Fusion Engine... 634
Menghan Li, Sheng Lu, Jun Han

Hardware Implementation of High Speed Fault-Tolerant Parallel Accelerator 637
Wenzhe Ma

Composite Filter-Based Bicubic Interpolation Method and FPGA Implementation 640
Li Zhang, Jingjing Liu, Yujie Zhu, Jianhua Zhang

MTJ Based Temperature Tracking Read/Write Assist for High Speed SRAM Bitcell...................... 643
Yongliang Zhou, Chengxing Dai, Jingxue Zhong, Yingxue Sun, Xin Li, Chunyu Peng

A 12V to 1V Tri-State DSD Hybrid Converter by Self-Balanced Dual Flying Capacitors with
0.3mV Output Ripple and 90.09% Peak Efficiency 646
Yixing Wang, Qianhui Liu, Yuhua Chen, Yizhe Yang, Yimeng Zhang, Yuming Zhang

System-Level Evaluation of AOS Gain Cell eDRAMs for Low-Power Normally-Off Computing................. 649
Long Chen, Yecheng Yang, Wei Li, Shao Hao Wang

A High Sigma Monte Carlo Analysis Solution Via Machine Learning for SRAM Margin Signoff 652
Amy Rao

Enhanced Multi-Bit Computation Using CIM SRAM Technology 655
Ruiyong Zhao, Yibo Hu, Zhipeng Ren, Yizhe Yin, Jing Chen

A Compute-In-Memory Macro Based on Complementary 2T2C FeRAM Cell for BNNs 658
Jinyu Li, Mingzhang Xie, Shisheng Xiong

A Novel High Speed Low Power Differential Circuit-Based FRAM Read Scheme 661
Qiuyu Tao, Jiabao Ye, Xuecheng Cui, Nan Jiang, Jiangtao Cao, Xibo Chen, Zhangbin Yang,
Daixiao Peng, Xi Cai, Xueguang Lian, Yong Ding, Jiuren Zhou, Bing Chen, Genquan Han

A 13-Bit, 1 MS/s Cyclic ADC, for High-Speed CMOS Image Sensor 664
Qi Lv, Rensheng Shen, Yu Cheng, Guoqiang Zhong, Yang Qu, Yuchun Chang

An Area-Efficient 16-Bit Four-Channel R-2R DAC Based on Switching On-Resistance Adaptive
Calibration Technique 667
Kejun Wu, Yuchen Liu, Yuhan Hu, Yu He, Zhen Yu, Ning Ning

A Background Calibration Method of Bandwidth Mismatch for Time-Interleaved ADCs Based on
Neural Network 670
Tianqi Yang, Longsheng Wang, Xin Zhao, Shubin Liu, Dengquan Li, Zhangming Zhu

A Second-Order Dual-Charge-Pump Passive Noise-Shaping SAR ADC for Medical Implant
Devices 673
Kangkang Sun, Xuanxin Ke, Haoning Sun, Yuchen Wang, Feng Yan, Jingjing Liu

A 114.4-DB DR, 26-KHz BW Discrete-Time Incremental Zoom ADC 676
Yuanhong Ding, Longjiang Jia, Jian Mei, Lei Deng, Rui Yin

A Multi-Phase Clock Self-Calibrating Circuit 679
Zhihuai Li, Li Jiang, Xueming Wei, Zilu Cai, Jiami Tang

A High-Resolution Low-Power Extended-Range Incremental $\sum\Delta$ ADC for Battery Management
System 682
Long Zhang, Quan Sun, Rui Liu, Zhe Zheng, Jingjing Zhang, Haitao Liu

An Infrared AFE Chip and System with Non-Invasive Blood Glucose Detection Output 685
Bin Li, Jiyuan Guo, Chengzhen Xie, Jian Mei, Lei Deng, Rui Yin

A 12-Bit 8GS/S Time-Interleaved Pipeline-SAR ADC with Calibration 688
Jie Pu, Jinda Yang, Yuanjun Cen, Jianwen Li, Rong Han, Xing Zhu, Lei Chen

An Ultra-High Frame Rate ROIC for Hyperspectral Detection 691
Angyang Li, Ningning Li, Jian Mei, Lei Deng, Rui Yin

A BackgroundDigital Calibration Method for DTCs Used in Digital PLL Employing Dual-Path
DTC 694
Renxuan Li, Xiaoyu Shan, Li Wang, Dongsheng Liu, Ang Hu

A High-Precision Sigma-Delta ADC for Battery Management System 697
Hao Xue, Liji Wu, Jing Hu, Zhiwei Li, Xiangmin Zhang

Multi-Sampling Mode CDAC Design for a 12-Bit 200MS/s Pipelined-SAR ADC 700
Tianyu Zhang, Fan Ye, Shunli Ma

DSP-PUF: A Software PUF Based on Digital Signal Processor for IoT Security 703
  Tengfei Yuan, Pengjun Wang, Yuejun Zhang, Mingze Ren, Shuang Hu

A Q/V Band 49.6-54.5GHz,3.53dB NF,45dB Gain,2.09° Phase Error,2-Way Phased-Array Receiver
for Satellite Application.. 706
  Congrui Li, Qi Zhao, Ruolan Chen, Shulan Chen, Yan Wang, Lei Zhang

A Fractional-N SPLL Using Space-Time Averaging and Phase Interpolator for Quantization Noise
Reduction 709
  Shengxiang Liu, Ke Sun, Chengyu Yang, Dongsheng Liu, Ang Hu

A 18V, 600mA Load Current, 22MHz High-Voltage Power Amplifier with Over-Temperature
Protection and Bidirection Enable Logic... 712
  Yuan Ren, Xin'An Wang

A 47μW Wake-Up Receiver with -77dBm Sensitivity Using a Mixer-First Architecture 715
  Weitao He, Yaxin Zeng, Bin Jia, Hao Min, Hao Xu, Na Yan

A Ka-Band CMOS Broadband Power Amplifier with 35.3% PAE for SATCOM Applications 718
  Zhiqing Liu, Yu Chu, Yuting Sun

RF Front-End Chip Design for Ku-Band with 130nm CMOS Technology................................... 721
  Huiquan Xie, Ziyu Wang, Tianrui Wang, Yifei Chen, Maliang Liu, Yintang Yang

Back-Gate Bias Assisting VCRO Design 724
  Chenglin Ye, Zheng Zhou, Xiaoyan Liu

A 3.2-To-7.1GHz Quad-Core Dual-Mode Oscillator Achieving 193.6 dBc/Hz Peak FoM........... 727
  Xiaoyu Shan, Renxuan Li, Mengming Zhang, Ang Hu, Dongsheng Liu

A 20.6 to 30.5 GHz Two Stage Cascode LNA in 40nm CMOS for Phase Array Tranceiver 730
  Lei Wang, Kefeng Han, Hao Xu, Rui Yin, Na Yan

A 12-32 GHz Power Amplifier with 32-DBm Psat and 25% PAE in 0.15μm GaN 733
  Xiangran Ni, Chunyue Bo, Tianyu Li, Qingyang Dong, Xin Jiang, Weijun Luo

A Source-Driven Push-Push Doubler with Wideband 2nd Harmonic Feedback........................... 736
  Yuyang Chen, Ao Zhang, Jianjun Gao, Jianjun Zhou

Low Power Processor for IoT Device 739
  Jincheng Li, Jiyuan Bai, Zelin Wang, Gengsheng Chen, Xiaofang Zhou

A Heterogeneous Integration System of Analog in-Memory Computing and Field-Programmable
Gate Array 742
  Hua Chen, Yiming Qu, Wenhao Wu, Yi Zhao

A 10-MHz Four-Phase Hysteretic Control DC-DC Converter with Inductor Current Self-Balancing............. 745
  Yushen Zhang, Ningning Li, Yibo Zhang, Yizhe Yang, Yimeng Zhang, Yuming Zhang

A High Precision −40°C to 150°C Bandgap Reference with Dual Temperature Compensation 748
  Yuhan Zhang, Jianzheng Li, Xiaomeng An, Lina Wang, Yajie Qin

A Biphasic Neural Stimulator with Adaptive Pulse-Width Modulation Charge Balancer.............. 751
  Hailong Tang, Wenxian Gu, Yifan Song, Hengchang Bi, Xing Wu, Liangjian Lyu

Assembly of Oxidized/Intrinsic 2D MXene Film for Improved Absorption Electromagnetic Shielding.. 754
    Yulin Guo, Siteng Li, Jiafeng Song, Yilin Sun, Zhifang Liu, Weijia Luo

Improved Channel Width and Morphology of Epi Silicon FinFET Via Low Thermal Budgets Fin Thinning Technology... 757
    Peng Wang, Yupeng Lu, Guanqiao Sang, Renjie Jiang, Lei Cao, Qingkun Li, Lianlian Li, Hang Zhang, Zhonrui Wang, Meihe Zhang, Qingzhu Zhang, Junfeng Li, Huaxiang Yin

Deep Investigation into Variability of Complementary Dopant Segregated Tunneling FET Based on Foundry Platform 760
    Rundong Jia, Jianfeng Hang, Kaifeng Wang, Yongqin Wu, Hongyan Han, Ye Ren, Weihai Bu, Runsheng Wang, Qianqian Huang, Ru Huang

Investigation of Common-Gate and Split-Gate Structures Based on CFET Standard Cells............................. 763
    Peishun Tang, Rongzheng Ding, Xiaona Zhu, Shaofeng Yu

Exploration of the Effect of Silver Impurity on the Minority Carrier Lifetime of Semiconductor.................... 766
    Xin Tian, Peizhi Zhao, Yudong Li, Jun Xu, Tianling Ren

Fabrication and Electrical Characterization of $Mo/Hf_xZr_{1-x}O_2/Mo$ Ferroelectric Capacitors............................ 768
    Chunsheng Jiang, Wencai Du, Qin Xie

Effect of Cascade Current Density and Plating Time on TSV Filling Effect in DC Power Supply.................. 771
    Weifeng Chen, Lijuan Peng, Xiaohui Wang, Fangzhou Wang, Guojian Ding, Qi Feng, Ping Yu, Peng Zuo, Feng Liu, Jiang Ma, Yang Wang, Haiqiang Jia, Hong Chen

High-Performance Carbon Nanotube Optoelectronic Transistors for Memory Applications........................... 774
    Shuang Liu, Heyi Huang, Yanqing Li, Yadong Zhang, Feixiong Wang, Yupeng Lu, Renjie Jiang, Jiali Huo, Huaxiang Yin

Investigation of the Channel Width Dependence of IGZO TFT by Experiment and TCAD Simulation 777
    Yanyu Yang, Yupeng Lu, Shuang Liu, Renjie Jiang, Jie Luo, Yunjiao Bao, Peng Wang, Gaobo Xu, Huaxiang Yin

A Test and Evaluation Platform for Quantitative Analysis of High-Reliability Designs................................. 780
    Yifeng Huang, Wenqing Wan, Chang Wu

Hot-Carrier Injection Characterization of n-LDMOS Transistors and Stress Tests in a Buck Converter Configuration.. 783
    Chun Yee Chu, Wai Tung Ng

Semimetal Alloy Contact with Low Resistivity and Enhanced Thermal Budget for MoS_2 FETs 786
    Kwok-Ho Wong, Mansun Chan

Automated Verification of Functional Interface Connections in Circuit Schematics 789
    Keli Long, Xingyu Gao, Lei Li, Jinxiang Wang, Fangfa Fu, Liangquan Qiao, Jinghan Zhou

Co-Optimization Design Method of Temperature Variation and Circuit Aging in Digital Circuits................. 792
    Songxuan He, Wangyong Chen, Ling Xiong, Linlin Cai

Boolean Matrix Factorization Algorithm Based on Error Shaping Technique and Its Application on Approximate Logic Synthesis 795
    Runhua Yang, Rensheng Shen

Automatically Device Sizing of Analog Circuit Through Sequential Model-Based Optimization with Circuit Recognition 798
Shun-Qi Dai, Xiao Wang, Yuan Lei, Bei-Ping Yan

Vanadium Oxide-Based Artificial Synapses for Construction of Artificial Neural System 801
Zhuoling Zhou, Libin Liang, Hongzhi Chen, Changjiu Teng, Shilong Zhao, Wenjun Chen

High Performance FeFET with α-IGZO Channel Enabled by Atomic-Layer-Deposited HfO_2 Interfacial Layer 804
Yinchi Liu, Hao Zhang, Xinlong Zhou, Dmitriy Anatolyevich Golosov, Chenjie Gu, Hongliang Lu, Shijin Ding, Wenjun Liu

A Simulation Study on Cell Scaling Impacts in 3D Charge-Trapping (CT) Flash Memory 807
Wanyu Li, Haitao Dong, Qianwen Wang, Yang Feng, Xuepeng Zhan, Jixuan Wu, Jiezhi Chen

Comprehensive Charaterizations on Read Disturbs in QLC Charge-Trap (CT) 3D NAND Flash 810
Shaoqi Yang, Xiaohuan Zhao, Peng Guo, Qianwen Wang, Guangkuo Yang, Xinyi Guo, Pengpeng Sang, Jixuan Wu, Xuepeng Zhan, Jiezhi Chen

Aspect Ratio Dependent Optimization and Comparison of Specific ON-Resistance of SJ and Hk MOSFETs with Extremely High Permittivity 813
Chenxing Wang, Zhentao Xiao, Zonghao Zhang, Zhenghao Jin, Zhiwan Liu, Zonglin Li, Zhemin An, Yunteng Jiang, Ruguan Li, Haimeng Huang, Hongqiang Yang

Copper Ion Migration in Van Der Waals $CuInP_2S_6$ Devices with Vertical and Lateral Structures 816
Jie Li, Yirong Guo, Pengying Chang

Simulation Study of the Impact of Split Gate on SiC DTMOS Short Circuit Withstand Capability 819
Zixun Chen, Jinping Zhang, Yang Liu, Xudong Ma, Bo Zhang

Improved Hall Mobility Measurement Distinguishing Interface Capturing Effect in 4H-SiC Inversion Channel 822
Xiangrui Fan, Hao Fu, Xinyu Zhang, Zilong Wu, Jiameng Sun, Jiaxing Wei, Siyang Liu, Weifeng Sun

A Superior SiC Lateral MOSFET with Patterned P-Bury Layer Made on N-Type Wafers 825
Xuke Yan, Junji Cheng, Xiaojun Fu, Bo Yi, Haimeng Huang, Hongqiang Yang

High Performance Termination Design and Fabrication for SiC MOSFET Device 828
Lei Huang, Junhou Cao, Chenlu Wang, Hao Fu, Jiaxing Wei, Siyang Liu, Weifeng Sun

Analysis of the Separation Degree for P-Pillar in SiC Super-Junction Structure Through "Multiple Epitaxy-Ion Implantation" Route 831
Hao-Bo Kang, Hao Yuan, Feng-Yu Du, Yu Zhou, Ke-Yu Liu, Xiao-Yan Tang, Chao Han, Qing-Wen Song, Yu-Ming Zhang

Numerical Analysis of the CIBL Effect on Short-Circuit Characteristics of DG-CSTBTs with Reduced Mesa Width 834
Zhengyu Lang, Jinping Zhang, Shiwei Zheng, Shuyang Huang, Haonan Deng, Bo Zhang

Innovations in GaN HEMT Design: Achieving Superior Power Output and Thermal Management 837
Shiming Li, Bowen Yang, Mei Wu, Ling Yang, Bin Hou, Meng Zhang, Xiaohua Ma, Yue Hao

An Enhanced RC-IGBT Incorporating Superjunction and Discontinuous Field Stop Layers for Improved Efficiency 841
Yiming Jia, Jieyu Long, Zhiwei Jing, Haimeng Huang, Hongqiang Yang

Simulation Study on 1200V CS-SemiSJ-IGBT for Reduced Switching Loss and Fast Switching 844
Luping Li, Zehong Li, Peng Chen, Yuzhou Wu, Qiansheng Rao, Ming Li, Haifeng Qin, Li Wan, Yang Yang, Wei Li, Min Ren

A Dual-Gate Trigger Thyristor for Reducing the Probability of False Triggering 847
Pengcheng Xing, Qingbo Wan, Jie Huang, Ruize Sun, Chao Liu, Wanjun Chen

Edge-Dependence of Threshold Voltage in MoS_2 Nanoribbon-Based 2D FETs 850
Zhirong Peng, Mansun Chan

Ultra Fast Diode Avalanche Shaper with Floating Junction 853
Zhen Yang, Yu Zhou, Xiao-Yan Tang, Chao Han, Qing-Wen Song, Yu-Ming Zhang

Silicon Carbide Diode Avalanche Shaper with Multi-Point Quasi-Uniform Triggering 856
Lin Cheng, Yu Zhou, Xiao-Yan Tang, Chao Han, Yu-Ming Zhang, Qing-Wen Song

Super Field Plate LIGBT with Improved Performance for Both Cell and Terminal Region 859
Weihao Lu, Jing Li, Jitong Wang, Chaoyang Peng, Chunwei Zhang

High-Temperature Oxidation of 4H-SiC and Gate Oxide Reliability Dependence on Oxidation
Temperature 862
Baoyan Feng, Xiaoyan Tang, Yi Bo Zhang, Chao Han, Hao Yuan, Qing Wen Song

Optimization for a High-Voltage Recessed-Gate β-Ga_2O_3 MOSFET by Gate and Drain Field Plate
Technology 865
Bo Yi, Yuan Qiao, Ming Dai, Fan Xu, Junji Cheng, Haimeng Huang, Moufu Kong, Xingli Jiang, Hongqiang Yang

A Novel Voltage Sensor with Composite Trench Structure for High Voltage IGBT 868
Yang Yang, Ze-Hong Li, Li-Hang Dong, Wei Li, Peng-Fei Jia, Zhi-Yu Yang, Li Wan, Yi-Shang Zhao, Tong-Yang Wang, Zi-Ming Xia

A Novel Triggered Voltage Sensing Structure for High Voltage IGBT 871
Yang Yang, Ze-Hong Li, Li-Hang Dong, Wei Li, Peng-Fei Jia, Zhi-Yu Yang, Li Wan, Yi-Shang Zhao, Lu-Ping Li, Zi-Ming Xia, Tong-Yang Wang

Investigation of Threshold Voltage Instability in p-GaN Gate HEMTs Under Surge Current Stress 874
Xiaoming Wang, Yu Shi, David Zhou, Haizhao Zhi, Yun Xia, Yuxi Wan, Ruize Sun, Xinghuan Chen, Wanjun Chen, Bo Zhang

A Novel Ga_2O_3 High-k Trench MOSFET with Improved Forward and Reverse Performance 877
Moufu Kong, Lewei Lyu, Haoran Wang, Zhaoyu Ai, Xinyang Chen, Fanxin Meng, Qiang Hu

A Nonlinear Behavioral Modeling Approach for Microwave Transistors Considering
Electrothermal-Aging Degradation 880
Lin Cheng, Hongliang Lu, Silu Yan, Junjun Qi, Jiantao Qiao, Yuming Zhang

Effect of Layer Thickness on the Transport Properties of ALD-Deposited ZnO/In_2O_3
Heterojunction Thin-Film Transistors 883
Zhenwei Li, Tiaoyang Li

Electrical and Thermal Analysis of CNT nTSV Applied to BS-PDN: A Modeling Study 886
Kai Ying, Baohui Xu, Jie Liang

A Unified Current-Voltage Compact Model for Organic Light-Emitting Diode 889
Wenbin Wang, Mingyu Ma, Wangjun Yang, Jianghao Ma, Hailong You, Cong Li

Threshold Voltage and Mobility Extraction of Negative Bias Temperature Instability in 22nm FD SOI MOSFETs 892

Yibo Hu, Hao Ge, Zhipeng Ren, Yizhe Yin, Ruiyong Zhao, Jing Chen

Modeling of Silicon Single-Photon Avalanche Diodes for Process and Design Optimization 895

Jing Fu, Anran Guo, Hongbo Zhang, Guowei Li, Huaping Ma, Ruizhi Li, Yuwei Chen

A Novel Modeling Method for BV Characteristics of ESD Protection Devices 898

Ke Zhang, Yang Wang, Xiangliang Jin

Analysis of the Impact of Parasitic Bipolar Amplification on Charge Sharing Based on Analytical Model 901

Yutao Zhang, Hongliang Lyu, Yuming Zhang, Ruxue Yao

Research on the Performance Degeneration of GGNMOS Under Total Ionizing Dose Radiation 904

Jiekai Feng, Ping Luo, Chengxin Li, Jiaxuan Hu, Peng Li, Pengfei Liao

The UIS Withstand Capability and Device Failure Mechanism of 650 V p-GaN Gate HEMTs 907

Qihang Huang, Luanxi Ma, Yanyu Nie, Shuting Huang, Jianggen Zhu, Yu Shi, Rongxin Du, David Zhou, Yuxi Wan, Bo Zhang, Qi Zhou

Time Dependent Dielectric Breakdown in n-MOSFETs Fabricated by Low-Temperature and Low-Pressure Mild Oxidation After Plasma Solidification 910

Qiao Teng, Yanning Chen, Fang Liu, Bo Wu, Yongfeng Deng, Dawei Gao

Simulation of BTI for GAA MOSFETs with Enhanced Parameters Extraction 913

Yongjia Wang, Yijiao Wang, Shuhan Wang, Jinghan Xu, Xiaoyan Liu

Foundamentals of Low-Resistive Indium-Violet Phosphorene Top Contact: An Ab-Initio NEGF Study 916

Huaipeng Wang, Sicheng Liu, Shuaihong Li, Zhifang Liu, Yilin Sun, Jianlong Xu, Dan Xie

Gold Thermocompression Wafer Bonding for Quartz MEMS Applications 919

Ting Yang, Dongxiang Han, Jun Xu, Tian-Ling Ren

An Adaptive Threshold Analog Front-End Circuit for Direct ToF LiDAR 922

Jianping Guo, Xiaoyang Zeng, Wenhong Li, Mingyu Wang

Design of Ultra-Broadband Metamaterial Absorber from Infrared to Terahertz 925

Xiangze Liu, Wenbin Zhou, Tiantian Shi, Yiming Liao, Feng Yan, Xiaoli Ji

Large Modulation Bandwidth Si-Based Avalanche Photodiode for Visible Light Communications 928

Jiabin Wu, Yidi Hu, Chiang Zhu, Zhichong Wang, Xiaona Zhu, Chao Shen

An Artificial Neuromuscular Synapse Based on a Ferroelectric $Pb(Zr_{1-x}Ti_x)O_3$/SiC Floating Gate Transistor 931

Yu Liu, Lin Lin, Xiang Wang, Chengyan Zhong, Junxiong Guo, Wen Huang, Yufeng Guo

Broadband Photodetectors Based on Graphene/Perovskite Hybrid Structure with Ferroelectric Gating 934

Zhongyang Liu, Shuangqi Dong, Mingjie Li, Huaipeng Wang, Dan Xie, Yilin Sun

Interconnection Design of Chiplet Technology 937

Ning Chen, Lei Shen, Chang Wu

Effects and Modeling Study on FDSOI MOSFETs at Cryogenic Temperature 940

Zhipeng Ren, Yibo Hu, Yizhe Yin, Ruiyong Zhao, Jing Chen

Author Index

IC SICT 2024

Proceedings of 2024 IEEE 17th International Conference on Solid-State & Integrated Circuit Technology (ICSICT)

Edited by   Fan Ye

                      Xiaona Zhu

                      Ting Ao Tang

◆IEEE Beijing Section

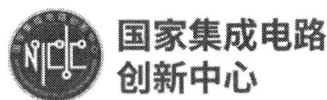

2024 IEEE 17th International Conference on Solid-State & Integrated Circuit Technology (ICSICT)

Patrons by

CXMT Corporation

GONGMO Semiconductor Inc.   Lingyange Semiconductor Inc.   ULVAC (Suzhou) Inc.

ACM Research (Shanghai), Inc.   XDL Technologies Inc.   Natural-Integration Adv. Semi. Tech. Co., Ltd.   Shanghai Dezhu Xinyuan Tech. Co., Ltd.

Welcome to ICSICT 2024

On behalf of the Conference, we are delighted and honored to extend a warm welcome to all attendees of ICSICT 2024.

ICSICT 2024 will take place in Zhuhai, China, from October 22nd to 25th, 2024. This marks the 17th edition of our conference series, which began in 1986. After two years of virtual meetings due to COVID-19, we are thrilled to return to an in-person format.

The conference is an international forum for VLSI circuit designers, system integrators, IC manufacturers, device engineers, and CAD/CAE tool developers to showcase the latest advancements, developments, and research findings in their respective fields. It also provides an excellent platform for academic and industry participants to share information and interests.

ICSICT 2024 has invited international experts to conduct tutorials on the first day of the conference. Additionally, eight world-renowned academic and industry leaders will deliver keynote speeches during the plenary sessions.

ICSICT has significantly impacted on both industry and academia, and we look forward to continuing this tradition with you at this year's in-person conference.

General Co-Chairs of ICSICT 2024

Jan Van der Spiegel
David Wei Zhang
Shaozhi Deng
Bin Zhao
Francois Rivet

Oct. 23rd, 2024

Conference Committee

Life Honorary Chair

Name	Affiliation	Country/Area
Yangyuan Wang	Peking University	China

General Co-Chairs

Name	Affiliation	Country/Area
Jan Van der Spiegel	University of Pennsylvania	USA
David Wei Zhang	Fudan University, NICIC	China
Shaozhi Deng	Sun Yat-Sen University	China
Bin Zhao	IEEE EDS	USA
Francois Rivet	University of Bordeaux	France
Jan Van der Spiegel	University of Pennsylvania	USA

Steering & Organizing Committee Co-Chairs

Jan Van der Spiegel	University of Pennsylvania	USA
Mengqi Zhou	IEEE Beijing Section	China
Ting-Ao Tang	Fudan University	China
Huihua Yu	Fudan University	China

Advisory Committee Co-Chairs

Ru Huang	Southeast University	China
Chenming Hu	UC Berkeley	USA

Cor Claeys	Proximus	Belgium
K.N.Tu	University of California, Los Angeles	USA
Hiroshi Iwai	Yang Ming Chiao Tung University	Japan

Technical Program Committee Co-Chairs

Shaofeng Yu	Fudan University/NICIC	China
Zhiyi Yu	Sun Yat-Sen University	China
Fan Ye	Fudan University	China
Ming Li	Peking University	China
Haruo Kobayashi	Gunma University	Japan
Bo Zhang	University of Electronic Science and Technology of China	China
Huaqiang Wu	Tsinghua University	China
Yong Lian	Shanghai JiaoTong University	China
Mansun Chan	Hong Kong University of Science and Technology	HongKong, China
Man-Kay Law	University of Macau	Macau, China
Andy Wu	Taiwan University	Taiwan, China
Yi Zhao	Zhejiang University	China

Publicity Co-Chairs

Rui Yin	Fudan University	China
Wei Xu	Fudan University	China
Junyu Wang	Fudan University	China
Jiting Sheng	Fudan University	China

Publication Committee Chair

| Mengqi Zhou | IEEE Beijing Section | China |

Secretary-General

| Xiaona Zhu | Fudan University | China |

Tutorial Session

Tuesday

Tuesday, October 22, 9: 00 – 18: 00

Tuesday, October 22, 9: 00 – 12: 15   Meeting Room 8
Tutorial Session T1   Sheraton Zhuhai Hotel 2nd Floor
Session Chair: Dr. Albert Li, Lingyange Semiconductor Inc., China
                  Dr. Y.K. Li, Zhuhai Fudan Innovation Institute, China

T1-1	**Low-frequency Noise Characterization as a Diagnostic Tool to Characterize Advanced Semiconductor Materials and Devices**
9: 00 ~10: 30	Prof. Cor Claeys, Proximus, Belgium
	Coffee Break
T1-2	**Power Super-junction Devices**
10: 45 ~12: 15	Prof. Wentong Zhang, University of Electronic Science and Technology of China, China

Tuesday, October 22, 13: 30 – 18: 30   Meeting Room 8
Tutorial Session T2   Sheraton Zhuhai Hotel 2nd Floor
Session Chair: Dr. Yuan Li, AzurEngine Technologies Inc., China

T2-1	**In-Memory Neuromorphic Computing Algorithm and Hardware**
13:30 ~15:00	Prof. Yufei Ma, Peking University, China
	Coffee Break
T2-2	**Artificial Intelligence for 6G: Implementations, Algorithms, and Optimizations**
15: 15 ~16: 45	Prof. Chuan Zhang, Southeast University, China
	Coffee Break
T2-3	**RF/mm-IC in silicon for wireless communication**
17: 00 ~18: 30	Prof. Hao Gao, Eindhoven University of Technology, The Netherlands

Technical Session

Wednesday

Wednesday, October 23, 8: 30 –9: 00

Wednesday, October 23, 8: 30 –9: 00	Grand Ball Room
Opening	Sheraton Zhuhai Hotel 1st Floor

Wednesday, October 23, 9: 00 –10: 30

Wednesday, October 23, 9: 00 –10: 30	Grand Ball Room
Keynote Session K1	Sheraton Zhuhai Hotel 1st Floor
Session Chair: Prof. Bin Zhao, IEEE EDS, USA	

K1-1	**Low-Power On-Device Computation for Future AI Expansion**
9: 00 ~ 9: 45	Dr. Paul Penzes, Vice President, Qualcomm, USA
K1-2	**Effective Deep Learning Models using Medical Images for Disease Diagnosis**
9: 45 ~10: 30	Prof. Myung Hoon Sunwoo, Ajou University, Korea
	Coffee Break

Wednesday, October 23, 10: 45– 12: 15

Wednesday, October 23, 10: 45–12: 15	Grand Ball Room
Keynote Session K2	Sheraton Zhuhai Hotel 1st Floor
Session Chair: Prof. Cor Claeys, Proximus, Belgium	

K2-1	**Piezotronics of the third- and fourth-generation semiconductors**
10: 45 ~11: 30	Prof. Zhong Lin Wang, Georgia Institute of Technology, USA
K2-2	**Atomic Layer Processing: Its Evolution, Diverse Applications, and Future Prospects**
11: 30 ~12: 15	Prof. Fred Roozeboom, University of Twente, The Netherlands

Wednesday, October 23, 13: 30 – 17: 15

Wednesday, October 23, 13: 30 – 17: 15   Meeting Room 1
Special Session: the Future of AI   Sheraton Zhuhai Hotel 2nd Floor
Session Chair: Prof. Jianguo Yang, Zhangjiang Laboratory, China

		Title
SS-1		**Transforming AI: The Impact of Computing-in-Memory on Future Technologies**
13:30 ~14:00		Tony Tae-Hyoung Kim *(Nanyang Technological University, Singapore)*
SS-2		**Ultra-low power multi-core hardware accelerators for AI on Edge**
14:00 ~14:30		Do Anh Tuan *(A*STAR, Singapore)*
SS-3		**Large language Model on Chip**
14:30 ~15:00		Hao Yu *(Southern University of Science and Technology, China)*
SS-4		**Progress and Challenges of Multi-physics Simulation EDA for Chiplet Packaging**
15:00 ~15:30		Wenliang Dai *(Xpeedic Corp., China)*
		Coffee Break
SS-5		**Emerging Non-volatile and Non-volatile/Volatile Fused Computing-in-memory Macros for Edge Inference and Learning**
15:45 ~16:15		Chunmeng Dou *(University of Chinese Academy of Sciences, China)*
SS-6		**0459: Memristor Crossbar's Design Technology for Improving PPA (Power-Performance-Area) of Neural Networks**
16:15 ~16:45		Kyeong-Sik Min *(Kookmin University, Korea)*
SS-7		**0462: Reliability of Memristor-based Neuromorphic Computing System**
16:45 ~17:15		Michiko Inoue *(Nara Institute of Science and Technology, Japan)*

Wednesday, October 23, 13: 30 – 15: 15   Meeting Room 2
Session B1: Analog Circuit I   Sheraton Zhuhai Hotel 2nd Floor
Session Chair: Prof. Wenning Jiang, Fudan University, China

	Title
B1-1	**0014: Design and Verification of Low-Temperature-Drift and Capacitor-Less LDO Based on 110nm Technology (invited)**
13:30 ~14:00	Yonggang Fu, Anping He, Zhaobin Wang *(Lanzhou University, China)*
B1-2	**0056: A Self-adaptive Gamma Voltage Regulation Circuit for AMOLED Displays**
14:00 ~14:15	Zhifeng Mao, Fei Gou, Bin Sheng, Jing Xie, Wenwei Xu, Wei Liu, Jun Xu *(Glenfly Tech Co., Ltd., China; Tsinghua University, China)*
B1-3	**0131: A Reconfigurable Thermoelectric Energy Harvesting Interface Based on OPDC and DSCT**
14:15 ~14:30	Peiyuan Fu, Jiabin Wang, Xufeng Liao, Lianxi Liu *(Xidian University, China; Xidian University Chongqing ICs Innovation Institute, China)*
B1-4	**0325: A Fixed-Peak-Current Single-Inductor-MultipleOutput DC-DC Converter Achieving 92.6% Peak Efficiency**
14:30 ~14:45	Fei Liu, Langyuan Wang, Shuyu Zhang, Hanlu Zhang, Na Yan *(Fudan University, China; Common Mode (GONGMO) Semiconductor Co., Ltd., China)*
B1-5	**0211: Buck-Boost Converter with Stable Transition Mode for Low Average Inductor Current**
14:45 ~15:00	Ningning Li, Yibo Zhang, Yushen Zhang, Yizhe Yang, Wenhao Yang, Yimeng Zhang, Yuming Zhang *(Xidian University, China)*

Wednesday, October 23, 13: 30 – 15: 15   Meeting Room 3
Session C1: EDA I   Sheraton Zhuhai Hotel 2nd Floor
Session Chair: Prof. Qi Wang, Zhuhai Fudan Innovation Institute, China

	Title
C1-1	**0281: Deep Learning Design-Flow with Static and Dynamic Optimizations (invited)**
13:30 ~14:00	Zhiqiang Que, Jose G. F. Coutinho, Wayne Luk *(Imperial College London, UK)*
C1-2	**0476:A New Era for Ai Processor Design Methodology with High Level Synthesis (invited)**
14:00	Yuan Li *(XDL Technologies Inc., China)*

~14:30	
C1-3	**0270: A QEMU-Based Virtual Platform of MPSoC**
14:30 ~14:45	Liangquan Qiao, Lei Li, Xingyu Gao, Jinxiang Wang, Fangfa Fu, Keli Long, Jinghan Zhou *(Harbin Institute of Technology, China; 58th Research Institute of China Electronics Technology Group Corporation, China)*
C1-4	**0193: A Parallel Harmonic Balance Method Based on GPU for Efficient Periodic Steady-State Analysis**
14:45 ~15:00	Zhengzhuo Wang, Yanliang Sha, Lingyun Ouyang, Quan Chen, Jianguo Hu *(Sun Yat-sen University, China; Southern University of Science and Technology, China)*
C1-5	**0245: Efficient Dynamic Memory Management for High Bandwidth Memory on FPGA**
15:00 ~15:15	Yuwei Qu, Yiqing Mao, Wenbo Yin, Lingli Wang *(Fudan University, China)*

Wednesday, October 23, 13: 30 – 15: 15   Meeting Room 4
Session D1: Novel Device I   Sheraton Zhuhai Hotel 2nd Floor
Session Chair: Prof. Zhengji Xu, Sun Yat-sen University, China

	Title
D1-1	**0312: Back End of Line (BEOL) Devices Using IGZO and P-type Oxides(Invited)**
13:30 ~14:00	John Robertson, Xuewei Zhang, Qingzhong Gui, Yuzheng Guo *(Cambridge University, UK; Wuhan University, China)*
D1-2	**0478: Miniaturization of High-speed GaN Based Laser Diodes(Invited)**
14:00 ~14:30	Junfei Wang, Chaowen Guan, Leihao Sun, Zhichong Wang, Chao Shen (Fudan University, China)
D1-3	**0122: Impact of Interfacial Layer on the Optoelectronic Performance of MoTe$_2$/Ge Heterojunction**
14:30 ~14:45	Wenyu Lei, Xiaokun Wen, Boyuan Di, Xinyue Xu, Haixin Chang, Wenfeng Zhang *(Huazhong University of Science and Technology, China)*
D1-4	**0098: MoS$_2$-WS$_2$ Heterostructure-enabled Optoelectronic Synaptic Diode**
14:45 ~15:00	Mingjie Li, Yingtao Ding, Jianzhi Hu, Hankun Zhao, Yilin Sun *(Beijing Institute of Technology, China)*
D1-5	**0472: Pseudo-Parallel Symmetrical and Crossed Perovskite Solar Cells for Bifacial Applications**
15:00	Guang-Wei Huang, Hsing-Mao Cheng, Jyi-Tsong Lin *(Sun Yat-Sen University,*

~15:15	*Taiwan, China)*

Wednesday, October 23, 13: 30 – 15: 15   Meeting Room 5
Session E1: Power Device I   Sheraton Zhuhai Hotel 2nd Floor
Session Chair: Prof. Shaofeng Yu, Fudan University, China

	Title
E1-1	**0282: Gate Driver ICs for Wide Bandgap Power Transistors (invited)**
13:30 ~14:00	Wai Tung Ng, Rophina Li, Wentao Cui, Jingyuan Liang *(University of Toronto, Canada)*
E1-2	**0291: Suppression of Back-Gating Effect by Integrated Substrate Termination Network for 200V Monolithic GaN Half-Bridge Power IC**
14:00 ~14:15	Mengyao Zhao, Yifei Zheng, Yanfeng Ma, Yuan Sun, Denggui Wang, Chuanqi Pan, Jianjun Zhou, Sheng Li, Siyang Liu, Long Zhang, Weifeng Sun *(Southeast University, China)*
E1-3	**0326: High Short-Circuit Capability and Low-Loss SOI-LIGBT with Double-Integrated NMOS**
14:15 ~14:30	Jialei Tan, Jie Wei, Jinlong Lu, Xindi Liu, Gaoqiang Deng, Wei Song, Pei Guo, Bo Zhang, Xiaorong Luo *(University of Electronic Science and Technology of China, China; Chengdu University of Information Technology, China)*
E1-4	**0328: Body Diode Degradation Mechanism Of 1200V SIC Power MOSFETs Under Gamma Rays Total Ionizing Dose Irradiation**
14:30 ~14:45	Yu Tian, Zhaoxu Song, Hao Fu, Jiaxing Wei, Siyang Liu, Weifeng Sun *(Southeast University, China)*
E1-5	**0361: Novel Heterojunction Field Plate β-Ga2O3 MOSFET with High Breakdown Voltage**
14:45 ~15:00	Xiangnan Li, Jie Wei, Kai Zhao, Linyao Hao, Xiaosong Peng, Yuxi Wei, Renkuan Liu, Wei Song, Pei Guo, Xiaorong Luo *(University of Electronic Science and Technology of China, China; Chengdu University of Information Technology, China)*

Wednesday, October 23, 13: 30 – 15: 15   Meeting Room 6
Session F1: Memory Device I   Sheraton Zhuhai Hotel 2nd Floor
Session Chair: Prof. Jian Huang, Sun Yat-sen University, China

	Title
F1-1	**0010: Sub-nanosecond Operation Speeds of Ferroelectric Domain Wall Memory (Invited)**
13:30 ~14:00	Anquan Jiang *(Fudan University, China)*
F1-2	**0096: Optimizing Flash Memory Towards Storage-Class Memory (SCM) Applications (Invited)**
14:00 ~14:30	Xinyi Guo, Yang Feng, Jing Liu, Junyu Zhang, Xuepeng Zhan, Jixuan Wu, Jiezhi Chen *(Shandong University, China; Institute of Microelectronics of Chinese Academy of Sciences, China; Neumem Co., Ltd, China)*
F1-3	**0020: Investigation of Reliability Characteristics of $Hf_xZr_{1-x}O_2$-Based FeFET and AFeFET Non-Volatile Memory**
14:30 ~14:45	Min Liao, Xianzhou Shao, Junshuai Chai, Xiaoqing Sun, Xiaoyu Ke, Hao Xu, Jinjuan Xiang, Xiaolei Wang, and Wenwu Wang *(Institute of Microelectronics, Chinese Academy of Sciences, China; Beijing Superstring Academy of Memory Technology, China)*
F1-4	**0066: Deep Understanding of Charge Trapping Phenomenon in n-FeFET and Endurance Improvement by Interlayer Engineering**
14:45 ~15:00	Saifei Dai, Hao Xu, Fengbin Tian, Xianzhou Shao, Xiaoqing Sun, Junshuai Chai, Xiaolei Wang, Wenwu Wang *(Institute of Microelectronics, Chinese Academy of Sciences, China; University of Chinese Academy of Sciences, China)*
F1-5	**0407: An FPGA-based Dual-mode SSD for Device-side Performance Optimization**
15:00 ~15:15	Xingyu Chen, Sirui Peng, Hankun Lv, Zhangbin Yang, Daixiao Peng, Xi Cai, Xueguang Lian, Yong Ding, Xiaoyong Xue *(Fudan University, China; University of Chinese Academy of Sciences, China; Institute of Electrical Engineering, Chinese Academy of Sciences, China; China Three Gorges Construction Engineering Corporation, China; Zhejiang University, China)*

Wednesday, October 23, 15: 30-17: 15

Wednesday, October 23, 15: 30-17: 15	Meeting Room 2
Session B2: Analog Circuit II	Sheraton Zhuhai Hotel 2nd Floor
Session Chair: Prof. Wenning Jiang, Fudan University, China	

	Title
B2-1	**0152: Dual-Loop Reference-less CDR with HLD for Wide Lock-in Range (invited)**

15:30 ~16:00	Chua-Chin Wang *(Sun Yat-Sen University, Taiwan, China)*
B2-2	**0089: A SET Sensitive Model of LC and Ring Voltage Controlled Oscillator in FinFET Technology**
16:00 ~16:15	Liu Heyuan, Yuan Hengzhou, Lu Jianzhuang, Chen Xiaowen, Sang Hao, Liu Jingtian, Guo Yang *(National University of Defense Technology, China; Academy of Military Sciences PLA China, China)*
B2-3	**0093: A Low Spur Wideband PLL in 65nm CMOS**
16:15 ~16:30	Zijun Wang, Biao Li, Teng Wang, Hong Li, Ruiting Niu, Jinpeng Lin *(Space Star Technology Limited Corporation, China)*
B2-4	**0217: A Low Power PLL Circuit with Signal 50% Duty Cycle Corrected in 180nm CMOS**
16:30 ~16:45	Bangtian Li, Xueke Li, Liying Chen, Chuantong Cheng, Jian Mei *(Tiangong University, China; Institute of Semiconductors, Chinese Academy of Sciences, China)*
B2-5	**0156: MTJ based Compensation for Charge Pump Temperature Drift**
16:45 ~17:00	Yongliang Zhou, Jingxue Zhong, Chengxing Dai, Yingxue Sun, Xin Li, Chunyu Peng *(Anhui University, China; Anhui Anxin Electronic Technology Co., Ltd, China）*
B2-6	**0213: A 112-Gb/s Coherent Receiver with a Novel Modulation Format**
17:00 ~17:15	Tianyuan Zhong, Boyang Zhang, Weixin Gai *(Peking University, China; Beijing Advanced Innovation Center for Integrated Circuits, China)*

Wednesday, October 23, 15: 30-17: 15	**Meeting Room 3**
Session C2: EDA II	Sheraton Zhuhai Hotel 2nd Floor
Session Chair: Prof. Qi Wang, Zhuhai Fudan Innovation Institute, China	

	Title
C2-1	**0292:An Improved Clock-Aware Global Placement Algorithm (invited)**
15:30 ~15:55	Ziang Ge, Pingqiang Zhou *(Shanghaitech University, China)*
C2-2	**0411: Analyzing Timing in Shorter Time: A Journey through Heterogeneous Parallelism for Static Timing Analysis (invited)**
15:55 ~16:20	Zizheng Guo, Yibo Lin, Runsheng Wang, Ru Huang *(Peking University, China)*
C2-3	**0216: TBPart-b: An Effective Hypergraph Partitioning Algorithm Considering Topological Order Balance for Processor-based Hardware Emulation**

16:20 ~16:34	Jing Tang, Shunyang Bi, Hailong You *(Xidian University, China)*
C2-4	**0068: FCE: A Fast CGRA Architecture Exploration Framework**
16:34 ~16:48	Sichao Chen, Yiqing Mao, Yuan Dai, Xuchen Gao, Wai-Shing Luk, Wenbo Yin, Lingli Wang *(Fudan University, China)*
C2-5	**0179: Research on Parametric Subthreshold Cell Delay Modeling Based on ANN**
16:48 ~17:02	Xuelian Zhang, Yuping Wu, Zhiqiang Li, Donglin Liu, Shushan Qiao *(Institute of Microelectronics of Chinese Academy of Sciences, China; University of Chinese Academy of Sciences, China)*
C2-6	**0393: A High-Performance Routing Architecture with 16 LUTs per CLB for Nanoscale FPGAs**
17:02 ~17:15	Sijing Yang, Jide Zhang, Hao Zhou, Lingli Wang *(Fudan University, China)*

Wednesday, October 23, 15: 30-17: 15   Meeting Room 4
Session D2: Novel Device II   Sheraton Zhuhai Hotel 2nd Floor
Session Chair: Prof. Zhengji Xu, Sun Yat-sen University, China

	Title
D2-1	**0450: Cryogenic and RF Modeling of On-Chip Passive Devices for Quantum Computer (Invited)**
15:30 ~16:00	Akira Tsuchiya *(The University of Shiga University, Japan)*
D2-2	**0457: Ferroelectric Transistors Based on Two Dimensional Materials (Invited)**
16:00 ~16:30	Wenwu Li *(Fudan university , China)*
D2-3	**0473: Comparison of Nanosheet and Fin Integration in Stacked Induced Tunnel Field-Effect Transistors**
16:30 ~16:45	Ruei-Cheng Tu, Chia-Yo Kuo, Jyi-Tsong Lin *(Sun Yat-Sen University, Taiwan, China)*
D2-4	**0160: Nonlinear Contact Behavior in MoS_2 Field Effect Transistors at Cryogenic Temperature**
16:45 ~17:00	Shihab Ahammed, Mansun Chan *(The Hong Kong University of Science and Technology, Hong Kong, China)*
D2-5	**0158: Experimental Verification of 1D Transport Model by Quantized Current**

	Spectrum of Si JNT Device
17:00 ~17:15	Zi-Meng Shang, Bo-Wei Wang, Wei-Hua Han *(Institute of Semiconductors,Chinese Academy of Sciences, China; University of Chinese Academy of Sciences, China)*

Wednesday, October 23, 15: 30-17: 15   Meeting Room 5
Session E2: Power Device II   Sheraton Zhuhai Hotel 2nd Floor
Session Chair: Prof. Shaofeng Yu, Fudan University, China

	Title
E2-1	**0401: Investigation of SiON Passivation for High Performance AlGaN/GaN HEMTs (invited)**
15:30 ~16:00	Difei Fan, Chenkai Deng, Jiming Zhang, Peiran Wang, Nick Tao, Qing Wang, Hongyu Yu *(Southern University of Science and Technology, China; Maxscend Microelectronics Company Limited, China)*
E2-2	**0009: 180nm BCD Technology Platform with 8V to 65V Isolated LDMOS**
16:00 ~16:15	Qi Ding, Renxiong Li, Ning Ning, Jun Huang, Yutuo Guo, Yu Wang, Kunqin He, Yaxin Li, Huaishan Wang, Juan Tang, Qiuyue Huo, Minghong Yuan, Pan Peng, Ming Qiao, Lulu Peng, Bo Zhang *(United Microelectronics Center Co., Ltd, China; University of Electronic Science and Technology of China, China;)*
E2-3	**0200: A Novel Insulated Gate-Triggered Thyristor with Integrated Super-Clamp Gate Transient Voltage Suppressor for Ultrahigh di/dt Pulse Switching**
16:15 ~16:30	Shiyu Deng, Yuxiao Yang, Xinqi Sun, Pengwei Zhou, Ruize Sun, Chao Liu, Wanjun Chen, Bo Zhang *(University of Electronic Science and Technology of China, China)*
E2-4	**0210: Device Instability in the Third Quadrant of Schottky-Type p-GaN Gate HEMTs: The Hole Defficiency & Trapping Effect**
16:30 ~16:45	Kuangli Chen, Shuting Huang, Jinggui Zhou, Ning Yang, Jianggen Zhu, Enchuan Duan, Bo Zhang, Qi Zhou *(University of Electronic Science and Technology of China (UESTC), China)*
E2-5	**0236: Static Characteristic Recovery Of SiC MOSFETs Induced By Dynamic Gate Stress After Total Ionizing Dose Irradiation**
16:45 ~17:00	Jiahao Hu, Xiaochuan Deng, Xing Zeng, Tao Xu, Haibo Wu, Xuan Li, Bo Zhang *(University of Electronic Science and Technology of China, China)*

Wednesday, October 23, 15: 30-17: 15   Meeting Room 6
Session F2: Memory Device II   Sheraton Zhuhai Hotel 2nd Floor
Session Chair: Prof. Jian Huang, Sun Yat-sen University, China

	Title
F2-1	**0141: A Simulation Comparison of Channel-All-Around and Gate-All-Around 3D Vertical Structure FeFET with IGZO Channel**
15:30 ~15:45	Xuebin Wang, Zhijian Guo, Yutao Li, Chengji Jin, Jixuan Wu, Guanhua Yang, Yuanxiao Ma, Masaharu Kobayashi, Fei Mo, Yeliang Wang *(Beijing Institute of Technology, China; The University of Tokyo, Japan; Institute of Microelectronics, Chinese Academy of Sciences, China; Shandong University, China; Xidian University, China)*
F2-2	**0174: Low Operating Voltage in HfO₂/ZrO₂ Superlattice Ferroelectric Capacitor Achieved by Thickness Scaling**
15:45 ~16:00	Dongya Li, Huan Liu, Peiyuan Du, Fei Yu, Chengji Jin, Xiao Yu, Yan Liu, Genquan Han, Yue Hao *(Xidian University, China; Zhejiang Lab, China; Hangzhou Institute of Technology, Xidian University, China)*
F2-3	**0257: Co-optimization of Oxide Semiconductor-based Ferroelectric Transistors Between Electrical Performance and Ambient Stability By Using TiO₂-IGZO Dual-Channel Layers**
16:00 ~16:15	Shangze Li, Xujin Song, Dijiang Sun, Xiaoyan Liu, Jinfeng Kang *(Peking University, China)*
F2-4	**0340: Enhancing Computational Precision in PLRAM-based In-memory Computing with High-Low Bit Concatenation**
16:15 ~16:30	Saike Zhu, Xiang Qiu, Yong Gong, Cimang Lu, Yi Zhao *(Zhejiang University, China; China Nanhu Academy of Electronics and Information Technology, China; East China Normal University, China; Flash Billion Semiconductor Co. Ltd., China)*
F2-5	**0062: FeFET based Logic-in-Memory Pipeline-Style Circuits**
16:30 ~16:45	Yang Li, Zhaohui Yang, Yinshui Xia *(Ningbo University, China)*
F2-6	**0172: Study of Vth Degradation Mechanism in FeFET with TiN/Al₂O₃/HfO₂/Al₂O₃/Hf₀.₅Zr₀.₅O₂/SiOₓ/Si Structure**
16:45 ~17:00	Runhao Han, Jia Yang, Tao Hu, Mingkai Bai, Yajing Ding, Xianzhou Shao, Saifei Dai, Xiaoqing Sun, Junshuai Chai, Hao Xu, Kai Han, Xiaolei Wang, Wenwu Wang, Tianchun Ye *(Institute of Microelectronics, Chinese Academy of Sciences, China; University of Chinese Academy of Sciences, China; Weifang University, China)*
F2-7	**0189: Random Number Generation from 3D-NAND Flash Memory Using Shallow Charge Trap Related Short-Term Retention Errors**
17:00 ~17:15	Ruibin Zhou, Jian Huang, Xianping Liu, Yuhan Wang, Xinrui Zhang, Yungen Peng, and Zhiyi Yu *(Sun Yat-sen University, China; Peng Cheng Laboratory, China)*

Wednesday, October 23, 17：30 – 18：30

Wednesday, October 23, 17: 30 –18: 30
Poster Session I   Sheraton Zhuhai Hotel 1st Floor

	Title
P1-1	**0007: A Fast-Response Current Source with High Impedance for Zero-Crossing-Based Circuits**
	Ruoyu Li, Xianglong Wang, Jianqiang Xu, Yintang Yang *(Xidian University, China）*
P1-2	**0028: Design of 12-bit Low-Power Single-Slope ADC with 2048 Columns for Infrared Focal Plane Array**
	Lixiang Han, Hao Li, Yihui Zhang, Liang Gao, Dongsheng Liu *(Huazhong University of Science and Technology, China; Hubei Optics Valley Laborator, China;)*
P1-3	**0040: A Low Power Narrow-Band Complex-Bandpass Filter Based on Feedforward Compensation Amplifiers for NB-IoT Applications**
	Xu Zhao, Ziqiang Wang, Jie Gan *(Beijing Smart-chip Microelectronics Technology Co.,Ltd, China; Tsinghua University ,China)*
P1-4	**0069: A 12V to 1V Tri-state DSD Hybrid Converter by Self-Balanced Dual Flying Capacitors with 0.3mV Output Ripple and 90.09% Peak Efficiency**
	Yixing Wang, Qianhui Liu, Yuhua Chen, Yizhe Yang, Yimeng Zhang, Yuming Zhang *(Xidian University, China;)*
P1-5	**0077: A Multi-Phase Clock Self-Calibrating Circuit**
	Zhihuai Li, Li Jiang, Xueming Wei, Zilu Cai, Jiami Tang *(Guilin University of Electronic , China)*
P1-6	**0090: A 18V, 600mA Load Current, 22MHz High-Voltage Power Amplifier with Over-Temperature Protection and Bidirection Enable Logic**
	Yuan Ren , Xin'an Wang *(Peking University, China)*
P1-7	**0153: A 10-MHz Four-Phase Hysteretic Control DC-DC Converter with Inductor Current Self-balancing**
	Yushen Zhang, Yibo Zhang, Yizhe Yang, Ningning Li, Wenhao Yang, Yimeng Zhang , Yuming Zhang *(Xidian University, China)*
P1-8	**0227: A High Precision -40 °C to 150 °C Bandgap Reference with Dual Temperature Compensation**

		Yuhan Zhang, Jianzheng Li, Xiaomeng An, Lina Wang, Yajie Qin *(Fudan University, China)*
P1-9	**0241: A Biphasic Neural Stimulator with Adaptive Pulse-Width Modulation Charge Balancer**	
	Hailong Tang, Wenxian Gu, Yifan Song, Hengchang Bi, Xing Wu, Liangjian Lyu *(East China Normal University, China)*	
P1-10	**0250: A PPG Analog Front-End With PVT-Insensitive High-Pass Frequency**	
	Zhaofeng Huang, Zepeng Huang, Hengchang Bi, Xing Wu, Liangjian Lyu *(East China Normal University, China)*	
P1-11	**0330: A Fully integrated FVF based low-noise voltage buffer for ADC reference**	
	Ikhwan Kim, Yajie Qin *(Fudan University, China)*	
P1-12	**0387: A Resistor-Free Grounded High-Frequency Memristor Emulator**	
	Xinying Su, Bingjun Xiong, Junjie Yu and Jingjing Liu *(Sun Yat-Sen University, China)*	
P1-13	**0027: An Ultra-Low-Leakage Current Sensing Interface for Wide Temperature Range**	
	Jinsheng Tang, Chun Zhao, Lin He *(Nanjing University of Posts and Telecommunications, China)*	
P1-14	**0029: A Global Threshold Voltage Finder Technology for the Readout Circuit of Event-based Vision Sensor**	
	Yanwen Su, Hao Li, Dongsheng Liu, Ang Hu, Kaiyue Li *(Huazhong University of Science and Technology, China; Hubei Optics Valley Laboratory, China)*	
P1-15	**0107: A Residue Amplifier With 72.27 dB Loop-Gain and 4.64 GHz Closed Loop Bandwidth consuming 6.4 mW for 12-Bit 1-Gsps Pipelined ADC**	
	Jiangbo Wei, Jin Liu, Wei Tian, Chao Wang, Maliang Liu *(Xi'an Microelectronics Technology Institute, China; Xidian University, China)*	
P1-16	**0124: A 180 mV–1.6 V Thermoelectric Energy Harvesting Converter with Low-Voltage Cold Start and Less than 1 μW Power Loss**	
	Chunlin Wang, Anzhi Yan, Tianyu Guo, Peng Wan, Houfang Liu, Yi Yang, Tianling Ren *(Tsinghua University, China)*	
P1-17	**0150: A Signal Conditioning ASIC With High Precision and Low Noise for MEMS Accelerometers**	
	Quan Sun, Rui Liu, Zhe Zheng, Lei Dong, Ji-jiang Wang *(Xi 'an Aerosemi Technology. Co., Ltd., China; Beijing Smart-chip Microelectronics Technology Co., China)*	

P1-18	**0185: Design of a high-precision self-calibration readout circuit for CMOS microbolometer**
	Qianhao Zhang, Jie Liu, Sheng Xu, Yiming Liao, Feng Yan, Xiaoli Ji *(Nanjing University, China; Nanjing University of Science and Technology, China)*
P1-19	**0297: A Smooth Two-Stage Soft Start Method for Current Mode Boost Converter**
	Yue Shi, Shi-dong Wang, Zekun Zhou, Bo Zhang, Zhigang Qin *(University of Electronic Science and Technology of China, China; Chengdu University of Information Technology, China; Saitama Institute of Technology, Japan)*
P1-20	**0327: A High-Voltage Smooth Self-Starting Reference Current Source Circuit**
	Dongyan Zhao, Jie Pan, Chenghao Zhang, Yidong Yuan, Yi Hu, Hongwei Shen，Zekun Zhou, Member, IEEE*(Beijing Smart-chip Microelectronics Technology Co., Ltd., China; University of Electronic Science and Technology of China, China)*
P1-21	**0276: A Temporal and Spatial Reuse Interpolation Hardware for VVC Motion Compensation**
	Huanxiang He, Shushi Chen, Leilei Huang, Yibo Fan *(Fudan University, China)*
P1-22	**0293: A Broadband Digital Beamforming Method Based on FPGA**
	Yiwen Tang, Guowen Jia, Zhen Zhang and Yue Zhang *(Sun Yat-Sen University, China)*
P1-23	**0305: Conditional cycle termination RANSAC**
	Tong Jiang, Yujie Huang, Liyuan Peng, Mingyu Wang, Wenhong Li,Minge Jing, Xiaoyang Zeng *(Fudan University, China; Shanghai ExploreX Technology Co., Ltd., China)*
P1-24	**0371: A multi-resolution propagation algorithm and pixel grouping storage strategy for PatchMatch Stereo**
	Kai Liu, Zhenyu Zhang, Haiwei Wang, Leilei Huang, Chunqi Shi, Long Xu, Runxi Zhang *(East China Normal University, China)*
P1-25	**0378: An XOR Arbiter PUF based on the IGZO TFT Devices**
	Xiang Chen, Yongliang Chen, Xiaole Cui *(Peking University Shenzhen Graduate School, China)*
P1-26	**0052: Design and Implementation of Hierarchical Storage Structure for MCCSIP-RAA**
	Longmei Nan, Yu Jin, Yiran Du, Tao Chen, Yanjiang Liu, Wei Li *(Institute of Information Science and Technology, China)*
P1-27	**0242: A Highly Scalable Hardware HEVC Encoder Based on FPGA**

	Guohao Xu, Chenlong He, Shiyan Yi, Leilei Huang, Xiaoyang Zeng, Yibo Fan *(Fudan University, China; East China Normal University, China)*
P1-28	**0259: A Hardware-friendly Fast Block Partition Decision Algorithm Based on Histogram of Oriented Gradient for AV1**
	Guohao Xu, Shiyan Yi, Zhijian Hao, Leilei Huang, Hao Zhang, Xiaoyang Zeng, Yibo Fan *(Fudan University, China; East China Normal University, China)*
P1-30	**0391: A High-Performance MTJ-LUT Circuit Using 4T1M Architecture**
	Yu Pan, Yuejun Zhang, Shuaicheng Guo, Yuanxin Tian, Bo Hong, Rui Fang, Liang Wen *(Ningbo University, China; China Coast Guard Academy, China)*
P1-31	**0261: Optimizing Communication Effciency of GNN Inference in Distributed System**
	Wenqian Zhou, Qiaosha Zou *(Fudan University, China; Zhejiang Lab, China)*
P1-32	**0295: SST: Simplified Space-Time Transformer based on Time-assisted Spatial MSA for 3D Human Pose Estimation**
	Sheng Lu, Qiyun Dong, Zhenyin Zhang, Gengsheng Chen, Yinna Zhu, Wei Xu *(Fudan University, China; Jiashan Fudan Institute, China)*
P1-33	**0173: SALTS: An Efficient and Flexible Self-Attention Accelerator with Long Token Support on FPGA**
	Kaiqi Chen, Xinhua Shi, Jun Han *(Fudan University, China)*
P1-34	**0222: RISC-V Neural Network Instruction Design and Simulation with Cache Scheduling via ROCC Interface**
	Siyao Dai, Zikang Zhou, Jun Han *(Fudan University, China)*
P1-35	**0226: Impact of external magnetic interference on the performance of MRAM-based neuromorphic computing**
	Yingtong He, Suihuan An, Yu Chen, Xue Zhou, Xihui Yuan, Weidong Zhang, Zheng Chai, Tai Min *(Xi'an Jiaotong University, China; Liverpool John Moores University, UK)*
P1-36	**0375: A Hardware Accelerator for Image Super Resolution with Algorithm Lightweighting and Custom Fusion Engine**
	Menghan Li, Sheng Lu, Jun Han *(Fudan University, China)*
P1-37	**0268: Hardware Implementation of High Speed Fault Tolerant Parallel Accelerator**
	Wenzhe Ma, Wenzhe Ma *(Fudan University, China)*

P1-38	0349: Composite Filter-based Bicubic Interpolation Method and FPGA Implementation
	Li Zhang, Jingjing Liu, Yujie Zhu, Jianhua Zhang *(Shanghai University, China)*
P1-39	0157: MTJ based Temperature Tracking Read/Write Assist for High Speed SRAM Bitcell
	Yongliang Zhou, Chengxing Dai, Jingxue Zhong, Yingxue Sun, Xin Li, Chunyu Peng *(Anhui University, China; Anhui Anxin Electronic Technology Co., Ltd, China)*
P1-40	0353: System-level Evaluation of AOS Gain Cell eDRAMs for Low-power Normally-off Computing
	Long Chen, Yecheng Yang, Wei Li, and Shao Hao Wang *(Fuzhou University, China)*
P1-41	0024: A High Sigma Monte Carlo Analysis Solution Via Machine Learning for SRAM Margin Signoff
	Amy Rao *(EBA Center, China)*
P1-42	0138: Enhanced Multi-bit Computation using CIM SRAM Technology
	Ruiyong Zhao, Yibo Hu, Zhipeng Ren, Yizhe Yin, Jing Chen *(Shanghai Institute of Microsystem and Information Technology, China)*
P1-43	0045: A Compute-in-Memory Macro Based on Complementary 2T2C FeRAM Cell for BNNs
	Jinyu Li, Mingzhang Xie, Shisheng Xiong *(Fudan University, China; China Resources Microelectronics Co., Ltd., China)*
P1-44	0057: A Novel High Speed Low Power Differential Circuit-Based FRAM Read Scheme
	Qiuyu Tao, Jiabao Ye, Xuecheng Cui, Nan Jiang, Jiangtao Cao, Xibo Chen, Jiuren Zhou, Bing Chen, Genquan Han *(Zhejiang University, China; Xidian University, China)*
P1-45	0023: A 13-bit,1 MS/s Cyclic ADC, for high-speed CMOS Image sensor
	Qi Lv, Rensheng Shen, Yu Cheng, Guoqiang Zhong, Yang Qu, Yuchun *(Dalian University of Technology, China)*
P1-46	0110: An Area-Efficient 16-bit Four-channel R-2R DAC Based on Switching On-resistance Adaptive Calibration Technique
	Kejun Wu, Yuchen Liu, Yuhan Hu, Yu He, Zhen Yu, Ning Ning *(University of Electronic Science and Technology of China, China)*
P1-47	0215: A Background Calibration Method of Bandwidth Mismatch for Time-Interleaved ADCs Based on Neural Network

	Tianqi Yang, Longsheng Wang, Xin Zhao, Shubin Liu, Dengquan Li, Zhangming Zhu *(Xidian University, China)*
P1-48	**0352: A Second-Order Dual-Charge-Pump Passive Noise Shaping SAR ADC for Medical Implant Devices**
	Kangkang Sun, Xuanxin Ke, Haoning Sun, Yuchen Wang, Feng Yan, Jingjing Liu *(Sun Yat-Sen University, China)*
P1-49	**0084: A 114.4-dB DR, 26-kHz BW Discrete-Time Incremental Zoom ADC**
	Yuanhong Ding, Longjiang Jia, Jian Mei, Lei Deng, Rui Yin *(Fudan University,China; National Integrated Circuit Innovation Center, China; Jiashan Fudan Institute, China)*
P1-50	**0087: A High-Resolution Low-Power Extended-Range Incremental ΣΔ ADC For Battery Management System**
	Long Zhang, Quan Sun, Rui Liu, Zhe Zheng, Jingjing Zhang, Haitao Liu *(Xi'an Aerosemi Technology Company Ltd., China; Beijing Smart-Chip Microelectronics Technology Company Ltd., China)*
P1-51	**0155: An Infrared AFE Chip and System with Non Invasive Blood Glucose Detection Output**
	Bin Li, Jiyuan Guo, Chengzhen Xie, Jian Mei, Lei Deng, Rui Yin *(Fudan University,China; National Integrated Circuit Innovation Center, China; Jiashan Fudan Institute, China)*
P1-52	**0183: A 12-Bit 8Gs/s Time-Interleaved Pipeline-SAR ADC with Calibration** Jie Pu, Jinda Yang, Jianwen Li, Rong Han, Xing Zhu, Lei Chen *(Chengdu Sino Microelectronics Technology Co., Ltd., China)*
P1-53	**0238: An Ultra-High Frame Rate ROIC for Hyperspectral Detection**
	Angyang Li, Ningning Li, Jian Mei, Lei Deng, Rui Yin *(National Integrated Circuit Innovation Center, China; Jiashan Fudan Institute, China; Fudan University, China)*
P1-54	**0273: A Background Digital Calibration Method for DTCs Used in Digital PLL Employing Dual-Path DTC**
	Renxuan Li, Xiaoyu Shan, Li Wang, Ang Hu, Dongsheng Liu *(Huazhong University of Science and Technology, China)*
P1-55	**0310: A High-Precision Sigma-Delta ADC for Battery Management System**
	Hao Xue, Liji Wu, Jing Hu, Zhiwei Li, Xiangmin Zhang *(Heilongjiang University, China: Tsinghua University, China: Beijing National Research Center for Information Science and Technology, China)*

P1-56	0413: Multi-Sampling Mode CDAC Design for a 12-bit 200MS/s Pipelined-SAR ADC
	Tianyu Zhang, Fan Ye, Shunli Ma *(Fudan University, China)*
P1-57	0356: DSP-PUF: A Software PUF Based on Digital Signal Processor for IoT Security
	Tengfei Yuan, Pengjun Wang, Yuejun Zhang, Mingze Ren, Shuang Hu *(Ningbo University, China; Wenzhou University, China)*
P1-58	0042: A Q/V Band 49.6-54.5GHz,3.53dB NF,45dB Gain,2.09° Phase Error,2-Way Phased-Array Receiver for Satellite Application
	Congrui Li, Qi Zhao, Ruolan Chen, Shulan Chen, Yan Wang, Lei Zhang *(Tsinghua University, China)*
P1-59	0265: A Fractional-N SPLL Using Space-time Averaging and Phase Interpolator for Quantization Noise Reduction
	Shengxiang Liu, Ke Sun, Chengyu Yang, Dongsheng Liu, Ang Hu *(University of Science and Technology, China)*
P1-60	0342: A 47 µW Wake-Up Receiver With -77dBm Sensitivity Using a Mixer-First Architecture
	Weitao He, Yaxin Zeng, Bin Jia, Hao Min, Hao Xu, Na Yan *(Fudan University, China; EPIC MEMS Corporation, China)*
P1-61	0346: A Ka-Band CMOS Broadband Power Amplifier with 35.3% PAE for SATCOM Applications
	Zhiqing Liu, Yu Chu, Yuting Sun *(Southwest China Institute of Electronic Technology, China)*
P1-62	0348: RF Front-End Chip Design for Ku-Band with 130nm CMOS Technology
	Huiquan Xie, Ziyu Wang, Tianrui Wang, Yifei Chen, Maliang Liu, Yintang Yang *(Xidian University, China)*
P1-63	0394: Back-gate Bias Assisting VCRO Design
	Chenglin Ye, Zheng Zhou, Xiaoyan Liu *(Peking University, China)*
P1-64	0034: A 3.2-to-7.1GHz Quad-Core Dual-Mode Oscillator Achieving 193.6 dBc/Hz Peak FoM
	Xiaoyu Shan, Renxuan Li, Mengming Zhang, Ang Hu, Dongsheng Liu *(Huazhong University of Science and Technology, China)*
P1-65	0301: A 20.6 to 30.5 GHz Two Stage Cascode LNA in 40nm CMOS for Phase Array Tranceiver

	Lei wang, Kefeng Han, Hao Xu, Rui Yin, Na Yan *(Fudan University, China; Jiashan Fudan Institute, China)*
P1-66	**0338: A 12-32 GHz Power Amplifier with 32-dBm Psat and 25% PAE in 0.15 μ m GaN**
	Xiangran Ni, Chunyue Bo, Tianyu Li, Qingyang Dong, Xin Jiang,Weijun Luo *(University of Chinese Academy of Sciences, China; Institute of Microelectronics of Chinese Academy of Sciences, China)*
P1-67	**0427: A Source-Driven Push-push Doubler with Wideband 2nd Harmonic Feedback**
	Yuyang Chen, Ao Zhang , Jianjun Gao, Jianjun Zhou *(Shanghai Jiao Tong University, China; Nantong University, China; East China Normal University, China)*
P1-68	**0367: Low Power Processor For IoT Device**
	Jincheng Li, Jiyuan Bai, Zelin Wang, GengSheng Chen, Xiaofang Zhou *(Fudan University, China; Jiashan Fudan Institute, China)*
P1-69	**0386: A Heterogeneous Integration System of Analog In Memory Computing and Field-Programmable Gate Array**
	Hua Chen, Yiming Qu, Wenhao Wu, Yi Zhao *(East China Normal University, China; China Nanhu Academy of Electronics and Information Technology, China; Zhejiang University, China)*

Thursday

Thursday, October 24, 9: 00 – 10: 30

Thursday, October 24, 9: 00 – 10: 30   Grand Ball Room
Keynote Session K3   Sheraton Zhuhai Hotel 1st Floor
Session Chair: Prof. Francois Rivet, University of Bordeaux, France

K3-1	Integrated Circuit Innovation in the Age of AI
9: 00 ~ 9: 45	Prof. Boris Murmann, University of Hawaii, USA
K3-2	**On-Chip ESD Protection: Methodologies, Challenges and Perspectives**
9: 45 ~10: 30	Prof. Albert Wang, University of California, Riverside, USA
	Coffee Break

Thursday, October 24, 10: 45 – 12: 15

Thursday, October 24, 10: 45 – 12: 15   Grand Ball Room
Panel Discussion   Sheraton Zhuhai Hotel 1st Floor

Session Chair: Prof. Jianguo Yang, Zhangjiang Laboratory, China
       Dr. Hailan Yi, Zhangjiang Laboratory, China

Opportunities and Challenges of Integrated Circuits in the AI Era

Tony Tae-Hyoung Kim *(Nanyang Technological University, Singapore)*, Xiaoyao Liang *(Shanghai Jiaotong University, China)*, Do Anh Tuan *(A*STAR, Singapore)*, Zhigang Ji *(Shanghai Jiaotong University, China)*, Chunmeng Dou *(University of Chinese Academy of Sciences, China)*

Thursday, October 24, 13: 30 – 15: 15

Thursday, October 24, 13: 30 – 15: 15   Meeting Room 1
Session A1: AI Circuit   Sheraton Zhuhai Hotel 2nd Floor
Session Chair: Prof. Yan Li, Fudan University, China

	Title
A1-1	**0468: Towards Efficient Computing Architecture and Chip For Embodied AI (invited)**
13:30 ~14:00	Hongbin Sun *(Xi'an Jiaotong University, China)*
A1-2	**0320: A High-Performance Multicore Testing Platform for Multi-Scenario Applications**
14:00 ~14:15	Zipeng Ling, Tianshu Zhuo, Zhuoyuan Yang, Jinhong Ye, JunHan, Jingtao Zhang *(State Key Laboratory of Integrated Chips and Systems, China; ZTE Corporation, China)*
A1-3	**0317: S-SIFT: A Simple SIFT Algorithm with High Efficiency**
14:15 ~14:30	Yixue Wang, Yujie Huang, Liyuan Peng, Mingyu Wang, Wenhong Li, Minge Jing, Xiaoyang Zeng *(Fudan University, China; Shanghai ExploreX Technology Co., Ltd., China)*
A1-4	**0345: Design of a High-Speed SAR Processor Based on the Chirp Scaling Algorithm**
14:30 ~14:45	Xianghe Cui, Yukun Song, Yurun Zhang, Jingyi Hu, Zhenmin Li, Duoli Zhang *(Hefei University of Technology, China)*
A1-5	**0196: Accelerating Matrix-Vector Multiplications of Large Language Models via Efficient Encoding**
14:45 ~15:00	Yongjin Tao, Wendi Sun, Song Chen, Yi Kang *(University of Science and Technology of China, China)*
A1-6	**0397: Flexible yet Efficient Transformer Acceleration with Unified Sparse Attention Support on FPGA**
15:00 ~15:15	Linfeng Zhong, Qingyu Guo, Runsheng Wang, Yuan Wang, Meng Li *(Peking University, China)*

Thursday, October 24, 13: 30 – 15: 15   Meeting Room 2
Session B3: Analog Circuit III   Sheraton Zhuhai Hotel 2nd Floor
Session Chair: Prof. Yubin Zhao, Sun Yat-sen University, China

	Title
B3-1	**0296: Beyond Bandwidth Trade-off: Simultaneous Wireless Power and Data Transfer System Design for Biomedical Implants (Invited)**
13:30 ~14:00	Quanrong Zhuang, Junyi Sun, Xusheng Zhang, Bo Li, Yi Shi, Hao Qiu *(Nanjing University, China)*
B3-2	**0011: A High Precision Operational Amplifier with Improved Bias Current Cancellation Circuit**
14:00 ~14:15	Zhili Zhang, Siyuan Yao, Hailong Wei *(Xi'an Microelectronics Technology Research Institute, China)*
B3-3	**0030: A 0.11-pJ/bit True Random Number Generator Based on a Clocked Current-Starved Inverter**
14:15 ~14:30	Kai Cheng, Chaowei Yang, Rui P. Martins, Pui-In Mak, Yong Chen *(University of Macau, Macao, China; Universidade de Lisboa, Portugal; Tsinghua University, China)*
B3-4	**0256: A Super-Mixed Current Decay Mode for Reducing the Angular Position Error in Stepper Motor**
14:30 ~14:45	Jian Fang, XuruiChen, Huajie Liu, Yuhan Jin *(University of Electronic Science and Technology of China, China)*
B3-5	**0065: A 109 dB 44-pArms Current Readout Circuit with Automatic Current Control for Multimodality Electrochemical Sensing**
14:45 ~15:00	Lina Wang, Jianzheng Li, Weiming Hu, Yajie Qin *(Fudan University, China)*
B3-6	**0080: A Low Temperature Coefficient Bandgap Reference For Temperature Sensor System**
15:00 ~15:15	Longjiang Jia, Yuanhong Ding, Jian Mei, Lei Deng, Rui Yin *(Fudan University, China; National Integrated Circuit Innovation Center, China; Jiashan Fudan Institute, China)*

Thursday, October 24, 13: 30 – 15: 15   Meeting Room 3
Session C3: RF Circuit I   Sheraton Zhuhai Hotel 2nd Floor
Session Chair: Prof. Gengzhen Qi, Sun Yat-sen University, China

	Title
C3-1	**0075: High-Efficiency Power Amplifier Design for Bluetooth Low Energy Applications (invited)**
13:30 ~14:00	Bharatha Kumar Thangarasu, Li Shuai, Yu Hongshi, Ge Wansi, Liu Yuqing, Nagarajan Mahalingam, Meng Fanyi, Kaixue Ma, Juin J. Liou, Bo Wang, Younan Hua, Xiaomin

	Li, Lu Zhenghao, and Kiat Seng Yeo *(Tianjin University, China; North Minzu University, China; Singapore University of Technology and Design, Singapore; Wintech Nano-Technology Services Pte Ltd, Singapore; Soochow University, China)*
C3-2	**0104: A 0.15-6.5GHz Stacked CMOS Power Amplifier With Low-Frequency Bandwidth Extension**
14:00 ~14:15	Shijiao Dong, Wei Li, Xingyu Ma, Fan Chen, Hongtao Xu *(Fudan University, China)*
C3-3	**0269: A 2-to-2.7GHz Class-G Switched-Capacitor PA with Cascode Switch-Reused Structure Achieving 25.92dBm Peak Power and 42% Efficiency**
14:15 ~14:30	Jie Deng, Gengzhen Qi *(Sun Yat-sen University, China)*
C3-4	**0294: A X-band High Linearity Tunable Bandpass Filter in 130nm CMOS**
14:30 ~14:45	Tianrui Wang, Ziyu Wang, Huiquan Xie, Yifei Chen, Haokun Lan, Maliang Liu, Yintang Yang *(Xidian University, China)*
C3-5	**0302: Analysis of Polar and Quadrature Digital Transmitters for Wi-Fi7 Applications**
14:45 ~15:00	Lixuan Cao, Yun Yin *(Fudan University, China)*

Thursday, October 24, 13: 30 – 15: 15   Meeting Room 4
Session D3: Novel Device III   Sheraton Zhuhai Hotel 2nd Floor
Session Chair: Prof. Mansun Chan, The Hong Kong University of Science and Technology, China

	Title
D3-1	**0458:Low-Dimensional Materials Enabled Wearable Circuits With Multi-level Detection and Wireless Communication Modules (invited)**
13:30 ~14:00	Li Tao *(Southeast University, China)*
D3-2	**0461: Memristive Circuits Based on Two-dimensional Layered Hexagonal Boron Nitride for Radiofrequency Applications (invited)**
14:00 ~14:30	Sebastian Pazos *(King Abdullah University of Science and Technology, Saudi Arabia)*
D3-3	**0452: Two-dimensional Ferroelectricity: Polarization Modulation and New Device (Invited)**
14:30	Fucai Liu *(University of Electronic Science and Technology of China, China)*

~15:00	
D3-4	**0475: Impact of Gate Overlap Length Modulation on Electrical Characteristics and Subthreshold Swing in Nanosheet TFETs with Varying Tunneling Mechanisms**
15:00 ~15:15	Zheng-Hong Zhong, Wei-Heng Tai, Jyi-Tsong Lin *(Sun Yat-Sen University, Taiwan, China)*

Thursday, October 24, 13: 30 – 15: 15   Meeting Room 5
Session E3: Power Device III   Sheraton Zhuhai Hotel 2nd Floor
Session Chair: Prof. Bo Zhang, University of Electronic Science and Technology of China (UESTC), China

	Title
E3-1	**0255: The Status of WBG Devices Towards Net-Zero Solutions (invited)**
13:30 ~14:00	Mikael Östling　*(KTH Royal Institute of Technology, Sweden）*
E3-2	**0121: Impact of the Resistive Silicon Base Wafer on Substrate Coupling in Power Integrated Circuits in GaN-on-Si Technology**
14:00 ~14:15	Zijin Jiang, Rui (Ray) Yao, Miao Cui, Zhao Wang, Sang Lam, Stephen Taylor *(Xi'an Jiaotong-Liverpool University, China; The University of Liverpool, UK)*
E3-3	**0128: A Novel Snapback-free Double-RESURF Reverse conducting LIGBT with Dual Conduction Paths**
14:15 ~14:30	Yun Xia, Yuxi Wan, Wei Zeng, Yu Shi, Xiaoping Wang, Wei Liu, Haizhao Zhi, Ziwei Zhou, Xixi Luo, Ruize Sun, Xiaoming Wang, Yan Wang, Wanjun Chen *(Shenzhen Pinghu Laboratory, China; University of Electronic Science and Technology of China, China; Tsinghua University, China)*
E3-4	**0168: Comparsion of SiC Planar and Trench Junction Barrier Schottky Diode With Surge Current Capability**
14:30 ~14:45	Ziming Zhao, Yancong Liu, Hao Yuan, Fengyu Du, Yu Zhou, Keyu Liu, Xiaoyan Tang, Qinwen Song, Yuming Zhang *(Xidian University, China)*

Thursday, October 24, 13: 30 – 15: 15   Meeting Room 6
Session F3: Memory Device III   Sheraton Zhuhai Hotel 2nd Floor
Session Chair: Prof. Anquan Jiang, Fudan University, China

	Title

F3-1	**0266: Simulation of Program/Erase Cycling and Retention Loss in 3-D CTF NAND Flash (invited)**
13:30 ~14:00	Anuj Kumar, Ravi Tiwari, Souvik Mahapatra (*Indian Institute of Technology Bombay, India*)
F3-2	**0275: Switch-off Mechanisms in GeAsTe Ovonic Threshold Switching Selector Device (invited)**
14:00 ~14:30	Zeyu Hu, Zheng Chai, Weidong Zhang, Jianfu Zhang (*Liverpool John Moores University, UK; Xi'an Jiaotong University, China*)
F3-3	**0334: Orthorhombic-I (Pbca) Phase: Origin of Antiferroelectricity in HfZrO Films (invited)**
14:30 ~15:00	Wei Liu, Zeping Weng, Jianguo Li, Wenchao Yan, Yiming Qu, Yi Zhao (*Zhejiang University, China; East China Normal University, China*)
F3-4	**0254: The Maximum Storage Capacity of Open-loop Written RRAM is Around 4 Bits**
15:00 ~15:15	Yongxiang Li, Shiqing Wang, Zhong Sun (*Peking University,China*)

Thursday, October 24, 15: 30 – 17: 15

Thursday, October 24, 15: 30 – 17: 15   Meeting Room 1
Session A2: Security   Sheraton Zhuhai Hotel 2nd Floor
Session Chair: Prof. Yan Li, Fudan University, China

	Title
A2-1	**0469: Hardware Security Linking Everything: from Lightweight PUF to Post-Quantum Cryptography Hardware (invited)**
15:30 ~16:00	Yijun Cui, Jiang Li, Jiansheng Chen, Fei Lyu, Chenghua Wang, Weiqiang Liu (*Nanjing University of Aeronautics and Astronautics, China*)
A2-2	**0219: Backward-edge Control Flow Integrity based on Return Address Encryption**
16:00 ~16:15	Fengshuo Tian, Kaixuan Wang, Jun Han (*Fudan University, China*)
A2-3	**0239: Stability Enhancement Technique for Monostable PUF Based on Hysteresis Effect of Schmitt Trigger**
16:15 ~16:30	Ruize Xu, Gang Li, Pengjun Wang, Hui Li, Xudong Wu (*Wenzhou University, China*)

A2-4	**0355: A Reliable Current Starved Inverter PUF Based on SRAM Memory Structure**
16:30 ~16:45	Mingze Ren, Pengjun Wang, Yuejun Zhang, Shutong Zhang, Zhikang Chen, Tengfei Yuan *(Ningbo University, China; Wenzhou University, China)*
A2-5	**0354: An Efficient Convolutional Neural Network Hardware IP for Epilepsy Detection**
16:45 ~17:00	Yi Gong, Yuejun Zhang, Jiangtao Tu, Rongxin Zou, Liang Wen *(Ningbo University, China)*
A2-6	**0082: TLBshield: A Low-cost Secure Reinforce on Translation Lookaside Buffer to Mitigate the Speculative Attacks**
17:00 ~17:15	Yuyang Liu, Runye Ding, Yujie Chen, Pujin Xie, Yao Liu, Zhiyi Yu *(Sun Yat-sen University, China)*

Thursday, October 24, 15: 30 – 17: 15   Meeting Room 2
Session B4: Mixed Signal I   Sheraton Zhuhai Hotel 2nd Floor
Session Chair: Prof. Yubin Zhao, Sun Yat-sen University, China

	Title
B4-1	**0038: Toward Unification of Digital Error Correction Algorithms for ADCs with Redundancy (invited)**
15:30 ~16:00	Haruo Kobayashi, Tomohiko Ogawa, Yutaro Kobayashi, Kentaroh Katoh, Jiangling Wei *(Gunma University, Japan; Fukuoka University, Japan; Yibin University, China)*
B4-2	**0482: A 1.2-V 2-GS/s Trimming-Free Input Buffer with Robust Output Common-mode Voltage (invited)**
16:00 ~16:30	Wei Zhang, Xizhu Peng, He Tang *(UESTC, China)*
B4-3	**0260: A 12-bit 1-MS/s SAR ADC Using V_{cm}-based Split MSB Switching and Segmented CDAC**
16:30 ~16:45	Zheng-Han Chen, Ya-Cong Zhang, Wen-Gao Lu, Zhong-Jian Chen *(Peking University, China)*
B4-4	**0333: A Simplified and Accelerated Opportunistic Bit Weight Calibration for High-Resolution ADCs**
16:45 ~17:00	Bingbing Ma, Wei Li, Hongtao Xu *(Fudan University, China)*

B4-5	0154: Background Calibration for Bit Weights in Pipelined SAR ADCs Using Split ADC Architecture
17:00 ~17:15	Zecheng Zhou, Longsheng Wang, Dongxian Ye, Yexin Zhu, Dengquan Li, Zhangming Zhu *(Xidian University, China)*

Thursday, October 24, 15: 30 – 17: 15   Meeting Room 3
Session C4: Sensor and MEMS I   Sheraton Zhuhai Hotel 2nd Floor
Session Chair: Prof. Luda Wang, Peking University, China
         Prof. Fan Ye, Fudan University, China

	Title
C4-1	0108: Atomically Thin Graphene Nanopore based MEMS Iontronic Devices for Sensing, Separation and Energy Applications (invited)
15:30 ~16:00	Luda Wang, Ruiyang Song, Ningran Wu *(Peking University, China)*
C4-2	0132: Smart Vision Chip (invited)
16:00 ~16:30	Liyuan Liu *(Institute Of Semiconductors, Chinese Academy Of Sciences, China)*
C4-3	0191: Systems-on-Chips for Invasive Brain-Computer Interfaces: Challenges and Opportunities (invited)
16:30 ~17:00	Jie Yang, Mohamad Sawan *(Westlake University, China)*
C4-4	0410: Multi-physics Simulation and Application of Ion Gel Based Triboelectric Nanogenerators
17:00 ~17:15	Chen Liu, Ruibo Wang, Ruiyi Gao, Yuming Zhang *(Xidian University, China; Air Force Engineering University, China)*

Thursday, October 24, 15: 30 – 17: 15   Meeting Room 4
Session D4: Novel Device IV   Sheraton Zhuhai Hotel 2nd Floor
Session Chair: Prof. Zhigang Ji, Shanghai Jiaotong University, China

	Title
D4-1	0264: Recent Progress in the Development of Complementary Field-Effect Transistors (invited)
15:30 ~15:57	Mansun Chan, Shengdong Zhang *(The Hong Kong University of Science and Technology, Hong Kong, China; Peking University, China)*

D4-2	0471: SC-CMOS: Revolutionizing Semiconductor Technology with High Electron Mobility Materials and Advanced Node Optimization (invited)
15:57 ~16:24	Jyi-Tsong Lin *(Sun Yat-Sen University, Taiwan, China)*
D4-3	0074: Metal-Oxide Thin-Film Transistors for Artificial Neural Networks (invited)
16:24 ~16:51	Yushen Hu, Tengteng Lei and Man Wong *(The Hong Kong University of Science and Technology, Hong Kong, China)*
D4-4	0203: Cryogenic Threshold Voltage and On-current Variability Analysis of GAA Nanosheet FETs at 4K
16:51 ~17:03	Zihao Liu, Tomoko Mizutani, Kiyoshi Takeuchi, Takuya Saraya, Hiroshi Oka, Takahiro Mori, Masaharu Kobayashi1 , Toshiro Hiramoto *(The University of Tokyo, Japan; National Institute of Advanced Industrial Science and Technology (AIST), Japan)*
D4-5	0188: Reverse-Biased PN Junction Isolation for Leakage Suppression and Strain Enhancement in Gate-All Around Nanosheet FETs
17:03 ~17:15	Boqian Shen, Chunlei Wu, Yumin Xu, Fei Zhao, Hanzhi Gu, Jian Ma, Yueyuan Yu, Yiming Xia, Qingqing Sun, David Wei Zhang *(Fudan University, China; Shanghai Integrated Manufacturing Innovation Center Co., Ltd, China; Jiashan Fudan Institute, China)*

Thursday, October 24, 15: 30 – 17: 15   Meeting Room 5
Session E4: Power Device IV   Sheraton Zhuhai Hotel 2nd Floor
Session Chair: Prof. Wentong Zhang, University of Electronic Science and Technology of China (UESTC), China

	Title
E4-1	0405: Comparative Study on Reliability of Conventional SiC MOSFET and JBS Integrated SiC MOSFET (invited)
15:30 ~16:00	Moufu Kong, Shurui Li, Hongfei Deng, Bo Yi, Hongqiang Yang, Sen Gong *(University of Electronic Science and Technology of China, China)*
E4-2	0445: Study on Single Event Effect of SiC MOSFET by Proton Irradiation
16:00 ~16:15	Wende Huang, Chengwen Fu, Yao Ma, Mingmin Huang , Xiaoping Dong, Qiang Yu *(Sichuan University, China; Sichuan Suining Lippxin Microelectronics Co., Ltd, China)*
E4-3	0341: Investigating Single-Event Burnout in 4H-SiC Inverters: Experiments and Simulations

16:15 ~16:30	Yong Gu, Yurui Yang, Hongyang Wen, Xiangyu Hou, Runhua Huang, Ao Liu, Bai Song, Jie Ma, Siayang Liu, Long Zhang, Weifeng Sun *(Southeast University, Nanjing China; Nanjing Electronic Device Institute, China)*

Thursday, October 24, 15: 30 – 17: 15	Meeting Room 6
Session F4: Memory Device IV	Sheraton Zhuhai Hotel 2nd Floor
Session Chair: Prof. Anquan Jiang, Fudan University, China	

	Title
F4-1	**DRAM Evolution and Challenges** （Invited）
15:30 ~16:00	Robert Liu *(CXMT Corporation, China)*
F4-2	**0335: Phase-Change Materials and Their Applications (invited)**
16:00 ~16:25	You Yin *(Gunma University, Japan)*
F4-3	**0449: High-Density and High-Reliability RRAM for Memory and Computing Applications (invited)**
16:25 ~16:50	Yimao Cai, Xiahong Zhou, Zongwei Wang, Lin Bao, Ling Liang, Cuimei Wang, Ru Huang *(Peking University, China)*
F4-4	**0343: Impact of Different MAC Schemes on Computing In Memory based on 1T1R Array**
16:50 ~17:03	Ruiqing Xie, Gaoqi Yang, Zongwei Wang, Linbo Shan, Jinshan Li, Chaoyi Ban, Lin Bao, Ling Liang, Cuimei Wang, Yimao Cai, Ru Huang *(Peking University, China)*
F4-5	**0426: Investigation of Gate Injection Charges Behavior on FeFETs with TiN/Al$_2$O$_3$/Hf$_{0.5}$Zr$_{0.5}$O$_2$/SiON/Si Structure by Analyzing ISPP/ISPE**
17:03 ~17:15	Jia Yang, Runhao Han, Saifei Dai, Tao Hu, Xianzhou Shao, Kanyi Li, Wenbo Fan, Xiaoqing Sun, Junshuai Chai, Hao Xu, Kai Han, Xiaolei Wang, Wenwu Wang, Tianchun Ye *(Key Laboratory of Fabrication Technologies for Integrated Circuits, Chinese Academy of Sciences, China; Institute of Microelectronics, Chinese Academy of Sciences, China; University of Chinese Academy of Sciences, China; Weifang University, China)*

Thursday, October 24, 17: 30 - 18: 30

Thursday, October 24, 17: 30 – 18: 30	
Poster Session II	Sheraton Zhuhai Hotel 1st Floor

	Title
P2-1	**0071: Assembly of Oxidized/Intrinsic 2D MXene Film for Improved Absorption Electromagnetic Shielding**
	Yulin Guo, Siteng Li, Jiafeng Song, Yilin Sun, Zhifang Liu, Weijia Luo *(Beijing Institute of Technology, China; Tsinghua University, China)*
P2-2	**0149: Semimetal Alloy Contact with Low Resistivity and Enhanced Thermal Budget for MoS$_2$ FETs**
	Kwok-Ho WONG, Mansun CHAN *(The Hong Kong University of Science and Technology, Hong Kong, China)*
P2-3	**0243: Copper Ion Migration in van der Waals CuInP$_2$S$_6$ Devices with Vertical and Lateral Structures**
	Jie Li, Yirong Guo, Pengying Chang *(Beijing University of Technology, China)*
P2-4	**0382: Edge-Dependent of Threshold Voltage in MoS$_2$ Nanoribbon-Based 2D FETs**
	Zhirong Peng, Mansun Chan *(The Hong Kong University of Science and Technology, Hong Kong, China)*
P2-5	**0438: Effect of Layer Thickness on the Transport Properties of ALD-deposited ZnO/In$_2$O$_3$ Heterojunction Thin-film Transistors**
	Zhenwei Li, Tiaoyang Li *(Fuzhou University, China)*
P2-6	**0417: Foundamentals of Low-Resistive Indium-Violet Phosphorene Top Contact: an ab-initio NEGF Study**
	Huaipeng Wang, Sicheng Liu, Shuaihong Li, Zhifang Liu, Yilin Sun, Jianlong Xu, Dan Xie *(Tsinghua University, China; Beijing Institute of Technology, China; Soochow University, China)*
P2-7	**0095: Broadband Photodetectors Based on Graphene/Perovskite Hybrid Structure with Ferroelectric Gating**
	Zhongyang Liu, Shuangqi Dong, Mingjie Li, Huaipeng Wang, Dan Xie, Yilin Sun *(Beijing Institute of Technology, China; Tsinghua University, China)*
P2-8	**0363: Interconnection Design of Chiplet Technology**
	Ning Chen, Chang Wu *(Fudan University, China)*
P2-9	**0102: Effects and Modeling Study on FDSOI MOSFETs at Cryogenic Temperature**
	Zhipeng Ren, Yibo Hu, Yizhe Yin, Ruiyong Zhao, Jing Chen *(Shanghai Institute of Microsystem and Information Technology, China)*

P2-10	**0247: Improved Channel Width and Morphology of Epi Silicon FinFET via Low Thermal Budgets Fin Thinning Technology**
	Peng Wang, Yupeng Lu, Guanqiao Sang, Renjie Jiang, Lei Cao, QingKun Li, Lianlian Li, hang zhang, zhonrui wang, meihe zhang, Qingzhu Zhang, Junfeng Li; Huaxiang Yin *(Institute of Microelectronics, China; University of Chinese Academy of Sciences, China)*
P2-11	**0364: Deep Investigation into Variability of Complementary Dopant Segregated Tunneling FET Based on Foundry Platform**
	Rundong Jia, Jianfeng Hang, Kaifeng Wang, Yongqin Wu, Hongyan Han2 , Ye Ren, Weihai Bu, Runsheng Wang, Qianqian Huang, Ru Huang *(Peking University, China; Seimiconductor Technology Innovation Center (Beijing), China)*
P2-12	**0379: Investigation of Common-Gate and Split-Gate Structures Based on CFET Standard Cells**
	Peishun Tang, Rongzheng Ding, Xiaona Zhu, Shaofeng Yu *(Fudan University, China)*
P2-13	**0187: Exploration of the effect of silver impurity on the minority carrier lifetime of semiconductor**
	Xin Tian, Peizhi Zhao, Yudong Li, Jun Xu, Tianling Ren *(Tsinghua University, China; Jiangsu Xinhua Semiconductor Technology Co., Ltd.; China)*
P2-14	**0267: Fabrication and Electrical Characterization of Mo/Hf$_x$Zr$_{1-x}$O$_2$/Mo ferroelectric capacitors**
	Chunsheng Jiang, Wencai Du, Qin Xie *(Guangxi Normal University; China)*
P2-15	**0116: Effect of Cascade Current Density and Plating Time on TSV Filling Effect in DC Power Supply**
	Weifeng Chen, Lijuan Peng, Xiaohui Wang, Fangzhou Wang, Guojian Ding, Qi Feng, Ping Yu, Peng Zuo, Feng Liu, Jiang Ma, Yang Wang, Haiqiang Jia, Hong Chen *(Songshan Lake Materials Laboratory, China; Shenzhen University, China)*
P2-16	**0396: High-Performance Carbon Nanotube Optoelectronic Transistors for Memory Applications**
	Shuang Liu, Heyi Huang, Yanqing Li, Yadong Zhang, Feixiong Wang, Yupeng Lu, Renjie Jiang, Jiali Huo, Huaxiang Yin *(Institute of Microelectronics, China; University of Chinese Academy of Sciences, China)*
P2-17	**0414: Investigation of the channel width dependence of IGZO TFT by experiment and TCAD simulation**
	Yanyu Yang, Yupeng Lu, Shuang Liu，Renjie Jiang, Jie Luo, Yunjiao Bao, Peng Wang, Gaobo Xu, Huaxiang Yin *(Institute of Microelectronics, China; University of Chinese Academy of Sciences, China)*

P2-18	**0120: A Test and Evaluation Platform for Quantitative Analysis of High-Reliability Designs**
	Yifeng Huang, Wenqing Wan, Chang Wu *(Fudan University, China)*
P2-19	**0171: Hot-Carrier Injection Characterization of n-LDMOS Transistors and Stress Tests in a Buck Converter Configuration**
	Chun Yee Chu, Wai Tung Ng *(University of Toronto, Canada)*
P2-20	**0044: Automated Verification of Functional Interface Connections in Circuit Schematics**
	Keli Long, Xingyu Gao, Lei Li, Jinxiang Wang, Fangfa Fu, Liangquan Qiao, Jinghan Zhou *(Harbin Institute of Technology, China; 58th Research Institute of China Electronics Technology Group Corporation, China)*
P2-22	**0214: Co-Optimization Design Method of Temperature Variation and Circuit Aging in Digital Circuits**
	Songxuan He, Wangyong Chen, Ling Xiong, Linlin Cai *(Sun Yat-sen University, China)*
P2-23	**0308: Boolean Matrix Factorization Algorithm based on Error Shaping Technique and its Application on Approximate Logic Synthesis**
	Botao Xiong, Runhua Yang, Yuchun Chang *(Dalian University of Technology, China)*
P2-24	**0148: Automatically Device Sizing of Analog Circuit through Sequential Model-Based Optimization with Circuit Recognition**
	Shun-Qi DAI, Xiao WANG, Yuan LEI, Bei-Ping YAN *(Hong Kong Applied Science and Technology Research Institute (ASTRI), Hong Kong, China)*
P2-25	**0229: Vanadium Oxide-Based Artificial Synapses for Construction of Artificial Neural System**
	Zhuoling Zhou, Libin Liang, Hongzhi Chen, Changjiu Teng, Shilong Zhao, Wenjun Chen *(Foshan University, China)*
P2-26	**0048: High performance FeFET with α-IGZO Channel Enabled by Atomic-Layer-Deposited HfO_2 Interfacial Layer**
	Yinchi Liu, Hao Zhang, Xinlong Zhou, Dmitriy Anatolyevich Golosov, Chenjie Gu, Hongliang Lu, Shijin Ding, and Wenjun Liu *(Fudan University, China; Research Institute of Fudan University in Ningbo, China; Belarusian State University of Informatics and Radioelectronics, Republic of Belarus; Ningbo University, China; Zhangjiang Fudan International Innovation Center, China)*

P2-27	0097: A Simulation Study on Cell Scaling Impacts in 3D Charge-trapping (CT) Flash Memory
	Wanyu Li, Haitao Dong, Qianwen Wang, Yang Feng, Xuepeng Zhan, Jixuan Wu, Jiezhi Chen (*Shandong University, China; Qingdao University of Science & Technology, China*)
P2-28	0390: Comprehensive Charaterizations on Read Disturbs in QLC Charge-Trap (CT) 3D NAND Flash
	Shaoqi Yang, Xiaohuan Zhao, Peng Guo, Qianwen Wang, Guangkuo Yang1, Xinyi Guo, Pengpeng Sang, Jixuan Wu, Xuepeng Zhan, Jiezhi Chen (*Shandong University, China; Shandong Sinochip Semiconductors Co., Ltd, China; Qingdao University of Science & Technology, China*)
P2-29	0047: Aspect Ratio Dependent Optimization and Comparison of Specific ON-Resistance of SJ and Hk MOSFETs with Extremely High Permittivity
	Chenxing Wang, Zhentao Xiao, Zonghao Zhang, Zhenghao Jin, Zhiwan Liu, Zonglin Li, Zhemin An, Yunteng Jiang, Ruguan Li, Haimeng Huang, Hongqiang Yang (*University of Electronic Science and Technology of China, China; GRG Metrology & Test Group Co., Ltd., China*)
P2-30	0092: Simulation Study of the Impact of Split Gate on SiC DTMOS Short Circuit Withstand Capability
	Zixun Chen, Jinping Zhang, Yang Liu, Xudong Ma, Bo Zhang (*University of Electronic Science and Technology of China, China; Weihai Singa Electronics CO.LTD, China*)
P2-31	0143: Improved Hall Mobility Measurement Distinguishing Interface Capturing Effect in 4H-SiC Inversion Channel
	Xiangrui Fan, Hao Fu, Xinyu Zhang, Zilong Wu, Jiameng Sun, Jiaxing Wei, Siyang Liu, Weifeng Sun (*Southeast University, China*)
P2-32	0163: A Superior SiC Lateral MOSFET with Patterned P-bury Layer Made on N-type Wafers
	Xuke Yan, Junji Cheng, Xiaojun Fu, Bo Yi, Haimeng Huang, Hongqiang Yang (*University of Electronic Science and Technology of China, China; The 24th Research Institute of China Electronics Technology Group Corporation, China*)
P2-33	0272: High Performance Termination Design and Fabrication For SiC MOSFET Device
	Lei Huang, Junhou Cao, Chenlu Wang, Hao Fu, Jiaxing Wei, Siyang Liu, Weifeng Sun (*Southeast University, China*)
P2-34	0422: Analysis of The Separation Degree For P-pillar in SiC Super-Junction Structure Through "Multiple Epitaxy-Ion Implantation" Route

	Hao-Bo Kang, Hao Yuan, Feng-Yu Du, Yu Zhou, ke-Yu Liu, Xiao-Yan Tang, Chao Han, Qing-Wen Song, Yu-Ming Zhang *(Xidian University, China; The Xidian-Wuhu Research Institute, China)*
P2-35	**0433: Numerical Analysis of the CIBL Effect on ShortCircuit Characteristics of DG-CSTBTs with Reduced Mesa Width**
	Zhengyu Lang, Jinping Zhang, Shiwei Zheng, Shuyang Huang, Haonan Deng, and Bo Zhang *(University of Electronic Science and Technology of China, China; Nanjing SilverMicro Electronics, China)*
P2-36	**0441: Innovations in GaN HEMT Design: Achieving Superior Power Output and Thermal Management**
	Shiming Li, Bowen Yang, Mei Wu, Ling Yang, Bin Hou, Meng Zhang, Xiaohua Ma,Yue Hao *(Xidian University, China)*
P2-37	**0446: An Enhanced RC-IGBT Incorporating Superjunction and Discontinuous Field Stop Layers for Improved Efficiency**
	Yiming Jia, Jieyu Long, Zhiwei Jing, Haimeng Huang, Hongqiang Yang *(University of Electronic Science and Technology of China, China)*
P2-38	**0105: Simulation Study on 1200V CS-SemiSJ-IGBT for Reduced Switching Loss and Fast Switching**
	Luping Li, Zehong Li, Peng Chen, Yuzhou Wu, Qiansheng Rao, Ming Li, Haifeng Qin, Li Wan, Yang Yang, Wei Li, Min Ren *(University of Electronic Science and Technology of China (UESTC), China; China Resources Microelectronics (Chongqing) Ltd., China; Shanghai Super Semiconductor Technology Company Ltd., China)*
P2-39	**0194: A Dual-Gate Trigger Thyristor for Reducing the Probability of False Triggering**
	Pengcheng Xing, Qingbo Wan, Jie Huang, Ruize Sun, Chao Liu, Wanjun Chen *(University of Electronic Science and Technology of China (UESTC), China)*
P2-40	**0202: Ultra Fast Diode Avalanche Shaper with Floating Junction**
	Zhen Yang, Yu Zhou, Xiao-Yan Tang, Chao Han, Qing-Wen Song, Yu-Ming Zhang *(Xidian University, China; Xidian-Wuhu Research Institute, China)*
P2-41	**0204: Silicon Carbide Diode Avalanche Shaper with Multi-Point Quasi-Uniform Triggering**
	Lin Cheng, Yu Zhou, Xiao-Yan Tang, Chao Han, Yu-Ming Zhang, Qing-Wen Song *(Xidian University China; Xidian-Wuhu Research Institute, China)*

P2-42	0209: Super Field Plate LIGBT with Improved Performance for Both Cell and Terminal Region
	Weihao Lu, Jing Li, Jitong Wang, Chaoyang Peng, Chunwei Zhang (*University of Jinan, China*)
P2-43	0248: High-temperature oxidation of 4H-SiC and gate oxide reliability dependence on oxidation temperature
	Baoyan Feng, Xiaoyan Tang, Yi bo Zhang, Chao Han, Hao Yuan, Qing wen Song (*Xidian University, China; Xidian-Wuhu Research Institute, China*)
P2-44	0299: Optimization for a High-voltage Recessed-gate β- Ga_2O_3 MOSFET by Gate and Drain Field Plate Technology
	Bo Yi, Yuan Qiao, Ming Dai, Fan Xu, JunJi Cheng, HaiMeng Huang, MouFu Kong, XingLi Jiang, HongQiang Yang (*University of Electronic Science and Technology of China, China; Chongqing Institute of Microelectronics Industry Technology, China; Chengdu Semi-Future Technology Co., Ltd, China*)
P2-45	0315: A Novel Voltage Sensor with Composite Trench Structure for High Voltage IGBT
	Yang Yang, Ze-Hong Li, Senior Member, IEEE, Li-Hang Dong, Wei Li, Peng-Fei Jia, Zhi-Yu Yang, Li Wan, Yi-Shang Zhao, Tong-Yang Wang, Zi-Ming Xia (*University of Electronic Science and Technology of China, China; China Resources Microelectronics (Chongqing) Limited, China; Chongqing Institute of Microelectronics Industry Technology, China*)
P2-46	0316: A Novel Triggered Voltage Sensing Structure for High Voltage IGBT
	Yang Yang, Ze-Hong Li, Senior Member, IEEE, Li-Hang Dong, Wei Li, Peng-Fei Jia, Zhi-Yu Yang, Li Wan, Yi-Shang Zhao, Lu-Ping Li, Zi-Ming Xia, and Tong-Yang Wang (*China Resources Microelectronics (Chongqing) Limited, China; University of Electronic Science and Technology of China, China; Chongqing Institute of Microelectronics Industry Technology, China*)
P2-47	0374: Investigation of Threshold Voltage Instability in p-GaN Gate HEMTs under Surge Current Stress
	Xiaoming Wang, Yu Shi, Chunhua Zhou, Haizhao Zhi, Yun Xia, Ruize Sun, Xinghuan Chen, Wanjun Chen, Bo Zhang (*University of Electronic Science and Technology of China, China; Shenzhen Pinghu Laboratory, China; China Electronic Product Reliability and Environmental Testing Research Institute, China*)
P2-48	0408: A Novel Ga_2O_3 High-k Trench MOSFET with Improved Forward and Reverse Performance
	Moufu Kong, Lewei Lyu, Haoran Wang, Zhaoyu Ai, Xinyang Chen, Fanxin Meng, Qiang Hu (*University of Electronic Science and Technology of China, China;*

	Chengdu High-tech Development Co.Ltd, China; Chengdu Semi-Future Technology Co. Ltd, China)
P2-49	**0054: A Nonlinear Behavioral Modeling Approach for Microwave Transistors Considering ElectrothermalAging Degradation**
	Lin Cheng, Hongliang Lu, Silu Yan, Junjun Qi, Jiantao Qiao, Yuming Zhang *(Xidian University, China)*
P2-50	**0344: Electrical and Thermal Analysis of CNT nTSV Applied to BS-PDN: A Modeling Study**
	Kai Ying, Baohui Xu, Jie Liang *(Shanghai University, China)*
P2-51	**0357: A Unified Current-Voltage Compact Model for Organic Light-Emitting Diode**
	Wenbin Wang, Mingyu Ma, Wangjun Yang, Jianghao Ma, Hailong You, Cong Li *(Xidian University, China)*
P2-52	**0162: Threshold Voltage and Mobility Extraction of Negative Bias Temperature Instability in 22nm FD SOI MOSFETs**
	Yibo Hu, Hao Ge, Zhipeng Ren, Yizhe Yin, Ruiyong Zhao, Jing Chen *(Shanghai Institute of Microsystem and Information Technology, China)*
P2-53	**0428: Modeling of Silicon Single-Photon Avalanche Diodes for Process and Design Optimization**
	Jing Fu, Anran Guo, Hongbo Zhang, Guowei Li, Huaping Ma, Ruizhi Li, Yuwei Chen *(National Key Laboratory of Integrated Circuits and Microsystems, China; CETC No.44 Research Institute, China)*
P2-54	**0046: A Novel Modeling Method for BV Characteristics of ESD Protection Devices**
	Ke Zhang, Yang Wang, Xiangliang Jin *(Hunan Normal University, China; Peking University, China)*
P2-55	**0070: Analysis of The Impact of Parasitic Bipolar Amplification on Charge Sharing Based on Analytical Model**
	Yutao Zhang, Hongliang Lyu, Yuming Zhang, Ruxue Yao *(Xidian University, China)*
P2-56	**0123: Research on the performance degeneration of GGNMOS under total ionizing dose Radiation**
	Jiekai Feng, Ping Luo, Chengxin Li, Jiaxuan Hu, Peng Li, Pengfei Liao *(Univ. of Elec. Sci. and Technol. Of China, China; Chongqing Institute of Microelectronics Industry Technology, China; The 24th Research Institute of China Electronics Technology Group Corporation, China)*

P2-57	**0208: The UIS Withstand Capability and Device Failure Mechanism of 650 V p-GaN Gate HEMTs**
	Qihang Huang, Luanxi Ma, Shuting Huang, Yanning Nie, Jianggen Zhu, Yu Shi, Rongxin Du, David Zhou, Yuxi Wan, Bo Zhang, Qi Zhou *(University of Electronic Science and Technology of China, China; Shenzhen Pinghu Laboratory, China)*
P2-58	**0100: Time Dependent Dielectric Breakdown in n-MOSFETs Fabricated by Low-Temperature and Low-Pressure Mild Oxidation After Plasma Solidification**
	Qiao Teng, Yanning Chen, Fang Liu, Bo Wu, Yongfeng Deng, Dawei Gao *(Zhejiang University, China; Beijing Smart-chip Microelectronics Technology Co., Ltd, China)*
P2-59	**0228: Simulation of BTI for GAA MOSFETs with Enhanced Parameters Extraction**
	Yongjia Wang, Yijiao Wang, Shuhan Wang, Jinghan Xu, Xiaoyan Liu *(Peking University, China; Beihang University, China)*
P2-60	**0033: Gold Thermocompression Wafer Bonding for Quartz MEMS Applications**
	Ting Yang, Dongxiang Han, Jun Xu, Tian-Ling Ren *(Tsinghua University, China; Nanjing University of Aeronautics and Astronautics, China)*
P2-61	**0083: An Adaptive Threshold Analog Front-End Circuit for Direct ToF LiDAR**
	Jianping Guo, Xiaoyang Zeng, Wenhong Li, Mingyu Wang *(Fudan University, China)*
P2-62	**0184: Design of Ultra-Broadband Metamaterial Absorber from Infrared to Terahertz**
	Xiangze Liu, Wenbin Zhou, Tiantian Shi, Yiming Liao, Feng Yan, Xiaoli Ji *(Nanjing University, China; Nanjing University of Science and Technology, China)*
P2-63	**0398: Large modulation bandwidth Si-based avalanche photodiode for visible light communications**
	Jiabin Wu, Yidi Hu, Chiang Zhu, Zhichong Wang, Xiaona Zhu, Chao Shen *(Fudan University, China)*
P2-64	**0400: An Artificial Neuromuscular Synapse Based on a Ferroelectric $Pb(Zr_{1-x}Ti_x)O_3$/SiC Floating Gate Transistor**
	Yu Liu, Lin Lin, Xiang Wang, Chengyan Zhong, Junxiong Guo, Wen Huang, Yufeng Guo *(Nanjing University of Posts and Telecommunications, China; University of Electronic Science and Technology of China, China; Chengdu University, China)*

Friday

Friday, October 25, 9: 00 – 10: 30

Friday, October 25, 9: 00 – 10: 30	Grand Ball Room
Keynote Session K4	Sheraton Zhuhai Hotel 1st Floor
Session Chair: Prof. Rui Yin, National Integrated Circuit Innovavtion Center, China	

K4-1	CMOS Digital Radiography
9: 00 ~9: 45	Prof. Youngcheol Chae, Yonsei University, Korea
K4-2	**High-Frequency and Wideband RF Filters for 6G and Wi-Fi 7**
9: 45 ~10: 30	Prof. Chengjie Zuo, University of Science and Technology of China, China
	Coffee Break

Friday, October 25, 10: 45– 12: 15

Friday, October 25, 10: 45 – 12: 15	Meeting Room 1
Session A3: Digital & Memory Circuit	Sheraton Zhuhai Hotel 2nd Floor
Session Chair: Prof. Fan Ye, Fudan University, China	

	Title
A3-1	**0451: One-Step Circuit Analysis Based on LCA for Sparse Coding (invited)**
10:45 ~11:15	Hanxi Xu, Zirui Chen, Qi Chen, Xiangshui Miao, Yuhui He *(Huazhong University of Science and Technology, China; Politecnico di Milano, Italy)*
A3-2	**0369: A Hybrid-Logic Scheme for High-Performance and Low-Power Decoders in 7nm Process**
11:15 ~11:27	Donghao Xia, Yuejun Zhang, Mengfan Xu, Liang Wen, Yiting Guo *(Ningbo University, China; China Coast Guard Academy, China)*
A3-3	**0049: Ternary Logic Units Design Based on the TDDFETs**
11:27 ~11:39	Hua Qiang, Haoran Lu, Xiaotao Liu, Linlin Xing, Bin Lu *(Shanxi Normal University, China)*
A3-4	**0474: Enhancing SRAM Cell Stability Through Single-Carrier CMOS Latch Integration**
11:39 ~11:51	Yuan-Yu Chuang, Pei-Zhang Xie, and Jyi-Tsong Lin *(Sun Yat-Sen University, Taiwan, China)*
A3-5	**0199: An RRAM based 9T1R NVSRAM for Low-Power Computing in Memory**
11:51 ~12:03	Huimeng Guo, Yujia Li, Tingrui Ren, Chenge Dong, Liang Wang, Yuanfu Zhao, Yanlong Zhang *(Hangzhou Dianzi University, China; Beijing Microelectronics Technology Institute, China; Beihang University, China)*
A3-6	**0377: A High-Resistance SOT Device Based Computing In-Memory Macro with High Sensing Margin and Multi-Bit MAC Operations for AI Edge Inference**
12:03 ~12:15	Junzhan Liu, Jinyao Mi, Haiyan Qin, He Zhang, Wang Kang *(Beihang University, China; Hangzhou International Innovation Institute, China)*

Friday, October 25, 10: 45 – 12: 15	Meeting Room 2
Session B5: Mixed Signal II	Sheraton Zhuhai Hotel 2nd Floor
Session Chair: Prof. Fan Ye, Fudan University, China	

	Title
B5-1	**0167: When Time Interleaving encounters Oversampling in ADC (invited)**
10:45	Mingqiang Guo, Dongyang Jiang, Shulin Zhao, Sai-Weng Sin, Rui P. Martins

~11:15	*(University of Macau, Macao, China; Universidade de Lisboa, Portugal)*
B5-2	**0322: A 0.000355mm² 4.6μm-Pitch 5.75fJ/Conv 6-bit SAR ADC for High Throughput Parallel Readout of Analog SRAM Computing-In-Memory**
11:15 ~11:30	Lin Wu, Lichen Feng, Hongwei Shan, Zhangming Zhu *(Xidian University, China)*
B5-3	**0051: A 250MS/s, 12 Bit Pipeline-SAR ADC Using Coarse-fine Ring Amplifier**
11:30 ~11:45	Linghao Liu, Junyan Ren, Fan Ye *(Fudan University, China)*
B5-4	**0420: A 0.71pJ/b 16Gb/s Equalizer with Inverter_based CTLE and 4-Tap Speculative DFE**
11:45 ~12:00	Huihong Zhang, Chuangao Yan, Luo Peng, Maliang Liu *(Xidian University, China)*
B5-5	**0350: A Digital Foreground Calibration Method for Pipeline SAR ADCs Using Extended Kalman Filter**
12:00 ~12:15	Dayan Zhou, Yuguo Xiang, Junyan Ren, Fan Ye *(Fudan University, China)*

Friday, October 25, 10: 45 – 12: 15   Meeting Room 3
Session C5: Sensor & MEMS II   Sheraton Zhuhai Hotel 2nd Floor
Session Chair: Prof. Zhanfeng Huang, Sun Yat-sen University, China

	Title
C5-1	**0099: A Flexible Graphene Acoustic Sensor for Sound Signal Acquisition and Spike Neural Network Recognition**
10:45 ~11:00	Lu-Yu Zhao, Hao-Yuan Shen, Yi-Wen Wu, Lu-Lu Zhang, Yu-Tao Li, Tian-Ling Ren *(Beijing Institute of Technology, China; Beijing University of Chemical Technology, China; Tsinghua University, China)*
C5-2	**0058: 0.15μm BCD Platform with High Sensitivity Hall Device and Low Noise CMOS for Sensor IC Applications**
11:00 ~11:15	Guiqiang Zheng, Qingyin Zhong, Jie Ma, Nannan Cheng, Yichen Li, Yongjia Li, Siyang Liu, Xiaofeng Sun, Dejin Wang, Sen Zhang, Long Zhang, Weifeng Sun *(Southeast University, China; CSMC Technologies Corporation, China)*
C5-3	**0166: A High Dynamic Range Pixel with Inverse Proportional Response**
11:15 ~11:30	Yuchen Wang, Wenji Mo, Haoning Sun, Jingjing Liu *(Sun Yat-sen University, China)*

C5-4	**0285: Enhancement of Image Sensor Pixel Performance through Ring-Shaped Vertical Transfer Gate Structure**
11:30 ~11:45	Shuang Yan, Shuai Yuan, Haoping Zheng, Yudi Zhao, Gang Du, Junchen Dong, Kai Zhao *(Beijing Information Science and Technology University, China; Peking University, China; HT-tech Jiangsu Co., Ltd., China)*

Friday, October 25, 10: 45 – 12: 15   Meeting Room 4
Session D5: Process I   Sheraton Zhuhai Hotel 2nd Floor
Session Chair: Prof. Heng Wu, Peking University, China

	Title
D5-1	**0251: Studies on Selective Deposition of SiO₂ by Rapid Atomic Layer Deposition (Invited)**
10:45 ~11:15	Sicong Shao, Jin Yan, Wang Li, Kun Cao , Rong Chen *(Huazhong University of Science and Technology, China)*
D5-2	**0404: Resistance Dependence of Cobalt on Line Width in Advanced Interconnects: First-Principles Modelling**
11:15 ~11:30	Kang Wang, Menglin Huang, Shiyou Chen *(Fudan University, China)*
D5-3	**0418: Reducation of Specific Contact Resistivity by Employing Pre-amorphization Implantation and In situ Steam Generation Oxidation**
11:30 ~11:45	Chang Liu, Xu Chen, Jinbiao Liu, Yanping He, Wenjuan Xiong, Weibing Liu, Mingshan Liu, Zhe Liu, Yaoqi Dong, Jeffrey Xu, Jing Xu, Jun Luo *(Institute of Microelectronics, Chinese Academy of Sciences, China. University of Chinese Academy of Sciences, China. Huawei Technologies Company limited, China)*
D5-4	**0432: High-performance Ultrathin ITO Thin Film Transistor With Ultralow Subthreshold Swing**
11:45 ~12:00	Yanheng Liu, Tiaoyang Li *(Fuzhou University, China)*

Friday, October 25, 10: 45 – 12: 15   Meeting Room 5
Session E5: Reliability I   Sheraton Zhuhai Hotel 2nd Floor
Session Chair: Prof. Ming Xiao, Sun Yat-sen University, China

	Title
E5-1	**0453: Challenges of Design for Reliability in Advanced CMOS Technology: From Single-mode to Mixed-mode Mechanisms (invited)**

10:45 ~11:15	Zixuan Sun, Lining Zhang, Ru Huang, Runsheng Wang *(Peking University, China)*
E5-2	**0232: Frequency-dependent Time-dependent Dielectric Breakdown (TDDB) Behavior and Physical Study in Gate Oxides (invited)**
11:15 ~11:45	Wei Liu, Chu Yan, Xinwei Yu, Yiming Qu, Wenchao Yan, Yi Zhao *(Zhejiang University, China; East China Normal University, China; Zhejiang Li-ryder Technologies Co. LTD, China)*
E5-3	**0181: Lightning Protection Stacked TVS Structure Based on a Novel Total-Ionizing-Dose Radiation-hardened Technology**
11:45 ~12:00	Zhao Qi, Hongquan Chen, Yirui Jia, Nailong He, Zhili Zhang, Sen Zhang, Ming Qiao, Bo Zhang *(University of Electronic Science and Technology of China, China; CSMC Technologies Corporation, China)*

Friday, October 25, 10: 15 – 12: 15   Meeting Room 6
Session F5: Device Modeling I   Sheraton Zhuhai Hotel 2nd Floor
Session Chair: Prof. Ming Li, Peking University, China

	Title
F5-1	**0466: Electric-Thermal Characteristics of Bottom P-i-N Isolated Nanosheet Gate-All-Around FETs (invited)**
10:45 ~11:15	Chunlei Wu, Hanzhi Gu, Jian Ma, Boqian Shen, Fei Zhao, Yueyuan Yu, Yiming Xia, Qingqing Sun, David Wei Zhang *(Fudan University, China; Shanghai Integrated Menufacturing Innovation Center Co., Ltd, China; Jiashan Fudan Institute, China)*
F5-2	**0180: Surface Potential-Based Compact Model for ITO Thin-film Transistors with Ultra-thin Channel**
11:15 ~11:30	Wenting Xu, Xinxin Shen, Zuoxu Yu, Tingrui Huang, Yuzhen Zhang, Weifeng Sun, Wangran Wu *(Southeast University, China)*
F5-3	**0186: A Continuous Full Channel Potential Model for Accurate Synthetic Electricfield Calculating in Gate-All-Around Devices**
11:30 ~11:45	Fei Zhao, Chunlei Wu, Yumin Xu, Boqian Shen, Jian Ma, Hanzhi Gu, Yueyuan Yu, Yiming Xia, Qingqing Sun, David Wei Zhang *(Fudan University, China; Shanghai Integrated Manufacturing Innovation Center Co., Ltd, China; Jiashan Fudan Institute, China)*
F5-4	**0271: Deep Learning and Adaptive Pattern Search Based BSIM-CMG Parameter Extraction Applicable to Process Migration**
11:45 ~12:00	Xingyu Li, Wangyong Chen, Linlin Cai *(Sun Yat-sen University, China)*

Friday, October 25, 13: 30 – 15: 15

Friday, October 25, 13: 30 – 15: 15   Meeting Room 1
Session A4: Processor   Sheraton Zhuhai Hotel 2nd Floor
Session Chair: Prof. Yao Liu, Sun Yat-sen University, China

	Title
A4-1	**0201: RISC-V Domain-Specific Processor for Accelerating SPHINCS+ on Multi-Core Architecture**
13:30 ~13:45	Shengnan Zhang, Yifan Zhao, Xinglong Yu, Jun Han *(Fudan University, China)*
A4-2	**0113: Design of an Out-of-Order Superscalar Processor with Improved Register Alias Table Recovery Method**
13:45 ~14:00	Wu Yang, Jun Zhang *(Central South University, China)*
A4-3	**0220: An SDPF RISC-V Processor with Two-stage Pseudo-pipelined Architecture for IoT Applications**
14:00 ~14:15	Wenji Mo, Yuchen Wang, Haoning Sun, Jingjing Liu *(Sun Yat-sen University, China)*
A4-4	**0366: A Unified Verification Scheme for the Acceleration of RISC-V Processor Design**
14:15 ~14:30	Zixiang Chen, Jiyuan Bai, Yueru Yu, Gengsheng Chen, Xiaofang Zhou *(Fudan University, China; Jiashan Fudan Institute, Jiaxing, China)*
A4-5	**0370: Asynchronous Arbitration Circuit Optimization for Multicore Neuromorphic Processors**
14:30 ~14:45	Jiajie Guo, Guangyao Lin, Bohan Wang, Zhiyi Yu, Shanlin Xiao *(Sun Yat-sen University, China)*

Friday, October 25, 13: 30 – 15: 15   Meeting Room 2
Session B6: Mixed Signal III   Sheraton Zhuhai Hotel 2nd Floor
Session Chair: Prof. Shuo Li, Fudan University, China

	Title
B6-1	**0289: A Second-Order Charge Pump Noise-Shaping SAR ADC (invited)**

13:30 ~14:00	Haoning Sun, Yuchen Wang, Wenji Mo, Kangkang Sun, Jingjing Liu *(Sun Yat-Sen University, China)*
B6-2	**0385: Computing in Memory for Accelerating Light Weighted On-Chip Learning in IoT Devices (Invited)**
14:00 ~14:30	Zhiwang Guo, Xiaoyong Xue, Jun Han, Peng Zhou, Xiaoyang Zeng *(Fudan University, China; Shaoxin Laboratory, China)*
B6-3	**0314: A Novel Beamforing Receiver Archiercture Combining MASH SDM and BSP**
14:30 ~14:45	Tao Zhong, Yuekang Guo, Jing Jin, Jianjun Zhou *(Shanghai Jiao Tong University, China)*

Friday, October 25, 13: 30 – 15: 15   Meeting Room 3
Session C6: RF Circuit Ⅱ   Sheraton Zhuhai Hotel 2nd Floor
Session Chair: Prof. Gengzhen Qi, Sun Yat-sen University, China

	Title
C6-1	**0424: A 10-GHz Low Power Class-C VCO with Long-Term Reliability and Tunable Performance in 28 nm FD-SOI for Satellite Communications (invited)**
13:30 ~14:00	Yann Deval, Henrique Iha Taguti, Ayoub Ait Ihda, Herve Lapuyade, Stephane Rochette, Francois Rivet *(Univ. Bordeaux, France; Thales Alenia Space, France)*
C6-2	**0313: A 191-GHz Harmonic Oscillator with Self-Feeding Line and Return-Path Gap Coupler Structure in 65nm CMOS**
14:00 ~14:15	Xiaohan Shen, Chen Jiang *(Fudan University, China)*
C6-3	**0278: A Blocker-Tolerant High-Linear Receiver Employing Baseband Noise-Cancelling and Bottom-Plate Switched-Capacitor Techniques**
14:15 ~14:30	Chenxiang Cai, Gengzhen Qi *(Sun Yat-sen University, China)*
C6-4	**0412: A High Sensitivity Series-Parallel Rectifier with Pre-Bias for RF Energy Harvesting Systems**
14:30 ~14:45	HaiQin Wu, Dejian Li, Xin Jin, Xufeng Liao, Lianxi Liu *(Xidian University, China; Beijing Smart-Chip Microelectronic Technology Co., Ltd, China)*
C6-5	**0103: A 24.3-43.7 GHz Variable-Gain Low-Noise Amplifier With Phase Self-Compensation**
14:45 ~15:00	Yue Wu, Wei Li, Mohan Zhou, Hongtao Xu *(Fudan University, China)*

C6-6	**0480: A Broadband Active Variable Attenuator With Phase Compensation Technique**
15:00 ~15:15	Zhiying Xia, Zhiqun Li, Bofan Chen, Xiaowei Wang *(Southeast University, China)*

Friday, October 25, 13: 30 – 15: 15   Meeting Room 4
Session D6: Process II   Sheraton Zhuhai Hotel 2nd Floor
Session Chair: Prof. Xiaona Zhu, Fudan University, China
              Prof. Fan Ye, Fudan University, China

	Title
D6-1	**0456:Advanced Logic Devices' DTCO Beyond 3nm Process Technology Node (invited)**
13:30 ~14:00	Xiaona Zhu, Hongliang Lu, Shaofeng Yu, David Wei Zhang *(Fudan University, China)*
D6-2	**0483:Defect-Centric Insights into Flicker Noise in UltraScaled FETs: From Physics to Compact Model for Circuit Level Simulation (invited)**
14:00 ~14:30	Chenyang Zhang, Yu Xiao, Pengpeng Ren, Shiyu Xia, Sheng Yang, Lining Zhang, Runsheng Wang, Zhigang Ji *(Shanghai Jiaotong University, China; Peking University, China)*
D6-3	**0454: A Modified Virtual-Source Model for Ballistic Transport Characterization of FinFETs at Cryogenic Temperature**
14:30 ~14:45	Hongbo Wang, Zirui Wang, Zixuan Sun, Runsheng Wang, Ru Huang *(Peking University, China)*
D6-4	**0114: Investigation on Asymmetric HfO_2-ZrO_2-HfO_2 Superlattice Gate Stacks with Ultra-low EOT for Advanced Transistors**
14:45 ~15:00	Haiyuan Lyu, Kun Zhong, Zhaohao Zhang, Huaxiang Yin *(Institute of Microelectronics, Chinese Academy of Sciences, China; University of Chinese Academy of Sciences, China)*
D6-5	**0119: Evaluation of Contact Resistance with the 'L' Kelvin Test Structure and the Modified Kelvin Test Structure**
15:00 ~15:15	Gui Chen, Yun-Hao Shao, Xin-Ping Qu *(Fudan University, China)*

Friday, October 25, 13: 30 – 15: 15   Meeting Room 5
Session E6: Reliability II   Sheraton Zhuhai Hotel 2nd Floor

Session Chair: Prof. Wangyong Chen, Sun Yat-sen University, China

	Title
E6-1	**0221: Characterization and Modeling of Non-conducting RF Hot Carrier Stress in FinFETs (invited)**
13:30 ~14:00	G. Niu, X. Ding, H. Zhang, W. Wang, K. Imura, F. Dai *(Auburn University, USA; Maxlinear Inc., USA)*
E6-2	**0164:Predictive Modelling of Hot Carrier Degradation (invited)**
14:00 ~14:30	James Brown , Kean Hong Tok , Rui Gao , Zhigang Ji , Weidong Zhang, Jian Fu Zhang *(Liverpool John Moores University, UK; No. 5 Electronics Research Institute of the Ministry of Industry and Information Technology, China; Shanghai Jiaotong University, China)*
E6-3	**0142: New Insights into the Saturation Behavior of the Hot Carrier Degradation in STI-based N-type LDMOS**
14:30 ~14:45	Zhuoqing Yu, Dan Gao, Yongsheng Sun, Junlin Huang *(Hisilicon, China)*
E6-4	**0081: The TID Response and HCI Degradation for multi-Vt nFinFETs**
14:45 ~15:00	Ruxue Yao, Hongliang Lu, Yuming Zhang, Yutao Zhang *(Xidian University, China)*

Friday, October 25, 13: 30 – 15: 15   Meeting Room 6
Session F6: Device Modeling II   Sheraton Zhuhai Hotel 2nd Floor
Session Chair: Prof. Xiaoyan Liu, Peking University, China

	Title
F6-1	**0059: Vertical Channel Transistor (VCT) for Advanced Logic and Memory Applications (invited)**
13:30 ~14:00	Mingmin Shi, Ran Bi, Ming Li *(Peking University, China; Beijing Advanced Innovation Center for Integrated Circuits, China)*
F6-2	**0423: High Precision I-V Characteristics SPICE Model for Silicon Carbide MOSFET**
14:00 ~14:15	Jinhong Shi, Yongxi Li, Jincheng Shi, Ruguan Li, Haimeng Huang, Hongqiang Yang *(University of Electronic Science and Technology of China, China; GRG Metrology & Test Group Co., Ltd., China)*
F6-3	**0440: An Analytical Model for Characterizing Density of States of Oxide Transistors**

14:15 ~14:30	Siyuan Hu, Chuanlin Sun, Junchen Dong, Zhensong Li, Kai Zhao, Dedong Han, Xing Zhang *(Peking University, China; Beijing Information Science & Technology University, China; Peking University Shenzhen Graduate School, China)*
F6-4	**0225: A Physics-Informed Neural Network Model for Body Potential Distribution in MOSFETs down to 50 K**
14:30 ~14:45	Honglin Wu, Fangxing Zhang, Xinyue Zhang, Baokang Peng, Wu Dai, Lining Zhang, Runsheng Wang, Ru Huang *(Peking University, China)*

Friday, October 25, 15: 30 – 17: 15

Friday, October 25, 15: 30 – 17: 15   Meeting Room 1
Session A5: FPGA Based Design   Sheraton Zhuhai Hotel 2nd Floor
Session Chair: Prof. Mingyu Wang, Sun Yat-sen University, China

	Title
A5-1	**0207: A Run-time Temperature Monitoring with Adaptive Duty Cycle Control for FPGA Applications**
15:30 ~15:45	Weizhi Li, Wangyong Chen, Haifeng Chen, Haoyu Zhang, Linlin Cai *(Sun Yat-sen University, China)*
A5-2	**0118: An FPGA-Based Top-K Gradient Compression Accelerator for Distributed Deep Learning Training**
15:45 ~16:00	Ziyao Wang, Jiayu Zhang, Kunyue Li, Jialei Sun, Feng Dong, Ke Chen, Yong Qiao, Jianfei Jiang *(National Key Laboratory of Advanced Micro and Nano Manufacture Technology, China; Beijing iQIYI Science & Technology Co.. Ltd.., China)*
A5-3	**0258: Dynamic-Matrix-Encryption Based Secure Strong PUF for Device Authentication Protocols**
16:00 ~16:15	Liangxiao Zhao, Gang Li, Pengjun Wang, Xuejiao Ma, Ziyu Zhou *(Wenzhou University, China; Wenzhou University of Technology, China; Ningbo University, China)*
A5-4	**0240: A Low Latency and High Throughout Hardware Design of Random Matrix Number Generator for FrodoKEM**
16:15 ~16:30	Shengfei Gu, Jiahao Lu, Tianze Huang, Jiaming Zhang, Kai Li, Cheng Wu, Mingbo Wang, Xianqi Mei, Ang Hu, Dongsheng Liu *(Huazhong University of Science and Technology, China; JinYinHu Laboratory, China)*
A5-5	**0465: A 4K60fps Ultra-Low-Latency Light Compression Encoder for Bandwidth-Constrained Scenarios**

16:30 ~16:45	Yanzhong Li, Leilei Huang, Yibo Fan *(Fudan University, China; East China Normal University, China)*
A5-6	**0086: Layer Pipelined Neural Network Accelerator Design on 2.5D FPGAs**
16:45 ~17:00	Mengxuan Wang, Chang Wu *(Fudan University, China)*
A5-7	**0237: Fast and Accurate Partial-Zoom Depth Estimation for SPAD LiDAR Readout on FPGA**
17:00 ~17:15	Lichen Feng, Hongwei Shan, Rundong Cai, Zhangming Zhu *(Xidian University, China)*

Friday, October 25, 15: 30 – 17: 15	Meeting Room 2
Session B7: Chip Test	Sheraton Zhuhai Hotel 2nd Floor
Session Chair: Prof. Shuo Li, Fudan University, China	

	Title
B7-1	**0135: In Situ Localization Techniques of Defects in Advanced Semiconductor Devices from Macroscale to Atomistic-scale (invited)**
15:30 ~16:00	Jialu Huang, Jingming Zhou, Zuoyuan Dong, Runsheng Wang, Junhao Chu, Xing Wu *(East China Normal University, China; Peking University, China)*
B7-2	**0402: Wafer-Level Characterization of Ring-Oscillators Frequency Degradation in FinFET Technology**
16:00 ~16:15	Hao Chang, Dan Gao, Yongsheng Sun, Junlin Huang *(Hisilicon Technologies Co., LTD, China)*
B7-3	**0246: Exhaustive Application-Dependent Testing for FPGA Interconnect Resources**
16:15 ~16:30	Wenwei Chen, Xinyu He, Tongshu Ding, Jian Wang, Jinmei Lai *(Fudan University, China)*
B7-4	**0303: A Comprehensive and Efficient Instruction-level Testing Method for Processor**
16:30 ~16:45	Zixin Yang, Zhichao Wei, Huanlin Luo, Jian Wang, Jinmei Lai *(Fudan University, China; Shanghai Academy of Spaceflight Technology, China)*
B7-5	**0006: Thermal Effect and Calibration for High Precision On-Wafer Analog IC Probe Testing**
16:45 ~17:00	Daisuke Iimori, Takayuki Nakatani, Shogo Katayama, Gaku Ogihara, Shuhei Yamamoto, Misaki Takagi, Yujie Zhao, Jianglin Wei, Anna Kuwana, Keno Sato, Takashi Ishida, Toshiyuki Okamoto, Tamotsu Ichikawa, Kentaroh Katoh, Kazumi

	Hatayama, Haruo Kobayashi *(Gunma University, Japan; Shenyang University of Chemical Technology, China; Yibin University, China; ROHM Semiconductor, Japan; Fukuoka University, Japan)*

Friday, October 25, 15: 30 – 17: 15   Meeting Room 3
Session C7: Sensor & MEMS III   Sheraton Zhuhai Hotel 2nd Floor
Session Chair: Prof. Zhanfeng Huang, Sun Yat-sen University, China

	Title
C7-1	**0460: Waterproof and Wearable Power Sources (invited)**
15:30 ~16:00	Sixing Xiong; Kenjiro Fukuda, Takao Someya *(RIKEN, Japan; The University of Tokyo, Japan)*
C7-2	**0146: A CMOS Pixel Sensor for Precise Track and Charge Measurement of Cosmic Ray Nuclei**
16:00 ~16:15	Ruikai Zhang, Wen He, Shanqiang Yang, Min Luo, Chenxu Wang, Cunfeng Feng, Meng Wang, Liang Zhang, Anqing Wang, Jianing Dong, Dong Liu, Yan Niu, Yang Zhou, Yuehong Gong, Xiaoli Wang, Shucheng Shi *(Harbin Institute of Technology, China; Shandong University, China; Institute of High Energy Physics Chinese Academy of Sciences, China; ShanDong JiaoTong University, China)*
C7-3	**0429: $Sc_{0.096}Al_{0.904}N$-Based Bimorph Piezoelectric Micro Machined Ultrasonic Transducers**
16:15 ~16:30	Ziye Zhai, Wenjuan Liu, Chengliang Sun *(Wuhan University, China)*

Friday, October 25, 15: 30 – 17: 15   Meeting Room 4
Session D7: 3D Integration   Sheraton Zhuhai Hotel 2nd Floor
Session Chair: Prof. Ming Li, Peking University, China

	Title
D7-1	**0283: HISIM: Design Exploration of 2.5D/3D Heterogeneous Integration for AI Computing (Invited)**
15:30 ~16:00	Zhenyu Wang, Pragnya Sudershan Nalla, Jingbo Sun, A. Alper Goksoy, Sumit K. Mandal, Jae-sun Seo, Vidya A. Chhabria, Jeff Zhang, Chaitali Chakrabarti, Umit Y. Ogras, Yu Cao *(Arizona State University, USA; University of Minnesota, USA; University of Wisconsin-Madison, USA; Indian Institute of Science, India; Cornell Tech, USA)*
D7-2	**Analysis of Current Status and Trends in Microsystem Integration Technology Based on TSV Advanced Packaging (Invited)**

16:00 ~16:30	Hua Yao *(Natural-Integration Advanced Semiconductor Technology Co., Ltd., China)*
D7-3	**0431: Flip 3D (F3D): A Novel 3D Integration Technology with Dual-side Integration Capabilities (invited)**
16:30 ~17:00	Heng Wu, Haoran Lu, Runsheng Wang, Ming Li, Yibo Lin, Weihai Bu, Jin Kang, Ru Huang *(Peking University, China)*
D7-4	**0035: Modeling and Simulation of A Conical 3D Monopole Antenna Embedded in Substrate for WNoC**
17:00 ~17:15	Junhao Wang, Ziyu Liu, Zhiyuan Zhu, Lin Chen, Qingqing Sun, Wei David Zhang *(Southwest University, China; Fudan University, China; Jiashan Fudan Institute, China)*

Friday, October 25, 15: 30 – 17: 15   Meeting Room 5
Session E7: Reliability III   Sheraton Zhuhai Hotel 2nd Floor
Session Chair: Prof. Zhigang Ji, Shanghai Jiaotong University, China

	Title
E7-1	**0463: Modeling and Parameter Extraction of Semiconductor Devices for Simulation and Design Optimization of ESD Protection Circuits on BCD Technologies for Automobile and Industry Applications (invited)**
15:30 ~16:00	Yuhua Cheng, Wei-wei Yu, Eugene Worley *(Peking University, China; Silicon Crossing, LLC, USA)*
E7-2	**0224: A Novel Double-zener Process and Multiplex Design for High-power Surge and High-speed ESD Devices Development**
16:00 ~16:15	Zhao Qi, YiRui Jia, Hongquan Chen, Ming Qiao, Zhaoji Li, Bo Zhang *(University of Electronic Science and Technology of China (UESTC), China;)*
E7-3	**0206: The Non-monotonic Instability of V_{TH} and $R_{ds,on}$ in P-GaN Gate HEMTs Under Repetitive Short Circuit Stress: The Role of Electric-field & Selfheating Effect**
16:15 ~16:30	Long Wang, Ning Yang, Shuting Huang, Jianggen Zhu, Kuangli Chen, Chao Feng, Haolin Hu, Wei Zeng, David Zhou, Yuxi Wan, Bo Zhang, and Qi Zhou *(University of Electronic Science and Technology of China (UESTC), China; Shenzhen Pinghu Laboratory, China)*

Friday, October 25, 15: 30 – 17: 15   Meeting Room 6
Session F7: Device Modeling III   Sheraton Zhuhai Hotel 2nd Floor
Session Chair: Prof. Chunlei Wu, Fudan University, China

	Title
F7-1	**0448: Neural Network Assisted Mosfets Model Development (invited)**
15:30 ~16:00	Xiaoyan Liu *(Peking University, China)*
F7-2	**0421:Modeling the Transient Characteristics with Trap Behaviors in LTPS-TFTs**
16:00 ~16:15	Haolin Li, Zheng Zhou, Xiaoyan Liu *(Peking University, China; Beijing Advanced Innovation Center for Integrated Circuits, China)*
F7-3	**0277: A Novel β-Ga$_2$O$_3$-Based Enhancement-Mode Transistor Combining Heterojunction Gate and Fin shaped Gate**
16:15 ~16:30	Yu Shao, Yunlong He, Xiaoli Lu, Songyao Wang, Xuefeng Zheng, Xiaohua Ma, Yue Hao *(Xidian University, China)*
F7-4	**0358: Electrical Characteristics and Thermal Reliability Investigation of TreeFET, FishboneFET, CombLikeFET and NSFET**
16:30 ~16:45	Mingyu Ma, Wenbin Wang, Haokun Li, Shujun Gao, Hailong You, Cong Li *(Xidian University, China)*

ICSICT 2024 Technical Sessions Overview

Date	Time	Meeting Room 1	Meeting Room 2	Meeting Room 3	Meeting Room 4	Meeting Room 5	Meeting Room 6
Oct.22	9:00-12:15	Tutorial Session T1 (Meeting Room 8)					
	13:30-18:30	Tutorial Session T2 (Meeting Room 8)					
	8:30-9:00	Opening (Grand Ball Room)					
	9:00-10:30	Keynote Session K1 (Grand Ball Room)					
	10:45-12:15	Keynote Session K2 (Grand Ball Room)					
Oct.23	13:30-15:15	Special Session the Future of AI	Session B1 Analog Circuit I	Session C1 EDA I	Session D1 Novel Device I	Session E1 Power Device I	Session F1 Memory Device I
	15:30-17:15		Session B2 Analog Circuit II	Session C2 EDA II	Session D2 Novel Device II	Session E2 Power Device II	Session F2 Memory Device II
	17:15-18:30	Poster Session I (1st Fl.)					
	19:00-21:00	Reception					
	9:00-10:30	Keynote Session K3 (Grand Ball Room)					
	10:45-12:15	Panel Discussion (Grand Ball Room)					
Oct.24	13:30-15:30	Session A1 AI Circuit	Session B3 Analog Circuit III	Session C3 RF Circuit I	Session D3 Novel Device III	Session E3 Power Device III	Session F3 Memory Device III
	15:30-17:15	Session A2 Security	Session B4 Mixed Signal I	Session C4 Sensor & MEMS I	Session D4 Novel Device IV	Session E4 Power Device IV	Session F4 Memory Device IV
	17:15-18:30	Poster Session II (1st Fl.)					
	9:00-10:30	Keynote Session K4 (Grand Ball Room)					
Oct.25	10:45-12:15	Session A3 Digital & Memory Circuit	Session B5 Mixed Signal II	Session C5 Sensor & MEMS II	Session D5 Process I	Session E5 Reliability I	Session F5 Device Modeling I
	13:30-15:30	Session A4 Processor	Session B6 Mixed Signal III	Session C6 RF Circuit II	Session D6 Process II	Session E6 Reliability II	Session F6 Device Modeling II
	15:30-17:15	Session A5 FPGA Based Design	Session B7 Chip Test	Session C7 Sensor & MEMS III	Session D7 3D Integration	Session E7 Reliability III	Session F7 Device Modeling III
	19:00-21:00	Closing & Banquet					

A High-Performance Multicore Testing Platform for Multi-Scenario Applications

Zipeng Ling [1*], Tianshu Zhuo [1], Zhuoyuan Yang [1], Jinhong Ye[1], JunHan[1*], Jingtao Zhang[2]

[1] State Key Laboratory of Integrated Chips and Systems, Shanghai 201203
[2] ZTE Corporation, Shenzhen 518057

* Email: 22212020111@m.fudan.edu.cn, junhan@fudan.edu.cn

Abstract—Cloud computing and high-performance data centers, as key pillars of computing infrastructure, face the challenges of rapidly growing network loads and high concurrency requests. Traditional data centers use modern processors to handle high bandwidth concurrent requests, and for the vast majority of network loads, do not require complex cache optimization strategies and adopt streaming processing, so there is no issue of cache consistency. This article proposes a high throughput multi-core platform for network processing, which includes input and output engines to complete data transmission and reception, as well as independent memory space and scheduler modules for each core. For the processing of network packets, due to the frequent switching of processor cores in different scenarios, the multi-core platform designed in this article can complete the scheduling and adaptation of processor cores for various application testing scenarios. The final experimental results show that the multi-core platform proposed in this article can efficiently process network packet processing requests in different scenarios simultaneously, achieving high data throughput.The platform that integrates four input/output engines and four PE cores can achieve a data throughput of 1.97Gbit/s in the scenario of forwarding packets.

Keywords—*Network packet processing, Multi-core platform, Multi-scene switching*

I. INTRODUCTION

Today, cloud computing and high-performance data centers have become key pillars of network infrastructure. They are composed of clusters connected by high-speed networks, where has a PB level data flow exchange or transmission every second. Emerging online services such as video communication, streaming, and online collaboration have increased the exchange traffic of cloud computing and high-performance data centers.

These requirements make network switching centers urgently need new hardware devices to handle such a large number of concurrent requests in parallel, among which network packet processing capability is a key indicator for handling concurrent requests. In the past, data centers used modern hyper threaded superscalar processors to handle high bandwidth concurrent requests, but due to the failure of Moore's Law, the performance of modern processors has grown to a bottleneck.

Therefore, current research hotspots have focused on dedicated accelerators and multi-core integrated networks. There are already many software and hardware technologies used in academia and industry to accelerate network packet processing[1]. In order to improve universality, decoupling, and scalability, the design of network packet processing accelerators tends to accelerate simple primitives. Di Girolamo et al. proposed PsPIN, a network computing accelerator based on RISC-V architecture general-purpose processor, which can be integrated into smart network cards

to accelerate packet processing[2]. The author proposes four indicators for network packet processing accelerators: locality(i.e. whether itcan be directly processed in network devices), programmability, programming granularity, and accessibility (i.e. whether user programs or library functions are allowed to directly configure network devices). At the same time, in order to improve processing bandwidth, multiple cores are often integrated. However, due to the complex cache consistency protocol and bus optimization of multi-core systems[3,5], it will increase a lot of unnecessary hardware overhead. For multi scenario switching in network processing, traditional multi-core platforms cannot efficiently switch corresponding processor cores for different scenarios, affecting processing bandwidth.

Based on these circumstances, this article designs and implements a multi-core platform with a configurable number of accelerators. Due to the fact that the processor core processes network packets in a streaming manner, there will be no data consistency issues with other cores. Therefore, multi-core platforms provide independent memory space for each core, reducing the overhead of caching in maintaining consistency protocols. For different processing scenarios, the scheduler of multi-core platforms can switch based on the processing requests in the data packet. The scheduler wakes up the PE to complete the corresponding processing by setting the CSR register of the processor core and using external interrupts.. At the same time, the multi-core platform has input and output engine , which can achieve simultaneous data transmission and processing between platform and processor, and improve throughput through FIFO and data crossover switches. The platform that integrates four input/output engines and four PE cores can achieve a data throughput of 1.97Gbit/s in the scenario of forwarding packets.

II. MULTI-CORE PLATFORM DESIGN

A. Overall architecture

The overall architecture of the multi-core platform implemented in the article is shown in the Fig.1 The multi-core platform mainly consists of IngressEngine 、 EgressEngine 、 CommandUnit 、 Program Loader and Context Loader.

The multi-core platform implemented in this article integrates four PE cores, each of which can run independently in different scenarios. PE is not within the scope of this article's design, it only needs to have network packet scene processing capabilities. The platform operation process is as follows:

1. IngressEngine reads and parses the input information, it sends the request information to the scheduler.

2. Schedule selects the corresponding idle PE core based on the parsing information.

3. When there are callable cores, the scheduler will send the core information to IgressUnit and the crossbar.

4. After receiving the response signal from the scheduler, IgressUnit will send the data to the input FIFO of the corresponding PE core through a crossover switch.

5. IngressEngine will give the scheduler a start signal after the handling is completed.

6. Then the scheduler will write the mcause register of the PE core. After that, the PE will exit the WFI state, run the interrupt processing function, and read the value of mcause to switch processing scenarios.

7. After PE completes data processing, the interrupt handling function will read the Credit register of EgressEngine to determine if there is enough space to store the results. If there is, it will write the corresponding output FIFO and set the high EOP signal to notify EgressEngine for transfer after completing the transmission.

8. The entire platform will cycle the above process until all packages are processed.

In addition, the platform also implemented a sharememory, which PE can access through atomic instructions to achieve data sharing between PEs. For scenarios such as sliding window algorithms that require maintenance of public variables, the Host CPU can pre-set the values of the corresponding variables, and finally the PE can access and update them atomically.

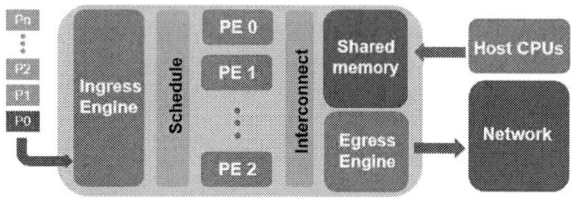

Fig. 1 Architecture diagram of testing platform

B. IngressEngine Design

IngressEngine is a module designed on network packet processing scenarios. which parses packet header information after receiving a network packet. The IngressEngine, as shown in Fig 2, consists of Igressunit, Mux, Cross Bar, and the Input FIFO for each PE. The IgressUnit will parse packet header information after receiving a network packert, then pass the parsed task information to the scheduler. After successfully sending a scheduling request, it will pause until the scheduler returns a response message. When IgressUnit receives the response message from schedule, it will load the remaining data of the packet into the input FIFO of the assigned PE in response message through the crossover switch. After all the data processed in a request is loaded into Input FIFO, IgressUnit will send a start signal to the scheduler, marking the completion of a task request. Subsequently, IgressUnit will read new packets from the bus again and repeat the above process. Each IgressUnit can independently process packet information and the quantity can be configured.

Through schedulers and cross switches, data can flow between any IgressUnit and PE, greatly improving processing bandwidth.

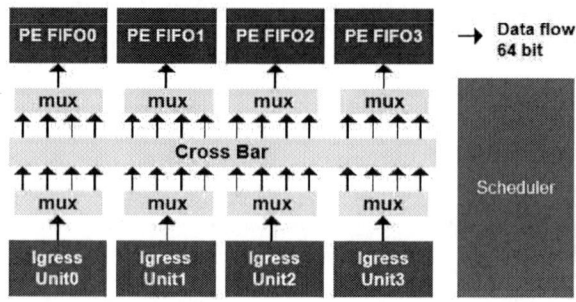

Fig. 2 Architecture diagram of Ingress Engine

The number of IgressUnit s can be configured. If the PE performs well in processing network packets and the input throughput is insufficient, a ratio of 2:1 or higher between IgressUnits and PE numbers can be achieved to enable PE to run at full capacity and improve network processing efficiency.

C. EgressUnit Design

EgressUnit is responsible for sending data processed by PE in the format of network packets . It needs to meet the characteristics of configurable quantity and high throughput. Internally, it consists of an output FIFO and a data handler. Each EgressUnit maintains a register named Credit, which represents the number of empty table entries in the FIFO.. After the PE completes the task processing, it will read all the EgressUnits' Credit to determines which EgressUnit to use. After confirmation, the PE will write the data to the corresponding EgressUnit's FIFO through a cross switch. When all data is transmitted, an EOP signal will be issued. Once the EgressUnit receives the EOP, it will start reading the data in the FIFO and send the data in the format of the network packet according to the requirements of the response network protocol.

Due to the decoupling structure of PE and EgressUnit, data forwarding and processing will not affect each other. And each PE can use all the EgressUnits on the platform, which can greatly improve the data transmission throughput.

D. Scheduler Design

Scheduler is designed based on the number of PEs and the type of work processing scenario, which maintains a status register for each PE. The status register is used to determine whether the PE is idle to receive new task requests.In order to improve scheduling efficiency and ensure that each PE has an equal probability of being scheduled, the scheduler adopts the Round Robin strategy to achieve scheduling.

When IgressUnit sends task request information, the scheduler will send the scheduled PE information back to IgressUnit and record its task type. When any IgressUnit sends a start signal, it will write the previously recorded corresponding task type into the mcause register of the PE, and then the PE will leave the WFI (wait for interrupt) state and start processing the task.

979-8-3503-6184-1/24 $31.00 © 2024 IEEE

III. EXPERIMENT RESULTS

In network processing, commonly used scenarios include loopback and sliding window algorithms. In order to test the data throughput of multi-core platforms in these scenarios, We added RAM to IgressUnit and EgressUnit respectively to simulate the packet sending and receiving process. We use scripts to generate test data packets in advance and store them in the RAM of IgressUnit. We also generate the correct result obtained from processing the test data packets in different scenarios and stored them in the RAM of EgressUnit. When EgressUnit forwards data to its own RAM, it can compare it with the correct results pre existing in RAM to determine the correctness of platform functionality. During the actual testing on this platform, a network packet processing engine with a dual emission six level pipeline and RISCV instruction set design (RV64I) was used as the PE.

We tested the throughput of platforms with integrated PE numbers ranging from 1 to 4 in the following scenarios: sliding window algorithm[6], ping-pong[2], histogram[7], and median calculation. The parameters of each component of the platform during testing are shown in Table 1. In order to test the impact of processor verification on platform throughput, we also tested the throughput of multi-core platforms in mixed scenarios and simply forwarding data packets. The test results are shown in the Fig.2.

Table.1 The parameters of platform

PE num	1~4
IgressUnit num	4
EgressUnit num	4
ShareMemory	32KB

From the Fig.2, it can be seen that the through-put of the multi-core platform is linked to the complexity of the testing scenario. The sliding window algorithm is the most complex computational scenario, with the lowest through-put. How-ever, transmit only requires sending packets from Ingress-Unit to EgressUnit, which is a simple algorithm. Therefore, the through-put can reach 1.97Gbit/s.In addition, we can see in the histogram that the throughput is significantly enhanced with the increase of the number of cores in each scenario. This is because the platform's ability to process data has increased, resulting in an increase in the amount of data it can receive. As the PE core requires a certain amount of time for initialization and wake-up, it is still the main limit.

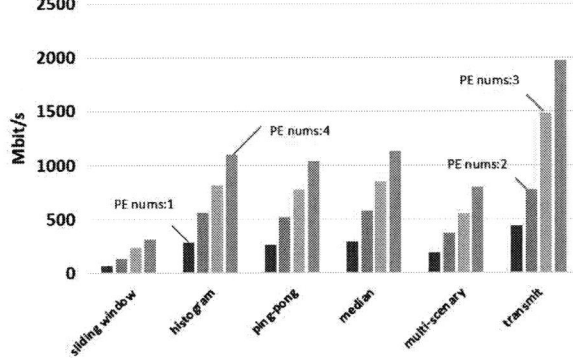

Fig. 3 Throughput of each scenario

IV. SUMMARY

This article designs a multi-core platform for network multi scenario processing, with configurable input and output engines and the ability to switch according to application scenarios. The scheduler can allocate idle PE for data packets that require different processing scenarios, improving the platform's operating bandwidth. By using data crossover switches to interconnect multiple PEs with IgressUnit and EgressUnit, the platform's input and output throughput is greatly improved. Under the influence of processor performance, the forwarding throughput can reach 1.97Gbit/s.

ACKNOWLEDGMENT

This work was supported by the National Natural Science Foundation of China under Grant 61934002 and 62234008.

REFERENCES

[1] Daniel Firestone, Andrew Putnam, Sambhrama Mundkur, et al. 2018. Azure accelerated networking: SmartNICs in the public cloud. In Proceedings of the 15th USENIX Conference on Networked Systems Design and Implementation (NSDI'18) [C]// USENIX Association, USA, 51–64.

[2] S. Di Girolamo et al., "A RISC-V in-network accelerator for flexible high-performance low-power packet processing," 2021 ACM/IEEE 48th Annual International Symposium on Computer Architecture (ISCA) [C]// Valencia, Spain, 2021, pp. 958-971, doi: 10.1109/ISCA52012.2021.00079.

[3] J. M. M. Junior, T. Khamvilai, L. Sutter and E. Feron, "Test platform for autopilot system embedded in a model of multi-core architecture using X-Plane flight simulator," 2019 IEEE/AIAA 38th Digital Avionics Systems Conference (DASC), San Diego, CA, USA, 2019, pp. 1-6, doi: 10.1109/DASC43569.2019.9081788.

[4] M. O. Gharan and G. N. Khan, "A Novel Virtual Channel Implementation Technique for Multi-core On-chip Communication," 2012 Third Workshop on Applications for Multi-Core Architecture, New York, NY, USA, 2012, pp. 36-41, doi: 10.1109/WAMCA.2012.12.K. Elissa, "Title of paper if known," unpublished.

[5] Y. Pan, J. Qiu, L. Chen, X. Gu, J. Chen and Y. Chen, "The research of multi-core parallel technology," 2012 8th International Conference on Natural Computation, Chongqing, China, 2012, pp. 1056-1059, doi: 10.1109/ICNC.2012.6234619.

[6] H. Lim, Y. Kim and K. Cheun, "An efficient sliding window algorithm using adaptive-length guard window for turbo decoders," in Journal of Communications and Networks, vol. 14, no. 2, pp. 195-198, April 2012, doi: 10.1109/JCN.2012.6253068.

[7] Torsten Hoefler. Distributed join algorithms on thousands of cores.Proceedings of the VLDB Endowment, 10(5), 2017

S-SIFT: A Simple SIFT Algorithm with High Efficiency

Yixue Wang[1,2], Yujie Huang*[1,2], Liyuan Peng[1,2], Mingyu Wang*[1], Wenhong Li[1],
Minge Jing[1], Xiaoyang Zeng*[1]

[1]State Key Lab of ASIC & System, Fudan University, 825 Zhangheng Rd, 200433, Shanghai, China
[2]Shanghai ExploreX Technology Co., Ltd., 188 Shengrong Rd, 200120, Shanghai, China

* Email: { mywang, huangyj19, xyzeng}@fudan.edu.cn

Abstract—Finding distinctions and connections between multiple visual targets through the detection of keypoints has become one of the research hot-spots in the field of computer vision. SIFT has received wide recognition and attention for its powerful performance in image match. However, the original SIFT has high complexity and time-consuming problems. In this paper, we propose Simple SIFT (S-SIFT), an improved SIFT algorithm in which the construction of Gaussian pyramid and the order of detection steps have been changed. Through the collaborative improvement and optimization on the scale space and algorithm framework, a more efficient local keypoints detecting method is achieved. Experiments demonstrat that S-SIFT guarantees satisfactory accuracy with a significant reduction in the computational overhead. Meanwhile, S-SIFT is robust to common image deformations such as rotation, scale change and affine.

Keywords—SIFT, S-SIFT, keypoints detection, image match, computer vision

I. INTRODUCTION

Keypoint detection is a basic and critical process in the field of vision, which perceives and connects two image targets with the same or similar attributes. The underlying layer of important applications such as simultaneous localization, maping, image stitching, and object recognition etc., are all based on keypoints being reliably extracted and matched across images.

SIFT algorithm [1] has been proposed by Canadian professor David G. Lowe in 1999 and been perfected in 2004 [2], which maintains good stability to image transformation. Along the SIFT algorithm, Bay first proposed the SURF algorithm in 2006 [3]. SURF alleviates the shortcomings of SIFT's high computational complexity and time-consuming. However, the increase in speed is accompanied by the decline in performance especially in scale and rotation changes. ORB proposed in 2011 [4], locates the corner points by FAST algorithm and uses gray scale center of mass method to obtain the orientation. The descriptor uses a 128-bit binary more concisely than sift. These greatly speed up the keypoints description computation and matching, but the false match rate increases at the same time. Keypoint features of images have been widely used due to their excellent properties, and subsequently many improved algorithms [5-8] have emerged, such as the GLOH algorithm [5], the Daisy algorithm [6], the AB-SIFT algorithm [7], and the PCA-SIFT algorithm [8], etc.

Methods based on deep learning, such as D2Net[9] and Patch2Pix[10], show excellent performance in keypoints detection. Deep learning largely avoids the requirement for human experience and prior knowledge. However, the image information involved in deep learning such as gradient is still at low level, and they are largely dependent on the framework and dataset. At the same time, the computational overhead is high, and most deep learning solutions run in poor real-time and require GPU acceleration.

SIFT for keypoints detection has been proven to be successful in many applications and is still one of the most popular and powerful algorithms today. Due to the complexity of the algorithm, the speed of operation has been limited. Some processes in the sift algorithm were found to be unnecessary in our experiments, and framework is still worth investigating. This paper will address the executability of some processes as well as propose new frameworks and algorithms based on SIFT. The aim is to simplify the computation as much as possible without compromising its performance and improve the execution efficiency of the algorithm.

II. REVIEW OF THE SIFT ALGORITHM

SIFT (Scale Invariant Feature Transform), which transforms image data into scale invariant coordinates with the local features. The computational stages for generating image keypoints are as follows.

A. Scale-space extrema detection

Detecting scale-space extremes involves first calculating image locations on all scales by constructing Gaussian pyramid and difference of Gaussian(DOG) pyramid.

The Gaussian pyramid of an image is defined as the function $L(x, y, \sigma)$, which is generated by the convolution of a variable scale Gaussian function $G(x, y, \sigma)$ with the input image $I(x, y)$:

$$L(x, y, \sigma) = G(x, y, \sigma) * I(x, y) \qquad (1)$$

For each octave of the Gaussian pyramid, images are repeatedly convolved with the Gaussian function to produce the set of gray scale space images shown in Fig.1(a). After the last layer of current octave is generated, the Gaussian image is downsampled by a factor of 2 to cte a new octave and the process is repeated until the build is completed.

Point $D(x, y, \sigma)$ in DOG pyramid can be calculated from the difference between two neighboring Gaussian images.

$$D(x, y, \sigma) = \big(G(x, y, k\sigma) - G(x, y, \sigma)\big) * I(x, y)$$
$$= L(x, y, k\sigma) - L(x, y, \sigma) \qquad (2)$$

As shown in Fig.1(b), neighboring Gaussian images are subtracted to obtain the DOG image on the right.

Compare the middle point with the other 3×3×3-1 points within neighborhood of the same scale and the corresponding

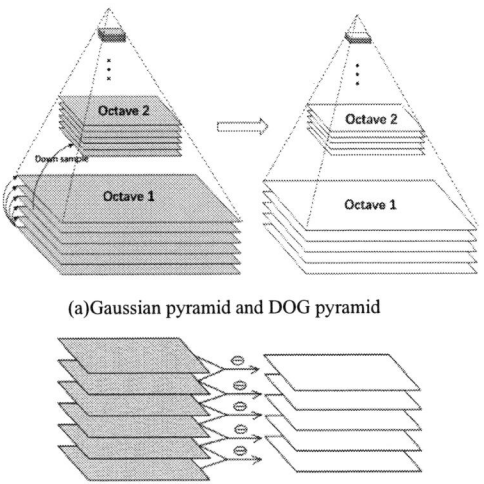

(a)Gaussian pyramid and DOG pyramid

(b)Gaussian pyramid construction

Fig. 1.　The Construction of Scale Space

neighborhoods of adjacent scales to successfully obtain the extreme points in the 2D image space and scale space, so that the keypoints detected through the initial localization have both positional coordinates and scale coordinates.

B. Keypoint localization

When the extreme points are detected, a problem can be found that the extreme points are not accurate enough as the Gaussian difference pyramid is discrete. It is likely that the true extreme point is nearby, and the Taylor expansion needs to be utilized in order to find the extreme point with higher sub-pixel positional accuracy. Points with low contrast and on edges are determined as the unstable extreme points which should be inhibited.

C. Orientation assignment

Orientations are assigned to each keypoint location based on the local image gradient orientation. All future operations are performed on image data that is transformed relative to the orientation, scale, and position specified for each feature.

D. Keypoint descriptor

The keypoint descriptor is created by calculating the gradient magnitude and direction for each image sample point in the region surrounding the keypoint location.

III. ALGORITHM

A. Construction of Scale Space

The process of constructing the scale space for SIFT algorithm can be summarized as Gaussian kernel convolution and downsampling. The purpose is to simulate the degree of blurring and size of objects in vision to mitigate the effect of scale on keypoints extraction. David G. Lowe upsamples the original image to have better results. However the process of up-sampling is not essential for normal images. Up-sampling using bilinear interpolation expands the width and height of the image, resulting in an image that is not precise enough to extract keypoints. At the same time, it creates a lot of duplication of keypoints with the original image scale, which leads to a dramatic increase in the number of keypoints in the up-sampled image. The repetition of keypoints results in a sharp drop in the final match rates. For computational vision,

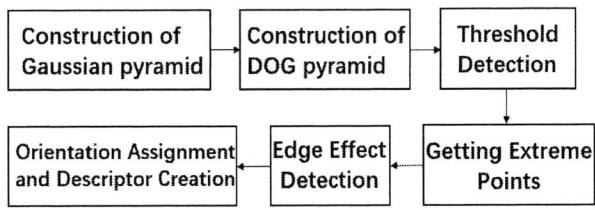

Fig. 2.　S-SIFT block diagram

higher accuracy of matching and more uniform distribution of keypoints are far more important than the number of keypoints. Up-sampling brings a huge amount of computation and increased time consuming, but it does not bring better matching accuracy and efficiency.

For the top level of the pyramid, the image resolution is too small to extract matchable keypoints. It is clearly unnecessary to bring in calculations to get more layers of the pyramid. Thus, The number of Gaussian pyramid groups O is set as:

$$O = \log_2 min(M, N) - 4 \qquad (1)$$

where M, N are the rows and columns of the original image. Make the top level image width and height must be at least 32 pixels.

B. Algorithm Framework

Caculating by Taylor expansion after getting extreme points leads to more accurate location of keypoints, but it also causes a significant computation consumption. The stable extreme points obtained in the Gaussian pyramid are actually sufficient to some extent for the alignment of the two images.

In order to obtain the stable keypoints, it is still need to do further refinement upon extreme points, which mainly includes inhibition of low contrast points and removal of edge response points. Since there is no need to position the keypoints with Taylor expansion, points with low contrast should be removed out in advance to reduce subsequent calculations, which can be done before getting extreme points in the DOG pyramid.

Unlike the original SIFT, S-SIFT changes the order in detecting keypoints. Fig 2 shows the S-SIFT block diagram. After constructing the DOG pyramid, the points $D(x, y, \sigma)$ less than the contrast threshold can be filtered away. Since low contrast points have been inhibited, unnecessary calculations in the subsequent process are effectively diminished. Changing the order of algorithm steps not only improves the speed of algorithms in software, but also suppresses redundant signal activity in hardware designs. Then extreme points and edge effect detection are carried out in the remaining points. Orientation assignment and keypoint descriptor is consistent with the original SIFT.

IV. EXPERIMENTS

In our experiments, compared methods include SIFT and SURF. Meanwhile, we use the code and the same set of parameters provided by the OpenCV to obtain the results for comparison. In S-SIFT, we set the contrast threshold to 1 and the rest of parameters are the same as in the original sift. Three groups of images(Fig.3) are used for our tests, where the right reference image is obtained by transforming the left one. In the transformations, we added rotations, scale scaling and affine to test. The resolution of the picture is 1920×1080, 3840×2160 and 2000×1500 in sequence. We use KNN [11]

Fig. 3.   Images Used for Testing

TABLE I.   THE PERFORMANCE COMPARISION

	Time	Keypoints num	Matches	Match correctness
Group1+SIFT	4.566s 7.108s	7731 11052	4234	45.08%
Group1+SURF	1.61s 2.464s	13713 18189	4959	34.95%
Group1+S-SIFT	1.283s 1.974s	3202 4657	1961	49.90%
Group2+SIFT	12.554s 17.849s	10606 9903	3210	31.46%
Group2+SURF	4.957s 6.504s	30829 33823	6902	21.35%
Group2+S-SIFT	3.094s 3.935s	5710 5964	1996	34.20%
Group3+SIFT	10.223s 13.269s	24505 29973	9833	36.10%
Group3+SURF	5.049s 5.485s	57397 52222	5587	10.19%
Group3+S-SIFT	2.891s 3.255s	8429 8901	3802	43.88%

TABLE II.   KNN TIME CONSUMPTION

	SIFT	SURF	S-SIFT
Group1	2.732s	3.464s	0.373s
Group2	2.863	14.19s	0.783s
Group3	21.687s	43.146s	1.704s

for matching, which is the most widely used algorithm. Since the transformation matrix of each image is already known, we can get the correct matching point by calculation. It is considered that the matching points are correct if the detected point is within 5 pixels of the correct match.

The proposed S-SIFT, the original SIFT and SURF method are respectively adopted to conduct matching experiment,. The results are shown in TABLE I and TABLE II. Match correctness is calculated as the number of matching points divided by the average of keypoints number in two original images.

Although SURF can get a large number of keypoints in a short time, the accuracy is much smaller than other algorithms,

which is less than a third of SIFT accuracy in Group3. At the same time, the KNN matching time increases dramatically due to the large number of keypoints. In applications such as image stitching and object recognition, a few hundred keypoints are enough to achieve good results. While the number of keypoints is reduced, S-SIFT's computation and matching times drop significantly, and its accuracy rises by few percentage points.

V. CONCLUSION

Aiming at high computation complexity and difficulty for practical use of the original SIFT , we propose a simpler and more efficient algorithm S-SIFT for keypoint detection. Through experiments we find that the upsampling process is actually unnecessary for generally large image resolutions. Meanwhile, S-SIFT remove the step of Taylor expansion and changed the sequence of steps. Through collaborative improvement and optimization on construction of scale space and algorithm framework, the performance of the algorithm has been improved significantly. Experiments demonstrate that the matching accuracy of S-SIFT is improved by few percentage points compared with sift, and S-SIFT is more robust to image transformations. While ensuring satisfactory accuracy, the computation time for extracting keypoints is reduced to less than a quarter of the original, and the KNN matching time can even be dropped substantially to 10% in Group3. However, the work in this paper leaves something to be desired. The generation of descriptors requires the utilization of information around keypoints, which is bound to demand a large amount of time consuming and storage. Further optimization can be done for the generation of keypoint descriptors.

REFERENCES

[1] Lowe D G. Object Recognition from Local Scale-Invariant Features[C]. Proceedings of the International Conference on Computer Vision,1999.

[2] Lowe D G. Distinctive Image Features from Scale-Invariant Keypoints [J]. International Journal of Computer Vision, 2004, 60(2):91-110.

[3] Bay H, Ess A, Tuytelaars T and Van Gool L. Speeded-Up Robust Features (SURF)[J]. Computer Vision and Image Understanding, 2008, 110(3):346-359.

[4] Rublee E, Rabaud V, Konolige K and Bradski G. ORB: an efficient alternative to SIFT or SURF [C]. International Conference on Computer Vision(ICCV),2011: 2564-2571.

[5] Mikolajczyk K, Schmid C. A Performance Evaluation of Local Descriptors[J]. IEEE Transactionson Pattern Analysis and Machine Intelligence,2005,27(10):1615-1630.

[6] Tola E, Lepetit V, Fua P. Daisy:An Efficient Dense Descriptor Applied to Wide Baseline Stereo[J]. IEEE Transactions on Software Engineering,2010,32(5):815-830.

[7] Sedaghat A, Ebadi H. Remote Sensing Image Matching Based on Adaptive Binning SIFT Descriptor[J]. IEEE Transactions on Geoscience and Remote Sensing,2015,53(10):5283-5293.

[8] Ke N Y, Sukthankar R. PCA-SIFT:A More Distinctive Representation for Local ImageDescriptors[C]. Proceedings of the 2004 IEEE Computer Society Conference on Computer Visionand Pattern Recognition,2004.

[9] Dusmanu M et al. D2-Net: A Trainable CNN for Joint Detection and Description of Local Features. Proceedings of the IEEE Conference on Computer Vision and Pattern Recognition(CVPR), 2019.

[10] Zhou Q, Sattler T and Leal-Taixe L. Patch2pix: Epipolar-guided pixel-level correspondences. Proceedings of the IEEE/CVF conference on computer vision and pattern recognition, 2021:4669-4678.

[11] Abeywickrama T, Cheema MA and Taniar D. k-Nearest Neighbors on Road Networks: A Journey in Experimentation and In-Memory Implementation. Proceedings of the VLDB Endowment,2016(9):492-503.

Design of a High-Speed SAR Processor Based on the Chirp Scaling Algorithm

Xianghe Cui, Yukun Song, Yurun Zhang, Jingyi Hu, Zhenmin Li, Duoli Zhang*

Department of Microelectronics, Hefei University of Technology, Hefei 230061, China

* Email: zhangduoli@hfut.edu.cn

Abstract—**This paper investigates the data stream structure in the imaging process based on the Chirp Scaling Algorithm (CSA) and proposes a pipelined high-performance Synthetic Aperture Radar (SAR) imaging processor architecture. Firstly, a multi-channel pipelined data flow structure that can match different operating frequencies and storage bandwidths is designed to reduce storage dependencies with full exploitation of pipeline efficiency. Secondly, based on the unified model of compensation functions, this paper implements a compensation function generation unit to save computational resources while ensuring calculation accuracy. Thirdly, a double matrix block data storage strategy is proposed, which improves the average read and write efficiency by 1.36 times compared to ordinary block strategies. This processor supports variable imaging size ranging from 1K to 32K, with the speed of processing 32K×32K, 16K×8K, and 2K×2K size images in 14.1s, 1.8s, and 0.062s respectively.**

Keywords—synthetic aperture radar (SAR), heterogeneous architecture, pipeline organization

I. INTRODUCTION

Synthetic Aperture Radar (SAR) is an active microwave imaging radar widely used in earth remote sensing and military reconnaissance [1]. Currently, SAR is developing towards high real-time performance and equipment miniaturization [2], while also having stringent requirements for observation range, observation resolution, and power consumption. Traditional CPU, DSP, and GPU are not capable of handling large-scale real-time SAR imaging processing tasks well, necessitating the design of specialized high-performance SAR processors.

Dong et al. [3] utilized multi-core DSP for SAR imaging, offering advantages of high precision and configurability. However, when facing large-scale and high-precision tasks, it consumes excessive power and has relatively slow imaging speed. Chan et al. [4] developed a SAR processor based on the Range-Doppler algorithm (RDA) on FPGA. This processor stores miscellaneous calculations in the RDA on-chip after MATLB computation, reducing the design complexity. Cao et al. [5] designed a SAR processor based on the Chirp Scaling algorithm (CSA) on FPGA, which reconstructs the algorithm into dynamic and static parts, saving resources and reducing power consumption, but it is slow in imaging large-scale data. The FPGA-based designs in [4] and [5] do not align with the trend towards miniaturization of carriers.

In response to the development trends of SAR imaging and the limitations of existing SAR imaging processors, this paper proposes a high-performance SAR processor using the CSA. This architecture features a multi-channel pipelined storage and computation structure, with the number of channels determined by the target frequency and external storage bandwidth. Additionally, a dual matrix block storage strategy is proposed to fully utilize DDR4 bandwidth. Furthermore, a simplified model for the phase compensation function is introduced to improve computational efficiency and conserve storage resources.

II. CS IMAGE-PROCESSING ALGORITHM

The CSA, illustrated in Fig. 1, stands as one of the more mature algorithms currently available. Employing Fast Fourier Transform (FFT) and complex multiplication for frequency domain modulation, it achieves signal scaling and shifting, thereby addressing the limitations of the RDA, which requires interpolation calculations for range cell migration correction. Not only does the CSA effectively process echo signals, but it also enhances computational efficiency [6].

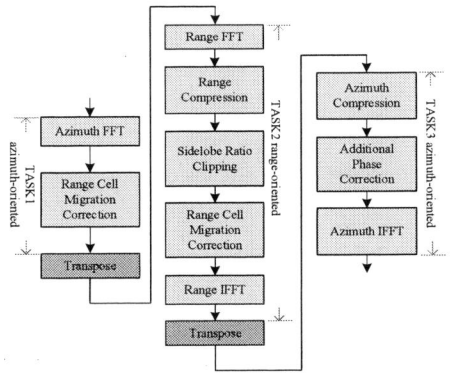

Fig. 1. CSA processing flow

III. SYSTEM ARCHITECTURE

A. Overall Architecture

Fig. 2 presents the heterogeneous CSA accelerator architecture designed in this paper. The system consists of a task control unit, computation unit, array storage unit, and storage controller. Through the task control unit, the system configures information such as data length and computation type for the computation unit and storage unit, thereby achieving acceleration of the CSA.

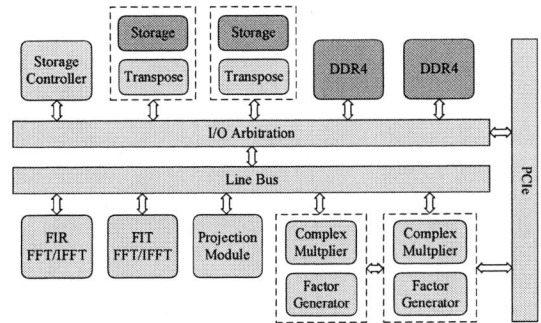

Fig. 2. SAR processor architecture

979-8-3503-6184-1/24 $31.00 © 2024 IEEE

The overall system structure adopts a pipelined architecture. In each task, data is read from the source into array storage. Once ready, the data flows to the computation units. After task completion, it is transferred back to array storage before being written to the destination, without passing through storage units in between. Both storage and computation units have pipelined structures, allowing data to enter the next unit directly after output from the previous one, without requiring reorganization, thus improving task computation efficiency. Fig. 3 illustrates the space-time diagram of this structure. The computational performance of this structure depends on the longest processing stage, therefore, it is necessary to balance the processing time of each module to fully utilize pipeline efficiency.

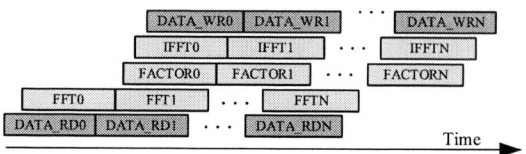

Fig. 3. Space-Time diagram of SAR processor

B. Computational Unit Design

1) FFT Unit Design：

In CSA, as shown in Table I, the performance of FFT determines the performance of the entire imaging system. Therefore, this paper designs a butterfly unit based on the Single-path Delay Feedback (SDF) structure. Compared to traditional butterfly units, it can start the next level of computation without waiting for the completion of each level, offering advantages of high parallelism and high throughput, while also saving storage units. Fig. 4 presents a multi-channel FFT structure composed of multiple sets of 12-level butterfly units and one base-N butterfly unit. This structure can achieve maximum bandwidth utilization and supports FFT operations from 1K to 32K points.

TABLE I. COMPUTATIONAL SCALE

Calculation type	Computational load ratio
FFT/IFFT	90%
Factor Generation	6.4%
Factor Generation	3.6%

Fig. 4. 8-channel SDF FFT

In some tasks, both FFT and Inverse Fast Fourier Transform (IFFT) operations need to be processed simultaneously. If the same FFT computation unit is used, the data order needs to be changed between the two computations. To save time consumed by data transformation, this paper designs two types of computation units: frequency-decimation-based FFT and time-decimation-based FFT. This method allows the next FFT task to start without the need for data reordering, significantly reducing task occupation time.

2) Factor Computation Unit Design：

This paper analyzes the phase compensation functions of various algorithms and discovers that there are many miscellaneous operations that account for a small proportion of the total computational load. These operations are diverse in nature. If a direct calculation approach is adopted, it would require a large number of addition, subtraction, multiplication, division, and trigonometric function units. On the other hand, storing all phase compensation functions on-chip would consume substantial storage resources. Therefore, this paper unifies the phase compensation functions into a single function model. The miscellaneous factors are calculated by a general processor, while the operations that account for a larger proportion of the computational load are implemented on-chip.

$$f(x, y, a, b) = \exp[j \times (x \times y - a)^n \times b] \quad (1)$$

Fig. 5 shows a factor computation unit that can implement phase compensation functions using a unified hardware module, thereby simplifying hardware design complexity. In practical operation, the task scheduler uploads the task mode, and the general processor transfers the pre-calculated phase vectors x, y, a, and the constant factor b into the storage units of the factor computation module. The address control module controls data retrieval based on the direction and number of current echo vectors. It then calculates the phase compensation vector corresponding to the current echo vector. When encountering special functions, this module supports direct transmission of phase compensation functions from general processors to result storage. To match data bandwidth, multiple sets of storage units and computation modules are combined to form factor computation module.

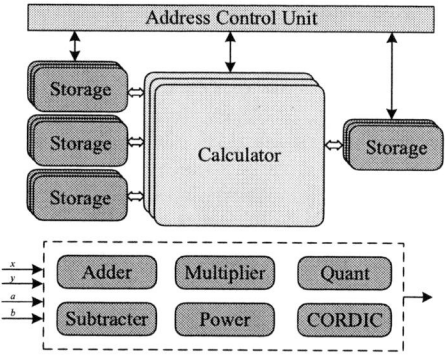

Fig. 5. Factor computation unit

C. High-efficiency Transpose Storage Unit

This paper uses a 64-bit wide DDR4 with a burst length of 8. The optimal read-write strategy for this DDR4 is to perform a burst, then switch Bank Group for another burst, followed by switching Bank for another burst. The DDR4 we use consists of 2 Bank Groups and 4 Banks. Therefore, one cycle comprises 64 sets of 64-bit data. Based on this, the paper divides the echo data into 8×8 data blocks.

If the matrix is divided into 8×8 blocks, the data read from DDR has a bandwidth of 90Gb/s. However, when transposing and writing the data, it requires continuous cross-row writing, resulting in the lowest DDR efficiency of about 10Gb/s, which causes a severe mismatch in read and write bandwidths. Fig. 6 illustrates a double matrix blocking method that divides the matrix into 64×64 blocks, and then further divides each 64×64

979-8-3503-6184-1/24 $31.00 © 2024 IEEE

matrix block into 8×8 sub-blocks to partition the echo matrix. This approach ensures that the amount of data read and write in each row is equal, balancing read and write efficiency. Fig. 7 describes the pseudocode logic for generating read and write addresses.

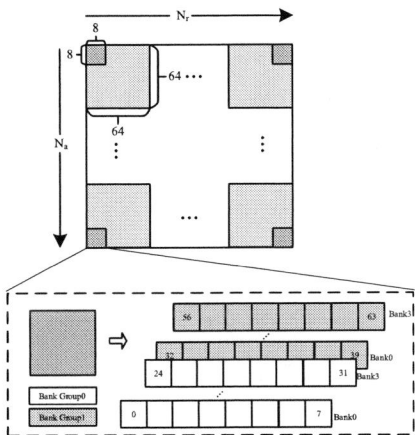

Fig. 6. Data partitioning and storage strategy

Algorithm 1 RD_Addr_Gen

```
1:  rd_addr ← 0;                              ▷ Initialize rd_addr to 0
2:  while 1 do                                ▷ Infinite loop
3:      for count1 = 1 to 64 do
4:          if count1 = 64 then
5:              rd_addr ← rd_addr + 64;
6:          else
7:              rd_addr ← rd_addr + 8;
8:          end if
9:      end for
10: end while
```

Algorithm 2 WR_Addr_Gen

```
1:  wr_addr ← 0;                              ▷ Initialize wr_addr to 0
2:  count3 ← 0;                               ▷ Initialize count3 to 0
3:  while 1 do                                ▷ Infinite loop
4:      if data_valid = 1 then
5:          for count2 = 1 to 8 do
6:              if count2 = 8 then
7:                  count3 ← count3 + 1;
8:                  if count3 = 7 then
9:                      wr_addr ← wr_addr + column_distance;
10:                 else
11:                     wr_addr ← wr_addr + 448;
12:                 end if
13:             else
14:                 wr_addr ← wr_addr + 8;
15:             end if
16:         end for
17:     end if
18: end while
```

Fig. 7. DDR read-write address

IV. EXPERIMENTS AND ANALYSIS

The system's functionality was verified on a Xilinx UltraScale+ XCVU13P. The system operates at a clock frequency of 150 MHz. Based on the system frequency and DDR4 bandwidth, the number of channels is set to 8. The processing speed for imaging echoes of different sizes is shown in Table 2, and the resource utilization is presented in Table 3. The imaging speed of this accelerator outperforms general-purpose processors and existing SAR imaging processors.

TABLE II. ECHO IMAGING TIME

Echo Size	Imaging Time(s)
32K×32K	14.1
16K×8K	1.8
2K×2K	0.062

TABLE III. RESOURCE UTILIZATION TABLE

Resource	Estimation	Utilization(%)
LUT	918200	53.14
FF	782115	22.63
DSP	2566	20.88
BRAM	1301	48.40
URAM	64	5.00

Comparing the images generated by this SAR imaging accelerator with those produced by MATLAB, the calculation between the two images yields a PSNR of 34.3dB. With the PSNR value exceeding 20dB, it can be concluded that the image quality is not affected. Due to data limitations, the large-scale imaging functionality is verified using point target images. The imaging results based on real data are shown in Fig. 8.

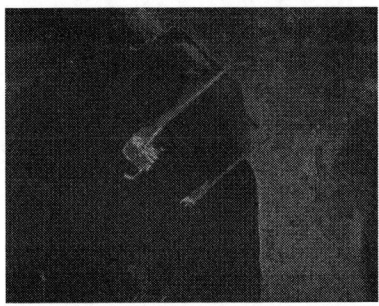

Fig. 8. SAR imaging results

V. CONCLUSION

This paper proposes a high-performance heterogeneous SAR processing architecture based on the CSA and verifies it on FPGA. First, the imaging process of the CSA is introduced, based on which a pipelined acceleration structure is designed. The multi-channel SDF structure FFT unit, factor generation unit, and data storage transposition unit are described in detail. Finally, the correctness of the system is verified through imaging on a hardware platform. This technology aims to address the real-time requirements of SAR systems by designing a high-performance heterogeneous SAR imaging structure, laying the foundation for high-performance real-time SAR imaging.

REFERENCES

[1] A. Moreira, "A golden age for spaceborne SAR systems," 2014 20th International Conference on Microwaves, Radar and Wireless Communications (MIKON), Gdansk, Poland, 2014, pp. 1-4.

[2] E. Schreiber, A. Heinzel, M. Peichl, M. Engel and W. Wiesbeck, "Advanced Buried Object Detection by Multichannel, UAV/Drone Carried Synthetic Aperture Radar," 2019 13th European Conference on Antennas and Propagation (EuCAP), Krakow, Poland, 2019, pp. 1-5.

[3] L. Dong, X. Meng and D. Zhu, "High-Squint SAR Imaging Technique Based on Multi-Chip DSP," 2021 7th Asia-Pacific Conference on Synthetic Aperture Radar (APSAR), Bali, Indonesia, 2021, pp. 1-5.

[4] Chan, Yee Kit, Yung Chong Lee, and Voon Chet Koo, "Design and Implementation of Synthetic Aperture Radar (SAR) Field-Programmable Gate Array (FPGA)-Based Processor" Appl. Sci. 2022, 12, 1808.

[5] Y. Cao, S. Jiang, S. Guo, W. Ling, X. Zhou and Z. Yu, "Real-Time SAR Imaging Based on Reconfigurable Computing," in IEEE Access, vol. 9, pp. 93684-93690, 2021.

[6] M. Y. Jin, F. Cheng and Ming Chen, "Chirp scaling algorithms for SAR processing," Proceedings of IGARSS '93 - IEEE International Geoscience and Remote Sensing Symposium, Tokyo, Japan, 1993, pp. 1169-1172 vol.3.

979-8-3503-6184-1/24 $31.00 © 2024 IEEE

Accelerating Matrix-Vector Multiplications of Large Language Models via Efficient Encoding

Yongjin Tao, Wendi Sun, Song Chen*, Yi Kang

School of Microelectronics, University of Science and Technology of China, Hefei 230026, China

*Email: tyj2018@mail.ustc.edu.cn, songch@ustc.edu.cn

Abstract—**Large language models (LLMs) have demonstrated remarkable success across various tasks. However, model inference remains a significant challenge due to frequent memory access and low compute-to-memory ratio. Model compression and hardware acceleration are promising approaches to address this issue. In this paper, we propose CORO for accelerating matrix-vector multiplications (MVMs) in LLMs, that adopts a non-uniform quantization method which processes weights with variable-length data representation. Our approach leverages the concentration of two adjacent quantized values into small ranges, allowing us to pair them together and represent them with fewer bit-widths via encoding. This method, based on characteristics of data distribution, can significantly reduce memory requirements and can be integrated into the existing accelerators such as coarse-grained reconfigurable arrays (CGRAs), which combine flexibility and energy efficiency. As a result, our accelerator achieves an average speedup of $1.7\times$ compared to prior work while maintaining superior model accuracy.**

Keywords—**Large language models, Matrix-vector multiplications, Quantization, Coarse-grained reconfigurable arrays**

I. INTRODUCTION

Decoder-only large language models, such as GPT4 [1], LLaMA2 [2] and GLM [3], have demonstrated superior performance in recent years. They are designed to generate context sequentially in an autoregressive manner by generating token one by one, taking into account the preceding generated tokens. The success of LLMs comes with extremely large model size. For instance, LLaMA2 has up to 70 billion parameters, leading to over 140GB memory requirements. Additionally MVMs contribute significantly to latency during edge inference, resulting in low hardware utilization. As shown in Table I, LLaMA2-7B has more than $78\times$ parameters than the encoder-based ViT-B-16 [4], while arithmetic intensity per parameter is much lower than that of ViT-B-16. Thus the performance of LLMs inference suffers from memory bandwidth constraints.

Model compression is a promising approach to alleviate the bandwidth demands imposed by LLMs. Quantization stands out as one of the most effective methods to reduce memory requirements by employing lower precision formats such as 8-bit integers to represent model weights and activations. SmoothQuant [5], for instance, is a post-training quantization technique that mitigates the challenges of quantization, shifting the focus from activations to weights in LLMs. To further enhance compression ratios, SPARK [6] utilizes variable-

TABLE I
PARAMETER AND OPERATION/PARAMETER RATIO COMPARISON OF ViT AND LLMS

Model	ViT-B-16	LLaMA2-7B	ChatGLM-6B
Params(B)	0.086	6.74	6.24
FLOPs/Params	390	1.97	1.93

length data representation to accommodate different value ranges, while Olive [7] prunes normal values to create space for outliers. However, they adapte systolic arrays for matrix-matrix multiplications, which suffer from low hardware utilization in accelerating MVMs. Additionally, errors introduced by coding can significantly impact the accuracy of LLMs.

In this paper, we introduce CORO for accelerating MVMs, adopting a non-uniform quantization based on a novel data representation called *mix*. The length of data in *mix* format varies among diverse value ranges. For instance, an 8-bit integer ranging from -8 to 7 can be encoded as *mix3* ("3" denotes the data length).

We make the following contributions in this paper:

- We introduce a novel data representation called *mix* to save bit-widths, which retains fine values and encodes every three coarse values as their intermediate value due to non-uniform margins of errors.
- We propose a non-uniform quantization method, which is based on *mix* format. We pair two adjacent values if they could be represented as *mix3* and *mix4* respectively. Otherwise, the values are encoded as *mix7*. This method achieves memory alignment and significantly reduces the memory requirements with negligible accuracy loss.
- We develop an efficient CGRA architecture and implement the decoding process as a configuration instruction to support our encoding scheme. We demonstrate that our architecture outperforms prior work in hardware utilization and performance.

II. QUANTIZATION

In this section, we introduce the encoding and decoding processes of *mix*. Furthermore, we employ this data representation to quantize the weights of LLMs, thereby reducing memory requirements while maintaining superior model accuracy.

979-8-3503-6184-1/24 $31.00 © 2024 IEEE

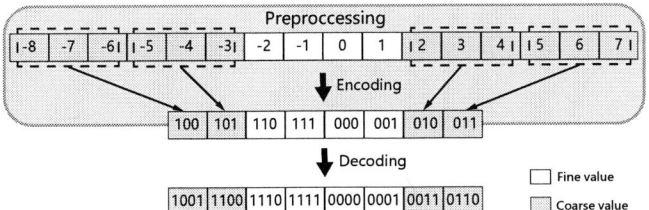

Fig. 1. The encoding and decoding processes for *mix3* format.

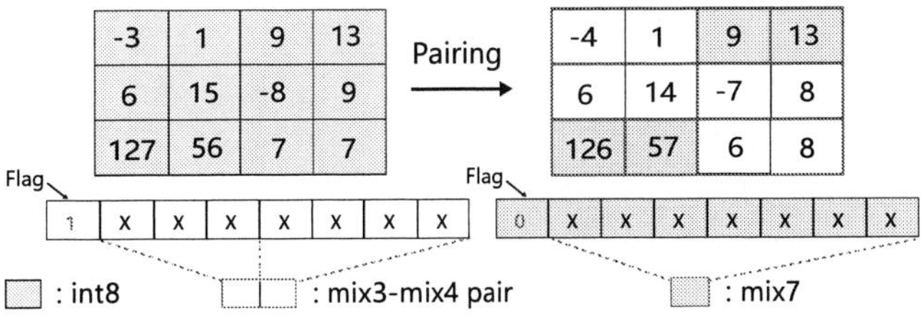

Fig. 2. The variable-length quantization framework.

A. Data representation

The weight tensors of LLMs often exhibit a Gaussian-like distribution [8], where the majority of values fall within a small range, and vary significantly in importance. As illustrated in Fig. 1, using *mix3* as an example, we categorize the *int4* values into fine and coarse distinctions based on value ranges. One-fourth of the values in the center range are defined as fine values, while the rest are considered coarse values. The model accuracy is particularly sensitive to errors in fine values. With a substantial proportion of fine values, it is crucial to consider error accumulation. In contrast, coding errors in coarse values are minimal.

During preprocessing, we retain fine values and encode every three coarse values as their intermediate value. This encoding scheme saves 1 bit per data, which serves as a flag, as discussed in section II-B.

Eq. 2 details how an element is encoded as *mix*, where the encoded value length is N. For a binary integer value X in the fine category, we remove the highest bit which is redundant. Else, we cast X into an unsigned number, then apply linear transformation and rounding to obtain the first N-bits of the encoded value.

Eq. 3 outlines the decoding process of the encoded value X. The XOR result between the two highest bits of X indicates whether it is classified as a coarse value. Fine values are converted into signed numbers, where the highest bit is filled with the sign bit. Coarse values are decoded through inverse transformation of the encoding process.

$$B = 2^{N-1} - 1 \tag{1}$$

$$Y_{encode} = \begin{cases} X_{N-1:0}, & X \in [-2^{N-2}, 2^{N-2}-1] \\ \left\lceil \dfrac{X_{unsigned} + B}{3} \right\rceil_{N-1:0}, & else \end{cases} \tag{2}$$

$$Y_{decode} = \begin{cases} X_{signed}, & X_{N-1} \oplus X_{N-2} = 0 \\ ((X \times 3 - B)_{N:0})_{signed}, & else \end{cases} \tag{3}$$

B. Quantization framework

The W8A8 quantization method [5] that mitigates the challenges of quantization by shifting the focus from activations to weights in LLMs. Following this, We propose a non-uniform quantization scheme, which utilizes the *mix* format.

Our key insignt is that allocate fewer bits for small values and more bits for large values. As illustrated in Fig. 2,

considering two adjacent elements, if the former requires 4 valid bits (ranging from -8 to 7) and the latter requires 5 valid bits, we pair them together and encode them as *mix3* and *mix4* respectively. If these conditions are not satisfied, the elements are encoded as *mix7*. Additionally, we use 1-bit flag to distinguish *mix3-mix4* pair and *mix7*. With this allocation strategy, we achieve memory alignment and conserve a substantial amount of parameter storage.

III. ARCHITECTURE

This section presents the hardware decoder designed for the *mix* format. Following this, we explore methods to accelerate MVMs using CGRA architecture.

A. Decoder

We encode weights into *mix* format offline to reduce memory requirements. To achieve efficient decoding, we design a decoder, which can be easily embedded into existing accelerators. Initially, our objective is to simplify the expression in Eq. 3. For illustrative purposes, we take N=3 for an example where $\{a, b\}$ denotes the concatenation of a and b. The resulting expression is as follows:

$$\begin{aligned} X \times 3 - B &= X + (X << 1) - 3 \\ &= \{0, X\} + \{X, 0\} + (1101)_2 \\ &= \{0, X\} + (1100)_2 + \{X, 0\} + (0001)_2 \\ &= \{\sim X_2, \sim X_2, X_{1:0}\} + \{X, 1\} \end{aligned} \tag{4}$$

The decoder for *mix3* is shown in Fig. 3b, it utilizes an adder, multiplexers and basic logic gates. Similarly, we implement the decoders of *mix4* and *mix7*.

B. CGRA architecture

The combination of word-level reconfigurability and energy efficiency of CGRA makes them ideal for edge inference. The CGRA architecture is designed for domain-specific acceleration as shown in Fig. 3a, comprises a grid of interconnected PEs. Each PE combines arithmetic and logical unit (ALU) for cumputing, a local register file (LRF), a weight register file (WRF), a crossbar for connectivity, and a configuration memory. The ALU is capable of performing decoding, multiply-accumulate operations (mac), and more, as detailed in [9].

Our achitecture enables dataflow excution of application kernels. As illustrated in Fig. 4, the dataflow graph (DFG) of MVMs consists of two nodes, which facilitating efficient processing. Activations are initially loaded into the WRF.

979-8-3503-6184-1/24 $31.00 © 2024 IEEE

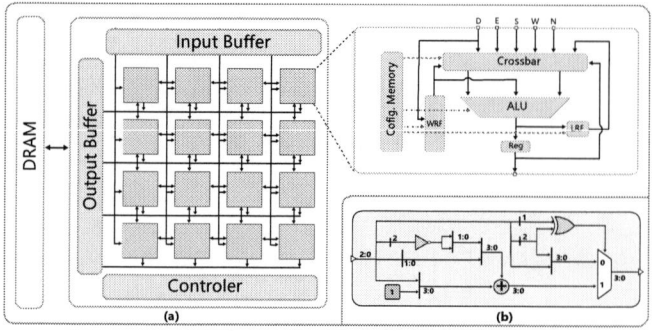

Fig. 3. (a) The CGRA architecture. (b) The decoder for mix3 format.

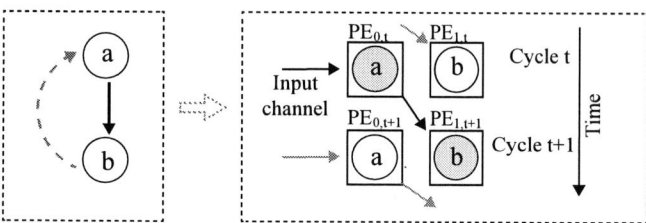

Fig. 4. DFG excution of MVMs on a 2×1 CGRA, where "a" and "b" denote decoding and mac operations, respectively. $PE_{i,t}$ denotes a PE instance at time t located at position i.

Subsequently, weights from the input channel (labeled as D port in Fig. 3a) drive the computation. During excution, PE0 decodes the data into integers at time t, whereas PE1 update the sum by adding the product of the decoded weights from the PE0 and the activations from the WRF at time t+1.

Moreover, non-linear functions are approximated and quantized in a hardware-friendly manner [10]. Subsequently, they are mapped onto our architecture in a reconfigurable manner. Due to limited space, we refrain from discussing the specific details.

IV. EVALUATION

We evaluate our quantization method on LLaMA2-7B using the LAMBADA dataset [11]. By comparing compression ratio and accuracy loss against previous works without fine-tuning, Table II illustrates that our approach reduces memory requirements by 42.4% with an approximate accuracy loss of 0.26%, outperforming SPARK. While Olive achieves 50% compression, it results in significant accuracy deterioration.

Our CGRA arranges PEs in a 4×4 mesh network, with each PE containing a 128B WRF. Compared to prior work [9], which is designed for accelerating deep learning networks with W8A8. Our work demonstrates an increase in hardware utilization from 25% to 94.8% and an average performance improvement of $1.7\times$ speedup.

We evaluate a 4×8 array with a total of 4096B WRF, which is capable of storing the entire activation vector of LLaMA2-7B. Our results demonstrate a $2\times$ speedup over the 4×4 array.

TABLE II
COMPRESSION RATIO AND ACCURACY LOSS FOR LLAMA2 ON LAMBADA DATASET.

	SPARK	Olive	**CORO**
Comp. ratio	31.0%	50.0%	**42.4%**
Acc. loss	1.78	3.93	**0.26**

V. CONCLUSION

In this work, we propose CORO to accelerate MVMs in LLMs based on a non-uniform quantization method. Notably, the matrix-matrix multiplications can also be decomposed into MVMs. Due to the sparsity at the bit level, we encode quantized values with variable lengths. Our evaluation demonstrates significant reductions in memory requirements while maintaining superior accuracy with our encoding scheme. To support our data representation, we develop the decoder circuit and integrate it into the CGRA as a configuration instruction. The result shows our architecture outperforms the prior work in both performance and hardware utilization.

ACKNOWLEDGMENT

This work was supported in part by the Strategic Priority Research Program of Chinese Academy of Sciences, Grant No. XDB44000000, in part by the University Synergy Innovation Program of Anhui Province under grant No. GXXT-2023-003, and in part by CAS Project for Young Scientists in Basic Research under grant No. YSBR029.

REFERENCES

[1] J. Achiam, S. Adler, S. Agarwal, L. Ahmad, I. Akkaya, F. L. Aleman et al., "GPT-4 technical report," arXiv preprint arXiv:2303.08774, 2023.

[2] H. Touvron, L. Martin, K. Stone, P. Albert, A. Almahairi, Y. Babaei et al., "Llama 2: Open foundation and fine-tuned chat models," arXiv preprint arXiv:2307.09288, 2023.

[3] A. Zeng, X. Liu, Z. Du, Z. Wang, H. Lai, M. Ding et al., "GLM-130B: An open bilingual pre-trained model," arXiv preprint arXiv:2210.02414, 2022.

[4] A. Dosovitskiy, L. Beyer, A. Kolesnikov, D. Weissenborn, X. Zhai, T. Unterthiner et al., "An image is worth 16x16 words: Transformers for image recognition at scale," arXiv preprint arXiv:2010.11929, 2020.

[5] G. Xiao, J. Lin, M. Seznec, H. Wu, J. Demouth, and S. Han, "SmoothQuant: Accurate and efficient post-training quantization for large language models," in International Conference on Machine Learning. PMLR, 2023, pp. 38 087–38 099.

[6] F. Liu, N. Yang, H. Li, Z. Wang, Z. Song, S. Pei et al., "SPARK: Scalable and precision-aware acceleration of neural networks via efficient encoding," in 2024 IEEE International Symposium on High-Performance Computer Architecture (HPCA). IEEE, 2024, pp. 1029–1042.

[7] C. Guo, J. Tang, W. Hu, J. Leng, C. Zhang, F. Yang et al., "Olive: Accelerating large language models via hardware-friendly outlier-victim pair quantization," in Proceedings of the 50th Annual International Symposium on Computer Architecture, 2023, pp. 1–15.

[8] Y. Li, X. Dong, and W. Wang, "Additive powers-of-two quantization: An efficient non-uniform discretization for neural networks," arXiv preprint arXiv:1909.13144, 2019.

[9] USTC, "Coarse-grained reconfigurable array systems and computing methods for deep learning," Patent CN202210798554.5, 2022.

[10] Y. Wu, Z. Wang, and W. D. Lu, "PIM-GPT: A hybrid process-in-memory accelerator for autoregressive transformers," arXiv preprint arXiv:2310.09385, 2023.

[11] D. Paperno, G. Kruszewski, A. Lazaridou, Q. N. Pham, R. Bernardi, S. Pezzelle et al., "The LAMBADA dataset: Word prediction requiring a broad discourse context," arXiv preprint arXiv:1606.06031, 2016.

Flexible yet Efficient Transformer Acceleration with Unified Sparse Attention Support on FPGA

Linfeng Zhong [1], Qingyu Guo [2], Runsheng Wang [245], Yuan Wang [25], Meng Li [324]*

[1]School of Electronic and Computer Engineering, [2]School of Integrated Circuits, [3]Institute for Artificial Intelligence &[4]Institute of Electronic Design Automation, Peking University, China
[5]Advanced Innovation Center for Integrated Circuits, China

* Email: zhonglinfeng@stu.pku.edu.cn, meng.li@pku.edu.cn

Abstract—The quadratic complexity of attention computation poses a challenge for traditional Transformers, which the window attention mechanism aims to mitigate. However, the diverse applications of Transformer architectures require custom network designs with various sparse attention patterns to meet specific task demands. To address this, we developed an FPGA-based accelerator that flexibly supports different window attention mechanisms across different heads. Our design employs a pipeline architecture and a reconfigurable LineBuffer module to enable efficient, low-resource consumption. Experimental results show that our accelerator achieves 1.9× GOPS/DSP and 3.9× GOPS/LUT computational efficiency compared to the FPGA baseline.

Keywords—*FPGA accelerator, window attention mechanism, sparse attention pattern, pipeline architecture*

I. INTRODUCTION

Transformers [1] have revolutionized natural language processing and computer vision by using attention mechanisms that prioritize relevant data segments, enhancing model performance. However, traditional full-attention operations associate each sequence element with every other, resulting in quadratic computational complexity. This becomes especially problematic as input sizes grow, causing a bottleneck in resource-intensive tasks like high-resolution image analysis and large-scale video processing. Fig. 1 shows the increase in computational demands, measured in floating-point operations (FLOPs), as input length grows. To address the quadratic complexity in Transformer models, sparse attention is a widely adopted solution [2].

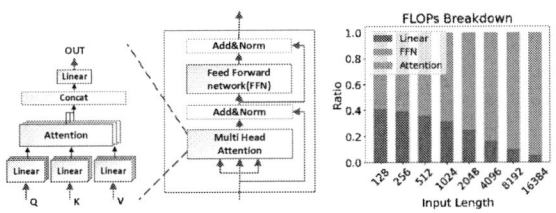

Fig. 1. The computation of vanilla attention mechanism and Floating point operations (FLOPs) breakdown for different batch sizes

The diverse applications of Transformer architectures require customized network designs with various sparse attention patterns to meet specific task demands [3]. This diversity highlights the need for adaptive solutions to manage these variations efficiently. Conventional hardware platforms with fixed computational architectures often have problems handling evolving demands.

In response, we develop an FPGA-based attention accelerator that supports a broad range of sparse attention patterns. This innovation enhances the adaptability of Transformer models to different computational environments, which include a pipeline architecture that facilitates concurrent processing of Transformer layers and a reconfigurable LineBuffer module for efficient handling of various sparse patterns. By optimizing the processing of various sparse patterns, our solution reduces the computational load, enabling faster and more efficient processing of transformer-based models across diverse tasks and data scales.

II. BACKGROUND AND RELATED WORKS

Sparse attention techniques fall into two categories: dynamic methods [4] and static methods. Dynamic methods can adapt in real time to the data being processed but often generate irregular sparsity patterns that are inefficient and difficult to optimize on hardware platforms. This inefficiency has led to a preference for static methods, which offer predictable and structured sparsity patterns better suited for hardware implementation.

Static sparse attention includes various approaches, such as block-wise, sliding window, dilated window, and hybrid forms like global combined with sliding window attention. Longformer [5] and Vision Longformer [6] use sliding and dilated window attention to manage larger contexts efficiently, while Swin Transformer [7] employs shifted window attention to facilitate interactions across shifted input regions.

TABLE I. EXISTING STATIC SPARSE ATTENTION ACCELERATOR.

Accelerator	static sparsity pattern					
	sliding widow	dilated window	global	block wise	strided	butterfly
SWAT	√		√			
SALO	√		√		√	
Butterfly						√
Our work	√	√	√	√	√	√

However, current hardware solutions typically support only a limited selection of these static patterns, limiting their adaptability and effectiveness across different Transformer architectures. In contrast, as shown in TABLE I. , our accelerator supports a wider range of sparse attention patterns. This allows for flexible adaptation to the specific needs of different heads, layers, model architectures, etc. significantly enhancing computational efficiency and model versatility in diverse settings. This extensive adaptability marks a substantial advancement over existing hardware solutions, which often exhibit limited scope and flexibility.

III. ARCHITECTURE AND IMPLEMENTATION

The architecture diagram of our accelerator is shown in Fig. 2(a). We propose a pipeline architecture to facilitate concurrent processing of different Transformer layers, which

979-8-3503-6184-1/24 $31.00 © 2024 IEEE

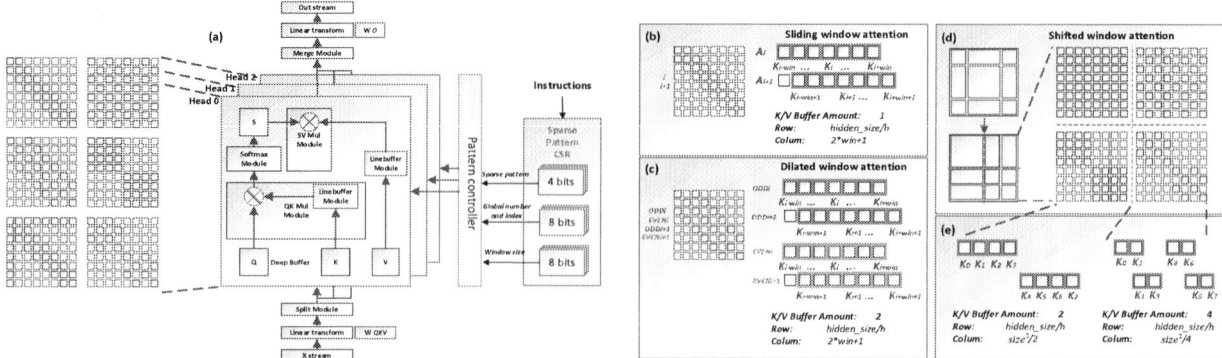

Fig. 2.(a) The Hardware View; (b) Support for Sliding Window Attention by LineBuffer module; (c) Support for Dilated Window Attention by LineBuffer module; (d) Sparse Patterns Corresponding to Shifted Window Attention Mechanism; (e) Support for Shifted Window Attention by LineBuffer module.

enables memory-efficient and low-latency attention acceleration. To ensure adequate throughput, specialized hardware has been developed for each component of the multi-head attention computation graph. Each head or module is instantiated separately without using time-division multiplexing. This design supports various sparse patterns in a unified manner while enabling specific configurations for different heads without extra hardware overhead.

To minimize hardware changes when switching between different patterns, we developed a pattern controller and a reconfigurable LineBuffer module. The pattern controller adjusts the settings of the LineBuffer module based on the configuration information for each head, which is provided by the sparse pattern control and status register (CSR). This information includes the sparse pattern, global token number and index, and window size, enabling the accelerator to effectively support various sparse patterns. Furthermore, due to different computational demands of these sparse patterns, we have deepened the FIFO in the merge module to align the data flow between the heads.

A. Support for Various Window Attentions

In both the sliding window and dilated window attention mechanisms, the QK multiplication module, SoftMax module, and SV multiplication module share a common sparsity pattern, as depicted in Fig. 3. For a specific input row of Q, denoted as Q_i, and the subsequent row vector Q_{i+1}, significant data reuse is observed in the attended rows of K (columns of K^T) and V. The most effective way to harness this data reuse is by adopting a row-major dataflow. We use fixed-size on-chip line buffers for the K and V inputs, while the Q input updates for each new row. This setup, illustrated in Fig. 2(b), inserts a new row of K at the bottom of the buffer and discards the top row of K each cycle, ensuring that data is loaded exactly once, thereby achieving 100% off-chip memory transfer efficiency.

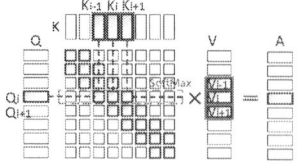

Fig. 3 Computation Pattern of Sliding Window Attention

In the case of dilated window attention, the sparse pattern is similar to that of sliding window attention, but with interleaved K and V corresponding to odd and even rows of

Q_i. To accommodate this, we use two line buffers: one for the odd rows of Q_i and another for the even rows of Q_i. The update pattern for these line buffers remains similar, effectively managing interleaved data access and ensuring efficient processing, as shown in Fig. 2(c).

In shifted window attention, the input image is divided into nine windows. These windows undergo a unique shift and mask process within the Swin Transformer, resulting in four types of window attention sparse patterns: full attention, stride attention, block-wise attention, and butterfly attention, as shown in Fig. 2(d). For these patterns, the sparsity patterns of K and V no longer require frequent updates with changes in Q_i. Consequently, the line buffers are configured as standard buffers, which do not need real-time updates, as illustrated in Fig. 2(e). For block-wise attention, only two line buffers are needed to store half of K each, fulfilling the computational requirements. Similarly, for butterfly attention, only four line buffers are required, which store K in an interleaved manner.

Global attention, which assigns important global tokens to be attended by all input tokens, is typically used in classification tasks to focus on the CLS token. This type of attention is similar to full attention and necessitates the storage of the entire K and V. We introduce a global index i to specify that the ith token requires full attention, as illustrated in Fig. 4. Within the line buffers for K and V used in sliding window attention, we additionally store the corresponding K_i and V_i. Subsequently, buffers storing the entire K and V are maintained until the computation for Q_i is completed, after which they are discarded.

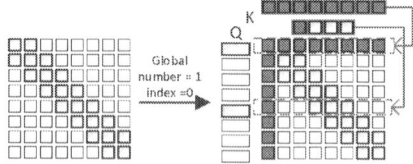

Fig. 4 Introducing Global Token into Locality Attention

B. Pipeline design and balance.

The design of our pipeline, illustrated in Fig 5, addresses the data dependencies inherent in the computation of Q, which requires access to the complete K and V matrices. To manage this, we structure the generation process such that the K and V matrices are produced first. This ensures that each row of Q can be computed in compliance with the necessary data dependencies. This method optimizes the computation process, allowing for efficient sequential processing of Q once

K and V have been established. Maintaining pipeline balance is crucial for the accelerator's performance. The Initiation Interval (II) of the entire pipeline is determined by the maximum II of all pipeline stages. As depicted in Fig. 5 (1), imbalanced cycles generate bubbles. Therefore, by allocating more computing resources to the QK Mul module, we achieve a balanced pipeline, as shown in Fig. 5 (2). By adjusting parallelism, we aim to equalize the II across all pipeline stages.

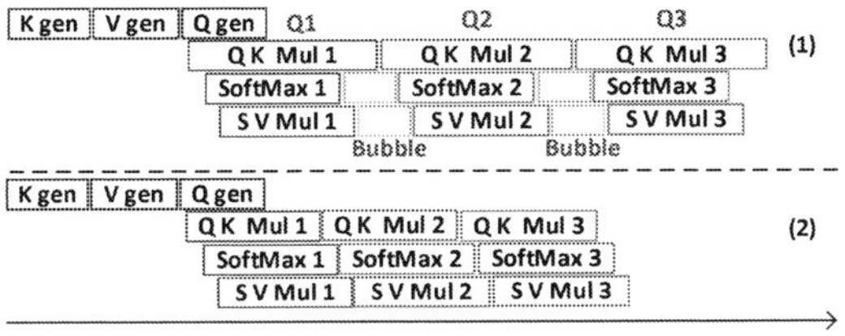

Fig. 5 Eliminating imbalance-induced bubble in pipeline.

IV. EXPERIMENTS AND RESULTS

Experimental Setup: this experiment employs the VCK190 FPGA platform and uses Vitis HLS 2023.2 for hardware synthesis. We implemented a baseline according to the SWAT [8] approach, which provides robust support for sliding window attention and global attention. However, it does not support the other sparse patterns that our design supports, necessitating additional hardware for these patterns. To ensure a fair comparison, we also designed linear module to generate Q,K and V for the baseline.

Resource Consumption Comparison: under a main frequency of 400 MHz, with a token sequence length of 128, a hidden size of 128, a window size of 32, and INT8 precision, we compared the resource consumption of the baseline and our attention module. The results, shown in TABLE II. , indicate that our design achieved 1.8× DSP efficiency, 2.0× LUT efficiency, and 2.6× BRAM efficiency compared to the baseline, considering only the attention module. These results highlight the superior efficiency of our FPGA-based design in terms of hardware resource utilization. The reduction in resource consumption is primarily attributed to the reconfigurable LineBuffer module, which demonstrates our design's flexible support for various sparse attention patterns.

TABLE II.   RESOURCE CONSUMPTION FOR DIFFERENT PATTERNS

Sparse Pattern	DSP	LUT	BRAM
sliding window+global	96	34594	256
dilated window	64	28234	192
block-wise	34	26570	256
butterfly	36	30954	384
strided	38	31620	264
ALL	**268**	**151972**	**1352**
Our work	**147**	**77075**	**512**

Computational Efficiency Comparison: the comparative analysis in this section is based on three different workloads, each derived from a single attention layer from the original models. We selected one attention layer from Longformer, one from the second stage of ViL, and one from the second stage of Swin Transformer (Swin-T). The key parameters of all three types of attention layers are summarized in TABLE III. .

TABLE III.   KEY PARAMETERS OF ATTENTION LAYERS

Parameter	Sequence length	Window size	Hidden size	Global Token
Longformer	4096	512	768	1
ViL-stage2	28×28	15×15	384	1
Swin-T stage2	56×56	7×7	192	0

To conduct a comprehensive evaluation, we implemented these attention layers on the FPGA and compared their performance with the baseline. The performance metrics included GOPS (Giga Operations Per Second) per DSP and GOPS per LUT. The results are shown in Fig. 6, demonstrating that our design achieved an average of 1.9× GOPS/DSP and 3.9× GOPS/LUT across the three workloads. These improvements underscore the enhanced computational efficiency of our FPGA-based attention accelerator.

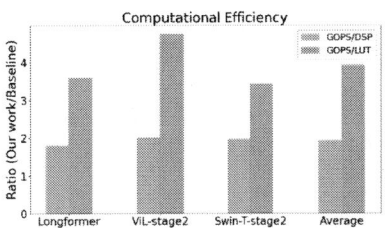

Fig. 6 Computational efficiency ratio by workload

V. CONCLUSION AND FUTURE WORK

We proposed an FPGA-based attention accelerator that can flexibly switch between different attention sparse patterns based on task requirements. This design is built on a pipeline architecture with a reconfigurable LineBuffer module at its core. Our accelerator supports a wide range of mechanisms, including sliding window attention, dilated window attention, and shifted window attention. Experimental results show an average of 1.9× GOPS/DSP and 3.9× GOPS/LUT computational efficiency over the baseline. This innovation improves scalability and adaptability, reducing computational load and enabling faster processing across diverse tasks. Future work will focus on enhancing adaptability to new Transformer architectures and extending applications to other domains.

REFERENCES

[1] Vaswani, Ashish, et al. "Attention is all you need." Advances in neural information processing systems 30 (2017).

[2] Child, Rewon, et al. "Generating long sequences with sparse transformers." arXiv preprint arXiv:1904.10509 (2019).

[3] Tay, Yi et al. "Efficient Transformers: A Survey." ACM Computing Surveys 55 (2020): 1 - 28.

[4] Zheng Qu et al. 2022. Dota: detect and omit weak attentions for scalable transformer acceleration. In ASPLOS-27. 14–26.

[5] Beltagy, Iz, Matthew E. Peters, and Arman Cohan. "Longformer: The long-document transformer." arXiv preprint arXiv:2004.05150 (2020).

[6] Zhang, Pengchuan, et al. "Multi-scale vision longformer: A new vision transformer for high-resolution image encoding." In the IEEE/CVF international conference on computer vision. 2021.

[7] Liu, Ze, et al. "Swin transformer: Hierarchical vision transformer using shifted windows." In the IEEE/CVF international conference on computer vision. 2021.

[8] Zhenyu Bai, Pranav Dangi, Huize Li, and Tulika Mitra, "SWAT: Scalable and Efficient Window Attention-based Transformers Acceleration on FPGAs," 2024, arXiv:2405.17025.

Backward-edge Control Flow Integrity based on Return Address Encryption

Fengshuo Tian[1], Kaixuan Wang[1], Jun Han[1]

[1] State Key Laboratory of Integrated Chips and Systems, Fudan University, Shanghai 200433, China
Email: junhan@fudan.edu.cn

Abstract

Backward-edge control flow integrity (CFI) defends against return-oriented programming (ROP) by protecting return addresses. In this paper, we implement a backward-edge control flow integrity by encrypting and decrypting return addresses using the Advanced Encryption Standard (AES). We utilize the gem5 simulator for modeling and evaluation, design an AES hardware accelerator, and integrate this accelerator into gem5 system using gem5 + RTL technique. The AES accelerator is implemented under TSMC 28nm technology, which can work at 1GHz, with an area of $10045\mu m^2$ and a power consumption of 1.31mW. Experimental results indicate that the performance overhead of the backward-edge CFI scheme is less than 0.1%.

Keywords

Control-flow integrity (CFI), advanced encryption standard (AES), return-oriented programming (ROP).

1. Introduction

ROP is an advanced memory attack technique that enables the execution of malicious code in an environment with Data Execution Prevention (DEP) enabled [1]. Attackers select instruction fragments, known as gadgets, from existing libraries or executable files. Gadgets typically end with *ret* instruction, and each gadget is responsible for performing a specific task, such as executing shellcode or elevating privileges. Finally, these gadgets will be chained together to construct malicious code, allowing attackers to perform arbitrary operations on the system.

Backward-edge CFI prevents attackers from tampering with return addresses to defend against ROP. Return address encryption is a mainstream backward-edge CFI scheme. When a *call* instruction executes, the return address is encrypted before being pushed onto the stack. When a *ret* instruction executes, the encrypted return address is popped from the stack, decrypted, and then the decrypted address is used for returning. If an attack attempts to tamper with return addresses, the default decryption process will still occur when *ret* executes. This unexpected decryption typically invalidates the attacker's jump addresses, which may ultimately result in system errors or exceptions.

Previous work, LEA-AES [2], utilizes CPU-built-in AES for encrypting and decrypting return addresses. This approach requires the ISA supporting AES, and the cost of AES calculation is tied to specific AES instructions. In this work, we design an independent AES hardware accelerator to reduce performance overhead. However, simulating RTL models individually cannot precisely evaluate their behavior within a SoC environment, and RTL simulation for SoC can be prohibitively slow. To address these challenges, we employ the gem5+RTL [3] method. Gem5+RTL is a flexible framework that facilitates the integration of RTL models with the full-system gem5 simulator. The basic idea is to compile the RTL Accelerator into Electronic System Level (ESL) to enable interaction. Gem5+RTL combines the advantages of both ESL model and RTL model, offering fast modeling speed, flexible model modification, rapid simulation speed, and accurate PPA evaluation.

In this work, the AES accelerator is integrated with gem5 O3 CPU. AES accelerator decouples AES operation from AES instructions, which further reduces the system's performance overhead and enhances its applicability. We evaluate our system on Dhrystone, Coremark, and SPEC CPU2006 benchmarks. Compared to LEA-AES, whose performance overhead is less than 4%, the performance overhead of this work is less than 0.1%, almost negligible.

The remainder of this paper is organized as follows. Section 2 is the implementation of our work. Section 3 discusses the security of this scheme. Section 4 presents the experimental results. Section 5 makes the summary.

2. Implementation

In this section, we describe the implementation of our backward-edge CFI scheme. We first present the overall architecture of the system. Then we introduce how to integrate AES accelerator with gem5. Finally, we explain the workflow of this scheme.

2.1 Overall architecture

Figure 1 illustrates the overall framework of our backward-edge CFI scheme based on return address encryption. The encryption-decryption unit (EDU) is designed to interact with gem5 O3 CPU. EDU consists of five main submodules: AES_Calculate_Module, AES_Schedule_Module, EDU_Inst_Queue, AES_Result_Buffer, and KEY_CFI.

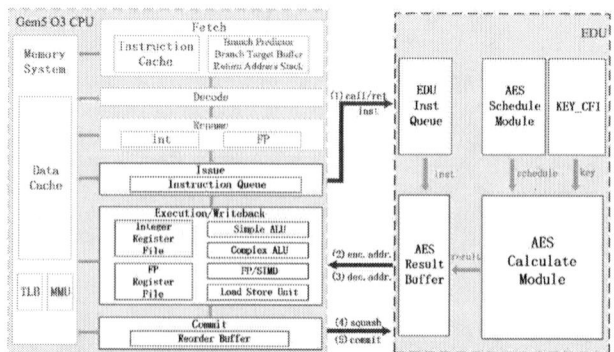

Figure 1. Overall architecture.

979-8-3503-6184-1/24 $31.00 © 2024 IEEE

In this work, the counter (CTR) mode of AES is adopted, in which the counter and the key are taken as the input of AES. The resulting AES output is then XORed with the plaintext to produce the ciphertext. The key is generated by the operating system and stored in the dedicated register KEY_CFI to prevent key leakage. The EDU_Inst_Queue maintains the order of *call* and *ret* instructions, allocates a counter for them, and restores the counter in the event of instruction squashing. The AES_Calculate_Module can instantiate multiple AES calculation units as needed to perform AES calculations in advance. The results will be stored in the AES_Result_Buffer. When a *call* or *ret* instruction executes, the buffer can be indexed using the previously allocated counter to retrieve the corresponding AES result, thereby reducing performance overhead. AES_Schedule_Module monitors idle AES calculation units and assigns computing tasks accordingly.

2.2 Integration of AES accelerator and gem5

AES [4] is a block encryption standard adopted by the Federal Government of the United States and is widely utilized across various fields. In this paper, we employ the AES-128 CTR mode, which consists of 11 rounds of computation. Figure 2 shows the workflow. AES calculation is primarily based on four fundamental operations, with corresponding operations executed in each round. In the design of AES hardware, each fundamental operation is completed within a pipeline cycle. AES workflow takes 40 cycles, resulting in significant performance overhead. To mitigate this, we compute the AES results in advance and store them in AES_Result_Buffer. Subsequent instructions only need to look up the buffer according to their counters. In this way, the performance overhead of AES calculation will be covered.

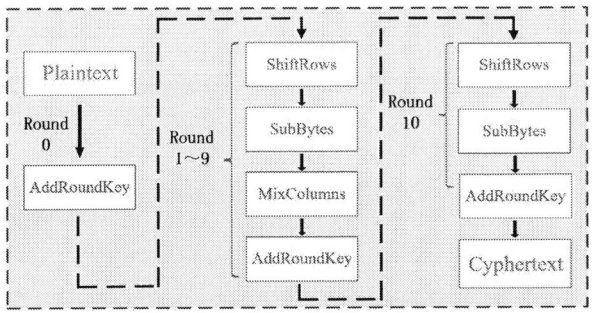

Figure 2. Workflow of AES accelerator.

Figure 3 shows the integration of AES accelerator and gem5. First, we utilize Verilator to compile the RTL code of AES into C++ code. The top file defines the AES_top class. Then we implement the Wrapper_AES class to encapsulate the C++ model of AES accelerator and provide two main functions, *tick* and *reset*. The tick function evaluates and updates signals at the beginning of each clock cycle. The reset function is responsible for sending a reset signal to initialize and reset the system. Additionally, Wrapper_AES class offers external interfaces for invoking the C++ model, applying drive signals, and receiving return signals. The wrapper and the C/C++ model of AES will be compiled into a shared library and integrated with gem5. This shared library allows the RTL code and gem5 to be compiled independently, meaning modifications to the RTL model do not necessitate recompiling the SoC in gem5. Finally, the rtl_AES class is provided, which implements all the functionalities required for interaction with gem5. The external interface of rtl_AES is responsible for interacting with other modules in Gem5, obtaining drive signals for AES_top from Gem5, and returning output signals from AES_top. The internal interface handles interactions with the wrapper, calling the tick function, applying drive signals to AES_top through the wrapper, obtaining output signals from AES_top.

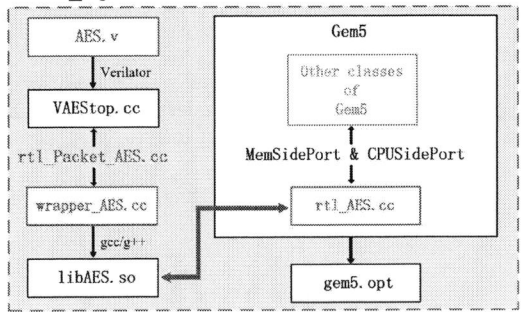

Figure 3. Integration and interaction of AES and gem5.

2.3 EDU workflow

To achieve backward-edge CFI, the EDU module is designed to interact with gem5 O3 CPU for encrypting and decrypting return addresses. The original pipeline of O3 CPU is also modified in certain logical details to support the behavior of EDU. The workflow is shown in Figure 4.

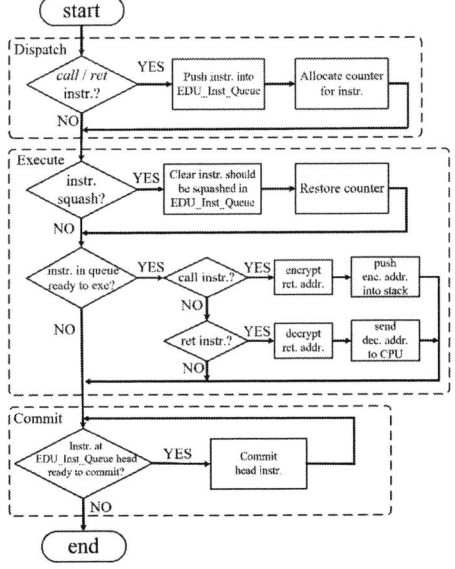

Figure 4. EDU workflow.

In the dispatch stage, if an instruction is *call* / *ret*, the instruction will be pushed into EDU_Inst_Queue and a counter will be allocated for it. The AES counter starts at 0, increments when a call instruction executes, and decrements when a ret instruction executes. This approach ensures that each pair of *call-ret* gets the same counter.

In the execute stage, EDU checks for instruction squashing. If squashing occurs, EDU clears the relevant instructions in EDU_Inst_Queue and restores AES counter to the state before speculative execution. Subsequently, EDU verifies whether an instruction in the queue is ready to execute by checking whether the AES result corresponding to the instruction's counter has been prepared. Typically, these AES results are calculated in advance and stored in AES_Result_Buffer to minimize performance overhead.

In the commit stage, the behavior of EDU_Inst_Queue is similar to ROB. It checks whether the head instruction of the queue is ready to commit. If the instruction is ready, it is popped from the queue and the check continues. Otherwise, the checking process is halted.

3. Security Evaluation

In an unsafe system, the return address is pushed directly onto stack when *call* executes. When *ret* executes, the return address is popped from the stack and used for the jump. In this work, when *call* executes, EDU will encrypt the return address, and then the encrypted return address is pushed onto the stack. When *ret* executes, the encrypted return address will be popped from the stack and sent to the EDU for decryption. The program will jump according to the decrypted address.

If attackers tamper with return addresses, EDU will still perform the default decryption during the return process. The unexpected decryption alters the tampered return addresses, which may result in system errors or exceptions. An example is shown in Figure 5.

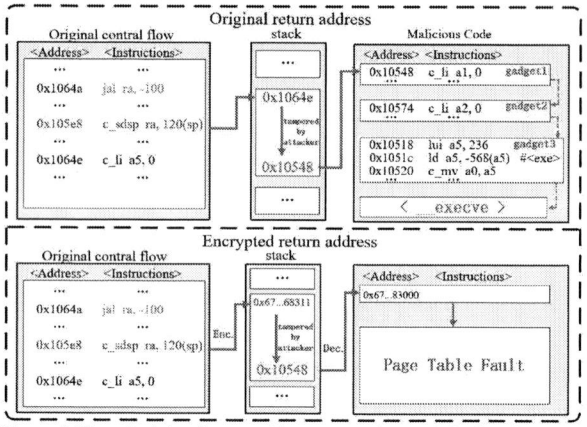

Figure 5. The principle of return address encryption to defend against ROP.

4. Experimental results

AES accelerator is synthesized using Synopsys Design Compiler (DC) on TSMC 28nm technology. The synthesis results show that AES accelerator can operate at a frequency of 1GHz, with an area of $10045\mu m^2$ and a power consumption of 1.31mW. The performance overhead of the entire backward-edge CFI system is evaluated using Dhrystone, Coremark and SPEC CPU2006 benchmarks, as detailed in Table 1. The simulation results are presented in Figure 6. Among all test sets, only *mcf* incurs a performance overhead of 0.084%, which is negligible.

Table 1. Parameters of processor and AES.

Parameters	Configuration
Processor	
Architecture	RISC-V GCB ISA, 1 core
Working Frequency	**1GHz**
Other Details	8-decode, 8-rename, 8-issue, 8-commit, 32 LQ entries, 32 SQ entries
AES	
Instantiated Units	1
Working Frequency	**1GHz**
Precomputation depth	32

However, what is unexpected is that the performance overhead of h264ref, omnetpp, and specrand is negative, which indicates that the system with return address encryption scheme has a performance improvement over the original system in these test sets. This phenomenon may be attributed to the operation of the EDU_Inst_Queue during instruction squashing. When a squash occurs, EDU takes the squashed instruction sent by the CPU as a reference. All instructions in EDU_Inst_Queue that are younger than the reference will be set as *Squashed* at that time. This action by EDU may occur earlier than in original system, leading to improved overall system performance.

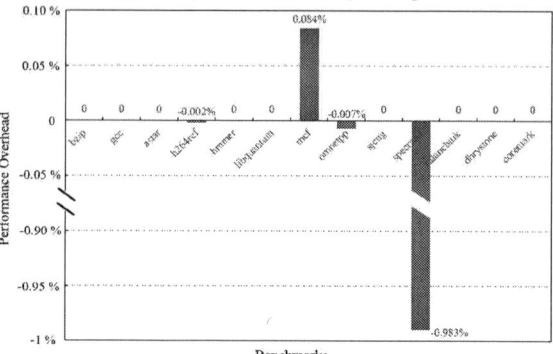

Figure 6. Performance overhead of backward-edge CFI based on return address encryption. The baseline is original system.

5. Summary

In this paper, a backward-edge CFI based on return address encryption is implemented. Instead of using AES instruction set, we design an independent AES accelerator to reduce performance overhead which incurs an additional area of $10045\mu m^2$ and an extra power consumption of 1.31mW. Experimental results from benchmark tests indicate that the performance overhead of this backward-edge CFI is less than 0.1%.

Acknowledgments

This work was supported by the National Natural Science Foundation of China under Grant 61934002 and 62234008.

References

[1] Hovav Shacham. The geometry of innocent flesh on the bone: Return-into-libc without function calls (on the x86). In Proceedings of the 14th ACM Conference on Computer and Communications Security (CCS), Alexandria, VA, October– November 2007.

[2] P. Qiu, Y. Lyu, J. Zhang, D. Wang and G. Qu, "Control Flow Integrity Based on Lightweight Encryption Architecture," in IEEE Transactions on Computer-Aided Design of Integrated Circuits and Systems, vol. 37, no. 7, pp. 1358-1369, July 2018

[3] Guillem López-Paradís, Adrià Armejach, and Miquel Moretó. "Gem5 + rtl: A Framework to Enable RTL Models Inside a Full-System Simulator. " In Proceedings of the 50th International Conference on Parallel Processing (ICPP '21). Association for Computing Machinery, New York, NY, USA, Article 29, 1–11. 2021.

[4] S. Gueron, Intel Advanced Encryption Standard (AES) New Instructions Set, Intel Corporat., Santa Clara, CA, USA, 2010.

Stability Enhancement Technique for Monostable PUF Based on Hysteresis Effect of Schmitt Trigger

Ruize Xu, Gang Li, Pengjun Wang*, Hui Li, Xudong Wu

College of Electrical and Electronic Engineering, Wenzhou University, Wenzhou 325000, China

* Email: wangpengjun@wzu.edu.cn

Abstract—Monostable physical unclonable function (PUF) plays an important role in the field of information security. In order to reduce the instability of monostable PUF due to environmental changes, a stability enhancement technique for monostable PUF based on the hysteresis effect of Schmitt trigger is proposed through the study of its working mechanism. This technique can realize efficient screening of unstable PUF cells by adding a screening unit based on Schmitt trigger in the monostable PUF circuit and utilizing the hysteresis effect of Schmitt trigger. Based on TSMC 65nm process, a fully customized layout is designed for PUF arrays with screening function. Experimental results show that the PUFs added to the screening unit maintain good randomness, uniqueness and reliability, while reducing the PUF bit error rate (BER) by more than 88% over the voltage range of 1.0 V to 1.4 V and the temperature range of −40 °C to 120 °C.

Keywords—Monostable PUF, Schmitt trigger, hysteresis effect, stability, screening

I. INTRODUCTION

With the wide adoption of Internet of Things (IoT) devices and the rapid development of cloud computing, there is a growing need for reliable hardware security solutions [1]. Physical unclonable function (PUF) has emerged, which gives each integrated circuit (IC) a unique "fingerprint" based on the unpredictable physical differences inherent in the manufacturing process of ICs [2]. These physical differences can be small variations in device parameters, such as the threshold voltage, resistance value, or capacitance value of a transistor. Based on these small physical differences, the PUF generates a series of specific response patterns to distinguish between different ICs, thus providing a unique identifier for the hardware device. The emergence of PUFs marks a shift from traditional key storage and management methods to more dynamic and secure key generation and authentication mechanisms [3].

PUFs can be categorized into monostable PUFs [4] and bistable PUFs [5] based on the number of output states. Monostable PUFs are commonly used in low-power devices and resource-constrained application environments. However, monostable PUFs encounter stability problems when facing voltage and temperature fluctuations in the operating environment [6], affecting their performance and security in critical applications such as key generation and authentication. Usually, techniques such as temporal majority voting (TMV) and error correction code (ECC) are employed to improve the stability of PUFs [7]. However, these schemes often require significant testing time. In view of this, this paper proposes a stability enhancement technique based on Schmitt trigger, which utilizes the hysteresis effect of Schmitt trigger to achieve efficient screening of unstable bits.

This work was supported in part by the National Natural Science Foundation of China (62234008, 62374117), the Zhejiang Provincial Natural Science Foundation of China (LY22F040004), the China Postdoctoral Science Foundation (2023M731776)

II. THEORETICAL FOUNDATION

A. Monostable PUF

The monostable PUF has only one stable output state, and its typical structure is based on an inverter, consisting of two stages with the same inverter structure and parameters, as shown in Fig. 1. The first stage acts as a voltage generator and generates the voltage V_M, which is normally distributed, by shorting the input and output of the inverter, as shown in Fig. 2(a). The second stage acts as an amplifier and consists of a chain of four-stage subthreshold inverters for entropy amplification and extraction. In the second stage amplifier, the inverter switching voltage V_T determines the output level as shown in Fig. 2(b), where V_{IL} and V_{IH} are the input low and input high threshold voltages of the inverter, respectively. Since the two stages have the same structure and parameters of the inverter, theoretically $V_M = V_T$. However, due to unavoidable process deviations, there will be a difference between V_M and V_T, which will be amplified by the amplifier. It is worth noting that V_M is between V_{IL} and V_{IH}, and the output logic value of the PUF unit is susceptible to flip by voltage and temperature, and these PUF units with unstable outputs are defined as unstable bits.

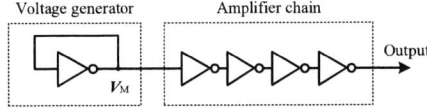

Fig. 1. Monostable PUF structure

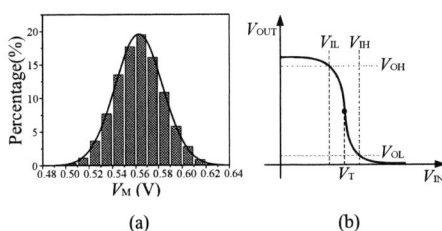

Fig. 2. Schematic diagram of monostable PUF operation (a) probability distribution of V_M (b) inverter voltage transfer characteristic curve

III. PROPOSED PUF CIRCUIT

A. Overall Structure

The overall structure of the proposed PUF circuit with screening function is shown in Fig. 3, which contains an input register (IR), timing control, row&column decoder, screening unit and an array of PUF units. An array of size 16×32 is realized. The PUF unit consists of a voltage generator and a single-stage inverter. The output of each PUF unit is transmitted through a transmission gate to avoid threshold loss. A three-stage inverter, a 2-to-1 multiplexer and a Schmitt trigger serve as the screening unit. In the screening unit, the timing controller manages the transmission of the 2-to-1 multiplexer, which can perform readout operations while detecting.

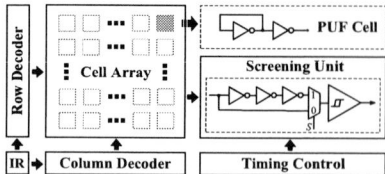

Fig. 3. Overall structure of the proposed PUF circuit

B. Unstable Bit Screening Circuit

To efficiently find out the potential unstable bits, this paper proposes an unstable bit screening circuit based on the hysteresis effect of Schmitt trigger. On the basis of the original PUF unit, a 2-to-1 multiplexer and a Schmitt trigger are added, as shown in Fig. 4(a). The screening technique works as follows: the voltage distribution in the PUF cell after amplification by a one-stage inverter is shown in Fig. 4(b). For a stable PUF unit, the amplified voltage of the Stage1 inverter is distributed near the low or high level, while for an unstable bit, the amplified voltage of the Stage1 inverter is still distributed near the inverter switching voltage. By choosing the appropriate size of the Schmitt trigger MOS tube, specific threshold voltages V_{T+} and V_{T-} are obtained so that the voltage distribution range of the unstable bit falls into its hysteresis interval. The control signal $S=1$ makes the output of the PUF unit first pass to the Schmidt trigger. Since the output of the PUF unit has only two possibilities: high or low level, one transition process of the Schmitt trigger will be selected. Then, the signal $S=0$ passes the amplified voltage from the Stage1 inverter to the Schmidt trigger. For the stable bit, the output voltage of this stage is distributed near the high or low level, avoiding the hysteresis interval, and because of the odd number of stages of inverters separated from the output of the PUF unit, the two are located in different level intervals, which will make the output of the Schmitt trigger changed. For the unstable bit, the output voltage of this stage falls into the hysteresis interval, and the output of the Schmitt trigger remains unchanged. The stability detection flow for the PUF unit is shown in Fig. 5.

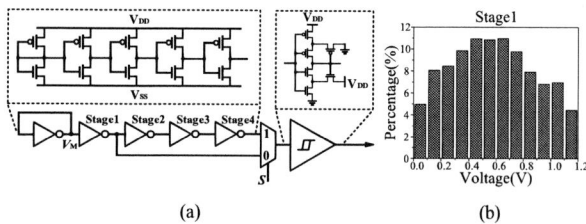

(a)   (b)

Fig. 4. Schematic diagram of screening circuit working principle (a) proposed screening circuit structure (b) voltage distribution after amplification by a one-stage inverter

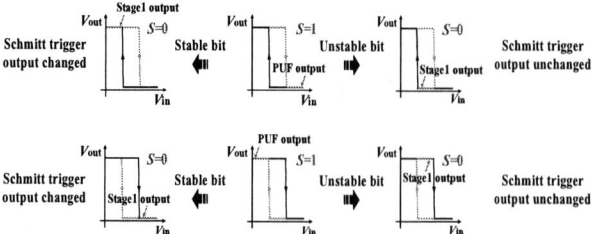

Fig. 5. Unstable bit screening flow chart

IV. EXPERIMENTAL RESULTS AND ANALYSIS

In this experiment, TSMC 65nm CMOS process rules were used to design the layout of the PUF array. The layouts of the PUF unit and the screening unit are shown in Fig. 6.

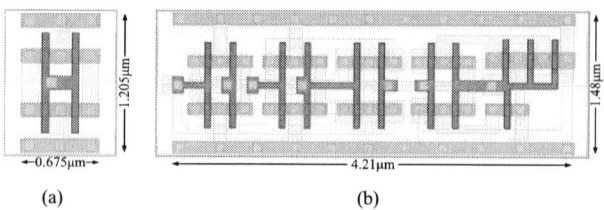

(a)   (b)

Fig. 6. Layout design (a) PUF unit (b) screening unit

A. Randomness, Uniqueness, and Reliability

Randomness is essential to ensure that PUF responses are unpredictable. Uniqueness measures the degree to which a PUF is distinct from another PUF. Reliability measures the ability of a PUF to produce the same response to the same challenge under noisy conditions. 20 Monte Carlo simulations of the PUF array were performed under golden condition (27°C and 1.2V). As shown in Fig. 7, the probabilities of the PUF circuits outputting "0" and "1" are 49.92% and 50.08% respectively, with an average inter Hamming Distance (HD) of 0.5050 and an average intra HD of 0.0035. This reflects that the PUFs with the addition of the screening unit still have good randomness, uniqueness and reliability.

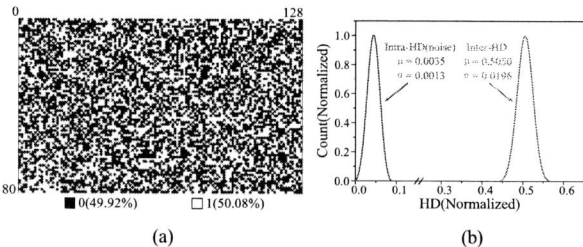

(a)   (b)

Fig. 7. PUF performance chart (a) 2D grayscale (b) intra and inter HD

B. Unstable Bit Screening Performance

Monte Carlo simulations were performed for 20 PUF arrays with screening function under the voltage range of 1.0V~1.4V and temperature range of −40°C~120°C. The output values of the PUFs at golden condition were taken as the standard values, and the output values of the PUFs generated at other voltages and temperatures are compared with the standard value, and those bits whose output values never change are regarded as stable bits, and vice versa are regarded as unstable bits. After screening, the screened unstable bit cells are masked and not counted in the bit error rate (BER). The enhancement of the PUF BER by the screening cells is expressed through BER_{Imp}, which is calculated by (1):

$$BER_{Imp} = \frac{BER_{native} - BER_{masked}}{BER_{native}} \quad (1)$$

The screening unit screens the unstable bits at golden condition. The native and masked BERs under different voltage and temperature conditions and the BER enhancement under extreme conditions are shown in Fig. 8. Under the 1.2V, the enhancement of the PUF BER reaches more than 94.74% in the temperature range of −40°C to 120°C, under the two extreme temperature conditions, −40°C and 120°C, the worst-

979-8-3503-6184-1/24 $31.00 © 2024 IEEE   20

case scenarios, the post-screening BER is reduced by 92.31% and 91.67%, respectively. At 1.0V, 1.1V, and 1.3V, the enhancement of BER for PUF in the temperature range of −40°C to 120°C is more than 87.5%, more than 95%, and more than 88.46%, respectively. In the three extreme conditions of 1.0V and −40°C, 1.1V and −40°C, and 1.3V and 120°C, the worst-case scenarios, the screening reduced the BER by 88.89%, 92.31%, and 87.5%, respectively. This shows that the screening unit significantly improves the stability of the PUF under extreme conditions. However, the worst-case enhancement of the BER of the PUF is only 84% under 1.4V in the range of −40°C to 120°C, under the temperature condition of 120°C, the BER is only reduced by 82.61%. Compared to other voltage conditions, the screening effect is poor. From the native BER situation, it can be seen that the PUF suffers the most interference under this voltage condition, which makes the PUF output voltage jump greatly, exceeding the screening interval and leading to missed screening. Overall, the BER is reduced by more than 91.89% in the 1.0V to 1.4V voltage range and 0°C to 100°C temperature range, and by more than 88% in the more extreme −40°C to 120°C temperature range. Screening of all unstable bits is achieved in the best case. The average masking ratio is 28.51%, and the minimum and maximum masking ratios are 26.36% and 31.45%, respectively. The hysteresis interval of the Schmitt trigger can be further expanded to achieve 100% realization of screening out all unstable bits, but the number of incorrectly screened stable bits will also increase, which will reduce the screening efficiency while improving the screening performance.

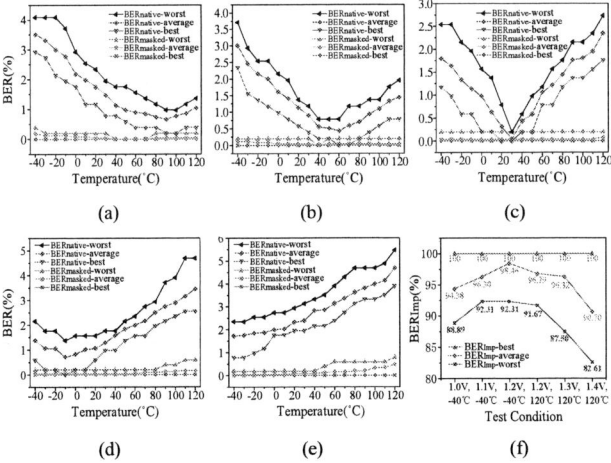

Fig. 8. Screening performance chart (a) 1.0V (b) 1.1V (c) 1.2V (d) 1.3V (e) 1.4V (f) BER improvement under extreme conditions

C. Performance Comparison

The performance of the proposed screening technique is compared with the traditional TMV method and existing screening techniques, as shown in Table I. The TMV method requires multiple sampling of the same PUF output value, while the proposed screening technique can complete the screening at the same time of reading the PUF output value, which significantly improves the screening performance while reducing the detection time. Compared with the existing bias screening techniques, the proposed technique is more effective in improving the stability of PUFs over a wider voltage and temperature range.

TABLE I. COMPARISON OF SCREENING PERFORMANCE

	Error detection technique	Test condition	Count of detections	Masking ratio	BER improvement
JSSC'20 [8]	TMV11	Noisy	11	—	48.3%
JSSC'20 [9]	V$_{SS}$ Bias	0.8V, −20°C~60°C	2	12.7%	84.1%
ISSCC'24 [10]	Reconfigurable current mirror	0.7V~1.0V, 0°C~90°C	2	23.2%	77.09%
	2b Current DAC			27.7%	87.18%
This work	Hysteresis effect	1.0V~1.4V, 0°C~100°C	2	31.45%	91.89%
		1.0V~1.4V, −40°C~120°C			88%

V. CONCLUSION

This paper proposes a stability improvement technique for monostable PUF based on the hysteresis effect of the Schmitt trigger. Within the voltage range of 1.0V to 1.4V and the temperature range of −40°C to 120°C, the PUF BER was reduced by more than 88%. The average masking ratio of the screening was 28.51%, with the maximum masking ratio being 31.45%. The proposed technique significantly reduces time cost while improving screening performance and demonstrates its effectiveness over a wide range of voltages and temperatures.

REFERENCES

[1] Z. Zhou, P. Wang and G. Li, "Bagua Protocol: A Whole-Process Configurable Protocol for IoT Sensing Devices Security Based on Strong PUF," in *IEEE Internet of Things Journal*, vol. 11, no. 1, pp. 805-819, 1 Jan.1, 2024.

[2] C. Herder, M.-D. Yu, F. Koushanfar, and S. Devadas, "Physical unclonable functions and applications: A tutorial," Proc. IEEE, vol. 102, no. 8, pp. 1126–1141, Aug. 2014.

[3] J. W. Lee, D. Lim, B. Gassend, G. E. Suh, M. van Dijk and S. Devadas, "A technique to build a secret key in integrated circuits for identification and authentication applications," in Proc. 2004 Symposium on VLSI Circuits (VLSI), pp. 176–179, 2004.

[4] G. Li, P. Wang, X. Ma, Y. Shi, B. Chen and Y. Zhang, "A Multimode Configurable Physically Unclonable Function With Bit-Instability-Screening and Power-Gating Strategies," in *IEEE Transactions on Very Large Scale Integration (VLSI) Systems*, vol. 29, no. 1, pp. 100-111, Jan. 2021.

[5] G. Li, P. Wang, X. Ma, J. Lian, J. Shu and Y. Zhang, "A 215-F² Bistable Physically Unclonable Function With an ACF of <0.005 and a Native Bit Instability of 2.05% in 65-nm CMOS Process," in *IEEE Transactions on Very Large Scale Integration (VLSI) Systems*, vol. 28, no. 11, pp. 2290-2299, Nov. 2020.

[6] S. Taneja, A. B. Alvarez and M. Alioto, "Fully Synthesizable PUF Featuring Hysteresis and Temperature Compensation for 3.2% Native BER and 1.02 fJ/b in 40 nm," in *IEEE Journal of Solid-State Circuits*, vol. 53, no. 10, pp. 2828-2839, Oct. 2018.

[7] Y. He, D. Li, Z. Yu and K. Yang, "ASCH-PUF: A "Zero" Bit Error Rate CMOS Physically Unclonable Function With Dual-Mode Low-Cost Stabilization," in *IEEE Journal of Solid-State Circuits*, vol. 58, no. 7, pp. 2087-2097, July 2023.

[8] D. Li and K. Yang, "A Self-Regulated and Reconfigurable CMOS Physically Unclonable Function Featuring Zero-Overhead Stabilization," in *IEEE Journal of Solid-State Circuits*, vol. 55, no. 1, pp. 98-107, Jan. 2020.

[9] K. Liu, Y. Min, X. Yang, H. Sun and H. Shinohara, "A 373-F² 0.21%-Native-BER EE SRAM Physically Unclonable Function With 2-D Power-Gated Bit Cells and V$_{SS}$ Bias-Based Dark-Bit Detection," in *IEEE Journal of Solid-State Circuits*, vol. 55, no. 6, pp. 1719-1732, June 2020.

[10] S. S. Kudva, M. E. Sinangil, S. Tell, N. Nedovic, S. Song and B. Zimmer "High-Density and Low-Power PUF Designs in 5nm Achieving 23× and 39× BER Reduction After Unstable Bit Detection and Masking," *2024 IEEE International Solid-State Circuits Conference (ISSCC)*, San Francisco, CA, USA, 2024.

A Reliable Current Starved Inverter PUF Based on SRAM Memory Structure

Mingze Ren [1], Pengjun Wang *[2], Yuejun Zhang *[1], Shutong Zhang [1], Zhikang Chen[1], Tengfei Yuan[1]

[1] Faculty of Electrical Engineering and Computer Science, Ningbo University, Ningbo, 315211, China
[2] College of Electrical and Electronic Engineering, Wenzhou University, Wenzhou, 325035, China

* Email: wangpengjun@wzu.edu.cn, zhangyuejun@nbu.edu.cn

Abstract—**This paper presents a current starved inverter physically unclonable function (PUF) based on static random memory (SRAM) structure. By improving the SRAM cell structure, the PUF cell is biased in the subthreshold region to form a current starved inverter PUF, and the DIBL effect amplifies the voltage distribution of the switching threshold in current starved inverter PUF. Meanwhile, the proposed PUF is monostable, which provides higher reliability compared to the conventional SRAM PUF and retains the storage function of conventional SRAM. We simulate and verify the proposed PUF circuit at 65nm process. The simulation results show that the proposed PUF exhibits superior uniqueness and randomness with an average inter-chip Hamming distance of 50.17% and a standard deviation of 0.0639, which passes the NIST 800-22 randomness test. The worst BER is 3.27% and 2.05 % over the temperature variation range of -20°C-80°C and the voltage variation range of 1V-1.4V, respectively.**

Keywords—physically unclonable function, current starved inverter, SRAM, subthreshold.

I. INTRODUCTION

As a fundamental hardware security element, the physical unclonable function (PUF) plays a crucial role in the realms of device verification and key derivation [1]. SRAM PUF (Static Random Memory PUF) [2] is the most commercially promising PUF because it utilizes the existing SRAM cell of the IoT device as the basic building block of the PUF, which eliminates the need for additional circuits and thus reduces the cost. However, conventional SRAM PUFs have low reliability due to the variation of voltage slope during power-up and the effect of noise. To solve these problems, scholars have proposed some monostable PUFs, such as the enhancement-enhancement (EE) SRAM PUF [3] and the inverter chain PUF [4]. Although EESRAM PUF improves the reliability by reducing the gain, it greatly reduces the performance of SRAM as a cache and is not compatible with conventional SRAM memory. Inverter chain PUFs have better reliability than conventional SRAM PUFs due to monostable characteristics, but they require additional circuitry to generate the PUF response compared to SRAM PUFs, which significantly increases the hardware cost.

In order to better trade-off between the reliability of the PUF and the additional hardware cost, this paper proposes a current-starved inverter PUF based on the SRAM storage structure, where the PUF cell is equivalent to a conventional SRAM in SRAM storage mode and can read and write data. In PUF mode, the bias voltage clamp causes the cell to operate in the subthreshold region, and the cross-coupled inverters are connected in parallel and the inputs and outputs are short-circuited to form a monostable current starved inverter PUF, which improves the reliability compared to the conventional SRAM PUF.

II. THE PROPOSED PUF DESIGN

A. CSI PUF Cell Based on SRAM Memory

The structure of the current starved inverter PUF based on the SRAM memory structure is shown in Fig. 1. Fig. 1(a) and (b) show the shared bias voltage generator and the shared mode switching circuit, respectively. Fig. 1(c) shows the proposed PUF Cell. P0, N0, N3 are mode configuration transistors controlled by the mode configuration circuit for mode switching. P1, P2, N1, N2 form a pair of cross-coupled inverters. N4, N5 are access transistors used to read the entropy source information out to the bit line BL (BLB).

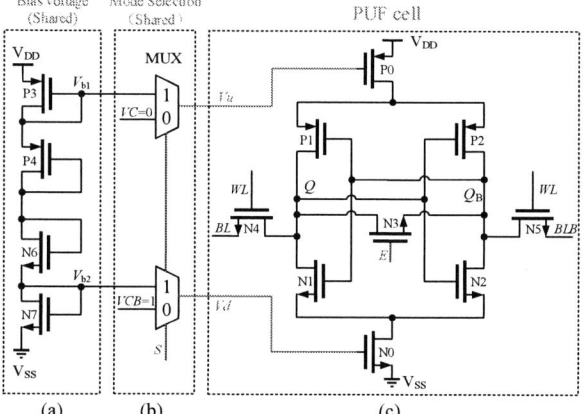

Fig. 1. (a) Bias voltage generation. (b) Mode switching circuit (c) PUF cell.

When the mode switching signal $S=E=0$, P0 conducts with N0 and N3 turns off. The PUF cell is structurally equivalent to a conventional 6T SRAM, retaining the SRAM cache function, and the conventional SRAM PUF response can be extracted.

When the mode switching signal $S=E=1$, as shown in Fig. 2, P0 and N0 are biased by the bias voltage, N3 conducts, and all the transistors enter into the subthreshold operating region, and the PUF cell is structurally equivalent to a current starved inverter PUF, whose switching threshold, V_M, is the entropy source of the PUF cell, and in order to derive the equation for the relationship of V_M, the *I-V* characteristics of the transistors operating in the subthreshold state are first written:

$$I_N = I_{SN} \exp\left(\frac{V_{GS} - V_{TH0,N} + \lambda V_{DS}}{m_n V_T}\right) \qquad (1)$$

$$I_P = I_{SP} \exp\left(\frac{V_{SG} - |V_{TH0,P}| + \lambda V_{SD}}{m_p V_T}\right) \qquad (2)$$

In (1)(2), $I_S = \mu C_{ox} \frac{W}{L}(m-1)V_T^2$, is the subthreshold intrinsic current, μ is the transistor mobility, C_{ox} is the gate oxygen capacitance, W and L are the width and length of the transistor channel, respectively, m is the subthreshold slope factor, V_T is the thermal voltage, V_{TH0} is the zero-bias threshold voltage at room temperature, and λ is the DIBL effect coefficient.

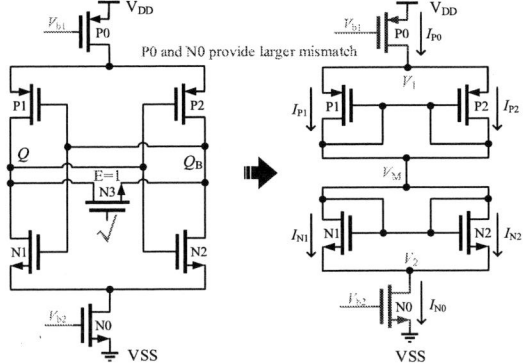

Fig. 2. Current starved inverter PUF based on SRAM memory structure.

To simplify the model, it is assumed that all transistors in Fig. 2 have the same λ, V_T, in a real design, their values are not equal. and thus the expressions for the currents flowing through P0-P2, N0-N2, and their relationships are given by (3)-(8):

$$I_{P0} = I_{SP0} \exp\left(\frac{(1+\lambda)V_{DD} - \lambda V_1 - V_{b1} - |V_{TH0,P0}|}{m_p V_T}\right) \quad (3)$$

$$I_{P1} + I_{P2} = I_{SPE} \exp\left(\frac{(1+\lambda)V_1 - (1+\lambda)V_M - |V_{TH0,P1}|}{m_p V_T}\right) \quad (4)$$

$$I_{SPE} = I_{SP1} + I_{SP2} \exp\left(\frac{|V_{TH0,P1}| - |V_{TH0,P2}|}{m_p V_T}\right) \quad (5)$$

$$I_{N0} = I_{SN0} \exp\left(\frac{V_{b2} + \lambda V_2 - V_{TH0,N0}}{m_n V_T}\right) \quad (6)$$

$$I_{N1} + I_{N2} = I_{SNE} \exp\left(\frac{(1+\lambda)V_M - (1+\lambda)V_2 - V_{TH0,N1}}{m_n V_T}\right) \quad (7)$$

$$I_{SNE} = I_{SN1} + I_{SN2} \exp\left(\frac{V_{TH0,N1} - V_{TH0,N2}}{m_n V_T}\right) \quad (8)$$

From the topology of the equivalent circuit shown in Fig. 2, it can be seen that The current at P0 equals the combined current at P1 and P2, and The current at N0 equals the combined current at N1 and N2, and thus the expressions for the matching voltages V_1 and V_2 can be obtained by associating (3)(4)(5) with associating (6)(7)(8) as shown in (8)(9), respectively:

$$V_1 = \frac{1}{1+2\lambda} \times [(1+\lambda)(V_{DD} + V_M) - V_{b1} - |V_{TH0,P0}| + |V_{TH0,P1}| + \quad (9)$$
$$m_p V_T \ln\left(\frac{I_{SP0}}{I_{SP1}}\right) - m_p V_T \ln\left(1 + \left(\frac{I_{SP2}}{I_{SP1}}\right) \exp\left(\frac{|V_{TH0,P1}| - |V_{TH0,P2}|}{m_p V_T}\right)\right)]$$

$$V_2 = \frac{1}{1+2\lambda} \times [(1+\lambda)V_M - V_{b2} + V_{TH0,N0} - V_{TH0,N1} - \quad (10)$$
$$m_n V_T \ln\left(\frac{I_{SN0}}{I_{SN1}}\right) + m_n V_T \ln\left(1 + \left(\frac{I_{SN2}}{I_{SN1}}\right) \exp\left(\frac{V_{TH0,N1} - V_{TH0,N2}}{m_n V_T}\right)\right)]$$

Similarly, The total current flowing through points P1 and P2 is equivalent to the combined current through points N1 and N2, so equating (4) with (7) and bringing in (9)(10) yields the expression for V_M as in (11):

$$V_M = \frac{m_n V_{DD}}{m_n + m_p} + \frac{1}{\lambda} \times \left[\frac{m_n V_{DD}}{m_n + m_p} + \frac{m_p (V_{TH0,N0} - V_{TH0,N1})}{m_n + m_p} - \right.$$
$$\frac{m_n (|V_{TH0,P0}| - |V_{TH0,P1}|)}{m_n + m_p} - \frac{m_n V_{b1} + m_p V_{b2}}{m_n + m_p} + m^* V_T \ln\left(\frac{I_{SP0} I_{SNE}}{I_{SN0} I_{SPE}}\right) \right] \quad (11)$$
$$+ \frac{1+2\lambda}{\lambda(1+\lambda)} \times \left[\frac{m_p V_{TH0,N1} - m_n |V_{TH0,P1}|}{m_n + m_p} + m^* V_T \ln\left(\frac{I_{SPE}}{I_{SNE}}\right) \right]$$

In (11), m^* is the subthreshold integrated effective slope factor, it can be seen that V_M is related to the process

parameters of all transistors of the equivalent circuit of the PUF cell in Fig. 2, and $1/\lambda$ is used as the mismatch amplification factor, which is generally $\lambda \ll 1$, so that the mismatches of P0 and N0 are amplified significantly resulting in a wide voltage distribution range for V_M. In addition, the weak positive temperature coefficient property of the V_M is amplified due to λ, and the V_M rise is more pronounced with increasing temperature.

B. Over Architecture of Proposed CSI PUF

The top-level structure of the proposed PUF are shown in Fig. 3. which includes the PUF array, decoder and driver circuits, bias voltage generator, and mode selection module, precharge module, voltage comparator, reference voltage generation and PUF response latch circuit. The array size is 32 × 8 and each bit line (BL/BLB) shares a voltage comparator. The reference voltage VREF is a positive temperature coefficient to properly compensate for the positive temperature coefficient of the V_M, which helps to reduce the BER caused by temperature variations. The voltage comparator compares the entropy source signal on the bit line with the reference voltage VREF and digitizes it to 0 or 1 and latches it.

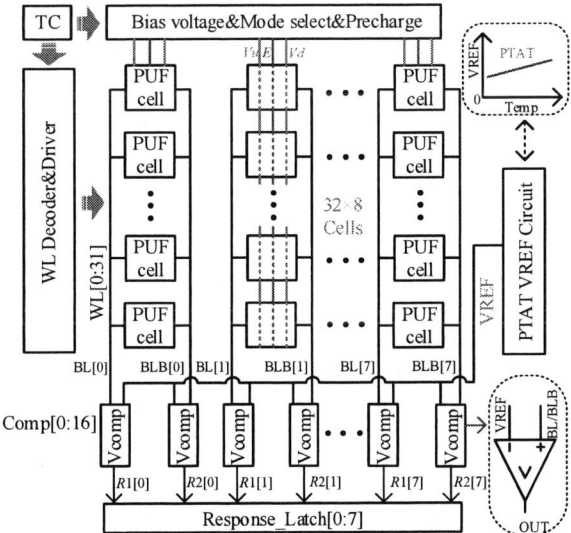

Fig. 3. Top-level array structure of the proposed PUF.

III. SIMULAITON RESULTS AND DISCUSSION

The proposed current starved inverter PUF based on SRAM memory structure is designed using Cadence virtuoso software and simulated using Spectre simulator to evaluate its performance. The proposed PUF is evaluated in terms of uniqueness, randomness and reliability.

A. Uniqueness

The uniqueness characterizes the differentiation between different PUF entities and is assessed by inter hamming distance (HD) between PUFs, which ideally should be 50%. Under nominal conditions, Monte Carlo simulation was used to simulate 40 times and collect 10240 bits of PUF responses to evaluate the uniqueness of the proposed PUFs using a block word length of 64. The results are shown in Fig. 4. The mean value of the uniqueness of the proposed PUF is 50.17% and the standard deviation is 0.0639, which indicates that the proposed PUF has a good degree of differentiation.

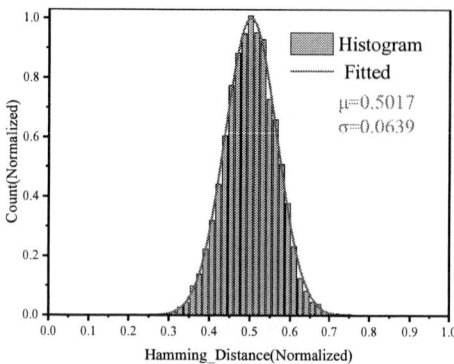

Fig. 4. Inter hamming distance.

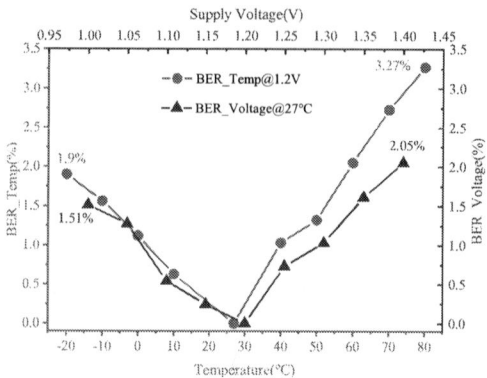

Fig. 5. BER with temperature and voltage variation.

B. Randomness

In order to better evaluate the randomness of the PUF proposed in this paper, a NIST 800-22 randomness test was performed on the collected 10240 bits responses, and the results of the test are shown in Table I, in which the suggested PUF successfully meets the NIST test criteria with an elevated rate of approval and a robust average *p*-value., which indicates that the proposed PUF has excellent randomness.

TABLE I.   NIST 800-22 TEST RESULTS

NIST test	Stream length(n)	No. of Runs	Average p-value	Pass(%)
Frequency	256	40	0.4326	97.5%
Block Frequency	256	40	0.4871	95%
Runs	256	40	0.5616	100%
Longest Runs	256	40	0.4199	100%
FFT	256	40	0.4092	97.5%
Non Overlapping Template	256	40	0.5173	100%
Serial	256	40	0.5204	98.75%
Approximate Entropy	256	40	0.5495	100%
Cumclative Sums	256	40	0.4493	96.25%

C. Reliability

Reliability in PUF circuits is characterized by their capacity to remain functional amidst varying environmental conditions. different temperature and voltage can lead to changes in the characteristics of the PUF circuits, which can cause the response to flip-flop resulting in a BER. In this experiment, eight PUF chips were simulated from a temperature range of -20°C to 80°C and a voltage range of 1 V to 1.4V. The average BER was calculated as shown in Fig. 5, and the worst BER caused by temperature was 3.27%, and the worst BER caused by voltage was 2.05%, which indicates that the proposed PUF has excellent reliability.

TABLE II.   COMPARISON WITH RELATED WORK

	JSSC[5]	TCASII[6]	TCASII[7]	This work
Tech.(nm)	130	28	40	65
Type	SRAM	SRAM	SRAM	CSI base on SRAM
Uniqeness(%)	50.13	49.57	49.64	50.17
NIST 800-22 pass	NA	NA	YES	YES
Native BER in temp&Voltage(%)	>6	>6	>10	3.27
Temp Range(°C)	0-80	-20-80	20-80	-20-80
Voltage Range(V)	0.9-1.2	0.7-1.1	0.7-1.2	1-1.4

IV. CONCLUSIONS

In this paper, we propose a current starved inverter PUF based on the SRAM memory structure, which benefits from its monostable structure and improves the reliability compared to the traditional SRAM without destroying the original SRAM memory structure thus ensuring its wide range of application scenarios. Table II summarizes the comparison of this work with other advanced work on traditional SRAM, and it can be seen that the reliability of the PUF proposed in this paper is better than that of the conventional SRAM PUF without any reliability improvement measures., and the uniqueness is comparable to it, which makes it very suitable for resource-constrained IoT devices.

ACKNOWLEDGMENT

This work is supported by the National Natural Science Foundation of China (62234008, 62174121, 62134002), the Science and Technology Innovation 2025 Major Project of Ningbo (2022Z203), Ningbo University and Ningbo Yongxin Microelectronics Technology Co., LTD. Digital Integrated Circuit Design Joint Laboratory (XQ2022000005), Ningbo University Graduate Education Practice Base (YJD202305), Zhejiang Xinmiao Talents Program (2024R405C105).

REFERENCES

[1] Z. Zhou, P. Wang and G. Li, "Bagua Protocol: A Whole-Process configurable protocol for IoT sensing devices security based on strong PUF," IEEE Internet of Things Journal, 2024, 11(1): 805-819.

[2] L. Ni and J. Zhang, " S2RAM PUF: An ultra-low power sub threshold SRAM PUF with zero bit error rate," ProceeDSngs of the 61th ACM/IEEE Design Automation Conference, 2024: 23-27.

[3] K. Liu, Y. Min, X. Yang, et al., "A 373-F² 0.21%-Native-BER EE SRAM physically unclonable function with 2-D Power-Gated bit cells and V_{SS} Bias-Based Dark-Bit detection," Journal of Solid-State Circuits, 2020, 55(6): 1719-1732.

[4] D. Li and K. Yang, "A Self-Regulated and reconfigurable CMOS physically unclonable function featuring Zero-Overhead stabilization, " Journal of Solid-State Circuits, 2020, 55(1): 98-107.

[5] Y. Su, J. Holleman and B. P. Otis, "A digital 1.6 pJ/bit chip identification circuit using process variations, " Journal of Solid-State Circuits, 2008, 43(1): 69-77.

[6] S. Park, M. Jeong, J. Kim, et al., "A 6T-SRAM-Based physically unclonable function with low BER through automated maximum mismatch detection, " Transactions on Circuits and Systems II: Express Briefs, 2024, 71(7): 3493-3497.

[7] L. Lu and T. Kim, "A high reliable SRAM-Based PUF with enhanced challenge response space," Transactions on Circuits and Systems II: Express Briefs, 2022, 69(2): 589-593.

An Efficient Convolutional Neural Network Hardware IP for Epilepsy Detection

Yi Gong [1], Yuejun Zhang*[1], Jiangtao Tu [1], Rongxin Zou [1], Liang Wen [2]

[1] Faculty of Electrical Engineering and Computer Science, Ningbo University, Ningbo 315211, China

* Email: zhangyuejun@nbu.edu.cn

Abstract—**Epilepsy is a common neurological disease which is sudden and unpredictable. In response to the current problems of low efficiency and high false detection rate in epilepsy detection, this paper proposes an efficient convolutional neural network (CNN) hardware intellectual property (IP) design scheme for epilepsy detection. According to the epilepsy pathology, the oversampling technique is used to solve the sample imbalance. Filters out the noise by using band-pass filters, and improves the model by using Z-score normalization. Then, the structure of convolutional neural network is optimized to build an end-to-end network model and train the corresponding weights as well as biases. A single convolutional kernel is used to lighten the network model and reduce the parameters. The 22-channel EEG data are processed in a parallelized manner to fully extract the correlation between the channels. The 3-layer convolutional and max pooling module is designed by combining with pipelining technique. Validated on the CHB-MIT dataset, the epilepsy detection accuracy is reached 97.45%. Under TSMC 65 nm process, the IP core area is measured as 1.625 mm^2, power consumption is measured as 4.225 mW, single detection delay is 0.01935 ms, and single detection energy consumption is 0.237 µJ/class.**

Keywords—*epilepsy detection, feature extraction, CNN, hardware IP*

I. INTRODUCTION

Epilepsy is a chronic brain disease in which sudden abnormal neuronal discharges in the brain lead to temporary impairment of brain function [1]. Existing epilepsy detection methods mainly include magnetoencephalography magnetic resonance imaging and electroencephalography (EEG). Among them, EEG is the most widely used epilepsy detection method, in which neurologists manually analyze EEG recordings to diagnose whether epilepsy has occurred. Due to the time-consuming and inefficient visual inspection of a huge amount of EEG, it imposes a great burden on healthcare workers. In order to protect the health of epilepsy patients and alleviate the workload of medical staff, there is an urgent need to design an efficient epilepsy detection device.

With the development of artificial intelligence, automatic epilepsy detection systems based on deep learning have gradually matured. Epilepsy detection based on deep learning mainly includes two types of methods. The first type is classified by manually extracting features and then inputting them into a neural network. For example, reference [2] utilizes multi-scale wavelet analysis to decompose the input EEG, obtaining components of different frequency bands. The decomposed multi-scale EEG are then input into CNN for classification through the attention mechanism. The second type is applying deep learning directly to the original EEG data. In contrast to the manual feature extraction approach, reference [3] uses CNN to extract feature from the original EEG data to distinguish between preictal, interictal, and ictal periods for the purpose of epilepsy detection. Compared with manual feature extraction, the deep learning method is able to

directly learning more distinct and robust feature from the original EEG signals, and outperforms the traditional method based on manual feature extraction [4].

Although deep learning algorithms have shown excellent classification performance for epileptic detection, it still faces great computational challenge in practical application, particularly in the hardware realization of the algorithms. In view of this, we will design an efficient convolutional neural network hardware IP for epilepsy detection.

II. EPILEPSY DETECTION MODEL

The structure of the one-dimensional CNN model for epilepsy detection is shown in Fig. 1, including three convolutional layers, three max pooling layers, two fully connected layers, and a sigmoid activation function layer. During the model design stage, the model is lightweighted. A single convolutional kernel is used to reduce the network size and computational burden, as well as to minimize classification performance loss. The sampling frequency of the CHB-MIT dataset is 256 Hz [4], while epilepsy related signals are mainly concentrated within 1~50 Hz. The original data are band-pass filtered from 1~50 Hz, and then the filtered data are normalized by Z-score. Finally, the feature data in the dimension 256 × 22 are extracted with a time window of 1 s. In order to better extract the internal spatial relationships between multi-channel EEG, the first layer of convolution is applied to all channels simultaneously, with a convolution kernel size of 3 × 22. While preserving the phase relation between channels, the data after the first layer convolution is turned into a single channel, which reduces the computational complexity. Then, the convolved features are made to do max pooling to reduce the number of features. In order to extract deeper level features, they are sequentially processed through the second and third convolutional pooling layers to obtain an 8 × 1 dimensional feature map, which is flattened and inputted into the two-layer fully connected layer. Finally, it is then input to the activation function sigmoid layer to obtain the classification result.

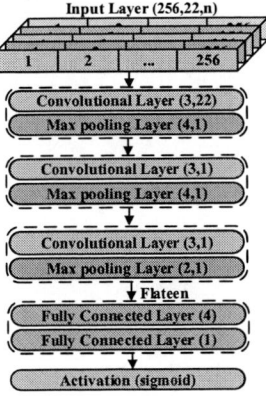

Fig. 1. Structure of epilepsy detection model.

III. EPILEPSY DETECTION HARDWARE IP DESIGN

The epilepsy detection hardware IP architecture is illustrated in Fig. 2 and includes five main parts: convolution module, max pooling module, fully connected module, EEG data, weight bias storage unit and sigmoid module. To ensure the calculation accuracy, the data type adopts IEEE-754 format. In IEEE-754, the half-precision floating-point type data include 1 bit sign, 5 bits exponent and 10 bits mantissa.

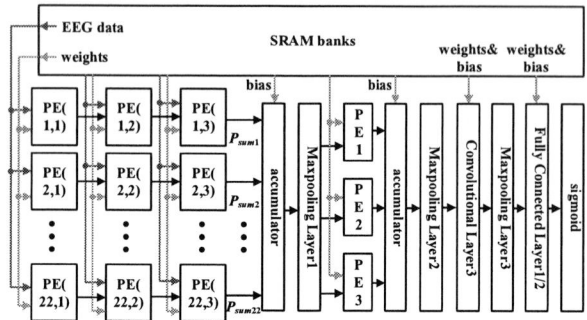

Fig. 2. General architecture of epilepsy detection hardware IP.

A. Convolution Module

Convolutional layer is the key component of CNN, which extract feature through convolutional operation. The convolution kernel slides the input data and calculates the weighted sum of each local region to generate the feature map. As shown in Fig. 3, taking the three processing elements (PE) of channel 1 as an example, PE (1, 1) performs the multiplication operation of the first data with the first weight. The result of the multiplication passes through a register and then is input into the adder of PE (1, 2) to accumulate with the multiplier output of PE (1, 2). The result is then input to the adder of PE (1, 3) through a register and accumulated with the output of the multiplier of PE (1, 3) to obtain the partial sum P_{sum1}, completing the convolution operation of channel 1. 22 channels are parallelized to obtain 22 partial sums, which are then input into the accumulator and accumulated with bias to get the first convolution output. The convolution kernel sliding is implemented through two registers with a sliding step of 1 and 0-padding to ensure that the data after convolution has the same dimension as the input data.

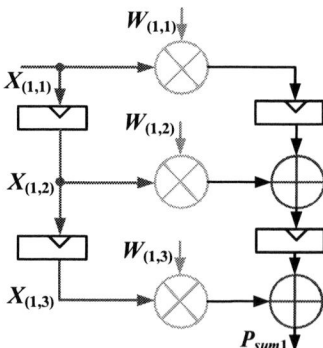

Fig. 3. Hardware circuit of convolution module.

B. Max Pooling Module

After the convolutional layer will obtain the features with large dimensions, the features are divided into a number of regions, and the maximum value is taken to get the new smaller dimensions features. Taking the first layer of max pooling as an example, the size of the first layer of max pooling is 4 × 1. The hardware design is implemented by two comparators, two registers, and two division clocks. Firstly, the input data are divided into two ways by the main frequency clock. One way is input to the first level comparator through register reg1, and the other way is directly input to the first level comparator. Then, the comparison result is read by a 2-division clock, and also divided into two ways to input into the second level comparator. Finally, the comparison result is input to the next convolutional layer via a 4-division clock. The max pooling size of the second and third layers are 4 × 1 and 2 × 1, respectively, and the hardware implementation is the same as the first layer max pooling.

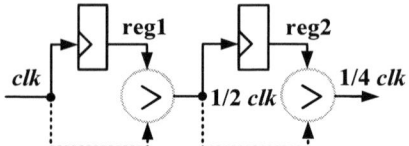

Fig. 4. Hardware circuit of max pooling module.

C. Fully Connected Module

It is the responsibility of the fully connected layer to map the extracted features to the network output. After the previous three layers of convolution and max pooling, the dimension of the output data is 8 × 1, which is flattened and input into the fully connected layer. The first layer adopts the method of time-division multiplexing to complete the calculation of four nodes. As shown in Fig. 5, for the first 8 clock cycles, the input data are sequentially input into the 8 multipliers through 7 registers respectively to multiply with the corresponding weights. The result after multiplication is input into the next level of adder to complete the accumulation operation. The partial sum obtained after multiplying and accumulating is then added to the bias of the first node to get the result of the first node. The next 8 clock cycles complete the weight update to get a new partial sum which is then added with the bias of the second node to get the result of the second node. Similarly, after 32 clock cycles, the calculation of four nodes is completed. These four nodes are input to the second fully connected layer to get the final result of the network.

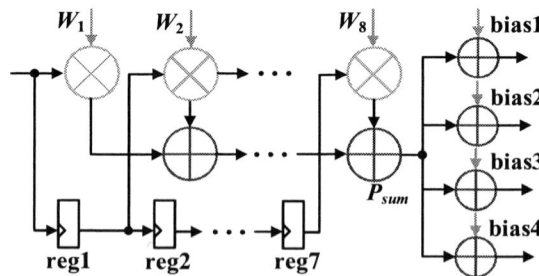

Fig. 5. Hardware circuit of fully connected module.

D. Sigmoid Module

The sigmoid function is often used as the output layer for binary classification problems in deep learning. The sigmoid function can map an arbitrary input value x to [0,1], which enables neural networks to learn more complex decision boundaries. Considering the hardware overhead, a piecewise second-order function is used to approximate the sigmoid function [5], whose mathematical form is shown in (1).

$$y = \begin{cases} \dfrac{1}{2}\left(\dfrac{1}{4}|x|-1\right)^2 , & -4 < x < 0 \\[3mm] 1-\dfrac{1}{2}\left(\dfrac{1}{4}|x|-1\right)^2 , & 0 \le x < 4 \end{cases} \qquad (1)$$

The sigmoid hardware circuit is designed according to (1), as shown in Fig. 6. Firstly, take the absolute value of the output data from the second fully connected layer. After taking the absolute value of the data multiplied by 0.25 and then subtracted by 1, the result is input to the multiplier as both the multiplier and the multiplied number to complete the squaring operation, and then input to the next level of the multiplier to multiply by 0.5. Finally, according to the sign bit positive or negative output the final result.

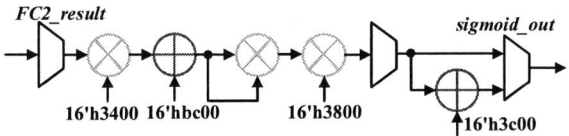

Fig. 6.   Hardware circuit of sigmoid module.

IV. EXPERIMENTAL RESULTS AND ANALYSIS

This study conducted experiments on a laptop equipped with AMD Ryzen 5 3500U 2.10 GHz processor. Epilepsy detection model is trained in Anaconda3, Tensorflow2.1, Python3.12 (CPU version) environment. The hardware IP is designed using the simulation tool VCS, the logic synthesis tool Design Compiler, and the physical layout tool IC Compiler.

The functional validation result of the epilepsy detection IP is illustrated in Fig. 7. *FC1_result*, *FC2_result*, and *sigmoid_out* denote the outputs of the first and second fully connected layers as well as the outputs of the sigmoid module, respectively. *Label* denotes the detection result, which is 0 under normal conditions, which inputs the EEG data of the seizure period, after 387 clock cycles to complete a single detection, the *Label* becomes 1.

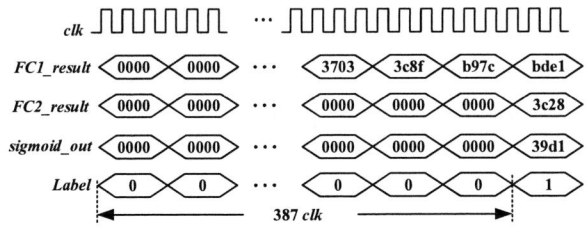

Fig. 7.   Functional validation results for epilepsy detection IP.

The accuracy and loss curves of the training and validation sets of the model are illustrated in Fig. 8. After 100 iterations, the proposed epilepsy detection model achieves 97.45% and 97.67% accuracy on the training and validation sets, respectively. The epilepsy detection IP layout as well as the key parameters are shown in Fig. 9. Under TSMC 65 nm process, the IP core area is measured as 1.625 mm², power consumption is measured as 4.225 mW, single detection delay is 0.01935 ms, and single detection energy consumption is 0.237 μJ/class.

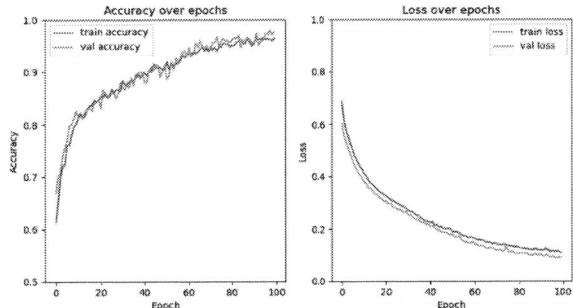

Fig. 8.   Accuracy and loss curves for epilepsy detection model.

Technology	TSMC 65nm 1P9M
Supply voltage	1V
Frequency	20MHz
Power	4.225mW
Accuracy	97.45%
Detection latency	0.01935ms
Energy/Classification	0.237μJ/class
Area	1.625mm²

Fig. 9.   Epilepsy detection IP layout and key parameters.

V. CONCLUSION

In this paper, we propose an efficient convolutional neural network hardware IP core for epilepsy detection. Validated on the CHB-MIT dataset, the epilepsy detection accuracy is reached 97.45%. Under TSMC 65 nm process, the IP core area is measured as 1.625 mm², power consumption is measured as 4.225 mW, single detection delay is 0.01935 ms, and single detection energy consumption is 0.237 μJ/class.

ACKNOWLEDGMENT

This work is supported by the National Natural Science Foundation of China (62174121, 61871244, 62134002), the Science and Technology Innovation 2025 Major Project of Ningbo (2022Z203), Ningbo University and Ningbo Yongxin Microelectronics Technology Co., LTD. Digital Integrated Circuit Design Joint Laboratory (XQ2022000005), Ningbo University Graduate Education Practice Base (YJD202305). the Ningbo University Student Research and Innovation Program (2024SRIP1306, 2024SRIP1310).

REFERENCES

[1] J. Yang and M. Sawan, "From seizure detection to smart and fully embedded seizure prediction engine: a review," IEEE Transactions on Biomedical Circuits and Systems, vol. 14, pp. 1008–1023, October 2020.

[2] Q. Xin, S. Hu, S. Liu, L. Zhao and Y. D. Zhang, "An attention-based wavelet convolution neural network for epilepsy EEG classification," IEEE Transactions on Neural Systems and Rehabilitation Engineering, vol. 30, pp. 957–966, April 2022.

[3] M. Zhou, C. Tian, R. Cao, *et al*, "Epileptic seizure detection based on EEG signals and CNN," Frontiers in Neuroinformatics, vol. 12, pp. 95, December 2018.

[4] A. L. Goldberge, L. A. N. Amaral, L. Glass, *et al*, "PhysioBank, PhysioToolkit, and PhysioNet: components of a new research resource for complex physiologic signals," Circulation 101 (23) (2000) 215–220.

[5] M. Zhang, S. Vassiliadis and J. G. Delgado-Frias, "Sigmoid generators for neural computing using piecewise approximations," IEEE Transactions on Computers, vol. 45, pp. 1045–1049, September 1996.

TLBshield: A Low-cost Secure Reinforce on Translation Lookaside Buffer to Mitigate the Speculative Attacks

Yuyang Liu, Runye Ding, Yujie Chen, Pujin Xie, Yao Liu*, Zhiyi Yu

School of Microelectronics Science and Technology, Sun Yat-sen University, Zhuhai, China

* Email: liuyy529@mail2.sysu.edu.cn, liuyao25@mail.sysu.edu.cn

Abstract—**Although hardware defenses against speculative attacks have been widely studied, the security of Translation Lookaside Buffer has seldom been studied. In this paper, we propose TLBshield to mitigate Spectre attack on TLB, and evaluates it on a XuanTie C910 RISC-V SOC on FPGA. Linux OS including filesystem and shell is also mounted in the RISC-V core. TLBshield automatically switches on and off the writeback channel from Page Table Walker to the second-level TLB, and increase the security level with almost no performance loss. The performance overhead of TLBshield under SPEC2017 benchmark is negligible, where the maximum performance loss is only 1.77%. Compared with the original RISC-V core, the success rate for attack is reduced from 100% to 55.7%.**

Keywords—*Translation Lookaside Buffer, Spectre attack, high-performance processor, FPGA*

I. INTRODUCTION

Speculative types of attacks have gained enormous attentions in the field of computer security in recent years, even since they have been discovered, for example, Spectre [1], Meltdown [2], Foreshadow [3], and so on. Compared with previous passive timing-based attacks with relatively weak attack ability [4], as the other types of timing-based attacks, speculative attacks are capable of actively stealing secrets directly by exploiting the security vulnerabilities of high-performance microarchitectural components.

The typical Spectre attack is mainly divided into two parts [5]. First, trigger the processor into a microarchitectural temporal state that can temporally touch secret data prohibited to the attacker. Second, reveal the secret data with timing traces via typical covert channels, and transmit the secret data via typical measures. Take Spectre V1 variant as an example, out-of-order execution and branch predictor are utilized as the leakage source. The timing information of cache is usually exploited as the covert channel to retrieve the data. The flush-reload attack strategy is applied throughout the whole attack process [6].

Adopted from the above idea, numerous variants of speculative attacks have been discovered, which involve different microarchitectural components and attack approaches. The corresponding works on defense have been also widely studied [7]. However, it is hard to propose a universal architecture that can eliminate all types of speculative attacks, because they are highly correlated with the typical microarchitecture and the defense may introduce significant performance loss. Previously, most of the attack and defense works focus on cache, because cache is usually considered to be the most vulnerable and practical covert channel.

Translation Lookaside Buffer (TLB) also bears the similar cache-like structure, so that timing-based attack is also applicable to TLB attacks. For example, Gras et al. use prime-probe strategy to gain the cryptographic key from the RSA with a 92% success rate [8]. Despite the similarity, TLB attack also bears obvious differences. The timing-based attack on TLB is triggered by memory translation instead of memory value in cache. Besides, the granularity and logic are different. In general, TLB involves more complicated logic than cache to support various memory page sizes. Due to these differences, a defending strategy against cache attack does not protect TLB.

At present, only a small number of works focus on TLB security. Deng et al. systematically study the security issue on TLB and propose a defense of TLB partition [9]. The idea is to split the TLB into two areas, where attacker and the victim can read and write on only one part. To further alleviate the high-performance penalty associated with partition design, Deng et al. also propose to reduce the performance overhead by randomizing reads and writes without sacrificing security.

We note that the TLB security especially under speculative attacks is of great importance and has seldom been studied. In this paper, we propose TLBshield that enhances the security of Memory Management Unit (MMU), and mitigates speculative attacks with negligible overhead on hardware and performance. We verify the idea of TLBshield on XuanTie C910 RISC-V core [10], mounted with the ported Linux operation system (OS) on Xilinx VCU128 FPGA development board. SPEC2017 is employed to verify the performance overhead.

II. THE SPECTRE ATTACK ON TLB

A. Spectre V1 attack mechanism

In the Spectre attack model, an attacker is not allowed to access the sensitive information due to the privilege isolation, but is able know the address of these sensitive information by attempts. During the Spectre V1 attack, the attacker triggers the processor to encounter a branch statement (such as an if-else statement); the speculative execution module predicts the instructions to be executed next, and then loads and executes those instructions before the branch target address is decided. If this prediction is wrong, the processor needs to squash these preloaded instructions. However, before the squash, sensitive data that has already been loaded into covert channels, for example, cache. The attacker crafts the attack program by encoding the sensitive data with address. If cache is utilized as the covert channel, the attacker can determine the encoded address value, that is the cache index, by measuring the timing difference due to cache miss and hit, and further infers the secret data associated with the index.

979-8-3503-6184-1/24 $31.00 © 2024 IEEE

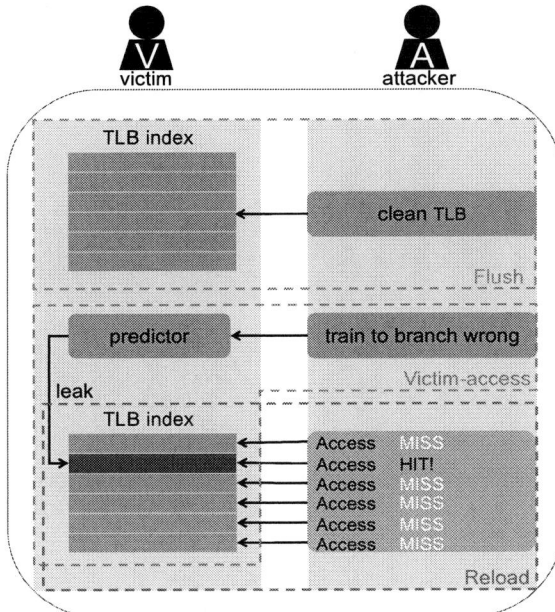

Fig. 1. Flush-reload attack steps of Spectre attack on TLB.

B. TLB-based Spectre V1 attack

TLB-based Spectre attack exploits TLB as the covert channel. The flush-reload strategy is utilized [6]. This approach is divided into three main stages, as shown in Fig. 1. The first stage is the flush phase. The attacker needs to clear the TLB page table from the TLB to get the initial state. The second stage is the victim-access phase. The sensitive content accessed by the victim is reloaded into the TLB. The final stage is the reload phase. The attacker revisits these TLB page table addresses and measures the access time. If the access time is short, it is highly probably that the address is related to a sensitive data.

III. TLBSHIELD DESIGN

A. XuanTie C910 Memory Management Unit Structure

XuanTie C910 is a processor of the RISC-V instruction set architecture. As shown in Fig. 2, The C910 MMU (*Memory Management Unit*) mainly contains *uTLB*, *jTLB*, *TLB Reg* and *PTW* (Page Table Walker) modules.

uTLB is the first-level TLB, which is divided into instruction TLB and data TLB. The data uTLB uses a 17-way fully associative structure. The 16-way data TLB is mixed storage of 4KB and 2MB size page table translations, and the other one only stores 1GB size page table translations. *uTLB* is not fit for attack, because the size is too small to meet the accuracy requirements of the attack.

jTLB is the second-level TLB, which is an instruction and data page table translation hybrid storage TLB. *jTLB* uses 4-way 256 index group association structure. *jTLB* starts addressing only when *uTLB* misses or *TLB Reg* requests. *jTLB* acts as the primary attack target, since it satisfies the accuracy requirements of the attack.

PTW is the module that translates a virtual address into a physical address. Page-walk occurs only if both *uTLB* and *jTLB* miss at the same time. *PTW* is connected to the (Load-Store Unit) LSU module through special channels to make up to three address accesses to convert virtual addresses into

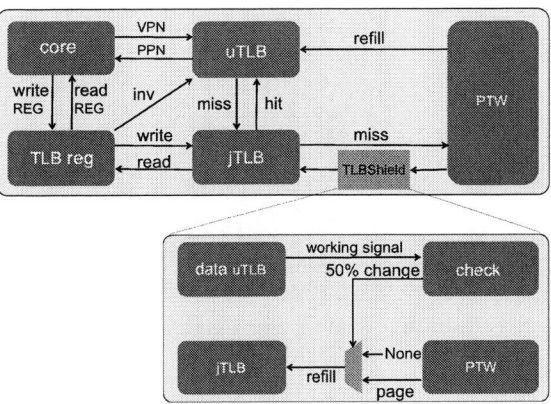

Fig. 2. XuanTie C910 MMU with TLBshield.

physical addresses. Both *uTLB* and *jTLB* are backfilled after obtaining the physical address.

TLB Reg module is designed to allow the operating system to read and write the TLB directly from the superuser mode status register.

B. TLBShield design

The straightforward idea to prevent speculative attack on TLB is to turn off the writeback channel between *PTW* and *jTLB*. However, this approach brings significant performance loss, which is considered impractical, as shown in Fig. 4.

We propose TLBshield, a low-cost secure mechanism on TLB, as shown in Fig. 2. TLBshield can balance the performance and security by temporally switching on and off the writeback channel from *PTW* to *jTLB*, as shown in Fig. 3. We add a module to convert the writeback channel on and off by identifying the working signal of the data *uTLB*.

When the data *uTLB* generates a working signal, a virtual address and a physical address translation is about to take place, which also indicates that a load or store instruction has been carried out. We create a new 1-bit register index, the 0 or 1 of which indicates the status of the writeback channel. Every time the *uTLB* working signal is high, the register index is continuously inverted at every clock edge, so that successful rate of writeback is only 50%.

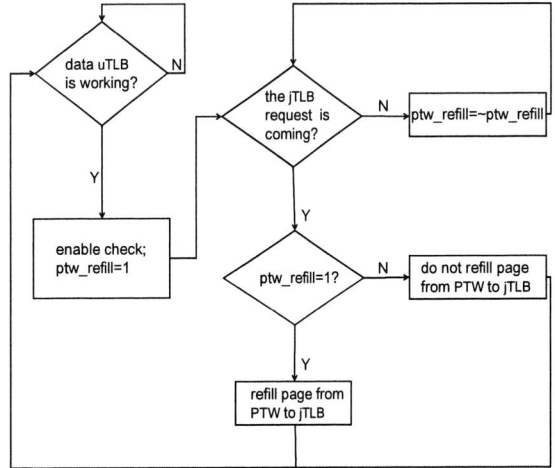

Fig. 3. The operation mechanism of TLBshield.

979-8-3503-6184-1/24 $31.00 © 2024 IEEE   29

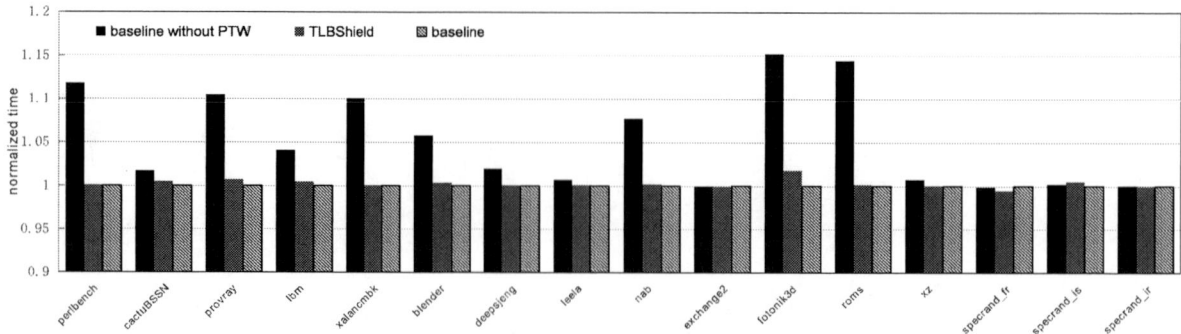

Fig. 4. The performance on SPEC2017 benchmark with the normalized time on baseline. The baseline means the unmodified TLB. The baseline without PTW means PTW cannot refill jTLB. The TLBShield means PTW can refill jTLB with 50% probability.

We believe that by frequently switching writeback channel, the TLB-based Spectre attack success rate can be significantly reduced with almost no performance impact. Since there is only a 50% chance of writeback from *PTW* to *jTLB*, the attack success rate of TLB-based Spectre is reduced to 50%.

The impact on performance is negligible due to the following reasons. From the hardware point of view, although there is the write channel from *PTW* is of 50% possibility to be turned off in TLBshield, the read is not influenced. Compared with disabling *PTW*, the performance is significantly improved. From the software point of view, the data read and write in the software program has locality, that is, the data address in the same region is read and written repeatedly. The smallest page table size is the address in the range of 4KB. The same virtual address translation is therefore revisited frequently. For N times of virtual address translation access, the probability that no virtual address translation write occurs is shown in (1). Since the probability is exponentially reduced with N, the final possibility of miss is low. The probability is reduced to less than one percent when N is greater than 6.

$$\text{Prob(no\_refill)} = 0.5^N \qquad (1)$$

IV. EXPERIMENTAL RESULTS AND DISCUSSION

In order to fairly evaluation TLBshield, we first implement a single XuanTie C910 core and the corresponding system on chip (SOC) on Xilinx VCU128 development board as the baseline. After power-on, the processor is externally mounted by the external PC via JTAG with OpenSBI, U-Boot and Linux OS. UART interface supports Linux shell in user mode.

Based on the baseline, we disable the writeback channel from *PTW* to *jTLB* as the control group. Together with TLBshield, we systematically evaluate the performance under SPEC2017 benchmark with the normalized time on baseline, as shown in Fig. 4. For the most obvious testcase **fotonik3d**, the performance overhead of TLBshield is 1.77%, while the performance on baseline without *PTW* degenerates dramatically on several testcases involved with TLB.

We also list the Coremark test result and the attack success rate in TABLE 1. Unlike SPEC2017, Coremark is a lightweight test and does not involve frequent address translations. Therefore, the Coremark performances on these three architectures are almost the same. The baseline without *PTW* completes disables the covert channel, so that it can eliminate the Spectre attack on TLB, in spite of the significant

performance loss in SPEC2017 test. After 350 rounds of attacks, the proposed TLBshield mitigates the success rate for attacks from 100% to 55.7%, which accords to the 50% on/off rate for writeback channel.

V. SUMMARY

In this paper, we propose TLBShield, a hardware defense architecture on XuanTie C910 core that is able to mitigate the speculative attacks on TLB. The performance lost under SPEC2017 is no more than 1.77% for all the testcases, and the success rate for attacks is reduced from 100% to 55.7%.

TABLE I. COMPARISON OF DIFFERENT TLB ARCHITECTURES

TLB architectures	*baseline without PTW*	*TLBshield*	*baseline*
Coremark (iterations/second)	420.83	420.71	420.66
Successful rate under 350 rounds of attacks	0 (0%)	195 (55.7%)	350 (100%)

*Coremark does not involve with frequent address translations, so the performance on Coremark is of no difference.

REFENCE

[1] P. Kocher, J. Horn, A. Fogh, D. Genkin, D. Gruss, W. Haas, et al. Spectre attacks: Exploiting speculative execution[J]. Communications of the ACM, 2020, 63(7): 93-101..

[2] M. Lipp, M. Schwarz, D. Gruss, T. Prescher, W. Haas, S. Mangard, et al. Meltdown[J]. arXiv preprint arXiv:1801.01207, 2018.

[3] J. Van Bulck, M. Minkin, O. Weisse, D. Genkin, B. Kasikci, F. Piessens, et al. Foreshadow: Extracting the keys to the intel {SGX} kingdom with transient {Out-of-Order} execution[C]//27th USENIX Security Symposium (USENIX Security 18). 2018: 991–1008.

[4] Daniel J Bernstein. 2005. Cache-Timing Attacks on AES.

[5] W. Xiong, J. Szefer. Survey of transient execution attacks and their mitigations[J]. ACM Computing Surveys (CSUR), 2021, 54(3): 1-36.

[6] Y. Yarom, K. Falkner. Flush + reload: A high resolution, low noise, L3 cache side-channel attack. In USENIX Security Symposium, 2014.

[7] X. Lou, T. Zhang, J. Jiang, Y. Zhang. A survey of microarchitectural side-channel vulnerabilities, attacks, and defenses in cryptography[J]. ACM Computing Surveys (CSUR), 2021, 54(6): 1-37.

[8] B. Gras, K. Razavi, H. Bos, C. Giuffrida. 2018. Translation Leak-aside Buffer: Defeating Cache Side-channel Protections with TLB Attacks. In USENIX Security Symposium. USENIX, 955–972.

[9] S. Deng, W. Xiong, S. Jakub. 2019. Secure TLBs. In Proceedings of the 46th International Symposium on Computer Architecture (pp. 346-359).

[10] C. Chen, X. Xiang, C. Liu, Y. Shang, R. Guo, D. Liu, et al. "Xuantie-910: A commercial multi-core 12-stage pipeline out-of-order 64-bit high performance RISC-V processor with vector extension: Industrial product." 2020 ACM/IEEE 47th Annual International Symposium on Computer Architecture (ISCA). IEEE, 2020.

One-Step Circuit Analysis Based on LCA for Sparse Coding

Hanxi Xu[1], Zirui Chen[1,2], Chenqi[1], Xiangshui Miao[1] and Yuhui He[1]

[1]School of Integrated Circuits, Huazhong University of Science and Technology, Wuhan, China.

[2]Dipartimento di Elettronica e Informazione and IU.NET, Politecnico di Milano, Milano, Italy

* Email: chenqi_whu@hust.edu.cn

Abstract

This paper proposes a novel sparse coding method based on the local competition algorithm, implemented using a closed-loop circuit. This approach can solve the sparse coding problem in a single step, significantly reducing time and energy consumption. Through simulations, we demonstrate the impact of different thresholds on reconstruction performance, selecting appropriate thresholds to balance image reconstruction quality and the number of activated neurons, thus achieving energy efficiency. Furthermore, this paper explores the effects of different activation functions on reconstruction performance and the number of activated neurons, further validating the effectiveness and flexibility of this method.

Keywords:

Sparse coding, Locally competitive algorithm, One-step circuit.

1. Introduction

Natural images can be well approximated by a small subset of elements from an overcomplete dictionary. The core of the sparse coding problem lies in how to select the appropriate elements and their corresponding coefficients from the dictionary to represent the signal[1]. Sparse coding has wide-ranging applications in fields such as image processing and compression, signal processing, machine learning and so on[2,3].The fundamental idea is to express the input data as a linear combination of a few non-zero activation values and their corresponding basis functions, thereby achieving data compression and dimensionality reduction (Figure 1). Sparse representation reduces the complexity of the input signal, making data processing and storage more efficient[4,5]. The solution of spare coding can be expressed as equation (1).

$$\min_{a}(|x - Da^T|_2 + \lambda|a|_0) \qquad (1)$$

Where $|\cdot|_2$ and $|\cdot|_0$ are the L_2- and L_0- norm, respectively. The first term represents the accuracy of the solution, the second term represents the sparsity, and λ represents the regulator between the two. Here, a denotes the activation values of the neurons, x represents the input signal, and D represents the dictionary elements.

This paper employs the Local Competition Algorithm (LCA) to address this issue, offering advantages in encoding spatiotemporal signals and biological plausibility[6].

In LCA, the membrane potential of output neurons is determined by the input, leakage term, and inhibition term[7]. The inhibition term ensures sparsity in the output by preventing multiple neurons with similar receptive fields from being activated simultaneously. Mathematically, it can be demonstrated that lateral inhibition among neurons can be achieved by iteratively removing the reconstructed signal from the network input, as shown in Equations (2) and (3).

$$\frac{du}{dt} = \frac{1}{\tau}(-u + (x - \hat{x})^T D + a) \qquad (2)$$

$$a_i = T_\lambda(u_i) \qquad (3)$$

where u represents the membrane potential of the neuron, $-u$ represents the leakage term, τ is a time constant, x is $M \times 1$ input, D is $M \times N$ dictionary, a is $1 \times N$ neuron activation value, $T_\lambda(\cdot)$ is the thresholding function with a threshold λ.

As the neuron's threshold increases, the solution of linear regression problems can obtain higher sparsity, while the accuracy will decrease.

Several hardware systems based on LCA has been developed[8-11], and in our previous work[12], we constructed a one-step circuit with $HfWO_x/VO_y$ neural component for its advantages in ultra-fast solution time. This study will replace the functionality of the neural component with operational amplifiers and simulate the circuit.

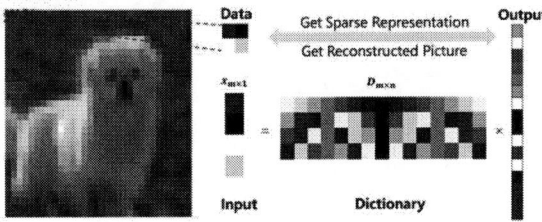

Figure 1. Sparse coding solution architecture

979-8-3503-6184-1/24 $31.00 © 2024 IEEE

2. LCA one-step circuit

Figure 2 shows that in some previous solutions to the sparse coding problem, algorithms have been mapped onto circuits to accelerate the solution. However, this is usually an iterative process[13,14] where intermediate data needs to be transferred back and forth between the FPGA and the RRAM array[7], consuming a significant amount of time and energy.

According to Equation (2), this paper proposes a one-step closed-loop circuit based on LCA and demonstrates its application to the sparse coding problem. The basic structure of the circuit is shown in Figure 3.

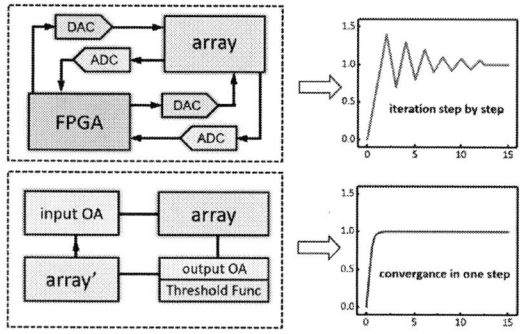

Figure 2.Comparison between previous work and ours

In the input unit, x is mapped to the input voltage waveform. The forward R array is the same as the backward R array, which serve as dictionary elements. The reconstructed signal can be obtained by reading the voltage from the output. In the reconstruction process, the more similar the dictionary elements are to the input voltage, the larger the corresponding neuron's charging current. Moreover, whether a neuron is activated largely depends on the threshold voltage (V_{th}). The larger the Vth, the fewer neurons are activated, resulting in higher sparsity but lower accuracy.

The most critical part is the threshold function in the output unit. Typically, the threshold function can be either a hard threshold (4-1) or a soft threshold function (4-2) , corresponding to different neuron activation methods and mathematical forms, namely L_0 and L_1 regularization, respectively. Among them, L_0 regularization performs better in terms of sparsity, meaning we can achieve reconstruction with fewer elements. Their mathematical forms are shown in Equation (4).

$$T_\lambda(u_i) = \begin{cases} u_i, & u_i > \lambda \\ 0, & u_i \leq \lambda \end{cases} \qquad (4-1)$$

$$T_\lambda(u_i) = \begin{cases} u_i - \lambda, & u_i > \lambda \\ 0, & u_i \leq \lambda \end{cases} \qquad (4-2)$$

In the circuit simulation of this paper, the hard threshold function is implemented using two operational amplifiers, while the soft threshold function is implemented using other methods from previous work[15].

With this closed-loop circuit, the reconstructed result can be obtained in one step, without the need for multiple iterations. The solving time primarily depends on the settling time of the operational amplifier.

3. Simulation results

The circuit simulation in this paper is based on TINA-TI. A 32×32 grayscale image from the CIFAR-10 dataset was selected for demonstration. The 32×32 image was divided into 256 2×2 images, making the circuit input 4×1. Input two sets of 2×1 voltage signals. A complete DCT dictionary was selected, with a size of 2×4, [1,0,1,1;0,1,1,1]. A normalized DCT dictionary was then mapped onto the resistor array. Two closed pixels in the chosen picture were selected to be the voltage input, (0.254,0.25).

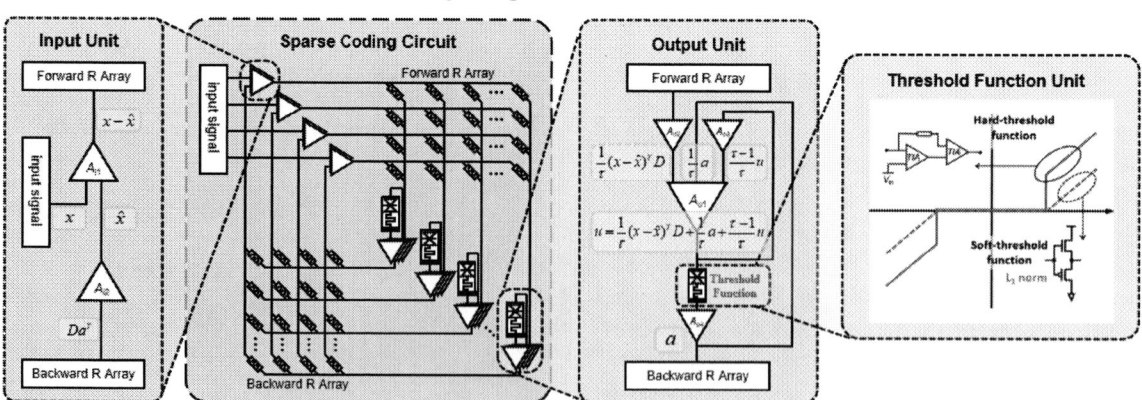

Figure 3. LCA closed-loop circuit schematic

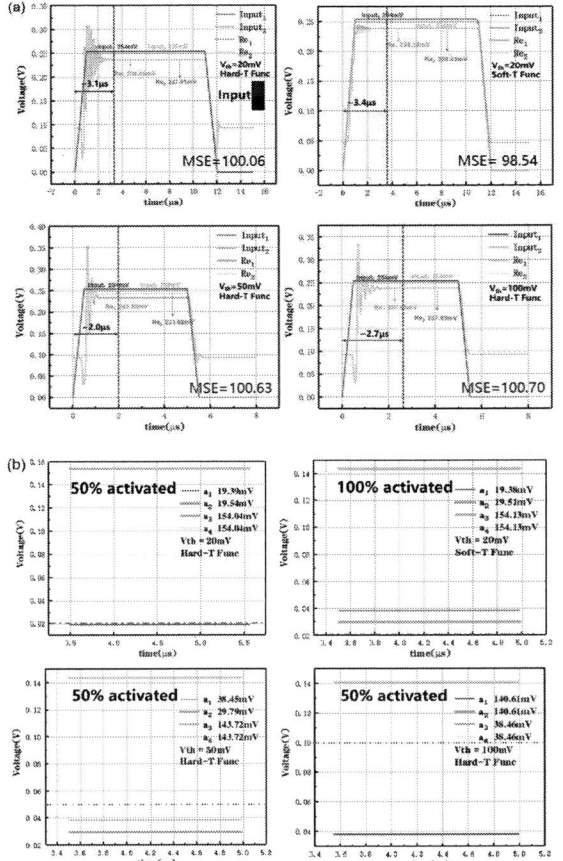

Figure 4. (a) Reconstructed results and (b) neurons' activation with different active functions and thresholds (MSE is in the units of $\times 10^{-5}$).

Firstly, we compare the impact of different threshold functions on the results under the same conditions, as shown in Figure 4(a). With a threshold of 20mV, when the activation function uses a hard threshold function, the reconstruction error MSE = $100.06e^{-5}$, with two neurons activated. When using a soft threshold function, MSE = $98.54e^{-5}$, with four neurons activated. In these two cases, the reconstruction accuracy does not differ significantly, but the hard threshold function performs better in terms of sparsity and convergence time. Considering the overall time and energy consumption, the hard threshold function is selected for the subsequent simulations.

Next, three different thresholds (V_{th}=20/50/100mV) were selected to explore the impact of the threshold on reconstruction accuracy and neuron activation. It can be observed that the higher the threshold, the less neurons 3

and 4 are activated, resulting in lower energy consumption.

In summary, we can utilize this system to explore the selection of corresponding thresholds under different solving requirements (focusing more on accuracy or sparsity), helping to achieve the same tasks with lower energy consumption. In practical situations, we can build larger-scale circuits to achieve more efficient reconstruction.

4. Conclusion

In this work, a sparse coding solution solver is designed. The architecture is established based on the LCA, where the activation function is mapped as a neural component. By combining operational amplifiers and other components, an iterative process of sparse coding is constructed with a closed-loop circuit. Based on this one-step convergence architecture, the impact of using different threshold functions on reconstruction accuracy and sparsity was explored. It is demonstrated that the hard threshold function can achieve reconstruction with fewer neurons while maintaining comparable accuracy. In terms of energy efficiency, the hard-T function outperforms the soft-T function.

In addition, selecting the right threshold is a crucial step in solving the problem. We need to choose an appropriate threshold to achieve the best balance between accuracy and sparsity while obtaining the desired results with minimal energy consumption. This circuit helps to conduct multiple explorations in a shorter time.

Moreover, in practical applications, we can build larger-scale circuits to achieve more efficient reconstruction.

Acknowledgments

The authors gratefully acknowledge the financial support by the National Key Research and Development Program of China (No. 2023YFB4502200), National Science Foundation of China (NSFC) with grant No. 62374063 and 92164204 and Interdisciplinary Research Program of HUST 2024JCYJ008.

References

[1] Olshausen, B. A. & Field, D. J. Emergence of simple-cell receptive field properties by learning a sparse code for natural images. Nature 381, 607–609 (1996).

[2] J. Zhang, D. Zhao, and W. Gao, "Group-Based Sparse Representation for Image Restoration," IEEE TRANSACTIONS ON IMAGE PROCESSING, vol. 23, no. 8, pp. 3336–3351, Aug. 2014, doi: 10.1109/TIP.2014.2323127.

[3] F. Luo, L. Zhang, X. Zhou, T. Guo, Y. Cheng, and T. Yin, "Sparse-Adaptive Hypergraph Discriminant Analysis for Hyperspectral Image Classification," IEEE GEOSCIENCE AND REMOTE SENSING LETTERS, vol. 17, no. 6, pp. 1082–1086, Jun. 2020, doi: 10.1109/LGRS.2019.2936652.

[4] Wright, J. et al. Sparse representation for computer vision and pattern recognition. Proc. IEEE 98, 1031–1044 (2010).

[5] Olshausen, B. A. & Field, D. J. Sparse coding with an overcomplete basis set: a strategy employed by V1? Vision Res. 37, 3311–3325 (1997).

[6] C. J. Rozell, D. H. Johnson, R. G. Baraniuk, and B. A. Olshausen, "Sparse Coding via Thresholding and Local Competition in Neural Circuits," Neural Computation, vol. 20, no. 10, pp. 2526–2563, Oct.2008,doi:10.1162/neco.2008.03-07-486.

[7] P. M. Sheridan, F. Cai, C. Du, W. Ma, Z. Zhang, and W. D. Lu, "Sparse coding with memristor networks," Nature Nanotech, vol. 12, no. 8, pp. 784–789, Aug. 2017, doi: 10.1038/nnano.2017.83.

[8] P. Maechler et al., "VLSI Design of Approximate Message Passing for Signal Restoration and Compressive Sensing," IEEE J. Emerg. Sel. Topics Circuits Syst., vol. 2, no. 3, pp. 579–590, Sep. 2012, doi: 10.1109/JETCAS.2012.2214636.

[9] J. Seo et al., "On-Chip Sparse Learning Acceleration with CMOS and Resistive Synaptic Devices," IEEE Trans. Nanotechnology, vol. 14, no. 6, pp. 969–979, Nov. 2015, doi: 10.1109/TNANO.2015.2478861.

[10] D. Zhang, Y. Hou, L. Zeng, and W. Zhao, "Hardware Acceleration Implementation of Sparse Coding Algorithm With Spintronic Devices," IEEE Trans. Nanotechnology, vol. 18, pp. 518–531, 2019, doi: 10.1109/TNANO.2019.2916149.

[11] X. Ji, X. Hu, Y. Zhou, Z. Dong, and S. Duan, "Adaptive sparse coding based on memristive neural network with applications," Cogn Neurodyn, vol. 13, no. 5, pp. 475–488, Oct. 2019, doi: 10.1007/s11571-019-09537-w.

[12] Z. Chen, Y. Zhou, et al. An Ultrafast (< 200 ns) Sparse Solution Solver made by HfWOx/VOy Threshold Tunable Neurons. 2023 International Electron Devices Meeting.

[13] K. K. Herrity, A. C. Gilbert, and J. A. Tropp, "Sparse Approximation Via Iterative Thresholding," in 2006 IEEE International Conference on Acoustics Speed and Signal Processing Proceedings, Toulouse, France: IEEE, 2006, p. III-624-III–627. doi: 10.1109/ICASSP.2006.1660731.

[14] Lee, H., Battle, A., Raina, R. & Ng, A. Y. in Proceedings of the 19th International Conference on Neural Information Processing Systems 801–808 (MIT Press, 2006).

[15] S. Wang, Y. Luo, P. Zuo, L. Pan, Y. Li, and Z. Sun, "In-memory analog solution of compressed sensing recovery in one step," Sci. Adv., vol. 9, no. 50,p.eadj2908,Dec.2023,doi:10.1126/sciadv.adj2908.

A Hybrid-Logic Scheme for High-Performance and Low-Power Decoders in 7nm Process

Donghao Xia [1], Yuejun Zhang[1]*, Mengfan Xu[1], Liang Wen[2]*, Yiting Guo[1]

[1] Faculty of Electrical Engineering and Computer Science, Ningbo University, Zhejiang, 315211, China
[2] Department of Electronic Technology, China Coast Guard Academy, Ningbo, 315211, China

* Email: zhangyuejun@nbu.edu.cn, lwen13@fudan.edu.cn

Abstract—Advancing memory circuitry has heightened the performance criteria for memory in diverse chip applications. Decoders, serving as peripheral circuits to memory devices, have be a prominent area of academic interest. The new NAND/NOR gate circuits are proposed using pass-transistor logic (PTL) and transmission gate logic (TGL). The two hybrid-logic design schemes of high-performance and low-power (HPLP) and HPLP inverting (HPLPI) utilizing PTL and TGL gate in combination with static CMOS logic for decoders are proposed, while achieving fewer transistors, faster speed, lower power dissipation as compared to conventional circuits. Experiments are conducted in TSMC 7nm FinFET process for hybrid-logic decoders. The simulation results indicate that, in comparison to conventional decoders, the proposed hybrid-logic 3-8 decoder demonstrates improvements, with the HPLP design reducing delay by over 19% and power by over 71%, and the HPLPI design reducing delay by over 17% and power by over 48%.

Keywords—hybrid-logic, decoder, HPLP, HPLPI, delay, power

I. INTRODUCTION

Static COMS logic with low power dissipation and good noise immunity is used as the dominant design methodology for digital integrated circuits. The static CMOS logic circuits consisting of NMOS and PMOS can realize arbitrary logic functions with superior performance and drive capability[1]. Initially, CMOS logic was composed of MOSFETs, with the evolution of integrated circuit scale and the advancement of fabrication technology, digital integrated circuits demanded higher and higher performance from semiconductor devices, and MOSFET were gradually replaced by FinFET, which are basically used in today's advanced process circuit design [2].

Digital integrated circuit design extensively incorporates memory, Memory is an integral part of digital circuit design. Decoders constitute a significant component of the peripheral circuitry surrounding SRAMs, In SRAM design, the decoder is a key part that affects its performance [3]. Typically, the outputs of the decoder are connected to the address lines of SRAM to access specific memory cells within the SRAM. As the scale of the memory array expands[4], the size of the decoder correspondingly increases. SRAM performance is affected by decoder decoding speed and power. In order to design higher performance SRAM, the design of the decoder should be directed toward high speed and low power consumption[5]. Conventional decoders are implemented using CMOS logic, a hybrid-logic 3-8 decoder circuit using PTL, TGL and inverter is proposed to realize a high-speed and low-power decoder. This design approach realizes full swing and also reduces transistors. Initially, Logic gate of PTL and TGL are designed. Subsequently, the hybrid-logic 3-8 decoders using our designed logic gate circuits and their layout scheme are proposed, the performance of the designed decoders are tested through a variety of experiments.

II. CONVENTIONAL CMOS LOGIC DECODER

Decoder is a combinational logic circuit consisting of n inputs and m outputs, where m = 2ⁿ. The decoder can be categorized into high-level outputs and low-level outputs depending on the type of outputs [6].

A 3-8 decoder has 3 inputs corresponding to 8 outputs, and conventional 3-8 decoders are composed of CMOS logic. Conventional decoders are categorized according to output as CMOS decoders and CMOS inverting (CMOSI) decoders. The CMOS 3-8 decoder comprise of 3 inverters and 8 3-input NORs, and its effective output level is 1, while the CMOSI 3-8 decoder comprise of 3 inverters and 8 3-input NANDs, and its effective output level is 0.

III. NEW HYBIRD-LOGIC DECODER

A. PTL gate circuits and TGL gate circuits

Conventional CMOS logic decoders are composed of simple logic gates. TGL and PTL can effectively enable the construction of diverse gate circuits, catering to the needs of decoder functionalities. PTL minimizes the transistor count by allowing the original to drive gate and source-drain. The difference between PTL and CMOS logic is that CMOS only allows the raw input to drive the gate side of the transistor. The 3-input AND/NAND using PTL logic are proposed, which both consume 5 transistors. PTL logic is challenged by voltage loss, weak NMOS pass 1 capability, weak PMOS pass 0 capability. Fig. 1(a) shows our proposed 3-input PTL AND, where we solve the problem of poor PMOS pass 0 capability by using an NMOS with gate connected to A' as a pull-down transistor. Fig. 1(b) shows our proposed 3-input PTL NAND, where we solve the problem of poor NMOS through-1 capability by using a PMOS with gate connected to A as a pull-up transistor. The addition of transistors enables our proposed gate circuits to provide full-swing in the realization of efficient logic.

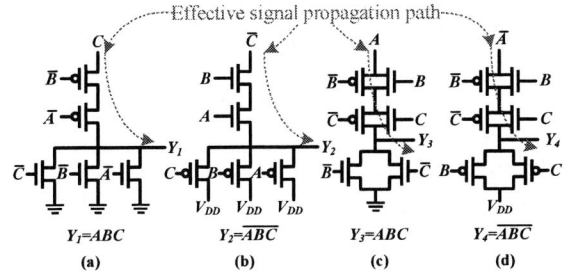

Fig. 1. PTL and TGL gate circuits. (a)PTL AND, (b)PTL NAND, (c)TGL AND, (d)TGL NAND.

TGL offers the advantages of low propagation delay, low power dissipation and full level transfer capability. The 3-input AND/NAND are proposed using TGL logic which

979-8-3503-6184-1/24 $31.00 © 2024 IEEE 35

both consume 6 transistors. Fig. 1(c) shows our proposed 3-input TGL AND, Fig. 1(d) shows our proposed 3-input TGL NAND.

B. Hybrid-logic design of 3-8 decoder

Traditional decoders are composed of inverters and NAND/NOR, for a 3-8 decoder with an effective signal of 1. A new high-performance and low-power (HPLP) hybrid-logic (HL) 3-8 decoder is proposed based on this design idea, Fig. 2 shows the schematic of the HPLP 3-8 decoder. For the

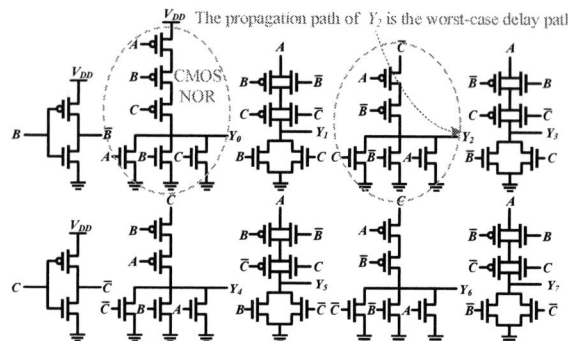

Fig. 2.   The schematic of HPLP 3-8 decoder using hybrid-logic.

HPLP 3-8 decoder, a CMOS NOR is used to propagate the Y_0, which reduces the complementary A as a propagation signal. Four TGL AND are used to propagate the Y_1, Y_3, Y_5 and Y_7, three PTL AND are used to propagate the signals Y_2, Y_4, and Y_6. The propagation path of signal Y_2 is the worst-case delay path for this design methodology. This design methodology consumes 49 transistors, which reduces the consumption of one inverter by a total of 5 transistors compared to a conventional 54T CMOS NOR 3-8 decoder.

For a 3-8 decoder with an effective signal of 0, a new design approach is proposed using hybrid-logic, Fig. 3 shows the schematic of the proposed HPLP inverting (HPLPI) 3-8 decoder. For the HPLPI 3-8 decoder, four TGL NAND are used to propagate the Y_0, Y_2, Y_4 and Y_6, and four PTL NAND are used to propagate the signals Y_1, Y_3, Y_5 and Y_7. The propagation path of signal Y_0 is the worst-case delay path for this design approach. This design approach consumes 48 transistors, which reduces the consumption of one inverter by a total of 6 transistors compared to a conventional 54T CMOS NAND 3-8 decoder.

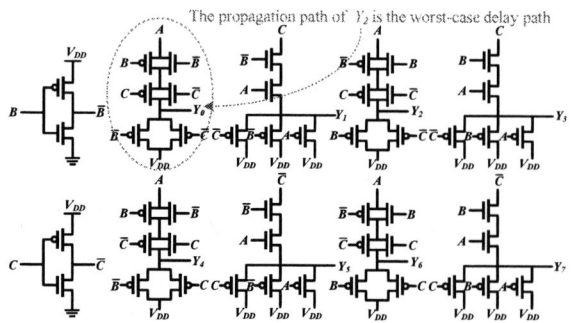

Fig. 3.   The schematic of HPLPI 3-8 decoder using hybrid-logic.

Eliminating inverters lessens the quantity of transistors, decreases logical effort, and diminishes the overall switching activity within the circuits, which in turn leads to a reduction in power consumption. These new hybrid-logic decoders offer significant improvements in both power dissipation and propagation delay.

IV. SIMULATION RESULTS

A. Simulation steps

In this section the Spectre is used to perform various simulations at the schematic level. All simulations are performed in the TSMC 7nm FinFET process, and in order to effectively compare data, we have utilized unit-size FinFET in our analysis.. The load capacitance size of the 3-8 decoder simulations is set to 0.5fF. All the circuits are simulated at 0.5 GHz, simulated with a process corner of TT and a temperature of 27°C. Our test time for both decoders is 16 ns.

B. Analysis of results

We first functionally verify the 3-8 HPLP and HPLPI decoders as a way of demonstrating that the designed 3-8 decoder has full voltage swing capability. Fig. 4(a) shows the functional waveforms of the HPLP 3-8 decoder and Fig. 4(b) shows the functional waveforms of the HPLPI 3-8 decoder to display their voltage with full swing capability.

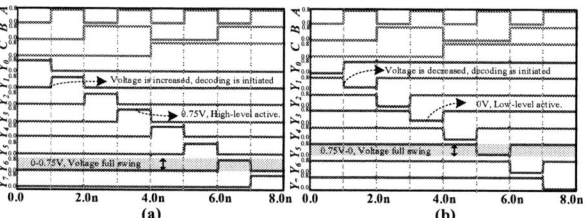

Fig. 4.   Simulated waveforms of the hybrid-logic 3-8 decoder.(a)HPLP, (b)HPLPI.

Evaluations of delay propagation, power dissipation, and Power Delay Product (PDP) are conducted on a range of decoders, including conventional CMOS, CMOSI, HPLP, and HPLPI. The evaluation results regarding delay, power and PDP are display in Fig.5(a), (b), (c) respectively. Power is evaluated by the average power dissipation, expressed in nw; delay is evaluated by the worst-case delay, expressed in ps; The PDP is calculated as the product of power and delay, expressed in eV. Delay, Power and PDP are simulated with varying voltage (0.55, 0.65, 0.75, 0.85, 0.95V). Fig. 5 demonstrates that the delay and power of the hybrid-logic 3-8 decoder outperforms the conventional CMOS decoder regardless of voltage.

Table I lists a variety of data comparisons between different 3-8 decoders, including the propagation delay, the power dissipation at 0.75 V, the number of transistors consumed, and the area of the two proposed hybrid-logic decoders. At 27°C, 0.75V, TT process corner, compared to CMOS and CMOSI, the HPLP 3-8 decoder demonstrate 77% and 71% decrease in power dissipation, 19% and 23% in propagation delay and 81% and 78% in PDP, respectively. The HPLPI 3-8 decoder demonstrate 58% and 48% decrease in power dissipation, 17% and 20% in propagation delay and 65% and 59% in PDP, respectively, compared to CMOS and CMOSI.

The propagation delay and power of the proposed hybrid-logic 3-8 HPLP decoder and HPLPI decoder are analyzed with multiple samples of 800 Monte Carlo simulation,

979-8-3503-6184-1/24 $31.00 © 2024 IEEE   36

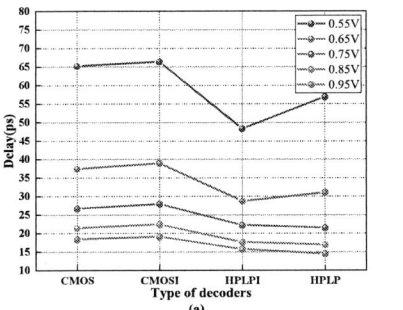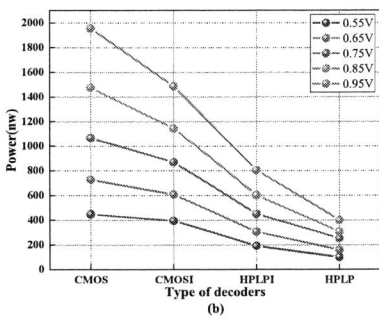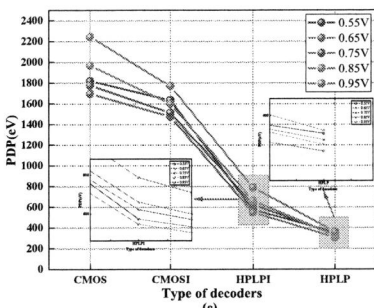

Fig. 5. Comparison of different 3-8 decoders at multiple voltages. (a) The results of delay, (b)the results of power, (c)the results of PDP.

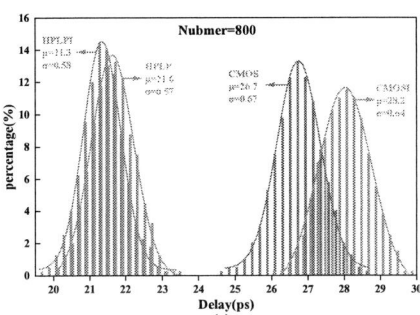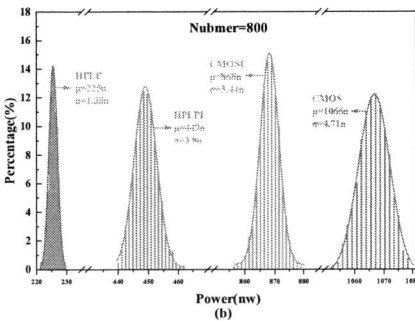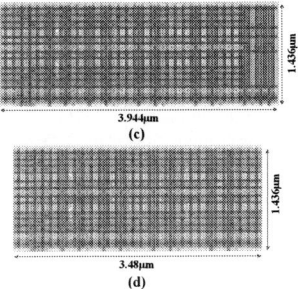

Fig. 6. Comparison of Monte Carlo simulation results of 3-8 decoders and the layout of 3-8 decoders. (a)The results of propagation delay, (b)the results of power dissipation, (c)the layout of HPLP, (d)the layout of HPLPI.

TABLE I.   COMPARISON OF DATA FOR DIFFERENT 3-8 DECODERS

3-8 Decoder	Comparison of data				
	Voltage (V)	Delay (ps)	Power (nw)	Transistors	Area (μm²)
CMOS	0.75	26.7	1066	54	--
CMOSI	0.75	27.9	868	54	--
HPLP	0.75	21.5	250	49	5.66
HPLPI	0.75	22.2	449	48	5.00

the simulation results are shown in Fig. 6(a), (b) respectively. Fig. 6(a), (b) represents the results of 800 process deviation simulations, from which we can see that both the mean value μ and the standard deviation σ of our proposed hybrid-logic decoder are smaller, and thus we can find that the performance of our proposed decoder is more stable during the process manufacturing process. The layouts of hybrid-logic decoders are designed in the TSMC 7nm FinFET process. Fig. 6(c) displays the layout of the HPLP 3-8 decoder and Fig. 6(d) displays the layout of the HPLPI 3-8 decoder. Compared to conventional CMOS logic decoders, Both HPLP and HPLPI decoders have reduced silicon area overhead.

CONCLUSION

This paper introduces a new hybrid-logic design scheme for 3-8 decoder. Initially, the logic gate circuits for PTL and TGL are designed, the 5T three -input AND/NAND using PTL and the 6T three-input AND/NAND using TGL are proposed. Subsequently, the hybrid-logic design approach for HPLP and HPLPI 3-8 decoder are proposed using PTL and TGL gate circuits, compared to conventional CMOS decoders, hybrid-logic HPLP and HPLPI 3-8 decoders have full voltage swing capability while reducing transistor count

and improving power and delay performance. The proposed hybrid-logic 3-8 decoder layout is designed in TSMC 7nm FinFET process, our simulation results in the process show that the hybrid-logic decoder outperforms the conventional CMOS logic decoder in all cases.

ACKNOWLEDGMENT

This work is supported by the National Natural Science Foundation of China (62174121, 61871244, 62134002), the Science and Technology Innovation 2025 Major Project of Ningbo (2022Z203), Ningbo University and Ningbo Yongxin Microelectronics Technology company, LTD. Digital Integrated Circuit Design Joint Laboratory (XQ2022000005), Ningbo University Graduate Education Practice Base (YJD202305).

REFERENCES

[1] D. Balobas, N. Konofaos, "Design of low-power high-performance 2–4 and 4–16 mixed-logic line decoders," IEEE Transactions on Circuits and Systems II: Express Briefs 64 (2) (2016): 176–180.

[2] Leavline, Epiphany Jebamalar, and Somasekaran Sujitha, "Design of FinFET based low power, high speed hybrid decoder for SRAM," Microelectronics Journal 126 (2022): 105499.

[3] N.S. Sumana, B. Sahana, A. Deshapande, "Design and implementation of low power-high performance mixed logic line decoders, RTEICT, 2019, pp. 529–534, Bangalore, India.

[4] Do A T, Zeinolabedin S M A, Kim T T H, "Energy-efficient data-aware SRAM design utilizing column-based data encoding," IEEE Transactions on Circuits and Systems II: Express Briefs, 2019, 67(10): 2154-2158.

[5] T. Singh, V. Prakash, S. S. Anwer and G. Ahmad, "Analyzing the performance of 6T SRAM cell and 64×64 memory array at lower technology nodes for low power design, CCPIS, 2023, pp. 1-6, Bhubaneswar, India.

[6] B. Joseph, G. C. Reddy and R. K. Kavitha, "Energy Efficient Memory Decoder for SRAM Based AI Accelerator," PCEMS, 2023, pp. 1-4, Nagpur, India.

Ternary Logic Units Design Based on the TDDFETs

Hua Qiang [1], Haoran Lu [1], Xiaotao Liu [1], Linlin Xing [1], Bin Lu*[1]

[1] School of Physics and Information Engineering, Shanxi Normal University, Taiyuan 030031, China

* Email: lubinsxnu@sina.cn

Abstract—This paper introduces the novel tunneling-drift-diffusion Field-Effect Transistors (TDDFETs) which exhibits two distinct current conduction mechanisms, namely the band-to-band tunneling (BTBT) and carrier drift-diffusion (DD). This mixed conduction mechanism makes the device can present additional state between the on and off states, and thus the device is very suitable for the design of ternary logic circuits. Based on the proposed TDDFETs, the ternary logic cells, such as the standard ternary inverter (STI), positive ternary inverter (PTI), negative ternary inverter (NTI), ternary NAND (T-NAND) and ternary NOR (T-NOR) are designed by the aid of Verilog-A. Additionally, the 3-to-1 ternary encoder and 1-to-3 ternary decoder are also designed using these basic units.

Keywords—Ternary Logic, STI, PTI, NTI, T-NAND, T-NOR.

I. INTRODUCTION

The latest advancements in artificial intelligence (AI) and the Internet of Things (IoT) have led to an exponential growth in data processing requirements [1]. The traditional binary systems are increasingly showing their limitations when dealing with the complex data and algorithms. By contrast, the ternary logic offers distinctive advantages, such as higher information density, reduced gate count requirements, and simplified computational processes [2], [3]. The ternary logic enables the conveyance of increased volume of information across a specified set of conductors, or the storage of greater quantity of data within a defined register length [4], [5]. As a result, it diminishes the chip complexity associated with the interconnectivity and the spatial requirements, enhancing the efficiency with which the hardware resources are harnessed [6]. Various ternary logic devices, such as Carbon Nanotube FET (CNTFET) and Quantum Dot Gate Field Effect Transistor (QDGFET), have been the research subject for an extended period [7], [8]. Despite their promising characteristics, the integration of these devices into the manufacturing of large-scale integrated circuits has been impeded by their incompatibility with the commercial CMOS platform [9], [10]. As for this issue, the TDDFET combining the BTBT and DD mechanisms is proposed for the suitability for ternary logic design and the compatibility with the conventional CMOS platform. In addition, based on the TDDFETs, the ternary logic cells are designed and used to design the ternary encoder and decoder.

II. DEVICE STRUCTURE AND OPERATION PRINCIPLE

Fig. 1(a) presents the schematic structure of the n-type TDDFET. The device presents a distinctive source consisting of two parts with different doping types. Notably, the device has two asymmetrical gates positioned at the top and bottom

This work was supported by the Basic Research Plan of Shanxi Province under Grant 202403021211225.

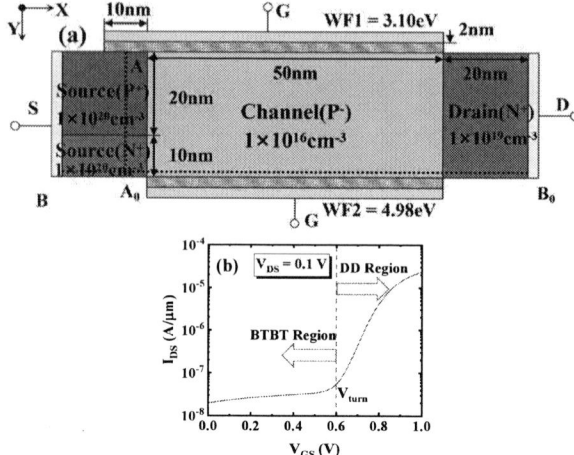

Fig. 1 (a). The structure and (b) the transfer curve of the proposed n-type TDDFET.

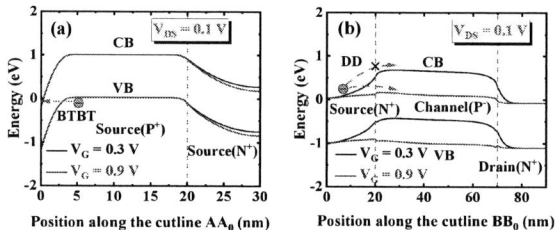

Fig. 2. The energy band diagrams along the cut-lines (a) AA_0 and (b) BB_0.

sections. The top gate is extended to cover a segment of the P+ source while the bottom gate just covers the channel body. The cutline AA_0 is perpendicular to the channel direction and located at the midpoint of the overlap region between the top gate and the P+ source. The BB_0 is parallel to the channel direction and 3 nm above the bottom gate.

The 2-D TCAD tools are used to study the device performance. The dynamic nonlocal BTBT model is adopted to consider the influence of the non-uniform electric field on the tunneling efficiency. The doping dependent mobility, high electric field saturation, band gap narrowing and the Shockley-Read-Hall models are also included.

Fig.1 (b) illustrates the transfer characteristics of the n-type TDDFET with drain-source voltage V_{DS} of 0.1V. Interestingly, this curve exhibits distinct behavior compared to conventional MOSFETs or TFETs. Upon closer inspection, a notable turning point, labeled as V_{turn}, becomes apparent. Below Vturn, the drain-source current I_{DS} exhibits a gradual increase with gate-source voltage V_{GS}, displaying a relatively smooth upward. However, beyond V_{turn}, I_{DS} experiences a significant increase followed by saturation as V_{GS} increases further. Remarkably, the behavior of I_{DS} for $V_{GS} > V_{turn}$ resembles

TABLE I. THE TRUTH TABLE OF THE STI, NTI AND PTI

Input	Output		
	STI	NTI	PTI
0	2	2	2
1	1	0	2
2	0	0	0

TABLE II. THE TRUTH TABLE OF T-NAND AND T-NOR

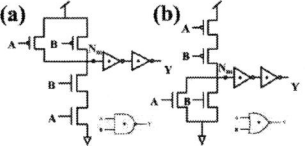

T-NAND		B			T-NOR		B		
		0	1	2			0	1	2
A	0	2	2	2	A	0	2	1	0
	1	2	1	1		1	1	1	0
	2	2	1	0		2	0	0	0

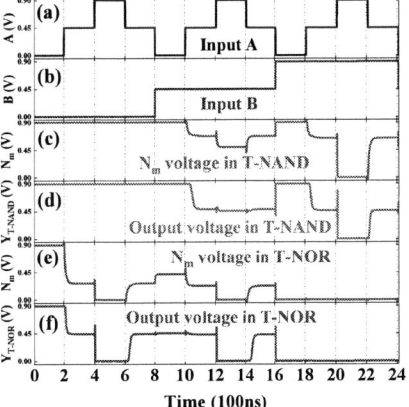

Fig. 5. The circuits of the (a) T-NAND and (b) T-NOR.

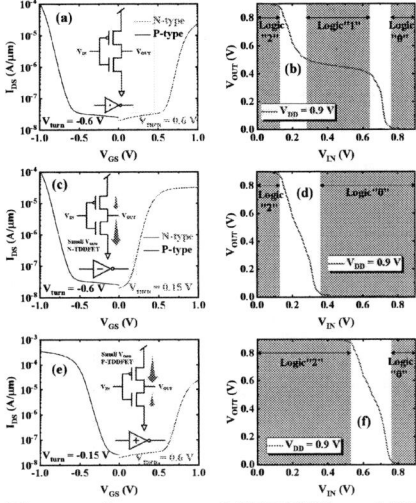

Fig. 3. The transfer curves of TDDFETs and VTCs for the (a) (b) STI, (c) (d) NTI and (e) (f) PTI, respectively.

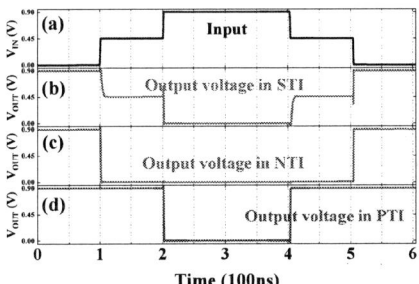

Fig. 4 .The (a) input signal and corresponding output signal for (b) STI, (c) NTI, (d) PTI.

Fig. 6. The input signal (a) A, (b) B, (c) Nm voltage and (d) output voltage for T-NAND, (e) Nm voltage and (f) output voltage for T-NOR.

that of MOSFETs. These distinct trends before and after V_{turn} is due to the two different conduction mechanisms.

Fig. 2 (a) illustrates the band diagrams along the cutline AA_0. It is evident that irrespective of whether $V_{GS} = 0.3$ V or 0.9 V, the conduction band near the interface between the top gate and the P+ source remains lower than the valence band farther away from the interface. Consequently, electrons in the region distant from the top gate can tunnel into the region near the top gate, resulting in the formation of the BTBT current near the top gate. The energy bands along the cutline BB_0 are depicted in Fig. 2(b). Evidently, at a lower voltage $V_{GS}=0.3$ V, a substantial energy barrier exists between the N+ source and the channel body, hindering the injection of electrons from the N+ source into the channel. Consequently, the BTBT current predominates the device current. However, at $V_{GS}=0.9$ V, this barrier significantly diminishes, enabling electrons from the N+ source to penetrate into the channel and subsequently be collected by the drain. Considering the number of electrons surpassing the barrier far exceeds that of the tunneling electrons along AA_0, the device current in this scenario is primarily governed by the DD current. Finally, this results in distinct turning voltage on the transfer

curve and makes the device suitable for the ternary logic design.

III. THE DESIGN OF THE TERNARY LOGIC

In this part, the TCAD tools are only used to design the TDDFETs and to generate look-up tables containing the drain current (I-V) and terminal capacitance (C-V) data for a defined range of V_{GS} and V_{DS}. Then the device is defined as a black-box device by the table based Verilog-A model and simulated by SPICE simulations.

The ternary inverter includes three different types, namely the Standard Ternary Inverter (STI), the Negative Ternary Inverter (NTI) and the Positive Ternary Inverter (PTI). The truth tables of the inverters are shown in Table I. The difference of these inverters is the output signal when the input signal is logic 1. The STI, NTI and PTI inverses input 1 as I, 0 and 2, respectively. The I-V curves and the voltage transfer curve (VTC) for the inverters are given in Fig. 3. The ternary inverters based on the TDDFET consist of two devices (NMOS and PMOS), which is the same as the CMOS-based binary inverter. For STI, the V_{turn} of the n-type and p-type TDDFETs is fixed as symmetrical, keeping the BTBT current comparable. In this way, when the input is near $V_{DD}/2$ (logic 1), the output locates near the $V_{DD}/2$ (logic 1). However, for NTI, when the input is near $V_{DD}/2$ (logic 1), the output should be near zero (logic 0). To obtain this, the Vturn is designed asymmetrically to make the n-

979-8-3503-6184-1/24 $31.00 © 2024 IEEE   39

TABLE III. THE TRUTH TABLE OF T-ENCODER

X_2	X_1	X_0	Y
0	0	1	0
0	1	0	1
1	0	0	2

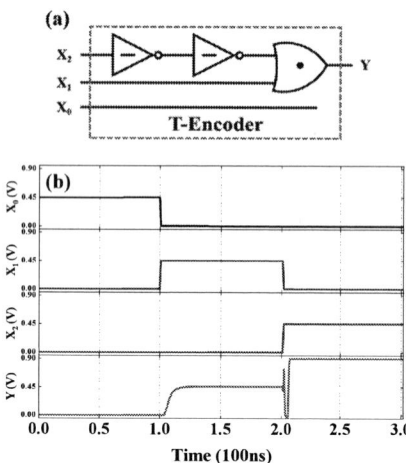

TABLE IV. THE TRUTH TABLE OF T-DECODER

X	Y_2	Y_1	Y_0
0	0	0	2
1	0	2	0
2	2	0	0

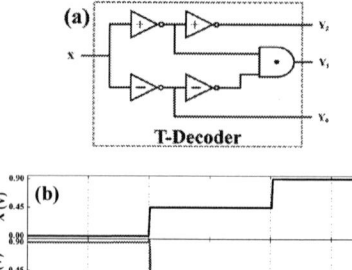

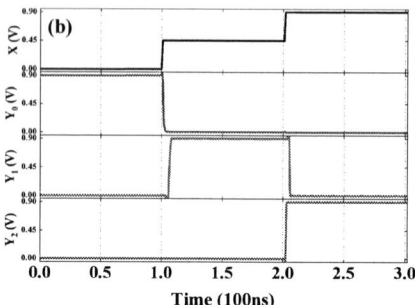

Fig. 7. (a) The circuit and (b) the transient performance of the T-Encoder.

Fig. 8. (a) The circuit and (b) the transient performance of the T-Decoder.

type BTBT larger than the p-type BTBT. The contrary situation is for PTI. The inverter transient performance is shown in Fig. 4.

The truth tables for T-NAND and T-NOR are given in Table II and the Fig. 5 presents the circuits. Obviously, the core of the circuit is the same with the binary logic. In the binary NAND, the device resistance can be simply considered as very large in off-state and very small in on-state. In this way, the output voltage is either close to V_{DD} or close to 0V. However, for ternary logic, the device can be in the BTBT region where the device resistance cannot be considered as very large or very small. This causes the output voltage corresponding to logic 1 to deviate from the desired value ($V_{DD}/2$). For example, when the input A=B=1 for the T-NAND, all the devices are in the BTBT region and assuming the same resistance, the voltage at node Nm equals to $4V_{DD}/5$, instead of the wanted value of $V_{DD}/2$, as shown in Fig. 6. To make the output voltage as close to the wanted value, two STIs are in series between the Nm and the output. Of course, the T-AND can be obtained if only one STI is in series. The analysis for the T-NOR is similar.

At last, the 3-to-1 ternary encoder (T-Encoder) capable of translating input signals into a ternary code and the 1-to-3 ternary decoder (T-Decoder) with opposite function of an encoder, are also designed in Fig. 7 and Fig. 8, respectively.

IV. CONCLUSION

This paper proposes the novel TDDFETs combining the BTBT and DD mechanisms. This mixed conduction mechanism makes the device can present additional state between the on and off states, and very suitable for the ternary logic design. Based on the TDDFETs, the ternary logic units including the STI, PTI, NTI, T-NAND and T-NOR are designed by the aid of Verilog-A description language. Moreover, based on the basic units, the 3-to-1

ternary encoder and 1-to-3 ternary decoder are also designed, indicating that the proposed TDDFET would be helpful for further ternary logic design.

REFERENCES

[1] J. Ko, J. Kim, T. Jeong, J. Jeong, and T. Song, "Exploration of Ternary Logic Using T-CMOS for Circuit-Level Design," *IEEE Trans. Circuits Syst. I*, vol. 70, no. 9, pp. 3612–3624, Sep. 2023, doi: 10.1109/TCSI.2023.3287274.

[2] X.-Y. Wang, C.-T. Dong, Z.-R. Wu, and Z.-Q. Cheng, "A review on the design of ternary logic circuits*," *Chinese Phys. B*, vol. 30, no. 12, p. 128402, Dec. 2021, doi: 10.1088/1674-1056/ac248b.

[3] M. Huang, X. Wang, G. Zhao, P. Coquet, and B. Tay, "Design and Implementation of Ternary Logic Integrated Circuits by Using Novel Two-Dimensional Materials," *Applied Sciences*, vol. 9, no. 20, p. 4212, Oct. 2019, doi: 10.3390/app9204212.

[4] A. Doostaregan and A. Abrishamifar, "Evaluating a Methodology for Designing CNFET-Based Ternary Circuits," *Circuits Syst Signal Process*, vol. 39, no. 10, pp. 5039–5058, Oct. 2020, doi: 10.1007/s00034-020-01400-2.

[5] J. W. Jeong *et al.*, "Tunnelling-based ternary metal–oxide–semiconductor technology," *Nat Electron*, vol. 2, no. 7, pp. 307–312, Jul. 2019, doi: 10.1038/s41928-019-0272-8.

[6] S. J. Basha and P. Venkatramana, "Design of Ternary Logic Circuits Using GNRFET and RRAM," *Circuits Syst Signal Process*, vol. 42, no. 12, pp. 7335–7356, Dec. 2023, doi: 10.1007/s00034-023-02445-9.

[7] A. Hazra and R. Goswami, Eds., *Carbon Nanomaterial Electronics: Devices and Applications*. in Advances in Sustainability Science and Technology. Singapore: Springer Singapore, 2021. doi: 10.1007/978-981-16-1052-3.

[8] S. Karmakar, "Ternary Logic Flip-Flops Using Quantum Dot Gate Field Effect Transistor (QDGFET)," *Silicon*, vol. 14, no. 18, pp. 12553–12565, Dec. 2022, doi: 10.1007/s12633-022-01949-4.

[9] F. Daneshvar, H. Chen, K. Noh, and H.-J. Sue, "Critical challenges and advances in the carbon nanotube–metal interface for next-generation electronics," *Nanoscale Adv.*, vol. 3, no. 4, pp. 942–962, 2021, doi: 10.1039/D0NA00822B.

[10] A. Naderi and B. A. Tahne, "Review—Methods in Improving the Performance of Carbon Nanotube Field Effect Transistors," *ECS J. Solid State Sci. Technol.*, vol. 5, no. 12, pp. M131–M140, 2016, doi: 10.1149/2.0021612j

Enhancing SRAM Cell Stability Through Single-Carrier CMOS Latch Integration

Yuan-Yu Chuang, Pei-Zhang Xie, and Jyi-Tsong Lin[*]

Department of Electrical Engineering, National Sun Yat-Sen University, Kaohsiung, 80424, Taiwan, ROC

*Email: jtlin@ee.nsysu.edu.tw

Abstract

In this paper, we propose the use of a single-carrier CMOS (SC-CMOS) latch to enhance the stability of SRAM cells. We consider three key parameters that affect the static noise margin (SNM) of SRAM: supply voltage, cell ratio, and pull-up ratio. Our results demonstrate that SC-CMOS SRAM exhibits superior SNM compared to conventional CMOS latches ($W_p/W_n=1$) across hold, read, and write modes. This highlights the excellent noise immunity and stable circuit performance of SC-CMOS SRAM.

1. Introduction

With the continuing miniaturisation of CMOS technology, devices are becoming increasingly susceptible to interference from noise sources. System-on-Chips (SoCs) and other integrated circuits are composed of a multitude of nanoscale devices, which can potentially give rise to noise within the circuit. Such noise can have a considerable effect on the overall operation of the system. For example, SRAM is especially vulnerable to noise interference due to its numerous nanoscale devices that are highly susceptible to external influences[1]. Stability has always been one of the major issues in designing SRAM cells. The sensitivity of the memory to process tolerances and operating conditions depends on cell stability. It must operate correctly in the presence of noisy signals. Therefore, we propose to use SC-CMOS latch to SRAM to improve the stability of the circuit. We have mentioned the performance comparison of SC-CMOS latch with conventional CMOS (Wp/Wn=1) latch in our previous articles[2]. SC-CMOS latch is superior to conventional CMOS(Wp/Wn=1) in terms of noise margin, voltage gain and delay time. This facilitates the application of SC-CMOS to SRAM.

Fig 1 shows the SC-CMOS used in a 6T SRAM cell. The cross-coupled SC-CMOS inverter remains bistable and its output node retains the data stored in the cell. However, as the input signal source increases, the stability of the cell decreases due to fluctuations in the node voltages. The SNM represents the allowable level of these noise voltages, thus quantifying the ability of these latch to maintain their state in the presence of noise. We measured the effects of supply voltage, cell ratio, and pull-up ratio on the SC-CMOS SNM applied to a 6T SRAM cell and compared it to a conventional CMOS (Wp/Wn=1) inverter applied to a 6T SRAM.

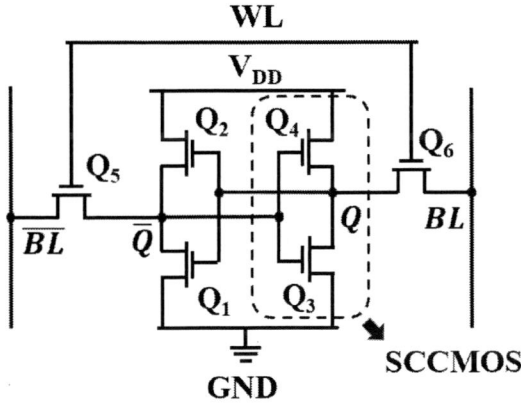

Fig 1 Schematic of SC-CMOS for 6T SRAM Cells.

2. Static Noise Margin Calculation Method

2.1 Butterfly curve

The SNM of the SRAM cell is obtained by plotting the voltage transfer curves (VTCs) of two cross-coupled latch. The VTC of one of the latch is transformed into a butterfly curve by a rotation relative to the line y = x[1]. The SNM is obtained by calculating the length of the square inside the two curves.

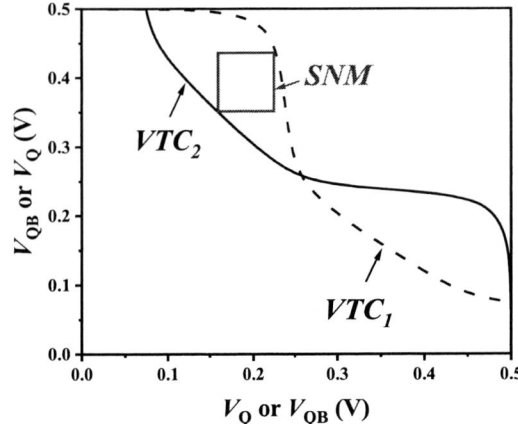

Fig 2 Schematic diagram of the maximum square edge length of SNM.

2.2 N-curve

An alternative method for measuring the noise margin of a SRAM cell is to utilise N-curves[3]. These are determined by clamping the word and bit lines to VDD, in a manner analogous to a read operation. A voltage sweep is applied at one of the nodes, and the current supplied by the source is subsequently measured[4]. The circuits and curve diagrams used to measure the N-curve are shown in Fig 3 and Fig 4. The N curve can be employed to quantify the read margin

979-8-3503-6184-1/24 $31.00 © 2024 IEEE

(SNM, as illustrated in Fig 4) and the write margin (WTV or write trip voltage). The positive peak current is the SINM, or static current noise margin, which is defined as the maximum current that can be injected into the cell in order to effect a change in its state. The negative peak current is the WINM, or write-in current noise margin, and it defines the amount of current that is required in order to write to the cell with both bit lines charged to V_{DD}[4].

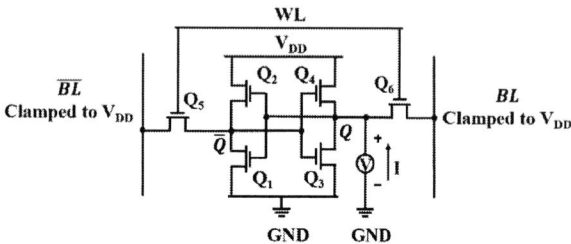

Fig 3 Schematic diagram of measuring N-curve circuit.

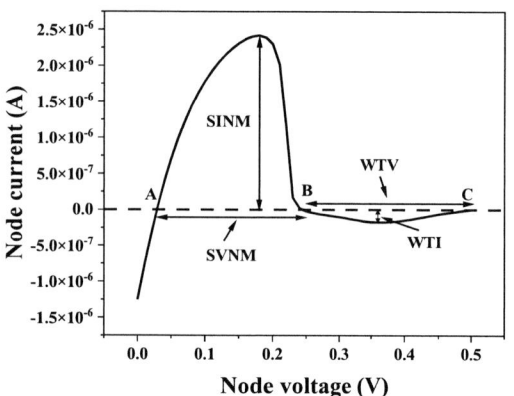

Fig 4 Plot of N-curve.

3. Simulation of SRAM Static Noise Margin

3.1 Effect of supply voltage

In order to reduce the power consumption of digital circuits, power supply voltage scaling is frequently employed as a means of achieving this objective. The noise margin of the SRAM cell is also a function of the supply voltage, as demonstrated in the model presented in [5]. Fig 5 shows the effect of different supply voltages on the noise margin of SC-CMOS SRAM cells. The supply voltage was subjected to analysis over a range of 0.1V to 0.5V. The noise margins for hold mode, read mode and N-curve are found to be proportional to the supply voltage. This outcome is anticipated, given that as the supply voltage declines, the impact of the noise present intensifies. Consequently, the circuit becomes more compact, and the voltage scaling is constrained to a level where the noise margin remains larger than the noise anticipated in the circuit.

3.2 Effect of cell ratio

In addition to supply voltage scaling, SNM is also determined by the cell ratio and pull-up ratio according to the SNM equation [5]. SNM in read and write mode is more susceptible to these ratios. The cell ratio is defined as the ratio between the size of the pull down transistor (Q_1, Q_3) of the SC-CMOS inverter and the size of the access transistor (Q_5, Q_6) as shown in Fig 1. Assuming that the data at node

Q is "0" and BL is pre-charged to VDD, i.e. operating in read mode. The size of the pull down transistors (Q_3) determines the strength of node Q to pull BL to GND. Fig 6 shows the effect of cell ratio on the noise margin of SC-CMOS SRAM cells. The value of the cell ratio is 0.5, 1, 1.5, 2 and 2.5. The read noise margin increases with cell ratio. It can be observed that an increase in the cell ratio results in a expansion of the pull down network. This phenomenon demonstrates that nodes storing a value of 0 exhibit robustness to noise introduced by the *BL* precharge voltage during read mode. The robust pull down network ensures that the node voltage remains constant. Furthermore, N-curve measurements indicate that as the cell ratio rises, the read noise current rises in tandem, thereby necessitating a greater current to change the cell state.

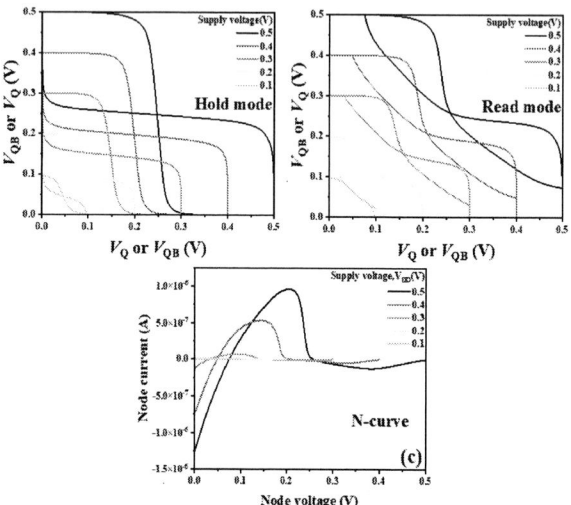

Fig 5 Effect of different supply voltages on the noise margin of SC-CMOS SRAM cell.

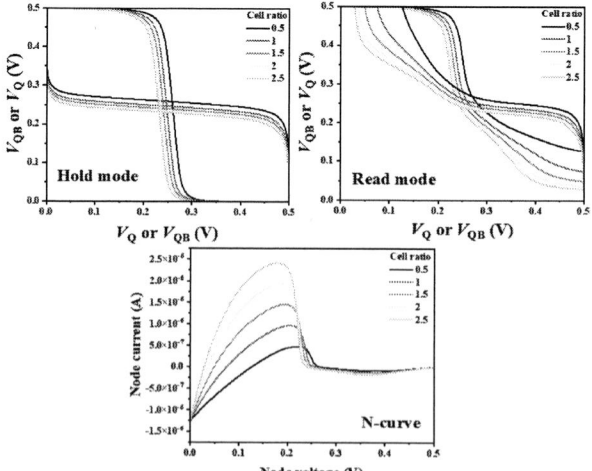

Fig 6 Effect of cell ratio on the noise margin of SC-CMOS SRAM cell.

3.3 Effect of pull up ratio

The pull up ratio is defined as the ratio between the size of the pull up transistors (Q_2, Q_4) of the SC-CMOS inverter and the size of the access transistors (Q_5, Q_6), as shown in Fig 1. The strength of the pull up transistor determines the difficulty of writing data or flipping the state of cell. Fig 7 shows the effect of pull-up ratio on the noise margin of SC-CMOS SRAM cell. The values of the pull up ratio are 1, 2, 3, 4 and 5. The strength of the pull up ratio determines how strongly the cell retains its stored data. Assuming that the data at $\bar{Q}$ is "1" and $\overline{BL}$ is connected to GND, i.e. write mode. As the pull up ratio increases, the difficulty of writing or pulling the node $\bar{Q}$ to GND also increases. This results in a reduction in both the write margin and the write trip voltage, which in turn leads to an increase in the read margin. Furthermore, the N-curve demonstrates that the write current noise margin is enhanced in proportion to the increase in pull up ratio. Consequently, a greater current is required to affect the state transition of the cell.

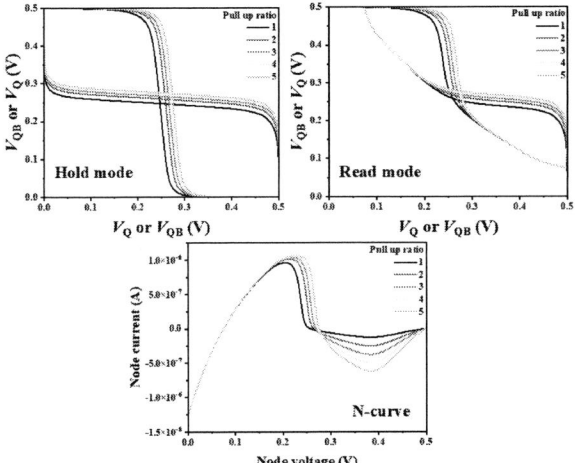

Fig 7 Effect of pull up ratio on the noise margin of SC-CMOS SRAM cell.

Table 1 SNM for butterfly curve measurement.

		Butterfly curve			
		SCCMOS SRAM		Conv. CMOS(W_p/W_n=1) SRAM	
	Metric	Worst Value	Best Value	Worst Value	Best Value
Supply voltage	Hold	17.2 mV	192 mV	14.7 mV	177 mV
	Read	6.23 mV	81.9 mV	1.68 mV	61.9 mV
Cell ratio	Hold	185 mV	191 mV	192 mV	196 mV
	Read	65.1 mV	105 mV	28.9 mV	150 mV
Pull up ratio	Hold	185 mV	192 mV	188 mV	197 mV
	Read	81.9 mV	100 mV	86.3 mV	102 mV

Table 2 SNM for N-curve measurement.

		N-curve			
		SCCMOS SRAM		Conv. CMOS(W_p/W_n=1) SRAM	
	Metric	Worst Value	Best Value	Worst Value	Best Value
Supply voltage	SVNM	26 mV	184 mV	24 mV	176 mV
	SINM	6.01×10^{-17} A	9.60×10^{-7} A	4.72×10^{-12} A	9.59×10^{-8} A
	WTV	25 mV	241 mV	28 mV	236 mV
	WTI	-3.47×10^{-12} A	-1.24×10^{-7} A	-2.64×10^{-12} A	-1.09×10^{-7} A
Cell ratio	SVNM	143 mV	215 mV	81 mV	221 mV
	SINM	4.73×10^{-7} A	2.41×10^{-6} A	4.19×10^{-8} A	2.75×10^{-7} A
	WTV	230 mV	256 mV	225 mV	225 mV
	WTI	-8.72×10^{-8} A	-8.72×10^{-8} A	-3.75×10^{-8} A	-1.88×10^{-7} A
Pull up ratio	SVNM	201 mV	184 mV	191 mV	173 mV
	SINM	1.06×10^{-6} A	9.60×10^{-7} A	1.18×10^{-7} A	9.59×10^{-8} A
	WTV	224 mV	241 mV	223 mV	241 mV
	WTI	-6.62×10^{-7} A	-1.24×10^{-7} A	-5.38×10^{-7} A	-1.09×10^{-7} A

Table 1 shows the best and worst measurements for SC-CMOS SRAM and conventional CMOS (W_p/W_n=1)

SRAM. With supply voltage scaling, SC-CMOS SRAM have higher SNM than conventional CMOS (Wp/Wn=1) SRAM in both hold and read modes, which means SC-CMOS SRAM have better noise immunity and stable circuit performance. Table 2 shows the measurements obtained from the N-curve. The N-curve also shows the same trend as the butterfly curve.

4. Conclusion

This study confirms that the SC-CMOS latch significantly enhances SRAM cell stability compared to conventional CMOS latches. SC-CMOS SRAM cells exhibit higher noise margins in hold and read modes, making them more resistant to noise and improving overall circuit performance. The noise margin increases with supply voltage and cell ratio, while adjusting the pull-up ratio improves the write margin. These findings suggest that SC-CMOS technology is a viable solution for enhancing the reliability of nanoscale SRAM cells in advanced CMOS technology.

Acknowledgments

This work was financially supported by the Ministry of Science and Technology of Taiwan, under Grant No. MOST109-2221-E-110-018-MY3.

References

[1] C. D. C. Arandilla et al., 2011 UkSim 13th International Conference on Computer Modelling and Simulation, pp. 534-539, 2011.

[2] J.-T. Lin et al., Discover Nano, p. 113, 2024.

[3] C. Wann et al., IEEE VLSI-TSA International Symposium on VLSI Technology, 2005. (VLSI-TSA-Tech). pp. 21-22, 2005.

[4] E. Grossar et al., IEEE Journal of Solid-State Circuits, pp. 2577-2588, 2006.

[5] E. Seevinck et al., IEEE Journal of Solid-State Circuits, pp. 748-754, 1987.

A RRAM based 9T1R NVSRAM for Low-Power Computing in Memory

Huimeng Guo[1,2], Yujia Li[2], Tingrui Ren[2,3], Chenge Dong[2,3], Liang Wang*[2],
Yuanfu Zhao[2], Yanlong Zhang[2]

[1] School of Electronic and Information, Hangzhou Dianzi University, Hangzhou 310018, China
[2] Beijing Microelectronics Technology Institute, Beijing 100076, China
[3] School of Integrated Circuit Science and Engineering, Beihang University, Beijing 100191, China
*Email: first_author:guohm163@163.com, corresponding_author:wangliang150200@163.com

Abstract—**With the rapid advancement of AI technology, the amount of data has been exponential growth, necessitating memory solutions that offer high speed, low energy consumption, non-volatility, and high reliability in memory cell. In this brief, we propose a 9T1R non-volatile SRAM (NVSRAM) based on Resistive Random Access Memory (RRAM) for low-power computing in memory (CIM). This cell combines the advantages of SRAM and RRAM to efficiently execute CIM. Additionally, it can power down during idle periods and back up data to RRAM, reducing SRAM standby power consumption. We provide a detailed introduction to the operating principles of three distinct modes: SRAM mode, NVSRAM mode, and CIM mode. Circuit simulations for each mode are presented to validate the functionality and performance of the design. Comprehensive performance evaluation tests reveal that the proposed unit achieves excellent results in terms of read static noise margin, energy consumption, and operation latency, underscoring its potential for advanced memory applications.**

Keywords—NVSRAM, Computing in Memory, RRAM

I. INTRODUCTION

As artificial intelligence (AI) rapidly develops, tremendous data exchanges between memory and processing units can result in latency and inefficiency when dealing with extensive datasets and intricate algorithms as depicted in Fig.1(a). This situation increasingly highlights the bottlenecks of von Neumann architecture [1]. Consequently, there is an urgent need to explore new storage solutions to overcome the limitations of current architecture and meet the higher demands for computational efficiency in future AI applications.

Computing-in-Memory (CIM) integrates computing capabilities directly within memory, thereby reducing data transmission latency and energy consumption. This innovation effectively mitigates the inherent bottlenecks of the traditional von Neumann architecture [2]. SRAM is indispensable in semiconductor memory due to its low power consumption and high speed advantage. Chen et al. proposed an 8T SRAM architecture with additional differential PMOS access transistors, enhancing speed and reliability in memory computing architecture and enables more complex Boolean logic operations [3]. Among the emerging memory technologies, RRAM is highly anticipated by its high density, non-volatile, and compatibility with existing advanced CMOS logic processes. Sing *et al.* proposed a 3D RRAM to achieve the low impedance matching (LIM) through a efficient vertical structure in CMOS logic compatible process [4].

NVSRAM combines the non-volatility of RRAM with the high speed and durability of SRAM, offering a brand new solution for building high-efficiency and large-capacity memory. However, the 6T2R structure is prone to short-circuit current issues [5], while the 8T2R structure exhibits high power consumption [6]. The CIM architecture based on

NVSRAM is shown in Fig. 1(b). The high efficiency and low power consumption of NVSRAM make it an ideal choice for CIM, and this new architecture has great potential in big data and AI applications. However, there is limited research on memory and computing integrated unit based on NVSRAM. This brief proposes a 9T1R NVSRAM-CIM cell based on RRAM for Boolean logic operations, which can reduce system power consumption and improve reliability in CIM.

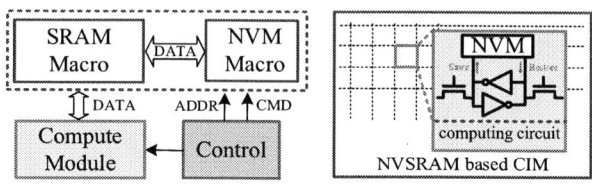

Fig. 1. (a) Conventional von Neumann architecture. (b) Computing-in-memory architecture based on NVSRAM.

This brief is organized as follows: Section II introduces the characteristics of RRAM. Section III presents the working principle and simulation verification of SRAM, NVSRAM and CIM mode, respectively. In Section IV, performance comparison is carried out. Finally, the brief conclusion is given in Section V.

II. CHARACTERISTICS OF RRAM

The RRAM device features a sandwich structure comprising a top electrode (TE), an intermediate transition metal oxide layer, and a bottom electrode (BE). The TE and BE are used to apply external stimuli. The resistance switching mechanism of RRAM involves applying different external voltage stimuli to the intermediate resistive switching layer, forming or disrupting conductive filaments (CF) to toggle between high and low resistance states. As depicted in Fig. 1(a), when a positive voltage (V_{set}) is applied at the TE, the CF forms, placing the RRAM device in a low resistance state (LRS). Conversely, when a negative voltage (V_{reset}) is applied at the BE, the conductive filament breaks, placing the RRAM device in a high resistance state (HRS). The transition from HRS to LRS is termed SET, whereas the transition from LRS to HRS is termed RESET.

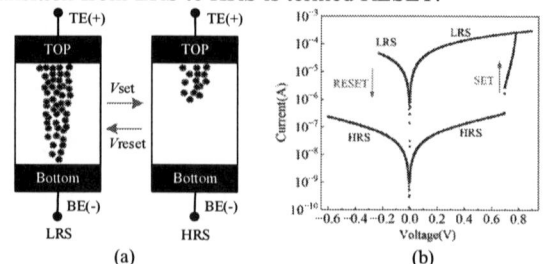

Fig. 2. (a) RRAM structure. (b) *V–I* curve of RRAM model.

979-8-3503-6184-1/24 $31.00 © 2024 IEEE

The RRAM compact model applied in this study is implemented in Verilog-A [7]. Fig. 2(b) presents the *V-I* characteristic curves of the switching behavior of the RRAM model. To highlight the switching characteristics, the current coordinate axis is in the log scale. The RRAM model has a V_{set} of about 0.78 V and a V_{reset} of about −0.25 V. The resistance values of the HRS and LRS are about 2.4MΩ and 3kΩ, respectively. When the initial resistance state of the RRAM is HRS, a scanning voltage ranging from 0 to 0.9V is applied to the TE terminal of the RRAM, then scanned back from 0.9V to 0V, with the BE terminal held at 0V. This completes the Set process, switching the RRAM to the LRS. Subsequently, the TE terminal is set to 0V, and a scanning voltage from 0 to 0.6V is applied to the BE terminal, then scanned back from 0.6V to 0V. This completes the Reset process, switching the RRAM back to the HRS, thus obtaining a complete voltage-current curve of the RRAM.

III. RRAM Based 9T1R CIM Cell

The CIM cell of 9T1R is mainly composed of SRAM module (M1-M6), RRAM module (M7, RRAM) and logic operation module (M8-M9), as shown in Fig. 3.

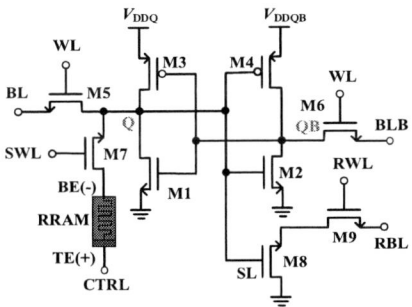

Fig. 3. Circuit schematic of 9T1R CIM cell

A. SRAM Mode

When the control signals SWL, RWL and SL are low, transistors M7-M9 are turned off, and the memory cell operates in the conventional 6T SRAM mode. The two inverters are cross coupled to store data, with M5 and M6 serving as access transistors. When WL is high, data is written to the storage nodes Q and QB from BL and BLB through the access transistors, and latched in the SRAM cell. However, when V_{DDQ} and V_{DDQB} are powered off, the data will be lost.

B. NVSRAM Mode

In order to make the storage unit non-volatile, a 1T1R structure is added to the SRAM storage node. When the power supply voltage drops or is turned off, the system switches to data storage mode and stores the data to the RRAM. When the voltage of SRAM is restored, the data in RRAM will be restored to the SRAM storage node. The process of storing and recovering data is shown schematically in Fig. 4, where the HRS represents Q=1 and the LRS represents Q=0.

The time-domain waveforms for the store and restore data Q=0 operation are shown in Fig. 5. During the writing phase, setting BL=0, BLB=1, WL is set high and open M5/M6 to write data "0" into the SRAM data node Q. In the store phase, firstly, SWL and CTRL are set to 1V, causing a positive potential drop on RRAM that changes the state from HRS to LRS. Then, setting CTRL at 0V to retain RRAM's state. During the power-off phase, V_{DDQ}, V_{DDQB}, SWL and CTRL

are all kept low to ensure the retention of data retention in RRAM. Finally, during the restore phase, V_{DDQ} is powered up to $1/2 V_{\text{DD}}$, then V_{DDQB} is powered up to V_{DD}, and M7 shuts down the pull-down path, initializing the Q node to a weak "1" and QB to "0". A pulse is applied to SWL to turn on M7 for a short period of time. V_{DDQ} then continues to power up to V_{DD} If RRAM is at HRS, the Q node is maintained at V_{DDQ}; if RRAM is at LRS, the Q node discharges to 0V.

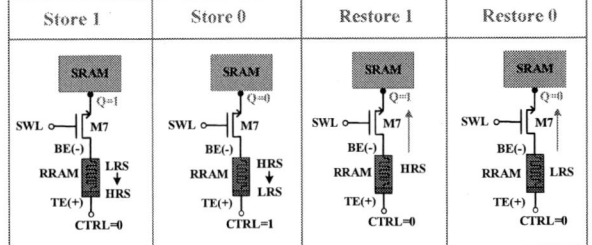

Fig. 4. Store and restore data in NVSRAM

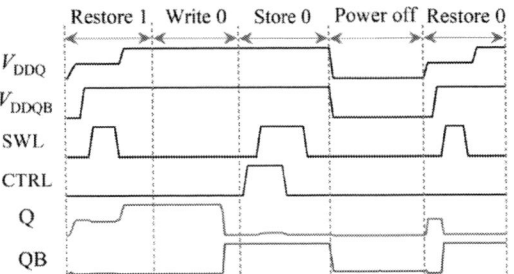

Fig. 5. Simulation results of store and restore operation in NVSRAM

C. CIM Mode

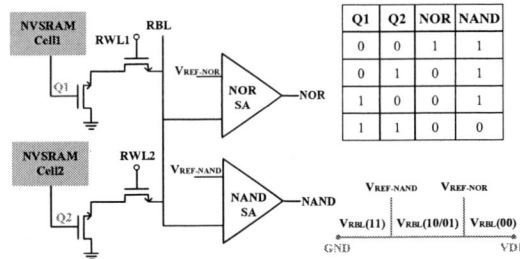

Fig. 6. Schematic of NOR and NAND operation

For NOR operation, the RBL of the NVSRAM cells is connected to the NOR sense amplifier (SA) and controlled using RWL1/RWL2, as shown in Fig. 6. In this operation, the NOR SA is designed with a reference voltage that outputs logic 1 when the input voltage is greater than the reference voltage, and logic 0 otherwise. For cell-1 and cell-2 with a value of 00, the RBL will be at V_{DD}, and the NOR SA will output logic 1. For the 01/10 case, by controlling RWL, the RBL voltage (0.36V) will be discharged below the NOR SA's reference voltage which can result in a NOR logic value of 0. For the 11 case, the RBL will also discharge from V_{DD} to a lower voltage (0.05V), resulting in a NOR logic value of 0 in Fig 7.

For NAND operation, a NAND SA is designed to connect to the RBL. Compared with the NOR operation, the reference voltage of the NAND SA is lower, resulting in potentially different outcomes when discharging to the same voltage. For cell-1 and cell-2 with a value of 00, the RBL will be V_{DD}, and the SA will produce a logic 1. For the 01/10 case, as shown in Fig. 7, the NAND logic value will be 1 due to the lower reference voltage of the SA during the NAND operation. For

979-8-3503-6184-1/24 $31.00 © 2024 IEEE

the 11 case, the RBL will discharge from V_{DD} to 0.05V, resulting in a NAND logic value of 0.

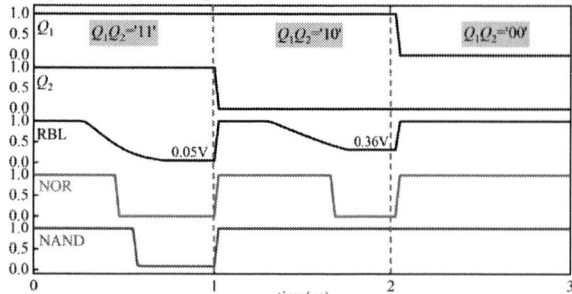

Fig. 7. Simulation results of nor and nand operation

IV. PERFOMANCE EVALUATION

In this section, we validate the performance of the proposed 9T1R NVSRAM-CIM cell in terms of stability, energy efficiency and latency. All the simulations are performed based on 180nm CMOS process and HSPICE.

In conventional SRAM, there is a conflict between RSNM and read current. Increasing the read current speed often comes at the expense of reducing RSNM. Fig. 8 shows the RSNM comparison between 6T SRAM and 9T1R NVSRAM cell at the same voltage. The 9T1R NVSRAM cell, as a basic computing unit, has a storage node separated from the read path, preventing read disturb issues. Consequently, it exhibits a larger RSNM compared to the traditional 6T SRAM, thereby enhancing read stability.

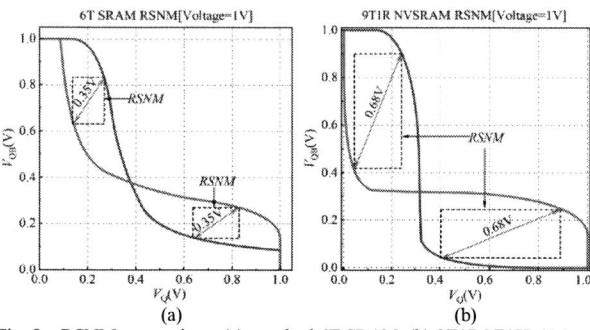

Fig. 8. RSNM comparison: (a) standard 6T SRAM; (b) 9T1R NVSRAM

Since NVSRAM backs up data before power-down and restores it upon power-up, it does not consume energy during data retention, only during the store and restore processes.

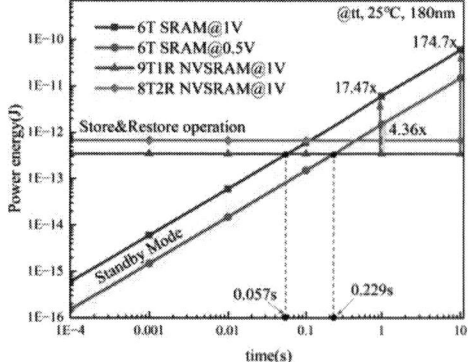

Fig. 9. Power energy comparison: 6T SRAM, 8T2R and 9T1R NVSRAM

In contrast, 6T SRAM continues to consume static leakage power during data retention. Fig. 9 compares the static retention power consumption of the proposed 9T1R

NVSRAM-CIM cell with 6T SRAM at different voltages. The static power consumption of 6T SRAM increases over time, while the total energy consumption of NVSRAM remains unchanged. As the retention time extends, the low power advantage of NVSRAM becomes significant. The figure shows that at 0.057s of data retention, the static power of traditional 6T SRAM at 1V is the same as that of the 9T1R. And at 1s, the static power consumption of 6T SRAM is 17.47 times that of the 9T1R.

Table I summarized the average energy consumption and latency of the proposed 9T1R NVSRAM cell across different operating modes. The results show the efficiency and speed of the NVSRAM cell in various scenarios, which can exhibit vast potential for low-power CIM computing applications.

TABLE I. AVERAGE ENERGY CONSUMPTION AND LATENCY DURING THE VARIOUS OPERATIONAL MODES OF PROPOSED NVSRAM CELL

	Avg Energy Consumption (aJ)	Avg Latency (ns)
Write Operation	5.97e3	0.1
Store Operation	2.29e5	20
Restore Operation	6.32e4	20
CIM Operation	Nor: 11.2 Nand: 20.1	Nor: 0.402 Nand: 0.613

V. CONCLUSION

In this paper, we present a low-energy 9T1R NVSRAM CIM cell. The proposed NVSRAM cell features a power-off data retention scheme and a single-ended sensing structure, which can effectively reduce CIM power consumption. And the cell provides higher robust performance for NAND and NOR operation, exhibiting high RSNM and energy efficiency during read and write operations. The results presented in the work indicate that this technique is a promising candidate for future low-energy on-chip CIM, offering significant improvements in both power efficiency and operational stability.

ACKNOWLEDGMENTS

This work was supported by the National Key Research and Development Program of China under Grant 2022YFB4400401.

REFERENCES

[1] C. -J. Jhang, C. -X. Xue, J. -M. Hung, F. -C. Chang and M. -F. Chang, "Challenges and Trends of SRAM-Based Computing-In-Memory for AI Edge Devices," IEEE Trans. Circuits Syst. vol. 68, pp. 1773-1786, May 2021.

[2] Y. Chen, J. Mu, H. Kim, L. Lu and T. T. -H. Kim, "BP-SCIM: A Reconfigurable 8T SRAM Macro for Bit-Parallel Searching and Computing In-Memory," IEEE Trans. Circuits Syst. vol. 70, pp. 2016-2027, May 2023.

[3] J. Chen, W. Zhao, Y. Wang, Y. Shu, W. Jiang and Y. Ha, "A Reliable 8T SRAM for High-Speed Searching and Logic-in-Memory Operations," IEEE Transactions on Very Large Scale Integration (VLSI) Systems, vol. 30, pp. 769-780, June 2022.

[4] S. Sing et al., "A New High Density 3D Stackable Via RRAM for Computing-in-Memory SOC Applications," IEEE Trans. Electron Devices, vol. 71, pp. 2399-2403, April 2024.

[5] W. Wang et al., "Nonvolatile SRAM Cell," 2006 International Electron Devices Meeting, San Francisco, CA, USA, 2006, pp. 1-4.

[6] P. Chiu et al., "A low store energy, low VDDmin, nonvolatile 8T2R SRAM with 3D stacked RRAM devices for low power mobile applications," 2010 Symposium on VLSI Circuits, Honolulu, HI, USA, 2010, pp. 229-230.

[7] J. Reuben, D. Fey and C. Wenger, "A Modeling Methodology for Resistive RAM Based on Stanford-PKU Model With Extended Multilevel Capability," IEEE Trans. Nanotechnol, pp. 647-656, 2019.

A High-Resistance SOT Device Based Computing-In-Memory Macro with High Sensing Margin and Multi-Bit MAC Operations for AI Edge Inference

Junzhan Liu [1,2], Jinyao Mi [1,2], Haiyan Qin [1,2], He Zhang [*1,2], Wang Kang [*1,2]

[1] School of Integrated Circuit Science and Engineering, Beihang University, Beijing 100191, China
[2] National Key Laboratory of Spintronics, Hangzhou International Innovation Institute, Beihang University, Hangzhou 311115, China

*liujunzhan@buaa.edu.cn, zhanghe@buaa.edu.cn, wang.kang@buaa.edu.cn

Abstract—Computing-in-memory (CIM) offers a promising solution to the memory wall issue. Magnetoresistive random-access memory (MRAM) is a favored medium for CIM due to its non-volatility, high speed, low power, and technology maturity. However, MRAM has continuously encountered the challenge of an insufficient high-resistance state (HRS) to low-resistance state (LRS) ratio, which affects the result accuracy of CIM. In this paper, based on SOT devices, we propose a 5T2M bit-cell structure that increases the high-to-low current ratio by modulating the subthreshold operation region. Besides, by jointly using high-resistance devices (MΩ level), the power consumption of the bit-cell array can be significantly reduced. Simultaneously, we have designed a compatible multi-bit implementation and macro architecture to support AI edge inference acceleration. This work was simulated under a 40-nm foundry process and a physically verified SOT-MTJ model. The results show that under the same high-to-low resistance ratio, a 52.6× high-to-low current ratio can be achieved, along with a 38.6%-98% bit-cell array power reduction.

Keywords—Computing-in-memory, SOT-MRAM, HRS/LRS ratio, multi-bit, artificial intelligence

I. INTRODUCTION

The rapid advancement of artificial intelligence (AI) has significantly increased the demand for high-performance and energy-efficient computing systems. Large amounts of unstructured data lead to significant memory access overheads, with computing system performance being constrained by memory performance, a phenomenon known as the memory wall. Computing-in-memory (CIM) has emerged as a promising solution to the memory wall challenge by integrating memory and computing units within the same framework. This approach minimizes data transfer, enhances computing efficiency, and meets the growing needs of AI applications. As shown in Fig. 1 (a), due to its efficiency, physical computing based on Ohm's law and Kirchhoff's current law has received widespread attention in accelerating multiply-and-accumulation (MAC) operations. Various media, such as SRAM, DRAM, Flash, ReRAM, MRAM, PCM, FeFET, etc. [1]-[5], have already been researched and applied in CIM. Among the various memory media, magnetoresistive random-access memory (MRAM) is a favored medium for CIM due to its non-volatility, high speed, low power consumption, and relatively mature technology. However, as illustrated in Fig. 1 (b), MRAM-based CIM faces a significant challenge: an insufficient high-resistance state (HRS) to low-resistance state (LRS) ratio, which leads to a limited sensing margin. This limitation hampers computing accuracy, posing a barrier to its widespread adoption.

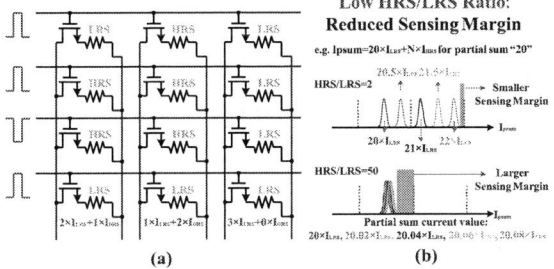

Fig. 1. (a) Conventional 1T1M bit-cell array for computing-in-memory; (b) The challenge of low HRS/LRS ratio leads to reduced sensing margin and current errors for partial sum results.

To address this challenge, we propose a novel 5T2M bit-cell structure based on spin-orbit torque (SOT) devices, which modulates the subthreshold operation region to significantly improve the high-to-low current ratio. Furthermore, by jointly using high-resistance devices (MΩ level), we have managed to reduce the power consumption of the bit-cell array. In addition, we have designed a compatible multi-bit implementation scheme and macro architecture, ensuring that our proposed structure can effectively support AI edge inference tasks.

II. PROPOSED HIGH SENSING MARGIN MULTI-BIT CIM

A. Proposed Architecture

As shown in Fig. 2, the proposed macro architecture consists of a 63-row × 64-column bit-cell CIM array, multi-bit pulse width input units, sub-array selectors & drivers, WWL &RWL decoders & drivers, WBL & WSL read/write drivers, multi-bit output units, and single slope ADCs. Each bit-cell consists of a 5T2M (5 transistors and 2 magnetic tunnel junctions, i.e., MTJ) structure. This structure fully leverages complementary bit-cells and the high gain of the subthreshold operation region to enhance the high-to-low current ratio. In the memory mode, the WWL decoders & drivers select specific rows, enabling the WBL &WSL read/write drivers to write data to or read data from specific 5T2M bit-cells. Weights are written to the two MTJs of each 5T2M bit-cell in complementary states through only a single write operation. In the CIM mode, the 63-row bit-cell CIM array is divided into three sub-arrays, with one 21-row × 64-column sub-array being activated at a time using sub-array selectors & drivers and RWL decoders & drivers. Activation signals are fed by the multi-bit pulse width input units to multi-bit output units. The analog output is then converted into digital signals through single slope ADCs.

979-8-3503-6184-1/24 $31.00 © 2024 IEEE   47

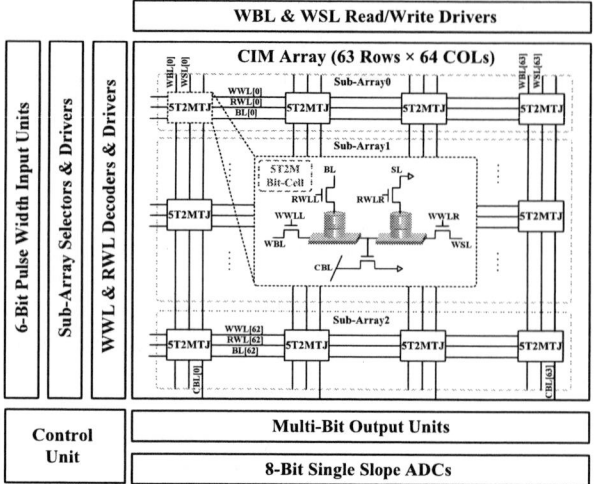

Fig. 2. The proposed macro architecture consists of a 63-row × 64-column bit-cell CIM array, multi-bit pulse width input units, sub-array selectors & drivers, WWL &RWL decoders & drivers, WBL & WSL read/write drivers, multi-bit output units, and single slope

B. 5T2M Bit-Cell Structure

Fig. 3 (a) shows the proposed 5T2M bit-cell structure. This bit-cell includes two complementary MTJs (MTJ1 and MTJ2), which are connected to the computing transistor T3 between their bottom electrodes. T3 operates in the subthreshold region, leveraging its high gain to achieve an increased high-to-low current ratio. Multi-bit configuration can be achieved by adjusting the number of T3 multipliers, with values of 1, 2, and 4 in our work, corresponding to 3-bit accuracy. Additionally, four transistors, T1-T2 and T4-T5, are used for switching between read/write and CIM modes. Fig. 3 (b) details the operation waveforms. When MODE=1, it is in the memory mode. During a write operation, when WWLL and WWLR are set to 1, the write voltage ($|V_{write1}-V_{write2}|$) is applied between WBL and WSL according to the data to be written. This completes the data writing to MTJ1 and MTJ2 in a single operation. During a read operation, taking the reading of MTJ1 as an example, when WWLL and RWLL are set to 1, the read voltage is applied between BL and WBL ($|V_{read1}-V_{read2}|$). The external read circuit completes the data reading from MTJ1. When MODE=0, it is in the CIM mode. At this time, RWWL and RWLR are set to 1, and BL and CBL are set to specific voltages (see simulation results section for details) for a period according to the pulse width input corresponding to the activation data conversion.

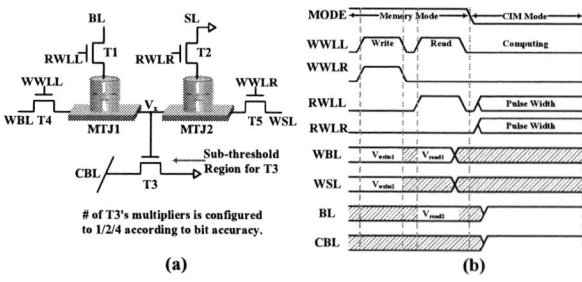

Fig. 3. (a) The proposed 5T2M bit-cell structure; (b) The operation waveforms of the proposed 5T2M bit-cell: divided into storage mode and CIM mode. In storage mode, data can be read and written, while in CIM mode, inputs are fed through pulse widths of RWLL and RWLR.

C. Multi-Bit Implementation

Multi-bit MAC operations are widely needed in AI edge inference applications. This work achieves 6-bit input, 3-bit weights, and 8-bit output based on the following methods. Fig. 4 illustrates the proposed multi-bit implementation using a single column of 5T2M bit-cells as an example. The bit-cell COL consists of 63 5T2M bit-cells. The computing transistor T3 (see Fig. 3 (a)) in every three adjacent 5T2M bit-cells has multiplier values of 1, 2, and 4, respectively, to represent 3-bit weight accuracy.

Each 6-bit input is divided into three 2-bit pulse width inputs. Each 2-bit pulse width input operates on every row of the 5T2M bit-cell column. The summation of different 2-bit inputs is controlled by the multi-bit current modulation unit. When the lowest 2-bit input is executed, SW0 is on while SW1 and SW2 are off. At this time, the current charges the capacitor at the IN_ADC terminal with a 1× multiplier. When the middle 2-bit input is executed, SW1 is on while SW0 and SW2 are off. At this time, the current charges the capacitor at the IN_ADC terminal with a 4× multiplier. When the highest 2-bit input is executed, SW2 is on while SW0 and SW1 are off. At this time, the current charges the capacitor at the IN_ADC terminal with a 16× multiplier. The summation of these three scaled currents results in the analog MAC output of the 6-bit input and 3-bit weights.

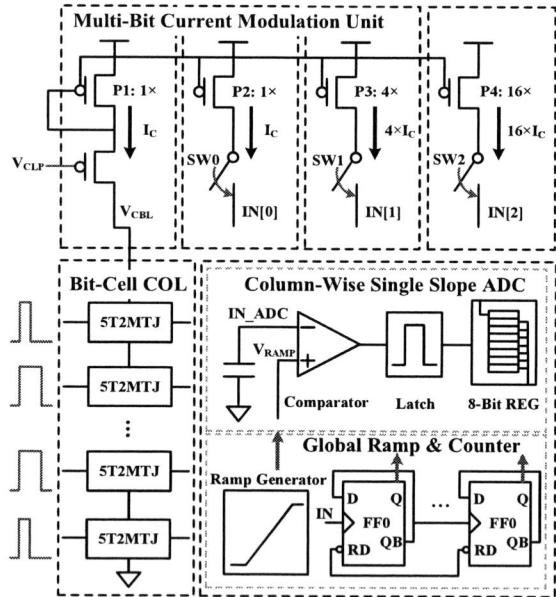

Fig. 4. The proposed multi-bit implementation, mainly including the bit-cell COL, multi-bit current modulation unit, and column-wise single slope ADC.

Then, the per-column single slope ADC converts the obtained analog MAC output into digital form. Each 8-bit single slope ADC consists of a comparator, a latch, and an 8-bit register. All columns share a single global RAMP generator and an 8-bit counter. As the RAMP generator produces a ramp waveform, the 8-bit counter begins counting. When the ramp waveform exceeds the voltage at the IN_ADC terminal capacitor, the comparator flips, and the count value is stored in the 8-bit register. Due to the simplicity and small area overhead of the single slope ADC, it is highly suitable for highly parallel CIM applications.

979-8-3503-6184-1/24 $31.00 © 2024 IEEE

III. SIMULATIONS AND DISCUSSIONS

We designed this macro using a 40nm foundry's process and physical verified SOT device model. Table I lists the relevant simulation parameters. The low and high resistance values of the MTJ are 1MΩ and 2MΩ, respectively, with a TMR of 100% at zero bias. The use of high-resistance devices significantly reduces power consumption, thereby enhancing parallelism. The 5T2M bit-cell with complementary dual MTJs employed in this work is equivalent to a series connection of one high-resistance and one low-resistance MTJ, which further reduces power consumption.

TABLE I
SIMULATION PARAMETERS OF THE PROPOSED MACRO

Parameters	Value
CMOS process	40nm
Bit-cell structure	5T2M
Supply voltage	1.1V
Temperature	300K
Array size	63 rows×64 columns
R_{LRS} of MTJ	1MΩ
TMR ratio at zero bias	100%

Fig. 5 (a) shows the current of the computing transistor (T3 in Fig. 3(a)) in the 5T2M bit-cell as a function of the voltage at point V_x. It can be observed that when V_x is ~0.15-0.4V, the MOS transistor operates in the subthreshold region, where the change in I_{CBL} is steeper. This characteristic is utilized in this work to amplify the high-to-low current ratio. With V_{CBL} fixed, Fig. 5 (b) shows the variation of I_{CBL} with respect to V_{BL}. When V_{BL} is approximately 0.6V, the ratio of I_{LRS} to I_{HRS} reaches its maximum, about 52.6, with I_{LRS} being 185nA and I_{HRS} being 3.52nA.

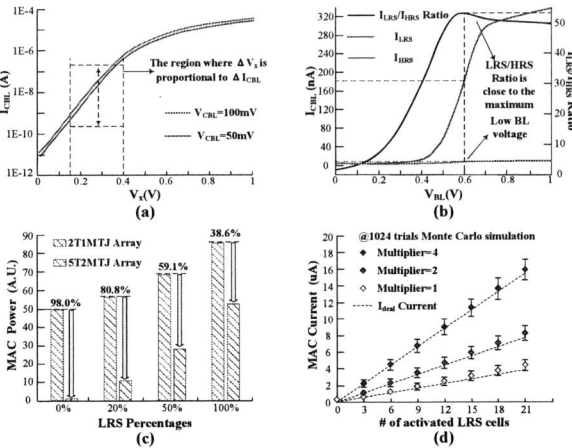

Fig. 5. (a) Current of the computing transistor in the 5T2M bit-cell as a function of the voltage at point V_x; (b) Variation of I_{CBL} with respect to V_{BL} with V_{CBL} fixed. At VBL≈0.6V, the ratio of I_{LRS} to I_{HRS} reaches its maximum of approximately 52.6; (c) Comparison of power consumption between the traditional 2T1M bit-cell and the proposed 5T2M bit-cell using high-resistance devices; (d) 1024 Monte Carlo simulation results of MAC current under different

Fig. 5 (c) compares the power consumption between the traditional 2T1M bit-cell and the proposed 5T2M bit-cell, both using high-resistance devices. The 5T2M bit-cell proposed in this work demonstrates power savings ranging from 38.6% to 98%, depending on the ratio of low-resistance states. This improvement is primarily attributed to the complementary design of the 5T2M bit-cell using dual MTJs, which further enhances the bit-cell resistance. Additionally, the computing transistors operating in the subthreshold region exhibit significantly lower computing current. Fig. 5 (d) presents the results of 1024 Monte Carlo simulations of MAC current under different multiplier conditions (m=1, 2, 4). Optimized high-to-low current ratio contributes to favorable sensing margins between different states. Fig. 6 illustrates the overall simulation waveform. In this simulation, a 6-bit input is divided into three 2-bit pulse width inputs, sequentially applied to V_{RWLL} and V_{RWLR}. SWX is used to control the multi-bit current modulation unit (see Fig. 4). The SS-ADC's sampling capacitor accumulates current information from these inputs, which is subsequently compared against a RAMP waveform to yield precise digital output results.

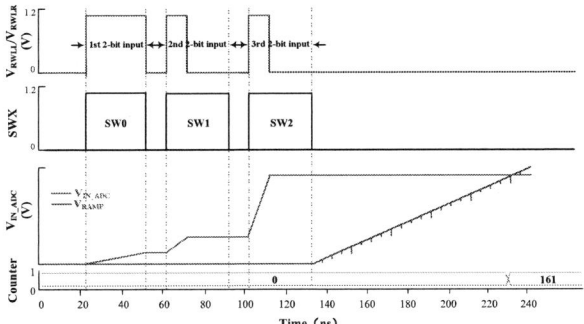

Fig. 6. The waveform of overall simulation.

IV. CONCLUSIONS

In this work, we propose a SOT CIM bit-cell with a significantly improved high-to-low current ratio, achieving up to 52.6× due to operation in the subthreshold region. Introducing high-resistance devices and a complementary MTJ structure reduces array power consumption by 38.6% to 98%. Based on these bit-cell characteristics, we design a multi-bit implementation scheme with 6-bit inputs, 3-bit weights, and 8-bit outputs. This approach is suitable for AI edge inference applications, ensuring efficient computing.

ACKNOWLEDGMENT

This work was supported by the Natural Science Foundation of China (62274008), and Beijing Natural Science Foundation (L223004).

REFERENCES

[1] H. Zhang, L. Jiang, J. Wu, et al., "CP-SRAM: Charge-Pulsation SRAM Marco for Ultra-High Energy-Efficiency Computing-In-Memory," Proceedings of the 59th ACM/IEEE Design Automation Conference (DAC), pp. 109-114, 2022.

[2] S. Kim, S. Um, W. Jo, et al., "Scaling-CIM: eDRAM In-Memory-Computing Accelerator with Dynamic-Scaling ADC and Adaptive Analog Operation," IEEE Journal of Solid-State Circuits (JSSC), 2024.

[3] E. Choi, I. Choi, V. Lukito, et al., "A 333TOPS/W Logic-Compatible Multi-Level Embedded Flash Compute-In-Memory Macro with Dual-Slope Computation," 2023 IEEE Custom Integrated Circuits Conference (CICC), 2023.

[4] D. Chen, Z. Guo, J. Fang, and X. Xue, "A Dual-Mode ReRAM CIM Macro for Low Power Memory-Augmented Neural Networks," 2022 IEEE 16th International Conference on Solid-State & Integrated Circuit Technology (ICSICT), 2022.

[5] H. Cai, Z. Bian, Y. Hou, et al., "33.4 A 28nm 2Mb STT-MRAM Computing-in-Memory Macro with a Refined Bit-Cell and 22.4-41.5TOPS/W for AI Inference," 2023 IEEE International Solid-State Circuits Conference (ISSCC), 2023.

RISC-V Domain-Specific Processor for Accelerating SPHINCS+ on Multi-Core Architecture

Shengnan Zhang[1], Yifan Zhao[1], Xinglong Yu[1], Jun Han[1]

[1]State Key Laboratory of Integrated Chips and Systems, Fudan University, Shanghai 200433, China

Email: junhan@fudan.edu.cn

Abstract—**SPHINCS+ is a hash-based signature scheme selected for post-quantum cryptography(PQC) standardization announced by the U.S. National Institute of Standards and Technology (NIST) in 2022. However, its slow computation time hinders its practical application, making it crucial to enhance its speed. The computations of SPHINCS+ require the use of tweakable hash functions, which can be selected from various hash operators. Among them, SHA-3 is chosen as a widely recognized and standardized hash function by NIST. This work proposes a coprocessor that includes a SHA-3 accelerator and its peripheral structure. Based on the RISC-V instruction set extension, we construct seven custom instructions for this coprocessor, achieving software-hardware co-acceleration. At the same time, the parallelizability of the FORS and WOTS+ within the SPHINCS+ is analyzed. We develop thread-level parallelism inherent in SPHINCS+ through multi-core programming, thereby achieving performance improvement. Finally, synthesis is conducted using the TSMC 28-nm CMOS technology at 800MHz. Compared to the benchmark results of the ARM Cortex-M4, the single-core acceleration for an overall speedup of SPHINCS+ reaches 23.1×, and the multi-core acceleration achieves an additional speedup of 3.4× for the verification process compared to the single-core.**

Index Terms—**Post-Quantum Cryptography(PQC), SPHINCS+, SHA-3, RISC-V instruction extension, parallel computing**

I. INTRODUCTION

The concept of quantum computers was proposed in the 1980s and has seen rapid development in recent decades. Traditional public-key cryptographic algorithms, such as Rivest-Shamir Adleman (RSA), and elliptic curve cryptography (ECC), are based on certain mathematical problems. However, these problems can be solved in polynomial time by quantum computers equipped with Shor's [1] and Grover's [2] algorithms. To address the threat posed by quantum computers to information security, the concept of Post-Quantum Cryptography(PQC) has emerged.

SPHINCS+ is a stateless hash-based signature scheme selected as one of the NIST PQC standardization schemes. As a stateless scheme, it has large signature sizes and long computation times, making research to improve its computational speed highly necessary.

SHA-3 [3] is a cryptographic hash function selected as the winner of the NIST hash function competition and became a standard in 2015. It offers high security and is suitable for hardware implementation; thus, it is chosen as the tweakable hash function used in SPHINCS+ among various hash operations. In SPHINCS+, a huge amount of hashes need to be computed, making the acceleration of SHA-3 one of the primary focuses of this work. Besides, the computation of WOTS+ is performed at the leaf nodes of each layer of the MXSS tree in the SPHINCS+ hypertree structure, while FORS is mainly used to process the input message. They both play significant roles in SPHINCS+ and accelerating them can effectively enhance the speed of SPHINCS+ verification.

In this paper, a dual approach is adopted. First, a SHA-3 accelerator is proposed, primarily targeting the acceleration of the keccak permutation in SHA-3. It is encapsulated into a coprocessor, along with an RISC-V extended custom instruction set to invoke this coprocessor, achieving the first stage of acceleration. Subsequently, we analyze

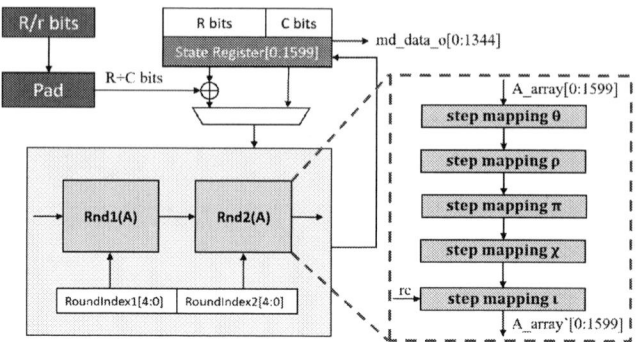

Fig. 1. SHA-3 Accelerator Framework

the parallelizability of WOTS+ and FORS within the SPHINCS+. A multi-core architecture is built on the gem5 [4] and Chipyard [5] platforms, and multi-core programming is implemented to achieve thread-level parallelism for the algorithm, resulting in the second stage of acceleration. The contributions of this work are as follows:

- A novel SHA-3 accelerator is proposed, with each call requiring only 12 cycles.
- Seven custom RISC-V instructions are defined to utilize the built coprocessor, achieving hardware-software co-acceleration.
- Thread-level parallel programming for WOTS+ and FORS in SPHINCS+ is implemented, with configurable core count.
- Multi-core development platforms are built on gem5 and Chipyard, and a combined software-hardware simulation approach is used for validation.
- The single-core implementation achieves an overall speedup of SPHINCS+ by 23.1×, while the eight-core implementation achieves an additional speedup of 3.4 × for the verification process compared to the single-core.

The rest of this paper is organized as follows. Section 2 describes the SHA-3 accelerator and the coprocessor architecture and explains the function of the seven custom instructions. Section 3 analyzes the parallelizability of WOTS+ and FORS, implements thread-level parallel programming, and constructs multi-core SoC architecture. Section 4 compares the simulation results. Section 5 concludes the paper.

II. SHA-3 ACCELERATOR AND COPROCESSOR

A. Design of SHA-3 Accelerator

Figure 1 shows the SHA-3 accelerator constructed in this work. In this diagram, *R* bits refers to the *rate* width, which varies depending on the SHA-3 instantiation type. *r* bits represents the remaining bit length of the input information. *C* bits refers to the capacity bits in the sponge function. *R+C* bits is the *state* width of the *Keccak-p* permutation, which is 1600 bits in total.

979-8-3503-6184-1/24 $31.00 © 2024 IEEE

TABLE I
FUNCTION OVERVIEW OF THE CUSTOM INSTRUCTION SET

Instruction	Function
keccak_init imm	Initialization of s
keccak_vl rd, rs1	Set operation data length
keccak_mode rd, rs1	Set SHA-3 instantiation type
keccak_load rs1	Transfer data of length vl from memory to the accelerator
keccak_store rs1	Transfer data of length vl from the accelerator to memory
keccak_absorb imm	Perform one absorb operation
keccak_squeeze imm	Perform one squeeze operation

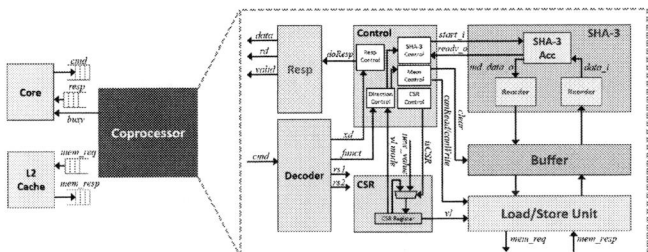

Fig. 2. Coprocessor architecture

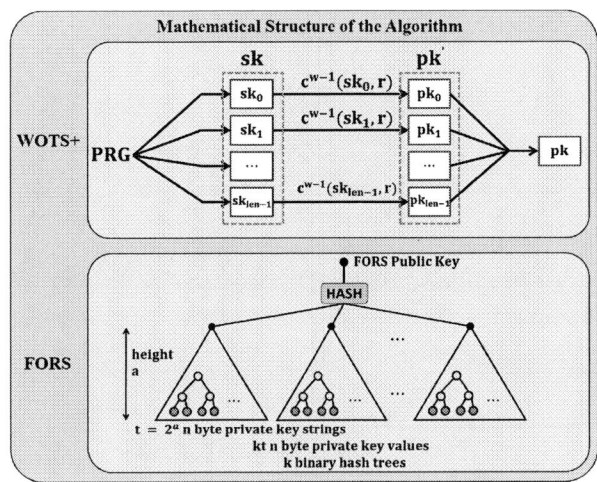

Fig. 3. Demonstration of Algorithm Parallelism

Rnd(A) denotes the five mapping steps required for one iteration. The necessary control signals are provided to *Rnd1(A)* and *Rnd2(A)*. The output of *Rnd1(A)* does not return to the state register but serves as the input to *Rnd2(A)*, while the output of *Rnd2(A)* is stored in the state register, awaiting input to *Rnd1(A)* for the next iteration. Due to the reuse of the *Rnd(A)* module, the runtime of a single call can be halved.

B. Construction of Coprocessor

Figure 2 shows the coprocessor architecture constructed in this paper. Inside the coprocessor, there are several important modules: the Decoder module, Control module, and CSR module, collectively referred to as the decode control module; the SHA-3 module; the Buffer and Load/Store Unit, collectively referred to as the memory access module; and the Resp module. The Decoder module decodes instructions, breaking them down into meaningful bit-field information, which is then passed to the Control module to generate control signals for other modules in the coprocessor. The SHA-3 module invokes the SHA-3 accelerator, handling computation tasks. The Buffer acts as a cache block, storing information transferred from and to memory, as well as providing input and output buffers for the SHA-3 accelerator. The Load/Store Unit manages the information exchange between the coprocessor and memory, storing incoming data in the Buffer according to a specific pattern and retrieving data from the Buffer to memory. Finally, the Resp module generates response information for the core.

C. RISC-V Custom Instruction Set

In this work, we extend the RISC-V instruction set by customizing seven instructions, with their functions detailed in Table I.

The custom instruction set is used to operate the coprocessor as described in the previous subsection. Additionally, we implement the modeling of this custom instruction set on gem5 for subsequent simulations.

III. MULTI-THREAD ACCELERATION

In SPHINCS+, the input information is first processed using FORS, and each leaf node of MXSS trees at every layer of the SPHINCS+ hypertree structure is a WOTS+ key pair. Therefore, they are critical

components of the algorithm during the computation of SPHINCS+. Especially in the verification process, the computations of FORS and WOTS+ chain hashing account for a larger proportion of the total runtime. In this work, we primarily analyze the parallelizability of WOTS+ and FORS within the SPHINCS+ and implement parallel acceleration, thereby accelerating SPHINCS+, which calls these two algorithms.

A. Analysis of Thread-Level Parallelism

Both WOTS+ and FORS possess a natural structure that enables parallel acceleration, which can be exploited for thread-level parallelism.

1) WOTS+: The public key of WOTS+ is derived through a chain of hashes. The computation processes for signing and verification are similar to key generation, where calculations between the elements that compose them are independent, as shown in Figure 3. Therefore, each element can undergo independent chain hashing operations, enabling parallel execution.

2) FORS: The signature of FORS is composed of the private keys at the leaf nodes of the trees and authentication paths. The public key is the hash of all tree roots. The signing process and public key generation occur concurrently in computation and are similar to the verification process. Calculations between trees are independent. Thus, the tree hashing process in each tree can proceed independently, enabling parallel execution, as shown in Figure 3.

B. Multi-core Architecture

A multi-core SoC architecture for parallel acceleration is constructed, as illustrated in Figure 4. Each tile is primarily composed of a core and a coprocessor. Multiple tiles are then connected via the system bus to form the multi-core architecture.

C. Multi-core Programming

In this work, multi-core programming is implemented on gem5 and Chipyard platforms with configurable core counts. For gem5, multithreaded programming is used, where each thread corresponds to a core, and the parallelizable program is distributed across the cores. For Chipyard, tasks are distinguished by the different core IDs, achieving the same program distribution as in gem5. Additionally, this work supports configurable core counts, with options for 2, 4, 8 cores.

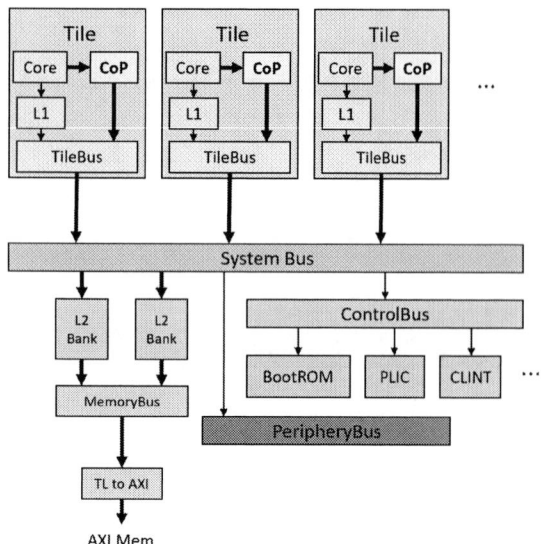

Fig. 4. Multi-core Architecture

TABLE II
ACCELERATION RESULTS OF THE CUSTOM INSTRUCTION SET FOR THE
ALGORITHM

Work	Algorithm	Cycle		
		Key Generation	Sign	Verify
This Work	WOTS+	439718	203008	245946
	FORS	5604758	5604758	262997
	SPHINCS+	3535966	87685367	5235453
Cortex-M4 [6]	SPHINCS+	65652221	2071760810	86340855

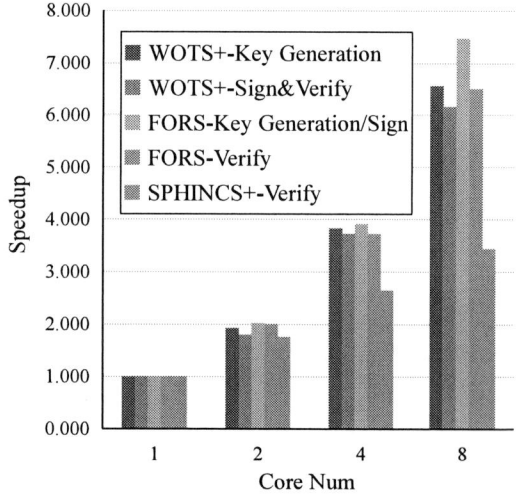

Fig. 5. Multi-core Acceleration Test Results

IV. EXPERIMENT RESULT AND COMPARATIVE ANALYSIS

In this work, we test SPHINCS+-SHAKE256-128f. We use Chipyard and gem5 for collaborative simulation and cross-validation. On gem5 platform, we take advantage of high-speed and large-scale capacity characteristics of software simulation to conduct preliminary tests, exploring the effects of custom instruction sets and multi-core programming on the computation speed. Subsequently, we perform hardware simulations on the Chipyard platform to verify and obtain accurate data, enabling better research and analysis. We evaluate the acceleration results of the custom instruction set for key generation, signing, and verification processes of WOTS+, FORS, and SPHINCS+. These results are compared with the benchmark results of the ARM Cortex-M4 [6], as shown in Table II. The results show speedups of 18.6x, 23.6x, and 16.5x for the key generation, signing, and verification processes, respectively, and a total time speedup of 23.1x.

Next, Figure 5 shows the results of multi-core parallel acceleration, which further improves the running speed based on the initial acceleration. The acceleration is more pronounced for WOTS+ and FORS due to their inherently parallel structures. In SPHINCS+, the verification process sees the most benefit from acceleration because the computations of WOTS+ and FORS occupy a larger proportion of the runtime during verification. A speedup of $3.4\times$ is achieved with eight cores acceleration. Improving the speed of the verification process is crucial for practical applications, such as the TLS 1.3 protocol [7], which is widely adopted to secure internet communication on all types of devices. In such applications, it only requires the verification of all signatures in the certificate chain issued by the Certificate Authority (CA).

Finally, we synthesize the coprocessor under the TSMC 28-nm CMOS technology at 800MHz. The SHA-3 accelerator uses 54.1k equivalent logic gates, while the entire coprocessor uses 93.7k equivalent logic gates.

V. SUMMARY

In this work, we propose an acceleration approach for the SPHINCS+ algorithm. Firstly, a SHA-3 accelerator is introduced, along with the development of a custom RISC-V instruction set and the construction of a corresponding coprocessor, enabling the acceleration through software-hardware co-design. Subsequently, by analyzing WOTS+ and FORS in SPHINCS+, multi-core programming is implemented. Finally, we compare the accelerated results with the benchmark results of the ARM Cortex-M4, achieving significant speedup. The multi-core acceleration for the SPHINCS+ verification process is also remarkable. Last but not least, the synthesis of the coprocessor is implemented.

VI. ACKNOWLEDGEMENTS

This work was supported by the National Natural Science Foundation of China under Grant 61934002 and 62234008.

REFERENCES

[1] Shor, Peter W. "Algorithms for quantum computation: discrete logarithms and factoring." Proceedings 35th annual symposium on foundations of computer science. Ieee, 1994.
[2] Grover, Lov K. "A fast quantum mechanical algorithm for database search." Proceedings of the twenty-eighth annual ACM symposium on Theory of computing. 1996.
[3] Dworkin, Morris J. "SHA-3 standard: Permutation-based hash and extendable-output functions." (2015).
[4] Binkert, Nathan, et al. "The gem5 simulator." ACM SIGARCH computer architecture news 39.2 (2011): 1-7.
[5] Amid, Alon, et al. "Chipyard: Integrated design, simulation, and implementation framework for custom socs." IEEE Micro 40.4 (2020): 10-21.
[6] Kannwischer, Matthias J., et al. "pqm4: Testing and Benchmarking NIST PQC on ARM Cortex-M4." (2019).
[7] Karl, Patrick, et al. "Post-quantum signatures on RISC-V with hardware acceleration." ACM Transactions on Embedded Computing Systems 23.2 (2024): 1-23.

979-8-3503-6184-1/24 $31.00 © 2024 IEEE

Design of an Out-of-Order Superscalar Processor with Improved Register Alias Table Recovery Method

Wu Yang [1], Jun Zhang *[1]

[1] College of Automation, Central South University, Changsha 410006, China

* Email: 3167018755@qq.com, junzhang@csu.edu.cn

Abstract—This paper implements a two-way out-of-order superscalar processor based on the RISC-V instruction set, which supports the RV32IM instruction set. The processor implements out-of-order memory execution to improve performance and uses arbitration of physical register file read ports to reduce resource consumption. Besides, this paper proposes a register alias table (RAT) recovery method based on reorder buffer marking, which can achieve rapid recovery of the RAT with a relatively low resource cost. When running CoreMark and Dhrystone, the average number of cycles the pipeline stalls due to RAT recovery is 0.056 and 0.175, respectively, significantly lower than that of the WALK algorithm. The CoreMark and Dhrystone benchmark scores for the design are 4.41 CoreMark/MHz and 2.29 DMIPS/MHz, respectively.

Keywords—RISC-V, out-of-order, superscalar, recovery of register alias table

I. INTRODUCTION

ARM and x86 dominate the main market of instruction set architectures (ISAs), yet using these two ISAs both involve intellectual property issues, requiring the payment of expensive patent fees, and they only support standardized design. RISC-V is an open-source ISA based on the reduced instruction set computer (RISC) principle, proposed by the University of California, Berkeley in 2010. It offers the advantage of modularity and customizable expansion, enabling refined system development tailored to specific needs. As the RISC-V ecosystem matures, it is gaining popularity in both academia and industry. Superscalar method is a type of pipeline technology that fetches and processes multiple instructions at a time; its essence is the ability to execute instructions independently in different pipelines[1]. Out-of-order execution improves instruction-level parallelism (ILP) by eliminating false data dependencies (write after write (WAW) and write after read (WAR)) and issuing instructions out of program order. Since the proposal of RSIC-V, many RISC-V out-of-order superscalar processors with different objectives and configurations have emerged. BOOM[2], developed by UC Berkeley, is a parameterized out-of-order superscalar processor that supports the RV64GC instruction set. It offers superior performance and occupies a smaller area compared to the similarly configured Cortex-A9 processor. RSD[3] is an FPGA-friendly two-way out-of-order superscalar processor that supports the RV32IM instruction set and is specially optimized to reduce FPGA resource consumption.

To meet the increasing performance demands of processors, this paper proposes a two-way out-of-order superscalar processor based on the RISC-V instruction set. It supports the RV32IM instruction set and implements out-of-order memory access. In addition, this paper proposes a

reorder buffer (ROB) marking based register alias table (RAT) recovery method, which can achieve rapid recovery of the RAT in the event of mispredictions with a relatively small resource overhead.

II. IMPLEMENTATION OF THE PROCESSOR

The processor architecture proposed in this paper is shown in Fig. 1, which has a total of 7 pipeline stages, including instruction fetch (IF), instruction decode (ID), dispatch (DP), issue (IS), register read (RR), execution (EX), and write-back (WB). In the IF stage, instructions are fetched from memory, and branch prediction is performed using the PC-Gshare algorithm[4]. In the ID stage, two decoders are set up for decoding. The tag generator module assigns a branch tag to each instruction to identify which instructions need to be cleared upon misprediction. The DP stage and IS stage form the scheduling module, which selects the instructions to be executed. In the RR stage, instructions receive source operands from the physical register file (PRF) or the bypass network. The execution stage is where the specific functions of the instructions are carried out by various execution units. In the write-back stage, the instruction results are written back to the PRF. The ROB stores instructions following renaming and commits an instruction once it is executed correctly and becomes the oldest instruction.

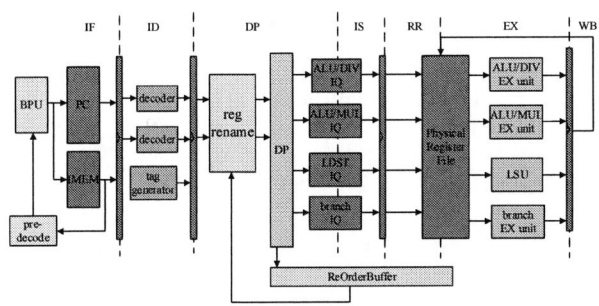

Fig. 1. Microarchitecture of the processor

A. Scheduling Modules

The scheduling module extracts ILP through register renaming and out-of-order issuing, and its structure is shown in Fig. 2. In the DP stage, register renaming is first performed. The freelist allocates a physical register to the destination register and updates the corresponding mapping in the RAT. Source registers read the corresponding physical register identifiers through the RAT, and the busytable is used to determine whether the corresponding physical registers are ready. After register renaming, WAW and WAR dependencies are removed. The dispatch module generates write enable signals to the corresponding issue queue based on

---

The National Natural Science Foundation of China(62274185).

979-8-3503-6184-1/24 $31.00 © 2024 IEEE

the type of the instruction, and the allocation unit searches for free entries in the issue queue to determine the write address.

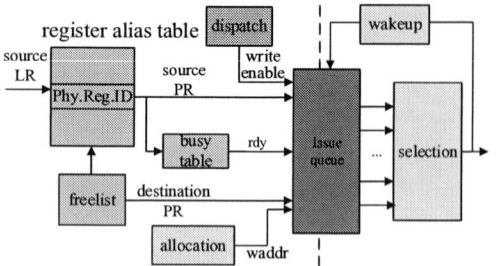

Fig. 2.   Structure of scheduling module

The issue queue (IQ) holds instructions waiting to be executed. Four distributed issue queues are implemented for storing instructions of different types. When the source operands of an instruction are ready and the corresponding execution unit is available, the instruction requests to be issued. The selection module selects an instruction to be sent to the corresponding execution unit according to a specific algorithm. The branch issue queue issues instructions in order, while other issue queues adopt the oldest-first algorithm. The function of wakeup module is to set the dependent source operands ready when a producer instruction is about to produce a result. Single-cycle instructions perform the wakeup operation in the issue stage, while multi-cycle instructions, including multiplication, division, and load, perform the wakeup operation two cycles before the result is obtained. This work implements the oldest-first algorithm based on the age matrix, and the wakeup logic is implemented by comparing the physical register identifier.

B.   Out-of-order Memory Instruction Execution

The execution efficiency of memory instructions is one of the key factors affecting the performance bottleneck of a processor. To enhance performance, this design implements out-of-order memory instruction execution. A special data dependency exists between memory instructions that access the same address, called memory dependency. The memory dependency also includes three types: WAW, WAR, and read after write (RAW). By writing data to memory in order after store instructions committed, WAW and WAR dependencies can be resolved. Therefore, only RAW dependency needs to be checked when executing memory instructions.

The microarchitecture of the load store unit (LSU) is shown in Fig. 3, where the store queue (STQ) stores the access address, data, and other information of store instructions, and the load queue (LDQ) stores the address and other information of load instructions. The execution of memory instructions requires two cycles. In the EX1 stage, the memory access address is calculated, and the STQ, LDQ, and Dcache are accessed. The load instruction writes its access address into the LDQ, and simultaneously searches for the youngest store instruction in the STQ that has the same access address and is older than it. If such a store instruction is found, data forwarding is performed. This is because there is a RAW data dependency between these two instructions. The store instruction writes its address and data into the STQ, and simultaneously searches for the youngest load instruction in the LDQ that has the same access address and is older than it. If such a load instruction is found, mark it in the ROB; flush the pipeline when that instruction commits because it violates the RAW data dependency, rendering the result incorrect. In

the EX2 stage, the results of the memory instructions are obtained.

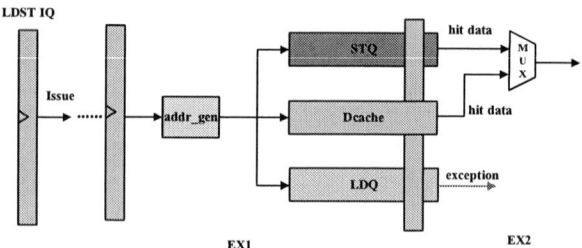

Fig. 3.   Microarchitecture of load store unit

When performing data forwarding and ordering violation detection, there is a need for age information to track relative age among instructions. Reference [5] proposes an age information mechanism for arbitration in the issue queue, which we have applied in LSU. Age information is composed of a sorting bit and a ROB pointer. In the DP stage, each memory instruction is assigned an ROB pointer and a 1-bit sorting bit. The sorting bit is initially set to 1, but when the ROB pointer wraps around from the maximum value back to 0, the sorting bits of all memory instructions in the pipeline are set to 0. Consequently, the memory instruction with the greater age information is considered newer.

C.   Reorder Buffer Marking Based RAT Recovery Method

When a misprediction occurs, it is necessary to recover the RAT to its correct state. Traditional recovery methods include checkpointing and the WALK method. However, checkpointing method has the issue of high resource consumption, and WALK method has the problem of high recovery penalty. This paper proposes a ROB marking based RAT recovery method, which can achieve rapid recovery of the RAT with small resource consumption.

The WALK method is carried out as follows: Begin at the newest instruction in the ROB, reading the previously mapped physical register identifiers one by one, and then write these identifiers back into the RAT. Continue until the instruction corresponding to the misprediction is reached, allowing the RAT to be recovered in the reverse order of modification. However, not every instruction's previously mapped identifier is valid for recovery. Some instructions leave the mapping unaltered, rendering their identifiers ineffective for recovery. Additionally, some instructions correspond to the same logical register, and their modifications to the mapping table will be overwritten by each other during the recovery process. Only the previously mapped physical register identifiers of instructions that initially changed the mapping of a logical register are valid. The RAT recovery method proposed in this paper is implemented based on this principle.

The specific implementation is as follows: In the DP stage, an ARF_GC and a ROB_GC are allocated for each branch instruction, and these resources are reclaimed in the EX stage. Each bit in the ARF_GC is associated with a logical register, and each bit in the ROB_GC corresponds to an entry in the ROB. The ARF_GC is used to mark logical registers whose mappings have been changed, initially set to all zeros. In the DP stage, when a destination register is renamed, the bit in the ARF_GC designated by destination register is set to 1. The ROB_GC is used to mark the ROB entries corresponding to the instructions that first changed the mapping of a logical register. The principle for determining whether an instruction

979-8-3503-6184-1/24 $31.00 © 2024 IEEE

meets this condition is whether the corresponding bit in the ARF_GC is 1.

This design includes 64 physical registers and a ROB with 44 entries. If the checkpointing method is adopted, each checkpoint requires 31 times 6 equals 186 flip-flops (FFs), while the method proposed in this design requires only 31 plus 44 equals 75 FFs, resulting in a saving of 111 FFs.

D. Arbitration of Physical Register File Read Ports

The processor has a total of 4 issue queues, with each queue issuing one instruction per cycle. To ensure that instructions are not stalled due to a lack of access ports, it would be necessary to set up 8 read ports for the PRF. However, the resources consumed increase with the number of ports. The likelihood of actually needing 8 read ports is quite low for several reasons: 1) Not every cycle will see all 4 issue queues issuing instructions. 2) Some operands are immediates. 3) Some register operands are read from the bypass network rather than the PRF. Therefore, this design only sets up 4 read ports, with read port arbitration occurring at the issue stage. Instructions that do not obtain enough ports are stalled, and the priority of read port arbitration is shown in TABLE I. This paper employs the I-LVT algorithm proposed in [6] for implementing the PRF.

TABLE I. PRIORITY OF READ PORT ARBITRATION

read port	Priority		
	first	*second*	*third*
1	rs1 of LDST	rs1 of ALU/DIV	rs2 of ALU/DIV
2	rs2 of LDST	rs1 of ALU/DIV	rs2 of ALU/DIV
3	rs1 of branch	rs1 of ALU/MUL	rs2 of ALU/MUL
4	rs2 of branch	rs1 of ALU/MUL	rs2 of ALU/MUL

The principle for determining whether the register operands will be read from the bypass network is whether the instruction is issued within two cycles after the operand is woken up. This is because, under this condition, when the dependent instruction is in the RR stage, the producer instruction is in the EX or WB stage.

III. RESULT

When running the CoreMark and Dhrystone benchmarks, the recovery of the RAT is performed using both the WALK method and the recovery method proposed in this paper separately. The number of mispredictions and the number of cycles the pipeline stalls due to the recovery of the RAT are recorded. The average number of cycles the pipeline stalls per misprediction is calculated, and the results are shown in TABLE II. The experimental results demonstrate that the misprediction recovery penalty of the method proposed in this paper is significantly less than that of the WALK method.

TABLE II. STATISTICS OF RECOVERY PENALTY

Recovery method	*program*	*misprediction*	*stall*	*Average*
WALK	coremark	7971	7501	0.941
WALK	dhrystone	740	884	1.195
proposed	coremark	8144	453	0.056
proposed	dhrystone	742	130	0.175

The comparison of CoreMark and Dhrystone scores between this design and some other similarly configured superscalar processors is shown in TABLE III. The comparison results indicate that this design has superior performance.

TABLE III. COMPARISON OF BENCHMARK

Processor	Out-of-order	*Dhrystone (DMIPS/MHz)*	*CoreMark (CoreMark/MHz)*
Reference[7]	N	1.06	3.84
BOOM	Y	——	3.91
RSD	Y	2.04	——
SweRV_EH1[8]	Y	——	4.34
This work	Y	2.29	4.41

The resource consumption is evaluated using the Xilinx XC7Z020-CLG484-1. The comparison of resource consumption between this design and other processors is shown in TABLE IV, and the resource consumption of BOOM is from[3]. The results show that this design has a low FPGA resource overhead, slightly lower than that of RSD. The maximum achievable frequency with this FPGA is 75 MHz.

TABLE IV. COMPARISON OF FPGA RESOURCE

Processor	*LUT*	*FF*
Reference[7]	30125	8532
BOOM	>40000	>20000
RSD	15379	8584
SweRV_EH1[8]	27989	13897
This work	15186	5926

IV. CONCLUSION

This paper implements a 2-way out-of-order superscalar processor based on the RISC-V instruction set, supporting the RV32IM instruction set. The experimental results show that this processor achieves high performance. To achieve rapid recovery of the RAT, this paper proposes a ROB marking based RAT recovery method, which serves as a compromise between WALK and checkpointing methods, providing much lower recovery latency compared to WALK and smaller resource consumption compared to checkpointing.

REFERENCES

[1] Y. He and X. Chen, "Survey and Comparison of Pipeline of Some RISC and CISC System Architectures," *2023 8th International Conference on Computer and Communication Systems (ICCCS)*, Guangzhou, China, pp. 785-790, April. 2023.

[2] C. Celio, D A. Patterson, and K. Asanovic, "The berkeley out-of-order machine (boom): An industry-competitive, synthesizable, parameterized risc-v processor," University of California at Berkeley Berkeley United States, Tech. Rep, 2015.

[3] S.Mashimo, A.Fujita, R.Matsuo, et al, "An open source FPGA-optimized out-of-order RISC-V soft processor," 2019 International Conference on Field-Programmable Technology (ICFPT), pp.63-71 ,Dec.2019.

[4] H. K. Kim, H. S. Kim, C. M. Eun, H. H. Cho and O. H. Jeong, "A high-performance branch predictor design considering memory capacity limitations," 2017 International Conference on Circuits, System and Simulation (ICCSS), London, UK, pp.49-53, Jul. 2017.

[5] A. Buyuktosunoglu, A. El-Moursy, and D.H. Albonesi, "An oldest-first selection logic implementation for non-compacting issue queues," 15th Annual IEEE International ASIC/SOC Conference, pp.31-35, Sep.2002.

[6] A.M.S. Abdelhadi, and G G.F. Lemieux, "Modular multi-ported SRAM-based memories," Proceedings of the 2014 ACM/SIGDA international symposium on Field-programmable gate arrays, pp.35-44, Feb.2014.

[7] G. T, A. Muraleedharan, and K. Varghese, "Design of a 32-bit, dual pipeline superscalar RISC-V processor on FPGA," 2020 23rd Euromicro Conference on Digital System Design (DSD), pp. 340-343, Aug.2020.

[8] Western Digital."VeeR EH1 RISC-V Core." github. Accessed: Jun. 20, 2024. [Online.] Available: https://github.com/chipsalliance/Cores-VeeR-EH1

An SDPF RISC-V Processor with Two-stage Pseudo-pipelined Architecture for IoT Applications

Wenji Mo, Yuchen Wang, Haoning Sun, and Jingjing Liu*

School of Electronics and Communication Engineering, Sun Yat-sen University, Shenzhen, China

* Email: mowj28@mail2.sysu.edu.cn, liujj77@mail.sysu.edu.cn

Abstract—Internet of Things (IoT) nodes are required to execute lightweight tasks for sensing information and simple signal processing. Thus low-power processors serve as crucial components of highly integrated IoT smart sensors. This paper proposes a low-power RISC-V instruction set architecture (ISA) based RV32I processor for embedded IoT nodes, which follows the serial data path. To enhance the performance of the serial data path followed (SDPF) processor, a pseudo-pipelined architecture is proposed. By partitioning and combining a portion of the instruction lifecycle tasks into two stages, a two-stage pseudo-pipelined structure is achieved, reducing the number of cycles per instruction (CPI) of the SDPF processor. The proposed processor is designed based on Verilog HDL, the implementation results on FPGA demonstrate that the performance is improved by 58% compared to the conventional SDPF RV32I processor. Besides, the synthesis is performed using a standard 0.18 μm CMOS process technology. Post-layout simulation results demonstrate that, the System on Chip (SoC), which consists of the proposed processor, a 2 kB IMEM and a 4 kB DMEM, achieves an average CPI of 36, an area of 0.988 mm², and an average power consumption of 182 μW/MHz.

Keywords—RISC-V, serial data path, pseudo pipelined, low power, IoT

I. INTRODUCTION

Embedded IoT nodes employed for information sensing and signal processing are required to operate continuously for extended periods in energy-constrained environments [1]. These devices have relatively low performance requirements and are highly sensitive to power consumption. Consequently, a low-power processor is necessary to manage the operation of the entire system. The RISC-V, an emerging open-source ISA [2], offers flexibility that allows users to remove redundant hardware and implement the instruction set in a modular manner. This enables the design of low-power processors specifically tailored for IoT sensor nodes. In the low-power design of RISC-V processors, retaining only the RV32I integer instruction set is an important approach. Moreover, power consumption and area can be reduced through various architectural designs. The single cycle design can effectively decrease power consumption but may lead to timing criticality in critical paths [3]. An alternative strategy is to apply a multi-cycle non-pipelined design, which reduces dynamic power by decreasing the toggle rate [4]. However, this approach introduces more registers, thereby increasing the chip area.

Traditional processors follow parallel data transfer techniques, but they face limitations in reducing power consumption and area. On the other hand, the processors based on serial data paths offer an attractive alternative [5]. By processing instructions in a serial data path with reduced register utilization, this approach can achieve lower power consumption and smaller area compared to parallel path processors. However, converting 32-bit wide data into serial form and transmitting it to the serial path require at least 32 cycles, which makes the implementation of a conventional pipeline architecture challenging, leading to a further decrease in core performance. This paper proposes an RV32I processor that follows a serial data path and features a pseudo-two-stage pipelined structure. By employing additional shift registers (SRs), the processor divides and combines a portion of the tasks within the instruction cycle into two stages, simulating a two-stage pipeline structure and reducing the CPI of the proposed SDPF processors. While delivering superior performance, the proposed design maintains low power consumption and a small area footprint. This paper is organized as follows: Section II introduces the pipeline simulation scheme and the data path design. Section III discusses the simulation results and Section IV concludes the paper.

II. DESIGN OF THE PROPOSED SDPF RISC-V PROCESSOR

A. Two-Stage Pseudo-Pipelined Structure

In previous work, a RISC-V processor adhering to a serial data path was first introduced in SERV, which is an open-source RV32I processor. In SERV, multiple SRs are employed to convert the data to be processed into a serial format and transmit it to other modules for serial execution. The core computes one bit of data per clock cycle, and the results of operations are written back to the register file in a serial manner. Consequently, in the RV32 baseline, the core requires at least 32 clock cycles to complete the computation of data or address. The architectural paradigm poses significant challenges in implementing a conventional pipelined processor structure within a serial data path. Furthermore, the overall architectural design of SERV resembles that of a traditional multi-cycle, non-pipelined processor. This implies that SERV requires the complete execution of the current instruction before fetching a new one, which leads to a significant degradation in performance. In order to solve this problem, this paper introduces a two-stage pseudo pipeline architecture aimed at optimizing the core implementation. The proposed architecture is illustrated in Fig. 1. The structures A and B represent two pseudo-pipeline architectures employed for processing different instructions in RV32I instructions set. The instruction processing flows of both the A and B structures commence with the *Instructions Fetch* (IF) stage. Each IF and *Instruction Decode* (ID) occupy one clock cycle respectively. Based on the decoding results, the core determines the subsequent execution flow. If the current instruction is a conditional-dependent instruction, it is processed according to the B structure. Otherwise, it follows the structure A for processing. In the structure A, the core determines the subsequent execution order based on the decoding results. If the current instruction is a non-source-

---

This work is supported by National Natural Science Foundation of China with project number 6217418.

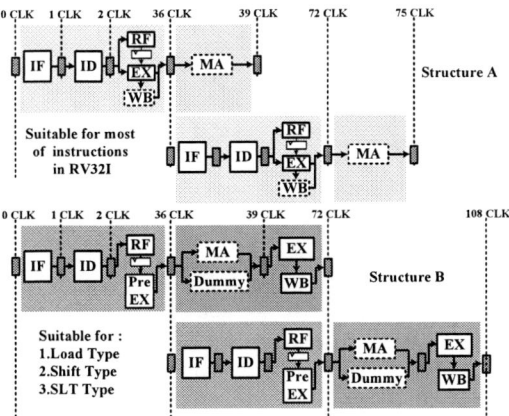

Fig. 1. The proposed two-stage pseudo-pipelined structure.

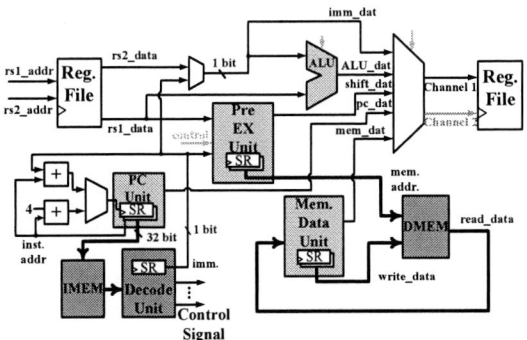

Fig. 2. The data path of the proposed SDPF RISC-V processor.

register-dependent instruction, it proceeds directly to the *Execution* (EX) stage; otherwise, *Register Files* (RF) reading need to be performed first. The RF, EX, and write-back (WB) are embedded within the same time frame. The register file transmits data bit-by-bit starting from the least significant bit (LSB), and it takes two cycles for the first bit of data to reach the EX module. During this phase, the core performs bit-wise computations on the serial input data in each clock cycle and continuously writes the new execution results back to the register file bit-by-bit. This entire process requires 32 clock cycles. After completing the WB, the core transitions into the second stage. The second stage of the structure A involves only *Memory Access* (MA), which is executed exclusively when processing S-type instructions. The core spend two cycles accessing the memory via the bus and writes data to the memory in the third cycle.

The structure B is a pseudo-pipelined architecture that is followed when processing load type (L-type), shift type and the *Set Less Than* (SLT) type instructions. Unlike the structure A, after reading the register files in the first stage, Pre-Execution (Pre–EX) is required, and the EX and WB is scheduled in the second stage. The Pre-EX is responsible for calculating the data that will be used in the EX. This stage requires 32 clock cycles to compute and determine the conditions necessary for the instruction. During this stage, if processing L-type instructions, the core calculates the address for MA. If processing SLT-type instructions, the core compares the two specified numbers. For shift-type instructions, although condition evaluation is not required, 32 clock cycles are needed to retrieve the data to be shifted and store it in the SR before performing the subsequent specified shift operations in the second stage. The second stage pseudo-pipeline encompasses three tasks: MA, EX, and WB. Among the three types of instructions involved in Structure B, only L-type instructions require the inclusion of the MA stage. Similar to S-type instructions, L-type instructions also necessitate three clock cycles to complete data retrieval. Furthermore, when processing the remaining two types of instructions, three dummy clock cycles are inserted during the original MA time slot to align with the timing of L/S-type instructions, thereby maintaining the consistency of the pipeline structure. The fundamental difference between processing instructions using Structure A or Structure B lies in handling condition-dependent instructions. Structure A is unable to simultaneously perform condition evaluation and the calculations of write-back data within the serial data path. To maintain the same pipeline structure, the two stages that

consume the most clock cycles are placed separately in the two stage of the pseudo-pipeline, resulting in two distinct methods of instruction processing.

Since the SERV processor executes instructions in either one or two stage and its design resembles a multi-cycle non-pipelined architecture [5], the CPI of SERV can be represented as Equation (1).

$$CPI_{SERV} = \frac{37f_1 + 71f_2}{f_1 + f_2} = 37 + \frac{34}{1 + \frac{f_1}{f_2}} \tag{1}$$

where f_1 denotes the number of one stage instructions, while f_2 represents the number of two stage instructions. The proposed processor, on the other hand, has a CPI of 36. This shows that the proposed processor have higher performance by reducing the CPI.

B. Data Path of the Proposed Processor

The data path architecture based on the aforementioned pseudo-pipelined structure is illustrated in Fig. 2, modules with a blue background indicate that data will be processed and computed serially within these modules, while modules with a red background signify that data is stored or computed in a 32-bit format. The thick lines in the figure represent data transmission in a 32-bit parallel manner, whereas the thin lines denote data transmission in a 1-bit serial manner. The instructions are first fetched from the instruction memory (IMEM) and then sent to the Decode Unit for decoding, which subsequently generates a series of control signals to drive the operation of the remaining core modules. Besides, Decode Unit converts the immediate value contained in the instruction signal via the SR and serially transmits it to the data path for subsequent module usage. All data that needs to be written back to the RF is connected to a multiplexer, which is controlled by the control signals generated by the Decode Unit. Based on the instruction type, the multiplexer selects the required data and serially writes it back to the register file. The MA operation is mainly handled by the Pre-EX Unit and the MEM Data Unit. The address for MA is generated by the SR in the Pre-EX Unit, and the data required for S-type instructions is generated by the SR in the MEM Data Unit. The core sends address signals, data signals, and handshake signals to the data memory (DMEM) via the bus. Unlike cache, the DMEM consists of an SRAM directly connected to the bus. Ideally, it takes three clock cycles from the core initiating the handshake to the DMEM returning a response signal. Additionally, when processing L-type instructions, the data returned from the DMEM is loaded into the SR of the MEM data Unit and converted to serial output, waiting to be written back to the register file.

979-8-3503-6184-1/24 $31.00 © 2024 IEEE

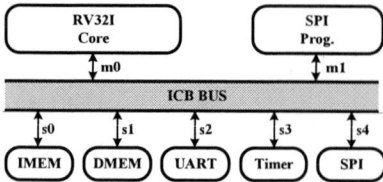

Fig. 3. Block diagram of SoC implementation for processor measurement.

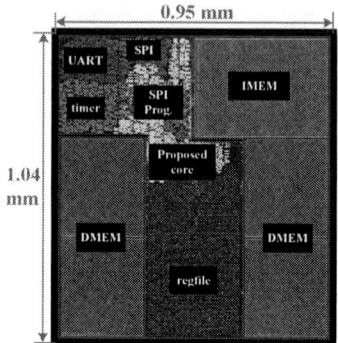

Fig. 4. The layout of the minimalist SoC including the proposed processor.

TABLE I. Experimental Results of RISC-V Minimalist SoC Implementation on FPGA.

SoC	LUTs	FFs	Static Power (mW)	Dynamic Power (mW)	Dhrystone	Coremark
SERV	493	533	157	114	0.034	0.023
This work	903	795	158	115	0.053	0.037

processor occupies an area of 0.988 mm². The post-layout simulation results indicate that, under an operating condition of 1.8 V, its average dynamic power consumption is 182 µW/MHz. Compared to the parallel data path followed (PDPF) RV32I processors, such as [3], the proposed processor reduces area by 33% and dynamic power consumption by 37%, and in comparison to [7], the proposed processor reduces area by 25.7% and dynamic power consumption by 73%. In comparison to SERV, the proposed processor exhibits an advantage in terms of average CPI. Based on the Coremark and Dhrystone results, it can be estimated that the actual CPI of SERV is approximately 62. The proposed processor, on the other hand, has a CPI of 36, which indicates that the proposed processor has a higher instruction processing efficiency. These characteristics make the proposed processor well-suited for operation in low-power IoT nodes.

III. IMPLEMENTATION RESULTS AND ANALYSIS

A. FPGA Implementation Result

To obtain meaningful experimental results, the proposed SDPF RISC-V processor has been tested within a minimalist SoC. As shown in Fig. 3, the SoC contains a 2 kB IMEM and a 4 kB DMEM, which are enough for the target application. Several communication interfaces are employed to transmit the results to the host computer. The internal interconnect of the SoC adopts the Internal Chip Bus (ICB) [6] from the Hummingbird E203 processor, as highlighted in blue. This is a custom bus suitable for low power processor cores with only two independent handshake channels, and users can customize priorities to implement multi master and multi slave peripheral mounting. SERV, an open-source RISC-V processor that also follows a serial data path, will be used as a suitable benchmark for comparison. The SERV core will be placed in an identical SoC and configured with its default settings.

Table I presents the test results obtained from a testing environment based on the Xilinx KC705 device. Dhrystone and CoreMark are two benchmark programs whose scores are positively correlated with performance. Compared to the SERV based SoC, the proposed SoC based on the SDPF RISC-V processor consumes more LUTs and FFs. This can be attributed to the additional SRs used in the pseudo-pipeline structure for jump address calculation and data computation in the Pre-EX module, which necessitate additional control logic. Consequently, this leads to higher static and dynamic power consumption compared to the SERV. However, compared to SERV based SoC, the proposed processor based SoC achieves an average performance improvement of 58%. Therefore, the additional power consumption and area overhead are acceptable in light of the performance improvements achieved.

B. Area and Power of the Proposed Processor

The proposed SDPF RISC-V processor were synthesized and placed-and-routed using standard 0.18 µm CMOS process with 1.8 V standard cell library. Fig. 4 shows the layouts of the minimalist SoC. The SoC incorporating the proposed

IV. CONCLUSION

This paper proposes a method to improve the performance of RISC-V processors that follow a serial data path. A two-stage pseudo-pipeline structure is proposed to reduce the CPI of the SDPF processors. The implementation result shows that the proposed processor improves computational performance by 58% compared to the SERV, while reducing dynamic power consumption by at least 33% compared to the PDPF processors. This validate the effectiveness of the pseudo-pipelined structure. The proposed SDPF RISC-V processor exhibits a high degree of suitability for IoT smart sensor applications.

REFERENCES

[1] C. Seok, M. M. Mahmud, M. Kumar, O. J. Adelegan, F. Y. Yamaner, and Ö. Oralkan, "A Low-power Wireless Multichannel Gas Sensing System Based on a Capacitive Micromachined Ultrasonic Transducer (CMUT) Array," *IEEE Internet of Things Journal*, vol. 6, no. 1, pp. 831–843, Feb. 2019.

[2] A. Waterman, Y. Lee, R. Avizienis, D.A. Patterson, and K. Asanovic, The RISC-V Instruction Set Manual Volume II: Privileged Architecture Version 1.9, Tech. Rep. UCB/EECS–2016–129, EECS Department, University of California, Berkeley, 2016.

[3] S. Shukla, P. K. Jha, and K. C. Ray, "An Energy-efficient Single-cycle RV32I Microprocessor for Edge Computing Applications," *Integration*, vol. 88, pp. 233–240, 2023.

[4] J. Saussereau, C. Leroux, J. B. Begueret, and C. Jego, "AsteRISC: A Size–Optimized RISC-V Core for Design Space Exploration". *2023 IEEE International Symposium on Circuits and Systems*, Monterey, CA, USA, 2023.

[5] M. Sarmiento, K. Nguyen, C. Duran, and R. Serrano, "Systems on a Chip With 8 and 32 Bits Processors in 0.18–µm Technology for IoT Applications," *IEEE Transactions on Circuits and Systems II: Express Briefs*, vol. 69, no. 5, pp. 2438–2442, May 2022.

[6] "Open-source Hummingbirdv2 E203 RISC-V processor core and SoC," github.com. [Online]. Available: https://github.com/riscv-mcu/e203_hbirdv2.

[7] F. F. Nascimento, R. N. Wuerdig, A. F. Ponchet, and B. Sanches, "RISC-V SoC Physical Implementation in 180 nm CMOS with a Quark Core Based on FemtoRV32," *2023 IEEE Seventh Ecuador Technical Chapters Meeting*, Ambato, Ecuador, 2023, pp. 1–6.

A Unified Verification Scheme for the Acceleration of RISC-V Processor Design

Zixiang Chen[1], Jiyuan Bai[1], Yueru Yu [1], Gengsheng Chen[1,2], Xiaofang Zhou[*,1]

[1] School of Microelectronics, Fudan University, Shanghai 201203, China
[2] Jiashan Fudan Institute, Jiaxing, Zhejiang Province, China

* Email: xiaofangzhou@fudan.edu.cn

Abstract—**This paper presents a unified verification scheme for design and verification of large-scale complex digital systems. The new scheme integrates the software simulation environment and the hardware accelerating resources to undertake a unified and automated verification of large-scale digital circuit systems, encompassing test generation, result comparison, and coverage analysis. The non-synthesizable UVM test bench is coupled with the hardware emulator to constitute a generalizable and flexible verification platform. By mapping RTL code to the emulator, the whole simulation process receives a significant acceleration. Besides, a thorough error analysis and coverage assessments validate the effectiveness of the scheme, providing a substantial feedback for designer's successive optimization. We apply this scheme to the verification task of a RISC-V processor design, achieving an up to 5.69x speedup and a comprehensive coverage rate of 93.80%, 18.92% increase over the conventional methods.**

Keywords—UVM, Palladium Z1, co-simulation, RISC-V

I. INTRODUCTION

Over the years, various technologies have been proposed to improve the verification process, evolving from Verilog/ VHDL testbenches to System Verilog testbenches and then to UVM (Universal Verification Methodology)[1] testbenches. Kabilan et al.[2] conducted a comparative study on these kinds of testbenches, pointing out that UVM Testbench offers an approximately 8 times faster speed with a higher coverage.

With the continuous growth in both size and complexity of modern circuit systems, researchers spend numerous efforts on accelerating the design & verification using UVM. CORE-V-VERIF[3], an open-source verification project supported by OpenHW, takes the approach of leveraging the modular and reusable advantages of UVM to construct a configurable verification platform for the validation of RISC-V processors with diverse features. Despite its remarkable flexibility and generalizability, there still leaves a big room for optimization in error detection and result presentation. Ehrlich et al. [4] introduces a hardware-in-the-loop verification method which runs UVM-SystemC for stimulus generation during real-time verification. Their method has a nonnegligible limitation in circuit scale and debugging visibility, making it difficult to identify errors during the verification process. Mohammad Ismael et al.[5] propose their AUTG tool (Automatic UVM Testbench Generator) for automatic generation of UVM test platforms, however, with a limited test coverage. Although the aforementioned approaches have optimized simulation speed to a certain extent, the traditional software simulation methods are obviously inadequate to meet the high demands from the escalating complexity of chip design on both the processing speed and test coverage. As a result, more recent works have been raised to study the joint use of hardware emulators in their automatic verification platforms. Ideally, the testbench and the DUT should be placed both on the hardware side. But

due to the non-synthesizable nature of UVM, designers have to leave their testbenches on the software side. To address this issue, Ruan et al.[6] propose a method which synthesizes all the behavioral components of the testbench, enabling both the testbench and the DUT to be executed at a high frequency on the hardware simulator. Ruan's method helps to reach a much higher verification speed. But it requires a substantial time and effort to synthesize all the behavioral parts of the testbench.

On the other hand, due to its inherent open-source nature, RISC-V architecture has received a wide acceptance in both AIot and cloud computing. In the past decade, various kinds of RISC-V based processors are under development with high quality pursuit and short development cycle. Confronted with diverse and intricate demands, design diversity emerges as a big challenge, necessitating the implement of efficient and rapid verification strategies for large-scale complex digital designs.

To address the above challenges, in this paper, by using a RISC-V based processor as the target under test, we propose a unified verification scheme to accelerate the verification process with enhanced test coverage. Our main contributions are as follows.

- We propose a unified verification scheme to accelerate the design verification. By using Palladium Z1 (PZ1)[7] as the hardware emulator, the new verification scheme significantly reduces the simulation verification time, achieving a maximum acceleration of up to 5.69 times compared to software simulation.

- We apply the proposed verification scheme for the test of a self-developed RISC-V based processor, achieving a comprehensive coverage of 93.80%.

II. HARDWARE ACCELERATION

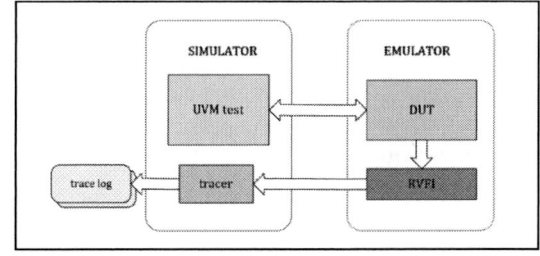

Fig. 1. Simulation accelerate mode

Fig. 1 is a typical simulation work flow using hardware emulator. The synthesizable portion of the RTL code (DUT and RVFI) is placed and conducted on the hardware emulator for acceleration, and the non-synthesizable parts are simulated on software simulator. In such a scenario, due to the existence of the communication overhead between the software side and

979-8-3503-6184-1/24 $31.00 © 2024 IEEE

the hardware side, the overall verification speed is severely limited by the software simulation speed and the message exchange. The efficiency can be improved by reducing the communication and the data dependencies between software and hardware. Since UVM is a class library of SystemVerilog, it is inherently non-synthesizable.

A. Compilation Optimization

As shown in Fig. 2, under the SA (Simulation Acceleration) mode, the RTL code to be deployed on hardware is initially compiled through the Cadence vlan process, followed by a further compilation using Cadence ixcom which sets the top-level module of the hardware deployment and specifies the hierarchy between hardware and software. An EMU database is then generated for simulation acceleration (xeDebug). To achieve a precise control over the compilation process, a step-by-step compiling strategy is applied to make the compilation process more controllable, facilitating the subsequent tasks of compilation management and error debugging.

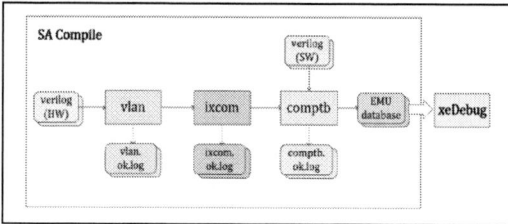

Fig. 2. SA mode compilation proscess

B. Communication Optimization

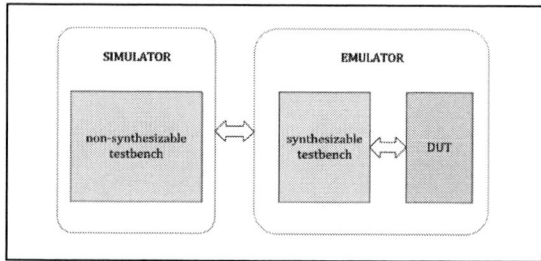

Fig. 3. The segmentation of testbench

As shown in Fig. 3, to reduce the communication overhead between hardware and software, the testbench is divided into synthesizable and non-synthesizable components. Those non-synthesizable are executed at the software side, leveraging its flexibility and processing capabilities, while the synthesizable portion is mapped to the hardware accelerator, harnessing the high-speed execution prowess of hardware. This segregation fosters a more efficient utilization of hardware accelerating resources by optimizing the communication through reduced unnecessary data transfers.

For instance, in traditional testing methodologies, clock signals are generated manually on the Testbench (software). However, for a joint simulation on PZ1, this approach is not recommended. Generating clock signal in software requires the hardware to retrieve it from the software environment for every cycle, resulting in a dramatic decrease in performance [8]. To enhance the performance, in this work, clock signals are generated on the hardware side, eliminating the need to retrieve them from the software side. To further reduce the software-hardware communication, we eliminate unnecessary

information printing, thereby minimizing the communication traffic between software and hardware.

III. VERIFICATION ENVIRONMENT

A. Basic Verification Scheme

Fig. 4 illustrates a comprehensive verification scheme for the design of a RISC-V processor. To apply an integrated and unified verification strategy that combines formal verification and UVM verification, we design a standard interface (RVFI) for formal verification. This provides a standardized method for designers to trace the instruction's execution status of the processor during the formal verification tests.

The scheme integrates an open-source random instruction generator (riscv-dv)[9] as the core of the random verification strategy. The test cases are randomly generated by riscv-dv and then forwarded to the golden model Spike[10] to generate the reference logs. The scheme compares the trace logs and the reference logs, checking the differences between the processor's actual performance and its expected behaviors.

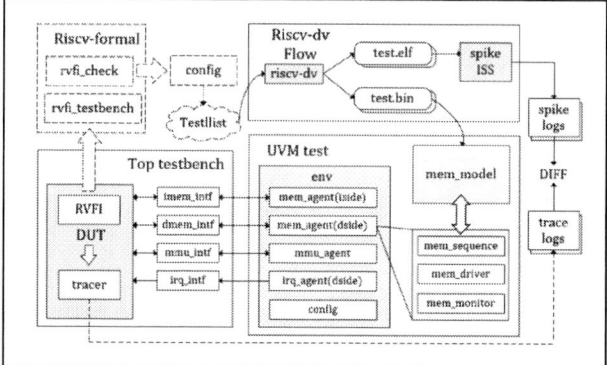

Fig. 4. Basic verification scheme

B. Integration of RISC-V Formal

RVFI (RISC-V Formal Interface) is a dedicated interface for the design verification of RISC-V processors, providing a standardized approach for describing and examining the target behaviors during its design and verification. Through RVFI, the actual behaviors of the DUT are collected and compared with their specifications to ensure that all the implements are completely consistent with their specifications.

We integrate riscv-formal [11] to the verification scheme via RVFI, using its output results to guide riscv-dv on creating test cases. The iteration count of test cases can be dynamically adjusted based on the results of formal verification.

C. Co-simulation System with ISS

The co-simulation platform for processor design, built on the aforementioned unified verification scheme, incorporates the integration of an ISS (instruction set simulator) with RTL-level simulation. Using a scoreboard to compare the behaviors between RTL simulation and ISS simulation, this platform significantly enhances verification efficiency and accuracy in problem probing. In the case when the execution results differ between hardware and software, the verification platform promptly halts the simulation and reports the error occurrence. Compared with the prior approach in which trace logs are analyzed after program execution, this method dramatically accelerates the simulation with instant feedbacks and a precise positioning of the errors.

979-8-3503-6184-1/24 $31.00 © 2024 IEEE

D. Automation Scripts and Scheme Reuse

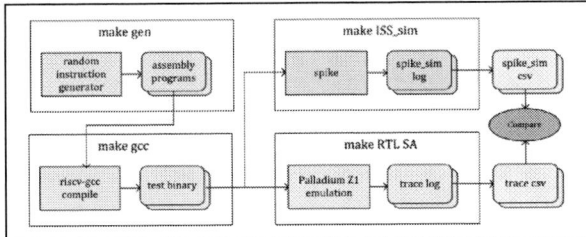

Fig. 5. Automation script flowcharts

As illustrated in Fig. 5, all the aforementioned tools are automatically executed through scripts to generate test reports. We develop a configuration file to inform the test platform about the features of the processors under test, which enables the whole verification process for the processors with different features to be simplified by a simple manipulation of editing the configuration file. The whole procedure applies not only to the design and verification of RISC-V based processors, but also to the other large-scale digital systems.

IV. EXPERIMENTAL RESULTS

A. Performance Comparison

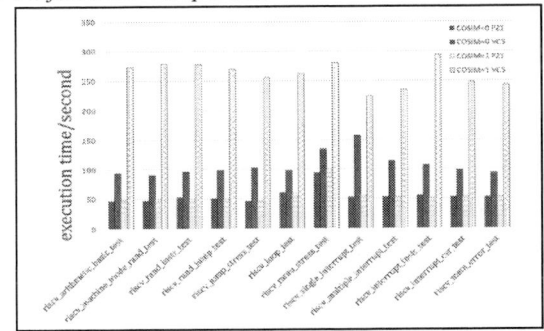

Fig. 6. Comparison of PZ1 and software simulation times

We use this unified verification scheme on our design of a RISC-V based processor. The basic simulation version and the co-simulation version of a total of 12 test cases are executed by using software simulator and hardware accelerator PZ1 respectively. Fig. 6 gives the running time result. In terms of overall runtime, against the basic version, PZ1 achieves an average acceleration of 1.96 times in comparison with simulation, with a maximum acceleration of 2.97 times and a minimum acceleration of 1.43 times. For the co-simulation version, PZ1 achieves an average acceleration of 4.86 times in comparison with simulation, with a maximum acceleration of 5.69 times and a minimum times of 2.81 times.

B. Coverage

Fig. 7 illustrates a statistical analysis of the test coverage for the target processor, using the conventional targeted testing and the random instruction testing using the new verification scheme proposed in this paper. The comprehensive coverage achieved with the conventional targeted testing (test suites: rv32ui-v, rv32um-v) is 74.88%, while the random instruction testing achieves a much high coverage of 93.80%, indicating

a prominent improvement of 18.92%. The random instruction testing method significantly enhances the coverage of various code metrics (lines, branches, states, toggles) and functional aspects (assertions, groups) of the complex circuits under test.

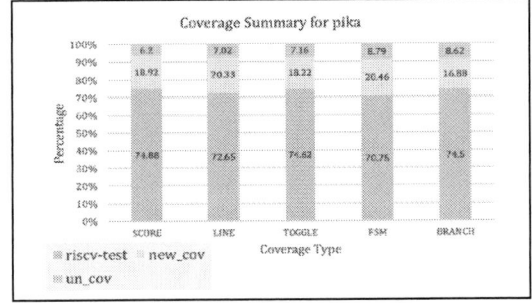

Fig. 7. Comparison of our work with riscv-tests coverage effectiveness

V. CONCLUSION

This paper presents a unified verification scheme providing a robust solution to the challenges posed by the increasing complexity of digital systems. By leveraging hardware emulation and optimizing communication between hardware and software, combined with the use of UVM, the scheme significantly enhances both the speed and the coverage of the verification process. This makes it an invaluable method and a generalizable approach as well as for the design and validation of the modern large-scale digital designs.

REFERENCES

[1] Accellera Systems Initiative. Universal Verification Methodology (UVM) 1.2 Specification. [EB/OL]. [2024-5-21]. Retrieved from https://www.accellera.org/downloads/standards/uvm

[2] K. T. Kai Xian and N. Kumar Thulasiraman, "An Automatic VHDL Testbench Generator for Medium Complexity Design," 2021 IEEE 19th Student Conference on Research and Development (SCOReD), Kota Kinabalu, Malaysia, 2021, pp. 113-118,

[3] OpenHW Group. CORE-V-VERIF. [EB/OL]. [2024-5-21]. Retrieved from https://github.com/openhwgroup/core-v-verif

[4] P. Ehrlich, T. Nguyen and T. Vörtler, "UVM-SystemC based hardware in the loop simulations for accelerated Co- Verification," in Design and Verification Conference and Exhibition (DVCon Europe), 2015.

[5] M. Ismael, A. Hroub and A. Abu-Issa, "AUTG: An Automatic UVM-based TestBench Generator for VLSI Chip Design Verification," 2023 International Conference on Microelectronics (ICM), Abu Dhabi, United Arab Emirates, 2023, pp. 162-167,

[6] Haocheng Huang et al., "A new event driven testbench synthesis engine for FPGA emulation," 2011 9th IEEE International Conference on ASIC, Xiamen, 2011, pp. 373-376,

[7] Cadence.(2024). Palladium Z1. [EB/OL]. [2024-5-21]. Retrieved from https://www.cadence.com/en_US/home/tools/system-design-andverification/acceleration-and-emulation/palladium-z1.html

[8] S. Kapoor, K. N. S. Batta and J. Nagpal, "Emulation: Accelerating Simulation for Rapid Verification of Modern Processor-based Subsystems," 2023 3rd International Conference on Intelligent Technologies (CONIT), Hubli, India, 2023, pp. 1-8,

[9] Google. riscv-dv. [EB/OL]. [2024-5-21] Retrieved from https://github.com/chipsalliance/riscv-dv.git.

[10] riscv-software-src. Spike, a RISC-V ISA Simulator.[EB/OL]. [2024-5-21]. Retrieved from https://github.com/riscv-software-src/riscv-isa-sim

[11] Clifford Wolf. riscv-formal. [EB/OL]. [2024-5-21]. Retrieved from https://github.com/SymbioticEDA/riscv-formal

Asynchronous Arbitration Circuit Optimization for Multicore Neuromorphic Processors

Jiajie Guo [1], Guangyao Lin [1], Bohan Wang [1], Zhiyi Yu [1,2], Shanlin Xiao*[1,2]

[1] School of Microelectronics Science and Technology, Sun Yat-sen University
[2] Guangdong Provincial Key Laboratory of Optoelectronic Information Processing Chips and Systems

* Email: guojj63@mail2.sysu.edu.cn, xiaoshlin@mail.sysu.edu.cn

Abstract—Spiking Neural Networks (SNNs), with their significant advantages of low power consumption and low latency, are a promising approach for edge computing and have attracted considerable attention in the scientific community. Multicore neuromorphic processors have been demonstrated as an effective solution for constructing large-scale SNN models. However, as the scale of these networks increases, the resource and power demands of inter-core communication escalate significantly. The inter-core arbiter represents an important component of the Power-Performance-Area (PPA) bottleneck. Compared to synchronous circuits, asynchronous circuits feature lower power consumption and enhanced robustness, making them more suitable for inter-core communication. We propose an asynchronous Quasi-Delay Insensitive (QDI) arbitration circuit based on hybrid encoding method and the NCL_X pipeline architecture, achieving a 22% improvement in speed and a 23% reduction in power consumption.

Keywords—*asynchronous circuit, multicore neuromorphic processors, arbitration*

I. INTRODUCTION

Spiking Neural Networks (SNNs) possess immense potential in the field of edge-computing, offering advantages of low power consumption and low latency. Multicore neuromorphic processors, using event-driven architectures that combine in-memory computing and brain-inspired principles [1], are an effective method for building large-scale SNN systems [2]. However, as the scale of SNN networks expands, the resource and power consumption demands of inter-core interfaces increase dramatically.

The Address-Event Representation (AER) communication protocol is the most common communication protocol in multicore neuromorphic processors [3]. Fig. 1 shows how AER transmits spikes from source neurons to target neurons. Each core includes a neuro-synaptic array and an inter-core interface. The inter-core arbiter, as a critical component of the communication interface, resolves conflicts from concurrent neuron requests and allocates the data bus for target neurons. It is one of the bottlenecks in optimizing the power, performance, and area (PPA) of SNN systems.

Traditional synchronous circuits are controlled by a global clock signal. Due to clock skew, the clock signal reaches different registers at different times. To ease this issue, clock tree networks are introduced in synchronous circuit design to balance the clock signal distribution. However, this method increases the chip area and power consumption. For example, in synchronous processors, the clock network consumes 40%

This work was supported in part by the Key-Area Research and Development Program of Guangdong Province under Grant 2023B0-303030004; in part by the National Natural Science Foundation of China (NSFC) under Grant 62334014. (Corresponding authors: Shanlin Xiao.)

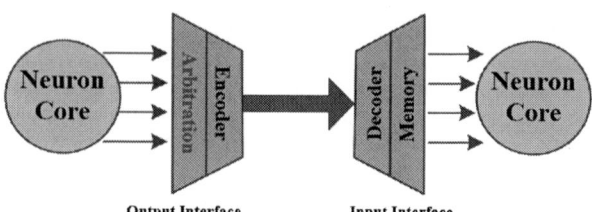

Fig. 1. Address-Event Representation Communication Protocol: the arbiter resolves request conflicts to obtain the target neuron address. The encoder then encodes this address and transmits it over the shared bus.

of the chip power [4]. Asynchronous circuits eliminate the dependence on a global clock and instead achieve data exchange between modules through local handshake protocol. This effectively reduces the power consumption, more suitable for the interconnect needs of multicore neuromorphic processors. The following are the paper's contributions.

- We propose an asynchronous Quasi-Delay Insensitive (QDI) arbitration circuit based on hybrid encoding method and the NCL_X pipeline architecture.

- A comparative analysis is performed on FPGA platform between the proposed asynchronous arbitration circuit and the traditional synchronous arbitration circuit.

II. PROPOSED HIERARCHICAL ARBITER

A. QDI Handshake Protocol

QDI circuits are a common type of asynchronous circuit that meets the condition of isochronous fork. QDI circuits have the advantages of loose constraints and high robustness. QDI circuits typically use a four-phase handshake protocol to replace the global clock and achieve signal communication between modules. Fig. 2 shows a diagram of the asynchronous QDI four-phase handshake protocol. The handshake signals consist of the Req and Ack lines. When these signals are active, they indicate that the data has been sent by the sender and received by the receiver, respectively. As shown in Fig. 2(b),

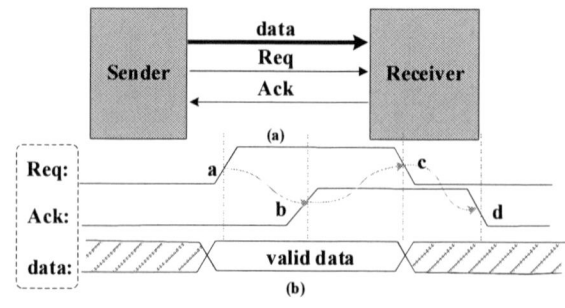

Fig. 2. (a) Asynchronous communication mode; (b) Four-phase handshake protocol.

a complete handshake cycle is achieved through the transitions of the four edges: a-b-c-d. Considering the low power consumption and high robustness requirements of the communication interface for SNNs, this paper adopts asynchronous QDI circuits and four-phase handshake protocol.

B. NCL_X Pipeline Architecture

NCL_X is one of the mainstream methods for implementing QDI circuits. As an expansion of the NULL Convention Logic (NCL) architecture, NCL_X has the advantages of low area cost [5]. In this experiment, we constructed the QDI circuit on FPGA platform using the NCL_X pipeline architecture.

Fig. 3 shows the NCL_X pipeline architecture. It consists of three components: NCL registers, combinational logic, and completion detection (CD) blocks. The CD blocks are used to avoid orphan hazards in QDI circuits [5]. Orphan Hazard refers to a situation in asynchronous QDI circuits where signal delays cause certain signals to become isolated or unrecognized at unintended times. This phenomenon typically occurs in complex logic networks where signal arrival times are inconsistent, potentially leading to incorrect outputs. This paper proposes two optimizations based on the NCL_X pipeline architecture. First, the traditional NCL_X pipeline architecture typically uses dual-rail encoding, which can result in high power consumption due to the high toggle rate when processing complex neuronal networks. To mitigate this problem, we propose a hybrid encoding method that combines one-of-four encoding with dual-rail encoding, effectively reducing power consumption. The optimization of the encoding method not only aligns with the spike request signal encoding mode from neurons but also reduces power consumption by lowering the circuit's toggling rate. Second, for simple QDI circuits, orphan hazards are less likely to occur, thus part of the CD block can be removed to reduce resource consumption.

C. Asynchronous Pipeline Circuit

Fig. 4 shows the hierarchical arbitration mechanism using 64 neurons as an example [6]. Initially, these neurons are divided into four high-level neuron clusters, each containing 16 neurons. Each high-level cluster is further subdivided into four mid-level neuron clusters. The division into low-level neuron clusters follows the same method. Arbitration starts with the high-level neuron clusters, and the active neurons in the cluster can transfer the request to the next level arbiter only when the arbiter grants processing permission to that cluster[6]. This rule also applies between mid-level and low-level arbiters. This arbitration method ensures orderly processing, reduces potential conflicts, and optimizes the efficiency of communication between neurons.

Inspired by the NCL_X pipeline architecture [5], the

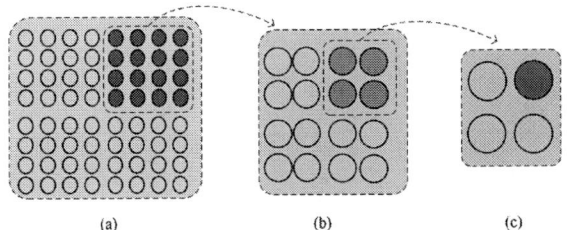

Fig. 4. Hierarchical arbitration mechanism scheme. (a) High level neuron cluster; (b) Medium level neuron cluster; (c) Low level neuron cluster.

asynchronous four-phase QDI circuit designed in this paper is depicted in Fig. 5. The workflow is divided into three stages.

The first stage is "decoupling stage", responsible for decoupling the long-term handshake protocol in neuron from the quick handshake protocol in arbiter. The subsequent CD blocks are used to check whether the low-level or mid-level neuron clusters still have active requests, if so, they do not release the grants of higher level. This implies that the arbiter does not require encoding higher-level requests each time it processes lower-level data, thereby enhancing energy efficiency.

The second stage is "arbitration stage". As shown in Fig. 6, each level of neuron clusters utilizes a four-input QDI tree arbiter [7]. This four-input arbitration tree employs a random arbitration method, meaning that when multiple requests are simultaneously active, the arbiter randomly selects one to grant. This approach ensures fairness in the arbitration process and enhances the circuit's scalability. The output of this stage is in one-hot encoding, providing the authorized address for subsequent stages.

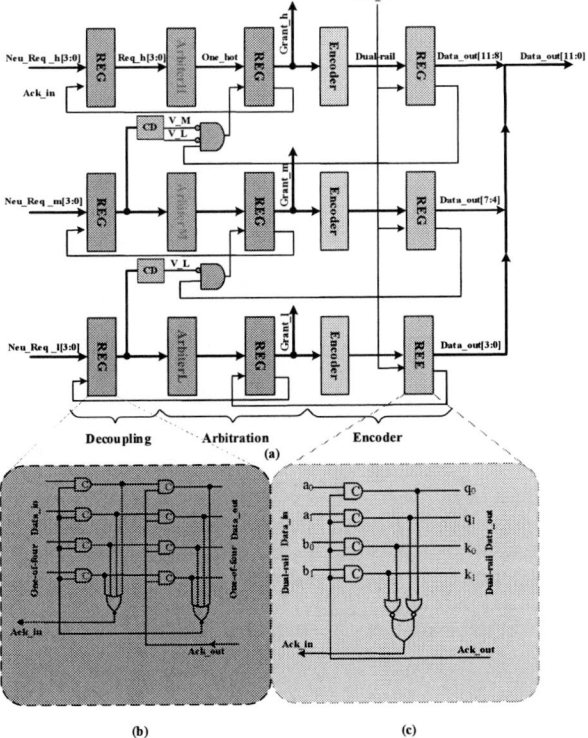

Fig. 5. Asynchronous four-phase QDI arbitration circuits. (a) Asynchronous pipelined hierarchical arbitration circuit; (b) NCL register for one-of-four encoding; (c) NCL register for dual-rail encoding.

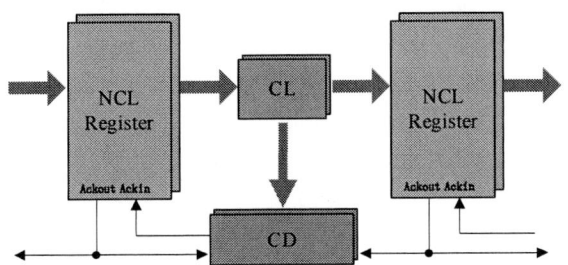

Fig. 3. NCL_X pipeline architecture.

979-8-3503-6184-1/24 $31.00 © 2024 IEEE

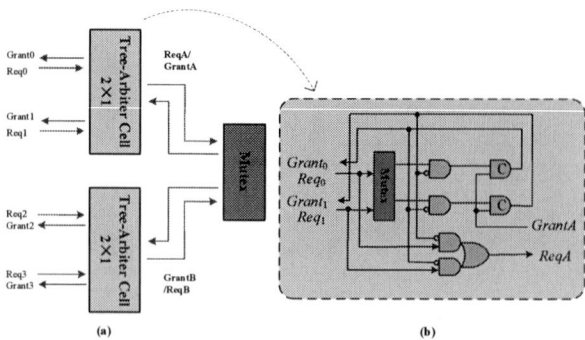

Fig. 6. QDI arbiter. (a) 4-input tree arbiter; (b) Tree arbiter cell design.

The third stage is "encoding stage". To align with the pulse request signals from neurons, the "decoupling stage" and "arbitration stage" use the one-of-four encoding method, which is then converted to dual-rail encoding after "encoding stage". The NCL registers corresponding to these two encoding methods are shown in Fig. 5(b) and Fig. 5(c). Eventually, the encoded data are merged into a complete packet as the output, representing the 12-bit neuron address.

III. EXPERIMENT RESULT

A. Experimental Conditions

This experiment was conducted on the Vivado 2018.3 platform, utilizing the ZCU102 development board. After ensuring that the timing constraints were met, synthesis and implementation were performed to obtain the experimental data.

B. Performance Comparison

For this experiment, a synchronous arbitration circuit with the same architecture was constructed to evaluate the performance advantages of the asynchronous hierarchical arbitration circuit. The comparison results are summarized in Table 1. In terms of speed, the maximum operating frequency of the asynchronous circuit is 22% higher than that of the synchronous circuit. With the input-to-output maximum delay being approximately the same, the normalized dynamic power consumption of the asynchronous arbitration circuit is reduced by 23% compared to the synchronous arbitration circuit. Due to the inherent characteristics of QDI circuits, the number of Lookup Tables (LUTs) is approximately 2.6 times that of the synchronous circuit. The reduction in power consumption is primarily attributed to the use of handshake signals for data transmission in asynchronous circuits, which saves the power typically consumed by the clock network. In the overall neural system processor, routing occupies only 15% of the resources [8], and the arbitration part within the routing consumes even less. Therefore, even with a 2.6-fold increase in resources, the advantages gained in power consumption and speed are more meaningful for interconnects. Moreover, asynchronous QDI circuits utilized in interconnect circuits possess an enhanced robustness advantage [9]. It means that the circuit can maintain stable performance in the face of Process, Voltage,

TABLE I. COMPARISON OF ASYNCHRONOUS HIERARCHICAL ARBITRATED CIRCUIT AND SYNCHRONOUS HIERARCHICAL ARBITRATED CIRCUIT ON FPGA

Arbitration circuit	Performance comparison			
	Limit freq[c] (MHz)	Area (LUTs)	Working on the same freq	
			Dyn.power (mW)	Dyn.power (mW/MHz)
ASYNC[a]	125	98	6	0.064
SYNC[b]	97	38	8	0.083

[a]. Asynchronous hierarchical arbitration circuit we proposed

[b]. Synchronous hierarchical arbitrated circuits with the same architecture

[c]. QDI circuits do not have clocks, so we use the delay's reciprocal as the operating frequency.

and Temperature (PVT) variations, which is essential for the reliability of large-scale integrated circuits. Therefore, from the perspectives of power consumption, speed, and robustness, the asynchronous QDI hierarchical arbiter is more suitable for multicore neuromorphic processors.

IV. CONCLUSION

A new asynchronous hierarchical arbitration circuit based on hybrid encoding method mechanism and the NCL_X pipeline architecture is proposed. Experimental results have validated the advantages of this asynchronous hierarchical arbitration circuit compared to traditional synchronous arbitration circuits, achieving a 22% improvement in speed and a 23% reduction in normalized power consumption. These results demonstrate that the asynchronous QDI hierarchical arbitration circuit is more suitable for use in multicore neuromorphic processors.

REFERENCES

[1] Chicca E, Stefanini F, Bartolozzi C, et al. Neuromorphic electronic circuits for building autonomous cognitive systems[J]. Proceedings of the IEEE, 2014, 102(9): 1367-1388.

[2] Leite V R C, Su Z, Whatley A M, et al. Cortical-inspired placement and routing: Minimizing the memory resources in multi-core neuromorphic processors[C]//2022 IEEE Biomedical Circuits and Systems Conference (BioCAS). IEEE, 2022: 364-368.

[3] Event-based neuromorphic systems[M]. John Wiley & Sons, 2014.

[4] Gronowski P E, Bowhill W J, Preston R P, et al. High-performance microprocessor design[J]. IEEE Journal of Solid-State Circuits, 1998, 33(5): 676-686.

[5] Cheng F C, Chen C. Can QDI combinational Circuits be implemented without C-elements?[C]//2013 IEEE 19th International Symposium on Asynchronous Circuits and Systems. IEEE, 2013: 134-141.

[6] Su Z, Hwang H, Torchet T, et al. Core interface optimization for multi-core neuromorphic processors[C]//2023 28th IEEE International Symposium on Asynchronous Circuits and Systems (ASYNC). IEEE, 2023: 89-98.

[7] Miorandi G, Bertozzi D, Nowick S M. Increasing impartiality and robustness in high-performance N-way asynchronous arbiters[C]//2015 21st IEEE International Symposium on Asynchronous Circuits and Systems. IEEE, 2015: 108-115.

[8] Lattard D, Beigne E, Bernard C, et al. A telecom baseband circuit based on an asynchronous network-on-chip[C]//2007 IEEE International Solid-State Circuits Conference. Digest of Technical Papers. IEEE, 2007: 258-601.

[9] Zou Q, Cui X, Zhong Y, et al. A fully asynchronous QDI mesh router based on 28nm standard cells[C]//2021 IEEE 14th International Conference on ASIC (ASICON). IEEE, 2021: 1-4.

979-8-3503-6184-1/24 $31.00 © 2024 IEEE

A Run-time Temperature Monitoring with Adaptive Duty Cycle Control for FPGA Applications

Weizhi Li, Wangyong Chen*, Haifeng Chen, Haoyu Zhang, Linlin Cai*

School of Microelectronics Science and Technology, Sun Yat-sen University, Guangzhou, 510275, China
*Email: chenwangy@mail.sysu.edu.cn, caillin3@mail.sysu.edu.cn

Abstract—To assist dynamic thermal management in FPGA based prototype system, this paper proposes a run-time temperature monitoring solution utilizing ring oscillators to monitor the temperature on different modules of FPGA projects and construct visual thermal maps of the full chips in HOST. To reduce the thermal overhead of the sensors based on ring oscillators, this paper proposes a feedback-driven adaptive duty cycle control technique and provides an indicator to quantify the cooling performance of this technique. Furthermore, the performance of monitoring system in terms of the thermal overhead, resolution and sensitivity of the configurable sensor network is comprehensively analyzed by experiments.

Keywords—*FPGA, ring oscillator, thermal monitoring, Pblock, duty cycle, sensor network.*

I. INTRODUCTION

Field Programmable Gate Arrays (FPGA) offer the flexibility to reconfigure the hardware for specific projects or requirements, making them widely used in hardware acceleration, prototyping, parallel computing cases. As technology scales, FPGAs achieves higher integration densities and performance improvements, but also generate more heat than the package can dissipate, which make the real-time temperature monitor crucial especially in the compute-intensive systems[1]-[3]. Improper thermal management in FPGAs may cause malfunction and irreversible damage[4][5].

To address this issue, there is a significant demand for precise, real-time thermal mapping of chips to support dynamic thermal management (DTM) techniques[5]. Some FPGAs tackle this challenge by integrating analog sensors, like system monitors in Xilinx FPGAs, which employ Analog-to-Digital Converters (ADC) and diodes positioned at fixed locations, typically in the middle of the chips[6]. However, due to the fixed positions and limited number of thermal-sensitive diodes, obtaining an on-chip thermal distribution remains challenging. Apart from embedded sensing, an alternative approach involves using external devices for temperature measurement. In [7], an infrared camera is utilized to gauge the thermal distribution of FPGAs. While this method saves internal FPGA resources, it relies on costly and cumbersome test equipment to measure surface temperatures. In recent years, soft sensors constructed from programmable FPGA resources have gained increased attention as an additional method. These soft sensors offer greater flexibility in terms of both placement and quantity, leveraging FPGA programmability and eliminating the need for external test devices.

In this work, we implement a soft sensor design involves creating Ring Oscillators (ROs) using Configurable Logic Blocks (CLBs) within FPGAs and reconstruct the run-time temperature map from the configurable thermal sensor network. Key features of the proposed system include leveraging Pblock tool for modular project management and employing feedback-driven adaptive duty cycle control technique to mitigate the thermal overhead caused by the sensor network itself.

II. THERMAL MONITORING SYSTEM ARCHITECTURE

This paper introduces a solution that uses programmable logic resources in FPGA to build a temperature monitoring system based on configurable RO sensor network, which suggests a configuration method of thermal sensor network in modular FPGA projects and dynamically reduce the thermal contribution from sensors. The holistic system architecture is shown in Fig.1.

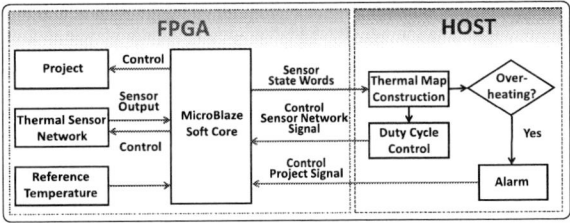

Fig.1 Block diagram of the system architecture.

The system includes the FPGA part and the HOST part. In the FPGA part, the Project is the users' FPGA projects and the thermal sensor network constructed by RO monitors the temperature of the Project. The reference temperature is acquired from the system monitor in Xilinx FPGAs for sensor calibration, which is based on the built-in thermal diode[6]. The MicroBlaze Soft Core runs the code written in C for reading the sensor outputs from the thermal sensor network. It sends the sensor state words containing the working states, outputs, and positions of sensors to the HOST part by serial port. The MicroBlaze Soft Core also controls the thermal sensor network and the Project based on control signals from the HOST part.

In the HOST part, the system constructs visual thermal maps according to the sensor state words. After that, the system generates control signals based on the thermal maps to control the duty cycle of the sensor network activation adaptively. Simultaneously, the system monitors peak temperatures within these maps to detect overheating conditions. Upon detection, users are alerted, and control signals are dispatched from the HOST to the MicroBlaze Soft Core. Upon receiving these signals, the MicroBlaze Soft Core takes action, either powering down the Project or initiating voltage and frequency scaling strategy to prevent overheating.

To monitor the temperature of the Project, the thermal sensor network is placed into the Project in the implementation process. About the layout of the thermal sensor network and the Project, this paper proposes a method that utilizes the Pblock tool in Vivado to partition all modules of the Project and then places the sensors on

979-8-3503-6184-1/24 $31.00 © 2024 IEEE

these modules. The method is shown in Fig.2. A thermal sensor based on RO is constructed conventionally using an odd number of inverters in a loop, an AND gate for enabling and a counter. Because of the correlation between oscillation frequency and temperature, the output of the counter is related to temperature. This method can get the temperature of different modules of the Project and this temperature information can assist users in optimizing high power consumption modules at the design level to reduce overall power consumption.

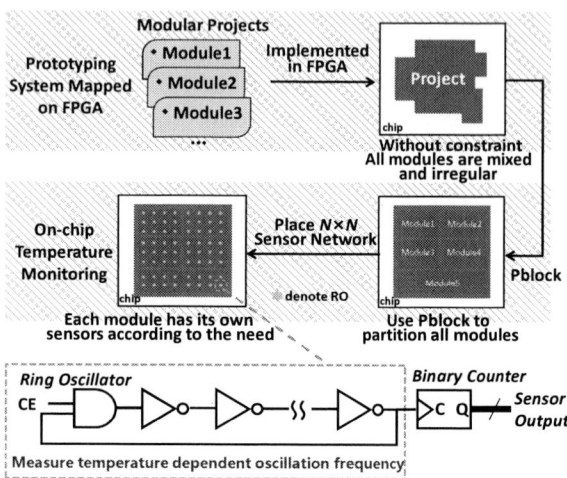

Fig.2 Placement of the thermal sensor network and the project

III. TEMPERATURE READING PROCESS

A. Getting temperatures from sensors

To get temperatures from sensors, firstly the MicroBlaze Soft Core needs to read the sensor outputs from the thermal sensor network. The connection between the thermal sensor network and the MicroBlaze Soft Core is shown in Fig.3(a). The MicroBlaze Soft Core sends three signals, i.e., CLR, CE, and address, to the thermal sensor network. CLR acts as a synchronous reset signal, forcing the counter output to a low state when activated. The CE is an active-high enable signal. When the CE is low, the RO and the counter remain idle, as shown in Fig.3(c). The address is a signal to choose a row of the thermal sensor network and the sensor outputs of this row are outputted into N 16-bit registers correspondingly. The code written in C language controls the MicroBlaze Soft Core to read the data from the registers.

The timing of a reading process is shown in Fig.3(b). The CLR signal comes to a high level at first to reset the counter outputs. Then the CLR comes back to a low level and the CE comes to a high level, so the RO starts oscillating and the counter starts counting. After a period of time t_s, the CE comes back to a low level and the counting stops. Finally, the address is added from 1 to N to output all counting results of the thermal sensor network line by line. The oscillation frequency of each RO can be obtained by dividing the sensor output n by t_s.

After getting the sensor outputs, the MicroBlaze Soft Core will send them to the HOST part, where the sensor outputs are converted to temperatures. According to relevant research [8], the RO oscillation frequency and

temperature have an approximate linear relationship. However, although at the same temperature, the oscillation frequencies of ROs at different positions may have significant differences due to process variations, which makes it impossible to establish a unified relationship between sensor outputs and temperature. To solve this problem, we conduct a calibration step before converting the frequency to temperature to eliminate the impact caused by process variation [9], where the outcome of the system monitor in Xilinx FPGAs serves as reference temperature to calibrate the RO based thermal sensors.

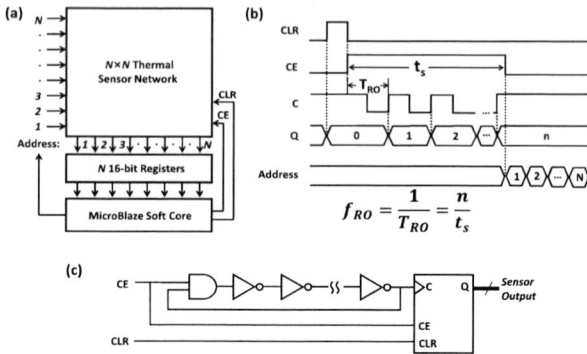

Fig.3 (a)The connection between a $N \times N$ thermal sensor network and the MicroBlaze Soft Core. (b)The timing of a reading process. (c)The input of CE and CLR in a sensor.

B. Constructing thermal maps

To observe the change of temperature, we read successive 40 thermal maps and chooses reference temperature and sensor temperatures from 3 representative positions, as shown in Fig.4(a). There are slightly more than 50 μs between every two maps with $t_s = 50$ μs in Fig.3(b), so approximately, 40 thermal maps last 2ms. In Fig.4(a), before about 0.4ms, the Project is on a working mode, which thermal map is shown in Fig.4(b). And between about 0.4ms to 1.5ms, the Project switches to another working mode, whose thermal map is shown in Fig.4(c). After about 1.5ms, the Project returns to the initial working mode.

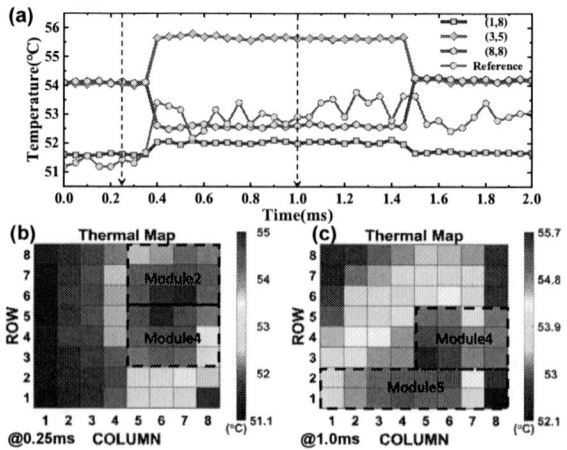

Fig.4 (a) The output temperatures of 3 sensors in different positions and the reference temperature during the time. (b)The thermal map when Time=0.25ms. (c)The thermal map when Time=1.0ms.

979-8-3503-6184-1/24 $31.00 © 2024 IEEE

IV. ADAPTIVE DUTY CYCLE CONTROL

Due to the high oscillation frequency of RO, the sensor network based on RO causes a non-negligible thermal overhead, especially with a large number of sensors. To decrease this overhead, the system dynamically turns off sensors that are at the lower temperature, because in practice, these sensor outputs are paid less attention with.

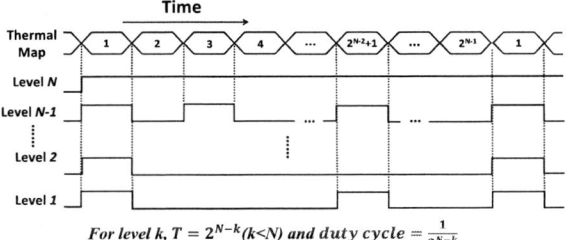

For level k, $T = 2^{N-k}(k<N)$ and duty cycle $= \frac{1}{2^{N-k}}$

Fig.5 Adaptive duty cycle control for the CE signals of sensors at different levels

In the HOST part, according to the thermal map constructed, the system divides the temperature between the maximum temperature and the minimum temperature for N levels equally and divides the sensors into their corresponding temperature levels. Then it generates control signals and sends them to the MicroBlaze Soft Core in the FPGA part to reduce the duty cycle of the CE signals of the sensors based on their temperature levels. By controlling the CE signals, the next 2^{N-1} thermal maps will only open parts of all sensors, as shown in Fig.5. After the 2^{N-1}th thermal map, the system opens all sensors and divides temperature levels and sensors again.

V. THE SYSTEM PERFORMANCE

To measure the thermal overhead of the thermal sensor network, we select a sensor S_{ref} located at the center of the sensor network and uses its output as a reference to observe temperature changes. The sensors except S_{ref} are denoted as S_{test} and the number of S_{test} is denoted as N_{test}. The closing percentage of S_{test} is denoted as φ. Obviously, the temperature obtained from S_{ref} is relevant to φ, which is denoted as $T(\varphi)$. Then the thermal overhead of a sensor can be obtained by:

$$T_{OH} = \frac{T(0) - T(100\%)}{N_{test}} \tag{1}$$

To reflect the temperature reduction caused by the increase in φ, define $\rho(\varphi)$ as:

$$\rho(\varphi) = \frac{T(0) - T(\varphi)}{T(0) - T(100\%)} \times 100\% \tag{2}$$

As φ increases, because of the reduction in thermal overhead of S_{test}, $T(\varphi)$ decreases and $\rho(\varphi)$ increases. The slope $(k_{\rho-\varphi})$ of $\rho(\varphi)$ versus φ reflects the efficiency of reducing thermal overhead by turning off sensors. We measure the slope $k_{\rho-\varphi}$ through the linear regression analysis of $\rho(\varphi)$ and φ. The key parameters are summarized in Table 1, where the resolution means the minimum temperature change that can cause a change in the sensor outputs and sensitivity denotes the decrease in

sensor outputs for every 1°C increase. Reducing the RO length will optimize sensitivity and resolution, but it will cause higher thermal overhead. Furthermore, there is no obvious correlation between $k_{\rho-\varphi}$ and RO length. The greater value of $k_{\rho-\varphi}$ reflects the better efficiency of the adaptive duty cycle control in reducing the thermal overhead of sensor network.

Table 1. The key system parameters

RO Length	3	5	7
$T(0)$ (℃)	45.4708	45.2098	44.6061
$T(100\%)$ (℃)	43.1124	43.4759	43.0983
T_{OH} (℃)[a]	0.0374	0.0275	0.0239
$k_{\rho-\varphi}$	0.97953	0.98779	0.97702
Resolution (℃)	0.00655	0.00974	0.01310
Sensitivity (1/℃)	152	102	76

a. The N_{test} in Eq.(1) is 63.

VI. CONCLUTION

To assist dynamic thermal management, this paper proposes a temperature monitoring system to construct thermal maps of the chips. The feature of proposed system is using Pblock tool to divide users' projects to monitor the temperatures of different modules of the projects and adaptively reducing the duty cycle of low-temperature area sensors through feedback to save thermal overhead of the sensor network. $k_{\rho-\varphi}$ of at least 0.97702 indicates that the adaptively duty cycle control is effective.

ACKNOWLEDGMENT

This work was supported in part by the National Natural Science Foundation of China under Grants 62304263, 62204269 and in part by the Guangdong Basic and Applied Basic Research Foundation under Grants 2023A1515011418, 2024A1515010349.

REFERENCES

[1] Y. Wang, et al., "Measurement of High-Temperature Delay Performance Degradation in FPGA Based on Distributed Temperature Sensor Networks," 2023 3rd International Conference on Computer Science, Electronic Information Engineering and Intelligent Control Technology (CEI), Wuhan, China, 2023.

[2] B. You, et al., "Realization of Parallel Distributed Temperature Sensor Network Based on Field Programmable Gate Array," 2023 5th International Conference on Electronic Engineering and Informatics (EEI), Wuhan, China, 2023.

[3] Heat Management in Integrated Circuits: On-chip and System-level monitoring and cooling, by Seda Ogrenci-Memik, ISBN: 9781849199346.

[4] A. Lesea, et al., Powering xilinx fpgas. Application Notes, 2002.

[5] G. Korkian, et al., "Exploration of ring oscillator based temperature sensors network accuracy on FPGA," International Symposium on Computer Architecture and Digital Systems, 2017, pp. 1-6.

[6] I. Xilinx, 7 Series FPGAs and Zynq-7000 SoC XADC Dual 12-Bit 1 MSPS Analog-to-Digital Converter User Guide, UG480 (v1.11) June 13, 2022.

[7] N. Abdullah, et al., "Thermal and power characterization of field-programmable gate arrays," ACM/SIGDA International Symposium on FPGA, 2011.

[8] N. Rahmanikia, et al., "Performance evaluation metrics for ring-oscillator-based temperature sensors on FPGAs: A quality factor," Integration, the VLSI Journal, vol 57, pp.81–100, March 2017.

[9] S. Xie, et al., "A low power all-digital self-calibrated temperature sensor using 65nm FPGAs," 2013 IEEE International Symposium on Circuits and Systems (ISCAS), Beijing, China, 2013.

979-8-3503-6184-1/24 $31.00 © 2024 IEEE

An FPGA-Based Top-K Gradient Compression Accelerator for Distributed Deep Learning Training

Ziyao Wang[1], Jiayu Zhang[1], Kunyue Li[1], Jialei Sun[1], Feng Dong[2], Ke Chen[2], Yong Qiao[2], Jianfei Jiang*[1]

[1] National Key Laboratory of Advanced Micro and Nano Manufacture Technology, Shanghai, China
[2] Beijing iQIYI Science & Technology Co.. Ltd.. Shanghai, China

* Email: Jiangjianfei@sjtu.edu.cn

Abstract—In distributed neural network training with multiple machines and devices, communication limitations often create efficiency bottlenecks due to the frequent exchange of model parameters and gradient information between computing nodes. This paper proposes an FPGA-based accelerator leveraging the gradient compression algorithm Top-K sparsification, which enhances performance by offloading computationally intensive compression operations to the FPGA. Experimental results demonstrate that the FPGA compression accelerator designed in this study achieves superior computing performance and compression efficiency compared to compression algorithms implemented on CPUs and GPUs. Specifically, the FPGA compresses the same amount of data 3.3-3.7 times faster than parallel solutions on CPUs and 1.3-1.8 times faster than on GPUs.

Keywords—*distributed neural network training, gradient compression, Top-K sparsification, FPGA-based accelerator*

I. INTRODUCTION

With the proliferation of large-scale datasets, distributed neural network training has become increasingly crucial, accelerating the training speed of large-scale deep models. By increasing the number of training nodes and leveraging data parallelism, significant reductions in computation time for forward and backward propagation on equivalent training data scales are achievable. However, distributed training requires frequent exchanges of gradient information between different computation nodes. The reduction in computation time achieved through parallel training may be insufficient to offset the increased cost of communication time, making communication a significant bottleneck in scaling distributed training [1].

Gradient compression, which reduces the volume of gradient data in distributed training, has emerged as an effective solution to alleviate the communication burden. Top-K sparsification is a prominent gradient compression algorithm that retains only the top K most significant absolute values in the gradient vector [2]. Previous research indicates that transmitting only 0.1% of the gradient can achieve accuracy comparable to training without compression [3]. Nonetheless, the computational overhead of compression algorithms is non-negligible, constituting a significant proportion of the overall training time. In specific scenarios, this additional computational burden might surpass the benefits of reduced communication overhead [4].

FPGAs offer several advantages in applying gradient compression algorithms, particularly in parallel computing, power efficiency, and real-time performance, thus providing more efficient hardware support for distributed deep learning [5]. This paper proposes an efficient design and implementation of an FPGA gradient compression accelerator, utilizing a high-performance Top-K gradient compression module based on the Bitonic Sort algorithm. A parallel extension scheme is presented to achieve linear improvements in processing capacity.

II. PROPOSED ACCELERATION SCHEME

A. Overall accelerator design

Figure 1 illustrates the accelerator's top-level architecture. The Control Status Register (CSR) module controls the interaction with the host. The Controller module defines a state machine that controls the compression engine. The compression engine performs data compression computations using a parallel pipeline method. Finally, the Ethernet FIFO buffers the data transmitted and received through the Ethernet port. The state machine also controls the data transfer between the modules and the host memory.

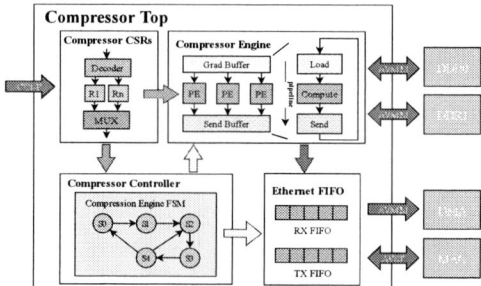

Fig. 1. Overall design framework diagram of accelerator

The CSR module implements a set of registers to make the design more general and capable of handling compression tasks with different data volumes and compression ratios. In a heterogeneous system, the host can configure these registers. Then, the CSR module sends the configured register information to other functional modules to determine the compression ratio, the base address of the data in DDR, the data length, and other information. Some readable registers indicate the current working status of the accelerator. The host can control the execution flow of the entire heterogeneous system by reading these registers. The register mapping information and the corresponding function descriptions are shown in Table Ⅰ.

TABLE I.   REGISTER MAP FOR ACCELERATOR CONFIGURATION

Address	Read/Write Type	Bit Width	Functional Description
18`h1_0000	R/W	32	Read address
18`h1_0002	R/W	64	Batch size
18`h1_0004	R/W	64	Round size
18`h1_0006	W	1	Enable signal
18`h1_0008	R	1	Completion status

979-8-3503-6184-1/24 $31.00 © 2024 IEEE

B. PE unit design

The core of the Top-K compression algorithm execution is the PE module, which sorts the input gradient data and retains the top K most significant gradients as the final output. The execution efficiency of the PE unit directly impacts the overall performance of the accelerator. Given the limited logic and storage resources of FPGA, it is essential to design an efficient parallel sorting method for FPGA.

This study observes that for two vectors of length K, P = [P_0, P_1, ..., P_{k-1}] and Q = [Q_0, Q_1, ..., Q_{k-1}] where P is sorted in ascending order and Q in descending order, the top K most significant data from both vectors can be represented as $V_i = \max\{P_i, Q_i\}$.

Based on this observation, a streaming execution sorting method is designed. For a gradient vector of size N to be sorted, it is first divided into multiple groups of size K. In each calculation round, a new set of data is inputted. The calculation module sorts the newly input vector in ascending order (SortInc) and the previously obtained result vector in descending order (SortDec), comparing the corresponding elements' sizes to obtain a new round of calculation results. The result buffer stores the outcome of each computation round and is used for the subsequent round. The specific hardware design implementation is shown in Figure 2.

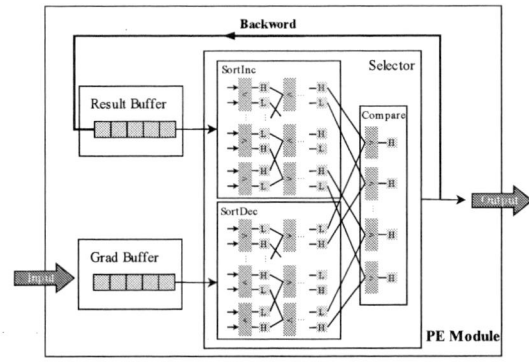

Fig. 2. PE Module Design Framework Diagram

This design employs the Bitonic Sort algorithm for the ascending and descending sorting processes. The algorithm is based on sorting networks. Compared with traditional sorting algorithms such as quicksort and max-heap sorting, sorting networks can utilize hardware to process multiple comparators simultaneously, significantly increasing the sorting speed. Consequently, it can be executed efficiently on parallel processing hardware such as FPGA. Additionally, we have pipelined this sorting network to fully utilize computing resources and enhance the throughput of the sorting network. The pipelined sorting network is shown in Figure 3.

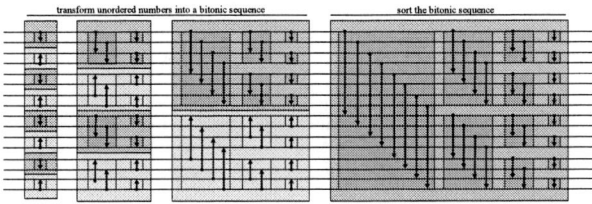

Fig. 3. Pipelined Bitonic Sort Networks

C. Compression Engine design

The compression engine module is responsible for reading and compressing gradient data from local memory and sending the compressed data to the Ethernet module. To fully utilize the hardware resources on the FPGA and further enhance the compression speed, multiple PE processing units are instantiated. The gradient data is split and input into different processing units to run in parallel. The parallel expansion scheme is shown in Figure 4.

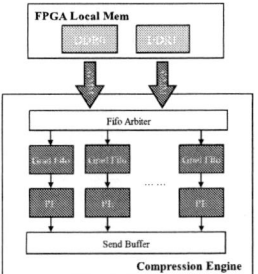

Fig. 4. Parallel Expansion Scheme

The compression engine begins executing its task upon receiving the start signal from the Control module. The FIFO Arbiter module initiates a read request to the DDR bus to access gradient data and employs round-robin arbitration to select different Grad Buffers for writing. Once the current Grad Buffer is prepared, the PE unit reads and compresses the data while the arbiter points to the next Grad Buffer for subsequent data reading. This overlap of data reading and compression across various processing modules enables pipeline execution. Although this approach facilitates the parallel processing of multiple gradient batches, it also results in a linear increase in resource overhead due to the replication of PE units.

Reflecting on the PE unit design, each calculation round necessitates waiting for the current computation to complete and for the results to be written back to the result buffer before new data can be processed. The interval between two data inputs is $\sum_{i=1}^{\log_2 n} i$ clock cycles. Therefore, each clock cycle in the computing process can be viewed as a set of computing resources. Inserting a Time Arbiter Module between the PE unit and Grad Buffers for clock cycle management allows for the flexible selection of multiple data sets for computation during different idle cycles. Consequently, multiple computing tasks can be executed simultaneously on a single PE unit. The optimized design scheme is illustrated in Figure 5.

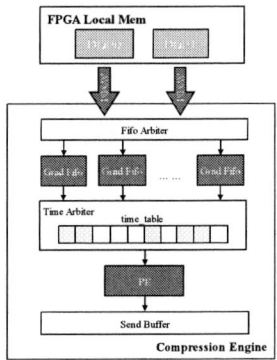

Fig. 5. Optimized Parallel Expansion Scheme

III. EXPERIMENTAL RESULTS

A. Hardware implementation results

We implemented the design proposed in this paper using a programmable accelerator card (PAC) equipped with an Intel Arria 10 GX FPGA as the device endpoint of a heterogeneous computing system. The implementation achieved an operating frequency of 204 MHz, with resource utilization detailed in Table II.

TABLE II.   ACCELERATOR RESOURCE UTILIZATION

Scheme	Parallelism	Logic	Memory	DSP
Before Opt.	1	49449	2737	0
	2	98898	5474	0
	4	197796	10948	0
	8	395592	21896	0
After Opt.	1	49917	2737	0
	2	**50578**	5473	0
	4	**53595**	10945	0
	8	**57966**	21889	0

B. Performance comparison results

In this experiment, we tested the performance of the accelerator proposed in this paper. We compared it with other FPGA-based Top-K hardware implementations to evaluate the system's acceleration performance. The results are shown in Table III, where N represents the data volume.

TABLE III.   COMPARISON OF THE IMPLEMENTATION RESULTS OF VARIOUS TOP-K SORTERS

Architecture	K	F(MHz)	Total Latency(ns)
[6]	8	383	20.9+2.61N
	128	382	335+2.62N
[7]	8	403	19.9+2.48N
	128	377	340+2.65N
Proposed (P=2)	8	204	30+1.88N
	128	204	140+1.09N
Proposed (P=4)	8	204	30+0.94N
	128	204	140+0.55N
Proposed (P=8)	**8**	**204**	**30+0.47N**
	128	**204**	**140+0.27N**

The accelerator proposed in this paper exhibits significant acceleration. This acceleration effect is particularly pronounced with increasing parallelism (P) and larger K values in the Top-K tasks. Compared to the solution proposed in [6], the accelerator presented in this paper achieves acceleration factors of 2.5 and 9.7 for acceleration tasks with a total data volume of 1 GB.

To further verify the superior performance of our accelerator, we compared the FPGA compressor designed in this paper with compression algorithms implemented on the CPU and GPU. The experiment utilized the OpenMP library on the CPU to implement a parallel Top-K compression method across multiple CPU cores. We used the highly optimized Top-K calculation operator in the PyTorch framework for the GPU tests executed on the Nvidia Tesla V100 GPU.

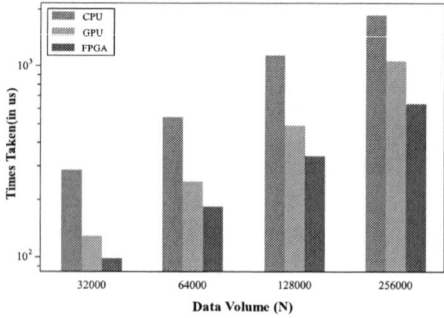

Fig. 6.   Comparison of Compression Time of FPGA, CPU And GPU

The experimental results are shown in Figure 6. In all scenarios, the compression ratio was set to 1%. Due to the significant differences in data, the vertical axis of the figure uses a logarithmic scale. The results indicate that, for the selected gradient data volumes, the FPGA compression accelerator designed in this paper consumes less time than the CPU and GPU. Specifically, the FPGA compression speed is 3.3 to 3.7 times faster than the parallel scheme on the CPU. Even compared to the high-performance GPU implementation of the parallel Top-K algorithm, the FPGA accelerator achieves a speedup of 1.3 to 1.8 times.

IV. SUMMARY

This paper detailed an FPGA-based Top-K gradient compression accelerator design. The core innovation is a high-performance Top-K algorithm based on a pipeline-optimized Bitonic Sort algorithm. The accelerator's hardware implementation and parallel extension have been completed, showcasing its advantages of high parallelism and low storage requirements. Finally, a heterogeneous acceleration system was constructed, and the accelerator was deployed. The proposed work addresses the communication bottleneck problem in distributed training.

REFERENCES

[1] Z. Zhang, C. Chang, H. Lin, Y. Wang, R. Arora, and X. Jin, "Is network the bottleneck of distributed training?," in Proceedings of the Workshop on Network Meets AI & ML, 2020, pp. 8-13.

[2] S. Shi, X. Chu, K. C. Cheung, and S. See, "Understanding top-k sparsification in distributed deep learning," arXiv preprint arXiv:1911.08772, 2019.

[3] Y. Lin, S. Han, H. Mao, Y. Wang, and W. J. Dally, "Deep gradient compression: Reducing the communication bandwidth for distributed training," arXiv preprint arXiv:1712.01887, 2017.

[4] S. Shi et al., "Towards scalable distributed training of deep learning on public cloud clusters," Proceedings of Machine Learning and Systems, vol. 3, pp. 401-412, 2021.

[5] E. Wang et al., "Deep neural network approximation for custom hardware: Where we've been, where we're going," ACM Computing Surveys (CSUR), vol. 52, no. 2, pp. 1-39, 2019.

[6] T. Chen, W. Li, F. Yu, and Q. Xing, "Modular serial pipelined sorting architecture for continuous variable-length sequences with a very simple control strategy," IEICE TRANSACTIONS on Fundamentals of Electronics, Communications and Computer Sciences, vol. 100, no. 4, pp. 1074-1078, 2017.

[7] C.-S. Lin and B.-D. Liu, "Design of a pipelined and expandable sorting architecture with simple control scheme," in 2002 IEEE International Symposium on Circuits and Systems (ISCAS), 2002, vol. 4: IEEE, pp. IV-IV.

Dynamic-Matrix-Encryption Based Secure Strong PUF for Device Authentication Protocols

Liangxiao Zhao [1], Gang Li *[1], Pengjun Wang[1], Xuejiao Ma [2,3], Ziyu Zhou [3]

[1] The College of Electrical and Electronic Engineering, Wenzhou University, Wenzhou 325000, China
[2] School of Data Science and Artificial Intelligence, Wenzhou University of Technology, Wenzhou, 325035, China
[3] Faculty of Electrical Engineering and Computer Science, Ningbo University, Ningbo, 315211, China

* Email: ligang@wzu.edu.cn

Abstract—**Strong physical unsolvable function (PUF) has a wide range of applications in internet of things (IoT) security. However, it is vulnerable to machine learning (ML) modeling attacks. This paper proposes a strong PUF anti-ML attack method based on dynamic-matrix encryption. It obfuscates the challenge by multiplying the dynamic encryption matrix (DEM) with the challenge matrix synchronizing initial values during the registration phase of the device security authentication protocol prevents our obfuscation from being invalidated when the algorithms and structures motivating the obfuscation are made public. Each round of DEM is updated with respect to each previous round of challenge which is similar to the properties of sequential logic circuits in that the current response does not depend only on the current challenge but is related to previous challenge. Experimental results show that even if an attacker collects 1 million challenge response pairs, the prediction accuracies of the several ML attacks we use are below 53%, with negligible effects on randomness, reliability, and uniqueness.**

Keywords—PUF, Machine Learning, Dynamic Matrix Encryption, Sequential Logic

I. INTRODUCTION

As the largest enabling factor for digital transformation, the Internet of things (IoT) has a wide range of application scenarios and huge market demand and is playing an important role in manufacturing, energy, transportation, agriculture, medical, and other industries. However, the rapid development of the IoT industry is also facing serious security problems. Devices located in the IoT awareness layer are difficult to implement complex security methods due to limited hardware resources, and therefore, lightweight and efficient security solutions are needed to protect these devices from attacks. The physical unclonable function (PUF) technique [1] provides an effective solution. PUF takes advantage of tiny inconsistencies in the chip manufacturing process to provide a unique identity for each device. With the advantages of low cost, low power consumption, and no need for additional storage space, it has a wide range of application prospects in the field of IoT security.

The practical application of PUF is generally in the form of a device security authentication protocol [2], which is mainly divided into registration and authentication phases, PUFModel based device authentication protocols diagram as shown in Fig. 1. In the registration phase, In order to reduce server storage space, instead of storing the actual challenge response pairs (CRPs), the PUFModel can be stored. The

machine learning (ML) is used to model the PUF and the model data is defined as a PUFModel[3] stored in the server. The server then synchronizes the generated encrypted information to the device and the registration phase is complete. In the authentication phase, when a device initiates an authentication request, the server randomly generates a certain number of groups of challenge C. The challenge C_{IS} and input it into PUFModel to obtain the R_{IS}. The R_{IS} is transmitted to the device side and compared with the R_D generated by the PUF for server authentication. After the server passes the authentication, the R_{dev} generated by the PUF is transmitted to the server side to compare with the R_{ser} generated by the PUFModel, and if the FHD $(R_{dev}, R_{ser}) < \tau$, the authentication is successful.

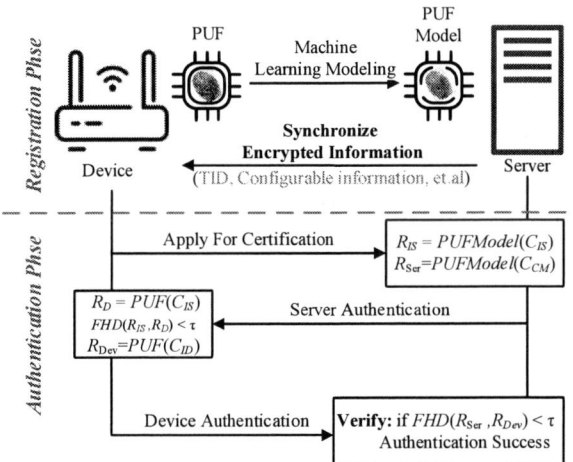

Fig. 1. PUFModel-based device authentication protocols diagram

In recent years, many counter-attack techniques have been proposed to defend against ML modeling attacks. These techniques defend against ML attacks by obfuscating the challenge or/and response, complicating the mapping relationship between the challenge and the response, and thus preventing the attacker from collecting valid CRPs to defend against ML modeling attacks. XOR-APUF [4], MPUF [5], PUF-FSM [6], and iPUF [7] are typical examples. However, the current strong PUF anti-machine learning attack techniques have certain limitations. For obfuscating challenge, when the obfuscating algorithm and structure are disclosed, the attacker can directly establish the relationship between the intermediate value and the response by obtaining the obfuscating intermediate value, so as to make the obfuscating invalid. For obfuscating responses, the stability of PUF will be reduced and the utility of PUF will be affected. Moreover, both the obfuscating challenge and the obfuscating response

---

This work was supported in part by the National Natural Science Foundation of China (62374117, 62234008), the Zhejiang Provincial Natural Science Foundation of China (LY22F040004), the China Postdoctoral Science Foundation (2023M731776), the Wenzhou Basic Scientific Research Projects (G20220005).

are determined by the current challenge, which is similar to the characteristics of combinatorial logic circuits, and the complexity of obfuscating is limited.

To solve the limitations of prior art, this paper propose dynamic matrix encryption (DME) based strong PUF for device authentication protocol. It has the following features: 1) Sequential logic-like obfuscating, dynamic updating of the encryption matrix by challenges so that the current response does not depend only on the current challenge, but is related to previous challenges, similar to the characteristics of sequential logic circuits. 2) Universality, the DME and corresponding authentication protocol can be used for all strong PUFs to resist ML attacks. 3) No effects on reliability, obfuscating only the challenge does not reduce the reliability of PUFs.

II. PUF CHALLENGE OBFUSCATION BASED ON DME

The accuracy of ML modeling is related to the complexity of the mapping relationship between challenge responses. The more complex the mapping relationship, the lower the modeling accuracy. In this paper, the challenge is mapped to a new value by multiplying the random challenge with a matrix in the dynamic encryption matrix, so that attackers cannot access the real challenges and are unable to build an accurate model of strong PUF.

DME based secure strong PUF for device authentication protocols is to obfuscate the challenge of strong PUF and can overcome the shortcomings, as shown in Fig. 2. In the registration phase of the authentication protocol to synchronize the initial Rubik's cube matrix M_0 and other encrypted information to the device before obfuscating challenge. Then the obfuscated challenge C' is put into the strong PUF to obtain the final response R. In each dynamic matrix encryption, the Rubik's cube matrix M dynamically updates the encryption matrix under challenge C and updates the Rubik's cube matrix as input for the next round of dynamic matrix encryption.

The detailed process of dynamic matrix encryption is as follows: Step1, the hamming weigh(HW)t is calculated for the c_0 to c_4 of challenge C, and one of the six faces of the Rubik's cube is selected for operation according to the HW. HW of 0 to 5 correspond to one of the six faces. Step2, after the face of the Rubik's cube is determined, the rotation direction of each face is determined according to the k_0 to k_5 of the Rubik's cube matrix of surfaces: '0' represents clockwise rotation of the corresponding face, '1' represents counterclockwise rotation of the corresponding face, according to the selected rotation direction, the corresponding face is rotated, the position of the elements in the Rubik's cube permutation group is adjusted, and the encryption matrix is generated. Step3, the first bit of the encryption matrix is used to determine the direction of matrix multiplication. If the k_0' is '0', the encryption matrix and the challenge matrix are postmultiplication otherwise are prepostmultiplication. Step4, for elements in the multiplied matrix, if they are odd, they are converted to '1' otherwise it is converted to '0'. The elements of the matrix are taken out in order for a 64-bit challenge C' is re-generated as input to the PUF, and the cube matrix is saved for use in the next round of encryption. The Rubik's cube matrix is permutation encrypted with a new challenge C for each round so that the cube matrix is related to the previous challenge C and the encryption operations for each round are related to each other, similar to the characteristics of sequential logic circuits.

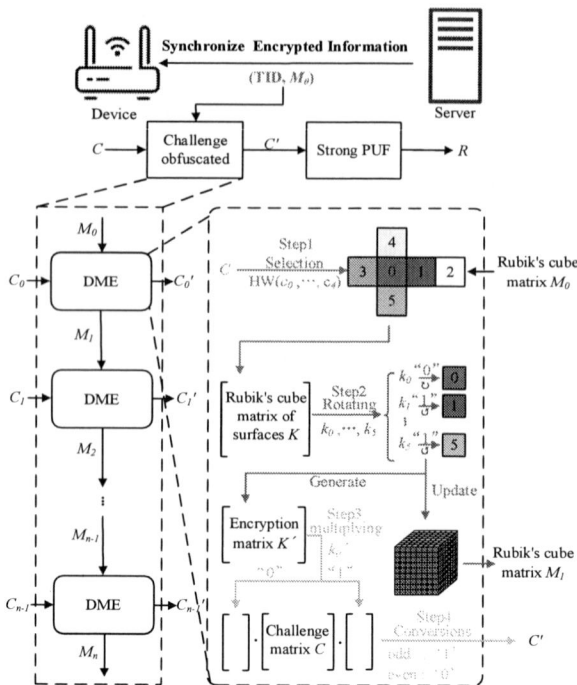

Fig. 2. PUF challenge obfuscation based on DME

III. EXPERIMENTAL RESULTS AND ANALYSIS

This paper chose to obfuscate APUF (the weakest of the most common strong PUFs against ML attacks) to evaluate the resistance of DME-based strong PUFs (DME-PUF) against ML attacks. Implemented PUF on Xilinx Artix-7 FPGA, the test platform is shown in Fig. 3. The randomness, uniqueness, reliability, and anti-ML capability of APUF and DME-PUF are compared. Finally, the performance of this obfuscating method is compared with several common strong PUF obfuscating methods.

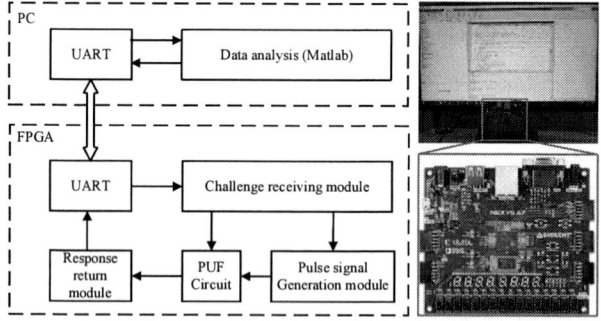

Fig. 3. Test platform

A. Resistance to ML attacks

We used logistic regression (LR), support vector machine (SVM), artificial neural nnetwork (ANN), and Light Gradient Boosting Machine (LGBM) algorithms respectively for modeling. We randomly divided the CRPs data set into the training set and test set. The number of test sets was fixed at 200,000 groups, and the number of test sets increased successively so that a total of 1 million groups of CRPs were collected. The attack prediction accuracy of the four attack methods was measured, as shown in Fig. 4.

979-8-3503-6184-1/24 $31.00 © 2024 IEEE

An APUF with a 64-bit challenge model prediction accuracy can exceed 95% when 650 CRPs are randomly selected for training, and when the number of CRPs is increased to 18050, the model prediction accuracy is as high as 99.9%[8]. For APUF obfuscated by DME even if an attacker collected 1 million CRPs, the accuracy was less than 53%. It is shown that the proposed DME can effectively defend against ML modeling attacks.

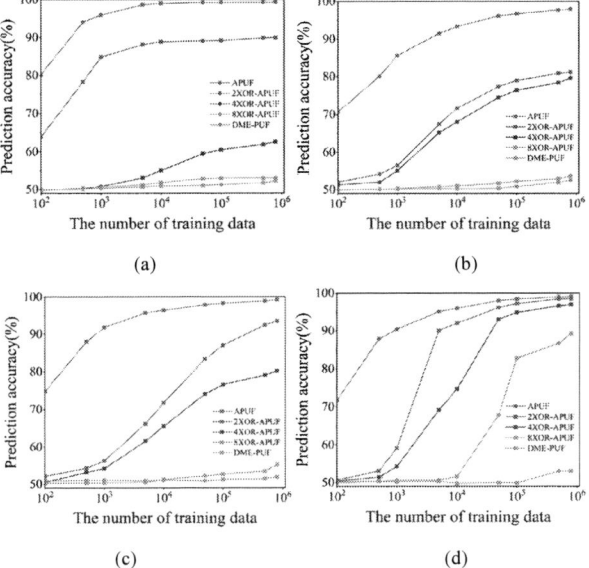

Fig. 4. Modeling attacks on strong PUF, (a) LR, (b) LGBM, (c)SVM, (d) ANN

B. Randomness, Reliability, and Uniqueness

The ratios of '0' and '1' in the PUF response are used to estimate randomness (Ran). Reliability (Rel) is used to measure the PUF's ability to reproduce the same response. Uniqueness (Uni) indicates the difference in response between different PUF instances[9]. In this study, we adopted the randomness test suite (SP 800-22) proposed by the National Institute of Standards and Technology (NIST) for randomness. DME-PUF passed 10 tests (Frequency, BlockFrequency, Runs, LongestRun, Rank, FFT, OverlappingTemplate, Serial-1, Serial-2, LinearComplexity) and the APUF passed only 6 tests. The Hamming distance (HD) can measure reliability and uniqueness[10], as shown in Fig. 5. The mean and standard deviation of the inter-PUF HD are 49.84% and 0.000435, respectively. The mean and standard deviation of the intra-PUF HD are 0.295% and 0.000458, indicating that the

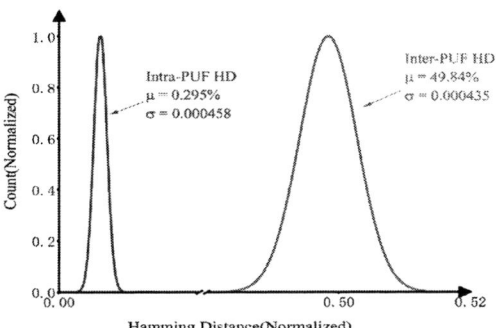

Fig. 5. Intra and Inter PUF HD

confusion has a negligible reduction in the reliability of PUF. The performance of DME-PUF and other strong PUF circuits is shown in Table I. DME-PUF improves the resistance of strong PUF to ML attacks while having a negligible effect on the other properties of PUF.

TABLE I. THE PERFORMANCE OF STRONG PUF

Type	ML attack prediction (%)				Ran	Uni	Rel
	LR	LGBM	SVM	ANN			
APUF	99.24	97.94	99.22	99.08	50.12	50.05	99.89
8-XOR-APUF	52.31	53.69	55.21	89.27	49.95	50.36	90.67
(2,1)-MPUF	70.52	80.37	91.51	97.49	50.56	50.23	97.25
(3,3)-iPUF	55.59	60.53	62.09	93.42	50.87	50.15	91.06
DME-PUF	52.57	51.83	51.11	52.76	50.05	49.84	99.71

IV. CONCLUSION

This paper proposed DME based secure strong PUF for device authentication protocols. This scheme produces the obfuscated challenges by multiplying the dynamic encryption matrix with challenge. The synchronization of the initial values is done in the registration phase of the device security authentication protocol to prevent our obfuscation from being invalidated when the algorithms and structures motivating the obfuscation are made public. The experiment results show that the proposed DME-PUF effectively resists modeling attacks of LR, LGBM, SVM, and ANN with a maximum prediction accuracy of 53% while remaining randomness, reliability, and uniqueness.

REFERENCES

[1] R Mae. "Physically Unclonable Functions: Constructions, Properties and Applications" [M]. Springer Publishing Company, Incorporated, 2013.

[2] Y Wang, X Mei, Z Chang, et al. "A Lightweight Authentication Protocol Against Modeling Attacks Based on a Novel LFSR-APUF,"[J] IEEE Internet of Things Journal, vol. 11, no. 1, 2024, pp. 283-29.

[3] Z Zhou, P Wang, G Li. "Bagua Protocol: A Whole-Process Configurable Protocol for IoT Sensing Devices Security Based on Strong PUF," [J]Internet of Things, vol. 11, no. 1, 2024, pp.805-819.

[4] G E Suh, S Devadas. "Physical Unclonable Functions for Device Authentication and Secret Key Generation"[C] 44th ACM/IEEE Design Automation Conference, 2007, pp. 9-14

[5] D Sahoo, D Mukhopadhyay, R Chakraborty, et al. "A Multiplexer-Based Arbiter PUF Composition with Enhanced Reliability and Security"[C], IEEE Transaction on Computers. 2018, pp.1-1.

[6] S J Wang, Y S Chen, S M Li. "Modeling Attack Resistant PUFs Based on Adversarial Attack against Machine Learning"[J]. IEEE Journal on Emerging and Selected Topics in Circuits and Systems, 2021, 11(2), pp.306-318.

[7] P Nguyen, D Sahoo, Jin C, et al. "The Interpose PUF: Secure PUF Design Against State-of-The-Art Machine Learning Attacks"[J]IACR Trans. Cryptographic Hardware and Embedded Systems, CHES, 2019, pp.243–290.

[8] X Ma, P Wang, G Li, Z Zhou." Machine Learning Attacks Resistant Strong PUF Design Utilizing Response Obfuscates Challenge with Lower Hardware Overhead"[J]. Microelectronics Journal, vol. 142,2023.

[9] G Li, P Wang, X Ma, Y. Shi, Y Zhang. "A Multimode Configurable Physically Unclonable Function with Bit-Instability-Screening and Power-Gating Strategies"[J]. IEEE Transactions on Very Large Scale Integration (VLSI) Systems, 2021, 29(01), pp.100-11.

[10] G Li, P Wang, X Ma, et al. "A 0.67-μm²/Bitcell Two-Transistor Leakage-Based Physically Unclonable Function with Native Bit-Instability of 0.89% at 65 Nm"[J]. Electronics Letters, 2020, 56(23): 1237-1239.

A Low Latency and High Throughout Hardware Design of Random Matrix Number Generator for FrodoKEM

Shengfei Gu[1], Jiahao Lu[1], Tianze Huang[1], Jiaming Zhang[1], Kai Li[1], Cheng Wu[1], Mingbo Wang[1], Xianqi Mei[1], Ang Hu[1,2], Dongsheng Liu*[1,2]

[1]School of Integrated Circuits, Huazhong University of Science and Technology, Wuhan, China
[2]JinYinHu Laboratory, WuHan 430040, China

* Email: gushengfei_2002@163.com, dsliu@mail.hust.edu.cn

Abstract—**FrodoKEM is a lattice-based key encapsulation mechanism，based upon the hardness of the Learning With Errors (LWE) problem. It takes great advantage in security by avoiding specific structures within lattices and is under consideration for standardization by International Organization for Standardization (ISO). In this paper, we propose a low latency and high throughout random matrix number generator for FrodoKEM. A pipelined and FrodoKEM-specified Secure Hash Algorithm 3 (SHA-3) structure is designed to reach high throughout and cost relatively little resource. This paper also designs a multi-data-flow fast sampler to save the cycle consumption of the sampling process. The design is implemented on XILINX Zynq UltraScale+ MPSoCs hardware platform. It reaches a max frequency of 400MHz and makes an improvement of 120× on the cycle number when comparing with the HW/SW codesign for FrodoKEM which still runs software and only operates basic operations in hardware.**

Keywords—*SHA-3,Random Matrix Number Generator，FrodoKEM, FPGA*

I. INTRODUCTION

Ensuring information security is crucial for maintaining stability across various fields, especially in sensitive areas such as banking, e-commerce, and the military. Typically, sensitive information in these domains is encrypted by cryptographic functions. However, Shor proposed a novel quantum algorithm capable of solving the prime factorization of large primes and the discrete logarithm problem in polynomial time, which made traditional public-key cryptosystems insecure under quantum computing[1].

NIST has launched an open selection process to identify standardized public-key algorithms that can resist quantum computing attacks. The process used the Keccak function as the SHA-3. In the third round of process, FrodoKEM was chosen as an additional algorithm[2]. It is under consideration for standardization by ISO[3]. The FrodoKEM algorithm is primarily used for key exchange processes, enabling secure communication channels between parties. This algorithm avoids specific patterns or structures within lattices, providing security advantages and high configurability. However, in recent years, researches on accelerating the hardware implementation of FrodoKEM were rare.

In this paper, we use FPGAs to accelerate the hardware implementation of the random matrix number generator for FrodoKEM. By inserting pipeline in SHA-3 module and optimizing combinational logic calculation, our design reaches high throughout and relatively low resource consumption. The cycle number of the sampling process is reduced through designing a multi-data-flow fast sampler.

This design achieves an average improvement of 120× on the cycle number when comparing with the HW/SW codesigns for FrodoKEM and reached a max frequency of 400MHz. Our design utilizes 8405 LUTs, 6822 FFs, and 1648 CLBs in F-1344 security mode.

The structure of this paper is as follows: Section II, we first briefly describe the framework of random matrix number generation module, and then detail the design of SHA-3 function and sampler. In Section III, we implement our design on hardware platform and make some discussion on its performance. A brief summary of our work is given in section IV.

II. ARCHITECTURE DESIGN

A. Principle of FrodoKEM Algorithm

FrodoKEM is a key encapsulation mechanism designed based on the Learning with Errors (LWE) problem. During the encryption process, it generates large amounts of ciphertexts and keys. These ciphertexts and keys are produced from a series of small pseudorandom number seeds through the extensible SHA-3 encryption function and a sampler.

FrodoKEM features three security modes: F-640, F-976, and F-1344. The higher the number, the higher the security level. The security parameter configurations of FrodoKEM in different modes vary mainly in terms of matrix size, seed length, the SHA-3 function used for generating random numbers, and the discrete symmetric error distribution.

The sponge construction shown in Fig. 1 is the key component of the SHA-3 function. It performs the computation of the input bitstream through a specific computational framework and outputs the required bitstream. The function that the construction produces from these components is called a sponge function, denoted by $Sponge[f, pad, r](N, d)$. A transformation function on fixed-length strings is denoted by f. First, the bitstream N is divided into groups of length r and "absorbed" into the structure. Once N is fully "absorbed" into the structure, an arbitrary number of output bits are "squeezed" out of its state, r bits every time.

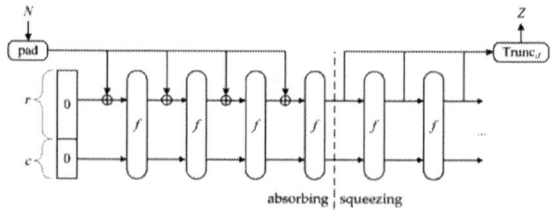

Fig. 1. The sponge construction [4]

B. Overall Architecture

According to the standard of the SHA-3 function, its main states include data padding, transformation, output truncating, and sampling. The overall structure is shown in Fig. 2. After initiating SHA-3, the random matrix number seed and model variable come into the registers. Depending on the specific model, the absorption data, padding rules, number of output rounds, and other configurations for the SHA-3 module are set accordingly.

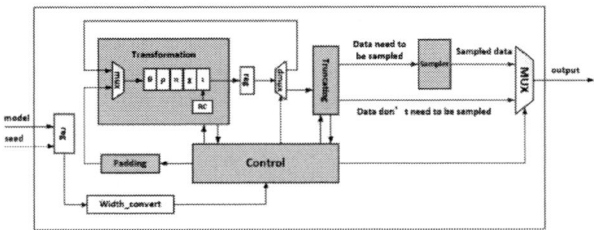

Fig. 2. Overall architecture

C. SHA-3 Structure

The transformation round, which maps the input data, is a critical part in SHA-3. A round of transformation consists of five step mappings, which are denoted by θ, ρ,π, χ, ι. Since the five step mappings correspond to bitwise AND, NOT, XOR, and cyclic shift operations, the most straightforward implementation of the transformation round is through a combinational logic design. In this design, one pipeline register is inserted between the step θ and ρ. In this way, combinational logic delay can be reduced and computation efficiency can be improved. At the same time, this approach doesn't increase the consumption of register resources and the number of computation cycles excessively. The structure is shown in Fig. 3.

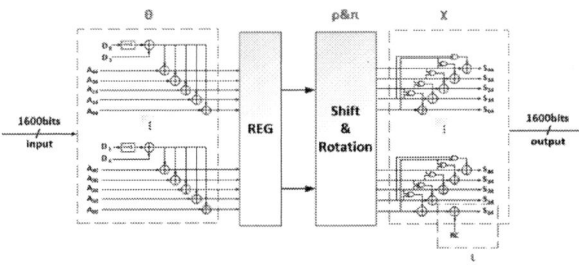

Fig. 3. Circuit of a transformation round

The final circuit diagram of a transformation round is shown in Fig. 4. In this process, XOR gates are primarily used for in θ step. The ρ and π steps only involve shifts and rotations of the bits, which do not consume gate circuits. The χ step mainly utilize NOT gates, AND gates, and XOR gates. Each ι calculation uses only one XOR gate.

D. Multi-data-flow Fast Sampler

The multi-data-flow fast sampler executes sampling from the error distribution χ of FrodoKEM. An error distribution table has been predefined in the standard. The sampling algorithm generates a sampling value by comparing data bits from the input random bit string with data from the error distribution table. This sampling value is then adjusted based on the sign bit and ultimately used as the output of the algorithm. This process ensures the randomness of the

samples and the accuracy of the distribution. The sampling algorithm is shown in Algorithm 1.

Algorithm 1 Sampling from the error distribution

Input: random bit string $r = (r_1, r_2, \ldots, r_{15})$
Output: sampling value e
1. Set $t \leftarrow (r_1, r_2, \ldots, r_{15})$,interpreted as a nonnegative integer
2. $e \leftarrow 0$
3. For $i = 0$ to $s - 1$ do
   3.1. (In constant time) If $t > T_\chi(i)$ then $e \leftarrow e + 1$
4. $e \leftarrow (-1)^{r_0} \cdot e$
5. Output e

Since the sampling process involves bitwise operations such as comparison and negation, the sampling process is also implemented by using combinational logic. One pipeline register is added to the sampling module, where the sampling values obtained by comparing data bits with the error distribution table are stored in eight 7-bit registers. In the next clock cycle, the sample values are adjusted for their sign and then output. This allows the comparison of the next set of data bits to occur while the current cycle's data is being output, reducing latency, and improving efficiency. The structure of sampler is shown in Fig. 4.

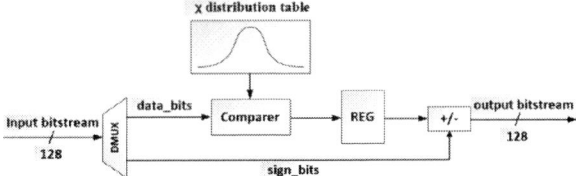

Fig. 4. The structure of sampler

III. HARDWARE IMPLEMENTATION AND DISCUSSION

Our design uses Vivado 2020.2 for simulation, synthesis and implementation. The hardware verification is conducted on the XILINX Zynq UltraScale+ MPSoCs hardware platform. We use the ILA probes to obtain the encrypted results from the hardware platform. The implementation condition is shown in Fig. 5.

Fig. 5. Implementation of the design

Our design utilizes 8405 LUTs, 6822 FFs, and 1648 CLBs in F-1344 security mode. The resource used by random matrix

---

979-8-3503-6184-1/24 $31.00 © 2024 IEEE

number generator in different security modes are shown in Table I.

TABLE I. Resource used by generator in different security modes

Security mode	LUT	FF	CLB	BRAM
F-640	8405	6822	1648	9.5
F-976	8185	6582	1680	9.5
F-1344	8101	6584	1681	9.5

To evaluate the performance of this design, we compared it with the related work presented in design [5]. The comparison results are shown in Table II.

TABLE II. Cycle counts for generating A matrix

Mode	Operation	Design[5]	Our design
F-640	Generate A Matrix	18,010,240	151,040
F-976		36,432,128	335,744
F-1344		64,926,288	612,864

The design [5] is implemented on a RISC-V hardware platform according to the FrodoKEM algorithm, which use loop structure in software algorithm to generate the random matrix number. The cycles of generating the A matrix were counted in their work. Our design optimizes the matrix generation process in the SHA-3 function and sampling algorithm by using pipeline technology and combinational logic design.

Through comparison, our design costs 151,040 cycles, 335,744 cycles and 612,864 cycles, in F-640 security mode, F-976 security mode and F-1344 security mode respectively. This design achieves an average improvement of 120× on the cycle number, which significantly reduces the cycle consumed in generating the A matrix. The low cycle consumption, combined with high frequency, leads to its low latency.

TABLE III. Performance of different works

Design	Operation	Platform	Throughput(Gbps)
Our design F-1344 security mode	SHA-3 function, sampling, data processing	Ultrascale+	18.54
[6]	SHA-3 function	Virtex-6	16.51
[7]	SHA-3 function	Virtex-7	16.58
[8]	SHA-3 function	Arria-10	22.60

The design can reaches a maximum frequency of 400MHz, with a peak throughput of 18.54Gbps and an efficiency of 10.87Mbps/Slice. It achieves relatively high throughput when compared with related works. Those works only operate the process of SHA-3 function. Among which, design[6] proposes a design with the least area consumption, and design[7] insert pipelining in π and χ process ,but their throughput and efficiency are not high. Design[8] has a similar SHA-3

function structure with ours and realizes higher throughout due to its efficient design for inputting data, but our design can realize more operation. The comparison results are shown in Table III.

IV. SUMMARY

This design focuses on the current lack of research on hardware acceleration for the FrodoKEM algorithm and uses an FPGA hardware platform to accelerate the random matrix number generation part of the FrodoKEM algorithm. The design has been validated on the UltraScale+ hardware platform. It utilizes 8405 LUTs, 6822 FFs, and 1648 CLBs in F-1344 security mode. By using multi-data-flow fast sampler and pipelined SHA-3 structure, the required cycles are significantly reduced, achieving a 120× improvement when comparing with the HW/SW codesign for FrodoKEM. The design can reaches a maximum frequency of 400MHz, with a peak throughput of 18.54Gbps and an efficiency of 10.87Mbps/Slice, showing good performance in terms of latency and throughput. In the future, based on the accelerated random matrix number generation part, the overall acceleration of the FrodoKEM algorithm can be completed.

V. ACKNOWLEDGMENT

This work is supported by the National Key Research and Development Program of China (No. 2021YFA0715502, No. 2023YFB4502100), the National Natural Science Foundation of China (No.62374064，No. 62104076, No. 62134002), the National Key Analog Integrated Circuit Laboratory Project of China. (No. JCKY2021210C004)，the Innovation Project of JinYinHu Laboratory. (Grant No. 2024JYH011401)

REFERENCES

[1] P.W.Shor, "Algorithms for quantum computation: Discrete logarithms and factoring," in *Proc. 35th Annu. Symp. Found. Comput. Sci.*,Santa Fe, NM, USA, Nov. 1994, pp. 124–134.

[2] G. Alagic, Status Report on the Third Round of the NIST Post-Quantum Cryptography Standardization Process. Gaithersburg, MD, USA: U.S.Department of Commerce, NIST, 2022.

[3] Intl. Organization for Standardization, "FrodoKEM: Learning with err ors key encapsulation preliminary draft standard," 2023. [Online].Ava ilable:https://frodokem.org/files/FrodoKEM-ISO 20230314.pdf.

[4] G. Bertoni, J. Daemen, M. Peeters, and G. V. Assche, ''Keccak,'' in *Proc. Annu. Int. Conf. Theory Appl. Cryptograph. Techn.* Springer, 2013,pp. 313–314.

[5] P. Karl, T. Fritzmann, and G. Sigl, "Hardware Accelerated FrodoKEM on RISC-V," in 2022 *25th International Symposium on Design and Diagnostics of Electronic Circuits and Systems (DDECS)*, 2022, pp. 154–159.

[6] V. Arribas, "Beyond the limits: SHA-3 in just 49 slices," in *Proc. 29th Int. Conf. Field Program. Logic Appl. (FPL)*, 2019, pp. 239–245.

[7] F. Kahri, H. Mestiri, B. Bouallegue, and M. Machhout, "High speed FPGA implementation of cryptographic Keccak hash function crypto-processor," *J. Circuits, Syst. Comput.*, vol. 25, no. 4, 2016,Art. no. 1650026.

[8] A. Sideris, T. Sanida, and M. Dasygenis, "High throughput pipelined implementation of the SHA-3 cryptoprocessor," in *Proc. 32nd Int. Conf. Microelectron. (ICM)*, 2020, pp. 1–4..

A 4K60fps Ultra-Low-Latency Light Compression Encoder for Bandwidth-Constrained Scenarios

Yanzhong Li[1], Leilei Huang[2]*, Yibo Fan[1]*

[1]State Key Laboratory of Integrated Chips and Systems, Fudan University, Shanghai 200433, China
[2]Institute of Microelectronic Circuits and Systems, East China Normal University, Shanghai 200241, China

* Email: llhuang@cee.ecnu.edu.cn, fanyibo@fudan.edu.cn

Abstract—Light compression standards are developed for low latency video transmission requirements. However, the compression ratio (CR) of most of the light compression standards is not enough for bandwidth-constrained scenarios. The limited CR leads to an increase in cache, which in turn affects latency and area. In this paper, we address to propose a high efficient light compression algorithm with a CR of 20. We implemented the proposed design on Xilinx ZCU102 FPGA. As a result, the codec system has an end-to-end latency of 250.76us with a 150MHz clock.

Keywords—*Light compression, low latency, rate control, limited bandwidth, visual lossless, HEVC, JPEG-XS.*

I. INTRODUCTION

The deep compression standards, like HEVC[1], VVC[2], and AV1[3], have significant compression performance. However, the complexity and latency make it difficult to embed them into the low latency system. On the other hand, the light compression standards like DSC[4], JPEG-XS[5], comes at the cost of compression efficiency in exchange for lower latency.

Among the existing light compression standards, DSC, JPEG-XS, and VC-2 standards are widely used. In [6], a DSC encoder based on FPGA was implemented, which can encode 1080P videos with a latency of 0.015ms at 60fps, with a maximum CR of less than 4. The JPEG-XS encoder proposed in [7] is implemented using HLS and can provide a slice-level latency of about 0.08ms, with a maximum CR about 10. The VC-2 standard proposed by [8] is an intra-frame compression standard used for professional applications, and its tools may be fit for low latency video encoding. The encoding efficiency of VC-2 is similar with JPEG-XS, but it lacks hardware implementation work.

As depicted in Fig. 1, the existing standards can meet the scenarios of high CR and frame-level latency, as well as low CR and subline-level latency. However, there is a gap between the two. Namely, a standard with middle CR and lower latency is required.

This paper is to propose a high efficient light compression standard to balance the latency and the CR. The issue we aim to address is three-folded:

- An optimized prediction algorithm and a context-independent entropy encoding are adopted to achieve higher throughput.
- A line-level stream parallel architecture is proposed to reduce the system latency.
- A Block History Reference (BHR) algorithm is proposed to achieve higher CR.

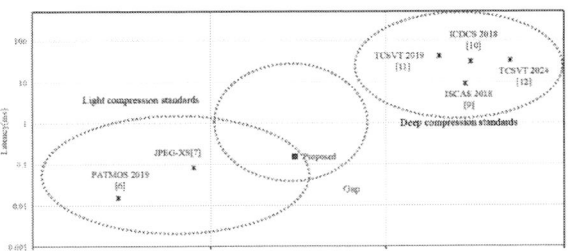

Fig. 1. The CR-Latency statistics of compression standards.

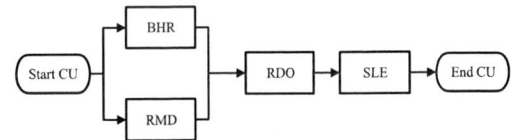

Fig. 2. The process flow of the proposed algorithm.

The proposed design is implemented on hardware, and an end-to-end latency of 250.76us of the codec system is achieved. For the 4K test sequence, the CR is 20 on average. With Xilinx ZCU102 FPGA, the codec system performs with 4K@60fps at a clock of 150MHz.

II. PROPOSED ALGORITHM

The conventional encoding algorithms employ small Coding Units (CUs) to enhance throughput, yet this approach often undermines compression efficiency. To address this, our proposed algorithm utilizes the intra prediction tools in on HEVC[12], with trims on fixed 8x8 CUs and less preditcion modes. This approach, combined with BHR, enhances efficiency for both screen and natural content sequences compared to existing light compression methods.

The process flow of our proposed algorithm is shown as Fig. 2. Rough Mode Decision (RMD) selects 3 candidate prediction modes for luma according to origin pixels. BHR searches the history information, and chooses a History Index (HI) for each 2x2 cube in the CU. Rate-Distortion Optimization (RDO) selected the best mode from the candidate list for luma and from a fixed set for chroma, and the selected modes will be further compared with BHR. Semi-fixed Length Encoding (SLE) performs the coefficients or HIs provided by RDO, and generate the final bitstream. This streamlined approach not only improves encoding efficiency but also reduces latency, making it well-suited for real-time applications.

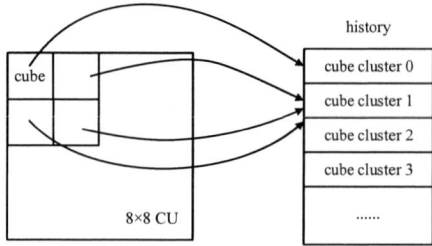

Fig. 3. Cube-history mapping in BHR.

A. Block History Reference

Deep compression standards have introduced complex prediction tools, such as Intra Block Copy (IBC) and Palette, to improve efficiency, especially for screen content coding. However, these tools are too expensive to light compression encoders, not only in complexity but also on latency. The extensive search processes they require lead to high cache overhead, increased computation time, and greater resource consumption.

In screen content coding (SCC) scenarios, abrupt changes on pixel values result in numerous high-frequency coefficients, complicating further compression. To address these issues, we propose the BHR algorithm. BHR leverages a Least Recently Used (LRU) cache to store historical 2x2 cubes and builds a history list. During encoding, pixels are partitioned into cubes and compared with this history. By using HIs for prediction instead of traditional methods, BHR reduces computational complexity and improves encoding efficiency.

As depicted in Fig. 3, each cube is abstracted into a cluster based on the luma and chroma features, which are mapped to the most closed HI within the LRU. Different from Platte mode, BHR maintains a history of clusters, eliminating the escape mode. In SCC scenarios, most of the cubes in a CU may be mapped to the same HI, and merge mode is supported to consolidate redundant HIs into efficient 1-bit data.

After the reconstruction of each CU, the resulting pixels are segmented into cubes. These cubes are then categorized based on luma and chroma features respectively, and assigned to different clusters. BHR employs a 32-entry LRU to store the clusters. Similar cubes are merged into the same cluster, and the representative nature of cubes in the LRU facilitates direct generation of CUs by assembling cube patterns akin to a puzzle.

B. Semi-fixed Length Encoding

Variable length coding (VLC) is a widely used method for low-cost lossless data coding. However, the VLC algorithms are usually with high complexity and data dependency. In that case, the latency of the encoding procedure is not benefit for low-latency design.

Our proposed entropy encoding algorithm is based on SLE in [13]. The CUs are divided into 2x2 cubes, and the coefficients in the same group is encoded with the same bit length. A syntax element of a cube is composed of two parts: the former provides the size of the coefficients, and the latter contains the symbols corresponding to the coefficients. Since the size of the signed coefficients is limited at 8 bits, 3 bits are allocated to represent the length of the coefficients in the cube.

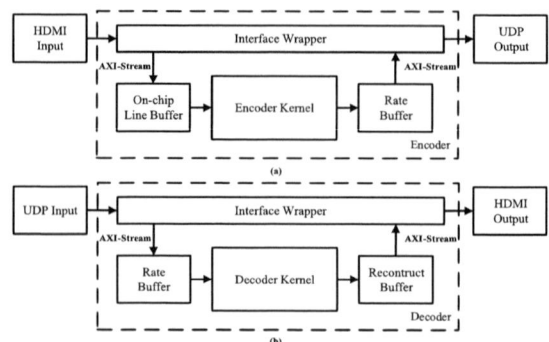

Fig. 4. Hardware architecture of the proposed (a) encoder; (b) decoder.

We tested the SLE proposed in [13] based on the VVC test sequence. Through the statistics of the coefficients, we noticed that the coefficients trend to be quantized into 0 with the increment of CR. However, it still requires 3 bits to transmit the bitplane information for all-zero cubes. We treat the 4x4 region consisting of four adjacent cubes as a coding unit (CU), and for the all-zero information of the cubes in each CU, our statistics are shown in Table I.

TABLE I.   ALL-ZERO CUBES STATISTICS

All-zero cube distribution	Proportion	Guide code
All cubes in CU	47.02%	1
Except the first one	17.04%	01
The last cube	11.27%	001
Others	24.67%	000

A variable-length guide code of up to 3 bits is applied before each syntax element of the CU as shown in Table I, which is used to skip some of the syntax elements of the cube. With the guide code, the 12-bit all-zero CU can be reduced to 1 bit, which means better compression efficiency at higher QP. The results show that our proposed algorithm can reduce the bitplane indicator code of the all-zero cube from 12 bits to 6.38 bits on average. This means that the compression efficiency of the customized algorithm is about 1.3058 times that of [13].

III. HARDWARE IMPLEMENTATION

Our proposed codec system achieves remarkably minimal latency, equivalent to the time required to process a single line of CUs at 60 fps. This achievement is facilitated by the effective reduction on inter-line dependencies and the hardware architecture optimization. As shown in Fig. 4, our encoder and decoder operate efficiently without relying on DDR caching by AXI-Stream bus. This design approach eliminates the DDR bus access latency, ensuring consistent low latency in our system, which is crucial for real-time video transmission applications.

IV. EXPERIMENTAL RESULTS

Experiments are conducted to test the performance of the proposed low latency encoder. We aim to prove that the proposed encoding framework has higher compression ratio and quality, which can better adapt to low latency applications. And the performance is compared with the HEVC encoder Kvazaar[14] and the JPEG-XS[7] encoder.

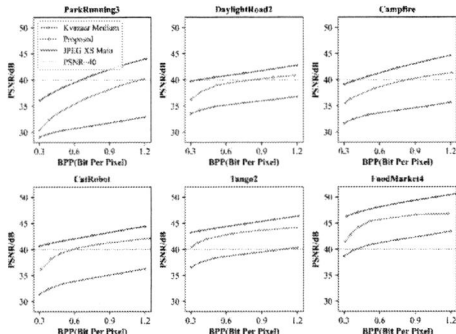

Fig. 5. The PSNR graph for 4K sequences.

A. Encoding Efficiency

We tested coding efficiency on the 6 sequences in constant QP mode on the VVC 4K test sequence, and the CR when the PSNR was over 40 is recorded in Table II. The simpler the sequence texture, the higher the compression ratio.

The encoding performance of the proposed algorithm at CBR mode is shown in Fig. 5. The encoding quality of the proposed algorithm is between Kvazaar Medium and JPEG-XS Main, and the performance is closer to Kvazaar. Due to the configured bandwidth constraint, the encoder gains a stable rate at the expense of PSNR.

TABLE II.   ENCODING EFFICIENCY

Sequence	PSNR	CR
ParkRunning3	40.22	10.05
DaylightRoad2	40.45	12.03
Campfire	40.67	12.09
CatRobot	40.33	19.13
Tango2	40.02	41.29
FoodMarket4	40.37	42.4

B. Hardware Performance

We implemented our proposed design on Xilinx ZCU102 FPGA, and the proposed baseline encoder IP costs 96837 LUT at a 150MHz clock. As shown in Table III, the equivalent cost of the JPEG-XS encoder at 60fps with a 150MHz clock is 88816 LUT. And the hardware cost of our proposed design is closed to the JPEG-XS encoder with equivalent performance, which is much lower than Kvazaar.

TABLE III.   PROPOSED ENCODER PERFORMANCE

Encoder	Kvazaar[14]	JPEG-XS[7]	Proposed
FPGA	Intel Arria 10 GX	Xilinx Alveo U50	Xilinx ZCU102
Throughput	4K@60fps	4K@152fps	4K@60fps
Clock	190MHz	196MHz	150MHz
Encoding delay	16.67ms	80us	126.19us
ALM/LUT	377649 ALM	172196 LUT	96837 LUT
DSP	1468	24	181

In our proposed design, the encoder and the decoder are parallelized in slices, and each slice consists of 8 pixel lines. With a clock of 150MHz, the encoding and decoding latency of our proposed design are 126.19us and 124.57us separately, which is closed to JPEG-XS with the same performance. The total end-to-end latency is 250.76us.

V. CONCLUSION

In this paper, we have presented a light compression encoding algorithm with higher CR. The implemented encoder is able to process 4K60fps video with ultralow latency. Our design offers a hardware cost close to the JPEG-XS, with higher encoding efficiency.

ACKNOWLEDGMENT

This work was supported in part by the National Key R&D Program of China (2023YFB4502802), in part by the National Natural Science Foundation of China (62031009), in part by Alibaba Innovative Research (AIR) Program, in part by Alibaba Research Fellow (ARF) Program.

REFERENCES

[1] G. J. Sullivan, J. -R. Ohm, W. -J. Han and T. Wiegand, "Overview of the High Efficiency Video Coding (HEVC) Standard," in IEEE Transactions on Circuits and Systems for Video Technology, vol. 22, no. 12, pp. 1649-1668, Dec. 2012, doi: 10.1109/TCSVT.2012.2221191.

[2] B. Bross et al., "Overview of the Versatile Video Coding (VVC) Standard and its Applications," in IEEE Transactions on Circuits and Systems for Video Technology, vol. 31, no. 10, pp. 3736-3764, Oct. 2021, doi: 10.1109/TCSVT.2021.3101953.

[3] J. Han et al., "A Technical Overview of AV1," in Proceedings of the IEEE, vol. 109, no. 9, pp. 1435-1462, Sept. 2021, doi: 10.1109/JPROC.2021.3058584.

[4] F. G. Walls and A. S. MacInnis, "VESA Display Stream Compression for Television and Cinema Applications," in IEEE Journal on Emerging and Selected Topics in Circuits and Systems, vol. 6, no. 4, pp. 460-470, Dec. 2016, doi: 10.1109/JETCAS.2016.2602009.

[5] A. Descampe et al., "JPEG XS—A New Standard for Visually Lossless Low-Latency Lightweight Image Coding," in Proceedings of the IEEE, vol. 109, no. 9, pp. 1559-1577, Sept. 2021, doi: 10.1109/JPROC.2021.3080916.

[6] N. Kefalas and G. Theodoridis, "Implementing VESA Display Stream Compression Encoder in FPGAs," 2019 29th International Symposium on Power and Timing Modeling, Optimization and Simulation (PATMOS), Rhodes, Greece, 2019, pp. 35-40, doi: 10.1109/PATMOS.2019.8862082.

[7] Yang, D., Chen, L. (2023). FPGA-Based Hardware Implementation of JPEG XS Encoder. In: Zhai, G., Zhou, J., Yang, H., Yang, X., An, P., Wang, J. (eds) Digital Multimedia Communications. IFTC 2022. Communications in Computer and Information Science, vol 1766. Springer, Singapore. https://doi.org/10.1007/978-981-99-0856-1_14

[8] VC-2 Video Compression. SMPTE ST 2042-1,2022.

[9] P. Sjövall, V. Viitamäki, J. Vanne, T. D. Hämäläinen and A. Kulmala, "FPGA-Powered 4K120p HEVC Intra Encoder," 2018 IEEE International Symposium on Circuits and Systems (ISCAS), Florence, Italy, 2018, pp. 1-5, doi: 10.1109/ISCAS.2018.8351873.

[10] M. Viitanen, J. Vanne, T. D. Hämäläinen and A. Kulmala, "Low Latency Edge Rendering Scheme for Interactive 360 Degree Virtual Reality Gaming," 2018 IEEE 38th International Conference on Distributed Computing Systems (ICDCS), Vienna, Austria, 2018, pp. 1557-1560, doi: 10.1109/ICDCS.2018.00168.

[11] S. Yang, B. Li, Y. Song, J. Xu and Y. Lu, "A Hardware-Accelerated System for High Resolution Real-Time Screen Sharing," in IEEE Transactions on Circuits and Systems for Video Technology, vol. 29, no. 3, pp. 881-891, March 2019, doi: 10.1109/TCSVT.2018.2809690.

[12] G. Xu et al., "A High Compression Efficiency Hardware Encoder for Intra and Inter Coding with 4K@30fps Throughput," in IEEE Transactions on Circuits and Systems for Video Technology, doi: 10.1109/TCSVT.2024.3417382.

[13] D. Zhou et al., "A 530 Mpixels/s 4096x2160@60fps H.264/AVC High Profile Video Decoder Chip," in IEEE Journal of Solid-State Circuits, vol. 46, no. 4, pp. 777-788, April 2011, doi: 10.1109/JSSC.2011.2109550.

[14] Sjovall, Panu & Lemmetti, Ari & Vanne, Jarno & Lahti, Sakari & Hämäläinen, Timo. (2022). High-Level Synthesis Implementation of an Embedded Real-Time HEVC Intra Encoder on FPGA for Media Applications. ACM Transactions on Design Automation of Electronic Systems. 27. 1-34. 10.1145/3491215.

Layer Pipelined Neural Network Accelerator Design on 2.5D FPGAs

Mengxuan Wang, Chang Wu*

School of Microelectronics, Fudan University, Shanghai, China
{20112020079,wuchang}@fudan.edu.cn

Abstract—The 2.5D FPGA integrates 3 or 4 FPGA dies together for larger design capacity and higher computing power. In this paper, we propose an advanced large neural network accelerator design on 2.5D FPGAs. Layer pipeline is considered in our design by making use of the dies for multiple accelerator cores with individual high bandwidth DDRs. Our experimental results show 237.93 FPS for ResNet18 on the Alveo U250 board, which is much better than the existing design using layer pipelining on the same board [1].

Index Terms—Neural Network Accelerator, Layer Pipeline, 2.5D FPGA

I. Introduction

Neural network algorithms are now capable of addressing practical problems across many fields, especially convolutional neural networks (CNN), which are increasingly applied in areas such as defect detection and object recognition. As CNNs continually improve accuracy, they also become increasingly complex and deeper, leading to higher computational demands. Consequently, neural network accelerators have become a research hotspot.

FPGA is used for neural network acceleration due to its low power consumption and reconfigurability, whether in data centers or at the edge. More and more companies are utilizing FPGA accelerators for their workloads. For example, large data centers like Amazon [2], Microsoft [3], and Huawei [4] have launched FPGA-based cloud acceleration services. At the edge, the cost advantage of FPGAs is even more pronounced, and there are already many FPGA-based CNN accelerator solutions. Sharma et al. [5] proposed an accelerator named DNNWeaver based on a PE/PU architecture, which enhances accelerator performance by adjusting the PE/PU configuration. You et al. [6] suggested using inter-layer pipelining to reduce data movement and decrease bandwidth requirements. However, most existing FPGA-based accelerator solutions are designed for traditional FPGAs, the advent of new-generation 2.5D FPGA devices offering substantial computational power and high bandwidth, these solutions have not been optimized for 2.5D FPGAs. Blevec et al. [1] uses multiple layer-based hardware accelerated cores to build a scalable heterogeneous pipelined architecture. They attempted to leverage the multi-resource advantages of 2.5D FPGAs, but the disordered data transfer between

different dies increased pipeline startup latency, resulting in reduced performance.

The 2.5D FPGA utilizes 2.5D stacked silicon interconnect (SSI) technology to package multiple Super Logic Regions (SLRs) into a single FPGA chip. The 2.5D FPGA enables FPGAs to meet the twin demands of higher logic capacity and heterogeneity. and it also permits lower latency communication between die and die. Since each die provides an independent external DDR interface, 2.5D FPGAs feature multiple DDR interfaces and high bandwidth. When considering a 2.5D FPGA as a single large FPGA for CNN accelerator design, the inter-die connection delays significantly reduce the accelerator's clock frequency. If we design it as multiple FPGAs, the computation dependencies between dies reduce parallelism, wasting bandwidth and computational power. Despite the many advantages of 2.5D FPGAs, these issues pose significant challenges in deploying CNN accelerators on 2.5D FPGAs.

In this paper, we propose a layer-pipelined neural network accelerator design for 2.5D FPGAs. We divide the convolution layer into several convolution blocks, and use block-based computational data flow, leveraging the high bandwidth of the multiple DDR interfaces in 2.5D FPGAs to reuse input feature maps and repeatedly load weights. This allows the next layer to commence computation with only a small amount of feature map data. Benefiting from the high bandwidth of 2.5D FPGAs, our layer-pipelined scheme achieves 237FPS on the Alveo U250, which is a 10.5-fold improvement compared to other layer-pipelined schemes.

II. Computing Architecture

We use a 2.5D FPGA with 4 dies as an example to introduce the layer-pipelined convolution neural network accelerator. The 2.5D FPGA with 4 SLRs integrates a DDR4 controller on each SLR. Adjacent SLRs are connected by thousands of Super Logic Links (SLLs), and the latency of SLL is 4 to 8 times that of internal SLR connections.

As shown in Figure 1a, we construct a configurable accelerator core on each SLR. For each accelerator core as shown in Figure 1b, We store the input feature maps and weights in the DDR of the current SLR, and the

979-8-3503-6184-1/24 $31.00 © 2024 IEEE

output feature maps in the DDR of the neighboring SLR. Specifically, we store the output feature maps from the accelerator core of the last SLR3 in the DDR0 of SLR0. Consequently, the input and output of the accelerator cores on the four dies are connected end-to-end, forming a ring structure. This ring structure is the hardware foundation that enables us to implement a layer pipeline. The reason for using this ring structure is that each accelerator core within a die has relatively low bandwidth requirements for output feature maps. Implementing the accelerator core within each die fully utilizes the resources of the SLR, avoiding the drawbacks of the high latency of SLLs.

We can access the data in any DDR through the PCIe bus. When deploying CNN models, convolutional layers are sequentially assigned to each die. Each accelerator core runs only one convolutional layer at a time, loading data for the next layer only after the current layer's computation is complete.

Each accelerator core consists of a set of processing engines (PEs), an input feature map buffer, a set of weight buffers, an output buffer, and a control module. All PEs share the input feature map (IFM) and output feature map (OFM) buffers and are controlled by an instruction controller. Each PE has an independent weight buffer. Within each PE, we implement convolution, activation, and pooling unit, forming a three-stage pipeline.

In our design, The number of PEs in each accelerator kernel, the size of each buffer, and the number of DSPs within each PE are all adjusted based on the characteristics of the convolutional layers running on each accelerator kernel. This approach balances the varying computational demands of different convolutional layers, thereby mitigating DSP idle time issues.

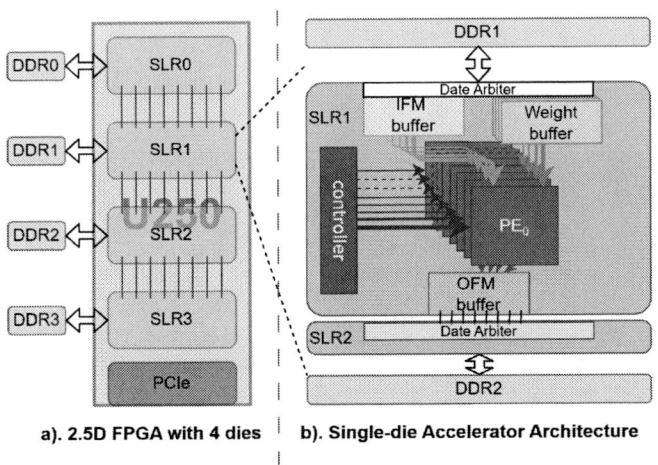

a). 2.5D FPGA with 4 dies   b). Single-die Accelerator Architecture

Fig. 1. The architecture of pipeline-based CNN accelerator. The U250 has four SLRs, each with an independent DDR4 memory. We implement an accelerator core on each SLR, writing the output results to the DDR of the adjacent SLR. This ultimately forms a ring structure.

III. Computing Dataflow

In our design, we propose using block convolution to handle data dependencies between die and die. The basic idea of block convolution is to split the feature maps and weight of convolutional layers into independent blocks. Block convolution employs a split-concat computing mechanism, there is no data dependency between the convolution results of adjacent spatial blocks.

We divide the convolution layer into several convolution blocks according to the process shown in Algorithm 1, and then perform computations. Specifically, we pre-divide the IFM into several data blocks of size $b_c \times b_h \times b_w$ and the weights into serveal blocks of size $b_n \times b_c \times k_y \times k_x$, based on the parameters H, W, C, N, k_x, k_y of each convolutional layer. Using these data blocks as units, we partition the entire convolution layer into multiple convolution blocks. Each time, we compute one convolution block and switch to the next block upon completion. and we cannot load all the parameters required for the entire convolution layer into the buffer. Dividing into data blocks significantly reduces the demand for buffer space. Additionally, computing convolutions in blocks allows each accelerator core to start with only a portion of the feature map data.

We follow the computation priority order of $H \to W \to N \to C$. This is because the entire result along the C direction is accumulated to produce an output in the N direction, allowing the next layer to begin computation. In other words, after loading an IFM data block, we repeatedly load the weights related to this data block. By leveraging the high bandwidth of the 2.5D FPGA, we ensure the pipeline can run continuously.

When the computation begins, the IFM buffer reads a data block of size $b_c \times b_h \times b_w$, and the weight buffer loads the corresponding weights of size $b_n \times b_c \times k_y \times k_x$. The computation proceeds while keeping the $b_h \times b_w$ window fixed, switching to a new b_c' until the entire C direction is computed. Then, The weights are switched to a new b_n' block and reload the IFM block, and the output feature map data in the output buffer is written to the DDR. Once the computation in the N direction is complete, the input feature map window moves in the order of W first and then H, repeating the process described above. At this point, the convolution core on the adjacent SLR can start computing the next convolutional layer. This repeated process forms a block-based layer pipeline.

IV. Evaluation Results

A. Experimental Setup

Our accelerator design is implemented with Vivado 2023.2 on the AMD Alveo™ U250 Data Center accelerator card. Its working frequency is 200 MHz, and the computation control software running on an Intel(R) Core(TM) i7-7700K CPU @ 4.20GHz.

Algorithm 1 Partition the convolution layer into several convolution blocks

---

for $h = 0$; $h < H$; $h = h + b_h$; do
  for $w = 0$; $w < W$; $w = w + b_w$; do
    for $n = 0$; $n < N$; $n = n + b_n$; do
      for $c = 0$; $c < C$; $c = c + b_c$; do
        ▷ Partition IFM data blocks and load to IFM buffer;
        $ifm[b_{c_{ifm}}][b_{h_{ifm}}][b_{w_{ifm}}] \leftarrow IFM[C][H][W]$;
        ▷ Partition weight data blocks and load to weight buffer, each weight buffer loads a data block;
        $weight[b_{n_{weight}}][b_{c_{weight}}][k_x][k_y] \leftarrow weight[N][C][k_x][k_y]$;
        ▷ We set b_n equal to the number of PEs. All PE units share the IFM block;
        $ofm[b_{n_{ofm}}][b_{h_{ofm}}][b_{w_{ofm}}] = conv(ifm, weight)$;
        if $c == C$ then
          ▷ When the C dimension computation is complete, the OFM is stored in the adjacent SLR's DDR;
          $OFM[N][H_{ofm}][W_{ofm}] \leftarrow ofm[b_{n_{ofm}}][b_{h_{ofm}}][b_{w_{ofm}}]$;
        else
          ▷ Otherwise, it is stored as a partial sum in the OFM buffer;
          $ofm[b_{n_{ofm}}][b_{h_{ofm}}][b_{w_{ofm}}] \leftarrow ofm[b_{n_{ofm}}][b_{h_{ofm}}][b_{w_{ofm}}]$;
        end if
      end for
    end for
  end for
end for

B. Experimental Results

To evaluate our design, we deployed ResNet18 and VGG16 on the Alveo U250 accelerator card. The experimental results are shown in Table I. For ResNet18, which has 18 convolutional layers, our design runs an average of 4 layers per die, with nearly equal computational loads for all accelerator cores. Consequently, the resource consumption of each accelerator core is nearly identical. This setup achieved an actual frame rate of 237.93 FPS. Compared to the existing design [1] using the U250, our inference speed is 10.5 times faster.

We also implemented the VGG16 network, which has 16 convolutional layers. Due to the structure of VGG16, the computational load on each accelerator core varied significantly. Our DSE algorithm adjusted the configuration of each accelerator core based on the model characteristics to maintain consistent pipeline wait times as much as possible. This approach resulted in achieving a frame rate of 98.4 FPS.

V. Conclusion

This paper presents a layer-pipelined neural network accelerator design for 2.5D FPGAs. We leverage the multi-die characteristics of 2.5D FPGAs to implement an accelerator core on each die, forming a ring structure. By combining this with a block-based data flow, we achieve a layer pipeline accelerator. Deploying ResNet18 on the Alveo U250 achieves 237.93 FPS, making it a highly significant attempt.

References

[1] H. L. Blevec, M. Léonardon, H. Tessier, and M. Arzel, "Pipelined Architecture for a Semantic Segmentation Neural Network on FPGA," in 2023 30th IEEE International Conference on Electronics, Circuits and Systems (ICECS), Dec. 2023, pp. 1–4.

[2] D. Rankin, J. Krupa, P. Harris, M. A. Flechas, B. Holzman, T. Klijnsma, K. Pedro, N. Tran, S. Hauck, S.-C. Hsu, M. Trahms, K. Lin, Y. Lou, T.-W. Ho, J. Duarte, and M. Liu, "Fpgas-as-a-service toolkit (faast)," in 2020 IEEE/ACM International Workshop on Heterogeneous High-performance Reconfigurable Computing (H2RC), 2020, pp. 38–47.

[3] A. M. Caulfield, E. S. Chung, A. Putnam, H. Angepat, J. Fowers, M. Haselman, S. Heil, M. Humphrey, P. Kaur, J.-Y. Kim et al., "A cloud-scale acceleration architecture," in 2016 49th Annual IEEE/ACM international symposium on microarchitecture (MICRO). IEEE, 2016, pp. 1–13.

[4] J. Guo, L. Zhang, J. Romero Hung, C. Li, J. Zhao, and M. Guo, "Fpga sharing in the cloud: a comprehensive analysis," Frontiers of Computer Science, vol. 17, no. 5, p. 175106, 2023.

[5] H. Sharma, J. Park, D. Mahajan, E. Amaro, J. K. Kim, C. Shao, A. Mishra, and H. Esmaeilzadeh, "Dnnweaver:fromhigh-leveldeepnetworkmodelstofpgaacceleration," in 2016 49th Annual IEEE/ACM International Symposium on Microarchitecture (MICRO), 2016, p. 1–12, citation Key: sharmaDNNWEAVER2016.

[6] W. You and C. Wu, "RSNN: A Software/Hardware Co-optimized Framework for Sparse Convolutional Neural Networks on FPGAs," IEEE Access, vol. PP, pp. 1–1, Dec. 2020.

TABLE I
omparison between our design and existing FPGA designs

	Blevec et al. [1]	Our design	Our design
Models	ResNet18	ResNet18	VGG16
Platform	Alveo U250	Alveo U250	Alveo U250
Process	16nm	16nm	16nm
Frequency	152 MHz	200 MHz	200MHZ
DSP	1043	5261	6343
LUT	490265	118361	119750
BRAM	2130	2206	2274
FPS	22.65	237.93	98.4

979-8-3503-6184-1/24 $31.00 © 2024 IEEE

Fast and Accurate Partial-Zoom Depth Estimation for SPAD LiDAR Readout on FPGA

Lichen Feng[1], Hongwei Shan[1], Rundong Cai[1], Zhangming Zhu*[1]

[1]Key Laboratory of Analog Integrated Circuits and Systems (Xidian University), Ministry of Education, School of Integrated Circuits, Xidian University, Xi'an 710071, P. R. China

* Email: lcfeng@xidian.edu.cn; zmyh@263.net

Abstract—Laser detection and ranging (LiDAR) based on single-photon avalanche diode (SPAD) has emerged as a widely applicable technology. The full-histogram method based on time-to-digital converter (TDC) obtains high accuracy at the cost of large memory. The compact full-zoom method with an asynchronous counter eliminates the use of TDC and full resolution histogram has been realized. However, it increases estimation error and latency. This paper proposes a fast and accurate partial-zoom method. The cause of increased depth error of full-zoom method is analyzed mathematically for the first time, and the accuracy is improved by introducing two auxiliary asynchronous counters. The speed is improved without accuracy degradation by reducing the number of zoom stages. Compared to the full-zoom method, the proposed partial-zoom method needs less laser cycles to obtain the better accuracy, achieving the reductions of 42% laser cycles and 58% RMSE. Furthermore, the proposed partial-zoom method is implemented on FPGA, demonstrating 92% memory, 80% power and 55% latency reduction compared to full-histogram method.

Keywords—SPAD LiDAR, depth estimation, asynchronous counters, partial zoom

I. INTRODUCTION

Laser detection and ranging (LiDAR) based on single-photon avalanche diode (SPAD) is effective in environment detection with depth information. As shown in Fig. 1 (a), besides the SPAD array used for converting photons into current, corresponding processing and memory units can also be integrated in the system to form an all-solid-state LiDAR. In general, the multi-channel time-to-digital converters (TDCs) with high time-resolution are working continuously to obtain the picosecond-level photon's direct time of flight (dToF). These dToFs are counted and stored in the memory to obtain the histogram. To alleviate the power and area requirements, many researches have been proposed [1]-[3], as shown in Fig. 1 (b). Among these solutions, zoom-based method obtains the best area efficiency, since only one asynchronous counter instead of the memory bank is need for an SPAD pixel. However, there is no explicit theoretical analysis for the degraded depth accuracy, and the large processing delay remains to be solved (e.g. 1024 bins needs 9 iteration stages). To find a balance between depth accuracy, latency and area, this paper proposes a fast and accurate partial-zoom method to address the above challenges. The major contributions of this paper are summarized below:

- Based on the probability distribution of the zooming stages, we quantitatively express the relationship between the probability of successful zooming (the range reduction of coarse-bins) and the number of laser cycles for the first time.

This work is supported in part by the National Key R&D Program of China (2022ZD0118903), in part by the National Natural Science Foundation of China under Grant U22A2013, Grant 62104175.

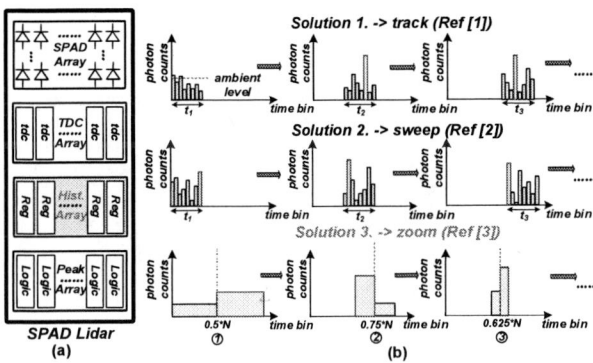

Fig. 1. (a) The system architecture of the SPAD LiDAR; (b) Three common solutions for alleviating the high memory demand.

- We proposed the fast and accurate partial-zoom method. The estimation accuracy and speed is further improved by counting edge bins at the last iteration stage and introducing two auxiliary counters. Compared to [3], the proposed method achieves 58% reduction on RMSE and 42% reduction on laser cycles.

- Implemented on FPGA, the proposed partial-zoom method achieves 41.6us@50MHz, consuming only 843 LUT, 1721 Flip Flops and 0.8kb BRAMs per channel.

II. THE FAST AND ACCURATE PARTIAL-ZOOM METHOD

A. The effectiveness of the counter-based zoom method

We assume that dToFs of the difference between transmitting and receiving photon sequence follows Gaussian distribution [4] with parameter μ (real time interval, $1 <= \mu <= T$) and σ as:

$$P_{Gau}(t = x) = \frac{1}{\sqrt{2\pi}\sigma} e^{\frac{-(x-\mu)^2}{2\sigma^2}} \quad (1)$$

Then we design an enabled double counter working from t_1 to t_2, which acts as an incrementing counter (called state *A*) from t_1 to $0.5*(t_2-t_1)$ and decrementing counter (called state *B*) from $0.5*(t_2-t_1)$ to t_2. That is, when the received photon arrived at t_1 to $0.5*(t_2-t_1)$, the counter adds one, on the contrary, the counter substrate one. Since each received photon is an independent event and follows Gaussian distribution, we can calculate the probability of state *A* and *B* for counter in one laser cycle as:

$$P(A) = \int_{t1}^{0.5*(t2-t1)} P_{Gau} \, d(t) \quad (2)$$

$$P(B) = \int_{0.5*(t2-t1)}^{t2} P_{Gau} \, d(t) \quad (3)$$

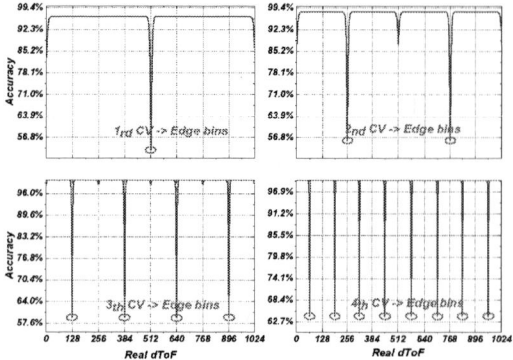

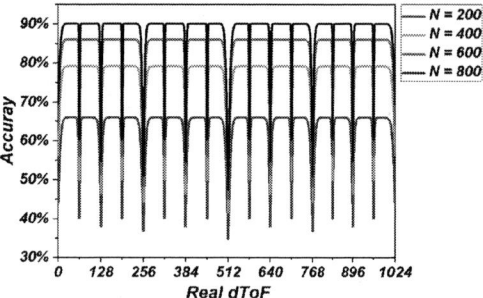

Fig. 2. Accuracies of four iteration stages versus different bins. CV is the control value of the each stage, that is the middle value of coarse-bins.

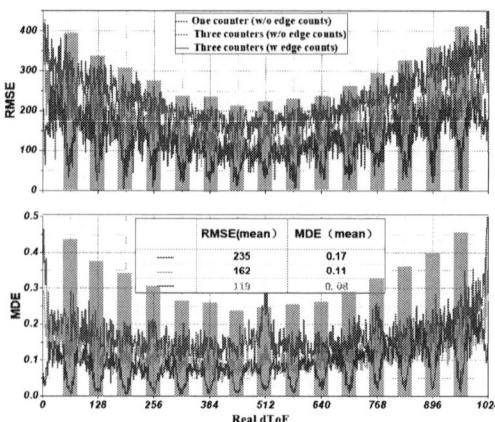

Fig. 3. Overall successful probability under different values of N.

Assuming that the length of received photon sequence is N, we can calculate the probability of the number X of state A being k (initial value is set as zero) after N laser cycles based on the binomial distribution as:

$$P_{Bin}(X = k) = C_N^k \times P(A)^k \times P(B)^{N-k} \qquad (4)$$

According to the basic principle of binary search in zoom [3] method, we are ought to calculate the probability of k>=0 (called state C, means that the number of occurrences of state A is larger than $N/2$) and k<0 (called state D, means that the number of occurrences of state A is shorter than $N/2$) as:

$$P(C) = \int_{N/2}^{N} P_{Bin} \, d(k) \qquad (5)$$

$$P(D) = \int_{1}^{N/2} P_{Bin} \, d(k) \qquad (6)$$

Based on the above definition and derivation, we generate a dataset and perform a probability simulation with the following parameters: $T=1024$, $N=1000$, $\sigma=8$. If $P(C) > P(D)$, the enable time of the counter would be adjusted to t_1 to $0.5*(t_2-t_1)$, on the contrary, it is $0.5*(t_2-t_1)$ to t_2. We repeat this process four times to reduce the range of enable time gradually, that is, the range of target bins (called coarse bins). The correct probability in each zooming (maximum value of $P(C)$ and $P(D)$) is shown in Fig. 2. We can conclude that the counter-based zoom method has a high accuracy in searching for the coarse position of target bin, thus it is effective.

B. The proposed partial-zoom histogram method

Ref. [3] proposed full-zoom method to reduce the number of bins, which needs 9 iteration stages for 1024 bins. This method decreases memory size due to the bin size is only one, however, a sufficient number of laser cycles is required to achieve better accuracy. Thus, we develop a fast and accurate partial-zoom method based on full-zoom method.

Fig. 4. The effectiveness of introducing two auxiliary counters and counting edge bins in the last iteration stage.

For a depth estimation task with 1024 bins, we use four rather than nine iteration stages to reduce the range of bins, that is, reduced the range from 1024 to 64. Then, in the fifth iteration, TDCs are utilized to calculate the dToFs of the transmitting and receiving photon, and store them in the memories. The index of maximum number of memories is the desired value. The proposed method leverages the advantages of both full-zoom and full-histogram, achieving a small memory size while enhancing accuracy and latency.

Now we define the parameter Acc as the probability that four zooming are all correct, and simulate under different values of N, as shown in Fig. 3. We can draw two conclusions. First, the edge bins (in this case, it is the multiple of 16) are more prone to errors, because $P(A)$ is nearly to $P(B)$, thus the difference between $P(C)$ and $P(D)$ is smaller. Second, the successful accuracy increases as the value of N increases, because larger N can amplify the difference between $P(A)$ and $P(B)$, however, it leads to a decrease in speed due to more laser cycles are needed. Based on the two conclusions, we propose two optimization strategies. First, edge bins are also counted besides 16 coarse-bins in the fifth iteration stage. Second, two more auxiliary counters (called C_1 and C_2) are utilized to work with main counter (called C_0) together. For example, if the range of coarse-bins of C_0 is from t_1 to t_2, the coarse-bins of C_1 and C_2 is from t_1 to $0.5*(t_2-t_1)$ and $0.5*(t_2-t_1)$ to t_2, respectively. In the next iteration stage, the initial value of C_0 is not zero but the counting result of C_1 or C_2 in the previous iteration, depending on the polarity of the counting result of C_0. In this way, the number of laser cycles for every iteration stage (except first) becomes $2*N$ rather N, which can increase accuracy without sacrificing speed.

C. Accuracy verification

To test the fast and accurate partial-zoom method, we generate a dataset, which is consisted of $T=1024$ categories and $n=1000$ samples per category. Each sample is generated based on Gaussian distribution with $N=400$ laser cycles. The performance metrics are relative mean expected absolute depth error (RMDE) and root mean square depth error (RMSE) [6], defined as:

$$\begin{cases} \text{RMDE} = \frac{1}{T} \sum_{j=1}^{T} \frac{1}{n} \sum_{i=1}^{n} |\mu_{e,ij} - \mu_{o,ij}| \\ \text{RMSE} = \frac{1}{T} \sum_{j=1}^{T} \sqrt{\frac{1}{n} \sum_{i=1}^{n} (\mu_{e,ij} - \mu_{o,ij})^2} \end{cases} \qquad (7)$$

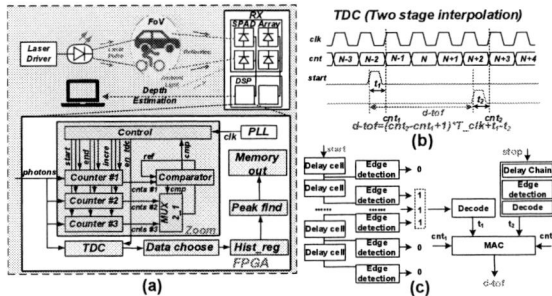

(a)

(b)

(c)

Fig. 5 (a) Circuit block diagrams of the partial-zoom method. (b) The computation of TDC (c) The circuit design of TDC.

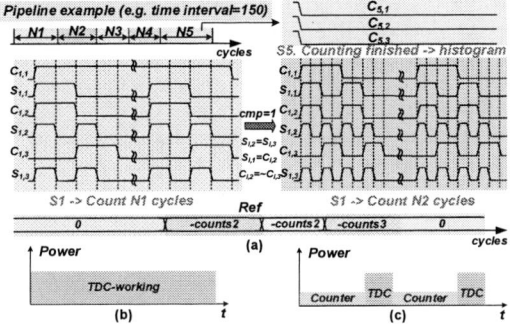

(a)

(b)

(c)

Fig. 6. (a) Timing diagram of the partial-zoom system. (b) Power analysis of full-histogram method. (c) Power analysis of patrial-zoom method.

where $\mu_{e,ij}$ and $\mu_{o,ij}$ are the time interval of the estimated and actual for the i_{th} sample of j_{th} category, respectively.

The simulated results of these two metrics are illustrated in Fig. 4. It can be seen the two optimization strategies proposed are effective, reducing the RMSE and RMDE globally and locally (edge bins), respectively. The RMSE and RMDE decreases 49.4% and 52.9% compared to the method of one counter without counting edge bins.

III. HARDWARE IMPLEMENTATION ON FPGA

The circuit block diagram including the TDC design and timing diagram of the proposed fast and accurate partial-zoom method are shown in Fig. 5 and 6.

In the block diagrams, the **control** module can generate different enable signals (**start** and **end** signal to control the counter operate in **A** or **B**) to derive three counters. When one iteration stage is completed, the count result of **counter1** would be compared with the **ref** signal. The comparison result determines not only the enable signal in the next iteration, but also the new **ref** signal in the next comparison. To be specific, when compared result is positive, new **ref** signal is **-cnt2**, on the contrary, it is the **-cnt3**. When four iterations are all completed, the **zoom** module stops working and received photons would be sent to **TDC** module to calculate the estimated dToF, which is stored in **Hist-ram** module if it is located in the desired coarse-bins. After five iteration stages (each stage with **N** cycles of laser pulse), **Peak-find** module is activated to find the index of maximum counts, and send it to the **Memory-out** module as the result of depth estimation. For the design of TDCs circuit, we employ a second-order interpolation structure and the second stage is based on Tapped delay line (TDL) architecture [5].

In the timing diagrams, we take **t=150** as example to explain. The entire pipeline operation can be divided into five

TABLE I
PERFORMANCE COMPARISONS WITH
PREVIOUS METHODS USED FOR DEPTH ESTIMATION

Algorithm	[3] full-zoom	[7] full-histogram	This work partial-zoom
RMSE/MDE	**188/17.1%**	**8/0.4%**	**119/8.6%**
Utilization* on FPGA	24 LUT 41 FF	813 LUT 1661 FF **10kb BRAM**	843 LUT 1721 FF **0.8kb BRAM**
Latency* @ 50MHz	**72.0us**	**92.0us**	**41.6us**
Memory Friendly?	Yes	No	**Yes**

*: Three algorithms are all re-implemented on the same FPGA;

stages, with the first four stages using counter-based zoom operation and the final stage using TDC-based histogram operation. $C_{j,i}$ represents the enable working time (state **A** and state **B**) of the j_{th} counter in the i_{th} iteration, $S_{j,i}$ represents the enable incrementing time (only state **A**)of the j_{th} counter in the i_{th} iteration. The comparison result is most likely positive when the first iteration is finished, thus, the $C_{1,2}$ is adjusted to $S_{1,1}$ to start the second iteration stage. When the first four iteration stages are all finished, $C_{j,5}$ are all pulled to ground to turn off three counters. Then the **TDC** module begins to calculate dToF and send it to the **Hist-ram** module.

Our proposed model has been implemented on Xilinx xc7z020 FPGA. The performance and comparison with previous works are summarized in Table I. The proposed partial-zoom method reduces 58%/98% RMSE/RMDE and 42% latency compared to [3]. And it reduces 92% memory, 55% latency and nearly 80% power compared to [7]. As shown in Fig. 6 (b) and (c), the power is decreased due to the use of low-power counter.

IV. CONCLUSION

In this paper, a fast and accurate partial-zoom method is proposed for the depth estimation of SPAD LiDAR, which acquires significant speed and accuracy improvement compared to the full-zoom method and remarkable reduction in resource consumption compared to the full-histogram method. The corresponding compact circuit system design is implemented on FPGA to verify the performance in circuit.

REFERENCES

[1] I. Gyongy et al., "A 200k FPS 256 × 128 SPAD dToF sensor with peak tracking and smart readout", *Proc. Int. Image Sensor Workshop*, pp. 85-88, 2021.

[2] D. Stoppa et al., "A reconfigurable QVGA/Q3VGA direct time-of-flight 3D imaging system with on-chip depth-map computation in 45/40 nm 3D-stacked BSI SPAD CMOS", *Proc. Int. Image Sensor Workshop*, pp. 53-56, 2021.

[3] B. Kim et al., "A 48×40 13.5 mm depth resolution flash LiDAR sensor with in-pixel zoom histogramming time-to-digital converter", *IEEE Int. Solid-State Circuits Conf. (ISSCC)*, pp. 108-110, Feb. 2021..

[4] V. Poisson et al., "A 2-stage EM algorithm for online peak detection an application to TCSPC data", *IEEE Trans. Circuits Syst. II Exp. Briefs*, vol. 69, no. 9, pp. 3625-3629, 2022.

[5] M. Arredondo-Velázquez et al., "Trimmed-TDL-Based TDC Architecture for Time-of-Flight Measurements Tested on a Cyclone V FPGA," *IEEE Trans. Instrum. Meas.*, vol. 72, pp.1-9, 2023.

[6] F. Gutierrez-Barragan et al., "Compressive single-photon 3D cameras," *IEEE Conf. Comp. Vision and Pattern Recogn. (CVPR)*, pp. 17833-17843, 2022.

[7] H. Seo et al., "Direct TOF Scanning LiDAR Sensor With Two-Step Multievent Histogramming TDC and Embedded Interference Filter," *IEEE J. of Solid-State Circuits*, vol. 56, no. 4, pp. 1022-1035, 2021.

979-8-3503-6184-1/24 $31.00 © 2024 IEEE

A Self-adaptive Gamma Voltage Regulation Circuit for AMOLED Displays

Zhifeng Mao [1,2], Fei Gou [1], Bin Sheng [1], Jing Xie [1], Wenwei Xu [1], Wei Liu [1], Jun Xu*[2]

[1] Glenfly Tech Co., Ltd., Shanghai 201203, China
[2] School of Integrated Circuits, Tsinghua University, Beijing 100084, China

* Email: JasonMao@glenfly.com, JunXu@tsinghua.edu.cn

Abstract—**To compensate the effect of AMOLED panel power supply voltage (ELVDD) fluctuation on luminance, in this paper, an analog circuit for gamma voltage regulation is proposed, which has self-adaptive tracking and simple structure. A source driver IC by using this design was fabricated in a 0.11-μm high voltage CMOS technology, and silicon measurement results show that this design can track the panel power supply voltage changes effectively by regulating gamma system, and feedback to the panel with a compensated source driver output voltage. When a fluctuation amount of ±0.2V is applied on the ELVDD, the maximum luminance change can be reduced from 108.6% to 1.5%.**

Keywords—panel power supply voltage fluctuation, voltage detection, gamma voltage, self-adaptive regulation, compensation circuit

I. INTRODUCTION

Active-matrix organic light emitting diode (AMOLED) has become a very popular display technology, and compares with liquid crystal display (LCD), AMOLED panel has many advantages and has been widely used in high imaging quality and high performance applications. AMOLED requires a driver IC to control the line driver circuit (emission driver) and the scanning circuit (scan driver), as shown in Fig. 1, to realize the switches of the thin-film transistor (TFT) pixel circuit to turn on and off in a line by line sequence. And the driver IC converts the imaging data to analog voltage (V_{DATA}), which is sent to the data line of the pixel circuit to charge the storage capacitor and generate current to make the diode light-emitting. Due to fluctuating changes in the panel power supply voltage (ELVDD), the variations on the light-emitting current will be introduced in. In order to eliminate the impact of fluctuations in ELVDD on the luminance of the AMOLED panel, the panel design or external driver IC need to make the appropriate compensation. Traditional compensation is to measure panel brightness and gamma curve then using the algorithm to correct. However, this approach requires specific calibration steps and on chip algorithm, so it is more complex, time-consuming, and high cost. In this paper, a self-adaptive gamma voltage regulation circuit is proposed, which detects ELVDD voltage of the panel in real time, and compares the measured voltage value with the target voltage value, then obtains the voltage difference, ΔELVDD, and provides this voltage difference to the gamma high voltage (VGMP) and the gamma low voltage (VGSP),

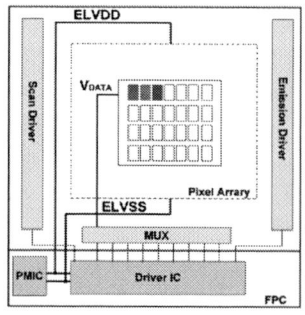

Fig. 1. AMOLED panel and module

which are generated by the gamma circuit and used as the reference of gray level. Proposed circuit can dynamically adjust the V_{DATA} voltage level and achieve a fixed voltage difference between ELVDD and V_{DATA} to keep the driving current and the luminance without changes. As the traditional methods need to measure the panel gamma curve and use the algorithm to calibrate the input image data, this design can adaptively adjust the gamma voltage and source driver output voltage directly, the procedure is simple and effective.

II. ARCHITECTURE

A. Pixel Circuit

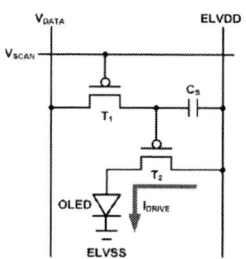

Fig. 2. 2T1C pixel circuit

The simple scheme of AMOLED pixel circuit unit is shown in Fig. 2. A 2T1C (Two transistors and one capacitor) architecture is used to control and assist the light emission, where T_1 and T_2 are the TFT devices. T_1 is used as the switching transmitter, and T_2 is used as the OLED light emission driver, and C_S is the charge storage capacitor. The current value I_{DRIVE} that dominates the brightness of the light emitting device is calculated as:

$$I_{DRIVE} = \frac{1}{2}\mu_p C_{ox}\left(\frac{W}{L}\right)(ELVDD\text{-}V_{DATA}\text{-}|V_{TH}|)^2 \quad (1)$$

Where μ_p represents the carrier mobility of the TFT device T_2, C_{ox} is the unit capacitance value of the T_2 gate oxide, W and L are the width and length of T_2, and V_{TH} refers to the T_2 threshold voltage. From (1), it can be seen that the driving current of OLED is strongly correlated with the TFT device intrinsic carrier mobility and threshold voltage. These device parameters are related to the manufacturing process, material, etc., and the value will be followed by variation in those factors, which causes issues with uniformity of OLED brightness. Different pixel circuits have been proposed to compensate for these nonideal factors, such as V_{TH} [1-2], and carrier mobility [3-4]. External IC compensation methods have also been studied in [5-6]. Basically those work can make the light-emitting current I_{DRIVE} independent of V_{TH} or carrier mobility. However, the influence of ELVDD still exists, so new pixel circuit and algorithm have been proposed to compensate for the internal power line IR-drop of AMOLED panel [7-10], which only focused on the systematic and static deviations. In this paper, a gamma voltage dynamic adaptive circuit is designed to further eliminate the influence of ELVDD variations on I_{DRIVE} by external chip compensation and no need to change the panel pixel circuit.

979-8-3503-6184-1/24 $31.00 © 2024 IEEE

B. Gamma Voltage Regulation Scheme

The panel power supply voltage ELVDD comes from the power management IC (PMIC), and due to the presence of load current and line resistance, ELVDD suffers the phenomenon of IR-drop, and shows a certain variation along the routing path. The diagram of the regulation and compensation circuit system is shown in Fig. 3, where the ELVDD terminal node of the driver IC detects the power supply voltage level on the panel through a low-pass filter, R_1 and C_1, and feed the detected ELVDDI into the subtraction module after a clamping circuit. ELVDDI is used to compare with a reference voltage ELVDDR, and then the difference of ΔELVDD between these two voltages is calculated. To trace ΔELVDD in the gamma reference voltage, we use the adder unit to sum ΔELVDD with VGMPR or VGSPR respectively, which is generated by the reference block, and the output VGMPD or VGSPD goes to next module to generate gamma reference VGMP or VGSP, where the values of ELVDDI, VGMPD, and VGSPD are defined respectively as:

$$ELVDDI=\Delta ELVDD+ELVDDR \qquad (2)$$

$$VGMPD=VGMPR+\Delta ELVDD \qquad (3)$$

$$VGSPD=VGSPR+\Delta ELVDD \qquad (4)$$

The image pixel data from application processor is converted to analog voltage V_{DATA} by the gamma gray level conversion, and the target V_{DATA} is a voltage level between, or equal to, VGMPR and VGSPR. Due to AMOLED gamma curve is approximated to linear, V_{DATA} can be expressed as:

$$V_{DATA}=\alpha VGMPR+(1-\alpha)VGSPR \qquad (5)$$

where $0\leq\alpha\leq1$, combined with (1), the target driving current I_{DRIVER} can be calculated as:

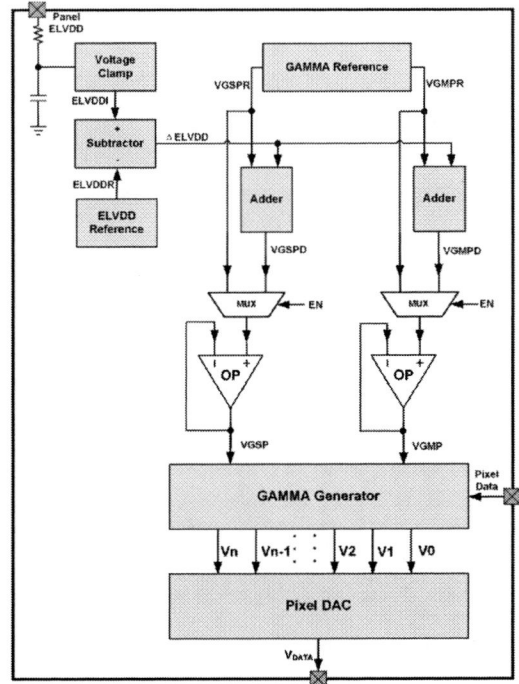

Fig. 3. Gamma self-adaptive circuit diagram

$$I_{DRIVER}=k[ELVDDR-\alpha VGMPR-(1-\alpha)VGSPR]^2 \qquad (6)$$

Where $k=\frac{1}{2}\mu_p C_{ox}\left(W/L\right)$, and we consider the drive current independent of V_{TH} with improved pixel circuits. Similarly, the actual driving current I_{DRIVEI} is:

$$I_{DRIVEI}=k[ELVDDI-\alpha VGMPD-(1-\alpha)VGSPD]^2 \qquad (7)$$

Substituting (2), (3), (4) to (7), and simplified it obtains:

$$I_{DRIVEI}=k[ELVDDR-\alpha VGMPR-(1-\alpha)VGSPR]^2 \qquad (8)$$

It can be seen from (6) and (8) that I_{DRIVER} and I_{DRIVEI} are equal. So in the case of deviations in ELVDD on the panel relative to the target ELVDDR, the light-emitting drive current are keep same as the expected value, thus eliminating current fluctuations caused by variations in the voltage of ELVDD.

III. CIRCUIT DESIGN

A. Operational amplifier with chop function

In the signal path of the dynamic tracking ELVDD, VGMPR and VGSPR go through the adder unit and buffer to generate VGMP and VGSP, but mismatches in the transistor and resistor, which is introduced in fabrication, causes offset voltage of the final output VGMP and VGSP, then reduces the accuracy of the gamma voltage. To eliminate the offset voltage, an operational amplifier architecture with chop function is shown in Fig. 4. A chop mux is inserted before amplifier input stage and intermediate stage, which alternately switches corresponding connections between the external input voltage and the amplifier input stage by control of the chop clock, so that the DC distortion and the low frequency noise are modulated to the high frequency, and the last stage of the buffer acts as a low pass filter to filter out the noise and the distortion voltage at the high frequency.

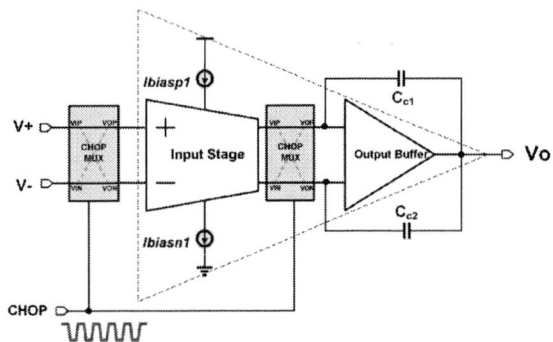

Fig. 4. Operational amplifier with chop function

B. ELVDD voltage clamp circuit

Usually, the ELVDD voltage comes from the power management chip, and it has abnormal large fluctuation occasionally. If the VGMP and VGSP still respond to this abnormal fluctuation and follow the ELVDD, the gamma voltage will also soar beyond its normal range. To avoid this situation, a voltage detection and clamp circuit is presented to limit the amplitude of VGMP and VGSP within a preset reasonable range. The voltage clamp circuit shown in Fig. 5, it detects the external ELVDD and feeds the divided voltage ELVDDF in two comparators with the threshold voltage of VRP and VRN. The output of comparators DH<1:0> control three switches to pass VRP, VRN or ELVDDF to a voltage

---

979-8-3503-6184-1/24 $31.00 © 2024 IEEE

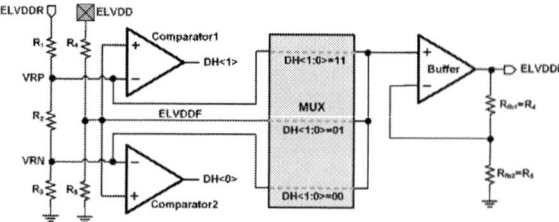

Fig. 5.   ELVDD voltage clamp circuit

buffer. The buffer regulates output voltage ELVDDI within an expected range, while its upward or downward limited level, ΔV_U or ΔV_D, depends on the value of VRP or VRN.

IV. SILICON RESULTS

Base on the development of a 0.11µm OLED driver IC product, we implemented the proposed self-adaptive gamma voltage regulation circuit in. Silicon results of ELVDD and V_{DATA} transient waveform is shown in Fig. 6, from the test waveforms, we can see that the final output voltage V_{DATA} to the panel of the source driver accurately follows ELVDD changes by the dynamic tracking circuit to suppress and compensate the brightness variations of the AMOLED panel effectively. In addition, in order to avoid overshooting or undershooting, a voltage clamp circuit is implemented, and the clamp amplitude ΔV_U and ΔV_D for upward and downward level can be adjusted. Test results show that when the clamp function is turned on, while ELVDD fluctuates more than the pre-set value, the V_{DATA} is clamped and devices operate within an expected safe voltage range. Fig. 7 shows the corresponding luminance outputs of the panel versus different ΔELVDD values at four gray levels. We can see that there is almost no change in brightness when the tracking mode is turned on compared to the drastic change in brightness when the tracking mode is turned off.

V. CONCLUSION

Traditional AMOLED compensation methods mostly focus on the electrical parameter variations of TFT transistor device. In this paper, we proposed a dynamic tracking and gamma regulating circuit to compensate the panel power supply ELVDD fluctuations. By monitoring the ELVDD changes in real time, the source driver output voltage V_{DATA} also takes this amount of change in to sustain a same voltage difference and panel luminance unchanged. For the ELVDD dynamic tracking mode turn off and turn on, with different gray level image data, the panel brightness deviations for the panel power supply voltage ELVDD changes are organized as Table I. When ELVDD tracking circuit is not enabled, the actual luminance change ratio of the panel reaches 32.2%~108.6% at 4 fixed gray level outputs. After ELVDD dynamic tracking circuit is turned on, the actual luminance change ratio of the panel at 4 fixed gray level outputs is only 0.5%~1.5%. Therefore, for the change of drive current I_{DRIVE} caused by the change of power supply voltage ELVDD, the

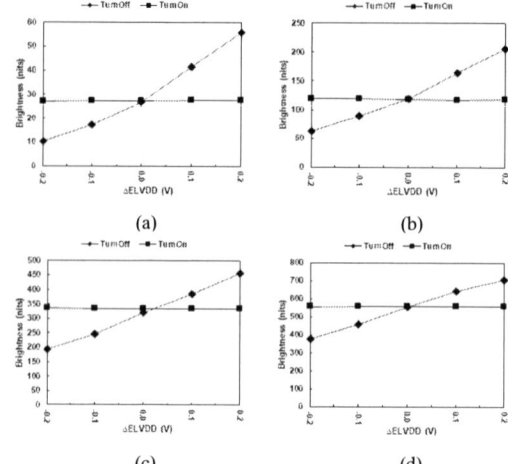

(a)   (b)

(c)   (d)

Fig. 7.   Luminance variations versus ΔELVDD changes. (a) 64 gray level. (b) 128 gray level. (c) 196 gray level. (d) 255 gray level.

proposed ELVDD dynamic tracking circuit can effectively reduce it, which improves the uniformity of the image quality.

TABLE I.   BRIGHTNESS VARIATIONS (ELVDD TRACKING MODE: TURN OFF VERSUS TURN ON)

Mode	Gray Level	Δ ELVDD (Volt)	Initial (nits)	Deviation (nits)	Deviation ratio
OFF	64	+/-0.2	26.7	29.0	108.6%
	128	+/-0.2	119.5	86.2	72.1%
	196	+/-0.2	320.1	137.6	43.0%
	255	+/-0.2	555.6	178.8	32.2%
ON	64	+/-0.2	27.4	0.4	1.5%
	128	+/-0.2	117.6	1.7	1.4%
	196	+/-0.2	333.3	2.1	0.6%
	255	+/-0.2	557.2	2.7	0.5%

REFERENCES

[1] Dawson, R. M. A., et al. "The impact of the transient response of organic light emitting diodes on the design of active matrix OLED displays." International Electron Devices Meeting 1998. Technical Digest (Cat. No.98CH36217) IEEE, 2002.

[2] Mo, Yeon Gon, et al. "Amorphous-oxide TFT backplane for large-sized AMOLED TVs." Journal of the Society for Information Display 19.1(2012):16-20.

[3] Lee, Jae Hoon, et al. "A new current scaling pixel circuit for AMOLED." Electron Device Letters IEEE 25.5(2004):280-282.

[4] Yamamoto, Tetsuro, et al. "Novel Pixel Circuit and Driving Method of AM-OLED for mobile application." Conference on Organic Light Emitting Materials and Devices 2006.

[5] In, Hai Jung, and O. K. Kwon. "External Compensation of Nonuniform Electrical Characteristics of Thin-Film Transistors and Degradation of OLED Devices in AMOLED Displays." IEEE Electron Device Letters 30.4(2009):377-379.

[6] Kwon, Jaewook, et al. "Design Considerations for External Compensation Approaches to OLED Display Degradation." IEEE International Symposium on Circuits and Systems IEEE, 2020.

[7] Kim, Yang Wan, et al. "40 Inch FHD AM-OLED Display with IR Drop Compensation Pixel Circuit." SID International Symposium: Digest of Technology Papers 1(2009):15.

[8] Alexander, et al. "Smart Power-Saving Driving Scheme for AMOLEDs Using Dynamic Power Rail Control." SID International Symposium: Digest of Technology Papers 42.1(2011):183-185.

[9] Lee, Soo-Yeon, et al. "A novel LTPS TFT pixel circuit for compensating IR drop of large area AMOLED display." Ecs Transactions 50.8 (2013): 269.

[10] Yum, Joohyuk, et al. "A Novel Method to Reduce Luminance Variation Due to IR-drop in Active Matrix OLED Displays." 2020 IEEE International Conference on Consumer Electronics (ICCE) IEEE, 2020.

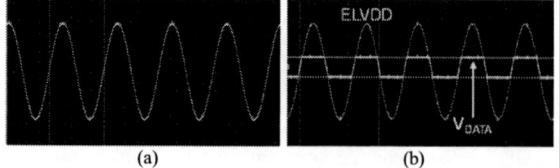

(a)   (b)

Fig. 6. V_{DATA} follows ELVDD transient waveform. (a) Clamp function turn off. (b) Clamp function turn on

A Reconfigurable Thermoelectric Energy Harvesting Interface Based on OPDC and DSCT

Peiyuan Fu[1], Jiabin Wang[1], Xufeng Liao[1,2], and Lianxi Liu*[1,2]

[1] Key Laboratory of AICAS (Ministry of Education), School of ICs, Xidian University, Xi'an 710071, China
[2] Xidian University Chongqing ICs Innovation Institute, Xiyong Microelectronics Park, Chongqing, 401331, China

* Email: lxliu@mail.xidian.edu.cn

Abstract—This paper presents a reconfigurable thermoelectric energy harvesting interface based on ordered power distributive control (OPDC) and digital self-calibration technology (DSCT). The proposed interface not only reduces the output ripple but also improves the efficiency of energy harvesting. When the output voltage reaches the ripple limit, the proposed OPDC strategy immediately switches the inductor current path from the load to the battery within one single cycle by changing the reconfigurable topology. As a result, the output ripples can be limited to a very small range, and the energy harvesting efficiency can be improved. In addition, in the situation of the battery supplied individually, the on-time can be adjusted adaptively by using DSCT. This not only ensures lower ripple but also ensures that the load receives the right amount of energy and avoids excessive waste of battery energy. The test results show that the peak tracking efficiency of TEG is 99.6%, the output power's range is 1μW-10mW, the peak efficiency can reach 87%, and the ripple is between 35mV.

Keywords—*ordered power distributive control, digital self-calibration technology, reconfigurable topology, thermoelectric energy harvesting, thermal electric generator (TEG)*

I. INTRODUCTION

Thermoelectric energy is very suitable for supplying small electronic devices with low power consumption due to its relatively stable energy and easy capture[1]. To recover and reuse the excess thermoelectric energy, a hybrid TEG and battery energy supply system is required[2]. Previously published studies have adopted TMC strategies to manage the hybrid TEG and battery energy supply system[2-4]. However, the harvesting and recovery of thermoelectric energy cannot be flexibly converted within one single cycle, resulting in large output ripples and low energy recovery efficiency. In this paper, a reconfigurable thermoelectric energy harvesting architecture based on OPDC is proposed to limit the output ripple to a very small range and manage the energy more efficiently. Besides, in the situation of the battery supplied individually, the constant on-time (COT) technique is typically employed[5]. However, there is a trade-off between output ripple and efficiency with the traditional COT technology. Therefore, this paper proposes DSCT to realize the adaptive adjustment of the on-time. This can break the trade-off, and increase the efficiency while reducing ripples.

II. SYSTEM ARCHITECTURE

Fig.1 shows the proposed system architecture. It mainly consists of the power stage and control stage circuitry. The control stage includes OPDC logic and gate driver circuitry, maximum power point tracking (MPPT) circuit, digital self-calibration on-time generation circuitry, and load monitoring circuit. The input sources of this hybrid energy supply system are the battery (BAT) and the thermal electric generator (TEG). To achieve a hybrid energy supply, the power stage adopts a reconfigurable topology. And the operating mode can be adjusted according to the input and output voltages under the control of OPDC logic, to realize the recovery and supply of energy in time. MPPT circuit is used to track TEG's maximum power point to improve the efficiency of thermoelectric energy harvesting. In addition to this, digital self-calibration on-time generation circuitry is only to realize DSCT in the situation of the battery supplied individually.

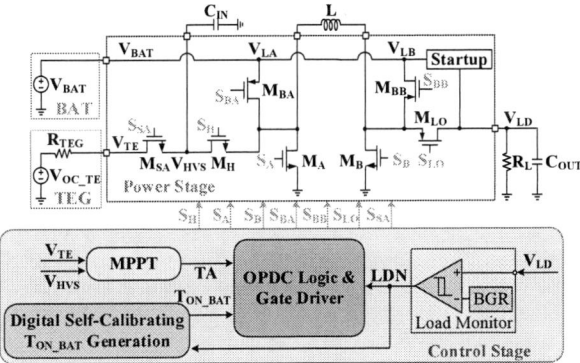

Fig. 1. The proposed reconfigurable TEG architecture.

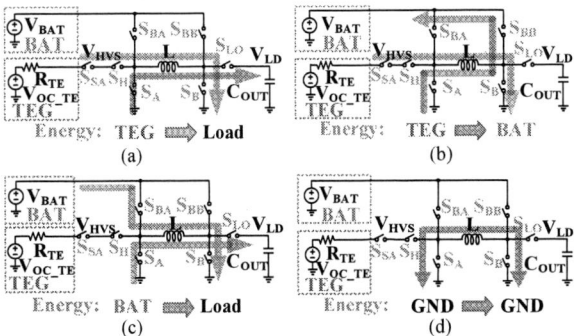

Fig. 2. The flow path of the energy (a) Harvest mode (b) Recycle mode (c) Backup mode (d) FW mode

The power stage adopts a reconfigurable topology to achieve different energy flow paths with minimal device cost. By varying the power stage topology, the energy can be supplied to the load or recycled to the battery for reuse in time, which avoids the waste of energy. Fig.2 shows the energy flow paths in four different operating modes. The red path represents the charge of the inductor current, and the blue path represents the discharge of the inductor current. Among them, the harvest mode is the stage in which the load obtains energy from the TEG. The recycle mode shows the TEG energy recovery stage, which can recover excess energy when the load is light. Backup mode represents the phase that the battery supplied individually. When the TEG energy is not

979-8-3503-6184-1/24 $31.00 © 2024 IEEE

enough to sustain the load, this mode will be active. In addition, the FW mode is freewheeling.

III. PROPOSED OPDC AND DSCT

A. OPDC and load monitoring

Fig.3(a) shows the waveform of the traditional TMC strategy. I_L is the inductor current, and V_{LD} is the output voltage. When TEG charges the inductor, the inductor current will rise. Then it will be released to the load or battery in a single cycle. Obviously, the energy is burst to the load or battery cycle by cycle. If the energy supplied for one cycle exceeds the energy required by the load, the maximum value of V_{LD} will be too high, which cannot be reduced until the next cycle. That results in a large V_{LD} ripple.

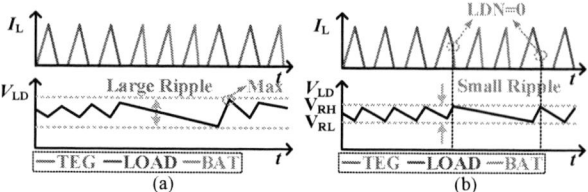

Fig. 3. The waveforms of I_L and V_{LD} under (a) TMC and (b) OPDC

Different from the TMC strategy, the charged inductor current will be released to the load and then to the battery in a single cycle. As shown in Fig.3(b), V_{RH} and V_{RL} are the upper and lower limits of the load monitoring circuit shown in Fig.1. When V_{LD} does not touch V_{RH} or V_{RL}, the charged inductor current will be released to the load. In other words, OPDC logic will keep the power stage in harvest mode. However, if V_{LD} touches V_{RH}, the load monitoring circuit will send a message LDN=0 to the OPDC logic circuit. Then OPDC logic will control the power stage switch to recycle mode. That means the excess inductive energy will be immediately recycled to the battery, even during one single cycle. Conversely, the power stage will switch the recycle mode to harvest mode once V_{LD} touches V_{RL}. Under the OPDC strategy above, the ripple voltage can be strictly limited to a reasonable range, and the excess energy can be recovered immediately. That improves the efficiency of energy acquisition, reduces the output ripple, and reduces energy waste. Moreover, a hysteresis comparator is used in the load monitoring circuit shown in Fig.1 to make sure that the V_{LD} stabilizes between V_{RH} and V_{RH}.

B. On-time generation circuitry based on DSCT

In the backup mode, the traditional COT technology has been used to deliver the energy from BAT to load. Conventional COT converter requires power devices with a fixed on-time in the inductive charging path. To meet the heavy load, a long on-time is required to store enough energy. However, the excess energy will be delivered to the C_{OUT} in the light load in one single cycle, leading to large output voltage ripples. There is a trade-off between the max load current and the ripple. This is easy to happen for WSN node applications with a wide load range. In this paper, the on-time control circuit based on DSCT is used to generate an adaptive on-time T_{ON_BAT}. Since the on-time T_{ON_BAT} can be adjusted according to the load current, the right amount of energy can be delivered to the load, so the ripple will be limited to a fixed range.

As shown in Fig.4, the on-time generation circuitry consists of an on-time control circuit based on DSCT and a

clock generator. The load monitor shown in Fig.1 uses one more comparator to detect the output voltage V_{LD}. When the load suddenly increases, the output voltage will inevitably drop. Until V_{LD} touches the lower limit which equals V_{RL}, LD_need=1. If V_{LD} successively touches V_{RL} in harvest mode, it means that the mode needs to be switched to backup mode, or even the on-time adjustment in backup mode is required. The details are as follows. When LD_need=1 occurs consecutively over two cycles, the LD_hungry jumps to high, indicating that the load is in urgent need of energy and the system enters backup mode. If LD_hungry=1 in two consecutive cycles, that means the current T_{ON_BAT} needs to be increased. Then, the LD_alarm jumps to high, and the counter starts to perform an addition operation. Then the output signal of the counter LD_ind<0:4> will be plus one from the previous output state. Finally, the LD_ind<0:4> signal is actually used to adjust the time interval generated by the subsequent clock generators.

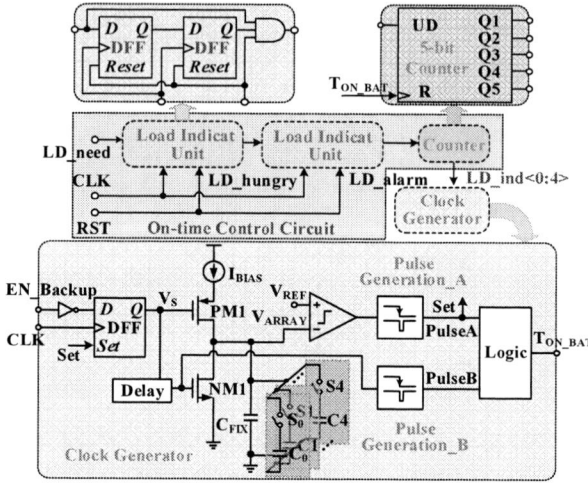

Fig. 4. The proposed on-time generation circuitry.

In the clock generator block, the switch S<0:4> is controlled by the counter's output LD_ind<0:4 >. The interval between Pulse A and Pulse B can be dynamically adjusted, which is exactly T_{ON_BAT}. The expression for T_{ON_BAT} is

$$T_{ON\_BAT} = \frac{C_{ARRAY}V_{REF}}{I_{BIAS}} \quad (1)$$

where V_{REF} is the reference voltage at the positive input of the comparator, I_{BIAS} is the bias current provided by the constant current source, and C_{ARRAY} is the total capacitance of the variable capacitance array. C_{ARRAY} consists of a fixed capacitor C_{FIX} and a variable capacitor C<0:4> shown in Fig. 4. Then Fig.5 shows the timing diagram of the clock generator. When V_S jumps to low, C_{ARRAY} will be charged, and Pulse B will be generated at the same time. When V_{ARRAY} on C_{ARRAY} reaches V_{REF}, Pulse A will be generated immediately. Then we can get an adaptive T_{ON_BAT} by adjusting C<0:4>.

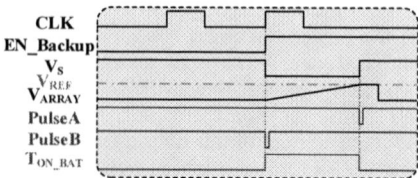

Fig. 5. Clock generator timing diagram

IV. TEST RESULTS AND DISCUSSION

The proposed circuit is implemented by a 0.18μm standard CMOS process, the core area is 0.91mm×0.61mm.

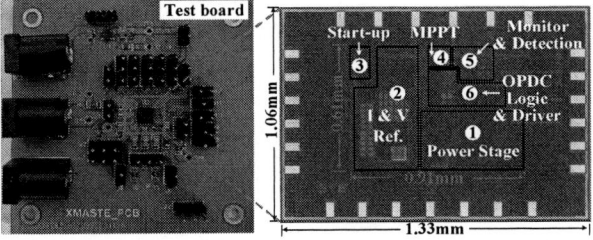

Fig. 6. The layout of the proposed circuit.

Fig.7 shows the measured waveforms of the control signals of the main power FET in different operating modes. The designed circuit can realize mode switching according to the load conditions and the input and output voltage.

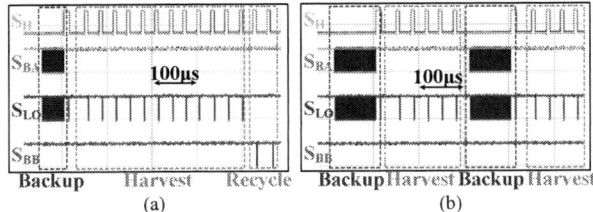

Fig. 7. Mode switching test results (a) light load (b) heavy load.

Since the output is strictly monitored by comparators, the output voltage ripple is kept within the comparison threshold. Fig.8 shows that the on-time T_{ON_BAT} can be adjusted adaptively using DSCT when the load current drops in backup mode. The output ripple is limited to 35mV.

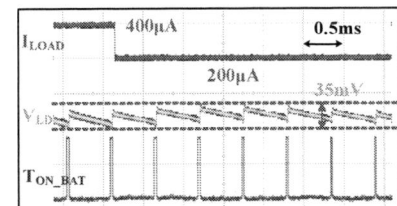

Fig. 8. Transient response and ripple.

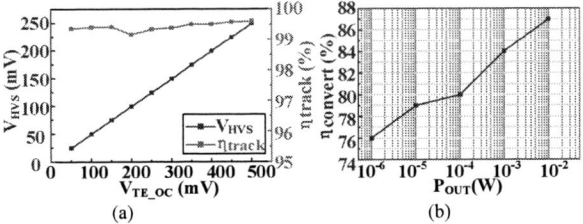

Fig. 9. (a) tracking curve and tracking efficiency curve (b) energy conversion efficiency curve.

Fig.9(a) shows that when the open-circuit voltage V_{OC_TE} of TEG is 500 mV, the tracking efficiency η_{track} peaks at 99.616%. Fig.9(b) shows that when the output power P_{OUT} changes from 1μW to 10mW, the energy conversion efficiency $\eta_{convert}$ is higher than 75% and peaks at 87%.

The performance comparison of the proposed thermoelectric energy harvesting interface with the published papers is given in Table I. By employing OPDC and DSCT, this work can achieve a very considerable reduction in ripples based on the relatively high efficiency.

TABLE I.   PERFORMANCE SUMMARY AND COMPARISON

Item	JSSC'19[4]	JSSC'22[4]	TCAS-I'23[2]	This Work
Process	0.18μm	65nm	0.18μm	**0.18μm**
Energy Source	TEG	TEG	TEG+PZT	**TEG**
Strategy	TMC+COT	TMC	TMC	**OPDC +DSCT**
Output Voltage	0.8V	0.9V~1.1V	1.8V	**1.2V**
Ripple	N/R	210mV @Max.	N/R	**35mV**
Peak Eff.	84%	93.3%	86.67%	**87%**

a. N/R: not reported.

V. CONCLUSION

Based on the 0.18μm standard CMOS process, we propose a reconfigurable TEG harvesting architecture with OPDC and DSCT, which improves the efficiency of thermoelectric energy acquisition and reduces the output ripples significantly. The test result shows the peak tracking efficiency is 99.6%. When the output power ranges from 1μW to 10mW, the peak energy conversion efficiency is 87%, and the ripple is between 35mV.

ACKNOWLEDGMENT

This work was supported in part by the National Natural Science Foundation of China under Grant 62131010, 62204183; in part by the Natural Science Basic Research Program of Shaanxi under Grant 2022JC-39; in part by the Key R&D Project in Shaanxi Province under Grant 2023-YBGY-189; and in part by the Chongqing Talents Program under Grant CQYC-2021030136; Natural Science Foundation of Chongqing (cstc2021jcyj-msxm3684). The authors would like to thank the Xidian University Chongqing Integrated Circuits Innovation Institute for their technical support.

REFERENCES

[1] R. Wang, Y. Liang and S. Du, "A 10-mV-Startup-Voltage Thermoelectric Energy Harvesting System With a Piezoelectric Starter," in *2022 IEEE International Symposium on Circuits and Systems (ISCAS)*, Austin, TX, USA, 2022, pp. 1482-1486.

[2] L. Liu, Y. Yu, X. Liao, J. Yin, J. Ma, and X. Wang, "MPPT Multiplexed Hybrid Energy Harvesting Interface With Adaptive Switching Cycle and Single-Cycle Sampling for Wearable Electronics," *IEEE Trans. Circuits Syst. I, Reg. Papers*, vol. 70, no. 8, pp. 3187-3197, Aug. 2023.

[3] Y. -S. Noh, J. -I. Seo, H. -S. Kim and S. -G. Lee, "A Reconfigurable DC-DC Converter for Maximum Thermoelectric Energy Harvesting in a Battery-Powered Duty-Cycling Wireless Sensor Node," *IEEE Journal of Solid-State Circuits*, vol. 57, no. 9, pp. 2719-2730, Sept. 2022.

[4] P. -H. Chen, H. -C. Cheng and C. -L. Lo, "A Single-Inductor Triple-Source Quad-Mode Energy-Harvesting Interface With Automatic Source Selection and Reversely Polarized Energy Recycling," *IEEE J. Solid-State Circuits*, vol. 54, no. 10, pp. 2671-2679, Oct. 2019.

[5] Y. Qian, H. Zhang, Y. Chen, Y. Qin, D. Lu, and Z. Hong, "A SIDIDO DC-DC Converter With Dual-Mode and Programmable-Capacitor-Array MPPT Control for Thermoelectric Energy Harvesting," IEEE Trans. Circuits Syst. II, Exp. Briefs, vol. 64, no. 8, pp. 952-956, Aug. 2017.

A Fixed-Peak-Current Single-Inductor-Multiple-Output DC-DC Converter Achieving 92.6% Peak Efficiency

Fei Liu [1], Langyuan Wang [2], Shuyu Zhang [2], Hanlu Zhang [2], Na Yan*[1]

[1] State Key Laboratory of Integrated Chip and Systems, Fudan University, Shanghai 200433, China
[2] Common Mode (GONGMO) Semiconductor Co., Ltd.

* Email: yanna@fudan.edu.cn

Abstract—A fixed-peak-current single-inductor-multiple-output DC-DC converter (SIMO) for wearable health-care devices is presented in this paper. It adopts burst mode control techniques to improve light-load efficiency, and limits the output ripple with peak current comparators (IPEAKs). The zero-crossing current comparators (IZCs) ensure that the system operates in discontinuous conduction mode (DCM), so the cross-modulation is suppressed. The proposed converter is fabricated in a standard 180 nm CMOS process. It can convert 1-1.6V or 3.3-5V battery voltage into stable output voltages of 1.8 V and 1.3V, and offers 100μA to 30mA load current with the efficiency over 80%. The peak efficiency is 92.6% with a supply voltage of 3.7 V, and the peak-to-peak output ripple is less than 25mV.

Keywords—SIMO converter, light load, efficiency

I. INTRODUCTION

In recent years, with the prosperity of portable device and wearable medical device market, the trend of miniaturization, multi-functionality, and long endurance has posed challenges to the power supply system. Figure 1 shows a power management system composed of a single-inductor-multiple-output DC-DC converter (SIMO), cascaded with linear low-dropout regulators (LDOs) and charge pumps. The architecture is highly favored, due to its good balance of efficiency, chip area, and output noise. In this system, SIMO serves as the first conversion circuit after the chemical battery, thus its conversion efficiency and output ripple dominate the performance of the whole system.

However, under the light load condition where the system operates most of the time, the efficiency of the inductive DC-DC converter falls significantly due to the increasing proportion of power losses, especially the quiescent current loss. Some recent studies attempt to reduce losses by employing hybrid control techniques [1], but the complex control methods increase chip area and consume additional power.

Burst mode (or pulse-skip mode) [2] is a control method that can effectively improve the converter's light-load efficiency. The control loop is relatively simple, and it enjoys the nature of good stability and fast load response. Furthermore, since the converter operates in DCM, there is no intermodulation between different output channels caused by the inductor current [3]. However, traditional ripple-based burst control often suffers from large output ripple. Based on burst mode control techniques, this paper proposes a single-inductor-dual-output DC-DC converter with fixed peak inductor current, which limits the maximum output ripple while retaining the advantages of traditional burst mode control method.

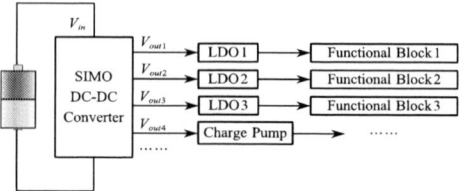

Fig. 1. Typical architecture of power management system

II. CIRCUIT IMPLEMENTATION

A. System Design

The block diagram of the converter proposed in this paper is shown in Figure 2. The input could be either V_{in1} (3.3-5V) or V_{in2} (1-1.6V), and the two output voltages V_{out1} (1.8V) and V_{out2} (1.3V) could be enabled simultaneously, each of which has a load capacity of 30mA.

The power stage includes MOSFET switches, output resistor digital-to-analog converters (RDACs), current comparators (IPEAKs and IZCs), off-chip inductor, and output capacitors. When V_{in1} is input, SW1, SW2, SW5, and SW6 are active, and both output channels operate step-down conversion; when V_{in2} is input, SW3, SW4, SW5, SW6, and SW7 are active, and output channel 1 and 2 operates in boost and buck-boost mode respectively.

The system operates under burst mode control, which means that there is switching activity only when the output feedback voltage V_{fb} is lower than V_{ref}. During the rest of the time, the current comparators and other modules are turned off by sleep signal in order to further save stand-by power consumption, with only the voltage comparators remain alive.

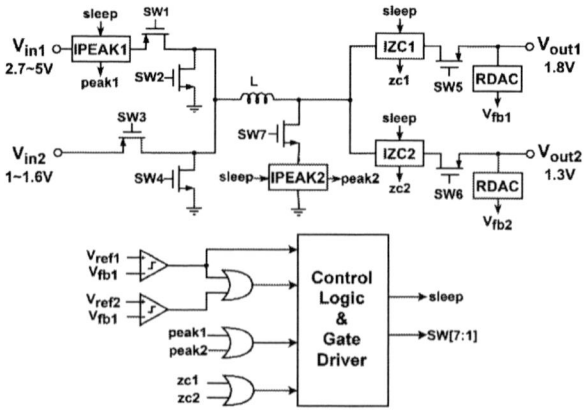

Fig. 2. Proposed SIMO DC-DC converter

979-8-3503-6184-1/24 $31.00 © 2024 IEEE

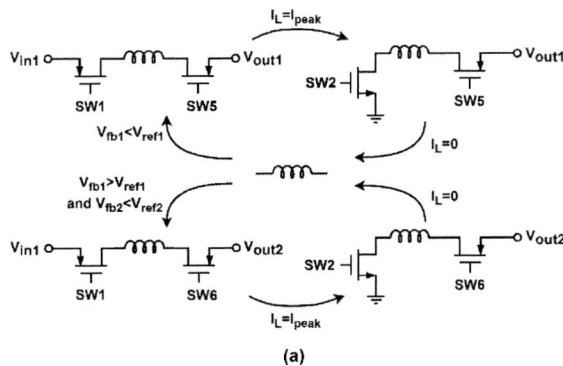

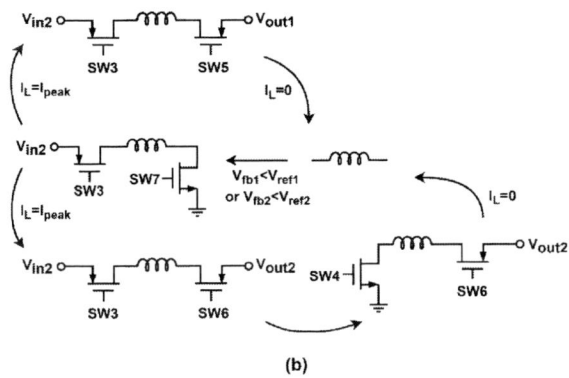

Fig. 3. Proposed control logic of SIMO converter (a) Vin1 (b) Vin2

In the power stage, the MOSFET switches are driven by the peak and zc signals generated by the current comparators instead of a system clock. The switching sequence of the proposed converter with different inputs is shown in Figure 3. For fixed input and output voltages, the conduction time of the input-side PMOS and NMOS is approximately constant, so the output ripple is limited by IPEAK threshold. For example, for buck conversion, the maximum ripple can be expressed as

$$V_{ripple,buck} = \frac{L}{2C} \frac{V_{in}}{V_{out}(V_{in} - V_{out})} (I_{peak} - I_{load})^2 \quad (1)$$

where V_{out} is the average value of the output voltage of the studied channel, I_{peak} is the threshold of the current peak comparator, and I_{load} is the load current of the output. The condition for reaching the theoretical maximum ripple is that $I_{load} \rightarrow 0$. The IZC current comparators ensure that the current flows to the load is always positive, thus avoids the loss caused by the reverse output current.

The switching period of each output can be expressed as

$$t_S = \frac{\int_{t_s} i_{L,out} dt}{I_{load}} \quad (2)$$

where $i_{L,out}$ is the inductor current that flows to the studied output.

The burst mode control technique is based on ripple, so the traditional small-signal analysis method that ignores the inductor current and output voltage ripple cannot be employed. Adopting the discrete-sampling modeling method proposed in [4], we can easily prove that the converter is inherently stable, thus no frequency compensation is required.

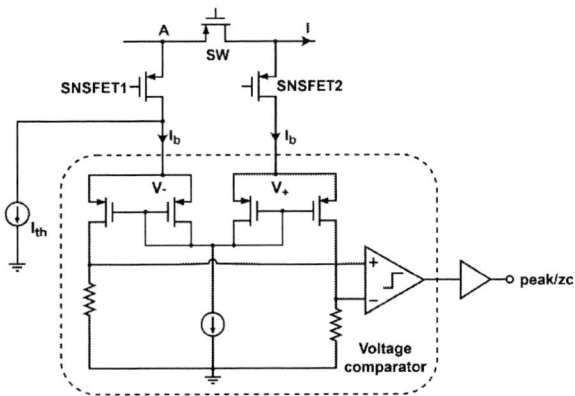

Fig. 4. Proposed peak/zc current comparator

B. Peak/ZC Current Comparator

The current comparator consists of a pair of sensing MOSFETs (SNSFETs), a bias current mirror, and a voltage comparator with a common-gate input stage, as shown in Figure 4. The SNSFETs and bias current mirror serves to convert the current information of the measured switch (SW) into voltage information, and the SNSFETs scales down the measured current, in order to reduce the power consumption of the bias current mirror.

Denote the size of SW as W/L, the current as I, and the size of SNSFETs are designed as $\frac{1}{K}W/L$. So if the conduction resistance of SW is R, then that of SNSFET is KR.

As shown in Figure 4, the bias current I_{th} is pulled down at the source of SNSFET1. Denote the voltage at point A as V_A, then the voltages at the positive and negative inputs of the voltage comparator V_+, V_- can be expressed as

$$V_+ = V_A - IR - KI_b R \quad (3)$$
$$V_- = V_A - KI_{th}R - KI_b R \quad (4)$$

where I_b is half of the bias current of the first stage of the comparator.

Then the threshold of the comparator can be simplified as

$$I = KI_{th} \quad (5)$$

which indicates that SNSFETs scale down the measured current by K times. When K is large, the current of SNSFET can be ignored compared to the current of SW, thus the current of the measured switch is approximately equal to the current of the main inductor.

In the layout, SNSFETs are placed contiguous to SW to reduce the mismatch. The routing resistances from SNSFETs to the voltage comparator will not affect the current threshold, as long as they are equal for SNSFET1 and SNSFET2.

III. SIMULATION RESULTS AND CONPARISIONS

Figure 5 shows the load transient simulation results. I_{load1} jumps from 1mA to 20mA at t=3.5ms in (a). I_{load2} jumps from 1mA to 24mA at t=6ms in (b). The effect of load transition on the output voltage is less than 1mV, submerged in the ripple. In addition, no cross-modulation is observed. The result is similar for downward load transition.

979-8-3503-6184-1/24 $31.00 © 2024 IEEE

Figure 6 (a) shows the power efficiency of the converter versus load. Figure 6 (b) is the steady state waveforms of the converter when it achieves the peak efficiency 92.6% under 3.7 V supply voltage, at this time the load current is 2 mA for each output channel. The peak-to-peak output ripple is 9mV and 25mV for V_{out1} and V_{out2}.

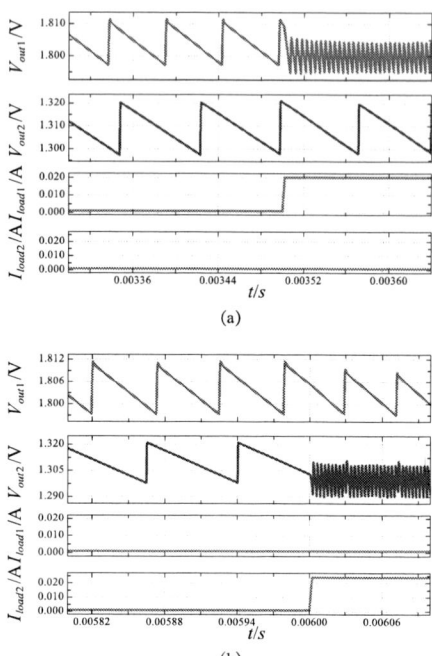

(a)

(b)

Fig. 5. Load transient simulation results (a) I_{load1} (b) I_{load2}

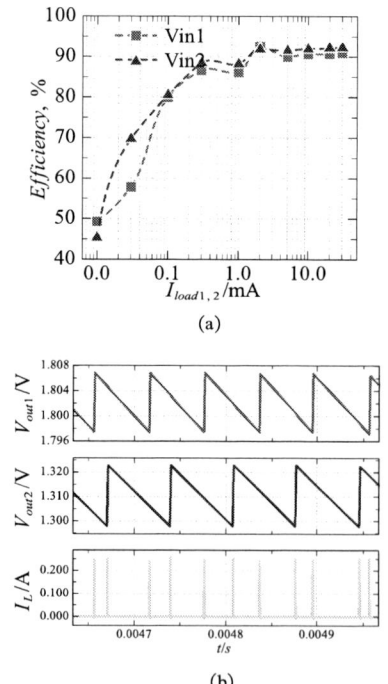

(a)

(b)

Fig. 6. (a) Efficiency v.s. load (b) steady state waveform at peak efficiency

TABLE I. PERFORMANCE COMPARISION

	[5]	[6]	This work[a]
Process	65nm CMOS	180nm BCD	180nm BCD
V_{in} (V)	1	3-5	1-1.6, 3.3-5
V_{out} (V)	0.4-0.8	0.8, 1.8, 12	1.3, 1.8
Inductor (μH)	82	4.7	2.2
# of output	10	3	2
Peak Efficiency	83.4%	86%	92.6%
V_{ripple} (mV)	80	50	25
Dynamic V_{droop} @ load transition	45mV @ 5 to 70μA	90mV @ 0 to 1mA	<1mV @ 1 to 20mA
Cross Regulation	0	0	0
Iq (μA)	42.8	0.2	4.2

[a.] Post Simulation Result

Table I summarizes the performance of the SIMO converter and gives a detailed comparison with state-of-the-art works. The proposed SIMO DC-DC converter shows significantly higher efficiency and lower ripple.

IV. CONCLUSION

This work proposes a power-efficient SIMO converter for wearable healthcare systems. Based on fixed-peak-current burst mode control, the converter achieves high efficiency under light load condition, and the output ripple is limited by the peak inductor current. The peak efficiency of the converter is 92.6%, with output ripple less than 25mV, and stand-by current less than 4.2μA. No cross regulation or voltage drop caused by load transition is observed.

ACKNOWLEDGMENT

This work is sponsored by Common Mode (GONGMO) Semiconductor Co., Ltd. Corresponding authors: Hao Xu, Na Yan.

REFERENCES

[1] T. -H. Yang et al., "A 94.3% Peak Efficiency Adaptive Switchable CCM and DCM Single-Inductor Multiple-Output Converter With 0.03 mV/mA Low Crosstalk and 185 nA Ultralow Quiescent," in IEEE Journal of Solid-State Circuits, vol. 57, no. 9, pp. 2731-2740, Sept. 2022

[2] F. Reverter and M. Gasulla, "Optimal Inductor Current in Boost DC/DC Converters Operating in Burst Mode Under Light-Load Conditions," in IEEE Transactions on Power Electronics, vol. 31, no. 1, pp. 15-20, Jan. 2016.

[3] D. S. Ma, W. H. Ki, C. Y. Tsui and P. K. T. Mok, "Single-inductor multiple-output switching converters with time-multiplexing control in discontinuous conduction mode", IEEE J. Solid-State Circuits, vol. 38, no. 1, pp. 89-100, Jan. 2003.

[4] R. Redl and J. Sun, "Ripple-Based Control of Switching Regulators—An Overview," in IEEE Transactions on Power Electronics, vol. 24, no. 12, pp. 2669-2680, Dec. 2009.

[5] D. Kim et al., "A 10-Output, Single-Inductor-Multiple-Output DC–DC Buck Converter With Integrated Output Capacitors for a Sub-mW System-on-Chip," in IEEE Solid-State Circuits Letters, vol. 4, pp. 56-59, 2021.

[6] K. -S. Yoon, S. Jung, J. -H. Lee, S. J. Kim, H. -S. Kim and G. -H. Cho, "A Single-Inductor–Multiple-Output (SIMO) 0.8-V/1.8-V/12-V Step-Up/Down Converter With Low-Quiescent Current for Implantable Electroceutical SoCs," in IEEE Solid-State Circuits Letters, vol. 4, pp. 182-185, 2021.

Buck-Boost Converter with Stable Transition Mode for Low Average Inductor Current

Ningning Li[1], Yushen Zhang[1], Yibo Zhang[1], Yizhe Yang[1], Wenhao Yang[1], Yimeng Zhang*[1,2], Yuming Zhang[1,2]

[1] School of Microelectronics,Xidian University
[2]Shaanxi Key Lab of Integrated Circuits and Systems,Xidian University

* Email: zhangyimeng@xidian.edu.cn

Abstract—**A DC-DC Buck-Boost converter with stable transition mode is proposed. The proposed transition mode control enables the loop operate with fast load transient and steady linear regulation when the input voltage is very close to the output voltage. The converter detects the input voltage and the output voltage considering the power transistor parasitic resistance to accurately switch the operating modes: buck/T-buck/T-boost/boost. In addition, transition mode with lower average inductor current is realized, which is reduced by 44.5% compared with the conventional control. Meanwhile, multi-mode adaptive on/off time control is adopted for faster transient response capability and 6.5% error is achieved at 2MHz switching frequency of the converter over the full input range. The circuit is realized in 0.18um BCD process.**

Keywords—Buck-Boost converter, stable transition mode, multi-mode adaptive COT/FOT control

I. INTRODUCTION

In recent years, consumer electronics such as portable devices are developing rapidly, and they are usually powered by lithium batteries. Wide input range four-switching buck-boost converter (FSBB) has received more and more attention because it can widen the effective power supply range of Li-ion batteries and prolong the usage time[2][3][4].

Buck-boost topology single-mode control has higher average inductor current compared to buck topology and boost topology, so it is usually designed for dual/multi-mode control: when the input voltage is less than the output voltage, the converter operates in boost mode; when the input voltage is greater than the output voltage, the converter operates in buck mode; when the input voltage is close to the output voltage, the converter operates in buck-boost transition mode. [1] adopts a dual-mode design, but the mode transition is accomplished by pulse frequency modulation, which has the problems of unstable duty cycle control and serious electromagnetic interference. [2] adopts a three-mode design, where the transition mode inserts a fixed duty cycle of the direct-connected phase compared to the conventional control, but the use of this approach requires a compromise between the average inductor current and transient response. [3] adopts a mixed peak/valley current feedback over the full input range, but there is the problem of unstable loop control due to the uncertainty of peak/valley current feedback in the transition mode. Hysteresis current control is used in [4] to adaptively complete the mode transition according to the shape of the inductor current of the direct-connected phase, but there are problems of random control and uncertain switching frequency in the transition mode.

In order to achieve high stability and low average inductor current in the transition mode, a four-mode buck-boost converter with buck loop valley current feedback/boost loop peak current feedback is proposed. In the following, the overall system description and design idea are in the Part II. The specific circuit implementation is in Part III. Simulation results verify the design in Part IV. Finally, the whole paper is summarized in Part V.

II. PROPOSED BUCK-BOOST CONVERTER

A. Top-Level Circuit

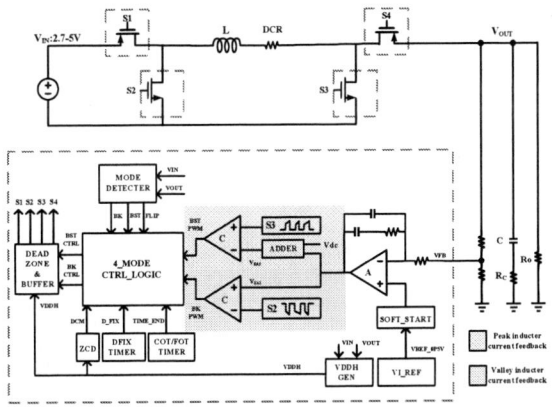

Fig. 1. System block diagram of the proposed buck-boost converter

The block diagram of the proposed four-mode buck-boost converter system is shown in Fig. 1. Two PWM comparators independently generate the PWM duty cycle information for the boost and buck loops, while the peak current feedback is used for the boost loop and valley current feedback is used for the buck loop, and a common loop compensator is used. In order to achieve stable switching between modes, the PI compensation output V_{EA} is superimposed with a DC level to control the boost loop. The mode detector accurately detects the operating mode according to the input voltage, output voltage and load current. A multi-mode adaptive timing circuit is used to achieve constant-on/fixed-off time (COT/FOT) control with constant switching frequency over the full input range.

B. Transition Mode

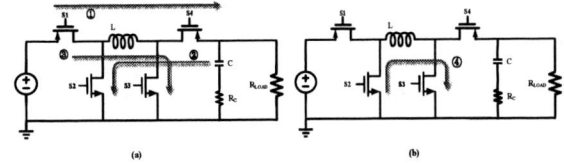

Fig .2. (a)Proposed transition mode, (b)DCM anti-ringing mode.

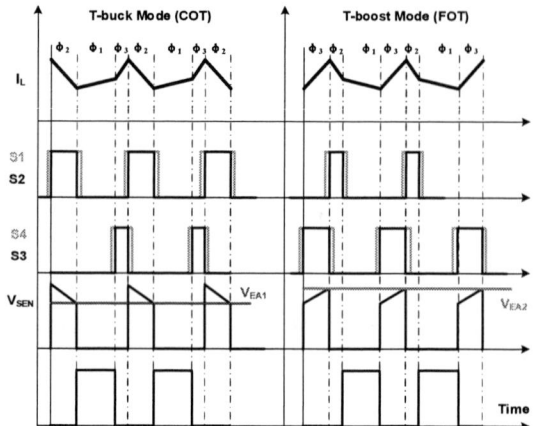

Fig. 3. inductor current ripple and control method of four-mode

When input voltage is close to output voltage, the proposed buck-boost converter operated in transition mode as shown in Fig. 2(a). Fig. 3 shows the inductor current waveforms and control method of the transition mode which is divided into T-buck and T-boost mode. In the T-buck mode, when the inductor current of phase 2 drops to the valley, path1 conducts and starts timing. When the timing is finished, phase 3 with fixed duty cycle D_{FIX} is inserted to charge the inductor. In the T-boost mode, when the inductor current of phase 3 rises to the peak, phase 2 with fixed duty cycle D_{FIX} is inserted to discharge the inductor. After that path 1 conducts and starts timing for a fixed time and then repeats the previous cycle. Fig. 2(b) shows the anti-ringing phase of the DCM mode to store inductor energy and reduce power loss.

In this design, the fixed duty cycle D_{FIX} is designed to be 10%, according to the buck-boost converter ideal conversion ratio equation:

$$\frac{V_{OUT}}{V_{IN}} = \frac{t_{s1}}{t_{s4}} = \frac{1 - t_{s2}}{1 - t_{s3}} \quad (1)$$

When converter operates in T-buck mode,

$$V_{IN} > V_{OUT}, t_{s2} > t_{s3} = 10\%T_s$$

The switching frequency is designed to be 2MHz, therefore the converter has an inductor current sampling time greater than 50ns, which is sufficient for the Sense-FET circuit to accurately sample the peak/valley current. Thus a stable loop control of the converter could be achieved when the input voltage is very close to the output voltage. The same is true for the T-boost mode.

Sampling the valley of the inductor current in phase2 in T-buck mode and the peak in phase3 in T-boost mode can effectively solve the unstable transition mode control problem caused by 1) uncertainty of the peak/valley value of the inductor current in phase 1, and 2) inaccuracy of the inductor current sampling in the small duty cycle, so that the converter can achieve a stable conversion ratio of 1:1. When considering the power transistor parasitic resistance present in the topology, the actual conversion ratio formula will be distorted, only the mode detection module is required to accurately detect the switching boundary, which will be mentioned later.

At the same time, the average inductor current in the transition mode is significantly reduced, by 44.5% compared to conventional buck-boost converter.

$$I_{Laverage}$$
$$= \begin{cases} \dfrac{I_o}{1 - D_{FIX}} & (Transition\ mode) \\ I_o\left(1 + \dfrac{V_{OUT}}{V_{IN}}\right) & (Conventional\ Buck - Boost) \end{cases} \quad (2)$$

$$\frac{I_{Laverage\_transition\_mode}}{I_{Laverage\_conventional}} = \frac{1}{(1 - D_{FIX})(1 + \frac{V_{OUT}}{V_{IN}})} = 55.5\%$$

III. CIRCUIT IMPLEMENTATION

The critical circuit modules of the converter include an accurate mode detector and a multi-mode adaptive on/off time circuit, which are described in the following subsections.

A. Accurate Mode Detector

In order to ensure that the transition mode sampling phase has a duty cycle greater than 10%, the actual voltage drop V_{SW} of a power transistor is sampled here to equate the effect of the parasitic resistance of the switching transistor in the topology, and the V_{SW} variable is added to the transition ratio calculation formula so that the switching from T-buck mode to T-boost mode occurs approximately synchronously with the flip-flop of the potential at the ends of the inductor.

According to the principle of inductive volt-second balance, the conversion ratio formula for the T-buck mode is calculated as:

$$V_{IN} - \frac{2V_{SW}}{1 - D_{bk}} = \frac{V_{OUT}(1 - D_{FIX})}{1 - D_{bk}} \quad (3)$$

In the T-boost mode, it is calculated as:

$$V_{IN} - \frac{2V_{SW}}{1 - D_{FIX}} = \frac{V_{OUT}(1 - D_{bst})}{1 - D_{FIX}} \quad (4)$$

The transition boundary between T-buck mode and T-boost mode is shown in Eq.(5)

$$V_{IN} - \frac{2V_{SW}}{1 - K_{bk}} = \frac{V_{OUT}(1 - K_{bst})}{1 - K_{bk}} = V_{OUT} \quad (5)$$

The mode detector is shown in Fig. 4, which is divided into two modules: 1) the conversion boundary detector generates the mode conversion boundary calculated according to the input voltage and the output voltage; 2) the phase detector makes the mode switching occur in the appropriate operating phase, which can ensure the bi-directional stable switching between modes.

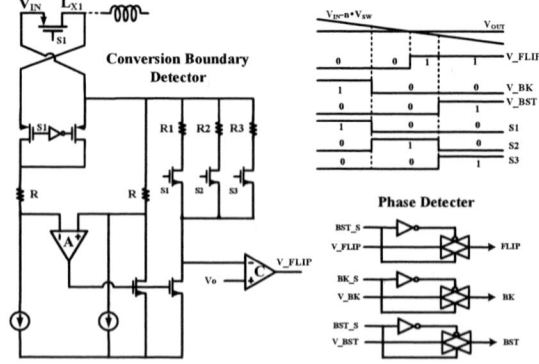

Fig. 4. The accurate mode detector of proposed buck-boost converter

B. Multi-mode Adaptive COT/FOT Circuit

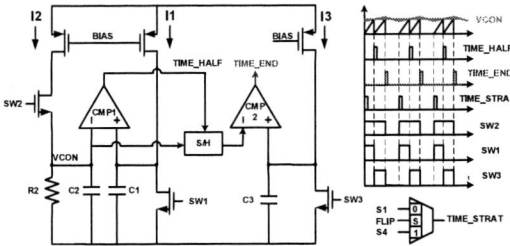

Fig. 5. Multi-mode Adaptive COT/FOT Circuit

The loop employs constant on-time/fixed off-time control. When a conventional COT/FOT timing circuit is used, the on-time of phase1 of the transition mode is calculated as:

$$T_s = \frac{k_2 RC}{k_1} \cdot \frac{1-D}{(1-D_{fix})(1-D-D_{FIX})} \quad (6)$$

It can be seen that a constant switching frequency cannot be obtained since the switching period is duty cycle dependent in the transition mode. Based on design ideas from [5], a new timing circuit shown in Fig.5 is proposed to achieve the fixed switching frequency with lower area cost compared with original structure. Multi-mode control is achieved by selecting COT or FOT control based on the mode detection signal FLIP. The switching frequency is calculated as:

$$\frac{1}{T_s} = \frac{I_1}{I_2} \frac{1}{2R_2 C_1} = F_{SW} \quad (7)$$

Also this new timing circuit does not need to be designed for slope compensation [5], thus providing high bandwidth design capability.

IV. SIMULATION RESULTS

As shown in Fig. 6, the converter starts up through four control modes at an input voltage of 2.7V. As the input voltage changes linearly from 2.7V to 5V, the output voltage hardly fluctuates and seamless transition between adjacent cycles in different control modes is achieved. The minimum switching frequency is 1.87MHz with a 6.5% frequency error within full input range.

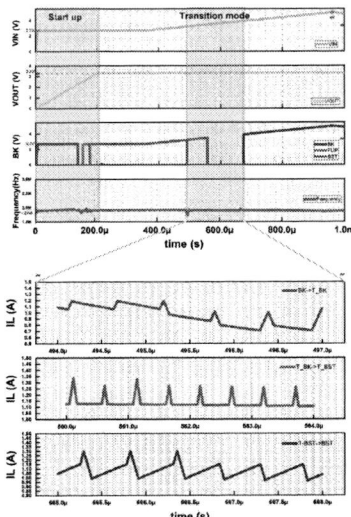

Fig. 6. Line regulation @Vin=2.7V->5V,Vout=3.3V,Io=1A

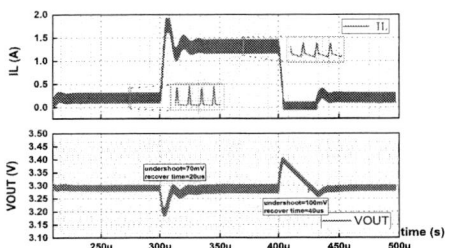

Fig.7. Transient Response in T-boost Mode (@Vin=3.3V, Vout=3.3V, Io=50mA->1A)

The load jump between 50mA and 1A at both input and output voltages of 3.3V is shown in Fig. 7. As shown in Table 1, the proposed converter can achieve large load, fast transient response and stable switching frequency in the transition mode compared with other papers.

TABLE I. THE PERFORMANCE COMPARISON TABLE

Target at $V_{IN} \approx V_{OUT}$	This work	[3]	[4]
Max I_{LOAD}	1A	100mA	400mA
Switching freq.	2MHz	1MHz	Varied
ΔV_O@load transient	70mV/950mA	none	50mV/400mA
$t_{recover}$	20us	none	70us
Peak Efficiency	90.1%	78%	98.1%

V. CONCLUSION

By designing specific current feedback in the transition mode and considering the effect of power transistor parasitic resistance to switch between T-buck and T-boost modes accurately, stable control and 44.5% lower average inductor current compared to the conventional control are achieved. Moreover, a multi-mode fixed-frequency COT/FOT control is used to achieve a high bandwidth of the loop. As a result, the proposed buck-boost converter supports full input and wide load with fast transient response, especially has good stable output capability when the conversion ratio is close to 1:1.

ACKNOWLEDGMENT

This work was supported by the National Natural Science Foundation of China (Grant No. 62234010).

REFERENCES

[1] P. Rajarshi and D. Maksimovic, "Smooth transition and ripple reduction in 4-switch non-inverting buck-boost power converter for WCDMA RF power amplifier," *ISCAS*, Seattle, WA, USA, 2008, pp. 3266-3269.

[2] M. Singh and A. A. Fayed, "A 1-A 6-MHz Digitally Assisted Buck–Boost Converter With Seamless Mode Transitions and Fast Dynamic Performance for Mobile Devices," in *IEEE Transactions on Power Electronics*, vol. 36, no. 4, pp. 4338-4351, April 2021.

[3] C. Xu and L. Liu, "A Four Modes and Smooth Transition Non-inverting Buck-Boost Converter," *2021 IEEE 14th International Conference on ASIC (ASICON)*, Kunming, China, 2021, pp. 1-4.

[4] X. -E. Hong, J. -F. Wu and C. -L. Wei, "98.1%-Efficiency Hysteretic-Current-Mode Noninverting Buck–Boost DC-DC Converter With Smooth Mode Transition," in *IEEE Transactions on Power Electronics*, vol. 32, no. 3, pp. 2008-2017, March

[5] G. Gritti, "Novel adaptive pulse width modulator provides quasi-fixed switching frequency in constant on/off-time controlled regulators," *APEC*, San Antonio, TX, USA, 2018, pp. 760-766

A SET Sensitive Model of LC and Ring Voltage-Controlled Oscillator in FinFET Technology

Liu Heyuan[1], Yuan Hengzhou[1*], Lu Jianzhuang[1#], Chen Xiaowen[1], Sang Hao[1], Liu Jingtian[2], Guo Yang[1]

[1]Key Laboratory of Advanced Microprocessor Chips and Systems, National University of Defense Technology, China
[2]Academy of Military Sciences PLA China, Beijing, China

* Email: niesaitong@163. com, # Email: lujz1977@163.com

Abstract—**In this study, a SET model is proposed to evaluate the sensitivity of LC and ring VCO. This paper analyzes the impact of SET on noise performance when it occurs on the cross-coupled transistors of an LCVCO and on the inverters of a ring VCO. It uses Cl ions with LET of 13.4 MeV*cm²/mg, Al ions with LET of 8.92 MeV*cm²/mg, and Ta ions with LET of 86.11 MeV*cm²/mg to inject into the LCVCO and ring VCO and observe the occurrence of SET. Ion experiments show that the SET model is effective and that the ring VCO has lower sensitivity to SET.**

Keywords—Voltage-Controlled Oscillator, phase-locked loop, single-event transient

I. INTRODUCTION

Advancements in space technology demand high-performance Serializer and Deserializer (SerDes) circuits, necessitating robust Clock and Data Recovery (CDR) systems. Phase-Locked Loops (PLLs), especially the Voltage-Controlled Oscillator (VCO), are critical for generating accurate clock signals. In aerospace, a key challenge is mitigating Single Event Transients (SET) in VCOs, which affect phase noise and power consumption. SETs can disrupt the output clock phase, increasing the Bit Error Rate (BER) of SerDes.

LCVCO and ring VCO are two VCO types used in PLLs. Ring VCO are easy to integrate but have inferior phase noise, while LC VCO provide better phase noise but require more chip area. Recent research on LCVCOs aims to improve frequency tuning ranges and phase noise. To realize wider frequency tuning ranges, multi-mode and multi-core VCOs [1] have emerged. A compact ring VCO based on CMOS inverters proposed in paper [2] exhibits good phase noise and radiation resistance, it introduces a high-frequency multibiased multiphase VCO (MBM-VCO) for clock and data recovery. It reduces VCO sensitivity to bias through multi-bias and interleaved bias techniques.

This paper presents the design of an LCVCO and a ring VCO, both at 6 GHz, analyzing SET impact on noise performance and validating models through heavy-ion experiments, highlighting ring VCO advantages in FinFET technology.

II. ANALYSIS OF LCVCO

A. Design of LCVCO

Fig. 1 illustrates the overall circuitry of the LCVCO. The designed LCVCO comprises cross-coupled NMOS transistors, coarse-tuning capacitor arrays, fine-tuning capacitor arrays, inductors, and tail current sources. To achieve VCO frequency tuning, the capacitor array consists of 8 sets of coarse-tuning capacitor banks and 6 sets of fine-tuning capacitor banks.

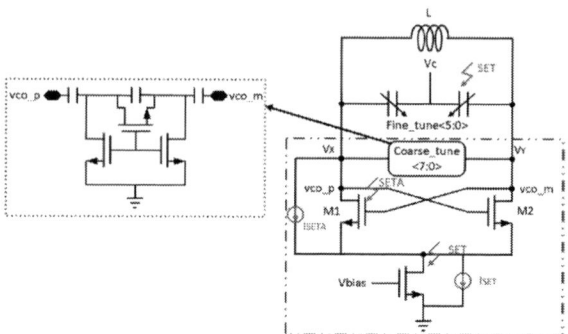

Fig. 1. LCVCO circuit structure.

B. SET Modeling and Analysis

Fig. 1. illustrates the occurrence of SET in the LCVCO. SET events occur at various nodes within the VCO, resulting in fluctuations in the oscillation signal output. When SETA occurs near the cross-coupled transistors, additional current flows from the drain to the substrate. Using a dual exponential current source to simulate the current when SET occurs, the function expression of the dual exponential current source is shown in equation (1):

$$I_{exp}(t) = I_0 \left(e^{-\frac{t}{t_f}} - e^{-\frac{t}{t_r}} \right) (0 < t < T) \qquad (1)$$

In the equation, t_r represents the rise time of the current, t_f represents the fall time of the current, and I_0 represents the peak current. Performing a Fourier transform on this function yields:

$$I(f) = I_0 \left(\frac{-\frac{1}{t_f} e^{-\frac{T}{t_f}} + \frac{1}{t_f}}{\frac{1}{t_f^2} + \omega^2} - \frac{-\frac{1}{t_r} e^{-\frac{T}{t_r}} + \frac{1}{t_r}}{\frac{1}{t_r^2} + \omega^2} \right)$$
$$- jI_0 \left(\frac{-\omega e^{-\frac{T}{t_f}} + \omega}{\frac{1}{t_f^2} + \omega^2} - \frac{-\omega e^{-\frac{T}{t_r}} + \omega}{\frac{1}{t_r^2} + \omega^2} \right) \qquad (2)$$

According to Equation (2), we can obtain the power spectral density (PSD) of the current over one period T:

$$\overline{I_{SET}^2} = \frac{1}{T} |I(f)|^2 \qquad (3)$$

The impact varies depending on the timing of the current injection into the LCVCO. Within the LCVCO, the switches of the cross-coupled transistors are dynamic, they continuously toggle their states as the output oscillation signal changes. Therefore, the effects of SET on the circuit differ depending on the timing of its occurrence.

The states of the cross-coupled transistors are dependent on the gate voltages, as illustrated in Fig. 2., where the voltage

979-8-3503-6184-1/24 $31.00 © 2024 IEEE

difference between the gates of the two cross-coupled transistors is denoted as V_{XY}. When $|V_{XY}| > \sqrt{2}(V_{GS} - V_{TH})$, where $(V_{GS} - V_{TH})$ denotes the overdrive voltage, one MOSFET turns off while the other turns on. Conversely, when $|V_{XY}| < \sqrt{2}(V_{GS} - V_{TH})$, both MOSFETs can be considered to be simultaneously on. The impact of SET on the circuit differs when both cross-coupled MOSFETs are simultaneously on compared to when only one is on. We will discuss these situations separately.

When $|V_{XY}| < \sqrt{2}(V_{GS} - V_{TH})$, indicating that both MOSFETs are simultaneously on, as depicted by the time T_1 in Fig. 2., V_{XY} reaches its minimum difference, and both MOSFETs are conducting. SET current flows through the resonant tank in this situation.

Within one period T, the time spent in this state is $2T_1$. In the model incorporating the dual-exponential current source as depicted in Fig. 3., the currents passing through M1 and M2 are considered approximately equal. At this time, the current flowing through the resonant tank is: $I_{Nort} \cong -\frac{I_{SET}}{2}$.

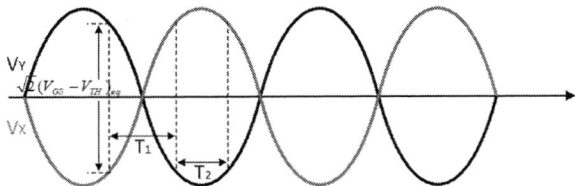

Fig. 2. Differential output signal of LCVCO.

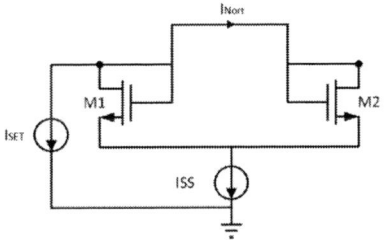

Fig. 3. Simplified LCVCO model with dual-exponential current source.

Paper [3] provides the spectral equation for phase noise:

$$S_{\Phi n}(f) = \overline{I_n^2} \frac{1}{4C_a^2 (2\pi f)^2} \frac{1}{V_0^2} \qquad (4)$$

Where $\overline{I_n^2}$ represents the PSD of the noise current, $\frac{1}{4C_a^2 (2\pi f)^2}$ represents the square of the impedance of the resonant tank. The noise current affects both the amplitude and phase of the VCO output signal, and multiplying by $\frac{1}{V_0^2}$ represents the effect of the noise current on the phase of the output signal. Therefore, the spectral density of the SET current noise is given by:

$$S_{\Phi n}(f) = \frac{\overline{I_{SET}^2}}{4} \frac{1}{4C_a^2 (2\pi f)^2} \frac{1}{V_0^2} \qquad (5)$$

When $|V_{XY}| < \sqrt{2}(V_{GS} - V_{TH})$ and if the gate voltage of M2 is greater, the SET current flows through the resonant tank and exits from M2, resulting in the maximum impact on noise. As depicted in Fig. 4., during one period T, there is a duration T_2 in this state. At this time, the current flowing through the

resonant tank is: $I_{Nort} = -I_{SET}$. In this situation, all the current flows through the resonant tank, resulting in the maximum impact on the circuit, and the noise effect on the circuit is:

$$S_{\Phi n}(f) = \overline{I_{SET}^2} \frac{1}{4C_a(2\pi f)^2} \frac{1}{V_0^2} \qquad (6)$$

When M1 is greater, the SET current cannot flow through the resonant tank, thus having minimal impact on the oscillation signal output of the LCVCO.

Considering the occurrence of SET within one period T and integrating the aforementioned three situations, the noise equation is obtained as follows:

$$S_{\Phi n}(f) = \left(\overline{I_{SET}^2} \frac{T_2}{T} + \frac{\overline{I_{SET}^2}}{4} \frac{2T_1}{T} \right) \frac{1}{4C_a(2\pi f)^2} \frac{1}{V_0^2} \qquad (7)$$

To obtain the values of T_1 and T_2 in Equation (7), let's assume the equation for the output signal as follows: $V_{out} = V_1 \sin\omega t$, where $V_1 = \frac{4}{\pi} R_p I_{SS}$. By setting $V_{out} = V_1 \sin\omega\Delta T = \sqrt{2}(V_{GS} - V_{TH})eq$, we can solve for the value of ΔT. Then, we can calculate $T_1 = 2\Delta T$ and $T_2 = \frac{T - 2T_1}{2}$.

III. Analysis Of Ring VCO

A. Design of Ring VCO

As illustrated in Fig. 4., each stage of the inverter introduces a signal propagation delay, which is caused by the charging and discharging of the load capacitance. The delay caused by the charging of the load capacitance is t_{dP}, while the delay caused by the discharging is t_{dN}. Let's assume that the delays caused by both charging and discharging are equal, denoted as $t_d = t_{dN} = t_{dP}$. Therefore, the clock period of the output oscillation signal for the ring VCO is $T_{OSC} = 6t_d$.

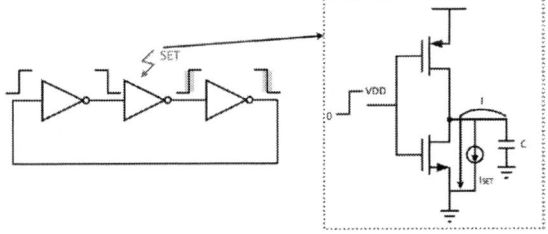

Fig. 4. SET occurrence in ring VCO.

B. SET Modeling and Analysis

When SET occurs in the ring VCO, it alters the inverter's capacitance charging and discharging time, affecting the output signal frequency. As shown in Fig. 4, SET on the NMOS transistor in the inverter causes extra current to flow from the NMOS drain to the substrate. At this point, the equation for capacitor discharge can be expressed as:

$$\int_0^{t_d'} \frac{I_N + I_{SET}}{C} dt = \frac{VDD}{2} \qquad (8)$$

Where I_N represents the current of the NMOS transistor in saturation, and I_{SET} is the current generated when SET occurs. Due to the addition of I_{SET}, the propagation delay becomes t_d'. The change in clock period before and after the occurrence of SET is $\Delta T = 6t_d - (5t_d + t_d')$. The corresponding phase variation is:

$$\phi_n = \frac{2\pi\Delta T}{T_{OSC}} \qquad (9)$$

The corresponding frequency variation is:

$$f_0 - \phi_n/(2\pi \cdot T_{OSC}) = f_0 - f_0\phi_n/(2\pi) \quad (10)$$

We have obtained the phase and frequency variations caused by SET within one period. Assuming SET occurs at the same position in each period, it results in a frequency variation of $f_0\phi_n/(2\pi)$ Hz per period. Therefore, the frequency noise of the closed-loop ring VCO is:

$$\phi_{n,osc}(t) = \int 2\pi \frac{f_0}{2\pi}\phi_n(t)dt \quad (11)$$

The corresponding spectrum of the noise is:

$$S_{\phi n,osc}(f) = \frac{f_0^2}{4\pi^2 f^2}S_{\phi n}(f) \quad (12)$$

IV. ANALYSIS AND EXPERIMENT RESULTS

A. Comparison of Noise between LCVCO and Ring VCO

Combining the spectra of noise when SET occurs in the LC VCO and ring VCO obtained from Chapters II and III, respectively, we calculate the ratio of noise magnitude when SET occurs in the LC VCO to that in the ring VCO based on the parameters used in practical design.

The designed oscillation frequency of the LCVCO is approximately $f = 6GHz$. Based on simulation results of the LCVCO, we can obtain $\sqrt{2}(V_{GS} - V_{TH}) \approx 212.13mV$, and the output oscillation signal amplitude is 355 mV. From this, we can derive T_1 and T_2, where $T_1 = 34ps$ and $T_2 = 98ps$.

The designed Ring VCO oscillates at a frequency of 6 GHz, with $\frac{W}{L} = 220$. We set the rise time of the dual-exponential current source $t_r = 20ps$, the fall time $t_f = 120ps$, and $I_0 = 800\mu A$.

Based on the parameters provided above, we can estimate the ratio of the noise PSD when SET occurs on the cross-coupled transistors in the LCVCO to the noise PSD when SET occurs on the NMOS transistors in the ring VCO.

$$\frac{S_{\Phi n,LCVCO}(f)}{S_{\phi n,ring\,vco}(f)} \approx 2.3 \quad (13)$$

Based on the above analysis, we can conclude that LCVCO is more sensitive to SET than ring VCO.

B. Heavy-Ion Experiment

The heavy ion experiment was conducted at the China Institute of Atomic Energy (CIEA) and Harbin Institute of Technology (HIT). Oscilloscope is used to capture the output signals after frequency division, with a trigger threshold set at $10ns \pm 400ps$.

Fig. 5. depicts the photograph of the test chip, where the ring VCO is integrated within a PLL. Any occurrence of SET in other parts of the PLL will also affect the output signal. Based on the relative sizes of the three-stage inverter in the ring VCO and the overall PLL area, we can estimate the probability of SET occurrence that caused by inverter. The PLL area is approximately $0.0399mm^2$, while the area occupied by the three-stage inverter is $0.000078mm^2$, accounting for 0.195% of the PLL area.

Similarly, we are interested in SET induced by the cross-coupled transistors in the LCVCO. The total area of the LCVCO is $0.055mm^2$, and the area occupied by the inductors

is $0.0207mm^2$. Excluding the area occupied by the inductors, the LCVCO area is $0.0343mm^2$ (since inductors do not experience SET that affects LCVCO oscillation). The area of the cross-coupled transistors is $0.000102mm^2$, which accounts for 0.297% of the LCVCO area (excluding inductors).

Then we can obtain the effective trigger counts of SET on the LCVCO cross-coupled transistors, denoted as t_1, and the effective trigger counts of SET on the ring VCO, denoted as t_2.

$$t_1 = \frac{area\ of\ LCVCO}{area\ of\ Cross - coupled\ transistor} \times times_{LCVCO} \quad (14)$$

$$t_2 = \frac{area\ of\ ring\ VCO}{area\ of\ PLL} \times times_{ring\,vco} \quad (15)$$

According to Tabel I, it can be concluded that LCVCO is more sensitive to SET than ring VCO, with $t1 > t2$. The test results are consistent with the model derivation.

Fig. 5. Test chip photograph.

TABLE I. CHARACTERISTIC OF IONS USED IN TEST AND THE RATIO OF EFFECTIVE TRIGGER COUNT

Ions	Energy [MeV]	LET [MeV · cm²/mg]	Fluence [cm⁻²]	Range [μm]	t_1/t_2
Cl	150	13.4	1×10^7	42.8	22.3
Al	110	8.92	1×10^7	45	3.04
Ta	500	86.11	6.2×10^6	37.9	1.39

V. ANALYSIS AND EXPERIMENT RESULTS

A model for the sensitivity to SET was developed for two different types of VCOs, analyzing the correlation between VCO design and SET. Experimental results confirmed that the model provides a rapid method for assessing VCO sensitivity. In FinFET technology, the ring VCO exhibits lower sensitivity to SET. Therefore, for applications below DDR and PCIE 3.0, the utilization of ring VCO is preferred.

REFERENCES

[1] Y. Shu, Z. Deng and X. Luo, "8.3 A 28GHz Scalable Inter-Core-Shaping Multi-Core Oscillator with DM/CM-Configured Coupling Achieving 193.3dBc/Hz FoM and 205.5dBc/Hz FoMA at 1MHz Offset," 2023 IEEE International Solid-State Circuits Conference (ISSCC), San Francisco, CA, USA, 2023, pp. 150-152, doi: 10.1109/ISSCC42615.2023.10067826.

[2] Y. Hengzhou et al., " A SET-Tolerant High-Frequency Multibiased Multiphase Voltage-Controlled Oscillator for Phase Interpolator-Based Clock and Data Recovery, " IEEE Trans. Nucl. Sci., vol. 69, no. 7, pp. 1725–1732, Jul. 2022.

[3] Behzad Razavi, "LC Oscillator Design," in Design of CMOS Phase-locked Loops. Cambridge, United Kingdom: Cambridge University Press, 2020, pp. 144-153.

A Low Spur Wideband PLL in 65nm CMOS

Zijun Wang*, Biao Li, Teng Wang, Hong Li, Ruiting Niu, Jinpeng Lin

Integrated Circuits Design Center, Space Star Technology Limited Corporation

Haidian District, Beijing 100086, P. R. China

* Email: kid20081018@126.com

Abstract—This paper presents a targeted approach to developing a low-spur phase-locked loop (PLL) circuit that operates within a wide frequency range. By controlling the pulse timing of the phase and frequency detector, the periodic ripple in the charge pump output voltage is reduced. Additionally, a voltage-controlled dynamic current compensation circuit is employed to mitigate current mismatch issues in the charge pump. Furthermore, a multi-stage VCO parallel circuit architecture is used to reduce VCO gain while ensuring the desired output frequency range, thereby achieving low-spur and low-noise performance. The proposed PLL circuit is fabricated in 65nm CMOS process, and experimental results indicate that the output frequency range of the PLL covers 30MHz to 4GHz. At an output frequency of 4GHz, the measured phase noise at a 1MHz offset from the center frequency reaches as low as -117.37dBc/Hz. Furthermore, the reference spur is significantly suppressed, reaching level as low as -71.01dBc.

Keywords—PLL, Charge Pump, Low-spur, VCO

I. INTRODUCTION

In the pursuit of improved spectrum efficiency and enhanced performance, contemporary communication systems, including mobile and satellite communications, have imposed more rigorous requirements on transmit and receive channels. The phase-locked loop (PLL), serving as a crucial component within these channels, must deliver a high-purity local oscillator signal with minimal spurious content to effectively meet the demanding criteria of high adjacent channel suppression and advanced modulation schemes like OFDM.

This paper analyzes the causes behind reference spur in PLL and proposes a targeted approach to minimize their occurrence. By optimizing the lock-in phase and frequency detector(PFD) switch pulse width, reducing the charge pump(CP) current mismatch, and implementing a low-gain voltage-controlled oscillator(VCO) in a parallel configuration, a wideband low-spur PLL circuit is introduced.

II. THEORETICAL ANALYSIS

Reference spur signals are induced by the undesirable elements of the circuits, it is important to analyze and mitigate their effects in CP-PLL circuits. These spur signals manifest as energy peaks in the frequency spectrum at frequencies corresponding to the reference clock's offset from the center frequency, as well as at the harmonic frequencies of the reference clock. The characterization and analysis of these reference spur signals are essential to ensure the desired performance of the PLL circuit.

Assuming an initial phase of 0 for the output signal, the phase of the VCO output can be represented as follows when the loop is locked:

$$\varphi_{out}(t) = \int_0^t \omega_{out}(\tau)d\tau = \int_0^t K_{vco}V_{ctrl}(\tau)d\tau = K_{vco}\int_0^t V_{ctrl}(\tau)d\tau \quad (1)$$

In scenarios where the phase deviation of the output signal caused by ripple is minimal, the output voltage of the VCO can be expressed as follows:

$$V_{out}(t) = V_0 \sin(\omega_0 t + \varphi_{out}) \approx V_0 \sin\omega_0 t + [V_0 K_{vco}\int_0^t V_{ctrl}(\tau)d\tau]\cos\omega_0 t$$

$$= V_0 \sin\omega_0 t + \frac{V_0 \Delta v K_{vco}}{2\omega_{ref}}[\cos(\omega_0 + \omega_{ref})t + \cos(\omega_0 - \omega_{ref})t] \quad (2)$$

It is evident that the occurrence of spur signals at the offset center frequency of ω_{ref} Hz can be observed, characterized by a peak amplitude of:

$$P_{spur} = 20\log\frac{\Delta v K_{vco}}{2\omega_{ref}} \quad (3)$$

The equation above reveals that the occurrence of reference spur is dependent on various factors such as the reference clock frequency (ω_{ref}), the fluctuation amplitude of the control voltage (Δv), and the gain of the VCO (K_{vco}). A narrower pulse width, higher pulse consistency, and reduced mismatch in the charge pump current result in a smaller value of Δv, consequently leading to a decrease in reference spur[1]. The Kvco exhibits a correlation with the gain of the VCO, whereby a lower gain results in a reduced level of reference spur signals[2].

In conclusion, this paper proposes a methodology for optimizing the low-spur circuit design of the PLL. The approach involves reducing the pulse width of the PFD's output, minimizing the mismatch in charge pump current, and decreasing the Kvco, while simultaneously ensuring a wide operating bandwidth. By employing these techniques, the proposed design aims to achieve enhanced performance in the PLL system.

III. CIRCUIT DESIGN

The overall architecture of the low-spur PLL circuit designed in this paper is illustrated in the figure below. It primarily comprises a phase-frequency detector (PFD), charge pump (CP), loop filter (LPF), voltage-controlled oscillator (VCO), divider (DIV), and output frequency division chain.

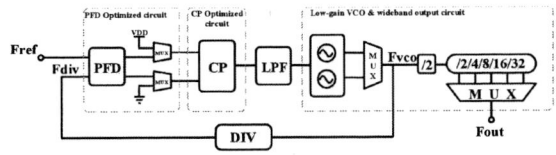

Fig. 1. Structure of PLL system

In this design, a PFD circuit with configurable reset pulse width and output pulse delay alignment is implemented to optimize the pulse width of the output control signal and ensure consistent delay among the various switch control signals. By utilizing the dynamic current compensation technique, the discrepancy between the charging and discharging currents in the CP is mitigated. Additionally, the VCO gain is minimized through the parallel connection of VCOs and the utilization of an output programmable division chain architecture, while ensuring a wide operating bandwidth[3].

A. Phase Frequency Detector (PFD)

In this design, the PFD is comprised of two distinct circuit components: the control signal generation circuit and the control signal processing circuit, as delineated in Figure 2 depicted below. In the control signal generation circuit ,to balance the pulse width to prevent aggravated the CP's current mismatch and caused the PFD and CP circuits falling into a dead zone, this design incorporates a configurable delay mechanism in the reset link, offering flexibility in controlling the reset duration[4]. In addition, there is an advanced MUX selection circuit is integrated to guarantee a constant level of the control signal in the locked state of the loop. As illustrated in Figure 3, by switching the UPO and DNO signals to a constant level during the locked state, the generation of this unwanted ripple can be completely avoided, thereby optimizing the overall spurious characteristics of the PLL. The control signal processing circuit generates the four essential pulse control signals UP, UN, UPN, and DNN, which are required for the subsequent stage of the CP. This meticulous process ensures that these four control signals are perfectly synchronized and exhibit no delay deviation.

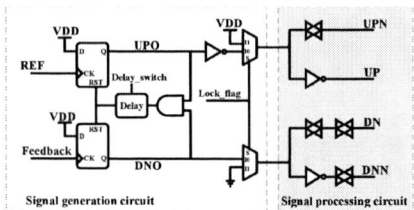

Fig. 2. Optimized PFD circuit

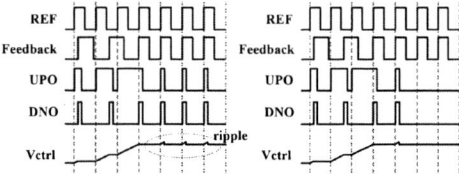

Fig. 3. Signal timing diagram of PFD

B. Charge Pump （CP）

To optimize the mismatch between the charge and discharge currents of the CP, a voltage-controlled current compensation circuit was meticulously devised. In order to address the non-ideal effect of charge sharing in the charge pump, a novel configuration utilizing a parallel switch structure at the drain terminal was introduced. This configuration not only enhances the switching speed, but also effectively mitigates the undesired charge sharing phenomenon. The CP circuit is depicted in Figure 4.

This paper presents a circuit design that enables dynamic tracking of Vctrl in order to dynamically adjust the charging and discharging currents (Iup and Idn), with the aim of minimizing current mismatch. As illustrated in Figure 5, the charging and discharging currents of CP are composed of a combination of fixed and dynamic components. The relationship between the dynamic currents and the output

voltage is mathematically described by formula 4, Notably, the greater the deviation of Vctrl from Vref, the larger the dynamic compensation current, thereby facilitating reduced current mismatch across a wide range of voltage control. The simulation results, illustrated in Figure 5,clearly indicate that the voltage-controlled current compensation CP circuit(red lines) offers an enhancement in the level of charge and discharge current matching, as compared to the conventional CP circuit(black lines).

$$I_{up} = I_{up1} + I_{up2}$$
$$I_{dn} = I_{dn1} + I_{dn2}$$
$$I_{up2} = I_1\,[V_{ctrl}\,/\,(V_{ref} + V_{ctrl})]$$
$$I_{dn2} = I_1\,[V_{ref}\,/\,(V_{ref} + V_{ctrl})] \tag{4}$$

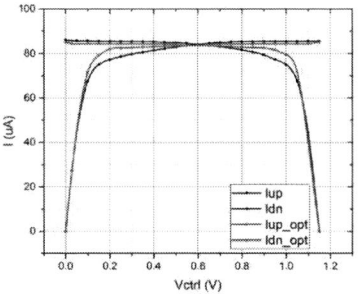

Fig. 4. Voltage-controlled current compensation CP circuit

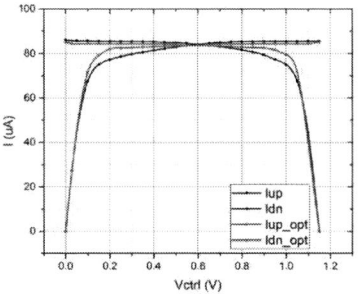

Fig. 5. Sumulation results of charging and discharging currents

The control switches employ a complementary parallel structure at the drain side. This configuration ensures a smooth flow of charging and discharging current, thereby enhancing the switching speed of the switches. Additionally, a unity gain amplifier is employed to clamp the output voltage Vctrl to the parallel switch voltage Vx. This mechanism guarantees that the output voltage remains stable and avoids any floating state, thereby preventing charge sharing during switch transitions.

C. Voltage-controlled oscillator （VCO）

This design incorporates an LC-VCO to enhance the noise performance of the phase-locked loop. By utilizing three parallel-connected VCOs followed by a frequency division chain, a wide range of operation is achieved, enabling the implementation of a broad bandwidth, low-gain VCO circuit.

The VCO circuit consists of two main components: a complementary differential negative resistance transistor pair and an LC resonant tank. Compared to a single-ended differential pair, the complementary differential pair allows for a larger swing, given the same bias conditions. By adjusting the size parameters of the NMOS and PMOS transistors, the output signal can be made more symmetrical,

979-8-3503-6184-1/24 $31.00 © 2024 IEEE

thereby improving the phase noise performance of the VCO. Additionally, biasing the NMOS transistor substrate in a reverse manner reduces the depletion layer charge, resulting in a lower transistor threshold voltage and facilitating low-voltage and low-power design.

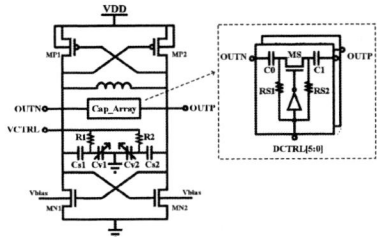

Fig. 6. Optimized VCO circuit

The capacitor C in the LC resonant cavity is constructed by the parallel connection of a coarse-tuned switch capacitor array and a fine-tuned variable capacitor. DCTRL controls the coarse tuning circuit, as shown in Figure 6. During the switch-off phase, the resistor connects the drain and source terminals of the switch to a high potential, preventing current noise caused by forward conduction between the drain and source terminals and the substrate PN junction. The capacitor fine-tuning circuit utilizes short-channel MOS variable capacitors (Cv1 and Cv2) to reduce the VCO's conversion gain (Kvco). The parallel capacitance value inside the resonant cavity is controlled by VCTRL. With the combined action of DCTRL and VCTRL, the VCO output achieves a wide bandwidth range, low conversion gain, and high phase noise performance.

IV. MEASUREMENT RESULTS

This paper implements the design, simulation, and fabrication of a low-spurious PLL chip based on the CMOS 65nm process. The micrograph and layout of the chip are depicted in figure 7, showcasing a phase-locked loop area measuring 1600um × 680um. Based on the photograph of the chip, it is evident that the PLL is situated at the central position of the overall transceiver chip. To mitigate voltage drops in the power supply, a parallel arrangement of wide top metal lines is employed for efficient power and ground routing. Furthermore, meticulous ground shielding is implemented on the upper and lower layers of the reference clock routing to ensure minimal interference with other crucial signals.

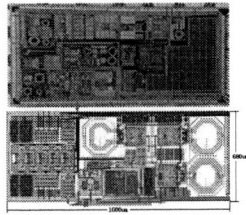

Fig. 7. Layout and micrograph of the chip

In this paper the PLL system produces pure frequency source signals ranging from 30MHz to 4GHz. The reference spur was tested at output frequencies of 1.38GHz and 4GHz, respectively. As depicted in figure 8, at an output frequency of 4GHz, with a 20MHz reference frequency offset, the reference spur was attenuated to -71.01dBc. As depicted in figure 9, the phase noise measured at 1MHz offset from 4GHz carrier is -117.37dBc/Hz.

By conducting a comprehensive comparison with other relevant references, it becomes apparent that the PLL designed in this paper possesses remarkable advantages across various performance metrics, including operating frequency range, spurious emissions, and phase noise characteristics.

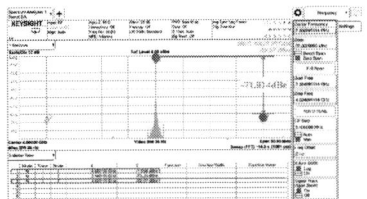

Fig. 8. Measured spectra with Ref. Spur at output frequency of 4GHz

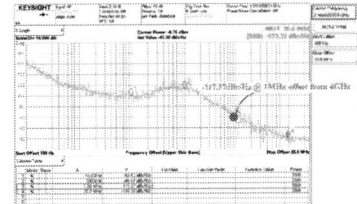

Fig. 9. Measured PN spectra at center frequency of 4GHz

TABLE I.   PERFORMANCE COMPARISON

	[5]	[6]	This work
Freq.(GHz)	1.6	3.2-3.8	0.03-4
Ref. spur(dBc)	-63	-69	-71.01
PN(dBc/Hz)	-118 (1MHz offset)	-107 (1MHz offset)	-117.37 (1MHz offset from 4GHz)

In conclusion, this paper analyzes the mechanism responsible for the generation of reference spur in a PLL system and presents the design, simulation, chip fabrication and measurement of a low reference spur PLL circuit based on CMOS 65nm process. The results demonstrate that the designed PLL chip effectively suppresses reference spur while exhibiting excellent performance with wide bandwidth.

REFERENCES

[1] J.Zu, X.Xing and H.Feng, "A Charge Pump with perfect current matching applied to phase-locked loop in 65nm CMOS," IEEE International Conference on ASIC(ASICON), p.1-4(2021)

[2] D.Cai et al. "A Dividerless PLL with low power and low reference spur by Aperture-Phase Detector and Phase-to Analog Converter," IEEE Transactions on Circuits and System I:Regular papers, vol.60, no.1, p.37-50(2013)

[3] N.Kim and Y.Moon, "A study on wide-band frequency synthesizer for advanced wireless communication," International SoC Design Conference, p.227-230(2011)

[4] A.Ghaemnia and M.B.Ghaznavi-Ghoushchi, "QD-PFD:Quasi dynamic dead-zone/blind-zone free PFD with 23nW-38 μ W for 2MHz-5GHz range and 150-ns setting time PLL applications," IEEE Transactions on Circuits and System II:Express Briefs, vol.71, no.1, p.91-95(2024)

[5] D.A.Pontes, H.D.Hernandez, D.Reyes and L.E.Rueda, "A 1.6-GHz low-jitter, low reference-spur single-loop Type-I PLL,"International Conference on Electrical, Computer and Energy Technologies (ICECET), p.1-4(2021)

[6] M. Osada, Z. Xu, T, Iizuka, "A 3.2-to-3.8GHz harmoic-mixer-based dual-feedback fractional-N PLL achieving -65dBc in-band fractional spur,"IEEE Solid-State Circuits Letters,vol.3, p.534-537(2020)

979-8-3503-6184-1/24 $31.00 © 2024 IEEE

A Low Power PLL Circuit with Signal 50% Duty Cycle Corrected in 180nm CMOS

Bangtian Li [1,2], Xueke Li [3], Liying Chen [1,2*], Chuantong Cheng [4*], Jian Mei [5]

[1] School of Electronics and Information Engineering, Tiangong University, Tianjin 300387, China
[2] Tianjin Key Laboratory of Photoelectric Detection Technology and System, Tianjin 300387, China
[3] College of Arts, Tiangong University, Tianjin 300387, China
[4] Institute of Semiconductors, Chinese Academy of Sciences, Beijing 100083, China
[5] National Integrated Circuit Innovation Center, Shanghai, China
* Corresponding Authors Email: nkchenliy@163.com, chengchuantong@semi.ac.cn

Abstract

A phase-locked loop (PLL) circuit with low power consumption and 50% duty cycle regulation has been implemented using the SMIC 180nm BCD process. The circuit fulfils the clocking requirements within the system on chip (SoC) and employs a four-stage differential loop oscillator to generate the clock phase with a minimal footprint, facilitating straightforward integration. The duty cycle corrector is a single-ended structure. The incorporation of a gain-enhanced charge pump enhances the performance of the loop, concomitantly reducing voltage ripple. This results in improved accuracy and an extended input duty cycle range and operating frequency range. The results of the post-simulation analysis demonstrate that the duty-cycle corrector achieves a 50% output duty cycle over a wide input duty cycle range of 10 MHz to 1 GHz. The PLL lock time is less than 7 μs, the phase noise performance reaches -100.57 dBc/Hz@1 MHz, and the overall power consumption is 1.628 mW, with a core chip area of 0.059mm^2.

1. Introduction

The duty cycle is of paramount importance in the design of any circuit system. In practice, different design solutions have varying requirements for pulse output. Large-scale SOCs typically necessitate a 50% duty cycle of the clock signal, while specific applications may require a 40% duty cycle or even less[1]. A general clock signal generated through a crystal oscillator or PLL may not meet the specific requirements of a given application. In such cases, the clock signal should be shaped to ensure that the output signal duty cycle meets the necessary requirements[2]. However, the duty cycle of the clock signal output directly from a phase-locked loop often deviates from the 50% mark, necessitating the use of an additional, specialized duty cycle adjustment circuit to adjust the clock signal's duty cycle.

In order to address the limitations of the conventional PLL, a novel digital signal duty cycle corrector circuit has been proposed[3]. The digital signal duty cycle corrector circuit exhibits robust immunity to loop noise and a brief lock time[4]. Nevertheless, the digital control circuit and delay element impose constraints on the operational frequency range. Analogue signal duty cycle corrector circuits are well-suited to high-speed applications, offering numerous hardware implementations[5]. However, loop stability constraints restrict their overall performance. The issue of stability is of paramount importance in analogue duty cycle modulated circuits, given the presence of feedback loops within the circuitry. Stability exerts a significant influence on the operating frequency range, input duty cycle range, and jitter performance[6]. Conventional signal duty cycle corrector circuits employ a three-stage ring oscillator to generate a reference voltage. In the event of process variations in the oscillator or an inverter with an inappropriate width-to-length

ratio, an incorrect reference voltage may be generated, which may result in unstable loop operation. The interrelated duty cycle corrector circuit employs a differential scheme to generate complementary signals for the bias voltage that are shifted 180° out of phase. This approach improves stability and reduces voltage ripple at the transconductor output. However, the bandwidth remains limited.

This paper presents a PLL circuit with a 50% duty cycle of the output signal, designed for low power and small area, and suitable for application on low voltage SoCs. The work proposes a signal duty cycle corrector circuit based on a charge pump structure, with an output duty cycle of 50% and a small area with low power consumption. It can receive signals with a wide input duty cycle range and high bandwidth with high stability. Section 2 presents the circuit structure design. Section 3 presents the results of the circuit simulation and an analysis thereof. Section 4 provides a conclusion.

2. Circuit Design

2.1 PLL overall circuit structure

The fundamental configuration of a PLL is a charge pump phase-locked loop. As illustrated in Figure 1, its primary components include a phase frequency detector, a charge pump (CP), a low pass filter, a divider, a voltage control ring oscillator (VCRO), a duty cycle corrector of 50%, and a current bias circuit.

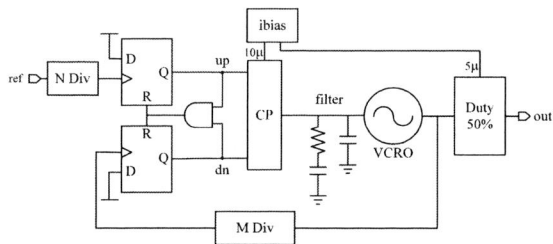

Figure 1. Proposed PLL structure with 50% duty cycle correction

Two principal categories of low-pass filters are distinguished: active filters, which exhibit a deficiency in noise performance and a proclivity for high power consumption[7]. Still offer the benefit of a compact configuration. The other is a passive filter, which is the antithesis of an active filter, exhibiting superior noise performance but a considerable footprint. The charge pump output, which comprises high-frequency and clutter components, can be filtered out by the loop filter. The output current of the charge pump flows through the loop filter after the formation of the control voltage, which is used to regulate the oscillation frequency of the VCRO. To reduce the lock time, the loop bandwidth can be increased in an appropriate manner. However, this may result in the loop becoming unstable and permanently losing lock. Alternatively, the locking time can be reduced by increasing the frequency of the PFD[8].

979-8-3503-6184-1/24 $31.00 © 2024 IEEE

2.2 Principle and design of duty cycle corrector

Figure 2 illustrates the circuit structure of the 50% duty cycle regulation of the signal proposed in this paper. The control stage is responsible for adjusting the duty cycle of the input signal CKin. The feedback control voltage Vctrl is generated by a charge pump, which is used to regulate the duty cycle of the input signal. A buffer is employed to drive the output load. The use of a gain-enhanced charge pump structure serves to reduce the ripple of the control voltage Vctrl, thereby improving current matching.

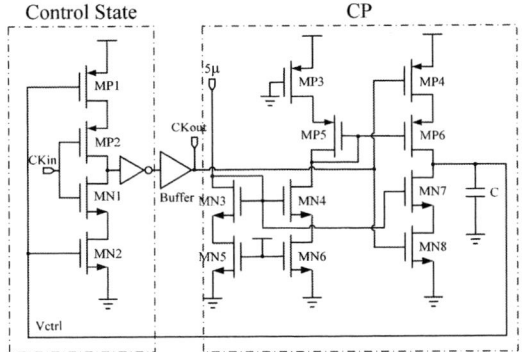

Figure 2. Proposed 50% duty cycle correction circuit

As illustrated in Figure 2, the Control State employs a push-pull configuration to expedite control operations. This structure oversees the regulation of both charging and discharging pathways, utilising an upper PMOS and lower NMOS as the switching mechanism. This results in a reduction in the time required for duty cycle correction, thereby increasing the speed of the lock process. The input signal for any duty cycle is represented by the variable CKin. The Vctrl voltage represents the charging voltage from the charge pump to the capacitor. As the voltage Vctrl decreases, the charging current increases and the discharging current decreases, resulting in an increase in the output duty cycle. Conversely, when the voltage Vctrl rises, the charging current decreases and the discharging current increases, leading to a decrease in the output duty cycle. Furthermore, the control stage lacks a direct current path from VDD to ground, allowing for a reduction in power consumption.

The control voltage Vctrl is of critical importance in the duty cycle adjustment of the signal and affects the jitter and duty cycle of the output signal. Therefore, the voltage jitter provided by the CP must be sufficiently small. In order to reduce the mismatch, the source switching structure charge pump is selected. This is located between the source of the charge/discharge transistor and VDD/GND, and the switching speed is faster. As the switching tube is not directly connected to the output port, the non-ideal characteristics brought about by the switching process (mainly charge sharing, charge injection and clock feedthrough) have a relatively small impact on the source switching structure.

Figure 2 illustrates a CP with a current source connected in the output circuit. The MP3, MP5, and MN3-MN6 components provide bias. The PMOS and NMOS current sources, designated MP6 and MN7, respectively, are connected to the outputs. MP4 and MN8 are, respectively, PMOS and NMOS switches. When the signal CKout is at a high level, the transistor MP4 is deactivated, while the transistor MN8 is activated. MN7 serves as the discharge path for the output capacitor. When the output capacitor is discharged, the PMOS transistor MP4 is activated, while the

NMOS transistor MN8 is deactivated. MP6 serves as the charge path for the output capacitor. This circuit is suitable for low-voltage designs, as stacking is no longer required.

3. The simulation result and discussion

The proposed PLL is implemented using a SMIC 180nm CMOS process with a supply voltage of 1.8V. Figure 3 shows the circuit layout, which occupies a total area of $0.059mm^2$.

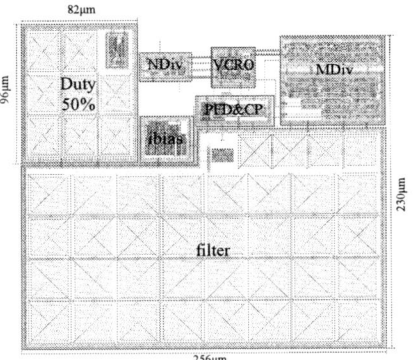

Figure 3. Layout of the proposed PLL

Figure 4 illustrates the input and output of the duty cycle corrector circuit in the form of a clock signal. The resulting output duty cycle is 50.8% and 49.87% when the input clock frequency is 1 GHz and the duty cycle is 1% and 99%, respectively. The 180nm CMOS device model is based on a Cadence SPICE simulation. Due to the constraints of the manufacturing process, the device is unable to function correctly when the pulse width is below 10 ps. Consequently, the operation of smaller and larger duty cycles at very high frequencies is constrained by the limitations of the process and circuit performance. Figure 5 illustrates the variation in the duty cycle corrector results in relation to the input signal frequency for distinct duty cycle inputs. The findings indicate that the output signal duty cycle spans a range of 47.39% to 50.96% across the frequency spectrum of 10 MHz to 1000 MHz.

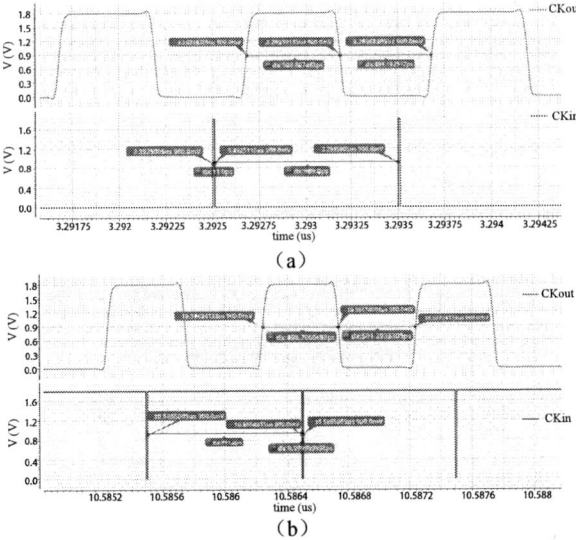

Figure 4. Post-simulated input and output clocks at 1GHz. (a) Input clock with 1% duty cycle and output clock with 50.8% duty cycle. (b) Input clock with 99% duty cycle and output clock with 49.87% duty cycle.

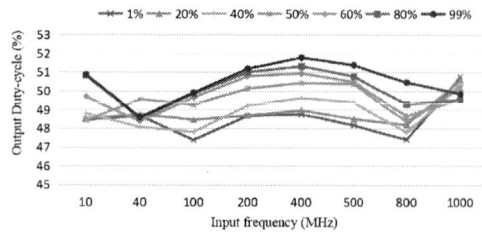

Figure 5. Post-simulated summary diagram of the input duty cycle of 1%, 20%, 40%, 50%, 60%, 80%, 99% corresponds to the output duty cycle.

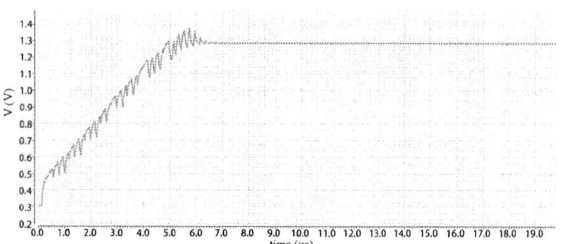

Figure 6. Tran simulation results of VCRO control voltage

The post-simulation results of the PLL demonstrates satisfactory performance. Figure 6 illustrates that the PLL lock time is less than 7 μs, which is sufficient to meet the requirement for fast clock locking in SoC systems. The wide frequency adjustment range of the PLL can be enhanced by utilising a 40 MHz signal as the external reference clock and subsequently adjusting the frequency after inputting the frequency divider NDiv before providing it to the PFD. At 432 MHz oscillation, a phase noise performance of -100.57 dBc/Hz is achieved at 1 MHz offset. The total power consumption is 1.628 mW, while the power contribution of each module of the PLL is shown in Table 1.

Table 1. The distribution of power in PLL circuits

	Power Consumption	Ratio
ibias	77.46μW	5.08%
NDiv	52.79μW	3.47%
MDiv	540μW	35.44%
PFD	6.93μW	0.45%
CP	23.05μW	1.51%
VCRO	719.28μW	47.21%
Duty50%	104.09μW	6.83%

4. Conclusion

This paper proposes a low-power PLL circuit with 50% duty cycle corrector and a lock time of less than 7 μs. The circuit has a core area of 0.059mm², which facilitates integration, and can meet the clocking requirements of SoC systems. The duty cycle corrector circuit performs well over a wide input duty cycle range and high bandwidth conditions. The operating frequency at which 1%~99% input duty and 50% output duty cycle can be achieved ranges from 10 MHz to 1 GHz.

Acknowledgments

This work was supported by the Youth Innovation Promotion Association of Chinese Academy of Sciences (Grant No. 2022109).

References

[1] K. H. Cheng, C. W. Su, and K. F. Chang, "A high linearity, fast-locking pulsewidth control loop with digitally programmable duty cycle correction for wide range operation", IEEE J Solid-State Circuits, vol. 43, no. 2, pp. 399–412, (2008)

[2] X. Jin, W. Park, D. S. Kang, Y. Ko, K. W. Kwon, and J. H. Chun, "A 4-GHz Sub-Harmonically Injection-Locked Phase-Locked Loop with Self-Calibrated Injection Timing and Pulsewidth", IEEE J Solid-State Circuits, vol. 55, no. 10, pp. 2724–2733, (2020)

[3] C.-C. Chung, D. Sheng, and S.-E. Shen, "High-Resolution All-Digital Duty-Cycle Corrector in 65-nm CMOS Technology", IEEE Trans Very Large Scale Integr VLSI Syst, vol. 22, no. 5, pp. 1096–1105, (2014)

[4] Y. J. Wang, S. K. Kao, and S. I. Liu, "All-digital delay-locked loop/pulsewidth-control loop with adjustable duty cycles", IEEE J Solid-State Circuits, vol. 41, no. 6, pp. 1262–1274, (2006)

[5] J. H. Lim et al., "A Delay Locked Loop with a Feedback Edge Combiner of Duty-Cycle Corrector with a 20%-80% Input Duty Cycle for SDRAMs", IEEE Transactions on Circuits and Systems II: Express Briefs, vol. 63, no. 2, pp. 141–145, (2016)

[6] W. M. Lin and H. Y. Huang, "A low-jitter mutual-correlated pulsewidth control loop circuit", IEEE J Solid-State Circuits, vol. 39, no. 8, pp. 1366–1369, (2004)

[7] Y. Yang, B. Lyu, F. Ye and J. Ren, "A 3GHz Phase-Locked Loop Design for SerDes Application," IEEE 16th International Conference on Solid-State & Integrated Circuit Technology (ICSICT), pp. 1-3, (2022)

[8] W. Deng et al., "A Self-Adapted Two-Point Modulation Type-II Digital PLL for Fast Chirp Rate and Wide Chirp-Bandwidth FMCW Signal Generation", IEEE J Solid-State Circuits, vol. 57, no. 4, pp. 1162–1174, (2022)

MTJ based Compensation for Charge Pump Temperature Drift

Yongliang Zhou*†, Jingxue Zhong*, Chengxing Dai*, Yingxue Sun*, Xin Li*, Chunyu Peng*

*School of integrated circuits, Anhui University, HeFei, China
† Anhui Anxin Electronic Technology Co., Ltd
* Email: zhouyongliang@ahu.edu.cn

Abstract

This research focus on the temperature compensation between CMOS and MTJ devices, for maintaining the consistent current output of wide-temperature charge pump. The investigation is based on the utilization of 150 nm CMOS technology, a cascaded of CMOS and MTJ is employed in the design, and a thorough analysis of temperature variations is conducted. The observed current mismatch of the developed charge pump within the voltage range of 1.5V to 3.5V is found to be below 0.3%. Low mismatch charge pump used in Phase-Locked Loop (PLL) circuit to reduce reference stray within the temperature range of -150°C to 125°C.

Key words: MTJ, charge pump, temperature drift, PLL

1. Introduction

In highly specialized operational settings like space and subterranean exploration, it is crucial for the phase-locked loop (PLL) to maintain synchronization consistently across varying temperature ranges. Consequently, the oscillator must achieve superior levels of frequency precision and temperature stability.

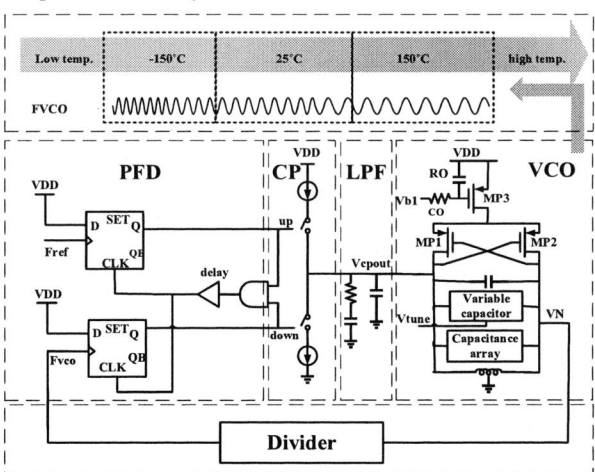

Figure 1. The conventional CPPLL architecture.

Figure 1 depicts the traditional architecture of PLL, comprising components such as the frequency and phase discriminator (PFD), charge pump (CP), low-pass filter (LPF), voltage-controlled oscillator (VCO), and divider. However, the VCO frequency demonstrates variability over a broad temperature range, with a tendency to decrease as temperature rises. This temperature-dependent characteristic may present difficulties for achieving PLL frequency and phase synchronization under significant temperature fluctuations. The charge pump's charge current raises the

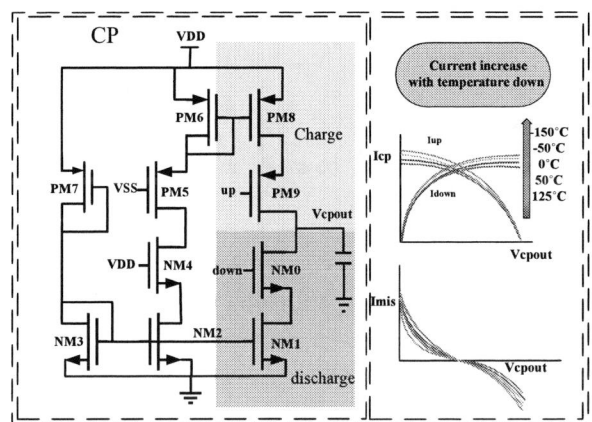

Figure 2. The traditional charge pump architecture.

VCO's control voltage, while the discharge current lowers it. By aligning the charge and discharge current values, the VCO's control voltage can effectively manage the VCO's frequency, ensuring stable alignment with the PLL circuit's reference frequency.This synchronization facilitates successful locking of the PLL to both phase and frequency.

Figure 2 illustrates the circuit schematic of a traditional charge pump, consisting of PM8, PM9, NM0, and NM1 components forming the charge pump circuit. The circuit's left section incorporates a current mirror, while the switching signals "up" and "down" regulate the charge and discharge operations of the charge pump. An important non-ideal characteristic of the charge pump stems from current mismatch induced by the channel length modulation effect in the MOS transistors. The output of PLL will experience non-ideal effects due to the presence of input jitter and reference spurious signals. Simulation findings from the conventional charge pump reveal that the matched current is subject to fluctuations based on temperature and charge pump output voltage, with the current magnitude increasing at lower temperatures.

2. Analysis of conventional charge pump

The simulation results depicted in Figure 3 illustrate the alignment of charge and discharge currents within the conventional charge pump under varying temperature conditions. It is apparent that enhancing the dimensions of the MOS tubes can enhance the current matching capabilities of the charge pump across diverse output voltage levels, thereby mitigating the impact of channel length modulation. Nevertheless, when subjecting the charge pump to a wide temperature range analysis, fluctuations in the matching current value are observed. The simulations indicate that lower temperature correspond to elevated matching current values. Consequently, substantial temperature fluctuations

979-8-3503-6184-1/24 $31.00 © 2024 IEEE

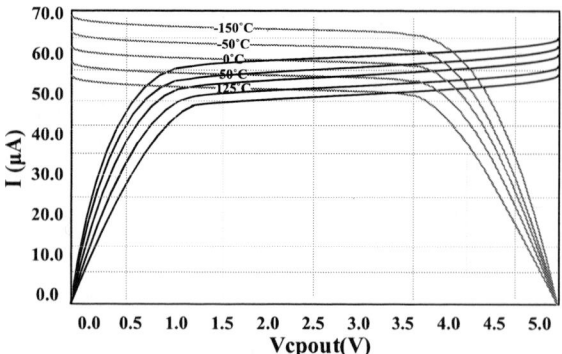

Figure 3. The matching current variation of traditional charge pumps at different voltages and temperatures.

may lead to an erratic control voltage for the VCO, influencing the frequency of VCO.

Figure 4 shows the results of a wide temperature scan of the traditional charge pump and voltage-controlled oscillator . As the temperature increases, the frequency of the VCO decreases. Therefore, temperature directly influences the frequency of VCO by adjusting the charge and discharge current values of the charge pump.

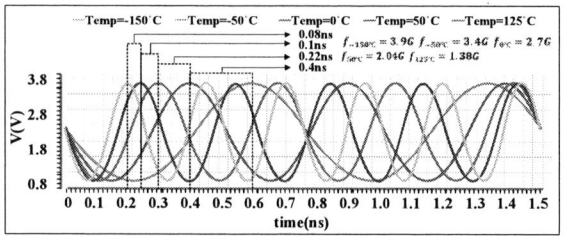

Figure 4. The output of VCO with traditional architecture at different temperatures.

3. Analysis of proposed charge pump

In CMOS technology, it is recognized that the transistor mobility and threshold voltage exhibit temperature-dependent variations, as depicted in Figures 5(a) and 5(b). The mobility of electrons and holes diminishes with increasing temperature, alongside a decrease in the threshold voltage of NMOS and PMOS transistors. Notably, the decline in mobility is more pronounced compared to the

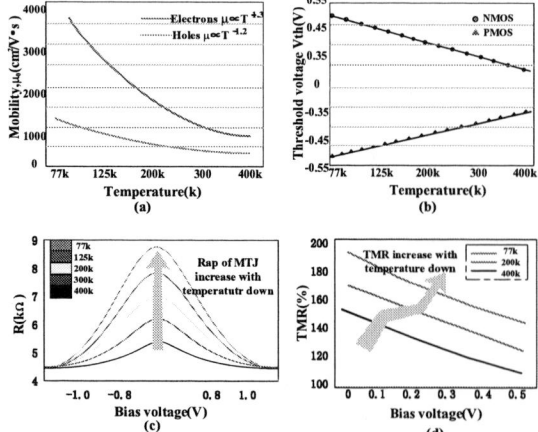

Figure 5. The wide temperature characteristics of CMOS and MTJ.

threshold voltage, leading to an escalation in MOS leakage current as temperature decreases. Figures 5(c) and 5(d) depict the temperature characteristics of the Magnetic Tunnel Junction (MTJ), which consists of an ultra-thin insulating layer sandwiched between two ferromagnetic sheets. The MTJ exhibits two resistance states -low resistance (R_P) and high resistance (R_{AP}). Notably, R_P remains relatively stable with temperature variations, whereas R_{AP} increases as temperature decreases. In the structure of MTJ, the Tunneling Magnetoresistance (TMR) effect is a spin-related tunnelling phenomenon that also increases with decreasing temperature.The TMR can be quantified as:

$$TMR= (R_{AP}-R_P)/R_P \qquad (1)$$

Figure 6 shows an improved charge pump where MTJs are connected in series in the charge and discharge branches. The MOSFET exhibits a positive temperature coefficient, implying that the on-resistance (Ron) of the MOSFET increases with the transistor temperature. In contrast, MTJ has a negative temperature coefficient, with their high resistance increasing as the temperature decreases. Initially, the MTJ is set to a high resistance state.By leveraging the temperature characteristics of CMOS and MTJ, a temperature compensation circuit is implemented in the proposed design to address the varying current values due to temperature fluctuations.

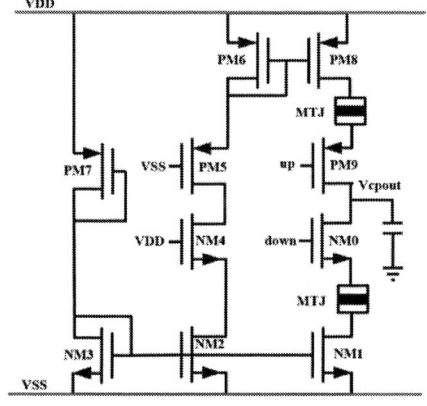

Figure 6. The proposed schematic of the charge pump.

Figure 7 shows the matlab data fitting according to the equations of [2] , [3], the R_{on} is expressed as:

$$R_{on}= 1/\mu Cox \frac{W}{L}(V_{GS}-V_{TH}) \qquad (2)$$

Where $\mu=T^{-m}$, $V_{TH} = aT + b$, The range of m is 1 to 2, a is the temperature coefficient of the threshold voltage and is found to be -0.68mV/°C, b is found to be about 0.567 V.

The fitted data is shown in figure 5(a). the R_{MTJ} is expressed as:

$$TMR(T)= (TMR_0+1)/(1+2Q\beta_{AP} \ln\left(\frac{k_BT}{E_c}\right))+1 \qquad (3)$$

$$R_{MTJ}=(1+TMR(T))R_P \qquad (4)$$

where TMR_0 is the TMR ratio at zero temperature, T = 0K, and Q describes the probability of a magnon involved in the tunneling process that will be used as a fitting parameter. $\beta_{AP} = Sk_BT/E_m$, where S is the spin parameter, k_B is Boltzmann's constant, E_m is related to the Curie temperature $E_m = Sk_BT/S + 1$of the ferromagnetic electrodes,and E_C is

979-8-3503-6184-1/24 $31.00 © 2024 IEEE   108

the magnon energy cutoff energy. The fitted data is shown in figure 7(b),figure 7(c) represents $Rtall = R_{MTJ} + R_{on}$. Figure 7(d) displays the histogram of Rtall in relation to temperature. It is apparent that Rtall remains nearly constant and exhibits minimal correlation with temperature. Therefore, achieving temperature compensation can be accomplished by leveraging the temperature characteristics of CMOS and MTJ.

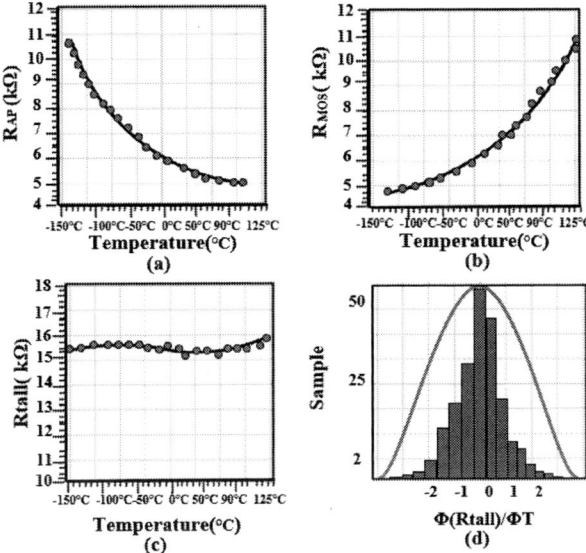

Figure 7. The equivalent resistance of charge pump simulated using Matlab.

In Figure 8, Figures 8(a), 8(b), and 8(c) represent the mismatch rates of the structures in references [4], [5], and [6], which are 1%, 1%, and 0.5% respectively. On the other hand, the improved charge pump structure achieved a mismatch rate of less than 0.3%, significantly enhancing the temperature stability of the charge pump.

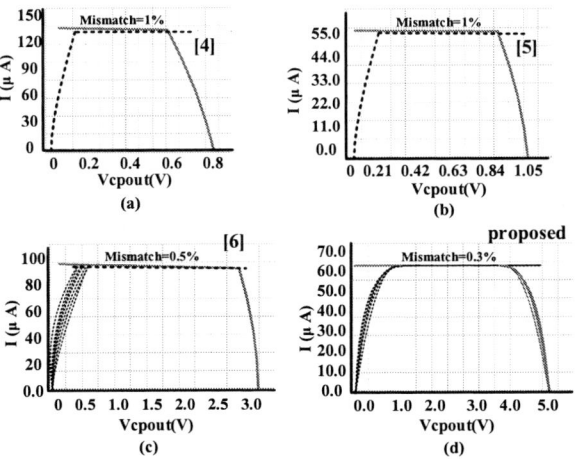

Figure 8. A comparison of the mismatch rate between the improved structure and other structures.

As illustrated in Figure 9, the proposed charge pump ensures a stable output voltage through the LPF, which results in a consistent control voltage for the VCO. As a result, the VCO's frequency remains relatively unchanged across a broad temperature range, facilitating easy frequency and phase locking for the PLL. This enhanced performance leads to significant improvements in the operational efficiency of PLL circuits.

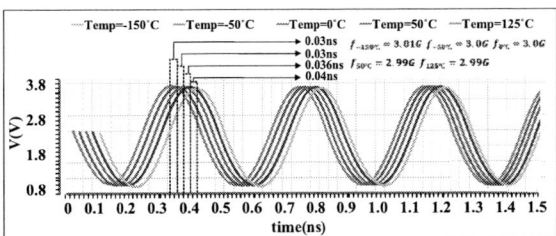

Figure 9. The output of VCO with proposed architecture at different temperatures.

4. Summary

In this paper, the temperature characteristics of MTJ are combined to compensate the current mismatch problem of conventional charge pump circuits over a wide temperature range. The results can achieve a current mismatch of less than 0.3% in the range of -150 ℃ ~125 ℃.

Table I. Comparison of parameters

REF	[4]	[5]	[6]	proposed
Supply(V)	0.8	1.05	3.3	5.0
Technolog	28nm	28nm	180nm	150nm
Matched range(V)	0.3~0.8	0.1~0.95	0.4~3.1	1.5~3.5
Temperature range(℃)	-40~85	/	-40~175	-150~125
Current Mismatched ratio	<1%	<1%	<0.5%	<0.3%

The performance of the proposed charge pump structure was compared with other literature based on the parameters summarised in Table I. It is clear that the current mismatch is significantly reduced over a wide temperature range.

ACKNOWLEDGMENT

This work is supported by Anhui Provincial Natural Science Foundation 2308085QF214, Natural Science Foundation of the Higher Education Institutions of Anhui Province under Grant 2023AH040011 and National Natural Science Foundation of China under Grant 62274001.

References

[1] T. Liu, X. Wang, R. Wang, G. Wu, T. Zhang and P. Gui, "A Temperature Compensated Triple-Path PLL With KVCO Non-Linearity Desensitization Capable of Operating at 77 K," in IEEE Transactions on Circuits and Systems I: Regular Papers, vol. 64, no. 11, pp. 2835-2843, Nov. 2017.

[2] Y. Zhou et al., "A CFMB STT-MRAM-Based Computing-in-Memory Proposal With Cascade Computing Unit for Edge AI Devices," in IEEE Transactions on Circuits and Systems I: Regular Papers, vol. 71, no. 1, pp. 187-200, Jan. 2024.

[3] K. Cao et al., "Low-Temperature Performance of Nanoscale Perpendicular Magnetic Tunnel Junctions With Double MgO-Interface Free Layer," in IEEE Transactions on Magnetics, vol. 55, no. 3, pp. 1-4, March 2019.

[4] L. Liu, Y. Ji, X. Liao, Z. Qin and H. Liang, "A 0.8-V, 2.55-GHz, 2.62-mW Charge-Pump PLL With High Spectrum Purity," in IEEE Transactions on Very Large Scale Integration (VLSI) Systems, vol. 30, no. 2, pp. 113-122.

[5] C. Zhang and Q. Wang, "A Low Current Mismatch High Swing Charge Pump for High Speed Phase Locked Loop," 2021 IEEE 3rd International Conference on Circuits and Systems (ICCS), Chengdu, China, 2021, pp. 126-129.

[6] J. Wang, Y. Jin and Q. Yuan, "A Temperature-Drift Suppressed Charge Pump for CPPLL Applied in MEMS Oscillator," 2023 IEEE 3rd International Conference on Electronic Technology, Communication and Information (ICETCI), Changchun, China, 2023, pp. 408-412.

A 112-Gb/s Coherent Receiver with a Novel Modulation Format

Tianyuan Zhong [1,3], Boyang Zhang [2,3], Weixin Gai* [2,3]

[1] School of Software and Microelectronics, Peking University, Beijing, China
[2] School of Integrated Circuits, Peking University, Beijing, China
[3] Beijing Advanced Innovation Center for Integrated Circuits, Beijing, China

* Email: wgai@pku.edu.cn

Abstract—This paper presents a 112-Gb/s coherent receiver (RX) based on a novel modulation format. The proposed coherent four-level pulse amplitude modulation (PAM-4) makes the phase detection of eliminating frequency offset simpler, and the chromatic dispersion (CD) equalizer can utilize a simplified-tap feed-forward equalizer (FFE) without affecting the equalization performance. And 4-way interleaving sampling relaxes the bandwidth requirements. Simulation shows the RX is capable of equalizing 112-Gb/s signals with 68 ps/nm CD and 200 MHz frequency offset and achieves an energy efficiency of 4.4 pJ/b in 28-nm CMOS technology with 1 V supply voltage.

Keywords—coherent, modulation, frequency offset, chromatic dispersion (CD), receiver (RX)

I. INTRODUCTION

As global cloud services grow at a rapid rate, demand for high bandwidth interconnects between data centers also grows. Coherent transmission plays a key role in the next data center interconnect (DCI) networks for higher data rates and longer distances [1]. However, at the same time, some non-ideal factors appear to affect the signals during transmission, such as frequency offset and chromatic dispersion (CD). To deal with these problems, an integrated complementary polarization-diversity coherent receiver (RX) with digital signal processing (DSP) was presented in [2], but DSP consumes much power. In [3], an all electrical Costas loop was presented to eliminate frequency offset, but the CD equalization was post-processing. An analog RX with equalization for both non-ideal factors was presented in [4], but the CD equalizer employed many taps, resulting in extra power consumption and difficult convergence. The phase detection was complex in [3], [4], [5]. In this paper, a novel modulation format based on coherent four-level pulse amplitude modulation (PAM-4) is proposed to eliminate frequency offset and CD by simpler phase detection and simplified taps CD equalizer respectively.

The rest of this article is organized as follows. Section II discusses the strategy of eliminating frequency offset and CD and introduces the architecture of the RX. Section III describes the circuit details of the RX. Section IV presents the simulation results, and Section V summarizes this article.

II. PROPOSED ARCHITECTURE

The proposed modulation format utilizes the same PAM-4 signals for both in-phase (I-phase) and quadrature-phase (Q-phase) throughout coherent transmission, and this can simplify phase detection and CD equalizer. The constellation diagrams of the proposed modulation format and non-ideal factors are shown in Fig. 1. In order to eliminate the frequency offset caused by transmission and demodulation, phase

This work was supported in part by the Beijing Major Science and Technology Project (Grant No. Z221100007722019).

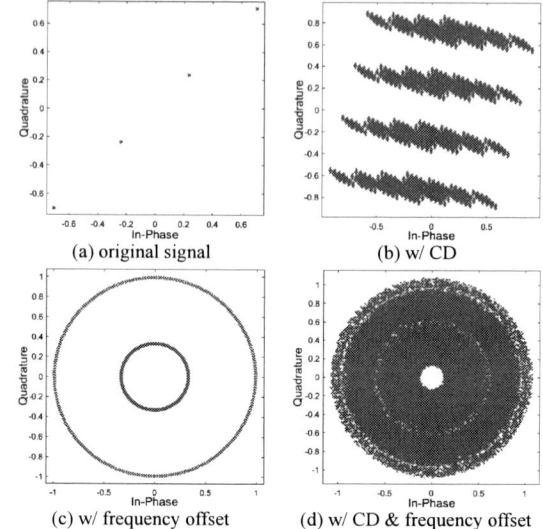

(a) original signal   (b) w/ CD

(c) w/ frequency offset   (d) w/ CD & frequency offset

Fig. 1. Constellation diagrams of the proposed modulation format

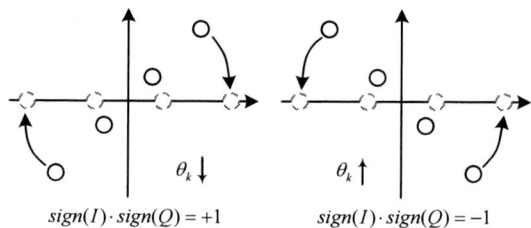

$sign(I) \cdot sign(Q) = +1$ $sign(I) \cdot sign(Q) = -1$

Fig. 2. Principle of phase detection of the proposed modulation format

detection is needed to track the frequency deviation. The principle of the phase detection for our modulation format is shown in Fig.2. Instead of restoring the constellation to its original position as Fig.1(a), we rotate the target constellation to the I-axis as the red circles show in Fig.2. If the current constellation is located in the first or third quadrant, where the signs of the I-phase and Q-phase signals are identical, the phase θ_k should be reduced, and vice versa. Therefore, only sign information is needed to recognize the angle, and the updating equation for θ_k using least mean square (LMS) method can be written as:

$$\theta_{k+1} = \theta_k - \Delta\theta \cdot [sign(I) \cdot sign(Q)] \qquad (1)$$

where $\Delta\theta$ is the phase step size and $sign(\cdot)$ is the sign of the I-phase or Q-phase signal. Compared with Costas loop-based phase detection in [3] and amplitude detection in [4] and mixing units in [5], the proposed modulation format and sign-

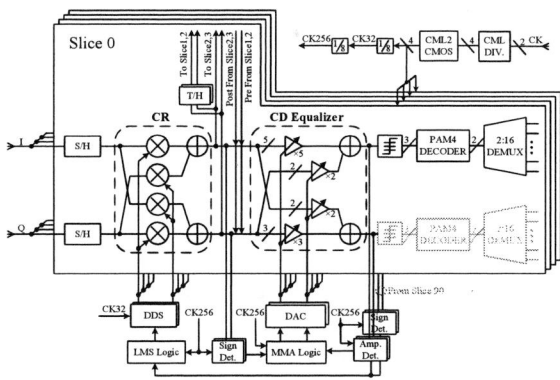

Fig. 3. Block diagram of the proposed Receiver

TABLE I. BER OF DIFFERENT EQUIVALENT LENGTHS

CD equivalent lengths(km)		3	4	5	6
BER	*Full Taps*	<1e-6	<1e-6	<1e-6	2.3e-3
	Simplified Taps	<1e-6	<1e-6	<1e-6	2.8e-3

based phase detection principle can significantly lower the difficulty of the frequency offset cancellation. Additionally, by rotating the target constellation to the I-axis, the eye height of the I-phase signal can be increased by 41%, and the Q-phase signal is only used for phase detection, therefore the equalization standards for Q-phase can be relaxed to further reduce the circuits complexity.

According to the analysis above, the block diagram of the proposed RX is shown in Fig. 3. On the left, I and Q are the same 56-Gbaud I-phase and Q-phase signals. They are electrical signals through demodulation and photoelectric conversion. The RX first samples the input signals using 4-way interleaving S/H to reduce clock frequency and bandwidth requirements. Then the sampled signals are sent to the Carrier Recovery (CR) to eliminate frequency offset initially. The output signals of CR are sent to the CD Equalizer. Considering the proposed modulation format and rotation strategy, the taps of feed-forward equalizer (FFE) can be simplified without affecting the overall CD equalization performance, and its structure is depicted in the middle of Fig. 3. Instead of all FFEs employing full 5 taps, the equalization of I-phase consists of 5 taps I-to-I and 2 taps Q-to-I. Meanwhile, the Q-phase equalization consists of 3 taps Q-to-Q and 2 taps I-to-Q. In the case of CD only which the dispersion coefficient equals 17 ps/(nm•km), the bit error rate (BER) for different equivalent distances corresponding to the two different modes of full taps and simplified taps is shown in Table I. It can be found that from 6km onwards, the BER of all taps and simplified taps begins to increase at a similar extent, and no obvious penalty is introduced by the simplified equalizer. After CD Equalizer, the equalized signals are sent to slicers, PAM4 decoders, and DEMUXs. These modules in Q-phase can be omitted because the constellation is rotated to the I-axis. The input clock signal CK is 28-GHz and passes through CML divider and CML2CMOS. The CMOS (rail-to-rail) clock signals are allocated for data path and algorithm modules.

The coefficients of CR are adjusted by LMS Logic through direct digital synthesizers (DDSs). To get a cleaner decision symbol and a better convergence of CR, the deviation information is obtained after CD Equalizer via amplitude

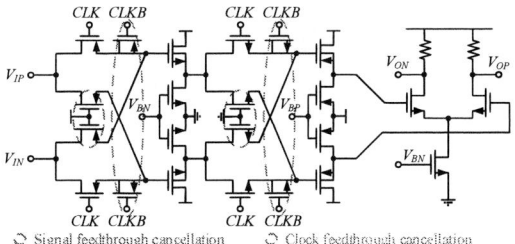

Fig. 4. Schematic of the 4-way Interleaving S/H

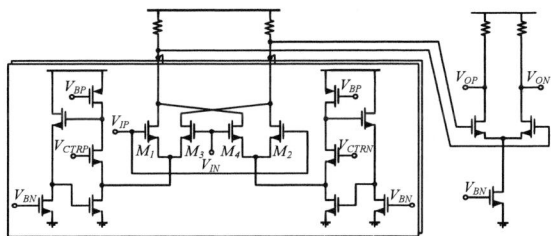

Fig. 5. Schematic of the CR

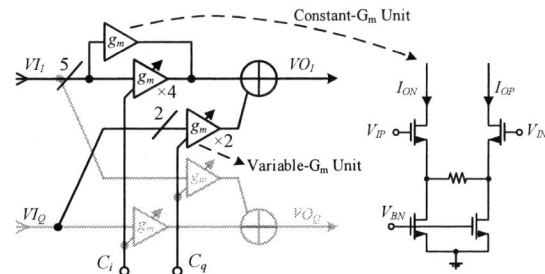

Fig. 6. Block diagram of the Chromatic Dispersion Equalizer

detectors in slice-0 and sign detectors in slice-90 as shown in Fig. 3. At the same time, the deviation information of CD equalizer is obtained only by amplitude detectors in slice-0. Multimodulus algorithm (MMA) [6] and DACs update the coefficients of CD equalizer. In order to reduce the power consumption, all the detectors work at 1/256 of the band rate.

III. CIRCUITS OF KEY BLOCKS

A. 4-way Interleaving S/H

To relax the bandwidth requirements, time interleaving sampling is implemented rather than active delay lines [7]. In this proposed RX, the FFE in CD equalizer has a maximum of 5 taps. To cover this span, the interleaving S/H consists of 2 stages of T/H and a common-source buffer as shown in Fig. 4. Each T/H switch utilizes cross-coupled transistors in the red circle for input feedthrough cancellation caused by parasitic capacitors C_{ds}, and a dummy transistor in the green circle for clock feedthrough cancellation. Every T/H is followed by a source follower which source and substrate are shorted together to ensure moderate gain. Finally, a buffer is applied to stabilize common-mode voltage and improve gain.

B. Carrier Recovery

CR consists of linearity-optimized multipliers which are shown in Fig. 3, and the I-phase and Q-phase have the same structure. For clarity, only one phase schematic of CR is shown in Fig. 5. V_{IP} and V_{IN} are the differential input signals

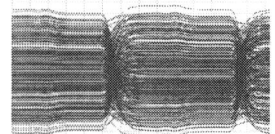
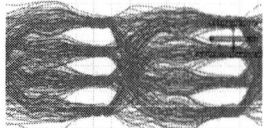

(a) S/H output (b) CD Equalizer output

Fig. 7. Eye diagrams of the in-phase

with frequency offset. The control signals V_{CTRP} and V_{CTRN} are sine and cosine waves with a certain frequency. The four transistors M_1, M_2, M_3, and M_4 multiply the gate voltage and the source voltage according to the current characteristics in the saturation region. Series resistors add the I-phase and Q-phase signals together. The source followers level-shift the control signals to a low common-mode voltage. This will bias the four middle transistors in the saturation region. Then the negative feedback loop ensures that the outputs follow the control signals precisely.

C. Chromatic Dispersion Equalizer

For CD equalizer, according to the analysis above, it consists of FFE with different tap numbers. At the same time, in order to ensure a maximum 5-UI span, extra T/Hs are added after CR to satisfy the time requirement as shown on the top of Fig. 3. The equalization of I-phase consists of 5 taps I-to-I and 2 taps Q-to-I. Meanwhile, the Q-phase equalization consists of 3 taps Q-to-Q and 2 taps I-to-Q. The block diagram of CR equalizer is shown in Fig. 6 which mainly focuses on I-phase. In I-to-I or Q-to-Q, the main tap is implemented by a constant-G_m unit as shown on the right of Fig. 6. While the remaining variable-G_m units utilize the same analog multiplier structure in CR.

IV. SIMULATION RESULTS

The 112-Gb/s RX with the novel modulation format and simplified phase detection and CD equalizer is implemented in 28-nm CMOS technology. In the case of 68 ps/nm (equivalent 4km-fiber) CD and 200 MHz frequency offset, the eye diagram of I-phase is shown in Fig. 7 (a) and (b) which are the output signals after S/H and CD equalizer respectively. The average eye height of equalized signals is 25 mV. The power breakdown of the RX is presented in Fig. 8. It consumes 500 mW at 1 V supply voltage in the simulation.

Table III presents the simulation performance and the comparison with prior works. The proposed PAM4-based modulation format realizes two bits per symbol the same as quadrature phase shift keyed (QPSK). The RX has a lower power efficiency at a 112-Gb/s data rate, compared with [3] and [7]. Although [3] can eliminate both frequency offset and CD, the equalization is post-processing and not integrated on the chip. Despite [5] can eliminate more severe frequency offset, it has no ability to eliminate CD.

V. CONCLUSION

In this article, a novel modulation format is proposed and makes the RX eliminate frequency offset more easily. The CD equalizer can also employ simplified taps FEE rather than full taps to reduce power consumption. The FFE consists of constant and variable Gm units. 4-way interleaving S/H relaxes bandwidth requirements. Simulation shows the circuits are capable of equalizing 112-Gb/s signals with 68

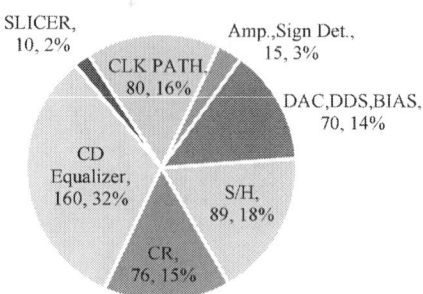

Fig. 8. Power breakdown of the RX

TABLE II. SIMULATION PERFORMANCE COMPARISION

	This work	TIM'24 [3]	ISSCC'23 [5]	JLT'20 [7]
Technology	28nm CMOS	130nm BiCMOS	28nm CMOS	130nm BiCMOS
Modulation	PAM4-based	QPSK	QPSK	DP-QPSK
Data rate (Gb/s)	112	50	24	40
Power consumption (mW)	500	480	76.8	2.5
Energy efficiency (pJ/bit)	4.4	9.6	3.2	62.5
Frequency offset (MHz)	200	250	600	N/A
Equalization	5-tap FFE[a]	N/A[b]	N/A	2-tap FFE
Transmission distance (km)	4	10	N/A	10

[a.] Only I-to-I employs 5-tap FFE

[b.] Post-processing equalization for CD

ps/nm CD and 200 MHz frequency offset. The RX achieves an energy efficiency of 4.4 pJ/b in 28-nm CMOS technology.

REFERENCES

[1] D. Tauber et al., "Role of Coherent Systems in the Next DCI Generation," in Journal of Lightwave Technology, vol. 41, no. 4, pp. 1139-1151, 15 Feb.15, 2023.

[2] H. Ji et al., "Photonic Integrated Self-Coherent Homodyne Receiver Without Optical Polarization Control for Polarization-Multiplexing Short-Reach Optical Interconnects," in Journal of Lightwave Technology, vol. 41, no. 3, pp. 911-918, 1 Feb.1, 2023.

[3] R. Ashok, S. Naaz, P. Jain, R. Kamran and S. Gupta, "All Electrical Costas Loop for Carrier-Forwarded Analog Coherent DCI Receivers: System Integration, Characterization, and Measurements," in IEEE Transactions on Instrumentation and Measurement, vol. 73, pp. 1-8, 2024.

[4] K. Sheng et al., "A 4.6-pJ/b 200-Gb/s Analog DP-QPSK Coherent Optical Receiver in 28-nm CMOS," in IEEE Journal of Solid-State Circuits, vol. 58, no. 1, pp. 45-56, Jan. 2023.

[5] A. E. Abdelrahman, M. G. Ahmed, M. A. Khalil, M. B. Younis, K. -S. Park and P. K. Hanumolu, "A Carrier-Phase-Recovery Loop for a 3.2pJ/b 24Gb/s QPSK Coherent Optical Receiver," 2023 IEEE International Solid-State Circuits Conference (ISSCC), San Francisco, CA, USA, 2023, pp. 1-3.

[6] Jenq-Tay Yuan and Kun-Da Tsai, "Analysis of the multimodulus blind equalization algorithm in QAM communication systems," in IEEE Transactions on Communications, vol. 53, no. 9, pp. 1427-1431, Sept. 2005.

[7] N. Nambath et al., "All-Analog Adaptive Equalizer for Coherent Data Center Interconnects," in Journal of Lightwave Technology, vol. 38, no. 21, pp. 5867-5874, 1 Nov.1, 2020.

979-8-3503-6184-1/24 $31.00 © 2024 IEEE

Beyond Bandwidth Trade-off: Simultaneous Wireless Power and Data Transfer System Design for Biomedical Implants

Quanrong Zhuang, Junyi Sun, Xusheng Zhang, Bo Li, Yi Shi, and Hao Qiu[*]

School of Electronic Science and Engineering, Nanjing University, China
Email: haoqiu@nju.edu.cn

Abstract—In this work, we presented a simultaneous wireless power and data transfer (WPDT) system utilizing the fundamental and harmonic components of the bridge inverter to transfer power and data, respectively. The conventional full bridge rectifier (FBR) introduces the interference voltage and could cause the problem of data flipping, which is solved by the proposed interference-free rectifier (IFR). Additionally, a tapped coil three capacitor (TL3C) topology was proposed to maximize the data channel gain without affecting the power channel gain. A 6.78 MHz system was implemented with the IFR fabricated in 180 nm CMOS process. It supported simultaneous 82mW load power (P_{Load}) and 4.0 Mb/s forward data rate (DR) at 52.6 % end-to-end efficiency (η_{E2E}).

Keywords—Biomedical implants, bandwidth trade-off, harmonic communication, interference-free rectifier (IFR), tapped coil three capacitor (TL3C) topology, simultaneous wireless power and data transfer (WPDT).

I. INTRODUCTION

The use of an inductive link for simultaneous WPDT in implantable devices has grown dramatically over the last decade. Well known examples are cochlear implants and visual prostheses. The requirement for both high η_{E2E} [1-3] and forward DR makes the system design challenging. One obvious method is to use two separate inductive links to transfer power and data respectively, since each link can be optimized independently. However, this method requires a large form factor and also suffers from cross-talk issue between these two links.

To reduce the cost and volume, another method is through a single inductive link. There are mainly three types methods : power carrier modulation, high-frequency data carrier injection and harmonic-utilization. About the power carrier modulation, in which schemes of on-off keying (OOK) [4-5], frequency-shift keying (FSK) [6] were used for data telemetry. The power carrier in OOK reduces the transmission power. FSK achieves constant output power during data transmission, but the modulation and demodulation circuits for FSK are complex and consume a lot of power. Besides power carrier modulation, high-frequency data carrier injection and harmonic utilization have also received more attention from scholars. They use an additional modulated signal source for data transmission. External signal sources require additional coupled inductors to transmit and extract the signal from the power channel, which significantly increases the cost and volume. Moreover, these schemes trade off η_{E2E} and DR on the basis of coils' quality factors (Qs). A high Q is requested for a high η_{E2E} whereas a low Q is preferred to a high DR. Additionally, the scheme in [6] can only apply to the short-distance (d) regime.

Due to their smaller size, cost and power consumption, single link WPDT system is more suitable for biomedical applications. To mitigate the trade-off between η_{E2E} and DR without adding extra cost, we presented a WPDT system employing the fundamental and harmonic components [7] of

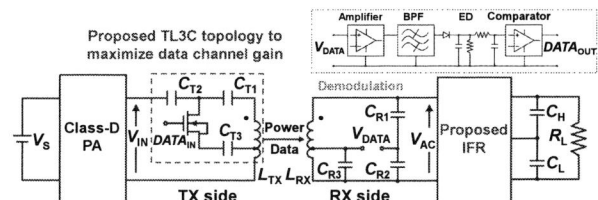

Fig. 1. Diagram circuit of proposed simultaneous WPDT system.

the output voltage of the class-D power amplifier (PA) on the TX side for power and data transfer, respectively. This harmonic communication method can achieve a high DR without increasing the working frequency of the bridge inverter. The conventional FBR on the RX side is attractive for its simplicity. However, its distorted V_{AC} waveform introduces interference with data transfer. To solve this problem, an IFR was proposed. Moreover, a TL3C topology was proposed to maximize the data channel gain without affecting the power channel gain.

II. WIRELES POWER AND DATA TRANSFER SYSTEM

Fig. 1 shows the overall system architecture, consisting of a Class-D PA, tapped TX and RX coils [8], the proposed IFR, and the load resistance (R_{L}). The fundamental (f_0) and 3rd harmonic ($3f_0$) components of the PA's V_{IN} are used to transfer power and data, respectively. The bandwidths of the data channel could be adjusted by selecting the inductance of the low side of tapped coil.

A. Proposed TL3C topology

Fig. 2 shows circuit schematics and channel gain for equivalent power and data channels. For the power channel at f_0, C_{T2}, C_{T1} and L_{TX} resonate in series, and C_{R1}, C_{R2}, and L_{RX} resonate in parallel, which is commonly used for implanted biomedical devices. For data channel at $3f_0$, two resonance loops improve the data channel gain. One is C_{T2}, C_{T3}, and L_{T2} in series, and the other is L_{R2} and C_{R3} in parallel. Data is modulated as $DATA_{\text{IN}}$ on the TX side and V_{DATA} is received on the RX side and then demodulated as $DATA_{\text{OUT}}$. The impedance of C_{T1} and $L_{\text{T1}}+M_{\text{T1T2}}$ is designed to resonant at f_0 so that C_{T3} and switch could be ignored at f_0.

Regardless of whether the switch is on or not, the gain of power channel is exactly equal to that of series-parallel topology at f_0. For the data channel, the circuit can be equivalent to magnetic resonance coupling (S-P) circuits too.

When the switch is closed, the current of data carrier on the L_{T2} increases and generate V_{DATA} at RX. In addition, by setting the C_{R1} over C_{R2} equal to L_{R1} over L_{R2}. The fundamental component of V_{AC} on L_{R2} is equal to C_{R2}. Crosstalk at fundamental frequency is eliminated by extracting V_{DATA} from these two nodes. The frequency characteristic response of TL3C topology is shown in Fig. 3. The gain of power and data channel is equal to $V_{\text{AC}}/V_{\text{IN}}$ and $V_{\text{DATA}}/V_{\text{IN}}$ respectively.

979-8-3503-6184-1/24 $31.00 © 2024 IEEE

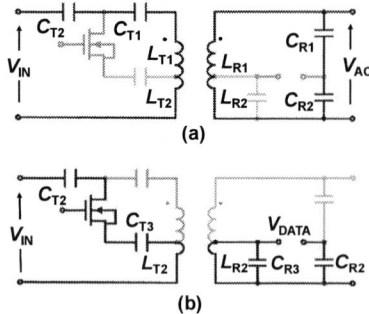

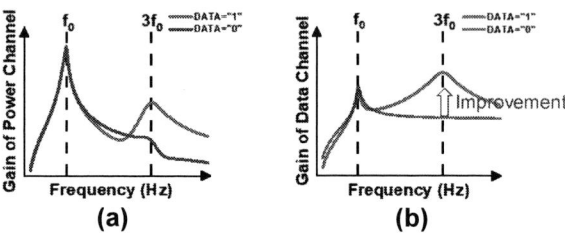

Fig. 2. Equivalent circuit schematics and channel gain at different frequencies (a) power channel at f_0 (b) data channel at $3f_0$.

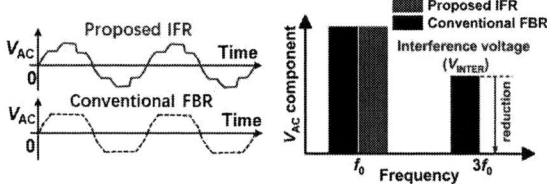

Fig. 3. Channel gain of (a) power channel and (b) data channel.

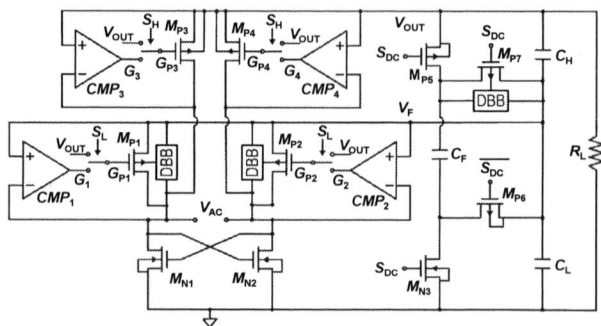

Fig. 5. Circuit schematic of proposed IFR.

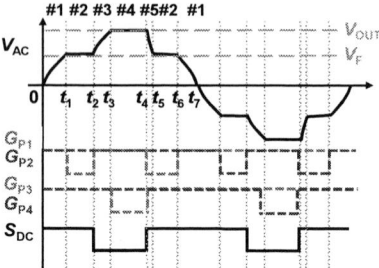

Fig. 6. Control signals.

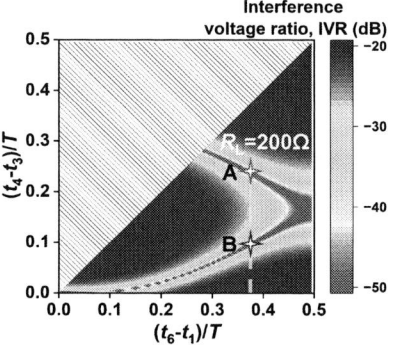

Fig. 4. Input V_{AC} waveform comparisons of the conventional FBR and proposed IFR.

B. Proposed Interference-Free Rectifier

Fig. 4 compares the input V_{AC} waveform of conventional FBR and proposed IFR. Taking the commonly used full bridge rectifier as an example, the non-linearity of the FBR typically results in V_{AC} deviates from the sine waveform, with its harmonic components illustrated in Fig. 4. V_{AC}'s component at $3f_0$ is defined as the interference voltage (V_{INTER}) and the ratio of V_{INTER} and V_{AC}'s component at f_0 is defined as interference voltage ratio (IVR). With the proposed IFR as shown in Fig. 5, a four-level stepped V_{AC} waveform is generated and V_{INTER} can be eliminated. This is of critical importance to data transfer.

Because V_{DATA} and V_{INTER} are different phases, the V_{INTER} in the conventional FBR can degrade the signal-to-noise ratio (SNR). At small distance, the gain of data channel is large, the data can be demodulated correctly. As d increases, data channel gain decreases and V_{INTER} can even dominate. As a result, the phenomenon of data flipping happens. On the other hand, V_{INTER} can be eliminated in the proposed IFR, which guarantees a high SNR at a small d and eliminates the problem of data flipping at a large d.

To generate the four-level stepped V_{AC} waveform, Fig. 5 shows the proposed 5-phase IFR, which is suitable for parallelly-resonated receiver. The IFR consists a pair of cross-couple NMOS and two pairs of PMOS active-diode to generate two voltage levels of V_{AC}. The charge pump circuit

Fig. 7. Design space.

is used to transfer the excess charge on C_L to C_H. In the steady state, the output voltage of IFR and voltage across C_F are equal to V_{OUT} and V_F, respectively. In this work, V_F is set as $V_{OUT}/2$. Due to the large capacity of C_L and C_F, two steps of V_{AC} can be generated by proper control of the two pairs of PMOS. During the positive half cycle, the operation sequence is #1→ #2→ #3→ #4→ #5→ #2→ #1 and corresponding control signals are shown in the left of Fig. 6.

To eliminate V_{INTER}, a careful design of V_{AC} waveform is requested. Fig. 7 shows the calculated IVR surface under different combinations of (t_4-t_3) and (t_6-t_1). (t_6-t_1) is correlated with R_L. Take $R_L = 200$ Ω, corresponding to (t_6-t_1)/T = 0.375 as an example, both designs at A and B can satisfy the IVR requirement. Considering the rectifier efficiency (η_{REC}) at A is greater than that at B, we chose the design at A in this work.

III.   EXPERIMENTAL RESULTS

The frequency of power carrier (f_0) is 6.78 MHz. The Class-D PA and the proposed IFR is fabricated in a 180 nm CMOS process. The ASK driver and demodulation is built with discrete operational amplifier.

979-8-3503-6184-1/24 $31.00 © 2024 IEEE

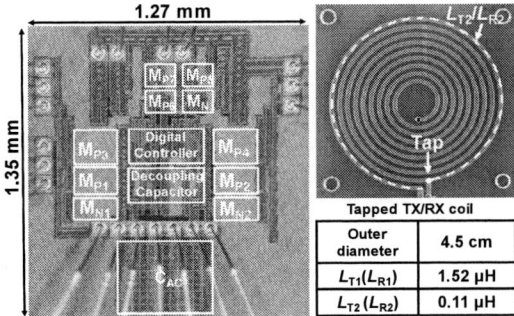

Fig. 8. Photos of IFR IC and tapped coils.

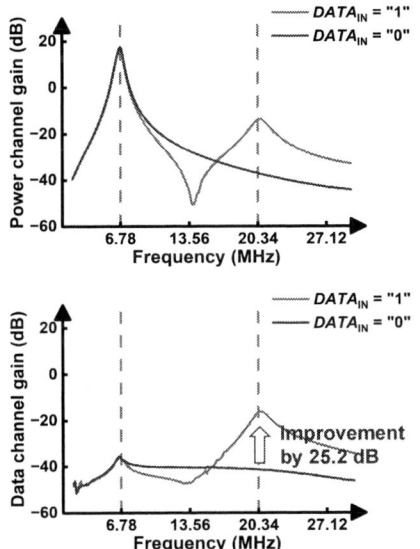

Fig. 9. Measured power and data channel gains.

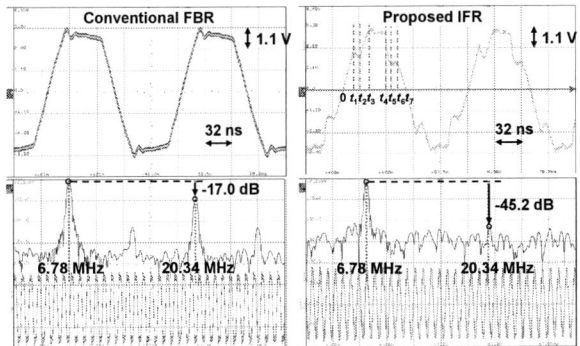

Fig. 10. Measured V_{AC} waveform comparisons.

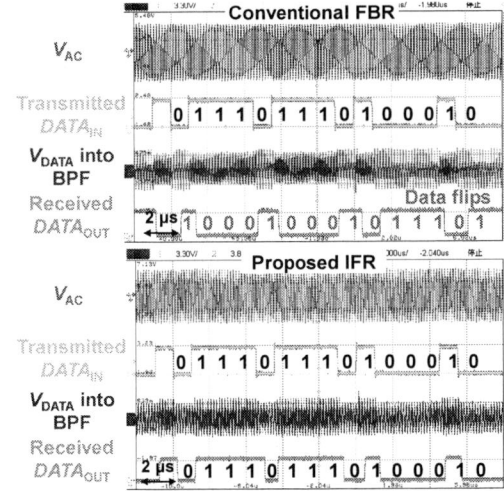

Fig. 11. Measured data transfer waveforms with DR = 1 Mb/s at d=25 mm.

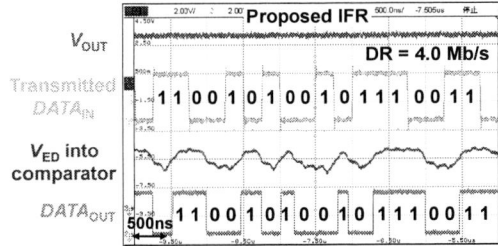

Fig. 12. Measured data rate of 4 Mb/s.

The chip photo of IFR is shown in the left of Fig. 8 and the IFR occupies a chip area of 1.71 mm². The TX/RX coil shown in the right of Fig. 8 is 9-turn PCB coil with a single tap. The inner coil is L_{T1} or L_{R1}, and its inductance is 1.52 µH, and the outer coil is L_{T2} or L_{R2}, and its inductance is 0.11 µH. The outer single turn of TX coil, used as data channel, is small and thus its power loss can be neglected.

Owing to the proposed TL3C topology, the data channel gain at $3f_0$ is improved by 25.2 dB at d = 18 mm. The switch between $DATA_{IN}$ "0" and "1" does not affect the power channel gain, which is demonstrated in Fig. 9.

Fig. 10 shows the measured V_{AC} waveforms. The asymmetry between two shoulders of V_{AC} waveform in the proposed IFR is ascribed to V_{ds} drop of MP1 and MP2. However, compared with the conventional FBR, its IVR is reduced from -17.0 dB to -45.2 dB.

Owing to the effective reduction of IVR, the problem of data flipping in the conventional FBR is solved by using the proposed IFR, which is demonstrated in Fig. 11. In the large distance, the gain of data link is low and V_{INTER} from FBR is larger than the V_{DATA} received from TX. Therefore, when the transmitter sends a signal representing '1', the receiver demodulates it and incorrectly interprets it as '0'. The IFR reduces the V_{INTER} component in V_{AC} significantly and the data flipping is avoided.

To further verify the simultaneous WPDT using the proposed IFR, a 4.0 Mb/s signal is applied as $DATA_{IN}$ [Fig. 12]. The received data can be correctly demodulated on RX side without affecting V_{OUT} (corresponding P_{Load} = 82 mW). To prove the performance of crosstalk cancellation, the BER were tested under different load conditions. The BER is as low as 1×10^{-9} for P_{Load} ranging from 8.8 to 82 mW [Fig. 13], which shows the crosstalk is well suppressed. Measured η_{REC} reaches 80.2 % at a P_{Load} of 61 mW is shown in Fig. 14 and the peak value of η_{E2E} reaches 52.6 % is shown in Fig. 15.

Table I shows a comparison table, in which this work reports the highest DR of 4.0 Mb/s with BER of 1×10^{-9}, and lowest IVR down to -45.2 dB owing to the proposed IFR. Two figures of merit (FoMs) (FoM₁ for DR and FoM₂ for η_{E2E}) are used for comparison, which are both highest compared to prior works supporting simultaneous WPDT.

979-8-3503-6184-1/24 $31.00 © 2024 IEEE   115

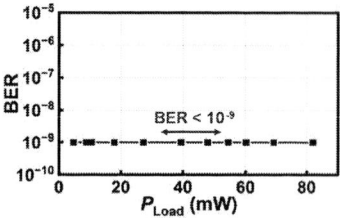

Fig. 13. BER varies with P_{LOAD} at d = 18 mm.

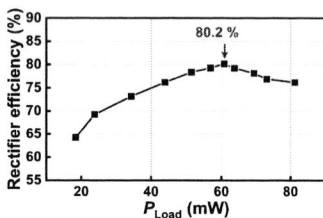

Fig. 14. Measured η_{REC} varies with output power.

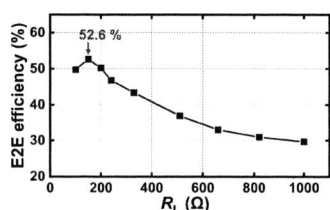

Fig. 15. Measured η_{E2E} varies with load resistance at $d = 10$ mm.

TABLE I

COMPARISON WITH PRIOR ART

	[4]	[5]	[6]	[9]	This work
Technology (nm)	130 CMOS	65 CMOS	180 BCD	180 CMOS	180 CMOS
Data direction	Downlink	Downlink	Downlink	Downlink	Downlink
Data transfer method	CWM	ASK	FSK	ESK	Harmonic communication
Number of coils	2	2	2	2	2
Distance (mm)	10	-	5	15	18
Coil diameter (mm)	43	-	30	6	45
Inverter frequency (MHz)	10	13.56	6.5 - 7.5	13.56	6.78
E2E efficiency	30% (Sim.) (d=10mm)	-	56.7% (d=5mm)	-	52.6% (d=10mm)
Load power (mW)	29.7	9.2	115	10.1	82
Data rate (Mb/s)	1.66	0.1 - 0.15	2.5	0.1	4.0
Bit error rate	-	10^{-3}	4×10^{-7}	-	1×10^{-9}
Interference voltage ratio (dB)	-	-	-	-	-45.2
FoM$_1$ (%)	17	0.92	36	0.74	59
FoM$_2$ (%)	7.0	-	9.5	-	12

FoM$_1$ is defined as DR/(Inverter frequency); FoM$_2$ is defined as $\eta_{E2E} \times d/(D_{TX}D_{RX})^{1/2}$, where D_{TX} and D_{RX} are respectively diameters of TX and RX coils.

IV. CONCLUSION

WPDT system is considered as an effective method to provide power and data communication for implantable electronic devices, with the advantages of small size, low power consumption and low cost. Targeting the trade-off between power and data transfer, we developed a 6.78 MHz harmonic communication system. In the system, the fundamental and harmonic components of the class-D PA's output voltage are used to transfer power and data, respectively. The tapped-coil three capacitor (TL3C) topology and interference-free rectifier (IFR) were proposed to enhance data channel gain and eliminate crosstalk between power and data channel. A high data rate of 4.0 Mb/s was achieved at a simultaneous load power of 82 mW.

ACKNOWLEDGMENT

This work was financially supported by the National Natural Science Foundation of China (62341408, 62374082, T2221003), the National Natural Science Foundation of China for Excellent Young Scholars (Overseas), the Engineering Research Center of Opto-Electro Materials and Chip Techniques, Nanjing University, Nanjing 210023, and the Interdisciplinary Research Center for Future Intelligent Chips (Chip-X), Nanjing University, Suzhou 215163, China.

REFERENCES

[1] H. Qiu, T. Sakurai, and M. Takamiya, "A 6.78 MHz multiple-transmitter wireless power transfer system with efficiency maximization by adaptive magnetic field adder IC," *IEEE J. Solid-State Circuits*, vol. 57, no. 8, pp. 2390–2403, Aug. 2022.

[2] H. Qiu, T. Sai and M. Takamiya, "A 6.78 MHz wireless power transfer system enabling perpendicular wireless powering with efficiency increase from 0.02% to 48.2% by adaptive magnetic field adder IC integrating shared coupling coefficient sensor," *Proc. IEEE Symp. VLSI Circuits*, pp. 1–2, Jun. 2020.

[3] H. Qiu, X. Zhang, J. Chen, M. Takamiya and Y. Shi, "A 6.78 MHz wireless power transfer system for simultaneous charging of multiple receivers with maximum efficiency using adaptive magnetic field distributor IC," *Proc. IEEE Symp. VLSI Circuits*, pp. 1–2, Jun. 2021.

[4] A. Trigui et al., "Generic wireless power transfer and data communication system based on a novel modulation technique," *IEEE Trans. Circuits and Systems I: Regular Papers*, vol. 67, no. 11, pp. 3978–3990, Nov. 2020.

[5] D. Ye, Y. Wang, Y. Xiang, L. Lyu, H. Min and C.-J. R. Shi, "A wireless power and data transfer receiver achieving 75.4% effective power conversion efficiency and supporting 0.1% modulation depth for ASK demodulation," *IEEE J. Solid-State Circuits*, vol. 55, no. 5, pp. 1386–1400, May 2020.

[6] Y. Park et al., "A frequency-splitting-based wireless power and data transfer IC for neural prostheses with simultaneous 115 mW power and 2.5 Mb/s forward data delivery," *IEEE Int. Solid-State Circuits Conf. (ISSCC)*, Feb. 2021, pp. 472–474.

[7] P. Wu, S. -P. Gao, Y. -D. Chen, Z. H. Ren, P. Yu and Y. Guo, "Harmonic-based integrated rectifier–transmitter for uncompromised harvesting and low-power uplink," *IEEE Trans. Microwave Theory and Techniques*, vol. 71, no. 2, pp. 870–880, Feb. 2023.

[8] P. P. Mercier and A. P. Chandrakasan, "Rapid wireless capacitor charging using a multi-tapped inductively-coupled secondary coil," *IEEE Trans. Circuits and Systems I: Regular Papers*, vol. 60, no. 9, pp. 2263–2272, Sep. 2013.

[9] S. -W. Hong, "A 13.56 MHz current-mode wireless power and data receiver with efficient power extracting controller and energy-shift keying technique for loosely coupled implantable devices," *IEEE Int. Solid-State Circuits Conf. (ISSCC)*, Feb. 2020, pp. 486–488.

A High Precision Operational Amplifier with Improved Bias Current Cancellation Circuit

Zhili Zhang [1], Siyuan Yao [1], Puyang Liu [1], Cheng Li [1], Lu You [1], Hailong Wei *[1]

[1] Xi'an Microelectronics Technology Research Institute, Xi'an, Shaanxi, 710048, China

* Email: 1712590048@qq.com

Abstract—**This brief presents a four-channel amplifier featured by low-offset voltage and low input bias current, which is mainly used in high precision instruments. The amplifier consists of four modules: biasing, input amplifier, bias current cancellation circuit and output stage. The biasing network adopts cross-coupled structure to improve the precision of mirror current and the offset voltage and drift are greatly reduced by using the trimming resistor. Besides, the input bias current is greatly reduced by using the improved bias current cancellation circuit (IBCCC). This device is fabricated using a standard bipolar IC process. The test results show that the input bias current is less than 1.0nA, the typical values of offset voltage and drift is 3.4µV and 0.027µV/℃ respectively.**

Keywords—*high precision, bias current cancellation, reduced input bias current, low offset voltage and drift*

I. INTRODUCTION

With the development of sensor technology, the need for precision in Active filter, switched capacitor circuits and signal processing has increased. When used in high-precision instruments, the input bias current (I_B) will cause errors in the sensitivity of the instrument. In order to keep the detection error below the specified tolerance level, it is necessary to select operational amplifier (op amp) with low input bias current (I_B), low input offset voltage (V_{OS}) and drift (TCV_{OS}) [1].

Op amp regularly has the disadvantages of high bias current, high offset voltage, low gain and high offset current, that's why the ordinary op amp cannot meet the needs for precise simulation and accurate calculation of weak signals in signal detection, acquisition and measurement. As a result, high-precision, low-offset integrated op amp has emerged [2].

Ideally, the input bias current of the op amp is zero because of the virtual open. But in a real circuit based on BJT process, the bias current cannot be zero because the base current of BJT is not equal to zero, which will lead to the reduction for DC input equivalent resistance (R_i) of amplifier. When it's used in multistage amplifiers, it will pose a negative impact on the performance of the preceding circuit. So, the input bias current is a very important index to measure the performance of the op amp [3].

In this thesis, the traditional bias current compensation structure is optimized to further decrease I_B, V_{OS} and TCV_{OS}.

II. CIRCUIT IMPLEMENTATION OF THE PROPOSED OP AMP

A. Traditional Technique of Bias Current Cancellation

The designer generally will use a bias current cancellation circuit (BCCC) to minish I_B and increase R_i [4]. The schematic of conventional BCCC is shown in Fig. 1. As shown on the left, the input pair consists of quad-connection transistors (Q_{1A}, Q_{1B}, Q_{2A}, Q_{2B}), whose base current required by Q_{11}/Q_{12}. As we can see, the voltage clamping structure (VC) consisting of transistors Q_2, Q_3, Q_{11}, Q_{12} and Q_{22} makes the collector-emitter voltage (V_{CE}) of Q_{11}/Q_{12} equal to V_{CE} of Q_2 plus a base-emitter voltage (V_{BE}), so Q_{11}/Q_{12} can provide a dc base current for the input transistor [5]. But in this scheme, the V_{CE} of Q_{11} and Q_{12} is not completely equal to V_{CE} of Q_2, that is, due to the early voltage, the current compensated by Q_{11}/Q_{12} cannot completely meet the base current needed by the input transistor. Hence, the accuracy of compensation current is low. In order to solve this drawback, the improved bias current cancellation circuit (IBCCC) is used to improve the current compensation accuracy, which is shown in the right-hand of Fig. 1.

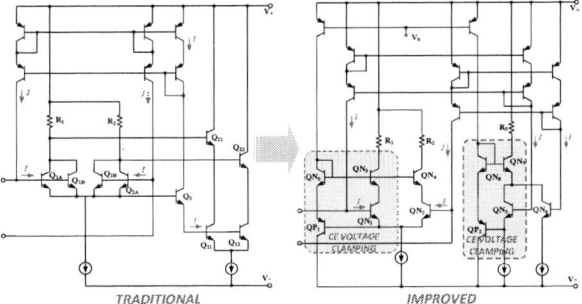

Fig. 1. Input bias current cancellation circuit

As shown in the right of Fig. 1, the V_{CE} of input pair transistors must be stable and its operational condition should be consistent with the bias current sampling structure in order to improve the bias current cancellation accuracy. The CE voltage clamping (CEVC) consisting of transistors QN_1, QP_1, QN_9, QN_3 clamps V_{CE} of QN_1 into a base-emitter voltage, and a differential circuit with the same CEVC is used to generate the base current of the input transistors (QN_1, QN_2), then the current is returned to the input terminal by the current mirror sampling circuit to realize the input bias current cancellation. In this structure, QN_5 and QN_6 have the same CE bias voltage as QN_1 and QN_2, so it can generate the more accurate current needed by the input pair transistors, thus effectively reducing the input bias current of the circuit.

B. Complete Schematic

Fig. 2 shows schematic diagram of high precision op amp in its entirety. The amplifier consists of a biasing circuit, an input bias current cancellation, a two-stage differential amplifier and an output stage.

979-8-3503-6184-1/24 $31.00 © 2024 IEEE

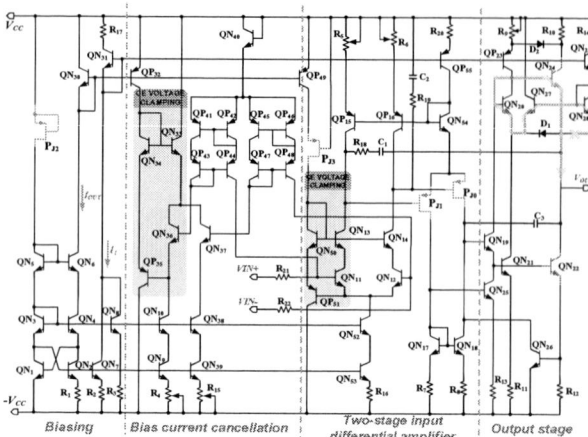

Fig. 2. Complete schematic of the designed Op amp

C. Biasing

As shown in Fig. 2, the leftmost part of which provides the op amp with necessary bias current, QN_1 and QN_2 form a cross-coupled pair of transistors. Start-up current is provided by PJFET, which can be cut off at a lower power supply voltage because of its small pinch-off voltage, therefore, a relatively stable start-up bias current can be provided over a wide range of supply voltages. The structure generates two types of bias currents, one is a PTAT current called I_{OUT}, the other has a low temperature coefficient, called I_1, generated by the addition of PTAT and CTAT [6].

$$I_1 = \frac{V_T \ln\left(\frac{I_{S8}}{I_{S4}}\frac{I_{C4}}{I_{C8}}\right) + V_{BE1}}{R_3} + \frac{V_T \ln\left(\frac{I_{S3}I_{S2}}{I_{S1}I_{S4}}\frac{I_{C4}I_{C1}}{I_{C3}I_{C2}}\right)}{R_1} \quad (1)$$

$$I_{OUT} = \frac{V_T \ln\left(\frac{I_{S3}I_{S2}}{I_{S1}I_{S4}}\frac{I_{C4}I_{C1}}{I_{C3}I_{C2}}\right)}{R_1} \quad (2)$$

where $V_T = kT/q$ is the thermal voltage at temperature T(K), $I_{S,n}$ ($n = 1,2,3,4,8$) are the transistor saturation currents, $I_{C,m}$ ($m = 1,2,3,4,8$) are the transistor collector currents.

Considering V_T has a positive temperature coefficient, and V_{BE} has a negative one, so currents, I_{OUT} and I_1, with positive and approximately zero coefficient temperature respectively can be obtained by weighting the transistor parameters I_S, R_3 and R_1 reasonably.

D. Two-stage Differential Op Amp

The two-stage differential op amp provides the required voltage amplifying gain for the design, as shown in Fig. 2.

In order to improve the output impedance, the first stage adopts the NPN with common-emitter–common-base (CE-CB) configuration as the input transistors. Its bias current is a PTAT current provided by the biasing, and the voltage gain is further enhanced by an active load consisting of QP_{16} and R_6 as well as QP_{15} and R_5, among which R_5/R_6 is termed emitter negative feedback resistor.

The second differential stage adopts PJFET as the input devices, and the bias current mirrored by I_1 is a low temperature coefficient current. PJFET used in the input

terminal of second-stage can effectively increase the input impedance, thus reducing the gain loss as well as input current and voltage offset of the first-stage [7].

E. Input Bias Current Cancellation

In analog op amp, reducing input bias current can improve the linearity and stability of the amplifier. In this design, the bias current cancellation structure is used to reduce the bias current and keep the high input impedance of the transistor.

To compare the performance of IBCCC and BCCC, Cadence virtuoso was used to simulate their bias currents. Fig. 3 shows the variation of input bias current IB follows with temperature range from -55℃ to 125℃, When T = 25℃, V_{CC} =15V, the curve of I_B with respect to power supply (V_{CC}) was obtained, as shown in Fig. 4.

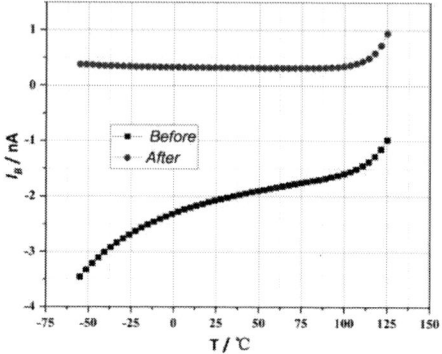

Fig. 3. Plot of I_B versus Temperature

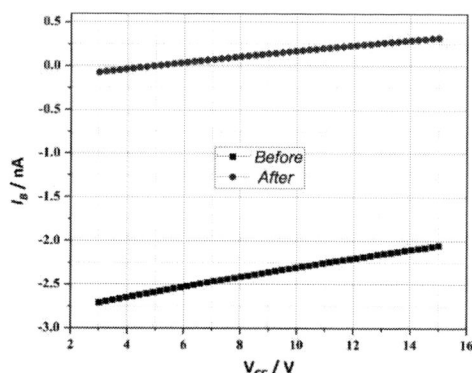

Fig. 4. Plot of I_B versus power supply

F. Output Stage

The Class AB output stage is shown in Fig. 2, in which uses high-power NPN transistors in both halves. QP_{23} provide bias current, which is mirrored by I_1.

In order to increase the gain of the output stage, the output stage adopts a unit gain positive feedback structure to increase the equivalent output impedance, as shown by the blue mark Line in Fig. 2. QN_{26} and R_{12} consists of over-temperature protection circuits used to protect output transistor QN_{22}. The specific process is: when the collector current of QN_{22} increase, it will produce voltage drop on R_{12}, which is greater than the on-voltage of QN_{26}, then, the collector of QN_{26} pulls current from the base of QN_{19} to reduce the base current of QN_{22}, and it prevents the burn-out of the sinking-current power transistor QN_{22}.

979-8-3503-6184-1/24 $31.00 © 2024 IEEE

G. the Design of Low Offset

Due to the two-stage differential amplifier structure with the first stage being fully differential, the input offset voltage caused by the conversion of the second stage from differential to single-ended can be almost ignored. From the analysis of the circuit structure, it can be seen that the main factors affecting the input offset voltage are the matching errors of symmetric devices.

The devices in the first-stage differential structure that have a significant impact on offset voltage are the matching of the input NPN transistors and the matching of the load PNP transistors. The input offset voltage can be expressed as follows

$$V_{OS1} \approx V_T \left(-\frac{\Delta R_C}{R_C} - \frac{\Delta I_S}{I_S} \right) \tag{3}$$

Where R_C represents the passive resistor in the structure, I_S is the transistor saturation current, ΔR_C is the offset of R_C, and similarly ΔI_S is the offset of I_S.

Simulation analysis shows that the input offset voltage and drift caused by the mismatch of the differential input transistors can be essentially satisfied by adjusting the initial value of the offset voltage simultaneously. The influence of the load PNP transistors on the input offset voltage is relatively minor, primarily affecting the input offset voltage drift.

The primary factor affecting input offset in the second-stage is the mismatch caused by PJFET transistors. The input offset voltage of the PJFET is given by

$$V_{OS2} = V_{TH} \sqrt{\frac{I_D}{I_{DSS0}}} \left(\sqrt{\frac{L_1}{W_1}} - \sqrt{\frac{L_0}{W_0}} \right) \tag{4}$$

From the equation, it is evident that the size mismatch of the PJFET determines the initial value of the offset voltage, while the threshold voltage (V_{TH}), unit size saturation current (I_{DSS0}), and DC bias current (I_D) collectively determine the temperature coefficient of the offset voltage [8].

III. SIMULATION RESULTS

Pre-simulation of the main parameters of this circuit was conducted under the following test conditions: R_L =2kΩ, V_{CC} =15V. Simulation curves for some parameters are shown in Fig. 5 and Fig. 6, and the simulation results for the main parameters are listed in TABLE I.

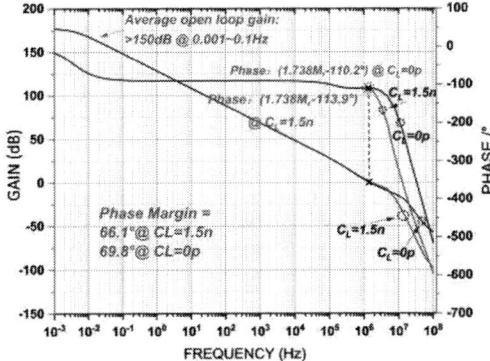

Fig. 5. Plot of Bode

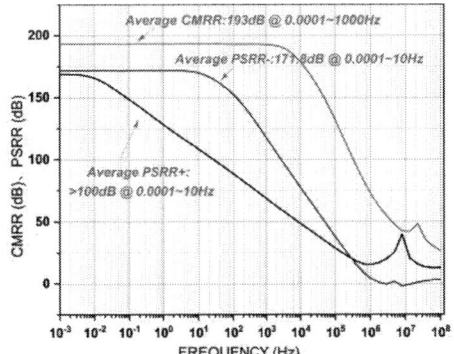

Fig. 6. PSRR、CMRR

The phase margin of the circuit obtained from simulation waveforms is greater than 60°, indicating good system stability, along with high common-mode rejection ratio (CMRR) and power supply rejection ratio (PSRR).

TABLE I.   SIMULATION RESULTS OF CIRCUIT PARAMETERS

Parameter	Condition	Result	Unit
Input offset voltage	T_A=25°C	≤ ± 0.1	μV
Input offset voltage drift	T_A=25°C	≤ ± 0.01	μV/°C
Input bias current	T_A=25°C	≤ ± 0.1	nA
CMRR	T_A=25°C	184	dB
Open loop voltage gain	R_L=10kΩ	158	dB
Unit gain bandwidth	T_A=25°C	1.01	MHz
Short-circuit Current	T_A=25°C	33	mA
Static power consumption	T_A=25°C，Io=0	789	μA

The simulation results of the circuit are shown in TABLE I, where all parameters meet the specified requirements. Parameters such as input offset voltage and drift, input bias current, and input offset current exhibit highly satisfactory simulation results.

IV. MEASUREMENT RESULTS

The circuit in this thesis is designed using a 40V BIPOLAR process and fabricated on-chip. During the circuit layout design, a large-sized four-transistor cross-coupled structure is employed for the PJFET, significantly improving matching to reduce input offset voltage. The die photograph is shown in the left image of Fig. 7, while the packaged chip is shown in the right image of Fig. 7, utilizing a DIP20 ceramic package.

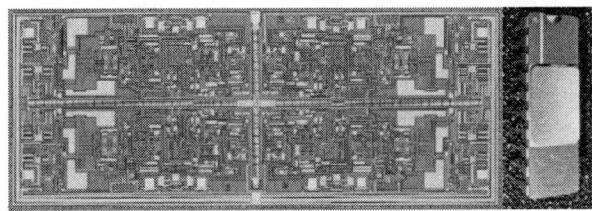

Fig. 7. The photo of die and package

At T=25°C and V_{CC}=15V, the bias currents of both positive and negative input terminals of multiple chips were tested. The data were plotted to form a distribution graph of input bias current, as shown in Fig. 8. It can be observed from the figure that the test results of input bias currents for each channel of multiple chips are consistently less than 1nA.

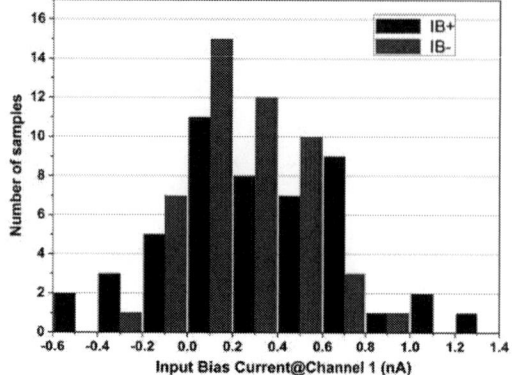

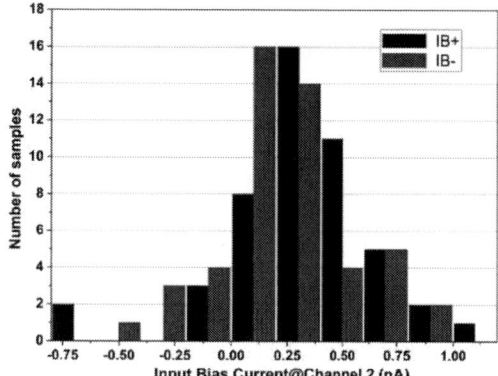

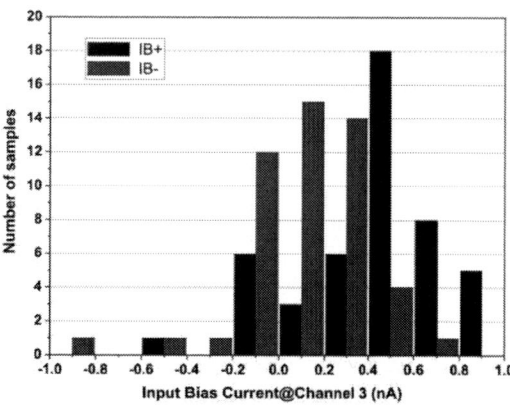

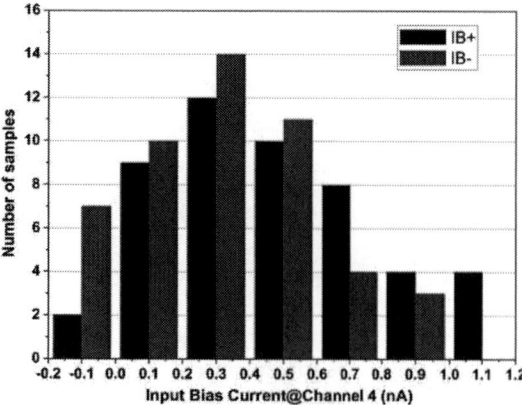

Fig. 8. Histogram of the measured input bias current

When V_{CC}=15V, the input offset voltages of multiple samples were tested. The test results were plotted as a histogram, as shown in Fig. 9 It can be seen that the typical value of input offset voltage drift is less than 0.027μV/℃.

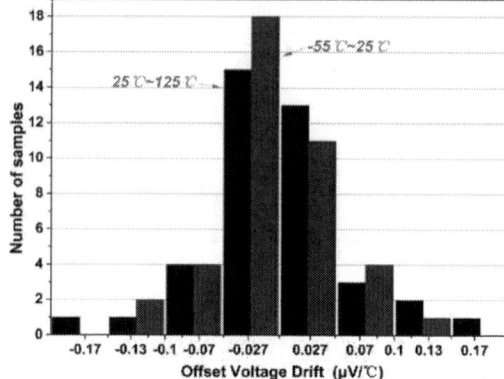

Fig. 9. Histogram of the measured input offset voltage drift at full temperature range

TABLE II presents the test results of various conventional parameters for the post-process chips, indicating good performance of the circuit. TABLE III summarizes the test results of key performance parameters and compares them with other op amps. Through comprehensive comparison, the indicators such as input bias current, input offset voltage and drift in this paper have been improved compared to devices in [9] and [10], making this circuit more suitable for high-precision applications.

TABLE II.   SUMMARY OF MEASURED PERFORMANCE

Parameter	Result	Unit
Input offset voltage	3.4	μV
Input offset voltage drift	0.027	μV/°C
Input bias current	1.0	nA
CMRR	162	dB
Open loop gain	149	dB
Unit gain bandwidth	1.94	MHz
PSRR	151	dB

TABLE III.   PERFORMANCE COMPARISON: SUMMARY OF MEASURED PARAMETERS

Parameter	This thesis	[9]	[10]	Unit
Offset voltage	3.4	100	30	μV
Voltage drift	0.027	0.4	0.3	μV/°C
Input bias current	1.0	6	±1.2	nA
Open loop gain	149	124.6	114	dB
Unit gain bandwidth	1.94	1	0.6	MHz

V. CONCLUSION

This thesis presents a high-precision op amp with low input bias current, low input offset voltage drift. By improving the bias current cancellation circuit, the input bias current is greatly reduced. Additionally, the circuit introduces adjustable resistors to the differential amplifier structure, fully optimizing the input offset voltage and drift. The chip has been successfully fabricated and corresponding test results have been obtained. The test results demonstrate that the circuit has achieved its design goals, with significant optimization of parameters such as input bias current, input offset voltage drift.

979-8-3503-6184-1/24 $31.00 © 2024 IEEE

REFERENCES

[1] V. Ivanov and M. Shaik, "5.1 A 10MHz-bandwidth 4μs-large-signal-settling 6.5nV/√ Hz-noise 2μV-offset chopper operational amplifier," *2016 IEEE International Solid-State Circuits Conference (ISSCC)*, San Francisco, CA, USA, 2016, pp. 88-89.

[2] Sun Jingwen, Ma Kui, Yang Fashun. "Overview of precision integrated operational amplifier,"(in Chinese), *Intelligent Computer and Applications*, vol. 10, no. 8, pp. 67-70, 2020.

[3] Chin K, Ohsawa M, Kitajima A, et al. "Input Bias Current Reduction Technique for Operational Amplifier in a Standard CMOS Technology," *Electron Comm Jpn*. 2020; 103: 30–36.

[4] H. L. Zeng, X. Xiao , L. You et al., "An improved instrumentation amplifier with low input bias current," (in Chinese), *Microelectronics and science*, vol. 39, no. 11, pp. 95-101, 2022.

[5] G. Erdi, "Amplifier techniques for combining low noise, precision and high-speed performance," *IEEE Journal of Solid-State Circuits*, vol. 16, no. 6, pp. 653-661, Dec. 1981.

[6] Paul R.Gray, Paul J.Hurst and Stephen H.Lewis. Analysis and Design of Analog Integrated Circuits (four edition) [M]. *John Wiley & Sons Inc*, 2009.

[7] K. Fukahori, Y. Nishikawa and A. R. Hamade, "A high precision micropower operational amplifier," in *IEEE Journal of Solid-State Circuits*, vol. 14, no. 6, pp. 1048-1058, Dec. 1979.

[8] W. Y. Li, W. C. Li, H. F. Jian et al., "Design of a High Precision and Low Noise Operational Amplifier," (in Chinese), *Microelectronics*, vol. 53, no. 5, pp. 800-806, 2023.

[9] OP470 Data Sheet, Analog Device Inc，2010. [Online]. Available: https://pdf.elecfans.com/LINER/OP-470.html#pdf

[10] OP07 Data Sheet, TEXAS INSTRUMENTS,2015. [Online]. Available: https://www.doc88.com/p-7038690183024.html

A 0.11-pJ/bit True Random Number Generator Based on a Clocked Current-Starved Inverter

Kai Cheng [1], Chaowei Yang [1], Rui P. Martins [1,2], Pui-In Mak [1], and Yong Chen [1,3]

[1] State-Key Laboratory of Analog and Mixed-Signal VLSI and IME/ECE-FST, University of Macau, Macao, China
[2] Instituto Superior Técnico, Universidade de Lisboa, 1049-001 Lisboa, Portugal
[3] (Now) Department of Electronic Engineering, Tsinghua University, Beijing 100084, China
Email: ychen_ee@tsinghua.edu.cn

Abstract—**This paper presents a clocked current-starved inverter-based true random number generator (TRNG) in 65-nm CMOS. The design of the TRNG uses inverter noise amplification and quantification through a sense latch. We add two extra transistors to a regular inverter to construct a current-starved inverter, used to remove the bias of the raw TRNG sequence generated by charge injection and SL offset. An adaptive calibration algorithm in conjunction with successive approximation register logic provides the biasing to the clocked current-starved inverter. Our TRNG occupies an area of only 0.0026 mm² and scores an energy efficiency of 0.11 pJ/bit for a 10 Mb/s random sequence with a 0.5 V supply. It can operate efficiently across a 0.5–1.4 V. Our TRNG is able to pass all the National Institute of Standards and Technology tests without requiring post-processing.**

Keywords—**CMOS, entropy, clocked current-started inverter, true random number generator (TRNG).**

I. INTRODUCTION

Current cryptographic systems widely use true random number generators (TRNGs) due to their unpredictability. In Fig. 1 we observe several types of reported TRNGs [1]−[3]. Although SAR analog-to-digital converter (ADC)-based TRNGs do not require post-processing, the capacitive digital-to-analog converter (CDAC) mismatch and the comparator offset easily affect their randomness; in addition, the output 1-bit random sequence requires multiple conversion cycles, which influence the efficiency of the random sequence generation [1],[7]. The latch-based TRNG has a simple structure and can generate 1-bit data with only one-time conversion, has low power, and a small area. However, the mismatch of the latch requires a very complex calibration loop for compensation and it is necessary to use post-processing algorithms to eliminate offset and correlation issues from the raw sequence [2]. An inverter-based TRNG avoids these problems with mismatch and offset [3], although multiple noise amplifications reduce the data throughput, besides, it requires calibration for clock feedthrough and charge injection. In this work, we propose a current starved inverter-based TRNG. On the other hand, we avoid the impact of charge injection and clock feedthrough through timing optimization, and we use adaptive calibration algorithms to eliminate the offset-induced data bias in the proposed TRNG structure.

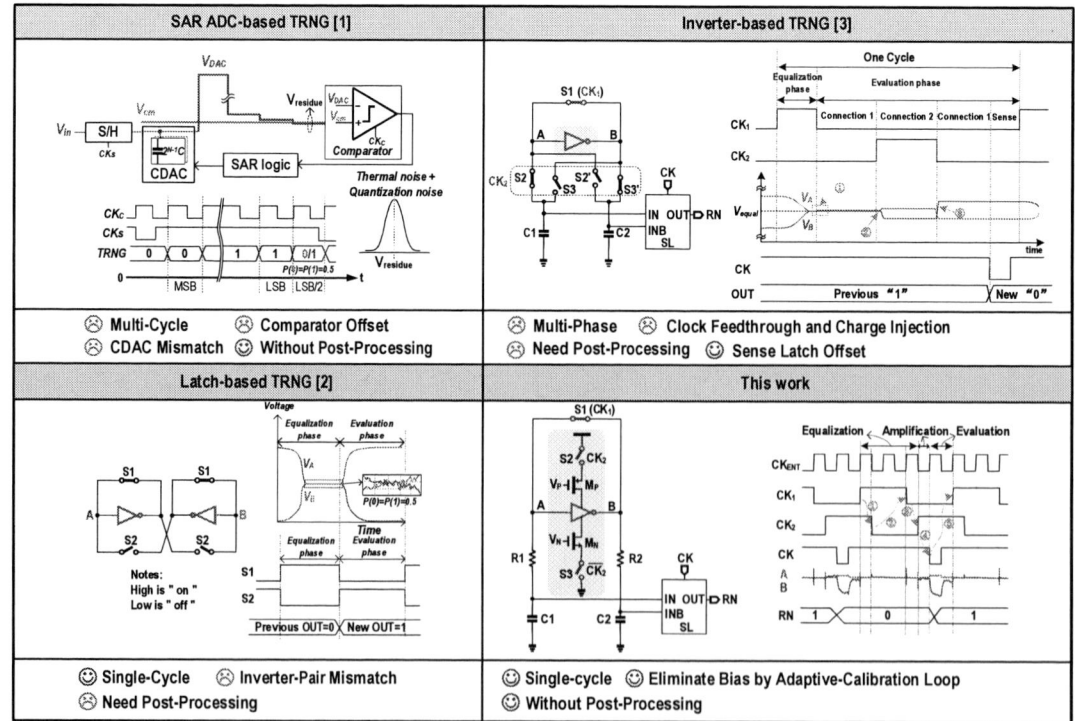

Fig. 1. Summary of the state-of-the-art CMOS TRNGs .

979-8-3503-6184-1/24 $31.00 © 2024 IEEE

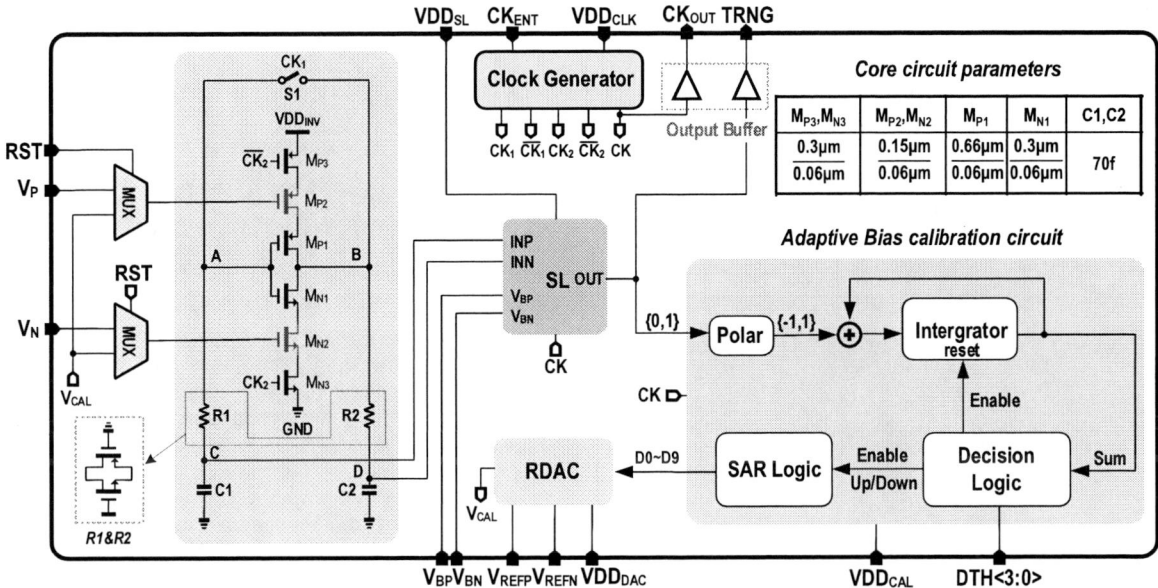

Fig. 2. Top architecture of the proposed TRNG.

II. TOP ARCHITECTURE OF THE PROPOSED CSI-TRNG

From Fig. 1, we can observe the proposed CSI-TRNG core block that includes a current-starved inverter with two switches, S2 and S3. We use the capacitors C1 and C2 to hold the voltage of node A (V_A) and node B (V_B) during the equalization phase, and the resistors R1 and R2 serve as the equivalent of two normally open switches. The implementation of the switches uses a transmission gate. Each operation cycle of the core block has three working phases: the equalization phase, the amplification phase, and the evaluation phase. In the equalization phase, S1, S2, and S3 are on, with the input and output of the inverter connected. Following this, with S2 and S3 opened, C1 and C2 held the voltage at V_A and V_B, respectively. After 3/2 clock cycles (the clock is CK_{ENT}, generated off-chip, 4x higher than the data rate), S1 opens, while the activation of S2 and S3 happens after a 1/2 clock cycle, marking the beginning of the amplification phase. CK goes low following a 1/2 clock cycle, and SL operates to generate the random data to finish the evaluation. This timing design prevents clock feedthrough and charge injection effects after S1 opens. M_P and M_N can also isolate the clock feedthrough effect caused by S2 and S3.

Fig. 2 depicts the top architecture of the proposed CSI-TRNG. The TRNG consists of a core block with a current-starved inverter and two sampling capacitances, an SL, a clock generator block, and an adaptive bias calibration block with an R2R DAC. The designed TRNG has two operational modes: "analog mode" and "calibration mode". When the reset signal RST is "low", the V_P and V_N voltages are supplied externally, the TRNG operates in analog mode. And the calibration mode starts when the RST is "high", meanwhile, V_{CAL} provides on-chip bias voltage instead of V_P and V_N. Ideally, with switch S1 turned off, thermal noise affects node A, amplified at the inverter's output and quantized into random bits through SL. In practical applications, the output of random sequences exhibits offsets caused by charge injection and

SL offset. In the current design, we introduce a current-starved structure to eliminate the offset voltage generated in the TRNG. By adjusting the gate voltages (V_{CAL}) of M_{P2} and M_{N2} we can effectively adjust the random sequence bias. Due to the influence of a series of offset voltages generated during the generation process and the introduction of a self-calibration loop, the current design can effectively resist the impact of process, voltage, and temperature (PVT) changes.

A. Entropy Generator

Fig. 3 shows the results of 100 transient noise simulations for node D (V_D). By taking SL to compare the voltage at a certain moment before analysis, we can obvserve that V_D follows a Gaussian distribution at this instant (μ = 0.439 V, σ = 2.1 mV). When we do not consider the influence of SL imbalance, the output of SL at this instant is P(0) = P(1) = 0.5.

B. Adaptive Calibration Algorithm

The basic requirement for whether a random sequence is random is that the proportions of "0" and "1" bits in the sequence are close to 50%. The proposed self-calibration algorithm therefore outputs the "0" bits in the data as polarity conversion to "−1", with the 40 data length summed by an integrator. If the sum obtained is in the range −3 ≤ sum ≤ 3, the current random sequence has no

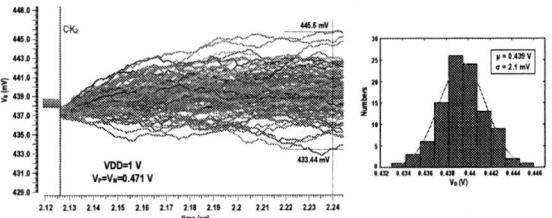

Fig. 3. Results of a noise amplification entropy source simulation.

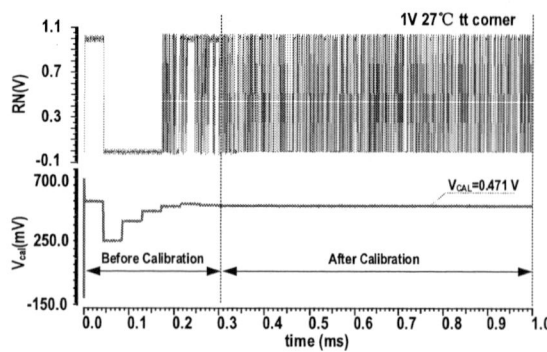

Fig. 4. Simulation results for the proposed adaptive bias calibration loop.

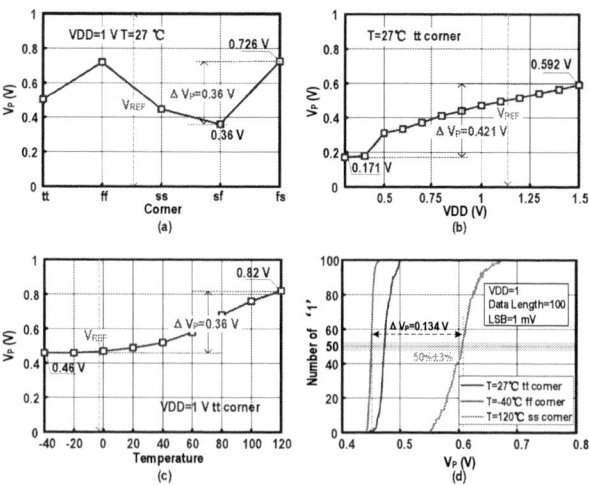

Fig. 5. Simulation results for (a) V_P versus corner; (b) V_P versus supply voltage; (c) V_P versus temperature; (d) V_P versus PVT.

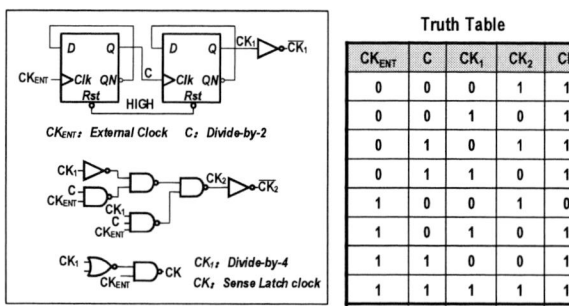

Fig. 6. Clock generation circuit and truth table.

obvious bias. As mentioned above, adjustments to V_{CAL} can eliminate the data bias. To expedite the search for V_{CAL}, we implemented a SAR logic and used it to regulate the 10-bit R-2R DAC to acquire V_{CAL}. If the supply voltage and reference voltage (V_{REFP}) are 1 V, the corresponding resolution accuracy is approximately 1 mV. As shown in Fig. 4, the transient simulation results for the calibration loop show that the SAR logic converges in the seventh search cycle, and the output sequence becomes random, corresponding to $V_{CAL} = 0.471$ V after calibration.

To verify the robustness of the proposed calibration scheme, some simulations of the PVT characteristics are

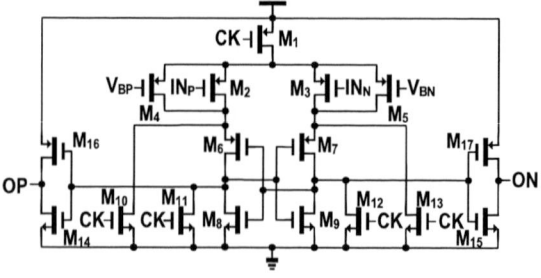

Fig. 7. Sense latch circuit.

presented. Fig. 5(a) shows that $\Delta V_P = \Delta V_N = 0.36$ V under different corners. From Fig. 5(b), we see that $\Delta V_P = \Delta V_N = 0.421$ V for various supply voltages (0.3–1.5 V). Fig. 5(c) shows that $\Delta V_P = \Delta V_N = 0.36$ V over a range of temperatures (−40–120 °C), and Fig. 5(d) shows that $\Delta V_P = \Delta V_N = 0.134$ V under the best, worst, and normal PVT conditions. All of the simulation results indicate that the proposed structure can compensate for the PVT effect by tuning the V_P (V_N) voltage.

C. Implementation of the Key Circuit

Fig. 6 shows the timing generation circuit for the proposed TRNG. CK_{ENT} is an off-chip clock. We can obtain the divide-by-2 C and the divide-by-4 CK_1 through the frequency divider. Subsequently, CK_{ENT}, CK_1, and C generate CK_2 and SL clock CK, with the specific logic implementation shown in the truth table. The proposed clock generation circuit mainly prevents the clock feedthrough problem caused by the disconnection of switch CK_1. The synchronization design of the off-chip clock can enable this design to operate across a wide range of frequencies.

Fig. 7 depicts the P-type SL circuit. When CK is low, SL completes the quantized output of the entropy source. To verify the impact of the SL offset on the random sequence offset, we introduce M4 and M5. Tuning of M4 and M5 can compensate for the offset voltage caused by process mismatch in SL.

III. MEASUREMENT RESULTS

Fig. 8 shows microphotographs of the proposed CSI-TRNG chip fabricated in 65 nm CMOS. The total circuit area of the TRNG is $42 \times 61 = 2562$ µm², and the core circuit area is $22 \times 16 = 352$ µm². Table I shows measured p-values for the 15 NIST randomness statistical tests; for tests with two or more sub-tests, cumulative sums, and serial tests we show the average value, while for the other tests, we show the minimum value.

Fig. 9(a) plots the measured autocorrelation on 1M consecutive bits with a lag of 1 to 5000 and displays the autocorrelation factor (ACF) within the 95% confidence bounds of a Gaussian distribution ($\mu = 0$, $\sigma = \pm 0.0019$). Fig. 9(b) shows the speckle pattern of 1M bits (1024 bits × 1024 bits), with no apparent color shift. From Fig. 9(c), it can be seen that in analog tuning mode, when $V_P = V_N = 0.38$ V, the entropy of the raw output is close to one. Fig. 9 (d) shows that when $V_{BN} = 0.1$ V and $V_{BP} = 0.113$ V, the entropy of the raw output is close to one, and we can eliminate the bias caused by the SL offset. Fig. 9(e) shows that the proposed CSI-TRNG has wide voltage (0.5–1.4 V)

979-8-3503-6184-1/24 $31.00 © 2024 IEEE

and throughput (1 Kb/s–10 Mb/s) ranges. Fig. 9(f) shows the proposed CSI-TRNG energy efficiency versus supply voltage at different data rates (1 Kb/s–10 Mb/s), where the energy range is 0.11 pJ/bit (0.5 V, 10 Mb/s) to 252 pJ/bit (1.4 V, 1 Kb/s).

Table II shows a comparison of our TRNG with state-of-the-art TRNGs with similar entropy sources. The

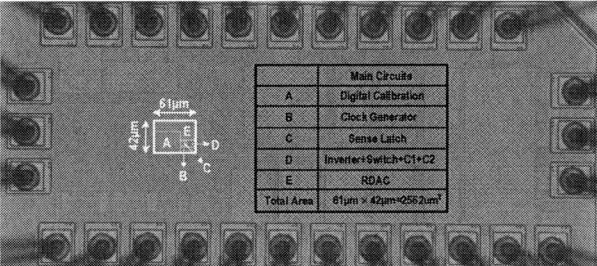

Fig. 8. Chip photograph.

TABLE I. Results of NIST PUB 800-22 tests

NIST TEST	P-value	Proportion
Frequency	0.448253	100/100
Block Frequency	0.374538	99/100
*Cumulative Sums	0.256305	99/100
Runs	0.334192	100/100
Longest Run of Ones	0.055361	99/100
Rank	0.251961	100/100
Discrete Fourier Transform	0.275709	98/100
*Non-Overlapping Template	0.474986	100/100
Overlapping Template	0.202268	98/100
Universal Statistical	0.175264	100/100
Approximate Entropy	0.772139	100/100
*Random Excursions	0.354210	100/100
*Random Excursions Variant	0.686532	100/100
*Serial	0.262249	98/100
Linear Complexity	0.816537	96/100

*For tests with two or more sub-tests, cumulative sums, and serial tests, we show the average value; for other tests, we show the minimum value.

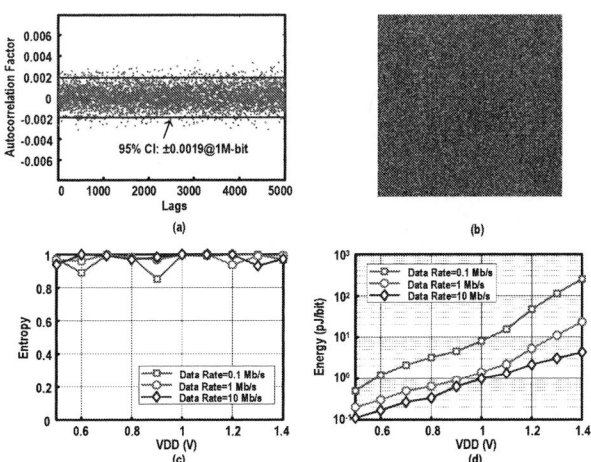

Fig. 9. (a) Measured autocorrelation of 1M consecutive bits; (b) bitmap of TRNG; (c) entropy versus supply voltage; (d) energy efficiency versus supply voltage for different data rates.

TABLE II. Performance summary and comparison with prior TRNGs

	This Work	JSSC'22 [2]	MWSCAS'20 [3]	JSSC'24 [4]
CMOS (nm)	65	130	130	28
Entropy Source	Noise Amplification	Meta-stability	Noise Amplification	Meta-stability
TRNG Topology	Clocked CS-Inverter	Latch	Inverter	StrongARM Latch
VDD (V)	0.5	0.3	0.7	0.7
VDD Range (V)	0.5-1.4	0.3-1	0.7-1	0.7-1.3
Throughput (Mb/s)	10	0.00787	0.4456	10000
Energy Efficiency (pJ/bit)	0.11	0.186	0.6585	0.121
Core Area (μm²)	2562	5561	1495	2254
Need Calibration ?	Yes	No	Yes	Yes
Need Post-Processing ?	No	Yes	Yes	Yes

proposed CSI-TRNG has a lower energy than the alternatives. The proposed TRNG, which does not use post-processing, also has an area of 2562 μm² and a wider voltage operating range and throughput.

IV. CONCLUSION

This paper reported an area- and power-efficient CSI-TRNG in 65 nm CMOS. We circumvented the issue of charge injection and clock feedthrough through the use of timing optimization. The application of an adaptive calibration algorithm eliminated the bias of the raw random sequence without the need for post-processing. With a relatively small footprint of 0.0026 mm² and high energy efficiency of 0.11 pJ/bit at a 0.5 V supply voltage and 10 Mb/s bit rate, the proposed scheme has a wide operating voltage range (0.5–1.4 V) and throughput (1 Kb/s–10 Mb/s), with strong robustness. A random number generated with our scheme passed all NIST tests.

REFERENCES

[1] A. Jayaraj, N. N. Gujarathi, I. Venkatesh, and A. Sanyal, "0.6–1.2 V, 0.22 pJ/bit true random number generator based on SAR ADC," *IEEE Transactions on Circuits and Systems II: Express Briefs*, vol. 67, no. 10, pp. 1765-1769, Oct. 2020.

[2] R. Zhang, X. Wang, K. Liu, and H. Shinohara, "A 0.186-pJ per bit latch-based true random number generator featuring mismatch compensation and random noise enhancement," *IEEE Journal of Solid-State Circuits (JSSC)*, vol. 57, no. 8, pp. 2498-2508, Aug. 2022.

[3] X. Wang, H. Liu, R. Zhang, K. Liu, and H. Shinohara, "An inverter-based true random number generator with 4-bit Von-Neumann post-processing circuit," in *2020 IEEE 63rd International Midwest Symposium on Circuits and Systems (MWSCAS)*, pp. 285-288, Aug. 2020.

[4] J. Kim and H. Chae, "A 10-Gb/s true random number generator using ML-resistant middle square method," *IEEE Journal of Solid-State Circuits (JSSC)*, pp. 1-9, Jan. 2024.

[5] Y. Cao, X. Zhao, W. Zheng, Y. Zheng, and C. H. Chang, "A New Energy-Efficient and High Throughput Two-Phase Multi-Bit per Cycle Ring Oscillator-Based True Random Number Generator," *IEEE Transactions on Circuits and Systems I: Regular Papers*, vol. 69, no. 1, pp. 272-283, Jan. 2022.

[6] M. Kim, U. Ha, K. J. Lee, Y. Lee, and H. J. Yoo, "A 82-nW chaotic map true random number generator based on a sub-ranging SAR ADC," *IEEE Journal of Solid-State Circuits (JSSC)*, vol. 52, no. 7, pp. 1953-1965, May 2017.

[7] C. Kai, Y. Chen, C. P. Stefano, R. P. Martins, and P.-I. Mak. "A 0.012-mm² 0.244-pJ/bit successive approximation register analog-to-digital converter-based true random number generator for Internet of Things applications in a 65-nm complementary metal–oxide–semiconductor," *Int J Circ Theor Appl.* 2024;1-28.

A Super-Mixed Current Decay Mode for Reducing the Angular Position Error in Stepper Motor

Jian Fang[*1], Xurui Chen[1], Huajie Liu[1], Yuhan Jin[1]

[1] Key Laboratory of Electronic Thin Films and Integrated Devices,
University of Electronic Science and Technology of China Chengdu, 610054, China

* Email: fjuestc@uestc.edu.cn

Abstract— **In stepper motors, discrepancies between the actual and setting currents of two-phase windings lead to errors in the rotor's angular position, resulting in performance degradation, increased power consumption, and system instability. Therefore, precise current control is crucial for accurate angular rotation with each step. In this paper, a super-mixed current decay mode is proposed to improve current accuracy. This mode ensures that the average actual current I_{MS} matches the setting current I_{TRIP} theoretically. We constructed a circuit and performed simulations to validate our approach. Implemented using a 0.18μm BCD process, our simulations demonstrate that this decay mode achieves stable operation and accurate current control, with variations in I_{MS} reduced to 0.04%.**

Keywords—stepper motor driver; current decay mode; chop PWM;

I. INTRODUCTION

Stepper motors have been extensively used in various industrial applications, including robotics and medical equipment, owing to their precise control capabilities in open-loop systems [1]. In two-phase stepper motors operating in half-step or micro-step modes, one crucial factor influencing angular positioning accuracy is the current error of winding. In each step, the desired angular position $\beta = \arctan(i_b/i_a)$ illustrated in Fig.1, is determined by the currents of the two windings: A and B phase. Discrepancies in these currents introduce errors in the rotor's angular position.

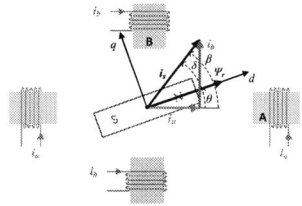

Fig. 1. Relationship between angular and current phases

Usually, the torque current remains constant, while the phase current i_a and i_b vary according to the angular steps. Due to the $i_b = i_s \sin(\beta)$ and $i_a = i_s \cos(\beta)$, the setting phase current I_{TRIP} should follow a sinusoidal-like waveform shown in Fig.2.

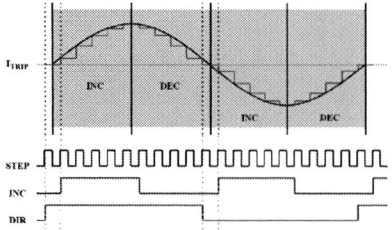

Fig. 2. Setting current waveform.

Stepper motor drivers commonly adopt chop PWM control with various decay modes. Several advanced decay modes have been proposed [2-3], including the Auto current decay mode introduced by TI (Texas Instruments) [4] and the SpreadCycle™ decay mode [5]. However, the actual phase winding current will deviate from the setting target current. This discrepancy is influenced by multiple factors, making it challenging to compensate for. To solve the above problem, this paper proposes a super-mixed current decay mode. The super-mixed decay mode ensures that the average winding current equals the setting current under the ideal case.

II. PRINCIPLE OF PROPOSED MODE

A typical chop PWM cycle comprises three distinct stages: charge, fast decay, and slow decay, as shown in Fig.3. During the charge stage, MH1 and ML2 turn on, allowing the winding current to rise until it reaches a predefined threshold. In the fast decay stage, ML1 and MH2 turn on, and the freewheeling current decays rapidly (to ensure that the current direction does not change). In the slow decay stage, ML1 and ML2 turn on, forming a freewheeling loop, and the current decays slowly. In the stabilized state, the average current of a chop PWM cycle I_{MS} is the average winding current.

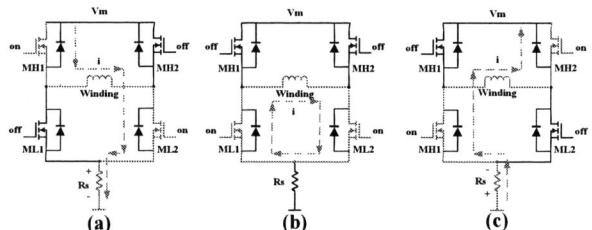

Fig. 3. (a) Charge mode. (b) Slow decay mode. (c) Fast decay mode

A. Winding Current Error of Preliminary Work

Stepper motor control often adopts two primary current decay strategies: charge-slow and charge-fast sequences. The charge-fast sequence exhibits a steep slope of decay current compared to the charge-slow sequence, resulting in increased winding current error and ripple [6]. Conversely, adopting the charge-slow sequence minimizes both ripple and error in winding currents to a greater extent. However, even with slow decay, some residual current error remains present.

The advanced decay mode incorporates a mechanism known as Hysteresis Current Control (HCC), which differs from standard HCC by integrating an additional slow decay stage during the decay process. Two distinct types of advanced decay modes are shown in Fig.4. Type 1 follows a charge-fast-slow sequence, while Type 2 follows charge-slow-fast. The primary objective of these modes is to minimize ripple and enhance current accuracy. Notably, the

979-8-3503-6184-1/24 $31.00 © 2024 IEEE

slow decay stage is pivotal in influencing the magnitude of the current error, with some residual error persisting. Moreover, the fast decay method is applied consistently throughout the chop cycle to enhance the overall responsiveness of the current.

The Auto current decay mode introduced by TI Co. Ltd, depicted in Fig.4(c), closely resembles Advanced decay mode Type 2. By employing a charge-fast-slow sequence, Auto decay minimizes the impact of t_{blank} time. During current charge steps, a single slow decay mode ensures minimal current ripple, while the fast-slow mode in decay steps facilitates rapid current reduction. The innovative technology pioneered by Trinamic and now under Analog Devices Co. Ltd., is highlighted in Fig.4(d). Its distinctive feature involves a charge-slow-fast-slow decay sequence within a single cycle. This method significantly reduces ripple and enhances current accuracy, especially in zero-crossing steps.

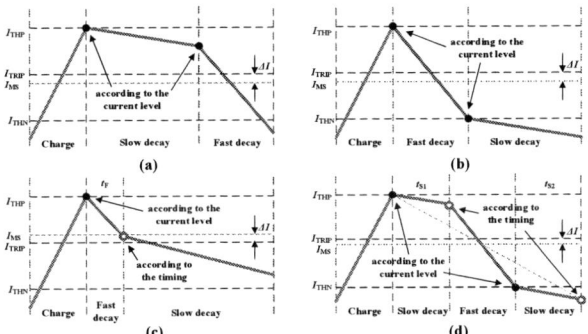

Fig. 4. (a) Advanced decay Type 1. (b) Advanced decay Type 2. (c) TI Auto current decay mode (t_F=30% t_{OFF}). (d) SpreadCycle™ decay ($t_{S1}=t_{S2}$)

B. Proposed Decay Mode and Its Principle

The proposed super-mixed current decay mode is shown in Fig.5. Unlike the SpreadCycle™ decay mode in a normal step, it primarily varies during the fast decay stage. This mode incorporates three defined current thresholds: I_{TRIP}, I_{THP} ($I_{\text{THP}}=I_{\text{TRIP}}+\Delta I$), and I_{THN} ($I_{\text{THN}}=I_{\text{TRIP}}-\Delta I$). The Charge stage concludes once the winding current reaches I_{THP}, initiating the first slow decay stage, which lasts for a predetermined duration t_{S1}. Following this, the fast decay stage begins concurrently with the controller recording the duration t_{F1} until the current decreases to I_{TRIP}. Upon reaching I_{TRIP}, the fast decay stage continues for an additional duration of t_{F2} equal to t_{F1}. Subsequently, the system transitions to the second slow decay stage, t_{S2}, which matches the duration of the first slow decay period ($t_{S2}=t_{S1}$). It is easy to prove that the average winding current I_{MS} over one chop cycle ideally equals I_{TRIP}, thereby minimizing current error.

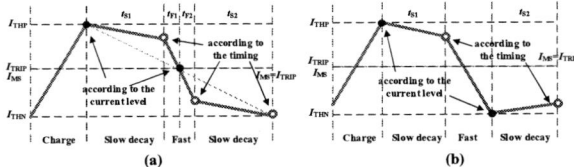

Fig. 5. Proposed super-mixed current decay mode (a) at a standard step ($t_{S1}=t_{S2}$ and $t_{F1}=t_{F2}$). (b)at the zero crossing step ($t_{S1}=t_{S2}$)

In the previously proposed decay mode, a residual current error occurs during the current zero-crossing step. Therefore, a strategy like the SpreadCycle™ strategy is adopted at the zero-crossing step. During the step change process, two distinct strategies are used, each tailored for increasing and decreasing current cases, as depicted in Fig.6. When the current increases, the strategy is the same as that shown in Fig. 5(a). Conversely, when the current decreases, the Charge stage is ignored, and the system transitions directly into the fast decay stage. Here, the fast decay continues until the current reaches I_{THN}. Following, the system enters a slow decay stage, which persists for a predetermined duration t_{S1}.

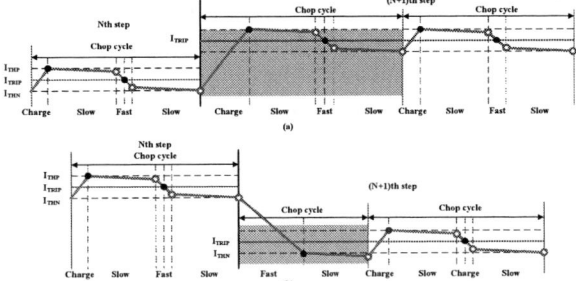

Fig. 6. (a) Proposed super-mixed current decay (Increasing current steps). (b) Proposed super-mixed current decay (Decreasing current steps)

III. CIRCUIT IMPLEMENTATION AND SIMULATION RESULT

A. Circuit Architecture of the Proposed Decay Mode

The circuit principle diagram is shown in Fig.7. The digital module receives STEP, INC, DIR, and ZeroSTEP signals and generates H1, H2, L1, and L2 signals to control H bridge switches. This module determines the motor's operational state (normal step, change step, or zero-crossing step) and rotation. Additionally, the winding current is compared with the setting currents by three comparators to realize the proposed strategy. The Comp1 compares the motor current with the maximum current I_{THP}, generating the end signal of charge and starting the first slow decay stage. The Timing module marks the end of the first slow decay stage (endslow1) and starts the fast decay stage. The Comp2 and TDS modules mentioned above realize comparison and the charging capacitors processes. Ultimately, this process produces the end of the fast decay signal (endfast) and ends the second slow decay signal (endslow2). The whole circuit architecture also includes the sin DAC and Indexer blocks.

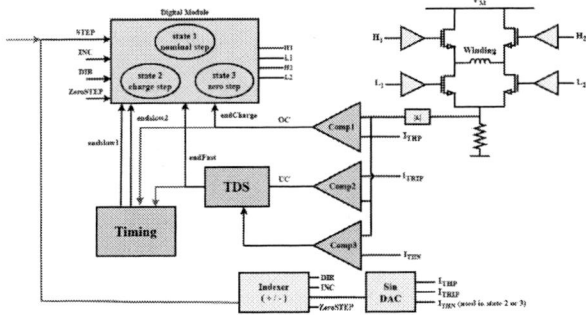

Fig. 7. The circuit architecture of the proposed decay mode

B. Implementation of Timing and Delay Switch Module

The principle diagram of the timing and delay switch module (TDS) is shown in Fig.8, to realize $t_{F1}=t_{F2}$. The circuit comprises two branches, including two switches and capacitors with the same charging current. After the first slow decay stage, switch Φ_1 opens, and capacitor C_1 will be charged to V_{CAP1} over a charging period of t_{F1}. When the current I_{MS}

979-8-3503-6184-1/24 $31.00 © 2024 IEEE   127

reaches the setting current I_{TRIP}, switch Φ_1 closes, and switch Φ_2 opens, initiating the charging of capacitor C_2. Upon V_{CAP2} reaching V_{CAP1}, the comparator activates, defining a charging period of t_{F2}. Due to the identical currents and capacitors charging to the same voltage, their charging durations will be equal, $t_{F1}=t_{F2}$. At the end of the fast decay stage, impulse Φ_{OFF} triggers the discharge of both capacitors in preparation for the next cycle. The relevant technical solutions have been reported previously [7].

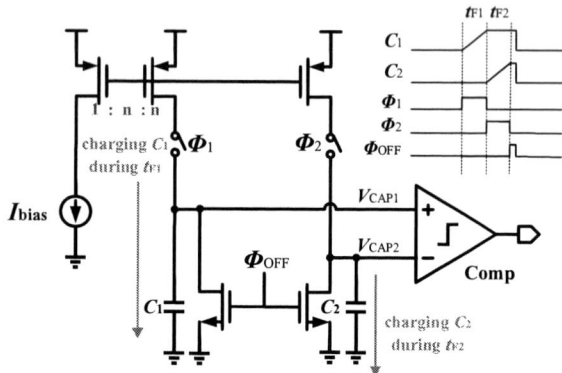

Fig. 8. The circuit diagram of TDS module (make $t_{F1}=t_{F2}$)

C. Simulation Verification Results

The proposed decay mode has been implemented in a 0.18μm BCD process. Fig.9 to Fig.11 show the simulation waveforms of winding current I_{MS} and crucial signals.

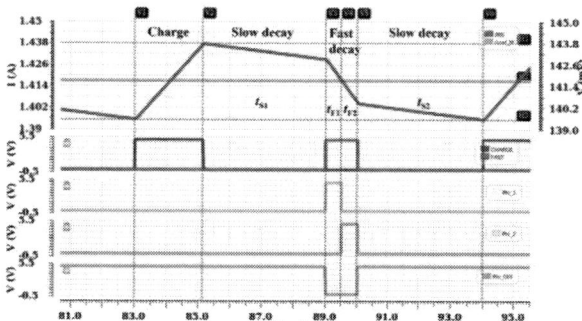

Fig. 9. Proposed decay mode simulation results

Fig. 9 shows the operational cycle of the proposed super-mixed decay mode, which follows a charge-slow-fast-slow sequence. Following the charge reaching I_{THP}, the current enters a slow decay stage. Subsequently, the TDS module starts timing, alternating switches Φ_1 and Φ_2 to ensure $t_{F1}=t_{F2}$. Then, the Φ_{OFF} switches on, concluding the fast decay stage, and the current returns to slow decay until reaching I_{THN}. In one PWM cycle, the discrepancy between the winding average current I_{MS} and the setting current I_{TRIP} is 0.04%, demonstrating precise current control and accurate motor angular positioning.

Fig.10 and Fig.11 demonstrate the robust performance of this super-mixed decay mode throughout the entire cycle, quickly adapting to achieve a steady state during both current increase and decrease. Furthermore, in zero crossing and changing steps, the proposed strategy also ensures that the winding average current I_{MS} matches the setting current I_{TRIP}.

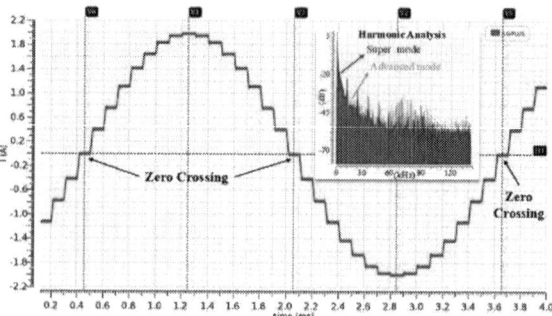

Fig. 10. Proposed decay mode simulation in a whole operating cycle

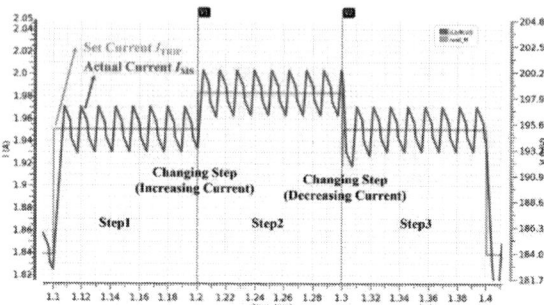

Fig. 11. Simulation results of proposed decay mode in changing steps

IV. CONCLUSION

A super-mixed decay mode is suggested in this paper. This mode follows a charge-slow-fast-slow sequence, with a fast decay stage divided into two periods. Because the charging current of two capacitors is the same, it ensures that the durations of the two fast decay stages are equal. This average actual current I_{MS} equals the setting current I_{TRIP} within each chop PWM cycle, reducing angular position errors in stepper motors. Circuit verification under the 0.18μm BCD process also meets design goals. And the current error is < 0.04%.

REFERENCES

[1] R. Rustam and M. I. Al Khoory, "FPGA-based stepper motor controller: An FPGA-based design project to improve level of the complexity in project-based learning", in Advances in Science and Engineering Technology International Conferences (ASET), Abu Dhabi, United Arab Emirates, 2018, pp. 1-5.

[2] -, "PWM Constant-Current Control Stepper Motor Driver", LV8729 datasheet, ON Semiconductor, 2014.

[3] -, "PWM chopper-type bipolar stepping motor driver IC", TB6564 datasheet, TOSHIBA, 2011.

[4] -, "35-V, 1-A bipolar stepper motor driver with integrated current sensing & 1/256 micro-stepping", DRV8428 datasheet, Texas Instruments, 2020 [Revised 2022].

[5] Bernhard Dwesteg, "Method and Circuit Arrangement for Controlling Current in Motors", U.S. Patent No. 9,030,150 B2, 2015.

[6] P. P. Acarnley and A. Hughes, "Machine/drive circuit interactions in small variable-reluctance stepping and brushless DC motor systems", IEEE Tran.on Industrial Electronics, vol. 35, pp. 67-74, Feb. 1988.

[7] Rongxing Lai, Jian Fang. et al. "Design and Analysis of The Average Current-Detection Method for Wide Input Voltage Range Constant-Current Lighting LED Driver", in International Symposium on Power Semiconductor Devices and ICs, Vienna, Austria, 2020, pp.266-269.

[8] Yoo, JW., Kim, JH., Kim, JH. et al. "Design of a Variable Reference Current Controller for Micro-stepping Motor Based on Vibration and Missing Step Characteristic Data". Int. J. Precis. Eng. Manuf., vol.24, pp.877–886, 2023.

A 109 dB 44-pArms Current Readout Circuit with Automatic Current Control for Multimodality Electrochemical Sensing

Lina Wang, Jianzheng Li, Weiming Hu, and Yajie Qin*

School of Information Science and Technology, Fudan University, Shanghai 200433, China

* Email: lnwang22@m.fudan.edu.cn, yajieqin@fudan.edu.cn

Abstract—This paper reports a high-resolution and wide dynamic range current readout circuit for multimodality electrochemical sensing in 0.11 μm standard CMOS process. The design utilizes a bidirectional current sensing potentiostat circuit to detect the oxidation-reduction current, which is beneficial for cyclic voltammetry (CV) and fast-scan cyclic voltammetry (FSCV). An automatic amplitude control (AAC) is employed to detect the amplitude of the input current and provides feedback to adjust the current digital to analog converter, which enables a wide dynamic range for the input current. This design achieves 109 dB current dynamic range and 44 pA current resolution in the detection current range of ±12.8 μA with an R^2 linearity of 0.99998. It attains 0.007% nonlinearity, offering minimal distortion The system consumes 12 μA static current from a 1.5 V supply and the power efficiency of the readout interface is 0.711.

Keywords—*readout circuit, potentiostat, high-resolution, wide-range*

I. INTRODUCTION

Biomarkers are essential biological parameters encompassing both biophysical and biochemical aspects that are critical for predicting the onset and progression of diseases, as well as for the development of efficacious therapeutic drug [1]. Current readout circuit can convert the concentration of the target biomarker into a corresponding electrical signal, and measures it to obtain information about the analyte. For different biomarkers, concentrations, and sensor sizes, the oxidation-reduction current can vary from pA to μA. Therefore, achieving low noise and high linearity within a wide input current range poses a challenge for the current readout circuit. Additionally, for wearable devices, low power consumption is also needed.

The input current is converted into voltage by a transimpedance amplifier (TIA) [2], or into time by a current-to-frequency converter (I-to-F) [3]. Previous research [3] achieves a current resolution of 8 nA within the current range of -7 μA to 10 μA using a current convey combined with an I-to-F converter. However, the comparator offset and time delay of the time-to-digital converter (TDC) will lead to decrease frequency accuracy and stability when operating at higher frequencies. The minimum detectable time of 0.5 μs in [3] partially limits its resolution. Therefore, compared to the I-to-F conversion, the current-to-voltage converter (I-to-V) design can achieve higher resolution. Reference [2] employes the capacitive feedback TIA (CTIA) to attain the input referred current noise of 21.6-pArms for measuring 50 nA current. A current buffer serves to reduce power

This work was supported by the National Natural Science Foundation of China (82227803).

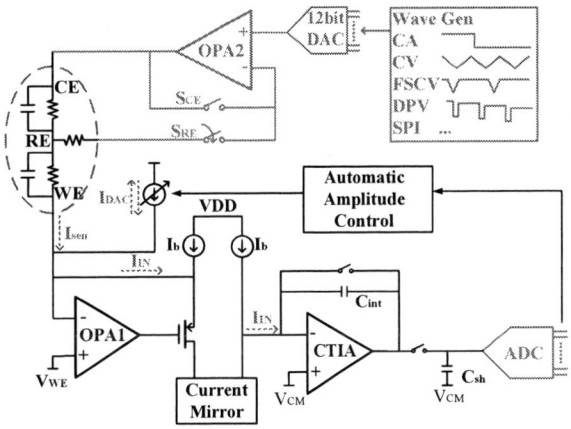

Fig. 1.   Architecture the proposed current readout circuit.

consumption, but the output impedance is relatively low, resulting in significant distortion. Moreover, the design uses offset current to detect the bidirectional current, which requires manual adjustment, making automatic current control unachievable.

This paper proposes a high-resolution and wide dynamic range current readout circuit with automatic current control that can adapt to a variety of electrochemical detection methods. A cascode current mirror is used as a current buffer to realize the continuous measurement of bidirectional current and achieve lower distortion. Combined with the closed-loop automatic amplitude control (AAC) , the input current range can be extended to ±12.8 μA. This design can flexibly address the detection needs of different biomarkers, while the input referred current noise is 44-pArms.

The paper is organized as follows. Section II describes the architecture of the proposed system. The design of current readout circuit is discussed in Section III. Section IV shows the post-layout simulation results, and a conclusion is addressed in Section V.

II. SYSTEM ARCHITECTURE

Fig. 1 shows the architecture of the proposed current readout circuit, which consists of a bidirectional current sensing potentiostat, a CTIA and a current digital to analog converter (IDAC) with automatic current control loop. In electrochemical sensing, depending on the excitation signal waveform, there are various measurement methods including chronoamperometry (CA), cyclic voltammetry (CV), fast scan cyclic voltammetry (FSCV), differential pulse voltammetry (DPV), etc. In this design, a fixed potential is applied to the working electrode (WE), while the potential of the reference electrode (RE) or counter electrode (CE) is varied to achieve multiple excitation signals. OPA1 and M_f form the potentiostat, providing a fixed voltage to the WE through the negative feedback loop. The cascode current mirror isolates the CTIA from the WE, enabling the architecture to accommodate various electrode models and electrochemical measurement methods while achieving bidirectional current sensing. The oxidation-reduction current is converted into a voltage signal by the CTIA, and ultimately digitized through an analog-to-digital converter (ADC). An AAC is employed to continuously monitor the input current range in real-time by comparing the output voltage to a threshold value. When the input current exceeds the specified range, AAC will automatically adjusts the IDAC to maintain the input current within the range of 200 nA. The automatic current control enables a wide dynamic range. Additionally, the IDAC can be used to eliminate the large background current from FSCV.

III. CIRCUIT DESIGN

A. Bidirectional Current Sensing Potentiostat Circuit

The potentiostat circuit provides a fixed voltage to the WE through the negative feedback composed of OPA1 and M_f as shown in Fig. 2(a). OPA1 adopts the folded cascode architecture to achieve high gain. The cascode current mirror consisting of M_f serves as current buffer isolating the electrode from the CTIA, which ensures the charge injection generated by the reset switch of CTIA does not affect the electrochemical charge balance at the electrode-electrolyte interface. In this design, the cascode current mirror can be used to achieve over 200 nA bidirectional input current range, with the static current I_b of 600 nA. Compared with [2], employing the cascode current mirror as the current buffer results in a lager output impedance. Moreover, the current flowing through the current buffer will not fluctuate excessively with the change of the input current, which is beneficial to reduce distortion.

As the first stage of the whole system, the bidirectional current sensing potentiostat circuit contributes the majority of the input referred current noise. Due to the high loop gain formed by OPA1 and M_f, the noise generated by M_f can be disregarded. Transistors $M_{n3,4}$ and $M_{p3,4}$ have negligible noise contribution as cascode transistors [4]. Fig. 2(b) shows the simplified noise model of the circuit. Therefore, the input referred current noise is given by:

$$i_{n,in}^2 = \left(\frac{1}{Z_{cell}}\right)^2 e_n^2 + I_{n,M_{n1}}^2 + I_{n,M_{n2}}^2 + I_{n,M_{p1}}^2 + I_{n,M_{p2}}^2 \quad (1)$$

where Z_{cell} is the equivalent impedance of the electrode, e_n denotes the equivalent input voltage noise of OPA1, and $I_{n,(x)}^2$ represents the current noise generated by transistor x. The noise of OPA1 need to be careful deigned from (1). In this design, the gm/id methodology is adopted to reduce noise and power consumption. Given the relatively large I_b of the cascode current mirror and the low power consumption of OPA1, the noise is mainly limited by the cascode current mirror. The current noise of M_{n1} can be expressed as:

$$I_{n,M_{n1}}^2 = \left(\frac{8kT}{3g_{m,M_{n1}}} + \frac{K_f}{f(WL)_{M_{n1}}}\right) g_{m,M_{n1}}^2 \quad (2)$$

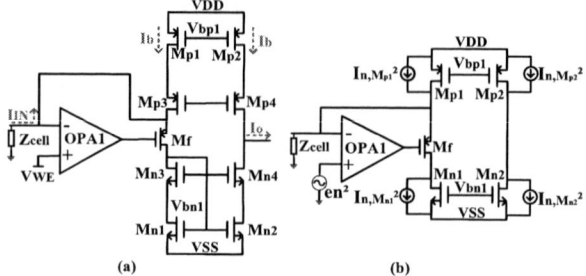

Fig. 2. (a) Schematic of bidirectional current sensing potentiostat. (b) The simplified noise model.

$$g_{m,M_{n1}} = \sqrt{\mu_n C_{ox} I_b \left(\frac{W}{L}\right)_{M_{n1}}} \quad (3)$$

where k is the Boltzmann constant, T is the operating temperature in Kelvin, $g_{m,M_{n1}}$ is the transconductance of M_{n1}, and K_f is a constant for flicker noise calculation, μ_n is the electron mobility, C_{ox} is the gate oxide capacitance per unit area. Therefore, by increasing the length of $M_{n1,2}$ and $M_{p1,2}$, the noise of the circuit is reduced.

B. CTIA Design

CTIA is a common circuit topology employed to convert the current input into a voltage output, facilitating subsequent analysis and processing. Since the input signal frequency in electrochemical sensing applications is close to direct current (DC), this design uses a discrete-time CTIA, which employs a reset switch to periodically reset the capacitor's charge [5]. The discrete-time CTIA using switch reset has high performance with low noise and low power consumption within a narrow bandwidth. The output voltage is given by

$$V_{out} = \frac{t_{int}}{C_{int}} I_{in} \quad (4)$$

where t_{int} is the integration time and C_{int} is the integration capacitor. In this design, $t_{int} = 16$ μs, and $C_{int} = 8$ pF. The subsequent stage is a switched capacitor sampling circuit, where the sampling capacitor is 1 pF.

C. Automatic Current Control Loop

A closed-loop AAC is implemented to regulate IDAC for preprocessing the input current through monitoring the output digital code continuously. The AAC utilizes SAR logic to quantize the output of the ADC to obtain the correct IDAC digital code. Preprocessing the input current through the IDAC enables a wide dynamic range. FSCV requires the handling of background current. Reference [6] introduces a region of interest (ROI) technique to remove a large portion of the non-informative background current. During this design, the IDAC removes most of the background current from the ROI. This design uses the 6-bit IDAC to ensure a current control range of ±12.6 μA with a step size of 200 nA.

IV. SIMULATION RESULTS

This system is designed in a 0.11 μm standard CMOS process and the following results are from the post-layout simulation. The total static current consumption is 12 μA, and the area is 0.06mm². The layout of the circuit is shown in Fig. 3. The simulation uses the electrical model for the electrodes shown in Fig. 4. In this model, $C_{dl,CE}$, $C_{dl,WE}$ are double-layer capacitances, R_{CE}, R_{WE} are the parallel resistances at the electrode-electrolyte interface, and $R_{S,CE}$,

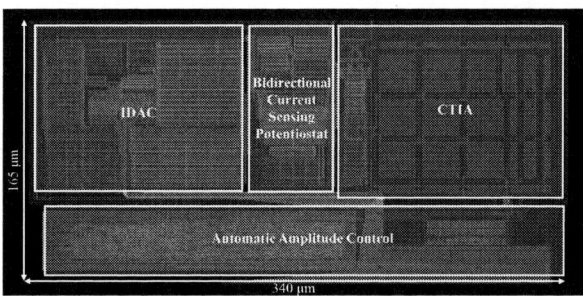

Fig. 3. Layout of the system in 0.11 µm technology.

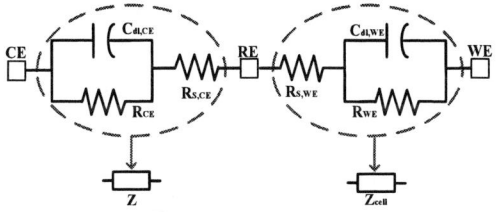

Fig. 4. The electrical model of the electrodes.

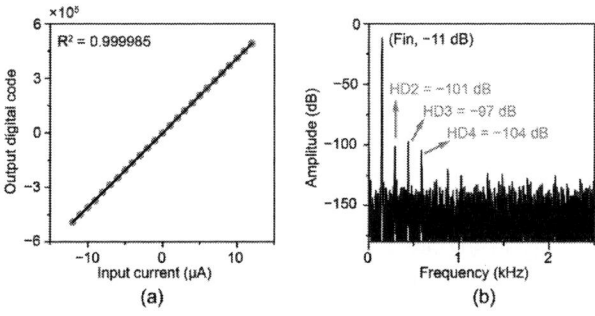

Fig. 5. Post-layout simulation results. (a) The output digital code with the input current range from -12 µA to 12 µA. (b) The spectrum of the output voltage with a 150 Hz 200 nApp current input.

$R_{S,WE}$ are the solution resistances. Different electrode model parameters are used to verify various sensing methods. For FSCV, the electrode uses model parameters of 1.1 nF, 500 MΩ, and 180 kΩ for $C_{dl,WE}$, R_{WE}, and $R_{S,WE}$, respectively [6]. To simulate electrodes for other measurement methods, model parameters of 500 nF, 500 kΩ, and 1 kΩ are used for $C_{dl,WE}$, R_{WE}, and $R_{S,WE}$ [7]. Simulation results indicate that different electrode model parameters do not affect the performance of the circuit.

Fig. 5(a) shows the relationship between the input current and the output digital code with an R^2 of 0.99998, using an ideal ADC. The linearity of the circuit is measured with a 150 Hz 200 nApp sinusoidal input signal. The spectrum of the output voltage is shown in Fig. 5(b). The total harmonic distortion (THD) is 84 dB, which is equivalent to 0.007% nonlinearity. Table I summarizes the performance of the current readout architecture proposed in this design and compares it with state-of-the-art designs. This current readout circuit achieves lower input referred noise of 44-pArms within 1 kHz bandwidth and a wider dynamic range of 109 dB compared to other designs. Our work is also among the lowest in distortion while providing ±12.8 µA input current range.

TABLE I. PERFORMANCE SUMMARY AND COMPARISONS

	TBioCAS'16 [2]	*TBioCAS'21 [3]*	*This work*
Technology	0.18 µm	0.18 µm	**0.11 µm**
Supply Voltage	1.8 V	1.2 V	**1.5 V**
Power Diss.	12.1 µW	19 µW	**18 µW**
Current Range	±200 pA-50 nA	-7~10 µA	**±12.8 µA**
Resolution	21.6 pA	87 pA	**44 pA**
Dynamic Range	104 dB	99 dB	**109 dB**
Power Efficiency	0.015	0.894	**0.711**
Input Referred Noise	21.6 pArms (DC-250 Hz)	87 pArms (~DC)	**44 pArms (DC-1 kHz)**
Linearity	N/A	0.999	**0.99998**
Distortion	0.8%	N/A	**0.007%**
Area	0.03 mm²	3.17 mm²	**0.06 mm²**

V. CONCLUSION

This paper introduces a bidirectional potentiostat-based current readout circuit with ACC for multimodality electrochemical Sensing. It utilizes a cascode current mirror circuit to achieve bidirectional current sensing and employs an ACC for automatic closed-loop control of the input current. The detection current range is ±12.8 µA while the R^2 linearity is above 0.99998. Circuit characterization results show that circuit achieves 44-pArms input referred current noise in 1 kHz bandwidth while only consuming 12 µA static current and 0.06 mm² area. This design achieves 0.007% nonlinearity representing the lowest distortion. With these merits, the proposed architecture demonstrates great potential for wearable electrochemical sensing application that require high-resolution conversion, multimodal capabilities and low-power dissipation.

REFERENCES

[1] J. R. Sempionatto, J. A. Lasalde-Ramírez, K. Mahato, J. Wang, and W. Gao, "Wearable Chemical Sensors for Biomarker Discovery in the Omics Era." Nat. Rev. Chem., vol. 6, no, 12, pp. 899-915, 2022.

[2] J. Guo, W. Ng, J. Yuan, S. Li, and M. Chan, "A 200-Channel Area-Power-Efficient Chemical and Electrical Dual-Mode Acquisition IC for the Study of Neurodegenerative Diseases," IEEE Trans. Biomed. Circuits Syst., vol. 10, no. 3, pp. 567-578, 2016.

[3] S. Y. Lu, and Y. T. Liao, "A 19 µW, 50 kS/s, 0.008-400 V/s Cyclic Voltammetry Readout Interface With a Current Feedback Loop and On-Chip Pattern Generation," IEEE Trans. Biomed. Circuits Syst., vol. 15, no. 2, pp. 190-198, 2021.

[4] H. Li, S. Parsnejad, E. Ashoori, C. Thompson, E. K. Purcell, and A. J. Mason, "Ultracompact Microwatt CMOS Current Readout With Picoampere Noise and Kilohertz Bandwidth for Biosensor Arrays," IEEE Trans. Biomed. Circuits Syst., vol. 12, no. 1, pp. 35-46, 2018.

[5] C. Chen, I. Kim, Y. Jiang, J. Zhang, R. Guo, Y. Ma, et al., "A 17.7µW CDS-CTIA for Wireless-powered Wearable Electrochemical Sweat Sensors," 44th Annu. Int. Conf. IEEE Eng. Med. Biol. Soc. U.K., pp. 4622-4625, July 2022.

[6] B. Nasri, T. Wu, A. Alharbi, K. D. You, M. Gupta, S. P. Sebastian, et al., "Hybrid CMOS-Graphene Sensor Array for Subsecond Dopamine Detection," IEEE Trans. Biomed. Circuits Syst., vol. 11, no. 6, pp. 1192-1203, 2017.

[7] Y. C. Chen, S. Y. Lu, and Y. T. Liao, "A Microwatt Dual-Mode Electrochemical Sensing Current Readout With Current-Reducer Ramp Waveform Generation," IEEE Trans. Biomed. Circuits Syst., vol. 13, no. 6, pp. 1163-1174, 2019.

A Low Temperature Coefficient Bandgap Reference For Temperature Sensor System

Longjiang Jia [1], Yuanhong Ding [1], Jian Mei [2], Lei Deng [2], Rui Yin*[1,3]

[1] Institute of Microelectronics, State Key Laboratory of Integrated Chip and Systems, Fudan University, Shanghai, China
[2] National Integrated Circuit Innovation Center, Shanghai, China
[3] Jiashan Fudan Institute, Jiaxing, China

* Email: ljjia23@m.fudan.edu.cn, yinrui@fudan.edu.cn

Abstract—**This paper describes the design of a bandgap reference used in temperature sensor system with low temperature coefficient and lowly sensitive to manufacturing process deviations. Also, the temperature-dependent voltage generated in this bandgap reference can be quantified by the ADC to obtain current temperature information. In order for bandgap reference performance to satisfy the requirements of temperature sensor systems. This paper employs a higher order temperature compensation technique based on a conventional structure, and the trimming technique is increased to reduce the influence of deviations generated during tape-out on the temperature coefficient. The bandgap reference is based on 180 nm CMOS process. The post-simulation results demonstrate that in the case of a supply voltage of 5V. The output reference voltage is 1.185 V, the temperature coefficient (TC) is 6.108 ppm/°C within the temperature range of -40°C to 125°C, and the PSRR is -68.15 dB at 1000 Hz.**

Keywords—trimming, high-order compensation, bandgap, temperature coefficient, temperature sensor

I. INTRODUCTION

Information from nature such as sound and temperature are analog signals that need to be converted to digital signals because only digital signals can be processed by computers. Temperature sensor systems are widely used in industrial control. The stability of the production process is ensured by monitoring the environmental temperature. With the rapid development of science and technology, the industrial control field for the temperature sensor system requirements are also increasing. Therefore it is necessary to design a high performance temperature sensor system.

The presents chip architecture of temperature sensor system is shown in Fig. 1, and consists of high precision bandgap reference, buffer, mux, ADC, clock and digital processing. The PTAT voltage of the bandgap reference is used as the temperature sensor part. Its performance affects the accuracy of the entire temperature sensor system, so the design of a high-performance bandgap reference is necessary.

In order to achieve high precision temperature measurements, high demands are placed not only on the performance of the ADC, but also on the bandgap reference that offers the supply voltage for the ADC. Therefore, the TC of the reference source should be less than 10 ppm/°C in -45 °C to 125 °C temperature range. Also, to improve the integration of the chip, the temperature information of the current environment can be provided directly by the PTAT voltage of the bandgap reference.

High order compensation is required for low temperature coefficient bandgap references. In this paper, high order compensation is incorporated into the conventional voltage mode bandgap reference[1]. The bandgap reference achieves low temperature coefficient. At the same time, trimming technique is added to reduce the bias due to manufacturing process. The accuracy after tape-out of the chip is improved.

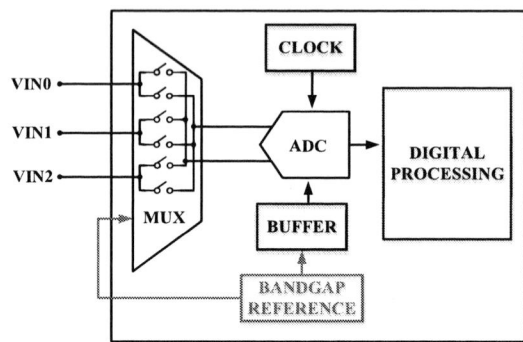

Fig.1. Temperature sensor system block diagram

The following is a description of the structure of this paper: Section II is dedicated to the bandgap reference theory and proposed structure of bandgap. Section III presents the post-simulation results. Finally, the conclusion of this paper in Section IV.

II. STRUCTURE OF BANDGAP REFERENCE

In order to guarantee that the bandgap reference (BGR) has a low temperature coefficient, the BGR of this design incorporates high-order compensation in the high temperature range. Furthermore, since the chip has deviation during the manufacturing process, which leads to a discrepancy between the tape-out experimental results and the post-simulation results, trimming is incorporated into the resistor part, allowing for minor adjustments to the resistor during the chip test process in order to obtain a lower temperature coefficient.

A. The Theory of Bandgap Reference

The critical characteristic of a bandgap reference is the ability to maintain a constant output voltage at different temperatures. Given that most process parameters fluctuate with temperature, a temperature-independent reference will usually exhibit process-independent characteristics as well. Combining two voltages with opposite temperature relationships with appropriate weights, the result will show zero temperature coefficient.

$$V_{ref} = \alpha V_1 + \beta V_2 \qquad (1)$$

The base-emitter voltage of a transistor is negatively temperature dependent. When two transistors are operated at

979-8-3503-6184-1/24 $31.00 © 2024 IEEE

different current values, the difference in base-emitter voltages is positively temperature dependent. Using the negative temperature coefficient and positive temperature coefficient voltages obtained above, we can design a reference with zero temperature dependence. As shown in Fig. 2, the voltage Vref can be defined as

$$V_{ref} = V_{BE2} + [V_T ln(N)][1+(R_2/R_3)] \qquad (2)$$

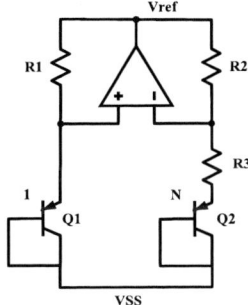

Fig.2. conventional bandgap reference

B. Proposed Structure of Bandgap Reference

In practice, high-order compensation of the bandgap reference is often required to obtain a lower temperature coefficient. The base-emitter voltage of a transistor drops faster with increasing temperature. Therefore, this design circuit compensates for the reference voltage in the high temperature range. Fig. 3 shows the schematic of the designed BGR.

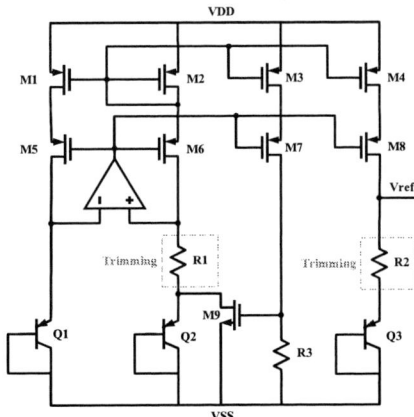

Fig.3. Schematic of proposed bandgap reference

The theory of high-order temperature compensation is accomplished through the utilization of M3,M7,M9 and R3. Because M9 draws only a small portion of the current, it operates in the subthreshold region. When the temperature increases, the current flowing through Q3 increases and the base-emitter voltage of Q3 decreases. At the same time the current drawn by M9 increases with the temperature, then the current flowing through Q3 decreases, slowing down the voltage drop of Q3. Thus, the high-order compensation of the bandgap reference is achieved.

The design should choose a suitable ratio for R2 and R1, when the ratio of R2/R1 is low, the peak value of output voltage Vref will move to the low temperature range. When the ratio of R2/R1 is high, the peak value of output voltage

Vref will move to the high temperature range. as shown in Fig. 4(a). The output voltage of the bandgap reference after adding higher order compensation is shown in Fig. 4(b).

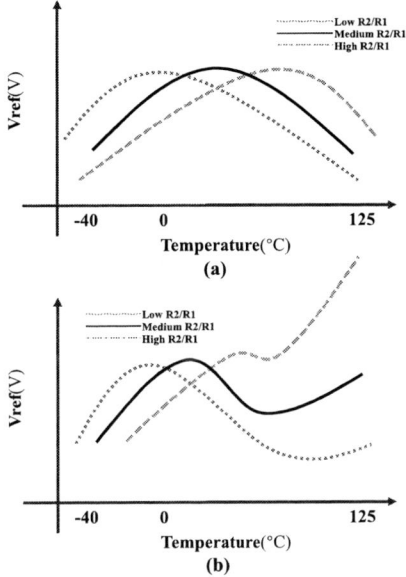

Fig.4. (a) high-order compensation not added (b) with the addition of high-order compensation

It can be observed that for different ratios of R2/R1, the waveform of the output reference voltage Vref is different. A reasonable design of the R2/R1 ratio is needed to obtain the minimum temperature coefficient. In this design, R2/R1=8.5.

At the same time, the chip in the tape-out process will inevitably produce manufacturing bias, which will lead to differences between the chip test results and post-simulation results. For the bandgap reference, the manufacturing brought about by the biggest impact of the manufacturing bias is the temperature coefficient. Considering this problem, this design adopts the trimming technique for R1 and R2, which ensures that the chip after the tape-out. The lowest temperature coefficient can be achieved by trimming.

III. LAYOUT DESIGN AND SIMULATION RESULT

The bandgap reference is based on 180 nm CMOS process. With a supply voltage of 5V,Post-simulation results indicate that the average value of the output voltage is 1.185V. The circuit layout is illustrated in Fig. 5. The total area of the layout is 0.039 mm².

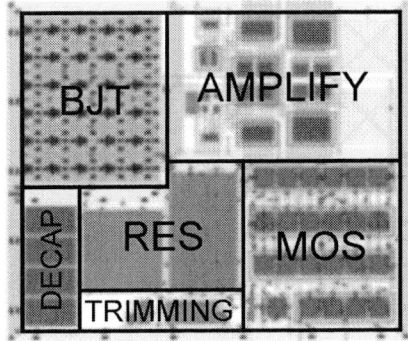

Fig.5. Layout of bandgap reference

979-8-3503-6184-1/24 $31.00 © 2024 IEEE

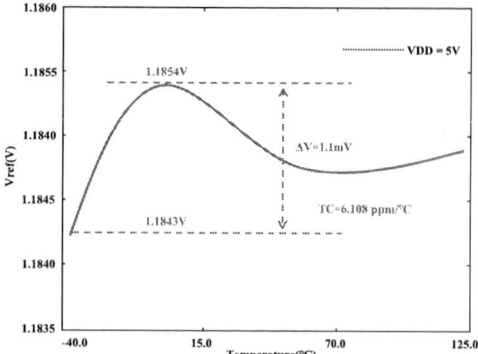

Fig.6. Simulation results of temperature coefficient

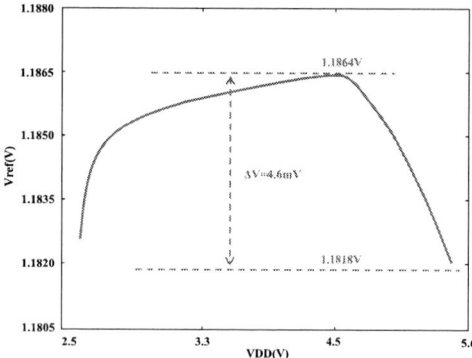

Fig.7. Simulation results of Vref as a function of power supply voltage

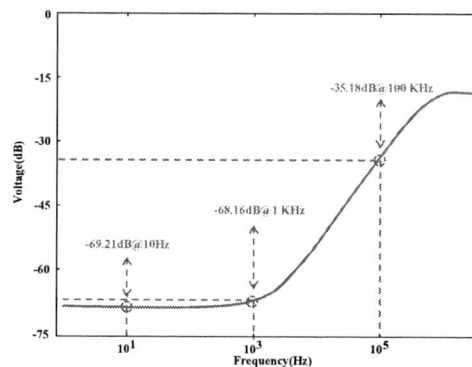

Fig.8. Simulation results of PSRR

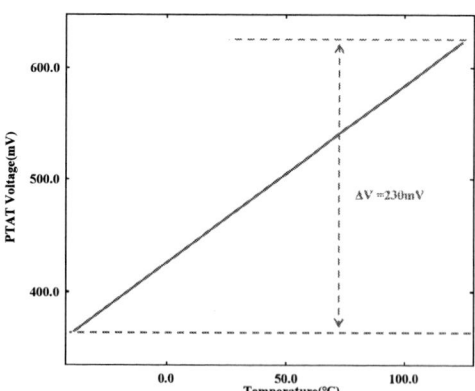

Fig.9. Simulation results of PTAT voltage

Fig. 6 illustrates the temperature coefficients of the designed bandgap reference. Within the temperature range of -40 °C to 125 °C, Vmax is 1.1854 V, Vmin is 1.1843 V, and ΔV is 1.1 mV. Based on these values, the temperature coefficient can be calculated is 6.108 ppm/°C.

Fig.7 illustrates the variation in output reference voltage in relation to the supply voltage, spanning a range from 2.6V to 5.5V. It can be seen that the average value of the output voltage is 1.185V and the output voltage fluctuation range ΔV is less than 4.6mV.

Fig.8 illustrates the PSRR of the reference source. The PSRR was observed to reach -69.21 dB at DC frequency and -68.16 dB at 1000 Hz.

Fig.9 shows the PTAT voltage of the bandgap reference. It can be seen that the voltage varies by 230mV over the temperature range of -40 to 125. this voltage information is quantified by the ADC to get the temperature information.

IV. CONCLUSION

In accordance with the requirements of the temperature sensor system, this paper describes the design of a reference source with temperature compensation. This structure employs a trimming technique to mitigate the impact of manufacturing process deviations and incorporates high-order temperature compensation. The designed bandgap reference has better performance and meets the fulfils the design requirements of the temperature sensor system, with a value of TC=6.108 ppm/°C and PSRR=-68.16 dB at Low frequency. Table I is a description of the differences between this work and other works:

TABLE I.   COMPARISON RESULTS WITH OTHER WORKS

Parameter	[2]	[3]	[4]	This Work
Process (nm)	180	45	600	180
Supply Voltage(V)	4	1.05	0.98	5
Output Voltage(V)	1.2	0.5	0.603	1.185
Temperature Range(°C)	-40~125	-40~125	0~100	-40~125
Temperature Coefficient(ppm/°C)	7.72	24.4	15	6.108
PSRR(dB)	-62.5	-60	-	-68.16

REFERENCES

[1] B. Razavi, "The Bandgap Reference [A Circuit for All Seasons]," in IEEE Solid-State Circuits Magazine, vol. 8, no. 3, pp. 9-12, Summer 2016.

[2] A. Hamouda, R. Arnold, O. Manck and N. -E. Bouguechal, "7.72 ppm/°C, ultralow power, high PSRR CMOS bandgap reference voltage," 2013 IFIP/IEEE 21st International Conference on Very Large Scale Integration (VLSI-SoC), Istanbul, Turkey, 2013, pp. 364-367.

[3] R. Nagulapalli, R. K. Palani and S. Bhagavatula, "A 24.4 ppm/°C Voltage Mode Bandgap Reference With a 1.05V Supply," in IEEE Transactions on Circuits and Systems II: Express Briefs, vol. 68, no. 4, pp. 1088-1092, April 2021.

[4] Ka Nang Leung and P. K. T. Mok, "A sub-1-V 15-ppm//spl deg/C CMOS bandgap voltage reference without requiring low threshold voltage device," in IEEE Journal of Solid-State Circuits, vol. 37, no. 4, pp. 526-530, April 2002.

Toward Unification of Digital Error Correction Algorithms for ADCs with Redundancy

Haruo Kobayashi *[1], Tomohiko Ogawa [1], Yutaro Kobayashi [1], Kentaroh Katoh [2], Jiangling Wei [3]

[1] Gunma University, 1-5-1 Tenjin-cho, Kiryu, Gunma 376-8515, Japan
[2] Fukuoka University, 8-19-1 Nanakuma, Jonan-ku, Fukuoka 814-0180, Japan
[3] Yibin University, Yibin-City, Sichuan 644000, China

* Email: koba@gunma-u.ac.jp

Abstract—This paper attempts to unify digital error correction algorithms in ADCs with some redundancy; non-binary SAR ADC, multiple-comparator SAR ADC, folding/interpolation ADC and pipeline ADC. We found that they have common structure of "correction of higher-bit errors using lower-bit information". This feature can lead to the relaxation of the ADC design requirement, and enables the ADC to be faster, lower power and reliable.

Keywords—ADC, digital error correction, redundancy, self-calibration, unification theory, digital assist analog technology

I. INTRODUCTION

As CMOS technology continues to evolve, addressing challenges related to transistor performance and variability is crucial for designing efficient ADCs. Digital error correction and self-calibration techniques play a vital role in achieving accurate and reliable AD conversion.

This paper focuses on the digital error correction for ADCs. While various algorithms and methods for digital error correction and self-calibration have been proposed for individual ADCs, there have been few attempts to classify and systematize them in a unified theory. Here, we explore the common structure of digital error correction algorithms for some ADCs; non-binary SAR ADC with one comparator, multiple-comparator SAR ADC, folding/ interpolation ADC and pipeline ADC. All these algorithms share the common structure of "correcting higher-bit judgment errors using lower-bit information." In other words, higher bits can tolerate some degree of errors, while lower bits must be accurately generated. This feature can lead to the design of fast, low-power and accurate ADCs.

II. DIGITAL ERROR CORRECTION AND SELF-CALIBRATION

There are two methods for the digital-assisted ADCs.

(i) Digital Error Correction: It involves redundant circuits or operations that allow the ADC to produce correct output despite non-ideal factors. These non-ideal and error factors are not directly measured.

(ii) Self-Calibration: It measures the non-ideal factors of the ADC within its circuit and stores the results in a memory automatically. Based on these values, the ADC output data is corrected and output during normal operation. There are two types of self-calibration: foreground calibration, which stops normal operation for calibration, and background calibration, which performs calibration concurrently without interrupting normal operation.

Here we focus on the digital error correction method.

III. SAR ADC AND DIGITAL ERROR CORRECTION

We consider two types of SAR ADCs with different redundancy features and generalize their algorithms.

A. SAR ADC with Single Comparator

Fig. 1 illustrates an SAR ADC with only one comparator.

(i) Binary search SAR ADC operation: Fig. 2 (a) depicts the operation of a binary search SAR ADC, which achieves N-bit resolution in N-comparison steps. While this approach minimizes the step count, any erroneous behavior can impact the accuracy of the overall ADC output.

(ii) Non-binary SAR ADC operation: Consider a non-binary SAR ADC that employs redundancy by using M comparisons (where N < M). Each step allows for a certain degree of comparator decision error. Fig. 2 (b) shows its example. For a given ADC binary output value, there can be multiple corresponding sets of comparator outputs. This redundancy accounts for cases where incorrect comparator decisions still yield the correct ADC output.

Let us define p(k) as the SAR ADC weight at k-th step. The ratio of p(k)/p(k+1) must be between 1 (unary) to 2 (binary).

$$1 \leq \frac{p(k)}{p(k+1)} \leq 2 \qquad (1)$$

For example, in 5-step binary search, p(1)=16, p(2)=8, p(3)=4, p(4)=2, p(5)=1. In a 5-bstep non-binary case, p(1)=8, p(2)=6, p(3)=3, p(4)=2, p(5)=1.

Also let us define an error correctable range q(k). The red region in Fig. 3 represents the analog input voltage range where the comparator output can be either 1 or 0, yet the correct ADC output is obtained. As the step count increases, this range narrows. Consequently, higher-order bits tolerate decision errors, while lower-order bits are less forgiving of such errors. Even if the comparator result is wrong at k-th step, we can obtain the correct output as long as $|V_{in}-V_{ref}(k)| < q(k)$ (where $V_{ref}(k)$ is the DAC output at k-th step) is satisfied. We have obtained the following relationship [1]:

$$q(k) = -p(k+1) + 1 + \sum_{i=k+2}^{M} p(i). \qquad (2)$$

Next, we consider the incomplete settling of the SAR ADC circuits such as the DAC and the sample-hold (S/H) circuit by assuming a first-order system with time constant τ. To satisfy the condition of $|V_{in}-V_{ref}(k)| < q(k)$, we have its required settling time $T_{settle}(k)$ at k-th step as follows [1]:

$$T_{settle}(k) < \tau \ln\left(\frac{p(k)+q(k-1)}{q(k)}\right). \qquad (3)$$

979-8-3503-6184-1/24 $31.00 © 2024 IEEE

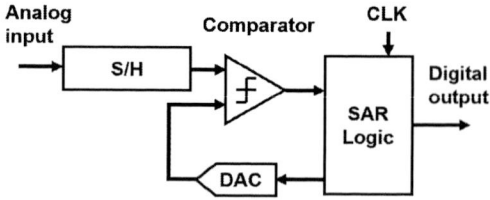

Fig. 1. SAR ADC with one comparator.

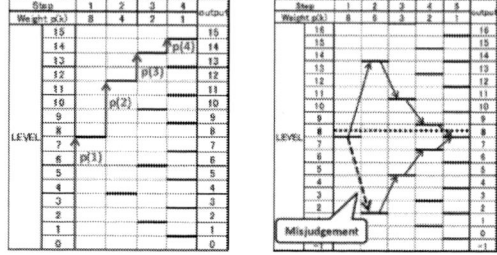

Fig. 2. (a) Binary SAR ADC algorithm; (b) Non-binary SAR ADC algorithm for correct and incorrect judgments.

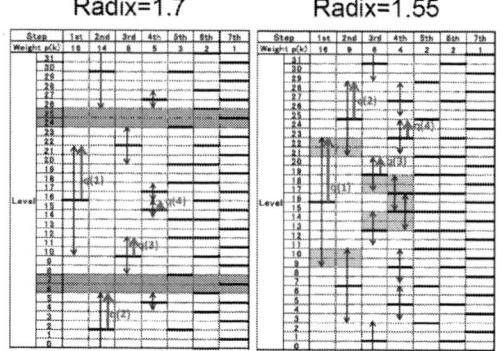

Fig. 3. Examples of p(k), q(k) of non-binary SAR ADCs.

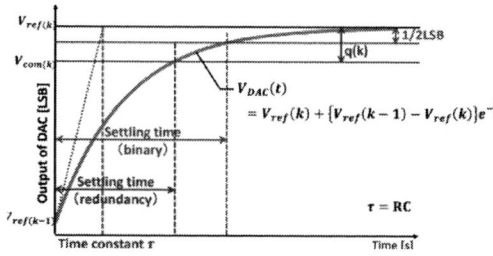

Fig. 4. Incomplete settling time of the internal DAC.

B. SAR ADC with Three Comparators

We have explored an SAR ADC using three comparators and binary search, which enables digital error correction (Fig. 5) [2]. We see in Fig. 5 (b) the error correctable range narrows as moving to the lower bits. Some degree of comparator misjudgment is acceptable for the higher bits, but accurate decision-making becomes crucial for the lower bits. The error correction range q(k) is given by

$$p(k+1) = -q(k) + 2 + 3\sum_{i=k+2}^{M} p(i) \qquad (4)$$

This research is conducted with the support of the Grant-in-Aid for JSPS KAKENHI Grant Number 21K04190.

Let $d(k)$ be the encoded output of the three comparators, and then the reference voltages of $V_{\text{refh}}(k)$, $V_{\text{refm}}(k)$, $V_{\text{refl}}(k)$ for the upper, middle, lower comparators generated by the corresponding DACs in Fig. 5 (a) are given as follows:

$$V_{\text{refm}}(k) = V_{\text{refm}}(k-1) + 3p(k) \quad (\text{for } d(k-1) = 3)$$
$$V_{\text{refm}}(k-1) + p(k) \quad (\text{for } d(k-1) = 2)$$
$$V_{\text{refm}}(k-1) - p(k) \quad (\text{for } d(k-1) = 1)$$
$$V_{\text{refm}}(k-1) - 3p(k) \quad (\text{for } d(k-1) = 0).$$

$$V_{\text{refh}}(k) = V_{\text{refm}}(k) + 2p(k+1).$$

$$V_{\text{refl}}(k) = V_{\text{refm}}(k) - 2p(k+1).$$

$$V_{\text{refh}}(M) = V_{\text{refm}}(M) + 1, \quad V_{\text{refl}}(M) = V_{\text{refm}}(M) - 1.$$

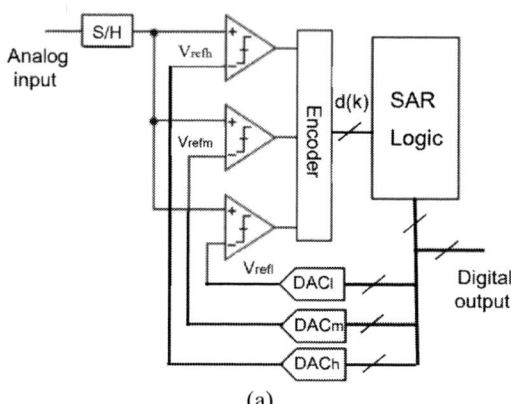

(a)

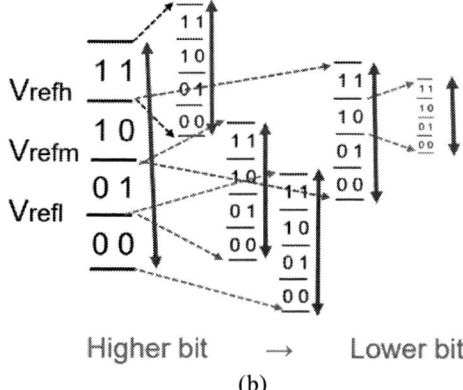

(b)

Fig. 5. SAR ADC with three comparators. (a) Configuration; (b) Operation.

Suppose that at k-th step, a comparator makes wrong decision. When the followings are satisfied, the correct AD conversion result can be obtained.

In case $d(k) = 3$, $V_{\text{refh}}(k) - V_{\text{in}} < q(k)$.

In case $d(k) = 2$, $V_{\text{in}} - V_{\text{refh}}(k) < q(k)$ and
$$V_{\text{refm}}(k) - V_{\text{in}} < q(k).$$

In case $d(k) = 1$, $V_{\text{in}} - V_{\text{refm}}(k) < q(k)$ and
$$V_{\text{refl}}(k) - V_{\text{in}} < q(k).$$

In case $d(k) = 0$, $V_{\text{in}} - V_{\text{refl}}(k) < q(k).$

C. SAR ADC with Non-Binary Search and Multiple Comparators

Now we generalize the redundant SAR ADC algorithms. The error correctable range at the k-th step with M steps (M>N) and L comparators (L =1, 3, 5, 7, …) is given by

$$q(k) = -p(k+1) + (L+1)/2 + L \sum_{i=k+2}^{M} p(i). \quad (5)$$

Equation (5) is generalization of (2) and (4). Then we can derive

$$q(k) \geq q(k+1). \quad (6)$$

We see that q(k) decreases as k increases.

Proof of (6):

It follows from (1) and (5)

$$q(k) - q(k+1) = -p(k+1) + 2\,p(k+2) \geq 0.$$

Then (6) is valid.   (QED)

Remark: (i) The redundant SAR ADC increases the number of steps compared to the binary one. But its well design can lead to fast AD conversion, thanks to the required time reduction for each step by DAC settling requirement relaxation.

(ii) A low power SAR ADC was proposed with two dynamic comparators: a low-power high-noise comparator for the early conversion stages, and a second comparator with lower noise but higher power consumption for the later stages (Fig. 6 (a)). However, the offset mismatch between two comparators degrades its linearity, and hence their offset analog calibration is required [3]. Then we investigated using a digitally-error-correcting non-binary SAR ADC algorithm for errors caused by their offset mismatch, which eliminates the need for the analog calibration [4].

(iii) Fibonacci sequence weight SAR ADC can be realized by choosing p(k) as the Fibonacci sequence. In 10-bit case, p(1)=55, p(2)=34, p(3)=21, p(4)=13, p(5)=8, p(6)=5, p(7)=3, p(8)=2, p(9)=1, p(10)=1. There the minimum settling time T_{settle_min} (k)for k-th step is equal for each k as follows [5]:

$$T_{settle\_min}(k) = \tau \ln(2\varphi + 1). \quad (\varphi = 1.62.... \text{ golden ratio})$$

This can be derived from (3) and Fibonacci sequence feature.

(iv) Pseudo silver ratio weight SAR ADC can be realized by choosing p(k) as the Pseudo silver ratio sequence. In 10-bit case, p(1)=16, p(2)=8, p(3)=8, p(4)=4, p(5)=4, p(6)=2, p(7)=2, p(8)=1, p(9)=1, p(10)=1. There the minimum settling time T_{settle_min} (k) for k-th step has two cases [6]:

$$T_{settle\_min}(k) = \tau \ln(2) \text{ or } \tau \ln(3).$$

This can be derived from (3) and silver ratio feature.

(v) According to our experiences, the overall figure-of-merits of the SAR ADC with non-binary search and single comparator would be better that those of the SAR ADC with binary search and three comparators.

(vi) Some testing related research results of a non-binary SAR ADC are discussed in [7].

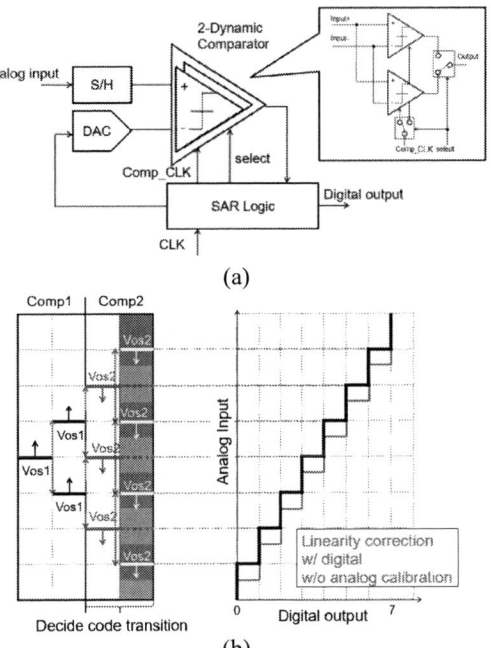

(a)

(b)

Fig. 6. (a) SAR ADC with a low-power comparator for higher bits and a high-power one for lower bits [3]; (b) Explanation of mismatch tolerance between two comparator offsets (Vos1, Vos2) by redundant steps [4].

IV. FOLDING-INTERPOLATION ADC AND DIGITAL ERROR CORRECTION

The folding-interpolation ADC maintains sampling speed equivalent to that of the flash ADC while significantly reducing the number of comparators and the complexity of digital encoder, and hence the power.

This ADC, for example in Fig. 7 [8], generates the higher 3-bit Gray codes (G5, G4, G3) using folding circuits and the lower bits (G2, G1, G0) using the interpolation circuits. The interpolation circuits generate the cyclic codes (C7, …., C0), which produce G2, G1, G0 by XOR logic (Fig. 8):

$$G2 = C4, \quad G1 = C2 \oplus C6, \quad G0 = C1 \oplus C3 \oplus C5 \oplus C7$$

The folding circuits are simple while the interpolation circuits are complex. Hence, when the folding circuits sample an analog input signal (Vin(t)), the interpolation circuits do slightly later at (Vin(t+Δt)). Notably, when Vin(t) is high-frequency or has a large amplitude (in other words, ΔVin(t)/Δt is large), the delay (Δt) can significantly impact the ADC accuracy.

Fig. 7 (b) shows that the transition points of C0 with respect to Vin and those of G5, G4, G3 are ideally equal. However, in practice, the folding circuits for G5, G4, G3 and the circuit for C0 have different delays, leading to deviations in the transition points. To address this, we perform digital error correction near the transition points of G5, G4, G3, aligning them with the transition point of C0. This digital error correction can be achieved through simple calculation of G5, G4, G3, C7, …, C0.

Here again the lower bits correct the higher bits, and the error correctable range is shown in [9].

979-8-3503-6184-1/24 $31.00 © 2024 IEEE   137

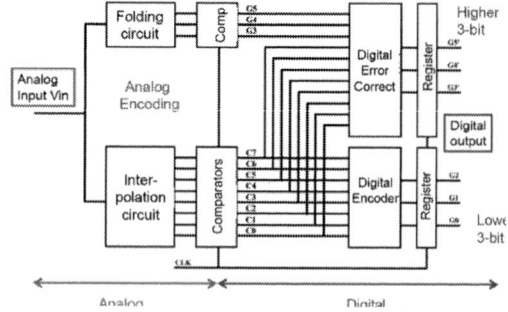

Fig. 7. 6-bit folding/interpolation ADC in [8].

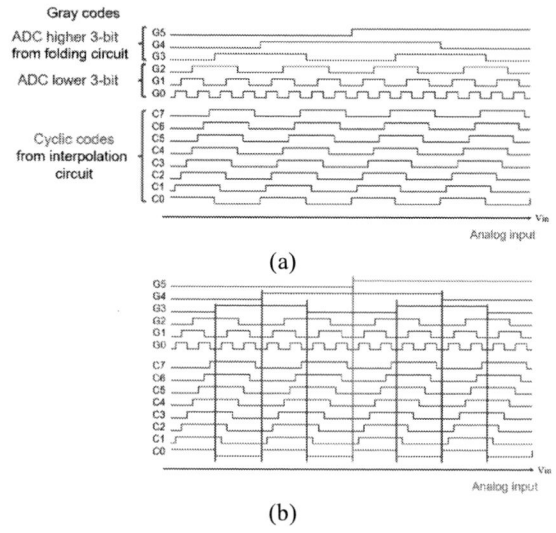

(a)

(b)

Fig. 8. (a) ADC output with respect to its input in Fig. 7; (b) Redundancy among G5, G4, G3 and C0.

V. PIPELINE ADC AND DIGITAL ERROR CORRECTION

Fig. 9 shows configuration of a 2-step pipeline ADC.

ADC1 generates the higher bits and the DAC produces its corresponding analog output. The subtractor and the residue amplifier produce the residue signal and ADC2 generates the lower bits. The DAC, subtractor and residue amplifier are often combined and designed as a multiply DAC (MDAC). Then ADC2 generates the lower bits.

Suppose that MDAC and ADC2 are ideal but ADC1 is not. Then even if ADC1 produces an incorrect output, it can be corrected and the overall correct ADC output can be obtained by widening the input range of ADC2, in other words, $N < N1 + N2$ where N is the overall ADC resolution, N1 is the ADC1 resolution, and N2 is the ADC2 one; thus, redundancy is provided. As long as ADC2 produces the correct output, the entire ADC system will yield the accurate digital output.

This structure effectively compensates for misjudgments in the higher bits using information from the lower bits. The wider the ADC2 input range is, the more tolerable against ADC1 errors the overall ADC is (i.e., the wider the error

correctable range is). This statement holds even for multi-step pipeline ADCs.

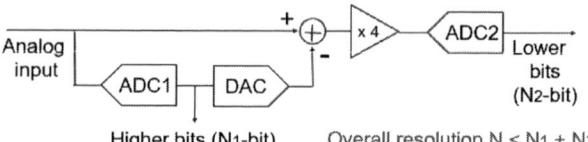

Fig. 9. Configuration of a 2-step pipeline ADC.

VI. CONCLUSION

This paper has tried to unify digital error correction algorithms for some classes of the ADCs with redundancy. There the error correctable range is an important concept.

In the case of the four types of redundant ADCs generated separately for higher and lower bits, a common structure where "higher-bit decision errors are corrected using lower-bit information" is observed. There the raw lower bits generation has to be accurate but the raw higher bits one does not have to be so accurate. Notice that for the ADC without redundancy, the higher bits have to be generated very accurately and this is very different from the ADC with redundancy.

These redundancies can lead to the ADC design requirement relaxation, speed improvement and power reduction.

REFERENCES

[1] T. Ogawa, H. Kobayashi, M. Hotta, Y. Takahashi, H. San, N. Takai, "SAR ADC algorithm with redundancy," IEEE APCCAS, Macao, China (Dec. 2008).

[2] M. Hotta, M. Kawakami, H. Kobayashi, H. San, N. Takai, T. Matsuura, A. Abe, K. Yagi, T. Mori, "SAR ADC architecture with digital error correction," IEEJ Trans. Electrical and Electronic Engineering, vol. 5, no. 6, pp. 651-659 (Nov. 2010).

[3] V. Giannini, P. Nuzzo, V. Chironi, A. Baschirotto, G. Van der Plas, J. Craninckx, " An 820 μ W 9b 40MS/s noise-tolerant dynamic-SAR ADC in 90nm digital CMOS," IEEE ISSCC, San Francisco, CA (Feb. 2008).

[4] T. Ogawa, H. Kobayashi, N. Takai, M. Hotta, H. San, T. Matsuura, A. Abe, K. Yagi, T. Mori, "Non-binary SAR ADC with digital error correction for low power applications," IEEE APCCAS, Kuala Lumpur, Malaysia (Dec. 2010).

[5] Y. Kobayashi, S. Shibuya, T. Arafune, S. Sasaki, H. Kobayashi, "SAR ADC design using golden ratio weight algorithm," IEEE ISCIT, Nara, Japan (Oct. 2015).

[6] Y. Kobayashi, T. Arafune, S. Shibuya, H. Kobayashi, "SAR ADC algorithm with redundancy using pseudo-silver-ratio," IEEJ Trans Electronics, Information and Systems, vol. 137 , no. 2, pp. 222-228 (Feb. 2017).

[7] T. Ogawa, H. Kobayashi, Y. Tan, S. Ito, S. Uemori, N. Takai, K. Niitsu, T. J. Yamaguchi, T. Matsuura, N. Ishikawa, "SAR ADC that is configurable to optimize yield," IEEE APCCAS, Kuala Lumpur, Malaysia (Dec. 2010).

[8] H. Kobayashi, T. Mizuta, K. Uchida, H. Matsuura, A. Miura, T. Yakihara, S. Oka, D. Murata, "A high-speed 6-bit ADC using SiGe HBT," IEICE Trans. Fundamentals, vol. E81-A, no. 3, pp.389-397 (Mar. 1998).

[9] H. Kobayashi, H. Sakayori, T. Tobari, H. Matsuura, "Error correction algorithm for folding/interpolation ADC," IEEE ISCAS, Seattle, WA (May 1995).

979-8-3503-6184-1/24 $31.00 © 2024 IEEE

A 1.2-V 2-GS/s Trimming-Free Input Buffer with Robust Output Common-Mode Voltage

Wei Zhang [1], Xizhu Peng [1], and He Tang *[1]

[1] School of Integrated Circuit Science and Engineering, University of Electronic Science and Technology of China, Chengdu, China

* Email: 202211022722@std.uestc.edu.cn, tanghe@uestc.edu.cn

Abstract—This paper presents a 1.2-V 2-GS/s trimming-free input buffer with robust output common-mode voltage. The proposed input buffer employs a push-pull structure to achieve high energy efficiency and a large input swing equal to the supply voltage. Utilizing the voltage stabilization technique, it has a robust output common-mode voltage without increasing the output load, which guarantees a wide swing and high linearity over PVT variations. Furthermore, the robust output voltage makes it suitable for driving backend ADCs, particularly those based on top-plate sampling techniques. The proposed 2-GS/s input buffer achieves 85.5-dB and 75.2-dB SFDR at low-frequency and near-Nyquist inputs respectively in 28-nm CMOS. The power consumption is only 8.7mW.

Keywords—*input buffer, analog-to-digital converter (ADC), , high speed, trimming-free, robustness, common-mode voltage*

I. INTRODUCTION

High-speed high-resolution analog-to-digital converter (ADC) finds wide-ranging applications in instruments, 5G communication, and satellite navigation systems [1]. With increasing sampling rates, the front-end circuit has become a bottleneck as its performance dominates the linearity and power consumption of the ADC [2]. The input buffer, serving as a crucial component of the front end, is currently attracting a lot of attention from the research community.

Traditionally, the high-linearity input buffers need 4 to 6 transistors in series between the supply rails, thus requiring the high supply voltage to ensure all transistors operate in saturation. Furthermore, due to large changes in the threshold voltage of transistors over process, voltage, and temperature (PVT), the output common-mode (CM) voltage of the input buffer differs from the ideal value, which reduces the differential peak-to-peak swing and is unfavorable for backend ADCs, especially those based on top-plate sampling technique. Reference [3] utilizes a feed-forward replica capacitor which copies the load current and injects it into the output node of the input buffer. This approach reduces the current variations of the input transistors, allowing for higher linearity. However, it still requires a 2.5-V supply, resulting in larger power consumption, additional power supply pins, and the risk of overvoltage, particularly when the circuit design is based on core devices. Additionally, the replica capacitor will introduce feed-forward current which can deteriorate the linearity of input signal. To avoid the drawbacks of replica capacitor, some transistors are introduced to bootstrap the drain voltage of input devices for higher linearity [4], [5]. However, it still necessitates high supply voltage and additional negative voltage modules, resulting in increased chip are and power consumption. A CM feedback circuit is employed to achieve robust output CM voltage [4], which utilizes resistances to detect the output common-mode voltage, thus lowering the gain, bandwidth, and swing of the buffer.

Due to many stacked transistors, it cannot allow a differential peak-to-peak input swing equal to the supply voltage. With the reduced number of stacked transistors, the flipped voltage follower (FVF) employs negative feedback to attain lower output impedance and high linearity [6]. However, it typically needs extra compensation capacitor to ensure the feedback loop to be stable and it is also unable to afford a differential swing of VDD.

We introduce a high-speed input buffer, which possesses robust output common-mode voltage, excellent linearity and a large swing over PVT. This paper is organized as follows. Section II analyzes the limitations of typical input buffers, and provides a detailed description of the proposed techniques and key designs, along with performance comparisons. Section III presents circuit simulation results. Section IV concludes the paper.

II. THE PROPOSED INPUT BUFFER

A. Overview of Typical Input Buffers

The input buffer serves to isolate the input signal from the backend ADC, reducing the kick-back effect of switch capacitor on the input signal [7]. Its significance becomes particularly pronounced when dealing with parasitic inductances and capacitances of the package's bond wires. Additionally, the input buffer has low output impedance, enabling it to drive switch capacitor loads effectively. Fig. 1 shows three low-supply input buffers, including the source follower (SF), the push-pull source follower (PPSF), and the self-adaptive current-compensation push-pull source follower (SACC-PPSF).

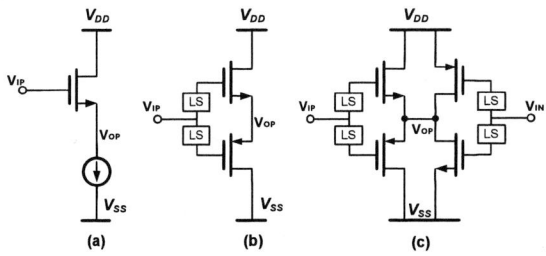

Fig. 1. Typical low-supply input buffers (single-ended for simplicity): (a) source follower, (b) push-pull source follower, (c) self-adaptive current-compensation push-pull source follower proposed in [8].

The source follower is a basic building block in analog integrated circuits as shown in Fig. 1(a). Due to high input impedance, large voltage swing, and simple structure, it is a commonly preferred option for buffer applications. However, the output impedance of the input transistor is modulated by the drain-source potential difference, leading to a highly nonlinearity. Besides, the output dc voltage (VOUT) is PVT-sensitive, which is related to the threshold voltage VTH and the mobility of charge carriers μn, as shown in (1).

$$V_{OUT} = V_{IN} - V_{TH} - \sqrt{\dfrac{2I_D}{\mu_n C_{ox}\left(\dfrac{W}{L}\right)}} \qquad (1)$$

To address the current inefficiency of the basic source follower, a push-pull architecture can be used, as shown in Fig. 1(b). It has a higher transconductance-to-current ratio, which results in roughly twice the efficiency compared to the SF. However, similar to the SF, the linearity of PPSF is still relatively poor because of the non-linear output impedance. The PPSF can be linearized by a self-adaptive current-compensation cell, as shown in Fig. 1(c). But its output dc voltage (2) will deviate from the design value over PVT, especially in processes where there is a significant difference between the VTH of NMOS and PMOS, such as in fast-NMOS-slow-PMOS (fs) corner. The PVT-sensitive output common-mode voltage leads to a narrow swing and large nonlinearity over PVT variations, and makes it unsuitable for driving backend ADCs, particularly those based on top-plate sampling techniques.

$$V_{OUT} = \dfrac{mV_{GN} + V_{GP} - mV_{THN} + |V_{THP}|}{1+m} \qquad (2)$$

$$m = \sqrt{\dfrac{\mu_n (W/L)_N}{\mu_p (W/L)_P}} \qquad (3)$$

B. The Proposed Robust Input Buffer

To meet the demands of low-power, high-speed, and high-precision ADCs, we require an input buffer with the following characteristics. First, the supply voltage of the input buffer should be as low as possible. This can reduce the risk of overvoltage, the power consumption, and the chip area [9]. Second, the input buffer should maintain a robust output common-mode voltage, which can ensure a wide swing even over PVT variations, and make it suitable for driving backend ADCs, particularly those based on top-plate sampling techniques. Third, the input buffer should maintain high linearity even under a low supply voltage. Thus, we propose a 1.2-V trimming-free input buffer with robust common-mode voltage, as shown in Fig. 2, which is based on the topology reported in Fig. 1(c). It has robust output common mode, ensuring high linearity and large swing even over PVT variations.

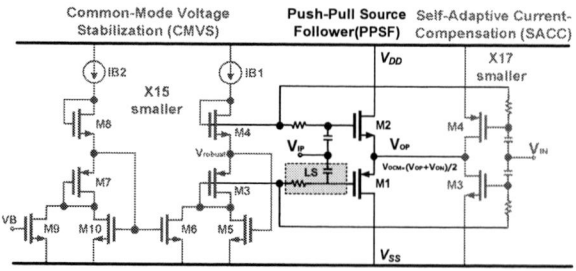

Fig. 2. The proposed input buffer with robust output common-mode voltage (single-ended for simplicity).

The common mode voltage stabilization (CMVS) shown in Fig. 2 is proposed to reduce variations of the output CM voltage over PVT. The sizes of transistors M3-6 and M7-10

are identical, respectively. Since the gates of M6 and M10 are tied together, they have the same gate voltage, and their drain current is approximately the same as long as they are both in saturation. The bias current IB1 is set close to IB2, so the drain currents of M5 and M9 are very close. Therefore, the gate voltage Vrobust of M5 will automatically follow the desired output common-mode voltage VB which can be generated by a bandgap reference or provided externally. The bias voltage of the main input buffer can be replicated from the gate voltages of M3 and M4, resulting in an almost constant output common-mode voltage, VOCM.

The CMVS circuit has a small size which is 1/15th that of the main buffer, and almost negligible increase in power consumption. In comparison to [4], which employs additional resistors at the output to detect CM voltage and adjust input bias, the proposed technique eliminates the need for additional output loads, thus not limiting the gain, bandwidth, and swing of the buffer. The simulated output common-mode voltage versus PVT is plotted in Fig. 3. For a comparison, the SACC push-pull source follower without CMVS, is also designed and simulated in 28-nm CMOS with similar total power. As shown in Fig. 3, the SACC-PPSF utilizing common mode stabilization technique exhibits a robust output common-mode voltage without trimming, which only varies by 17 mV over PVT. This represents an 83% reduction in common-mode variations compared to the SACC-PPSF, while power consumption remains nearly the same. Fig. 4 shows 500 samples of output common-mode voltage. As can be seen, the output common mode of the PPSF with CMVS exhibits relatively small CM voltage variations, similar to the SACC-PPSF.

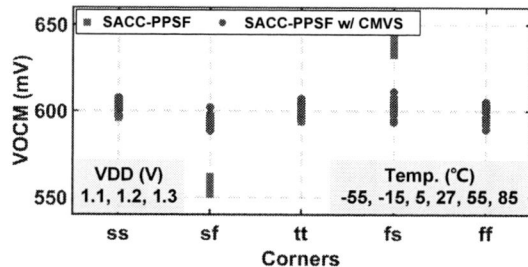

Fig. 3. Simulated output common mode vs. PVT.

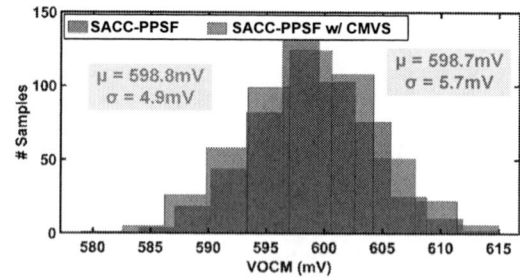

Fig. 4. 500-iteration Monte Carlo simulation results of output CM voltage.

It should be noted the SACC cell must be appropriately biased to effectively suppress the nonlinearity of the main buffer. However, transistor parameters exhibit significant variation over PVT, particularly the threshold voltage. Therefore, the SACC cell cannot maintain a proper dc operating point under all PVT conditions. The CMVS mentioned in Fig. 2 not only helps achieve a robust output common mode but also can automatically provide the SACC cell with a proper bias point over PVT, without the need for trimming. The performance of the SACC buffer with/without the CMVS under different corners is demonstrated in Fig. 5. It can be observed that the SACC-PPSF with CMVS shows better PVT robustness, particularly in ss and ff corners.

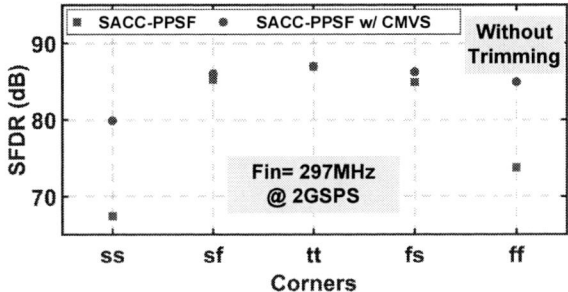

Fig. 5. Simulated SFDR vs. corners.

III. SIMULATION RESULTS

The proposed buffer is evaluated with circuit simulations in 28-nm CMOS with a 2-GHz sampling rate, 1.2-V differential input swing, 500-fF sampling capacitor, and 1.2-V supply voltage. Fig. 6 shows that the proposed input buffer achieves high SFDR without trimming at low-to-mid-frequency inputs over PVT variations. Furthermore, the worst performance is observed in the ss corner, due to the largest RC and lowest power consumption in this corner. However, the proposed buffer still maintains a satisfactory degree of linearity. Fig. 7 displays the SFDR of the proposed input buffer in a 500-run Monte Carlo simulation with different input frequencies. The results show a slight variation in the SFDR, with a small deviation of σ/μ= 1.6% at a 982-MHz input.

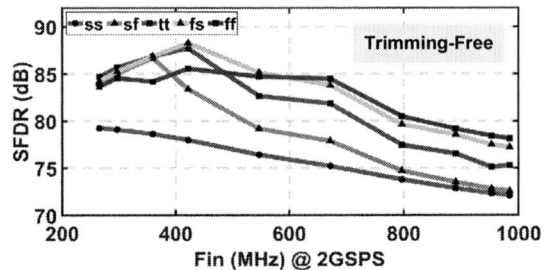

Fig. 6. Simulated SFDR vs. input frequency.

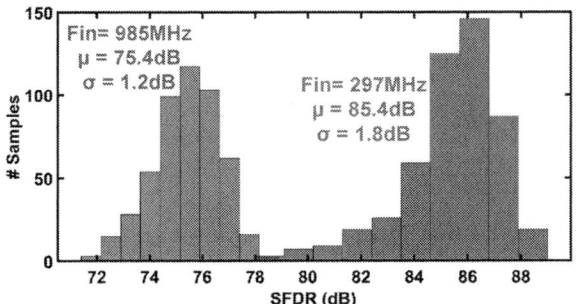

Fig. 7. 500-iteration Monte Carlo simulation results of SFDR at low-frequency and near-Nyquist inputs respectively.

Table I provides a performance summary and comparison with other recent works, all based on simulation results in modern CMOS technologies. It can be seen that the proposed input buffer achieves high linearity with lower power consumption, and it has a robust output common-mode voltage without increasing the output load, which guarantees a wide swing over PVT variations. Besides, the robust output voltage makes it suitable for driving backend ADCs, particularly those based on top-plate sampling techniques.

TABLE I.   PERFORMANCE SUMMARY AND COMPARISON

	This work	*TCAS-2 2022 [2]*	*VLSI 2023 [4]*	*TVLSI 2024 [10]*
Technology [nm]	28	28	28	65
Fs [GS/s]	2	8	2	4
Supply [V]	1.2, 0	2.5, -1	1.65, -1	2.5, -
Robust output CM?	Yes	No	Yes, but increasing load	No
Power [mW]	8.7	136.5	33	27
Area [mm²]	0.0065	0.133	0.285	0.0155
SFDR [dB] @ 297MHz	85.5[a]	-	77[a]	-
SFDR [dB] @ 985MHz	75.2[a]	76[a]	71.8[a]	72.5[a]
Load[fF]	500	600	256	600

a. Simulated SFDR

IV. CONCLUSION

This paper introduces a 1.2-V 2-GS/s 8.7-mW trimming-free input buffer with robust output common-mode voltage. Based on a push-pull architecture, the proposed input buffer achieves high energy efficiency and allows a differential peak-to-peak swing equal to VDD. The PVT variations of the output common-mode voltage is suppressed through the voltage stabilization module with only a 6.7% increase in area. The proposed input buffer is well-suited for use in the front end of ADCs due to its robust output voltage, large swing and high linearity.

REFERENCES

[1] A. Ramkaj, M. Perrott, B. Haroun and B. Murmann, "High-Linearity High-Bandwidth (>20GHz) T&H Front Ends Using Active Bootstrapping and Heterogeneous SiGe/CMOS Circuit Co-Design," 2023 IEEE International Symposium on Circuits and Systems (ISCAS), Monterey, CA, USA, 2023, pp. 1-5, doi: 10.1109/ISCAS46773.2023.10181490.

[2] Z. Huang et al., "A 6-GHz Bandwidth Input Buffer Based on AC-Coupled Flipped Source Follower for 12-bit 8-GS/s ADC in 28-nm CMOS," in IEEE Transactions on Circuits and Systems II: Express Briefs, vol. 69, no. 10, pp. 4163-4167, Oct. 2022, doi: 10.1109/TCSII.2022.3188534.

[3] A. M. A. Ali et al., "A 16-bit 250-MS/s IF Sampling Pipelined ADC With Background Calibration," in IEEE Journal of Solid-State Circuits, vol. 45, no. 12, pp. 2602-2612, Dec. 2010, doi: 10.1109/JSSC.2010.2073194.

[4] L. Ricci et al., "A 2GS/s 11b 8x Interleaved ADC with 9.2 ENOB and 69.9dB SFDR in 28nm CMOS," 2023 IEEE Symposium on VLSI Technology and Circuits (VLSI Technology and Circuits), Kyoto, Japan, 2023, pp.1-2, doi: 10.23919/VLSITechnologyandCir57934.2023.10185370

[5] A. M. A. Ali et al., "A 12-b 18-GS/s RF Sampling ADC With an Integrated Wideband Track-and-Hold Amplifier and Background Calibration," in IEEE Journal of Solid-State Circuits, vol. 55, no. 12, pp. 3210-3224, Dec. 2020, doi: 10.1109/JSSC.2020.3023882.

[6] R. G. Carvajal et al., "The flipped voltage follower: a useful cell for low-voltage low-power circuit design," in IEEE Transactions on Circuits and Systems I: Regular Papers, vol. 52, no. 7, pp. 1276-1291, July 2005, doi: 10.1109/TCSI.2005.851387.

[7] A. M. A. Ali et al., "A 14-bit 125 MS/s IF/RF Sampling Pipelined ADC With 100 dB SFDR and 50 fs Jitter," in IEEE Journal of Solid-State Circuits, vol. 41, no. 8, pp. 1846-1855, Aug. 2006, doi: 10.1109/JSSC.2006.875291.

[8] Y. Cao, M. Zhang, Y. Zhu, R. P. Martins and C. -H. Chan, "22.1 A 12GS/s 12b 4× Time-Interleaved Pipelined ADC with Comprehensive Calibration of TI Errors and Linearized Input Buffer," 2024 IEEE International Solid-State Circuits Conference (ISSCC), San Francisco, CA, USA, 2024, pp. 388-390, doi: 10.1109/ISSCC49657.2024.10454350.

[9] X. Pan, B. Rui, Y. Cao, Y. Zhu, C. -H. Chan and R. P. Martins, "A 12b 1GS/s ADC with Lightweight Input Buffer Distortion Background Calibration Achieving >75dB SFDR over PVT," 2023 IEEE Custom Integrated Circuits Conference (CICC), San Antonio, TX, USA, 2023, pp. 1-2, doi: 10.1109/CICC57935.2023.10121262.

[10] D. Li, T. Feng, J. Ding, Y. Shen, S. Liu and Z. Zhu, "A Wideband Input Buffer Based on Cascade Complementary Source Follower," in IEEE Transactions on Very Large Scale Integration (VLSI) Systems, doi: 10.1109/TVLSI.2024.3349564.

A 12-bit 1-MS/s SAR ADC Using V_{cm}-based Split MSB Switching and Segmented CDAC

Zheng-Han Chen, Ya-Cong Zhang*, Wen-Gao Lu, Zhong-Jian Chen

School of Integrated Circuits, Beijing Advanced Innovation Center for Integrated Circuits, Peking University, Beijing 100871, China

* Email: zhenghantju@163.com, zhangyc@pku.edu.cn

Abstract—**This paper designs a 12-bit SAR ADC in 0.18-μm CMOS technology. To save power and suppress the effect of capacitor mismatch, V_{cm}-based split MSB switching technology is used. To reduce the capacitance of CDAC and achieve better matching, a novel layout for segmented CDAC is proposed, which significantly reduces the parasitic capacitor at the top plate of sub CDAC. At a 3.3-V supply and a sample rate of 1 MS/s, the SAR ADC has an ENOB of 11.2 bits without missing any code.**

Keywords—SAR ADC, V_{cm}-based split MSB switching, novel segmented CDAC layout

I. INTRODUCTION

As an important part of sensors, analog-to-digital converter (ADC) is responsible for converting natural analog signals into digital signals that computers can easily process. Nowadays, successive approximation register (SAR) ADC has played an important role in medium-speed and low-speed applications. Thanks to the process scaling, SAR ADC has also shown great potential in high-speed applications.

Capacitive digital to analog converter (CDAC) switching takes up a significant part of SAR ADC power. To reduce it, monotonic switching [1], bi-directional single-side switching [2], V_{cm}-based switching [3], split-capacitor-based switching [4], charge-average switching [5] and V_{cm}-based split MSB switching [6] have been proposed. They consume 19%, 17%, 12%, 31%, 25%, 8% of the conventional switching power, respectively. Switching the less capacitors to opposite directions, V_{cm}-based split MSB switching has the best DNL and INL among the above technologies. The capacitance of CDAC increases exponentially as the number of SAR ADC bits increases. Segmented CDAC can be used to interrupt the exponential growth. In traditional CDAC layout, the bottom plate of the unit capacitor wraps the top plate totally to achieve better matching [1]. However, a large parasitic capacitor will be introduced to the top plate of sub CDAC when traditional CDAC layout is used in segmented CDAC.

In this paper, a 12-bit SAR ADC using V_{cm}-based split MSB switching and segmented CDAC is proposed. The main CDAC is 9-bit and the sub CDAC is 3-bit. To reduce parasitic capacitance at the top plate of sub CDAC, a novel layout of the unit capacitor is proposed. Every two unit capacitors share a top plate trace. The top plate is routed between two unit capacitors while the bottom plates are routed at both sides. Using the above design methods, the SAR ADC proposed in this paper achieves 11.2-bit ENOB and consumes 135.3 μW with a sample rate of 1 MS/s at a 3.3-V supply.

The rest of this paper is organized as follows: the second part introduces the circuit implementation, the third part gives the measurement results, and the fourth part summarizes the whole paper.

II. STRUCTURE AND IMPLEMENTATION

A. CDAC shematic

The CDAC schematic is shown in Fig. 1. The complete CDAC is differential while only one side is shown for simplicity. To reduce the capacitance, segmented CDAC with 9-bit main CDAC and 3-bit sub CDAC is used. For 12-bit accuracy, bottom-plate sampling is used. If using V_{cm}-based switching, the MSB capacitor will be $C_7=2^7C_u$. But in V_{cm}-based split MSB switching whose DNL and INL are better, C_7 is split into the same capacitor array as the rest of main CDAC. The sub CDAC uses V_{cm}-based switching because its mismatch contributes less to DNL and INL [7]. After sampling, the bottom plate of main CDAC is disconnected from V_{in} and connected to V_{cm}. Then, the comparator performs comparison and the first bit B_1 is decided. According to B_1, C_7 is switched to V_{RH} or V_{RL} and B_2 is decided. If B_2 and B_1 are the same, C_6 will be switched to the same voltage as C_7. Otherwise, C_{6m} will be switched backed to V_{cm}. The same process will be repeated until B_9 is determined. If B_n and B_1 are the same, $C_{(8-n)}$ will be switched to the same voltage as C_7. Otherwise, $C_{(8-n)m}$ will be switched backed to V_{cm}.

The 12-bit SAR ADC using the above CDAC is modeled and simulated in MATLAB. For comparison, V_{cm}-based switching with 31-bit thermometer codes is also involved. According to the process used, C_u is 10 fF with a mismatch standard deviation of 0.55%.

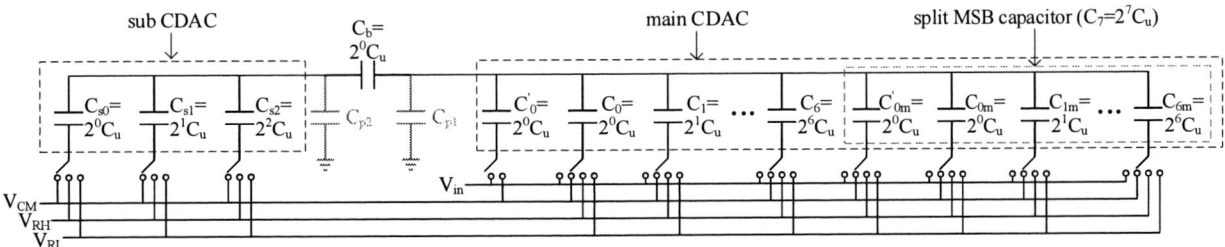

Fig. 1. CDAC schematic

Only considering the effect of CDAC mismatch, 100000 Monte Carlo simulations is performed in MATLAB. The standard deviation of DNL and INL is shown in Fig. 2. The probability of ENOB is shown in Fig. 3. V_{cm}-based split MSB switching has better ENOB while using obviously fewer logic gates. Moreover, 99.98% of simulation results have an ENOB more than 11 bits, which is sufficient for design.

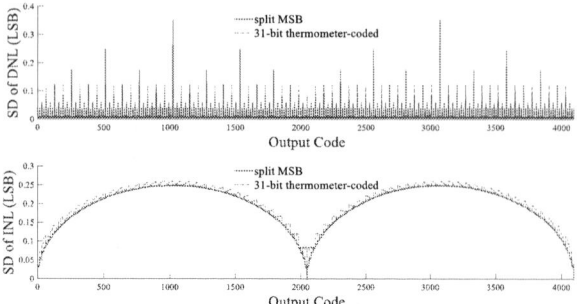

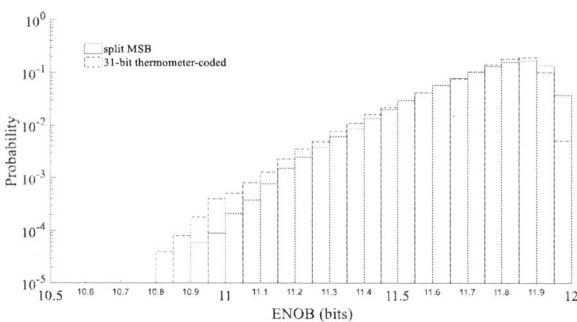

Fig. 2.　Standard deviation of DNL and INL

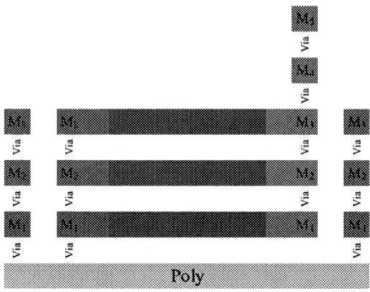

Fig. 3.　Probability of ENOB

B. CDAC layout

The traditional layout of C_u is a sandwich capacitor in which the bottom plate encloses the top plate to minimize the mismatch from parasitic capacitance [1]. In Fig. 1, C_{p1} doesn't affect the weight, whereas C_{p2} does. Therefore, the top plate of C_b should be connected to sub CDAC while the bottom plate of C_b should be connected to main CDAC. Moreover, a dummy C_u should be inserted between C_b and sub CDAC. Otherwise, there will be significant parasitic capacitance between the bottom plate of sub CDAC and the bottom plate of C_b which is also the top plate of main CDAC. In this case,

the top plate of C_b passes through the dummy C_u and connects to sub CDAC, which means a large C_{p2}.

In order to curb the above problem, a novel layout for segmented CDAC is proposed. The top view and cross-section view of C_u layout are shown in Fig. 4 and Fig. 5, respectively.

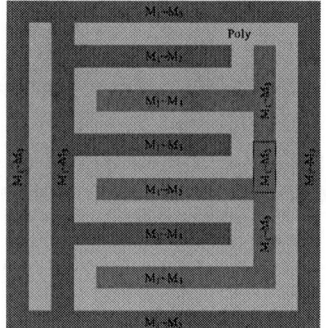

Fig. 4.　Top view of C_u layout

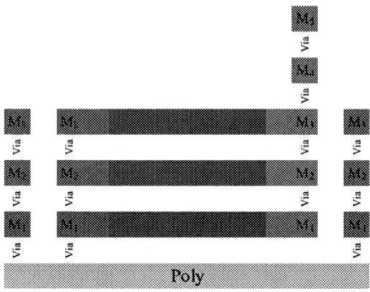

Fig. 5.　Cross-section view of C_u layout

The red part is the top plate, the blue part is the bottom plate, and the purple part is the overlapping area of them. The bottom and top plates are in an interdigitated manner. The bottom plate wraps the top plate except the upper surface. The top plate is led out through high-layer metal and ready to route. The overall layout of CDAC is shown in Fig. 6. The red line is the top plate of main CDAC, the green line is the top plate of sub CDAC, and the blue line is the bottom plate of each C_u. Every two C_u share a top plate trace. In this way, a smaller C_{p2} is obtained at the expense of a larger C_{p1}. Fortunately, C_{p1} doesn't contribute to the weight. According to the parasitic extraction from Calibre, C_{p2} is 0.9 fF in the novel layout while 2.4 fF in the traditional one.

Fig. 6.　CDAC layout

979-8-3503-6184-1/24 $31.00 © 2024 IEEE

III. Measurement

The circuit is fabricated using a standard 0.18-μm CMOS process with a 3.3-V supply and a core area of 540μm×280μm. The chip and PCB is shown in Fig. 7. In order to shield against external interference, the die is coated with black glue.

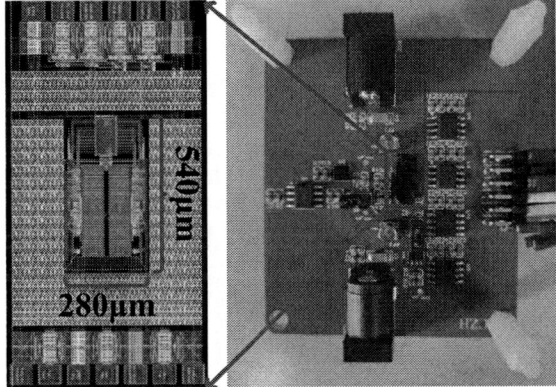

Fig. 7. Chip and PCB

The measured differential nonlinearity (DNL) and integral nonlinearity (INL) of the SAR ADC are shown in Fig. 8. Although segmented CDAC is used, there is no code loss because of the clever switch timing and CDAC layout. The peak DNL and INL are 1.29/-0.44 LSB and 0.98/-0.90 LSB, respectively. DNL and INL are not strictly symmetrical, which may be caused by the non-ideal V_{cm} that is not strictly equal to 1.65V during the test.

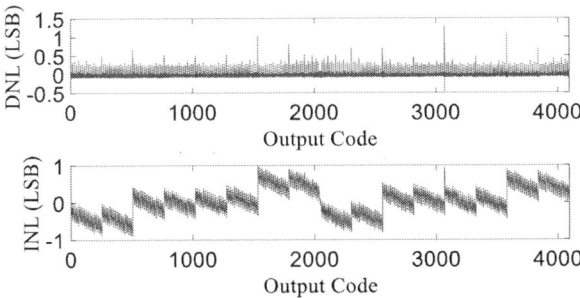

Fig. 8. Measured DNL and INL

The measured ENOB at different sample rates is shown in Fig. 9. At the target sample rate of 1 MS/s, the SAR ADC achieves an ENOB of 11.2 bits with a power of 135.3 μW. Thanks to the design margin, the circuit works well even at a sample rate slightly higher than 1 MS/s. A measured FFT spectrum at 1 MS/s is shown in Fig. 10.

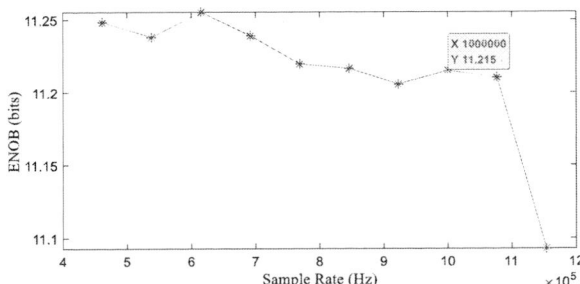

Fig. 9. Measured ENOB versus sample rate

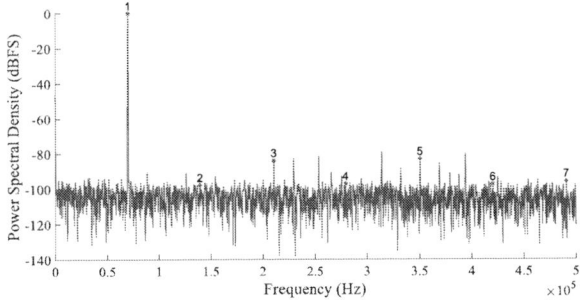

Fig. 10. Measured FFT spectrum at 1 MS/s

IV. Conclusion

In this paper, a 12-bit SAR ADC is designed. The segmented CDAC with V_{cm}-based split MSB switching contributes to small total capacitance and good matching. According to the simulation results of MATLAB, the ENOB is better than that of CDAC with 31-bit thermometer codes. In order to solve the thorny problem in sub CDAC layout, a novel CDAC layout is proposed. According to the extraction results of Calibre, the parasitic capacitor at the top plate of sub CDAC is much smaller than that of traditional CDAC layout. At a 3.3-V supply and a sample rate of 1 MS/s, the SAR ADC achieves a 11.2-bit ENOB with a power consumption of 135.3 μW. The experiment results demonstrate the effectiveness of the proposed SAR ADC.

Acknowledgment

This work is supported by STIC under QYJS-2022-1502-B.

References

[1] C. C. Liu, S. J. Chang, G. Y. Huang and Y. -Z. Lin, "A 10-bit 50-MS/s SAR ADC with a monotonic capacitor switching procedure," in IEEE Journal of Solid-State Circuits, vol. 45, no. 4, pp. 731-740, April 2010

[2] A. Sanyal, and S. Nan, "SAR ADC architecture with 98% reduction in switching energy over conventional scheme." Electronics Letters 49.4 (2013): 248-250.

[3] Y. Zhu, C. H. Chan, U. F. Chio, S. W. Sin, S. P. U, R. P. Matins, F. Maloberti, "A 10-bit 100-MS/s reference-free SAR ADC in 90-nm CMOS," in IEEE Journal of Solid-State Circuits, vol. 45, no. 6, pp. 1111-1121, June 2010

[4] B. P. Ginsburg and A. P. Chandrakasan, "500-MS/s 5-bit ADC in 65-nm CMOS with split capacitor array DAC," in IEEE Journal of Solid-State Circuits, vol. 42, no. 4, pp. 739-747, April 2007

[5] C. Y. Liou and C. C. Hsieh, "A 2.4-to-5.2fJ/conversion-step 10b 0.5-to-4MS/s SAR ADC with charge-average switching DAC in 90nm CMOS," 2013 IEEE International Solid-State Circuits Conference Digest of Technical Papers, San Francisco, CA, USA, 2013, pp. 280-281.

[6] Z. M. Zhu, Y. Xiao, L. Liang, L. X. Liu and Y. T. Yang, "A 3.03 μW 10-bit 200 KS/s SAR ADC in 0.18-μm cmos." Journal of Circuits, Systems and Computers 22.04 (2013): 1350026.

[7] W. Tsutomu, H. X. Li, and K. Murase, "Statistical analysis on the effect of capacitance mismatch in a high-resolution successive-approximation ADC." IEEJ Transactions on electrical and electronic engineering 6.S1 (2011): S89-S93.

A Simplified and Accelerated Opportunistic Bit-Weight Calibration for High-Resolution ADCs

Bingbing Ma, Wei Li*, Hongtao Xu

State Key Laboratory of Integrated Chips and Systems, Fudan University, Shanghai 201203, P.R.China

* Email: 19112020089@fudan.edu.cn, w-li@fudan.edu.cn

Abstract—This paper presents a background bit-weight calibration (BWC) for high-resolution pipelined successive-approximation-register (P-SAR) analog-to-digital converters (ADCs) based on comparator metastability, in which the conventional pseudo-random number generator (PNG) is simplified as a ÷2 divider, requiring only simple D-flip-flops (DFFs). The simplified PNG completely eliminates the systematic errors inherent in the conventional PNGs, which is essential for high-resolution ADCs. By updating all the bit weights simultaneously, the proposed calibration algorithm converges twice as fast as the conventional method does.

Keywords—ADC, calibration, bit weight, metastability, high resolution, pseudo-random number generator

I. INTRODUCTION

As superhigh-data-rate communications develop, e.g., the 5G/6G communications, high-speed and high-resolution analog-to-digital converters (ADCs) are becoming more and more essential. To achieve high speed and high dynamic range under an acceptable power consumption, the trend in ADC designs have shifted to the dual 'compactness + calibration' mind [1][2]. On the 'compactness' side, the ADC sampling capacitors should be as small as possible, thus increasing bandwidth and reducing power consumption. Nowadays, the sampling capacitors have been downsized to the kT/C noise limit, as with the pipelined successive-approximation-register ADC (P-SAR) [3][1]. On the calibration side, mismatches among small unit capacitors require calibration to mitigate harmonic distortion, which have been a standard and a must [4].

Among different calibrations, the one based on comparator metastability (MTS) is probably the most feasible in terms of complexity [3]. However, there are drawbacks with this method. Firstly, the digital cost is high. In a 12-bit ADC, the bit-weights are represented as $\geq$ 16-bit long words, requiring a pseudo-random number generator (PNG) of greater than 16 bits. Moreover, if each of the N_{bc} bit weights to be calibrated requires an individual PNG, then N_{bc} such PNGs are required, incurring considerable digital cost. Secondly, errors in each PNG accumulate to higher bit weights. In an L-bit PNG, the sum of all the $(2^L -1)$ sequence values $\{1,-1\}$ is always 1 or -1 rather than 0, introducing error in the calibration result. As shown in Fig.1, in the conventional calibration [3], the higher bit weights are calibrated based on the lower weights, e.g., W_{12} = Ave($D_{BE} \times PN$)+W_{11}+W_{10}+W_9+W_8, the errors introduced by PNGs in W_8 through W_{11} are accumulated into W_{12}. This systematic error cannot be eliminated, limiting the calibration accuracy. Thirdly, there is waiting time for each bit weight calibration before all the lower weights are calibrated. Owing to the fact that the calibration process is carried out from the least significant bit (LSB) to the most significant bit (MSB),

This paper is supported by National Key R&D Program of China under Grant 2023YFB4403802, and by the Fundamental Research Funds for Central Universities in China.

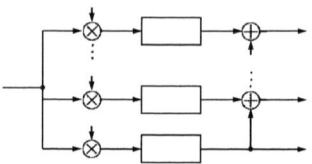

Fig. 1. The conventional bit-weight calibration algorithm [3] based on comparator MTS.

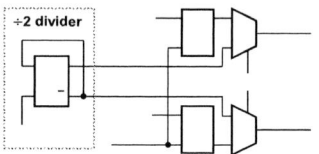

Fig. 2. The proposed simplified PNG implementation.

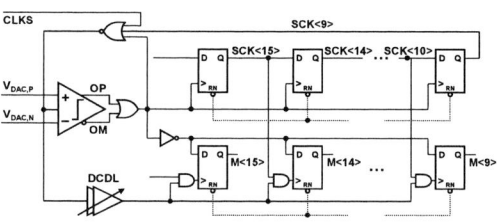

Fig. 3. The comparator VALID and metastability extraction circuit;

only after W_8 is fully calibrated to its final value, then can W_9 begin to be calibrated, and only after W_9 fully calibrated, then can W_{10} begin, and so on. This waiting time is undesirable where the system startup time is important.

To address the aforementioned issues, this work proposed a calibration in which the PNG is simplified as a ÷2 divider, and all the weights are calibrated simultaneously.

II. THE PROPOSED CALIBRATION

A. The Simplified Pseudo-random Number Generator

As shown in Fig. 2, in the proposed calibration, the PNG is simplified as a ÷2 divider, which is merely a D-flip-flop (DFF) with its inverted output connected back to its data input. This ÷2 divider is clocked by metastability signals (M<k>). As shown in Fig. 3, whenever a comparator metastability occurs, the VALID signal is still low (VALIDB is high) when the delayed comparator clock (CLKCD) triggers the MTS DFFs to output the corresponding M<k> as 'high'. This M<k> in turn triggers the ÷2 divider to generate a new simplified pseudo-random number, SPN<k>, to overwrite the comparator output (OP/OM). After VALID goes to high, the capacitor switching clock (SCK<k>) chooses either OM/OP or SPN<k> to switch the capacitor. The MTS window size can be adjusted by the digitally controlled delay line (DCDL).

979-8-3503-6184-1/24 $31.00 © 2024 IEEE

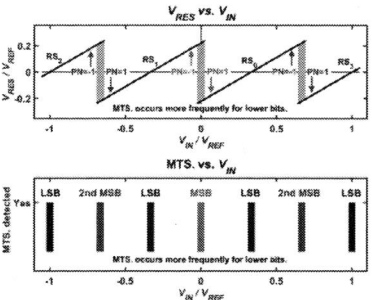

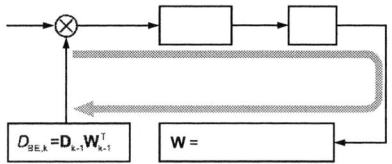

Fig. 4. MTS probability are much higher for the LSBs than the MSBs.

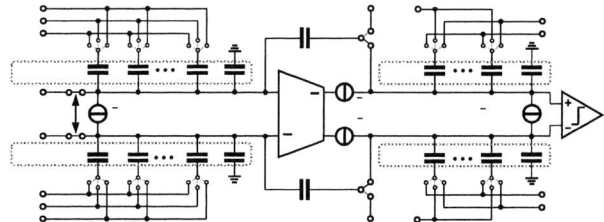

Fig. 6. The P-SAR model to be calibrated.

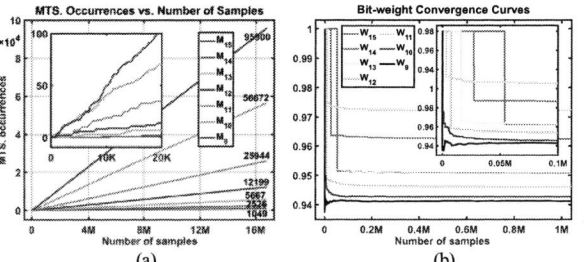

Fig. 5. The proposed bit weight updating algorithm.

(a)   (b)

Fig. 7. (a) MTS occurrences vs. number of samples; (b) Convergence curves of bit weights.

B. Simultaneous Updating of All Bit Weights

Fig. 4 shows a typical residue voltage curve as a function of the input amplitude. It is easily seen that there is only one decision level for the MSB, while there are 2^{N-n} decision levels for the MSB-n bit in an N-bit SAR ADC. Consequently, the comparator MTS probabilities for every two adjacent bits exhibit a 1:2 relationship. In an N-bit SAR ADC, the LSB MTS occurs 2^N times as frequently as the MSB does [5].

The above observation leads to the proposed bit-weight updating algorithm shown in Fig. 5, where all the weights are updated simultaneously. Specifically, when W_k is being updated, its back-end digital code $D_{BE,k}$ is calculated as

$$D_{BE,k} = \boldsymbol{D}_{k-1}\boldsymbol{W}_{k-1}^T \quad (1)$$

where $\boldsymbol{D}_{k-1}=[D_{k-1}, D_{k-2},\dots, D_0]$ and $\boldsymbol{W}_{k-1}=[W_{k-1}, W_{k-2},\dots, W_0]$ are the vectors of all the lower bits and weight values to the current bit D_k and weight W_k, with k = N-1, N-2, … N-K being the index of the K weights to be calibrated. The current weight W_k is calculated as the average of $D_{BE,k}$ times ($-SPN_k$)

$$W_k = Ave\left(-SPN_k \times D_{BE,k}\right) \quad (2)$$

where SPN_k = 1 if it is high or -1 if it is low. In (2), ($-SPN_k \times D_{BE,k}$) are averaged only when the comparator metastability M<k> has occurred for an even number of times, thus completely eliminating the systematic error of the conventional PNG mentioned in Section I. To accomplish this, the averaging engine is clocked by SPN<k>, which is the output of the ÷2 divider.

III. CIRCUIT MODEL DESCRIPTION

The proposed calibration is verified by a two-stage P-SAR model, as shown in Fig. 6. This ADC consists of a 7-bit SAR as the front end (FE) and a 9-bit SAR as the back end (BE), bridged by a residue amplifier (RA). The digital output code is $D_{15}:D_0$, with $D_{15}:D_9$ being the FE bits and $D_8:D_0$ the BE bits. The FE sampling capacitor (CDAC1) is 6.67 pF (single-side) to ensure 86-dB SNR, which is arranged in a [64, 32, …, 1]-units array. The mismatch among these unit capacitors is σ(ΔC/C) = 3%, which is somehow exaggerated to demonstrate the proposed calibration. The BE sampling capacitor (CDAC2) is 520 fF to obtain a BE SNR of 65 dB.

Capacitors in CDAC2 are assumed to exhibit no mismatch, since a high-gain RA greatly relaxes the linearity requirement of the back end. The rms noise voltages of the FE and BE comparators are respectively 1 mV and 0.1 mV, as in the high-resolution P-SAR, decision errors due to the FE comparator's noise is well compensated by the interstage redundancy [6], while the noise of the BE comparator should be as low as possible to achieve a high SNR. The RA's open loop gain is designed to be 1000 and its feedback factor is approximately 1/64 to introduce 1-bit interstage redundancy. Other nonidealities such as sampling switch distortion, reference buffer noise, and incomplete settling of CDAC reference and the RA, are not included in the simulated model, since they are not essentially pertinent in validating the effectiveness of the proposed calibration.

IV. SIMULATION RESULTS

In the model simulations, an ideal sine wave of 0.99 full-scale was applied, and it was sampled at Nyquist rate. A total number of 2^{24} samples were taken for calibration.

Fig. 7(a) shows the number of MTS occurrences increasing with the number of samples taken. After all the 2^{24} samples were converted, there were 1049, 2526, 5667, … and 95900 times of MTS detected for D_{15} (M_{15}), D_{14} (M_{14}), …, and D_9 (M_9), respectively, which validated the fact that the MTS probability of the LSB is much higher than that of the MSB. Owing to Nyquist sampling and the very small MTS window size of 50uV (WS1), the numbers of MTS exhibit a relation of approximately 1:2 between every two adjacent bits, i.e., 1049: 2526: 5667: ⋯: 95900 ≈ 1: 2: 4: ⋯: 64.

In Fig. 7(b), it can be seen that all the bit weights of the first stage (W_{15} to W_9) were calibrated simultaneously, with the lower weights converging to their final values much faster than the higher weights. Specifically, W_9 begins to jump to its final value after less than just several thousand samples, thanks to the frequent MTS of D_9 (M_9 in Fig. 7 (a)). After 0.05 million samples, its variation is no longer remarkable while the first great jump of W_{15} has just begun. The convergence speeds of other weights lie in between, which is consistent with intuition.

979-8-3503-6184-1/24 $31.00 © 2024 IEEE   147

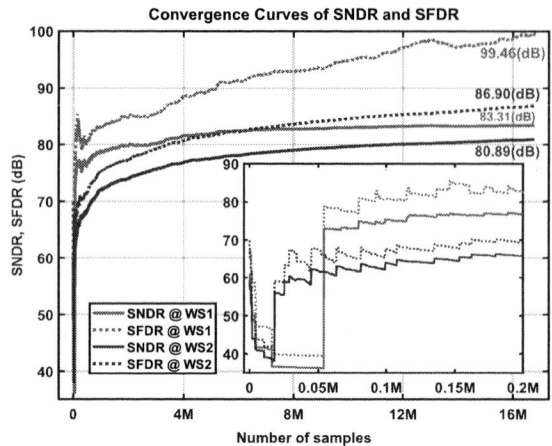

Fig. 8. Convergence curves of SNDR and SFDR.

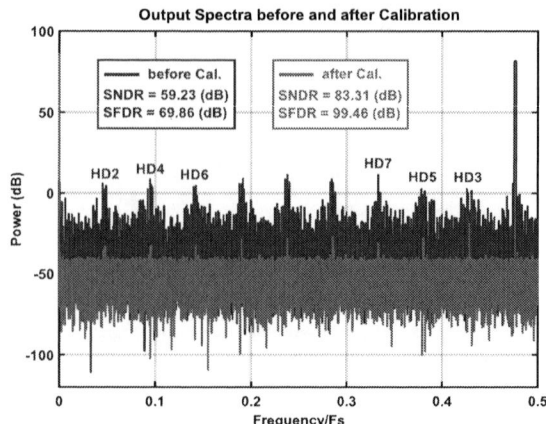

Fig. 9. Output spectra before and after calibration.

Fig. 8 shows the convergence curves of SNDR and SFDR under MTS window sizes of 50µV and 100µV, respectively. On the one hand, the numbers of MTS occurrences with WS2 = 100µV are twice those under WS1 = 50µV. As a matter of fact, with WS2 = 100µV, there are 2212, 5052, 11307, 24317, 52037, 113067, and 191843 times of MTS detected for W_{15} through W_9, respectively, with all other simulation conditions being the same. Therefore, the convergence speed is twice higher as that with WS1 = 50µV. Specifically, in the sub-graph it is clearly seen that the bule lines (WS2 = 100 µV) begin to jump as twice earlier as the red lines (WS1 = 50 µV). On the other hand, smaller MTS window size would generate more accurate band-end digital codes (D_{BE}), resulting in more accurate calibration results, that is why the final values (SNDR = 83.31 dB, SFDR = 99.46 dB) of the red lines are remarkably higher than those (SNDR = 80.89 dB, SFDR = 86.90 dB) of the blue lines.

The contrast of output spectra before and after calibration (under WS1 = 50 µV) is shown in Fig. 9. It is seen that with calibration, the harmonics are greatly suppressed, with the SNDR improved from 59.23 dB to 83.31 dB and the SFDR boosted by 29.6 dB to 99.46 dB, validating the effectiveness of the proposed calibration in high-resolution applications.

The main aspects of the proposed calibration are summarized and compared with other works in Table I. Owing to the fact that the proposed algorithm calibrates both capacitor mismatches as well as interstage gain error, it is clarified as Type III in terms of calibration depth [4].

V. CONCLUSIONS

This work proposed a background bit-weight calibration in which DFFs have been proven superior to conventional PNGs due to their simplicity. Unlike the conventional PNG which is primarily composed of LFSR, the simplified PNG introduces no systematic errors, which is essential in high-resolution scenarios. By simultaneously updating all the bit weights, the proposed calibration algorithm achieved a bit-weight convergence speed that is theoretically twice as fast as that of the conventional method, which is favorable in applications where the startup time is important. The idea of replacing the PNG with simple DFFs can be extended to other calibrations wherever the PN is injected in the vicinity of comparator threshold, simplifying the designs.

TABLE I
COMPARISON WITH OTHER CALIBRATION METHODS

	This work	JSSC'10 [1]	JSSC'11 [2]	JSSC'15 [3]	TVLSI'24 [4]
Cal. depth	III	II	I	III	I
Operation mode	BG	BG	BG	BG	FG
Architecture	SAR/ P-SAR	Pipeline	SAR	SAR/ P-SAR	SAR
PNG	**DFFs**	None	LFSR	LFSR	None
Analog overhead to the core	DCDL	Auxiliary ADC	Cap., CMP.	DCDL	None
Digital overhead to the core	MTS detector		Inner product blocks	PNG, MTS detector	None
Post-processing	ACC, AVE	LMS	LMS	ACC, AVE	Cordic
Number of samples required	10M	0.04M	1M	20M	0.5M

Note: ACC = accumulation, AVE = averaging, BG = background, FG = foreground, LFSR = linear feedback shift register, M = million.

REFERENCES

[1] A. M. A. Ali et al., "A 16-bit 250-MS/s IF Sampling Pipelined ADC With Background Calibration," in *IEEE Journal of Solid-State Circuits*, vol. 45, no. 12, pp. 2602-2612, Dec. 2010.

[2] W. Liu, P. Huang and Y. Chiu, "A 12-bit, 45-MS/s, 3-mW Redundant Successive-Approximation-Register Analog-to-Digital Converter With Digital Calibration," in *IEEE Journal of Solid-State Circuits*, vol. 46, no. 11, pp. 2661-2672, Nov. 2011.

[3] Y. Zhou, B. Xu and Y. Chiu, "A 12 bit 160 MS/s Two-Step SAR ADC With Background Bit-Weight Calibration Using a Time-Domain Proximity Detector," in *IEEE Journal of Solid-State Circuits*, vol. 50, no. 4, pp. 920-931, April 2015.

[4] B. Ma, W. Li and H. Xu, "Analysis and Calibration of Bit Weights in SAR and Pipelined SAR ADCs Based on Code Distribution," in *IEEE Transactions on Very Large Scale Integration (VLSI) Systems*, vol. 32, no. 6, pp. 977-990, June 2024.

[5] D. Bankman, A. Yu, K. Zheng and B. Murmann, "Understanding Metastability in SAR ADCs: Part I: Synchronous," in *IEEE Solid-State Circuits Magazine*, vol. 11, no. 2, pp. 86-97, Spring 2019.

[6] V. Giannini, P. Nuzzo, V. Chironi, A. Baschirotto, G. Van der Plas and J. Craninckx, "An 820µW 9b 40MS/s Noise-Tolerant Dynamic-SAR ADC in 90nm Digital CMOS," 2008 *IEEE International Solid-State Circuits Conference - Digest of Technical Papers*, San Francisco, CA, USA, 2008, pp. 238-610.

979-8-3503-6184-1/24 $31.00 © 2024 IEEE

Background Calibration for Bit Weights in Pipelined SAR ADCs Using Split ADC Architecture

Zecheng Zhou, Longsheng Wang, Dongxian Ye, Yexin Zhu, Dengquan Li[*], and Zhangming Zhu

Key Laboratory of Analog Integrated Circuits and Systems (Ministry of Education), School of Integrated Circuits, Xidian University, China

* Email: dqli@xidian.edu.cn

Abstract—**A background calibration method of bit weights for pipelined successive-approximation-register (SAR) analog-to-digital converters (ADCs) is presented. The method corrects the errors from capacitor mismatch and inter-stage gain error. The ADC is split into two independent converters, each converting the same input signal. The dither window is constructed by adding opposite polarity offsets to each comparator of the converters to obtain the bit weights. Simulation results show that after 2×10^5 cycles, the signal-to-noise and distortion ratio (SNDR) and the spurious-free dynamic range (SFDR) are improved from 54.1dB to 69.9dB and 63.8dB to 95.6dB, respectively. The proposed calibration method has the advantages of high accuracy, fast convergence speed and running in background.**

Keywords—*pipelined SAR ADC, background calibration, bit weight, split ADC, offset, signal-dependent PN injection.*

I. INTRODUCTION

High-speed and high-precision analog-to-digital converters (ADCs) with low power are the key blocks for image processing, wireless communication, and instrumentation equipment. Combining the advantages of high speed of pipelined ADCs and low power of successive-approximation-register (SAR) ADCs, pipelined SAR ADC is a promising architecture. It is much more energy-efficient than conventional pipelined ADCs. However, several challenges are encountered. The first issue is that the ADC achievable linearity is limited by capacitor mismatch in the coarse-stage SAR ADC. In addition, with the decrease of the intrinsic gain, achieving a large open-loop gain becomes increasingly challenging, which is crucial for maintaining accurate inter-stage gain of close-loop RAs. Although open-loop RAs exhibit high energy-efficiency and high-speed, their downside is the high process, voltage, and temperature (PVT) sensitivity. The inaccurate inter-stage gain limits the performance of the ADC. In [1], dither window is constructed by adding opposite polarity offsets to the paired comparators. DAC-based injection is employed to alleviate the swing increment of the residue voltage. However, it still needs about 2 million samples to converge. In [2], the split ADC architecture is first introduced to solve the issue of slow convergency speed in conventional statistics-based approaches. Nonetheless, a complicated algorithm is adopted to calibrate the offset and gain mismatch between the two split ADCs.

In this paper, a bit weights background calibration method for pipelined SAR ADCs is proposed. To obtain bit weights, opposite polarity offsets are applied to comparators of each half of the split ADC. By adding dither capacitors, the residue swing increment due to dither window is alleviated. With the split ADC architecture, the calibration method requires only 200K samples to achieve convergence. This paper is

This work was supported by the National Key R&D Program of China (2023YFB4403303), and National Natural Science Foundation of China (62274129, 62021004).

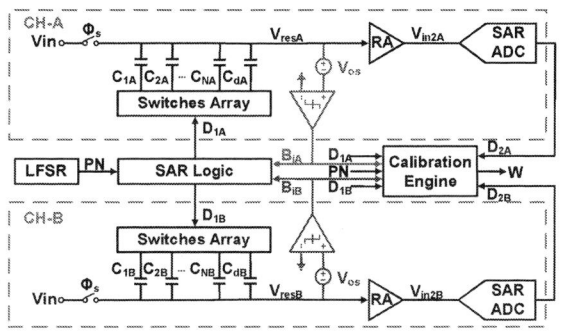

Fig. 1. Block diagram of the proposed pipelined SAR ADC

organized as follows. Section II introduces the proposed calibration method. Simulation results are shown in Section III. Finally, Section IV concludes this paper.

II. PROPOSED CALIBRATION METHOD

A. PN Injection

Fig. 1 shows the block diagram of the ADC. For simplicity, a two-stage ADC is presented to illustrate the calibration method, which can be applied to multi-stage ADCs. The ADC is split into two identical channels CH-A and CH-B. Each channel converts the same input signal. Opposite polarity offsets (V_{OS} and $-V_{OS}$) are inserted in the comparators of two channels. To neutralize voltage swing increasement due to the offsets of comparators, dither capacitors ($C_{dA,B}$) are added in the DAC. After dither injection, the output sequences of the backend SAR ADCs are employed to calculate bit weights. The capacitor ratio and inter-stage gain in pipelined SAR ADCs consist of a bit weight, which is

$$W_i = GV_{ref} \times \frac{C_i}{C_t} \tag{1}$$

where G is the inter-stage gain, V_{ref} is the reference voltage, C_i is the i-th capacitance and C_t is the total capacitance of the capacitor digital to analog converter (CDAC) in the first stage SAR ADC. Note that capacitor mismatch and inter-stage gain are lumped to the bit weights deviation. The goal of calibration is to search the optimal bit weights.

Conventionally, to obtain the optimal bit weights, a pseudo number (PN) signal is inserted to the first stage ADC. The injected PN signal travels through the residual path and gets quantized by the back-end stage. By correlating the outputs of the back-end ADC with the input PN signal, accurate weights can be calculated [3]. To illustrate the process of the dither injection, a 5-bit first stage SAR ADC is presented, as shown in Fig. 2. A dither window is constructed by opposite polarity offsets (V_{os} and $-V_{os}$) of the comparators in the two channels. A XOR gate is employed to monitor the outputs of the two comparators, which indicates whether and where the residual

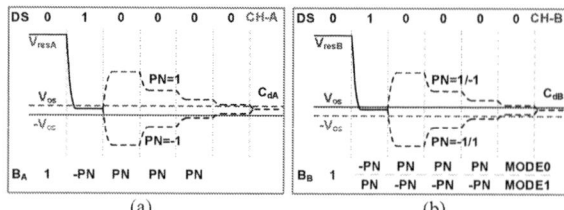
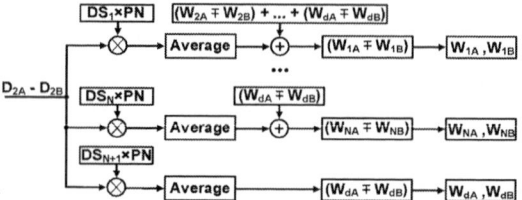

Fig. 2. The process of the dither injection for a 5-bit first stage SAR ADC: (a) channel A (b) channel B

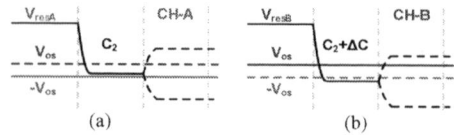

Fig. 3. Schematic for calculating the bit weights

voltage ($V_{resA,B}$) falls into the window. The detection results are recorded in a register array (DS). During the first cycle, $V_{resA,B}$ is out of the dither window, and the corresponding bit of DS is 0. In the next cycle, as $V_{resA,B}$ falls into the window, DS_2 becomes high and thus the corresponding bit $B_{iA,B}$ is overwritten by the PN. During the last cycle, $C_{dA,B}$ is switched according to PN. When the V_{OS} is set within a reasonable range, $V_{resA,B}$ will fall into the window no more than once. Thus, the rest bits of DS are set to 0. If the $V_{resA,B}$ does not fall into the windows in any of the cycles, DS_6 will be set to 1.

B. Two Injection Modes

In order to obtain the bit weights for channels A and B, it is necessary for one channel to have two modes of dither injection. Assuming the $V_{resA,B}$ is within the window during the j-th cycle and falls outside of the window in all subsequent cycles, the input of the second stage SAR ADC ($V_{in2A,B}$) can be expressed as

$$V_{in2A} = G_A \cdot V_{in} - \left(\sum_{i=1}^{j-1} B_{iA} \cdot W_{iA} \right) + PN \cdot W_{jA} - \left(\sum_{i=j+1}^{N} PN \cdot W_{iA} \right) - PN \cdot W_{dA} \quad (2)$$

$$V_{in2B} = G_B \cdot V_{in} - \left(\sum_{i=1}^{j-1} B_{iB} \cdot W_{iB} \right) \pm PN \cdot W_{jB} \mp \left(\sum_{i=j+1}^{N} PN \cdot W_{iB} \right) \mp PN \cdot W_{dB} \quad (3)$$

where B_{iA} and B_{iB} are the outputs of the comparators in the i-th cycle. Channel B has two modes of dither injection. In mode 0, the bits falling within the window are rewritten as -PN, while in mode 1, they are rewritten as PN. After the quantization, channel B transitions to another dither injection mode.

The dual-mode characteristic of channel B results in the fact that the determination of whether $V_{resA,B}$ falls into the dither window cannot merely rely on the XOR of the results from the two comparators. When channel B operates in mode 1, the polarity of the dither injected in the two channels is opposite. Therefore, the comparison results of the remaining bits are also opposite. In this case, even though the outputs of the comparators are opposite, it does not mean that $V_{resA,B}$ falls into the window. Thus, when $V_{resA,B}$ falls into the window and the corresponding bit of DS is set to 1, the remaining bits of DS will be forced to 0, and no further dither injection will occur afterwards.

C. Calibration of Bit Weights

The input of the second stage ADC ($V_{in2A,B}$), which contains information of all the bit weights, is digitalized as $D_{2A,B}$. The bit weights are calculated with the correlation algorithm in the digital domain. Fig. 3 shows the schematic of the calibration algorithm, where DS, PN, and $D_{2A,B}$ are from the ADC and serve as inputs to the calibration engine. The difference between the outputs D_{2A} and D_{2B} of the second stage SAR ADC is multiplied by the PN and separately summed and averaged according to the DS_i and the injection mode of channel B. Assuming $V_{resA,B}$ falls into the window

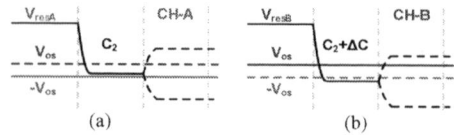

Fig. 4. Effect of capacitor mismatch on residual voltage: (a) channel A (b) channel B

during the j-th cycle for M times. After being correlated with PN for M times, $D_{2A} - D_{2B}$ is the summed and averaged to be

$$\frac{\sum_{k=1}^{M} \left(D_{2A}(k) - D_{2B}(k) \right) \times PN}{M} = \delta + PN^2 \times \left(W_{jA} \mp W_{jB} \right)$$
$$- \sum_{i=j+1}^{N} PN^2 \times \left(W_{iA} \mp W_{iB} \right) - PN^2 \times \left(W_{dA} \mp W_{dB} \right) \quad (4)$$

$$\delta = \frac{PN \times \sum_{k=1}^{M} \left[\left(G_A - G_B \right) \times V_{in}(k) - \sum_{i=1}^{j-1} B_i(k) \times \left(W_{iA} - W_{iB} \right) \right]}{M} \quad (5)$$

where δ contains the input signal and is turned to be noise. If M is large enough, the magnitude of δ will approach zero. Thus, the convergence speed is determined by the magnitude of δ. If a small dither window is used to decrease the amplitude of δ, fewer input samples will fall in the comparator threshold. In this case, although W_d can be calculated faster, the slower convergence of other bit weights results in a decreased speed of the entire calibration algorithm. In this work, by subtracting the outputs D_{2A} and D_{2B}, the amplitude of δ is significantly reduced. Thus, there is no need to accumulate a great number of samples to remove the input signal. Since PN^2 is always 1, the combination of bit weights is obtained. Likewise, other combinations of bit weights can be calculated, as shown in Fig. 3. The injection mode of Channel B determines whether the obtained result is the sum or difference of the bit weights of the two channels. With two injection modes, the actual bit weights of the two channels can be calculated separately, which are employed to generate the final output.

D. Analysis of Nonidealities

The offset voltage may deviate from the ideal value due to mismatch between the comparators. As long as the window size is smaller than one LSB voltage of the first stage, $V_{resA,B}$ will not fall into the window again in the following cycles. However, under a larger offset voltage, the window size may exceed the maximum allowable range, which would deteriorate the calibration algorithm. Therefore, an offset calibration scheme is employed to generate V_{os}, while also eliminating the mismatch of the comparators [1].

Capacitor mismatch results in the discrepancy between two channels, which may lead to an increase of the residual swing. A 3-bit SAR ADC is presented in Fig. 4. In the second cycle, V_{resB} falls out of the dither window due to capacitor mismatch. However, it is still lower than V_{os}, while V_{resA} is higher than $-V_{os}$. Thus, the outputs of two comparators are opposite and the second bit is rewritten by PN, which may increase the residual swing of channel B. As a result, when

979-8-3503-6184-1/24 $31.00 © 2024 IEEE

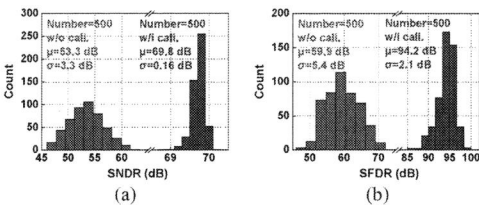

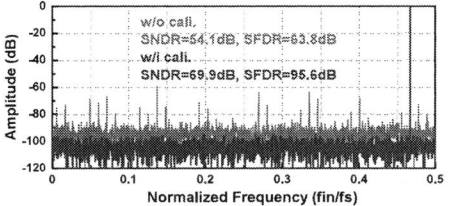

Fig. 5. Histograms of (a) SNDR and (b) SFDR before and after calibration

Fig. 6. Spectrum of the ADC output before and after calibration

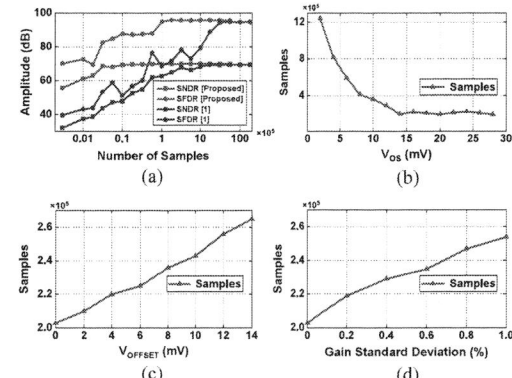

Fig. 7. (a) Comparison of SNDR/SFDR convergence results using different methods. Convergence speed versus (b) Vos, (c) offset and (d) gain errors

TABLE I.   COMPARISON WITH RELATED WORKS

	This Work*	[1]*	[4]	[5]
Approach	Background			Foreground
Architecture	Pipeline SAR			Pipeline
Resolution (bit)	12	12	12	13
fS (MS/s)	400	100	160	260
ΔSNDR (dB)	15.8	17.5	19.3	10.6
ΔSFDR (dB)	31.8	39.5	35.9	21.9
Convergence speed (Msamples)	0.2	2	10-30	0.08

* simulation results

comparison results with other related works. The proposed method operates background with fast convergence speed.

choosing V_{os}, some redundancy range should be reserved.

For the conventional split ADC calibration scheme, a challenge is encountered when offset and gain errors in each channel are taken into consideration. The basic concept of conventional split ADC is that the difference of two split channels outputs is zero if both channels are calibrated. However, even if the ADC is correctly calibrated, any offset and gain mismatches between the ADC channels can result in a non-zero difference, which will disrupt the calibration process. In the proposed method, after correlated with PN, the unknown signal containing offset and gain errors is averaged to be zero. Therefore, this method is insensitive to the mismatch between channels, avoiding the complex calibration of offset and gain errors as in [2].

III. SIMULATION RESULTS

A 5b+8b 400MS/s pipelined SAR ADC model with 1-bit redundancy is employed to verify the calibration method. The unit capacitance C_u is 16fF to meet the kT/C noise requirement. The capacitor array of the first stage is [64, 32, 16, 8, 4, 2, 1, 1] C_u. The V_{os} is chosen to $1/64V_{ref}$, and C_d is $0.5C_u$ to realize the neutralization of the voltage swing increment. The standard deviation of the capacitor mismatch and inter-stage gain error is 3%. The kT/C noise of the sampling capacitor, the residue amplifier noise and the second stage comparator noise are all included in the model.

Histograms of SNDR and SFDR of 500-run Monte-Carlo simulation are shown in Fig. 5. On average, the SNDR and SFDR are improved by 16.5dB and 34.3dB. Fig. 6 shows the ADC output spectrum before and after calibration for a full-scale sine wave input at a frequency of 188MHz. After calibration, the SNDR and SFDR are improved from 54.1dB and 63.8dB to 69.9dB and 95.6dB, respectively. Fig. 7(a) illustrates the comparison of SNDR and SFDR convergence results using the conventional method [1]. After about 200K cycles, the calibration converges. It can be seen that the convergence speed of the proposed method is 9 times faster than the conventional method [1]. Fig. 7(b) presents the samples used to iterate versus the V_{os}. Fig. 7(c) and Fig. 7(d) illustrate the relationship between convergence speed and mismatch of the two channels. The amplitude of measurement noise δ is increased by the mismatch of two channels, leading to a reduction in calibration speed. Table I shows the

IV. CONCLUSION

A bit weight calibration for pipelined SAR ADC using split ADC architecture is proposed. When the input residues fall into the windows formed by the offset voltages applied to the comparators in each channel, dither signal is injected to the ADC. The actual bit weight can be obtained by calibration algorithm. By subtracting the outputs of the two channels, the calibration speed is significantly increased. It takes about 200K cycles to obtain the calibrated results. Simulation results of the 12-bit pipelined SAR ADC show that SNDR and SFDR are improved by 15.8dB and 31.8dB, respectively.

REFERENCES

[1] J. Sun, M. Zhang, L. Qiu, J. Wu and W. Liu, "Background calibration of bit weights in pipelined-SAR ADCs using paired comparators," *IEEE Trans. Very Large Scale Integr. (VLSI) Syst.*, vol. 28, no. 4, pp. 1074-1078, Apr. 2020.

[2] J. McNeill, M. C. W. Coln, and B. J. Larivee, "'Split ADC' architecture for deterministic digital background calibration of a 16-bit 1-MS/s ADC," *IEEE J. Solid-State Circuits*, vol. 40, no. 12, pp. 2437–2445, 2005.

[3] E. Siragusa and I. Galton, "A digitally enhanced 1.8 V 15 b 40 MS/s CMOS pipelined ADC," *IEEE J. Solid-State Circuits*, vol. 39, no. 12, pp. 2126–2138, Dec. 2004.

[4] Y. Zhou, B. Xu, and Y. Chiu, "A 12 bit 160 MS/s two-step SAR ADC with background bit-weight calibration using a time-domain proximity detector," *IEEE J. Solid-State Circuits*, vol. 50, no. 4, pp. 920–931, Apr. 2015.

[5] A. W. Hassan, D. Zhou and J. Silva-Martinez, "Matrix-Based Digital Calibration Technique for High-Performance SAR and Pipeline ADCs," in *IEEE Transactions on Circuits and Systems I: Regular Papers*, vol. 71, no. 1, pp. 20-28, Jan. 2024.

979-8-3503-6184-1/24 $31.00 © 2024 IEEE

When Time Interleaving encounters Oversampling in ADC

Mingqiang Guo, Dongyang Jiang, Shulin Zhao, Sai-Weng Sin, and Rui P. Martins[1]

State-Key Laboratory of Analog and Mixed-Signal VLSI, Institute of Microelectronics and Department of ECE / Faculty of Science and Technology, University of Macau, Macao, China
[1] Instituto Superior Técnico, Universidade de Lisboa, Portugal
Email: mqguo@um.edu.mo

Abstract— **Time-interleaved ADC architecture is widely used in high-speed applications because this structure can increase the speed of ADC linearly by adding the number of channels. On the other hand, the oversampling technique, especially the technique that combines oversampling and noise shaping, can significantly improve the accuracy of the ADC through a trade-off between speed and resolution. So, what kind of sparks will be produced when time interleaving meets oversampling? This paper will be based on the three works of our research group in recent years, combined with other excellent papers, and try to discuss how to combine the time-interleaved and oversampling to solve the mismatch between the channels of the time-interleaved ADC and break the speed bottleneck through the traditional sigma-delta modulators or noise-shaping SAR ADC.**

Keywords – Time interleaving, Noise Shaping, ADC, Oversampling.

I. INTRODUCTION

In the past few decades, Time-Interleaving (TI) has been widely used in high-speed and wide-bandwidth Analog-to-Digital Converters (ADCs). This technology can linearly increase the sampling speed/bandwidth of the ADC by increasing the number of channels [1]. For example, as shown in Fig. 1, the overall sampling frequency of the TI ADC will become n-times of a single channel. However, there are mismatches between different channels. From the source of the error, we can divide it into offset, gain, and timing skew mismatch. Compared with offset and gain, the error caused by timing mismatch is not only related to the amplitude of the input signal but also to the frequency of the input signal [2], as shown in Fig. 2.

On the other hand, oversampling techniques, at the expense of speed/bandwidth, can certainly reduce the noise floor of the ADC within the effective bandwidth of interest, resulting in a higher signal-to-noise ratio (SNR). Fig. 3 [3] shows the processing chain: Step 1 shows the sampled analog signal was limited in 0-f_B by a 1/N (where N is oversampling ratio (OSR)) times of Nyquist band f_N; Step 2 presents quantization error is spread over the entire Nyquist band just like noise, and most of this item is outside f_B; Step 3 exhibits the impaction of the digital filter to cancel the noise outside f_B; Step 4 shows the final outputs after decimation, the noise in band occupies a small Nyquist interval. Actually, combined with the techniques of oversampling and noise shaping, it becomes more effective

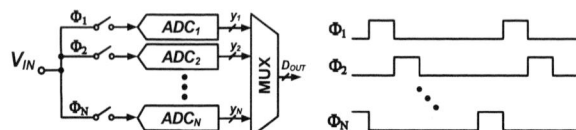

Figure 1. (a) A N-channel TI ADC, and (b) timing diagram.

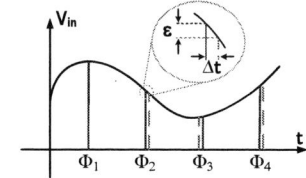

Figure 2. The timing mismatch error in a 4-channel TI ADC.

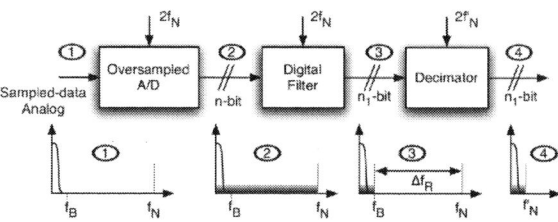

Figure 3. The application of the oversampling technique in an ADC [3].

because the noise spectrum is much lowered in the signal band [3].

Time-interleaving can enhance speed/bandwidth but suffers from poor accuracy due to mismatches between different channels. The oversampling can enhance the accuracy of the ADC but at the expense of speed/bandwidth. So, it seems that these two technologies can be combined to make up for each other's shortcomings and obtain a better compromise under certain specifications of the ADC. We have sorted out three different ideas, which will be discussed in detail in sections II, III, and IV.

II. APPLING OVERSAMPLING IN TIME-INTERLEAVED ADC

The Split structure was originally proposed to solve the linearity problem of cyclic ADC [4] or Pipelined ADC [5]. The basic idea of this technique is based on the structure of using the Reference ADC to calibrate the main ADC. Now, we split

979-8-3503-6184-1/24 $31.00 © 2024 IEEE

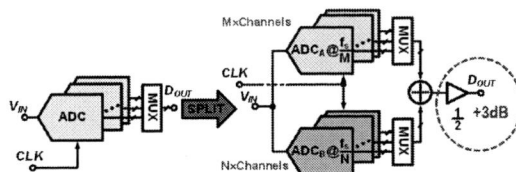

Figure 4. The split Time-Interleaved architecture.

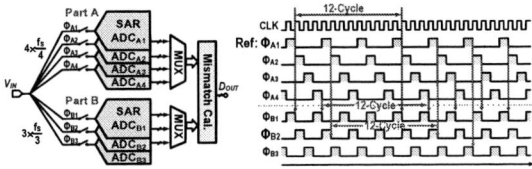

Figure 5. The structure and timing diagram of a 3/4-channel Split Time-interleaved ADC.

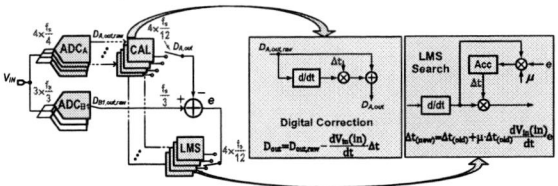

Figure 6. The timing mismatch calibration of a 3/4-channel Split Time-interleaved ADC [6].

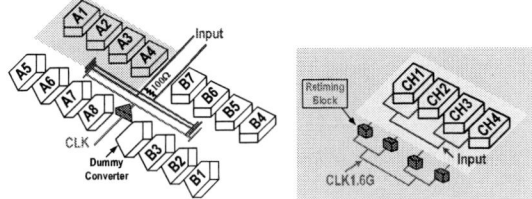

Figure 7. The circuit implementation of a 7/8-channel Split Time-interleaved SAR ADC [6].

the original ADC into two almost identical parts, A and B. They are independent of each other but have exactly the same ideal signal transfer function. Due to non-ideal factors, there is a deviation between A and B, and we then use the difference between the two functions for calibration. The final output result is to average the output of parts A and B to obtain a 3dB SNR improvement relative to A or B itself. Therefore, the split structure is also a special application of the oversampling technique. Following this idea, we proposed a split time-interleaved structure to solve the mismatch between different channels [6].

Fig.4. presents the basic idea of split time-interleaved; now, a TI ADC is split into two parts, A (*M* channels) and B (*N* channels), with the same entire sampling rate but different interleaving channels [6]. *M* and *N* must be mutual prime numbers to keep one of the sub-ADC$_{A(or\ B)}$ as the reference of sub-ADC$_{B(or\ A)}$. Then, Fig. 5. shows how to implement and operate a split TI ADC. ADC$_A$ in the A part and ADC$_B$ in the B part sample the same input signal V_{in} at the falling edge of its sampling clock (Φ_{A1}-Φ_{A4}, Φ_{B1}-Φ_{B3}). And ADC$_{A1/A2/A3/A4}$ and ADC$_{B1/B2/B3}$ will meet each other after every 12 clock periods. In this way, we can use the constantly changing method of the difference between the A and B channels to calibrate the channel mismatch.

Fig. 6 shows the details of the timing mismatch calibration in the split time-interleaved ADC, we use a fully digital method to correct the timing skew error in the digital domain. A least-mean-square (LMS) search method is used to update the timing skew Δt by minimizing the e, which is the difference between two parts (A and B).

Benefitting from the proposed split time-interleaved architecture, we obtain a fast convergence calibration in channel mismatch calibration without the addition of power and hardware, and it also avoids the problem of impedance change of the reference-assisted calibration method. This work can be regarded as a trade-off between accuracy and speed; we use a very small OSR (OSR=2) to obtain a fast and simple calibration in time-interleaved ADC. Fig. 7 presents the circuit implementation of the split TI ADC. We chose a 7/8 channel SAR ADC to verify this architecture, and it obtains a 54.2dB

SNDR at Nyquist rate with a 1.6GS/s sampling rate while it consumes 12.2 mW at 28nm CMOS.

III. APPLING TIME-INTERLEAVING IN ΔΣ MODULATOR

Combined with the technique of oversampling and noise shaping, we can obtain a ΔΣ modulator. As shown in Fig. 8(a), the analog input enters the loop filter, then it is digitized by the quantizer, and a DAC to finish the ΔΣ feedback loop. The noise Transfer Function (NTF) brings the noise-shaping feature, making the ΔΣ modulator as good choice for a high-resolution application. When we try to put an N-way ΔΣ modulator in timing interleaving to enhance the bandwidth. We find it will take a problem, as shown in Fig. 8(b), due to the change of topology, the order of NTF from z^{-1} to z^{-N}, and it makes a bandpass feature other than the high pass we need. Therefore, the straight parallel is not a common solution. So how do we apply Time-Interleaving to ΔΣ modulator?

One solution is called extrapolating TI ADC; some papers also call it prediction ADC. Fig. 9 shows a transformation procedure from a single-channel 1st-order ΔΣ to a 2-channel 1st-order case: the traditional TI will build the digital block filter for multi-rate systems. By those cross-coupled paths, we can keep the same NTF. However, the whole scheme is very complex, and those extra paths greatly reduce the feedback factor for integrators. In general, a de-MUX is needed at the input of the modulator, and if we try to equal the inputs X_1 and X_2, the TI input sampler can be eliminated (Fig. 10). It results in a modified signal transfer function equals $1 + z^{-1}$ for 2 channels, but it doesn't affect NTF. Therefore, it will not degrade the signal-to-quantization noise ratio (SQNR). Once we write the time-domain transfer function of two nodes, U_1 and U_2, we can see each node contains the information of X, Y_1, and Y_2. As a result, it is possible to use one channel to get another channel. Here, we make U_2 unchanged and try to use X, U_2, and Y_2 to estimate U_1 and Y_1.

As shown in Fig. 10(b), the summation will be done in front of the quantizer by an analog adder, while the subtraction is

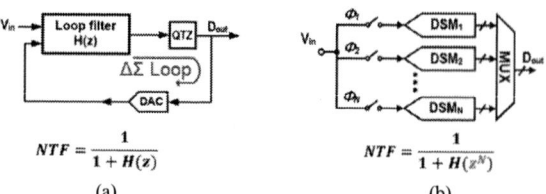

$$NTF = \frac{1}{1 + H(z)}$$

(a)

$$NTF = \frac{1}{1 + H(z^N)}$$

(b)

Figure 8. From (a) single-channel to (b) Direct N-channel time-interleaved in a ΔΣ modulator.

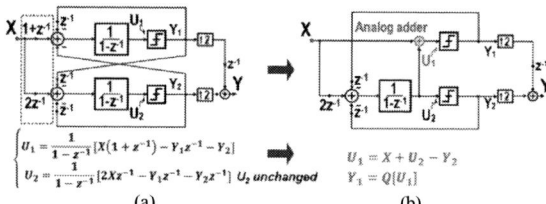

Figure 9. A digital block filter 2-channel time-interleaved in a ΔΣ modulator.

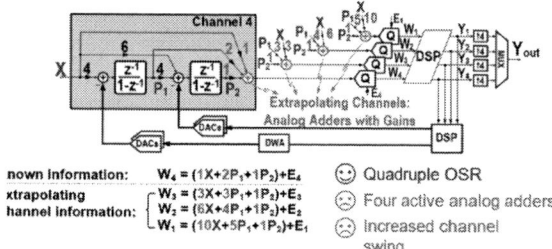

$$U_1 = \frac{1}{1 - z^{-1}}[X(1 + z^{-1}) - Y_1 z^{-1} - Y_2]$$

$$U_2 = \frac{1}{1 - z^{-1}}[2Xz^{-1} - Y_1 z^{-1} - Y_2 z^{-1}] \quad U_2 \text{ unchanged}$$

$$U_1 = X + U_2 - Y_2$$
$$Y_1 = Q[U_1]$$

(a)   (b)

Figure 10. From (a) input extrapolation to (b) output extrapolation in a 2-channel time-interleaved in a ΔΣ modulator.

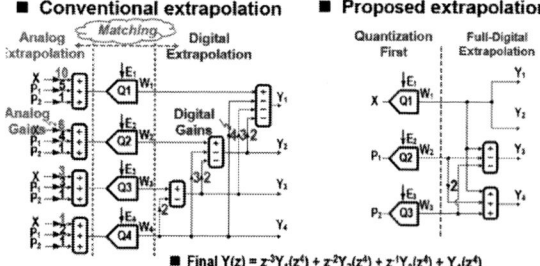

nown information: $W_4 = (1X + 2P_1 + 1P_2) + E_4$ ☺ Quadruple OSR

xtrapolating $\begin{cases} W_3 = (3X + 3P_1 + 1P_2) + E_3 \\ W_2 = (6X + 4P_1 + 1P_2) + E_2 \\ W_1 = (10X + 5P_1 + 1P_2) + E_1 \end{cases}$ ☹ Four active analog adders

hannel information: ☹ Increased channel swing

Figure 11. A possible solution to extension to 2nd-order, 4×TI.

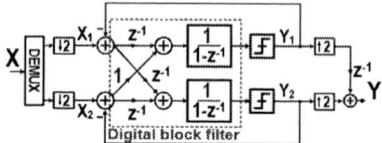

■ Final $Y(z) = z^{-3}Y_1(z^4) + z^{-2}Y_2(z^4) + z^{-1}Y_3(z^4) + Y_4(z^4)$

Figure 12. Comparison to conventional extrapolation and proposed extrapolation [7].

done in the digital domain. Finally, with the one analog channel Y_2, we successfully extrapolate Y_1, and build the overall 2-channels TI ΔΣ modulator.

Fig. 11 shows the transformation from a conventional 2nd order Feedforward ΔΣ modulator to a 4-channel TI extrapolating case.

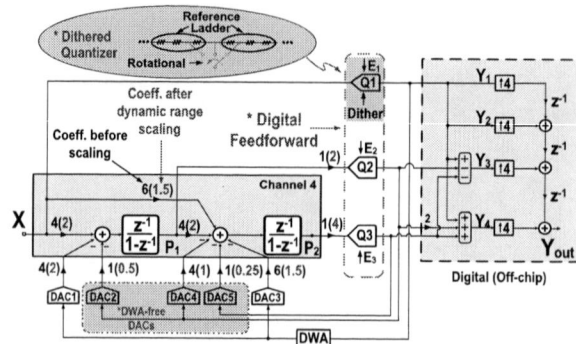

Figure 13. The implementation of the 4-Paths 2nd order TI ΔΣ modulator [7].

with one channel information input X, first integrator output P_1 and second integrator P_2, the fourth channel is unchanged with $W_4 = (X + 2P_1 + P_2)$, the outputs of the remaining 3 channels can be extrapolated by their time-domain equations. All analog channels consist of X, P_1 and P_2, based on different weights. The ΔΣ is a recursive operation, we will find the channel swing will be accumulated, the most serious channel is the outmost channel $W_1 = 10X + 5P_1 + P_2$. For such a 4-paths 2nd-order extrapolating ΔΣ modulator, the pros are the quadruple OSR and single channel realization. However, the disadvantages are also obvious: first, it needs four active analog adders to construct the channel information; second, the increased swing easily saturates the circuit, making the design impractical.

Conventional extrapolation has 3 steps: analog extrapolation to accumulate the channel information, quantizer to get the data, and digital extrapolation to recover the channel output. The final output could be derived using poly-phase decomposition. During the process, we have increased analog and digital gains, the mismatch may produce an inaccurate NTF. This is another problem.

Let's look at our proposed scheme [7]. The basic idea is as follows: since the summation of X, P_1 and P_2 contributes to each channel, we can directly quantize these three pieces of information first, shift the summation to the digital domain, and finally make a full-digital extrapolation (Fig. 12). Thus, we get three benefits: (a) all the extrapolating gains are now in the digital domain, no matching problem; (b) all the analog adders are removed; (c) the input swings before Q1-3 significantly reduce.

Fig. 13 shows the final implementation of the 2nd-order ΔΣ modulator using digital feed-forward extrapolation, it achieves equivalent OSR=208 with a practical OSR=52, from silicon results, we get an 86.1dB SNDR 5 MHz Bandwidth with 23.1mW @fs=520MS/s with 28nm CMOS.

IV. COMBINED WITH NOISE-SHAPING SAR AND TIME-INTERLEAVING

In recent years, noise-shaping has become a popular research direction in high-precision applications because of its excellent energy efficiency ratio. Combining time-interleaved and noise-shaping can achieve larger bandwidth while maintaining a high energy efficiency ratio [8][9]. [8] presents a

979-8-3503-6184-1/24 $31.00 © 2024 IEEE

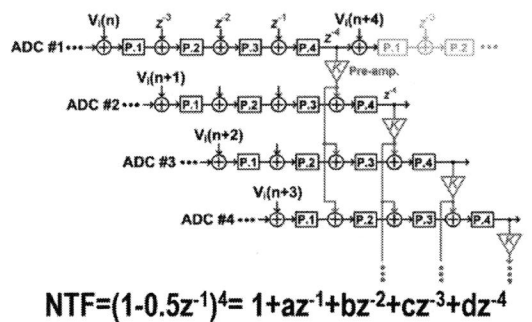

$$NTF=(1-0.5z^{-1})^4= 1+az^{-1}+bz^{-2}+cz^{-3}+dz^{-4}$$

Figure 14. The operation of the 4-way 4-th order TI NS SAR ADC with midway error feedback. [8]

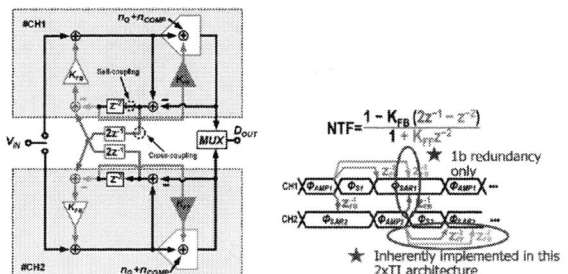

Figure 15. The 2-way TI NS SAR ADC with noise-shaping enhancement [9].

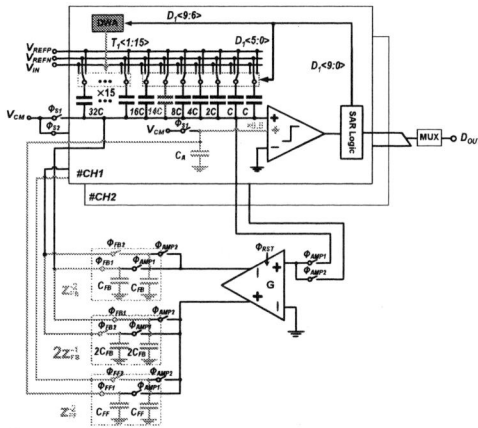

Figure 16. The circuit implementation of the 2-way TI NS SAR ADC with shared residue amplifier based on error feedback and error feedforward [9].

4-channel time-interleaved (TI) noise-shaping (NS) SAR ADC with the midway error feedback to achieve a 4-th order noise shaping using the inherent delay between 4 channels (Fig. 14). However, the 4-phase midway EF leads to an extra redundant bit to avoid overload of the SAR, which degrades the SAR's conversion efficiency. And the NTF must be mild to avoid instability.

To overall the limitation of previous TI NS, we proposed a two-way TI NS SAR [9]. As shown in Fig. 15, due to the inherent delay between 2 channels, the residue from one channel cross-coupling to the other channel generates z^{-1}.

Because the amplified residue is unavailable at the beginning of the SAR conversion, the midway error feedback is still necessary. The residue self-coupling to its own channel in the next phase of SAR conversion generates z^{-2}, the midway error feedback is adopted to avoid a high swing of signal on the CDAC array during MSB of SAR conversion. Moreover, we proposed a noise-shaping enhancement, a 2nd-order error feedforward with z^{-2} forms transfer function in the denominator of the Noise-Transfer-Function (NTF), which makes the NS more aggressive.

Fig. 16 presents the implementation of this TI NS SAR ADC. It uses a 9-bit asynchronous split monotonic SAR with bottom-plate sampling, and a partial DWA used to cancel the impaction of capacitor mismatch. It achieves a SNDR 73.5dB and DR 74.7dB @ fs=330MS/s, OSR=5.5, BW=30MHz while it consumes 3.07mW at 28nm CMOS.

V. CONCLUSION

This paper reviews published articles to explore three possibilities of combining Time-Interleaved and Oversampling techniques: 1. Used a special oversampling – Split architecture into a Nyquist TI ADC to address channel mismatch issues. 2. Used a time-interleaved technique into a DSM, to enhance the equivalent OSR, thus improving the accuracy of ADC. 3. Optimize the TI NS SAR architecture, proposed an NS enhancement technique by 2nd-order error feedforward to achieve less NTF peaking and DR loss.

REFERENCES

[1] B. Razavi, "Design Considerations for Interleaved ADCs," *IEEE Journal of Solid-State Circuits*, vol. 48, no. 8, pp. 1806-1817, Aug. 2013.

[2] M. Guo, S. -W. Sin, L. Qi, D. Xu, G. Wang and R. P. Martins, "Background Timing Mismatch Calibration Techniques in High-Speed Time-Interleaved ADCs: A Tutorial Review," *IEEE Transactions on Circuits and Systems II: Express Briefs*, vol. 69, no. 6, pp. 2564-2569, June 2022.

[3] F. Maloberti, Data Converters. Dordrecht, The Netherlands: SpringerVerlag, 2007.

[4] J. McNeill, M. C. W. Coln and B. J. Larivee, ""Split ADC" architecture for deterministic digital background calibration of a 16-bit 1-MS/s ADC," *IEEE Journal of Solid-State Circuits*, vol. 40, no. 12, pp. 2437-2445, Dec. 2005.

[5] J. Mao, M. Guo, S. -W. Sin and R. P. Martins, "A 14-Bit Split-Pipeline ADC With Self-Adjusted Opamp-Sharing Duty-Cycle and Bias Current," *IEEE Transactions on Circuits and Systems II: Express Briefs*, vol. 65, no. 10, pp. 1380-1384, Oct. 2018.

[6] M. Guo, *et al*, "A 1.6-GS/s 12.2-mW Seven-/Eight-Way Split Time-Interleaved SAR ADC Achieving 54.2-dB SNDR With Digital Background Timing Mismatch Calibration," *IEEE Journal of Solid-State Circuits*, vol. 55, no. 3, pp. 693-705, March 2020.

[7] D. Jiang, L. Qi, S. -W. Sin, F. Maloberti and R. P. Martins, "A Time-Interleaved 2nd-Order ΔΣ Modulator Achieving 5-MHz Bandwidth and 86.1-dB SNDR Using Digital Feed-Forward Extrapolation," *IEEE Journal of Solid-State Circuits*, vol. 56, no. 8, pp. 2375-2387, Aug. 2021.

[8] L. Jie, B. Zheng, and M. P. Flynn, "A Calibration-Free Time-Interleaved Fourth-Order Noise-Shaping SAR ADC," *IEEE Journal of Solid-State Circuits*, vol. 54, no. 12, pp. 3386-3395, Dec. 2019.

[9] S. Zhao et al., "A 3.07 mW 30 MHz-BW 73.2 dB-SNDR Time-Interleaved Noise-Shaping SAR ADC With Self-Coupling Second-Order Error-Feedforward," *IEEE Journal of Solid-State Circuits*, vol. 58, no. 10, pp. 2722-2732, Oct. 2023.

A 0.000355mm² 4.6μm-Pitch 5.75fJ/Conv 6-bit SAR ADC for High Throughput Parallel Readout of Analog SRAM Computing-In-Memory

Lin Wu[1], Lichen Feng*[1], Hongwei Shan[1], Zhangming Zhu[1]

[1]Key Laboratory of Analog Integrated Circuits and Systems (Xidian University), Ministry of Education, School of Integrated Circuits, Xidian University, Xi'an 710071, P. R. China

* Email: lwu_2@stu.xidian.edu.cn, lcfeng@xidian.edu.cn

Abstract—Numerous SRAM-based analog Computing-In-Memory (CIM) macros have been verified in silicon to show great energy efficiency improvement. However, the large area of the existing analog-digital-converters (ADCs), especially the wide pitches between the ADCs, limits the number of computing results that can be quantized and readout in parallel. In this paper, we propose a 4.6μm-pitch 5.75fJ/conv 6-bit SAR ADC in 65nm CMOS process. The area and power consumption of the complete ADC are significantly decreased by employing a pre-comparator and the unit-length DAC technology with Vcm-based switching scheme. The proposed ADC occupies only 0.000355mm² with the width of 4.6μm, which can be aligned with four bitcell columns, showing superior potential for improving both throughput and energy efficiency of analog CIM macros.

Keywords—Computing-In-Memory, SAR ADC, Unit-length DAC, Pitch

I. INTRODUCTION

Computing-In-Memory (CIM) has garnered significant attention from both academia and industry in recent years, particularly SRAM-based ones. Analog SRAM-based CIM macro typically is comprised of an SRAM array with computing units, and analog-digital-converters (ADCs) for quantizing the computing results [1]. Conventional SAR ADCs have large areas due to the large capacitor arrays, which fail to align with each bitcell column of the SRAM. As a result, a layout involving multiple bitcell columns sharing a single ADC is commonly employed in practice, significantly impeding the throughput.

The 8-bit SAR ADC proposed in [1] utilizes a customized sandwich-like unit capacitor to achieve an area of 0.0025mm². In [3], the 13μm-pitch is obtained by precisely tuning the layout of the SAR/SS ADC with partial bit quantization technique, which sacrifices the accuracy. The same technique of using SS ADCs to reduce area in [4] achieves a 5μm pitch, but this leads to the increased length of 190μm and the decreased sampling rate of 20MS/s. An exploration into the layout of differential thin-shaped capacitors is proposed in [5], which leverages the differential input to achieve a unit capacitance value of 125aF. However, the pitch of this ADC is still 36μm, and the switching logic remains to be improved.

In this paper, a 65nm 0.000355mm² 4.6μm-pitch 6-bit SAR ADC is proposed, which can be aligned with as few as 4 SRAM bitcell columns. By employing a pre-comparator and the unit-length DAC technology with Vcm-based switching scheme, the Walden FoM of the proposed ADC is as low as 5.75fJ/conv at the conversion speed of 50MS/s.

This work is supported in part by the National Key R&D Program of China (2022ZD0118903), in part by the National Natural Science Foundation of China under Grant U22A2013, Grant 62104175.

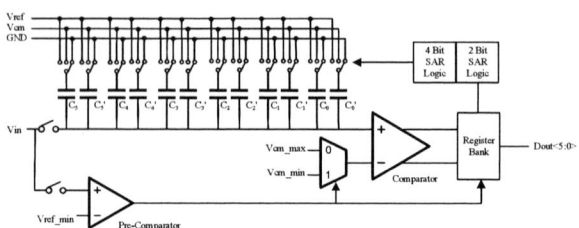

Fig. 1. The architecture of the proposed SAR ADC.

The remainder of this paper is organized as follows. Section II presents the overall architecture of the proposed SAR ADC with a pre-comparison stage. Section III describes the particular implementation and circuit configuration of the key blocks. Section IV shows the layout and simulation results of the proposed SAR ADC. Finally, the summary and conclusions are provided in Section V.

II. PROPOSED SAR ADC ARCHITECTURE

Figure 1 depicts the main building blocks of the proposed SAR ADC: a pre-comparator, a main-comparator, a register bank, a SAR logic stack, and a differential capacitor array for digital-analog-conversion (DAC). The main-comparator is based on a two-stage dynamic latch. The SAR ADC is single-end sampling due to that the SRAM array with computing units passes computing results in single-end form.

The pre-comparator is the first device the input signal flows through, which detects whether the signal amplitude is larger than the predefined threshold voltage. For the signal lower than the threshold, the ADC merely activates the least significant 4 bits for quantization; the signal larger than the threshold activates the 6 bits of quantization. Correspondingly, the negative input of the main-comparator is selected differently for high-level and low-level input signals. The extra pre-comparison causes the comparator to compare once more for high-level signal and once less for low-level signal as compared to the conventional 6-bit ADC. It appears that this is not advantageous in terms of area, speed, and power consumption. However, in practical applications of neural networks, where the inputs to the ADC tend to be concentrated in the range of less than 1/2 or even 1/4 of the full swing range, the effect of the pre-comparator in reducing power consumption is remarkable.

III. SYSTEM AND CIRCUIT DETAILS

A. Pre-comparator

As shown in Fig. 2, the pre-comparator is a customized inverter, whose transfer characteristic curve is finely tuned by

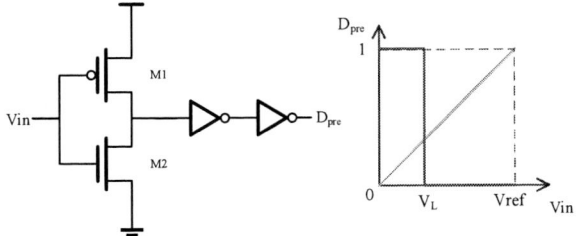

Fig. 2. The pre-comparator and its transfer characteristic curve.

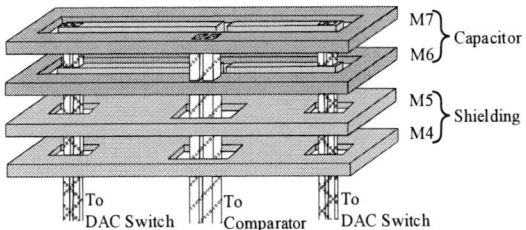

Fig. 4. The 3D layout of a DAC unit capacitor pair.

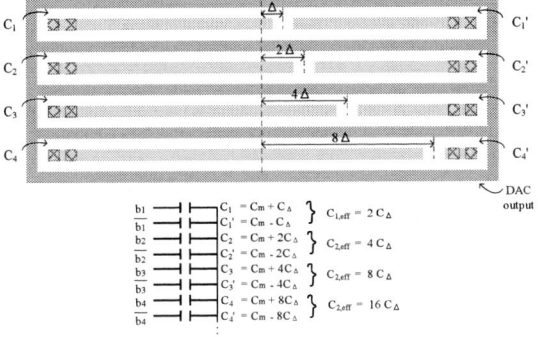

Fig. 3. The differential capacitor array.

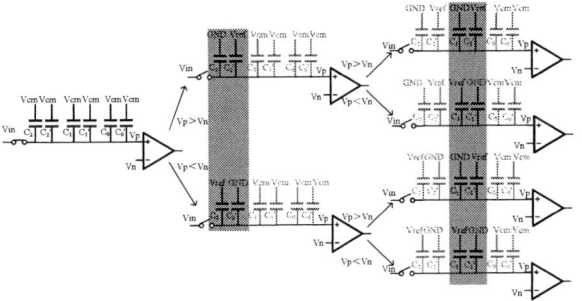

Fig. 5. The Vcm-based switching scheme for 3-bit DAC.

finely regulating the width-to-length ratio of NMOS and PMOS.

In order to leave enough redundant bits for the 6-bit quantization to further guarantee the accuracy, the threshold voltage of the inverter is controlled to be approximately 220mV (corresponding to 13 of the 6-bit digital code), which is slightly smaller than the full-scale quantization value of 15 of the 4-bit ADC. The 2-LSB redundant bits (15-13) are set to prevent the pre-comparator's mis-comparisons from causing input signals that should go through 6 bit logic to incorrectly select 4 bit logic. This means that even if there is a 2-LSB offset in the pre-comparator, the input signal is guaranteed to be correctly quantized, which ensures the accuracy.

B. Unit-length DAC Layout Technique

As shown in Fig. 3, the unit-length DAC [5] with differential strip capacitors has an elongated metal layer with a binary progression of the length difference between two adjacent metal strips. In the case of a differential voltage input to the two metal strips, this length difference can correspond to a capacitance difference, which is of very low capacitance per unit and ultra-small area. For a DAC of unit length, n bits are quantized for n strip capacitors, which achieves significant area reduction compared to the 2^n unit capacitors of conventional DACs.

The total capacitance is considered as the sum of all capacitances, while the effective capacitance that generates the power dissipation is the sum of the difference between a set of capacitances, which is the key for this technique to avoid parasitic capacitance disturbances while guaranteeing small unit capacitance values. C and C' are characterized as larger and smaller capacitances, respectively, with a small value binarized difference between them.

As shown in Fig. 4, the proposed SAR ADC is designed in 65nm CMOS technology with two metal layers, M6 and

M7, used to form the two plates that make up the capacitors, and the M4 and M5 layers connected to GND to mask the effects of the underlying metal and the active region. Tthe other modules (comparator, logic gates, registers, etc.) use M1-3 to complete the routing. Due to the large spacing between the two plates, it is not difficult to predict that this capacitor is reasonably well-matched. Accuracy is ensured by the differential nature of this capacitor layout.

There are still some problems with this technology, however, and its flaw lies in the loss of output swing. It can be deduced that for a certain Vref, the output swing S_{out} of an N-bit DAC degrades as follows.

$$S_{out} = \frac{2^{N-2}C_\Delta}{NC_m}\text{Vref} \tag{1}$$

where S_{out} represents the output swing, C_Δ (0.5fF) and C_m (4.6fF) are the capacitance difference and common mode capacitance, respectively, which lead to a 30% degradation of the output swing. This degradation can be compensated for by equally scaling the Vcm and Vref.

C. Vcm-based Switching Scheme for Unit-length DAC

To further improve the energy efficiency, the Vcm-based switching scheme for unit-length DAC is proposed. An example of 3-bit DAC is shown in Fig. 5, and described in detail as follows.

During the sampling stage, the bottom plates of all capacitors are connected to Vcm. For the output of the first comparison cycle, if Vp > Vn, the bottom plate of C_{n-1} (the larger) of the first largest differential strip capacitor pairs switches to GND and C_{n-1}' (the smaller) to Vref; if Vp < Vn, the switching is reversed. And then, the second comparison cycle begins and the result directs the voltage switching of bottom plates of the second largest capacitor pair. The switching process is repeated until the completion of the n-th comparison period for the n-bit DAC.

979-8-3503-6184-1/24 $31.00 © 2024 IEEE   157

Fig. 6. The layout of 32-column SAR ADCs and the DAC

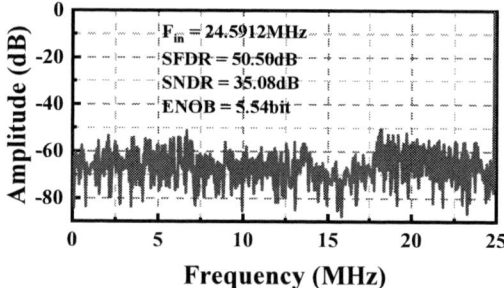

Fig. 7. The layout of 32-column SAR ADCs and the DAC

The effective unit capacitance unit-length DAC cannot be designed too small because of the low bit count of the SAR ADC proposed in this paper; otherwise, the parasitic capacitance effect would be too large to ensure robustness of accuracy.

IV. LAYOUT AND EXPERIMENTAL RESULTS

A. Layout

The layout of 32 proposed ADC routed in parallel columns is shown in Fig. 6, where the capacitors used for a single ADC occupy only 4.12μm in width. The area of a single SAR ADC is $4.635 \times 76.58 \mu m^2$. The small-to-limit width allows each ADC to be aligned with four SRAM bitcell columns, which means that a CIM macro with 32 columns of a 128-column SRAM array can be quantized and readout simultaneously, leading to a significant throughput improvement.

B. Simulation Results

As shown in Fig. 7, the measured SFDR, SNDR and ENOB are 50.50dB, 35.08dB and 5.54bit with a Nyquist-rate input, respectively.

For a full-swing input signal, the total power consumption is 13.37μW, only <1.5% of which is caused by the Vref and DAC block and 2.2 μW of which comes from the comparators. Predictably, in specific CIM applications, the overall power consumption is expected to be reduced by another 20% or so, as the input signal is mostly less than 1/4 of a range. In this case, almost all of the input signals will trigger only 4-bit quantization, reducing comparator and register bank power consumption by nearly 25%.

TABLE I.   PERFORMANCE SUMMARY AND COMPARISON

	ICICM'23 [2]	AICAS'23 [3]	ISCAS'22 [4]	This work
Tech(nm)	110	55	65	65
Pitch(μm)	49.6	13	5	**4.6**
Area(μm²)	2550	4290	950	**355**
Supply(V)	1.2	1.2	1.2	1.0
Conver. speed	70 MS/s	350 MS/s	20 MS/s	50 MS/s
ENOB(bit)	7.4@8bit	6.51@8bit	7.87@8bit	5.54@6bit
FoM (fJ/conv)	47.26	44.8	10.2	**5.75**

Table I shows a performance summary and comparison against other low-area and power-efficient ADCs for CIM [1]-[4], which demonstrates quite significant advantages of this work in terms of area and energy efficiency. This work achieves a 10.7× and 2.8× reduction in pitch compared to [2] and [3], and 8.2×, 7.8× and 1.8× reduction in the Walden FoM against [2], [3] and [4]. Besides, in terms of chip size, 7.2×, 12.1×, and 2.7× reduction compared to [2], [3] and [4] are demonstrated. In this work, area-efficient capacitor arrays couple with customized mapping layout schemes to realize circuits with an ultra-small pitch for front-end SRAM arrays. The high energy efficiency also benefits from the setting of the pre-comparator and tightly controlled transistor sizes.

V. CONCLUSION

SRAM-based analog CIM macros have demonstrated significant energy efficiency increasement in silicon. The throughput of which is limited by the huge size, particularly the wide pitches of the ADCs. In this paper, a power-efficient and ultra-small-area 6-bit SAR ADC is proposed with the Walden FoM of 5.75fJ/conv and an area of 0.000355mm² by using by using a pre-comparator and the unit-length DAC technique. A single ADC proposed can be aligned with a 4-column SRAM array with the 4.6μm pitch, demonstrating its great ability to improve the throughput and energy efficiency for analog CIM macros.

REFERENCES

[1] C. -J. Jhang, C. -X. Xue, J. -M. Hung, F. -C. Chang and M. -F. Chang, "Challenges and Trends of SRAM-Based Computing-In-Memory for AI Edge Devices," IEEE Transactions on Circuits and Systems I: Regular Papers, vol. 68, no. 5, pp. 1773-1786, May 2021.

[2] X. Xu, Y. Shui and A. Wang, "A 0.0025mm2 8-bit 70MS/s SAR ADC with a Linearity-Improved Bootstrapped Switch for Computation in Memory," 2023 8th International Conference on Integrated Circuits and Microsystems (ICICM), Nanjing, China, 2023, pp. 412-416.

[3] Y. Li, L. Du and Y. Du, "A Column-Parallel Time-Interleaved SAR/SS ADC for Computing in Memory with 2-8bit Reconfigurable Resolution," 2023 IEEE 5th International Conference on Artificial Intelligence Circuits and Systems (AICAS), Hangzhou, China, 2023, pp. 1-5.

[4] W. Fan, Y. Li, L. Du, L. Li and Y. Du, "A 3-8bit Reconfigurable Hybrid ADC Architecture with Successive-approximation and Single-slope Stages for Computing in Memory," 2022 IEEE International Symposium on Circuits and Systems (ISCAS), Austin, TX, USA, 2022, pp. 3393-3397.

[5] P. Harpe, "A Compact 10-b SAR ADC With Unit-Length Capacitors and a Passive FIR Filter," IEEE Journal of Solid-State Circuits, vol. 54, no. 3, pp. 636-645, March 2019.

A 250MS/s, 12 Bit Pipeline-SAR ADC Using Coarse-Fine Ring Amplifier

Linghao Liu, Junyan Ren and Fan Ye*

State Key Laboratory of ASIC and System
Department of Microelectronics, Fudan University, Shanghai, China

* Email: fanye@fudan.edu.cn

Abstract—The ring amplifier is an energy effective choice for high performance pipeline ADC. However, sensitivity to process, supply voltage and temperature (PVT) variations and complex stability mechanism make it less practical. This brief introduces a coarse-fine ring amplifier (CFRingAmp) that allows individual designs of large-signal and small-signal processes. With a large and a smaller dead-zones switched by the self-detect-and-switch block, both stability and accuracy are well designed over corners. Furthermore, a 250MS/s, 12-bit pipeline-SAR ADC is implemented adopting the proposed CFRingAmp. Simulated in 28nm CMOS, the ADC achieves an SNDR of 64.1 dB and an SFDR of 74.9 dB over corners at Nyquist frequency while consuming 4.5mW, resulting a Walden figure-of merit (FoM) of 12.7 fJ/conversion step.

Keywords—Analog-to-digital converter (ADC), closed loop, pipeline successive approximation register (SAR), interstage amplifier, coarse-fine ring amplifier (CFRingAmp).

I. INTRODUCTION

To achieve high performance pipeline ADCs, one of the key bottlenecks is the performance of residue amplifier (RA), which is done by operational transconductance amplifier (OTA) in traditional design. However, it has been more difficult to design a cost-effective OTA with lower supply voltage and intrinsic gain in scaled CMOS technologies. To meet the growing demand, open-loop amplifier is becoming more popular for its simplicity and low power consumption, such as inverter-based RA and gm-R based RA. While circumventing the drawbacks of closed-loop system, these architectures suffer from limited linearity, and thus rely on calibration or compensation techniques.

Another attractive alternative is the ring amplifier [1], [2], [3]. By adopting oscillator-like typology, ring amplifier can reach high gain through three-stage amplification even with low intrinsic gain. By inserting a dead-zone to bias the third stage at subthreshold area or even shut it down, the dominant pole is generated at the output with little quiescent current. Nevertheless, the stability mechanism dependent on dead-zone makes it sensitive to process, supply voltage and temperature (PVT) variations.

This work introduces a coarse-fine ring amplifier (CFRingAmp) that divides the amplification procedure into two parts by adopting a self-detect-and-switch block to change the dead-zone size. This division allows individual consideration of large signal effect and final settling process [3], which helps choose a more coordinated operation point while ensuring both accuracy and stability over corners. Furthermore, a 12-bit, 250MS/s pipeline-SAR ADC is implemented employing the proposed interstage amplifier.

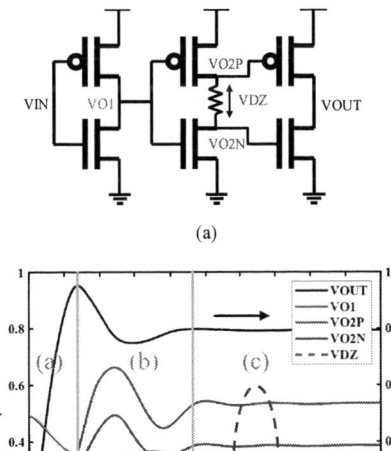

(a)

(b)

Fig. 1. (a)Basic structure and (b) settling mechanism of ring amplifier.

II. PROPOSED COARSE-FINE RING AMPLIFIER

A. Analysis of Stability and accuracy of Ring Amplifier

The settling mechanism of ring amplifier can be divided into three stages: (a) initial ramping, (b) oscillation and (c) stable state [1] as shown in Fig. 1. (b). Practical design usually demands the number of oscillations to be smaller than two, and this large-signal stability can be tuned in accordance to the expression [1]:

$$\frac{t_d \cdot SlewRate}{\psi} \leq \frac{1}{2A_1}\left(\frac{V_{DD}-V_{SS}}{A_2}-2V_{DZ}\right) \tag{1}$$

where t_d is the time delay from V_{IN} to V_{O2} and ψ is the scaling factor from V_{OUT} to V_{IN}. Notice that A_1 and A_2 are negative values.

A large dead-zone can force the output stage to cut off and thus avoids unexpected oscillations. However, the final settled V_{IN} value is limited to $V_{DZ}/(A_1A_2)$ in this case. Another working point applies only weak-zone to ensure weak-inversion at stable state [4]. In this situation, the ringamp works as traditional closed-loop amplifier with its error decided by open-loop gain and bandwidth, which is more favorable for high accuracy design.

979-8-3503-6184-1/24 $31.00 © 2024 IEEE

Coarse-fine ring amplifier

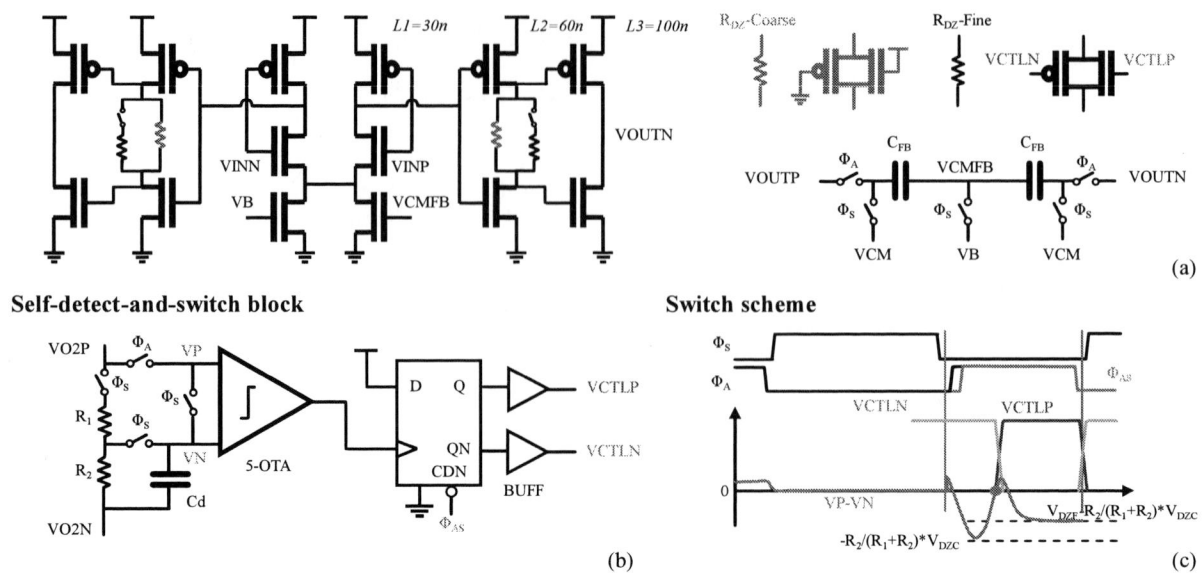

Fig. 2.   Structure and operation of proposed coarse-fine ring amplifier. (a) Circuit of CFRingAmp. (b) Self-detect-and-switch block. (c) Switch scheme of self-detect-and-switch block.

B. Structure and operation of the CFRingAmp

Fig. 2 presents the circuit and operation scheme of the proposed coarse-fine ring amplifier. The fully differential input-stage offers better common-mode rejection and harmonic suppression. The dead-zone is generated by self-bias resistors composed of R_{DZC} and R_{DZF}.

The switch action between the coarse-fine amplification mode is done by the self-detect-and-switch block as shown in Fig. 2. (b). During Φ_S phase, a storage capacitor Cd keeps a portion of dead-zone, $R_2/(R_1+R_2)*V_{DZ}$. During Φ_A phase, VP and VN are connected to VO2P and the top plate of Cd, respectively, where the value kept at the inputs of comparator before switching is given by:

$$V_P(t)\text{-}V_N(t)=V_{DZ}(t)\text{-}\frac{R_2}{R_1+R_2}V_{DZC} \tag{2}$$

where V_{DZC} is the dead-zone value generated by R_{DZC}. As Fig. 1. (b) shows, outputs of the 2nd stage of traditional ringamp are close to each other at rail level during initial ramping and finally go back to a preset dead-zone size, thus the value of $V_{DZ}(t)$ first decreases and then increases. After the level-shift of $V_{DZ}(t)$ explained in (2), there is a rising edge of the inputs of the comparator passing through zero point as the red curve shown in Fig. 2. (c). Then the output of the comparator triggers the following DF and connects R_{DZF} to circuit parallel with R_{DZC}, which forms a smaller dead-zone and completes the switch action between coarse and fine amplification mode.

C. Key Advantages

In high performance design, a critical problem is that the dead-zone should be chosen to ensure large-signal stability over PVT variations, which degenerates the bandwidth and causes over damping in fast corner. For the proposed CFRingAmp, these two factors can be designed separately by employing different dead-zone during coarse and fine phases, breaking the trade-off between stability and speed.

TABLE I.   STABILITY AND BANDWIDTH SIMULATION RESULT.

		tt	ff	ss	fs	sf
Coarse	V_{DZC} (mV)	295	354	249	307	273
	PM (deg)	87.9	89.9	84.6	89.0	86.5
	GBW (Hz)	205M	131M	238M	183M	257M
Fine	V_{DZF} (mV)	106.5	117.5	95.7	106.5	93.9
	PM (deg)	71.0	81.5	54.5	74.9	64.4
	GBW (Hz)	1.35G	1.05G	1.43G	1.26G	1.55G
C-F	SFDR (dB)	69.6	59.0	69.1	68.6	68.1
	A_V (dB)	76.3	68.9	81.7	75.2	76.9

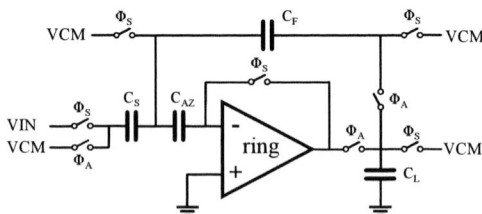

Fig. 3.   Switch capacitor circuit in single-ended form.

Besides, the CFRingAmp saves power of the 2nd stage. Generally, the gain is mainly provided by the last stage with minimal length applied to the first two stages to ensure convergence from oscillations. For CFRingAmp, it is allowed to increase the gain of the 2nd stage and apply a larger V_{DZC} to maintain the large-signal stability while the final settling accuracy is decided by fine phase only. In this case, the secondary pole is more dependent on intrinsic parasitism and allows smaller current of the 2nd stage.

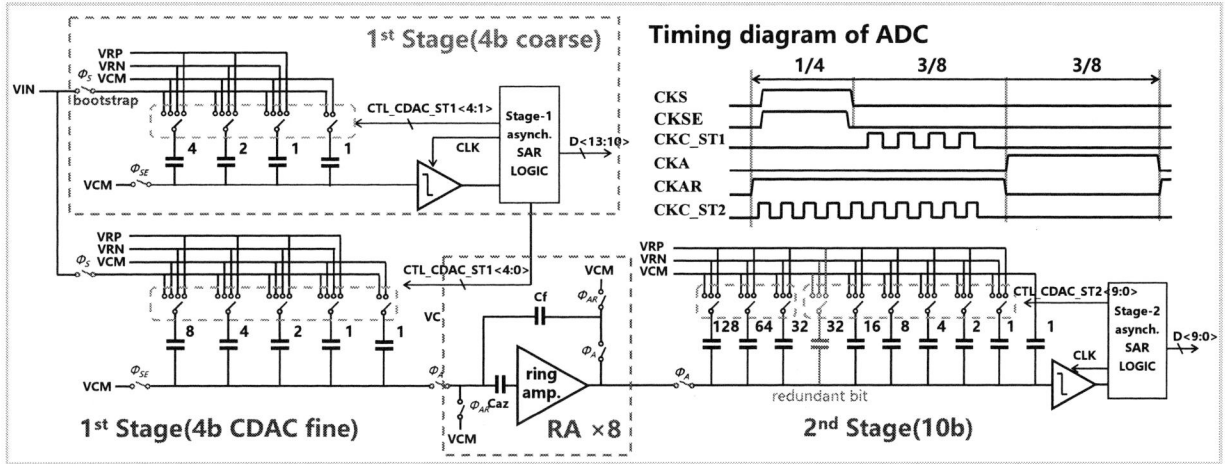

Fig. 4. Architecture and timing diagram of the ADC.

D. Simulation Result

The switched capacitor (SC) circuit in single-ended form is shown in Fig. 3 with a closed-loop gain of 8, where C_S=512fF, C_F=64fF, C_L=128fF and a large C_{AZ}=1pF is chosen to ensure stability during auto-zero phase. The lengths of the three stages are L_1=30n, L_2=60n, L_3=100n, respectively, where allocates a 43.1dB gain to first two stages and achieves a total dc gain of 76.3dB. As described in Table. 1, a large V_{DZC} around 300mV is chosen to ensure large-signal stability over different corners and a small V_{DZF} around 100mV prominently enhances the bandwidth by 7 times on the premise of maintaining reasonable phase margin. The power consumption is 2.64mW with 1.84mW allocated to the first stage for an input equivalent noise of 169µV. It is worth noting that only 201µW is consumed by the 2nd stage thanks to the superiorities of CFRingAmp analyzed before.

To simulate the dynamic characteristics, a 119.14MHz sine wave is exerted to the SC circuit and generate full-range output with 250MHz clock and 1.5ns amplification. Notice that no noise was added during this simulation. As Table.1 shows, the spurious free dynamic range (SFDR) are larger than 59.0dB over the five corners, which is equivalent to 9.5-bit linearity for subsequent pipeline stages.

III. ADC ARCHITECTURE AND IMPLEMENTATION

Fig. 4 presents the architecture of the 12-bit, 250MS/s pipeline-SAR ADC composed of a 4-bit coarse-fine SAR with eight-fold scaling as the 1st stage, and a 128fF, 10-bit SAR 2nd stage with 1-bit redundancy. The blue color switches are bootstraps performing bottom-plate sampling and top-plate sampling for the 1st stage and 2nd stage, respectively. The aperture error between coarse and fine ADCs and CDAC building error are accommodated by an interstage redundant bit.

The ADC is simulated in 28nm CMOS and 0.9V supply voltage. The SNDR and SFDR achieved at Nyquist frequency are larger than 64.1dB and 74.9dB over corners, with 64.8dB and 76.3dB at tt corner, respectively. The power consumption is 4.5mW, where the CFRingAmp consumes 2.4mW and the clock block consumes 1.4mW. Table 2. summaries the performance and compares this design with prior arts having similar specification.

TABLE II.   PERFORMANCE AND COMPARISON.

	This work[1]	ISCAS−21 Chen[1]	ISSCC−14 [3][2]	ISSCC−22 Zhan[2]
Resolution[b]	12	10	10.5	13
ADC Architecture	Pipe-SAR	SAR	1.5b/stage	Pipe-SAR
Residue Amp	CFRing	\	Ring	Ring
Sampling Rate[MS/S]	250	500	100	200
Process	28nm	40nm	65nm	28nm
Supply Voltage[V]	0.9	1.1	1.2	0.9
SNDR@Nyq.(dB)	64.8	60.3	56.6	66.7
SFDR@Nyq.(dBc)	76.3	69.0	64.7	87.2
Power[mW]	4.5	4.16	2.46	1.3
FoMw@Nyq.(fJ/c.-s)	12.7	9.87	44.5	3.7

$* FoM_W = Power/(2^{ENOB} \cdot 2 \cdot BW)$.

[1] Result from simulation. [2] Result from measurement.

IV. CONCLUSION

This brief introduces a coarse-fine ring amplifier. With the help of a self-detect-and-switch block, a large dead-zone for stability and a small dead-zone for accuracy are employed successively, which ensures dc gain of 68.9 dB, GBW of 1.05GHz and SFDR of 59.0 dB over corners. Adopting the proposed interstage amplifier, a 250MS/s, 12-bit pipeline-SAR is implemented in 28nm CMOS. The simulation result achieves 64.1dB SNDR and 74.9dB SFDR over corners at Nyquist frequency while consuming 4.5mW, resulting a Walden figure-of-merit of 12.7 fJ/c.-s.

ACKNOWLEDGMENT

This work was supported by the National Natural Science Foundation of China under grant 62074038.

REFERENCES

[1] B. Hershberg, S. Weaver, K. Sobue, S. Takeuchi, K. Hamashita and U. -K. Moon, "Ring Amplifiers for Switched Capacitor Circuits," in IEEE JSSC, vol. 47, no. 12, pp. 2928-2942, Dec. 2012.

[2] Y. Lim and M. P. Flynn, "26.1 A 1mW 71.5dB SNDR 50MS/S 13b fully differential ring-amplifier-based SAR-assisted pipeline ADC," 2015 IEEE ISSCC, San Francisco, CA, USA, 2015, pp. 1-3.

[3] Y. Lim and M. P. Flynn, "11.5 A 100MS/s 10.5b 2.46mW comparator-less pipeline ADC using self-biased ring amplifiers," 2014 IEEE ISSCC, San Francisco, CA, USA, 2014, pp. 202-203.

[4] B. Hershberg and U. -K. Moon, "A 75.9dB-SNDR 2.96mW 29fJ/conv-step ringamp-only pipelined ADC," 2013 Symposium on VLSI Circuits, Kyoto, Japan, 2013, pp. C94-C95.

A 0.71pJ/b 16Gb/s Equalizer with Inverter_based CTLE and 4-Tap Speculative DFE

Huihong Zhang [1], Chuangao Yan [2], Peng Luo [3], Maliang Liu*[1]

[1] Xidian University, Xi'an, 710068, China

* Email: 1824024358@qq.com, mlliu@xidian.edu.cn

Abstract—This paper presents a SerDes receiver equalizer with small area at 16Gb/s, which consists of a two-stage adjustable continuous time linear equalizer (CTLE) and a half-rate speculative decision feedback equalizer (DFE). The CTLE adopts an inverter-based structure to reduce area and maintain low power assumption. The DFE makes use of proposed summers and DFFs to increase feedback timing margin. Compensating channel loss of 23.5dB at 8GHz, the equalizer achieves a jitter of 11ps and a timing margin of 44ps (0.7UI) when BER<10^{-12}. It consumes 11.4mW power, equivalent to 0.7125pJ/b, and occupies an area of 138um*85um designed in 65nm CMOS technology.

Keywords—equalizer, inverter-based, small area, low power consumption

I. INTRODUCTION

With the increasing amount of data, area and power consumption are important consideration points as more chips are applied to a system. Larger area and power consumption mean more cost and heat dissipation. Thus, compensating the channel loss of 23.5dB at 8GHz, this paper aims to realize an equalizer with small area as well as low power consumption.

In order to expand bandwidth, passive inductors or T-coils are usually applied to the conventional CTLE, as shown in Fig. 1. However, this method results in large area overhead. It has been demonstrated that inverter-based CTLE has potential to realize area reduction as well as low power consumption[1]. This work needs to compensate a high channel loss. As a result, an equalizer of a 2-stage inverter-based CTLE and a DFE is proposed. This structure occupies much smaller area as well as low power consumption.

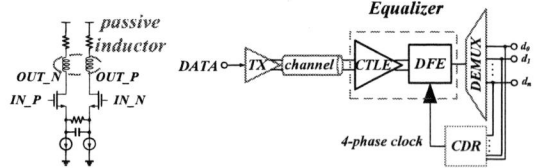

Fig. 1 CTLE with passive inductors   Fig. 2 The equalizer in SerDes

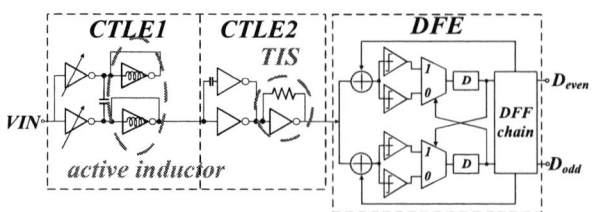

Fig. 3 Diagram of equalizer

II. OVERALL STRUCTURE OF THE EQUALIZER

Fig. 2 and Fig. 3 present the equalizer structure. It consists of a 2-stage inverter-based CTLE and a half-rate speculative DFE with 4 taps. The core of the CTLE measures 54um*51um, and consumes 4.4mW, which shows an advantage compared to conventional CTLE. Meanwhile, the DFE measures 81um*85um, and consumes 7.0mW.

III. DETAILS ABOUT THE PROPOSED EQUALIZER

A. Inverter-based CTLE

[1] mentions that inverter-based structure occupies a smaller area and can also consume low power. In order to compensate a high channel loss at 8GHz, a 2-stage CTLE is proposed. CTLE1 adopts the structure in [1] and it can adjust gain by switchable gm cells. CTLE2 uses a branch with a capacitor and TIS to implement compensation. A bias structure and common mode feedback circuits are used to ensure the high-frequency input and output are biased at the appropriate point, which are not in the figure.

Fig. 4 shows AC response of 2-stage CTLE. The 2-stage CTLE achieves a high-frequency gain of 22.6 dB at 8GHz. Compared to the CTLE in [1], which achieves a high-frequency gain of 6 dB at 28GHz, this one achieves higher gain at the needed frequency. Also, the proposed CTLE measures 54um*51um, much smaller than the CTLE with passive inductors in [3] measuring 800um*900um.

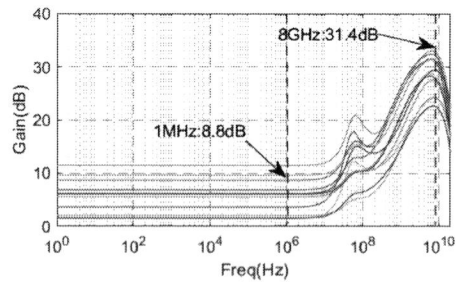

Fig. 4 AC response of 2-stage CTLE(all corners)

B. Speculative DFE

Due to inter-symbol interference, each symbol affects the values of adjacent symbols. The influence of each symbol on subsequent symbols determines the tap coefficients of the DFE:

$$S[n] = D[n] - \alpha_1 D[n-1] - \cdots - \alpha_4 D[n-4] \quad (1)$$

979-8-3503-6184-1/24 $31.00 © 2024 IEEE

The DFE in this structure adopts a half-rate speculative structure for tap1, as shown in Fig. 3. This structure can reduce the data rate of the subsequent stages, making the design less complex and lower power consumption. Also, it increases the timing margin of tap1.

The structure of the summer is shown in the Fig. 5. The summer is implemented using the structure in [4]. This work adds active inductors and RC degradation network to expand bandwidth.

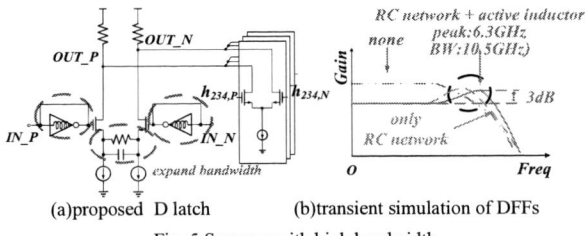

(a)proposed D latch   (b)transient simulation of DFFs

Fig. 5 Summer with high bandwidth

The slicer adopts the structure in Fig. 6(a)[5], which is 50% faster than StrongArm slicer. In order to eliminate disorder between current MOSFETs in the first stage, the differential input structure is used. TH_N is replaced by GND, which can reduce the use of DAC. And it can also realize speculation feature. Meanwhile, in Monte Carlo analysis, the slicer's standard deviation σ of input offset voltage is approximately 20mV. This work scales up the MOSFET whose input is TH_P and calibrate input offset voltage by adjusting TH_P. This means the TH_P MOSFET undertakes two tasks of speculation and calibration. All of these changes are shown in Fig. 6(b).

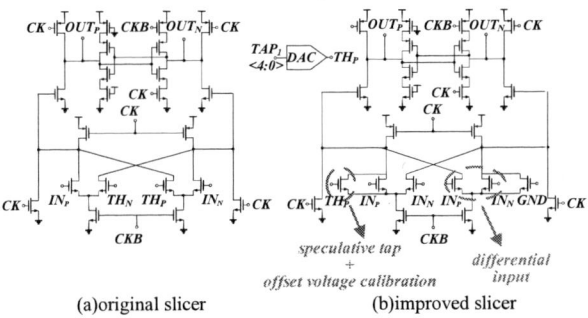

(a)original slicer   (b)improved slicer

Fig. 6 Slicer

This work improves a DFF with the dual ports in [6] to a single-end DFF, as shown in Fig. 7(c), which reduces the signal and clock load. The smaller inverters in Fig. 7 have smaller size. Comparison between different DFFs is shown in Fig. 8 and Fig. 9. The inverter-based DFF is 27% faster than the transmission-gate-based one. Fig. 7(d) shows this one can double driving capability by doubling the size of the second D latch, which can reduce the use of buffers.

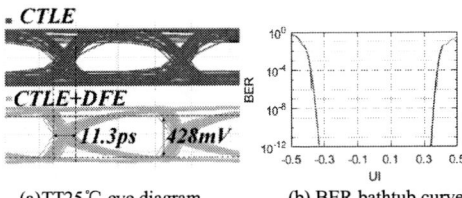

(a)TG-based DFF1   (b)TG-based DFF2

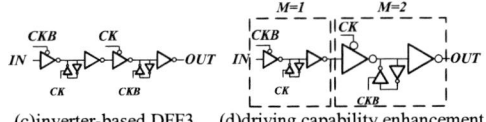

(c)inverter-based DFF3   (d)driving capability enhancement

Fig. 7 DFF structure

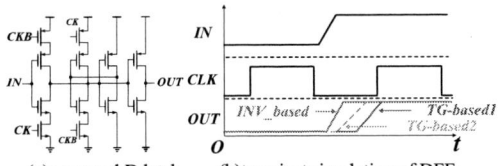

(a)proposed D latch   (b)transient simulation of DFFs

Fig. 8 Details of proposed structure and transient simulation of DFFs

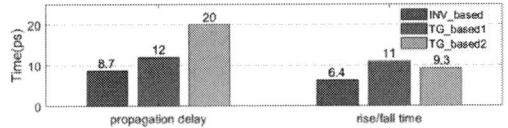

Fig. 9 Comparison between DFFs

IV. SIMULATION AND ANALYSIS

The simulation results show that the designed equalizer has a power consumption of 11.4mW under the TT corner, 1V, and 25°C condition, equivalent to 0.7125pJ/b. Power consumption in other corners is shown in Fig. 10, ranging from 9mW to 15.8mW. The CTLE and DFE consume 4.4mW and 7.0mW respectively. The area of the core is 138um*85um, including the CTLE of 54um*51um and the DFE of 81um*85um.

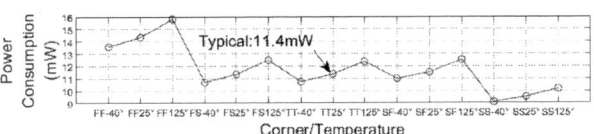

Fig. 10 Power consumption of the equalizer(all corners)

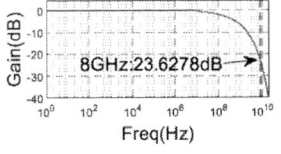

(a) channel S21 curve

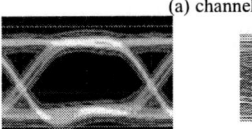

(a) eye diagram from TX   (b) eye diagram after channel

Fig. 11 Signals before and after channel

(a)TT25°C eye diagram   (b) BER bathtub curve

Fig. 12 Eye diagram and BER curve after equalization

979-8-3503-6184-1/24 $31.00 © 2024 IEEE   163

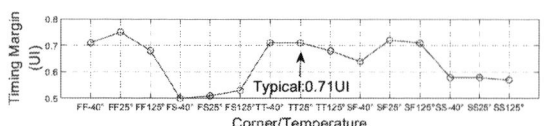

Fig. 13 Timing margin(UI) when BER<10⁻¹² (all corners)

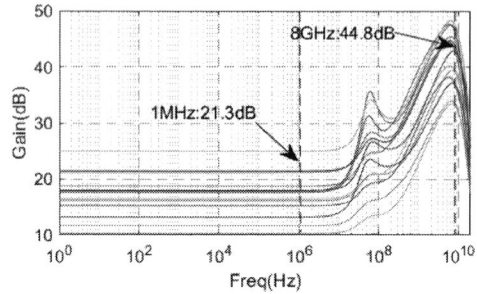

Fig. 14 AC response of the equalizer(all corners)

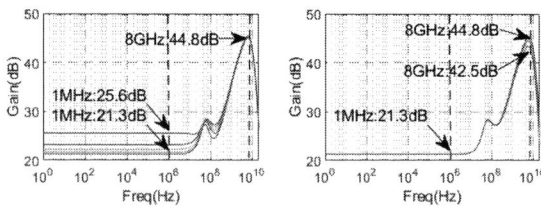

(a)gain adjustment at low frequency   (b)gain adjustment at high frequency

Fig.15 Adjustable AC response (TT, 25°C)

Fig.16 Photo of chip containing the proposed equalizer

Fig. 11 shows the channel has an attenuation of 23.5dB at 8GHz. The eye diagram after the channel is completely closed. After passing through the equalizer, the eye diagram of the signal significantly opens up, as shown in Fig. 12. Under TT corner, 1V, and 25°C condition, the timing margin of the signal is 44ps (0.7UI) when BER<10⁻¹² and the eye height is 428mV. Fig. 13 shows that among all corners, the worst timing margin is 31ps (0.5UI) and the best timing margin is 47ps (0.74UI). The simulation result confirms that this equalizer works well in all corners.

The amplitude-frequency response curve in Fig. 14 shows that the high-frequency gain of CTLE+DFE is approximately 23.5dB higher than the low-frequency gain. The residual ISI is canceled by DFE. Fig. 15 shows the gain can be adjusted slightly from 21.3dB to 25.6dB at low frequency and from 42.5dB to 44.8dB at high frequency.

Compared to the work in [1], which also uses an inverter-based CTLE structure, although the proposed equalizer works at lower frequency, it equalizes much higher channel attenuation. As for other similar process-based equalizers in TABLE I, the area of the proposed one is smaller and the power efficiency is low, too. The chip photo is in Fig. 16.

V. CONCLUSION

This work presents a gain-adjustable equalizer with an area of 138um*85um, which is much smaller than structure with passive inductors as well as low power consumption (0.71pJ/b). It has a high-frequency gain from 16.9dB to 23.5dB at 8GHz. The CTLE is implemented using inverters to reduce area and keep low power consumption. The summer, slicer and DFF in DFE are improved to extend bandwidth, calibrate the input offset voltage and increase timing margin. Simulation results in all corners show good eye width(>0.5UI) and eye height($\approx$400mV), confirming the robustness and potential of the proposed structure.

TABLE I. COMPARISON BETWEEN EQUALIZERS

	This work	[1]	(7)	(3)	(8)
Technology	65nm	16nm	65nm	65nm	65nm
Equalization	Inverter-based CTLE + DFE	Inverter-based CTLE	CTLE + 12tap DFE	CTLE (LFE) + 2-tap DFE	CTLE + 1-Tap DFE
Data rate (Gb/s)	16	56	20	28	16.6
Channel loss (dB) at Nyq.	23.5	8	20	20	19
Eye width (UI) @BER< 10⁻¹²	0.70	>0.24	0.24	n/a	0.73
Power (mW)	11.4	6	15.5	24	24.1
Energy Efficiency(pJ/b)	0.71	0.11	0.72	0.86	1.43
Area (mm2)	0.0118	0.0003	0.34	0.72	n/a
Measurement	Post Layout	Yes	Yes	Post Layout	Yes

REFERENCES

[1] K. Zheng, Y. Frans, K. Chang and B. Murmann, "A 56 Gb/s 6 mW 300 um2 inverter-based CTLE for short-reach PAM2 applications in 16 nm CMOS," 2018 IEEE Custom Integrated Circuits Conference (CICC), San Diego, CA, USA, 2018, pp. 1-4

[2] Q. Pan, Y. Wang and C. P. Yue, "A 42-dB Ω 25-Gb/s CMOS Transimpedance Amplifier With Multiple-Peaking Scheme for Optical Communications," in IEEE Transactions on Circuits and Systems II: Express Briefs, vol. 67, no. 1, pp. 72-76, Jan. 2020

[3] M. Kim, J. Bae, U. Ha and H. -J. Yoo, "A 24-mW 28-Gb/s wireline receiver with low-frequency equalizing CTLE and 2-tap speculative DFE," 2015 IEEE International Symposium on Circuits and Systems (ISCAS), Lisbon, Portugal, 2015, pp

[4] Dengbao Liu, Lin He, Yu-Kai Chou and Fujiang Lin, "A low-power 10-Gb/s receiver with merged CTLE and DFE summer," 2016 13th IEEE International Conference on Solid-State and Integrated Circuit Technology (ICSICT), Hangzhou, China, 2016, pp. 1579-1581

[5] K. -C. Chen, W. W. -T. Kuo and A. Emami, "A 60-Gb/s PAM4 Wireline Receiver with 2-Tap Direct Decision Feedback Equalization Employing Track-and-Regenerate Slicers in 28-nm CMOS," 2020 IEEE Custom Integrated Circuits Conference (CICC), Boston, MA, USA, 2020, pp. 1-4

[6] A. E. Abdelrahman, M. G. Ahmed, M. A. Khalil, M. B. Younis, K. -S. Park and P. K. Hanumolu, "12.3 A Carrier-Phase-Recovery Loop for a 3.2pJ/b 24Gb/s QPSK Coherent Optical Receiver," 2023 IEEE International Solid-State Circuits Conference (ISSCC), San Francisco, CA, USA, 2023, pp. 1-3

[7] X. Wu, Z. Wang, Z. Zhao, C. Zhang and Z. Wang, "A 20Gbuad NRZ/PAM4 Receiver Frontend in 65nm CMOS," 2022 IEEE 16th International Conference on Solid-State & Integrated Circuit Technology (ICSICT), Nanjing, China, 2022, pp. 1-3

[8] Y. -H. Kim, Y. -J. Kim, T. Lee and L. -S. Kim, "A 21-Gbit/s 1.63-pJ/bit Adaptive CTLE and One-Tap DFE With Single Loop Spectrum Balancing Method," in IEEE Transactions on Very Large Scale Integration (VLSI) Systems, vol. 24, no. 2, pp. 789-793, Feb. 2016

979-8-3503-6184-1/24 $31.00 © 2024 IEEE

A Digital Foreground Calibration Method for Pipeline SAR ADCs Using Extended Kalman Filter

Dayan Zhou [1], Yuguo Xiang [1], Junyan Ren [1] and Fan Ye [1]

[1] State-Key Laboratory of Integrated Chips and Systems,
Fudan University, Shanghai 201203, China

Email: fanye@fudan.edu.cn

Abstract—This article presents a digital foreground calibration algorithm based on sinewave fitting to mitigate mismatches and nonlinearities in analog-to-digital converters (ADCs). The ground truth of ADCs is obtained through sinewave fitting, and the weights in the ADCs output code reconstruction block are modified by the Least Mean Square (LMS) algorithm. An Extended Kalman Filter (EKF) is employed in sinefit block to reduce hardware complexity and facilitate iterative calculation. A modeling 12-bit pipeline successive-approximation-register (SAR) ADC is used to verify the performance of proposed calibration scheme. The simulation results show that SFDR and SNDR are improved by 32.3 dB and 19.2 dB after calibration, respectively.

Keywords—Analog-to-digital converter (ADC), Digital Calibration, Sinefit, Kalman Filter, Least Mean Square (LMS)

I. INTRODUCTION

The signal transmission speed in modern communication systems is continuously improving, which poses higher precision and speed requirements for ADCs. For pipeline successive approximation-register(SAR) ADCs, the errors caused by capacitor mismatch and amplifier nonlinearity are the bottlenecks that constrain the accuracy of the ADCs. To relax the design constraints of ADCs in the analog field, digital calibration has been a widely used technique for improving ADCs' performance.

Digital calibration can be background or foreground, based on whether it will interrupt the normal operation of the ADCs during calibration. Although background calibration can handle errors caused by PVT changing and device aging, it usually requires additional analog circuit units, resulting in more area and power consumption overhead. Foreground calibration uses the ideal output as a reference to calibrate the ADC. Due to the ease of generating sine signals, fitting the sine output is often used in practice to obtain the reference signal. There are currently multiple methods available for implementing sine fitting algorithms in software, such as fast Fourier transform (FFT) [1] algorithm and nonlinear least square (NLS) algorithm [2]. However, these algorithms require significant computing and storage resources, resulting in difficulties to be implemented on hardware.

In this paper, we propose a hardware-friendly digital foreground calibration scheme for SAR ADCs. The proposed calibration scheme uses an Extended Kalman filter (EKF) to implement the sine fitting block, followed by a least mean square (LMS) block to modify the weights of ADC reconstruction formula. The proposed EKF-based sinefit algorithm can reduce the computation resources in sine fitting, which saves hardware overhead and enables iterative calculation. The rest of this paper is organized as follows. Section II introduces the proposed calibration scheme. Section III describes the simulation results. Finally, the conclusion is drawn in Section IV.

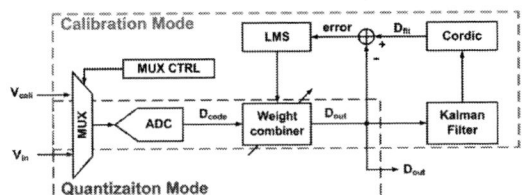

Fig. 1. The sinefit-based digital calibration scheme

II. THE PROPOSED CALIBRATION SCHEME

A. The System Architecture

Fig .1 shows the proposed sinefit-based digital calibration scheme. While in calibration mode, the single-tone sine signal V_{cali} is sent to ADC, generating binary output D_{code}. The variable weight combiner converts D_{code} to a decimal value D_{out}, which is a distorted sine signal. The Kalman Filter followed by a Coordinate Rotation Digital Compute (Cordic) module is used to generate a ground truth D_{fit} through D_{out}. The LMS block modifies the weights in the combiner according to the error between D_{fit} and D_{out}, making D_{out} converge to D_{fit} iteratively. Once the error is lower than the given threshold, the mode can be switched to quantization mode. After passing through the trained weight combiner, the performance of the ADC output will be improved.

B. The EKF-based Sinefit Algorithm

The EKF involves estimation of the state of a discrete-time nonlinear dynamic system [3],

$$x_{i+1} = F(x_i) + w_i \tag{1}$$

$$z_i = H(x_i) + v_i \tag{2}$$

where x_i represents the unobserved state of the system and z_i is the only observed signal. w_i and v_i are the process noise and the observation noise, respectively. The system dynamic functions F and H are assumed to be known.

Specifically regarding the sinefit problem, x_i equals [A_i ω_i ϕ_i d_i]^T, which is the parameter vector to be estimated. Here A_i, ω_i and d_i are the amplitude, digital frequency, and DC offset of sine signal, respectively. $\phi_i = \phi_{i-1} + \omega_{i-1}\Delta t$ represents the real-time phase of the sine signal. The real-time phase instead of the initial phase to omit the sequence number i in the expression of z_i. We assume that parameters of input signal are invariant during the foreground calibration process and no process noise is introduced. Therefore (1) can be rewritten as

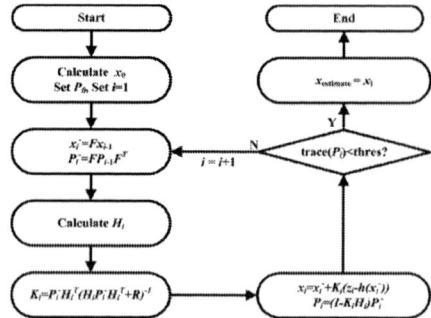

Fig. 2. The procedure of Extended Kalman Filter

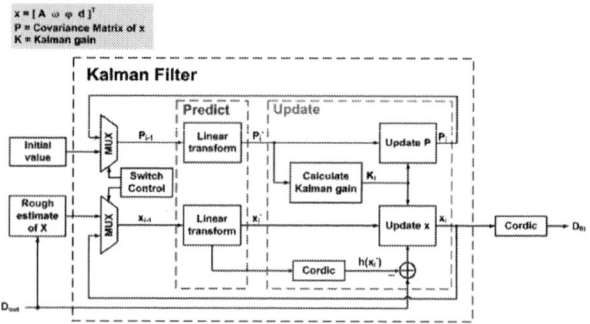

Fig. 3. The proposed sinefit algorithm based on Extended Kalman Filter

$$x_{i+1} = Fx_i = \begin{bmatrix} 1 & 0 & 0 & 0 \\ 0 & 1 & 0 & 0 \\ 0 & 1 & 1 & 0 \\ 0 & 0 & 0 & 1 \end{bmatrix} \begin{bmatrix} A_i \\ \omega_i \\ \phi_i \\ d_i \end{bmatrix} \quad (3)$$

In (2), z_i represents the ADC output code. v_i represents the nonideal factors in the conversion, such as kT/C noise, quantization noise, and harmonics caused by nonlinearities. $H(x_i)$ is the ideal code of input signal, which equals $A_i\sin(\phi_i)$+ d_i. It is worth noting that H is a nonlinearity function and a first-order Tayler expansion is used to approximate it in the EKF solution, which is

$$H_i = \frac{\partial H(x_i)}{\partial x_i} = [\sin(\phi_i) \quad 0 \quad A_i\cos(\phi_i) \quad 1] \quad (4)$$

After obtaining the parameter vector x_i for the current state, F and H together provide an estimate of predicted sine signal value for the next state, which is represented as prior estimate z_i. The EKF gives the optimal estimate based on the weighted sum of prior estimate and measured value z_i^-. The weight is Kalman gain K_i derived from P_i, the covariance matrix of x_i. P_i and x_i are predicted and updated together, causing P_i to decrease with every iteration, eventually converging x_i to the ideal value. Fig. 2 shows the procedure of the Kalman filter. This can be summarized in the following six steps:

1) Initialization. This is a rough estimate block, which provides an initial value $x_0 = [\ A_0 \quad \omega_0 \quad \phi_0 \quad d_0\ ]^T$ of the iterative algorithm. A_0 and d_0 can be obtained from the root mean square and mean of the sequence, respectively. Using the three-point method [4], the rough estimates of digital frequency ω_0 and phase ϕ_0 are obtained:

$$\omega_0 = \arccos(\frac{z_1 + z_3}{2z_2}) \quad (5)$$

$$\phi_0 = \arctan(\frac{z_1 \sin(\omega_0)}{z_2 - z_1 \cos(\omega_0)}) \quad (6)$$

The trigonometric and inverse trigonometric functions can be calculated using Cordic algorithm. P is the covariance matrix of x, and its initial value P_0 is set as a 4×4 identity matrix multiplied by a constant ε_0. The value of ε_0 is related to the accuracy of the initial value x_0.

2) Predict. Based on the system state equation, the prior estimate x_i^- and P_i^- are derived from (7) and (8),

$$x_i^- = Fx_{i-1} \quad (7)$$

$$P_i^- = FP_iF^T \quad (8)$$

3) Calculate matrix H_i based on (4).

4) Calculate Kalman gain K_i using (9), where R is the variance of v_i.

$$K_i = P_i^- H_i^T (H_iP_i^- H_i^T + R)^{-1} \quad (9)$$

5) Update. Using Kalman gain as the weight, the posterior estimates x_i and P_i are given by (10) and (11).

$$x_i = x_i^- + K_i(z_i - h(x_i^-)) \quad (10)$$

$$P_i = P_i^- - K_iH_iP_i^- \quad (11)$$

6) Calculate the trace of matrix P_i. The trace of P_i represents the variance of estimated parameters. If it is smaller than the given threshold, which means that the parameter has converged to the expected accuracy. Therefore the iterative algorithm is finished. If not, return to step 2.

Fig. 3 illustrates the sine fitting algorithm based on EKF. Initially, the ADC output value D_{out} is sent to the rough estimate block, which is the initialization step mentioned previously, to provide the initial value of x_0. The EKF then uses the initial value of x_0 and P_0 to generate estimates of x and P through the predict step and the update steps. The linear transform in Fig. 3 corresponds to (7) and (8), the calculation of Kalman gain corresponds to (9), and the update of x and P corresponds to (10) and (11), respectively. From the second iteration onward, the Kalman filter utilizes the previous posterior estimates x_i and P_i as the initial values for the current calculation, which is implemented by a multiplexer. The Kalman filter then iteratively calculates the four parameters of the sine signal. When the trace of P_i is smaller than the given threshold, the parameters are considered as converged, and the parameters x_i are sent to a Cordic block. The sine fitted code is obtained through the Cordic block to generate the reference signal D_{fit}.

C. The Weight Combiner and LMS algorithm

The weight combiner is based on the prior knowledge of the ADC. Taking pipeline SAR ADCs as an example, the nonidealities include the capacitor mismatch, in-stage amplifier offset and nonlinearity, and the nonlinearity caused by sample and hold circuits. The mismatches can be corrected by linear matrix multiplication, and nonlinearity can be approximatively corrected by power function.

The LMS block modifies the weights in the combiner according to the error given by Kalman Filter, and its iteration equation is given by

979-8-3503-6184-1/24 $31.00 © 2024 IEEE

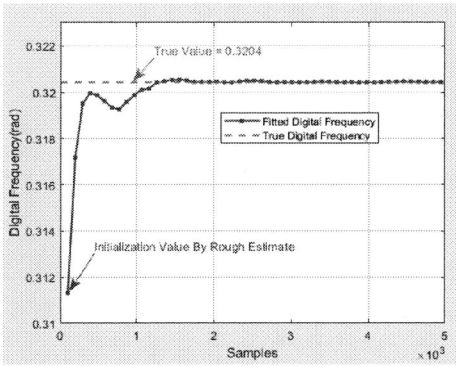

Fig. 4. Convergence process of digital frequency in EKF-based sinefit

$$W_{i+1} = W_i + 2\mu\varepsilon_i \frac{\partial \varepsilon_i}{\partial W_i} \qquad (12)$$

where W_i is the weights, ε_i is the error.

III. THE SIMULATION RESULTS

The EKF-based sinefit is iterative. Fig. 4 shows the convergence process of digital frequency in EKF-based sinefit algorithm. The initial value of digital frequency is set by three-point method. Then the proposed sinefit algorithm will calculate the parameters including digital frequency iteratively. By around 3×10^3 samples, the fitted digital frequency is close to its true value.

To verify the performance of the calibration algorithm, a modeling ADC is used to generate test data. The modeling ADC is a two-stage pipeline SAR ADC working at a sampling frequency of 100MHz. The first stage is 5 bit and the second stage is 8 bit, with 1 bit redundance.

Fig. 5 shows the FFT plot of output signal of the modeling ADC before and after the calibration, respectively. The SFDR and SNDR of the modeling ADC are 51.5 dB and 50.9 dB. Owing to the proposed calibration scheme, the SFDR and SNDR are improved to 83.8 dB and 70.1 dB.

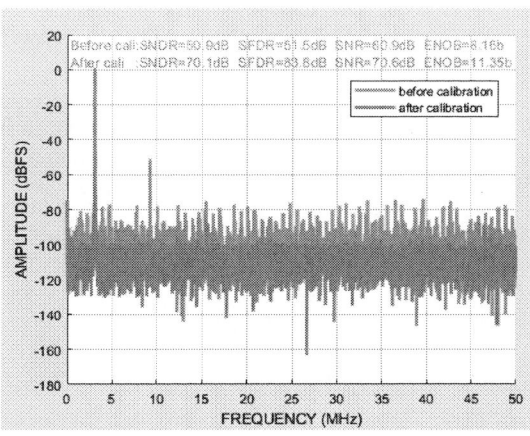

Fig. 5. The FFT spectrum of ADC output before and after calibration

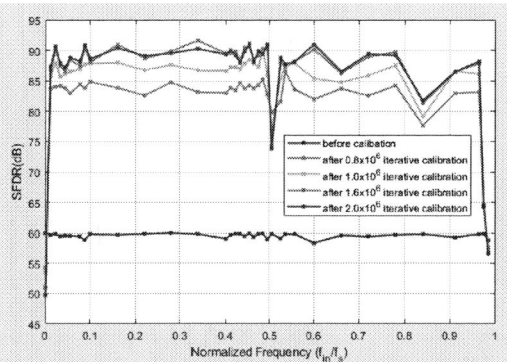

Fig. 6. The SFDR versus input frequencies with different numbers of iterations

Fig. 6 shows the variation of SFDR of the output code after calibration with the input signal frequencies under different numbers of iterations. At most input frequencies, as the number of iterations increases, the SFDR of the output code also gradually improves. After approximately 1.6×10^6 iterations, SFDR reached its peak (85dB~90dB), indicating that the proposed correction algorithm has reached its limit. At some input frequencies, the results show a degradation of SFDR performance. This is because when f_{in} approaches $\pm nf_s/2$ ($n = 0, 1, 2 \dots$), the output signal no longer conforms to the sine signal model, leading to the mis-convergence of sinefit block. Therefore the frequency of calibration signal f_{cali} must be chosen to be coprime with f_s, to ensure optimal calibration performance.

IV. CONCLUSION

A digital foreground calibration scheme is presented. The calibration uses EKF-based sinefit algorithm to generate the ground truth of ADC output. Compared to previous sinefit algorithms, this method implements an iterative computation, and the output of ADC does not need to be stored. Therefore a relatively lower consumption of calculation and storage resources is achieved.

REFERENCES

[1] D. Zhai et al., "High-Speed and Time-Interleaved ADCs Using Additive Neural-Network-Based Calibration for Nonlinear Amplitude and Phase Distortion," in IEEE Transactions on Circuits and Systems I: Regular Papers, vol. 69, no. 12, pp. 4944-4957, Dec. 2022, doi: 10.1109/TCSI.2022.3201016.

[2] "IEEE Std. 1057–2007", IEEE Standard for Digitizing Waveform Recorders.

[3] E. A. Wan and R. Van Der Merwe, "The unscented Kalman filter for nonlinear estimation," Proceedings of the IEEE 2000 Adaptive Systems for Signal Processing, Communications, and Control Symposium (Cat. No.00EX373), Lake Louise, AB, Canada, 2000, pp. 153-158, doi: 10.1109/ASSPCC.2000.882463.

[4] Saber M, Elkenawy, "Design and implementation of accurate frequency estimator depend on deep learning, " International Journal of Engineering & Technology, 2020, 9(2).

[5] Q. Chen, "Design and implementation of time-interleaved calibration algorithm based on sine fitting, " Southeast University, 2017.

A Second-Order Charge Pump Noise-Shaping SAR ADC

Haoning Sun, Yuchen Wang, Wenji Mo, Kangkang Sun and Jingjing Liu*

Sun Yat-Sen University, Shenzhen, China

* sunhn3@mail2.sysu.edu.cn, liujj77@mail.sysu.edu.cn

Abstract—In this paper, a low-power, high-precision passive charge pump noise-shaping (NS) successive approximation register (SAR) analog-to-digital converter (ADC) is proposed for ultrasonic sensor applications in the Internet of Things (IoT). A residual voltage integration technique is introduced to realize a second-order noise transfer function (NTF), thereby achieving a trade-off between the power consumption and accuracy. To address the charge loss issue during the noise-shaping integration process, a charge pump (CP) voltage-multiplying principle is employed, which compensates for the charge loss to a certain extent. The proposed NS SAR ADC is implemented using a standard 180-nm CMOS process. Simulation results demonstrate that the circuit consumes 90.4μW under a 1.2V supply voltage at a sampling rate of 10MS/s. The proposed second-order NS SAR ADC achieves a signal-to-noise and distortion ratio (SNDR) of 87.55dB, an effective number of bits (ENOB) of 14.25bits, a Schreier figure of merit (FoMs) of 178dB.

Keywords—Analog-to-digital converter, charge pump, noise-shaping, successive approximation register, three-input-pair comparator

I. INTRODUCTION

The next-generation Internet of Things (IoT) has become the dominant trend in the future development of the electronics industry [1]. As a critical component, ADC determines the core competitiveness of the entire system. In the IoT application domain, various sensor devices have a strong demand for low-power, high-precision ADCs. With the continuous advancement of technology, the system requirements for ADC accuracy are constantly increasing [2]. Ultrasonic sensors have extensive applications in the IoT, enabling detection and measurement using ultrasound, and they normally require low-power, high-precision (12-14 bits) ADCs [3].

NS SAR ADC is one of the most promising hybrid ADC architectures. Similar to the SAR ADC, it features low power and low cost, while its high signal-to-noise ratio is comparable to that of the Δ-Σ ADC. The first modern NS SAR ADC is implemented using a cascaded integrator feed-forward (CIFF) structure [4]. A loop filter is used and consumes relatively high power. The ADC has limited accuracy due to first-order noise-shaping. Liu incorporates a dynamic amplifier into the NS SAR ADC architecture [5]. It achieves an improved first-order noise-shaping performance. However, the power consumption remains relatively high due to the additional circuitry introduced by the dynamic amplifier and the associated filtering components. Chen employs a capacitor-based charge pump voltage doubler to compensate for the voltage loss during the residue voltage integration process [6]. This design achieves second-order noise-shaping, but only the second-order integration process is compensated.

This work is supported by National Natural Science Foundation of China with project number 62174181.

Consequently, the noise shaping performance has a potential for further improvement.

This paper presents a low-power, high-accuracy noise-shaping SAR ADC suitable for ultrasonic sensors in IoT applications. The objective is to enhance the accuracy and reduce power consumption while achieving more efficient noise shaping with a lower raw bit count. This work employs a passive NS technique using residue voltage integration and utilizes a three-input-pair comparator to sum the input voltage and the NS integration voltage. Furthermore, this work leverages the capacitive charge pump voltage-multiplying principle to compensate for the voltage loss during the NS integration process, thereby enhancing the NS performance and improving the accuracy of ADC. By eliminating the need of additional active amplifiers and relying solely on switched-capacitor circuits to implement the noise-shaping functionality, the ADC's power consumption is reduced. The structure of this paper is as follows. Section II presents the proposed NS SAR ADC architecture and its operating principles. The ADC simulation results are provided in Section III. Finally, Section IV concludes the paper.

II. THE PROPOSED CP NS SAR ADC DESIGN

A. Voltage-multiplying charge pump

The voltage-multiplying charge pump can achieve the effect of multiplying the input voltage through periodic charging and discharging processes. As illustrated in Fig. 1, the operating principle of a voltage-multiplying charge pump is demonstrated. The four capacitors, possessing identical capacitance values, are configured with their upper plates connected to V and their lower plates grounded.

When the charge pump is in the voltage acquisition stage, the four capacitors form a parallel connection through switch switching, as shown in Fig. 1(a). After charging to a voltage equal to V across the four capacitors, the charge pump is ready to enter the voltage-multiplying stage. When the charge pump is in the voltage-multiplying stage, the four capacitors are switched to series connection, as shown in Fig. 1 (b). The

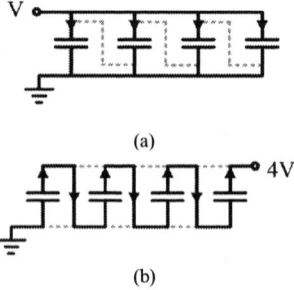

Fig. 1 Voltage-multiplying charge pump working principle, (a) voltage acquisition mode, (b) voltage-multiplying mode.

lower plate of the end capacitor is grounded, and the upper plate of each capacitor is sequentially connected to the lower plate of the next level capacitor. The upper plate of the last capacitor serves as the output terminal. Due to the fact that each capacitor is charged to V during the acquisition mode, when the lower plate of the higher-level capacitor is connected to the upper plate of the lower-level capacitor, this lower plate potential will be raised. And each level of capacitor potential is raised by V. From this, it can be seen that the voltage-multiplying charge pump composed of four capacitors can approximately obtain four times the voltage at the output end. Therefore, this voltage-multiplying charge pump can amplify the input voltage by approximate four times.

B. Second-order CP NS SAR ADC

Fig. 2. illustrates the structure of the designed passive second-order CP NS SAR ADC. For simplicity, the diagram shows the single-ended positive input section of the NS SAR ADC, though the actual design is differential. The structural details of second-order CP NS circuit and the timing diagrams of control signals are depicted in Fig. 3, where $C_{res}=C_{DAC}=2C_{int1}=4C_{int2}$. The C_{DAC} is the total capacitance of CDAC. CLK_S denotes the ADC sampling clock and CLK_C represents the control clock for the ADC comparator. Φ_{res} controls the integration capacitor C_{res} to sample the residual voltage V_{res}. Φ_1 controls the charge sharing between C_{res} and C_{int1}. Φ_2 controls the charge sharing between C_{res} and C_{int2}. Φ_{CP1} controls the first-stage integration voltage V_{int1} amplification via the charge pump configuration with C_{int1}. Φ_{CP2} controls the second-stage integration voltage V_{int2} amplification via the charge pump configuration with C_{int2}. Φ_{rst} controls the reset operation of the C_{res} voltage.

During one sampling period, the ADC undergoes three phases, including the sampling and second-order NS phase, comparison phase and residue V_{res} collection phase. In the sampling and second-order NS phase, the signal is sampled and sent to the comparator for comparison, and the NS structure performs the integration operation. During the comparison phase, the comparator outputs the comparison result to the SAR logic, which controls the DAC capacitor array C_{DAC} to switch the top plate voltage levels. This process performs a successive approximation and outputs a digital code, completing the quantization task. After the comparison phase, the residual voltage V_{res} on the top plate of the capacitor array is sampled onto capacitor C_{res} via switch Φ_{res}. Due to the voltage loss from charge sharing, the voltage collected on C_{res} is approximately $0.5V_{res}$, which is used as the integration voltage for noise shaping in the subsequent sampling cycle.

At the beginning of the next sampling cycle, Φ_1 is high, while Φ_{CP1} maintains low. At this moment, the two capacitors

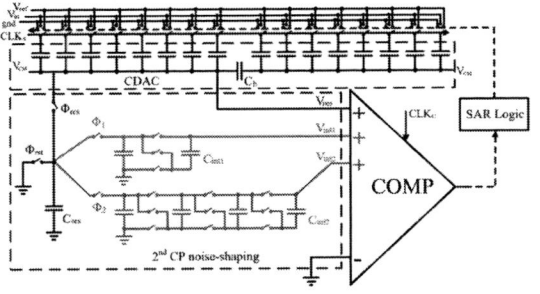

Fig. 2 The architecture of the proposed second-order CP NS SAR ADC.

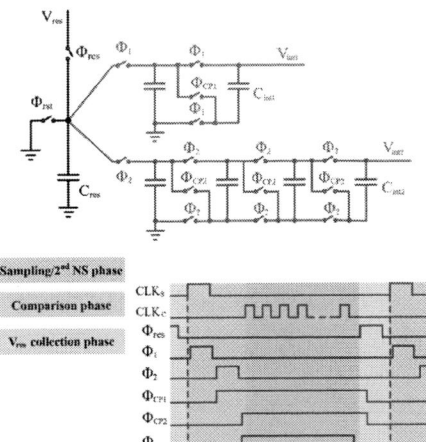

Fig. 3 The architecture of the proposed second-order CP NS circuit and the timing diagrams of control signals.

C_{int1} are connected in parallel, resulting in a total capacitance value of $2C_{int1}$. According to the principle of charge conservation, the following can be obtained:

$$0.5V_{res}(z)C_{res} + 2V_{int1}(z)C_{int1}z^{-1} = V_{int1}(z)(C_{res} + 2C_{int1}) \quad (1)$$

Subsequently, Φ_1 switches to low, and the voltage on C_{res} is V_{int1}. Φ_2 is switched to high, while Φ_{CP2} remains low. At this moment, the four capacitors C_{int2} are connected in parallel, resulting in a total capacitance value of $4C_{int2}$. According to the principle of charge conservation, (2) can be obtained:

$$V_{int1}(z)C_{res} + 4V_{int2}(z)C_{int2}z^{-1} = V_{int2}(z)(C_{res} + 4C_{int2}) \quad (2)$$

Simultaneously, after Φ_1 is at a low level, Φ_{CP1} is at a high level, transitioning C_{int1} capacitors from the integration phase to the amplification phase. At this point, the two capacitors are connected in series, providing the first-order integration voltage $V'_{int1} = 2 V_{int1}$ to the three-input-pair comparator. Similarly, after Φ_2 is at a low level and Φ_{CP2} is at a high level, transitioning C_{int2} capacitors from the integration phase to the amplification phase. The four capacitors are also connected in series, providing the second-order integration voltage $V'_{int2} = 4 V_{int2}$ to the three-input-pair comparator. Additionally, after Φ_2 is at a low level, Φ_{rst} is at a high level, performing the reset operation on capacitor C_{res}, which means clearing the voltage on C_{res}.

During the comparison phase, the comparator sums and compares the residual voltage V_{res}, first-order integration voltage V'_{int1} and second-order integration voltage V'_{int2}, thereby achieving the second-order noise shaping . Due to the quantization noise Q introduced during the comparator quantization phase, the ADC output D_{out} can be expressed as follows:

$$D_{out}(z) = V'_{int1}(z) + V'_{int2}(z) + V_{in}(z) + Q(z) \quad (3)$$

Based on (1), (2) and (3), the following relationship can be derived:

$$D_{out}(z) = V_{in}(z) + (1-0.5z^{-1})^2 Q(z) \quad (4)$$

It can be observed that this CP NS SAR ADC achieves a second-order noise-shaping function. The NTF is given by

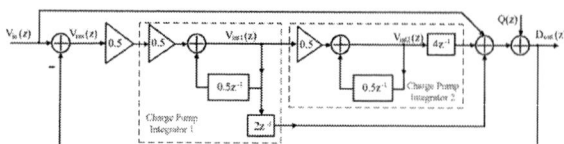

Fig. 4 The signal flow diagram of the proposed CP NS SAR ADC.

$(1-0.5z^{-1})^2$. Fig. 4 shows the signal flow diagram of the proposed CP NS SAR ADC.

III. SIMULATION RESULTS AND DISCUSSION

The proposed second-order CP NS SAR ADC is implemented using a standard 180nm CMOS process. This ADC consumes 90.4μW at a supply voltage of 1.2V. The power consumption breakdown of each block is illustrated in Fig. 5. The spectral density of the proposed ADC is shown in Fig. 6. When the input signal is a sine signal of 38kHz and the sampling rate of ADC is 10MS/s, the output spectrum density shows that SNDR is 87.55dB, corresponding to an ENOB of 14.25bit. The performance comparison with previous noise-shaping ADCs is given in Table I. It can be observed that the proposed CP NS SAR ADC circuit achieves a high precision within an appropriate bandwidth. The obtained SNDR of 87.55dB outperforms other ADCs in the table except [7]. It employs active amplifiers to achieve fourth-order noise shaping in a 28nm process, resulting in high precision but also significant power consumption. The FoM$_s$

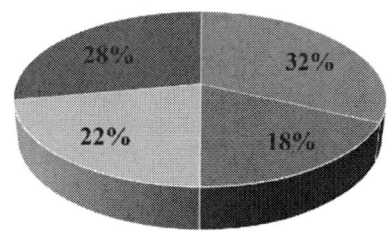

■ CDAC ■ Noise Shaping ■ Sar Logic ■ SH/Comparator

Fig. 5 The power consumption breakdown for each ADC block.

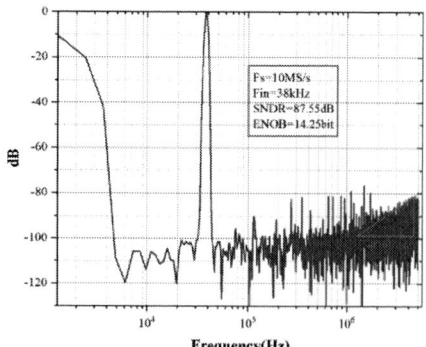

Fig. 6. Simulated spectral density of the proposed ADC for a 38kHz input signal sampled at 10MS/s.

TABLE I.   PERFORMANCE SUMMARY AND COMPARISON

Specifications	[2]	[5]	[6]	[7]	This work
Technology (nm)	65	40	65	28	**180**
Supply (V)	0.8	1.1	1	1	**1.2**
Sampling rate (S/s)	128k	8.4M	64M	2M	**10M**
Bandwidth (kHz)	4	262	8000	100	**100**
Original bit	14	8	8	8	**8**
SNDR(dB)	79.1	80	65	87.6	**87.55**
ENOB(bit)	12.8	13	10.5	14.3	**14.25**
Power(μW)	1.37	143	252.9	120	**90.4**
FoM$_s$(dB)	173.8	173	169.9	177	**178**

of the proposed CP NS SAR ADC is the best in the table. This indicates that the proposed CP NS SAR ADC achieves a better trade-off among precision, speed, and power consumption.

IV. CONCLUSION

This paper proposes a second-order CP NS SAR ADC. By using an 8-bit V$_{CM}$-based SAR ADC, a 14.25-bit ADC suitable for ultrasonic sensors is realized. To achieve high energy efficiency, this work employs a passive NS technique using residue voltage integration and utilizes a three-input-pair comparator to sum the input voltage and the NS integration voltages. The capacitive charge pump voltage-multiplying principle is used to compensate the voltage loss during the NS integration process, thereby enhancing the NS performance and improving the accuracy of the ADC. The FoM$_s$ of the proposed second-order NS SAR ADC is 178dB. It consumes 90.4μW from a 1.2V supply at 10MS/s. The SNDR is 87.55dB, corresponding to an ENOB of 14.25bit. Simulation results demonstrate that the proposed NS SAR ADC achieves a favorable trade-off between conversion accuracy and power consumption.

REFERENCES

[1] Y. Zhang and J. Zhang, "A 20 MHz Bandwidth 79 dB SNDR SAR-Assisted Noise-Shaping Pipeline ADC With Gain and Offset Calibrations," *IEEE Journal of Solid-State Circuits*, vol. 57, no. 3, pp. 745-756, Mar. 2022.

[2] P. Harpe and E. Cantatore, "An oversampled 12/14b SAR ADC with noise reduction and linearity enhancements achieving up to 79.1dB SNDR," in *IEEE International Solid-State Circuits Conference (ISSCC)*, Mar. 2014, pp. 194-195.

[3] D. Chen and X. Cui, "A Survey on Analog-to-Digital Converter Integrated Circuits for Miniaturized High Resolution Ultrasonic Imaging System," *Micromachines*, vol. 13, no. 1:114, Jan. 2022.

[4] J. A. Fredenburg and M. P. Flynn, "A 90-MS/s 11-MHz-bandwidth 62-dB SNDR noise-shaping SAR ADC," *IEEE Journal of Solid-State Circuits*, vol. 47, no. 12, pp. 2898-2904, Dec. 2012.

[5] C. Liu and M. Huang, "A 0.46mW 5MHz-BW 79.7dB-SNDR noise-shaping SAR ADC with dynamic-amplifier-based FIR-IIR filter," in *IEEE International Solid-State Circuits Conference (ISSCC)*, Feb. 2017, pp. 466-467.

[6] Z. Chen, M. Miyahara, and A. Matsuzawa, "A 2nd order fully-passive noise-shaping SAR ADC with embedded passive gain," in *IEEE Asian Solid-State Circuits Conference (A-SSCC)*, Nov. 2016, pp. 309-312.

[7] Y. Shu and L. Kuo, "An Oversampling SAR ADC With DAC Mismatch Error Shaping Achieving 105 dB SFDR and 101 dB SNDR Over 1 kHz BW in 55 nm CMOS," *IEEE Journal of Solid-State Circuits*, vol. 51, no. 12, pp. 2928-2940, Dec. 2016.

979-8-3503-6184-1/24 $31.00 © 2024 IEEE

Computing in Memory for Accelerating Light-Weighted On-Chip Learning in IoT Devices

Zhiwang Guo [1], Xiaoyong Xue *[1,2], Qiang Zhou *[1,2], Xiaoyang Zeng [1]

[1] State Key Laboratory of Integrated Chips and Systems, School of Microelectronics, Fudan University, Shanghai 201203, China
[2] Transcputing Tech Co., LTD, Shanghai 200135, China

* Email: xuexiaoyong@fudan.edu.cn, Alex.zhou@transcputing.com

Abstract—**Computing in memory (CIM) is a promising approach to achieve on-chip inference of neural networks (NNs) with high energy efficiency for IoT devices. However, CIM accelerators without supporting on-chip training, will lack environmental adaptability and incur data security issues. Light-weighted on-chip learning has shown great potential in IoT devices. In this paper, several circuits with on-chip training techniques are reviewed and discussed. First, the innovations of the weight memory array including memory cell and weight mapping for on-chip training are explained. Second, several algorithms achieving light-weighted on-chip learning based on the CIM scheme are discussed. Third, the non-ideality aware training is considered, including pulse-to-pulse variability (PPV), noise, non-linearity, and quantization accuracy loss.**

Keywords—*AI, neural networks, computing-in-memory, on-chip learning, IoT device*

I. INTRODUCTION

With the development of artificial intelligence (AI) edge hardware, an increasingly serious security issue is the leakage of the weights and the neural network (NN) models trained from the cloud and then stored in edge devices, considering the expensive resources invested in the training process and the private user data. Some AI chips have already encrypted some weights and personal information on chip, such as embedded XNOR cipher for lightweight encryption [1]. However, user data still needs to be transmitted to the cloud, which incurs security concerns of the consumers. Moreover, IoT devices barely adapt to the environmental changes without local training. Therefore, on-chip training has become an indispensable function in future IoT devices.

Computing in memory (CIM) is considered to be one of the mainstream AI edge hardware architectures, which can break down the Von Neumann bottleneck achieving higher energy efficiency and lower latency [2]. Researchers are currently exploring floating-point (FP) CIM to achieve AI applications with higher precision especially on-chip training [1,2]. While today's typical NNs using FP usually exceed on-chip memory capacity. Moreover, the IoT devices are targeted at light-weighted and portable application scenarios. Therefore, it is urgent for resource-constrained edge devices to achieve light-weighted on-chip learning based on CIM.

On-chip training calls for more operations to be performed than that only for the inference. As early as 2018, related work [3] in ISSCC had involved on-chip training, but it support inference classifier using CIM and stochastic gradient descent (SGD) training without CIM in two separate circuits. Light-weighted on-chip learning tries to achieve inference-training using CIM in one memory array. On-chip training for NNs is generally composed of four steps, feed-forward (FF), error calculation (EC), gradient (ΔW) calculation (GC), and weight

update (WU) [4], while inference only has the FF process. Using the DNN with batch gradient descent (BGD) as an example, the details of the training are shown in Algorithm-1. Given the structure of a DNN, the number of layers L, activation function $\sigma(z)$, loss function $J(W,b,x,y)$, iteration step size α, iteration number I, epochs, batch_size, and the number of samples m, weight W and bias b to be updated after training can be calculated.

Algorithm 1
Input:
L, $\sigma(z)$, $J(W,b,x,y)$, α, I, datasets, epoch, batch_size, m, $\{(x_1,y_1),(x_2,y_2),...,(x_m,y_m)\}$.
Output:
W, b
1 Initialize parameters of DNN, W, and b.
2 for epoch:
for i=1 to I:
for l=2 to L:
FF $a^{i,l} = \sigma(z^{i,l}) = \sigma(W^l a^{i,l-1} + b^l)$
for l=L-1 to 2:
BP $\delta^{i,L} = \dfrac{\partial J(W,b,x,y)}{\partial z^{i,L}} = (a^{i,L} - y) \odot \sigma'(z^{i,L})$
for l=2 to L:
GC $W^l = W^l - \alpha \sum\limits_{i=1}^{batch} \delta^{i,l}(a^{i,l-1})^T$
GC $b^l = b^l - \alpha \sum\limits_{i=1}^{batch} \delta^{i,l}$
WU W^l and b^l
3 Return W^l and b^l

According to the Algorithm-1, the AI accelerator will require additional circuits to support BP, GC, WU processes for on-chip training. Several circuits with on-chip training techniques are reviewed in the rest of this article. In Section II, the innovation of the CIM array is given. In Section III, recent circuits for on-chip training are discussed. The novel algorithms are given in Section IV. Finally, Section V draws the non-ideality aware training.

II. ARRAY INNOVATION

The different data flow and the topological structure between FF and BP call for a transpose of network weights, as shown in Algorithm-1. It will increase one data transpose and load process each time between FF and BP, which cannot be accepted for light-weight on-chip training. More researchers focus on achieving in-situ training which stores and updates the weights directly in the memristors and performs

979-8-3503-6184-1/24 $31.00 © 2024 IEEE

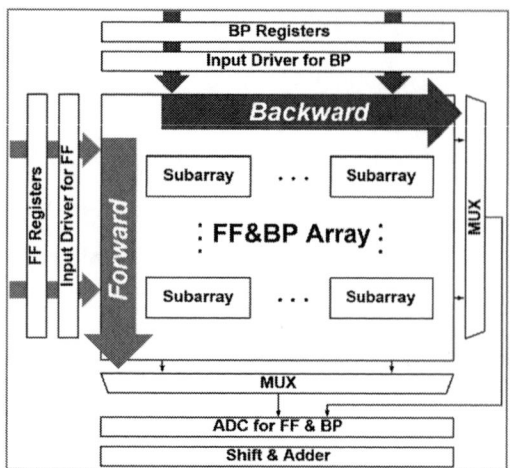

Fig. 1. The structure for FF and BP processes achieving on-chip training.

computations at the original place [5]. Direct weight access in the vertical and horizontal directions can decrease the memory activities without the rewrite process of weights, as shown in Fig. 1. Urgent challenge for on-chip training is to solve the transpose read in the cell structure.

A. SRAM

In recent years, SRAM-based CIM on-chip training researches focus on 6T SRAM with local computing unit (LCC) [2,6,7,8] and novel cell structures including, 4T1T SRAM [9], 7T SRAM [10,11], 8T SRAM [11], 9T SRAM [12,13]. All of them support transpose read which is necessary for on-chip training based on CIM and can achieve higher energy efficiency and lower latency. 6T SRAM with a local

computing unit can solve SRAM write-disturb and optimize process variation. Related work [2] proposed the 6T SRAM with two-way transpose multiply cell (TWT-MC) which can support FF and BP acceleration, as shown in Fig. 2(a). 4T1T SRAM in work [9] comprises two pull-up transistors (T1, T2), a pull-down transistor (T3), a read transistor (T4), and a write transistor (T5) shown in Fig. 2(b). 4T1T SRAM achieves normal read and transpose read by applying different voltages. In normal (transpose) read, the RBLs (RWLs) are fixed to be HIGH (LOW), and the RWLs (RBLs) are clamped to be LOW (HIGH) represented as 1 (0). In works [10,11] shown in Fig. 2(c), increasing a transpose read transistor in 7T SRAM achieves the accumulation of partial sums in the BP, on the basis of the 6T SRAM cell. In Fig. 2(d), two transistors read two states of Q and QB separately in the 8T SRAM cell with a high symmetrical structure [11]. 9T SRAM cell in researches [12,13] provides less discharging current variation and improves the output voltage linearity compared with the 7T SRAM cell. In Fig. 2(e), T6 is a discharging transistor, and T7 or T8 is a cascode transistor in the FF or BP symmetrically.

B. ReRAM

Recently, the structures of transpose 2T1R ReRAM (T-ReRAM) [14], parallel-connected 2T1R [15], serial-connected 2T1R [16], and 1T1R with shared-path transpose read [17] were proposed to achieve transpose read. The three 2T1R cells add a redundant transistor to form a decoupled path or a balanced reversed path while they double bit-cell area at least. The transpose 2T1R cell uses the added transistor to amplify the voltage between the selector and the ReRAM resistor with higher read performance but more vulnerable to PVT variations, as shown in Fig. 3(a) [14]. In Fig. 3(b), the parallel-connected 2T1R takes the added transistor as another selector in parallel with the existing one [15]. The serial-connected 2T1R cell in Fig. 3(c) may degrade the normal and the transpose read performance for large parasitic path resistance [16]. The shared-path 1T1R in work [17] employs a new voltage biasing scheme to eliminate the influence of the body effect on the transpose read, improving both read margin and speed, as shown in Fig. 3(d). 2T2R is composed of two 1T1R cells which store the positive and negative weights, as shown in Fig. 3(e) [18]. In Fig. 3(f), 4T2R uses two access transistors (T3 and T4) and two ReRAMs (R1 and R2) for normal access operation, and two additional computing transistors (T1 and T2) are added for a ternary weight inference operation [19]. However, 2T2R and 4T2R cannot achieve transpose read, the weight needs to be remapped between FF and BP processes.

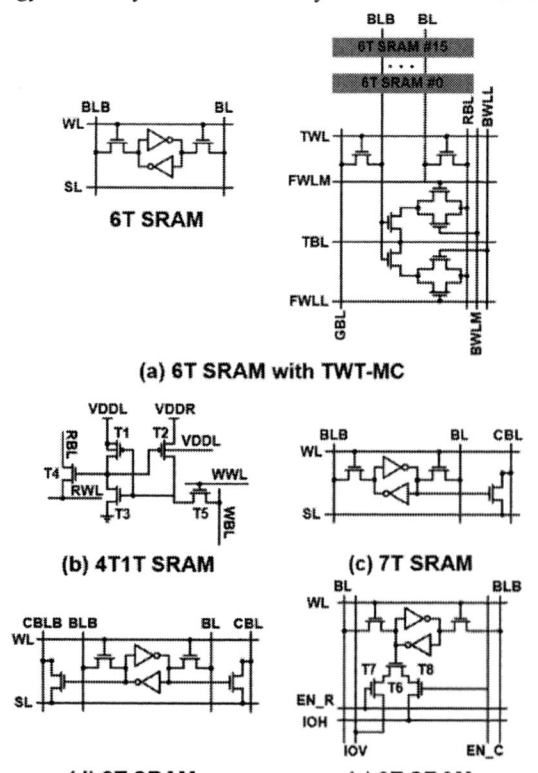

Fig. 2. The schematics of the different SRAM cells.

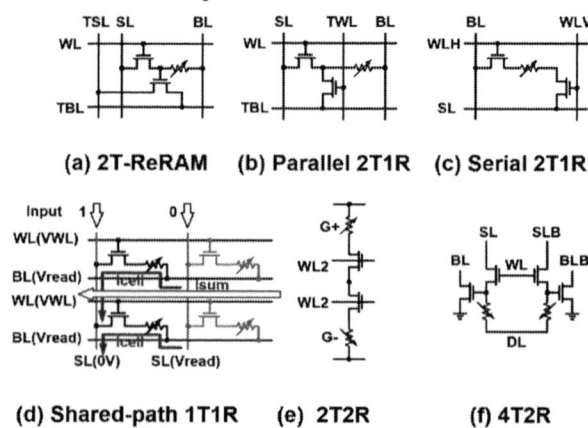

Fig. 3. The schematics of the different ReRAM cells.

979-8-3503-6184-1/24 $31.00 © 2024 IEEE

172

C. MRAM, PCM, Fe-FET, FeRAM, and Flash

CIM researches based on MRAM [20], PCRAM [21], Fe-FET [22], FeRAM [23], and Flash [24] are unique in their own potential and are widely investigated for on-chip AI accelerator training. MRAM, PCRAM, and FeRAM have fast industrial development. Besides, MRAM and PCRAM have remarkable endurance, especially for the frequent weight update process in on-chip training. Due to the technological progress of Flash and its indispensability, Flash memory can be more compatible with the current system structure.

III. CIRCUIT FOR TRAINING

A. Weight Mapping

The bits of different weights are usually adjacently arranged for the convenience of inference. In the BP process, the accumulation of different bits cannot be directly computed in a row. The bit-interleaving weight mapping achieves row accumulation and has been widely used in recent works [9,17]. The different bits of weights stored in different subarrays can lead to huge speed improvement and energy saving.

B. Hybrid Memory

There is a trend towards combining the advantages of two types of memories in recent works for on-chip training based on CIM. A hybrid precision synapse combining ReRAM with capacitor is proposed in works [5,16], which leverages the symmetric and fast update in the volatile capacitor, as well as the non-volatility and large dynamic range of the NVM. Another research uses a plastic cell array that combines Flash and SRAM, and it allows for the storage and the computation of both long-term and short-term information in training [24].

C. ADC and SA

There are some researchers propose novel ADCs aimed at training. In works [6,8], a small-offset gain-enhancement sense amplifier (SOGE-SA) is proposed to tolerate a small read margin. An energy efficiency 6-bit programmable single-slope SAR (PSS-SAR) ADC is used in works [12,13]. A differential merged-into-array ADC (DMA-ADC) saves the power of training [24].

IV. ALGORITHM FOR TRAINING

This Section reviews recent algorithms for light-weighted training. Sparse optimization algorithm is one of the most important methods reducing computational complexity due to the sparsity of NNs [25,26]. A sparsity-optimized CIM macro with a high bit-level-sparsity encoding scheme is designed to reduce the power consumption of one MAC operation [26]. Meanwhile, some circuit designs of zero skipping also utilize the sparsity of NNs. The tensor-train (TT) algorithm in works [26] is a tensor decomposition method, which decomposes entire DNN computation to small-size matrix multiplications. What's more, the mixed-precision training algorithm is a novel light-weighted training method to improve energy efficiency, which usually used with hybrid memory together [16,25]. BNN (binary neural network) realizes ultimate computing compression by binarizing the parameters of the NNs and the activation values to 1 and -1, but requests higher accuracy (full precision, 32-bit FP) when training [27,28]. One-shot learning or few-shot learning shows great potential in scenarios lacking training samples or light-weighted training [29,30]. Such as most facial recognition applications, IoT devices need to identify a person through a single image or a few facial samples. Research [31] combines offline training (using previous data generation networks) with online training (real-time data training) regarded as a promising method, achieving 5.7× energy efficiency improvement.

V. NON-IDEALITY AWARE TRAINING

Recent researches have mainly adopted the strategies of modeling and compensation for non-idealities aware training including pulse-to-pulse variability (PPV) [32], device noise aware training [33], random telegraph noise (RTN) [34], non-linearity of devices and circuits [35], and quantization accuracy [36]. Data-driven framework in [37] uses gaussian distribution to analytically model the device level non-idealities to improve the accuracy of training. Insight and modeling of PPV for training aims to decrease the variability and curve fluctuation in analog resistive memory impact on training [32]. Two-stage training framework in [33] injects noises of devices or peripheral circuits aware training to improve accuracy. Research [34] is trained on the artificially created dataset with assumed fluctuation patterns of the RTN. A fully analog CIM is proposed in work [35] avoiding the utilization of the ADCs, and a circuit-algorithm co-design scheme is necessary to solve much noise in the fully analog circuit. CIMQ focuses on the accuracy loss during the quantization process in [36]. An end-to-end benchmarking framework is proposed introducing nonlinearity, asymmetry, device-to-device and cycle-to-cycle variation in work [38].

ACKNOWLEDGMENTS

This work was supported in part by the National Key R&D Program under Grant 2023YFB4404700, in part by the National Natural Science Foundation of China under Grant 62274038, in part by the Science and Technology Commission of Shanghai Municipality under Grant 21TS1401200 and Grant 22ZR1407100, and in part by State Key Laboratory of Integrated Chips and Systems under Grant SKLICS-Z202315.

REFERENCES

[1] W. Li, S. Huang, X. Sun, H. Jiang and S. Yu, "Secure-RRAM: A 40nm 16kb Compute-in-Memory Macro with Reconfigurability, Sparsity Control, and Embedded Security," 2021 IEEE Custom Integrated Circuits Conference (CICC), Austin, TX, USA, 2021, pp. 1-2, doi: 10.1109/CICC51472.2021.9431558.

[2] N. Pan, X. Cui, X. Qiao, K. Xiao, Q. Guo and Y. Wang, "A 28nm 64Kb SRAM based Inference-Training Tri-Mode Computing-in-Memory Macro," 2022 IEEE International Symposium on Circuits and Systems (ISCAS), Austin, TX, USA, 2022, pp. 2561-2565, doi: 10.1109/ISCAS48785.2022.9937705.

[3] S. K. Gonugondla, M. Kang and N. Shanbhag, "A 42pJ/decision 3.12TOPS/W robust in-memory machine learning classifier with on-chip training," 2018 IEEE International Solid-State Circuits Conference - (ISSCC), San Francisco, CA, USA, 2018, pp. 490-492, doi: 10.1109/ISSCC.2018.8310398.

[4] H. Jiang et al., "A Two-way SRAM Array based Accelerator for Deep Neural Network On-chip Training," 2020 57th ACM/IEEE Design Automation Conference (DAC), San Francisco, CA, USA, 2020, pp. 1-6, doi: 10.1109/DAC18072.2020.9218524.

[5] Y. Luo and S. Yu, "Accelerating Deep Neural Network In-Situ Training With Non-Volatile and Volatile Memory Based Hybrid Precision Synapses," in IEEE Transactions on Computers, vol. 69, no. 8, pp. 1113-1127, 1 Aug. 2020, doi: 10.1109/TC.2020.3000218.

[6] J. -W. Su et al., "15.2 A 28nm 64Kb Inference-Training Two-Way Transpose Multibit 6T SRAM Compute-in-Memory Macro for AI Edge Chips," 2020 IEEE International Solid-State Circuits Conference - (ISSCC), San Francisco, CA, USA, 2020, pp. 240-242, doi: 10.1109/ISSCC19947.2020.9062949.

[7] H. Jiang et al., "A Two-way SRAM Array based Accelerator for Deep Neural Network On-chip Training," 2020 57th ACM/IEEE Design

Automation Conference (DAC), San Francisco, CA, USA, 2020, pp. 1-6, doi: 10.1109/DAC18072.2020.9218524.

[8] J. -W. Su et al., "Two-Way Transpose Multibit 6T SRAM Computing-in-Memory Macro for Inference-Training AI Edge Chips," in IEEE Journal of Solid-State Circuits, vol. 57, no. 2, pp. 609-624, Feb. 2022, doi: 10.1109/JSSC.2021.3108344.

[9] C. Zhao et al., "A 28-nm 36 Kb SRAM CIM Engine With 0.173 um2 4T1T Cell and Self-Load-0 Weight Update for AI Inference and Training Applications," in IEEE Journal of Solid-State Circuits, doi: 10.1109/JSSC.2024.3399615.

[10] K. Bong, S. Choi, C. Kim, S. Kang, Y. Kim and H. -J. Yoo, "14.6 A 0.62mW ultra-low-power convolutional-neural-network face-recognition processor and a CIS integrated with always-on haar-like face detector," 2017 IEEE International Solid-State Circuits Conference (ISSCC), San Francisco, CA, USA, 2017, pp. 248-249, doi: 10.1109/ISSCC.2017.7870354.

[11] H. Jiang, X. Peng, S. Huang and S. Yu, "CIMAT: A Compute-In-Memory Architecture for On-chip Training Based on Transpose SRAM Arrays," in IEEE Transactions on Computers, vol. 69, no. 7, pp. 944-954, 1 July 2020, doi: 10.1109/TC.2020.2980533.

[12] X. Zhang, Y. Jo, J. Liu, J. Zhou, Y. Zheng and T. T. -H. Kim, "A Local Transpose 9T SRAM Compute-In-Memory Macro with Programmable Single-Slope SAR ADC," 2022 IEEE Asian Solid-State Circuits Conference (A-SSCC), Taipei, Taiwan, 2022, pp. 6-8, doi: 10.1109/A-SSCC56115.2022.9980672.

[13] Y. -J. Jo, X. Zhang, J. Liu, J. Zhou, Y. Zheng and T. T. -H. Kim, "Transposable 9T-SRAM Computation-In-Memory for on-Chip Learning With Probability-Based Single-Slope SAR Hybrid ADC for Edge Devices," in IEEE Solid-State Circuits Letters, vol. 6, pp. 81-84, 2023, doi: 10.1109/LSSC.2023.3260090.

[14] L. Wang et al., "A 14nm 100Kb 2T1R Transpose RRAM with >150X resistance ratio enhancement and 27.95% reduction on energy-latency product using low-power near threshold read operation and fast data-line current stabling scheme," 2021 Symposium on VLSI Technology, Kyoto, Japan, 2021, pp. 1-2.

[15] S. Kim et al., "NVM neuromorphic core with 64k-cell (256-by-256) phase change memory synaptic array with on-chip neuron circuits for continuous in-situ learning," 2015 IEEE International Electron Devices Meeting (IEDM), Washington, DC, USA, 2015, pp. 17.1.1-17.1.4, doi: 10.1109/IEDM.2015.7409716.

[16] Y. Luo and S. Yu, "Benchmark non-volatile and volatile memory based hybrid precision synapses for in-situ deep neural network training," in Proc. Asia South Pac. Design Autom. Conf. (ASP-DAC), 2020, pp. 422–427.

[17] Z. Guo et al., "An Emerging NVM CIM Accelerator With Shared-Path Transpose Read and Bit-Interleaving Weight Storage for Efficient On-Chip Training in Edge Devices," in IEEE Transactions on Circuits and Systems II: Express Briefs, vol. 70, no. 7, pp. 2645-2649, July 2023, doi: 10.1109/TCSII.2023.3240193.

[18] Y. Geng et al., "An On-chip Layer-wise Training Method for RRAM based Computing-in-memory Chips," 2021 Design, Automation & Test in Europe Conference & Exhibition (DATE), Grenoble, France, 2021, pp. 248-251, doi: 10.23919/DATE51398.2021.9473931.

[19] K. Zhou et al., "A 28 nm 81 Kb 59–95.3 TOPS/W 4T2R ReRAM Computing-in-Memory Accelerator With Voltage-to-Time-to-Digital Based Output," in IEEE Journal on Emerging and Selected Topics in Circuits and Systems, vol. 12, no. 4, pp. 846-857, Dec. 2022, doi: 10.1109/JETCAS.2022.3196678.

[20] H. Wang, Y. Zhao, C. Li, Y. Wang and Y. Lin, "A New MRAM-Based Process In-Memory Accelerator for Efficient Neural Network Training with Floating Point Precision," 2020 IEEE International Symposium on Circuits and Systems (ISCAS), Seville, Spain, 2020, pp. 1-5, doi: 10.1109/ISCAS45731.2020.9181003.

[21] C. Zhou et al., "ML-HW Co-Design of Noise-Robust TinyML Models and Always-On Analog Compute-in-Memory Edge Accelerator," in IEEE Micro, vol. 42, no. 6, pp. 76-87, 1 Nov.-Dec. 2022, doi: 10.1109/MM.2022.3198321.

[22] W. Shim and S. Yu, "Ferroelectric Field-Effect Transistor-Based 3-D NAND Architecture for Energy-Efficient on-Chip Training Accelerator," in IEEE Journal on Exploratory Solid-State Computational Devices and Circuits, vol. 7, no. 1, pp. 1-9, June 2021, doi: 10.1109/JXCDC.2021.3057856.

[23] Y. Luo, Y. -C. Luo and S. Yu, "A Ferroelectric-Based Volatile/Non-Volatile Dual-Mode Buffer Memory for Deep Neural Network

Accelerators," in IEEE Transactions on Computers, vol. 71, no. 9, pp. 2088-2101, 1 Sept. 2022, doi: 10.1109/TC.2021.3122872.

[24] L. Wang et al., "34.9 A Flash-SRAM-ADC-Fused Plastic Computing-in-Memory Macro for Learning in Neural Networks in a Standard 14nm FinFET Process," 2024 IEEE International Solid-State Circuits Conference (ISSCC), San Francisco, CA, USA, 2024, pp. 582-584, doi: 10.1109/ISSCC49657.2024.10454372.

[25] S. -H. Sie et al., "MARS: Multimacro Architecture SRAM CIM-Based Accelerator With Co-Designed Compressed Neural Networks," in IEEE Transactions on Computer-Aided Design of Integrated Circuits and Systems, vol. 41, no. 5, pp. 1550-1562, May 2022, doi: 10.1109/TCAD.2021.3082107.

[26] R. Guo et al., "TT@CIM: A Tensor-Train In-Memory-Computing Processor Using Bit-Level-Sparsity Optimization and Variable Precision Quantization," in IEEE Journal of Solid-State Circuits, vol. 58, no. 3, pp. 852-866, March 2023, doi: 10.1109/JSSC.2022.3198413.

[27] Y. Fujiwara and T. Kawahara, "BNN Training Algorithm with Ternary Gradients and BNN based on MRAM Array," TENCON 2023 - 2023 IEEE Region 10 Conference (TENCON), Chiang Mai, Thailand, 2023, pp. 311-316, doi: 10.1109/TENCON58879.2023.10322327.

[28] Y. S. Chong, W. L. Goh, Y. S. Ong, V. P. Nambiar and A. T. Do, "Recovering Accuracy of RRAM-based CIM for Binarized Neural Network via Chip-in-the-loop Training," 2022 IEEE International Symposium on Circuits and Systems (ISCAS), Austin, TX, USA, 2022, pp. 2958-2962, doi: 10.1109/ISCAS48785.2022.9937271.

[29] Y. Li et al., "Monolithic 3D Integration of Logic, Memory and Computing-In-Memory for One-Shot Learning," 2021 IEEE International Electron Devices Meeting (IEDM), San Francisco, CA, USA, 2021, pp. 21.5.1-21.5.4, doi: 10.1109/IEDM19574.2021.9720534.

[30] D. Reis, A. F. Laguna, M. Niemier and X. S. Hu, "A Fast and Energy Efficient Computing-in-Memory Architecture for Few-Shot Learning Applications," 2020 Design, Automation & Test in Europe Conference & Exhibition (DATE), Grenoble, France, 2020, pp. 127-132, doi: 10.23919/DATE48585.2020.9116292.

[31] N. Cao et al., "A 65 nm Wireless Image SoC Supporting On-Chip DNN Optimization and Real-Time Computation-Communication Trade-Off via Actor-Critical Neuro-Controller," in IEEE Journal of Solid-State Circuits, vol. 57, no. 8, pp. 2545-2559, Aug. 2022, doi: 10.1109/JSSC.2022.3159473.

[32] Z. Yu, Z. Wang, S. Bao, Y. Ling, Y. Cai and R. Huang, "A New Insight and Modeling of Pulse-to-Pulse Variability in Analog Resistive Memory for On-Chip Training," in IEEE Transactions on Electron Devices, vol. 69, no. 6, pp. 3100-3104, June 2022, doi: 10.1109/TED.2022.3164630.

[33] H. -W. Kuo, R. -H. Wang, Z. Li, S. -T. Lin, M. -F. Chang and K. -T. Tang, "A Two-stage Training Framework for Hardware Constraints of Computing-in-Memory Architecture," 2022 IEEE Asia Pacific Conference on Circuits and Systems (APCCAS), Shenzhen, China, 2022, pp. 30-34, doi: 10.1109/APCCAS55924.2022.10090308.

[34] A. Yamada, N. Misawa, C. Matsui and K. Takeuchi, "ReRAM CiM Fluctuation Pattern Classification by CNN Trained on Artificially Created Dataset," 2023 IEEE International Reliability Physics Symposium (IRPS), Monterey, CA, USA, 2023, pp. 1-6, doi: 10.1109/IRPS48203.2023.10118305.

[35] K. Zhou et al., "An Energy Efficient Computing-in-Memory Accelerator With 1T2R Cell and Fully Analog Processing for Edge AI Applications," in IEEE Transactions on Circuits and Systems II: Express Briefs, vol. 68, no. 8, pp. 2932-2936, Aug. 2021, doi: 10.1109/TCSII.2021.3065697.

[36] J. Bai, S. Sun, W. Zhao and W. Kang, "CIMQ: A Hardware-Efficient Quantization Framework for Computing-In-Memory-Based Neural Network Accelerators," in IEEE Transactions on Computer-Aided Design of Integrated Circuits and Systems, vol. 43, no. 1, pp. 189-202, Jan. 2024, doi: 10.1109/TCAD.2023.3298705.

[37] M. -G. Lin et al., "D-NAT: Data-Driven Non-Ideality Aware Training Framework for Fabricated Computing-In-Memory Macros," in IEEE Journal on Emerging and Selected Topics in Circuits and Systems, vol. 12, no. 2, pp. 381-392, June 2022, doi: 10.1109/JETCAS.2022.3171268.

[38] X. Peng, S. Huang, H. Jiang, A. Lu and S. Yu, "DNN+NeuroSim V2.0: An End-to-End Benchmarking Framework for Compute-in-Memory Accelerators for On-Chip Training," in IEEE Transactions on Computer-Aided Design of Integrated Circuits and Systems, vol. 40, no. 11, pp. 2306-2319, Nov. 2021, doi: 10.1109/TCAD.2020.3043731.

979-8-3503-6184-1/24 $31.00 © 2024 IEEE

A Novel Beamforing Receiver Architecture Combining MASH SDM and BSP

Tao Zhong [1], Yuekang Guo [1], Jing Jin*[1], Jianjun Zhou [1]

[1] Department of Micro/Nano Electronics, Shanghai Jiao Tong University, Shanghai, China

* Email: jinjing@sjtu.edu.cn

Abstract—**This paper proposes a millimeter-wave phased array receiver architecture based on multi-stage noise-shaping continuous-time bandpass sigma-delta modulator and digital-domain beamforming techniques. This architecture ensures the signal bit-width of each bit-stream processing channel is the same as that of using single-loop sigma-delta modulator, thereby minimizing the power consumption and design complexity of the bit-stream processing module. The feasibility of the proposed architecture is verified through system-level modeling and simulation of a 9-antenna phased array receiver.**

Keywords—5G, beamforming, millimeter-wave, CTSDM, BSP

I. INTRODUCTION

With the booming development of millimeter-wave applications in recent years, such as 5G applications, beamforming technology is receiving increasing attention. The system bandwidth of 5G technology is much higher than earlier technologies, enabling higher data transmission speeds. However, the transmission in the millimeter-wave band is limited by factors such as the output power of power amplifiers, transmission path losses, and more pronounced shadowing losses, imposing strict requirements on link budgets. Benefiting from the shorter wavelength compared to traditional technologies, millimeter-wave applications allow for the use of larger antenna arrays within the same area. Large-scale antenna arrays, combined with beamforming technology, can effectively compensate for transmission path losses, thereby making up for the shortcomings of millimeter-wave wireless transmission [1].

To achieve high integration and accurate beamforming in large-scale antenna arrays, the performance requirements and power constraints of beamforming receivers are very stringent. The architecture of beamforming receivers can be divided into three categories based on the implementation of signal down-conversion and phase shifting, as shown in Fig. 1: analog-domain beamforming (ABF) [2], hybrid beamforming (HBF) [3], and digital-domain beamforming (DBF) [4]. The performance of ABF is limited by the high insertion loss and limited phase resolution of the phase shifters in the RF signal path, making it difficult to achieve high-precision beamforming control in large-scale antenna arrays. The drawback of the HBF architecture is the complex LO network in the design of large-scale antenna arrays and results in higher system power consumption. These drawbacks make the HBF architecture more suitable for applications that are less sensitive to system integration and power consumption. The DBF architecture implements both RF signal down-conversion and phase shifting in the digital domain. The advantages of digital-domain down-conversion and phase shifting circuits compared to analog-domain circuits lie in their higher integration and reliability. With the advancement

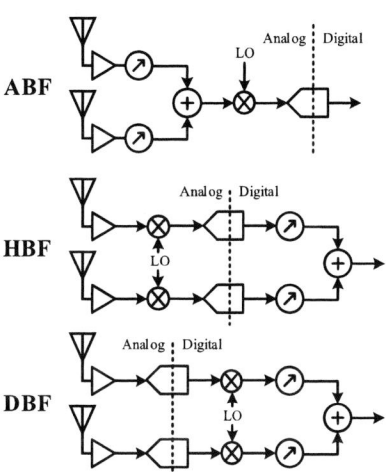

Fig. 1. The architecture of beamforming receivers.

of CMOS technology, its energy efficiency and integration can be further improved, making it very suitable for high-integration large-scale millimeter-wave phased array designs.

This paper proposes a DBF receiver system architecture that combines multi-stage noise-shaping (MASH) sigma-delta modulator (SDM) and bit-stream processing (BSP). This architecture improves the signal dynamic range without increasing the signal bit-width of each BSP channel, thereby ensuring that the complexity of BSP design is not increased, and the power consumption and area of the system are minimized. The structure of this paper is as follows. The next section provides a brief introduction to the existing beamforming receiver systems based on continuous-time (CT) bandpass (BP) SDM and BSP. Section III describes the proposed architecture. Section IV presents the simulation results of system-level modeling.

II. DBF RECEIVER USING CTBPSDM AND BSP

A. Basic Principle

Fig. 2 shows the block diagram of traditional DSP and BSP. In the traditional DSP processing, the output signal of SDM first goes through an decimation filter, and then the low-speed, high-resolution digital signal is processed by the DSP module. However, the larger bit-width results in the inevitable presence of complex multipliers in the DSP. In contrast, BSP directly processes the low-resolution output signal of SDM and uses simple MUXs to implement down-conversion and phase shifting of the output signal. Compared to traditional DSP, it has been demonstrated in [4] and [5] that BSP greatly optimizes the power consumption and area of digital circuits.

This work was supported by the Chinese National Natural Science Foundation under Grant 62122051.

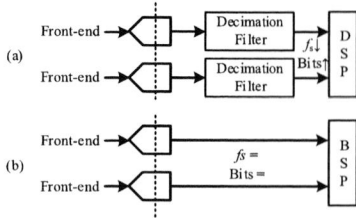

(a) ... (b) ...

Fig.2. Beaming forming receiver using (a)DSP, (b)BSP.

The interleaved bit-stream processing (IL-BSP) technique shown in Fig. 3(a) [5] reduces the power consumption and area of digital circuits to only 20% of the traditional DSP implementation.

The sampling rate of CTBPSDM in the IL-BSP architecture is selected as 4 times the IF frequency. This simplifies the NTF design of CTBPSDM and enables the digital-domain down-conversion (DDC) to be implemented using a simple 2:1 MUX, as shown in the Fig. 3(b). The output voltage levels of the quantizer in SDM include 5 levels, as mentioned in [6]. The 5-level output is a cost-effective choice, which allows the phase shift operation in BSP to be implemented using a simple 5:1 MUX. The coefficients in the 5:1 MUX include $\pm 2k$, $\pm 1k$, and 0. If we use the values stored in a set of registers as the coefficient k, we only need to perform a shift read on the values in the registers to obtain the coefficient $2k$. This eliminates the need for complex arithmetic logic or additional registers in the digital circuit to express the coefficient $2k$.

B. Improve System Dydnamic Range

To further improve the dynamic range of the system and relax the stringent front-end circuit link budget requirements in millimeter-wave receiver, enhancing the dynamic range of CTBPSDM is an effective approach. Generally, there are three ways to improve the ideal SQNR of SDM: increasing the oversampling ratio (OSR), increasing the number of quantizer output voltage levels, and increasing the order of noise shaping. In the BSP-based architecture, the sampling rate of CTBPSDM is fixed at 4 times the IF frequency, and the optimal solution for the number of quantizer output voltage levels is 5. Attempting to increase the order of noise shaping, is an intuitive and cost-effective solution. However, in existing designs, single-loop SDM has been used, and higher-order noise shaping would compromise the stability of the SDM, significantly increasing the complexity and reducing the reliability of the system. To ensure system reliability, using MASH or sturdy-MASH (SMASH) SDM is a feasible solution.

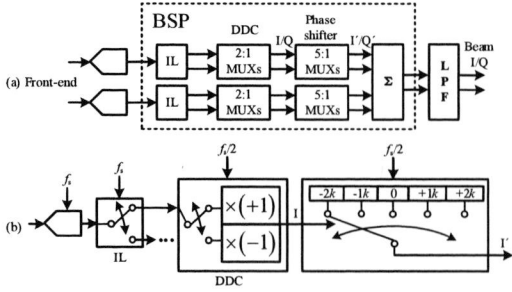

Fig.3. (a)IL-BSP, (b)DDC and phase shifer using MUXs.

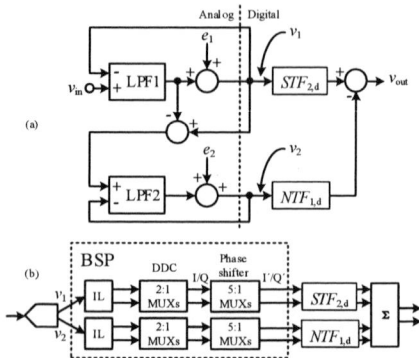

Fig.4. Block diagram: (a)MASH SDM, (b) MASH SDM with BSP.

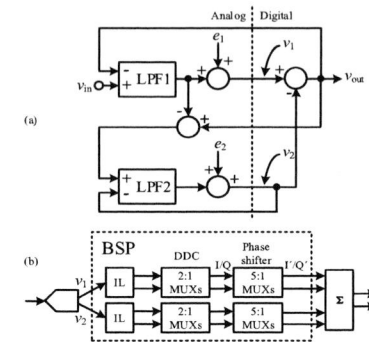

Fig.5. Block diagram: (a)SMASH SDM, (b) SMASH SDM with BSP.

III. DBF Receiver Using MASH SDM and BSP

A. MASH SDM with BSP

Fig. 4(a) shows a basic MASH SDM architecture. Under ideal conditions, the quantization noise e_1 is eliminated in the output, leaving only the quantization noise e_2, which is shaped by noise shaping. The MASH SDM's final output v_{out}, passes through digital filtering and digital summation circuits. Consequently, the bit-width of this digital output will be significantly larger than the bit-width of the quantizer output. Applying DDC and digital domain phase shift to this signal would result in unacceptable power consumption and area for digital circuits.

In order to combine MASH SDM with BSP, we propose the receiver chain architecture shown in the Fig. 4(b). When using MASH SDM, the digital filter used for noise cancellation is moved after the BSP processing. Although the BSP path will be doubled in this way, the design of the BSP path remains the same, and by simply using two sets of the original single-loop SDM and BSP circuits, the order of quantization noise shaping can be doubled. Since the new digital filter used for quantization noise cancellation is located after DDC, its transfer function also needs to be changed to match the STF and NTF after DDC.

B. SMASH SDM with BSP

A common issue with MASH SDM is that the cancellation of quantization noise e_1 relies on the matching between the analog-domain and digital-domain filters. To overcome this issue, SMASH SDM can be used, as shown in Fig. 5(a), instead of cancel out e_1, the output of SMASH SDM contains two quantizers' noises shaped by noise shaping.

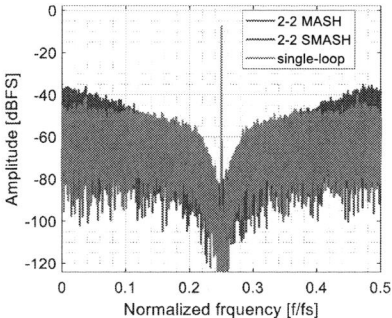

Fig.6. Output spectrum of CTBPSDMs.

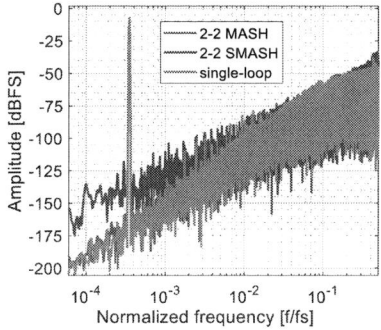

Fig.7. Output spectrum of I signal in one single channel.

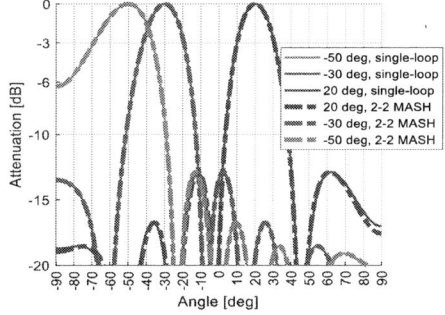

Fig.8. Simulated beam patterns with proposed beamforning receiver.

For the case of using SMASH SDM, we cannot remove this digital summing circuit because the feedback of the first loop is based on the final output v_{out} of the SDM. One intuitive approach is to use a SMASH SDM architecture with a quantizer output level of 3. In this case, the final output v_{out} of the SDM still only has 5 levels, but this approach weakens the benefits brought by the noise shaping order. We propose the architecture shown in the Fig. 5(b), the outputs of the two quantizers are connected to two BSP paths respectively, and the digital summation circuit is still retained for the feedback path of the SMASH SDM. An additional summation circuit is added to the digital outputs of two BSP paths to cancel the additional quantization noise present in the BSP signal.

IV. SIMULATION RESULTS

To validate the proposed architecture of combining MASH SDM with BSP, a system-level model of a

beamforming receiver system was constructed, which consists of a 9-element linear array of antennas. In order to compare with the case of using single-loop SDM, the MASH and SMASH SDM architectures adopted a 2-2 structure, while the single-loop SDM adopted a 4th-order noise-shaping architecture. All quantizers in the SDMs produce a 5-level output.

A. Single Receiver Chain

Fig. 6 shows the output spectrum of the SDM models. It can be seen that the ideal 2-2MASH SDM and the 4th-order single-loop SDM have consistent output results, the output of the 2-2 SMASH SDM shows more quantization noise. Fig. 7 compares the final output spectrum of the I path in a single channel by using different beamforming receiver architecture. The results in the figure demonstrate the feasibility of the proposed architecture. Moving the digital filter used for noise elimination in the MASH SDM to the BSP does not affect the final output results. SMASH exhibits greater quantization noise when the loop filter order is the same.

B. 9-antenna Beamforming Receiver

Fig. 8 presents the simulation results of system-level modeling for a 9-element antenna array. The beamforming angles are set at -50, -30, and 20 degrees. Comparing the proposed architecture combining MASH and BSP with the traditional architecture, the beamforming performance remains unaffected, further validating the feasibility of the proposed architecture in this paper.

V. SUMMARY

The paper proposes a receiver system architecture for DBF that combines MASH SDM and BSP. The feasibility of achieving high integration and precise beam control in large-scale mm-wave phased array design is demonstrated through simulation results. By using the proposed architecture, exponential growth of quantization noise shaping effect can be achieved at the cost of linear increase in area and power consumption.

REFERENCES

[1] F. W. Vook, A. Ghosh and T. A. Thomas, "MIMO and beamforming solutions for 5G technology," 2014 IEEE MTT-S International Microwave Symposium (IMS2014), Tampa, FL, USA, 2014, pp. 1-4.

[2] I. Ahmed et al., "A Survey on Hybrid Beamforming Techniques in 5G: Architecture and System Model Perspectives," in IEEE Communications Surveys & Tutorials, vol. 20, no. 4, pp. 3060-3097, Fourthquarter 2018.

[3] Y. A. Atesal, B. Cetinoneri, K. M. Ho and G. M. Rebeiz, "A Two-Channel 8–20-GHz SiGe BiCMOS Receiver With Selectable IFs for Multibeam Phased-Array Digital Beamforming Applications," in IEEE Transactions on Microwave Theory and Techniques, vol. 59, no. 3, pp. 716-726, March 2011.

[4] J. Jeong, N. Collins and M. P. Flynn, "A 260 MHz IF Sampling Bit-Stream Processing Digital Beamformer With an Integrated Array of Continuous-Time Band-Pass ΔΣ Modulators," in IEEE Journal of Solid-State Circuits, vol. 51, no. 5, pp. 1168-1176, May 2016.

[5] S. Jang, J. Jeong, R. Lu and M. P. Flynn, "A 16-element 4-beam 1GHz-IF 100MHz-bandwidth Interleaved Bit-Stream digital beamformer in 40nm CMOS," 2017 IEEE Radio Frequency Integrated Circuits Symposium (RFIC), Honolulu, HI, USA, 2017, pp. 124-127.

[6] D. A. Johns and D. M. Lewis, "Design and analysis of delta-sigma based IIR filters," in IEEE Transactions on Circuits and Systems II: Analog and Digital Signal Processing, vol. 40, no. 4, pp. 233-240, April 1993.

979-8-3503-6184-1/24 $31.00 © 2024 IEEE

In Situ Localization Techniques of Defects in Advanced Semiconductor Devices from Macro-scale to Atomistic-scale

Jialu Huang[1], Jingming Zhou[2], Zuoyuan Dong[1], Runsheng Wang[*2], Junhao Chu[*1], Xing Wu[*1]

[1] In Situ Devices Center, School of Integrated Circuits, East China Normal University, Shanghai, China
[2] Institute of Electronic Design Automation, Peking University, Wuxi, Jiangsu, China

* Email: xwu@cee.ecnu.edu.cn

Abstract—As manufacturing processes for advanced semiconductor devices become more complex, precise defect localization and characterization are crucial for device failure analysis and reliability improvement. This review summarizes *in situ* defects localization methods across different scales. Specifically, it covers defect detection techniques from macroscopic level (such as EFA) to microscopic level (such as FIB-SEM and CR technique), even down to atomic level (typically TEM). Also, corresponding research findings are presented. These techniques provide significant support and assistance for developing advanced semiconductor devices.

Keywords—*In situ* defect localization, advanced semiconductor devices, infrared detector

I. INTRODUCTION

Moore's Law has long been the driving force behind the semiconductor industry, enabling continuous shrinking device sizes and increases in transistor density [1, 2]. However, as technology nodes to the nanoscale, manufacturing processes have become increasingly complex, where even minor defects in materials and structures can severely impact device performance and reliability . Therefore, precise localization and characterization of defects in advanced semiconductor device manufacturing have become crucial steps to ensure device quality and performance [3–5].

In situ localization of defects provides a method for real-time monitoring and analysis of defects during device fabrication and testing. This approach can detect and analyze defects at various scales, ranging from macroscopic to atomic levels, employing a range of advanced techniques including optical microscopy, electron microscopy, scanning probe microscopy, X-ray microscopy, and atomic probe tomography [6–8].

This review introduces *in situ* localization of defects in advanced semiconductor devices, exploring various methods and their applications across different scales, as shown in Fig. 1. We firstly introduce defect localization techniques at the macroscopic scale, followed by investigations into microscopic defect analysis, and finally discuss the methods for atomic-scale characterization. By using these multiscale analytical tools, comprehensive insights into the defect formation mechanisms and the impact of these defects on device reliability, thereby providing strong basis for the optimization of advanced devices.

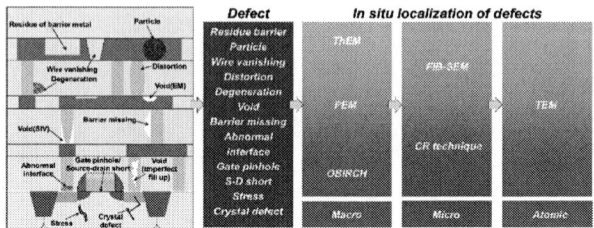

Fig. 1. Defects in intergrated circuits and *in situ* defects localization methods at different scales.

II. MACRO-LEVEL DEFECT DETECTION TECHNIQUES

Based on optical imaging systems, electrical failure analysis (EFA) techniques are used to detect defects at macroscopic level. These techniques locate defects by utilizing luminescent and heat-generating properties at the failed position. The most widely used techniques include thermal emission microscopy (ThEM), photon emission microscopy (PEM) and optical beam induced resistance change (OBIRCH) [9–11].

ThEM pinpoints the fault locations by analyzing the thermal radiation distribution generated by defects, utilizing heat conduction and thermal radiation principles. As shown in Fig. 2a, while appropriate bias is applied on the circuits, joule heating generate at the detect area, then thermal emission is finely captured by the infrared detector. Minor irregularities on the sample can cause localized temperature variations, reflecting differences in microstructure, components, or electrical properties. ThEM collect the information and indicate possible failed points. Fig. 2b depicts the outcome of localization using this technique. However, it is worth mentioning that its resolution is limited (approximately 5 μm), making it just suitable for primarily wafer-level defect localization.

PEM is a high-resolution technique used to localize defects and faults in integrated circuits by detecting the emission of photons generated at the fault point [12]. As shown in Fig. 2c, driven by appropriate voltage on the device, the defect emits photons of specific wavelengths due to accelerated carrier scattering or electron-hole pair recombination, usually in the infrared wavelength. By recording the positions of these photons, a signal map can be grabbed. This signal map is then compared with infrared photos to pinpoint the potential failure location. Using this technology, the commercially produced instrument is known as the emission microscope (EMMI) (Fig. 2d) .

The fundamental principle of OBIRCH is depicted in Fig. 2e: in bias condition, the surface of the device under test is scanned by a laser beam. By recording these laser induced changes of the resistance or current in bias circuits, defects or failure locations can be precisely identified. Fig. 2f shows an actual OBIRCH example. Compared to ThEM, OBIRCH offers higher accuracy (1 μm), making it suitable for defect localization at the chip level [13].

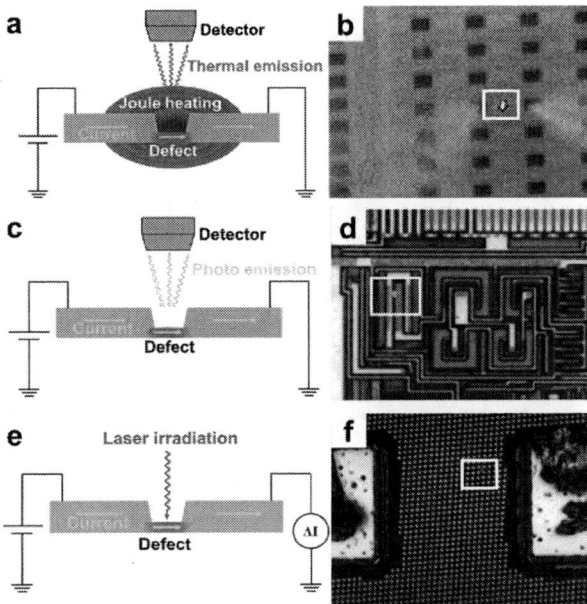

Fig. 2. The mechanism diagram of (a) ThEM, (c) PEM and (e) OBIRCH. (b) ThEM, (d) PEM and (f) OBIRCH examples [13].

III. MICRO-LEVEL DEFECT DETECTION TECHNIQUES

EFA techniques provide a rough estimate of defect locations. Advanced micro-level defect detection techniques such as focused current-ratio (CR) and technique ion beam-scanning electron microscopy (FIB-SEM) are required for precise determination.

A. Current-ratio (CR) technique

Current-ratio (CR) technology is a defect localization method based on electrical properties. As shown in Fig. 3a, the drain and source currents are measured before and after breakdown [14]. The breakdown location along the channel length is then calculated using the following parameters [15]:

$$s_{BD} = \frac{I_{D-BD}}{I_{D-BD} + I_{S-BD}}$$

Through the collection and analysis of experimental results, it was found that when the ratio s approaches 0, the breakdown point is located at the source side in the channel; when the ratio s approaches 1, the breakdown point is located at or near the drain side (Fig. 3b) [16]. Based on these findings, the on-state time dependent dielectric breakdown (TDDB) position of the fin field-effect transistor (FinFET) was determined. In consideration of the influence of the drain voltage (V_{ds}), the equation has been revised and its revised expression is as follows:

$$s_{BD} = \frac{(I_{initial} - I_D)V_G}{I_S(V_G - V_{ds}) + V_{ds}I_{initial} - I_D V_G}$$

As illustrated in Fig. 3c, the figure depicts the statistical distribution of the breakdown locations along the channel direction in the on-state TDDB mode and contrasts it with the breakdown locations in the gate-only TDDB mode [17]. The

results demonstrate that the breakdown locations of the on-state TDDB are predominantly situated near the source (red bars), whereas the breakdown locations of the gate-only TDDB are primarily concentrated in the central region between the source and the drain (green bars). This suggests that drain bias causes a shift in the breakdown location of the gate dielectric layer, moving it from the middle of the channel towards the source side.

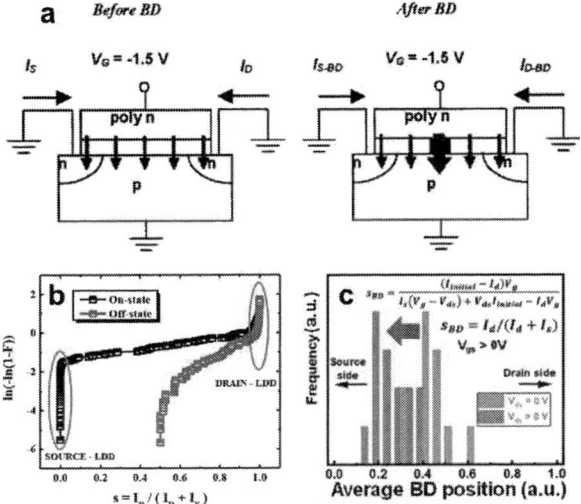

Fig. 3. (a)Bias configuration is used to evaluate the source and drain current before stress and after the breakdown occurrence in the case of an nMOSFET [14]. (b) Weibit distribution plots for sorted ratios show multi-modal breakdown behavior in the LDD diffusion (vertical) and channel (central) regions [16]. (c) Breakdown localizations of FinFET on-state TDDB [17].

B. FIB-SEM

FIB-SEM integrates the capabilities of a focused ion beam system and a scanning electron microscope to achieve precise material removal and high-resolution imaging [18–20]. The FIB utilizes a finely focused ion beam, typically composed of gallium, to mill or sputter material at the microscopic scale, facilitating site-specific sample preparation. Concurrently, the SEM provides detailed imaging and compositional analysis by scanning the sample with a focused electron beam. As shown in Fig. 4a, the FIB-SEM dual-beam system has been extensively used to prepare TEM samples for various devices, including memory devices, logic devices, and sensors [21]. This synergistic combination allows for micro-level pinpointing of defects in advanced semiconductor devices and enables the precise and detailed study of surface and sub-surface features.

However, as device structures become increasingly complex, traditional FIB-SEM sample preparation methods struggle to meet the demand for high-quality TEM sample preparation. Consequently, new methods offering enhanced efficiency and superior sample quality have been developed for advanced devices. Fig. 4b and 4c illustrate a refined FIB-SEM sample preparation technique designed to improve experimental efficiency [22]. The method further refines the specimen after removing the amorphous layer on the surface. Subsequently, the film is sectioned into multiple narrower parts, experiments on these parts can be carried out simultaneously. This approach eliminates variations in experimental conditions, thereby enhancing the consistency and reliability of testing.

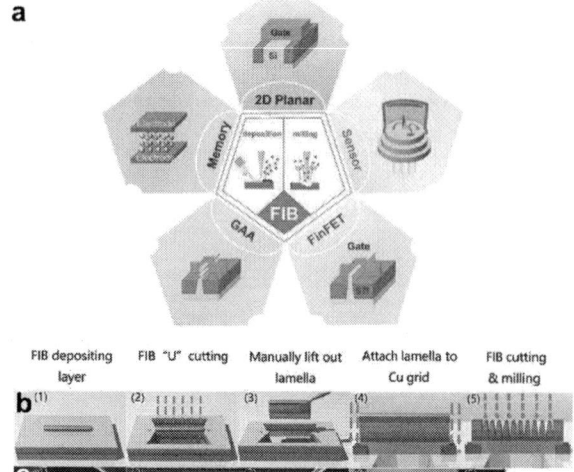

Fig. 4. (a) Overview of the TEM sample preparation technique of devices based on the FIB–SEM system [21]. (b) The series of images (1)–(5) show the samples prepared by FIB for *in situ* TEM analysis. (c) The corresponding SEM images of the process [22].

IV. ATOMIC LEVEL DEFECT DETECTION TECHNOLOGY

Transmission electron microscopy (TEM) stands as a pivotal technology for detecting atomic-level defects, utilizing a focused electron beam to reveal the microstructure and defects of materials at the nanoscale [23, 24]. This advanced microscopy technique surpasses the limitations of traditional optical microscopes, offering sub-angstrom resolution to observe defects and crystal structures at the atomic level directly [25, 26].

Fig. 5 presents the results of the scanning TEM (STEM) defect analysis of the enhanced modified lateral silicon-controlled rectifier (EMLSCR) [13]. By comparing the TEM bright-field images of the device before and after failure, as shown in Fig. 5a and 5b, it can be observed that the central part of the EMLSCR device contains a damaged active region. Further examination at higher magnification, as shown in Fig. 5c and 5d, reveals the damaged area in the SCR section. The nickel silicide layer has melted and infiltrated into the silicon, exhibiting uneven damage. Comparing Fig. 5c and 5d, the cathode N^+ region shows the most severe damage, with noticeable melting traces in the active region. Energy-dispersive X-ray spectroscopy (EDX) mapping was conducted to confirm the chemical nature of the failure. The mapping image in Fig. 5e shows that almost all nickel silicide has penetrated into the active silicon region. Fig. 5e-i indicates that the initial thickness of the nickel silicide layer was 22 nm. As shown in Fig. 5e-ii, in the anode P^+ region, the depth of nickel penetration into the silicon layer is uneven, ranging from approximately 30 nm to 226 nm. Fig. 5e-iii illustrates that the nickel silicide in the cathode N^+ region is the most severely damaged part of the device, with a nickel penetration depth of up to 600 nm. The damage depth in the cathode active region is approximately four times that of the anode.

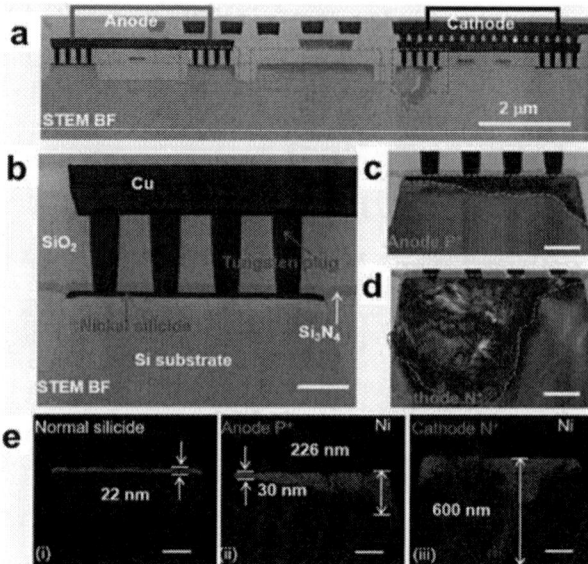

Fig. 5. (a) STEM BF image of the complete view of the EMLSCR device after breakdown, with dashed lines marking abnormal regions. (b) STEM BF image of the complete view of the fresh EMLSCR device. (c) TEM images of (c) anode P^+ and (d) cathode N^+ regions after breakdown. (e) Ni mapping of (i) normal silicide, (ii) anode P^+, and (iii) cathode N^+. The scale bar is 200 nm [13].

The aforementioned study primarily utilized low-resolution characterization techniques via TEM. Low-resolution characterization typically offers a broader field of view, facilitating an understanding of the overall morphology and key features of the sample, as well as identifying target areas for further high-resolution analysis. Through high-resolution characterization, it is possible to observe the sample's details in greater depth, obtain higher resolution images, and acquire more precise structural information [27–29]. This enables the revelation of the sample's internal microstructure, crystal defects, and interfacial characteristics with finer detail.

The high-resolution TEM characterization results of a single fin of FinFET device are shown in Fig. 6 [17]. Fig. 6a illustrates the overall structure of a single fin, revealing a height of 46 nm and a width of 10 nm, resulting in a height-to-width ratio close to 5:1, indicative of a tall and narrow fin structure. Fig. 6b displays the arrangement of silicon atoms inside the fin, exhibiting clear double-atomic "dumbbell" structures. The upper part of the fin corresponds to the (100) silicon plane, with a nearest neighbor spacing of 1.3 Å; the sidewalls correspond to the (110) silicon plane, with a nearest neighbor spacing of 1.9 Å. Fig. 6c details the structure of the gate dielectric layer, which includes approximately 1 nm of SiO_2 as an interlayer and about 2 nm of HfO_2, forming the typical high-k metal gate (HKMG) stack structure. Fig. 6d and 6e present a comparative atomic-scale STEM characterization of TDDB in gate dielectrics for devices subjected to V_{gs}-only voltage ($V_{ds} = 0$) and on-state TDDB ($V_{ds} > 0$) soft breakdown. Following soft breakdown, both V_{gs}-only TDDB and on-state TDDB modes reveal dielectric breakdown induced epitaxy (DBIE) in the gate dielectric layer. This is attributed to defect accumulation during breakdown, leading to increased leakage current and localized positive thermal feedback. Compared to V_{gs}-only TDDB, the DBIE formation point (breakdown initiation point)

979-8-3503-6184-1/24 $31.00 ©2024 IEEE

in on-state TDDB shifts from the root to the middle of FinFET fins (Fig. 6d and 6e). Additionally, strain analysis of silicon lattice was conducted for devices before and after on-state breakdown. The interface of intact devices appears smooth with a uniform gradient in silicon lattice (Fig. 6f), whereas devices after soft breakdown exhibit significant protrusions at the interface accompanied by strain-induced stretching (Fig. 6g).

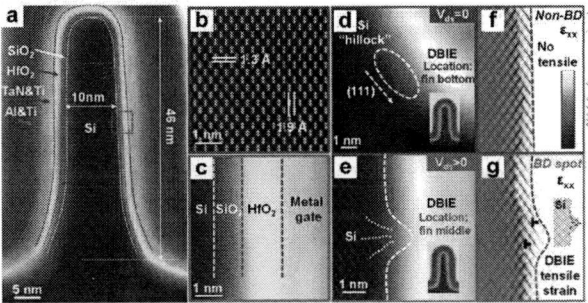

Fig. 6. (a)-(c) HRSTEM image of virgin fin and its atomistic structure. (d)-(g) HRSTEM images of DBIE mophologies compared with Vgs-only TDDB and on-state TDDB [17].

With the advancement of scientific research, researchers are no longer satisfied with static analysis but are increasingly focusing on the dynamic responses of materials, such as crystal growth, phase transitions, electrochemical reactions, and mechanical deformations. This shift has sparked significant interest, leading to the development of *in situ* TEM . This specialized technique is designed for real-time observation and analysis of materials' microstructure and dynamic behaviors under various environmental and experimental conditions [30–32]. In contrast to traditional TEM methods, *in situ* TEM allows researchers to simulate specific operating conditions, such as high temperatures, pressures, different atmospheres, or electrical fields, enabling real-time observation and analysis of the evolution of material devices during experiments [33].

Fig. 7 illustrates the *in situ* TEM images of the gate structure under a constant voltage of 1.9 V, along with its corresponding high-resolution transmission electron microscopy (HRTEM) images [22]. Fig. 7a-d depict a series of low-magnification TEM images taken during stress loading, where areas marked by green dashed ellipses illustrate the expansion of defects in both vertical and horizontal directions. Fig. 7e-g present high-magnification images. The cross-sectional HRTEM image in Fig. 7e reveals the $ZrO_2/Al_2O_3/InGaAs$ layer, with Al_2O_3 in an amorphous state. Initially, the sample is in pristine condition with no observable defects and low gate current. Experimental findings indicate an extremely low leakage current density (10^{-8} A cm^{-2}) under low bias voltage. Upon applying a constant voltage of 1.9 V, the gate current sharply rises to 3×10^{-7} A within 60 s, accompanied by an order of magnitude increase observed through electrical monitoring. Simultaneously, the contrast of the ZrO_2 layer undergoes a sudden change, revealing defects such as bit errors and other structural features. After breakdown, new grain boundary distortions are detected in the ZrO_2 layer, absent in the initial sample. These distortions coincide with a hillock-like morphology appearing on the InGaAs substrate. The fast Fourier transform (FFT) analysis in Fig. 7f-g confirms a phase transition to cubic zirconia in the region after breakdown, whereas the ZrO_2 phase before breakdown is monoclinic.

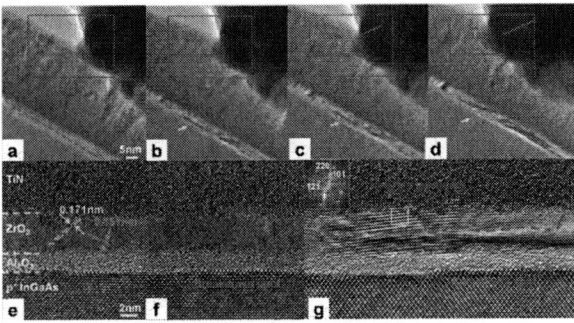

Fig. 7. *In situ* TEM images of the gate stack under a constant stressing voltage of 1.9 V and corresponding HRTEM images [22].

V. CONCLUSION

This paper reviews *in situ* localization of defects in advanced semiconductor devices, discussing various methods and their applications at different scales. At the macro-scale, techniques like EFA swiftly identify and assess major defects in devices. At the micro-scale, FIB-SEM and CR techniques are widely used for detailed analysis of defect structures and morphology. At the atomic-scale, TEM enables precise localization and compositional analysis of individual atoms, revealing the nature and formation mechanisms of defects. Multiscale analysis tools provide comprehensive insights into the defect formation mechanisms and the impact on device reliability, thereby providing strong basis for the optimization of advanced devices. In the future, *in situ* localization of defects techniques will continue to evolve alongside technological advancements, aiding the semiconductor industry in addressing coming challenges.

REFERENCES

[1] Mack CA, "Fifty years of Moore's Law," IEEE Transactions on Semiconductor Manufacturing, vol. 24, pp. 202–207, 2011.

[2] Schaller RR. "Moore's law: past, present and future," IEEE Spectrum, vol. 34, pp. 52-59, 1997.

[3] Z. Chen et al., "Research on the application of microscopic analysis technology in PCB inspection and failure analysis," International Conference on Electronic Packaging Technology (ICEPT), IEEE: Shihezi City, China, pp. 1–5, 2023.

[4] H. Choi et al., "High resolution short defect localization in advanced FinFET device using EBAC and EBIRCh," IEEE 24th International Symposium on the Physical and Failure Analysis of Integrated Circuits (IPFA), IEEE: Chengdu, pp. 1–4, 2017.

[5] A. Graff et al., "Physical failure analysis methods for wide band gap semiconductor devices," IEEE International Reliability Physics Symposium (IRPS), IEEE: Burlingame, CA, USA, pp. 3B.2-1-3B.2-8, 2018.

[6] X. Wu et al., "Atomic scale modulation of delf-rectifying resistive switching by interfacial defects,". Advanced Science, vol. 5, pp. 1800096, 2018.

[7] X. Wu et al., "Role of oxygen vacancies in HfO2-based gate stack breakdown," Applied Physics Letters, vol. 96, pp. 172901, 2010.

[8] X. Wu et al., "Electrode material dependent breakdown and recovery in advanced high-κ gate stacks," Applied Physics Letters, vol. 96, pp. 202903, 2010.

[9] K. Ouyang et al., "Fault localization of functional failure by using dynamic EMMI analysis technique," International Conference on Electronic Packaging Technology (ICEPT), IEEE: Dalian, China, pp. 1–3, 2022.

[10] Z. Sun et al., "PEM/OBIRCH in failure localization of flip-chip," International Conference on Electronic Packaging Technology (ICEPT), IEEE: Wuhan, China, pp. 1272–1274, 2016.

[11] Y. Li et al., "A comprehensive review for micro/nanoscale thermal mapping technology based on scanning thermal microscopy," Journal of Thermal Science, vol. 31, pp. 976–1007, 2022.

[12] L. Tian et al., "Failure analysis of damaged dielectric on resistor and capacitor with EMMI and IR-OBIRCH," IEEE International Symposium on the Physical and Failure Analysis of Integrated Circuits, IEEE: Singapore, Singapore, pp. 1–4, 2012.

[13] X. Chen et al., "Direct visualization of breakdown-induced metal migration in enhanced modified lateral silicon-controlled rectifiers," IEEE Transactions on Electron Devices, vol. 68, pp. 1378–1381, 2021.

[14] F. Crupi et al., "A comparative study of the oxide breakdown in short-channel nMOSFETs and pMOSFETs stressed in inversion and in accumulation regimes," IEEE Transactions on Device and Materials Reliability, vol. 3, pp. 8, 2003.

[15] F. Crupi et al., "A novel methodology for sensing the breakdown location and its application to the reliability study of ultrathin Hf-silicate gate dielectrics," IEEE Transactions on Electron Devices, vol. 52, pp. 1759–1765, 2005.

[16] T. Garba-Seybou et al., "Location of oxide breakdown events under off-state TDDB in 28nm N-MOSFETs dedicated to RF applications," 2023 IEEE International Reliability Physics Symposium (IRPS), pp. 1–8, 2023.

[17] Z.Y. Dong et al., "Catching the missing EM consequence in soft breakdown reliability in advanced FinFETs: impacts of self-heating, on-state TDDB, and layout dependence," 2023 IEEE Symposium on VLSI Technology and Circuits (VLSI Technology and Circuits), IEEE: Kyoto, Japan, pp. 1–2, 2023.

[18] V. Brogden et al., "Material sputtering with a multi-ion species plasma focused ion beam," Advances in Materials Science and Engineering, pp. 1-9, 2021.

[19] Y. Liu et al., Direct-writing of 2D diodes by focused ion beams," Advanced Functional Materials, vol. 31, pp. 2102708, 2021.

[20] M. Peng et al., "Time-resolved focused ion beam microscopy: modeling, estimation methods, and analyses," IEEE Transactions on Computational Imaging, vol. 7, pp. 547-561, 2021.

[21] Z.J. Zhang et al., "The trends of in situ focused ion beam technology: toward preparing transmission electron microscopy lamella and devices at the atomic scale," Advanced Electronic Materials, vol. 8, pp. 2101401, 2022.

[22] X. Wu et al., "Probing and manipulating the interfacial defects of InGaAs dual‐layer metal oxides at the atomic scale," Advanced Materials, vol. 30, pp. 1703025, 2018.

[23] C. Luo et al., "Probing gate dielectrics for two-dimensional electronics at atomistic scale using transmission electron microscope," IEEE Trans Electron Devices. vol. 70, pp. 1499–1508, 2023.

[24] X. Wu and L.T. Sun, "Advanced methodologies for atomic-scale nanofabrication and dynamic characterization," Proceedings of the 20th IEEE International Symposium on the Physical and Failure Analysis of Integrated Circuits (IPFA); IEEE: Suzhou, China, pp. 393–399, 2013

[25] X.Q. Chen et al., "Nanoscale analysis of breakdown induced crack propagation in DTSCR devices," IEEE International Reliability Physics Symposium (IRPS), IEEE: Dallas, TX, USA, pp. P48-1-P48-5, 2022.

[26] S.J. Pennycook et al., "Material structure, properties, and dynamics through scanning transmission electron microscopy," Journal of Analytical Science and Technology, vol. 9, pp. 11, 2018..

[27] S. Zheng et al., "Electron-beam-induced current (EBIC) imaging technique to quicken polysilicon defect localization in MOSFETs," Microelectronics Reliability, vol. 128, pp. 114432, 2022.

[28] X. Yang et al., "Transmission line pulse induced breakdown of FinFETs."

[29] X. Yang et al., "Metal migration induced breakdown from gate contact in bulk FinFET devices," IEEE International Symposium on the Physical and Failure Analysis of Integrated Circuits (IPFA), IEEE: Singapore, Singapore, pp 1–4, 2021.

[30] X. Yang et al., "A review of in situ transmission electron microscopy study on the switching mechanism and packaging reliability in non-volatile memory," Journal of Semiconductors, vol. 42, pp. 013102, 2021.

[31] G.H. Ryu et al., "*In-situ* atomic-scale dynamics of thermally driven phase transition of 2D few-layered 1T PtSe$_2$ into ultrathin 2D nonlayered PtSe crystals. Chemistry of Materials, vol. 31, pp. 9895–9903, 2019.

[32] F.C. Shen et al., "Atomic-scale investigation of electromigration with different directions of electron flow into high-density nanotwinned copper through *in situ* HRTEM," Acta Materialia, vol. 219, pp. 117250, 2021.

[33] Y.W. Zhang et al., "Review of electrical stimulus methods of in situ transmission electron microscope to study resistive random access memory," Nanoscale, vol. 14, pp. 9542–9552, 2022.

Wafer-Level Characterization of Ring-Oscillators Frequency Degradation in FinFET Technology

Hao Chang[1]*, Dan Gao[1], Yongsheng Sun[1], and Junlin Huang[1]

[1] COT Department, Hisilicon Technologies Co., LTD, Shenzhen 518129, China

* Email: changhao18@huawei.com

Abstract—**In this paper, frequency degradation of ring-oscillator (RO) is systematically investigated in advanced FinFET technology. To link the time evolutions of RO circuit degradation with the device parameter shift, a high-resolution wafer-level frequency characterization unit is introduced using the measure-stress-measure method. Both dynamic and static aging modes are compared in aspects of time exponent, accelerate factor and recovery behavior. The different contribution of hot carrier injection (HCI) and bias temperature instability (BTI) can explain the behavioral differences between the two aging modes. This work is helpful for establishing correlation between device and circuit degradation.**

Keywords—ring-oscillator (RO), bias temperature instability (BTI), hot carrier injection (HCI)

I. INTRODUCTION

Bias temperature instability (BTI), hot carrier injection (HCI) and time dependent dielectric breakdown (TDDB) are considered as main degradation mechanisms effecting CMOS logic circuit functionality [1]. These mechanisms introduce defects that degrade device performance, such as threshold voltage (V_T) mobility (μ) drift and idsat degradation, resulting in circuit performance drop and even long-term functional failures [2]. Based on above device-level degradation mechanisms, conventional circuit aging models are established [3]. To find the correlation between discrete device and circuit degradation, RO has been widely researched considering the consistency of its frequency degradation with product Fmax degradation, and its measurability under wafer level situations [4].

As a basic unit of the RO and digital logic circuit, the inverter suffers different reliability mechanisms in static and dynamic working states (shown in Fig. 1). BTI/TDDB dominates the static aging period while HCI mainly occurs in dynamic level switch phase [5]. Therefore, long-term RO frequency degradation introduced by BTI depends on the duty cycle of stress signal; while degradation due to HCI is in direct proportion to the circuit operating frequency [6]. Besides, AC BTI recovery effect should also be considered [7].

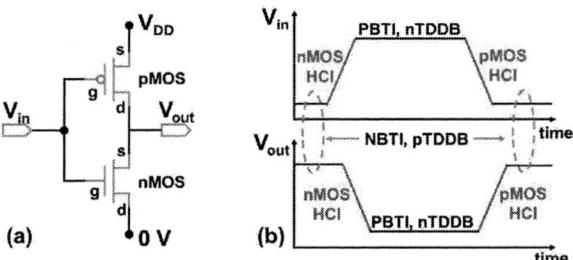

Fig. 1. (a) CMOS inverter and (b) domainnent device degradation modes during different work states.

With the introduction of FinFET structure into CMOS mainstream manufacturing [8], circuit aging modeling faces challenges due to complex degradation mechanisms [9]. In this paper, a systematic study of RO frequency degradation is developed in high-scaled FinFET technology. Based on the wafer-level frequency measurement units (FMUs), different aging and recovery behaviors in dynamic and static modes are founded and explained.

II. EXPERIMENTAL SETUP

The RO used in this work are fabricated with advanced high-k metal-gate FinFET technology. The schematic diagram is shown in Fig. 2, a NAND gate followed by 100 stages of inverters end-to-end forms the main RO structure. After the level shift structure, the RO signal is sent to a 2048-stage divider. To avoid the impact of frequency division structure aging on frequency measurement, the common drain voltage (VDD) terminals of RO and divider are individually connected to different supplies, and the leakage current are recorded with SMUs. The divider output signal is captured by the FMUs for frequency measurement.

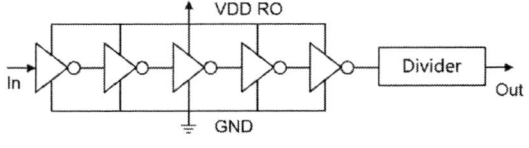

101 Stages Inverter

Fig. 2. Simplized schematic diagram of RO in this work. The NAND gate works as an enable gate and the first stage of RO. For dynamic aging mode, $V_{EN} = VDD_{RO} = V_{STR}$; for static aging mode, $V_{EN} = 0V$, $VDD_{RO} = V_{STR}$.

The frequency measurement process is followed by the traditional measure-stress-measure (MSM) method. The RO is biased under stress voltage (V_{STR}) for a certain period, then reduced to operating voltage during measurement period. The FMUs record the divider output, and extract the frequency through fast Fourier transform (FFT). The FFT program converts the time-domain output signal into frequency-domain signal, which effectively eliminate the impact of noise on frequency measurement. Under appropriate sample frequency and sample step setting, the test unit can detect a frequency change over 0.1%, thereby ensuring data accuracy. All measurements are repeated three times, and the averaged results are shown in this paper.

III. RESULTS AND DISCUSSION

A. Measure Delay Influence

The first issue to verify is the impact of measure delay on characterization results. As shown in Fig. 3, the RO is stressed in dynamic mode for 1ks with varying measure delays. The degradation of the RO frequency (Δf) shows a power-law distribution, with the time slope decreasing from 0.32 to 0.26

979-8-3503-6184-1/24 $31.00 © 2024 IEEE

as the measurement delay is reduced from 40.88s to 0.092s. The BTI recovery effect is suppressed with the reduction in measure delay, resulting in a decrease in time slope [4]. However, when the measure delay is below 0.488s, the time slope remains at 0.26, which indicates that the discharge of fast traps is already saturated. It also should be noticed that at longer stress time, the impact of measure delay clearly diminishes, and the total frequency degradation tends to be the same [10]. In this study, measure delay of 0.488s is chosen unless otherwise specified.

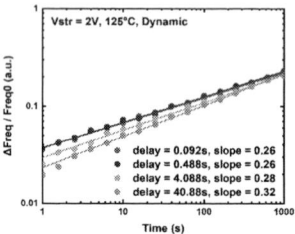

Fig. 3. Time evolutions of frequency degradation in dynamic mode with meausre delay from 0.092s to 40.88s. Stress condition: Vstr = 2 V, 125°C.

B. Degradation Behaviors in Dyanmic and Static Stresses

The RO is then stressed under different V_{STR} in dynamic mode. It can be seen from Fig. 4 (a) that, no obvious change in time slope is observed across the V_{STR} range. For further verification, the output current of VDD_{RO} (I_{VDD}) is recorded by SMUs. The I_{VDD} comes from the short-circuit current generated when the N/PMOS of each inverter are conducting simultaneously, so it also reflects the electric performance degradation of transistors during stress. As shown in Fig. 4 (b), the I_{VDD} exhibits power-law degradation versus stress time, and a consistent time slope is seen across different V_{STR}, which indicates that the degradation mechanism has not changed.

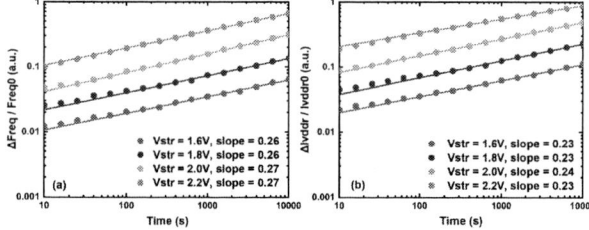

Fig. 4. Time evolutions of (a) frequency and (b) I_{VDD} degradations in dynamic mode with different V_{STR} from 1.6 V to 2.2 V at 125°C.

For comparison, the RO is also stressed in static mode. During the static aging mode, the enable signal is set to zero, and each level inverter of the RO keeps at a constant output. As shown in Fig. 5 (a), at V_{STR} of 2V, the frequency degradation in static mode is 1.85 times of that in dynamic mode. The time slope of Δf is 0.23 for static aging and 0.27 for dynamic aging. More intuitively, the time slope of I_{VDD} degradation in static mode is 0.18 (Fig. 5 (b)), which is close to NBTI time slope of discrete device (around 0.167 derived by Reaction-Diffusion model) [11]. The total Δf after 10ks dynamic and static stresses are plotted as a function of V_{STR} in Fig. 5 (c), the voltage accelerate factors (VAF) are extracted through power-law fitting. The lower VAF in static mode (~ 6.14) can also be contributed to the higher fraction of BTI, which is close to NBTI VAF (~ 5) of discrete device [13]. While the higher time slope of 0.27 and higher VAF of 7.33 in dynamic mode can be attributed to more HCI contribution

[10]. It also should be noted that, considering the DC TDDB effect, we don't recommend applying a relatively high V_{STR} at static mode. At a static stress of 2.2V, the time slope of Δf shows a two-stage distribution, and I_{VDD} increases abnormally at later stage of stress (not shown here), which indicates the unwilling oxide damage.

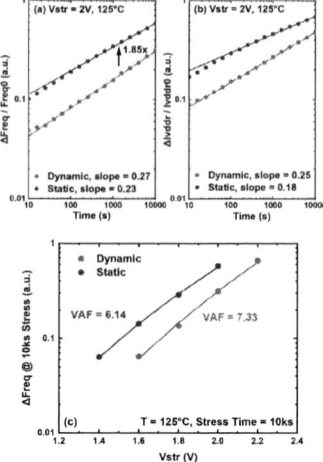

Fig. 5. Comparisons of (a) frequency and (b) I_{VDD} degradations in dynamic and static modes at V_{STR} of 2V. (c) Total frequency degradation as a function of V_{STR} in dynamic and static modes.

As shown in Fig. 6, the temperature dependence is also discussed. The activation energy (Ea) of 0.17 eV extracted in static stress is consistent with the Arrhenius model which describes kinetic degradation caused by NBTI [14]. While a relatively low Ea of 0.12 eV is observed in dynamic stress, which is more close to the parameters of HCI in discrete device. Interestingly, dynamic Δf degradation exhibit two-stage power-law distribution at temperature of 75°C and 100°C. For stress time below 1ks, a BTI-like time slope of 0.22 is seen. As stress time exceeds 1ks, HCI gradually takes the dominant position, and raise the time slope to 0.27. This also explains the phenomenon observed in Fig. 3 where the long-term degradation gradually tends to be the same at different delays. As the proportion of unrecoverable HCI degradation increases, the recovery effect due to measurement delay diminishes.

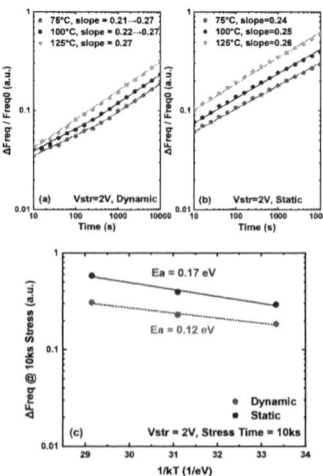

Fig. 6. Time evolutions of frequency degradation at different temperatures ranging from 75°C to 125°C in (a) dynamic and (b) static mode. Vstr = 2V.

C. Recovery Behaviors after Dynamic and Static Stresses

The recovery behaviors of RO in dynamic and static modes are also compared. As shown in Fig. 7, a group of ROs are subjected to 1ks stress followed by 1ks recovery cycle. The V_{STR} and V_{REC} are 2V and 0V, respectively. The static stress exhibits a 2.16 times of Δf at the end of stress cycle compared to the dynamic mode. However, after the recovery cycle, the recovery ratio in static mode reaches 28.3%, compared to the 11.4% recovery ratio in dynamic mode. The different recovery ratios also corresponds to the BTI fraction.

As explained in Fig. 8, in static aging mode, only half of PMOS transistors is subjected to DC NBTI stress, but the total stress time is over 2 times than dynamic mode. While in dynamic mode, PMOS transistors are subjected to AC NBTI stress (~GHz), a large part of NBTI degradation has already been recovered during stress. In addition, HCI, the other dominant mechanism in dynamic mode, has no significant recovery effect by itself [10]. As a result, the dynamic mode shows a lower recovery ratio compared to static mode.

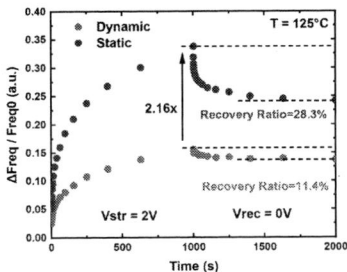

Fig. 7. Time evolutions of frequency degradation in dynamic and static modes for 1ks stress (V_{STR}=2V) followed by a 1ks recovery cycle (V_{REC}=0V).

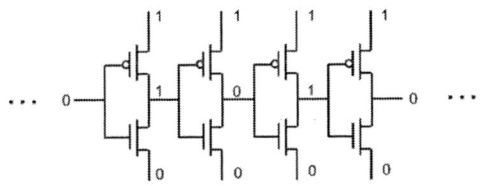

| PMOS | NBTI | Off-HCI | NBTI | Off-HCI |
| NMOS | Off-HCI | PBTI | Off-HCI | PBTI |

Fig. 8. Main degradation mechanisms of N/PMOS of each stage of inverters in static aging mode.

IV. CONCLUSION

In this paper, a systematic study of dynamic and static RO frequency degradation is discussed from accurate wafer-level characterization data. A higher NBTI contribution leads to the observed lower time slope (~0.23), lower VAF (~6.14), higher Ea (0.17 eV) and higher recovery ratio (~28.3%). These aging parameters based on circuit-level degradation are compared and show good consistency with discrete device models. As for dynamic mode, HCI dominates in long-term degradation over 1ks stress time, and NBTI degradation is suppressed due to high frequency level switch (~GHz). These results bring a comprehensive perspective in the correlation between RO frequency degradation with discrete device aging, which shows guiding significance for circuit reliability design.

REFERENCES

[1] T. Nigam, B. Parameshwaran, and G. Krause, "Accurate product lifetime predictions based on device-level measurements," *2009 IEEE International Reliability Physics Symposium*, Montreal, QC, Canada: IEEE, 2009, pp. 634–639.

[2] F. Cacho, P. Mora, W. Arfaoui, X. Federspiel, and V. Huard, "HCI/BTI coupled model: The path for accurate and predictive reliability simulations," *2014 IEEE International Reliability Physics Symposium*, Waikoloa, HI, USA: IEEE, Jun. 2014, p. 5D.4.1-5D.4.5.

[3] J. Peng, D. M. Huang, G. F. Jiao, and M. F. Li, "A reliability model for CMOS circuit based on device degradation," *2012 IEEE 11th International Conference on Solid-State and Integrated Circuit Technology*, Xian, China: IEEE, Oct. 2012, pp. 1–3.

[4] A. Kerber, T. Nigam, P. Paliwoda, and F. Guarin, "Reliability Characterization of Ring Oscillator Circuits for Advanced CMOS Technologies," *IEEE Trans. Device Mater. Relib.*, vol. 20, no. 2, pp. 230–241, Jun. 2020.

[5] C. Schlunder *et al.*, "HCI vs. BTI? - Neither one's out," *2012 IEEE International Reliability Physics Symposium (IRPS)*, Anaheim, CA, USA: IEEE, Apr. 2012, p. 2F.4.1-2F.4.6.

[6] D. Sengupta and S. S. Sapatnekar, "Estimating Circuit Aging Due to BTI and HCI Using Ring-Oscillator-Based Sensors," *IEEE Trans. Comput.-Aided Des. Integr. Circuits Syst.*, vol. 36, no. 10, pp. 1688–1701, Oct. 2017.

[7] J. Diaz-Fortuny, P. Saraza-Canflanca, E. Bury, M. Vandemaele, B. Kaczer, and R. Degraeve, "A Ring-Oscillator-Based Degradation Monitor Concept with Tamper Detection Capability," *2022 IEEE International Reliability Physics Symposium (IRPS)*, Dallas, TX, USA: IEEE, Mar. 2022, pp. 1–7.

[8] J.-W. Han, H. Y. Wong, D.-I. Moon, N. Braga, and M. Meyyappan, "Stringer Gate FinFET on Bulk Substrate," *IEEE Trans. Electron Devices*, vol. 63, no. 9, pp. 3432–3438, Sep. 2016.

[9] R. Wang, Z. Yu, J. Zhang, Z. Sun, Z. Zhang, and R. Huang, "Understanding Hot Carrier Degradation and Variation in FinFET Technology," *2020 IEEE 15th International Conference on Solid-State & Integrated Circuit Technology (ICSICT)*, Kunming, China: IEEE, Nov. 2020, pp. 1–4.

[10] A. Kerber, X. Wan, Y. Liu, and T. Nigam, "Fast Wafer-Level Stress-and-Sense Methodology for Characterization of Ring-Oscillator Degradation in Advanced CMOS Technologies," *IEEE Trans. Electron Devices*, vol. 62, no. 5, pp. 1427–1432, May 2015.

[11] H. Kufluoglu and M. A. Alam, "A Generalized Reaction–Diffusion Model With Explicit H–H2 Dynamics for Negative-Bias Temperature-Instability (NBTI) Degradation," *IEEE Transactions on Electron Devices*, vol. 54, no. 5, pp. 1101-1107, May 2007.

[12] Y. Kim, H. Shim, M. Jin, J. Bae, C. Liu and S. Pae, "Investigation of HCI effects in FinFET based ring oscillator circuits and IP blocks," *2017 IEEE International Reliability Physics Symposium (IRPS)*, Monterey, CA, USA, 2017, pp. 4C-2.1-4C-2.4,

[13] M. Jin *et al.*, "Reliability characterization of 10nm FinFET technology with multi-VT gate stack for low power and high performance," *2016 IEEE International Electron Devices Meeting (IEDM)*, San Francisco, CA, USA: IEEE, Dec. 2016, pp. 15.1.1-15.1.4.

[14] A. Kerber et al., "Device reliability metric for end-of-life performance optimization based on circuit level assessment," *2017 IEEE International Reliability Physics Symposium (IRPS)*, Monterey, CA, USA, 2017, pp. 2D-3.1-2D-3.5.

Exhaustive Application-Dependent Testing for FPGA Interconnect Resources

Wenwei Chen [1], Xinyu He, Tongshu Ding, Jian Wang, Jinmei Lai *

[1] State Key Laboratory of ASIC and System, School of Microelectronics, Fudan University, Shanghai 201203, China

* Email: chenww23@m.fudan.edu.cn, jmlai@fudan.edu.cn

Abstract—Up to now, in the field of application-dependent testing for FPGA interconnect resources, the most widely used method is single-term function method, and the most accurate fault model is asymmetric bridging fault model. In this paper, a new fault model called Boolean bridging fault model is proposed, which is a superset of asymmetric bridging fault model. Based on single-term function method, we minimize the number of test configurations to the greatest extent possible, and theorems in set theory ensure that the result is optimal for both asymmetric bridging fault model and Boolean bridging fault model. Moreover, for special cases called feedback bridging faults, the study demonstrates that repeated sampling is a simple method that can achieve high coverage.

Keywords—*FPGA, application-dependent test, bridging fault, single-term function method, interconnect resource*

I. INTRODUCTION

Field programmable gate arrays (FPGAs) can implement arbitrary user-specific designs. In some scenarios, users don't need the entire FPGA to be completely fault-free, and they only care about whether a specific design can work correctly. Thus, application-dependent testing (ADT) has emerged.

The single-term function method [1, 2, 3, 4, 5, 6, 7, 8] is the most widely used in the ADT for FPGA interconnect resources, which will generate a series of partially modified configurations by reconfiguring the functions of look-up tables (LUTs) and the initial values of the flip-flops (FFs). By testing those modified configurations, interconnect resource faults in the original configuration can be detected in an interpretable way.

The exhaustive fault list for FPGA interconnect resources contains stuck-at faults (SAFs) of all nets and bridging faults (BFs) between all net pairs. If the number of nets in the circuit is M, the number of potential BFs is M(M-1)/2. Up to now, the most accurate fault model is asymmetric BF model [9], which is adopted by [6, 7, 8]. Moreover, the work in [6] points out that if asymmetric BFs form combinational loops, then oscillations may happen in the circuit. These special cases are called feedback BFs.

The work of this paper includes:

- Boolean BF model. Based on the internal structure of FPGA interconnect resources, there are many inverse nodes in the nets, which can lead to different fault behaviors. Boolean BF model, which is a superset of asymmetric BF model, includes these situations.

- Sperner Codes. For exhaustive asymmetric BF list, we minimized the number of configurations required for single-term function method, by generating a special set of net codes called Spener codes. The set theory theorem in [10] proves that the result is optimal.

- Katona Codes. For exhaustive Boolean BF list, we minimized the number of configurations required for single-term function method, by generating a special set of net codes called Katona codes. The set theory theorem in [11] proves that the result is optimal. Compared with Sperner codes, Katona codes can cover Boolean BFs at the cost of 1-2 configurations.

- Repeated sampling method. For feedback BFs, the study shows some limitations of existing approaches, and demonstrates that repeated sampling is a simple method that can achieve high coverage.

The paper is structured as follows. Section II introduces the single-term function method and existing fault models briefly. Section III describes Boolean BF model. In Section IV, we present the simple steps to generate Sperner codes and Katona codes, and propose the repeated sampling methods. In Section V, the experimental results are compared with previous works.

II. BACKGROUND

A. Single-Term Function Method

The prerequisite of the single-term function method is that the circuit is composed of nets, LUTs and FFs in the FPGA. All the FFs are synchronized by the same clock signal. The circuit does not contain any other components, such as carry chains, BRAMs and DSPs, or combination loops, such as latches and ring oscillators. An example is shown in Fig. 1.

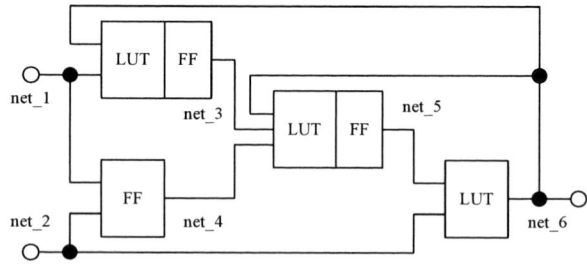

Fig. 1. An example design in FPGA

The exhaustive fault list contains stuck-at faults (SAFs) of all nets and bridging faults (BFs) between all net pairs. Meanwhile, the custom fault list is a subset of the exhaustive fault list, which depends on the topology of the circuit and the user-defined fault list strategy. In this paper, the exhaustive fault list is adopted.

The core of single-term strategy is generating logical vectors, which outputs an M×N logical matrix, where M is the number of nets in the circuit, and N presents the number of logical vectors. Each element in the matrix is a Boolean value.

Once the logical matrix is determined, each logical vector will be transformed into a modified test configuration in an interpretable and automatic way [2,3,4]. Thus, the number of configurations is also equal to N.

In each modified configuration, the function of each LUT is configured to the single-term function corresponding to the values of its inputs and output, and the initial value of each FF is configured to the value of its output. Moreover, the test vector is also determined by the values of all primary inputs.

If the error value caused by the fault is sampled, it will definitely propagate to observable outputs, which is promised by the property of single-term functions. Therefore, the target is to minimize the number of logical vectors while satisfying the sensitization conditions of the fault list.

B. Existing Fault Models

SAF model includes stuck-at-0 and stuck-at-1.

The sensitization condition of the exhaustive SAF list is that all nets need to be applied 1 and 0 at least once.

Symmetric BF model include wired-AND and wired-OR.

The sensitization condition of the exhaustive symmetric BF list is that all net pairs need to be applied (0, 1) or (1, 0) at least once.

Asymmetric BF model include dominant, dominant-AND and dominant-OR [9].

The sensitization condition of the exhaustive asymmetric BF list is that all net pairs need to be applied (0, 1) and (1, 0) at least once.

For single-term function method, if all asymmetric BFs are covered completely, then all SAFs and symmetric BFs will also be covered completely. Thus, asymmetric BF model is the most accurate fault model.

In most cases, a fault will result in a stable error value. when the error value is sampled, it will propagate to outputs, and the required number of test clock cycles is not greater than the maximum sequential depth of the circuit. However, the work in [6] points out that if asymmetric BFs form combinational loops, then oscillations may happen in the circuit, and the sampling results will be uncertain. These special cases are called feedback BFs.

III. BOOLEAN FAULT MODEL

In the field of ADT for FPGA interconnect resources, nets are regarded as simple equipotential bodies, and their internal structure is neglected, which deviates from reality. In modern FPGA, routing is implemented by SRAM-controlled buffered multiplexers [12], which are also known as programmable interconnect points (PIPs) [13], as shown in Fig. 2.

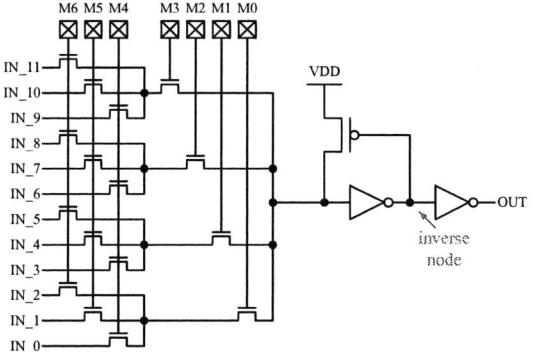

Fig. 2. The structure of programmable intereconnect point (PIP) in FPGA

Each buffer consists of two inverters cascaded together, so there is an inverse node in each buffer. BFs related to inverse nodes may lead to different phenomena, called Boolean BF

model. The characteristic is that even if the logical values of two nets are the same, the Boolean BF can still cause error results. Thus, the sensitization condition of the exhaustive Boolean BF list is that all net pairs need to be applied (0, 1), (1, 0), (0, 0), (1, 1) at least once.

IV. EXHUASTIVE ADT FOR INTERCONNECT RESOURCES

A. Limitations of Existing Approaches

The existing approaches have the following limitations:

- The EXH technique in [6] can generate logical vectors for exhaustive asymmetric BF list, but it is not optimal, which will be solved by Sperner codes in IV.B.

- Exhaustive Boolean BF list has not been researched, which will be solved by Katona codes in IV.C.

- The FDB technique in [6] is presented to test feedback BFs, but it is only applicable to combinational circuits, or sequential circuits without feedback signals, such as pipelines, and it does not work on sequential circuits with feedback signals, such as state machines. Besides, feedback BFs caused by Boolean BFs can't be covered. Repeated sampling in IV.D will solve these problems.

The framework of our work is shown in Fig. 3.

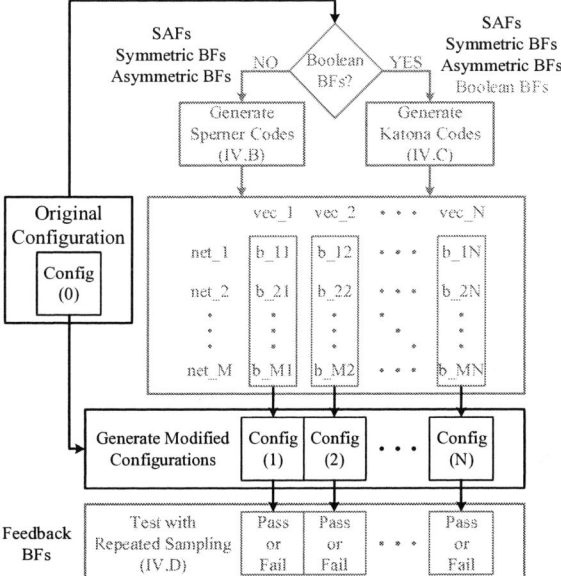

Fig. 3. The framework of exhaustive ADT for interconnect resources

B. Sperner Codes

The optimal logical vectors for exhaustive asymmetric BF list, called Sperner codes, can be generated as follows:

- Calculate N based on M, so that N is the smallest positive integer that satisfies $C(N, \lfloor N/2 \rfloor) \geq M$.

- Generate all N-bit binary codes with $\lfloor N/2 \rfloor$ bits of 1 and $\lceil N/2 \rceil$ bits of 0. Obviously, there are $C(N, \lfloor N/2 \rfloor)$ codes meeting the requirement. Select any M codes from them as row vectors to form an M×N matrix, then column vectors are the desired optimal logical vectors.

The set theory theorem in [10] proves the logical vectors generated by Sperner codes is optimal for asymmetric BFs.

C. Katona Codes

The optimal logical vectors for exhaustive Boolean BF list, called Katona codes, can be generated as follows:

- Calculate N based on M, so that N is the smallest positive integer that satisfies $C(N-1, \lfloor N/2 \rfloor - 1) \geq M$.

- Generate all (N-1)-bit binary codes with $\lfloor N/2 \rfloor - 1$ bits of 1 and $\lceil N/2 \rceil$ bits of 0. There are $C(N-1, \lfloor N/2 \rfloor - 1)$ codes meeting the requirement. Select any M codes from them as row vectors to form an M×(N-1) matrix, and add an all-1 column to form an M×N matrix, then column vectors are the desired optimal logical vectors.

The set theory theorem in [11] proves the logical vectors generated by Katona codes is optimal for Boolean BFs.

Compared with Sperner codes, Katona codes can cover Boolean BFs at the cost of 1-2 configurations.

D. Repeated Sampling Method

For feedback BFs, oscillations may happen, and the result of a single sampling may be just the same as the fault-free case, which prevents the fault from being detected.

Increasing test clock cycles and sampling repeatedly is a simple and effective method to capture error values. In the scheme, the testing time for each configuration is set to 1ms, which is much shorter than the typical bitstream downloading time. Assuming the test frequency is 50MHz, approximately 50000 test clock cycles can be added to capture the potential oscillations. The periods of oscillations caused by BFs cannot reach the millisecond scale. Therefore, the sampling results come from different oscillation periods. Moreover, the jitter of the oscillation will accumulate, causing the sampling results to exhibit randomness on the millisecond scale. Once there is one sampling value that is different from the expected value, the feedback BF is detected. Thus, high coverage of feedback BFs can be achieved by using repeated sampling.

V. RESULTS AND COMPARISONS

The approaches are applied to ISCAS'89 large-scale benchmark circuits implemented on FPGA xc7vx690t, and the experiments are listed in Table I. Sperner codes can cover exhaustive asymmetric BF list, which is the same as the EXH technique in [6], and the number of configurations can be reduced significantly. Meanwhile, Katona codes can cover exhaustive Boolean BF list, which is more complete, and the number of configurations is still much lower.

TABLE I.   EXPERIMENT RESULTS FOR ISCAS'89 BENCHMARKS

Circuit	#nets	#configurations		
		This work		[6]
		Sperner codes	Katona codes	EXH
s1238	522	12	13	18
s1423	674	12	13	18
s1488	661	12	13	18
s5378	2814	14	15	22
s9234	5633	15	16	24
s13207	8013	16	17	24
s15850	9849	16	17	26
s35932	16100	17	18	26
s38417	22207	17	18	28
s38584	19291	17	18	28

Both repeated sampling method and the FDB technique in [6] aim at covering feedback BFs, and the comparisons are listed in Table II. Repeated sampling method has advantages. Firstly, it doesn't need additional configurations. Secondly, feedback BFs caused by Boolean BFs can be covered. Thirdly, it can still work when there are feedback signals from FFs.

TABLE II.   TEST METHODS FOR FEEDBACK BFS

Comparisons	This work	[6]
	Repeated sampling	FDB
additional configurations	0	2
feedback BFs caused by asymmetric BFs	YES	YES
feedback BFs caused by Boolean BFs	YES	NO
combinational circuits	YES	YES
sequential circuits without feedback signals	YES	YES
sequential circuits with feedback signals	YES	NO

ACKNOWLEDGMENT

This work was supported in part by the National Natural Science Foundation of China under Grant 62074101G.

REFERENCES

[1] M. B. Tahoori, E. J. McCluskey, M. Renovell and P. Faure, "A multi-configuration strategy for an application dependent testing of FPGAs," *22nd IEEE VLSI Test Symposium, 2004. Proceedings.*, Napa Valley, CA, USA, 2004, pp. 154-159.

[2] M. B. Tahoori, "Using Satisfiability in Application-Dependent Testing of FPGA Interconnects," Proceedings 2003. Design Automation Conference (IEEE Cat. No.03CH37451), 2003, pp. 678-681.

[3] M. B. Tahoori, "Application-Dependent Testing of FPGA Interconnects," Proceedings 18th IEEE Symposium on Defect and Fault Tolerance in VLSI Systems, 2003, pp. 409-416.

[4] M. B. Tahoori, "Application-Dependent Testing of FPGAs," in IEEE Transactions on Very Large-Scale Integration (VLSI) Systems, vol. 14, no. 9, pp. 1024-1033, Sept. 2006.

[5] T. N. Kumar and F. Lombardi, "A Novel Heuristic Method for Application-Dependent Testing of a SRAM-Based FPGA Interconnect," in IEEE Transactions on Computers, vol. 62, no. 1, pp. 163-172, Jan. 2013.

[6] A. Cilardo, "New Techniques and Tools for Application-Dependent Testing of FPGA-Based Components," in IEEE Transactions on Industrial Informatics, vol. 11, no. 1, pp. 94-103. 2015.

[7] S. Banik and S. Roy, "Application Dependent Testing of FPGA Interconnect Using Satisfiability Modulo Theory," 2018 3rd International Conference for Convergence in Technology (I2CT), 2018, pp. 1-5.

[8] S. Banik, S. Roy and B. Sen, "Application-Dependent Testing of FPGA Interconnect Network," in IEEE Transactions on Very Large-Scale Integration (VLSI) Systems, vol. 27, no. 10, pp. 2296-2304, Oct. 2019.

[9] J. M. Emmert, C. E. Stroud and J. R. Bailey, "A new bridging fault model for more accurate fault behavior," 2000 IEEE Autotestcon Proceedings. IEEE Systems Readiness Technology Conference. Future Sustainment for Military Aerospace (Cat. No.00CH37057), 2000, pp. 481-485.

[10] E. Sperner, "Ein Satz über Untermengen einer endlichen Menge," Math Z 27, 544–548 (1928).

[11] G.O.H. Katona, "Two applications (for search theory and truth functions) of Sperner type theorems," Periodica Mathematica Hungarica 3.1(1973):19-26.

[12] A. Boutros and V. Betz, "FPGA Architecture: Principles and Progression," in *IEEE Circuits and Systems Magazine*, vol. 21, no. 2, pp. 4-29, Secondquarter 2021.

[13] US Patent 7,202,698 B1, Xilinx Inc, Integrated Circuit Having a Programmable Input Structure with Bounce Capability, 2007-4-10.

A Comprehensive and Efficient Instruction-level Testing Method for Processor

Zixin Yang [1], Zhichao Wei [1], Huanlin Luo [2], Jian Wang [1], Jinmei Lai*[1]

[1] State Key Laboratory of ASIC and System, School of Microelectronics, Fudan University, Shanghai 201203, China
[2] Shanghai Academy of Spaceflight Technology, Shanghai 201109, China

* Email: jmlai@fudan.edu.cn

Abstract—Testing of the processor is crucial, which necessitates a comprehensive test set to ensure an efficient and thorough coverage. We herein propose an Equivalence Class Partitioning and Boundary Value Analysis (ECP-BVA) constrained instruction stream generation method on coverage-guided fuzzing (CGF). The updated ECP-BVA constraints mutate the instruction field to maximize the coverage metrics. The accelerated coverage-guided fuzzing skipes redundant instructions to improve the testing efficiency. An efficient test set is built by fuzzing and the experimental results indicate that our method can maximize coverage metrics close to 100% in a short time, thereby ensuring a comprehensive and efficient coverage for processor testing.

Keywords—processor testing, instruction stream, coverage-guided fuzzing

I. INTRODUCTION

Highly efficient methods for processor testing are crucial to find hardware vulnerabilities existing in processor, which can be described as undocumented instructions [1] and instruction flaws [2] at the instruction-level. They require strong test set generation techniques to ensure a comprehensive testing process. Due to the continuous expansion of processors function, manually writing testcases is time-consuming and pure random generation provides only very limited coverage.

Tradionally, instruction stream generation used for processor testing can be achieved using model-based approach, which is separated from the architecture description. The generator uses format specifications to guide instruction generation and is capable of integrating constraints generated by algorithms. An abstract constraint satisfaction problem (CSP) based insturction generation framework [3] was proposed to reduced generation fail rate. Since the model may not fully captures all functional characteristics of the processor, the test coverage is insufficient.

New methods have been developed to ensure a thorough testing process, such as coverage-guided fuzzing (CGF) technique [4]-[5]. A CGF approach has been proposed to improve the integrated functional coverage by utilizing constraint-based mutation [5]. However, the custom mutation suffer from limitations as no modification in registers and immediates of the instructions, resulting in an incomplete coverage. Besides, the fuzzing process is complex and time-consuming, as the fixed constraints.

In this study, we proposed an ECP-BVA constrained instruction stream generation method on coverage-guided fuzzing for processor testing. The main contributions of this work are: (1) The updated ECP-BVA constraints tailored for instruction generation can mutate the instruction field towards specific cases, which effectively maximizes the functional coverage metrics. (2) The coverage-guided fuzzing skips a considerable number of redundant instructions and upgrades the ECP-BVA constraints to accelerate the fuzzing process, resulting in the improvement of testing efficiency. (3) The method that enables a comprehensive and efficient instruction-level coverage can also be used to test the applications with related Instruction Set Architecture (ISA), such as disassemblers, ISA simulators, CPU emulators.

II. BACKFROUND

A. Functional coverage metrics

We define the functional coverage metrics for register (Rx) and immediate (Ix) which are suitable for a large set of ISAs [5]. In the condition when an instruction with one destination register (Rd) and one source register (Rs), the coverage metric R_1 is covered when $Rd = Rs$ and $Rd \neq Rs$ are each satisfied at least once. The coverage metric R_2 extends R_1 to the case where the instruction has one destination register (Rd) and two source registers ($Rs1$, $Rs2$). R_2 is covered when each of the following four conditions [5] occurs at least once.

- $Rs1 = Rd \wedge Rs2 = Rd$
- $Rs1 = Rd \wedge Rs2 \neq Rd$
- $Rs1 \neq Rd \wedge Rs2 = Rd$
- $Rs1 \neq Rd \wedge Rs2 \neq Rd \neq Rd \wedge Rs1$

The metrics $V(Rx)$ and $V(Ix)$ necessitate that the Rx and Ix satisfy each value from the set $\{min, -1, 0, 1, max\}$ at least once, where min and max represent the smallest and largest possible values of Rx and Ix. For immediates interpreted as unsigned values (e.g. shift amounts), negative values are excluded from the set. For an instruction including the Rx and Ix parameters, all related metrics should be applied to the corresponding parameters, respectively.

B. Instruction format of ARM ISA

The testing ARM cortex-M3 processor [6] featuring a Harvard bus architecture and a 3-stage pipeline, implements the ARMv7-M ISA. In order to create valid instructions, we adhere to the constraints of the machine instruction format of ARMv7-M ISA [7], which includes functional field of condition opcode (*cond*), instruction opcode (*opcode*) and non-functional field of destination register (*Rd*), register that contains the first operand (*Rs*), shifter operand.

III. INSTRUCTION GENERATION FOR PROCESSOR TESTING

A. Testing Framework

Our insturction-level processor testing method is shown in Fig. 1. First, a test set is generated by ECP-BVA constrained instruction stream generation on coverage-guided fuzzing

979-8-3503-6184-1/24 $31.00 © 2024 IEEE

(Step 1). Then, the test set is used to perform functional testing of the processor at instruction-level (Step 2) by comparing the results with the reference Instruction Set Simulator (ISS).

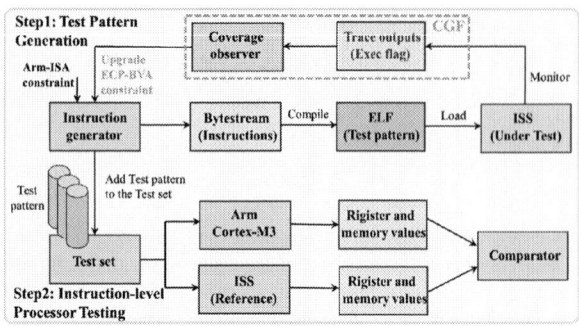

Fig. 1. Testing framework of ECP-BVA constrained instruction stream generation on coverage-guied fuzzing.

B. ECP-BVA constrained instruction stream generation

The instruction generator (Fig. 1) are added with specific constraints to guide the generation towards interesting cases. An ADD instruction is initially formed by injecting *cond* and *opcode*, while maintaining randomization of the non-functional fields adhere to ARM instruction format (Fig. 2a).

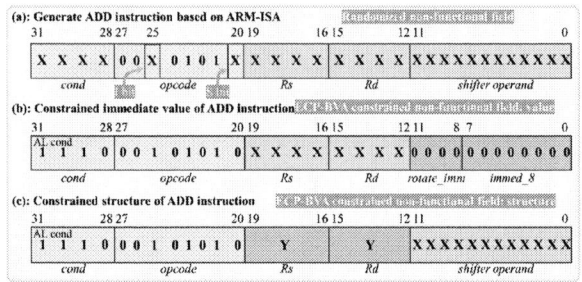

Fig. 2. Illustration of the constraints for instruction generation.

We then change the non-functional field based on ECP-BVA constraints. We have developed a set of rules for constraining the values of the instructions. For example, the 32-bit fixed-point addition data can be divided as: positive, zero and negative based on ECP. Then according to BVA, the corresponding boundary values are selected: the smallest and largest positive number (0x00000001, 0x7FFFFFFF); the zero number (0x0); the smallest and largest negative number (0x80000000, 0xFFFFFFFF). Besides, 0x80000001 is added to get 6 test points, because its operation with 0xFFFFFFFF results in the smallest negative number. Fig. 2b shows an example where a legal 32-bit zero number encoded as an 8-bit (*immed_8*) and 4-bit (*rotate_imm*) immediate according to Data-processing operands [7] with a corresponding *cond* and *opcode*, which satisfies one condition of the V(*Ix*) of ADD instruction. We have also added rules constraining register structure. For registers, general purpose registers are divided into low registers (*R0 – R7*) and high registers (*R8 – R12*), where *R0* is hardwired and thus a special case. The constraints include: set *Rd* to *R0*, set *Rd* to the value of *Rs1* or *Rs2*, and mutate *Rs1* to match *Rs2*. Fig. 2c demonstrates an example where the register values of *Rd* and *Rs* being randomized but kept equal in an ADD instruction. Moreover, the ECP-BVA constraints are updated with the coverage (Section III-C).

In addition to the generation of individual instructions, the instruction sequence is also considered, which comprises a

series of instructions tailored to accomplish a specific task (e.g. loading large immediates, computational chain, etc.). A new instruction sequence will be generated with a probability of 10%. Otherwise (no existing or new instruction sequence), a single randomized instruction will be generated with a 20% probability of ECP-BVA constraints, as shown in Fig. 3.

Input: An instruction sequence *seq*, random probability *rp*
Output: A generated instruction *ins*

1. **if** *seq* ≠ nill and *seq*.next() **then**
2.   *ins* ← *seq*.next()
3. **end if**
4. **if** *rp* **then**
5.   *seq* ← sequence generator()
6.   *ins* ← *seq*.start()
7. **else**
8.   *ins* ← Randomized non-functional field(*ins*)
9.   **if** *rp* **then**
10.   *ins* ← ECP-BVA constrained value(*ins*)
11.   **end if**
12.   **if** *rp* **then**
13.   *ins* ← ECP-BVA constrained structure(*ins*)
14.   **end if**
15. **end if**

Fig. 3. Instruction generation algorithm.

C. Coverage-guided fuzzing

As shown in Fig. 1, after instructions generation, the instruction stream which is consist of a series of instructions, is compiled and linked by GCC compiler. Then the executable file (ELF), serving as a test pattern, is loaded into the ISS (Under Test) by a script. The ISS is initialized to ensure that all executions start in the same state. Subsequently, we observe the execution of the test pattern in ISS by tracing the Exec flag, which includes a series of internal state information (e.g. time, program counter, memory address, opcode, etc.) after one instruction is completed. Fig. 4 shows our algorithm for coverage-guided fuzzing. After capturing the flag, we analyze its functional coverage information (Line 1-9), which illustrates two specific cases (Fig. 2b-c) as an example.

Input: An Exec flag *exec*, a test pattern *tp*
Output: Coverage *cove*, Constraints *ECP-BVA*, testset *set*

1. **while** *exec* ≠ nill **do**
2.   **if** *exec*.opcode == ADD **then**
3.   **if** *exec*.Rd == *exec*.Rs **then**
4.   *cove*.ADD.match(1)
5.   **end if**
6.   **if** *exec*.Imm == 0 **then**
7.   *cove*.ADD.match(2)
8.   **end if**
9.   **end while**
10. **if** *cove*.new.match ≠ nill **then**
11.   *set* ← Testset generation(*tp*)
12.   **if** *cove*.new.match in *ECP-BVA* **then**
13.   *ECP-BVA* ← Constrain_upgrade(*cove*.new.match)
14.   **end if**
15. **end if**

Fig. 4. Algorithm of coverage-guided fuzzing for the coverage observer.

If the execution of a test pattern can increase the functional coverage information, then the test pattern will be added to the test set (Line 11). In addition, the ECP-BVA constraints will be updated based on the functional coverage information. If the newly satisfied coverage metric exists within the ECP-BVA constraints, then the corresponding constraint will be removed (Line 12-13).

IV. EVALUATION

In this section, we present the evaluation results of the proposed method. We built the test set extracted from the ISS Under Test on ARM Cortex-M3 processor and ISS Reference, respectively. The goal of our experiments is to evaluate the coverage and efficiency of our approach for processor testing.

A. Test setup

As Device-under-test (DUT), we used an ARM Cortex-M3 core based STM32F103C8T6 MCU. As ISS reference and under test, we implemented the ARM ISA in Gem5 simulator. The instruction generator and coverage observer were all written using Python. All experiments were conducted on Ubuntu 20.04 LTS operating system.

B. The Improvement of Functional Coverage

Table I shows the evaluation results where the coverage was measured on the ISS Under Test. The second column of the table shows the required time to generate the entire test set. The third column shows the functional coverage obtained by different methods.

Compared with randomized instruction generation (Ran.), ECP-BVA constraints can significantly improve the functional coverage. It is noteworthy that our method (CGF) is capable of maximizing most coverage metrics to 100% and no bug is found in the DUT, while only the metric V(Rd) is 85.6%. The coverage is better than the previous report [5], which was experimented on an RISC-V core. The ARM and RISC ISA have similarities in functionality (e.g. load-store architecture) and instruction structure (Rd, Rs, shifter operand, etc.). Additionally, ARM ISA has a larger number of instructions but the same data bit width, making the comparison of metrics meaningful due to the same level of traversal complexity. The reason for a low V(Rd) metric is that the value of Rd relies on the operand values and the operation. Certain results may be impossible for some operations. Meanwhile, V(Rd) can be improved by adding test cases manually. Our proposed method efficiently produces a comprehensive test set comprising 4506 test patterns (method to obtain it in Section III-C), varying in complexity with a mere 1 instruction to the most intricate one with 29 instructions, with an average of 3 instructions per test pattern, ensuring a broad coverage for functional testing.

TABLE I.   FUNCTIONAL COVERAGE MEASURED AT ISS UNDER TEST

Gene rator	Time (s)	Functional coverage (%)					
		R₁	R₂	V(Rd)	V(Rs1)	V(Rs2)	V(Imm)
[5]	32492	100	100	81.1	98.2	100	100
Ran.	36213	89.1	77.6	64.8	67.4	70.6	69.3
CGF	13949	100	100	85.6	100	100	100

C. The Improvement of Testing Efficiency

We leveraged the Ratio of Legal Instructions (RLI) [8] evaluation metric to evaluate the efficiency of the test set. The formula of RLI is shown in equation (1), where the legal instructions refers to an instruction increases the coverage.

$$\frac{Number\ of\ legal\ instructions\ in\ test\ set}{Number\ of\ total\ instructions\ in\ test\ set} \times 100\% \qquad (1)$$

We demonstrate the effectiveness of our proposed method by comparing with the no fuzzing process (No fuzz.). Fig. 5 shows the experimental results measured at the six different coverage metrics, where the vertical axis is the number of the required instructions in test set when increasing 10% coverage. The average instruction number has been reduced from about 1145 to 229 as illustrated in Fig. 5 and the RLI has been increased from 2.2% (No fuzz.) to 11.5% (CGF) using (1). The results indicate that CGF can significantly improve the test efficiency because the fuzzing process skips a considerable number of redundant instructions.

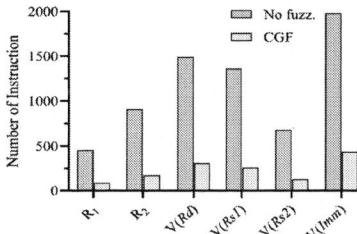

Fig. 5.   Comparison of the number of the required instructions tested by no fuzzing method and CGF.

Morever, Table I indicates that our CGF approach takes significantly less time to achieve similar functional coverage compared with the previous study [5]. This is because the ECP-BVA constraints are updated with the functional coverage information to accelerate the fuzzing process, which furthur improves the efficiency of the testing.

V. SUMMARY

Test coverage and efficiency need to be improved in instruction generation. We propose an ECP-BVA constrained instruction stream generation on CGF for processor testing. The updated ECP-BVA constraints effectively maximize most coverage metrics to 100%. The CGF skips redundant instructions and is accelerated by the updated constraints to improve the testing efficiency. Taken together, our proposed approach takes less time when generating the test set up to a better coverage compared to the previous report, serves as an effective instruction-level method for processor testing.

REFERENCES

[1]   X. Li, Z. Wu, Q. Wei, and H. J. I. A. Wu, "Uisfuzz: An efficient fuzzing method for cpu undocumented instruction searching," IEEE Access, vol. 7, pp. 149224-149236, 2019.

[2]   C. J. B. H. Domas, "Breaking the x86 ISA," vol. 1, pp. 1-6, 2017.

[3]   Y. Katz, M. Rimon, and A. Ziv, "Generating instruction streams using abstract CSP," in 2012 Design, Automation & Test in Europe Conference & Exhibition, 2012, pp. 15-20.

[4]   N. Bruns, V. Herdt, D. Große, and R. Drechsler, "Efficient cross-level processor verification using coverage-guided fuzzing," in Proceedings of the Great Lakes Symposium on VLSI 2022, 2022, pp. 97-103.

[5]   V. Herdt, D. Große, H. M. Le, and R. Drechsler, "Verifying instruction set simulators using coverage-guided fuzzing," in 2019 Design, Automation & Test in Europe Conference & Exhibition, 2019, pp. 360-365.

[6]   "Cortex™ -M3 Technical Reference Manual," https://documentation-service.arm.com/static/5e8e107f88295d1e18d34714

[7]   "ARM Architecture Reference Manual," https://documentation-service. arm.com/static/5f8dacc8f86e16515cdb865a.

[8]   G. Wang, Z. Zhu, X. Cheng, and D. Meng, "A High-Coverage and Efficient Instruction-Level Testing Approach for x86 Processors," IEEE Transactions on Computers, vol. 72, no. 11, pp. 3203-3217, 2023.

Thermal Effect and Calibration for High Precision On-Wafer Analog IC Probe Testing

Daisuke Iimori [1], Takayuki Nakatani [1], Shogo Katayama [1], Gaku Ogihara [1], Shuhei Yamamoto [1], Misaki Takagi [1], Yujie Zhao [2], Jianglin Wei [3], Anna Kuwana [1], Keno Sato [4], Takashi Ishida [4], Toshiyuki Okamoto [4], Tamotsu Ichikawa [4], Kentaroh Katoh [5], Kazumi Hatayama [1], and Haruo Kobayashi*[1]

[1] Gunma University, Japan 376-8515
[2] Shenyang University of Chemical Technology, China 110142, [3] Yibin University, China 644000
[4] ROHM Semiconductor, Japan 222-8537, [5] Fukuoka University, Japan 814-0180

* Email: koba@gunma-u.ac.jp

Abstract—This paper describes issues and countermeasures of thermal effects on high precision on-wafer probe testing of analog ICs. Due to recent demands for high reliability and low power of IoT systems, μV-order DC voltage and nA-order DC current in analog ICs need to be tested with good linearity at low cost for mass production. There thermo-electromotive force (TEF) has to be taken care of for high precision measurement using micro-probes. As its countermeasure, a de-embedding calibration for the thermal effects is proposed. Our experiments show that the TEF fluctuation depends on the probe materials and it is proportional to the temperature. Also, it depends on the contact pressure between the probes and the PAD. Our experiments show that it can be calibrated down to about 1 μV by the de-embedding method with the constant probe contact pressure, in the temperature range from 30 °C to 130 °C; this is sufficient in most cases of IoT application analog IC testing.

Keywords—*high precision analog IC testing, on-wafer probe testing, thermo-electromotive force, DC-AC conversion, calibration*

I. INTRODUCTION

Wafer probe test costs of analog ICs at their mass production stage are much less than the corresponding test costs of assembled packages [1, 2]. There, we recognize that the thermal effect or thermo-electromotive force (TEF) of the device under test (DUT) with prober (Fig. 1) is an error factor in high-precision DC signal testing, and recently this becomes serious for IoT device testing. The TEF is the voltage arising from the difference in the temperature at two points along the circuit, known as Seebeck effect. For on-wafer IC testing using automatic test equipment (ATE), the temperature difference between the probe card and the DUT on the wafer can be often higher than 100 ℃ (Fig. 2). However, to the best of our knowledge, their intensive research results have not been reported, though some tutorials for thermal effects and small DC voltage measurements are in [3, 4, 5].

This paper investigates the thermal effect and proposes its calibration method for accurate DC characteristics testing of analog IC at the mass production stage. (i) We found that the thermal effect (or TEF) depends on the probe materials (copper, tungsten) and the contact pressure between the probes and the PAD. These were investigated with experiments. (ii) Then, as their calibration, the de-embedding method is proposed: first, measure the TEF using "load" status and "short" status circuits under the test in the experimental test system environment, and next calibrate by subtracting the output voltage of the circuits under the test in "load" status

from the one in "short" status. Experiment was conducted to show the effectiveness of our proposed method quantitatively.

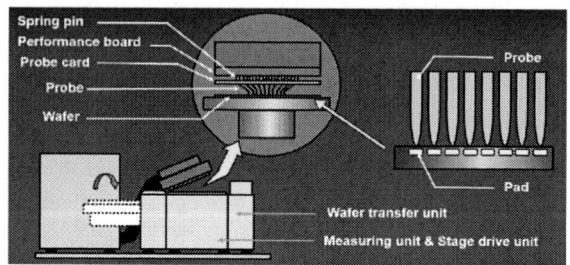

Fig. 1. On-wafer IC testing system.

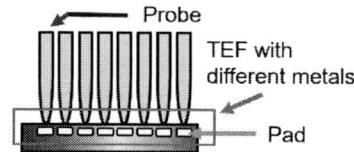

Fig. 2. Temperature effect causing part between probes and DUT in on-wafer IC testing system.

II. ANALOG IC TESTING WITH ON-WAFER PROBING

Probes are used to test each die before it is cut from the wafer (Fig. 1). When conducting temperature tests using a prober for high-precision analog IC tests that handle μV order, its TEF affects the measurement accuracy.

We modeled the TEF in the wafer probe testing (Fig. 3).

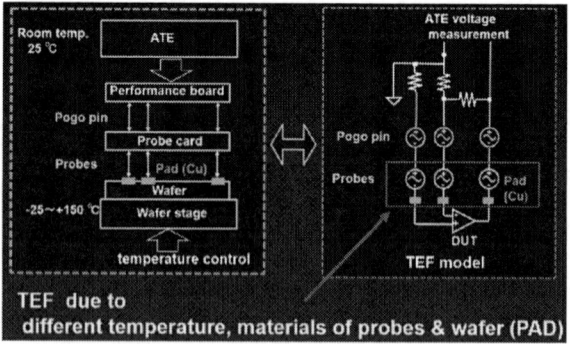

Fig. 3. TEF model in on-wafer testing system using a prober.

The temperature test at the wafer stage was performed in the range from 30 °C to 150 °C. The TEF is caused by the

979-8-3503-6184-1/24 $31.00 © 2024 IEEE

temperature difference between the wafer and the probe card (Fig. 2), and also caused by different metal connections.

III. TEF MEASUREMENT AND CALIBRATION METHODS

A. High Precision Measurement by DC-AC Conversion

Here, high-precision measurement by DC-AC conversion is explained. Measurement of μV-order DC signal with a digital multi-meter (DMM) suffers from low frequency noise or system noise and hence the DC-AC measurement method was employed [6, 7, 8]. The TEF voltages were measured in load status and short status on the order of several μV accuracy using DC-AC conversion measurements (Fig. 4). The prober was realized equivalently using a hot plate and a jack; the jack was moved up and down to make the probe and the PAD to be into contact.

Also, the combination of probes in load status and short status is as follows: in load status, three probes are used, such as tungsten, copper and copper (3 probes). In short status, two probes are used, such as tungsten and copper (2 probes).

Figs. 5 (a) (b) show the DC-AC conversion measurement systems in load status and short status, respectively.

Fig. 6 shows the actual measurement environment setup. The TEF generated by the probes was converted to an electronic AC signal, which was analyzed by FFT in the LabVIEW. The board was covered with Styrofoam to prevent radiation and thermal convection from the hot plate.

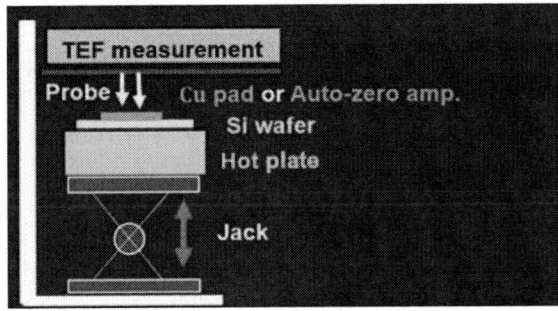

Fig. 4. Probe TEF measurement using hot plate and jack.

B. Calibration Method

Proposed de-embedding calibration consists of measurement of the TEF using "load" status and "short" status circuits and calibration by subtracting V_{short} from V_{load} (Fig. 7).

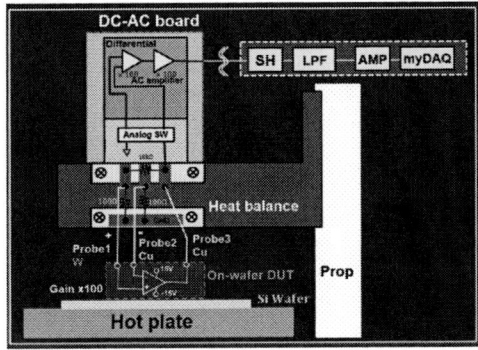

(a) Load status.

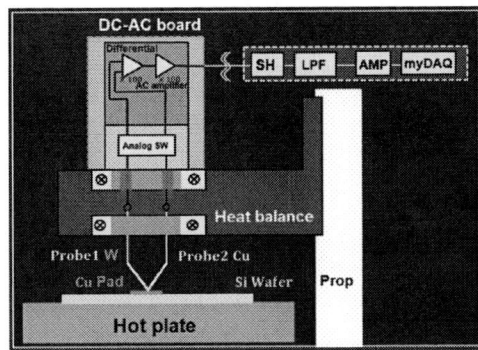

(b) Short status.

Fig. 5. TEF measurement system using DC-AC conversion with different combinations of probe materials, using in the whole system in Fig. 4.

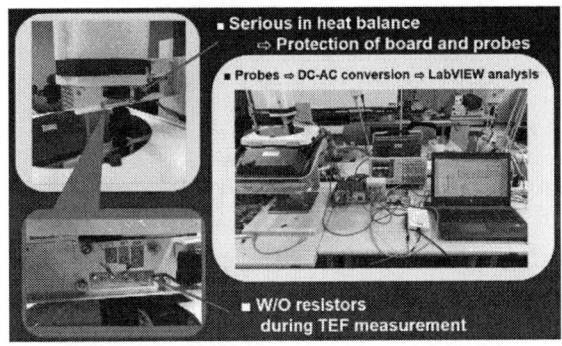

Fig. 6. Measurement environments with different combinations probe materials.

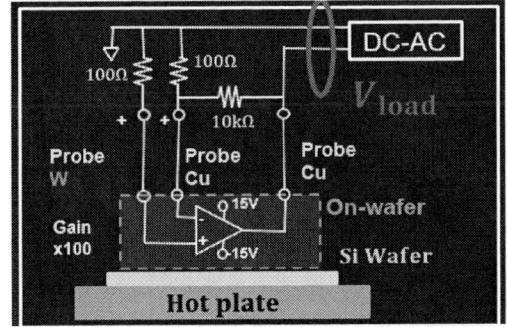

(a) Load status.

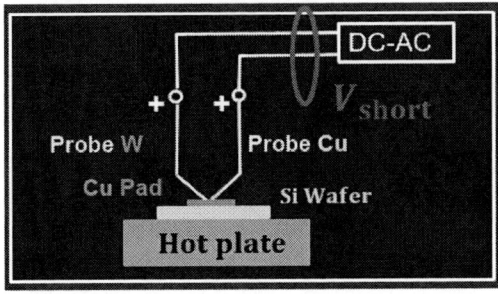

(b) Short status.

Fig. 7. V_{load} and V_{short} in the temperature effect calibration using the proposed de-embedding method.

IV. EXPERIMENTAL RESULTS

The thermal effect depending on the probe contact pressure was measured in two ways. First, the experiment was

979-8-3503-6184-1/24 $31.00 © 2024 IEEE   193

performed at room temperature to investigate the TEF due to the contact pressure. A wafer is placed directly on the scale meter. Next, the experiment was performed to measure TEF variation due to temperature change using the hot plate. It was placed on the top of the scale meter, and the wafer was placed on it for measurement.

Fig. 8 shows the experiment result of V_{short} at room temperature when the contact pressure was changed; there is a proportional relationship between the contact pressure and V_{short}. Also, it depends on the probe materials; in case of two probes of "tungsten, copper" (different materials), V_{short} is much larger than the one in case of "tungsten, tungsten" (the same materials). Different probe materials cause larger TEF.

Fig. 9 shows the experimental result of V_{short} at several temperatures when the contact pressure changes in case of two probes "tungsten, copper". There is a proportional relationship between the contact pressure and V_{short}; if the contact pressure increases by 0.2 kg, V_{short} increases by 14.5 µV at 100 °C.

Fig. 10 shows the experiment result of V_{load} at several temperatures when the contact pressure was changed in case of three probes "tungsten, copper, copper". There are proportional relationships between the TEF and the contact pressure and between the TEF and the temperature.

Fig. 11 summarizes the measurement errors for V_{load} and V_{short} with the proposed calibration. As the contact pressure changes, the TEF also changes; if the contact pressure is kept constant using a scale meter, the measurement error between V_{load} and V_{short} is as small as 1 µV. The reason for the TEF change due to the contact pressure change would be the changes of the contact area and resistance.

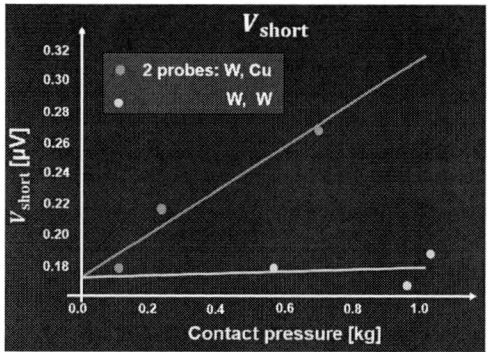

Fig. 8. V_{short} measurement for the contact pressure effect of two probes with W, Cu and two probes with W, W at room temperature in short status.

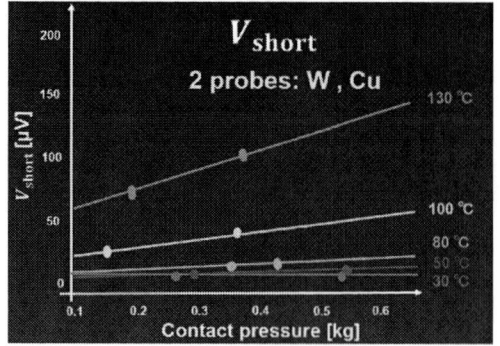

Fig. 9. V_{short} measurement for the contact pressure effect of two probes with W, Cu at varous temperatures in short status.

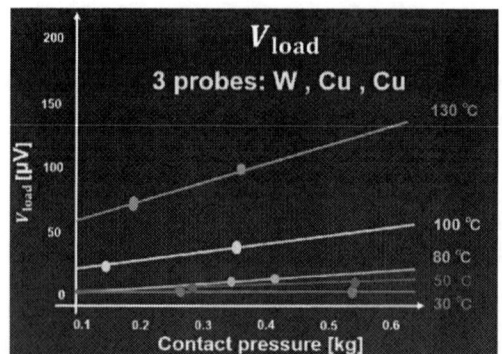

Fig. 10. V_{load} measurement for the contact pressure effect of three probes with W, Cu, Cu at various temperatures in load status.

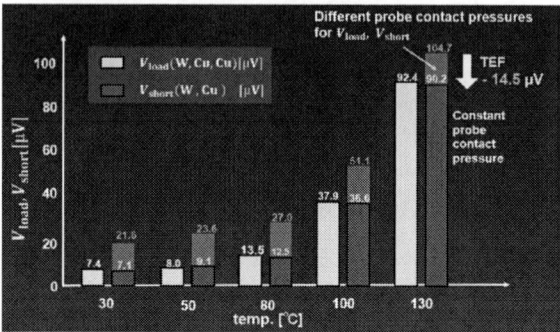

Fig. 11. Summary of TEF measurement errors in load status and short status when the proposed calibration is applied in cases of the same and different contact pressures between both statuses.

V. CONCLUSION

We have investigated the thermal effect or TEF in the on-wafer analog IC testing system and its calibration method. It was found by experiments that the TEF is proportional to the temperature, and it depends on the probe materials and the contact pressure between the probe and the PAD. The thermal effect voltage measurement was performed by DC-AC conversion to avoid the low frequency noise or system noise. We have also proposed its calibration method using the de-embedding method; Our experiments show that the TEF fluctuation of the probe can be calibrated down to about 1 µV by using a scale meter for the constant contact pressure in the temperature range from 30 °C to 130 °C.

REFERENCES

[1] W. R. Mann, et. al., "The leading edge of production wafer probe test technology," IEEE International Test Conference (Oct. 2004).

[2] S. Bhattacharya and A. Chatterjee, "High coverage analog wafer-probe test design and co-optimization with assembled-package test to minimize overall test cost," IEEE VTS (April 2003).

[3] B. Dobki and J. Williams (Editors), Analog Circuit Design-A Tutorial Guide to Application and Solutions, Linear Tech. (Sept. 2011).

[4] Low Level Measurements Handbook, 7th Edition, Precision DC Current, Voltage and Resistance Measurement, Keithley.

[5] K. Blake, "Op amp precision design: PCB layout techniques," MicroChip Application Notes, AN1258.

[6] Y. Sasaki, et. al., "Accurate and fast testing technique of operational amplifier DC offset voltage in µV-order by DC-AC conversion," IEEE ITC Asia (Sept. 2019).

[7] K. Sato, et. al., "Accurate testing of precision voltage reference by DC-AC conversion," IEEE ATS (Nov. 2020).

[8] G. Ogihara, et. al., "Evaluation of high-precision nano-ampere current measurement method for mass production," IEEE ICECS (Nov. 2021).

Deep Learning Design-Flow with Static and Dynamic Optimizations

Zhiqiang Que, Jose G. F. Coutinho, Wayne Luk

Department of Computing, Imperial College London, UK. {z.que, jgfc, w.luk}@imperial.ac.uk

Abstract—This paper presents a novel co-optimization framework that integrates both top-down and bottom-up flows through an automated, cross-stage design methodology to address the inefficiencies of current FPGA-based deep neural network (DNN) accelerator optimization. Our approach addresses two main optimization challenges: the design of multi-level optimization strategies that span software and hardware domains, and the development of effective search mechanisms for fine-tuning these strategies statically and dynamically. By employing a library of customizable tasks for optimization, transformation, and control, our framework facilitates significant reductions in DSP and LUT usage across various DNN architectures without requiring extensive human intervention or deep domain expertise. Results show our approach reduces DSP and LUT usage of two neural networks respectively by up to 16.7 and 6.2 times while maintaining accuracy and requiring little human effort or domain expertise, demonstrating the benefits and potential of the proposed co-optimization framework.

I. INTRODUCTION

Recent advances in deep learning demand efficient, high-performance applications, highlighting the importance of FPGA-based DNN accelerator design and optimization. The challenge lies in combining machine learning expertise with a deep understanding of hardware architecture to balance high inference accuracy with the power consumption, latency, and throughput constraints of FPGA devices. This optimization is often managed currently without systematic considerations, leading to inefficiencies in information flow between application needs and hardware capabilities.

To address this problem, we propose a novel co-optimization framework that integrates both top-down and bottom-up information flows through an automated, cross-stage design methodology. Our work addresses two main optimization challenges: (i) the design of multi-level optimization strategies covering software and hardware domains and the (ii) development of effective search mechanisms for fine-tuning these strategies. Specifically, we present the following contributions:

1) A co-optimization framework for FPGA-based DNN accelerators that enables rapid development of customized cross-stage design flows, automating the design iteration process (Section III).
2) A library of reusable static and dynamic optimization, transformation and control tasks supported by our co-optimization framework (Section IV).
3) A comprehensive evaluation of our framework through diverse benchmarks and optimization strategies (Section V).

II. RELATED WORK

The development of FPGA-based DNN accelerators has benefited significantly from foundational work [1, 2] which pioneered the integration of algorithmic and hardware optimizations. While these efforts introduced the concept of co-design, the need for dynamic feedback mechanisms to continually refine optimization strategies was not fully addressed until later. Recent advances [3] have begun to fill this gap by integrating feedback mechanisms into the co-design process. However, these methodologies typically support only limited dynamic adjustments post-deployment.

Platforms like Xilinx's Vitis AI [4] and Intel's Open-VINO [5] have facilitated the deployment of DNNs on FPGAs but offer limited flexibility for introducing new optimizations beyond their offering. On the other hand, frameworks such as FINN [6] and HLS4ML [7] provide essential building blocks for DNN accelerator construction but lack support for dynamic optimizations. Our framework addresses these limitations by providing an automated, customizable design flow that supports the integration and improvement of optimization strategies using reusable tasks.

Furthermore, while some current frameworks allow developers to describe and customize DNN optimization strategies, their scope is often limited. For instance, ScaleHLS [8] and SOTA-OPT [9] focus on HLS-based optimization strategies and hardware designs, but their optimization scope is restricted to those based on MLIR. TVM [10] is a general-purpose DNN compilation framework that offers performance portability across different types of devices, but optimizations occur only at the graph (IR) level. Our approach extends beyond these methods by facilitating the design and continuous refinement of optimization techniques that span across various domains, from software to hardware.

This paper builds upon our previous work [11] by introducing **control** capabilities within our co-optimization framework. Key enhancements include: (i) the introduction of branch mechanisms to facilitate the selection of static and dynamic optimization paths tailored to specific hardware vendors, enabling a customized approach to DNN accelerator design; (ii) the implementation of feedback loops which integrate insights from lower stages of the design process, thereby influencing and refining decision-making at higher levels of our design approach. These mechanisms support the development of new strategies and allow precise fine-tuning of the optimization process.

979-8-3503-6184-1/24 $31.00 © 2024 IEEE

III. APPROACH

This section outlines our co-optimization approach for creating customizable design-flows that optimize FPGA-based DNNs. A design-flow is a multi-stage pipeline converting high-level specifications into hardware designs, refining model abstractions at each stage based on target device features. For instance, a DNN model in TensorFlow can be converted to a C++ HLS model using HLS4ML, and subsequently to an RTL model using Vivado HLS. Feedback allows optimizations to benefit from insights obtained from runtime profiling.

Fig. 1(a) illustrates our design flow architecture, which integrates pipe tasks and a meta-model. This architecture is structured as a directed cyclic graph, where pipe tasks serve as nodes performing functions such as optimization, and dependencies are represented as edges. Tasks communicate through a unidirectional data flow, becoming active when triggered by the scheduler and remaining idle otherwise. The meta-model functions as a central hub divided into two sections: the configuration (CFG) which contains the parameters for every design-flow task, and the model space (M-SPACE) which stores and manages models generated by O-tasks.

This architecture enables tasks to interconnect and share data through the meta-model, facilitating cross-stage optimizations involving multi-level optimization strategies that encompass both software and hardware domains. Each task includes adjustable parameters that can be dynamically adapted and tuned throughout the design flow.

IV. IMPLEMENTATION

Fig. 1(b) presents a set of pre-built pipe tasks in our customizable framework. Our framework can be further extended to support custom tasks tailored to specific needs. Each pipe task has unique characteristics, including **multiplicity**—the number of input and output channels managed, which influences how tasks can be combined; **parameters**—which allow task behavior to be tuned, such as accuracy tolerance; and **role**, which determines the task type. We consider three types of roles:

1) K**-tasks**: These are generic tasks that control the top-down and bottom-up flows introduced in this paper. Examples include: **JOIN**, which merges multiple paths into one; and **STOP**, which terminates the design flow;

2) O**-tasks**: These are self-contained optimization tasks that enhance deep learning models based on specific objectives and constraints. Our current pipe task repository includes **PRUNING** and **SCALING**, which are implemented using the Keras API (version 2.9.0), and **QUANTIZATION**, which is performed using C++ source-to-source transformations via the Artisan framework [12].
(a) A PRUNING task automatically finds the highest pruning rate that keeps accuracy loss within a specified tolerance ($\leq \alpha_p$), using a binary search method. This ensures efficiency while maintaining performance within acceptable limits. (b) A SCALING O-task systematically

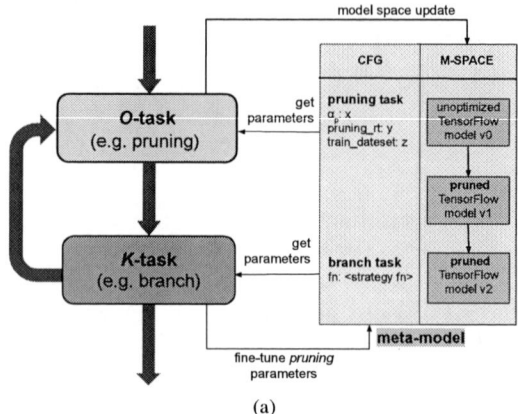

(a)

Type	Role	Multiplicity	Parameters
JOIN	K	many-to-1	-
BRANCH	K	1-to-many	fn: meta-model → bool
STOP	K	1-to-0	fn: meta-model → output
HLS4ML	λ	1-to-1	default_precision IOType FPGA_part_number clock_period test_dataset
VIVADO-HLS	λ	1-to-1	project_dir
INTEL-HLS	λ	1-to-1	project_dir
KERAS-MODEL-GEN	λ	0-to-1	train_en train_test_dataset train_epochs
PRUNING	O	1-to-1	tolerate_acc_loss (α_p) pruning_rate_thresh (β_p) train_test_dataset train_epochs
SCALING	O	1-to-1	default_scale_factor tolerate_acc_loss (α_s) scale_auto max_trials_num train_test_dataset train_epochs
QUANTIZATION	O	1-to-1	tolerate_acc_loss (α_q) train_test_dataset

(b)

Fig. 1. (a) An example connecting an O-task and a K-task. The O-task primarily updates the model space with new model variations, while the K-task handles control flow management. Each connection represents a unidirectional stream. (b) Pipe tasks implemented (the parameter list is simplified due to space constraints).

reduces the DL model size by scaling down layer dimensions, ensuring the model fits within FPGA resource constraints while monitoring whether accuracy loss exceeds a set threshold (α_s). The process halts when accuracy loss surpasses α_s. (c) A QUANTIZATION O-task optimizes HLS C++ DNN model translations by automatically adjusting bitwidths to minimize hardware used and unintended side effects, while ensuring layer dependencies constrain the quantization options and maintain accuracy within a predefined loss threshold (α_q). This process iteratively refines precision settings in the C++ kernel, validated through co-design simulation to keep accuracy loss under the acceptable limit. Additional details about these three optimization tasks are available [11].

3) λ**-tasks**: These tasks perform transformations on the model space, such as synthesis. Examples include **HLS4ML**, which translates a DNN model into an HLS C++ model, and **Vivado-HLS** and **Intel-HLS**, which translate an HLS C++ model into an RTL model.

V. EVALUATION

This section evaluates our approach with the control mechanisms introduced in this paper to support two specific scenarios: **(i)** branching to support path selection for specific FPGA targets (Section V-A), and **(ii)** feedback loops to refine the strategy of the optimization process (Section V-B).

Experiments were conducted using Python 3.9 on benchmark workloads from typical Deep Neural Network (DNN) applications, including jet tagging [13] from high energy physics domain, and time series analysis using Long Short-Term Memory (LSTM) networks [14]. The jet tagging application targets FPGA-based triggers at CERN's LHC, involving an input rate of 40 MHz and response latency under 1 microsecond. Default clock frequencies are set at 100 MHz for AMD Zynq 7020, and at 200 MHz for AMD KU115 and for Intel A10 1150. The HLS4ML library provides a λ-task with 18-bit fixed-point precision, including 8 integer bits.

A. Branching Flow

Fig. 2(a) illustrates a design flow that performs pruning for two alternate targets using the BRANCH K-task, for AMD/Xilinx FPGAs and Intel FPGAs. A user-defined selection function is supplied as a parameter to the BRANCH K-task which encodes a strategy determining the path forward for the design.

Fig. 3 presents the results of applying this design flow to an LSTM model on the MNIST dataset. Specifically, Fig. 3(a) illustrates the pruning rate and accuracy at each step using the PRUNING O-task. The tolerance is set to less than α_p (2%) in this design. Figs 3(b) and (c) show the resource utilization of the LSTM design after each pruning step respectively on an AMD KU115 FPGA and on an Intel A10 1150 FPGA. The DSP consumption is reduced from 6011 (108%) to 2101 (38.1%) on the AMD FPGA after the final pruning rate is optimized to be 71.9%. Compared to AMD's HLS compiler, which prefers DSP blocks, Intel's HLS compiler tends to favor the use of soft multipliers for implementation. As shown in Fig 3(c), most of the computation kernels are implemented using logic resources rather than DSP blocks.

While this design flow currently supports two types of FPGAs, it can be extended to include additional paths such as GPU, CPU and ASIC technologies. Moreover, this evaluation underscores the flexibility of our approach in utilizing the same software optimization task, specifically PRUNING, across multiple hardware targets.

B. Feedback Flow

Fig. 2(b) illustrates a more complex use of the BRANCH K-task for the jet tagging model on an AMD Zynq 7020 FPGA. This design flow supports multiple optimization strategies in each loop iteration: no compression (baseline), pruning, quantization, and combined pruning and quantization. Each strategy is progressively more aggressive in compression and requires more time. For instance, quantization, a hardware optimization, takes longer than pruning, a software optimization.

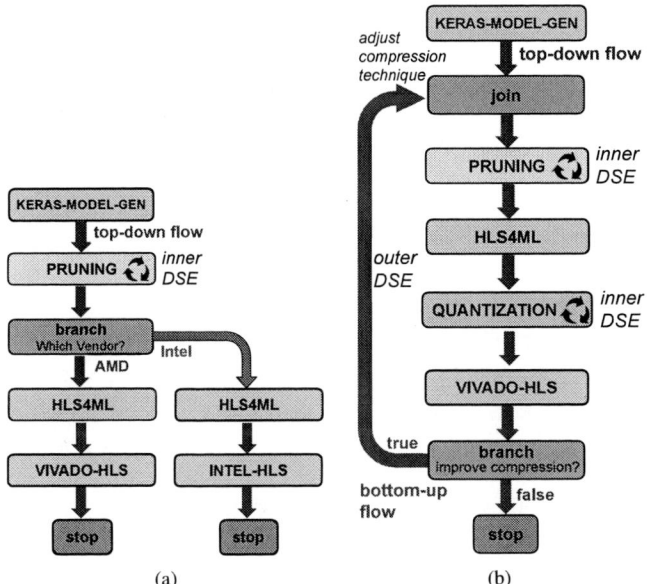

Fig. 2. (a) Pruning optimization targeting different vendors using the BRANCH K-task; (b) Strategy that integrates PRUNING and QUANTIZATION O-tasks.

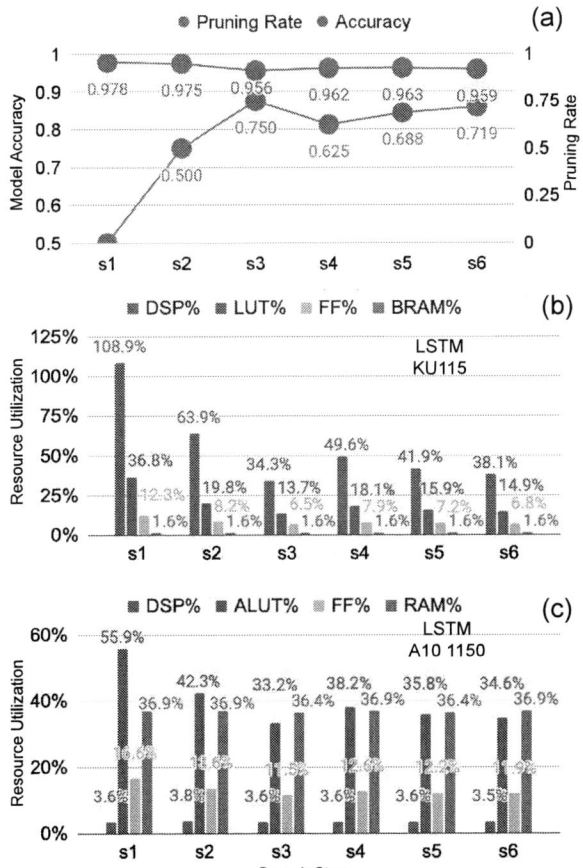

Fig. 3. LSTM model optimization on two FPGA platforms: (a) pruning rate and model accuracy using the PRUNING O-task; (b) resource utilization on an AMD KU115 FPGA; (c) resource utilization on an Intel A10 1150 FPGA.

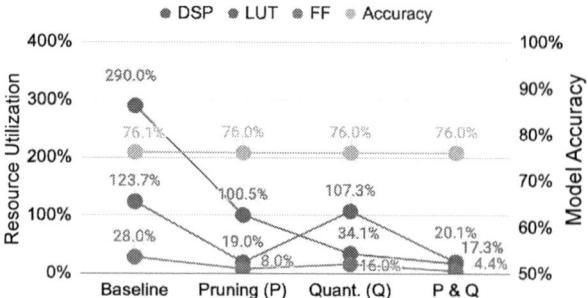

Fig. 4. Combining different compression strategies based on PRUNING and QUANTIZATION O-tasks for jet tagging on an AMD Zynq 7020 FPGA.

The combined approach of pruning and quantization takes the longest time. So optimization time can itself be optimized.

In this scenario, the BRANCH K-task employs a user supplied function that implements our optimization strategy —- either halting if the design meets the resource requirements based on hardware synthesis reports, or activating a more aggressive compression strategy in the next loop iteration. This is managed by modifying the configuration (CFG) section of the meta-model, shared across all tasks in the design flow, which enables or disables specific O-tasks. Fig. 4 shows that, when employing the maximum compression strategy, DSP usage decreases significantly from 290% to 17.3% and LUT usage from 123.7% to 20.1%, achieving reductions respectively of up to 16.7 and 6.2 times.

Alternatively, a user-defined function may be provided as a parameter to the BRANCH K-task, designed to refine the O-task parameters during each iteration of a Design Space Exploration (DSE) loop. This includes adjusting accuracy tolerance parameters α_p and α_q to balance accuracy and resource utilization effectively.

VI. CONCLUSIONS

This paper presents a novel co-optimization framework tailored for the automation of FPGA-based DNN accelerator design and optimization. This framework enhances current design methods by implementing a cross-stage automated design flow that integrates both top-down and bottom-up optimization strategies. It offers a library of customizable optimization, transformation, and control tasks that can be dynamically adjusted across the design process, facilitating significant reductions in resource utilization without compromising accuracy. In particular, results show that our approach reduces DSP and LUT usage of two neural networks respectively by factors of up to 16.7 and 6.2 times, while maintaining accuracy and requiring little human effort or domain expertise.

Moreover, the flexible architecture of our framework enables incorporation of new optimization tasks and strategies, which supports adaptability to evolving hardware and software landscapes. Current and future work includes supporting more diverse types of neural networks, including those targeting graph [15] and transformer [16] models, and incorporating more optimization strategies such as balancing initiation interval [17, 18] and utilizing emerging FPGA resources such as AI Engines [19] and AI Tensor blocks [20].

ACKNOWLEDGEMENT

The support of the United Kingdom EPSRC (grant numbers EP/X036006/1, EP/V028251/1, EP/L016796/1, EP/N031768/1, EP/P010040/1, and EP/S030069/1), CERN, Intel, AMD and SRC is gratefully acknowledged.

REFERENCES

[1] Y. Yang *et al.*, "Synetgy: Algorithm-hardware co-design for convnet accelerators on embedded FPGAs," in *Proceedings of the ACM/SIGDA International Symposium on Field-Programmable Gate Arrays*, 2019.

[2] C. Hao and D. Chen, "Deep neural network model and FPGA accelerator co-design: Opportunities and challenges," in *2018 14th IEEE International Conference on Solid-State and Integrated Circuit Technology (ICSICT)*. IEEE, 2018, pp. 1–4.

[3] H. Fan *et al.*, "Algorithm and hardware co-design for reconfigurable cnn accelerator," in *27th Asia and South Pacific Design Automation Conference (ASP-DAC)*. IEEE, 2022.

[4] V. Kathail, "Xilinx Vitis unified software platform," in *Proceedings of the 2020 ACM/SIGDA International Symposium on Field-Programmable Gate Arrays*, 2020, pp. 173–174.

[5] "OpenVINO toolkit," https://github.com/openvinotoolkit/openvino, 2023.

[6] Y. Umuroglu *et al.*, "FINN: A framework for fast, scalable binarized neural network inference," in *Proceedings of the 2017 ACM/SIGDA International Symposium on Field-Programmable Gate Arrays*, 2017.

[7] F. Fahim *et al.*, "hls4ml: an open-source co-design workflow to empower scientific low-power machine learning devices," in *Research Symposium on Tiny Machine Learning*, 2020.

[8] H. Ye *et al.*, "ScaleHLS: A new scalable high-level synthesis framework on multi-level intermediate representation," in *International Symposium on High-Performance Computer Architecture (HPCA)*. IEEE, 2022.

[9] N. B. Agostini *et al.*, "An MLIR-based compiler flow for system-level design and hardware acceleration," in *Proceedings of the 41st IEEE/ACM International Conference on Computer-Aided Design*, 2022.

[10] T. Chen *et al.*, "TVM: An automated end-to-end optimizing compiler for deep learning," in *13th USENIX Symposium on Operating Systems Design and Implementation (OSDI 18)*, 2018.

[11] Z. Que *et al.*, "MetaML: Automating customizable cross-stage design-flow for deep learning acceleration," in *2023 30th International Conference on Field-Programmable Logic and Applications (FPL)*. IEEE.

[12] J. Vandebon, J. G. F. Coutinho, W. Luk, E. Nurvitadhi, and T. Todman, "Artisan: A meta-programming approach For codifying optimisation strategies," in *2020 IEEE 28th Annual International Symposium on Field-Programmable Custom Computing Machines (FCCM)*, 2020.

[13] J. Duarte *et al.*, "Fast inference of deep neural networks in FPGAs for particle physics," *Journal of Instrumentation*, vol. 13, no. 07, 2018.

[14] Z. Que *et al.*, "Optimizing reconfigurable recurrent neural networks," in *2020 IEEE 28th Annual International Symposium on Field-Programmable Custom Computing Machines (FCCM)*, pp. 10–18.

[15] Z. Que, H. Fan, H. Loo, H. Li, M. Blott, M. Pierini, A. Tapper, and W. Luk, "LL-GNN: Low latency graph neural networks on FPGAs for high energy physics," *ACM Transactions on Embedded Computing Systems*, vol. 23, no. 2, pp. 1–28, 2024.

[16] F. Wojcicki *et al.*, "Accelerating transformer neural networks on fpgas for high energy physics experiments," in *International Conference on Field-Programmable Technology (ICFPT)*. IEEE, 2022.

[17] Z. Que *et al.*, "Accelerating recurrent neural networks for gravitational wave experiments," in *International Conference on Application-specific Systems, Architectures and Processors (ASAP)*. IEEE, 2021.

[18] M. Rognlien *et al.*, "Hardware-aware optimizations for deep learning inference on edge devices," in *International Symposium on Applied Reconfigurable Computing*. Springer, 2022.

[19] "Xilinx AI Engines and Their Applications," in *WP506(v1.1)*, July 10, 2020.

[20] M. Langhammer, E. Nurvitadhi, B. Pasca, and S. Gribok, "Stratix 10 NX architecture and applications," in *The 2021 ACM/SIGDA International Symposium on Field-Programmable Gate Arrays*, 2021, pp. 57–67.

979-8-3503-6184-1/24 $31.00 © 2024 IEEE

A QEMU-Based Virtual Platform of MPSoC

Liangquan Qiao [1], Lei Li [2], Xingyu Gao [2], Jinxiang Wang [1], Fangfa Fu*[1], Keli Long [1], Jinghan Zhou [1]

[1] Department of Microelectronics Center, Harbin Institute of Technology, Harbin 150000, China
[2] 58th Research Institute of China Electronics Technology Group Corporation, Wuxi 214000, China

* Email: 23b920138@stu.hit.edu.cn, fff1984292@hit.edu.cn

Abstract—The MPSoC represents a promising direction for embedded systems to improve the performance and energy efficiency. Virtual platforms such as QEMU enable us to discover the interactive design space exploration at early stage in the design. Unfortunately, QEMU has no full ability of multiprocessor co-simulation. In this paper, we tackle the challenging problem of supporting MPSoC on QEMU by using multiple QEMUs to co-simulate. The virtual platform achieves scalability while maintaining a unified and coherent memory address space by flexibly using the remote ports and the memory agents. Furthermore, a neural network case and a FFT case are used to show co-simulation of MPSoC in the virtual platform.

Keywords—co-simulation, MPSoC, shared memory, unified memory address, QEMU

I. INTRODUCTION

The recent remarkable progress in modern microelectronic technology enabled implementation of highly complex multiprocessor systems on single chips (MPSoCs) and created a big push towards development of various kinds of high-performance embedded systems. The parallel processing featured by MPSoC improves the performance and energy efficiency of embedded systems to handle more complex applications[1].

To develop such complex computing systems and ensure their reliability and efficiency, virtual platforms are often used at early design stages to facilitate hardware-software co-design and function verification. In the past, virtual platforms focused on software hardware co-design. The interface between QEMU and SystemC for hardware modeling has been achieved in [2]. Such virtual platform integrated one QEMU process is useful for SoC that integrated a processor and several accelerators but not for MPSoC. As far as we know, few works have used multiple QEMUs to simulate MPSoC. The dual-kernel simulator mentioned in [5] manages the shared memory through socket API and system bus modeling by SystemC. Reference [6] concurrently runs two heterogeneous CPU architectures in a QEMU process. However, the virtual platform in question does not support more than two processors.

In this paper, we consider constructing a QEMU-based virtual platform of MPSoC. We tackle this challenging problem by considering such strategy that multiple QEMUs co-simulate. This strategy maintains a unified, coherent memory space for MPSoC. In other words, any processor can directly access the entire memory address space via a socket-based device, known as the memory agent (MA). It also can easily combine with other simulation platforms such as SystemC for flexible and modular co-simulation.

The rest of this paper is organized as follows. Section II presents the system design of the proposed simulation platform. The experimental results are shown in section III. Section IV concludes the paper.

II. ARCHITECTURE

In this section, our virtual platform for MPSoC is presented, as shown in Fig. 1, which consists of multiple processors (i.e., P_i) and shared memory (SM) that are achieved by multiple QEMU process. The SM is composed of a memory agent (MA) server, a memory backend, and several MA clients. Additionally, we specify the QEMU process that supports SimCtrl as the master QEMU process and the remaining processes as slave QEMU processes.

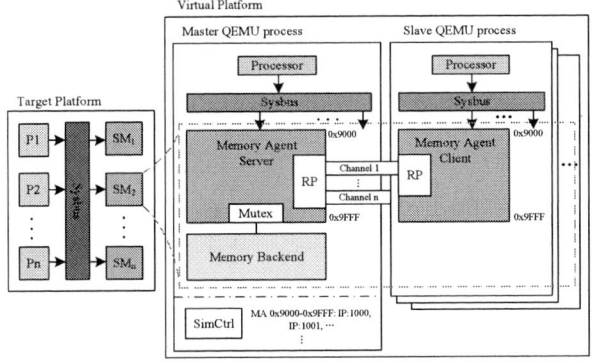

Fig. 1. Overview of proposed virtual platform architecture.

A. Preparation of the virtual platform

In the virtual platform, any SM is composed of the MA and the memory backend. These components are used to maintain a unified memory address space, enabling multiple processors to access the remote shared memory simulated by other QEMU processes.

- SM is a memory in the target platform. In the target platform, every processor can access any SM. In other words, SM serves a physical memory in the MPSoC. However, SM is merely a virtual object in the virtual platform, which comprises a memory agent and a memory backend. When a processor accesses SM, it actually accesses the memory backend through the memory agent within the virtual platform.

- Memory backend is the real memory in the QEMU process. Memory backend and SM correspond to each other, so the size of memory backend is equal to the size of SM.

- MA is the agent of SM in the QEMU process. In our virtual platform, a processor can access remote memory instanced in other QEMU processes through the MA. When a QEMU process has a memory backend, the MA services as an MA Server, and the MAs of the rest of the processes are called MA Clients. The MA Server is the only device that can

---

This work is supported by National Key Research and Development Program of China (Grant No. 2023YFB4403502).

979-8-3503-6184-1/24 $31.00 © 2024 IEEE

actually access the memory backend among all the processes. So, the MA Server has a mutex mechanism that guarantees that only one processor can access the memory backend at any one time. RP establishes the channel between the MA Server and MA Client and is responsible for passing memory access requests and returning results between the MA Server and MA Client.

B. Boot sequence

SimCtrl serves as the central coordination mechanism for the boot sequence of multiple QEMU instances, as shown in Fig. 1. It maintains a comprehensive table that outlines the established connections. Prior to initiating the QEMU boot process, each QEMU instance transmits the specific port numbers utilized for establishing these connections to SimCtrl. Once all the connections in the table are successfully established and validated, SimCtrl proceeds to trigger the boot of all QEMU processes at this moment.

C. Multiprocessor co-simulation

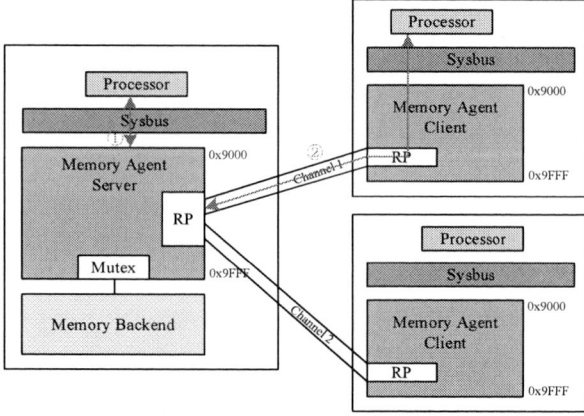

Fig. 2. Communication between multiple QEMU processes

Fig. 2 illustrates that multiple QEMU processes communicate with each other through the MAs and RPs to achieve multiprocessor co-simulation.

- Processor directly accesses memory backend through the MA Server. If there is an MA Server in the current process, the processor can send the memory access request directly to the MA Server, as indicated by the red line 1 in Fig. 2.

- Processor accesses memory backend via the MA Client. As indicated by red line 2 in Fig. 2, if there is an MA Client in the current process, the processor can only send the memory access request to the MA Server through the channel established by the RP between the MA Server and MA Client. The processor then waits for the reply from the MA Server.

- The MA Server serves as the sole intermediary between the processor and the memory backend, facilitating exclusive access to the latter. It maintains an organized queue of memory access requests, executing reads and writes to the memory backend in accordance with the specifics of the requests, and subsequently returns the outcomes of these operations to the appropriate requestor.

III. SIMULATION AND DISCUSSION

A. Experiment Environment

In our simulation platform, processor is simulated by QEMU. The experimental environment is shown in Table I.

TABLE I.　　EXPERIMENTAL ENVIRONMENT

Environment		
Guest OS	None	None
Processor Core	Cortex-A9 MPCore	Cortex-A53 MPCore
Instruction Set	ARM v7-a	ARM v8-a
Benchmark	MNIST and FFT	
Guest simulator	QEMU[a]	
Host OS	VMWare Ubuntu 20.04.6 LTS	
Host hardware	AMD Ryzen 7 7840HS	

[a.] Based on Xilinx QEMU-v2022.2

B. Scalability

One of the most typical features of the virtual platform is scalability. Ideally, our virtual platform can run any number of QEMU processes by adding remote port devices and memory agents to each process.

For example, if we want to construct a MPSoC that integrates four processors in our virtual platform, we need to configure four QEMU processes, where each process contains an MA Server, three MA Clients, and four RPs. Then we need to specify which one is the master process among the four QEMU processes. In this way, we can achieve a four processors virtual platform.

C. Simulation Speed

Another of the most typical features of the virtual platform is that it maintains a unified memory address space. Thus, a processor from one QEMU process can directly access the host memory of another QEMU process. A unified memory address space represents a new trend for MPSoC, aiming to provide high-bandwidth and low-latency connectivity, which helps maintain cache consistency between host and attached devices.

For assessing the simulation speed of our virtual platform, we use three methods to retrieve image data from the host memory, which were taken from the MNIST dataset. The first method involves one processor reading data from host memory and writing it to device memory. The second method utilizes DMA to transfer image data from host memory to device memory. The third method allows the processor to access the SM through a unified memory address (UMA). Fig. 3 depicts the simulation speed in our virtual platform. Undoubtedly, the first method is the slowest, and the CPU cannot perform other tasks during this process. The speed of the third method is slower than that of the second method, because it is limited by the performance of BSD sockets, leading to higher communication overhead. But a unified memory space offers unparalleled advantages in real systems as it can maintain cache consistency between the host and devices.

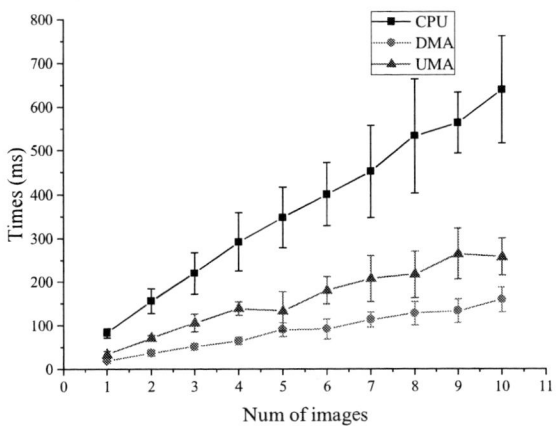

Fig. 3. Remote memory access speed.

```
X[0]=(136.0000,0.0000i)
X[1]=(-8.0000,40.2187i)
X[2]=(-8.0000,19.3137i)
X[3]=(-8.0000,11.9728i)
X[4]=(-8.0000,8.0000i)
X[5]=(-8.0000,5.3454i)
X[6]=(-8.0000,3.3137i)
X[7]=(-8.0000,1.5913i)
X[8]=(-8.0000,-0.0000i)
X[9]=(-8.0000,-1.5913i)
X[10]=(-8.0000,-3.3137i)
X[11]=(-8.0000,-5.3454i)
X[12]=(-8.0000,-8.0000i)
X[13]=(-8.0000,-11.9728i)
X[14]=(-8.0000,-19.3137i)
X[15]=(-8.0000,-40.2187i)
```

Fig. 4. The result of FFT case.

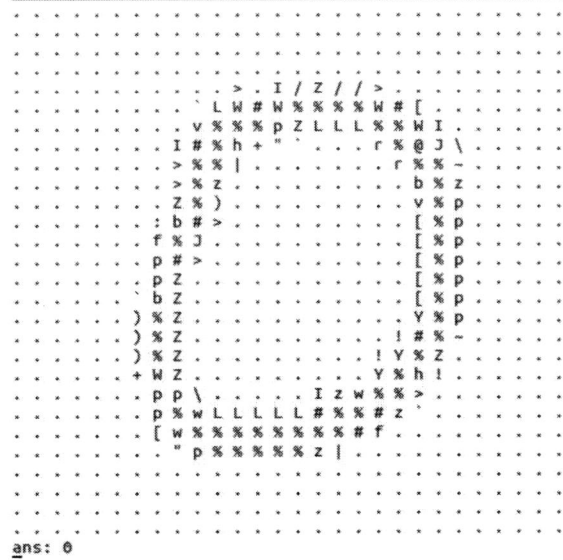

Fig. 5. The result of MNIST recognition.

D. Co-simulation case

In order to demonstrate the co-simulation capability of the virtual platform, we utilized a configuration comprising one ARM Cortex-A53 MPCore and two ARM Cortex-A9 MPCores to execute two cases: FFT and MNIST handwriting recognition.

In the FFT case, the Cortex-A53 is considered as the main controller that assigns task to two Cortex-A9 processors. The Cortex-A9 processors read data from the host memory and write the results back to the host memory. Together, the two Cortex-A9 processors collaborate to complete FFT calculations. Fig. 4 illustrates the FFT results.

For MNIST recognition in this study, we design a model consisting of five convolutional cores, one convolutional layer, one pooling layer, and a fully connected layer. Following training, the model achieved an accuracy of 83.5% on the training set and 86.67% on the test set. The predicted results on our virtual platform using the test set were statistically consistent with those obtained in previous tests. Fig. 5 shown a recognition case of MNIST.

IV. CONCLUSION

In this paper, we propose an easy and convenient method for constructing a virtual platform for MPSoC with QEMU. The virtual platform can build the architecture for co-simulation of multiple processors by using the remote ports and the memory agents. The virtual platform is scalable by using multiple QEMU to co-simulate. The virtual platform maintains a unified and coherent memory address space to provide quick simulation speed. In the experimental results, we demonstrate the scalability and simulation speed of the virtual platform, and we show two co-simulation cases running on the virtual platform.

REFERENCES

[1] W. Wolf, A. A. Jerraya and G. Martin, "Multiprocessor system-on-chip (MPSoC) technology", IEEE Trans. Comput.-Aided Design Integr. Circuits Syst., vol. 27, no. 10, pp. 550-561, Oct. 2008.

[2] E. Díaz, R. Mateos and E. Bueno, "Virtual Platform of FPGA based SoC for Power Electronics Applications," 2019 IEEE 28th International Symposium on Industrial Electronics (ISIE), Vancouver, BC, Canada, 2019, pp. 1371–1376.

[3] M. C. Chiang, T. C. Yeh and G. F. Tseng, "A QEMU and SystemC-Based Cycle-Accurate ISS for Performance Estimation on SoC Development," IEEE Trans. Comput.-Aided Des. Integr. Circuits Syst. vol. 30, no. 4, pp. 593–606, April 2011.

[4] F. Cucchetto, A. Lonardi and G. Pravadelli, "A common architecture for co-simulation of SystemC models in QEMU and OVP virtual platforms," 2014 22nd International Conference on Very Large Scale Integration (VLSI-SoC), Playa del Carmen, Mexico, 2014, pp. 1–6.

[5] C. S. Peng, L. C. Chang, C. H. Kuo, and B. D. Liu, "Dual-core virtual platform with QEMU and SystemC," 2010 International Symposium on Next Generation Electronics, Kaohsiung, Taiwan, 2010, pp. 69–72.

[6] I. H. Chen, C. T. King, Y. H. Chen, and J. -M. Lu, "Full System Emulation of Embedded Heterogeneous Multicores Based on QEMU," 2018 IEEE 24th International Conference on Parallel and Distributed Systems (ICPADS), Singapore, 2018, pp. 771–778.

A Parallel Harmonic Balance Method Based on GPU for Efficient Periodic Steady-State Analysis

Zhengzhuo Wang[1], Yanliang Sha[2], Lingyun Ouyang[2], Quan Chen[2], *Jianguo Hu[1,3]

[1]*School of Microelectronics Science and Technology, Sun Yat-sen University, Zhuhai, China*
[2]*School of Microelectronics, Southern University of Science and Technology, Shenzhen, China*
[3]*Shenzhen Research Institute of Sun Yat-Sen University, Shenzhen, China*

*Email: wangzhzh57@mail2.sysu.edu.cn, hujguo@mail.sysu.edu.cn

Abstract—**The analysis of radio-frequency (RF) circuits, including periodic steady-state (PSS) simulation, has traditionally been highly computationally demanding, particularly with a substantial quantity of frequency harmonic components. This paper presents a parallel harmonic balance (HB) method that leverages graphical processing unit (GPU) to facilitate efficient PSS analysis. The method focuses on parallelizing the construction of the nonlinear Jacobian matrix, the most time-intensive component, while combining GPU to enhance simulation performance. Following an outline of the HB Jacobian matrix structure, this paper elaborates on the development process based on a computational platform equipped with CUDA-enabled GPUs. Numerical experiments indicate that the proposed parallel HB method achieves a threefold increase in speed compared to the conventional HB method, while GPU utilization results in almost 10x acceleration for the corresponding section.**

Keywords—**harmonic balance method, GPU, periodic steady-state analysis, RF simulation**

I. INTRODUCTION

PSS analysis is an important and fundamental simulation methodology which plays a vital role in the design for analog and RF systems. This analysis is essential to solve the periodic static operating points of RF systems and to calculate the frequency-domain response with other characteristics [1], [2]. Typically, there are three approaches to numerically find the PSS solution [3]–[5]: the finite difference method (FDM), the shooting-Newton method and the HB method. Compared to the first two methods, the HB method is commonly the choice for RF systems when the nonlinear elements have weak nonlinear behavior [6], where the FDM and shooting-Newton method are more suitable for strong nonlinear systems. This paper only focuses on the HB method because it is faster than the other two methods when accuracy tolerance is acceptable, which means that only RF systems with weak nonlinearity are concerned.

The HB method characterizes systems through nonlinear equations to address the PSS problem, wherein waveforms are expressed as Fourier series and the unknowns correspond to their Fourier coefficients. Consequently, an increase in the number of unknowns escalates both the complexity of solving these nonlinear equations and the duration required for

PSS analysis. Generally, achieving accurate system analysis necessitates a sufficient quantity of unknowns within the HB method, which significantly amplifies computational complexity [7]. Numerical experiments have indicated that constructing and factorizing HB Jacobian matrices consumes a substantial portion of time in this methodology, where these matrices tend to be considerably larger and denser [8]. Furthermore, alternative approaches [9], [10] have been proposed to optimize the construction and factorization processes of the HB Jacobian matrix with an aim to mitigate computational complexity. While these efforts do alleviate some computational burdens, they inherently lack parallelism.

Simultaneously, the full development and widespread use of GPUs provides high-quality hardware support for performing massively parallel tasks. A common GPU contains several hundred cores, while the high-end ones have tens of thousands of cores [11], which is an inherent advantage to be used for massively parallel data processing. This paper focuses on combining GPU with conventional HB method, aiming to expedite PSS analysis with enhanced accuracy and efficiency. Our primary contributions include:

1) Implementation of conventional HB method for PSS analysis and explanation for the cause of time consuming in HB process.
2) Parallelization of nonlinear Jacobian matrix construction on hybrid GPU platforms and development of the HB-GPU method.

Our objective with these advancements is to establish HB-GPU as an efficient method for the PSS analysis of large-scale systems.

II. BACKGROUND

For a PSS problem, the modified nodal analysis (MNA) yields the following differential equation:

$$\mathbf{G}\boldsymbol{x}\left(t\right) + \mathbf{C}\frac{\mathrm{d}\boldsymbol{x}\left(t\right)}{\mathrm{d}t} + \boldsymbol{f}\left(\boldsymbol{x}\left(t\right)\right) = \boldsymbol{b}\left(t\right) \tag{1}$$

where $\mathbf{G} \in \mathbb{R}^{N \times N}$ includes resistors and conductors, $\mathbf{C} \in \mathbb{R}^{N \times N}$ includes capacitors and inductors, while $\boldsymbol{b}\left(t\right) \in \mathbb{R}^{N}$ is the vector of set of input sources and N is the total number

979-8-3503-6184-1/24 $31.00 © 2024 IEEE

of generalized nodes which contains unknown currents in inductors and independent voltage sources.

To compute the PSS response in the period T with the input stimuli, both the unknown response vector $\boldsymbol{x}(t)$ and the input source vector $\boldsymbol{b}(t)$ should be extended to Fourier series as follows:

$$\boldsymbol{x}(t) = X_0 + \sum_{k=1}^{K} \left[X_k^C \cos(k\omega_0 t) + X_k^S \sin(k\omega_0 t) \right] \quad (2)$$

$$\boldsymbol{b}(t) = B_0 + \sum_{k=1}^{K} \left[B_k^C \cos(k\omega_0 t) + B_k^S \sin(k\omega_0 t) \right] \quad (3)$$

where ω_0 is the angular frequency in rad/sec, given by $\omega_0 = 2\pi/T$; $X_0, X_k^{C,S} \in \mathbb{R}^N$ and $B_0, B_k^{C,S} \in \mathbb{R}^N$ are respectively the unknown and known coefficients of the PSS response and input source, and K is the highest order of harmonic components selected.

Besides, $\boldsymbol{f}(\boldsymbol{x}(t)) \in \mathbb{R}^N$, the current vector of nonlinear elements, should also be extended to Fourier series in the HB method as follows:

$$\boldsymbol{f}(\boldsymbol{x}(t)) = F_0(\overline{\boldsymbol{X}}) + \sum_{k=1}^{K} \left[F_k^C(\overline{\boldsymbol{X}}) \cos(k\omega_0 t) + F_k^S(\overline{\boldsymbol{X}}) \sin(k\omega_0 t) \right] \quad (4)$$

where the frequency response transforms:

$$\overline{\boldsymbol{X}}^T = \left[X(0\omega_0) \ X(\omega_0) \ \cdots \ X(K\omega_0) \ X(-K\omega_0) \ \cdots \ X(-\omega_0) \right] \quad (5)$$

in which $X(k\omega_0)$ and $X(-k\omega_0)$ respectively represents the cosine and sine components of Fourier coefficients.

Solving the extended equations using the Newton–Raphson method yields the HB Jacobian matrix:

$$\mathcal{J} = \mathcal{J}_l + \mathcal{J}_{nl} = \mathbf{G} \otimes \mathbf{I}_H + \mathbf{C} \otimes \mathbf{K}_H + \frac{\partial \overline{\boldsymbol{F}}(\overline{\boldsymbol{X}}^{(j)})}{\partial \overline{\boldsymbol{X}}^{(j)}} \quad (6)$$

where j is the iteration index, $\mathbf{I}_H$ is an identity matrix and $\mathbf{K}_H$ is a diagonal matrix defined as:

$$\mathbf{K}_H = \begin{bmatrix} 0 & 0 & 0 & 0 & 0 & 0 & 0 \\ 0 & z\omega_0 & 0 & 0 & 0 & 0 & 0 \\ \vdots & \vdots & \ddots & \vdots & \vdots & \vdots & \vdots \\ 0 & 0 & 0 & zK\omega_0 & 0 & 0 & 0 \\ 0 & 0 & 0 & 0 & -zK\omega_0 & 0 & 0 \\ \vdots & \vdots & \vdots & \vdots & \vdots & \ddots & \vdots \\ 0 & 0 & 0 & 0 & 0 & 0 & -z\omega_0 \end{bmatrix} \quad (7)$$

where z is the unit of an imaginary number ($z^2 = -1$).

III. NEW PARALLEL HB METHOD

A. Structure of nonlinear HB Jacobian Matrix

In constructing the nonlinear HB Jacobian matrix, the Fourier inverse and Fourier transforms are successively applied to the nonlinear function, as shown in Algorithm 1:

$$\overline{\boldsymbol{F}}(\overline{\boldsymbol{X}}^{(j)}) = \overline{\Gamma} \boldsymbol{f}^{(j)}(\overline{\Gamma}^{-1} \overline{\boldsymbol{X}}^{(j)}) \quad (8)$$

where $\overline{\Gamma}, \overline{\Gamma}^{-1}$ are Fourier and inverse Fourier operators respectively. Substituting (8) into (6) yields:

$$\mathcal{J}_{nl} = \frac{\partial \overline{\boldsymbol{F}}(\overline{\boldsymbol{X}}^{(j)})}{\partial \overline{\boldsymbol{X}}^{(j)}} = \begin{bmatrix} \overline{\Gamma \left[\frac{\partial f_1}{\partial x_1}\right] \Gamma^{-1}} & \cdots & \overline{\Gamma \left[\frac{\partial f_1}{\partial x_N}\right] \Gamma^{-1}} \\ \vdots & \ddots & \vdots \\ \overline{\Gamma \left[\frac{\partial f_N}{\partial x_1}\right] \Gamma^{-1}} & \cdots & \overline{\Gamma \left[\frac{\partial f_N}{\partial x_N}\right] \Gamma^{-1}} \end{bmatrix} \quad (9)$$

where $\Gamma, \Gamma^{-1} \in \mathbb{R}^{N \times N}$ are matrix representation for the Fourier and inverse Fourier operators respectively, and all partial derivatives at each single time step are written in the sub-matrices $\overline{\left[\frac{\partial f_i}{\partial x_j}\right]} \in \mathbb{R}^{(2K+1) \times (2K+1)}$. Furthermore, $\boldsymbol{f}(\boldsymbol{x}(t))$ is algebraically nonlinear and its value is only related to the value of $\boldsymbol{x}(t)$ at a single time step, which transforms all sub-matrices $\overline{\left[\frac{\partial f_i}{\partial x_j}\right]}$ into the following diagonal matrix form:

$$\overline{\left[\frac{\partial \boldsymbol{f}_i}{\partial \boldsymbol{x}_j}\right]} = \begin{bmatrix} \frac{\partial f_i(x_j(t_0))}{\partial x_j(t_0)} & 0 & \cdots & 0 \\ 0 & \frac{\partial f_i(x_j(t_1))}{\partial x_j(t_1)} & \cdots & 0 \\ \vdots & \vdots & \ddots & \vdots \\ 0 & 0 & \cdots & \frac{\partial f_i(x_j(t_{2K}))}{\partial x_j(t_{2K})} \end{bmatrix} \quad (10)$$

Algorithm 1 Parallel Nonlinear Harmonic Balance Method

Data: $\mathbf{G}$, $\mathbf{C}$, $\boldsymbol{b}$ from the Parser, initial guess $\boldsymbol{x}(0)$ from the transient analysis, the convergence tolerance

Result: Final frequency response $\overline{\boldsymbol{X}}^{(j+1)}$

RecyclingIndex $j, i = 0$;
  Compute the conductivity matrix $\overline{\boldsymbol{Y}} = \mathbf{G} \otimes \mathbf{I}_H + \mathbf{C} \otimes \mathbf{K}_H$;
  Slice $\mathbf{FFT}(\boldsymbol{b}(0))$ to obtain $\overline{\boldsymbol{B}}$;
  Slice $\mathbf{FFT}(\boldsymbol{x}(0))$ to obtain $\overline{\boldsymbol{X}}^{(0)}$;

while $\|\Phi(\overline{\boldsymbol{X}}^{(j+1)})\| > tol$ **do**
  if $j = 0$ **then**
    | $\overline{\boldsymbol{X}}^{(j)} = \overline{\boldsymbol{X}}^{(0)}$
  else
    | $\overline{\boldsymbol{X}}^{(j)} = \overline{\boldsymbol{X}}^{(j+1)}$
  end
  Insert $\mathbf{IFFT}(\overline{\boldsymbol{X}}^{(j)})$ equally spaced to obtain $\boldsymbol{x}(j)$;
  for $i < P$ **do**
    | Calculate Jacobian sub-matrices in the time domian $(i++)$;
  end
  $\mathbf{FFT}(\dots)$;
  Construct nonlinear HB Jacobian matrix $\mathcal{J}_{nl}$ in parallel ;
  Compute the HB Jacobian matrix $\mathcal{J} = \mathcal{J}_l + \mathcal{J}_{nl}$;
  LU decomposition to obtain LU factors ;
  Update $\overline{\boldsymbol{X}}^{(j+1)}$ and solve $\Phi(\overline{\boldsymbol{X}}^{(j+1)})$; $j++$
end

B. Parallelization by GPU

GPU is equipped with numerous cores that enable simultaneous processing of matrices, unlike CPU which handles multiple matrices sequentially. In this work, the computation of the P Jacobian sub-matrices in Algorithm 1 is distributed across P GPU cores for parallel processing, where each core computes a single Jacobian sub-matrix at the current time step. The resulting Jacobian sub-matrix undergoes FFT and is subsequently placed by its corresponding GPU core into the appropriate location within $\mathcal{J}_{nl}$, executed concurrently by all P cores.

PyCUDA enables Python to use CUDA to control GPU cores, facilitating the above parallelization on GPU. Besides, considering GPU vendors provide corresponding libraries for mathematical operation and their usage can already achieve an improvement of one order of magnitude or more when compared to CPU, the relevant mathematical libraries have been used.

This approach significantly reduces computation time compared to traditional serial processing on CPU. However, the CPU on the hybrid platform is tasked with managing other computational operations that are not amenable for parallelization, including moving and scaling of matrix blocks. Since utilizing GPU for these tasks does not yield any reduction in computation time and cannot be optimized algorithmically, it constrains further improvements to the HB-GPU method.

IV. NUMERICAL RESULTS

We developed the novel parallel HB method in Python. The experiments were conducted on Linux network servers equipped with an Intel(R) Core(TM) i7-10700 CPU (2.9 GHz and 128 GB memory) and a Tesla V100 GPU (5120 CUDA cores and 32 GB memory). Initially, we assessed the accuracy of the conventional HB method in comparison to the shooting-Newton method. Subsequently, we investigated the parallel runtime during the construction of the nonlinear Jacobian matrix. Finally, we demonstrated the overall impact of utilizing GPU acceleration for enhancing the HB method's performance in periodic steady-state analysis, as illustrated in Fig. 1.

A varying amount of harmonic components were selected for each test case to differentiate the computational scale, specifically the Jacobian matrix of size N. Furthermore, simulation times for the HB-GPU method are contrasted with those of both the conventional HB method and the shooting method, thereby illustrating the speedup achieved through parallelism in enhancing the efficiency of the HB method.

V. CONCLUSION

The accuracy and efficiency of periodic steady-state analysis are crucial for modern RF systems design. Our main contribution in this work is to extend the conventional HB method, leveraging GPU for parallel the intricate and computationally expensive step among it. We derive the detailed formulations for the structure of nonlinear Jacobian matrices, rendering GPU for their parallel construction to speed up effectively. Numerical results indicate that the parallel HB method effectively speeds up the PSS analysis threefold, and demonstrate an increasing acceleration ratio as the matrix size increases.

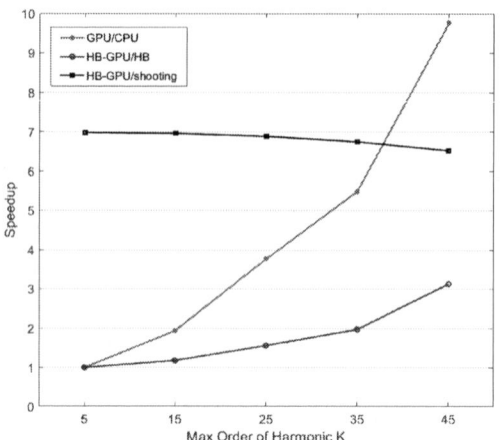

Fig. 1. Speedup Ratio after GPU was Used

ACKNOWLEDGMENT

This work is supported in part by National Science and Technology Major Project of China (No.2021ZD0114600), in part by the Department of Science and Technology of Guangdong Province (No.2021B1101270003), and in part by Shenzhen Science and Technology Program (No.JCYJ20220818102002005).

REFERENCES

[1] J. Hu, D. Wang, and J. Wu, "A 2 kbits low power eeprom for passive rfid tag ic," *Chinese Journal of Electronics*, vol. 31, no. 1, pp. 18–24, 2022.

[2] X. Guan and Q. Wu, "Irs-enabled spectrum sharing: Interference modeling, channel estimation and robust passive beamforming," *ZTE Communications*, vol. 20, no. 1, pp. 28–35, 2022.

[3] T. Aprille and T. Trick, "Steady-state analysis of nonlinear circuits with periodic inputs," *Proceedings of the IEEE*, vol. 60, no. 1, pp. 108–114, 1972.

[4] K. Kundert, J. White, and A. Sangiovanni-Vincentelli, *Steady-State Methods for Simulating Analog and Microwave Circuits*, ser. The Springer International Series in Engineering and Computer Science. Springer US, 2013.

[5] N. Li, S. Liu, Q. Chen, *et al.*, "A fast electromagnetic field and radio frequency circuit co-simulation approach for strongly coupled coil array in magnetic resonance imaging," *IEEE Transactions on Magnetics*, vol. 54, no. 11, pp. 1–5, 2018.

[6] R. Telichevesky, K. S. Kundert, and J. K. White, "Efficient steady-state analysis based on matrix-free krylov-subspace methods," in *32nd Design Automation Conference*, 1995, pp. 480–484.

[7] Z. Xiong, P. Zhao, J. Fan, *et al.*, "Mixed electric and magnetic coupling design based on coupling matrix extraction," *ZTE Communications*, vol. 21, no. 4, pp. 85–90, 2023.

[8] L. Han, X. Zhao, and Z. Feng, "An efficient graph sparsification approach to scalable harmonic balance (hb) analysis of strongly nonlinear rf circuits," in *2013 IEEE/ACM International Conference on Computer-Aided Design (ICCAD)*, 2013, pp. 494–499.

[9] K. Mayaram, D. Lee, S. Moinian, *et al.*, "Computer-aided circuit analysis tools for rfic simulation: algorithms, features, and limitations," *IEEE Transactions on Circuits and Systems II: Analog and Digital Signal Processing*, vol. 47, no. 4, pp. 274–286, 2000.

[10] O. Nastov, R. Telichevesky, K. Kundert, *et al.*, "Fundamentals of fast simulation algorithms for rf circuits," *Proceedings of the IEEE*, vol. 95, no. 3, pp. 600–621, 2007.

[11] P. Boncz, W. Lehner, and T. Neumann, "Special issue: Modern hardware," *The VLDB Journal*, vol. 25, pp. 623–624, 2016.

Efficient Dynamic Memory Management for High Bandwidth Memory on FPGA

Yuwei Qu*, Yiqing Mao*, Wenbo Yin, and Lingli Wang[†]

State Key Laboratory of Integrated Chips and Systems, Fudan University
[†]llwang@fudan.edu.cn

Abstract—**Efficient memory allocation is essential for high-performance computing and neural networks implemented on FPGA, particularly in scenarios demanding rapid data processing and substantial bandwidth. Traditional DDR memory often fails to meet these requirements. To address this, a dynamic memory allocator on FPGA utilizing High Bandwidth Memory (HBM) is developed. Our design includes *Data Dispatch*, *Linear Allocator*, and *HBM Driver* modules to optimize memory allocation and data writing processes, respectively. The *Linear Allocator* minimizes resource usage while reducing clock cycles required for allocation, and the *Data Dispatch* module enhances allocation efficiency by multiple AXI4 interfaces. Experimental results on the Alveo U280 Card show that our design achieves a 52.60% reduction in LUT usage and a 28.99% reduction in DFF usage compared to the HeapAllocator. Additionally, using multiple AXI4 interfaces tackling 300 variable-sized allocations, the quad-port configuration achieves a 36.21% reduction of clock cycles averagely, equivalent to a 1.5× efficiency improvement over the DDR memory.**

Index Terms—**Dynamic Memory Allocation, High Bandwidth Memory (HBM), Data Dispatch**

I. INTRODUCTION

In the realm of high-performance computing and neural network acceleration on FPGA, efficient dynamic memory allocation is critical. Compared to DDRs, HBMs offer a significantly higher bandwidth and lower latency, which is an ideal choice for data-intensive applications such as machine learning and complex data analysis. Meanwhile, the implementation of dynamic memory management functions such as *malloc()* and *free()* at the hardware level is crucial, which can facilitate efficient allocation and deallocation of memory resources, optimizing the utilization of available memory and ensuring that computational tasks on FPGA-based systems run smoothly. This capability becomes even more critical when leveraging HBM's capabilities, as it allows for faster data access and processing, thereby enhancing overall system performance.

Several related works have attempted to address dynamic memory allocation but have notable limitations. A C library called "Libmem" for memory management has been proposed by Nicholas V. Giamblanco et al., targeting BRAM resources only without considering DDR or HBM [1]. A Dynamic On-Chip Memory Management Unit (DOMMU) has been introduced by Ghada Dessouky et al., also focusing solely on BRAM resources [2]. "SysAlloc," a hardware dynamic

memory allocation scheme, is presented by Zeping Xue et al., but it requires improvements in memory utilization and allocator performance [3]. An HLS-based dynamic memory allocation method is proposed by Argyris Kokkinis et al. to minimize memory allocation failures in multi-accelerator Dynamic Memory Management (DMM) frameworks, but the garbage collection mechanism incurs performance loss [4].

To address these limitations, we propose an architecture to optimize memory management for FPGA-based high-performance computing applications. The architecture consists of three main components: Data Dispatch, Linear Allocator, and HBM Driver. The hardware-based dynamic memory allocator handles C functions like *malloc()* and *free()* with low resource usage. The data dispatch module enhances multi-port writes to HBM. At the same time, the driver module ensures efficient memory operations. The above components aim at leveraging HBM's high bandwidth and low latency. This architecture is verified on the Alveo U280 platform with the following contributions:

- **Parallelism**: An innovative *Data Dispatch* module segments incoming data and assigns tasks to multiple AXI4 interfaces, enabling concurrent multi-port writing to HBM and maximizing its high bandwidth.
- **Efficiency**: The *Linear Allocator* is based on a bookkeeping strategy, remarkably minimizing resource usage while reducing allocation clock cycle number.
- **Design Automation**: A push-button workflow is developed to generate the whole memory management system.

The paper is organized as follows: Section II provides the background. Section III outlines the proposed methodology, including the three main design components. Section IV details resource utilization and experimental results. Section V discusses future work and concludes the paper.

II. BACKGROUND

A. Alveo U280 HBM2 Architecture

The Alveo U280 Card features two HBM2 stacks, each with 4GB capacity and 16 Pseudo Channels (PCs) providing 256MB of memory per channel (Fig. 1). Theoretically, each PC has a maximum bandwidth of 14.375 Gb/s, totaling 230 Gb/s for a single HBM subsystem [5]. While each PC's performance is lower than the 19 Gb/s of a DDR4 channel [6], performance can be improved by efficiently integrating

---

*Both authors contributed equally to this research.

979-8-3503-6184-1/24 $31.00 © 2024 IEEE

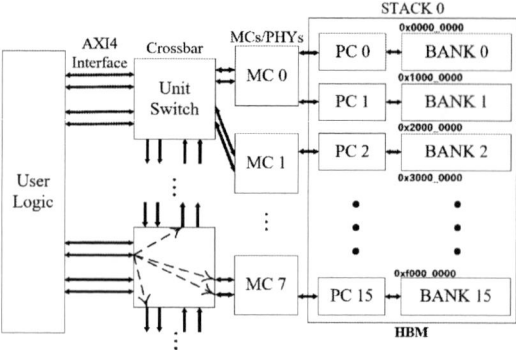

Fig. 1. U280 HBM2 Architecture

multiple AXI masters into the HBM subsystem. In addition to HBM2, the U280 includes DDR4 memory configured as single rank, 64-bit with error correcting code (ECC) [6].

The FPGA's programmable logic provides 16 HBM AXI4 interfaces, allowing access to any memory location via a built-in switch network. However, simultaneous access to the same PC by two interfaces can lead to memory bank conflicts, limiting bandwidth scalability and posing a challenge for fully utilizing HBM's high throughput.

III. DESIGN METHODOLOGY

The dynamic memory allocator (DMA) optimizes memory allocation by HBM2, featuring both serial (Fig. 2a) and parallel (Fig. 2b) access architectures.

The parallel design employs multiple AXI4 interfaces for concurrent data writing. The Data Dispatch module in Section III-A segments incoming data and assigns tasks to available AXI4 interfaces. The Linear Allocator in Section III-B then coordinates with multiple HBM Drivers in Section III-C to store data efficiently across different HBM addresses. This approach enhances memory allocation efficiency and system performance by leveraging HBM's high bandwidth.

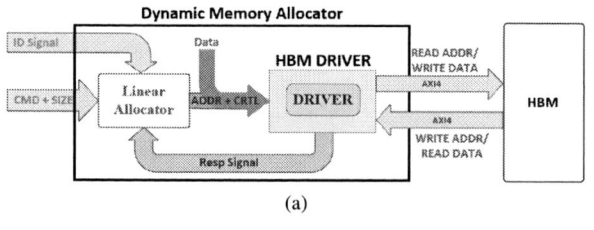

(a)

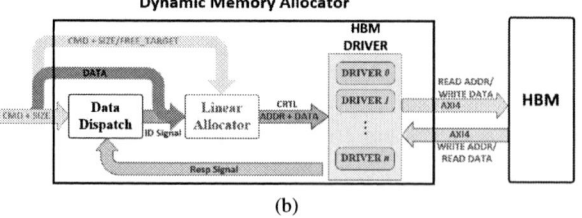

(b)

Fig. 2. The hardware architecture of the dynamic memory allocator. (a) Serial Design. (b) Parallel Design.

A. Data Dispatch

The Data Dispatch module is designed to optimize the efficiency of writing data to HBM by multiple AXI4 Slaves, without increasing the number of allocator modules or resource usage. The module operates by segmenting the input data stream based on the number of enabled AXI4 interfaces in the HBM. It assigns write tasks to the idle AXI4 interfaces according to the write completion feedback signals (*write_finish*) from each corresponding HBM Driver module in Section III-C. Additionally, it outputs an ID signal to inform the allocator module about which base address in the HBM should be prioritized for allocation as shown in Alg. 1. This approach ensures a pipeline-like data writing process into the HBM, preventing writing conflicts among multiple AXI4 interfaces and ultimately enhancing the overall allocation efficiency by effectively utilizing multiple HBM AXI4 interfaces.

Algorithm 1: Data Dispatch

Input: Respond Signal $write\_finish$ from HBM
      Driver, Data Input Stream D_{in}
Output: Data Output D_{out}, ID Signal ID

1 Reset all outputs and internal states;
2 slice D_{in} into $n = number\_of\_Channels$ segments;
3 **for** $j = 1; j \leq n$ **do**
4   **while** $write\_finish[j] == 1$ **do**
5   Assign $D_{in}[j]$ to the idle AXI4 interface;
6   **return** $ID \leftarrow j, D_{out} \leftarrow D_{in}[j]$;
7   **end**
8   **if** *all AXI4 interfaces are busy* **then**
9   Wait for $write\_finish[j]$ signal;
10   **end**
11 **end**

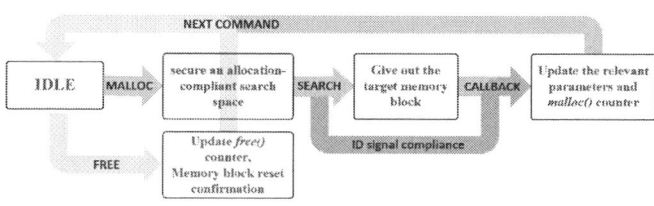

Fig. 3. Linear Allocator Workflow

B. Linear Allocator

The Linear Allocator module manages HBM memory using a simple book-keeping strategy, dividing memory into equal-sized blocks matching the number of enabled AXI4 interfaces. The value of the ID signal corresponds one-to-one with the memory blocks divided. In this way, the ID signal can prioritize memory allocation from specific blocks, simplifying the *malloc()* command cycle count and interfacing with the Data Dispatch module to prevent bandwidth loss from multiple AXI4 interfaces accessing the same pseudo channel.

The Linear Allocator processes two commands as shown in Fig. 3: *malloc()* and *free()*. For *malloc()*, it selects available

979-8-3503-6184-1/24 $31.00 © 2024 IEEE   206

blocks based on the ID signal, updates the address offset according to the request size, returns the starting address to the user, and records the remaining space and *malloc()* execution count. For *free()*, it updates the *free()* execution count and resets the address offset when the *free()* count matches the *malloc()* count. Implemented in Verilog, the module executes *malloc()* within 3-4 clock cycles and *free()* within 2 clock cycles.

C. HBM Driver

The HBM Driver module, based on the AXI4 bus protocol, is designed according to HBM's characteristics and AXI4 communication methods. It includes an interface that is compliant with the AXI4 protocol for address, data, and control channels connected to the HBM controller. An address mapping module maps AXI4 bus addresses to HBM storage units. A data transfer controller manages data transfers between the AXI4 bus and HBM, handling initiation, termination, and flow control according to the AXI4 protocol. The module uses FIFO IP cores to simplify timing logic.

IV. EXPERIMENTAL RESULTS

The experimental platform used is the Alveo U280 Card, with all modules individually and systematically verified. The evaluation includes resource utilization, performance, and on-board experimental data.

A. Resource Utilization and Performance

Based on the analysis of the provided table (Table. I), the Linear Allocator proposed shows significantly lower resource consumption. The "*Cycles*" column in the table refers to the average number of clock cycles spent on a single allocation operation. Compared to other examples, the allocator requires fewer clock cycles to complete an allocation operation and can operate at a higher clock frequency.

More specifically, it demonstrates a 52.60% reduction in LUT usage and a 28.99% reduction in DFF usage compared to the most resource-efficient HeapAllocator [4]. Additionally, the BRAM usage of the proposed allocator is 0.

TABLE I
ON-BOARD RESOURCE CONSUMPTION AND PERFORMANCE OF MEMORY ALLOCATORS

Allocators	LUTs	DFFs	Cycles	Frequency
This Paper	1411	1112	≈5	400MHz
SysAlloc (256blocks) [3]	4446	3894	≈100	150MHz
DOMMU [2]	≈11750	≈2073	-*	140MHz
HeapAllocator [4] (Heap Depth 512)	2977	1566	≈10^4	360MHz
DMM [7]	6075	2295	≈20	175.26MHz

*the corresponding reference does not provide specific figures.

B. On-Board Experimental Data

This section compares the performance of single-port, dual-port, and quad-port configurations for processing three different cases, each involving 300 variable-sized allocations. The experimental results for the single-port serial design are consistent with using the DDR memory as external memory. According to the bar chart (Fig. 4), the dual-port configuration reduces the clock cycles required to complete the allocation process by 15.26% on average compared to single-port serial access, while the quad-port configuration achieves a 36.21% reduction in clock cycles, equivalent to a 1.5× efficiency improvement over the DDR memory.

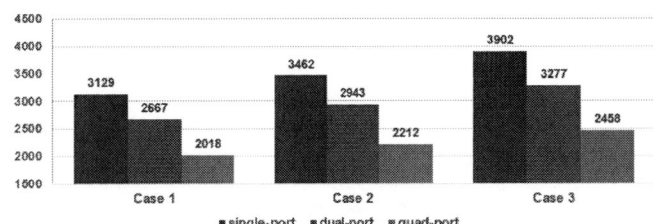

Fig. 4. Cycle Latency under specific dataset

V. FUTURE WORK AND CONCLUSION

Addressing memory fragmentation caused by high allocation efficiency remains a significant challenge. Future work will focus on developing techniques to minimize fragmentation while maintaining or improving allocation performance.

Our proposed dynamic memory allocator for FPGA-based systems using HBM improves resource efficiency and allocation performance. The Linear Allocator module reduces LUT and DFF usage, and the Data Dispatch module accelerates data writing through parallelism. This design provides a robust solution for high-performance computing, achieving better resource utilization and system efficiency. Experimental results show significant reductions in resource consumption and improvements in allocation efficiency.

ACKNOWLEDGMENT

This work is supported by the National Natural Science Foundation of China under grant 62174035.

REFERENCES

[1] Giamblanco, Nicholas V., and Jason H. Anderson. "A dynamic memory allocation library for high-level synthesis." 2019 29th International Conference on Field Programmable Logic and Applications (FPL). IEEE, 2019.

[2] Dessouky, Ghada, et al. "Adaptive Dynamic On-chip Memory Management for FPGA-based reconfigurable architectures." 2014 24th International Conference on Field Programmable Logic and Applications (FPL). IEEE, 2014.

[3] Xue, Zeping, and David B. Thomas. "SysAlloc: A hardware manager for dynamic memory allocation in heterogeneous systems." 2015 25th International Conference on Field Programmable Logic and Applications (FPL). IEEE, 2015.

[4] Kokkinis, Argyris, Dionysios Diamantopoulos, and Kostas Siozios. "Dynamic Heap Management in High-Level Synthesis for Many-Accelerator Architectures." 2022 32nd International Conference on Field-Programmable Logic and Applications (FPL). IEEE, 2022.

[5] Xilinx, A. M. D. "Vitis Unified Software Platform Documentation, Application Acceleration Development, UG1393 (v2022. 2), 7 December 2022." (2023).

[6] Xilinx, A. M. D. "Alveo U280 Data Center Accelerator Card Data Sheet: Alveo U280 Card Data Sheet, DS963 (v1.7), 23 June 2023." (2023).

[7] Ozer, C. (2014). A dynamic memory manager for FPGA applications (Master's thesis, Middle East Technical University).

979-8-3503-6184-1/24 $31.00 © 2024 IEEE

An Improved Clock-Aware Global Placement Algorithm

Ziang Ge and Pingqiang Zhou
School of Information Science and Technology
Shanghaitech University, Shanghai, P. R. China
{geza2022, zhoupq}@shanghaitech.edu.cn

Abstract—Traditional very-large-scale integrated (VLSI) circuits physical design flows that optimize clock networks after placement are limited by the quality of register placement. Prior research on clock-aware placement shows effectiveness in minimizing clock-net wirelength during placement but suffers from large runtime overhead due to iterative clock-tree construction process. In this paper, we propose to accelerate clock-aware placement with a fast clock-tree synthesis method and a preconditioner for clock wirelength gradient. Experimental results show that the proposed method can save 30% runtime over prior work with only 1% placement quality degradation.

Index Terms—Placement, clock-network synthesis, nonlinear optimization, low power.

I. INTRODUCTION

Due to their frequent switching and large capacitance, clock networks often account for over 30% of total power consumption in synchronous sequential designs [1]. Total wirelength is a critical optimization objective in clock network synthesis, since larger clock wirelength results in greater clock capacitance and thus require more power for distribution of clock signals [2].

Fig. 1 illustrates the steps of VLSI physical design. Clock network synthesis is usually performed after placement, which determines the locations of the sequential logics such as registers. Considering that the quality of synthesized clock network is greatly affected by register locations, such design flow limits the design space for clock network synthesis. To overcome such limitation, previous works propose to optimize the total switching power of clock networks by minimizing clock wirelength during placement stage [1], [4], [5]. This is achieved by adding additional force on registers during global placement. The direction of the additional force is estimated by constructing a virtual clock-tree at each global placement iteration. However, the runtime overhead for iterative virtual clock-tree construction become non-negligible over the many iterations of global placement. [5] proposes a grid-based algorithm to accelerate DME-based virtual clock-tree construction, at the cost of a 20% runtime overhead.

To further accelerate clock-aware global placement, we apply a much faster clock-tree synthesis method for virtual clock-tree construction. The key insight is that the purpose of virtual clock-tree construction is to estimate the direction of force

This work was supported in part by the National Natural Science Foundation of China under the Grant No. 62074100.

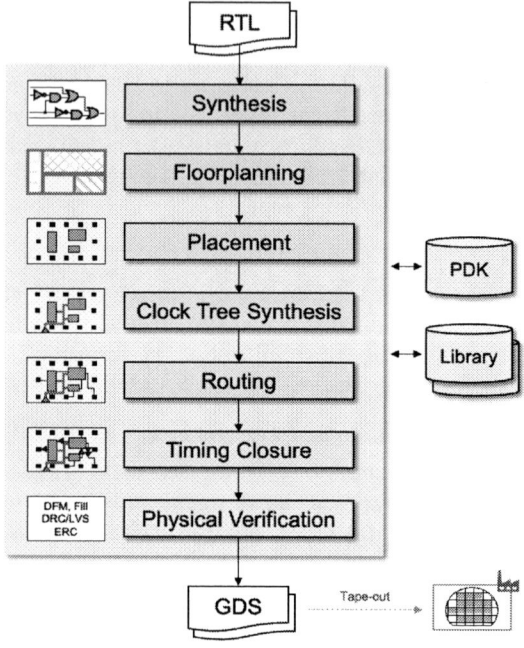

Fig. 1. VLSI physical design flow [3].

to pull registers together, rather than obtaining a high-quality clock-tree. Thus, we can sacrifice virtual clock-tree quality for lower runtime. Furthermore, for the first time, we propose a preconditioner for clock wirelength gradient to balance the optimization of clock-net wirelength and signal wirelength. The proposed technique is integrated into ePlace [6]. Experimental results on eight CLKISPD'05 benchmarks [1] show that, compared to the state-of-the-art clock-aware placement algorithm [5], our methods can reduce the runtime of clock-aware global placement by 30% on average.

II. RELATED WORKS

In this section, we present an overview of the state-of-the-art clock-aware global placement algorithm [5]. RePlAce [7] is chosen as the placement backbone, in which the global placement problem is formulated as an unconstrained optimization problem

$$\min_{\mathbf{v}} f(\mathbf{v}) = W(\mathbf{v}) + \lambda N(\mathbf{v}) \tag{1}$$

979-8-3503-6184-1/24 $31.00 © 2024 IEEE

where $\mathbf{v} = (\mathbf{x}, \mathbf{y})^T = (x_1, \ldots, x_n, y_1, \ldots, y_n)^T$ denotes a solution to accommodate all n objects and $W(\mathbf{v})$ is the wirelength function. $N(\mathbf{v})$ is the density function, which is modeled by eDensity [6], and λ is the penalty factor for adjusting the ratio between wirelength and density.

[5] extends the original optimization problem to consider clock-tree wirelength and the new objective function is defined as

$$f(\mathbf{v}) = W_u(\mathbf{v}) + \eta W_c(\mathbf{v}) + \lambda N(\mathbf{v}) \qquad (2)$$

where $W_u(\mathbf{v})$ denotes signal-net wirelength and $W_c(\mathbf{v})$ is the clock-net wirelength. η is the weight used to scale the clock-net wirelength. The nonlinear optimization problem in Eq. (2) is then solved using Nesterov's method [6]. Solving such problem using Nesterov's method involves efficiently computing the gradient function. The gradient function $f(v)$ over one variable x_i can be derived as

$$\frac{\partial f(\mathbf{v})}{\partial x_i} = \frac{\partial W_u(\mathbf{v})}{\partial x_i} + \eta \frac{\partial W_c(\mathbf{v})}{\partial x_i} + \lambda \frac{\partial N(\mathbf{v})}{\partial x_i} \qquad (3)$$

where the density gradient is computed using fast Fourier transform (FFT) [6]. The gradient of signal-net wirelength is obtained by differentiating the WA wirelength model [8]. The weight η is used to explore the optimization tradeoff between signal-net wirelength and clock-net wirelength. It should be noted that in the practice of nonlinear optimization, the gradient vector ∇f is often multiplied by a preconditioner [6] to accelerate its convergence process.

Since the existing wirelength models for signal-nets do not offer accurate estimation of clock-net wirelength [1], to obtain the gradient of clock-net wirelength, a virtual clock-tree is constructed at each global placement iteration. Then the gradient of clock-net wirelength for each register is estimated by accumulating gradients along the clock-tree. Specifically, the constructed virtual clock-tree is decomposed into a set of two-pin pseudo-nets and the clock-net gradient on leaf nodes (registers) is estimated by accumulating the wirelength gradients of the two-pin nets along each root-to-leaf path. Thus, the clock-net gradient function is defined as

$$\frac{\partial W_c(\mathbf{v})}{\partial x_i} = \sum_{t_j \in \mathcal{T}} w_{l_j} \frac{\partial W_{t_j}(\mathbf{v})}{\partial x_i} \qquad (4)$$

where $\mathcal{T}$ is the set of all two-pin pseudo-nets from a register r_i to the root, and w_{l_j} denotes the weight of gradient of level l_j at which the two-pin pseudo-net t_j is located in the clock-tree. Fig. 2 gives an example of arboreal clock-net gradient.

The runtime overhead of constructing the virtual clock-tree using nearest-neighbor graph (NNG) and deferred-merge embedding (DME) at every global placement iteration can be significant. For example, the complexity of naive NNG construction is $O(n^3)$ for n registers. Therefore, a grid-based algorithm for NNG construction is proposed in [5] to accelerate virtual clock-tree construction. The NNG has a time complexity of $O(n^2 \log n + k^2 n)$, where k is the average number of sinks in a grid.

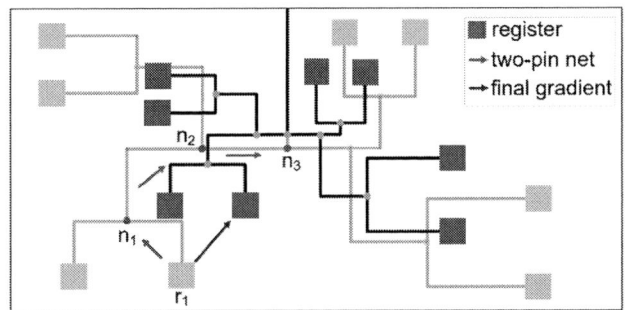

Fig. 2. Arboreal gradient accumulation in [5].

III. METHODOLOGY

Although the grid-based algorithm can accelerate virtual clock-tree construction, its runtime overhead is still non-negligible (20% as reported in [5]). More importantly, the effort to build a well-optimized virtual clock-tree with NNG and DME algorithm can be wasteful, because

- DME-based algorithm tries to minimize clock skew while our work targets the minimization of clock-tree wirelength.
- In each iteration of global placement, previous virtual clock-tree is discarded and a new one is built from scratch with updated register locations. Therefore, over-optimizing a virtual clock-tree with current register locations can be meaningless especially during the early stage of global placement, when the registers are far from their final positions.

For efficient and effective virtual clock-tree construction, we propose to replace NNG and DME with the method of means and medians (MMM) [9], which is much faster and adequately effective to guide the movement of registers.

A. Virtual Clock-Tree Construction with MMM

We adopt ePlace [6] as our placement backbone and use the objective function defined in Eq. (2). We apply the MMM algorithm to construct a virtual clock-tree with current register locations in one global placement iteration and follow the arboreal clock-net gradient accumulation to obtain clock-net wirelength gradient. The overall computation flow is shown in Fig. 3. MMM is a classic algorithm for clock-tree synthesis. It recursively partitions the set of sinks (registers) into two subsets of equal cardinality (median). Then, the center of mass of the set is connected to the centers of mass of the two subsets (mean). The MMM algorithm has a time complexity of only $O(n \log n)$. Notice that the effectiveness of MMM depends heavily on the choice of partition directions for median computation, in this work, we alternate the cut direction during a single MMM run because all benchmark circuits we use in this work have an approximate square shape. The main drawback of MMM is that it only minimizes clock skew heuristically. But this is acceptable because we only need to optimize clock wirelength during global placement.

979-8-3503-6184-1/24 $31.00 © 2024 IEEE

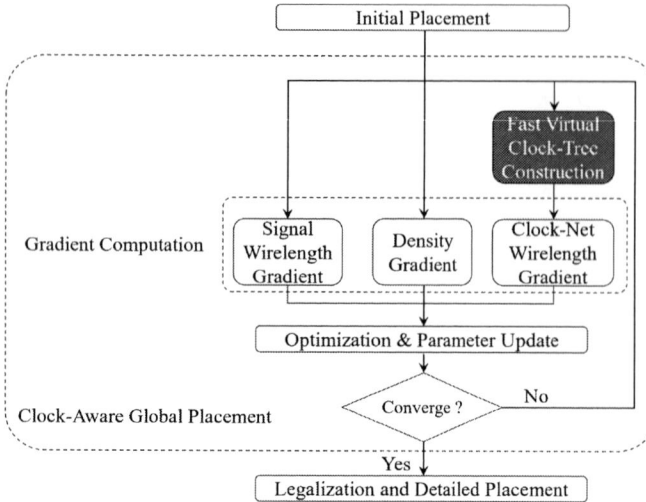

Fig. 3. Overall flow of our algorithm.

The strategy for adjusting η in [5] is adopted and η is defined as

$$\eta = k_1 e^{k_2(k_3 - \tau)} \qquad (5)$$

where k_1, k_2, k_3 are chosen to be $6, 10, 0.1$ respectively in our experiments, and τ is the global density overflow, which is defined as the total overflow area in all placement bins over the area of total movable blocks [10] and it usually starts from around 1.0 and ends with 0.1. We observe that η is very small during early stage of global placement (when the global density overflow is fairly high) and $\eta \frac{\partial W_c(\mathbf{v})}{\partial x_i}$ is thus close to 0. Therefore, we only conduct virtual clock-tree construction when $\tau < 0.6$ and clock-net gradient is set to 0 when $\tau \geq 0.6$.

The weight of gradient at level i is calculated as

$$w_{l_i} = \frac{1}{\sqrt{2\pi}\sigma} e^{\left(-\frac{l_i^2}{2\sigma^2}\right)} \qquad (6)$$

where l_i is defined as

$$l_i = 3\left(1 - \frac{i}{L_t}\right) \qquad (7)$$

Here L_t is the total level count of the clock-tree. We follow [5] for the definition of σ in Eq. (6)

$$\sigma = p - \frac{1}{e^{\tau+q}} \qquad (8)$$

where p and q are set to 10.35 and -1.68 respectively.

B. Clock-Net Wirelength Preconditioner

Preconditioning helps reduce the condition number of a problem and has been verified to be effective for accelerating the nonlinear global placement [6]. Signal wirelength preconditioner and density preconditioner have been used in [6] for signal wirelength optimization.

In this work, we propose a clock-net wirelength preconditioner to 1) speed up the convergence of clock-net wirelength optimization, and 2) relieve the imbalance between object

gradients in Eq. (3). We approximate $\mathbf{H_f}$ of Eq. (2) to be a positive definite diagonal matrix $\widetilde{\mathbf{H}}_\mathbf{f}$

$$\mathbf{H_f} \approx \widetilde{\mathbf{H}}_\mathbf{f} = \begin{pmatrix} \widetilde{\mathbf{H}}_{\mathbf{f_{x,x}}} & 0 \\ 0 & \widetilde{\mathbf{H}}_{\mathbf{f_{y,y}}} \end{pmatrix} \qquad (9)$$

where

$$\widetilde{\mathbf{H}}_{\mathbf{f_{x,x}}} = \begin{pmatrix} \frac{\partial^2 f}{\partial x_1^2} & 0 & \cdots & 0 \\ 0 & \frac{\partial^2 f}{\partial x_2^2} & \cdots & 0 \\ \vdots & \vdots & \ddots & \vdots \\ 0 & 0 & \cdots & \frac{\partial^2 f}{\partial x_n^2} \end{pmatrix} \qquad (10)$$

is the x part of the precondioner, and the y part $\widetilde{\mathbf{H}}_{\mathbf{f_{y,y}}}$ can be expressed in a similar way. Then we use $\widetilde{\mathbf{H}}_\mathbf{f}$ as the preconditioner and calculate $\nabla f_{\text{pre}} = \widetilde{\mathbf{H}}_\mathbf{f}^{-1} \nabla f$ in nonlinear optimization.

To obtain $\widetilde{\mathbf{H}}_{\mathbf{f_{x,x}}}$, according to Eq. (2), we have

$$\frac{\partial^2 f(\mathbf{v})}{\partial x_i^2} = \frac{\partial^2 W_u(\mathbf{v})}{\partial x_i^2} + \eta \frac{\partial^2 W_c(\mathbf{v})}{\partial x_i^2} + \lambda \frac{\partial^2 N(\mathbf{v})}{\partial x_i^2} \qquad (11)$$

We then need to compute or estimate $\partial^2 W_u(\mathbf{v})/\partial x_i^2$, $\partial^2 W_c(\mathbf{v})/\partial x_i^2$ and $\partial^2 N(\mathbf{v})/\partial x_i^2$ separately. The readers are referred to [6] for the estimations of the terms $\partial^2 W_u(\mathbf{v})/\partial x_i^2$ and $\partial^2 N(\mathbf{v})/\partial x_i^2$. For the second term $\partial^2 W_c(\mathbf{v})/\partial x_i^2$, after differentiating Eq. (4) with respect to x_i, we can get the second-order gradient of the clock-net wirelength function as

$$\frac{\partial^2 W_c(\mathbf{v})}{\partial x_i^2} = \sum_{t_j \in \mathcal{T}} w_{l_j} \frac{\partial^2 W_{t_j}(\mathbf{v})}{\partial x_i^2} \qquad (12)$$

Here we use the degree of two-pin pseudo-nets, 2, as the estimation of $\partial^2 W_{t_j}(\mathbf{v})/\partial x_i^2$, so we get

$$\frac{\partial^2 W_c(\mathbf{v})}{\partial x_i^2} = 2 \sum_{t_j \in \mathcal{T}} w_{l_j} \qquad (13)$$

IV. EXPERIMENTAL RESULTS

Our algorithm is implemented in C++ and ran on a Linux machine with Intel i7 12700 2.1GHz CPU and 16GB memory. We conduct experiments on the CLKISPD'05 benchmarks [1], which include register lists generated based on the ISPD2005 benchmarks [11]. 15% of standard cells are selected to be registers in each benchmark. Table II shows the details of the benchmarks.

For fair comparison, we implement the algorithm [5] in C++ and run on the same machine. Notice that although RePlAce is chosen as the placement backbone in [5], it shares the core algorithm with ePlace. Therefore we choose ePlace as the placement backbone of our algorithm. We propose two techniques, MMM and preconditioner for clock wirelength, to speed up the clock-aware global placement algorithm. To show their effectiveness, we design two algorithms for comparison. One is *Our-MMM*, which only includes MMM for virtual clock-tree construction, and the other is *Ours-MMM-Pre* which includes both techniques.

979-8-3503-6184-1/24 $31.00 © 2024 IEEE

TABLE I
RESULTS ON THE CLKISPD'05 BENCHMARK SUITE

Bench	[5]			Ours-MMM			Ours-MMM-Pre		
	ClkWL	HPWL	Time (s)	ClkWL	HPWL	Time (s)	ClkWL	HPWL	Time (s)
clkad1	1.37	77.47	97.37	1.34	77.92	78.1	1.37	77.89	82.64
clkad2	1.53	85.79	179.53	1.48	89.81	150.79	1.50	88.31	137.68
clkad3	3.08	209.29	349.62	3.13	207.93	253.22	3.19	206.90	245.31
clkad4	3.17	193.66	377.9	2.91	204.24	303.16	3.29	199.29	268.51
clkbb1	1.65	94.05	202.55	1.58	95.11	151.58	1.60	94.73	154.88
clkbb2	3.40	158.02	388.5	3.69	155.52	230.42	3.64	157.56	238.02
clkbb3	5.47	393.53	1314.02	5.12	470.51	1132.72	5.99	379.96	641.22
clkbb4	12.04	879.42	4113.9	9.93	951.90	3293.8	11.51	909.78	3011.31
Avg	1.00×	1.00×	1.00×	0.96×	1.05×	0.77×	1.01×	1.02×	0.70×

TABLE II
STATISTICS OF CLKISPD'05 BENCHMARKS.

Name	Cells	Regs	Nets	Macros
clkad1	210K	32K	221K	56
clkad2	255K	38K	266K	177
clkad3	451K	68K	466K	721
clkad4	494K	74K	516K	1329
clkbb1	278K	42K	284K	30
clkbb2	535K	84K	577K	923
clkbb3	1095K	165K	1123K	666
clkbb4	2169K	327K	2230K	639

We use NTUplace3 [10] as our legalizer and skip detailed placement because it optimizes signal wirelength only and may harm the result of clock-net wirelength. After legalization, we build the final zero-skew clock-tree for evaluation using DME-based method [1], [5]. The wirelength of signal nets is estimated with HPWL, while the wirelength of the clock net is obtained by summing the Manhattan length of all edges in the final clock-tree. Since global placement might easily diverge, we record the result of best overflow for each run.

Experimental results are shown in Table I with wirelength in $\times 10^6$ and CPU time in seconds. Both Ours-MMM and Ours-MMM-Pre are faster than [5] with 23% and 30% lower runtime respectively. Compared to [5], Ours-MMM achieves 4% lower clock-net wirelength and 5% higher signal wirelength while Ours-MMM-Pre achieves a more balanced result with only 1% and 2% degradation on clock-net wirelength and signal wirelength respectively. Since we use the same placement backbone here, experimental results show that our method can effectively alleviate the overhead brought by iterative virtual clock-tree construction while maintaining high quality of placement solution. Furthermore, after applying the preconditioner for clock-net wirelength, our method can achieve faster convergence speed and better balance between the optimizations of signal wirelength and clock-net wirelength.

V. CONCLUSION

In this work, we propose an improved clock-aware placement methodology for electrostatics-based nonlinear placer to minimize runtime overhead. Experiments show that our method can achieve 30% lower runtime with negligible quality

degradation compared with previous methods on the CLK-ISPD'05 benchmarks.

REFERENCES

[1] D.-J. Lee and I. L. Markov, "Obstacle-aware clock-tree shaping during placement," *IEEE Transactions on Computer-Aided Design of Integrated Circuits and Systems*, vol. 31, no. 2, pp. 205–216, 2012.

[2] T.-H. Chao, Y.-C. Hsu, J.-M. Ho, and A. Kahng, "Zero skew clock routing with minimum wirelength," *IEEE Transactions on Circuits and Systems II: Analog and Digital Signal Processing*, vol. 39, no. 11, pp. 799–814, 1992.

[3] A. B. Kahng, J. Lienig, I. L. Markov, and J. Hu, *VLSI Physical Design: From Graph Partitioning to Timing Closure*, 1st ed. Springer Publishing Company, Incorporated, 2011.

[4] Y. Wang, Q. Zhou, X. Hong, and Y. Cai, "Clock-tree aware placement based on dynamic clock-tree building," in *IEEE International Symposium on Circuits and Systems*, 2007, pp. 2040–2043.

[5] J. Ding, L. Lu, Z. Fu, J. Ma, M. Gong, Y. Qi, and W. Yu, "Clock aware low power placement," in *IEEE/ACM International Conference on Computer Aided Design*, 2023, pp. 01–08.

[6] J. Lu, P. Chen, C.-C. Chang, L. Sha, D. J.-H. Huang, C.-C. Teng, and C.-K. Cheng, "ePlace: Electrostatics-based placement using fast fourier transform and nesterov's method," *ACM Transactions on Design Automation of Electronic Systems*, vol. 20, no. 2, pp. 1–34, 2015.

[7] C.-K. Cheng, A. B. Kahng, I. Kang, and L. Wang, "RePlAce: Advancing solution quality and routability validation in global placement," *IEEE Transactions on Computer-Aided Design of Integrated Circuits and Systems*, vol. 38, no. 9, pp. 1717–1730, 2019.

[8] M.-K. Hsu, V. Balabanov, and Y.-W. Chang, "TSV-aware analytical placement for 3-D IC designs based on a novel weighted-average wirelength model," *IEEE Transactions on Computer-Aided Design of Integrated Circuits and Systems*, vol. 32, no. 4, pp. 497–509, 2013.

[9] M. Jackson, A. Srinivasan, and E. Kuh, "Clock routing for high-performance ICs," in *ACM/IEEE Design Automation Conference*, 1990, pp. 573–579.

[10] T.-C. Chen, Z.-W. Jiang, T.-C. Hsu, H.-C. Chen, and Y.-W. Chang, "NTUplace3: An analytical placer for large-scale mixed-size designs with preplaced blocks and density constraints," *IEEE Transactions on Computer-Aided Design of Integrated Circuits and Systems*, vol. 27, no. 7, pp. 1228–1240, 2008.

[11] G.-J. Nam, C. J. Alpert, P. Villarrubia, B. Winter, and M. Yildiz, "The ISPD2005 placement contest and benchmark suite," in *International Symposium on Physical Design*, 2005, pp. 216–220.

Analyzing Timing in Shorter Time: A Journey through Heterogeneous Parallelism for Static Timing Analysis

Zizheng Guo[1,2], Yibo Lin[1,2,3]*, Runsheng Wang[1,2,3], Ru Huang[1,2,3]

[1]School of Integrated Circuits, Peking University [2]Institute of Electronic Design Automation, Peking University
[3]Beijing Advanced Innovation Center for Integrated Circuits

Abstract—This paper reviews recent work on the acceleration of static timing analysis (STA), with a special focus on parallel and heterogeneous computing techniques. Timing analysis is one of the most critical tasks in circuit design. The ever-increasing size and complexity of modern circuit design has asked for unprecedented STA runtime speed-up which has to be achieved through CPU-GPU heterogeneous computing. GPU-accelerated STA is however difficult due to its nature of irregular computation and memory access patterns. We demonstrate and analyze the algorithm design and scheduling considerations in recent works targeting various different STA stages, and discuss future directions of STA acceleration as well as the future of timing optimization in heterogeneous circuit design flow.

Index Terms—Static timing analysis, Heterogeneous computing

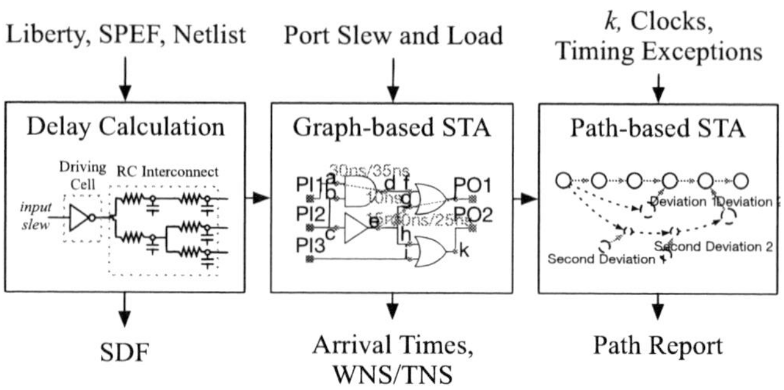

Fig. 1: Overview of STA steps, inputs, and outputs.

I. INTRODUCTION

The analysis and optimization of timing is one of the most critical tasks in circuit design as it is directly tied to the chip's correctness and runtime performance. With the ever-increasing size and complexity of modern circuit design as well as the ever-rising difficulty to model physics in advanced nodes, the runtime it takes to complete a design cycle has become much longer. This runtime burden is worsened by the increasing demand to explore and search the design space more thoroughly for power, performance, and area (PPA) improvements. Within this runtime burden, static timing analysis (STA) has become a remarkable bottleneck.

During the circuit design flow, STA will be invoked for hundreds to thousands of times because most design steps like synthesis, placement, and routing are iterative. As a result, the performance of STA is critical for efficient design turnaround. STA consists of graph algorithms running on large circuit graphs with millions of nodes and edges. An efficient STA engine thus calls for massive parallelism that is beyond the reach of traditional multi-core CPUs.

In this paper, we survey recent works on GPU-accelerated STA. These works focus on different STA stages including graph-based analysis, path-based analysis, delay modeling,

*Corresponding author: Yibo Lin (yibolin@pku.edu.cn).

etc. We review various techniques on GPU-friendly algoithms, data structures, scheduling algorithms, and other engineering efforts to overcome the inherent difficulty in STA parallelization such as irregular graph computation workload and memory access patterns.

The rest of the paper is organized as follows. Section II introduces the background of STA in a heterogeneous computing perspective. Then, Sections III, IV, and V reviews recent work categorized into different STA stages including delay modeling, graph-based analysis, and path-based analysis respectively. Finally, Section VI concludes the review and discuss future directions.

II. PRELIMINARIES

STA takes as input a circuit represented as a directed acyclic graph (DAG) with nodes indicating pins and edges indicating signal connections. Along with the circuit, STA also takes physical information such as cell library in process design kit (PDK) as well as parasitics information for circuit interconnects. With these inputs and further clock and path settings, STA analyzes design timing and outputs worst negative slack (WNS), total negative slack (TNS), and a set of critical logic paths [1].

STA is usually divided into 3 steps, as shown in Figure 1:

979-8-3503-6184-1/24 $31.00 © 2024 IEEE

1) Delay calculation. This step makes use of parasitics information to derive a compact modeling of the metal interconnects.
2) Graph-based analysis. This step propagates voltage slew and signal arrival time of pins throughout the circuit graph in topological order. After this step, WNS and TNS as well as pin slacks can be derived.
3) Path-based analysis. This step searches for top-k most critical signal paths with k given by designers. Optionally, the path slacks are recalculated based on path-specific conditions for better accuracy.

The performance of STA is critical as it is frequently used in circuit design. The runtime of delay calculation and graph-based STA is proportional to the scale of circuit DAG, which can exceed millions or even more pins and arcs. The runtime of path-based STA is further multipled by the number of paths k requested, which may range from tens to thousands. A typical STA run on a million-sized design can take tens of minutes to hours and become a bottleneck in the physical design flow.

Parallel computation is the key to accelerating STA due to its problem scale. Current major STA engines like Open-STA [2], PrimeTime [3], and OpenTimer [4] all supports parallel STA using multi-core CPUs. It has been widely observed, however, that CPU-based parallel STA cannot scale beyond 8–16 CPU threads [4], and the performance may even degrade after that threshold.

Heterogeneous computing using general-purpose graphics computing units (GPGPU) has been shown to provide unprecedented speed-up on a variety of tasks, with machine learning a notable example. However, successful heterogeneous application requires balanced and regular patterns in computation and memory access, which is hardly the case for STA. Delay calculation requires solving interconnect equations in a long-tailed net size distribution, which incurs workload imbalance between working threads. Graph-based and path-based analyses work on highly irregular circuit graph with induced task dependencies and irregular memory access. These make GPU-accelerated STA quite challenging.

III. Delay Modeling

Delay modeling or delay calculation is the first step in STA and it determines the analysis accuracy when choosing from different delay models. This section introduces GPU-accelerated delay calculation works arranged inside a brief review of different delay models. One can refer to [5], [6] for a more comprehensive introduction of delay models.

There are usually two separate delay models inside a STA engine for net and cell delay calculation, respectively. Net delay model analyzes the voltage response of metal interconnects represented in resistors and capacitors (RC) networks. Cell delay model characterizes the standard cell voltage response

into compact forms like look-up tables conditioned on the environment of the cell. Voltage response consists of signal delay and transition time.

A. Heterogeneous net delay model

One widely used and simple net delay model is the Elmore delay model [7]. Elmore delay approximates net delay and slew by summing up RC products along the tree path from the source to every sink. In reality, this process is implemented as a tree-based dynamic programming algorithm [4]. Guo Z. et al [8], [9] present a GPU-accelerated Elmore delay calculator on top of OpenTimer. As GPU has very limited call stack, they propose a sweeping-based algorithm to simulate the dynamic programming on GPU. An Elmore delay speed-up of $2.54\times$ and overall speed-up of $3.69\times$ over OpenTimer has been reported.

As process technology continues to develop, high-order voltage effects including resistive shielding have evolved in sub-14nm nodes. This calls for more accurate interconnect modeling than the Elmore approximation. To this end, Arnoldi models and other model order reduction techniques have been proposed [10], [11], [12]. They construct the linear system equation of each RC interconnect, and then reduce the system order using algebraic tricks. Guo Z. et al [13] present a GPU-accelerated Arnoldi delay calculator based on the coordinate-transformed Arnoldi algorithm [12] and reported $7.27\times$ speed-up over PrimeTime. Advanced net delay models face more severe workload imbalance challenge because both the interconnect size variation and the time complexity of interconnect modeling are higher. They tackled this problem by splitting the algebraic computations (e.g., sparse LU decomposition) into multiple stages with different parallelism to exploit.

B. Heterogeneous cell delay model

The most widely used cell model is called non-linear delay model (NLDM) which is often combined with the Elmore model to derive delays for both nets and cells. NLDM models cell timing arcs as linearly-interpolated look-up tables (LUTs) indexed by input voltage transition and output capacitive load. A look-up table query needs to access two index arrays to locate interpolation point, and then access the result matrix 4 times. The CASTA timer presented by Wang H. et al [14] propose to optimize such memory access by placing indices and values close to each other (table-index remapping) and inside texture memory. They reported up to $14.89\times$ speed-up on NLDM calculation.

As transistors shrink to nanoscale, current-source models (CSM) begin to replace NLDM for its better accuracy especially in signoff scenarios. CSM does not directly model the delay and transition of cell arcs, but instead models a cell as a time- and voltage-controlled current source (for composite current source model, CCS [15], [16]) or voltage

979-8-3503-6184-1/24 $31.00 © 2024 IEEE   213

source (for effective current source model, ECSM [17]). Lin S. et al [18] propose a GPU-accelerated CCS model calculator. They propose an efficient matrix inverse precomputation technique making use of the similarity of conductance matrices and achieve up to 3.4× speed-up in 2% error.

IV. GRAPH-BASED ANALYSIS

A delay calculator is itself not an end-to-end STA engine without graph-based analysis. End-to-end STA engines are more difficult to accelerate on GPU due to the synchronization and data transfer overhead between CPU and GPU. Prior works like [14], [19] can achieve high speed-up ratios when measuring only the kernel runtime. However, their end-to-end performance may even be 0.9× inferior to a CPU flow. As a result, a GPU-accelerated graph-based STA engine must incorporate efficient task-scheduling strategies to overcome the overhead. Guo Z. et al [8] overlaps memory transfer and independent computations using CUDA streams. They make choices on CPU/GPU task placement based on a runtime breakdown to avoid over-optimization. In their journal extension [9], they extend this framework to multi-corner graph-based STA by placing the corner-level parallelism at the GPU thread-level to achieve better scalability up to 25.67×.

Arrival time propagation in graph-based STA is another challenge due to its dependency constraints. To sort out the dependency, levelization is widely used [8], [14] as a preprocessing step to make sure nodes within each topological level can run in parallel. However, levelization itself is shown to take a significant amount of runtime, so GPU-accelerated topological sorting and levelization is designed [8] that accelerates it up to 4.51×. For traditional CPU-based task parallelism, a better partition can also improve the performance by reducing scheduling cost. Zhang B. et al [20] show that such partition can be efficiently generated with the help of GPU.

Besides parallelism within a circuit graph, prior works also explore the application of GPUs or FPGAs in Monte-Carlo-based statistical static timing analysis (SSTA) [21], [22]. SSTA using Monte Carlo provides another parallelism across different independent simulation runs which fits nicely with heterogeneous platforms.

V. PATH-BASED ANALYSIS

Path searching is the ultimate step in STA flow. The search of top-k critical paths often relies on prefix-suffix tree algorithms [23] or its improvements [24], [25]. The algorithm behaves like a A* search that relies on a priority heap and a first-in-first-out (FIFO) queue. One challenge is that A* algorithms are inherently sequential because only the best current solution (in our formulation, current most-critical path) can be used to expand solution space. Fortunately, it turns out that this A* algorithm constraint can be carefully relaxed and turned into an equivalent iterative-pruning algorithm. Guo G.

et al [26], [27] propose the above technique. Their evaluation generates up to 1 million paths with up to 45× shorter runtime than a saturated CPU parallel STA engine. Later, they propose a path-search-oriented graph partitioning strategy [28] to overcome the single-GPU memory limitation.

Path-based analysis is also a stage where various user-defined constraints and exceptions are applied. These constraints, settings, and exceptions often create unique challenges for accelerating STA. One notable example is common path pessimism removal (CPPR)[1]. CPPR requires adjustment of path criticality based on the common clock driving paths between its launching and capturing registers. A brute-force algorithm for CPPR needs to enumerate all pairs of registers which make the path search extremely slow for large designs. Recently, better algorithms called depth-based CPPR have been proposed [29], [30], [31] that clusters registers into groups based on a batch of depth-relevant criticality adjustments. Depth-based CPPR is also implemented on GPU in the HeteroCPPR framework by Guo Z. et al [32]. They make use of the independence between clustered groups to scale their algorithm to multiple GPUs and achieve up to 16× speed-up.

Besides CPPR, other timing exceptions like false paths, multi-cycle paths, path margins, and set delays have been more and more frequently used in modern chip designs. These rules are written in a design constraints file. Each rule gives a path pattern (including from, through, and to) and actions on the paths. For simple inclusive rules in final timing report, Guo G. et al [33] propose constrained subgraph scanning and sub-forest expansion techniques making use of the topological structure. For advanced exclusive rules like false paths, multi-cycle paths, etc., Guo Z. et al propose the HeteroExcept framework [34] that solves most of the common timing exceptions. They prove the NP-hardness of general exclusive rules handling. Practically, their GPU-accelerated exception-aware STA achieves 6.84× speed-up over PrimeTime by various techniques like GPU exception footprinting and copy-on-write algorithms.

VI. CONCLUSION AND FUTURE CHALLENGES

This paper reviews the current state-of-the-art heterogeneous algorithms and frameworks for STA. Heterogeneous CPU/GPU parallelism have introduced runtime benefit to all major STA stages including delay calculation, graph-based analysis, and path-based analysis. While each stage has its unique challenge, the common challenges are irregular computation and memory patterns as well as imbalanced workload. Prior works introduce various algorithm transformation, preprocessing, and scheduling techniques to achieve an end-to-end speed-up ranging from 3× to 45× on different workloads.

---

[1] also known as clock reconvergence pessimism removal (CRPR).

With these advancements in heterogeneous STA, there are a number of new challenges and opportunities.

- *Efficient inter-GPU partitioning for large circuits.* Modern system on a chip (SoC) design contains more than tens to hundreds of millions of circuit elements. This scale is beyond the capacity of a single GPU node, or even a classical CPU node. Automatic partitioning and hierarchical STA techniques can help in scaling the STA task to multiple GPUs and multiple nodes, which is vital for the successful application of heterogeneous STA.

- *Advanced manufacturing technology.* The development of process node continues to require even more complex circuit delay models, taking into account aging, multi-input switching, electromigration, advanced variation, etc. Specifically, the introduction of 3D ICs create challenge in multi-corner STA due to the explosion of corner count.

- *Timing-driven optimization.* The ultimate goal of timing analysis is to optimize circuit performance based on the evaluation. This not only calls for effective use of fine-grained timing analysis results through heterogeneous timing-driven design algorithms, but also calls for cross-stage prediction of timing results to enable early feedback for both RTL and physical design.

ACKNOWLEDGEMENT

This work was supported in part by the Natural Science Foundation of Beijing, China (Grant No. Z230002) and the 111 project (B18001).

REFERENCES

[1] J. Bhasker and R. Chadha, *Static Timing Analysis for Nanometer Designs: A Practical Approach*, 1st ed. Springer Publishing Company, Incorporated, 2009.

[2] "OpenSTA," https://github.com/The-OpenROAD-Project/OpenSTA.

[3] "Synopsys PrimeTime," http://www.synopsys.com.

[4] T. Huang, G. Guo, C. Lin, and M. D. F. Wong, "OpenTimer v2: A New Parallel Incremental Timing Analysis Engine," *IEEE TCAD*, vol. 40, no. 4, pp. 776–789, 2021.

[5] J. Croix and D. Wong, "Blade and razor: cell and interconnect delay analysis using current-based models," in *Proc. DAC*, 2003, pp. 386–389.

[6] U. Baur, P. Benner, and L. Feng, "Model order reduction for linear and nonlinear systems: a system-theoretic perspective," *Archives of Computational Methods in Engineering*, vol. 21, no. 4, pp. 331–358, 2014.

[7] W. C. Elmore, "The transient response of damped linear networks with particular regard to wideband amplifiers," *Journal of applied physics*, vol. 19, no. 1, pp. 55–63, 1948.

[8] Z. Guo, T.-W. Huang, and Y. Lin, "Gpu-accelerated static timing analysis," in *Proc. ICCAD*. ACM, 2020.

[9] ——, "Accelerating static timing analysis using cpu-gpu heterogeneous parallelism," *IEEE TCAD*, pp. 1–1, 2023.

[10] A. Odabasioglu, M. Celik, and L. T. Pileggi, "Prima: Passive reduced-order interconnect macromodeling algorithm," *IEEE TCAD*, vol. 17, no. 8, p. 645, 1998.

[11] C. L. Ratzlaff and L. T. Pillage, "Rice: Rapid interconnect circuit evaluation using awe," *IEEE TCAD*, vol. 13, no. 6, pp. 763–776, 1994.

[12] L. Miguel Silveira, M. Kamon, I. Elfadel, and J. White, "A coordinate-transformed Arnoldi algorithm for generating guaranteed stable reduced-order models of RLC circuits," in *Proc. ICCAD*, Nov. 1996, pp. 288–294.

[13] Z. Guo, T.-W. Huang, Z. Jin, C. Zhuo, Y. Lin, R. Wang, and R. Huang, "Heterogeneous static timing analysis with advanced delay calculator," in *Proc. DATE*, 2024.

[14] H. H.-W. Wang, L. Y.-Z. Lin, R. H.-M. Huang, and C. H.-P. Wen, "Casta: Cuda-accelerated static timing analysis for VLSI designs," in *Proc. ICPP*. IEEE, 2014, pp. 192–200.

[15] S. Simoglou, I. Lilitsis, N. Blias, and C. Sotiriou, "Full Stage Delay Calculation Using Full Waveform Propagation and Standard Library CCS Model," in *Proc. ISQED*. San Francisco, CA, USA: IEEE, Apr. 2024, pp. 1–8.

[16] D. Garyfallou, S. Simoglou, N. Sketopoulos, C. Antoniadis, C. P. Sotiriou, N. Evmorfopoulos, and G. Stamoulis, "Gate delay estimation with library compatible current source models and effective capacitance," *IEEE TVLSI*, vol. 29, no. 5, pp. 962–972, 2021.

[17] Cadence, "ECSM Library Format." [Online]. Available: https://www.cadence.com/en_US/home/alliances/standards-and-languages/ecsm-library-format.html

[18] S. Lin, G. Guo, T.-W. Huang, W. Sheng, E. F. Young, and M. D. Wong, "GCS-Timer: Gpu-accelerated current source model based static timing analysis," in *Proc. DAC*, 2024.

[19] K. E. Murray and V. Betz, "Tatum: Parallel timing analysis for faster design cycles and improved optimization," in *Proc. FPT*. IEEE, 2018, pp. 110–117.

[20] B. Zhang, D.-L. Lin, C. Chang, C.-H. Chiu, B. Wang, W. L. Lee, C.-C. Chang, D. Fang, and T.-W. Huang, "G-PASTA: Gpu-accelerated partitioning algorithm for static timing analysis," in *Proc. DAC*, 2024.

[21] K. Gulati and S. P. Khatri, "Accelerating statistical static timing analysis using graphics processing units," in *Proc. ASPDAC*. IEEE, 2009, pp. 260–265.

[22] J. Cong, K. Gururaj, W. Jiang, B. Liu, K. Minkovich, B. Yuan, and Y. Zou, "Accelerating Monte Carlo based SSTA using FPGA," in *Proc. FPGA*, 2010, pp. 111–114.

[23] T.-W. Huang and M. D. Wong, "OpenTimer: A high-performance timing analysis tool," in *Proc. ICCAD*. IEEE, 2015, pp. 895–902.

[24] K. Zhou, Z. Guo, T.-W. Huang, and Y. Lin, "Efficient critical paths search algorithm using mergeable heap," in *Proc. ASPDAC*, 2022.

[25] C. Chang, T.-W. Huang, D.-L. Lin, G. Guo, and S. Lin, "Ink: Efficient incremental k-critical path generation," in *Proc. DAC*, 2024.

[26] G. Guo, T.-W. Huang, Y. Lin, and M. Wong, "Gpu-accelerated path-based timing analysis," in *Proc. DAC*. ACM, 2021.

[27] G. Guo, T.-W. Huang, Y. Lin, Z. Guo, S. Yellapragada, and M. D. F. Wong, "A gpu-accelerated framework for path-based timing analysis," *IEEE TCAD*, pp. 1–1, 2023.

[28] G. Guo, T.-W. Huang, and M. Wong, "Fast sta graph partitioning framework for multi-gpu acceleration," in *Proc. DATE*, 2023, pp. 1–6.

[29] Z. Guo, T.-W. Huang, and Y. Lin, "A provably good and practically efficient algorithm for common path pessimism removal in large designs," in *Proc. DAC*. ACM, 2021.

[30] Z. Guo, M. Yang, T.-W. Huang, and Y. Lin, "A provably good and practically efficient algorithm for common path pessimism removal in large designs," *IEEE TCAD*, pp. 1–1, 2021.

[31] T. Sun and C. Feng, "Dac-cppr: A fast and accurate approach for common path pessimism removal with divide and conquer on the clock tree." IEEE, 2023, pp. 263–268.

[32] Z. Guo, T.-W. Huang, and Y. Lin, "HeteroCPPR: Accelerating common path pessimism removal with heterogeneous cpu-gpu parallelism," in *Proc. ICCAD*. ACM, 2021.

[33] G. Guo, T.-W. Huang, Y. Lin, and M. Wong, "Gpu-accelerated critical path generation with path constraints," in *Proc. ICCAD*, 2021, pp. 1–9.

[34] Z. Guo, Z. Zhang, W. Li, T.-W. Huang, X. Shi, Y. Du, Y. Lin, R. Wang, and R. Huang, "HeteroExcept: A CPU-GPU heterogeneous algorithm to accelerate exception-aware static timing analysis," in *Proc. ICCAD*, 2024.

TBPart-b: An Effective Hypergraph Partitioning Algorithm Considering Topological Order Balance for Processor-based Hardware Emulation

Jing Tang[1], Shunyang Bi[2], Hailong You*

Xidian University, Xi'an 710071, China

Email: tangjing2022@stu.xidian.edu.cn, hlyou@mail.xidian.edu.cn

Abstract—As the scale of circuit designs continues growing, processor-based emulation plays a crucial role in design verification. To fully leverage the parallel capabilities of emulation, we need to consider the topological order (TB) relationships between gates when partitioning large-scale circuits into processors. However, existing advanced partitioners, like hMETIS and KaHyPar, do not currently take this issue into consideration. In this paper, we propose a partitioning algorithm TBPart considering topological balance to utilize parallel capabilities fully. TBPart improves the topological balance of the partitioning results by considering the gain of topological order balance and cut-size brought by vertices movement. We evaluate TBPart using the ISPD98 benchmark tests, and experimental results demonstrate its effectiveness in improving the topological order balance. On average hand, 0.2×cut-size loss can result in 0.54×TB metric, and runtime can be reduced by half compared to the segment tree data structure.

Keywords—Hypergraph partitioning, topological order balance, hardware emulation, VLSI verification

I. INTRODUCTION

As the scale of VLSI design rapidly expands, Processor-Based Hardware Emulation (PBE) has been widely adopted in chip verification. In a typical compilation flow of PBE, several steps are performed: starting from the RTL description, synthesizing into a gate-level netlist, partitioning the netlist, and finally, mapping the partition to processors and scheduling. In dealing with the huge scale circuits, partitioning plays an ever more important role in system performance [1]. When partitioning a netlist, we typically transform it into a hypergraph partitioning problem.

After partitioning, the processors will parallelly emulate the netlist within the correct time order in the scheduling stage. To fully enhance the parallel potential, an appropriate schedule should fit the hierarchical structure of the netlist, which denotes the topological ordering of the netlist. Therefore, considering the topological order of the netlist in the partitioning stage is necessary.

A toy example is given in "Fig. 1" to demonstrate the importance of this topological order-balanced partitioning. There is a Directed Acyclic Hypergraph (DAH) containing 6 vertices and 5 hyperedges. The first vertex of each hyperedge is its source, and the remaining vertices are drains.

Topological order: $\{(v_1), (v_2, v_3), (v_4, v_5), (v_6)\}$.

Clearly, if we place gates with the same topological order on the same processor, even though the communication cost

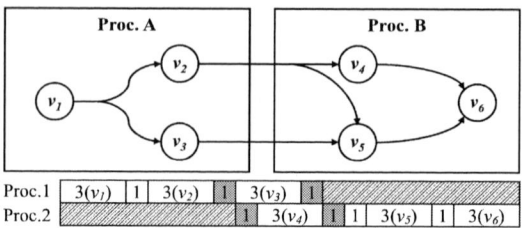

(a) Partitioning without considering topological order.

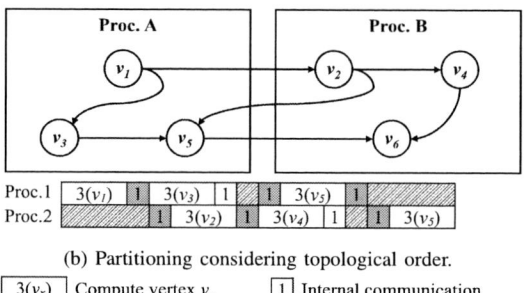

(b) Partitioning considering topological order.

| $3(v_x)$ | Compute vertex v_x | 1 | Internal communication |
| Idle state | | 1 | External communication |

Fig. 1: Partitioning considering different objectives.

is reduced, it will ultimately lead to an increase in emulation time.

However, to the best of our knowledge, existing research has not focused on the topological order balance(TB) of partitioning results. In existing research, partitioners such as hMETIS (1998) [2], KaHyPar (2017) [3], and SpecPart (2022) [4], only aim to minimize cut-size. TritonPart (2023) aims to reduce cut-size on timing-critical paths [5]. In 2020, P. Sanders et al. proposed a partitioning algorithm for directed acyclic hypergraphs [6], but its goal is to ensure that there are no cyclic relationships between partitions, which does not align with our goal. In summary, existing partitioning research for hypergraphs cannot address the TB partitioning problem.

In this work, we propose a hypergraph partition algorithm TBPart to minimize the cut-size and TB metric. The main contributions of this paper are summarized as follows:

- We propose a k-way hypergraph partitioning algorithm TBPart that optimizes the TB of partition result with high quality, while keeping the cut-size in small size.

979-8-3503-6184-1/24 $31.00 © 2024 IEEE

- We utilize a new data structure named block, which is more efficient than the segment tree structure.

The remainder of this paper is organized as follows. Section II defines the topological order balance partitioning problem and presents the constraints and objectives. Section III introduces our partitioning algorithm. Section IV provides details on the experimental results, demonstrating the effectiveness and efficiency of TBPart. Finally, Section V concludes the paper.

II. PRELIMINARIES

We map the gate-level netlist of the circuit into a *Directed Acyclic Hypergraph* (DAH), which effectively represents the connectivity and dependency relationships between gates.

Given a directed acyclic weighted hypergraph $H(V, E, c, w)$ consists of a set of vertices $V = \{v_1, v_2, \cdots, v_n\}$ and a set of hyperedges $E = \{e_1, e_2, \cdots, e_m\}$. Here, $c(v_i)$ and $w(e_j)$ respectively denote the weights function of vertices and hyperedges. $I(v)$ is defined as the set of hyperedges incident to v. The vertices in a hyperedge e are called its pins and denoted as $pins(e)$.

A k-way partition of a hypergraph H is the process of placing vertices into k disjoint partitions. It is defined as $\Pi = \{V_1, V_2, \cdots, V_k\}$, $\cup_{i=1}^{k} V_i = V$ and $V_i \neq \emptyset$ for $i = 1, 2, \cdots, k$. A k-way partition Π is ε-balanced if each partition $V_i \in \Pi$ satisfies the balance constraint $c(V_i) \leq L_{max} := (1 + \varepsilon) \lceil c(V)/k \rceil$, where $c(V_i) := \sum_{v \in V_i} c(v)$.

A. Cut-size

The connectivity set of a hyperedge e is denoted by $\Lambda(e, \Pi) := \{V_i \in V | pins(e) \cap V_i \neq \emptyset\}$ which means the number of different partitions that the vertices $pins(e)$ on hyperedge e were distributed into. The connectivity is denoted by $\lambda(e, \Pi) := |\Lambda(e, \Pi)|$. We consider e is cut if $\lambda(e, \Pi) > 1$. A very important metric, cut-size, is defined as $\sum_{e \in E} (\lambda(e, \Pi) - 1) w(e)$.

B. TB Metric

The topological order of the vertices is denoted by $\tau(v)$. A very significant parameter, the set of vertices distributed in partition i with the topological order of t, is defined as $P_i(t)$. Suppose the operation of computing the standard deviation for a set of values X is denoted as $std(X)$. For partition i, we define the standard deviation of the number of distinct topological order vertices as $std(P(t)) := std(\{P_1(t), P_2(t), \cdots, P_k(t)\})$. In this paper, another very important metric, TB metirc, is denoted as $K := \sum_{t=0}^{max(\tau)} std(P(t))/|P(t)|$, where $P(t)$ means all vertices with the topological order of t. The smaller this value, the more balanced the topological order.

Therefore, we can set our objective function as "(1)":

$$min \sum_{e \in E} (\lambda(e, \Pi) - 1) w(e) + \alpha K \qquad (1)$$

In "(1)", α represents the coefficient of the TB parameter.

III. METHODOLOGY

A. Improvement of cut-size

Fiduccia-Mattheyses (FM) algorithm [7] is a heuristic bi-partitioning algorithm proposed by C. M. Fiduccia and R. M. Mattheyses in 1982. Its objective is to minimize cut costs, and it has a linear time complexity. The FM algorithm calculates the gain of moving each vertex and selects the best move in each iteration. Once a vertex is moved, it is locked for the current iteration and cannot be moved again. Additionally, the FM algorithm considers the weight balance of partitions.

B. Improvement of TB Metric

Similar to the FM algorithm, we determine which vertex to move by calculating the gain in improving TB due to vertex movement. However, in the case of TB, gain represents the improvement of a group of vertices.

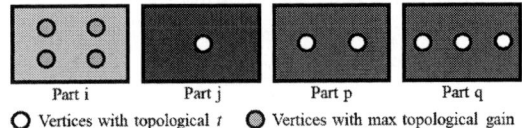

○ Vertices with topological t ◎ Vertices with max topological gain

Fig. 2: Distribution of vertices with a topological order of t.

As shown in the figure, with 4 partitions, we need to calculate the gain for vertices with a topological order of t. The TB gain calculation equation is as "(2)".

$$g_j^i(t) = (|P_i(t)| - |P_j(t)| - 1)/|P(t)| \qquad (2)$$

$g_j^i(t)$ means the gain of moving a vertex with topological order t from partition i to partition j. By calculating, we can find that the value of $g_j^i(t)$ is maximal, indicating a preference for moving a vertex from partition i to partition j.

Theorem 1: Let vertex $v \in P_i(t)$ be moved from partition i to j, the gains of other vertices with topological order t can be updated quickly in $O(1)$.

Proof: Let other partition q which $q \neq i \neq j$. After the movement, $|P_i(t)| - = 1$ and $|P_j(t)| + = 1$. By "(2)", the new gain $g_j^i(t) - = 2/|P(t)|$, $g_q^i(t) - = 1/|P(t)|$, $g_i^j(t) + = 2/|P(t)|$, $g_q^j(t) + = 1/|P(t)|$, $g_i^q(t) + = 1/|P(t)|$, $g_j^q(t) - = 1/|P(t)|$. ∎

C. Data Structure

It is noticeable that when updating gains, all vertices in $P_i(t)$ are simultaneously increased or decreased by $1/|P(t)|$ or $2/|P(t)|$. Given this characteristic, as shown in "Fig. 3", we utilize a new structure named block to store gains.

Meanwhile, a segment tree is also a suitable data structure because it excels at performing operations on a range of data.

Suppose there are k partitions and a total of N vertices. Using block can achieve an update operation in $O(\log(max(\tau) \times k^2) + \log(N))$ time. In contrast, the update time complexity of a segment tree is typically $O(k\log(kN))$. Clearly, if N is significantly larger than $max(\tau)$, the segment tree will have a noticeable efficiency advantage.

979-8-3503-6184-1/24 $31.00 © 2024 IEEE   217

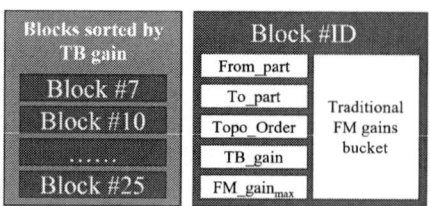

Fig. 3: Block data structure. #ID can be calculated by $Topo\_Order \times k^2 + From\_part \times k + To\_part$.

IV. EXPERIMENTAL RESULTS

Table I describes the 17 hypergraphs used in our experiments, which are from the ISPD 98 VLSI Circuit Benchmark Suite. The ibm15 hypergraph did not successfully convert to DAH and therefore does not appear in Table I. All experiments were conducted on a server running Ubuntu 22.04.1 LTS, equipped with an Intel(R) Xeon(R) Gold 6248R CPU @ 3.00GHz processor and 256GB RAM.

TABLE I: Case Information

Case	$|V|$	$|E|$	$|Pin|$	$max(\tau)$	cut-size	K
ibm01	12752	6637	35672	28	580	6.10
ibm02	19601	9500	61502	27	629	5.97
ibm03	23136	12425	68366	25	1678	3.94
ibm04	27507	14988	78742	15	1570	1.99
ibm05	29347	15127	91803	45	3432	6.25
ibm06	32498	17197	100592	22	1601	2.98
ibm07	45926	23570	132136	21	2001	3.46
ibm08	51309	25401	153010	26	1824	4.39
ibm09	53395	27148	158398	23	1476	4.13
ibm10	69429	35106	212798	43	2309	10.50
ibm11	70558	35894	201496	24	2176	5.55
ibm12	71076	35270	225395	16	3259	1.10
ibm13	84199	43513	257400	22	1788	3.31
ibm14	147605	76883	418228	15	3495	1.70
ibm16	183484	90971	568703	40	3662	7.04
ibm17	185495	89804	609092	30	5574	5.77
ibm18	210613	102662	621807	24	3294	3.43

We used the results from hMETIS as the initial partitioning, with parameters set to the number of partitions $k = 4$ and other parameters set to default values. Then, we use TBPart for secondary partitioning, with the constraint that the cut-size loss should be $0.2\times$ initial partitioning results.

For the baseline, due to the absence of the related existing work, we develop the segment tree data structure as baseline which is easy to follow. Note that the segment tree version employs the same topological order balancing strategy presented in Section III.

The experimental results are as shown in the "Fig. 4". Compared to the baseline, the block version achieves $0.94\times$TB metric in only $0.53\times$runtime. This result demonstrates that the block data structure has superior performance than the segment tree version when employing the same strategy. On average hand, an additional $0.2\times$times cut-size loss can achieve $0.54\times$TB metric, which proves that our strategy is effective.

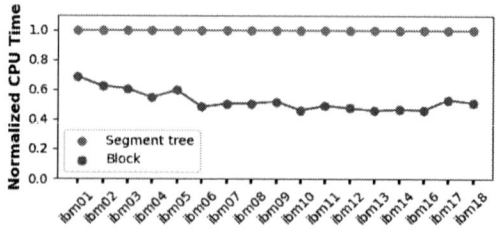

(a) Average runtime comparison over 5 runs.

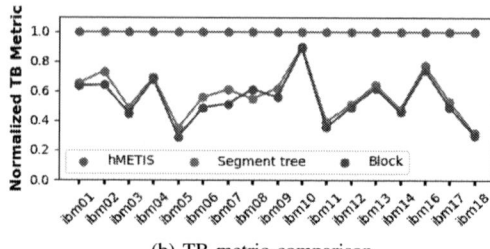

(b) TB metric comparison.

Fig. 4: Comparisons based on ISPD98 benchmarks for $k = 4$ and $\varepsilon = 10\%$ when achieving $0.2\times$cut-size loss.

V. CONCLUSION

In this paper, we investigated a hypergraph partitioning algorithm TBPart applied to VLSI hardware emulation and proposed a partitioning result improvement strategy that considers topological order balance. TBPart calculates gains based on vertices movements, aiming to maximize TB gains with minimal cut-size loss. Through experimental validation, we found that typically, a $0.2\times$cut-size loss results in $0.54\times$TB metric. Compared to the segment tree version, runtime can be reduced by half. In the future, we aim to apply TBPart in a multi-level partitioning paradigm, which will further reduce runtime and enhance quality.

REFERENCES

[1] S.-J. Chen and C.-K. Cheng, "Tutorial on VLSI partitioning," VLSI design, vol. 11, no. 3, pp. 175–218, 2000.

[2] G. Karypis, R. Aggarwal, V. Kumar, and S. Shekhar, "Multilevel hypergraph partitioning: application in VLSI domain," in Proceedings of the 34th annual conference on Design automation conference - DAC '97, Anaheim, California, United States: ACM Press, 1997, pp. 526–529.

[3] Y. Akhremtsev, T. Heuer, P. Sanders, and S. Schlag, "Engineering a direct k -way Hypergraph Partitioning Algorithm," in 2017 Proceedings of the Ninteenth Workshop on Algorithm Engineering and Experiments (ALENEX), Society for Industrial and Applied Mathematics, Jan. 2017, pp. 28–42.

[4] I. Bustany, A. B. Kahng, I. Koutis, B. Pramanik, and Z. Wang, "SpecPart: A Supervised Spectral Framework for Hypergraph Partitioning Solution Improvement," in Proceedings of the 41st IEEE/ACM International Conference on Computer-Aided Design, in ICCAD '22. New York, NY, USA: Association for Computing Machinery, Dec. 2022, pp. 1–9.

[5] I. Bustany, G. Gasparyan, A. B. Kahng, I. Koutis, B. Pramanik, and Z. Wang, "An Open-Source Constraints-Driven General Partitioning Multi-Tool for VLSI Physical Design," in 2023 IEEE/ACM International Conference on Computer Aided Design (ICCAD), Oct. 2023, pp. 1–9.

[6] D. Seemaier, S. Schlag, and P. D. D. C. Schulz, "Acyclic n-Level Hypergraph Partitioning," PhD Thesis, Karlsruher Institut für Technologie (KIT), 2020. Accessed: Feb. 08, 2024.

[7] C. M. Fiduccia and R. M. Mattheyses, "A linear-time heuristic for improving network partitions," in Papers on Twenty-five years of electronic design automation, ACM, Jun. 1988, pp. 241–247.

979-8-3503-6184-1/24 $31.00 © 2024 IEEE

FCE: A Fast CGRA Architecture Exploration Framework

Sichao Chen*, Yiqing Mao*, Yuan Dai, Xuchen Gao, Wai-Shing Luk, Wenbo Yin, and Lingli Wang[†]

State Key Laboratory of Integrated Chips and Systems, Fudan University
20112020050@fudan.edu.cn, {maoyq22, daiy21, xcgao22}@m.fudan.edu.cn, {luk, wbyin, llwang}@fudan.edu.cn

Abstract—**Conventional design space exploration (DSE) flows of Coarse-grained reconfigurable arrays (CGRAs) are based on black-box optimization methods, which are slow and ineffective. This paper proposes FCE, a fast CGRA DSE framework that presents the CGRA design space to the application data flow graph (DFG) mapper and directly employs it to explore the architecture in the mapping process. FCE achieves over 1600× speedup and better result quality than conventional DSE methods, which boosts the CGRA DSE process.**

Index Terms—**CGRA, design space exploration, DFG mapping**

I. INTRODUCTION

Coarse-grained reconfigurable arrays (CGRAs) bring hardware designs with high flexibility, reconfigurability, and energy efficiency, making them attractive for specific application domains, such as cryptographic processing [1], [2], neural network acceleration [3], [4], and scientific computing [5], [6]. During the CGRA development, developers should evaluate and tune the design to optimize the CGRA architecture. This process is called design space exploration (DSE).

Modern CGRA frameworks [7]–[11] consist of automatic DSE flows with different algorithms. PRAD [10] and THRAM [7] utilize the Bayesian optimization (BO) algorithm, while AURORA [8] and REVAMP [9] use other heuristics algorithms, including simulated annealing (SA). These DSE flows are based on black-box optimization methods, which include iterative data flow graph (DFG) mapping tests, making the DSE process time-consuming. Furthermore, they are ineffective because they contain many invalid iterations. CDE [11] utilizes the graph analysis algorithm in DSE flow to avoid the iterative mapping tests. However, the generality of CDE-explored architectures is deficient since they are specific to the applications used during the DSE.

We propose FCE, a fast CGRA architecture exploration framework with a novel DSE method. FCE presents the CGRA design space to the DFG mapping tool and directly utilizes it to explore the architecture in the mapping procedure. Table I compares the features of our DSE tool with other tools. Our contributions include:

- We propose a fast CGRA DSE flow based on the DFG mapping algorithm. Compared to conventional black-box DSE tools, our DSE tool can speed up the architecture exploration process by over 1600× with better results.

* Both authors contributed equally to this research.
† Corresponding author.

- We implement a new cost function for our DSE flow, achieving the same architecture generality as conventional black-box DSE tools.
- We develop an accurate area estimation model based on the ASAP 7nm [12] library, which can quickly estimate the CGRA area with a low error (<1.06%) compared with the commercial synthesis results.

The rest of this paper is organized as follows: Section II provides an overview of FCE. Section III introduces the CGRA DSE method. Section IV presents the experimental results in comparison with existing CGRA DSE tools. Section V concludes the paper with future work.

TABLE I: Feature comparison of different DSE tools.

DSE Tool	DSE Time	Hardware Evaluation Model	Architecture Generality
AURORA [8]	$\sim 10^3$ s	✓	✓
REVAMP [9]	$\sim 10^4$ s	-	✓
PRAD [10]	$\sim 10^3$ s	✓	✓
THRAM [7]	$\sim 10^3$ s	✓	✓
HETA [13]	$\sim 10^4$ s	✓	✓
CDE [11]	< 10s	✓	-
FCE (this work)	< 10s	✓	✓

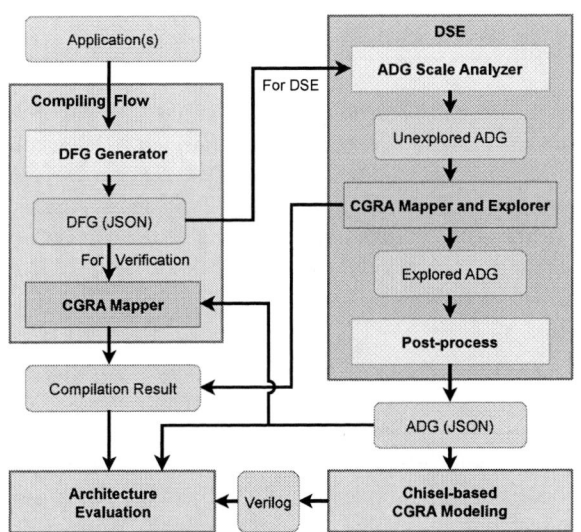

Fig. 1: FCE framework.

II. FCE FRAMEWORK

As shown in Fig. 1, the FCE framework includes the CGRA software development and hardware optimization flow. The applications are converted to DFGs by the DFG generator in the compiling flow. The DFG files have two usages: ① Being sent to the DSE flow for architecture exploration. ② Being sent to the CGRA mapper for verification. The DSE flow explores the CGRA architecture according to the application DFGs and generates the CGRA architecture description graph (ADG). The ADG files can be used for CGRA mapping in the compiling flow and Verilog generation in the Chisel-based CGRA modeling tool. According to the CGRA Verilog codes and application compilation results, synthesis and simulation tools can evaluate and verify the CGRA architecture. We also propose an accurate area estimation model to assess the CGRA hardware area on the ASAP 7nm [12] according to the ADG.

III. DSE METHOD

Our CGRA DSE framework fuses the hardware architecture exploration procedure and the application DFG mapping test. Algorithm 1 shows our DSE flow.

Algorithm 1: FCE DSE Flow

Input: DFG set S, CGRA architecture design constraints C
Output: Explored ADG H, PnR results Ps and Rs

1   $\{Ps, Rs\} = \{\varnothing, \varnothing\}$;
2   $H = analyzeScale(S, C)$;
3   $queDFG = sort(S)$;
4   **foreach** G **in** $queDFG$ **do**
5     $\{P, R\} = \{\varnothing, \varnothing\}$;
6     $MapandExplore(G, H, P, R)$;
7     $Ps.insert(P)$;
8     $Rs.insert(R)$;
9   $PostProcess(H, C)$;
10   **return** H, Ps, Rs;

Firstly, the *analyzeScale()* function in line 2 of Algorithm 1 outputs the minimum suitable least-connected ADG according to the design constraints C and the DFGs. For a DFG set S consisting of p DFGs $G_k, k \in \{1, 2, \ldots, p\}$, let $n_{i,k}$, $n_{o,k}$, and $n_{c,k}$ be the input, output, and computing vertex number of G_k, respectively. The minimum suitable scale of the CGRA ADG is defined as follows:

$$Scale_{min} = minLegal(n_{i,k,max}, n_{o,k,max}, n_{c,k,max}, C)$$

Then, the *sort()* function in line 3 sorts the DFGs by their vertex numbers in descending sequence. Next, the function *MapandExplore()* in line 6 maps the sorted DFGs on the CGRA architecture and explores the design. We modify the source codes of the CGRA DFG mapping tool to implement *MapandExplore()*. The modification is applicable for arbitrary spatial CGRA mapping tools, which contains two changes:

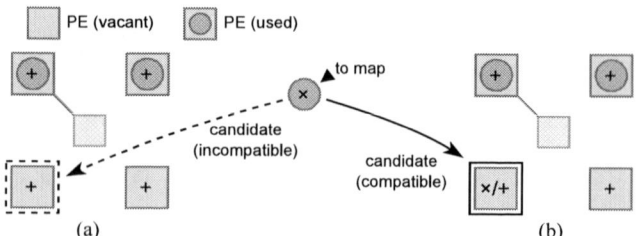

Fig. 2: Example of PE operation configuration. (a) ADG before the PE operation configuration. (b) ADG after the PE operation configuration.

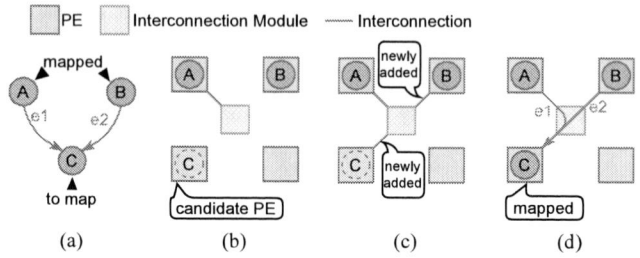

Fig. 3: Example of interconnection exploration. (a) DFG to map. (b) Original ADG. (c) Adding necessary interconnections to ADG. (d) Routing result.

(1) When the proper PE operation and connection resources for placement and routing are absent, the *MapandExplore()* function can append them to the architecture. Fig. 2 and Fig. 3 show these procedures. In the original mapping flow, the ADGs are read-only, so the mapper directly returns the empty result in this situation.

(2) The cost function of *MapandExplore()* is different from the original CGRA mapper. When the *MapandExplore()* function can find existing hardware resources to place a DFG vertex or route a DFG edge, it should consider the routing latency and decide whether to use the existing resources or add new resources. When adding the new hardware resources, the *MapandExplore()* function should strike a balance between the latency and the ADG symmetry to guarantee the generality of the explored architecture.

When the *MapandExplore()* function completes execution, the explored architecture is regularized by the *PostProcess()* function in line 9. This process enhances the symmetry and generality of the architecture. At last, the DSE flow returns the architecture exploration, placement, and routing results.

IV. EXPERIMENTAL RESULTS

To verify the efficiency of FCE, we utilize it and THRAM [7] to explore the GIB-connected [14] CGRAs separately. The baseline architecture is the optimal 8×8 homogeneous CGRA architecture. We use the DFG files provided by the benchmark suite in THRAM's source code directory to make fair comparisons.

979-8-3503-6184-1/24 $31.00 © 2024 IEEE

A. Hardware Area and Power

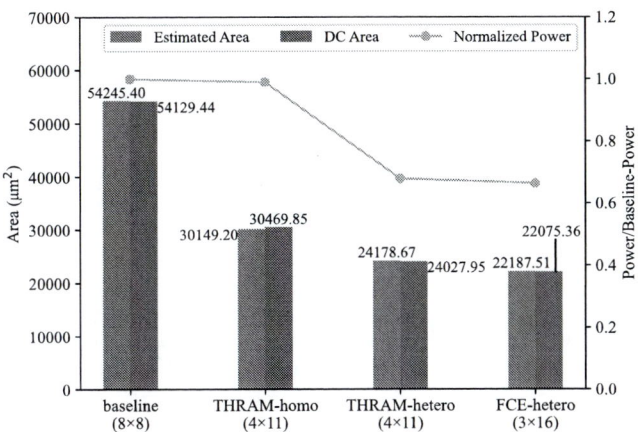

Fig. 4: DSE results comparison.

Fig. 4 shows the DSE results. The *cosine1*, *arf*, *stencil3d*, *ewf*, *fir*, *fir1*, and *resnet2* benchmarks are used for DSE. According to the evaluation results of Synopsys DC, the FCE-explored heterogeneous CGRA lowers the area by 59.2% and reduces the power by 33.5% from the baseline. Compared with the THRAM-explored heterogeneous CGRA, it accomplishes 8.1% area reduction and 2.2% power reduction. The area estimator keeps the error <1.06% in the tests.

B. Exploration Speed

Table II shows the DSE running time of different tools on an i9-12900K CPU. FCE can explore better CGRA architectures much quicker than THRAM. The speedup is more than 1600×.

TABLE II: Exploration times of different DSE flows.

DSE Flow	Exploration Time (s)	Speedup
THRAM-homo [7]	3053.61	1.53×
THRAM-hetero [7]	4672.07	1.00×
FCE-hetero (this work)	2.882	1621.12×

C. Architecture Generality

To verify the generality of the explored CGRA architectures, we test their mapping compatibility with several DFGs not used in DSE, including *cosine2*, *fft*, *resnet1*, *centro-fir*, and *md*. Table III shows the results. The CGRA architectures explored by FCE and THRAM reach the same generality in the test. All the applications except *md* can be successfully mapped on the architectures, for the scale of *md* is too large.

TABLE III: Generality of the explored architectures.

Application DFG	Mapping Compatiblity	
	THRAM-hetero [7]	FCE-hetero
cosine2	✓	✓
fft	✓	✓
resnet1	✓	✓
centro-fir	✓	✓
md	-	-

V. Conclusion

We propose FCE, a fast CGRA architecture exploration framework with a novel DSE method. Our DSE flow accomplishes more than 1600× speedup, better result quality, and the same architecture generality compared with conventional DSE tools. At present, our DSE algorithm is applicable for arbitrary spatial CGRAs. In future work, we will extend the fast DSE algorithm for temporal CGRAs.

Acknowledgment

This work is supported by the National Natural Science Foundation of China under grant 62174035.

References

[1] G. Sayilar and D. Chiou, "Cryptoraptor: High throughput reconfigurable cryptographic processor," in *2014 IEEE/ACM International Conference on Computer-Aided Design (ICCAD)*. IEEE, 2014, pp. 155–161.

[2] T. Qu, Z. Dai, Y. Liu, and L. Chen, "A high flexible shift transformation unit design approach for coarse-grained reconfigurable cryptographic arrays," *Electronics*, vol. 11, no. 19, 2022. [Online]. Available: https://www.mdpi.com/2079-9292/11/19/3144

[3] T. Geng, C. Wu, C. Tan, B. Fang, A. Li, and M. Herbordt, "Cqnn: a cgra-based qnn framework," in *2020 IEEE High Performance Extreme Computing Conference (HPEC)*, 2020, pp. 1–7.

[4] R. Kazerooni-Zand, M. Kamal, A. Afzali-Kusha, and M. Pedram, "Memristive-based mixed-signal cgra for accelerating deep neural network inference," *ACM Trans. Des. Autom. Electron. Syst.*, may 2023, just Accepted. [Online]. Available: https://doi.org/10.1145/3595638

[5] M. Emani, V. Vishwanath, C. Adams, M. E. Papka, R. Stevens, L. Florescu, S. Jairath, W. Liu, T. Nama, and A. Sujeeth, "Accelerating scientific applications with sambanova reconfigurable dataflow architecture," *Computing in Science & Engineering*, vol. 23, no. 2, pp. 114–119, 2021.

[6] G. Charitopoulos, I. Papaefstathiou, and D. N. Pnevmatikatos, "Creating customized cgras for scientific applications," *Electronics*, vol. 10, no. 4, 2021. [Online]. Available: https://www.mdpi.com/2079-9292/10/4/445

[7] J. Li, Y. Qiu, G. Zhu, Q. Zhu, W. Yin, and L. Wang, "Thram: A template-based heterogeneous cgra modeling framework supporting fast dse," in *2023 IEEE International Symposium on Circuits and Systems (ISCAS)*, 2023.

[8] C. Tan, C. Xie, A. Li, K. J. Barker, and A. Tumeo, "Aurora: Automated refinement of coarse-grained reconfigurable accelerators," in *2021 Design, Automation & Test in Europe Conference & Exhibition (DATE)*, 2021, pp. 1388–1393.

[9] T. K. Bandara, D. Wijerathne, T. Mitra, and L.-S. Peh, "Revamp: A systematic framework for heterogeneous cgra realization," in *Proceedings of the 27th ACM International Conference on Architectural Support for Programming Languages and Operating Systems*, ser. ASPLOS '22. New York, NY, USA: Association for Computing Machinery, 2022, p. 918–932. [Online]. Available: https://doi.org/10.1145/3503222.3507772

[10] B. Peng, S. Sun, Y. Dai, J. Li, Y. Qiu, K. Wang, W. Yin, and L. Wang, "Prad: A bayesian optimization-based dse framework for parameterized reconfigurable architecture design," in *2023 IEEE 31st Annual International Symposium on Field-Programmable Custom Computing Machines (FCCM)*, 2023, pp. 226–226.

[11] S. Chen, Y. Dai, J. Zhang, H. Kuang, X. Gao, W.-S. Luk, W. Yin, and L. Wang, "Cde: A novel cgra development environment with fast design space exploration framework," in *2024 2nd International Symposium of Electronics Design Automation (ISEDA)*, 2024, pp. 772–772.

[12] L. T. Clark, V. Vashishtha, L. Shifren, A. Gujja, S. Sinha, B. Cline, C. Ramamurthy, and G. Yeric, "Asap7: A 7-nm finfet predictive process design kit," *Microelectronics Journal*, vol. 53, pp. 105–115, 2016.

[13] Y. Dai, J. Li, Q. Zhu, Y. Qiu, Y. Hu, W. Yin, and L. Wang, "Heta: A heterogeneous temporal cgra modeling and design space exploration via bayesian optimization," *IEEE Transactions on Very Large Scale Integration (VLSI) Systems*, vol. 32, no. 3, pp. 505–518, 2024.

[14] K. Shi, X. Zhou, H. Zhou, and L. Wang, "An optimized gib routing architecture with bent wires for fpga," *ACM Trans. Reconfigurable Technol. Syst.*, vol. 16, no. 1, dec 2022. [Online]. Available: https://doi.org/10.1145/3519599

Research on Parametric Subthreshold Cell Delay Modeling Based on ANN

Xuelian Zhang[1,2], Yuping Wu *[1,2], Zhiqiang Li[1,2], Donglin Liu[1,3], Shushan Qiao[1]

[1] Institute of Microelectronics of Chinese Academy of Sciences, Beijing, China
[2] Beijing Key Laboratory of Three-Dimensional and Nanometer Integrated Circuit Design Automation Technology, Beijing, China
[3] University of Chinese Academy of Sciences, Beijing, China

Email: zhangxuelian@ime.ac.cn, wuyuping@ime.ac.cn

Abstract—With the operating voltage decreasing to subthreshold, cell delay distribution tends to be a flatten, non-Gaussian distribution. which makes timing analysis and optimization become more important for the integrated circuit (IC) design. Although SPICE-based Monte Carlo (MC) analysis is very accurate, it is timing-consuming and thus impractical for the complex subthreshold IC design. Fast and accurate cell delay modeling is more significant. This paper proposes a parameterized cell delay modeling to predict the statistical mean(μ) and variance(σ) of subthreshold cell delay by Artificial Neural Network (ANN) model. The parameter includes both geometric parameters (i.e., transistor width and length) and operating conditions (i.e., operating voltage, temperature, input signal slew and output load capacitance). Experimental results demonstrate that the root mean square errors of μ and σ at 0.3V±10% is 3.31% and 4.34% respectively with 1100× speedup, compared with SPICE-based MC analysis. A parameterized cell delay model can predict cell delay quickly and accurately based on geometries, which is not mentioned in the prior works.

Keywords—*parametric cell delay modeling, subthreshold voltage, Artificial Neural Network*

I. INTRODUCTION

With continuous CMOS technology scaling, accurate timing analysis and optimization play an ever-increasing role in the complex integrated circuit (IC) design. In addition, the ultra-low voltage design reduces the operating voltage below the threshold. As a result, the impact of process variation increases rapidly, as shown in Fig.1, resulting in unpredictable circuit delay. Although SPICE-based Monte Carlo (MC) analysis can offer golden accuracy results, the simulation takes a huge execution time. Every path in the circuit consists of several cells and interconnects. Static timing analysis computes the delay of each cell by reading the library cell, which stores the cell delays of a set of input signal slews and output loads base on the look-up tables (LUTs) for each standard cell. As the number of LUTs increases, there is an unacceptable analysis and storage overhead.

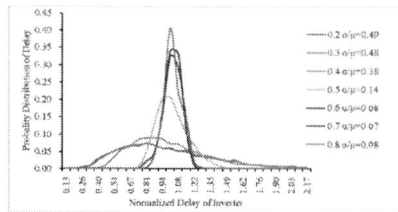

Fig. 1.   Statistical delay distribution of different voltages

Cell delay models can be divided into two categories: physical analysis model and empirical model. The physical analysis model provides the relationship between cell delay and current or output voltage. [1] presented an iterative methodology, which computes the driver output waveform. [2] proposed a non-iterative analytical approximation method for the surface potential replace solved iteratively. [3] proposed an analytical delay model at near-threshold domain by taking input slew. [4] proposed an analytical delay model at sub-threshold domain with the inverse Gaussian distribution function. The empirical model constructs the relationships between inputs and outputs with training data. For example, the error propagation technique and Response Surface Methodology[5], Random Forest model[6] and a novel deep learning waveform-delay model[7] focused on input slew and output load, Aadam[8] considered the transistor width/length ratio. However, the transistor width and length is neglected in the empirical model, which limiting the applicability for timing analysis and optimization.

This paper focuses on parametric ANN-based cell delay by considering both geometric parameters and operating conditions. The contributions can be summarized as follows:

- In the cell characterization extraction, one cell model replaces multiple cell models of the same type with different driving capabilities, to accelerate the characterization extraction.

- In the phase of timing analysis and optimization, parametric cell delay replaces SPICE-based MC analysis to accelerate timing analysis and optimization.

- Considering the transistor width and length, the process porting of IC design can be accelerated.

II. RELATED WORK

We proposed a compact, physics-based current model for fully depleted silicon on-insulator (FDSOI) MOSFETs and applies it to delay variability analysis[2].

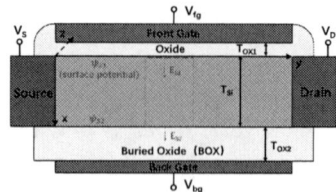

Fig. 2.   Schematic diagram of FDSOI MOSFET[2].

A. Current Model

Assume the back gate to be in weak inversion, the back surface potential is approximately a linear expression of the front surface potential. We used a non-iterative analytical approximation method to obtain an accurate initial solution for the surface potential. Then, the expression of ideal current is derived based on the surface potential and accurate charge density model.

$$I_0 = \mu \frac{W}{L} \left\{ \frac{Q_i(L) + Q_i(0)}{2} [\psi_{S1}(L) - \psi_{S1}(0)] + \frac{kT}{q} [Q_i(0) - Q_i(L)] \right\} \quad (1)$$

Note: μ is carrier mobility.

B. Cell Delay Model

We considered variations in the threshold voltage as the main factor on the propagation delay, and derived the physical analytic expressions of the statistical mean and variance of the cell delay in the case of multiple input signals[4].

$$Td = \begin{cases} \frac{C_{tot} mV_T}{I_0 \cdot e^{\frac{-Vthb - Vth}{mV_T}} \lambda} e^{\frac{-Vdd}{mV_T}} \left(e^{\frac{\lambda Vdd}{2mV_T}} - e^{\frac{\lambda Vout(\tau)}{mV_T}} \right) + \frac{\tau}{2}, t > \tau \\ k0 \frac{mV_T \tau}{Vdd} ln[\frac{Vdd \cdot C_{tot}}{I_0 \cdot e^{\frac{-Vthb - Vth}{mV_T}} \lambda \tau} (e^{\frac{-\lambda Vdd}{2mV_T}} - e^{\frac{\lambda Vdd}{mV_T}}) + 1] - \frac{\tau}{2}, 0 \leq t \leq \tau \end{cases} \quad (2)$$

Note: C_{tot} is the sum of the load capacitance and the coupling capacitance, m is sub-threshold slope, V_T is thermal voltage, λ represents Drain Induced Barrier Lowering effect coefficient, Vth is the threshold voltage at zero bias, and $Vthb$ means the increment of the threshold voltage caused by the body effect.

We can compute the statistical mean and variance of basic logical cell use (1) and (2). The complex cells can be regarded as path.

III. ANN-BASED CELL DEALY MODEL

Implementing physics-based analytical models requires development to accurately analyze and understand specific models. The development process is time-consuming. The empirical model instead of physical analytical model, can establish the delay model for each type of cell. Based on the physical analysis model and device modeling[9], this paper adopts the ANN to train empirical model. The flow is shown in Fig.3.

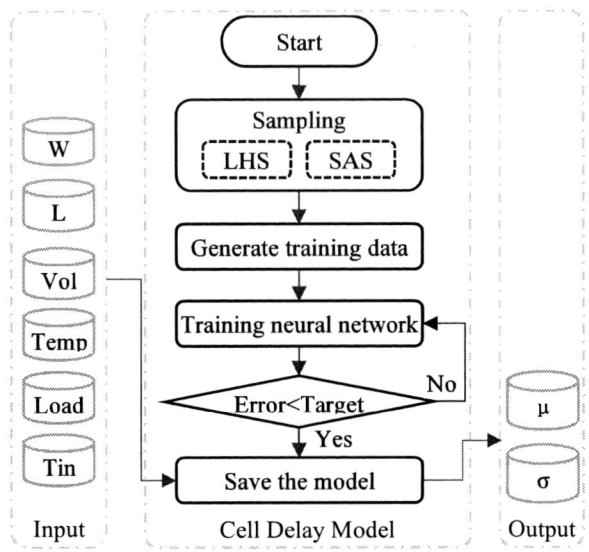

Fig. 3.   The flow of the cell delay model

The input of ANN-based delay model includes both geometric parameters and operating conditions. The operating conditions of each cell mainly includes the operating voltage(Vol), temperature(Temp), input slew(Tin) and output load capacitance(Load). Geometric parameters include the width(W) and length(L) of each transistor in the cell. Considering mirror, symmetry and other factors, the transistor parameter of the same size in the cell is set to the same parameter variable.

The output of ANN-based delay model are statistical mean(μ) and variance(σ) of cell delay due to process variation.

A. Sampling

Latin Hypercube Sampling(LHS) is mainly used to perform MC simulations. Comparing with other sampling methods, because of the Latin hypercube properties, the advantage of LHS is that much smaller sample size is required in order to obtain specific accuracy.

Because of the difference in the structure of different types of the cell, the input number of different cell delay model is also different. The waveform of cell delay variation with any of the input parameters can be fitted as a smooth curve, as shown in Fig.4. Thus, we can also use self-adaptive sampling(SAS) to generate the training data.

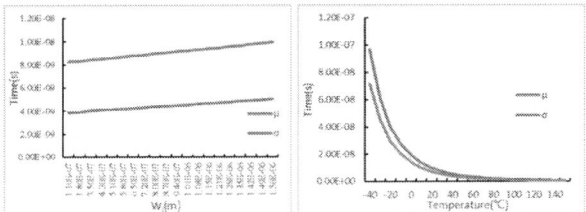

Fig. 4.   The delay waveform with width and temperature

Assume that the cell delay of input combinations P_1, P_2 and P_i are D_1, D_2 and D_i respectively. If Error is less than the specified value, P_1 and P_2 are the sampling points. Since the input dimension is not less than six. the spatial dimension sampling complexity of adaptive is large. This paper performs SAS in the form of points, lines and surfaces.

$$D_{pre} = \frac{D_2 - D_1}{P_2 - P_1}(P_i - P_1) + D_1 \quad (3)$$

$$Error = \frac{|D_i - D_{pre}|}{D_i} \quad (4)$$

B. Training the ANN-Based Model

The model leverages the backpropagation (BP) algorithm for training, and employs the adaptive momentum estimation algorithm to accelerate the convergence.

$$log - sigmoid = \frac{1}{1 + e^{-x}} \quad (5)$$

Activation functions commonly used in neural networks includes Sigmoid, tanh, ReLU, Leaky Relu, Swish, Softmax and so on. The derivative of Sigmoid types activation function is smooth, which can meet the requirements of high order derivatives in IC simulation. The common Sigmoid functions are log-sigmoid and tanh-sigmoid function. Combined with the physical analytic expression of current (1) and delay (2), log-sigmoid is selected as the activation function, as in (5).

The Root-Mean-Square-Error (RMSE) of the relative error is defined as the loss function, which is defined as follows:

$$RMSE = \sqrt{\frac{1}{N} \sum_{i=1}^{n} \left| \frac{D_{pre}^i - D_{ref}^i}{D_{ref}^i} \right|^2} \quad (6)$$

Note: D_{ref}^i is the golden standard data, D_{pre}^i is the predicted value of the delay model.

IV. EXPERIMENTAL RESULTS

We use ANN to build the parameterized cell delay modeling. ANN consists of an input layer, three hidden layers and an output layer. In order to validate the parameterized cell delay model, it was applied to different cell and different operating conditions based on 20nm-40nm FDSOI technology. The operating voltage is 0.3V±10%.

In the sampling phase, the sampling number of SAS is related to the structure of the cell. The sampling number of LHS needs to be tested. The delay of cell operating at subthreshold region is more sensitive to process variation. So the sampling number of the MC simulation corresponding to each set of input is at least 2000. Taking AND as an example, geometric parameters include L, W1,W2 and W3, the sample number of LHS is shown in Table Ⅰ.

TABLE I.   SELECTION OF LHS SAMPLING NUMBER

Sampling number	Hidden layer	Error
2000	[48,36,12]	11.27%
5000	[48,36,12]	6.37%
10000	[48,36,12]	4.43%
	[44,36,28]	3.49%

A. Comparison of Accuracy

Using the statistical analysis results of the conventional MC simulation as the golden standard data, the proposed model is implemented with different parameterized cells. Table II shows the RMSE for different cells.

TABLE II.   RMSE OF DIFFERENT CELL

Cell	μ_{error}	σ_{error}	Cell	μ_{error}	σ_{error}
INV	3.31%	4.34%	BUF	2.53%	3.78%
NAND2	3.14%	4.15%	XOR2	2.51%	3.77%
NOR2	3.21%	3.98%	XNOR2	1.61%	2.32%
AND2	3.11%	3.49%	MUX2	3.26%	3.87%

Taking AND as an example, the transistor lengths are 20nm, 28nm and 40nm respectively, the predicted result is shown in Fig.5. And the cell delay prediction of different driving capabilities is shown in Fig.6 and Fig.7, the length is 28nm, three sets of widths are X1(180nm, 150nm, 180nm), X2(180nm,360nm, 150nm) and X3(540nm,1080nm,450nm).

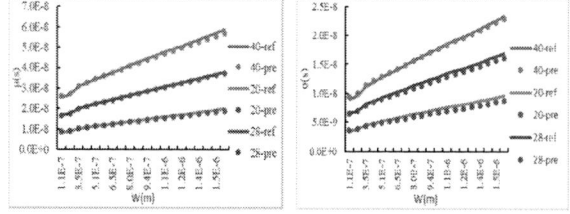

Fig. 5.   Prediction results for different feature size

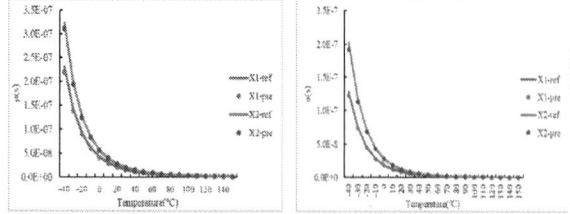

Fig. 6.   Prediction results for different temperature

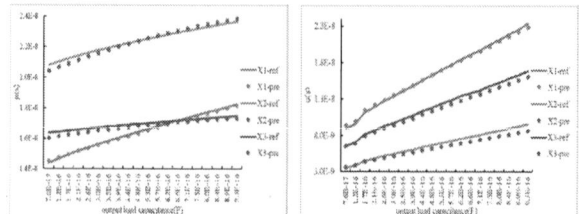

Fig. 7.   Prediction results for different load

B. Comparison of Efficiency

Table III shows the acceleration of different cell. Note that MC sampling for a single set of inputs is not less than 2000 times. The trained model can predict the cell delay by inputs.

TABLE III.   COMPARISION OF RUNNING TIME FOR DELAY PREDICTION

Cell	MC(s)	Ours(s)	Speed-up
INV	35.26	0.032	1100×
BUF	73.51	0.034	2193×
NAND2	119.21	0.063	1897×
NOR2	131.67	0.062	2122×
XOR2	945.10	0.066	14318×

V. CONCLUSIONS

In order to build a fast and accuracy subthreshold cell delay model to accelerate the timing analysis and optimization, this paper proposes a parameterized cell delay model based on ANN, which predicts cell delay by transistor geometries. At 0.3V±10%, the RMSE of statistical mean and variance of the cell delay predicted is 3.31% and 4.34% respectively with 1100× speedup, comparing to conventional MC simulation. In the future, we plan to develop algorithms to accelerate timing analysis and optimization with our proposed model.

REFERENCES

[1] D. Garyfallou, S. Simoglou, N. Sketopoulos, et al., "Gate Delay Estimation With Library Compatible Current Source Models and Effective Capacitance," in IEEE Transactions on Very Large Scale Integration (VLSI) Systems, vol. 29, no. 5, pp. 962-972, May 2021

[2] Z. Mao, Y. Wu, L. Chen, X. Zhang, "A Computationally Efficient Model for FDSOI MOSFETs and Its Application for Delay Variability Analysis," Appl. Sci. 2022,12, 5167.

[3] H. Jiang, B. Xu, P. Cao and H. Cai, "Analytical Delay Model in Near-Threshold Domain Considering Transition Time," 2021 IEEE International Conference on Integrated Circuits, Technologies and Applications (ICTA), Zhuhai, China, 2021, pp. 234-235.

[4] J. Wang, Y. Wu, X. Zhang, Z. Li, L. Chen, "Analytical Delay Modeling for a Sub-Threshold Cell Circuit with the Inverse Gaussian Distribution Function," Electronics 2023, 12, 1387.

[5] L. Brusamarello, G. I. Wirth, P. Roussel, M. Miranda, "Fast and accurate statistical characterization of standard cell libraries," Microelectronics Reliability,Vol. 51, Iss. 12, 2011, pp. 2341-2350.

[6] Y. Attaoui, M. Chentouf, Z. Alaoui Ismaili, A. Mourabit, "Machine learning application for cell delay accuracy improvement at post-placement stage: A case study for combinational cells," Integration, Vol. 90, 2023, pp. 261-270.

[7] W. Raslan, Y. Ismail, "Deep-learning cell-delay modeling for static timing analysis," Ain Shams Engineering Journal,Vol. 14, Issue 1,2023.

[8] S. M. Ebrahimipour, B. Ghavami, H. Mousavi, M. Raji, Z. Fang and L. Shannon, "Aadam: A Fast, Accurate, and Versatile Aging-Aware Cell Library Delay Model using Feed-Forward Neural Network,"2020 IEEE/ACM International Conference On Computer Aided Design (ICCAD), San Diego, CA, USA, 2020, pp. 1-9.

[9] H. Kang, Y. Wu, L. Chen, X. Zhang, "Research on Device Modeling Technique Based on MLP Neural Network for Model Parameter Extraction," Appl. Sci. 2022, 12, 1357.

A High-Performance Routing Architecture with 16 LUTs per CLB for Nanoscale FPGAs

Sijing Yang, Jide Zhang, Hao Zhou, Lingli Wang*

State Key Laboratory of Integrated Chips and Systems, Fudan University, Shanghai, China

*Email: yangsj20@fudan.edu.cn, llwang@fudan.edu.cn

Abstract—Traditional island-style FPGAs commonly feature 8 LUTs per CLB based on the CB-SB model, which may not effectively capture the complexities of modern commercial FPGAs. In contrast, contemporary commercial FPGA designs have increased the number of LUTs per CLB to enhance performance, reflecting advancements in technology nodes. To address these challenges and reduce the critical path delay, we propose a high-performance routing architecture with 16 LUTs per CLB based on the General Routing Block (GRB) model. This architecture uses multi-LUT structures to explore the performance advantages in nanoscale FPGAs. Our design space exploration demonstrates significant benefits: compared to the traditional 8-LUT routing architecture, the 16-LUT routing architecture reduces the critical path delay by 14.90% and area-delay product by 14.01% respectively.

Keywords—*FPGA Routing Architecture, LUT, Design Space Exploration, General Routing Block*

I. INTRODUCTION

The routing architecture of Field Programmable Gate Array (FPGA) is profoundly affected by the internal configuration of CLBs and the topology of global segments. These factors play a critical role in determining the critical path delays, routing area, and flexibility of the FPGA. Academic research has explored various FPGA routing architectures to meet diverse application requirements. Architectures based on 8 LUTs per CLB and a length-4 CB-SB routing scheme have shown superior performance [1].

With advancements in the IC process, the scaling rates of metal wires and logic units differ. Architectures that once excelled are no longer optimal for nanoscale FPGAs, which is evident in the latest commercial FPGA designs. For example, Xilinx has introduced the Versal architecture at the 7nm technology node [2], integrating 32 LUTs within a CLB to effectively enhance FPGA performance. This evolution includes redesigned internal CLB structures with enhancements such as high-speed interconnects, multiplexers (MUXes), and carry chains.

In this paper, we propose a new FPGA routing architecture with 16 LUTs per CLB at the academic 7nm technology [7] to verify that integrating more LUTs within a CLB can effectively enhance FPGA performance, which aims to bridge the gap between academic research and commercial FPGA architecture design. Our key contributions include:

• We propose a 16-LUT routing architecture based on the GRB model for modern FPGAs, utilizing various segment types, bend patterns and two-level MUXes with output sharing to improve performance.

• Compared to the baseline architecture, our design achieves a 14.90% reduction in the critical path delay and a 14.01% reduction in the area-delay product. For large benchmarks

This work is supported by the National Natural Science Foundation of China under grant 62174035.

with over 10,000 netlist primitives in VTR 8.0 [6], the reduction is up to 18.03% and 18.17% respectively.

II. BACKGROUND AND RELATED WORK

A. General Routing Block Model

The 16-LUT routing architecture is based on the General Routing Block (GRB) model [3], which comprises three fundamental components: the Input Connection Box (ICB), the Output Connection Box (OCB), and the General Switch Box (GSB). It presents the two-level MUXes [4] with output sharing, which means the outputs of the first-level MUXes are used as the inputs of multiple second-level MUXes. Besides, the routing segments in the GRB model support bent patterns, which can connect the horizontal and vertical channels directly.

B. GRAEBO

The GRB architecture exploration via Bayesian Optimization (GRAEBO) is proposed in [5]. GRAEBO leverages this method to explore the design space of the GRB model, achieving an effective balance between exploration and exploitation. Additionally, GRAEBO incorporates pruning rules to improve the efficiency of the Design Space Exploration (DSE) process.

GRAEBO accounts for two categories of parameters: the number and bend patterns of each segment type, as well as the driving relationships among CLB pins, ICBs, OCBs, GSBs, and global segments.

III. 16-LUT FPGA ROUTING ARCHITECTURE DESIGN

To design a 16-LUT FPGA routing architecture, we decompose the existing 8-LUT routing architecture model [3] based on the architecture file labels, expand each module sequentially, and then combine them. During Bayesian optimization, we parameterize the GRB model to select critical parameters impacting performance. We carefully design the parameter space to avoid excessive runtime, which can hinder Bayesian convergence and prolong iteration times.

A. Hyperparameter Space

The hyperparameter space θ for the 16-LUT routing architecture encompasses three categories: segment distribution, driving relationship, and feedback signal distribution.

The segment lengths in the 16-LUT routing architecture range from 1 to 12. The probability of each interconnect wire bending is represented by $bendProb = [pb_1, pb_2, \cdots, pb_{12}]$. The position of each interconnect wire bend is represented by $bendPos = [ps_1, ps_2, \cdots, ps_{12}]$. The direction of each interconnect wire bend is represented by $bendType = [t_1, t_2, \cdots, t_{12}]$.

For the ICB's MUX driving sources, let $i = [i_{OCB}, i_{CLB}]$ represent the proportions of OCB and CLB outputs, respectively. For the GSB's MUX driving sources, let $g = [g_{OCB}, g_{CLB}]$ represent the proportions of OCB and CLB outputs respectively.

The distribution of fast feedback signals in the MUX driving of the ICB and GSB is detailed as follows:

For the ICB, let fb_{ICB} denote the number of feedback signals in the second-level MUX of each ICB connected to the same LUT, and p_{ICB} represents the proportion of such LUTs among the total number of LUTs.

For the GSB, depending on the reuse condition r:

• If r is True, fb_{GSB} represents the number of feedback signal sources in the first-level MUX of the GSB. p_{GSB} denotes the proportion of first-level MUXes with feedback signals in the total number of first-level MUXes.

• If r is False, fb_{GSB} stands for the number of feedback signals in the second-level MUX of the GSB. p_{GSB} represents the proportion of second-level MUXes with feedback signals in the total number of second-level MUXes.

B. Global Segment Design

The transition from an 8-LUT routing architecture to a 16-LUT routing architecture involves replicating all global segments and LUT resources, which doubles the numbers of segments and LUTs while keeping their types and lengths unchanged, effectively doubling the channel width. Additionally, the numbers of CLB inputs and outputs are also doubled. The first-stage MUX in the OCB is doubled in both number and size, and corresponding adjustments are made to the GLB-MUX in the ICB to accommodate the expanded MUX count.

Resources from the original 8-LUT routing architecture (including LUTs, MUXes, and segments) are categorized as Group A, while the newly duplicated resources form Group B. To facilitate signal transmission between Groups A and B and complete the architecture design, segment rules for the ICB, GSB, and OCB are needed to design and optimize.

As shown in Fig. 1, driving sources for the second-stage MUXes in the ICB and GSB are evenly divided between Groups A1, A2, B1, and B2. Subsets A1 and B2 connect to the second-stage MUX in Group A, while subsets A2 and B1 connect to the second-stage MUX in Group B, enabling efficient signal exchange between the two groups.

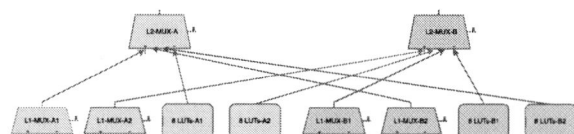

Fig. 1. Driving sources of the second-stage MUXes exchange.

Moreover, the first-stage MUX in the OCB has its driving sources doubled, mirroring the original driving sources from CLB outputs in Group A to those in Group B. This results in identical driving sources for the first-stage MUX in both Groups A and B, distinguished only by their sequential numbering.

This approach ensures seamless integration of duplicated resources in Group B with existing Group A resources,

enhancing the scalability and performance of the FPGA routing architecture.

C. Local Segment Design

Within the ICB, MUXes are classified into two types based on their feedback connections: feedback signals connecting to the first-stage MUX and feedback signals directly connecting to the second-stage MUX. Our study primarily investigates the design and distribution of feedback signals for the second-stage MUX to minimize delays in the first-stage MUX. Key aspects of this architectural exploration include determining the number of second-stage MUXes driven by feedback and the extent of feedback integration into these MUXes.

Similarly, within the GSB, feedback signals are categorized into multiplexed and non-multiplexed modes: multiplexed signals connect to the first-stage MUX for selection and distribution, while non-multiplexed signals directly connect to the second-stage MUX to reduce delays in the first-stage MUX. Our architectural exploration focuses on whether to employ multiplexing, determining the number of feedback-driven MUXes, and optimizing the degree of feedback integration within these MUXes. These parameters are critical in enhancing architecture efficiency and are refined by Bayesian optimization methods.

D. Objective Function

In the evaluation of architectures across various VTR 8.0 benchmarks [6], we utilize a cost function to assess their performance relative to a baseline:

$$\text{cost}_0(\theta) = \frac{1}{k} \sum_{b=1}^{k} \left(\frac{a_{b,\theta}}{a_{b,base}} \right)^{\alpha} \left(\frac{d_{b,\theta}}{d_{b,base}} \right)^{\beta} \tag{1}$$

where k is the number of benchmarks. $a_{b,\theta}$ and $d_{b,\theta}$ are the area and delay of the architecture parameterized by θ on benchmark b. $a_{b,base}$ and $d_{b,base}$ are the area and delay of the baseline on benchmark b. Parameters α and β control the tradeoff between area and delay. We primarily investigate the impact of multi-LUT structures on critical path delay, thus setting $\alpha = 0$ and $\beta = 1$.

IV. Experiment Result

A. Baseline Introduction

Parameters such as area and delay for segments and MUXes are derived using the transistor sizing tool COFFE 2 [7], specifically calibrated for the 7nm technology node. The exploration of the 16-LUT routing architecture utilizes the optimized FPGA routing architecture based on the GRB [3]. The architecture information includes segment distribution and driving relationships. The channel width of the baseline architecture is 204.

TABLE I. BASELINE SEGMENT DISTRIUTION

Name	Length	Number	Bend Pattern[a]
L1	1	14	ST
L2	2	8	ST
L4	4	4	- U -
L8	8	4	ST
L12	12	2	ST

[a.] "ST" indicates straight, "U" indicates a 90-degree counterclockwise bend, "-" indicates no bend.

TABLE II. BASELINE DRIVING RELATIONSHIP

Sink	Source
L1	L1, L2, L8, CAS-MUX output, CLB output

Sink	Source
L2	L1, L2, L8, L12, OCB output, CLB output
L4	L1, L2, L4, OCB output
L8	L1, L2, L8
L12	L1, L2, L4, L12

B. Best 16-LUT Routing Architecture

Through 1800 iterations, the detailed information of best 16-LUT routing architecture is listed in TABLE III and TABLE IV.

TABLE III.   BEST 16-LUT ARCH SEGMENT DISTRIBUTION

Name	Length	Number	Bend Pattern[a]
L1	1	24	ST
L2	2	10	ST
L3	3	6	ST
L4	4	8	- U -
L5	5	6	- - D -
L6	6	12	ST

[a] "D" indicates a 90-degree clockwise bend.

In the best 16-LUT routing architecture, the channel width is 392. $fb_{ICB} = 2$, $p_{ICB} = 0.644$, $fb_{GSB} = 2$, $p_{GSB} = 0.959$, $r = True$. Approximately 10 LUTs in the ICB feature 2

feedback inputs, enhancing connectivity to their second-level MUXes. In the GSB, 245 first-level MUXes incorporate 2 feedback inputs.

TABLE IV.   BEST 16-LUT ARCH DRIVING RELATIONSHIP

Sink	Source
L1, L2, L3, L4, L5, L6	L1, L2, L3, L4, L5, L6, OCB output, CLB output

Compared to the baseline 8-LUT routing architecture, detailed in TABLE V, the 16-LUT routing architecture design achieves significant performance gains: an average 14.90% reduction in the critical path delay and an average 14.01% reduction in the area-delay product. Since each CLB in the 16-LUT routing architecture contains twice the number of logic units compared to the baseline, the area and area-delay product used to calculate the ratio are doubled relative to the actual routing area of the baseline architecture. These enhancements capitalize on doubled routing resources relative to the baseline, as reported by VTR 8.0, highlighting the effectiveness of scaling resources to optimize FPGA performance metrics for advanced applications.

TABLE V.   COMPARISON OF 8-LUT BASELINE AND BEST 16-LUT ROUTING ARCHITECTURE

Benchmark	Routing Area (MWTA)			Critical Path Delay (ns)			Area-Delay Product		
	Baseline	16-LUT Arch	Ratio	Baseline	16-LUT Arch	Ratio	Baseline	16-LUT Arch	Ratio
arm_core	18609.9	36505.6	-1.92%	38.83	25.99	-33.05%	722583.3	948908.3	-34.34%
bgm	19096.0	37699.3	-1.29%	32.34	25.71	-20.49%	617572.3	969384.7	-21.52%
blob_merge	18019.6	34886.8	-3.20%	12.92	13.80	6.86%	232795.2	481598.3	3.44%
boundtop	15221.7	30718.0	0.90%	3.55	2.82	-20.64%	54031.1	86529.8	-19.93%
ch_intrinsics	14143.3	28465.2	0.63%	3.05	3.12	2.12%	43171.6	88731.2	2.77%
diffeq1	16876.4	34111.7	1.06%	30.55	27.24	-10.84%	515616.2	929226.6	-9.89%
diffeq2	16876.4	34111.7	1.06%	24.22	21.99	-9.20%	408773.4	750235.7	-8.23%
LU8PEEng	19116.2	39277.7	2.73%	122.92	118.63	-3.49%	2349763.3	4659395.7	-0.85%
mkDelayWorker32B	18609.9	37612.8	1.06%	12.93	9.09	-29.68%	240693.7	341938.7	-28.94%
mkPktMerge	17256.8	34886.8	1.08%	6.69	5.26	-21.44%	115509.6	183456.4	-20.59%
mkSMAdapter4B	16394.9	32696.7	-0.28%	8.13	6.76	-16.86%	133232.7	220898.6	-17.10%
or1200	18019.6	36401.5	1.01%	19.79	21.46	8.44%	356638.5	781227.2	9.53%
raygentop	17726.3	35771.5	0.90%	9.33	6.71	-28.08%	165391.7	240051.4	-27.43%
sha	16876.4	32303.7	-4.29%	20.18	19.36	-4.06%	340638.3	625532.1	-8.18%
stereovision0	18609.9	37926.5	1.90%	6.51	5.21	-20.09%	121227.3	197425.3	-18.57%
stereovision1	19054.9	38574.1	1.15%	15.32	9.52	-37.89%	291932.5	366827.3	-37.17%
stereovision3	12486.8	21674.4	-13.2%	3.17	3.60	13.81%	39535.8	78104.8	-1.22%
AVERAGE		**-0.63%**			**-14.90%**			**-14.01%**	

V. CONCLUSION

In this paper, we propose a new FPGA routing architecture with 16 LUTs per CLB based on the GRB model at the academic 7nm technology. We plan the topology rules for both global and local segments, focusing on investigating the driving sources of second-level MUXes and the distribution of feedback signals within the ICB and GSB. Our results demonstrate that integrating more LUTs within a CLB and designing optimized routing channels effectively enhances the performance of modern FPGA routing architectures. Moreover, we plan to utilize VTR 9.0 from Github to reduce runtime for large designs and explore the 32-LUT routing architecture to narrow the gap with the Versal architecture.

REFERENCES

[1] O. Petelin and V. Betz, "The speed of diversity: Exploring complex FPGA routing topologies for the global metal layer," 2016 26th International Conference on Field Programmable Logic and Applications (FPL), Lausanne, Switzerland, 2016, pp. 1-10.

[2] Gaide, Brian, Dinesh D. Gaitonde, Chirag Ravishankar and Trevor Bauer. "Xilinx adaptive compute acceleration platform: Versal™ architecture." Proceedings of the 2019 ACM/SIGDA International Symposium on Field-Programmable Gate Arrays, 2019, 10 pages.

[3] J. Qian, Y. Shen, K. Shi, H. Zhou and L. Wang, "General routing architecture modelling and exploration for modern FPGAs," 2021 International Conference on Field-Programmable Technology (ICFPT), Auckland, New Zealand, 2021, pp. 1-9.

[4] Y. Shen, J. Qian, K. Shi, L. Wang, and H. Zhou, "Two-level MUX design and exploration in FPGA routing architecture," in International Conference on Field-Programmable Logic and Applications (FPL), 2021, pp. 234–241.

[5] S. Zheng, J. Qian, H. Zhou and L. Wang, "GRAEBO: FPGA General Routing Architecture Exploration via Bayesian Optimization," 2022 32nd International Conference on Field Programmable Logic and Applications (FPL), Belfast, United Kingdom, 2022, pp. 282-286.

[6] Murray, O. Petelin, S. Zhong, J. M. Wang, M. ElDafrawy, J.-P. Legault, E. Sha, A. G. Graham, J. Wu, M. J. P. Walker, H. Zeng, P. Patros, J. Luu, K. B. Kent and V. Betz, "VTR 8: high performance CAD and customizable FPGA architecture modelling", ACM TRETS, 2020.

[7] Sadegh Yazdanshenas and Vaughn Betz. 2019. "COFFE 2: Automatic modelling and optimization of complex and heterogeneous FPGA architectures", ACM Trans. Reconfigurable Technol. Syst. 12, 1, Article 3 (March 2019), 27 pages.

High-Efficiency Power Amplifier Design for Bluetooth Low Energy Applications

Bharatha Kumar Thangarasu *[1], Li Shuai [1], Yu Hongshi [1], Ge Wansi [1], Liu Yuqing [1], Nagarajan Mahalingam [1], Meng Fanyi [1], Kaixue Ma [1], Juin J. Liou [2], Bo Wang [3], Younan Hua [4], Xiaomin Li [4], Lu Zhenghao [5], and Kiat Seng Yeo [1,3]

[1] School of Microelectronics, Tianjin University, No. 92, Weijin Road, Nankai District, Tianjin, China, 300027
[2] School of Electrical and Information Engineering, North Minzu University, Ningxia, Yinchuan, China, 750021
[3] Singapore University of Technology and Design, 8 Somapah Road, Singapore, 487372
[4] Wintech Nano-Technology Services Pte Ltd, 10 Science Park Rd, Singapore Science Park II, Singapore, 117684
[5] Soochow University, 50 E Ring Rd, Gusu District, Suzhou, Jiangsu, China, 215006

* Email: kumar@tju.edu.cn, yeokiatseng@tju.edu.cn, kiatseng_yeo@sutd.edu.sg

Abstract—This paper presents a differential common source fixed gain high efficiency power amplifier with performance determined by using chip-on-board PCB measurements. The amplifier is based on inductive load without degeneration which is designed using 40nm CMOS process. The proposed amplifier design achieves power gain of 10 dB, operating frequency in the 2.4 GHz ISM band, power consumption of 24.2 mW using 1.1 V supply voltage, a power added efficiency of over 60%, an output third order intermodulation intercept point of +15 dBm and the output referred 1-dB gain compression point of +10.4 dBm. The designed amplifier occupies a core area of 650 μm x 450 μm.

Keywords—Bluetooth-low energy (BLE), CMOS, fully differential amplifier, high efficiency, low power, power amplifier, transmitter (TX).

I. INTRODUCTION

Normally, a power amplifier is the most power consuming block in the RF transceiver [1]. However, for Bluetooth low-energy (BLE) application, the standard targets to reduce the overall power consumption and imposes a challenge for the power amplifier (PA) design. In recent years the wired connections between the devices such as phone, speakers, headsets, laptops, etc. which was replaced using Bluetooth standard evolved further for low power applications as the BLE standard. This standard is targeted for very low power operation and is to be operated in the 2.4 GHz unlicensed industrial, scientific and medical (ISM) band which is limited within 2.402 – 2.480 GHz. BLE standard suggests a Gaussian frequency shift keying (GFSK) modulation scheme with a point-to-point (including piconet) communication topology that specifically targets industries where the demand is for ultra-low power at the expense of low throughput. A typical

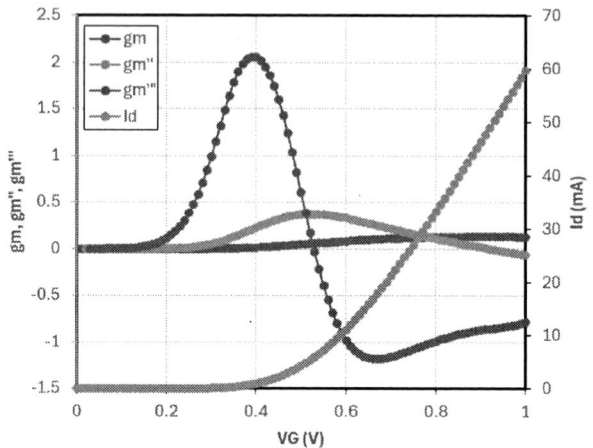

Fig. 2. Bias point selection.

BLE protocol includes periodically turning on the RF radio, transferring (transmitter) or receiving (receiver) a few bytes to kilobytes of data, and then turning off after data transfer and going back to the sleep mode [2], [3].

Generally, the PA designed for achieving a better linearity performance becomes the most power consuming circuits and additionally with a limited efficiency the power consumption

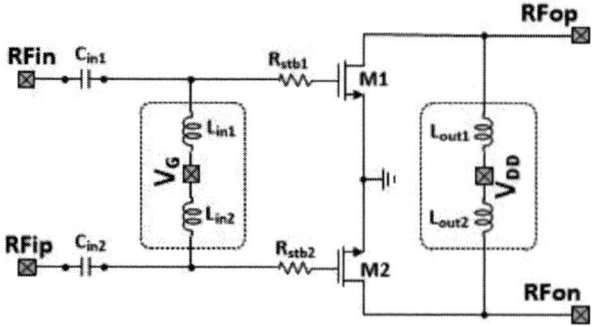

Fig. 1. Proposed differential PA schematic.

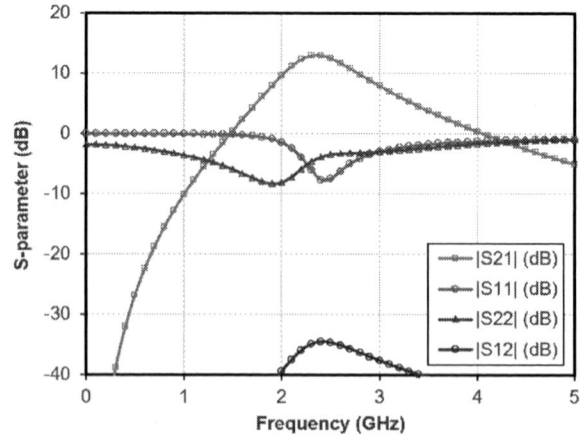

Fig. 3. Simulated 4-port S-parameters plots.

979-8-3503-6184-1/24 $31.00 © 2024 IEEE   228

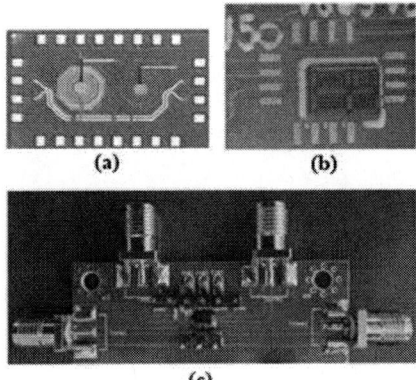

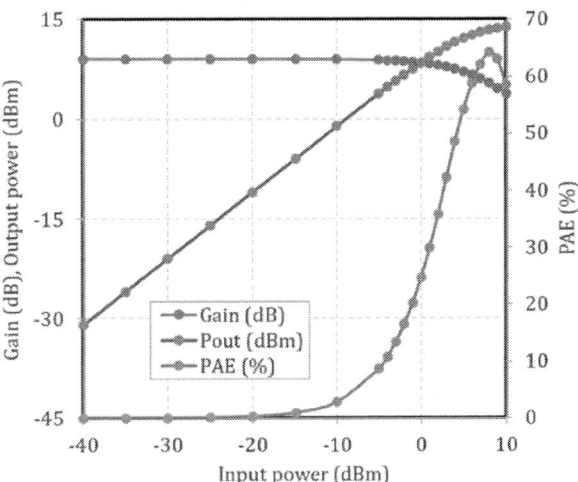

Fig. 4. (a) Microphotograph, (b) CoB wire-bonding and (c) assembled PCB photograph.

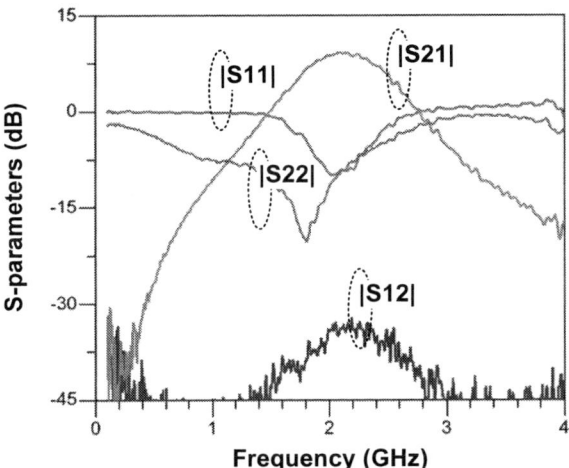

Fig. 5. Measured 4-port S-parameters plots.

Fig. 6. Measured gain, output power and PAE against the input power.

is increased multifold. This becomes a restriction for its usage in the BLE transmitters and insufficient attention was given to this. The BLE radios by using the GFSK modulation scheme which is a fixed amplitude modulated signal can afford to have the linearity performance relaxed [4].

From this work, we propose a method to improve the efficiency of the PA by compromising the linearity performance. The proposed design is a fully differential power amplifier by transition of bias point from class AB towards close to class B bias point is selected to improve the power added efficiency (PAE) with the tradeoff of linearity. The details of proposed design are provided in subsequent section.

II. CIRCUIT DESIGN

The proposed amplifier is a fully differential amplifier as shown in Fig. 1 with both input and output matched to 100-Ω differential impedance. The amplifier is based on common source (CS) amplifier with inductive load. To achieve the specifications, the proposed design is a single stage amplifier, without any degeneration inductors. In the proposed design, the dc feed bias inductors L_{in1} and L_{in2} are tuned to work as the L-section input matching network with the input capacitors C_{in1} and C_{in2}, respectively to achieve better input matching. Similarly, the load inductors L_{out1} and L_{out2} also function as the dc feed inductors at the drain terminals as well as works as the

output matching network to achieve the required output matching and efficiency against the linearity performance.

From the S-parameters simulation of the transistors M1 and M2 suggested conditional instability at the operating frequency. To mitigate this instability and to minimize the effect on the efficiency and linearity performance of the proposed PA design, the gate resistance R_{stb1} and R_{stb2} included improves the stability factors such that the overall amplifier becomes unconditionally stable for any input and output impedances.

From the application of this PA to be used in BLE transmitters, which is based on GFSK modulation scheme, the signal amplitude is fixed and the information is carried using the variable frequency. Hence, it suggests the linearity performance of the design can be relaxed and the efficiency of the PA can be improved as a design tradeoff. In addition, by incorporating minimum passive components in the overall design, the power losses can be reduced significantly. With this, the proposed design is targeted to achieve improved efficiency as required for the low power BLE applications. Furthermore, to improve the efficiency, the gate bias V_G is reduced to change the PA operating mode from class AB towards close to class B.

The bias condition of the CMOS based PA is mainly set by the gate voltage VG. From this gate voltage against the drain current, the first order transconductance gm, second derivative gm'' and the third derivative gm''' will be obtained as shown in Fig. 2. This plot indicates the effect of non-linearity that will be introduced in the drain current due to the different VG bias voltage conditions. The desirable transconductance gm component is the first order term for which an improved linearity performance can be achieved. The higher order terms mainly the second order term will be cancelled out in a differential amplifier and hence the most critical is the third order term (gm''') which has positive value for low bias voltage and negative for high bias voltage. Most of the time the high order terms will result in high order intermodulation products that will affect the spectrum purity. In addition to the transconductance and its derivatives, for the proper bias voltage selection we also need to consider the increasing drain current with increase in the bias voltage as shown in Fig. 2. Hence, if the bias voltage is appropriately

979-8-3503-6184-1/24 $31.00 © 2024 IEEE   229

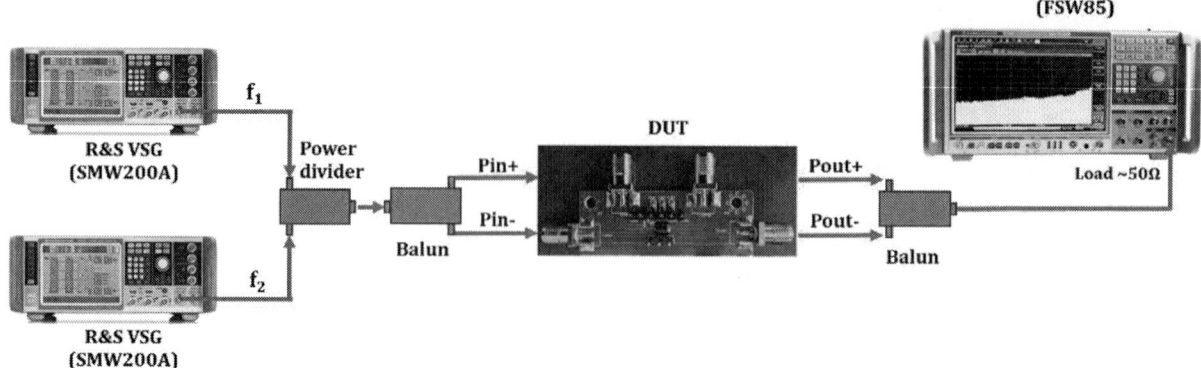

Fig. 7. Test setup for 2-tone measurement.

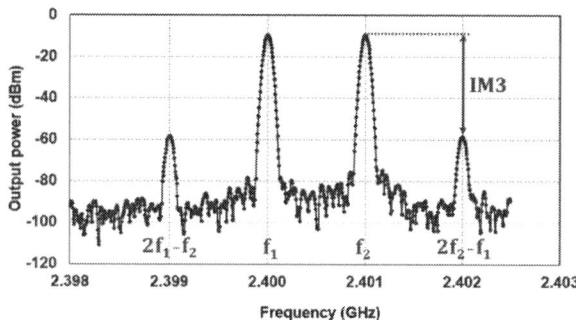

Fig. 8. Output spectrum plot with input 2-tone for IP3 measurement.

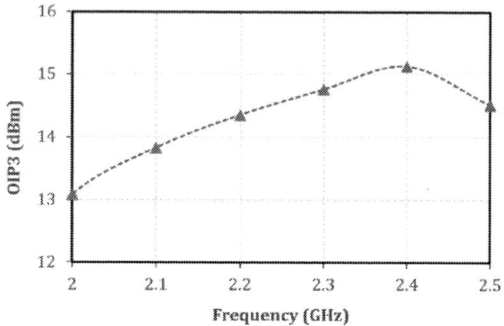

Fig. 9. Measured OIP3 plot against input frequency.

selected it will determine the overall PA performance and for the proposed design the bias is mainly considered for an improved efficiency with a tradeoff against the linearity. This design statement is verified by the simulated S-parameters shown in Fig. 3 and also further by the measurement results in the subsequent section III.

III. EXPERIMENTAL RESULTS

The proposed differential PA is designed using TSMC 40nm CMOS process. The PA design consumes a dc power of 24 mW from a 1.1 V supply voltage. The die microphotograph is shown in the Fig. 4 (a) with a die area of 1.4 mm × 0.7 mm. The design is fabricated and sub-diced to be integrated as chip-on-board using wire-bonding on the printed circuit board (PCB) as in Fig. 4 (b). The test FR4 based PCB photo is shown in Fig. 4 (c). The proposed PA is measured using R&S FSW85 signal analyzer, R&S SMW200A vector signal analyzer, and R&S ZVA43 vector network analyzer.

A. S-parameter measurements

The PA is a fully differential amplifier and the S-parameter is measured using 4-port vector network analyzer. The calibration is based on 4-port balanced SOLT technique using the SMA cables. To enable S-parameter against input power for linearity measurements, a power calibration using power meter is performed against the input power range and the input frequency range. From this setting, the 4-port S-parameters are measured and the results are plotted as shown in Fig. 5. The design has an overall better return loss at both input and output ports. The design also shows a gain of 10.1 dB in the interested frequency range with very low reverse isolation.

B. Gain compression and power added efficiency (PAE)

The measurement results as shown in Fig. 6 are obtained which includes the gain, and output power obtained against the input power. From this result, the 1-dB compression point is measured for the input power as IP_{1dB} and output power OP_{1dB}, for which the gain drops from the linear characteristics by 1 dB. This value provides a metric of the PA linearity performance and is expected to be as high as possible. For the proposed design, the measured OP_{1dB} is +10.4 dBm and the saturated output power is +14 dBm at 2.45 GHz frequency. The Fig. 6 also includes the PAE results against the input power which has the maximum PAE of 64.1 % indicating a higher efficiency with improved linearity performance.

C. Third intermodulation intercept point using 2-tone test:

The 2-tone test bench setup is shown in Fig. 7. The test is performed using a wideband power divider and balun modules interfaces to single ended equipment such as the two R&S SMW200A signal generator, each for one tone combined using power divider module followed by a balun to generate the differential input to the DUT. The DUT's differential output is combined using an output balun provided to the single ended spectrum analyzer to plot the PA output spectrum as shown in Fig. 8. From the output plot as shown in Fig. 8 for an input with 2-tone of equal amplitude of -10 dBm with a frequency offset $(f_1 - f_2)$ of 1 MHz, the IM_3 deviation is obtained as 50.1 dB. The total loss of the test setup including the RF cables, power divider and balun is 12 dB. The OIP_3 as determined by the test setup is plotted against the frequency and peaks within the 2.4 GHz band as shown in Fig. 9.

979-8-3503-6184-1/24 $31.00 © 2024 IEEE   230

TABLE I.   PERFORMANCE COMPARISON TABLE OF THE PROPOSED PA AGAINST THE STATE-OF-THE-ART WORKS.

Parameter	[5]	[6]	[7]	[8]	This work
CMOS Technology	40nm	180nm	180nm	90nm	**40nm**
Frequency (GHz)	2.4/5	2.4	2.4	2.4	**2.4**
Gain Max (dB)	17.5	17*	12.7	17*	**10.1**
PAE Max (%)	25	35	30	33	**64.1**
PAE at OP_{1dB} (%)	11*	24	28*	32*	**40**
OP_{1dB} (dBm)	24.4	14	16	24*	**10.4**
P_{sat} (dBm)	25.4	26	16.8	26.3	**14**
OIP_3 (dBm)	-	-	27	-	**15.2**
VDD (V)	2.5	3.3	1.8	2	**1.1**
DC current (mA)	-	-	-	64.3	**22**
Chip area (mm^2)	1.14	3.4	1.8	1.875	**1**

* Estimated from the measurement results

The measured results obtained from the evaluation PCB of the proposed differential PA is summarized and compared against the state-of-the-art works in Table I. From the results it is observed that the proposed PA design covers the operating frequency range in the 2.4 GHz ISM band. Though, the small signal gain of this work is lower as compared to other works, the power added efficiency is better than the state-of-the-art works which is also desirable of this PA design and become more suitable for the BLE applications. Overall the proposed design have comparable linearity performance demonstrated by the measured output gain compression point and the third order intermodulation intercept point even with a low supply voltage of 1.1V with low dc power consumption. This also explains the reason for the improved efficiency performance. Furthermore, the proposed PA design is using a completely integrated single stage differential amplifier design with a compact die area. Hence, facilitating a seamless integration in a BLE transmitter.

IV. CONCLUSION

This paper proposes a single stage differential power amplifier designed to achieve improved efficiency along with the overall acceptable gain and linearity performance. The proposed amplifier is implemented using a 40nm CMOS process technology and finally targeted as the chip-on-board technology for measurement purpose. This amplifier achieves an improved efficiency and adequate linearity suitable for low power BLE applications.

REFERENCES

[1] M. Babaie et al., "A fully integrated bluetooth low-energy transmitter in 28 nm CMOS with 36% system efficiency at 3 dBm," *IEEE J. Solid-State Circuits*, vol. 51, no. 7, pp. 1547-1565, July 2016.

[2] A. H. Masnadi Shirazi, H. M. Lavasani, M. Sharifzadeh, Y. Rajavi, S. Mirabbasi and M. Taghivand, "An ultralow-power current-reused direct-conversion Bluetooth-low-energy receiver front-end in 40-nm CMOS," *IEEE Trans. Microw. Theory Techn.*, vol. 69, no. 5, pp. 2697–2711, May 2021.

[3] W. Fang, Z. Li, Z. Li, X. Wang and J. Du, "High-efficiency digital power amplifier design based on 22-nm CMOS process," *8th Int. Conf. Integrated Circuits Microsystems (ICICM)*, Nanjing, China, December 2023, pp. 5–9.

[4] J. Li, Y. Wu and X. Lv, "Study on a 0.13−μm CMOS class-E 2.4 GHz adjustable power amplifier for IoT application," *Int. Conf. Microw. Millimeter Wave Technology (ICMMT)*, Harbin, China, January 2023, pp. 1–3.

[5] B. Liu, R. Zhou and Z. Zhu, "Reconfigurable 2.4/5.0-GHz dual-band CMOS power amplifier for WLAN 802.11ax," *IEEE Trans. Circuits Syst. I, Reg. Papers* (Early Access).

[6] J. Ho and H. -W. Tsao, "CMOS power amplifier with novel transformer power combiner," *IEEE Trans. Circuits Syst. II: Express Briefs*, vol. 70, no. 7, pp. 2365–2369, July 2023.

[7] K. -C. Lin, H. -K. Chiou, P. -C. Wu, W. -H. Chen, C. -L. Ko and Y. -Z. Juang, "2.4-GHz complementary metal oxide semiconductor power amplifier using high-quality factor wafer-level bondwire spiral inductor," *IEEE Trans. Components, Packaging Manufacturing Technology*, vol. 3, no. 8, pp. 1286–1292, August 2013.

[8] E. Kaymaksut and P. Reynaert, "Transformer-based uneven Doherty power amplifier in 90 nm CMOS for WLAN applications," *IEEE J. Solid-State Circuits*, vol. 47, no. 7, pp. 1659–1671, July 2012.

A 0.15-6.5GHz Stacked CMOS Power Amplifier With Low-Frequency Bandwidth Extension

Shijiao Dong, Wei Li*, Xingyu Ma, Fan Chen, Hongtao Xu

State Key Laboratory of Integrated Chips and System, Fudan University, Shanghai 201203, P.R.China

* Email: 22212020060@m.fudan.edu.cn, w-li@fudan.edu.cn

Abstract—**This paper presents a wideband power amplifier (PA) for sub-7 GHz wireless applications. The PA employs a stacking topology with resistive shunt–shunt feedback for high output power, good linearity, and simultaneous input and output matchings. A negative feedback network is proposed to extend low-frequency bandwidth. The proposed PA has been implemented in 40nm CMOS technology for verification. The PA operates from 3 V power supply and achieves OP_{1dB} of 18.1 ~ 20.3 dBm with power added efficiency of 26.6% ~ 45.6% within 616% fractional bandwidth from 0.15 to 6.5 GHz. It also demonstrates better than -10dB return loss at both input and output. This PA core area is only 0.01 mm².**

Keywords—Broadband power amplifier (PA), stacked PA, shunt–shunt negative feedback

I. INTRODUCTION

The sub-7 GHz frequency band is emphasized by the emerging 5G communication standard for various wireless applications such as Wi-Fi, Bluetooth, and cellular. Consequently, wideband and efficient power amplifiers (PA), allowing to support multiple services across a broad frequency range, are becoming increasingly important.

The main challenge to design a broadband PA is to maintain wideband gain frequency response while delivering sufficient saturated output power (P_{sat}) across multiple octaves with good power-added efficiency (PAE). Among the various design approaches, several notable techniques are transformer-based wideband matching networks [1], stacked configuration [2],[3],[4] and resistive negative feedback technique [5],[6], respectively.

As the frequency decreases to sub- 1GHz, the inductor quality factor decreases and the required area increases dramatically, making transformers-based method difficult for broadband PA design. The stacked structures can withstand higher supply voltages by stacking multiple transistors, and R_{opt} can be increased to reduce the matched impedance transformation ratio, therefore, the stacking structures are widely used in low-frequency broadband PA designs. Resistive negative feedback allows impedance matching over a wide bandwidth, but introduces deterioration in gain and efficiency.

In this work, a broadband PA with novel negative feedback network to extend the low-frequency-bandwidth while maintaining highly flat gain is proposed. The design employs a stacking topology with the proposed resistive shunt–shunt feedback to successfully achieve 0.15~6.5GHz wide bandwidth with ±0.4dB flat gain and good efficiency performance. This paper is organized as follows. Section II discusses the circuit design of the proposed PA. Section III shows the post-simulation results and a comparison with state-of-the-art works. Conclusion is given in Section IV.

This paper is supported by National Key R&D Program of China under Grant 2023YFB4403802, and by the Fundamental Research Funds for Central Universities in China.

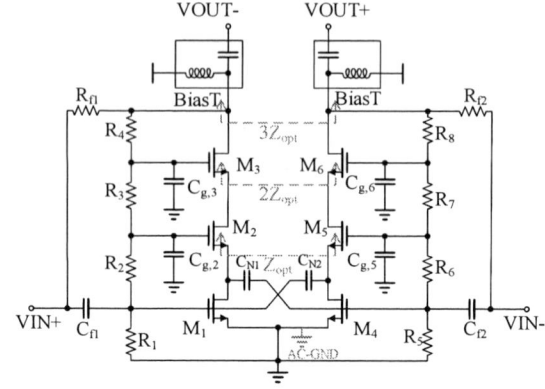

Fig. 1. Schematic of the proposed PA

II. CIRCUIT DESIGN

The configuration of the proposed PA is shown in Fig. 1, which is based on the transistor-stacking technique to provide high output power and wideband gain. Such differential structure intends to provide a differential PA outputs to establish a duplicated self-interference signal for a full duplex receiver. As depicted in Fig. 1, this structure employs a three transistor stacking to work at 3V power supply, while maintaining the output impedance at differential 100 ohms to achieve a wideband output matching. The negative feedback (R_{f1}/R_{f2}, C_{f1}/C_{f2}) are introduced to achieve wideband input matching and enhance linearity. We also introduce neutralization capacitors (C_{N1}/C_{N2}) in the first transistor to enhance stability and compensate for phase variations caused by gate-drain capacitance. Due to the absence of inductors, the differential structure can double the output power while occupying extra 0.005 mm² area. Additionally, it help form an AC ground to mitigate the degradation caused by bonding inductance shown in Fig. 1.

Differing from conventional stacked structures, we introduce bias capacitors ($C_{g,3}/C_{g,6}$, $C_{g,2}/C_{g,5}$) at the gates, where the gate capacitances and C_{gs} capacitances form a couple of capacitance divider network respectively. These allow direct coupling of the output of each transistor to the input of the next, forming a three transistor series amplification of the signal, resulting in a large output swing. Simultaneously, the gate capacitance value can be utilized to adjust the load impedance of the preceding transistor to achieve maximum output power. The gate capacitance of the k-th transistor ($C_{g,k}$) and the optimal load impedance [$(k-1)Z_{opt}$] required by the (k-1)-th transistor should satisfy (1). The relationship between gate capacitance and OP_{1dB} is depicted in Fig 2. In order to achieve the maximum OP_{1dB}, we choose 2.6 pF for $C_{g,2}$ and $C_{g,5}$, 1.3 pF for $C_{g,3}$ and $C_{g,6}$.

979-8-3503-6184-1/24 $31.00 © 2024 IEEE

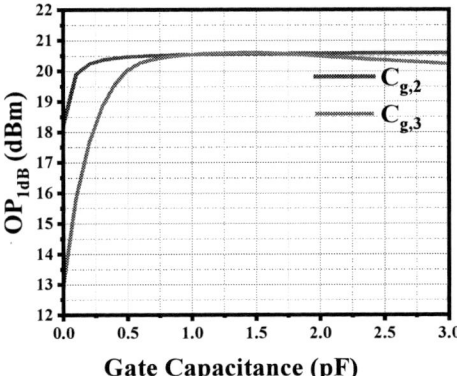

Fig. 2. OP$_{1dB}$ of the PA versus Cg,2 and Cg,3

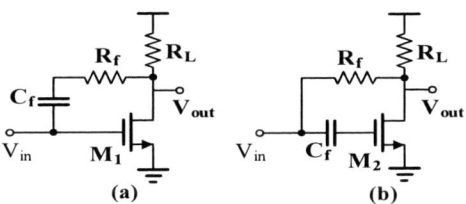

Fig. 3. (a) Conventional feedback structure and (b) Proposed feedback structure

$$C_{g,k}=\frac{C_{gs,k}+C_{gd,k}\left(1+g_{m,k}Z_{opt}\right)}{(k-1)g_{m,k}Z_{opt}-1},k=2,\ 3 \qquad (1)$$

Where, kZ_{opt} represents the optimal load impedance of the k-th transistor. Considering output matching, we set $3Z_{opt}$ to be approximately equal to differential 100 ohms. $C_{gs,k}$ is the gate-source parasitic capacitance of transistor M_k. $C_{gd,k}$ is the gate-drain parasitic capacitance of transistor M_k. $g_{m,k}$ is the transconductance of transistor M_k.

As shown in Fig. 3(a), the feedback resistor (R_f) can be utilized to provide input impedance matching. Due to the high supply voltage of the power amplifier, capacitors (C_f) are needed for DC voltage blocking. Unlike typical resistor-capacitor series structure introduced between input and output, we proposed the feedback resistor between input and output, with capacitors placed between the input and the gate of transistor M_2 shown in Fig. 3(b). This arrangement provides better gain flatness while avoiding excessively large capacitor values.

For the conventional feedback structure, it can be seen from its gain expression (2) that a large feedback capacitance is required at low frequencies in order to maintain a flat gain.

$$\frac{V_{out}}{V_{in}}=\frac{1/\left(R_f+(1/j\omega C_f)\right)-g_m}{1/\left(R_f+(1/j\omega C_f)\right)+(R_L+r_o)/(R_Lr_o)} \qquad (2)$$

The proposed feedback architecture gains are as follows:

$$\frac{V_{out}}{V_{in}}=\frac{1/R_f-g_mC_f/\left(C_f+C_{gs}\right)}{1/R_f+(R_L+r_o)/(R_Lr_o)} \qquad (3)$$

where, R_f and C_f are the feedback resistor and feedback capacitor, respectively; C_{gs} is the transistor gate parasitic capacitance. g_m is the transistor transconductance; r_o is the characteristic impedance; and R_L is the load impedance.

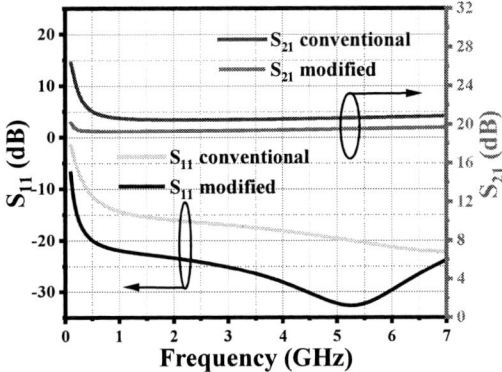

Fig. 4. S$_{21}$ and S$_{11}$ comparisons of PA under conventional and proposed feedback configurations (pre-simulation)

From (3), it can be seen that the proposed structure decouples the feedback capacitor (C_f) and frequency (ω), allowing for flat gain at low frequencies. At the same time, this structure reduces the capacitive impedance at the input,

which facilitates input impedance matching. The input impedance is calculated as:

$$Z_{in}=\frac{1+R_f/(R_L//r_o)}{1/(R_L//r_o)+g_mC_f/\left(C_f+C_{gs}\right)}//\frac{C_f+C_{gs}}{j\omega C_fC_{gs}} \qquad (4)$$

As shown in Fig. 4, the proposed feedback structure exhibits flatter gain and better input matching at low frequencies.

III. SIMULATION RESULTS

The proposed ultra-wideband power amplifier was implemented in 40nm CMOS with 3V power supply. Fig. 5(a) shows the layout of the PA with a size of 0.31mm^2 including pads, and the core area is only 0.01 mm^2. The PA operates from 0.15 to 6.5 GHz.

Fig. 5(b) shows the post-simulation results of the S-parameters. The maximum S$_{21}$ is 18.5 dB, and the gain variation within the operating bandwidth is less than 0.8 dB. Both S$_{11}$ and S$_{22}$ are less than -10 dB.

Fig. 5(c) shows the post-simulated saturated output power (P$_{sat}$), OP$_{1dB}$ and Peak PAE as a function of the frequency. From 0.15 to 6.5 GHz, P$_{sat}$ is 20.8 ~ 22.7 dBm, OP$_{1dB}$ is 18.1 ~ 20.3 dBm. The output power is limited by the gate widths and the numbers of the stacked transistor. The maximum PAE ranges from 26.6% to 45.6%. The PAE performance degradation with frequency is due to the pad parasitic capacitance. This can be mitigated by further adjusting the output matching network. The AM–AM and AM–PM distortions at 3 GHz are plotted in Fig. 5(d). The PA demonstrates a gain variation of 0.3 dB at the input power of 0 dBm, and the phase variation is about 2° in the input power range of −20 ~ 0 dBm.

A summary of the PA performance and comparison is shown in Table I. It is demonstrated that the proposed PA achieves an excellent OP$_{1dB}$ with much higher PAE within 616% fractional bandwidth from 0.15 to 6.5 GHz. The core area is only 0.01mm^2. It is verified by the above results that the proposed feedback structure can successfully extend low-frequency bandwidth, which will be applied in the future work.

979-8-3503-6184-1/24 $31.00 © 2024 IEEE

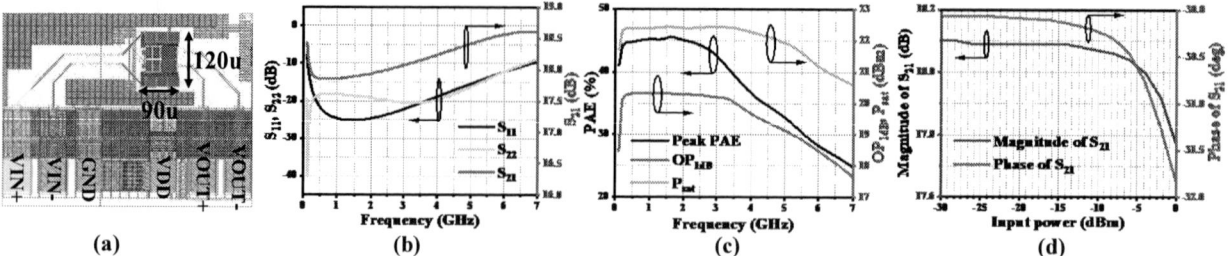

Fig. 5.(a) The layout of the proposed PA. (b) Post-simulated S parameters. (c) Post-simulated PAE and saturated output powers versus frequency. (d) Post-simulated AM–AM and AM–PM distortions at 3 GHz.

Table I. Performance Summary & Comparison With State-Of-The-Art CMOS PAs

Ref.	This work*	[1]-ISSCC 2015	[2]-TCAS II 2016	[5]-TCAS II 2021	[3]-ICICM 2022*	[6]-IMWTL 2023	[4]-ICMIMT 2023
Architecture	1-stage/ 3-stack	Class AB cascode	2-stage/ 3-stack	Wideband Pre-Distortion	2-stage/ 3-stack	Cascode+resistive feedback	2-stage/ 4-stack
On chip inductor	No	Yes	No	Yes	Yes	Yes	No
Freq. (GHz)	0.15-6.5	2-6	0.1-6.5	0.8-3.3	2-6	0.33-2.5	0.1-3
ΔBW (%)	616.6	115.5	793.8	153.9	115.5	238.9	529.5
Gain (dB)	18.1±0.4	23.6±0.8	18.4±1.5	12-15	19.2±0.5	19.5-22.5	30±1.2
P_{sat} (dBm)	20.8-22.7	20.1-22.4	22	23-24	21	19.5-21.5	27
OP_{1dB} (dBm)	18.1-20.3	17.8-20.7	18-19	20	17.8-18.6	16.3-17.6	-
Peak PAE (%)	26.6-45.6	19-28.4	13-20	31	19.6-27.8	35-52.4	23
S_{11} (dB)	<-10	<-11	<-11	<-10	<-7.3	<-10	<-11
S_{22} (dB)	<-10	<-5	<-5.1	<-10	<-8.8	<-10	<-11
Supply (V)	3	1.8	6	3.3	6	3.3	7.2
Core Area (mm²)	0.01	0.75	0.42	1	0.36	0.42	0.54
Process (nm)	40	65	180	180	130	65	180

*: post-simulation

IV. CONCLUSIONS

In this paper, we propose a feedback network that achieves low-frequency bandwidth extension by decoupling frequency and capacitance value. The designed PA employs a three-transistor stacking technique and the proposed feedback network to achieve a relative bandwidth of 616% from 0.15 to 6.5 GHz. The post-simulation results show a gain of 18.1 ± 0.4 dB and less than -10 dB input and output return loss performance over the operating frequency. With a core area of only 0.01 mm², OP_{1dB} of 20.3 dBm and a peak efficiency of 45.6% are achieved. These results confirm that the proposed feedback structure is the efficient approach to extend the low-frequency bandwidth.

REFERENCES

[1] W. Ye, K. Ma and K. S. Yeo, "2.5 A 2-to-6GHz Class-AB power amplifier with 28.4% PAE in 65nm CMOS supporting 256QAM," 2015 IEEE International Solid-State Circuits Conference - (ISSCC) Digest of Technical Papers, San Francisco, CA, USA, 2015, pp. 1-3, doi: 10.1109/ISSCC.2015.7062914.

[2] H. -F. Wu, Q. -F. Cheng, X. -G. Li and H. -P. Fu, "Analysis and Design of an Ultrabroadband Stacked Power Amplifier in CMOS Technology," in IEEE Transactions on Circuits and Systems II: Express Briefs, vol. 63, no. 1, pp. 49-53, Jan. 2016, doi: 10.1109/TCSII.2015.2504926.

[3] L. Wei, Y. Zhang, X. Tang, Y. Sun and F. Huang, "Broadband Power Amplifier Design in 130nm CMOS," 2022 7th International Conference on Integrated Circuits and Microsystems (ICICM), Xi'an, China, 2022, pp. 322-325, doi: 10.1109/ICICM56102.2022.10011257.

[4] Q. Lin, P. -F. Zhao, L. -N. Jia, R. -L. Yang and H. -F. Wu, "A 0.1-3 GHz 0.5W two-stage quadruple-stacked CMOS Power PA," 2023 International Conference on Microwave and Millimeter Wave Technology (ICMMT), Qingdao, China, 2023, pp. 1-3, doi: 10.1109/ICMMT58241.2023.10277570.

[5] S. Mariappan, J. Rajendran, H. Ramiah, P. -I. Mak, J. Yin and R. P. Martins, "An 800 MHz-to-3.3 GHz 20-MHz Channel Bandwidth WPD CMOS Power Amplifier For Multiband Uplink Radio Transceivers," in IEEE Transactions on Circuits and Systems II: Express Briefs, vol. 68, no. 4, pp. 1178-1182, April 2021, doi: 10.1109/TCSII.2020.3035758.

[6] N. Ginzberg and E. Cohen, "A Wideband CMOS Power Amplifier With 52% Peak PAE Employing Resistive Shunt Feedback for Sub-6 GHz 5G Applications," in IEEE Microwave and Wireless Technology Letters, vol. 33, no. 2, pp. 192-195, Feb. 2023, doi: 10.1109/LMWC.2022.3209822.

979-8-3503-6184-1/24 $31.00 © 2024 IEEE

A 2-to-2.7GHz Class-G Switched-Capacitor PA with cascode Switch-Reused Structure Achieving 25.92dBm Peak Power and 42% Efficiency

Jie Deng [1], Gengzhen Qi [1*]

[1] School of Microelectronics Science and Technology, Sun Yat-sen University, Zhuhai, China
*Email: qigzh@mail.sysu.edu.cn

Abstract—**This article proposes a 2-to-2.7GHz Class-G Switched-Capacitor power amplifier (SCPA). It features a cascode Switch-Reused structure in the switching unit and a transformer-based impedance matching network. The cascode Switch-Reused structure improves the drain efficiency and reduce the parasitic capacitance effectively at the output of the switching unit. The transformer-based matching network saves 78.5% output matching network layout area when compared to traditional inductive matching network [2]. For the first time we propose a method for controlling the switching unit without using a decoder eliminating the time delay 31ps between bits and the decoding circuit is simplified. In the case of 1.4V and 2.8V supply voltage operation the maximum output power of this work achieves 25.92dBm and the SCPA operates in a frequency range of 2.0 to 2.7GHz when the maximum system efficiency is greater than 30% which is suitable for 2.4GHz Wi-Fi Indoor Access Point (AP) (≤20 dBm) and Outdoor AP (≤27 dBm). It achieves 42% system efficiency without using a decoder. The average output power is 17.1dBm and the average system efficiency is 21.8%.**

Keywords—*Class-G, Switched-Capacitor Power Amplifier (SCPA), decoder, On-Chip matching network, polar transmitter*

I. INTRODUCTION

Radio Frequency (RF) power amplifiers are the main energy-consuming device in mobile wireless terminals. Wireless communication standards always use spectrum efficient OFDM modulation to encode signal information of amplitude and phase, such as Wi-Fi applications. The employment of such non-constant envelope modulation requires a high-linear power amplifier (PA), operating at a less-than-peak signal level to realize higher linearity and inherently reduced efficiency. Consequently, many efforts prefer more efficient switching amplifiers with linearization techniques, such as the pulse-width modulation (PWM) [2], out-phasing [2] and the envelope elimination and restoration (EER) [1]. The PWM PA is limited by the transmitted output power, thus the dynamic range of PA is relative to the minimum pulse width. The minimum output power of the out-phasing PA is limited by the loading mismatch that occurs when the external phase angle is large. Another drawback of out-phasing PA is the need for two PAs and also an area-consuming passive power-combining network. Compared to PWM and out-phasing PAs, the EER PA offers the best performance balance among the linearity, output power and efficiency. Yet, it sacrifices power consumption (25mW), especially the power-hungry analog power modulator [1].

This work was supported in part by the National Natural Science Foundation of China (NSFC) under Grant 62104263, and in part by GuangDong Basic and Applied Basic Research Foundation under Grant 2022A1515-012295. (Corresponding author: Gengzhen Qi)

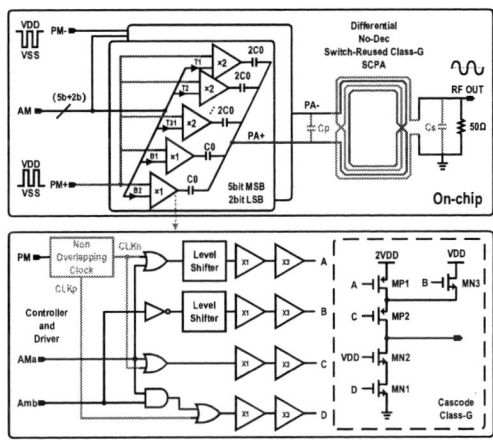

Fig. 1. The block diagram of the proposed Switch-Reused Class-G SCPA.

The Switched-Capacitor power amplifier (SCPA) architecture modulates envelope signals digitally, and it extends EER to the digitally modulated switching topology for high average efficiency and output power [1]. Herein, we propose a cascode Switch-Reused structure and a transformer-based impedance matching network for the Class-G SCPA to improve the drain efficiency and reduce the parasitic capacitance effectively at the output of the switching unit, also saving die area when compared to traditional inductive output matching network. Besides, a proposed method for controlling the switching unit without using a decoder is introduced to eliminate the time delay between bits and the decoding circuit is simplified. In 65nm CMOS process, the maximum output power achieves 25.92dBm, with 42% system efficiency. The average output power is 17.1 dBm and the average efficiency is 21.8%.

II. IMPLEMENTATION DETAILS

The block diagram of the cascode Switch-Reused Class-G SCPA is shown in Fig. 1. In which the digital phase (PM) and envelope amplitude (AM) signal are injected to the SCPA. The output power resolution is designed with 5-bit MSBs and 2-bit LSBs. The digital codeword representing the AM component is the sampled value of the envelope amplitude. The PM signal is converted into a non-overlapping differential switch waveform and applied to selected drivers to drive the Class-G switch unit. The output signal from the PA array is finally filtered through a transformer matching network to generate an RF signal at a 50Ω impedance.

To optimize the output power and system efficiency (SE) of our Class-G SCPA, we can analyze their dominant factors first. The output power can be expressed in (1), in which V_{DD}

is the voltage switched at the bottom plates and $2/\pi$ is the first coefficient of the Fourier series. R_{opt} is the optimum termination resistance ($R_{opt} = 9.7\Omega$) for the desired output

$$P_{out} = \frac{2}{\pi^2}\left(\frac{n}{N}\right)^2 \frac{V_{DD}^2}{R_{opt}} \quad (1)$$

$$\eta_{ideal} = \frac{4n^2}{4n^2 + \frac{\pi n(N-n)}{Q_{loaded}}} \quad (2)$$

power. The drain efficiency (DE) is defined in (2), in which Q_{loaded} is the loading quality factor. Normally, the typical On-Chip Q values (e.g., 10-15) limit Q_{loaded} to ~2-3 for fully integrated CMOS implementations. The Thévenin voltage is the source voltage scaled by (n/N), where n is the number of unit capacitors whose bottom plates are switched between V_{GND} and V_{DD}. N is the total number of capacitors. N determines the scaling of the output power resolution, and we choose N = 31. When n is larger, the efficiency and power of the output are larger [1]. The transformer is recommended to reduce output matching network layout area by 78.5% compared with three inductors in Fig. 4, although the output power is reduced by 12.3%. The matching network transforms 50Ω into R_{opt}. The primary inductance (L_{pri}) of the transformer and C_P are used to tune the total capacitance $4(N+1) \cdot C_0$ at the carrier frequency. $C_0 = 111fF$; $Z_{in2} = 9.7 + 19.3i$; $Cp = 1pF$; $Cs = 963.54pF$; $L_{pri} = 1.33nH$; $L_{sec} = 1.68nH$; $k = 0.75$.

A. 5-bit MSB Non-Decoder Design Method

The conventional 5-bit decoder a combinational logic circuit which is used to decode AM signal from binary to thermometer [6] has nearly 70 logic gates (e.g., NOR, NAND, INV) in Fig. 2(a). In order to simplify the decoding circuit our proposed without the decoder scheme the gating control method of the binary AM code on the PA array is shown in Fig. 2(b). It only uses 10 inverters as drivers, which greatly simplifies the decoding circuit. In addition, there is signal competition and risk in combinational logic circuit, so that the decoded signals of certain channel(e.g., T_{16} and T_{31}) have time difference. Input 1MHz square wave analog AM signal, the two channels T_{16} and T_{31} in the traditional scheme have 31ps delay, while the delay of our proposed scheme is 0ps, which solves the problem of delay after decoding AM signal.

B. cascode Switch-Reused Class-G SCPA

The enhanced-efficiency Class-G SCPA [1] boosts the average drain efficiency by introducing an additional efficiency peak in the power back-off (PBO) region without any discontinuity in efficiency and linearity. To maximize the efficiency, we use a cascode Switch-Reused Class-G SCPA for area saving and efficiency improvement [6], as shown in Fig. 3(b). The M_{P3} switches at carrier frequency when M_{N3} is conducting, and the output voltage is V_{DD} to V_{GND}. The proposed switching architecture removes the M_{P3} from the conventional switch and reuses the existing common-source common-gate transistor M_{P2} as the switching device in V_{DD} mode to reduce the parasitic capacitance at the switching output. When $|V_{GSMP3}| = V_{DD}$ and M_{P1} switches at carrier frequency, the output voltage is $2 V_{DD}$ to V_{GND}. By changing the AM code to switch the three modes of 2VDD, V_{DD} and V_{GND}. The effect of capacitor charge leakage when the chip bit is switched to V_{GND} is reduced, so the SCPA always worked at the saturation power output and the system efficiency was improved. The sizes of the MOSFETs are $W_P/L_P = $

$75um/60nm$ and $W_N/L_N = 30um/60nm$ with two in parallel, while the MOSFET output resistor $R_{on} = 16.8\Omega$.

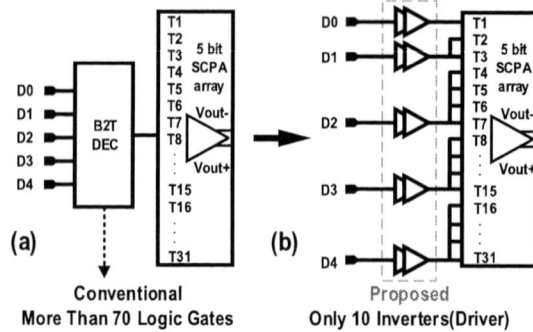

Fig. 2. (a) The conventional 5-bit decoder has nearly 70 logic gates. (b) our proposed scheme only uses 10 inverters as drivers

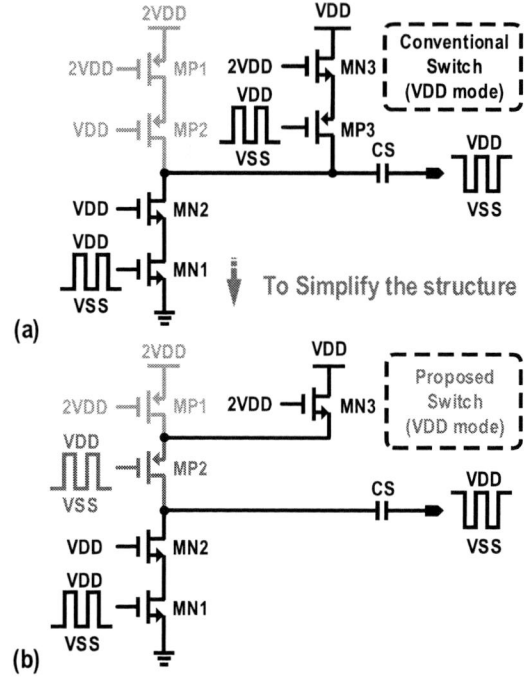

Fig. 3. (a) Traditional Class-G switch scheme. (b) The proposed cascode Switch-Reused Class-G structure for reducing the usage of MOSFET.

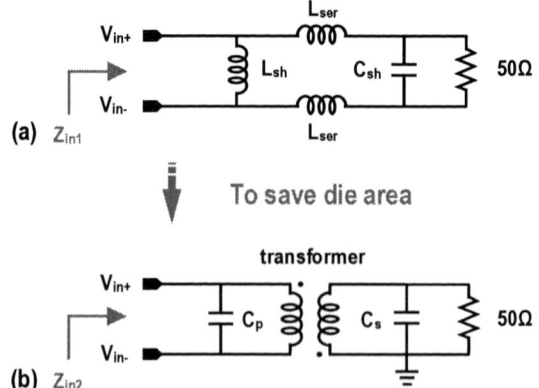

Fig. 4. output matching network with (a) the area of [2] is 1000um×270um. (b) the area our proposed is only 282um×206um.

979-8-3503-6184-1/24 $31.00 © 2024 IEEE

TABLE I. COMPARISON WITH STATE-OF-THE-ART

	This Work	JSSC'13 [1]	JSSC'18 [2]	JSSC'17 [3]	JSSC'21[4]	JSSC'23[5]	JSSC'19[6]	JSSC'21[7]
Topology	Simplified Class-G No-Dec	Class-G	C-2C	Class-G Doherty	Hybrid-Doherty	Multi-mode DTX	Quadrature Class-G	Quadrature SCPA/Hybrid Doherty
Supply (V)	1.4 / 2.8	1.4 / 2.8	1.2 / 2.4	1.2 / 2.4	1.1	0.95	1.2 / 2.5	1.2 / 2.4
Freq (GHz)	2 - 2.7	2.15	1.8	3.5	3.3	2.4	2.2	2.4
Peak Pout(dBm)	25.92	24.3	24	25.3	23.6	23.18	30.1	30.3
Peak η (%)	42.0 (SE)	44.0 (SE)	40.0 (SE)	30.4 (SE)	34.7	52.59	37.0 (SE)	36.5
AVg.Pout(dBm)	17.1	16.8	18.9	17.1	22.78	12.23	19.5	20.4
AVg.SE (%)	21.8	33	21.2	18	16.9	18.81	14.7	22.6
Technology	65nm CMOS	65nm CMOS	65nm CMOS	65nm CMOS	40nm CMOS	40nm CMOS	65nm CMOS	40nm CMOS

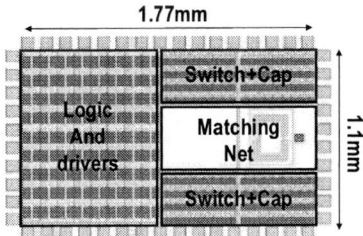

Fig. 5. The layout of our proposed Class-G SCPA.

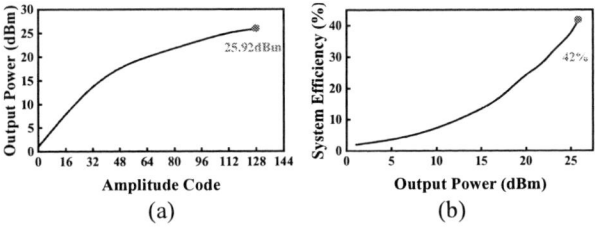

(a)   (b)

Fig. 6. (a) Measured output power versus codeword. (b) Measured SE versus output power.

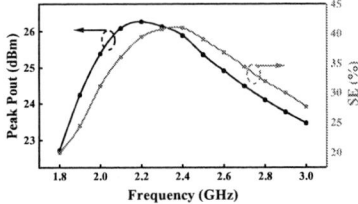

Fig. 7. Measured Peak Pout and System Efficiency versus Frequency.

III. SIMULATION RESULTS

The input PM is a 2.4GHz square-wave single tone signal, and the AM signal is a binary encoding of the high (V_{GND} = 0V - $2V_{DD}$ = 2.8V) and low (V_{GND} = 0V - V_{DD} = 1.4V) level of the switching array gate. The circuit is simulated and the conclusion is drawn. Fig. 6(a) plots the relationship between output power and envelope amplitude code with a maximum output power of 25.92dBm, and Fig. 6(b) depicts the relationship between SE and output power, with a maximum system efficiency of 42%. By changing the encoding of the AM signal, as expected by (1), the output power rises with the increase of the envelope amplitude AM, while efficiency rises with the increase of the output power as predicted by (2). The input PM signal was changed to SCPA frequency sweep from 1.8 to 3GHz. the maximum system efficiency and peak output power are also measured for each frequency. The maximum

output power peaks at 2.2GHz, the maximum system efficiency peaks at 2.4GHz, and the SCPA operates in a frequency range of 2.0 to 2.7GHz when the maximum system efficiency is greater than 30% as shown in Fig. 7.

IV. CONCLUSIONS

We propose a cascode Switch-Reused Class-G SCPA with 1.4 and 2.8V power supply operating in a frequency range of 2.0 to 2.7GHz. For the first time, we propose a method for controlling the switching unit without using a decoder eliminating the time delay 31ps between bits and the decoding circuit is simplified. The transformer saves 78.5% output matching network layout area when compared to [2]. The work achieves 25.92dBm peak output power and 42% system efficiency at 2.4GHz. The average output power is 17.1dBm and the average system efficiency is 21.8%. Our proposed Class-G SCPA is suitable for the 2.4GHz Wi-Fi application. The die area of the proposed SCPA is 1.96mm² implemented in 65nm CMOS process.

REFERENCES

[1] S. -M. Yoo et al., "A Class-G Switched-Capacitor RF Power Amplifier," in IEEE Journal of Solid-State Circuits, vol. 48, no. 5, pp. 1212-1224, May 2013, doi: 10.1109/JSSC.2013.2252754.

[2] Z. Bai, A. Azam, D. Johnson, W. Yuan and J. S. Walling, "Split-Array, C-2C Switched-Capacitor Power Amplifiers," in IEEE Journal of Solid-State Circuits, vol. 53, no. 6, pp. 1666-1677, June 2018, doi: 10.1109/JSSC.2018.2805872.

[3] V. Vorapipat, C. S. Levy and P. M. AsbeckIEEE, "A Class-G Voltage-Mode Doherty Power Amplifier," in IEEE Journal of Solid-State Circuits, vol. 52, no. 12, pp. 3348-3360, Dec. 2017, doi: 10.1109/JSSC.2017.2748283.

[4] H. J. Qian, B. Yang, J. Zhou, H. Xu and X. Luo, "A Quadrature Digital Power Amplifier With Hybrid Doherty and Impedance Boosting for Complex Domain Power Back-Off Efficiency Enhancement," in IEEE Journal of Solid-State Circuits, vol. 56, no. 5, pp. 1487-1501, May 2021, doi: 10.1109/JSSC.2021.3059113.

[5] M. Beikmirza, Y. Shen, L. C. N. de Vreede and M. S. Alavi, "A Wideband Energy-Efficient Multi-Mode CMOS Digital Transmitter," in IEEE Journal of Solid-State Circuits, vol. 58, no. 3, pp. 677-690, March 2023, doi: 10.1109/JSSC.2022.3222028.

[6] S. -W. Yoo, S. -C. Hung and S. -M. Yoo, "A Watt-Level Quadrature Class-G Switched-Capacitor Power Amplifier With Linearization Techniques," in IEEE Journal of Solid-State Circuits, vol. 54, no. 5, pp. 1274-1287, May 2019, doi: 10.1109/JSSC.2019.2904209.

[7] B. Yang, H. J. Qian and X. Luo, "Quadrature Switched/Floated Capacitor Power Amplifier With Reconfigurable Self-Coupling Canceling Transformer for Deep Back-Off Efficiency Enhancement," in IEEE Journal of Solid-State Circuits, vol. 56, no. 12, pp. 3715-3727, Dec. 2021, doi: 10.1109/JSSC.2021.3113511.

A X-band High linearity Tunable Bandpass Filter in 130nm CMOS

Tianrui Wang , Ziyu Wang ,Huiquan Xie , Yifei Chen , Haokun Lan, Maliang Liu*, Yintang Yang

School of Microelectronics, Xidian University, Xi'an 710071, China

* Email: a15802423645@163.com, mlliu@xidian.edu.cn

Abstract—This paper presents the design of a tunable bandpass filter using 130nm CMOS technology. The designed X-band high-linearity tunable bandpass filter covers a frequency range of 6GHz to 12GHz, with tunable bandwidth and center frequency. It achieves a maximum bandwidth greater than 4GHz and provides out-of-band suppression of over 50dB. The typical insertion loss in the passband is 5dB, with return loss greater than 10dB. The filter's IP1dB is greater than 25dBm and employs an integrated design approach combined with full-wave electromagnetic (EM) simulation for optimal performance. The bare die is packaged using advanced flip-chip packaging techniques, with dimensions of 2.698mm×1.331mm.

Keywords—Tunable, High linearity, Bandpass Filter , X-band

I. INTRODUCTION

With the advancement of technology, the speed and accuracy of Analog-to-Digital Converters (ADCs) continue to improve, enabling the possibility of RF direct sampling receivers. They can be widely utilized in wireless communication, broadcasting, radar, and various other fields to efficiently capture and process RF signals. The X-band, covering from 8GHz to 12GHz, finds applications in radar detection, satellite communication and so on. Figure 1 illustrates the structure of a widely used RF direct sampling receiver. Filter can improve the receiver's sensitivity and interference resistance by selectively choosing the desired signals and reject the undesired ones.

With the development of miniaturized, tunable and highly integrated circuit systems, the traditional architecture based on multi-chip integration on PCB clearly cannot meet the requirements of the system for the performance of the RF front end[1]. To adapt to the trend of integration, LTCC technology and MEMS technology have replaced PCB technology[2]. However, due to the complexity of circuits and limitations in technology, it still suffers from large size constraints[3].

This paper presents a tunable bandpass filter operating in the X-band, implemented using 130nm CMOS technology. The designed filter features high linearity, ultra-wide bandwidth, high roll-off speed and fully digital continuous tunability[4], offering a dynamically tunable solution for advanced communication applications.

II. CIRCUIT ANALYSIS AND IMPLEMENTATION

The overall structure of the tunable filter from 6GHz to 12GHz is shown in the Figure 2. It consists primarily of a 7th-order elliptic high-pass and low-pass filter connected in series. The capacitors in both the high-pass and low-pass filters are tunable respectively. By utilizing a lvshift circuit, the control signals are converted to +2.5V and -2.5V levels, thereby enhancing the linearity of the switch. Input and output are connected to the pads and digital control signals are followed by an inverter buffer for driving, with ESD protection implemented on all connections.

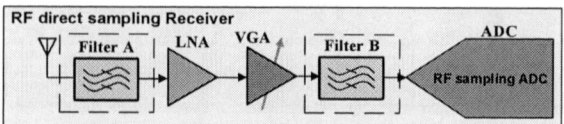

Fig. 1. RF direct sampling Receiver Block diagram

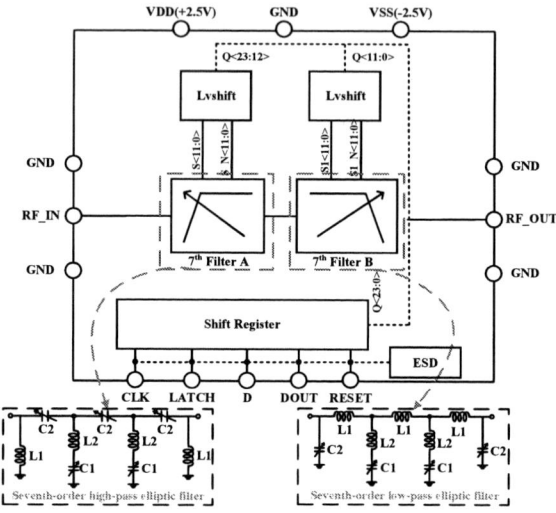

Fig. 2. Block diagram of a proposed tunable filter

The tunable bandpass filter proposed is composed of a high-pass and low-pass filter. Both the high-pass and low-pass filters utilize elliptical structures due to their fast roll-off rates. The frequency response of the low-pass elliptical filter is shown in Figure 3 and its expression is given by:

$$|H(j\omega)| = \frac{1}{\sqrt{1 + \varepsilon^2 R_n(\omega^2)}} \qquad (1)$$

$R_n(\omega^2)$ an n-th order elliptic rational function equation,

$$R_n(\omega) = \begin{cases} \dfrac{\omega(\omega_1{}^2 - \omega^2)\cdots(\omega_k{}^2 - \omega^2)}{(1 - \omega_1{}^2\omega^2)\cdots(1 - \omega_k{}^2\omega^2)}, n = 2k+1 \\ \dfrac{(\omega_1{}^2 - \omega^2)\cdots(\omega_k{}^2 - \omega^2)}{(1 - \omega_1{}^2\omega^2)\cdots(1 - \omega_k{}^2\omega^2)}, n = 2k \end{cases} \qquad (2)$$

where ε is the parameter related to in-band ripple, ω_k is the zero of $R_n(\omega)$. Each zero provides an attenuation of 20dB/decade, and the transition narrows as zeros increases.

The 6-bit variable capacitor uses MOSFETs as RF switches, connected in series to achieve variable capacitance. The schematic of 6bit Switched Capacitor Array is shown in Figure 4. To ensure optimal RF performance, MOSFETs in the RF switches are fabricated with minimum gate lengths. In

979-8-3503-6184-1/24 $31.00 © 2024 IEEE   238

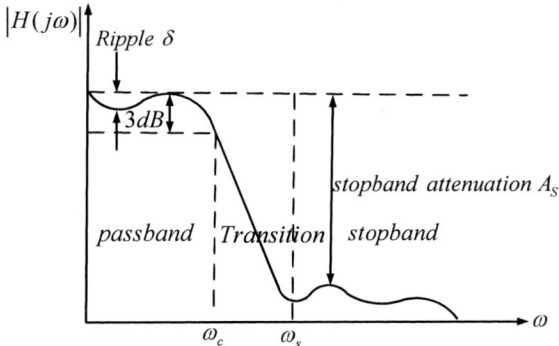

Fig. 3. Frequency response curve of a elliptic low-pass filter

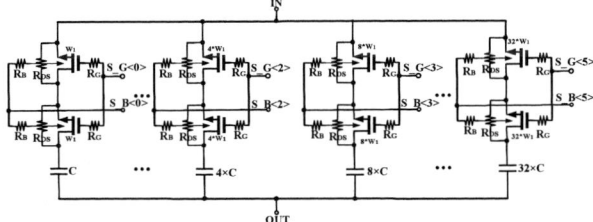

Fig. 4. The schematic of 6bit Switched Capacitor Array C1

addition, R_G and R_B are connected in series to the gate and substrate of a MOSFET, respectively. R_G effectively reduces the impact of control voltage switching on the RF signal and increase the linearity. While R_B can reduce signal leakage and crosstalk through the substrate. However, if R_G is too large,

it can increase the switching time of the MOSFET. Considering the speed and linearity, R_G and R_B are chosen to be 22kΩ. Furthermore, R_{DS} is placed between the source and drain to ensure uniform voltage distribution across the source-drain during off-state, which is 75kΩ.

The gate width of the MOSFET determines both its on-resistance (R_{ON}) when conducting and its off-capacitance (C_{OFF}), as depicted in the Figure 5. They determine the insertion loss and isolation between channels of the switch. Due to the inverse relationship between R_{ON} and C_{OFF}, as gate width increases, the R_{ON} rapidly decreases initially, with the rate of decrease gradually slowing down. Increasing the width of gate does not significantly contribute to the improvement of $R_{ON} \times C_{OFF}$. Considering layout constraints, a gate width of 20μm is chosen for the smallest RF switch capacitor.

Additionally, the optimization of input-output matching is required to reduce insertion loss. Figure 6 illustrates the ZIN Smith chart of the HPF with frequency and state variations. $Z = R + i(\omega L - 1/\omega C)$, The Smith chart ZIN matching can be adjusted by tuning the size of inductors, capacitors, and the width of MOS switches, as shown in Figure 6.

The values of capacitance and inductance can be obtained by extracting the Y-parameters and can be calculated using

$$Cap = \frac{imag(Y_{11})}{2\pi f}, Leff = \frac{imag(1/Y_{11})}{2\pi f} \quad (3)$$

$$Q_c = \frac{imag(Y_{11})}{real(Y_{11})}, Q_L = \frac{imag(1/Y_{11})}{real(1/Y_{11})} \quad (4)$$

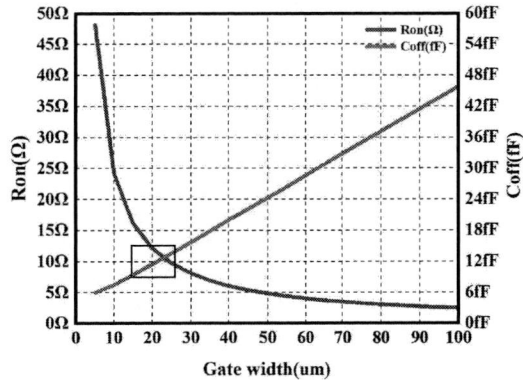

Fig. 5. Effect of Gate Width Variation on R_{ON} and C_{OFF}

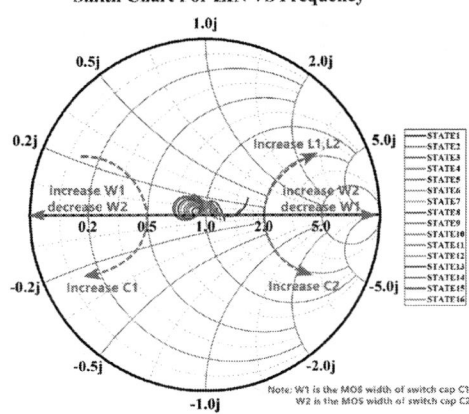

Fig. 6. Filter Smith Chart for ZIN Variation on Frequency and State

TABLE I. DEVICE VALUES IN THE CIRCUIT

Device values	HPF 6G~12G	LPF 6G~12G
C1(pF)	0.15~0.57	0.08-0.91
C2(pF)	0.3-2.6	0.04-0.16
L1(nH)	0.7	0.66
L2(nH)	0.54	1.2

Both the high-pass and low-pass filter are composed of two tunable capacitors and two fixed inductors, with the values of the capacitors and inductors specified in Table I.

III. SIMULATION RESULTS

The layout of the design is shown in Figure 7. The layout interconnects introduce parasitic resistance and capacitance, requiring the use of EM simulation to extract parasitic parameters and optimize them based on the Smith chart.

The results of the post-layout simulation are shown in Figure 8,9 and 10. The frequency and bandwidth can be continuously adjusted from 6GHz to 12GHz, with input and output return losses exceeding 10dB. Group delay characterizes the flatness of the phase response of a filter. It is less than 0.6ns within the frequency range.

Additionally, the linearity of the filter in the front-end of a system is crucial for ensuring optimal system performance. The Input P1dB is over 25dBm, while the Input IP3 is over 38dBm, which is shown in Figure 10.

979-8-3503-6184-1/24 $31.00 © 2024 IEEE

TABLE II.   COMPARISON OF THE PROPOSED TUNABLE INSET FILTERS WITH SIMILAR STATE-OF-THE-ART DESIGNS

Ref	Architecture	Tunable Range	Order	Roll-off Speed[1]	Insertion loss(dB)	BW range(Hz)
[6]	Waveguide	8.7GHz-9.6GHz	N/A	6%	4.1	76M-80M
[7]	GeTe switches	7.4GHz-8.1GHz	3	13.3%	3.2	500M
[8]	Varactor	1.1GHz-1.88GHz	3	13.3%	4.3-6.8	90M
[9]	Suspended stripline	3.7GHz-6GHz	2	25%	1.9-2.5	90M-130M
[10]	Ferrite	3.8GHz-5.2GHz	N/A	40%	1.6-3	500M
This Work	**Switched cap**	**6GHz-12GHz**	**7**	**10%**	**5**	**500M-4000M**

[1]Note: Roll-off speed is defined as: $\left|\frac{f_{Cutoff}-f_{3dB}}{f_{center}}\right|$. Cutoff frequency is defined as f_{Cutoff},

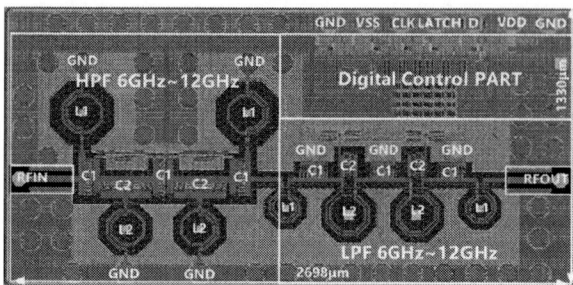

Fig. 7. Layout of the proposed 6GHz-12GHz tunable BPF

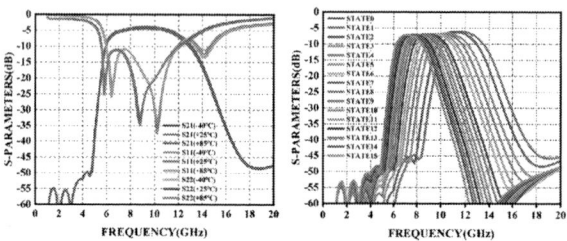

Fig. 8. EM simulated S21,S11,S22 of the proposed tunable BPF

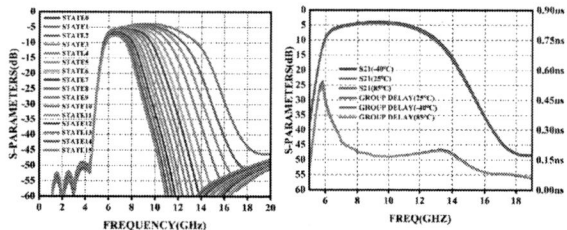

Fig. 9. EM simulated S21 and Group Delay of the proposed tunable BPF

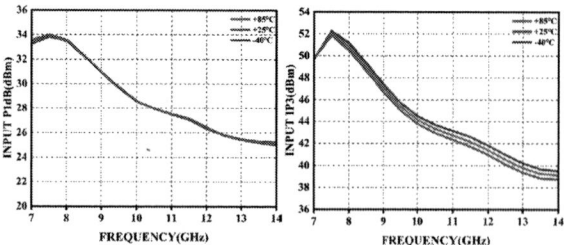

Fig. 10. EM simulated Input P1dB and IP3 of the proposed tunable BPF

A comparison between similar designs is shown in Table II. The proposed tunable bandpass filter in this paper offers significant advantages in terms of bandwidth, frequency range and roll-off speed, albeit at the expense of insertion loss.

IV. CONCLUSION

This paper proposes a design of a tunable bandpass filter based on 130nm CMOS, and compares the performance with those of filters of different architectures mentioned in papers of recent years. This filter achieves a higher frequency band with a maximum bandwidth of 4G at the cost of an insertion loss of about 5dB, a tunable range covering 6GHz-12GHz with flexible tunable of bandwidth and center frequency point, and high linearity and fast roll-off speed, and it can serve as a compact alternative to X-band filter arrays and find applications in satellite communications, radar, and so on.

REFERENCES

[1]   Y. Yang, H. Liu, Z. J. Hou, X. Zhu, E. Dutkiewicz and Q. Xue, "Compact On-Chip Bandpass Filter With Improved In-Band Flatness and Stopband Attenuation in 0.13-μm (Bi)-CMOS Technology," in IEEE Electron Device Letters, vol. 38, no. 10, pp. 1359-1362, Oct. 2017

[2]   Y. Guo, Q. Wang and J. Chen, "S-band Band-pass Filter by Using Integrated Passive Device Technology," 2022 23rd International Conference on Electronic Packaging Technology (ICEPT), Dalian, China, 2022

[3]   Zheng and W. Sheng, "Compact Lumped-Element LTCC Bandpass Filter for Low-Loss VHF-Band Applications," in IEEE Microwave and Wireless Components Letters, vol. 27, no. 12, pp. 1074-1076, Dec. 2017

[4]   P. R. Sriram, R. Bhaskaran, D. Bharathi, P. Srinivasan, N. Krishnan and H. U. Habiba, "Reconfigurable X-band bandpass filter using SIR with variable capacitor," 2017 Sixth International Conference on Future Generation Communication Technologies (FGCT), Dublin, Ireland, 2017, pp. 1-4

[5]   Deliyannis T, Sun Y, Fidler J K. Continuous-Time Active Filter Design [M]. Boca Raton: CRC Press, 1998.

[6]   B. A. Belayev, K. V. Lemberg and A. M. Serzhantov, "An X-band magnetically tunable bandpass filter based on novel waveguide cavity resonator," 2016 Asia-Pacific Microwave Conference (APMC), New Delhi, India, 2016

[7]   M. Wang, F. Lin and M. Rais-Zadeh, "An X-band reconfigurable bandpass filter using phase change RF switches," 2016 IEEE 16th Topical Meeting on Silicon Monolithic Integrated Circuits in RF Systems (SiRF), Austin, TX, USA, 2016, pp. 38-41

[8]   Z. Zhao, J. Chen, L. Yang and K. Chen, "Three-Pole Tunable Filters With Constant Bandwidth Using Mixed Combline and Split-Ring Resonators," in IEEE Microwave and Wireless Components Letters, vol. 24, no. 10, pp. 671-673, Oct. 2014,

[9]   C. -C. Cheng and G. M. Rebeiz, "High- Q 4–6-GHz Suspended Stripline RF MEMS Tunable Filter With Bandwidth Control," in IEEE Transactions on Microwave Theory and Techniques, vol. 59, no. 10, pp. 2469-2476, Oct. 2011

[10] H. Lin et al., "Integrated non-reciprocal dual H- and E-Field tunable bandpass filter with ultra-wideband isolation," 2015 IEEE MTT-S International Microwave Symposium, Phoenix, AZ, USA, 2015, pp. 1-4.

Analysis of Polar and Quadrature Digital Transmitters for Wi-Fi7 Applications

Lixuan Cao, Yun Yin*

State Key Laboratory of Integrated Chips and Systems, School of Microelectronics, Fudan University, Shanghai 201203, China

*Email: yiny@fudan.edu.cn

Abstract—This work compares digital transmitters (DTX) of polar and quadrature architectures for Wi-Fi7 applications in the aspects of digital front-end (DFE) and digital power amplifier (DPA). The main conclusions are as follows: (1) The Coordinated Rotation Digital Computer (CORDIC) module of polar DFE can only be placed after the up-sampling module, leading to higher power consumption. (2) Digital predistortion of the quadrature DFE is more complex. (3) The average output power of polar DPA is about 2.5 dB larger than that of the quadrature counterpart with the same EVM requirement. (4) The average efficiency of polar DPA is about 8% higher. (5) Polar DPA suffers from more severe memory effect.

Keywords—Wi-Fi7, polar, quadrature, digital transmitter.

I. INTRODUCTION

Wi-Fi7 supports extremely high data communication rate, which means the use of more complex signal modulation schemes (4096-QAM) and a wider bandwidth (320-MHz). Therefore, there is a pressing demand for broadband, high efficiency, and high linearity RF transmitters. A SCPA (switched-capacitor power amplifier)-based digital RF transmitter presents a viable solution [1][2][3][4]. In the realm of digital RF transmitters, polar and quadrature architectures stand as two mainstream options. Compared to the quadrature architecture, the polar digital power amplifier (DPA) achieves higher output power and average efficiency, but it necessitates a more intricate digital front-end (DFE) comprising additional components such as Coordinated Rotation Digital Computer (CORDIC) module, phase modulators, and AM/PM (amplitude modulation/phase modulation) synchronization circuitry. Given that selecting an appropriate architecture is a crucial step preceding RF transmitter circuit design, it holds significant research significance to evaluate the choice of RF transmitter architecture to best fit Wi-Fi7 protocol.

This work aims to compare the polar and quadrature transmitter architectures in two aspects: DFE and DPA. For the comparison of DFE, MATLAB Simulink is utilized to construct behavior-level models of both architectures for Wi-Fi7 application. As for the comparison of DPA, a reconfigurable SCPA circuit capable of operating in both polar and quadrature modes is designed to evaluate the performance. The results can serve as a valuable reference for architecture selection and offer some insights into the design of the Wi-Fi7 transmitter. Designers can take consideration of their past design experience to select the most appropriate transmitter architecture.

II. COMPARISON OF DFE

In this section, MATLAB Simulink is used to design the DFE models of polar and quadrature architectures for Wi-Fi7 applications. The architectures, quantization bit width sand sampling rates are compared. The analysis in this work mainly focuses on the 5925MHz ~ 7125MHz band.

A. Behavior-Level Model Design

1) Quadrature DFE

Assuming that there are only two sampling rates for the DFE input signal, $f_s = 160\text{MHz}$ for the signals with 20-MHz, 40-MHz and 80-MHz bandwidths and $f_s = 640\text{MHz}$ for the signals with 160-MHz and 320-MHz. The input signal sampling rate of DPA (also the output signal sampling rate of DFE) should be $1/N$ of the carrier frequency, where N is an integer. Considering that the DPA input signal sampling rate should not be too low or too high, it is selected as 1975MHz-2375MHz with $N=3$.

Accordingly, the implementation of up-sampling process can be designed. Firstly, the low-bandwidth signal is up-sampled by a factor of four to achieve a 640-MHz sampling rate equivalent to that of the high-bandwidth signal. Subsequently, all five bandwidth signals go through a Farrow fractional up-sampling module to generate the final output signal. This process is shown in Fig. 1(a). Noting that DFE of the quadrature architecture only does the upsampling and filtering, so Fig. 1(a) also shows the DFE of the quadrature architecture.

2) Polar DFE

Fig. 1(b)(c) shows the DFE of polar architecture, where the phase modulator module is omitted. What should be considered is the placement of CORDIC module. In Wi-Fi6 or earlier wireless communication protocols, CORDIC can be placed in the middle of the DFE, afterwards only the AM signal is up-sampled, while the PM signal is kept at low frequency, as shown in Fig. 1(b). Such a structure allows the CORDIC to work at low frequency and it only needs one up-sampling branch, reducing the overall power consumption.

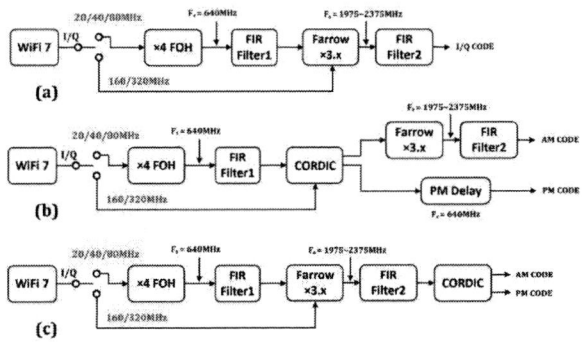

Fig. 1. The DFE of (a) quadrature, (b) middle-connected-CORDIC polar, (c) last-connected-CORDIC polar Wi-Fi7 DTX.

This work was supported by the State Key Laboratory of Radio Frequency Heterogeneous Integration under Grant 2023005.

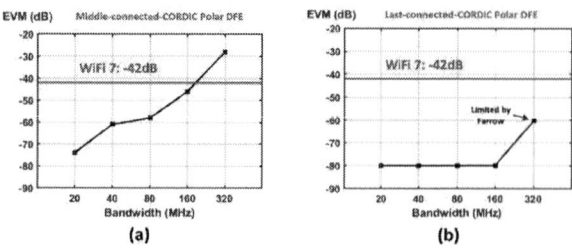

(a)   **(b)**

Fig. 2. EVM evaluation with 20-320MHz Wi-Fi7 4096QAM signals fed into (a) middle-connected-CORDIC and (b) last-connected-CORDIC DFE.

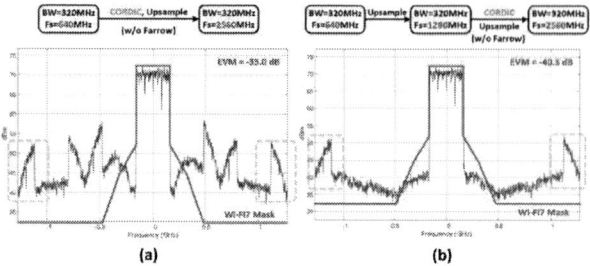

(a)   **(b)**

Fig. 3. Spectrum and EVM evaluation of the output signal of the polar DFE with CORDIC operating at (a) 640MHz and (b) 1280MHz, where the output sampling rate is set to 2.56GHz.

However, this structure cannot be applied to Wi-Fi7 signals. Fig. 2(a) shows the output EVM curve when the Wi-Fi7 4096-QAM signals with different bandwidths are fed into the DFE of Fig. 1(b). It can be seen that the output signal with 320-MHz bandwidth cannot meet the requirements of the Wi-Fi7 protocol (-42dB). Fig. 3 illustrates the spectrum and EVM of the output signal of the polar DFE with CORDIC operating at 640MHz and 1280MHz, where the output sampling rate is set to 2.56GHz to avoid additional distortion caused by the Farrow module. Although the performance is better after increasing the operating frequency of CORDIC to 1280MHz, the output signal spectrum still has large spur, which cannot meet the spectrum mask and EVM requirements. Meanwhile, in the actual design of DFE, it is indispensable to leave a certain margin for DPA distortion (e.g. EVM < -55dB for DFE). Therefore, it is necessary to further increase the operating frequency of CORDIC. Considering the complexity of the architecture, as well as the limitation of the operating frequency, it is the best choice to place CORDIC as the last stage and avoid upsampling the AM signals after CORDIC. Finally, the DFE of the polar architecture is shown in Fig. 1(c). Simulation result shows EVM of the output signal is less than -60dB, as in Fig. 2(b).

B. Comparison

In addition to the DFE architecture, other factors will be discussed below.

1) Quantization Bit Width

When the EVM of the output signals of polar DFE and quadrature DFE are equal, assuming -55dB, a quantization bit width of 12 is required for both of them. However, if the quantization bit width before CORDIC in polar DFE is increased to 16, the EVM will be reduced to -59dB, which indicates that the EVM of polar DFE is mainly limited by the quantization of the up-sampling part (i.e. the quadrature part). The results are presented in Fig. 4.

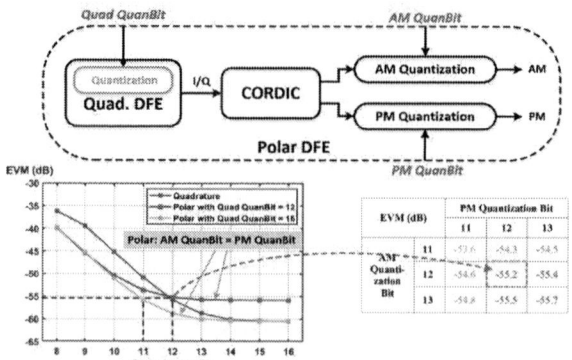

Fig. 4. The quantization process of the DFE and the EVM testing results.

2) Output Signal Sampling Rate

For DFE itself, the sampling rate does not limit its performance. However, due to the CORDIC module, the output signal bandwidth of polar DFE is much wider than that of quadrature DFE. Therefore, polar DFE needs a higher output signal sampling rate, which can improve the attenuation of sampling images.

3) Digital Pre-Distortion (DPD)

The Wi-Fi7 protocol supports very complex modulation schemes (4096-QAM), which necessarily requires the DPDs for linearization. Polar DFE usually needs 1-D DPDs, while quadrature DFE needs 2-D DPDs for good linearity, which is more complex.

III. Comparison of DPA

For the comparison of DPA in this section, a reconfigurable DPA is designed [1], which can work in polar or quadrature mode. Wi-Fi7 modulation signals are fed to this DPA, and the average output power, average efficiency, and EVM performance of the two modes are compared[2].

A. IQ-Reuse Reconfigurable SCPA

Fig. 5(a) illustrates the block diagram of the proposed 14 bits IQ-reuse reconfigurable SCPA in quadrature mode. While in polar mode, AM<13:0> = I<13:0> + Q<13:0> and the same PM-modulated-LO signals are sent to both LOI and LOQ ports. Its input code range is limited by $|I| + |Q| < 2^{13}$ for the quadrature mode and AM $< 2^{13}$ for the polar mode, as shown in Fig. 5(b).

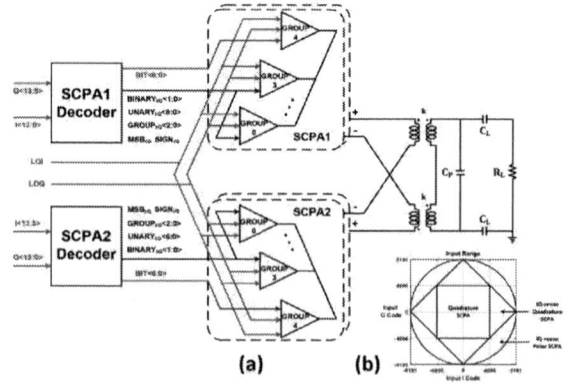

(a)   **(b)**

Fig. 5. (a) Block diagram of the proposed IQ-reuse reconfigurable SCPA. (b) Input range of quadrature, IQ-reuse-quadrature, polar SCPAs.

979-8-3503-6184-1/24 $31.00 © 2024 IEEE

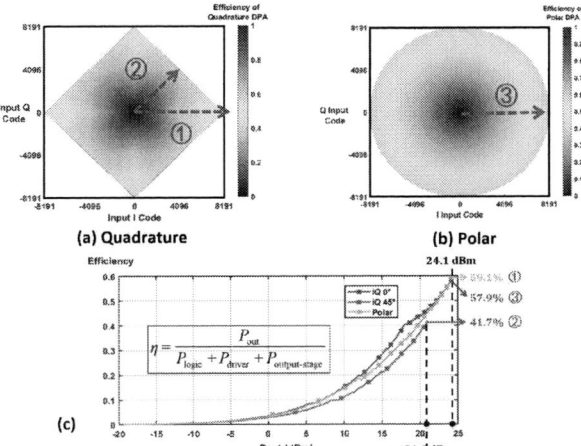

Fig. 6. Static efficiency map in IQ plane of (a) quadrature mode, and (b) polar mode. (c) Efficiency curve of direction ①~③.

B. Comparison

1) Static Efficiency

In order to capture the major differences in architecture comparison, only the power consumption of logic module, drivers and power stages is considered, as expressed in Fig. 6(c). Fig. 6 shows the static efficiency map of the SCPA. Input signals along the I/Q axis in the quadrature mode has similar static efficiency as the polar mode, while the static efficiency in the 45° direction is obviously reduced.

2) Average Pout, Average Efficiency and EVM

Using the static efficiency map, the average efficiency of dynamic modulation signals can be calculated. Here Wi-Fi7 MCS13 320-MHz modulation signal is fed into the SCPA. For the original signal, the output power and efficiency is quite low due to the high peak-to-average power ratio (PAPR). Thus, amplification and clipping for the signals is necessary. Fig. 7 shows the clipping operation on the input IQ plane, which can reduce the signal PAPR to increase the output power and efficiency, but it will bring distortion to the signals and deteriorate the EVM performance. Therefore, by amplifying the modulation signal with different multiples and clipping, different average output power, average efficiency and EVM performance can be obtained. The result is shown in Fig. 8, and some conclusions can be drawn:

- For the same -50dB EVM performance or the same 30% efficiency, the average output power gap between polar and quadrature SCPA is about 2.5dB. This gap is slightly different under different criteria.

- With the same 15-dBm P_{avg}, the efficiency of polar SCPA is 8% higher than that of quadrature SCPA, and the efficiency gap is larger for larger P_{avg}.

Therefore, polar DPA achieves better performance in terms of EVM, output power, and efficiency. However, the aforementioned experimental results are obtained by quasi-static calculation. In the practical design, due to the higher sampling rate and the need to process wider signal bandwidth (~GHz), polar DPA would suffer from more severe memory effect, its overall efficiency will be degraded.

IV. CONCLUSION

This work compares digital RF TX of polar/quadrature

TABLE I. COMPARISON OF WI-FI7 POLAR AND QUADRATURE DTX

Arch.	Polar	Quadrature
BB Arch.	CORDIC last-connected, complex	Simple
Quan. Bit	Same 12 bits, for -55dB EVM	
f_s	Higher, 3-5f_s	f_s
DPD	1-D DPDs	Complex 2-D DPDs
P_{avg}	2.5dB Higher (Same EVM and Eff.)	-
Eff.	About 8% Higher (P_{avg} = 15dBm)	-
EVM	Better	-
Others	Severe memory effect	IQ-mismatch

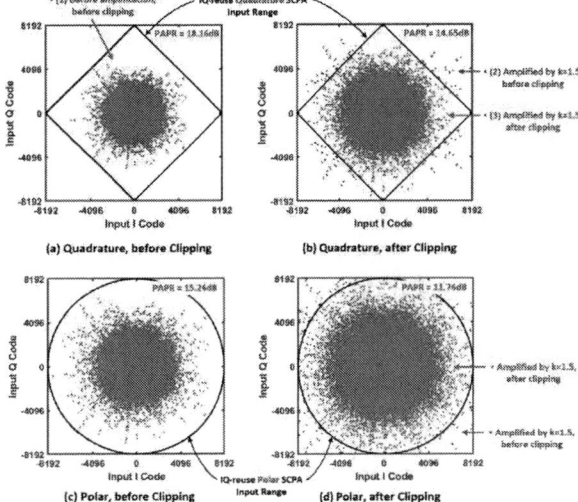

Fig. 7. Scatter diagrams of Wi-Fi7 MCS13 320-MHz signals for (a) quadrature SCPA w/o clipping and (b) w/clipping; (c) polar SCPA w/o clipping and (d) w/clipping.

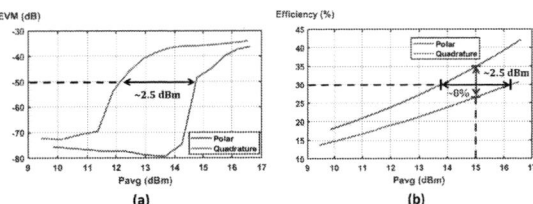

Fig. 8. (a) EVM vs. P_{avg} (b) Efficiency vs. P_{avg} of the IQ-reuse SCPA.

architectures for Wi-Fi7 applications. The main conclusions are as follows: (1) The CORDIC module of polar DFE can only be placed after the up-sampling module. (2) DPDs of the quadrature DFE is more complex. (3) The average output power of polar DPA is about 2.5 dB larger. (4) The average efficiency of polar DPA is about 8% higher. (5) Polar DPA would suffer from more severe memory effect. A summary of the comparison is presented in TABLE I.

REFERENCES

[1] Y. Li et al., "A 15-Bit Quadrature Digital Power Amplifier With Transformer-Based Complex-Domain Efficiency Enhancement," in IEEE Journal of Solid-State Circuits, vol. 57, no. 6, pp. 1610-1622, June 2022.

[2] Y. Li et al., "A Quadrature Digital Power Amplifier With Wide Efficiency Enhancement Coverage and High Dynamic Power Range," in IEEE Journal of Solid-State Circuits, vol. 59, no. 7, pp. 2133-2144, July 2024.

[3] Y. Yin, H. Xu, "Digital-intensive RFIC design techniques for transmitters in ISSCC 2023," Journal of Semiconductors, 44(4), 2023, 1-2.

[4] Chan C H, Cheng L, Deng W, et al. Trending IC design directions in 2022. J Semicond, 2022, 43, 071401.

Atomically Thin Graphene Nanopore based MEMS Iontronic Devices for Sensing, Separation, and Energy Applications

Luda Wang*[1,2,3], Ruiyang Song[1], Ningran Wu[2]

[1] National Key Laboratory of Advanced Micro and Nano Manufacture Technology,
School of Integrated Circuits, Peking University, Beijing 100871, P. R. China
[2] Academy for Advanced Interdisciplinary Studies, Peking University, Beijing, 100871, China
[3] Beijing Advanced Innovation Center for Integrated Circuits, Beijing 100871, P. R. China

* Email: luda.wang@pku.edu.cn

Abstract—**Inspired by the fine-tuned ionic transport across biological nanochannels, two-dimensional atomically thin graphene nanopores are designed as micro-electromechanical system (MEMS) iontronic devices to study the ionic transport under nanoconfinement. Understanding the nanofluidic transport process at such thinnest limit endows the ability to control ionic transport at the nanoscale, and paves ways for a range of practical applications, such as biosensors, membrane separation, and energy conversion. In this review, we focused on the applications of graphene nanopore-based MEMS iontronic devices, highlighting their potential in nanofluidic iontronics.**

Keywords—*2D material, nanopore devices, MEMS, ionic transport, nanofluidic iontronics*

I. INTRODUCTION

Biological ion channels demonstrate remarkable ionic transport capabilities facilitated by the exquisite structure and specific functional groups. These biological nanochannels possess a remarkable ability to precisely regulate the transport of diverse ions [1]. This capability forms the essential foundation for diverse physiological processes, encompassing crucial functions such as neural signaling, muscular contraction, cellular communication, etc.

However, the applications of biological nanochannels in phospholipid bilayers have been limited by their low mechanical strength and unstable physical and chemical properties. Inspired by the biological channels, the artificial solid-state nanochannels based on micro-electro-mechanical systems (MEMS) technology have been designed as iontronic devices to quantitatively study ionic transport at the nanoscale. Solid-state nanochannels offer advantages over biological counterparts owing to their stable physical and chemical properties, which are less affected by pH, temperature, and other environmental factors. Additionally, their pore size and surface functional groups can be precisely controlled, allowing for effective modulation of ionic transport. Therefore, silicon-based MEMS technology has enabled micro-nano fluidic devices into various applications, including biosensing and separation processes [2].

Despite much effort, achieving angstrom-scale sizes and transport behaviors comparable to biological nanochannels in artificial MEMS devices remains challenging. Atomically thin two-dimensional (2D) materials have been utilized to investigate ionic transport in nanofluidics owing to their compatibility with MEMS fabrication, ultrathin ionic pathways, and controllable angstrom-scale pore structures [3]. Among them, graphene is considered as a promising platform

for developing iontronic devices under extreme confinement because of ultralow flow resistance, stable physical and chemical properties, as well as the simplicity of structure and composition.

Hence, the atomically thin graphene nanopores with angstrom-scale structure and precisely-controlled functional groups are designed as MEMS iontronic devices. These nanopore devices can show excellent sensing accuracy and ion permeability owing to the atomic thickness nature. In this paper, we reviewed the applications of graphene nanopore based MEMS iontronic devices, particularly in the fields of biosensing, ionic separation, and energy conversion, highlighting their significant potential in nanofluidic iontronics (Figure 1).

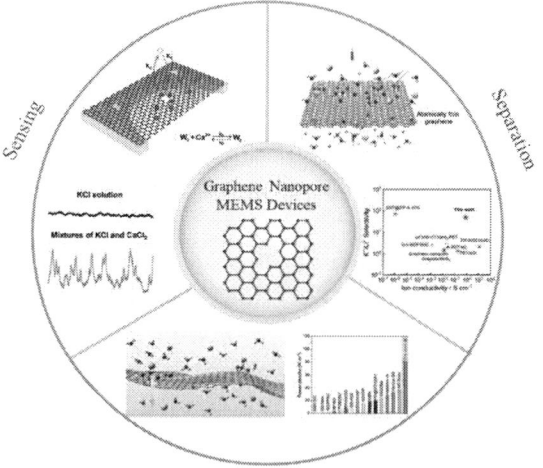

Fig. 1. Atomically thin graphene nanopore based MEMS iontronic devices for applications in sensing, separation, and energy conversion.

II. GRAPHENE NANOPORE DEVICE FOR SENSING

Atomically thin graphene nanopores demonstrate high spatial resolution and sensitivity in single-molecule sensing, and can be utilized in DNA sequencing. The working principle is based on measuring ionic current change due to the blockage of molecules. Achieving these capabilities relies on accurately detecting ionic transport through graphene nanopores. However, the low-frequency flicker noise in the current raises challenges to electrical read-out, and limits the signal-to-noise ratio of nanopore devices. Therefore,

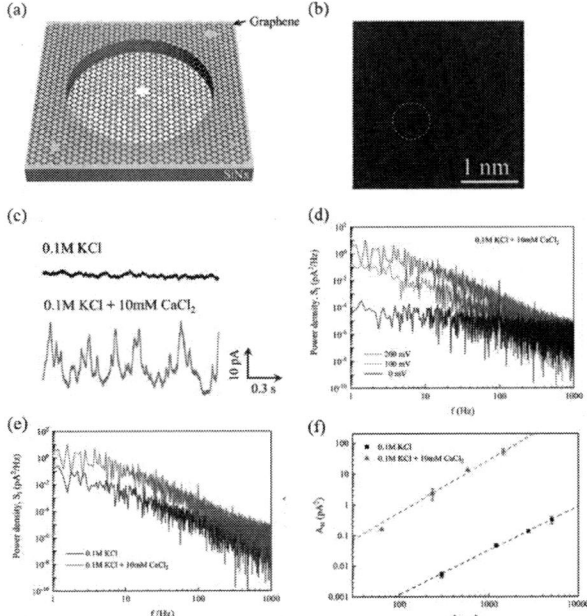

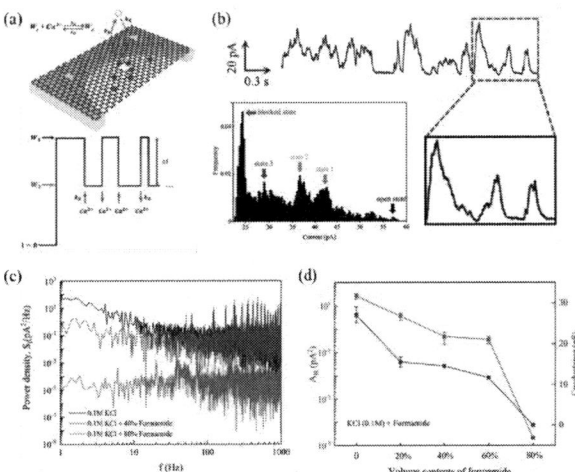

Fig. 1. The presence and modulation of low-frequency flicker noise in graphene nanopore. (a) Schematic of the statistically isolated intrinsic graphene nanopore on the SiNx substrate. (b) Annular dark-field scanning transmission electron microscopy image of the individual graphene nanopore. (c) Ionic current traces of the nanopore device in KCl solution and mixtures of KCl and CaCl₂. (d) The power spectra of current under different applied voltages in mixture solutions indicate the presence of low-frequency flicker noise. (e) The comparison of power spectra in 0.1 M KCl and mixture solutions. (f) Relationship between the amplitude of the noise power spectrum A_H and current (log-log scales). Adopted from Ref. [4].

comprehending the physical mechanisms behind low-frequency flicker noise and effectively regulating it are crucial for increasing the temporal resolution of advanced graphene nanopore devices in molecular sensing fields.

By MEMS techniques, the graphene nanopore iontronic devices were fabricated on the perforated silicon nitride (SiNx) substrate [4]. The intrinsic defects of graphene synthesized via chemical vapor deposition (CVD) were statistically isolated by SiNx substrate with a 60 nm aperture, as shown in Fig. 1. (a). The annular dark-field scanning transmission electron microscopy (STEM) was used to characterize the individual graphene nanopores (Fig. 1. (b)).

To investigate the presence and modulation of low-frequency flicker noise in ionic current through atomically thin nanopores, the ionic current was measured by the patch-clamp amplifier, and the power spectra were analyzed. The current traces in mixtures of KCl and CaCl₂ exhibited stronger fluctuations compared to those in KCl alone, as shown in Fig. 1. (c). Fig. 1. (d) illustrates the power spectra of current noise under various applied voltages in mixture solutions, confirming the presence of low-frequency noise in graphene nanopore iontronic devices. The Hooge's relation was applied to calculate the amplitude A_H of noise, and was expressed as

$$S(f) = \frac{A_H}{f^\gamma} = \frac{\alpha I^2}{f^\gamma} \quad (1)$$

The millimole concentration of calcium ions in mixture solutions can induce a larger amplitude of the power density in the low-frequency range, which is nearly 1 to 2 orders of magnitude higher than that in KCl, as shown in Fig. 1. (e-f).

Fig. 2. The mechanism and reduction of low-frequency flicker noise in graphene nanopore. (a) Schematic illustration of the reversible binding of calcium ions, causing the ionic current fluctuations and low-frequency noise. (b) The typical current traces and all-point histogram in mixture solutions, illustrating the transitions between open and blocked states. Peaks in the histogram represent distinct pore states. (c) The reduction of low-frequency noise in graphene nanopore devices based on the suppression of electrostatic interactions. The power spectra of the noise in 0.1 M KCl and mixtures of formamide. (d) The amplitude of the noise A_H and ionic conductance in 0.1 M KCl and the mixtures of formamide. Adopted from Ref. [4].

To illuminate the physical mechanisms of current fluctuations and flicker noise in nanopore devices, the kinetic analysis of transitions between nanopore states was analyzed, as shown in Fig. 2. (a). The multivalent ions were accumulated on the nanopore surface due to the electrostatic interaction. Then the reversible adsorption-desorption of calcium ions at surface ionizable sites caused the surface charge fluctuations and transitions between different conductance states. These transitions accounted for the fluctuating of ionic current and low-frequency noise observed experimentally. Fig. 2. (b) exhibited recorded current traces and all-point histogram in mixture solutions. Peaks in the histogram represent distinct pore states, illustrating the transitions between open and blocked states.

The flicker noise due to surface charge fluctuations can be suppressed by adding organic solvents with high dielectric constant into electrolyte solutions. Fig. 2. (c) showed the power spectra of the low-frequency noise in 0.1 M KCl and mixtures of formamide. The noise amplitude A_H decreased significantly as the formamide content increased from 0% to 80%, as shown in Fig. 2. (d). Notably, the presence of 40% formamide reduced the noise amplitude by two orders of magnitude compared to pure KCl solution, guiding the design of nanopore sensors with high signal-to-noise ratios by weakening the electrostatic interactions.

III. GRAPHENE NANOPORE DEVICE FOR IONIC SEPARATION

Atomically thin graphene is considered as a promising platform for ionic separation thanks to the ultra-low transport barrier and the tunable surface properties, which has the potential to mimic the ultrahigh ion permeability and selectivity of biological ion nanochannels. By the precisely control of pore size and design of charged functional groups, the graphene nanopore devices can be employed as iontronic filters to separate diverse ions with different sizes and charges efficiently.

979-8-3503-6184-1/24 $31.00 © 2024 IEEE

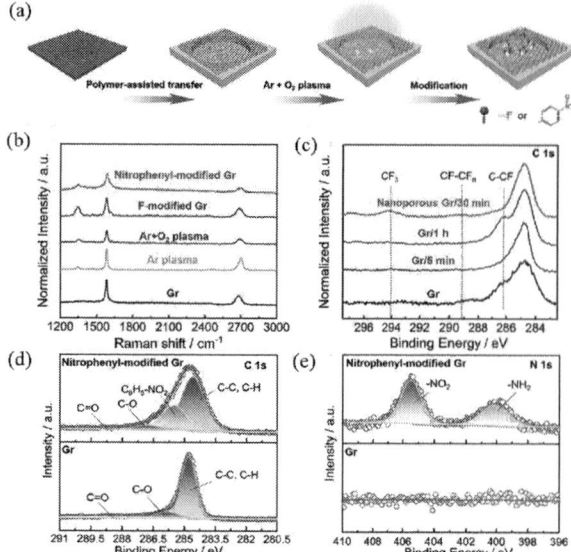

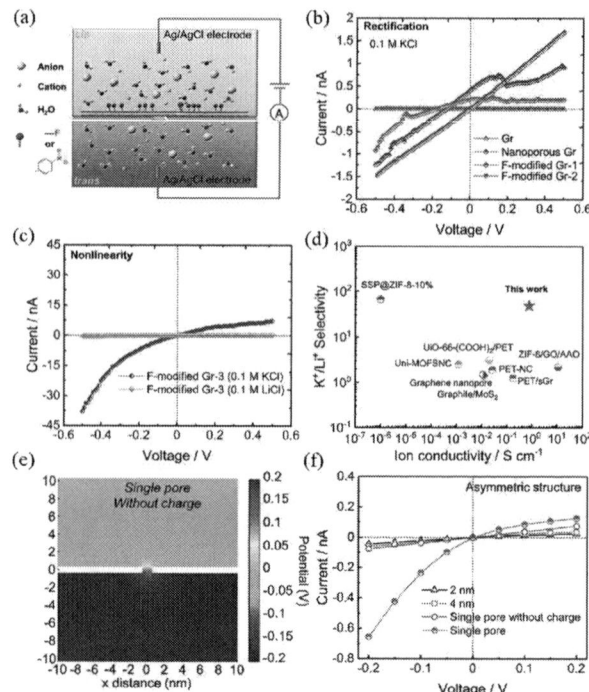

Fig. 3. The fabrication and characterization of functionalized graphene nanopore devices. (a) Fabrication processes of controlled graphene nanopores and modification of devices. (b) Raman spectra of pristine double-layer graphene and modified graphene. (c) X-ray photoelectron spectroscopy (XPS) of the C 1s core-level spectra of pristine double-layer graphene and graphene reacted with AgF. (d-e) XPS of the C 1s and N 1s core-level spectra of pristine double-layer graphene and nanoporous double-layer graphene functionalized by 4-nitrophenyldiazonium tetrafluoroborate (4-NBD). Adopted from Ref. [5].

In the conventional method, the graphene nanopore devices were fabricated by the direct Au ion bombardment, yielding that the K^+/Na^+ selectivity ratio was only nearly 1.1 while it is about 10^4 for biological channels. Through controllable etching and covalent functionalization, the modified graphene nanopore devices exhibited asymmetric ion transport behaviors and can effectively sieve monovalent ions, with a selectivity ratio of approximately 48.6 for K^+ over Li^+ [5].

The fabrication processes of functionalized graphene nanopore devices for iontronic separation are shown in Fig. 3. (a). The CVD graphene on copper foil was transferred onto SiNx using the polymer-assisted transfer method. Then the nanopore array was prepared by the coordinated approach involving "two-step" etching and covalent modification techniques. The defect sites were initially created in the graphene lattice by Ar plasma, which was subsequently enlarged to nanopores using O_2 plasma. At last, the covalent functionalization of nanopore devices was performed in solution through fluorination and nitrophenyl diazonium chemistry, respectively. Since the iontronics response of nanopore devices strongly depends on the atomic configuration, comprehensive structural characterizations were performed, including Raman spectra and X-ray photoelectron spectroscopy (XPS), to verify the formation of chemical bonds and the existence of functional groups, as shown in Fig. 3. (b)-(e).

The ionic transport and separation performance were characterized by the electrical measurement of ionic current in a custom-designed system consisting of two fluidic cells filled with electrolyte solution by an electrochemical analyzer, as shown in Fig. 4. (a). After modification, the ionic current exhibited pronounced ionic current rectification behavior (Fig. 4. (b)), arising from the asymmetric attachment

Fig. 4. Ionic transport and separation behavior of functionalized graphene nanopores. (a) The experimental setup of ionic transport measurements. (b) Current–voltage curves of graphene nanopore devices, showing the rectification behavior. (c) Current–voltage curves of graphene nanopore devices in KCl and LiCl solutions. (d) Comparison of the K^+/Li^+ selectivity with reported results. (e) The numerical calculations of electrostatic potential in graphene nanopores. (f) The calculated ionic current of asymmetrical graphene nanopores, showing the effect of charge. Adopted from Ref. [5].

of functional groups at the entrance of the graphene nanopores. The ionic conductance of the nanopore device in 0.1 M KCl is 20 nS. However, the conductance is extremely low at 0.43 nS in 0.1 M LiCl (Fig. 4. (c)). This demonstrated effective sieving capability for K^+/Li^+ with a selectivity ratio of ~48.6, exhibiting high selectivity and conductivity, as shown in Fig. 4. (d). Based on the theoretical simulation, the electrostatic potential and ionic current were calculated to show the key role of electrostatic interactions between ions and charged nanopores, as shown in Fig. 4. (e) and (f).

IV. GRAPHENE NANOPORE DEVICE FOR ENERGY CONVERSION

Besides ionic separation, the 2D material nanopore device is an advanced platform for osmotic energy conversion. The osmotic energy can be harnessed from the salinity gradient and electrochemical potential difference across an ion-selective permeable membrane separating seawater and freshwater. The precisely modified graphene nanopores can exhibit minimal flow resistance and high anion/cation selectivity, making them highly suitable for osmotic energy conversion technologies.

In conventional modification methods, there are three primary challenges of graphene nanopore device for osmotic power conversion. Firstly, the presence of random reaction sites can decrease the efficiency of targeting specific locations for modification. Secondly, stochastic defects arising during functionalization can compromise the ion selectivity. Lastly, the negatively charged nature of nanopores presented an energy barrier to attach more negatively charged functional groups.

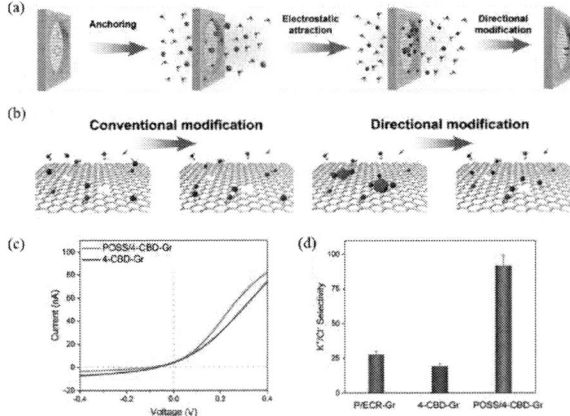

Fig. 5. The fabrication and characterization of directionally modified graphene nanopore devices. (a) Schematic diagram of the fabrication process of directionally modified graphene nanopores. (b) Comparison of functionalization processes between conventional modification and directional modification. (c) Current–voltage curves of directionally modified graphene nanopores and conventionally modified graphene nanopores at 100-fold concentration gradient. (d) Comparison of ion selectivity of two methods at 100-fold concentration gradient. Adopted from Ref. [6].

To overcome these obstacles, the electrostatic-directional method was designed to position functionalized groups near graphene nanopores effectively [6]. Initially, octa-ammonium-polyhedral-oligomeric-silsesquioxane (POSS, positively charged) was attracted to the graphene nanopores via electrostatic forces, marking their locations (pre-anchor process). Subsequently, the positively charged POSS facilitates the attraction of negatively charged 4-carboxybenzenediazonium tetrafluoroborate (4-CBD) into the nanoconfined space for subsequent functionalization. This approach ensures that the 4-CBD reactant accumulates primarily near the pores, enabling directional modification with enhanced efficiency and achieving a controllable surface charge density around the nanopore (Fig. 5. (a)). The comparison of functionalization processes between conventional modification and directional modification is shown in Fig. 5. (b). The current-voltage curves of directional modified devices (POSS/4-CBD-Gr) exhibited increased nonlinearity compared to conventionally modified membranes (4-CBD-Gr) (Fig. 5. (c)), indicating the increased modification efficiency and negative surface charge. Furthermore, the cation selectivity was 4 times larger than that of 4-CBD-Gr, as shown in Fig. 5. (d).

Based on this novel method, the performance of POSS/4-CBD-Gr was investigated in salinity gradient energy harvesting. The ionic conductance of POSS/4-CBD-Gr device showed non-linear behavior at low ionic concentrations, deviating from a linear relationship (Fig. 6. (a)). Additionally, it displayed nonlinear current-voltage curves across a wide range of KCl concentrations (Fig. 6. (b)), suggesting a high surface charge density and significant electrostatic interactions between nanopores and ions. The directionally functionalized graphene nanopore devices demonstrated significantly improved energy output, achieving a power density of 81.6 W/m^2 and an high energy conversion efficiency of 35.4%, as shown in Fig. 6. (c). The osmotic energy conversion performance of reported graphene-based devices are summarized in Fig. 6. (d). Remarkably, the directionally modified graphene nanopore devices demonstrated superior power density compared to other graphene-based devices, exhibiting the potential of modified

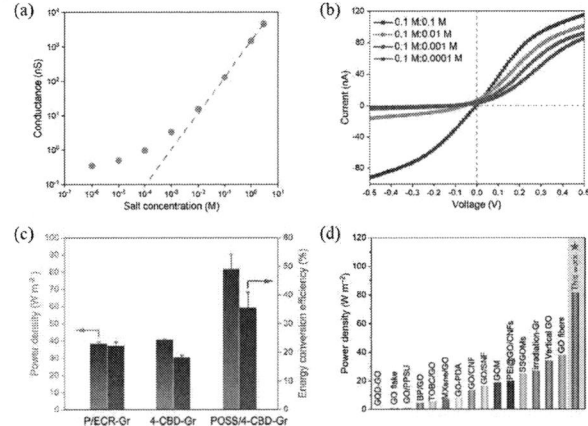

Fig. 6. Osmotic energy conversion by directionally modified graphene nanopore devices. (a) Transmembrane ionic conductance as a function of KCl concentration. (b) Current–voltage curves of directionally modified graphene nanopore devices at different concentration gradients. (c) Power density and energy conversion efficiency comparison between different methods at 100-fold concentration gradient. (d) Summary of the osmotic power performance of the state-of-the-art results. Adopted from Ref. [6].

graphene nanopores as iontronic devices for blue energy harvesting.

V. CONCLUSION

Thanks to the ultra-low flow resistance and tunable surface properties, graphene nanopore iontronic devices based on MEMS technology are extensively applied in high-resolution sensing, highly selective ionic separation, and efficient energy conversion. By design of nanopore structure and specific functional groups, the ionic transport behaviors under nano-confinement were controlled, which will promote the development of MEMS-based nanofluidic iontronics technologies.

ACKNOWLEDGMENTS

This work was financially supported by the National Natural Science Foundation of China (NSFC Nos. 62274004, 62004004, and T2188101) and the China Postdoctoral Science Foundation (No. 2022M710184).

REFERENCES

[1] Y. Hou, Y. Ling, Y. Wang, M. Wang, Y. Chen, X. Li, and X. Hou, "Learning from the brain: bioinspired nanofluidics," J. Phys. Chem. Lett., vol. 14, pp. 2891, 2023.

[2] M. Tilli, M. Paulasto-Krockel, M. Petzold, H. Theuss, T. Motooka, V. Lindroos, Handbook of Silicon Based MEMS Materials and Technologies (Third Edition), Elsevier, 2020.

[3] L. Wang, M. Boutilier, P. Kidambi, D. Jang, N. Hadjiconstantinou, R. Karnik, "Fundamental transport mechanisms, fabrication and potential applications of nanoporous atomically thin membranes," Nat. Nanotechnol., vol. 12, pp. 509, 2017.

[4] R. Song, H. Zeng, S. Zhang, Y. Wang, X. Han, X. Chen, L. Wang, "Low-frequency flicker noise in stochastic ionic transport across atomically thin graphene nanopores," Cell Rep. Phys. Sci., vol. 4 (1), pp. 101210, 2023.

[5] L. Guo, Y. Liu, H. Zeng, S. Zhang, R. Song, J. Yang, X. Han, Y. Wang, L. Wang, "Covalently functionalized nanopores for highly selective separation of monovalent ions," Adv. Mater., vol. 36, pp. 2307242, 2023.

[6] Y. Liu, S. Zhang, R. Song, H. Zeng, L. Wang, "Preanchoring enabled directional modification of atomically thin membrane for high-performance osmotic energy generation," Nano Lett., vol. 24(1), pp. 26, 2024.

Systems-on-Chips for Invasive Brain-Computer Interfaces: Challenges and Opportunities

Jie Yang, Member, IEEE, and Mohamad Sawan, Life Fellow, IEEE

CenBRAIN Neurotech, School of Engineering, Westlake University, Hangzhou, Zhejiang, China, 310030

Email: {yangjie, sawan}@westlake.edu.cn

Abstract—Invasive brain-computer interface (BCI) technology has recently gained significant attention due to its vast potential in treating neurodegenerative diseases and aiding in impairment rehabilitation. However, mainstream invasive BCI systems suffer from high power consumption, bulky setups, and inadequate channel capacity, which fail to meet the demands of implantable medical devices such as closed-loop neuromodulation, minimally invasive procedures, full implantation, and long-term operation. Systems-on-Chips (SoCs) play a critical role in BCI technology, being pivotal for the miniaturization and enhancement of BCI systems. This paper focuses on SoCs-based closed-loop neuromodulation and high-throughput neural recording, exploring the design challenges and opportunities. Also, we discuss the future perspectives and conclude the paper.

Keywords—*Brain-computer interface, System-on-chip, Integrated circuits, Neuromodulation, Neural recording, Neurostimulation.*

I. INTRODUCTION

Brain-computer interfaces (BCIs) create a direct, bidirectional communication channel between the brain and external devices. Invasive BCIs, due to their proximity to neurons, offer higher spatial and temporal resolution, making them more suitable for clinical applications compared to non-invasive BCIs. As shown in Fig. 1, these invasive systems can send stimuli into the brain to modulate neural activity for treating brain diseases [1] or provide sensory input to restore functions such as vision [2], hearing, and tactile sensation [3]. More importantly, invasive BCIs can record brain signals and decode human intentions, which is critical for functional rehabilitation in conditions such as spinal cord injuries [4] or speech impairments [5]. Typical applications of invasive BCI systems are illustrated in Fig 1. However, current invasive BCI systems are still bulky and power-hungry, primarily due to the limited degree of integration. Consequently, these systems are not widely available for a larger patient population and are often confined to laboratory environments.

Since 2010, the rapid development of BCI applications has driven significant efforts to integrate the circuit components of BCI systems onto a single silicon die, thereby reducing system size and power consumption. This integration has resulted in several notable advancements. For instance, system-on-chip (SoC) solutions have been developed that incorporate neural signal acquisition, microcontroller units (MCUs), and stimulation functions for closed-loop neural modulation aimed at epilepsy suppression [6]. Advancements, including high-throughput recording [7], data compression [8], [9], and low-power high-bandwidth wireless transmission techniques [10], have been achieved. As BCI systems continue to transition to chip-level integration, they are becoming increasingly smaller and more energy-efficient [11]. This evolution is critical for the development of miniaturized, long-lasting BCI devices. However, despite these advancements, numerous challenges need to be addressed to fully realize their potential.

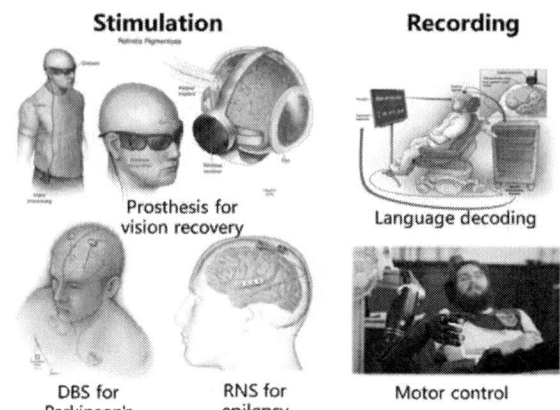

Fig. 1. Representative BCI applications.

We report in this paper ICs for closed-loop neuromodulation and high-throughput neural recording, explore the main design approaches, and highlight the necessary future adjustments and innovations. The remaining sections of this paper are organized as follows: Section II and Section III describes the design opportunities and challenges in closed-loop neural modulation systems, and high-throughput neural signal recording, respectively. Section IV give the future perspective and we will conclude the paper with a summary.

II. CHALLENGE AND OPPORTUNITY FOR NEUROMODULATION

Traditional neuromodulation devices, such as deep brain stimulation (DBS), vagus nerve stimulation (VNS) and responsive neural stimulation (RNS) are proven effective for various brain disorders as shown in Fig. 2(a), their implantation methods are illustrated in Fig. 2(b). These devices are surgically implanted in the chest and connected to electrodes placed near the target brain regions. They can deliver continuous or periodic stimulation to these regions to modulate abnormal neural activities, alleviating symptoms such as tremors, rigidity, and other related conditions. Despite their clinical success, these devices face significant challenge.

Their inability to adapt to dynamic neural activity or the changing needs of patients limits their capacity to provide personalized and optimal treatment outcomes. Moreover, as shown in Fig. 2(c), continuous nerve stimulation can lead to habituation and neurochemical changes, reducing treatment efficacy and raising safety concerns. Side effects such as dyskinesia have frequently been reported in epilepsy and Parkinson's patients due to continuous stimulation. Closed-loop neuromodulation, as illustrated in Fig. 2(d), which aims to establish a neural activity monitoring, signal processing, and stimulation loop is believed to provide adaptive and on-demand neurostimulation. A typical example is RNS for epileptic seizure detection [12]. It monitors brain waves at the

979-8-3503-6184-1/24 $31.00 © 2024 IEEE

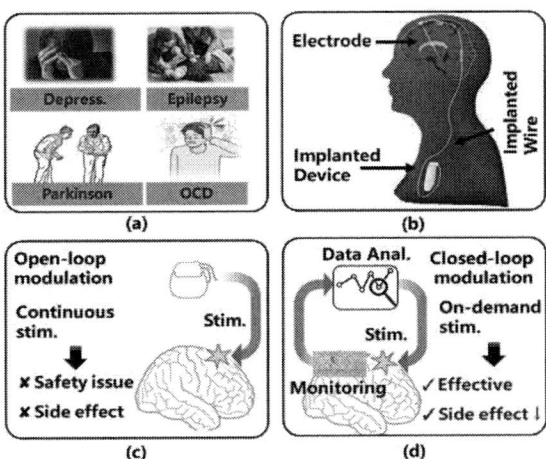

Fig. 2. Application, implantation form and modalities of neuromodulation.

TABLE I. CLOSED-LOOP NEUROMODULATION ICs

Work	JSSC '22[14]	JSSC '23[15]	JSSC '22[16]	TbioCAS '23[22]	ISSCC '24 [23]
Process (nm)	28	40	65	55	55
Area (mm²)	0.1	1.96	8	5.06	0.98
Classi-fier	LR	SVM	Neural-Tree	Corse:BNN Fine: CNN	CNN
Feature	Band power	Energy	Band power, phase	Raw-Data	Hybrid Feature
Accur-cy	97.9%, 98.2%	92% 99.1%	94%, 98.9%	99.2%, 98.1%	98.8% 94%
Power	1.5µW	2.31 mW	453µW	143µW	25.8µW

seizure focus and, once unusual electrical activity that can lead to a seizure is detected, stimulation is provided to help brain waves return to normal state, thereby easing or eliminating the symptoms. Closed-loop neuromodulation is regarded as the next-generation neuromodulation approach to address the abovementioned challenges.

To truly close the loop, the detection of individualized abnormal neural activity requires high sensitivity and specificity. This, in turn, necessitates the generalization of the classification model, making it power and computationally demanding compared to the capabilities of an implantable chip [13]. Table I lists recent efforts made to design ASICs to tackle this challenge. One approach is to implement biomarker extraction and classical machine learning classifiers, such as logistic regression (LR) [14], support vector machine (SVM) [15], decision trees [16], or their variants, in an application-specific manner to reduce the power consumption of on-device microprocessors and provide more computing power. Recently, neural networks have garnered significant attention in the field of closed-loop neuromodulation, as they are believed to offer enhanced accuracy and adaptability for individualized neuromodulation [17]. However, the increased power and computational demands of these networks exacerbate the conflict between the limited on-device power budget and the required computational load. Algorithm-hardware co-design is believed to be a promising methodology to address this issue [18] [19] [20]. Lightweight and compact neural networks that with reduced parameters and computations for various biomarker detection have been proposed [21], and low-power ASIC architectures that leverage the unique characteristics of electrophysiology signals have been implemented [22]. These advancements have reduced the power consumption of state-of-the-art biomarker detection is reduced to the level of hundreds of microwatts. Notably, low-shot learning and corresponding hardware is emerging in the field to better deal with patients specific model learning [23] [24].

The examination of neural activity during stimulation poses another challenge for closed-loop neuromodulation. The stimulation artifact exhibit as a short, high-amplitude peak followed by a slow, exponential decay which superimposed on the neural activity. It can saturate the front-end amplifier

and induce unwanted frequency bands to the band of interest for biomarker diagnosis. Front-end design including increasing dynamic range, artifact subtraction and fast recovery [25].

Despite substantial advancements in various AI algorithms and specialized low-power ASICs, the implementation of closed-loop neuromodulation in clinical practice has been slow. This is primarily due to the challenges posed by the lack of interpretability of artificial intelligence in biomedical devices, which complicates clinical application. Combining the latest advancements in model generalization with clinically accepted interpretability is thus a crucial focus area. It is essential to ensure that AI models not only perform effectively but are also comprehensible and trusted by clinicians to achieve successful implementation in neuromodulation.

III. CHALLENGE AND OPPORTUNITY FOR HIGH-THROUGHPUT NEURAL RECORDING

Advanced BCI decoding relies on high-throughput intracranial signals to increase the communication bandwidth between the brain and the external device. Early practice utilizes medical instruments with a few channels for directional cursor control [26], then dedicated BCI instruments of over hundred channels for complex motor control [27]. Recently, with over thousands of channels, Neuralink has demonstrated high efficiency and smooth control of complex games. Additionally, as neural decoding finds more applications in rehabilitation, there is an increasing demand for fully wireless implanted systems. Fig. 3 illustrates the trends of increasing channel count and decreasing system dimensions. However, the superposition of high-throughput and fully implantation pose two challenges for the design of neural recording ICs.

The first challenge brought by high throughput is the rapid increase in data rate for wireless transmission. For a typical 256-channel wireless recording systems with standard sampling, the data transmission rate will over 50Mbps which exceeds the capability of low-power transmission protocols such as Bluetooth and UHF [28]. On-chip data compression is a feasible approach to reduce the bandwidth requirement. Compressive sensing is widely used in neural recording [8] [29] which is able to compress the signal more than 10 times, however the reconstruction process is cumbersome, affecting

TABLE II. HIGH CHANNEL COUNT NEURAL RECORDING ICS

Work	VLSI '21 [7]	CICC '24 [30]	JSSC '20[32]	JSSC '18 [33]	JSSC '17[34]
Process (nm)	65	40	180	65	40
Archic-ture	TDM (CCIA-SAR)	Rapid Multi. (CCIA-ASC)	FDM	Direct-ADC (CTDSM)	VCO-based
Area per channel (mm²)	0.0062	0.0032	0.019	0.024	0.135
Power per channel (μW)	2.72	1.38	1.97	0.8	7
No. of channels	1024	32	15	16	4

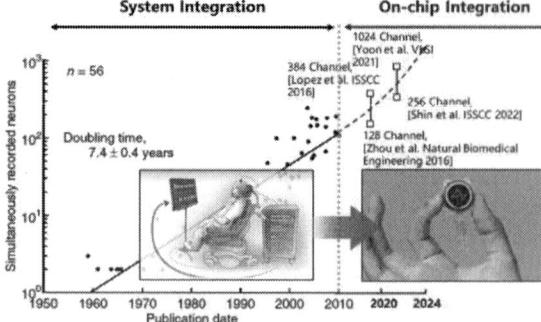

Fig. 3. The channel count of neural recording devices is increasing while the size of devices is become smaller.

the real-time process. In the application scenarios that focus on the action potential (AP), spike detection is utilized where an AP will be condensed to 1 bit signal. The achieved sample rate can be as large as thousand times [7], [30]. In addition to reducing data rate through data compression, another solution is adopting new wireless transmission methods that capable of transmitting few hundred Mb in cost of microwatt power. Impulse radio ultra-wideband (IR-UWB) is recognized as one of the most promising technologies for wireless BCI communication technology. By generating short pulses in a low-duty cycle, it supports high data rates while minimizing power consumption and maintaining simplicity in design [31]. Currently, IR-UWB transmitters is now able to achieve data rate of over hundreds of Mb at power cost range from μW to less than 10 mW.

Table II lists representative neural recording designs. Common neural recording utilizes time-division architecture that shares the analog-to-digital (ADC), but has independent signal conditioning path for the seek of the low-noise performance [7]. However, chip area grows linearly with the number of growing channels which will compromise the size of the chip. To realize high-density recording, rapid multiplexing is introduced where a single AFE circuit is used for multiple channels [30]. However, this brings challenges like DC-offset difference between channels can easily saturate amplifier and lower input impedance. Frequency division multiplexing modulate the signal to different frequency to share one front-end channel [32]. Direct digitalization aim to digitalize raw neural signals directly without a low-noise amplification stage to reduce power and area overhead. Low-noise performance is achieved by oversampling ADC with intrinsic noise shaping capabilities [33]. VCO-based direct digitization design can use more digital cells, which can take the benefit from advanced process node [34].

IV. FUTURE PERSPECTIVES

Current closed-loop neuromodulation chips often overstate the benefits of machine learning algorithms. In reality closed-loop control should support human data analysis by swiftly formulating modulation strategies, rather than excluding human involvement. Closed-loop neuromodulation with human-in-the-loop that emphasizes the interaction between control algorithms and clinical practitioners may be more practical for clinical use. Closed-loop systems also vary depending on the specific indication. Balancing the low-noise requirements for front-end analog

processes, the advanced node requirements for low-power signal processing, and the high voltage compliance needed for stimulation is a key consideration. Therefore, finding the optimal design among these factors is a crucial topic for future research. Currently, neuromodulation techniques are diversifying to include not only electrical methods but also ultrasound, heat, optical, and others. This trend toward multi-modality also provide new opportunities for advanced circuits designs.

Recent advancements in high-throughput neural recording have achieved rates as high as ten thousand channels. It remains to be studied whether this level meets the threshold for functional rehabilitation based on BCIs. If this limit has been reached, further increasing throughput may no longer be the primary focus of development in this field. Instead, the emphasis might shift back to satisfying the needs of neuroscience, assisting in the study of neural mechanisms, and exploring the mysteries of the brain. This shift would introduce new circuit design specifications and functions.

As a highly interdisciplinary research area, BCI involve not only chip design but also aspects strongly related to electrodes, packaging, system architecture, and application scenarios. Therefore, in addition to designing chips that meet current specifications, designers must also anticipate and address emerging demands from related interdisciplinary fields. These demands are driven by the evolving of applications, which continually introduce new requirements and challenges. Current BCI applications focused on long-term implantation, opportunities are still open for pre, intra and postoperative demands.

V. CONCLUSIONS

In this paper, we introduce the primary applications of BCIs, focusing on the treatment of brain diseases and rehabilitation. Driven by the diverse needs of these applications, we explore the design challenges of closed-loop neuromodulation and neural recording. These challenges include computational demands, the need for simultaneous recording and stimulation, and the high power and area costs associated with high channel count. We discuss current methodologies utilized to address these issues. Finally, we offer perspectives on the future directions of BCI technology, emphasizing the need to satisfy practical clinical usage and meet the demands of neuroscience research.

979-8-3503-6184-1/24 $31.00 © 2024 IEEE

ACKNOWLEDGMENT

The authors would like to acknowledge funding support from the "Pioneer" and "Leading Goose" R&D Program of Zhejiang (2024C03002), STI2030-Major Projects (2022ZD0208805), the Key Project of Westlake Institute for Optoelectronics (Grant No. 2023GD004), and Westlake University,

REFERENCES

[1] S. M. Won, E. Song, J. T. Reeder, and J. A. Rogers, "Emerging Modalities and Implantable Technologies for Neuromodulation," *Cell*, vol. 181, no. 1, pp. 115–135, Apr. 2020, doi: 10.1016/j.cell.2020.02.054.

[2] C. Wang, C. Fang, Y. Zou, J. Yang, and M. Sawan, "Artificial intelligence techniques for retinal prostheses: a comprehensive review and future direction," *J. Neural Eng.*, vol. 20, no. 1, p. 011003, Feb. 2023, doi: 10.1088/1741-2552/acb295.

[3] S. N. Flesher *et al.*, "A brain-computer interface that evokes tactile sensations improves robotic arm control," *Science*, vol. 372, no. 6544, pp. 831–836, May 2021, doi: 10.1126/science.abd0380.

[4] H. Lorach *et al.*, "Walking naturally after spinal cord injury using a brain–spine interface," *Nature*, vol. 618, no. 7963, pp. 126–133, Jun. 2023, doi: 10.1038/s41586-023-06094-5.

[5] C. Feng *et al.*, "Acoustic inspired brain-to-sentence decoder for logosyllabic language." Nov. 05, 2023. doi: 10.1101/2023.11.05.562313.

[6] J. Yang and M. Sawan, "From Seizure Detection to Smart and Fully Embedded Seizure Prediction Engine: A Review," *IEEE Trans. Biomed. Circuits Syst.*, vol. 14, no. 5, pp. 1008–1023, Oct. 2020, doi: 10.1109/TBCAS.2020.3018465.

[7] D.-Y. Yoon, S. Pinto, S. Chung, P. Merolla, T.-W. Koh, and D. Seo, "A 1024-Channel Simultaneous Recording Neural SoC with Stimulation and Real-Time Spike Detection," in *2021 Symposium on VLSI Circuits*, Kyoto, Japan: IEEE, Jun. 2021, pp. 1–2. doi: 10.23919/VLSICircuits52068.2021.9492480.

[8] X. Liu *et al.*, "A Fully Integrated Wireless Compressed Sensing Neural Signal Acquisition System for Chronic Recording and Brain Machine Interface," *IEEE Trans. Biomed. Circuits Syst.*, vol. 10, no. 4, pp. 874–883, Aug. 2016, doi: 10.1109/TBCAS.2016.2574362.

[9] D. Wu, Y. Shi, Z. Wang, J. Yang, and M. Sawan, "C^2SP-Net: Joint Compression and Classification Network for Epilepsy Seizure Prediction," *IEEE Trans. Neural Syst. Rehabil. Eng.*, vol. 31, pp. 841–850, 2023, doi: 10.1109/TNSRE.2023.3235390.

[10] W. Zou, R. Eskandari, X. Liu, J. Chen, J. Yang, and M. Sawan, "Wireless Data Transceivers for Brain-machine Interfaces," in *2023 IEEE Biomedical Circuits and Systems Conference (BioCAS)*, Toronto, ON, Canada: IEEE, Oct. 2023, pp. 1–5. doi: 10.1109/BioCAS58349.2023.10388570.

[11] M. Sawan *et al.*, "Emerging Trends of Biomedical Circuits and Systems," *Found. Trends® Integr. Circuits Syst.*, vol. 1, no. 4, pp. 217–411, 2021, doi: 10.1561/3500000005.

[12] T. L. Skarpaas, B. Jarosiewicz, and M. J. Morrell, "Brain-responsive neurostimulation for epilepsy (RNS® System)," *Epilepsy Res.*, vol. 153, pp. 68–70, Jul. 2019, doi: 10.1016/j.eplepsyres.2019.02.003.

[13] B. Zhu, U. Shin, and M. Shoaran, "Closed-Loop Neural Prostheses With On-Chip Intelligence: A Review and a Low-Latency Machine Learning Model for Brain State Detection," *IEEE Trans. Biomed. Circuits Syst.*, vol. 15, no. 5, pp. 877–897, Oct. 2021, doi: 10.1109/TBCAS.2021.3112756.

[14] A. Chua, M. I. Jordan, and R. Muller, "SOUL: An Energy-Efficient Unsupervised Online Learning Seizure Detection Classifier," *IEEE J. Solid-State Circuits*, vol. 57, no. 8, pp. 2532–2544, Aug. 2022, doi: 10.1109/JSSC.2022.3172231.

[15] Y.-Y. Hsieh, Y.-C. Lin, and C.-H. Yang, "A 96.2-nJ/class Neural Signal Processor With Adaptable Intelligence for Seizure Prediction," *IEEE J. Solid-State Circuits*, vol. 58, no. 1, pp. 167–176, Jan. 2023, doi: 10.1109/JSSC.2022.3218240.

[16] U. Shin *et al.*, "NeuralTree: A 256-Channel 0.227-μJ/Class Versatile Neural Activity Classification and Closed-Loop Neuromodulation SoC," *IEEE J. Solid-State Circuits*, vol. 57, no. 11, pp. 3243–3257, Nov. 2022, doi: 10.1109/JSSC.2022.3204508.

[17] Z. Wang, J. Yang, H. Wu, J. Zhu, and M. Sawan, "Power efficient refined seizure prediction algorithm based on an enhanced benchmarking," *Sci. Rep.*, vol. 11, no. 1, p. 23498, Dec. 2021, doi: 10.1038/s41598-021-02798-8.

[18] J. Yang, S. Zhao, J. Wang, S. Lin, Q. Hou, and M. Sawan, "Precise and low-power closed-loop neuromodulation through algorithm-integrated circuit co-design," *Front. Neurosci.*, vol. 18, p. 1340164, Mar. 2024, doi: 10.3389/fnins.2024.1340164.

[19] Y. Wei *et al.*, "A Review of Algorithm & Hardware Design for AI-Based Biomedical Applications," *IEEE Trans. Biomed. Circuits Syst.*, vol. 14, no. 2, pp. 145–163, Apr. 2020, doi: 10.1109/TBCAS.2020.2974154.

[20] S. Zhao, C. Fang, J. Yang, and M. Sawan, "Emerging Energy-Efficient Biosignal-Dedicated Circuit Techniques: A Tutorial Brief," *IEEE Trans. Circuits Syst. II Express Briefs*, vol. 69, no. 6, pp. 2592–2597, Jun. 2022, doi: 10.1109/TCSII.2022.3169004.

[21] S. Zhao, J. Yang, and M. Sawan, "Energy-Efficient Neural Network for Epileptic Seizure Prediction," *IEEE Trans. Biomed. Eng.*, vol. 69, no. 1, pp. 401–411, Jan. 2022, doi: 10.1109/TBME.2021.3095848.

[22] S. Zhao *et al.*, "A 0.99-to-4.38 uJ/class Event-Driven Hybrid Neural Network Processor for Full-Spectrum Neural Signal Analyses," *IEEE Trans. Biomed. Circuits Syst.*, vol. 17, no. 3, pp. 598–609, Jun. 2023, doi: 10.1109/TBCAS.2023.3268502.

[23] J. Liu *et al.*, "33.1 A High-Accuracy and Energy-Efficient Zero-Shot-Retraining Seizure-Detection Processor with Hybrid-Feature-Driven Adaptive Processing and Learning-Based Adaptive Channel Selection," in *2024 IEEE International Solid-State Circuits Conference (ISSCC)*, San Francisco, CA, USA: IEEE, Feb. 2024, pp. 542–544. doi: 10.1109/ISSCC49657.2024.10454405.

[24] C.-W. Tsai *et al.*, "SciCNN: A 0-Shot-Retraining Patient-Independent Epilepsy-Tracking SoC," in *2023 IEEE International Solid- State Circuits Conference (ISSCC)*, San Francisco, CA, USA: IEEE, Feb. 2023, pp. 488–490. doi: 10.1109/ISSCC42615.2023.10067518.

[25] A. Zhou, B. C. Johnson, and R. Muller, "Toward true closed-loop neuromodulation: artifact-free recording during stimulation," *Curr. Opin. Neurobiol.*, vol. 50, pp. 119–127, Jun. 2018, doi: 10.1016/j.conb.2018.01.012.

[26] M. J. Vansteensel *et al.*, "Fully Implanted Brain–Computer Interface in a Locked-In Patient with ALS," *N. Engl. J. Med.*, vol. 375, no. 21, pp. 2060–2066, Nov. 2016, doi: 10.1056/NEJMoa1608085.

[27] L. R. Hochberg *et al.*, "Reach and grasp by people with tetraplegia using a neurally controlled robotic arm," *Nature*, vol. 485, no. 7398, pp. 372–375, May 2012, doi: 10.1038/nature11076.

[28] B. Ji *et al.*, "Recent advances in wireless epicortical and intracortical neuronal recording systems," *Sci. China Inf. Sci.*, vol. 65, no. 4, p. 140401, Apr. 2022, doi: 10.1007/s11432-021-3373-1.

[29] M. Shoaran, M. H. Kamal, C. Pollo, P. Vandergheynst, and A. Schmid, "Compact Low-Power Cortical Recording Architecture for Compressive Multichannel Data Acquisition," *IEEE Trans. Biomed. Circuits Syst.*, vol. 8, no. 6, pp. 857–870, Dec. 2014, doi: 10.1109/TBCAS.2014.2304582.

[30] J. Chen *et al.*, "A Neuron-Inspired 0.0032mm^2 −1.38μW/Ch Wireless Implantable Neural Interface with Direct Multiplexing Front-End and Event-Driven Spike Detection and Transmission," in *2024 IEEE Custom Integrated Circuits Conference (CICC)*, Denver, CO, USA: IEEE, Apr. 2024, pp. 1–2. doi: 10.1109/CICC60959.2024.10529097.

[31] R. Eskandari and M. Sawan, "Challenges and Perspectives on Impulse Radio-Ultra-Wideband Transceivers for Neural Recording Applications," *IEEE Trans. Biomed. Circuits Syst.*, vol. 18, no. 2, pp. 369–382, Apr. 2024, doi: 10.1109/TBCAS.2023.3331049.

[32] J. H. Park *et al.*, "A 15-Channel Orthogonal Code Chopping Instrumentation Amplifier for Area-Efficient, Low-Mismatch Bio-Signal Acquisition," *IEEE J. Solid-State Circuits*, vol. 55, no. 10, pp. 2771–2780, Oct. 2020, doi: 10.1109/JSSC.2020.2991542.

[33] C. Kim, S. Joshi, H. Courellis, J. Wang, C. Miller, and G. Cauwenberghs, "Sub-μV rms -Noise Sub-μW/Channel ADC-Direct Neural Recording With 200-mV/ms Transient Recovery Through Predictive Digital Autoranging," *IEEE J. Solid-State Circuits*, vol. 53, no. 11, pp. 3101–3110, Nov. 2018, doi: 10.1109/JSSC.2018.2870555.

[34] W. Jiang, V. Hokhikyan, H. Chandrakumar, V. Karkare, and D. Markovic, "A ±50-mV Linear-Input-Range VCO-Based Neural-Recording Front-End With Digital Nonlinearity Correction," *IEEE J. Solid-State Circuits*, vol. 52, no. 1, pp. 173–184, Jan. 2017, doi: 10.1109/JSSC.2016.2624989.

Multi-physics Simulation and Application of Ion-Gel Based Triboelectric Nanogenerators

Chen Liu[1]*, Ruibo Wang[1], Ruiyi Gao[2], and Yuming Zhang[1]

[1] School of Microelectronics and the State Key Laboratory of Wide-Bandgap Semiconductor Devices and Integrated Technology, Xidian University, Xi'an 710071, China
[2] Fundamentals Department, Air Force Engineering University, Xi'an 710051, China

* Email: liuchen@xidian.edu.cn

Abstract—By combining the principles of contact electrification and electrostatic induction, triboelectric nanogenerators (TENGs) exhibit unique advantages in the fields of energy harvesting and sensing. Tribotronics is proposed as a field that couples triboelectricity with semiconductors, which imposes strict requirements on the output of TENGs to drive semiconductor devices more efficiently. The effect of the surface microstructure and motion frequency on the output performance of TENGs has been systematically investigated by numerical simulation. The theoretical analysis indicates that the output voltage increases by 17.1 times as the motion frequency rises from 0.25 Hz to 4 Hz, resulting in a four-fold increase in the peak output power. A new estimation method of the equivalent surface charge density is also proposed for TENGs with various morphologies. It can be noted that the open-circuit voltage and short-circuit charge of the devices with pyramid patterns on the contacting surfaces are approximately 2.2 and 1.8 times greater than a TENG without any patterns, respectively. The experimental output characteristics of the TENGs align well with the simulation results. Furthermore, a tactile-sensing element has been successfully demonstrated by coupling the ion-gel based TENG and the In_2O_3 based thin-film transistor (TFT).

Keywords—triboelectric nanogenerator (TENG), multi-physics modeling, surface microstructure, tactile sensing, thin-film transistor (TFT).

I. INTRODUCTION

Triboelectric nanogenerators (TENGs), firstly proposed by Wang in 2012, possess the properties of triboelectrification and charge transfer to efficiently convert mechanical energy from the environment into electrical energy [1]. As the power source, TENGs have great potential applications in smart biomedical sensors and wearable intelligent electronic devices [2].

Tribotronics is an emerging research field that couples the triboelectricity and semiconductors [3]. One of the most common approaches is the integration of TENGs with thin-film transistors (TFTs). TENGs serve as inputs to TFTs, utilizing triboelectrification effects to convert signals generated by external stimuli into electrical signals. These signals further modulate the channel conductivity of TFTs. It is noteworthy that the critical parameters of the front-end TENGs should be delicately designed to effectively modulate the output performance of the transistors.

The influence of the surface microstructure and motion frequency on the output characteristics of TENGs has been investigated in detail by numerical simulation in this work. Further, an ion-gel based TENG is combined with a TFT utilizing In_2O_3 as the channel layer to form the tactile-sensing

element, which enables the active mechanosensation and nanoscale tactile perception.

II. SIMULATION AND EXPERIMENTAL PROCESS

A. COMSOL Simulation

Contact-separation (CS) is currently the most widely used among the four basic operational modes of TENGs. An advanced TENG is proposed by utilizing polydimethylsiloxane (PDMS) and ion-gel as both the triboelectric layer materials and electrical double layers which demonstrate higher sensitivity and faster response time [4].

The modeling of the CS-mode TENG was carried out using COMSOL Multiphysics software. The electrostatics submodule was used in COMSOL to analyze the electrostatic force, electric field, and potential distribution. Meanwhile, moving meshes and electrical circuit physics models were selected for all simulations taking into account the frequency variation and load resistance.

The geometric dimension of the TENG was 6 mm × 3 mm, with the critical parameters summarized in TABLE I. The simulated results of three different patterned surface-based TENGs including line, hemisphere, and pyramid have been compared with that with the plain surface.

TABLE I. SIMULATION RELATED PARAMETERS

Name of the units	Parameter Value
Upper and Bottom Electrode	$Material: Copper; \varepsilon_{r0} = 1, d_0 = 50\ \mu m$
Upper Dielectric Layer	$Material: PDMS; \varepsilon_{r1} = 2.75, d_1 = 100\ \mu m$
Bottom Dielectric Layer	$Material: ion-gel; \varepsilon_{r2} = 14.8, d_2 = 20\ \mu m$
Distance Between Electrodes	$x_{min} = 0.5\ \mu m,\ x_{max} = 1\ mm$ $f = 0.25, 0.5, 1, 2, 4\ (Hz)$ $x(t) = \frac{x_{max}}{2}(1 + \sin(2\pi f t + 1.5\pi)) + x_{min}$
Surface Charge Density	$\sigma_0 = 50\ \mu C m^{-2}$
Load Resistance	$R = 10\ M\Omega \sim 1\ T\Omega$
Microstructures	$\Delta d = 5\ \mu m$; line, hemisphere, and pyramid

The surface modification of triboelectric layers can enlarge the effective contact area, thus increasing the output voltage of the TENG [5]. In terms of micropatterns, the surface charge density on one side of the triboelectric layer needs to be corrected according to the principles of

979-8-3503-6184-1/24 $31.00 © 2024 IEEE

electrostatic induction and charge conservation. The modified equations are as follows.

$$\sigma'_{Line} = \frac{l + 2nd_3}{l}\sigma_0 \qquad (1)$$

$$\sigma'_{Hemisphere} = \frac{l + n\pi d_3 - 2nd_3}{l}\sigma_0 \qquad (2)$$

$$\sigma'_{Pyramid} = \frac{l + n\sqrt{5}d_3 - nd_3}{l}\sigma_0 \qquad (3)$$

where l denotes the length of the TENG, n is the number of microstructures, and d_3 and σ_0 represent the thickness of the specific micropattern and the original surface charge density, respectively.

B. Experimental Process

The TENG was assembled by combining PDMS and ion-gel triboelectric layers with the copper electrodes by the adhesive tape. The CS motion was achieved using a linear motor as shown in Fig. 1(a). In_2O_3 based TFTs were fabricated on the glass substrate. First, the In_2O_3 channel layer was deposited on the substrate by radio frequency magnetron sputtering and the gate area was defined by thermal evaporation of the Al electrode. Then, the ion-gel polymer was spin-coated on the gate area by the screen-printing technique. Finally, the tactile-sensing element was formed by coupling the ion-gel based TENG with the In_2O_3 based TFT, in which the ion-gel served as both the electrolyte gate dielectric and triboelectric layer as illustrated in Fig. 1(c).

The open-circuit voltage and short-circuit charge of the TENG were recorded by the electrometer (Keithley 6514) and digital multimeter (Keithley DMM 6500). The output characteristics of the tactile sensor were assessed by the source meter (Keithley 2410) and digit multimeter (Keithley 34470A).

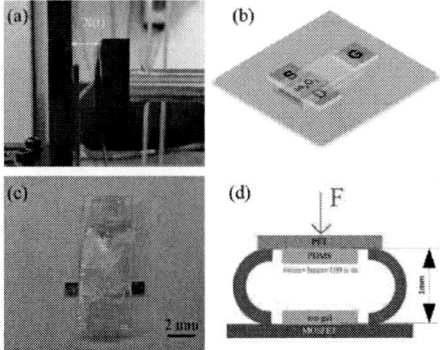

Fig. 1. (a) The CS motion of an ion-gel based TENG. (b) Schematic illustration of In_2O_3 based TFT. (c) Optical image of the tactile sensor. (d) Test principle of the tactile sensor.

III. RESULTS AND DISCUSSION

A. COMSOL Simulation Results

In this study, various motion frequencies were applied to the TENGs for comparison of the difference in the open-circuit voltage (V_{OC}), short-circuit charge (Q_{SC}), and the output voltage across the resistance. As shown in Fig. 2, the V_{OC} and Q_{SC} of the TENG are independent of the motion frequency, which is consistent with the V–Q–x relationship [6]. When an external load resistance of 300 MΩ is connected,

the V_{OC} increases by 17.1 times as the motion frequency rises from 0.25 Hz to 4 Hz. It can be noted from Fig. 3 that the peak output voltage is proportional to the load resistance, while the peak current demonstrates an inverse trend. Thus, the maximum output power increases by 4 times. The output charge is not significantly obstructed at a motion frequency of 0.5 Hz when the external resistance is relatively low as shown in Fig. 3(c), resulting in an output behavior similar to the short circuit condition. However, as the resistance further increases, the obstruction to the charge flow becomes more obvious, and this phenomenon is more pronounced at 2 Hz as illustrated in Fig. 3(d). The low currents generated at high resistances allow only a limited quantity of transferred charge.

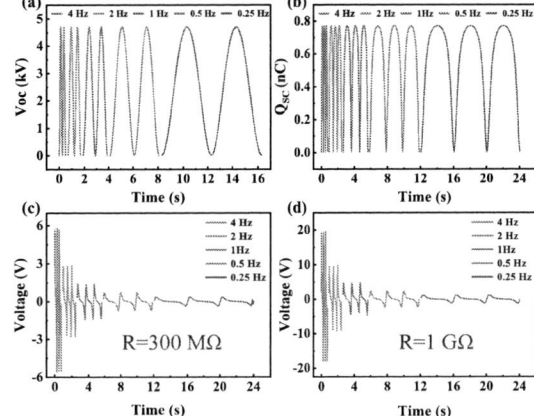

Fig. 2. The simulated (a) open-circuit voltage and (b) short-circuit transferred charge at different motion frequencies. The output voltage across a load resistance of (c) 300 MΩ and (d) 1 GΩ.

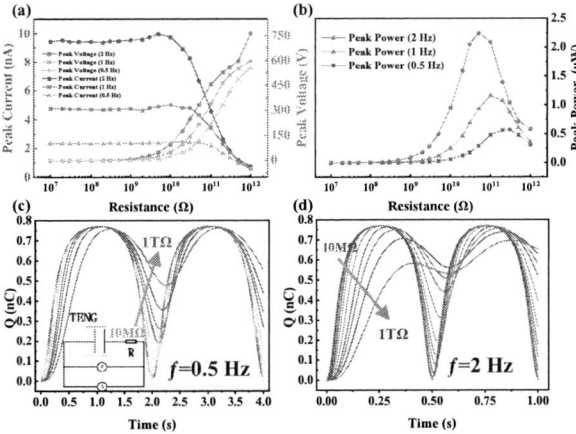

Fig. 3. (a) The peak voltage and current under different load resistances. (b) The peak power of the TENG at external loads. The output charge at the motion frequency of (c) 0.5 Hz and (d) 2 Hz.

Surface patterning is an effective technique for enhancing the output performance of TENGs. Four different surface microstructures, including plane, line, hemisphere, and pyramid, are depicted in Fig. 4a. It can be observed from Fig. 5 that the values of V_{OC} and Q_{SC} of the TENG with pyramid patterns on the contacting surfaces are approximately 2.2 and 1.8 times greater than that without any patterns, respectively. Moreover, the devices with the pyramid pattern show the highest output voltage compared with the other microstructures owing to its higher surface charge density [7].

979-8-3503-6184-1/24 $31.00 © 2024 IEEE   253

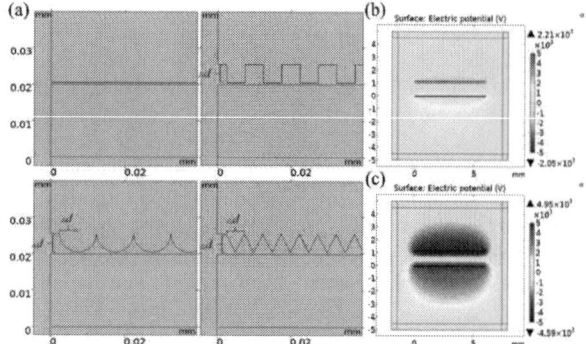

Fig. 4. (a) Ion-gel based TENGs with different surface micropatterns. Surface electric potential of the device (b) with plane pattern and (c) with pyramid pattern.

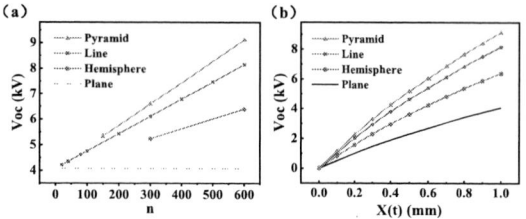

Fig. 5. The value of V_{OC} as a function of (a) the number of micropatterns and (b) the distance between electrodes.

B. Experimental Results

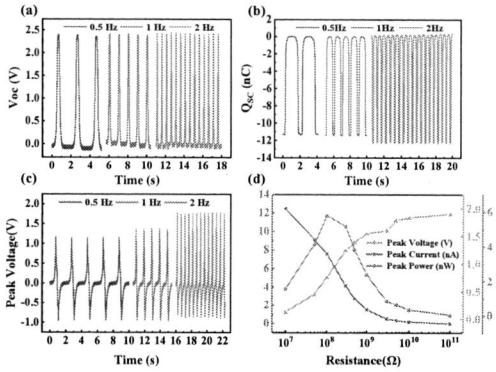

Fig. 6. The measured (a) V_{OC} and (b) Q_{SC} at different motion frequencies. (c) The measured peak voltage under a resistance of 300 MΩ for different motion frequencies. (d) The behaviour of the TENG power output at external loads.

The open-circuit voltage and output charge measurements were carried out by varying frequencies and loads. V_{OC} and Q_{SC} remain stable in spite of frequency as illustrated in Fig. 6. When the external load is added, the output voltage rises from 1.16 V to 1.78 V as the frequency increases from 0.5 Hz to 2 Hz. At the given 1 Hz motion profile, the output voltage is proportional to the resistance. However, the peak current exhibits an inverse trend, which is consistent with the simulation result. The peak power is predicted to be 5.8 nW under a load resistance of 100 MΩ.

The tactile sensations could be mimicked by mechanically pressing the TENG to generate the pulse voltage that modulates the output current of the In$_2$O$_3$ based TFT. As demonstrated in Fig. 7, there is a significant difference in the output current measured at a constant drain voltage of 0.5 V by varying pressing frequencies. The frequency of the obtained output current pulse is positively correlated with the pressing frequency, rendering the active tactile sensor as a promising mechanical frequency monitoring sensor.

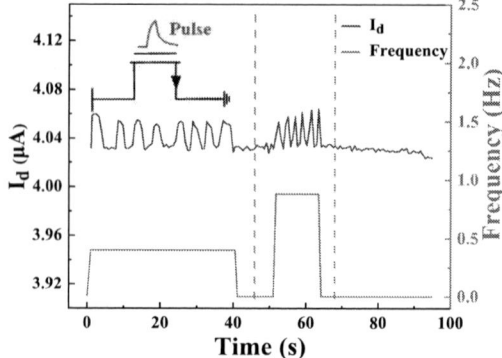

Fig. 7. Relationship between the output current and the pressing frequency of the tactile sensor.

IV. CONCLUSION

In this study, the influence of the surface microstructure and motion frequency on the output performance of ion-gel based TENGs has been investigated in detail by numerical simulation. Meanwhile, a new estimation method of the equivalent surface charge density is proposed and the open-circuit voltage and short-circuit charge of the devices with various surface micropatterns can be accurately predicted, which are consistent with the experimental results. Moreover, the ion-gel based TENG is coupled with the In$_2$O$_3$ based TFT to form a tactile-sensing element, which can precisely monitor the mechanical frequency demonstrating promising prospects in futuristic human machine interaction and intelligent sensing.

ACKNOWLEDGMENT

This work was supported by the Natural Science Basic Research Program of Shaanxi (Grant No. 2024JC-YBQN-0622), National Natural Science Foundation of China (Grant No. 92164202), Xidian University Specially Funded Project for Interdisciplinary Exploration (Grant No. TZJH2024023).

REFERENCES

[1] F. R. Fan, Z. Q. Tian, and Z. L. Wang, "Flexible triboelectric generator ! ," Nano Energy, vol. 1, pp. 328-334, January 2012.

[2] T. Cheng, J. Shao, and Z. L. Wang, "Triboelectric nanogenerators," Nature Reviews Methods Primers, vol. 3, 2023.

[3] C. Zhang, J. Zhao, Z. Zhang, T. Bu, G. Liu, and X. Fu, "Tribotronics: an emerging field by coupling triboelectricity and semiconductors," International Journal of Extreme Manufacturing, vol. 5, July 2023.

[4] Y. Liu, C. Zhao, Y. Xiong, J. Yang, H. Jiao, Q. Zhang, R. Cao, Z. L. Wang, and Q. Sun, "Versatile Ion-Gel Fibrous Membrane for Energy-Harvesting Iontronic Skin," Advanced Functional Materials, vol. 33, 2023.

[5] M. A. P. Mahmud, J. Lee, G. Kim, H. Lim, and K. B. Choi, "Improving the surface charge density of a contact-separation-based triboelectric nanogenerator by modifying the surface morphology," Microelectronic Engineering, vol. 159, pp. 102-107, 2016.

[6] S. Niu, S. Wang, L. Lin, Y. Liu, Y. S. Zhou, Y. Hu, and Z. L. Wang, "Theoretical study of contact-mode triboelectric nanogenerators as an effective power source," Energy & Environmental Science, vol. 6, 2013.

[7] S. Hasan, A. Z. Kouzani, S. Adams, J. Long, and M. A. P. Mahmud, "Comparative study on the contact-separation mode triboelectric nanogenerator," Journal of Electrostatics, vol. 116, 2022.

979-8-3503-6184-1/24 $31.00 © 2024 IEEE

A Flexible Graphene Acoustic Sensor for Sound Signal Acquisition and Spiking Neural Network Recognition

Lu-Yu Zhao[1, 2], Hao-Yuan Shen[1, 2], Yi-Wen Wu[1, 2], Lu-Lu Zhang[2], Yu-Tao Li*[1], Tian-Ling Ren[3]

[1] School of Integrated Circuits and Electronics, MIIT Key Laboratory for Low-Dimensional
Quantum Structure and Devices, Beijing Institute of Technology, Beijing 100081, China;
[2] School of Information Science and Technology, Beijing University of Chemical Technology,
Beijing 100029, China;
[3] School of Integrated Circuits and Beijing National Research Center for Information Science and Technology (BNRist),
Tsinghua University, Beijing 100084, China;

* Email: ytli@bit.edu.cn

Abstract: As artificial intelligence continues to develop and mature, sound sensing and recognition technology has played a crucial role in fields such as human-computer interaction. This article fabricated a microstructure-based graphene acoustic sensor and used Spiking Neural Network (SNN) to identify the collected data. By combining a micro-pyramid structure on a flexible substrate, the as-fabricated sensor can cover the main frequency range of human sound (200-3000 Hz), display excellent mechanical sensitivity (S= 10.9 kPa⁻¹) and fast response ability (5.8 ms), and can capture complex changes in sound. Converting sound signals into pulses can reduce losses during transmission, so a Spiking Neural Network is constructed to recognize sound datasets and an accuracy of 96.5% is achieved. This paper provides the possibility for new applications of carbon-based acoustic sensors in intelligent sound signal recognition systems.

Keywords—Microstructure, Spiking Neural Network, Graphene Acoustic Sensor, Flexible Sensor.

I. INTRODUCTION

In the rapidly developing information age, sound sensing technology plays a crucial role in fields such as human-computer interaction. Although traditional acoustic sensor can meet basic sound capture needs, with the popularity of wearable technology and smart homes, the demand for acoustic sensor with higher sensitivity, wider dynamic range, and better flexibility is increasing.[1, 2]

Graphene not only has high mechanical strength and flexibility, but can also produce ultra-thin sensor films. Its extremely high conductivity and fast response speed make it very suitable for processing sound signals. On this basis, introducing microstructures can enhance the performance of graphene acoustic sensor.[3] Gu[4] et al describes an innovative flexible electronic eardrum (EE) incorporating a microstructured pyramid array. The EE device demonstrates high sensitivity, a high signal-to-noise ratio (about 55 dB), and rapid response time (76.9 μs). The microstructure improves the performance of sensor, enabling them to more effectively capture and convert acoustic signals. Furthermore, converting sound data into pulses can reduce signal quality loss during transmission due to signal attenuation or noise interference. Therefore, choosing SNN to process the collected data is more reasonable.

In the field of sound signal recognition, SNN has shown significant potential, especially because they simulate the way biological neural systems process information, making them naturally suitable for processing time series data such as sound. SNN can effectively capture and encode the temporal dynamic characteristics of sound signals by simulating the pulse firing behavior of neurons, which is particularly important for identifying continuous sound streams and complex sound patterns.[5]

In this work, a microstructure-based graphene acoustic sensor was successfully developed, achieving efficient capture of sound signals. By constructing micro-pyramid structures on a PDMS flexible substrate and harnessing the excellent electrical properties of graphene films, the sensor exhibits high sensitivity and adaptability, enabling operation across diverse application scenarios. Additionally, the sensor's ability to perceive sound signals was tested, including its frequency, pressure, repeatability, and response time, confirming that the graphene acoustic sensor can capture distinct sound features and subtle variations in sound. Moreover, by constructing a SNN to recognize the collected sound signals, a recognition accuracy of 96.5% was achieved. This device holds tremendous potential in sound processing and intelligent sound signal recognition systems.

II. TEST AND RESULT

As shown in Fig 1 (a), the sensor uses graphene as the sensitive material. From the SEM image, it can be observed that the graphene surface is loose and porous, exhibiting a high surface area to volume ratio. Its internal cavities and channels effectively scatter and absorb sound waves, reducing the reflection and transmission of sound waves, thus enhancing the capture of sound signals. The SEM image of PDMS in Fig 1 (a) reveals its surface micro-pyramid structure. Using PDMS as the substrate ensures the sensor's flexibility to adapt to variously shaped surfaces. Traditional methods of microstructure acquisition may suffer from poor preparation precision and difficulty in controlling the size and distribution of microstructures.[6] Therefore, this paper utilizes photolithography techniques, as shown in Fig 1 (b), to etch micro-pyramid structures with a side length of 10 μm x 10 μm on silicon wafer. Pour PDMS into the mold to obtain a micro-pyramid structure, and drop-casting the graphene oxide solution to form a thin film. Then Laser scribing technology is used to obtain graphene thin films with improved conductivity. The micro-pyramid structures produced by this method exhibit good controllability and repeatability.

979-8-3503-6184-1/24 $31.00 © 2024 IEEE

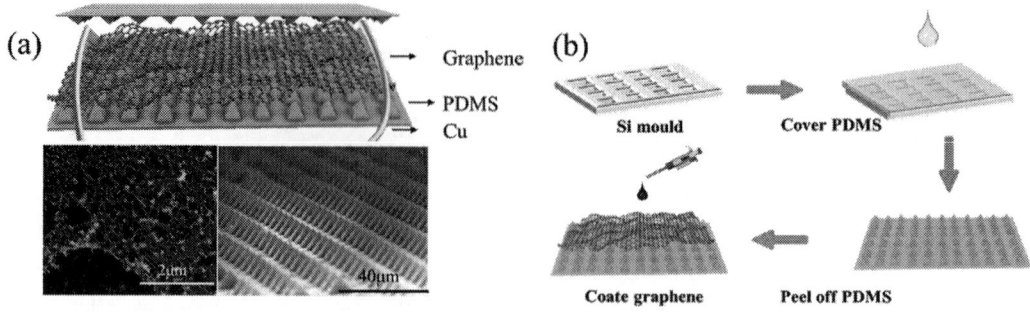

Fig. 1. (a) Schematic diagram of sensor structure. (b) Sensor preparation process diagram.

To demonstrate the performance of the graphene acoustic sensor, Fig 2 conducted corresponding tests from the perspectives of the sensor's perception of sound signals, frequency, pressure, repeatability, and response time. As shown in Fig 2 (a), it can be seen that as the decibel of the sound signal increases, the amplitude of the obtained signal also increases, thus distinguishing different intensities of sound signals. The frequency of human sound primarily ranges between 200-3000 Hz.[7] As shown in Fig 2 (b), the sensor effectively covers this range, peaking around 700 Hz. Sound signals primarily propagate through sound waves, which are relatively weak compared to vibrations caused by pressure. Only by being sensitive enough to distinguish small changes, can capture the sound signals. As a piezoresistive type of acoustic sensor, it exhibits excellent mechanical sensitivity (S= 10.9 kPa^{-1}). As shown in Fig 2 (c), the sensor can perceive very small sound signals and has good linearity.

Meanwhile, in Fig 2 (d), the repeatability of the sensor was tested, and it can be seen that even after multiple repetitions, the waveform of the signal can still remain consistent. Through the analysis of the characteristics of sound signals, it can be seen that the changes in sound signals are very fast. Therefore, sensor need to have a fast response time to better record all sound characteristics. As shown in Fig 2 (e), the response time of the sensor is 5.8 ms, and the recovery time is 6.8 ms, which can effectively recognize sound signals.

In order to further demonstrate the recognition ability of the sensor for sound signals, as shown in Fig 3 (a), sound signals of male and female sounds speaking the same sentence were recorded separately, and Fast Fourier Transformation (FFT) transformation and conversion were performed to spectrograms. It can be seen that the sound signals recorded by the graphene acoustic sensor clearly exhibit distinct features of the sound signals, and the sensor has a certain ability to differentiate between male and female voices. Fig 3 (b) and (d) respectively show the recognition performance of a scale and different tones, while Fig 3 (c) shows the recognition performance of two words with similar pronunciations. It can be seen that even for words with similar pronunciations, the sound signals recorded by the sensor are different (the red border section). Based on the above tests, it can be seen that the sensor meets the conditions for recognizing sound signals. Therefore, it can serve as a data source for neural network classification. From the perspective of maintaining signal quality, a SNN was constructed to convert the collected sound signals into pulses and perform recognition.

Converting continuous signal data into discrete pulse sequences is crucial for input processing in SNN. In this article, a special peak seeking function is constructed to detect the peaks in the original signal and convert these peaks into pulse form. Determine some local maximum values are considered as peaks based on the set prominence. Prominence ensures that only sufficiently significant peaks are selected.

As shown in Fig 4 (a), it can be seen that for two words with similar pronunciations, the pulse conversion results are not the same, indicating that the constructed pulse conversion function can effectively distinguish sound data. This neural network model consists of two fully connected linear layers and two Leaky integrated and fire (LIF) neuron layers (beta = 0.95) to simulate the behavior of biological neurons.

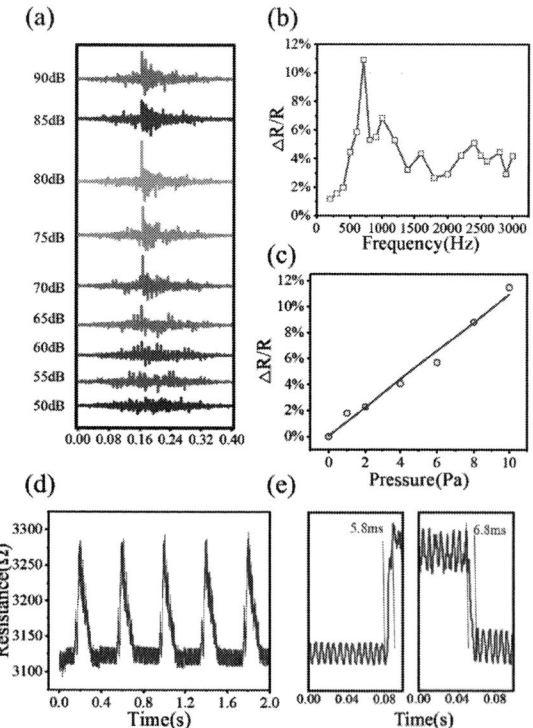

Fig. 2. Resistance change rate of graphene acoustic sensor at different (a) decibels, (b) frequencies and (c) pressures. (d) Repetitive testing (e)Response time

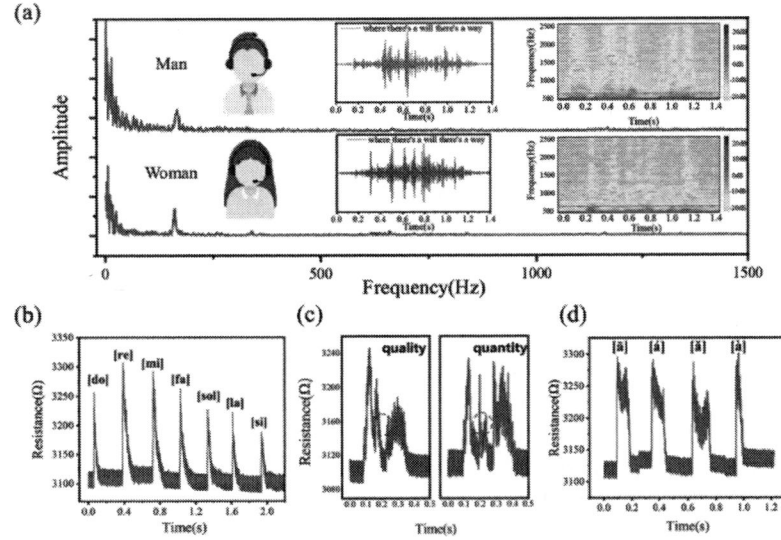

(a)

Fig. 3. (a) The spectrogram and FFT transformation of male and female phonetics. (b) The phonetic curve of a scale. (c) The curve of words with similar pronunciations. (d) Curve of different tones.

Using the prepared sensor, sound data was collected, obtaining a total of 1000 sound datasets containing ten English words from 0 to 9. Fig 4 (b) shows Test Loss and Test Accuracy, demonstrating the model's generalization ability on test data. Fig 4 (c) shows the confusion matrix of the recognition results, and the final recognition accuracy remains stable at 96.5%.

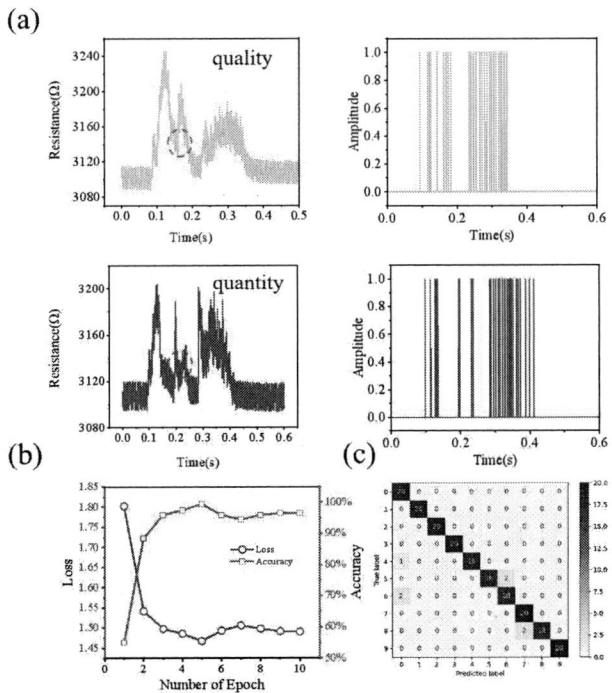

Fig. 4. (a) Pulse conversion results of words with similar pronunciations. (b) Test Loss and Test Acc of SNN. (c) Identification results of the 0-9 dataset.

III. CONCLUSION

This article developed a sound signal recognition method using a microstructure-based graphene acoustic

sensor integrated with SNN. By incorporating flexible micro-pyramid structures, the sensitivity and adaptability of the graphene acoustic sensor to different acoustic environments were enhanced. The sensor was systematically tested for its ability to detect sound signal intensity, frequency (200-3000 Hz), and pressure, demonstrating high sensitivity, rapid response time (5.8 ms), and excellent repeatability. An SNN was constructed to recognize the collected sound signals, achieving a high recognition accuracy of 96.5%. This not only shows the reliability of the sensor in capturing sound signals but also demonstrates its potential applications in advanced sound signal recognition systems, providing support for flexible sensor technology and neural network applications in sound processing.

ACKNOWLEDGMENTS

This research was supported by the National Natural Science Foundation of China (62201026, 61971026).

REFERENCES

[1] S. Lee et al., "A transparent bending-insensitive pressure sensor," Nature nanotechnology, vol. 11, no. 5, pp. 472-478, 2016.

[2] Tao, LQ., Tian, H., Liu, Y. et al. An intelligent artificial throat with sound-sensing ability based on laser induced graphene. Nat Commun 8, 14579 (2017).

[3] Y. Wang et al., "Ultra-sensitive graphene strain sensor for sound signal acquisition and recognition," Nano Research, vol. 8, pp. 1627-1636, 2015.

[4] Y. Gu, X. Wang, W. Gu, Y. Wu, T. Li, and T. Zhang, "Flexible electronic eardrum," Nano Research, vol. 10, no. 8, pp. 2683-2691, 2017/08/01 2017, doi: 10.1007/s12274-017-1470-1.

[5] J. Wu, E. Yılmaz, M. Zhang, H. Li, and K. C. Tan, "Deep spiking neural networks for large vocabulary automatic speech recognition," Frontiers in neuroscience, vol. 14, p. 513257, 2020.

[6] Z. Shi et al., "Morphological engineering of sensing materials for flexible pressure sensors and artificial intelligence applications," Nano-micro letters, vol. 14, no. 1, p. 141, 2022.

[7] K. Tong, Q. Zhang, J. Chen, H. Wang, and T. Wang, "Research on throat speech signal detection based on a flexible graphene piezoresistive sensor," ACS Applied Electronic Materials, vol. 4, no. 7, pp. 3549-3559, 2022.

0.15μm BCD Platform with High Sensitivity Hall Device and Low Noise CMOS for Sensor IC Applications

Guiqiang Zheng[1], Qingyin Zhong[1], Jie Ma[1], Nannan Cheng[1], Yichen Li[1], Yongjia Li[1], Xiaofeng Sun[2], Dejin Wang[2], Sen Zhang[2], Long Zhang[1*], Siyang Liu[1*], Weifeng Sun[1*]

[1] National ASIC System Engineering Research Center, Southeast University, Nanjing, China
[2] CSMC Technologies Corporation, Wuxi, China

* Email: longzh@seu.edu.cn; liusy2017@seu.edu.cn; swffrog@seu.edu.cn

Abstract—In this paper, a 40V 0.15μm Bipolar-CMOS-DMOS (BCD) platform with high sensitivity Hall device and low noise CMOS device is proposed. The lateral Hall device (LHD) with a current-related sensitivity (S_I) of 378V/AT and the vertical Hall device (VHD) with a S_I of 161V/AT are integrated in the platform without extra masks. The low noise CMOS with fluoride ion implantation shows a 10 times optimization in 1/ *f* noise and the static electric parameters almost unchanged. The proposed BCD platform presents a cost-effective solution in the design and fabrication of sensor ICs.

Keywords—*Bipolar-CMOS-DMOS (BCD), Hall device, Low noise device, LDMOS.*

I. INTRODUCTION

Sensor integrated circuits (ICs) such as Hall sensor ICs are widely used in magnetic field detection, position detection and current measurement [1]. Hall sensor ICs includes Hall device, amplifiers, modulators and other circuits. Hall device, CMOS device and LDMOS are three key components in Hall sensor ICs, which limit the applications and the accuracy of the detection [2]. In this work, a 0.15μm Bipolar-CMOS-DMOS (BCD) platform integrated with high sensitivity Hall device, 5V low noise CMOS device (LND), and 40V low on-resistance LDMOS is proposed with a compromise balance between cost and performance.

II. FEARURES OF THE 0.15UM BCD PLATFORM

Fig. 1 shows the modulars of the 0.15μm BCD platform. The low power 0.15μm CMOS technology platform is recognized as the standard platform. 1.8V and 5.5V CMOS are included in order to meet high speed and low power applications. Additionally, the several kinds of device options and wiring options are prepared. The device options are consisted of seven elements: 5V low noise CMOS (LND), LDMOS (7V-40V), Hall device (lateral Hall device (LHD) and vertical Hall device (VHD)), high-resistance Poly-Si resistor, MIM Capacitor, MTP and Flash. Required devices and metal options can be selected in modular so that the reduction of process cost is achieved.

The key process flow is shown in Fig. 2. The staring material is an 8-inch P-epi wafer on P⁺ substrate. N⁺ Buried Layer (NBL) is formed on it. NBL is used for Hall device, high-side LDMOS, and isolated devices. Then, the P-type epitaxial layer, with an appropriate doping concentration and

This work was supported by the National Key R&D Program of China (2023YFB4403700), the natural science foundation of Jiangsu Province (BK20231150, BK20232006), the Distinguished Young Scholars Program of Southeast University (2242022R40010) and the Technological Achievements of Jiangsu Province (No.BA2022005).

thickness, is grown on NBL to achieve the requirement of breakdown voltage capable up to 40V n/pLDMOS. There are two deep wells for isolations, which are DNWELL and DPWELL, respectively. After the definition of active region with shallow trench isolation (STI), there are two implantation, NM and PM, in drift regions for low resistance. The process offers up to five level metals and the top metal with a thickness of 3.3μm.

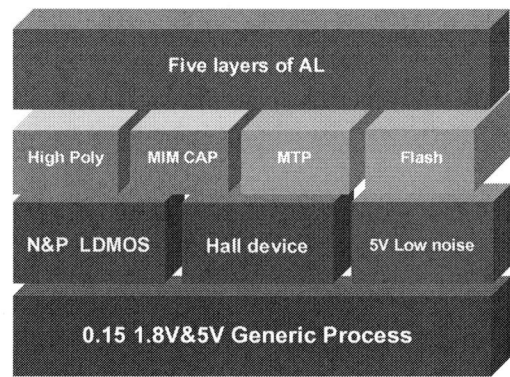

Fig. 1. Modulars of the 0.15um BCD platform.

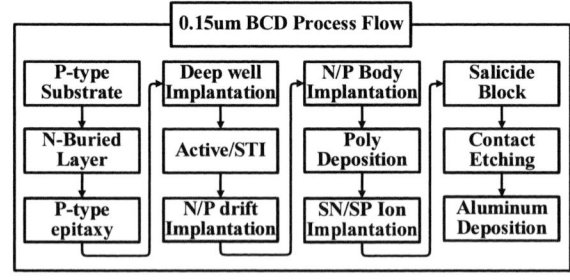

Fig. 2. Key process flow of 0.15um BCD platform.

III. KEY DEVICE DESIGN AND CHARACTERISTICS

A. Hall device

The LHD and VHD are designed and fabricated based on the 0.15μm BCD platform. As shown in Fig. 3(a), the octagonal NM well is used as magnetic-sensitive region in LHD, which is surrounded by bulk ring and isolation (ISO) ring. Four electrodes are paired into two groups: Hall electrodes and bias electrodes, which are set on the edge of the NM well. The LHD detects magnetic field that perpendicular

979-8-3503-6184-1/24 $31.00 © 2024 IEEE

to the chip surface. With the STI set in the surface of NM well, the effective thickness of NM is 0.6μm. The ISO ring, which includes the NX, DN, and BN wells, diminishes the junction field effect from P-sub and isolates the effect from other devices in the chip.

As shown in Fig. 4, electrodes are placed at the center of the device in parallel in VHD. The VHD senses the magnetic field that parallel to the chip surface. The DN well is used as magnetic-sensitive region in VHD. The 7μm depth of DN well can provide space for the rotation of bias current to improve the detection sensitivity.

Key parameters of the proposed Hall devices are shown in Table I. The LHD presents a current-related sensitivity (S_I) of 378V/AT and the VHD presents a S_I of 161V/AT. Fig. 7 shows the change of S_I of the proposed Hall device over a temperature range of 25°C to 125°C. S_I of the LHD increases to 405V/AT, while the SI of the VHD increases to 175 V/AT. The S_I changes lower than 10% present the temperature stability of the proposed Hall device.

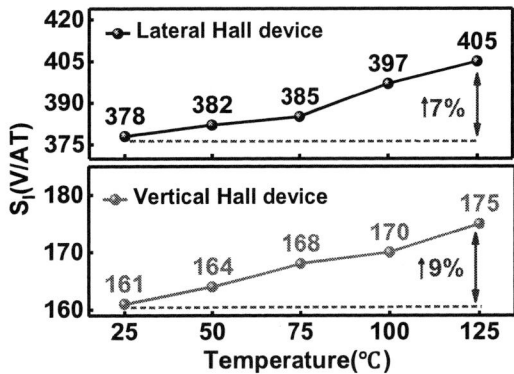

Fig. 5. S_I changes of the proposed Hall device over the temperature from 25 °C to 125 °C.

TABLE I. SUMMARY OF HALL DEVICE CHARACTERISTICS

Parameter	LHD	Parameter	VHD
Magnetic-sensitive well	NM well	Magnetic-sensitive well	DN well
L	90μm	L1	25μm
W	5μm	M	3μm
/	/	L2	2μm
S_I	378V/AT	S_I	161V/AT
Input Resistance	25.1kΩ	Input Resistance	10.4kΩ

B. CMOS

As shown in Fig. 6, the fluoride (F) ions are implanted in channel region to obtain the LND, based on the standard CMOS device (STD) process. The F atoms can diffuse into the gate oxide and locate near the Si/SiO₂ interface, which can minimize the interface traps in gate oxide and improve the $1/f$ noise [3].

Fig 7 shows the power spectrum density (PSD) of STD and LND at different gate voltages (V_{gs}). The $1/f$ noise is improved more than 10 times from 10Hz to 10kHz at different V_{gs}. The PMOS presents a better optimization in 1/f noise than NMOS. As shown in Table II, there seems no shift in Vth, I_{dlin}, I_{dsat}, I_{off} and BV_{off} of the LND with F ion implantation, which presents that LND can replace STD in circuits without other sacrifices.

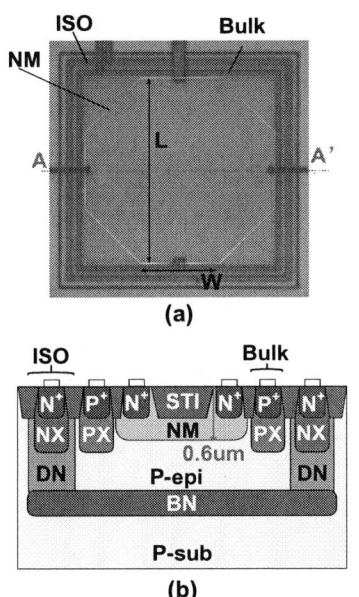

(a)

(b)

Fig. 3. (a) Photograph of the top view and (b) cross section view along the line A-A' of the proposed LHD.

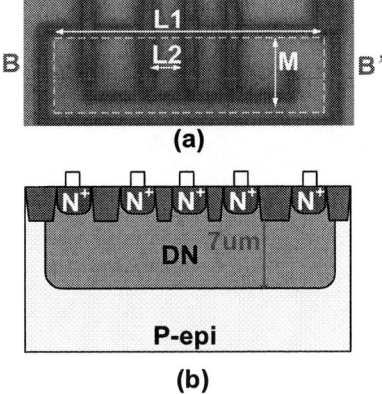

(a)

(b)

Fig. 4. (a) Photograph of the top view and (b) cross section view along the line B-B' of the proposed VHD

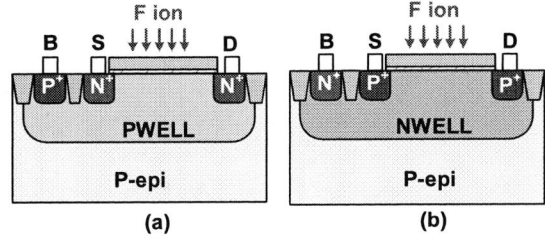

(a)   (b)

Fig. 6. Schematic of F ion implantation for (a) low noise NMOS and (b) low noise PMOS.

979-8-3503-6184-1/24 $31.00 © 2024 IEEE

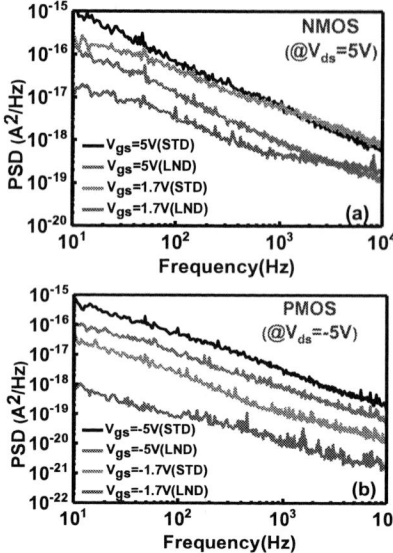

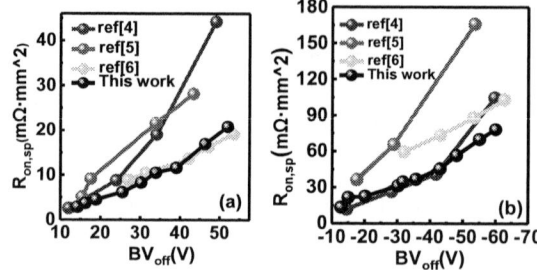

Fig. 9. Relationship between $R_{on,sp}$ and BV_{off} in this work and previous reports of (a) nLDMOS and (b) pLDMOS.

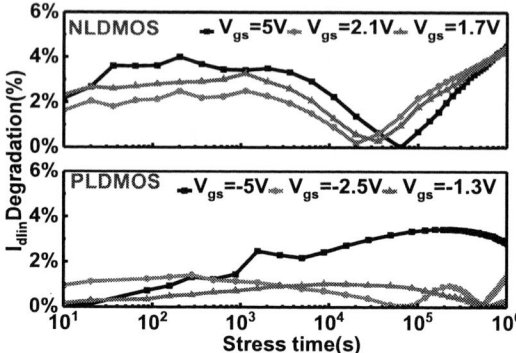

Fig. 7. The $1/f$ noise spectrum density of (a) NMOS and (b) PMOS at different V_{gs}.

TABLE II.   ELECTRONIC CHARACTERS OF STANDARD CMOS DEVICE (STD) AND LOW NOISE CMOS DEVICE (LND)

Parameter	NMOS		PMOS	
	STD	LND	STD	LND
$V_{th}(V)$	0.668	0.667	-0.776	-0.769
$I_{dlin}(\mu A/\mu m)$	65.15	65.13	-16.98	17.15
$I_{dsat}(\mu A/\mu m)$	599.1	595.7	-283.9	-286.8
$I_{off}(\mu A/\mu m)$	0.2	0.2	-0.2	-0.2
$BV_{off}(V)$	10.3	10.3	-9.6	-9.6

Fig. 10. Measured I_{dlin} degradations of 40V LDMMOS at Abs.V_{ds}=44V.

C. LDMOS

The cross-section views of 40V nLDMOS in the platform are shown in Fig. 8. The design of the salicide block (SAB) oxide field plate and NM for nLDMOS aims to achieve the tradeoff between off-state breakdown voltage (BV_{off}) and specific on-resistance ($R_{on,sp}$). Fig. 9 presents the trade-off performance between BV_{off} and Ron,sp for n/pLDMOS, respectively. The LDMOS in the 0.15μm platform shows good performance compared to other reports [4][5][6].

The linear current (I_{dlin}) degradation of LDMOS is measured under three gate voltage conditions (I_{submax}, low V_{gs} and max V_{gs}). From Fig. 10, the I_{dlin} degradation after 1×10^6 s stress time is proven to be lower than 5%, indicating high hot carrier injection reliability of the LDMOS in the platform.

IV. CONCLUSION

In conclusion, the proposed 40V 0.15μm BCD platform with high sensitivity Hall device module and low noise CMOS module presents a cost-effective solution in the design and fabrication of the sensor ICs.

REFERENCES

[1] Crescentini M, Marchesi M, Romani A, et al. A broadband, on-chip sensor based on Hall effect for current measurements in smart power circuits[J]. IEEE Transactions on Instrumentation and Measurement, 2018, 67(6): 1470-1485.

[2] Toh E H, Sun Y, Zheng P, et al. A modular three-dimensional Hall effect sensor for performance optimization[J]. IEEE Sensors Journal, 2021, 22(12): 11256-11263.

[3] Wright P J, Saraswat K C. The effect of fluorine in silicon dioxide gate dielectrics[J]. IEEE Transactions on Electron Devices, 1989, 36(5): 879-889.

[4] Riccardi D, Causio A, Filippi I, et al. BCD8 from 7V to 70V: a new 0.18μm Technology Platform to Address the Evolution of Applications towards Smart Power ICs with High Logic Contents[C]//Proceedings of the 19th International Symposium on Power Semiconductor Devices and IC's. IEEE, 2007: 73-76.

[5] Park I Y, Choi Y K, Ko K Y, et al. BD180-A new 0.18 μm BCD (Bipolar-CMOS-DMOS) Technology from 7V to 60V[C]//2008 20th International Symposium on Power Semiconductor Devices and IC's. IEEE, 2008: 64-67.

[6] Iwamoto K, Kori M, Terada C, et al. Advanced 300mm 0.13 μm BCD technology from 5V to 80V with highly reliable embedded Flash[C]//2014 IEEE 26th International Symposium on Power Semiconductor Devices & IC's (ISPSD). IEEE, 2014: 402-405.

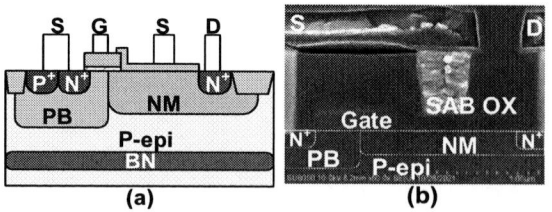

Fig. 8.   (a) Schematic and (b) SEM cross-section view of the 40V nLDMOS.

A High Dynamic Range Pixel with Inverse Proportional Response

Yuchen Wang, Wenji Mo, Haoning Sun, Jingjing Liu*

School of electronics and communication engineering, Sun Yat-sen University, Shenzhen 518107, China

* Email: wangych79@mail2.sysu.edu.cn, liujj77@mail.sysu.edu.cn

Abstract—This paper proposes a high dynamic range (HDR) pixel that combines linear response and inverse proportional response. This pixel achieves nonlinear compression of light intensity under inverse proportional response to improve dynamic range (DR), suitable for CMOS image sensors (CIS) with rolling shutter operation. The proposed pixel is composed of only 4 MOSFET. In HDR mode, it loads the output signal with brightness information dynamically onto the column signal bus. This pixel does not rely on the I-V characteristics of the CMOS subthreshold region and adopts a hard reset structure, overcoming the problems of poor low light SNR performance and image lagging in the traditional logarithmic pixels. The use of low threshold NMOS transistors in the pixel circuit results in the swing of the pixel output close to the power supply voltage, improving the low-voltage performance of the pixel circuit and bringing higher DR. Under a standard CMOS process, the pixel pitch is 6.6μm with a fill factor of 37.6%. The post simulation results indicate that the proposed pixel has good linear and inverse proportional responses to photocurrent. Compared to the linear mode, this pixel has a DR improvement of at least 31.9dB in the HDR mode.

Keywords—CMOS image sensors (CIS), high dynamic range (HDR), inverse proportional response

I. INTRODUCTION

Dynamic range (DR) of CMOS image sensors (CIS) is one of the important criteria for evaluating performance. Human eye can achieve a DR of about 130dB by adjusting the pupil size. However, the DR of current commercial cameras usually does not exceed 70dB. Many methods to expand the DR of CIS have been investigated.

The traditional HDR is achieved by imaging the same scene multiple times with different exposure times [1]. Other techniques that use linear pixels to extend DR include completing two different exposure times within one single frame [2], adding additional integration capacitors to the pixels, using high-resolution ADC for high supply voltages [3], and using pulse frequency pulse width modulation to obtain for digital pixel outputs for low supply voltages [4], etc.

In addition, logarithmic response pixels can also be used to improve the DR. The logarithmic pixels utilize the I-V characteristic of MOSFETs operating in the subthreshold region to achieve nonlinear compression of light intensity. However, operating in the subthreshold region makes them vulnerable to fixed pattern noise (FPN). Small pixel output swing, poor low light SNR performance, and image lagging are also issues with the logarithmic pixels [5-6].

This paper proposes a novel HDR pixel, which mimics the human eye and exhibits an inverse proportional response to light intensity. The pixel can also achieve the same linear response characteristics as traditional 3T pixels by reconfiguring the control signal. This pixel circuit consists of only 4 MOSFETs and is suitable for CIS with rolling shutter operation. Different from the traditional linear pixels that output stable voltage values, the proposed pixel dynamically loads the output signal with brightness information onto the column signal bus, thus obtaining larger DR. And unlike traditional logarithmic pixels, the proposed pixel does not have transistors operating in the subthreshold region, resulting in a smaller FPN due to the process variations. Hard reset structure is used in the pixel, so there is no image lagging. In addition, low threshold NMOS transistors are used within the pixel, achieving a swing of the pixel output close to the power supply voltage, resulting in better SNR performance for low supply voltages.

II. THE PROPOSED PIXEL STRUCTURE AND OPERATIONS

The structure of the proposed HDR pixel circuit and quantization circuit are shown in figure 1, and the schematic of a single pixel is shown in the dashed box. One pixel consists of one photodiode and four MOSFETs, requiring three control signals including a reset signal rst, gating signal sel and $\overline{sel}$. Among them, M1 and M3 are nominal threshold voltage PMOS, M2 is a native NMOS, and M4 is a medium threshold voltage NMOS. M1 uses PMOS to hard reset the photodiode to VDD, which can avoid image lagging caused by charge residue and also improve the swing of the pixel output signal. The native NMOS as the source follower and the transmission gate can also improve the output swing. The use of medium threshold voltage NMOS in M4 is a compromise between the performance and layout of the transmission gate under low power supply voltage. A larger swing provides a better SNR performance. The pixel output signal swing is only limited by the threshold voltage of the native NMOS M2 and the voltage drop of the source follower load.

The proposed high dynamic range pixels can operate in linear mode or HDR mode by changing the control signal and quantization circuit transmission path. The comparator compares the pixel output with a rising slope signal to obtain a pulse width modulation (PWM) signal containing the pixel brightness information. Then, the counter converts the PWM signal into a digital code to complete the quantization process of the pixel output.

A. HDR Mode

When the pixel operates in HDR mode, the quantization circuit keeps the switch HDR on and the switch LIN off. The key node voltage and control signal waveforms in the HDR mode are shown in Fig. 2. The maximum exposure time for the pixel is set to T. Firstly, before the time 0, M1 keeps on

---

This work is supported by National Natural Science Foundation of China with project number 62174181.

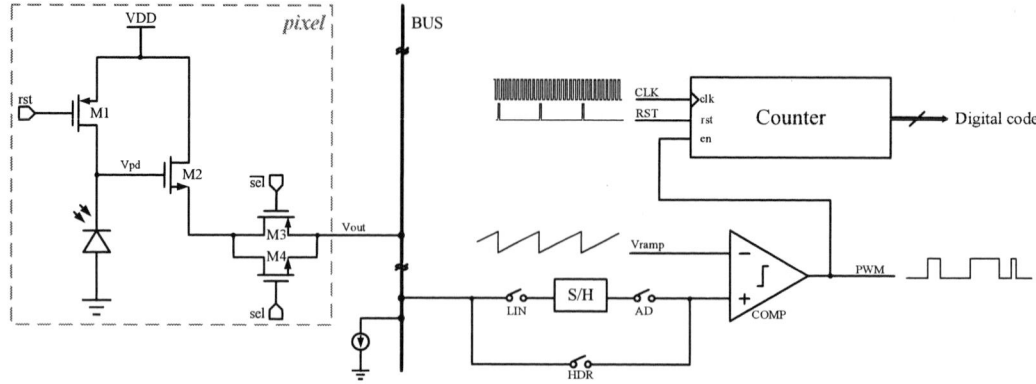

Fig. 1. The structure of the proposed HDR pixel circuit and quantization circuit.

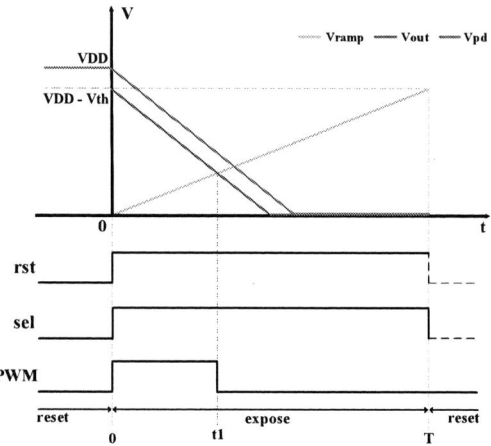

Fig. 2. Key node voltage and control signal waveforms in HDR mode.

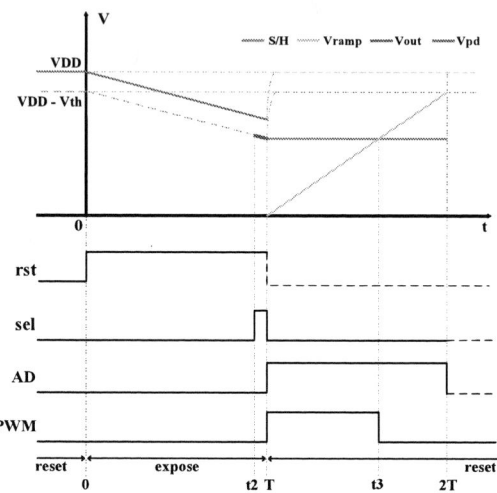

Fig. 3. Key node voltage and control signal waveforms in Linear mode.

and the photodiode node is reset to the power supply voltage VDD. At the time 0, M1 turns off and the cathode node voltage Vpd of the photodiode gradually decreases under the illumination. At the same time, the transmission gate composed of M3 and M4 turn on, and Vpd is amplified by M2 and transmitted to the column signal bus through the transmission gate as $Vout$. $Vout$ has a decrease of M2 threshold voltage compared to Vpd signals, which is given by (1):

$$Vout = Vpd - Vth = VDD - \frac{Ipd \cdot t}{C} - Vth \quad (1)$$

where t denotes any time within the range of 0 to T. Ipd represents the photocurrent generated by the photodiode. C is the total parasitic capacitance of the photodiode, drain of M1, and gate of M2. Vth is the threshold voltage of M2. In HDR mode, pixel exposure, read out, and quantization are performed at the same time. The output of proposed pixel is a dynamic waveform rather than a stable voltage, which is different from the traditional pixels.

Within $0{\sim}T$, the negative input terminal voltage of the comparator has an expression for the ramp signal $Vramp$ that rises from 0 to $VDD - Vth$:

$$Vramp = \frac{VDD - Vth}{T} \cdot t \quad (2)$$

$Vout$ is directly connected to the positive input of the comparator, and it gradually decreases from $VDD - Vth$ at the time 0. Therefore, within $0{\sim}T$, $Vout$ and $Vramp$ have only one intersection point. Let this moment be $t1$, at this

intersection, the comparator flips from high level to low level, that is, the comparator output converts the pixel output signal into a PWM signal with a pulse width of $t1$. Combine (1) and (2) to obtain the expression for pulse width:

$$Vout = VDD - \frac{Ipd \cdot t1}{C} - Vth$$
$$= Vramp = \frac{VDD - Vth}{T} \cdot t1 \quad (3)$$

$$PW_{HDR} = t1 - 0 = \frac{(VDD - Vth) \cdot C}{\frac{(VDD - Vth) \cdot C}{T} + Ipd} \quad (4)$$

where PW_{HDR} represents the pulse width of the comparator's output PWM signal in HDR mode. From (4), it can be seen that PW_{HDR} is inversely proportional to Ipd. When $Ipd = 0$, $PW_{HDR} = T$. And, when $Ipd = \infty$, $PW_{HDR} = 0$. Therefore, in an ideal scenario, the proposed pixel in the HDR mode would not experience saturation.

B. Linear Mode

When the pixel operates in the linear mode, the switch HDR is off and the key node voltage and control signal waveforms are shown in Fig. 3. The pixel is exposed within $0{\sim}T$. At the end of the exposure, $Vout$ satisfies (1) when $t = T$. The waveform of LIN is same as sel. At the time $t2{\sim}T$, the transmission gate conducts, and the pixel output $Vout$ is transmitted to the sample and hold (S/H) circuit through the column signal bus. Within $T{\sim}2T$, the positive input of the comparator is connected to the S/H, which maintains $Vout$

979-8-3503-6184-1/24 $31.00 © 2024 IEEE   262

voltage level at the time T, while the negative input of the comparator is connected to a ramp voltage $Vramp$ that rises from 0 to $VDD - Vth$. Within $T \sim 2T$, there is a unique intersection point between the S/H voltage and $Vramp$, which causes the comparator output to transition from high level to low level. Let this moment be $t3$, the expression of $t3$ regarding the photocurrent can be obtained:

$$S/H = VDD - \frac{Ipd \cdot T}{C} - Vth$$

$$= Vramp = \frac{VDD - Vth}{T} \cdot (t3 - T) \qquad (5)$$

The pulse width of the comparator output signal can be expressed as:

$$PW_{LIN} = t3 - T = T - \frac{T^2}{(VDD - Vth) \cdot C} \cdot Ipd \qquad (6)$$

where PW_{LIN} is the pulse width of the PWM signal output of the comparator in the linear mode. From (6), it can be seen that PW_{LIN} has a linear relationship with Ipd, and there is no difference from the traditional 3T pixels.

III. POST-LAYOUT SIMULATION RESULT

The proposed pixel layout adopts a stander 180nm technology, occupying an area of $6.6 \times 6.6\mu m$ and a filling factor of 37.6%, as shown in Fig. 4. Each pixel contains one photodiode, four MOSFETs, three horizontal metal wires and three vertical metal wires.

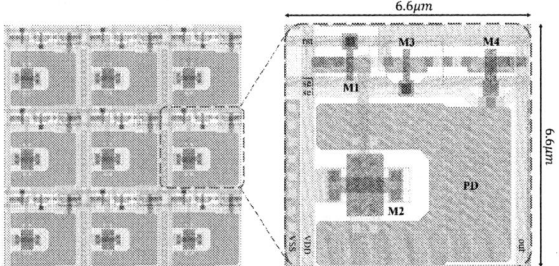

Fig. 4. Layout of proposed pixel.

The post-layout simulation uses a two-stage rail to rail operational amplifier circuit as a comparator to compare the pixel output signal $Vout$ with the ramp signal $Vramp$ and obtain the PWM signal. The parasitic parameters are extracted. The parasitic capacitance of the photodiode node is 1.6fF, and the threshold voltage of M2 is 0.91mV. Under a working voltage of 1V, set the exposure time T to 0.5ms, and simulate in linear mode and HDR mode, respectively. The curves of the relationship between the pulse duty cycle of the PWM signal and the photocurrent are shown in Fig. 5 and 6, respectively.

The post-layout simulation results show that the PWM signal in both operating modes have a duty cycle close to 100% when the photocurrent is low. They also exhibits a property of decreasing pulse width with increasing photocurrent. In the linear mode, the duty cycle of the PWM signal exhibits a good linearity within the photocurrent range of 1~6.5pA and is fully saturated at 7.6pA. In the HDR mode, the duty cycle of PWM signal exhibits an inverse relationship with the photocurrent. The response curve still shows downward trend even at a photocurrent of 300pA, which is approximately 39.5 times compared to the linear mode. Therefore it corresponds to a DR improvement of at least 31.9dB in the HDR mode respect to the linear mode.

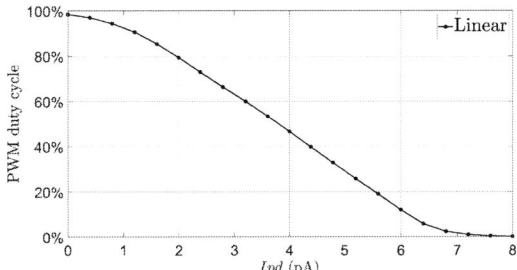

Fig. 5. Simulated response curve of the proposed pixel in linear mode.

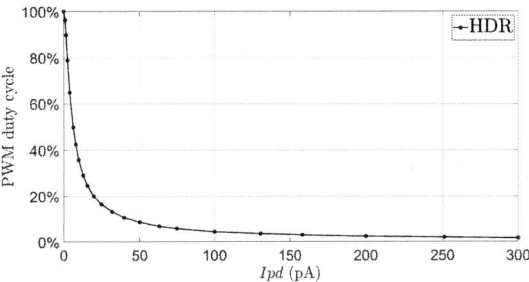

Fig. 6. Simulated response curve of the proposed pixel in HDR mode.

IV. CONCLUSION

This paper proposes an HDR pixel combined linear and inverse proportional response. The pixel circuit is composed of only 4 MOSFETs, achieves a 6.6μm pixel pitch and 37.6% fill factor under a stander 180nm CMOS technology. The use of hard reset structure and low threshold MOSFETs in the pixel circuit increases the swing of pixel output and improves the performance under the low supply voltage. In the HDR mode, the pixel output is dynamically read out onto the column bus and compared with the ramp voltage to obtain a PWM signal that responds inversely to the photocurrent, greatly expanding the dynamic range of the pixel. The post-layout simulation results show that the DR of the proposed pixel in the HDR mode is at least 31.9dB higher than that in the linear mode.

REFERENCES

[1] R. Wang, Y. Yin, L. Li, X. Wang, and Y. Chang, "A high dynamic range CMOS image sensor with dual charge transfer phase," *IEEE International Conference on Solid-State and Integrated Circuit Technology*, Hangzhou, China, 2016, pp. 1369-1371.

[2] S. Lou, Y. Qu, G. Zhong, Y. Zheng, B. Xiong, Q. Zhou, Y. Li, X. Wang, and Y. Chang, "An over 140 dB dynamic range CMOS image sensor combined DCG and logarithmic response," *IEEE Transactions on Electron Devices*, vol. 70, no. 9, pp. 4719-4724, Sep. 2023.

[3] M. Seo, T. Takasawa, K. Yasutomi, K. Kagawa, and S. Kawahito, "A low-noise high-sensitivity CMOS image sensor for scientific and industrial applications," *IEEE SENSORS*, Valencia, Spain, 2014, pp. 2163-2166.

[4] A. Chiou and C. Hsieh, "An ULV PWM CMOS imager with Adaptive-Multiple-Sampling linear response, HDR imaging, and energy harvesting," *IEEE Journal of Solid-State Circuits*, vol. 54, no. 1, pp. 298-306, Jan. 2019.

[5] N. Tu, R. Hornsey, and S. Ingram, "CMOS active pixel image sensor with combined linear and logarithmic mode operation," *IEEE Conference on Electrical and Computer Engineering*, Waterloo, Canada, 1998, pp. 754-757.

[6] H. Cheng and S. Collins, "A wide dynamic range integrating pixel with an accurately controlled logarithmic response," *IEEE International Conference on Sensing Technology*, Taipei, Taiwan, 2008, pp. 68-71.

979-8-3503-6184-1/24 $31.00 © 2024 IEEE

Enhancement of Image Sensor Pixel Performance through Ring-Shaped Vertical Transfer Gate Structure

Shuang Yan[1], Shuai Yuan*[2], Haoping Zheng[1], Yudi Zhao[1], Gang Du*[3], Junchen Dong[1], Kai Zhao*[4,1]

[1] Key Laboratory of Information and Communication Systems, Ministry of Information Industry,
Beijing Information Science and Technology University, Beijing, China
[2] School of Software and Microelectronics, Peking University, Beijing 100871, China
[3] School of Integrated Circuits, Peking University, Beijing 100871, China
4 HT-tech Jiangsu Co., Ltd., Nanjing 211806, China

* Email: shuai.yuan@stu.pku.edu.cn, gangdu@pku.edu.cn, zhaokai@bistu.edu.cn

Abstract—As the pixel size of image sensors enters the sub-micron level, the traditional transfer gate structure is difficult to meet the performance requirements of the new generation of products. In this paper, a new vertical transfer gate structure, Ring-Shaped Vertical Transfer Gate (RVTG), is proposed to enhance the gate control and increase the full well capacity (FWC). By comparing and analyzing with the conventional structure, this study shows that the RVTG structure reduces the potential barrier at the channel by 40%, which facilitates the transfer of electrons within the PD and thus reduces the image trailing. Meanwhile, the RVTG structure has a higher potential barrier at the turn-off moment, which effectively improves FWC.

Keywords—*CMOS Image Sensor, Transfer Gate, Ring-shaped Vertical Transfer Gate, Full Well Capacity*

I. INTRODUCTION

In the field of CMOS Image Sensors (CIS), with the growing demand for high-resolution and high-dynamic-range imaging, pixels become narrower and deeper[1][2]. However, this technological advancement has imposed more stringent requirements on the pixel's transfer gate structure. As the pixel pitch shrinks into the submicron region, the conventional Planar Transfer Gate (PTG)[3] encounters diminished efficiency in charge transfer and insufficiency in gate control capability, which limits its application in future higher resolution CIS. Vertical Transfer Gate (VTG)[4] was introduced to improve the electron transfer efficiency and save its footprint. Although VTG has made significant progress in improving electron transfer efficiency, there is still room for improvement in Full Well Capacity (FWC) and gate control accuracy.

This paper aims to further optimize the transfer gate structure, and propose a novel Ring-Shaped Vertical Transfer Gate (RVTG) structure. The RVTG structure further improves the concentration of the electron density and the potential profile of the pixels by its unique ring design, which leads to a more accurate gate control and a higher FWC.

II. DEVICE STRUCTURE

Fig. 1 demonstrates the pixel structures of PTG, VTG and RVTG , as well as their top views. The TG and FD footprints of the RVTG structure are smaller than those in the VTG and PTG structure. The structural and doping parameters of the RVTG are shown in Table 1, and the main parameters of the VTG are identical to those of the RVTG. The pixel size of both structures is set to 0.8μm, and the two structures are optimized with the same doping concentration and front deep trench isolation (FDTI) structure.

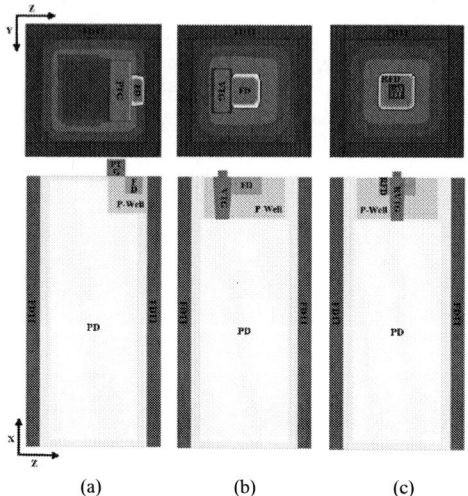

(a)   (b)   (c)
Fig. 1 Schematic structures of the three pixel units:
(a) PTG; (b) VTG; (c) RVTG

TABLE I.   STRUCTURAL PARAMETERS OF THE PROPOSED RVTG

Parameter (Unit)	Value
Pixel size(μm)	0.8
Silicon thickness(μm)	3.7
DTI depth(μm)	3.7
DTI width(nm)	100
RVTG depth(nm)	300
Channel doping(cm^{-3})	5×10^{16}
FD doping(cm^{-3})	5×10^{19}
PD doping(cm^{-3})	8×10^{16}
Well doping(cm^{-3})	$1 \times 10^{17} \sim 2.5 \times 10^{17}$

The VTG is a vertical gate structure that controls the conduction or cutoff between the photodiode (PD) and the Floating Diffusion (FD) region. The FD is typically a region that is heavily doped adjacent to PTG or VTG. The proposed RVTG structure introduces a Ring-shaped Floating Diffusion (RFD) region around the VTG's terminal to form a three-dimensional circular inversion layer at the ON state, and then to increase the effective gate width. The doping and volume of the FD region should be precisely set to accommodate charges equal to the pixel's full-well capacity, such that the RFD is narrower in the lateral dimension, as is shown in Fig.1 (b) and (c). The FD acts as a drain for a bulk MOSFET, thus having a significant impact on the channel electric potential.

Due to its smaller lateral dimensions, the RFD design should have better gate control capabilities compared to the FD.

In this study, the three-dimensional pixel structure was simulated using the TCAD tool Sentaurus[4], which employs the drift-diffusion model, the Philips uniform mobility model, the high-field saturation model, and the Shockley-Read-Hall (SRH)[6] to generate a composite model, and ray-tracing for the Optical simulation.

III. RESULTS AND DISCUSSION

By intercepting the 2D potential diagram of the RVTG structure at specific moments and extracting the potential profile along the path shown in Fig. 2(a), the potential profiles before and after reset and under different light intensities at an exposure time of 6μs are plotted in Fig. 2(b). As a high-level reset pulse is applied to the TG, an inversion layer is formed at the TG oxide surface, and the electrons in the PD are transferred to the FD. After reset, the potential inside the PD rises to 1.3V at X=0.8μm, and the PD is ready for exposure. At the exposure stage, the accumulation of photogenerated electrons leads to a decrease in the potential inside the PD, which is proportional to the light intensity.

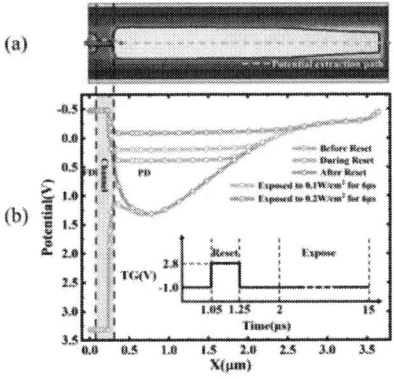

Fig. 2 Potential profiles of RVTG at different moments

To evaluate the effect of light intensity on the FWC of the RVTG pixels, its photo response under different light conditions is compared in Fig. 3. With the increase of light intensity, the photoelectron accumulation rate increases and the pixel reaches saturation faster. As the charge photogeneration rate equilibrates with the charge loss rate, the FWC of the pixel is determined. Due to the feedforward[7] effect, the FWC slightly increases with the growth of light intensity, and reached it maximum value of 7.8ke- when TG = -1.0V, and the light intensity is 0.3W/cm².

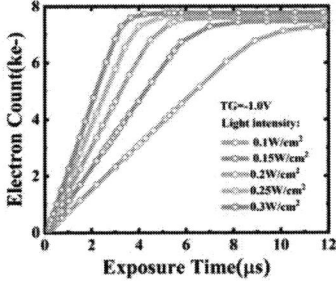

Fig. 3 Light response under different light intensities, TG=-1.0V

In order to analyze the FWC of VTG and RVTG structure pixels, the light intensity was fixed to 0.2W/cm² and the maximum number of electrons of the two structure pixels was compared at different negative TG voltages. As demonstrated in Fig. 4, the FWC of both structures increases accordingly as the turn-off voltage decreases gradually. The FWC of RVTG is significantly larger than that of VTG when the turn-off voltage is -0.6V~0V, while the FWC of RVTG is slightly lower than that of VTG when the turn-off voltage is -0.6V~-1.0V. When the turn-off voltage is -0.6V, the FWC of the two structures is basically equal, at around 7.5ke-.

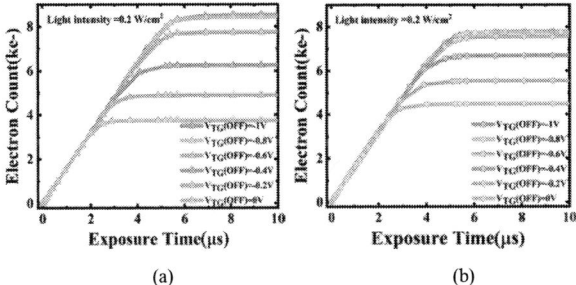

(a)   (b)

Fig. 4 Maximum number of electrons for (a)VTG structure pixel and (b)RVTG structure pixel at different turn-off voltages

The potential profiles of the VTG and RVTG structures at -0.2V, -0.6V, and -1.0V turn-off voltages were extracted along the path shown in Fig. 5 at a light intensity of 0.2W/cm², respectively. As shown in Fig. 6, a potential barrier exists at the channel position of X=0.4μm for both VTG and RVTG structures when the transfer gate is on. Compared to the VTG structure, the RVTG structure has a 40% lower potential barrier at the channel and the potential barrier is significantly smoother than that of the VTG structure, an advantage that facilitates the subsequent transfer of electrons within the PD and thus reduces image trailing.

At a TG voltage of -0.2V, the RVTG structure has a higher potential barrier and is able to accommodate more electrons compared to the VTG structure. With the decrease of TG voltage, both VTG and RVTG have increasing potential barriers at X=0.2μm, and the gap between the two potential barriers is gradually narrowing. At a turn-off voltage of -0.6V, the FWC of the two structures are essentially equal. As the turn-off voltage reaches -1.0V, although the potential barrier heights of the two structures are basically the same, the VTG structure is able to accommodate more electrons with the same barrier height attributing to its higher maximum potential inside the PD after reset, and thus has a slightly higher FWC than the RVTG structure.

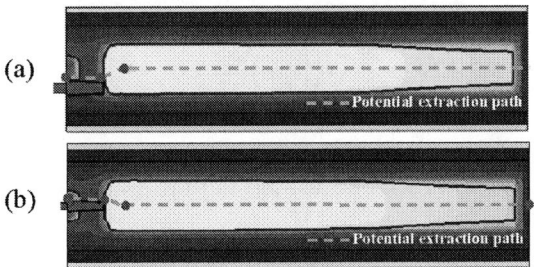

Fig. 5 Potential extraction paths for two pixel structures, (a)VTG, (b)RVTG

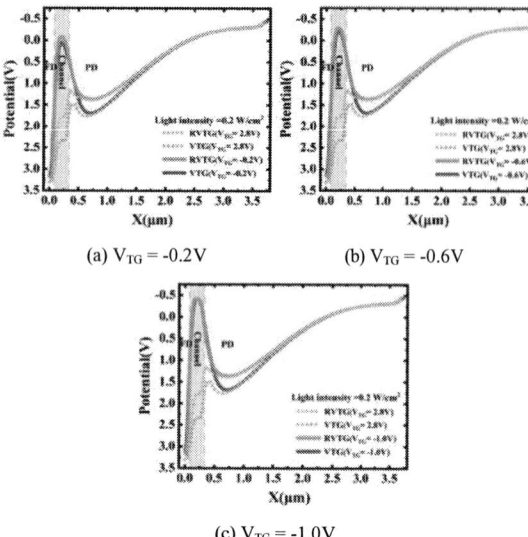

(a) $V_{TG} = -0.2V$ (b) $V_{TG} = -0.6V$

(c) $V_{TG} = -1.0V$

Fig. 6 Potential profiles of the two structures at different negetive transfer gate voltages

Fig. 8 illustrates the potential distribution from point A to point B along the path shown in Fig. 7 in both the ON and OFF states of the TG. As shown in Fig. 8(a), at the moment when the TG is turned on, the channel potential in the RVTG structure is significantly lower than that of the VTG structure, which results in a highest potential inside the PD of the VTG structure being higher than that of the RVTG structure. In the RVTG structure, electrons concentrate more closely to the gate oxide than that of the VTG structure, exhibiting superior gate control ability. Furthermore, the RVTG channel, with its ring-shaped channel layout, not only occupies a smaller footprint but also exerts less influence on the electric potential of the surrounding areas, which is beneficial for reducing read-out noise.

As shown in Fig. 8(b), the channel potential in the RVTG structure is much lower than that in the VTG structure at the moment of transfer gate turned off, which can improve the potential barrier of the RVTG structure at the channel and effectively enhance the control ability of the transfer channel, thus improving FWC. Moreover, the channel electron concentration of a RVTG structure is significantly lower than that of a VTG structure, and thus the leakage current along the channel at OFF state should be much lower for the RVTG structure.

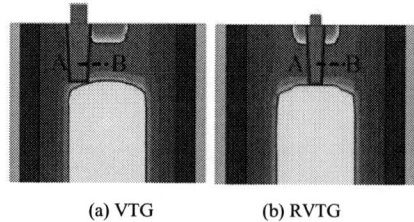

(a) VTG   (b) RVTG

Fig. 7 Potential extraction paths for two pixel structures

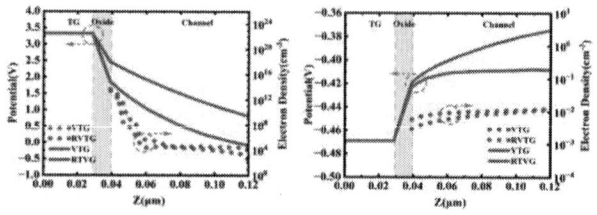

(a)   (b)

Fig. 8 Potential profiles of VTG and RVTG at the moment of (a)TG is ON, and (b)TG is OFF

IV. SUMMARY

A RVTG structure for submicron CIS pixel is proposed and evaluated in this paper. The FWC of the proposed RVTG structure is superior to that of VTG structure with the turn-off voltage is larger than -0.6V. The FWC gradually increases as the turn-off voltage decreases, and it also increases with the enhancement of light intensities. The RVTG structure has a lower potential in the channel than the VTG structure during OFF stage, which creates a higher potential barrier and better transfer gate control capability, thus increasing the FWC of the pixel structure. At the ON state, the RVTG structure has a lower potential barrier along the channel, which is more conducive to efficiently transferring electrons inside the PD, and reducing image trailing.

ACKNOWLEDGMENTS

This research was funded by the National Key R&D Program of China (Grant No. 2022YFB4402000), National Natural Science Foundation of China (Grant No. 92264105), Beijing Natural Science Foundation (Grant No. 4232067), and the R&D Program of Beijing Municipal Education Commission (Grant No. KM202211232007).

REFERENCES

[1] Y. Kim et al., "A 1/2.8-inch 24Mpixel CMOS image sensor with 0.9μm unit pixels separated by full-depth deep-trench isolation," 2018 IEEE International Solid-State Circuits Conference - (ISSCC), San Francisco, CA, USA, 2018, pp. 84-86.

[2] D. Park et al., "A 0.8 μm Smart Dual Conversion Gain Pixel for 64 Megapixels CMOS Image Sensor with 12k e- Full-Well Capacitance and Low Dark Noise," 2019 IEEE International Electron Devices Meeting (IEDM), San Francisco, CA, USA, 2019, pp. 16.2.1-16.2.4.

[3] S. Kim, J. H. Kim, U. Kwon, K. Lee and D. S. Kim, "Potential Engineering to Enhance Transfer Characteristics of Advanced CIS Pixel based on VTG - FDTI scheme," 2021 International Conference on Simulation of Semiconductor Processes and Devices (SISPAD), Dallas, TX, USA, 2021, pp. 297-300.

[4] J. H. Park et al., "Optimization of Photodiode Design Through Analysis of Full-Well Capacity and Image Lag in 0.5 μm CMOS Image Sensors With Vertical Transfer Gates," in IEEE Electron Device Letters, vol. 43, no. 10, pp. 1697-1700, Oct. 2022.

[5] Sentaurus Device Manual. Mountain View, CA, USA, 2013. Synopsys Inc.

[6] B. Ruch, M. Jech, G. Pobegen and T. Grasser, "Applicability of Shockley-Read-Hall Theory for Interface States," 2020 IEEE International Electron Devices Meeting (IEDM), San Francisco, CA, USA, 2020, pp. 22.1.1-22.1.4.

[7] M. Sarkar, B. Buttgen and A. J. P. Theuwissen, "Feedforward Effect in Standard CMOS Pinned Photodiodes," in IEEE Transactions on Electron Devices, vol. 60, no. 3, pp. 1154-1161, March. 2013.

979-8-3503-6184-1/24 $31.00 © 2024 IEEE

A 10-GHz Low Power Class-C VCO with Long-Term Reliability and Tunable Performance in 28 nm FD-SOI for Satellite Communications

Yann Deval[1], Henrique Iha Taguti[1], Ayoub Ait Ihda[1], Hervé Lapuyade[1], Stephane Rochette[2], and

François Rivet[1]

[1] IMS Lab., Univ. Bordeaux, Bordeaux INP, CNRS UMR 5218, Talence, France
[2] Thales Alenia Space, Toulouse, France
Email: yann.deval@ims-bordeaux.fr

Abstract—A VCO was designed in 28-nm FD-SOI for robustness against aging mechanisms such as Time Dependent Dielectric Breakdown (TDDB) and Hot Carrier Injection (HCI). A high-swing class-C oscillator featuring a control circuit was chosen as the base of the architecture. A cascode topology was adopted to improve reliability and a local capacitive feedback was used to improve transductance. A combination of design and data analysis on aging effects of the technology contributed to the design of a robust and suitable device for space applications with state-of-the-art performances. The VCO has a frequency range of 10-11.7 GHz, a tuning range of up to 16.4%, a power consumption down to 4.88 mw, 0.073 mm^2 of area and achieves a phase noise of down to -114 dBc/Hz, a FoM of 185 dBc/Hz and a FoM$_T$ of 190 dBc/Hz at 1 MHz offset, regardless of the two transistors used instead of just one to improve robustness.

Index Terms—FD-SOI, frequency synthesis, hot carrier injection (HCI), phase noise tuning, reliability, satellite communications, time dependent dielectric breakdown (TDDB), voltage-controlled oscillator (VCO).

I. INTRODUCTION

In this work, we propose a reliable and adaptable VCO with a performance trade-off [1]. The oscillator operates in class-C with a transformer-based resonator for efficient power consumption. The fully depleted silicon-on insulation (FD-SOI) short-channel devices time-dependent dielectric breakdown (TDDB) and time to breakdown (TBD) are significantly affected by the drain voltage, as discussed in [2]. A cascode structure in the active device is used mitigate HCI and improve TDDB by splitting electric field stress across two transistors. The reliability of the circuit is improved even under a large oscillation amplitude. FDSOI transistors allow variation of the threshold voltage through body biasing, enabling adjustments of the oscillator performance and avoiding the addition of lossy varactors. It allows flexible trade-offs to address each specific need accordingly: a better phase-noise at the cost of power consumption when stringent requirements are necessary; a lower power mode for relaxed and idle cases; or a calibration for achieving just the right level of performance, without lacking or being excessive without necessity.

II. VCO DESIGN FOR SPACE AND RELIABILITY

A. Understanding Hazards

Power yields to a significant constraint in space due to the challenges of heat dissipation in vacuum. Electronics rely on conduction and radiation for heat dissipation, as there is no air flow to carry heat away. Reducing power consumption and managing heat become crucial in extending the circuit lifespan.

Wear-out and aging effects are exacerbated by technology scaling, increasing electric fields and temperature. Large voltage fluctuations can further deteriorate reliability due to phenomena such as TDDB and HCI [3], [4].

TDDB is a major concern in IC reliability, which causes irreversible damage to the dielectric layer, and eventually leads to breakdown and failure of the device [3], [4]. HCI can shift V_{th} and decrease the mobility of the conducting carrier (i.e., degrade drain current, gain in transconductance, phase noise, voltage swing) [3].

B. Remarks and Considerations for the Design

VCOs are widely used in transceivers. As an application example of data links in satellites, we can mention two use cases [5]: the connection of the satellite with Earth base station, which requires an accurate and stable carrier frequency from the VCO; and a more relaxed and less stringent performance for connecting with other satellites in the same constellation – due to shorter distance, no atmospheric losses and less propagation losses.

C. Circuit Design

The class-C oscillator offers higher current efficiency compared to class-B due to its pulse-shaped current [6], though voltage efficiency remains similar. Tohidian et al. [7] propose a high swing class-C oscillator without a current source, enhancing voltage swing and efficiency. The control circuit addresses the trade-off between robust start-up and maximum amplitude.

The increased oscillation voltage centered on V_{DD}, however, creates reliability concerns due to accelerated gate oxide deterioration, particularly concerning TDDB.

979-8-3503-6184-1/24 $31.00 © 2024 IEEE

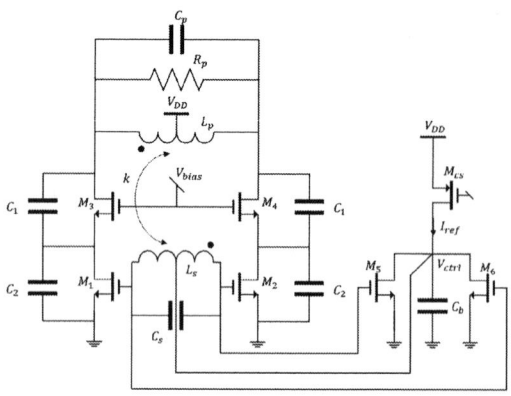

Fig. 1. Proposed class-C oscillator.

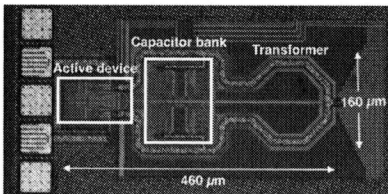

Fig. 2. Chip microphotograph.

Dedicated study carried out by STMicroelectronics [2] demonstrated a strong reduction of the breakdown time (TBD) for 28-nm FDSOI short-channel devices when increasing the drain voltage, especially for values higher than 0.6 V.

Fig. 1 presents the proposed high-swing class C oscillator with a cascode structure, which mitigates HCI and improves TDDB by splitting the electric field stress between two transistors. This modification of the architecture will provide a significant increase of TBD (i.e. increase its lifespan) as demonstrated in [2]. The back-gate of 28 nm FD-SOI transistors offers wide tuning range, allowing control over amplitude and phase noise.

It is important to note that, while improving long-term reliability, the cascode structure reduces output voltage headroom, and this degrades phase noise.

III. MEASUREMENTS

The proposed class-C VCO has been implemented in 28 nm FD-SOI technology and placed in a 48-pin QFN package.

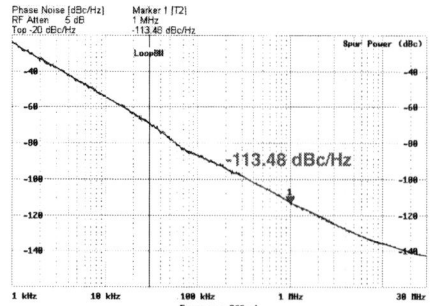

Fig. 3. Measured phase noise at 1 MHz for a frequency of 11.72 GHz.

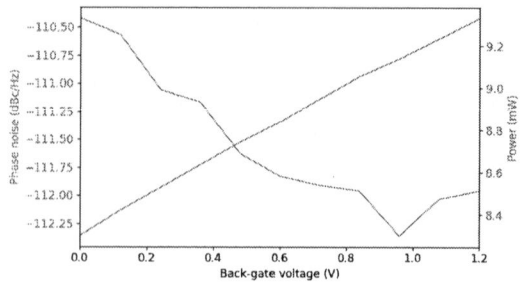

Fig. 4. Measured phase noise at 1MHz offset and power versus back-gate voltage.

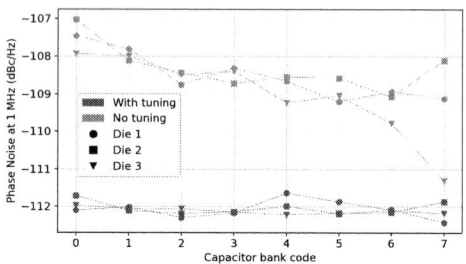

Fig. 5. Measurements of phase noise versus frequency tuning range for three different dies. Comparison between phase noise with tuning to achieve -112 dBc/Hz (blue curve), without tuning (orange curve).

The VCO core occupies an area of 160 μm x 460 μm (Fig. 2). The circuit includes an output buffer to drive the 50 Ω load of the measurement instrument and to isolate the circuit. The DC voltage is generated by a dedicated PCB using batteries as the power supply. The RF PCB has a socket for measuring different dies, paving the way for industrial robustness demonstration. The output spectrum and phase noise are measured with R&S FSUP26 spectrum analyzer.

The VCO consumes 8.8 mW from a 1.2 V supply and 4.88 mW from a 1 V supply. It operates from 10.00 GHz to 11.74 GHz, with a tuning range of 16.0% centered around 10.87 GHz at 1.2 V supply; at 1 V supply the VCO performs similarly, operating from 9.95-11.73 GHz, with a tuning range of 16.4%, with slightly higher phase noise (-113.86 dBc/Hz vs -111.83 dBc/Hz at 1 MHz of offset, respectively). All frequency bands overlap, ensuring continuous frequency tuning without blind spots.

Applying back-bias voltage and/or changing the supply voltage from 1 V to 1.2 V have little impact on the output frequency. Both operation modes demonstrated a similar overall FoM (around 185.5 dBc/Hz). This possibility allows an even larger range of performance tuning, while maintaining the specification frequency range.

Fig. 4 shows how the body bias voltage affects the performance of the VCO. It improves the phase noise at the cost of power consumption. Fig. 5 shows the low variation between different dies, validating industrial robustness, a key parameter for space applications. Moreover, the capability of employing body biasing allows phase noise tuning (e.g. -112 dBc/Hz). Figs. 4, 5 and 6 demonstrate the ability to optimize

979-8-3503-6184-1/24 $31.00 © 2024 IEEE

TABLE I
VCO COMPARISON

Ref.	Technology	Frequency (GHz)	TR (%)	Vdd (V)	PN @ 1MHz (dBc/Hz)	Power (mW)	Area (mm2)	FoM @ 1MHz	FoM$_T$ @ 1MHz
[8]	0.12 um SiGe BiCMOS	9.9-12.45	23	2	-122	54	0.8*	183	190
[9]	40 nm CMOS	11.74-12.63	7.3	1.1	-120	13.23	0.04	190.5	187.8
[10]	65 nm CMOS	7.76-8.62	10.5	0.6	-120	10.6	0.38	188.4	188.8
[11]	90 nm CMOS	9.5-11.7	10.6	1.2	-117	13.3	0.2	185.4	185.9
[12]	22 nm FDSOI	10.5	24.3	0.4	-97	0.57	0.16	179.9	187.9
[13]	22 nm FDSOI	6-13	70	1	-85.1	4.92	0.02	158.2	174.8
[14]	22 nm FDSOI	13.3-16.2	19.7	1	-119.7	35	0.05	187	192.9
This work	**28 nm FDSOI**	**10.00-11.74**	**16.0**	**1.2**	**-113.9**	**8.8**	**0.07**	**185.4**	**189.5**
		9.95-11.73	**16.4**	**1.0**	**-111.8**	**4.88**	**0.07**	**185.6**	**189.9**

* Including pads

$$FoM = -PN(\Delta\omega) + 20log_{10}(\tfrac{\omega_0}{\Delta\omega}) - 10log_{10}(P_{mW})$$

$$FoM_T = FoM + 20log_{10}(\tfrac{TN}{10})$$

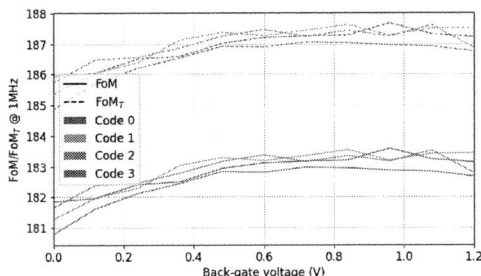

Fig. 6. Figure of Merit (FoM and FoM$_T$) versus back-gate voltage for different codes (calculated with measured values of each point).

the oscillator.

Table I summarizes the performance of the proposed VCO and compares it to state-of-the-art VCOs. Reliability comes at the cost of performance. Due to its application requirements and purpose, the VCO does not need to achieve a phase noise as low as the others at Table I. The main challenge in this work was to design a high-performance and low-cost VCO while being reliable at the same time.

IV. CONCLUSION

Electronics for space applications require reliability and state-of-the-art performance. Unfortunately, most electronics are designed for high performance and low cost, disregarding reliability. A high-swing class-C oscillator was chosen as the base of the architecture for achieving good phase noise and low-power. A cascode topology was used to improve reliability against TDDB and increase its lifespan, at the cost of power and phase noise. A combination of design and data from the foundry on aging effects, specifically addressing TDDB for the 28-nm FDSOI technology, allowed the design of a robust and adaptable VCO with state-of-the-art performance.

In this work, we presented a VCO designed in 28-nm FD-SOI robust against aging mechanisms such as TDDB and HCI. The VCO has a frequency range of 10-11.7 GHz, a tuning range of up to 16.4%, a power consumption of down to 4.88 mW, 1-to-1.2 V supply, 0.073 mm^2 of area, achieves a phase noise of down to -114 dBc/Hz, a FoM of 185.6 dBc/Hz and a FoM$_T$ of 189.9 dBc/Hz at 1 MHz offset.

REFERENCES

[1] A. Ait Ihda, *Low power, high-reliability class-C voltage-controlled oscillator for space application in 28 nm FD-SOI technology*. Ph.D. Thesis, Université de Bordeaux, Dec. 2022.

[2] X. Federspiel, M. Rafik, M. Arabi, A. Cros, and F. Cacho, "FDSOI Mosfet gate dielectric breakdown Vd dependancy," in *2018 International Integrated Reliability Workshop (IIRW)*, pp. 1–4, 2018.

[3] M. Babaie and R. B. Staszewski, "A study of RF oscillator reliability in nanoscale CMOS," in *2013 European Conference on Circuit Theory and Design (ECCTD)*, pp. 1–4, 2013.

[4] E. Maricau and G. Gielen, *Analog IC Reliability in Nanometer CMOS*. Springer New York, 2013.

[5] Y. Deval, A. Cathelin, R. Guillaume, A. A. Ihda, H. Lapuyade, and F. Rivet, "Benefits of an FD-SOI feature: Optimal power budget of wireless links through phase noise tuning," *2019 IFIP/IEEE International Conference on Very Large Scale Integration (VLSI-SoC)*, 2019.

[6] J. Chen, F. Jonsson, H. Olsson, L.-R. Zheng, and D. Zhou, "A current shaping technique to lower phase noise in LC oscillators," in *2008 15th IEEE International Conference on Electronics, Circuits and Systems*, pp. 392–395, 2008.

[7] M. Tohidian, A. Fotowat-Ahmadi, M. Kamarei, and F. Ndagijimana, "High-swing class-C VCO," in *2011 Proceedings of the ESSCIRC (ESSCIRC)*, pp. 495–498, 2011.

[8] E. Wagner, O. Shana'a, and G. M. Rebeiz, "A Very Low Phase-Noise Transformer-Coupled Oscillator and PLL for 5G Communications in 0.12 μ m SiGe BiCMOS," *IEEE Transactions on Microwave Theory and Techniques*, vol. 68, no. 4, pp. 1529–1541, 2020.

[9] X. Liu, J. Jin, C. Yang, Y. Liu, and J. Zhou, "A 12-GHz Transformer Feedback Class-F$_{2,3}$ Voltage-Controlled Oscillator Using Noise Circulating With FoM of 190.5 dBc/Hz," *IEEE Microwave and Wireless Components Letters*, vol. 31, no. 11, pp. 1231–1234, 2021.

[10] H. Jia, B. Chi, and Z. Wang, "An 8.2 GHz triple coupling low-phase-noise class-F QVCO in 65nm CMOS," in *ESSCIRC Conference 2015 - 41st European Solid-State Circuits Conference (ESSCIRC)*, pp. 124–127, 2015.

[11] P.-Y. Wang, M.-C. Chou, Y.-T. Chen, Y.-C. Chang, D.-C. Chang, and S. S. H. Hsu, "A Ku-band low-phase-noise transformer coupled VCO for satellite communications," in *2016 IEEE International Symposium on Radio-Frequency Integration Technology (RFIT)*, pp. 1–3, 2016.

[12] P. Kumar, J. Al-Eryani, B. T. Ulaşlı, E. Böhme, F. Vanselow, E. N. Isa, and L. Maurer, "A 400 mV, widest-tuning- VCO with a central Frequency of 10.5 GHz and FTR of 2.5 GHz, designed in 22 nm FDSOI CMOS technology," in *2019 Joint International EUROSOI Workshop and International Conference on Ultimate Integration on Silicon (EUROSOI-ULIS)*, pp. 1–4, 2019.

[13] L. Szilagyi, Z. Tibenszky, C. Carta, R. Henker, and F. Ellinger, "6-to-13 GHz Voltage Controlled Oscillator with 7 dBm Output Power in 22 nm FD-SOI," in *2021 IEEE 21st Annual Wireless and Microwave Technology Conference (WAMICON)*, pp. 1–4, 2021.

[14] S. Balamurali, G. Mangraviti, Z. Zhong, P. Wambacq, and J. Craninckx, "A 13-16 GHz Low-Noise Oscillator with Enhanced Tank Energy in 22-nm FDSOI," in *ESSCIRC 2023- IEEE 49th European Solid State Circuits Conference (ESSCIRC)*, pp. 125–128, 2023.

979-8-3503-6184-1/24 $31.00 © 2024 IEEE

A 191-GHz Harmonic Oscillator with Self-Feeding Line and Return-Path Gap Coupler Structure in 65nm CMOS

Xiaohan Shen, Chen Jiang*

State Key Laboratory of Integrated Chips and Systems (SKLICS), Fudan University, Shanghai, China

* Email: cjiang@fudan.edu.cn

Abstract—This paper presents a 191-GHz harmonic oscillator with high DC-to-RF efficiency and small physical size in a 65nm CMOS technology. This harmonic oscillator utilizes a novel structure to simultaneously maximize the fundamental oscillation and second harmonic extraction. The post-layout simulation indicates that, the 191-GHz oscillator achieves a peak DC-to-RF efficiency of 1.83%, with an output power of 0.13 dBm. The harmonic oscillator achieves a phase noise of -100.4 dBc/Hz at 1MHz offset as well as a FoM of -159.1 dBc/Hz, which shows promising applications in future THz radar systems and 6G wireless communication systems.

Keywords—Harmonic oscillator, THz, Self-Feeding Line, Return-Path Gap Coupler.

I. INTRODUCTION

The past decade has witnessed a rapid development in sub-millimeter and THz applications, such as imaging, sensing, spectroscopy and wireless communication. And they are widely used in many areas, such as security and industrial inspection, biomedical analysis, material science, space communication and so on. However, as the working frequency of the applications is close to f_{max} of the transistor, how to realize high-power THz sources remains to be a critical challenge.

There are two approaches to achieve an over-100-GHz signal on silicon: harmonic oscillators and multiplier chains. They are both based on the nonlinearity of the device. But compared with multiplier chains, harmonic oscillators don't need RF inputs. Moreover, they are more power efficient and compact.

From a systematic view, several performance metrics are desired for signal generation: high output power, wide tuning range, high DC-to-RF efficiency, small physical size, and low phase noise. High output power means high SNR for THz communication systems. Wide tuning range is crucial for spectroscopy, indicating large spectrum coverage. High DC-to-RF efficiency and small physical size can enhance battery life and reduce fabrication costs. Low phase noise indicates signal generation stability and measuring sensitivity.

This paper presents an efficient 191-GHz harmonic oscillator with self-feeding line and return-path gap coupler structure. With an optimized transistor layout structure, the power and efficiency of the oscillator is enhanced further. Designed in a 65nm CMOS technology, the post-layout simulation indicates that this oscillator achieves a peak DC-to-RF efficiency of 1.83%, as well as an output power of 0.13dBm.

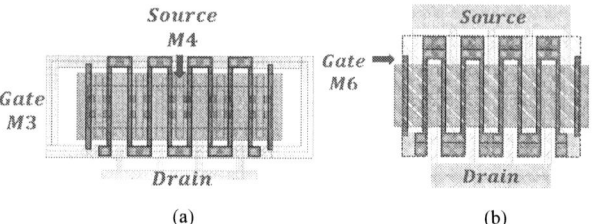

Fig. 1. The standard layout (a) and the optimized layout (b) of transistors

At 1MHz offset, the simulated phase noise is -100.4 dBc/Hz. All the above performances of the oscillator show promising applications in future integrated THz systems.

II. DESIGN OF HARMONIC OSCILLATOR

A. RF Device Investigation

In the high frequency oscillators, transistors are working near their activity boundary, which means high performance sensitivity for transistor embedding. Therefore, the bottom-up design method is used in order to achieve a better overall performance.

Generally speaking, a MOSFET transistor can be regarded as a two-port network, thus we can use small signal y-parameter to describe its properties. Moreover, as the working frequency is approached to f_{max}, it is more challenging to design an efficient oscillator. we use unilateral power gain U (also called Mason's U) [1] as an indicator of the transistor's performance.

The TSMC 65nm CMOS process has already provided nmos_rf model and standard layout for RF integrated circuit design, as shown in Fig. 1(a). However, there are some limitations of this model: only 2 vias between PO and M1 layer for gate; large drain-source (M2-M4) overlap; short distance for gate-source (M3-M4) and gate-drain (M3-M2). These factors can increase gate resistance and parasitic capacitance, which decrease the behavior of the transistor at high frequency. In order to enhance the performance of the transistor, we optimize the layout by adding 2 vias to connect PO and M1, separating drain from source to avoid overlap, using M6 for gate to increase the distance with source and drain [2]. Fig. 1(b) shows the optimized layout. Besides, we also investigate the optimal width per finger, optimal finger number and multiplier number for Mason's U. As a trade-off, optimized transistors with 4 multipliers, each has 8 fingers with 1μm width, are chosen for this oscillator design. As shown in Fig. 2, the post-layout simulation indicates that the Mason's U of the optimized layout is 7.33 at 95 GHz (V_g =

979-8-3503-6184-1/24 $31.00 © 2024 IEEE

0.96V, $V_d = 1.2V$), compared to 6.32 for the standard layout, with nearly 16% improvement for output power generation.

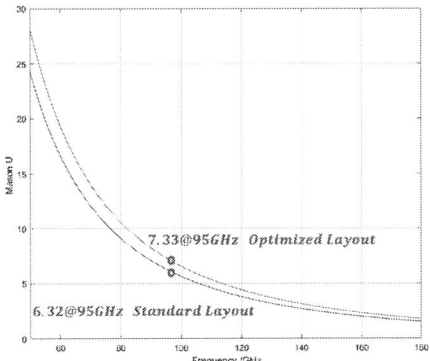

Fig. 2. Mason' U of the standard layout and the optimized layout

B. Self-Feeding Line and Return-Path Gap Coupler Structure

For the harmonic oscillators, the harmonic signal generation is based on the nonlinearity of the transistors. The nonlinearity is greater as the fundamental voltage swing gets larger. In order to maximize the output power, we need to meet the requirements for maximum fundamental oscillation [3]. Due to the existence of the parasitic capacitance C_{GDO} and gate resistance R_G, the optimal phase between gate and drain at 95 GHz is 164° rather than 180°. As a result, we need to use self-feeding lines to satisfy the phase condition. The self-feeding line is realized with grounded CPW lines. The signal trace is realized with *2μm* wide M9 metal layer, while the ground plane and ground wall are realized with M1-M3 metal overlap and M1-M9 metal stack, respectively. The distance between two ground walls is *12μm*. The simulated characteristic impedance of the self-feeding line is *42Ω*.

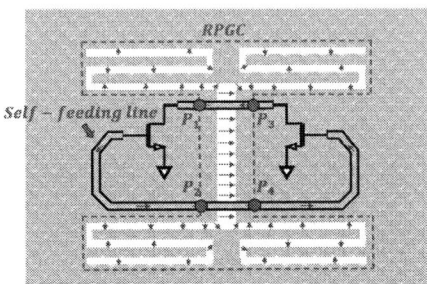

Fig. 3. Self-feeding line and return-path gap coupler structure

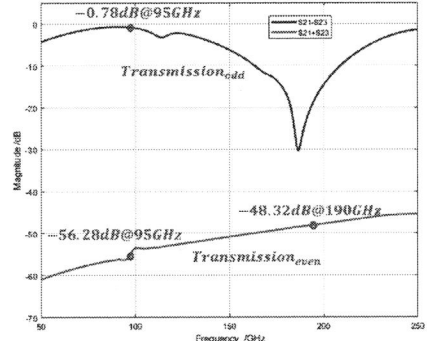

Fig. 4. the simulated results of transmission in odd and even mode of the return-path gap coupler

As for the harmonic output port, we desire to extract harmonic power and avoid fundamental signal. Besides, the half equivalent circuit for conventional push-push harmonic oscillator at harmonic frequency indicates that, the MOSFET is connected like a diode, which means most of the harmonic power is escaped through the channel and can't be collected through a certain port. This is called the self-power loading /cancellation effect [4]. In order to separate gate from drain at harmonic frequency, a return-path gap coupler (RPGC) structure is employed. There is a wide slot line in the middle, each side lies a transistor with self-feeding lines connected. Two λ/4 slot lines are placed both on the top and bottom of the structure, so standing waves can exist in the middle section of the slot line. In order to reduce the radiation loss, the two slot lines on the top and bottom are split into two, respectively, and bent to the horizontal direction, so that the generated far fields will cancel each other, as shown in Fig.3. The electromagnetic field simulation shows that the transmission in odd mode between P1(P3) and P2(P4) is -0.78 dB at 95 GHz, while the transmission in even mode is below -40 dB in all concerned range, as shown in Fig.4. As a result, the oscillator can only oscillate out-of-phase at the fundamental frequency. The generated second harmonic signal is in-phase. According to the electromagnetic field simulation results, the even mode transmission is blocked, which means isolation between gate and drain at the harmonic frequency is formed, eliminating the self-power loading/cancellation effect.

From the output port at the middle of the self-feeding line on the top, the fundamental signals generated by two transistors are out-of-phase, so they can cancel each other. While the harmonic signals are in-phase, the power is added at the port. In the end, we realize the maximum power generation and the harmonic signal extraction using self-feeding line and return-path gap coupler structure.

C. Probing Pad and Output Matching

Probing pad is the structure for chip testing. It is a passive network connected into a *50Ω* load. The pad is realized with *54μm × 60μm* M9 metal layer. According to HFSS simulation, the impedance of the probing pad structure is $Z_A = (56.8 - j11.3) \Omega$. But the optimal output impedance for the oscillator is $Z_B = (6.0 - j14.6) \Omega$. In this case, we need an impedance matching network to transform Z_A into Z_B.

In this paper, the T-type open-circuit transmission lines are utilized for impedance matching. The impedance matching scheme is shown in Fig.5.

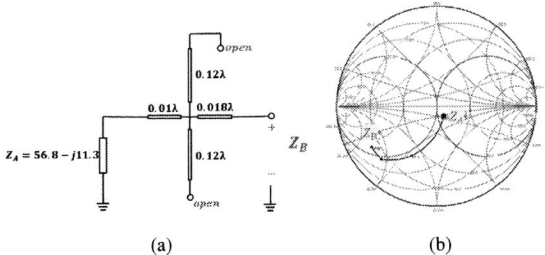

Fig. 5. T-type open-circuit transmission lines matching network (a) and the matching scheme on a Smith chart (b)

III. POST-LAYOUT SIMULATION RESULTS

The 191-GHz harmonic oscillator with self-feeding line and return-path gap coupler structure is designed using a TSMC 65nm CMOS process. The full layout area is *320μm×*

TABLE I
PERFORMANCE COMPARASION OF SOME SUB-MILLIMETER-WAVE OSCILLATORS

Ref.	Technology	Frequency (GHz)	Output Power (dBm)	DC-to-RF Efficiency	Phase Noise (dBc/Hz)	Tuning Range	DC Power (mW)	Area (mm²)	|FoM| (dBc/Hz)
ISSCC 2016[5]	130-nm SiGe	190.5	-2.1	0.21%	-102.6@10MHz	20.7%	183-294	0.64	139.9
TMTT 2013[6]	130-nm SiGe	212	-7.1	0.65%	-92@1MHz	2.8%	30	0.073	156.4
RFIC 2016[3]	130-nm SiGe	210	1.4	2.4%	-87.5@1MHz	10.6%	26-61	0.08	160.1
JSSC 2013[4]	65-nm CMOS	260	0.5	0.14%	-78@1MHz	1.45%	800	2.3	120.3
JSSC 2012[7]	65-nm CMOS	290	-1.2	0.23%	-78@1MHz	4.5%	325	0.36	132.3
This Work	65-nm CMOS	191	0.13	1.83%	-100.4@1MHz	0.63%	56.2	0.094	159.1

Calculated based on $FoM = L(\Delta f) - 20lg(\frac{f_0}{\Delta f}) + 10lg(\frac{P_{DC}}{1mW}\frac{1}{\eta_{osc}}\frac{10}{TR})$

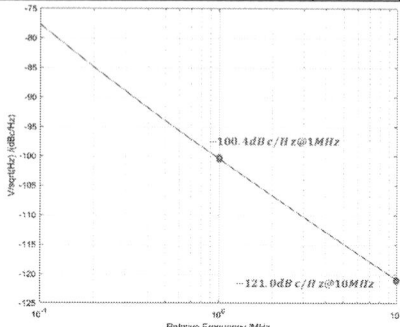

Fig. 8. Simulated phase noise of the oscillator

294μm, while the core area of the oscillator is *320μm× 92μm*. The layout of the harmonic oscillator and the 3-D view of the upper metal layer structure is shown in Fig.6, including the self-feeding lines, the return-path gap coupler structure, the impedance matching structure and the probing pad structure.

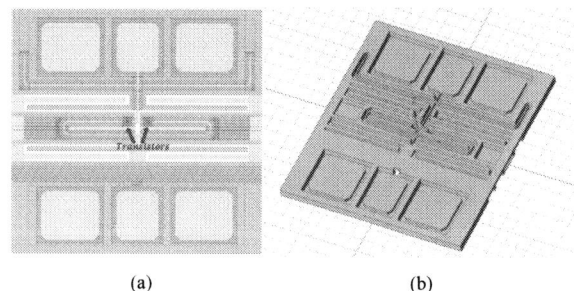

(a)   (b)

Fig. 6. Layout of the harmonic oscillator (a) and 3-D view of the upper metal layer structure (b)

According to post-layout simulation results, the DC power of the oscillator is 56.2mW with the gate and drain biases of 0.96V and 1.2V, respectively. The 191-GHz oscillator achieves a peak DC-to-RF efficiency of 1.83%. At 27°C, the output power of the oscillator is 0.13 dBm, while 0.94 dBm at -23°C and -0.90 dBm at 77°C, respectively. By setting the Gate bias V_g from 0.6V to 1.2V, the simulated oscillating frequency changes from 192.1 GHz (0.6V) to 190.9 GHz (at 0.92V), which means 0.63% of tuning range. The simulated results under different V_g are shown in Fig. 7.

Fig.8 shows the simulation result of the phase noise after considering the parasitic components. At 1MHz offset, the simulated phase noise reaches -100.4 dBc/Hz. Compared to other prior works in table I, this oscillator achieves a FoM of -159.1 dBc/Hz.

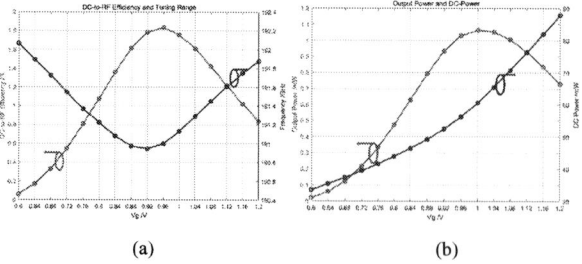

(a)   (b)

Fig. 7. Simulated results of the oscillator: (a) DC-to-RF efficiency and tuning range, (b) output power and DC power under different V_g

IV. CONCLUSION

In this paper, a 191-GHz harmonic oscillator is presented. By optimizing the MOSFET layout, the power generating performance is partly enhanced. With the self-feeding line and the return-path gap coupler structure, this oscillator achieves high DC-to-RF efficiency and small physical size, which makes it suitable for future THz radar systems and wireless communication systems.

REFERENCE

[1] Hamid Khatibi, Somayeh Khiyabani, Ehsan Afshari, "An Efficient High-Power Fundamental Oscillator Above fmax/2: A Systematic Design," *IEEE Trans. Microw. Theory Technol.*, vol. 65, no. 11, pp. 4176-4189, Nov. 2017.

[2] Chun Wang, Hans Herdian, Wenbin Zheng, Chenxin Liu, Jill Mayeda, Yuxuan Liu, Olivia Angel Yong, Wenqian Wang, Yuncheng Zhang, Carrel da Gomez, Abanob Shehata, Sena Kato, Ibrahim Abdo, Teruo Jyo, Hiroshi Hamada, Hiroyuki Takahashi, Hiroyuki Sakai, Atsushi Shirane, Kenichi Okada, "A 236-to-266GHz 4-Element Amplifier-Last Phased-Array Transmitter in 65nm CMOS", *in IEEE Int. Solid-State Circuits Conf. Dig.*, pp.416–417, Feb. 2024.

[3] Chen Jiang, Andreia Cathelin, and Ehsan Afshari, "An efficient 210 GHz compact harmonic oscillator with 1.4dBm peak output power and 10.6% tuning range in 130nm BiCMOS," *in 2016 IEEE Radio Frequency Integrated Circuits Symposium (RFIC)*, pp.194–197, May 2016.

[4] Ruonan Han and Ehsan Afshari, "A CMOS high-power broadband 260-GHz radiator array for spectroscopy," *IEEE J. Solid-State Circuits*, vol. 48, pp. 3090–3104, Dec. 2013.

[5] R. Kananizadeh and O. Momeni, "A 190.5GHz mode-switching VCO with 20.7% continuous tuning range and maximum power of-2.1dBm in 0.13μm CMOS," *in IEEE Int. Solid-State Circuits Conf. Dig.*, pp. 52–53, Feb. 2016.

[6] P. Chiang, O. Momeni and P. Heydari, "A 200-GHz inductively tuned VCO with-7dBm output power in 130-nm SiGe BiCMOS," *IEEE Trans. Microw. Theory Technol.*, vol. 61, pp. 3666–3673, Oct. 2013.

[7] Y. M. Tousi, O. Momeni and E. Afshari, "A novel CMOS high-power terahertz VCO based on coupled oscillators: theory and implementation," IEEE *J. Solid-State Circuits*, vol. 47, no. 12, pp. 3032–3042, Dec. 2012.

979-8-3503-6184-1/24 $31.00 © 2024 IEEE

A Blocker-Tolerant High-Linear Receiver Employing Baseband Noise-Cancelling and Bottom-Plate Switched-Capacitor Techniques

Chenxiang Cai [1], Gengzhen Qi [1*]

[1] School of Microelectronics Science and Technology, Sun Yat-sen University, Zhuhai, China
* Email: qigzh@mail.sysu.edu.cn

Abstract—**This article proposes a radio frequency (RF) wideband blocker-tolerant receiver (RX) that covers 0.5-to-2GHz. It features the baseband noise-cancelling technique and bottom-plate switched-capacitor (SC) N-path filter to improve the noise figure (NF) and linearity. Unlike [2] that achieves noise-cancelling at RF, our proposed BB noise-cancelling structure saves the number of passive switches (i.e., mixers), saving dynamic power, and extends the RF operation. Besides, the bottom-plate SC N-path filter is added at the RF input, effectively enhancing the out-of-band (OOB) rejection, thus improving the linearity. At baseband, a negative capacitance is introduced around the trans-impedance amplifier (TIA) to extend the passband bandwidth (BW). Implemented in a 65nm CMOS technology, the simulation results present that the receiver achieves a 51.8-to-52.7dB voltage gain and a 17.7-to-24.1MHz passband BW when covering a 0.5-to-2GHz local oscillator (LO) frequencies. It also exhibits a 2.3dB NF and a 39.8dBm OOB-IIP$_3$ at 80MHz offset. The power consumption is 62.1-to-90.3mW.**

Keywords—*bandpass filter, baseband, mixer-first, N-path filter, negative capacitance, noise-cancelling, noise figure, out-of-band linearity, single-ended.*

I. INTRODUCTION

With the continuously development of wireless communications, the growing demand for high transmission speed and signal leads to great challenges for the radio frequency (RF) transceivers, especially the RF receivers (RXs). Considering the off-chip RF filters (e.g., SAW filters) increase both the cost and the form factor of the system, the blocker-tolerant RXs are becoming the research hotspot to deal with both strong out-of-band (OOB) blockers and self-interference from the transmitter. However, the classical RXs can only achieve effective OOB blocker suppression for narrowband due to fixed external RF filtering, thus multiple RXs are needed to cover a large number of RF bands. Thus, a tunable wideband RX is desired to cover different RF bands. Meanwhile, high linearity is crucial for the signal quality, especially in a frequency-division duplexing (FDD) system. Besides, wide passband BW is also necessary to support high speed data transmission. Therefore, the RXs need to achieve wide passband BW, high linearity and low noise figure (NF). In [1], a high-linearity RX combining 2-stage N-path filtering with bottom-plate mixing is proposed, and it raises the OOB-IIP$_3$ up to 44dBm while sacrificing NF (10.3dB). In [2], the frequency-translational noise-cancelling receiver employs two separate down-conversion paths to achieve noise-

cancelling, which shows low NF (1.9dB) and comparable OOB-IIP$_3$ (13.5dBm), but sacrificing the passband BW (2MHz). To improve both the linearity and passband BW, a baseband noise-cancelling receiver is introduced in [3] and achieves 18dBm OOB-IIP$_3$ and 175MHz baseband BW, yet it pays too much power consumption (172mW).

Herein, we propose a wideband blocker-tolerant RX to optimize the linearity, NF and passband BW, by featuring the techniques of baseband noise-cancelling topology and bottom-plate switched-capacitor (SC) N-path filter. Besides, at baseband (BB) a negative capacitance is introduced around the trans-impedance amplifier (TIA) to extend the passband BW. Implemented in 65nm CMOS process, the simulated power consumption is 62.1-to-90.3mW under 1.2V power supply. The proposed RX achieves a 51.8-to-52.7dB voltage gain and a 17.7-to-24.1MHz passband BW at 0.5-to-2GHz RF range. Meanwhile, it exhibits a 2.3dB NF thanks to the baseband noise-cancelling technique and a 39.8dBm OOB-IIP$_3$ at 80MHz offset.

II. RECEIVER ARCHITECTURE AND IMPLEMENTATION

The schematic of our proposed RX is shown in Fig. 1, which is based on the bottom-plate SC *N*-path filter and the mixer-first topology. Our RX employs BB noise-cancelling and BW-extended techniques to improve both the NF and passband BW. The linearity (i.e., IIP$_3$) is relative to the OOB rejection of the gain response, especially at the RF input side. Thus, a bottom-plate SC *N*-path pre-filter is introduced, in which the switch is 180μm/60nm. The capacitor C_1 is sized 60pF to achieve high roll-off filtering. The BB resistor R_m (500Ω) are employed to achieve input impedance matching due to the frequency-translational property of N-path mixers. However, NF will degrade due to the thermal noise R_m, therefore the noise-cancelling technique can be a prospective solution. According to [3], the auxiliary noise-cancelling path uses a low-noise transconductance amplifier (LNTA) and a TIA to cancel the thermal noise of the matching resistors. Since a LNTA operating at the baseband achieves higher performance than that at RF frequency, our proposed receiver introduces this auxiliary path in BB and the auxiliary class-AB transconductance G_m is designed 100mS. In addition, a capacitive positive feedback path across the TIA can extend the channel BW, thanks to the negative capacitance (NC) created due to the Miller theory [5]. $C_{n,main}$ and $C_{n,aux}$ in the main and auxiliary paths are designed 50 and 100pF, respectively, to extend the BW. Finally, signals are recombined by properly scaling each path. R_{rec} is designed 500Ω to achieve NF optimization while not degrading the BB gain.

This work was supported in part by the National Natural Science Foundation of China (NSFC) under Grant 62104263, and in part by the GuangDong Basic and Applied Basic Research Foundation under Grant 2022A1515-012295. (Corresponding author: Gengzhen Qi)

979-8-3503-6184-1/24 $31.00 © 2024 IEEE

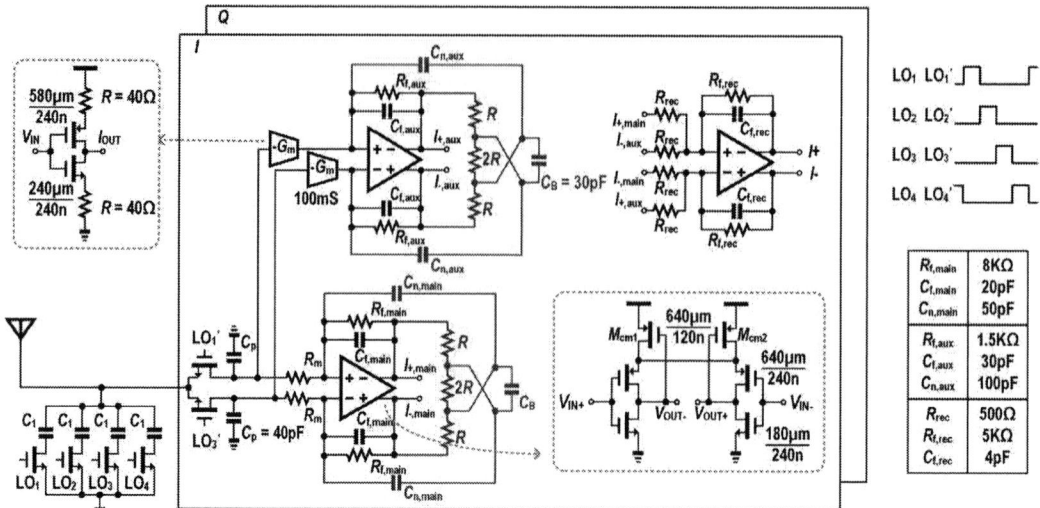

Fig. 1. Schematic of the proposed receiver.

A. Bottom-Plate SC N-path Filtering

The *N*-path filter is a passive SC *N*-path network. For classical top-plate mixing in Fig. 2(a), when the switch is closed, V_S is variable due to the source terminal connected to a capacitor, resulting in V_{GS} being a variable value. Therefore, V_{GS} generate the modulation due to the square law, resulting in the non-linearity. To resolve it, we employ the bottom-plate mixing technique in Fig. 2(b). V_S is connected to the ground instead of a capacitor [1]. When the switch is closed, V_D will be pulled to ground and V_{GS} will be a constant, thereby improving the linearity. Besides, reducing R_{SW} of the switches can improve the linearity further [4]. The OOB-IIP$_3$ (in voltage) can be computed to be [7]:

$$V_{IIP3} = \sqrt{\frac{4}{3} \frac{(1+\rho)^4}{\rho^3 [2g_2^2 - g_3(1+\rho)]}} \qquad (1)$$

where $\rho = R_{SW}/R_S$, $g_2 = -[2(V_{GS}-V_{TH})]^{-1}$ and $g_2 = -(2V_{SAT}^2)^{-1}$. From (1), it can be concluded that large mixer switches (320µm/120nm) can effectively improve IIP$_3$.

B. LNTA, Low-Noise TIA and Negative Capacitance

LNTA exploits an inverter-based amplifier with degenerated resistors. The transconductances of PMOS and NMOS are designed equally (50 mS) to improve the linearity [8]. A CMOS inverter can serve as a low-noise quite-linear BB TIA, and the stability concerns for the closed-loop operation are avoided in the single-stage amplifier. $M_{cm1,2}$ are used for the common-mode feedback, and they only generate common mode noise which will be cancelled at the differential output. Due to the differential architecture, $-A_a$ for the NC can simply be implemented by wire crossing shown in Fig. 1, while low-ohmic resistors R (40Ω) realize high-linearity with $A_a = 0.5$. And the performance is mainly determined by R-to-R and C-to-C ratios, hence, it is insensitive to process, voltage and temperature (PVT) variations.

C. 4-Phase 25%-Duty-Cycle LO Generator

Fig. 4 shows the proposed 25%-duty-cycle LO generator (LOGEN). The self-biased input buffers are employed to amplify the input signals and make 2LO$_P$ and 2LO$_N$ 50%-duty-cycle. After a divider-by-2 and "AND" logic, we obtain the 25%-duty-cycle LO waveforms (LO$_{1-4}$). The non-overlap

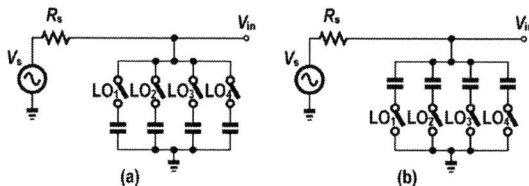

Fig. 2. (a) Top-plate and (b) bottom-plate mixing *N*-path filter.

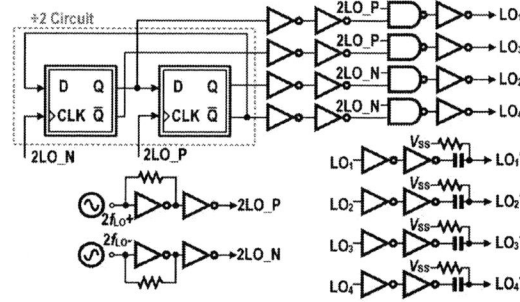

Fig. 3. Schematic of the four-phase 25%-duty-cycle LO generator.

LO waveforms can be obtained by optimizing the size of NMOS and PMOS transistors in the output buffer.

III. SIMULATED RESULTS AND COMPARISON

A. Simulated results

The power consumption is simulated 62.1-to-90.3mW at 0.5-to-2GHz RF range, in which the static power is 51.2mW due to BB LNTA and TIA cells, and the rest is the dynamic power due to the LO generator. In Fig. 4(a), the simulated RF to BB gain ranges from 51.8-to-52.7 dB with a 17.7-to-24.1MHz RF BW, and S_{11} is <−10dB. The OOB rejection at 80MHz offset reaches up to 62dB, which shows a sharp roll-off at OOB frequency to guarantee the effective suppression on the blocker signals. In Fig. 4(b), the IIP$_3$ profile is simulated at [f_{LO}+80MHz] and [f_{LO}+159 MHz], and the receiver achieves a high OOB-IIP$_3$ from 36.2-to-39.8dBm mainly due to the *N*-path filter and the large mixer switches. It also exhibits a 2.3-to-3.9dB NF, thanks to the noise cancellation and high BB

979-8-3503-6184-1/24 $31.00 © 2024 IEEE

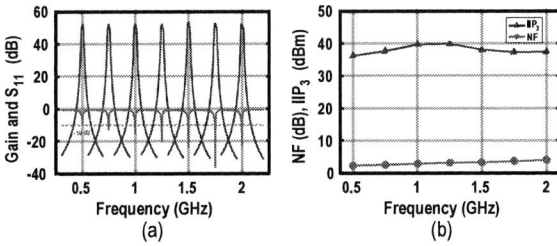

Fig. 4. Simulated (a) gain, BW and S_{11}, (b) NF and IIP3.

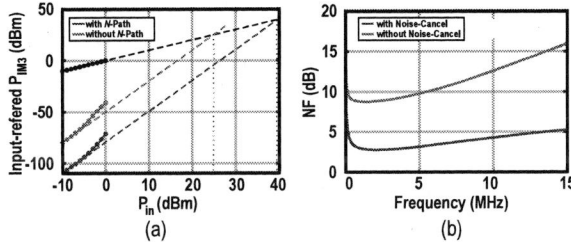

Fig. 5. (a) Simulated IIP3 with and without N-path filter. (b) Simulated NF with and without baseband noise-cancelling.

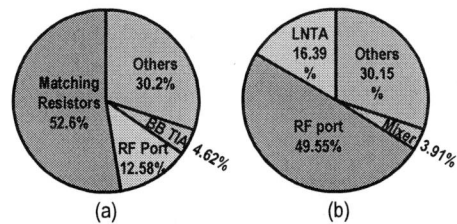

Fig. 6. Noise contribution break-up percentage of (a) without baseband noise-cancelling and (b) with noise-cancelling.

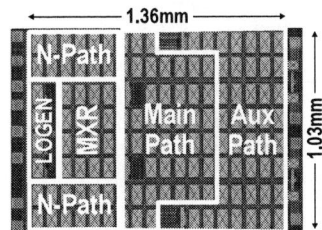

Fig. 7. Layout of our proposed RX in 65nm CMOS process.

voltage gain. Simulate at 1GHz, Fig. 5(a) shows that N-path filter improves OOB-IIP3 from 25-to-39.7dBm. Before introducing the noise-cancelling path, R_m is the most noise contributor as shown in Fig. 6(a). In Fig. 6(b), the RF port contributes most noise due to the BB noise-cancelling technique, which improve NF from 8.8 to 2.8dB. The layout of our proposed RX is show in Fig. 7 in 65nm CMOS process.

B. Comparison with the Prior Art

Table I benchmarks this work with respect to the state-of-the-art receivers. While the receiver in [1] achieves high IIP3, our receiver shows >30dB higher gain and >6dB lower NF at 2GHz. Compared with [2], our receiver achieves a much wider BW and >20dB larger OOB-IIP3. The receiver in [3] consumes the static power 162.4mW and the dynamic power 1.7mW/GHz under the 0.83V supply, our receiver achieves much less power consumption and >15dB larger OOB-IIP3. With comparable OOB-IIP3, our receiver achieves wider RF range and lower NF when compared with [9].

TABLE I. SUMMARY AND COMPARISON

Reference	[1]	[2]	[3]	This Work
Technology	28nm CMOS	40nm CMOS	22nm CMOS	65nm CMOS
RF Range (GHz)	0.1 to 2.0	0.08 to 2.7	1 to 6	0.5 to 2
Gain (dB)	16	72	22.4	51.8 to 52.7
BW (MHz)	13	4	175	17.7 to 24.1
DSB NF (dB)	4.1 to 10.3	1.9@2GHz	2.5 to 5	2.3 to 3.9
OOB-IIP3 (dBm)@Offset	44@80MHz	13.5@80MHz	18*@1GHz	39.8@80MHz
Power (mW)	34 to 96	35.1 to 78	172	62.1 to 90.3
Supply (V)	1.2/1.0	1.3	0.83	1.2

* Estimate from figure

IV. CONCLUSION

In this article, a baseband noise-cancelling wideband receiver is achieved. An N-path filter is introduced to pre-filter, the OOB interferes and improve the linearity. The main RX path exploits large mixer switches to improve linearity and NF, and linear baseband resistors for the input matching. To cancel the thermal noise of baseband resistors, the auxiliary noise-cancelling path is introduced. Finally, the NC extend BW. Implemented in a 65nm CMOS process, the receiver achieves a 51.8-to-52.7 dB gain and a 17.7-to-24.1MHz BW across 0.5-to-2 GHz LO frequencies. It also exhibits a 2.3dB NF and a 39.8dBm OOB-IIP3 at 80MHz offset. The power consumption is 62.1-to-90.3mW.

REFERENCES

[1] Y. -C. Lien, E. A. M. Klumperink, B. Tenbroek, J. Strange and B. Nauta, "High-Linearity Bottom-Plate Mixing Technique with Switch Sharing for N-path Filters/Mixers," *IEEE J. Solid-State Circuits*, vol. 54, no. 2, pp. 323-335, Feb. 2019.

[2] D. Murphy *et al.*, "A Blocker-Tolerant, Noise-Cancelling Receiver Suitable for Wideband Wireless Applications," *IEEE J. Solid-State Circuits*, vol. 47, no. 12, pp. 2943-2963, Dec. 2012.

[3] A. N. Bhat, R. A. R. van der Zee and B. Nauta, "A Baseband-Matching-Resistor Noise-Canceling Receiver with a Three-Stage Inverter-Only OpAmp for High In-Band IIP3 and Wide IF Applications," *IEEE J. Solid-State Circuits*, vol. 56, no. 7, pp. 1994-2006, July 2021.

[4] J. W. Park and B. Razavi, "Channel Selection at RF Using Miller Bandpass Filters," *IEEE J. Solid-State Circuits*, vol. 49, no. 12, pp. 3063-3078, Dec. 2014.

[5] Y. -C. Lien, E. A. M. Klumperink, B. Tenbroek, J. Strange and B. Nauta, "Enhanced-Selectivity High-Linearity Low-Noise Mixer-First Receiver with Complex Pole Pair Due to Capacitive Positive Feedback," *IEEE J. Solid-State Circuits*, vol. 53, no. 5, pp. 1348-1360, May 2018.

[6] C. Andrews and A. C. Molnar, "Implications of Passive Mixer Transparency for Impedance Matching and Noise Figure in Passive Mixer-First Receivers," *IEEE Trans. Circuits Syst. I, Reg. Papers*, vol. 57, no. 12, pp. 3092-3103, Dec. 2010.

[7] D. Yang, C. Andrews and A. Molnar, "Optimized Design of N-Phase Passive Mixer-First Receivers in Wideband Operation," *IEEE Trans. Circuits Syst. I, Reg. Papers*, vol. 62, no. 11, pp. 2759-2770, Nov. 2015.

[8] H. Kundur Subramaniyan, E. A. M. Klumperink, V. Srinivasan, A. Kiaei and B. Nauta, "RF Transconductor Linearization Robust to Process, Voltage and Temperature Variations," *IEEE J. Solid-State Circuits*, vol. 50, no. 11, pp. 2591-2602, Nov. 2015.

[9] G. Qi, B. van Liempd, P. -I. Mak, R. P. Martins and J. Craninckx, "A SAW-Less Tunable RF Front End for FDD and IBFD Combining an Electrical-Balance Duplexer and a Switched-LC N-Path LNA," *IEEE J. Solid-State Circuits*, vol. 53, no. 5, pp. 1431-1442, May 2018.

A High Sensitivity Series-Parallel Rectifier with Pre-Bias for RF Energy Harvesting Systems

HaiQin Wu [1], Dejian Li [2], Xin Jin [2], Xufeng Liao[1], and Lianxi Liu[1]*

[1] School of Integrated Circuits, Xidian University, Xi'an 710071, China.
[2] Beijing Smart-Chip Microelectronic Technology Co., Ltd, Beijing, 100192, China.

* Email: lxliu@mail.xidian.edu.cn

Abstract—This paper introduces a high-sensitivity rectifier for RF energy harvesting systems. It utilizes a series-parallel configuration to increase the output voltage and then optimize the sensitivity, without increasing the number of rectifier stages. To address the reverse leakage issue in CCDD structures during alternating MOSFET conduction, dynamic threshold voltage adjustment and gate with pre-bias are proposed, which balances forward conduction loss and reverse leakage, achieving optimal PCE. The proposed rectifier, implemented in 65nm technology, achieves a 1V output at -34.5dBm input power at 915MHz, with a peak PCE of 49.28% at -24.6dBm input power.

Keywords—*CMOS rectifier, high sensitivity, high power conversion efficiency, RF energy harvesting,*

I. INTRODUCTION

In recent years, RF energy harvesting technology has found increasing applications in IoT and wireless sensor networks. The main parameters for evaluating the performance of RF rectifiers include Power Conversion Efficiency (PCE) and sensitivity. To improve sensitivity, multi-stage series-connected CCDD rectifiers are typically used[1][4]. However, as the number of stages increases, parasitic effects significantly intensify, thereby reducing the PCE of the rectifier. Besides, PCE is mainly influenced by forward conduction loss and reverse leakage current. Many studies have employed various methods to reduce the threshold voltage to decrease forward conduction loss[2]. However, this also results in increased reverse leakage current. Therefore, it is essential to balance forward conduction loss and reverse leakage.

This paper proposes a series-parallel rectifier with gate pre-biasing. Without increasing the number of rectifier stages, a significant improvement in sensitivity was achieved. By utilizing the substrate modulation effect, the MOSFET threshold voltage is dynamically adjusted. Additionally, pre-bias is introduced to the MOSFETs' gates. This approach

balances forward conduction loss and reverse leakage current to improve PCE.

II. OVERALL DESIGN OF THE PROPOSED CIRCUIT

To effectively improve the sensitivity and PCE of the rectifier, the overall circuit design proposed in this work is shown in Figure 1. The rectifier adopts a 4-stage series-parallel configuration to achieve higher sensitivity. The substrate of the MOSFETs is connected to the gate, dynamically adjusting the threshold voltage. Additionally, gate biasing is introduced, with each NMOS gate biased at the midpoint of the previous stage's output and each PMOS gate biased at the midpoint output of the current stage. This effectively improves the sensitivity and PCE of the rectifier.

III. SERIES-PARALLEL CONFIGURED RECTIFIER FOR ENHANCED OUTPUT VOLTAGE

The traditional CCDD rectifier is shown in Figure 2. The input signals V_{RFP} and V_{RFN} are differential voltage signals. When V_{RFP} is higher than V_{RFN}, M_{P1} and M_{N1} are turned on, and the current flow is indicated by the blue arrows in Figure 2(a). Similarly, when V_{RFP} is lower than V_{RFN}, M_{P2}, and M_{N2} are turned on, and the current flow is as indicated by the red arrows in Figure 2(b).

The sensitivity of the rectifier is typically defined as the minimum input power needed to achieve an output of 1V when the load is either open-circuit or 10MΩ. To enhance the sensitivity of the rectifier, it is often necessary to increase the number of rectifier stages. Many studies have utilized N-stage voltage multiplier rectifiers. As shown in Figure 3(a), taking a 2-stage CCDD rectifier cascade as an example, each stage is connected in series to provide the required V_{OUT}. However, as the number of stages increases, parasitic effects significantly intensify, leading to a reduction in the rectifier's PCE.

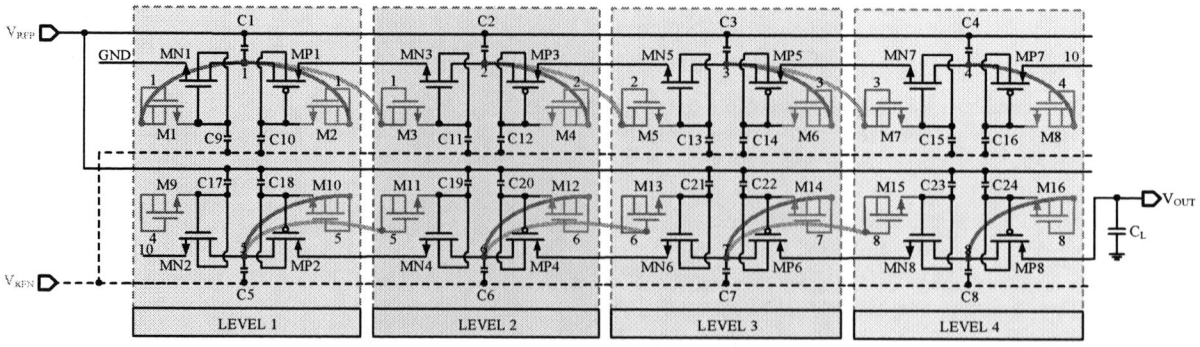

Fig. 1. Overall Design of the Proposed Circuit.

979-8-3503-6184-1/24 $31.00 © 2024 IEEE

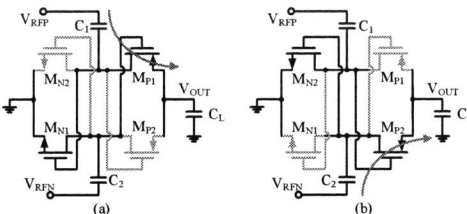

Fig. 2.   Operational Principle of the CCDD Rectifier

To address this issue, this design proposes a series-parallel configured rectifier, as shown in Figure 3(b). In traditional series rectifiers, the initial potential during charge transfer is always 0. The output voltage is increased by coupling the input signal and transmitting it stage by stage. However, in the series-parallel structure rectifier, the initial voltage of the lower half rectifier is no longer 0. Instead, it is connected to the output of the last stage in the upper half rectifier coupled to the V_{RFP} signal. This structure can double the output voltage of the traditional series configuration without adding extra circuitry. It significantly improves the rectifier's sensitivity without reducing conversion efficiency.

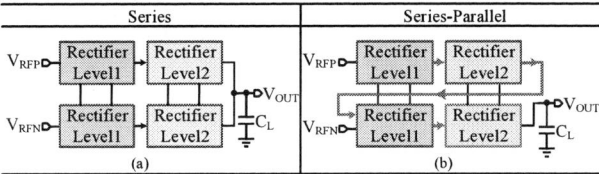

Fig. 3.   Schematic of Series and Series-Parallel Configured Rectifiers

IV. DYNAMIC THRESHOLD ADJUSTMENT AND PRE-BIAS

To address the reverse leakage issue in traditional CCDD rectifiers, this paper adopts a rectifier structure. This structure features shared gate and substrate potentials, as shown in Figure 4. By lowering the threshold voltage of the conducting MOSFETs while increasing the threshold voltage of the cutoff MOSFETs. This design reduces conduction loss and minimizes reverse leakage, thereby improving the rectifier's PCE.

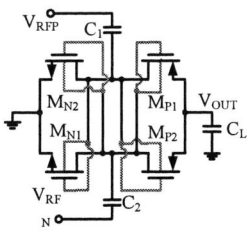

	$V_{RFP} + V_{RFN} -$		$V_{RFP} + V_{RFN} -$	
M_{N1}	on	V_{TH} ⤵	off	V_{TH} ⤴
M_{P1}	on	V_{TH} ⤵	off	V_{TH} ⤴
M_{N2}	off	V_{TH} ⤴	on	V_{TH} ⤵
M_{P2}	off	V_{TH} ⤴	on	V_{TH} ⤵

Fig. 4.   Schematic of Reverse Leakage Current in Traditional Structure

The rectifier performance is significantly impacted when the RF energy harvesting system operates at extremely low input power levels. Weak RF energy may not provide sufficient voltage to drive the circuit, leading to insufficient gate voltage for MOSFET conduction. To solve this problem, this design further optimizes the dynamic threshold voltage adjustment structure proposed in section III.A. Diode-connected MOSFET biasing is introduced at the gate of NMOS and PMOS. The gate voltage is supplied by both the introduced bias and the coupled input signal, providing sufficient conduction voltage for the MOSFETs.

The conduction resistance of MOSFETs in the linear and subthreshold regions is calculated as follows:

$$R_{ON} = \frac{1}{\mu_n C_{OX} \frac{W}{L}(V_{GS} - V_{TH})} \qquad (1)$$

$$R_{ON(subthreshold)} = \frac{V_T}{I_{D0} \exp(\frac{V_{GS} - V_{TH}}{\eta V_T})(1 - \exp(-\frac{V_{DS}}{V_T}))} \qquad (2)$$

where μ_n is the electron mobility, C_{OX} is the gate oxide capacitance per unit area, W is the gate width, L is the gate length, V_{GS} is the gate-source voltage, V_{TH} is the threshold voltage, I_{D0} is a factor related to process and dimensions, η is the subthreshold slope factor (typically between 1 and 2), and V_{DS} is the drain-source voltage. From the above calculations, it can be seen that the conduction resistance of the MOSFET is inversely proportional to V_{GS}. Thus, introducing a bias increases V_{GS}, reducing conduction resistance and improving MOSFET performance. However, this also lowers the conduction resistance during cutoff, worsening reverse leakage.

Therefore, to achieve optimal PCE, it is crucial to balance the effects of forward conduction loss and reverse leakage on the rectifier. Figure 5 shows the second stage coupled to the positive input terminal in the overall rectifier circuit of Figure 1. Taking NMOS as an example, the gate biases that can be introduced include pre-biasing by 1.5 stages, 1 stage, 0.5 stages, and 0 stages. Introducing a high gate bias reduces conduction loss but increases the reverse leakage from the source to the drain of the MOSFET. Conversely, introducing a low gate bias increases conduction loss but reduces reverse leakage. Therefore, in this design, the NMOS gate is biased at 1 stage ahead and the PMOS gate is biased at the output of the current stage, as shown by the solid line in Fig.5.

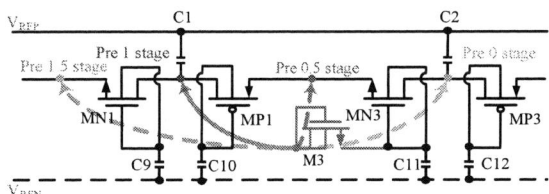

Fig. 5. Schematic of NMOS gate bias at different positions

V. SIMULATION RESULTS

The proposed series-parallel rectifier with pre-biasing is designed and implemented using a 65nm standard process. The layout is shown in Figure 6. The chip area for the RF energy harvesting system is 1.08mm × 1.4mm, with the core rectifier area being 389μm × 235μm. The external circuit includes a balun and L-match impedance matching network that provides the differential RF input signal needed to drive the rectifier. As shown in Figure 7, the circuit achieves impedance matching at the operating frequency of 915MHz. In this design, NMOS transistors use a deep N-well process, which dynamically adjusts the threshold voltage. This enhances the rectifier's sensitivity and efficiency.

Figure 8 shows the simulation results for different biasing methods. It can be observed that the rectifier achieves the highest efficiency when the NMOS gate is biased at 1 stages ahead and the PMOS gate is biased at the output of the current stages.

979-8-3503-6184-1/24 $31.00 © 2024 IEEE

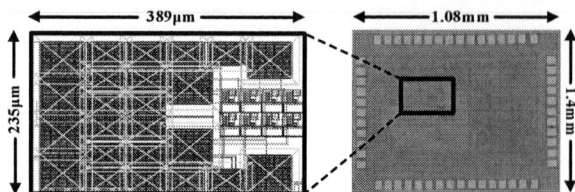

Fig. 6. Layout of the proposed series-parallel rectifier with pre-biasing

Simulations were conducted for the proposed rectifier, the traditional 4 stages series-CCDD rectifier, the traditional 8 stages series-CCDD rectifier, and the rectifier without biasing to compare their PCE under different input power P_{IN} levels. The simulation results are shown in Figure 9. The results indicate that the proposed rectifier maintains a high PCE at input power levels below -22dBm. Additionally, at an input power of -34.5dBm, the rectifier can output a voltage of 1V. At an input power of -24.6dBm, the PCE_{MAX} reaches 49.28%. Compared to other structures, the proposed rectifier demonstrates superior performance in sensitivity and PCE under low input power conditions.

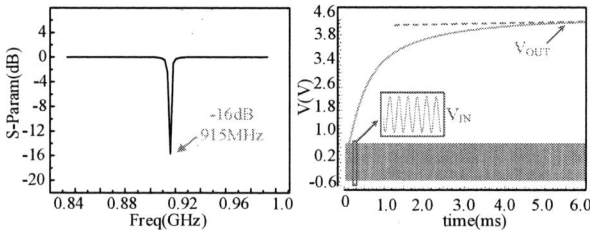

Fig. 7. Impedance Matching and Circuit Transient Simulation

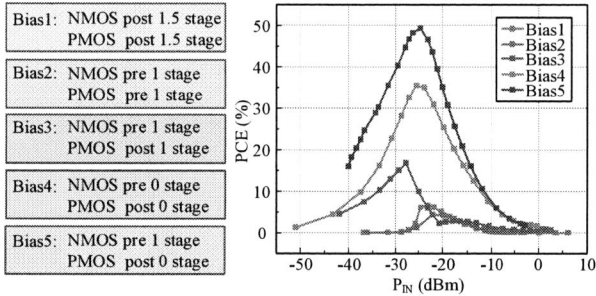

Fig. 8.Relationship between PCE and P_{IN} for Rectifiers with Different Biasing Positions

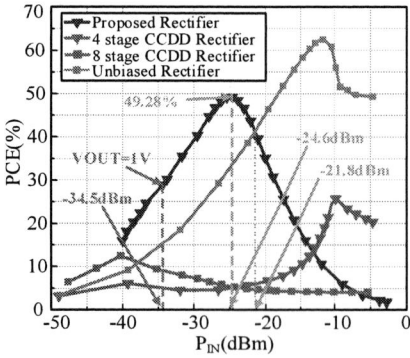

Fig. 9.Simulation of the Relationship between PCE and P_{IN} for the Proposed Rectifier, Series CCDD Rectifier, and Unbiased Rectifier

Table 1 summarizes the performance of the proposed rectifier and compares this work with related relevant papers in recent years. Compared with other works, this work uses a series-parallel structure, dynamic threshold adjustment, and the introduction of gate over-bias. The highest sensitivity and PCE are achieved.

TABLE I. PERFORMANCE COMPARISON

References	*This work*	*TCASII*[2]	*TCASI*[3]	*JSSC*[4]
Technology(nm)	65	180	180	90
Frequency(Hz)	915M	433M	2.4G	868M
Stage	4	1	1	5
Sensitivity(dBm)	-34.5	-17	-26.7	-27
PCE$_{max}$	49.28%	51.5%	32.3%	40%

VI. CONCLUSION

This paper proposes a series-parallel rectifier with pre-biasing to achieve high efficiency and high sensitivity at a frequency of 915MHz. This design utilizes a series-parallel structure. The output voltage is increased without increasing the number of rectifier stages, which in turn optimizes the sensitivity. To address the reverse leakage issue in CCDD rectifiers. The forward conduction loss and reverse leakage are balanced by dynamic threshold voltage adjustment and gate pre-bias. Under a 10MΩ load resistance, the circuit can output 1V at an input power of -34.5dBm. Additionally, the peak PCE of the system reaches 49.28%.

ACKNOWLEDGMENT

Supported by Laboratory Specialized Scientific Research Projects of Beijing Smart-chip Microelectronics Technology Co., Ltd. (SGSC0000SJQT2400925). The authors also would like to thank the Xidian University Chongqing Integrated Circuits Innovation Institute for their technical support.

REFERENCES

[1] G. Papotto, F. Carrara and G. Palmisano, "A 90-nm CMOS Threshold-Compensated RF Energy Harvester," in *IEEE Journal of Solid-State Circuits*, vol. 46, no. 9, pp. 1985-1997, Sept. 2011.

[2] M. H. Ouda, W. Khalil and K. N. Salama, "Self-Biased Differential Rectifier With Enhanced Dynamic Range for Wireless Powering," in *IEEE Transactions on Circuits and Systems II: Express Briefs*, vol. 64, no. 5, pp. 515-519, May 2017.

[3] S. Nagaveni, P. Kaddi, A. Khandekar, and A. Dutta, "Resistance Compression Dual-Band Differential CMOS RF Energy Harvest er Under Modulated Signal Excitation." in *IEEE Transactions on Circuits and Systems I: Regular Papers*, vol. 67, no. 11, pp. 40 53-4062, Nov. 2020.

[4] M. Stoopman, S. Keyrouz, H. J. Visser, K. Philips and W. A. Serdijn, "Co-Design of a CMOS Rectifier and Small Loop Ante nna for Highly Sensitive RF Energy Harvesters," in *IEEE Journ al of Solid-State Circuits*, vol. 49, no. 3, pp. 622-634, March 2 014.

A 24.3-43.7 GHz Variable-Gain Low-Noise Amplifier With Phase Self-Compensation

Yue Wu, Wei Li*, Mohan Zhou, Hongtao Xu

State Key Laboratory of Integrated Chips and System, Fudan University, Shanghai 201203, P.R.China

* Email: 23212020164@m.fudan.edu.cn, w-li@fudan.edu.cn

Abstract—This paper presents a variable-gain low-noise amplifier (VG-LNA) with a gain bandwidth of 24.3-43.7 GHz in 28-nm CMOS technology. A co-design methodology for the first two stages of LNA and the high-order matching network are applied to implement the input matching and low noise figure (NF) in a broadband. The gate-inductive gain-peaking technique is adopted to optimize the gain flatness in high frequency. The proposed phase self-compensation technique without introducing extra devices helps reduce the phase error and realize lower insertion loss. The proposed VG-LNA achieves a stimulated peak gain of 16.4 dB and NF of 3.59-4.38 dB across 24.3 to 43.7 GHz with RMS gain error of less than 0.45 dB and RMS phase error of less than 3.9°.

Keywords—5G, low noise amplifier (LNA), broadband, millimeter-wave (mm-wave), attenuator

I. INTRODUCTION

With the rapid development of 5G communication, the wideband phased-array system covering multiple frequency bands (24/28/37/39/43GHz) plays an important role. Accordingly, millimeter-wave (mm-wave) broadband variable-gain low-noise amplifier (VG-LNA) has gradually become a research hotspot, as one of the crucial components of millimeter wave phased-array system, millimeter wave broadband VG-LNA not only needs to amplify weak antenna signals with low noise figure and high gain, but also provide different gain modes to adapt the signal amplitude coming from the phase array antenna.

There are two key points in the design of mm-wave wideband VG-LNA: one is to simultaneously achieve input impedance matching and noise matching over a wide frequency range. The other is to achieve precise gain control. In previously reported wideband mm-wave VG-LNA [1]-[4], transformers are often used to achieve input impedance and noise matching, but it is still very challenging to achieve both broadband and low noise. The gain control mechanisms proposed in these studies include tunable load, current steering, current slicing, etc. But these gain control mechanisms will have a great impact on the input and output impedance matching, and the gain tuning accuracy is limited, and a large phase error will be introduced. Another way of gain control mechanism is the switched attenuator, which can achieve accurate gain control, but requires a phase compensation mechanism to reduce the phase error in different gain states. Based on which, the capacitive compensation [2][5] and inductive compensation technique [4] have been widely used in the published works, where the parallel capacitor can realize the phase lag compensation for attenuation state. However, the insertion of additional devices increases the insertion loss of attenuation.

In order to address the above challenges, a broadband VG-LNA in 28-nm CMOS is proposed in this paper. The high-

This paper is supported by National Key R&D Program of China under Grant 2023YFB4403802, and by the Fundamental Research Funds for Central Universities in China.

order matching network of shunt-series resonator is used to realize both input impedance matching and noise matching, in which the weak coupling of the transformer can tune the value of the inductor and optimize the noise figure (NF). In order to reduce the phase error under different gain states and reduce the insertion loss of the attenuator, the phase self-compensation technique is proposed without introducing extra circuit. Section II presents the detailed design of the proposed VG-LNA. Section III shows the post-simulation results and conclusions are given in Section IV.

II. CIRCUIT DESIGN

The schematic of the proposed VG-LNA is shown in Fig.1. The LNA consists of two-stage common-source amplifier (CS) and one-stage cascode amplifier, and is cascaded by the attenuator (ATT). The ATT has 4 gain control units that are implemented by simplified T-type attenuators and π-type attenuators with phase self-compensation.

A. Broadband Input Impedance Matching and Co-Design Methodology for Low NF and Gain Flatness

In LNA design, the CS amplifier is always adopted in input stage to reduce noise, and cascode amplifier is adopted to provide high gain. In the mm-wave band, the parasitic capacitance of the transistor cannot be ignored, which will have a large impact on the input impedance and high frequency gain. The shunt-series RLC resonator is adopted in the first CS stage, which contains a ladder-based transformer L_P, L_{G1}. L_P and L_{G1} are coupled through the coupling coefficient k_1. Through this high-order impedance matching network, two different poles can be generated at the input, which can not only eliminate the influence of parasitic capacitance C_{gs1}, but also absorb the pad capacitance. The small signal analysis of the input matching network is shown in Fig.2, and in order to simplify the analysis, here we assume that the coupling coefficient k_1=0. The input impedance can be calculated as follows

$$Z_{in}\big|_{k_1=0} = \cfrac{1}{\cfrac{1}{sL_P} + sC_1 + \cfrac{1}{\cfrac{g_{m1}L_{S1}}{C_{eq}} + s(L_{G1}+L_{S1}) + \cfrac{1}{sC_{eq}}}} \quad (1)$$

Where, $C_{eq}=C_{gs1}+A\times C_{gd1}$, From (1), it can be observed that the shunt resonant pole $f_1=1/2\pi\sqrt{L_PC_1}$ and series resonant pole $f_2=1/2\pi\sqrt{(L_{G1}+L_{S1})C_{eq}}$. Hence, by properly designing the position of the poles, impedance matching can be achieved over a wide bandwidth.

The total NF of the LNA is mostly contributed by the first two stages. In order to achieve the flatness of the gain and the low NF in the wideband, the co-design methodology of the first and the second stage is adopted. Due to the influence of

979-8-3503-6184-1/24 $31.00 © 2024 IEEE

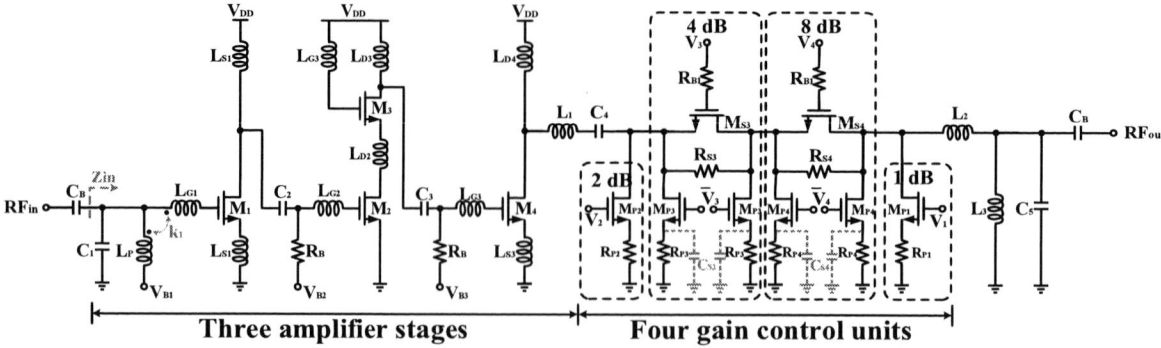

Fig. 1.   Schematic of the proposed broadband VG-LNA.

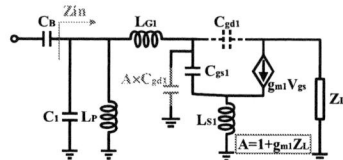

Fig. 2.   Equivalent circuit of the input matching network.

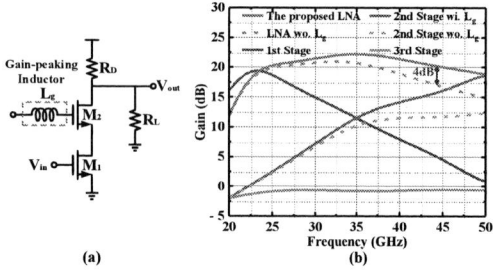

Fig. 3.   (a) Cascode amplifier with gate-inductive gain-peaking technique. (b) Post-simulated small-signal gains of the three stages LNA under high gain state.

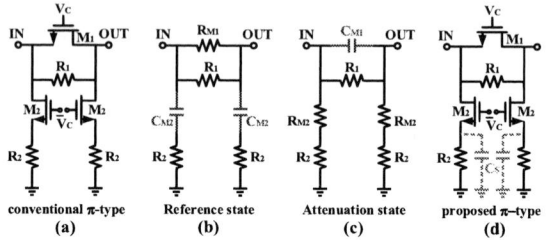

Fig. 4.   (a) Conventional π-type attenuator. (b) Equivalent circuit model of the conventional π-type attenuator in the reference state. (c) Equivalent circuit model of the conventional π-type attenuator in the attenuation state. (d) Proposed π-type attenuator with phase self-compensation methodology.

the attenuator load, the third-stage common source amplifier provides little amplification. In order to deal with the problem that the gain decreases rapidly at high frequency due to the parasitic capacitance, the second stage of cascode amplifier is designed to be high gain at high frequency. Here a gate-inductive gain-peaking technique [6] is adopted to further compensate for the high frequency gain as shown in Fig.3(a). By means of such technique, and the peak gain of the LNA is designed at different frequencies, the gain at high frequency is significantly improved, i.e. 4dB gain improved at 44 GHz, gain flatness is improved as well. Obviously, as the gain of the first two stages increases, the NF of the whole circuit is suppressed to a lower level. Based on the above co-design method, the gain flatness and low NF in a broadband can be realized simultaneously.

B. Phase Self-Compensation Technique with Low Insertion Loss

The schematic of the switched π-type attenuator is shown in Fig.4(a). This architecture generally has a low insertion loss and a small area, which makes it suitable for large-scale applications in phased-array systems. In order to analyze the source of the phase error of the ATT, the simplified small-signal equivalent circuits for the reference state and the attenuation state of the π-type ATT are depicted in Fig.4(b)(c) respectively. R_{M1} and R_{M2} are the ON-state resistances of the switch transistors M_1 and M_2 respectively, and C_{M1} and C_{M2} are

the parasitic capacitances of the switch transistors M_1 and M_2 respectively for an OFF-state. Due to the parasitic capacitance of the switch transistor in the OFF-state, the zero-pole of the transfer function is introduced. The zeros and poles of the conventional attenuator cells under a reference state and an attenuation state are derived in [4].

The difference in frequency between the zero pole of the attenuation state and the reference state ultimately results in the phase error of the two states. [5] proposed a phase compensation method by adding capacitors parallel to the resistor R_2, which will worsen the insertion loss of ATT. The discussion above only considers the parasitic capacitance C_{gd}, but in fact, the parasitic capacitance of the transistor is also a high-order model, based on which we can utilize the parasitic effect of the transistor itself to reduce the phase error under different gain states and reduce the insertion loss of the attenuator. As is shown in Fig.4 (d) the phase self-compensation technique is proposed in this paper without introducing additional phase compensation circuit. By properly designing the size of the switch transistor, the source of the transistor contributes intrinsic 5.44 fF and 6.12 fF of capacitance respectively in 4 dB and 8 dB π-type attenuator unit. Thus, we can control the position of the zero-pole of the different state to be relatively consistent, and then realize the phase correction.

III. SIMULATION RESULTS

The proposed VG-LNA with phase self-compensation method was implemented in 28-nm CMOS with a supply voltage of 1.2V. Fig. 5. shows the layout of the VG-LNA with a core size of 0.16 mm². The VG-LNA consumes 20.1 mA. Fig. 6. shows the post-simulated results of S-parameters and noise figure. The simulated bandwidth is up to 19.4 GHz, which ranges from 24.3 to 43.7 GHz, covering the K- band

TABLE I. PERFORMANCE SUMMARY AND COMPARISON WITH STATE-OF-THE-ART WORKS

	This Work*	MWCL 2018[1]	APMC 2017[2]	RFIC 2019[3]	IMS 2021[4]	TMTT 2021[7]	ASICON 2023[8]	JSSC 2021[9]
Process	**28-nm CMOS**	40-nm CMOS	65-nm CMOS	65-nm CMOS	90-nm CMOS	55-nm CMOS	40-nm CMOS	28-nm CMOS
BW (GHz)	**24.3-43.7**	26-33	33.5-39	20-43	26-30.5 33.8-40.6	6.5-12	4.7-18	22.9-38.2
Peak Gain (dB)	**16.4**	26	21	14.5	21.4	20.7	19.3	14.5
Gain Tuning Range (dB)	**15**	8	31	21.5	9.8	18	8.8	N/A
NF (dB)	**3.59-4.38**	3.4-4.3	4-4.2	5.5-8.2	4.7-7.8	3.26-5.0	5-6.4	2.65-4.62
RMS Gain Error (dB)	**<0.45**	<0.38	0.8	N/A	N/A	<0.6	N/A	N/A
RMS Phase Error (deg.)	**<3.9**	<5.8	5.4	N/A	1.8/4.32	<4.5	N/A	N/A
VDD (V)	**1.2**	1.1	1	1.1	1	1.3	1.1	N/A
Pdc (mW)	**20.1**	31.4	28	30.87	17.9	75	66	18.9
Core Size (mm²)	**0.16**	0.26	0.52**	0.34	0.45	0.98	0.31	0.16
FoM#	**22.98**	10.12	1.27	2.09	3.29	0.44	1.84	15.06

* post-simulated results. **estimated from die photo. # $FoM = \dfrac{Gain[abs.] \times BW[GHz]}{(F-1) \times P_{dc}[mW] \times Area[mm^2]}$

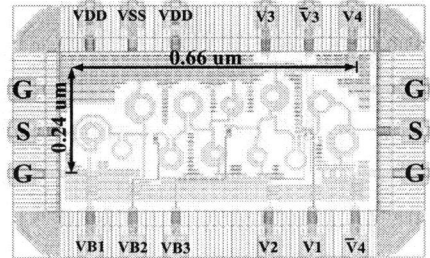

Fig. 5. The layout of the proposed VG-LNA.

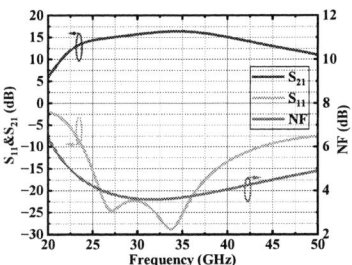

Fig. 6. Post-stimulation results of S-parameters and NF.

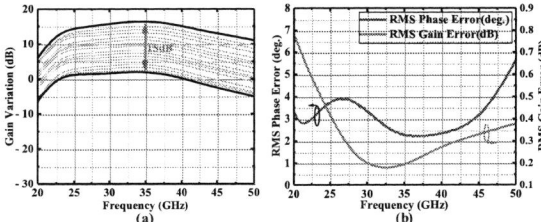

Fig. 7. (a) Post-stimulated VG-LNA gain for all 16 gain states. (b) the corresponding RMS gain error and the corresponding RMS phase error.

and the whole *Ka*-band. The simulated peak gain is 16.3 dB at 34 GHz, and the NF remains $\leq$ 4.38 dB over the entire frequency range. Fig. 7(a)(b) show the simulated VG-LNA gain for all 16 gain states, and the corresponding RMS gain error are less than 0.45 dB and the RMS phase error are less than 3.9°.

A summary of the VG-LNA performance and comparison is shown in Table I. It is demonstrated that the proposed VG-LNA is effective in reducing RMS phase error with lower NF over a wider frequency band simultaneously.

IV. CONCLUSIONS

This paper presents a 24.3-43.7 GHz broadband VG-LNA with phase self-compensation technique in 28-nm CMOS process. The gain-peaking inductor technique is adopted to achieve gain flatness in a wide band. At the maximum gain state, the proposed VG-LNA has a peak gain of 16.4 dB over 24.3-43.7 GHz, and the NF is 3.59-4.38 dB under good input matching. The VG-LNA achieves a dB-linear gain tuning range of 15 dB with RMS gain error of less than 0.45 dB and RMS phase error of less than 3.9 °.

REFERENCES

[1] M. Elkholy, S. Shakib, J. Dunworth, V. Aparin and K. Entesari, "A Wideband Variable Gain LNA With High OIP3 for 5G Using 40-nm Bulk CMOS," in IEEE Microwave and Wireless Components Letters, vol. 28, no. 1, pp. 64-66, Jan. 2018.

[2] Z. Jiang et al., "A 33.5–39 GHz 5-bit variable gain LNA with 4 dB NF and low phase shift," 2017 IEEE Asia Pacific Microwave Conference (APMC), Kuala Lumpur, Malaysia, 2017, pp. 1200-1202.

[3] T. Wu, C. Zhao, H. Liu, Y. Wu, Y. Yu and K. Kang, "A 20 ~ 43 GHz VGA with 21.5 dB Gain Tuning Range and Low Phase Variation for 5G Communications in 65-nm CMOS," 2019 IEEE Radio Frequency Integrated Circuits Symposium (RFIC), Boston, MA, USA, 2019, pp. 71-74.

[4] K. -C. Chang, Y. Wang and H. Wang, "A Broadband Variable Gain Low Noise Amplifier Covering 28/38 GHz bands with Low Phase Variation in 90-nm CMOS for 5G Communications," 2021 IEEE MTT-S International Microwave Symposium (IMS), Atlanta, GA, USA, 2021, pp. 764-767.

[5] Z. Zhang et al., "A DC–Ka-Band 7-Bit Passive Attenuator With Capacitive-Compensation-Based Bandwidth Extension Technique in 55-nm CMOS," in IEEE Transactions on Microwave Theory and Techniques, vol. 69, no. 8, pp. 3861-3874, Aug. 2021.

[6] Y. -H. Yu, Y. -S. Yang and Y. -J. E. Chen, "A Compact Wideband CMOS Low Noise Amplifier With Gain Flatness Enhancement," in IEEE Journal of Solid-State Circuits, vol. 45, no. 3, pp. 502-509, March 2010.

[7] H. Gao et al., "A 6.5–12-GHz Balanced Variable-Gain Low-Noise Amplifier With Frequency-Selective Gain Equalization Technique," in IEEE Transactions on Microwave Theory and Techniques, vol. 69, no. 1, pp. 732-744, Jan. 2021.

[8] S. Han, X. Wu, W. Li, Y. Wang, Y. Lin and H. Xu, "A 4.7-to-18-GHz Ultra-Wideband Variable-Gain Balun-LNA Using 3rd-Order-Band-Pass Input Matching in 40-nm CMOS," 2023 IEEE 15th International Conference on ASIC (ASICON), Nanjing, China, 2023, pp. 1-4.

[9] Z. Deng, J. Zhou, H. J. Qian and X. Luo, "A 22.9–38.2-GHz Dual-Path Noise-Canceling LNA With 2.65–4.62-dB NF in 28-nm CMOS," in IEEE Journal of Solid-State Circuits, vol. 56, no. 11, pp. 3348-3359, Nov. 2021.

979-8-3503-6184-1/24 $31.00 © 2024 IEEE

A Broadband Active Variable Attenuator With Phase Compensation Technique

Zhiying Xia[1,2,3], Zhiqun Li[1,2,3,*], Bofan Chen[1,2,3], and Xiaowei Wang[1,2,3],

1 Institute of RF-& OE-ICs, Southeast University, 210096 Nanjing, China;
2 State Key Laboratory of Millimeter Waves, 210096 Nanjing, China;
3 Engineering Research Center of RF-& OE-ICs, Jiangsu Province, 210096 Nanjing, China;
* Correspondence: zhiqunli@seu.edu.cn

Abstract—This paper presents a 8–16 GHz active variable attenuator with high gain resolution and low phase variation in TSMC 40-nm CMOS technology for wideband active phased-array applications. The proposed circuit employs a phase compensation method by cascading the current-steering stage and the current-tuning stage, which exhibit opposite phase variation trends during gain tuning, achieving an Root-Mean-Square (RMS) phase error of less than 1.5 ° within 8-16 GHz. The proposed active attenuator exhibits a simulated attenuation tuning range of 16 dB with 0.2 dB gain resolution, while the RMS amplitude error over 8-16 GHz is better than 0.13 dB. In addition, the input 1 dB compression point (IP$_{1dB}$) @ maximum gain state is better than -9.4 dBm with 22.3 mW power consumption.

Keywords—attenuator, phase compensation, transformer

I. INTRODUCTION

With the increasing demands of modern wireless communication systems, the phased array technique attracts widespread attention from universities and industries. Compared to conventional phased arrays based on passive components, active phased arrays offer advantages in terms of the effective isotropic radiated power of transmitters and the signal-to-noise ratio of receivers. As a gain control circuit, the active attenuator is one of the key building blocks of an active phased array[1],[2]. And Gain control range along with gain control resolution are two important parameters of the attenuators. Additionally, the associated phase variation during gain tuning is another crucial metric, which should be reduced to ease calibration difficulty.

In the previous works [8]-[13], several gain control circuits with low phase variation have been reported. A distributed attenuator with 96.5% relative bandwidth is realized in [3] while poor performance on the amplitude error and phase error is exhibited. An inductive phase compensation network is employed in [4] while the gain control range is only 8.2 dB. A two-stage phase-invariant structure combining with the current-steering stage and current-splitting stage is adopted in [5], but the bandwidth is limited. The current-steering stage and bias-adjusting mechanism are employed in [6], where the achieved relative bandwidth is less than 15%. Moreover, a current-tuning structure is adopted in [7] and the bandwidth is extended, although the phase variation is limited.

In this paper, we present a two-stage active variable attenuator working at 8-16 GHz, utilizing a novel phase-invariant technique. The design features a current-steering stage and a current-tuning stage, which exhibit opposing phase variation trends during gain tuning. By cascading these two stages, the maximum phase variation is effectively reduced. Consequently, the proposed active variable attenuator achieves an RMS phase error of less than 1.5 ° over

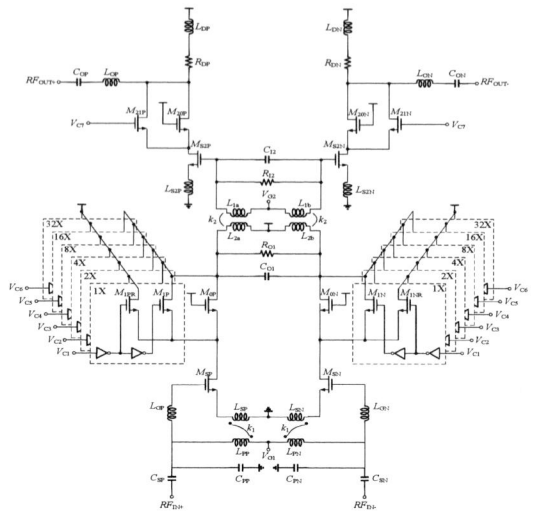

Fig.1. Schematic of proposed active variable attenuator

8-16 GHz, with a gain control range of 16 dB and a gain resolution of 0.2 dB.

The paper is divided as follow. Section II presents the circuit design while Section III provides Electromagnetic (EM) simulation results with a comparison to the recent reported works. Finally, Section IV offers the conclusions.

II. CIRCUIT DESIGN

A. Proposed Phase-Invariant Technique

In this paper, a two-stage active attenuator is proposed to minimize the phase variation during gain tuning. The design concept from [5] is adopted here, which involves using a two-stage circuit for phase compensation. The phase variations of these two stages versus gain change in opposite directions, resulting in a canceling effect when cascaded. The traditional current steering VGA is based on a cascode structure (M_S and M_i) with a replica common gate transistor M_{iR}, which is shown in Fig.2(a). Transistor M_i and M_{iR} will switch between on-state and off-state during attenuation tuning, and the corresponding parasitic capacitance is different. Then the relative phase variation will be introduced with positive value.

In order to minimize the phase variation, a circuit with an opposite insertion phase variation trend to that of conventional current-steering VGA is adopted. As presented in Fig.2, the concept of the proposed technique combining the current-steering stage with positive phase variation and the current-tuning stage with negative phase variation is briefly illustrated. Note that the differential circuits are drawn in the single-ended

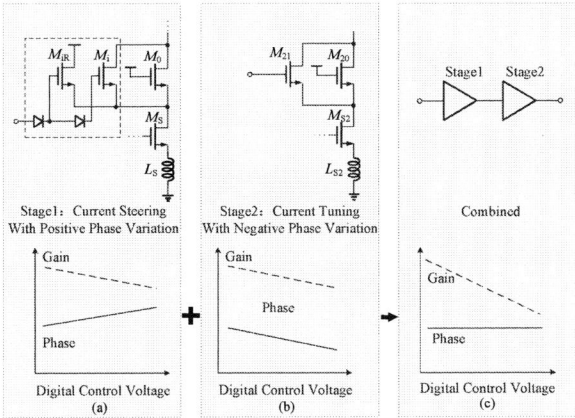

Fig.2. Concept of the proposed phase-invariant technique

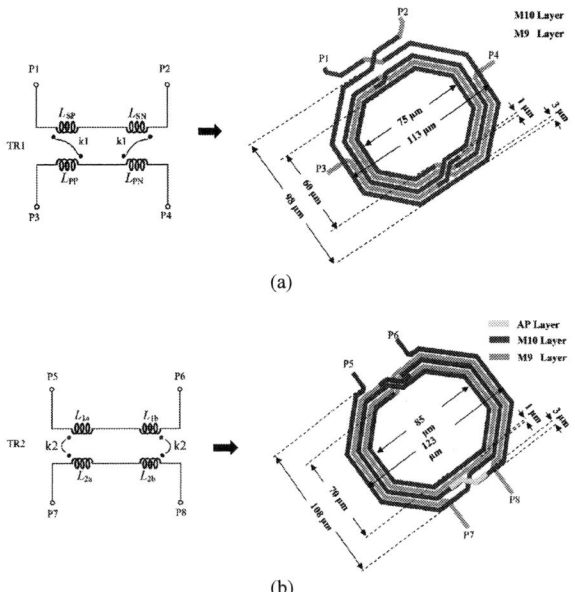

Fig.3. (a) TR1 of the attenuator in the layout and (b) TR2 of the attenuator in the layout

form to simplify the representation. As the gain increases, the insertion phases of the first-stage circuit and the second-stage circuit change in opposite directions, leading to a reduction in phase variation after cascading. According to the EM simulation results, an RMS phase variation of less than 1.5 ° within 8-16 GHz can be achieved during 16 dB gain tuning.

B. Gain Control Mechanism And Bandwidth Extension

Attenuation variation is achieved by the digitally controlled common-gate transistors with different sizes [8]. Specifically, there are 6-bit control signals (V_{C1}-V_{C6}) used in the first stage to achieve 8 dB attenuation and 0.2 dB gain resolution, while another control signal (V_{C7}) is utilized in the second stage to further extend the attenuation by an additional 8 dB. Consequently, an attenuation range of 16 dB with a gain resolution of 0.2 dB is obtained.

The improvement in bandwidth performance is realized by the combination of the transformer-based feedback network at the input stage and a shunt peaking technique at the output stage. Specifically, a shunt–series transformer TR1 combined with the gate inductor and a parallel capacitor is employed for input matching as shown in Fig.1. The layout of TR1 is shown in Figure.3(a), while the layout of its transformer TR2 is shown in Figure.3(b). Additionally, the inductor shunt peaking technology is applied to the second stage to compensate for the gain difference within the band, and a 3-dB bandwidth is extended.

III. SIMULATION RESULTS

This paper proposed a transformer-based active variable attenuator with high gain resolution and low phase variation in TSMC 40-nm CMOS technology. The layout of the proposed circuit is shown in Fig.4, the core area is 530 μm × 750 μm, and the power consumption P_{DC} is 22.3 mW.

According to the EM simulation results in Fig.5, the insertion loss is 0.2-3.2 dB in the maximum gain mode, while S11 and S22 are better than -9.5 dB in all attenuation modes. Meanwhile, 81 desired attenuation modes with a 16 dB gain control range and 0.2 dB gain resolution are selected from 128 attenuation modes. Consequently, the RMS amplitude and phase errors of the selected states over 8–16 GHz are superior to 0.13 dB and 1.5 °, respectively, as depicted in Figure.6 and Figure.7. In addition, the simulated IP$_{1dB}$ ranges from -9.4 dBm to -7.4 dBm, while the Noise Figure (NF) ranges from 6.6 dB to 8.2 dB across the 8-16 GHz frequency band, as shown in Fig.8. Additionally, Table I presents a comparison

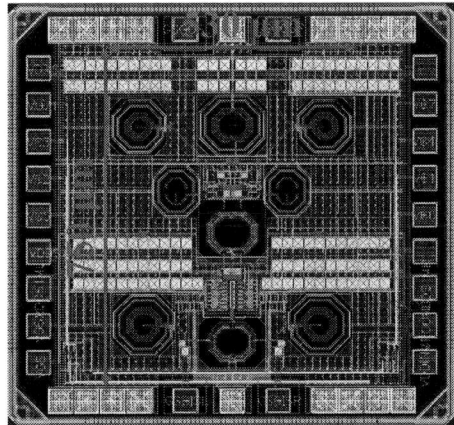

Fig.4. The layout of the proposed active variable attenuator.

between this work and the recent reported gain control circuits, and a better performance on gain resolution and phase variation is shown in our work.

IV. CONCLUSION

The proposed 8–16 GHz active variable attenuator with 16 dB gain control range and 0.2 dB gain resolution is designed on TSMC 40-nm CMOS technology. The current-steering stage with positive phase variation and the current-tuning stage with negative phase variation are cascaded in the proposed circuit, achieving an RMS phase error of less than 1.5 ° over 8-16 GHz. The transformer-based feedback network is employed for the input stage and the simulated S11 better than -9.5 dB is realized. Meanwhile, the utilization of MCR in the interstage along with the shunt peaking technique in the output stage are combined to enhance the 3-dB bandwidth. The proposed active variable attenuator exhibits ultralow gain and phase error within a larger bandwidth.

979-8-3503-6184-1/24 $31.00 © 2024 IEEE

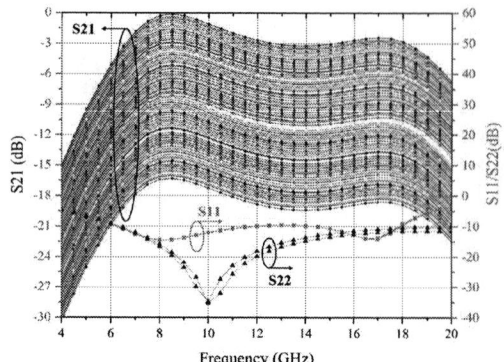

Fig.5. Simulated S-parameters characteristics of the active attenuator

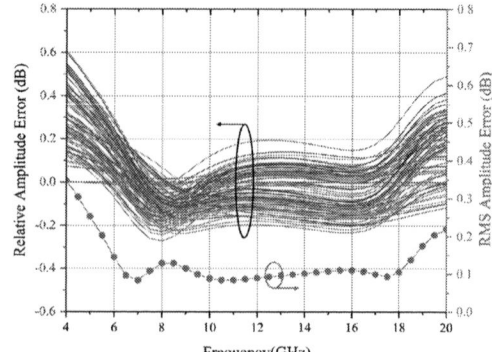

Fig.6. Simulated relative amplitude error and RMS amplitude error

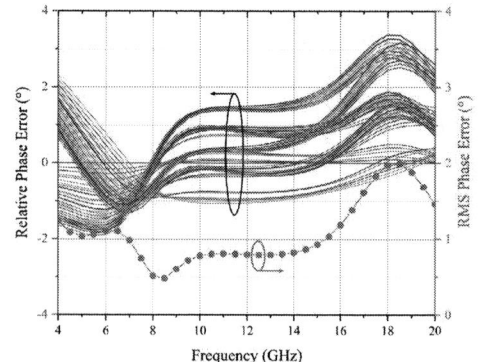

Fig.7. Simulated relative phase error and RMS phase error

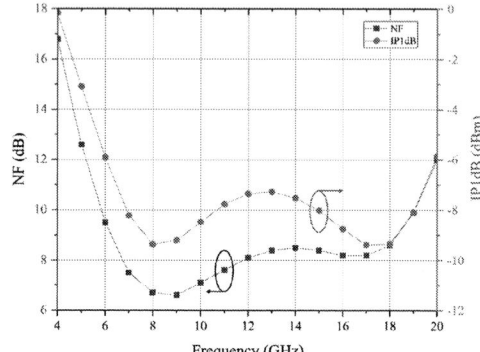

Fig.8. Simulated NF and IP$_{1dB}$ of the active variable attenuator.

TABLE I.   PERFORMANCE COMPARISON OF THE PUBLISHED GAIN CONTROL CIRCUIT

Reference	Tech.	Fre. (GHz)	Peak Gain (dB)	ΔGain (dB)	Gain Resolution (dB)	RMS Amp. Err. (dB)	RMS Pha. Err. (°)	IP1dB (dBm)	P$_{DC}$ (mW)	Core Area (mm²)
[1] MWCL'2016	0.25-μm SiGe	6-12.5	-10.2	16.5	0.26	< 0.26	< 3.5	12.5@ 10GHz	0	0.29
[3] TMTT'2020	65-nm CMOS	15-43	-2.9	14	1	< 1.7	< 6	14 @ 35GHz	0	0.29
[2] EuMIC'2022	130-nm SiGe	8-18	13.2	30	0.5	< 0.49	< 12.3	> -1.2	23.4	0.24
[8] MWCL'2022	130-nm SiGe	16~24	15	15	1	< 0.3	< 15	N/A	14.1	0.19
[7] TCAS II'2024	40-nm CMOS	6-15.3	-1.7	15.5	0.5	< 0.3	< 2.95	> -1.1	29.6	0.075
This Work*	**40-nm CMOS**	**8~16**	**-0.2**	**16**	**0.2**	**< 0.13**	**< 1.5**	**> -9.4**	**22.3**	**0.4**

Simulation Results

ACKNOWLEDGMENT

The authors thank the Jiangsu Provincial Key R&D Program under Grant BE2022052-2 for their support.

REFERENCES

[1] M. Davulcu, C. Caliskan, I. Kalyoncu, M. Kaynak, and Y. Gurbuz, "7-Bit SiGe-BiCMOS Step Attenuator for X-Band Phased-Array RADAR Applications," IEEE Microwave and Wireless Components Letters, vol. 26, no. 8, pp. 598-600, 2016, doi: 10.1109/LMWC.2016.2585565.

[2] K. Altintas, T. A. Ozkan, M. Yazici, M. Kaynak, and Y. Gurbuz, "A 8–18 GHz Low Noise Variable Gain Amplifier with 30 dB Gain Control Range," in 2022 17th European Microwave Integrated Circuits Conference (EuMIC), 26-27 Sept. 2022 2022, pp. 304-307, doi: 10.23919/EuMIC54520.2022.9923440.

[3] K. Park, S. Lee, and S. Jeon, "A New Compact CMOS Distributed Digital Attenuator," IEEE Transactions on Microwave Theory and Techniques, vol. 68, no. 11, pp. 4631-4640, 2020, doi: 10.1109/TMTT.2020.3017820.

[4] S. H. Kim, T. H. Jang, J. H. Kim, and C. S. Park, "A Wideband 120-GHz Variable Gain Amplifier With Multistage Phase Compensation,"

IEEE Transactions on Microwave Theory and Techniques, vol. 68, no. 6, pp. 2419-2427, 2020, doi: 10.1109/tmtt.2020.2980518.

[5] D.-S. Siao, J.-C. Kao, and H. Wang, "A 60 GHz Low Phase Variation Variable Gain Amplifier in 65 nm CMOS," IEEE Microwave and Wireless Components Letters, vol. 24, no. 7, pp. 457-459, 2014, doi: 10.1109/lmwc.2014.2316253.

[6] J.-L. Kuo et al., "60-GHz Four-Element Phased-Array Transmit/Receive System-in-Package Using Phase Compensation Techniques in 65-nm Flip-Chip CMOS Process," IEEE Transactions on Microwave Theory and Techniques, vol. 60, no. 3, pp. 743-756, 2012, doi: 10.1109/tmtt.2011.2176508.

[7] B. Chen, Z. Li, Z. Xia, Z. Fang, and D. Zhou, "A 0.075-mm2 6-15.3 GHz Active Digital Step Attenuator With Novel Current-Tuning Topology for Phased-Array Radar System," IEEE Transactions on Circuits and Systems II: Express Briefs, pp. 1-1, 2024, doi: 10.1109/tcsii.2024.3375860.

[8] Z. Xu et al., "A 16–24-GHz SiGe Decibel-Linear Low-Gain-Error Digitally Controlled High-Efficiency Variable Gain Amplifier," IEEE Microwave and Wireless Components Letters, vol. 32, no. 6, pp. 543-546, 2022, doi: 10.1109/lmwc.2021.3138845.

Waterproof and wearable power sources

Sixing Xiong
1. RIKEN
2-1, Hirosawa, Wako, Saitama, 351-0198, Japan
sixing.xiong@riken.jp

Kenjiro Fukuda
1. RIKEN
2-1 Hirosawa, Wako, Saitama 351-0198, Japan
kenjiro.fukuda@riken.jp

Takao Someya
1: RIKEN
2: The University of Tokyo
7-3-1 Hongo, Bunkyo-ku, Tokyo 113-8656, Japan
someya@g.ecc.u-tokyo.ac.jp

Abstract—**Wearable electronics can be effortlessly incorporated into human skin or fabrics for continuous physiological health monitoring. A reliable energy supply system is vital for ensuring the long-term operation of these devices. Ultraflexible organic photovoltaics have gained attention as a promising power source for wearable electronics due to their stretchable and lightweight characteristics. However, achieving waterproofing for ultraflexible organic electronics without compromising mechanical flexibility and conformability has been a considerable challenge. This study underscores advancements in the development of waterproof, ultra-flexible solar cells designed to power wearable electronics and sensors. We demonstrate the fabrication of waterproof, ultraflexible organic photovoltaics by employing in-situ growth of a hole-transporting layer, which enhances adhesion between the active layer and the anode. The resulting 3 μm-thick organic photovoltaics retain 89% and 96% of their initial performance after 4 hours of water immersion and after undergoing 300 stretching and releasing cycles at 30% strain underwater, respectively. Additionally, these ultraflexible devices withstand machine washing test with an exceptionally thin encapsulation layer, a feat not previously reported.**

Keywords—*wearable electronics, energy supply, ultra-flexible organic photovoltaics, waterproof devices*

I. INTRODUCTION

With the rapid expansion of the Internet of Things (IoTs), the demand for portable energy sources to power a myriad of wearable sensors and smart devices is increasing [1-3]. Organic photovoltaics (OPVs) have emerged as a promising solution to this demand. OPVs possess the ability to convert light energy from both natural sunlight and artificial sources into electrical energy without needing additional converters or connection wires [4]. This makes them exceptionally versatile and convenient for various applications.

The potential applications of OPVs in wearable technology are extensive. For instance, they can power wearable devices like smartwatches and fitness trackers, providing a sustainable and portable energy source that enhances the functionality and convenience of these gadgets. Additionally, OPVs can be embedded in clothing to power sensors that monitor various health metrics, offering a seamless way to integrate health monitoring into daily life [5, 6].

As wearable sensors continue to develop, emphasizing increased flexibility and stretchability, there is a rising need for power sources that share these mechanical attributes [7, 8]. Traditionally, flexible OPVs are produced on polymer substrates with thicknesses ranging from 100 to 300 μm. These substrates, however, fail to offer optimal conformability or stretchability for integration with ultrathin wearable bioelectronics or textile devices. By reducing substrate thickness to just a few micrometers, ultra-flexible OPVs with a thickness of less than 10 μm can be effectively incorporated into irregular and stretchable surfaces, including

human skin and fabrics [9-11]. Achieving this level of integration is challenging with conventional flexible OPVs [12, 13]. The novel combination of ultra-flexible solar cells with wearable electronics creates new opportunities for self-sustaining sensor systems embedded in daily items.

Ultra-flexible OPVs offer numerous benefits; however, they also face certain drawbacks. A primary concern is the relatively high water vapor transmission rate of the flexible substrate and encapsulation layer. There is a trade-off between achieving mechanical flexibility and ensuring the waterproofness of these devices, which presents a significant challenge for their long-term performance and reliability [14-16].

In this study, we introduce an ultraflexible OPV incorporating the in-situ growth of an AgOx hole transporting layer, , which ensures substantial waterproofness even under mechanical stress by reinforcing the adhesion between the electrode and the active layer. The in-situ growth process involves directly depositing Ag onto the active layers, followed by annealing in air at 85 °C for 24 hours. Consequently, our 3 μm-thick OPVs exhibit significantly improved waterproofness, retaining 89% and 96% of the initial efficiency after 240 minutes of immersion and 300 stretching/releasing cycles at 30% strain underwater, respectively. These features enable our ultraflexible OPVs to withstand machine washing processes. With a champion efficiency of 14.3% under one sun illumination, our developed OPVs outperform existing waterproof OPVs [17-20]. The strategy significantly improves the waterproofness from the fundamental structure, enhancing the waterproofness of OPVs without compromising mechanical durability and conformability.

II. RESULTS AND DISCUSSION

A. Comparsion of the conventional and waterproof OPVs structure

Fig. 1a illustrates the conventional OPVs structure and the individual layer thickness. The waterproof devices are shown in Fig. 1b. For the waterproof devices structure, we employed the in-situ growth of a hole-transporting layer to strengthen interface adhesion between the active layer and anode. Both conventional and waterproof devices were fabricated on transparent polyimide (tPI) substrate to block ultraviolet (UV) light, thus efficiently enhancing the light stability under 1-sun illumination. To fabricate the tPI substrate, the tPI precursor was spin-coated onto the glass and thereafter transferred to an N_2-filled oven at 250 °C for 8 hours to cure a uniform the tPI substrate. Indium tin oxide (ITO) was used as a transparent electrode. The ITO electrode was sputtered onto the tPI and patterned by photolithography. Here, we used PEI-Zn (ethoxylated polyethyleneimine (PEIE) chelated with Zn^{2+}) as the electron transporting layer, which demonstrates better

979-8-3503-6184-1/24 $31.00 © 2024 IEEE

environmental stability than the ZnO electron transporting layer. To form uniform PEI-Zn layer, the substrate was treated with oxygen plasma. Then, the PEI-Zn precursor solution (70 mg zinc acetate dehydrate dissolved in 1 ml 1 wt% PEIE 2-methoxyethanol solution was spin-coated onto the ITO electrode at 3500 rpm, followed by thermal annealing at 180 °C for 30 min in air. After annealing, the samples were transferred into the N_2-fulled glovebox to fabricate the active layer. PM6:Y6 was selected as the active layer due to its high power conversion performance. The active layer solution (PM6:Y6 = 7 mg: 9 mg dissolved in 1 mL mixed solvent of chloroform:1-chloronaphthalene) was spin-coated in the glovebox at 3500 rpm and then annealed at 110 °C for 10 min. For the conventional device, a MoO_x hole-transporting layer (7.5 nm thick) and an Ag electrode (thickness of 100 nm) were sequentially deposited onto the active layer. The effective area of the solar cells was measured to be 4 mm². For the waterproof device, only a 100-nm-thick Ag electrode was deposited onto the active layer. After that, the devices were annealed at 85 °C for 24 h in the air to form an efficient hole transporting layer. For better connections, an external wiring consisting of Cr (3 nm thick)/Au (100 nm thick) electrodes was attached to the cathode and anode of the OPVs by anisotropic electrically conductive adhesive tape. Finally, a 1-μm-thick layer of parylene was deposited through chemical vapor deposition to encapsulate the devices for better environment stability. Then, the ultra-flexible organic solar cells devices were delaminated from the supporting glass substrate and used as free-standing devices.

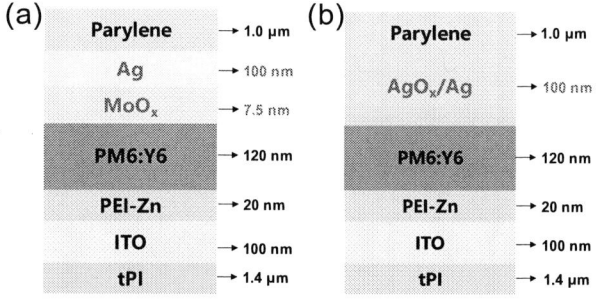

Fig. 1 The device structure of (a) conventional organic photovoltaic device and (b) waterproof organic photovoltaic device.

B. Characterization of waterproof OPVs

Fig. 2a illustrates the photograph of the waterproof and ultraflexible OPVs. The current density–voltage (J–V) characteristics were measured by a Keithley 2400 Source Meter under 1 sun illumination. The light source simulator was XES-40S3, SAN-EI ELECTRIC (AM 1.5 global spectrum). A silicon reference diode was used to calibrate the light density of the light density. Before annealing, the fabricated OPVs showed open-circuit voltage (V_{OC}) of 0.05 V, short-circuit current density (J_{SC}) of 15.8 mA cm^{-2}, fill factor (FF) of 0.27, and PCE of 0.2% (Fig. 2b), and an average PCE of 0.2% (Table 1). After 24 h of annealing, the performance improved to V_{OC} of 0.77 V, J_{SC} of 26.5 mA cm^{-2}, FF of 0.71, and PCE of 14.3%. The average PCE of 30 devices was 13.6%. Fig. 2c displays the PCE distribution of 30 waterproof solar cell devices after 24 h of annealing at 85 °C in the air.

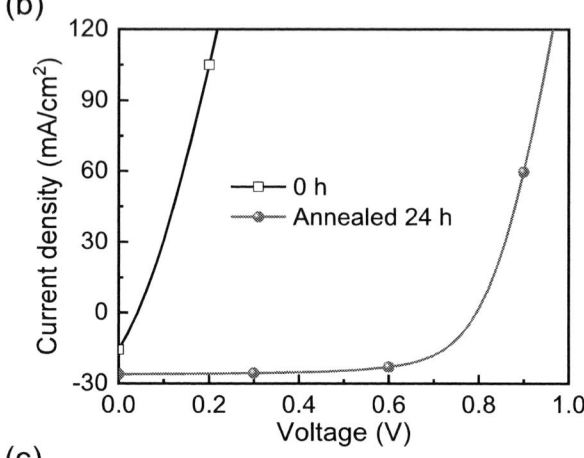

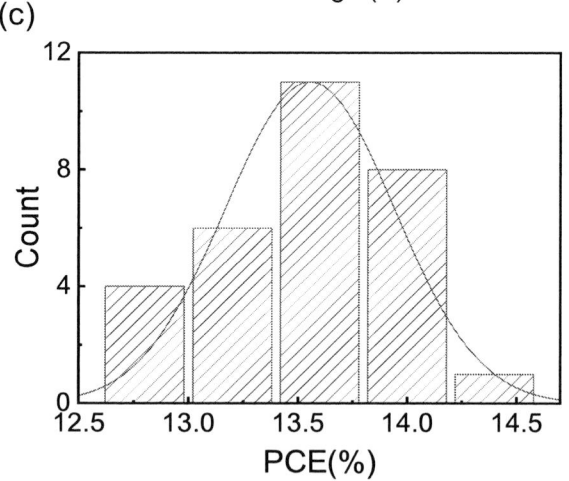

Fig. 2 (a) Photograph of the ultraflexible and waterproof OPVs (tPI/ITO/PEIZn/PM6:Y6/AgOx/Ag/Parylene) immersing in water. Scale bar: 1 cm. (b) J–V characteristics of devices with 0 h and 24 h annealing treatment at 85 °C in air. (c) PCE distribution of 30 devices after 85 °C annealing treatment for 24 h in the air.

The OPV devices exhibit remarkable mechanical durability due to their ultrathin structure. A compression-stretching test was conducted to assess their mechanical durability. As demonstrated in our previous research [21], the ultra-flexible OPVs were transferred onto a pre-stretched elastomer. Upon releasing the elastomer, the ultra-flexible

OPV experienced compression. The degree of deformation was quantified as the ratio of the change in length upon release to the initial length of the device.

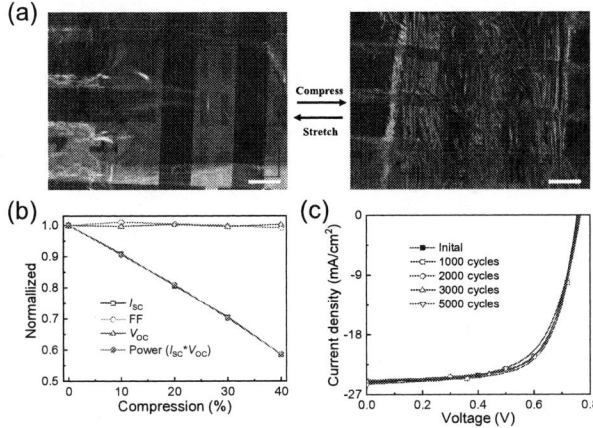

Fig. 3 (a) Schematic illustrations of pre-stretching and compressing. Scale bar: 2 mm. (b) Evolution of V_{OC}, J_{SC}, FF, and power of the ultra-flexible OPV compressions of 0%, 10%, 20%, 30% and 40%.(c) J–V characteristics of the free-standing ultraflexible devices under cyclic compressing–stretching deformation with 30% compression.

Fig. 3a shows the optical images of ultra-flexible OPV device in initial state and compressed state, respectively. Additionally, we evaluated the performance of the device under different levels of compression. We assessed the short circuit current (I_{SC}) of the ultra-flexible OPV under various levels of compression due to difficulties in confirming the effective area during compression. Fig. 3b displays the V_{OC}, I_{SC} and FF of the ultra-flexible OPV device of the ultra-flexible OPV device under different compression strain (0%, 10%, 20%, 30% 40%). Notably, the V_{OC} and FF of the ultra-flexible OPV device exhibited negligible changes even under 40% compression. Due to the area compressed, the I_{SC} decreased with the increase of the compression value. The I_{SC} and the output power changes displayed linearity with the compression value. These results demonstrate the extraordinary mechanical durability of the ultra-flexible OPV device, highlighting its potential for applications in flexible wearable electronics.

C. Waterproofness of OPVs

Furthermore, we investigated the waterproofness of the ultra-flexible OPV devices. First, we immersed the devices into deionized water to evaluate their long-term static waterproofness. The OPV devices exhibited a 11% reduction in PCE after 4 h of water immersion (Fig. 4a). The OPV device also remained operational when fully immersed in water under light illumination. Fig. 4b displays the J_{SC} of the device under light illumination in water. The current of the waterproof OPV remained above 91% of its initial efficiency after operation in water for 64 min. Moreover, the ultra-flexible OPV devices can be integrated with textiles. Fig. 4c illustrates the machine-washing test results for free-standing OPVs without detergent. The OPVs devices remained 90% of their original PCE after two washing cycles in a washing machine. Notably, the concept of ultraflexible OPVs enduring

machine washing has not been reported in previous literature. Furthermore, these findings indicate that OPVs with AgOx/Ag electrode can sustain continuous operation underwater, thereby providing a long-running power source for underwater autonomous systems and sensors.

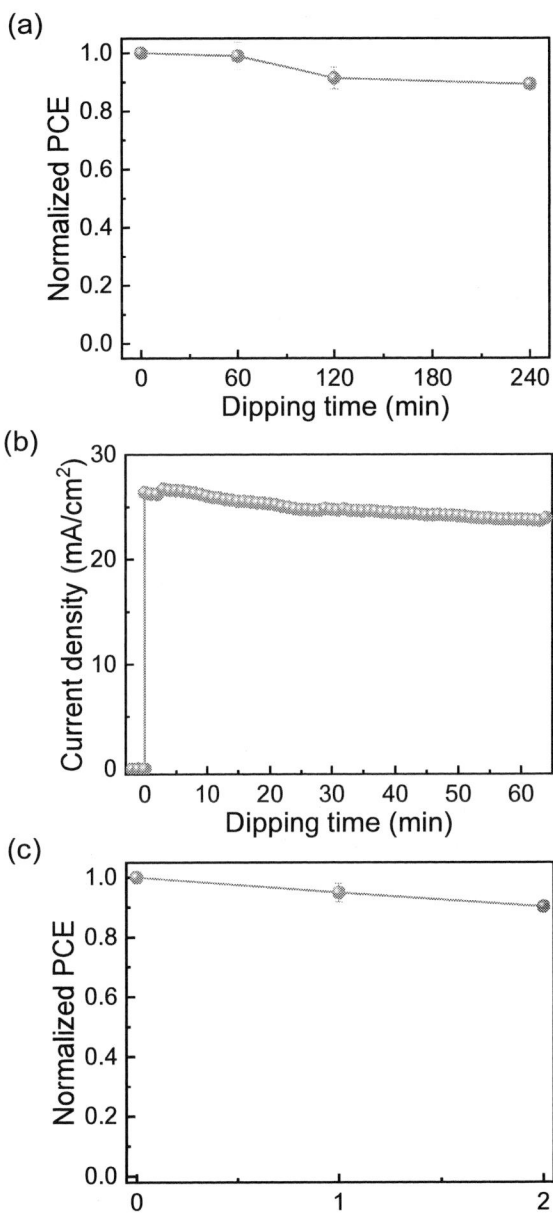

Fig. 4 (a) The evolution of PCE for waterproof devices with different immersion times in water. (b) The evolution of J_{SC} for waterproof devices as a function of dipping time in water under light illumination. (c) The evolution of PCE for waterproof devices under washing process by washing machine.

III. SUMMARY

In this study, we have successfully developed waterproof and ultraflexible OPVs through the in-situ growth of an AgOx hole-transporting layer. These OPVs exhibit exceptional stretchability and waterproofing despite their thin structure, making them highly suitable for integration into wearable electronics. Notably, t the devices exhibited only a 10% reduction in efficiency even after enduring two 66-minute cycles in a washing machine. The successful implementation of these waterproof organic OPVs holds promising implications for the advancement of self-powered wearable electronics. Furthermore, it paves the way for the development of underwater electronics for the Internet of Things in the near future.

ACKNOWLEDGMENT

The authors thank Prof. Y. Zhou of Huazhong University of Science and Technology (China), Dr. S. Lee, Prof. K. Tajima, and Dr. K. Nakano of Center for Emergent Matter Science, RIKEN (Japan) and Prof. T. Yokota of The University of Tokyo (Japan) for fruitful discussions and technical support.

REFERENCES

[1] S. Saifi, X. Xiao, S. Cheng, H. Guo, J. Zhang, and Peter Müller-Buschbaum, et al. An ultraflexible energy harvesting-storage system for wearable applications. Nat Commun 15, 6546 (2024).

[2] R. Liu, Z. Wang, K. Fukuda, and T. Someya, Flexible self-charging power sources. Nat. Rev. Mater. 7, 870–886 (2022)

[3] S. Park, S. Heo, W. Lee, D. Inoue, Z. Jiang, and K. Yu et al. Self-powered ultra-flexible electronics via nano-grating-patterned organic photovoltaics. Nature, 561, 516–521 (2018)

[4] R. Liu, M. Takakuwa, A. Li, D. Inoue, D. Hashizume, and K. Yu et al. An efficient ultra-flexible photo-charging system integrating organic photovoltaics and supercapacitors. Adv. Energy Mater., Adv. Energy Mater. 10, 2000523 (2020)

[5] S. Hashemi, S. Ramakrishna, and A. Aberle, Recent progress in flexible–wearable solar cells for self-powered electronic devices. Energy Environ. Sci., 13, 685-743 (2020)

[6] X. Xu, K. Fukuda, A. Karki, S. Park, H. Kimura, and H. Jinno, et al., Nobuhiro Watanabe, Shuhei Yamamoto, Satoru Shimomura, Daisuke Kitazawa, Thermally stable, highly efficient, ultraflexible organic photovoltaics. Proc. Natl. Acad. Sci. U.S.A., 115 (18) 4589-4594 (2018)

[7] K. Fukuda, K. Yu, and T. Someya, The future of flexible organic solar cells. Adv. Energy Mater., 10, 2000765 (2020)

[8] J. Wang, Y. Ochiai, N. Wu, K. Adachi, D. Inoue, and D. Hashizume et al., Intrinsically stretchable organic photovoltaics by redistributing strain to PEDOT:PSS with enhanced stretchability and interfacial adhesion. Nat. Commun., 15, 4902 (2024)

[9] F. Qin, W. Wang, L. Sun, X. Jiang, L. Hu, and S. Xiong, et al., Robust metal ion-chelated polymer interfacial layer for ultraflexible non-fullerene organic solar cells. Nat. Commun., 11, 4508 (2020)

[10] H. Jinno, T. Yokota, M. Koizumi, W. Yukita, M. Saito, and I. Osaka et al., Self-powered ultraflexible photonic skin for continuous bio-signal detection via air-operation-stable polymer light-emitting diodes. Nat. Commun., 12, 2234 (2021)

[11] S. Cheng, Z. Lou, L. Zhang, H. Guo, Z. Wang, and C. Guo et al., Ultrathin hydrogel films toward breathable skin-integrated electronics. Adv. Mater. 35, 2206793 (2023)

[12] K. Yu, S. Rich, S. Lee, K. Fukuda, T. Yokota, and T. Someya, Organic photovoltaics: Toward self-powered wearable electronics, Proc. IEEE 107 (10), 2137-2154 (2019)

[13] M. Kaltenbrunner, M. White, E. Głowacki, T. Sekitani, T. Someya, amd N. Sariciftci et al., Ultrathin and lightweight organic solar cells with high flexibility. Nat. Commun., 3, 770 (2012)

[14] Q. Burlingame, M. Ball, and Y. Loo, It's time to focus on organic solar cell stability. Nat. Energy, 5, 947–949 (2020)

[15] L. Duan, A. Uddin, Progress in stability of organic solar cells. Adv. Sci., 7, 1903259 (2020)

[16] J. Han, F. Bao, D. Huang, X. Wang, C. Yang, and R. Yang, et al., A universal method to enhance flexibility and stability of organic solar cells by constructing insulating matrices in active layers. Adv. Funct. Mater., 30, 2003654 (2020)

[17] H. Zhen, K. Li, C. Chen, Y. Yu, Z. Zheng and Q. Ling, Water-borne foldable polymer solar cells: one-step transferring free-standing polymer films onto woven fabric electrodes. J. Mater. Chem. A 5, 782-788 (2017).

[18] E. Jeong, Y. Jeon, S. Cho, and K. Choi, Textile-based washable polymer solar cells for optoelectronic modules: toward self-powered smart clothing. Energy Environ. Sci. 12, 1878-1889 (2019).

[19] H. Jinno, K. Fukuda, X. Xu, S. Park, Y. Suzuki, M. Koizumi, et al. Stretchable and waterproof elastomer-coated organic photovoltaics for washable electronic textile applications. Nat. Energy 2, 780-785 (2017).

[20] D. Lv, Q. Jiang, Y. Shang, and D. Liu, Highly efficient fiber-shaped organic solar cells toward wearable flexible electronics. npj Flex. Electron. 6, 38 (2022)

[21] S. Xiong, K. Fukuda, S. Lee, K. Nakano, X. Dong, and T. Yokota, et al., Ultrathin and efficient organic photovoltaics with enhanced air stability by suppression of zinc element diffusion. Adv. Sci., 9, 2105288 (2022)

A CMOS Pixel Sensor for Precise Track and Charge Measurement of Cosmic Ray Nuclei

Ruikai Zhang[1], Wen He[1], Shanqiang Yang[1], Min Luo[1], Chenxu Wang[1*], Cunfeng Feng[3], Meng Wang[3], Liang Zhang[3], Anqing Wang[3], Jianing Dong[3], Dong Liu[3], Yan Niu[3], Yang Zhou[4],Yuehong Gong[5], Xiaoli Wang[2] and Shucheng Shi[2]

[1]Harbin Institute of Technology, Weihai, 264209, China
[2] Shandong University, Qingdao, 266237, Shandong, China
[3] Shandong University, Weihai, 264209, China
[4]Institute of High Energy Physics Chinese Academy of Sciences, 19B Yuquanlu, 100049, Beijing, China
[5] ShanDong JiaoTong University, Weihai, 264200, China

*Email: ruikaizhang2000@hotmail.com, *chenxuwang@hit.edu.cn

Abstract—This paper presents a fully functional 10-bit digital output CMOS pixel sensor for precise track and charge measurement of cosmic ray nuclei. Charge measurement is achieved by quantization of e/h pairs generated by ionization energy lost dE/dX of incident particle on the 4mm×2mm silicon detecting area. The pixel sensor chip operates in rolling shutter sequence at 52KHz frame rate, a spatial resolution better than 15μm is achieved by 50μm-pitch pixel array, dynamic range for charge measurement up to 23000 electrons is achieved on 25 electron resolution, which is adequate for discrimination of nuclei up to ferric and above.

Keywords—CMOS pixel sensor, particle detection, ionization energy lost, nuclei charge measurement

I. INTRODUCTION

CMOS pixel sensors has been equipped widely in the inner tracker of particle colliders and shown good performance in terms of spatial resolution, detect efficiency and other parameters. However, binary-output structure is mainly used in these ground-based experiments, make charge measure a shortcoming of these sensors. For this reason, silicon micro-strip is now a favorable technique for spatial cosmic ray detect experiment, such detectors are selected by major international experiment such as AMS and DAMPE, and reported to be well-functioning. However, micro strip detectors are suffering from several shortcomings in detect accuracy, fabrication, material budget and other concerned indexes. CMOS pixel sensors, based on the 2-D pixel array, is capable of incident particle track direct measurement, and have the potential for nuclei charge discrimination based on deposit energy measurement, which, with a lower material budget and easier fabrication, makes it a better potential detector type.

Based on the XFAB 0.18um CIS process, this paper designs a fully functional pixel sensor that can perform both track and charge measurement of incident particle. This article is organized as follows. The second section introduces the principle of charge measurement and relevant design requirements. The third section introduces the chip design. Simulation results are presented in the fourth section, and finally conclusion drown in the fifth section.

II. CHARGE MEASURE PRINCIPLE AND REQUIREMENTS

Charge measurement of the incident particle is based on quantization of its ionization energy loss on the silicon chip. As cosmic ray nuclei hit the pixel sensor, ionization occurs and leads to energy deposition in silicon, According to Bath formula, ionization energy loss of the incident particle is proportional to the square of its nuclear charge, while loga-rithmic with the energy of incident particle. Thus, by quanti-fying e/h pairs generated in the pixels, discrimination of in-cident particle charge can be performed.

The main requirements for CMOS pixel sensor to meas-ure energy deposition of incident particle include a high sen-sitivity in detection, a low noise floor and a uniform back-ground. High sensitivity is required to maximize the signal in the presence of noise, while low noise is required to extend the lower detection limit of the sensor for low Z particle inci-dent, and uniform background is also critical to prevent a signal of interest overwhelmed by spatial noise.

III. CIRCUIT DESIGN

A. Chip Architecture

The overall diagram of the sensor chip is shown in Figure 1, which is comprised of the control circuit, pixel matrix size of 32×64, readout circuit for each column, and a data compression circuit, Auxiliary and driving circuit are also implemented to build a fully functional chip. The pixel sensor adopts distributed front-end readout circuit to lower spatial noise, column end full-differential CDS-PGA to provide high sensitivity, column level 10-bit SAR-ADC for high precision quantization, and data compression circuit to prevent data stacking under high frame rate.

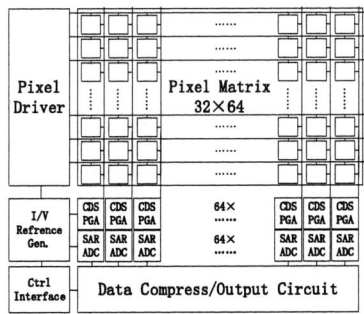

Figure 1 Chip overall structure diagram

After particle hit, electrons generated is firstly collected in situ by the diodes in the pixel matrix and form a voltage drop, pixels are then selected row by row by the control circuit for readout to the column end. Each column of pixels connects to a full-differential PGA which allows for single to double CDS readout, and an ADC for quantization of the signal. As most pixels are expected not to be fired during a readout cycle, data compress based on channel fire discrimination is then performed to reduce scale of the output data. 16-bit width parallel digital output is finally performed using custom defined data format.

B. Pixel Design

The pixel circuit includes a sensing diode and its front-end readout circuit.

The diode normally works in integrating mode to collect the electrons generated inside the pixel and forms a voltage drop upon its barrier capacitor. After a certain period, front-end readout circuit collects the voltage and reset the diode. To ensure best collecting capacity, the diode is designed according to simulation results of the physical process, two different diodes are designed for comparison of charge collecting performances.

The main difficulty in pixel amplifier design is the exclusive use of NMOS transistors inside the pixel area due to the parasitic collecting diode introduced by n-well boxes of PMOS transistors, as all e/h pairs generated by the incident particle are expected to be collected by the diode.[1]

Output of traditional particle detect sensors are mostly binary, thus saturation and linearity are not critical for those applications, one typical pixel structure is shown in Figure 2(a)[1]. For better performances on dynamic range and readout accuracy, a distributed amplifier shown in Figure 2(b) [2] is implemented, whose PMOS transistors are placed at the column-end and reused by all pixels in the same column. The amplifier is connected in close-looped configuration to provide a stable gain as well as a lower FPN, feedback structures with different gain factors are studied and simple gain-one buffer without DC level shift is implemented due to best spatial noise reduction and gain consistency throughout the pixel matrix, further signal processing is performed at column-end of the matrix.

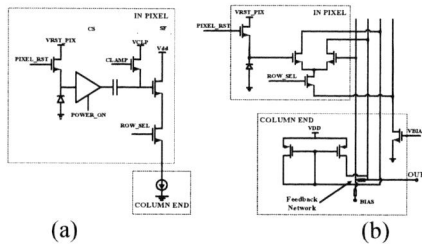

(a)   (b)

Figure 2 (a) Traditional in-pixel circuit of binary-output sensors
(b) Pixel circuit with distributed amplifier structure

The Correlated Double Sampling (CDS) working timing is used for the pixel. During the readout cycle (600ns), two voltages are transmitted to the column end respectively: the first one is the voltage on the diode after the integrating cycle, and the second is the reset voltage of the pixel, by subtracting these two voltages, the net voltage drop is delivered from the pixel, while low frequency correlated noise can also be eliminated.

C. Column-end CDS-PGA

The CDS-PGA circuit prototype is shown in (a) (b)   (c)

Figure 3(a). The circuit uses two parallel S/H stages to store the two voltages collected from the pixels, then signal subtracting is performed using the op-amp.

To provide a larger output dynamic range as well as a better anti-interference performance, a CDS-PGA with single-to-double technique is adopted. The circuit, shown in (a) (b)   (c)

Figure 3(b), is mainly consists of two parts: DC level shifter and full differential signal amplifier. [3]

The DC level shifter is based on switched capacitor array, which is divided into two sub-arrays of different sizes, with their upper plate connected together to sample the input signal. A reference voltage V_{ref}, other than ground voltage, is put on the bottom plate of the smaller sub-array before signal sampling. After signal sampling, the bottom plates of the smaller array are discharged from V_{ref} to ground, thus form a certain DC level shift. The output voltage is:

$$V_{out} = V_{in} - \frac{C_1}{C_1 + C_3} V_{ref} \quad (1)$$

Where V_{out} is the output voltage of the level shifter, V_{in} is the input voltage from the pixel, C_1 is the smaller capacitor, and C_3 is the bigger capacitor, shown in (a)   (b) (c)

Figure 3(b).By putting different V_{ref} on the two S/H stages, a DC level shift among the two voltages can be performed.

$$V_{shift} = - \frac{C_{1,2}}{C_{1,2} + C_{3,4}} (V_{reftop} - V_{refbot}) \quad (2)$$

The DC level shift technique can introduce an offset due to parasitic capacitor, while large signal can overwhelm this offset easily, measurement of small signal input may suffer a non-negligible downgrade. Thus, a configuration changeable structure is designed, shown in (a)   (b) (c)

Figure 3(c). The input signal firstly goes through a programmable threshold discriminator, the PGA then use the result to decide its configuration: for small input, the PGA uses a traditional configuration without level-shifting to improve accuracy, while for large input the PGA uses the single to double configuration to enlarge the output dynamic range.

The aforementioned two CDS-PGA structures are implemented on column-end 1-48 and 49-64 of the pixel matrix respectively.

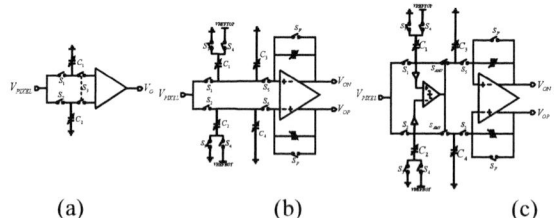

(a)   (b)   (c)

Figure 3 (a) CDS-PGA diagram
(b) single-to-double CDS-PGA
(c) configuration changeable CDS-PGA

D. Column-level SAR-ADC

SAR structure ADC is adopted in this sensor chip considering both frequency, accuracy, power and area demanding for column level integration. The ADC mainly consist of three parts: a capacitor array sub-DAC, a dynamic comparator with asynchronous clock generation circuit and a SAR logic control circuit. Using full dynamic structure, the ADC has no static power dissipation.

The CDAC architecture is shown in Figure 4. A centered voltage bias is added as the initial state of the capacitors. During transition, the top and bottom sub-DACs flip in opposite direction to form a full-scale swing, thus providing better anti-interference performance as well as a faster settling time. For column integration, number of the capacitor is minimized using centered bridge capacitor.[4]

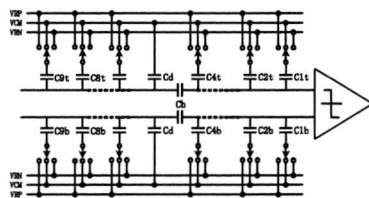

Figure 4 Architecture of the CDAC

The core comparator uses full-dynamic Lewis-Gray type structure with dynamic pre-amp for kick-back noise reduction, key node parasitic capacitance is optimized according to simulation results. Asynchronous gated clocks are generated based on delay-line to eliminate the ADC from external high-frequency clock, a low-frequency external clock is used to switch the ADC between sampling and conversion process, circuit architecture shown in Figure 5.[5]

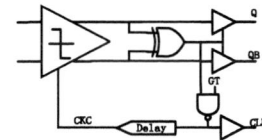

Figure 5 Comparator with asynchronous clock generation

The SAR control logic circuit is mainly composed of shift registers and CDAC bit control logic. Monotonous switching timing is used to minimize the number of switches between voltage references.

E. Data Compression and Output Circuit

Data compression circuit is shown in Figure 1. The compression method is based on fired channel filtering. Output of all ADCs first go through a digital threshold comparator for fire discrimination, then the fired channels are picked out using polling pick method, the ADC values of these channels are then encoded with their row and column indexes to provide track and dE/dX information, while signals in other channels are dumped. For picking fired channels in less clock periods, the channels are divided into several parallel banks. After data encapsulation, output through a 16-bit width parallel interface is finally performed.

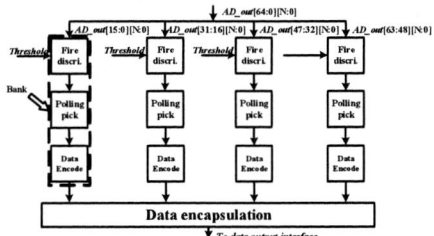

Figure 6 Data compression circuit diagram

IV. SIMULATION RESULTS

The simulation results of the front-end readout circuit shows good results, CVF of the front-end circuit is

$16\pm0.2\mu V/e^-$, the dynamic range of the circuit is 23000 electrons, which indicates a large detect range and output linearity.

The open-looped LF gain of the CDS-PGA core amplifier is 66.8dB, unity-gain bandwidth of the core amplifier is 68.2MHz, which ensures low gain offset of the CDS-PGA in close loop configuration. Pre-layout and post-layout simulation result of the SAR-ADC are shown in Table I.

Table I Performance of the column level SAR-ADC

Performance index	*Pre-layout*	*Post-layout*
Sampling Rate(MS/s)	10	10
Dynamic Range(V)	2.4	2.4
Max DNL(LSB)	0.0625	0.4325
Max INL(LSB)	0.125	1.0625
SNDR(dBFS)@Low Freq.	60.76	55.34
FoM(fJ/Conv.step)	87.06	189.39

The overall chip size is 5mm×5mm. The control interface is located on the bottom right corner, while digital and analog test output pad are located on the bottom and top side respectively, testing pads are located nearby the respective modules where they are used.

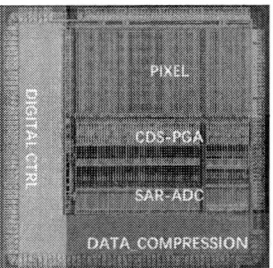

Figure 7 Chip layout

ACKNOWLEDGMENT

This work was mainly supported by the NSF project of China with granted No. U2106202 and 12075142. The research presented in this paper is also partially supported by Major scientific and technological innovation projects of Shandong Province of China(Grant No.2020CXGC010705, 2021ZLGX05, and 2022ZLGX04) and Shandong Provincial Natural Science Foundation (Grant ZR2023MA074).

REFERENCES

[1] M. Gelin et al., "Intermediate Digital Monolithic Pixel Sensor for the EUDET High Resolution Beam Telescope," IEEE Transactions on Nuclear Science, vol. 56, no. 3, pp. 1677-1684, June 2009

[2] M. Beiderman, T. Tam, A. Fish, G. A. Jullien and O. Yadid-Pecht, "A Low-Light CMOS Contact Imager With an Emission Filter for Biosensing Applications," in IEEE Transactions on Biomedical Circuits and Systems, vol. 2, no. 3, pp. 193-203, Sept. 2008

[3] Z. Li, "Research on the Key Techniques of High Performance Column Readout Circuits for CMOS Image Sensors", PHD thesis, Zhejiang Univ, 2020.

[4] J. Zhao, Z. Huang and X. Hou, "A 10-bit 50-MS/s SAR ADC with Binary-Scaled Recombination Weighting Capacitor Array," 2022 7th International Conference on Integrated Circuits and Microsystems (ICICM), Xi'an, China, 2022, pp. 135-139

[5] P. Mounika, D. Verma and K. -Y. Lee, "An Improved Dynamic Latch Comparator with Low Power Consumption for SAR ADC Applications," 2022 19th International SoC Design Conference (ISOCC), Gangneung-si, Korea, Republic of, 2022, pp. 43-44

979-8-3503-6184-1/24 $31.00 © 2024 IEEE

Sc$_{0.096}$Al$_{0.904}$N-based Bimorph Piezoelectric Micro Machined Ultrasonic Transducers

Ziye Zhai[1], Wenjuan Liu[2,*], Chengliang Sun[2]

[1] The Hongyi honor college of Wuhan University, Wuhan, China
[2] The Institute of Technological Sciences Wuhan University, Wuhan, China

* Email: lwjwhu@whu.edu.cn

Abstract—The paper presents a Sc$_{0.096}$Al$_{0.904}$N-based bimorph piezoelectric micromachined ultrasonic transducers (PMUT) and its liner array with 100 elements for medical imaging applications. The PMUT is fabricated using the Sc$_{0.096}$Al$_{0.904}$N-based bimorph with thickness of 500nm, respectively. The diameter of PMUT is 120μm, of which the center frequency is 1.9MHz. The 3D models are simulated to analysis the resonant frequency, the central displacement, the electrical impedance and the crosstalk response in frequency domain. The active area of the PMUT liner array has a dimension of 150μm in both width and length. The phased liner array is simulated using 3D FEM simulation models where the focus depth can be adjusted by the excited signals. Our works validate the bimorph PMUT and its array, thereby accelerating the development of the phased array in medical imaging.

Keywords—PMUT, bimorph, scandium-doped aluminum nitride (ScAlN), FEM

I. INTRODUCTION

With the rapid development of micromachined ultrasonic transducer (MUT), the advantages such as miniaturization, low cost, and highly integrated drivers, have become increasingly prominent. The emergence of MUTs has solved a series of shortcomings of traditional ultrasonic transducers, such as large size, serious acoustic impedance mismatch, and insufficient integration. Piezoelectric micromachined ultrasonic transducer (PMUT) arrays have occupied an increasing market share in the fields of industry, medical care, and automobiles. PMUTs have been widely used in fingerprint recognition, ranging positioning, gesture recognition, and imaging [1]-[3].

Common piezoelectric materials include PZT, aluminium nitride (AlN) and zinc oxide. AlN with the advantages of low dielectric loss, low thermal noise and easy compatibility with the CMOS process, the thin film is a good material for making the PMUT piezoelectric layer. However, the low piezoelectric coefficient of the AlN limits the noise resolution, sensitivity, and signal-to-noise ratio of the AlN-based PMUT. It is found that the d33 of the piezoelectric membrane increased by 400% after doping with Sc, Liu et al [4]. Thus, the researchers turn their eyes on the piezoelectric receiving sensors based on scandium-doped AlN (ScAlN) [5].

In addition, structural optimization of PMUT is another important method to improve its acoustic performance and main characteristics. The traditional PMUT structure is a sandwich-type planar diaphragm structure. In order to improve the sensitivity, electromechanical coupling factor and transmission capacity of PMUT, Sina Akhbari et al. at the University of California, Berkeley designed and manufactured a curved membrane AlN-based PMUT that is compatible with CMOS circuits and achieves more efficient acoustic coupling and higher sound pressure [6]. However,

this structure will introduce residual stress, resulting in a decrease in the service life of the device. In 2015, Tao Wang et al. from the National University of Singapore used etched holes to obtain a PMUT with a piston-like vibration mode, which can be used for ultrasonic imaging [7]. However, the etched holes limit the use scenarios of PMUT. He et al. proposed a double-layer PMUT array to improve the normal displacement modulus and bandwidth [8]. Compared with the single-layer PMUT array, the normal displacement modulus at the center of the double-layer PMUT array is doubled. Therefore, this work proposes a ScAlN-based bimorph PMUT linear array with higher energy efficiency and sensitivity. Simulation results show that the center displacement of the ScAlN-based bimorph PMUT is 6% higher than that of the AlN-based bimorph PMUT and 43% higher than that of the unimorph PMUT.

II. STRUCTURE AND DESIGN

This study proposes a bimorph PMUT linear array based on ScAlN. Compared with the unimorph piezoelectric layer on Si or SiO$_2$ device layer, the bimorph diaphragm has the advantages of low bending stiffness, large electrostatic capacitance, and high output sensitivity, but the manufacturing process is more complicated [9]. Since the ratio of the piezoelectric coefficient e31 to the relative dielectric constant ε33 and r of Sc$_{0.096}$Al$_{0.904}$N is greater than that of PZT and AlN, this better piezoelectric material is used to obtain higher sensitivity.

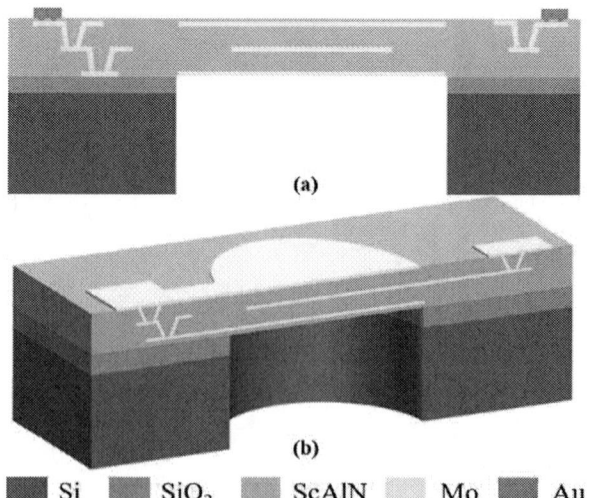

Fig. 1. Cross-section and axonometric view of an element of the designed bimorph PMUT based on ScAlN.

This study presents a bimorph PMUT array to achieve ultrasonic transmission and reception. Its circular diaphragm consists of Sc$_{0.096}$Al$_{0.904}$N bimorph films and three Mo electrodes. Fig.1 shows the structure of the bimorph PMUT element based on $Sc_{0.096}Al_{0.904}N$, and its resonant frequency is designed to be 1.9MHz. The actuation layer consists of two $Sc_{0.096}Al_{0.904}N$ diaphragms with a thickness of 500nm and a diameter of 120μm. The electrodes are made of molybdenum material with a thickness of 100nm to provide electrical connections.

III. FABRICATION AND CHARACTERIZATION

The proposed bimorph PMUT is prepared using a $Sc_{0.096}Al_{0.904}N$ process platform on a silicon substrate. Fig. 2. shows a process which is used to fabricate the testing device. At first, an 8-in Si wafer is used, on which 500nm SiO_2, 25nm AlN seed layer, and 100m lower Mo electrode is deposited sequentially, and then the lower electrode pattern is formed [Fig.2(a)]. The next step is the deposition of the first ScAlN (500nm) film and the lower Mo via [Fig.2(b)]. The middle Mo electrode (100nm) is deposited and patterned [Fig.2(c)]. The second ScAlN (500nm) film is deposited and the hole of the middle Mo electrode is opened. After that, the upper Mo electrode (100nm) is deposited and processed graphically [Fig.2(d)]. The lower Mo electrode is connected to the upper Mo electrode through the lower Mo extraction to form an electrical port, and the middle Mo electrode is used as another electrical port. A passivation layer of 25nm AlN is deposited to inhibit the oxidation of the Mo electrode and to open the through-hole of the upper Mo electrode [Fig.2(e)]. Au electrode pads (1μm) are formed by the lift-off process. Before the back side processing, a 500nm SiO_2 protective layer is deposited on the front side of the wafer. The bottom surface of the wafer is thinned to 400 μm and polished by chemical-mechanical polishing (CMP). Then, a 2μm SiO_2 is deposited on the back side and etched to use as a hard mask [Fig.2(f)]. The posterior cavity is a deep-made reactive ion etching (DRIE). Finally, the silica is released by a buffer oxide etching (BOE) solution [Fig.2(g)].

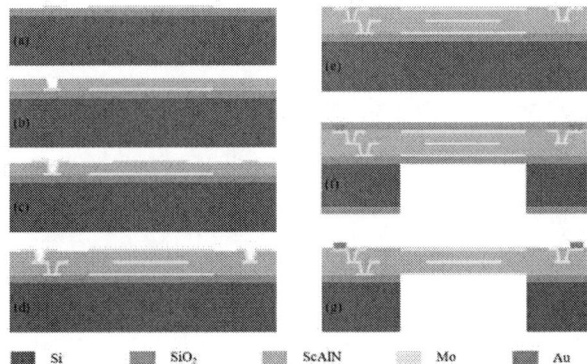

Fig. 2. Schematic illustration of the ScAlN-based bimorph PMUT fabrication process flow. (a) Lower electrode. (b) Lower via. (c) Middle electrode. (d) Middle via and upper electrode. (e) Upper via. (f) Au pad and bottom release. (g) Oxide release.

Fig.3(a) shows a scanning electron microscope (SEM) image of a 10×10 bimorph PMUT linear array. The radii of the bimorph diaphragm and the middle Mo electrodes are 60μm and 40μm, respectively. The thickness of the upper ScAlN and lower ScAlN layers are designed to be 500μm, and the neutral plane of the piezoelectric stack film is still located on the middle Mo electrode, which is the basis for the bimorph to enhance the electrical response [Table I and Fig.3(b)].

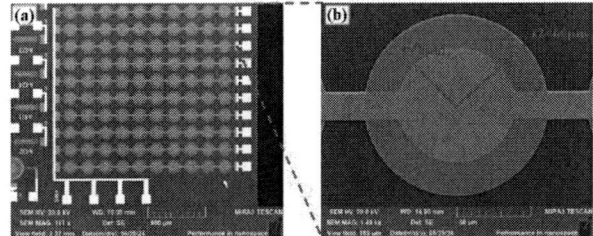

Fig. 3. The array (a) and an array element (b) SEM images of the ScAlN-based bimorph PMUT

TABLE I. BASIC PARAMETERS OF THE DEVICE

BASIC PARAMETERS OF THE DESIGNED ScAlN-based bimorph PMUT	
Parameters	*Value*
Radius of the upper Mo electrode	60 μm
Radius of the middle Mo electrode	40 μm
Radius of the lower Mo electrode	60 μm
Thickness of the Mo electrode	100 μm
Thickness of the ScAlN	500 μm
Pitch of the PMUT liner array	150 μm
Size of the array	150×150 μm²

IV. MEASUREMENT AND DISCUSSION

In this study, finite element simulations are performed on ScAlN-based bimorph PMUT linear arrays and phased arrays. The main characteristics of the ScAlN-based bimorph PMUTs are calculated and analyzed using three-dimensional models.

A. PMUT cell and liner array

In this study, a transducer array model is constructed to evaluate the frequency characteristics and crosstalk characteristics of the transducer array. For the substrate side boundary of the 3×3 parallel array, a 10V excitation signal is applied to the unit, and the original perfectly matched layer domain is replaced by a free boundary condition to simulate the actual vibration of the chip after cutting at the periphery.

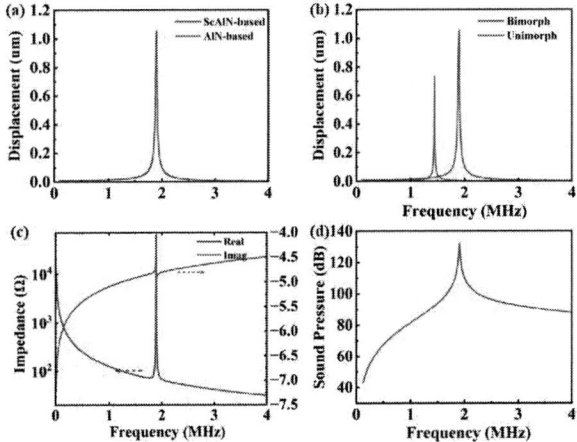

Fig. 4. Simulation results: (a) Center displacement response of ScAlN-based bimorph PMUT cell and AlN-based bimorph PMUT cell. (b) Center displacement response of ScAlN-based bimorph PMUT cell and ScAlN-

based unimorph PMUT cell. (c) Electrical impedance response of the cell. (d) Sound pressure of ScAlN-based bimorph PMUT cell at 2mm in air.

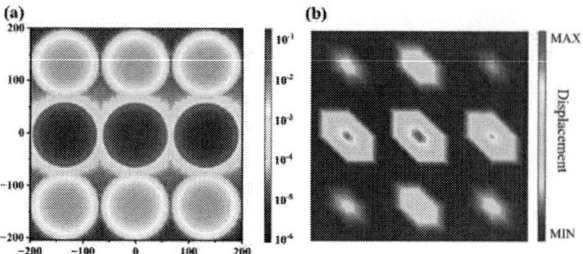

Fig. 5. 3 × 3 PMUT array displacement obtains from (a) finite element simulation and (b) experiment.

The finite element simulation results of a bimorph PMUT unit are shown in Fig.4. Fig.4(a) shows the difference between the center displacement of the bimorph PMUT units based on ScAlN and AlN. Since the ratio of the piezoelectric coefficient e31 to the relative dielectric constant ε 33 and r of $Sc_{0.096}Al_{0.904}N$ is greater than that of AlN. The center displacement of the ScAlN-based bimorph PMUT element increased by 6% compared to the AlN-based transducer. Compared with the single-crystal PMUT element, the center displacement of the bimorph PMUT element increased by 43%, which means that the ScAlN-based bimorph PMUT has better electromechanical conversion capability [Fig.4(b)]. The ScAlN-based bimorph PMUT has good electrical impedance at the resonant frequency [Fig.4(c)], and its sound pressure at 2mm in the air domain can reach 130dB [Fig.4 (d)].

Fig.5(a) and Fig.5(b) are comparisons of the displacement results of the 3×3 ScAlN-based bimorph PMUT array when only the second row of units are excited by the finite element simulation and experiment. The experimental results of the crosstalk response are similar to the simulation results, thus confirming the correctness of the finite element simulation.

B. PMUT phased array

Phased array technology enables ultrasound to be focused and deflected for scanning, further improving the sensitivity and signal-to-noise ratio of the PMUT transmission and reception process. Therefore, this study constructs a finite element simulation model of a 7×7 ScAlN-based bimorph PMUT phased array and obtains the ultrasound focusing results through transient analysis.

The ultrasound waves are emitted from the central O_i of each transducer element. Each wave shall have a specific phase shift so that all waves synchronously reach the focus F. For each PMUT element, the Euclidean distance $|| O_iF ||$ between O_i and F is calculated. The phase shift is calculated by:

$$\varphi_i = 2\pi(|| O_iF || \% \lambda) \tag{1}$$

where % is the modulo operation (the remainder of the division).

The bottom electrode and top electrode of each unit are grounded, and the middle electrode is independently controlled. The phase shift of each array element excitation signal is calculated based on the acoustic path difference at each position and formula (1). A 1V sine wave signal with different phase shifts is applied to each PMUT unit, and transient analysis is performed.

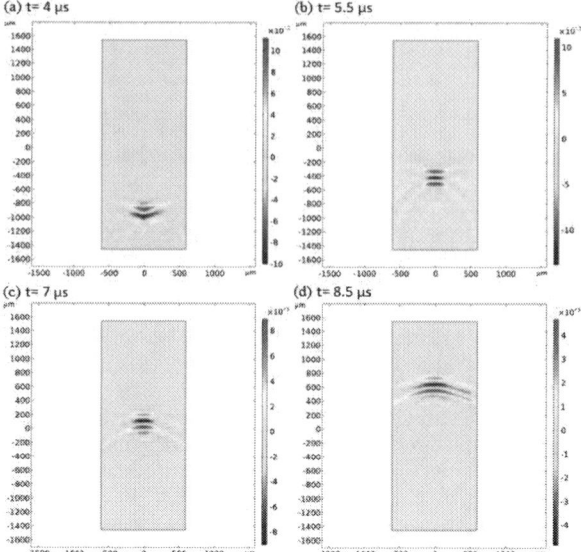

Fig. 6. Ultrasonic-focused acoustic pressure results were achieved by a PMUT phase array. (A) The sound beam travels 4μs, and the sound wave is unfocused. (b) The sound beam travels 5.5μs, and the sound wave is focused. (c) (d) The sound beam travels 7μs and 8.5μs.

Fig.6. shows the finite element simulation results of the ScAlN-based bimorph PMUT phased array, and the acoustic wave is successfully focused. The performances of the ultrasonic sound pressure in the air domain within 10μs are shown in Fig.6(a) to Fig.6(d). Fig.6(a) shows the sound pressure performance before the acoustic wave is focused at t=4μs. Fig.6(b) shows the sound pressure performance when the acoustic wave is focused at t=5.5μs, and the propagation of the acoustic wave after focusing is shown in Fig.6(c) and 6(d). The simulation results show that when the sound wave does not reach the specified focal point, the sound pressure sinks downward, and when it reaches the specified focal position, it is successfully focused into a beam knot. After focusing, when it continues to propagate, the sound pressure first sinks outward and then gradually becomes flat.

V. CONCLUSION

This study aims to evaluate the performance, crosstalk, and implementation of ScAlN-based bimorph PMUT elements and arrays to enable acoustic focusing of PMUT phased arrays. Using finite element method, the models of the bimorph PMUT single unit, the PMUT liner array and the PMUT phased array are analyzed, respectively. The analytical simulations have shown that the energy efficiency and sensitivity of the ScAlN-based bimorph PMUTs can be 6% higher than AlN-based PMUTs, and 43% higher than unimorph PMUTs. Experimental results are consistent with simulation results. The further phased array will be implemented in the future, which is suitable for medical imaging applications.

REFERENCES

[1] Drinkwater B, Wilcox D. Ultrasonic arrays for non-destructive evaluation: A review [J]. NDT & e International, 2006, 39(7): 525-541.

[2] Tajik A, Seward J, Hagler D, Mair D, Lie J. Two-dimensional real-time ultrasonic imaging of the heart and great vessels. Technique, image orientation, structure identification, and validation [C]// Mayo Clinic Proceedings. 1978: 271-303.

[3] Lu Y, Tang H, Fung S, Boser B, Horsley D. Short-range and high-resolution ultrasound imaging using an 8 MHz aluminum nitride PMUT array [C]// 2015 28th IEEE International Conference on Micro Electro Mechanical Systems (MEMS). 2015: 140-143

[4] WANG C H, CHEN X Y, WANG L, et al. Bioadhesive ultrasound for long-term continuous imaging of diverse organs [J]. Science, 2022, 377(6605): 517-+.I. S. Jacobs and C. P. Bean, "Fine particles, thin films and exchange anisotropy," in Magnetism, vol. III, G. T. Rado and H. Suhl, Eds. New York: Academic, 1963, pp. 271–350.

[5] K. Elissa, "Title of paper if known," unpublished.Y. Kusano, I. Ishii, T. Kamiya, A. Teshigahara, G.-L. Luo, and D. A. Horsley, "High-SPL air-coupled piezoelectric micromachined ultrasonic transducers based on 36% ScAlN thin-film," IEEE Trans. Ultrason., Ferroelectr., Freq. Control, vol. 66, no. 9, pp. 1488–1496, Sep. 2019, doi: 10.1109/TUFFC.2019.2921983.

[6] Akhbari S, Sammoura F, Shelton S, et al. Highly responsive curved aluminum nitride pMUT[C]//2014 IEEE 27th International Conference on Micro Electro Mechanical Systems (MEMS). IEEE, 2014: 124-127

[7] Wang T, Sawada R, Lee C. A piezoelectric micromachined ultrasonic transducer using piston-like membrane motion[J]. IEEE Electron Device Letters, 2015, 36(9): 957-959.

[8] L. -M. He, W. -J. Xu, W. -J. Liu, X. -B. Wang, J. Zhou and J. -Y. Ren, "Performance and Crosstalk Evaluation of 2-D Array Piezoelectric Micromachined Ultrasonic Transducer with 3-D Finite Element Simulation," 2019 IEEE International Ultrasonics Symposium (IUS), Glasgow, UK, 2019, pp. 792-795,doi: 10.1109/ULTSYM.2019.8925676.

[9] C. Yang et al., "Bimorph piezoelectric MEMS microphone with tractive structure," in Proc. IEEE Int. Ultrason. Symp. (IUS), Oct. 2022, pp. 1–4, doi: 10.1109/IUS54386.2022.9957653.

979-8-3503-6184-1/24 $31.00 © 2024 IEEE

TeO$_2$ as Amorphous P-type Transistor for Back-End-of-Line Applications

John Robertson [1*], Xuewei Zhang [1], Qingzhong Gui [2], Yuzheng Guo [2]

[1] Engineering Dept, Cambridge University, Cambridge CB3 0FA, UK
[2] School of Electrical Engineering, Wuhan University, Wuhan 430072, China

* Email: jr214@cam.ac.uk, yguo@whu.edu.cn

Abstract—**Electronic devices would benefit from low-cost amorphous (a-) bipolar oxide semiconductors for back-end-of-line CMOS devices. InGaZn oxide is a good n-type amorphous oxide, but p-type oxides are quite rare. Those presently known suffer from various deficiencies, such as high leakage currents (SnO), having a layer structure that is incompatible with being amorphous, needing high-temperature refractory metals (ZnRh$_2$O$_4$) or forming deep non-doping acceptor-states (ZnO). We show by ab-initio molecular dynamics that a-TeO$_2$ can have shallow Sb$_{Te}$ acceptor centers due to stronger covalent bonding.**

Keywords— P-type BEOL, oxide, semiconductor

I. INTRODUCTION

New processing techniques have allowed back-end-of-line (BEOL) oxides to play a role in new device designs. These devices need bipolar oxides grown by atomic layer deposition and processable at ~400C. So far there are many suitable n-type oxides such as InGaZnO$_4$ (IGZO)[1] or In$_2$O$_3$ [2], but no satisfactory p-type oxide. Overall, their band states lie too deep below the vacuum level, typified by nitrogen doped ZnO, whose acceptor states form deep levels [3].

P-type oxides do exist but so far they each suffer from various problems. SnO is a valuable p-oxide due to its Sn 5s-like upper valence band character, but its small 0.7 eV band gap causes high leakage currents [4]. There are various p-type transparent conducting oxides like CuAlO$_2$ [5] but their local layer structure is incompatible with isotropic amorphous bonding. Amorphous p-type oxides such as the spinel Zn$_2$RhO$_4$ do exist, but this uses the costly refractory metal Rh [6] incompatible with a need for low processing temperatures. Finally, data-based searches have found other oxides like SnTa$_2$O$_6$ but these may suffer from dopant self-compensation by intrinsic defects [7,8]. Thus, there is still no fully satisfactory p-type oxide.

Bulk TeO$_2$ is a glassy oxide known for its large non-linear optical coefficient [9]. It has a relatively low glass transition temperature compared to SiO$_2$ [10]. It has several crystalline polymorphs α, β and γ, which are non-planar layer compounds [11]. It is a compound of two group 6 elements, but unusually, the Te site has a coordination of 4 while the O site has a coordination of 2. TeO$_2$ could be hole dopable, unlike ZnO, because its bonding has a more covalent character.

In 1996, Hosono et al [12] noted that amorphous materials could have high electron mobility similar to their crystalline counterparts if the conduction band edge states have isotropic s-like character (like SnO$_2$). This was applied to n-type oxides. On the other hand, if the oxide has an s-like valence band maximum it could have a high hole mobility. For this to occur, the oxide should possess filled s-like lone-pair states.

Guo et al [13] noted that, theoretically, TeO$_2$ had a low effective hole mass and high hole mobility, due to this filled Te 5s orbitals. Zavabeti et al [14] experimentally measured various properties of TeO$_2$ such as its hole mobility of 146 cm^2/V.s an effective hole mass of 0.51 and a weakly p-like character of its undoped films. Shi et al [15] produced β-TeO$_2$ by pulsed laser and sol-gel deposition, but again it was undoped, perhaps due to safety issues. Thus, a major problem affecting many p-type oxides was the possibility of compensation of doping by intrinsic defects [16] which remains unresolved. Substitutionally doped a-TeO$_2$ films must be deposited and shown to be fully conductive, not compensated, ideally experimentally. Substitutional doping of the Te site of TeO$_2$ is simpler in that it avoids the polaron problem, but this must use a group V element like As, Sb or perhaps Bi, which has not occurred yet possibly because of safety concerns. Instead of experimental confirmation, we use ab-initio molecular dynamics (AIMD) here to study doping.

The amorphous phase of TeO$_2$ is similar to that of β-TeO$_2$, the most stable of its phases. Here we study the electronic structure of β-TeO$_2$, then a-TeO$_2$ by AIMD, and finally the extrinsic doping of a-TeO$_2$ by AIMD of both the Te site and the O site. Interestingly, undoped TeO$_2$ has already been studied by AIMD by Bernasconi et al [17,18].

II. METHODS

We calculate the electronic structure of β-TeO$_2$ by density functional theory (DFT) in the generalized gradient approximation (GGA) and then use HSE hybrid functionals [19] to correct the underestimate of the band gap and any error of the energies referred to the vacuum level. To find the vacuum level, layers of the crystalline phase are placed in a supercell separated by a 15Å vacuum gap from the next cell.

III. RESULTS

The various phases of TeO$_2$ are fundamentally derived from the parent tetragonal phase of TeO$_2$ in which Te is 6-fold

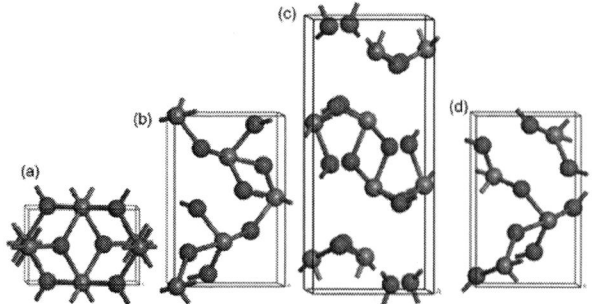

Fig. 1(a-d). The lattice structure of ideal rutile, α, β and γ-TeO$_2$ respectively. Orange balls=Te, red balls =O.

979-8-3503-6184-1/24 $31.00 © 2024 IEEE

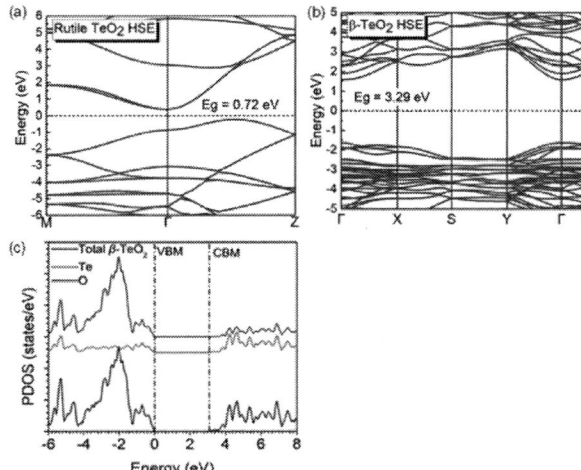

Fig. 2(a,b). HSE band structure of the idealized rutile structure and the β polytype of TeO₂. (c) the partial density of states of β-TeO₂ in HSE.

bonded and oxygen is 3-fold bonded [20]. This phase has only a small 0.7 eV band gap (Fig 1a). This bonding distorts into various layered phases in which the Te site becomes 4-fold coordinated but with 2 additional weak bonds, and the O site becomes 2-fold coordinated. The lattice adopts various rumpled layered structures with band gaps ranging from 2.6 to 3.29 eV depending on the phase. These phases differ only slightly in total energy per formula unit. The β-phase has a direct band gap (Fig 1b) whereas the α- and γ- phases have slightly indirect gaps [21].

Fig 1(c) shows the calculated partial density of states (PDOS) of β-TeO₂. The PDOS of the valence band maximum shows Te s states which cause the high hole mobility of TeO₂.

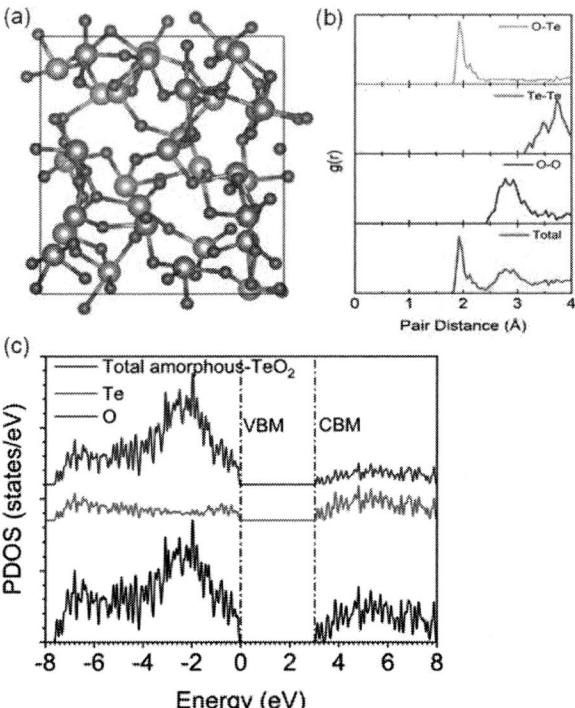

Fig. 3(a). Structure of amorphous TeO₂ by ab-initio molecular dynamics, (b) pair distribution function of a-TeO₂, (c) calculated (HSE) PDOS of a-TeO₂.

These states partially mix with the O 2p states which is why the hole mobility of p-type amorphous oxides is lower than that of the crystalline phase.

We then continue to calculate the properties of the amorphous TeO₂ phase by AIMD. A-TeO₂ resembles glassy SiO₂ in the more covalent quality of its bonding whereas a typical metal oxide like IGZO has a more ionic bonding. Here we form an amorphous TeO₂ phase by AIMD using a relatively small 96 atom model. A cubic supercell of TeO₂ is heated to 2000K for 4 ps, then cooled to 300K in 20 ps and then the network is relaxed at 300K for 5 ns.

The resulting structure is shown in Fig 3(a). The radial distribution function (rdf) and partial rdfs of this structure are shown in Fig 3(b). All the nearest neighbor bonds are Te-O bonds, as in β-TeO₂. The weaker Te-Te bonds are slightly longer in the amorphous phase. The main Te-O bonds lie at 1.96Å. The second peak of O-O neighbors lies at around 2.8 Å. The results resemble those earlier of Pietrucci et al [18].

The electronic structure is calculated by GGA and then corrected by hybrid functionals in Fig 3(c). It is interesting that the disorder has partially washed out the difference between the upper Te s-like DOS peak and the peak at -2 eV due mainly to O 2p states. This is similar to the earlier results [17] that used GGA only.

We now continue and directly calculate the electronic structure of doped a-TeO₂ by AIMD. We examine the substitutional acceptor doping on both the Te and the O sites. We choose to replace two Te or two O sites with As or N respectively, so as not to deal with spin polarisation.

The doped a-TeO₂ network with As on Te sites is shown in Fig. 4(a). The two acceptor sites are each 4-fold coordinated. The PDOS is shown using the hybrid functional calculations in Fig 4(b). The Fermi energy of the doped network is seen to lie below the top of the valence band, and that it is doped by shallow acceptors. This lattice shows the acceptors are shallow centers and they do not form deep polaron states separated by a gap from the main valence band as is seen in for example ZnO or Ga₂O₃. Nor are the acceptors

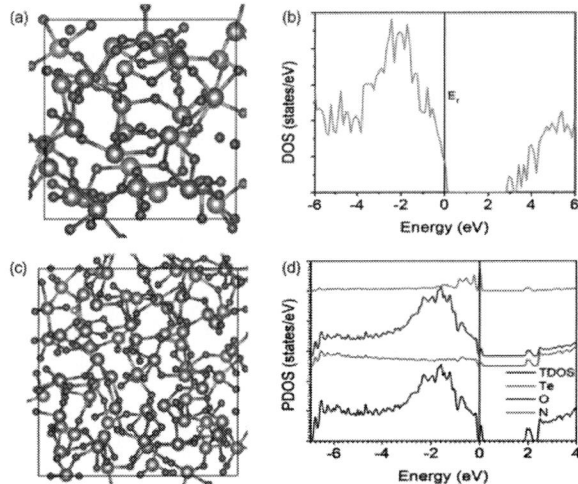

Fig. 4(a). AIMD structure of a-TeO₂ with two As$_{Te}$ substitutional sites, (b) calculated HSE DOS of a-TeO₂ with two As$_{Te}$ showing Fermi level lying in the top of the valence band, (c) two substitutional two N$_O$ sites, (d) calculated HSE PDOS for a-TeO₂ containing two N$_O$ sites.

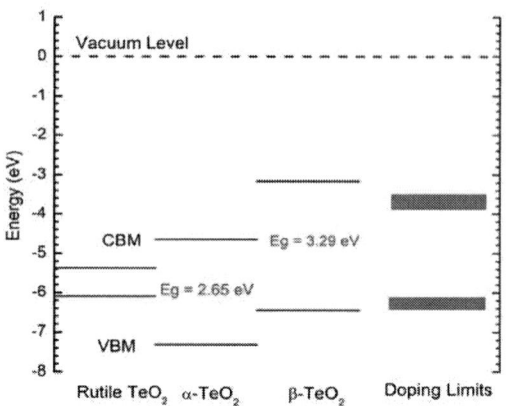

Fig. 4. Calculated HSE band edge energies referred to the vacuum level for different phases of TeO_2, compared to the approximate doping limits [16].

self-compensated in the network by the formation of native defect levels higher in the gap. This result is simpler than the previous indirect method of considering self-compensation as a separate issue [21].

We also considered the replacement of two O sites by two N atoms to form two N_O sites. The resulting network is shown in Fig 4(c) using a larger cell. The analogous partial density of states is shown in Fig 4(d). The response in this case is more complex. The lattice disproportionates to having sites with different coordinations, a negatively charged under-coordinated site and a positively charged over-coordinated site. One filled site is at the valence band maximum and the other is state is empty and near the conduction band minimum.

The calculated valence band edge energy (ionisation potential) of the β-TeO_2 lies just within the doping limits guidelines set by comparison with other oxides [8,16,22]. It is possible due to the unusual layer-like structure of the amorphous phase that the internal ionization potential of a-TeO_2 lies above that of β-TeO_2, in the same way that this may occur in a-$SnTa_2O_6$ [23].

We note that our proposed doping mechanism differs from that proposed recently by Liu et al [24]. Our mechanism uses conventional substitutional doping by acceptors, that of Liu

uses more complex processes which might be incompatible with scaling to small BEOL device dimensions.

IV. CONCLUSION

In summary, this paper has shown by ab-initio molecular dynamics that amorphous TeO_2 can be substitutionally doped p-type that is not compensated by intrinsic defects. We found that substitutional As or Sb atoms at Te sites will lead to the Fermi energy lying below the valence band maximum. If these results are experimentally confirmed, this could open the way to CMOS back-end-of-line oxide-based devices.

V. REREFERENCES

[1] H Hosono, J Non-Cryst Solids **352** 851 (2006)

[2] M Si, Y Hu, Z Lin, X Sun, A Charnas, D Zheng, X Lyu, H Wang, K Cho, P Ye, Nanolett **21** 500 (2021)

[3] J L Lyons, A Janotti, C G van de Walle, Appl Phys Lett **95** 252105 (2009)

[4] J B Varley, A Schliefe, A Janotti, C G van der Walle, Appl Phys Lett **103** 082119 (2013)

[5] H Kawazoe, M Yasukawa, H Hyodo, M Kurita, H Yanagi, H Hosono, Nature **389** 930 (1997)

[6] T Kamiya, et al, Adv Funct Mater **15** 968 (2005)

[7] M Barone et al, J Phys Chem C **126** 3764 (2022)

[8] Z Zhang, Y Guo, J Robertson, Chem Mater **34** 643 (2022)

[9] M BenYahia et al, J Phys Chem B **112** 10777 (2008)

[10] N S Tagiara, et al, J Non-Cryst Solids **457** 116 (2017)

[11] Y Li, W Fan, H G Sun, X P Cheng, P Li, X Zhao, J Appl Phys **107** 093506 (2010)

[12] H Hosono et al, J Non-Cryst Solids **203** 334 (1996)

[13] S Guo et al, Nanoscale **10** 8397 (2018)

[14] A Zavabeti et al, Nat Electronics **4** 277 (2021)

[15] J Shi, Z Sheng, I Zhu, X Xu, Y Gao, D Tang, K H L Zhang, App Phys Lett **122** 101901 (2023)

[16] J Robertson, S J Clark, Phys Rev B **83** 075205 (2011)

[17] M Ceriotti, F Pietrucci, M Bernasconi, Phys Rev B **73** 104304 (2006)

[18] F Pietrucci, S Caravati, M Bernasconi, Phys Rev B **78** 064203 (2008)

[19] J Heyd, G E Scuseria, M Erzenhof, J Chem Phys **120** 7274 (2004)

[20] J Robertson, J Phys C **12** 4767 (1979)

[21] J Robertson, X Zhang,Q Gui, Y Zheng, Appl Phys Lett **124** 212101 (2024)

[22] H Hosono, Jpn J Appl Phys **52** 09001 (2013)

[23] Y Hu, D Schlom, S Datta, KJ Cho, Appl Surf Sci **613** 155981 (2022)

[24] A Liu, et al, Nature **629** 798 (2024)

Miniaturization of high-speed GaN based laser diodes

Junfei Wang, Chaowen Guan, Leihao Sun, Zhichong Wang, and Chao Shen

School of Information Science and Technology, Fudan University, 200433, Shanghai, China

chaoshen@fudan.edu.cn

Abstract—In this work, we design and fabricated a GaN based mini-laser featured with 1.8 µm width ridge and 300 µm length resonant cavity. -3 dB bandwidth of the mini-laser is around 4.8 GHz, and the -10 dB is around 5.8 GHz.

Keywords—GaN, mini-laser, visible light communication

I. INTRODUCTION

Visible light communication (VLC) is an optical wireless communication technology whereby the transmitted signal is loaded on the light with wavelength between 375 and 780 nm[1,2]. Compared with the existed wireless communication technology in the radio frequency range, VLC is characterized by the plentiful spectrum (400-800 THz), electromagnetic immunity and license-free. These above advantages make the VLC be regarded as an essential component of the next-generation 6G wireless network[3].

To date, light emitting diode (LED) based VLC system has demonstrated great potential in these applications. Plentiful investigations are related to the improvement of the communication behaviors of the LED. Miniaturization of the active region area of the LED can be effective method to improve the modulation response of the LED. For instance, a record 7.91 Gbps data transmission rate based on a micro-LED was accomplished. The increased demand of the data rate puts forward higher request on the modulation bandwidth of the light source. However, modulation bandwidth of the LED is limited by the carrier lifetime which is on the ns level, maximum achievable bandwidth of LED is limited in 1 GHz. In contrast to the LED, GaN based laser diode (LD) behaves high spectral purity and high electrical-optical conversion efficiency at relative high injection level. More importantly, bandwidth of the LD is not limited by the carrier lifetime. Some related researches have demonstrated the GaN based blue LD is featured with the modulation bandwidth exceeding 1 GHz. Miniaturization of the LD structure leads to the increase of the carrier concentration which results in the higher material gain and lower mode loss during the prograation of the photon in the resonant cavity.

In this work, we design and fabricated the GaN based blue mini-laser which is featured with the 1.8 µm width ridge and 300 µm length resonant cavity. The narrow ridge waveguide is prepared by the UV lithography and the ICP etch. The resonant cavity is formed by the thinning of the epitaxial wafer and the cleavage.

II. DESIGN AND RESULTS

A. Design

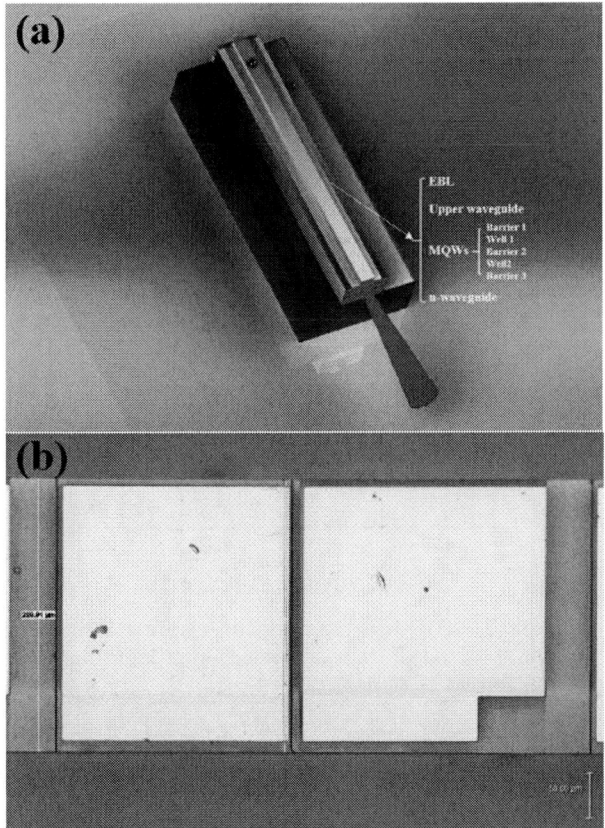

Figure 1. Schematic of the designed GaN mini-laser (a) Three-dimensional structure schematic (b) Microscope of the fabricated mini-laser.

Fig.1 (a) shows a three-dimensional structure schematic of the designed GaN mini-laser. Active region of the epitaxial structure is shown in the picture. The active region includes two 3 nm thickness quantum well and three 5 nm thickness quantum barrier. Fig.1 (b) shows the microscope photo of the fabricated mini-laser. The mini-laser is featured with 1.8 µm width ridge and 300 µm length resonant cavity. Golden rectangular in the picture with the mission corner is the n-electrode and the other is the p-electrode.

979-8-3503-6184-1/24 $31.00 © 2024 IEEE

B. Results

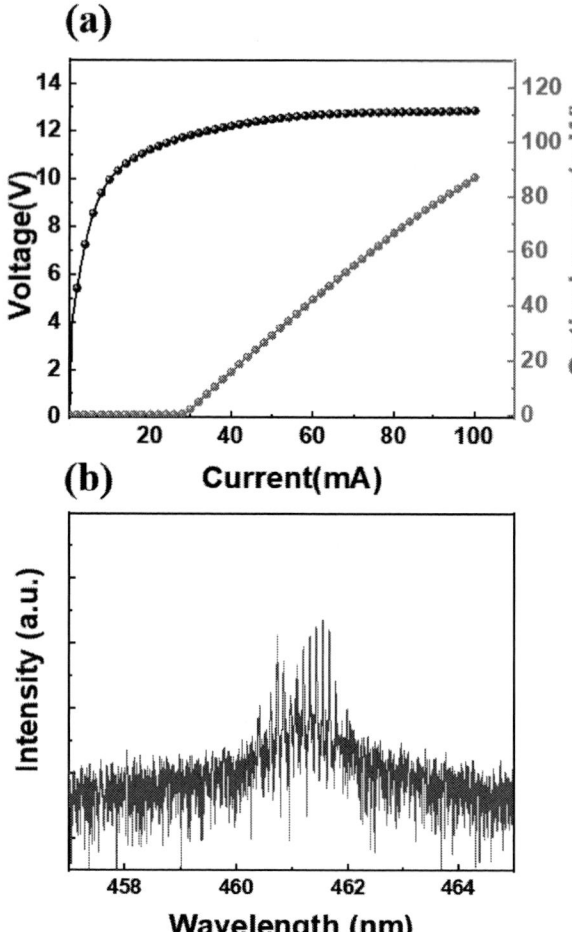

(a)

(b)

Figure 2. (a) L-I-V characteristic of the mini-laser (b) Emission spectra of the fabricated mini-laser under the injection current of 87 mA.

Fig.2 (a) shows the L-I-V character of the fabricated mini-laser. At the cw electrical condition, threshold current of the mini-laser is around 29 mA and the corresponding threshold current density is 5.37 kA/cm^2. Threshold voltage of the mini-laser is around 12 V. The relative high threshold voltage arises from that the n-electrode is deposited on the n-cladding layer. Compared with the n-contact layer, the n-cladding layer is characterized by lower doping concentration which can result in higher contact resistivity. Slope efficiency of the mini-laser is around 1.24 W/A. Fig.2 (b) shows emission spectra of the mini-laser under room temperature at the injection current of 87 mA which is 3 times threshold current. Central wavelength of the emission spectra is between 460 nm to 462 nm.

Fig.3 shows the intrinsic small signal modulation response of the fabricated mini-laser under the injection current of 87 mA. There is an apparent roll-off at low frequency due to the lager series resistance. Therefore, we build up the equivalent circuit model of the mini-laser to extract the intrinsic modulation response. -3 dB bandwidth of the mini-laser is around 4.8 GHz, and the -10 dB is around 5.8 GHz.

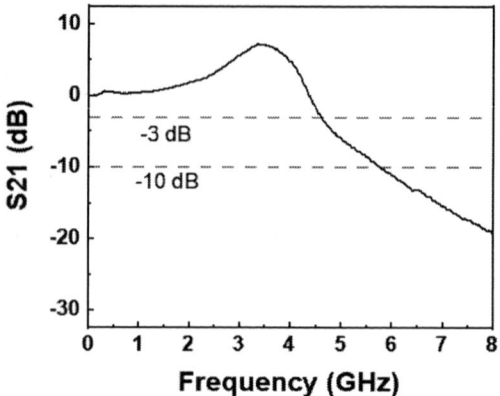

Figure 3. Extracted intrinsic small signal modulation response of the mini-laser.

III. SUMMARY

In this work, we design and fabricate the GaN based mini-laser featured with 1.8 μm width ridge and 300 μm length resonant cavity. Minimization of the characteristic of the laser improves the small signal modulation response of the LD. As a result, -3 dB bandwidth of the mini-laser is around 4.8 GHz, and the -10 dB is around 5.8 GHz.

ACKNOWLEDGMENT

This research is partially funded by National Key Research and Development Program of China (2022YFB2802803), Natural Science Foundation of China Project, grant numbers 61925104, 62031011, 62274042; Natural Science Foundation of Shanghai, grant number 21ZR1406200; The Key Research and Development Program of Jiangsu Province (BE2021008-5).

REFERENCES

[1] Junfei Wang, Junhui Hu, Chaowen Guan, Yuqi Hou, Zengyi Xu, Leihao Sun, Y Wang, Yuning Zhou, Boon S Ooi, and Jianyang Shi, "High-speed GaN-based laser diode with modulation bandwidth exceeding 5 GHz for 20 Gbps visible light communication," Photon. Res, no. 2024.

[2] Junhui Hu, Zeyuan Guo, Jianyang Shi, Xiong Jiang, Qinmiao Chen, Hui Chen, Zhixue He, Qinghai Song, Shumin Xiao, and Shaohua Yu, "A metasurface-based full-color circular auto-focusing Airy beam transmitter for stable high-speed underwater wireless optical communications," Nature Communications vol. **15**, no. 2041-1723 pp. 2944 2024.

[3] Yuqi Hou, Zhichong Wang, Meixin Liu, Shulan Yi, Xiaoqian Wang, Liang Xia, Guangyi Liu, Jianyang Shi, Ziwei Li, Junwen Zhang, Nan Chi, and Chao Shen, "Wide field-of-view laser-based white light transmitter for visible light communications," Optics Letters vol. **49**, no. pp. 2805 2024. doi:http://doi.org/10.1364/OL.522667

979-8-3503-6184-1/24 $31.00 © 2024 IEEE

Impact of Interfacial Layer on the Optoelectronic Performance of MoTe₂/Ge Heterojunction

Wenyu Lei[1,2], Xiaokun Wen[1,2], Boyuan Di[1,2], Xinyue Xu[1,2], Haixin Chang[1,2], and Wenfeng Zhang[1,2]*

[1] Key Laboratory of Material Processing and Die & Mould Technology, School of Materials Science and Engineering, Huazhong University of Science and Technology, Wuhan 430074, PRC.
[2] Shenzhen R&D Center of Huazhong University of Science and Technology, Shenzhen 518000, PRC.

* Email: 308563494@qq.com, wfzhang@hust.edu.cn

Abstract—In this paper, three types of ultra-thin interfacial oxide layers including GeO_2, Al_2O_3, and SiO_2, are designed to inserted into the interface of MoTe₂/Ge heterojunction to investigate its impact on the photoelectric performance. The device with GeO_2 interfacial layer exhibits the maximum responsivity and detectivity as well as the relatively large rectification ratio, reaching values of 1.38 A/W, 1.32×10^{13} Jones, and 140, respectively. For the device with Al_2O_3 interfacial layer, the maximum rectification ratio and the relatively high detectivity are achieved, reaching values of 807 and 1.56×10^{11} Jones, respectively. While for the MoTe₂/Ge heterojunction with SiO_2 interfacial layer, which shows an ideality factor of 2.14 and the lowest rectification ratio together with the detectivity. These results indicate that both the interface quality and bandgap of ultrathin interfacial layer would be crucial for the photoelectronic performance of MoTe₂/Ge heterojunction.

Keywords—interfacial layer, MoTe₂/Ge, optoelectronic, oxides, heterojunction

I. INTRODUCTION

In recent years, the heterojunction fabricated by combining two-dimensional (2D) materials with traditional semiconductors attracts increasing attention due to their distinctive optoelectronic characteristics[1]. The optoelectronic performance of heterojunction is dependent on the light harvesting properties of the absorber layers and the charge separation at the interface, while the strong interface recombination impact the performance of heterojunction-based optoelectronic devices seriously[2]. In this regard, inserting an ultrathin interfacial passivation layer to reduce interface recombination would be an effective strategy[3]. Additionally, the interface layer also has a significant impact on the band alignment, carrier injection and collection, rectifying behavior, as well as photovoltaic efficiency of the heterojunction, while is rarely discussed.

Herein, we have designed three types of ultrathin interfacial oxide layers including GeO_2, Al_2O_3, and SiO_2, to insert into the interface between MoTe₂ and Ge, in order to investigate their impact on the optoelectronic performance of MoTe₂/Ge heterojunction. The MoTe₂/Ge heterojunction with ultrathin GeO_2 layer (ideality factor of 1.21) shows the maximum responsivity of 1.38 A/W and detectivity of 1.32×10^{13} Jones, and the MoTe₂/Ge heterojunction with Al_2O_3 layer (ideality factor of 1.16) show the maximum rectification ratio of 807. In comparison, the MoTe₂/Ge heterojunction with SiO_2 layer (ideality factor of 2.14) exhibits the lowest rectification ratio and the detectivity. These results demonstrate that good interface quality and the relatively narrow bandgap of ultrathin interface layers are crucial for the optoelectronic performance of as-constructed MoTe₂/Ge heterojunction.

II. EXPERIMENTAL DETAILS

Fig. 1(a) presents the schematic diagram of the MoTe₂/Ge heterojunction photodetector. The devices comprises of an ultra-thin oxide layer (GeO_2, Al_2O_3, and SiO_2) sandwiched between MoTe₂ and Ge. The fabrication of device with ultrathin GeO_2 interfacial layer started with the HCl-lasted chemical cleaned n-Ge substrate, followed with the transfer of MoTe₂ film by water-soluble PVP/PVA double-layer mediator, and accomplish by the eelectrodes deposition. For such process, the unavoidable GeO_2 layer (~3 nm) was formed due to thermal oxidation of the Ge substrate during the removal step of PVP/PVA stamp in 70 ℃ water, which was confirmed in our previous investigation[4]. Similarly, the fabrication of device with ultrathin Al_2O_3 interfacial layer is processed with MoTe₂ film transfer, only differed by the Al_2O_3 was formed by the spontaneous oxidation of a 2 nm thick Al layer at room temperature. Meanwhile, The device with ultrathin SiO_2 interfacial layer was fabricated by transferring the as-synthesized Ge flakes on Si/SiO_2 grown substrate onto the MoTe₂ film using PMMA stamp. The residual ultrathin SiO_2 was observed during the NaOH-assisted wet transfer process of Ge flakes. To better understand the critical role of the ultrathin interfacial layer, the device without any interfacial layer was also fabricated by the NaOH-assisted wet transferring of the MoTe₂ film onto the Ge substrate. The detailed fabrication processes for the four types of devices without and with different types of interfacial layer are further depicted in **Fig. 1(b)**.

III. RESULTS AND DISCUSSION

Fig. 1 (c) shows the Raman spectra of the as-constructed MoTe₂/Ge heterojunction, MoTe₂ film, and Ge, respectively. The Raman peaks located at 172 and 233 cm⁻¹ correspond to the out-of-plane A_{1g} mode and in-plane E^1_{2g} mode of 2H-MoTe₂, while an additional peak located at 300 cm⁻¹ corresponds to Ge[3]. For the MoTe₂/Ge heterojunction, all the above peaks can be clearly observed, thus verifies the heterojunction formation.

Next, cross-sectional TEM observation for the device with ultrathin SiO_2 interfacial layer is further performed, as shown in **Fig. 2(a)**. An ultrathin interfacial layer of SiO_2 sandwiched between MoTe₂ and Ge was clearly observed, and its corresponding elemental map with well identified Mo, Te, Ge, and Si elements is shown in the inset of **Fig. 2(a)**. The atomic-scale HAADF image further confirmed ~4 nm thickness of SiO_2 layer, without any atomic intermixing or damage (**Fig. 2(b)**). Similarly, the cross-sectional TEM characterization for the device with the GeO_2 interfacial layer was also performed in our previous research[4]. The Al_2O_3 interfacial layer formed by the spontaneous oxidation of the Al layer has also been confirmed in previous study[2]. Oppositely, the NaOH-assisted

979-8-3503-6184-1/24 $31.00 © 2024 IEEE

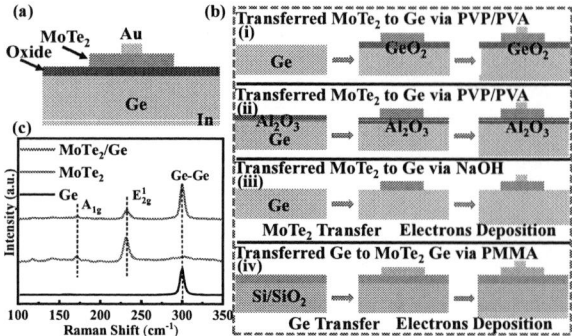

Figure 1. (a) The schematic diagram of the MoTe₂/Ge heterojunction with ultra-thin interfacial layers (GeO₂, Al₂O₃, and SiO₂). (b) The detailed fabrication process of four types of MoTe₂/Ge heterojunction without/with different types of interfacial layers. (c) Raman spectra of the Ge, MoTe₂ film, and MoTe₂/Ge heterojunction under 532 nm laser excitation.

wet transfer of the MoTe₂ film at room temperature inhibits the oxidation of the Ge substrate effectively, thus no interfacial layer can be observed for the device fabricated with such process.

Fig. 3(a) illustrates the *I-V* curves of the MoTe₂/Ge heterojunction with and without interfacial layer. The reverse current was observed to decrease by two, three, and five orders of magnitude after the insertion of GeO₂, Al₂O₃, and SiO₂ interfacial layer, respectively. Meanwhile, the forward current with the insertion of GeO₂ and Al₂O₃ is reduced by less than one order of magnitude, which is much smaller than that with SiO₂ insertion (~five orders of magnitude). Therefore, **Fig. 3(b)** further summarized the rectification ratio between −1 and +1 V for different interfacial layer insertion. The rectification ratio of the device with GeO₂ and Al₂O₃ interfacial layer can be as high as 140 and 807, respectively, which is much higher than that with SiO₂ and without interfacial layer (less than 10). These results indicate that the insertion of GeO₂ and Al₂O₃ favors to improve the rectifying characteristic of the MoTe₂/Ge heterojunction. Moreover, the ideality factor (n), which characterizes the deviation from an ideal p-n heterojunction, can be calculated using the following equation based on the *I-V* curves:

$$n= \frac{q}{k_B T} \frac{dV}{dln\ I} \qquad (1)$$

where q, k_B, and T denote the electronic charge, Boltzmann's constant, and Kelvin temperature, respectively. The ideality factors for the MoTe₂/Ge heterojunction with GeO₂, Al₂O₃, SiO₂, and without interfacial layer are determined as 1.21, 1.16, 2.14 and 1.92, respectively, revealing a more ideal diode behavior by inserting the appropriate interfacial oxide layer to passivate the interfacial defects. Notably, the ideality factor of the MoTe₂/Ge heterojunction with GeO₂ and Al₂O₃ is more close to 1 for an ideal p-n heterojunction, indicating that GeO₂ and Al₂O₃ are more suitable as interfacial layer for the MoTe₂/Ge heterojunction relative to SiO₂.

To interpret the impact of interfacial layer on charge separation and transport mechanism, the theoretical band alignment of the MoTe₂/Ge heterojunction with different interfacial oxide layer (GeO₂, Al₂O₃, and SiO₂) was shown in **Fig. 3(c)**. The E_C values of MoTe₂, Ge, GeO₂, Al₂O₃ and SiO₂ are 3.8, 4.13, 2.24, 1.31 and 0.9 eV, respectively, while E_V values are 4.8, 4.8, 7.59, 8.01 and 9.8 eV, respectively. **Fig. 3(d)** illustrates the band diagram of the MoTe₂/Ge heterojunction with/without interfacial layer at the

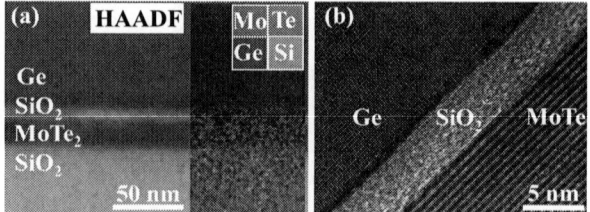

Figure 2. (a) Typical HAADF image of the interface for the MoTe₂/Ge heterojunction with ultrathin SiO₂ interfacial layer, the insert shows the corresponding EDS elemental mapping image. (b) The corresponding HR-TEM cross-section image of the MoTe₂/Ge heterojunction with SiO₂ interfacial layer.

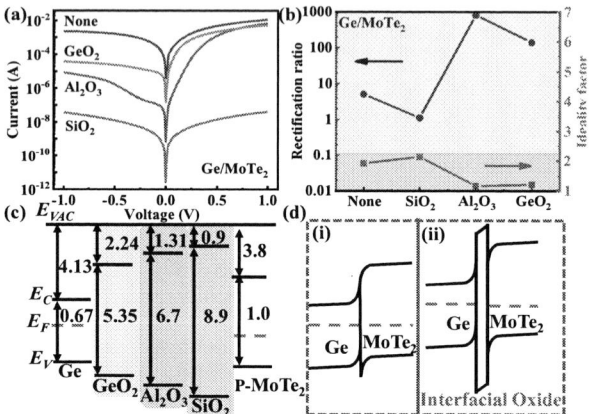

Figure 3. (a) *I-V* characteristics of the MoTe₂/Ge heterojunction with and without interfacial layer were measured in the dark. (b) Rectification ratio and ideality factor of the MoTe₂/Ge heterojunction with and without interfacial oxide layer. (c) Band alignment of the MoTe₂/Ge heterojunction with GeO₂, Al₂O₃, and SiO₂ layer. (d) Band diagram of the MoTe₂/Ge heterojunction without/with interfacial layer at the equilibrium state.

equilibrium state, where the diffusion and drift motion of charge carriers are in balance. When V_bias is < 0 V, the width of the depletion region is increased and the device current is dominated by drift motion of the minority carriers, theoretically resulting in a low dark current. Notably, the small ΔE_C (0.33 eV) as well as negligible ΔE_V between MoTe₂ and Ge favor to the carrier transport, thus leading to a non-negligible reverse current under the reverse bias. In this regard, the inserted interfacial layer functions as barrier for the transport of carriers, thus reducing the reverse current significantly. As the bandgap of the inserted layer increases from 5.35 eV (GeO₂) to 6.7 eV (Al₂O₃) and up to a maximum of 8.9 eV (SiO₂), the barrier for minority carriers drift increases simultaneously, thus resulting in a reduced reverse current. When V_bias is > 0 V, the depletion region is narrowed and the device current is dominated by majority carrier diffusion. The forward current is also reduced after the insertion of interfacial layer, as the oxide layer can also block the transport of carriers under the applied forward bias. Meanwhile, due to the high bandgap of SiO₂, a significant reduction of forward current was observed.

Then, a systematic evaluation of device performance was further conducted on MoTe₂/Ge heterojunction with/without interfacial layer. **Fig. 4(a)** shows the *I-V* characteristics measured under 1550 nm light illumination with a light

979-8-3503-6184-1/24 $31.00 ©2024 IEEE

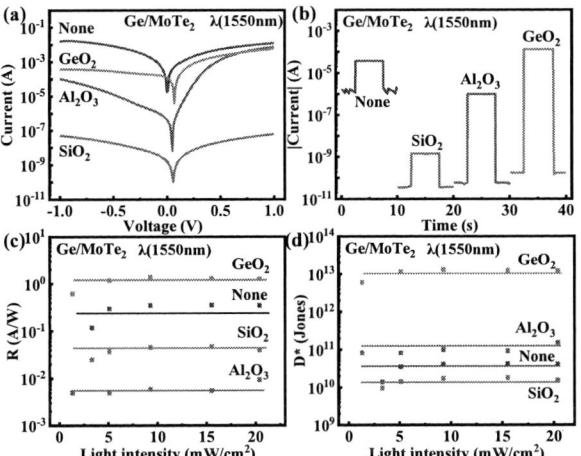

Figure 4. (a) *I-V* characteristics and (b) photoresponse of the MoTe₂/Ge heterojunction with/without interfacial layer when measured under 1550 nm light illumination with a light intensity of 20.37 mW/cm². (c) Photoresponsivity and (d) specific detectivity of the MoTe₂/Ge heterojunction with/without interfacial layer under 1550 nm light illumination with varied light intensity at V_{bias} = 0 V.

intensity of 20.37 mW/cm². It can be observed that the minimum current of the *I-V* curves shift towards the forward voltage, which can be attributed to the photovoltaic effect, indicating the heterojunction can operate without an external power source. **Fig. 4(b)** further presents the corresponding temporal response of the *I-V* curves at zero bias. First, the dark current (I_{dark}) of the MoTe₂/Ge heterojunction without interfacial layer is higher than that of the MoTe₂/Ge heterojunction with interfacial layer, which is consistent with the *I-V* curves in **Fig. 3(a)**. Second, the photocurrent of the MoTe₂/Ge heterojunction with GeO₂, Al₂O₃, and SiO₂ was observed to be gradually decreased, revealing that the ultrathin interfacial barrier would inhibit the transport of photo-generated charge carriers, thus the formation of photocurrent. The photocurrent on/off modulation ratio (I_{ON}/I_{OFF}) of the MoTe₂/Ge heterojunction with GeO₂, Al₂O₃, SiO₂, and without interfacial layer was determined as 7.7×10⁵, 1.6×10⁴, 40, and 21, respectively, indicating the crucial impact of interfacial oxide layer on device performance.

Quantitatively, the responsivity (R) to evaluate the optoelectronic performance can be calculated using the following equation:

$$R = \frac{(I_{light} - I_{dark})}{PS} \quad (2)$$

where P is the illumination power, S is the effective light area, and I_{light} and I_{dark} are the current under illumination and without illumination, respectively. As shown in **Fig. 4(c)**, the calculated responsivity at different light intensity for the MoTe₂/Ge heterojunction with GeO₂, Al₂O₃, and SiO₂ interfacial layer exhibits maximum R value of ~1.38, ~0.01, and ~0.05 A/W, respectively. The GeO₂ interfacial layer shows the highest responsivity among the as-investigated structures, indicating its superior effect to improve device performance. In comparison, the larger bandgap of Al₂O₃ and SiO₂ may impede the transport of photogenerated carriers, resulting in lower responsivity than that of the device without interfacial layer. Meanwhile, the Specific detectivity (D*), which is another figure of merit for photodetector and can be calculated with the following equation:

$$D^* = R\frac{S^{1/2}}{(2qI_{dark})^{1/2}} \quad (3)$$

where q is the unit charge, was also evaluated. **Fig. 4(d)** shows the light intensity-dependent detectivity of the MoTe₂/Ge heterostructure with/without interfacial layer, indicating that the highest detectivity up to 1.32×10^{13} Jones with an ultrathin GeO₂ layer. Due to the effective suppression of dark current, the detectivity with an ultrathin Al₂O₃ layer can reach up to 1.56×10^{11} Jones, which is higher than that of device without an interfacial layer (1.82×10^{10} Jones).

IV. SUMMARY

In summary, we have investigated the impact of three types of ultrathin interfacial layers including GeO₂, Al₂O₃, and SiO₂, on the photoelectronic performance of the MoTe₂/Ge heterojunction. For the devices with GeO₂ and Al₂O₃ layer, where the ideality factor approaches 1, high responsivity and detectivity, and large rectification ratio can be achieved. While for the MoTe₂/Ge heterojunction with SiO₂ interfacial layer, it displays a high ideality factor of 2.14 with the low detectivity, and rectification ratio. These results indicate that both good interface quality and the relatively narrow bandgap of the interface layer are crucial for the optoelectronic performance of MoTe₂/Ge heterojunction.

ACKNOWLEDGMENT

The authors acknowledge support from the National Natural Science Foundation of China (Grant No. 62074061 and 52272152), the Natural Science Foundation of Hubei Province of China (Grant No. 2022CFA031), Interdisciplinary Research promotion of Huazhong University of Science and Technology (Grant No. 2023JCYJ040 and 2023JCYJ007), and Shenzhen Science and Technology Program (Grant No. JCYJ20230807143714029). Analytical Center of Huazhong University and Center of Micro-Fabrication and Characterization of Wuhan National Laboratory for Optoelectronics are acknowledged for access to facilities.

REFERENCES

[1] Z Lu et al., "Ultrahigh Speed and Broadband Few-Layer MoTe₂/Si 2D-3D Heterojunction-Based Photodiodes Fabricated by Pulsed Laser Deposition," *Adv. Funct. Mater.*, vol. 30, p.1907951, Feb. 2020.

[2] D. Wu et al., "Ultrabroadband and High-Detectivity Photodetector Based on WS₂/Ge Heterojunction through Defect Engineering and Interface Passivation," *ACS Nano*, vol. 15, no. 6, pp. 10119-10129, May 2021.

[3] W. Chen et al., "Ultrahigh sensitive near-infrared photodetectors based on MoTe₂/germanium heterostructure," *Nano Res.*, vol. 13, pp. 127-132, Jan 2020.

[4] W. Lei et al., "Vertical MoTe₂/Ge Heterojunction Photodiode for 1550-nm Near-Infrared Photodetection," *IEEE Electron Device Lett.*, vol. 69, pp. 6825-6829, Dec. 2022.

MoS₂-WS₂ Heterostructure-enabled Optoelectronic Synaptic Diode

Mingjie Li[1], Yingtao Ding[1], Jianzhi Hu[1], Hankun Zhao[1], Yilin Sun*[1]

[1] School of Integrated Circuits and Electronics, Beijing Institute of Technology, Beijing 100081, CHINA

* Email: mingjieli@bit.edu.cn, sunyl@bit.edu.cn

Abstract—We report an optoelectronic synaptic diode based on a vertically stacked MoS₂-WS₂ van der Waals heterostructure. This diode demonstrates outstanding rectification characteristic, achieving a rectification ratio of up to 4.2×10^3. A remarkable persistent photocurrent is observed under white light illumination, attributed to the presence of charged dielectric surface trap states. Benefitting from such photo-induced memory behavior, optoelectronic synaptic plasticity such as short-term and long-term plasticity can be emulated by as-fabricated MoS₂-WS₂ heterostructure under various light stimuli. Moreover, a high linearity of synaptic weight updating is achieved with increasing light power, which further demonstrates its great potential in artificial neural networks for image processing.

Keywords—Two-dimensional materials, Van der Waals heterostructure, Synaptic plasticity, Diode, Band alignment

I. Introduction

Recent advances in two-dimensional (2D) transition metal dichalcogenides (TMDs) have facilitated the development of various promising technologies, including nanoelectronics, photonics, sensing and opto-electronics [1-2]. These advancements are attributed to its natural atomic-scale thickness, layer-dependent tunable energy band, high carrier mobility and extraordinary compliance with flexible substrates [3-5]. The extensive material database allows for the convenient identification of suitable materials for designing and stacking heterostructures, thereby enhancing the performance of electronic devices [6-7].

Neuromorphic synaptic devices, as an innovative branch that may transcend the von Neumann architecture, have arisen significant attention [8]. The essence of an artificial synaptic device is to establish a non-volatile and tunable conductance state, resulting in various bio-inspired information technologies. Seunghwan Seo et al. proposed an artificial synapse with electrically tunable conductance using WSe₂ and MoS₂ hybrid channels. Their approach involved CF₄ plasma treatment on h-BN flakes to create a floating gate layer [9]. Additionally, Ehnho Lee et al. utilized a sputtering process to bring in sulfur defect, thereby enabling a large charge memory window [10]. Van der Waals (vdWs) heterostructures constructed by stacking layers of different 2D materials provide the possibility to design and fabricate neuromorphic devices, which can fully utilize the electrical and optical properties of 2D materials as well as the band structure of vdWs heterojunctions.

In our work, a few layers of MoS₂ and WS₂ were vertically stacked to form a vdWs heterostructure. This junction exhibited a rectification ratio of up to 4.2×10^3 due to the strong built-in electric field at the interfaces. The photosensitivity of the heterostructure was confirmed by its ability to absorb visible light, as indicated in the current-voltage *(I-V)* curve. The persistent photocurrent suggested a charge-trapping process at the interface between gate dielectric and MoS₂, which contributed to the synaptic behaviors obtained under varying durations, incident light power and pulse numbers. The post-synaptic current (PSC) exhibited strong linearity with increasing incident light intensity, highlighting the advantageous performance of our fabricated synaptic diode in high-performance neural networks. Finally, the underlying charge memory mechanism was discussed.

II. Experimental

MoS₂ and WS₂ few-layer samples were mechanically exfoliated from bulk polycrystalline materials obtained commercially, using blue adhesive plastic film. After carefully selecting materials with appropriate thickness and designing the heterostructure stack, the MoS₂-WS₂ heterostructure was assembled via an optical dry-transfer platform assisted by polyvinyl alcohol (PVA), thereby forming a vdWs interface [11-12]. The contact area between the two-dimensional materials and the metal was defined using laser direct writing technology, utilizing a 2 mm diameter lens, 80 W power intensity, and a 1% grating size. Finally, Cr/Au with a thickness of 10/50 nm was deposited on the SiO₂/Si substrate via electron beam evaporation.

Fig. 1(a) describes the schematic diagram of MoS₂-WS₂ heterostructure, with an inset displaying an optical microscopic image of as-fabricated device. The surface morphology of the heterostructure was examined using atomic force microscopy (AFM, Oxford Cypher S), revealing thicknesses of 76.1 nm for MoS₂ and 7.0 nm for WS₂, as shown in Fig. 1(c). Raman spectra were collected using the Horiba-Jobin-Yvon Raman system under 532 nm laser excitation, illustrating the in-plane and out-of-plane vibration modes of MoS₂ and WS₂ (Fig. 1(d)). The electrical properties of the optoelectronic diode were measured using a B1500A semiconductor parameter analyzer (Keysight). White light illumination with controllable incident intensity and duration was provided by the microscope light source integrated in the probe station system. The incident power to the device was calibrated by a light power meter.

III. Results and Discussions

Fig. 1(b) shows the rectification behavior of the MoS₂-WS₂ vdWs heterostructure. In our electrical performance measurements, the MoS₂ is grounded while a DC voltage is applied to the WS₂. When a positive voltage is applied, forward bias generates an electric field that opposes to the existing built-in electric field, therefore weakening the overall electric field strength in the barrier region. As a result, the diffusion current dominated by the majority carriers increases sharply, indicating the ON state of the diode [13]. Conversely, the reverse bias enhances the electric field in the depletion region, leading to a low conductance OFF state. This rectification behavior is line with the characteristics of traditional PN junction diodes, owing to the alignment of the MoS_2-WS_2 energy bands [14-15]. Although both materials are n-type semiconductors, the difference in electron affinity contributes to the formation of an interfacial built-in electric field, facilitating the junction characteristics with a rectification ratio of up to 4.2×10^3 in the dark. Subsequently, light illumination is applied during the sweep to investigate the photosensitive behavior of our MoS_2-WS_2 heterostructure diode. The photocurrent in the *I-V* measurement is observed with the increasing incident light power, particularly notable in the reverse bias region with low dark current (Fig 1. (b)).

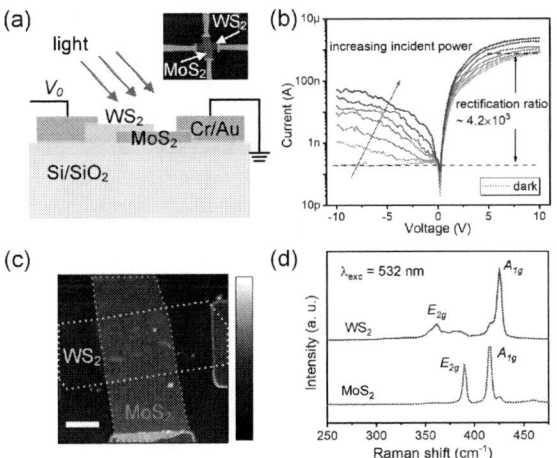

Figure 1. (a) Schematic diagram and optical microscopic graph of MoS_2-WS_2 heterostructure diode; (b) Current-voltage characteristics of the MoS_2-WS_2 heterostructure in dark and white light; (c) AFM image of MoS_2-WS_2 vdWs heterostructure; (d) Raman spectrum of MoS_2 and WS_2.

Current-time (*I-t*) measurements are carried out to investigate the optical response of the heterostructure diode. In the following tests, we bias the diode at -6 V to keep a low and coincident dark current. Persistent photocurrents are observed even after the light is removed, involving a specific charge storage mechanism, which will be discussed later. When the light pulse irradiates to the junction under a certain bias, the conductance of the channel changes and maintains after light removed with sufficient strength. This photo-induced memory behavior makes it possible to emulate plasticity of synaptic structures in living organisms.

In the biological nervous system, adjacent neurons transmit biological information through synaptic structure. Fig. 2(a) depicts the forward propagation of the stimuli between neurons, where the pre-synaptic neuron releases neurotransmitters and the post-synaptic neuron receive those peculiar by the acceptor. In our artificial synapse, the applied white light acts as a pre-synaptic terminal to provide reliable stimuli, while the current flowing through the two-terminal diode represents the post-synaptic current. The change in post-synaptic current (ΔPSC) reflects the change in the state of the channel conductance. This tunable synaptic weight strategy, known as plasticity, determines the ability of neural networks to process complex information [17].

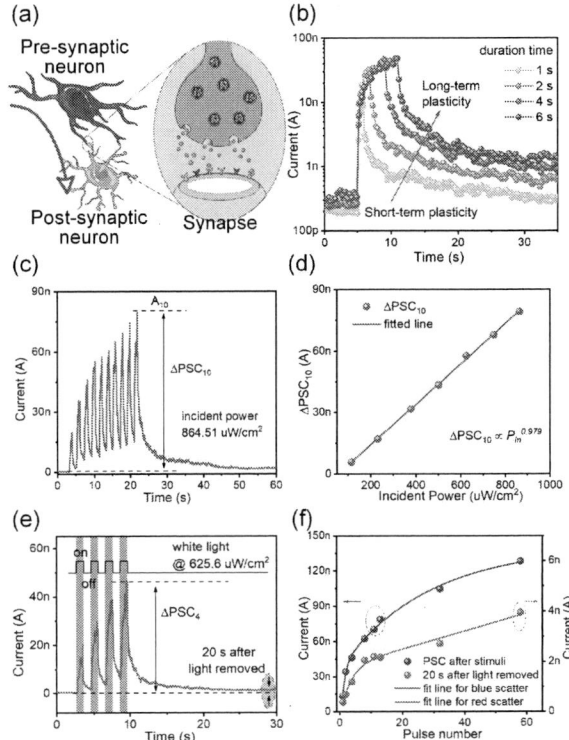

Figure 2. (a) Schematic of the process of the biological information transport by neurotransmitter between adjacent neurons; (b) Conversion from STP to LTP controlled by the light duration time; (c) Evolution of ΔPSC under ten pulses of consecutive stimulation; (d) The extracted ΔPSC_{10} varied with the incident power; (e) Four pulse of white light-caused change in channel conductance state and the retention behavior; (f) Different pulse number-induced ΔPSC and the memory performance of synaptic diode.

To further investigate the artificial synaptic plasticity of the stacked heterostructure diode, light pulses with varying sequences and power, simulating various pre-synaptic stimulus conditions, are applied to the junction region. In Fig. 2(b), under a light duration of 1 s, the PSC changes weakly and quickly returns to its initial state, indicating a weak connection. As the irradiation time increases, ΔPSC evolves significantly and maintains a certain level of conductance, demonstrating the transition from short-term relaxation recovery to long-term stable memory. Besides, we conduct multiple-pulse stimuli on the junction to examine the photoresponse of the synaptic diode. As shown in Fig. 2(c), ten successive pulses with a duration/interval time of 1 s/1 s and incident power of 864.51 μW/cm^2 are applied to the heterostructure diode. The PSC variation with the number of stimuli indicates a linear response to light pulse of the heterostructure diode, which is conductive to emulating the weight updating rules in neural networks.

The ten-pulse test is carried out under varying incident powers, and the variation of ΔPSC_{10} with incident light power is extracted and fitted, as shown in Fig. 2(d). The ΔPSC_{10} fit well to the incident power (P_{in}) with a relationship of $\Delta PSC_{10} \sim P_{in}^{\theta}$, where θ determines the response of the photocurrent to light intensity. Here, the fitted curve shows a θ of 0.979, verifying the strong proportional relationship and contributing to a controllable long-term potentiation synaptic plasticity.

We then evaluate the long-term potentiation plasticity triggered by light pulses in our artificial heterostructure diode. Fig. 2(e) depicts ΔPSC triggered by four light pulses with an incident intensity of 625.6μW/cm². The photocurrent remains a high level of 1.19 nA after the light is removed for 20 s. This rapid recovery may be related to the natural built-in electric field of PN junction. Fig. 2(f) shows the long-term potentiation synaptic weights with the increasing light pulse number. The conductance state, after the light is removed for 20 s, illustrates a similar growth trend with the ΔPSC. This indicates a weight mapping rule without crosstalk.

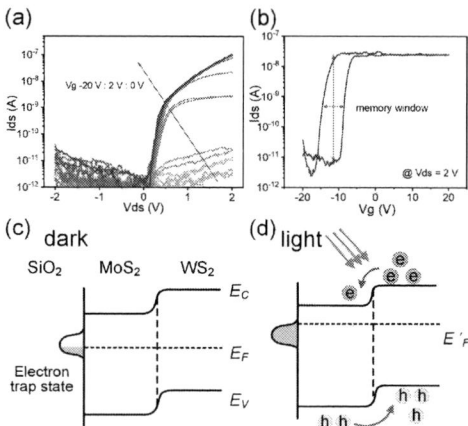

Figure 3. Double sweep of (a) output curve and (b) transfer curve. The schematic diagram of the energy band alignment of MoS₂-WS₂ heterostructure under (c) dark and (d) illumination.

It is evident that, unlike traditional PN junctions, the fabricated MoS₂-WS₂ heterostructure diode exhibits a memory capacity in response to light pulse. Thus, we further investigate the origin of the memory mechanism in the heterostructure through the double sweep of the output and transfer curve. Fig. 3(a) and (b) depict the three-terminal behavior of this heterostructure, where the introduction of gate terminal causes the hysteresis. The gate-tunable hysteresis behaviors in both output and transfer curves indicate the possible charge trapping/de-trapping process through the electron trap states at the interface between MoS₂ and silicon dioxide. Fig. 3(c) and (d) illustrate the possible charge storage mechanism in this heterostructure diode. As depicted in Fig. 3(c), upon contact between MoS₂ and WS₂ in the dark, band alignment is accomplished through electrons flow, with some electron trap states partially occupied at the gate dielectric interface.

When light irradiates the junction region, photons with enough energy generate electron-hole pairs, dramatically increasing the photocurrent. And photo-generated electrons are gradually trapped by the interface defect states. After illumination, the recombination of photo-generated pairs was inhibited due to the limited de-trapping process of the trapped photo-generated electrons, resulting in a persistent photocurrent even after the light is removed.

IV. CONCLUSIONS

MoS₂-WS₂ vdWs heterostructure-based optoelectronic synaptic diode is fabricated on the SiO₂/Si substrate. This diode exhibits good rectification ratio of up to 4.2×10³ due to the built-in electric field. Different from the traditional PN junction devices, the natural trap state at the surface of silicon dioxide provide enough storage space for the capture of photogenerated electrons, thereby leading to a persistent photocurrent. This charge memory mechanism contributes to a linear conductance change with the light pulse, emulating the synaptic plasticity. This strategy of constructing charge capture mechanism in heterostructure diode provides reference for high performance neural network.

ACKNOWLEDGMENT

This work was supported by the National Natural Science Foundation of China (Nos. 62104017 and 92373105) and the Beijing Institute of Technology Research and Innovation Promoting Project (Grant No. 2023YCXY032).

REFERENCES

[1] Q. H. Wang, K. K. Zadeh, A. Kis, J. N. Coleman and M. S. Strano, "Electronics and optoelectronics of two-dimensional transition metal dichalcogenides," Nature Nanotechnology, vol. 7, pp. 699–722, Novermber 2012.

[2] H. Qiu, Z. Yu, and T. Zhao et al., "Two-dimensional materials for future information technology: status and prospects," Science China Information Sciences, vol. 67, June 2024.

[3] W. Choi, N. Choudhary, and G. H. Han et al., "Recent development of two-dimensional transition metal dichalcogenides and their applications," Materials Today, vol. 20, pp. 116–130, April 2017.

[4] R. Cheng, S. Jiang, and Y. Chen et al., "Few-layer molybdenum disulfide transistors and circuits for high-speed flexible electronics," Nature Communications, vol 5. October 2014.

[5] D. Jiang, Z. Liu, and Z. Xiao et. al., "Flexible electronics based on 2D transition metal dichalcogenides," Journal of Materials Chemistry A, vol. 10, pp. 89–121, 2022.

[6] Y. Liu, X. Duan, Y. Huang, and X. Duan, "Two-dimensional transistors beyond graphene and TMDCs," Chemical Society Reviews, vol. 47, pp. 3288–3409, August 2018.

[7] Y. Meng, J. Feng, and S. Han et al., "Photonic van der Waals integration from 2D materials to 3D nanomembranes," Nature Reviews Materials, vol. 8, pp. 498–517, April 2023.

[8] K. He, C. Wang, Y. He, J. Su, and X. Chen, "Artificial neuron devices," Chemical Reviews, vol. 123, pp. 13796–13865, November 2023.

[9] S. Seo, B. -S. Kang, and J. -J. Lee et al., "Artificial van der Waals hybrid synapse and its application to acoustic pattern recognition," Nature Communications, vol 11. August 2020.

[10] E. Lee, J. Kim, and J. Park et al., "Realizing electronic synapses by defect engineering in polycrystalline two-dimensional MoS₂ for neuromorphic computing," ACS Applied Materials & Interfaces, vol. 15, pp. 15837–15847, March 2023.

[11] Y. Li, S. Weng, and R. Niu et al., "Poly(vinyl alcohol)-assisted exfoliation of van der Waals materials," ACS Omega, vol. 7, pp. 38774–38781, October 2022.

[12] M. Onodera, Y. Wakafuji, and T. Hashimoto et al., "All-dry fip-over stacking of van der Waals junctions of 2D materials using polyvinyl chloride," Scientific Reports, vol. 12, December 2022.

[13] L. Yuan, N. Nerngchamnong, and L. Cao et al., "Controlling the direction of rectification in a molecular diode," Nature Communications, vol. 6, March 2015.

[14] A. Balapure, J. R. Dutta, and R. Ganesan, "Recent advances in semiconductor heterojunctions: a detailed review of the fundamentals of photocatalysis, charge transfer mechanism and materials," RSC Applied Interfaces, vol. 1, pp. 43–69, 2024.

[15] H. Xia, M. Luo, and W. Wang et al., "Pristine PN junction toward atomic layer devices," Light: Science & Applications, vol. 11, June 2022.

[16] L. F. Abbott and Wade G. Regehr, "Synaptic computation," Nature, vol. 431, pp. 796-803, October 2004.

[17] M. Li, Z. Liu, and Y. Sun et al., "Tailoring Neuroplasticity in a Ferroelectric-Gated Multi-Terminal Synaptic Transistor by Bi-Directional Modulation for Improved Pattern Edge Recognition," Advanced Functional Materials, vol. 33, November 2023.

Pseudo-Parallel Symmetrical and Crossed Perovskite Solar Cells for Bifacial Applications

Guang-Wei Huang [1], Hsing-Mao Cheng[1], Jyi-Tsong Lin [1]

[1]Department of Electrical Engineering

National Sun Yat-Sen University, 80424 Kaohsiung, Taiwan

Email: jtlin@ee.nsysu.edu.tw

Abstract

In this paper, we propose the development of pseudo-parallel symmetrical and crossed perovskite solar cells, evaluated through simulation using Silvaco TCAD Atlas under various albedo conditions. Our pseudo-parallel symmetrical perovskite solar cells achieved a short-circuit current density (Jsc) of 50.899 mA/cm², an open-circuit voltage (Voc) of 0.804 V, and a conversion efficiency (η) of 34.624% under a 30% albedo environment. At 100% albedo, the symmetrical cells reached a Jsc of 78.818 mA/cm² and a conversion efficiency of 54.087%, while the crossed cells exhibited a Jsc of 78.794 mA/cm², a Voc of 0.816 V, and a conversion efficiency of 54.259%. Compared to SHJ and PERC solar cells, our pseudo-parallel designs demonstrated superior efficiency, with energy boosts of 105.599% for symmetrical cells and 106.042% for crossed cells under 100% albedo. Even at 30% albedo, the energy boost remained above 31% for both structures. These results highlight the potential of pseudo-parallel symmetrical and crossed perovskite solar cells in enhancing bifacial solar cell performance, particularly in environments with varying albedo levels

.*Index Terms—Bifacial, perovskite, pseudo-parallel, solar cells.*

I. Introduction

In recent years, perovskite solar cells (PSCs) have emerged as a leading photovoltaic technology due to their exceptional power conversion efficiency (PCE) and potential for low-cost production. Since their inception in 2009, PSCs have improved efficiency from 3.8%[1] to over 31.25% in 2024, surpassing many traditional thin-film solar cells like CdTe and CIGS [2][3]. This progress is attributed to unique properties like earth-abundant elements, low processing temperatures, tunable band gaps, and high absorption coefficients [4][5].

Recent research on tandem multi-junction structures has enabled PSCs to surpass the theoretical single-junction Shockley-Queisser limit of 33.7% [6]. Despite these advancements, PSCs face challenges related to stability under environmental conditions. Techniques such as spin coating, deposition, and passivation layers reduce interface traps and enhance carrier extraction, thereby improving both efficiency and stability . [7].

Lead-free double halide perovskite absorbers are explored for better environmental friendliness and stability, achieving simulated PCEs around 20.39% [8][9]. Bifacial PSCs, which capture light from both sides, demonstrate reduced recombination losses and improved performance. Advanced materials like C60 for electron transport layers (ETLs) and Spiro-OMeTAD for hole transport layers (HTLs) enhance PSC efficiency and stability [10][11][12].

Many studies focus on tandem p-i-n or n-i-p structures, optimizing perovskite composition. Single-crystal perovskites, with lower defect density and higher carrier diffusion lengths, improve conversion efficiency over polycrystalline films [13]. Interdigitated back contact (IBC) solar cells, known for high efficiency, have been adapted to bifacial applications, improving short-circuit current density and overall efficiency [14][15].

To address issues like minority carrier recombination and thermal degradation, pseudo-parallel symmetrical and crossed structures have been proposed. These structures improve output and reduce working temperature, benefiting from decreased infrared absorption [16]. By applying perovskite material into pseudo-parallel structures, we aim to decrease solar cell thickness and increase Voc, fill factor, and conversion efficiency [17].

In this work, we propose pseudo-parallel symmetrical and crossed perovskite solar cells using single-crystal CH3NH3PbI3 and C60 as ETL. C60, known for reducing trap states, enhances F.F. and suppresses hysteresis. Spiro-OMeTAD, a widely used HTL, further improves conversion efficiency .[18]

Integrating the advantages of perovskite solar cells and pseudo-parallel symmetrical structures, our proposed cells will be thinner than silicon solar cells while maintaining high conversion efficiency. The pseudo-parallel structure facilitates carrier transport, increasing current and reducing recombination. Thus, pseudo-parallel perovskite solar cells combine the advantages of both pseudo-parallel structure and perovskite solar cells.

II. Simulation & Parameter

In this paper, our pseudo-parallel symmetrical perovskite solar cells were modeled as operating under a global standard solar spectrum (AM 1.5G) illumination with a total incident power density of 100 mW/cm² and a light intensity calculated for wavelengths from 300 nm to 1200 nm . Simulations software were performed using Silvaco Atlas with the ray-tracing model.

The pseudo-parallel perovskite solar cells materials parameters are shown in Table I. By using these parameters, all of our extended simulation is based on, and the measurements are the same as the calibration reference. Fig. 1 compares the measured I-V curve with perovskite solar cells reference [19], the close agreement of two I-V curves in Fig. 1 ensures the accuracy of our simulation.

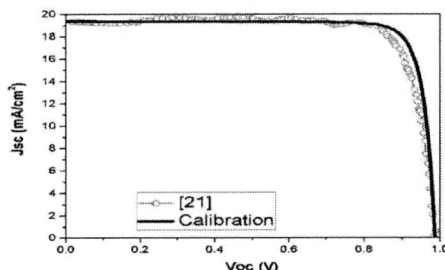

Fig. 1. Compared of the physically measured I-V curve of the perovskite solar cells [19] and the calibrated simulation curve of that cell.

TABLE I
Device parameter used in all our simulations.

Parameters	CH₃NH₃PbI₃	Spiro-OMeTAD	C₆₀
Energy gap (eV)	1.5	3.2	1.7
Donor Doping Density (cm⁻³)	4×10^{16}		1×10^{19}
Acceptor Doping Density (cm⁻³)		1×10^{19}	
Electron mobility (cm²/Vs)	50	2×10^{-4}	1.6
Hole mobility (cm²/Vs)	50	2×10^{-4}	1.6
Effective conduction band density (cm⁻³)	2.5×10^{20}	2.5×10^{18}	9.44×10^{19}
Effective valence band density (cm⁻³)	2.5×10^{20}	1.8×10^{19}	6.72×10^{19}
Permittivity (F/m)	18	2.1	4.25
Affinity (eV)	3.9	3.2	2.66

III. Results and Discussion

After calibrating our model with reference perovskite solar cells, we applied the parameters to our pseudo-parallel symmetrical cells to assess the impact of the ETL/HTL width ratio on performance. A 3% ETL width ratio was found to be optimal, resulting in a Jsc of 33.087 mA/cm², Voc of 0.805 V, F.F. of 78.52%, and an efficiency of 22.16% under 30% albedo. Adding an antireflection layer further improved performance, increasing the Jsc to 34.572 mA/cm² and efficiency to 23.117% under 0% albedo (Fig. 2). We also examined different passivation strategies under bifacial conditions, finding that monofacial passivation significantly enhanced performance, with a Jsc of 76.385 mA/cm² and an efficiency of 51.713% at 100% albedo(TABLE II). These results highlight the importance of optimizing both the ETL/HTL ratio and passivation to minimize recombination and improve efficiency in bifacial solar cells.

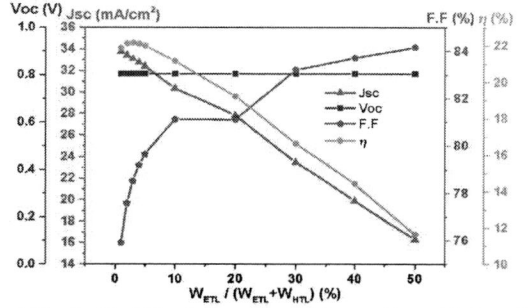

Fig. 2. The J_{sc}, V_{oc}, F.F and conversion efficiency with modulating width ratio between ETL and HTL under 0 % albedo environment.

TABLE II

The parameter of adding passivation in pseduo-parallel symmetrical perovskite solar cells.

	Albedo (%)	Jsc (mA/cm²)	Voc (V)	F.F. (%)	η (%)	Energy Boost(%)
Without passivation	0	34.572	0.853	78.301	23.117	11.948
	100	42.476	0.856	71.116	25.879	
Monofacial passivation	0	37.713	0.805	83.953	25.485	102.915
	100	76.385	0.824	82.129	51.713	
Bifacial passivation	0	38.599	0.582	81.055	18.216	103.732
	100	77.239	0.601	80.003	37.112	

We choose an appropriate thickness, 25 μm, to obtain the maximum benefit under albedo 100% environment. As for the optimal thickness, the short-circuit current density, J_{sc} reaches 78.818 mA/cm², open-circuit voltage, V_{oc} reaches 0.797 V, fill factor reaches 84.812 % and the conversion efficiency reaches 54.087 %, under albedo 100% environment. Using Si₃N₄ as passivation layer make the energy boost of our cells highest compared with other materials. Although Si₃N₄ passivation has low conversion efficiency under albedo 0% environment, but it has more benefit under bifacial environment.The refractive index of passivation or antireflection layer can be defined by [20]:

$$N_{passivation} \sqrt{N_{perovskite} \times N_{ETM\&HTM}} \qquad (1)$$

where $n_{perovskite}$ is 2.59, n_{ETL} is 2.2 and n_{HTL} is 1.7, substituting into (1), we can calculate the refractive index of $N_{passivation}$ is 2.38 and 2.11. Therefore, we can know that the refractive index of Si₃N₄ is 2.022, this material is most suitable in our cells.

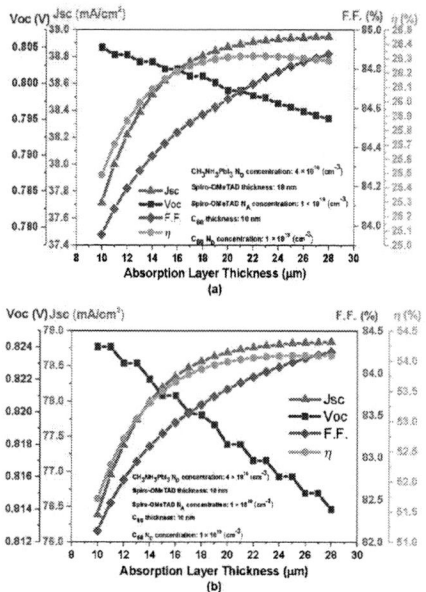

Fig. 3. The performance of J_{sc}, V_{oc}, F.F. and conversion efficiency with modulating the thickness of absorption layer (a) under albedo 0 % environment. (b) under albedo 100% environment.

Table VI compared the other SHJ[9] [21] or PERC [10][22] pseudo-parallel symmetrical and crossed structure solar cells, the SHJ structure has lower energy boost than PERC solar cells and our cells. But SHJ structure has higher conversion efficiency in albedo 0% environment. All of the structure in Table VI have surrouding eletronic field so that all of them have better benefit than other structure, because the carrier have lower probability being recombined. Furthermore, the structure have benefit in bifacial environment. The psuedo-parallel strcture allows the holes and electrons to move only half as far to reach the anode or cathode.

Fig. 4 shows the structure of our pseduo-parallel crossed perovskite solar cells

Table VII presents the Jsc, Voc, F.F. and conversion efficiency of our pseudo-parallel symmetrical and crossed perovskite solar cells under albedo 0%, 30% and 100% .

TABLE VI
Comparsion of different pseudo-parallel symmetrical and cross structure.

	Albedo (%)	Jsc (mA/cm²)	Voc (V)	F.F. (%)	η(%)	Energy Boost (%)
Symmetrical SHJ [9]	0	36.94	0.75	84.35	23.50	N/A
	30	46.41	0.76	84.35	29.81	26.8
	100	68.53	0.77	84.31	44.69	90.1
RSC-SHJ [11]	0	37.24	0.767	83.24	23.74	N/A
	30	46.93	0.772	83.96	30.42	28.14
	100	71.33	0.781	84.96	47.30	99.24
EWT [22]	0	39.45	0.695	83.6	22.92	N/A
	30	49.47	0.699	83.42	28.86	25.91
	100	72.85	0.705	83.0	43.69	90.6
Crossed PERC [10]	0	40.1	0.67	81.2	22.0	N/A
	30	52.16	0.68	81.19	28.91	31.7
	100	80.3	0.69	81.3	45.4	106.42
symmetrical PSCs Fig. 5 (b)	0	38.934	0.797	84.812	26.307	N/A
	30	50.899	0.804	84.618	34.624	31.615
	100	78.818	0.816	84.126	54.087	105.599
Crossed PSCs Fig. 11	0	38.922	0.797	84.919	26.334	N/A
	30	50.883	0.804	84.805	34.691	31.735
	100	78.794	0.816	84.415	54.259	106.042

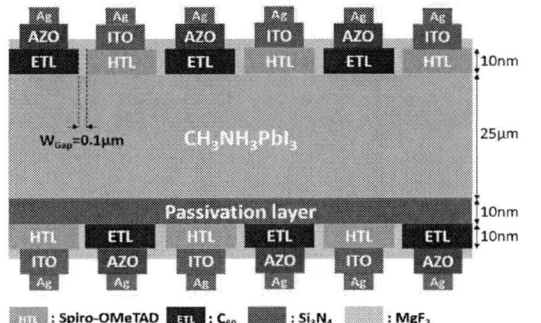

Fig. 4. Our pseudo-parallel crossed perovskite solar cells.

TABLE VII
Parameter of our symmetrical and crossed structure.

	Albedo (%)	Jsc (mA/cm²)	Voc (V)	F.F. (%)	η(%)	Energy Boost (%)
Symmetrical PSCs	0	38.934	0.797	84.812	26.307	N/A
	30	50.899	0.804	84.618	34.624	31.615
	100	78.818	0.816	84.126	54.087	105.599
Crossed PSCs	0	38.922	0.797	84.919	26.334	N/A
	30	50.883	0.804	84.805	34.691	31.735
	100	78.794	0.816	84.415	54.259	106.042

IV. Conclusions

In this study, we developed pseudo-parallel symmetrical and crossed perovskite solar cells, achieving a Jsc of 78.818 mA/cm², Voc of 0.816 V, F.F. of 84.126%, and a conversion efficiency of 54.087% for the pseudo-parallel symmetrical perovskite solar cells. Similarly, the pseudo-parallel crossed perovskite solar cells reached a Jsc of 78.794 mA/cm², Voc of 0.816 V, F.F. of 84.415%, and a conversion efficiency of 54.259%. Under more realistic albedo conditions of 30%, the energy boost for the symmetrical structure remained at 31.615%, while the crossed cells achieved 31.735%. These findings demonstrate that pseudo-parallel symmetrical and crossed designs significantly enhance the efficiency of bifacial solar cells, particularly when optimized with advanced materials and structural configurations. Future research should focus on further refining these designs and exploring their scalability for practical applications.

References

[1] A. Kojima, K. Teshima, Y. Shirai, and T. Miyasaka, "Organometal Halide Perovskites as Visible-Light Sensitizers for Photovoltaic Cells," *Journal of the American Chemical Society*, vol. 131, no. 17, pp. 6050–6051, Apr. 2009. DOI: 10.1021/ja809598r.

[2] D. Luo, R. Su, W. Zhang, Q. Gong, and R. Zhu, "Minimizing Non-Radiative Recombination Losses in Perovskite Solar Cells," *Nat. Rev. Mater*, vol. 5, pp. 44–60, Feb. 2020, DOI: 10.1038/s41578-019-0141-2.

[3] J.-Y. Jeng, et al., "CH₃NH₃PbI₃ perovskite/fullerene planar-heterojunction hybrid solar cells," *Adv. Mater*, vol. 25, pp. 3727–3732, Jul. 2013, DOI: 10.1002/adma.201300153.

[4] X. Zheng, et al., "Managing Grains and Interfaces via Ligand Anchoring Enables 22.3%-Efficiency Inverted Perovskite Solar Cells," *Nat. Energy*, vol. 5, pp. 131–140, Jan. 2020, DOI: 10.1038/s41560-019-0501-5.

[5] S. Mariotti, et al., "Interface Engineering for High-Performance Triple-Halide Perovskite–Silicon Tandem Solar Cells," *Science*, vol. 381, pp. 63–69, Apr. 2023, DOI: 10.1126/science.abq1847.

[6] X. Y. Chin, K. W. Tan, R. Singh, E. Köhnen, G. Li, and N. Mathews, "Interface Passivation for 31.25%-Efficient Perovskite/Silicon Tandem Solar Cells," *Science*, vol. 381, pp. 59–63, Mar. 2023, DOI: 10.1126/science.adb3782.

[7] Q. Tan, J. Wu, et al., "Inverted Perovskite Solar Cells Using Dimethylacridine-Based Dopants," *Nature*, vol. 620, pp. 545–551, Jul. 2023, DOI: 10.1038/s41586-023-05996-2.

[8] S. M. Park, N. Ahn, M. Choi, and S. Jang, "Engineering Ligand Reactivity Enables High-Temperature Operation of Stable Perovskite Solar Cells," *Science*, vol. 381, pp. 209–215, May. 2023, DOI: 10.1126/science.adi2374.

[9] F. Li, C. Ma, H. Wang, J. Hu, W. Huang, and L. Wang, "Hydrogen-Bond-Bridged Intermediate for Perovskite Solar Cells with Enhanced Efficiency and Stability," *Nat. Photon*, vol. 17, pp. 478–484, Jun. 2023, DOI: 10.1038/s41566-023-01072-4.

[10] S. Yu, J.-W. Lee, H.-S. Kim, and N.-G. Park, "Homogenized NiOx Nanoparticles for Improved Hole Transport in Inverted Perovskite Solar Cells," *Science*, vol. 382, pp. 1399–1404, Jul. 2023, DOI: 10.1126/science.adi5648.

[11] E. Aydin, et al., "Enhanced Optoelectronic Coupling for Perovskite/Silicon Tandem Solar Cells," *Nature*, vol. 623, pp. 732–738, Sep. 2023, DOI: 10.1038/s41586-023-05980-w.

[12] C. Liu, U. W. Paetzold, and M. Ba, "Bimolecularly Passivated Interface Enables Efficient and Stable Inverted Perovskite Solar Cells," *Science*, vol. 382, pp. 810–815, Aug. 2023, DOI: 10.1126/science.adj2389.

[13] A. J. Knight, E. M. Tennyson, and T. A. White, "Electronic Traps and Phase Segregation in Lead Mixed-Halide Perovskite," *ACS Energy Lett*, vol. 4, pp. 75–84, Feb. 2019, DOI: 10.1021/acsenergylett.8b01852.

[14] Z. Xu, B. Chen, Y. Zhao, and S. Tan, "Halogen Redox Shuttle Explains Voltage-Induced Halide Redistribution in Mixed-Halide Perovskite Devices," *ACS Energy Lett*, vol. 8, pp. 513–520, Mar. 2023, DOI: 10.1021/acsenergylett.2c02493.

[15] J. T. DuBose and P. V. Kamat, "Hole Trapping in Halide Perovskites Induces Phase Segregation," *Acc. Mater. Re.*, vol. 3, pp. 761–771, Dec. 2022, DOI: 10.1021/am2021.07348.

[16] J. Yoon, et al., "Al2O3-TiO2 Hybrid Passivation Layers for High-Performance Perovskite Solar Cells," *Adv. Mater*, vol. 35, 2105984, Jan. 2023, DOI: 10.1002/adma.202105984.

[17] Y. Huang, et al., "Low-Temperature Processed Perovskite Solar Cells with Enhanced Stability," *ACS Appl. Mater. Interfaces*, vol. 15, pp. 10012–10019, Apr. 2023, DOI: 10.1021/acsami.3c02493.

[18] W. Chen, et al., "Efficient and Stable Perovskite Solar Cells with Tailored Interfaces via Molecular Engineering," *Joule*, vol. 7, pp. 1108–1122, Mar. 2023, DOI: 10.1016/j.joule.2023.03.019.

[19] S. O. Kasap, *Optoelectronics and Photonics: Principles and Practices (2nd ed.)*, Prentice Hall, 2001.

[20] W.-H. Chen, H.-C. Chen, and C.-Y. Wu, "Recessed Back-Surface-Field Crystalline Silicon Solar Cells with Heterojunction for Bifacial Application," in Proceedings of the *14th IEEE International Conference on Solid-State and Integrated Circuit Technology*, Nov. 2018, pp. 1–3. DOI: 10.1109/ICSICT.2018.8564906.

[21] Y.-Y. Hu, "Both Side Collection on Emitter Wrap Through Bifacial Solar Cells," M.S. thesis, Dept. Electr. Eng., National Sun Yat-Sen Univ., Kaohsiung, Taiwan, 2017.

[22] M. P. Brennan, A. L. Abramase, R. W. Andrews, and J. M. Pearce, "Effects of Spectral Albedo on Solar Photovoltaic Devices," *Solar Energy Materials and Solar Cells*, vol. 124, pp. 111–116, 2014. DOI: 10.1016/j.solmat.2014.01.046.

Cryogenic and RF Modeling of On-Chip Passive Devices for Quantum Computer

Akira Tsuchiya

Dept. Electronic Systems Engineering, The University of Shiga University
tsuchiya.a@e.usp.ac.jp

Abstract—This paper discusses modeling of on-chip passive devices in cryo-CMOS for quantum computers. To solve scaling problem of quantum computer, frontend IC working at cryogenic environment is a promising technology. There are many reports of frontend IC in 4-K stage in dilution refrigerator, but cryogenic device modeling still has some unclear points. For future fine-tuned circuit design, it is important to establish proper device model in cryogenic environment. This paper focuses on passive devices which are key for analog/RF blocks. In cryogenic environment, the resistivity of metal wire is no longer a constant. Numerical simulations reveals that we need size-dependent and frequency-dependent resistivity for accurate modeling of wire resistance.

Index Terms—quantum computer, cryo-CMOS, passive device, resistivity

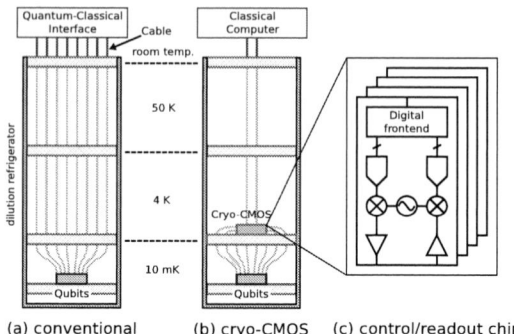

Fig. 1. Quantum Computer and Cryo-CMOS.

I. Introduction

Quantum computer has a potential to make a big change in computing. After 2010's, development of quantum computer has been a highly competitive area. In 2024, the number of qubits is around 100. However, we need to scale-up the number of qubits to realize quantum supremacy. One of the challenges is cabling bottleneck [1]. Some types of qubits work at extremely low temperature, for example, superconducting qubits at 10 mK and semiconductor spin qubits at 100 mK. To realize such cryogenic environment, dilution refrigerator is necessary. For control/readout of qubits, several cables per each are required to connect the inside and the outside of the dilution refrigerator. Due to the spatial and thermal capacity of dilution refrigerator, the number of cables is already near the limit.

One promising solution for this problem is cryo-CMOS [2], [3]. If we can put interface ICs inside the dilution refrigerator, the number of cables is reduced drastically. Many cryo-CMOS implementations have been reported [4]–[8]. 4-Kelvin stage is widely used, but also milli-Kelvin IC and assembly have been developed [9]–[11]. However, we cannot extrapolate the temperature dependency around room temperature to deep cryogenic region because there are some particular phenomena. This paper focuses modeling of metal resistivity in cryogenic environment. Temperature coefficient model of resistivity is no longer valid in cryogenic, so we need to consider size-dependent resistivity.

This work is supported by JST Moonshot R&D Program (Grant Number JPMJMS226A).

II. Overview of Cryo-CMOS

This section introduces cryo-CMOS for quantum computer and related works on device modeling.

A. Overview of Cryo-CMOS for Quantum Computer

As mentioned in Introduction, one of the problems in scaling quantum computer up is the number of cables through dilution refrigerator. A simplified figure is shown in Fig. 1. In case of superconducting qubits, the qubits are at 10 mK stage in dilution refrigerator. Conventionally, the interface between qubits and the classical computing system is in room temperature. So, as shown in Fig. 1 (a), many cables from room temperature to cryogenic environment are needed for quantum-classical connection. The space in the dilution refrigerator is limited, and the thermal inflow through the cables will exceed the cooling capacity. Large capacity dilution refrigerator is in development [12], however it is not a fundamental solution. One solution is cryo-CMOS as shown in Fig. 1 (b). If we implement the frontend of quantum-classical interface, we can reduce the number of cables drastically. Fig. 1 (c) shows a typical architecture of frontend circuit. Modulated RF pulse is used to control/readout of qubits. RF part (local oscillator, mixer, and amplifiers) transmit/receive RF pulse, and it is controlled by digital part via DAC and ADC. Since the cooling capacity of dilution refrigerator is 1–2 W, very low-power implementation is required, for example, about 10 mW/qubit for 1,000-qubit system. To achieve such performance, we need fine-tuning of all circuits. Thus, accurate device modeling is essential.

979-8-3503-6184-1/24 $31.00 © 2024 IEEE

B. Modeling of Active Devices

Transistor characteristics has been discussed for not only quantum computer but also other applications such as aerospace. As the temperature goes down, the threshold voltage rises but the circuit performance improves compared to room temperature. So, CMOS ICs can work in cryogenic environment without major modification. There are many reports of CMOS working at 4 K [4]–[8], and at 100 mK [9]. However, there are some cryogenic particular behaviors, for example kink effect [13] and career freeze-out. Also, Ref. [14] reported that self-heating effect is significant. Thus, device model which reflects cryogenic physics is necessary and many groups are working on device modeling [15]–[20].

C. Modeling of Passive Devices

Since cryo-CMOS frontend has analog and RF blocks, passive devices play important roles. Poly-resistor is a widely used device for precision resistance in analog/RF circuits. However, Ref. [21] reported that temperature dependency becomes complex in cryogenic region. Resistance of metal wire decreases as the temperature goes down, but wire width dependency and self-heating effect is significant [22]. Another important device is on-chip inductor, which is used for LC oscillator, mixer and so on. Refs. [23], [24] shows measurement results and they show that on-chip inductor has better quality-factor in cryogenic environment because of low parasitic resistance. However, accurate wire resistivity and substrate characteristics are needed to predict the inductor characteristics accurately. These results indicate that accurate modeling and cryogenic-specific design strategy is needed.

III. RESISTIVITY IN CRYOGENIC ENVIRONMENT

We focus on the resistivity of metal wire. As discussed in Ref. [22], the resistivity depends on wire width. We discussed the cause of this behavior by numerical evaluation [25].

A. Resistivity in Cryogenic Region

The root of resistivity is scattering of electron. When current flows in metal, free electrons run along electric field. If the free electron hit something, the free electron loses kinetic momentum. This loss is the root of resistivity. Fig. 2 shows cause of scattering. In room temperature, the dominant factor is atoms of metal. This is called phonon scattering. As the temperature becomes lower, the effective radius of phonon shrinks. Then, the probability of scattering decreases and it leads low resistivity. This temperature dependency appears as the temperature coefficient of resistivity. The other causes are surface of metal (surface scattering), grain boundary, and impurity, but these are negligible in many cases.

At deep cryogenic region, the probability of hitting atom becomes very low, and the phonon scattering is not the dominant factor of the resistivity [26]. Thus, at deep cryogenic region, the resistivity is determined by the dimension and impurity of metal [27].

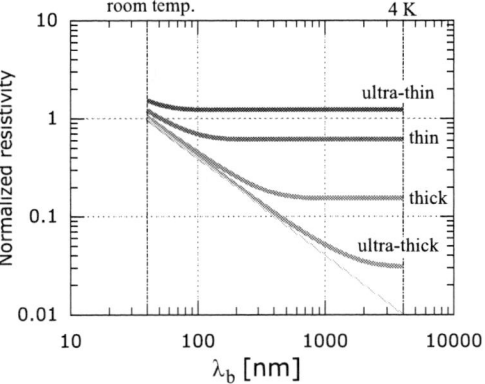

Fig. 2. Free electron scattering in metal (Cu).

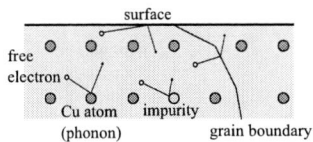

Fig. 3. The mean free path of electron versus normalized resistivity.

B. Numerical Evaluation

When the main cause of resistivity changes from phonon to other factors, wire resistance model should be changes. As shown in Ref. [22], size dependency becomes significant in cryogenic environment. On-chip wire has a wide variety of width and thickness. So, we proposed a method to model the resistivity considering surface scattering [25]. The method evaluates the effective mean free path of electron numerically. Fig. 3 shows the resistivity normalized by the value at the room temperature. The horizontal axis is the mean free path of electron λ_b, so the left side is high temperature, and the right side is lower temperature. The types of wires, ultra-thin, thin, thick, ultra-thick have the thickness of 50 nm, 100 nm, 400 nm, and 4000 nm, respectively. The results shows that the resistivity differs about 40 times by the dimension of wire, even the material is the same. Also, surface scattering affects the frequency dependency of the resistance as called "anomalous skin effect". Fig. 4 shows the impact of anomalous skin effect evaluated by the proposed method. In ultra-thick metal, the impact of anomalous skin effect is about 10% at 10 GHz. These results imply that new modeling and design strategy are needed in cryo-CMOS. It might impact on not only passive devices but also signal wire and power distribution network.

IV. CONCLUSION

This paper discussed device modeling of cryo-CMOS for quantum computer. To achieve scaling of qubits, frontend circuits at 4 K stage in dilution refrigerator is a promising solution. Cryo-CMOS is required to achieve high-precision RF pulse generation with very low power, so more accurate device models are necessary. In this paper, we focus on modeling of

979-8-3503-6184-1/24 $31.00 © 2024 IEEE

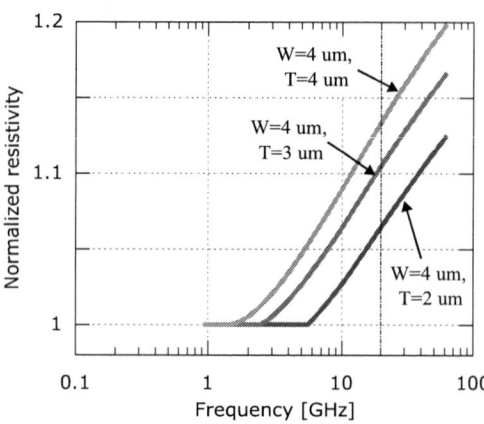

Fig. 4. The mean free path of electron versus normalized resistivity.

metal resistivity. Since the main cause of resistivity changes, we need resistivity model considering electron scattering at the surface and/or impurity. Our numerical modeling revealed that the resistivity strongly depends on the size of wire, and the difference is over 40 times at maximum. We have to develop accurate modeling and design strategy for future improvement of cryo-CMOS frontend ICs.

REFERENCES

[1] J. C. Bardin, D. H. Slichter, and D. J. Reilly, "Microwaves in quantum computing," *IEEE Journal of Microwaves*, vol. 1, no. 1, pp. 403–427, 2021.

[2] E. Charbon, "Cryo-CMOS electronics for quantum computing: Bringing classical electronics closer to qubits in space and temperature," *IEEE Solid-State Circuits Magazine*, vol. 13, no. 2, pp. 54–68, 2021.

[3] R. B. Staszewski, I. Bashir, E. Blokhina, and D. Leipold, "Cryo-CMOS for quantum system on-chip integration: Quantum computing as the development driver," *IEEE Solid-State Circuits Magazine*, vol. 13, no. 2, pp. 46–53, 2021.

[4] J. Park, S. Subramanian, L. Lampert, T. Mladenov, I. Klotchkov, D. J. Kurian, E. Juarez-Hernandez, B. P. Esparza, S. R. Kale, A. B. K. T., S. P. Premaratne, T. F. Watson, S. Suzuki, M. Rahman, J. B. Timbadiya, S. Soni, and S. Pellerano, "A fully integrated cryo-CMOS SoC for state manipulation, readout, and high-speed gate pulsing of spin qubits," *IEEE Journal of Solid-State Circuits*, vol. 56, no. 11, pp. 3289–3306, 2021.

[5] K. Kang, D. Minn, J. Lee, H.-J. Song, M. Lee, and J.-Y. Sim, "A cryogenic controller IC for superconducting qubits with DRAG pulse generation by direct synthesis without using memory," in *2023 IEEE International Solid-State Circuits Conference (ISSCC)*, pp. 33–35, 2023.

[6] L. L. Guevel, C. Wang, and J. C. Bardin, "A 22nm FD-SOI <1.2mW/active-qubit AWG-free cryo-CMOS controller for fluxonium qubits," in *2024 IEEE International Solid-State Circuits Conference (ISSCC)*, vol. 67, pp. 1–3, 2024.

[7] B. Prabowo, O. Pietx-Casas, M. A. Montazerolghaem, G. Scappucci, L. M. K. Vandersypen, F. Sebastiano, and M. Babaie, "A cryo-CMOS receiver with 15K noise temperature achieving 9.8dB SNR in $10\mu s$ integration time for spin qubit readout," in *2024 IEEE International Solid-State Circuits Conference (ISSCC)*, vol. 67, pp. 474–476, 2024.

[8] Y. Guo, Q. Liu, W. Huang, Y. Li, T. Tian, N. Wu, S. Zhang, T. Li, Z. Wang, N. Deng, Y. Zheng, and H. Jiang, "A cryo-CMOS quantum computing unit interface chipset in 28nm bulk CMOS with phase-detection based readout and phase-shifter based pulse generation," in *2024 IEEE International Solid-State Circuits Conference (ISSCC)*, vol. 67, pp. 476–478, 2024.

[9] B. C. Paz, L. Le Guevel, M. Cassé, G. Billiot, G. Pillonnet, A. G. M. Jansen, R. Maurand, S. Haendler, A. Juge, E. Vincent, P. Galy, G. Ghibaudo, M. Vinet, S. de Franceschi, T. Meunier, and F. Gaillard, "Variability evaluation of 28nm FD-SOI technology at cryogenic

temperatures down to 100mK for quantum computing," in *2020 IEEE Symposium on VLSI Technology*, pp. 1–2, 2020.

[10] H. Oka, H. Asai, T. Inaba, S. Shitakata, H. Yui, H. Fuketa, S. Iizuka, K. Kato, T. Nakayama, and T. Mori, "Milli-kelvin analysis revealing the role of band-edge states in cryogenic MOSFETs," in *2023 International Electron Devices Meeting (IEDM)*, pp. 1–4, 2023.

[11] M. Taguchi, T. Okidono, T. Miki, and M. Nagata, "Si interposer with Cu TSVs on Cu substrate thermally and electrically anchoring qubit chips in millikelvin assembly," in *2024 IEEE 74th Electronic Components and Technology Conference (ECTC)*, pp. 447–450, 2024.

[12] M. I. Hollister, R. C. Dhuley, and G. L. Tatkowski, "A large millikelvin platform at Fermilab for quantum computing applications," *IOP Conference Series: Materials Science and Engineering*, vol. 1241, p. 012045, may 2022.

[13] T. Lu, Y. Zhang, Y. Zhang, J. Xu, G. Guo, and C. Luo, "Cryogenic modeling of MOSFET device based on BSIM and EKV models," 2021.

[14] P. A. T Hart, M. Babaie, A. Vladimirescu, and F. Sebastiano, "Characterization and modeling of self-heating in nanometer bulk-CMOS at cryogenic temperatures," *IEEE Journal of the Electron Devices Society*, vol. 9, pp. 891–901, 2021.

[15] A. Beckers, F. Jazaeri, A. Ruffino, C. Bruschini, A. Baschirotto, and C. Enz, "Cryogenic characterization of 28 nm bulk CMOS technology for quantum computing," in *2017 47th European Solid-State Device Research Conference (ESSDERC)*, pp. 62–65, 2017.

[16] H. Homulle, L. Song, E. Charbon, and F. Sebastiano, "The cryogenic temperature behavior of bipolar, MOS, and DTMOS transistors in standard CMOS," *IEEE Journal of the Electron Devices Society*, vol. 6, pp. 263–270, 2018.

[17] Z. Tang, Z. Wang, A. Guo, L. Liu, C. Cao, X. Luo, W. Wu, Y. Guo, Z. Zhi, Y. Hu, Y. Cao, G. Shang, L. Yu, S. Hu, S. Chen, Y. Zhao, and X. Kou, "Cryogenic CMOS RF device modeling for scalable quantum computer design," *IEEE Journal of the Electron Devices Society*, vol. 10, pp. 532–539, 2022.

[18] Q. Berlingard, J. Lugo-Alvarez, M. Bawedin, T. Mota-Frutuoso, C. Durand, D. Gloria, P. Galy, and M. Cassé, "Capacitance RF characterization and modeling of 28 FD-SOI CMOS transistors down to cryogenic temperature," in *2023 18th European Microwave Integrated Circuits Conference (EuMIC)*, pp. 37–40, 2023.

[19] J. Pérez-Bailón, M. Tarancón, S. Celma, and C. Sánchez-Azqueta, "Cryogenic measurement of CMOS devices for quantum technologies," *IEEE Transactions on Instrumentation and Measurement*, vol. 72, pp. 1–7, 2023.

[20] M. Shintani, T. Iwasaki, and T. Sato, "Gaussian process-based device model toward a unified current model across room to cryogenic temperatures," in *2024 IEEE 36th International Conference on Microelectronic Test Structures (ICMTS)*, pp. 1–5, 2024.

[21] J. Marqués-García, J. Pérez-Bailón, S. Celma, and C. Sánchez-Azqueta, "Characterization of 65-nm CMOS integrated resistors in the cryogenic regime," *IEEE Transactions on Instrumentation and Measurement*, vol. 73, pp. 1–3, 2024.

[22] K. Okamoto, T. Tanaka, M. Miyamura, H. Ishikuro, K. Uchida, T. Sakamoto, and M. Tada, "Cryogenic CMOS performance analysis including BEOL characteristics at 4K for quantum controller application," in *2022 IEEE International Interconnect Technology Conference (IITC)*, pp. 139–141, 2022.

[23] B. Patra, M. Mehrpoo, A. Ruffino, F. Sebastiano, E. Charbon, and M. Babaie, "Characterization and analysis of on-chip microwave passive components at cryogenic temperatures," *IEEE Journal of the Electron Devices Society*, vol. 8, pp. 448–456, 2020.

[24] Q. Berlingard, J. Lugo-Alvarez, L. Contamin, C. Durand, P. Galy, A. Juge, S. De Franceschi, M. Vinet, T. Meunier, and M. Cassé, and F. Gaillard, "RF performances at cryogenic temperatures of inductances integrated in a FDSOI technology," in *2021 Joint International EU-ROSOI Workshop and International Conference on Ultimate Integration on Silicon (EuroSOI-ULIS)*, pp. 1–4, 2021.

[25] A. Tsuchiya, "Mean-free-path-based evaluation of size effect and anomalous skin effect in on-chip interconnects under cryogenic environment," in *2024 IEEE 28th Workshop on Signal and Power Integrity (SPI)*, pp. 1–4, 2024.

[26] A. B. Pippard, *The Dynamics of Conduction Electrons*. Gordon and Breach Science Publishers, 1965.

[27] R. P. Reed and R. P. Mikesell, *Low Temperature Mechanical Properties of Copper and Selected Copper Alloy*. National Bureau of Standards, United States,, 1967.

Comparison of Nanosheet and Fin Integration in Stacked Induced Tunnel Field-Effect Transistors

Ruei-Cheng Tu[1], Chia-Yo Kuo[2], Jyi-Tsong Lin[3]

[1] Department of Electrical Engineering, National Sun Yat-Sen University, Kaohsiung 80424, Taiwan, R.O.C.

Email: 2012.ray.just.2@gmail.com[1], jeff5035@gmail.com[2], jtlin@ee.nsysu.edu.tw[3]

Abstract—**Nanosheet transistors are poised to become the next-generation choice for smaller dimensions. To meet future high-performance computing demands and low-power applications, this study proposes a vertically stacked nanosheet structure with a high I_{ON}/I_{OFF} ratio, combined with Schottky iTFET. By utilizing SiGe, which has three times the carrier mobility of Si, and leveraging line tunneling mechanisms, we achieved superior Band-to-Band characteristics, resulting in improved switching performance and lower Subthreshold Swing (*SS*).**

Additionally, we compared the horizontal stacking of Fin structures with our Nanosheet. The results indicate that our device not only achieved an outstanding I_{ON}/I_{OFF} ratio of 2.7 × 10^8, but also demonstrated a 779% increase in I_{ON} in the horizontal stacking of Fin iTFET when comparing 5-layer stacking to a single Fin layer. This further proves that the horizontal stacking of Fin iTFET can effectively enhance I_{ON}, making it a leading candidate for future low-power devices.

Keywords—Nanosheet induced tunnel field-effect transistor (NS iTFET), Fin induced tunnel field-effect transistor (Fin iTFET), subthreshold swing (*SS*), line tunneling, band-to-band tunneling (BTBT), SiGe.

I. INTRODUCTION

As Moore's Law progresses, in the pursuit of faster switching speeds, lower power consumption, and smaller sizes, transistor technology has evolved from traditional Planar FETs, through Nanowire and FinFETs, to next-generation Nanosheet technologies. As CMOS technology continues to scale down, new components for enhancing performance have been integrated into recent generations. For instance, the gate dielectric has evolved to high-K/metal gate, and the conducting channel has transitioned to a Fin structure[1]. Nanosheets exhibit remarkable design flexibility: the channel can be widened to increase current or narrowed to limit power consumption. Vertically stacked nanosheet transistors have been identified as the primary component structure for the 3-nanometer technology node and beyond[2].

However, Nanosheets come with their own set of challenges, such as the trade-offs between switching speed, power consumption, process complexity, and cost. These trade-offs are closely related to the effective channel width (*W*eff). A wider channel means more current can be driven, enabling faster transistor switching, but it also results in a more complex and expensive manufacturing process.

II. DEIVCE OF STRUCTURE AND SIMULATION METHOD

We conducted a stacked comparison between Nanosheet iTFET and Fin iTFET, as shown in Fig 1. By replacing traditional Si with SiGe, which has three times the carrier

mobility of Si, we achieved superior Band-to-Band tunneling [3]. Utilizing Schottky contacts with different work functions, a PN junction can be formed. To enhance the reliability of the comparison, we used the same overall structure, with dimensions of 55 × 55 × 65. With the lateral stacking of Fin iTFET, the width of the devices also increased, as shown in Fig 2, resulting in dimensions of 55 × 75 × 65. As the number of lateral stacks of Fin iTFET increased, the width continued to increase. Compared to MOSFET devices, since iTFET primarily utilizes a line tunneling mechanism, its tunneling current is proportional to the width (*W*) and length (*L*). Therefore, a larger volume allows our iTFET devices to exhibit superior circuit characteristics[4, 5].

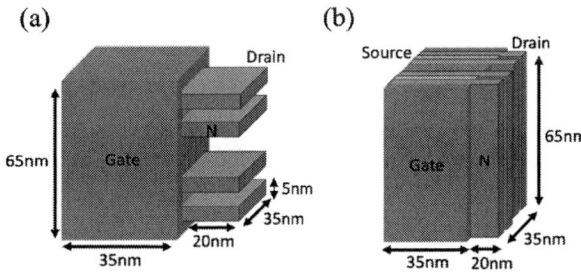

Fig. 1. SiGe Body width is 35 nm of (a) Nanosheet iTFET (b) Fin iTFET.

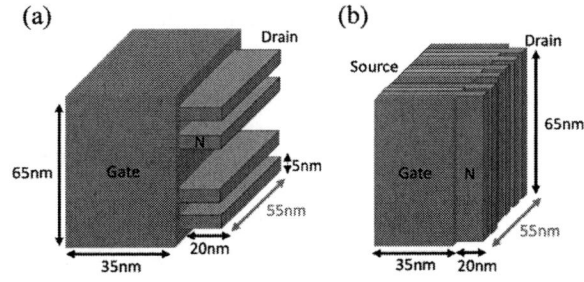

Fig. 2. SiGe Body width is 55 nm of (a) Nanosheet iTFET (b) Fin iTFET.

TABLE I.
The SiGe Body width is 35 nm / 55 nm for
DEVICE PARAMETERS.

Parameter	Value
Body Thickness (t_{Body})	5 nm
Gate Oxide Thickness (t_{ox})	5 nm
Gate channel Length (L_G)	35 nm
Total length of Nanosheet iTFET and Fin iTFET (*L*)	55 nm
Total width of Nanosheet iTFET and Fin iTFET (*W*)	45 nm / 65 nm
Total height of Nanosheet iTFET and Fin iTFET (*H*)	65 nm
Body Doping in Nanosheet iTFET (N_D)	1×10^{18} cm^{-3}
Body Doping in Fin iTFET (N_D)	1×10^{18} cm^{-3}
Schottky Barrier Hight (φb)	0.8 eV

979-8-3503-6184-1/24 $31.00 © 2024 IEEE   313

In this paper, we utilized Sentaurus TCAD to simulate the electrical characteristics of three different Tunnel Field-Effect Transistor (TFET) structures that we proposed, considering various parameter variations. For precise tunneling current calculations, we employed the Dynamic Nonlocal Path Band-to-Band Tunneling Model. We also incorporated several non-ideal device effects, including the Shockley-Read-Hall recombination (SRH) model, Bandgap narrowing, High-field saturation mobility models, and Auger recombination model, along with quantum confinement effects to account for minute fabrication details. To validate the accuracy and feasibility of our simulations, we used experimentally fabricated Si/SiGe heterojunctions and considered TFETs that exhibit both line and point tunneling simultaneously. The calibration of our models for these simulations is depicted in Fig 3.

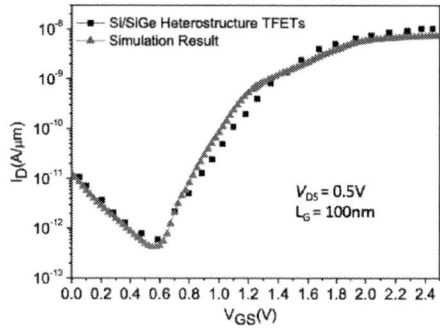

Fig. 3. TCAD model calibration using experimental data[6].

III. ELECTRICAL CHARACTERISTICS DISCUSSION

Fig 4 presents the $I_D V_G$ curves for Nanosheet iTFET and Fin iTFET with a SiGe body width of 35nm. Both Nanosheet iTFET and Fin iTFET, which primarily utilize line tunneling mechanisms, exhibit small subthreshold swings of 20.67 mV/dec and 23.92 mV/dec, respectively. The I_{ON}/I_{OFF} ratios are 2.7×10^8 and 2.14×10^7, respectively. The lower I_{ON}/I_{OFF} ratio is due to the inherently lower I_{ON} of TFETs. We can increase the I_{ON} of the device by either widening the Nanosheet or stacking the Fin horizontally.

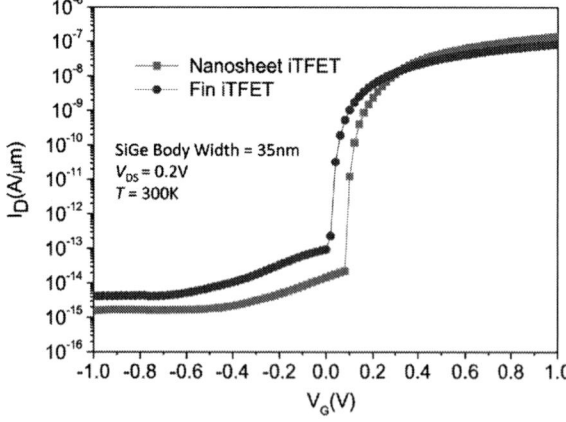

Fig. 4. I_D-V_G characteristic curves of the Nanosheet iTFET and Fin iTFET, SiGe Body Width set to 35nm.

Fig 5 and Fig 6 present the current variations of Nanosheet iTFET and Fin iTFET with different SiGe Body widths, different Nanosheet widths, and different Fin layers. We observed that with the increase in the number of Fin layers, the I_{ON} current for both Nanosheet iTFET and Fin iTFET shows an upward trend, as shown in Fig 7. During the horizontal stacking process, the increase in I_{ON} for Fin iTFET is greater than that for Nanosheet iTFET when adjusting the width. For Nanosheet iTFET, the increase in I_{ON} ranges from 184% to 397%, while for Fin iTFET, it ranges from 161% to 779%. Theoretically, the current should be proportional to the number of stacking layers. Therefore, for a horizontal stacking number of 5 layers, the theoretical value should be 500%. However, due to the line tunneling mechanism we used, the increase in current reached 779%.

Consequently, Fin iTFET horizontal stacking is more effective in enhancing I_{ON} than the width adjustment of Nanosheet iTFET. This demonstrates that by using stacking methods, we can effectively improve the relatively low I_{ON} of TFET while achieving lower subthreshold swing and lower power consumption.

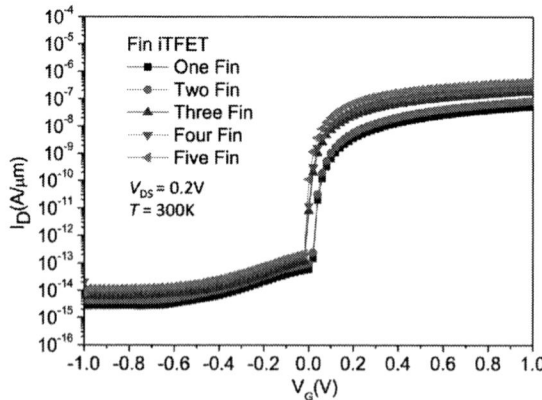

Fig. 5. I_D-V_G characteristic curves for different widths of the Nanosheet iTFET.

Fig. 6. I_D-V_G characteristic curves for different number of fins of the Fin iTFET.

979-8-3503-6184-1/24 $31.00 © 2024 IEEE

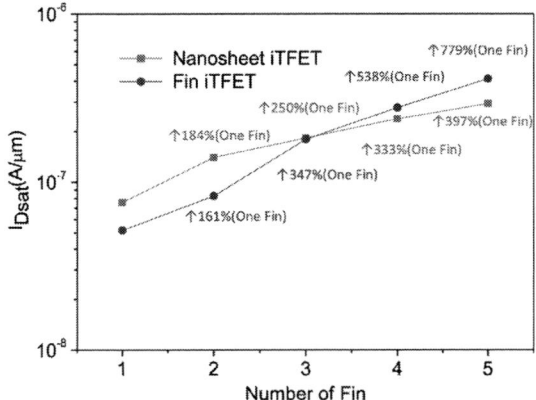

Fig. 7. I_{ON} for different number of fins of the Nanosheet iTFET and Fin iTFET.

As the number of Fin iTFET lateral stacking layers increases, the I_{ON} current shows an upward trend, but the I_{OFF} current in the off-state also increases. When the stacking layers increase from 1 to 5, the I_{OFF} rises by nearly an order of magnitude, indicating the need to consider the I_{ON}/I_{OFF} ratio. We compared the stacked Fin iTFET with the Nanosheet iTFET, as shown in Fig 8. The I_{ON}/I_{OFF} of the stacked Fin iTFET remains constant, while for the Nanosheet iTFET, as the width increases, the I_{OFF} does not increase as much, resulting in a gradual rise in the I_{ON}/I_{OFF} ratio.

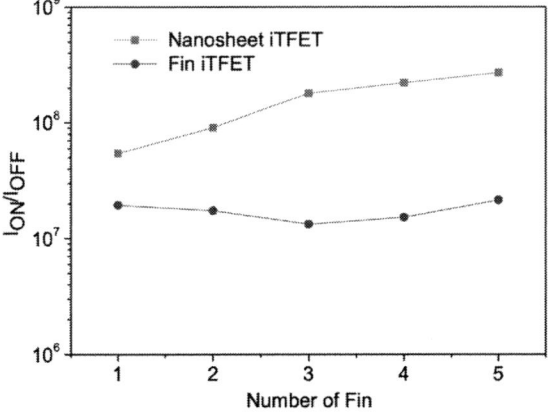

Fig. 8. I_{ON}/I_{OFF} for different number of fins of the Nanosheet iTFET and Fin iTFET.

CONCLUSION

In this study, we selected SiGe with a carrier mobility three times that of Si, employed a line tunneling mechanism, and applied traditional nanosheet stacking combined with iTFET technology. This approach successfully achieved superior Band to Band characteristics, a lower subthreshold swing, and higher I_{ON}. We also explored the horizontal stacking of Fin iTFET using the same line tunneling mechanism. By comparing the efficiency of horizontally stacked Fin iTFET with vertically stacked Nanosheet iTFET at the same volume, we found that the horizontal stacking of Fin iTFET more effectively enhanced I_{ON}.

When the number of horizontal stacking layers reached 5, the theoretical increase in current should have been 500%, but due to the line tunneling mechanism, the actual increase reached 779%.

In future IoT and AI applications, the demand for higher voltage has become an unavoidable issue in power consumption for upcoming devices. When supply voltage decreases, maintaining high performance and a steep subthreshold swing becomes crucial. Whether through horizontal stacking of Fin iTFET or vertical stacking of Nanosheet iTFET, I_{ON} can be effectively increased while maintaining a good I_{ON}/I_{OFF} ratio. Based on these findings, we believe that Nanosheet iTFET and Fin iTFET will emerge as the preferred components for low-power and fast-switching applications in the future.

REFERENCE

[1]   D. Guo *et al.*, in *2016 IEEE Symposium on VLSI Technology*, pp. 1-2, 2016.

[2]   J. Yang *et al.*, in *2022 China Semiconductor Technology International Conference (CSTIC)*, pp. 1-3, 2022.

[3]   W. Vandenberghe *et al.*, in *2008 International Conference on Simulation of Semiconductor Processes and Devices*, pp. 137-140, 2008.

[4]   Ashita *et al.*, in *2019 IEEE SOI-3D-Subthreshold Microelectronics Technology Unified Conference (S3S)*, pp. 1-3, 2019.

[5]   J. T. Lin *et al.*, *IEEE Transactions on Electron Devices*, pp. 769-775, 2018.

[6]   X. Liu *et al.*, *IEEE Journal of the Electron Devices Society*, pp. 976-980, 2020.

Nonlinear Contact Behavior in MoS₂ Field Effect Transistors at Cryogenic Temperature

Shihab AHAMMED* and Mansun CHAN

Department of Electronic and Computer Engineering, The Hong Kong University of Science and Technology, Clear Water Bay, Kowloon, Hong Kong

* Email: sahammed@connect.ust.hk

Abstract— **The performance of MoS₂-based devices depends on contact resistance, influenced by the Schottky barrier. Typically, metal-contacted MoS₂ devices show non-ohmic behavior around 77 K due to excessive Schottky barrier height and wide depletion region, limiting carrier injection. This study explores n-type MoS₂-based Ni-contacted field effect transistors from 300 K to 4 K to find out the behavior of carrier injection. We observe ohmic conduction from 300 K to 77 K due to the minimized Schottky barrier. The low SBH justifies the enhanced carrier injection, vital for reliable low-temperature operation. However, the transition from linear to nonlinear behavior below 77 K indicates the domination of contact resistance. This investigation provides critical insights into carrier injection at various temperatures, advancing the understanding of metal-TMD interfaces.**

Keywords— *MoS₂, 2-D transistor, cryogenic, contact, Schottky barrier*

I. INTRODUCTION

Effective carrier injection, essential for optimizing device performance, hinges on the Schottky barrier height (SBH) and width (SBW) [1]. Reducing SBH and SBW lowers contact resistance, but lowering of SBH is difficult due to the Fermi level pinning near the conduction band of MoS₂ [2, 3]. Room temperature ohmic contacts can be achieved through work function engineering of various metals, yet these contacts often become Schottky at cryogenic temperatures [4]. Recent research suggests using semimetals for ohmic contacts at low temperatures, but their low melting points present reliability challenges during the back end of line (BEOL) process [5]. Therefore, further investigation into contact engineering is necessary to achieve stable ohmic contacts with metal contacts at both room and cryogenic temperatures.

Schottky barrier engineering could solve the carrier injection problems at the metal/2D-semiconductor interface. Few studies have explored methods to thin the Schottky barrier width and lower the Schottky barrier height to improve carrier injection [1, 6]. One of the approaches involves doping the semiconductor at the contact region [7]. However, doping MoS₂ during fabrication often introduces defects, which increase contact resistance and hinder carrier injection [8]. Therefore, new techniques are needed to develop these devices and characterize their properties at various temperatures, especially at cryogenic temperatures.

In this study, we propose MoS₂-based Ni-contacted field-effect transistors (FETs) doped with Rhenium (Re) to address the doping challenges during fabrication. This method aims to achieve a minimized Schottky barrier which in turn improves the carrier injection. Consequently, n-type MoS₂ FETs with metal contacts are expected to achieve higher on-current and linear contact behavior at both room temperature and cryogenic temperatures.

II. FABRICATION AND CRYOGENIC CHARACTERIZATION OF N-TYPE MoS₂ TRANSISTOR

To investigate the electrical properties of n-type MoS₂-based FETs, we fabricated three batches of devices. The fabrication process started with the deposition of 30 nm Al_2O_3 layer at 300 °C by plasma-enhanced ALD, which serves as a high-k gate dielectric. This layer is grown on a heavily doped p-type Si substrate, which functions as the back gate. In this work, we utilized Re-doped MoS₂ synthetic crystal from '2Dsemiconductors'. After mechanical exfoliation of MoS₂ on top of Al_2O_3 layer, the samples are cleaned with acetone, rinsed with IPA/DI water, and then baked on a hotplate to eliminate any remaining organic contaminants. To prevent oxidation of the MoS₂ layer, the O_2 plasma process is conducted at an RF power of 50 W. The top source/drain (S/D) contacts are subsequently defined using photolithography. The channel size of all of the devices is maintained as 1 μm. To protect the integrity of the MoS₂ flake, we used slow-rate evaporation. Specifically, we deposited 15 nm of Ni at a rate of 0.5 Å / s and 50 nm of Au at a rate of 1 Å / s under a high vacuum of 3×10^{-7} Torr, which minimized the defect formation in MoS₂. After doing the liftoff process the device was then wire bonded to go for the cryogenic measurement.

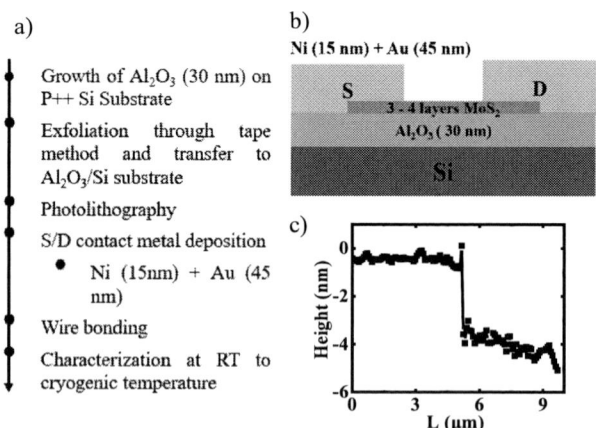

Fig. 1. a) The process flow of the fabrication and characterization of MoS₂ FET b) 2D Schematic illustration of back-gated MoS₂ FET c) The height of the MoS₂ channel (~ 3.2 nm) measured With Atomic Force Microscopy.

The physical layout of the FET and their manufacturing process flow are depicted in Fig. 1. The Cryogenic Fluid Management (CFM) system (for cooling down from RT to 4K), interfaced with semiconductor device parameter analyzer has been used for the characterization of the n-type MoS₂ FETs.

979-8-3503-6184-1/24 $31.00 © 2024 IEEE

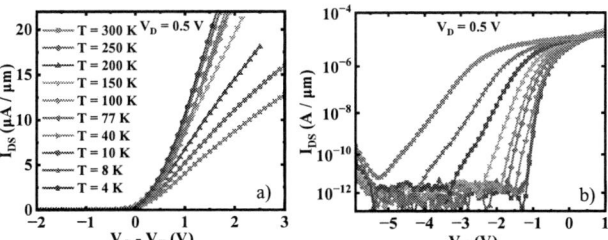

Fig. 2. I_D - V_G graph with a) linear and b) log scale. The temperature is varied from 300 K down to 4 K with a fixed drain voltage, V_D = 0.5 V.

The variation of drain current with gate voltage at different temperatures is depicted in Fig. 2. From the three batches of devices, all exhibited comparable electrical performance. The data presented here are from a device with a ~ 3.2 nm thick MoS₂ sheet. Transfer characteristics were measured by varying the gate voltage from – 6 V to 1 V, with temperatures ranging from room temperature down to 4 K. Initially, the temperature was reduced by 50 K steps. Below 100 K, smaller steps were used for detailed analysis.

The gate voltage was converted to overdrive voltage to compare the drain current, as the threshold voltage shifts significantly with temperature changes. Fig. 2(a) shows that the drain current increases as the temperature decreases, indicating significant improvement in MoS₂-based devices. This improvement is attributed to the efficient carrier injection in the Re-doped MoS₂/Ni interface. These kinds of characteristics suggest the dominance of field emission (direct tunneling of carriers) through the Schottky barrier, unlike conventional MoS₂ FETs, where carrier injection is primarily through thermionic emission. However, the current increase rate diminishes at deep cryogenic temperatures due to the saturation of carrier mobility indicating Coulomb scattering and the dominant effect of the contact resistance.

To investigate the contact characteristics of our devices, we characterized the I_D – V_D graph across a temperature range from RT down to 4 K. Fig. 3 illustrates the output characteristics of the device at 300 K, 77 K, 40 K, and 4 K. Our findings indicate that the contacts remain purely ohmic down to 77 K, a significant improvement compared to previous Ni-based MoS₂ FETs [5]. From the perspective of

charge transport mechanisms, thermionic emission over the Schottky barrier is more temperature sensitive compared to field emission. Given the consistent linear behavior from 300 K to 77 K, we infer that carrier injection might be dominated by field emission, which is less affected by temperature variations than thermionic emission. However, a slight non-linear behavior emerges as the temperature further decreases to 4 K. We believe this linear to non-linear transition in the I_D – V_D graph is attributable to the impact of contact resistance at deep cryogenic temperatures.

III. EXTRACTION OF KEY PARAMETERS AND DISCUSSION

To further understand the carrier injection mechanism, we investigated the effective SBH, resistance, and mobility of our devices. To extract effective SBH, The temperature-dependent transfer characteristics were analyzed using the thermionic current equation and replotted in an Arrhenius fashion [9], as shown in Fig. 4(a). The Arrhenius curves were fitted linearly to determine the effective barrier for current flow and can be expressed as,

$$\phi_{B,eff}(V_G) = \frac{K_B}{q}\left[\frac{\Delta \ln(I_D(V_G)/T^{\frac{3}{2}})}{\Delta T^{-1}}\right] \tag{1}$$

Where $\phi_{B,eff}$ is the effective SBH also known as activation energy, K_B is the Boltzmann constant, q is the elemental charge and I_D, V_G, and T are the drain current, gate voltage and temperature respectively. The final step involved plotting the SBH as a function of the applied gate voltage (V_G) which is shown in Fig 4(b). Near the depletion region, the effective ϕ_B linearly decreases with increasing gate voltage when thermally activated transport dominates. The slope changes when field-emission transport becomes significant at the Schottky barrier. The crossover point occurs when the band flattens, where the effective ϕ_B equals the Schottky barrier height. To determine the V_G at which the band flattens, we plotted two linear lines on the data points in Fig. 4(b). The curves intersect at $V_G \approx -1$ V, indicating the flat band voltage (V_{FB}), and we estimated SBH ≈ 18 meV for the Ni/MoS₂ contact. This low ϕ_B suggests that the carrier transport across the Ni/MoS₂ contact has been improved. Previous studies on conventional Ni-based MoS₂ devices have shown that the effective SBH is around 30 meV, which is significantly higher than that of our n-type MoS₂ FET [5]. Despite a lower SBH, we assume carrier injection is dominated by field emission. This is due to the constant linear behavior from RT to 77 K, as 18 meV should cause nonlinearity at low temperatures if the carrier injection is mainly dominated by thermionic emission. Additionally, the increasing on-current with decreasing temperature also suggests field emission dominance, especially at lower temperatures.

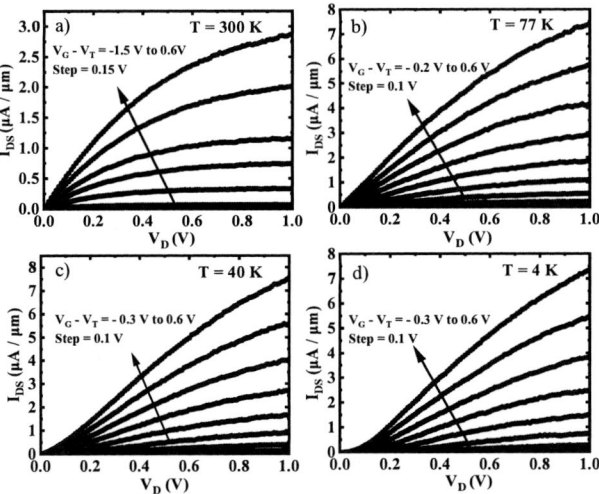

Fig. 3. Output characteristics of the n-type MoS₂ FET with Ni contact at a) 300 K, b) 77 K, c) 40 K, and d) 4 K.

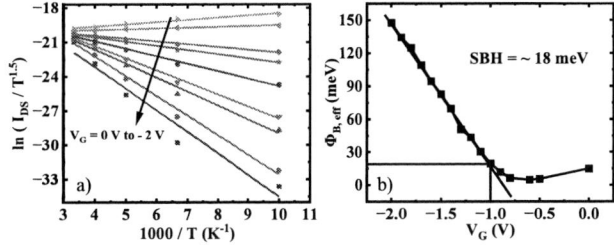

Fig. 4. a) Richardson plot of the transfer curves to extract the SBH b) Effective SBH as a function of gate voltage to extract the effective Schottky barrier height.

979-8-3503-6184-1/24 $31.00 © 2024 IEEE

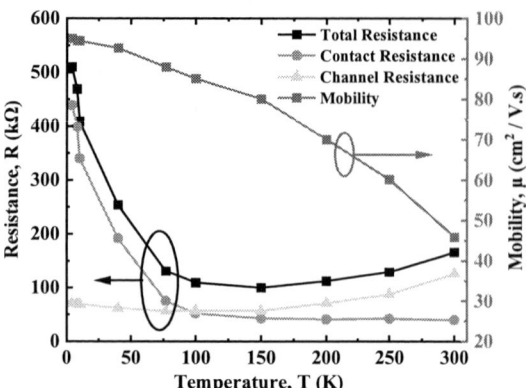

Fig. 5. Extracted total resistance and the mobility of the MoS₂ FET with the variation of temperatures.

To further understand the behavior of the carrier injection at different temperatures we investigated the resistance and the mobility of the carrier. To calculate the total resistance of our devices, we performed linear fitting of the output graph at very low drain voltages (0 to 0.1 V) [10]. By maintaining a consistent overdrive voltage $V_G - V_T = 0.5$ V across all temperature points from 300 K to 4 K, we ensured comparability of the measurements. We extracted the contact resistance by the Y functional method by considering the mobility degradation for the external factors and the channel resistance is extracted by subtracting the contact resistance from the total resistance [11]. We also investigated carrier mobility across varying temperatures and plotted it alongside the resistance in Fig. 5. The effective mobility of the device is extracted using the following equation [12], where g_m represents the transconductance, C_{OX} denotes the gate capacitance, and W and L are the geometric parameters of the channel.

$$\mu = \frac{g_m \cdot L}{V_d \cdot W \cdot C_{OX}}, \text{ and } g_m = \frac{\partial I_D}{\partial V_G} \qquad (2)$$

This plot reveals that the initial variation in total resistance with temperature is not significantly impacted by contact resistance. Notably, the channel resistance decreases with decreasing temperature from 300K to 150 K, likely due to increased carrier mobility which in turn decreases the overall resistance. However, under 150K, the overall resistance starts to increase due to the dominant effect of coulomb scattering and the hindrance of carrier injection at the contact.

Interestingly, at temperatures below 77 K, the rate of increase in total resistance becomes significant, likely due to a substantial rise in contact resistance which becomes the dominant component. It has been observed from Fig. 3 that the variation in I_D at cryogenic temperatures (77 K to 4 K) is not significant at high drain voltages as the mobility of the carriers got saturated in this region. This phenomenon suggests that carriers still have sufficient energy to overcome the Schottky barrier height when the drain voltage is higher, and the contact resistance has a very negligible impact. However, we observe a slight non-linear behavior at low drain voltages below 77 K, likely due to reduced thermal energy or carrier freezeout effects, which result in increased contact resistance and greater difficulty in carrier injection.

IV. Conclusion

These findings highlight the potential of Schottky barrier engineering as well as the carrier injection mechanism in metal/2D-semiconductor interfaces at various temperatures. The variation in channel resistance with temperature is not very significant compared to overall resistance. However, contact resistance becomes a serious issue at low temperatures even if it is a n-doped MoS₂, which primarily affects carrier injection. This increased contact resistance at low temperatures also results in non-linear contact behavior. Future work should focus on optimizing the contact engineering process to achieve stable ohmic contacts across a wider temperature range, particularly at deep cryogenic temperatures. This study offers essential insights into carrier injection and the transition from linear to nonlinear behavior at deep cryogenic temperatures, shedding new light on metal-TMD interfaces.

Acknowledgment

This work was supported by the General Research Fund (GRF) from the Research Grants Council (RGC) of Hong Kong under grant 16203222. We acknowledge the Nanosystem Fabrication Facility (CWB) of the HKUST for the device/system fabrication.

References

[1] A. Rai, H. C. P. Movva, A. Roy, D. Taneja, S. Chowdhury, and S. K. Banerjee, "Progress in Contact, Doping and Mobility Engineering of MoS2: An Atomically Thin 2D Semiconductor," *Crystals*, vol. 8, no. 8, Art. no. 8, Aug. 2018, doi: 10.3390/cryst8080316.

[2] W.-C. Cheng, X. Liu, and M. Chan, "Carrier Injection Mechanism of Metal-MoS2 Ohmic Contact in MoS2 FETs," in *2018 IEEE International Conference on Electron Devices and Solid State Circuits (EDSSC)*, Jun. 2018, pp. 1–2. doi: 10.1109/EDSSC.2018.8487111.

[3] "A Fermi-Level-Pinning-Free 1D Electrical Contact at the Intrinsic 2D MoS2–Metal Junction - Yang - 2019 - Advanced Materials - Wiley Online Library." Accessed: Mar. 09, 2024. [Online]. Available: https://onlinelibrary.wiley.com/doi/full/10.1002/adma.201808231

[4] X. Liu *et al.*, "Low temperature carrier transport study of monolayer MoS2 field effect transistors prepared by chemical vapor deposition under an atmospheric pressure," *J. Appl. Phys.*, vol. 118, no. 12, p. 124506, Sep. 2015, doi: 10.1063/1.4931617.

[5] P.-C. Shen *et al.*, "Ultralow contact resistance between semimetal and monolayer semiconductors," *Nature*, vol. 593, no. 7858, pp. 211–217, May 2021, doi: 10.1038/s41586-021-03472-9.

[6] X. Zhang, L. Zhang, and M. Chan, "Doping enhanced barrier lowering in graphene-silicon junctions," *Appl. Phys. Lett.*, vol. 108, no. 26, p. 263502, Jun. 2016, doi: 10.1063/1.4954799.

[7] Z. Ma, L. Zhang, C. Zhou, and M. Chan, "High Current Nb-Doped P-Channel MoS₂ Field-Effect Transistor Using Pt Contact," *IEEE Electron Device Lett.*, vol. 42, no. 3, pp. 343–346, Mar. 2021, doi: 10.1109/LED.2021.3056178.

[8] S. Das *et al.*, "Transistors based on two-dimensional materials for future integrated circuits," *Nat. Electron.*, vol. 4, no. 11, pp. 786–799, Nov. 2021, doi: 10.1038/s41928-021-00670-1.

[9] S. B. Mitta *et al.*, "Electrical characterization of 2D materials-based field-effect transistors," *2D Mater.*, vol. 8, no. 1, p. 012002, Nov. 2020, doi: 10.1088/2053-1583/abc187.

[10] C. J. Estrada, Z. Ma, L. Zhang, and M. Chan, "Threshold Voltage Model for 2-D FETs With Undoped Body and Gated Source," *IEEE Trans. Electron Devices*, vol. 70, no. 5, pp. 2575–2580, May 2023, doi: 10.1109/TED.2023.3255833.

[11] F. Liao *et al.*, "MoS2 dual-gate transistors with electrostatically doped contacts," *Nano Res.*, vol. 12, no. 10, pp. 2515–2519, Oct. 2019, doi: 10.1007/s12274-019-2478-5.

[12] C. J. Estrada, Z. Ma, and M. Chan, "Complementary Two-Dimensional (2-D) FET Technology With MoS₂/hBN/Graphene Stack," *IEEE Electron Device Lett.*, vol. 42, no. 12, pp. 1890–1893, Dec. 2021, doi: 10.1109/LED.2021.3124823.

979-8-3503-6184-1/24 $31.00 © 2024 IEEE

Experimental Verification of 1D Transport Model by Quantized Current Spectrum of Si JNT Device

Zi-Meng Shang[1,2], Bo-Wei Wang[1,2], Wei-Hua Han*[1,2]

[1] Engineering Research Center of Semiconductor Integrated Technology, Institute of Semiconductors,
Chinese Academy of Sciences, Beijing 100083, P. R. China
[2] Center of Materials Science and Optoelectronics Engineering, University of Chinese Academy of Sciences,
Beijing 100049, P. R. China
*Corresponding author's Email: weihua@semi.ac.cn

Abstract — The nonlinear electron transport in silicon junctionless nanowire transistor (JNT) was characterized by the one-dimensional (1D) transport spectrum at the temperature of 8 K. We demonstrate the essential physical picture of the steady-state 1D current-voltage curves, which are relevant to the operating mechanism under controlling of the gate voltage and source-drain bias voltage. The results show that the nonlinear conductance plateaus are quantized by different heights due to $\Delta 2$ and $\Delta 4$ valley splitting in the linear-response regime as the source-drain bias voltage increases.

Keywords — *silicon junctionless nanowire transistor, one-dimensional transport, operating mechanism, subband spacing*

I. INTRODUCTION

As the feature size of transistors continues to shrink, the quantum effect caused by the channel size constraint becomes more obvious, leading to the vigorous development of quantum devices. Junctionless nanowire transistors (JNTs), in which the channel is formed along the center axis [1], are attractive for studying quantum properties. Compared with conventional devices, the channel energy levels in quantum devices are discrete due to quantum confinement effects, resulting in subband splitting [2]. It is represented as current steps and differential conductance oscillations in transfer and output curves [3]. The electrical characteristics can be regulated by the gate voltage and source-drain bias voltage respectively.

Herein, we fabricated an n-type silicon JNT device with an 18 nm-diameter channel and observed the one-dimensional transport spectrum. We demonstrate the essential physical picture of the steady-state current-voltage curves, which are relevant to the operating mechanism under controlling of the gate voltage and source-drain bias voltage. The results show that the nonlinear conductance plateaus are quantized by different heights due to $\Delta 2$ and $\Delta 4$ valley splitting in the linear-response regime as the source-drain bias voltage is increased.

II. EXPERIMENT

Fig. 1(a) shows the structural diagram of silicon JNT device, which is fabricated on a (100) silicon-on-insulator (SOI) wafer with 55 nm thick top silicon layer. In order to reduce interface damage, the top silicon is thermally oxidized for 40 minutes to form a buffer layer. The SOI wafer was then uniformly doped and rapidly annealed to reach the concentration of 2×10^{18} cm^{-3}. The desired nanowire pattern is transferred to the oxide layer of substrate by electron beam lithography and inductively coupled plasma (ICP) etching. Then we obtained the 25 nm-diameter silicon nanowires with flat edges by silicon ICP etching and again reduced the etching damage by dissolving the sacrificed oxide layer in a 5% HF solution. After thermally oxidizing the SOI wafer to form the gate oxide layer with 6 nm, we deposited 230 nm-thick polysilicon layer. The polysilicon gate with the length of 485 nm was prepared in a similar manner using EBL and ICP. The top-view SEM image of the nanowire and gate structure is shown in Fig.1(b). Lastly, the device was finished after evaporating Ni/Al and forming contact pads by conventional lithography and lift-off process.

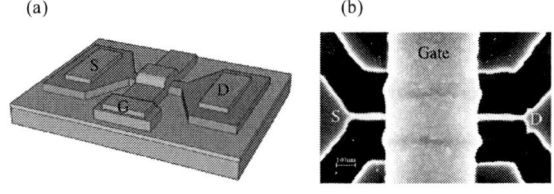

Figure 1. (a) Structural diagram and (b) top-view SEM image of silicon JNT device.

III. RESULT AND DISCUSSION

Fig. 2 shows the transfer characteristics under different source-drain bias voltages at the temperature of 8 K. The potential barrier of the channel decreases as the gate voltage increases. Three operation mechanisms including subthreshold, saturation and linear regimes are formed successively, as shown in the inset of Fig. 2(a). We define $V_{g\text{-sat}}$ as the gate voltage entering saturation state, which is equal to the summation of source-drain bias voltage V_{ds} and threshold voltage V_{th}, i.e. $V_{g\text{-sat}} = V_{ds} + V_{th}$.

The threshold voltage can be extracted by linear extrapolation [4] in transconductance curves. The threshold voltage decreases slightly with the increase of the source-drain bias voltage, which originates from drain induced barrier lowering (DIBL) effect. The source-drain bias voltage, threshold voltage and saturation gate voltage are displayed in table 1. From both drain current curves and transconductance curves, the saturation gate voltage $V_{g\text{-sat}}$, which is shown by the red dash line in Fig. 2, tends to move to the right side with the increasing V_{ds} bias voltage. When $V_{gs} < V_{th}$, the conductive channel has not yet formed, so the device is off. When $0 < V_{gs} - V_{th} < V_{ds}$, the device operates in the saturation regime. The Fermi energy level of channel $E_{channel}$ is between source Fermi level E_{fs} and drain Fermi level E_{fd}. The drain current is saturated due to the energy level window between E_{fs} and $E_{channel}$. When $0 < V_{ds} < V_{gs} - V_{th}$, the device operates in the linear region and the drain current increases linearly. All the subbands in E_{fs}-$E_{channel}$ level window can participate in transport of electron. Morever, the conductive channel of JNTs is formed along

This work was supported by the National Natural Science Foundation of China under Grant No.62374157.

979-8-3503-6184-1/24 $31.00 © 2024 IEEE

the center axis, and the effective channel width increases as the gate voltage increases until the effective width is comparable to the physical scale of the nanowires.

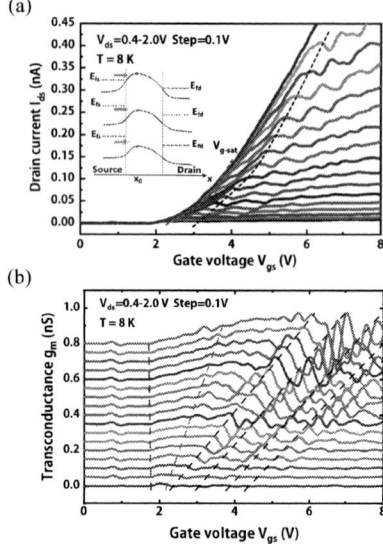

Figure 2. (a) Transfer characteristic curves under the source-drain bias voltages from 0.4 V to 2.0 V at a temperature of 8 K. The inset shows energy band diagram for the three operation regimes varying with gate voltage. (b) Transconductance characteristics at the temperature T=8K.

TABLE I.   THE THRESHOLD VOLTAGE AND SATURATION GATE VOLTAGE WITH DIFFERENT SOURCE-DRAIN BIAS VOLTAGE

$V_{ds}(V)$	$V_{th}(V)$	$V_{g\text{-}sat}(V)$
0.40	1.800	2.200
0.50	1.792	2.292
0.60	1.784	2.384
0.70	1.776	2.476
0.80	1.776	2.576
0.90	1.760	2.66
1.00	1.760	2.76
1.10	1.752	2.852
1.20	1.752	2.952
1.30	1.744	3.044
1.40	1.744	3.144
1.50	1.744	3.244
1.60	1.736	3.336
1.70	1.736	3.436
1.80	1.736	3.536
1.90	1.728	3.628
2.00	1.728	3.728

The black dash line in Fig. 2(a) distinguishes the drain current, for which the effective channel width is equal to the physical size. A wide effective channel will lead to a small subband energy level spacing which may increase the number of subbands involved in electron transport. When the scale of conducting channel reaches that of the physical channel, we can see groups of current oscillations (the right side of black dash line in Fig. 2(a)) and transconductance peaks (black dash line in Fig. 2(b)), which indicate electron transport behavior in one-dimensional subbands. The

subband spacing can be calculated using the experimental formula:

$$\Delta E_{\exp} = \frac{\pi \hbar^2 C_{ox}}{2m^* e} \cdot \Delta V_{gs} \qquad (1)$$

where ΔE_{exp} represents the spacing of discrete subband level, the gate capacitance per unit area C_{ox} is about 5.74×10^{-3} F/m² with 6 nm-thick oxide layer. Thus, we can calculate the ΔE_{exp} is 17.03 meV, 16.85 meV, 10.78 meV respectively according to the gate voltage spacing ΔV_{gs} with the measured data of 0.75V, 0.77V, 0.44V.

The subband spacing can also be calculated using the theoretical formula:

$$E_{m,n} = \frac{\hbar^2 \pi^2}{2m^*} \cdot \left(\frac{m^2}{l_x^2} + \frac{n^2}{l_y^2} \right) \qquad (2)$$

where $E_{m,n}$ represents the discrete level of the subband, l_x and l_y represent the width and height of the nanowires, the effective mass of electron $m^* = 0.19m_0$, m or n is 1, 2, ···. Here, the diameter of silicon nanowire is about 18 nm. Assuming that $l_x = l_y = 18$ nm, we can figure out that the theoretical ΔE_{theory} is 18.21meV, 18.21meV, 12.14meV, respectively. Therefore, the two sets of data ΔE_{exp} and ΔE_{theory} are consistent, showing that carrier transport depends on subbands under one-dimensional quantum confinements.

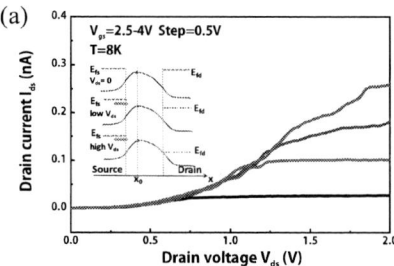

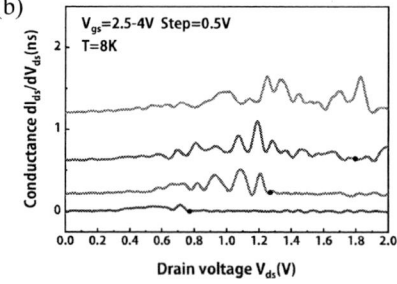

Figure 3. (a) The output characteristics of the device at a temperature of 8K. The inset shows energy band for low and high V_{ds}. (b) The differential conductance characteristics at T=8K.

Fig. 3 shows the quantized drain current in the output characteristics of the device at the temperature of 8 K. The gate potential barrier is stable at a constant gate voltage setup that V_{gs} is higher than V_{th}. With the increasing V_{ds}, the Fermi energy level of drain E_{fd} gradually declines to make the drain current I_{ds} go through subthreshold, linear and saturation regimes as shown in the inset of Fig. 3(a). At lower V_{ds}, the device operates in the linear regime, in which the gate potential is below E_{fs} and E_{fd}. The subband successively participates in the electron transport through the widening E_{fs}-$E_{channel}$ window. In Fig. 3(b), we can observe the oscillations of differential conductance curves damping in a

979-8-3503-6184-1/24 $31.00 © 2024 IEEE   320

similar way under different V_{gs}. It results from the drain current saturation when the drain Fermi level is aligned with the top of gate potential barrier. We define the saturation source-drain bias voltage as $V_{d\text{-sat}}$, at which the drain current I_{ds} starts approaching a stable state in the output characteristic curves. The first subband of the channel is always aligned with E_{fd}. The saturation bias voltage $V_{d\text{-sat}}$ tends to move to the right side with the increasing gate voltages V_{gs}. The number of subbands participating in the electron transport will increase due to the gate potential barrier decreasing at the same time. The saturation bias voltages $V_{d\text{-sat}}$ at different gate voltages are represented to be 0.78 V, 1.25 V, 1.77 V ⋯by the black dot signs in Fig.3(b). The read data from output characteristic curves are consistent with those from transfer characteristic curves in Table 1. It indicates that the saturation voltages regulated by V_{gs} and V_{ds} can be well matched.

Fig. 4(a) shows a schematic energy range diagram for subband current flow through linear and saturation regimes. In order to quantify the energy range, we provide the V-shape diagram of transconductance with source-drain bias voltages and gate voltages in Fig. 4(b). The decrease of drain Fermi level ΔE_{fd} with the increasing bias voltage V_{ds} can be calculated by the formula

$$\Delta E_{fd} = \alpha \cdot (1 - \beta) \cdot \Delta V_{ds}, \quad (3)$$

where $\alpha = C_g / C_\Sigma$ is the coupling parameter of gate capacitance and β is barrier shape parameter [5]. According to the slopes $k_1 = C_g / (C_g + C_d) = 0.332$ and $k_2 = -C_g / C_s = -21.59$ in the V-shape diagram, we can obtain the value $\alpha = 0.33$. The barrier shape parameter $\beta = 0.75$, which is gotten by the ratio of the gate-drain distance to the total channel length. Fig. 4 (c) presents the magnified output characteristic curve at $V_{gs} = 3$ V. Each current step corresponds to an additional subband involving in electron transport. The internals of bias voltage ΔV_{ds} for each current step are 0.168 V, 0.080 V, 0.152 V, 0.140 V, 0.100 V, respectively corresponding to ΔE_{fd} = 13.86 meV, 6.6 meV, 12.54 meV, 11.55 meV, 8.25 meV.

Furthermore, we clearly observed the heights of those current steps are different, which originates from the sixfold degenerate Δ valleys splitting into two sets of subbands with degeneracy of 2 ($E^{\Delta 2}$) and 4 ($E^{\Delta 4}$) [6]. The degenerate subband spacing can be calculated theoretically by the equation (2). Here, the effective channel diameter is about 12 nm according to the linearly approximate expression $ch_{eff} / ch_{phy} = V_{gs}^{ch_{eff}} / V_{gs}^{ch_{phy}}$, where ch_{eff} is the channel effective size and ch_{phy} is the channel physical size. The electron effective mass of the $E^{\Delta 2}$ subband is equal to the transverse effective mass m_t (=0.19m_0) of the electron, while the electron effective mass of the $E^{\Delta 4}$ subband is related to both the transverse effective mass m_t and the longitudinal effective mass m_l (=0.98m_0) and is equal to (m_t+m_l)/2 [7]. Because of the difference in effective mass, the energy level interval of the $E^{\Delta 4}$ subband is smaller than that of the $E^{\Delta 2}$ subband, as shown in the inset of Fig.4(c). Thus, we can figure out that $E_{1,1}^{\Delta 2}$ = 27.32 meV, $E_{1,2}^{\Delta 2}$= 68.30 meV and $E_{1,1}^{\Delta 4}$ = 8.88 meV, $E_{1,2}^{\Delta 4}$ = 22.20 meV, $E_{2,2}^{\Delta 4}$ = 35.52 meV, $E_{1,3}^{\Delta 4}$ = 44.40 meV, $E_{2,3}^{\Delta 4}$ = 54.72 meV. These subbands with different degeneracy can be arranged in order $E_{1,1}^{\Delta 4}$, $E_{1,2}^{\Delta 4}$, $E_{1,1}^{\Delta 2}$,

$E_{2,2}^{\Delta 4}$, $E_{1,3}^{\Delta 4}$, $E_{2,3}^{\Delta 4}$ and marked by arrows in Fig. 4(c). Then we can calculate the degenerated subband spacing is 13.32 meV, 5.12 meV, 8.20 meV, 8.88 meV, 10.32 meV, respectively. The degenerated subband spacing is in agreement with the experimental spacing of drain Fermi level we calculated before.

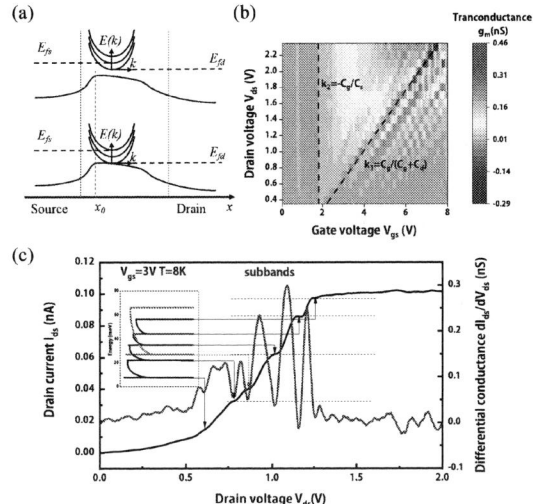

Figure 4. (a) Physical diagram of 1D transport with subbands at linear and saturation regimes. (b) The diamond diagram of transconductance with source-drain bias voltages and gate voltages. (c) Drain current and transconductance characteristics at the temperature of 8 K.

IV. SUMMARY

We fabricated silicon junctionless nanowire transistors with 18 nm-diameter channel on SOI substrates. We experimentally verified 1D transport model by quantized current spectrum of silicon JNT device. Through the energy level diagram, we analyze the operating mechanism under regulation by gate voltage and source-drain bias voltage. By calculating the theoretical and experimental energy level spacing of subbands, we confirmed that the different current steps and differential conductance oscillations in the linear regime originate from 1D transport through the subbands of splitting degenerate valleys.

[1] J. P. Colinge, C. W. Lee, A. Afzalian, N. D. Akhavan, R. Yan, I. Ferain *et al.*, "Nanowire transistors without junctions," *Nature Nanotechnology*, vol. 5, pp. 225-229, March 2010.

[2] S. C. Rustagi, N. Singh, Y. F. Lim, G. Zhang, S. Wang, G. Q. Lo *et al.*, "Low-Temperature Transport Characteristics and Quantum-Confinement Effects in Gate-All-Around Si-Nanowire N-MOSFET," *IEEE Electron Device Letters*, vol. 28, pp. 909-912, October 2007.

[3] J. T. Park, J. Y. Kim, C. W. Lee, and J. P. Colinge, "Low-temperature conductance oscillations in junctionless nanowire transistors," *Applied Physics Letters*, vol. 97, pp. 172101, October 2010.

[4] A. Ortiz Conde, F. J. García Sánchez, J. J. Liou, A. Cerdeira, M. Estrada, and Y. Yue, "A review of recent MOSFET threshold voltage extraction methods," *Microelectronics Reliability*, vol. 42, pp. 583-596, April 2002.

[5] H. Kothari, A. Ramamoorthy, R. Akis, S. M. Goodnick, D. K. Ferry, J. L. Reno *et al.*, "Linear and nonlinear conductance of ballistic quantum wires with hybrid confinement," *Journal of Applied Physics*, vol. 103, pp. 013701, January 2008.

[6] A. Rahman, M. S. Lundstrom, and A. W. Ghosh, "Generalized effective-mass approach for n-type metal-oxide-semiconductor field-effect transistors on arbitrarily oriented wafers," *Journal of Applied Physics*, vol. 97, February 2005.

[7] R. Kim, and M. S. Lundstrom, "Characteristic Features of 1-D Ballistic Transport in Nanowire MOSFETs," *IEEE Transactions on Nanotechnology*, vol. 7, pp. 787-794, November 2008.

Impact of Gate Overlap Length Modulation on Electrical Characteristics and Subthreshold Swing in Nanosheet TFETs with Varying Tunneling Mechanisms

Zheng-Hong Zhong, Wei-Heng Tai, Jyi-Tsong Lin*

Department of Electrical Engineering, National Sun Yat-Sen University, Kaohsiung 80424, Taiwan, ROC

* Email: jtlin@ee.nsysu.edu.tw

Abstract— This paper explores the impact of gate (overlap) length modulation on the electrical characteristics and subthreshold swing of nanosheet TFETs, using Sentaurus 3D TCAD simulations. We compare conventional point tunneling TFETs (NS-TFET) with line tunneling TFETs (NS-LTFET) under various gate lengths (L_G) and overlap lengths (L_{OV}). Our results demonstrate that the NS-LTFET, dominated by line tunneling, achieves higher I_{ON} and superior subthreshold characteristics compared to the NS-TFET. Increasing L_{OV} in NS-LTFETs significantly enhances the ON-state current and reduces the subthreshold swing, highlighting the potential of line tunneling structures for future low-power applications.

Keywords—Nanosheet, Tunnel Field-Effect Transistor (TFET), Line Tunneling, Point Tunneling, Subthreshold Swing (SS), Sentaurus TCAD, Overlap Length.

I. INTRODUCTION

As device dimensions continue to shrink, maintaining the validity of Moore's Law has driven rapid advancements in electronic device technology. Traditional Metal-Oxide-Semiconductor Field-Effect Transistors (MOSFETs) have long been the backbone of the semiconductor industry due to their mature fabrication techniques and excellent performance characteristics. However, as MOSFET dimensions reach nanoscale critical dimensions, they face numerous challenges, including severe short channel effects (SCEs), increased leakage current, velocity saturation, and drain-induced barrier lowering (DIBL) [1]. These issues degrade device performance and increase power consumption.

To address these challenges, researchers have explored alternative devices offering better channel control and reduced leakage current. The Tunnel Field-Effect Transistor (TFET) has emerged as a promising candidate. Unlike MOSFETs, which rely on thermally excited carriers [2], TFETs use band-to-band tunneling (BTBT) as the primary carrier injection mechanism, enabling them to achieve sub-60 mV/decade subthreshold swings (*SS*) and making them highly suitable for low-power applications[3].

However, TFETs face numerous challenges in development and design due to various non-ideal effects in practical operation and increased fabrication costs. Early TFET designs suffered from low ON-state current (I_{ON}) due to limited tunneling regions and high tunneling resistance. Defects in semiconductor materials could also promote Trap-Assisted Tunneling (TAT), leading to increased leakage current [4]. Additionally, the ambipolar effect could further increase leakage current and reduce the ON/OFF current ratio [5]. These issues result in subpar subthreshold swing and overall performance of TFETs at room temperature. Recent advancements in TFET research have focused on overcoming these challenges through various innovative approaches.

One key to enhancing TFET performance is the introduction of heteromaterial structures, particularly using narrow bandgap materials for the source and wide bandgap materials for the drain [6]. This approach shortens the tunneling distance, enhances band-to-band tunneling efficiency, and reduces leakage current. In traditional TFETs, band-to-band tunneling (BTBT) mainly occurs at the high electric field region between the source and channel, known as point tunneling. This type of tunneling is independent of the gate electric field direction and is deeply influenced by the local electric field. However, the electric field within the channel is typically non-uniform, leading to a smaller tunneling region and significantly limiting the switching performance of the device. In contrast, line tunneling, which aligns with the gate electric field direction and is perpendicular to the channel, has a tunneling carrier quantity proportional to the effective length of the overlap between the source and the channel. This provides a controllable and scalable tunneling region [5]. Further optimization of the I_{ON} and switching performance (I_{ON}/I_{OFF}) allows TFETs to overcome existing performance limitations and achieve more efficient applications.

In this paper, we utilize Sentaurus 3D TCAD for simulations, focusing on optimizing I_{ON} and subthreshold characteristics. We discuss the potential of structural modulation of TFETs under different tunneling mechanisms. We also combine line tunneling TFETs with GAA nanosheet structures, demonstrating their structural potential and developmental prospects.

II. CALIBRATION AND PROPOSED DEVICE STRUCTURE

A. Simulation Setup and Calibration

Fig 1 shows the calibration results of the simulation. We selected the Heterojunction GAA TFET by R. Rooyackers *et al.* [7] for calibration and utilized the Dynamic non-local path band-to-band tunneling (BTBT) model to calculate the tunneling behavior. Trap-assisted tunneling (TAT) has been activated by the Shockley-Read-Hall (SRH) recombination model. Additionally, the Thin-Layer Mobility model has been enabled to ensure high accuracy and reliability when simulating semiconductor devices with nanoscale thickness. To account for band gap narrowing in semiconductor materials at high doping concentrations or high carrier concentrations, the Band Gap Narrowing model was adopted. Auger recombination, Fermi statistics, and the Density Gradient Quantum Correction model were also incorporated to achieve more precise simulation results.

979-8-3503-6184-1/24 $31.00 © 2024 IEEE

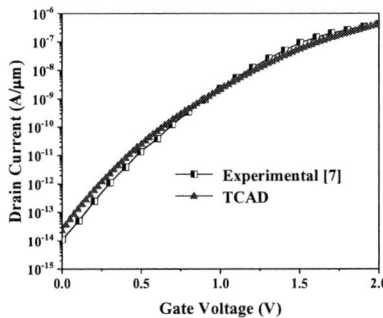

Fig 1. Calibration results of Si/Ge heterojunction GAA TFET.

B. Proposed Devise Structures

Fig 2 illustrates our proposed nanosheet TFET structure. We use Germanium as the source material and Silicon for the channel and drain. As previously mentioned in the introduction, using a narrow bandgap material in the TFET source can achieve larger tunneling currents. Fig 2 (a) and Fig 2 (c) depict the conventional point tunneling Nanosheet TFET (NS-TFET). Fig 2 (b) and Fig 2 (d) show the Nanosheet Line Tunneling TFET (NS-LTFET) with current characteristics entirely dominated by line tunneling. In the complete line tunneling structure, the source region fully overlaps with the gate region, and an SiO2 isolation layer is used between the source and drain to prevent point tunneling. This design ensures that tunneling occurs only in the gate-controlled region, thereby enhancing device performance. Both structures have a gate length of 50 nm and a gate work function of 4.15 eV.

Fig 3 presents the transfer characteristics of the NS-TFET and NS-LTFET. It can be observed that, under the same drive voltage ($V_{DS} = 0.4$ V), the NS-LTFET, with line tunneling-dominated current characteristics, exhibits a higher I_{ON} and a lower average subthreshold swing (SS_{AVG}). This indicates superior switching characteristics.

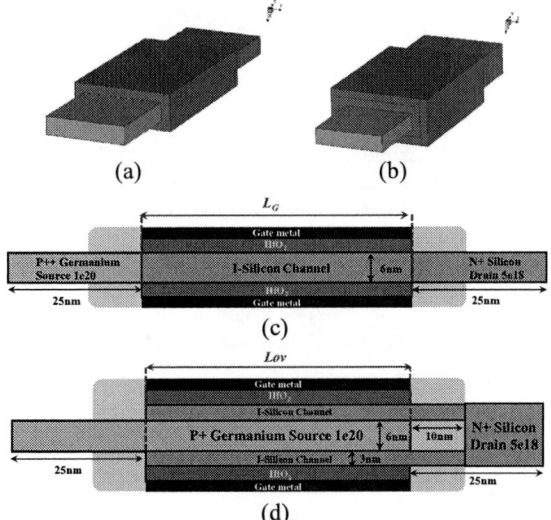

(a)   (b)

(c)

(d)

Fig 2. (a) 3D-TCAD of NS-TFET, (b) 3D-TCAD of NS-LTFET, (c) Cross-section of the NS-TFET, (d) Cross-section of the NS-LTFET.

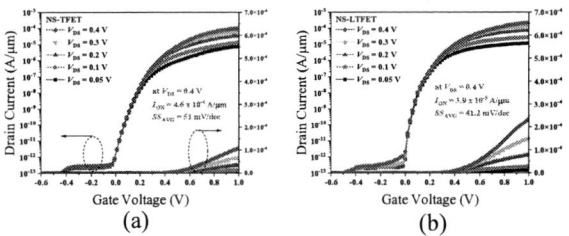

Fig 3. (a) Transfer characteristics of the NS-TFET, (b) Transfer characteristics of the NS-LTFET.

III. RESULTS AND DISCUSSION

A. Impact of Gate and Channel Length on Band to Band Generation Behavior

In this section, we define the gate length of the NS-TFET and the intrinsic channel as L_G, and the overlap length of the gate and the Germanium source in the NS-LTFET as L_{OV}. L_G and L_{OV} are illustrated in Fig 2 (c) and (d). We adjusted these lengths and compared the band-to-band generation distribution characteristics of the two device structures.

As shown in Fig 4 (a), with an increase in L_G, the distance between the source and drain increases, causing the tunneling carriers in the NS-TFET to become more concentrated within the channel. This enhances the gate's control over the carriers in the channel. As a point tunneling device, the tunneling region of the NS-TFET is located at the heterojunction interface and the high electric field area of the gate. This characteristic remains unchanged with variations in L_G, as observed in the figure.

In the NS-LTFET shown in Fig 4 (b), as L_{OV} increases from 30 nm to 50 nm, the primary band-to-band generation region in the channel extends, allowing for more tunneling carriers. In other words, in the NS-LTFET, which utilizes complete line tunneling, the number of tunneling carriers is proportional to the overlap region L_{OV} between the gate and the source in the channel. The generation of more tunneling carriers per unit time also implies a higher on-state current and superior subthreshold characteristics.

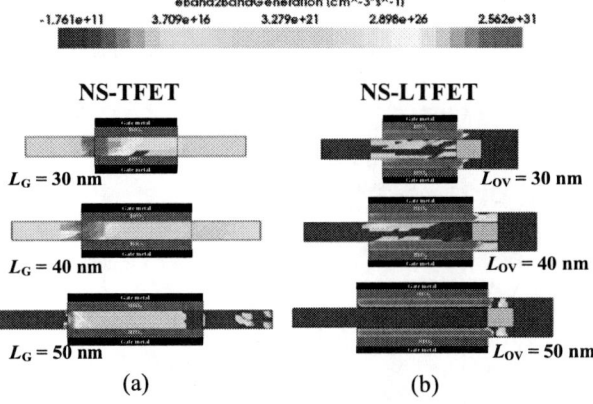

Fig 4. (a) The band-to-band generation distribution of NS-TFET with varying L_G, (b) The band-to-band generation distribution of NS-LTFET with varying L_{OV}.

979-8-3503-6184-1/24 $31.00 © 2024 IEEE   323

B. Impact of Gate and Channel Lengths on the Comparison of Different Tunneling Mechanisms

Fig 5 shows the transfer characteristics of the NS-TFET when adjusting L_G from 30 nm to 50 nm at $V_{DS} = 0.4$ V. As seen in the figure, as L_G increases, both I_{ON} and I_{OFF} of the NS-TFET slightly increase, but the change is minimal, and the subthreshold characteristics remain almost unchanged. Fig 6 illustrates the transfer characteristics of the NS-LTFET when adjusting L_{OV} from 30 nm to 50 nm at $V_{DS} = 0.4$ V. The figure shows that the length of L_{OV} significantly affects the performance of the NS-LTFET. Both I_{ON} and I_{OFF} of the NS-LTFET increase substantially, with the changes being much more pronounced than in the NS-TFET. The subthreshold characteristics also improve with the increase in L_{OV}. It is evident from Fig 6 that the longer the L_{OV}, the steeper the slope of the transfer curve, indicating a lower subthreshold swing.

In summary, the NS-TFET, which is dominated by point tunneling, and the NS-LTFET, which is dominated by line tunneling, exhibit significant differences in their response to changes in device length. Due to the localized tunneling nature of point tunneling, adjusting the device length has a limited impact on the current increase in the NS-TFET. In contrast, the tunneling amount and current magnitude in the line tunneling-dominated NS-LTFET are proportional to the device length. As shown in Fig. 8(a), the I_{ON} of the NS-LTFET increases significantly with longer gate length compared to the NS-TFET. When L_{OV} increases from 30 nm to 50 nm, the I_{ON} of the line tunneling structure increases by 884%, with no significant change in I_{OFF}, as shown in Fig. 8(b). This makes the line tunneling structure highly advantageous for increasing I_{ON}, as its tunneling current is proportional to the overlap region between the source and gate within the channel. Therefore, the on-state current can be increased by extending the length and width of the channel in the nanosheet line tunneling TFET structure.

Table I presents the average subthreshold swing (SS_{AVG}) of the NS-TFET and NS-LTFET under different L_G, L_{OV} values. It can be observed that as L_G increases, the SS_{AVG} of the NS-TFET remains almost unchanged. In contrast, the NS-LTFET achieves a lower SS_{AVG} as L_{OV} increases. This indicates that line tunneling TFETs not only have an advantage in on-state current but also possess greater potential for optimization in subthreshold characteristics.

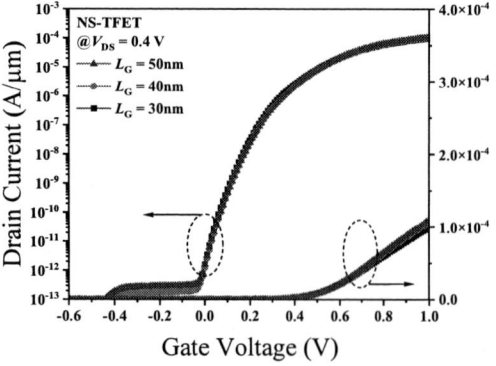

Fig 5. Transfer characteristics of the NS-TFET with different L_G.

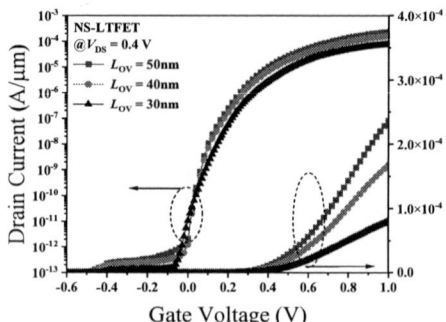

Fig 6. Transfer characteristics of the NS-LTFET with different L_{OV}.

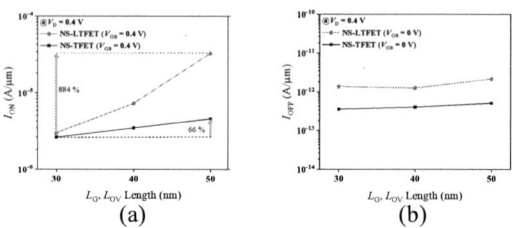

Fig 7. (a) The impact of varying L_G, L_{OV} on the ON-state current of NS-TFET and NS-LTFET, (b) The impact of varying L_G, L_{OV} on the OFF-state current of NS-TFET and NS-LTFET.

Table I
SS_{AVG} OF NS-TFET AND NS-LTFET WITH DIFFERENT L_G; L_{OV}

L_G; L_{OV}	30 nm	40 nm	50 nm
SS_{AVG} (NS-TFET)	51.7 mV/dec	51.5 mV/dec	51 mV/dec
SS_{AVG} (NS-LTFET)	50.4 mV/dec	44.5 mV/dec	41.2 mV/dec

IV. CONCLUSION

This study examines the effects of gate length (L_G) and overlap length (L_{OV}) on nanosheet TFETs using Sentaurus 3D TCAD simulations. Nanosheet Line tunneling TFETs (NS-LTFET) show superior performance compared to traditional point tunneling Nanosheet TFETs (NS-TFET), with significant improvements in ON-state current (I_{ON}) and subthreshold swing (SS) as L_{OV} increases. These results highlight the advantages of Nanosheet line tunneling structures for future low-power electronics, providing a foundation for optimizing TFET designs for enhanced efficiency and scalability.

REFERENCES

[1] J. Wu *et al.*, IEEE Transactions on Electron Devices, pp. 3019-3024, 2015.
[2] Y. H. Liao *et al.*, IEEE Electron Device Letters, pp. 1860-1863, 2019.
[3] A. M. Ionescu *et al.*, Nature, pp. 329-337, 2011.
[4] K. S. Singh *et al.*, IEEE Transactions on Device and Materials Reliability, pp. 404-412, 2020.
[5] J. T. Lin *et al.*, IEEE Transactions on Electron Devices, pp. 6049-6056, 2023.
[6] T. Krishnamohan *et al.*, in 2008 IEEE International Electron Devices Meeting, 2008.
[7] R. Rooyackers *et al.*, in 2013 IEEE International Electron Devices Meeting, 2013.

979-8-3503-6184-1/24 $31.00 © 2024 IEEE

Recent Progress in the Development of Complementary Field-Effect Transistors

Mansun Chan[1] and Shengdong Zhang[2]

[1]Department of Electronic and Computer Engineering, The Hong Kong University of Science and Technology,
Clear water bay, Kowloon, Hong Kong
[2]School of Integrated Circuits, Peking University, Beijing 100871, China

Email: mchan@ust.hk, zhangsd@pku.edu.cn

Abstract— **The continuous scaling of semiconductor devices has driven significant advancements in the industry, but also introduced challenges such as short channel effects and higher metal resistance. This paper explores the development of Complementary Field-Effect Transistors (CFETs), which stack n-type and p-type FETs to enhance performance and reduce power consumption. The historical evolution, technical challenges, and future potential of CFETs are discussed, highlighting innovative processes like back-side power rail integration and the use of 2D materials for improved device efficiency.**

Keywords — 2D transistor, CFET, stacked transistors, nano-sheet (key words)

I. INTRODUCTION

The continuous scaling of devices has significantly contributed to the rapid growth and innovation in the semiconductor industry. However, as technology progresses with each new process node, the challenges associated with device scaling have become more apparent. The pursuit of reducing gate length and metal pitch, combined with an escalation in device density, and more effective interconnect has encountered challenges such as short channel effects, stronger parasitic effects, and higher metal resistance. To address these challenges and optimize overall circuit performance, the development of new device architectures that effectively balance these factors is essential [1].

The most advanced manufacturing solution to produce highly scaled device is the utilization of nanosheet transistors, which offer a wider channel compared to finFETs. However, these transistors face constraints due to the minimum pMOS/nMOS separation. Recent advancements in semiconductor technology have introduced stacked Complementary Field Effect Transistors (CFETs) [2] as a promising avenue for enhancing integrated circuit performance. These innovative transistor structures leverage the stacking of n-type and p-type FETs, resulting in a compact footprint and reduced power consumption. CFETs are at the forefront of addressing the scaling challenges of traditional CMOS technology, achieving superior switching speeds and drive currents while contributing to the miniaturization of electronic components and maintaining high efficiency.

The concept of CFETs is not new and was demonstrated by our team almost two decades ago using double silicon-on-insulator (SOI) layers [3,4]. While this early demonstration showcased the potential of CFETs, persistent technical

challenges continue to hinder their widespread manufacturing. In light of the renewed interest in CFETs, there exists an opportunity to integrate emerging processes such as back-side power rail to enhance the design optimization of CFETs. This paper delves into the historical perspective of the challenges and opportunities presented by these advanced CFET processes, leveraging our innovative contributions in this field while projecting their limitations and future potentials.

II. THE EARLY CFET ARCHITECTURE

While many researchers have compared CFET with 3D chip architecture, CFET is not strictly 3D because the nFET and pFET are tightly stacked on top of each other without any interconnect wires between them. Additionally, a single-gate feature is used to control both devices, aiming to reduce fabrication costs at the expense of layout flexibility. Therefore, it will primarily benefit circuits that require complementary logics such as inverters or SRAM cells. The potential advantages of CFET for generic functions like pass-transistor logic (PTL) have yet to be explored.

Earlier CFET designs employed a layer-by-layer approach where the bottom transistors were fabricated before stacking the top transistor on top of them. This technology has been utilized to construct very compact SRAM cells with excellent soft-error immunity. The main challenge of the layer-by-layer process lies in forming high-quality single crystal or large grain size polysilicon films for the upper layer transistor. Through techniques such as laser recrystallization and metal-induced lateral crystallization, the grain size of polysilicon can be significantly enhanced to achieve near single crystal quality. Stacked CFET circuits utilizing grain-enhanced films were first demonstrated, as shown in Fig. 1.

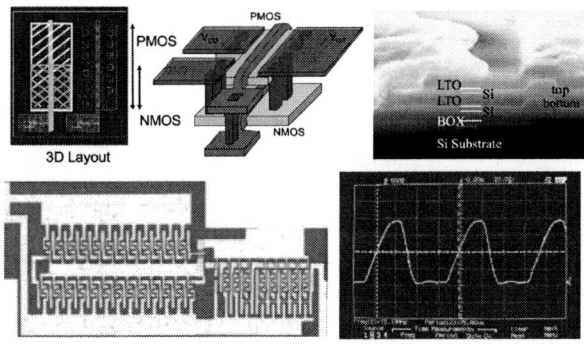

Fig. 1: Demonstration of stacked CMOS using SOI film as the bottom and re-crystallized large grain polysilicon as the top active layer including output of 25 stage inverter [7]

This work was supported by the General Research Fund (GRF) from the Research Grants Council (RGC) of Hong Kong under Grant number 16203222.

While using recrystallized film remains an option to form stacked CFET structures today with the introduction of new processing techniques and new materials such as IGZO [8], the performance of the upper layer transistors remains unsatisfactory for high-performance circuits. The first demonstration of a monolithic Stacked CFET structure with truly single-crystal silicon utilized both the top silicon film and the substrate of an SOI wafer to form the stacked transistors, as depicted in Fig. 2 [9]. This structure featured a double-gate pFET on top of a single-gate nFET to ensure matching output currents for transistors occupying the same footprint. The double-gated pFET resembled the structure of a nano-sheet transistor today.

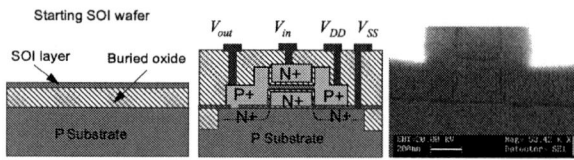

Fig. 2:CFET process using both top SOI film and bottom substrate to form the top and bottom transistors in the CFET stack [9]

The continuous scaling process necessitates the use of a double-gate or even multi-gate structure to control short channel effects. The planar nFET on the SOI substrate is no longer effective in managing leakage current for highly scaled transistors. This has prompted the need to form multiple stacked single crystal silicon films to develop the stacked CFET. The initial demonstration involved a double oxygen implant process to create double layers of SOI films [3, 4]. Stacked FinFETs were then fabricated on the double-layer SOI films, showcasing various circuit building blocks like inverters and logic gates, as illustrated in Fig. 3. This demonstration marks one of the earliest examples resembling the modern stacked CFET structures currently under development.

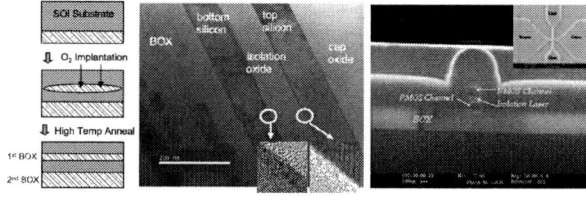

Fig. 3: Illustration on the formation of double-SOI film and the final structure of the CFET based on stacked FinFET based on [3, 4]

III. THE FORMATION OF MULTILAYER STACKED Si FILMS

To enhance the current density of nano-transistors, stacking nanowires or nano-sheets has become a crucial approach. The previously mentioned method, which involved multiple oxygen implants to generate two distinct single crystal silicon films, is insufficient for this purpose. Two innovative techniques have been introduced to produce stacked multiple single crystal silicon films. One approach involves utilizing an SiGe interleave structure, with the SiGe layer serving as a sacrificial layer [10]. By removing the SiGe films located between the silicon layers, isolated single crystal silicon films can be achieved. This method has now become the primary manufacturing process for stacked multilayer nanosheet transistors and the recently developed stacked CFET structures.

The second approach entails utilizing Inductive Coupled Plasma (ICP) dry etching followed by Bosch cycle treatment to create vertical silicon ridges with non-uniform cross-sections, as illustrated in Figure 4 [11]. This process has yielded highly favorable transistor characteristics [12]. However, in contrast to the multilayer Si/SiGe method, the Bosch etching process has limitations in controlling the lateral dimension and aspect ratio of the nanowires, thereby restricting its applicability in production.

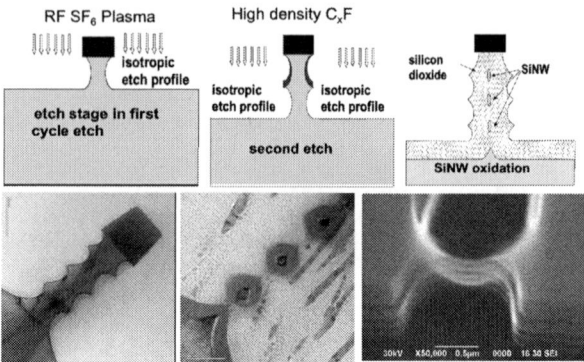

Fig. 4:The Bosch etching and stress limited oxidation process together with the SEM images of the cross-sections of the nanowires [12]

In contrast to the manufacturing process of multilayer nanosheet transistors, the Bosch process offers certain advantages for producing stacked CFET structures. This is because the latter necessitate an insulator to separate the top and bottom transistors, a task that proves challenging within the Si/SiGe multilayer structure. However, by employing Bosch etching on an SOI wafer, it becomes feasible to achieve two distinct layers of single crystal silicon films separated by an insulator, as depicted in Fig. 5.

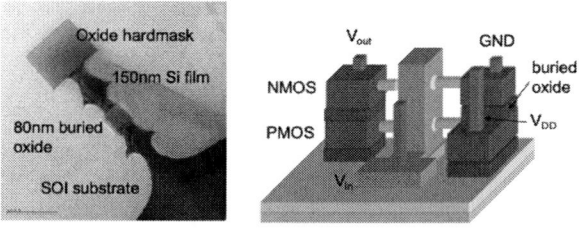

Fig. 5:The formation of distinct multilayer silicon film for constructing nanowire stacked CFETs.

A common method for fabricating stacked CFETs involves utilizing the Si/SiGe multilayer nano-sheet with SiGe as sacrificial layers. The process begins with constructing the channel region stack, where the thickness of the central SiGe layer is increased to create a wider gap between the upper and lower clusters of nanosheets. Following this, the channel region is defined with a dummy gate, and the area outside the channel region of the top FET is etched. The source and drain region of the top FET is removed, while the source/drain region of the bottom FET remains intact. Subsequently, a second spacer is employed to protect the top FET, and the source/drain region of the bottom FET is passivated. During this step, the passivating insulator also fills the void left by the thick middle SiGe layer, effectively separating the top and bottom FETs. Source/drain epitaxy is then conducted to grow the source/drain region of the top FET. The next steps involve removing the dummy gate to release the silicon channel by eliminating the SiGe interlayer. The actual gate is deposited in place of the dummy gate, with the choice of a single or dual gate material based on process complexity considerations. Finally, vias are opened, and metal is deposited to establish the interconnects. A simplified process flow is illustrated in Fig. 6 [14].

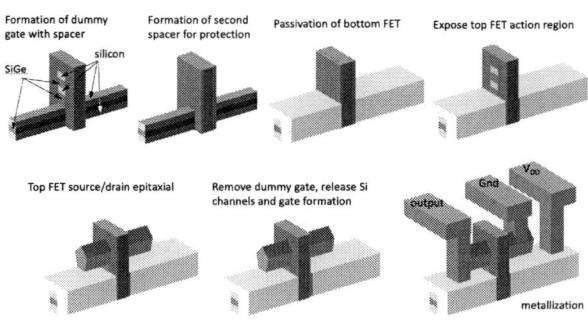

Fig. 6: CFET process with bottom FET passivation before S/D epitaxial of the top transistor [14]

The fabrication process described is not perfect with many intricate challenges. Initially, the top epitaxial source/drain regions often exhibit small and irregular shapes, leading to variations in contact formation. Additionally, achieving the accurate threshold voltage poses another obstacle, particularly due to the complexities involved in need to use distinct materials for the gate dielectric and material, forming different gate-stack configurations between the top and bottom FETs to ensure optimal performance and functionality.

IV. THE ROUTING OF STACKED CFETs

The conventional stacked CFETs were designed with an architecture where all the interconnects are positioned on top of the chips. To enable power supply and signal extraction from the bottom transistors, the nFET and pFET cannot be entirely stacked on top of each other. A via region connecting to the bottom FET is essential to enable the metal line to reach the bottom FET from the top, as depicted in Fig. 6. Even though the anticipated area savings compared to the planar structure are expected to exceed 50% due to the elimination of the space between the pFET and nFET, the actual area savings are slightly less than 50% due to the additional via landing area required for the bottom FET. Fig. 7 presents a comparison of simple logic circuits constructed using planar FinFET and stacked CFET, demonstrating the potential area reduction.

In order to fully leverage the benefits of the stacked CFET structure, recent research [16] indicates that 3D-stacked CMOS systems will require interconnects positioned both above and below the devices. This can be achieved through various methods. A recent advancement known as buried power rail technology involves relocating the interconnects responsible for supplying power to logic cells - but not involved in data transmission - to the silicon layer beneath the transistors. This approach can significantly reduce the primary overhead in standard cell design by eliminating the necessity for additional contact areas for power supply lines when stacking transistors. Ultimately, there is an expectation that signal lines can also be routed beneath the active devices, resulting in split structures with fully double-sided

interconnectivity. The cross-section of different CFET configurations with different interconnect schemes and their corresponding area savings is depicted in Fig. 8 to compare their compactness together with the estimated area saving.

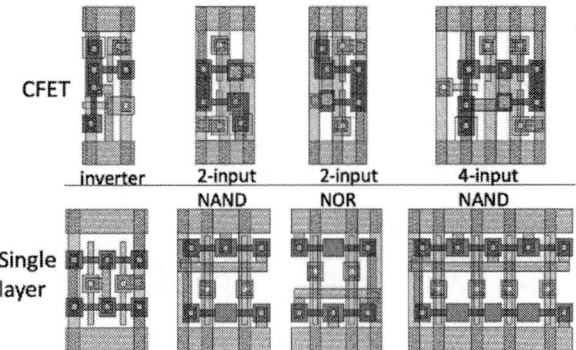

Fig. 7: Layouts of some of the share-gate single sided interconnect CFET standard cells compared with the conventional 2-D layout with a single layer of transistors [15]

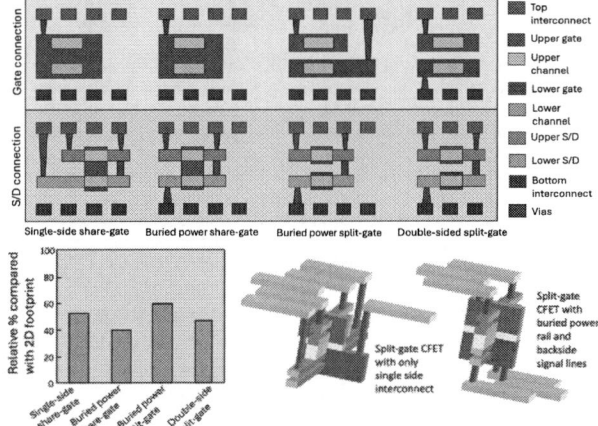

Fig. 8: Different CFET routing methods including single-sided shared-gate structure with buried power rail, split-gate structure with buried power rail and split-gate structure with fully double side interconnect. The bottom picture shows the relative area saving of different routing scheme compared with planar structure together with a 3D schematic of the typical structures.

V. THE NEXT GENERATION STACKED CFET

The semiconductor industry has identified 2D material as a potential alternative of replacing the Si channel for the nano-sheet transistor era. Anticipated superior electrostatic control and high channel mobility of the transition metal dichalcogenide (TMD) are expected to enhance the stacked CFET performance. More recently, TSMC has demonstrated a process to form stacked nano-sheets [17] even though transistor characteristics are only extracted from a single layer device. With the structural compatibility between the nanosheet transistors and stacked CFET, it is very natural to extend the nanosheet transistor structure to form stacked CFET structures. A possible process to fabricated stacked CFET with 2D channel material is shown in Fig. 9.

The fabrication of stacked CFETs utilizing 2D transistors presents several challenges that need to be addressed. Firstly, the lack of a reliable process to form high-quality monolayer 2D materials in a controlled location hinders the

979-8-3503-6184-1/24 $31.00 © 2024 IEEE

manufacturing process. Additionally, 2D materials lack out-of-plane bonds on their surface, complicating the deposition of high-k dielectrics. The amorphous nature of common high-k dielectrics may lead to poor interfaces with the 2D channel materials, impacting device performance. Furthermore, the impact of edge termination on the transistor performance remains inadequately studied, posing a potential obstacle. Lastly, establishing contact between the metal interconnect and the 2D material with low contact resistance is a challenging aspect of the fabrication process that requires further exploration and development.

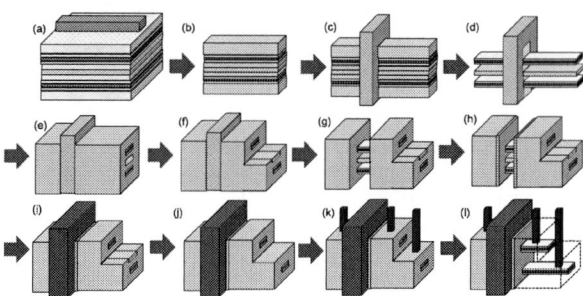

Fig. 9: Fabrication steps of a stacked CFET with 2D nano-sheet transistors including (a) formation of gate hardmask on the channel stacked materials with an isolation layer between the top and bottom FETs; (b) channel region etching; (c) formation of dummy gate; (d) 2D film release outside the gate stack; (e) passivation and support for the nanosheet; (f) expose the bottom S/D contact region; (g) removal of dummy gate; (h) inner gate sidewall formation; (i) gate electrode material deposition and definition; (j) removal of isolation layer on top of the bottom FET contact; (k) contact to the source and drain region; (i) a more detail view of the source/drain contact

The fabrication of stacked CFETs utilizing 2D transistors presents several challenges that need to be addressed. Firstly, the lack of a reliable process to form high-quality monolayer 2D materials in a controlled location hinders the manufacturing process. Additionally, 2D materials lack out-of-plane bonds on their surface, complicating the deposition of high-k dielectrics. The amorphous nature of common high-k dielectrics may lead to poor interfaces with the 2D channel materials, impacting device performance. Furthermore, the impact of edge termination on the transistor performance remains inadequately studied, posing a potential obstacle. Lastly, establishing contact between the metal interconnect and the 2D material with low contact resistance is a challenging aspect of the fabrication process that requires further exploration and development.

VI. Conclusion

In conclusion, the advancement of CFET technology holds the promise of transforming the semiconductor device landscape by extending Moore's Law and achieving high performance, low power consumption, and increased density. Although the fabrication process for CFETs remains complex, major technology firms have made significant strides and introduced their stacked CFET solutions [18-19]. Looking forward, exploring the fabrication of nano-sheet transistors

using 2D materials presents an intriguing research avenue. However, this direction also introduces new challenges, such as establishing proper contact with the source and drain regions and effectively passivating the edge termination of the materials employed. While obstacles persist, ongoing research and development in CFET technology utilizing 2D materials open up exciting possibilities for the future of semiconductor devices, potentially offering enhanced performance and more energy-efficient electronics.

References

[1] R. Chau, "Process and Packaging Innovations for Moore's Law Continuation and Beyond," 2019 IEDM, San Francisco, 2019, pp. 1.1.1-1.1.6, doi: 10.1109/IEDM19573.2019.8993462.

[2] J. Ryckaert et al., "The Complementary FET (CFET) for CMOS scaling beyond N3," 2018 IEEE Symposium on VLSI Technology, 2018, pp.141-142, doi: 10.1109/VLSIT.2018.8510618.

[3] X. Wu, P. C. H. Chan, S. Zhang, C. Feng, and M. Chan, "Stacked 3-D Fin-CMOS Technology", *IEEE Electron Device Letter*, Vol. 26, No. 6, pp. 416 - 418, June 2005.

[4] X. Wu, P. C. H. Chan, S. Zhang, C. Feng, and M. Chan, "A Three-Dimensional Stacked Fin-CMOS Technology for High Density ULSI Circuits", *IEEE Trans. on Electron Devices*, Vol. 52, No. 9, pp. 1998-2003, September 2005

[5] T. Yamanaka, et. al., "A 20μm², new poli-Si PMOS load (PPL) SRAM cell having excellent soft error immunity", 1988 IEDM, pp. 48-51.

[6] A. Gat, et. al., "CW Laser Anneal of Polycrystalline Silicon: Crystalline Structure, Electrical Properties", APL, 1978, pp. 54-55

[7] V. W.-C. Chan, P. -C.-H. Chan and M. Chan, "Three Dimensional CMOS Integrated Circuits on Large Grain Polysilicon Films", 2000 IEDM Tech. Dig., pp. 161-164.

[8] S. W. Chang et. al., "First Demonstration of Heterogeneous IGZO/Si CFET Monolithic 3D Integration with Dual Workfunction Gate for Ultra Low-power SRAM and RF Applications", 2021 IEDM, pp. 733-736

[9] S. Zhang, R. Han, X. Lin, X. Wu, and M. Chan, "A Stacked CMOS Technology on SOI Substrate", *IEEE Electron Device Letters*, Vol. 25, No. 9, pp. 661-663, September 2004

[10] L. K. Bera et. al., "Three Dimensionally Stacked SiGe Nanowire Array and Gate-All-Around p-MOSFETs", 2006 IEDM

[11] R. M. Y. Ng et. al., "A new Approach to Fabricate Vertically Stacked Single-Crystalline Silicon Nanowires", 2007 IEEE EDSSC

[12] R. M. Y. Ng et. al., "A new Approach to Fabricate Vertically Stacked Single-Crystalline Silicon Nanowires", IEEE EDL, 2009, pp. 520.

[13] C.-Y. Huang et al., "3-D Self-aligned Stacked NMOS-on-PMOS Nanoribbon Transistors for Continued Moore's Law Scaling," 2020 IEDM, pp. 20.6.1-20.6.4, doi: 10.1109/IEDM13553.2020.9372066.

[14] X. Lin and M. Chan, "Stacked Complementary Field-Effect Transistors: Promises and Challenges", 2024 EDTM

[15] M. Chan, Stacked CMOS Technology, Ch. 3 in Wafer Level 3-D ICs Process Technology, Springer, 2008

[16] L. Liebmann, et. al., "CFET Design Options, Challenges, and Opportunities for 3D Integration," 2021 IEDM, pp. 3.1.1-3.1.4, doi:10.1109/IEDM19574.2021.9720577.

[17] Y.-Y. Chung et. al., "Monolayer-MoS2 Stacked Nanosheet Channel with C-type Metal Contact", 2023 IEDM

[18] C.Y. Huang, et. al. "3-D Self-aligned Stacked NMOS-on-PMOS Nanoribbon Transistors for Continued Moore's Law Scaling", 2020 IEDM.

[19] S. Liao, et. al., "Complementary Field-Effect Transistor (CFET) Demonstration at 48nm Gate Pitch for Future Logic Technology Scaling", 2022 IEDM

Metal-oxide thin-film transistors for artificial neural networks

Yushen Hu, Tengteng Lei and Man Wong

Abstract—**The technology of thin-film transistors (TFTs) enables the construction of both single- and dual-gate devices. When built on metal oxide (MO) semiconductors with large energy bandgaps, TFTs with exceptionally low leakage current can be realized. This combination of materials, structures and electrical characteristics enables the deployment of MO TFTs in a wide range of applications. Presently reviewed are the deployment of a dual-gate TFT as a biomimetic electronic synapse and its application to the construction of artificial neural networks (ANNs).**

Index Terms—**Thin-film transistor (TFT), metal oxide (MO) semiconductor, dual-gate (DG), artificial neural network (ANN), spiking neural network (SNN).**

I. INTRODUCTION

Metal-oxide (MO) semiconductors, such as indium gallium zinc oxide (IGZO) or indium tin zinc oxide (ITZO), are being deployed as active channels in thin-film transistors (TFTs) [1]. Both single-gate (SG) and dual-gate (DG) TFTs have been reported [2]. A DG TFT with at least a portion of its channel sandwiched between two gate electrodes allows simultaneous modulation of its channel potential by the bias applied on the electrodes. When MO semiconductors with relatively large energy bandgaps were deployed, TFTs with exceptionally low leakage current could be obtained. Despite the moderate field-effect mobility, this unique combination of materials, structures and electrical characteristics enables the deployment of MO TFTs in a wide range of applications (Fig. 1), including memory [3], analog/digital circuits [4], sensors [5], displays [6], flexible electronic systems [7], and neuromorphic circuits [8]. The present discourse is focused on the application of MO TFTs to the realization of neuromorphic computation systems.

The simultaneous modulation of the drain current I_d of a DG TFT by the "input" and "weight" signals placed on its two gate electrodes enables the deployment of such a TFT as a biomimetic artificial synapse. The high impedance of a gate electrode placed above an insulating gate dielectric allows coupling of an input signal generated by a source with limited driving capability to a larger number of post-synaptic artificial neurons. Furthermore, the ultra-low leakage current

This work was supported by the National Key R&D Program 2022YFB3607100, Fundamental and Applied Fundamental Research Fund of Guangdong Province 2021B1515130001 and in part by Hong Kong Innovation and Technology Commission Grant GHP/018/21SZ.

Y. Hu, T. Lei, and M. Wong are with the Department of Electronic and Computer Engineering, The Hong Kong University of Science and Technology, Hong Kong, M. Wong is also with Guangzhou HKUST Fok Ying Tung Graduate School, Guangzhou, China (e-mail: eemwong@ust.hk).

($<10^{-18}$ A/μm [9]) of a MO TFT enables the use of a simple capacitor, featuring ease of construction, ease of operation and high reliability, as a memory element. These unique capabilities of MO TFTs have allowed the construction of in-memory-computation systems built on the basic units of DG TFTs as artificial synapses and capacitors as memory elements. Presently reviewed is the MO TFT technology and its application to the realization of artificial neural networks (ANNs), including spiking neural networks (SNNs).

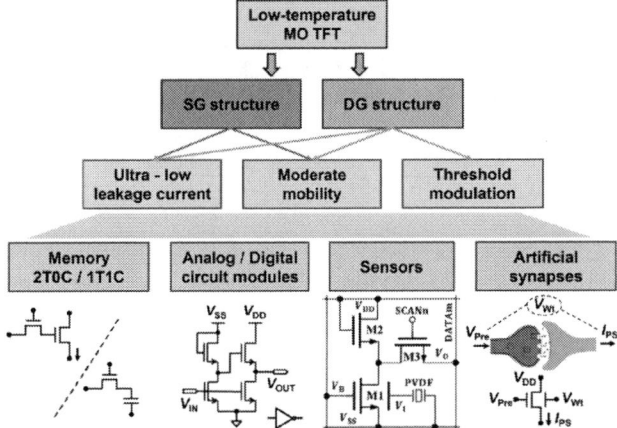

Fig. 1. Hierarchical schematics linking the unique combination of materials, structures and electrical characteristics of MO TFTs to their deployment in a variety of applications.

II. TFT CONFIGURATIONS AND CHARACTERISTICS

Shown in Fig. 2 are the schematic cross-sections of monolithically integrated DG TFT, SG TFT, and storage capacitor, all based on ITZO [8].

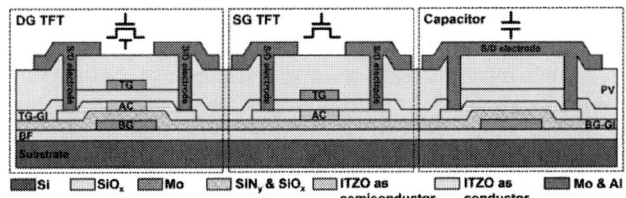

Fig. 2. (a) Schematic cross-sections of a parallel DG TFT (Left), a bottom SG TFT (Center), and a storage capacitor (Right). Legends: BG (bottom gate), GI (gate insulator), AC (active channel), TG (top gate) and S/D (source/drain).

For these TFTs, the I_d *vs.* gate voltage V_g transfer characteristics of an SG TFT with channel width W and length L both of 10 μm and the I_d *vs.* top-gate voltage V_{tg} characteristics of a DG TFT ($W/L = 10/10$ μm) are shown respectively in Figs. 3a and 3b. The turn-on voltage V_{on} of the SG TFT is about -2.5 V. It is obvious that the V_{on} of the DG TFT can be modulated (Fig. 3b) by the voltage V_{bg} applied on the bottom gate electrode.

979-8-3503-6184-1/24 $31.00 © 2024 IEEE

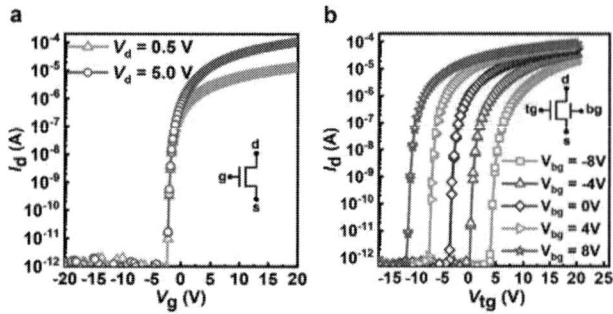

Fig. 3. Measured transfer characteristics of (a) SG and (b) DG ITZO TFTs.

III. IMPLEMENTATION OF ANNs

3.1 Conventional ANN

A biological neural network is composed of a collection of neurons that are extensively connected through synapses. A signal reaching the pre-synaptic terminal triggers the release of neurotransmitters that bind to specific receptors on a post-synaptic neuron, thus opening the corresponding ion channels that leads to the generation of either an excitatory post-synaptic current (EPSC) or an inhibitory post-synaptic current (IPSC) in the post-synaptic neuron (Fig. 4a) [10].

Shown in Fig. 4b is an ANN circuit unit implementing only EPSC. The unit is composed of a SG TFT M_E for accessing a storage capacitor C_E storing the weight signal V_{Wt} and a DG TFT D_E responsible for the generation of an EPSC that contributes to an accumulated output current I_{ACU}. Demonstrating the utility of such a unit, a 6×4 ANN has been reported [11] for recognizing the Tetris-like patterns labeled #1, #2, #3, and #4. These are coded and applied to training the network using a feed-forward gradient-descent algorithm. It can be seen from the inference results (Fig. 4c) that the designated patterns are properly recognized but other patterns not in the training set are properly rejected.

It is well-recognized that the use of only EPSC or IPSC but not both results in a unidirectional I_{ACU} that prevents the network from accomplishing non-monotonic classification tasks. Such limitation has been overcome by applying the input voltage modulation or the differential method to the two- or three-port computation elements popularly deployed in conventional ANNs. One common issue with these implementations is that the finite impedance of the elements limits the size of the resulting ANNs for signal sources with a finite current-driving capability. For the differential method, additional circuits are required to implement the logic function "NOT" for the two current outputs, thus increasing the complexity of the peripheral circuitry.

Illustrated in Fig. 4d is an artificial synapse incorporating both EPSC and IPSC, achieved through the integration of two DG TFTs D_E and D_I [12]. Each DG TFT is associated with a "1-access TFT-1-storage capacitor" memory circuit. The storage electrode of the capacitor is connected to the top gate of the corresponding DG TFT. The process of writing a weight signal V_{WtE} onto capacitor C_E is accomplished by activating the access TFT M_E through the application of V_{ScanE}, while deactivating the access TFT M_I by applying V_{ScanI}. Similarly, the activation of M_I and deactivation of M_E can be employed to write weight signal V_{WtI} onto capacitor

C_I. The input signal X is applied to the bottom-gate electrodes of both TFTs D_E and D_I.

The relationship between the X and the EPSC flowing through D_E is investigated, while ensuring D_I is deactivated. Similarly, the dependence on X of IPSC passing through D_I is examined with D_E turned off. With the I_{ACU} readout line fixed at 2.6 V, typical EPSC at different V_{WtE} values and IPSC at various V_{WtI} values are presented in Fig. 4e. This configuration has been applied to the realization of the complete set of 2-input logic functions [12].

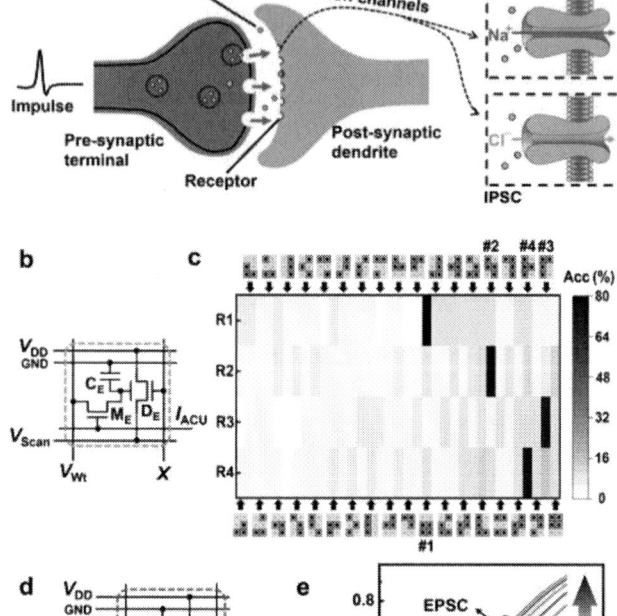

Fig. 4. (a) Schematic cartoon of a biological synapse. (b) Schematic diagram of a single ANN circuit unit implementing only EPSC. (c) Grey-scale map showing the inference outcome of Tetris-like patterns containing 3 or 4 solid shapes. (d) Schematic diagram and (e) measured characteristics of a single ANN circuit unit implementing both EPSC and IPSC.

3.2 SNN

SNN is a type of ANN that more vividly mimics the behavior of a biological neural system in terms of signal encoding, learning mechanism, and neuron operation. Signals are encoded and transmitted across a synapse in the form of action potential "spikes". The pre-synaptic neuron fires a voltage spike V_{Pre}, while the post-synaptic neuron generates a spike V_{Post}. Affecting the magnitude of the post-synaptic current triggered by V_{Pre}, the strength V_{Wt} of the synaptic connection is modified through spike-timing-dependent plasticity (STDP) [13]. This modification depends on the relative timing of V_{Pre} and V_{Post}, resulting in either enhancement or reduction of V_{Wt}.

t_{Pre} and t_{Post} are denoted as the instances in time at which spikes are fired by a pair of pre- and post-synaptic neurons. It is worth noting that the condition $(t_{Post} - t_{Pre}) > 0$, which may reflect a cause-effect correlation, may not hold true for every pair of pre- and post-synaptic neurons in a large network of neurons. During a learning or training process, correlation is tested within "learning windows" with window margin denoted as τ_{Pre} and τ_{Post}. Consequently, the weight V_{Wt} of a synaptic connection governed by the STDP mechanism is adjusted. Specifically, V_{Wt} is increased ($\Delta V_{Wt} > 0$) for $\tau_{Pre} > (t_{Post} - t_{Pre}) > 0$, indicating a positive correlation, or it is decreased ($\Delta V_{Wt} < 0$) for $\tau_{Post} > (t_{Pre} - t_{Post}) > 0$, indicating a negative correlation.

Illustrated in Fig. 5a is the schematic representation of a "decay circuit" designed to generate stretched waveforms V_{Pre}^* and V_{Post}^*, in response to the respective input spike waveforms V_{Pre} and V_{Post} [13]. Determined by capacitor C_{STDP} and the channel resistance of the drive TFT T_{STDP} under the influence of a gate bias V_B, the respective time constants of V_{Pre}^* and V_{Post}^* are designed to match the desired τ_{Pre} and τ_{Post}. Shown in Fig. 5b is the dependence of V_{Pre}^* on V_B when triggered by a 50-µs wide V_{Pre}. It is evident that a more negative V_B results in a higher resistance of T_{STDP}, consequently leading to a larger τ_{Pre}.

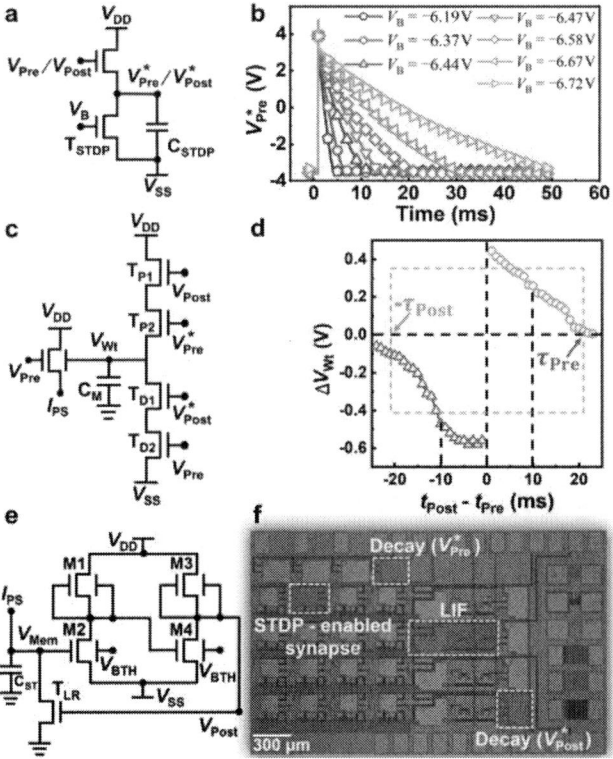

Fig. 5. (a) Schematic diagram of a decay circuit. (b) Dependence on V_B of V_{Pre}^* waveforms triggered by a 50-µs V_{Pre}. (c) Schematic diagram of an STDP-enabled SNN synapse circuit. (d) Measured STDP learning window. (e) Schematic circuit of a LIF circuit. (f) Schematic of a 4 × 4 SNN.

Depicted in Fig. 5c is a circuit for implementing the STDP mechanism. The upper pair of serially connected TFTs T_{P1} and T_{P2} are activated when $\tau_{Pre} > (t_{Post} - t_{Pre}) > 0$. They are responsible for charging the weight capacitor C_M. For the duration of V_{Post}, ΔV_{Wt} is made > 0, thereby

facilitating "long-term potentiation". Similarly, the lower pair of TFTs T_{D1} and T_{D2} are activated when $\tau_{Post} > (t_{Pre} - t_{Post}) > 0$. They are responsible for discharging C_M. For the duration of V_{Pre}, ΔV_{Wt} is made < 0, resulting in "long-term depression". The gate bias values V_B used to generate the V_{Pre}^* and V_{Post}^* signals are -6.5 and -6.7 V, respectively. By adjusting the timing difference $t_{Post} - t_{Pre}$ and recording the corresponding changes in V_{Wt}, an STDP learning window is obtained and displayed in Fig. 5d.

Emulating a post-synaptic neuron, a leaky-integrate-and-fire (LIF) circuit made of a capacitor (C_{ST}) and a leaky/reset TFT T_{LR} is constructed to accumulate a train of post-synaptic current spikes I_{PS}. This results in the generation of a post-synaptic voltage V_{Mem} (Fig. 5e). V_{Mem} is also the input to a subsequent "comparator" responsible for the simultaneous generation of V_{Post} and resetting of C_{ST}. A 4 × 4 SNN (Fig. 5f) has been reported for pattern recognition [13].

IV. CONCLUSION

The utilities of MO TFTs exhibiting ultra-low leakage current and moderate mobility are reviewed. Besides conventional SG TFTs, DG TFTs allowing modulation of the threshold voltage have also been reported. Leveraging these properties, neuromorphic computing elements based on DG TFTs and capacitive memory units have been realized. These are combined and applied to realize ANNs and their more biomimetic counterparts, SNNs.

V. ACKNOWLEDGMENT

Micro-fabrication was carried out at the Nanosystem Fabrication Facility of The Hong Kong University of Science and Technology.

REFERENCES

[1] Shi, Y *et al.*, *Applied Physics Letters*, vol. 121, no. 21, Nov. 2022, doi: 10.1063/5.0123253.

[2] Lei, T *et al.*, *IEEE Trans. Electron Devices*, vol. 69, no. 6, pp. 3186-3191, Jun. 2022, doi: 10.1109/TED.2022.3167940.

[3] Ryu, S *et al.*, *Advanced Materials Technologies*, May. 2024, doi: 10.1002/admt.202302209.

[4] Liao, C *et al.*, *Microelectronics Journal*, vol. 46, no. 10, pp. 923-927, Oct. 2015, doi: 10.1016/j.mejo.2015.08.003.

[5] Geng, D *et al.*, *IEEE Sensors Journal*, vol. 17, no. 3, pp. 585-586, Feb. 2017, doi: 10.1109/JSEN.2016.2639525.

[6] Hara, Y *et al.*, *Journal of the Society for Information Display*, vol. 26, no. 3, pp. 169-177, Apr. 2018, doi: 10.1002/jsid.648.

[7] Xiao, X *et al.*, *ACS applied materials & interfaces*, vol. 10, no. 31, pp. 25850-25857, Dec. 2017, doi: 10.1021/acsami.7b13211.

[8] Hu, Y *et al.*, *IEEE Electron Device Letters*, vol. 43, no. 5, May. 2022, doi: 10.1109/LED.2022.3164684.

[9] Wang, Y *et al.*, *SID Symposium Digest of Technical Papers*, vol. 52, pp. 413-416, Feb. 2021, doi: 10.1002/sdtp.14505.

[10] Pan, E *et al.*, *Eneuro*, vol. 9, no. 1, Dec. 2021, doi: 10.1523/ENEURO.0375-21.2021.

[11] Hu, Y *et al.*, *IEEE Transactions on Electron Devices*, vol. 69, no. 10, Oct. 2022, doi: 10.1109/TED.2022.3201836.

[12] Hu, Y *et al.*, *IEEE Transactions on Circuits and Systems I: Regular Papers*, vol. 71, no. 4, Apr. 2024, doi: 10.1109/TCSI.2023.3347773.

[13] Hu, Y *et al.*, *IEEE Transactions on Circuits and Systems II: Express Briefs*, doi: 10.1109/TCSII.2024.3372886.

979-8-3503-6184-1/24 $31.00 © 2024 IEEE

Cryogenic Threshold Voltage and On-current Variability Analysis of GAA Nanosheet FETs at 4K

Zihao Liu[1], Tomoko Mizutani[1], Kiyoshi Takeuchi[1], Takuya Saraya[1], Hiroshi Oka[2], Takahiro Mori[2], Masaharu Kobayashi[1], Toshiro Hiramoto[1]

[1] Institute of Industrial Science (IIS), The University of Tokyo, 4-6-1 Komaba Meguro-ku, Tokyo 153-8505, Japan
[2] National Institute of Advanced Industrial Science and Technology (AIST), Tsukuba, Ibaraki 305-8468, Japan

* Email: zh-liu@nano.iis.u-tokyo.ac.jp, hiramoto@nano.iis.u-tokyo.ac.jp

Abstract—The current and threshold voltage variability for 4nm to 7nm wide gate-all-around (GAA) Nanosheet (NS) FETs are analyzed and compared at cryogenic temperature (CT) and room temperatures (RT). It is found that linear region performance is strongly deteriorated, while saturation region performance is better than mature traditional devices. The reason for such phenomena is discussed.

Keywords—nanowire, nanosheet, cryogenic, variability analysis, threshold voltage, on-current, overdrive current

I. INTRODUCTION

Cryogenic device has attracted much attention from industry and academia alike due to emerging fields of applications such as quantum computing [1]. On the other hand, device scaling has also come to a point that traditional device architectures is no longer sufficient to catch up with the scaling and performance demands. Therefore, nanowire and consequently GAA nanosheet FETs (NSFETs) have been chosen as the next candidate to continue device scaling, while the interest on using NSFETs in peripheral circuits for deep cryogenic systems is also getting increasingly higher.

Nevertheless, while existing literatures and research did cover this topic to certain extent, due to the fabrication difficulties of nanowire and nanosheet devices, the size of those devices is typically large [2] and may not necessarily be comparable to highly scaled traditional devices. In this research, we carried out the analysis with extremely thin and narrow NSFETs down to 4nm width, and at the temperature of 4K.

II. DEVICE PREPARATION AND MEASUREMENT

The detailed preparation processes of the NSFETs using the university laboratory were described in [3]. The length of all NSFETs are 100nm, with height being 3nm. The NSFETs are not intentionally doped. The width are 4-7nm. 100 transistors for each width were measured both at RT and CT. For comparison purposes, a batch of bulk and FDSOI FETs were also measured. The bulk and FDSOI FETs were prepared at external fabs using 65nm technology. The width and length are 140nm and 60nm respectively.

In this work, there are a few key metrics of our interest, and they are defined as follows: (i) V_{thc}: threshold voltage defined by constant current of 1×10^{-7} W/L [A], where the effective perimeter width W_{eff} is used for NSFETs. (ii) I_{on}: I_d at 1.2V. (iii) I_{ov}: overdrive current defined as I_d at $V_g = V_{thc} + 0.5V$ for bulk and FDSOI FETs, and $V_g = V_{thc} + 0.8V$ for NSFETs.

This work is supported by JST SPRING, Japan, Grant Number JPMJSP2108.

III. RESULTS

A. I_d-V_g measurements

The I_d-V_g plots are shown as in Fig.1 (linear region) and Fig.2 (saturation region). It's obvious that as compared to RT, the variability in CT is much larger overall. While subthreshold region shows a slight variability increase in CT, the above-threshold suffers from much higher variability as compared to RT in the linear region. On the other hand, the variability performance in the saturation region is much better and the difference between RT and CT is not extremely large.

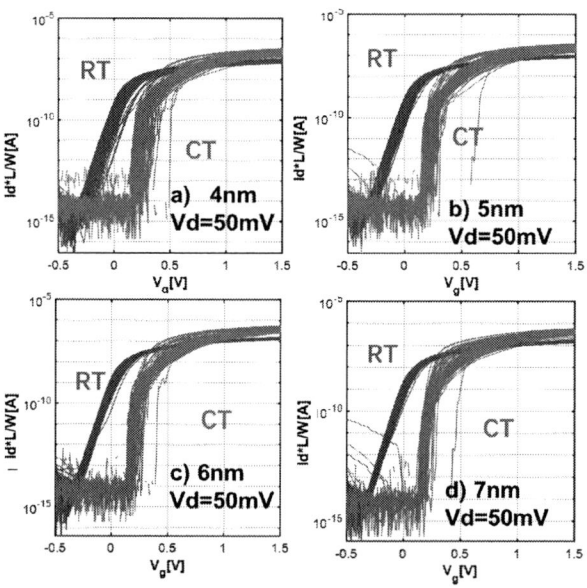

Fig.1. I_d-V_g plots in linear region for 4-7nm NSFET at RT and CT.

B. V_{thc} quantile plot and I_d-V_d measurements

The V_{thc} quantile plots are shown in Fig. 3. In both linear and saturation, V_{thc} increases under CT due to change in fermi level with temperature. The quantile plots indicated that in linear region, when changing from RT to CT, the slope of the higher V_{thc} distribution changed significantly, indicating that the V_{thc} variability in the higher V_{thc} region has severely increased.

To look into this issue, the I_d-V_d plots were plotted and shown as in Fig. 4. At low V_d situations at CT, the I_d-V_d characteristics show non-ohmic behaviors. Such phenomena didn't happen in RT, however. The reason for such behavior is quite complicated. Non-ohmic contact at CT and freeze-out in the extension region [4-6] are considered to be the most likely causes. Such phenomena also indicates that under CT at linear region, the threshold voltage analysis is quite difficult and typical methods, including extrapolation of threshold voltage, may not be reliable.

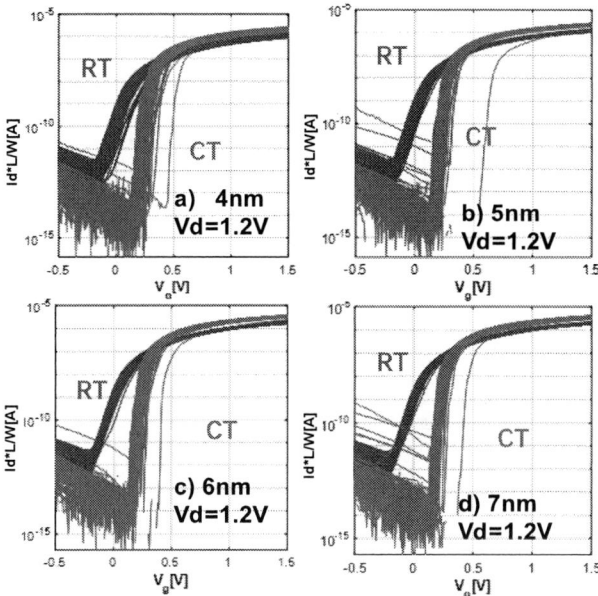

Fig.2. I_d-V_g plots in saturation region for 4-7nm NSFET at RT and CT.

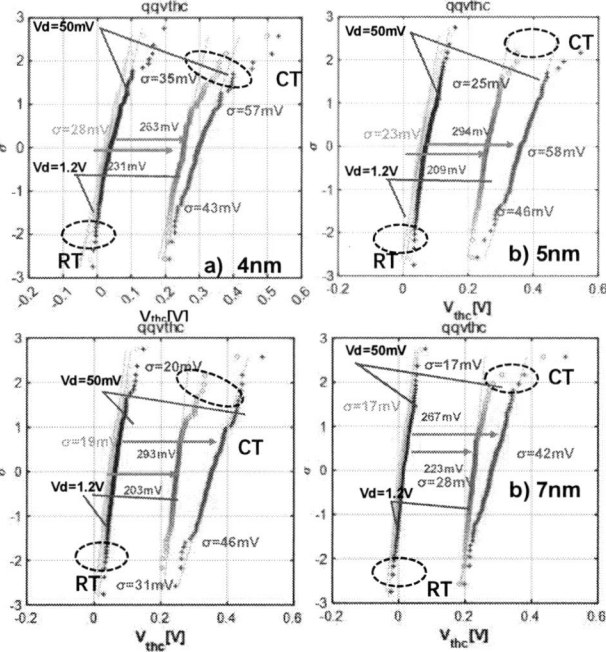

Fig.3. V_{thc} quantile plots for 4-7nm NSFET at RT and CT.

C. I_{on} and I_{ov} characteristics

The I_{on} and I_{ov} characteristics are shown as in Fig.5 and 6, respectively. For conciseness, only 4nm and 7nm results are shown. The mean of I_{on} increases as expected, due to reduced phonon scattering at CT and thus higher mobility. In linear region, the I_{on} variability significantly increased by 2 to 4 times with decreasing temperature, while in saturation region the variability increase is again very mild. The reason for such huge performance difference is considered to be the

resistance variability induced by the above mentioned non-ohmic contacts and the freeze-out effect whose impact is much more severe in linear region than in saturation region. While previous works [7] have demonstrated that one of the key limitations for the performance of such narrow NSFET is actually silicon thickness induced mobility fluctuation ($\mu_{fluctuation}$) [8], $\mu_{fluctuation}$ has very small temperature dependence, so the effect of $\mu_{fluctuation}$ degradation affect both RT and CT on-current.

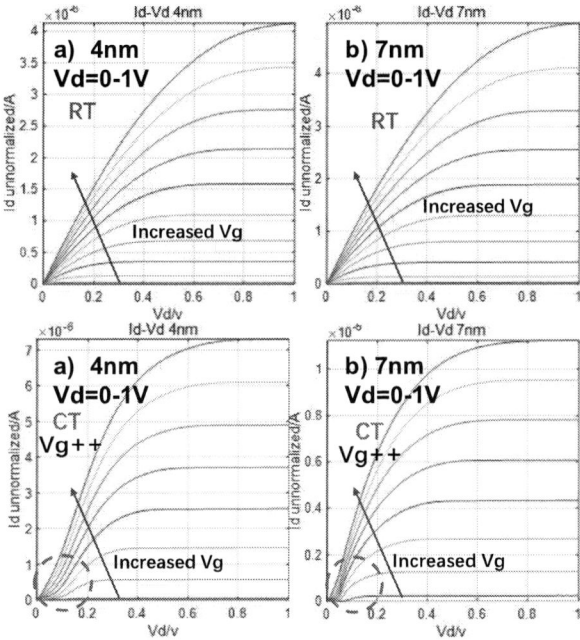

Fig.4. I_d-V_d plots for 4 and 7nm NSFET at RT and CT.

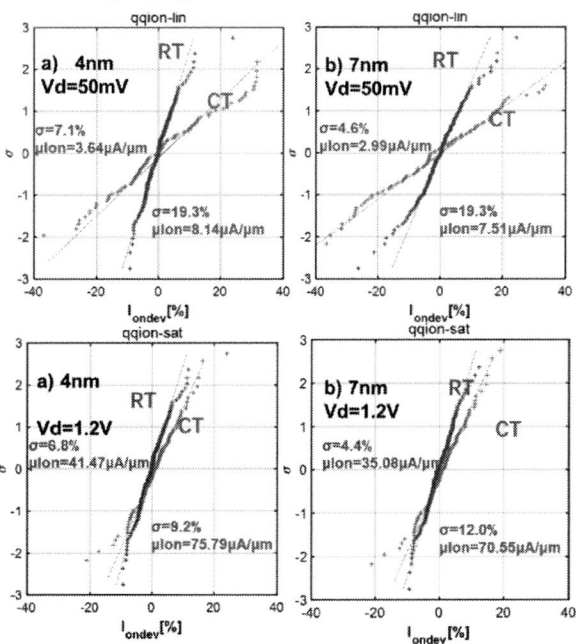

Fig.5. I_{on} quantile plots for 4 and 7nm NSFET at RT and CT.

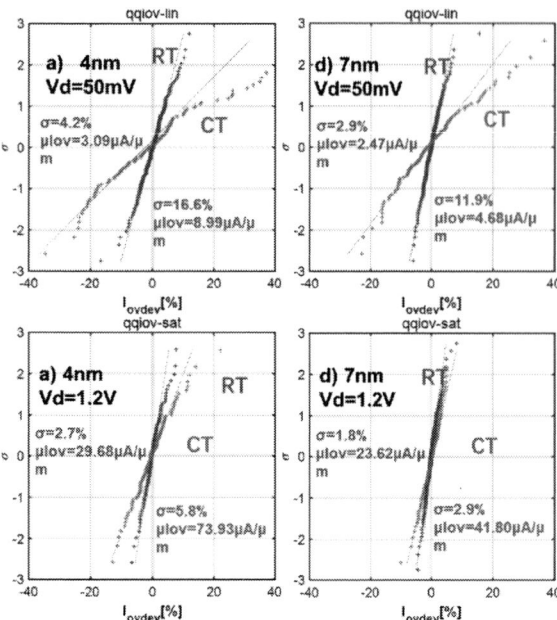

Fig.6. I_{ov} quantile plots for 4 and 7nm NSFET at RT and CT.

I_{ov} is plotted to remove the impact of threshold voltage. I_{ov} followed the same trend as I_{on} in general. However, it is noticed that 7nm device I_{ov} variability (2.9%) is reduced dramatically as compared to I_{on} (12.0%). This hints that V_{thc} may have contributed more than 50% to the overall variability and could have been the major contributor to the overall variability. This is different from room temperature, whose major contributor of variability was identified to be transconductance factors [7].

Furthermore, a similar analysis was carried out for I_{ov} using bulk and FDSOI devices mentioned before [9]. Comparably speaking, the I_{ov} variability of NSFETs is still small at CT. In absolute terms, NSFET has similar I_{ov} variability to bulk FETs. Given the much smaller size of NSFET, the I_{ov} variability is very good. Even in 4nm, the NSFETs I_{ov} variability in saturation region only doubled between CT and RT, and comparing the 7nm case, NSFETs has an even smaller change (61.1% of NSFET vs. 107.7% of FDSOI) [9], as shown in Table I, indicating the superior I_{ov} variability of NSFET due to better gate control even in CT.

TABLE I. I_{ov} VARIABILITY COMPARSION IN SATURATION REGION.

$\sigma I_{ov}(\%)$	Bulk	FDSOI	4nm NSFET	7nm NSFET
RT	2.9	1.3	2.7	1.8
CT	5.8	2.7	5.8	2.9
$\Delta\sigma I_{ov}(\%)$	100	107.7	114.8	61.1

D. Pelgrom plots

The Pelgrom plot is a typical effective method to understand how the variability changes with device size. If the variability of the devices is purely random, then the fitting line should actually pass through origin. The linear and saturation I_{on} and I_{ov} Pelgrom plots are thus plotted as in Fig. 7.

Under RT, the variability of device performs as expected for 4-7nm devices. However, in linear region under CT, the slope drastically increases, and the data are not on the linear line through the origin, indicating that non-random variability, most likely non-ohmic and carrier freeze out. In saturation region, however, the overall variability performance falls more nicely on the ideal line. Although the slope is a bit higher, the outcome is very similar to the FDSOI results reported in previous works [10], again indicating the superior performance of NSFET even under cryogenic conditions.

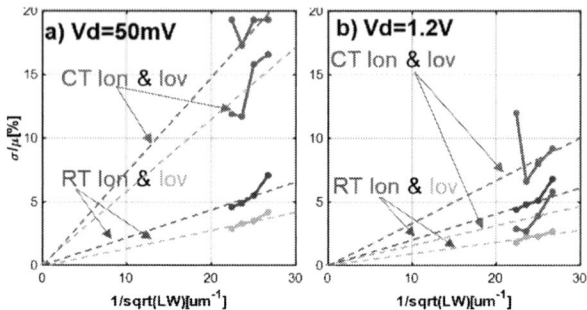

Fig.7. Pelgrom plot for I_{on} and I_{ov}.

IV. CONCLUSIONS

The threshold voltage and on-current variability for 4-7nm width GAA NSFETs under cryogenic condition are analyzed. It was found that the NWFETs suffer larger variability especially in linear region due to the non-ohmic contact and freeze out effects. Such impact is also present in saturation region although with a reduced extent. The performance characteristics is also similar to FDSOI devices. Although NSFETs still outperform tranditional devices in saturation region, threshold voltage engineering will be very important if NSFETs will be applied in future cryogenic systems.

ACKNOWLEDGMENT

The authors would like to thank Dr. Suzuki Ryota for the support with preparation of test devices.

REFERENCES

[1] IEEE IRDS 2023.

[2] J. Gu et al., Nanomaterials 2021, 11(2), 309.

[3] R. Suzuki et al., JJAP, vol. 52, 104001, 2011.

[4] T. Inaba et al., Applied Physics Express 15, 084004, 2022.

[5] K. Takeuchi et al., JJAP 63, SC1023, 2023.

[6] S. Aymeloglu et al., IEEE TED, Vol. 23, Issue 4, 1976.

[7] Z. Liu et al., JJAP 61, SC1006, 2022.

[8] K. Uchida et al., IEDM, 33.5, 2003.

[9] Z. Liu et al., Submitted to Int. Conf. Solid-State Devices and Materials (SSDM), 2024.

[10] T. Mizutani et al., SNW, p.71, 2012.

Reverse-Biased PN Junction Isolation for Leakage Suppression and Strain Enhancement in Gate-All-Around Nanosheet FETs

Boqian Shen[1], Chunlei Wu[1,2,3*], Yumin Xu[1], Fei Zhao[1], Hanzhi Gu[1], Jian Ma[1], Yueyuan Yu[1], Yiming Xia[1], Qingqing Sun[1,2,3], David Wei Zhang[1,2,3]

[1] School of Microelectronics, Fudan University, Shanghai 200433, China
[2] Shanghai Integrated Manufacturing Innovation Center Co., Ltd, Shanghai 201203, China
[3] Jiashan Fudan Institute, Jiaxing 314100, China

* Email: wuchunlei@fudan.edu.cn

Abstract—**In this work, a novel PN junction isolation (PN-I) scheme is proposed to suppress the parasitic channel leakage in Gate-All-Around (GAA) nanosheet (NS) FET. By introducing a reverse-biased PN junction underneath the stacked NS channels, PN-I scheme can reduce off leakage current to the same level as devices with buried dielectric isolation. Moreover, fabrication process flow has also been discussed based on TCAD process simulations. The results show that PN-I configuration can also achieve p-channel stress enhancement, owing to its excellent process compatibility with subsequent process steps including source/drain stressors epitaxy. The results suggest great potential of the proposed PN isolation scheme in prospective low power logic applications.**

Keywords—Gate-All-Around, nanosheet, sub-fin leakage, Reverse-biased PN junction, stress engineering

I. INTRODUCTION

Gate-All-Around (GAA) nanosheet FET has been widely recognized as the mainstream logic device structure for beyond 3 nm node applications, owing to its superior gate electrostatics[1]. To meet the requirement of low power applications, however, sub-fin leakage in GAA nanosheet FET is one of the most important problems to be addressed. Different from sub-fin leakage in FinFET, GAA nanosheet has a tri-gated sub-fin with much wider "fin width", the gated sub-fin region acts as a parasitic channel with poor gate control and accordingly undesired leakage problems. On the other hand, compared to FinFETs, the main transport orientation of lateral NS channel is changed from (110) to (100), resulting in lower hole mobility and degraded N/P balance in GAA NS FETs. Therefore, effective p-channel stress engineering is also of great importance for GAA CMOS technology development.

Multiple approaches have been discussed to address the sub-channel leakage issue, including punch through stoppers (PTS) [2], partial/full bottom dielectric isolation (BDI) [3] and narrowed sub-fin technique [4]. For aggressively scaled GAA nanosheet FETs, however, PTS suffers from increased process difficulty and high sensitivity to process variations. BDI scheme, which blocks the sub-channel leakage path with dielectrics, can achieve much lower leakage current. While introducing dielectric layers underneath the source /drain (S/D) regions also brings about increased challenges to subsequent processes such as S/D stressors epitaxial [5].

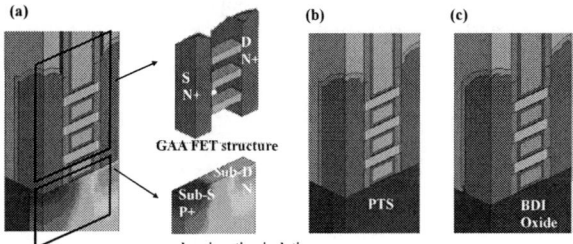

Fig. 1. Schematics of GAA FETs with different leakage control schemes, (a) proposed PN-I, (b) PTS and (c) BDI. (n-type FETs)

TABLE I.   PAREMETER USED IN SIMULATION

Parameter	Value
Gate length, L_G	16 nm
Nanosheet thickness, T_{NS}	6 nm
Nanosheet width, W_{NS}	20 nm
Sacrifice layer thickness, T_{SAC}	12 nm
Gate spacer length, L_{SP}	5 nm
Interficial layer thickness, T_{IL}	0.5 nm
High-k material thickness, T_{HK}	1.5 nm

In this work, we propose a novel PN junction isolation (PN-I) scheme for parasitic channel leakage control in GAA FETs. The TCAD simulation results show that GAA FETs with PN-I configuration can achieve suppressed sub-channel leakage and enhanced channel stress at the same time. Excellent process compatibility has also been validated.

II. DEVICE STRUCTURE AND PROCESS DESIGN

Fig. 1 a) shows the structure of the proposed PN-I GAA FET. Compared with PTS and BDI schemes, the proposed PN-I scheme has asymmetrically doped sub-source and sub-drain regions underneath the S/D regions respectively, so as to get a reverse-biased P-i-N or PN junction along the sub-channel. Taking n-type devices as example, the sub-source is formed by p-type implantation with peak concentration of $N_{SUBS} = 2 \times 10^{20}$ cm^{-3}, the sub-drain is formed by n-type implantation with peak concentration of $N_{SUBD} = 1 \times 10^{18}$ cm^{-3}. While the doping concentrations of the n+ doped source/drain is set as $N_{SD} =$

979-8-3503-6184-1/24 $31.00 © 2024 IEEE

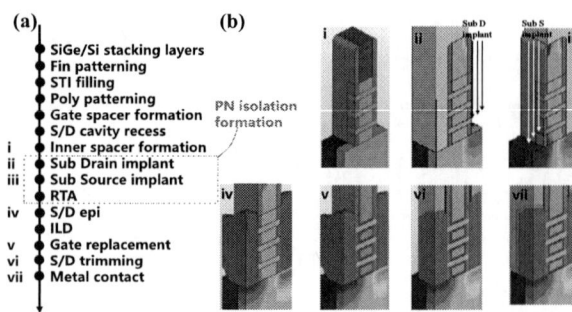

Fig. 2. Process flow for GAA FET with PN-I, (a) overall process sequence, (b) key process steps.

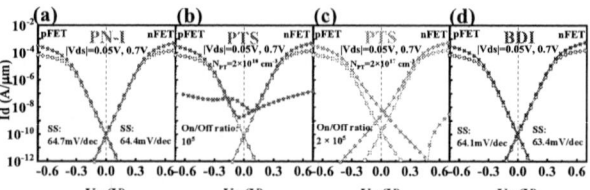

Fig. 3. Transfer characteristics of GAA FETs with (a) PN-I, (b) PTS of N_{PT} = 2×10^{18} cm^{-3}, (c) PTS of $N_{PT} = 2 \times 10^{17}$ cm^{-3}, and (d) BDI schemes. The gate work function is set as 4.490 eV for nFET and 4.814 eV for pFET to fix the OFF-current of GAA FET with full BDI (I_{off} = 100 pA/μm at V_{GS} = 0 V) for comparison.

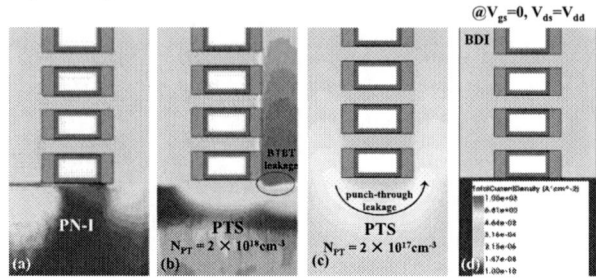

Fig. 4. Off-state (V_{GS} = 0 V) current density contour of n-type GAAFET with (a) PN-I, (b) PTS of N_{PT} of 2×10^{18} cm^{-3} (c) PTS of $N_{PT} = 2 \times 10^{17}$ cm^{-3}, and (d) BDI at V_{ds}=V_{dd}.

1.5×10^{20} cm^{-3}. As for PTS scheme, p+ doped PTS of N_{PT} = 2×10^{18} cm^{-3} and 2×10^{17} cm^{-3} is used for comparison. Device parameters used in TCAD simulation are listed in Table I, following 3 nm node design rules [6].

The fabrication flow of the PN-I GAA FET is the same with main stream GAA FETs only with two extra sub-source and sub-drain implantation steps after inner spacer formation, as shown in Fig. 2. Here, sub-source implantation of 3 keV, 5e15 cm^{-2} and sub-drain implantation of 3 keV, 5e12 cm^{-2} is used, followed with annealing for dopants activation and surface recovering. The epitaxial S/D is one of the most critical stressors in advanced CMOS fabrication. Here the epitaxial $Si_{0.6}Ge_{0.4}$ is considered in p-type GAA FETs and ideal defect-free S/D epitaxy is assumed.

All the process simulation and electrical characteristic are conducted based on Synopsys TCAD Sentaurus simulation tools [7]. The Self-consistent calculation is achieved though transport equation combined with Poisson equation and density-gradient quantum correction. The Auto-orientation Inversion-Accumulation-Layer mobility, High field saturation velocity models, and Dynamic nonlocal BTBT model, Shockley-Read-Hall recombination models are included. For the strain models, a multivalley electron and hold mobility model is used. Physical models used have been delicately calibrated with the experimental data [8][9].

III. RESULTS AND DISCUSSION

A. PN-I vs. PTS-Parasitic Leakage Suppression

Fig. 3 shows the simulated transfer characteristic of GAA FETs with PN-I, BDI and PTS respectively. It can be seen that the PN-I scheme can effectively reduce the leakage current to the same level as BDI scheme, achieving perfectly controlled subthreshold leakage and a near 60mV/dec SS. The excellent sub-channel leakage block capability of PN-I scheme, as shown in Fig.4 a), can be attributed to the small off-state leakage dictated by the reverse-biased PN diode, which is about 4 decades lower compared to a typical Si MOS channel [10].

While PTS scheme shows the highest off leakage current. A GIDL-like leakage tail and a punch-through leakage current is witnessed in PTS of $N_{PT} = 2 \times 10^{18}$ cm^{-3} and $N_{PT} = 2 \times 10^{17}$ cm^{-3} respectively. As shown in Fig.4 b) and c), the high sub-channel leakage in PTS scheme can be attributed to both punch-though current and band-to-band tunneling current at the drain junction. For PTS with higher $N_{PT} = 2 \times 10^{18}$ cm^{-3}, the drain region and PTS channel are both heavily doped, practically forming a steep PN junction at drain side.

Thus, significant band-to-band tunneling occurs at drain junction under large drain voltage V_{DS}=V_{DD}, resulting in the GIDL-like leakage tail shown in Fig. 3 b). In PN-I scheme, however, the undoped or lightly doped sub-channel together with the lightly-implanted sub-drain forms a graded junction, hence band-to-band tunnelling at drain junction is negligible. Meanwhile, the PTS doping cannot be too low either, or the sub-channel would suffer from severe punch through leakage (Fig. 3 c) and Fig. 4 c)).

B. PN-I vs. BDI-Channel strain enhancement

In GAA NS devices, the epitaxial growth of S/D stressors may have to start from multiple seed surfaces of the stacked NS tips and the bottom Si substrate. One of the most concerned issues of BDI scheme is the absence of the bottom Si substrate, leading to more severe dislocations and defects in following S/D epitaxy and significant channel stress loss [5]. PN-I scheme, however, shows great process compatibility with S/D stressors epitaxy, owing to the intact bottom Si substrate in S/D cavity.

To give insight into the impacts of PN-I structure on the channel stress, longitudinal channel stress evolution of critical process steps of p-type GAA FETs with BDI and PN-I have been simulated, as shown in Fig. 5. Here the epitaxial $Si_{0.6}Ge_{0.4}$ is used as S/D stressors, other stressors including sacrificial layers, STI, metal gate and ILD layers are also taken into account [11].

As can be seen in Fig.5 b), the S/D region in GAA pFET with BDI suffers severe stress loss in S/D epitaxy step. Consequently, the NS channels get a small tensile stress transferred from the strained sacrificial layer during the S/D carve step, instead of the desired compressive stress for hole mobility improving. In GAA pFET with PN-I, however, the stored stress in epitaxial S/D can be effectively transferred to the NS channels during the S/D epitaxy and channel release steps (Fig.5 a)). An average compressive stress of -1.95GPa has been obtained (Fig.6), indicating great potential of the proposed PN-I scheme in prospective GAA

979-8-3503-6184-1/24 $31.00 © 2024 IEEE

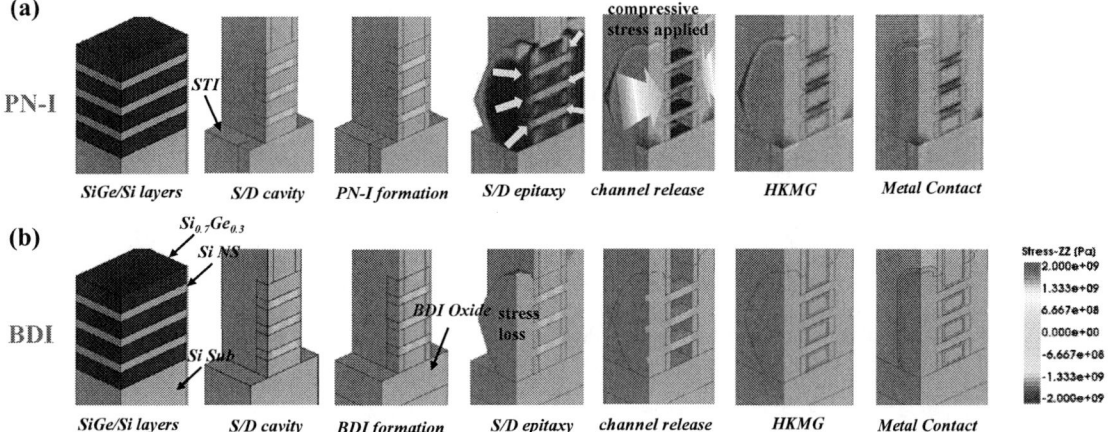

Fig. 5. Longitudinal channel stress evolution of critical steps in the process of p-type GAA FETs with (a) PN-I, and (b) BDI scheme respectively.

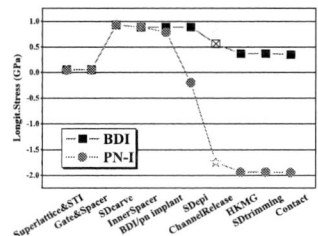

Fig. 6. Average longitudinal channel stress evolution of p-type GAA FETs with PN-I and BDI.

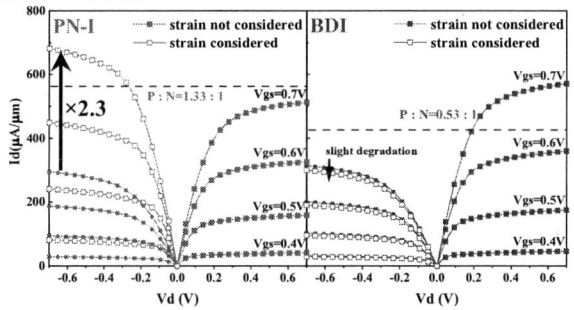

Fig. 7. Output characteristics of n/p GAA FETs with PN-I and BDI scheme.

FET optimization for p-channel stress engineering.

Fig.7 compares the output characteristics of n/p GAA FETs with PN-I and BDI. As can be seen in the figure, p-type GAA FET with PN-I has achieved ~2.3X higher on-current and improved N/P current matching, attributing to enhanced p-channel stress.

IV. CONCLUSION

By introducing a reverse-biased PN junction underneath the stacked NS channels, the proposed PN-I scheme in GAA NS FETs can achieve not only full BDI-comparable parasitic channel leakage but also enhanced NS channel stress. An high compressive stress of -1.95GPa and ~2.3X on current boost has been obtained in PN-I based GAA pFET, achieving a great P/N current matching ratio of 1.33 : 1. The results would provide insights regarding the exploration of GAA FETs optimization for low power logic applications.

ACKNOWLEDGMENT

This work was supported in part by NSFC under Grant 62304048, in part by the Science and Technology Commission of Shanghai Municipality under Grant 21TS1400800.

REFERENCES

[1] G. Bae et al., "3nm GAA Technology featuring Multi-Bridge-Channel FET for Low Power and High Performance Applications," 2018 IEEE International Electron Devices Meeting (IEDM), San Francisco, CA, USA, 2018, pp. 28.7.1-28.7.4.V.

[2] Jegadheesan, K. Sivasankaran and A. Konar, "Optimized Substrate for Improved Performance of Stacked Nanosheet Field-Effect Transistor," in IEEE Transactions on Electron Devices, vol. 67, no. 10, pp. 4079-4084, Oct. 2020.

[3] J. Zhang et al., "Full Bottom Dielectric Isolation to Enable Stacked Nanosheet Transistor for Low Power and High Performance Applications," 2019 IEEE International Electron Devices Meeting (IEDM), San Francisco, CA, USA, 2019, pp. 11.6.1-11.6.4.

[4] J. Gu et al., "Narrow Sub-Fin Technique for Suppressing Parasitic-Channel Effect in Stacked Nanosheet Transistors," in IEEE Journal of the Electron Devices Society, vol. 10, pp. 35-39, 2022

[5] M. Saleh, A. M. Bayoumi and H. Abdelhamid, "Impact of Bottom Dielectric Isolation of Si-Stacked Nanosheet Transistor on Stress and Self-Heating at 3-nm Node and Beyond," in IEEE Transactions on Electron Devices, vol. 70, no. 11, pp. 5535-5542, Nov. 2023.

[6] IRDS 2023, [Online]. https://irds.ieee.org/

[7] TCAD Sentaurus User Guide Version O-2018.06, June 2018, Synopsys, Inc., MountainView CA, USA, 2018..

[8] L. Knoll et al., "Inverters With Strained Si Nanowire Complementary Tunnel Field-Effect Transistors," in IEEE Electron Device Letters, vol. 34, no. 6, pp. 813-815, June 2013

[9] N. Loubet et al., "Stacked nanosheet gate-all-around transistor to enable scaling beyond FinFET," in Proc. Symp. VLSI Technol., Jun. 2017, pp. T230–T231.

[10] S. O. Koswatta, M. S. Lundstrom and D. E. Nikonov, "Performance Comparison Between p-i-n Tunneling Transistors and Conventional MOSFETs," in IEEE Transactions on Electron Devices, vol. 56, no. 3, pp. 456-465, March 2009.

[11] G. Eneman et al., "Stress simulations for optimal mobility group IV p- and nMOS FinFETs for the 14 nm node and beyond," 2012 International Electron Devices Meeting, San Francisco, CA, USA, 2012, pp. 6.5.1-6.5.4.

979-8-3503-6184-1/24 $31.00 © 2024 IEEE   337

Studies on Selective Deposition of SiO_2 by Rapid Atomic Layer Deposition

Sicong Shao , Jin Yan , Wang Li , Kun Cao , Rong Chen*

School of Mechanical Science and Engineering, Huazhong University of Science and Technology, Wuhan 430074, China

* Email: rongchen@mail.hust.edu.cn

Abstract—**With the rapid development of the semiconductor industry, the traditional process is facing the challenge of insufficient manufacturing accuracy due to the increasingly small size of microelectronic devices. Atomic layer deposition (ALD), as a bottom-up self-limiting nanofilm preparation process, has great application prospects in the field of semiconductors. However, due to its slow growth rate, it cannot meet the needs of industrial production, so rapid atomic layer deposition (RALD) was developed. In this work, in order to study the selectivity of SiO_2 grown by RALD on patterned substrates, the basic process and electrical properties of RALD were first explored, and some basic properties of SiO_2 deposited by RALD were verified. After that, the Cu and Si substrates were selected, and the template method was used to carry out the experiment of area-selective atomic layer deposition (ASD) on RALD. Finally, the results of selective deposition experiments were discussed by AES and other characterizations.**

Keywords—*rapid atomic layer deposition, area-selective atomic layer deposition, self-assembled monolayers, metal-dielectric patterns*

I. INTRODUCTION

With the rapid development of the semiconductor industry, the feature size of microelectronic devices has been gradually reduced to less than 5nm, followed by more and more complex device structures. This has also caused the current top-down subtractive manufacturing methods, such as lithography, etching, etc., often face the challenge of edge placement error (EPE) [1]. Therefore, a bottom-up additive manufacturing method with self-alignment characteristics is needed. As an emerging and self-limiting nano-film preparation technology, atomic layer deposition (ALD) can achieve high-precision and good shape-preserving film growth near the atomic scale, which well meets the needs of reducing the feature size. At the same time, its extended Area-Selective atomic layer deposition (ASD) can realize the fully self-aligned via (FSAV) on the patterned structure [2], which effectively solves the etching error caused by EPE, and has broad application prospects in the fields of microelectronics, optoelectronic devices and catalysts [3,4].

At present, there are two main methods to achieve selective deposition. One is to use the inherent selectivity between the two substrates to achieve thin film deposition in the growth area (GA) and non-deposition in the non-growth area (NGA) [5]. The other is to deposit inhibitors on NGA before the ALD step, such as SAMs [6], to block the deposition of the film on it. Due to the less inherent selectivity between different substrates, the second selective deposition method is mainly used. However, for the widely used dielectric material SiO_2, it often requires high temperature or Plasma-assisted deposition, which will lead to the rapid failure of the inhibitor and the inability to achieve selective growth. Moreover, due to the characteristics of

ALD layer-by-layer nano-scale growth, the rate of SiO_2 deposition in the general process is only 0.05-0.2nm/cycle, which is approximately equivalent to 0.1-12nm/min. In the semiconductor industry, the processing time of a wafer is required to be less than 5min, and the thickness of each wafer should reach 50nm [7]. Therefore, the traditional ALD process cannot meet the needs of industrial growth of SiO_2. ALD process with faster growth rate and area-selectivity is needed. Previous study has proposed a new concept of ALD process called Rapid Atomic Layer Deposition (RALD) [8]. This process can achieve a high deposition rate of SiO_2, up to 12-35nm/cycle, which is hundreds of times faster than the traditional ALD process, while maintaining the self-limiting characteristics of ALD. Many literatures have discussed the reaction mechanism of SiO_2 growth by RALD. The growth kinetics of rapid SiO_2 ALD can be understood in terms of the temperature dependence of nucleation and cross-linking and the pressure dependence of the siloxane polymerization rate. At present, the mechanism of RALD is relatively mature, but there are relatively few reports on the use of rapid ALD for area-selective deposition.

Therefore, based on the basic process research of RALD, this paper will discuss the influence of different aluminum sources on the surface morphology and electrical properties of RALD. Referring to the area-selective deposition method in other studies, the template method was used to explore the area-selectivity of RALD on Si (GA) and Cu (NGA) substrates, and the patterned substrate was prepared to further verify the fully self-aligned via on the patterned structure. This is of great significance to the cost reduction and efficiency increase of the semiconductor industry in the future.

II. EXPERIMENTAL METHOD

A. Preparation of Materials and Experimental Steps

Trimethylaluminum (TMA, > 99.9 %) or Dimethylaluminium i-propoxide(DMAI, > 99.9 %) was used as aluminum precursor source, tris(tert-butoxy)silanol (TBS, > 99.9 %) was used as silicon precursor source, N_2 was used as carrier gas and cleaning gas, and n-octadecanethiol (ODT, > 97 %) was used as an inhibitor to achieve selective deposition on lightly doped Si (natural oxide layer about 1.7nm) and ODT-treated Cu substrates. The three precursor sources were stored in a separate source bottle. The preparation method of Cu substrate is to evaporate a layer of chromium on the Si substrate and then plate copper. In order to enhance the binding force of Cu on the Si substrate, 10nm chromium was first evaporated as the seed layer, and then 100nm Cu was grown on it as the substrate. The 200μm patterned Cu/Si substrate was prepared by AZ5214 photoresist lithography. After the lithography process is completed, a patterned copper surface is prepared using the same method as above. Subsequently, the

979-8-3503-6184-1/24 $31.00 © 2024 IEEE   338

photoresist, together with any remaining metal chromium and copper, was ultrasonically removed in acetone for 15 minutes. Before the ALD process, ODT was grown on the surface of the Cu substrate. After the ALD process, acetic acid (> 99.9 %) was used for post-treatment in order to remove the SAMs on Cu and the physically adsorbed SiO_2 film.

B. SAMs Functionalizations

In order to grow ODT on the surface of Cu substrate, the substrate evaporated Cu should be soaked in anhydrous ethanol (> 99.9 %) and then taken out. The residual ethanol on the surface was purged with N_2 to remove organic impurities and inorganic dust. Then, the substrate was exposed to ultraviolet light for 10 min, and the surface was ultrasonically cleaned with absolute ethanol for 2 min to achieve oxidation and hydroxylation of the Cu surface. Then, it was immersed in 50mM ODT ethanol solution, heated to 50°C and soaked for 48h. After removing the substrate, the substrate was ultrasonically cleaned with anhydrous ethanol for 2min, then purged with N_2, and stored in a glove box in a N_2 environment.

C. RALD Parameters

The cavity temperature of the RALD process for the growth of SiO_2 thin film is 150°C. The temperature of the TMA precursor source bottle is kept at room temperature (25°C) without additional heating. The DMAI source bottle is heated to 80°C and kept warm. The TBS source bottle is heated to 95°C and then kept warm. The temperature of the pipeline throughout the deposition process is 110°C. A complete typical cycle of the process is carried out in the form of ACBC. First, 0.1s TMA or DMAI is introduced, followed by 20s N_2 purge, 0.5s TBS, and 20s N_2 purge.

D. SAMs Removal

After completing the RALD process, the substrate was ultrasonically washed with acetic acid for 10min, and then the sample was immersed in deionized water to remove excess acetic acid. After removing the excess acetic acid, it was dried with N_2.

E. Characterization Methods

The film thickness at 65° (SE, M-200X, J.A.Woollam Co., Inc.) was measured by ellipsometry, and the results were obtained by fitting the Cauchy model of alumina. The surface roughness of the film was measured by atomic force microscopy (AFM, SPM9700). The electrical properties of the samples were measured by a semiconductor characteristic analyzer (4200-SCS, Keithley). Auger electron spectroscopy (AES, JAMP9510F, JEOL) was used to scan the surface composition and elements of Si and Cu substrates, respectively.

III. RESULT AND DISCUSSION

A. The Basic Process of RALD

Figure 1 (a) and (b) are the growth rate of RALD at different temperatures, the change of film thickness with the number of RALD cycles and the growth rate at different TMA and TBS pulse times. Figure 1 (c) and (d) are the AFM test results. The process parameters are the same as the typical process described above except for the corresponding variables in the experiment shown in Figure 1 (a), (b).

Figure 1 (a) shows that the growth rate of the RALD process on the equipment in this paper reaches the highest at 175°C, which is 1.27nm/cycle, and the appropriate growth temperature window is 150-175°C. However, compared with the growth rate of more than 12nm/cycle reached in previous studies, it is greatly reduced. It is speculated that the condensation of some TBS precursor sources leads to relatively small pulse pressurization, which requires greater pulse pressure to achieve higher growth rate. The small figure in Figure 1 (a) clearly suggests that the growth rate under different cycles is well maintained on a fitting line, reflecting the linear growth and self-limiting characteristics of RALD. The large and small graphs in Figure 1 (b) are the saturated pulse curves of TBS and TMA, respectively. Relatively speaking, the pulse time of TMA between 0.1-0.4s has little effect on the growth rate of RALD. It can be considered that under the TBS pulse of 0.5s, the TMA pulse of more than 0.1s has reached saturation. With the increase of TBS pulse time, the growth rate is also increasing. Even when the TBS pulse reaches 1.5s, there is no sign of rate saturation, which also confirms the reason why the highest growth rate is lower than previous studies. The roughness of RALD films with TMA and DMAI as aluminum sources measured in Figure 1 (c) and (d) is Ra = 0.293nm and Ra = 0.232nm, respectively, which is a small value, indicating the uniformity of the deposited films.

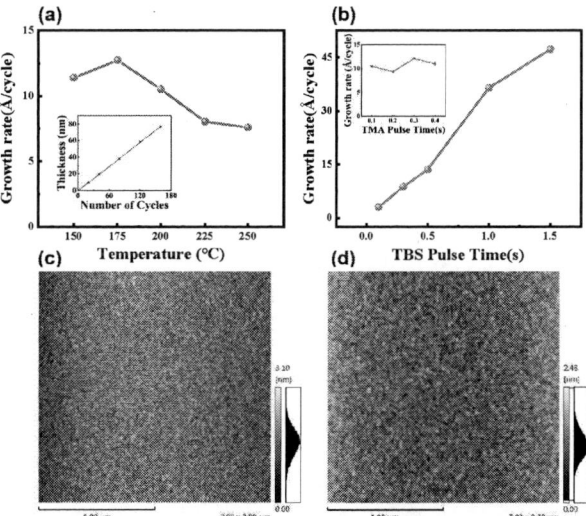

Fig. 1. (a) Temperature window and linear growth. (b) The saturated growth of TBS and TMA. Surface roughness of SiO_2 thin films grown by RALD using (c) TMA and (d) DMAI as aluminum sources.

B. The Electrical Properties of RALD

In this work, we compared the CV and IV characteristics of RALD films grown with TMA and DMAI as aluminum sources, as shown in Figure 2. Thin films with thickness of 7.38nm and 7.69nm were deposited by using TMA and DMAI as aluminum sources, respectively.

Figure 2 (a), (b) show the CV characteristics of the films respectively. Figure 2 (c), (d) show the IV characteristics of the two films. The dielectric constants (k values) of the two films calculated from CV curves are 4.55 and 4.12 respectively, which is close to the dielectric constant of the intrinsic SiO_2 of 3.97. It indicate that no matter what kind of aluminum source is used, most of the film components are Si, which is consistent with the description of the mechanism that Al only acts as a catalyst.

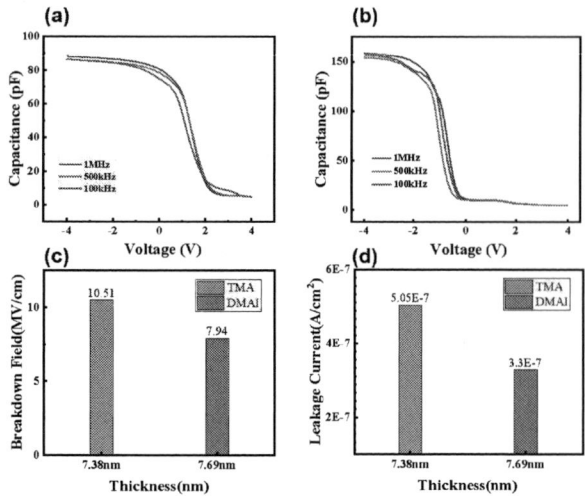

Fig. 2. The CV curve of (a) 7.38nm film grown with TMA as aluminum source and (b) 7.69nm film grown with DMAI as aluminum source. (c) Breakdown voltage of two films. (d) Leakage current of two films.

Figure 2 (c), (d) are the breakdown voltage and leakage current results obtained from the IV curve after calculation. The Large breakdown voltage (>7.94MV/cm) and lower leakage current (<5.05E-07A/cm²@2MV/cm) indicate the good electrical properties of RALD grown SiO_2 thin films. The small dispersion between CV curves at different frequencies also proves the conclusion above. In summary, the above results show that different aluminum sources have no significant effect on the electrical properties of RALD grown SiO_2.

C. The Area-Selectivity of RALD

In order to verify the selectivity of RALD on ODT-Cu and Si, A 6nm SiO_2 film was deposited on the Si substrate and an ODT-Cu sample was also placed in the cavity at the same time. The aluminum source and silicon source used were TMA and TBS, respectively. After the completion of the ALD process, the ODT-Cu and Si sample were subjected to AES test. After that, the same ODT-Cu sample and Si sample were subjected to acetic acid ultrasonic cleaning post-processing to remove ODT on the Cu surface and verify that the SiO_2 film on the Si surface is not affected by acetic acid cleaning. The AES test was performed again, and the results were shown in Figure 3.

Figure 3 (a), (b) are the broad-spectrum scanning results of Cu surface and Si surface before and after pickling, and (c) are the AES surface scanning images of the patterned structure. Figure 3 (a) shows that there is no obvious Al peak and Si peak on the Cu substrate before and after acetic acid pickling, indicating that ODT completely blocks the SiO_2 deposition of RALD on Cu, and acid pickling removes ODT on Cu. Figure 3 (b) shows that there are Al peaks and Si peaks on the Si substrate before and after pickling, indicating the normal growth on the Si substrate, and the acid pickling has no effect on the RALD growth film. Figure 3 (c) is the AES surface scan energy spectrum image of the acid-washed U-shaped patterned substrate of Cu, Al and Si elements. The U-shaped pattern is the Cu substrate, and the rest is the Si substrate. In the detection of Al and Si elements, the Cu

substrate shows a very small amount of Al and Si elements, indicating that the Al and Si on the Cu substrate are basically removed. The results are consistent with the above-mentioned broad-spectrum scanning results, indicating that ODT achieves good barrier performance on both light sheets and structured patterns.

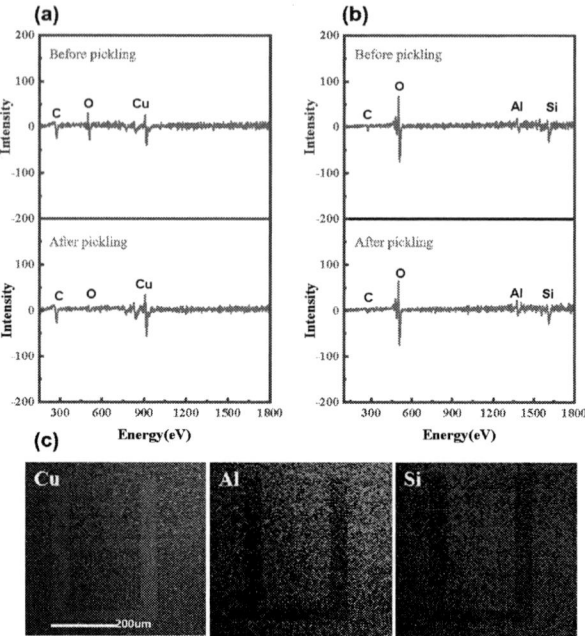

Fig. 3. The results of AES wide scan on the surface of (a) ODT-Cu and (b) Si before and after pickling. (c) U-shaped structure patterning AES spectral images.

IV. CONCLUSION

In this work, it is found that SiO_2 RALD reaches the maximum growth rate at 175°C and is greatly affected by the pulse size and time of TBS, while the pulse amount of TMA has little effect on the growth rate. AFM verified that the surface of the film grown by RALD was very smooth. The same electrical test was carried out on SiO_2 grown by RALD with different aluminum sources TMA and DMAI, and approximate results were obtained. It was indirectly verified that the film element composition was mostly Si and a small part of Al. The results of CV and IV characteristics show that the films grown by RALD have excellent electrical properties. When verifying the barrier property of ODT to RALD grown SiO_2, it was found that AES showed good results. The Al and Si signals are not shown on the ODT-Cu substrate in the AES broad spectrum before and after acid pickling, showing good barrier performance of ODT. The AES surface scan energy spectrum image also shows that good area-selectivity is achieved on the 200 micron structure chip.

ACKNOWLEDGMENT

This work is supported by the National Key R&D Program of China (2022YFF1500400), the National Natural Science Foundation of China (52350349, 52273237) and Tencent foundation. The authors would also like to acknowledge the support from the Analytic Testing Center and the Flexible Electronics Research Center of the HUST.

REFERENCES

[1] Chen R, Li Y C, Cai J M, Cao K, and Lee H B R. Atomic level deposition to extend Moore's law and beyond[J]. International Journal of Extreme Manufacturing, 2020, 2(2): 022002.

[2] Pasquali M, Brady-Boyd A, Lesniewska A, et al. Area-Selective Deposition of AlO x and Al-Silicate for Fully Self-Aligned Via Integration[J]. ACS Applied Materials & Interfaces, 2023, 15(4): 6079-6091.

[3] Lausecker C, Muñoz-Rojas D, Weber M. Atomic layer deposition (ALD) of palladium: from processes to applications[J]. Critical Reviews in Solid State and Materials Sciences, 2023: 1-23.

[4] Zhang J, Li Y, Cao K, et al. Advances in atomic layer deposition[J]. Nanomanufacturing and Metrology, 2022, 5(3): 191-208.

[5] Liu T L, Bent S F. Area-selective atomic layer deposition on chemically similar materials: Achieving selectivity on oxide/oxide patterns[J]. Chemistry of Materials, 2021, 33(2): 513-523.

[6] Choi Y, Kim H J, Kim E, et al. Molecular Mechanism of Selective Al_2O_3 Atomic Layer Deposition on Self-Assembled Monolayers[J]. ACS Applied Materials & Interfaces, 2023, 15(34): 41170-41179.

[7] Won S J, Kim J R, Suh S, et al. Effect of Catalyst Layer Density and Growth Temperature in Rapid Atomic Layer Deposition of Silica Using Tris (tert-pentoxy) silanol[J]. ACS Applied Materials & Interfaces, 2011, 3(5): 1633-1639.

[8] Hausmann D, Becker J, Wang S, et al. Rapid vapor deposition of highly conformal silica nanolaminates[J]. Science, 2002, 298(5592): 402-406.

Resistance Dependence of Cobalt on Line Width in Advanced Interconnects: First-Principles Modelling

Kang Wang[1], Menglin Huang[1*], Shiyou Chen[1*]

[1] School of Microelectronics and Key Laboratory of Computational Physical Sciences (MOE), Fudan University, Shanghai, China

* Email: menglinhuang@fudan.edu.cn, chensy@fudan.edu.cn

Abstract—The tremendous potential of the cobalt metal in advanced interconnect is systematically investigated from a theoretical perspective. Based on first-principles calculation, we propose that although copper is dominant in bulk resistivity and grain boundary scattering, when considering adhesion and diffusion barrier layers, cobalt possesses lower actual line resistance than copper for line width below about 20 nm. These features demonstrate the ability of cobalt metal to continue the interconnect pitch scaling.

Keywords—cobalt interconnect, NEGF, grain boundary, Mayadas-Shatzkes model

I. INTRODUCTION

The introduction of dual damascene copper (Cu) in back end of line (BEOL) has successfully realized the interconnect pitch scaling in the past generations of advanced process nodes [1]. However, as the continuous reduction is reaching the physical limit of tens of nanometers, Cu interconnect technology has suffered from a sharp increase of metal resistance and severe electromigration (EM) concerns, making it ineligible to continuously lead the pitch miniaturization [2-3]. To mitigate the adverse effects, multiple candidate metals have been proposed as the alternative of the next-generation interconnect materials, including cobalt (Co), Iridium (Ir), ruthenium (Ru) and so on [4-5].

Among these metals, Co has attracted numerous attention on account of its high EM reliability, enhanced manufacturability at nanoscale dimensions and lower line resistance compared to Cu with a thicker liner [6]. For example, Co interconnect technology has been implemented at the lowest two interconnect layers in Intel 10-nm technology node, showing a 2x reduction in via resistance and a 5-10x improvement in EM [7]. These features indicate that Co metal can serve as a suitable candidate for the next-generation interconnect technology, however, the verification of which in theory is greatly needed.

Here, we systematically investigate the potential of the Co metal for advanced interconnect from a theoretical perspective. Although it is predicted that the Cu interconnect exhibits lower line resistance due to lower bulk resistivity and grain boundary scattering, when considering adhesion and diffusion barrier layers, Co has advantage over Cu in terms of lower actual line resistance when the line width is below about 20 nm.

II. COMPUTATIONAL METHODS

The explorations on structural relaxation and electronic properties are implemented by the Vienna Ab initio Simulation Package (VASP) within the framework of the density functional theory (DFT) [8]. The exchange correlation energy is described by the Perdew-Burke-Ernzerhof (PBE) functional [9] and dense Gamma-centered k meshes are adopted for structural optimization and self-consistent calculations.

The transport properties are calculated by utilizing a combination of the DFT and the non-equilibrium Green's functions (NEGF) method [10]. The SG-15 pseudopotentials are used and the real-space mesh cutoff is set to 155 Hartree. The k-points density of the channel and electrode regions are 6×6×1 and 6×6×200 Gamma-centered grids for transport calculations, respectively.

The drain current for per spin channel at the specific gate voltage (V_G) and drain voltage (V_D) is an integration of the transmission coefficients T utilizing the Landauer-Büttiker formula:

$$I = \frac{e}{h} \int_{-\infty}^{+\infty} \{T(E, V_D, V_G)[f_S(E - \mu_S) - f_D(E - \mu_D)]\} dE \qquad (1)$$

For metal interconnects, the low-bias and zero-temperature conductance G is required to evaluate the line resistance, which can be directly calculated from the simplification of the Landauer-Büttiker formula:

$$G = \frac{e^2}{h} T(E_F) \qquad (2)$$

$$\gamma = \frac{A}{G} \qquad (3)$$

where $T(E_F)$ represents the transmission coefficient at Fermi energy, and γ is the area normalized specific resistivity [11-12].

III. RESULTS AND DISCUSSIONS

Co metal possesses two common polymorphs, the face-centered-cubic (FCC) and hexagonal close packed (HCP) phase. The spin-polarized band structures and density of states (DOS) for two phases are illustrated in Fig. 1(a) and 1(b), respectively. For bulk Co, the HCP phase is more stable at room temperature, while the FCC phase becomes thermodynamically more stable above 450 °C [13]. However, due to the increasingly significant impact of surface energy, the confined structures may form a polymorph or metastable phase differing from the stable bulk phase. From the perspective of the cohesive energy (the energy difference between the average energy of an isolated atom and the energy of an atom in a solid), we can find the HCP phase is energetically favorable by merely 0.018 eV/atom, much less than that of Ru metal (0.111 eV/atom). This energy barrier can easily be overcome with the increased surface energy triggered by the fine structure and the extra energy provided by the high temperature (300-400 °C) in the BEOL process, indicating the coexistence of two phases in the Co interconnect. Hence, the estimation of line resistance in Co

979-8-3503-6184-1/24 $31.00 © 2024 IEEE

interconnects requires a comprehensive consideration of the two phases. Moreover, the cohesive energy of Co exceeds about 40% compared to that of bulk Cu, showing the more solid bonds, which strengthens the ability to resist EM phenomena.

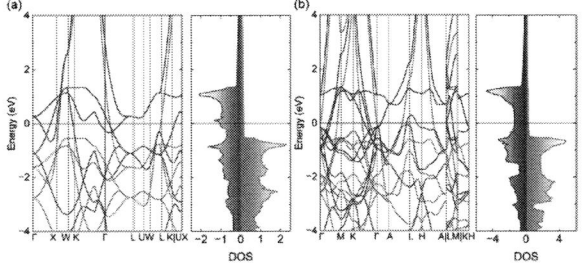

Fig. 1. Spin-polarized band structures and DOS of (a) FCC Co and (b) HCP Co. The spin-up and spin-down bands are depicted by magenta and blue lines, respectively.

We subsequently investigate the area normalized specific resistivity of representative orientations for both phases, as shown in Table I. Due to the intrinsic magnetism in bulk Co, the transmission coefficients contributed by different spin channels exhibit prominent difference, i.e., the spin-down state dominates. The specific resistivity between different orientations shows little anisotropy. Moreover, the specific resistivity of FCC phase shows generally a slight decline compared to that of HCP phase, for instance, the 30.173×10^{-12} $\Omega \cdot cm^2$ for [111] orientation in FCC Co and 34.555×10^{-12} $\Omega \cdot cm^2$ for [0001] orientation in HCP Co.

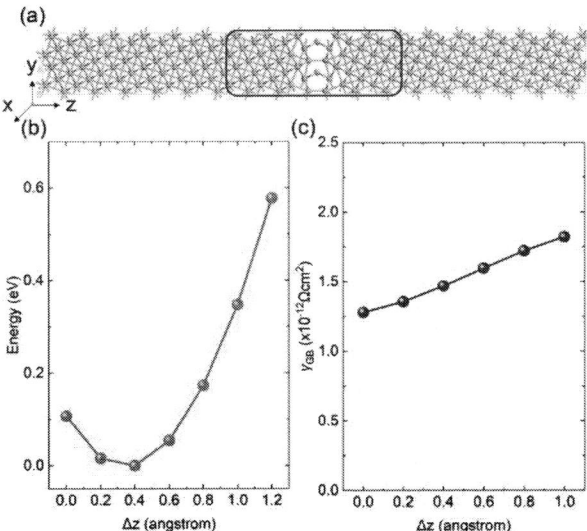

Fig. 2. (a) The atomic-scale representation of a supercell with grain boundary for transport calculation. The blue block denotes the structure needed to be relaxed. (b) The optimized energy curve implemented by VASP. (c) The specific resistivity of Cu Σ5 grain boundary using different relaxed structures.

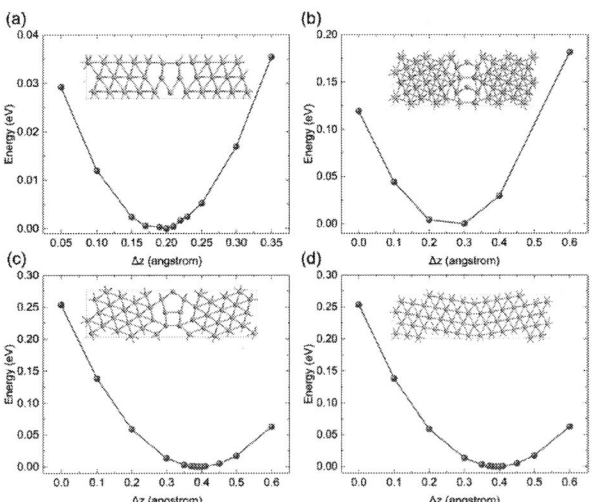

Fig. 3. The relaxed structures of representative coincident site lattice symmetric tilt grain boundaries (a) Σ3, (b) Σ5, (c) Σ9 and (d) Σ11 in FCC Co.

TABLE I.   Values of specific resistivity (in units of 10^{-12} $\Omega \cdot cm^2$) for different orientations

Lattice	Orientation	Area (Å²)	T_{UP}	T_{DOWN}	γ
FCC	[100]	42.799	6.294	15.110	24.865
	[110]	69.890	9.169	20.094	28.655
	[111]	42.799	5.201	12.367	30.173
	[112]	15.132	1.928	4.430	29.078
	[113]	20.488	2.651	6.109	28.604
	[120]	27.627	3.644	8.167	28.301
	[122]	18.532	2.350	5.320	29.347
HCP	[0001]	42.998	4.671	10.281	34.555
	[110]	34.713	4.387	5.279	37.397
	[100]	20.041	2.381	3.048	38.702
	[120]	26.512	3.311	4.043	37.593

Then we turn to the electron transmission across the grain boundary. Fig. 2 shows the method of structural relaxation, where several layers of atoms close to grain boundary are freely optimized, while the remaining atoms are only allowed to move along the z direction integrally when stretching or compressing the length of z axis. Using the Cu Σ5 grain boundary as an example, the optimal structure relaxed by VASP is 0.4 Å stretch along the z direction, whose specific resistivities of grain boundary is calculated as 1.468×10^{-12} $\Omega \cdot cm^2$, which is in good agreement with the prior result, 1.49×10^{-12} $\Omega \cdot cm^2$ [14]. Hence, the VASP will be used to optimize the structure.

We choose the representative coincident site lattice (CSL) grain boundaries in FCC cobalt, i.e., Σ3, Σ5, Σ9 and Σ11, whose relaxed structures are illustrated in Fig. 3. The specific resistivity of grain boundary γ_{GB} and reflection coefficient r can be calculated utilizing the whole specific resistivity with (γ_R) and without ($\gamma_{[hkl]}$) grain boundary [11]:

$$\gamma_{GB} = \gamma_R - \gamma_{[hkl]} \qquad (4)$$

$$r = 1 - \frac{\gamma_{[hkl]}}{\gamma_R} \qquad (5)$$

979-8-3503-6184-1/24 $31.00 © 2024 IEEE   343

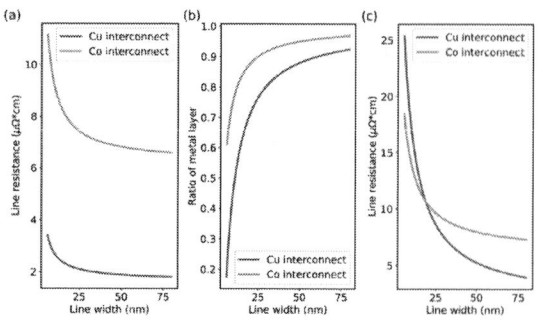

Fig. 4. (a) The line resistance of Cu and Co interconnects using the Mayadas-Shatzkes model. (b) The ratio of metal layer to the interconnect volume and (c) the actual line resistance accounting for adhesion and diffusion barrier layers.

TABLE II. Values of specific resistivity and reflection coefficient for representative coincident site lattice (CSL) grain boundaries in FCC Co

CSL	Area (Å^2)	T_{UP}	T_{DOWN}	γ_{GB}	r
Σ3	15.132	1.740	2.014	12.766	0.305
Σ5	27.627	3.376	2.845	17.886	0.387
Σ9	18.532	2.149	2.626	11.132	0.275
Σ11	20.488	2.508	3.084	9.632	0.252

Table II demonstrates the values of specific resistivity and reflection coefficient for representative grain boundaries in FCC cobalt. It's worth noting that the transmission coefficients of different spin channels have different attenuation degrees, with the spin-down channel being more seriously affected by grain boundary scattering. For example, the transmission coefficients decrease by 5.4% for spin-up channel and 49.5% for spin-down channel after the presence of grain boundary Σ11.

To quantify the impact of grain boundary scattering, Mayadas-Shatzkes empirical model is implemented to predict the line resistance of Co interconnect [11]:

$$\alpha = \frac{\lambda}{d}\frac{r}{1-r}, \gamma = \frac{\gamma_0}{1 - 3/2\alpha + 3\alpha^2 - 3\alpha^3 \ln(1+1/\alpha)} \quad (6)$$

where the γ_0 represents the bulk resistivity and γ is the resistivity accounting for grain boundary. We assume the line width is equal to the average grain size, and r refers to the average reflection coefficient of the four representative grain boundaries. Some data about bulk resistivity can be referred to [15].

Fig. 4(a) indicates Cu interconnect possesses a dominant position in low line resistance than Co interconnect, despite the nonnegligible size effect with the scaling of line width. However, the thickness of adhesion and diffusion barrier layers does not decreases proportionally with the line width, resulting in the increasing occupancy of liner in the interconnect volume, as shown in Fig. 4(b). Assuming the thinnest liner [11], we find that Co demonstrates lower actual line resistance than Cu for line width below about 20 nm, demonstrating the tremendous potential of Co metal in advanced interconnect, as plotted in Fig. 4(c).

IV. SUMMARY

Based on first-principles calculation, we systematically investigate the potential of the Co metal in advanced interconnect. Our results show Cu interconnect possesses a dominant position in bulk resistivity and grain boundary scattering than Co interconnect in spite of nonnegligible size effect with the scaling of line width. However, when considering adhesion and diffusion barrier layers, Co possesses lower actual line resistance than Cu for line width below about 20 nm, demonstrating the ability of Co metal to continue the interconnect pitch scaling.

ACKNOWLEDGMENT

This work was supported by Science and Technology Commission of Shanghai Municipality (Explorer project, 21TS1401000), National Natural Science Foundation of China (NSFC) under grant Nos. 12174060 and 12334005, the National Key Research and Development Program of China (2022YFA1402904), and Project of MOE Innovation Platform.

REFERENCES

[1] C.-C. Yang et al., "Line Resistance Reduction in Advanced Copper Interconnects," IEEE Electron Device Letters, vol. 38, no. 11, pp. 1579-1582, November 2017.

[2] A. Pyzyna et al., "Resistivity of copper interconnects at 28 nm pitch and copper cross-sectional area below 100 nm²," 2017 IEEE International Interconnect Technology Conference (IITC), Hsinchu, Taiwan, 2017, pp. 1-3.

[3] N. A. Lanzillo, U. Bajpai, I. Garate, and C. T. Chen, "Size-Dependent Grain-Boundary Scattering in Topological Semimetals," Physical Review Applied, vol. 18, pp. 034053, September 2022.

[4] D. Gall, "The search for the most conductive metal for narrow interconnect lines," Journal of Applied Physics, vol. 127, pp. 050901, February 2020.

[5] D. Gall, A. Jog, and T. Zhou, "Narrow interconnects: The most conductive metals," International Electron Devices Meeting (IEDM), San Francisco, CA, USA, 2020, pp. 32.3.1-32.3.4.

[6] T. Nogami et al., "Advanced BEOL Materials, Processes, and Integration to Reduce Line Resistance of Damascene Cu, Co, and Subtractive Ru Interconnects," 2022 IEEE Symposium on VLSI Technology and Circuits (VLSI Technology and Circuits), Honolulu, HI, USA, 2022, pp. 423-424.

[7] C. Auth et al., "A 10nm high performance and low-power CMOS technology featuring 3rd generation FinFET transistors, Self-Aligned Quad Patterning, contact over active gate and cobalt local interconnects," 2017 IEEE International Electron Devices Meeting (IEDM), San Francisco, CA, USA, 2017, pp. 29.1.1-29.1.4.

[8] G. Kresse, and J. Hafner. "G. Kresse and J. Hafner," Physical Review B, vol. 47, pp. 558, January 1993.

[9] J. P. Perdew, K. Burke, and M. Ernzerhof, "Generalized Gradient Approximation Made Simple," Physical Review Letters, vol. 77, pp. 3865, October 1996.

[10] M. Brandbyge, J. L. Mozos, P. Ordejón, J. Taylor, and K. Stokbro, "Density-Functional method for nonequilibrium electron transport," Physical Review B, Condensed matter and materials physics, vol. 65, pp. 165401, March 2002.

[11] T. M. Philip et al., "First-Principles Evaluation of fcc Ruthenium for its use in Advanced Interconnects," Physical Review Applied, vol. 13, pp. 044045, April 2020.

[12] N. A. Lanzillo, P. Bhosale, C. Lavoie, D. J. Dechene, R. R. Robison, and K. Choi, "Spin-dependent electron scattering in cobalt interconnects," Journal of Physics D: Applied Physics, vol. 52, pp. 495302, September 2019.

[13] X. Ma et al., "Guiding Synthesis of Polymorphs of Materials Using Nanometric Phase Diagrams," Journal of the American Chemical Society, vol. 140, pp. 17290–17296, November 2018.

[14] M. César, D. Liu, D. Gall, and H. Guo, "Calculated Resistances of Single Grain Boundaries in Copper," Physical Review Applied, vol. 2, pp. 044007, October 2014.

[15] D. Gall, "Electron mean free path in elemental metals," Journal of Applied Physics, vol. 119, pp. 085101, February 2016.

Reduction of Specific Contact Resistivity by Employing Pre-amorphization Implantation and In-situ Steam Generation Oxidation

Chang Liu[1,2,3], Xu Chen[1,2,3], Jinbiao Liu[1,2], Yanping He[1,2,3], Wenjuan Xiong[1,2], Weibing Liu[1,2], Mingshan Liu[4], Zhe Liu[4], Yaoqi Dong[4], Jeffrey Xu[4], Jing Xu*[1,2,3], Jun Luo[1,2,3]

[1] Key Laboratory of Fabrication Technologies for Integrated Circuits, Chinese Academy of Sciences, Beijing 100029.
[2] Institute of Microelectronics, Chinese Academy of Sciences, Beijing 100029, China.
[3] School of Integrated Circuits, University of Chinese Academy of Sciences, Beijing 100049, China.
[4] Huawei Technologies Company limited, Shenzhen 518100, China.

* Email: xujing@ime.ac.cn, liuchang2023@ime.ac.cn

Abstract—Specific contact resistivity (ρ_c) at the transistor source/drain has become a bottleneck for further improvement of modern Si CMOS. The increase of surface dopant concentration (Ns) and the reduction of Schottky barrier height (SBHs) are considered as one of the most effective approaches to reduce ρ_c. In this paper, the combining effect of pre-amorphization implantation (PAI) and *in-situ* steam generation (ISSG) oxidation on ρ_c is studied. The results show that ρ_c can be significantly reduced by the germanium (Ge) or arsenic (As) PAI and ISSG oxidation after phosphorus (P) implantation. When the Ge implantation dose is set at 1×10^{15} cm^{-2} and ISSG is performed, the ρ_c achieves a value of 2.82×10^{-8} $\Omega\cdot$cm^2. When the implantation dose of As is 6×10^{14} cm^{-2} and ISSG is performed, the ρ_c reaches 1.65×10^{-8} $\Omega\cdot$cm^2. In comparison with the Ge PAI and ISSG, the segregation of P and As at the interface by As PAI and ISSG processes provide the higher carrier concentrations to reduce ρ_c.

Keywords—Ge/As PAI, ISSG oxidation, specific contact resistivity, dopant segregation

I. INTRODUCTION

As device miniaturization advances, the relative contribution of source/drain resistance to the overall resistance of the transistor escalates significantly. Particularly at the 32nm technology node in CMOS fabrication, the source/drain parasitic resistance (R_{para}) surpasses the channel resistance ($R_{channel}$), adversely impacting device performance enhancement[1], [2]. The contact resistance (R_c) has become the dominant factor of the R_{para}. Notably, the proportion of source/drain contact resistance increases as the contact area diminishes, making it a critical factor in limiting further device size reduction[3]. Moreover, with the transition of transistors from planar to 3D FinFET architectures, current flow predominantly occurs in the vertical direction[4]. Therefore, selecting metal silicides that form low contact resistance with the source and drain is crucial. To decrease the contact resistance, strategies involve either increasing the contact area or reducing the ρ_c. As device size decreases, the effective contact area becomes smaller, making the reduction of ρ_c particularly important[5], [6], [7].

Germanium (Ge) Pre-amorphization Implantation (PAI) is a widely used method for reducing specific contact resistivity[8]. Before metal deposition, a certain dose of Ge is implanted on the Si surface to form a layer of amorphous silicon. This increases the free energy of the system, promotes Si diffusion, and enhances the Ti-Si reaction during subsequent annealing [9], [10]. Alternatively, arsenic (As) PAI has been proposed as a substitute for Ge PAI. Besides causing amorphization, As also reduces the barrier height[11], [12].

Dopant segregation is another effective method to reduce ρ_c. By dopant segregation, impurity concentration at the Si surface increases, thus improving the carrier penetration rate and reducing ρ_c. The *in-situ* steam generation (ISSG) oxidation process forms a thin oxide layer on the Si surface, which redistributes impurities within Si substrate. ISSG facilitates impurity segregation, enhances the surface doping concentration, and reduces ρ_c[13].

Given that PAI promotes Si diffusion and enhances Ti-Si bonding, and ISSG can increase the surface doping concentration and reduce the barrier height, we combined the PAI and ISSG processes in this study to explore the peak of N_s value. In order to quantify the performance of the method, refined transmission line model (RTLM) structures were fabricated and characterized in terms of ρ_c while dopant profile and the N_s value were characterized using secondary ion mass spectrometry (SIMS) and spreading resistance probe (SRP).

II. EXPERIMENTAL PROCEDURE

In this study, RTLM structures with varied spacing from 2 μm to 5 μm were used to extract the ρ_c. The structure is shown as Fig. 1 while the schematic diagram of the process is presented in Fig. 2. The 8-inch wafers with a resistivity of 0.5-100 $\Omega\cdot$cm were used as the substrate. The natural oxide layer on the surface was removed using diluted HF solution. After forming a P-well, P were implanted with dose of 3×10^{15} cm^{-2} to form an N-type heavy doping layer. Subsequently, As PAI or Ge PAI with varying doses as detailed in Table 1 was carried out in the contact areas. Then the thin silicon dioxide (SiO$_2$) layer was grown by ISSG oxidation at 600 ℃, and the thickness of the oxide layer was 3 nm while a control wafer skipped PAI and ISSG as the reference. To activate dopants, all wafers experienced a spike annealing step at 1050 °C. After this step, mesas of active area were formed by the first photolithography and reactive ion etching (RIE). Following the removal of the residual oxide layers, a multilayered structure comprising 250 Å of SiO$_2$, 500 Å of Si$_3$N$_4$, and 2000 Å of SiO$_2$ – collectively referred to as the ONO layer – was sequentially deposited to serve as the masking layer for the second photolithography process. After the second lithography, the contact areas were formed. Subsequently, 5 nm Ti, 3 nm TiN and 400 nm W were deposited. Ti silicide was formed after rapid thermal annealing at 550 °C for 60 s. Finally, the metal Pad was formed by the third lithography.

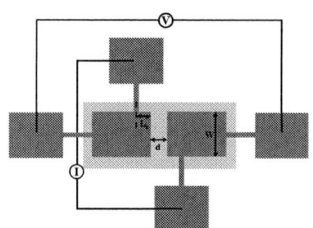

Fig. 1.   Schematic of RTLM structure

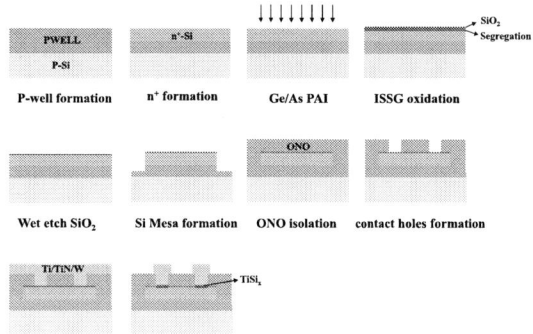

Fig. 2.   Schematic diagram of the process for the preparation of the RTLM structure

TABLE I.   EXPERIMENTAL CONDITIONS OF THE SAMPLES

Wafer ID	Ge implantation dose(cm^{-2})	As implantation dose(cm^{-2})	SiO$_2$ thickness caused by ISSG（nm）
Ref.	/	/	/
Ge PAI1+ISSG	6x10^{14}	/	3
Ge PAI2+ISSG	8x10^{14}	/	3
Ge PAI3+ISSG	1x10^{15}	/	3
As PAI1+ISSG	/	6x10^{14}	3
As PAI2+ISSG	/	8x10^{14}	3
As PAI3+ISSG	/	1x10^{15}	3
ISSG	/	/	3

III. RESULTS AND DISCUSSION

As shown in Fig. 3, the ρ_c of the reference sample, which underwent impurity activation annealing directly after implantation, is 3.68×10^{-8} $\Omega\cdot$cm^2. For the samples with Ge PAI and ISSG, the ρ_c at a dose of 6×10^{14} cm^{-2} is 4.09×10^{-8} $\Omega\cdot$cm^2, which is higher than that of the reference wafer. As the Ge dose increases, ρ_c gradually decreases. The specific contact resistivity of sample with Ge dose of 1×10^{15} cm^{-2} decreases to 2.82×10^{-8} $\Omega\cdot$cm^2 which is 23% lower than the reference. For the samples of As PAI and ISSG, the ρ_c is lower than that of the reference. With increasing the As dose, ρ_c gradually increases. When the As dose is 6×10^{14} cm^{-2}, the ρ_c reaches its lowest value of 1.65×10^{-8} $\Omega\cdot$cm^2, which is 55% lower than the reference. In addition, the ρ_c of the sample with only ISSG is 3×10^{-8} $\Omega\cdot$cm^2. Though still lower than reference

value, the mere 18% reduction in ρ_c suggests its weaker effect than the combination of ISSG and PAI.

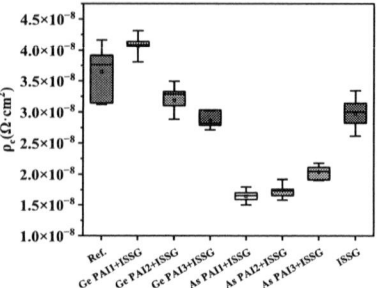

Fig. 3.   Extracted ρ_c of TiSi$_x$/n$^+$-Si contacts with PAI and ISSG

In order to analyze the reasons for the reduction of specific contact resistivity in PAI and ISSG oxidation, we used SIMS to characterize impurity concentrations in some samples before metal deposition. The effects of Ge PAI, As PAI and ISSG oxidation on the surface impurity concentration were observed. As shown in Fig. 4, all samples exhibit impurity segregation at the SiO$_2$/Si interface under ISSG oxidation, resulting in the accumulation of P atoms.

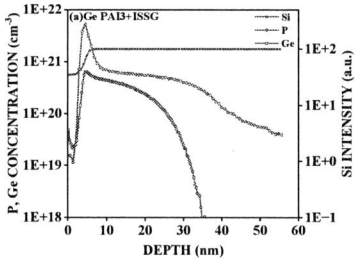

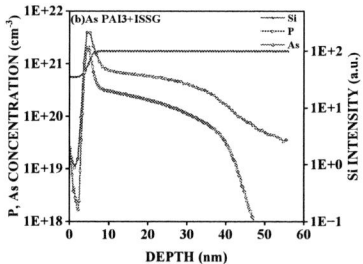

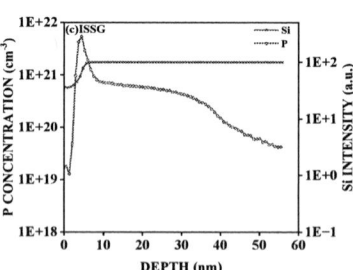

Fig. 4.   SIMS profiles of the samples: (a)Ge PAI3+ISSG, (b)As PAI3+ISSG and (c) ISSG.

Ge and As also accumulate on the surface of Si. It is worth noting that, by comparing Fig. 4(a) and (b), the interface morphology differs at the same implantation dose of Ge and As PAI. As forms a distinct peak at the Si/SiO$_2$ interface,

979-8-3503-6184-1/24 $31.00 © 2024 IEEE

while Ge does not show an obvious peak. It can be attributed to the different diffusion coefficient and segregation coefficient. The separation coefficient of As on Si/SiO_2 surface is much higher than 1(m>1), while the diffusion rate in silica is slow[14]. During the ISSG oxidation process, As accumulates on the surface of Si, similar to P. Specifically, As reaches a high concentration of 1.4×10^{21} cm^{-3} on the surface.

Due to the limitations of SIMS characterization, the carrier concentrations of two samples (#Ge PAI3+ISSG and #As PAI3+ISSG) were further analyzed by SRP. As shown in Fig. 5, the carrier concentration in As PAI sample is significantly higher than that in Ge PAI samples. The maximum carrier concentration is 3.82×10^{20} cm^{-3}, which is significantly higher than that of pure ISSG in the references[13]. Through the combination of As PAI and ISSG oxidation, P can be separated on the surface to improve the impurity concentration, while a higher concentration of As can be achieved to enhance the PAI effect and reduce the barrier height. As a result, the increase of N_s value leads to the reduction of ρ_c.

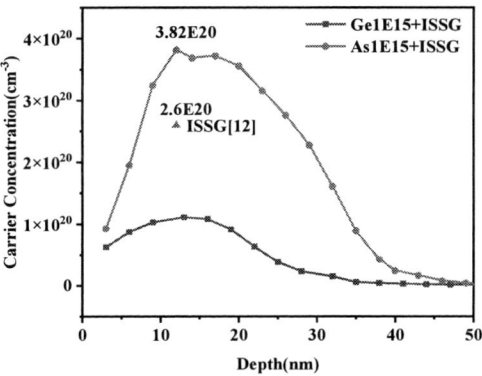

Fig. 5. Carrier Concentration of the samples before silicidation.

IV. CONCLUSION

Due to the enhancement of N_s by P and As dopant segregation , the combination of PAI and ISSG oxidation can effectively reduce the ρ_c value for TiSi$_x$/n$^+$-Si contacts. When the Ge implantation dose is 1×10^{15} cm^{-2} and ISSG is performed, the specific contact resistivity reaches 2.82×10^{-8} $\Omega \cdot cm^2$, which is 23% lower than the reference. When the As implantation dose is 6×10^{14} cm^{-2} and ISSG is performed, the specific contact resistivity reaches 1.65×10^{-8} $\Omega \cdot cm^2$, which is 55% lower than the reference. These reductions were 6% and 45%, respectively, compared to the sample with ISSG only. The results achieved in this work are beneficial for providing useful insights to the optimization of S/D contact technology in state-of-the-art CMOS technology.

ACKNOWLEDGMENT

We thank Dr. Mingshan Liu from Huawei Technologies Co., Ltd. for his team support. This work was supported in part by Guangdong Province Research and Development Program in Key Fields under Grant 2021B0101280002 and in part by the Strategic Priority Research Program of Chinese Academy of Sciences under Grant XDA0330300.

REFERENCES

[1] Topaloglu R O. 3-2-1 contact: an experimental approach to the analysisof contacts in 45 nm and below[C]//Proceedings of the 12th ACM/IEEE international workshop on System level interconnect prediction. 2010: 59-66.

[2] Noori A M, Balseanu M, Boelen P, Cockburn A, Demuynck S, et al. Manufacturable Processes for ≤32-nm-node CMOS Enhancement by Synchronous Optimization of Strain-Engineered Channel and External Parasitic Resistances[J]. IEEE transactions on electron devices, 2008, 55(5): 1259-1264.

[3] Kim S D, Park C M, Woo J C S. Advanced model and analysis of series resistance for CMOS scaling into nanometer regime. I. Theoretical derivation[J]. IEEE Transactions on Electron Devices, 2002, 49(3): 457-466.

[4] Yu H, Schaekers M, Peter A, Pourtois G, Rosseel E, et al. Titanium Silicide on Si: P With Precontact Amorphization Implantation Treatment: Contact Resistivity Approaching 1x10-9 Ohm-cm2[J]. IEEE Transactions on Electron Devices, 2016, 63(12): 4632-4641.

[5] Jacob A P, Xie R, Sung M G, Liebmann L, Lee RTP, et al. Scaling challenges for advanced CMOS devices[J]. International Journal of High Speed Electronics and Systems, 2017, 26(01n02): 1740001.

[6] Yu H, Schaekers M, Everaert J L, Horiguchi N, Meyer K D, et al. A snapshot review on metal–semiconductor contact exploration for 7-nm CMOS technology and beyond[J]. MRS Advances, 2022, 7(36): 1369-1379.

[7] Razavieh A, Zeitzoff P, Nowak E J. Challenges and limitations of CMOS scaling for FinFET and beyond architectures[J]. IEEE Transactions on Nanotechnology, 2019, 18: 999-1004.

[8] Mao S, Liu J, Wang Y, Liu W, Hu Y, et al. Investigation on Contacts Thermal Stability for 3D Sequential Integration[C]//2022 IEEE International Reliability Physics Symposium (IRPS). IEEE, 2022: P37-1-P37-4.

[9] Mao S, Wang G, Xu J, et al. Impact of Ge preamorphization implantation on both the formation of ultrathin TiSi x and the specific contact resistivity in TiSi x/n-Si contacts[J]. IEEE Transactions on Electron Devices, 2018, 65(10): 4490-4498.

[10] Yu H, Schaekers M, Rosseel E, Peter A, Lee JG, et al. 1.5× 10− 9 Ωcm2 Contact resistivity on highly doped Si: P using Ge pre-amorphization and Ti silicidation[C]//2015 IEEE International Electron Devices Meeting (IEDM). IEEE, 2015: 21.7. 1-21.7. 4.

[11] Mao S, Zhao C, Liu J, Wang G, Li M, et al. Specific contact resistivity improvement by as preamorphization implantation for ti-based ohmic contacts on n+-Si[J]. IEEE Transactions on Electron Devices, 2020, 67(4): 1726-1729.

[12] Mao S, Zhao C, Liu J, Wang G, Zhang Y, et al. Experimental investigation of as preamorphization implant on electrical property of Ti-based silicide contacts[J]. IEEE Transactions on Electron Devices, 2021, 68(4): 1835-1840.

[13] Zhang D, Zhao C, Xu J, Gao J, Liu J, et al. A Novel Method to Reduce Specific Contact Resistivity of TiSi x/n+-Si Contacts by Employing an In-Situ Steam Generation Oxidation Prior to Ti Silicidation[J]. IEEE Electron Device Letters, 2021, 42(7): 958-961.

[14] Suzuki K, Yamashita Y, Kataoka Y, Kataoka Y, Yamazaki K, Kawamura K. Segregation coefficient of boron and arsenic at polycrystalline silicon/SiO2 interface[J]. Journal of the Electrochemical Society, 1993, 140(10): 2960.

High-performance Ultrathin ITO Thin-Film Transistor With Ultralow Subthreshold Swing

Yanheng Liu[1], Tiaoyang Li*[1]

*[1]Fuzhou University-Jinjiang Joint Institute of Microelectronics and School of Physics, Information Engineering and Microelectronics, Fuzhou University, Fuzhou 350108, China

*Email: 241110016@fzu.edu.cn, tyl@fzu.edu.cn

Abstract—Indium tin oxide (ITO) has garnered significant attention for its transparency, wide bandgap, and high mobility. In this study, we fabricated high-performance thin-film transistors (TFTs) using semiconductor ITO as the active channel. By adjusting the oxygen partial pressure during the magnetron sputtering process and employing a 10 nm HfLaO gate dielectric, we achieved a field-effect mobility of up to 40 cm²/Vs. We then investigated the impact of ITO channel thickness on device performance, achieving a threshold voltage greater than 0 V and a subthreshold swing as low as 61 mV/dec when the ITO channel thickness was reduced to 5 nm. Notably, when the channel length was decreased to 70 nm, the on-state current reached 722 μA/μm, and the maximum transconductance reached 215 μS/μm with V_{ds} = 1 V. These results underscore the significant potential of ITO transistors for applications in high-performance, low-power electronic devices.

Keywords—Indium Tin Oxide, Thin Film Transistors, Field-effect Mobility, Subthreshold Swing

I. INTRODUCTION

As modern electronic devices continue to evolve towards portability, low power consumption, and high performance, the demand for material selection and performance optimization in electronic components is on the rise. Oxide semiconductors, characterized by a wide bandgap (E_g > 2 eV), suitable mobility (10 to 100 cm²/V·s), and low-temperature fabrication (T < 400 °C), are commonly used in the channel layer of TFTs[1]. These materials have been extensively applied in three-dimensional stacked memory devices[2] and post-process integration of silicon technology[3]. ITO, known for its high electrical conductivity and good transparency, has been widely used in display technology and solar cells[4]. Recent studies have shown that oxide semiconductors can overcome the short-channel effect [5]and achieve high-performance transistors, while also exhibiting extremely low static power consumption, making the fabrication of ultra-low power electronic devices possible. However, the application of ITO in the field of TFTs has been limited by its high carrier concentration and poor gate control performance[6]–[8]. Recent research indicates that oxide semiconductors can overcome the short-channel effect, enabling high-performance transistors with extremely low static power consumption. During the fabrication of TFTs, the control of oxygen content is one of the key factors in achieving high-performance devices. Oxygen vacancies in ITO thin films directly affect the concentration and type of carriers, thereby influencing the threshold voltage, subthreshold slope, and current on-off ratio of the transistor. This study focuses on the crucial role of the oxygen content ratio in ITO thin films and the impact of oxygen partial pressure during growth by magnetron sputtering on the electrical properties of TFTs. The study shows that as the oxygen partial pressure increases, the

oxygen content in ITO thin films gradually increases, effectively filling oxygen vacancies, thereby reducing the carrier concentration and achieving the transition from a metallic state to a semiconducting state. In terms of electrical properties, an appropriate oxygen partial pressure can significantly improve the current on-off ratio of the transistor, reduce the off-state leakage current, and enhance the gate's electrostatic control over the channel, thereby optimizing the subthreshold slope and mobility of the transistor. Through experimental data and theoretical analysis, the optimal oxygen partial pressure conditions were determined, providing important insights into the key factors in achieving high-performance ITO TFTs.

II. EXPERIMENTS

The fabrication process of the ITO TFT is encapsulated in Fig. 1. It commences with the selection of a highly doped, low-resistivity silicon substrate characterized by a resistivity

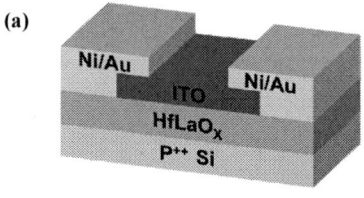

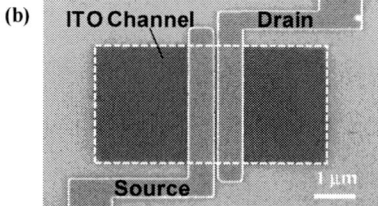

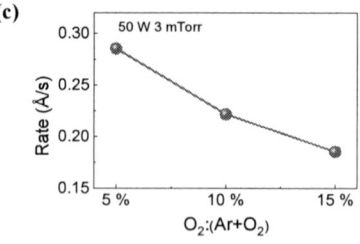

Fig. 1. (a) Device structure (b) SEM image of ITO TFT with W/L of 3.5 μm/70 nm and (c) The growth rate of ITO thin films under varying O₂/(Ar+O₂) ratios

of less than 0.005 Ω·cm. Utilizing Atomic Layer Deposition (ALD), the high-k dielectric HfLaOₓ is deposited at a temperature of 300 °C.Subsequently, an ITO thin film is applied via ultra-high vacuum magnetron sputtering under a base vacuum of less than 10⁻⁷ Torr. A specific ratio of Argon

979-8-3503-6184-1/24 $31.00 © 2024 IEEE

(Ar) and oxygen (O$_2$) is introduced into the chamber, with the pressure maintained at 50 mTorr. The RF power, set at 50 W, is engaged, and adjustments to the capacitance and inductance in the RF coupling circuit are made to keep the reflected power to within 1 W. For the isolation of the active layer, PMMA A4 is employed as a mask photoresist, with electron-beam lithography (EBL) being utilized to define the channel region pattern. This pattern is developed in a MIBK: IPA (1:3) solution for 50 seconds, followed by a rinse in IPA for 40 seconds. Etching is performed with a 5 % HCl solution for 3 minutes, and the PMMA A4 photoresist is eliminated by immersing in NMP at 80 °C for an hour. The substrate is then cleansed in IPA for 5 minutes. In the final stages of fabrication, for the formation of the source and drain metals, PMMA A3 is spin-coated onto the substrate as an EBL mask photoresist, ensuring a uniform layer at a rotation speed of 3000 rpm for 60 seconds. The source and drain contact metal regions are delineated using EBL. The sample undergoes physical vapor deposition (PVD) to deposit the Ni/Au metals for the source and drain contacts. Subsequently, the sample is immersed in acetone at 50 °C for 30 minutes, facilitating the dissolution of the PMMA, thereby stripping the metal from regions outside the source and drain electrodes.

A. Optimization of oxygen content concentration in indium tin oxide (ITO)

Fig. 2 presents the transfer and output electrical characteristics of a back-gated top-contact device incorporating a 6 nm ITO thin film at room temperature.

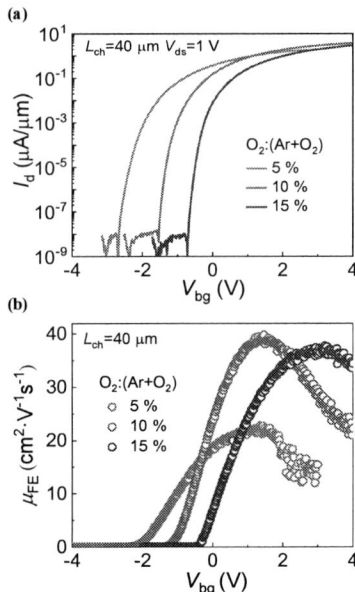

Fig. 2. (a) Transfer characteristics and (b) mobility characteristics at different O$_2$/(Ar+O$_2$) ratios

Characterized by a channel width and length of 40 µm, the thin-film transistor's electrical performance is significantly influenced by the modulation of oxygen partial pressure during the magnetron sputtering process. This process demonstrates a pronounced positive shift in the threshold voltage (V$_{th}$) with increased oxygen concentration. The device impressively achieves an on/off current ratio of 10^8 at a drain-source voltage (V$_{ds}$) of 1 V. Furthermore, the mobility within the linear region is meticulously extracted by the saturation equation.

$$I_{ds}=\mu_n C_{ox}\frac{W}{L}\left[\left(V_{gs}-V_{th}\right)V_{ds}-\frac{1}{2}V_{ds}^2\right] \quad (1)$$

$$\mu_{FE}=\frac{g_m L_{ch}}{W_{ch}C_{ox}V_{ds}} \quad (2)$$

Equation (2) is derived from the premise that (1) is applied to the linear operation region of a TFT with exceedingly small V$_{ds}$. In this context, g$_m$ denotes the transconductance, L represents the channel length, W signifies the channel width, and C$_{ox}$ refers to the gate oxide unit capacitance. Employing a 10 nm HfLaO$_x$ as the gate dielectric, C$_{ox}$ is quantified at 1.16 µC/cm^2. Observations reveal that at an oxygen partial pressure of 10 %, the ITO film attains its maximum mobility of 39.77 cm^2/V·s. Consequently, the optimal deposition environment for the ITO thin film is identified to be at an oxygen partial pressure of 10 %. Unbound metal ions predominantly source the carriers within the ITO thin film. A reduction in the concentration of oxygen vacancies effectively diminishes the carrier concentration, thereby transitioning the ITO thin film from a metallic to a semiconducting state and consequently achieving enhanced electronic transport capabilities.

B. The impact of indium tin oxide (ITO) thickness on transistor performance

This study examines the impact of the ITO active layer thickness on transistor performance, focusing on the reduction from 6 nm to 4 nm.

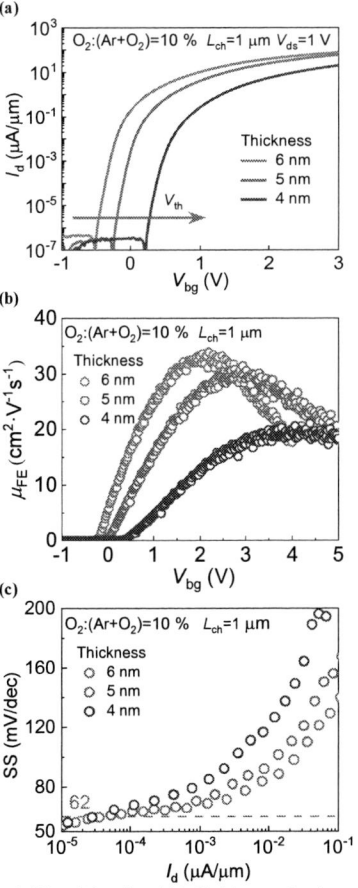

Fig. 3. At an O$_2$/(Ar+O$_2$) ratio of 10 %, (a) transfer characteristics and (b) mobility characteristics of ITO thin films of different thicknesses, and (c) subthreshold swing characteristics, where the lowest subthreshold swing (SS) of 61 mV/dec is achieved when the ITO film thickness (t$_{ITO}$) is 5 nm

979-8-3503-6184-1/24 $31.00 © 2024 IEEE

As illustrated in Fig. 3(a), at room temperature and an oxygen partial pressure of 10%, with a device aspect ratio of 3.5 μm/1 μm, there is a positive shift in the threshold voltage (V_{th}) with decreasing active layer thickness. Although an increase in the ITO film thickness reduces the subthreshold slope, the on/off current ratio remains relatively constant. Fig. 3(b) demonstrates that the mobility of the transistor reaches its peak of 31.46 cm²/V·s at a thickness of 5 nm.

$$SS = \frac{\partial V_{gs}}{\partial \log_{10} I_{sub}} \qquad (3)$$

Furthermore, Fig. 3(c) indicates that as the ITO thickness diminishes, the subthreshold swing (SS) is significantly reduced, achieving the lowest value of 62 mV/dec. This reduction in SS with decreasing thickness is reminiscent of observations from Silicon-On-Insulator (SOI) literature, where a thinner channel body leads to an increase in the number of electrons in orbits with a smaller effective mass due to subband energy level modulation, thereby enhancing mobility. This phenomenon is analogous to the behavior seen in ITO transistors[8].

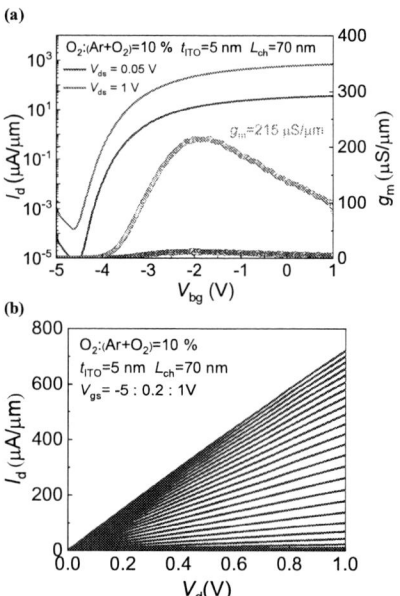

Fig. 4. (a) (b) Transfer and output characteristics of the channel with 70-nm and 5-nm thickness

Fig. 4(a) presents the transfer characteristics of the ITO transistor with a 70 nm channel length, showcasing a transconductance of 215 μS/μm. Figure 4(b) displays the output characteristics with a gate-source voltage (V_{gs}) sweep from -5 V to 1 V in increments of 0.2 V. At a drain-source voltage (V_{ds}) of 1 V, the transistor exhibits a maximum on-state current of 722 μA/μm, which fully demonstrates the advantages of short-channel effects and excellent contact performance characteristic of ITO transistors.

III. Summary

This article delves into the impact of the oxygen content ratio on the electrical characteristics of indium tin oxide (ITO) thin-film transistors. By precisely controlling the oxygen composition ratio during the magnetron sputtering process, we successfully transformed ITO from a metallic state to a semiconducting state, significantly reducing the carrier concentration of the thin film. Experimental results demonstrate that as the oxygen partial pressure increases, key electrical parameters such as the threshold voltage, subthreshold slope, and current on-off ratio of the ITO thin film are markedly optimized. Notably, when the oxygen partial pressure is at 10 %, the ITO thin film exhibits the highest electron mobility and the best overall electrical performance, confirming the importance of appropriate oxygen partial pressure in achieving high-performance ITO transistors.

Acknowledgment

We thank Mingwei Wang, Benhui Lin, and Guosheng Chen at Fuzhou University-Jinjiang Joint Institute of Microelectronics of Fuzhou University for their support in e-beam lithography and metal deposition. This work was funded by the National Natural Science Foundation of China (Grant No. 62204042), the Science and Technology Major Project of Fujian Province, China (Grant No. 2021HZ021027), and the Natural Science Foundation of Fujian Province, China(Grant No. 2021J05118).

References

[1] S. Li, M. Tian, Q. Gao, M. Wang, T. Li, Q. Hu, X. Li, and Y. Wu, "Nanometre-thin indium tin oxide for advanced high-performance electronics," *Nat. Mater.*, vol. 18, no. 10, pp. 1091–1097, Oct. 2019.

[2] X. Duan, K. Huang, J. Feng, J. Niu, H. Qin, S. Yin et al., "Novel Vertical Channel-All-Around(CAA) IGZO FETs for 2T0C DRAM with High Density beyond 4F2 by Monolithic Stacking," in *2021 IEEE International Electron Devices Meeting (IEDM)*, 2021, p. 10.5.1-10.5.4.

[3] J. Zhang, Z. Lin, Z. Zhang, K. Xu, H. Dou, B. Yang, A. Charnas, D. Zheng, X. Zhang, H. Wang, and P. D. Ye, "Back-End-of-Line-Compatible Scaled InGaZnO Transistors by Atomic Layer Deposition," *IEEE Trans. Electron Devices*, vol. 70, no. 12, pp. 6651–6657, Dec. 2023.

[4] T. Minami, "Present status of transparent conducting oxide thin-film development for Indium-Tin-Oxide (ITO) substitutes," *Thin Solid Films*, vol. 516, no. 17, pp. 5822–5828, Jul. 2008.

[5] M. Zhang, D. Xu, R. Duan, J. Shi, J. Dong, Y. Wang, D. Han, and X. Zhang, "P-1.12: Sub-200nm Nano-scale Indium-Zinc-Oxide Ultra-thin Channel Transistors," *Symp Digest of Tech Papers*, vol. 55, no. S1, pp. 664–666, Apr. 2024.

[6] H. Kim, A. Piqué, J. S. Horwitz, H. Mattoussi, H. Murata, Z. H. Kafafi, and D. B. Chrisey, "Indium tin oxide thin films for organic light-emitting devices," *Applied Physics Letters*, vol. 74, no. 23, pp. 3444–3446, Jun. 1999.

[7] I. Hamberg and C. G. Granqvist, "Evaporated Sn-doped In2O3 films: Basic optical properties and applications to energy-efficient windows," *Journal of Applied Physics*, vol. 60, no. 11, pp. R123–R160, Dec. 1986.

[8] K. Uchida, H. Watanabe, A. Kinoshita, J. Koga, T. Numata, and S. Takagi, "Experimental study on carrier transport mechanism in ultrathin-body SOI nand p-MOSFETs with SOI thickness less than 5 nm," in *Digest. International Electron Devices Meeting*, San Francisco, CA, USA, 2002, pp. 47–50.

Defect-Centric Insights into Flicker Noise in Ultra-Scaled FETs: From Physics to Compact Model for Circuit Level Simulation (Invited)

Chenyang Zhang[1], Yu Xiao[1], Pengpeng Ren[1], Shiyu Xia[1], Sheng Yang[1], Lining Zhang[2], Runsheng Wang[3,4], Zhigang Ji*[1,4]

[1]Department of Micro/Nano Electronics, SJTU, China, [2]ECE, PKU, Shenzhen, China, [3]School of Integrated Circuits, PKU, China, [4]Institute of EDA, PKU, Wuxi, China.

Email: zhigangji@sjtu.edu.cn

Abstract—Flicker noise has become a tricky problem in nano-scaled devices with the extensive use of analog/mixed-signal circuits. In this work, we proposed a new method to separate the intertwined noise sources. Based on this, it is found that in advanced nodes, oxide traps, channel scattering and parasitic access resistance can all cause flicker noise in nano-scaled devices. In addition, the noise of defects can be obtained through random telegraph noise (RTN), but a large amount of time will be consumed in testing and extraction. Thus, we developed automatic hidden Markov model (AHMM) technology that can automatically extract defect information based on measured RTN. In addition, we verified that the BTI defects extracted using discharge-based multi-pulse (DMP) technology are consistent with the RTN defects, which means that the defect distribution obtained through macroscopic relaxation spectrum technology can directly calculate the RTN and the noise it generates, greatly simplifying the difficulty of testing. By introducing equivalent circuits for simulation, the new compact model can be easily used with commercial simulators for circuit level analysis.

Keywords — FinFETs, flicker noise, RTN, compact model, automatic hidden Markov model (AHMM), discharge-based multi-pulse (DMP)

Introduction

Flicker noise (1/f) becomes as an urgent problem with the downscaling of CMOS technologies, due to the increasing demand for analog/mixed-signal circuits in mobile and AI applications [1]. The accurate modeling of device flicker noise directly affects the accuracy of circuit level evaluation. However, the conventional corner model tends to overestimate the impact of flicker noise due to severe device-to-device variations, it will lead to reduce the design margin. Moreover, the conventional understanding of device mobility fluctuation and/or carrier number fluctuation (CNF) mechanisms cannot be validated in nano-scaled devices in terms of bias dependence [2, 3], physical significance of extracted defect numbers, and other aspects. While the opinion prevailing in recent years attributes the origin of flicker noise to the superposition of random telegraph noise (RTN) from multiple traps, the limited number of traps on nanoscale devices fails to explain the nearly perfect 1/f slope observed in experimental results [4]. Therefore, the lack of a correct understanding of the origin of flicker noise in

nano-scaled devices hinders the establishment of accurate statistical modeling, and affects the accurate use of noise models in circuit simulations.

In this work, we reveal that in advanced technology, the oxide traps, channel scattering, and parasitic access resistance all contribute to the flicker noise in nano-scaled devices based on the proposed experimental separation method. In addition, the noise model caused by defects requires multiple parameters, including the impact of defects on current, capture time, emission time, and the number of defects, which increases the complexity of testing and modeling. Although the above information can be obtained through RTN, testing and extraction are very time-consuming. This article utilizes the developed automatic hidden Markov model (AHMM) to achieve automated batch extraction of device RTN, and verifies the consistency between RTN and BTI defects using discharge-based multi-pulse (DMP) technology, which will reduce a lot of time in testing and extracting defect noise. Based on this understanding, we present a physics based compact model that can be readily used for circuit-level simulations.

Method for Noise Source Separation

A. Decoupling method for intertwined noise sources

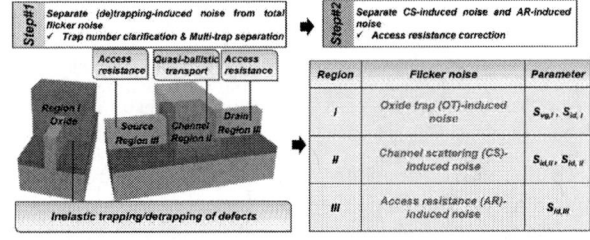

Fig. 1 Several nonideal effects in nano-scaled FinFET, including inelastic (de)trapping of oxide defects, quasi-ballistic transport and parasitic access resistance, will contribute to the total flicker noise. Separation method of different noise components from different device regions is illustrated.

In nano-scaled devices, flicker noise can originate from three sources: (de-)trapping in the dielectric layer for the oxide trap (OT), carriers scattering during transport for channel scattering (CS) and source/drain access resistance (AR) [5, 6]. As shown in Fig. 1, we define three regions of the FinFET device: gate oxide (Region I), channel (Region II), and source/drain extension (Region III) to discuss the corresponding noise sources. The induced flicker noise spectrum is denoted as $S_{id,I}$

979-8-3503-6184-1/24 $31.00 © 2024 IEEE

for OT-induced noise, $S_{id,II}$ for CS-induced noise, and $S_{id,III}$ for AR-induced noise, respectively. Advanced industrial-grade FinFETs with different sizes are characterized in this work.

Due to the fact that all three noise sources contribute to the total noise and can be independently modeled, a separation method is proposed to independently analyze and model the three noise sources. The proposed separation method has two steps. First, as the noise spectrum of trapping/detrapping of individual OTs presents as a Lorentzian form, while the spectrum of CS-induced noise and AR-induced noise all shows "1/ f" dependence of frequency, the total noise can be expressed as follows:

$$S_{id} = \frac{K}{f} + \sum_i \frac{A_i}{1+(2\pi f \tau_i)^2} \qquad (1)$$

where K is the coefficient representing the contribution from the CS and AR components, f is the frequency, i is the serial number of the trap, A_i is the i_{th} defect amplitude, and τ_i is the characteristic time constant. Due to the downscaling of device sizes and increased impact of OTs, the amplitude of the RTN induced by handful OTs is sufficient to cause the evident humps in the flicker noise. This provides the possibility to separate the OT-induced noise. With (1), the OT-induced noise can be separated from the total noise. After removing the trap-induced noise, the second step is to separate the CS-induced noise and AR-induced noise. The total resistance of the device can be simplified as a series connection of channel resistance and AR.

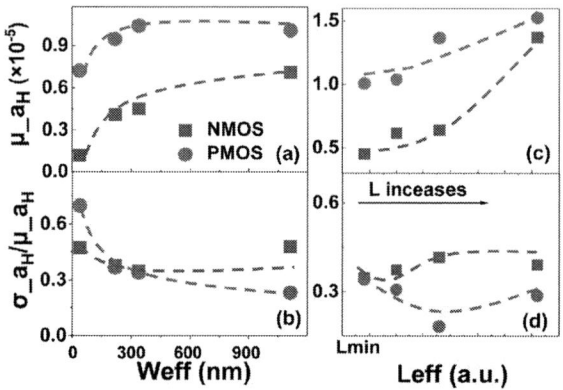

Fig. 2 (a) Mean and (b) normalized sigma values of extracted α_H with different channel widths under $L_{eff} = L_{min}$. (c) Mean and (d) normalized sigma values of extracted α_H with different channel lengths under $W_{eff} = 338$ nm.

With the above procedure, the three noise sources can be separated to model independently. Regarding the noise caused by CS, the normalized power spectral density can be modeled using the traditional Hooge model. Additionally, with the downscaling of device channel length, quasi-ballistic transport becomes remarkable, its impact is incorporated into Hooge parameter [2], which can be expressed as follows:

$$\alpha_H = \alpha_{H0} + b\mu^2 \qquad (2)$$

$$\mu^{-1} = \mu_0^{-1} + \mu_{ballistic}^{-1} \qquad (3)$$

$$\mu_{ballistic} = kL \qquad (4)$$

$$\frac{S_{id,II}}{I_D^2} = \frac{\alpha_H}{fN} = \frac{q\alpha_H \mu V_D}{fL^2 I_D} \qquad (5)$$

where α_H is the Hooge parameter and N is the number of channel carriers, α_{H0}, b, and k are constant parameters, μ is the effective mobility, μ_0 is the mobility for the long channel, and $\mu_{ballistic}$ is the equivalent ballistic mobility, which can be written as a linear relationship with channel length L.

For the AR-induced noise, normalized power spectral density manifests "1/ f" dependence [7], which can be expressed by:

$$\frac{S_{id,III}}{I_D^2} = \frac{K_r}{f} \qquad (6)$$

where K_r is the coefficient. Similar to the CS-induced noise, the variation of AR-induced noise can be characterized by the mean and sigma values of K_r.

Regarding the OT-induced noise, it behaves as RTN in the time domain. The noise spectral density in the frequency

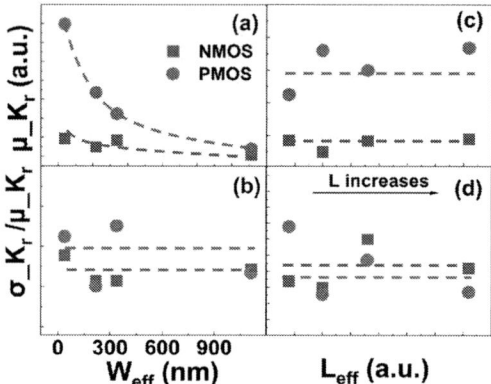

Fig. 3. (a) Mean and (b) normalized sigma values of extracted Kr in AR-induced noise with different channel widths under $L_{eff} = L$min. (c) Mean and (d) normalized sigma values of extracted Kr in AR-induced noise with different channel lengths under $W_{eff} = 338$ nm.

domain can be modeled as the superposition of different OTs [4, 8], which can be expressed by:

$$S_{vg,I} = \sum_{i=N} \frac{4(\Delta V_i)^2}{(\tau_{c,i}+\tau_{e,i})[(1/\tau_{c,i}+1/\tau_{e,i})^2 + (2\pi f)^2]} \qquad (7)$$

$$S_{id,I} = g_m^2 S_{vg,I} \qquad (8)$$

where N_t is the trap number, ΔV_i is the i_{th} trap amplitude, $\tau_{c,i}$ and $\tau_{e,i}$ are the i_{th} trap capture/emission time constants, and g_m is device transconductance.

Finally, the total flicker noise model can be constructed as follows:

$$S_{id} = S_{id,I} + S_{id,II/III} \qquad (9)$$

Since the model parameters can be extracted from the experiment after noise source separation, the model applies to different channel lengths, widths, and operating biases. In addition, the new mechanisms of trapping/detrapping of OTs and carrier transport are included in the modeling to achieve high accuracy.

B. Model parameter extraction

Due to the fact that the accuracy of the model depends on the validation of model parameters under different biases and device dimensions, the trends of the three noise sources need to be examined under various conditions. Regarding the CS-induced noise, **Fig. 2** shows the mean and sigma values of α_H

with different channel widths and lengths. It is clearly shown that the mean value has very weak dependence on channel width in PMOS, however, a decreasing trend can be observed in NMOS. Normalized sigma in both NMOS and PMOS shows a clear increasing trend with smaller channel width, since devices with narrower channels have a weaker averaging effect. For the length dependence, the mean value decreases with a shorter channel in both NMOS and PMOS, because the quasi-ballistic transport becomes more remarkable with a shorter channel length, leading to weaker CS [9].

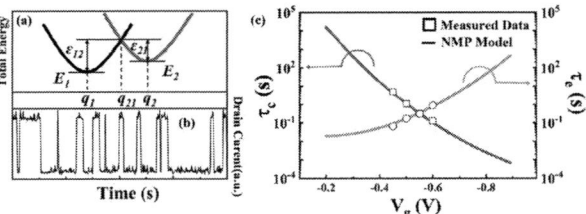

Fig. 4 (a) Trap state transitions can be well described by the two-state NMP model. **(b)** Measured *RTN* can be well described by NMP model.

For the AR-induced noise, the model parameters Kr in both NMOS and PMOS show a significant dependence on W_{eff}. This is because AR is constant for different L_{eff} values, and increases as W_{eff} narrows, resulting in a stronger impact of noise caused by AR (**Fig. 3**).

About the OT-induced noise, it represents itself as the random telegraph noise (RTN) in the time domain. The inelastic tunnelling based on multi-phonon-assisted non-radiative transition (NMP) theory can explain the behavior of OT under different bias (**Fig. 4a**). By simplifying RTN as the two-state transition the key physical parameters of each trap including the energy level, E_t, the spatial location, x_t and the relaxation energy, S (**Fig. 4b**), can be obtained from the emission and capture time constants under different gate biases [10].

Furthermore, this model is also suitable for fitting noise after different hot carrier degradation (HCD) stresses (**Fig. 5**), and the specific method will not be described in detail here.

Efficient Defect Extraction

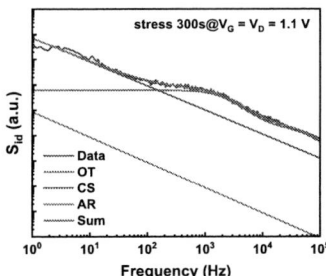

Fig. 5 Demonstration of separating the total noise after HCD stress.

As described in the previous section, defect extraction in noise can be obtained through RTN testing. However, due to the time-consuming nature of defect testing and extraction, it is not conducive to commercial applications [10].

In this work, based on the previous hidden Markov model (HMM) analysis method [11], aiming at the automatic extraction of complex multi-trap RTN, the automatic HMM

(AHMM) algorithm was proposed [12]. The flow chart is shown in **Fig. 6**, in which modules for automatically detecting RTN and judging the number of RTN traps are added, thus reducing the pressure of manual judgment.

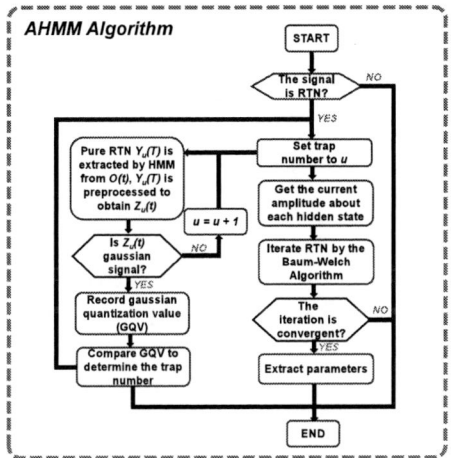

Fig. 6 The flow chart of accelerated AHMM algorithm.

Meanwhile, we recently identified two types of preexisting traps in PMOS [13] and one type of preexisting trap in NMOS [14]. By comparing the trap properties extracted from the discharge-based multi-pulse (DMP) technique and the atomic simulation using density functional theory [15], the origin of the traps can be identified, as shown in **Figs. 7 and 8**. When we compared the properties of the traps extracted from OT-induced noise, excellent agreement can be achieved for both Et and S in NMOS and PMOS [5]. The good agreement from three independent studies (DMP extraction, Ab initio calculation, and OT-induced noise extraction) strongly supports those traps account for the trap induced noise.

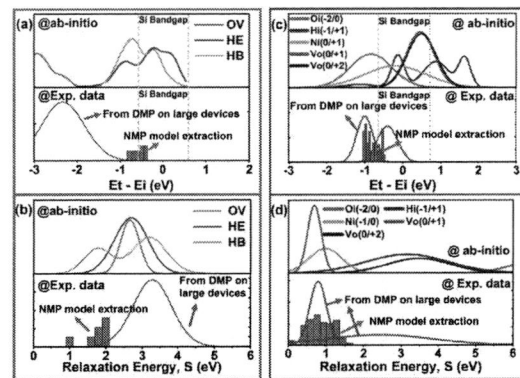

Fig. 7 Comparison of the ab-initio calculation, discharge-based multi-pulse (DMP) technique extraction and NMP model extraction for **(a)** and **(c)** energy level, **(b)** and **(d)** relaxation energy of Type-A in the IL layer and Type-B in the HK layer.

Compact Model and Circuit Simulation

Based on the statistical flicker noise model which we mentioned above, the model parameters can be extracted from the experiment after noise source separation. The model is applicable for different channel length, width and also for

979-8-3503-6184-1/24 $31.00 © 2024 IEEE

statistical simulation under any operating condition (**Fig. 9b**). The accuracy of this model is verified by comparing between the experimental results and model prediction on devices under different biases (**Fig. 9a&c**). The noise distribution is affected by the uneven energy level distribution of oxide traps.

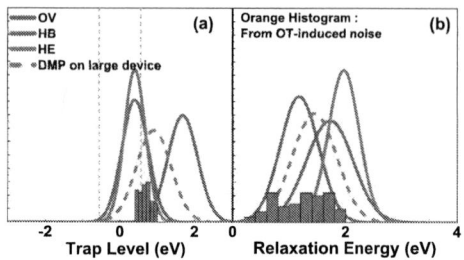

Fig. 8 Comparison between the ab-initio calculation, DMP technique extraction on large NMOS and NMP model extraction on nanoscale NMOS for **(a)** trap energy level and **(b)** relaxation energy of traps in the IL.

For noise analysis on circuit level, one voltage and one current noise source can be connected to each transistor (**Fig. 10a**). Ring Oscillator with transistors of different sizes is used for demonstration (**Fig. 10b**). Phase noise from oxide trap and channel scattering/access resistance are equally important, which increases with size shrinking (**Fig. 10c**). MC simulations are also conducted to show the circuit variation (**Fig. 10d**).

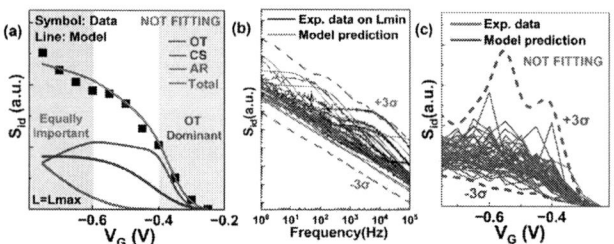

Fig. 9 Comparison between the proposed model prediction and experiment data **(a)** of S_{id} with different V_G on L_{max} device, **(b)** with varying frequency and **(c)** with varying V_G bias.

Conclusion

In this work, we have established a complete set of compact model with defect-centric insights suitable for circuit level simulation, based on our physical understanding of flicker noise. Flicker noise can originate from three sources: (de-)trapping in the dielectric layer for the oxide trap, channel scattering and source/drain access resistance. Our model can successfully separate complex intertwined noise sources. Among them, the defects noise can be obtained by measuring random telegraph noise (RTN), but due to the need for extensive testing and the complexity of extraction, this will increase the testing time. This article proposes a novel automatic hidden Markov model (AHMM) program that will greatly accelerate the extraction speed. In addition, the consistency of bias temperature instability (BTI) and RTN defects was verified through the discharge-based multi-pulse (DMP) method, which means that the distribution of defects can be obtained through relatively simple DMP testing, simplifying the repeated testing of RTN. The above methods can be used to accelerate circuit simulation and increase the accuracy of noise simulation in circuits. This

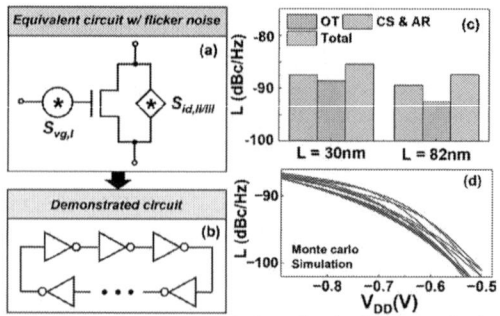

Fig. 10 (a-b) Illustration of the equivalent circuit of flicker noise for circuit simulation, demonstrated on ring oscillator. Simulated phase noise (@1Mhz) under different **(c)** channel length and **(d)** VDD.

article provides a more accurate and efficient process and model for simulating circuit noise.

Acknowledgments

This work was supported in part by the National Natural Science Foundation of China under Grant 92164205, 61927901, 62125401, 62027818, 61874034, and 11974320.

References

[1] P. Kushwaha *et al.*, *IEEE Electron Device Letters*, vol. 40, no. 6, pp. 985-988, 2019

[2] F. N. Hooge, *IEEE Transactions on Electron Devices*, vol. 41, no. 11, pp. 1926-1935, 1994

[3] K. K. Hung *et al.*, *IEEE Transactions on Electron Devices*, vol. 37, no. 3, pp. 654-665, 1990

[4] M. Liu *et al.*, in *2021 IEEE International Electron Devices Meeting (IEDM)*, 11-16 Dec. 2021 2021, pp. 18.6.1-18.6.4

[5] P. Ren *et al.*, *IEEE Transactions on Electron Devices*, vol. 71, no. 5, pp. 3377-3382, 2024

[6] J. Wu *et al.*, in *2023 International Electron Devices Meeting (IEDM)*, 9-13 Dec. 2023 2023, pp. 1-4

[7] L. K. J. Vandamme *et al.*, *IEEE Transactions on Electron Devices*, vol. 55, no. 11, pp. 3070-3085, 2008

[8] J. Wu *et al.*, *IEEE Transactions on Electron Devices*, vol. 70, no. 11, pp. 6078-6081, 2023

[9] W. Wu *et al.*, *IEEE Transactions on Electron Devices*, vol. 65, no. 6, pp. 2573-2577, 2018

[10] C. Zhang *et al.*, in *2024 International VLSI Symposium on Technology, Systems and Applications (VLSI TSA)*, 22-25 April 2024 2024, pp. 1-2

[11] H. Miki *et al.*, in *2012 International Electron Devices Meeting*, 10-13 Dec. 2012 2012, pp. 19.1.1-19.1.4

[12] Y. Xiao *et al.*, in *2024 IEEE International Reliability Physics Symposium (IRPS)*, 14-18 April 2024 2024, pp. P75.TX-1-P75.TX-4

[13] Y. Xue *et al.*, *IEEE Transactions on Electron Devices*, vol. 70, no. 9, pp. 4518-4524, 2023

[14] Y. Xue *et al.*, in *2024 IEEE International Reliability Physics Symposium (IRPS)*, 14-18 April 2024 2024, pp. 1-5

[15] Z. Ji *et al.*, in *2013 IEEE International Electron Devices Meeting*, 9-11 Dec. 2013 2013, pp. 15.6.1-15.6.4

A Modified Virtual-Source Model for Ballistic Transport Characterization of FinFETs at Cryogenic Temperature

Hongbo Wang[1], Zirui Wang[1], Zixuan Sun[1], Runsheng Wang[1*], Ru Huang[1]
[1]School of Integrated Circuits, Peking University, Beijing, China

*Email: r.wang@pku.edu.cn

Abstract—**In this paper, a modified cryo-physics based virtual-source model is developed to describe the quasi-ballistic transport in foundry level 16/14 nm FinFET at cryogenic temperatures (down to 15 K). For the first time, band-tail states and interface states, which have negligible effects on device electrostatics at cryogenic temperature, are incorporated in the virtual-source model. What's more, an efficient ballistic efficiency (B_{sat}) extraction method is also developed. The method not only eliminates the need for empirical parameters but also enables accurate extraction for a single device at any temperature. This work is helpful for the physical understanding and modeling of the transport in cryogenic CMOS.**

I. Introduction

Cryogenic CMOS is gathering significant attention due to its prospects in quantum computing and high-performance computing [1-5]. At cryogenic temperature, the transport and electrostatic properties of the device differ significantly from those at room temperature, posing substantial challenges for cryogenic modeling. Previous studies mainly focus on the electrostatic abnormalities introduced by band-tail states and interface states, causing the SS saturation and inflection [6-9]. However, due to the operation of cryogenic devices close to the ballistic limit, there have been few studies on their transport analyses, especially ballistic transport. Several works reported on ballistic transport of cryogenic devices either rely on fitting models which fail to reflect the physical mechanisms or require characteristics from multiple devices or temperatures, which makes the ballistic efficiency extraction rather complex. Furthermore, none of these works consider the effect of band-tail states and interface states, which may lead to inaccurate results.

In this paper, a modified MVS-2 model is proposed for transport analysis of cryogenic devices, considering the band-tail states and interface states. The model is validated at foundry-level 16/14 nm FinFET at different temperatures (15-300 K). A new ballistic efficiency (B_{sat}) extraction method balancing both accuracy and efficiency is proposed. This work provides new insights and deep understandings for cryogenic CMOS modeling.

II. Physics Based Modeling

As shown in **Fig.1**, the "virtual source" is developed based on the assumption that the current is the product of the charge density Q_{inv} and local carrier velocity v_{x0} at the top of the barrier, or "virtual source(VS)". Q_{inv} can be divided into three parts: carriers that are thermally injected from the source to the VS point, backscattered from the channel and transported from the drain to the VS point, given by equation (2). v_{x0} is the drift velocity at low V_{ds} and thermal velocity at high V_{ds}. Introducing the smooth function F_{sat} in (1), MVS-2 model describes I-V characteristics very well in SOI, FinFET and III–V transistors [10]-[11].

However, at cryogenic temperature, band-tail states and localized states significantly impact the carrier density [12], as shown in **Fig. 2**. The band-tail states are induced by the disorder, such as impurity, surface roughness, and so on, providing additional leakage path and acting as parasitic capacitance [13]. Interface states are induced by dangling bonds, degrading the subthreshold swing (SS) and threshold voltage (V_{th}). Many non-ideal phenomena are related to these two effects, such as SS-temperature relation saturation, I-V protrusion, $SS - I_{ds}$ inflection and so on. It can be found in **Fig. 3** that VS model without considering band-tail states cannot fit the experiments at cryogenic temperature. Therefore, it's necessary to incorporate these two effects in cryogenic VS model.

Since Q_0 only considers the role of conduction band carrier transport, Q_1 introduced by interface states and Q_2 introduced by band-tail states are considered to obtain total charge density Q_{total} at VS point. Interface states with gaussian distribution centered around the sharp conduction/valance band edge are considered in this work, and the charge occupied Q_1 can be computed in (3). The band-tail states are usually modeled as density of states in exponentially decaying form with a decay parameter W_t, and the occupied charge density Q_2 is computed by hyperfunction in [7][14], as shown in (4). Considering the 2-D electrostatics, the surface potential ψ_s at the VS point can be divided into three parts, potential from source-VS, gate-VS, drain-VS capacitors in (5). The Q_{total} equals to Q_0 in traditional models without considering the band-tail states and interface states. It must be noted that $\frac{c_\Sigma}{c_{g-vs}}$ is related to SS, and the Boltzmann limit needs to be replaced by $(Wt/q)ln10$ since dominate subthreshold current is transported by band-tail states [14]-[16]. In this case, we can solve for Q_{inv} and ψ_s by joining the equations above.

As shown in **Fig. 4**, the foundry-level 16/14 nm 1finger nFinFET is characterized at different temperatures from 15 K to 300 K. The proposed model is agreed well with the transfer and output characteristics in full temperature range, indicating the validity of our model.

To further determine the effect of band-tail states, the currents contributed by band-tail states and conduction band are separated as shown in **Fig. 5**. It can be found that the subthreshold current is dominated by band-tail states while on-state current is dominated by conduction band, which is what we expected.

979-8-3503-6184-1/24 $31.00 © 2024 IEEE

III. Ballistic Efficiency Extraction

The ballistic efficiency is crucial in ballistic transport analysis since it describes the portion of carriers transporting in the channel without scattering. Based on the back scattering theory, the ballistic efficiency is given in (7). L_c is the critical length of the low-field region near the source of the channel, which describes the length of back scattering. λ is the mean free path for backscattering.

In the past, many ballistic efficiency extraction methods are proposed, which can be divided into three fusions: empirical fitting of B_{sat} with temperature [17]-[19], direct calculation by injection velocity divided by thermal velocity [20]-[21], and linear extrapolation by apparent mobility and extracted injection velocity. But these methods either need more than one device or temperature, or too simple to capture the nature of ballistic transport.

Therefore, a new method is proposed based on the parabolic approximation of channel potential as shown in **Fig. 6**. Based on this assumption, L_c can be determined accordingly. After determining the mean free path λ in (8), the ballistic efficiency can be calculated accordingly. Compared with other methods, the new method only makes an assumption of parabolic approximation of channel potential which is agreed well with the real channel potential by TCAD in several kT drop in energy. The extraction can be achieved within single device at certain temperature.

Fig. 7 shows the extracted ballistic efficiency from 15 to 300 K. It can be found that B_{sat} increases with temperature decreasing. When the temperature is below 100 K, transistor is near ballistic transport with ballistic efficiency around 0.9. As the temperature increases above 200 K, the ballistic efficiency decreases rapidly, changing to quasi-ballistic or drift-diffusion modes.

IV. CONCLUSION

In this paper, the VS Transport Model is modified by introducing band-tail states and interface states to describe the ballistic transport for cryogenic devices. The model is validated successful in foundry level nFinFET. What's more, an efficient ballistic efficiency extraction method is proposed. This work provides great help for cryogenic transport analysis and cryogenic CMOS modeling.

Acknowledgment

This work was supported by NSFC (61927901, 62125401) and the 111 Project (B18001).

References

[1] H. L. Chiang et al., "Cold CMOS as a Power-Performance-Reliability Booster for Advanced FinFETs," 2020 IEEE Symposium on VLSI Technology, Honolulu, HI, USA, 2020, pp. 1-2.

[2] S. S. T. Nibhanupudi, S. R. Sundara Raman, M. Cassé, L. Hutin and J. P. Kulkarni, "Ultra-Low-Voltage UTBB-SOI-Based, Pseudo-Static Storage Circuits for Cryogenic CMOS Applications," in IEEE Journal on Exploratory Solid-State Computational Devices and Circuits, vol. 7, no. 2, pp. 201-208, Dec. 2021.

[3] R. Saligram, S. Datta and A. Raychowdhury, "CryoMem: A 4K-300K 1.3GHz eDRAM Macro with Hybrid 2T-Gain-Cell in a 28nm Logic Process for Cryogenic Applications," 2021 IEEE Custom Integrated Circuits Conference (CICC), Austin, TX, USA, 2021, pp. 1-2.

[4] F. Arute et al. "Quantum Supremacy Using a Programmable Superconducting processor," in Nature 574.7779, 2019, pp. 505-510.

[5] M. Steffen, D. P. DiVincenzo, J. M. Chow, T. N. Theis and M. B. Ketchen, "Quantum computing: An IBM perspective," in IBM Journal of Research and Development, vol. 55, no. 5, pp. 13:1-13:11, Sept.-Oct. 2011.

[6] A. Kamgar, "Subthreshold behavior of silicon MOSFETs at 4.2 K," Solid-State Electron., vol. 25, no. 7, pp. 537-539, Jul. 1982.

[7] H. Bohuslavskyi et al., "Cryogenic subthreshold swing saturation in FDSOI MOSFETs described with band broadening," IEEE Electron Device Lett., vol. 40, no. 5, pp. 784-787, May 2019.

[8] H. Bohuslavskyi, "Cryogenic electronics and quantum dots on siliconon-insulator for quantum computing," Ph.D. dissertation, Laboratoire d'Electronique et de Technologie de l'Information, Commun. Univ. Grenoble Alpes, Saint-Martin-d'Hères, France 2016.

[9] P. Sarangapani, Y. Chu, J. Charles, G. Klimeck, and T. Kubis, "Band-tail formation and band-gap narrowing driven by polar optical phonons and charged impurities in atomically resolved III-V semiconductors and nanodevices," Phys.Rev.A,Gen.Phys., vol. 12, Oct. 2019.

[10] S. Rakheja, M. S. Lundstrom and D. A. Antoniadis, "An Improved Virtual-Source-Based Transport Model for Quasi-Ballistic Transistors—Part I: Capturing Effects of Carrier Degeneracy, Drain-Bias Dependence of Gate Capacitance, and Nonlinear Channel-Access Resistance," in IEEE Transactions on Electron Devices, vol. 62, no. 9, pp. 2786-2793, Sept. 2015.

[11] S. Rakheja, M. S. Lundstrom and D. A. Antoniadis, "An Improved Virtual-Source-Based Transport Model for Quasi-Ballistic Transistors—Part II: Experimental Verification," in IEEE Transactions on Electron Devices, vol. 62, no. 9, pp. 2794-2801, Sept. 2015.

[12] A. Beckers, F. Jazaeri and C. Enz, "Inflection Phenomenon in Cryogenic MOSFET Behavior," in IEEE Transactions on Electron Devices, vol. 67, no. 3, pp. 1357-1360, March 2020.

[13] Y. Luo and A. J. Flewitt, "Understanding localized states in the band tails of amorphous semiconductors exemplified by a-Si:H from the perspective of excess delocalized charges," Phys. Rev. B, vol. 109, no. 104203, pp. 1-9, Mar. 2024.

[14] A. Beckers, F. Jazaeri and C. Enz, "Theoretical Limit of Low Temperature Subthreshold Swing in Field-Effect Transistors," in IEEE Electron Device Letters, vol. 41, no. 2, pp. 276-279, Feb. 2020.

[15] H. Bohuslavskyi et al., "Cryogenic subthreshold swing saturation in FDSOI MOSFETs described with band broadening," IEEE Electron Device Lett., vol. 40, no. 5, pp. 784–787, May 2019.

[16] H. Bohuslavskyi, "Cryogenic electronics and quantum dots on siliconon-insulator for quantum computing," Ph.D. dissertation, Laboratoire d'Electronique et de Technologie de l'Information, Commun. Univ. Grenoble Alpes, Saint-Martin-d'Hères, France 2016.

[17] M. J. Chen, H. T. Huang, K. C. Huang, P. N. Chen, C. S. Chang, and C. H. Diaz, "Temperature dependent channel backscattering coefficients in nanoscale MOSFETs," in Proc. IEDM, Dec. 2002, pp. 39-42.

[18] V. Barral, T. Poiroux, M. Vinet, J. Widiez, B. Previtali, P. Grosgeorges, et al., "Experimental determination of the channel backscattering coefficient on 10-70 nm-metal-gate double-gate transistors," Solid-State Electron., vol. 51, no. 4, pp. 537-542, Apr. 2007.

[19] R. Wang, H. Liu, R. Huang, J. Zhuge, L. Zhang, D. W. Kim, et al., "Experimental investigations on carrier transport in Si nanowire transistors: Ballistic efficiency and apparent mobility," IEEE Trans. Electron Devices, vol. 55, no. 11, pp. 2960-2967, Nov. 2008.

[20] V. Barral, T. Poiroux, J. Saint-Martin, D. Munteanu, J. L. Autran, and S. Deleonibus, "Experimental investigation on the quasi-ballistic transport: Part I—Determination of a new backscattering coefficient extraction methodology," IEEE Trans. Electron Devices, vol. 56, no. 3, pp. 408-419, Mar. 2009.

[21] V. Barral, T. Poiroux, D. Munteanu, J. L. Autran, and S. Deleonibus, "Experimental investigation on the quasi-ballistic transport: Part IIBackscattering coefficient extraction and link with the mobility," IEEE Trans. Electron Devices, vol. 56, no. 3, pp. 420-430, Mar. 2009.

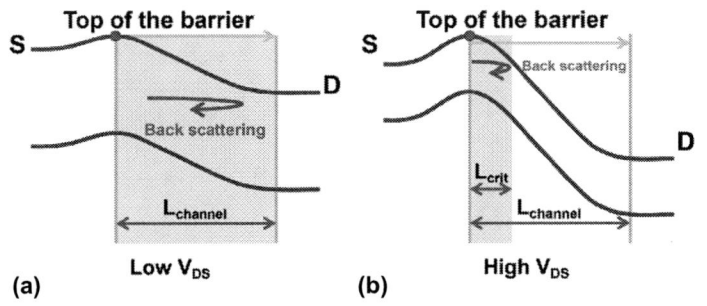

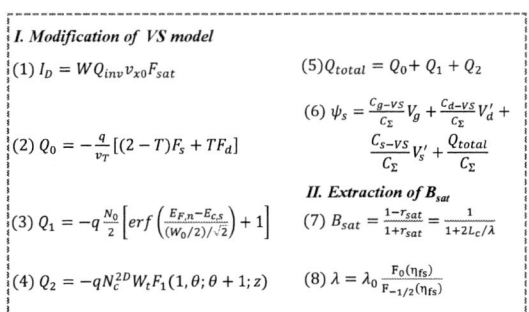

Fig. 1. Sketch of energy band and backscattering at different V_d biases.

Table I. Equations used in this work.

I. Modification of VS model

$$(1)\ I_D = WQ_{inv}v_{x0}F_{sat}$$

$$(2)\ Q_0 = -\frac{q}{v_T}[(2-T)F_s + TF_d]$$

$$(3)\ Q_1 = -q\frac{N_0}{2}\left[erf\left(\frac{E_{F,n}-E_{c,s}}{(W_0/2)/\sqrt{2}}\right)+1\right]$$

$$(4)\ Q_2 = -qN_c^{2D}W_tF_1(1,\theta;\theta+1;z)$$

$$(5)\ Q_{total} = Q_0 + Q_1 + Q_2$$

$$(6)\ \psi_s = \frac{C_{g-VS}}{C_\Sigma}V_g + \frac{C_{d-VS}}{C_\Sigma}V_d' + \frac{C_{s-VS}}{C_\Sigma}V_s' + \frac{Q_{total}}{C_\Sigma}$$

II. Extraction of B_{sat}

$$(7)\ B_{sat} = \frac{1-r_{sat}}{1+r_{sat}} = \frac{1}{1+2L_c/\lambda}$$

$$(8)\ \lambda = \lambda_0 \frac{F_0(\eta_{fs})}{F_{-1/2}(\eta_{fs})}$$

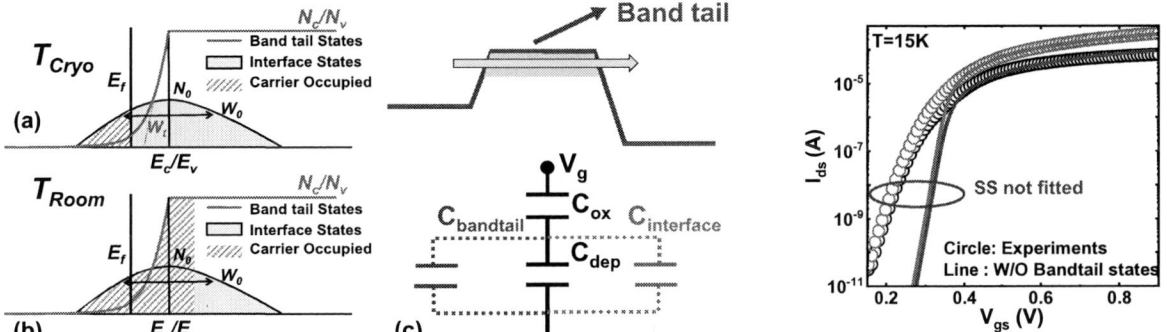

Fig. 2. Sketch of effect of band-tail states and interface states.

Fig. 3. VS model without band-tail states cannot fit experiments.

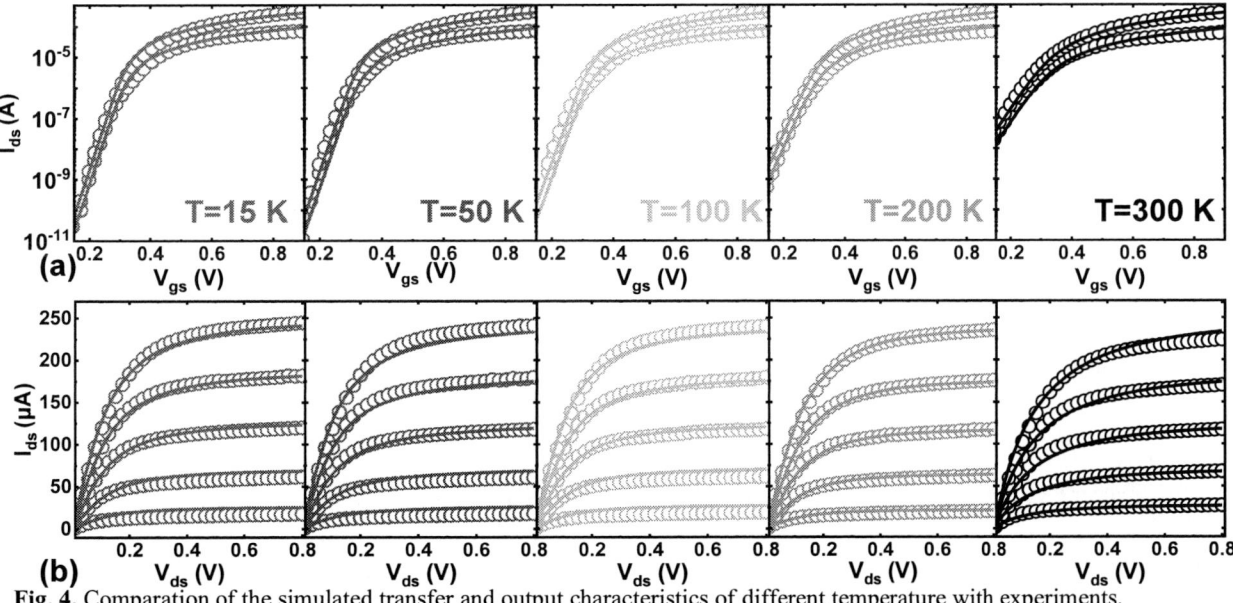

Fig. 4. Comparation of the simulated transfer and output characteristics of different temperature with experiments.

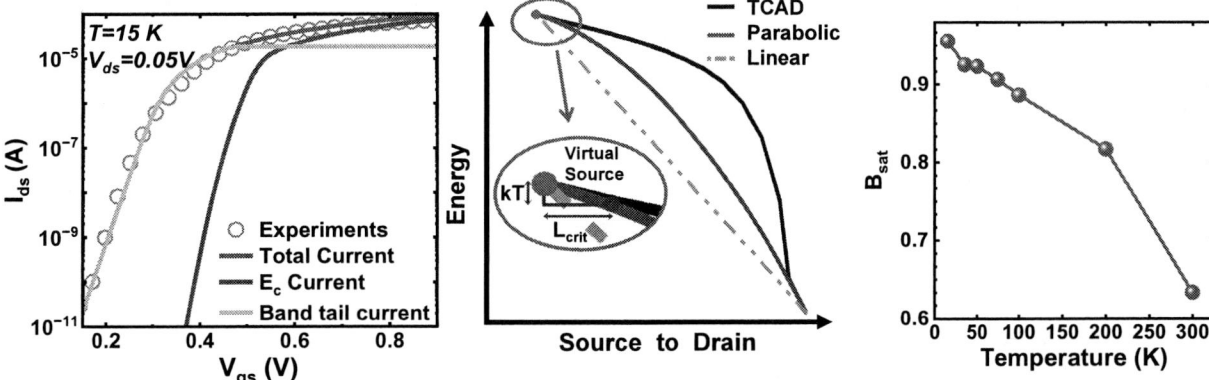

Fig. 5. Comparation of contributions of band-tail current and E_c current at cryogenic temperature.

Fig. 6. Comparison of the parabolic estimation and linear estimation of channel potential with TCAD simulations.

Fig. 7. Extracted ballistic efficiency of different temperatures. At cryogenic temperature, B_{sat} is approaching 1.

979-8-3503-6184-1/24 $31.00 © 2024 IEEE

Investigation on Asymmetric HfO_2-ZrO_2-HfO_2 Superlattice Gate Stacks with Ultra-low EOT for Advanced Transistors

Haiyuan Lyu [1,2,3], Kun Zhong [1,2,3], Zhaohao Zhang [1,2,3], Huaxiang Yin*[1,2,3]

[1] Key Laboratory of Fabrication Technologies for Integrated Circuits, Chinese Academy of Sciences, Beijing 100029.
[2] Institute of Microelectronics, Chinese Academy of Sciences, Beijing 100029, China.
[3] School of Integrated Circuits, University of Chinese Academy of Sciences, Beijing 100049, China.

* Email: lvhaiyuan23@mails.ucas.ac.cn, yinhuaxiang@ime.ac.cn

Abstract—In this work, equivalent oxide thickness (EOT) and leakage currents of HfO_2-ZrO_2-HfO_2 gate stack with symmetric (HZH) and asymmetric (HHZH or HZHH) structure were investigated based on metal oxide semiconductor capacitors (MOSCAPs). The results indicate that ultra-low EOT (< 0.8 nm) can be realized on both symmetric and asymmetric superlattice gate stacks. Furthermore, compared to the asymmetric stacks (0.73 nm EOT and 0.022 A/cm² for HHZH; 0.76 nm EOT and 0.016 A/cm² for HZHH), the symmetric HZH has the lowest EOT (0.68 nm) and the lowest leakage current (0.012 A/cm²). The results offer a design guide for the HZH superlattice gate stacks and their implementation into advanced transistors.

Keywords— *hafnium zirconium oxide, equivalent oxide thickness, leakage current, thickness ratios*

I. INTRODUCTION

Thinning equivalent oxide thickness (EOT) or inversion layer thickness (T_{inv}) in gate dielectric is capable of directly enabling improved electrostatics, enhanced drive current, and reduced power consumption in transistors. Previous technologies, such as interfacial layer (IL) scavenging or permittivity enhancement in high-κ film may cause mobility and reliability degradation [1-4].

Recently, the HfO_2-ZrO_2-HfO_2 (HZH) superlattice gate stack was proposed to reduce EOT in a great extent. The ultra-thin EOT originates from the tetragonal/orthorhombic (t/o-) phases mixed in the HZH thin-film system and obtaining the flattening of the ferroelectric double-well energy landscape by antiferroelectric inducing depolarization field energies [5,6]. Furthermore, a thickness-proportion controlled HfO_2-ZrO_2-HfO_2 (TPC-HZH) superlattice gate stack with improved thermal stability [7] was also reported for a compatible advanced back-end-of-line (BEOL) process. Nevertheless, the impact of the HZH structure with different layer thicknesses on its electrical properties still lacking.

In this work, by adjusting the thickness ratio among HfO_2-ZrO_2-HfO_2, three different HZH gate stacked (symmetric HZH and asymmetric HHZH, HZHH) based metal oxide semiconductor capacitors (MOSCAPs) were fabricated. Furthermore, the device's characteristics with the different structures, such as EOT, flat band voltage (V_{fb}), and leakage current were investigated. Although ultra-low EOT (< 0.8 nm) can be realized on both symmetric and asymmetric super-lattice gate stacks, the symmetric HZH has

the lowest EOT (0.68 nm) and the lowest leakage current (0.012 A/cm²).

II. DEVICE FABRICATION

The fabrication process of the MOS structure is shown in Fig. 1(a). Firstly, the natural oxide layer on the p-Si <100> substrate was removed with buffered oxide etch (BOE) treatment. Then, the SiO_2 interfacial layer was grown on the substrates with ozone treatment for 20 seconds. Subsequently, the HfO_2-ZrO_2-HfO_2 films with three different thickness ratios were deposited using the atomic layer deposition (ALD) process by varying the cycle ratios between HfO_2 and ZrO_2 (HHZH: 12: 21: 2, HZHH: 2: 21: 12, HZH: 7: 12: 7), as shown in Fig. 1(b). The deposition rates of the HfO_2 and ZrO_2 films were 0.75 and 0.64 Å/cycle, respectively. Then, TiN and tungsten films are deposited by physical vapor deposition (PVD). After that, the devices were annealed at 250 °C for 60 seconds to solidify the interface of the oxide layers.

Capacitance versus voltage (C-V) curves and the leakage current versus voltage curves were performed on the Keysight-1500 semiconductor parameter analyzer.

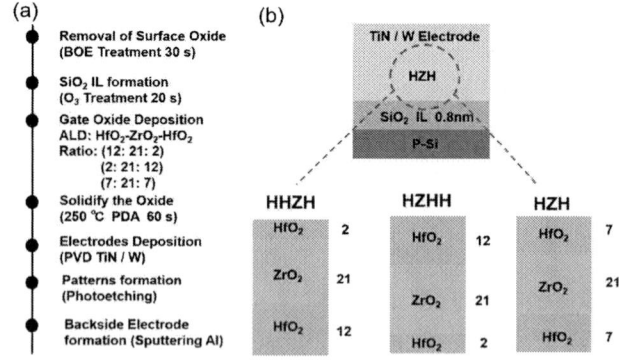

Fig. 1. (a) key Process flow and (b) the schematic of three different HZH (symmetric HZH and asymmetric HHZH, HZHH) films-based MOS capacitors.

III. RESULTS AND DISCUSSION

Fig. 2(a, b, c) shows the C-V curves of three MOS structures (HHZH, HZHH, HZH) at different frequencies (5 kHz, 10 kHz, 100 kHz, 500 kHz), respectively. Since the

979-8-3503-6184-1/24 $31.00 © 2024 IEEE

ultra-thin oxide layer thickness and thus large tunneling leakage current, the parallel C-G model is chosen to perform the C-V curves [8].

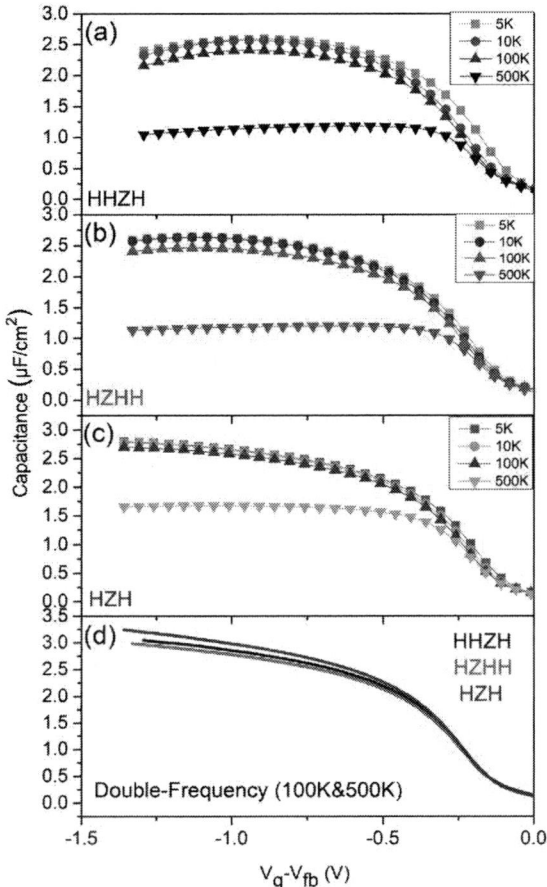

Fig. 2. Capacitance versus V_g-V_{fb} curves of (a) HHZH, (b) HZHH, and (c) HZH films-based MOS capacitors measured at four different frequencies and (d) C-V curve extracted by double frequency method.

The C-V curves shown in Fig. 2(d) were achieved at 100 kHz and 500 kHz to allow for the accurate parameter extraction for the frequency-dependent decreasing [7]. As the figures show, although the physical thickness of these HZH films are same, the capacitances of symmetric HZH are larger than that of asymmetric HHZH or HZHH.

Subsequently, corresponding EOT and V_{fb} were extracted using a capacitance-voltage simulator based on quantum mechanical effects (University of California, Berkeley) [7]. It can be seen from Fig. 3(a) that the EOTs of HZH with symmetric or asymmetric structures are all lower than the thickness of 0.8-nm IL which has been confirmed in our previous works.

Compared to the HHZH with an EOT of 0.73 nm and the HZHH with an EOT of 0.76 nm, the symmetric structure HZH exhibits the smallest EOT at 0.68 nm as shown in Fig. 3(a). Since the ultra-low EOT HZH structure features a coexistence of orthorhombic and tetragonal phases, the symmetric HZH structure is more likely to have a smaller grain size than asymmetric structures, as a decrease in the orthorhombic phase grain size within the HZH structure helps to reduce the EOT value [9,10]. V_{fb} of HHZH, HZHH,

and HZH films were also extracted and shown in Fig. 3(b). Compared with symmetric HZH, asymmetric structures have a larger V_{fb} value which indicates more oxygen vacancies in the films[11].

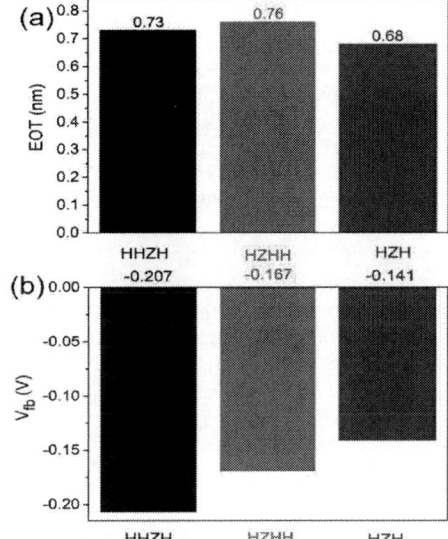

Fig. 3. (a) EOT and (b) V_{fb} of HHZH, HZHH and HZH films-based MOS capacitors (Extracted from Fig. 2(d)).

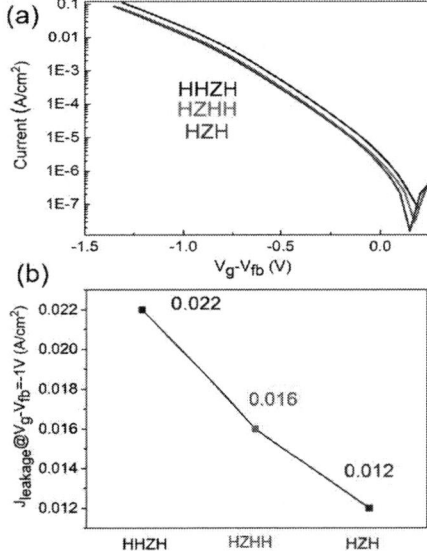

Fig. 4. (a) Leakage currents versus V_g-V_{fb} curves of HHZH, HZHH and HZH films-based MOS capacitors. (b) Leakage currents extracted at V_g-V_{fb} = -1V.

Due to the importance of leakage current on the reliability of gate stacks, the leakage currents were measured and shown in Fig. 4. As can be seen from Fig. 4(a), the symmetric HZH structure exhibits the smallest leakage current. By extracting the leakage current values at V_g-V_{fb} = -1V, it can be seen in Fig. 4(b) that the leakage current of HZH is the lowest, reaching 0.012 A/cm² compared to 0.022 A/cm² for HHZH and 0.016 A/cm² for HZHH. The reason for the leakage current increase might be the large oxygen vacancy concentration in the asymmetric HZH structure [12],

979-8-3503-6184-1/24 $31.00 © 2024 IEEE

which is consistent with that of the V_{fb} shifting analysis mentioned earlier.

IV. SUMMARY

In this work, we fabricate three HfO_2-ZrO_2-HfO_2 stacked MOS structures with different thickness ratios. By investigating the C-V and leakage current characteristics, we demonstrate that the symmetric HZH has the lowest EOT (0.68 nm) and the lowest leakage current (0.012 A/cm²), compared to the asymmetric structures. The research results validate the impact of different structures on the electrical properties of HZH stacks, which assists with further studies on the HZH structure and its underlying mechanisms.

ACKNOWLEDGMENT

The authors acknowledge support from the National Natural Science Foundation of China (Granted No. 92064003), and the Youth Innovation Promotion Association, Chinese Academy of Sciences under Grant 2023130.

REFERENCES

[1] J. Zhou et al., "Al-doped and Deposition Temperature-engineered HfO2 Near Morphotropic Phase Boundary with Record Dielectric Permittivity (~68)," 2021 IEEE International Electron Devices Meeting (IEDM), San Francisco, CA, USA, 2021, pp. 13.4.1-13.4.4.

[2] J Zhu, T L Li, B Pan, L Zhou and Z G Liu, "Enhanced dielectric properties of ZrO2 thin films prepared in nitrogen ambient by pulsed laser deposition," Journal of Physics D: Applied Physics, vol. 36, no. 36, pp. 389-393, January 2003.

[3] T. Ando et al., "Understanding mobility mechanisms in extremely scaled HfO2 (EOT 0.42 nm) using remote interfacial layer scavenging technique and Vt-tuning dipoles with gate-first process," 2009 IEEE International Electron Devices Meeting (IEDM), Baltimore, MD, USA, 2009, pp. 1-4.

[4] S. -i. Ohmi, M. Tanuma and J. -W. Shin, "Effect of SiO2 Interfacial Layer Reduction on MFSFET With 5 nm-Thick Ferroelectric Nondoped HfO2 by Deposition Rate Control," IEEE Transactions on Semiconductor Manufacturing, vol. 36, no. 4, pp. 553-557, Nov. 2023.

[5] Cheema S S et al., "Ultrathin ferroic HfO2–ZrO2 superlattice gate stack for advanced transistors," Nature, vol. 604, no. 7904, pp. 65-71, April. 2022.

[6] N. Shanker et al., "CMOS Demonstration of Negative Capacitance HfO2-ZrO2 Superlattice Gate Stack in a Self-Aligned, Replacement Gate Process," 2022 International Electron Devices Meeting (IEDM), San Francisco, CA, USA, 2022, pp. 34.3.1-34.3.4.

[7] K. Zhong et al., "Demonstration of EOT-Scaled FinFET Based on Thickness-Proportion Controlled HZH Superlattice Gate Stacks with Improved Thermal Stability (≥ 450 °C)," IEEE Electron Device Letters, in press.

[8] K. J. Yang, and C. Hu, "MOS capacitance measurements for high-leakage thin dielectrics," IEEE Transactions on Electron Devices, vol. 46, no. 7, pp. 1500-1501, Jul. 1999.

[9] M. Hoffmann, S. S. Cheema, N. Shanker, W. Li and S. Salahuddin, "Quantitative study of EOT lowering in negative capacitance HfO2-ZrO2 superlattice gate stacks," 2022 International Electron Devices Meeting (IEDM), San Francisco, CA, USA, 2022, pp. 13.2.1-13.2.4.

[10] J. Liao et al., "Grain Size Engineering of Ferroelectric Zr-doped HfO2 for the Highly Scaled Devices Applications," IEEE Electron Device Letters, vol. 40, no. 11, pp. 1868-1871, Nov. 2019.

[11] Takagi, Kensuke and Tomoya Ono. "First-principles study on leakage current caused by oxygen vacancies at HfO2/SiO2/Si interface." Japanese Journal of Applied Physics 57 (2018).

[12] B. Cui et al., "Back-End-of-Line Compatible HfO2/ZrO2 Superlattice Ferroelectric Capacitor With High Endurance and Remnant Polarization," IEEE Electron Device Letters, vol. 44, no. 6, pp. 1011-1014, June 2023.

Evaluation of contact resistance with the 'L' Kelvin test structure and the modified Kelvin test structure

Gui Chen, Yun-Hao Shao, Xin-Ping Qu*

School of Microelectronics, Fudan University, Shanghai 200433, China

*Email: 22112020074@m.fudan.edu.cn, xpqu@fudan.edu.cn

Abstract—**The 'L' Kelvin and modified Kelvin structures are widely used to extract contact resistance in 3D interconnect technology. The modified Kelvin structure has the advantage of not needing alignment, but we found the results of this structure are prone to negative. The 'L' Kelvin structure can extract a smaller contact resistance, but when the misalignment is significant, the extracted contact resistance is also negative. In this study, we use a simple method to simulate the bonding process and quickly analyze the reasons for the negative resistance of the two structures. The Finite-element method (FEM) is used to analyze the internal current flow and potential distribution of the two Kelvin structures. The possible causes of negative resistance are put forward to avoid negative resistance in the testing process. The influence of the bonded interface oxide layer on the contact resistance is also analyzed.**

Keywords—*'L' Kelvin structure, modified Kelvin structure, Finite-element method (FEM), specific contact resistance, negative contact resistance*

I. INTRODUCTION

Specific contact resistance is one of the critical indicators to evaluate the quality of the bonded metal interface in 3D hybrid bonding technology [1]. Anderson [2] proposed a four-terminal test structure to measure the contact resistance between PtSi-Si, which reduces the effect of parasitic resistance, including the resistance of metal lines and the probe-pad contacts on the test result of contact resistance. We now call this structure the 'L' Kelvin test structure [3] or the Cross-bridge Kelvin resistor (CBKR) structure[4]. Stucchi [5] pointed out that the measured value can even become negative for small resistors in resistance measurements of vertical interconnect elements by CBKR. In order to eliminate errors due to possible misalignments, K. N. Chen [6] proposed a simple test structure called a modified Kelvin structure.

In this study, we use a simple method to simulate the bonding process and quickly measure the contact resistance in different structures. The Finite-element method (FEM) is used to analyze the internal current flow and potential distribution of the two Kelvin structures. This work used Co-to-Co bonding [7] as an example to evaluate the contact resistance obtained with two test structures.

II. TEST STRUCTURES DESCRIPTION

Fig. 1 shows the illustration of the 'L' Kelvin structure and the modified Kelvin structure. The (100) p-type Si wafer with a 2 μm-thick silicon oxide film was used as a substrate. Then, we used the two-step PVD (Physical Vapor Deposition) 10 nm Mo / 200 nm Cu / 10 nm Co on the PVD 10 nm Co / 200 nm Cu surface with a vacuum break followed by annealing to simulate the real metal-metal bonding. The specific manufacturing process is reported in previous work [8].

The metal-metal contact resistance with the two test structures has been modeled using the simulation analysis

software Ansys Electronics Q3D. The simulation models of two test structures are shown in Fig. 2. For the modified structures, the metal line widths vary from 1, 5, 10, to 20 μm, and the contact areas vary from 1, 25, 100, to 400 μm². The ratio between h_{Cu} and metal width (W) is less than 1. The width of the 'L' Kelvin structure is fixed at 20 μm, with the h_{Cu} of 200 nm.

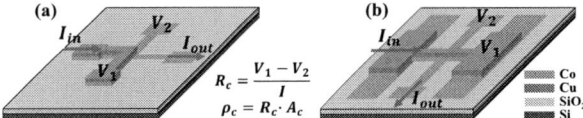

Fig. 1. Schematic illustration of (a) 'L' Kelvin structure and (b) Modified Kelvin structure for measuring the contact resistance between metals.

For both structures, a current of 1 A was applied to the bottom metal, and ground was set on the upper metal, resulting in a current flowing from the bottom metal to the upper metal through the contact area. The potential difference between metals V_1 and V_2 can be extracted from the simulation, enabling the calculation of the contact resistance (R_c). In order to reduce the error between the experimental value and the simulated value, the resistivity of the actual fabricated values is adopted in this work. The I-V curves of the fabricated structures were measured using Keithley 4200. A current of -100 ~100 mA was applied from the upper metal to the bottom metal, and the voltage drop across the contact area was measured.

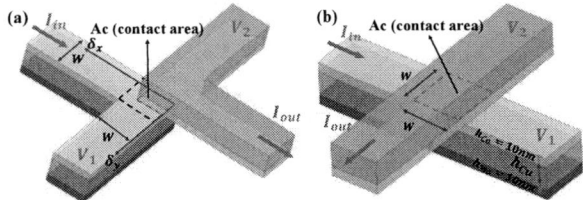

Fig. 2.   3-D model of (a) 'L' Kelvin structure and (b) modified Kelvin structure.

III. DISCUSSION

A. Modified Kelvin structure

The design contact area of the modified Kelvin test structure is 10 × 10 μm². Because of the lithographic and lift-off error, the final actual contact area is about 12 × 14 μm², and the h_{Cu} is about 200 ± 10 nm. Fig. 3 (a) shows the measured specific contact resistance distribution of the as-deposited samples. About 48% of the samples with modified Kelvin structures have negative contact resistance. After different annealing processes (pressures (0, 0.5, 1 MPa), annealing durations (0.5, 1, 2 h), and temperatures (200, 250, 300, 350 °C)), all values of contact resistance for the samples with the modified Kelvin structure turn negative and the

979-8-3503-6184-1/24 $31.00 © 2024 IEEE   361

summary results as shown in Fig. 3 (b). This means that the modified Kelvin structure is not suitable for the measurement of very small contact resistance.

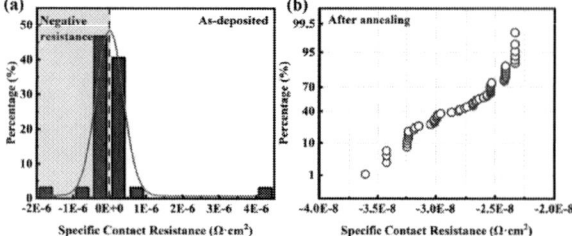

Fig. 3. The specific contact resistance distribution of the (a) as-deposited samples with modified Kelvin structure and (b) the ones after different annealing processes.

Fig. 4 shows the distribution of specific contact resistance with different ratios of h_{Cu}/W. The contact resistance of the annealed ones is about 18% lower when the resistance is positive. However, negative resistance will appear for the h_{Cu}/W value less than 0.6 for W=1 μm and h_{Cu}/W value less than 0.7 for W=5, 10, 20 μm. The smaller the value of h_{Cu}/W, the more negative the contact resistance tends to be. With the increase of h_{Cu}/W, the contact resistance increases. When the metal width is 10 μm, and the h_{Cu} is 200 nm, the obtained specific contact resistance according to the simulation is -1.63 × 10^{-8} Ω·cm².

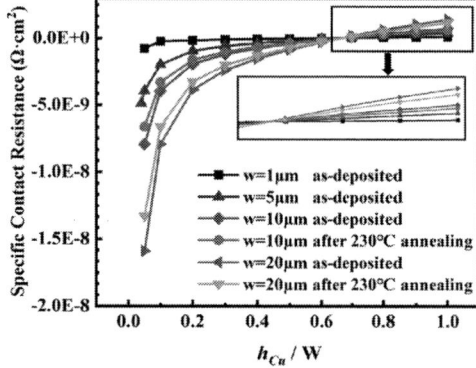

Fig. 4. The simulated specific contact resistance variation with the values of different ratios of h_{Cu}/W by using the modified Kelvin test structure for the as-deposited metal line and the ones after annealing at 230°C.

Fig. 5 shows the potential distribution and current density of samples with the modified Kelvin structure. When the current flows from the bottom metal to the upper metal, the current will turn 90° as shown in Fig.5. The current tends to flow to the path with the lowest resistance. When the ratio of h_{Cu}/W is less than 0.7, the current lowest resistance path includes the bottom metal, the left corner in contact with the upper metal, and the upper metal. The left corner of the bottom line is passed through by a small current, shown in the red circle in Fig. 5 (d), so the voltage drop along this metal line is small. Thus, the voltage from I_{in} is brought to the V_2 and higher than the V_1, which causes the negative contact resistance. When the ratio of h_{Cu}/W is larger than 0.7, the current density of the left and right of the upper and bottom metal contact is relatively uniform, as shown in the red and blue circles in Fig. 5 (e) (f).

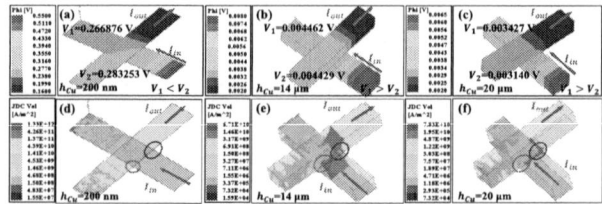

Fig. 5. (a-c) potential distribution and (d-f) current distribution during the simulation of contact resistance extraction of modified Kelvin structure with W=20 μm at different h_{Cu}.

B. 'L' Kelvin structure

The designed metal line width of the simulated 'L' Kelvin test structure is 20 μm, and the h_{Cu} is 200 nm. Because of the lithographic and lift-off error, the metal line width of the fabricated 'L' Kelvin test structure is about 20 ± 2 μm, and the h_{Cu} is about 200 ±10 nm. The illustration of the possible misalignments between the upper layer and fixed bottom layer with origin 0 is shown in Fig. 6. We define δ_x and δ_y as the lateral and vertical location that the upper layer aligned with the bottom layer. The minimum alignment area is 5 × 5 μm² contact area with δ_x and δ_y equal to 5, and the maximum alignment is 40 × 40 μm². The ideal contact without misalignment is 20 × 20 μm².

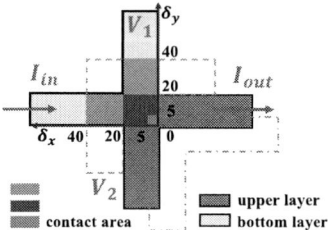

Fig. 6. The illustration of different misalignments of the 'L' Kelvin structure.

Table I presents the comparison between the simulated and experimental results of the specific contact resistance under different misalignment conditions. In the experiment, it was observed that with a metal line width of approximately 20 μm and a contact area of 15 × 15 μm², the specific contact resistance measured was 3.23 × 10^{-7} Ω·cm². Remarkably, a negative specific contact resistance of -2.7 × 10^{-7} Ω·cm² was observed when a misalignment of 30 in the directions of δ_x and δ_y between the upper and lower metals. This led to an in-depth discussion employing simulation to elucidate the rationale behind the observed negative resistance. As illustrated in Table I, the simulated contact resistance closely aligns with the experimental results. The difference between the measured and simulated values could be attributed to cobalt oxide at the actual interface due to air exposure between the two PVDs.

The simulated results demonstrate that the contact resistance decreases as the contact area increases. The symmetrical structure shows a minimal difference in resistance between contact areas of 5 × 20 μm² and 20 × 5 μm². However, for the structure with a contact area of 10 × 10 μm², the specific contact resistivity is 1.46 times higher. Figs. 7 (a-c) compares the potential distribution and current density of these three types of misalignments. It is observed that when the contact area is 10 × 10 μm², the current density at the contact point is significantly higher compared to 5 × 20 μm² and 20 × 5 μm², resulting in an increased voltage difference

979-8-3503-6184-1/24 $31.00 © 2024 IEEE

between the upper and lower metals and consequently higher contact resistivity. In the case of positive contact resistance, similar to Figs. 7 (a-c), the path of lowest resistance includes the bottom metal, the right side in contact with the upper metal, and the upper metal. A small current passes through the right side of the bottom metal, indicated by the blue circle in Figs. 7 (a-c), resulting in a minimal voltage drop along the same metal. Consequently, the voltage drop between the top and bottom metals is more easily measurable. Conversely, in the case of negative contact resistance, as shown in Fig. 7 (e), the path of lowest resistance includes the bottom metal, the contact edge between the upper and bottom metals, and the upper metal. So, the voltage from I_{in} is brought to V_2 and is higher than V_1, which causes the negative contact resistance. The ideal contact occurs when there is no misalignment, and the contact area is 20×20 μm², as shown in Fig. 7 (d). The simulated specific contact resistance is minimal, up to 4.80×10^{-10} Ω·μm², which is 1 to 2 orders of magnitude lower than the contact with misalignment. The current density distribution is symmetrical, with V_1 and V_2 measuring 1.26638 V and 1.26626 V, respectively. Therefore, to measure the voltage drop between the upper and bottom metals accurately, the resolution of the testing equipment should be less than 0.1 mV.

Based on the analysis above, it is evident that the 'L' Kelvin structure is significantly affected by the uneven current distribution. The specific contact resistance between the upper and bottom metals can only be accurately reflected in the absence of misalignment.

TABLE I.   THE SPECIFIC CONTACT RESISTANCE DISTRIBUTION OF DIFFERENT MISALIGNMENTS OF THE 'L' KELVIN TEST STRUCTURE

No.	Misalignment			Simulated Result	Experimental Result
	δ_x (μm)	δ_y (μm)	area (μm²)	ρ_c (Ω·cm²)	ρ_c (Ω·cm²)
#1	5	5	25	6.50×10^{-8}	/
#2	5	20	100	8.44×10^{-8}	/
#3	15	15	225	1.07×10^{-7}	3.23×10^{-7}
#4	10	10	100	1.24×10^{-7}	/
#5	20	20	400	4.80×10^{-10}	/
#6	20	5	100	8.48×10^{-8}	/
#7	30	30	700	-3.50×10^{-7}	-2.7×10^{-7}
#8	40	40	800	-7.82×10^{-7}	/

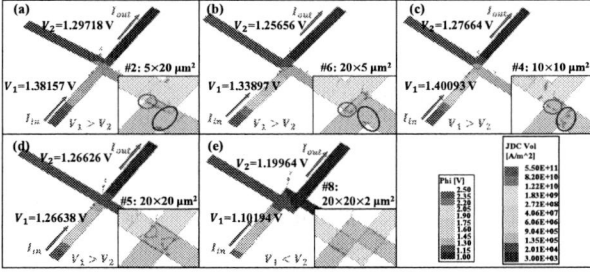

Fig. 7.   Potential distribution and current distribution during the simulation of contact resistance extraction of 'L' Kelvin structure with h_{Cu}=200 nm, W=20 μm.

In the actual bonding, oxidation of the interconnect metal will inevitably appear. Our previous work showed that the oxide at the bonding interface is non-uniform, with the oxidation area exceeding half of the bonding area [8]. Thus, the actual area through which the current flows is less than the bonding area. According to the results discussed above, the contact resistance will increase when the bonding area is smaller than the width of the metal line interconnect. However, the resistivity of a mixture of oxide and metal is difficult to determine. In some cases, the oxide is very thin, so it can cause tunneling; in other cases, the oxide is embedded in the metal layer. Due to the software limitation, the oxide layer is set to be uniform, and here we use "resistivity" to demonstrate its conductivity in the case of tunneling or embedding. The resistivity is in the range of $1 \times 10^{-5} \sim 1 \times 10^{-4}$ Ω·m in this work. Table II shows the effects of oxide layers with different resistivity on contact resistance extraction of the 'L' Kelvin structure. As the resistivity of the oxide layer increases, the contact resistance increases. When the resistivity of the oxide layer is 1×10^{-4} Ω·m, the contact resistivity increases by about 10 times.

TABLE II.   SPECIFIC CONTACT RESISTANCES OF 'L' KELVIN STRUCTURE (h_{Cu}=200 NM, W=20 MM) WITH DIFFERENT OXIDE LAYERS

No.	Metal Line Resistivity				Simulated result
	Mo (Ω·m)	Cu (Ω·m)	Co (Ω·m)	Oxide (Ω·m)	ρ_c (Ω·cm²)
#1	5.20×10^{-8}	1.69×10^{-8}	6.64×10^{-8}	No	5.28×10^{-10}
#2				1×10^{-5}	1.30×10^{-9}
#3				1×10^{-4}	6.60×10^{-9}

IV. CONCLUSION

Through a simple approach, the metal-metal contact resistance is measured. Also, the metal-metal contact resistance is simulated through the Finite-element method. The results show that for the interface with small contact resistance, a negative contact resistance value can be obtained by using the modified Kelvin structure. It was found that the misalignment would also result in negative resistance to the 'L' Kelvin structure. The contact resistance is affected by the bonding interface oxide layer. The 'L' Kelvin structure is suitable for the measurement of small specific contact resistance like 4.8×10^{-10} Ω·cm².

REFERENCES

[1] J Jourdon, S Lhostis, S Moreau, N Bresson, and H Fremont, IEEE Transactions on Electron Devices, vol. 66, pp. 2699-2703, 2019.

[2] R. M. Anderson, Journal of the Electrochemical Society, vol. 122, pp. 1337-1347, 1975.

[3] N Stavitskil, J. H. Klootwijk, HWV Zeijl, AY Kovalgin, and RAM Wolters, IEEE ICMTS, pp. 199-204, 2008.

[4] A. T. Schreyer, and C. K. Saraswat, IEEE Electron Device Letters, vol. 7, pp. 661-663, 1986.

[5] M Stucchi, F Fodor, and E. J. Marinissen, European Test Symposium, ISSN. 1530-1877, Tallinn, Estonia, 25-29 May, 2020.

[6] K. N. Chen, A. Fan, C. S Tan, IEEE Electron Device Letters, vol. 25, pp. 10-12, 2004.

[7] P Wang, Y. H Shao, Z. H Ni, CF Hu, XP Qu, AIP Advances, vol. 12, pp. 115101, 2022.

[8] Gui Chen, Yun-Hao Shao, Xin-Ping Qu, IITC, P25, San Jose, California, 3-6 June, 2024.

979-8-3503-6184-1/24 $31.00 © 2024 IEEE

Flip 3D (F3D): A Novel 3D Integration Technology with Dual-side Integration Capabilities

Heng Wu*, Haoran Lu, Runsheng Wang, Ming Li, Yibo Lin, Weihai Bu, Jin Kang, Ru Huang

School of Integrated Circuits, Peking University, Beijing, China

* Email: hengwu@pku.edu.cn

Abstract—In this work, we proposed the Flip 3D integration, a novel 3D technology with dual-side transistors and interconnects. Two layers of transistors and interconnects are formed on each side of wafer and back-to-back stacked, featuring a much more manufacturing-friendly process flow with low aspect ratio (AR). On transistor level, Flip-FET(FFET), a new stacking transistor architecture is enabled by F3D and experimentally demonstrated. Standard cell (STC) libraries with minimum 2.5 track height (2.5T) design are established, proving further scaling possibility and better intra-cell routability over CFET thanks to dual-side signal tracks. Due to smaller cell area and less parasitic, FFET outperforms CFET for both Fin and Nanosheet channels. New concepts of dual-side global interconnects are introduced and the P&R result of a RISCV32I core further validates the superiority of F3D integration. The F3D integration can be further extended to other complex circuits integration, such as memory stacking, memory to logic stacking, delivering great flexibility and extendibility for future's electronic applications.

Keywords—Flip3D Integration, stacked transistor, CFET, functional backside, dual-side interconnects, standard cell design

I. INTRODUCTION

With the conventional scaling coming to an end [1], hyper scaling with DTCO serves as a key enabler for future logic technology. New Methods such as fin depopulation and track reduction [2][3] are introduced, as shown in Fig.1. Meanwhile, BSPDN [4] and stacked transistors such as CFET [5], also attract hot discussions recently.

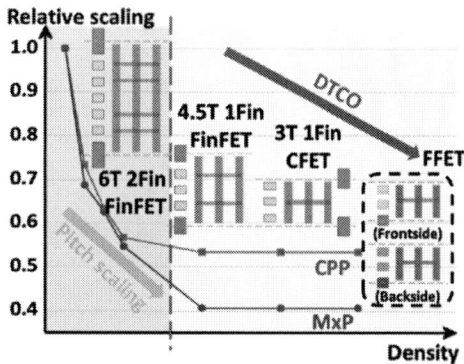

Figure 1: The scaling road of CMOS technology. With the ending of pitch scaling, hyper scaling with DTCO plays the key role in post Moore regime.

As given in Fig. 2, The industry is adapting BSPDN, with backside signal (BSS) and backside passive device (BSD) on the functional backside roadmap, showing a clear trend of more complex and functional integration on the wafer backside on the roadmap. In this work, as an extension, stacking transistors on wafer backside is proposed.

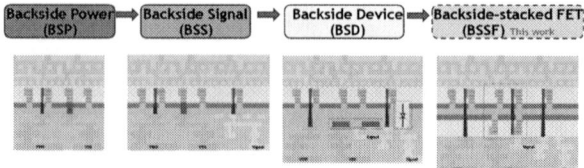

Figure 2: Roadmap of wafer backside integration. New concept of backside stacked FET is proposed in this work.

On the other hand, limited by intra-cell routing resources, CFET cell height reduction ends at 3T design with 3 frontside-only signal tracks [5] and 2 backside-only buried power rails (BPR) [6]. The CFET also faces great manufacturing challenges due to the high AR processes.

For here, to maximize the potential of functional backside and overcome the process complexity of CFET, we proposed a breakthrough transistor level 3D integration technology: Flip 3D (F3D) integration, as the possible ultimate form of functional backside, with backside-stacked active and dual-side interconnection capability. Based on this concept, a new transistor stacking architecture: Flip FET (FFET) is experimentally demonstrated for the first time and comprehensively studied. Key processes of this brand-new backside-stacked FET (BSSF) architecture, including the wafer bonding, thinning and self-aligned active, were successfully developed. From our comprehensive study in the aspects of process complexity, STC design flexibility, SRAM scalability, PPA analysis, and P&R assessment.

In this work, FFET is used as an example of F3D integration to show the great flexibility and extendibility. As given in Fig.3, the FFET has unique transistor stacking capabilities, with two layers of transistor back-to-back stacked. Below the conventional frontside transistor, another layer of transistor is stacked on the wafer backside, delivering much smaller standard cell height with better routability.

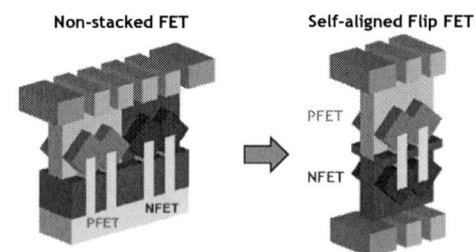

Figure 3: Concept of self-aligned Flip FET. The PFET and NFET are stacked back-to-back by F3D integration on each side of wafer.

II. PROCESS FLOW AND DEVELOPMENT

Fig. 4 summarizes the critical steps of the FFET flow, taking 3.5T dual fin STC as an example. The fin is etched

979-8-3503-6184-1/24 $31.00 © 2024 IEEE

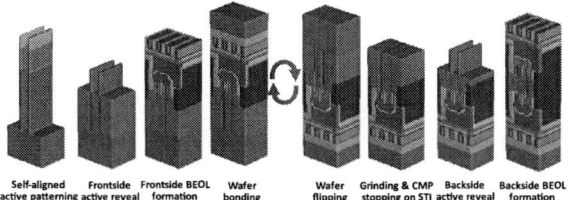

Self-aligned | Frontside | Frontside BEOL | Wafer | Wafer | Grinding & CMP | Backside | Backside BEOL
active patterning | active reveal | formation | bonding | flipping | stopping on STI | active reveal | formation

Figure 4: Process flow of Flip-FET proposed in this work. Two layers of transistors are formed on the same active and stacked in a self-aligned fashion on the wafer frontside and backside.

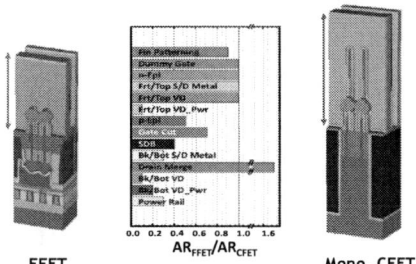

FFET | AR$_{FFET}$/AR$_{CFET}$ | Mono. CFET

Figure 5: Comparison of process complexity of FFET and Mono. CFET. The FFET shows much lower AR compared with CFET.

first, followed by the STI process. After the frontside (FS) FEOL & BEOL formation, a carrier wafer is bonded to it and flipped. Then, the Si substrate of the active wafer is thinned down with CMP stopping on STI. The backside (BS) fin is then revealed by STI recess in a self-aligned manner, followed by the BS FEOL & BEOL formation. Most importantly, the processes of both sides follow the standard FinFET/NSFET flow except those for inter-side connections. Its process implementation is far easier than Mono. CFET [5], as validated by the lower AR in Fig. 5. Only one process step in FFET has a higher AR that in the CFET. This process, the Drain Merge, is not that challenging compared with those complex processes with even higher AR in CFET considering that supervia is already widely used in CMOS technology.

The key processes of FFET were also experimentally demonstrated. Fig. 5(a) shows the SEM image of the frontside transistor before the backside processes. Fig. 5(b) depicts the partial backside thinning down with some Si remaining after the wafer bonding and flipping process. Fig. 5(c) shows the precise CMP stop on STI. Backside STI recess by wet etch to reveal the backside active is shown in Fig. 5(d). Considering the asymmetrical fin profile for top and bottom of the fin channel, additional fin trimming and smoothing can be used for profile improvement. Fig. 5(e) shows the backside gate formation with improved fin shapes. Note that, thanks to the separated gate process, split-gate structure can be easily realized by FFET, in contrast to CFET.

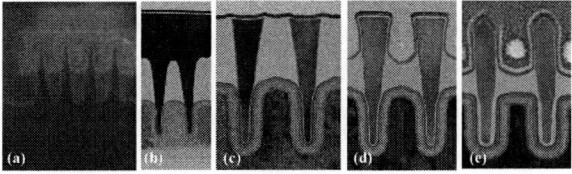

Figure 6: Key experimental process step TEMs of FFET. (a) Original frontside FinFET, (b) Flipped fin after wafer bonding, flipping and thinning, (c) wafer CMP stopped on STI, (d) STI recess with fin revealed, (e) Gate structure after RMG.

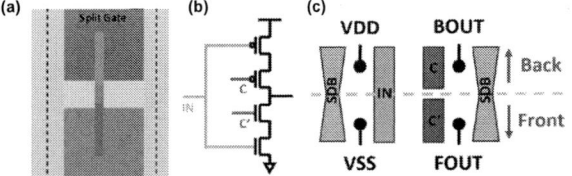

Figure 7: (a) Split-gate structure, (b) circuit diagram of C²MOS, complementary clock signals are connected separately to a NMOS and PMOS. (c) Simplified structure of C2MOS by FFET. FFET can support C²MOS without area penalty.

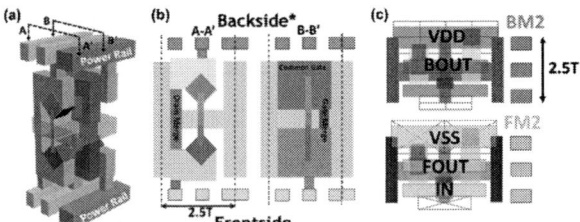

Figure 8: CMOS inverter based on FFET: (a)3D structure. (b) Cross sectional images of (a) in A-A' and B-B' directions. (c) Backside and Frontside layouts.

III. STANDARD CELL DESIGN

For FFET, we use nFET at FS and pFET at BS for the thermal budget constraint of pFET [7][8] in FFET STC design. Fig. 8(a) shows the 3D cartoon of a 2.5T FFET inverter (INV) and the cross sectional images at gate and SD are provided in Fig. 8(b). Fig. 8(c) shows the layout of the 2.5T INV cell.

The FFET has more design flexibility than the CFET in three aspects, as compared in Fig. 8. Firstly, with 2 signal tracks and 1 power rail shared with adjacent STCs placed in a mirror configuration on both sides, 2.5T FFET has 1 more signal track than 3T CFET [5], thus more intra-cell routing resources. Secondly, with similar structure like the non-stacked FET at each side, S/D and gate pins of FFET can be connected to any signal track, while the active structure of the top transistor blocks the connection between bottom contact and signal tracks in active region. Thirdly, thanks to the separated gate (Fig. 9(left)) process, FFET can support crucial sequential logic cells such as: transmission gate and C²MOS (Fig. 9(right)) without area penalty. While typical CFET is normally common-gated and has to waste extra CPPs for sequential logic design. To overcome this issue, SG CFET was recently proposed [9] to support split gate structure with quite challenging process flows. However, extra space of 0.5T cell height has to be sacrificed to leave space for the via connecting bottom gate and signal tracks.

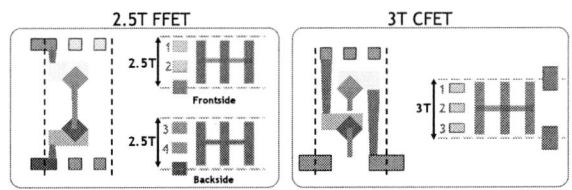

Figure 9: Comparison of FFET and CFET Inverter at source sides connecting power. FFET has smaller cell height and much simpler processes for power connections.

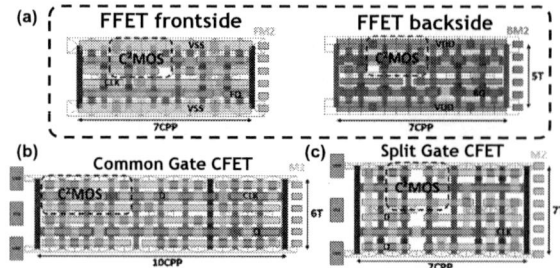

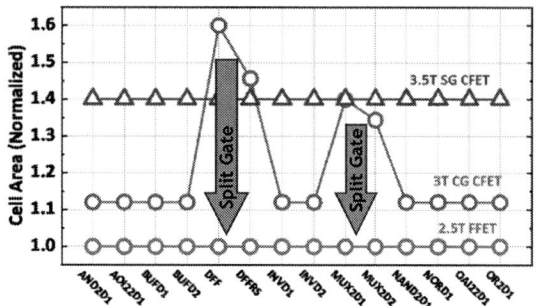

Figure 10: DFF layouts with folded 2-row design for (a) FFET, (b) CG CFET and (c)SG CFET. FFET has fewer CPPs than CG CFET and smaller cell height than both CG and SG CFET.

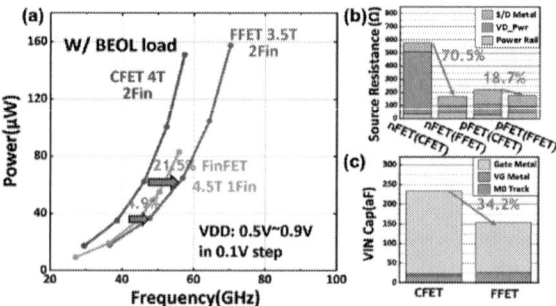

Figure 12: Power-frequency plots of the 4.5T single-fin FinFET, 4T dual-fin CFET and 3.5T dual-fin FFET with the same M0 pitch at similar footprint. A typical BEOL load is assumed here. (b) Reduced source resistance of FFET due to the shorter VD to power. (c) Reduced input capacitance of FFET due to shorter gate extension.

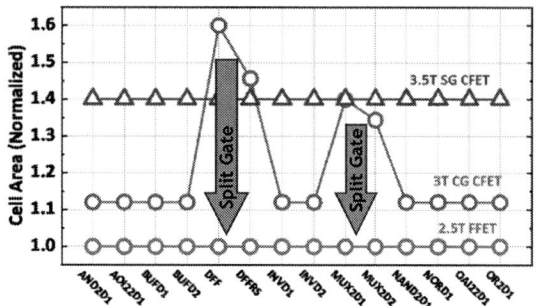

Figure 11: STC area comparison between 2.5T FFET, 3T CG CFET and 3.5T SG CFET. 2.5T FFET has the smallest cells. Thanks to the area-efficient separated gate nature of FFET, extra area benefits can be seen for cells requiring split gate.

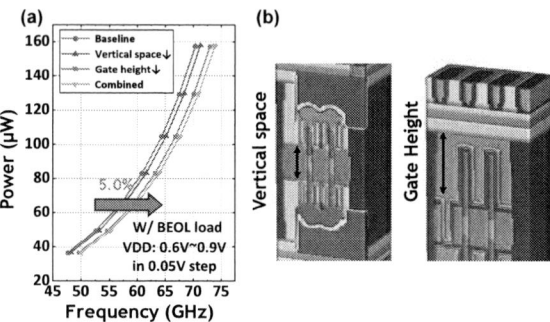

Figure 13: (a) Power-performance for FFET with DTCO elements. (b) Reduction of vertical space, which lowers the resistance of Drain Merge and Gate Merge, and reduction of gate height, which reduces the gate-to-drain capacitance.

Fig. 10 gives DFF layouts with folded 2-row design for the three stacked FETs above and FFET is the smallest due to reduced cell height and the area-efficient split-gate design. As discussed previously, CFET suffers from area penalties in cells requiring C^2MOS design. The area of key cells in our STC libraries are compared in Fig. 11, further proving the area benefits of FFET. Note that in every FFET STC, each output pin is dual-sided and delivers the same output signal. This is important for dual-side global interconnects.

transistors and gate height, 5.0% frequency gain at iso-power @VDD=0.7V can be realized in FFET, as given in Fig. 13.

DTCO knobs in the top-left corner include reduction of vertical space, which lowers the resistance of Drain Merge and Gate Merge, and reduction of gate height, which reduces the gate-to-drain capacitance.

IV. PPA ANALYSIS

For power-performance-area(PPA) evaluation, A 15-stage ring oscillator with FO3 and typical BEOL loads [12] was used. The device models were calibrated to ref [11, 13-15]. We assume the same intrinsic transistor performance for all the architectures studied.

Note that for fair footprint benchmark, we compared the 4.5T single-fin FinFET, the 4T dual-fin CG CFET and the 3.5T dual-fin FFET by using the same M0 pitch for all the stacked transistors studied, which results in smaller Mx pitch for CG CFET. As shown in Fig. 12(a), at iso-power @VDD=0.7V with typical BEOL load, frequency of FFET exceeds FinFET by 4.9% due to reduced BEOL parasitic RC from smaller cell size and exceeds CFET by 21.5% due to smaller parasitic within the cell and outside the cell. FFET INV also has 70.5% reduced source resistance (Fig. 12. (c)) due to the far longer VD to BPR for top device in CFET [5]. Furthermore, FFET has a wider gate cut thus shorter gate extension because FFET MOL is formed after the gate cut while the bottom tier of CFET [5] isn't. This leads to 34.2% reduction in input capacitance (Fig. 12(c)). By inserting more DTCO knobs such as the vertical space between stacked

V. DUAL-SIDE INTERCONNECTS

For logic gates, output pin is typically composed of common S/D of a pair of nFET and pFET. Similarly, in FFET STC, each output pin is linked by the Drain Merge (the via connecting FS and BS S/D, shown in Fig. 8(b)) and gets connected to FS and BS M0 to drive the next stage cell with

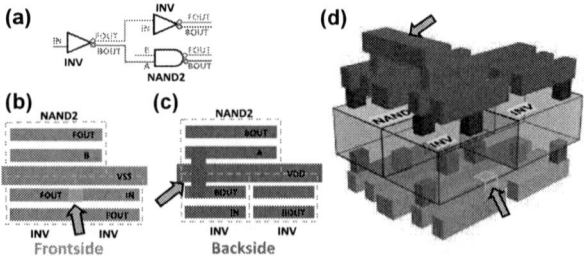

Figure 14: Principles of dual-side interconnects. Each output is linked by the Drain Merge. (a) A typical FFET circuit with a INV fanout to a INV and NAND2. Interconnects layout of (a) for frontside(b) and backside(c). (d) 3D schematic of the routing.

979-8-3503-6184-1/24 $31.00 © 2024 IEEE

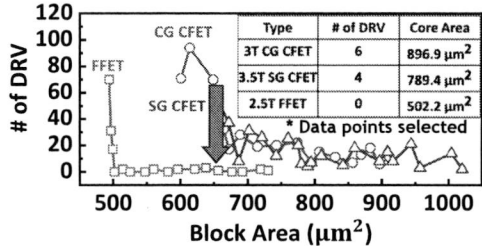

Figure 15: Block-level P&R results of a RISCV32I core for FFET, CG CFET and SG CFET. The FFET has much less design rule violations.

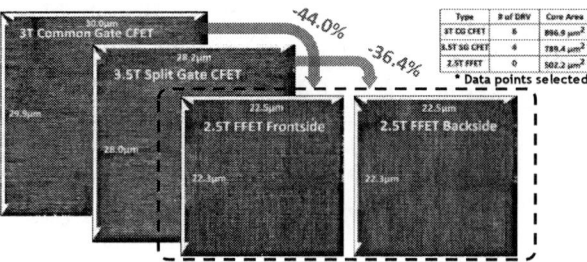

Figure 16: BEOL layout of CG CFET, SG CFET and FFET with similar DRV.

input pins on FS or BS. Fig. 14(a) shows an FFET INV with dual-side output (FOUT & BOUT) and input at either side (FIN or BIN). A typical FFET circuit consisted of an INV driving the next stage INV and NAND is illustrated in Fig. 14(b) and the dual-side global interconnects layout and the 3D schematic of routing is given explicitly in Fig. 14(c-d). Signals can also reach the other side (inter-side interconnects) in field region which is called the Signal Tap Cell by using the same processes as Drain Merge, as given in Fig. 8(b). Therefore, no extra process is needed for inter-side interconnects such as the nTSV.

Based on the new concept of interconnects above, block-level routability evaluation on FFET and CG & SG CFET W/ BSPDN was conducted by doing P&R with a RISCV32I core. FFET has much less design rule violation (DRV) than CG & SG CFET at the same chip area (Fig. 15) as the input pin number is halved at each side for FFET. With less area and better routability, FFET has 44% and 36.4% core area reduction against CG & SG CFET respectively, as in Fig. 16.

VI. F3D FOR FUTURE

The concept of Flip 3D integration can be divided into 4 categories from device to system level. First: the dual-side interconnects, enabling greatly improved circuit routability; Second: The FFET, with dual-side active and interconnects, significantly increasing transistor density; Third: the dual-side functionality with logic and memories on each side of wafer, delivering much higher memory bandwidth for better SOC performance; Fourth: the dual-side chip hybrid bonding, featuring much enhanced system connectivity.

Fig. 17 compares the proposed F3D integration with existing M3D integration. Limited by the single-sided process, the conventional M3D can only stack chips on the frontside. While the F3D features chip stacking on both sides of wafer, with much broader space for future's 3D integration.

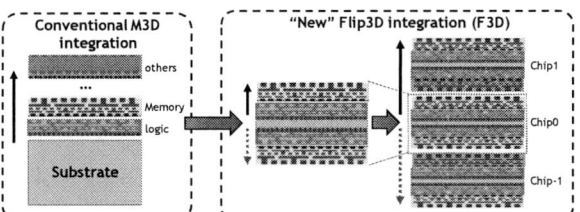

Figure 17: comparison of conventional M3D integration with the F3D integration proposed in this work. The F3D provides new insights to stack chip on both sides of wafer.

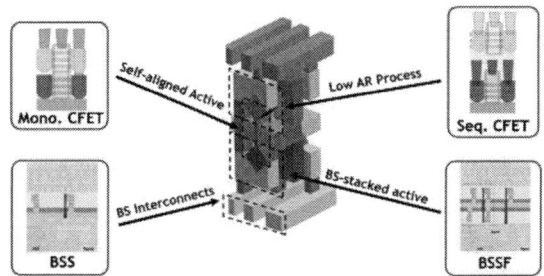

Figure 18: Key features of FFET: the low AR process from Seq. CFET, the self-aligned active process from Mono. CFET, the dual side interconnects from backside signal and backside stack active from backside stacked-FET.

VII. CONCLUSIONS

In this work, we proposed a novel concept of F3D integration, with dual-side integration capability. Based on F3D, a new transistor stacking method is demonstrated as the FFET, as shown in Fig. 18. The FFET takes advantages of self-aligned active and gate from Mono. CFET[16-17], manufacturing-friendly lower AR process from Seq. CFET, backside interconnects from BSS[18] and backside active from backside stacked FET (BSSF). With higher design flexibility of STCs and SRAM, smaller cell size and better routability, FFET shows great potential of scalability. This work provide new insights for future logic technology.

REFERENCES

[1] M. Radosavljević, et al., IEDM 2021.
[2] S. -Y. Wu, et al., IEDM 2022.
[3] J. Cai, VLSI 2021, Short Course.
[4] M. Kobrinsky, et al., IEDM 2023.
[5] J. Ryckaert, et al., VLSI 2018.
[6] A. Veloso, et al., IEDM 2023.
[7] J. Wang, et al., VLSI 2021.
[8] A. Vandooren, et al., IEDM 2018.
[9] L. Liebmann, et al., IEDM 2021.
[10] H. -H. Liu, et al., TED, p.883, 2023.
[11] G. Yeap, et al., IEDM 2019.
[12] S. Yang, et al., IEDM 2023.
[13] J. Jeong, et al., VLSI 2023.
[14] R. Li, et al., www.techinsights.com/blog/qualcomm-snapdragon-8-gen1-soc, 2022.
[15] Y. Lee, et al., IEEE Access, p. 80695, 2022.
[16] S. Liao, et al., IEDM 2023.
[17] B. Chehab, et al., SPIE 2021.
[18] W.Rachmady, et al., IEDM 2019.

Modeling and Simulation of A Conical 3D Monopole Antenna Embedded in Substrate for WNoC

Junhao Wang[1], Ziyu Liu[2,3*], Zhiyuan Zhu[1], Lin Chen[2], Qingqing Sun[2], and Wei David Zhang[2]

[1]School of Electronic Information Engineering, Southwest University, Chongqing China
[2]School of Microelectronics, Fudan University, Shanghai, China
[3] Jiashan Fudan Institute, Jiaxing, Zhejiang Province 314100, China
*Email: liuziyu@fudan.edu.cn

Abstract—In this paper, a three-dimensional (3D) monopole antenna applied for short-range communication wireless on-chip networks was proposed to increase the bandwidth and reduce the operating frequency. The 3D antenna used two conical via as the transmitting and receiving end. To verify the antenna performance, modeling and simulations were set up by changing the substrate materials. To further increase the bandwidth and lower the insertion loss, the number of grounding layer and middle grounding layer layout were compared based on the their influence on the S-parameters. The optimized antenna system had a bandwidth of 5.4 GHz between 22.37 and 27.77 GHz. Within the bandwidth, a minimum return loss of -39.39 dB and a insertion loss of -8.30 dB at 27.36 GHz was obtained. The 3D antenna could work at the minimum and maximum frequency of 17 and 37 GHz.

Keywords—Monopole, TSV, 3D integration，3D antenna

I. INTRODUCTION

As the complexity and integration density of electronic systems increase, improving the performance and reducing the power are the demands of integrated circuits (ICs). System-on-a-chip (SoC) and wireless network-on-a-chip (WNoC) have been proposed successively to reduce the latency and the power, and improve inter-chip communication performance [1]. However, the wireless communication performance between different chips in WNoC systems still need to be improved. 3D antennas have been proposed to solve these problems and also increase integration density [2].

In the literature, a 3D antenna was integrated into a silicon substrate [2] and a prototype based on through silicon via in the printed circuit board (PCB) was demonstrated to improve the insertion loss [3]. A 3D antenna based on a glass substrate was also proposed and simulations showed the resonance frequencies were between 70 and 90 GHz [4]. A 3D monopole antenna for WNoC was also evaluated [5-6]. All these antennas had narrow bandwidth and high resonance frequencies.

In this paper, a 3D monopole antenna based on a conical via structure is proposed for WNoC. The effect of substrate materials, grounding layer number and middle grounding layer layout on the return and insertion loss are simulated using high frequency structure simulator (HFSS) to reveal the effects of these parameters on the 3D antenna performance.

II. DESIGN

A. 3D Antenna Design

Figure 1 (a) shows the overall design of 3D antenna system with three grounding layers to improve the signal performance. The top, middle, and bottom ground layer (both 1 mil thick) are designed as the shielding structures. Two conical signal vias are used as the transmitting and receiving antennas, which has a transmission distance of 10 mm. A coplanar waveguide (CPW) is used for the feed, and the CPW has both linear and circular segments as shown in Fig. 1 (b). The linear segment has a length of 450 mil and a width of 17 mil. The circular segment has a radius of 13.5 mil and the space between the ground layer and CPW trajectory at the top layer is 10 mil.

The detailed structure of conical signal via is shown in Fig. 2, which has the upper radius of 4 mil, the lower radius of 7 mil, and the height of 34 mil. The air gap size is 8 mil, the disk size is 13.5 mil, and the metal thickness is 1 mil as shown in Fig. 2. The grounding via has a cylindrical structure with the radius of 10 mil and height of 35 mil as shown in Fig. 1 (a).

B. Substrate Material Design

Substrates with high resistance (HR) silicon, glass, and Rogers RO4003 are all commonly used in high-frequency application especially antenna due to the dielectric constant and low tangential loss. Thus, these three substrate materials are compared in this paper. HR-Silicon, glass, and Rogers RO4003 has a dielectric constant of 11.7, 5.5 and 3.55, and a tangential loss of 2.6×10^{-3}, 0, and 2.7×10^{-3}.

C. Number and Layout Design of Mid-grounding Layer

To optimize the performance of 3D antenna, different number of grounding layers and different layouts of middle grounding layer are designed. The two grounding layers in Fig. 3 (a) and three grounding layers in Fig. 3 (e) are compared. Figures 3 (a) to (d) presented four inward-extending designs including no middle grounding layer, one-third, one-half and two-thirds area in the middle grounding layer. Meanwhile, four inward-contracting designs containing large area (regular), one-third, one-half and two-thirds area of total area are also designed as shown in Figs. 3 (e) to (h). Figures 3 (e) and (i) also made a regular and irregular design.

III. SIMULATION RESULTS

In section A, the S parameters were compared for three different substrate materials. In section B, antennas with two and three grounding layers were compared. In section C, the effect of middle grounding layer layout was investigated.

A. Effect of Substrate Materials on S parameters

Figure 4 showed the S parameter comparison of three substrate materials, in which the green line was for HR-silicon, the blue line for the glass, and red line for Rogers RO4003.

- In the HR-silicon substrate, all the insertion loss was lower than -10 dB, which meant no bandwidth could

979-8-3503-6184-1/24 $31.00 © 2024 IEEE

be used. Thus, S parameter for glass and Rogers RO4003 was compared.

- In the glass substrate, the bandwidth was 1.18 GHz between 20.51 and 21.69 GHz. At 21.05 GHz, the minimum return loss is -17.08 dB, and the insertion loss is -8.29 dB.

- In Rogers RO4003 substrate, the bandwidth was 2.42 GHz between 25.17 and 27.59 GHz. In addition, there were also two smaller bandwidth of 1.24 GHz and 1.7 GHz. At 26.36 GHz, the minimum return loss was -24.48 dB, the insertion loss was -7.34 dB.

- Fig. 4 also showed the glass substrate reduced the operating frequency of 3D antenna by about 3.38 GHz. But the bandwidth of Rogers RO4003 was much larger than the glass, so Rogers RO4003 was used as the basis in the subsequent simulation.

B. Comparative of Grounding Layer Number

The number of grounding layers was studied in the substrate of Rogers RO4003. The specific results were shown in Fig. 5. The red and blue line stood for the three-layer and two-layer grounding structure, respectively.

- Insertion loss for minimum return loss: At 26.36 GHz, the three-layer structure had the minimum return loss of -24.5 dB and insertion loss of -7.47 dB. The two-layer structure had the minimum return loss of -36.15 dB and insertion loss of -8.34 dB at 27.36 GHz. The insertion loss of the two-layer structure was a little worse than that of three-layer.

- Bandwidth: the three-layer structure had three large bandwidths of 1.24, 2.42 and 1.7 GHz respectively. The two-layer structure had two large bandwidths of 3.44 and 2.45 GHz, and the maximum bandwidth was 30 percent higher than that of the three-layer structure.

C. Simulation of Middle Grounding Layer Layout

- Middle grounding layer extended inward as shown in Figs. 3 (a)-(d): Compared with the two-layer structure, when the area of middle ground layer gradually increased toward the ground via, the trend of S-parameter remained the same and the effective interval of return loss (<-10 dB) increased as shown in Fig. 6. The maximum effective interval of return loss was 9.56 GHz. The insertion loss increased a little with the maximum by 0.2 dB in the effective interval of return loss. The bandwidth was as large as 5.4 GHz for the designs with the 1/3, 1/2, 2/3 total area of middle ground layer.

- Middle grounding layer contracted inward as shown in Figs. 3 (e)-(h): The middle grounding layer contracted in the direction close to the grounding via and the area kept shrinking. Figure 7 showed the area contraction have little effect on the S11 and S21 parameter results, so the maximum bandwidth range remained around 2.42 GHz. Figure 7 showed small changes occurred at the frequency point but the maximum shift of return loss and insertion loss was separately smaller than 5 percent and 0.5 percent.

- Irregular middle grounding layer: The irregular design increased the middle grounding layer area by 26% compared with regular one as shown in Figs. 3 (a) and (i). Figure 8 presented the irregular layout had little effect on the S11 and S21. However, the excessive grounding layer area would waste substrate area, which can be embedded with more devices. Thus, large middle grounding area was not benefit for the shielding performance and proper design was especially important for 3D antenna.

IV. CONCLUSION

A conical 3D monopole antenna applied for WNoC was proposed and the antenna in different substrate materials was investigated. Rogers RO4003 material showed better performance in the three-layer structure. According to the analysis of grounding layer number, the antenna with three grounding layer had three smaller bandwidths of 1.24, 2.42 and 1.7 GHz and two grounding layer brought two large bandwidths of 3.44 and 2.45 GHz, for which insert loss had only a little difference. Based on the grounding layer layout design, large middle grounding layer area did not improve the performance but optimized layout could. It was found that the inward-extending layout for middle grounding layer could further optimize the bandwidth compared with two grounding layers. The effective interval of return loss increased to 9.56 GHz with the insertion loss worsening only by 0.2 dB and the bandwidth became as larger as 5.4 GHz. Finally, compared with other 3D antennas in the literatures [5-6], the operating resonance frequency of the proposed 3D antenna (mainly worked from 17 to 37GHz) was reduced by 25 GHz, the insertion loss was improved by about 32 dB. In addition, the optimized antenna had a bandwidth of 5.4 GHz with the minimum return loss of -39.39 dB at 27.36 GHz.

ACKNOWLEDGMENT

This work was supported by STI 2030—Major Projects (Grant 2022ZD0209200), and Key Project of Ministry of Industry and Information Technology of the People's Republic of China (TC230A076-2).

REFERENCES

[1] V. F. Pavlidis and E. G. Friedman, "Interconnect-Based Design Methodologies for 3D Integrated Circuits," in Proceedings of the IEEE, 2009: 123-140. .

[2] K. Kim and K. Ko, "Integrated dipole antennas on silicon substrates for intra-chip communication," IEEE-APS International Symposium. 1999 Digest. Orlando, FL, USA, 1999: 1582-1585.

[3] Pano V, Tekin I, Liu Y, "TSV-Based Antenna for On-Chip Wireless Communication," IET MICROW ANTENNA P, 2019, 14 (4).

[4] Hwangbo S, Rahimi A, Kim C, et al. Through Glass Via (TGV) disc loaded monopole antennas for millimeter-wave wireless interposer communication. IEEE, 2015.

[5] Wu J, Kodi A K, Kaya S, Monopoles Loaded with 3D-Printed Dielectrics for Future Wireless Intra-Chip Communications, IEEE Trans. Antennas Propag., 2017, 65 (12): 6838-6846.

[6] Tasolamprou A C, Mirmoosa M S, Tsilipakos O, Intercell Wireless Communication in Software-defined Metasurfaces, ISCAS IEEE, 2018: 1—5.

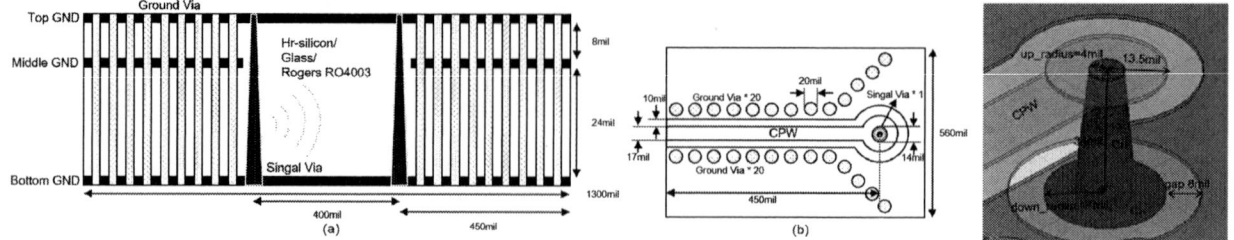

Fig. 1. Overall structure diagram of 3D antenna (a) Cross section (b) Top view (Half of thesymmetrical structure).   Fig. 2. Signal via structure diagram.

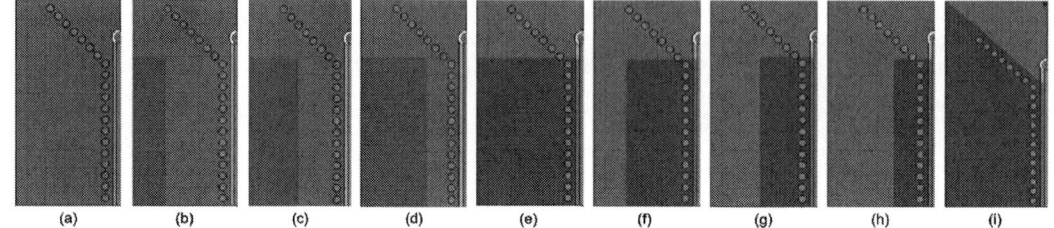

Fig. 3. Four inward-extending middle grounding layer designs including from (a) to (d); Four inward contracting middle grounding layer designs (e) to (h). Regular and irregular middle grounding layer design (e) and (i). (a) two grounding layers (b)1/3 area (c) 1/2 area (d) 2/3 aera (e) regular aera (f) 2/3 area (g)1/2 area (h)1/3 area (i) irregular area.

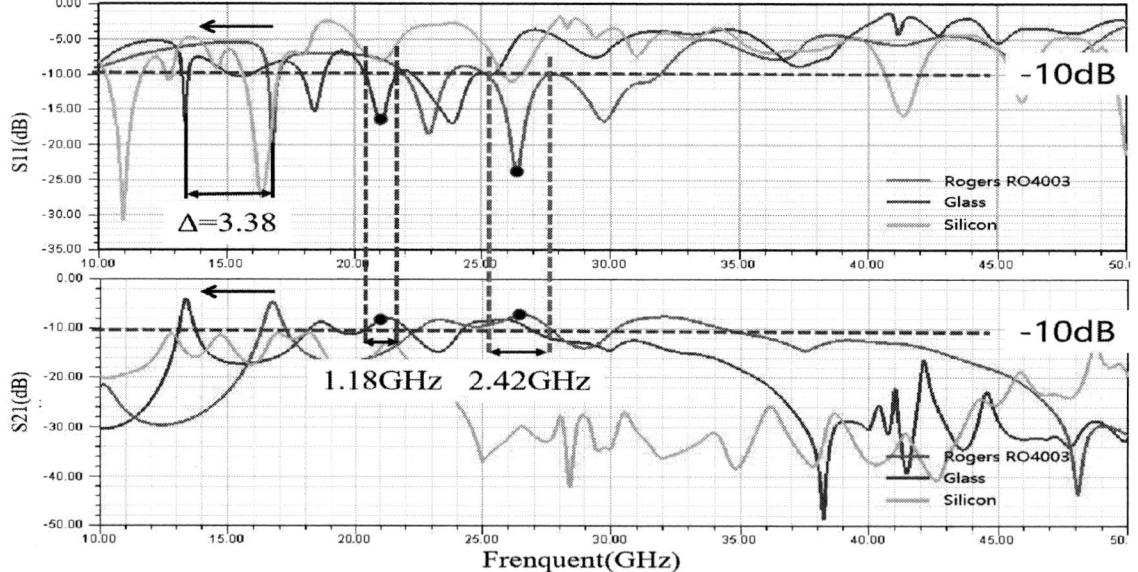

Fig. 4. Effect of substrate materials on S11 and S21

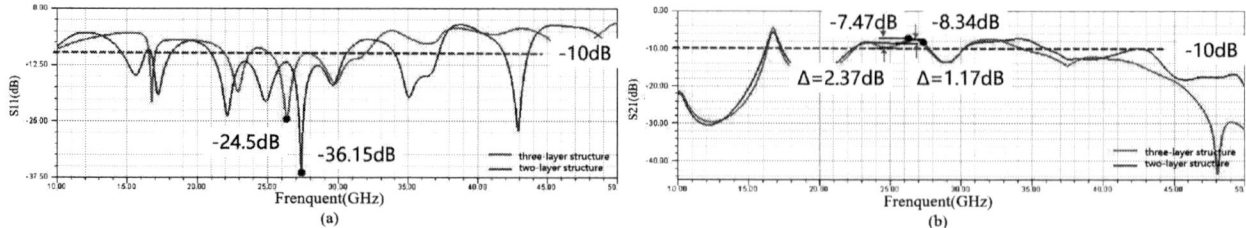

Fig. 5. Comparison of three-layer and two-layer structures in Rogers RO4003.(a) return loss (b) insertion loss.

979-8-3503-6184-1/24 $31.00 © 2024 IEEE   370

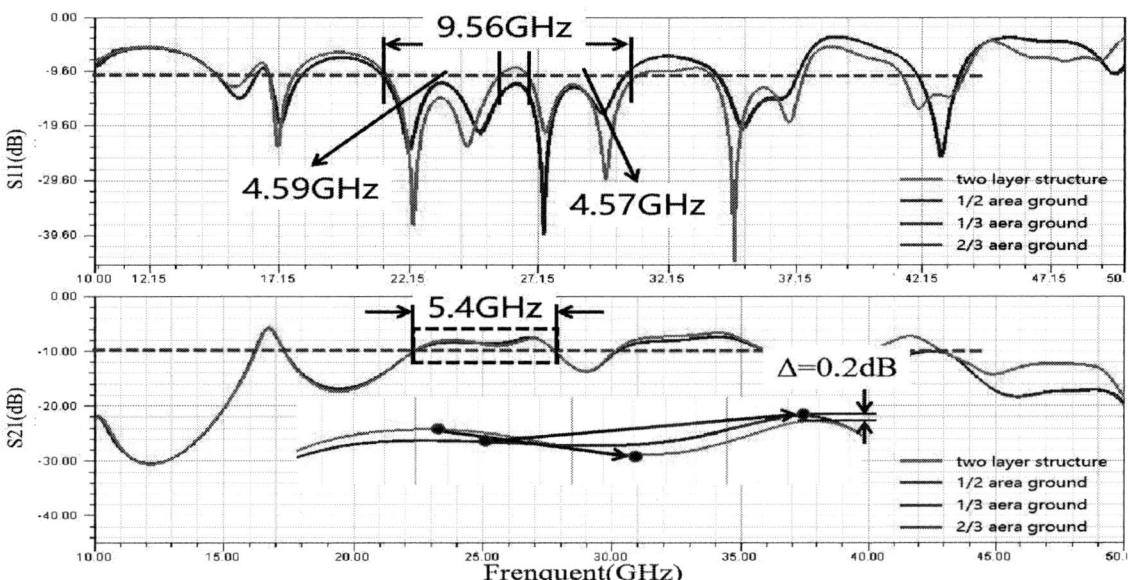

Fig. 6. Comparison of grounding layer extends inwards

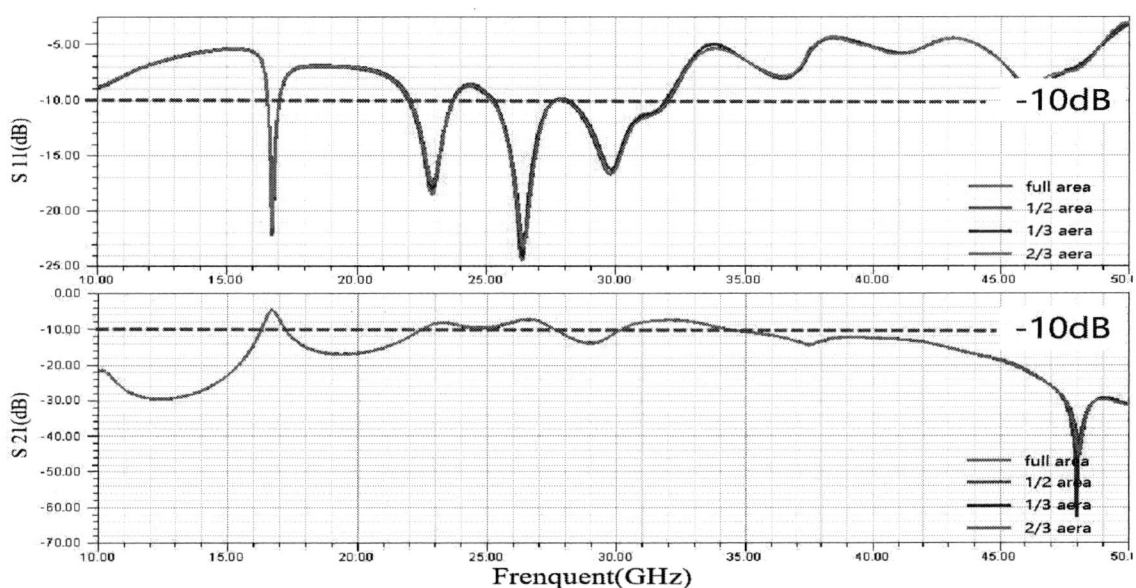

Fig. 7. Effect of grounding layer contracted inward (based on a three-layer structure).

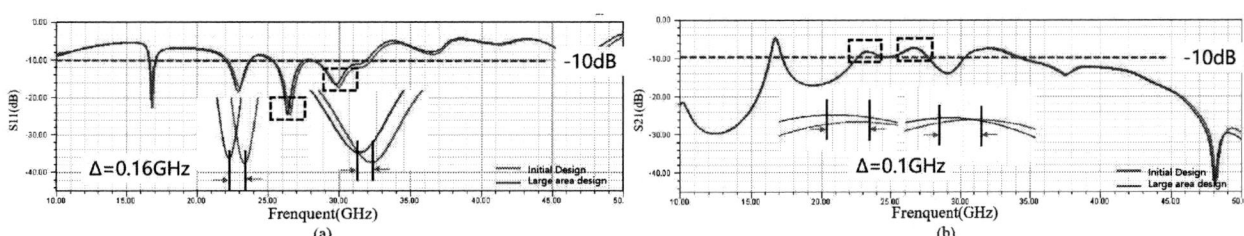

Fig. 8. Comparison of large grounding layer and regular design.(a) return loss (b) insertion loss.

Gate Driver ICs for Wide Bandgap Power Transistors

Wai Tung Ng*, Rophina Li, Wentao Cui, Jingyuan Liang

The Edward S. Rogers Sr. Dept. of Electrical & Computer Engineering
University of Toronto, 10 King's College Road, Toronto ON Canada M5S 3G4

* Email: ngwt@ece.utoronto.ca

Abstract—**Wide bandgap (WBG) power devices such as Gallium Nitride (GaN) and Silicon Carbide (SiC) power transistors are the workhorse of modern power electronics. Although these power semiconductor devices have MOS-like gate electrodes, turning them on or off quickly requires much more than just applying a high or low voltage. Recent trends for smart gate driver ICs are to integrate a variety of complex functions to provide better protection, monitoring, and local control of the switching behaviors of the power devices. This paper starts with a review of basic gate driving requirements. This is followed by the introduction of recent developments in smart integrated gate drivers that are specific to the stringent requirements for GaN and SiC power transistors. Smart gate driver ICs with innovative integrated features such as dynamic gate driving and dead-time correction to minimize EMI and switching losses will be discussed. Techniques to provide sub-nanosecond time resolutions to automate the determination of the dynamic gate drive profiles dedicated to WBG power devices will be described. Finally, new gate drive features such as aging detection and compensation for the SiC devices will also be presented.**

Keywords—Aging Detection, Deadtime Correction, Dynamic Gate Driving, GaN and SiC Power Transistors, Gate Driver ICs

I. Gate Drive Requirements

Power transistors are designed to have high voltage and current ratings. The MOS-like gate terminals of silicon-based power MOSFETs, Insulated Gate Bipolar Transistors (IGBTs), Wide bandgap (WBG) power devices such as Gallium Nitride (GaN) and Silicon Carbide (SiC) power transistors generally exhibit an appreciable amount of gate capacitance. Normally, these power transistors should be turned on and off at high speed to minimize switching losses. This can be achieved by using a gate driver circuit with very low output resistance to quick charge and discharge the input gate capacitance. However, the gate drive loop can be considered as an RLC circuit as shown in Fig. 1(a). $R_{G,H}$ and $R_{G,L}$ are the pull-up and pull-down equivalent output resistance of the gate driver. L_G is the parasitic from the connection between the gate driver and the power transistor. C is the input capacitance of the power transistor, C_{ISS}. A small gate resistance ($R_{G,H}$ or $R_{G,L}$) will lead to underdamp situation, resulting in excessive oscillation at the gate. This could lead to unwanted electromagnetic interference (EMI) and false turn-on (or off) if the ringing exceeds the threshold voltage, V_{TH} as shown in Fig. 1(b). A large gate resistance would result in overdamp and slow turn-on (or off).

Conventional wisdom would call for a very closed placement between the gate driver IC and the power transistor to minimize the parasitic inductance. Unfortunately, due to the inherently low on-resistance of the GaN and SiC power transistors. Their die size is usually quite small. In a typical half-bridge configuration as shown in Fig. 2, the spacing of the die placement has a significant impact on the thermal dissipation. A large spacing is preferred for lower thermal resistance as shown in Fig. 3, but this would lead to larger parasitic gate loop inductance. Therefore, gate driver circuits with dynamically adjustable gate resistance would be required to suppress possible ringing oscillation from different RLC loops.

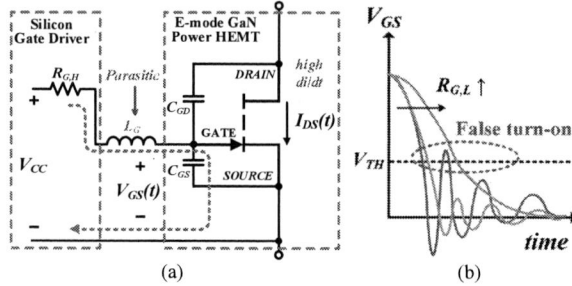

Fig. 1. (a) A typical RLC gate drive circuit loop. $R_{G,H}$ is the output resistance of the gate driver. (b) During turn-off (or turn-on), a low gate resistance will result in unwanted oscillation at gate terminal [1].

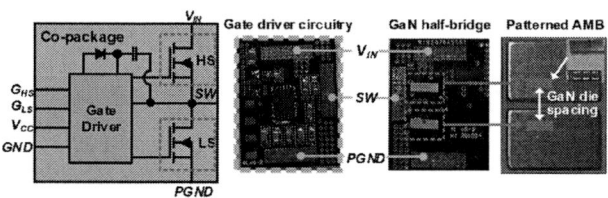

Fig. 2. A typical block diagram of a GaN half-bridge PCB module. The two GaN power transistor dies are flip-chip mounted onto the bottom of the PCB. The back side of the transistor dies are attached to the direct bonded copper (DBC) to heat dissipation [2].

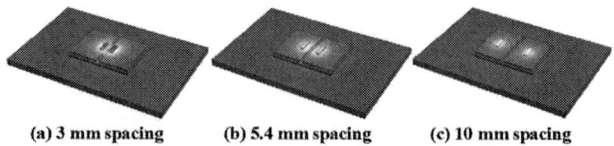

(a) 3 mm spacing   (b) 5.4 mm spacing   (c) 10 mm spacing

Fig. 3. Simulated temperature profiles of the GaN dies showing larger distance spacing would lead to lower thermal resistance [2].

In this paper, the design trends of smart gate driver ICs for GaN and SiC power transistors will be reviewed. In addition to dynamic gate driving strength with sub-nanosecond timing, other advanced features such as dead-time correction and aging detection will be discussed.

II. Dynamic Gate Drive

Dynamic gate driving is a new trend to adjust the output resistance of the gate driver IC on-the-fly. A very low output resistance is used initially during each switching event to provide fast transitions. After the power device is turned on

---

979-8-3503-6184-1/24 $31.00 © 2024 IEEE

(or off), the output resistance is changed quickly to a high value to damp any subsequent oscillation [3]. A conceptual implementation of such dynamic gate driver and the driving pattern is depicted in Fig. 4(a). The benefits of dynamic gate driving versus fixed gate resistances are quite apparent as can be seen in Fig. 5. A main obstacle for the adoption of dynamic gate driving is the sub-nanosecond timing precision required for the output resistance pattern. In addition, rapidly changing load conditions may also require constant adjustment of this timing. As a result, significant effort has been placed to simply the programming of these ICs. Zhang *et al.* has proposed an easy way to adjust a pre-defined driving pattern using a simple external bias resistor [1]. The timing of the pre-defined 2, 3, or 4 level driving patterns, as shown in Fig. 4(b), can be stretched or compressed by changing a bias current controlled by an external resistor. Cui *et al.* has demonstrated an effective way to implement an auto-timing scheme by detecting the onset of the Miller plateau [4]. A switched capacitor high-pass filter is applied to the gate signal to capture the beginning and the end of the Miller plateau commonly observed in power MOSEFTs. The continuous development of these techniques will help to promote the wide acceptance of dynamic gate drivers.

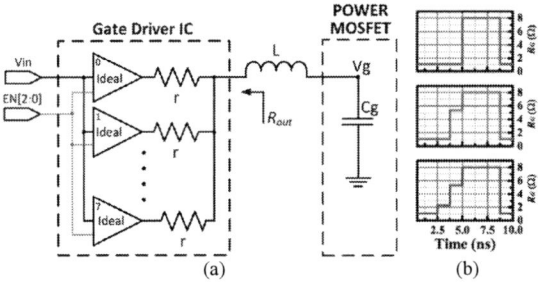

Fig. 4. Conceptual diagram on how a gate driver with multiple output stages in parallel and the driving patterns that can be created [1].

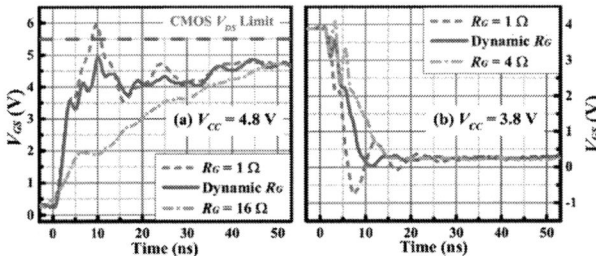

Fig. 5. Measure waveforms at the gate terminal of a GaN power transistor during turn-on and turn-off with fixed and dynamic gate resistance [1].

III. DEAD-TIME CORRECTION

The control signals for all switching power output stages require dead-time to prevent shoot-through current and body diode conduction, or reverse conduction in the case of GaN power transistors. These phenomena are illustrated in Fig. 6(a) and (c). Currently, most power topologies employ fixed dead-times between the high-side (HS) and low-side (LS) gate drive signals. However, this would only be practical if the load current is constant. For frequently changing loads, the dead-time must also be corrected continuously to minimize switching loss. To implement this function, it is best to start with a dead-time that is slightly longer than necessary. A smart gate driver IC can detect the duration of the unwanted body diode conduction or reverse conduction. This duration will then be subtracted from the current dead-time and applied to

the next switching cycle [5, 6]. In steady state operation, the dead-time correction algorithm can occasionally deviate from the optimized dead-time by deliberately increase the dead-time slight and correct it again to account for load current variations.

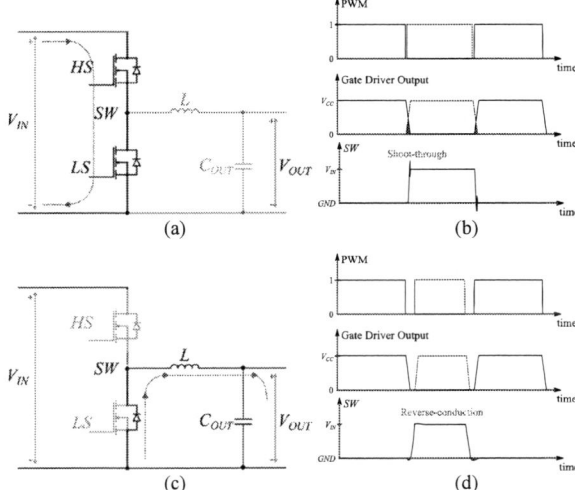

Fig. 6. Dead-time is an important consideration when driving a half-bridge. If the dead-time is too short, it will lead to shoot-through current as in (a) and (b). If it is too long, it will lead to body diode (or reverse conduction) as in (c) and (d).

One of the challenges for the dead-time correction circuit is the detect the duration of the reverse conduction by monitoring the voltage signal at the switching node (*SW*). Since the voltage at this output node can swing between the supply voltage and less than a few voltage below ground level, a clamping circuit is required to protect the sensing input node of the gate driver IC. A possible solution is illustrated in the circuit block diagram of a smart gate driver with dead-time correction as shown in Fig. 7. In this implementation, instead of using the SenseFET to mirror the current in the main power transistor, the gate voltage of this smaller device is connected to a DC voltage, $V_{G,sense}$. If the SW node is at a low or negative voltage, the SenseFET will act like a pass-transistor and signal is connected directly to the SENSE terminal of the IC. If the SW node is at a high voltage, the source terminal is clamped to a safe low voltage defined by $V_{G,sense}$.

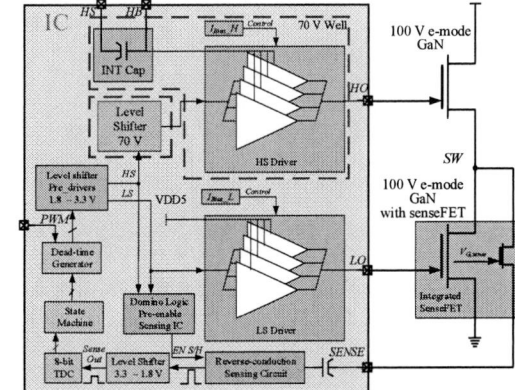

Fig. 7. Circuit block diagram of a gate driver IC with dynamic gate driving strength and dead-time correction. The SenseFET is used as a clamping circuit to prevent large voltage from damaging the SENSE input [6].

979-8-3503-6184-1/24 $31.00 © 2024 IEEE

The operation of this dead-time correction scheme is illustrated by the waveforms in Fig. 8. On the left, dead-time correction is disabled. The SW node shows a train of switching pulses with negative spikes, indicating the presence of reverse conduction. On the right hand size, dead-time correction is enabled. The negative spikes are eliminated, indicating the optimum dead-time is being used.

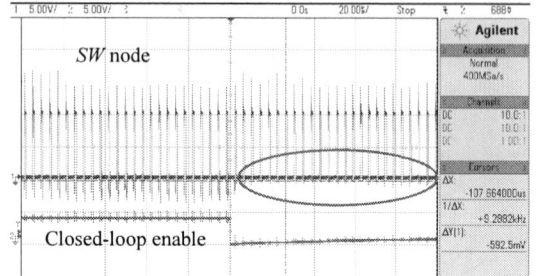

Fig. 8. The operation of the dead-time correction circuit. After activation, the negative pulses due to reverse conduction at the switching node is eliminated [6].

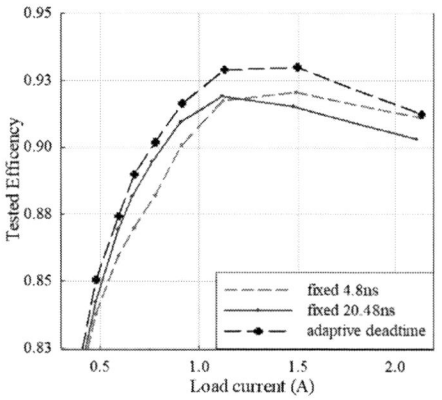

Fig. 9. Measured efficiency for different dead-times, at V_{IN} = 20 V, V_{OUT} = 4 V, f_{SW} = 250 kHz [6].

The benefit of continuous dead-time correction (black curve) can be seen in the measured power conversion efficiency for a 20 to 4 V buck converter as shown in Fig. 9. High efficiency can be maintained throughout the entire output current range. The red curve represents the situation with a short dead-time. The efficiency is reduced at low output current where the shoot-through current would be significant. At high output current, conduction loss dominates and the efficiency becomes closer to the optimum value. The blue curve is when an excessively long dead-time is used. Due to the absence of shoot-through current, the efficiency at low output current is closer to optimal, but the efficiency drops significantly at high output current due to reverse conduction.

IV. AGING DETECTION AND COMPENSATION

Although SiC power MOSFETs have superior on-resistance and voltage blocking capabilities when compared to their silicon-based counterparts, they are still relatively expensive. As a result, they are normally employed in high power applications requiring high reliability. It is important to monitor the health condition of these devices and flag their immanent failure. This would allow preventive maintenance before any catastrophic event takes place. Capturing the Miller plateau of the gate signal can reveal a lot of information including drain current level and V_{TH} shift [7].

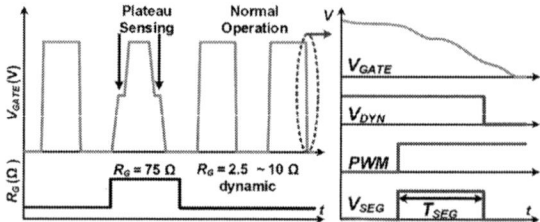

Fig. 10. Due to the excellent figure of merit for SiC and GaN power transistors, their die size can be much smaller, leading to very small gate capacitiance. The Miller plateau visible in silicon-based power devices are normall difficult to detect unless a sufficiently large R_G is used [7].

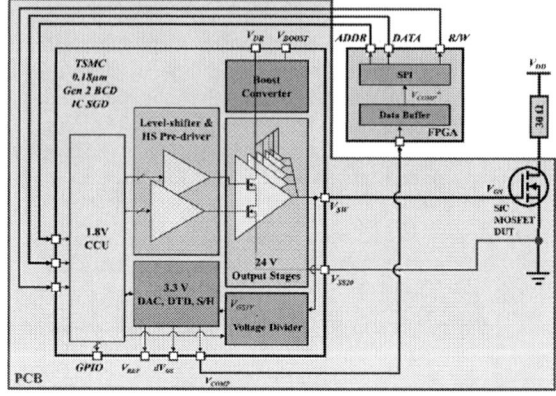

Fig. 11. Functional overview of a smart gate driver IC with aging detection and compensation capability. An FPGA is used to provide the digital interface. A SiC DUT resistive load of R_L = 30 Ω, The nominal gate drive supply voltage of V_{DR} = 15 V is provided by a built-in boost converter [7].

Aging stress applied at the gate of SiC power MOSFETs has been reported to result in non-recoverable V_{TH} shift by up to +1.5 V in planar structures and +3.5 V in trench gate devices [9, 10]. For SiC MOSFETs, the drain current (I_D) and on-resistance (R_{ON}) has a strong dependence on the gate drive voltage (V_{DR}) [11]. Most SiC power devices have very low specific on-resistance, a small device size is all it takes to carry a significant amount of current. As a result, the C_{ISS} for SiC power transistors can be much smaller that their silicon-based equivalent. This will cause the Miller plateau to be very difficult to observe for V_{TH} shift related aging detection purposes. Since aging occurs slowly over time, it is quite adequate to slow down the switching process with a more prominent Miller plateau by employing a large gate resistance on an occasional basis as shown in Fig. 10. This can be accomplished by using a smart gate driver IC with a segmented output stage to provide dynamic driving strength. The circuit block diagram of a smart gate driver IC with aging detection and compensation capability is shown in Fig. 11. In addition to the dynamic gate driver, this IC chip also includes an Miller plateau detection circuit and an internal boost converter to provide the appropriate V_{DR}.

Wang *et al.* has demonstrated the deterioration of drain current for a SiC power MOSFET stressed with a resistive load at 200 °C for up to 200 hours as shown in Fig. 12(a) [8]. This is mainly caused by an increased in V_{TH}. To compensate for this degradation, the smart gate driver IC can estimate the amount of V_{TH} shift and instruct the on-chip boost converter to increase V_{DR} to compensate for lowered drain current. As can be seen in Fig. 12 (b), an increase of +0.5 V in V_{DR} can effectively bring the current level back to its initial

979-8-3503-6184-1/24 $31.00 © 2024 IEEE

performance. At the same time, the smart gate driver IC can also raise a flag to indicate the severity of the aging effect, prompting the user to arrange for a shutdown to carry out preventive maintenance.

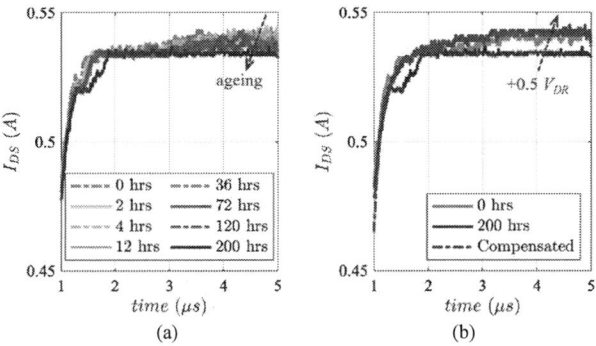

Fig. 12. (a) An aging SiC power MOSFET would normally exhibit in an increase in on-resistance, leading to reduced drain current. (b) This can be partially compensated by increasing the gate drive voltage by 0.5 V in this example [8].

V. Conclusions

In this paper we have discussed the different features and capabilities of modern smart gate driver IC designs dedicated to WBG power semiconductor devices. Gate drivers with parallel output stages can be programmed to provide various dynamic gate driving patterns with sub-nanosecond timing precision. This technique allows GaN and SiC power transistors to switch at high speed without incurring unnecessary gate ringing. Significant effort is underway to make dynamic gate driving as transparent as possible for the users. These include simplified programming and auto-timing methods. Continuous dead-time correction is another important feature required to maintain optimal power conversion efficiency under a wide range of loading conditions. Gate drivers with local sensing and dead-time correction can effectively suppress shoot through current and reverse (or body diode) conduction. Finally, aging detection and compensation techniques are especially important for SiC power MOSFETs where a small V_{TH} shift could result in significant increase in on-resistance. By monitoring the Miller plateau, and appropriate boost in V_{DR} can partially restore some of the lost performance.

As smart gate driver circuits continue to evolve, more features packed ICs will be able to provide high speed, high efficiency, and reliable operation for WBG power semiconductor devices.

Acknowledgment

The authors would like to thank Fuji Electric and NSERC Canada for financial support. Technical support from Fuji Electric and Taiwan Semiconductor Research Institute (TSRI) are also very much appreciated.

References

[1] W.J. Zhang, J.S. Yu, W.T. Cui, Y.H. Leng, J.Y. Liang, Y.T. Hsieh, H.H. Tsai, Y.Z. Juang, W.K. Yeh, and W.T. Ng, "A Smart Gate Driver IC for GaN Power HEMTs With Dynamic Ringing Suppression," *IEEE Trans. Power Electronics*, Vol. 36, No. 12, pp. 14119 – 14132, December 2021.

[2] W.J. Zhang, J. Liang, W.T. Cui, N. Kim, R. Li, A. Catuneanu, M. Birkett, J.G. Burgers, and W.T. Ng, "Compact GaN Power Modules with Direct Bonded Liquid-Cooled Heat Exchanger Suitable for EV

Applications," *IEEE 8th Workshop on Wide Bandgap Power Devices and Applications (WiPDA 2021)*, Nov. 7-11, 2021.

[3] A. Akhavan Fomani, A. Shorten and W.T. Ng, "An Integrated Segmented Gate Driver with Adjustable Driving Capability," *2010 IEEE Energy Conversion Congress and Exposition*, pp. 2430 – 2433, Atlanta, Georgia, USA, September 12-16, 2010.

[4] W.T. Cui, W.J. Zhang, J.Y. Liang, H. Nishio, H. Sumida, H. Nakajima, Y.-T. Hsieh, H.-H. Tsai, Y.-Z. Juang, W.-K. Yeh, and W.T. Ng, "A Dynamic Gate Driver IC with Automated Pattern Optimization for SiC Power MOSFETs," *Int'l Symposium on Power Semiconductor Devices and ICs (ISPSD 2022)*, Vancouver, Canada, May 22-25, 2022.

[5] G.H. Thuc, and C.J. Chen, "A Gate Driver IC for GaN-Based Synchronous Buck Converter with a Double-Sided Adaptive Dead-Time Generator," *2023 IEEE Applied Power Electronics Conference and Exposition (APEC)*, Florida, FL, 19-23 March 2023.

[6] W.J. Zhang, Y.H. Leng, J.S. Yu, Y.S. Lu, C.Y. Cheng and W.T. Ng, "An Integrated Gate Driver for E-Mode GaN HEMTs with Active Clamping for Reverse Conduction Detection," *The 31st Int'l Symposium on Power Semiconductor Devices and ICs (ISPSD 2019)*, pp. 83-86, Shanghai, May 19-23, 2019.

[7] M.Q. Wang, W.J. Zhang, W.T. Cui, J.Y. Liang, H. Nishio, H. Sumida, H. Nakajima, and W.T. Ng, "A Smart Gate Driver for SiC Power MOSFETs with Aging Compensation and Ringing Suppression," *Int'l Symposium on Power Semiconductor Devices and ICs (ISPSD 2021)*, Nagoya, Japan, May 30 – Jun 3, 2021.

[8] M.Q. Wang, J.P. Zhang, H. Nishio, M. Iwamoto, H. Sumida, W.T. Ng, "Application of a Smart Gate Driver to Detect Aging in SiC Power MOSFETs," *Int'l Symposium on Power Semiconductor Devices and ICs (ISPSD 2023)*, Hong Kong, China, May 28 - Jun 1, 2023.

[9] D. Ngwashi and L. Phung, "Recent review on failures in silicon carbide power MOSFETs", *Microelectronics Reliability*, vol. 123, p. 114169, 2021.

[10] H. Jiang, X. Zhong, G. Qiu, L. Tang, X. Qi, and L. Ran, "Dynamic Gate Stress Induced Threshold Voltage Drift of Silicon Carbide MOSFET," *IEEE Electron Device Letters*, vol. 41, no. 9, pp. 1284–1287, 2020.

[11] H. Li, X. Liao, Y. Hu, Z. Huang, and K. Wang, "Analysis of Voltage Variation in Silicon Carbide MOSFETs During Turn-On and Turn-Off," *Energies*, vol. 10, no. 10, p. 1456, 2017.

Suppression of Back-gating Effect by Integrated Substrate Termination Network for 200V Monolithic GaN Half-Bridge Power IC

Mengyao Zhao[1], Yifei Zheng[1], Yanfeng Ma[1], Yuan Sun[1], Denggui Wang[2], Chuanqi Pan[2], Jianjun Zhou[2], Sheng Li[1], Long Zhang[*1], Siyang Liu[*1], Weifeng Sun[*1]

[1] National ASIC System Engineering Research Center, School of Integrated Circuits, Southeast University, Nanjing, Jiangsu, China
[2] National Key Laboratory of Solid-State Microwave Devices and Circuits, Nanjing, Jiangsu, China

* Email: longzh@seu.edu.cn, liusy2017@seu.edu.cn, swffrog@seu.edu.cn

Abstract—We fabricate a 200V monolithic GaN half-bridge power IC featuring an integrated substrate termination network (ISTN). The ISTN utilizes two-dimensional electron gas resistors to dynamically control the substrate potential so that detrimental back-gating effects and dynamic on-resistance (R_{on}) degradation is effectively suppressed. The proposed GaN power IC scheme with ISTN demonstrates remarkable performance by minimizing the increase in dynamic R_{on} for both the high-side and low-side HEMT at power supply of 200V, respectively. The fabricated monolithic ISTN can minimize parasitic elements compared with the board-level substrate termination network in previous work, while no additional process complexity is required.

Keywords—GaN-on-Si, half-bridge, monolithic integration, back-gating effect

I. INTRODUCTION

GaN half-bridge circuit is an important candidate of power electronics comprising a high-side (HS) and a low-side (LS) High-Electron-Mobility-Transistor (HEMT) in series. Unfortunately, the potential difference between the conductive substrate and the source electrode (V_{BS}) can induce detrimental back-gating effect (also called substrate bias effect) to deteriorate the performance of circuit. It is summarized that on-resistance (R_{on}) is increased under static negative V_{BS} (static back-gating effect), or when a positive V_{BS} changes from a high value to a low value (dynamic back-gating effect) [1].

Several schemes have been proposed to eliminate the back-gating effect and related dynamic on-resistance (R_{on}) degradation. Engineered Bulk Silicon (EBUS) and Silicon On Insulator (SOI) substrates enable the source electrode of each HEMT to be connected to its "local electrical substrate", ensuring that V_{BS} is zero for both HS and LS transistors simultaneously [2], [3], [4], [5]. HEMTs with hybrid-drain and virtual-body structures take use of hole injection to accelerate the release of trapped electrons thus mitigating dynamic R_{on} degradation [6], [7], [8], [9]. Another approach is to establish resistor network to modulate substrate potential dynamically during switching cycles, which can also alleviate back-gating effect in half-bridge circuit [10], [11].

This work was supported by the National Key R&D Program of China (2023YFB4403700), the national natural science foundation of China (62274032), the natural science foundation of Jiangsu Province (BK20231150, BK20232006), the Technological Achievements of Jiangsu Province (No.BA2022005).

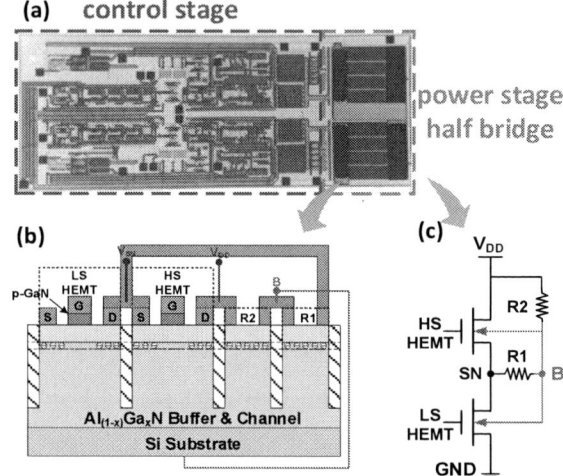

Fig. 1 (a) Photograph of the fabricated monolithic GaN-on-Si IC. (b) Cross-section of GaN-on-Si half-bridge circuit with ISTN (c) Schematic of half-bridge circuit with ISTN.

However, the resistor network in [10], [11] is established by means of external circuit at board-level. In order to further minimize parasitic components, it is beneficial to integrate the half-bridge circuit with resistor network on a single chip. In this work, we design and fabricate a 200V GaN-on-Si half-bridge power IC with Integrated Substrate Termination Network (ISTN), and observe only a slight R_{on} degradation in both HS and LS HEMTs. ISTN can also shorten the switching time due to a smaller V_{BS} voltage swing. Moreover, the incorporation of ISTN does not introduce process complexity.

II. DEVICE PARAMETERS AND ISTN DESIGN

Our monolithic GaN-on-Si IC consists of a control stage and a power stage (Fig.1(a)). The control stage with gate drivers, level shifters and other components has been reported in [12]. The power stage is a half-bridge circuit integrated with ISTN (a reference circuit without ISTN is fabricated at the same time). The epitaxial structure of the half-bridge circuit is composed of 5μm graded $Al_xGa_{(1-x)}N$ buffer layer, 300nm GaN channel layer, and 15nm $Al_{0.22}Ga_{0.78}N$ barrier layer and a 70nm Mg-doped p-GaN layer (Fig. 1(b)). The ohmic contact adopts Ti/Al/Ni/Au multi-layer metal structure, which is formed by rapid thermal annealing at 800℃ for 30 s. Sputtered W metal is used to form Schottky contact for the gate of HEMTs. The back metal Au is used for substrate contact.

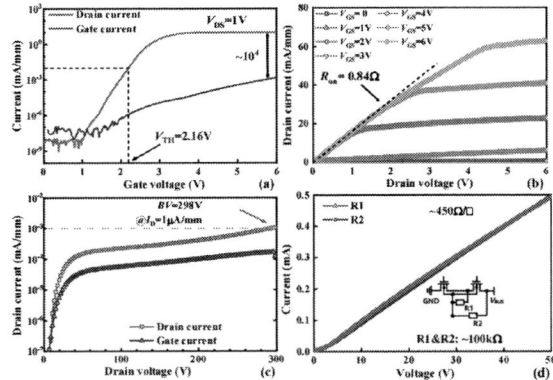

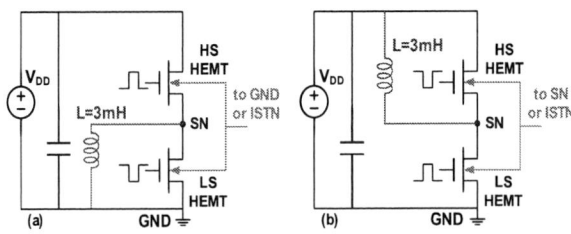

Fig. 5 (a) Transfer characteristics at V_{DS} = 1V. (b) Output characteristics at V_{GS} = 0V to 6V. (c) Off-state breakdown curve at V_{DS} = 0V.(d) Measured I-V curve of 2DEG resistors R1 and R2.

Fig. 6 Inductive clamp double-pulse switching test circuit. (a) Test circuit for HS, with the inductor parallel to LS and substrate connected to GND or ISTN. (b) Test circuit for LS, with the inductor parallel to HS and substrate connected to SN or ISTN.

The ISTN is formed by two 100kΩ Two-Dimensional-Electron-Gas (2DEG) resistors R1 and R2, where R1 is connected between the substrate (B) and switching node (SN), and R2 is connected between B and power supply (VDD) (Fig. 1(c)). Both LS and HS transistors are enhancement-mode HEMTs with a threshold voltage (V_{TH}) of 2.16V (at I_D = 10µA/mm and V_{DS} = 1V), static on-resistance (R_{on}) of 0.84Ω (at V_{GS} = 6V and V_{DS} = 1V), and off-state breakdown voltage (BV) of 298V (Fig. 5).

III. CIRCUIT PERFORMANCE AND DISCUSSIONS

Inductive clamp double-pulse switching test are adopted to evaluate the back-gating effect of half-bridge circuit (Fig. 6). The gate input signals are 0~5V square waves with a period of 20µs and duty cycle of 50%, and VDD varies from 50V to 200V with a step of 50V. The parallel inductor is set to 3mH, and the inductor current is driven to 0.33A, 0.67A, 1.00A and 1.33A when VDD = 50V, 100V, 150V and 200V, respectively (Fig. 2). There are three types of substrate connection: GND (connected to ground), SN (connected to switching node) and ISTN. The voltage drop across the LS and HS HEMT($V_{DS,LS}$ and $V_{DS,HS}$) are measured under the three substrate connection to characterize the dynamic R_{on}. The V_{DS} waveforms during two switching cycles are displayed in Fig. 3.

For HS, there is no dynamic R_{on} degradation under SN connection because $V_{BS,HS}$ remains 0V all the time. However, it faces a **static negative $V_{BS,HS}$** under GND connection during on-state (static back-gating effect), which can partially deplete 2DEG and significantly increase $V_{DS,HS}$ during on-state [1],

especially when VDD is higher than 150V (Fig. 3(a)). The increase of R_{on} is related to the value of negative $V_{BS,HS}$.

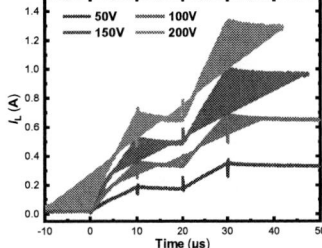

Fig. 2 Inductor current (I_L) in the double-pulse switching test at VDD from 50V to 200V.

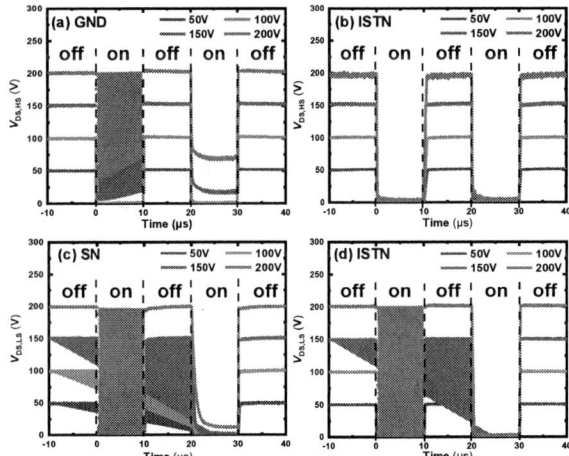

Fig. 3 V_{DS} waveforms under inductive clamp double-pulse switching test at VDD from 50V to 200V. (a) $V_{DS,HS}$ under GND connection. (b) $V_{DS,HS}$ under ISTN connection. (c) $V_{DS,LS}$ under SN connection. (d) $V_{DS,LS}$ under ISTN connection.

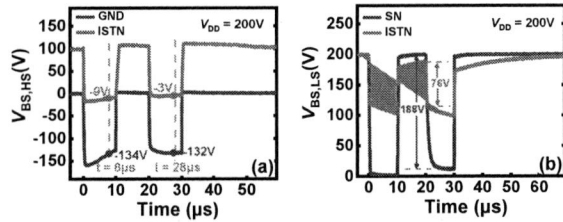

Fig. 4 V_{BS} waveforms of (a) HS and (b) LS in two switching cycles. ISTN shifts the average substrate potential and decreases V_{BS} swing.

In terms of LS, GND connection does not deteriorate its performance but SN connection results in dynamic R_{on} degradation. When LS is at off-state with SN connection, there is a **positive $V_{BS,LS}$** which induces electron to be injected into the buffer layer. Upon switched to on-state, electrons trapped in the buffer layer cannot be released immediately, and continue to partially deplete 2DEG (dynamic back-gating effect). Hence, $V_{DS,LS}$ exhibits a moderate increase at the beginning of the on-state (around 21µs), and gradually decreases along with the release of trapped electrons (Fig. 3(c)). The extent of R_{on} degradation is related to the swing of $V_{BS,LS}$ when switching from off-state to on-state.

When ISTN is integrated with the half-bridge circuit, V_{BS} can be dynamically modulated and both $V_{DS,HS}$ and $V_{DS,LS}$ remain nearly unaffected even at VDD = 200V (Fig. 3(b) and (d)). Fig. 4 plots the change of V_{BS} under the three types of

979-8-3503-6184-1/24 $31.00 © 2024 IEEE   377

substrate connections during two switching cycles. Compared

TABLE I. VBS OF HS HEMT AND ΔVBS OF LS HEMT UNDER VDD =150V AND 200V

Device	Substrate Connection	$V_{BS,HS}$ (t=28μs)		$\Delta V_{BS,LS}$ (from off-state to second on-state)	
		VDD =150V	VDD =200V	VDD =150V	VDD =200V
HS HEMT	GND	-130V	-132V	/	/
	ISTN	-2V	-3V	/	/
LS HEMT	GND	/	/	143V	188V
	ISTN	/	/	55V	76V

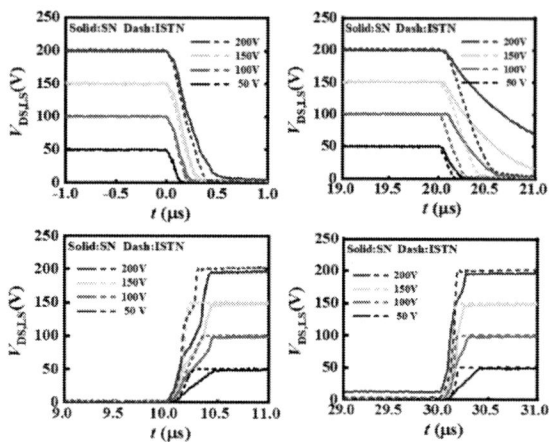

Fig. 7 $V_{DS,LS}$ waveforms at the switching instant. ISTN can accelerate the switching speed of LS.

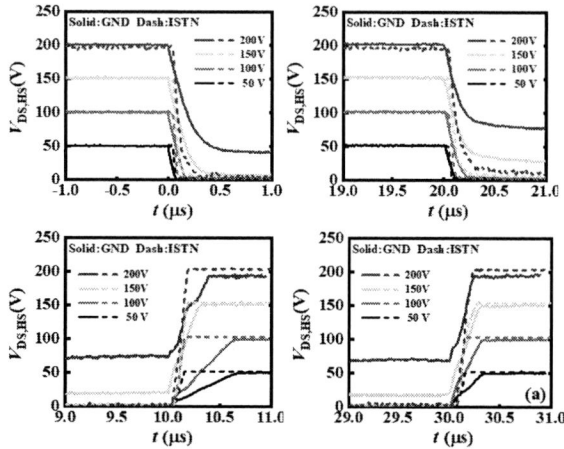

Fig. 8 $V_{DS,HS}$ waveforms at the switching instant. ISTN can accelerate the switching speed of HS.

with SN connection, ISTN connection significantly reduces the static negative $V_{BS,HS}$ during on-state. For instance, $V_{BS,HS}$ is -132V at 28μs under GND connection, and it is minimized to -3V under ISTN connection (Fig. 4(a)). On the other hand, the swing of $V_{BS,LS}$ ($\Delta V_{BS,LS}$) is also reduced under ISTN connection than SN connection when switching from off-state to on-state. For example, at the beginning of the second turn-on process (21μs), $\Delta V_{BS,LS}$ significantly decreases from 188V to 76V (Fig. 4(b)). Hence, the impact of dynamic back-gating effect is effectively alleviated. The comparison of V_{BS} (at 28μs) and ΔV_{BS} (at 21μs) is listed in Table.I.

The swing of V_{BS} not only cause dynamic back-gating effect but also delays the turn-on and turn-off process, due to

the charge and discharge of substrate-coupled capacitor. Hence, the smaller ΔV_{BS} under ISTN connection can also accelerate the switching speed. Fig. 7 and Fig. 8 focus on the V_{DS} waveforms in the vicinity of switching instant (i.e. 10, 20, 30 and 40 μs). For LS, its fall time (t_{fall}) under ISTN is 23% and 37% shorter than SN connection in the first and second turn-off process, respectively (Fig. 7 (a) and (c)). Its rise time in turn-on process (t_{rise}) is also reduced by 20% and 44% (Fig. 7 (b) and (d)). Similarly, t_{fall} and t_{rise} of HS are significantly shortened as well (Fig. 8), which leads to a higher frequency and efficiency of the circuit.

IV. CONCLUSION

In this study, a 200V monolithic GaN half-bridge power IC with ISTN is proposed and fabricated. The substrate is connected to SN and VDD through 2DEG resistors, allowing for on-chip dynamic control of the substrate potential. Unlike the monolithic half-bridge IC with GND or SN connection, the proposed GaN power IC scheme with ISTN shows minimal impact on the R_{on} of both the HS and LS HEMTs at VDD of 200V, and also exhibits a faster switching speed.

REFERENCES

[1] S. Yang, C. Zhou, S. Han, J. Wei, K. Sheng, and K. J. Chen, "Impact of Substrate Bias Polarity on Buffer-Related Current Collapse in AlGaN/GaN-on-Si Power Devices," IEEE Trans. Electron Devices, vol. 64, no. 12, Art. no. 12, Dec. 2017.

[2] J. Wei, M. Zhang, G. Lyu, and K. J. Chen, "GaN Integrated Bridge Circuits on Bulk Silicon Substrate: Issues and Proposed Solution," IEEE J. Electron Devices Soc., vol. 9, pp. 545–551, 2021.

[3] G. Lyu, J. Wei, T. Chen, J. Zhang, and K. J. Chen, "Substrate and Trench Design for GaN-on-EBUS Power IC Platform," IEEE Trans. Electron Devices, vol. 69, no. 7, Art. no. 7, Jul. 2022.

[4] X. Li, M. Van Hove, M. Zhao, K. Geens, W. Guo, S. You et al., "Suppression of the Backgating Effect of Enhancement-Mode p-GaN HEMTs on 200-mm GaN-on-SOI for Monolithic Integration," IEEE Electron Device Lett., vol. 39, no. 7, Art. no. 7, Jul. 2018.

[5] X. Li, M. Van Hove, M. Zhao, K. Geens, V. Lempinen, J. Jormunen et al., "200 V Enhancement-Mode p-GaN HEMTs Fabricated on 200 mm GaN-on-SOI With Trench Isolation for Monolithic Integration," IEEE Electron Device Letters, vol. 38, no. 7, pp. 918–921, Jul. 2017.

[6] E. Fabris et al., "Hot-Electron Trapping and Hole-Induced Detrapping in GaN-Based GITs and HD-GITs," IEEE Trans. Electron Devices, vol. 66, no. 1, pp. 337–342, Jan. 2019.

[7] S. Kaneko, M. Meneghini, C. De Santi, M. Borga, Y. Kinoshita, K. Tanaka et al., "Current-collapse-free operations up to 850 V by GaN-GIT utilizing hole injection from drain," in 2015 IEEE 27th International Symposium on Power Semiconductor Devices & IC's (ISPSD), in ISPSD. Hong Kong, China: IEEE, May 2015, pp. 41-44.

[8] J. Yang, J. Wei, M. Wang, M. Nuo, H. Yang, T. Li et al., "650-V GaN-on-Si Power Integration Platform Using Virtual-Body p-GaN Gate HEMT to Screen Substrate-Induced Crosstalk," 2023.

[9] J. Yang, J. Wei, Y. Wu, M. Nuo, Z. Chen, X. Yang et al., "Virtual-Body p-GaN Gate HEMT With Enhanced Ruggedness Against Hot-Electron-Induced Degradation," IEEE Electron Device Letters, vol. 45, no. 5, pp. 770–773, May 2024.

[10] S. Moench, R. Reiner, P. Waltereit, D. Meder, M. Basler, R. Quay et al., "Asymmetrical Substrate-Biasing Effects at up to 350V Operation of Symmetrical Monolithic Normally-Off GaN-on-Si Half-Bridges," in 2019 IEEE 7th Workshop on Wide Bandgap Power Devices and Applications (WiPDA), Raleigh, NC, USA: IEEE, Oct. 2019, pp. 28–35.

[11] B. Weiss, R. Reiner, V. Polyakov, P. Waltereit, R. Quay, O. Ambacher et al., "Substrate biasing effects in a high-voltage, monolithically-integrated half-bridge GaN-Chip," in 2017 IEEE 5th Workshop on Wide Bandgap Power Devices and Applications (WiPDA), in WiPDA. Albuquerque, NM: IEEE, Oct. 2017, pp. 265–272.

[12] Y. Zheng, B. Li, Q. Dong, Y. Ying, D. Song, J. Zhu et al., "A 200-V Half-Bridge Monolithic GaN Power IC With High-Speed Level Shifter and dV S /dt Noise Immunity Enhancement Structure," IEEE Trans. VLSI Syst., vol. 32, no. 3, pp. 542–551, Mar. 2024.

979-8-3503-6184-1/24 $31.00 © 2024 IEEE

High Short-Circuit Capability and Low-Loss SOI-LIGBT with Double-Integrated NMOS

Jialei Tan [1], Jie Wei*[1], Jinlong Lu [1], Xindi Liu [1], Gaoqiang Deng [1], Wei Song [1], Pei Guo [1],
Bo Zhang [1], and Xiaorong Luo*[1,2],

[1] State Key Laboratory of Electronic Thin Films and Integrated Devices,
University of Electronic Science and Technology of China, Chengdu, 610054, China
[2] College of Microelectronics,
Chengdu University of Information Technology, Chengdu, 610225, China
* Email: weijieuestc@uestc.edu.cn, xrluo@uestc.edu.cn

Abstract—A novel SOI lateral insulated gate bipolar transistor (LIGBT) featuring double-integrated NMOS (DNM) is proposed and investigated. The DNM (MOS1 and MOS2) self-adaptively controls the states of parasitic diode D_0 (P-well/N+ cathode). In the on state, the D_0 adaptively turns on to enhance the conductivity modulation effect to improve the on-state voltage drop (V_{on}). In the saturation and short circuit state with high anode voltage (V_{AK}), the D_0 is adaptively turned off and reverse-biased to reduce the saturation current (I_{sat}) and improve the latch-up immunity and achieve a better short-circuit withstanding time (t_{sc}). During turning-off period with increasing V_{AK}, the DNM LIGBT significantly reduces turnoff loss (E_{off}) due to the adaptively turn-on MOS1 providing a low resistance path to extract the stored carriers quickly. DNM LIGBT decreases the E_{off} by 76% at the same V_{on} compared with the SCM LIGBT and improves t_{sc} by 252% and 27% compared with those of Con. and SCM LIGBTs.

Keywords—SOI-LIGBT, on-state voltage drop (V_{on}), turn off loss (E_{off}), short-circuit, double-integrated NMOS

I. INTRODUCTION

The silicon-on-insulator (SOI) lateral insulated gate bipolar transistors (LIGBTs) are important components in smart power ICs due to their easy integration, high current capability, and low on-state voltage drop (V_{on}) [1-2]. The tradeoff between V_{on} and the turn-off loss (E_{off}) is an inevitable issue for LIGBTs [3-5]. This relationship can be improved by further reducing V_{on}. Common technologies include trench gates technology [6], carrier storage technology [7], and triggering the parasitic P-well/N+ diode at the cathode to enhance conductivity modulation [8-10]. However, these methods may lead to high E_{off} and large saturation current (I_{sat}), thus reducing the short-circuit withstanding time (t_{sc}). To further improve tradeoff between V_{on} and E_{off} and increase t_{sc}, a 300-V-rated novel SOI LIGBT with double-integrated NMOS (DNM LIGBT) is proposed and investigated by Sentaurus TCAD. The models adopted in simulation include Lackner avalanche generation, Auger recombination, SRH recombination Philips unified mobility, HighFieldSaturation mobility, and Enormal mobility.

II. STRUCTURE AND MECHANISM

Fig. 1 shows the schematic cross section view of the DNM LIGBT. It features double-integrated NMOS (MOS1 and MOS2), which are separated from the Con. LIGBT region by the oxide trench. The gate of the MOS1 (V_{G1}) is self-adaptively controlled by the floating P+ region above the N-drift in the Con. LIGBT region. The surface electric field

This work is supported by the National Natural Science Foundation of China under Grant 62004031, 62104030 and 62304076.

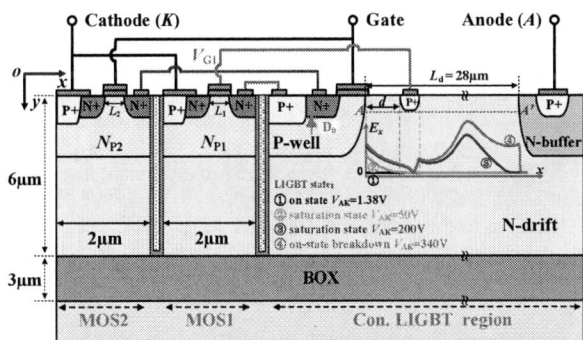

Fig. 1. Schematic cross section view of DNM LIGBT. The inset shows surface lateral electric field (E_x) distributions to form V_{G1} in the N-drift region for different states.

distributions along AA' to form the V_{G1} is labelled as the inset shown. The gates voltage of the MOS2 and Con. LIGBT region are controlled by main gate. The drains of the MOS1/MOS2 are shorted to the cathode P+/N+ part of Con. LIGBT region, respectively. A parasitic thyristor (T^*) composes of a PNP (P+ anode/N-drift/P-well) transistor and an NPN (N-drift/P-well/N+ cathode) transistor. N_{P1} and N_{P2} are the doping concentration of P-well in MOS1 and MOS2. L_1/L_2 and T_{G1}/T_{G2} are the channel length and gate oxide thickness of MOS1/MOS2, respectively. N_d and L_d are the doping concentration and length of N-drift, respectively. d is the space between floating P+ region and P-well. $N_{P1} = N_{P2} = 2 \times 10^{16}$ cm^{-3}, $L_{G1} = L_{G2} = 0.6\mu m$, $T_{G1} = T_{G2} = 100$nm, $L_d = 28\mu m$, $N_d = 2.6 \times 10^{15}$ cm^{-3}, and $d = 11\mu m$ are set in simulation expect for special statement.

Fig. 2(a)-2(d) show the equivalent circuits of the DNM LIGBT at different operation states. Fig. 2(e)-2(h) show electron and hole current flowlines and current density contours at the cathode side of DNM and Con. LIGBT at V_{AK}=1.3V and 3V, respectively. Fig. 3 shows the forward characteristic and hole density for DNM LIGBT. At the initial on-state stage with low anode voltage V_{AK} in Fig. 2(a) and the part-I of Fig. 3(a), $V_{G1} \approx V_{AK} < V_{th1}$ (the threshold voltage of MOS1). The channel of MOS1 and MOS2 is turned-off and turned-on, respectively. The electron current flowing through the N-channels of the MOS2 and Con. LIGBT region (I_{M0}) becomes the base current of the PNP transistor. The injected holes from the P+ anode would accumulate in the P-well in Con. LIGBT region to rise the potential of P-well, while the parasitic diode D_0 (P-well/N+ cathode, labelled in Fig. 1) is turned off with $V_{PN} < V_{bi}$ ($V_{PN} = V_P - V_N$, and V_{bi} is the built-in potential for D_0).

As V_{AK} increases to $V_{PN} > V_{bi}$ and $V_{G1} < V_{th1}$ in Fig. 2(b) and the part-II of Fig. 3(a), D_0 turns on and T^* is triggered. Thus,

979-8-3503-6184-1/24 $31.00 © 2024 IEEE

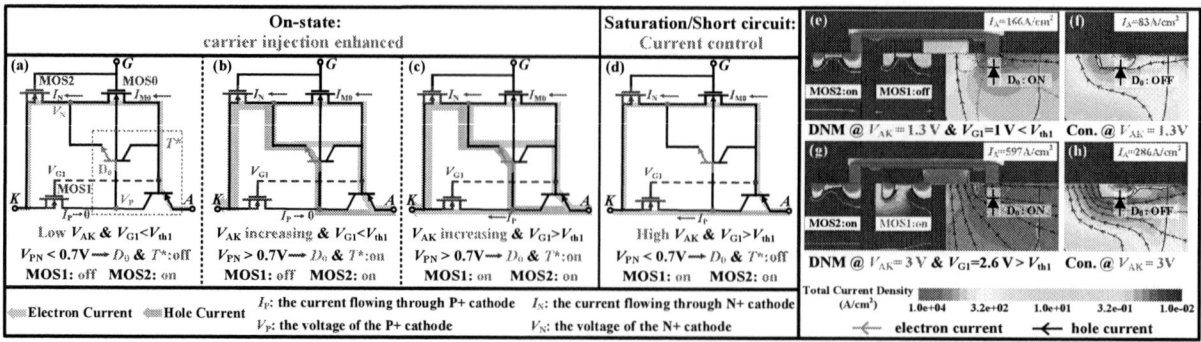

Fig. 2 Operation mechanisms for the DNM LIGBT. Equivalent circuits for DNM LIGBT at (a)-(c) on-state with V_{AK} increasing, (d) saturation / short circuit state. Electron/hole current flowlines and current density contours distribution at the cathode side for (e)/(g) DNM LIGBT and (f)/(h) Con. LIGBT at V_{AK} = 1.3V / 3V.

the DNM LIGBT enhances the conductivity modulation effect dramatically, and the hole density of DNM LIGBT at the cathode side increases by 7.7× compared to that of SCM LIGBT as shown in Fig. 3(b). Note that MOS1 is turned-off at $V_{G1} < V_{th1}$ and I_P is almost 0. As the simulated results shown in Fig. 2(e)-2(f) with V_{AK} = 1.3V, the I_A = 166 A/cm² of DNM LIGBT is higher than 83A/cm² of Con. LIGBT, wherein the electron current lines of DNM LIGBT not only flow through the inversion layer like the Con. LIGBT in Fig. 2(f), but also

flow through the turned-on D_0 in Fig. 2(e).

With the increasing of V_{AK} till $V_{G1} > V_{th1}$ in Fig. 2(c) and the part-III region of Fig. 3(a), MOS1 turns on adaptively and provides another hole current path, leading to a further enhancement in current capacity. As shown in Fig. 2(g)-2(h), at V_{AK} = 3V, I_A = 597 A/cm² for the DNM LIGBT is much higher than 286 A/cm² of Con. LIGBT.

In the saturation and short circuit state with high V_{AK} and $V_{G1} > V_{th1}$ in Fig 2(d) and part-IV of Fig. 3(a), the D_0 turns off adaptively and the saturation current (I_{sat}) of DNM LIGBT would decrease dramatically, as the relational expressions shown in (1)-(2):

$$\text{high } V_{AK} \xrightarrow{\text{turned-on}}_{\text{MOS1}} R_1 \downarrow \longrightarrow V_P - V_N = (I_P \times R_1) \downarrow - (I_N \times R_2) < V_{bi} \Big\} \begin{matrix} D_0 \text{ turns off \&} \\ \text{negative biased} \end{matrix} \quad (1)$$

$$\begin{matrix} V_P - V_N < 0V \xrightarrow{\text{bulk}}_{\text{effect}} V_{th}(V_P) \uparrow > V_{th0} \\ V_N = I_N \times R_2 > 0V \longrightarrow V_{GS}(V_N) \downarrow < V_{GS0} \end{matrix} \Big\} \longrightarrow I_{sat} \propto [V_{GS}(V_N) - V_{th}(V_P)]^2 \downarrow \quad (2)$$

Here R_1 and R_2 are the channel resistance of MOS1 and MOS2, respectively. $V_{th}(V_P)$ / V_{th0} and $V_{GS}(V_N)$ / V_{GS0} are the threshold voltage and gate-to-source-voltage of MOS0 in DNM / Con. LIGBT. According to expression (1)-(2), D_0 turns off with $V_{PN} < V_{bi}$ and gets gradually negative biased with increasing V_{AK}, and the I_{sat} of DNM LIGBT would be lower than that of Con. LIGBT because of the bulk effect and lower $V_{GS}(V_N)$. Both weaken the conductivity modulation effect and improve the latch-up immunity. It's worth noting that the depletion region expands with the increasing V_{AK}, and the V_{G1} remains much lower than V_{AK} to prevent gate breakdown of MOS1 at high V_{AK}.

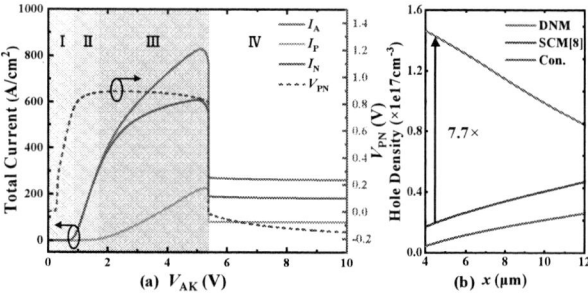

Fig. 3 (a) Forward I–V characteristic for DNM LIGBT. (b) Hole density comparison for different LIGBTs at V_{AK}=1.3V and y=3μm.

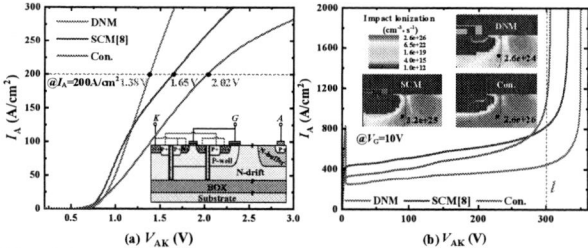

Fig. 4 Forward I–V characteristics: (a) linear region and (b) on-state breakdown characteristics of different LIGBTs. The inset in (a) shows the schematic view of SCM LIGBT and the inset in (b) shows impact ionization distribution of different LIGBTs at V_{AK}=300V.

III. RESULTS AND DISCUSSION

Fig. 4 shows the forward I–V characteristics of different LIGBTs with the same P+ anode doping concentration (N_A = 1×10^{18} cm⁻³). In Fig. 4(a), the DNM LIGBT achieves a lower V_{on} of 1.38V at I_A = 200A/cm² because of its superior electron injection efficiency as demonstrated in Fig. 2(e)-2(h). For the saturation state in Fig. 4(b), the DNM LIGBT obtains the lowest I_{sat} due to the turned off D_0 and bulk effect shown in (1)-(2). Accordingly, the impact ionization at the cathode side of DNM LIGBT is two orders of magnitude lower than that of Con. LIGBT at V_{AK} = 300V, which allows DNM LIGBT achieve the highest on-state breakdown voltage (BV_{on}) compared with SCM and Con. LIGBT. Furthermore, the lower I_{sat} of DNM LIGBT helps improve the t_{sc} to 21.5μs, increased by 27% and 252% compared with those of SCM and Con. LIGBT in Fig. 5.

Fig. 6(a) shows the turning-off waves of different LIGBTs.

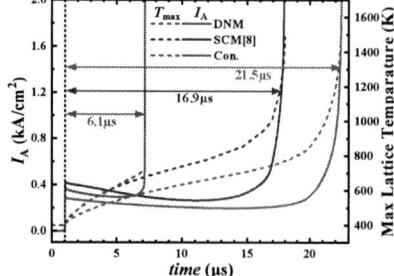

Fig. 5 Short-circuit characteristics of different LIGBTs.

979-8-3503-6184-1/24 $31.00 © 2024 IEEE

The DNM LIGBT achieves the shortest current off time t_{off} of 15 ns, much shorter than 90 ns of SCM LIGBT at the same $V_{on}=1.4V$ and 119ns of Con. LIGBT at $V_{on}=1.7V$. The DNM LIGBT decreases the E_{off} from 3.22 mJ/cm² to 1.36 mJ/cm² of the SCM LIGBT. Fig. 6(b) compares the holes distribution in the x-dimension from moment t_1 to t_6 labelled in Fig. 6(a). Before moment t_2, the DNM LIGBT extracts excess holes a little slowly than that of SCM LIGBT, because $V_{G1} < V_{th1}$ at low V_{AK} and DNM LIGBT can only removes holes through recombination. After moment t_2, the DNM LIGBT extracts excess holes faster. Because the adaptively turned-on MOS1 not only provides a low resistance hole extracting path to quickly remove the excess hole from N-drift, but also decrease the V_P to turn off the D_0 and decrease electron injection from cathode side. Note that the V_{on} of Con. LIGBT is larger than other two devices due to lack of D_0 to enhance the modulation effect, and it exhibits the longest t_{off} and highest E_{off}.

Fig. 7 shows the influences of N_{P1} and N_{P2} on V_{on} and E_{off}. A smaller N_{P2} (i.e., lower V_{th}) contributes to a lower V_{on}, because the lower resistance of MOS2 helps D_0 turn on earlier to enhance the conductivity modulation effect and decrease the V_{on}, and then lead to a higher E_{off}. E_{off} shows a slight increase with the increasing N_{P1} because the resistance of hole extracting path is slightly increased for high N_{P1}. Meanwhile, V_{on} is almost irrelevant to N_{P1} since the MOS1 is turned off at $I_A = 200A/cm^2$.

Fig. 8 shows the tradeoff relationship between V_{on} and E_{off} for different LIGBTs. The DNM LIGBT achieves a better V_{on}-E_{off} tradeoff than the other two LIGBTs, because its superior electron injection efficiency and low resistance hole extracting path provided by self-adaptively turn-on MOS1. Compared with the SCM LIGBT, the DNM LIGBT reduces E_{off} by 76% at the same V_{on} and decreases V_{on} by 22% at the same E_{off}. Compared with the Con. LIGBT, the DNM LIGBT reduces V_{on} by 46% at the same E_{off}.

IV. CONCLUSION

A novel DNM LIGBT with better V_{on}–E_{off} tradeoff and short circuit performance is proposed. Due to the enhanced conductivity modulation induced by the adaptively turn-on D_0, the DNM LIGBT improves V_{on} dramatically in the on-state. In the saturation and short circuit state with high V_{AK}, the D_0 adaptively turns off. Thus, DNM LIGBT not only increases the BV_{on} by 13.4% compared with the Con. LIGBT, but also increases the t_{sc} by 252% and 27% compared with the Con. and SCM LIGBTs, respectively. During turning off, the DNM LIGBT achieves an ultrafast t_{off} of 15 ns by self-adaptively turning on the MOS1 to extract the excess holes and turn off the D_0 to reduce electron injection. Therefore, the DNM LIGBT decreases the E_{off}/V_{on} by 76%/22% at the same V_{on}/E_{off} compared with the SCM LIGBT.

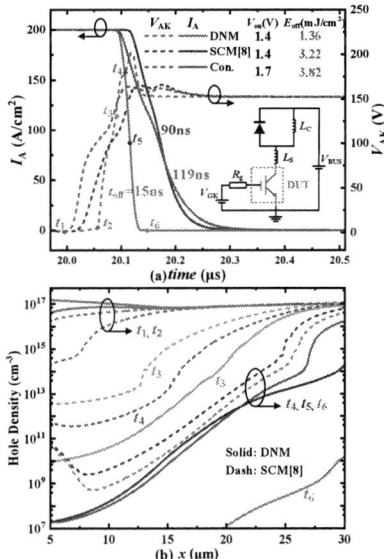

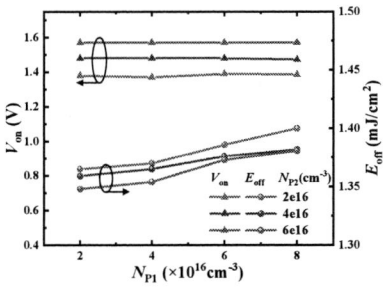

Fig. 6 (a) Turning-off waves of different devices. The inset is the inductive load circuit used in simulation ($L_C = 1\mu H$, $L_S = 5nH$, $R_g = 10\Omega$, and $V_{BUS} = 150V$). (b) Hole density distributions in the x-dimension from t_1 to t_6 period labelled in (a).

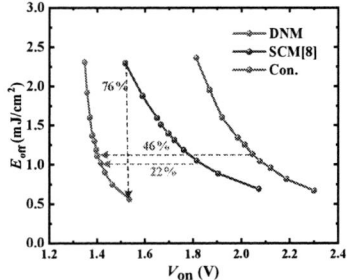

Fig. 7 Influences of N_{P1} and N_{P2} on V_{on} and E_{off}.

Fig. 8 Tradeoff relationship between V_{on} and E_{off} at $I_A=200A/cm^2$.

REFERENCES

[1] N. Iwamuro and T. Laska, "IGBT History, State-of-the-Art, and Future Prospects," IEEE Trans. Electron Devices, vol. 64, no. 3, pp. 741-752, March 2017.

[2] J. Wei, P. Zhu, and K. Yang, et al. "Ultralow Loss Lateral Insulated Gate Bipolar Transistor with U-shape Trench Anode," IEEE ICSICT, Nangjing, pp. 1-3, 2022.

[3] J. Zhu, L. Zhang, and W. Sun, et al. "Electrical Characteristic Study of an SOI-LIGBT With Segmented Trenches in the Anode Region," IEEE Trans. Electron Devices, vol. 63, no. 5, pp. 2003-2008, May 2016.

[4] K. Yang, W. Su, and J. Wang, et al. "High-Current and Short-Circuit Capability SOI-LIGBT With Double-Integrated Self-Adapted MOS-Resistors," IEEE Trans. Electron Devices, vol. 70, no. 2, pp. 667-674, Feb. 2023.

[5] J. Wei, P. Zhu, and K. Yang, et al. "Novel Ultrafast Low-Loss LIGBT With Reverse-Conduction Capability," IEEE Trans. Electron Devices, vol. 70, no. 5, pp. 2622-2626, May 2023.

[6] M. Harada, T. Minato, and H. Takahashi, et al. "600 V trench IGBT in comparison with planar IGBT—An evaluation of the limit of IGBT performance," IEEE ISPSD, May 1994, pp. 411–416.

[7] Shigeki, Akio, and Youichi, et al. "Carrier-storage effect and extraction-enhanced lateral IGBT (E2LIGBT): A super-high speed and low on-state voltage LIGBT superior to LDMOSFET," IEEE ISPSD, Bruges, pp. 393-396, 2012.

[8] W. Sun, Z. Yang, and J. Zhu, et al. "Investigation on Self-Adjust Conductivity Modulation SOI-LIGBT Structure (SCM-LIGBT) for Monolithic High-Voltage IC, " IEEE Trans. Electron Devices, vol. 64, no. 9, pp. 3762-3767, Sept. 2017.

[9] B. Zhang, M. Kong, and B. Yi, et al. "A novel low on-state voltage SOI LIGBT with enhanced conductivity modulation," IEEE Trans. Electron Devices, vol. 66, no. 11, pp. 4865–4869, Nov. 2019.

[10] W. Sun, J. Zhu, and Z. Yang, et al. "A composite structure named self-adjusted conductivity modulation SOI-LIGBT with low on-state voltage," IEEE ISPSD, Sapporo, pp. 85-88, 2017.

Body Diode Degradation Mechanism of 1200V SiC Power MOSFETs Under Gamma Rays Total Ionizing Dose Irradiation

Yu Tian, Zhaoxu Song, Hao Fu, Jiaxing Wei*, Siyang Liu, Weifeng Sun

National ASIC System Engineering Research Center, School of Integrated Circuits, Southeast University, Nanjing, 210096, China

* Email: jiaxingwei@seu.edu.cn

Abstract—The body diode (BD) degradation mechanism of 1200V SiC power MOSFETs under gamma rays total ionizing dose (TID) irradiation is investigated in details. According to the measured electrical characteristic results of the BD of SiC power MOSFETs before and after the TID irradiation, it demonstrates that the degradation mechanism of the BD is mainly derived from positive holes captured in the gate oxide layer during the irradiation, which can be reflected by the negative shift in C_g-V_g curves. Therefore, the channel of the devices under test (DUTs) after the irradiation are weakly open under the OFF state, and the total current density of the BD shows an increasing trend. It can be revealed by the Silvaco TCAD simulations, which is in agreement with the measured data. In addition, the reverse recovery test of the DUTs is also performed, which shows that the TID irradiation has no obvious impact on the dynamic characteristic of the BD.

Keywords—SiC power MOSFETs, total ionizing dose irradiation, body diode, degradation mechanism, positive holes

I. INTRODUCTION

Silicon carbide (SiC) power metal oxide semiconductor field effect transistors (MOSFETs) receive much attention from worldwide research institutions and vehicle industries because of higher mobility, faster switching speed, higher thermal conductivity, higher critical electric field, and lower energy losses compared to conventional silicon MOSFETs [1]. With SiC power MOSFETs showing high performance in many situations, it is found that they have good potential in airspace field [2], especially in suffering from total ionizing dose (TID) irradiation [3]. Therefore, it is necessary to investigate the irradiation reliability of SiC power MOSFETs under TID irradiation.

At present, there are many literatures that reported to explore the degradation mechanism of SiC power MOSFETs under the TID irradiation [4-7]. For example, *Kexin Gao* et al. studied the degradation behavior and mechanism of static electrical parameters of SiC MOSFETs at different gate voltage. *S. Liang* et al. exhibited the dynamic characteristics of SiC MOSFETs in switching performances or unclamped inductive switching (UIS) test. However, few of them investigate the influence of the TID irradiation on the body diode (BD) of SiC power MOSFETs. Although *Fu-Jen Hsu* et al. compared the forward current degradation of BD in JMOS and DMOS under the TID irradiation, they lacked the complete experiments tests and simulations verification before and after the irradiation. The BD degradation mechanism of SiC power MOSFETs under the TID irradiation is still urgent to discuss in details. In this work, a dose rate of 50 rad(Si)/s Co^{60} gamma rays is used to irradiate a kind of 1200V SiC power MOSFETs. The BD degradation mechanism of the devices under test (DUTs) is discussed comprehensively according to experiments and verified by simulations. The experimental results show good agreements with the simulations.

II. DEVICE STRUCTURE AND IRRADIATION CONDITION

The schematic cross section of the DUTs (C3M0075120K from CREE Corporation) is shown in Fig. 1. The length of channel is 0.4 μm, the width of cell is 5.8 μm, while the width of P+, N+, and JEFT region are 1 μm. Other dimensions are listed in Table I. In addition, there are three groups of samples (A0, A1, A2) suffering from different irradiation dosage to study the influence of the TID effect on DUTs before and after the irradiation. The irradiation conditions are listed in Table II.

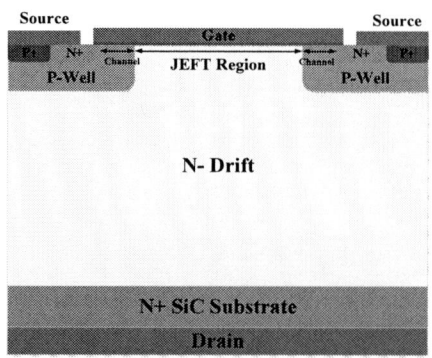

Fig. 1. Schematic cross section of a symmetric cell of the SiC power MOSFET investigated in this work.

TABLE I. DEVICE PARAMETERS FOR SIMULATION

Symbols	Device Parameters	Values
L_{ch}	Length of Channel	0.4 μm
W_{cell}	Width of Cell	5.8 μm
W_{N+}	Width of N+	1 μm
W_{P+}	Width of P+	1 μm
W_{JEFT}	Width of JEFT	1 μm
W_{OHM}	Width of OHM	1.3 μm
T_{oxide}	Thickness of OXIDE	35 nm
W_{gate}	Width of gate poly	2.8 μm

979-8-3503-6184-1/24 $31.00 © 2024 IEEE

TABLE II.	IRRADIATION CONDITIONS OF THE SAMPLES	
TID (krad (Si))	**Bias Conditions**
	$V_{gs} = 0V \& V_{ds} = 0V$
0	Sample A0
100	Sample A1
500	Sample A2

III. RESULTS AND DISCUSSIONS

The forward characteristics of the BD of the DUTs under different gate and source bias (V_{gs}) before and after the TID irradiation are shown in Fig. 2. It can be seen that the forward current of BD (I_{sd}) has an increased trend with TID increases. In addition, it is noted that the shift of I_{sd} is diminished when the V_{gs} becomes more negative. Especially when the channel is completely pinched OFF(V_{gs}=-4V), the TID irradiation rarely impacts on I_{sd}. It infers that the channel is influenced by the TID and produces current to make the I_{sd} higher.

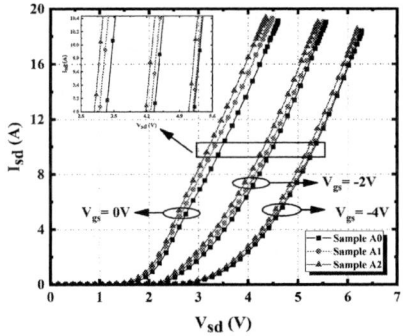

Fig. 2. I_{sd}-V_{sd} curves of the DUTs at different V_{gs} after the TID irradiation.

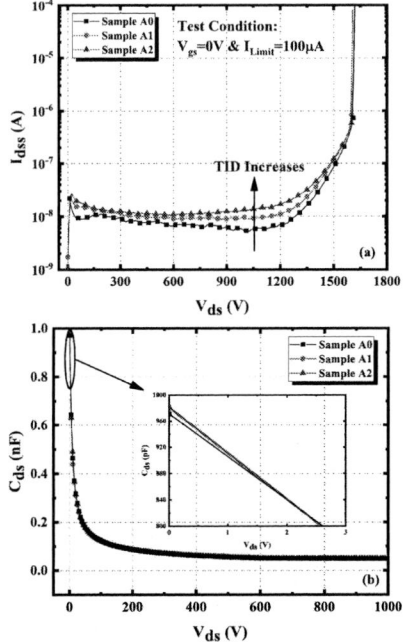

Fig. 3. Measured curves of (a) I_{dss}-V_{ds} and (b) C_{ds}-V_{ds} of the DUTs after the TID irradiation.

The blocking and the capacitance characteristics curves are shown in Fig. 3(a) and Fig. 3(b), respectively. A gradual increase trend in the drain leakage current (I_{dss}) and the stable breakdown voltage with the increased TID irradiation are found in Fig. 3(a). Meanwhile, Fig. 3(b) shows that the capacitance of the BD is unchanged, which means that the TID irradiation rarely damages the PN junction of the BD and the blocking characteristic. In fact, the TID irradiation makes the channel opened partially, causing the I_{dss} to become larger as same trend as the I_{sd}. The BD electrical characteristics of the DUTs are not broken down after the TID irradiation.

To figure out the degradation mechanism of the BD, the gate leakage current and the C_g-V_g curves are investigated in Fig. 4(a) and Fig. 4(b). As shown in Fig. 4(a), the gate leakage current (I_{gss}) becomes larger during the irradiation, which can prove that the gate oxide is damaged under the gamma rays. While Fig. 4(b) exhibits negative shift of C_g-V_g before and after the TID irradiation, it infers that the positive charges are accumulated along channel region and JFET region in the gate oxide [8]. They are mainly derived from the holes captured in the oxide trapped charges, which weakly opens the channel and causes the BD degradation.

Finally, the reverse recovery testing of the BD is performed with a double pulse test circuit, as shown in Fig. 5(a). The reverse recovery curves with the schematic inset of waveform used in this work are compared in Fig. 5(b) and several relevant parameters (T_a, T_b, T_{rr}, I_{rr}, and Q_{rr}) extracted from the curves before and after 500krad (Si) are listed in Table III. Although there is a slight negative shift of reserve recovery curves after 500krad (Si) irradiation, the parameters between samples A0 and A2 have no obvious change, which can illustrate that the dynamic characteristics of BD would not be influenced by the TID irradiation.

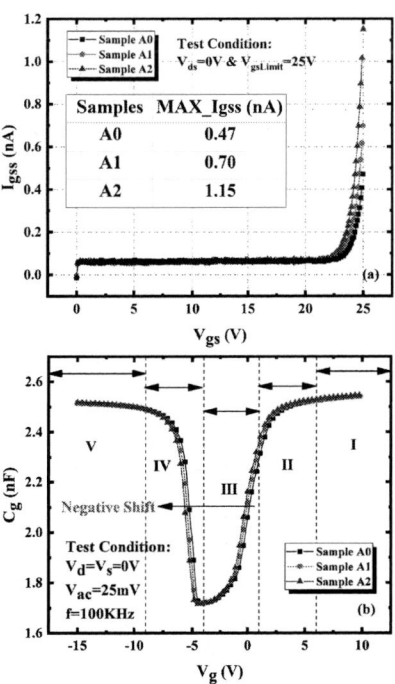

Fig. 4. Measured curves of (a) I_{gss}-V_{gs} and (b) C_g-V_{gs} of the DUTs after the TID irradiation.

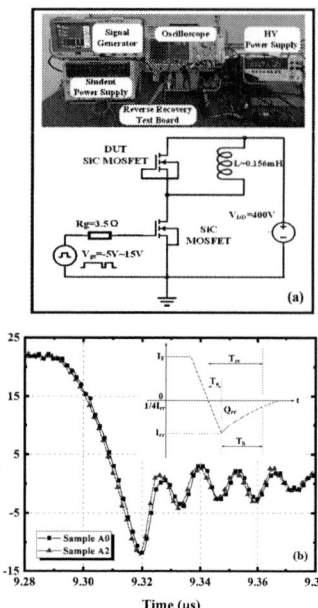

Fig. 5. Reverse recovery (a) testing system and (b) characteristics curves of the DUTs after 500krad (Si) TID irradiation.

TABLE III. PARAMETERS OF REVERSE RECOVERY

Samples	Parameters				
	T_a (ns)	T_b (ns)	T_{rr} (ns)	I_{rr} (A)	Q_{rr} (nC)
A0	6.8	4.4	11.2	12.2	87.05
A2	7.2	4.4	11.6	12.2	87.64

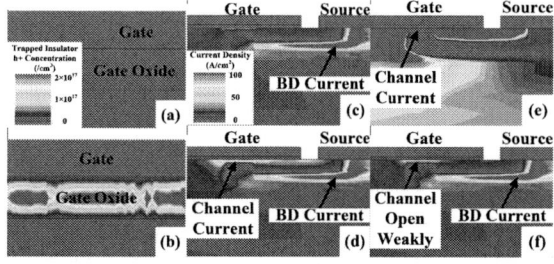

Fig. 6. Simulated holes concentration after (a) 0krad (Si), (b) 500krad (Si) irradiation, current density after (c) 0krad (Si), (d) 100krad (Si) irradiation at V_{gs}=0V, current density after 500krad (Si) irradiation at (e) V_{gs}=0V, and (f) V_{gs}=-4V.

In order to explore the degradation mechanism of the BD of DUTs, the Silvaco TCAD is used to simulate the changes in gate oxide and forward current density of the BD before and after the irradiation, as shown in Fig. 6. It can be seen that many holes are captured in gate oxide and current path gradually changes from the BD to the channel after the TID irradiation, which make the current density become larger as TID increases. Even when the channel is almost pinched OFF (V_{gs} = -4V), high dose irradiation can still open the channel slightly. These phenomena in simulations reflect the TID irradiation has an obvious impact on the gate oxide, causing the channel weakly opened and current density increased, which meet the results of experiments and further verify the BD degradation mechanism of SiC power MOSFETs.

IV. CONCLUSION

In this work, the 1200V SiC power MOSFETs are irradiated by a dose rate of 50 rad(Si)/s Co60 gamma rays to investigate the BD degradation mechanism under TID irradiation. Compared to the electrical characteristic of the BD before the irradiation, the forward current and the leakage current of the BD become higher with the increased TID. Furthermore, the gate leakage current and the CV curves of the BD exhibit there existing many holes captured in the gate oxide layer to weakly open the channel, which is the main reason of BD degradation of SiC power MOSFETs.

ACKNOWLEDGMENT

This work was supported in part by the National Natural Science Foundation of China under Grant 62174029, in part by the Fund for Transformation of Scientific and Technological Achievements of Jiangsu Province under Grant BA2023001, in part by the Research and Development Plan of Jiangsu Province under Grant BE2022073 and Grant BE2022048-3, in part by the Distinguished Young Scientists Foundation of Jiangsu Province under Grant BK20230025, in part by the Fundamental Research Funds for the Central Universities under Grant 2242024RCB0028, in part by the Important Special Project of Nanjing City under Grant 2021-11004, and in part by the Fund for Transformation of Scientific and Technological Achievements of Wuxi City under Grant C20231021.

REFERENCES

[1] K. Mikami, M. Kaneko, and T. Kimoto, "High-Mobility 4H-SiC p-Channel MOSFETs on Nonpolar Faces," in IEEE Electron Device Letters, vol. 45, no. 7, pp. 1113-1116, July 2024, doi: 10.1109/LED.2024.3401001.

[2] C. Peng, Z. Lei, Z. Zhang, Y. He, T. Ma, and Y. Chen, "Bias and Temperature Dependence of Radiation-Induced Degradation for SiC MOSFETs," IEEE Trans. Nucl. Sci., vol. 71, no. 5, pp. 1186–1193, May 2024, doi: 10.1109/TNS.2024.3384767.

[3] Y. Tang et al., "Synergistic Effect of Negative Bias Instability and Total Ionizing Dose on SiC MOSFETs," IEEE Trans. Nucl. Sci., vol. 70, no. 8, pp. 1990–1994, Aug. 2023, doi: 10.1109/TNS.2023.3273172.

[4] K. Gao, Y. Chen, S. Zheng, M. Liao, X. Xu, and M. Lu, "Degradation behavior and mechanism of SiC power MOSFETs by total ionizing dose irradiation under different gate voltages," in 2021 IEEE Workshop on Wide Bandgap Power Devices and Applications in Asia (WiPDA Asia), Wuhan, China: IEEE, Aug. 2021, pp. 46–50. doi: 10.1109/WiPDAAsia51810.2021.9656082.

[5] S. Liang et al., "Investigation on Dynamic Degradation of SiC MOSFETs after Total Ionizing Dose Radiation," in 2023 IEEE Energy Conversion Congress and Exposition (ECCE), Nashville, TN, USA: IEEE, Oct. 2023, pp. 5757–5762. doi: 10.1109/ECCE53617.2023.10362271.

[6] S. Liang et al., "Observations on Ruggedness Degradation of Planar-gate SiC MOSFETs after Total Ionizing Dose Radiation," in 2023 IEEE Energy Conversion Congress and Exposition (ECCE), Nashville, TN, USA: IEEE, Oct. 2023, pp. 5379–5384. doi: 10.1109/ECCE53617.2023.10362840.

[7] F.-J. Hsu et al., "Radiation Influence Comparison between SiC JMOS and DMOS," in 2020 32nd International Symposium on Power Semiconductor Devices and ICs (ISPSD), Vienna, Austria: IEEE, Sep. 2020, pp. 146–149. doi: 10.1109/ISPSD46842.2020.9170058.

[8] J. Wei et al., "Interfacial damage extraction method for SiC power MOSFETs based on C-V characteristics," 2017 29th International Symposium on Power Semiconductor Devices and IC's (ISPSD), Sapporo, Japan, 2017, pp. 359-362, doi: 10.23919/ISPSD.2017.7988992.

Novel Heterojunction Field Plate β-Ga$_2$O$_3$ MOSFET with High Breakdown Voltage

Xiangnan Li [1], Jie Wei [*1], Kai Zhao [1], Linyao Hao [1], Xiaosong Peng [1], Yuxi Wei [*1], Renkuan Liu [1], Wei Song [1], Pei Guo [*1], Xiaorong Luo[1,2]

[1] State Key Laboratory of Electronic Thin Films and Integrated Devices,
University of Electronic Science and Technology of China, Chengdu 610054, China
[2] College of Microelectronics, Chengdu University of Information Technology, Chengdu 610225, China
* Email: weijieuestc@uestc.edu.cn, yxwei@uestc.edu.cn, guopei6174@163.com

Abstract—A novel β-Ga$_2$O$_3$ metal–oxide–semiconductor field-effect transistor (MOSFET) featured heterojunction field plate (HFP) is proposed to enhance its breakdown voltage (BV) and reduce specific ON-resistance ($R_{ON,SP}$), and its mechanism is investigated by simulation. The HFP is composed of P-NiO above the drift and regrown Ga$_2$O$_3$ at drain side. The P-NiO and N-drift Ga$_2$O$_3$ work like the P/N pillars of superjunction. Thus, the HFP structure not only modulates the surface electric field (E-field) distribution to improve the BV in the off-state, but also introduces assisted depletion effect to improve the drift doping concentration to reduce the $R_{ON,SP}$. Based on the charge balance principle between the P-NiO and N-drift Ga$_2$O$_3$, the BV, power-figure-of-merit (PFOM = $BV^2/R_{ON,SP}$) and $R_{ON,SP}$ of the proposed HFP β-Ga$_2$O$_3$ MOSFET are improved by 40.5%, 230.8% and 40.3% compared with those of conventional β-Ga$_2$O$_3$ MOSFET at the same dimensions. Thus, this proposed device provides a new design strategy for high-power β-Ga$_2$O$_3$ MOSFETs.

Keywords— β-Ga$_2$O$_3$, P-NiO, breakdown voltage (BV), MOSFET, heterojunction field plate

I. INTRODUCTION

THE developments of power supplies, motor drives, and electric vehicles have imposed higher requirements for energy-saving electronic systems and efficient semiconductor power devices accordingly [1]. Among all the emerging semiconductor materials, β-Ga$_2$O$_3$ owns its unique advantages over SiC / GaN materials due to an ultrawide bandgap of 4.5~4.9 eV and a consequent high theoretical breakdown electric field strength (E_C) of 8 MV/cm [2]. With the progressive development of Ga$_2$O$_3$ power transistors, various electric field management strategies have been employed to take full advantage of the intrinsic E_C of Ga$_2$O$_3$ to achieve high breakdown voltage (BV), including vertical trenched-fin channel structure [3], field plate constructions [4]–[6], superjunction (SJ) technology [7] and heterojunction gate [8]. Although high BV of Ga$_2$O$_3$ MOSFETs has been obtained, the specific ON-resistance ($R_{ON,SP}$) would increase due to the tradeoff between $R_{ON,SP}$ and BV in power devices.

Aiming at the bottleneck problem of the tradeoff between the $R_{ON,SP}$ and BV, a novel β-Ga$_2$O$_3$ MOSFET with heterojunction field plate (HFP) is proposed and studied in this work. The HFP could not only modulate the electric field (E-field) distribution to improve the BV, but also assist to deplete the N-drift region to improve the doping concentration and thus reduces the $R_{ON,SP}$. Thus, the proposed HFP β-Ga$_2$O$_3$ MOSFET realizes a high power-figure-of-merit (PFOM = $BV^2/R_{ON,SP}$).

II. STRUCTURE AND MECHANISM

This work is supported by the National Natural Science Foundation of China under Grant 62374028, and by the Postdoctoral Fellowship Program of CPSF under Grant GZC20240199 and GZC20240200

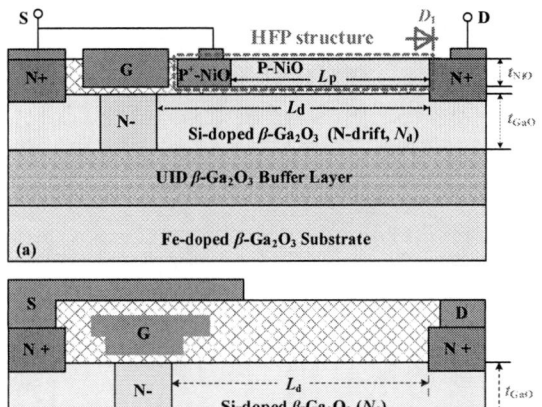

Fig. 1 Schematic cross section view of (a) proposed HFP β-Ga$_2$O$_3$ MOSFET and (b) conventional β-Ga$_2$O$_3$ MOSFET.

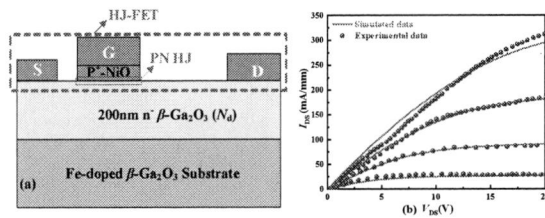

Fig. 2 (a) Schematic cross-section of β-Ga$_2$O$_3$ HJ-FET [10]. (b) Comparison of output characteristics of the simulated and fabricated HJ-FET.

Fig. 1(a) shows the schematic cross section view of the HFP β-Ga$_2$O$_3$ MOSFET. It features a heterojunction field plate (HFP) consisted of P-NiO above the N-drift surface and N+ Ga$_2$O$_3$ at the drain side, forming a heterojunction diode (D_1). The P-NiO and N-drift Ga$_2$O$_3$ region are separated by Al$_2$O$_3$ layer. The N- region under the gate is a local thermal oxidation region, and the doping concentration is lower than other parts of the drift region. The doping concentration of β-Ga$_2$O$_3$ N-drift region is N_d, except for N- region under the gate of 5e16 cm^{-3} formed by local thermal oxidation. The hole concentrations for P$^+$-NiO and P-NiO region is 1e19 cm^{-3} and N_p. The length (L_d) and thickness (t_{GaO}) of N-drift region are 10 μm and 200 nm. The length (L_p) and thickness (t_{NiO}) of P-NiO are 9 μm and 50 nm. The thickness of Al$_2$O$_3$ layer is 50nm. Fig. 1(b) shows the schematic cross section view of Con. β-Ga$_2$O$_3$ MOSFET.

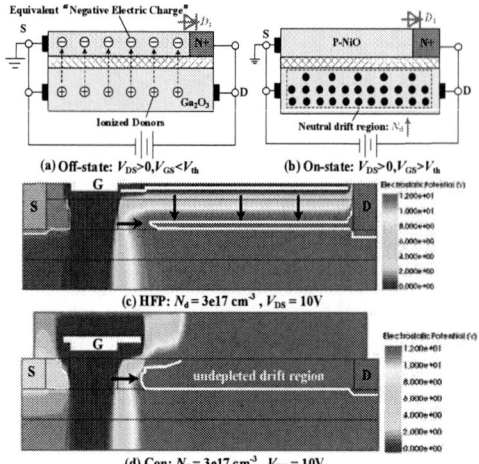

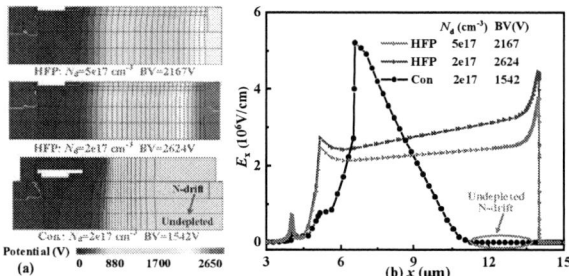

Fig. 3 Operation mechanism for the HFP β-Ga$_2$O$_3$ MOSFET. (a) Off-state: $V_{DS} > 0$, $V_{GS} < V_{th}$, and reverse biased D_1 sustains V_{DS}; (b) On-state: $V_{GD} > 0$ and $V_{GS} > V_{th}$. (c)-(d) Equipotential contour and depletion boundary distribution for HFP and Con. β-Ga$_2$O$_3$ MOSFET, wherein the arrows indicate depletion directions and the white lines are depletion boundaries.

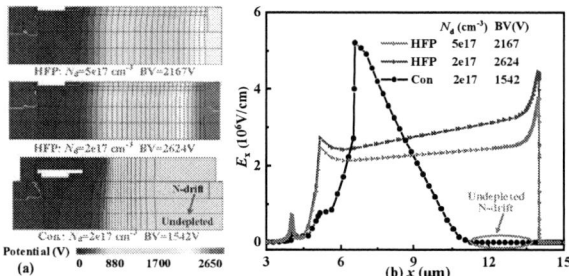

Fig. 4 (a) Potential lines distribution and (b) N-drift surface x-component of E-field ($y = 5$nm) for HFP and Con. β-Ga$_2$O$_3$ MOSFET at breakdown.

The electrical characteristics are investigated by the Sentaurus TCAD simulation in this work [9]. To ensure the simulation accuracy, the physical models employed in simulation are calibrated based on the reported measured output curves from β-Ga$_2$O$_3$ MOSFET with P-NiO heterojunction gate [10], as shown in Fig. 2(a). This includes refining the energy bandgap, mobility and recombination model, and so on. The electron mobility of Ga$_2$O$_3$ is set to 118 cm^2 / V·s [10]. As shown in Fig. 2(b), a good agreement between the experimental and simulation data demonstrates the accuracy of the physical models and the reliability of the simulation outcomes.

III. RESULTS AND DISCUSSION

Fig. 3 shows the operation mechanism for the HFP β-Ga$_2$O$_3$ MOSFET at different operation states. In the off-state with $V_{GS} < V_{th}$ and $V_{DS} > 0$ in Fig. 3(a), the P-NiO in HFP assists depleting the N-drift Ga$_2$O$_3$ as the vertical arrows shown, just equivalent of P/N pillars of superjunction structure, to modulate the electric field distribution to improve the BV. Meanwhile, the reverse biased diode D_1 helps suppress the leakage current in HFP. Therefore, the HFP allows the higher N_d in the on-state to reduce the $R_{ON,SP}$ in Fig. 3(b). Fig. 3(c)-3(d) compare the deletion boundary distributions of the HFP and Con. β-Ga$_2$O$_3$ MOSFET in the off-state with the same $V_{DS} = 10$ V and $N_d = 3$e17 cm^{-3}. The

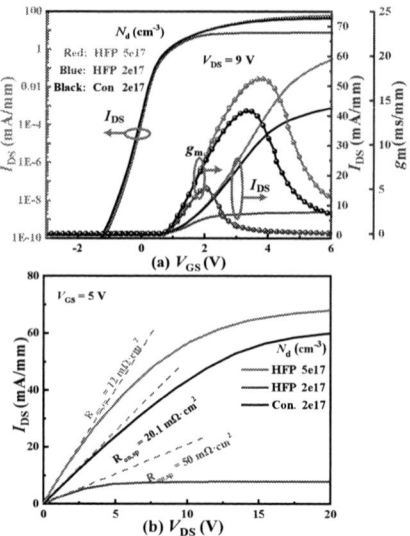

Fig. 5 (a) Simulated transfer characteristics in linear and semi-logarithmic scale. (b) Forward output characteristics.

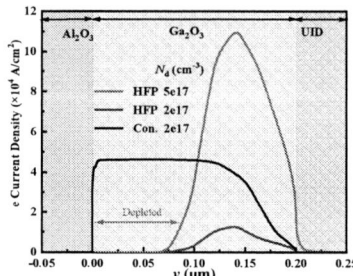

Fig. 6 On-state current-density distributions of HFP and Con. β-Ga$_2$O$_3$ MOSFET at $x = 7$ μm.

simulation results verify that the proposed device depletes the N-drift more quickly.

Fig. 4 depicts the equipotential lines and N-drift surface lateral electric field (E_x) distribution at breakdown for both devices. Obviously, the HFP β-Ga$_2$O$_3$ MOSFET depletes the whole N-drift region with $N_d = 2$e17 / 5e17 cm^{-3}, while the N-drift is NOT fully depleted for Con. β-Ga$_2$O$_3$ MOSFET even with $N_d = 2$e17 cm^{-3} shown in Fig. 4(a). Because the premature breakdown occurs around the source field plate for Con. β-Ga$_2$O$_3$ MOSFET as the E_x shown Fig. 4(b). Owing to the assisted depletion effect induced by the P-NiO in HFP, the E_x and equipotential lines distribution of HFP β-Ga$_2$O$_3$ MOSFET are more uniform, and the average E_x value is higher to improve the BV. Compared to BV of 1542V for Con. β-Ga$_2$O$_3$ with $N_d = 2$e17 cm^{-3}, the BV of 2624V / 2167V for HFP β-Ga$_2$O$_3$ MOSFET with $N_d = 2$e17 / 5e17 cm^{-3} is improved by 70.2% / 40.5%. Note that the average E_x for $N_d = 5$e17 cm^{-3} is lower than that of $N_d = 2$e17 cm^{-3}, because the higher N_d easily leads to vertically premature breakdown around the gate.

Fig. 5 shows the transfer and on-state output characteristics for both devices. Fig. 6 illustrates the current density distribution perpendicular to the channel direction in the on-state. Owing to local thermal oxidation treatment for the N-drift below the gate region, both devices achieve enhance-mode with threshold voltage of about 0.5V in Fig. 5(a). The maximum transconductance of 17.4mS/mm for HFP β-Ga$_2$O$_3$ MOSFET with $N_d = 5$e17 cm^{-3} is higher than 13.8mS/mm of Con. β-Ga$_2$O$_3$ MOSFET with $N_d = 2$e17 cm^{-3}.

979-8-3503-6184-1/24 $31.00 © 2024 IEEE

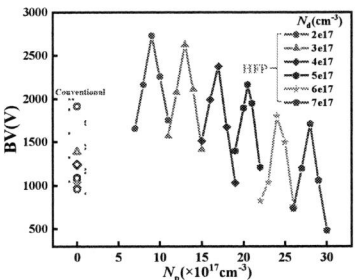

Fig. 7 Dependence of BV on the N_p and N_d.

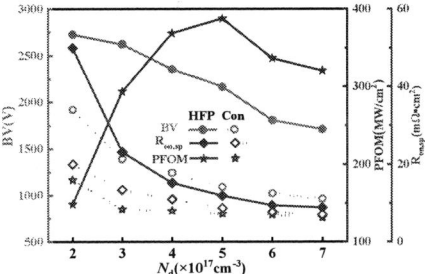

Fig. 8 Dependences of BV, $R_{ON,SP}$ and PFOM on the N_d for HFP-MOSFET and Con. MOSFET.

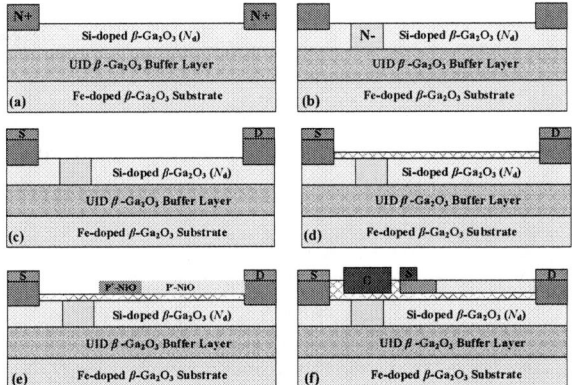

Fig. 9. Feasible process steps to fabricate HFP β-Ga$_2$O$_3$ MOSFET. (a) Regrown N+ epitaxy; (b) Local thermal oxidation treatment; (c) Ohmic contact metal deposition; (d) ALD for field oxide; (e) Magnetron sputtering to form P type NiO; (f) Gate and HFP electrode formation.

Fig. 5(b) compares the on-state output characteristics. With the same $N_d = 2e17$ cm^{-3}, the Con. β-Ga$_2$O$_3$ MOSFET owns higher output current and lower $R_{ON,SP}$ of 20.1 mΩ·cm^2 than HFP β-Ga$_2$O$_3$ MOSFET, because the P-NiO in HFP depletes the N-drift region and narrows the current path in the Ga$_2$O$_3$ N-drift as shown in Fig. 6. Owing to the assisted depletion effect induced by HFP structure, the HFP β-Ga$_2$O$_3$ MOSFET with higher $N_d = 5e17$ cm^{-3} achieves higher output current and lower $R_{ON,SP}$ of 12 mΩ·cm^2, which is decreased by 40.3% compared with Con. β-Ga$_2$O$_3$ MOSFET in Fig. 5(b).

Fig. 7 summarizes the dependence of BV on the doping concentration N_p and N_d of P-NiO and Ga$_2$O$_3$ N-drift. Based on the charge balance between P-NiO and Ga$_2$O$_3$ N-drift, it satisfies the formula of $N_d \times t_{GaO} = N_p \times t_{NiO}$ to realize the optimal BV value. Therefore, the BV firstly increases and then decreases with the increasing N_p for given N_d, t_{GaO} and t_{NiO} values. For low N_p value, the device prematurely breakdown in the N-drift around the gate, just like conventional device with N-drift NOT fully depleted. For high N_p value, the device breakdown at the heterojunction of P-NiO/N+ Ga$_2$O$_3$ (D1) at

the drain side. The maximum BV (BV_{max}) of different N_d for the proposed device is higher than that of Con. β-Ga$_2$O$_3$ MOSFET as shown in the blue dash rectangle region, because of the electric modulation effect induced by the HFP structure.

Fig. 8 summarizes the dependences of BV, $R_{ON,SP}$ and power-figure-of-merit (PFOM = $BV^2 / R_{ON,SP}$) on the N_d. The optimal BV and $R_{ON,SP}$ values for the both devices decrease with the increasing N_d, because high N_d easily cause premature breakdown around the gate. For the same N_d, the HFP β-Ga$_2$O$_3$ MOSFET achieves better BV than Con. β-Ga$_2$O$_3$ MOSFET. With the increasing N_d, the BV of Con. β-Ga$_2$O$_3$ MOSFET decreases quickly, which leads to a decreasing PFOM. However, the PFOM of HFP β-Ga$_2$O$_3$ MOSFET firstly increases and then decreases with increasing N_d, because the HFP not only modulate the electric field distribution to improve the BV, but also assists depletion the N-drift to increase the N_d and decrease the $R_{ON,SP}$. Compared with Con. β-Ga$_2$O$_3$ MOSFET, the maximum PFOM of HFP β-Ga$_2$O$_3$ MOSFET is improved by 230.8%.

Fig. 9 shows a feasible process steps to fabricate a prototype MOSFET. The key steps are to realize good selectively regrown N+ Ga$_2$O$_3$ by MOCVD in step (a) [8], and form hole density controllable P-NiO region above the drift region in step (e) [10].

IV. CONCLUSION

A novel high voltage β-Ga$_2$O$_3$ MOSFET with heterojunction field plate is presented and investigated by simulation. Based on the superjunction mechanism formed between the P-NiO and N-drift Ga$_2$O$_3$, the HFP structure not only modulates the lateral E-field distribution to improve the BV in the off-state, but also assists depleting the N-drift to improve the N_d and decrease $R_{ON,SP}$. Therefore, the proposed HFP β-Ga$_2$O$_3$ MOSFET achieves better tradeoff between the BV and $R_{ON,SP}$. In comparison with the Con. β-Ga$_2$O$_3$ MOSFET, this novel device improves the BV, $R_{ON,SP}$ and PFOM by 40.5%, 40.3% and 230.8%.

REFERENCES

[1] T. Onuma, S. Saito, K. Sasaki, et al. "Valence band ordering in β-Ga$_2$O$_3$ studied by polarized transmittance and reflectance spectroscopy," Japanese Journal of Applied Physics, vol 54, no. 11, pp. 112601,2015.

[2] M. Orita, H. Ohta, M. Hirano, et al. "Deep-ultraviolet transparent conductive β-Ga$_2$O$_3$ thin films," Applied Physics Letters, vol 77, no. 5, pp. 4166-4168, 2000.

[3] W. Li, K. Nomoto, Z. Hu, et al. "Single and multi-fin normally-off Ga$_2$O$_3$ vertical transistors with a breakdown voltage over 2.6 kV," IEEE IEDM, p. 12-14, 2019.

[4] S. Sharma, K. Zeng, S. Saha, et al. "Field-plated lateral Ga$_2$O$_3$ MOSFETs with polymer passivation and 8.03 kV breakdown voltage," IEEE Electron Device Letters, vol. 41, no. 6, pp. 836–839, Jun. 2020.

[5] K. Zeng, A. Vaidya, and U. Singisetti, "A field-plated Ga$_2$O$_3$ MOSFET with near 2-kV breakdown voltage and 520 mΩ·cm^2 on-resistance,"Appl. Phys. Express, vol. 12, no. 8, pp.081003. Aug. 2019.

[6] K. Tetzner, E.B. Treidel, O. Hilt, et al. "Lateral 1.8 kV β- Ga$_2$O$_3$ MOSFET with 155 MW/cm^2 power figure of merit," IEEE Electron Device Letters, vol. 40, no. 9, pp. 1503–1506, Sep. 2019.

[7] Y. Wang, H. Gong, X. Jia, et al. "Demonstration of β-Ga$_2$O$_3$ Superjunction-Equivalent MOSFETs," IEEE Transactions on Electron Devices, vol. 69, no. 4, pp. 2203-2209, Mar. 2022.

[8] X. Zhou, Q. Liu, W. Hao, et al. "Normally-off β-Ga$_2$O$_3$ power heterojunction field-effect-transistor realized by p-NiO and recessed-gate," IEEE ISPSD, pp. 101-104, 2022.

[9] X. Luo, L. Hao, Y. Wei, et al. "High breakdown voltage β-Ga$_2$O$_3$ Schottky barrier diode with fluorine-implanted termination," Microelectronics Journal, pp 106269, 2024.

[10] C. Wang, H. Gong, W. Lei, et al, "Demonstration of the p-NiOx /n-Ga$_2$O$_3$ heterojunction gate FETs and diodes with BV^2/Ron,sp figures of merit of 0.39 GW/cm^2 and 1.38 GW/cm^2," IEEE Electron Device Letters, vol. 42, no. 4, pp. 485–488, Apr. 2021.

Investigation of SiON Passivation for High Performance AlGaN/GaN HEMTs

Difei Fan [1], Chenkai Deng [1], Jiming Zhang [1], Peiran Wang [1], Nick Tao [2], Qing Wang*[1,3,4], Hongyu Yu*[1,3,4]

[1]School of Microelectronics, Southern University of Science and Technology, Shenzhen 518055, China
[2] Maxscend Microelectronics Company Limited, Wuxi 214072, China
[3] Engineering Research Center of Integrated Circuits for Next-Generation Communications,
Ministry of Education, Southern University of Science and Technology, Shenzhen 518055, China
[4] GaN Device Engineering Technology Research Center of Guangdong, Southern University of Science and
Technology, Shenzhen 518055, China

* Email: wangq7@sustech.edu.cn (Q.W.); yuhy@sustech.edu.cn (H.Y.)

Abstract—This study examines the impact of silicon oxynitride passivation composition (SiON) with refractive indices ranging from 1.5 to 1.8 on AlGaN/GaN high electron mobility transistors (HEMTs). The significant improvements in two dimensional electron gas (2DEG) concentration, mobility, and sheet resistance are observed at devices with a refractive index of 1.67 SiON passivation, due to it effectively reduces surface state defects. Compared to conventional SiN$_x$ passivation, SiON passivation exhibits lower parasitic capacitance, thereby reducing the dielectric constant. Therefore, enhancing the device's transconductance (g_m), cutoff frequency (f_T), and maximum oscillation frequency (f_{max}), providing an efficient passivation solution for future high-performance GaN radio frequency (RF) devices.

Keywords—AlGaN/GaN HEMTs; SiON passivation; RF

I. INTRODUCTION

AlGaN/GaN-based high electron mobility transistors (HEMTs) are recognized as promising candidates for next-generation radio frequency (RF) power amplifiers [1-3], due to their high saturated electron velocity, high critical breakdown field, and high-temperature stability [4-6]. However, the significant lattice mismatch between Si and GaN generates numerous defects during epitaxial growth, which are present in the bulk GaN and on the AlGaN surface. These defects ultimately lead to device performance degradation and eventual failure [7-9].

To deal with this issue, passivation treatment has been commonly utilized, with SiO$_2$ and SiN$_x$ being the most frequently used passivation materials. SiO$_2$ is an attractive passivation material for GaN due to its large bandgap energy and conduction band offset (E_c = 3.6 eV). However, it has a relatively lower dielectric constant (ε_r = 3.9) and weaker durability. On the other hand, SiN$_x$, widely used as a passivation layer for GaN electronic devices, has a higher dielectric constant (ε_r = 7) but a lower conduction band offset from GaN (E_c = 2.4 eV) [10]. As a tradeoff approach, silicon oxynitride (SiON) would be an alternative candidate to take good properties from both SiO$_2$ and SiN$_x$ [11,12]. V. Desmaris and J.Y. Shiu et al. confirmed the feasibility of using SiON passivation on GaN HEMTs as early as 2008 [13]. Liu et al. reported that SiON exhibits a small lattice mismatch with GaN, resulting in great interface quality and a large mean free path of electrons [14]. Compared to SiO$_2$, SiON is more effective at blocking the penetration of impurities [15,16]. Additionally, positive fixed charges have been shown to exist in the SiON passivation layer, which reduce the surface potential and expand the quantum well below the Fermi level.

This phenomenon increases the 2DEG carrier density and improves the switching performance of HEMTs [17]. Despite previous research suggesting the potential benefits of SiON passivation, there is still no unanimous consensus on the optimal preparation method for applying a SiON passivation layer to GaN HEMT devices.

In this work, we prepared SiON passivation materials with varying refractive indices and investigated their effects on electron mobility, 2DEG concentration, and sheet resistance. The DC and AC characteristics of GaN HEMTs with different refractive indices ratios of SiON passivation were compared to those with traditional SiN$_x$ passivation using Silvaco TCAD simulations. The results demonstrate that a reasonable adjustment of the stoichiometric ratio of SiON passivation material can significantly improve surface defects on AlGaN and enhance key device parameters such as transconductance (g_m), cutoff frequency (f_T), and maximum oscillation frequency (f_{max}).

II. DEVICE STRUCTURE AND FABRICATION

To optimize the deposition conditions of SiON films, we adjusted the flux of the reactant gases SiH$_4$, N$_2$O, NH$_3$, and N$_2$. The RF power, temperature, and pressure in the reaction chamber were fixed at 30 W, 350 ℃, and 500 mTorr, respectively, based on the standard values for SiO$_2$ and SiN$_x$ deposition processes as well as reference [18]. The influence of each variable on the refractive index was investigated independently to elucidate their relationships. As shown in Table I, the refractive index increased with the flux of SiH$_4$, NH$_3$, and N$_2$, while it decreased with the flux of N$_2$O. To date, we have achieved refractive indices range from 1.520 to 1.787.

TABLE I.   RECIPE OF PECVD DEPOSITION CONDITIONS

Sample	RF(W)	SiH$_4$ (sccm)	NH$_3$ (sccm)	N$_2$O (sccm)	N$_2$ (sccm)	Refractive Index
1	30	6	3	150	500	1.520
2	30	6	3	60	500	1.568
3	30	6	12	30	500	1.670
4	30	10	48	30	500	1.751
5	30	20	48	30	1000	1.787

Figure 1 illustrates the schematic diagram of the van der Pauw Hall measurement sample. Initially, the wafer underwent dicing into 1 cm × 1 cm sections, followed by

979-8-3503-6184-1/24 $31.00 © 2024 IEEE

solvent cleaning. To achieve ohmic contacts, a 20/110/40/50 nm Ti/Al/Ti/Au metal stack were deposited at the corners of each sample using electron-beam evaporation, followed by a rapid thermal annealing process conducted at 850°C for 45 seconds in a N_2 atmosphere. Then, SiON passivation layers with varying refractive indices were deposited using plasma-enhanced chemical vapor deposition (PECVD). Finally, the SiON passivation layer was etched at each corner of the sample to ensure the exposure of the underlying metal electrode stack. After fabrication, ellipsometry was employed to measure the thickness and refractive index of the passivation layer.

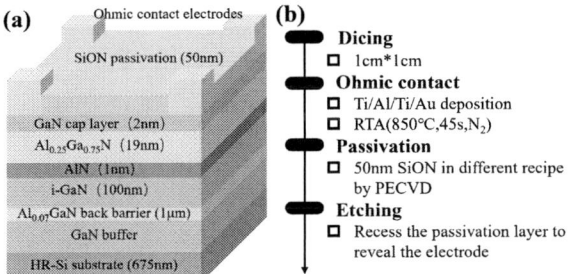

Fig. 1. (a) Schematic diagram and (b) process flow of van der Pauw hall measurement sample.

III. RESULTS AND DISCUSSION

A. Surface Morphology Performance

The surface topography and roughness of the SiON passivation layer with different refractive index were characterized using atomic force microscopy (AFM). As shown in Figure 2(f), within the refractive index range of 1.5 to 1.8, the root-mean-square (RMS) roughness of all samples remained consistently low, at less than 0.7 nm. The RMS roughness displayed a noticeable increase within the range of n = 1.670 – 1.751, followed by a decrease within the range of n = 1.751 – 1.787. This trend was closely related to the reactant gas concentrations.

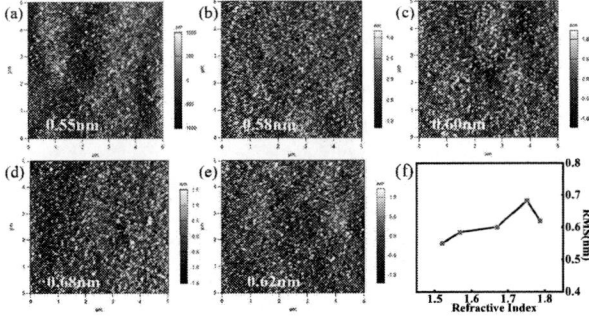

Fig. 2. (a) - (e) AFM images (5 × 5 μm²) for sample from n=1.520 ~ 1.787 and (f) n as a function of RMS.

B. Element Component Analysis

The X-ray photoelectron spectroscopy (XPS) analysis was performed to investigate the elemental composition of the passivation layer surface. The binding energies were calibrated using C 1s with a peak observed at 284.8 eV. Figures 3(b) and 3(c) depict the XPS spectra for N 1s and O 1s, respectively. The atomic concentrations of each element across all samples are summarized in Table II. Notably, an increase in refractive index correlates with a decrease in the

O component and an increase in the N component. Films with higher N content exhibit higher refractive indices, approaching that of SiN_x (n = 2.0). In contrast, passivation layers tend to resemble SiO_2 (n = 1.48). The Si component shows consistent behavior under different conditions. Figure 3(a) reveals a slight shift towards lower binding energy in the Si 2p peak of the XPS analysis as the refractive index increases. Specifically, the Si 2p binding energy of 103.3 eV for sample 1 corresponds to Si-O bonds, while the Si 2p binding energy of 102.1 eV for sample 5 corresponds to Si-N bonds, confirming variations in Si-bonding as described above.

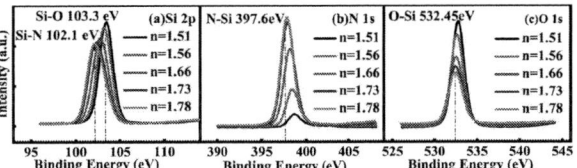

Fig. 3. XPS measurements of the passivation layer: (a) Si 2p, (b) N 1s and (c) O 1s peak deviation

TABLE II.   ATOMIC CONCENTRATION TABLE FROM XPS

Sample	C 1s (%)	N 1s (%)	O 1s (%)	Si 2p (%)
1	10.99	3.46	59.51	26.04
2	12.17	8.02	53.56	26.26
3	13.46	16.69	43.41	26.45
4	13.62	21.92	37.50	26.95
5	14.10	23.21	35.07	27.62

C. Hall effect measurement

After characterizing the surface properties and composition of the SiON material, van der Pauw Hall measurement was employed to assess its passivation efficacy on AlGaN/GaN heterostructures.

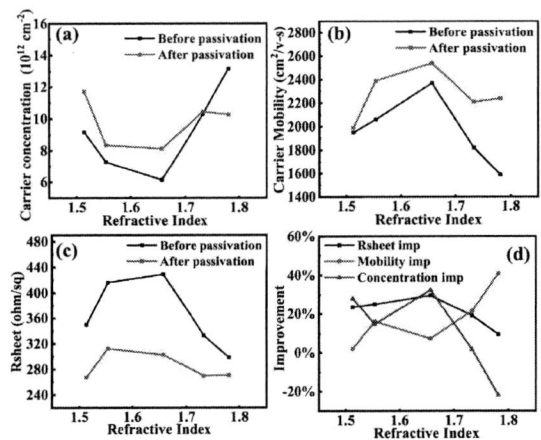

Fig. 4. Hall effect measurement of samples with n=1.520 to n=1.787: (a) – (c) R_{sheet}, carrier mobility and concentration of 2DEG and (d) their improvement in percentage.

The optimization of sheet resistance (R_{sheet}), carrier mobility, and carrier concentration under varying refractive indices before and after passivation is presented in Figure 4.

979-8-3503-6184-1/24 $31.00 © 2024 IEEE   389

In Figure 4(a), at n = 1.67, the carrier concentration increased by approximately 32% post-passivation, rising from 6.14×10^{12} cm^{-2} to 8.12×10^{12} cm^{-2}. This enhancement can be attributed to reduced positive fixed charges within the passivation layer, thereby lowering surface potential and extending the quantum well below the Fermi level. Figure 4(b) illustrates a general improvement trend in mobility, with enhancements exceeding 21% for refractive indices n $\geqslant$ 1.751, peaking at a 41% increase when n = 1.787. Additionally, Figure 4(c) indicates that R_{sheet} improvement initially increases and then decreases, with optimal enhancement observed in the refractive index range of n = 1.52 to 1.67, reaching its maximum at n = 1.67 before declining. Notably, variations in 2DEG carrier concentration significantly influence the observed trends in R_{sheet} development. These results suggest that deposition conditions yielding a thin film refractive index around 1.67 demonstrate effective passivation effects.

D. T-CAD Simulation Analysis

1) Device Structure Design and Simulation Setup:

As is shown in Figure 5, The AlGaN/GaN HEMTs structure implemented by TCAD simulation. The epitaxial structure consists of a 20 nm i-Al$_{0.25}$Ga$_{0.75}$N barrier layer with a 600 nm i-GaN channel layer, a 50 nm passivation layer and a Si-substrate. The reported devices have a gate length (L_g) of 0.25 μm, a gate-source distance (L_{gs}) of 0.8 μm and a gate-drain distance (L_{gd}) of 2.7 μm.

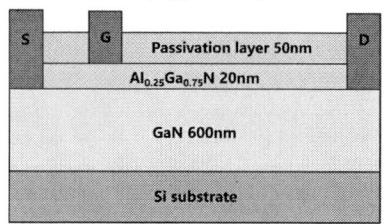

Fig. 5. Cross-section of the AlGaN/GaN HEMTs in TCAD simulation.

TABLE III. CALIBRATION PASSIVATED MATERIAL PARAMETERS FOR SIO$_2$, SION AND SIN

Passivated material		Refractive index	Dielectric constant
SiON	Sample 1	1.520	4.13
	Sample 2	1.568	4.39
	Sample 3	1.670	4.98
	Sample 4	1.751	5.48
	Sample 5	1.787	5.70
SiN$_x$		2.049	7.50

The equation for refractive index and dielectric constant is Equation (1), where n is refractive index, ε and μ are dielectric constant and magnetic permeability. In Table 3, n and ε of SiN$_x$ are shown. It can be calculated that μ is around 0.56. Set magnetic permeability of SiON μ = 0.56, The dielectric constant of each sample is calculated by different refractive index by Equation (1).

$$n = \sqrt{\varepsilon\mu} \qquad (1)$$

2) DC Characteristics:

Figure 6 illustrates the DC transfer characteristics and g_m curve of the HEMTs as a function of the refractive index of the passivation layer. From Figure 6(a), it is evident that the refractive index of the passivation layer has a negligible effect on the drain current and threshold voltage. The threshold voltage (V_{th}) of the SiN$_x$ passivated HEMTs measures -3.985 V, slightly higher than that of the SiON passivated HEMTs at -3.987 V. The maximum g_m values range from 362.19 to 361.46 mS/mm as the refractive index of the HEMTs passivation layer varies from n = 1.520 to n = 2.049. Compared to the devices with SiN$_x$ passivation, the SiON passivated HEMT exhibits higher transconductance. However, as the refractive index increases, the BV gradually degrades, from 30.324 to 28.294 V.

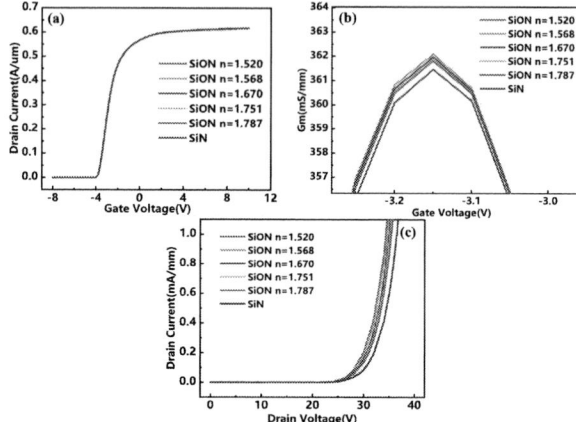

Fig. 6. (a) Transfer characteristics, (b) G$_m$ curves and (c) OFF-state breakdown characteristics of varying the refractive index of passivation layer HEMT.

3) RF Characteristics:

The small-signal characteristics of those GaN HEMTs with varying refractive indices SiON passivation and SiN$_x$ passivation are shown in Figure 7. The cut-off frequency f_T value is extracted from the extrapolation of the $|H_{21}|^2$ parameter, where its slope equals −20 dB/dec and reaches the gain of 0 dB. The maximum oscillation frequency f_{max} value is extracted from the unilateral power gain, where its slope equals −20 dB/dec and reaches the gain of 0 dB. As shown in Figure 7, all devices with SiON passivation exhibit higher f_T and f_{max} values compared with the devices with SiN$_x$ passivation. Additionally, as the refractive index decreases, the relative dielectric constant of the passivation layer also decreases, thereby achieving better small signal characteristics.

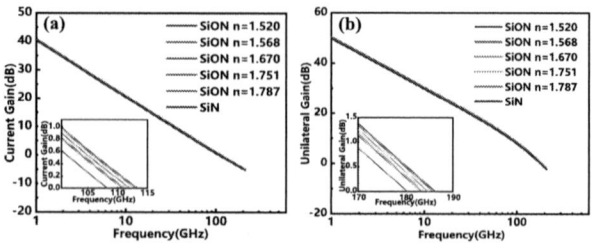

Fig. 7. (a) cut-off frequency f_T and (b) maximum oscillation frequency f_{max} of varying the refractive index of passivation layer HEMT.

IV. CONCLUSION

This work studied the influence of different refractive indices of SiON on the performance of GaN RF devices. We found that at a refractive index of 1.67, the devices demonstrated optimal performance. The two-dimensional electron gas in GaN HEMTs was enhanced by 32%, and the sheet resistance decreased by 29%. Compared with traditional SiN_x passivation, the devices with SiON passivation exhibited higher peak transconductance, f_T and f_{max}. Overall, this work demonstrates that SiON can be considered as a potential passivation layer material for high-performance AlGaN/GaN HEMTs.

REFERENCES

[1] Jones E A, Wang F, Ozpineci B. Application-based review of GaN HFETs[C]. 2014 IEEE Workshop on Wide Bandgap Power Devices and Applications, 2014: 24-29.

[2] Iucolano F, Boles T. GaN-on-Si HEMTs for wireless base stations[J]. Materials Science in Semiconductor Processing, 2019, 98: 100-105.

[3] He J, Cheng W C, Wang Q, et al. Recent Advances in GaN‐Based Power HEMT Devices[J]. Advanced Electronic Materials, 2021, 7(4): 2001045.

[4] Meng F, Zhang J, Zhou H, et al. Transport characteristics of AlGaN/GaN/AlGaN double heterostructures with high electron mobility[J]. Journal of Applied Physics, 2012, 112(2): 023707.

[5] Wośko M, Szymański T, Paszkiewicz B, et al. MOVPE growth conditions optimization for AlGaN/GaN/Si heterostructures with SiN and LT-AlN interlayers designed for HEMT applications[J]. Journal of Materials Science: Materials in Electronics, 2019, 30(4): 4111-4116.

[6] Zeng F, An J X, Zhou G, et al. A comprehensive review of recent progress on GaN high electron mobility transistors: Devices, fabrication and reliability[J]. Electronics, 2018, 7(12): 377.

[7] Green B M, Chu K K, Chumbes E M, et al. The effect of surface passivation on the microwave characteristics of undoped AlGaN/GaN HEMTs[J]. IEEE Electron Device Letters, 2000, 21(6): 268-270.

[8] Cheng K-Y, Wu S-C, Yu C-J, et al. Comparative study on performance of AlGaN/GaN MS-HEMTs with SiN_x, SiO_x, and SiNO surface passivation[J]. Solid-State Electronics, 2020, 170: 107824.

[9] Koley G, Tilak V, Eastman L F, et al. Slow transients observed in AlGaN/GaN HFETs: effects of SiN/sub x/passivation and UV illumination[J]. IEEE Transactions on Electron Devices, 2003, 50(4): 886-893.

[10] Hashizume T, Nishiguchi K, Kaneki S, et al. State of the art on gate insulation and surface passivation for GaN-based power HEMTs[J]. Materials science in semiconductor processing, 2018, 78: 85-95.

[11] Balachander K, Arulkumaran S, Egawa T, et al. A comparison on the electrical characteristics of SiO_2, SiON and SiN as the gate insulators for the fabrication of AlGaN/GaN metal–oxide/insulator–semiconductor high-electron mobility-transistors[J]. Japanese journal of applied physics, 2005, 44(7R): 4911.

[12] Balachander K, Arulkumaran S, Egawa T, et al. Demonstration of AlGaN/GaN metal-oxide-semiconductor high-electron-mobility transistors with silicon-oxy-nitride as the gate insulator[J]. Materials Science and Engineering: B, 2005, 119(1): 36-40.

[13] Desmaris V, Shiu J-Y, Rorsman N, et al. Influence of oxynitride (SiO_xN_y) passivation on the microwave performance of AlGaN/GaN HEMTs[J]. Solid-state electronics, 2008, 52(5): 632-636.

[14] Liu X, Wang X, Zhang Y, et al. Insight into the near-conduction band states at the crystallized interface between GaN and SiN_x grown by low-pressure chemical vapor deposition[J]. ACS applied materials & interfaces, 2018, 10(25): 21721-21729.

[15] Green M, Gusev E, Degraeve R, et al. Ultrathin (< 4 nm) SiO_2 and Si–O–N gate dielectric layers for silicon microelectronics: Understanding the processing, structure, and physical and electrical limits[J]. Journal of Applied Physics, 2001, 90(5): 2057-2121.

[16] Sun Z, Huang H, Wang R, et al. Effects of SiON/III-nitride interface properties on device performances of GaN-based power field-effect transistors[J]. Journal of Physics D: Applied Physics, 2020, 54(2): 025109.

[17] Liu S-C, Huang C-K, Chang C-H, et al. Effective passivation with high-density positive fixed charges for GaN MIS-HEMTs[J]. IEEE Journal of the Electron Devices Society, 2017, 5(3): 170-174.

[18] Han S-W, Park S-H, Kim H-S, et al. Normally-off AlGaN/GaN-on-Si MOS-HFET with a monolithically integrated single-stage inverter as a gate driver[J]. Electronics Letters, 2017, 53(3): 198-199.

180nm BCD Technology Platform with 8V to 65V Isolated LDMOS

Qi Ding[1,2], Renxiong Li[1], Ning Ning[1,2], Jun Huang[1], Yutuo Guo[1], Yu Wang[1], Kunqin He[1], Yaxin Liu[1], Huaishan Wang[1], Juan Tang[1], Qiuyue Huo[1], Minghong Yuan[1], Pan Peng[1], Ming Qiao[2,3,4]*, Lulu Peng[1]*, Bo Zhang[2]

[1] United Microelectronics Center Co., Ltd, Chongqing, P.R. China
[2] State Key Laboratory of Electronic Thin Films and Integrated Devices, University of Electronic Science and Technology of China, Chengdu, P.R. China
[3] Institute of Electronic and Information Engineering of UESTC in Guangdong, Dongguan, P.R. China
[4] Shenzhen Institute for Advanced Study, University of Electronic Science and Technology of China, Shenzhen, P.R. China

* Email: dingxiaoqi_0804@163.com, qiaoming@uestc.edu.cn, lulu.peng@cumec.cn,

Abstract—**This work presents a 180nm BCD technology platform with 1.8V/5V CMOS, BJT, 8-65V isolated LDMOS (Lateral Double-Diffused MOSFET) and other devices such as diodes, resistors, capacitors etc., which possesses competitive Ron,sp (Specific On-Resistance) of LDMOS and can meet different application requirements.**

Keywords—*BCD technology, LDMOS, Breakdown Voltage, Specific On-Resistance*

I. INTRODUCTION

In recent years, power IC market is increasing rapidly, and BCD (BJT, CMOS, DMOS) technology provides an excellent solution in circuits integration[1,2], such as automotive, industrial, mobile and communication applications. Bipolar devices enable precision analog and robust ESD solutions, CMOS enables logic and digital semiconductor IP, memory, analog and interface functions, and DMOS enable high-voltage analog and power delivery[3]. Low Ron,sp (Specific On-Resistance) of LDMOS (Lateral Double-Diffused MOSFET) is critical for conduction losses and die size shrinking in power IC.

II. 180NM BCD TECHNOLOGY FEATHERS

A. Platform feathers

The BCD platform is developed with 180nm process node, and Al BEOL interconnection with options of top metal thickness. The platform possesses comment logic and analog devices, such as 1.8V&5V CMOS, BJT, diode, resistor, and variable capacitors (MOM, MIM, Varactor), also provides 18V BJT and 8-65V LDMOS, as shown in Figure1. These devices can satisfy variable design requirements.

B. Key devices design

We use N type buried layer, silicon epitaxial layer and 'Pepi' ring as the main isolation method, to achieve isolated LDMOS on both vertical and horizontal directions; Simultaneously, we use poly field plate and RESURF (Reduced Surface Field) technology to balance the horizonal and vertical electric fields to achieve lower Ron,sp and higher BV (Breakdown Voltage). LOCOS (Local Oxidation of Silicon) and STI (Shallow Trench Isolation) are also utilized to improve the sustained voltage between drain and gate, and also could enhance reliability performance.

Figure 2 shows TEM cross section of 40V NLDMOS with LOCOS. Figure 3(a) and 3(b) show the schematic structures of LDMOS, 'PSUB' and 'Pepi' are both no doped P type area. 8V&15V LDMOS is STI-type isolated LDMOS, 20~65V LDMOS is LOCOS-STI type LDMOS.

As shown in Figure 3(a), 15V NLDMOS and PLDMOS use Pepi between bulk and drifting area to adjust horizontal BV, and use BN and DNW as isolation layers; Because of 15V devices drain drifting region use the same well with CMOS (IONW and PW), they are friendly for cost down.

30V and 40V LDMOS (Figure 3(b)) use LOCOS under gate and over drifting region instead of STI, this method not only can improve reliability performance, but also can optimize Ron,sp because of the shallow depth of oxide; HVNW and HVPW have two types implantation respectively with one mask, the deeper implantation type is opposite to the shallower drifting region implantation type and forming RESURF structure to optimize BV and Ron,sp; Pepi is added between Nring (N Type Ring)/Bulk and Nring/substrate to improve BV of Nring/Bulk and Nring/substrate, so that Nring will obtain good isolation performance when it is on high potential.

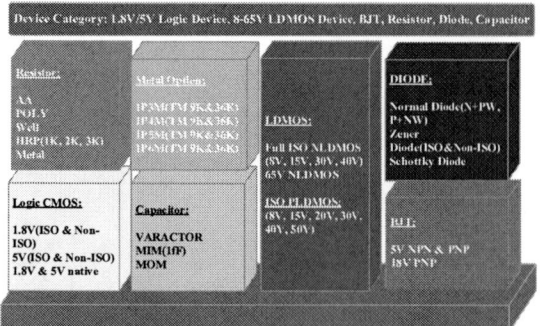

Fig. 1. The 180nm BCD technology platform device category

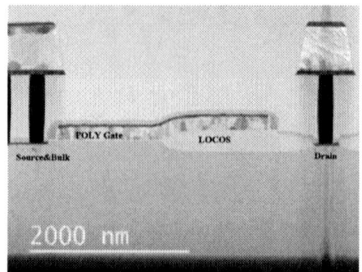

Fig. 2. TEM cross section of 40V NLDMOS

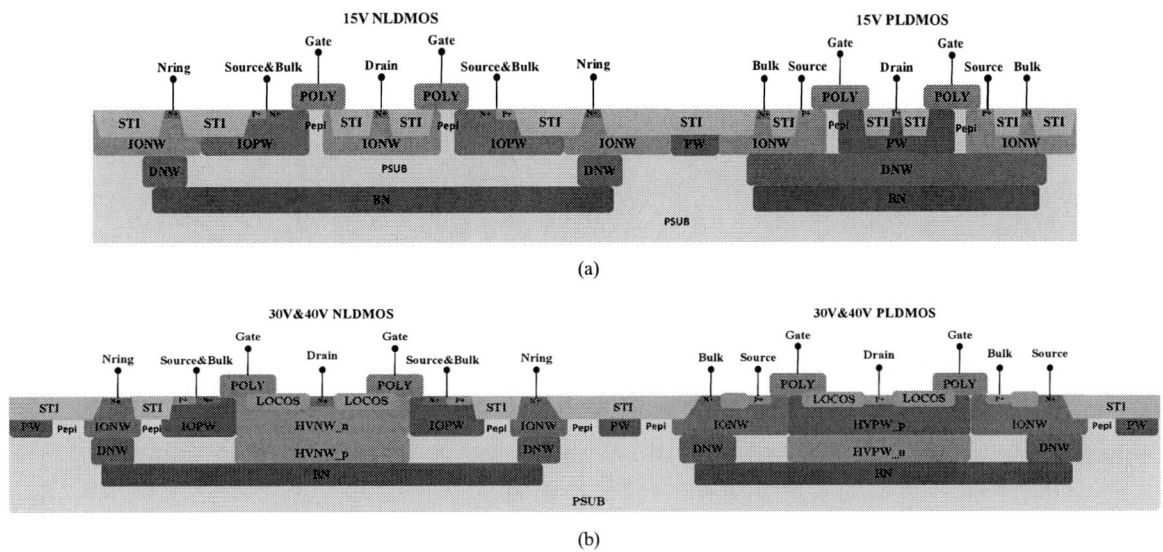

Fig. 3. (a) Schematic structure of 15V LDMOS, (b) Schematic structure of 30&40V LDMOS.

III. KEY DEVICE CHARACTERISTICS

Table1 shows the common electrical parameters of LDMOS in a wide range of 8V-65V. There are two types LDMOS of switch and analog for different applications. The two types of LDMOS have similar BV and Vth; Because of the different Lch(Channel Length), the switch devices have larger Idsat and under the same BV, switch devices have lower Ron,sp. The electrical parameters such as BV, Idsat, Vth, Ron,sp show good fullmap uniformity, as shown in Figure 4, mostly, the variation of these parameters is lower than 1% between every die.

Table I. Electrical parameters of LDMOS

		Lch	BV	Vth	Idsat	Ron,sp
	unit		V	V	μA/um	mΩ·mm²
NLDMOS	8V	switch	14.3	1.34	510	4.8
		analog	14.3	1.22	447	6
	15V	switch	20.5	1.12	444	10.3
		analog	20.5	1.11	374	13.9
	30V	switch	41.1	1.1	360	20.5
		analog	41.1	1.15	340	22.9
	40V	switch	54.3	1.1	420	39
		analog	54.3	1.15	380	44.6
	65V	analog	92	1.12	375	85

		Lch	BV	Vth	Idsat	Ron,sp
	unit		V	V	μA/um	mΩ·mm²
PLDMOS	8V	analog	-13.7	-0.75	-357	15.8
	15V	switch	-29.5	-0.71	-243	54.7
		analog	-29.6	-0.73	-196	69.5
	20V	switch	-31	-0.71	-220	46.9
		analog	-31.2	-0.76	-220	53.3
	30V	analog	-43.8	-0.78	-170	88.7
	40V	switch	-55.2	-0.7	-200	95.8
		analog	-55.2	-0.78	-170	119
	50V	switch	-61.3	-0.7	-190	114
		analog	-61.3	-0.78	-160	139

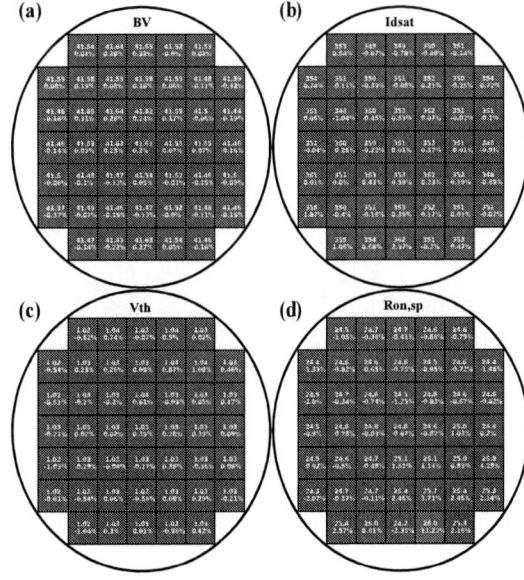

Fig. 4. 30V NLDMOS electrical full map (BV, Idsat, Vth, Ron,sp).

As presented in Figure 5, is electrical characteristics curves (Ids-Vgs and Ids-Vds) of 15V NLMOS and PLDMOS. Figure 5(b) and Figure 5(d) show wide SOA (Safe Operation Area) with drain voltage up to 30V and gate voltage from 0V to 5V.

Figure 6 shows the temperature dependent characteristics curves of 30V PLDMOS. Figure 6(a), (b), (c) are Ids-Vds curves under different temperatures, and Figure 6(d) is Vth temperature linear curve. The Ids-Vds curves show good SOA performance under different temperatures. The drain current decreases with the temperature increases and Vth also decreases linearly with temperature from -40°C to 125°C.

979-8-3503-6184-1/24 $31.00 © 2024 IEEE

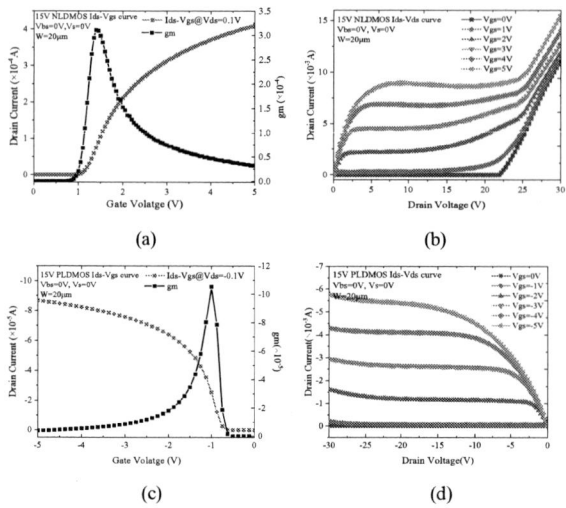

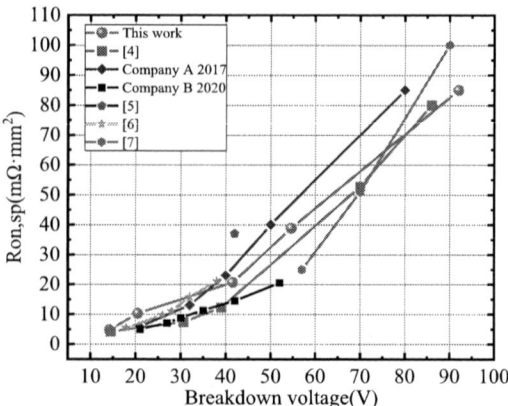

Fig. 7. The Ron,sp comparation with other works.

Fig. 5. 15V LDMOS electrical curve (a)15V NLDMOS Ids-Vgs curve, (b)15V NLDMOS Ids-Vds curve, (c)15V PLDMOS Ids-Vgs curve, (d) 15V PLDMOS Ids-Vds curve.

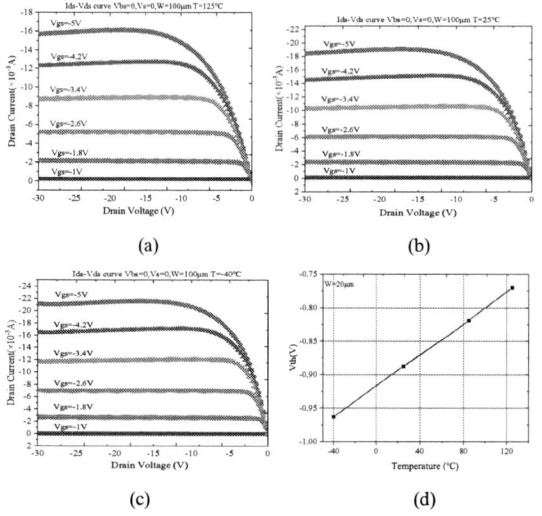

Fig. 6. 30V PLDMOS temperature characteristics (a) 30V PLDMOS Ids-Vds curve @125℃, (b) 30V PLDMOS Ids-Vds curve @25℃, (c) 30V PLDMOS Ids-Vds curve @-40℃, (d) 30V PLDMOS Vth under different temperature.

Figure 7 shows the Ron,sp comparation with other works, which Ron,sp is comparable with other works and shows advantage when BV is larger than 80V.

IV. CONCLUSION

The 180nm BCD platform we proposed has comprehensive device library and good uniformity of device performance. We use the method of poly field plate, LOCOS, RESURF etc. to optimize BV and Ron,sp; use N type buried layer and 'Pepi' to isolate the devices form substrate on vertical and horizontal directions. Finally, the Ron,sp is 85mΩ·mm² when BV is 92V. Ron,sp is comparable with other works and shows advantage when BV is larger than 80V.

V. REFERENCE

[1] Feng Jin, et al., "Best-in-Class LDMOS with Ultra-Shallow Trench Isolation and P-Buried Layer from 18V to 40V in 0.18μm BCD Technology," *Proc. of ISPSD*, 2017, pp 295-298.

[2] Yen-Ming Chen, et al., "A 150V Novel High-Voltage LDMOS in a 0.18um BCD Plug-In Process," *Proc. of ISPSD*, 2018, pp 331-334.

[3] François Hébert, et al., "Building Blocks of Past, Present and Future BCD Technologies," *Proc. of ISPSD*, 2021, pp 11-16.

[4] Hsueh-Liang Chou, et al., "0.18μm BCD Technology Platform with Best-in-Class 6 V to 70 V Power MOSFETs," *Proc. of ISPSD*, 2012, pp 401-404.

[5] Huihui Wang, et al., "A 0.35μm 600V Ultra-Thin Epitaxial BCD Technology for High Voltage Gate Driver IC," *Proc. of ISPSD*, 2018, pp 311-314.

[6] Ming Li, et al., "0.18μm BCD technology platform with performance and cost optimized fully isolated LDMOS," *Proc. of EDSSC*, 2015, pp 820-822.

[7] Y. Hao, et al., "A 0.18μm SOI BCD Technology for Automotive Application," *Proc. of ISPSD*, 2015, pp 177-180.

A Novel Insulated Gate-Triggered Thyristor with Integrated Super-Clamp Gate Transient Voltage Suppressor for Ultrahigh *di/dt* Pulse Switching

Shiyu Deng, Yuxiao Yang, Xinqi Sun, Pengwei Zhou, Ruize Sun, Chao Liu,
Pengcheng Xing, Xiaoming Wang, Wanjun Chen*, and Bo Zhang
University of Electronic Science and Technology of China, Chengdu, China

* Email: 202222021619@std.uestc.edu.cn, wjchen@uestc.edu.cn

Abstract—This study proposes an insulated gate-triggered thyristor integrated with a super-clamp transient voltage suppressor (SC-IGTT) applied to high *di/dt* pulse application, overcoming gate oxide breakdown caused by significant voltage oscillations due to parasitic inductance. The SC-IGTT integrates an NPN transistor composed of an N-well, a P-well, and an N+ region beneath the gate. Attributed to the super energy dispassion capability of the NPN transistor, the induced gate-cathode voltage would be clamped effectively, thus protecting the gate from gate oxide breakdown. Simulation results show that, in comparison to IGTT integrated with diode-clamp TVS (DC-IGTT), SC-IGTT achieves a 38.7% reduction in clamping voltage in transmission line pulse (TLP) testing. Furthermore, SC-IGTT clamped the induced gate-cathode voltage at 15.6V with ultrahigh *di/dt* of 12.7kA/μs, which is 35.3% lower than that of DC-IGTT, indicating an enhanced gate robustness.

Keywords—Pulse power, insulated gate triggered thyristor, di/dt, transient gate overvoltage

I. INTRODUCTION

With the continuous advancement of pulse power systems, solid-state devices have gradually replaced gas switches due to their advantages of long lifespan, high repeatability, compactness, and low cost [1], [2], [3], [4], [5]. Among solid-state devices, the insulated gate-triggered thyristor (IGTT) has been demonstrated to be an attractive choice that provides high surge current capabilities with a peak current of several kiloamperes and *di/dt* of tens of kiloamperes per microsecond. Furthermore, its gate-triggered characteristics simplify the driving process, eliminating the need for complex driving circuits [6], [7], [8]. However, with higher *di/dt*, gate oxide breakdown becomes serious, partly attributed to transient overvoltage between the gate and cathode. The presence of parasitic common-source-inductance (L_C) within the package leads to a high voltage difference between the gate and cathode during high *di/dt* operation, potentially causing gate oxide layer breakdown and device damage if the voltage surpasses the gate safe operating range [9]. Currently, there are several approaches to addressing this issue: the first method is utilizing a larger gate-source capacitance to enhance the gate-cathode coupling effect by constructing a dummy gate at the cathode side [10]. However, this approach results in increased device turn-on delay and sacrifices active conducting area, which results in decreased *di/dt* capability. The second approach is integrating a diode-based TVS between the gate and cathode, unluckily, the holding voltage of the TVS diode is higher than its breakdown voltage, compromising the gate driving voltage, and making it unable to achieve low resistance in dissipating energy [11].

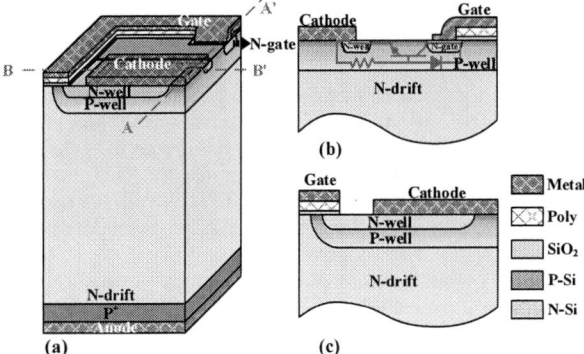

Fig. 1. The structure of the proposed SC-IGTT: (a) the diagram scheme of the device structure; (b) the cross-section of line AA'; (c) the cross-section of line BB'.

This paper presents an IGTT integrated with super-clamp TVS (SC-IGTT) for high *di/dt* applications. Firstly, the working mechanism of the structure is given, and the structure is verified by Sentaurus TCAD. Secondly, a comparison was made between SC-IGTT and the previously proposed IGTT integrated with Diode-Clamp TVS (DC-IGTT). At last, the gate protection capability of the structure was validated through simulation of pulse discharge at different voltages.

II. DEVICE STRUCTURE AND MECHANISM

Fig. 1(a) illustrates the proposed structure, wherein the gate of the proposed structure, as opposed to the conventional IGTT (C-IGTT), is a gate metal overlaps an additional injected N-well called N-gate, while the cathode metal covers the N-well and the P-well (cathode short region). In this manner, an NPN transistor is formed by utilizing the N-gate as the collector, the P-well as the base, and the N-well as the emitter, as illustrated in Fig. 1(b). At the same time, the cell structure was not affected, as shown in Fig. 1(c), indicating SC-IGTT has a similar surge capacity to that of C-IGTT. The specific composition of the super-clamp structure is illustrated in Fig. 1(b), where the N-gate and P-well form the collector of the NPN transistor, while also serving as the trigger avalanche diode for it. Considering the case where the gate-cathode voltage (V_{G-Ca}) is greater than zero; when the V_{G-Ca} is lower than the breakdown voltage of the avalanche diode, the avalanche diode is reverse biased, the NPN is not triggered, and there is no current path for the super-clamp TVS. When the applied voltage exceeds the breakdown voltage of the avalanche diode, the current is conducted through the P-well to the cathode. As the current increases, the base potential of the NPN transistor is elevated due to the parasitic resistance of the P-well. Until the emitter junction forms by the P-well

979-8-3503-6184-1/24 $31.00 © 2024 IEEE

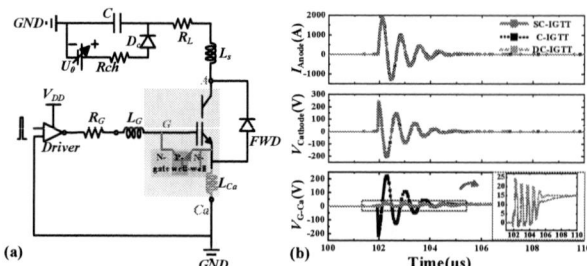

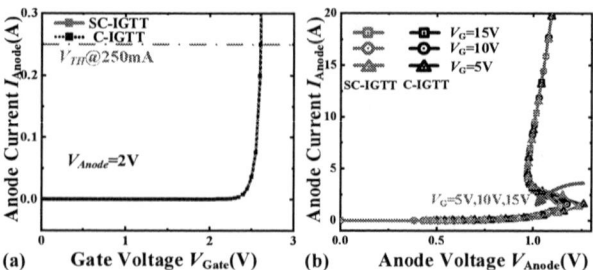

Fig. 2. (a) The high-*di/dt* pulse test circuit in simulation. In the figure, C, U_0, R_{Ch}, D_C are the storage capacitance, supplying voltage, charge resistance and charge diode. L_s, L_G, L_{Ca}, R_G and R_L are anode parasitic inductance, gate parasitic inductance, cathode parasitic inductance, gate resistance, and circuit load resistance. (b) The simulation results include anode current I_{Anode}, cathode voltage $V_{Cathode}$ and gate-cathode voltage V_{G-Ca}. In this simulation C=0.2μF, L_S=50nH, L_C=15nH, L_G=10nH, R_G=10Ω.

Fig. 3. The transfer characteristics. (a) The V_{TH} is measured at I_A = 250mA with V_A = 2V. (b) The I-V curves at V_G=5V,10V,15V.

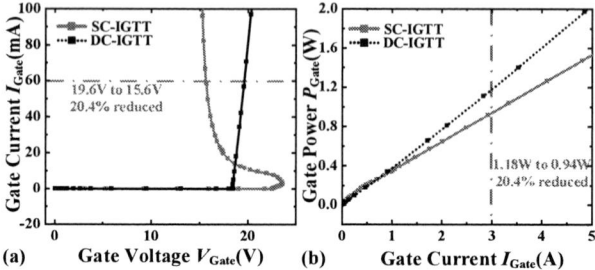

Fig. 4. Static gate reverse characteristic. (a) The gate breakdown voltage. (b) The power varies with the current.

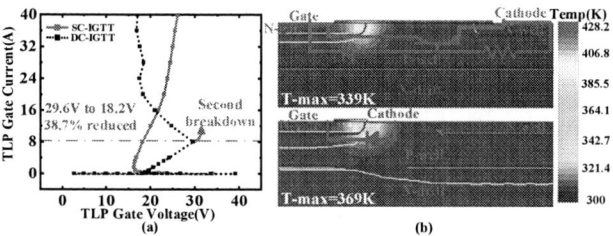

Fig. 5. TLP simulation results. (a) I-V curves. (b) Local heat map at 30ns after giving gate current=4A.

and N-well is opened, thereby causing the NPN transistor to conduct. The current generated by avalanche breakdown enters the base and becomes the base current, which is amplified by the gain β of the NPN transistor. Therefore, the NPN transistor of the super-clamp TVS is fully turned on, leading to a lower holding voltage after breakdown, and superbly clamp gate-cathode voltage. When V_{G-Ca} is under zero, the PN junction formed by P-well and N-well allows for direct forward conduction to release energy as the P-well is also in direct contact with the cathode. It is important to highlight that the modification of the implant layer layout and metal layer layout is sufficient to achieve the desired structure without introducing any additional process. Luckily, when N-gate and N-well are used at the same concentration, the breakdown voltage of the diode formed with P-well meets the requirements, allowing N-gate and N-well to be injected together.

$$V_{GC-max} \approx \frac{1}{\sqrt{1 + \left(\omega_0 r_g C_{gca}\right)^2}} \cdot \frac{di}{dt} \cdot L_C \quad (1)$$

Fig. 2(a) illustrates the circuit structure used in the simulation of pulsed discharge in the study. During the charging phase, the power supply U_0 charges the capacitor C through the charging resistor R_C, with the gate driver providing low-level output, ensuring that the IGTT remains non-conductive and withstands voltage. During the discharge phase, U_0 is turned off, and the gate driver gives set voltage to the gate, initiating the pulse discharge, and causing a rapid increase in current flow. At this juncture, the presence of packaged parasitic inductance between the cathode and ground, as well as circuit parasitic inductance (L_c), can lead to a significant increase in cathode voltage under high *di/dt* conditions. This can result in substantial voltage oscillation at the gate-cathode, as shown in Eq. (1), potentially leading to device damage [10]. However, for SC-IGTT, the super-clamp TVS at the gate will effectively clamp the gate-cathode voltage, thereby protecting the device as illustrated in Fig. 2(b).

III. SIMULATION RESULTS

In this section, a simulation comparison was conducted on the performance of SC-IGTT in comparison to C-IGTT and DC-IGTT. As shown in Fig. 3(a), the threshold voltage (V_{TH}) is 2.6V@ 250mA, consistent with C-IGTT. Fig. 3(b) illustrates that the designed structure does not affect cell design and the set breakdown voltage is above the gate operating voltage, resulting in minimal impact on the forward

characteristic curve. As a key point, a comparison was made between the gate reverse characteristics of the proposed SC-IGTT and DC-IGTT. During static simulation, the SC-IGTT exhibited a snapback IV curve after reaching breakdown voltage at the gate, showing lower holding voltage and a steeper I-V relationship compared to the DC-IGTT, which, due to its diode-clamp, could only dissipate current at a higher holding voltage. At I_{Gate}=60mA, the voltage for SC-IGTT was 15.6V, which was 20.4% lower than the 19.6V of DC-IGTT as illustrated in Fig. 4. These results indicate that compared to DC-IGTT, SC-IGTT dissipates energy at a lower resistance and posses a lower heat accumulation, providing superior clamp capability.

For the purpose of comparing SC-IGTT and DC-IGTT from transient voltage protection and heat accumulation more comprehensively, the transmission line pulse (TLP) test was conducted using simulation. It used a multi-pulse mode, simulating the response of the device at different current levels multiple times. The simulation employs the thermal model and transient simulation to observe the response of the device under sustained current stress within 100 nanoseconds. As the current increases, heat accumulation becomes more severe, ultimately leading to thermal breakdown and device damage. Due to the mentioned snapback IV characteristics of the gate, the SC-IGTT exhibits lower power, compared to the

979-8-3503-6184-1/24 $31.00 © 2024 IEEE

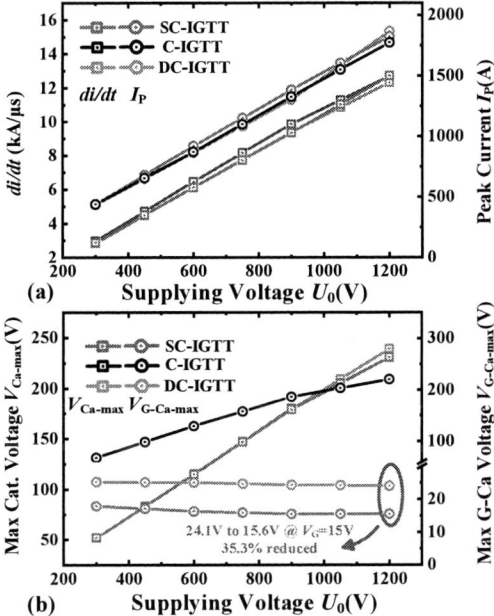

Fig. 6. Pulse discharge simulation results with different supplying voltage: (a) peak current I_{Peak} as well as di/dt; (b) max cathode voltage $V_{C\text{-}max}$ as well as max gate-cathode voltage $V_{GC\text{-}max}$.

lower holding voltage while maintaining a breakdown voltage similar to DC-IGTT. This enhances gate robustness and ensures optimal gate protection.

IV. CONCLUSION

This paper introduces a novel SC-IGTT for high di/dt power pulse applications. A super-clamp TVS based on an NPN transistor is integrated between the gate and cathode, protecting the gate from transient overvoltage under high di/dt conditions. This study focuses on the comparison of the proposed device and DC-IGTT on their reverse characteristics. In the static test, the SC-IGTT exhibits a gate voltage of 15.6V at 60mA, which is 20.4% lower than the 19.6V of the DC-IGTT, indicating lower resistance during energy dissipation. During the TLP test, the stable voltage of the SC-IGTT at 8A is 18.2V, which is 38.7% lower than the 29.6V of the DC-IGTT, and the peak temperature in the thermal map for the SC-IGTT is lower, demonstrating lower heat accumulation and a higher TLP level. Additionally, in the pulse discharge simulation at di/dt=12.7kA/µs, the SC-IGTT clamps $V_{G\text{-}Va\text{-}max}$ to 15.6V, which is 35.3% lower than the 24.1V of the DC-IGTT, demonstrating its superiority in gate robustness.

ACKNOWLEDGMENT

This work was supported in part by the National Natural Science Foundation of China under Grant U21A20499 and 62334003, the University-Industry Collaborative Education Program (231004866165126).

REFERENCES

[1] J. Mankowski and M. Kristiansen, "A review of short pulse generator technology," IEEE Transactions on Plasma Science, vol. 28, no. 1, pp. 102–108, 2000

[2] W. Jiang et al., "Compact pulsed power and its industrial applications," in 2009 IEEE Pulsed Power Conference, 2009, pp. 1-10.

[3] W. Jiang et al., "Compact solid-state switched pulsed power and its applications," Proceedings of the IEEE, vol. 92, no. 7, pp. 1180-1196, 2004.

[4] S. C. Glidden and H. D. Sanders, "Solid State Spark Gap Replacement Switches," in 2005 IEEE Pulsed Power Conference, 2005, pp. 923–926.

[5] J. A. VanGordon, S. D. Kovaleski, and G. E. Dale, "Characterization of power IGBTS under pulsed power conditions," in 2009 IEEE Pulsed Power Conference, 2009, pp. 280-282.

[6] W. Chen et al., "Design and Characterization of High di/dt CS-MCT for Pulse Power Applications," IEEE Transactions on Electron Devices, vol. 64, no. 10, pp. 4206-4212, 2017.

[7] W. Chen et al., "Experimentally demonstrate a cathode short MOS-controlled thyristor (CS-MCT) for single or repetitive pulse applications," in 2016 28th International Symposium on Power Semiconductor Devices and ICs (ISPSD), 2016, pp. 311-314.

[8] W. Chen et al., "High Peak Current MOS Gate-Triggered Thyristor With Fast Turn-On Characteristics for Solid-State Closing Switch Applications," IEEE Electron Device Letters, vol. 37, no. 2, pp. 205-208, 2016.

[9] C. Liu et al., "Transient overvoltage induced failure of MOS-controlled thyristor under ultra-high di/dt condition," in 2017 29th International Symposium on Power Semiconductor Devices and IC's (ISPSD), 2017, pp. 139-142.

[10] C. Liu et al., "Voltage Coupling Enhancement for Transient Gate Overvoltage Suppression of Insulated Gate Trigger Thyristor in Ultrahigh di/dt Pulse Applications," IEEE Transactions on Power Electronics, vol. 36, no. 3, pp. 3346-3353, 2021.

[11] C. Liu et al., "A Novel Insulated Gate Trigger Thyristor Integrated With Gate Transient Voltage Suppressor for Ultrahigh di/dt Pulse Switching," IEEE Electron Device Letters, vol. 44, no. 10, pp. 1676-1679, 2023.

DC-IGTT in TLP testing. At a TLP current of 8A, the voltage for the SC-IGTT after achieving stability is 18.2V, which is 38.7% lower than the 29.6V for the DC-IGTT. From the curves, it can be observed that the IV curve of the DC-IGTT exhibits a second snapback at this point, indicating thermal breakdown has occurred. In contrast, the SC-IGTT does not exhibit a second snapback even up to 40A., as shown in Fig. 5(a). Additionally, as depicted in Fig. 5(b), the local heat map after voltage stabilization during the TLP test at 4A shows that the transient maximum temperature for SC-IGTT was 339K, whereas for DC-IGTT it was 369K, indicating lower heat accumulation in SC-IGTT. The above experiments demonstrate that the SC-IGTT exhibits lower heat accumulation and a higher ESD level, thus enhancing gate robustness.

In order to demonstrate the gate protection capability of SC-IGTT in practical pulse applications, simulations of pulse discharge were conducted at different voltage levels. As depicted in Fig. 6(a), the pulse performance of SC-IGTT closely resembles that of C-IGTT and DC-IGTT. Taking the case of a supplying voltage of 1200V as an example, the peak current and di/dt for C-IGTT were 1825A and 12.32kA/µs, while they were 1775.1A, 12.72kA/µs for DC-IGTT, and for SC-IGTT, the peak current was 1867A with a di/dt of 12.7kA/µs, showing minimal performance differences between the three. However, under such a high di/dt condition, significant voltage oscillations were observed at the cathodes of all three devices, with the maximum cathode voltage for both above 230V. Furthermore, as shown in Fig. 6(b), the maximum gate-cathode voltage for C-IGTT was as high as 220.9V, posing a high risk of gate oxide breakdown and device damage. In contrast, SC-IGTT robustly clamped the gate-cathode voltage to 15.6V while DC-IGTT was only 24.1V, which was reduced by 35.3%. It indicates the SC-IGTT effectively protects the gate by suppressing gate-cathode voltage oscillation with its super-clamp TVS at a

979-8-3503-6184-1/24 $31.00 © 2024 IEEE

Device Instability in the Third Quadrant of Schottky-Type p-GaN Gate HEMTs: The Hole Defficiency & Trapping Effect

Kuangli Chen[1], Shuting Huang[1], Jinggui Zhou[1], Ning Yang[1], Jianggen Zhu[1], Enchuan Duan[1], Bo Zhang[1], and Qi Zhou †[1, 2]

[1] University of Electronic Science and Technology of China (UESTC), Chengdu
[2] Institute of Electronic and Information Engineering, UESTC, Dongguan
† Email: zhouqi@uestc.edu.cn

Abstract—**In this work, the third-quadrant characteristics instability of commercial 100 V Schottky p-GaN gate High Electron Mobility Transistors (HEMTs) is studied under OFF-state drain bias and negative gate bias stress. For zero voltage gate turn-off scenarios, the threshold voltage shifts up to +0.395 V in reverse conduction with drain stressed at 50 V. For negative voltage gate turn-off scenarios, the imposition of a negative gate bias stress appears to exacerbate the device instability. A substantially large threshold voltage shift of +0.716 V in reverse conduction is measured with drain stressed at 50 V and gate stressed at -4 V. The underlying mechanisms are further revealed based on intricate device physics. For lower drain voltage stress, the threshold voltage shift is predominantly influenced by the hole deficiency effect in the p-GaN gate. With the increase of drain voltage stress, impact ionization is initiated in the drift region and followed by hole trapping-induced asymmetric barrier lowering effect, which in turns to result in a decrement in threshold voltage shift.**

Keywords—*Device instability, high electron mobility transistors, p-GaN gate, the third quadrant, reverse conduction*

I. INTRODUCTION

Gallium nitride (GaN) devices have become as formidable contenders for power applications achieving high switching frequency, high power density, and high power conversion efficiency [1-3]. In recent years, the p-GaN gate enhancement-mode (E-mode) GaN HEMTs have been commercialized for fast charging and further penetrating for industrial/vehicle electronics such as power converters in data center and LiDAR [4-6]. GaN HEMTs exhibit inherent bidirectional conduction capability due to their symmetric lateral channel and the zero reverse recovery charge, enabling the typical freewheeling diode functionality in power conversion applications.

Taking the low-side power GaN HEMT in half-bridge for example, the device cyclically operates in OFF-state and reverse conduction mode. Normally, the E-mode GaN HEMT can be turned off at zero volt gate bias. However, the p-GaN gate HEMTs exhibit a low forward threshold voltage (e.g., < 2 V) [7]. In the OFF-state, to prevent the false turn-on caused by the ringing effect and gate control noise induced by the parasitic effect at high operation frequency, a negative gate bias is used to guarantee the fully turn-off of the device, albeit it may result in higher dead-time power losses [8-9]. Hence, it is of great interest to study the third-quadrant characteristics of p-GaN gate HEMTs under varied bias conditions of zero/negative gate bias $V_{GS, OFF}$ and high drain bias $V_{DS, OFF}$ to assess the device stability in reverse conduction mode. Despite

This work was supported in part by the National Natural Science Foundation of China under Grant 62174019, in part by the Guangdong Basic and Applied Basic Research Foundation, China, under Grant 2021B1515140039, 2024A1515012139 and in part by Sichuan Science and Technology Program under Grant 2023YFG0138.

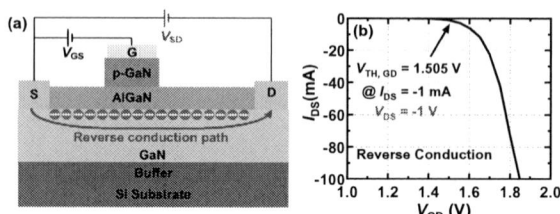

Fig. 1. (a) The schematic of reverse conduction current flows and (b) the third-quadrant curve of the Schottky-type p-GaN gate HEMT.

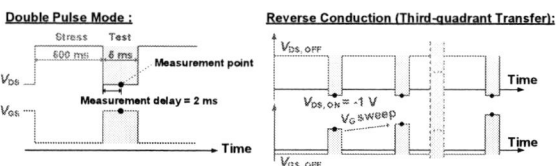

Fig. 2. The waveforms of gate and drain bias during the quasi-static measurement for reverse conduction in the third quadrant.

the $V_{TH,GS}$ stability in forward conduction of p-GaN gate HEMTs has been widely studied [10-12], the stability of the inherent $V_{TH,GD}$ and the third-quadrant characteristics are scarcely reported up to date.

In this work, a quasi-static measurement approach was proposed to evaluate the third-quadrant characteristics stability of 100 V commercial p-GaN gate HEMTs. Significant V_{TH} instability both in forward ($V_{TH, GS}$) and reverse ($V_{TH, GD}$) conduction is observed. Particularly, a substantial threshold voltage shift, denoted as $\Delta V_{TH,GD}$ of +0.395 V and $\Delta V_{TH, GS}$ of +0.436 V, is observed for zero voltage turn-off scenarios (i.e. @ $V_{DS,OFF}$=50 V, $V_{GS,OFF}$=0 V). The V_{TH} instability becomes more pronounced for negative voltage turn-off scenarios (i.e. @ $V_{DS, OFF}$ = 50 V and $V_{GS, OFF}$ = -4 V), with the maximum $\Delta V_{TH, GD}$=+0.716 V. Intriguingly, the dependence of $\Delta V_{TH, GD}$ with the increasing $V_{DS, OFF}$ exhibits a prominent divergence from that in $\Delta V_{TH, GS}$. For lower $V_{DS, OFF}$ (i.e. $V_{DS, OFF} \leq 50$ V), the threshold voltage shift is governed by hole deficiency effect in p-GaN gate. For higher $V_{DS, OFF}$ (i.e. $V_{DS, OFF} > 50$ V), it is dominated by hole trapping-induced asymmetric barrier lowering effect, which results in unbalanced barrier height respectively for forward and reverse conduction. The insights into device instability behaviors and the underlying physics delineated in this study are helpful for application engineers to understand the power performance fluctuation of the converter and further carry out circuit optimization.

II. EXPERIMENTAL RESULTS

The devices used in this work are 100 V/24 mΩ commercial Schottky-type p-GaN gate HEMTs [13]. Fig. 1 (a)

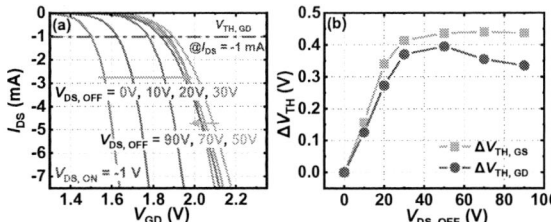

Fig. 3. (a) The pulsed reverse transfer curves using quasi-static measurement with $V_{DS, OFF} = 0 \sim 90$ V and $V_{GS, OFF} = 0$ V. (b) The variation of reverse $\Delta V_{TH, GD}$ compared with forward $\Delta V_{TH, GS}$.

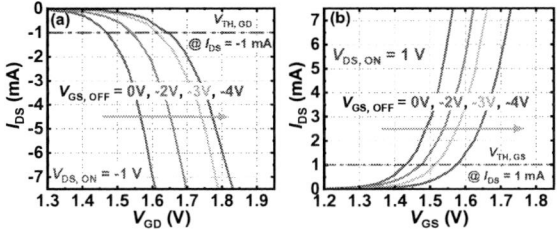

Fig. 4. (a) The reverse and (b) forward pulsed transfer curves using quasi-static measurement with varied $V_{GS, OFF}$ under $V_{DS, OFF} = 0$ V.

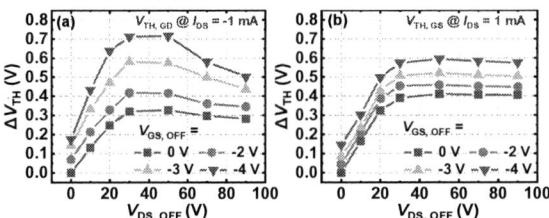

Fig. 5. The extracted (a) $\Delta V_{TH, GD}$ and (b) $\Delta V_{TH, GS}$ under OFF-state stress $V_{DS, OFF} = 0 \sim 90$ V & $V_{GS, OFF} = 0 \sim -4$ V.

illustrates the cross-sectional structure and its lateral reverse conduction current flows along 2DEG channel. The intrinsic transfer curve in the third quadrant is also shown in Fig.1 (b), in which the reverse conduction threshold voltage $V_{TH,GD}$ defined at $I_{DS} = -1$ mA is 1.505 V. Furthermore, the OFF-state drain bias $V_{DS, OFF}$ and negative gate bias stress $V_{GS, OFF}$ are applied to the device under test (DUT) utilizing the semiconductor parameter analyzer Agilent B1505A. As shown in Fig. 2, the proposed quasi-static measurement approach is conducted in double-pulse mode to avoid the self-heating effect and the associated V_{th} instability [14]. Two gate bias scenarios were used: a) zero gate bias of $V_{GS}=0$ V & b) negative gate bias of $V_{GS}=-2, -3, -4$ V.

The pulsed reverse transfer curves of the DUT under drain stress voltage $V_{DS, OFF}$ ranging from 0 to 90 V are shown in Fig. 3 (a). A prominent positive threshold voltage shift with the increasing $V_{DS, OFF}$ is observed in reverse conduction. As summarized in Fig. 3 (b), the reverse threshold voltage shift $\Delta V_{TH, GD}$ exhibit a rapid increase under lower $V_{DS, OFF}$ (i.e., $\leq$ 50 V), which is the same as the $\Delta V_{TH, GS}$ in forward conduction. For $V_{DS, OFF} = 50$ V, the $\Delta V_{TH, GD}$ and $\Delta V_{TH, GS}$ reach the maximum value of +0.395 V and +0.436 V, respectively. Such a substantial $\Delta V_{TH, GD}$ in reverse conduction may cause higher unfavorable dead-time power losses of the circuit, thus posing a significant obstacle to achieving even higher power efficiency of the system. However, a distinct instability trend between $\Delta V_{TH, GS}$ and $\Delta V_{TH, GD}$ is observed for $V_{DS, OFF}>50$ V. After the initial increase, the positive $\Delta V_{TH, GS}$ gradually saturated with the increased $V_{DS,OFF}$. In contrast, a decline in the negative shift in $\Delta V_{TH, GD}$ is observed for $V_{DS, OFF}$ increased above 50 V.

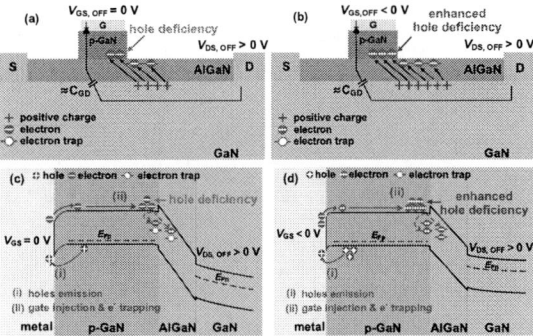

Fig. 6. The schematic cross section of the charge distribution under OFF-state drain bias stress (a) without and (b) with negative $V_{GS, OFF}$. The band diagrams are also illustrated in (c) and (d), respectively.

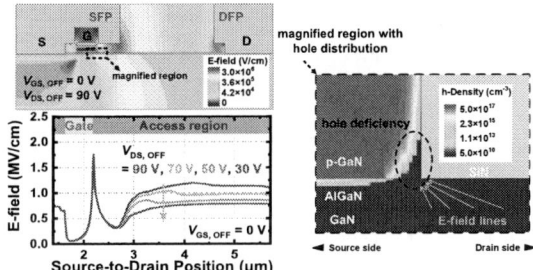

Fig. 7. The simulated E-field distribution along the channel with varied $V_{DS, OFF}$ at $V_{GS, OFF} = 0$ V. Inset: The hole distribution and E-field vectors at the drain-side gate edge.

Besides the zero gate voltage turn-off scenario, the negative gate voltage turn-off scenario is also evaluated to explore the device instability dependence on negative gate bias $V_{GS, OFF}$. As shown in Fig. 4 (a) and (b), the pulsed transfer curves show even more severe positive threshold voltage shift in both reverse and forward conduction when negative $V_{GS, OFF}$ is applied. As summarized in Fig.5 (a) and (b), the maximum $\Delta V_{TH, GD}$ and $\Delta V_{TH, GS}$ are 0.716 V and 0.595 V at $V_{DS, OFF} = 50$ V and $V_{GS, OFF} = -4$ V, respectively. More importantly, the different variation trend between $\Delta V_{TH, GD}$ and $\Delta V_{TH, GS}$ become more conspicuous with a higher negative $V_{GS, OFF}$. It can be seen that the $\Delta V_{TH, GD}$ reaches the maximum value at $V_{DS, OFF} = 50$ V and turns into a decline trend for $V_{DS, OFF} > 50$ V, which shows significant difference as that in $\Delta V_{TH, GS}$. The distinct V_{TH} instability behavior respectively in reverse and forward conduction indicates that the device instability is governed by sophisticated device physics.

III. DEVICE INSTABILITY MECHANISMS AND DISCUSSION

For the GaN HEMT with a Schottky p-GaN gate, the p-GaN layer is an electrically isolated "floating region". As the device is turned off, in the drain-side access region adjacent to the gate edge, the fixed positive charges are presented with the depletion of the 2DEG as illustrated in Fig. 6 (a). Correspondingly, the negative charges in the p-GaN gate layer are ionized acceptors since the holes flow out of the p-GaN layer through the forward-biased gate metal/p-GaN Schottky junction under a high $V_{DS, OFF}$ [15]. However, as the device is turned on, the ionized acceptors generated in the OFF-state stress scenario cannot be completely neutralized due to the reverse-biased metal/p-GaN Schottky junction. Hence, a portion of the stored net negative charges remain in the p-GaN layer, which causes the hole deficiency effect [16]. In this manner,

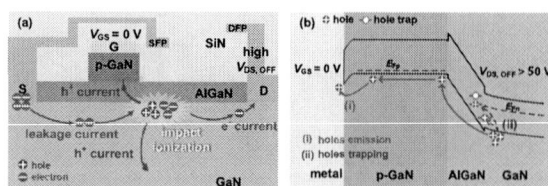

Fig. 8. (a) The schematic cross section to illustrate the process of holes generation by impact ionization. (b) The schematic band diagram and holes trapping under high $V_{DS, OFF}$ at $V_{GS, OFF}$ = 0 V.

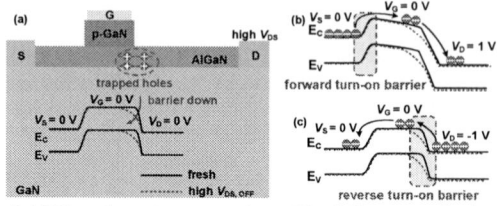

Fig. 9. (a) The schematic cross section and band diagrams under the gate with $V_{DS, OFF}$. The impact of trapped holes on the process of (b) forward conduction and (c) reverse conduction.

additional gate bias is required to offset the residual negative charges, resulting in a positive shift in V_{TH}. As shown in Fig. 6 (b), when a negative $V_{GS, OFF}$ is further applied, the hole deficiency effect is enhanced, leading to a larger positive ΔV_{TH} than that in the device solely stressed by $V_{DS, OFF}$ with V_{GS} = 0 V.

Furthermore, to investigate the notable discrepancies emerge for $V_{DS, OFF}$ > 50 V, TCAD simulations were conducted as shown in Fig. 7. As $V_{DS, OFF}$ increases, due to the E-field modulation effect of the field plate, the E-field peak transfers from the edge of the source field plate (SFP) to the drain electrode, particularly for $V_{DS, OFF} \geq$ 50 V. Under high E-field, the impact ionization could be initialized by the leakage current flows from the source as shown in Fig. 8 (a). A portion of the impact ionization induced holes flow to the low-potential substrate, while more holes flow vertically towards the gate which is driven by the vertical E-field in the p-GaN/AlGaN gate terminal. Consequently, as illustrated in Fig. 9 (a), the positively charged traps reduce the energy barrier height at the drain-side gate edge, while the energy barrier at the source-side gate edge remains unchanged.

Such an asymmetric barrier lowering effect (ABLE) ultimately results in the observed ΔV_{TH} dispersion in forward and reverse conduction modes. As shown in Fig. 9 (b), in forward conduction, the unchanged energy barrier at the source-side gate edge requires a sufficiently higher $V_{TH, GS}$ to turn on the 2DEG channel, leading to stable $\Delta V_{TH, GS}$ for $V_{DS, OFF}$ > 50 V. On the contrary, in reverse conduction, as shown in Fig. 9 (c), the occupied hole traps induced energy barrier lowering at the drain-side gate edge facilitates to turn on the 2DEG channel, while a lower $V_{TH, GD}$ is able to turn on the device. In this manner, the positive $\Delta V_{TH, GD}$ exhibits a decrease for $V_{DS, OFF}$ > 50 V. These results and in-depth mechanisms can be related to reverse conduction loss and deadtime loss, which is critical for the low-side power device in half-bridge circuits.

IV. CONCLUSION

In this study, the instability of third-quadrant characteristics in the of p-GaN gate HEMT under OFF-state stress is comprehensively investigated. For zero gate voltage turn-off scenario, the positive threshold voltage shift is observed in both reverse and forward conduction. At $V_{DS, OFF}$ = 50 V, the $\Delta V_{TH, GD}$ and $\Delta V_{TH, GS}$ reach +0.395 V and +0.436 V, respectively. For negative gate voltage turn-off scenario, the V_{TH} shows even severe instability. At $V_{DS, OFF}$ = 50 V and $V_{GS, OFF}$ = -4 V, the $\Delta V_{TH, GD}$ and $\Delta V_{TH, GS}$ increase up to 0.716 V and 0.595 V, respectively. Furthermore, notable discrepancies of ΔV_{TH} emerge for $V_{DS, OFF}$ > 50 V between reverse and forward conduction. The observed device instability is attributed to the hole deficiency in the p-GaN gate and hole trapping induced asymmetric barrier lowering effect.

REFERENCES

[1] L. Li and A. Wakejima, "Polarization-engineered quaternary barrier InAlGaN/AlGaN heterostructure field-effect transistors toward robust high-frequency power performance in AlGaN channel electronics," *IEEE Trans. Electron Devices*, vol. 68, no. 11, pp. 5535–5540, Nov. 2021.

[2] J. Wang, *et al.*, "Vertical GaN-on-GaN p-n diodes with 10-A forward current and 1.6 kV breakdown voltage," in *Proc. 76th Device Res. Conf. (DRC)*, Jun. 2018, pp. 1–2.

[3] K. J. Chen, *et al.*, "GaN-on-Si power technology: Devices and applications," *IEEE Trans. Electron Devices*, vol. 64, no. 3, pp. 779–795, Mar. 2017.

[4] K. Kumar, S. Banerjee, and R. R. Kumar, "Efficiency analysis of a 7.4 kW dynamic wireless charging system for electric vehicle through GaN devices," in *Proc. IEEE Int. Conf. Power Electron., Drives Energy Syst. (PEDES)*, Dec. 2020, pp. 1–6.

[5] Y.S. Ma *et al.*, "29.6 A digital-type GaN driver with current-pulse balancer technique achieving sub-nanosecond current pulse width for high-resolution and dynamic effective range LiDAR system," in *IEEE Int. Solid-State Circuits Conf. (ISSCC) Dig. Tech. Papers*, Feb. 2019, pp. 466–468.

[6] H. Matsumori, T. Kosaka, K. Sekido, K. Kim, T. Egawa, and N. Matsui, "Isolated DC-DC converter utilizing GaN power device for automotive application," in *Proc. IEEE Appl. Power Electron. Conf. Expo. (APEC)*, Mar. 2019, pp. 1704–1709.

[7] J. O. Gonzalez, B. Etoz, and O. Alatise, "Characterizing threshold voltage shifts and recovery in Schottky gate and ohmic gate GaN HEMTs," in *Proc. IEEE Energy Convers. Congr. Expo. (ECCE)*, Oct. 2020, pp. 217–224.

[8] R. Li, J. Zhu and M. Xie, "Parasitic Parameter Effects on the dv/dt-induced Low-side MOSFET False Turn-on in Synchronous Buck Converters," 2019 *IEEE Applied Power Electronics Conference and Exposition (APEC)*, Anaheim, CA, USA, 2019, pp. 2247-2254.

[9] Q. Zhu and M. Xie, "A New Analytical Model for Predicting dv/dt-Induced Low-Side MOSFET False Turn ON in Synchronous Buck Converters," *IEEE Trans. Power Electronics*, vol. 34, no. 6, pp. 5500-5512, June 2019.

[10] Y. Shi, *et al.*, "Bidirectional threshold voltage shift and gate leakage in 650 V p-GaN AlGaN/GaN HEMTs: The role of electron-trapping and hole-injection," 2018 *IEEE 30th International Symposium on Power Semiconductor Devices and ICs (ISPSD)*, May 2018, pp. 96-99.

[11] N. Yang, *et al.*, "Study of the Short-Circuit Capability and Device Instability of p-GaN Gate HEMTs by Repetitive Short Circuit Stress," *IEEE Trans. on Power Electronic*, vo. 39, no. 2, pp. 2247-2257, Feb. 2024.

[12] L. Sayadi, *et al.* "Threshold Voltage Instability in p-GaN Gate AlGaN/GaN HFETs," *IEEE Transactions on Electron Devices*, vol. 65, no. 6, pp. 2454–2460, Apr. 2018.

[13] EPC2007C, Efficient Power Convers., USA, 2022. [Online]. Available: https://epc-co.com/

[14] K. J. Chen, *et al.*, "The Device Instability of p-GaN Gate HEMTs Induced by Self-Heating Effect Investigated by on-State Drain Current Injection (DCI) Technique," in *IEEE Transactions on Electron Devices*, vol. 69, no. 10, pp. 5496-5502, Oct. 2022.

[15] L. Sayadi, G. Iannaccone, S. Sicre, O. Häberlen, and G. Curatola, "Threshold voltage instability in p-GaN gate AlGaN/GaN HFETs," *IEEE Trans. Electron Devices*, vol. 65, no. 6, pp. 2454–2460, Jun. 2018.

[16] J. Wei, *et al.*, "Charge Storage Mechanism of Drain Induced Dynamic Threshold Votlage Shift in p-GaN Gate HEMTs," *IEEE Electron. Dev. Lett.*, vol. 40, no. 4, pp. 526-529, Apr. 2019.

Static Characteristic Recovery Of SiC MOSFETs Induced By Dynamic Gate Stress After Total Ionizing Dose Irradiation

Jiahao Hu [1], Xiaochuan Deng*[1, 2], Xing Zeng [1], Tao Xu [1], Haibo Wu [1], Xuan Li [1], Bo Zhang [1]

[1] School of Integrated Circuit Science and Engineering, University of Electronic Science and Technology of China, Chengdu 611731, China
[2] Institute of Electronic and Information Engineering in Guangdong, University of Electronic Science and Technology of China, Dongguan 523808, China

* Email: xcdeng@uestc.edu.cn

Abstract—**More defects at the SiC/SiO₂ interface of SiC MOSFETs cause the decrease in the static parameters to be particularly significant when the SiC MOSFETs gate oxide is subjected to a certain extent of stress. In this paper, the variation in the static characteristics of SiC MOSFETs with total ionizing dose (TID) irradiation and dynamic gate stress is fully studied. After TID irradiation with a 15 V gate voltage of 1 Mrad, the breakdown voltage of the devices decreased by 70%, the threshold voltage for negative drifting decreased by 1.79 V, and the on-resistance decreased by 7.5%. However, when dynamic gate stress is applied to SiC MOSFETs after TID irradiation, the breakdown voltage, threshold voltage and on-resistance recover completely. Further analysis indicated that this phenomenon is related to the amount of oxide trapped charges in the gate oxide.**

Keywords—*static characteristic, total ionizing dose, dynamic gate stress, oxide trapped charges.*

I. INTRODUCTION

Silicon carbide (SiC) metal–oxide–semiconductor field-effect) have become popular power semiconductor devices in high-power applications such as photovoltaic inverters and new energy vehicles due to their high breakdown voltage, high operating frequency and low switching loss [1]. However, the weaker SiC/SiO₂ interface quality of SiC MOSFETs can lead to significant gate oxide reliability issues. Currently, with the rapid development of the aerospace industry, research on the application of power semiconductor devices in the space environment has become a hot topic. Among them, total ionizing dose (TID) irradiation can cause degradation of SiC MOSFETs gate oxide and reduce the reliability of its application. The degradation phenomenon caused by TID irradiation mainly manifests as the degradation of static characteristics such as the threshold voltage (V_{th}), on-resistance (R_{on}), breakdown voltage (BV) and gate leakage current (I_{gss}).

It has been reported that the main reason for the decrease in the static parameters of SiC MOSFETs caused by TID irradiation is the trapping of holes by interface charges and oxide trapped charges at the SiC/SiO₂ interface, resulting in negative drift of V_{th} and R_{on} [2]. However, in recent years, more studies have focused on the impact of TID irradiation on the gate oxide of SiC MOSFETs under constant gate bias conditions. Since SiC MOSFETs often operate in high-frequency environments, the degradation of static parameters such as V_{th} and R_{on} caused by dynamic gate stress has also become an important factor affecting the gate oxide reliability of SiC MOSFETs [3]. Therefore, it is very important to further study the effect of dynamic gate stress on the reliability of the gate oxide of SiC MOSFETs after TID irradiation.

In this paper, a commercial 1200 V planar SiC MOSFET was subjected to 1 Mrad dose of TID irradiation and dynamic gate stress, and its static characteristics were subsequently evaluated. In addition, the mechanism of static characteristic variations was studied by capacitance–voltage (C–V) measurements, TCAD simulations and subthreshold defect charge extraction. The results show that the static parameters of SiC MOSFETs decrease significantly after the TID experiment, but they recover completely after the dynamic gate stress test. Through the analysis of defect charges at the SiC/SiO₂ interface, it can be seen that the recovery of static characteristics is mainly related to the amount of oxide trapped charges.

II. EXPERIMENTAL DETAILS

The device under test (DUT) used in this experiment was a commercial 1200 V planar SiC MOSFET with the device model C3M0160120D, manufactured by Cree/Wolfspeed. The gate oxide was composed of 40-nm SiO₂, and the package type of the DUTs was TO-247-3. Fig. 1 shows a cross-sectional view of the n-channel conventional planar SiC MOSFETs. The irradiation source was ⁶⁰Co-γ, with a dose rate of 20 rad/s and a total dose of 1 Mrad (Si). The total duration of the irradiation was 14 h, and a constant gate-source bias voltage of $V_{gs} = 15$ V and $V_{gs} = -5$ V was applied to the DUTs using a power supply. After the irradiation experiment, dynamic gate stress was applied to the DUTs, with a gate bias stress of +15/-5 V. The duty ratio of the square wave voltage was 50%, and the frequency was 100 kHz. The total duration of dynamic gate stress was 14 h, which is consistent with the irradiation time. On the other hand, considering the recovery of interface charges during the time delay between experiments and static characteristic testing, to accurately evaluate the static properties of SiC MOSFETs, we used a preconditioning process to eliminate recoverable charges at the SiC/SiO₂ interface. Preconditioning pulses of +15/-5 V lasting 12 s were applied to the initial DUTs and after TID irradiation and dynamic gate stress experiments. Among them, the preconditioning process includes a positive gate bias of 15 V for 5 s, a zero gate bias for 2 s and a negative gate bias for 5 s, and subsequent gates and sources are shortened for more than 10 minutes[4-5]. An Agilent B1505A semiconductor device analyzer was used to measure the static parameters, and the C-V curve was obtained with a Keithley 4200A-SCS instrument. To study the mechanism of BV degradation after TID irradiation and dynamic gate stress, TCAD simulation

---

979-8-3503-6184-1/24 $31.00 © 2024 IEEE   401

was used in this paper. The size of the SiC MOSFETs and the thickness of the gate oxide layer for simulation were determined by using the process data obtained from physical analysis of the C3M0160120 device and measurement of the static parameters.

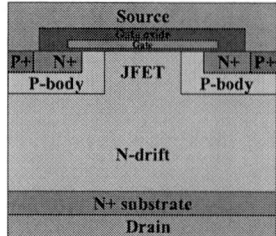

Figure 1. Cross-sectional view of the planar SiC MOSFET

III. RESULTS AND DISCUSSION

The ΔV_{th} of the DUTs after TID irradiation and dynamic gate stress experiments are shown in Fig. 2. Here, 0–14 h is the TID irradiation time, and 15–28 h is the dynamic gate stress time. After TID irradiation of 1Mrad with bias voltages of 15 V and -5 V, the V_{th} values of the DUTs decrease by 1.79 V and 0.67 V, respectively. This is mainly attributed to the trapping of holes by the trapped charges in the gate oxide [2]. On the other hand, after 14 h of dynamic gate stress, the V_{th} of the DUTs fully recovers, and there is a slight positive drift. The literature shows that the dynamic gate stress can result in electrons being injected into the gate oxide layer [4]. Therefore, the recovery of V_{th} may be related to the negative charges injected into the gate oxide by applying dynamic gate stress.

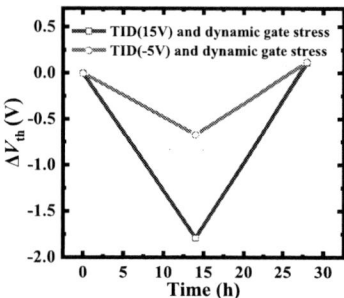

Figure 2. ΔV_{th} of the DUTs after TID irradiation and dynamic gate stress

Fig. 3 shows the variation in R_{on} of DUTs after TID irradiation and dynamic gate stress. The variation trends of R_{on} are consistent with those of V_{th}.

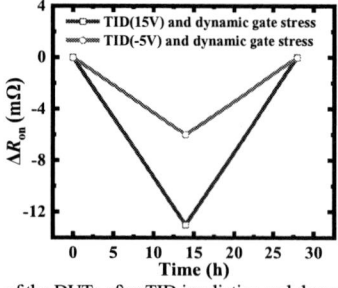

Figure 3. ΔR_{on} of the DUTs after TID irradiation and dynamic gate stress

Fig. 4 indicates that the I_{gss} is always stable below 5 nA, which shows that the quality of the gate oxide is intact and not affected by TID irradiation or dynamic gate stress. In addition, the breakdown voltage variations of the DUTs are shown in Fig. 5. After TID irradiation with a gate voltage of 15 V, the BV of the DUTs decreases to 557 V and recovers to more than

1600 V with dynamic gate stress. However, the BV of the DUTs has no significant variation when the gate voltage is -5 V. To explain this phenomenon, a TCAD simulation is used to illustrate the leakage current change in the DUTs with TID irradiation and dynamic gate stress.

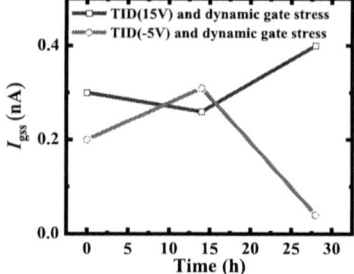

Figure 4. I_{gss} of the DUTs after TID irradiation and dynamic gate stress

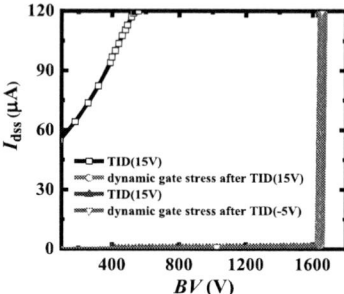

Figure 5. BV of the DUTs after TID irradiation and dynamic gate stress

Fig. 6 (a) shows that when V_{th} is strongly negatively affected after irradiation, the leakage current in the channel region increases sharply under BV testing. This could be resulted from the channel of the SiC MOSFETs is in a subthreshold state when V_{th} is seriously negative drift. Therefore, even if V_{gs} is 0 V during BV testing, an increase in the leakage current can result in a decrease in the breakdown voltage. Fig. 6 (b) shows that after dynamic gate stress, the leakage current of the DUTs is mainly concentrated at the PN junction. After V_{th} is recovered, the channel is no longer in a subthreshold state.

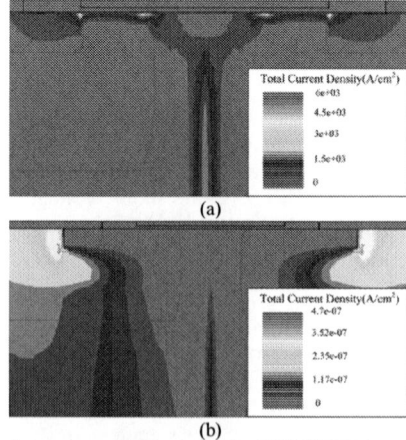

Figure 6. Leakage current distribution of the SiC MOSFETs under BV testing with (a) TID irradiation, (b) dynamic gate stress

Fig. 7 shows the $C\text{-}V$ curves of the SiC MOSFETs with respect to the initial time, TID irradiation and dynamic gate stress. During the measurement, the drain and source are grounded, while the gate bias scans from -15 V and 15 V. The frequency is 1 MHz, and a 25 mV AC voltage small signal is loaded on the gate to detect capacitance values. It can be seen that in the range of 0-5 V and -10 to -5 V in Fig. 7(a) and (b), after TID irradiation and dynamic gate stress, the curves first drift negatively and then drift positively. This can be explained by the fact that positive charges are injected into the gate oxide above the channel region and JFET region upon irradiation, and then negative charges are injected into the gate oxide after dynamic gate stress.

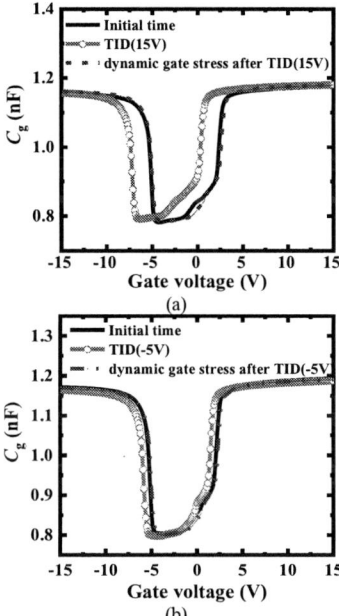

Fgure 7. *C-V* curves of the DUTs with an initial time, TID irradiation and dynamic gate stress of (a) 15 V, (b) -5 V

To further analyze the type of defect charges in the gate oxide, the interface traps and oxide trapped charges were extracted from the subthreshold curves. Since the degradation of the threshold voltage is mainly caused by the interface traps and the oxide trapped charges, the relationships are shown as[6]

$$\Delta V_{\text{th}} = \Delta V_{\text{it}} + \Delta V_{\text{ot}} \tag{1}$$

$$\Delta V_{\text{ot}} = \frac{q \Delta N_{\text{ot}}}{C_{\text{ox}}} \tag{2}$$

$$\Delta V_{\text{it}} = \frac{q \Delta N_{\text{it}}}{C_{\text{ox}}} \tag{3}$$

where ΔV_{it} is the voltage change caused by the interface charges and ΔN_{it} is the change in the interface trapped charge density. Similarly, ΔV_{ot} is the voltage variation caused by the oxide trapped charges, and ΔN_{ot} is the change in the oxide trapped charge density. C_{ox} is the gate oxide capacitance per unit area. According to equations (1), (2) and (3), the change trends of the SiC/SiO₂ interface defect charges are obtained, as shown in Fig. 8. Regardless of whether the gate voltage is 15 V or -5 V in the TID, the number of oxide trapped charges is much greater than the number of interface charges after TID irradiation. This indicates that the decrease in the static characteristics of the DUTs is mainly caused by the trapping of holes in the oxide traps, and the contribution of the interface traps to the decrease in the static parameters is limited. After dynamic gate stress, the densities of the two kinds of charges

are almost zero, which further verifies the recovery of the static parameters.

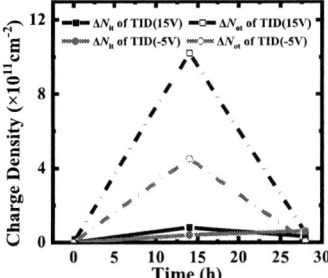

Figure 8. Variations in the interface charge (N_{it}) and oxide charge density (N_{ot}) in SiC MOSFETs after irradiation and AC gate stress

IV. CONCLUSION

This paper analyzes the effects of TID irradiation under different gate bias conditions and dynamic gate stresses on the static characteristics of SiC MOSFETs. The research results indicate that the threshold voltage, on-resistance and breakdown voltage of SiC MOSFETs are severely decreased by TID irradiation, mainly because of the trapping of holes by SiC/SiO₂ interface traps in SiC MOSFETs under irradiation. The degradation of the breakdown voltage is mainly attributed to the presence of a subthreshold current in the channel after TID irradiation. In addition, the static characteristics of SiC MOSFETs can be recovered completely by using dynamic gate stress. This phenomenon results from the decrease of the number of defect charges at the SiC/SiO₂ interface, which mainly include oxide trapped charges. In summary, the study of the degradation and recovery of static parameters under ionizing irradiation and dynamic gate stress presented in this work is highly important for understanding the gate oxide degradation mechanism and reliability design of SiC MOSFETs.

Acknowledgments

This work was supported in part by National Science Foundation of China under Grant 62334004 and the Guangdong Basic and Applied Basic Research Foundation under Grant 2023A1515010538.

REFERENCES

[1] Shi B, Ramones A I, Liu Y, et al. "A review of silicon carbide MOSFETs in electrified vehicles: Application, challenges, and future development," IET Power Electronics, vol. 16, pp. 2103-2120, May. 2023.

[2] Tan H, Zhang L, Tan R, et al. "A Deep Insight Into the Ionizing Radiation Effects and Mechanisms on the Dynamic Characteristics of SiC MOSFETs," IEEE Transactions on Electron Devices, vol. 71, pp. 1145-1152, Feb 2024.

[3] Zhong X, Jiang H, Qiu G, et al. "Bias temperature instability of silicon carbide power MOSFET under AC gate stresses," IEEE Transactions on Power Electronics, vol. 37, pp. 1198-2008, Feb 2022.

[4] Jiang H, Zhong X, Qiu G, et al. "Dynamic gate stress induced threshold voltage drift of silicon carbide MOSFET," IEEE Electron Device Letters, vol. 41, pp. 1284-1287, Sep 2020.

[5] Zhao H, Li X, Wu Y, et al. "Investigation of Inrush Current Induced Trench Gate Degradation inside SiC MOSFET by New Fowler-Nordheim Localization Methodology." IEEE Transactions on Power Electronics, vol. 39, pp. 4947-4951, May 2024.

[6] P. J. McWhorter, P. S. Winokur, "Simple technique for separating the effects of interface traps and trapped-oxide charge in metal-oxidesemiconductor transistors," Appl. Phys. Lett., vol. 48, pp. 133–135, Jan. 1986.

The Status of WBG Devices Towards Net-Zero Solutions

Mikael Östling

KTH Royal Institute of Technology, School of Electrical Engineering and Computer Science, Electrum 229, 164 40 Kista-Stockholm

Email: mostling@kth.se

Abstract— The world is facing a dramatic increase in the need for electricity and needs to battle the ever-increasing carbon dioxide emissions. New, more efficient, wide bandgap power electronics devices are developed to enable the urgent plans towards net-zero loss in electricity production, distribution and consumption. These devices will be very important pieces in the worldwide puzzle to enable fossil free electricity production, distribution and consumption. This paper discusses the status of wide bandgap device technology in the perspective of maturity of the devices for roll-out and implementation for industrial commercial use.

Keywords—**Wide bandgap devices, Device performance, Efficiency, Device applications.**

I. INTRODUCTION

The world is being electrified at a fast pace. The main driving force is to reduce the CO2 footprint. However, the energy mix is still not aggressive enough in generating fossil free electricity. The trend is, however, positive. In 2024 the energy mix is based on 61% fossil energy, 11% nuclear and 30% renewable. The fossil dependance is down by 7% units over the past 10 years. The global electricity consumption was 29480 TWh in 2023 [1]. What can we do to reduce electricity consumption from a semiconductor perspective? Quite a lot in fact.

If we assume that the world is doubling electricity consumption during the next 20 years, there is a total consumption of nearly 60000 TWh/year in 2045. By improving just 1% in the energy efficiency of the power devices in the whole chain from power generation, power distribution and power consumption, we can save about 600 TWh /year. In the textbook by Baliga [2] we learn that more than 50% of the electric power is controlled by power devices the total energy savings would then be 300 TWh/year. This corresponds to about 90 medium sized fossil fueled power plants (about 3-3,5 TWh/year). In practice the improved efficiency potential is much larger than 1%. The introduction of more efficient power semiconductors is speeding up. The two main wide bandgap (WBG) semiconductor materials silicon carbide (SiC) and gallium nitride (GaN) are rapidly finding commercial usage in a variety of important applications.

II. DEVICE APPLICATION AREAS FOR WBGS

A. SiC Device Applications

The most common application areas for SiC are:

Electric Vehicles (EVs): SiC MOSFETs are widely used in the powertrain of electric vehicles. Their ability to handle high voltages and currents with minimal losses makes them ideal for inverters, which convert DC from the battery to AC for the motor. Companies like Tesla and Toyota have integrated SiC MOSFETs in their EV systems to enhance efficiency, reduce cooling requirements, and increase driving range.

Renewable Energy Systems: SiC MOSFETs are employed in solar inverters and wind turbine converters to improve the efficiency of converting and managing power from renewable sources. Their high switching frequency and low losses reduce the size and cost of passive components like inductors and capacitors.

Industrial Power Supplies: SiC MOSFETs are used in industrial power supplies and uninterruptible power supplies (UPS) due to their high efficiency and compact size. These applications benefit from reduced cooling requirements and improved power density.

Data Centers: SiC MOSFETs are increasingly used in power conversion systems for data centers. Their efficiency and ability to handle high power densities make them suitable for power supplies and server power management units (PMUs), helping to reduce energy consumption and cooling costs.

Aerospace and Defense: The high-temperature and radiation-resistant properties of SiC make it suitable for aerospace and defense applications.

B. GaN Device Applications

For GaN devices we find these most frequent application areas:

Fast Chargers: GaN technology is widely used in fast chargers for smartphones, tablets, and laptops. These chargers are more efficient and compact than traditional silicon-based chargers, providing faster charging times and better thermal management.

Power Adapters: GaN power adapters are used in a variety of consumer electronics due to their smaller size and higher efficiency.

Telecommunications 5G Base Stations: GaN transistors are used in RF amplifiers for 5G base stations due to their high frequency and power capabilities.

Satellites and RF Communications: GaN devices are employed in satellite communication systems and RF transmitters because of their efficiency and ability to handle high power levels.

Power Supplies in Data Centers: GaN-based power supplies in data centers improve efficiency and reduce the physical footprint, leading to lower operational costs and improved thermal management.

979-8-3503-6184-1/24 $31.00 © 2024 IEEE

Electric Vehicles (EVs) On-Board Chargers: GaN devices are used in on-board chargers for electric vehicles to achieve higher efficiency and faster charging times.

III. WBG Materials Issues and opportunities

Both SiC and GaN are compound materials which means that the materials quality in the growth process is a challenge compared to silicon. Over the past 35 years we have seen a steady improvement in the materials quality of SiC wafers. The production of SiC wafers is today mainly at 150 mm wafer size but in a few years the bulk of the production will be at 200 mm wafer size. This is a rapid and impressive progress from wafers of poor quality at < 50 mm diameter in the early 90's. However, even to date, the materials quality is limiting the device performance and yield in some applications.

The table below summarizes some of the key advantages of using SiC and GaN versus Si. Most important advantage of the WBG materials is the high critical electric field, and for SiC also the high thermal conductivity.

Material Property	Symbol	Si	4H-SiC	GaN
Bandgap (eV at 300 K)	E_g	1.12	3.2	3.4
Critical electric field (MV/cm)	E_C	0.25	2.2	3
Electron mobility (cm²/V·cm)	μ_n	1350	1020	400
Hole mobility (cm²/V·cm)	μ_p	480	120	30
Saturation electric velocity (cm/s)	V_{sat}	1	2	2.5
Thermal conductivity at 300 K (W/cm·K)	λ	1.5	4.5	1.3
Relative permittivity	ε_r	11.9	10	9.5

Table 1: Materials parameters for Si, SiC and GaN [2-5].

The main benefit in device performance utilizing SiC and GaN for power devices is the possibility to use the almost 10 times higher electric field strength than in Si technology. By doping the drift layer in the device about 100 times higher than for the corresponding Si device, a nearly 100 times lower on-resistance can be achieved in a vertical MOSFET design.

A. SiC Materials Issues

Bipolar degradation during operation is still problematic for 4H-SiC power devices such as PiN diodes and transistors, including MOSFETs [6-8]. One major issue to overcome is the stepwise increase in the forward voltage drop during the forward bias operation [9]. This degradation is caused by expansion of stacking faults (SFs) during operation where a basal plane dislocation (BPD) is dissociated into a pair of partial dislocations (PDs) with a stacking fault in between [10].

Generally, most BPDs in SiC substrates can be converted to harmless threading edge dislocations (TEDs), at the epi/substrate interface by a proper epitaxial growth technology [11].

However, the conversion point where BPD converts to TED is critical. At high current density, the minority-carrier concentration (holes) can be high and reach the n+ substrate region [6,7,12]. This recombination process provides energy for growth of SFs at BPDs or at BPD/TED conversion points [6].

Therefore, the quality of the substrate as well as the design of epilayers are important issues.

The latest generations of SiC devices have been improved, for instance by utilizing thin highly doped layers with short minority carrier lifetime to prevent holes reaching the substrate in order not to exceed the threshold for SF formation [13].

It has also been confirmed that bipolar degradation can be suppressed by using novel high-quality substrates with inherent low BPD density [6]. However, there is presently a shortage of such "BPD free" substrates.

A key development breakthrough to yield BPD free substrates is highly desired and longed-for. Possibly, a new player in Sweden, the company Kiselkarbid i Stockholm AB (KISAB), is emerging with 4H-SiC wafers with very low or BPD free substrates [14].

B. GaN Materials Issues

When growing Gallium Nitride on Silicon and Silicon Carbide substrates, several critical material issues arise, impacting device performance, reliability, and manufacturing efficiency. An orientational summary of the most urgent material issues for each substrate type follows.

Growing GaN on Si: GaN and Si have a significant lattice mismatch (~17%), which leads to high dislocation densities in the GaN layer, which severely will impact on the device performance of both lateral and vertical GaN power devices. The thermal expansion coefficient mismatch between GaN and Si can cause thermal stress during cooling from the growth temperature, leading to cracking and warping of the GaN layer, and may result in too high dislocation densities and defects. The lower thermal conductivity provided by the Si substrate impairs the thermal management of the devices. On the positive side the cost reduction by growing GaN on Si is substantial.

Growing GaN on SiC: The lattice mismatch between GaN and SiC is much smaller (~3.4%), resulting in fewer dislocations compared to GaN on Si. However, dislocations still need to be managed to optimize device performance. SiC has high thermal conductivity making it suitable for high-power and high-frequency applications where efficient heat dissipation is critical. SiC substrates are more expensive and less readily available compared to Si. This can increase the overall cost of GaN-based devices and limit their widespread adoption.

The GaN epitaxial quality is in general better for GaN on SiC than GaN on Si, and is still the preferred choice for the sensitive RF power transistors with respect to harmonic distortion. Recent reviews and references on GaN on Si and GaN on SiC are provided [15-17].

IV. WBG Device Status

A. Commercial Devices

Power SiC devices are commercially available from several vendors in the voltage range 1.2 kV to 1.7 kV. The main device type is n-MOSFETs. These devices are often sold in single device packages of type TO-247 or similar. Most vendors are also producing rectifying diodes to be used together with the transistors. The diodes are of type Schottky Barrier Diodes (SBD) or Junction Barrier Schottky Diodes (JBS). In the references [18-26] some major companies' products can be found. A few examples of higher voltage MOSFETs are available. For example, GeneSiC Semiconductor (part of Navitas) has recently released a commercial 3.3 kV MOSFET [27].

Commercial GaN devices are mostly lateral High Electron Mobility Transistors (HEMT) [28-32], and the main products are in a lower voltage range than for SiC MOSFETs. The major offering from the commercial vendors is for

979-8-3503-6184-1/24 $31.00 © 2024 IEEE

applications from 48V to 650V. The most frequent applications are mobile electronics power supplies, EV charging, and 5G telecom RF power.

B. Research Devices for High Voltage

Worldwide, large efforts are spent on research and development of SiC devices for voltages between 5-20 kV. These applications are of particular interest for traction, power grid and other industrial applications. One trend is to utilize bipolar devices such as bipolar junction transistors (BJT) or insulate gate bipolar transistors (IGBT), and Gate Turn-off Transistors (GTO).

Many of the challenges in designing high-voltage devices are focusing on the junction termination extension (JTE), which controls and reduces the high electric surface field. The design is often a trade-off for cost since the JTE takes up a large area with no current flow. In figure 1, a 15 kV BJT is shown as a cross section depicting both the vertical design and the lateral JTE design. This transistor was optimized to yield a minimum JTE size.

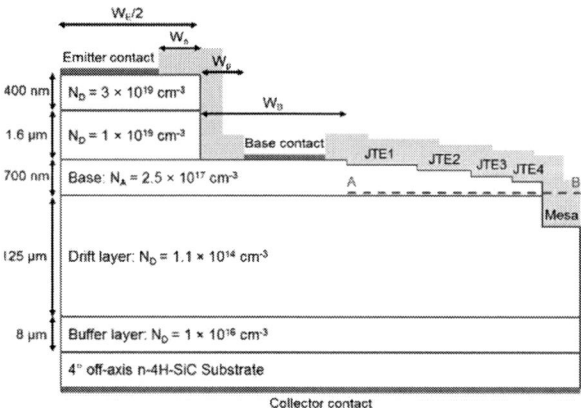

Figure 1. Cross-sectional view of a 15 kV BJT with optimized JTE [33].

Some of the recent device research papers are listed in the following references [34-38]. In Europe a new initiative to launch a pilot line for research devices in WBG materials and power modules has been granted [39].

V. SUMMARY AND CONCLUSIONS

Power efficiency is a key feature in the battle to reduce loss in electricity production, distribution and consumption. WBG devices are an efficient way to move. Massive improvements are under way in the WBG domain, with a market expanding by un-precedent speed. The WBG materials progress is very positive and production capability is quickly expanding. In a few years most WBG device production will be done on 200 mm wafer technology. The establishment of new fabrication facilities are launched in Europe, Asia and USA.

ACKNOWLEDGMENT

The author is grateful for the discussions and input by Margareta Linnarsson and Mattias Ekström and Carina Zaring.

REFERENCES

[1] https://ourworldindata.org/electricity-mix

[2] B Jayant Baliga. "Advanced High Voltage Power Device Concepts". Springer, Science & Business Media, 2011 https://doi.org/10.1007/978-1-4614-0269-5

[3] Hossein Elahipanah, "Design and Realization of 4H-SiC Bipolar Junction Transistors", PhD Thesis, 2017, KTH, ISBN 978-91-481-8

[4] R. Rupp, "Silicon Carbide Power Devices: Present and Future," *IEEE Journal of Emerging and Selected Topics in Power Electronics*, vol. 4, no. 3, pp. 523-529, Sep. 2016.

[5] T. Kimoto, "Material science and device physics in SiC technology for high-voltage power devices," *Japanese Journal of Applied Physics*, vol. 54, no. 4, p. 040103, 2015.

[6] K. Konishi, R. Fujita, K. Kobayashi et al.,), "Nucleation sites of expanding stacking faults detected by in operando x-ray topography analysis to design epitaxial layers for bipolar-degradation-free SiC MOSFETs", AIP Advances **12**, 035310 (2022)

[7] S.Torimi, Y.Obiyama, M.Tsukuda and I. Omura, "Numerical study of 4H-SiC PiN diod to enable forward bias degradation prediction considering BPD-TED conversion position in the SiC epitaxial layer", 2019 International Conference on Solid State Devices and Materials (SSDM2019).

[8] T. Neyer, M. Domeij, H. Das and S. Sunkari, "Is there a perfect SiC MOSFETs Device on an imperfect crystal?," *2021 IEEE International Reliability Physics Symposium (IRPS)*, Monterey, CA, USA, 2021, pp. 1-6, doi: 10.1109/IRPS46558.2021.9405098.

[9] T. Kimoto and H. Watanabe, "Defect engineering in SiC technology for high-voltage power devices", Appl. Phys. Express 13 (2020) 120101.

[10] T. Kimoto *et al.*, "Understanding and reduction of degradation phenomena in SiC power devices," *2017 IEEE International Reliability Physics Symposium (IRPS)*, Monterey, CA, USA, 2017, pp. 2A-1.1-2A-1.7

[11] S. Ha, P. Mieszkowski, M. Skowronski and L.B. Rowland, "Dislocation conversion in 4H silicon carbide epitaxy" Journal of Crystal Growth 244 (2002) 257.

[12] S. Palanisamy, T. Basler, J. Lutz, C. Künzel, L. Wehrhahn-Kilian and R. Elpelt, "Investigation of the bipolar degradation of SiC MOSFET body diodes and the influence of current density," *2021 IEEE International Reliability Physics Symposium (IRPS)*, Monterey, CA, USA, 2021, pp. 1-6, doi: 10.1109/IRPS46558.2021.9405183.

[13] A. Iijima and T. Kimoto, "Estimation of the critical condition for expansion/contraction of single Shockley stacking faults in 4H-SiC PiN diodes", Appl. Phys. Lett. 116 (2020) 092105, doi:10.1063/1.5143690

[14] https://kisabsemi.com/n-type-sic/

[15] Matteo Meneghini, et al.; GaN-based power devices: Physics, reliability, and perspectives. *J. Appl. Phys.* 14 November 2021; 130 (18): 181101. DOI: 10.1063/5.0061354

[16] M. Meneghini, G. Meneghesso, and E. Zanoni, "Review and Outlook on GaN and SiC Power Devices: Industrial State-of-the-Art and Future Directions," IEEE Trans. Electron Devices, vol. 64, no. 10, pp. 4055-4064, Oct. 2017. DOI: 10.1109/TED.2017.2730939.

[17] Atsunori Tanaka, et al.; Structural and electrical characterization of thick GaN layers on Si, GaN, and engineered substrates. *J. Appl. Phys.* 28 February 2019; 125 (8): 082517.https://doi.org/10.1063/1.5049393

[18] https://www.wolfspeed.com/products/power/sic-mosfets/#product-table

[19] https://www.wolfspeed.com/products/power/sic-schottky-diodes/

[20] https://www.st.com/en/power-transistors/stpower-sic-mosfets/products.html

[21] https://www.onsemi.com/products/discrete-power-modules/silicon-carbide-sic/silicon-carbide-sic-diodes

[22] https://www.onsemi.com/products/discrete-power-modules/silicon-carbide-sic/silicon-carbide-sic-mosfets

[23] https://www.coherent.com/materials/wide-bandgap-electronics/sic-power-devices/sic-mosfets

[24] https://www.rohm.com/products/sic-power-devices/sic-mosfet

[25] https://www.rohm.com/products/sic-power-devices/sic-schottky-barrier-diodes'

[26] https://www.infineon.com/cms/en/product/power/mosfet/

979-8-3503-6184-1/24 $31.00 © 2024 IEEE

[27] https://genesicsemi.com/sic-mosfet/

[28] https://www.infineon.com/cms/en/product/power/mosfet/

[29] https://www.qorvo.com/products/discrete-transistors/gan-hemts

[30] https://www.rohm.com/products/gan-power-devices#productGF

[31] https://www.transphormusa.com/en/products/

[32] https://navitassemi.com/gan-power-ics/

[33] Arash Salemi, "Silicon Carbide Technology for High- and Ultra-High-Voltage Bipolar Junction Transistors and PiN Diodes", PhD Thesis 2017, KTH, SBN 978-91-7729-183-1

[34] A. Salemi, H. Elahipanah, K. Jacobs, C. -M. Zetterling and M. Östling, "15 kV-Class Implantation-Free 4H-SiC BJTs With Record High Current Gain," in *IEEE Electron Device Letters*, vol. 39, no. 1, pp. 63-66, Jan. 2018, doi: 10.1109/LED.2017.2774139

[35] van Brunt, Edward, et al. "27 KV, 20 A 4H-SiC n-IGBTs." Materials Science Forum, vol. 821–823, Trans Tech Publications, Ltd., June 2015, pp. 847–850. doi: 10.4028/www.scientific.net/msf.821-823.847

[36] Hidenori Kitai *et al*, "A superior 15 kV SiC MOSFET with current spreading layer for high-frequency applications", 2019 *Jpn. J. Appl. Phys.* 58 SBBD16 DOI 10.7567/1347-4065/ab01d1

[37] N. Yun *et al.*, "Developing 13-kV 4H-SiC MOSFETs: Significance of Implant Straggle, Channel Design, and MOS Process on Static Performance," in *IEEE Transactions on Electron Devices*, vol. 67, no. 10, pp. 4346-4353, Oct. 2020, doi: 10.1109/TED.2020.3017150

[38] N. Watanabe, H. Okino, H. Shimizu and A. Shima, "Power Loss Reduction of N-Channel 10-kV SiC IGBTs With Box Cell Layout," in *IEEE Transactions on Electron Devices*, vol. 70, no. 7, pp. 3768-3773, July 2023, doi: 10.1109/TED.2023.3279799

[39] WBG Pilot Line, Chips Joint Undertaking, https://www.chips-ju.europa.eu/pilot-lines-detail/

Impact of the Resistive Silicon Base Wafer on Substrate Coupling in Power Integrated Circuits in GaN-on-Si Technology

Zijin Jiang[1], Rui (Ray) Yao[1,3], Miao Cui[1], Zhao Wang[2], Sang Lam*[1], and Stephen Taylor[3]

[1] Department of Electrical and Electronic Engineering; [2] Department of Communications and Networking; School of Advanced Technology, Xi'an Jiaotong-Liverpool University (XJTLU), Suzhou 215123, Jiangsu Province, China
[3] Department of Electrical Engineering and Electronics, The University of Liverpool, Liverpool L69 3GJ, UK

* Email: Zijin.Jiang20@student.xjtlu.edu.cn, Rui.Yao22@student.xjtlu.edu.cn, s.lam.cn@ieee.org

Abstract—We report a computational investigation into substrate coupling in power integrated circuits (ICs) in GaN-on-Si technology. Three-dimensional (3D) electromagnetic (EM) simulations have been performed to solve EM fields numerically. S-parameter results in addition to the electric field intensity distribution have been obtained to investigate the impact of the resistive silicon (Si) base wafer on the substrate coupling in GaN-on-Si technology. Various parameters including the conductivity and thickness of the resistive Si substrate as well as the GaN buffer layer have been adopted in the 3D EM simulations. The substrate coupling is found to be relatively small below 10 MHz but become non-negligible at high frequencies (with $|S_{21}|$ = -53 dB at 100 MHz). The EM coupling is mainly through the top GaN buffer layer rather than the resistive Si substrate underneath. The use of resistive Si base wafers of different thickness and conductivity values provide only slight improvement in the substrate coupling. Decreasing the conductivity of the GaN buffer layer is more helpful in the reduction of the substrate coupling by about 20 dB to 60 dB for $|S_{21}|$ at all simulated frequencies from 1 to 100 MHz.

Keywords—*substrate coupling, gallium nitride (GaN), GaN-on-Si technology, resistive silicon substrate, power integrated circuits (ICs), electromagnetic (EM) simulation*

I. INTRODUCTION

Gallium nitride (GaN) as a wide bandgap semiconductor material [1] has emerged as a promising solution for making power electronic devices and circuits, which are essential for various important applications in power conversion systems [2]. Compared with power transistors based on bulk silicon (Si) technology, GaN-based technology has the device option of high electron mobility transistors (HEMTs) which can allow performance breakthrough in high frequency power electronics [3]. There has been development of GaN-based power integrated circuits (ICs) operating at high frequencies from 2 MHz to 50 MHz and beyond [4]-[7]. Despite all the advantages of the monolithic solution, substrate coupling and crosstalk can be problematic [8]-[13] in GaN power ICs with high-voltage and low-voltage devices sharing the same substrate . This can be a particular acute issue in GaN-on-Si technology in which the base wafer is resistive Si as a low-cost substrate with relatively good thermal conductivity.

Over the past decade, both theoretical and experimental research has been reported on substrate coupling and crosstalk in GaN-on-Si technology. However, the research studies mainly focus on voltage or current distortions [9]-[10] about the substrate coupling. Little research has been done to investigate the substrate coupling using S-parameters, especially when the frequency reaches multi-MHz and beyond. In this work, we perform electromagnetic (EM) investigation to provide a systematic analysis of the influence of the resistive Si base wafer on the substrate coupling in GaN-based power ICs. The analysis of S-parameter results as well as the electric field intensity distribution can provide insight into the substrate coupling up to 100 MHz beyond the equivalent circuit models [8] can tell.

II. GAN-ON-SI STRUCTURE

In our EM investigation of the resistive Si base wafer's impact on substrate coupling in GaN power ICs, we adopted the structure consisting of three layers as shown in Fig. 1. There are one metallic base layer for grounding the IC at the bottom, a p-type Si base wafer in the middle, and a GaN buffer layer above. Two aluminium (Al) strips are in contact with the GaN buffer layer to represent terminals (e.g. drain and gate) of two separate HEMTs sharing the same substrate with certain lateral distance (700 μm) apart from each other. In the actual fabricated devices [14], there was a 25-nm AlGaN barrier layer between the Al strips and the GaN buffer layer. Since the AlGaN layer is too thin and ohmic contacts are formed between the Al strips and the GaN layer, the effect of this AlGaN layer is negligible. The two Al strips have equal area and serve as two 50-ohm signal ports (one with voltage excitation of 1.0 V and the other for detection) in our EM coupling investigation. Therefore, any electric field computed at port 2 compared with that at port 1 will reveal the EM coupling in the structure. In our investigated structure, the resistive Si base wafer's thickness varies from 50 μm to 200 μm with different conductivity values, and the GaN buffer layer's thickness from 1 μm to 5 μm also with different conductivity values. This to find out the impact of especially the resistive Si base wafer on the substrate coupling.

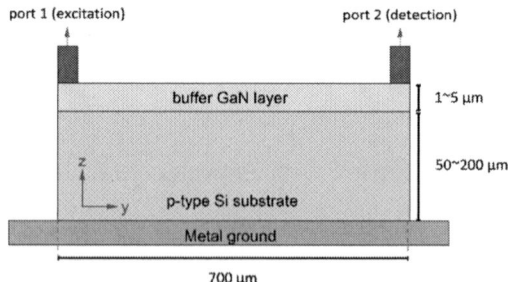

Fig. 1. Schematic cross-sectional diagram of the GaN-on-Si structure for investigating the substrate coupling

III. 3D EM SIMULATION & S-PARAMETER RESULTS

In our computational investigation, three-dimensional (3D) EM simulation was conducted using Ansys high-frequency structure simulator (HFSS). The signal frequencies span from

979-8-3503-6184-1/24 $31.00 © 2024 IEEE

1 MHz to 100 MHz which are typical operation frequencies of GaN-based power ICs [4]-[7]. As for the materials, the buffer GaN layer is set with $\varepsilon_{rGaN} = 8.9$ and $\sigma_{GaN} = 2000$ S/m; and the p-type Si base wafer is set with $\varepsilon_{rSi} = 11.9$ and $\sigma_{Si} = 2000$ S/m. HFSS uses the finite element method (FEM) to subdivide structures into many smaller subsections and solve the Maxwell's equations across inter-element boundaries between the finite elements, thus finding a solution for the EM fields at different spatial points in the simulated structure.

Fig. 2 shows the electric (E) field intensity (magnitude) plot of the structure's cross-section in the front view with logarithmic scale to better observe the changes. It is obvious that the E-field magnitude exhibits a non-linear decrease along both the y-axis and z-axis from the excitation port. However, a significant drop in E-field magnitude occur at the boundary between the GaN buffer layer and the resistive Si substrate.

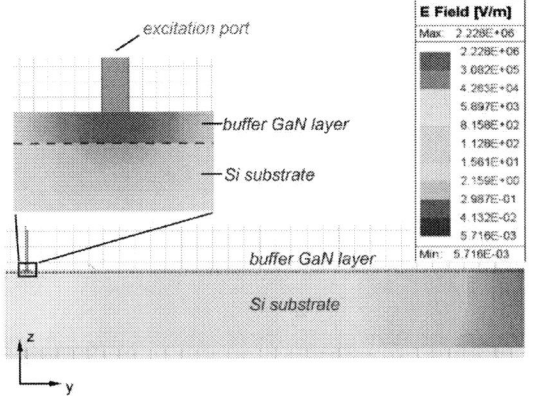

Fig. 2. Schematic cross-sectional view with electric field intensity distribution (on logarithmic scale) of the GaN-on-Si structure

Fig. 3 shows E-field magnitude at the increasing lateral distance away from the excitation port. It drops rapidly within the first 20 µm away from the excitation port and decreases much slower after. The E-field seems to weaken inversely with the distance. In varying the resistive Si substrate thickness from 100 µm, the E-field magnitude increases considerably at large lateral distance beyond 200 µm away from the excitation port. With HFSS, we can directly compute S-parameters using EM fields and we concentrate on $|S_{21}|$ which is a good measure of the substrate coupling from the excitation to detection ports.

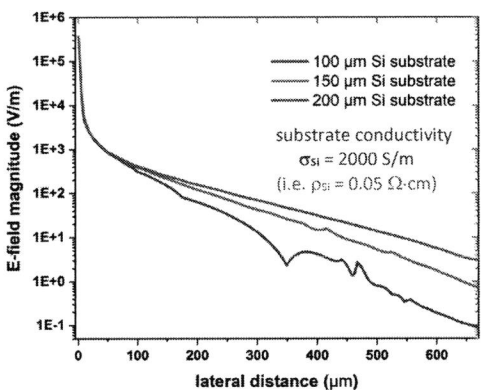

Fig. 3. E-field magnitude curves at 5 MHz along lateral line in the middle of the GaN buffer layer, starting from the excitation to detection ports, for the Si substrate thickness of 100 µm, 150 µm and 200 µm

Fig. 4 shows the $|S_{21}|$ results from 1 MHz to 100 MHz with the resistive Si substrate thickness of 50 µm, 100 µm, 150 µm and 200 µm. At relative high frequencies beyond about 10 MHz, the frequency dependence of $|S_{21}|$ has the same trend: increasing $|S_{21}|$ as the frequency increases. The resistive Si substrate of 200-µm thickness results in the largest $|S_{21}|$ value (meaning the worst EM coupling among the four Si substrate thickness cases), with $|S_{21}| = -70$ dB at 1 MHz while it is 20 dB lower (i.e. $S_{21}| = -91$ dB for a 100-µm thick Si substrate. The thicker Si substrate (200 µm or 150 µm) is likely to allow a larger conduction current flow in the resistive Si even though the current density ($J_{cond} = \sigma_{Si}E_{Si}$) might be relatively small related to the E-field. Such $|S_{21}|$ results are consistent with the E-field magnitude plot (Fig. 3), for a weaker E-field yields lower $|S_{21}|$ levels. As for a 50-µm Si substrate, the E-field magnitude is expected to increase considerably because of the overall much thinner structure. Then a relatively higher level of $|S_{21}|$ is resulted (Fig. 4). Neither a very thick (200 µm) nor a very thin (50 µm) Si substrate is good. These results imply that an optimum Si substrate thickness should be adopted to minimise the substrate coupling.

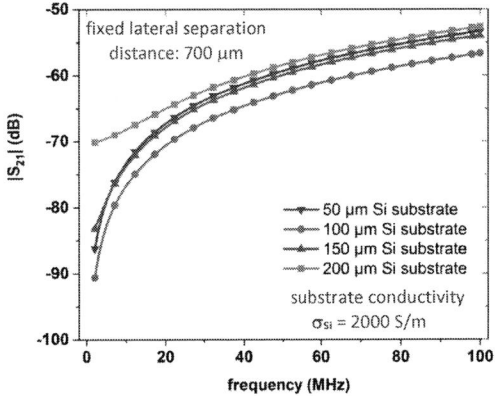

Fig. 4. $|S_{21}|$ results of the GaN-on-Si device structure from 1 to 100 MHz for different Si substrate thickness (with $\sigma_{Si} = 2000$ S/m, i.e. $\rho_{Si} = 0.05$ Ω·cm)

Fig. 5 shows the influence of GaN buffer layer thickness on the $|S_{21}|$ results. It can be observed that a thicker GaN buffer layer (up to 6 µm) contributes to somewhat smaller $|S_{21}|$ values, indicating weaker substrate coupling. However, the improvement in $|S_{21}|$ is not significant (only -0.5 dB/µm in this case of the GaN buffer layer thickness). As it is expensive to deposit a thicker GaN buffer layer, it is not cost effective to use a thick GaN buffer layer to suppress the substrate coupling.

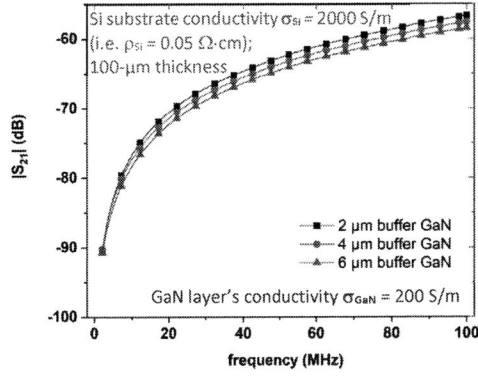

Fig. 5. $|S_{21}|$ results of the GaN-on-Si structure from 1 to 100 MHz for different thickness of the GaN layer (with $\sigma_{GaN} = 200$ S/m, i.e. $\rho_{GaN} = 0.5$ Ω·cm)

979-8-3503-6184-1/24 $31.00 © 2024 IEEE

Fig. 6 shows that if the GaN buffer layer has a smaller conductivity (σ_{GaN}) value, the substrate coupling would be strongly suppressed: $|S_{21}|$ can be massively lowered by 70 dB if the GaN buffer layer's conductivity decreases from 200 S/m to 2 S/m. This implies that the EM coupling is predominantly through the GaN buffer layer. The smaller σ_{GaN} of the GaN buffer layer would mean the EM coupling through the displacement current rather than the conduction current (J_{cond}).

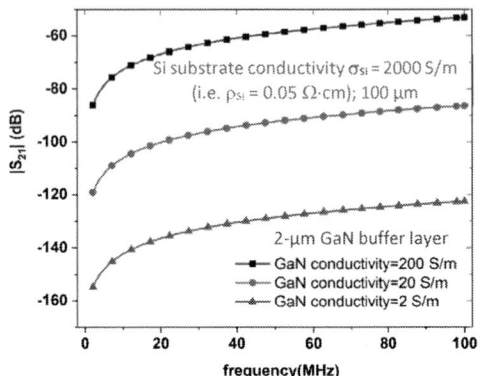

Fig. 6. $|S_{21}|$ results of the GaN-on-Si structure from 1 to 100 MHz with different conductivity (σ_{GaN}) values of the GaN buffer layer, while the Si substrate conductivity is fixed at $\sigma_{Si} = 2000$ S/m (i.e. $\rho_{Si} = 0.05\ \Omega \cdot$cm)

Fig. 7 shows that a smaller Si substrate conductivity (σ_{Si}) also results in somewhat weaker substrate coupling, but the improvement is not significant. This indicates that the influence of the resistive Si base wafer on the substrate coupling is not strong. It also implies the flexibility of using low- or medium-resistivity Si base wafers ($\rho_{Si} = 1$ to 50 m$\Omega \cdot$cm for fabricating power devices [15]), as far as minimising the substrate coupling is concerned in GaN-on-Si power ICs. $|S_{21}|$ can be lowered with $\rho_{Si} = 500$ m$\Omega \cdot$cm or beyond for the Si substrate. However, the use of high resistivity Si base wafers have other issues such as wafer bow and warpage for development of power ICs, in addition to the relatively higher wafer cost.

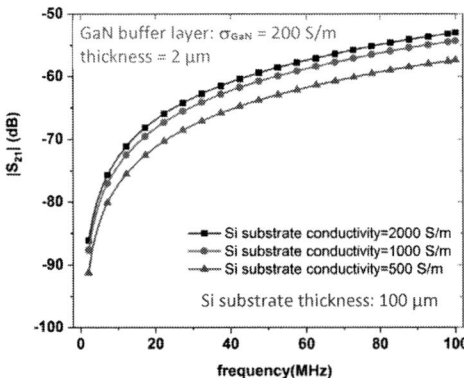

Fig. 7. $|S_{21}|$ results of the GaN-on-Si structure with different conductivity values of the Si substrate, while $\sigma_{GaN} = 200$ S/m for a 2-μm GaN layer

IV. CONCLUSION

We have used FEM-based 3D EM simulation to investigate the substrate coupling in GaN-on-Si technology from 1 MHz to 100 MHz, looking at particularly the influence of different conductivity values and thickness of the resistive Si base wafer. It is found that the substrate coupling is predominantly through the GaN buffer layer. The impact of the resistive Si base wafer on the substrate coupling only come into place when there can be no more reduction of the coupling through the GaN buffer layer. At 100 MHz, the substrate coupling gives $|S_{21}| = -50$ dB or below. Meanwhile, increasing the thickness of the resistive Si base wafer provides little help in suppressing the substrate coupling. Decreasing the resistive Si base wafer's conductivity is not as helpful as in decreasing the GaN buffer layer's conductivity to vastly lower $|S_{21}|$. These results are helpful for consideration of low- or medium-resistivity Si base wafers in the design and manufacturing power ICs in GaN-on-Si technology, to avoid signal interference via substrate coupling.

ACKNOWLEDGMENT

This work is supported in part by PGRS funding (FOSA2406036) of XJTLU. Both R. Yao and S. Lam acknowledge various help from Prof. Alex M. H. Wong of City University of Hong Kong, especially for R. Yao's training in the use and access to Ansys HFSS within CRAE Laboratory.

REFERENCES

[1] T. J. Flack et al., "GaN technology for power electronic applications: a review," J. Electronic Materials, 45(6), pp. 2673-2682, June 2016.

[2] N. Mohan, T. Undeland, and W. Robbins. Power Electronics: Converters, Applications, and Design, editor B. Zobrist (Hoboken: Wiley, 2003).

[3] Yijie Wang et al., "A review of high frequency power converters and related technologies," IEEE Open Journal of the Industrial Electronics Society, vol. 1, pp. 247-260, September 2020.

[4] S. Ujita, Y. Kinoshita, H. Umeda et al., "A fully integrated GaN-based power IC including gate drivers for high-efficiency DC–DC Converters," Proc. of IEEE Symposium on VLSI Circuits, June 2016, pp. 1–2.

[5] Y. Zhang, M. Rodriguez, and D. Maksimovic, "Very high frequency PWM buck converters using monolithic GaN half-bridge power stages with integrated gate drivers," IEEE Transaction on Power Electronics, vol. 31, no. 11, pp. 7926-7942, November 2016.

[6] Tz-Wun Wang, Yu-Yung Kao, Sheng-Hsi Hung et al., "Monolithic GaN-based driver and GaN switch with diode-emulated GaN technique for 50-MHz Operation and sub-0.2-ns deadtime control," IEEE Journal of Solid-State Circuits, vol. 57, no. 12, pp. 3877-3888, December 2022.

[7] Yifei Zheng, Boyu Li, Qianheng Dong et al., "A 200-V half-bridge monolithic GaN power IC with high-speed level shifter and dVS/dt noise immunity enhancement structure," IEEE Tran. on Very Large Scale Integration (VLSI) Systems, vol. 32, no. 3, pp. 542-551, 2024.

[8] S. Aamir Ahsan et al.,, "Analysis and modeling of cross-coupling and substrate capacitances in GaN HEMTs for power-electronic applications," IEEE Tran. Electron Devices, 64(3), 816-823, Mar. 2017.

[9] J. Wei, M. Zhang, G. Lyu and K. J. Chen, "Substrate effects in GaN-on-Si integrated bridge circuit and proposal of engineered bulk silicon substrate for GaN power ICs," 2020 IEEE Workshop on Wide Bandgap Power Devices & Applications in Asia, Suita, Japan, 2020, pp. 1-4.

[10] Q. Jiang, Z. Tang, C. Zhou, S. Yang and K. J. Chen, "Substrate-coupled cross-talk effects on an AlGaN/GaN-on-Si smart power IC platform," IEEE Tran. on Electron Devices, 61(11), 3808-3813, Nov. 2014.

[11] R. Sun, J. Lai, W. Chen et al., "Crosstalk suppression in monolithic GaN devices based on inverted E-field decoupling," IEEE Transactions on Electron Devices, vol. 68, no. 4, pp. 1542 - 1549, April 2021.

[12] D. Pagnano, G. Longobardi et al., "Suppression of substrate coupling in GaN high electron mobility transistors (HEMTs) by hole injection from the p-GaN gate " Appl. Phys. Lett. 115, 203502 (2019), November 2019.

[13] M. Cui and S. Lam, "Use of DC probes for multi-MHz measurements of crosstalk and substrate coupling in gallium nitride power integrated circuits," 2024 IEEE 36th International Conference on Microelectronic Test Structures (ICMTS), Edinburgh, United Kingdom, 2024, pp. 1-5.

[14] M. Cui, Y. Cai, S. Lam et al., "Characterization of transient threshold voltage shifts in enhancement- and depletion-mode AlGaN/GaN metal-insulator-semiconductor (MIS)-HEMTs," 2018 IEEE Intl. Conf. on Electron Devices & Solid State Circuits, Shenzhen, China, 2018, pp. 1-2.

[15] N. Machida, "Si wafer technology for power devices: A review and future directions," 2018 IEEE 30th Intl. Symposium on Power Semiconductor Devices and ICs (ISPSD), Chicago, IL, USA, 2018, pp. 12-1.

A Novel Snapback-free Double-RESURF Reverse-conducting LIGBT with Dual Conduction Paths

Yun Xia*[1], Yuxi Wan [1], Wei Zeng [1], Yu Shi [1], Xiaoping Wang [1], Wei Liu [1], Haizhao Zhi [1], Ziwei Zhou [1],
Xixi Luo [1], Ruize Sun [2], Xiaoming Wang [2], Yan Wang [3], Wanjun Chen [2].

[1] Shenzhen Pinghu Laboratory, Shenzhen 518111, China.
[2] State Key Laboratory of Electronic Thin Films and Integrated Devices, University of Electronic Science and Technology of China, Chengdu 610054, China
[3] School of Integrated Circuits, Tsinghua University, Beijing 100084, China

* Email: xiayun@phlab.com.cn

Abstract—To achieve snapback-free and reduce switching loss, a novel double-RESURF reverse-conducting lateral insulated gate bipolar transistor with dual conduction paths (DCP RC-LIGBT) is proposed and investigated. By separating the device into the LIGBT part and the LDMOS part with an oxide layer, the proposed device exhibits dual conduction paths. In the forward conduction state, unipolar conduction only exhibits in the LDMOS part, and bipolar conduction only exhibits in the LIGBT part, thus no snapback phenomenon happens. Besides, due to its bipolar conduction is not influenced and part of the forward conduction current being unipolar current, a better tradeoff relationship between forward conduction voltage (V_F) and turn-off loss (E_{OFF}) is obtained. In the reverse conduction state, the reverse conduction current is solely exhibited in the LDMOS part, the narrowed reverse conduction area results in a significantly reduced reverse recovery charge (Q_{RR}). As a result, the performance of DCP RC-LIGBT is significantly improved. The simulation results show that compared to the double-RESURF RC-LIGBT with trench barrier and shorted anode (TBSA RC-LIGBT), the proposed device achieves snapback-free forward conduction performance, a 45.0% reduction in turn-off loss at forward conduction voltage (V_F) of 1.32V, and achieves a 69.8% reduction in Q_{RR}.

Keywords—RC-LIGBT, double-RESURF, turn-off loss, snapback phenomenon, reverse recovery charge.

I. INTRODUCTION

Reverse conducting lateral insulated gate bipolar transistor (RC-LIGBT) integrates the freewheeling diode and LIGBT in one single chip by introducing the N^+ anode in the anode region of the LIGBT, which enables it with reverse conducting capability [1-6]. However, the N^+ anode causes a parasitic laterally diffused metal oxide semiconductor (LDMOS) in conventional RC-LIGBT. As the N^+ anode and the P^+ anode are shorted, the turning on of the conventional RC-LIGBT is switched from unipolar conduction mode (LDMOS mode) to the bipolar conduction mode (LIGBT mode), and the rapid decrease of on-state resistance results in an undesirable snapback problem, which hinders its application [1-6]. Suppressing the conduction of the N^+ anode, i.e. hindering the unipolar conduction is used to relieve this problem, for example, the RC-LIGBT with trench barrier and shorted anode (TBSA RC-LIGBT) introduced additional anode resistance (R_{anode}) between the N^+ anode and P^+ anode, which successfully suppress the snapback phenomenon. However, its reverse conduction performance has to be compromised [1]. Thus, there is a tradeoff relationship between forward conduction performance and reverse conduction performance and this tradeoff relationship will get worse when double-RESURF technology is used [7].

To solve this, a novel double-RESURF RC-LIGBT with dual conduction paths (DCP RC-LIGBT) is proposed and investigated in this paper. By separating the device into the LIGBT part and the LDMOS part with an oxide layer, the proposed device exhibits dual conduction paths. In the forward conduction state, unipolar conduction only exhibits in the LDMOS part, and bipolar conduction only exhibits in the LIGBT part, thus no snapback phenomenon happens, and the bipolar conduction is not undermined. As a result, the performance of the proposed DCP RC-LIGBT can be significantly improved.

II. DEVICE STRUCTURES AND MECHANISMS

Fig. 1(a) shows the schematic cross-sectional views of the proposed DCP RC-LIGBT. For comparison, the double-RESURF LDMOS, double-RESURF LIGBT, and the TBSA RC-LIGBT with double-RESURF technology are also studied in this paper, as shown in Fig. 1(b), 1(c), and 1(d), respectively.

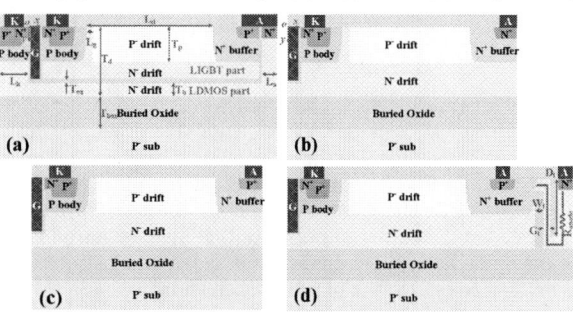

Fig. 1. Schematic views of (a) proposed DCP RC-LIGBT, (b) double-RESURF LDMOS, (c) double-RESURF LIGBT, and (d) double-RESURF TBSA RC-LIGBT.

As shown in Fig. 1(a), the proposed device is divided into two parts: the LIGBT part and the LDMOS part by an oxide layer. Both parts contribute to the conduction of the forward conduction current while the reverse conduction current is only conducted in the LDMOS part.

In the forward conduction state, the LDMOS part turns on first. With the increase of anode voltage, the LIGBT part gradually turns on. The conduction of both parts does not interact with each other, thus no snapback phenomenon occurs and the bipolar conduction is not undermined by the shorted N^+ anode. While in the TBSA RC-LIGBT, as depicted in Fig. 1(d), its forward performance is greatly influenced by the R_{anode}. A high R_{anode} is needed to suppress the conduction of the N^+ anode. If a low R_{anode} is used, the conduction of the P^+ anode will be influenced by the N^+ anode, which may cause a snapback phenomenon, and undermine the conductivity

979-8-3503-6184-1/24 $31.00 © 2024 IEEE

modulation in the drift region, leading to a poor tradeoff relationship between forward conduction voltage (V_F) and turn-off loss (E_{OFF}).

In the reverse conduction state, the current of the DCP RC-LIGBT is solely conducted in the LDMOS part. While in the TBSA RC-LIGBT, the current is conducted in the whole drift region. Compared with the TBSA RC-LIGBT, the reverse conduction path in the DCP RC-LIGBT is narrowed, which induces reduced hole carriers stored in the drift region as well as reduced reverse recovery charge (Q_{RR}).

III. RESULTS AND DISCUSSION

Device simulation and mixed-mode simulation results are obtained by using MEDICI TCAD. Calibrated models including IMPACT.I, CONSRH, AUGER, BGN, CONMOB, ANALYTIC, FLDMOB, and SRFMOB2 are all considered [1]. Optimized key device parameters are listed in Table i. If it is not specified, the carrier lifetime of 1 µs is set for the studied devices.

Table I. Key device parameters

Parameters	LIGBT.	LDMOS.	TBSA.	DCP.
T_{box} (µm)	3	3	3	3
T_d (µm)	4	4	4	4
L_d (µm)	14	14	14	14
T_p (µm)	2	2	2	2
L_g (µm)	0.5	0.5	0.5	0.5
$L_a/L_k/T_{ox}/T_b$ (µm)	/	/	/	1/1/0.05/0.5
$D_t/W_t/G_t$ (µm)	/	/	3.5/0.2/0.5	/
N⁻ drift doping ($\times 10^{15} cm^{-3}$)	10.3	10.3	10.3	10.3
P⁻ drift doping ($\times 10^{15} cm^{-3}$)	7	7	7	7

The forward-blocking characteristic of the studied devices is shown in Fig. 2. The DCP RC-LIGBT exhibits a similar blocking characteristic as the other studied devices. As shown in the inset of Fig. 2, both the LIGBT part and LDMOS part of the DCP RC-LIGBT are depleted and the breakdown point is located at the corner of the N⁺ buffer, thus the introduction of the oxide layer between the LIGBT part and LDMOS part is not detrimental to the blocking characteristic.

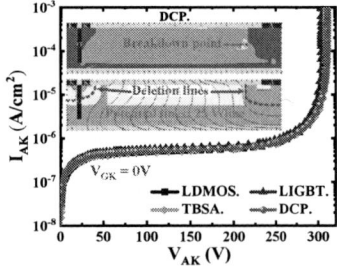

Fig. 2. Forward-blocking characteristic.

The forward conduction characteristic of the studied devices is shown in Fig. 3(a). The snapback phenomenon occurs in the TBSA RC-LIGBT, while the proposed device exhibits a snapback-free curve. The current line distribution of different current densities in Fig. 4(a) shows that, with the increase of current density, the conduction mode of TBSA RC-LIGBT changes from the unipolar conduction mode to bipolar mode, which causes a snapback phenomenon. While

in the proposed DCP RC-LIGBT, unipolar conduction is limited in the LDMOS part, and bipolar conduction is limited in the LIGBT part, thus the conduction of these two parts is independent as shown in Fig. 3(b) and Fig. 4(a), consequently, snapback-free is achieved. Since its bipolar conduction is not influenced by its unipolar conduction, thus the proposed DCP RC-LIGBT obtains a 10.8% reduction in V_F at $I_{AK} = 150$ A/cm² compared with the TBSA RC-LIGBT, as shown in Fig. 3(a).

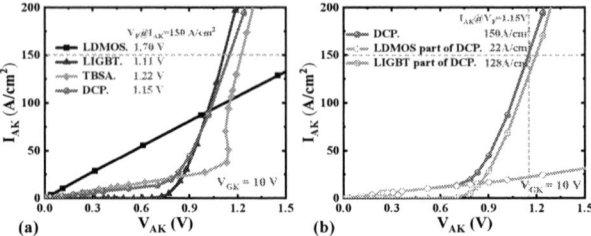

Fig. 3. (a) Forward conduction characteristic. (b) Forward conduction current distribution of the DCP RC-LIGBT.

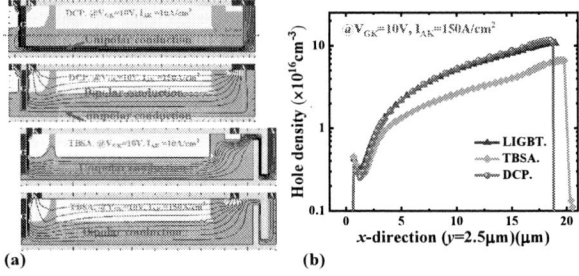

Fig. 4. (a) Forward conduction current lines distribution at different current densities. (b) Hole density distribution.

The turn-off characteristics of the studied devices are shown in Fig. 5. The proposed DCP RC-LIGBT exhibits a faster turn-off transient compared with the LIGBT and TBSA RC-LIGBT, so its E_{OFF} can be reduced. The improved hole density distribution as shown in Fig. 5(b) accounts partly for the reduced E_{OFF}. As shown in Fig. 5(b), due to the conductivity modulation in the DCP RC-LIGBT is not weakened by the shorted N⁺ anode, the DCP RC-LIGBT achieves a similar hole distribution as the LIGBT, which helps for the extraction of hole carriers during the turn-off period compared with the TBSA RC-LIGBT. Besides, since nearly 15% current of the DCP RC-LIGBT is unipolar current which is conducted in the LDMOS part, its turn-off performance is improved compared with the LIGBT.

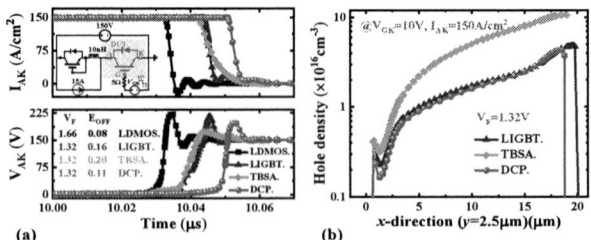

Fig. 5. (a) Turn-off characteristic. The V_F of the LIGBT, the TBSA RC-LIGBT, and the DCP RC-LIGBT are set to 1.32V by adjusting the doping of the P⁺ anode. (b) The hole density distribution at the beginning of turn-off.

The tradeoff relationships between the V_F and E_{OFF} of the studied devices are shown in Fig. 6. The proposed device obtains a better tradeoff relationship at both T = 300 K and

400 K than the LIGBT and the TBSA RC-LIGBT. With the same V_F of 1.32 V, the E_{OFF} at T = 300 K of the proposed device reduces by 31.3% and 45.0% in comparison with the LIGBT and TBSA RC-LIGBT. Though the LDMOS shows the lowest E_{OFF}, however, due to the lack of bipolar conduction, its V_F is greatly influenced by the temperature, while the V_F of the proposed device shows less temperature dependence because its bipolar conduction is dominant.

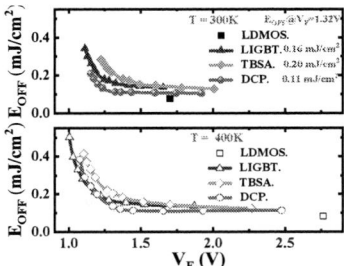

Fig. 6. V_F-E_{OFF} tradeoff relationship.

The reverse conduction curve of the LDMOS, the TBSA RC-LIGBT, and the DCP RC-LIGBT is shown in Fig. 7(a). The proposed device achieves a smaller reverse conduction voltage (V_R) than TBSA RC-LIGBT even though its reverse conduction path is significantly narrowed. This is because a high R_{anode} is used in the TBSA RC-LIGBT to relieve the snapback phenomenon in the forward conduction state, while its reverse conduction current needs to go through the R_{anode}, thus a high R_{anode} results in a high V_R. As depicted in Fig. 7(b), the conduction path of the DCP RC-LIGBT is limited in the LDMOS part, while in the LDMOS and the TBSA RC-LIGBT, their drift region is fully conducted, which causes a great number of hole carriers stored in their drift region.

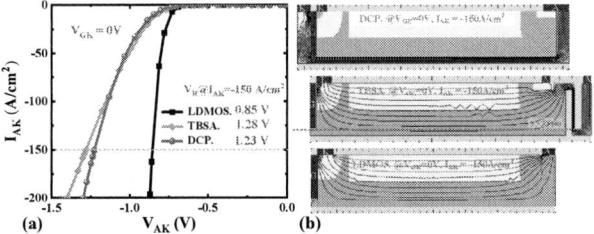

Fig. 7. (a) Reverse conduction characteristic. (b) Reverse conduction current lines distribution.

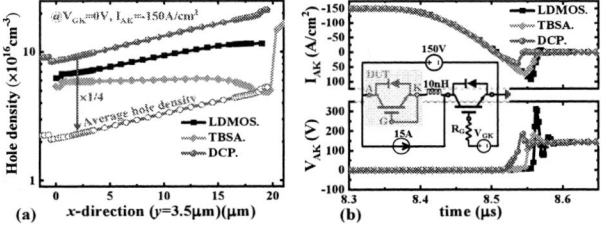

Fig. 8. (a) Reverse recovery characteristic. (b)The hole density distribution at the beginning of the reverse recovery process

With the thickness of the reverse conduction path dramatically shrunk by about 3/4, the average hole density in the drift region of the DCP RC-LIGBT is dramatically reduced as shown in Fig. 8(a). This results in a significantly reduced Q_{RR}, as shown in Fig. 8(b). The Q_{RR} of the proposed device is 0.83 µC/cm², reducing by 65.1% and 69.8%

compared to the 2.38 µC/cm² of the LDMOS and the 2.68 µC/cm² of the TBSA RC-LIGBT, respectively.

Fig.9 shows the influence of P⁻ drift doping and N⁻ drift doping on the performance of DCP RC-LIGBT. The V_F and V_R are less influenced by the P⁻ drift doping and the N⁻ drift doping, this is because its reverse conduction current and major forward conduction current is bipolar current. Its breakdown voltage (BV) shows high tolerance to the variation of the P⁻ drift doping and the N⁻ drift doping. Its E_{off} can be improved by increasing the P⁻ drift doping. Its Q_{RR} can be reduced by reducing the P⁻ drift doping and the N-drift doping.

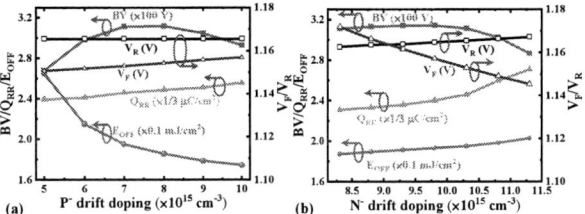

Fig. 9. The influence of (a) P⁻ drift doping and (b) N⁻ drift doping on the performance of DCP RC-LIGBT

IV. CONCLUSIONS

A novel double-RESURF RC-LIGBT with dual conduction paths is proposed in this paper. With an oxide layer introduced in the drift region, the proposed DCP RC-LIGBT is divided into the LDMOS part and the LIGBT part, unipolar conduction in the forward conduction state only exhibits in the LDMOS part, while bipolar conduction only exhibits in the LIGBT part, thus no snapback phenomenon happens, and the bipolar conduction is not influenced, which results in a better V_F-E_{OFF} tradeoff relationship. Compared to the double-RESURF TBSA RC-LIGBT, the proposed device achieves snapback-free performance and a 45.0% reduction in E_{OFF} at V_F of 1.32 V. In the reverse conduction state, the current in the proposed device is only exhibited in the LDMOS part. Because its reverse conduction path is about 1/4 the width of that of the double-RESURF TBSA RC-LIGBT, thus it achieves a 69.8% reduction in Q_{RR}. Consequently, the proposed device achieves snapback-free conduction performance and dramatically reduces switching losses.

REFERENCES

[1] Y. Xia, W. Chen, C. Liu, R. Sun, Z. Li and B. Zhang, "An Ultralow Loss Reverse-Conducting LIGBT With Embedded P-P-N Diode in Oxide Trench," in IEEE Transactions on Electron Devices, vol. 69, no. 12, pp. 6956-6962.

[2] J. K. O. Sin and S. Mukherjee, "Lateral insulated-gate bipolar transistor (LIGBT) with a segmented anode structure," in IEEE Electron Device Letters, vol. 12, no. 2, pp. 45-47, Feb. 1991.

[3] J. Zhu et al., "Electrical characteristic study of an SOI-LIGBT with segmented trenches in the anode region", IEEE Trans. Electron Devices, vol. 63, no. 5, pp. 2003-2008, May 2016.

[4] L. Sun, B. Duan and Y. Yang, "Novel Snapback-Free SOI LIGBT With Shorted Anode and Trench Barriers," in IEEE Transactions on Electron Devices, vol. 68, no. 5, pp. 2408-2413, May 2021.

[5] M. R. Simpson, "Analysis of negative differential resistance in the I-V characteristics of shorted-anode LIGBT's," in IEEE Transactions on Electron Devices, vol. 38, no. 7, pp. 1633-1640, July 1991.

[6] Jung-Hoon Chul, Dae-Seok Byeon, Jae-Keun Oh, Min-Koo Han and Yearn-Ik Choi, "A fast-switching SOI SA-LIGBT without NDR region," 12th IEEE ISPSD, Toulouse, France, 2000, pp. 149-152.

[7] V. Loong Choo et al., "Reverse Recovery and Carrier Lifetime in Body Diodes of LDMOS Transistors," 2021 33rd IEEE ISPSD, Nagoya, Japan, 2021, pp. 307-310

979-8-3503-6184-1/24 $31.00 © 2024 IEEE

Comparsion of SiC Planar and Trench Junction Barrier Schottky Diode with Surge Current Capability

Zi-Ming Zhao [1], Yan-Cong Liu [1], Hao Yuan *[1], Feng-Yu Du [1], Yu Zhou [1], Ke-Yu Liu [1], Xiao-Yan Tang [1], Chao Han [1,2], Qing-Wen Song [1,2], Yu-Ming Zhang [1,2]

[1] Key Laboratory of Wide Band Gap Semiconductor Materials and Devices, School of Microelectronics, Xidian University, Xi'an, China
[2] The Xidian-Wuhu Research Institute, wuhu 241000, China

* Email: zimingzhao@stu.xidian.edu.cn, haoyuan@xidian.edu.cn

Abstract—The surge capabilities of SiC planar junction barrier schottky diodes and trench junction barrier schottky diodes are analyzed and compared by experiment and simulation in this paper. Though experimental comparison, the maximum surge current of SiC trench JBS is larger than SiC planar JBS, improved by 8.5%. The leakage current of the trench structure is approximately one order of magnitude lower than that of the planar structure at voltages below 1300 V, sacrificing only a small forward voltage drop. Thanks to the deep P+ region design, the p-n junction turn-on voltage in SiC trench JBS diodes is substantially decreased, and the results of both the experiment and simulation studies demonstrate that the trench structure exhibits superior surge capability.

Keywords—junction barrier schottky (JBS) diode, silicon carbide (SiC), surge current, planar, trench

I. INTRODUCTION

Featuring lower forward voltage drop, higher breakdown voltage, lower switching and lower reverse leakage current and superior thermal characteristics, SiC JBS diodes will become the core devices in power electronics. In some industrial application, SiC JBS diodes are required to withstand more than 10 times rated current. High surge reliability is critical for the more application scenarios of SiC JBS diodes. There are considerable investigations on the surge ruggedness of SiC devices in recent years, but most of them mainly focused on the comparison of surge current capability of SiC diodes [1], and the locations and causes of surge failures [2]. In some new structures, toroidal and hexagonal structures have been investigated as potential solutions to the current congestion of JBS diodes [3]. It has been demonstrated that JBS diodes with a "saddle" finger layout can lead to a reduction in peak temperature [4]. It is evident that process conditions exert a considerable influence on the surge current capability of SiC MPS/JBS diodes [5]. Nevertheless, there is a paucity of research investigating the SiC trench JBS diode surge current capability and the means of enhancing the SiC JBS diode surge current capability. In this paper, the surge current capability of SiC JBS diodes is evaluated by experiments, and is analyzed by the TCAD simulations.

II. EXPERIMENT SETUPS AND SMULATION STUDY

A. Experimental Test Setup

The surge experiment is conducted at 25℃. Fig.1 show the surge experiment test bench. A 10ms half sinwave is generated by the ATP-ST-1000A test system developed by ALLTOPLUS. The waveforms captured by the current clamps and differential voltage probes are displayed on the oscilloscope.

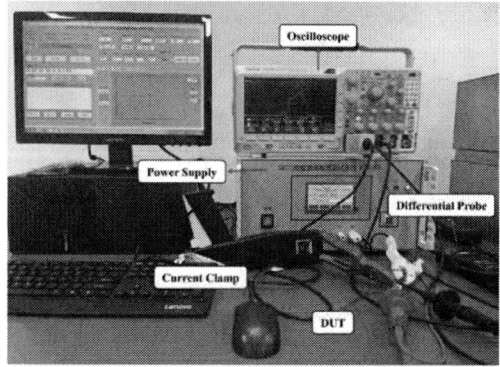

Fig. 1 Surge experiment test bench

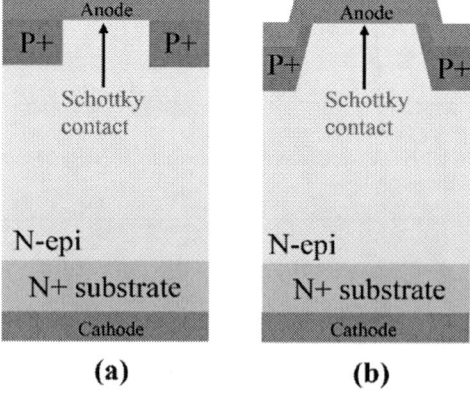

Fig. 2 Cross-sectional views of (a) planar SiC JBS diode structure and (b) trench JBS diode structure

SiC JBS diodes contain P+ regions and Schottky regions. The cross-sectional view of a planar SiC JBS diode structure and a trench SiC JBS diode structure are shown in Fig. 2(a) and Fig. 2(b) respectively. The pin-areas are implemented doping of 1×10^{19} cm^{-3} and epitaxial wafers are available in 10um thickness. The n- drift region is doped with the density of approximately 6×10^{15} cm^{-3}. The depth of P+ areas is 1 μm and the depth of trench is also 1μm.

979-8-3503-6184-1/24 $31.00 © 2024 IEEE

B. Experimental results

Prior to examining the surge current capability of the device, it is essential to consider its static characteristics.

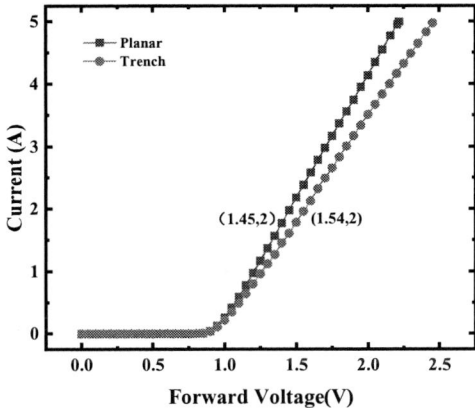

Fig. 3 Experimental results of static forward I–V characteristics of SiC planar JBS diodes and SiC trench JBS didoes

The measured forward I–V characteristics of planar JBS diodes and trench JBS didoes are shown in Fig. 3. As illustrated in the inset, the forward voltage drop of the trench JBS diode is greater than that of the planar JBS diode at a rated current of 2A, increasing by 0.09V.

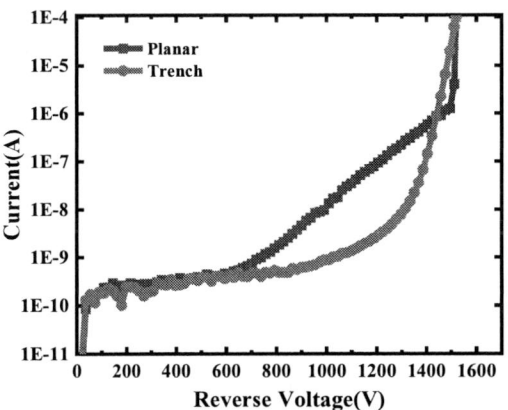

Fig. 4 Experimental results of static reverse I–V characteristics of SiC planar JBS diodes and SiC trench JBS didoes

Fig.4 depicts the reverse I–V characteristics of planar JBS diodes and trench JBS didoes. The findings indicate that the leakage current of the trench structure is approximately one order of magnitude lower than that of the planar structure at voltages below 1300 V. After the static characterizations, single pulse surge current tests are carried out on the fabricated diodes to analyze and compare the surge current capability of the diodes with different designs. During each test, such current pulse is applied on the device while the device voltage and current waveforms are recorded. After each test, the device is cooled down for several minutes to ensure that its temperature drops down to the room temperature. The current increases gradually from 5A until the point of failure of the device is reached.

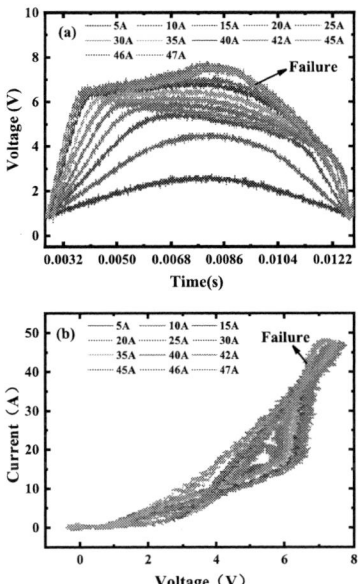

Fig. 5 (a)Voltage waveforms of the SiC planar JBS diodes with varied peak current amplitude up to 47 A and (b) corresponding I–V curves

Fig. 5(a) extracts the Time-Voltage curve of the SiC planar JBS diodes under the surge current stress condition. Fig. 5(b) shows the corresponding I–V curves. Under higher surge current conditions, the waveforms show a collapse at 47A. This represents the occurrence of the device failure.

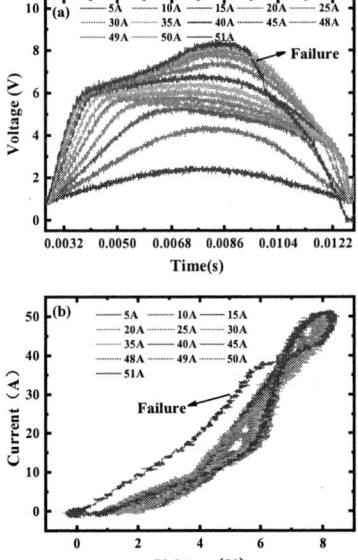

Fig. 6 (a)Voltage waveforms of the SiC trench JBS diodes with varied peak current amplitude up to 47 A and (b) corresponding I–V curves

Similarly, Fig. 6(a) extracts the Time-Voltage curve of the SiC trench JBS diodes under the surge current stress condition. Fig. 6(b) shows the corresponding I–V curves. With the increase of surge current, the waveforms show a collapse at 51A. The device may malfunction or even deteriorate in performance. It is evident that the trench JBS diode exhibits greater resistance to surge current, with an improvement of 8.5%.

C. Simulation results

Simulation studies of planar JBS diodes and trench JBS diodes based on the SENTAURUS software are carried out. The device active area is setup as 1 mm² too, which is consistent with fabricated devices. The surge current is set to 55A by applying a half sine current pulse with the a duration of 10ms.

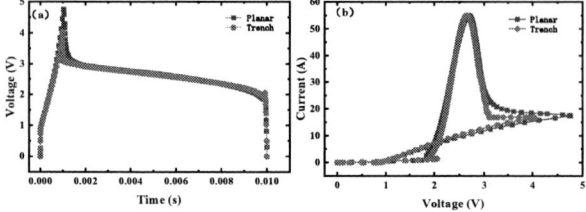

Fig. 7 Simulated I–V characteristics of SiC planar JBS diodes and SiC trench JBS didoes at 51A (a) voltage waveforms with single pulse surge current and (b) corresponding I–V curves

Fig. 7(a) shows the voltage waveforms for planar JBS diodes and trench JBS didoes, and Fig. 7(b) shows the corresponding I–V curves. As shown in Fig. 7(a) and Fig. 7(b), the trench JBS diode is characterised by a lower dropout and a smaller bipolar on-state voltage. A corresponding shift to the left of I-V curve for the SiC trench diode structure is observed, as shown in Fig. 7(b).

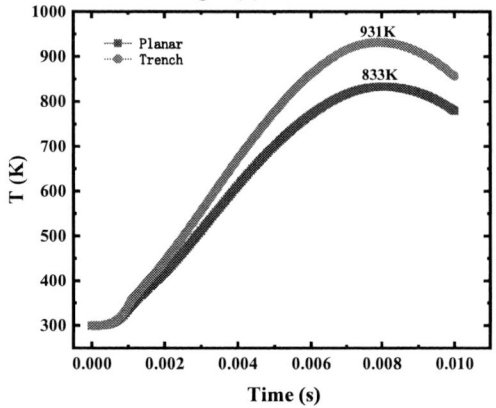

Fig. 8 Simulated maximum lattice temperature curve with time of SiC planar JBS diodes and SiC trench JBS didoes at 51A

Fig.8 depicts maximum lattice temperature curve with time of SiC planar JBS diodes and SiC trench JBS didoes at 51A. At the beginning, lattice temperature changes slowly at first. When the surge current increases, lattice temperature increases dramatically. Lattice temperatures of the planar JBS diode and the trench JBS didoe reach 931k and 833k, respectively. It has been demonstrated that the SiC trench JBS diode exhibits a lower maximum lattice temperature for a given surge current.

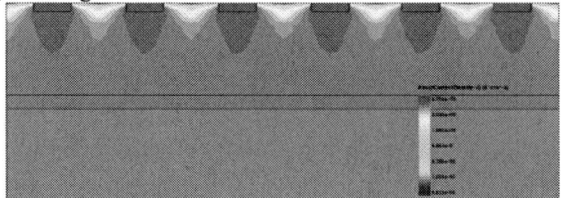

Fig. 9 Simulated hole density current distribution of SiC planar JBS diodes at 5ms

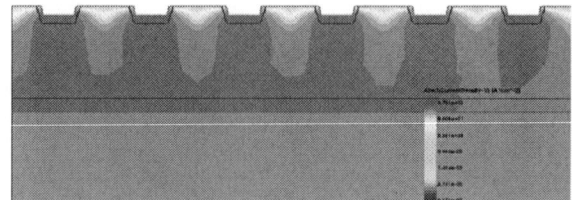

Fig. 10 Simulated hole density current distribution of SiC trench JBS diodes at 5ms

Fig. 9 and Fig. 10 illustrate simulated hole density current distribution of SiC planar JBS diodes and SiC trench JBS diodes at 5ms respectively. There is higher hole current injection for the SiC trench JBS diode at 5ms respectively. The area of concentrated heat build-up is clearly visible.

D. Discussion

The results of experimental and simulation studies exhibit a consistent pattern. The SiC trench JBS diode has a higher surge current capability under the same surge current condition. Concurrently, it exhibits favourable static characteristics. The experimental results demonstrate an 8.5% enhancement in the trench structure relative to the planar structure. Similarly, the results by simulation are consistent with the experiment. The trench structure exhibits a smaller bipolar on-state voltage and a reduced capacity for heat accumulation.

III. SUMMARY

In this work, the surge ruggedness in SiC planar JBS diodes and SiC trench JBS didoes is investigated and compared by experiments and simulations. The experimental results indicate that the maximum surge current in the SiC trench JBS diode (51A) is larger than that of the SiC planar JBS diode (47A), which is increased by 8.5%. The trench structure exhibits a deeper P+ region, which results in a smaller bipolar on-state voltage. The localised congestion in currents has been alleviated. As a conclusion, in maximum surge current aspect, the surge capability of the trench JBS didoe is better than the planar JBS didoe under the same condition.

ACKNOWLEDGMENT

This work was supported in part by the National Key R&D Program of China under Grant 2021YFB3601800 and the Major Projects of Shanxi Province (Grant no. 202101030201001)，in part by the Natural Science Basic Research Program of Shaanxi under Grants 2024JC-YBMS-474 and 2024JC-YBQN-0647.

REFERENCES

[1] S. Palanisamy, J. Kowalsky, J. Lutz, T. Basler, R. Rupp and J. Moazzami-Fallah, ISPSD, pp. 367-370(2018).

[2] J. Wu, N. Ren, H. Wang and K. Sheng, IEEE Journal of Emerging and Selected Topics in Power Electronics,,pp. 1496-1504(2019).

[3] V. Banu, M. Berthou, J. Montserrat, X. Jordà and P. Godignon, CAS, pp. 147-150(2017).

[4] A. Shimbori and A. Q. Huang, WiPDA,, pp. 17-21(2022)

[5] Xu, H., Ren, N., Wu, J., Zhu, Z., Guo, Q., Sheng, K., Materials, pp. 663(2021).

Comparative Study on Reliability of Conventional SiC MOSFET and JBS Integrated SiC MOSFET

Moufu Kong*, Shurui Li, Hongfei Deng, Bo Yi, Hongqiang Yang, Sen Gong

State Key Laboratory of Electronic Thin Films and Integrated Devices of China, University of Electronic Science and Technology of China, Chengdu, China,611731

* Corresponding author: Moufu Kong, Email: kmf@uestc.edu.cn

Abstract—The reliability of the conventional silicon carbide (SiC) MOSFET and the junction barrier Schottky (JBS) diode integrated SiC MOSFET (JMOS) is investigated and compared in this article. The JMOS exhibits a slightly higher specific on-resistance ($R_{on,sp}$) compared to the conventional SiC MOSFET. However, it demonstrates a significantly lower on-state voltage drop (V_{SD}) of the diode due to the integrated JBS diode, in contrast to the body PN diode of the conventional SiC MOSFET. The reliability experiments including high temperature reverse bias (HTRB), high temperature gate bias (HTGB) and high humidity and high temperature reverse bias (H3TRB) for the conventional SiC MOSFET and JMOS both are conducted. The results indicate that the SiC MOSFET with integrated JBS diode exhibits similar reliability to the conventional SiC MOSFET, with both demonstrating good reliability.

Keywords—JBS, SiC MOSFET, reliability, HTRB, H3TRB, HTGB

I. INTRODUCTION

Silicon carbide (SiC) is a wide band gap material with high critical breakdown electric field and high thermal conductivity, therefore, it has been extensively researched and employed in power devices [1]-[2]. And the SiC power MOSFET is now widely utilized due to its high operating frequency, low power loss, and low noise. However, the body diode of conventional the SiC power MOSFET has high on-state voltage drop, poor reverse recovery performance, and the presence of bipolar degradation effect, and the junction barrier Schottky (JBS) diode integrated SiC MOSFET (JMOS) is developed to improve the reverse recovery and suppress the bipolar degradation effect of the device. However, the reliability of the SiC MOSFET and JMOS in practical applications has always been a challenge [3]-[6]. And in particular, minimal research has been conducted on the reliability of JMOS. In this paper, the high temperature reverse bias (HTRB) test, high humidity high temperature reverse bias (H3TRB) test and high temperature gate bias (HTGB) test are conducted for the conventional SiC MOSFET (C-MOS) and JMOS. The reliability of the two devices is evaluated by comparing the threshold voltage (V_{th}), breakdown voltage (BV), leakage current, on-resistance (R_{on}) and other data before and after the test.

II. DEVICE STRUCTUREs AND CHARACTERISTICs

SiC MOSFET is usually used with anti-paralleled SiC JBS diode in practical applications, one is because the SiC PN body diode has a higher on-state voltage drop(>3V), which leads to a high on-loss of the body diode, the other is because the body diode is a bipolar device, which introduces a bipolar degradation effect, leading to performance degradation of the device [7]. The new SiC MOSFET structure with integrated JBS diode in the cell, which can be fabricated under the existing process steps, and the integration of JBS diode in the

This work was supported in part by the Key R & D project of science and technology plan of Sichuan province (Grant No. 2023YFG0005).

cell can simplify the package and reduce the volume and cost of the JMOS compared to the external anti-parallel SiC JBS in the package of C-MOS. And Fig. 1 (a) and (b) show the proposed SiC JMOS cell and SiC C-MOS cell structures, respectively. And Fig. 1 (c) and (d) demonstrate the JMOS square cell layout and the C-MOS square cell layout, respectively. And from cell the layout, it is indicated that the JBS diode occupies only a very small part of the entire cell.

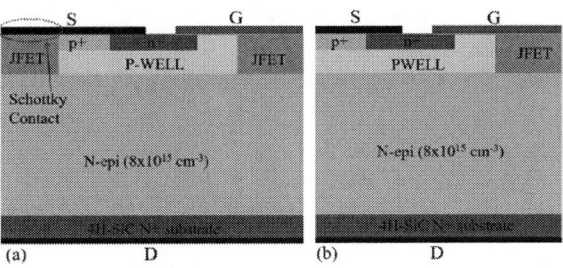

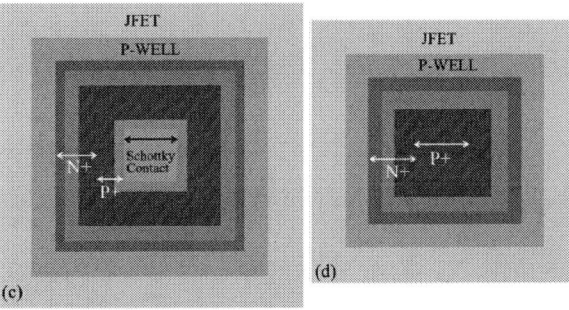

Fig.1 (a) device structure of proposed SiC JMOS, (b) structure of SiC C-MOS, (c) JMOS square cell layout, (d) C-MOS square cell layout.

The SiC JMOS and C-MOS both are fabricated and packaged with a TO-247-3 package. And Fig.2 (a) and (b) reveal the micrographs of the SiC JMOS chip and the packaged JMOS device, respectively. Table 1 also lists the electrical performance test results of the two devices. Fig.3 (a) and (b) also show the output I-V curves (b) the diode I-V curves of the two devices, respectively.

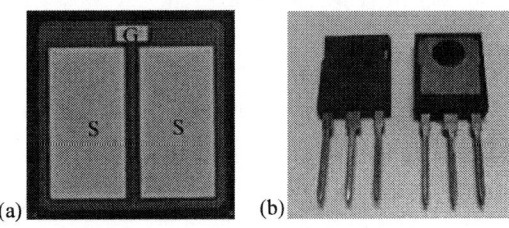

Fig.2 (a) Micrographs of the SiC JMOS chip, (b) the packaged JMOS device

As can be seen from Table 1, since the JBS takes up some part of the JMOS area, in the case that the total chip area of the two structures is equal, the actual MOSFET active area of the JMOS structure is smaller than that of the conventional structure, so the R_{on} of the JMOS is slightly higher than that of C-MOS , but the V_{SD} of JMOS (at $I_F = 10A$) is reduced by 60% from 3.8V to 2.28V, compared with that of the C-MOS.

TABLE I COMPARISON OF ELECTRICAL RROPERTIES OF TWO STRUCTURES

Parameters	Conditions	JMOS	C-MOS
$I_{DSS\text{-}HV}$	$V_{DS}=1200V$	9.44 μA	8.68 μA
I_{GSS+}	$V_{GS}=20V$	30.92 nA	34.43 nA
I_{GSS-}	$V_{GS}=-5V$	8.29 nA	9.03 nA
V_{th}	$V_{GS}=V_{DS}, I_{DS}=5mA$	3.40 V	3.21 V
R_{ON}	$V_{GS}=20V, I_{DS}=20A$	59.3 mΩ	51.1 mΩ
BV	$I_{DS}=100μA$	1329 V	1358 V
V_{SD}	$I_F=10A$	2.28 V	3.8 V

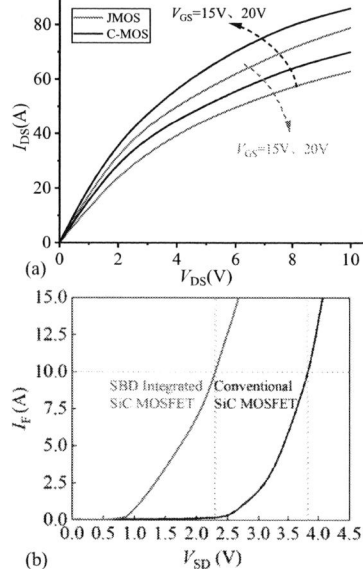

Fig.3 (a) output I-V curves, and (b) the diode I-V curves of the two devices

III. EXPERIMENTAL RESULT AND DISCUSSION

The HTRB, HTGB and H3TRB reliability test for the packaged SiC JMOS and C-MOS devices are carried out, and there are 45 samples in each test with no devices failed during the entire test. The electrical parameters (average value) of the samples before and after the reliability test are investigated and compared to judge the reliability of the devices.

A.HTRB

The HTRB test is conducted to verify the leakage current of the chip over extended periods to ensure long-term stability [8]. During this test, the gate-source is shorted, and the drain is subjected to 80% of the rated breakdown voltage (BV) of the device, specifically 960V for a 1200V-class device. The test duration is 1000 hours at 175°C. Fig. 3(a) and 3(b) illustrate the comparison results and change rates of electrical parameters of the SiC JMOS and C-MOS before and after the HTRB test, respectively. The SiC JMOS and C-MOS both demonstrate a slightly reduction in V_{th} and R_{on} after testing, with reductions of approximately -4.8% and -1.1% for JMOS, and -1.3% and -3.6% for C-MOS, respectively. And the high blocking voltage leakage current ($I_{DSS\text{-}HV}$) and BV are almost unchanged. The gate leakage current (I_{GSS+} and I_{GSS-}) of both JMOS and C-MOS changes slightly with a change rate of less than 12.5%. These results demonstrate that JMOS exhibits

excellent reliability in the HTRB test, showing no significant difference compared to C-MOS.

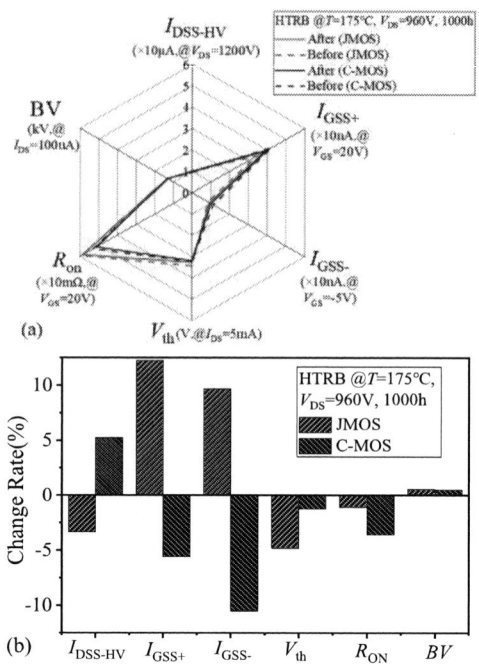

Fig.3. (a)electrical parameters of JMOS and C-MOS before and after HTRB test，(b)change rate of the electrical parameters

B.H3TRB

The wet and thermal environments can induce physical and chemical effects on devices, leading to moisture absorption, expansion, insulation resistance decline, rust on tubes and shells, electric leakage, and potential short circuits. Moisture and electrical stress can also cause electrochemical migration of metal materials. The H3TRB test primarily assesses the device's ability to withstand electrical, thermal, and humidity stresses. During the test, samples are exposed to conditions with a relative humidity (RH) of 85% and a temperature of 105°C. The gate-source is shorted, and the drain is subjected to 80% of the rated BV for 1000 hours [9]. Fig. 4(a) and (b) illustrate the comparison results and change rates of electrical parameters of the JMOS and C-MOS before and after the HTRB test, respectively. And the results indicate minimal changes in V_{th}, R_{on}, and BV, about -6.9%、-3.8%、0.08% for JMOS, and -3.9%, -7.0% and 0.23% for C-MOS respectively, after the H3TRB test. Although there are significant changes in $I_{DSS\text{-}HV}$, they remain within a reasonable order of magnitude and do not significantly impact reliability.

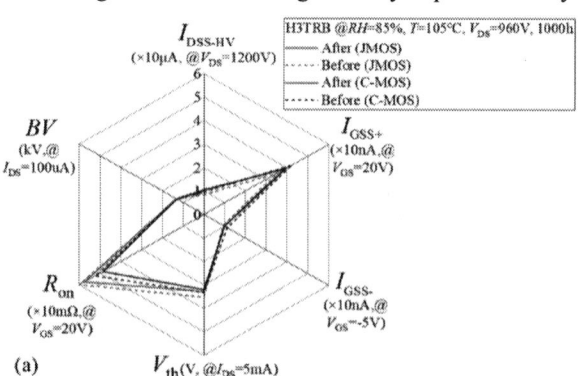

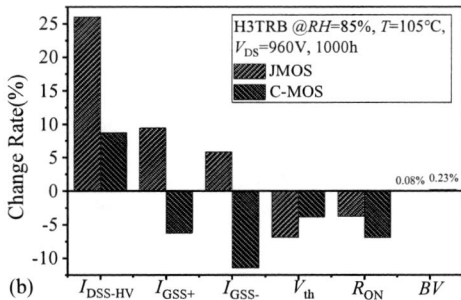

Fig.4. (a)electrical parameters of JMOS and C-MOS before and after H3TRB test, (b)change rate of the electrical parameters

C.HTGB

The HTGB test evaluates the reliability of a device's gate oxide layer by subjecting it to high temperature stress with two variations: high temperature gate forward bias and high temperature gate reverse bias [10]. During this test, samples are exposed to 175°C, with one group subjected to a +20V gate to source voltage (V_{GS} = +20V) and another to -5V gate to source voltage (V_{GS} = -5V). The duration of each test is 1000 hours. The results of the reverse bias and forward bias tests are illustrated in Fig. 5 and Fig. 6 respectively. Under forward bias, electrons from the inversion layer on the channel surface become trapped by defects at the SiC/SiO$_2$ gate oxide interface, causing a forward shift in V_{th}. Conversely, under reverse bias, holes from the channel surface are injected into the SiC/SiO$_2$ gate oxide layer, resulting in a reverse shift in the V_{th}, as depicted in the figures. Under forward bias, the change rates of V_{th} for JMOS and C-MOS are 32.8% and 30.9%, respectively. Under reverse bias, the change rates of V_{th} for JMOS and C-MOS are -17.2% and -9.4%, respectively. Furthermore, the HTGB forward bias test notably affects I_{DSS-HV}. However, these changes remain within a reasonable range, indicating no significant impact on reliability.

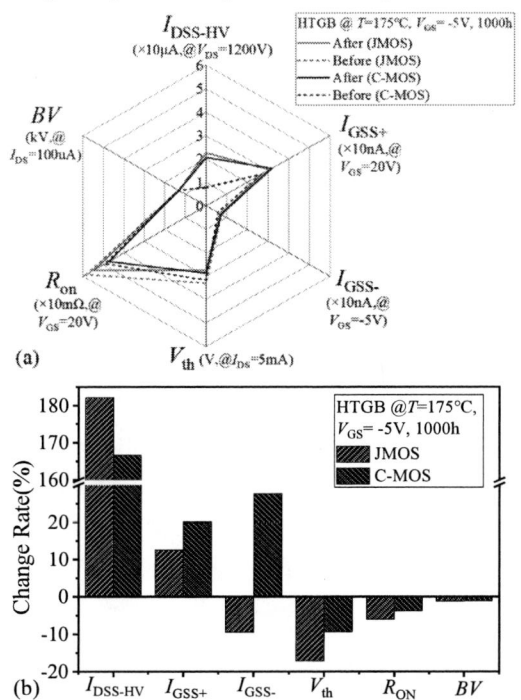

Fig.5 (a) electrical parameters of JMOS and C-MOS before and after HTGB (reverse bias) test，(b)change rate of the electrical parameters

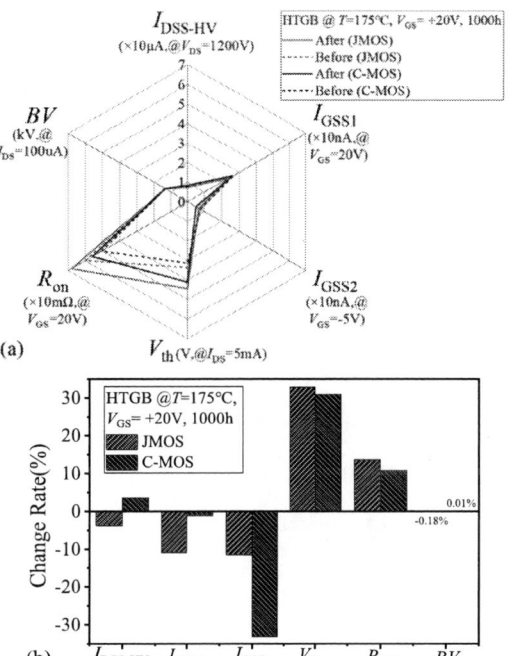

Fig.6 (a)electrical parameters of JMOS and C-MOS before and after HTGB (forward bias) test，(b)change rate of the electrical parameters

IV. CONCLUSION

In this study, a new JBS diode-integrated SiC MOSFET (JMOS) and a conventional SiC MOSFET were both designed and fabricated on the same process platform. It was observed that the JMOS exhibited improved performance in the reverse on-state compared with the conventional MOSFET. After packaging, both devices underwent a 1000-hour reliability test. By comparing the electrical parameters before and after reliability testing, we conclude that there is no significant difference in reliability between the JMOS and conventional MOSFET, both devices demonstrate high reliability.

REFERENCES

[1] Francesco La Via, et.al., "Emerging SiC Applications beyond Power Electronic Devices", Micromachines 2023, pp. 1-37.

[2] B. Jayant Baliga, "Silicon Carbide Power Devices: Progress and Future Outlook", IEEE Journal of Emerging and Selected Topics in Power Electronics, vol. 11, no. 3, JUNE 2023,pp. 2400-2411.

[3] B. Hull, et.al., "Reliability and stability of SiC power MOSFETs and Next-Generation SiC MOSFETs", IEEE Workshop on Wide Bandgap Power Devices and Applications,2014, pp. 139-142.

[4] D. A. Gajewski, et al. "SiC power device reliability", IEEE International Integrated Reliability Workshop (IIRW), 2016,pp.29-34.

[5] A. J. Lelis, et.al., Habersat, "SiC MOSFET reliability and implications for qualification testing", in proc. IEEE IRPS,2017, pp. 1-4.

[6] Y. Wu, et.al., "Comparative Study on Reliability and Application Features of SiC MOSFET", IEEE 2nd PEAS,2023, pp. 1650-1654.

[7] X. Wei, et.al., "Design and Fabrication of 1.2kV SBD Integrated 4H-SiC MOSFET for Reliable Reverse Conduction", the 4th International Conference on Power and Energy Technology, 2022,pp. 166-169.

[8] F. Hoffmann, et.al., "Long Term High Temperature Reverse Bias (HTRB) Test on High Voltage SiC-JBS-Diodes", in proceedings of the 30th ISPSD,2018, pp. 435-438.

[9] In-Hwan Ji, et.al., "High Temperature and High Humidity Reliability Evaluation of Large-Area 1200V and 1700V SiC Diodes", in proc. IEEE IRPS, 2023, pp. 81-84.

[10] D. B. Habersat, et.al, "Permanent and Transient Effects of High Temperature Bias Stress on Room-Temperature V_{th} Drift Measurements in SiC Power MOSFETs", in proc. IEEE IRPS, 2019, pp. 1-4.

Study on Single Event Effect of SiC MOSFET by Proton Irradiation

Wende Huang [12], Chengwen Fu [1], Yao Ma*[1], Mingmin Huang*[1], Xiaoping Dong[1], Qiang Yu [3]

[1] College of Physics, Sichuan University, Chengdu 610065, China
[2] Key Laboratory of Microelectronics of Sichuan Province, Sichuan University, Chengdu 610065, China
[3] Sichuan Suining Lippxin Microelectronics Co., Ltd, Suining 629099, China

* Email: 1753783441@qq.com, mayao@scu.edu.cn, mmhuang@scu.edu.cn

Abstract—**The single event burnout (SEB) of SiC MOSFET by proton irradiation is studied based on experiments of SiC MOSFETs irradiated by 300 MeV proton. The SEB voltages of 650 V and 1.2 kV SiC MOSFETs are about 400 V and 900 V, respectively. The SEB of SiC MOSFETs is caused by secondary particles induced by proton. Geant4 is used to calculate linear energy transfer (LET) distributions of secondary particles, and mechanism of SEB induced by secondary particles is studied by TCAD simulations. Besides, it is interested to find that the single event leakage current (SELC) does not occur in SiC MOSFETs irradiated by protons, which is different from the case irradiated by heavy ions.**

Keywords—*SiC MOSFET, proton irradiation, single event burnout (SEB), single event leakage current (SELC)*

I. INTRODUCTION

Since silicon carbide (SiC) has advantages of wide band gap, high critical electric field, high thermal conductivity, and high melting point, the SiC MOSFET has great advantages in power applications and also has broad prospect in aerospace applications [1]. Single event effect is the most important failure phenomenon of SiC MOSFETs in harsh irradiation environments, which can be caused by heavy ions, neutrons and protons [2]-[3]. Previous work has mainly concerned the single event effect of SiC MOSFETs induced by high-energy heavy ions, mechanism of which is that carriers (electrons and holes) generated by heavy ions will redistribute the electric field in the n-drift and form highly localized electric field to generate localized high temperature, and the parasitic npn transistor will aggravate this process [4]. Besides, the single event leakage current (SELC) can be observed in SiC MOSFETs irradiated by heavy ions with a reverse bias voltage lower than that for SEB (SEB voltage) [5].

Although the SEB of SiC MOSFET by protons is less sensitive than that by heavy ions, the fluence of protons is usually much higher than that of heavy ions in space, i.e., the probability of devices suffering protons can be higher than suffering heavy ions. Hence, it is also important to study SEB of SiC MOSFET by protons. However, there are few related works. Recently, a work reported that proton-generated secondary particles may be the reason of SEB of SiC MOSFETs irradiated by protons [6]. More works need be carried out to study the experiment phenomenon of SEB of SiC MOSFET by protons and its mechanism.

This paper studies the phenomena and mechanism of SEB of 650 V and 1.2 kV SiC MOSFETs irradiated by 300 MeV protons. Experiments details and results are introduced in Section II, where the different phenomenon between proton irradiation and heavy ion irradiation of SiC MOSFET are discussed. The mechanism of SEB by proton irradiation is studied by TCAD simulations in Section IV. Conclusion is presented in Section V.

II. EXPERIMENTS DETAILS AND RESULTS

Experiments were carried out at the Space Environment Ground Simulation Installations (SESRI), Harbin Institute of Technology. The energy of protons was 300 MeV, the fluence rate was 2.5×10^8 p·cm^{-2}·s^{-1}, and there was a 3 s output of protons in a period of 12 s. Two types of SiC MOSFETs were used, i.e., C3M0350120D (1.2 kV, 7.6 A) and C3M0120065D (650 V, 22 A). The Keithley 2470 source meter was used to provide biased voltages at the drain and monitor the drain current. The Keithley 2450 source meter was used to detect gate current with zero bias between gate and source. The irradiation was set to be stopped if the SEB occurs or the fluence reaches about 2.5×10^{10} p·cm^{-2}. The current limits of the two source meters were both set to be 1 mA.

Table I lists experiment conditions and results of samples, where samples #1 and #2-1 were used to evaluate the SEB by a stepped increased drain voltage, and samples #2-2 and #2-3 were used to evaluate the SELC by a constant drain voltage.

Fig. 1 shows waveforms of drain voltage (V_{DS}), drain current (I_D) and gate current (I_G) of samples #1 and #2-1. The SEB voltages of samples #1 and #2-1 are about 900 V and 400 V, respectively. For sample #1, SEB occurred with a sudden increase of both I_D and I_G after the drain voltage keeping at 900 V for about 45 s. For sample #2, SEB occurred with an immediate increase of I_D but I_G unchanged at the moment of the drain voltage increasing from 300 V to 400 V.

TABLE I. EXPERIMENT CONDITIONS AND RESULTS

Sample	Rated Voltage	Drain Voltage	SEB	Fluence (cm^{-2})
#1	1.2 kV	800→900 V	Yes	1.85×10^{10}
#2-1	650 V	300→400 V	Yes	1.32×10^{10}
#2-2	650 V	350 V	No	2.26×10^{10}
#2-3	650 V	300 V	No	2.47×10^{10}

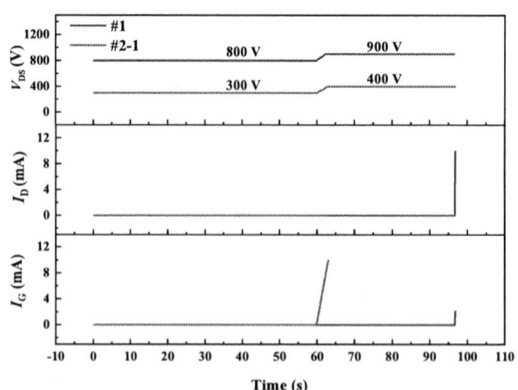

Fig. 1. Waveforms of V_{DS}, I_D and I_G of samples #1 and #2-1.

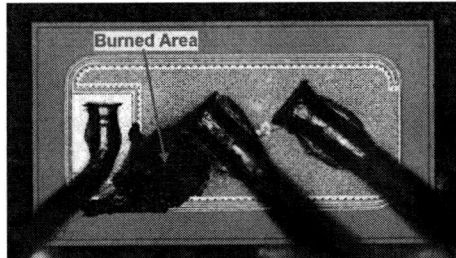

Fig. 2. Photo of sample #1 by the metallography microscope.

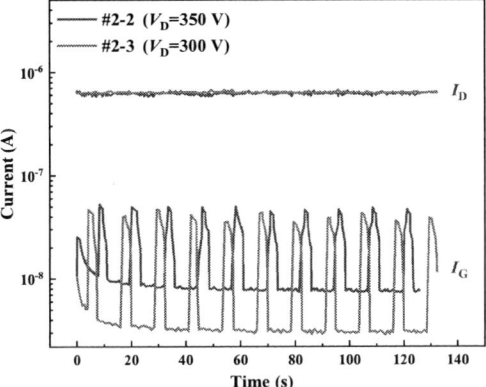

Fig. 3. Waveforms of I_D and I_G of samples #2-2 and #2-3.

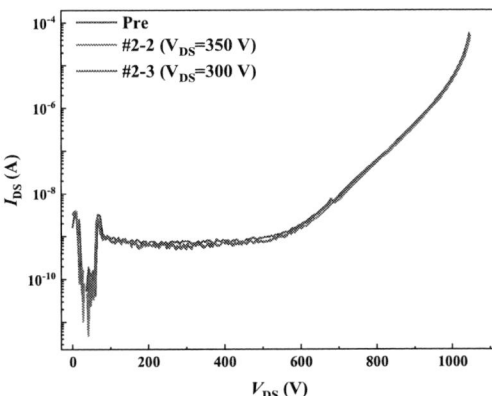

Fig. 4. Reverse I-V curves of samples #2-2 and #2-3.

After irradiation, sample #1 has failed with its three electrodes being shorted. Fig. 2 shows the photo of sample #1 by the metallography microscope, where a large burned area covers both the gate and source pads. However, the sample #2 has not failed after irradiation. This may be because that the current limiting protection by the source meters can be timely for the case with only I_D increasing suddenly.

Fig. 3 shows waveforms of I_D and I_G of samples #2-2 and #2-3. There are gate current pulses with pulses of protons, but I_D and I_G are nearly unchanged after every pulse of protons. Fig. 4 shows reverse I-V curves of samples #2-2 and #2-3 before and after irradiation. Both samples #2-2 and #2-3 have no degradation in leakage current, which means that there is no SELC for the SiC MOSFET irradiated by protons, which is something different from the case of the SiC MOSFET irradiated by heavy ions [7].

The secondary particles generated by protons in SiC MOSFETs play an important role in the response of the SiC

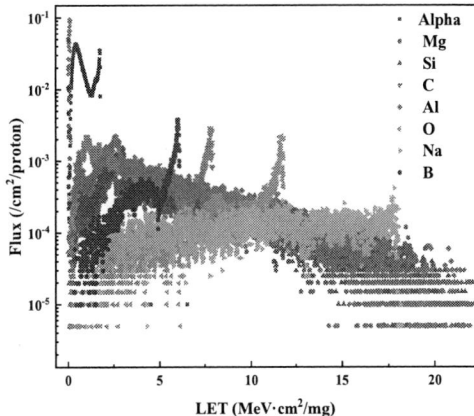

Fig. 5. LET distributions of secondary particles produced by 300 MeV proton irradiation in the SiC MOSFET.

TABLE II. KEY PARAMETERS OF SiC MOSFET USED IN SIMULATIONS

Parameters	Values
cell width	10 μm
p-base width	3.2 μm
oxide thickness	50 nm
n-drift doping & thickness	2×10^{15} cm^{-3}, 10 μm
channel doping & length	2×10^{17} cm^{-3}, 1 μm

MOSFET during proton irradiation. Fig. 5 shows linear energy transfer (LET) distributions of secondary particles produced by 300 MeV proton irradiation in the SiC MOSFET, which are calculated by the Geant4 software. Generally, a high LET is likely to induce SEB. Hence, it is quite possible for secondary particles of Si, Mg, and Al with a relative high LET to induce SEB in samples #1 and #2-2. As for the SELC, it can be manifested until enough damages has been accumulated in the SiC MOSFET, which needs many secondary particles with high enough LET. It is reported that heavy ions with LET ≥ 9.3 MeV·cm^2/mg may be able to induce SELC [8]. It can be calculated by Fig. 9 that the total fluence of secondary particles with LET ≥ 9 MeV·cm^2/mg is $< 1\times10^4$ cm^{-2} (i.e., < 300 particles) for samples #2-2 and #2-3. Such a low number of particles is not easy to accumulate enough damages. Hence, there is no SELC in samples #2-2 and #2-3.

III. TCAD SIMULATIONS AND ANALYSIS

In order to investigate the mechanism of SEB of the SiC MOSFET by proton irradiation, a two-dimensional TCAD simulation is conducted by Sentaurus software. Simulation results of reverse and transfer I-V curves have been calibrated with measurement results, where the breakdown voltage is 1500 V and the threshold voltage is 3.5 V. Key parameters of the device used in simulations are listed in Table II.

According to LET distributions of secondary particles in Fig. 5, the influence of secondary particles with LET = 20 MeV·cm^2/mg (equivalent to 0.13 pC/μm) is studied by simulations. It can be calculated by TRIM software that the track length of the secondary particle is 4.86 μm. Thus, the length of electron-hole pairs area generated by the secondary particle is set to be 4.86 μm. As shown in Fig.6, four incident locations of protons ($x = 0.5, 1.8, 2.8,$ and 5 μm) and six center depths that generates the secondary particle ($y = 0, 2, 4, 6, 8,$ and 10 μm) are considered in simulations. The drain voltage is set to be 900 V, i.e., the SEB voltage as depicted in Fig. 1.

979-8-3503-6184-1/24 $31.00 © 2024 IEEE

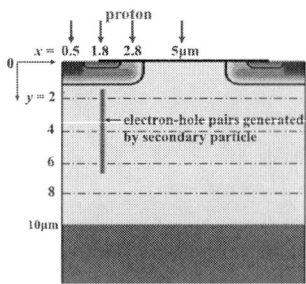

Fig. 6. The incident position of the secondary particles in simulation.

TABLE III. PEAK TEMPERATURES DURING IRRADIATION TRANSIENT

Peak Temperature	$x = 0$ μm	$x = 1.8$ μm	$x = 2.8$ μm	$x = 5$ μm
$y = 0$ μm	304 K	308 K	310 K	306 K
$y = 2$ μm	373 K	452 K	uncontrol	700 K
$y = 4$ μm	uncontrol	uncontrol	uncontrol	uncontrol
$y = 6$ μm	uncontrol	uncontrol	uncontrol	uncontrol
$y = 8$ μm	507 K	uncontrol	uncontrol	315 K
$y = 10$ μm	320 K	327 K	328 K	311 K

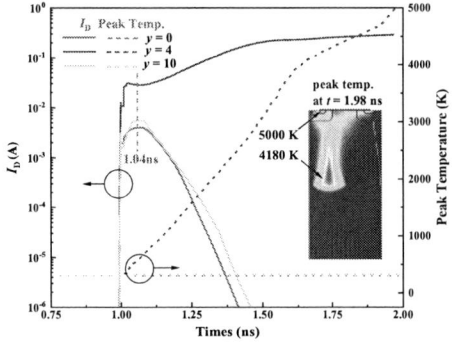

Fig. 7. Peak temperature and I_D of three cases given in Table III.

Table III presents simulation results of peak temperatures of the device during the irradiation transient in the cases with the four incident locations (x = 0.5, 1.8, 2.8, and 5 μm) and six center depths that generates the secondary particle (y = 0, 2, 4, 6, 8, and 10 μm). It can be found that the SEB is most likely to occur if the proton strikes through the channel region (x = 2.8 μm) and the secondary particle is generated at the middle of the n-drift region (y = 4 ~ 6 μm).

Fig.7 shows the peak temperature and I_D of the three cases given in Table III, i.e., center depths are y = 0, 4, and 10 μm with incident location of x = 2.8 μm. The I_D in the case with the center depth of y = 4 μm fails to fall to zero, and then the peak temperature becomes uncontrollable, where the high temperature occurs at the source contact and bottom of the n-drift. However, I_D in the other two cases can fall to zero within 0.5 ns, which avoids the high temperature spot in the device.

Fig. 8 shows electric field distributions along x = 2.8 μm of the three cases at time of 1.04 ns in Fig. 7, where electron current distributions are given in the inset. In the case with the center depth of y = 4 μm, carriers generated in the depleted n-drift are much more than the other two cases. Thus, the peak electric field in the former can be much higher than the latter two, resulting in a higher hole current generated by impact ionization to keep the paratactic npn transistor on and inducing SEB in the former.

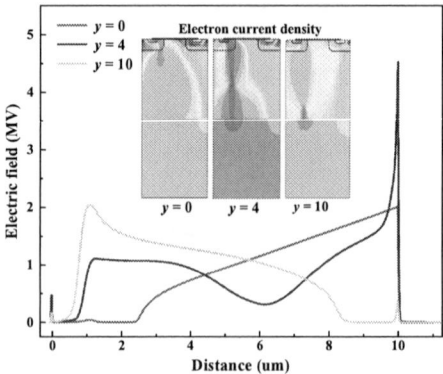

Fig. 8. Electric field distribution along x = 2.8 μm and electron current distributions of three cases at time of 1.04 ns in Fig. 7.

IV. CONCLUSION

The SEB of SiC MOSFET by 300 MeV proton irradiation is investigated through experiments and simulations. The SEB voltages of 650 V and 1.2 kV SiC MOSFETs are about 400 V and 900 V, respectively. The SEB is most likely to occur if the proton strikes through the channel region and the secondary particle is generated at the middle of the n-drift. If the hole current generated by the impact ionization is high enough, the paratactic npn transistor can be kept on to induce SEB. Besides, the SELC does not occur, which may be due to that number of secondary particles with high LET is too small.

ACKNOWLEDGMENT

This work was supported by Sichuan University – Suining Strategic Cooperation Project under Grant No. 2023CDSN-13. The authors would like to thank SESRI, Harbin Institute of Technology for supporting the proton irradiation experiments.

REFERENCES

[1] J. A. Cooper, M. R. Melloch, R. Singh, A. Agarwal and J. W. Palmour, "Status and prospects for SiC power MOSFETs," IEEE Transactions on Electron Devices, vol. 49, no. 4, pp. 658-664, April 2002.

[2] E. Mizuta, S. Kuboyama, H. Abe, Y. Iwata and T. Tamura, "Investigation of Single-Event Damages on Silicon Carbide (SiC) Power MOSFETs," in IEEE Transactions on Nuclear Science, vol. 61, no. 4, pp. 1924-1928, Aug. 2014.

[3] A. Akturk, J. M. McGarrity, R. Wilkins, A. Markowski and B. Cusack, "Space and Terrestrial Radiation Response of Silicon Carbide Power MOSFETs," 2017 IEEE Radiation Effects Data Workshop (REDW), New Orleans, LA, USA, 2017, pp. 1-5.

[4] D. R. Ball et al., "Ion-Induced Energy Pulse Mechanism for Single-Event Burnout in High-Voltage SiC Power MOSFETs and Junction Barrier Schottky Diodes," in IEEE Transactions on Nuclear Science, vol. 67, no. 1, pp. 22-28, Jan 2020.

[5] Peng, C., Lei, Z., Chen, Z., Yue, S., Zhang, Z., He, Y., Huang, Y., "Experimental and simulation studies of radiation - induced single event burnout in SiC - based power MOSFETs. " IET Power Electronics 14, 1700-1712, April 2021.

[6] C. Peng et al., "Mono-Energetic Proton Induced Damages in SiC Power MOSFETs," in IEEE Transactions on Device and Materials Reliability, vol. 23, no. 1, pp. 64-71, March 2023.

[7] C. Martinella et al., "Current Transport Mechanism for Heavy-Ion Degraded SiC MOSFETs," in IEEE Transactions on Nuclear Science, vol. 66, no. 7, pp. 1702-1709, July 2019.

[8] J. -M. Lauenstein, M. C. Casey, R. L. Ladbury, H. S. Kim, A. M. Phan and A. D. Topper, "Space Radiation Effects on SiC Power Device Reliability, " 2021 IEEE International Reliability Physics Symposium (IRPS), Monterey, CA, USA, 2021, pp. 1-8.

979-8-3503-6184-1/24 $31.00 ©2024 IEEE

Investigating Single-Event Burnout in 4H-SiC Inverters: Experiments and Simulations

Yong Gu[1], Yurui Yang[1], Yawen Xv, Hongyang Wen[1], Xiangyu Hou[1], Runhua Huang[2], Ao Liu, Song Bai[2], Jie Ma[1], Long Zhang*[1], Siyang Liu*[1], Weifeng Sun*[1]

[1] National ASIC System Engineering Research Center, Southeast University, Nanjing China
[2] Nanjing Electronic Device Institute, Nanjing 100048, China

* Email: longzh@seu.edu.cn, liusy2017@seu.edu.cn, swffrog@seu.edu.cn

Abstract—In this paper, the failure mechanism of single-event burnout (SEB) in 4H-SiC inverter is studied by experiments and simulations. The most sensitive region to SEB in 4H-SiC inverter circuit has been identified by conducting pulsed laser simulations of single-event experiments. Experimental evidence demonstrates that the maximum Linear Energy Transfer (LET) value tolerated by 4H-SiC inverter circuit falls between 64.07MeV·cm²/mg and 92.25MeV·cm²/mg. Sentaurus TCAD is employed to reveal the failure mechanism. After heavy ion irradiation, electron-hole pairs generated at the edge of the N-well cause the opening of the P+/N-WELL/P-epi bipolar junction transistor, followed by the activation of the N+/P-epi/N-well transistor, leading to circuit latching and eventual failure.

Keywords—*4H-SiC, Integrate, SiC Inverter, Single-event burnout (SEB), Pulsed laser, Heavy-ion*

I. INTRODUCTION

The chips utilized in deep space exploration equipment exhibit numerous degradation effects following exposure to cosmic ray radiation [1], [2]. Such as single event burnout (SEB) and single-event gate rupture (SEGR) [2], [4]. 4H-Silicon carbide (4H-SiC) exhibit a wider bandgap, higher critical breakdown electric field, and superior thermal conductivity compared to silicon materials [5]. As a result, the radiation tolerance of SiC devices is much stronger than that of silicon devices. Studies has been conducted to assess how single-event irradiation affects the performance of SiC power Metal Oxide Semiconductor Field Effect Transistors (MOSFETs) [6], [7]. The failure mechanisms of SiC MOSFETs when exposed to single-event irradiation have also been examined [8], [9]. Upon heavy ion impact on the SiC MOSFET, activation of the parasitic n-p-n bipolar junction transistor (BJT) occurs, causing concentrated transient energy dissipation at the drift/substrate junction and resulting in notable SEB failures.

4H-SiC integrated circuits，which currently experiencing rapid development，demonstrates an excellent radiation resistance. It holds potential for application in aerospace, space exploration, nuclear power plants, and other related fields. However, SiC integrated circuits possess a more intricate process flow and architecture compare to SiC MOSFETs. This increased complexity significantly facilitates the activation of parasitic BJTs following heavy ion irradiation. Nonetheless, there is limited documentation regarding the failure mechanisms of SiC integrated circuits subsequent to heavy ion impacts.

In this paper, 4H-SiC CMOS inverters process platform is developed and fabricated. Picoseconds-level pulsed laser is utilized to study the sing-event effect. Test results demonstrate that the maximum LET value that 4H-SiC CMOS inverters can withstand, and the failure mechanism, are revealed by Sentaurus TCAD.

II. PROCESS AND EXPERIMENTS

4H-SiC CMOS devices are fabricated on an N+ substrate (N-Sub) with double P-epitaxy (P-epi) layers. Fig. 1 illustrates the schematic cross-sectional view of the 4H-SiC CMOS process platform. The doping concentrations for N-Sub, P-epi1, and P-epi2 are 1.0×10^{19} cm^{-3}, 1.0×10^{17} cm^{-3} and 1.0×10^{16} cm^{-3}, respectively. N-well formation utilizes nitrogen ion implantation, followed by Nitrogen and Aluminum ion implantation for N+ and P+ ohmic contact formation, respectively. Subsequently, a 1650°C annealing process is performed on the wafer. A 45 nm gate oxide layer is grown using a wet oxidation process, followed by deposition and etching processes to complete fabrication. Polysilicon doping employs phosphorus. For the nMOS and pMOS designs in the 4H-SiC inverter, W/L ratios of 3 μm/6 μm and 3 μm/38 μm are used, respectively. The layout of CMOS device adopts a multi-fingered structure. The VDD voltage is 20V, and the input voltage ranges from 0 to 20V. Fig. 2 illustrates the input/output timing curve of the fabricated 4H-SiC inverter. Pulsed lasers are extensively used in the study of SEE[10]. The pulsed laser provide energy to the irradiated lattice, generating electron-hole pairs, which is similar to heavy ion impacts on semiconductors. In this study, a picosecond-level pulsed laser is employed to simulate heavy ion strikes on a 4H-SiC inverter.

The layout of the inverter circuit is segmented into six regions, each potentially sensitive to single-event irradiation, as illustrated in Fig. 3. These regions are individually scanned

This work was supported by the National Key R&D Program of China (2023YFB4403700), the natural science foundation of Jiangsu Province (BK20231150, BK20232006), the Distinguished Young Scholars Program of Southeast University (2242022R40010), the Technological Achievements of Jiangsu Province (No.BA2022005), and the Distinguished Young Scientists Foundation of Jiangsu Province (BK20230025).

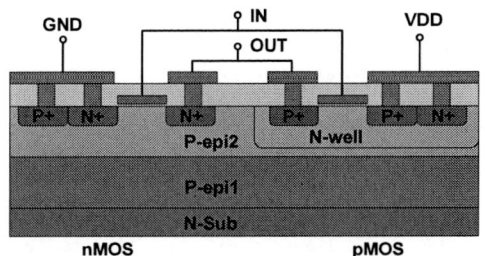

Fig. 1. Schematic diagram of the inverter circuit fabrication platform.

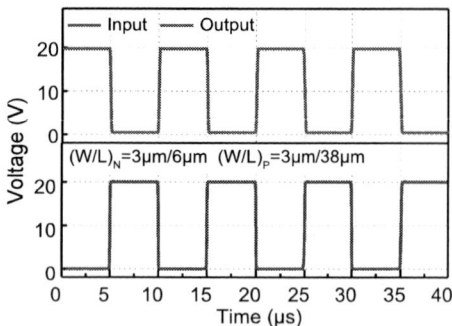

Fig. 2. Input-Output curve of 4H-SiC inverter circuit.

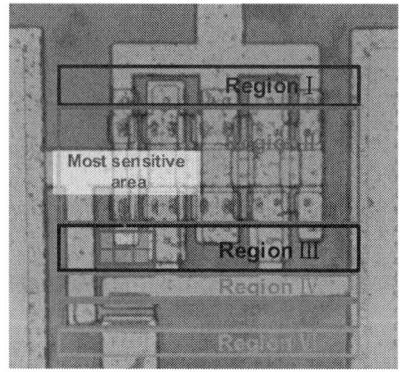

Fig. 3. Schematic diagram delineating regions potentially sensitive to single-event irradiation by the inverter.

using pulsed lasers with equivalent LET values setting at 2.57MeV·cm²/mg, 23.07MeV·cm²/mg. 64.07MeV·cm²/mg, 92.25 MeV·cm²/mg and 125.56MeV·cm²/mg. The inverter operates under 0V and 20V input voltage conditions. Table I summarizes the maximum tolerable LET values and the minimum LET values causing failure for regions I to VI in the 4H-SiC inverter circuit. The test outcomes identify region III as the most susceptible within the 4H-SiC inverter circuit, failing at LET = 92.25 MeV·cm²/mg.

Fig. 4 and Fig.5 illustrate the output voltage curves of the inverter following pulsed laser irradiation at LET values of 64.07 MeV·cm²/mg and 92.25 MeV·cm²/mg in region III, with input voltages of 0V and 20V, respectively. When region III is irradiated with a pulsed laser below 64.07 MeV·cm²/mg, the inverter's output voltage exhibits noticeable fluctuations, though individual outputs can recover. However, when a pulsed laser with LET= 92.25MeV·cm²/mg irradiates region III, the output voltage undergoes an irreversible transition from its normal level of 20V to 0V.

TABLE I
LET VALUES SUMMARY FOR IRRADIATION TOLERANCE
ACROSS VARIOUS REGIONS.

Regions	The maximum laser pulse energy for recoverable output waveform	The minimum laser pulse energy for irrecoverable output waveform
Region I	92.25 MeV·cm²/mg	125.56 MeV·cm²/mg
Region II	92.25 MeV·cm²/mg	125.56 MeV·cm²/mg
Region III	64.07 MeV·cm²/mg	92.25 MeV·cm²/mg
Region IV	92.25 MeV·cm²/mg	125.56 MeV·cm²/mg
Region V	92.25 MeV·cm²/mg	125.56 MeV·cm²/mg
Region VI	>125.56 MeV·cm²/mg	--

Pulsed laser testing is conducted across every point in region III to identify its most sensitive area. This investigation pinpointed the area marked by the red box in Fig. 3, situated at the edge of the N-well between the N+ and P+ electrodes, as the region of highest sensitivity. When subjected to pulsed laser irradiation in this area, the 4H-SiC inverter exhibited failure at a minimum LET value of 92.25 MeV·cm²/mg. Conversely, outside the marked region in region III, the minimum LET required for failure of the 4H-SiC inverter exceeded 125.56 MeV·cm²/mg.

III. MECHANISM AND ANALYSIS

Sentaurus TCAD is employed to elucidate the failure mechanism of the inverter. The simulation process platform is meticulously calibrated to ensure accuracy. The 3D structure of the most sensitive area, highlighted in red in Fig. 3, is modeled using Sentaurus TCAD. Fig. 6 illustrates the variation in total leakage current over time at LET = 92.25 MeV·cm²/mg. Initially, the circuit demonstrates minimal leakage current. Following exposure to heavy ions, there is a sudden and persistent increase in leakage current, suggesting the formation of a persistent new current path within the circuit. After impact of heavy ions, current density distribution at 1ns of the red-boxed marked area, which is under the SiC surface 0.2μm, are observed and illustrated in Fig. 7(a) (0V input) and Fig. 7(b) (20V input). Evidenced from Fig. 7, a significant current flow is established from N-well to N+ electrode. With the input signal set to 0, the temporal evolution of current density along section D-D1 following heavy ion irradiation is shown in Fig. 8. Before heavy ion impact (Fig. 8(a)), there's negligible leakage from VDD to GND. During irradiation, a substantial quantity of electron-hole pairs is generate along the ion's trajectory (Fig. 8(b)). Electrons are drawn to the P+ region by the electric

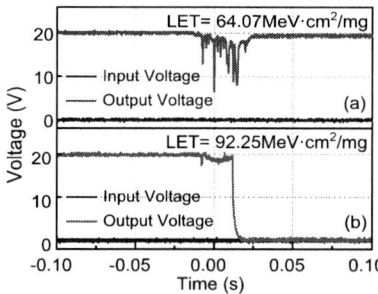

Fig. 4 Input and output curves of the 4H-SiC inverter at heavy ion incidence at (a) LET= 64.07MeV·cm²/mg and (b) LET= 92.25MeV·cm²/mg (input voltage 0V).

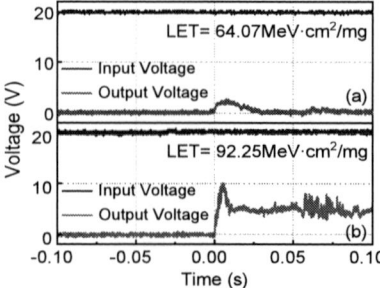

Fig. 5 Input and output curves of the 4H-SiC inverter at heavy ion incidence at (a) LET= 64.07MeV·cm²/mg and (b) LET= 92.25MeV·cm²/mg (input voltage 20V).

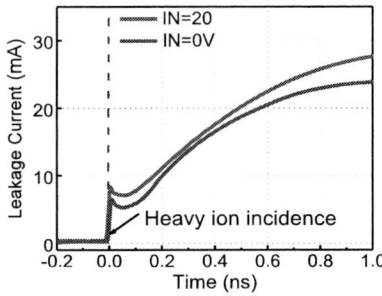

Fig. 6 Variation in total leakage current and as the function of time at LET=92.25MeV·cm²/mg.

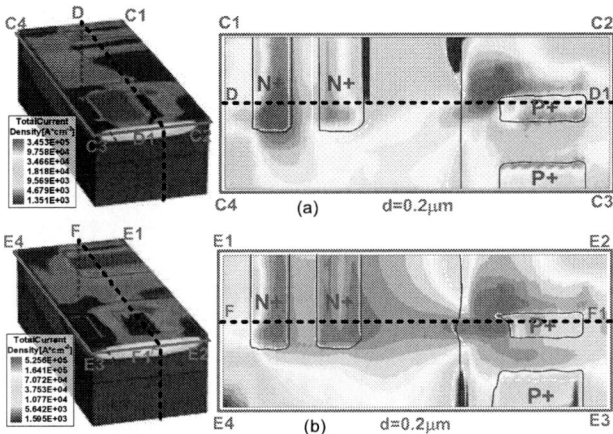

Fig. 7 Current density schematic diagram of the of the 3-D structure in the red box area and the cross-section 0.2um away from the semiconductor surface. (a)input signal = 0V, (b)input signal = 20V.

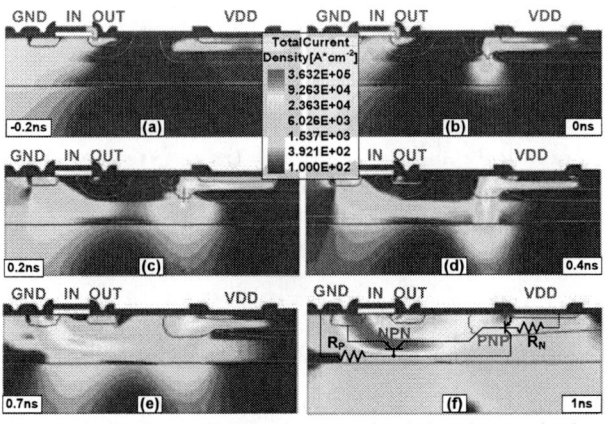

Fig. 8 Schematic diagram depicting the temporal variation of current density at section D-D1 before and after heavy ion irradiation (input signal is 0V).

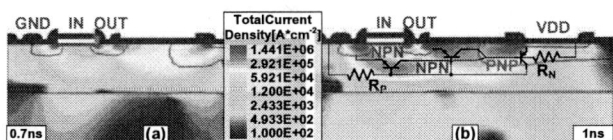

Fig. 9 Schematic diagram depicting the temporal variation of current density at section F-F1 before and after heavy ion irradiation (input signal is 20V).

field (Fig. 8(c)). The P+/N-well junction is forward-biased, activating the parasitic p-n-p transistor (Fig. 8(d)). Thus,

there's an influx of holes from P+ electrode to the P-epi. These carriers subsequently drift to the N+ electrode at GND side (Fig. 8(e)), elevating the potential of P-epi, which makes the p-n junction forward-bias. The n-p-n parasitic transistor is on. The device burnout is due to the positive feedback, resulting in a surge in the device's leakage current (Fig. 8(f)). The failure mechanism under an input voltage of 20V mirrors that at 0V. However, there is a slight difference between t=0.7ns and t=1ns. As shown in Fig.9(a) and Fig.9(b), NMOS is turned on at this state, most of the current flows to the drain of the NMOS first, and then flows to GND through the channel of the NMOS.

IV. CONCLUSION

The failure mechanism of 4H-SiC inverter struck by heavy ions is explored through experiments and simulations. The chips are scanned by pulsed laser to determine the most vulnerable location. Sentaurus TCAD is utilized to reveal the failure mechanism: the collision of heavy ions with the semiconductor results in the generation of substantial electron-hole pairs. Driven by the electric field, these carries drift, causing the activation of parasitic p-n-p and n-p-n transistors. Thus, the circuit failure attributes to latch-up effect in the circuit.

REFERENCES

[1] Y. Lei, J. Fang, Q. Shi, S. Li, L. Shi, X. Xiong, X. Luo and B. Zhang, "Thermal and Electrical Study of Single-Event Burnout and Hardening in 600 V Lateral DMOSFETs With Optimized Trench Drain," in *IEEE Transactions on Electron Devices*, vol. 70, no. 10, pp. 5294-5299, Oct. 2023.

[2] L. Zhang, Y. Gu, J. Ma, X. Hou, H. Wen, J. Hong, S. Liu, A. Liu, R. Huang, S. Bai and W. Sun, "Leakage Performance of 4H-SiC CMOS Logic Circuits After Gamma Irradiation," in *IEEE Electron Device Letters*, vol. 45, no. 4, pp. 542-545, April 2024.

[3] Z. Lei, C. Zhang, M. Wu, X. Zou, J. Yan, and Z. Chen, "Comprehensive study of the radiation effects on the LDMOS transistors," *Microelectron. Rel.*, vol. 139, Dec. 2022, Art. no. 114793.

[4] G. H. Johnson, J. H. Hohl, R. D. Schrimpf and K. F. Galloway, "Simulating single-event burnout of n-channel power MOSFET's," in *IEEE Transactions on Electron Devices*, vol. 40, no. 5, pp. 1001-1008, May 1993.

[5] K. Hamada, M. Nagao, M. Ajioka and F. Kawai, "SiC—Emerging Power Device Technology for Next-Generation Electrically Powered Environmentally Friendly Vehicles," in *IEEE Transactions on Electron Devices*, vol. 62, no. 2, pp. 278-285, Feb. 2015.

[6] R. A. Austin, B. D. Sierawski, R. A. Reed, R. D. Schrimpf, K. F. Galloway, D. R. Ball and A. F. Witulski, "Inclusion of Radiation Environment Variability for Reliability Estimates for SiC Power MOSFETs," in *IEEE Transactions on Nuclear Science*, vol. 67, no. 1, pp. 353-357, Jan. 2020.

[7] C. Abbate, G. Busatto, D. Tedesco, A. Sanseverino, F. Velardi and J. Wyss, "Gate Damages Induced in SiC Power MOSFETs During Heavy-Ion Irradiation—Part II," in *IEEE Transactions on Electron Devices*, vol. 66, no. 10, pp. 4243-4250, Oct. 2019.

[8] X. Zhou, Y. Jia, D. Hu and Y. Wu, "A Simulation-Based Comparison Between Si and SiC MOSFETs on Single-Event Burnout Susceptibility," in *IEEE Transactions on Electron Devices*, vol. 66, no. 6, pp. 2551-2556, June 2019.

[9] X. Zhou, Y. Tang, Y. Jia, D. Hu, Y. Wu, T. Xia, H. Gong and H. Pang, "Single-Event Effects in SiC Double-Trench MOSFETs," in *IEEE Transactions on Nuclear Science*, vol. 66, no. 11, pp. 2312-2318, Nov. 2019.

[10] C. Liang, R. Ma, K. Li, Y. Su, H. Gong, K. L. Ryder, P. Wang, A. L. Sternberg, E. X. Zhang, M. L. Alles, R. A. Reed, S. J. Koester, D. M. Fleetwood and R. D. Schrimpf, "Laser-Induced Single-Event Transients in Black Phosphorus MOSFETs," in *IEEE Transactions on Nuclear Science*, vol. 66, no. 1, pp. 384-388, Jan. 2019.

979-8-3503-6184-1/24 $31.00 © 2024 IEEE

Challenges of Design for Reliability in Advanced CMOS Technology:
From Single-mode to Mixed-mode Mechanisms

Zixuan Sun[1,3], Lining Zhang[2,3], Ru Huang[1,3], Runsheng Wang[1,3*]

[1]School of Integrated Circuits, Peking University, Beijing 100871, China. [2]School of Electronic and Computer Engineering, Peking University Shenzhen Graduate School, Shenzhen 518055, China. [3]Institute of Electronic Design Automation, Peking University, Wuxi 214000, China

[*]Email: r.wang@pku.edu.cn

Abstract

As CMOS devices scale down, reliability issues are gradually becoming one of the major challenges for circuit applications. Key factors such as hot carrier degradation (HCD), bias temperature instability (BTI), off-state stress degradation (OSD), time-dependent dielectric breakdown (TDDB) and electromigration (EM) are critical bottlenecks that constrain reliability. In practical circuit operations, voltage bias and workloads are complex and variable, and devices often experience the combined effects of multiple reliability factors. Traditional device characterization and modeling that only consider a single-mode reliability factor, are no longer sufficient to meet the circuit design requirement for high reliability, necessitating the consideration of mixed-mode reliability effects. Therefore, in this talk, our recent efforts on the mixed-mode reliability issues will be summarized in three parts: mixed-mode reliability issues under conventional on-state stress, under complex on-state stress, and under off-state stress, respectively.

First, for the mixed-mode issues under conventional on-state stress, we reveal the key roles of self-heating effects (SHE) in the deterioration of HCD, on-state TDDB and EM in FinFETs. Then, we systematically study the mixed-mode degradation under two types of complex on-state stress conditions: body-biased on-state stress, and alternating on-state stress. Finally, we conduct detailed research on the mechanism and modeling of mixed-mode reliability issues under off-state stress conditions, focusing on FinFET devices and advanced DRAM peripheral circuit devices. These works provide an in-depth understanding of mixed-mode reliability in nanoscale CMOS devices, which is important for the reliability evaluation of advanced technology and the reliability design for circuits.

Acknowledgments

This work was partly supported by NSFC (62125401, 61927901) and the 111 Project (B18001).

References

[1] Z. Sun et al., "The Understanding and Compact Modeling of Reliability in Modern Metal–Oxide–Semiconductor Field-Effect Transistors: From Single-Mode to Mixed-Mode Mechanisms," in Micromachines, vol. 15, no.1, pp. 127, 2024.

[2] Z. Sun et al., "Investigation of Interplays between Body Biasing and Hot Carrier Degradation (HCD) in Advanced NMOS FinFETs," in IEEE IRPS, Grapevine, TX, USA, 2024, pp. P72.TX-1-P72.TX-5.

[3] Z. Sun et al., "Transient Self-Heating Effects on Mixed-Mode Hot Carrier and Bias Temperature Instability in FinFETs: Experiments and Modeling," in IEEE Transactions on Electron Devices, vol. 70, no. 11, pp. 5528-5534, 2023.

[4] Z. Dong et al., "Catching the Missing EM Consequence in Soft Breakdown Reliability in Advanced FinFETs: Impacts of Self-heating, On-State TDDB, and Layout Dependence," in IEEE VLSI Technology and Circuits, Kyoto, Japan, 2023, pp. 1-2.

[5] Z. Sun et al., "Investigation of the Off-State Degradation in Advanced FinFET Technology—Part II: Compact Aging Model and Impact on Circuits," in IEEE Transactions on Electron Devices, vol. 70, no. 3, pp. 921-927, 2023.

[6] Z. Sun et al., "Investigation of the Off-State Degradation in Advanced FinFET Technology—Part I: Experiments and Analysis," in IEEE Transactions on Electron Devices, vol. 70, no. 3, pp. 914-920, 2023.

[7] R. Wang et al., "Understanding Hot Carrier Reliability in FinFET Technology from Trap-based Approach," in IEEE IEDM, San Francisco, CA, USA, 2021, pp. 31.2.1-31.2.4.

979-8-3503-6184-1/24 $31.00 © 2024 IEEE

Frequency-dependent Time-dependent Dielectric Breakdown (TDDB) Behavior and Physical Study in Gate Oxides

Wei Liu [1], Chu Yan [1], Xinwei Yu [1], Yiming Qu *[2], Wenchao Yan [3], Yi Zhao [1,4]

[1] College of Electronic Engineering and Information Science, Zhejiang University, Hangzhou 310027, China
[2] School of Integrated Circuits, East China Normal University, Shanghai 200241, China
[4] Zhejiang Li-ryder Technologies Co. LTD, Hangzhou 311300, China
[3] State Key Laboratory of Silicon and Advanced Semiconductor Materials, Zhejiang University, Hangzhou 310027, China

* Email: ymqu@ic.ecnu.edu.cn

Abstract—In this work, AC time-dependent dielectric breakdown (TDDB) was systematically investigated with considerable experimental data using various stress patterns. It is confirmed that TDDB lifetime could be improved at GHz. By analyzing the electrical parameter degradation during low-frequency AC stress, the TDDB lifetime deterioration under low frequency is physically attributable to the charges accumulated at the high-*k* (HK)/interfacial layer (IL) interface due to Maxwell-Wagner instability that generates additional defects in IL during the low frequency AC TDDB stress.

Keywords—oxide reliability, CMOS, TDDB, AC/DC, frequency dependance

I. INTRODUCTION

In modern integrated circuits, time-dependent dielectric breakdown (TDDB) is still considered a major reliability issue [1]. Transistors serving in consumer products and automotive ICs are both required to have a higher temperature tolerance, lower failure rate, and longer lifetime, which leads to reliability margins shrinking and requires more accurate lifetime evaluation [2-3]. To meet TDDB requirements, it has already been realized in the 1990s that only DC reliability studies are insufficient to predict circuit reliability [4]. Therefore, AC TDDB has always been a focus point through the technology nodes developing from bulk planar MOSFETs to tri-gate FinFETs, experiencing dielectric innovations from silicon dioxides to high-*k* materials [5-9]. It is widely accepted that high-frequency AC TDDB stress conditions bring lifetime gain owing to a lower defect generation rate or recovery effect under high-frequency operation conditions [10]. As for low-frequency, AC stress leads to a decay of TDDB lifetime compared with DC stress [11].

Except for frequency, more waveform factors are schematically shown in Fig. 1. The duty factor, i.e. the stress time (T_s) and relaxation time (T_r), directly determines defect generation and accumulation rate through the capture/emission time constant. Moreover, unipolar and bipolar bias conditions also influence, which is represented by low voltage level V_{low} [12]. To our best knowledge, there is still no clear conclusion if AC TDDB is worse or better than DC TDDB. Note that all TDDB data are strongly related to the gate-stack process, especially the nitrogen treatment and thermal annealing. The frequency-dependent TDDB behavior can be different in an optimized gate stack [13]. Since it is desirable to use the results of DC reliability studies to predict circuit reliability, we must clearly investigate what effect waveform has on TDDB.

II. TDDB CHARACTERIZATION

In this study, the foundry-level transistors on 300 mm wafers have been used. Typical constant voltage stress (CVS) was applied on the gate electrode with the source/drain/substrate electrode grounded. All experiments were performed in the MOSFET inversion regime at 85°C. Traditional CVS evitable interrupts the stress to make measurements. It is reported that a clear effect of stress interruption is evidenced on TDDB [14]. To overcome this limitation, the on-the-fly (OTF) technique was used to determine the time to breakdown (T_{BD}) for both uninterrupted DC and AC stress modes, as shown in Fig. 1. Different stress pulse widths (T_s) and relaxation pulse widths (T_r) were applied for physical mechanism analysis.

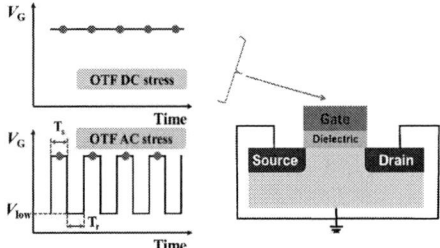

Figure 1. Schematic of the OTF monitoring methodology without sense for both uninterrupted DC and AC stresses.

III. TDDB AT GHz HIGH FREQUENCY

Fig. 2 shows the adopted ultra-fast measurement system, mainly including an arbitrary waveform generator (AWG), a digital phosphor oscilloscope (DPO) [15]. AWG is used to generate ultra-fast waveform. Here the AWG used can generate waveform with less than 1 ns period. which is necessary for our ultra-fast tests. DPO is used to monitor the real-time waveforms.

Fig. 3(a) shows the stress current traces (log-log scale) of SOI MOSFETs for DC and AC TDDB, respectively. The

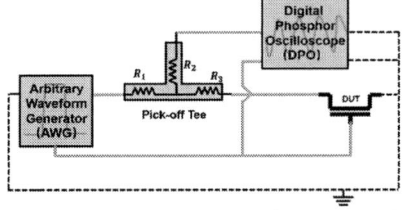

Figure 2. Ultra-Fast measurement architecture. AWG used for both gate and drain driving voltage. DPO used as voltage monitor to calculate current.

This work was supported in part by the National Key Research and Development Program of China under Grant 2020AAA0109001, National Natural Science Foundation of China under Grant 62204086, and Chenguang Program of Shanghai Education Development Foundation and Shanghai Municipal Education Commission under Grant 23CGA35.

979-8-3503-6184-1/24 $31.00 © 2024 IEEE

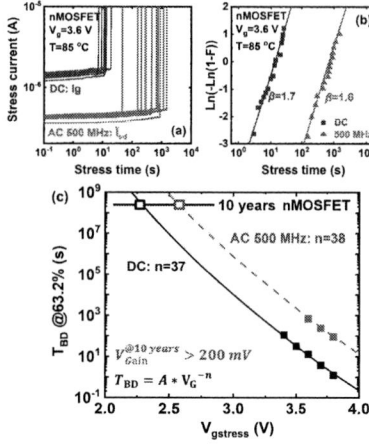

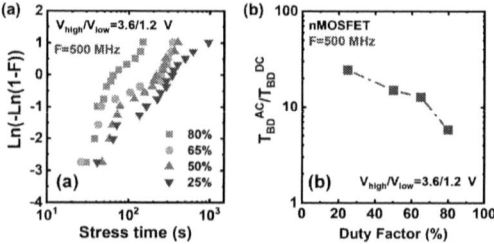

Figure 3. (a) Typical stress current traces until breakdown measured under DC and AC 500 MHz stress. (b) Weibull distributions of T_{BD} under DC and AC 500 MHz stress. No significant change in β is observed. (3) DC/AC TDDB voltage acceleration and lifetime extrapolation.

TBD of 500 MHz AC stress is much longer than that of DC stress. The cumulative distribution function (CDF) of each T_{BD} (dots) are calculated and plotted in Fig. 3(b), along with the cell-based statistical analysis model is also applicable to the AC TDDB. And the AC 500 MHz results show no significant change in Weibull slope β. Fig. 3(c) compares the voltage acceleration and lifetime extrapolation of DC/AC TDDB. It is confirmed that both T_{BD} are Power-law dependent on $V_{gstress}$, and the voltage acceleration factor (VAF) η_{AC} is comparable to the η_{DC}. It implies that the failure mechanism of AC TDDB is not out of the voltage-induced breakdown regime. And the slightly improved VAF increases the margin of reliability at nominal voltage. Compared to DC, AC 500 MHz has a voltage gain (more than 200 mV) at 10 years. Compared to DC, AC 500 MHz has a voltage gain (more than 200 mV) at 10 years.

Furthermore, the effects of different AC stress patterns on TDDB regarding frequency, V_{low} and duty factors are analyzed to explore the intrinsic reasons for the AC TDDB gain. Fig. 4 shows both the AC and DC Weibull distributions of T_{BD}. Pre-existing defects in the gate oxide are the main reason for T_{BD} deviation from Weibull distribution in the low CDF, which will lead to the non-random generation of defects during subsequent TDDB degradation. Overall, there is no significant change in the Weibull slope for different frequencies. Fig. 4(b) shows the AC/DC lifetime ratio (T_{BD}^{AC}/T_{BD}^{DC}) versus frequency, and T_{BD} extracted from Fig. 3(a) at 63.2% CDF. The AC TDDB lifetime is lower than the DC TDDB at low frequencies. And the T_{BD}^{AC}/T_{BD}^{DC} gradually increases with increasing frequency until GHz,

Figure 5. (a) Weibull distributions of T_{BD} for various duty factor conditions at 500 MHz. (b) Lifetime ratio (T_{BD}^{AC}/T_{BD}^{DC}) of nMOSFET with respect to duty factor.

which still does not have a saturation trend. This GHz frequency threshold around tenths of MHz, postulated by [8], from which TDDB increase should saturate.

To explore the contribution of the AC stress and relaxation stage for TDDB degradation in the time dimension, AC TDDB behaviors of different duty factor are shown in Fig. 5. It is found that the Weibull distribution of T_{BD} becomes distorted and the T_{BD}@63.2% has significant dependence on duty factor. Recovery effect is the intrinsic reason for AC TDDB lifetime increasing.

IV. TDDB At Hundred Hz Low Frequency

Although the low-frequency stress cannot mimic real-circuit operations like the GHz stress shown in section III, its comprehensive study is worth understanding the physical mechanism behind it. Therefore, this section focuses on the defect generation under AC stress below 1 kHz to clarify the TDDB lifetime deterioration.

Fig. 6 compares time-to-failure (TTF) results of FinFETs under DC and AC TDDB stresses with a 50% duty factor (DF), where TTF was defined by the hard breakdown. A typical Weibull distribution was observed in the range of T_s from 100 ms to 1 µs and TTF increases as the frequency increases. However, in the low-frequency range (T_s>1 ms), devices suffer worse degradation compared to the DC stress condition. Similar results were also reported in previous studies for FinFET TDDB [8] and usually this behavior was attributed to the charging and discharging of bulk oxide traps under low frequency stresses. Because the defect generation rate can be accelerated due to the energy release during discharging [14].

According to our previous work, two two pre-conditions are required for the AC TDDB lifetime deterioration to occur:

1) T_s is long enough for the full occupation of bulk oxide traps. (T_s>1 µs under stress voltage condition V_{stress})

2) T_r is long enough for the discharging of bulk oxide traps. (T_r>10 ms under relaxation voltage condition $V_{relaxation}$)

AC stress condition that satisfy the above pre-

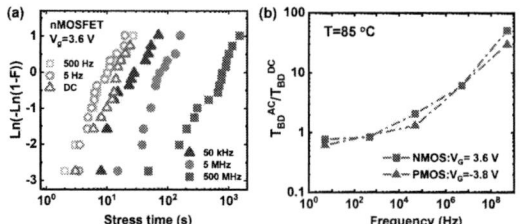

Figure 4. (a) Weibull distributions of T_{BD} under various frequency conditions. (b) Frequency dependent failure time ratio (T_{BD}^{AC}/T_{BD}^{DC}) of NMOS/PMOS.

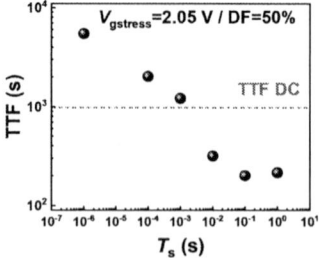

Figure 6. TDDB lifetime vs pulse width of AC stresses. When T_s<100 ms, TTF increases as frequency increases. However, TTF saturates to a value worse than DC for higher T_s.

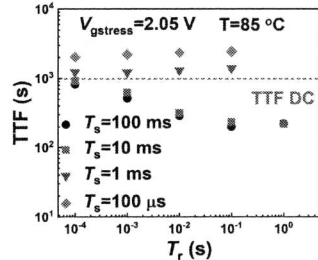

Figure 7. AC TDDB lifetime under various stress patterns. For stress patterns with $T_s > 10$ ms, TTF increases as T_r decreases. For stress patterns with $T_s < 1$ ms, TTF slightly increases as T_r increases.

conditions should introduce a shorter TDDB lifetime compared to the DC stress condition. As shown in Fig. 7, stress patterns with different T_s and T_r were applied to the device to verify the above proposed theory. Evidently, two trends can be observed in the results. The TTF increases with decreasing T_r in the stress patterns of $T_s = 100$ ms and 10 ms. This can be explained by the suppression of the oxide trap discharging due to the T_r reduction. However, for the stress patterns of $T_s = 1$ ms and 100 µs, the TTF decreases slightly with T_r decreasing, which is contrary to the results derived from the aforementioned theory. T_s is long enough for the full occupation of bulk oxide traps and T_r is also long enough for the discharge that accelerates the defect generation rate. But the TTF under such stress patterns is still higher than that under the DC condition.

Based on the above measurements results, the decay of the TDDB lifetime under low-frequency AC stress could be attributed to generation of extra interface defects. On the other hand, this order of time scale (~ms) is relatively long for the behavior of traditional traps, therefore triggering us to consider dielectric related phenomena which usually performs in longer time scale compare to the traps de-trapping.

Maxwell-Wagner (MW) instability of the bilayer gate stack could be used to explain the above results [16]. Fig. 8 explains the mechanism of TDDB lifetime decay under low frequency. First, when stress voltage is applied to the gate electrode, charges start to accumulate at the interface of two dielectric layers due to the different conductivities. Second, after a certain period of time (~10 ms), the charge density at the interface will be saturated and the MW instability reaches the equilibrium state. Then the charges rush out after the stress bias is removed, leading to the damage to the interfacial layer. Finally, after the stress bias is removed for long enough time, the charges could be totally released and the system reaches equilibrium again. The above process will be repeated throughout the low frequency AC TDDB stresses

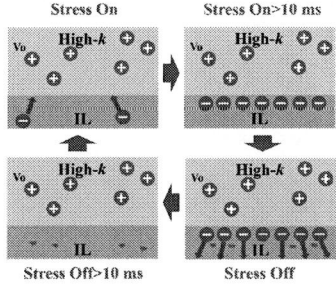

Figure 8. Schematic describes the triggering of the physical mechanism responsible for low-frequency TTF deterioration.

and ultimately results in the lifetime deterioration. As for the AC stresses with $T_s < 10$ ms, the charges at the interface cannot be completely filled in, thus the damage to interface layer is suppressed when the stress is removed. Similarly, for the AC stresses with $T_r < 10$ ms, the degradation of interface layer is suppressed because the charges cannot be completely released.

V. Conclusion

The frequency dependence of AC TDDB has been extensively investigated up to GHz in order to understand the reliability behaviors closer to real product operation. And more than 10× lifetime gain is observed when operation coming to GHz. The effect of frequency, V_{low}, and duty factor on AC TDDB, are concluded to study the related failure mechanisms. In addition, the low-frequency TDDB behavior has been systematically studied by using OTF monitoring methodology. Based on the experiment results, it has been found that the trapping/de-trapping of traps could not fully explain the low-frequency TDDB lifetime deterioration. The Maxwell-Wagner instability was proved to be responsible for it.

References

[1] J. W. McPherson, "Physical model for the frequency dependence of time-dependent dielectric breakdown (TDDB)," AIP Advances, vol. 13, no. 5, pp. 055217, 2023.

[2] Failure Mechanism Based Stress Test Qualification for Integrated Circuits, document AEC-Q100, Automotive Electronics Council, 2014.

[3] Y. Zhao, "Design of higher-k and more stable rare earth oxides as gate dielectrics for advanced CMOS devices," Materials, vol. 5, no. 8, pp. 1413-1438.

[4] E. Rosenbaum, Z. Liu, and C. Hu, "The effects of oxide stress waveform on MOSFET performance," in Proc. Internat. Electron Dev. Meet. (IEDM), pp. 719–722, 1991.

[5] M.S. Liang, S. Haddad, W. Cox, and S. Cagnina, "Degradation of very thin gate oxide MOS devices under dynamic high fieldicurrent stress," in Proc. Internat. Electron Dev. Meet., pp. 394–397,1986.

[6] M. Kerber et al., "Influence of charge trapping on AC reliability of high-k dielectrics," in Proc. Int. Rel. Phys. Symp. (IRPS), pp. 585–588 , 2004.

[7] K. Lee et al., "Frequency dependent TDDB behaviors and its reliability qualification in 32nm High-k/Metal gate CMOSFETs," in Proc. Int. Rel. Phys. Symp. (IRPS), pp. 2A.3.1–2A.3.5, 2011.

[8] I. Hirano et al., "Time-dependent dielectric breakdown (TDDB) distribution in n-MOSFET with HfSiON gate dielectrics under DC and AC stressing," Micro. Relia., vol. 13, no. 5, pp. 1868–1874, 2013.

[9] R. Ranjan, Y. Liu, T. Nigam, A. Kerber, and B. Parameshwaran, "Impact of AC voltage stress on core NMOSFETs TDDB in FinFET and planar technologies," in Proc. Int. Rel. Phys. Symp. (IRPS), pp. DG.10.1–DG.10.5, 2017.

[10] X. Yu, C. Yan, Y. Ding, Y. Qu, and Y. Zhao, "GHz AC to DC TDDB modeling with defect accumulation efficiency model," in Proc. Int. Rel. Phys. Symp. (IRPS), pp. 4C.3.1–4C.3.6, 2023.

[11] C. Yan, Y. Ding, X. Yu, Y. Qu, and Y. Zhao, "Physical study of low-frequency TDDB lifetime deterioration in advanced FinFETs," in Proc. Int. Rel. Phys. Symp. (IRPS), pp. 2A.3.1–2A.3.6, 2024.

[12] A. Kerber, E. Cartier, B.P. Linder, S.A. Krishnan, and T. Nigam, "TDDB failure distribution of metal gate/high-k CMOS devices on SOI substrates," in Proc. Int. Rel. Phys. Symp. (IRPS), pp. 505–509 , 2009.

[13] C.L. Chen et al., "The physical mechanism investigation of AC TDDB behavior in advanced gate stack," in Proc. Int. Rel. Phys. Symp. (IRPS), pp. 5B.5.1–5B.5.5, 2014.

[14] E.Y. Wu, D.P Ioannou, and C.B. Larow, "Influence of charge trapping on failure detection and its distributions for nFET high-k stacks," in Proc. Internat. Electron Dev. Meet. (IEDM), pp. 433–436, 2011.

[15] https://www.liryder.com/en/

[16] S. Knebel et al., "Influence of Frequency Dependent Time to Breakdown on High-K/Metal Gate Reliability," IEEE Transactions on Electron Devices, vol. 60, no. 7, pp. 2368–2371, 2013.

Lightning Protection Stacked TVS Structure Based on a Novel Total-Ionizing-Dose Radiation-hardened Technology

Zhao Qi*[1,2], Hongquan Chen[1,2], Yirui Jia[1], Nailong He[3], Zhili Zhang[3], Sen Zhang[3], Ming Qiao[1,4,5] and Bo Zhang[1]

[1] State Key Laboratory of Electronic Thin Films and Integrated Devices, University of Electronic Science and Technology of China, Chengdu, P. R. China
[2] Chongqing Institute of Microelectronics Industry Technology, UESTC, Chongqing, P. R. China
[3] CSMC Technologies Corporation, Wuxi, China.
[4] Institute of Electronic and Information Engineering of UESTC in Guangdong, Dongguan, P. R. China.
[5] Shenzhen Institute for Advanced Study, UESTC, Shenzhen, China

* Email: qizhao@uestc.edu.cn

Abstract—To mitigate the total ionizing dose (TID) effects of the transient voltage suppressor (TVS), a novel radiation-hardened technology that includes hardening process with oxide etch-back technology and low-voltage (LV) stacked TVS structure is proposed. In this work, the high breakdown voltage (BV) is achieved by employing a series of LV devices to replace sensitive low concentration drift-region in traditional high voltage devices. Furthermore, a novel low-trap-density oxide layer is developed by the high-temperature local oxidation of silicon (LOCOS) etch-back technology and subsequently low-temperature secondary thin oxidation, which greatly reduces the density of hole traps. By combining these two improvements, the TID-hardened lateral TVS with ultra-low clamping voltage (V_{CL}) and high transient power density for three voltage levels are achieved.

Keywords—*TVS, TID, radiation-hardened, hole traps*

I. INTRODUCTION

Transient Voltage Suppressor (TVS) is an important type of semiconductor protection device characterized by its fast response, accurate breakdown voltage (BV), and flexible design [1]. However, semiconductor devices exposed to radiation environments are vulnerable to Total Ionizing Dose (TID) effects, which may lead to performance degradation and even functional failures. High voltage (HV) devices are more sensitive than Low-voltage (LV) ones because of their low concentration drift region and thick oxide layer upon the drift-region [2,3].

The mechanism of device failure induced by TID effects is illustrated in Fig.1(a). Within a radiation environment, electron-hole pairs are generated by ionization within semiconductor devices. Electrons, due to their high mobility, drift out of the oxide layer. A portion of the holes recombine with electrons, while unrecombined holes migrate towards the Si-SiO$_2$ interface under the influence of the electric field. These unpaired holes are subsequently captured by trap centers, leading to the generation of interface states at the Si-SiO$_2$ interface. Such trapped charges can cause degradation to key parameters of a semiconductor device, such as the threshold voltage, radiation-induced leakage current, as shown in Fig. 1(b), and degenerated power consumption [4-7]. The performance of irradiated devices decreases with

This work was supported in part by the China Postdoctoral Science Foundation under Grant 2021M700684, and in part by the Chongqing Natural Science Foundation under Grant cstc2021jcyj-msxmX1023.

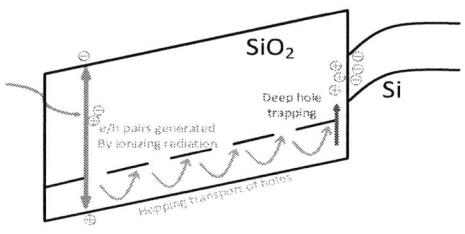

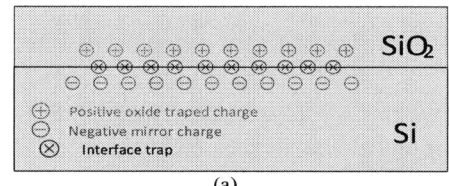

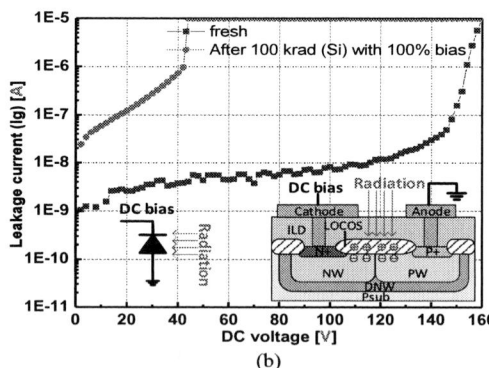

Fig. 1. (a) Diagram of the mechanism of device failure caused by TID effects. (b) Example of BV degradation of HV TVS diode fabricated by traditional HV LOCOS process.

increasing TID, and a sufficiently large TID can lead to permanent failure. Identify applicable funding agency here. If none, delete this text box.

In this paper, innovative designs are proposed from both structure and process aspects. In terms of the device structure, a LV stacked technique is employed to avoid thick SiO$_2$ upon the drift-region and the low concentration of the drift-region in traditional HV TVS. On the other hand, in order to reduces the density of Si/SiO$_2$ interface hole traps, a novel radiation hardening process with the local oxidation of silicon (LOCOS)

979-8-3503-6184-1/24 $31.00 © 2024 IEEE

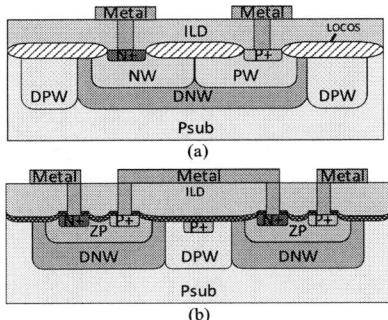

Fig. 2 Cross-sectional of the (a) conventional lateral HV TVS, and (b) lateral LV stacked TVS.

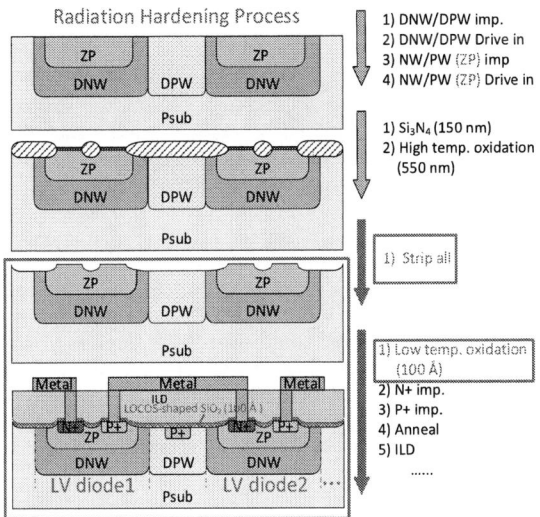

Fig. 3 Diagram of the flow for the radiation-hardening back-end technology.

Etch-Back and low-temperature secondary oxidation technology is proposed and developed.

II. RADIATION-HARDENED TVS

A. Low-Voltage Stacked TVS Structure

The cross-sectional view of conventional HV TVS and the low-voltage stacked TVS structure developed in this paper are illustrated in Fig. 2. Conventional HV devices employ thicker field oxide layer on its voltage-withstanding layer (NW and PW), which make them more susceptible to TID. To avoid this issue, this paper proposes a new low-voltage stacked TVS structure. The new lightning-protection diode cancels the NW and PW voltage-withstanding layers and adopts high-concentration Zener-P-well (ZP) and LV diode serial connection structure to against TID, while achieving high voltage tolerance, as shown in Fig. 2(b). Meanwhile, a P+ under metal (PPUM) was implanted into the DPW to block the potential leakage channel. By employing this stacked structure, different voltage levels of HV TVS can be achieved by varying the number of stacked LV diode.

B. Novel Radiation-Hardening Process

Based on the low-voltage stacked structure, the thick thermal oxide layer in the traditional LOCOS process will be eliminated and replaced by a thin low-temperature oxide layer with fewer hole traps. The proposed novel radiation-hardened

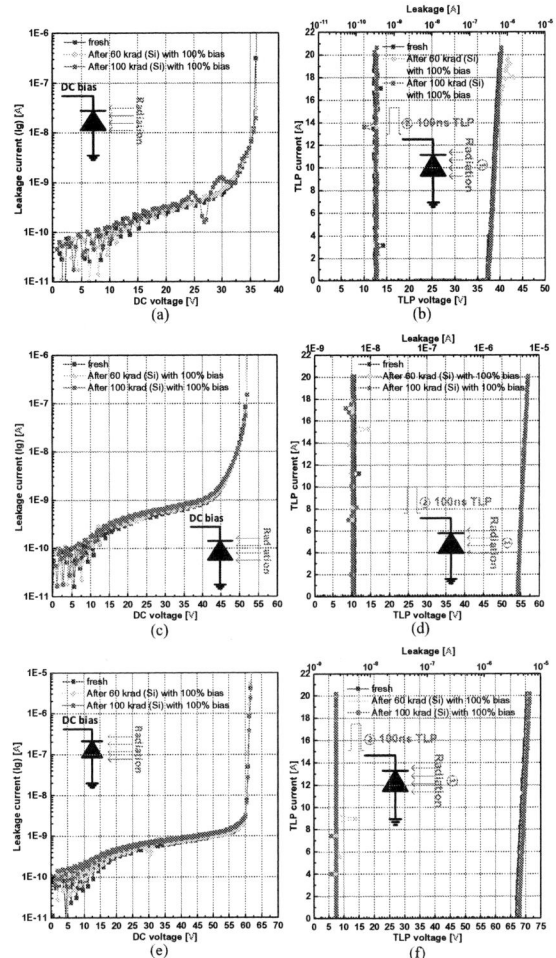

Fig. 4 Measured I-V characteristic curves at fresh, after 60 krad(Si) irradiation, and after 100 krad(Si) of (a) 28 V TVS, (c) 48 V TVS and (e) 60 V TVS, and the measured TLP characteristic curves at fresh, after 60 krad(Si) irradiation, and after 100 krad(Si) of (b) 28 V TVS, (d) 48 V TVS and (f) 60 V TVS.

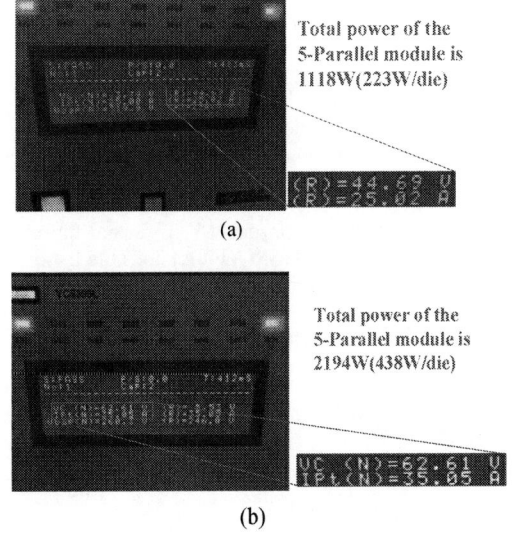

Fig. 5 (a) 10/1000-μs lightning test of the proposed 28V TVS 5-parallel module and (b) proposed 48V TVS 5-parallel module.

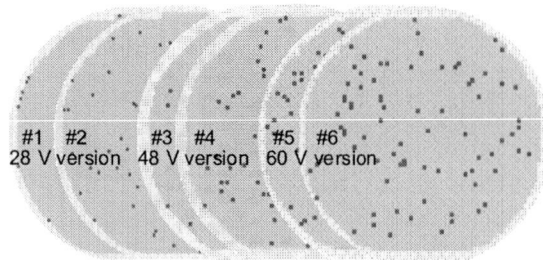

Fig. 6 The wafer yield distribution maps of the proposed stacking TVS in three versions of 28V, 48V, and 60 V.

TVS fabrication process is illustrated in Fig. 3. Following the conventional LOCOS process flow, all high-temperature oxide layers on the silicon surface are etched away, and then a low-temperature secondary oxidation is performed at 850°C to form a locos-shaped thin SiO_2 layer of approximately 10 nm thick. In this way, the hole trap density introduced by the thick oxide layer will be greatly reduced, and the radiation resistance of the novel TVS will be significantly improved.

III. RESULT AND DISCUSSION

To investigate the TID effects on the proposed TVS, samples of three voltage levels of 28 V, 48 V and 60 V are irradiated in ^{60}Co source with a dose rate of 50 rad(Si)/s under 100% bias. Fig. 4(a) presents the static BV test curves for a 28V TVS (with BV of 36V) under no irradiation, after 60 krad(Si) irradiation, and after 100 krad(Si) irradiation. It shows negligible degradation in BV and leakage current after exposed to 100 krad(Si) under 100% bias. Additionally, Fig. 4(b) exhibits the TLP characteristics, proving that the proposed TVS can still ensure ESD capability after irradiation. The same test was performed on 48 V and 60 V TVS, the test results are shown in Fig. 4(c)-(f), demonstrating a significant radiation hardening effect.

To verify the surge current capability of proposed TVS, a standard 10/1000 µs lightning pulse test was performed on a 28V TVS module (5 dies in parallel), as shown in Fig. 5(a), exhibiting its excellent lightning peak current (I_{PP}) of 25 A with only the V_{CL} of 44.69 V (>1 kW). This indicates that each die withstands approximately 5 A of lightning strike current while achieving a V_{CL} of only 44.69 V, despite having a chip area of just 1 mm^2 (1 mm*1 mm). Further tests on the 48V TVS module containing 5 parallel chips (as shown in Fig. 5b) revealed that it endured a lightning strike power of 2194W, with an I_{PP} reaching an astonishing 35 A and a V_{CL} of only 62.61 V. This indicates that a single 1.5 mm*1.5 mm 48 V TVS chip passed a lightning strike current of approximately 7 A and possessed extremely low V_{CL}. A comparison of the key indicators between the proposed 48V TVS and the existing vertical anti-lightning TVS (SMAJ48A) in Table I shows that under similar die size, the lateral TVS proposed in this paper exhibits higher I_{PP} and lower V_{CL} compared to traditional vertical TVS, indicating that the proposed TVS device has better overcurrent protection and clamping performance.

TABLE I. COMPARISON OF KEY INDICATORS BETWEEN THE PROPOSED 48V TVS AND THE TRADITIONAL SMAJ48A

	Stand off Voltage (V)	BV@1 mA (V)	V_{CL} (V)	I_{PP} (A)	Power (W)
SMAJ48A	48	56.1	77.4	5.2	400
Proposed	48	53.3	62.6	7	430

Additionally, the 3σ yields for 28 V, 48 V, and 60 V wafers are presented in Fig. 6 and summarized in Table II, demonstrating the technical stability.

TABLE II. CHIP PROBING TEST RESULT SUMMARY

Voltage level	Wafer ID	Yield (%)	Leakage current[a]	BV[b]
28 V	#1	99.56	0.1-60 nA	35.8-36.3 V
28 V	#2	99.24		
48 V	#3	99.30	0.1-60 nA	51.5-52.5 V
48 V	#4	98.72		
60 V	#5	98.02	0.1-60 nA	64.8-65.9 V
60 V	#6	97.39		

[a.] Leakage are tested in 28 V for 28 V TVS, 48 V for 48 V TVS and 60 V for 60 V TVS.

[b.] The voltage at 10 mA current.

IV. SUMMARY

A novel radiation-hardened technology that includes a novel hardening process and the new LV stacked structure is proposed, aiming to enable TVS to withstand TID radiation and ensure their lightning protection capability in irradiation environment. The TID-hardened TVS for three voltage levels of 28 V, 48 V and 60 V are fabricated and tested. Under standard 10/1000 µs pulse testing, TID-hardened lateral TVS exhibited a high transient power density with I_{PP} of 7 A for 48 V TVS in the case of a small area of 1.5 mm*1.5 mm, and possessed extremely low V_{CL} of 62.61 V, which demonstrates a clear superiority over traditional TVS devices. Besides, compared to the pre-irradiation test data, the proposed TVS exhibited negligible degradation in both I-V and TLP characteristics after 100 krad(Si) irradiation, proving that the proposed TVS can still ensure excellent overcurrent protection capability after irradiation.

REFERENCES

[1] Zhao Qi, Ming Qiao, Jingqi Wei, et al., "Novel Multifunctional Transient Voltage Suppressor Technology for Modular EOS/ESD Protection Circuit Designs," IEEE International Symposium on Power Semiconductor Devices and ICs, 2023, Hong Kong, China

[2] Zhangyi'an Yuan, Ming Qiao, Xinjian Li, et al., "Improved Model on Buried-Oxide Damage Induced by Total-Ionizing-Dose Effect for HV SOI LDMOS," IEEE Transactions on Electron Devices, vol. 68, NO. 4, pp. 2064-2070, 2021

[3] Zhuojun Chen, Wenzhao Lu, Ming Wu, et al., "Comparative Study of Total Ionizing Dose Effects onthe Silicon-Controlled Rectifier Devices for HV and LV ESD Protections," IEEE 26th International Symposium on Physical and Failure Analysis of Integrated Circuits, 2019, Hangzhou, China

[4] T. Borel, S. Furic, E. Leduc, et al., "Total Ionizing Dose Effect in LDMOS Oxides and Devices," IEEE Transactions on Nuclear Science, vol. 66, NO. 7, 2019

[5] Ming Wu, Chenchen Zhang, Wei Peng, et al., "A Radiation-Hardened Dual-Direction SCR Based on LDMOS for ESD Protection in the Extreme Radiation Environment," IEEE Transactions on Nuclear Science, vol. 67, NO. 4, 2020

[6] Wei Liang, Kostas Alexandrou, Maxim Klebanov, et al., "Characterization of ESD Protection Devices under Total Ionizing Dose Irradiation," IEEE 24th International Symposium on Physical and Failure Analysis of Integrated Circuits, 2017, Chengdu, China

[7] M. Young, Zhong Daohong, Ma Zhen, "Breakdown Voltage Shift Mechanism Analysis of Bi-directional ESD Protection Device After Total Ionizing Dose Test," International EOS/ESD Symposium on Design and System, 2022, Chengdu, China

Characterization and Modeling of Non-conducting RF Hot Carrier Stress in FinFETs

G. Niu[1], X. Ding[1], H. Zhang[2], W. Wang[2], K. Imura[2], and F. Dai[1]

[1]Alabama Micro/Nano Electronics Science and Technology Center,
Electrical and Computer Engineering Department, Auburn University, Auburn, AL, USA
[2]Maxlinear Inc., Carlsbad, CA, USA

Abstract—**A significant reliability concern in RF power amplifiers is non-conducting RF hot carrier stress. We present experimental characterization and lifetime modeling of non-conducting RF hot carrier stress in modern FinFETS and demonstrate the importance of near-threshold stress. The relationship between RF and DC stresses is examined, and the quasi-static approximation is discussed. Die-to-die threshold voltage variation makes the measurement and modeling of V_{ds} dependence difficult. A solution using the source current is proposed and demonstrated.**

Index Terms—**hot carrier stress, RF stress, non-conducting stress, FinFET, power amplifier**

I. Introduction

An important reliability concern in RF power amplifiers (PA) is non-conducting (NC) or off-state RF hot carrier stress, which occurs when the transistor experiences a high V_{ds} when the V_{gs} goes below the threshold. While the channel is not significantly conducting, or in "off-state", hot carriers produced by the gate-induced-drain-leakage (GIDL) and channel current still cause significant degradation [1] [2] [3] [4]. NC stress exists in CMOS logic gates, DRAM and SRAMs as well [5]. The situation is worsened in modern CMOS due to limited supply voltage and breakdown voltage.

NC stress is historically evaluated at $V_{gs} = 0$ V, the off-state condition in CMOS inverters. For RF circuits, however, this is not sufficient, as the transistor V_{gs} dynamically changes and have a high probability of being around the threshold voltage. An example of V_{gs} and V_{ds} waveforms, along with the probability density of the dynamic operating point in the V_{gs} and V_{ds} plane, are shown in Fig. 1 (a) and (b), respectively, for a digitally modulated RF PA at peak power. The probability of V_{gs} being around the V_{th} is high, at which the V_{ds} is near its maximum. Hot carrier stress models in design kits are typically based on DC stress data at V_{gs} well above the threshold voltage, which is not sufficient for RF PAs. Therefore, it is essential to experimentally characterize RF NC stress at $V_{gs} = V_{th}$ for PAs.

In another design, the minimum V_{gs} can reach 0 V or even become negative. Experiments should be conducted for all possible V_{gs} values. For the devices tested, however, the degradation at $V_{gs} = V_{th}$ is found to be much more significant than at $V_{gs} = 0$ V, even after considering the smaller V_{ds} for $V_{gs} = V_{th}$ in the simulated waveforms. We therefore briefly discuss $V_{gs} = 0$ V and mainly discuss $V_{gs} = V_{th}$ in this work.

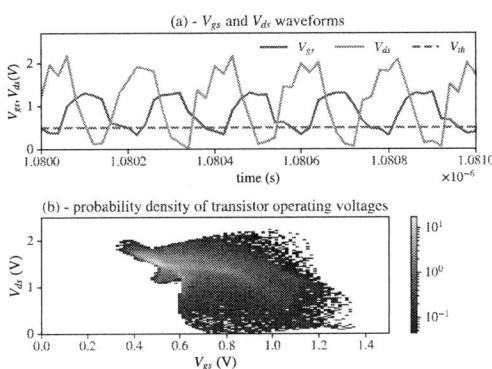

Fig. 1: (a) Simulated V_{gs} and V_{ds} waveforms; and (b) corresponding probability density of RF operating point in a digitally modulated RF PA at peak power.

II. Experiments

$L = 135$ nm n-channel IO devices used in RF PA design are fabricated using a 14/16-nm FinFET technology from a major foundry. A 5 GHz RF NC stress is used due to its relevance for our PA. For selected experiments, 2 GHz RF stress was also used. The nominal V_{dd} is 1.8 V. The devices have hundreds of fins, and no dependence of the stress response on the number of fins is observed.

The source and body are tied to the ground in the RF test structures. The gate and drain are probed using GSG probes, and voltages are supplied through bias tees. The gate is given a DC V_{gs} below or equal to V_{th}, and a 50 Ω RF termination to avoid reflections. The drain is given a DC bias and RF power that needs to be carefully chosen so that the resulting net V_{ds} waveform is relevant for circuit operation.

We set the drain DC bias and RF power so that the V_{ds} minimum is zero, and the V_{ds} maximum is at a desired level. Such a waveform is not only relevant for RF PAs, but also suppresses potential unintentional reverse mode on-state stress as detailed below. S_{22} is used to estimate the drain impedance, which is then used to calculate the required RF source power. Circuit simulations using a calibrated transistor model, including pad parasitics, are performed to confirm the required RF power for a desired $V_{ds,max}$. Simulated V_{gs} and V_{ds} waveforms under NC RF stresses of various $V_{ds,max}$ are shown in Fig. 2 for a gate DC bias of 0 V. As $|S_{12}|$ is small

979-8-3503-6184-1/24 $31.00 © 2024 IEEE

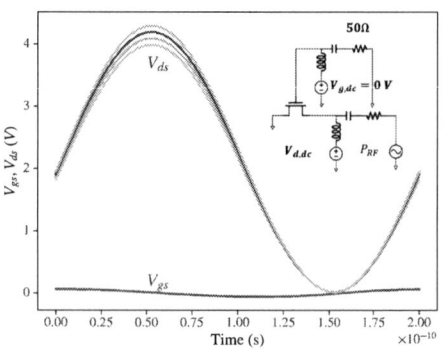

Fig. 2: Simulated V_{gs} and V_{ds} waveforms under RF NC stresses of various $V_{ds,max}$ at $V_{gs,dc} = 0$ V.

for low V_{gs}, the RF swing of V_{gs} is negligible, making V_{gs} practically DC despite the high V_{gs} required for accelerated testing. Lifetime modeling is simplified as only one voltage is time-varying.

The stress is periodically interrupted to measure I-V and s-parameters, from which g_m, C_{gs}, C_{gd} characteristics are extracted. The constant switching between applying RF stress and measuring s-parameters is thus required and achieved using a Keysight N5242B PNA-X capable of high RF output power. Separate channels are set up for RF stress and s-parameter measurement. The cable and probe losses are significant at RF. They must be accounted for using a power meter to obtain accurate RF voltage at the drain terminal, which is critical for determining the RF V_{ds} dependence of degradation. An HP4155 is used for DC bias and I-V measurement.

One issue we encountered was unintentional reverse mode on-state stress, which occurs when the RF power is applied before DC bias is established due to the inevitable latency between the DC and RF sources. Consider a gate voltage of 0 V. Before the drain DC bias is established, the drain sees a large negative voltage, e.g., -2 V at its negative peak for a 2 V amplitude, and a large amount of reverse current flows under high V_{ds}, which degrades the device fast. In our waveform design, the DC bias is kept relatively low, and the degradation is negligible. Therefore, we can apply the RF power after the DC bias is well established by a safe delay without unintentional DC stress. Similarly, we turn off the DC bias after RF power is turned off by another delay.

The stress voltages are chosen so that the measurements for one stress condition can be completed within 30 hours, the limit of typical laboratory on-wafer RF measurement.

III. RF stress vs DC stress

The percentage change of I_{dsat}, the I_d at $V_{gs} = V_{ds} = V_{dd}$, is chosen below to represent degradation denoted as D, a somewhat arbitrary but popular choice. As full bias range I-V and S-parameters are also measured [4], we have a complete

picture of the device degradation corresponding to each I_{dsat} degradation level. Lifetime τ is defined as the stress time for a 10% I_{dsat} degradation, a widely used industry standard. Our goal is to determine the RF V_{ds} maximum allowed for a τ of 10 years.

As RF stress experiments are much more complex, costly, and time-consuming compared to DC stress, we are interested in finding out if the RF stress is quasi-static, or loosely speaking, quasi-dc, meaning that the rate of degradation at any time t, dD/dt, is the same as that measured under DC stress at a voltage equal to the instantaneous voltage, in our NC stress case, $V_{ds}(t)$. If true, degradation under RF stress can be calculated from DC stress measurements using quasi-static approximation (QSA).

Both the DC and RF stress $D(t)$ show a typical power law time dependence, i.e., $D(t) = At^n$, with A being a function of the stress voltage. For QSA to possibly hold, n must be the same for different stress voltages and must be the same for DC and RF stresses. If both are true, QSA formulation can be used to calculate RF stress degradation from DC stress, which is then compared to RF stress measurements. If a good agreement is achieved, we can further use QSA to determine the highest allow RF stress by extrapolation.

At $V_{gs}=0$ V, we found that the NC RF stress is not quasi-static at all [3]. GIDL dominates the drain current, and the hot carriers are likely related to or initiated by the GIDL, as was also reported for NC DC stress [1] [3] [5] [6].

However, at $V_{gs}=V_{th}=0.5$ V, QSA approximately holds, provided that the die-to-die variation of the hot carrier initiating subthreshold channel current is correctly accounted for [4], as detailed below.

IV. V_{gs}=0 V Stress

Fig. 3 compares the DC and RF stresses at $V_{gs}=0$ V. DC stress $V_{ds}=4.1, 4.2,$ and 4.3 V. and RF stress $V_{ds,max}=3.9, 4.0, 4.1,$ and 4.2 V. The DC stresses approximately share one slope (n) parameter, while the RF stresses share another, suggesting a non-quasi-static mechanism.

Fortunately, utilizing the shared n for all RF stresses, we can directly model the RF stress lifetime as a function of $V_{ds,max}$. A linear relation is observed between $\ln(\tau_{RF})$ and $1/V_{ds,max}$, as shown in Fig. 4 [3]. The allowed $V_{ds,max}$ for a 10-year lifetime is then found by extrapolating the linear fit to 10 years, 3.32 V, sufficient for our PA design if we only consider NC RF stress occurring at V_{gs} of 0 V. However, the NC RF stress experienced when V_{gs} is above 0 V, particularly around V_{th}, can cause much more degradation, even after considering on average a smaller V_{ds} than at $V_{gs} = 0$ V, which we address below.

V. Near-threshold Stress

As shown in Fig. 1, the stacked transistor in the PA operates near the threshold voltage, $V_{gs}=0.5$ V, with a high probability.

979-8-3503-6184-1/24 $31.00 © 2024 IEEE

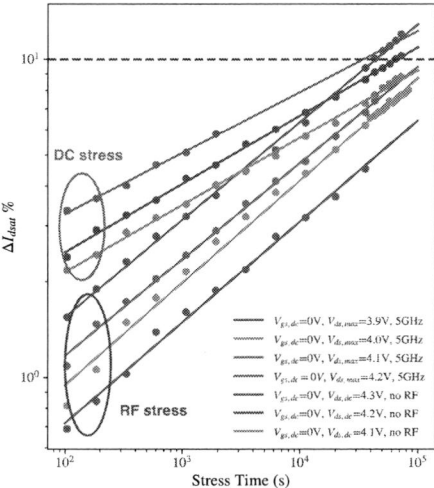

Fig. 3: ΔI_{dsat} versus stress time for V_{gs}=0 V DC and RF stresses.

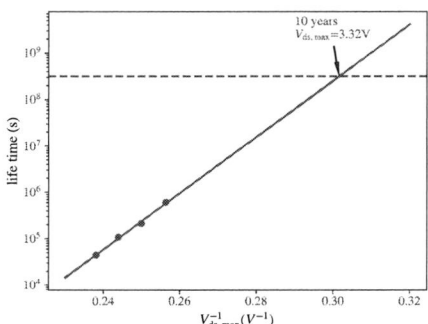

Fig. 4: RF stress lifetime versus $V_{\mathrm{ds,max}}^{-1}$. V_{gs}=0 V.

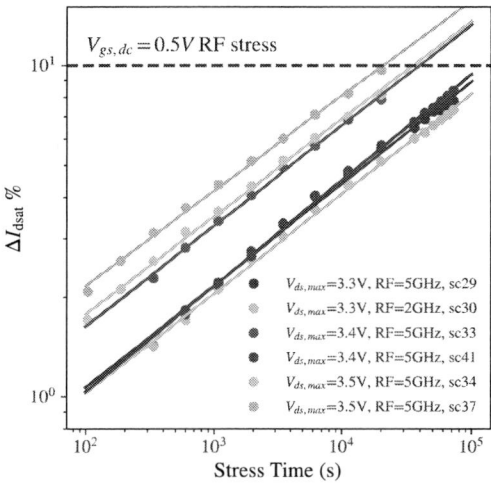

Fig. 5: Measured and fitted ΔI_{dsat} versus stress time for RF stresses at V_{gs}=0.5 V with varying $V_{\mathrm{ds,max}}$.

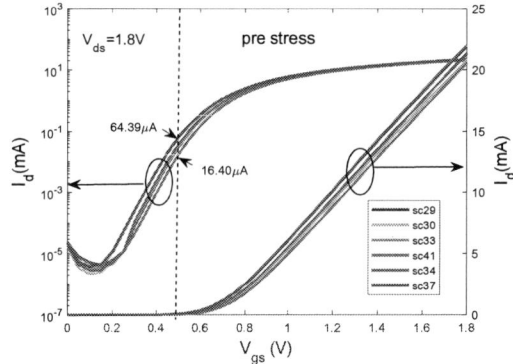

Fig. 6: Die-to-die variation of pre-stress $I_{\mathrm{d}} - V_{\mathrm{gs}}$.

Therefore, we make DC and RF stress measurements at V_{gs}=0.5 V. V_{ds} is reduced compared to V_{gs}=0 V.

Identifying the $V_{\mathrm{ds,max}}$ dependence of stress response, however, is quite tricky, as shown in Fig. 5 using a set of RF stress measurements with varying $V_{\mathrm{ds,max}}$. For example, the two $V_{\mathrm{ds,max}}$=3.5 V 5 GHz RF stresses, denoted as sc34 and sc37, show a significant difference, which is a result of die-to-die variation of the subthreshold current, as shown by the pre-stress $I_{\mathrm{d}} - V_{\mathrm{gs}}$ in Fig. 6. The slope parameter n varies from 0.29 to 0.32, centering around 0.3, and similar to the dc stress n, making QSA possible. Lifetime is extracted for all RF stress measurements in Fig. 5, and used to generate the $1/\tau_{\mathrm{RF}}$ versus $V_{\mathrm{ds,max}}$ plot in Fig. 7 as "+" symbols.

Analysis of measurement data shows that the primary source of die-to-die variation of the stress response is the subthreshold current that initiates the hot carriers. We thus use the source current instead of V_{gs} in modeling the DC degradation function

[1] [4]:

$$D_{\mathrm{DC}}(t) = \left(K \frac{I_{\mathrm{s}}}{N_{\mathrm{fin}}} V_{\mathrm{ds}}^m t \right)^n, \qquad (1)$$

where I_{s} is the source current, N_{fin} is the number of fins, K, m, and n are parameters determined from DC stress data. The t for a $D_{\mathrm{DC}}(t) = C$, with $C = 10\%$ is by definition τ_{DC}, the inverse of which is given by [4]:

$$\frac{1}{\tau_{\mathrm{DC}}} = C^{-\frac{1}{n}} \cdot K \cdot \frac{I_{\mathrm{s}}}{N_{\mathrm{fin}}} \cdot V_{\mathrm{ds}}^m. \qquad (2)$$

Note that I_{s} also depends on V_{ds}. Applying QSA, we can model τ_{RF} through integrating $1/\tau_{\mathrm{DC}}$ over one period T [7]:

$$\tau_{\mathrm{RF,QS}} = \frac{T}{\int_0^T \frac{1}{\tau_{\mathrm{DC}}(V_{\mathrm{ds}}(t))} dt}. \qquad (3)$$

For sine waves, the result is frequency-independent. The modeled $1/\tau_{\mathrm{RF,QS}}$ are shown in Fig. 7 as circles. A reasonable agreement between measurement and modeling is achieved. At

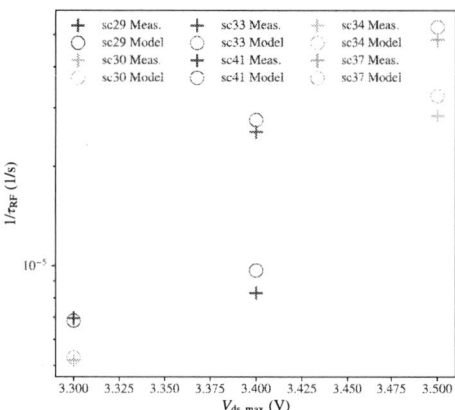

Fig. 7: $1/\tau_{\mathrm{RF}}$ versus $V_{\mathrm{ds,max}}$ from measurements and QSA modeling.

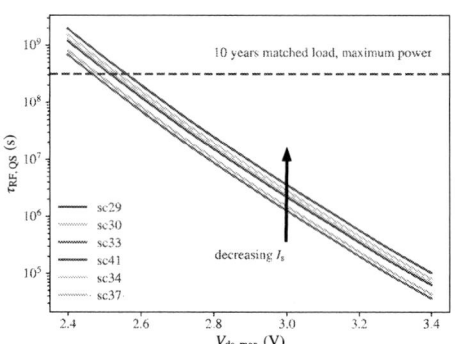

Fig. 8: QSA modeled τ_{RF} versus $V_{\mathrm{ds,max}}$. $V_{\mathrm{gs,dc}}$=0.5 V.

the same $V_{\mathrm{ds,max}}$, the $1/\tau_{\mathrm{RF,QS}}$ difference between different dies is entirely due to I_{s} difference. The allowed $V_{\mathrm{ds,max}}$ for a 10-year lifetime is then found by calculating $\tau_{\mathrm{RF,QS}}$ as a function of $V_{\mathrm{ds,max}}$, using the I_{s} of the devices in Fig 5, as shown in Fig. 8. The $V_{\mathrm{ds,max}}$ varies from 2.46 to 2.56 V, which is sufficient for our PAs that see less than 2 V. Measurement and modeling of the $V_{\mathrm{ds,max}}$ dependence of τ_{RF} would be difficult if not impossible without considering the die-to-die variation of I_{s}.

During measurement, die-to-die variation of the hot carrier inducing subthreshold current can be approximately accounted for by plotting the degradation as a function of $I_{\mathrm{s}}(V_{\mathrm{ds,max}}) \cdot t$, with I_s being the source current at $V_{\mathrm{ds,max}}$, as shown in Fig. 9. The degradation data for the same $V_{\mathrm{ds,max}}$ from different dies are then much closer than in Fig. 5, making the identification and subsequent modeling of $V_{\mathrm{ds,max}}$ dependence much easier.

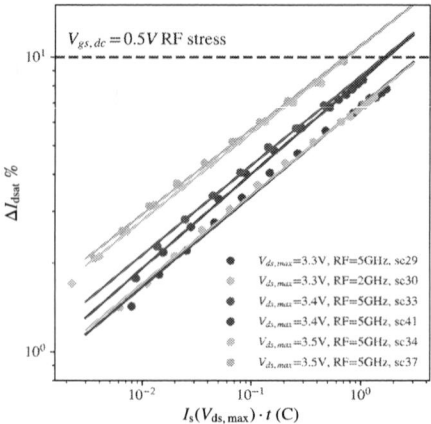

Fig. 9: Measured (symbols) and fitted (lines) ΔI_{dsat} versus $I_{\mathrm{s}}(V_{\mathrm{ds,max}}) \cdot t$ for the same RF stresses in Fig. 5.

VI. Conclusion

We have experimentally characterized non-conducting RF hot carrier stress on FinFETs used for RF PAs from a 14/16-nm technology. The near-threshold stress is shown to be much more significant than traditional off-state stress at V_{gs}=0 V for PA designs. At V_{gs}=0 V, the RF stress degradation is not quasi-static, and lifetime can only be modeled from RF stress measurement. At $V_{\mathrm{gs}} = V_{\mathrm{th}}$, the RF stress degradation is approximately quasi-static, and the lifetime can be modeled using QSA, provided that the die-to-die variation of the subthreshold current is accounted for. A simple method to account for die-to-die variation is demonstrated, which is critical for modeling the NC RF stress lifetime.

References

[1] A. Cattaneo, S. Pinarello, J.-E. Mueller, and R. Weigel, "Impact of DC and RF non-conducting stress on nMOS reliability," in *IEEE International Reliability Physics Symposium*, 2015, pp. XT.4.1–XT.4.4.

[2] L. Negre, D. Roy, F. Cacho, P. Scheer, S. Jan, S. Boret, D. Gloria, and G. Ghibaudo, "Reliability characterization and modeling solution to predict aging of 40-nm MOSFET DC and RF performances induced by RF stresses," *IEEE Journal of Solid-State Circuits*, vol. 47, no. 5, pp. 1075–1083, 2012.

[3] X. Ding, G. Niu, H. Zhang, W. Wang, K. Imura, and F. Dai, "Impact of non-conducting RF and DC hot carrier stresses on FinFET reliability for RF power amplifiers," in *IEEE RFIC Symposium*, 2022, pp. 199–202.

[4] X. Ding, G. Niu, H. Zhang, W. Wang, K. Imura, and F. F. Dai, "Nonconducting rf and dc hot carrier stresses in 14/16-nm finfets for rf power amplifiers," *IEEE Transactions on Electron Devices*, vol. 70, no. 8, pp. 4028–4035, 2023.

[5] K. Hofmann, S. Holzhauser, and C. Kuo, "A comprehensive analysis of nfet degradation due to off-state stress," in *IEEE International Integrated Reliability Workshop Final Report, 2004*, 2004, pp. 94–98.

[6] N.-H. Lee, D. Baek, and B. Kang, "Effect of off-state stress and drain relaxation voltage on degradation of a nanoscale nmosfet at high temperature," *IEEE Electron Device Letters*, vol. 32, no. 7, pp. 856–858, 2011.

[7] A. J. Scholten, D. Stephens, G. D. Smit, G. T. Sasse, and J. Bisschop, "The relation between degradation under DC and RF stress conditions," *IEEE Transactions on Electron Devices*, vol. 58, no. 8, pp. 2721–2728, 2011.

979-8-3503-6184-1/24 $31.00 © 2024 IEEE

Predictive modelling of hot carrier degradation

James Brown[1], Kean Hong Tok[1], Rui Gao[2], Zhigang Ji[3], Weidong Zhang[1], and Jian Fu Zhang[1*]

[1]School of Engineering, Liverpool John Moores University, Liverpool L3 3AF, UK.
[2]No. 5 Electronics Research Institute of the Ministry of Industry and Information Technology, China.
[3]School of Microelectronics, Shanghai Jiaotong University, Shanghai 200240, China.
* Email: j.f.zhang@ljmu.ac.uk

Abstract

Hot carrier degradation is limiting the lifetime of MOSFETs at present, and it plays a major role in determining the maximum operation voltage and speed of advanced CMOS technologies. Traditionally, the ageing kinetics follow a power law and lifetime can be predicted by extrapolating the straight line on a log-log plot. For advanced CMOS nodes, however, the hot carrier ageing does not always follow a power law, making the linear extrapolation inapplicable. The challenge is how to predict device lifetime accurately in this case. The aim of this work is to provide a defect framework and based on it, to develop a methodology for accurate device lifetime prediction. This framework consists of four different types of traps during hot carrier ageing: as-grown electron traps, generated electron traps, as-grown hole traps, and generated hole traps. To enable accurate device lifetime prediction, the kinetics of each type of trap must be modelled separately. This requires separating the measured ageing into the contributions of these four types of traps. A new experimental technique is developed to enable this separation. It will be shown that the combination of the contribution of these four types of traps can predict device ageing when it is highly non-linear on the log-log plot.

1. Introduction

In 1980s, the operation voltage for digital circuit was fixed at 5 V [1], when the device sizes were downscaled by following the Moore's law. This leads to an increase in the electrical field within the device, which accelerates charge carriers in the conduction channel and makes them 'hot'. These hot carriers bombard the device, generate defects, and cause device ageing. The hot carrier ageing (HCA) was the dominant ageing mechanism, limiting the device lifetime [2-5].

In the two decades between 1990s and 2000s, the operation biases were reduced progressively for smaller devices, partially mitigating the HCA. The other ageing mechanisms, such as time-dependent-dielectric breakdown [6], negative [7-10] and positive [11,12] bias temperature instabilities were limiting device lifetime and random telegraph noise also becomes a major source of instability [13-16]. Recently, however, the HCA

comes back as the lifetime limiting mechanism [3-5]. This is because, for advanced CMOS nodes, the operation voltage is downscaled at slower rate than the physical sizes of the device, resulting in an increase of electrical field within the device. As the silicon bandgap cannot be downscaled, the operation voltage must be adequate to induce an inversion channel, limiting its downscaling to 0.6~0.7 V, as shown in Fig. 1 [1].

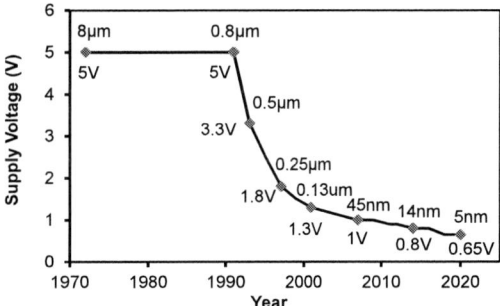

Fig. 1. The operation voltage for different CMOS technologies. [1].

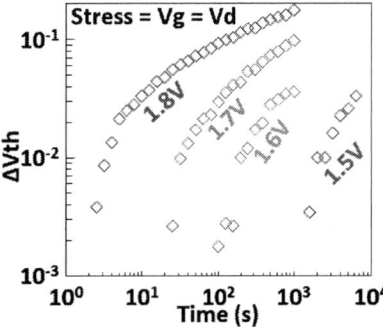

Fig. 2. The kinetics of hot carrier ageing under different biases, which does not follow a simple power law. [5].

The device lifetime is typically defined as the time for the degradation of a given device property to reach a certain level, such as a reduction of drain current by 10% [1,2] or a threshold voltage increase by 50 mV [17,18]. The typically required device lifetime is 10 years [1,17,18]. Since it is not practical for an ageing test to last years, device lifetime prediction becomes essential.

979-8-3503-6184-1/24 $31.00 © 2024 IEEE

When the ageing follows a power law, a common prediction method is to plot the ageing against time on a log-log scale and then linearly extrapolate the kinetics. In reality, however, the ageing kinetics do not always follow a perfect power law, as shown in Fig. 2 [5]. This invalidates the prediction method by linear extrapolation and the challenge is how to reliably predict device lifetime for the case in Fig. 2. The objective of this work is to develop a methodology to overcome this challenge, based on a proposed defect framework.

2. Devices and Experiments

Devices used in this work were nFINFETs with a channel length of 28 nm and a channel width of 1 μm. Both HCA and measurements were carried out at 125 °C. To monitor the HCA, the stress was periodically interrupted, and the Id versus Vg were recorded by using a 10 μs gate pulse under a drain bias of 0.1 V [19]. The detailed measurement methods are given in section 3.

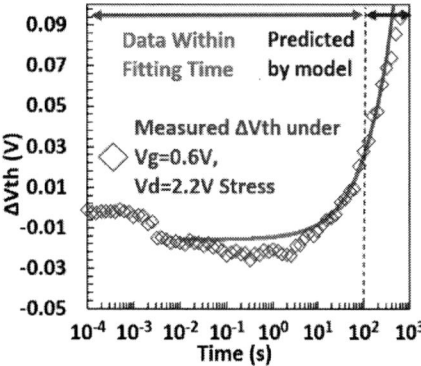

Fig. 3. The negative shift of Vth during the initial stage of the stresses. [5].

3. Defect framework and separation technique

The HCA is recorded after 1 sec in Fig. 2. To understand why the ageing kinetics does not follow a power law, the HCA at shorter time was recorded in Fig. 3. In contrast with the positive shift at longer stress time, it can be clearly seen that Vth is shifted in the negative direction initially. This indicates that positive charges were formed by filling hole traps first during the stress.

To support the claim above, Vg=-0.7 V and Vd=+0.7 V were applied in Fig. 4, which is in favor of hole injection. Vth was indeed shifted negative and then saturated. When hole-detrapping condition (Vg=0 V and Vd= -0.7 V) was applied, Vth recovered to its fresh level, i.e. ΔVth=0. The same saturation level was reached again after reapplying Vg=-0.7 V and Vd=+0.7 V. This supports the presence of as-grown hole traps (AH), which can be efficiently charged and discharged [20,21].

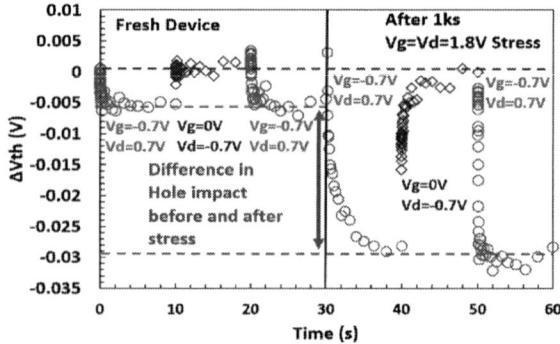

Fig. 4. The presence of as-grown hole traps (left panel) and the increased hole trapping after stresses (right panel). [5].

To separate the contribution of AH, a test technique shown in Fig. 5 was developed, where ΔVth were measured both before and after discharging AH. After discharging AH, the symbol 'o' in Fig. 6 shows that the ΔVth is no longer negative, confirming the neutralization of trapped holes. The contribution of AH can be evaluated from the difference in the ΔVth before and after discharging AH, i.e. 'x'-'o' in Fig. 6. The result is given in Fig. 7(a). It can be seen that filling AH reached a saturation level, S_{AH}, quickly. For longer stress time, however, hole trapping starts increasing beyond S_{AH}, so that new hole traps are generated [22,23]. By subtracting S_{AH}, Fig. 7(b) shows that the generated hole traps (GH) follow a power law. To support that new hole traps are generated, the panel on the right of Fig. 4 shows that there are indeed more hole traps after stresses.

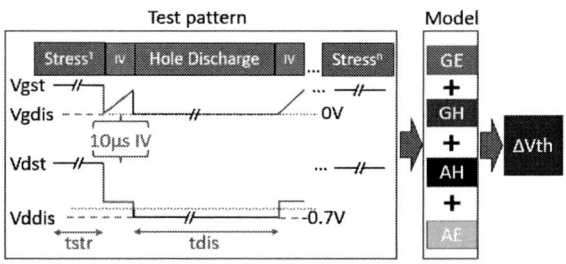

Fig. 5. The test procedure for separating the contribution of hole traps to the measured ΔVth. The ΔVth was recorded both before and after discharging hole traps [5].

After removing the contribution of hole traps, the contribution of electron traps can be assessed from the symbol 'o' in Fig. 6. Similar to hole traps, Fig. 8(a) shows that filling as-grown electron traps (AE) [24,25] reaches its saturation level, S_{AE}, and then rises again by generating new electron traps (GE) [26,27]. After subtracting S_{AE}, GE also follows a power law.

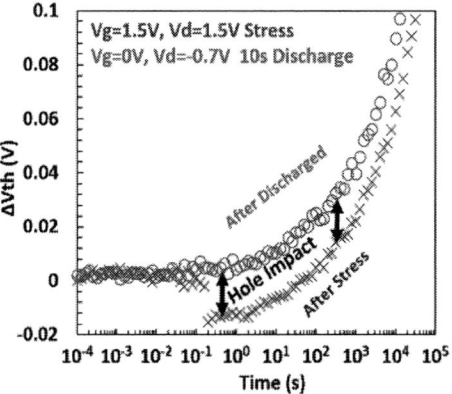

Fig. 6. A comparison of the measured ΔVth before ('x') and after ('o') neutralizing hole traps. [5].

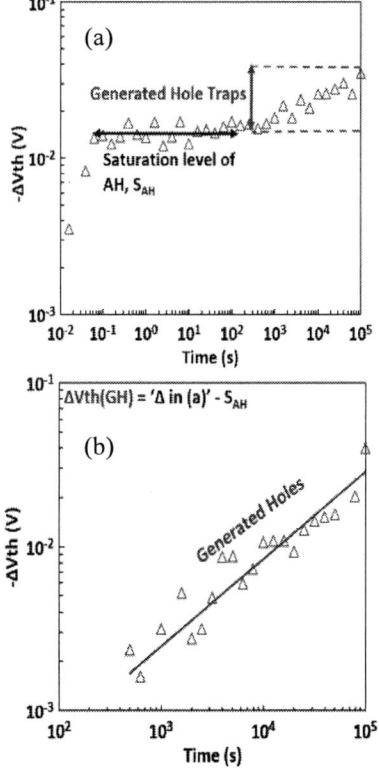

Fig. 7. (a) The contribution of hole trapping to ΔVth, evaluated from the two sets of data in Fig. 6. (b) The generated hole traps after subtracting S_{AH} in (a). [5].

In summary, the defect framework consists of four different types of traps: As-grown hole traps (AH), Generated hole traps (GH), As-grown clcctron traps (AE), and Generated electron traps (GE). This lays the foundation for developing a methodology to predict device lifetime.

4. Results and Discussions

According to the proposed defect framework, the measured ΔVth consists of contributions from four different types of traps,

$$\Delta Vth = C_{GE}t^{n_{GE}} + C_{GH}t^{n_{GH}} + S_{AE} + S_{AH} \quad (1)$$

The next question is how to determine the six parameters in eq.(1): C_{GE}, n_{GE}, C_{GH}, n_{GH}, S_{AE}, and S_{AH}. A common methodology for extracting model parameters is to fit the model with the test data. If we follow this methodology, we can extract these parameters simultaneously and Fig. 9 shows that the model agrees well with test data within the fitting range of 100 sec. When these fitted parameters were used to predict future ageing, however, Fig. 9 shows that they failed. As a result, a good fit with test data must not be used as the criterion to validate a model and it is essential to verify its capability to predict future ageing.

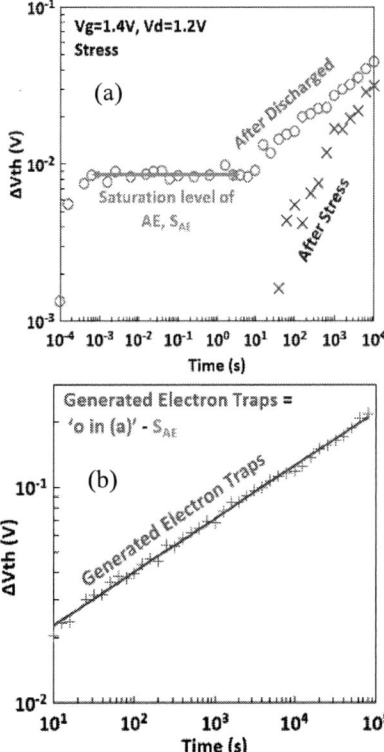

Fig. 8. (a) The contribution of electron traps to ageing. (b) The generated electron traps after subtracting the as-grown electron traps, S_{AE} in (a). [5].

When eq.(1) is fitted with test data, we are extracting six parameters simultaneously based on the least square error criterion and Fig. 9 shows that their true values cannot be obtained by this extraction methodology. To determine the values of these parameters, we must limit the number of parameters to be extracted. This can be achieved by determining the kinetics for each type of traps separately, as shown in Figs. 7 and 8, where the fitting parameters are always limited at two. The eq.(1) is then used to combine the contribution of these four types of traps. Fig. 9 shows that the parameters extracted by this methodology not only agree well with test data in the fitting range, but also can predict future ageing.

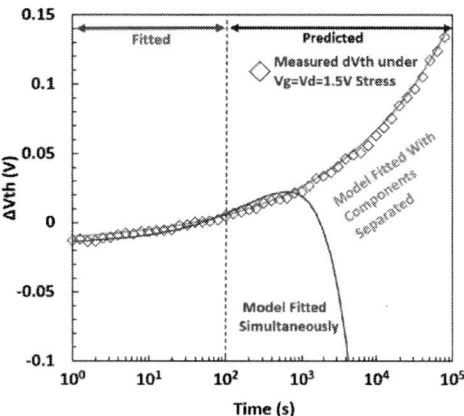

Fig. 9. The model parameters were extracted from test data before 100 sec. They were then used to predict the following ageing. When all parameters were simultaneously extracted by fitting with eq.(1), the prediction fails. When parameters were extracted separately for each type of traps, the prediction agrees well with the test data. [5].

5. Summary

A framework has been proposed for defects during stresses and it consists of four types of traps: As-grown hole traps, Generated hole traps, As-grown electron traps, and Generated electron traps. Each of them has different ageing kinetics, resulting in non-linear log-log plot of the measured ΔVth against stress time. A technique has been developed to separate their contributions. It should be emphasized that the model parameters for these four types of traps must be extracted separately, rather than simultaneously by fitting the measured total ΔVth. Good prediction, rather than fitting, should be used to validate a model.

Acknowledgments

The authors would like to thank IMEC for supplying test samples. This work is supported by the EPSRC of UK under the grant no. EP/T026022/1.

References

[1] J. F. Zhang et al, Electronics, 11(9), 2022, https://doi.org/10.3390/electronics11091420.
[2] C. Hu et al, IEEE Trans. Electron Devices, vol. ED-32, pp. 375–385, 1985.
[3] M. Duan et al, iedm, p.547, 2015. doi:10.1109/IEDM.2015.7409742.
[4] M. Duan et al, IEEE Trans. Electron Devices, 2017. doi: 10.1109/TED.2017.2691008.
[5] J. Brown et al, IEEE Access, 2023. doi:10.1109/ACCESS.2023.3329077
[6] J. Ma et al, IEEE Trans. Electron Devices, 2019. doi: 10.1109/TED.2018.2881294.
[7] Z. Ji et al, iedm, 2013. doi:10.1109/IEDM.2013.6724638
[8] R. Gao et al, IEEE Trans. Electron Devices, 2017. doi: 10.1109/TED.2017.2742700.
[9] M. Duan et al, IEEE Trans. Electron Devices, 2014. doi: 10.1109/TED.2014.2335053
[10] Z. Ji et al, VLSI Tech., 2015 doi:10.1109/VLSIT.2015.7223693
[11] R. Gao et al, iedm, 2016. doi:10.1109/IEDM.2016.7838520
[12] R. Gao et al, 2018. doi:10.1109/TED.2018.2857000
[13] R. Wang et al., IEDM, p.388, 2018.
[14] K. H. Tok et al, IEEE Trans. Electron Devices, 2022 doi:10.1109/TED.2022.3176585
[15] K. H. Tok et al, IEEE Trans. Electron Devices, 2022 doi: 10.1109/TED.2022.3195690
[16] K. H. Tok et al, IEEE Trans. Electron Devices, 2023 doi:10.1109/TED.2023.3253665
[17] J. F. Zhang, in BTI for Devices and Circuits. doi:10.1007/978-1-4614-7909-3
[18] J. F. Zhang et al, Microelectronics Reliability, 2018. https://doi.org/10.1016/j.microrel.2017.11.026
[19] X. F. Zheng et al, IEEE Trans. Electron Devices, 2009. doi: 10.1109/TED.2009.2035193.
[20] J. F. Zhang et al, IEEE Electron Dev Lett, 2006. doi:10.1109/LED.2006.882566.
[21] J. F. Zhang et al, IEEE Trans. Electron Devices, 2000. doi: 10.1109/16.822284.
[22] J. F. Zhang et al, IEEE Electron Dev Lett, 2007. doi: 10.1109/LED.2007.893219.
[23] J. F. Zhang et al, IEEE Electron Dev Lett, 2008. doi: 10.1109/LED.2008.2006288.
[24] I. S. Goh et al, Semicod.Sci.Tech., 1995. doi:10.1088/0268-1242/10/6/013.
[25] M. Duan et al, IEEE Trans. Electron Devices, 2016. doi:10.1109/TED.2016.2590946.
[26] W. D. Zhang, IEEE Electron Dev Lett, 2006. doi: 10.1109/LED.2006.873384.
[27] M. H. Chang et al, IEEE Trans. Electron Devices, 2006. doi:10.1109/TED.2006.874155.

New Insights into the Saturation Behavior of the Hot Carrier Degradation in STI-based N-type LDMOS

Zhuoqing Yu[1*], Dan Gao[1], Yongsheng Sun[1], and Junlin Huang[1]

[1] Team of Design For Reliabilty, Hisilicon, Xinjinqiao Road 2222, Shanghai, P. R. China

* Email: yuzhuoqing@hisilicon.com

Abstract—The saturation behavior of Hot Carrier Degradation (HCD) kinetics in a Shallow Trench Isolation (STI)-based N-type Laterally Diffused MOS (NLDMOS) is studied which is found mainly due to the decrease of impact ionization rate rather than the interface precursor limitation. A compact model is proposed and verified with experiments over different bias and temperature conditions. The results provide new insights to NLDMOS HCD physics and are helpful to the design-for-reliability (DFR) of circuit.

Keywords—*LDMOS, hot carrier, impact ionization, saturation.*

I. INTRODUCTION

Laterally Diffused MOS (LDMOS) is widely adopted in High-voltage (HV) applications like RF communication and smart power management integrated circuit, owing to its compatibility with the standard CMOS manufacturing process flow [1-4]. The typical working condition of LDMOS involves high drain voltage and high current density, which makes hot carrier degradation (HCD) in LDMOS a major reliability concern in circuit design [3,6,7]. Therefore, understanding the HCD in LDMOS physical mechanisms and its accurate modeling is important for reliability-aware circuit design, especially for HCD induced current degradation which directly relates to the ON-state resistance (Ron) [5] degradation impacting the circuit energy efficiency [7,8].

Conventionally, the HCD in LDMOS is understood to be caused by the hot carrier generated by impact ionization in the device substrate [2,5]. The generates hot carriers react with precursors (i.e. Si-H or Si-O [9,10]) at the interface, causing bond-isolation and generate interface traps. Usually, the kinetics of LDMOS HCD is observed to show a saturating behavior in its time dependence, which is understood to originate in the limited interface precursor amount as described in the Bond-Dispersion (B-D) model [2]. However, in this work, the saturation behavior of a N-type LDMOS (NLDMOS) HCD kinetics is found to originate in the decrease of impact ionization rate rather than the limitation of interface precursors. A compact model is proposed and verified to describe the observations. The understandings provide new insights into HCD physics in NLDMOS and are helpful to the circuit design-for-reliability (DFR) practice.

II. DEVICES AND EXPERIMENTS

The device utilized in this work is a foundry-level STI-based n-type LDMOS (NLDMOS). The basic structure of the device is illustrated in Fig. 1(a). Well-calibrated TCAD simulations are applied to LDMOS under stress conditions, as shown in Fig. 1(b), which shows single peak of impact ionization rate at the STI corner, indicating that most of the interface traps are generated around this region. Similar results can be observed under other stress conditions used in this work (not shown here).

Conventional Measure-Stress-Measure (MSM) method is adopted with I_{ds}-V_{gs} curves sensed periodically during stress

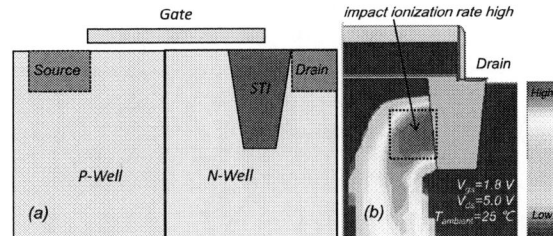

Fig. 1. (a) The schematic cross section of NLDMOS in this work. (b) The typical results of calibrated TCAD simulation of impact ionization rate in NLDMOS under a HCD stress condition. Single high rate region is found near drain-side STI.

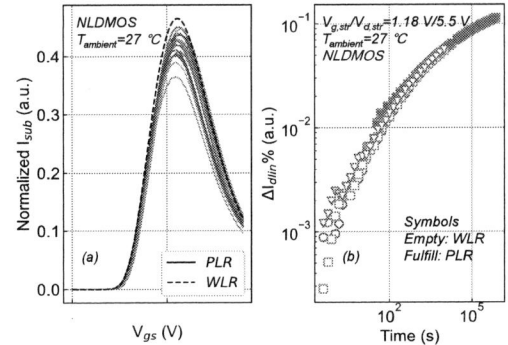

Fig. 2. Typical experimental results from both WLR and PLR of (a) I_{sub}-V_{gs} curves in pristine devices. (b) HCD degradation traces in NLDMOS. The results from both method agree well with each other.

time. The reliability test are performed in both wafer level (WLR) and package level (PLR) in order to capture both short-term (~1E4 s) and long-term (~1E6 s) degradation. The experiments from WLR and PLR show good consistency in pristine devices' I_{sub}-V_{gs} curves (Fig. 2(a)) and HCD degradation traces (Fig. 2(b)), indicating that both methods are applicable in HCD tests. The LDMOS devices are stressed under HCD worst-case conditions with different $V_{d,str}$ and $T_{ambient}$, which corresponds to the $V_{g,str}$ at each corresponding I_{sub} maximum conditions.

III. EXPERIMENTAL RESULTS AND MODELING

A. The Saturation Behavior of HCD Kinetics

The typical results of NLDMOS HCD degradation are shown in Fig. 3. As shown in Fig. 3(a), linear-region current (I_{dlin}) is observed to degrade with stress time, which follows a saturation trend instead of a single power law. Meanwhile, negligible degradation is observed in linear-region threshold voltage (V_{thlin}) and saturation current (I_{dsat}). The results indicate that the traps generated during HCD are mainly localized in the drain region that can decrease carrier mobility while has less impacts on V_{thlin} and I_{dsat}. The saturation behavior of HCD kinetics indicates a possible pessimistic prediction of I_{dlin} degradation (or lifetime) if the conventional

979-8-3503-6184-1/24 $31.00 © 2024 IEEE 441

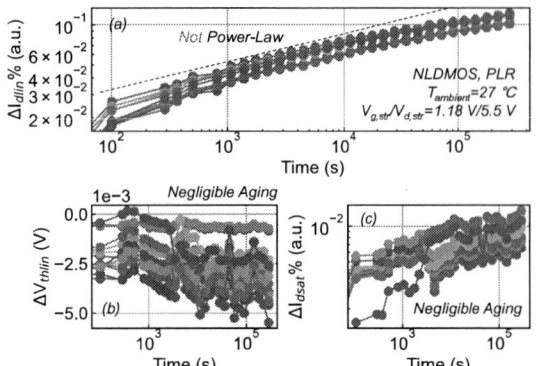

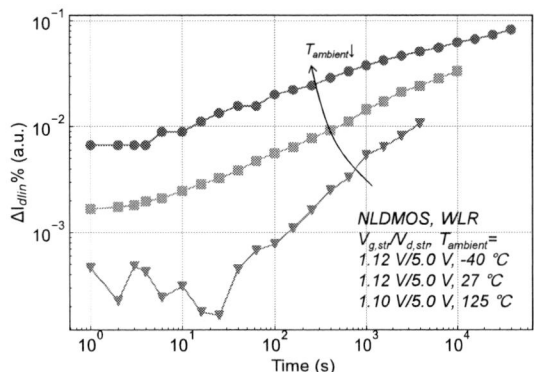

Fig. 3. The typical experimental results of HCD in NLDMOS of (a) I_{dlin} degradation, (b) V_{thlin} degradation and (c) I_{dsat} degradation. The I_{dlin} degradation kinetics shows saturating behavior in its time dependence.

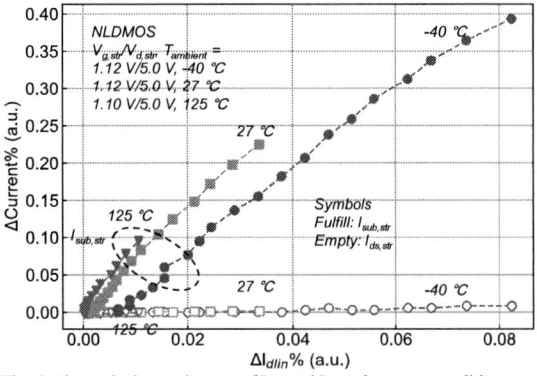

Fig. 4. The experimental results of the NLDMOS HCD temperature dependence. The degradation increases with decreasing temperature indicating a negative temperature dependence.

single power law is applied with short-term degradation experiments. The experiment results of the HCD temperature dependence are shown in Fig. 4. HCD in NLDMOS shows a negative temperature dependence, which indicates the degradation originates in the impact ionization generated hot carriers [11]. Since impact ionization rate can be indicated by the ratio I_{sub}/I_{ds} according to the Lucky Election Model (LEM) [12], the measurement of I_{sub} characteristics can be adopted as a good approach to monitor the HCD aging rate [1].

Conventionally, the saturation behavior of HCD kinetics is understood to originate in the limitation of interface precursors [10,13]. However, in the NLDMOS utilized in this work, I_{sub} under stress conditions is observed to decrease with

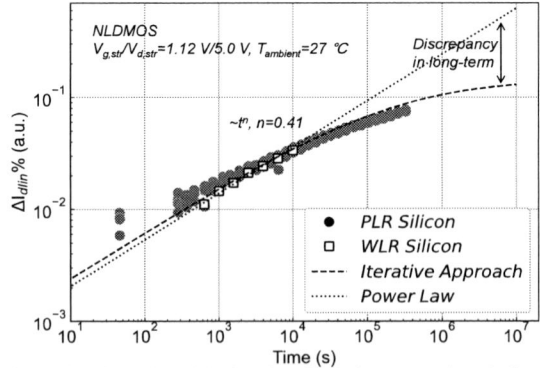

Fig. 5. The typical experiments of I_{sub} and I_{ds} under stress conditions over HCD. I_{sub} under stress condition is observed to degrade with aging.

aging amount, while I_{ds} under stress condition remains approximately the same, as shown in Fig. 5. The results indicate that impact ionization rate decreases with stress time that can contribute to the saturation behavior. The decrease of the impact ionization rate may originate in the modification of substrate drift-region electric field or carrier energy distribution that caused by the generated traps [13].

B. Iterative Simulation Approach to Saturation Behavior

To further analyze the HCD saturation behavior, an iterative simulation approach is utilized to mimic the HCD degradation traces.

First, to extract initial aging rate, the HCD is modeled with short-term WLR experiments following LEM framework as

$$\Delta I_{dlin}\% = A \times I_{ds,str} \times (I_{sub,str}/I_{ds,str})^m \times t^n \quad (1)$$

where A, m and n are fitting parameters. $I_{ds,str}$ and $I_{sub,str}$ represent corresponding current under stress conditions, which are obtained from experiments.

Then, to reflect the decrease of aging rate, I_{sub} degradation is modeled with an empirical model based on short-term experiments in Fig. 5 as

$$\Delta I_{sub}\% = k \times \Delta I_{dlin}\% \quad (2)$$

where k is a fitting parameter.

Last, based on (1) and (2), the degradation trace in long-term is obtained by iteratively simulate degradation and update degradation rate (by updating degraded I_{sub}) in each time step with history effect considered [14]. The typical experiments and simulation results are demonstrated in Fig. 6. The long-term prediction results of iterative approach simulations agree well with the PLR experiments, which indicates the saturation behavior mainly originates in the decrease of impact ionization rather than interface precursor limitation. Moreover, the traditional simulation method without considering aging rate decrease is shown to cause a large discrepancy in long-term degradation prediction, which results in a much more pessimistic degradation prediction.

Fig. 6. The illustration of the iterative simulation approach to the long-term HCD degradation. The prediction results with the iterative approach agree well with the long-term degradation experiments.

The Saturation Model of HCD and Extracted Parameters		
Total Degradation	$\Delta I_{dlin}\% = 1/k(T) \times (1-exp(-AR_0 \times k(T)*t^n))$	n=0.30
Saturating Mechanism	$k(T)=k_0*T^a$	α=1.2488
Aging Rate	$AR_0 = A \times exp(-b/V_d) \times exp(-E_a/(k_BT/q))$	b=46.24
		E_a=-0.074

Table 1. The compact model of NLDMOS HCD considering saturating behavior and the extracted parameters.

C. Compact Modeling of HCD Kinetics

Based on the understandings above, a compact model of NLDMOS HCD kinetics considering saturating behavior is proposed as Table. 1. Contrary to the traditional model that based on precursor limiting assumption [10,13], the total degradation in this work considered the decrease of impact ionization rate and can be derived as follows. According to (1), the degradation of each small time period can be expressed as

$$\Delta(\Delta I_{dlin}\%) = AR \times \Delta(t^n) \tag{3}$$

where n is a fitting parameter. Considering the aging rate (AR) decreases with degradation amount, an empirical model of AR is given as

$$AR = AR_0 \times (1 - k(T) \times \Delta I_{dlin}\%) \tag{4}$$

Combining (3) and (4), the total degradation in long-term can be expressed as

$$\Delta I_{dlin}\% = 1/k(T) \times (1 - \exp(-AR_0 \times k(T) \times t^n)) \tag{5}$$

The fitting and predicting results of the models are illustrated in Fig. 7 and 8. The models can well describe the short-term degradation and their prediction results agree well with experiments in long-term degradation for different voltage and ambient temperature conditions. The values of the extracted model parameters (as shown in Table. 1) are within

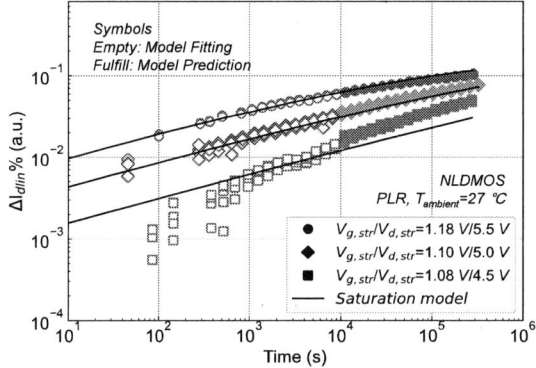

Fig.7. The fitting and prediction results of the model in bias dependence. The simulation results agree well with the experiments.

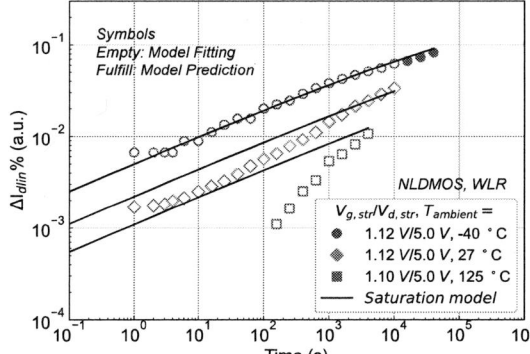

Fig. 8. The fitting and prediction results of the model in temperature dependence. The simulation results agree well with the experiments.

typical value ranges. The temperature dependence of aging rate is negative ($E_a < 0$) while that of the saturating mechanism is positive ($\alpha > 0$), which indicates that though impact ionization rate is higher under low ambient temperature, the generated traps will have higher impacts on the decreasing of electric field or carrier energy distribution thus limit the aging rate in long-term.

IV. SUMMARY

In this work, the saturating behavior of HCD kinetics in a N-type Shallow Trench Isolation (STI)-based LDMOS is found to mainly originate in the decrease of impact ionization rate rather than interface precursor limitation during stress. A compact model is proposed based on the physical understandings which shows good agreements with experiments over different bias and ambient temperature conditions. The results provide new insights into LDMOS HCD physics and modeling which are helpful to the design-for-reliability of circuit.

ACKNOWLEDGMENT

The author would like to address special thanks to Dr. Min Zhang in Hisilicon for the helpful discussions and TCAD simulation supports.

REFERENCES

[1] S. Reggiani et al., "Characterization and Modeling of High_Voltage LDMOS Transistors," *Hot Carrier Degradation in Semiconductor Devices*, Springer, 2015, pp.309-339.

[2] B.K. Mahajan et al., "An Analytical Model of Hot Carrier Degradation in LDMOS Transistors Rediscovery of Universal Scaling," TED, pp.3923-3929, 2021.

[3] S. Cimino et al., "Optimized LDMOS Offering for Power Management and RF Applications," IRPS, pp.P57-1-P57-5, 2022.

[4] S.E. Liu et al., "High Voltage Tolerant Design with Advanced Process for TV Application," IRPS, pp.1-4, 2019.

[5] J.F. Chen et al., "On-Resistance Degradation Induced by Hot-Carrier Injection in LDMOS Transistors With STI in the Drift Region," EDL, pp.1071-1073, 2008.

[6] J.F. Chen et al., "Convergence of Hot-Carrier-Induced Saturation Region Drain Current and On-Resistance Degradation in Drain Extended MOS Transistors," TED, pp. 2843-2847, 2009.

[7] C. Schlunder, "Circuit Reliability: Hot-Carrier Stress of MOS Transistors in Different Fields of Application," *Hot Carrier Degradation in Semiconductor Devices*, Springer, 2015, pp.445-476.

[8] S.J.C.H. Theeuwen and J.H. Qureshi, "LDMOS Technology for RF Power Amplifiers," TMTT, pp. 1755-1763, 2012.

[9] M.A. Alam et al., "Hot carrier Degradation in Classical and Emerging Logic and Power Electronic Devices: Rethinking Reliability for Next-Generation Electronics," EDTM, pp. 1-3, 2021.

[10] P. Moens et al., "Towards a Universal Model for Hot Carrier Degradation in DMOS Transistors," ISPSD, pp. 61-64, 2010.

[11] A. Bravaix et al., "Hot-Carrier Degradation in Decananometer CMOS Nodes: From an Energy-Driven to a Unified Current Degradation Modeling by a Multiple-Carrier Degradation Process," *Hot Carrier Degradation in Semiconductor Devices*, Springer, 2015, pp.57-103.

[12] C. Hu, "LUCKY-ELECTRON MODEL OF CHANNEL HOT ELECTRON EMISSION," IEDM, pp.22-25, 1979.

[13] Y.M. Randriamihaja et al., "Microscopic scale characterization and modeling of transistor degradation under HC stress," MR, pp. 2513-2520, 2012.

[14] S. Guo et al., "Towards Reliability-Aware Circuit Design in Nanoscale FinFET Technology:–New-Generation Aging Model and Circuit Reliability Simulator," ICCAD, pp. 780-785, 2017.

AUTHOR INDEX

Ahammed, Shihab...........316	Chen, Jing.............655, 892, 940
Ai, Zhaoyu...........877	Chen, Kaiqi...........624
An, Suihuan...........630	Chen, Ke...........68
An, Xiaomeng...........748	Chen, Kuangli...........398, 454
An, Zhemin...........813	Chen, Lei...........688
Bai, Jiyuan...........59, 739	Chen, Lin...........368
Bai, Mingkai...........484	Chen, Liying...........104
Bai, Song...........423	Chen, Long...........649
Ban, Chaoyi...........508	Chen, Ning...........937
Bao, Lin...........504, 508	Chen, Peng...........844
Bao, Yunjiao...........777	Chen, Quan...........202
Bi, Hengchang...........551, 751	Chen, Rong...........338
Bi, Ran...........526	Chen, Ruolan...........706
Bi, Shunyang...........216	Chen, Shiyou...........342
Bo, Chunyue...........733	Chen, Shulan...........706
Brown, James...........437	Chen, Shushi...........587
Bu, Weihai...........364, 760	Chen, Sichao...........219
Cai, Chenxiang...........273	Chen, Song...........10
Cai, Linlin...........65, 523, 792	Chen, Tao...........602
Cai, Rundong...........83	Chen, Wangyong...........65, 523, 792
Cai, Xi...........466, 661	Chen, Wanjun...........395, 411, 847, 874
Cai, Yimao...........504, 508	Chen, Weifeng...........771
Cai, Zilu...........679	Chen, Wenjun...........801
Cao, Jiangtao...........661	Chen, Wenwei...........186
Cao, Junhou...........828	Chen, Xiang...........599
Cao, Kun...........338	Chen, Xibo...........661
Cao, Lei...........757	Chen, Xinghuan...........874
Cao, Lixuan...........241	Chen, Xingyu...........466
Cen, Yuanjun...........688	Chen, Xinyang...........877
Chai, Junshuai...........460, 463, 484, 511	Chen, Xu...........345
Chai, Z. 494	Chen, Xurui...........126
Chai, Zheng...........630	Chen, Yanning...........910
Chan, Mansun...........316, 325, 786, 850	Chen, Yifei...........238, 721
Chang, Haixin...........301	Chen, Yong...........122
Chang, Hao...........183	Chen, Yongliang...........599
Chang, Pengying...........816	Chen, Yu...........630
Chang, Yuchun...........664	Chen, Yuhua...........646
Chen, Bing...........661	Chen, Yujie...........28
Chen, Bofan...........282	Chen, Yuwei...........895
Chen, Fan...........232	Chen, Yuyang...........736
Chen, Gengsheng...........59, 620, 739	Chen, Zheng-Han...........143
Chen, Gui...........361	Chen, Zhikang...........22
Chen, Haifeng...........65	Chen, Zhong-Jian...........143
Chen, Hong...........771	Chen, Zirui...........31
Chen, Hongquan...........430, 451	Chen, Zixiang...........59
Chen, Hongzhi...........801	Chen, Zixun...........819
Chen, Hua...........742	Cheng, Chuantong...........104
Chen, Jiezhi...........457, 807, 810	Cheng, Hsing-Mao...........307

Cheng, Junji	825, 865
Cheng, Kai	122
Cheng, Lin	856, 880
Cheng, Nannan	258
Cheng, Yu	664
Cheng, Yuhua	447
Chenqi	31
Chu, Chun Yee	783
Chu, Junhao	178
Chu, Yu	718
Chuang, Yuan-Yu	41
Coutinho, Jose G. F.	195
Cui, Miao	408
Cui, Wentao	372
Cui, Xianghe	7
Cui, Xiaole	599
Cui, Xuecheng	661
Dai, Chengxing	107, 643
Dai, F.	433
Dai, Ming	865
Dai, Saifei	463, 484, 511
Dai, Shun-Qi	798
Dai, Siyao	627
Dai, Wu	536
Dai, Yuan	219
Deng, Chenkai	388
Deng, Gaoqiang	379
Deng, Haonan	834
Deng, Hongfei	417
Deng, Jie	235
Deng, Lei	132, 676, 685, 691
Deng, Shiyu	395
Deng, Xiaochuan	401
Deng, Yongfeng	910
Deval, Yann	267
Di, Boyuan	301
Ding, Guojian	771
Ding, Qi	392
Ding, Rongzheng	763
Ding, Runye	28
Ding, Shijin	804
Ding, Tongshu	186
Ding, X.	433
Ding, Yajing	484
Ding, Yingtao	304
Ding, Yong	466, 614, 661
Ding, Yuanhong	132, 676
Dong, Chenge	44
Dong, Feng	68
Dong, Haitao	807
Dong, Jianing	289
Dong, Junchen	264, 533

Dong, Lei	572
Dong, Li-Hang	868, 871
Dong, Qingyang	733
Dong, Qiyun	620
Dong, Shijiao	232
Dong, Shuangqi	934
Dong, Xiaoping	420
Dong, Yaoqi	345
Dong, Zuoyuan	178
Du, Feng-Yu	414, 831
Du, Gang	264
Du, Peiyuan	472
Du, Rongxin	907
Du, Wencai	768
Du, Yiran	602
Duan, Enchuan	398
Fan, Difei	388
Fan, Wenbo	511
Fan, Xiangrui	822
Fan, Yibo	77, 587, 605, 608
Fang, Jian	126
Fang, Rui	614
Fanyi, Meng	228
Feng, Baoyan	862
Feng, Chao	454
Feng, Cunfeng	289
Feng, Jiekai	904
Feng, Lichen	83, 156
Feng, Qi	771
Feng, Yang	457, 807
Fu, Chengwen	420
Fu, Fangfa	199, 789
Fu, Hao	382, 822, 828
Fu, Jing	895
Fu, Peiyuan	89
Fu, Xiaojun	825
Fukuda, Kenjiro	285
Gai, Weixin	110
Gan, Jie	611
Gao, Dan	183, 441
Gao, Dawei	910
Gao, Jianjun	736
Gao, Liang	581
Gao, Rui	437
Gao, Ruiyi	252
Gao, Shujun	545
Gao, Xingyu	199, 789
Gao, Xuchen	219
Ge, Hao	892
Ge, Ziang	208
Golosov, Dmitriy Anatolyevich	804
Gong, Sen	417

Gong, Yi 25
Gong, Yong 478
Gong, Yuehong 289
Gou, Fei 86
Gu, Chenjie 804
Gu, Hanzhi 335, 514, 520
Gu, Shengfei 74
Gu, Wenxian 751
Gu, Yong 423
Guan, Chaowen 299
Gui, Qingzhong 296
Guo, Anran 895
Guo, Huimeng 44
Guo, Jiajie 62
Guo, Jianping 922
Guo, Jiyuan 685
Guo, Junxiong 931
Guo, Mingqiang 152
Guo, Pei 379, 385
Guo, Peng 810
Guo, Qingyu 13
Guo, Shuaicheng 614
Guo, Tianyu 569
Guo, Xinyi 457, 810
Guo, Yirong 816
Guo, Yiting 35
Guo, Yuekang 175
Guo, Yufeng 931
Guo, Yulin 754
Guo, Yutuo 392
Guo, Yuzheng 296
Guo, Zhijian 469
Guo, Zhiwang 171
Guo, Zizheng 212
Han, Chao 414, 831, 853, 856, 862
Han, Dedong 533
Han, Dongxiang 919
Han, Genquan 472, 661
Han, Hongyan 760
Han, Jun 1, 16, 50, 624, 627, 634
Han, Kai 484, 511
Han, Kefeng 730
Han, Lixiang 581
Han, Rong 688
Han, Runhao 484, 511
Han, Wei-Hua 319
Hang, Jianfeng 760
Hao, Linyao 385
Hao, Sang 98
Hao, Yue 472, 542, 837
Hao, Zhijian 608
Hatayama, Kazumi 192

He, Chenlong 605
He, Huanxiang 587
He, Kunqin 392
He, Lin 560
He, Nailong 430
He, Songxuan 792
He, Weitao 715
He, Wen 289
He, Xinyu 186
He, Yanping 345
He, Yingtong 630
He, Yu 667
He, Yuhui 31
He, Yunlong 542
Hengzhou, Yuan 98
Heyuan, Liu 98
Hiramoto, Toshiro 332
Hong, Bo 614
Hongshi, Yu 228
Hou, Bin 837
Hou, Xiangyu 423
Hu, Ang 74, 563, 694, 709, 727
Hu, Haolin 454
Hu, Jiahao 401
Hu, Jianguo 202
Hu, Jianzhi 304
Hu, Jiaxuan 904
Hu, Jing 697
Hu, Jingyi 7
Hu, Qiang 877
Hu, Shuang 703
Hu, Siyuan 533
Hu, Tao 484, 511
Hu, Weiming 129
Hu, Yi 584
Hu, Yibo 655, 892, 940
Hu, Yidi 928
Hu, Yuhan 667
Hu, Yushen 329
Hu, Z. 494
Hua, Younan 228
Huang, Guang-Wei 307
Huang, Haimeng 530, 813, 825, 841, 865
Huang, Heyi 774
Huang, Jialu 178
Huang, Jian 487
Huang, Jie 847
Huang, Jun 392
Huang, Junlin 183, 441
Huang, Lei 828
Huang, Leilei 77, 587, 596, 605, 608
Huang, Menglin 342

Huang, Mingmin	420	Jing, Minge	4, 593
Huang, Qianqian	760	Jing, Zhiwei	841
Huang, Qihang	907	Jingtian, Liu	98
Huang, Ru	212, 355, 364, 426, 504, 508, 536, 760	Kang, Hao-Bo	831
Huang, Runhua	423	Kang, Jin	364
Huang, Shuting	398, 454, 907	Kang, Jinfeng	475
Huang, Shuyang	834	Kang, Wang	47
Huang, Tianze	74	Kang, Yi	10
Huang, Tingrui	517	Katayama, Shogo	192
Huang, Wen	931	Katoh, Kentaroh	135, 192
Huang, Wende	420	Ke, Xiaoyu	460
Huang, Yifeng	780	Ke, Xuanxin	673
Huang, Yujie	4, 593	Kim, Ikhwan	554
Huang, Zepeng	551	Kobayashi, Haruo	135, 192
Huang, Zhaofeng	551	Kobayashi, Masaharu	332, 469
Huo, Jiali	774	Kobayashi, Yutaro	135
Huo, Qiuyue	392	Kong, Moufu	417, 865, 877
Ichikawa, Tamotsu	192	Kumar, Anuj	490
Ihda, Ayoub Ait	267	Kuo, Chia-Yo	313
Iimori, Daisuke	192	Kuwana, Anna	192
Imura, K.	433	Lai, Jinmei	186, 189
Ishida, Takashi	192	Lam, Sang	408
Ji, Xiaoli	575, 925	Lan, Haokun	238
Ji, Zhigang	351, 437	Lang, Zhengyu	834
Jia, Bin	715	Lapuyade, Hervé	267
Jia, Guowen	590	Lei, Tengteng	329
Jia, Haiqiang	771	Lei, Wenyu	301
Jia, Longjiang	132, 676	Lei, Yuan	798
Jia, Peng-Fei	868, 871	Li, Angyang	691
Jia, Rundong	760	Li, Bangtian	104
Jia, Yiming	841	Li, Biao	101
Jia, Yirui	430, 451	Li, Bin	685
Jiang, Chen	270	Li, Bo	113
Jiang, Chunsheng	768	Li, Cheng	117
Jiang, Dongyang	152	Li, Chengxin	904
Jiang, Jianfei	68	Li, Cong	545, 889
Jiang, Li	679	Li, Congrui	706
Jiang, Nan	661	Li, Dejian	276
Jiang, Renjie	757, 774, 777	Li, Dengquan	149, 670
Jiang, Tong	593	Li, Dongya	472
Jiang, Xin	733	Li, Gang	19, 71
Jiang, Xingli	865	Li, Guowei	895
Jiang, Yunteng	813	Li, Hao	563, 581
Jiang, Zijin	408	Li, Haokun	545
Jianzhuang, Lu	98	Li, Haolin	539
Jin, Chengji	469, 472	Li, Hong	101
Jin, Jing	175	Li, Hui	19
Jin, Xiangliang	898	Li, Jianguo	498
Jin, Xin	276	Li, Jianwen	688
Jin, Yu	602	Li, Jianzheng	129, 748
Jin, Yuhan	126	Li, Jie	816
Jin, Zhenghao	813	Li, Jincheng	739

Li, Jing 859
Li, Jinshan 508
Li, Jinyu 658
Li, Junfeng 757
Li, Kai 74
Li, Kaiyue 563
Li, Kanyi 511
Li, Kunyue 68
Li, Lei 199, 789
Li, Lianlian 757
Li, Luping 844
Li, Lu-Ping 871
Li, Meng 13
Li, Menghan 634
Li, Ming 364, 526, 844
Li, Mingjie 304, 934
Li, Ningning 95, 691, 745
Li, Peng 904
Li, Qingkun 757
Li, Renxiong 392
Li, Renxuan 694, 727
Li, Rophina 372
Li, Ruguan 530, 813
Li, Ruizhi 895
Li, Ruoyu 548
Li, Shangze 475
Li, Sheng 376
Li, Shiming 837
Li, Shuaihong 916
Li, Shurui 417
Li, Siteng 754
Li, Tianyu 733
Li, Tiaoyang 348, 883
Li, Wang 338
Li, Wanyu 807
Li, Wei 146, 232, 279, 602, 649, 844, 868, 871
Li, Weizhi 65
Li, Wenhong 4, 593, 922
Li, Xiangnan 385
Li, Xiaomin 228
Li, Xin 107, 643
Li, Xingyu 523
Li, Xuan 401
Li, Xueke 104
Li, Yang 481
Li, Yanqing 774
Li, Yanzhong 77
Li, Yichen 258
Li, Yongjia 258
Li, Yongxi 530
Li, Yongxiang 501
Li, Yudong 766

Li, Yujia 44
Li, Yu-Tao 255
Li, Yutao 469
Li, Zehong 844
Li, Ze-Hong 868, 871
Li, Zhaoji 451
Li, Zhenmin 7
Li, Zhensong 533
Li, Zhenwei 883
Li, Zhihuai 679
Li, Zhiqiang 222
Li, Zhiqun 282
Li, Zhiwei 697
Li, Zonglin 813
Lian, Xueguang 466, 661
Liang, Jie 886
Liang, Jingyuan 372
Liang, Libin 801
Liang, Ling 504, 508
Liao, Min 460
Liao, Pengfei 904
Liao, Xufeng 89, 276
Liao, Yiming 575, 925
Lin, Guangyao 62
Lin, Jinpeng 101
Lin, Jyi-Tsong 41, 307, 313, 322
Lin, Lin 931
Lin, Yibo 212, 364
Ling, Zipeng 1
Liou, Juin J. 228
Liu, Ao 423
Liu, Chang 345
Liu, Chao 395, 847
Liu, Chen 252
Liu, Dong 289
Liu, Donglin 222
Liu, Dongsheng 74, 563, 581, 694, 709, 727
Liu, Fang 910
Liu, Fei 92
Liu, Feng 771
Liu, Haitao 682
Liu, Houfang 569
Liu, Huajie 126
Liu, Huan 472
Liu, Jie 575
Liu, Jin 566
Liu, Jinbiao 345
Liu, Jing 457
Liu, Jingjing 56, 168, 261, 557, 640, 673
Liu, Junzhan 47
Liu, Kai 596
Liu, Ke-Yu 414, 831

Liu, Lianxi	89, 276
Liu, Linghao	159
Liu, Maliang	162, 238, 566, 721
Liu, Mingshan	345
Liu, Puyang	117
Liu, Qianhui	646
Liu, Renkuan	385
Liu, Rui	572, 682
Liu, Shengxiang	709
Liu, Shuang	774, 777
Liu, Shubin	670
Liu, Sicheng	916
Liu, Siyang	258, 376, 382, 423, 822, 828
Liu, Wei	86, 411, 427, 498
Liu, Weibing	345
Liu, Wenjuan	292
Liu, Wenjun	804
Liu, Xiangze	925
Liu, Xianping	487
Liu, Xiaotao	38
Liu, Xiaoyan	475, 539, 724, 913
Liu, Xindi	379
Liu, Yan	472
Liu, Yan-Cong	414
Liu, Yang	819
Liu, Yanheng	348
Liu, Yanjiang	602
Liu, Yao	28
Liu, Yaxin	392
Liu, Yinchi	804
Liu, Yu	931
Liu, Yuchen	667
Liu, Yuyang	28
Liu, Zhe	345
Liu, Zhifang	754, 916
Liu, Zhiqing	718
Liu, Zhiwan	813
Liu, Zhongyang	934
Liu, Zihao	332
Liu, Ziyu	368
Long, Jieyu	841
Long, Keli	199, 789
Lu, Bin	38
Lu, Cimang	478
Lu, Haoran	38, 364
Lu, Hongliang	444, 804, 880
Lu, Jiahao	74
Lu, Jinlong	379
Lu, Sheng	620, 634
Lu, Weihao	859
Lu, Wen-Gao	143
Lu, Xiaoli	542

Lu, Yupeng	757, 774, 777
Luk, Wai-Shing	219
Luk, Wayne	195
Luo, Huanlin	189
Luo, Jie	777
Luo, Jun	345
Luo, Min	289
Luo, Peng	162
Luo, Ping	904
Luo, Weijia	754
Luo, Weijun	733
Luo, Xiaorong	379, 385
Luo, Xixi	411
Lv, Hankun	466
Lv, Qi	664
Lyu, Haiyuan	358
Lyu, Hongliang	901
Lyu, Lewei	877
Lyu, Liangjian	551, 751
Ma, Bingbing	146
Ma, Huaping	895
Ma, Jian	335, 514, 520
Ma, Jiang	771
Ma, Jianghao	889
Ma, Jie	258, 423
Ma, Kaixue	228
Ma, Luanxi	907
Ma, Mingyu	545, 889
Ma, Shunli	700
Ma, Wenzhe	637
Ma, Xiaohua	542, 837
Ma, Xingyu	232
Ma, Xudong	819
Ma, Xuejiao	71
Ma, Yanfeng	376
Ma, Yao	420
Ma, Yuanxiao	469
Mahalingam, Nagarajan	228
Mahapatra, Souvik	490
Mak, Pui-In	122
Mao, Yiqing	205, 219
Mao, Zhifeng	86
Martins, Rui P.	122, 152
Mei, Jian	104, 132, 676, 685, 691
Mei, Xianqi	74
Meng, Fanxin	877
Mi, Jinyao	47
Miao, Xiangshui	31
Min, Hao	715
Min, Tai	630
Mizutani, Tomoko	332
Mo, Fei	469

Mo, Wenji	56, 168, 261
Mori, Takahiro	332
Nakatani, Takayuki	192
Nan, Longmei	602
Ng, Wai Tung	372, 783
Ni, Xiangran	733
Nie, Yanyu	907
Ning, Ning	392, 667
Niu, G.	433
Niu, Ruiting	101
Niu, Yan	289
Ogawa, Tomohiko	135
Ogihara, Gaku	192
Oka, Hiroshi	332
Okamoto, Toshiyuki	192
Ostling, Mikael	404
Ouyang, Lingyun	202
Pan, Chuanqi	376
Pan, Jie	584
Pan, Yu	614
Peng, Baokang	536
Peng, Chaoyang	859
Peng, Chunyu	107, 643
Peng, Daixiao	466, 661
Peng, Lijuan	771
Peng, Liyuan	4, 593
Peng, Lulu	392
Peng, Pan	392
Peng, Sirui	466
Peng, Xiaosong	385
Peng, Xizhu	139
Peng, Yungen	487
Peng, Zhirong	850
Pu, Jie	688
Qi, Gengzhen	235, 273
Qi, Junjun	880
Qi, Zhao	430, 451
Qiang, Hua	38
Qiao, Jiantao	880
Qiao, Liangquan	199, 789
Qiao, Ming	392, 430, 451
Qiao, Shushan	222
Qiao, Yong	68
Qiao, Yuan	865
Qin, Haifeng	844
Qin, Haiyan	47
Qin, Yajie	129, 554, 748
Qin, Zhigang	578
Qiu, Hao	113
Qiu, Xiang	478
Qu, Xin-Ping	361
Qu, Yang	664

Qu, Yiming	427, 498, 742
Qu, Yuwei	205
Que, Zhiqiang	195
Rao, Amy	652
Rao, Qiansheng	844
Ren, Junyan	159, 165
Ren, Min	844
Ren, Mingze	22, 703
Ren, Pengpeng	351
Ren, Tian-Ling	255, 919
Ren, Tianling	569, 766
Ren, Tingrui	44
Ren, Ye	760
Ren, Yuan	712
Ren, Zhipeng	655, 892, 940
Rivet, François	267
Robertson, John	296
Rochette, Stephane	267
Sang, Guanqiao	757
Sang, Pengpeng	810
Saraya, Takuya	332
Sato, Keno	192
Sawan, Mohamad	248
Sha, Yanliang	202
Shan, Hongwei	83, 156
Shan, Linbo	508
Shan, Xiaoyu	694, 727
Shang, Zi-Meng	319
Shao, Sicong	338
Shao, Xianzhou	460, 463, 484, 511
Shao, Yu	542
Shao, Yun-Hao	361
Shen, Boqian	335, 514, 520
Shen, Chao	299, 928
Shen, Hao-Yuan	255
Shen, Hongwei	584
Shen, Lei	937
Shen, Rensheng	664, 795
Shen, Xiaohan	270
Shen, Xinxin	517
Sheng, Bin	86
Shi, Chunqi	596
Shi, Jincheng	530
Shi, Jinhong	530
Shi, Mingmin	526
Shi, Shucheng	289
Shi, Tiantian	925
Shi, Xinhua	624
Shi, Yi	113
Shi, Yu	411, 874, 907
Shi, Yue	578
Shuai, Li	228

Sin, Sai-Weng 152
Someya, Takao 285
Song, Jiafeng 754
Song, Qing-Wen 414, 831, 853, 856, 862
Song, Ruiyang 244
Song, Wei 379, 385
Song, Xujin 475
Song, Yifan 751
Song, Yukun 7
Song, Zhaoxu 382
Su, Xinying 557
Su, Yanwen 563
Sun, Chengliang 292
Sun, Chuanlin 533
Sun, Dijiang 475
Sun, Haoning 56, 168, 261, 673
Sun, Jialei 68
Sun, Jiameng 822
Sun, Junyi 113
Sun, Kangkang 168, 673
Sun, Ke 709
Sun, Leihao 299
Sun, Qingqing 335, 368, 514, 520
Sun, Quan 572, 682
Sun, Ruize 395, 411, 847, 874
Sun, Weifeng 258, 376, 382, 423, 517, 822, 828
Sun, Wendi 10
Sun, Xiaofeng 258
Sun, Xiaoqing 460, 463, 484, 511
Sun, Xinqi 395
Sun, Yilin 304, 754, 916, 934
Sun, Yingxue 107, 643
Sun, Yongsheng 183, 441
Sun, Yuan 376
Sun, Yuting 718
Sun, Zhong 501
Sun, Zixuan 355, 426
Taguti, Henrique Iha 267
Tai, Wei-Heng 322
Takagi, Misaki 192
Takeuchi, Kiyoshi 332
Tan, Jialei 379
Tang, Hailong 751
Tang, He 139
Tang, Jiami 679
Tang, Jing 216
Tang, Jinsheng 560
Tang, Juan 392
Tang, Peishun 763
Tang, Xiao-Yan 414, 831, 853, 856
Tang, Xiaoyan 862
Tang, Yiwen 590

Tao, Nick 388
Tao, Qiuyu 661
Tao, Yongjin 10
Taylor, Stephen 408
Teng, Changjiu 801
Teng, Qiao 910
Thangarasu, Bharatha Kumar 228
Tian, Fengbin 463
Tian, Fengshuo 16
Tian, Wei 566
Tian, Xin 766
Tian, Yu 382
Tian, Yuanxin 614
Tiwari, Ravi 490
Tok, Kean Hong 437
Tsuchiya, Akira 310
Tu, Jiangtao 25
Tu, Ruei-Cheng 313
Wan, Li 844, 868, 871
Wan, Peng 569
Wan, Qingbo 847
Wan, Wenqing 780
Wan, Yuxi 411, 454, 874, 907
Wang, Anqing 289
Wang, Bo 228
Wang, Bohan 62
Wang, Bo-Wei 319
Wang, Chao 566
Wang, Chenlu 828
Wang, Chenxing 813
Wang, Chenxu 289
Wang, Chunlin 569
Wang, Cuimei 504, 508
Wang, Dejin 258
Wang, Denggui 376
Wang, Fangzhou 771
Wang, Feixiong 774
Wang, Haiwei 596
Wang, Haoran 877
Wang, Hongbo 355
Wang, Huaipeng 916, 934
Wang, Huaishan 392
Wang, Jiabin 89
Wang, Jian 186, 189
Wang, Ji-Jiang 572
Wang, Jinxiang 199, 789
Wang, Jitong 859
Wang, Junfei 299
Wang, Junhao 368
Wang, Kaifeng 760
Wang, Kaixuan 16
Wang, Kang 342

Wang, Langyuan 92
Wang, Lei 730
Wang, Li 694
Wang, Liang 44
Wang, Lina 129, 748
Wang, Lingli 205, 219, 225
Wang, Long 454
Wang, Longsheng 149, 670
Wang, Luda 244
Wang, Meng 289
Wang, Mengxuan 80
Wang, Mingbo 74
Wang, Mingyu 4, 593, 922
Wang, Peiran 388
Wang, Peng 757, 777
Wang, Pengjun 19, 22, 71, 703
Wang, Qianwen 807, 810
Wang, Qing 388
Wang, Ruibo 252
Wang, Runsheng 13, 178, 212, 351, 355, 364, 426, 536, 760
Wang, Shao Hao 649
Wang, Shi-Dong 578
Wang, Shiqing 501
Wang, Shuhan 913
Wang, Songyao 542
Wang, Teng 101
Wang, Tianrui 238, 721
Wang, Tong-Yang 868, 871
Wang, W. 433
Wang, Wenbin 545, 889
Wang, Wenwu 460, 463, 484, 511
Wang, Xiang 931
Wang, Xianglong 548
Wang, Xiao 798
Wang, Xiaohui 771
Wang, Xiaolei 460, 463, 484, 511
Wang, Xiaoli 289
Wang, Xiaoming 395, 411, 874
Wang, Xiaoping 411
Wang, Xiaowei 282
Wang, Xin'An 712
Wang, Xuebin 469
Wang, Yan 411, 706
Wang, Yang 771, 898
Wang, Yeliang 469
Wang, Yijiao 913
Wang, Yixing 646
Wang, Yixue 4
Wang, Yongjia 913
Wang, Yu 392
Wang, Yuan 13
Wang, Yuchen 56, 168, 261, 673

Wang, Yuhan 487
Wang, Zelin 739
Wang, Zhao 408
Wang, Zhengzhuo 202
Wang, Zhichong 299, 928
Wang, Zhonrui 757
Wang, Zijun 101
Wang, Ziqiang 611
Wang, Zirui 355
Wang, Ziyao 68
Wang, Ziyu 238, 721
Wang, Zongwei 504, 508
Wansi, Ge 228
Wei, Hailong 117
Wei, Jiangbo 566
Wei, Jianglin 192
Wei, Jiangling 135
Wei, Jiaxing 382, 822, 828
Wei, Jie 379, 385
Wei, Xueming 679
Wei, Yuxi 385
Wei, Zhichao 189
Wen, Hongyang 423
Wen, Liang 25, 35, 614
Wen, Xiaokun 301
Weng, Zeping 498
Wong, Kwok-Ho 786
Wong, Man 329
Worley, Eugene 447
Wu, Bo 910
Wu, Chang 80, 780, 937
Wu, Cheng 74
Wu, Chunlei 335, 514, 520
Wu, Haibo 401
Wu, Haiqin 276
Wu, Heng 364
Wu, Honglin 536
Wu, Jiabin 928
Wu, Jixuan 457, 469, 807, 810
Wu, Kejun 667
Wu, Liji 697
Wu, Lin 156
Wu, Mei 837
Wu, Ningran 244
Wu, Wangran 517
Wu, Wenhao 742
Wu, Xing 178, 551, 751
Wu, Xudong 19
Wu, Yi-Wen 255
Wu, Yongqin 760
Wu, Yue 279
Wu, Yuping 222

Wu, Yuzhou...844
Wu, Zilong...822
Xia, Donghao..35
Xia, Shiyu...351
Xia, Yiming..........................335, 514, 520
Xia, Yinshui...481
Xia, Yun...411, 874
Xia, Zhiying...282
Xia, Zi-Ming......................................868, 871
Xiang, Jinjuan...460
Xiang, Yuguo...165
Xiao, Shanlin..62
Xiao, Yu...351
Xiao, Zhentao..813
Xiaowen, Chen...98
Xie, Chengzhen...685
Xie, Dan...916, 934
Xie, Huiquan.......................................238, 721
Xie, Jing...86
Xie, Mingzhang...658
Xie, Pei-Zhang..41
Xie, Pujin..28
Xie, Qin...768
Xie, Ruiqing...508
Xing, Linlin..38
Xing, Pengcheng....................................395, 847
Xiong, Bingjun...557
Xiong, Ling..792
Xiong, Shisheng..658
Xiong, Sixing..285
Xiong, Wenjuan...345
Xu, Baohui...886
Xu, Fan..865
Xu, Gaobo..777
Xu, Guohao...605, 608
Xu, Hanxi...31
Xu, Hao...................460, 463, 484, 511, 715, 730
Xu, Hongtao........................146, 232, 279
Xu, Jeffrey..345
Xu, Jianlong...916
Xu, Jianqiang..548
Xu, Jing...345
Xu, Jinghan..913
Xu, Jun..86, 766, 919
Xu, Long...596
Xu, Mengfan..35
Xu, Ruize...19
Xu, Sheng..575
Xu, Tao..401
Xu, Wei..620
Xu, Wenting..517
Xu, Wenwei..86

Xu, Xinyue...301
Xu, Yumin...335, 520
Xue, Hao...697
Xue, Xiaoyong.....................................171, 466
Xv, Yawen..423
Yamamoto, Shuhei.......................................192
Yan, Anzhi...569
Yan, Bei-Ping..798
Yan, Chu...427
Yan, Chuangao..162
Yan, Feng.........................575, 673, 925
Yan, Jin...338
Yan, Na..........................92, 715, 730
Yan, Shuang..264
Yan, Silu..880
Yan, Wenchao......................................427, 498
Yan, Xuke..825
Yang, Bowen..837
Yang, Chaowei..122
Yang, Chengyu..709
Yang, Gaoqi..508
Yang, Guangkuo...810
Yang, Guanhua..469
Yang, Guo...98
Yang, Hongqiang...........417, 530, 813, 825, 841, 865
Yang, Jia...484, 511
Yang, Jie..248
Yang, Jinda..688
Yang, Ling...837
Yang, Ning...398, 454
Yang, Runhua...795
Yang, Shanqiang..289
Yang, Shaoqi...810
Yang, Sheng..351
Yang, Sijing...225
Yang, Tianqi...670
Yang, Ting...919
Yang, Wangjun..889
Yang, Wenhao..95
Yang, Wu...53
Yang, Yang........................844, 868, 871
Yang, Yanyu..777
Yang, Yecheng..649
Yang, Yi...569
Yang, Yintang.....................238, 548, 721
Yang, Yizhe.......................95, 646, 745
Yang, Yurui..423
Yang, Yuxiao...395
Yang, Zhangbin....................................466, 661
Yang, Zhaohui..481
Yang, Zhen...853
Yang, Zhi-Yu......................................868, 871

Yang, Zhuoyuan 1
Yang, Zixin 189
Yao, Rui Ray 408
Yao, Ruxue 444, 901
Yao, Siyuan 117
Ye, Chenglin 724
Ye, Dongxian 149
Ye, Fan 159, 165, 700
Ye, Jiabao 661
Ye, Jinhong 1
Ye, Tianchun 484, 511
Yeo, Kiat Seng 228
Yi, Bo 417, 825, 865
Yi, Shiyan 605, 608
Yin, Huaxiang 358, 757, 774, 777
Yin, Rui 132, 676, 685, 691, 730
Yin, Wenbo 205, 219
Yin, Yizhe 655, 892, 940
Yin, Yun 241
Ying, Kai 886
You, Hailong 216, 545, 889
You, Lu 117
Yu, Fei 472
Yu, Hongyu 388
Yu, Junjie 557
Yu, Ping 771
Yu, Qiang 420
Yu, Shaofeng 763
Yu, Wei-Wei 447
Yu, Xiao 472
Yu, Xinglong 50
Yu, Xinwei 427
Yu, Yueru 59
Yu, Yueyuan 335, 514, 520
Yu, Zhen 667
Yu, Zhiyi 28, 62, 487
Yu, Zhuoqing 441
Yu, Zuoxu 517
Yuan, Hao 414, 831, 862
Yuan, Minghong 392
Yuan, Shuai 264
Yuan, Tengfei 22, 703
Yuan, Xihui 630
Yuan, Yidong 584
Yuqing, Liu 228
Zeng, Wei 411, 454
Zeng, Xiaoyang 4, 171, 593, 605, 608, 922
Zeng, Xing 401
Zeng, Yaxin 715
Zhai, Ziye 292
Zhan, Xuepeng 457, 807, 810
Zhang, Ao 736

Zhang, Bo 379, 392, 395, 398, 401, 430, 451, 454,
.... 578, 819, 834, 874, 907
Zhang, Boyang 110
Zhang, Chenghao 584
Zhang, Chenyang 351
Zhang, Chunwei 859
Zhang, David Wei 335, 514, 520
Zhang, Duoli 7
Zhang, Fangxing 536
Zhang, H. 433
Zhang, Hang 757
Zhang, Hanlu 92
Zhang, Hao 608, 804
Zhang, Haoyu 65
Zhang, He 47
Zhang, Hongbo 895
Zhang, Huihong 162
Zhang, Jiaming 74
Zhang, Jian Fu 437
Zhang, Jian 494
Zhang, Jianhua 640
Zhang, Jiayu 68
Zhang, Jide 225
Zhang, Jiming 388
Zhang, Jingjing 682
Zhang, Jingtao 1
Zhang, Jinping 819, 834
Zhang, Jun 53
Zhang, Junyu 457
Zhang, Ke 898
Zhang, Lei 706
Zhang, Li 640
Zhang, Liang 289
Zhang, Lining 351, 426, 536
Zhang, Long 258, 376, 423, 682
Zhang, Lu-Lu 255
Zhang, Meihe 757
Zhang, Meng 837
Zhang, Mengming 727
Zhang, Qianhao 575
Zhang, Qingzhu 757
Zhang, Ruikai 289
Zhang, Runxi 596
Zhang, Sen 258, 430
Zhang, Shengdong 325
Zhang, Shengnan 50
Zhang, Shutong 22
Zhang, Shuyu 92
Zhang, Tianyu 700
Zhang, W. 494
Zhang, Wei David 368
Zhang, Wei 139

Zhang, Weidong 437, 630
Zhang, Wenfeng 301
Zhang, Xiangmin 697
Zhang, Xing 533
Zhang, Xinrui 487
Zhang, Xinyu 822
Zhang, Xinyue 536
Zhang, Xuelian 222
Zhang, Xuewei 296
Zhang, Xusheng 113
Zhang, Ya-Cong 143
Zhang, Yadong 774
Zhang, Yanlong 44
Zhang, Yi Bo 862
Zhang, Yibo 95, 745
Zhang, Yihui 581
Zhang, Yimeng 95, 646, 745
Zhang, Yue 590
Zhang, Yuejun 22, 25, 35, 614, 703
Zhang, Yuhan 748
Zhang, Yu-Ming 414, 831, 853, 856
Zhang, Yuming 95, 252, 444, 646, 745, 880, 901
Zhang, Yurun 7
Zhang, Yushen 95, 745
Zhang, Yutao 444, 901
Zhang, Yuzhen 517
Zhang, Zhaohao 358
Zhang, Zhen 590
Zhang, Zhenyin 620
Zhang, Zhenyu 596
Zhang, Zhili 117, 430
Zhang, Zonghao 813
Zhao, Chun 560
Zhao, Dongyan 584
Zhao, Fei 335, 514, 520
Zhao, Hankun 304
Zhao, Kai 264, 385, 533
Zhao, Liangxiao 71
Zhao, Lu-Yu 255
Zhao, Mengyao 376
Zhao, Peizhi 766
Zhao, Qi 706
Zhao, Ruiyong 655, 892, 940
Zhao, Shilong 801
Zhao, Shulin 152
Zhao, Xiaohuan 810
Zhao, Xin 670
Zhao, Xu 611
Zhao, Yi 427, 478, 498, 742
Zhao, Yifan 50
Zhao, Yi-Shang 868, 871
Zhao, Yuanfu 44

Zhao, Yudi 264
Zhao, Yujie 192
Zhao, Zi-Ming 414
Zheng, Guiqiang 258
Zheng, Haoping 264
Zheng, Shiwei 834
Zheng, Xuefeng 542
Zheng, Yifei 376
Zheng, Zhe 572, 682
Zhenghao, Lu 228
Zhi, Haizhao 411, 874
Zhong, Chengyan 931
Zhong, Guoqiang 664
Zhong, Jingxue 107, 643
Zhong, Kun 358
Zhong, Linfeng 13
Zhong, Qingyin 258
Zhong, Tao 175
Zhong, Tianyuan 110
Zhong, Zheng-Hong 322
Zhou, David 454, 874, 907
Zhou, Dayan 165
Zhou, Hao 225
Zhou, Jianjun 175, 376, 736
Zhou, Jinggui 398
Zhou, Jinghan 199, 789
Zhou, Jingming 178
Zhou, Jiuren 661
Zhou, Mohan 279
Zhou, Pengwei 395
Zhou, Pingqiang 208
Zhou, Qi 398, 454, 907
Zhou, Qiang 171
Zhou, Ruibin 487
Zhou, Wenbin 925
Zhou, Wenqian 617
Zhou, Xiahong 504
Zhou, Xiaofang 59, 739
Zhou, Xinlong 804
Zhou, Xue 630
Zhou, Yang 289
Zhou, Yongliang 107, 643
Zhou, Yu 414, 831, 853, 856
Zhou, Zecheng 149
Zhou, Zekun 578, 584
Zhou, Zheng 539, 724
Zhou, Zhuoling 801
Zhou, Zikang 627
Zhou, Ziwei 411
Zhou, Ziyu 71
Zhu, Chiang 928
Zhu, Jianggen 398, 454, 907

Zhu, Saike 478
Zhu, Xiaona 763, 928
Zhu, Xing 688
Zhu, Yexin 149
Zhu, Yinna 620
Zhu, Yujie 640
Zhu, Zhangming 83, 149, 156, 670
Zhu, Zhiyuan 368
Zhuang, Quanrong 113
Zhuo, Tianshu 1
Zou, Qiaosha 617
Zou, Rongxin 25
Zuo, Peng 771

IEEE
445 Hoes Lane
Piscataway, NJ 08854-4141

ISBN 979-8-3503-6184-1

2024 IEEE 17th International Conference on Solid-State & Integrated Circuit Technology (ICSICT 2024)

Zhuhai, China
22-25 October 2024

Pages 444-942

IEEE Catalog Number: CFP24829-POD
ISBN: 979-8-3503-6184-1

2024 IEEE 17th International Conference on Solid-State & Integrated Circuit Technology (ICSICT 2024)

Zhuhai, China
22-25 October 2024

Pages 444-942

IEEE Catalog Number: CFP24829-POD
ISBN: 979-8-3503-6184-1

**Copyright © 2024 by the Institute of Electrical and Electronics Engineers, Inc.
All Rights Reserved**

Copyright and Reprint Permissions: Abstracting is permitted with credit to the source. Libraries are permitted to photocopy beyond the limit of U.S. copyright law for private use of patrons those articles in this volume that carry a code at the bottom of the first page, provided the per-copy fee indicated in the code is paid through Copyright Clearance Center, 222 Rosewood Drive, Danvers, MA 01923.

For other copying, reprint or republication permission, write to IEEE Copyrights Manager, IEEE Service Center, 445 Hoes Lane, Piscataway, NJ 08854. All rights reserved.

*** This is a print representation of what appears in the IEEE Digital Library. Some format issues inherent in the e-media version may also appear in this print version.

IEEE Catalog Number: CFP24829-POD
ISBN (Print-On-Demand): 979-8-3503-6184-1
ISBN (Online):   979-8-3503-6183-4
ISSN: 2835-7612

Additional Copies of This Publication Are Available From:

Curran Associates, Inc
57 Morehouse Lane
Red Hook, NY 12571 USA
Phone: (845) 758-0400
Fax: (845) 758-2633
E-mail: curran@proceedings.com
Web: www.proceedings.com

TABLE OF CONTENTS

A High-Performance Multicore Testing Platform for Multi-Scenario Applications 1
Zipeng Ling, Tianshu Zhuo, Zhuoyuan Yang, Jinhong Ye, Jun Han, Jingtao Zhang

S-SIFT: A Simple SIFT Algorithm with High Efficiency 4
Yixue Wang, Yujie Huang, Liyuan Peng, Mingyu Wang, Wenhong Li, Minge Jing, Xiaoyang Zeng

Design of a High-Speed SAR Processor Based on the Chirp Scaling Algorithm 7
Xianghe Cui, Yukun Song, Yurun Zhang, Jingyi Hu, Zhenmin Li, Duoli Zhang

Accelerating Matrix-Vector Multiplications of Large Language Models Via Efficient Encoding 10
Yongjin Tao, Wendi Sun, Song Chen, Yi Kang

Flexible Yet Efficient Transformer Acceleration with Unified Sparse Attention Support on FPGA 13
Linfeng Zhong, Qingyu Guo, Runsheng Wang, Yuan Wang, Meng Li

Backward-Edge Control Flow Integrity Based on Return Address Encryption 16
Fengshuo Tian, Kaixuan Wang, Jun Han

Stability Enhancement Technique for Monostable PUF Based on Hysteresis Effect of Schmitt Trigger 19
Ruize Xu, Gang Li, Pengjun Wang, Hui Li, Xudong Wu

A Reliable Current Starved Inverter PUF Based on SRAM Memory Structure 22
Mingze Ren, Pengjun Wang, Yuejun Zhang, Shutong Zhang, Zhikang Chen, Tengfei Yuan

An Efficient Convolutional Neural Network Hardware IP for Epilepsy Detection 25
Yi Gong, Yuejun Zhang, Jiangtao Tu, Rongxin Zou, Liang Wen

TLBshield: A Low-Cost Secure Reinforce on Translation Lookaside Buffer to Mitigate the Speculative Attacks 28
Yuyang Liu, Runye Ding, Yujie Chen, Pujin Xie, Yao Liu, Zhiyi Yu

One-Step Circuit Analysis Based on LCA for Sparse Coding 31
Hanxi Xu, Zirui Chen, Chenqi, Xiangshui Miao, Yuhui He

A Hybrid-Logic Scheme for High-Performance and Low-Power Decoders in 7nm Process 35
Donghao Xia, Yuejun Zhang, Mengfan Xu, Liang Wen, Yiting Guo

Ternary Logic Units Design Based on the TDDFETs 38
Hua Qiang, Haoran Lu, Xiaotao Liu, Linlin Xing, Bin Lu

Enhancing SRAM Cell Stability Through Single-Carrier CMOS Latch Integration 41
Yuan-Yu Chuang, Pei-Zhang Xie, Jyi-Tsong Lin

A RRAM Based 9T1R NVSRAM for Low-Power Computing in Memory 44
Huimeng Guo, Yujia Li, Tingrui Ren, Chenge Dong, Liang Wang, Yuanfu Zhao, Yanlong Zhang

A High-Resistance SOT Device Based Computing-In-Memory Macro with High Sensing Margin and Multi-Bit MAC Operations for AI Edge Inference 47
Junzhan Liu, Jinyao Mi, Haiyan Qin, He Zhang, Wang Kang

RISC-V Domain-Specific Processor for Accelerating SPHINCS+ on Multi-Core Architecture 50
Shengnan Zhang, Yifan Zhao, Xinglong Yu, Jun Han

Design of an Out-Of-Order Superscalar Processor with Improved Register Alias Table Recovery
Method 53
Wu Yang, Jun Zhang

An SDPF RISC-V Processor with Two-Stage Pseudo-Pipelined Architecture for IoT Applications 56
Wenji Mo, Yuchen Wang, Haoning Sun, Jingjing Liu

A Unified Verification Scheme for the Acceleration of RISC-V Processor Design 59
Zixiang Chen, Jiyuan Bai, Yueru Yu, Gengsheng Chen, Xiaofang Zhou

Asynchronous Arbitration Circuit Optimization for Multicore Neuromorphic Processors 62
Jiajie Guo, Guangyao Lin, Bohan Wang, Zhiyi Yu, Shanlin Xiao

A Run-Time Temperature Monitoring with Adaptive Duty Cycle Control for FPGA Applications 65
Weizhi Li, Wangyong Chen, Haifeng Chen, Haoyu Zhang, Linlin Cai

An FPGA-Based Top-K Gradient Compression Accelerator for Distributed Deep Learning Training 68
*Ziyao Wang, Jiayu Zhang, Kunyue Li, Jialei Sun, Feng Dong, Ke Chen, Yong Qiao, Jianfei
Jiang*

Dynamic-Matrix-Encryption Based Secure Strong PUF for Device Authentication Protocols 71
Liangxiao Zhao, Gang Li, Pengjun Wang, Xuejiao Ma, Ziyu Zhou

A Low Latency and High Throughout Hardware Design of Random Matrix Number Generator for
FrodoKEM 74
*Shengfei Gu, Jiahao Lu, Tianze Huang, Jiaming Zhang, Kai Li, Cheng Wu, Mingbo Wang,
Xianqi Mei, Ang Hu, Dongsheng Liu*

A 4K60fps Ultra-Low-Latency Light Compression Encoder for Bandwidth-Constrained Scenarios 77
Yanzhong Li, Leilei Huang, Yibo Fan

Layer Pipelined Neural Network Accelerator Design on 2.5D FPGAs 80
Mengxuan Wang, Chang Wu

Fast and Accurate Partial-Zoom Depth Estimation for SPAD LiDAR Readout on FPGA 83
Lichen Feng, Hongwei Shan, Rundong Cai, Zhangming Zhu

A Self-Adaptive Gamma Voltage Regulation Circuit for AMOLED Displays 86
Zhifeng Mao, Fei Gou, Bin Sheng, Jing Xie, Wenwei Xu, Wei Liu, Jun Xu

A Reconfigurable Thermoelectric Energy Harvesting Interface Based on OPDC and DSCT 89
Peiyuan Fu, Jiabin Wang, Xufeng Liao, Lianxi Liu

A Fixed-Peak-Current Single-Inductor-Multiple-Output DC-DC Converter Achieving 92.6% Peak
Efficiency 92
Fei Liu, Langyuan Wang, Shuyu Zhang, Hanlu Zhang, Na Yan

Buck-Boost Converter with Stable Transition Mode for Low Average Inductor Current 95
*Ningning Li, Yushen Zhang, Yibo Zhang, Yizhe Yang, Wenhao Yang, Yimeng Zhang, Yuming
Zhang*

A SET Sensitive Model of LC and Ring Voltage-Controlled Oscillator in FinFET Technology 98
*Liu Heyuan, Yuan Hengzhou, Lu Jianzhuang, Chen Xiaowen, Sang Hao, Liu Jingtian, Guo
Yang*

A Low Spur Wideband PLL in 65nm CMOS 101
Zijun Wang, Biao Li, Teng Wang, Hong Li, Ruiting Niu, Jinpeng Lin

A Low Power PLL Circuit with Signal 50% Duty Cycle Corrected in 180nm CMOS 104
Bangtian Li, Xueke Li, Liying Chen, Chuantong Cheng, Jian Mei

MTJ Based Compensation for Charge Pump Temperature Drift 107
Yongliang Zhou, Jingxue Zhong, Chengxing Dai, Yingxue Sun, Xin Li, Chunyu Peng

A 112-Gb/s Coherent Receiver with a Novel Modulation Format 110
Tianyuan Zhong, Boyang Zhang, Weixin Gai

Beyond Bandwidth Trade-Off: Simultaneous Wireless Power and Data Transfer System Design for
Biomedical Implants 113
Quanrong Zhuang, Junyi Sun, Xusheng Zhang, Bo Li, Yi Shi, Hao Qiu

A High Precision Operational Amplifier with Improved Bias Current Cancellation Circuit 117
Zhili Zhang, Siyuan Yao, Puyang Liu, Cheng Li, Lu You, Hailong Wei

A 0.11-PJ/Bit True Random Number Generator Based on a Clocked Current-Starved Inverter 122
Kai Cheng, Chaowei Yang, Rui P. Martins, Pui-In Mak, Yong Chen

A Super-Mixed Current Decay Mode for Reducing the Angular Position Error in Stepper Motor 126
Jian Fang, Xurui Chen, Huajie Liu, Yuhan Jin

A 109 dB 44-PArms Current Readout Circuit with Automatic Current Control for Multimodality
Electrochemical Sensing 129
Lina Wang, Jianzheng Li, Weiming Hu, Yajie Qin

A Low Temperature Coefficient Bandgap Reference for Temperature Sensor System 132
Longjiang Jia, Yuanhong Ding, Jian Mei, Lei Deng, Rui Yin

Toward Unification of Digital Error Correction Algorithms for ADCs with Redundancy 135
Haruo Kobayashi, Tomohiko Ogawa, Yutaro Kobayashi, Kentaroh Katoh, Jiangling Wei

A 1.2-V 2-GS/s Trimming-Free Input Buffer with Robust Output Common-Mode Voltage 139
Wei Zhang, Xizhu Peng, He Tang

A 12-Bit 1-MS/s SAR ADC Using Vcm-Based Split MSB Switching and Segmented CDAC 143
Zheng-Han Chen, Ya-Cong Zhang, Wen-Gao Lu, Zhong-Jian Chen

A Simplified and Accelerated Opportunistic Bit-Weight Calibration for High-Resolution ADCs 146
Bingbing Ma, Wei Li, Hongtao Xu

Background Calibration for Bit Weights in Pipelined SAR ADCs Using Split ADC Architecture 149
Zecheng Zhou, Longsheng Wang, Dongxian Ye, Yexin Zhu, Dengquan Li, Zhangming Zhu

When Time Interleaving Encounters Oversampling in ADC 152
Mingqiang Guo, Dongyang Jiang, Shulin Zhao, Sai-Weng Sin, Rui P. Martins

A 0.000355mm^2 4.6µm-Pitch 5.75fJ/Conv 6-Bit SAR ADC for High Throughput Parallel Readout
of Analog SRAM Computing-In-Memory 156
Lin Wu, Lichen Feng, Hongwei Shan, Zhangming Zhu

A 250MS/s, 12 Bit Pipeline-SAR ADC Using Coarse-Fine Ring Amplifier 159
Linghao Liu, Junyan Ren, Fan Ye

A 0.71pJ/b 16Gb/s Equalizer with Inverter_Based CTLE and 4-Tap Speculative DFE 162
Huihong Zhang, Chuangao Yan, Peng Luo, Maliang Liu

A Digital Foreground Calibration Method for Pipeline SAR ADCs Using Extended Kalman Filter 165
Dayan Zhou, Yuguo Xiang, Junyan Ren, Fan Ye

A Second-Order Charge Pump Noise-Shaping SAR ADC 168
Haoning Sun, Yuchen Wang, Wenji Mo, Kangkang Sun, Jingjing Liu

Computing in Memory for Accelerating Light-Weighted On-Chip Learning in IoT Devices 171
Zhiwang Guo, Xiaoyong Xue, Qiang Zhou, Xiaoyang Zeng

A Novel Beamforing Receiver Architecture Combining MASH SDM and BSP 175
Tao Zhong, Yuekang Guo, Jing Jin, Jianjun Zhou

In Situ Localization Techniques of Defects in Advanced Semiconductor Devices from Macro-Scale
to Atomistic-Scale 178
Jialu Huang, Jingming Zhou, Zuoyuan Dong, Runsheng Wang, Junhao Chu, Xing Wu

Wafer-Level Characterization of Ring-Oscillators Frequency Degradation in FinFET Technology 183
Hao Chang, Dan Gao, Yongsheng Sun, Junlin Huang

Exhaustive Application-Dependent Testing for FPGA Interconnect Resources 186
Wenwei Chen, Xinyu He, Tongshu Ding, Jian Wang, Jinmei Lai

A Comprehensive and Efficient Instruction-Level Testing Method for Processor 189
Zixin Yang, Zhichao Wei, Huanlin Luo, Jian Wang, Jinmei Lai

Thermal Effect and Calibration for High Precision On-Wafer Analog IC Probe Testing 192
*Daisuke Ilmori, Takayuki Nakatani, Shogo Katayama, Gaku Ogihara, Shuhei Yamamoto,
Misaki Takagi, Yujie Zhao, Jianglin Wei, Anna Kuwana, Keno Sato, Takashi Ishida, Toshiyuki
Okamoto, Tamotsu Ichikawa, Kentaroh Katoh, Kazumi Hatayama, Haruo Kobayashi*

Deep Learning Design-Flow with Static and Dynamic Optimizations 195
Zhiqiang Que, Jose G. F. Coutinho, Wayne Luk

A QEMU-Based Virtual Platform of MPSoC 199
Liangquan Qiao, Lei Li, Xingyu Gao, Jinxiang Wang, Fangfa Fu, Keli Long, Jinghan Zhou

A Parallel Harmonic Balance Method Based on GPU for Efficient Periodic Steady-State Analysis 202
Zhengzhuo Wang, Yanliang Sha, Lingyun Ouyang, Quan Chen, Jianguo Hu

Efficient Dynamic Memory Management for High Bandwidth Memory on FPGA 205
Yuwei Qu, Yiqing Mao, Wenbo Yin, Lingli Wang

An Improved Clock-Aware Global Placement Algorithm 208
Ziang Ge, Pingqiang Zhou

Analyzing Timing in Shorter Time: A Journey Through Heterogeneous Parallelism for Static
Timing Analysis 212
Zizheng Guo, Yibo Lin, Runsheng Wang, Ru Huang

TBPart-B: An Effective Hypergraph Partitioning Algorithm Considering Topological Order
Balance for Processor-Based Hardware Emulation 216
Jing Tang, Shunyang Bi, Hailong You

FCE: A Fast CGRA Architecture Exploration Framework 219
  Sichao Chen, Yiqing Mao, Yuan Dai, Xuchen Gao, Wai-Shing Luk, Wenbo Yin, Lingli Wang

Research on Parametric Subthreshold Cell Delay Modeling Based on ANN 222
  Xuelian Zhang, Yuping Wu, Zhiqiang Li, Donglin Liu, Shushan Qiao

A High-Performance Routing Architecture with 16 LUTs Per CLB for Nanoscale FPGAs 225
  Sijing Yang, Jide Zhang, Hao Zhou, Lingli Wang

High-Efficiency Power Amplifier Design for Bluetooth Low Energy Applications 228
  *Bharatha Kumar Thangarasu, Li Shuai, Yu Hongshi, Ge Wansi, Liu Yuqing, Nagarajan
  Mahalingam, Meng Fanyi, Kaixue Ma, Juin J. Liou, Bo Wang, Younan Hua, Xiaomin Li, Lu
  Zhenghao, Kiat Seng Yeo*

A 0.15-6.5GHz Stacked CMOS Power Amplifier with Low-Frequency Bandwidth Extension 232
  Shijiao Dong, Wei Li, Xingyu Ma, Fan Chen, Hongtao Xu

A 2-To-2.7GHz Class-G Switched-Capacitor PA with Cascode Switch-Reused Structure Achieving
25.92dBm Peak Power and 42% Efficiency 235
  Jie Deng, Gengzhen Qi

A X-Band High Linearity Tunable Bandpass Filter in 130nm CMOS 238
  Tianrui Wang, Ziyu Wang, Huiquan Xie, Yifei Chen, Haokun Lan, Maliang Liu, Yintang Yang

Analysis of Polar and Quadrature Digital Transmitters for Wi-Fi7 Applications 241
  Lixuan Cao, Yun Yin

Atomically Thin Graphene Nanopore Based MEMS Iontronic Devices for Sensing, Separation, and
Energy Applications 244
  Luda Wang, Ruiyang Song, Ningran Wu

Systems-On-Chips for Invasive Brain-Computer Interfaces: Challenges and Opportunities 248
  Jie Yang, Mohamad Sawan

Multi-Physics Simulation and Application of Ion-Gel Based Triboelectric Nanogenerators 252
  Chen Liu, Ruibo Wang, Ruiyi Gao, Yuming Zhang

A Flexible Graphene Acoustic Sensor for Sound Signal Acquisition and Spiking Neural Network
Recognition 255
  Lu-Yu Zhao, Hao-Yuan Shen, Yi-Wen Wu, Lu-Lu Zhang, Yu-Tao Li, Tian-Ling Ren

0.15μm BCD Platform with High Sensitivity Hall Device and Low Noise CMOS for Sensor IC
Applications 258
  *Guiqiang Zheng, Qingyin Zhong, Jie Ma, Nannan Cheng, Yichen Li, Yongjia Li, Xiaofeng Sun,
  Dejin Wang, Sen Zhang, Long Zhang, Siyang Liu, Weifeng Sun*

A High Dynamic Range Pixel with Inverse Proportional Response 261
  Yuchen Wang, Wenji Mo, Haoning Sun, Jingjing Liu

Enhancement of Image Sensor Pixel Performance Through Ring-Shaped Vertical Transfer Gate
Structure 264
  Shuang Yan, Shuai Yuan, Haoping Zheng, Yudi Zhao, Gang Du, Junchen Dong, Kai Zhao

A 10-GHz Low Power Class-C VCO with Long-Term Reliability and Tunable Performance in 28
nm FD-SOI for Satellite Communications 267
  *Yann Deval, Henrique Iha Taguti, Ayoub Ait Ihda, Hervé Lapuyade, Stephane Rochette,
  François Rivet*

A 191-GHz Harmonic Oscillator with Self-Feeding Line and Return-Path Gap Coupler Structure in 65nm CMOS.. 270
Xiaohan Shen, Chen Jiang

A Blocker-Tolerant High-Linear Receiver Employing Baseband Noise-Cancelling and Bottom-Plate Switched-Capacitor Techniques 273
Chenxiang Cai, Gengzhen Qi

A High Sensitivity Series-Parallel Rectifier with Pre-Bias for RF Energy Harvesting Systems 276
Haiqin Wu, Dejian Li, Xin Jin, Xufeng Liao, Lianxi Liu

A 24.3-43.7 GHz Variable-Gain Low-Noise Amplifier with Phase Self-Compensation 279
Yue Wu, Wei Li, Mohan Zhou, Hongtao Xu

A Broadband Active Variable Attenuator with Phase Compensation Technique.. 282
Zhiying Xia, Zhiqun Li, Bofan Chen, Xiaowei Wang

Waterproof and Wearable Power Sources... 285
Sixing Xiong, Kenjiro Fukuda, Takao Someya

A CMOS Pixel Sensor for Precise Track and Charge Measurement of Cosmic Ray Nuclei............................ 289
Ruikai Zhang, Wen He, Shanqiang Yang, Min Luo, Chenxu Wang, Cunfeng Feng, Meng Wang, Liang Zhang, Anqing Wang, Jianing Dong, Dong Liu, Yan Niu, Yang Zhou, Yuehong Gong, Xiaoli Wang, Shucheng Shi

$Sc_{0.096}Al_{0.904}N$-Based Bimorph Piezoelectric Micro Machined Ultrasonic Transducers................................. 292
Ziye Zhai, Wenjuan Liu, Chengliang Sun

TeO_2 as Amorphous P-Type Transistor for Back-End-Of-Line Applications.. 296
John Robertson, Xuewei Zhang, Qingzhong Gui, Yuzheng Guo

Miniaturization of High-Speed GaN Based Laser Diodes.. 299
Junfei Wang, Chaowen Guan, Leihao Sun, Zhichong Wang, Chao Shen

Impact of Interfacial Layer on the Optoelectronic Performance of $MoTe_2$/Ge Heterojunction...................... 301
Wenyu Lei, Xiaokun Wen, Boyuan Di, Xinyue Xu, Haixin Chang, Wenfeng Zhang

MoS_2-WS_2 Heterostructure-Enabled Optoelectronic Synaptic Diode 304
Mingjie Li, Yingtao Ding, Jianzhi Hu, Hankun Zhao, Yilin Sun

Pseudo-Parallel Symmetrical and Crossed Perovskite Solar Cells for Bifacial Applications........................... 307
Guang-Wei Huang, Hsing-Mao Cheng, Jyi-Tsong Lin

Cryogenic and RF Modeling of On-Chip Passive Devices for Quantum Computer 310
Akira Tsuchiya

Comparison of Nanosheet and Fin Integration in Stacked Induced Tunnel Field-Effect Transistors 313
Ruei-Cheng Tu, Chia-Yo Kuo, Jyi-Tsong Lin

Nonlinear Contact Behavior in MoS_2 Field Effect Transistors at Cryogenic Temperature............................. 316
Shihab Ahammed, Mansun Chan

Experimental Verification of 1D Transport Model by Quantized Current Spectrum of Si JNT Device.. 319
Zi-Meng Shang, Bo-Wei Wang, Wei-Hua Han

Impact of Gate Overlap Length Modulation on Electrical Characteristics and Subthreshold Swing in Nanosheet TFETs with Varying Tunneling Mechanisms 322

Zheng-Hong Zhong, Wei-Heng Tai, Jyi-Tsong Lin

Recent Progress in the Development of Complementary Field-Effect Transistors 325

Mansun Chan, Shengdong Zhang

Metal-Oxide Thin-Film Transistors for Artificial Neural Networks 329

Yushen Hu, Tengteng Lei, Man Wong

Cryogenic Threshold Voltage and on-Current Variability Analysis of GAA Nanosheet FETs at 4K 332

Zihao Liu, Tomoko Mizutani, Kiyoshi Takeuchi, Takuya Saraya, Hiroshi Oka, Takahiro Mori, Masaharu Kobayashi, Toshiro Hiramoto

Reverse-Biased PN Junction Isolation for Leakage Suppression and Strain Enhancement in Gate-All-Around Nanosheet FETs 335

Boqian Shen, Chunlei Wu, Yumin Xu, Fei Zhao, Hanzhi Gu, Jian Ma, Yueyuan Yu, Yiming Xia, Qingqing Sun, David Wei Zhang

Studies on Selective Deposition of SiO_2 by Rapid Atomic Layer Deposition 338

Sicong Shao, Jin Yan, Wang Li, Kun Cao, Rong Chen

Resistance Dependence of Cobalt on Line Width in Advanced Interconnects: First-Principles Modelling 342

Kang Wang, Menglin Huang, Shiyou Chen

Reduction of Specific Contact Resistivity by Employing Pre-Amorphization Implantation and In-Situ Steam Generation Oxidation 345

Chang Liu, Xu Chen, Jinbiao Liu, Yanping He, Wenjuan Xiong, Weibing Liu, Mingshan Liu, Zhe Liu, Yaoqi Dong, Jeffrey Xu, Jing Xu, Jun Luo

High-Performance Ultrathin ITO Thin-Film Transistor with Ultralow Subthreshold Swing 348

Yanheng Liu, Tiaoyang Li

Defect-Centric Insights into Flicker Noise in Ultra-Scaled FETs: From Physics to Compact Model for Circuit Level Simulation (Invited) 351

Chenyang Zhang, Yu Xiao, Pengpeng Ren, Shiyu Xia, Sheng Yang, Lining Zhang, Runsheng Wang, Zhigang Ji

A Modified Virtual-Source Model for Ballistic Transport Characterization of FinFETs at Cryogenic Temperature 355

Hongbo Wang, Zirui Wang, Zixuan Sun, Runsheng Wang, Ru Huang

Investigation on Asymmetric HfO_2-ZrO_2-HfO_2 Superlattice Gate Stacks with Ultra-Low EOT for Advanced Transistors 358

Haiyuan Lyu, Kun Zhong, Zhaohao Zhang, Huaxiang Yin

Evaluation of Contact Resistance with the 'L' Kelvin Test Structure and the Modified Kelvin Test Structure 361

Gui Chen, Yun-Hao Shao, Xin-Ping Qu

Flip 3D (F3D): A Novel 3D Integration Technology with Dual-Side Integration Capabilities 364

Heng Wu, Haoran Lu, Runsheng Wang, Ming Li, Yibo Lin, Weihai Bu, Jin Kang, Ru Huang

Modeling and Simulation of a Conical 3D Monopole Antenna Embedded in Substrate for WNoC 368

Junhao Wang, Ziyu Liu, Zhiyuan Zhu, Lin Chen, Qingqing Sun, Wei David Zhang

Gate Driver ICs for Wide Bandgap Power Transistors 372
    Wai Tung Ng, Rophina Li, Wentao Cui, Jingyuan Liang

Suppression of Back-Gating Effect by Integrated Substrate Termination Network for 200V
Monolithic GaN Half-Bridge Power IC 376
    *Mengyao Zhao, Yifei Zheng, Yanfeng Ma, Yuan Sun, Denggui Wang, Chuanqi Pan, Jianjun
    Zhou, Sheng Li, Long Zhang, Siyang Liu, Weifeng Sun*

High Short-Circuit Capability and Low-Loss SOI-LIGBT with Double-Integrated NMOS 379
    *Jialei Tan, Jie Wei, Jinlong Lu, Xindi Liu, Gaoqiang Deng, Wei Song, Pei Guo, Bo Zhang,
    Xiaorong Luo*

Body Diode Degradation Mechanism of 1200V SiC Power MOSFETs Under Gamma Rays Total
Ionizing Dose Irradiation 382
    Yu Tian, Zhaoxu Song, Hao Fu, Jiaxing Wei, Siyang Liu, Weifeng Sun

Novel Heterojunction Field Plate β-Ga$_2$O$_3$ MOSFET with High Breakdown Voltage 385
    *Xiangnan Li, Jie Wei, Kai Zhao, Linyao Hao, Xiaosong Peng, Yuxi Wei, Renkuan Liu, Wei
    Song, Pei Guo, Xiaorong Luo*

Investigation of SiON Passivation for High Performance AlGaN/GaN HEMTs 388
    Difei Fan, Chenkai Deng, Jiming Zhang, Peiran Wang, Nick Tao, Qing Wang, Hongyu Yu

180nm BCD Technology Platform with 8V to 65V Isolated LDMOS 392
    *Qi Ding, Renxiong Li, Ning Ning, Jun Huang, Yutuo Guo, Yu Wang, Kunqin He, Yaxin Liu,
    Huaishan Wang, Juan Tang, Qiuyue Huo, Minghong Yuan, Pan Peng, Ming Qiao, Lulu Peng,
    Bo Zhang*

A Novel Insulated Gate-Triggered Thyristor with Integrated Super-Clamp Gate Transient Voltage
Suppressor for Ultrahigh Di/Dt Pulse Switching 395
    *Shiyu Deng, Yuxiao Yang, Xinqi Sun, Pengwei Zhou, Ruize Sun, Chao Liu, Pengcheng Xing,
    Xiaoming Wang, Wanjun Chen, Bo Zhang*

Device Instability in the Third Quadrant of Schottky-Type p-GaN Gate HEMTs: The Hole
Defficiency & Trapping Effect 398
    *Kuangli Chen, Shuting Huang, Jinggui Zhou, Ning Yang, Jianggen Zhu, Enchuan Duan, Bo
    Zhang, Qi Zhou*

Static Characteristic Recovery of SiC MOSFETs Induced by Dynamic Gate Stress After Total
Ionizing Dose Irradiation 401
    Jiahao Hu, Xiaochuan Deng, Xing Zeng, Tao Xu, Haibo Wu, Xuan Li, Bo Zhang

The Status of WBG Devices Towards Net-Zero Solutions 404
    Mikael Ostling

Impact of the Resistive Silicon Base Wafer on Substrate Coupling in Power Integrated Circuits in
GaN-On-Si Technology 408
    Zijin Jiang, Rui Ray Yao, Miao Cui, Zhao Wang, Sang Lam, Stephen Taylor

A Novel Snapback-Free Double-RESURF Reverse-Conducting LIGBT with Dual Conduction
Paths 411
    *Yun Xia, Yuxi Wan, Wei Zeng, Yu Shi, Xiaoping Wang, Wei Liu, Haizhao Zhi, Ziwei Zhou, Xixi
    Luo, Ruize Sun, Xiaoming Wang, Yan Wang, Wanjun Chen*

Comparsion of SiC Planar and Trench Junction Barrier Schottky Diode with Surge Current
Capability 414
    Zi-Ming Zhao, Yan-Cong Liu, Hao Yuan, Feng-Yu Du, Yu Zhou, Ke-Yu Liu, Xiao-Yan Tang,
    Chao Han, Qing-Wen Song, Yu-Ming Zhang

Comparative Study on Reliability of Conventional SiC MOSFET and JBS Integrated SiC MOSFET 417
    Moufu Kong, Shurui Li, Hongfei Deng, Bo Yi, Hongqiang Yang, Sen Gong

Study on Single Event Effect of SiC MOSFET by Proton Irradiation 420
    Wende Huang, Chengwen Fu, Yao Ma, Mingmin Huang, Xiaoping Dong, Qiang Yu

Investigating Single-Event Burnout in 4H-SiC Inverters: Experiments and Simulations 423
    Yong Gu, Yurui Yang, Yawen Xv, Hongyang Wen, Xiangyu Hou, Runhua Huang, Ao Liu, Song
    Bai, Jie Ma, Long Zhang, Siyang Liu, Weifeng Sun

Challenges of Design for Reliability in Advanced CMOS Technology: From Single-Mode to
Mixed-Mode Mechanisms.. 426
    Zixuan Sun, Lining Zhang, Ru Huang, Runsheng Wang

Frequency-Dependent Time-Dependent Dielectric Breakdown (TDDB) Behavior and Physical
Study in Gate Oxides.. 427
    Wei Liu, Chu Yan, Xinwei Yu, Yiming Qu, Wenchao Yan, Yi Zhao

Lightning Protection Stacked TVS Structure Based on a Novel Total-Ionizing-Dose Radiation-
Hardened Technology 430
    Zhao Qi, Hongquan Chen, Yirui Jia, Nailong He, Zhili Zhang, Sen Zhang, Ming Qiao, Bo
    Zhang

Characterization and Modeling of Non-Conducting RF Hot Carrier Stress in FinFETs 433
    G. Niu, X. Ding, H. Zhang, W. Wang, K. Imura, F. Dai

Predictive Modelling of Hot Carrier Degradation 437
    James Brown, Kean Hong Tok, Rui Gao, Zhigang Ji, Weidong Zhang, Jian Fu Zhang

New Insights into the Saturation Behavior of the Hot Carrier Degradation in STI-Based N-Type
LDMOS.. 441
    Zhuoqing Yu, Dan Gao, Yongsheng Sun, Junlin Huang

The TID Response and HCI Degradation for Multi-Vt nFinFETs 444
    Ruxue Yao, Hongliang Lu, Yuming Zhang, Yutao Zhang

Modeling and Parameter Extraction of Semiconductor Devices for Simulation and Design
Optimization of ESD Protection Circuits on BCD Technologies for Automobile and Industry
Applications... 447
    Yuhua Cheng, Wei-Wei Yu, Eugene Worley

A Novel Double-Zener Process and Multiplex Design for High-Power Surge and High-Speed ESD
Devices Development.. 451
    Zhao Qi, Yirui Jia, Hongquan Chen, Ming Qiao, Zhaoji Li, Bo Zhang

The Non-Monotonic Instability of V_{TH} and $R_{ds,on}$ in P-GaN Gate HEMTs Under Repetitive Short-
Circuit Stress: The Role of Electric-Field & Self-Heating Effect.. 454
    Long Wang, Ning Yang, Shuting Huang, Jianggen Zhu, Kuangli Chen, Chao Feng, Haolin Hu,
    Wei Zeng, David Zhou, Yuxi Wan, Bo Zhang, Qi Zhou

Optimizing Flash Memory Towards Storage-Class Memory (SCM) Applications 457
    Xinyi Guo, Yang Feng, Jing Liu, Junyu Zhang, Xuepeng Zhan, Jixuan Wu, Jiezhi Chen

Investigation of Reliability Characteristics of $Hf_xZr_{1-x}O_2$-Based FeFET and AFeFET Non-Volatile Memory 460
 Min Liao, Xianzhou Shao, Junshuai Chai, Xiaoqing Sun, Xiaoyu Ke, Hao Xu, Jinjuan Xiang, Xiaolei Wang, Wenwu Wang

Deep Understanding of Charge Trapping Phenomenon in n-FeFET and Endurance Improvement by Interlayer Engineering 463
 Saifei Dai, Hao Xu, Fengbin Tian, Xianzhou Shao, Xiaoqing Sun, Junshuai Chai, Xiaolei Wang, Wenwu Wang

An FPGA-Based Dual-Mode SSD for Device-Side Performance Optimization 466
 Xingyu Chen, Sirui Peng, Hankun Lv, Zhangbin Yang, Daixiao Peng, Xi Cai, Xueguang Lian, Yong Ding, Xiaoyong Xue

A Simulation Comparison of Channel-All-Around and Gate-All-Around 3D Vertical Structure FeFET with IGZO Channel 469
 Xuebin Wang, Zhijian Guo, Yutao Li, Chengji Jin, Jixuan Wu, Guanhua Yang, Yuanxiao Ma, Masaharu Kobayashi, Fei Mo, Yeliang Wang

Low Operating Voltage in HfO_2/ZrO_2 Superlattice Ferroelectric Capacitor Achieved by Thickness Scaling 472
 Dongya Li, Huan Liu, Peiyuan Du, Fei Yu, Chengji Jin, Xiao Yu, Yan Liu, Genquan Han, Yue Hao

Co-Optimization of Oxide Semiconductor-Based Ferroelectric Transistors Between Electrical Performance and Ambient Stability by Using TiO_2-IGZO Dual-Channel Layers 475
 Shangze Li, Xujin Song, Dijiang Sun, Xiaoyan Liu, Jinfeng Kang

Enhancing Computational Precision in PLRAM-Based In-Memory Computing with High-Low Bit Concatenation 478
 Saike Zhu, Xiang Qiu, Yong Gong, Cimang Lu, Yi Zhao

FeFET Based Logic-In-Memory Pipeline-Style Circuits 481
 Yang Li, Zhaohui Yang, Yinshui Xia

Study of V_{th} Degradation Mechanism in FeFET with $TiN/Al_2O_3/HfO_2/Al_2O_3/Hf_{0.5}Zr_{0.5}O_2/SiO_x/Si$ Structure 484
 Runhao Han, Jia Yang, Tao Hu, Mingkai Bai, Yajing Ding, Xianzhou Shao, Saifei Dai, Xiaoqing Sun, Junshuai Chai, Hao Xu, Kai Han, Xiaolei Wang, Wenwu Wang, Tianchun Ye

Random Number Generation from 3D-NAND Flash Memory Using Shallow Charge Trap Related Short-Term Retention Errors 487
 Ruibin Zhou, Jian Huang, Xianping Liu, Yuhan Wang, Xinrui Zhang, Yungen Peng, Zhiyi Yu

Simulation of Program/Erase Cycling and Retention Loss in 3-D CTF NAND Flash 490
 Anuj Kumar, Ravi Tiwari, Souvik Mahapatra

Switch-Off Mechanisms in GeAsTe Ovonic Threshold Switching Selector Device 494
 Z. Hu, Z. Chai, W. Zhang, Jian Zhang

Orthorhombic-I (Pbca) Phase: Origin of Anti-Ferroelectricity in HfZrO Films 498
 Wei Liu, Zeping Weng, Jianguo Li, Wenchao Yan, Yiming Qu, Yi Zhao

The Maximum Storage Capacity of Open-Loop Written RRAM is Around 4 Bits 501
 Yongxiang Li, Shiqing Wang, Zhong Sun

High-Density and High-Reliability RRAM for Memory and Computing Applications 504
Yimao Cai, Xiahong Zhou, Zongwei Wang, Lin Bao, Ling Liang, Cuimei Wang, Ru Huang

Impact of Different MAC Schemes on Computing in Memory Based on 1T1R Array 508
Ruiqing Xie, Gaoqi Yang, Zongwei Wang, Linbo Shan, Jinshan Li, Chaoyi Ban, Lin Bao, Ling Liang, Cuimei Wang, Yimao Cai, Ru Huang

Investigation of Gate Injection Charges Behavior on FeFETs with TiN/Al$_2$O$_3$/Hf$_{0.5}$Zr$_{0.5}$O$_2$/SiON/Si Structure by Analyzing ISPP/ISPE 511
Jia Yang, Runhao Han, Saifei Dai, Tao Hu, Xianzhou Shao, Kanyi Li, Wenbo Fan, Xiaoqing Sun, Junshuai Chai, Hao Xu, Kai Han, Xiaolei Wang, Wenwu Wang, Tianchun Ye

Electric-Thermal Characteristics of Bottom P-I-N Isolated Nanosheet Gate-All-Around FETs 514
Chunlei Wu, Hanzhi Gu, Jian Ma, Boqian Shen, Fei Zhao, Yueyuan Yu, Yiming Xia, Qingqing Sun, David Wei Zhang

Surface Potential-Based Compact Model for ITO Thin-Film Transistors with Ultra-Thin Channel 517
Wenting Xu, Xinxin Shen, Zuoxu Yu, Tingrui Huang, Yuzhen Zhang, Weifeng Sun, Wangran Wu

A Continuous Full Channel Potential Model for Accurate Synthetic Electricfield Calculating in Gate-All-Around Devices 520
Fei Zhao, Chunlei Wu, Yumin Xu, Boqian Shen, Jian Ma, Hanzhi Gu, Yueyuan Yu, Yiming Xia, Qingqing Sun, David Wei Zhang

Deep Learning and Adaptive Pattern Search Based BSIM-CMG Parameter Extraction Applicable to Process Migration 523
Xingyu Li, Wangyong Chen, Linlin Cai

Vertical Channel Transistor (VCT) for Advanced Logic and Memory Applications 526
Mingmin Shi, Ran Bi, Ming Li

High Precision I-V Characteristics SPICE Model for Silicon Carbide MOSFET 530
Jinhong Shi, Yongxi Li, Jincheng Shi, Ruguan Li, Haimeng Huang, Hongqiang Yang

An Analytical Model for Characterizing Density of States of Oxide Transistors 533
Siyuan Hu, Chuanlin Sun, Junchen Dong, Zhensong Li, Kai Zhao, Dedong Han, Xing Zhang

A Physics-Informed Neural Network Model for Body Potential Distribution in MOSFETs Down to 50 K 536
Honglin Wu, Fangxing Zhang, Xinyue Zhang, Baokang Peng, Wu Dai, Lining Zhang, Runsheng Wang, Ru Huang

Modeling the Transient Characteristics with Trap Behaviors in LTPS-TFTs 539
Haolin Li, Zheng Zhou, Xiaoyan Liu

A Novel β-Ga$_2$O$_3$-Based Enhancement-Mode Transistor Combining Heterojunction Gate and Fin Shaped Gate 542
Yu Shao, Yunlong He, Xiaoli Lu, Songyao Wang, Xuefeng Zheng, Xiaohua Ma, Yue Hao

Electrical Characteristics and Thermal Reliability Investigation of TreeFET, FishboneFET, CombLikeFET and NSFET 545
Mingyu Ma, Wenbin Wang, Haokun Li, Shujun Gao, Hailong You, Cong Li

A Fast-Response Current Source with High-Impedance for Zero-Crossing-Based Circuits 548
Ruoyu Li, Xianglong Wang, Jianqiang Xu, Yintang Yang

A PPG Analog Front-End with PVT-Insensitive High-Pass Frequency 551
Zhaofeng Huang, Zepeng Huang, Hengchang Bi, Xing Wu, Liangjian Lyu

A Fully Integrated FVF Based Low-Noise Voltage Buffer for ADC Reference............................... 554
Ikhwan Kim, Yajie Qin

A Resistor-Free Grounded High-Frequency Memristor Emulator.. 557
Xinying Su, Bingjun Xiong, Junjie Yu, Jingjing Liu

An Ultra-Low-Leakage Current Sensing Interface for Wide Temperature Range.......................... 560
Jinsheng Tang, Chun Zhao, Lin He

A Global Threshold Voltage Finder Technology for the Readout Circuit of Event-Based Vision
Sensor 563
Yanwen Su, Hao Li, Dongsheng Liu, Ang Hu, Kaiyue Li

A Residue Amplifier with 72.27 dB Loop-Gain and 4.64 GHz Closed-Loop Bandwidth Consuming
6.4 mW for 12-Bit 1-Gsps Pipelined ADC 566
Jiangbo Wei, Jin Liu, Wei Tian, Chao Wang, Maliang Liu

A 180 mV–1.6 V Thermoelectric Energy Harvesting Converter with Low-Voltage Cold Start and
Less than 1 μW Power Loss 569
Chunlin Wang, Anzhi Yan, Tianyu Guo, Peng Wan, Houfang Liu, Yi Yang, Tianling Ren

A Signal Conditioning ASIC with High Precision and Low Noise for MEMS Accelerometers 572
Quan Sun, Rui Liu, Zhe Zheng, Lei Dong, Ji-Jiang Wang

Design of a High-Precision Self-Calibration Readout Circuit for CMOS Microbolometer 575
Qianhao Zhang, Jie Liu, Sheng Xu, Yiming Liao, Feng Yan, Xiaoli Ji

A Smooth Two-Stage Soft Start Method for Current Mode Boost Converter 578
Yue Shi, Shi-Dong Wang, Zekun Zhou, Bo Zhang, Zhigang Qin

Design of 12-Bit Low-Power Single-Slope ADC with 2048 Columns for Infrared Focal Plane Array 581
Lixiang Han, Hao Li, Yihui Zhang, Liang Gao, Dongsheng Liu

A High-Voltage Smooth Self-Starting Reference Current Source Circuit 584
Dongyan Zhao, Jie Pan, Chenghao Zhang, Yidong Yuan, Yi Hu, Hongwei Shen, Zekun Zhou

A Temporal and Spatial Reuse Interpolation Hardware for VVC Motion Compensation 587
Huanxiang He, Shushi Chen, Leilei Huang, Yibo Fan

A Broadband Digital Beamforming Method Based on FPGA... 590
Yiwen Tang, Guowen Jia, Zhen Zhang, Yue Zhang

Conditional Cycle Termination RANSAC 593
*Tong Jiang, Yujie Huang, Liyuan Peng, Mingyu Wang, Wenhong Li, Minge Jing, Xiaoyang
Zeng*

A Multi-Resolution Propagation Algorithm and Pixel Grouping Storage Strategy for PatchMatch
Stereo... 596
Kai Liu, Zhenyu Zhang, Haiwei Wang, Leilei Huang, Chunqi Shi, Long Xu, Runxi Zhang

An XOR Arbiter PUF Based on the IGZO TFT Devices 599
Xiang Chen, Yongliang Chen, Xiaole Cui

Design and Implementation of Hierarchical Storage Structure for MCCSIP-RAA 602
  Longmei Nan, Yu Jin, Yiran Du, Tao Chen, Yanjiang Liu, Wei Li

A Highly Scalable Hardware HEVC Encoder Based on FPGA 605
  Guohao Xu, Chenlong He, Shiyan Yi, Leilei Huang, Xiaoyang Zeng, Yibo Fan

A Hardware-Friendly Fast Block Partition Decision Algorithm Based on Histogram of Oriented
Gradient for AV1 608
  Guohao Xu, Shiyan Yi, Zhijian Hao, Leilei Huang, Hao Zhang, Xiaoyang Zeng, Yibo Fan

A Low Power Narrow-Band Complex-Bandpass Filter Based on Feedforward Compensation
Amplifiers for NB-IoT Applications 611
  Xu Zhao, Ziqiang Wang, Jie Gan

A High-Performance MTJ-LUT Circuit Using 4T1M Architecture 614
  Yu Pan, Yuejun Zhang, Shuaicheng Guo, Yuanxin Tian, Bo Hong, Rui Fang, Liang Wen, Yong Ding

Optimizing Communication Efficiency of GNN Inference in Distributed Systems 617
  Wenqian Zhou, Qiaosha Zou

SST: Simplified Space-Time Transformer Based on Time-Assisted Spatial MSA for 3D Human
Pose Estimation 620
  Sheng Lu, Qiyun Dong, Zhenyin Zhang, Gengsheng Chen, Yinna Zhu, Wei Xu

SALTS: An Efficient and Flexible Self-Attention Accelerator with Long Token Support on FPGA 624
  Kaiqi Chen, Xinhua Shi, Jun Han

RISC-V Neural Network Instruction Design and Simulation with Cache Scheduling Via ROCC
Interface 627
  Siyao Dai, Zikang Zhou, Jun Han

Impact of External Magnetic Interference on the Performance of MRAM-Based Neuromorphic
Computing 630
  Yingtong He, Suihuan An, Yu Chen, Xue Zhou, Xihui Yuan, Weidong Zhang, Zheng Chai, Tai Min

A Hardware Accelerator for Image Super-Resolution with Algorithm Lightweighting and Custom
Fusion Engine 634
  Menghan Li, Sheng Lu, Jun Han

Hardware Implementation of High Speed Fault-Tolerant Parallel Accelerator 637
  Wenzhe Ma

Composite Filter-Based Bicubic Interpolation Method and FPGA Implementation 640
  Li Zhang, Jingjing Liu, Yujie Zhu, Jianhua Zhang

MTJ Based Temperature Tracking Read/Write Assist for High Speed SRAM Bitcell 643
  Yongliang Zhou, Chengxing Dai, Jingxue Zhong, Yingxue Sun, Xin Li, Chunyu Peng

A 12V to 1V Tri-State DSD Hybrid Converter by Self-Balanced Dual Flying Capacitors with
0.3mV Output Ripple and 90.09% Peak Efficiency 646
  Yixing Wang, Qianhui Liu, Yuhua Chen, Yizhe Yang, Yimeng Zhang, Yuming Zhang

System-Level Evaluation of AOS Gain Cell eDRAMs for Low-Power Normally-Off Computing 649
  Long Chen, Yecheng Yang, Wei Li, Shao Hao Wang

A High Sigma Monte Carlo Analysis Solution Via Machine Learning for SRAM Margin Signoff 652
Amy Rao

Enhanced Multi-Bit Computation Using CIM SRAM Technology 655
Ruiyong Zhao, Yibo Hu, Zhipeng Ren, Yizhe Yin, Jing Chen

A Compute-In-Memory Macro Based on Complementary 2T2C FeRAM Cell for BNNs 658
Jinyu Li, Mingzhang Xie, Shisheng Xiong

A Novel High Speed Low Power Differential Circuit-Based FRAM Read Scheme 661
Qiuyu Tao, Jiabao Ye, Xuecheng Cui, Nan Jiang, Jiangtao Cao, Xibo Chen, Zhangbin Yang,
Daixiao Peng, Xi Cai, Xueguang Lian, Yong Ding, Jiuren Zhou, Bing Chen, Genquan Han

A 13-Bit, 1 MS/s Cyclic ADC, for High-Speed CMOS Image Sensor 664
Qi Lv, Rensheng Shen, Yu Cheng, Guoqiang Zhong, Yang Qu, Yuchun Chang

An Area-Efficient 16-Bit Four-Channel R-2R DAC Based on Switching On-Resistance Adaptive
Calibration Technique 667
Kejun Wu, Yuchen Liu, Yuhan Hu, Yu He, Zhen Yu, Ning Ning

A Background Calibration Method of Bandwidth Mismatch for Time-Interleaved ADCs Based on
Neural Network 670
Tianqi Yang, Longsheng Wang, Xin Zhao, Shubin Liu, Dengquan Li, Zhangming Zhu

A Second-Order Dual-Charge-Pump Passive Noise-Shaping SAR ADC for Medical Implant
Devices 673
Kangkang Sun, Xuanxin Ke, Haoning Sun, Yuchen Wang, Feng Yan, Jingjing Liu

A 114.4-DB DR, 26-KHz BW Discrete-Time Incremental Zoom ADC 676
Yuanhong Ding, Longjiang Jia, Jian Mei, Lei Deng, Rui Yin

A Multi-Phase Clock Self-Calibrating Circuit 679
Zhihuai Li, Li Jiang, Xueming Wei, Zilu Cai, Jiami Tang

A High-Resolution Low-Power Extended-Range Incremental $\sum\Delta$ ADC for Battery Management
System 682
Long Zhang, Quan Sun, Rui Liu, Zhe Zheng, Jingjing Zhang, Haitao Liu

An Infrared AFE Chip and System with Non-Invasive Blood Glucose Detection Output 685
Bin Li, Jiyuan Guo, Chengzhen Xie, Jian Mei, Lei Deng, Rui Yin

A 12-Bit 8GS/S Time-Interleaved Pipeline-SAR ADC with Calibration 688
Jie Pu, Jinda Yang, Yuanjun Cen, Jianwen Li, Rong Han, Xing Zhu, Lei Chen

An Ultra-High Frame Rate ROIC for Hyperspectral Detection 691
Angyang Li, Ningning Li, Jian Mei, Lei Deng, Rui Yin

A BackgroundDigital Calibration Method for DTCs Used in Digital PLL Employing Dual-Path
DTC 694
Renxuan Li, Xiaoyu Shan, Li Wang, Dongsheng Liu, Ang Hu

A High-Precision Sigma-Delta ADC for Battery Management System 697
Hao Xue, Liji Wu, Jing Hu, Zhiwei Li, Xiangmin Zhang

Multi-Sampling Mode CDAC Design for a 12-Bit 200MS/s Pipelined-SAR ADC 700
Tianyu Zhang, Fan Ye, Shunli Ma

DSP-PUF: A Software PUF Based on Digital Signal Processor for IoT Security 703
   Tengfei Yuan, Pengjun Wang, Yuejun Zhang, Mingze Ren, Shuang Hu

A Q/V Band 49.6-54.5GHz,3.53dB NF,45dB Gain,2.09° Phase Error,2-Way Phased-Array Receiver
for Satellite Application 706
   Congrui Li, Qi Zhao, Ruolan Chen, Shulan Chen, Yan Wang, Lei Zhang

A Fractional-N SPLL Using Space-Time Averaging and Phase Interpolator for Quantization Noise
Reduction 709
   Shengxiang Liu, Ke Sun, Chengyu Yang, Dongsheng Liu, Ang Hu

A 18V, 600mA Load Current, 22MHz High-Voltage Power Amplifier with Over-Temperature
Protection and Bidirection Enable Logic 712
   Yuan Ren, Xin'An Wang

A 47µW Wake-Up Receiver with -77dBm Sensitivity Using a Mixer-First Architecture 715
   Weitao He, Yaxin Zeng, Bin Jia, Hao Min, Hao Xu, Na Yan

A Ka-Band CMOS Broadband Power Amplifier with 35.3% PAE for SATCOM Applications 718
   Zhiqing Liu, Yu Chu, Yuting Sun

RF Front-End Chip Design for Ku-Band with 130nm CMOS Technology 721
   Huiquan Xie, Ziyu Wang, Tianrui Wang, Yifei Chen, Maliang Liu, Yintang Yang

Back-Gate Bias Assisting VCRO Design 724
   Chenglin Ye, Zheng Zhou, Xiaoyan Liu

A 3.2-To-7.1GHz Quad-Core Dual-Mode Oscillator Achieving 193.6 dBc/Hz Peak FoM 727
   Xiaoyu Shan, Renxuan Li, Mengming Zhang, Ang Hu, Dongsheng Liu

A 20.6 to 30.5 GHz Two Stage Cascode LNA in 40nm CMOS for Phase Array Tranceiver 730
   Lei Wang, Kefeng Han, Hao Xu, Rui Yin, Na Yan

A 12-32 GHz Power Amplifier with 32-DBm Psat and 25% PAE in 0.15µm GaN 733
   Xiangran Ni, Chunyue Bo, Tianyu Li, Qingyang Dong, Xin Jiang, Weijun Luo

A Source-Driven Push-Push Doubler with Wideband 2nd Harmonic Feedback 736
   Yuyang Chen, Ao Zhang, Jianjun Gao, Jianjun Zhou

Low Power Processor for IoT Device 739
   Jincheng Li, Jiyuan Bai, Zelin Wang, Gengsheng Chen, Xiaofang Zhou

A Heterogeneous Integration System of Analog in-Memory Computing and Field-Programmable
Gate Array 742
   Hua Chen, Yiming Qu, Wenhao Wu, Yi Zhao

A 10-MHz Four-Phase Hysteretic Control DC-DC Converter with Inductor Current Self-Balancing 745
   Yushen Zhang, Ningning Li, Yibo Zhang, Yizhe Yang, Yimeng Zhang, Yuming Zhang

A High Precision −40°C to 150°C Bandgap Reference with Dual Temperature Compensation 748
   Yuhan Zhang, Jianzheng Li, Xiaomeng An, Lina Wang, Yajie Qin

A Biphasic Neural Stimulator with Adaptive Pulse-Width Modulation Charge Balancer 751
   Hailong Tang, Wenxian Gu, Yifan Song, Hengchang Bi, Xing Wu, Liangjian Lyu

Assembly of Oxidized/Intrinsic 2D MXene Film for Improved Absorption Electromagnetic Shielding............754
Yulin Guo, Siteng Li, Jiafeng Song, Yilin Sun, Zhifang Liu, Weijia Luo

Improved Channel Width and Morphology of Epi Silicon FinFET Via Low Thermal Budgets Fin Thinning Technology............757
Peng Wang, Yupeng Lu, Guanqiao Sang, Renjie Jiang, Lei Cao, Qingkun Li, Lianlian Li, Hang Zhang, Zhonrui Wang, Meihe Zhang, Qingzhu Zhang, Junfeng Li, Huaxiang Yin

Deep Investigation into Variability of Complementary Dopant Segregated Tunneling FET Based on Foundry Platform............760
Rundong Jia, Jianfeng Hang, Kaifeng Wang, Yongqin Wu, Hongyan Han, Ye Ren, Weihai Bu, Runsheng Wang, Qianqian Huang, Ru Huang

Investigation of Common-Gate and Split-Gate Structures Based on CFET Standard Cells............763
Peishun Tang, Rongzheng Ding, Xiaona Zhu, Shaofeng Yu

Exploration of the Effect of Silver Impurity on the Minority Carrier Lifetime of Semiconductor............766
Xin Tian, Peizhi Zhao, Yudong Li, Jun Xu, Tianling Ren

Fabrication and Electrical Characterization of Mo/Hf$_x$Zr$_{1-x}$O$_2$/Mo Ferroelectric Capacitors............768
Chunsheng Jiang, Wencai Du, Qin Xie

Effect of Cascade Current Density and Plating Time on TSV Filling Effect in DC Power Supply............771
Weifeng Chen, Lijuan Peng, Xiaohui Wang, Fangzhou Wang, Guojian Ding, Qi Feng, Ping Yu, Peng Zuo, Feng Liu, Jiang Ma, Yang Wang, Haiqiang Jia, Hong Chen

High-Performance Carbon Nanotube Optoelectronic Transistors for Memory Applications............774
Shuang Liu, Heyi Huang, Yanqing Li, Yadong Zhang, Feixiong Wang, Yupeng Lu, Renjie Jiang, Jiali Huo, Huaxiang Yin

Investigation of the Channel Width Dependence of IGZO TFT by Experiment and TCAD Simulation............777
Yanyu Yang, Yupeng Lu, Shuang Liu, Renjie Jiang, Jie Luo, Yunjiao Bao, Peng Wang, Gaobo Xu, Huaxiang Yin

A Test and Evaluation Platform for Quantitative Analysis of High-Reliability Designs............780
Yifeng Huang, Wenqing Wan, Chang Wu

Hot-Carrier Injection Characterization of n-LDMOS Transistors and Stress Tests in a Buck Converter Configuration............783
Chun Yee Chu, Wai Tung Ng

Semimetal Alloy Contact with Low Resistivity and Enhanced Thermal Budget for MoS$_2$ FETs............786
Kwok-Ho Wong, Mansun Chan

Automated Verification of Functional Interface Connections in Circuit Schematics............789
Keli Long, Xingyu Gao, Lei Li, Jinxiang Wang, Fangfa Fu, Liangquan Qiao, Jinghan Zhou

Co-Optimization Design Method of Temperature Variation and Circuit Aging in Digital Circuits............792
Songxuan He, Wangyong Chen, Ling Xiong, Linlin Cai

Boolean Matrix Factorization Algorithm Based on Error Shaping Technique and Its Application on Approximate Logic Synthesis............795
Runhua Yang, Rensheng Shen

Automatically Device Sizing of Analog Circuit Through Sequential Model-Based Optimization with Circuit Recognition 798
  Shun-Qi Dai, Xiao Wang, Yuan Lei, Bei-Ping Yan

Vanadium Oxide-Based Artificial Synapses for Construction of Artificial Neural System 801
  Zhuoling Zhou, Libin Liang, Hongzhi Chen, Changjiu Teng, Shilong Zhao, Wenjun Chen

High Performance FeFET with α-IGZO Channel Enabled by Atomic-Layer-Deposited HfO$_2$ Interfacial Layer 804
  Yinchi Liu, Hao Zhang, Xinlong Zhou, Dmitriy Anatolyevich Golosov, Chenjie Gu, Hongliang Lu, Shijin Ding, Wenjun Liu

A Simulation Study on Cell Scaling Impacts in 3D Charge-Trapping (CT) Flash Memory 807
  Wanyu Li, Haitao Dong, Qianwen Wang, Yang Feng, Xuepeng Zhan, Jixuan Wu, Jiezhi Chen

Comprehensive Charaterizations on Read Disturbs in QLC Charge-Trap (CT) 3D NAND Flash 810
  Shaoqi Yang, Xiaohuan Zhao, Peng Guo, Qianwen Wang, Guangkuo Yang, Xinyi Guo, Pengpeng Sang, Jixuan Wu, Xuepeng Zhan, Jiezhi Chen

Aspect Ratio Dependent Optimization and Comparison of Specific ON-Resistance of SJ and Hk MOSFETs with Extremely High Permittivity 813
  Chenxing Wang, Zhentao Xiao, Zonghao Zhang, Zhenghao Jin, Zhiwan Liu, Zonglin Li, Zhemin An, Yunteng Jiang, Ruguan Li, Haimeng Huang, Hongqiang Yang

Copper Ion Migration in Van Der Waals CuInP$_2$S$_6$ Devices with Vertical and Lateral Structures 816
  Jie Li, Yirong Guo, Pengying Chang

Simulation Study of the Impact of Split Gate on SiC DTMOS Short Circuit Withstand Capability 819
  Zixun Chen, Jinping Zhang, Yang Liu, Xudong Ma, Bo Zhang

Improved Hall Mobility Measurement Distinguishing Interface Capturing Effect in 4H-SiC Inversion Channel 822
  Xiangrui Fan, Hao Fu, Xinyu Zhang, Zilong Wu, Jiameng Sun, Jiaxing Wei, Siyang Liu, Weifeng Sun

A Superior SiC Lateral MOSFET with Patterned P-Bury Layer Made on N-Type Wafers 825
  Xuke Yan, Junji Cheng, Xiaojun Fu, Bo Yi, Haimeng Huang, Hongqiang Yang

High Performance Termination Design and Fabrication for SiC MOSFET Device 828
  Lei Huang, Junhou Cao, Chenlu Wang, Hao Fu, Jiaxing Wei, Siyang Liu, Weifeng Sun

Analysis of the Separation Degree for P-Pillar in SiC Super-Junction Structure Through "Multiple Epitaxy-Ion Implantation" Route 831
  Hao-Bo Kang, Hao Yuan, Feng-Yu Du, Yu Zhou, Ke-Yu Liu, Xiao-Yan Tang, Chao Han, Qing-Wen Song, Yu-Ming Zhang

Numerical Analysis of the CIBL Effect on Short-Circuit Characteristics of DG-CSTBTs with Reduced Mesa Width 834
  Zhengyu Lang, Jinping Zhang, Shiwei Zheng, Shuyang Huang, Haonan Deng, Bo Zhang

Innovations in GaN HEMT Design: Achieving Superior Power Output and Thermal Management 837
  Shiming Li, Bowen Yang, Mei Wu, Ling Yang, Bin Hou, Meng Zhang, Xiaohua Ma, Yue Hao

An Enhanced RC-IGBT Incorporating Superjunction and Discontinuous Field Stop Layers for Improved Efficiency 841
  Yiming Jia, Jieyu Long, Zhiwei Jing, Haimeng Huang, Hongqiang Yang

Simulation Study on 1200V CS-SemiSJ-IGBT for Reduced Switching Loss and Fast Switching 844
Luping Li, Zehong Li, Peng Chen, Yuzhou Wu, Qiansheng Rao, Ming Li, Haifeng Qin, Li Wan, Yang Yang, Wei Li, Min Ren

A Dual-Gate Trigger Thyristor for Reducing the Probability of False Triggering 847
Pengcheng Xing, Qingbo Wan, Jie Huang, Ruize Sun, Chao Liu, Wanjun Chen

Edge-Dependence of Threshold Voltage in MoS_2 Nanoribbon-Based 2D FETs 850
Zhirong Peng, Mansun Chan

Ultra Fast Diode Avalanche Shaper with Floating Junction 853
Zhen Yang, Yu Zhou, Xiao-Yan Tang, Chao Han, Qing-Wen Song, Yu-Ming Zhang

Silicon Carbide Diode Avalanche Shaper with Multi-Point Quasi-Uniform Triggering 856
Lin Cheng, Yu Zhou, Xiao-Yan Tang, Chao Han, Yu-Ming Zhang, Qing-Wen Song

Super Field Plate LIGBT with Improved Performance for Both Cell and Terminal Region 859
Weihao Lu, Jing Li, Jitong Wang, Chaoyang Peng, Chunwei Zhang

High-Temperature Oxidation of 4H-SiC and Gate Oxide Reliability Dependence on Oxidation
Temperature 862
Baoyan Feng, Xiaoyan Tang, Yi Bo Zhang, Chao Han, Hao Yuan, Qing Wen Song

Optimization for a High-Voltage Recessed-Gate β-Ga_2O_3 MOSFET by Gate and Drain Field Plate
Technology 865
Bo Yi, Yuan Qiao, Ming Dai, Fan Xu, Junji Cheng, Haimeng Huang, Moufu Kong, Xingli Jiang, Hongqiang Yang

A Novel Voltage Sensor with Composite Trench Structure for High Voltage IGBT 868
Yang Yang, Ze-Hong Li, Li-Hang Dong, Wei Li, Peng-Fei Jia, Zhi-Yu Yang, Li Wan, Yi-Shang Zhao, Tong-Yang Wang, Zi-Ming Xia

A Novel Triggered Voltage Sensing Structure for High Voltage IGBT 871
Yang Yang, Ze-Hong Li, Li-Hang Dong, Wei Li, Peng-Fei Jia, Zhi-Yu Yang, Li Wan, Yi-Shang Zhao, Lu-Ping Li, Zi-Ming Xia, Tong-Yang Wang

Investigation of Threshold Voltage Instability in p-GaN Gate HEMTs Under Surge Current Stress 874
Xiaoming Wang, Yu Shi, David Zhou, Haizhao Zhi, Yun Xia, Yuxi Wan, Ruize Sun, Xinghuan Chen, Wanjun Chen, Bo Zhang

A Novel Ga_2O_3 High-k Trench MOSFET with Improved Forward and Reverse Performance 877
Moufu Kong, Lewei Lyu, Haoran Wang, Zhaoyu Ai, Xinyang Chen, Fanxin Meng, Qiang Hu

A Nonlinear Behavioral Modeling Approach for Microwave Transistors Considering
Electrothermal-Aging Degradation 880
Lin Cheng, Hongliang Lu, Silu Yan, Junjun Qi, Jiantao Qiao, Yuming Zhang

Effect of Layer Thickness on the Transport Properties of ALD-Deposited ZnO/In_2O_3
Heterojunction Thin-Film Transistors 883
Zhenwei Li, Tiaoyang Li

Electrical and Thermal Analysis of CNT nTSV Applied to BS-PDN: A Modeling Study 886
Kai Ying, Baohui Xu, Jie Liang

A Unified Current-Voltage Compact Model for Organic Light-Emitting Diode 889
Wenbin Wang, Mingyu Ma, Wangjun Yang, Jianghao Ma, Hailong You, Cong Li

Threshold Voltage and Mobility Extraction of Negative Bias Temperature Instability in 22nm FD SOI MOSFETs 892

Yibo Hu, Hao Ge, Zhipeng Ren, Yizhe Yin, Ruiyong Zhao, Jing Chen

Modeling of Silicon Single-Photon Avalanche Diodes for Process and Design Optimization 895

Jing Fu, Anran Guo, Hongbo Zhang, Guowei Li, Huaping Ma, Ruizhi Li, Yuwei Chen

A Novel Modeling Method for BV Characteristics of ESD Protection Devices 898

Ke Zhang, Yang Wang, Xiangliang Jin

Analysis of the Impact of Parasitic Bipolar Amplification on Charge Sharing Based on Analytical Model 901

Yutao Zhang, Hongliang Lyu, Yuming Zhang, Ruxue Yao

Research on the Performance Degeneration of GGNMOS Under Total Ionizing Dose Radiation 904

Jiekai Feng, Ping Luo, Chengxin Li, Jiaxuan Hu, Peng Li, Pengfei Liao

The UIS Withstand Capability and Device Failure Mechanism of 650 V p-GaN Gate HEMTs 907

Qihang Huang, Luanxi Ma, Yanyu Nie, Shuting Huang, Jianggen Zhu, Yu Shi, Rongxin Du, David Zhou, Yuxi Wan, Bo Zhang, Qi Zhou

Time Dependent Dielectric Breakdown in n-MOSFETs Fabricated by Low-Temperature and Low-Pressure Mild Oxidation After Plasma Solidification 910

Qiao Teng, Yanning Chen, Fang Liu, Bo Wu, Yongfeng Deng, Dawei Gao

Simulation of BTI for GAA MOSFETs with Enhanced Parameters Extraction 913

Yongjia Wang, Yijiao Wang, Shuhan Wang, Jinghan Xu, Xiaoyan Liu

Foundamentals of Low-Resistive Indium-Violet Phosphorene Top Contact: An Ab-Initio NEGF Study 916

Huaipeng Wang, Sicheng Liu, Shuaihong Li, Zhifang Liu, Yilin Sun, Jianlong Xu, Dan Xie

Gold Thermocompression Wafer Bonding for Quartz MEMS Applications 919

Ting Yang, Dongxiang Han, Jun Xu, Tian-Ling Ren

An Adaptive Threshold Analog Front-End Circuit for Direct ToF LiDAR 922

Jianping Guo, Xiaoyang Zeng, Wenhong Li, Mingyu Wang

Design of Ultra-Broadband Metamaterial Absorber from Infrared to Terahertz 925

Xiangze Liu, Wenbin Zhou, Tiantian Shi, Yiming Liao, Feng Yan, Xiaoli Ji

Large Modulation Bandwidth Si-Based Avalanche Photodiode for Visible Light Communications 928

Jiabin Wu, Yidi Hu, Chiang Zhu, Zhichong Wang, Xiaona Zhu, Chao Shen

An Artificial Neuromuscular Synapse Based on a Ferroelectric $Pb(Zr_{1-x}Ti_x)O_3/SiC$ Floating Gate Transistor 931

Yu Liu, Lin Lin, Xiang Wang, Chengyan Zhong, Junxiong Guo, Wen Huang, Yufeng Guo

Broadband Photodetectors Based on Graphene/Perovskite Hybrid Structure with Ferroelectric Gating 934

Zhongyang Liu, Shuangqi Dong, Mingjie Li, Huaipeng Wang, Dan Xie, Yilin Sun

Interconnection Design of Chiplet Technology 937

Ning Chen, Lei Shen, Chang Wu

Effects and Modeling Study on FDSOI MOSFETs at Cryogenic Temperature 940

Zhipeng Ren, Yibo Hu, Yizhe Yin, Ruiyong Zhao, Jing Chen

Author Index

The TID Response and HCI Degradation for multi-Vt nFinFETs

Ruxue Yao[1], Hongliang Lu[1]*, Yuming Zhang[1], Yutao Zhang[1]

[1]School of Microelectronics, Xidian University, Xi'an 710071, China
*E-mail: hllv@mail.xidian.edu.cn

Abstract

The total ionizing dose (TID) response and hot carrier injection (HCI) degradation of multi-Vt bulk nFinFETs are investigated in this paper. X-ray experiments show that multi-Vt nFinFETs are insensitive to TID irradiation. Stress experiments show significant HCI damage in multi-Vt nFinFETs and the most dramatic degradation in low-Vt nFinFETs. 3D TCAD simulation indicates that high-doping concentration in the channel stop region improves the TID tolerance. With the same electrical stress applied, the low-Vt nFinFET has the largest channel electric field, exacerbating the HCI degradation.

1. Introduction

FinFETs are of great interest in many application areas, such as ASIC design, Internet of Things, memory design, etc. Low power consumption and high transistor densities are the primary requirements for realizing these applications, which require multi-Vt transistor options [1]. Low-Vt transistors are mainly used for high-speed circuits with the disadvantage of high leakage current. On the other hand, high-Vt transistors have better leakage characteristics but poor switching speed performance. Standard-Vt transistors play a balancing role between power consumption and efficiency. To enable innovations ranging from power-sensitive mobile SoCs to speed-driven high-performance computing (HPC), multi-Vt transistor options are needed to achieve design flexibility that optimizes power consumption and performance [2].

When using multi-Vt transistors, its reliability needs to be taken into account. On the one hand, with the growth of the aerospace industry, FinFETs will be a potential choice for high-performance integrated circuits in aerospace applications. Total Ionizing Dose (TID) is one of the radiation effects that needs to be evaluated and studies have shown that the shallow trench isolated oxide (STI) is a critical location for TID damage [3]. On the other hand, multi-Vt nFinFETs face their own HCI aging effects. Research has shown that the HCI degradation mechanism of short-channel devices changes [4]. In this paper, the TID response and the Vt-dependence of HCI degradation for multi-Vt nFinFETs are investigated by X-ray irradiation and accelerated stress experiments.

This article is organized as follows. Section 2 provides device information and experimental details. Section 3 shows the TID and HCI experimental results for multi-Vt nFinFETs. Section 4 adds TCAD simulations to analyze the above results. Finally, Section 5 summarizes the conclusions.

2. Devices and experimental details

The devices under test (DUTs) are multi-Vt core-nFinFETs fabricated in 14nm technology. The low-Vt/standard-Vt/high-Vt devices with threshold voltages of 0.2V, 0.3V, and 0.37V, respectively. The transistors have a channel length of 14 nm and a nominal operating voltage (Vdd) of 0.8 V. The cross-section of the device along the channel width is shown in Fig. 1(a). nFinFETs utilize a high K metal gate (HKMG) stack consisting of HfO_2 and SiO_2 interlayers, and achieve different V_{th} through gate work function (WF) engineering with multiple metal layers.

The experimental arrangement is shown in Fig. 1(b). The DUTs were divided into two groups: one for TID irradiation and the other for HCI testing. TID irradiation experiments were performed on an on-line wafer-level X-ray irradiation platform at the Xinjiang Institute of Physics and Chemistry, Chinese Academy of Sciences. nFinFETs were cumulatively irradiated up to 2 Mrad(Si) at a dose rate of 611.985 rad(Si)/s, with the ON-bias during irradiation (where Vg = 0.8 V, the other terminals grounded). The stress conditions for the DUTs were Vg = Vd = 1.5 V with the other terminals grounded. The stress duration was 1000s and was interrupted to measure the transistor characteristics. I-V characteristics of the DUTs were measured immediately before and after irradiation, pre- and post-stress per step length. All experiments were performed at room temperature.

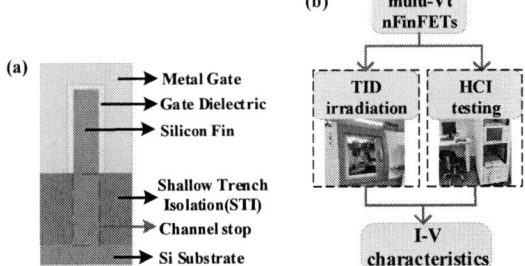

Fig. 1: (a)The cross-section of the nFinFET along the channel width, (b) The experimental arrangement.

3. Experimental results

This section demonstrates the DC response of multi-Vt nFinFETs after X-ray irradiation and the degraded behavior after hot carrier stress experiments.

3.1 TID Response

Fig. 2 shows the linear region transfer characteristic curves of the multi-Vt nFinFETs before and after ON-bias irradiation. As can be seen in Fig. 2(a), after 2 Mrad (Si) irradiation, low-Vt/standard-Vt/high-Vt devices show a slightly threshold voltage decrease (< 5mV) and on-state current rise ($<2\%$). In Fig. 2(b), the low-Vt device has the largest intrinsic off-state current (before irradiation). No post-irradiation leakage current rise was observed in any of the multi-Vt devices, which is usually the critical TID response of n-type transistors. The experimental results indicate that multi-nFinFETs have better TID irradiation tolerance. Since the multi-Vt nFinFETs have only a slight TID response, no significant dependence of TID irradiation on the threshold voltage is observed.

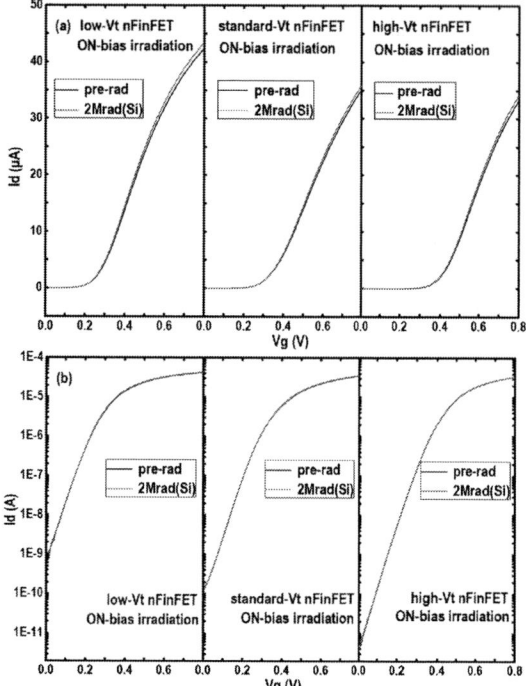

Fig. 2 Id-Vg curves in the linear region (Vd = 0.05V) before and after irradiation: (a)Linear current coordinates, (b)Logarithmic current coordinates.

3.2 HCI Degradation

Fig. 3 illustrates the HCI degradation behavior of low-Vt nFinFET:(a) Id-Vg@Vd=0.05V and (b) transconductance shift (Gm-Vg curve). As shown in Fig. 3(a), after 1000s of hot carrier stress, the threshold voltage of the nFinFETs shows a significant increase accompanied by a decrease in drain current. Negligible changes were observed for the subthreshold swing. In Fig. 3(b), the peak transconductance magnitude is almost constant, but the Gm-Vg curve shifts to the right with the increase of stress time. This suggests that under stress conditions, a large number of hot electrons are injected into the gate oxide to form negatively oxide trapped charges, thus shortening the device lifetime.

Fig. 4 illustrates the sensitive DC parameters shift as a function of stress time for multi-Vt nFinFETs: (a) threshold voltage shift ΔV_{th}, and (b) percentage change in on-state current ΔI_{dlin}. The threshold voltage is obtained by the constant current method ($I_{th}=1\mu A$) and the on-state current is the drain current at Vg=0.8V&Vd=0.05V. From Fig. 4(a), it can be seen that after 1000s stress, the V_{th} increase is 102.6mV, 75.7mV and 56.5mV for low-Vt/standard-Vt/high-Vt nFinFETs, respectively. In Fig. 4(b), the percentage of I_{dlin} degradation is 13.9%, 11.5% and 9.1% for low-Vt/standard-Vt/high-Vt nFinFETs, respectively. With the largest threshold voltage shift and current degradation percentage, low-Vt nFinFET exhibits the most sensitivity to HCI damage.

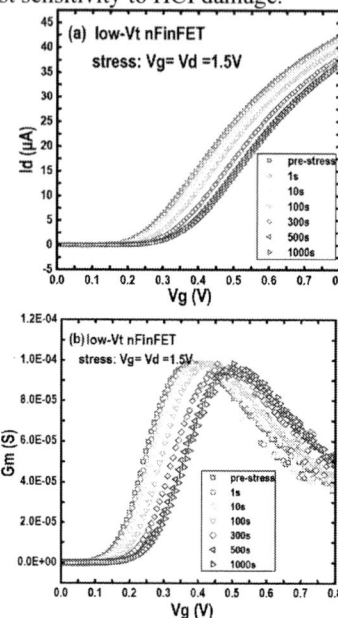

Fig. 3 HCI degradation behavior of low-Vt nFinFET: (a) Id-Vg@Vd=0.05V and (b) transconductance shift (Gm-Vg curve).

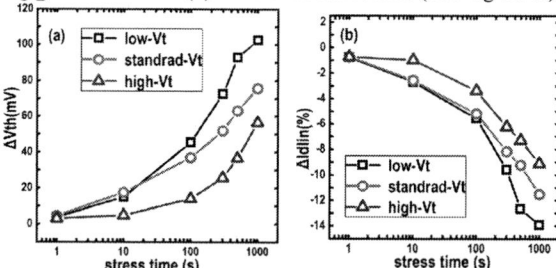

Fig. 4 The sensitive DC parameters shift as a function of stress time for multi-Vt nFinFETs: (a) threshold voltage shift ΔV_{th}(mV), (b) percentage change in on-state current ΔI_{dlin} (%).

4. Simulation and Discussion

3D simulations were implemented using Sentaurus TCAD to explore the TID response mechanism and different degrees of HCI degradation of multi-Vt core-nFinFETs. The device structure was built from a typical FinFET process flow. The TCAD model is calibrated by fine-tuning the material parameters, and the reference values are provided by the test results.

The TID damage of the gate oxide is neglected due to

its ultra-thin thickness. The TID response of the nFinFET is analyzed by adding uniform positively trapped charges to the STI. As shown in Fig. 5, the Id-Vg simulation results before and after irradiation obtained using the calibrated low-Vt nFinFET TCAD model are in good agreement with the experimental results. This indicates that the slightly decrease in the V_{th} and the rise in the I_{dlin} of the nFinFET are caused by the irradiation-induced accumulation of trapped charges in the STI. Experimental and simulation results show that the positive charges in the STI do not lead to a rise in leakage current, which is related to the high-doping concentration in the channel stop region (sub-fin). Channel stop is a region located below the fin, with STI present on both sides of the fin (the rectangular area surrounded by dashed line in Fig. 1(a)). In calibrating the TCAD model we used high-doping concentrations ($N_{channel\text{-}stop}$=2.8E18cm^{-3}) to reproduce punch-through stop layer (PTSL) [5]. In short-channel core-nFinFETs, high-doping concentration in sub-fin region suppresses the formation of drain-to-source parasitic paths induced by positive trapped charges in the STI. This improves the TID tolerance of multi-Vt nFinFETs.

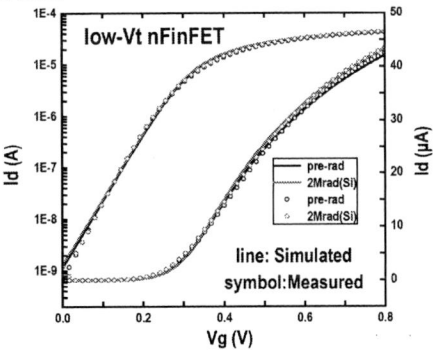

Fig. 5 Pre- and post-irradiation Id-Vg simulation results obtained with the calibrated low-Vt nFinFET TCAD model compared with measured data. (N_{ot} (trapped charge) density in the STI region is 7E17cm^{-3}).

According to the experimental results, the HCI degradation of multi-Vt nFinFETs is dominated by the injection of hot electrons into the gate oxide to form negatively oxide trapped charges, resulting in an increase in the threshold voltage and a decrease in the drain current. Experiments show that low-Vt nFinFETs have the worst HCI degradation. The metal work function (MWF) of high-Vt device is higher than that of low-Vt device. When a positive gate voltage is applied, it causes the energy band to bend and confines electrons at the oxide/semiconductor interface. Therefore high-Vt device requires a higher gate voltage to invert the channel, implying that the threshold voltage of the high-Vt device is higher than that of the low-Vt device. Low-Vt/standard-Vt/high-Vt nFinFETs are achieved by tuning the MWF and applying the same stress (Vg=Vd=1.5V). The

simulation results are shown in Fig. 6, where the channel longitudinal electric field is largest for the low-Vt nFinFET, with an electric field spike near the gate dielectric. Thus, more hot electrons are injected into the gate oxide during the stress time and the HCI degradation of the low-Vt device is most intense. In other words, when tuning the MWF individually to achieve low-Vth, it results in a larger electric field on the gate dielectric for a given gate voltage. Therefore, when using low-Vt devices, more attention needs to be paid to reliability issues to ensure chip lifetime.

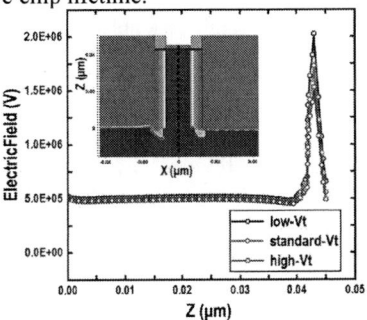

Fig. 6 Electric field distribution of multi-Vt nFinFETs along the Z direction of the channel (inset is a cross-section of the simulated structure along the length of the channel).

5. Conclusion

Experimental and TCAD simulation results show that the slightly TID response of multi-Vt nFinFETs is caused by the radiation-induced positively trapped charges in the STI. High-doping concentration in the channel stop region suppresses the radiation-induced leakage current increase and improves the TID tolerance of the core-nFinFETs. The HCI damage of multi-Vt nFinFETs is dominated by hot electron injection into the gate oxide. At the same electrical stress, low-Vt nFinFET has the largest gate dielectric electric field and thus the worst HCI degradation. Therefore, more consideration needs to be given to the aging reliability of low-Vt in the selection of multi-Vt nFinFETs.

Acknowledgments

This work was supported by the National Natural Science Foundation of China. Grant No. 62374120.

References

[1] Y. L. Chen, W. K. Yeh, and H. T. Hsu, Journal of Electronic Materials, 52(2), p.1391 (2023).

[2] A. Rahman, J. Dacuna, and P. Nayak, IEEE International Reliability Physics Symposium (IRPS), p.6F.4-1(2018).

[3] X. W. Wei, J. W. Cui, and D. Luo, IEEE Electron Device Letters, 44, p.1931 (2023).

[4] S. Mahapatra and R. Saikia, IEEE Transactions on Electron Devices, 65, p.3088 (2018).

[5] R. Li, Y. Liu, K. Zhang, C. Zhao, H. Zhu and H. Yin, IEEE International Conference on Solid-State and Integrated Circuit Technology (ICSICT), p.1(2014).

Modeling and Parameter Extraction of Semiconductor Devices for Simulation and Design Optimization of ESD Protection Circuits on BCD Technologies for Automobile and Industry Applications

Yuhua Cheng[1]*, Wei-wei Yu[1], and Eugene Worley[2]

[1] Shanghai Research Institute of Microelectronics, Peking University, Shanghai, China
[2] Silicon Crossing, LLC, Irvine, CA 92620, USA
* Email: chengyh@shrime-pku.org.cn

Abstract

In this paper, modeling and parameter extraction of semiconductor devices for simulation and design optimization of ESD protection circuits on BCD technologies were discussed. Concepts on developing optimal and advanced devices for cost effective and robust ESD protection capability were introduced, which can be executed with the support of the test chip design experience and on-wafer device characterization equipment environment. Turnkey solutions of ESD protection device development and debugging of ESD surge failure could be offered to circuit designers for designing efficient and robust ESD protection circuit blocks, which need to be simulated and optimized with the model library for the ESD protection devices.

1. Introduction

Growing IC applications for power management, power conversion and automotive chips require I/O interfaces typically between 12V and 100V. Most LCD/OLED display technologies also require driving voltages between 10 and 60V. Even cell phone devices and MEMS sensors need I/O interfaces above 10V. All those circuits require a high voltage interface with strong and robust ESD protection capability. Usually, bipolar-CMOS-DMOS (BCD) technologies are used with new options and features such as high voltage or bipolar modules for specific high voltage applications. Although these system applications represent fast-growth markets, the silicon process technologies cannot offer cost-efficient and high-performance ESD solutions. A good ESD protection solution is to provide a low cost but safe, robust current path while limiting the voltage drop below the critical voltage determined by the circuit to be protected. In BCD technologies, however, many different protection clamp types used in the industry are with significant performance and cost burdens such as high leakage current, large silicon area consumption, and extensive custom (trial-error-trial) development cycles for a new product design or for a mature product design to be transferred to a different process/fab. Also, to guide the design optimization of ESD protection device modules, circuit designers have long relied on the device models and process development libraries (PDK) provided by foundries, as well as ESD tools such as Spectre or HSpice. However, up to date, most of foundries if not all provide only basic ESD protection devices and do not provide models, related optimization design methods, and PDK solutions required for ESD devices. The designers have to optimize and improve the ESD protection modules through multiple chip fabrication iterations and independently develop ESD protection modules according to the circuit and system requirements. Furthermore, it is still a fact that most design companies lack a deep understanding of ESD devices and design experience for an optimized ESD protection circuit, often resulting in extended product development cycles and increased development costs due to ESD issues, causing huge economic losses to the IC design industry. Therefore, it is absolutely necessary to find methods, tools, and solutions on low cost and robust ESD devices and circuit blocks that can help design companies accelerate the development process of ESD protection design.

To address the long-standing issue of ESD protection and provide turnkey solutions, the efforts on the following research and capabilities are required: (1) Test chip design capability for ESD protection structures with separate control for the trigger and holding voltage; (2) Device modeling capability to create scalable devices models and extract the model parameters for circuit simulation and design optimization; (3) Test equipment capability to measure and characterize devices at DC/AC/RF/TLP levels: (4) PDK development capability to implement the model with extracted parameters with circuit simulators; (5) Circuit design capability for efficient and robust ESD protection blocks; (6) Failure analysis and debug capability for ESD related issues; (7) ESD design and optimization capability for turnkey solutions.

In this paper, we discuss modeling and parameter extraction of semiconductor devices for simulation and design optimization of ESD protection circuit blocks, based on designs and characterization of test structures of various ESD protection devices. Turnkey solutions with the optimized ESD protection devices could be offered for circuits and systems that need very efficient and robust ESD protection design, which can be modeled and simulated.

979-8-3503-6184-1/24 $31.00 © 2024 IEEE

2. Test Chip Design of ESD Protection Structures

The circuit ports to be protected from ESD surge are usually divided into the three types: an entire power domain (supply pin protection) or a single input, output, or I/O circuit (I/O pin protection). Circuit designers use different ESD protection devices to protect integrated circuits against ESD stress, according to the circuit/system requirements. The mainstream devices include Zener diodes, BigFET, PMOS/PNP, NMOS/NPN, SCR, and the above combined.

Zener-based protection devices are provided in many technology nodes. The breakdown and holding voltages are designed above the supply voltage level to ensure latch-up-immune ESD protection. The drawbacks of Zener diodes for ESD protection are large device sizes and too high holding voltages [1], which limits its applications. The BigFET is extensively used in low-voltage technology nodes for general-purpose I/O libraries. However, in high-voltage technologies, the application of BigFET devices is less popular due to the fact large area consumption is needed even for a low ESD HBM protection level [2, 3]. PMOS or PNP has a characteristic similar to the Zener device, however the robustness level per micron device width could be higher than the Zener device due to the bipolar PNP action. This results in a somewhat better performance for the leakage and silicon area consumption. NMOS or NPN is used because the breakdown robustness level is higher compared to Zener diodes, BigFET and PNP or PMOS devices, leading to smaller silicon area and reduced leakage. However, one of big drawbacks is the latch-up weakness due to a low holding voltage, typically much lower than the supply voltage [4]. Silicon Controlled Rectifier (SCR)-based protection devices were widely used to protect the high voltage signal bins [5, 6]. SCRs have good ESD characteristics with low leakage and low capacitance but are notorious for being slow to turn on and largely unresponsive to CDM. Also SCRs require careful design to control holding and trigger current to avoid latch-up problems [7]. Improved latch-up high-holding-current (HHI) SCR approach is reported [8], which should be done very carefully by designing and optimizing ESD protection structures for high voltage circuit applications.

The main purpose of the test chip design is to design various ESD protection structures to extract scalable model parameters and develop optimal and advanced devices with separate control for the trigger and holding voltages, that is, the key parameters should be set and adjust independently according to requirements of the circuit and system to be protected. Figure 1 shows an example of the test chip to characterize the on wafer ESD performance of several different ESD protect structures. More test structures are designed for model parameter extraction and advanced device optimization.

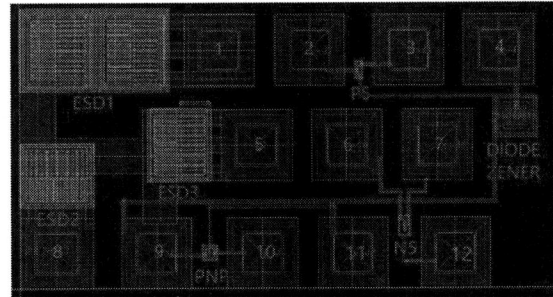

Figure 1. A test chip example for ESD device characterization.

3. On-Wafer Characterization of ESD Devices

ESD protection is typically qualified using human-body model (HBM), machine model (MM), and charged-device model (CDM) testers. For an in-depth analysis, Transmission Line Pulse (TLP) testers are used to characterize the ESD relevant performance parameters of the protection clamps as shown in Figure 2.

Figure 2. A TLP test system

TLP testers have been used widely to characterize the performance of ESD devices and to determine the optimal design of those devices, however, for a more robust and optimal design in terms of both ESD and other types of reliability issues such as latch-up and electro-thermal failure, additional characterization methods and equipment should be used together with a TLP tester to fully understand the ESD device behaviors. This is because TLP characteristic only presents averaged values of voltage and current versus time waveforms, and hence the information related to the time dependent behavior would be ignored in the TLP curves. Also, the TLP pulse width is typically limited to 100ns; that is enough for ESD failure analysis but not for EOS (electrical overstress) with a much longer timeframe, which needs additional characterization via different measurement equipment such as DC/AC meters and Oscilloscopes. Furthermore, TLP measurements are performed on 2 pins, leaving other pins floating. No bias is applied at V_{dd} so latch-up issues cannot be investigated. Therefore, in addition to standard TLP analysis, it is important for high voltage applications to look carefully into the full waveform information and to include longer pulse durations in the evaluation.

Other measurement approaches need to be used to

979-8-3503-6184-1/24 $31.00 © 2024 IEEE

verify transient latch-up susceptibility and thermoelectric effects. In our case, On-wafer Agilent Modeling Characterization System, as shown in Figure 3, is used. The system can perform DC、 AC and RF measurements at the chunk temperatures from -40 ℃ to 150ºC.

Figure 3. On-Wafer Agilent Modeling Characterization System with a 12 Inch Semi-auto Prober Station suable for DC-AC-RF measurements at different temperature conditions (from -40 ℃ to +200ºC).

4. Modeling and Parameter Extraction

Developing models for circuit simulation is highly desirable, especially for high performance circuits for ESD protection purposes, wherein a qualification test design target of meeting the ESD protection requirements is desired with the minimum chip area and hence the capacitance. The two requirements are that the ESD clamp could survive the ESD surge current and prevent damage to the I/O circuit. To perform circuit simulations for ESD protection design optimization, 3 model types are required: a tester model, a device failure model, and a high current ESD conduction model. Tester models must meet the IEC 61000-4-2, HBM, and CDM test specs. Failure models typically entail 2 types, electrothermal and gate dielectric TDDB. TDDB failure models are simple but electrothermal models are more complex. The high current conduction models require TLP data and may include transient effects.

For the models to be used for the simulation and optimization of the circuit design with ESD protection devices, all of the model parameters need to be extracted from the measured data. The methodology of designing and characterizing test structures as well as finding model parameters for various semiconductor devices is the key for these models to be inserted into a circuit simulator such as Spice for design simulation and optimization. Examples of model parameter extractions include a wide current range diode I-V model and transient model of diode conduction, electrothermal failure models of diodes and metal, and a snap-back NFET I-V characteristic model. For the models to be used for design optimization purpose, the key model parameters should be scalable with the geometry of the devices, and be extracted with the measured data from a set of test structures designed according to certain rules and constraints. Furthermore, for a model to better describe the electrical behaviors and enhance the model convergence and robustness in the whole operation

regimes, additional model terms are added in addition to the theoretical mathematic model expressions.

Diode devices, as an example in this paper, have been widely used for ESD protection in various kinds of circuit design. For DC diode conduction the exponential diode I-V model in series with fixed parasitic resistors works well for the ESD conduction regime. The diode conduction model provides the required offset voltage for the high current TLP I-V curve. However, for the low-level exponential region this conduction model does not work. To get the model to work over the entire conduction range, i.e., from the low-level exponential region to the high current

TLP linear region, a conductivity modulation term is added as seen in Eq. 1.

$$V_d = I_d \left(R_{fxd} + \frac{Ro}{1 + KI_d^\alpha} \right) + V_{therm} \ln \left(\frac{I_d}{I_s} + 1 \right) \quad \text{(Eq. 1)}$$

where I_s is the diode equation saturation current, R_{fxd} is the series fixed resistance, R_0 is the background diode channel resistance, I_d is the diode terminal current, V_d is the diode terminal voltage, V_{therm} is the diode thermal voltage constant, and K is the empirical conductivity modulation constant. The term with R_0 is the conductivity modulation equation. Normally, the parameter α is 1 but it was introduced as an empirical fitting parameter to see if more accuracy could be obtained.

Figure 4a shows a log I_{diode} vs. V_{diode} plot for the low current regime and Figure 4b is a corresponding linear plot in the high current regime. The low current regime was obtained from DC measurements and the high current regime can be obtained from TLP measurements. As seen in the figures, with the extracted model parameters, （Eq. 1） does a good job matching the measured I-V data ranging from a low of 10nA to 0.55A [9].

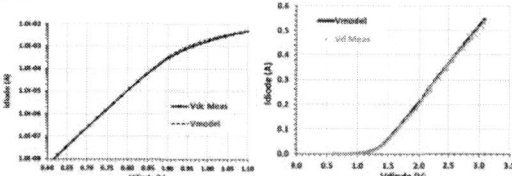

Figure 4a a log I_{diode} vs. V_{diode} plot for the low current regime; Figure 4b a corresponding linear plot in the high current regime.
Extracted parameters: R_{fxd} is 3.241 Ω , R_0 is 31.406 Ω , K is 148.69 and α is 1.16 [9].

STI diodes are known to have a considerable voltage overshoot for fast current rise times on the order of 100pS [10]. Since the STI diode voltage overshoot occurs on the leading edge of the current pulse the power is relatively constant and, therefore, is not a factor in electrothermal failures. However, gate dielectrics can respond to the voltage overshoot. Thus, for high performance I/O's that use LV MOSFETs the overshoot can lower ESD protection performance. To model voltage overshoot to a current pulse, the model assumes that the elements that are responsible for overshoot are resistances, one fixed, R_{fixed}, in series with a conductivity modulation resistor, R_{mod}, and a diffusion capacitance, C_{diff}, as shown in Figure 5. As with the conductivity modulation for the DC case,

$$R_{mod} = \frac{R_0}{1 + KI_{inj}^a}$$

(Eq. 2)

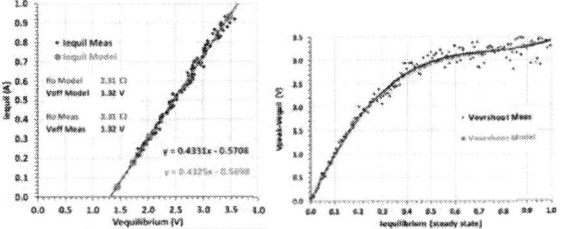

Figure 5 The equivalent circuit for STI Diode.

Figure 6a shows the results for the equilibrium or steady state I-V plot for an STI diode. The orange dots are the points optimized by solver and the blue dots are from measured TLP data. Least squares line fits for both data sets are shown along with the slope resistance and offset voltage which are the very nearly the same for the 2 data sets. Figure 6b shows a plot of the overshoot voltage, i.e. peak voltage minus the equilibrium voltage, versus equilibrium current. The orange dots are from model fitting and the blue dots from measured data.

Figure 6a Equilibrium I-V plots for measured and model data; Figure 6b Overshoot voltage versus equilibrium current.

As another example, model parameters were extracted for a snap-back NFET with a schematic diagram of the model shown in Figure 7. In this example, a grounded gate NFET with 2 regions, the pre-snap-back and post snap-back regions, must be fitted to extract these model parameters. The base-emitter junction is modeled as a diode with a series resistor, R_{bext}. The collector current is modeled as a constant β multiplied by the base current. A body-source shunt resistor, R_{body}, is used to model the effective body tie resistance. A constant leakage source is used to trigger the NFET from the blocking state to the conducting state. Using a constant current for the leakage trigger source is simpler than using a voltage dependent current. A drain resistor, R_{drain}, is used to model external resistances such as contact, silicide block, etc. on the drain.

An exponential function was used for the impact ionization avalanche model. The standard Miller function can lead to instabilities as when the breakdown voltage parameter is reached. The stable avalanche function used here is

$$M - 1 = K_m\left(e^{K_e(V_c - V_b)} - 1\right)$$

(Eq. 3)

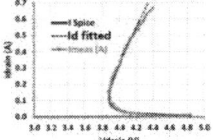

Figure 7 Snap-back NFET model with parameters to be extracted.

There are a total of 8 parameters (R_{bext}, β, R_{body}, I_s, I_{leak}, R_{drain}, K_m and K_e) to be optimized. Figure 8 shows the I-V plot results for the Spice simulation, fitting and measured data. The fit is quite good up to the point where heating effects start kicking in near failure current.

Figure 8 Grounded gate snap-back NFET I-V (Measured, simulated, and fitted).

5. Summary

In this paper, modeling and parameter extraction of ESD protection devices on BCD technologies were discussed. Test chip design experience and on-wafer device characterization equipment environment to develop optimal and advanced devices for cost effective and robust ESD protection capability are needed to support the execution. Service for a turnkey solution of ESD protection device development and debugging of ESD surge failure could be offered, in addition to providing circuit designers the model library for designing efficient and robust ESD protection circuit blocks for automobile and industry applications with strong ESD reliability requirements.

References

[1] Oleg Semenov et al., ESD Protection Device and Circuit Design for Advanced CMOS Technologies, Springer, 2008

[2] O. Quittard et al., "ESD protections for the High-Voltage CMOS Technology", EOS/ESD 2006

[3] G. Notermans et al., "Designing HV Active Clamps for HBM Robustness", EOS/ESD 2007.

[4] B. Keppens et al., "ESD Protection Solutions for High Voltage Technologies", EOS/ESD 2004.

[5] M. Mergens et al., "ESD Protection Considerations in Advanced High-Voltage Technologies for Automotive", EOS/ESD 2006.

[6] K. Reynders, "Design and Characterization of a High Voltage SCR with High trigger current", EOS/ESD 2005.

[7] Y. Fukuda et al., "Solving the problems with traditional Silicon Controlled Rectifier (SCR) approaches for ESD", RCJ 2008.

[8] B. Sorgeloos et al., "On-Chip ESD Protection with Improved High Holding Current SCR (HHISCR) Achieving IEC 8kV Contact System Level" IEW 2010 and EOS/ESD 2010.

[9] E. Worley, "Parameter Extraction methods and Circuit Simulation Models", 2024 Asia International ESD Workshop, Singapore, 2024.

[10] Z. Pan, et. al., "Understanding and modelling of diode voltage overshoots during fast transient ESD events", IEEE Trans. On Electron Devices, vol. 61, no. 8, p. 2682-268. 2014.

A Novel Double-zener Process and Multiplex Design for High-power Surge and High-speed ESD Devices Development

Zhao Qi[1,2*], YiRui Jia[1], Hongquan Chen[1,2], Ming Qiao[1,3,4], Zhaoji Li[1] and Bo Zhang[1]

[1] State Key Laboratory of Electronic Thin Films and Integrated Devices, University of Electronic Science and Technology of China (UESTC), Chengdu, P. R. China.
[2] Chongqing Institute of Microelectronics Industry Technology, UESTC, Chongqing, P. R. China.
[3] Shenzhen Institute for Advanced Study, UESTC, Shenzhen, P. R. China.
[4] Institute of Electronic and Information Engineering of UESTC in Guangdong, Dongguan, P.R. China.

* Email: qizhao@uestc.edu.cn

Abstract—In this work, a novel Double-zener process (DZP) for Transient Voltage Suppressor (TVS) was developed in 0.5 μm Bipolar CMOS DMOS (BCD) process for the design of Electrical Overstress (EOS) and Electrostatic Discharge (ESD) protection devices. The DZP is designed by adding two P-zener implantation (EP1 and EP2) with different energy. Moreover, multiplex units are designed for various applications, including a dual-directional (DD) low-capacitance silicon-controlled rectifier (DD-LCSCR), a dual-directional NPN (DD-NPN) clamp, and a high-power lightning protection array (LPA). The DD-LCSCR fabricated by EP2 step achieves a capacitance (C) of 0.5 pF and a surge peak current (I_{PP}) of 7 A. The DD-NPN clamp using an EP1 step to decrease the breakdown voltage (BV), achieving a holding voltage (V_h) of 5.5 V, a clamping voltage (V_{CL}) of 14 V@50 A, and an ESD level of ±30 kV. Using the EP1 step and stacking technology, the 6-stacked LPA shows the excellent power density and V_{CL} which is competitive with vertical TVS products.

Keywords—Electrostatic Discharge (ESD), Transient Voltage Suppressor (TVS), Double-Zener Process (DZP)

I. INTRODUCTION

Transient Voltage Suppressors (TVS) are widely used in electronic systems to protect the Integrated Circuit (IC) from Electrostatic Discharge (ESD) event for its excellent characteristics. However, different applications in different usage scenarios require TVS to have specific characteristics, which increases the complexity of the manufacturing process. For instance, high-speed interfaces necessitate TVS with high surge current (I_{PP}) and minimal capacitance (C) [1-3]. Power supply applications demand TVS with specific breakdown voltage (BV), high holding voltage (V_h) to prevent the latch-up effect (LU), and minimal clamping voltage (V_{CL}) [4,5]. Certain integrated circuits, such as those used in Controller Area Networks (CAN), require system-level electrostatic discharge protection[6,7]. Faced with this situation, the improvements of manufacturing process to meet the different requirements of TVS for different applications are necessary. In this work, a novel Double-zener process (DZP) based on 0.5 μm Bipolar CMOS DMOS (BCD) process was developed, which is deeply compatible with the standard BCD process for on-chip protection. Additionally, it is suitable for multiple applications with different requirements, including those requiring low capacitance (low-C), low clamping voltage (low-V_{CL}), and effective lightning protection, all while delivering excellent performance.

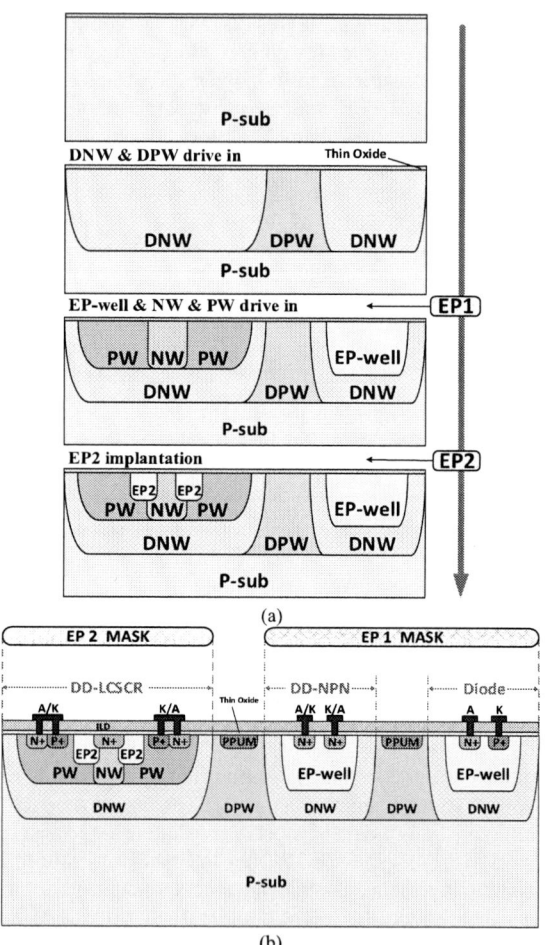

Fig. 1. (a) The key EP processes of the novel DZP, (b) Three main TVS cells, including DD-LCSCR, DD-NPN and Diode array, integrated by DZP.

II. APPROACH AND MECHANISM

The development of the TVS DZP focused on two essential aspects: the enhanced P-Zener implantation (EP1/2) and the implementation of advanced dielectric techniques. With different combinations of EP1/2 and advanced dielectric techniques, TVS with specific characteristics can be designed

---

This work was supported in part by the China Postdoctoral Science Foundation under Grant 2021M700684, and in part by the Chongqing Natural Science Foundation under Grant cstc2021jcyj-msxmX1023.

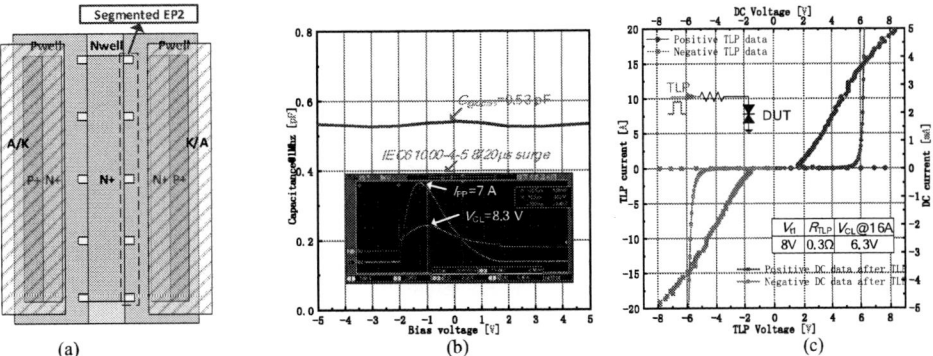

Fig. 2. (a) The Segmented EP2 design for point-triggered TVS, (b) the tested capacitance under 0-bias and the 8/20 μs I_{PP} (IEC 61000-4-5) test data, (c) the DD transmission line pulse (TLP) curves and DC curves after TLP of the DD-LCSCR.

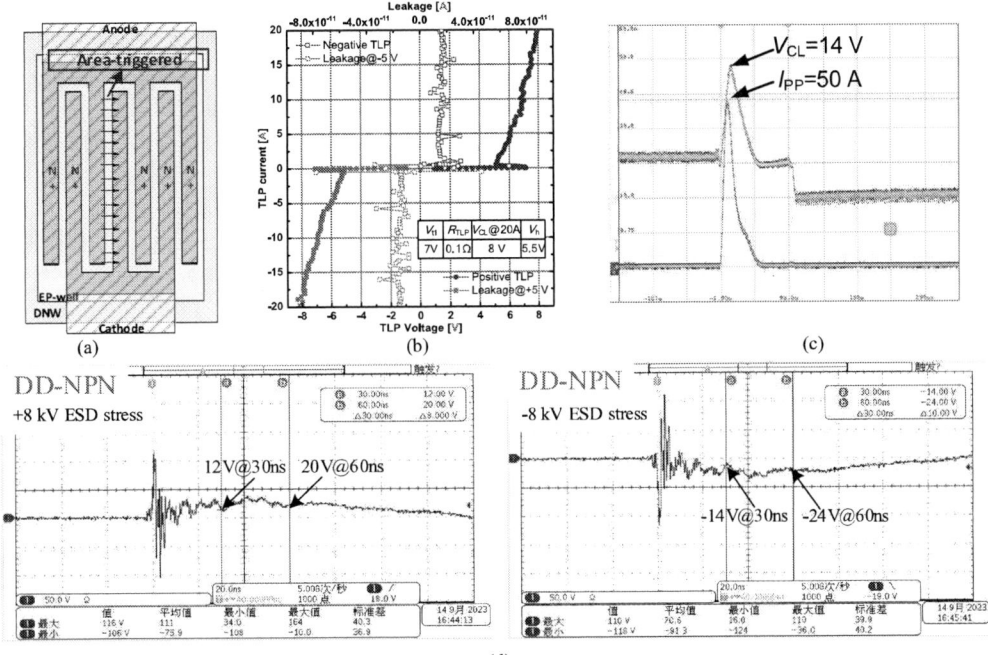

Fig. 3. (a) The area-triggered DD-NPN diagram, (b) the TLP curves of the DD-NPN with V_h of ±5.5 V, (c) the 8/20-μs surge test data of the DD-NPN and (d) the ±8 kV DD ESD contact test curve (IEC61000-4-2) of the DD-NPN.

for different applications. In this work, multiplex units were designed based on the EP1/2 for a range of different applications. These include a low-power dual-directional (DD) low-capacitance silicon-controlled rectifier (DD-LCSCR) unit, a dual-directional NPN clamp unit, and a high-power lightning protection array (LPA) unit, all demonstrating excellent performance and yield.

Fig. 1a illustrates the key enhancement of the DZP, which involves two EPs (EP1 and EP2). EP1 mask is used for EP-well implantation after the driving in process of DNW & DPW in the conventional process flow. EP-well drive in together with NW & PW to form EP1, which can provide a high-concentration and large-area base region for the high-current Bipolar TVS to restrain the high-injection effect. EP2 mask is used after the driving in for EP2 implantation. EP1 is designed for area-triggered devices that handle high power, while EP2 is tailored for point-triggered devices that operate at low power, as shown in Fig. 1b. From a functional perspective, EP1 is required to pass high current, while EP2 is solely for triggering purposes.

The DZP employs low dose DN/DP isolation and a dielectric combination of low-temperature thin oxide and inter-layer dielectric (ILD) to prevent boron absorption. Additionally, a P+ under metal (PPUM) layer is utilized to block the possible inverted channel between two DN regions for further reducing metal insulator semiconductor (MIS) structure leakage. The DZP process simply adds two additional masks, the rest of the process is fully compatible with the conventional process, allowing the devices to be used for both ESD and surge protection, combining the two functions together.

III. RESULTS AND SIGNIFICANCE

Fig. 1b shows the three integrable TVS units designed for different applications fabricated by DZP, including DD-LCSCR, DD-NPN clamp and LPA. Experimental results demonstrate excellent performance of the proposed DZP technology under Transmission Line Pulse (TLP) and surge testing, providing reliable and multiplex solutions for integrated and discrete protection devices.

979-8-3503-6184-1/24 $31.00 © 2024 IEEE

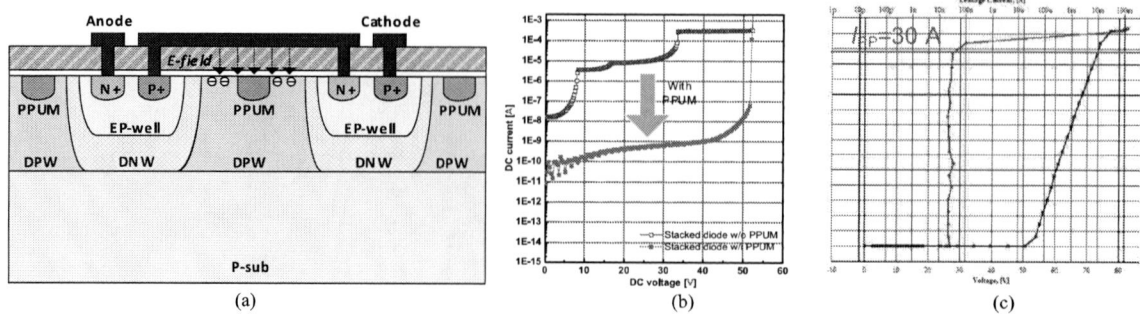

Fig. 4. (a) Stacked HAD, (b) the BV curve w/i or w/o PPUM and (c) the 8/20-μs test curve.

A. DD-LCSCR

Firstly, the DZP process, when combined with the EP2 mask, can be used to develop the dual-directional low-capacitance silicon-controlled rectifier (DD-LCSCR). The DD-LCSCR, as shown in Fig. 2a, features segmented EP2 regions to achieve a low V_{t1} and minimize C. Through an optimized layout, the DD-LCSCR achieves a C of 0.5 pF and a V_{t1} of 8 V, while supporting an I_{PP} of 7 A with a good V_{CL} of 8.3 V, as shown in Fig. 2b and 2c. It can be foreseen that, with the reduction of feature size, the LCSCR can achieve lower C and higher I_{PP}.

B. DD-NPN Clamp

When the DZP utilizes only the EP1 mask, it is suitable for developing high-current clamp devices. Fig. 3a illustrates the design of a DD-NPN clamp for ±5 V supply (Area:550 *450 um²), where the base region is formed using a high concentration (~1E18/cm³ level) EP-well to decrease the BV and restrain high-injection effect. Fig. 3b to 3d present transmission line pulse (TLP), surge, and ESD tested data for the DD-NPN, showing a V_h of 5.5 V, a V_{CL} of 14 V@50 A, and an ESD level of ±30 kV. Therefore, DDNPN and its scaled versions can be suitable for chip-level ESD protection and system-level surge protection under 5V without latch-up effects.

C. Stacked LPA

For high-power lightning protection, the DZP combining the EP1 mask and stacking technology can develop a high-voltage diode array that rivals the performance of vertical TVS. Fig. 4a shows a 6-stacked diode LPA (the BV of every unit is designed as 8.5 V), the body region of each diode is formed by EP-well for achieving low BV and low body resistance. Fig. 4b shows the BV curves of 6-stacked LPA (Area:1.4 *1.4 mm²) w/i or w/o PPUM, demonstrating the leakage blocking effect of the PPUM. The surge test (IEC61000-4-5 8/20 μs waveform) shown in Fig. 4c exhibits that the LPA can get BV of 52 V and handle an I_{PP} of 30 A with a low V_{CL} of 74V.

Table 1 presents a comprehensive comparison between the TVS manufactured using DZP and same-level products. It can be seen that the devices manufactured using DZP, especially the V_{CL}, are lower than those of the same level. Additionally, the devices in this article have the advantage of being able to integrate with BCD process ICs, making them highly suitable for integrated high-level IC ESD protection or discrete protection device design.

IV. CONCLUSION

In conclusion, the proposed DZP technology in this study has successfully achieved high performance in various types of protection devices, such as low C required for high-speed interface protection, low clamping and anti-latch-up capabilities for traditional supply applications, and high-power protection for high voltage applications, with better performance (especially in terms of V_{CL}) compared to the latest similar products. On the other hand, the advantage of DZP lies in its full compatibility with BCD processes, which allows the resulting devices to be integrated with ICs to meet specific integrated protection needs. It is a novel TVS technology that covers multi-purpose and multi-voltage applications with high performance.

REFERENCES

[1] Z. Qi, M. Qiao, F. Zhao, Z. Li and B. Zhang, "Novel Integrated Low Capacitance Transient Voltage Suppressor Array with Capacitance Equalization Technique for System-Level EOS/ESD Protection," 2021 33rd International Symposium on Power Semiconductor Devices and ICs (ISPSD), Nagoya, Japan, 2021, pp. 203-206.

[2] K. -H. Lo et al., "Ultralow Capacitance Transient Voltage Suppressor Design," in IEEE Transactions on Electron Devices, vol. 63, no. 8, pp. 3064-3068, Aug. 2016.

[3] Q. Cui, J. A. Salcedo, S. Parthasarathy, Y. Zhou, J. J. Liou and J. J. Hajjar, "High-Robustness and Low-Capacitance Silicon-Controlled Rectifier for High-Speed I/O ESD Protection," in IEEE Electron Device Letters, vol. 34, no. 2, pp. 178-180, Feb. 2013.

[4] Z. Qi et al., "Novel Silicon-Controlled Rectifier With Snapback-Free Performance for High-Voltage and Robust ESD Protection," in IEEE Electron Device Letters, vol. 40, no. 3, pp. 435-438, March 2019.

[5] K. -I. Do, B. -B. Song and Y. -S. Koo, "A Novel Dual-Directional SCR Structure With High Holding Voltage for 12-V Applications in 0.13-μm BCD Process," in IEEE Transactions on Electron Devices, vol. 67, no. 11, pp. 5020-5027, Nov. 2020.

[6] C. -H. Chuang and M. -D. Ker, "System-Level ESD Protection for Automotive Electronics by Co-Design of TVS and CAN Transceiver Chips," in IEEE Transactions on Device and Materials Reliability, vol. 17, no. 3, pp. 570-576, Sept. 2017.

[7] Y. Wang, Z. Zhong, X. Jin, Y. Peng, J. Luo and J. Yang, "A Novel Gate-Controlled Dual Direction SCR With Enhanced Failure Current for On-Chip ESD Protection of Industry-Level Controller Area Network Bus," in IEEE Journal of Emerging and Selected Topics in Power Electronics, vol. 10, no. 6, pp. 7615-7626, Dec. 2022

TABLE I. COMPARISON OF KEY INDICATORS BETWEEN THE PROPOSED DEVICES AND THE SAME-LEVEL PRODUCTS

	$BV@1 mA$ (V)	$V_{CL@TLP}$ (V)	I_{TLP} (A)	Cap (pF)	I_{PP} (A)	$V_{CL@Ipp}$ (V)
PESD5V0 C1BLS-Q	8	9	>20	0.35	6.5	/
LCSCR	6	8	>20	0.5	7	8.3
SDW05C-01FTG	6	10.5	>20	/	30	15
DD-NPN	7	8	>20	/	50	14

The non-monotonic instability of V_{TH} and $R_{ds,on}$ in P-GaN Gate HEMTs Under Repetitive Short-Circuit Stress: The role of electric-field & self-heating effect

Long Wang[1], Ning Yang[1], Shuting Huang[1], Jianggen Zhu[1], Kuangli Chen[1], Chao Feng[3], Haolin Hu[3], Wei Zeng[3], David Zhou[3], Yuxi Wan[3], Bo Zhang[1], and Qi Zhou†[1,2]

[1] School of Integrated Circuit Science and Engineering, University of Electronic Science and Technology of China (UESTC), Chengdu
[2] Institute of Electronic and Information Engineering, UESTC, Dongguan
[3] Shenzhen Pinghu Laboratory, Shenzhen, China

† Email: zhouqi@uestc.edu.cn

Abstract—In this work, the instability of threshold voltage (V_{TH}) and on-resistance ($R_{ds,on}$) in 650 V Schottky-type p-GaN gate HEMTs under repetitive short-circuit (SC) stress is investigated. The unique non-monotonic instability behavior of V_{TH} and $R_{ds,on}$ during SC stress is observed, which is found to be respectively dominated by electric-field and self-heating effect. Under mild repetitive SC stress, the V_{TH} and $R_{ds,on}$ exhibit a positive shift dynamics with the enhancing stress intensity, which is dominated by the electric-field induced electron-trapping in the gate stack and access region. Conversely, under more stringent repetitive SC stress, as more energy generated in the SC stress event, the thermally enhanced hole injection and suppressed hot electrons effect are identified as the key factors which in turns to lead to a negative shift kinetics in V_{TH} and $R_{ds,on}$. The findings provide a comprehensive understanding of the SC withstand capability and degradation mechanism of the device.

Keywords—non-monotonic instability, p-GaN gate high electron mobility transistors (HEMTs), short circuit (SC), hole injection.

I. INTRODUCTION

Gallium nitride (GaN) high electron mobility transistors (HEMTs) exhibit excellent properties in high power conversion efficiency, switching speed, and high power density. The short-circuit (SC) fault operation occurring in GaN HEMTs in power applications may induce substantially high electric-field and self-heating in the devices. This can lead to significant device instability or even catastrophic failure, hindering stable and safe operation of the overall power system.

Recent studies have analyzed the robustness of commercial 600V/650V p-GaN gate HEMTs under various SC conditions. The SC capability of Schottky-type p-GaN gate HEMTs primarily depends on $I_{g(on)}$ and I_{ds} has been reported [1-3]. Significantly thermal accumulation results in device failure during SC stress is also observed [4], [5]. A rapid device degradation under different SC conditions, including variation in V_{TH} and $R_{ds,on}$ is reported [6], but the underlying mechanisms remain unclear. A monotonic variation in V_{TH}, $R_{ds,on}$, and C_{oss} after repetitive SC stress with $V_{gs}= 6$ V, $V_{dc} = 100$ V, and $T_{SC} = 5$ us has been reported [7].

In this work, the SC ruggedness of commercialized 650 V Schottky-type p-GaN gate HEMTs is comprehensively

This work was supported in part by the National Natural Science Foundation of China under Grant 62174019, in part by the Guangdong Basic and Applied Basic Research Foundation, China, under Grant 2021B1515140039, 2024A1515012139 and in part by Sichuan Science and Technology Program under Grant 2023YFG0138.

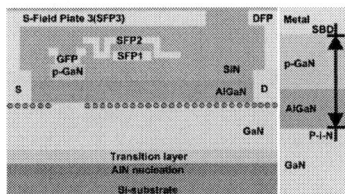

Fig. 1. Schematic cross-sectional structure of the studied Schottky-type p-GaN gate HEMT (650 V/100 mΩ) (left) and equivalent model of the p-GaN gate-stack (right).

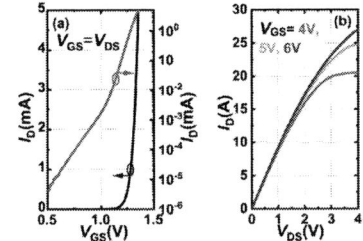

Fig. 2. (a) Transfer curves and (b) output curves of the fresh device.

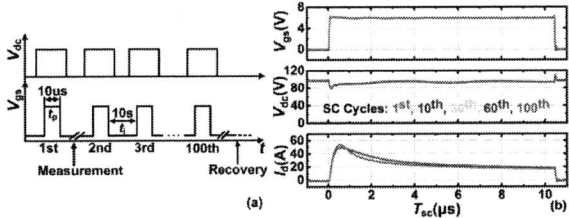

Fig. 3. (a) Modified timing diagrams of V_{dc} and V_{gs}. and (b) waveform under repetitive SC stress with $V_{gs} = 6$ V, $V_{dc} = 100$ V.

studied by using the repetitive SC test with various electrical conditions (e.g., V_{gs} and V_{dc}). The unique non-monotonic instability of V_{TH} and $R_{ds,on}$ during the repetitive SC stress is observed. Although both electric-field and SC induced self-heating are simultaneously presented in the device, the electron spill over and trapping driven by the electric-field is the root cause responsible for the positive shift in V_{TH} and $R_{ds,on}$, while the enhanced hole injection triggered by severe self-heating effect associated with suppressed hot electron generation take over to dominate the negative shift in V_{TH} and $R_{ds,on}$.

II. EXPERIMENTAL METHOD

The device used in this work is commercial 650 V/100 mΩ Schottky-type p-GaN gate HEMTs. Fig. 1 shows the schematic cross-sectional device structure and the equivalent model of the p-GaN gate-stack (comprising Schottky metal/p-GaN/AlGaN/GaN). Fig. 2(a) and (b) show the static transfer and output characteristics of the device, respectively.

The repetitive SC stress is applied using the ITC57250 test system, the test circuit is available in our previous report [8]. The SC stress and device parameters test schemes are plotted in Fig. 3(a). During the SC stress phase, a high voltage pulse V_{dc} together with a gate pulse voltage V_{gs} are simultaneously applied to the drain and gate terminal, respectively. Each SC event lasts for 10 μs with a 10 s interval to mitigate the thermal accumulation and associated device instability. The static *I-V* characteristics (e.g., transfer, output) of DUT are measured by Keysight B1505A after the 1st, 10th, 30th, 60th, and 100th SC events to record the device instability dynamics. The device under test (DUT) undergoes 100 SC stress cycles with varied V_{gs}=3, 4, 5, & 6 V, and V_{dc}=100, 200, & 300 V. Fig. 3(b) shows the typical SC waveforms under repetitive SC stress with V_{gs}=6 V & V_{dc}=100 V.

III. RESULTS AND DISCUSSION

Fig. 4(a) and (b) shows the transfer and output curves variation dynamics during repetitive SC stress with $V_{gs} = 4$ V, $V_{dc} = 100$ V. With the increasing number of SC cycles, V_{TH} (defined at $I_D = 1$ mA) shifts significantly from the initial value of 1.27 V in the fresh device to 1.95 V after the 100th cycle. Concurrently, $R_{ds,on}$ (extracted at $I_D = 4.5$ A) increases from 100.2 mΩ to 134.7 mΩ and the drain-source current (extracted at $V_{DS} = 4$ V) decreases from 26.63 A to 16.88 A. Notably, DUT exhibits a substantial V_{TH} shift of 0.40 V and an increased $R_{ds,on}$ of 11% after the 1st cycle. It can be inferred that such a substantial instability may cause power performance degradation (e.g. dead-time & conduction power loss) while the device suffers from a single SC event during applications. Fig. 4(c) and (d) shows the variations in V_{TH} and $R_{ds,on}$ of DUTs after the 1st and 10th SC events under SC stress with $V_{dc} = 100$ V and various V_{gs}. It can be seen that the positive ΔV_{TH} and $R_{ds,on}$ initially increases and then decreases as V_{gs} increases from 3 V to 6 V. It should be noted that the DUT exhibits the maximum ΔV_{TH} of +0.69 V and a 35%

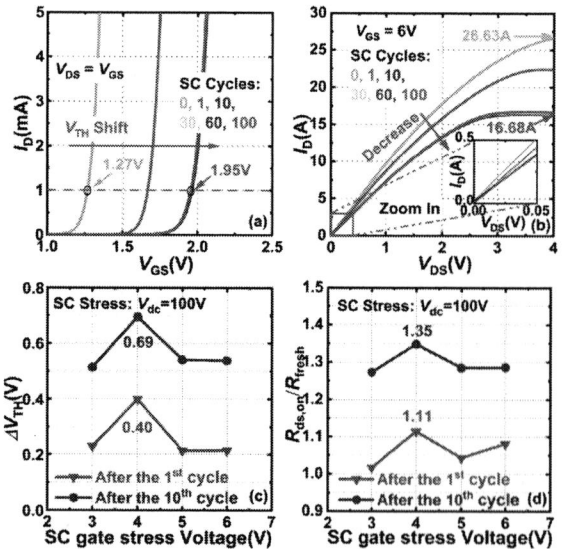

Fig. 4. (a) Transfer and (b) output curves before and after repetitive SC stress with $V_{gs} = 4$ V, $V_{dc} = 100$ V; (c)the variations in V_{TH} and (d) the variations in $R_{ds,on}$ after the 1st and 10th SC cycle.

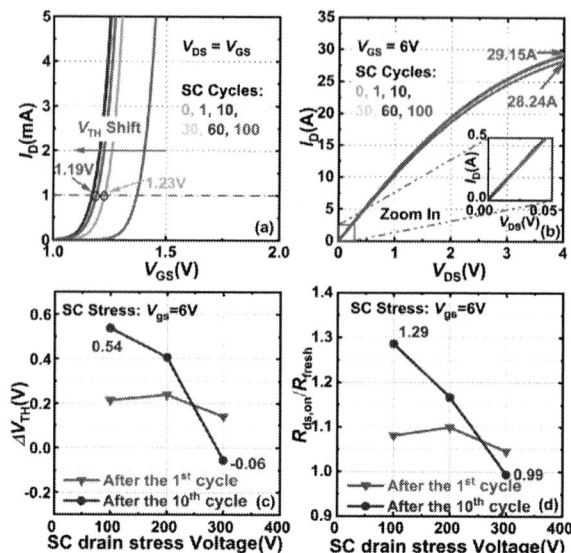

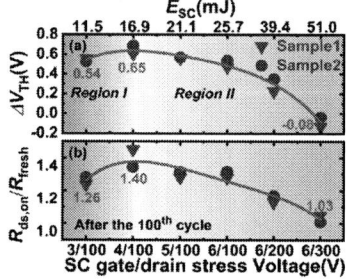

Fig. 5. (a) Transfer and (b) output curves before and after repetitive SC stress with $V_{gs} = 6$ V, $V_{dc} = 300$ V; (c)the variations in V_{TH} and (d) the variations in $R_{ds,on}$ after the 1st and 10th SC cycle.

Fig. 6. (a) The ΔV_{TH} and (b) $\Delta R_{ds,on}$ after 100 SC cycles of sample1 & sample2 *vs.* SC stress voltage and energy.

increase in $R_{ds,on}$ after 10 SC cycles stress under stress conditions of $V_{gs} = 4$ V, $V_{dc} = 100$ V.

Fig. 5(a) and (b) shows the transfer and output curves during repetitive SC stress with $V_{gs} = 6$ V, $V_{dc} = 300$ V. As the number of SC events increases, it is interesting to see that the V_{TH} exhibits non-monotonic shift. The V_{TH} shifts positively from 1.23 V to 1.37 V after the 1st cycle, and then turns into a negative shift to 1.19 V after the 100th cycle. Meanwhile, the changes in $R_{ds,on}$ and I_D of the device are almost negligible. Fig. 5(c) and (d) shows the variations in V_{TH} and $R_{ds,on}$ of DUTs under SC stress with $V_{gs} = 6$ V and various V_{dc}. Normally, with severer stress (e.g. higher V_{dc}), the instability of V_{TH} and $R_{ds,on}$ should become more pronounced. However, in this work the ΔV_{TH} and $R_{ds,on}$ exhibits unexpected decrement for V_{dc} further increased beyond 100 V @ $V_{gs} \geq 6$ V.

Fig. 6 presents a dual X-axis plot to illustrate the relationship between the ΔV_{TH} and $R_{ds,on}$ with SC stress gate/drain voltage & SC energy (E_{sc}). It can be seen that the instability of ΔV_{TH} and $R_{ds,on}$ versus SC stress exhibit two correlation region. In *Region I*, as SC energy escalates, the positive ΔV_{TH} and $R_{ds,on}$ increase become more pronounced. In this region, ΔV_{TH} & $\Delta R_{ds,on}/R_{fresh}$ reach the peak values of 0.65 V and 40%, respectively. Conversely, in *Region II*, the V_{TH} and $R_{ds,on}$ show prominent negative shift with the increasing SC stress energy. Notably, under the SC stress condition of $V_{gs} = 6$ V, $V_{dc} = 300$ V corresponds to $E_{sc} = 51$ mJ, the V_{TH} exhibits a negative shift of -0.08 V compared to the fresh device.

979-8-3503-6184-1/24 $31.00 © 2024 IEEE

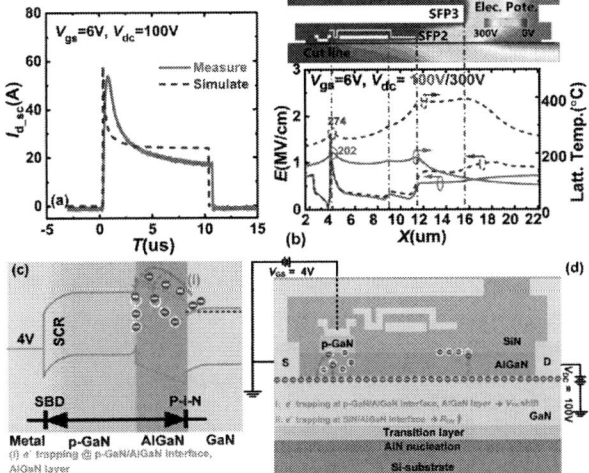

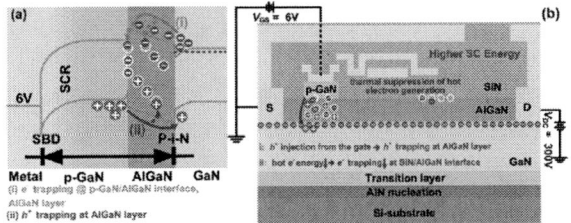

Fig. 7. (a) Simulated and measured I_{d_sc} waveform; (b) electrothermal simulation of lattice temperature and E-field at V_{gs} = 6 V, V_{dc} = 100 & 300 V; (c) energy band diagram in gate-stack and (d) charge distribution and trap-filling in p-GaN gate HEMTs under mild repetitive SC stress.

Accordingly, $\Delta R_{ds,on}/R_{fresh}$ shows a negligible variation of 3%. The distinctively non-monotonic device instability property along with the increasing SC energy suggests that the device degradation in *Region I & II* are respectively governed by different underlying physics.

The TCAD simulation is employed to reveal the device physics during the SC stress. The TCAD model is calibrated by the SC waveform as shown in Fig. 7(a). The electric field and lattice temperature distributions in 2DEG channel under SC stress are shown in Fig. 7(b). In the SC event, the forward gate voltage ($V_{gs} \leq 4$ V) tends to bend down the gate-stack energy-band together with a high-density 2DEG formed in the channel. Associated with the high peak electric-field at the drain-side gate edge (see Fig. 7(b)), electrons can be accelerated by the locally high electric-field, becoming hot electrons and spill over the channel. Partial of the electrons are trapped in the p-GaN/AlGaN gate-stack as shown in Fig. 7(c). Then the filled traps act as negative charges to deplete the 2DEG channel, which results in the positive ΔV_{TH} in *Region I*. Additionally, the high electric-field at source field plate (SFP) edge may also assists electrons transfer from 2DEG channel to be trapped at the residue dielectric/AlGaN interface states as shown in Fig. 7(d), which features the virtual gate effect and leads to an increased $R_{ds,on}$. For power applications, the positive ΔV_{TH} requires a higher gate drive voltage to fully turn-on the 2DEG channel, which may increase the dead-time power loss. Besides, the increased $R_{ds,on}$ is detrimental to the conduction power loss.

In *Region II*, with higher gate stress voltage (4 V < $V_{gs} \leq$ 6 V), the enhanced electric-field generates substantial hole injection as shown in Fig. 8. Such a hole injection effect is further enhanced by thermal activation, which is identified in the SC stress with higher SC energy. As shown in Fig. 7(b), as V_{dc} increases from 100 to 300 V, the resultant higher E_{SC} during stress increases the junction temperature T_j at the drain-side gate edge from 202 to 274°C. The elevated T_j enhances the ionization of acceptors in the p-GaN layer and then trigger prominent hole injection from the p-GaN to the GaN channel as depicted in Fig. 8. The electric-field & thermally enhanced hole injection and associated hole trapping as shown in Fig. 9 lead to reduced net negative charges stored in the gate region. Consequently, the DUT exhibits a negative shift in V_{TH}. Additionally, the significantly elevated device temperature at

Fig. 8. Hole carrier distribution with stress of V_{gs} = 6 V, (a) at V_{dc} = 100 V; (b) at V_{dc} = 300 V.

Fig. 9. (a) Energy band diagram in gate-stack and (b) charge distribution and trap-filling in p-GaN gate HEMTs under stringent repetitive SC stress.

the SFP2 and SFP3 edges suppresses the generation of hot electrons, ultimately mitigating the degradation of $R_{ds,on}$ as shown in Fig. 9(b).

IV. CONCLUSION

The SC ruggedness of 650 V Schottky-type p-GaN gate HEMTs under various repetitive SC stress is comprehensively investigated. The studied devices exhibit significant non-monotonic instability in both V_{TH} and $R_{ds,on}$ after repetitive SC stress. Detailed device characterization combined with TCAD simulation is carried out to reveal the degradation mechanisms. The electron-trapping (@ the gate-stack and residue dielectric/AlGaN interface states) triggered by locally high electric-field result in a prominent positive shift in V_{TH} and increase in $R_{ds,on}$ during SC stress. However, under stringent SC stress with increased SC energy, hole injection from the p-GaN layer to GaN channel and subsequent hole trapping is enhanced by the severe self-heating effect, which mitigates the positive V_{TH} shift and even results in an ultimate negative V_{TH} shift. Besides, the hot electron effect is also suppressed due to increased T_j that mitigates the $R_{ds,on}$ degradation.

REFERENCES

[1] M. Riccio, *et al.*, "Short circuit robustness analysis of new generation Enhancement-mode p-GaN power HEMTs," in *2018 IEEE ISPSD*, Chicago, IL, USA, 2018, pp. 104-107.

[2] T. Oeder, *et al.*, "Experimental study of the short-circuit performance for a 600V normally-off p-gate GaN HEMT," in *2017 IEEE ISPSD,*, Sapporo, Japan, 2017, pp. 211-214.

[3] M. Fernández, *et al.*, "P-GaN HEMTs Drain and Gate Current Analysis Under Short-Circuit," *IEEE Electron Device Letters*, vol. 38, no. 4, pp. 505-508, April 2017.

[4] C. Abbate, *et al.*, " Failure analysis of 650 V enhancement mode GaN HEMT after short circuit tests," *Microelectronics Reliability*, vol. 88-90, pp. 677-683, September 2018.

[5] J. Sun, *et al.*, "Short Circuit Capability Characterization and Analysis of p-GaN Gate High-Electron-Mobility Transistors Under Single and Repetitive Tests," *IEEE Trans. on Ind. Electronics*, vol. 68, no. 9, pp. 8798-8807, Sept. 2021.

[6] H. Li *et al.*, "Robustness of 650-V Enhancement-Mode GaN HEMTs Under Various Short-Circuit Conditions," *IEEE Trans. on Ind. Appl.*, vol. 55, no. 2, pp. 1807-1816, March-April 2019.

[7] S. Li *et al.*, "Understanding Electrical Parameter Degradations of P-GaN HEMT Under Repetitive Short-Circuit Stresses," *IEEE Trans. on Power Electronics*, vol. 36, no. 11, pp. 12173-12176, Nov. 2021.

[8] N. Yang, *et al.*, "Study of the Short-Circuit Capability and Device Instability of p-GaN Gate HEMTs by Repetitive Short-Circuit Stress," *IEEE Trans. on Power Electronics*, vol. 39, no. 2, pp. 2247-2257, Feb. 2024.

979-8-3503-6184-1/24 $31.00 © 2024 IEEE

Optimizing Flash Memory Towards Storage-Class Memory (SCM) Applications

Xinyi Guo[1], Yang Feng[1], Jing Liu[2], Junyu Zhang[3], Xuepeng Zhan[1], Jixuan Wu[1], and Jiezhi Chen*[1]

[1] School of Information Science and Engineering, Shandong University, Qingdao, P. R. China; [2] Key Laboratory of Microelectronic Devices and Integrated Technology, Institute of Microelectronics of Chinese Academy of Sciences, Beijing, P. R. China; [3] Neumem Co., Ltd, Hefei, P. R. China

* Email: chen.jiezhi@sdu.edu.cn

Abstract—**Aiming at the applications of flash memories as storage-class memory (SCM), we did a comprehensive study on the operation schemes of flash cells for fast Program/Erase (PE) cycling, high endurance and read stabilities. By adopting the channel-hot-electron injection (CHEI) and hot-hole injection (HHI) as PE schemes, as large as ~5V memory window (MW) can be realized and the endurance can reach ~10^6 PE cycles. As for 1-bit operation with sub-1V MW, the speed can be as fast as ~10ns and the endurance is over 10^9 PE cycles. Moreover, with the incremental-step-pulse-program (ISPP) method, 64 states (6-bit/cell) with high linearity and robust stabilities could be obtained. Our results demonstrate that there is a large flexibility to optimize flash cells as SCM, especially for the read-intensive applications and energy-efficient computing-in-memory (CiM) architectures.**

Keywords—*Flash memory, Storage-class memory, reliability, Computing-in-memory*

I. INTRODUCTION

Along with the dramatically increasing demands to deal with huge amounts of data, the strategies for high-efficient data processing systems attach much more attentions. A key point is to close the gap of capacity and speed between DRAM and flash memory by some novel concept memory products. In this background, storage class memory (SCM) was proposed as the next generation of memory and storage, and several non-volatile emerging memories have been well studied, including PCRAM, MRAM, RRAM, FRAM [1], etc. However, if taking the issues of economic, array capabilities and scaling vectors into account, there still exist a lot of challenges for emerging memories. For example, as the most suitable candidate to replace NAND flash memory, PCRAM-based 3D X-Point was a great success in technology, however, 3D X-point memory business was shut down after 7 years' development. Instead, over the years, flash memory continues dominating non-volatile memory (NVM) markets and is even accelerated after its turning from 2D to 3D-stacked structures for both NOR flash [2] and NAND flash [3]. Recently, an open industry-standard interconnect of "CXL" (compute express link) was proposed to provide high-efficient inter-connections between different memories [4]. So far, besides the reported ultra-high bit density of 28.5Gb/mm^2 and 3.2Gbps high-speed I/O rate in 280-layer 3D NAND [5], high-speed XL-Flash SCM was unveiled for ultra-high-end SSD (solid-state drives) [6]. All these new methods and products (CXL, high-density 3D NAND flash, high-performance XL-Flash, etc.) have demonstrated great potentials to provide flexible and feasible approaches to develop the solutions to construct high-efficient data processing systems.

In addition, Computing-in-memory (CiM) is another novel approach to achieve high-efficient data processing, and it has been intensively studied in non-volatile emerging memories [7-8]. In the viewpoint of processing stabilities and large-array capabilities, flash memory is a suitable candidate to construct large-array CiM architectures in general-purpose applications. In our previous work, the first NAND flash data-searching prototype was demonstrated [9] and a partial-differential-equation (PDE) solver with 32-bit floating point processing precision was proposed in flash-based CiM arrays [10-11]. Importantly, a novel and special point of flash CiM is that we can lower the MW and improve the speed and endurance to meet the requirements of CiM applications. However, so far, it is still unknown whether we can have further optimizations on flash cells to make it more suitable to design architectures than can satisfy the requirements of both SCM and CiM.

In this work, to provide CiM-compatible SCM solutions (Fig.1), the operation schemes of flash memory cell have been studied by adopting the methods of channel-hot-electron injection (CHEI) and hot-hole injection (HHI) for Program / Erase (PE). On the one hand, it is found that as large as ~5V memory window (MW) could be achieved by ~100ns CHEI program and the endurance can reach ~10^6 PE cycles; on the other hand, by minimizing MW to ~1V for 1-bit (Single-level Cell, SLC) processing, the speed can be improved to as high as ~10ns and over 10^9 PE cycles can be achieved. In addition, 64 states (6-bit/cell) with high linearity and robust stabilities can be obtained experimentally, which provides stable states that can be well utilized for flash-based CiM applications.

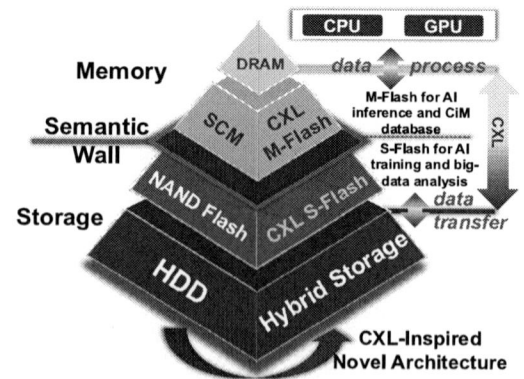

Fig. 1 AI-compatible pyramid architecture on the basis of versatile Flash memory technologies [12]. Memory-class Flash (M-Flash) and Storage-class Flash (S-Flash) are utilized for different purposes based on the system-required properties.

979-8-3503-6184-1/24 $31.00 © 2024 IEEE

II. CELL STRUCTURE AND CHARACTERIZATION METHODS

The TEM (Transmission Electron Microscope) figure of a flash cell and fundamental I-V curves are shown in Fig. 2. Normally, the threshold voltage (V_{th}) of a flash cell is adjusted via FN erasing and CHEI programming, which have been widely adopted in NOR flash memory. However, to satisfy the requirements of SCM, we choose to employ CHEI for programming while HHI for erasing, which can largely improve PE operation speed. Also, the separated word-line (WL) and bit-line (BL) in the memory array allow single-cell selectivity and individual PE operations. Also, the gate bias required in HHI operation is much lower than that to activate FN tunneling. The HHI scheme involves applying a positive voltage to the BL and a negative voltage to the WL, which promotes band-to-band tunneling (BTBT) and the generated electrons are collected at the drain side. Similarly, by applying the biases at the substrate bias of p-well and drain junction, the holes that flow towards the p-well can be injected into the floating gate (FG) under a high negative voltage of FG. During this process, holes injected into FG could recombine with the electrons stored in FG, thus lowering V_{th} value. HHI operation simplifies the peripheral circuit design by employing a zero bias at the substrate.

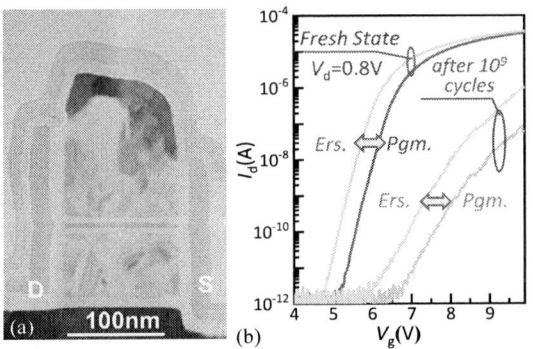

Fig.2. (a) TEM of a flash cell; (b) fundamental I_d-V_g curves. [13]

III. RESULTS AND DISCUSSIONS

By adopting CHEI-HHI scheme, the degradation can be mitigated compared to the conventional CHEI-FN scheme. In addition to the programming and erasing schemes, the trade-off between the memory window (MW) and endurance is characterized and shown in Fig.3. It is found that as large as 5V MW could be achieve with 10^6 PE cycling capability, indicating that our methods also provide a robust solution for the storage usage. Impressively, by designing a smaller MW, the flash cell can attain much higher endurance, exceeding 10^{10} cycles in 0.2V MW operations. This is enough for 1-bit/cell and 2-bit/cell operations in CiM applications. It should be noted, although there exists a decrease in endurance with a larger MW, we can still process 4-bit/cell operations (with a 1V MW) with a good endurance of 10^7 cycles. This is good enough for the large-scale neural networks because most on-chip trainings are still processed in the standard DRAM unit. The programming speed is also studied in Fig. 4. Utilizing CHEI method, as fast as 10ns programming speed is achieved for slight adjustments of V_{th}. In Fig.4(b), the HHI erasing speed is plotted, showcasing that V_{th} states can be finely tuned by HHI method.

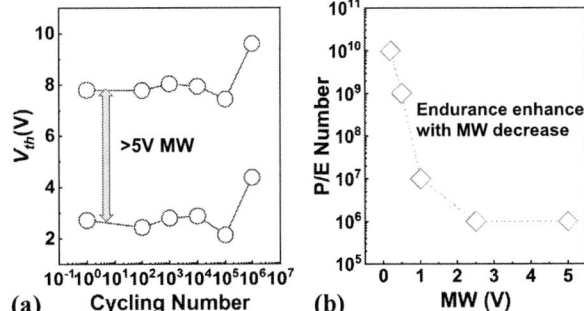

Fig.3. The endurance of cells programmed with (a) >5V MW and (b) the trade-off between MW and endurance (PE cycles).

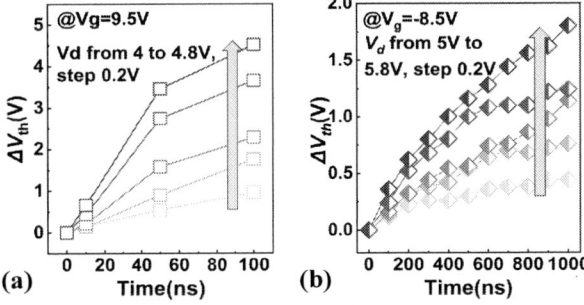

Fig.4. The programming and erasing speed of V_{th} modulations (absolute values), by (a) CHEI and (b) HHI at various drain biases.

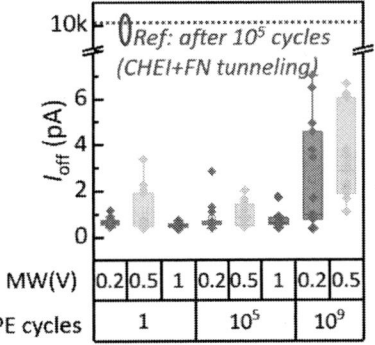

Fig.5. I_{off} of different MW and cycles compared with the traditional programming scheme, where in each box contains 15 flash memory cells. [13]

Next, leakage currents (I_{off}) after cycling are characterized to avoid deteriorated computing accuracy as the result of on/off ratio degradation. Although both the traditional FN tunneling and the HHI can degrade the sub-threshold swing (S.S.) [13], this is beneficial for precise device tunning in CiM, as it can enlarge MW between adjacent programming states, enabling more accurate weight-updating. So, the degradation of I_{off} seriously impacts computational performance due to the deteriorated on/off ratio. However, HHI method can suppress I_{off} to sub-10pA even after 10^9 cycles, as shown in Fig.5, which is much lower than the standard FN method. This can be understood because the traditional FN tunneling can cause serious degradation to gate dielectrics and the interface of flash cells, while HHI mainly degrades the interface.

979-8-3503-6184-1/24 $31.00 © 2024 IEEE

Benefiting from the fine-tuning capbility of HHI and the ultra-fast programming speed, the flash cell utilizing the proposed scheme could be tuned precisely. As shown in Fig.6, 64 states (6-bit/cell) with a high linearity and robust stabilities can be obtained precisely. Importantly, these tunned states are quite stable and it can be further adjusted by HHI method. Therefore, though the optimizations of cell structures and operation schemes, it is believed that well-controlled cell tunning could significantly broadens the potentials of flash-based CIM achitectures in various applications.

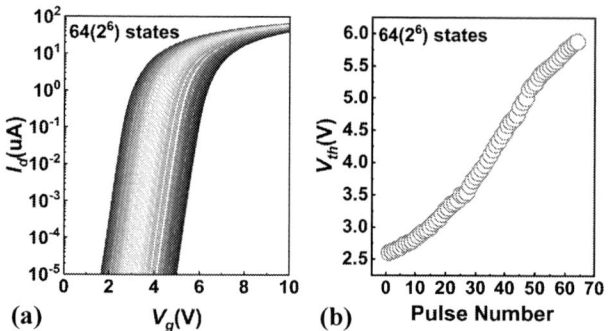

Fig.6. (a) The cells can be tuned into 64 distinct states for 6-bit/cell operation, using the proposed programming scheme of combined CHEI and HHI. (b) The adjusted V_{th} with varying pulse numbers.

IV. CONCLUSIONS

A comprehensive study on the optimizations of flash cell operations has been done in this work, aiming at fast PE operations, high endurance and robust read stabilities. By adopting CHEI and HHI methods, we plotted the trade-off between MW and the endurance. It is demonstrated that: 1) as large as ~5V MW can be realized with ~10^6 cycles; 2) lowering the MW to sub-1V help to achieve much higher endurance (10^9~10^{10} cycles); 3) 6-bit/cell operation with a high linearity and stabilities can be realized in flash cells. All these results indicate that the optimizations of flash memory can balance its properties and reliabilities, which can be used to construct high-speed memory/storage in SCM applications, as well as the design of CiM architectures to process general-purpose computing tasks.

ACKNOWLEDGMENT

This work was supported by China Key Research and Development Program under Grant (Nos. 2023YFB4402500, 2023YFB4402400), National Natural Science Foundation of China (Nos. 62034006, 92264201, U23B2040), and Natural Science Foundation of Shandong Province (ZR2023QF054, ZR2023LZH007, tsqn202306059).

REFERENCES

[1] M. A. Zidan, J. P. Strachan, and W. D. Lu, "The future of electronics based on memristive systems," *Nature Electron.*, vol. 1, no. 1, pp. 22–29, Jan. 2018.

[2] H.-T. Lue, et al., "3D AND: A 3D Stackable Flash Memory Architecture to Realize High-Density and Fast-Read 3D NOR Flash and Storage-Class Memory," IEEE International Electron Devices Meeting (IEDM), San Francisco, CA, USA, 2020, pp. 6.4.1-6.4.4.

[3] H. Tanaka et al., "Bit Cost Scalable Technology with Punch and Plug Process for Ultra High Density Flash Memory," IEEE Symposium on VLSI Technology, Kyoto, Japan, 2007, pp. 14-15.

[4] D. D. Sharma, "Compute Express Link," White Paper, Compute Express Link Consortium, Mar. 2019. https://docs.wixstatic.com/ugd/

[5] W. Jung et al., "13.3 A 280-Layer 1Tb 4b/cell 3D-NAND Flash Memory with a 28.5Gb/mm2 Areal Density and a 3.2GB/s High-Speed IO Rate," IEEE International Solid-State Circuits Conference (ISSCC), San Francisco, CA, USA, 2024, pp. 236-237.

[6] T. Shiozawa, et al., "Emerging Usage and Evaluation of Low Latency FLASH," 2020 IEEE International Memory Workshop (IMW), Dresden, Germany, 2020, pp. 1-4.

[7] A. Sebastian, et al, "Memory devices and applications for in-memory computing," Nature Nanotechnol., vol. 15, no. 7, pp. 529–544, July 2020.

[8] P. Yao et al., "Fully hardware -implemented memristor convolutional neural network," Nature, vol. 577, pp. 641–646, Jan. 2020.

[9] F. Wang, et al., "Implementation of Data Search in Multi-Level NAND Flash Memory by Complementary Storage Scheme," in IEEE Electron Device Letters, vol. 41, no. 8, pp. 1189-1192, Aug. 2020.

[10] Y. Feng, et al., "Design-Technology Co-Optimizations (DTCO) for General-Purpose Computing In-Memory Based on 55nm NOR Flash Technology," 2021 IEEE International Electron Devices Meeting (IEDM), San Francisco, CA, USA, 2021, pp. 12.1.1-12.1.4.

[11] Y. Feng, et al., "A Novel Array Programming Scheme for Large Matrix Processing in Flash-Based Computing-in-Memory (CIM) With Ultrahigh Bit Density," in IEEE Transactions on Electron Devices, vol. 70, no. 2, pp. 461-467, Feb. 2023.

[12] S. Ohshima, "Challenges & Opportunities of Flash-native Solutions Around Compute, Storage and AI," Flash Memory Summit (FMS), San Francisco, CA, USA, August, 2024.

[13] Y. Feng, et al., "Optimized operation scheme of flash-memory-based neural network online training with ultra-high endurance," in J. Semicond, vol. 45, no. 1, pp.012301, 2024.

Investigation of Reliability Characteristics of $Hf_xZr_{1-x}O_2$-Based FeFET and AFeFET Non-Volatile Memory

Min Liao[1,2], Xianzhou Shao[1,2], Junshuai Chai[1,2], Xiaoqing Sun[1,2], Xiaoyu Ke[1,2], Hao Xu[1,2], Jinjuan Xiang[3], Xiaolei Wang*[1,2], and Wenwu Wang[1,2]

[1] Key Laboratory of Fabrication Technologies for Integrated Circuits, Chinese Academy of Sciences, Beijing 100029.
[2] Institute of Microelectronics, Chinese Academy of Sciences, Beijing 100029, China.
[3] Beijing Superstring Academy of Memory Technology, Beijing 100176, China
* Email: liaomin@ime.ac.cn, wangxiaolei@ime.ac.cn

Abstract—In this work, we investigate the performance and reliability of $Hf_xZr_{1-x}O_2$-based field-effect transistors (FETs) with different gate stacks (FE, AFE, or DE/AFE). By introducing a top Al_2O_3 dielectric layer (DE), we demonstrate enhanced non-volatile memory performance and reliability of $Hf_{0.25}Zr_{0.75}O_2$-based AFeFET. Firstly, we investigate the memory window (MW) dependence on the write pulses. Secondly, we explore the reliability of the FeFET and AFeFET by monitoring the leakage evolution during cycling. The enhanced endurance with AFE is attributed to the suppression of gate degradation during electric field cycling. These findings contribute to understanding the devices with FE or AFE gate stacks.

Keywords—ferroelectric (FE), Antiferroelectric (AFE), reliability, non-volatile memory (NVM).

I. INTRODUCTION

Doped HfO_2 ferroelectric field-effect transistors (FeFETs) have attracted significant interest in non-volatile memory (NVM) applications due to their compatibility with CMOS processes [1]. Ferroelectricity can be induced by introducing different dopants (such as Si, Al, and Zr) into HfO_2 films [2]. Zr is a common dopant, by changing the Zr doping content, the ferroelectricity of the $Hf_{1-x}Zr_xO_2$ film can be changed. Increasing the Hf content makes the films paraelectric while increasing the Zr content induces antiferroelectric behavior [3]. Compared to ferroelectric (FE), antiferroelectric (AFE) materials offer several advantages including faster speed and higher endurance [4]. However, unlike ferroelectric (FE) materials, AFE does not exhibit remanent polarization (P_r) at zero fields and is therefore not suitable for use in NVMs. The characteristics of FeFET have been extensively studied [5], however, research on AFE-based devices is still lacking. Persic et al. demonstrated the introducing of built-in bias in the AFE stack using asymmetric electrodes with different work functions (WF) or by employing interface engineering approaches based on fixed charges or dipole formation [6]. These methods may provide feasible solutions for applying AFE materials to NVM applications.

In this work, the reliability of FE and AFE devices is discussed. The dependence of MW on the write pulse is quantified by using switching dynamics measurement. The reliability is compared by monitoring the gate current evolution during endurance degradation by direct current (DC) I_g-V_g. The underlying mechanism for the enhanced endurance of AFeFET is proposed. Additionally, the enhancement of Al_2O_3 on the performance and reliability of

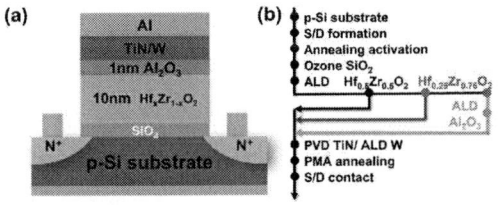

Fig. 1. (a) Schematic diagram of device structure. (b) process flow for the fabrication of the FE, AFE, and DE/AFE.

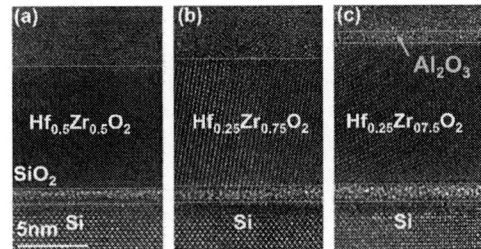

Fig. 2. TEM images for gate structure of fabricated devices.

AFeFETs is demonstrated and discussed.

II. DEVICE FABRICATION

Fig. 1(a)-(b) shows the schematic diagram and process flow of FeFET and AFeFET. The field-effect transistor (FET) was fabricated through a gate-last process. After source/drain formation, the gate stack was grown. First, a 0.7 nm SiO_2 interfacial layer was formed by ozone oxidation. 10 nm $Hf_{0.5}Zr_{0.5}O_2$ (FE) and $Hf_{0.25}Zr_{0.75}O_2$ (AFE) films were then deposited by atomic layer deposition (ALD) at 300 °C. The additional 1 nm Al_2O_3 dielectric layer was deposited on top of the AFE. After that, TiN and W were grown. The ferroelectric orthorhombic phase was then formed by annealing at 550 °C for 60 s in N_2 ambient. After this, the source/drain contacts were defined, and Al was used as the contact metal. Finally, forming gas annealing was performed in a 5% H_2/95% N_2 ambient at 550°C. We define these three devices as FE, AFE, and DE/AFE, respectively, with each FET having a gate length/gate width of 5/150 μm. Fig. 2 shows high-resolution TEM (HRTEM) images of the three device structures. We derived the final physical thicknesses of the FE, AFE, and DE/AFE devices to be 8.5 nm, 8.8 nm, and 9.7 nm, respectively.

We then investigated the switching dynamics of the three devices. The waveform of the switching dynamic

979-8-3503-6184-1/24 $31.00 © 2024 IEEE

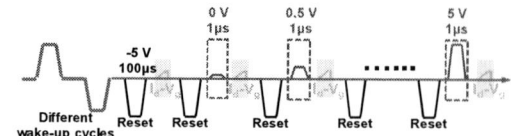

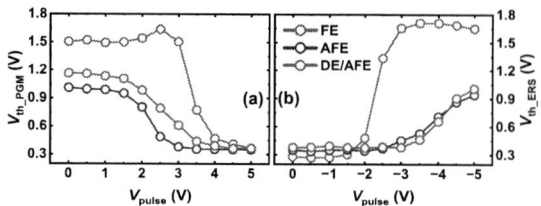

Fig. 3. The waveform diagram for switching dynamic measurement.

Fig. 4. (a) and (b) show the relationship between the V_{th} and V_{pulse} for the three devices.

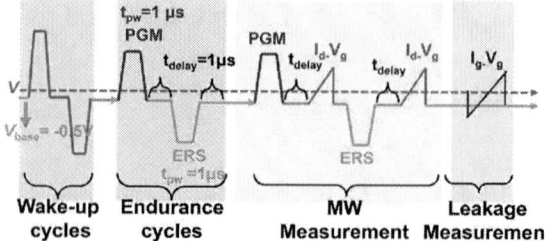

Fig. 5. Schematic of the waveform used for the MW measurement and leakage measurement for the devices.

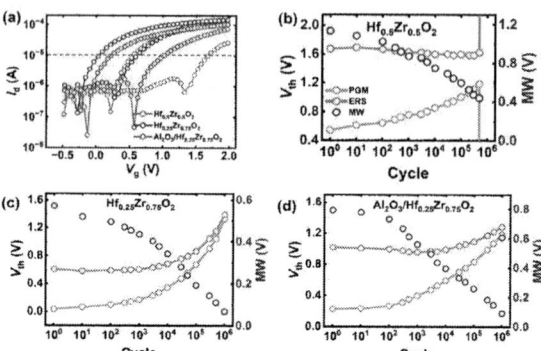

Fig. 6. Endurance characteristics of these three devices. (a) I_d-V_g curves of three devices. (b)-(d) V_{th} and MW of the three devices during cycling.

measurement is presented in Fig. 3. The SPGU module in Agilent Keysight B1500 is employed to generate voltage pulses for cycling during the wake-up and endurance processes. The B1530 WGFMU module is utilized for subsequent I_d-V_g tests. Different wake-up conditions are applied for these three devices. With bipolar electric cycling (E-cycling) wake-up conditions (±5.5V/100μs/rectangular wave/10cycles) for FE, (±6V/100μs/rectangular wave/100cycles) for AFE, and (±6.5V/100μs/rectangular wave/100cycles) for DE/AFE. All the devices are waked-up, and a reset voltage uses a rectangular pulse of +5 V/-5 V with a pulse width of 100 μs, while the write pulse width of pulse amplitude (V_{pulse}) is 1 μs. For FE, the base voltage (V_{base}) is 0 V, and for AFE and DE/AFE, V_{base} is -0.5 V. After the reset wave, continuously increasing write V_{pulse} is applied, and then the ramped I_d-V_g is used to monitor the threshold voltage (V_{th}), V_{th} is extracted by the constant current method ($I_d = 10$ μA).

Fig. 4 shows the corresponding V_{th} of three devices under different V_{pulse} conditions. The changes in V_{th} and V_{pulse} approximately conform to the sigmoid function. Fig. 4(a) demonstrates that for FE, V_{th_PGM} remains in a high threshold state when V_{pulse+} is below 3 V, but as V_{pulse+} exceeds 3 V, V_{th_PGM} begins to decrease. By reaching 4 V, V_{th_PGM} approaches the low threshold voltage of approximately 0.3 V. Similarly, in AFE and DE/AFE, V_{th_PGM} remains high when V_{pulse+} is below 2 V and 1.5V, respectively, but gradually decreases as V_{pulse+} exceeds these values. In Fig. 4(b), the changing trends of V_{pulse-} and V_{th_ERS} are depicted. FE can achieve a higher threshold voltage at -3 V, whereas AFE-based devices reach a higher threshold state at around 4.5V.

Based on the above analysis, the electrical conditions for endurance measurement were determined. For the FE, AFE, and DE/AFE a pulse amplitude of +4.5 V/-3.5 V, +3.5 V/-4.5 V, and +4.2 V/-4.5 V, respectively, the pulse width is 1 μs.

Fig. 5 shows the measurement of MW and leakage during cycling. During the endurance test, program (PGM) and erase (ERS) voltages were set respectively, and the pulse width was 1 μs, with rising and falling times of 20 ns. For FE, the I_d-V_g ramps up from 0V to 2 V with V_{base}=0 V, while for AFE and DE/AFE, the I_d-V_g ramps from -0.5 V to 2 V with V_{base}=-0.5 V. The leakage test procedure involves DC I_g-V_g, with the voltage range set from -2 V to 2 V.

Fig. 6 illustrates the endurance characteristics of the devices. In Fig. 6(a), the I_d-V_g curves of the three devices after

wake-up are presented. The MW of the three devices after wake-up are 1.1 V, 0.57 V, and 0.8 V respectively. Fig. 6(b)-(d) shows the endurance characteristics of the three devices. The MW of FE gradually decreases with cycling, reaching 0.6 V after 10^5 cycles, and the device breaks down after 5×10^5 cycles. For AFE and DE/AFE devices, there was no breakdown after 10^6 cycles. However, the MW is reduced to about 0.1 V. Based on these experiments, FE demonstrates superior MW characteristics compared to AFE. AFE and DE/AFE exhibit better breakdown characteristics. Moreover, the introduction of 1 nm Al_2O_3 on the top of AFE shows a 40% increase in MW after wake-up compared to AFE.

We then further explored the leakage characteristics of the devices during the cycling. Fig. 7(a)-(c) shows the leakage characteristics of the three devices. It is evident that for FE, the leakage current increases significantly with cycling, whereas for AFE devices, the leakage current remains stable within 10^5 cycles, with a slight increase observed beyond 10^6 cycles. Interestingly, for DE/AFE devices, the leakage current remains almost unchanged within the 10^6 cycles. Based on these leakage characteristics, we infer that, compared to FE devices, AFE devices experience lower degradation in gate leakage during cycling. This observation aligns with the endurance results presented in Fig. 6. Moreover, the incorporation of Al_2O_3 further reduces leakage and enhances breakdown characteristics.

III. SUMMARY

In this work, we investigate non-volatile memory (NVM) applications, focusing on FE, AFE, and DE/AFE devices. FeFETs based on $Hf_{0.5}Zr_{0.5}O_2$ exhibit a larger MW compared to $Hf_{0.25}Zr_{0.75}O_2$-based AFeFET, while AFeFET shows improved endurance and leakage characteristics. The introduction of a 1 nm Al_2O_3 layer on top of $Hf_{0.25}Zr_{0.75}O_2$

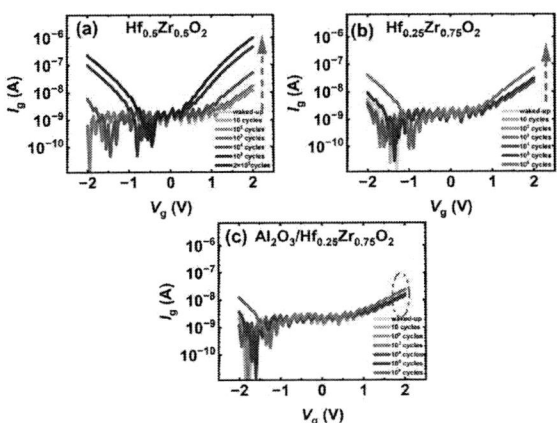

Fig. 7. Leakage characteristics of these three devices. (a)-(c) are FE, AFE, and DE/AFE-based FET.

enhances both the MW and endurance of AFE devices. The superior reliability of AFeEFT is attributed to lower gate degradation. This work provides insights into AFE-based devices for non-volatile storage.

ACKNOWLEDGMENT

This work was supported in part by the National Natural Science Foundation of China under Grant Nos. 92264104 and 52350195, in part by National Key Research and Development Program of China under Grant No. 2022YFB4400300, in part by R&D Program of Beijing Municipal Education Commission under Grant No.KZ202210009014, in part by the Young Elite Scientists Sponsorship Program under Grant No. BYESS2023033, and Supported by the Postdoctoral Fellowship Program of CPSF under Grant No. GZC20232925.

REFERENCES

[1] J. Müller, T. S. Böscke, S. Müller, E. Yurchuk, P. Polakowski, J. Paul, D. Martin, "Ferroelectric hafnium oxide: A CMOS-compatible and highly scalable approach to future ferroelectric memories," in *2013 IEEE International Electron Devices Meeting*, 9-11 Dec. 2013 2013, pp. 10.8.1-10.8.4, doi: 10.1109/IEDM.2013.6724605.

[2] M. Materano, P. D. Lomenzo, A. Kersch, M. H. Park, T. Mikolajick, and U. Schroeder, "Interplay between oxygen defects and dopants: effect on structure and performance of HfO$_2$-based ferroelectrics," *Inorganic Chemistry Frontiers*, vol. 8, no. 10, pp. 2650-2672, 2021, doi: 10.1039/d1qi00167a.

[3] Müller, Johannes, Böscke, Tim S, Schröder Uwe, Mueller, Stefan, Bräuhaus, Dennis, Böttger, Ulrich, "Ferroelectricity in Simple Binary ZrO$_2$ and HfO$_2$," *Nano Letters*, vol. 12, no. 8, pp. 4318-4323, 2012/08/08 2012, doi: 10.1021/nl302049k.

[4] X. Lyu, M. Si, X. Sun, M. Capano, H. Wang, and P. Ye, "Ferroelectric and anti-ferroelectric hafnium zirconium oxide: Scaling limit, switching speed and record high polarization density," in *2019 Symposium on VLSI Technology*, 2019: IEEE, pp. T44-T45, doi: 10.23919/VLSIT.2019.8776548.

[5] J. Chai, Hao Xu, Jinjuan Xiang, Yuanyuan Zhang, Shujing Zhao, Fengbin Tian, "Endurance Improvement of Si FeFET by a Fully CMOS-Compatible Process: Insertion of HfO$_x$ at Hf$_{0.5}$Zr$_{0.5}$O$_2$/SiO$_x$ Interface to Suppress Oxygen Vacancy Generation," *IEEE Transactions on Electron Devices*, vol. 69, no. 12, pp. 7156-7160, 2022, doi: 10.1109/ted.2022.3217997.

[6] M. Pešić, T. Li, V. Di Lecce, M. Hoffmann, M. Materano, C. Richter, "Built-in bias generation in anti-ferroelectric stacks: Methods and device applications," *IEEE Journal of the Electron Devices Society*, vol. 6, pp. 1019-1025, 2018, doi: 10.1109/JEDS.2018.2825360.

Deep Understanding of Charge Trapping Phenomenon in n-FeFET and Endurance Improvement by Interlayer Engineering

Saifei Dai[1,2,3], Hao Xu[1,2]*, Fengbin Tian[1,2], Xianzhou Shao[1,2,3], Xiaoqing Sun[1,2], Junshuai Chai[1,2], Xiaolei Wang[1,2]*, Wenwu Wang[1,2]

[1]Key Laboratory of Fabrication Technologies for Integrated Circuits, Chinese Academy of Sciences, Beijing, China
[2]Institute of Microelectronics, Chinese Academy of Sciences, Beijing, China
[3]University of Chinese Academy of Sciences, Beijing, China
*Email: xuhao@ime.ac.cn; wangxiaolei@ime.ac.cn

Abstract—**A comprehensive model is developed to study the charge trapping phenomenon in n-FeFET to pave the way for endurance improvement. The excess electron trapping and the coupling of ΔQ_t and ΔP_s are explained by trap occupancy. The MW degradation and endurance fatigue mechanism are explored, and new insight into endurance improvement is revealed by ascending the trap energy level (E_t). We also find that the N incorporation into the interlayer not only increases the dielectric constant but also can ascend the E_t and improve endurance.**

Index Terms—**Si-FeFET, Ferroelectric, Model, Memory window, Trapped charges**

I. INTRODUCTION

HfO$_2$-based ferroelectric FETs (FeFETs) have received significant attention as candidates for high-density, non-volatile, emerging memory applications due to their high speed, low voltage, non-destructive, scalability, and CMOS compatibility [1-2]. However, the main drawback lies in its poor endurance. The researchers found that endurance failure was closely related to charge trapping and trap generation [3-5]. The typical polarization charge of ferroelectric-HfO$_2$ (10-30 μC/cm^2), far exceeds the maximum charge density that can be supported by an insulator (e.g., SiO$_2$: 5 μC/cm^2) in metal-ferroelectric-insulator-semiconductor (MFIS) stack, leads to an excess electric field across the interlayer. This is considered as the reason for charge trapping at the ferroelectric (FE)/interlayer (IL) interface [6-8]. Due to the occurrence of charge trapping, most of the polarization in the FE is screened by the trapped charges. For example, the ferroelectric polarization (P_s) and charge distribution in the n-FeFET were monitored by the split quasi-static capacitance-voltage (QSCV) method and the trapped charge density (Q_t) was about 10^{14} cm^{-2} at gate voltage of 3.5 V, and most of the trapped charges remain after removing the gate voltage [9]. The severe charge trapping especially the excess electrons trapping was reported by other works [3, 10-11].ΔQ_t and ΔP_s between program (PGM) and erase (ERS) states were extracted by one-time current measurements and a fixed coupling between ΔP_s and ΔQ_t (~90%) was observed at different pulse conditions. The increase in $\Delta Q_t/\Delta P_s$ ratio was considered to be the dominant contributor to FeFET endurance [11-12]. However, the physical origin behind the excess electrons trapping ($\Delta Q_t/\Delta P_s > 1$) at $V_g = V_{PRG}$ (generally > 3 V) and the fixed coupling ratio of $\Delta Q_t/\Delta P_s$ at $V_g = 0$ V is still unclear. The guidelines for endurance improvement are not comprehensively understood yet.

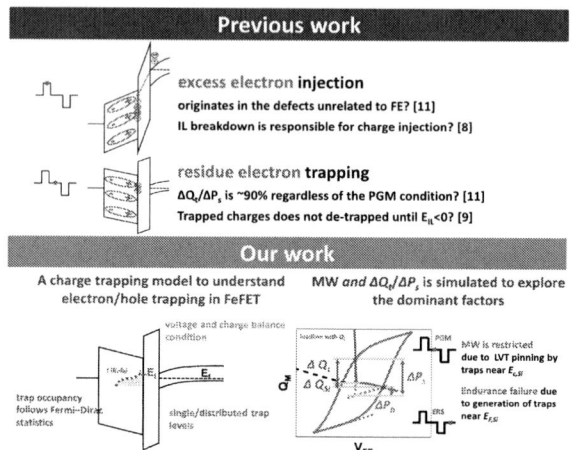

Fig. 1 An electron/hole trapping model considering trap occupancy is developed to understand the trapping phenomenon in n-FeFET. Furthermore, performance and endurance are discussed by simulating memory window (MW) and $\Delta Q_t/\Delta P_s$. The ERS state and PGM state could be determined by intersecting the loadline with Q_m-V_{FE} minor loop, as highlighted in the figure. Thus, ΔQ_t and ΔP_s could be obtained by defining the trapped charges and spontaneous polarization charge change between the ERS state and PGM state. The HVT and LVT are threshold voltage (V_{th}) for ERS state and PGM state, respectively. The V_{th} is extracted at surface potential $\psi_s = 2\phi_B$ [13].

Figure 1 compares our work with the previous ones [8,9,11]. In this work, a charge trapping model considering trap occupancy is developed to understand the trapping phenomenon in n-FeFET. The MW degradation and endurance mechanism are discussed by simulating the coupling ratio between ΔQ_t and ΔP_s. This model is also used to investigate interlayer optimization by combining experimental demonstration to improve endurance.

II. CHARGE TRAPPING MODEL OF FeFET OPERATION

A. Charge trapping model

The charge trapping model is developed by introducing trap occupancy into the classic FeFET operation model, as described in **Fig. 2(a)**. The classic model has been used for studying MW and the depolarization field by setting a fixed charge density [5, 13]. In this work, the trapped charge is assumed to be located at the FE/IL interface. The trap occupancy is assumed to comply with Fermi-Dirac statistics and the Fermi level at the FE/IL interface equals to Fermi level of the substrate in the stationary state [14]. Thus, the Q_t could be determined self-consistently at a given V_g. The MOS loadline including charge trapping shows an abrupt increase in charge density due to trap filling (blue line in **Fig. 2(b)**).

979-8-3503-6184-1/24 $31.00 © 2024 IEEE

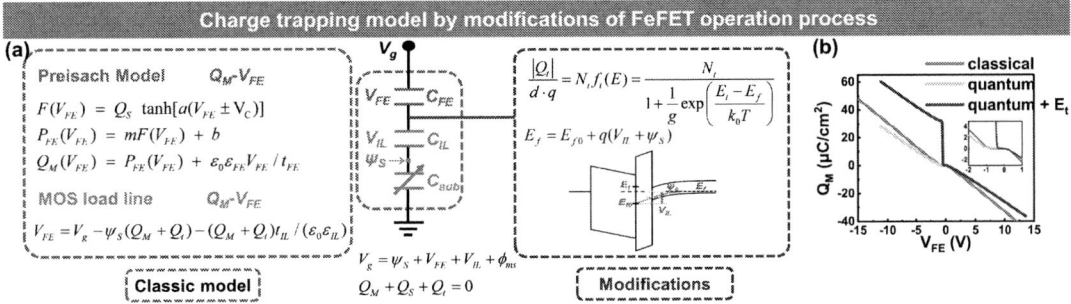

Fig. 2 (a) Simulation framework of charge trapping model by modifications of classic FeFET operation model. In the stationary state, the trap occupancy complies with Fermi-Dirac statistics. The thickness of interface (d) is considered as 0.2 nm. The degeneracy factor (g) is set to 2 for donor and 4 for acceptor, respectively. E_{f0} is the Fermi-level at FE/DE interface at flat-band condition. (b) Simulated MOS loadline with/without traps near $E_{c,Si}$. The charges in the MOS system are simulated by solving Schroedinger's and Poisson's equations self-consistently with the Fermi-Dirac distribution [15] (the classical simulated result is also given by the orange line). Charge trapping leads to a sharp increase in Q_m (in the blue line), and V_{MOS} is pinned.

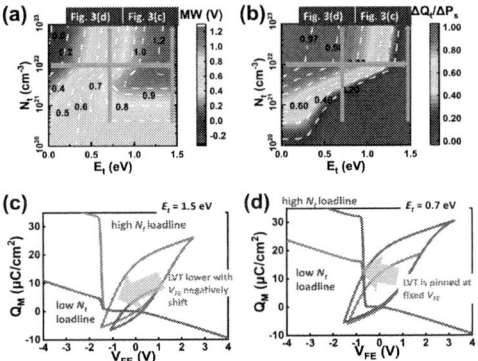

Fig. 3 The simulated dependence of N_t and E_t on (a) MW and (b) $\Delta Q_t/\Delta P_s$. (c) and (d) show the Q_m-V_{FE} plot for simulations of MW and $\Delta Q_t/\Delta P_s$ with high-level E_t (1.5 eV) and low-level E_t (0.7 eV), respectively. Simulation parameters: ε_{FE}=30, P_r=20 μC/cm^2, P_s=23 μC/cm^2, E_c=1.5 MV/cm, t_{FE}=10 nm, t_{IL}=1 nm.

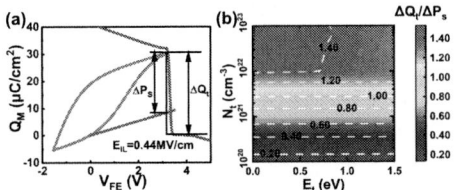

Fig. 4 (a) The Q_m-V_{FE} plot for V_g=4V. (a) The simulated dependence of N_t and E_t on $\Delta Q_t/\Delta P_s$. Excess Q_t injection is achieved due to large N_t.

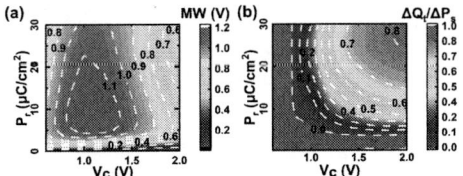

Fig. 5 The simulated dependence of remanent polarization and coercive voltage on (a) MW and (b) $\Delta Q_t/\Delta P_s$.

B. Simulation results

Firstly, the charge trapping model is used to investigate the dependence of trap density (N_t) and trap energy level (E_t) on MW and $\Delta Q_t/\Delta P_s$. The simulation results are shown in **Fig. 3(a)-(b)**. **Fig. 3(c)-(d)** shows the Q_m-V_{FE} plot for simulations with a high-level E_t (1.5 eV) and low-level E_t (0.7 eV, which means E_t near the Fermi-level of the substrate at the flat-band condition), respectively. For E_t = 1.5 eV, as N_t increases, the P_s tends to be saturated, thus the MW increases, while the ratio of $\Delta Q_t/\Delta P_s$ is always ~0. For E_t = 0.7 eV, as N_t increases, the PGM state (after PGM pulse) is pinned at fixed V_{FE} due to the trap filling process, as shown in **Fig. 3(d)**. Consequently, the MW maintains ~0.7 V, and the ratio of $\Delta Q_t/\Delta P_s$ increases from 0 to 90%. Besides, for N_t = 10^{22} cm^{-3} (horizontal arrow in **Fig. 3(a)-(b)**), as E_t ascends, most Q_t is detrapped when V_g sweeps back to 0 V, thus the MW becomes larger and the ratio of $\Delta Q_t/\Delta P_s$ decreases. From these results, the Q_t detrapping always happens only if V_g decreases due to Fermi-level lowering. And the $\Delta Q_t/\Delta P_s$ at V_g = 0 V is expected to be a fixed value in the stationary state, regardless of PGM pulse conditions. The root cause for this phenomenon is the high density of N_t and the low-level E_t distribution of the trap. And the LVT state is pinned at a fixed intersection point between the loadline and saturated hysteresis loop, leading to a fixed ratio of $\Delta Q_t/\Delta P_s$.

Secondly, the excess electron trapping is explained by the simulation. **Figure 4(a)** shows the Q_m-V_{FE} plot for V_g = 4 V. A ΔQ_t larger than ΔP_s is obtained. This phenomenon could be understood as electron injection into the traps, and the trapping process is determined by the Fermi-level at the FE/IL interface rather than P_s. In addition, due to the trap-filling process, the electric field in the interlayer (E_{IL}) is pinned at a fixed value much lower than the breakdown limitation. In **Fig. 4(b)**, $\Delta Q_t/\Delta P_s$ is expected larger than 100% when N_t > 3 × 10^{21} cm^{-3}.

Thirdly, the endurance mechanism is discussed. (i) Trap generation (N_t increases) with the low-level E_t is the dominant factor for endurance. Based on the above simulation results in **Fig. 3(a)-(b)**, trap generation with the low-level E_t will make the $\Delta Q_t/\Delta P_s$ ratio increase and the MW reduce to 0 V. However, trap generation with high-level E_t will only lead to a fixed MW and $\Delta Q_t/\Delta P_s$ ratio before the breakdown occurs. (ii) The P_s fatigue will lead to a sharp endurance failure when P_s is lower than 5 μC/cm^2. As shown in **Fig. 5(a)-(b)**, for a given coercive voltage with charge trapping (N_t = 10^{22} cm^{-3}, E_t = 0.7 eV), the MW is nearly unchanged as the P_s decreases when P_s > 5 μC/cm^2. After P_s degradation below 5 μC/cm^2, the MW sharply decreases to 0 V. Meanwhile, the $\Delta Q_t/\Delta P_s$ ratio is always decreased.

The above discussions indicate that the root cause for endurance failure of FeFET is trap generation with low-level E_t and P_s fatigue. However, considering that the increase of $\Delta Q_t/\Delta P_s$ ratio is reported [11-12], trap generation with low-level E_t is the dominant factor for endurance failure.

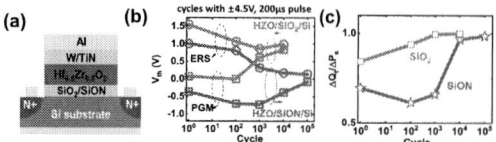

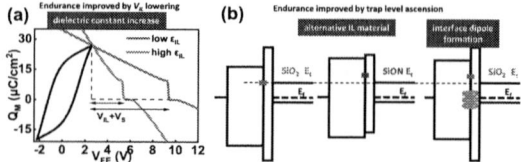

Fig. 6 (a) Schematic diagram of the Si FeFET gate structure of different interlayer (SiO₂ or SiON). (b) HVT and LVT evolution during endurance cycling. (c) $\Delta Q_t/\Delta P_s$ ratio evolution during cycling.

Fig. 8 Guidelines for endurance improvement in n-FeFET by interlayer engineering. (a) The increasing dielectric constant is an effective method to reduce E_{IL}. (b) By ascending E_t, the charge trapping could be suppressed significantly.

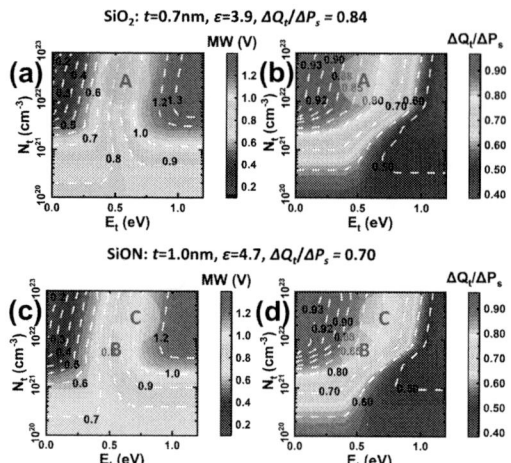

Fig. 7 The MW and $\Delta Q_t/\Delta P_s$ simulation results for samples. (a) and (b) SiO₂, (c) and (d) SiON. Here, in addition to the electron traps, in order to match the experimental results, another trap with $E_t = -0.2$ eV and $N_t = 10^{21}$ cm⁻³ was introduced in both samples to account for a small amount of hole trapping. The difference between measured MW and simulated MW is possible due to read-induced electron trapping.

III. EXPERIMENTAL RESULTS FOR DIFFERENT INTERLAYERS OF SiO₂ OR SiON

The interlayer optimization is a potential method for reliability improvement. The nitridation of interlayer provides a better endurance owing to the advantage of increasing the dielectric constant [16-18]. Here, we further investigate the N incorporated interlayer in FeFET with our developed model by combining experimental results.

The gate structure and process flow are schematically shown in **Fig. 6(a)**. The detailed experiments can be found in [18]. The endurance of the SiON sample is one order larger than the SiO₂ sample, as shown in **Fig. 6(b)**. The ΔQ_t and ΔP_s are measured based on our previous work [19], and the $\Delta Q_t/\Delta P_s$ ratio during endurance fatigue is shown in **Fig. 6(c)**. The $\Delta Q_t/\Delta P_s$ ratio increases to 100% during cycling, but with different initial ratios (~84% for SiO₂ and ~70% for SiON).

The comprehensive understanding of the endurance mechanism for both samples is investigated by simulations. The MW and $\Delta Q_t/\Delta P_s$ simulation results for SiO₂ and SiON samples are shown in **Fig. 7(b)-(e)**. The corresponding region to the experimental results (MW~0.9 V and $\Delta Q_t/\Delta P_s$ ratio~84% for SiO₂ and 70% for SiON) are highlighted as A for SiO₂ and B/C for SiON, respectively. Comparing the results of regions A and B, less N_t is expected for the SiON sample due to the smaller $\Delta Q_t/\Delta P_s$ ratio. The reason could be explained by the previous work as the suppression of trap generation and charge trapping due to lower E_{IL} [18].

Meanwhile, a new insight into endurance improvement is revealed by comparing with regions A and C, and a higher E_t is expected for the SiON sample. According to our previous work, the trap levels of the oxygen vacancy (V₀) trap are distributed in the mid-gap of Si, and the trap level in the Hf₀.₅Zr₀.₅O₂/SiON/Si structure is further away from the valence band maximum (VBM) of Si than that in Hf₀.₅Zr₀.₅O₂/SiO₂/Si structure [20, 21]. These results indicate that V₀ traps play a key role in the endurance failure of FeFET, and the introduction of N can ascend the trap level of V₀ away from the VBM of Si. This provides a possible origin for the E_t difference and reveals a potential direction for FeFET endurance improvement.

The guidelines for performance boost and endurance improvement in n-FeFET by interlayer engineering are summarized in **Fig. 8**. (i) Increasing the dielectric constant is still an effective method to reduce E_{IL}. Thus, the trap generation could be suppressed and endurance is expected to improve. (ii) The endurance will be improved by ascending the trap energy level. This could be accomplished by alternative IL materials or dipole layer formation between FE and IL.

IV. CONCLUSION

In summary, we investigate the charge trapping phenomenon in n-FeFET by developing a simulation framework considering trap occupancy. The excess electron trapping and the coupling of ΔQ_t and ΔP_s are comprehensively understood. The MW is a trade-off of competing mechanisms between P_s enhancement and Q_t increase. The dominant factor for endurance failure is trap generation with low-level E_t. Guidelines for interlayer engineering of n-FeFET are provided.

ACKNOWLEDGMENT

This work was supported in part by the National Natural Science Foundation of China under Grant Nos. 92264104 and 52350195, in part by National Key Research and Development Program of China under Grant No. 2022YFB4400300, in part by R&D Program of Beijing Municipal Education Commission under Grant No.KZ202210009014, in part by the Young Elite Scientists Sponsorship Program under Grant No. BYESS2023033, and supported by the Postdoctoral Fellowship Program of CPSF under Grant No. GZC20232925.

REFERENCES

[1] Dünkel, S. *et al.*, *IEDM*, 2017, pp. 19.7.1-19.7.4. [2] Beyer, S. *et al.*, *IMW*,2020, pp. 1-4. [3] Yurchuk, E. *et al.*, *TED*, 2016, pp. 3501-3507. [4] Gong, N. *et al.*, *EDL*, 2018, pp. 15-18. [5] Ni, K. *et al.*, *TED*, 2018, pp. 2461-2469. [6] Muller, J. *et al.*, *NVMTS*, 2016, pp. 1-7. [7] Deng, S. *et al.*, *EDL*, 2020, pp. 1348-1351. [8] Si, M. *et al.*, *TED*, 2021, pp. 5108-5113. [9] Toprasertpong, K. *et al.*, *IEDM*, 2019, pp. 23.27.21-23.27.24. [10] Zhou, H. *et al.*, *IEDM*, 2020, pp. 18.6.1-18.6.4. [11] Ichihara, R. et al., *VLSI*, 1-2, 2020. [12] Ichihara, R. *et al.*, *IEDM*, 2021, 6.3.1-6.3.4. [13] Wang, X. *et al.*, *TED*, 2020, pp.4500-4506. [14] Synopsys TCAD User manual, Sentaurus™ Device User Guide, 2019, pp. 467-498. [15] http://www-device.eecs.berkeley.edu/qmcv. [16] Ali, T. *et al.*, *TED*, 2018, pp. 3769-3774. [17] Tan, A. J. *et al.*, *EDL*, 2021, pp. 994-997. [18] Tian, F. *et al.*, *TED*, 2021, pp. 5872-5878. [19] Zhao, S. *et al.*, *TED*, 2021, pp. 1561-1567. [20] J. Chai *et al.*, *JAP.*, 2022, pp.105301. [21] F. Tian *et al.*, *TED*, 2024, pp.1040-1047.

An FPGA-based Dual-mode SSD for Device-side Performance Optimization

Xingyu Chen[1], Sirui Peng[1], Hankun Lv[1], Zhangbin Yang[2,3,4], Daixiao Peng[4], Xi Cai[4], Xueguang Lian[4], Yong Ding[5], Xiaoyong Xue*[1]

[1] State Key Laboratory of Integrated Chips and Systems, School of Microelectronics, Fudan University, Shanghai 201203, China
[2]University of Chinese Academy of Sciences, Beijing 100049, China
[3]Institute of Electrical Engineering, Chinese Academy of Sciences, Beijing 100190, China
[4]China Three Gorges Construction Engineering Corporation, Chengdu 610000, China
[5]College of Integrated Circuits, Zhejiang University, Hangzhou 310000, China,

* Email: xuexiaoyong@fudan.edu.cn

Abstract—This paper proposes a Dual-mode Solid State Drive(DM-SSD) based on FPGA that can switch modes between a traditional SSD and an open-channel SSD (OCSSD) according to users' needs. The DM-SSD is constructed through hardware and software co-design. The hardware design of the DM-SSD is mainly responsible for command fetch and data transmission while the software part is for command parsing and scheduling. Both SSD modes use the same hardware design. and users can switch the mode of SSD by flashing the firmware. Based on the traditional SSD system, the Open-Channel SSD features are added to the embedded software according to the Open-Channel SSD Specification 2.0. We have evaluated and analyzed the performance of the two SSD modes separately and comparatively.

Keywords—Solid state drive, open-channel SSD, embedded software design.

I. INTRODUCTION

Flash-based Solid-State Drives (SSDs) have gradually become the main form of storage in the field of consumer electronics and data centers in recent years due to its high performance, low power consumption and shock resistance, which make them an ideal replacement for Hard Disk Drives (HDDs) [1]. But the HDD has completely different characteristics including "one-dimensional" address space and in-place overwrites. SSDs need to adopt the block address interface with these characteristics for compatibility, which results in a mismatch between the host interface and NAND flash. To address this, SSDs introduce the Flash Translation Layer (FTL), which mainly plays the roles of address translation, garbage collection, wear leveling, bad block management and power-down recovery [2]. However, the functions of address translation and garbage collection, as well as the unpredictability of the I/O requests, together lead to several significant shortcomings of traditional SSDs [3]: (1) reduced lifetime and degraded performance duo to write amplification [4-5], (2) "tail-latency" problem from the unpredictability of requests [6-7] and (3) cost increasing caused by address mapping tables and over-provisioning.

Many solutions have been proposed to address these shortcomings. One of the more mainstream approaches is the Open-Channel SSD (OCSSD) which exposes structural characteristics of the NAND flash to the host and transfers all or part of the work of the FTL from the SSD to the host [8]. However, in some cases, this may produce undesirable results like software incompatibility, host resources occupation and I/O bandwidth reduction of the host interface [9].

To leverage the advantages of both traditional SSD and OCSSD, we propose a dual-mode SSD (DM-SSD) that allows users to select the mode of SSD between traditional SSD and OCSSD according to their needs by flashing the firmware of the embedded CPU. Therefore, users can choose the mode freely according to different applications and scenarios to maximize the performance of the DM-SSD.

The key contributions are summarized as follows:

- We propose an SSD controller architecture enabling the switch traditional SSD and OCSSD.

- We design the Non-Volatile Memory Express (NVMe) controller and Nand Flash controller along with a corresponding embedded software system based on FPGA.

- We evaluate the performance of DM-SSD's both modes in a real-world test platform.

The rest of the paper is organized as follows. We discuss the hardware design of DM-SSD in Section II. In Section III, we discuss the embedded software design. Section IV presents the implementation and evaluation results of the DM-SSD. Finally, we conclude the paper in Section V.

II. HARDWARE DESIGN

A. Overall Architecture of the SSD controller

The SSD controller includes NVMe Controller, NAND Flash Controller and Microblaze embedded CPU. The controller connects to the NAND Flash submodule through the FPGA Mezzanine Card (FMC) and completes the NVMe command interaction with the host through the Peripheral Component Interconnect Express (PCIe) interface. The hardware part is mainly responsible for command fetch and data transmission, which is irrelevant with any detailed command and address space implementation. Thus, it could enable mode switch. The overall architecture of the SSD controller is shown in Figure 1.

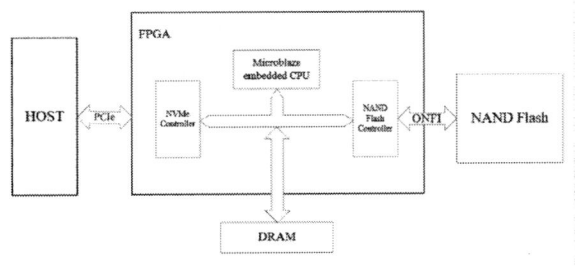

Fig. 1. Overall architecture of SSD controller

This work was supported in part by the National Key R&D Program under Grant 2023YFB4404700, in part by the National Natural Science Foundation of China under Grant 62274038, in part by the Science and Technology Commission of Shanghai Municipality under Grant 21TS1401200 and Grant 22ZR1407100, in part by the research project funding by China Three Gorges Construction Engineering Corporation （No.JGJD0323003) and China Three Gorges corporation (No.202203269).

979-8-3503-6184-1/24 $31.00 © 2024 IEEE

B. NVMe Controller Hardware Design

The NVMe controller use XDMA IP to interact with host memory data. XDMA IP provides four types of interfaces to the users: (1) the AXI-Lite interface is used to maintain registers between NVMe and the host. (2) the AXI-MM interface is for data and commands transmission. (3) the XDMA configuration interface enables embedded software to configure the XDMA and start PCIe transfers. (4) the interrupt interface sends interrupt requests to the host. Considering the above interface and protocol requirements, the following four modules are set up in the NVMe logic: IO logic, Admin logic, NVMe host-side registers, and embedded system PS-side registers. NVMe logical structure and overall system framework is shown in Figure 2.

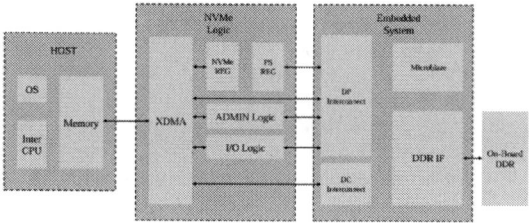

Fig. 2. NVMe logical structure and overall system framework.

Both the admin logic and the I/O logic implement the following six parts: the SQ command capture module, the SQ data capture module, the CQ message submission module, the CQ data submission module, the H2C descriptor submission module, and the C2H descriptor submission module. In addition, the I/O logic has a PRP chain table capture module separately. The NVMe host-side registers are configured according to the NVMe Specification and mapped to PCIe BAR space.

C. NAND Flash Controller Hardware Design

The NAND Flash Controller (NFC) uses NV-DDR1 (ONFI2.2) protocol to connect NAND Flash. The NFC hardware is divided into three layers: command classification layer, timing generation layer, and physical interface layer. The command classification layer supports seven commands: Reset (Async), Set Feature (Async), Program Page (Sync), Read Page (Sync), Get Feature (Sync), Erase Block (Sync) and Read Status (Sync). We introduce the workflow using the Read Page state as an example. The operation can be divided into six processes: sending the first Command (00h), sending the flash address of the data to be read (C1, C2, R1, R2, R3), sending the second Command (30h), waiting for the RB signal to be pulled low, waiting for the RB signal to be pulled high and the data readout section. The workflow of the read page is shown in Figure 3.

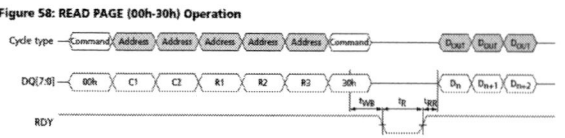

Fig. 3. The workflow of the read page.

The timing generation layer supports five command segments: Command (Async), Data output (Async), Command (Sync), Data output (Sync), Data input (Sync). The physical interface layer contains the following modules:

cross-clock domain module, data phasing logic, data detection logic and physical interface module for input, output and inout ports.

III. EMBEDDED SOFTWARE DESIGN

A. NVMe Software Design

The NVMe software use state machine to schedule the operations of NVMe controller with 4 states: DISABLE, ENABLE, RUNNING, and SHUT DOWN. It supports most of admin commands and all mandatory IO commands required by NVMe specification 1.2.

The processing flow of admin commands is mainly related to SQ opcodes, which is used to determine whether handling data through DMA is required.

The key part of the IO command is the Physical Region Page (PRP) mechanism, which is used for data transfer between the host and the SSD. Each PRP Entry describes the starting address of a contiguous segment of physical memory. If we need more than 2 PRP Entries, we need to use a PRP List, which is consisting of the PRP Entry.

B. Traditional SSD Software Design

The traditional SSD software is responsible for processing request data and controlling the behavior of the NFC to make the SSD functional as a logic block device. The traditional SSD software Consists of NVMe controller, FTL Layer and scheduler. The overall architecture is shown in Figure 4.

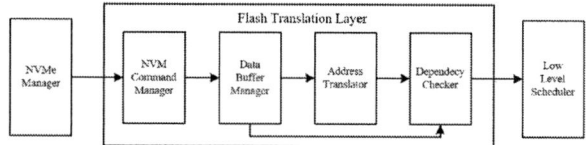

Fig. 4. Traditional SSD embedded software architecture

The FTL is used to perform request segmentation, data buffer management, address translation and dependency checking. The FTL will process the request from the host and generate a series of subsequent operations and finally drive the NVMe control and NFC to complete host requests. The Low-level scheduler determines the final execution order of operations generated by the FTL module. Low-level scheduler flowchart is shown in Figure 5.

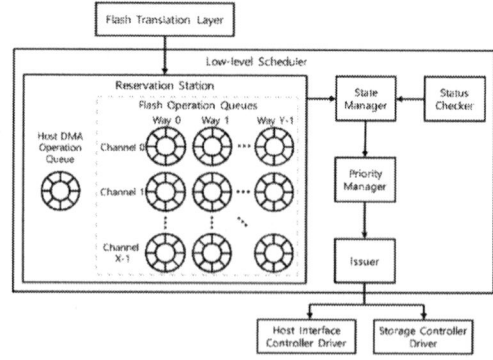

Fig. 5. Low-level scheduler flowchart

C. OCSSD Software Design

This design is based on Open-Channel SDD Specification 2.0. The embedded software design of the OCSSD is modified from the embedded software of the traditional SSD with the addition of Open-Channel features including Geometry to report all the hierarchies of address space and Get Log Page for host to acquire state of all chunks. Due to the requirement of host FTL (pblk), the software adds additional vector operations for read, write and reset. OCSSD software also supports the logic to maintain all chunks' metadata and check read and write constraints when processing IO commands.

D. Dual-Mode Design

All the hardware-software interface is shared between traditional SSD software and OCSSD software and all the hardware design is irrelevant with any detailed command and address space implementation. Thus, the embedded software can be changed without any cost.

Users can use the firmware update feature supported by NVMe specification and transfer different firmware image to the SSD to realize the switch between traditional SSD and OCSSD according to different requirements.

IV. IMPLEMENTATION AND EVALUATION

We have implemented DM-SSD using Xilinx KCU105 FPGA board and two Micron SLC NAND chips. The FPGA board connects to the host with PCIe 3.0 X8 interface and NAND chips operate in NV-DDR1 Mode 5.

The test platforms for DM-SSD are shown in Figure 6. We run FIO tests direct on Linux NVMe driver for traditional SSD and on pblk, which is the kernel-based OCSSD host-side FTL, for OCSSD.

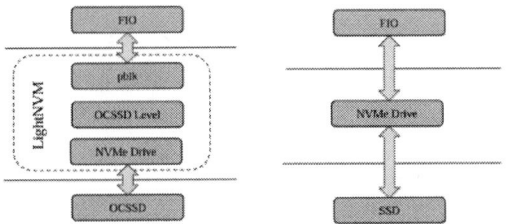

Fig. 6. OCSSD test platform (left) and SSD test platform (right)

Based on the above test platform, we used time-based tests driven by FIO to test the read and write performance of conventional SSD design and OCSSD design respectively. We tested the performance of bandwidth and average latency at different I/O depths and block sizes (bs). The specific test parameters and performance is shown in table 1 and Figure 7. Our experimental results show that OCSSD improves bandwidth by an average of 1.50×/8.11× and its average latency is 0.91×/0.09×compared to traditional SSD in write

and read performance respectively. Users are allowed to freely choose the mode of SSD according to their needs and ecosystem, thus maximizing the performance of the DM-SSD.

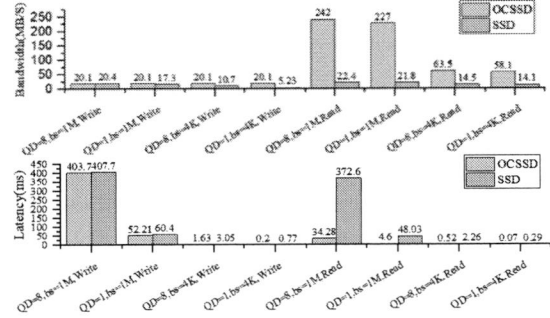

Fig. 7. Bandwidth performance(up) and latency performance(down)

V. SUMMARY

In this paper, we propose an DM-SSD architecture which enables the user to freely choose between two modes, traditional SSD and OCSSD. OCSSD has better bandwidth performance than SSD but is less compatible for historical reasons. OCSSD and traditional SSD share the same hardware system, and we implemented the software systems of OCSSD and traditional SSD separately, which makes users can fully utilize the performance of DM-SSD according to their needs.

REFERENCES

[1] C. Dirik and B. Jacob, "The Performance of PC Solid-State Disks (SSDs) as a Function of Bandwidth, Concurrency, Device Architecture, and System Organization,"in ISCA, PP. 279–289, June 2009.

[2] S.-P. Lim, S.-W. Lee, and B. Moon, "FASTer FTL for Enterprise-Class FlashMemory SSDs," in SNAPI, 2010, PP. 3-12.

[3] Bjørling M, Aghayev A, Holmberg H, et al. "ZNS: Avoiding the Block Interface Tax for Flash-based SSDs". 2021 USENIX Annual Technical Conference (USENIX ATC 21), 2021: 689-703.

[4] Lu Y, Shu J, Zheng W. "Extending the lifetime of flash-based storage through reducing write amplification from file systems". 11th USENIX Conference on File and Storage Technologies (FAST 13), 2013: 257-270.

[5] Yang J, Plasson N, Gillis G, et al. "Don't Stack Your Log On My Log". 2nd Workshop on Interactions of NVM/Flash with Operating Systems and Workloads (INFLOW 14), 2014.

[6] Dean J, Barroso L "A. The tail at scale. Communications of the ACM", 2013, 56(2): 74-80.

[7] González J, Bjørling M, Lee S, et al. "Application-driven flash translation layers on open-channel SSDs". Proceedings of the 7th non Volatile Memory Workshop (NVMW), 2016: 1-2.

[8] González J, Bjørling M. "Multi-tenant I/O isolation with open-channel SSDs". Nonvolatile Memory Workshop (NVMW), 2017.

[9] Zhang X, Zhu F, Li S, et al. "Optimizing Performance for Open-Channel SSDs in Cloud Storage System". 2021 IEEE International Parallel and Distributed Processing Symposium (IPDPS), 2021: 902-91

A Simulation Comparison of Channel-All-Around and Gate-All-Around 3D Vertical Structure FeFET with IGZO Channel

Xuebin Wang[1], Zhijian Guo[1], Yutao Li[1], Chengji Jin[5], Jixuan Wu[4], Guanhua Yang[3], Yuanxiao Ma[1], Masaharu Kobayashi[2], Fei Mo[1]*, Yeliang Wang[1]*

[1] School of Integrated Circuits and Electronics, Beijing Institute of Technology, Beijing, China.
[2] Institute of Industrial Science, The University of Tokyo, Tokyo, Japan.
[3] State Key Lab of Fabrication Technologies for Integrated Circuits, Institute of Microelectronics, Chinese Academy of Sciences, Beijing, China.
[4] School of Information Science and Engineering, Shandong University, Qingdao, China.
[5] Hangzhou Institute of Technology, Xidian University, Hangzhou, China.

* Email: mofei@bit.edu.cn, yeliang.wang@bit.edu.cn,

Abstract—We have compared the memory performance of vertical structure InGaZnO (IGZO) channel ferroelectric field effect transistors (FETs) with channel-all-around and gate-all-around structures by 3D TCAD simulation for high-density application. The memory window (MW), on current (I_{on}), and subthreshold swing (SS) are systematically studied and discussed in terms of ferroelectric film thickness, channel thickness, and diameter. It reveals that the channel-all-around structure has a larger MW and higher I_{on} due to the overlap region between the gate and drain/source based on this simulation, while the memory performance of the gate-all-around structure can be improved through shrinking the length of the underlap region.

Keywords—*In-Ga-Zn-O (IGZO) channel, vertical structure, ferroelectric field effect transistor (FeFET), TCAD simulation, channel-all-around, gate-all-around*

I. INTRODUCTION

In a highly digital and information-oriented society, more and more data has been generated, collected, and utilized for AI and IoT applications. Although 3D structure NAND memories have high density, their high power consumption and low operation speed become the bottlenecks. To load and store a tremendous amount of data, it is important to develop memories with higher density, high operation speed, and low power consumption.

In recent years, the HfO_2-based FET has attracted significant attention for its good CMOS process compatibility [1]. A 3D vertical HfO_2-FeFET has been demonstrated with a poly-Si channel [2]. However, poly-Si channel-based FeFETs suffer from a high thermal budget and low mobility. Particularly, a low-k interfacial layer is unavoidably formed between the ferroelectric layer and the poly-Si channel during the RTA process, leading to reliability degradation. Oxide semiconductor such as IGZO is a promising channel material for monolithic 3D structure memories thanks to its back-end-of-line (BEOL) process, high mobility, and nearly zero low-k interfacial layer between the channel and the ferroelectric layer [3-6]. The gate-all-around and channel-all-around vertical structure oxide semiconductor channel FETs have been successfully demonstrated [7,8]. Although, the memory operation of the gate-all-around (GAA) structure IGZO channel FeFET has been studied using simplified structures [9,10]. The memory operation of the channel-all-around (CAA) structure IGZO channel FeFET is not clear yet. It is

necessary to give a comprehensive study of CAA and GAA structures in erase operation with a more realistic structure.

In this work, we investigate the memory characteristics of CAA and GAA structure IGZO channel FeFETs by 3D TCAD simulation for 3D monolithic integration. The MW, I_{on}, and SS are compared regarding ferroelectric film thickness, diameter, and underlap length.

II. DEVICE STRUCTURE AND SIMULATION METHOD

The GAA and CAA 3D structure IGZO channel FeFETs are built as shown in Fig. 1 (a) and (b), Fig. 1 (c) and (d) illustrate the cross-sectional view of GAA and CAA structure, respectively.

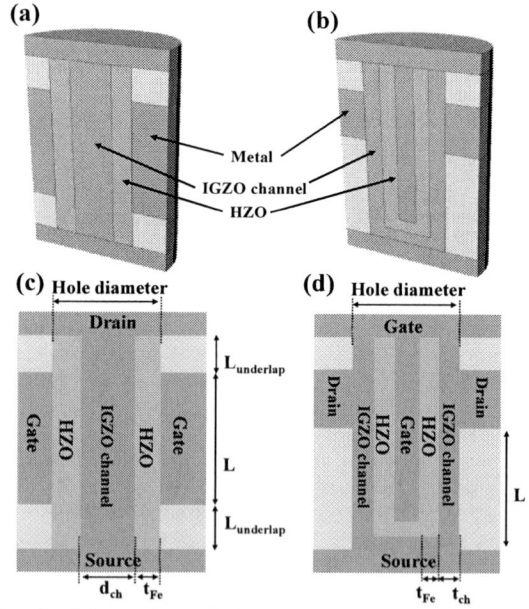

Fig. 1. Schematics of the device structures of IGZO channel FeFET (a) GAA structure and (b) CAA structure, cross-sectional view of the (c) GAA structure and (d) CAA structure.

The GAA and CAA IGZO channel FeFET are junctionless transistors with constant mobility (= 10 $cm^2/V \cdot S$) IGZO channel. The bandgap (E_g) and carrier concentration of the

IGZO channel are 3.2 eV and 10^{18} cm^{-3}. The remanent polarization charge density (P_r) of simulated HfZrO (HZO) material is 23 μC/cm, and the coercive field is 1.16 MV/cm. The GAA and CAA structures are designed with the same hole diameter and HZO thickness for fair comparison, as shown in Fig. 1. The IGZO channel thickness of the CAA structure is fixed at 10 nm. In contrast, the nanowire diameter (d_{IGZO}) of the GAA structure is dependent on the HZO thickness. The gate length of both structures is 40 nm. The default parameters are shown in Tab. 1.

Default Parameters	GAA Structure	CAA Structure
Hole Diameter	60 nm	60 nm
HZO Thickness	10 nm	10 nm
Length of Gate	40 nm	40 nm
Channel Thickness	-	10nm
Length of underlap	20 nm	-

Table 1. The default parameters for simulation.

III. RESULTS AND DISCUSSION

First, we compared GAA and CAA structures with default parameters. The simulated results of GAA and CAA structures of IGZO FeFET with 3 V bidirectional DC sweep gate voltage are shown in Fig. 2. Because the GAA structure has an underlap region, which causes a considerable channel resistance, the I_{on} of the GAA structure is lower and V_{th} of GAA structure is higher than CAA structure. In addition, since the large bandgap, a few minority carrier holes are generated during the erase operation in the IGZO channel. Polarization switching hardly occurs due to the body's potential floating. In the case of the CAA structure, benefiting from the overlap region, the electric field from the drain and source concentrates near the gate. The electric field can facilitate the switching of polarization. Thus, a large MW is observed because more polarization switching occurs during the erase operation. In the case of the GAA structure, the underlap region makes the gate far from the drain and source, and the electric field from the drain and source is sparse near the gate. It is difficult to induce polarization switching of the HZO layer in the gate region. Thus, the MW of the GAA structure is smaller than the CAA structure.

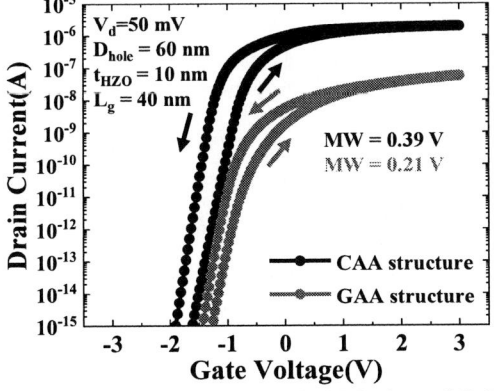

Fig. 2. Simulated I_d-V_g curves of IGZO channel FeFET with CAA and GAA structures using default parameters.

A. Effect of ferroelectric film thickness variation

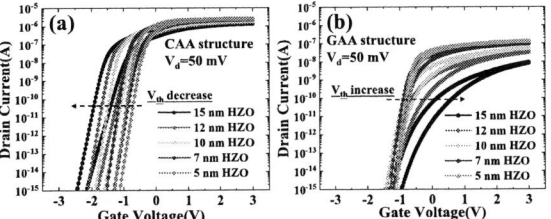

Fig. 3. The simulated I_d-V_g curves of (a) the CAA structure and (b) the GAA structure IGZO channel FeFET, with varying HZO thickness.

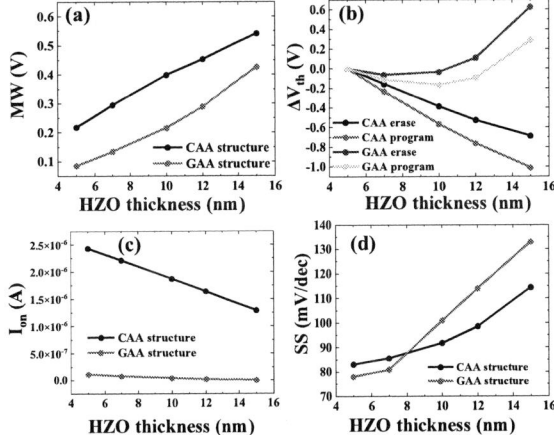

Fig. 4. The extracted (a) MW, (b) ΔV_{th} at the program and erase state, (c) I_{on} and (d) SS of CAA and GAA structure IGZO channel FeFETs.

The effects of ferroelectric film thickness are systematically investigated for CAA and GAA structures. The simulation results of I_d-V_g curves are shown in Fig. 3. The MWs, ΔV_{th} at the program and erase state, I_{on}, and SS are plotted against HZO thickness in Fig. 4. The MW of CAA structure is larger than the GAA structure, thanks to the overlap region as shown in Fig. 4 (a). The MW of both structures increases with increasing HZO thickness because the coercive voltage of HZO film is proportional to thickness, a thicker HZO film leads to a large coercive voltage, inducing a larger MW. In Fig. 4 (b), as HZO thickness increases, the ΔV_{th} of the CAA structure decreases because of a smaller gate capacitance. It notes that the ΔV_{th} of the CAA structure follows different trends. The ΔV_{th} decreases first and then increases, this is because the HZO thickness causes the IGZO channel diameter to decrease with constant hole diameter. At thin HZO film, gate capacitance is the primary factor for V_{th}. As HZO thickness increases, the ΔV_{th} decreases. At thick HZO film, the channel diameter is the primary factor for V_{th}. As HZO thickness increases, channel diameter decreases, and the ΔV_{th} increases. Fig. 4 (c) shows the I_{on} of the two structures decreasing with increasing HZO thickness because of the reduced IGZO channel diameter. The extracted SS at reverse V_g sweep is described in Fig. 4 (d). The two structures have comparable SS in thin HZO film, but the GAA structure has a larger SS in thick HZO film because of a more significant channel resistance in the underlap region.

B. Effect of Hole diameter variation

Next, we simulated the effect of hole diameter variation, as shown in Fig. 5. The hole diameter is reduced from 60 nm to 48 nm without changing HZO thickness. The MW of GAA and CAA structures increase with hole diameter decreasing because of electric field concentration in narrow hole diameter. The V_{th} of the CAA structure is slightly dependent on the hole diameter because of the unchanged HZO and channel thickness. However, the V_{th} of the GAA structure obviously shifts in a positive direction, because the channel diameter decreases caused by the hole diameter decreasing. Moreover, the narrow channel induces a larger resistance in the underlap region, leading to a low I_{on}.

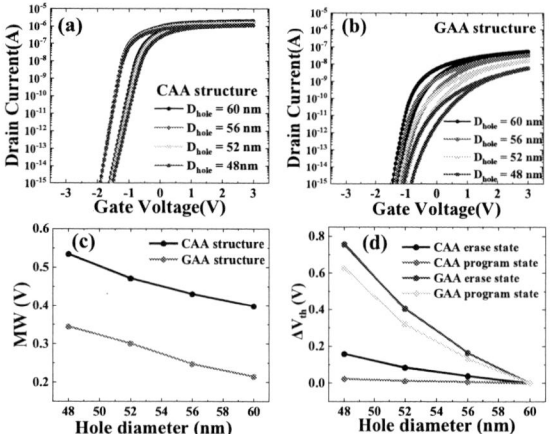

Fig. 5. Simulated I_d-V_g curves of IGZO channel FeFET with (a) CAA and (b) GAA structures with hole diameter variation, (c) extracted MW, and (d) ΔV_{th} at program and erase state.

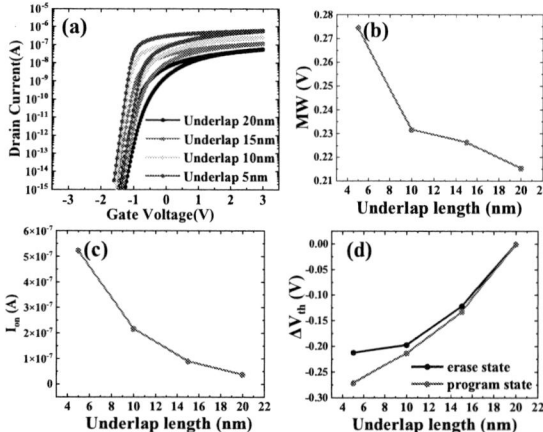

Fig. 6. Simulated (a) I_d-V_g curves of GAA structure IGZO channel FeFET with underlap length variation, extracted (b) MW, (c) I_{on}, and (d) ΔV_{th} at program and erase state.

C. Underlap region dependence of GAA structure

Finally, we studied the memory characteristics of GAA structures by reducing the length of the underlap region. The

I_d-V_g curves with different underlap lengths and extracted parameters are shown in Fig. 6. Due to the drain and source being closer to the gate, a large MW can be obtained by reducing the underlap length. Thanks to reduced extra resistance in the underlap region, the I_{on} becomes higher. However, the MW and I_{on} are still smaller than the CAA structure.

IV. CONCLUSION

In this work, we compared the memory characteristics of GAA and CAA structure IGZO FeFET by 3D TCAD. From simulation results, CAA structure IGZO FeFET shows a larger MW, higher I_{on}, and lower SS than GAA structures. It found that the underlap region of the GAA structure significantly affects memory performance. Although reducing the length of the underlap region can improve the GAA structure performance, unfortunately, the memory performance is still worse than the CAA structure. CAA structure IGZO channel FeFET shows a more promising candidate for 3D monolithic integration.

REFERENCES

[1] J. Müller et al., "Ferroelectricity in HfO2 enables nonvolatile data storage in 28 nm HKMG," 2012 Symposium on VLSI Technology (VLSIT), Honolulu, HI, USA, 2012, pp. 25-26.

[2] K. Florent et al., "Vertical Ferroelectric HfO2 FET based on 3-D NAND Architecture: Towards Dense Low-Power Memory," 2018 IEEE International Electron Devices Meeting (IEDM), San Francisco, CA, USA, 2018, pp. 2.5.1-2.5.4.

[3] F. Mo et al., "Experimental Demonstration of Ferroelectric HfO2 FET with Ultrathin-body IGZO for High-Density and Low-Power Memory Application," 2019 Symposium on VLSI Technology, Kyoto, Japan, 2019, pp. T42-T43.

[4] M. Zeng et al., "First Demonstration of Annealing-Free Top Gate La:HZO-IGZO FeFET with Record Memory Window and Endurance," 2023 International Electron Devices Meeting (IEDM), San Francisco, CA, USA, 2023, pp. 1-4.

[5] Z. Chen et al., "Improved MW of IGZO-channel FeFET by Reading Scheme Optimization and Interfacial Engineering," 2023 IEEE International Memory Workshop (IMW), Monterey, CA, USA, 2023, pp. 1-4.

[6] S. Hooda et al., "BEOL Compatible Extremely Scaled Bilayer ITO/IGZO Channel FET with High Mobility 106 cm2/V.s," 2023 7th IEEE Electron Devices Technology & Manufacturing Conference (EDTM), Seoul, Korea, Republic of, 2023, pp. 1-4.

[7] H. Fujiwara et al., "Surrounding Gate Vertical-Channel FET With a Gate Length of 40 nm Using BEOL-Compatible High-Thermal-Tolerance In-Al-Zn Oxide Channel," in IEEE Transactions on Electron Devices, vol. 67, no. 12, pp. 5329-5335, Dec. 2020.

[8] X. Duan et al., "Novel Vertical Channel-All-Around(CAA) IGZO FETs for 2T0C DRAM with High Density beyond 4F2 by Monolithic Stacking," 2021 IEEE International Electron Devices Meeting (IEDM), San Francisco, CA, USA, 2021, pp. 10.5.1-10.5.4.

[9] F. Mo et al., "A simulation study on memory characteristics of InGaZnO-channel ferroelectric FETs with 2D planar and 3D structures", Japanese Journal of Applied Physics, 61, SC1013 (2022), February 9, 2022.

[10] Z. Du et al., "Simulation for the Feasibility of IGZO Channel in 3D Vertical FeFET Memory Based on TCAD," 2023 IEEE International Conference on Integrated Circuits, Technologies and Applications (ICTA), Hefei, China, 2023, pp. 51-52.

Low Operating Voltage in HfO$_2$/ZrO$_2$ Superlattice Ferroelectric Capacitor Achieved by Thickness Scaling

Dongya Li [1,3], Huan Liu [1,2,3,*], Peiyuan Du[1], Fei Yu[1,3], Chengji Jin[1,2], Xiao Yu[1,3,*], Yan Liu[1], Genquan Han[1,3], Yue Hao[1]

[1] State Key Discipline Laboratory of Wide Band Gap Semiconductor Technology, Xidian University, Xi'an, 710071, China
[2] Research Center for Intelligent Chips, Zhejiang Lab, Hangzhou 311121, China
[3] Hangzhou Institute of Technology, Xidian University, Hangzhou 311231, China

* Corresponding Author's Email: Xin_pzh@163.com, yuxiao@ieee.org;

Abstract—In this study, the influence of thickness in HfO$_2$/ZrO$_2$ superlattice (SL_HZO) ferroelectric capacitors with 1 nm TiO$_2$ seed layer has been explored. Thanks to the relatively unchanged E_c (coercive field), as the SL_HZO film thickness decreases to 6 nm, the operating voltage is reduced to 0.8 V for one-shot operation, while maintaining a remnant polarization (P_r) above 6.2 µC/cm^2. Moreover, a $2P_r$ value of 20.4 µC/cm^2 at an operating voltage of 1 V for stable memory operation has been achieved. These voltages are within the range compatible with scaled silicon technologies.

Keywords—HfO$_2$-ZrO$_2$ superlattice, ferroelectric film, Scaling down, low operating voltage, TiO$_2$ seed layer.

I. Introduction

Since the discovery of ferroelectricity in Si-doped hafnium in 2011, hafnium-based thin films have been extensively researched for use in various emerging devices. Such as ferroelectric random access memory (FeRAM), ferroelectric field-effect transistors (FeFET), and ferroelectric tunnel junctions (FTJ). The widespread interest is due to their compatibility with complementary metal-oxide-semiconductor (CMOS) technologies, excellent scalability, and fast read and write operations [1-3]. However, for memory devices, low power consumption has become the primary focus. One of the crucial challenges is reducing their high operating voltage to satisfy the requirements for embedded FeRAM. Simply lowering the operating voltage may lead to reliability issues related to endurance and retention characteristics due to smaller electric field cycles, which is detrimental to data security in memory devices [4]. In addition, HfO$_2$-doped ferroelectric films demonstrate that the coercive field (E_c) remains constant regardless of the film thickness. This consistency is attributed to the minimal impact of the interface layer on the switching characteristics [5]. As a result, it is anticipated that by further reducing the film thickness, low-voltage operation of HfO$_2$-based FeRAM devices can be achieved [3]. To further optimize the characteristics of ferroelectric films, superlattices offer a promising alternative approach. Superlattices are periodic structures composed of two or more materials with layer thicknesses in the nanometer range [6].

In this work, we investigated the ferroelectric properties of HfO$_2$/ZrO$_2$ superlattice (SL_HZO) films with different thickness scales at various annealing temperatures.

Compared to a 10 nm SL_HZO film, slightly decreasing P_r with decreasing thickness to 6 nm is due to the smaller grain size of SL_HZO. The coercive voltage (V_c) also decreased with the reduction in thickness. The operating voltage of 6 nm SL_HZO is reduced to 0.8 V ($2P_r > 10$ µC/cm^2). This work will provide a reference for low-power applications in future non-volatile memory implementation.

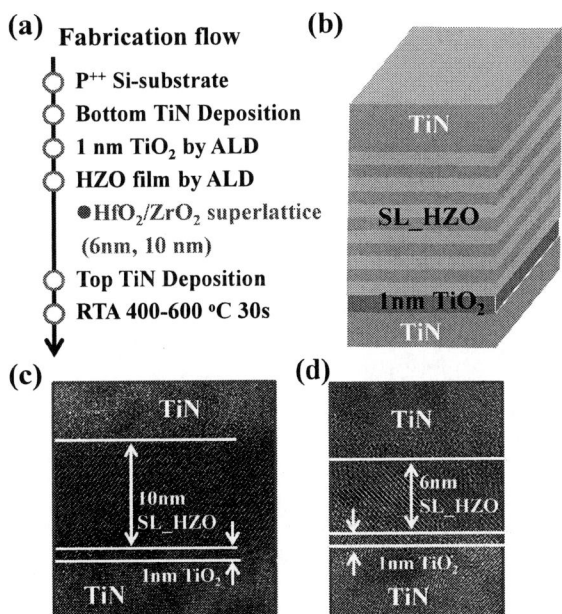

Figure 1. (a) Key fabrication flow for TiN/TiO$_2$/SL_HZO/TiN ferroelectric capacitors with different thicknesses of SL_HZO. (b) Schematic of SL_HZO device structure. The TEM image of MFM capacitors for SL_HZO with thickness of (c) 10 nm, and (d) 6 nm.

II. Device fabrication

TiN/TiO$_2$/SL_HZO/TiN ferroelectric capacitors were fabricated with the thickness scaling of SL_HZO from 10 nm down to 6 nm. Figure. 1 (a) shows the key fabrication process of devices. First, 100 nm TiN was deposited on the heavily doped p-type Si substrate by reactive sputtering after surface cleaning. Then 1 nm TiO$_2$ was deposited as the seed layer, and the HfO$_2$/ZrO$_2$ superlattice ferroelectric film with the thickness of 6, and 10 nm were deposited by atomic layer deposition system at 250 °C, respectively. Each HfO$_2$/ZrO$_2$ superlattice stack consisted of 8 cycles of HfO$_2$ and 8 cycles of ZrO_2. $TiCl_4$, TEMAHf, TEMAZr, and H_2O were used as the Ti, Hf, Zr, and oxygen sources, respectively. 40 nm TiN top electrode was formed by sputter and wet-etch. Finally, the samples were treated by post-metal annealing (PMA) at 400 to 600 °C for 30 s in a nitrogen atmosphere, respectively.

Figure 1 (b) shows the schematic of HfO_2/ZrO_2 superlattice MFM structures. The TEM images of SL_HZO with the thickness of 10 nm and 6 nm are shown in Figures 1 (c-d). Among them, the 10 nm and 6 nm SL_HZO samples exhibited good ferroelectric crystallization. In this study, the polarization-voltage (*P-V*) characteristics were measured with TF 3000. The capacitance-voltage (*C-V*) test was performed by the Keithley 4200 parameter analyzer.

III. Results and Discussion

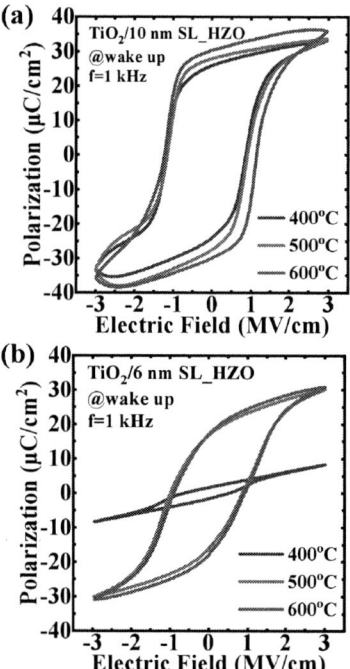

Figure 2. *P-V* loops of (a) 10 nm and (b) 6 nm SL_HZO devices annealed at 400-600 °C after being woken up.

Figure 2 illustrates the *P-V* characteristics of 10 nm and 6 nm SL_HZO devices with different annealing temperatures measured at 3 MV/cm. The period taken in the *P-V* measurement was fixed to 1 kHz. Wake-up pulses of 10000 cycles were applied before all measurements. The 10 nm SL_HZO devices after the wake-up process exhibited good ferroelectric characteristics with a wide range of annealing temperatures (400-600 °C). Compared to the stable ferroelectric hysteresis observed in the 6 nm SL_HZO annealed at 500 °C. Ferroelectric capacitors with 6 nm SL_HZO annealed at 400 °C exhibit weak ferroelectric properties ($1~\mu C/cm^2 < 2P_r < 10~\mu C/cm^2$). The reason why thinner SL_HZO films require higher annealing temperatures to exhibit ferroelectricity will be discussed in detail later. For SL_HZO devices annealed at 500 °C, the $2P_r$ of 10 nm SL_HZO and 6 nm SL_HZO is 55.6 and 33.1 $\mu C/cm^2$, respectively. The slightly decreasing P_r with decreasing thickness to 6 nm may be due to the smaller grain size of SL_HZO. Additionally, the E_c of 10 nm SL_HZO and 6 nm

SL_HZO is 1.0 MV/cm and 0.98 MV/cm, respectively. There is almost no change in the coercive field as the SL_HZO thickness decreases. This is consistent with previous reports in solid solution HZO film [5]. The constant E_c, independent of the HZO thickness, can be attributed to the polarization switching kinetics in thin HZO.

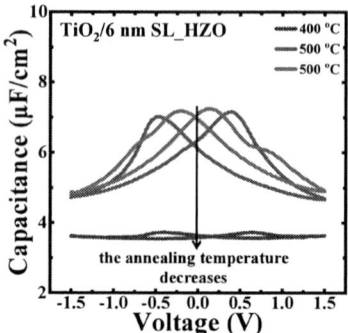

Figure 3. *C-V* characteristic of 6 nm SL_HZO device annealed at 400-600 °C, The relative dielectric constant decreases as the annealing temperature decreases.

Figure 3 illustrates the capacitance-voltage (*C-V*) characteristics of 6 nm HZO annealed at temperatures ranging from 400 to 600 °C. The relative dielectric constant decreases with decreasing annealing temperature. Specifically, a relatively low dielectric constant was observed in the sample annealed at 400°C, which corresponds well with the results shown in Figure 2. For the 6 nm SL_HZO device annealed at 400 °C, the insufficient annealing temperature results in a predominantly amorphous phase, with only a small portion crystallizing into the ferroelectric phase. This phenomenon has also been reported for HfO_2 and ZrO_2 films, where a larger portion of the amorphous phase is found in thinner films [7]. This can be attributed to the increased surface energy in thinner films, making crystallization more difficult as the thickness of ferroelectric film decreases [5]. As the thickness of SL_HZO decreases, the required annealing temperature for crystallization increases.

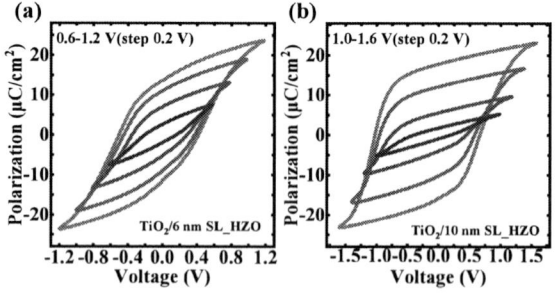

Figure 4. (a) The *P-V* loops of 6 nm SL_HZO device with different sweeping voltage (0.6-1.2 V) and (b) 10 nm SL_HZO device with different sweeping voltage (1.0-1.6 V).

Reducing the operating voltage, in addition to scaling dimensions, is crucial for decreasing power consumption. To further investigate the operating voltage characteristics, *P-V* tests were conducted on 10 nm and 6 nm SL_HZO devices with a small voltage sweep. Figure 4 shows the *P-V* loops of 6 nm SL_HZO at 0.6-1.2 V and 10 nm SL_HZO at 1.0-1.6 V after wake-up, with a step size of 0.2 V. Clearly, the 6 nm HZO exhibited larger remnant polarization at lower voltages. the coercive voltage Vc decreases with decreasing SL_HZO thickness. At ± 1 V, the 10 nm SL_HZO device does not exhibit $2P_r$ above 5 μC/cm². However, the 6 nm SL_HZO device achieves a high $2P_r$ of 20.4 μC/cm². The low operating voltage achieved in thin SL_HZO film is due to the unchanged Ec. The extracted $2P_r$ and E_c as a function of voltage are shown in Figure 5. For the 10 nm SL_HZO, a voltage of ± 1.2 V was required to achieve a $2P_r$ over 10 μC/cm², but the $2P_r$ reached 33.7 μC/cm² at 1.6 V with E_c of 0.76 MV/cm. The operating voltage of the 6 nm SL_HZO device can be further reduced. Ferroelectric hysteresis with $2P_r$ greater than 10 μC/cm² can still be observed even at an operating voltage as low as 0.8 V ($E_c = 0.44$ MV/cm), which is sufficiently low to be compatible with the operating voltage of the most advanced nodes in current logic technology. The excellent operating voltage performance is confirmed based on both 6 nm and 10 nm SL_HZO devices.

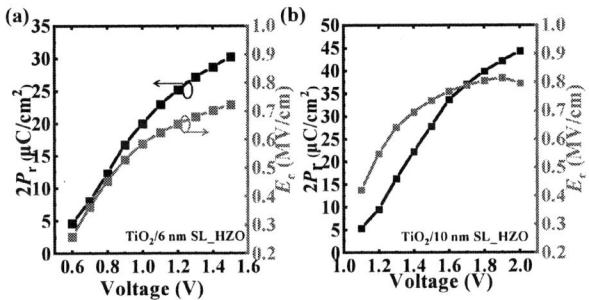

Figure 5. The extracted $2P_r$ and E_c versus voltage for (a) 6 nm SL_HZO device from 0.6 V to 1.5 V and (b) 10 nm SL_HZO device from 1.1 V to 2.0 V, respectively.

IV. Summary

The impact of the thickness of SL_HZO on the ferroelectricity of TiN/TiO₂/SL_HZO/TiN devices has been investigated. The thinner SL_HZO films require higher annealing temperatures to exhibit ferroelectricity. The lower operating voltage can be obtained by the thickness of SL_HZO scaling down to 6 nm. The operating voltage is reduced to 0.8 V for $2P_r$ above 10 μC/cm², and 1 V for $2P_r$ value of 20.4 μC/cm². This work will serve as a reference for low-power applications in future non-volatile memory implementations.

ACKNOWLEDGMENT

This work was supported by the National Natural Science Foundation of China (Grant No. 62204226, 62374151, 62025402), the National Key Research and Development Project (Grant No. 2023YFB4402301).

REFERENCES

[1] T. S. Boscke, J. Muller, D. Brauhaus, U. Schroder, and U. Bottger, IEEE International Electron Devices Meeting (IEDM), pp. 24.5.1-24.5.4 (2011).

[2] F. Huang, Y. Wang, X. Liang, Y. Zhang, X. Yuan, Z. Wang, B. Peng, L. Deng, Q. Liu, L. Bi, and M. Liu, IEEE Electron Device Letters (EDL), pp. 330–333 (2017).

[3] T. S. Boscke, J. Muller, D. Brauhaus, U. Schroder, and U. Bottger, IEEE International Electron Devices Meeting (IEDM), pp. 24.5.1-24.5.4 (2011).

[4] K. Tahara, K. Toprasertpong, Y. Hikosaka, K. Nakamura, H. Saito, M. Takenaka, and S. Takagi, IEEE Symposium on VLSI Technology and Circuits (VLSI Technology and Circuits), pp. 1-2 (2021).

[5] K. Toprasertpong, K. Tahara, Y. Hikosaka, K. Nakamura, H. Saito, M. Takenaka, and S. Takagi, ACS Applied Materials & Interfaces, pp. 51137-51148 (2022).

[6] Z. Zhao, Y. Chen, J. Wang, Y. Chen, J. Zou, Y. Lin, Y. Xing, C. W. Liu, and C. Hu, IEEE Electron Device Letters (EDL), pp. 553-556 (2022).

[7] X. Luo, K. Toprasertpong, M. Takenaka, and K. Takagi, Applied Physics Letters (APL), pp. 232904 (2021).

Co-optimization of Oxide Semiconductor-based Ferroelectric Transistors Between Electrical Performance and Ambient Stability By Using TiO$_2$-IGZO Dual-Channel Layers

Shangze Li [1], Xujin Song [2], Dijiang Sun [2], Xiaoyan Liu [2], Jinfeng Kang*[2]

[1] School of Software and Microelectronics, Peking University, Beijing 100871, China
[2] School of Integrated Circuits, Peking University, Beijing 100871, China

* Email: 2201210615@stu.pku.edu.cn, kangjf@pku.edu.cn

Abstract—Integrating oxide semiconductor (OS) channel with hafnium-based ferroelectric gate is one of the promising solutions to achieve high performance of ferroelectric thin-film transistors (FeFETs). In this study, we propose and demonstrate a new approach to realize both markedly enhanced electrical properties and good ambient stability of OS-FeFETs by adopting dual-channel TiO$_2$-IGZO and La-doped HfO$_2$ ferroelectric configurations. The dual-channel device exhibited improved performance, with the memory window of 1.5 V, no obvious degradation after two-month air exposure and an on-state current comparable to that of the IGZO single-channel. The proposed TiO$_2$-IGZO stacked channel offers a promising avenue to enhance the performance of oxide semiconductor FeFETs.

Keywords—FeFETs, HfO$_2$-based ferroelectric, TiO$_2$-IGZO dual-channel, memory window, ambient stability

I. INTRODUCTION

Hafnium oxide based ferroelectric field effect transistors (FeFETs) have attracted huge research interest, owing to its CMOS compatibility, high speed, low power operation and 3D integration capability [1], [2]. Oxide semiconductor (OS) channels such as Indium-Gallium-Zinc-Oxide (IGZO) or Indium-Zinc-Oxide (IZO), have been extensively employed due to their ability to reduce the growth of the interfacial layer during the annealing process [3], [4]. Furthermore, OS-FETs based on hafnium-based ferroelectric materials are promising candidates for 3D vertical NAND and monolithic 3D integration technologies. This is due to the fact that they are capable of attaining a high degree of conformality through a straightforward process, with the assistance of atomic layer deposition (ALD) [5]. However, as an n-type semiconductor for the majority of OS, the typical OS such as IGZO lacks a certain number of hole carriers, which results in the inability to effectively apply the erasing voltage to the hafnia-ferroelectric layer in OS-FeFETs [6]. Furthermore, the threshold voltage (V_{th}) of OS-FeFETs will shift in the near future, primarily due to the phenomenon of charge trapping or injection at the interface between the OS-ferroelectric layer. These problems need to be solved urgently to realize the wider application of ferroelectric transistors.

A series of studies has been conducted with the objective of developing a solution to the issue of OS-FeFETs being challenging to erase. For example, the insertion of an additional p-type CuOx layer between the n-type oxide semiconductor InZnOx and the ferroelectric surface has been

demonstrated to enhance the polarization switching [7]. Furthermore, the dual-gate control structure can be employed to achieve partial erasure [8]. Additionally, the alternating of the ferroelectric HfO$_2$ by a biased ZrO$_2$ anti-ferroelectric layer effectively reduces the ERS voltage [9]. In addition to the challenge of achieving reliable erasing, the ambient stability of the OS channel, which ensures the long-term retention of OS channel TFTs, has seldom been addressed in the context of OS-FeFETs.

This work addresses the issues of both atmospheric stability and erasure of IGZO-FeFETs. The results of the memory window and ferroelectric tests performed on the FeFETs are evaluated in order to assess their electrical performance. The environmental stability of the fabricated FeFETs was evaluated after two-month of air exposure. The incorporation of ultra-thin TiO$_2$ films has been demonstrated to significantly enhance the performance of IGZO-FeFETs, paving the way for the development of novel applications for 3D integrated stacking.

II. EXPERIMENTS

The structure of the dual-channel IGZO-TiO$_2$ FeFETs device and the process flow are illustrated in Fig. 1. The dimensions of the channel were 10 μm in length and 100 μm in width. A 30 nm TiN layer was deposited by radio frequency (RF) sputtering as the gate electrode. A 9 nm HfLaOX (HLO) layer was deposited using TDMAHf and La(thd)$_3$ precursors with ozone as an oxidant at 300°C by ALD. The oxide semiconductor TiO$_2$ layer was deposited using TiCl$_4$ as a precursor and ozone as an oxidant at 300°C by ALD. The sample was subjected to annealing under an N$_2$ environment at 600°C for 30 seconds using a rapid thermal annealing (RTA) process to crystallize the HLO film. Subsequently, Mo source/drain electrodes were formed by RF sputtering. The 20 nm IGZO channel layer was

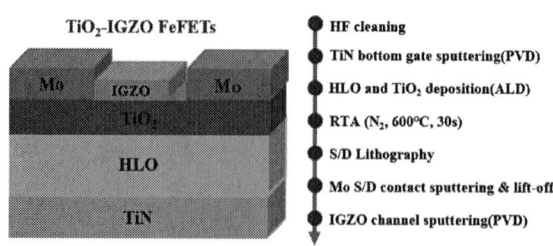

Fig. 1. Schematic and process flow for the TiO$_2$-IGZO FeFETs.

979-8-3503-6184-1/24 $31.00 © 2024 IEEE

deposited by RF sputtering with the atomic composition ratio In: Ga: Zn of 1:1:1.

The electrical characteristics of the FeFETs were quantified using a Keysight B1500A semiconductor parameter analyzer, which was equipped with a WGFMU. In order to assess the ambient stability, the freshly fabricated devices were subjected to a two-month exposure to room air and subsequently subjected to electrical test.

III. RESULTS AND DISCUSSION

A. DC Transfer Characteristics

By sweeping the gate voltage (V_g) from -4 V to 4 V in double sweep, the drain voltage was set at 1V, resulting in the transient I_d-V_g curve observed in Fig. 2. Notable memory window (MW) value of 1.5V was achieved for the TiO_2–IGZO FeFETs, which demonstrates the absence of the erase issue commonly observed in IGZO. This can be attributed to the high quality of the interface layer between ALD-TiO_2 and ALD-HLO. In contrast, the memory window was observed to be almost immeasurable in the IGZO-FeFETs, as illustrated in the figure. In TiO_2-IGZO FeFETs, a certain number of hole carriers in TiO_2 assist HLO in performing polarization switching [10]. This resolves the issue of the poor hole-carrying characteristics in IGZO, which renders the erasure operation impossible to perform. Meanwhile, the

on-state current of TiO_2-IGZO FeFETs was observed to reach a value close to 10 µA, while the on-state current of the 20 nm TiO_2-FeFETs was only 1 µA. Consequently, the conduction characteristic of the TiO_2-IGZO dual-channel FeFETs are demonstrably superior to those of TiO_2 single-channel FeFETs, but similar to those of IGZO-FeFETs.

B. Ambient Stability

In this part, we have evaluated the ambient stability of the devices, the topic that has seldom previously discussed in the research literature on OS-FeFETs. By exposing the devices to room air for a period of two months, we have conducted repeated tests of the transient I_d-V_g curve in comparison to the pristine devices. The results are presented in Fig. 3. It is evident that a notable shift of the V_{th} to a negative direction is observed in the IGZO-FeFETs, primarily due to the degradation of the HLO-IGZO interface, which may result from the oxygen vacancy related charge trapping and injection at the interface.

However, in TiO_2-IGZO FeFETs, a negative shift in V_{th} of less than 0.1 V occurs after exposure, which demonstrates the suitability of HLO for use with TiO_2 and indicates that the interface is not significantly degraded during two-month of atmospheric exposure.

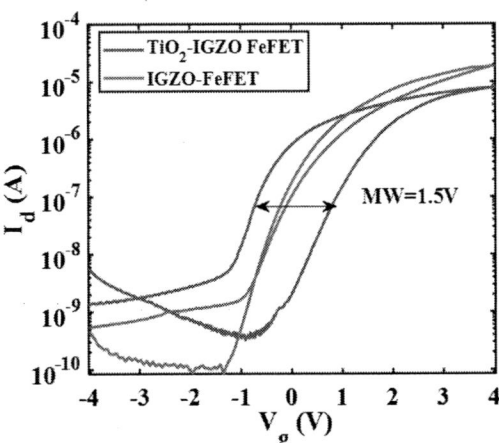

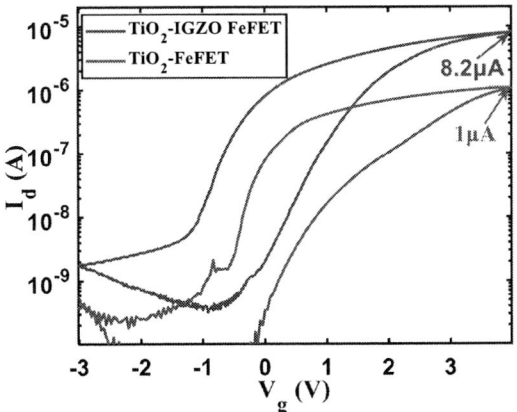

Fig. 2. Measured double sweep DC Id-Vg curves with Vds= 1 V

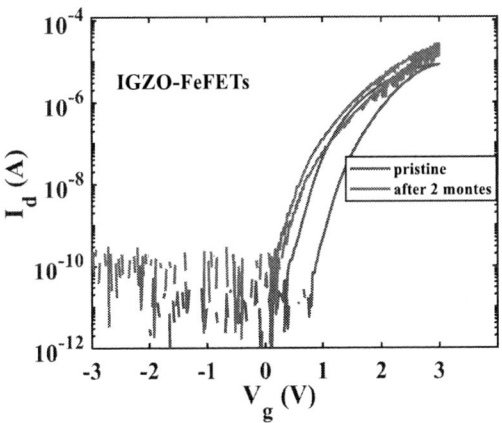

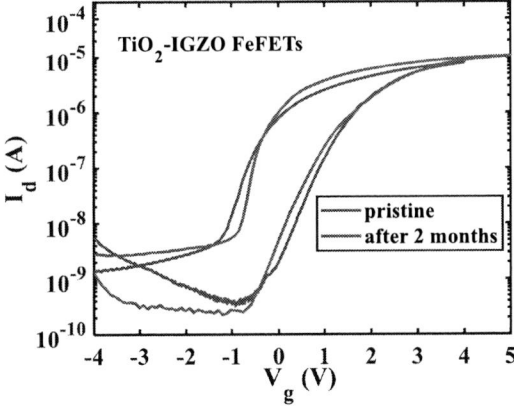

Fig. 3. Ambient stability test for TiO_2-IGZO FeFETs and IGZO-FeFETs after two-month exposed to room air.

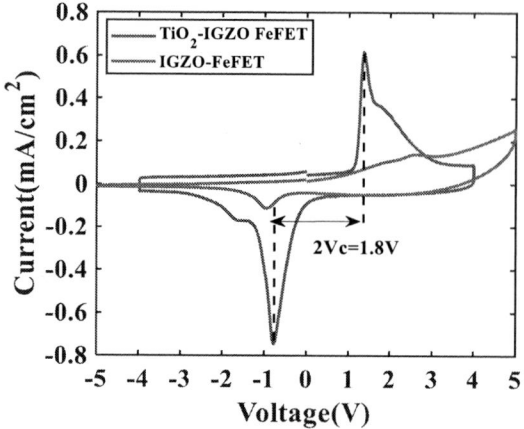

Fig. 4. Transient I-V under 4V, 1kHz triangular pulse .

C. Polarization Reversal Characteristic

As illustrated in Fig. 4, the transient I-V curve of the TiO_2-IGZO FeFETs exhibits a pronounced current peak with an ultra-low coercive voltage of $2Vc = 1.8V$, which is analogous to the low coercive field characteristics observed in the TiO_2-FeFETs [11]. This indicates that the TiO_2 layer inserted into the device plays a role in facilitating the HLO polarization switching. In contrast, in IGZO-FeFETs, the current peak is observed exclusively in the negative direction, with the positive direction being less pronounced due to the insufficient availability of hole carriers in IGZO to facilitate the polarization switching.

IV. CONCLUSION

In the study, we propose and demonstrate a new technical solution for the fabrication of high performance oxide semiconductor-based FeFETs by inserting a TiO_2 layer into the IGZO-FeFETs to form a TiO_2-IGZO dual-channel field effect transistor. The experimental results show that the TiO_2-IGZO dual-channel field effect transistor exhibits significantly enhanced ambient stability, a stable memory window of 1.5 V and comparable on-current behaviors compared to IGZO-channel FeFETs. The reported TiO_2-IGZO dual-channel FeFETs offers a promising avenue for the development of OS-FeFETs in 3D stacking applications.

ACKNOWLEDGMENT

These devices were manufactured in National Micro/Nano Fabrication Laboratory of Peking University.

REFERENCES

[1] H. Mulaosmanovic, E. T. Breyer, S. Dunkel, S. Beyer, T. Mikolajick, and S. Slesazeck, "Ferroelectric field-effect transistors based on HfO2: a review," Nanotechnology, vol. 32, no. 50, Sep 22 2021.

[2] G. Kim et al., "Design Guidelines of Thermally Stable Hafnia Ferroelectrics for the Fabrication of 3D Memory Devices," presented at the 2022 International Electron Devices Meeting (IEDM), 2022.

[3] I.-J. K. M.-K. Kim, and J.-S. Lee, "CMOS-compatible ferroelectric NAND flash memory for high-density, low-power, and high-speed three-dimensional memory," Sci. Adv., vol. 7, no. 3, Jan. 2021.

[4] F. Mo et al., "Low-Voltage Operating Ferroelectric FET with Ultrathin IGZO Channel for High-Density Memory Application," IEEE Journal of the Electron Devices Society, vol. 8, pp. 717-723, 2020.

[5] I.-J. Kim, M.-K. Kim, and J.-S. Lee, "Vertical ferroelectric thin-film transistor array with a 10-nm gate length for high-density three-dimensional memory applications," Applied Physics Letters, vol. 121, no. 4, 2022.

[6] J. S. Kim et al., "Investigating the Reasons for the Difficult Erase Operation of a Charge-Trap Flash Memory Device with Amorphous Oxide Semiconductor Thin-Film Channel Layers," physica status solidi (RRL) – Rapid Research Letters, vol. 15, no. 2, 2021.

[7] I.-J. Kim, M.-K. Kim, and J.-S. Lee, "Design Strategy to Improve Memory Window in Ferroelectric Transistors With Oxide Semiconductor Channel," IEEE Electron Device Letters, vol. 44, no. 2, pp. 249-252, 2023.

[8] M. Kobayashi, "IGZO channel ferroelectric memory FET," 2020 27th International Workshop on Active-Matrix Flatpanel Displays and Devices 2020, 2020.

[9] Z. Liang et al., "A Novel High-Endurance FeFET Memory Device Based on ZrO2 Anti-Ferroelectric and IGZO Channel," presented at the 2021 IEEE International Electron Devices Meeting (IEDM), 2021.

[10] M. K. Nowotny, T. Bak, J. Nowotny, and C. C. Sorrell, "Titanium vacancies in nonstoichiometric TiO2 single crystal," physica status solidi (b), vol. 242, no. 11, 2005.

[11] X. Song et al., "Ultrathin TiO2 Channel HfLaO FeFET with Low Operation Voltage," presented at the 2024 Conference of Science and Technology for Integrated Circuits (CSTIC), 2024.

Enhancing Computational Precision in PLRAM-based In-memory Computing with High-Low Bit Concatenation

Saike Zhu [1,2,3], Xiang Qiu [4], Yong Gong*[3], Cimang Lu [5], Yi Zhao*[1,2,3]

[1] College of Electronic Engineering and Information Science, Zhejiang University, Hangzhou 310027, China
[2] International Joint Innovation Center, Zhejiang University, Haining 314400, China
[3] China Nanhu Academy of Electronics and Information Technology, Jiaxing 314001, China
[4] School of Integrated Circuits, East China Normal University, Shanghai 200241, China
[5] Flash Billion Semiconductor Co. Ltd., Shanghai 201210, China

* Email: yizhao@zju.edu.cn.

Abstract—**Computing-in-memory (CIM) technology has emerged as a strong contender for edge computing, owing to its high performance and low power consumption. Programmable Linear Random Access Memory (PLRAM) shows significant potential in memory-intensive edge-computing applications. However, accuracy degradation occurs in PLRAM-based computing with larger array sizes due to programming limitations. In this work, we propose a hybrid precision computing architecture to improve inference accuracy in PLRAM-based chips. This is achieved by performing two forward propagations using the concatenation of high-bit and low-bit regions. Consequently, the correlation coefficients of the fully connected layers (FC1, FC2, FC3, and FC4) are improved from 0.9943, 0.9894, 0.9578, and 0.9214 to 0.9947, 0.9901, 0.9742, and 0.9807, respectively. Additionally, the classification error rate of an 11-keyword speech recognition model decreased by 11.76%, demonstrating the effectiveness of the proposed architecture in achieving high computational accuracy.**

Keywords—*Compute-In-Memory (CIM), Memristor, Flash memory, Hybrid precision, Analog compensation.*

I. INTRODUCTION

Computing-In-Memory (CIM) technology is increasingly attractive for data-intensive AI applications due to its ability to reduce data communication overhead between the processor and storage in traditional Von-Neumann computing architecture, thereby significantly enhancing computational efficiency [1-7]. Flash memory, utilizing floating-gate transistors, can store multi-bit data in a single cell by adjusting the charge level in the floating gate. In [8], Gao et al. proposed Programmable Linear Random Access Memory (PLRAM), a modified NOR-Flash structure. This modification improves linearity and reliability, making it more suitable for the CIM paradigm.

The PLRAM-based CIM architecture, depicted in Figure 1, includes PLRAM cells with 128 multi-levels, achieving 7-bit accuracy [9]. To mitigate the issue of weight drift in PLRAM devices, Gao et al. proposed a self-calibrating program/erase scheme. This approach measures the conductance of PLRAM cells prior to each programming cycle and adjusts the control voltage accordingly, ensuring high precision and stability in the programming/erase process. As demonstrated in Figure 2 (a), the 128*128 PLRAM array achieves a programming accuracy of 7 bits per cell.

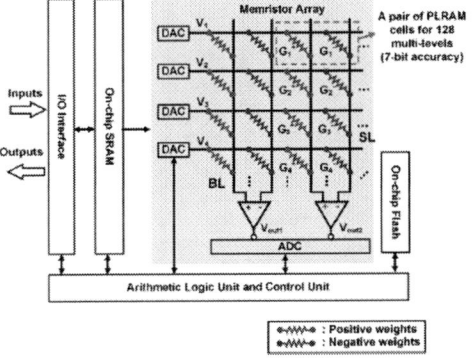

Fig. 1. System architecture of the PLRAM-based CIM array chip. Weights are split into positive and negative parts and written in adjacent BLs

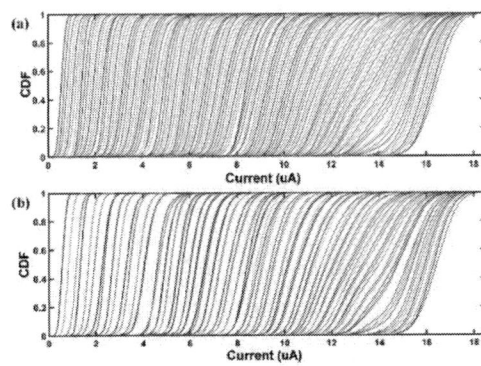

Fig. 2. The cumulative distribution function (CDF) of PLRAM cells under different array sizes. Device precision decreases with larger array sizes due to programming limitations. (a) All 16384 cells in a 128*128 array are programmed into 128 states. The tight distribution guarantees 7-bit precision in 1 sigma [9]. (b) 1 million cells in a 1024*4096 array are programmed into 128 states, approximately achieving 6-bit precision.

However, for larger PLRAM-based arrays, considering the trade-offs between circuit area and programming time, a block-based program/erase scheme was implemented. As illustrated in Figure 2 (b), this approach reduced the precision of PLRAM cells to 6 bits in the 1024*4096 array.

To address the precision challenge in CIM architecture [10], Chi et, al proposed a scheme for improving computational precision by combining high-bit and low-bit parts of weights and high-bit and low-bit parts inputs [11]. However, cumulative errors from multiple segmented

---

This work was supported by the National Key Research and Development Program of China (No. 2020AAA0109001), the "Pioneer" and "Leading Goose" R&D Program of Zhejiang Province (2023C01018) , the Industry-University-Research Project of Pudong New Area, Shanghai (PKX2021-D04) and Shanghai Science and Technology Funding Project (22DZ2205100).

979-8-3503-6184-1/24 $31.00 © 2024 IEEE

calculations can significantly affect the final precision, especially in deep neural networks where error accumulation is more pronounced. Yu et al. proposed storing high-bit weights in non-volatile memory (eNVM) and low-bit weights in capacitors to achieve higher numerical precision [12]. However, the nonlinearity of the low-bit capacitors can lead to a decrease in training accuracy.

In this study, we proposed a novel high-low bit concatenation architecture for in-memory computing to address the precision degradation observed in larger array sizes due to programming limitations. By performing two forward propagations using the concatenation of high-bit and low-bit regions, the inference accuracy of the CIM-based chip was significantly improved.

II. EXPERIMENTAL METHODS

A. Analog error compensation in PLRAM-based Array

In large-scale PLRAM-based arrays, programming limitations can introduce significant errors in weights, adversely impacting the accuracy of Multiply-Accumulate (MAC) operations. To mitigate this issue, an analog compensation scheme is implemented. The process begins by writing the trained neural network weight matrix into the Most Significant Bit (MSB) memristor array. Subsequently, the conductance matrix is measured by applying fixed voltages to the array. The target weight matrix is then compared to the measured conductance matrix to determine the error matrix. These measured errors are scaled by a specific factor and programmed as compensation weights into the Least Significant Bit (LSB) memristor array. By incorporating these compensation weights, the overall output error is significantly reduced, thereby enhancing the computational accuracy of the PLRAM-based array.

Figure 3 illustrates an example of the compensation process. The target weights in the first bit line (BL) are [8, 8, 9]^T, which yield an ideal output of 126. Due to programming limitations in larger arrays, the actual weights stored in the PLRAM-based array are [8.1, 7.6, 9.3]^T, The errors [−0.1, 0.4, −0.3]^T are scaled up by the scaling factor of 16 and programmed into the BL of the LSB array. Assuming the same variation as in the main array, the actual compensation weights become [−1.5, 6, −4.5]^T, resulting in a column sum of -33 in the compensation array. Finally, this output is scaled down by 1/16 and added to the main weight output, reducing the output error from 2.2 to 0.1.

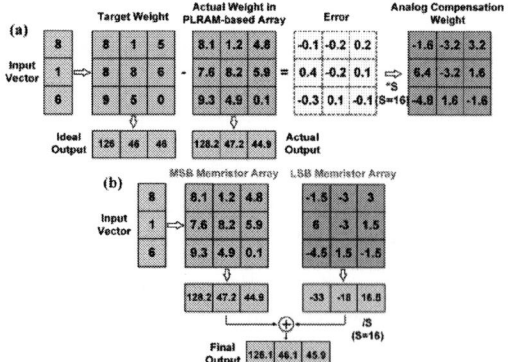

Fig. 3. (a) Compared to the ideal MAC operation (yellow), the actual weights stored in the PLRAM-based array (green) contain errors. The measured errors are scaled by a factor (S) to obtain compensation weights (blue). (b) The recovered MAC operation using the compensation scheme.

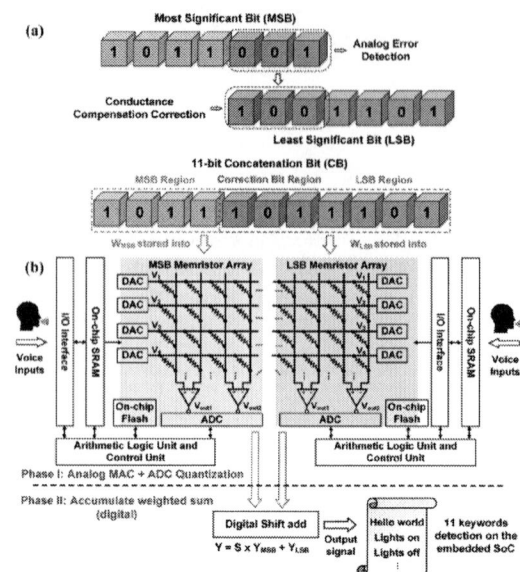

Fig. 4. High-Low Bit Concatenation Architecture in PLRAM-Based Chip. (a) Error Correction and High-Low Bit Concatenation. After measuring the analog error in the high-bit region, conductance compensation is performed in the low-bit region to improve its precision. (b) Computation Flow of High-Low Bit Concatenation. High-bit and low-bit weights undergo two MAC analog forward propagations. The results are then shifted and added in the digital shift-add unit. F represents the significance factor for the W$_{MSB}$.

B. Hybrid precision architecture

Due to the inevitable analog errors in memristor arrays, a hybrid precision architecture that concatenates the MSB and the LSB has been proposed, as shown in Figure 4. In this architecture, the MSB weights and the LSB weights are stored in two separate memristor arrays. Figure 4 illustrates an 11-bit Concatenation Bit (CB) design, consisting of 7-bit MSB and 7-bit LSB, with a 3-bit overlap forming the Correction Bit Region. This region is dynamically adjustable. If the overlap in the Correction Bit Region is 2 bit, the overall CB is 12-bit.

The weight shift-add process includes two phases: Phase I involves analog MAC and ADC operations, while Phase II involves digital shift-add operations between columns to accumulate the weighted sums from both LSB and MSB. The final output is expressed as $Y = S \times Y_{MSB} + Y_{LSB}$, where S is the scaling factor, set as an integer power of 2. The weights W_{MSB} and W_{LSB} are stored separately, and their partial sums are digitally accumulated on the peripheral circuit.

III. RESULTS AND DISCUSSIONS

To investigate the impact of high-low bit concatenation on computation accuracy, we deployed a commercial speech recognition neural network model on the CIM chip. In this study, the computation accuracy is quantified using the correlation coefficient (CC) between the chip outputs and the ideal model outputs. A higher CC indicates greater accuracy.

As shown in Figure 5, fully connected layers FC1 and FC3 achieve the highest correlation coefficients of 0.9946 and 0.9753, respectively, at 12-bit concatenation bit (CB) precision. Similarly, FC2 and FC4 attain the highest correlation coefficients of 0.9895 and 0.9809, respectively, at 13-bit CB precision. In other words, we set the scaling factor S to 32 for FC1 and FC3, and to 64 for FC2 and FC4. This hybrid precision architecture significantly enhances computation accuracy.

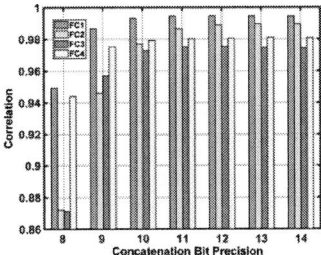

Fig. 5. Correlation Coefficients (CC) Across Four Fully Connected Layers (FC1, FC2, FC3, FC4) for Various Concatenation Bit Precision Levels.

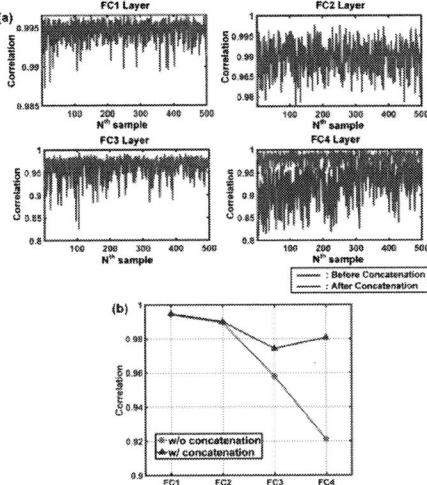

Fig. 6. Correlation Coefficient (CC) of Four Fully Connected Layers (FC1, FC2, FC3, FC4) Before and After Weight Concatenation. (a) Correlation coefficients between chip outputs and ideal model outputs for 500 different speech input samples. The red lines represent the CC before weight concatenation, and the blue lines represent the CC after weight concatenation. (b) Mean correlation of 500 different speech input samples across FC1 – FC4 layers.

As shown in Figure 6, the improvement becomes more pronounced with increasing layer depth due to the cumulative amplification effect of hardware errors. The hybrid precision architecture enhances the correlation coefficients (CC) across all layers, as demonstrated by the mean correlation of 500 different speech input samples before and after weight concatenation. Table I compares the accuracy before and after weight concatenation. The mean correlation coefficients for FC1, FC2, FC3, and FC4 before optimization are 0.9943, 0.9894, 0.9578, and 0.9212, respectively. After optimization, these coefficients improve to 0.9947, 0.9901, 0.9742, and 0.9807, respectively. The final classification accuracy increases from 96.6% to 97.0%, demonstrating the effectiveness of the proposed architecture in enhancing computation accuracy.

TABLE I.   COMPARISON OF ACCURACY BEFORE AND AFTER WEIGHT CONCATENATION

	Mean Correlation of 500 samples				Final Classification Accuracy
	FC1	FC2	FC3	FC4	
Before Optimization	0.9943	0.9894	0.9578	0.9212	96.6 %
After Optimization	0.9947	0.9901	0.9742	0.9807	97.0 %

IV. SUMMARY

In this study, we propose a novel method to enhance the computational precision of PLRAM-based Compute-in-Memory (CIM) arrays using high-low bit concatenation. This architecture mitigates precision degradation in larger arrays caused by programming limitations. By implementing a hybrid precision computing architecture that utilizes both the Most Significant Bit (MSB) and Least Significant Bit (LSB) regions, two forward propagations are performed to improve inference accuracy. To validate the effectiveness of this architecture, an 11-keyword commercial speech recognition neural network model was deployed on the CIM chip. This implementation significantly enhanced the inference accuracy of the network's intermediate layers, reducing the keyword recognition error rate by 11.76%.

REFERENCES

[1] Z. Dong, X. Ji, G. Zhou, M. Gao and D. Qi, "Multimodal Neuromorphic Sensory-Processing System With Memristor Circuits for Smart Home Applications," IEEE Transactions on Industry Applications, vol. 59, no. 1, pp. 47-58, 2023.

[2] Z. Dong, X. Ji, C. S. Lai, D. Qi, G. Zhou and L. L. Lai, "Memristor-Based Hierarchical Attention Network for Multimodal Affective Computing in Mental Health Monitoring," IEEE Consumer Electronics Magazine, vol. 12, no. 4, pp. 94-106, 2023.

[3] Z. Dong, C. Lai, Z. Zhang, D. Qi, M. Gao, S. Duan, "Neuromorphic extreme learning machines with bimodal memristive synapses," Neurocomputing, vol. 453, pp. 38-49, 2021.

[4] X. Ji, C. Lai, G. Zhou, Z. Dong, D. Qi and L. Lai, "A Flexible Memristor Model With Electronic Resistive Switching Memory Behavior and Its Application in Spiking Neural Network," IEEE Transactions on NanoBioscience, vol. 22, no. 1, pp. 52-62, 2023.

[5] X. Ji, D. Qi, Z. Dong, C. Lai, G. Zhou, and X. Hu, "TSSM: Three State Switchable Memristor Model Based on Ag/TiO x Nanobelt/Ti Configuration," International Journal of Bifurcation and Chaos, vol.31, no. 7, pp. 2130020, 2021.

[6] Z. Dong, X. Ji, C. S. Lai and D. Qi, "Design and Implementation of a Flexible Neuromorphic Computing System for Affective Communication via Memristive Circuits," IEEE Communications Magazine, vol. 61, no. 1, pp. 74-80, 2023.

[7] X. Ji, Z. Dong, Y. Han, C. S. Lai and D. Qi, "A Brain-inspired Hierarchical Interactive In-memory Computing System and its Application in Video Sentiment Analysis," IEEE Transactions on Circuits and Systems for Video Technology, 2023.

[8] S. Gao, G. Yang, X. Qiu, C. Yang, C Zhang, B. Li, C. Gao, H. Jiang, Z. Wang, J. Hu, J. Xiao, B. Zhang, C.H. Lee, Y. Zhao, and W. Kong, "Programmable Linear RAM: A New Flash Memory-based Memristor for Artificial Synapses and Its Application to Speech Recognition System," in 2019 IEEE International Electron Devices Meeting (IEDM), pp. 14.1.1- 14.1.4, 2019.

[9] S. Gao, Y. Cong, Z. Zhang, X. Qiu, C. Lee, and Y. Zhao, "Superior Data Retention of Programmable Linear RAM (PLRAM) for Compute-in-Memory Application," in 2020 IEEE International Reliability Physics Symposium (IRPS), pp. 1-5, 2020.

[10] J. He, Y. Huang, M. Lastras, T. T. Ye, C. -Y. Tsui and K. -T. Cheng, "RVComp: Analog Variation Compensation for RRAM-based In-Memory Computing," 2023 28th Asia and South Pacific Design Automation Conference (ASP-DAC), Tokyo, Japan, pp. 1-6, 2023.

[11] P. Chi, S. Li, C. Xu, T. Zhang, J. Zhao, Y. Liu, Y. Wang, and Y. Xie , "PRIME: a novel processing-in-memory architecture for neural network computation in ReRAM-based main memory," ACM SIGARCH Computer Architecture News, vol. 44, pp. 27–39, 2016.

[12] S. Yu, W. Shim, X. Peng and Y. Luo, "RRAM for Compute-in-Memory: From Inference to Training," in IEEE Transactions on Circuits and Systems I: Regular Papers, vol. 68, no. 7, pp. 2753-2765, 2021.

FeFET based Logic-in-Memory Pipeline-Style Circuits

Yang Li [1], Zhaohui Yang [2], Yinshui Xia*[1]

[1] Faculty of Electrical Engineering and Computer Science Ningbo University Ningbo 315211, China

* Email: 2311100017@nbu.edu.cn

Abstract—Emerging non-volatile memory technologies provides a solution to break through the limitations of the von-Neumann architecture and using these technologies to develop logic-in-memory circuits. In this paper, we propose a novel circuit design style based on hafnium oxide ferroelectric memory technology, using a accumulator as an example. This design method offers a new direction for the development of logic-in-memory circuits.Additionally, the method significantly increases data processing capability and provides greater flexibility for circuit design.

Keywords—logic-in-memory(LiM), ferroelectric, non-volatile, accumulator

I. INTRODUCTION

With the rapid growth of demand for new technologies like computing, and artificial intelligence (AI) , big data, etc. Computility becomes critical issues in today's digital systems. However, von-Neumann architecture based separation of computation and storage faces power wall and storage wall. To tackle these challenges, researchers are exploring emerging devices, new computing paradigms, and innovative circuit architectures.

The logic-in-memory (LiM) circuits mentioned in 1970 [1] provided a direction for addressing these challenges by reducing data transfer between system memory and computational cores. In recent years, with the discovery of emerging non-volatile (NV) memory devices such as resistive random-access memory (RRAM), phase-change memory (PCM), and ferroelectric technology, researchers have integrated these devices with logic itself to form LiM circuits. In comparison, FeFETs, serving as a three-terminal device, exhibits the switch ratio of up to 10^6 [2], enabling its operation as a switch rather than a resistor. Further, its triterminal configuration segregates the write and read paths, markedly improving circuit design flexibility and substantially decreasing the complexity of peripheral control circuits. Additionally, hafnium oxide-based FeFETs have been demonstrated to be compatible with standard CMOS process libraries [3].

This paper centers on the circuit design employing ferroelectric devices based on hafnium oxide [4]. Utilizing dynamic current (DL) mode as described in [6], we employ the concept of latched pipelines and their combination. Two modules controlled by asynchronous clocks are utilized for circuit design. This method not only mitigates data contention but also substantially improves data processing speed. We designed a DL Pipeline-style circuit based on FeFET and elucidate functionality through the design of

This work was supported in part by the National Natural Science-Foundation of China under Grants U23A20351 and 62304115; in part by the Natural ScienceFoundation of Zhejiang Province under Grant LDT23F04021F04. (Corresponding author: Yinshui Xia).

accumulator. The organization of the remaining sections of this paper is as follows: The second section delves into the foundational principles of FeFET devices, simulation models, and the writing and storage mechanisms of these devices. In the third section, we propose the methodologies of circuit design based on FeFET. The fourth section concludes with a summary.

II. PRELIMINARIES

In this section, we primarily discuss FeFET devices, simulation models, and the mechanism for writing logic states.

A. Device of FeFET

As shown in Fig. 1(a), a ferroelectric layer is introduced between the gate and metal layers to fabricate the FeFET device. Due to the negative capacitance character of the ferroelectric layer, the equivalent circuit diagram of the FeFET is shown in Fig. 1(b). When C_{FE}/C_{MOS} is large, the polarization state is not preserved, thus termed as negative capacitance field-effect transistors (NCFETs).

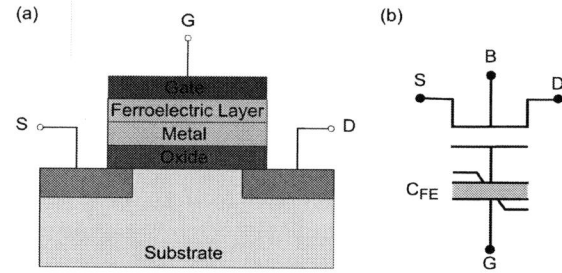

Fig. 1. (a) FeFET device structure (b) the equivalent circuit composed of a nonlinear ferroelectric capacitor in series with a MOSFET

However, when C_{FE}/C_{MOS} is sufficiently low, the polarization state of the ferroelectric layer can be retained [6], resulting in the hysteresis characteristics curve, as shown in Fig. 2 (a), thereby generating a storage window for the device [7].

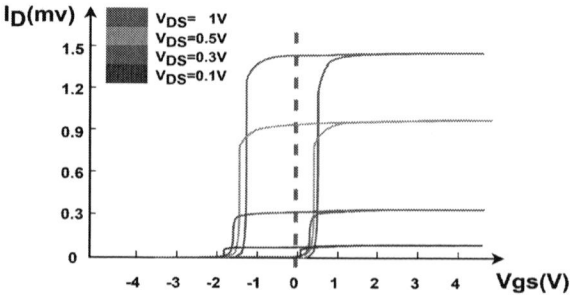

Fig. 2. FeFET I-V curves exhibiting hysteresis

979-8-3503-6184-1/24 $31.00 © 2024 IEEE

B. FeFET model

In this paper, we employ the Landau-Khalatnikov (LK) model. The L-K equation describes the relationship between the applied electric field (E) and the internal polarized charges within the gate. This equation is nonlinear [8]:

$$E - \rho \frac{dP}{dt} = \alpha P + \beta P^3 + \gamma P^5 \qquad (1)$$

Here, α, β, and γ serve as static parameters, while ρ indicates the dynamic coefficient, P is the internal polarization state. The parameter settings are specified below:

TABLE I. The L-K model parameter setting

α	$-1.05 \times 10^9 \text{m F}^{-1}$
β	$1 \times 10^7 \text{m}^5\text{F}^{-1}\text{C}^{-2}$
γ	$6 \times 10^{11}\text{m}^9\text{F}^{-1}\text{C}^{-4}$
ρ	$0.25 \ \Omega \ \text{m}$

C. Writing and storage of FeFET

As shown in Fig. 3(a), with $V_{GS} = V_{DD}$, the FeFET is set to logic 1, and it remains in the open state. With $V_{GS} = -V_{DD}$, the FeFET is set to logic 0, and it stays closed. At $V_{GS} = 0V$, the FeFET stores the previous state. In consideration of circuit overhead and the avoidance of negative voltage usage in subsequent cascades, we choose the second writing method outlined in [5] when writing the logic states.

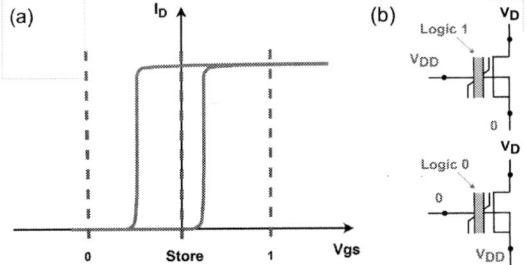

Fig. 3. (a) FeFET *I-V* curves with tunable hysteresis(b)The write method for FeFET without a negative power supply

As depicted in Fig. 3(b), when V_{DD} is applied to the gate and 0V to the source, the V_{GS} becomes V_{DD}, causing the FeFET to write logic 1. Conversely, when the source and gate are reversed for writing, the FeFET writes logic 0.

III. THE DESIGNE OF FEFET-BASED CIRCUIT

When circuits require both high performance and area reduction, CMOS DL logic emerges as an excellent choice. The recent proliferation of various NV devices such as MTJs and FTJs has been notable. However, operating as dual-terminal devices, they continue to exhibit considerable leakage current even in high-resistance states. Conversely, FeFET, as a novel NV three-terminal device, offers a switch ratio of up to 10^6 and a smaller threshold swing, making it better suited to capitalize on the advantages of CMOS DL logic.

A. DL logic of circuit based on FeFET

The circuit, as shown in Fig. 4, comprises an up-pull clock network, a down-pull clock network, and a logic block.

Additionally, the notion of designing a variety of basic gates has been put forward in [5]. Unlike traditional DL logic, a clock-controlled MOS transistor must be directly inserted between the up-pull clock network and the logic block to prevent a path between the write signal and V_{DD} when *CLK* is low during FeFET write operations.

The operational principle, akin to that described in [5], is as follows: According to Fig. 4, in the CLK module, CLK' functions as the write enable voltage. When CLK' is high, the FeFET acts as a storage unit, writing logic 1 with a high signal and logical 0 with a low signal. The CLK' module operates inversely. The data stored in the FeFET is non-volatile. We have designed two modules: the CLK module and the CLK' module, with detailed descriptions to follow in later sections.

B. FeFET based Storage Pipeline-style circuit

Traditional setups of such pipelines necessitate the integration of latches across various logic blocks to facilitate data propagation, thus enhancing operational efficiency. However, given FeFET's innate capacity for retention, it can supplant latches. Nonetheless, simultaneous FeFET writing and reading across modules during cross-module operation is infeasible. Hence, we introduce the concept of *CLK* and *CLK'* modules. Illustrated in Fig. 4, the latch-based pipeline encompasses two modules: the *CLK* and *CLK'* modules, each comprising FeFET storage element and NMOS working networks.

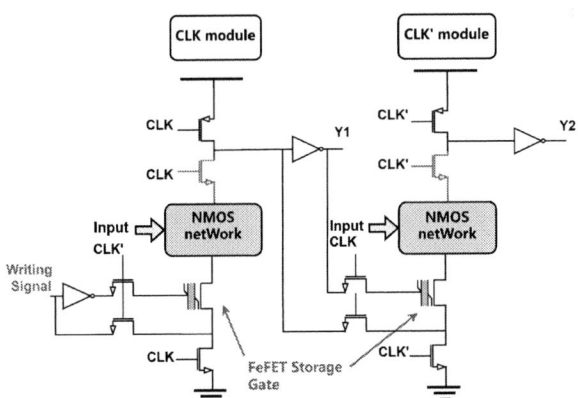

Fig. 4. The structure of DL circuit based on FeFET

This operation mode is similar to standard pipeline, as demonstrated in Fig. 5: When *CLK* is high, the CLK' module handles precharging, and the CLK module simultaneously performs calculations, storing the results in the following CLK' module. When *CLK* changes to low, the CLK module takes over pre-charging, and the CLK' module executes calculations, storing the results in the following CLK module.

C. FeFET based Storage Pipeline-style accumulator

To further establish that FeFETs can function dually as switches and storage devices in a pipeline configuration, we present a more sophisticated LiM circuit as a case study. As shown in Fig. 6(a), conventional accumulator comprises a full adder and D flip-flops for storage, whereas the FeFET-based accumuator, which operates on the pipeline principle, is depicted in Fig. 6(b). It includes CLK and CLK' modules. When *CLK* is high, the CLK' module conducts precharging, and the CLK module performs computation, storing the computed result in the FeFETs of the CLK' module. When *CLK* switches to low, the CLK module undertakes precharging, while the CLK' module performs computation, storing the result in the FeFETs of the CLK module. Shown in Fig. 7 is the schematic diagram of the accumulator we designed.

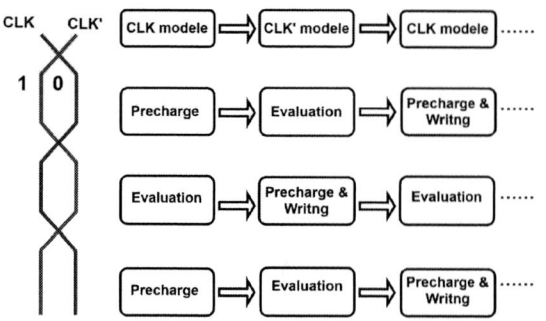

Fig. 5. The operational principle of DL pipelined circuit based on FeFET

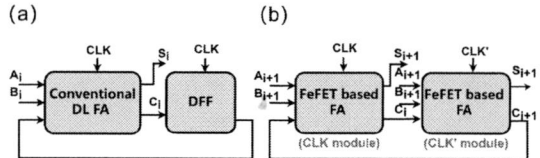

Fig. 6. Components of accumulator (a)conventional DL accumulator (b)FeFET based accumulator

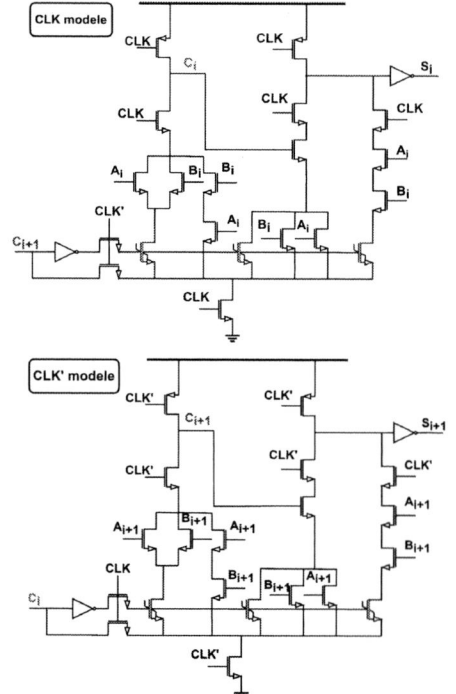

Fig. 7. The schematic of FeFET based Pipeline-style accumulator

The simulation waveforms for the FeFET-based pipeline-style accumulator, depicted in Fig. 8, indicate accurate functional performance. The dual-module design enhances the computation process via parallel processing, enabling more tasks to be executed within each clock cycle. Consequently, although the circuit area is larger, the reduced

number of clock cycles leads to lower overall power consumption. This design method is highly beneficial for applications focused on achieving high efficiency and low power usage.

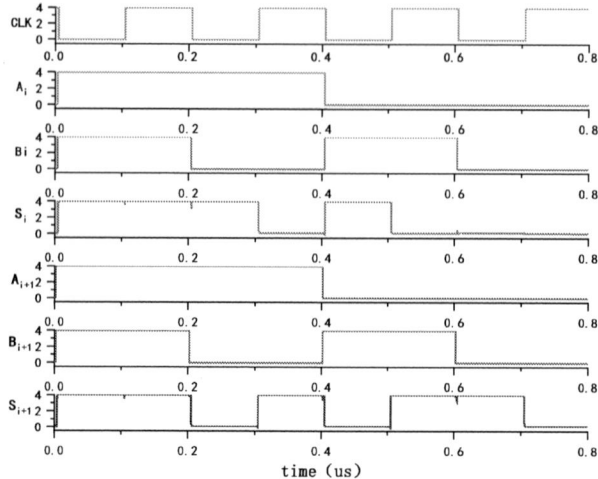

Fig. 8. Simulation waveforms of FeFET based Pipeline-style accumulator

IV. Conclusion

This paper describes an NV LiM circuit designed with FeFET in a DL Pipeline-style manner. The circuit maximizes the unique hysteresis characteristics of FeFET and proposes their use as high-switching-ratio NV devices. These devices utilize their hysteresis properties for storage and their high switching ratios for computation. A segmented pipeline design with two clock modules allows the circuit to operate interactively at different clock levels, greatly enhancing data processing efficiency. The output variables differ with clock levels, providing a novel direction for further research on FeFET-based circuits.

References

[1] Stone, Harold S, A Logic-in-Memory Computer[J]. IEEE Transactions on Computers, 1970, pp. 73-78.

[2] Yeung C W, Khan A I, Sarker A, et al. Low power negative capacitance FETs for future quantum-well body technology[C]//2013 International symposium on VLSI technology, systems and application (VLSI-TSA). IEEE, 2013, pp. 1-2.

[3] Müller J, Böscke T S, Müller S, et al. Ferroelectric hafnium oxide: A CMOS-compatible and highly scalable approach to future ferroelectric memories[C]//2013 IEEE International Electron Devices Meeting. IEEE, 2013, pp. 10-8.

[4] Breyer E T, Mulaosmanovic H, Slesazeck S, et al. Demonstration of versatile nonvolatile logic gates in 28nm HKMG FeFET technology[C]//2018 IEEE International Symposium on Circuits and Systems (ISCAS). IEEE, 2018, pp. 1-5.

[5] Yin X , Chen X , Niemier M ,et al. Ferroelectric FETs-Based Nonvolatile Logic-in-Memory Circuits[J]. IEEE Transactions on Very Large Scale Integration (VLSI) Systems, 2018, pp. 159-172.

[6] Salahuddin S, Datta S. Use of negative capacitance to provide voltage amplification for low power nanoscale devices[J]. Nano letters, 2008, pp. 405-410.

[7] Mulaosmanovic H, Breyer E T, Dünkel S, et al. Ferroelectric field-effect transistors based on HfO2: a review[J]. Nanotechnology, 2021, pp. 502002.

[8] S. Sivasubramanian, A. Widom and Y. Srivastava, Equivalent circuit and simulations for the Landau-Khalatnikov model of ferroelectric hysteresis[J], IEEE Transactions on Ultrasonics, Ferroelectrics, and Frequency Control, 2003, pp. 950-957.

[9] Wahab M A, Alam M A. A verilog-A compact model for negative capacitance FET[D], Purdue University, 2017.

Study of V_{th} Degradation Mechanism in FeFET with TiN/Al$_2$O$_3$/HfO$_2$/Al$_2$O$_3$/Hf$_{0.5}$Zr$_{0.5}$O$_2$/SiO$_x$/Si Structure

Runhao Han[1,2,3], Jia Yang[1,2,3], Tao Hu[1,2,3], Mingkai Bai[1,2,3], Yajing Ding[1,2,3], Xianzhou Shao[1,2,3], Saifei Dai[1,2,3], Xiaoqing Sun[1,2,3], Junshuai Chai[1,2,3], Hao Xu[1,2,3], Kai Han[4], Xiaolei Wang[1,2,3, *], Wenwu Wang[1,2,3] and Tianchun Ye[1,2,3]

[1] Key Laboratory of Fabrication Technologies for Integrated Circuits, Chinese Academy of Sciences, Beijing 100029.
[2] Institute of Microelectronics, Chinese Academy of Sciences, Beijing 100029, China.
[3] School of Integrated Circuits, University of Chinese Academy of Sciences, Beijing 100049, China.
[4] School of Physics and Electronic Information, Weifang University, Shandong, 261061, China

* Email: hanrunhao@ime.ac.cn, wangxiaolei@ime.ac.cn

Abstract—In this work, we investigate the degradation of the threshold voltage after erasing operation in HfO$_2$-based ferroelectric silicon channel field-effect transistors (HfO$_2$ Si-FeFETs) with a TiN/Al$_2$O$_3$/HfO$_2$/Al$_2$O$_3$/Hf$_{0.5}$Zr$_{0.5}$O$_2$/SiO$_x$/Si (MAHAFIS) structure. We found that the de-trapping of charges injected from the metal gate is the main factor for the degradation. Moreover, the concentrated distribution of trap energy levels exacerbates this issue. This differs from the FeFET with the TiN/Hf$_{0.5}$Zr$_{0.5}$O$_2$/SiO$_x$/Si (MFIS) structure, where the degradation is due to the increased trapped charges injected from the silicon substrate. Therefore, for FeFET with a top interlayer, increasing deep level traps and reducing charge de-trapping is crucial to suppress the degradation of the threshold voltage after erasing operation.

Keywords—FeFET, charge trapping/de-trapping, gate leakage, trap.

I. INTRODUCTION

The HfO$_2$ Si-FeFET-based V-NAND is considered a potential solution to address the scaling challenges of CTF-VNAND, owing to the low write voltage of FeFETs. A large MW of at least 6 V (TLC) within a 20 nm ferroelectric gate stack is necessary for multi-bit operation in the V-NAND. [1]However, the memory window (MW) of the HfO$_2$ Si-FeFETs typically ranges from 1 to 2 V, which limits its application in V-NAND. [2]

The conventional structure of HfO$_2$ Si-FeFETs is metal/ferroelectric/interlayer/Si (MFIS). Recently, many efforts have focused on inserting a top dielectric interlayer between the metal gate and the ferroelectric layer to achieve a higher MW. [3, 4] The gate structure is metal/top interlayer/ferroelectric/bottom interlayer/Si (MIFIS). The MIFIS structure can induce the charge injected from the metal gate. Lim et al. from Samsung Electronics and Yoon et al. from SK Hynix realized an MW of 5.5 V and 10.54 V, respectively, adopting the MIFIS structure. [3, 4] In our previous work, we used a top Al$_2$O$_3$ dielectric interlayer in FeFET and realized an MW of 4.19 V. [5] By further optimizing the gate stack structure, we realized a maximum MW of 10.04 V by inserting an Al$_2$O$_3$/HfO$_2$/Al$_2$O$_3$ (AHA) top dielectric interlayer. [6] However, the threshold voltage after the erase operation (V_{th_ERS}) of the TiN/Al$_2$O$_3$/HfO$_2$/Al$_2$O$_3$/Hf$_{0.5}$Zr$_{0.5}$O$_2$/SiO$_x$/Si (MAHAFIS) device degrades rapidly. The degradation of V_{th_ERS} during cycling leads to poor endurance in the MAHAFIS device.

There are two types of trapped charges in the MAHAFIS structure. The first is the trapped charges injected from the metal gate, and the second is injected from the silicon substrate. The former contributes to the V_{th_ERS}, while the latter causes degradation. [3] Two kinds of charges make the physical origin of the V_{th_ERS} degradation in the MAHAFIS structure more complex and unclear.

In this work, we discuss the physical origin of V_{th_ERS} degradation in the MAHAFIS structure. Three factors affecting V_{th_ERS} are discussed: (I) ferroelectric polarization, (II) charges injected from silicon substrate ($Q_{t\_bottom\_IL}$), and (III) charges injected from the metal gate ($Q_{t\_top\_IL}$). Our results indicate that compared to conventional MFIS FeFETs, the main factor for the V_{th_ERS} degradation in MAHAFIS FeFETs is the charge de-trapping of $Q_{t\_top\_IL}$.

II. SAMPLE PREPARATION

We designed the FeFET with two kinds of gate structures, i.e., MAHAFIS and MFIS structures, as schematically shown in Fig. 1(a). The two samples were fabricated using the gate-last process on an 8-inch p-type silicon wafer. The fabrication process flow of the two devices is shown in Fig. 1(b). For the MAHAFIS sample, 8.5 nm Hf$_{0.5}$Zr$_{0.5}$O$_2$, 1.5 nm Al$_2$O$_3$, 2.5 nm HfO$_2$, and 1.5 nm Al$_2$O$_3$ were successively deposited using atomic layer deposition (ALD) at 300 °C in the same chamber. The MFIS sample only underwent the deposition of 8.5 nm Hf$_{0.5}$Zr$_{0.5}$O$_2$. Next, 10 nm TiN and 75 nm W were grown. Then all samples were annealed at 400 °C in N$_2$ for 60 s to achieve the orthorhombic phase. The gate length/width (L/W) of the FeFET in this work is 5/150 μm. Fig. 1(c) shows the High-Resolution Transmission Electron Microscopy (HRTEM) images. The electrical properties were all measured by Keysight B1500A, and the V_{th} is defined by the constant current method.

III. RESULT AND DISCUSSION

Fig. 2 shows the electrical measurement waveform. Fig. 3(a) shows the I_d-V_g curves for both devices. The MAHAFIS device shows an MW of 8.02 V at +10/-9 V, while the MFIS device shows an MW of 0.95 V at +4.5/-4.5 V. The top AHA dielectric interlayer can effectively increase the MW due to the $Q_{t\_top\_IL}$. [3] Fig. 3(b) shows the endurance test results for the MAHAFIS and the MFIS devices. The V_{th_ERS} of the

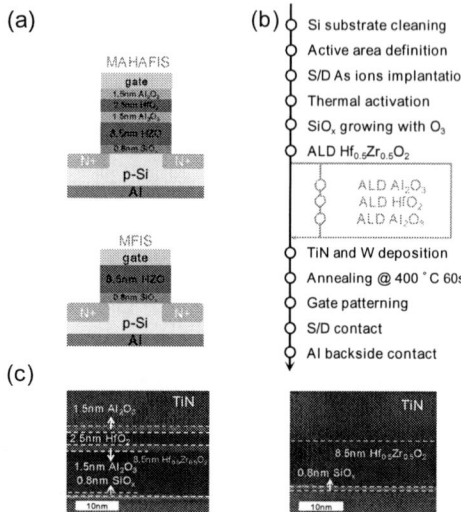

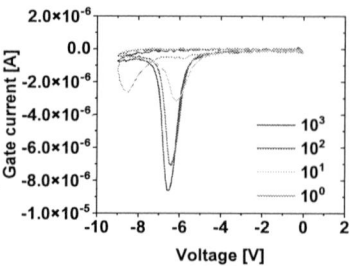

Fig. 4. Ferroelectric polarization switching current at +10/-9 V during 10^3 cycles.

B. $Q_{t\_bottom\_IL}$ injected from the silicon substrate

The charge trapping of $Q_{t\_bottom\_IL}$ leads to a decrease in V_{th_ERS}. Fig. 5(a) shows the endurance test results for the MAHAFIS device under -9 V unipolar stress. The V_{th_ERS} decreases from 6.34 V to 4.16 V after 10^2 cycles. Compared to bipolar stress, the deterioration rate of V_{th_ERS} for the case of unipolar stress has slowed down. Fig. 5(b) shows that the gate leakage current of the MAHAFIS device remains largely unchanged during -9 V unipolar stress cycling. The gate leakage current can reflect the trap density in the gate stack. [7] Therefore, no significant increase in trap density during the unipolar stress test.

We also conducted the unipolar stress test at -4.5 V on the MFIS device. Fig. 5(a) and (c) show the V_{th_ERS} and gate leakage current of the MFIS device during -4.5 V unipolar stress cycling, respectively. The MFIS device shows only a slight decrease in $V_{th,ERS}$, with the gate leakage current remaining largely unchanged. The leakage current reflects no significant increase in trap density, indicating the charge trapping of $Q_{t\_bottom\_IL}$ does not increase during -4.5 V unipolar stress cycling. Therefore, the V_{th_ERS} of the MFIS device remains relatively stable during -4.5 V unipolar stress cycling. Next, the electric field of the SiO_x bottom interlayer (E_{SiO_x}) was calculated using TCAD without considering $Q_{t\_top\_IL}$ and $Q_{t\_bottom\_IL}$. The simulation parameters are summarized in Table I. The simulated gate voltage is -9 V for the MAHAFIS device and -4.5 V for MFIS device. Here, the E_{SiO_x} in the MFIS device is higher than that in the MAHAFIS device. Even with a higher E_{SiO_x}, the V_{th_ERS} of the MFIS device does not degrade significantly. This means the degradation of V_{th_ERS} in the MAHAFIS device is not due to an increase in $Q_{t\_bottom\_IL}$.

Fig. 1. (a) The structure of two devices. (b) The fabrication process flow of two devices. (c) The HRTEM images of the MAHAFIS and MFIS structures.

MAHAFIS device decreases from 6.41 V to 1.93 V during 10^2 cycles, leading to a rapid decrease in MW. The V_{th} shift in the MAHAFIS device can be given as [3]:

$$\Delta V_{th} = \frac{\Delta(Q_{t\_top\_IL} - Q_{t\_bottom\_IL})}{C_{top\_IL}} + \frac{\Delta(P - Q_{t\_bottom\_IL})}{C_{FE}} \quad (1)$$

where $Q_{t\_top\_IL}$ means charges injected from the metal gate. $Q_{t\_bottom\_IL}$ means charges injected from the silicon substrate. C_{top_IL} and C_{FE} are the top interlayer capacitance and background ferroelectric layer capacitance, respectively. P is ferroelectric polarization. Three factors influence the V_{th_ERS}: ferroelectric polarization, $Q_{t\_top\_IL}$, and $Q_{t\_bottom\_IL}$, which are discussed below.

Fig. 2. The pulse sequence of electrical measurement.

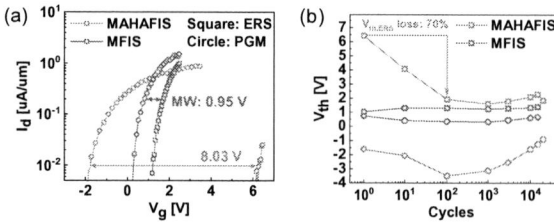

Fig. 3. (a) and (b) Measured I_d–V_g curves of maximum MW during endurance cycling and endurance test results for the MAHAFIS and MFIS devices, respectively.

TABLE I. SIMULATION PARAMETER

Name	Unit
Remnant Polarization	18 $\mu C/cm^2$
Saturation Polarization	20 $\mu C/cm^2$
Coercive Field	1.2 MV/cm
FE Thickness	8.5 nm
SiO_x Thickness	0.8 nm
AHA Thickness	1.5 nm, 2.5 nm, and 1.5 nm

A. ferroelectric polarization

Fig. 4 shows the ferroelectric polarization current at +10/-9 V during 10^3 cycles. The ferroelectric polarization current increases stem from the wake-up effect in the HfO_2-based ferroelectric materials. It improves the ferroelectric polarization and thus increases the V_{th_ERS} rather than its degradation.

979-8-3503-6184-1/24 $31.00 © 2024 IEEE

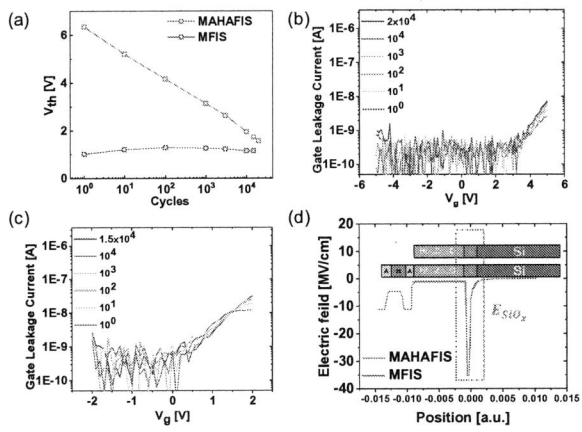

Fig. 5. (a) Endurance test results for the MAHAFIS and MFIS devices under unipolar stress. (b) and (c) Gate leakage current for the MAHAFIS and MFIS devices. (d) The electric field of each layer in gate stacks calculated by TCAD for MAHAFIS and MFIS devices.

C. $Q_{t\_top\_IL}$ injected from metal gate

Fig. 6(a) and (b) show the incremental step pulse programming (ISPP) characteristics and the ISPP slope for both devices. The MAHAFIS device shows a substantial improvement, with an ISPP slope of 5.83, which is 16 times higher than that of the MFIS device. The steep ISPP slope in the MAHAFIS device indicates that trap energy levels may concentrate at a narrow energy range. Such trap distribution enhances the charge trapping of $Q_{t\_top\_IL}$, thereby improving ISPP performance. However, this concentrated trap distribution can also increase the de-trapping of $Q_{t\_top\_IL}$. We deduce that the quality of the AHA top interlayer deteriorates under electric stress, leading to the de-trapping of $Q_{t\_top\_IL}$. Gonzalez *et al.* reported that when a highly negative voltage is applied on the Al_2O_3 films, new traps creation occurs. And the traps form a localized conduction path leading to an enhanced de-trapping of $Q_{t\_top\_IL}$. [8] Interestingly, it seems that the Al_2O_3 film can form local leakage paths without requiring a high trap density. This is evident from the lack of a significant increase in leakage current observed in the MAHAFIS device under unipolar stress. The charge de-trapping of $Q_{t\_top\_IL}$ leads to a decrease in the V_{th_ERS}. The concentrated distribution of trap energy levels exacerbates this situation, causing a significant V_{th_ERS} degradation in the MAHAFIS device during +10/-9 V bipolar stress cycling.

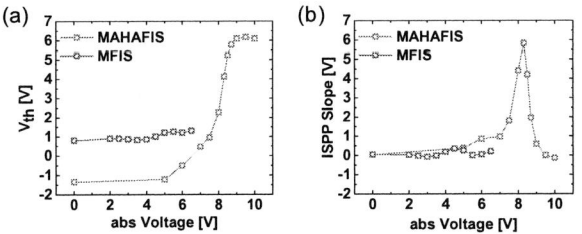

Fig. 6. (a) and (b) ISPP and ISPP slope for the MAHAFIS and MFIS devices, respectively.

The enhancement in data retention serves as additional evidence. Fig. 7 shows the retention characteristics of the MAHAFIS device improve after 10^2 cycles. This improvement can be understood as follows. After 10^2 bipolar stress cycling, there is an enhanced de-trapping of $Q_{t\_top\_IL}$,

leading to a reduction in remaining $Q_{t\_top\_IL}$. Therefore, the de-trapping of $Q_{t\_top\_IL}$ decreased during the retention test after 10^2 cycles. It has been reported that the $Q_{t\_bottom\_IL}$ increases during endurance tests, eventually leading to the degradation of V_{th_ERS} in the MFIS device. [9] However, in the MAHAFIS device, the main factor for the V_{th_ERS} degradation is the enhanced charge de-trapping of $Q_{t\_top\_IL}$.

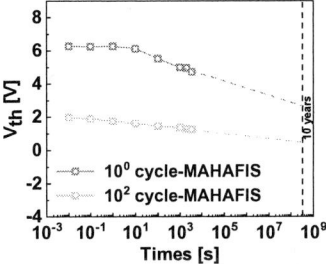

Fig. 7. Retention characteristics measured at room temperature after 10^0 and 10^2 cycles at +10/-9 V bipolar stress for the MAHAFIS device.

IV. CONCLUSIONS

The FeFET device shows a large MW and steep ISPP slope by inserting an AHA top interlayer between the metal gate and ferroelectric layer. However, the rapid degradation of V_{th_ERS} during bipolar stress cycling leads to the reduction of the MW. Unlike conventional MFIS FeFETs, the degradation of V_{th_ERS} in MAHAFIS FeFETs is caused by the de-trapping of charges injected from the metal gate. Bipolar electric stress exacerbates this degradation compared to unipolar stress. Increasing deep level traps in the top interlayer might provide a solution.

ACKNOWLEDGMENT

This work was supported in part by the National Natural Science Foundation of China under Grant Nos. 92264104 and 52350195, in part by National Key Research and Development Program of China under Grant No. 2022YFB4400300, and in part by the R&D Program of Beijing Municipal Education Commission under Grant No.KZ202210009014, in part by the Young Elite Scientists Sponsorship Program under Grant No. BYESS2023033, and Supported by the Postdoctoral Fellowship Program of CPSF under Grant No. GZC20232925.

REFERENCES

[1] D. Das, L. Fernandes, P. V. Ravindran, T. Song, C. Park, N. Afroze *et al.*, in 2024 IEEE International Memory Workshop (IMW), 2024, pp. 1-4.

[2] K. Florent, M. Pesic, A. Subirats, K. Banerjee, S. Lavizzari, A. Arreghini *et al.*, in 2018 IEEE International Electron Devices Meeting (IEDM), 2018, pp. 2.5.1-2.5.4.

[3] S. Lim, T. Kim, I. Myeong, S. Park, S. Noh, S. M. Lee *et al.*, in 2023 International Electron Devices Meeting (IEDM), 2023, pp. 1-4.

[4] S. Yoon, S. I. Hong, D. Kim, G. Choi, Y. M. Kim, K. Min *et al.*, in 2023 IEEE Symposium on VLSI Technology and Circuits (VLSI Technology and Circuits), 2023, pp. 1-2.

[5] T. Hu, X. Sun, M. Bai, X. Jia, S. Dai, T. Li *et al.*, IEEE Electron Device Letters, vol. 45, no. 5, pp. 825-828, 2024.

[6] R. Han, T. Hu, J. Yang, M. Bai, Y. Ding, X. Shao *et al.*, unpublished.

[7] F. Tian, X. Sun, S. Li, S. Xu, J. Chai, J. Xiang *et al.*, IEEE Transactions on Electron Devices, vol. 71, no. 2, pp. 1040-1047, 2024.

[8] M. B. Gonzalez, J. M. Rafi, O. Beldarrain, M. Zabala, and F. Campabadal, Microelectronic Engineering, vol. 109, pp. 57-59, 2013/09/01/ 2013.

[9] X. Shao, J. Chai, F. Tian, S. Zhao, J. Duan, X. Ke *et al.*, IEEE Transactions on Electron Devices, vol. 70, no. 6, pp. 3043-3050, 2023.

Random Number Generation from 3D-NAND Flash Memory Using Shallow Charge Trap Related Short-Term Retention Errors

Ruibin Zhou[1], Jian Huang[1,*], Xianping Liu[1,2], Yuhan Wang[1], Xinrui Zhang[1], Yungen Peng[1], and Zhiyi Yu[1]

[1]School of Microelectronics Science and Technology, Sun Yat-sen University, Zhuhai, 510275, China

[2]Peng Cheng Laboratory, Shenzhen, 518055, China

*E-mail: huangj573@mail.sysu.edu.cn

Abstract—Read noise and operation latency variation are major sources of entropy in existing flash memory-based True Random Number Generators (TRNGs). We proposed that the short-term retention errors caused by the charging and discharging of shallow traps in 3D-NAND flash memory can serve as entropy for TRNGs. By utilizing the back-pattern effect and bypassing the intrinsic determinacy of the design and process of 3D-NAND flash memory, we developed a method with only normal user operations for random number extraction. Using this method, we successfully extracted random bitstreams without post-debiasing process. We evaluated the randomness of the generated bitstream using the NIST SP 800-22 statistical test suite and it passed all 15 tests. This method can be conveniently implemented in electrical systems that use 3D NAND flash memory as a storage medium.

Index Terms—Flash Memory, True Random Number Generators, Shallow Charge Traps, Short-term Data Retention

I. INTRODUCTION

True Random Number Generators (TRNGs) can offer an unbreakable layer of security for different applications, such as secure communications and data encryption, by ensuring unpredictability.Flash memory is considered as a promising candidate for TRNGs due to high density, low power consumption and cost effectiveness [1]. And it typically comes in the form of commodity chips and is widely used in various applications. However, commodity chips typically offer limited usable operations to end-users, making entropy extraction challenging. Currently, only a limited number of reports have demonstrated true random numbers generation in flash memory, typically utilizing read noise or operation latency [2]. These approaches usually involve unconventional operations such as intended interrupt during normal erase/program, which makes practical application complex.

We find that by utilizing the short-term retention errors of 3D NAND flash, true random numbers can be generated through normal memory operations, achieving a bitstream generation throughput comparable to reported flash-based TRNGs. The retention errors are often used to generate true random numbers in volatile memory, such as DRAM [3]. In non-volatile memory, due to the rarity of retention errors within

a short period, they have not been considered for TRNGs. However, the unique short-term retention phenomenon in 3D NAND charge-trapping type flash makes this possible. The short-term retention effect is caused by unstable charges trapped in the shallow trap sites of the memory cells. These charges can detrap within a short time (within seconds), leading to retention errors [4]. It can serve as a potential entropy source for several reasons: 1) These trap sites predominantly result from material inconsistencies, interface anomalies, or defects related to wafer fabrication, making their locations within the cell unpredictable. 2) The charge injection and emission from a trap site are inherently stochastic [5]. 3) Charges can be trapped and subsequently detrapped at various locations within the cell, and the detrapping direction and mechanism may vary [6]. This leads to unpredictable threshold voltage (Vth) fluctuations in individual cells and cause unpredictable read errors.

In this article, we introduce the innovative use of discharging of the shallow charge traps in 3D NAND flash memory as an entropy source to generate true random numbers. By leveraging NAND flash's back pattern effect through routine erase, program and read operations in commercially available 3D NAND flash memory, we uncovered errors caused by the shallow charge traps related short-term retention effect and used the error location information to generate true random numbers.

II. METHOD

In commodity 3D NAND flash memory, the short-term retention issue is not readily apparent from the bit error rate due to its sufficient margin with the designed read reference voltage from manufacturer. To overcome this problem by using only normal user operations, we utilized the back pattern effect (BPE) of NAND flash memory. To enhance the BPE, we designed a special data pattern that programs all bits in a word line (WL) to the P6 Vth state. For TLC NAND flash, which stores three bits per cell, there are a total of eight voltage states (P0-P7), with the P6 Vth state being the second highest. The targeted blocks are programmed with this data pattern.

To get the short-term retention error bitmap in a page, the first read is performed right after the program and the second read is performed after one second. The two read results are

---

This work was supported by Guangdong Province Key Areas R&D Plan Project (Grant No. 2023B0303030004) and the 100 Talents Program of Sun Yat-sen University (Grant No. 76220-12230040).

979-8-3503-6184-1/24 $31.00 © 2024 IEEE

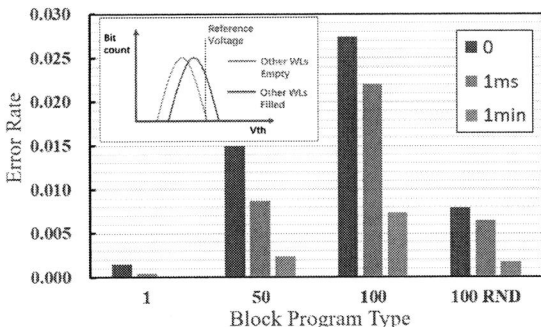

Fig. 1. The error rate of the first WL for different programming scenarios at different read delay time. The number 1, 50 and 100 represents the programmed percentages of a block with P6 state pattern. "100 RND" indicate the block is filled with random pattern after the first WL. The inset shows the schematic of the WL0 Vth distribution before and after the rest of the WLs are programmed.

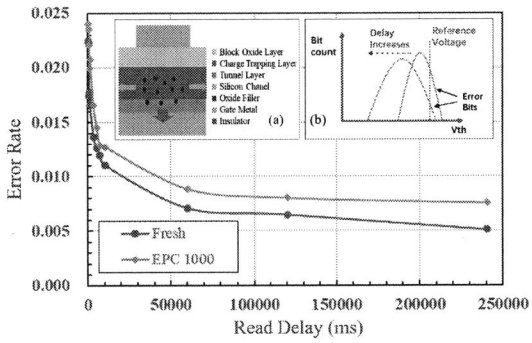

Fig. 2. The short-term retention effect of WL0 characterized by the error rate versus delay time. The inset (a) shows schematic of a memory cell. The Charge Trapping Layer is sandwiched by the oxide layers. The blue dots indicate the charges trapped in the memory cell. The arrows indicate the possible leakage direction of the charges trapped in the shallow traps, and inset (b) shows the schematic of the WL0 Vth distribution right after the program and after the delay time.

then compared and form the page level error bitmap. To extract bitstream from the error bitmap, we used an even-odd method instead of using the absolute address of the error. The even or odd error address read from the chip determines the binary state "0" or "1" of the bitstream. This way mitigates the issue of certain locations being more error-prone due to the process limitations, particularly along the long word lines in NAND flash memory. Considering the design and process systematic of 3D NAND flash [7], 4 bits are grouped together to use as an error address unit. By grouping 4 bits together, we only need to determine if the errors originate from the first half byte or the last half byte within one byte of data to generate the corresponding random bit. Furthermore, to avoid the weak locations in the WL induced by manufacturing processes, we pick locations that exhibit only one bit error within a 32-bit span.

Below is a summary of the bitstream generation process:

(1) Erase the targeted block.
(2) Program all the pages of the targeted block with a fixed data pattern.
(3) Read the pages right after program and then read the same pages after 1 second. XOR two read data to locate the error bits.
(4) For each word in the read page, if the 4-bit data unit address is odd, bit 0 is extracted. Otherwise, bit 1 is extracted.

III. RESULTS AND DISCUSSION

Fig. 1 characterize the BPE for the first WL under various programming scenarios. As the number of programmed WLs increases, more bit errors occur in the earliest programmed WL0. The error bit count from WL0 is highest when the entire block is programmed to the P6 state. Due to the upper tails of the P6 state move up and cross the read reference level between the P6 and P7 states, resulting in a significant number of error bits (Fig. 1 inset).

Fig. 2 inset (a) shows the structure of a memory cell, where the charge trapping layer is sandwiched between two insulator

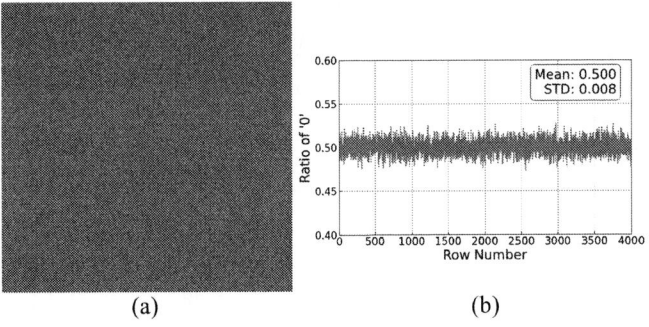

Fig. 3. (a) Two-dimensional bitmap image of a bitstream, measuring 4000 pixels by 4000 pixels. (b) The statistic of the ratio of "0" bit for the rows in the bitmap.

layers. The charges trapped in shallow charge traps may escape from vary directions. We characterized the short-term retention effect of WL0 in Fig. 2. The WL0 is read at different delay time after the whole block is fully programmed. Immediately following the programming, the readout data exhibits a very high error rate. As the waiting time increases, the error rate decreases rapidly. For the page we measured, approximately 65% of the errors disappeared within the 10 seconds for the fresh block, correlating with the fast charge detrapping induced Vth downshift (Fig. 2 inset (b)). This data suggests that most of the error bits observed shortly after program are short-term retention errors, which can be utilized as an entropy source.

Fig. 3(a) shows a two-dimensional bitmap image, measuring 4000 by 4000 pixels, which visualizes the random bitstream generated by our method. In this visualization, black represents '0' bit, while white represent '1' bit. The image reveals a nearly equal distribution of '1' bits and '0' bits throughout the generated bitstream. The uniformity is further confirmed by the ratio statistics in Fig. 3(b). The ratio of bit "0" of each row in the bitmap (consisting of 4000 bits) is calculated across 4000 rows. The mean ratio is 0.500 with a standard deviation of 0.008, indicating that the generated bitstream is bias-free

TABLE I
NIST 800-22 TEST SUITE RESULTS OF THE BITSTREAMS FROM THE PROPOSED METHOD

Test	P-value	Pass Rate	Pass/Fail
Freq.	0.003577	1.000	Pass
BlockFreq. (M = 128)	0.041438	1.000	Pass
CumSum	0.015065, 0.392456	1.000, 1.000	Pass
Runs	0.585209	1.000	Pass
LongRun	0.105618	1.000	Pass
Rank	0.186566	0.952	Pass
FFT	0.311542	0.952	Pass
NonOverlapTemp (m = 9)	0.002465 (min.)	0.995 (avg.)	Pass
OverlapTemp (m = 9)	0.186566	1.000	Pass
Univ.	0.689019	1.000	Pass
AppEntropy (m = 10)	0.186566	1.000	Pass
RandExc.	0.213309 (min.)	1.000 (avg.)	Pass
RandExcVar.	0.066882 (min.)	1.000 (avg.)	Pass
Serial (m = 16)	0.875539, 0.311542	1.000, 1.000	Pass
LinComplex. (M = 500)	0.585209	0.952	Pass

and does not require a de-biasing process as it's necessary with other flash memory TRNGs. This simplifies the process of random number generation and enhances actual throughput of the TRNG [8].

To validate the randomness of the bitstream, the proposed method was assessed using the NIST SP 800-22 statistical test suite. We generated 21 bitstreams, each consisting of 10^6 bits. As indicated in Table I, the bitstreams passed all 15 tests, confirming the randomness of the TRNG output. It should be noted that in some of the tests the proportion is less than 1, indicating that some bitstreams failed specific tests. This is common for the reported TRNGs [9], [10], and the test suite allows a permissible number of failures.

From the latency of the flash memory operation, we can estimate the throughput of the TRNG form the following equation:

$$\text{Throughput} = \frac{\text{Bit\_Count}}{n \times t_{\text{erase}} + m \times t_{\text{program}} + l \times t_{\text{read}}} \quad (1)$$

where Bit_Count is the number of generated bits, m, n and l represent the number of erase, program, and read operations, respectively. The latencies for erase, program and read operations are denoted as t_{erase}, $t_{program}$ and t_{read}, respectively. Since our method does not require a post de-biasing process, there is no loss of bits from de-biasing, allowing all generated bits to be utilized. Here, 2.1×10^7 bits in the Fig. 3(a) were generated from 30 cycles of erasing and programming across 60 blocks, with 30 page reads per block for each cycle. We measured an average read latency of 67 us/page, a programming latency of 530 us/page, a block erase

latency of 4800 us/block. The 1-second retention time can be largely concealed by the program latency through strategically sequencing the read and program operations across different blocks. From equation (1), the throughput is estimated to be 28.05 Kbit/s. We need to note that the speed is calculated from a single die with no parallel plane operation. In a 3D NAND flash memory system, multiple planes and dies operating in parallel can significantly increase the throughput.

IV. CONCLUSION

In this paper, we proposed that the shallow traps related short-term retention errors in 3D NAND flash memory can serve as a new kind of entropy source for the commercial flash memory based TRNGs. Our approach successfully generated the random bitstreams with a speed of about 28.05 Kbit/s without any post debiasing process. Notably, all memory chip operations involved are standard flash memory erase, program and read operations. The generated bitstreams are successfully validated by the NIST SP 800-22 statistical test suite. Our study demonstrate that the shallow traps related retention errors in 3D NAND flash memory can be implemented in commercial memory system for convenient true random number generation.

REFERENCES

[1] Y. Wang, W.-k. Yu, S. Wu, G. Malysa, G. E. Suh, and E. C. Kan, "Flash memory for ubiquitous hardware security functions: True random number generation and device fingerprints," in *2012 IEEE Symposium on Security and Privacy*. IEEE, 2012, pp. 33–47.

[2] H. Gordon, J. Edmonds, S. Ghandali, W. Yan, N. Karimian, and F. Tehranipoor, "Flash-based security primitives: Evolution, challenges and future directions," *Cryptography*, vol. 5, no. 1, p. 7, 2021.

[3] C. Keller, F. Gürkaynak, H. Kaeslin, and N. Felber, "Dynamic memory-based physically unclonable function for the generation of unique identifiers and true random numbers," in *2014 IEEE international symposium on circuits and systems (ISCAS)*. IEEE, 2014, pp. 2740–2743.

[4] M. Kim and H. Shin, "Analysis and compact modeling of fast detrapping from bandgap-engineered tunneling oxide in 3-d nand flash memories," *IEEE Transactions on Electron Devices*, vol. 68, no. 7, pp. 3339–3345, 2021.

[5] Y. V. Gomeniuk, R. Litovski, V. Lysenko, I. Osiyuk, and I. Tyagulski, "Current stochasticity of field emission of charge from traps in the transition layer of implanted mis structures," *Applied surface science*, vol. 59, no. 2, pp. 91–94, 1992.

[6] H. Mertens, R. Ritzenthaler, A. Hikavyy, M.-S. Kim, Z. Tao, K. Wostyn, S. A. Chew, A. De Keersgieter, G. Mannaert, E. Rosseel *et al.*, "Gate-all-around mosfets based on vertically stacked horizontal si nanowires in a replacement metal gate process on bulk si substrates," in *2016 IEEE symposium on VLSI technology*. IEEE, 2016, pp. 1–2.

[7] P. Nowakowski, M. Ray, P. Fischione, and J. Sagar, "Top-down delayering by low energy, broad-beam, argon ion milling—a solution for microelectronic device process control and failure analyses," in *2017 28th Annual SEMI Advanced Semiconductor Manufacturing Conference (ASMC)*. IEEE, 2017, pp. 95–101.

[8] Z. Zheng, Y. Zhang, M. Huang, Z. Chen, S. Yu, and H. Guo, "Bias-free source-independent quantum random number generator," *Optics Express*, vol. 28, no. 15, pp. 22 388–22 398, 2020.

[9] S. Chakraborty, A. Garg, and M. Suri, "True random number generation from commodity nvm chips," *IEEE Transactions on Electron Devices*, vol. 67, no. 3, pp. 888–894, 2020.

[10] G. Kim, J. H. In, Y. S. Kim, H. Rhee, W. Park, H. Song, J. Park, and K. M. Kim, "Self-clocking fast and variation tolerant true random number generator based on a stochastic mott memristor," *Nature communications*, vol. 12, no. 1, p. 2906, 2021.

979-8-3503-6184-1/24 $31.00 © 2024 IEEE

Simulation of Program/Erase Cycling and Retention Loss in 3-D CTF NAND Flash

Anuj Kumar, Ravi Tiwari and Souvik Mahapatra*

Department of Electrical Engineering, Indian Institute of Technology Bombay, Mumbai 400076, India

* Corresponding and Presenting author; Phone: +91-222-572-0408, Email: souvik@iitb.ac.in

Invited Paper

Abstract: **RDD model is implemented in Sentaurus TCAD and standalone modes to simulate trap generation in tunnel oxide of 3-D CTF NAND. Equivalence of trap generation kinetics is shown between P/E bipolar cycling, unipolar cycling and DC. TCAD generated few P/E bipolar cycling kinetics is extended to large number of P/E cycles (used in products) by standalone with unipolar pulses and also TCAD with equivalent-time DC. Generated trap density from RDD model is used in ABDWT model to simulate data retention loss after P/E cycling. Model results show good match with experimental data.**

Keywords: **3-D CTF NAND, P/E cycling, Data retention, RDD model, ABDWT model.**

I. INTRODUCTION

Data retention (DR) is an important qualification metric of a nonvolatile memory like 3-D Charge Trap Flash (CTF) NAND. It is well known that tunnel oxide (TO) traps play a crucial role in DR loss by triggering multiple mechanisms like Trap Assisted Tunneling (TAT) and Detrapping (DT) [1]-[5]. Program/Erase (P/E) cycling causes trap generation (TG) in TO and increases DR loss during post cycling bake. Although a product under actual use undergoes P/E cycling over its useful life, it is replicated with a distributed cycling scheme during qualification, where certain % of cycles are done in rapid succession at a given temperature (T), and the remaining % of cycles are done with inserted delay at the same or a different T [2], [6]. Recent results from 3-D CTF NAND show: (a) TG density has a power-law relation with cycle count, with power factor of 0.5, (b) there is no impact of inserted delay between cycles on TG density, and (c) TG density does not change during post-cycling retention bake [2]. The TAT process is directly related to TG density and demonstrate the above behavior. However, the DT process involves detrapping of trapped electrons in generated TO traps, and show impact of inserted delay and retention bake (but the density of traps does not change). It is important to understand the mechanism governing trap generation (TG) in TO during P/E cycling and subsequent DR, and develop a suitable framework for modeling these effects. Recently, the Activated Barrier Double Well Thermionic (ABDWT) model is successfully used to explain experimental DR loss after cycling [3]-[5]. It uses TO trap density as an input.

The Reaction-Diffusion-Drift (RDD) framework is used to simulate experimental TG features in Bias Temperature Instability (BTI), Stress Induced Leakage Current (SILC) and Time Dependent Dielectric Breakdown (TDDB) stress in logic devices [7]. Logic devices are subjected to unipolar stress, and the RDD model implementations show identical TG time kinetics between deterministic 1-D standalone and 3-D TCAD and stochastic 3-D standalone (mean) versions [7]. The 1-D standalone RDD model was used to simulate TG during distributed P/E cycling in NAND [8]. However, unipolar pulses were used and TG was assumed to occur in the E step. Recently, RDD model is implemented in TCAD under bipolar pulses, suitable for simulating TO-TG during P/E cycling in NAND [9], [10]. A basic setup under limited bipolar P/E cycling count (#20) is shown in [9]. The setup is enhanced in [10] to handle higher P/E cycle count (#100), and a matched standalone RDD model with unipolar pulses is used to extend P/E cycle count to ~10K used in products. ABDWT model with RDD generated traps is used to model DR loss data [10].

It is not possible to directly validate TO-TG with cycling, it is always indirectly validated using post-cycling DR loss data with ABDWT model (where trap density is an input) [3]-[5], [10]. Hence it is necessary to check the step-by-step correlation between logic-level unipolar pulses and NAND level bipolar pulses, to independently substantiate the RDD model under bipolar pulses. First, a thick logic-equivalent capacitor (LEC) structure is used to connect RDD unipolar and bipolar pulse results. Next, the model is used in a 2-D cross-sectional NAND string structure to link bipolar pulse results from LEC structure. To overcome the limitation of TCAD execution time (limiting P/E cycling count to ~100), similarity of bipolar pulse TG kinetics to that from unipolar pulse (standalone implementation, extended to 10K cycles) and DC (TCAD) is shown. TO-TG results obtained from RDD model is used in ABDWT model to match DR data.

II. BACKGROUND

Fig.1 shows the 2-D cross sectional schematic of a 3-D CTF NAND with different charge loss mechanisms during retention bake [1]. DR is divided into in-cell (TAT and DT) and intercell (lateral migration, LM) contributions, with the former impacting both solid pattern (SP) and checkerboard pattern (CP) data, and later impacting only CP data (in SP, all word-lines (WLs) are at same programmed state, while in CP, the WLs are in different programmed state). TO-TG during P/E cycling impact TAT and DR [1]-[5]. Fig.1 also shows measured cell V_T distributions (CVD) for different programmed state, before and after retention bake. A left-shift is observed due to DR, and a particular sigma value is usually relevant for product read-window budget.

Fig.2 shows the schematic of RDD model. Hydrogen (H) passivated defects (X-H) are dissociated at or near the poly-Si channel/TO interface during P and E pulses. Released H atoms diffuse and initiate subsequent reactions to dissociate other H passivated defects and Oxygen (O) related bonds (Y-H, Z-H/O) in the TO bulk. Released H_2 molecule or ion (H_2^+, OH^-) species diffuse or drift in the TO and beyond. In LEC structure, only one-sided diffusion is considered in the gate insulator and metal gate [7]. In NAND, the Nitride (N) in ONO based TO acts as a blocker, and it forces sidewise diffusion and drift (considered in bottom O of TO and poly-

Si channel) [9], [10]. Broken X-, Y- and Z- are generated traps, the reactions are solved at two different interfaces for simplicity, first at the poly-Si/channel (density N_{OT1}) and the subsequent one in the TO bulk (density N_{OT2}). In logic, $N_{OT1}=N_{OT2}$ as all H atoms from first reaction participate in the second, while in NAND, N_{OT2} is slightly less than N_{OT1} due to sidewise diffusion. The reaction (K_{F1}/K_{R1}) related to H release depends on channel electrons/holes, TO electric field (E_{OX}) and T. All other reactions and diffusivities are Arrhenius T activated. The parameter K_{F3} determines ion to molecule ratio and determine TG time kinetics. Two RDD models are simultaneously solved irrespective of unipolar or bipolar pulses. N_{OT1} and N_{OT2} are defined as electron and hole related (eTrap and hTrap) for $+V_G$ and $-V_G$ (in case of bipolar) or 0 (in case of unipolar) pulse.

Fig.2 also shows the schematic of ABDWT model [3]. A trap (E_1) is separated from the reservoir (channel, E_2) by a barrier (E_B), which is Normally distributed with T activated mean (E_{BM}) and spread (E_{BS}). The TO electric field during retention bake from a programmed state initiates over-the-barrier loss from E_1, which is used for TAT and DR [3]-[5].

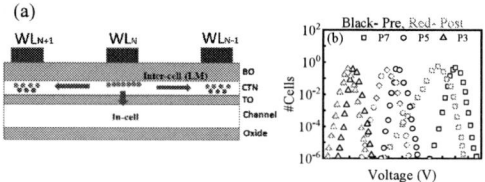

Fig.1. (a) 2-D cross section schematic of a CTF NAND showing different DR loss mechanisms, (b) CVD for different programmed states before and after DR bake.

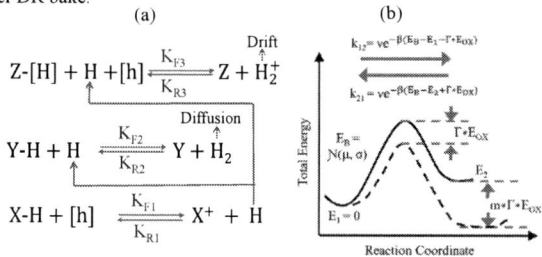

Fig.2. Schematic of (a) RDD model and (b) ABDWT model.

III. LOGIC EQUIVALENT CAPACITOR RESULTS

Step-by-step validation is done for DC, unipolar AC with $+V_G$ and bipolar AC with $+V_G/-V_G$ pulse. Fig.3 shows the Sentaurus TCAD generated structure and simulated H, H_2, H_2^+ and OH^- species from DC RDD model simulation.

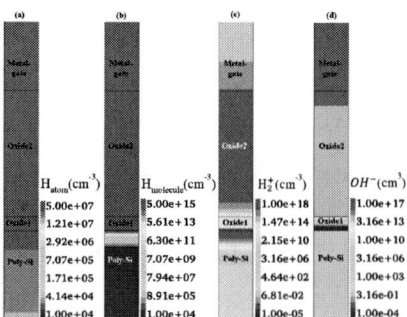

Fig.3. RDD model simulated density of (a) H, (b) H_2, (c) H_2^+ and (d) OH^- at the end of DC bipolar $+V_G/-V_G$ segments of duration 1Ks/1Ks.

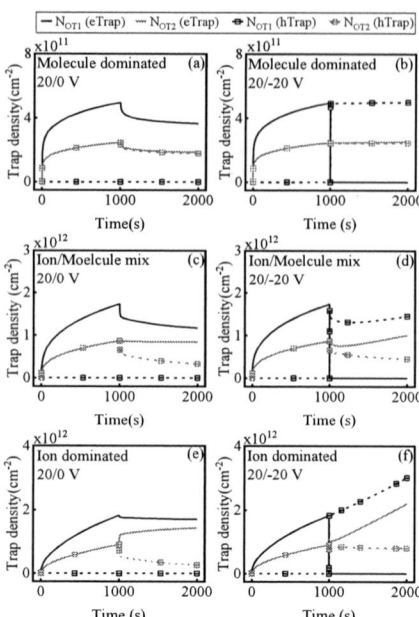

Fig.4. RDD model simulated eTrap and hTrap density for (left) unipolar and (right) bipolar DC sequences, for (top) molecule, (mid) molecule-ion mixed and (bottom) ion dominated cases.

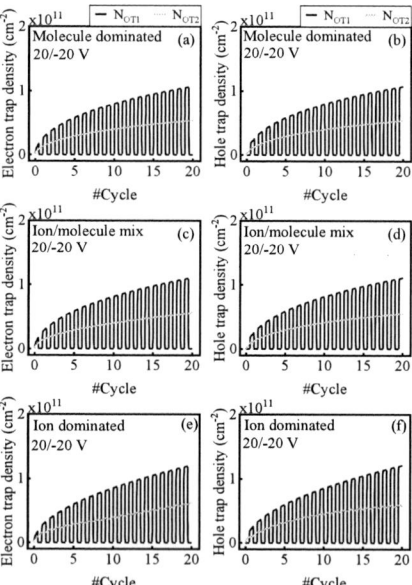

Fig.5. RDD model simulated (left) eTrap and (right) hTrap density related to N_{OT1} and N_{OT2} for bipolar AC pulses, (top) molecule, (mid) molecule-ion mixed and (bottom) ion dominated cases are shown.

Fig.4 plots the RDD model simulated trap time kinetics under DC unipolar (left panels) and bipolar (right panels) biasing, with molecule dominated, mixed molecule and ion, and ion dominated cases shown respectively in top, middle and bottom panels. Only eTrap is seen for N_{OT1}, while both for N_{OT2} for $+V_G/0$ as H from first reaction trigger second, while both eTrap and hTrap are seen for N_{OT1} and N_{OT2} for $+V_G/-V_G$. The time kinetics between unipolar $+V_G/0$ and bipolar $+V_G/-V_G$ cases are different for the second segment of simulation, and more importantly, a clear impact is seen depending on the relative dominance of molecules or ions.

If ions are present, OH^- is generated and are pulled towards gate (drift) at $+V_G$; they either diffuse or pushed back (drift) in second segment respectively at 0V (for unipolar) or $-V_G$ (for bipolar), and H_2^+ is generated and pulled towards gate for bipolar case. At long-time, TG kinetics is determined by diffusion/drift of molecule/ion species.

Fig.5 plots the RDD model simulated eTrap (left panels) and hTrap (right panels) density related to N_{OT1} and N_{OT2} versus P/E cycle count under bipolar AC pulses. Fig.6 plots the RDD model simulated total trap density (sum of N_{OT1} and N_{OT2}, sum of eTrap and hTrap for both N_{OT1} and N_{OT2}) for successive DC bipolar biasing (left panels) and bipolar AC pulse (right panels), for different bias combinations for $+V_G/-V_G$. All results are from TCAD, and molecule, mixed molecule-ion and ion dominated cases are plotted in top, middle and bottom panels respectively. Once again, the DC second segment results are very different depending on the relative contributions from molecules and ions, Fig.6 (a, c, e), similar to Fig.4, while the AC results appear to be very similar between different cases shown in Fig.5 and Fig.6 (b, d, f). Note, this apparent discrepancy is due to the time scale involved in DC (long) and AC (short, due to few P/E cycles). DC long-time data are governed by the diffusion/ drift phases, while AC short-time data are governed by the reaction phase, and therefore, DC results depend on, while the AC results do not depend on molecule to ion ratio. For DC, equal $+V_G/-V_G$ sequences result in continuous buildup of total traps, while unequal $+V_G/-V_G$ sequences result in recovery (if $+V_G > |-V_G|$) or buildup (if $+V_G < |-V_G|$). A net buildup is seen for AC irrespective of $+V_G/-V_G$ magnitude. Since equal pulse width and identical parameters are used for eTrap and hTrap, identical total trap density is seen for $+20V/-16V$ and $+16V/-20V$ simulations.

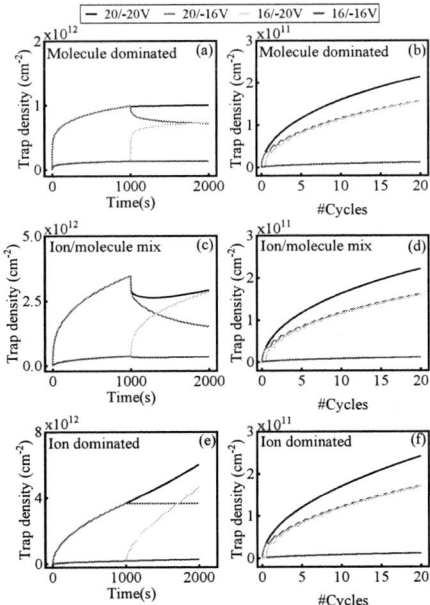

Fig.6. RDD model simulated total trap density under (left) consecutive DC and (right) AC pulses, different combinations of bipolar $+V_G/-V_G$ biases are used, for (top) molecule, (mid) molecule-ion mixed and (bottom) ion dominated cases.

Fig.7 plots the RDD model simulated total trap density under bipolar AC pulse, for different combinations of P/E bias (left panels) and T (right panels), from TCAD. Only a limited (#100) P/E cycle count is possible due to run-time

limitation in TCAD. The standalone RDD model is used under unipolar pulse ($+V_G/0$), and matched to replicate the bipolar TCAD results for all cases. The standalone model is used to extend the cycling to larger (10K) count. In both versions, simulations are done using molecule, mixed and ion dominated cases, shown respectively in top, middle and bottom panels. Larger timescales are involved due to higher P/E cycle count, and now the results are different between different cases and depends on the molecule to ion ratio. A fully molecule or ion dominated case result in power-law dependence on P/E cycle count with a slope of 0.17 and 0.5 respectively, while intermediate slopes can be obtained for the mixed molecule-ion case (0.33 in this case). Note that when distributed cycling is used, the recovery of TG in the time gap between cycles depends on the relative dominance of molecules or ions, with no recovery seen for the case of complete ion domination [8]. Finally, TCAD based RDD simulation under DC is used to replicate the results for all the three molecule/ion ratio cases, under all bias and T.

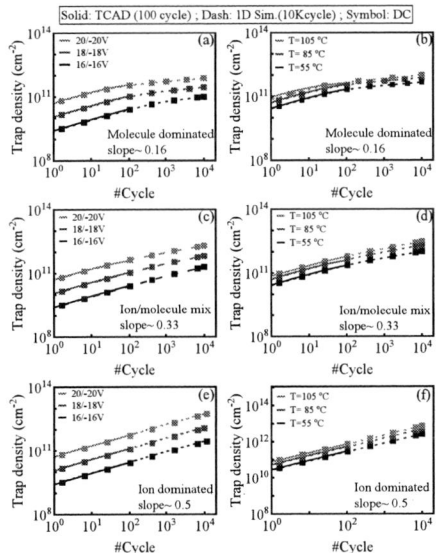

Fig.7. RDD model simulated total trap density under bipolar AC pulses from TCAD, for (left) different combinations of $+V_G/-V_G$ and (right) T, for (top) molecule, (mid) molecule-ion mixed and (bottom) ion dominated cases. Results from standalone RDD model under unipolar pulse and DC TCAD are also shown.

IV. NAND STRUCTURE RESULTS

Fig.8 shows the 2-D cross sectional cut of a 3-D NAND string used for TCAD simulation, with 3 consecutive WLs and Select FETs. The biasing scheme during P and E are shown, with electron and hole density in poly-Si channel respectively. The generated trap profiles along the NAND string at channel poly-Si/TO interface (N_{OT1}) and inside TO bulk (N_{OT2}) are plotted after 100 P/E cycles. Since P bias is applied to target WL while E bias to all WLs, N_{OT1} eTrap is only under target WL while N_{OT1} hTrap is under all WLs. However, N_{OT2} is spread out along the entire NAND string for both eTrap and hTrap, as this is caused by diffusion of H released from the first reaction.

Fig.9 plots the RDD model simulated total trap density under bipolar AC pulse, for different combinations of P/E bias (left panels) and T (right panels), from TCAD. Results from few cycles (#20) are shown in top panels, while in the bottom panels, results are shown up to #100 cycles. Once again, the standalone model is used under unipolar pulse, and calibrated with TCAD data and extended to 10K cycle counts for all cases. Results are also replicated using TCAD DC simulation. The ion dominated case is simulated, hence the resulting TO-TG density from TCAD and standalone versions show power law dependence on cycle count, with 0.5 slope, which holds for all P/E cycling bias and T. One of the cycling conditions is same as in [4], and reproduces the TO-TG density that model data retention after different P/E cycle counts.

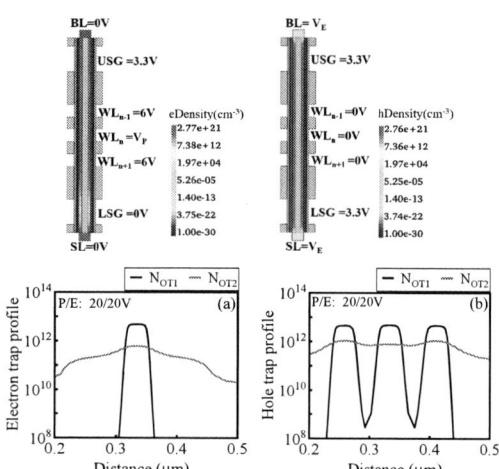

Fig. 8. 2-D schematic of 3-D NAND string under P and E bias conditions, electron and hole densities in the channel are shown (top). Simulated trap densities at the first and second interfaces are shown after 100 P/E cycles.

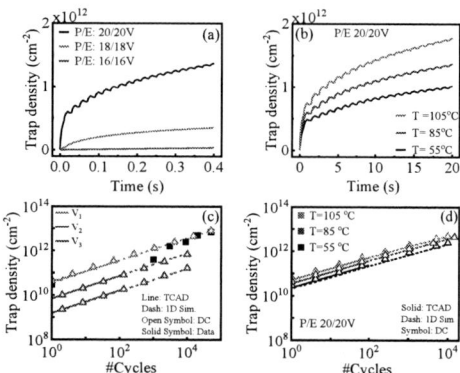

Fig. 9. RDD model simulated trap density under bipolar pulse from TCAD, for various combinations of (left) $+V_G/-V_G$ bias and (right) T, for (top) 20 cycles and (bottom) 100 cycles. Results from standalone RDD model under unipolar pulse and DC TCAD are also shown. All simulations are for ion dominated case. Symbols (bottom, left) are data for one case, which matches DR results (Fig. 10).

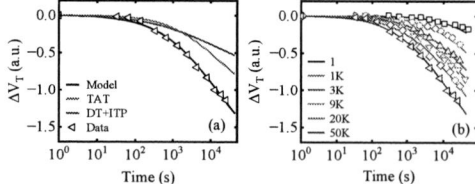

Fig. 10. ABDWT modeling of measured DR loss data, (a) after one P/E cycling case with DT and TAT subcomponents, and (b) after different P/E cycling cases, under SP mode.

RDD model generated TO traps versus P/E cycle for one case of Fig. 9 is used in ABDWT model to simulate DR loss characteristics during SP retention bake after different P/E cycles. Fig. 10 (a) plots the measured and modeled overall DR loss for one P/E cycle condition and the underlying DT and TAT subcomponents. Fig. 10 (b) plots the modeling of measured DR loss for different P/E cycling conditions.

V. CONCLUSION

A step-by-step validation of trap generation time kinetics from TCAD implemented RDD model is demonstrated, by using logic equivalent capacitor and NAND structures. DC unipolar and bipolar sequences are used to understand the behavior under AC unipolar and bipolar pulses. Identical time kinetics of generated traps is observed between bipolar and unipolar AC and DC, and also between standalone and TCAD implementations. This is valid across different bias and T conditions, and also for model parameter choices that determine power-law slope of generated traps versus cycle count. The run-time, and therefore cycle count limitation of TCAD bipolar cycling is overcome by using the standalone unipolar or TCAD DC simulation. RDD model generated traps are used in ABDWT model to reproduce DR loss data during SP measurements after different P/E cycle counts.

ACKNOWLEDGEMENT

Rashmi Saikia and Himanshu Rai for retention modeling. Mohit Bajaj of Synopsys for help with TCAD setup.

References

[1] C. Woo et al., "Modeling of Charge Failure Mechanisms during the Short Term Retention Depending on Program/Erase Cycle Counts in 3-D NAND Flash Memories," IEEE Int. Rel. Phys. Symp. (IRPS), 2020, pp. 1-6.

[2] G. Nicosia, N. Righetti and Y. Dong, "Distributed Cycling in Charge Trap-Based 3D NAND Arrays: Model and Qualification Tests Implications," IEEE Int. Mem. Works. (IMW), 2023, pp. 1-4.

[3] R. Saikia, A. Ansari and S. Mahapatra, "A Physics-based Model for Long Term Data Retention Characteristics in 3D NAND Flash Memory," IEEE Int. Rel. Phys. Symp. (IRPS), 2023, pp. 1-6.

[4] R. Saikia, H. Rai and S. Mahapatra, "Modeling of Post-Cycling Retention Bake in 3-D CTF TLC NAND Arrays," IEEE Int. Rel. Phys. Symp. (IRPS), 2024, pp. 1-6.

[5] K. Thakor et al., "Comprehensive physics-based modeling of post-cycling long-term data retention in 176L 3-D NAND Flash Memories," IEEE Int. Mem. Works. (IMW), 2024, pp. 1-4.

[6] N. Mielke, H. P. Belgal, A. Fazio, Q. Meng and N. Righos, "Recovery Effects in the Distributed Cycling of Flash Memories," IEEE Int. Rel. Phys. Symp. (IRPS), 2006, pp. 1-6.

[7] S. Mahapatra et al., "A Generic Trap Generation Framework for MOSFET Reliability—Part I: Gate Only Stress–BTI, SILC, and TDDB," in IEEE Trans. Electron Devices, vol. 71, no. 1, pp. 114-125, 2024,

[8] T. Samadder, S. Kumar, K. Thakor and S. Mahapatra, "A Theoretical Framework for Trap Generation and Passivation in NAND Flash Tunnel Oxide During Distributed Cycling and Retention Bake," IEEE Int. Rel. Phys. Symp. (IRPS), 2024, pp. 1-6.

[9] A. Kumar, R. Tiwari, M. Bajaj, D. Dolgos, L. Smith and S. Mahapatra, "Study of Trap Generation in NAND Flash Tunnel Oxide using TCAD," IEEE Electron Devices Tech. and Manf. (EDTM), 2024, pp. 1-3.

[10] A. Kumar, R. Saikia, A. S. Bisht, R. Tiwari, H. Rai and S. Mahapatra, "Modeling of Trap Generation in 3-D NAND Charge Trap Flash Memory," Int. Conf. on Sim. of Semiconductor Processes and Devices, 2024, in press.

[11] Sentaurus Device user guide, Version T-2022.03

Switch-off mechanisms in GeAsTe Ovonic Threshold Switching Selector Device

Z. Hu[1], Z. Chai[2], W. Zhang[1], Jian Zhang[1]

[1]School of Engineering, Liverpool John Moores University, Liverpool L3 3AF, UK,
[2]Xi'an Jiaotong University, China

Abstract

Understanding the switching process in ovonic threshold switching (OTS) devices is an important research topic. Less attention has been paid to the fast switching-off process in the past, especially in modern nanoscale OTS devices. In this work, the OTS switching-off process in 1S1Rs operations are investigate. The underlying mechanisms can be explained by the dynamic resistance of OTS induced by the transition of defect clusters, and the impact of series resistance value on the switching off process is revealed. The defect cluster shrinking stage and rupture stage can be measured and characterized, respectively, which correspond to two distinct switch-off mechanisms. This research sheds new light on OTS switching mechanism and its impact on 1S1Rs operation.

1. Introduction

Selector device plays a key role in suppressing the sneak path currents in the crossbar arrays that utilize the emerging non-volatile memory devices [1, 2], such as the phase change memory (PCRAM) [3], the Redox based and oxide based resistive switching memory (ReRAM/OxRAM) [4]. The selector is required to have fast volatile switching characteristics and is connected in series with the non-volatile memory device. When the memory device is not selected, it is biased at a half of the operation voltage, at which the selector should have very low off-state current and can suppress the overall leakage current and avoid the sneak current path. When the memory device is selected at the operation voltage, the selector is at the ON state, and should allow a sufficiently high current to pass through so that the memory device can be programmed. A large on-off current ratio is an essential requirement for the selector, therefore [5]. Among the competing technologies, OTS selectors have achieved fast switching speed in the order of ns, high ON-state current density larger than $20MA/cm^2$, high half-bias nonlinearity larger than 10^5, and excellent endurance $>10^{12}$ cycles [6-8].

Despite the recent progress, further detailed electrical experimental evidence is needed to investigate the switching off process. Several new observations in GeAsTe OTS switching are investigated in this work, including the full OTS quasi-static switching-off process and two distinct mechanisms of OTS switch-off: the OTS switches off at zero total impedance of 1S1Rs when the Rs value is small, and it switches off at the minimum OTS current when Rs is large. It is found that the switching off process is associated with the defect cluster shrinking kinetics, which becomes Rs independent after the normalization. On the other hand, the rupture criteria of the defect cluster are dependent on Rs. Significant impact of Rs on the 1S1Rs operations and parameters are also identified.

2. Devices and Experiments

A 20nm amorphous GeAsTe chalcogenide film is deposited by room temperature physical vapor deposition (PVD) and passivated with a low-temperature BEOL process. The TiN/GeAsTe/TiN selector uses a pillar (TiN) bottom electrode that defines the device size down to 50 nm and integrates with a serial resistor Rs in the range of 1.8 kΩ to 400 kΩ in a 300 mm process flow, as shown in **Fig.1a**. The I-V characterization was carried out by using a Keithley 4200A semiconductor analyzer either with the SMU for DC sweep or with the embedded 4225-PMU ultrafast I-V module for voltage sweep. A typical DC I-V of the 1S1R measured by a triangular pulse is shown in **Fig. 1b,** from which the OTS-only I-V can be obtained by subtracting the voltage across the series Rs, $V_{OTS}=V_{1S1R}-I\cdot Rs$, since the linear I-V at ON-state is dominated by Rs, as illustrated in **Fig.2a**, in which V_{th} is the threshold voltage for off-to-on switching, V_{hold} is the voltage where the on-to-off switching occurs. Note that these values are different for 1S1R and OTS-only. The threshold current and holding current, on the other hand, are the same for 1S1R and OTS-only due to the series connection. The I-V curves with various Rs values are shown in **Fig. 2b** for the 1S1R, and in **Fig. 2c** for the extracted OTS-only, respectively. The larger Rs value leads to lower on-state and holding current, and larger threshold and holding voltage for the 1S1R. However, the OTS-only I-Vs in **Fig. 2c** follow a similar trajectory during the switching off process, demonstrating the validity of this extraction method, as the OTS in these devices have the same configuration and parameters.

979-8-3503-6184-1/24 $31.00 © 2024 IEEE

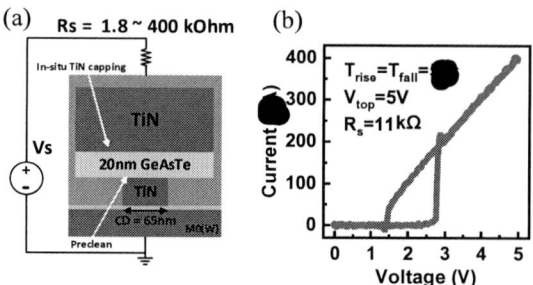

Fig. 1 Fig. 1 (a) Illustration of the OTS structure in series with a resistor Rs. (b) A typical I-V of 1S1Rs measured by a triangular pulse. Pulse conditions and Rs value are labelled. Device size (CD) is 65nm unless specified otherwise.

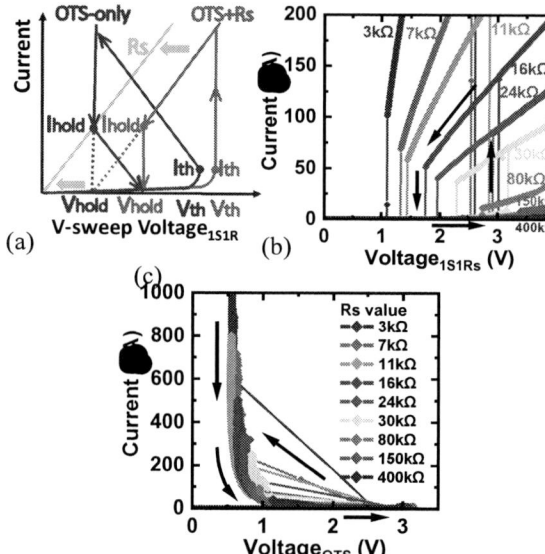

Fig. 2 (a) Illustration of OTS-only I-V extraction. (b) I-V during DC voltage sweep measured for 1S1Rs with 3kΩ ≤Rs ≤ 400kΩ, and (c) the extracted I-V for OTS-only.

3. Results and discussions

The switching process in OTS has been described by different models, including the earlier thermally induced instability model [9], the electronic injection induced space charge model [10], the impact ionization induced generation and recombination model [11], the thermally assisted charge hopping model [11], the field induced nucleation model [12], and more recently, the electric field induced local bond modification models [13-16]. In the latest models, the defects transition from the ground state to the excited state at a high field due to the local

bonds modification and become delocalized, so that the defect clusters are formed which lead to the off-to-on switching, and vice-versa for the on-to-off switching at a low electric field where the local bonds configuration recovered, and defects return to ground state and become localized again.

Furthermore, the operation of 1S1Rs in which the selector is connected with a series resistor (Rs) were also investigated [17, 18], and it was suggested that the Rs plays an important role in the switching-on and switching-off process. For example, it was found that Rs defines the load line for the selector and hence its operation points and the holding current, so that the selector is either switching volatilely between the static off-state and on-state, or oscillating in the transitional negative-differential-resistance (NDR) state [19-22]. Despite these early efforts, detailed experimental evidence is still lacking for characterizing the fast and sharp switching off process in modern OTS devices. For example, it is not clear how the switching process and parameters in nanoscale OTS devices are affected by the series resistance and what are the corresponding OTS switching mechanisms, which should provide important information for understanding the operation and modelling of the 1S1R circuit.

A detailed inspection on the OTS-only I-V curves can reveal more significant differences in the switching process. For smaller Rs values in the range below 30 kΩ, the I-V curves largely follow the same trajectory, as shown in Fig. 2c, agreeing with previous results [23]. However, as shown in Fig.3a, for larger Rs values at and above 80 kΩ, a significant non-linear ON state (NL-ON) region is observed, before the OTS switches to the OFF state, which deviates from the linear ON state that is defined by the Rs value. The origin of this NL-ON region needs to be investigated, as it could cause large errors if the Rs values in the 1S1Rs structure are read out from the current measured within this region. To rule out the possible contribution to this non-linear deviation by the test methods and speed, the results of DC voltage sweep, DC current sweep, and pulse voltage sweep with Rs = 80 kΩ are compared in Fig. 3b and Fig. 3c, respectively. The NL-ON region is observed and overlaps in all three cases, and similar results are also observed at other Rs values larger than 80kΩ. This confirms that the NL-ON is not caused by the test methods or speed used in our measurements, hence the instrument settings and the RC effects in the probe and cable connections are not responsible for the occurrence of NL-ON.

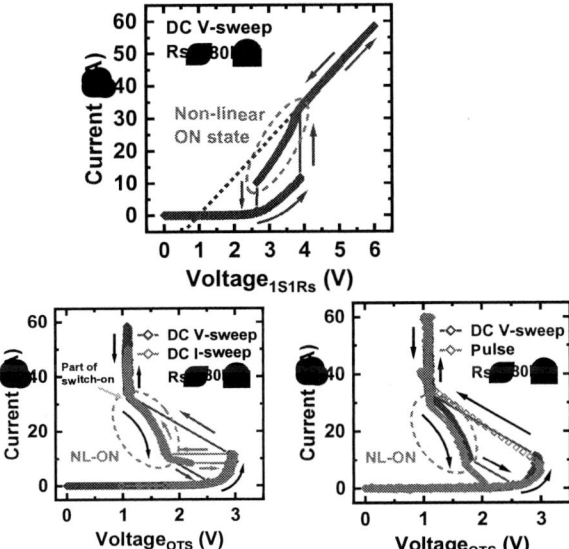

Fig. 3 (a) A significant non-linear ON state (NL-ON) region is observed before the OTS switches to OFF state with Rs≈80 kΩ. (b) Both DC V-sweep and I-sweep show NL-ON that overlap with each other during switch-off, and also overlap with a part of switch-on of DC V-sweep. (c) Good agreement is observed in DC and pulse V-sweep, supports that NL-ON is not caused by the measurement speed in our setup.

It is well known that the OTS device exhibits a negative differential resistance region during the switching process, as evident in **Fig. 3**. For the off-to-on voltage sweeps, the OTS current increases abruptly at V_{th}, and reaches the ON state at near the top end of, and in some cases just within the NL-ON region. For the on-to-off voltage sweeps, the OTS goes through the NL-ON region as the current reduces, until it reaches the holding point where the device switches off abruptly. The negative differential resistance in this large NL-ON region can be clearly seen, in which $dV_{OTS}/dI_{OTS} < 0$.

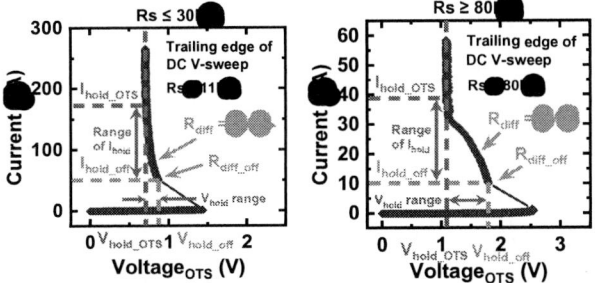

Fig. 4 Definition of Vhold_OTS, Ihold_OTS, Vhold_off and Ihold_off for OTS-only (a) when Rs≤30kΩ, and (b) when Rs≥80kΩ.

In contrast, when Rs is smaller than 80 kΩ, a much subtler

NL-ON region is observed, as shown in Fig. 4a for Rs =11 kΩ, where the subtler NL-ON only slightly deviates from the linear ON region. This confirms that this significant difference in the OTS switching off process can only be caused by the difference in Rs values. A part of the UP traces when reaching the ON state also falls within the NL-ON region and overlap with the DOWN traces in Fig.3, suggests that the NL-ON is a common feature for both the switch-on and -off processes.

As described in the latest models, the defects return from the delocalized excited state to the localized ground state during the switching-off process, so that the conductivity of OTS decreases. From the above analysis, this process starts whilst the OTS is still at ON-state and its negative differential resistance increases when the electric field is getting smaller. To characterize the NL-ON region in the switching-off process, $R_{diff\_OTS} = \delta V_{OTS}/\delta I_{OTS}$ is defined as the dynamic differential resistance of OTS-only, as shown in Figs. 4a&b, with smaller and larger Rs values, respectively. R_{diff_OTS} is negligible in the linear ON state, where the linear I-V is dominated by Rs. The I-V curves enter the NL-ON at $(I_{hold\_OTS}, V_{hold\_OTS})$, where the increase of $|R_{diff\_OTS}|$ is no longer negligible in comparison to Rs, due to the defect clusters shrinking with the decreasing current [14]. The OTS switches off at $(I_{hold\_off}, V_{hold\_off})$ when the defect clusters rupture, and this is the OTS holding point commonly used for 1S1R [13, 14, 24], where $R_{diff\_OTS} = R_{diff\_off}$. The switching-off process is clearly different for Rs≤30 kΩ and Rs≥80 kΩ as Fig.4 shows it is controlled by different mechanisms.

The impact of Rs on the boundary condition of the NL-ON region, i.e., $(I_{hold\_OTS}, V_{hold\_OTS})$ and $(I_{hold\_off}, V_{hold\_off})$, is further examined. **Fig. 5** shows that I_{hold_OTS} decreases inversely with Rs, agreeing with previous works that follows a linear line in the log-log scale [20]. At the onset point of the NL-ON region, the total differential resistance is dominated by Rs, so the inverse relation between I_{hold_OTS} and Rs is as expected, since the OTS operates with Rs as the load, and $(I_{hold\_OTS}, V_{hold\_OTS})$ is the on-set of defect cluster shrinking in the OTS when the differential resistance of OTS starts to become negative. I_{hold_off} also decreases inversely when Rs is small (Rs ≤ 30kΩ), as expected. But for larger Rs values (Rs ≥ 80kΩ), I_{hold_off} no longer decreases with Rs and remains at a constant minimum value just below 10 µA. This provides clear evidence that there are two distinct switching-off mechanisms, associated with small and large Rs values, respectively. As shown in **Fig.5b**, for the smaller Rs, OTS switches off when $|R_{diff\_OTS}|$ reaches Rs (≤30 kΩ), where the total impedance of the 1S1Rs circuit, $Rs+R_{diff\_OTS}$, becomes 0. For larger Rs, the maximum value of $|R_{diff\_OTS}|$, $|R_{diff\_max}|$, becomes

979-8-3503-6184-1/24 $31.00 © 2024 IEEE   496

saturated and can no longer reach Rs (80kΩ-400kΩ), hence the total impedance of 1S1Rs remains positive. The OTS current must decrease to a minimum holding level to trigger the switch-off, as shown in **Figs. 5a**.

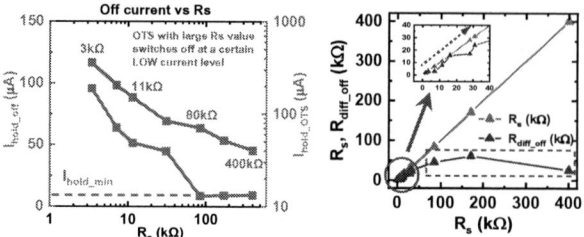

Fig.5 (a) I_{hold_OTS} and I_{hold_off} vs Rs for OTS-only. OTS switches OFF at a common I_{hold_min} when 80kΩ ≤Rs≤400kΩ. (b) R_{diff_off} is plotted vs Rs. For smaller Rs (inset), OTS switches off when Rs+R_{diff_OTS} =0. For larger Rs, the maximum R_{diff_OTS} is less than 80 kΩ, OTS switches OFF when it reaches the minimum I_{hold}.

As shown in **Fig. 6a**, the inverse of the slopes of normalized I-V curves in the NDR region has a peak in the middle of the NL-ON region, where the R_{diff_OTS} reaches the maximum value, for both the OTS-only and the 1S1R operations. Hence the shrinking speed of the defect cluster accelerates at the beginning of NL-ON. After the peak, the speed starts to reduce as the current decreases further and eventually the clusters rupture at the minimum current for Rs≥80kΩ.

It is observed in **Fig.6b** that both I_{OTS} and V_{OTS} show two regions in the NL-ON range, (I_{hold_OTS} - I_{hold_off}) and (V_{hold_off} -V_{hold_OTS}), and its Rs dependence. One is from ~2kΩ to 30 kΩ, and this is the region that $|R_{diff\_OTS}|$ can become large enough to reach the smaller Rs value, in which both I_{hold_off} and V_{hold_off} change slightly; and the other one is from 80kΩ to 400kΩ where the OTS switch off at the same I_{hold_min} level and at a much larger V_{hold_off}. Hence the full OTS switch-off characteristics and two different switch-off mechanisms are revealed.

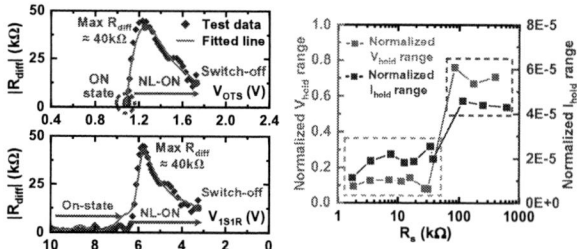

Fig.6 (a) I_{hold_OTS} and I_{hold_off} vs Rs for OTS-only. OTS switches OFF at a common I_{hold_min} when 80kΩ ≤Rs≤400kΩ. (b) R_{diff_off} is plotted vs Rs. For smaller Rs (inset), OTS switches off when Rs+R_{diff_OTS} =0. For larger Rs, the maximum R_{diff_OTS} is less than 80 kΩ, OTS switches OFF when it reaches the minimum I_{hold}.

When the OTS current reduces, the defect clusters start to shrink, and the defect number in the clusters reduces, so that $|R_{diff\_OTS}|$ increases and V_{OTS} increases. When $|R_{diff\_OTS}|$ reaches the Rs (≤30kΩ), the total impedance of 1S1Rs becomes 0 and the OTS is out of equilibrium and switches OFF at V_{hold_off}. For larger Rs (≥80kΩ), the maximum $|R_{diff\_OTS}|$ value can no longer reach Rs, the OTS is forced to operate at NL-ON state in a low current regime, where I_{OTS} keeps decreasing until it is below the minimum OTS holding current, then the conductive cluster is broken and the OTS switches OFF [14]. Hence the I-V trajectory in the NL-ON is determined by defect clusters.

4. Conclusions

Experimental evidence in this work shed new insights into the full switching-off process of GeAsTe OTS and its dependence on the series resistance value. When the Rs is small, the OTS switches off when the total differential resistance reaches zero. When the Rs value is large, the OTS switches off at a constant minimum OTS holding current level. The changes in $|R_{diff\_OTS}|$ during the switching-off process can be attributed to the defect cluster shrinking kinetics in the OTS and can be therefore measured and characterized. The resulted non-linear ON region and the dependence on the resistance level in the memory element have a significant impact on the parameters and simulation of 1S1R operation.

Acknowledgments

EPSRC UK grant EP/S000259/1 & EP/Y008235/1. The authors would like to thank colleagues at IMEC, Belgium, for test samples and fruitful discussions.

References

[1] G. Burr, et al, J Vac. Sci. & Tech. B, 2014. [2]. J. T. Zhou, et al, IEEE TED, 2014. [3]. H. Cheng, et al, IEDM, 2018. [4]. D. Robayo, et al, IMW, 2019. [5]. S. Jia, et al, Nat Commun, 2020. [6]. Y. Koo, et al, VLSI, 2016. [7]. B. Govoreanu, et al, VLSI 2017. [8]. F. Hatem, et al, Int. Elct. Dev. Meet, 2019. [9]. D. Eaton, et al, J. American Ceramic Society, 1964. [10]. N. Mott, Philosophical Magazine, 1971. [11]. D. Adler, J. Appl. Phys., 1980. 17. [12]. V. G. Karpov, Appl. Phys. Lett., 2007. [13]. S. Kabuyanagi, et al, VLSI 2020. [14]. R. Degraeve, IRPS, 2021. [15]. S. Clima, et al, Physica Status Solidi-Rapid Research Lett., 2020. [16]. P. Noe, et al, Sci Adv, 2020. [17]. J. M. Lopez et al, IMW2021. [18]. C. Wu, et al, VLSI, 2021. [19]. Pryor et al, J. Non-Crystalline Solids, 1972. [20]. A. J. Hughes, et al, J. Non-Crystalline Solids, 1975. [21]. Y. Yu, et al, ACS Appl. Electr. Mat., 2020. [22]. S. Lavizzariet et al, IEEE TED, 2010. [23]. D. Garbin, et al, IEDM, 2019. [24]. W. Devulder, et al, Thin Solid Films, 2022.

Orthorhombic-I (Pbca) Phase: Origin of Anti-ferroelectricity in HfZrO Films

Wei Liu[1], Zeping Weng[1, 2, *], Jianguo Li[1], Wenchao Yan[3], Yiming Qu[3], and Yi Zhao[1, 2, 3, *]

[1]College of Information Science and Electronic Engineering, Zhejiang University, Hangzhou 310027, China
[2]State Key Laboratory of Silicon and Advanced Semiconductor Materials, Zhejiang University, Hangzhou 310027, China
[3]School of Integrated Circuits, East China Normal University, Shanghai 200241, China

* Email: {yizhao, wengzp}@zju.edu.cn

Abstract—**In this work, we experimentally investigate the anti-ferroelectric behavior of $Hf_{1-x}Zr_xO_2$ (HZO) films, depending on the film thickness and Zr content. It was found that the orthorhombic-I phase is the fundamental source of anti-ferroelectric properties in HZO. Additionally, reversible transitions between the tetragonal and orthorhombic-I phases were observed to indirectly contribute to anti-ferroelectric behavior in HZO films. Moreover, through a comprehensive evaluation of polarization capability and degradation, HZO films with an Hf/Zr atomic ratio of 1:4 and thickness ranging from 6 to 10 nm have been verified to possess relatively robust anti-ferroelectric properties.**

Keywords—*anti-ferroelectric, HZO, orthorhombic-I, phase transition, tensile stress*

I. INTRODUCTION

Hafnium-based ferroelectric (FE) materials, particularly $Hf_{1-x}Zr_xO_2$ (HZO), have garnered significant attention in emerging memory applications due to complementary metal oxide semiconductor (CMOS) compatibility, excellent scalability, and potential for high-density 3D vertical integration [1, 2]. Despite these advantages, the endurance of FE HZO films remains inferior to that of traditional PZT-based ferroelectrics [3]. In contrast, Zr-rich HZO (including ZrO_2) anti-ferroelectric (AFE) films have attracted increasing interest due to their remarkable higher endurance [3]. However, challenges such as low polarization and undesirable wake-up and fatigue effects during cycling remain problematic and are closely related to film thickness and Zr content [4]. To date, there is a lack of systematic experimental studies addressing the effects of both film thickness (<10 nm) and Zr content on the AFE behavior of HZO.

In this study, the AFE properties of HZO thin films with varying thicknesses (5, 6, 8, 10, and 12 nm) and different Zr contents (Hf/Zr=1:1, 1:3, 1:4, 1:7, and pure ZrO_2) were systematically investigated. The experimental results indicate that HZO(1:4) films exhibit the most robust AFE performance at thicknesses of 6-10 nm among all AFE HZO films. Additionally, it was confirmed that the orthorhombic-I (o-I) phase directly contributes to the AFE properties of HZO films, while the tetragonal (t-) phase must first transform into the o-I phase to exhibit AFE characteristics indirectly. Furthermore, the reversible/irreversible phase transitions among the t-, o-I, and orthorhombic-III (o-III) phases are identified as the cause of the differences in AFE performance

across various AFE HZO films. This study also demonstrates that the asymmetry in AFE properties is related to the asymmetric tensile stress of the top and bottom electrodes.

II. EXPERIMENTS

Fig. 1 shows the detailed W/TiN/HZO/TiN/Si capacitor device fabrication process. HZO films were deposited on a heavily-doped p-type Si substrate coated with a TiN electrode using Plasma-ALD at a wafer temperature of 280 °C. The deposition rates of HfO_2 and ZrO_2 were both 0.1 nm per cycle. TEMAH, TEMAZ, and plasma O were used as the Hf, Zr, and O precursors, respectively. Both top and bottom 15 nm-thick TiN electrodes were deposited at 400 °C, using $TiCl_4$ and NH_3 as the Ti and N precursors. After the sputtering deposition of the 100 nm-thick W capping layer, the annealing process was performed for 60 s at 500 °C in an atmosphere of nitrogen using a rapid thermal annealing system. A conventional photolithography/etching process was used to fabricate metal-insulator-metal (MIM) capacitors with a diameter of 80 μm. 100 nm Al was deposited by thermal evaporation for the back contact. The different ratios of Hf/Zr were controlled by the ALD cycles. It was confirmed by X-ray photoelectron spectroscopy (XPS) that the atomic ratio of Hf/Zr is equal to the ratio of ALD cycles (data not shown). For convenience, we denote HZO films fabricated by different Hf:Zr atomic ratios (e.g., x:y) as HZO(x:y). The electrical characteristics of AFE capacitors were measured using the Keysight B1500A semiconductor device analyzer and the Liryder F3000 ferroelectric analyzer [5]. Cs-corrected scanning transmission electron microscopy (STEM) in annular brightfield (ABF) mode was also carried out for material characterization.

III. RESULTS AND DISCUSSION

A. Anti-ferroelectric properties of different HZO films

Fig. 2 shows the capacitance-voltage (C-V) curves measured at 100 kHz for different devices, illustrating that the AFE properties of HZO films are significantly influenced by film thickness and Zr content. Under an electric field of 3 MV/cm, reduction in film thickness suppresses negative

This work was supported in part by the National Key Research and Development Program of China under Grant 2020AAA0109001, and in part by the "Pioneer" and "Leading Goose" R&D Program of Zhejiang Province under Grant 2023C01018, and in part by the Shanghai Pujiang Program under Grant 22PJD019.

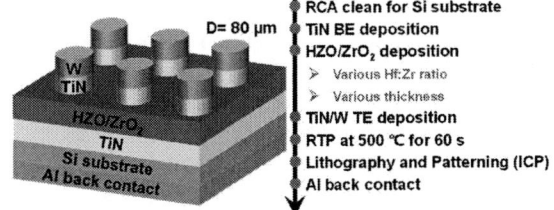

Fig. 1. Process flow of W/TiN/HZO/TiN/Si capacitor devices

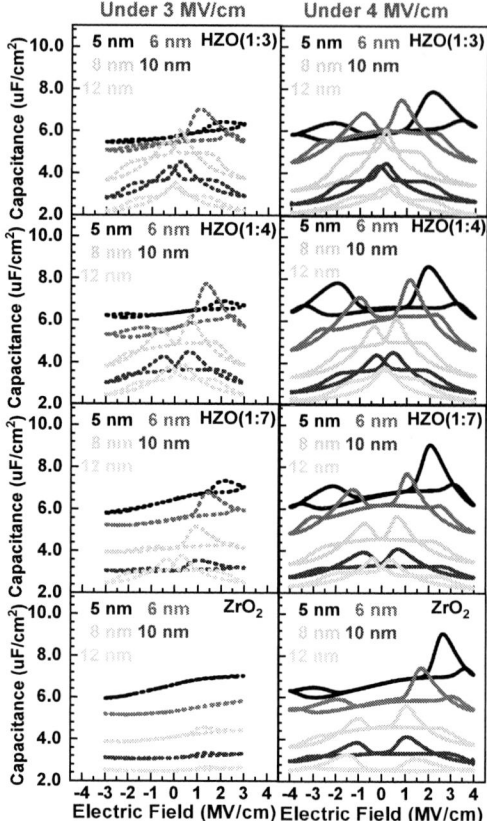

Fig. 2. The *C-V* curves of HZO capacitors with different thicknesses and Hf/Zr atomic ratios under 3 MV/cm and 4 MV/cm, respectively.

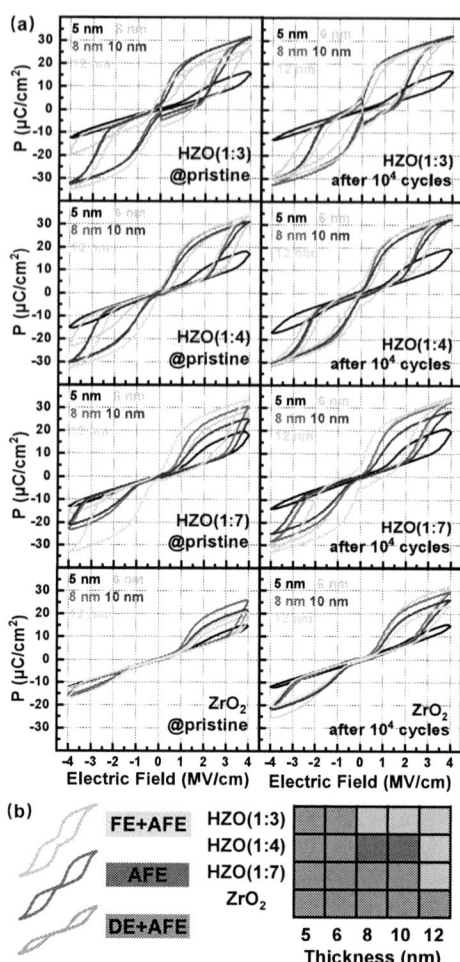

Fig. 3. (a) The *P-E* curves of HZO capacitors with different thicknesses and Hf/Zr atomic ratios. (b) The types of AFE behaviors in different HZO films.

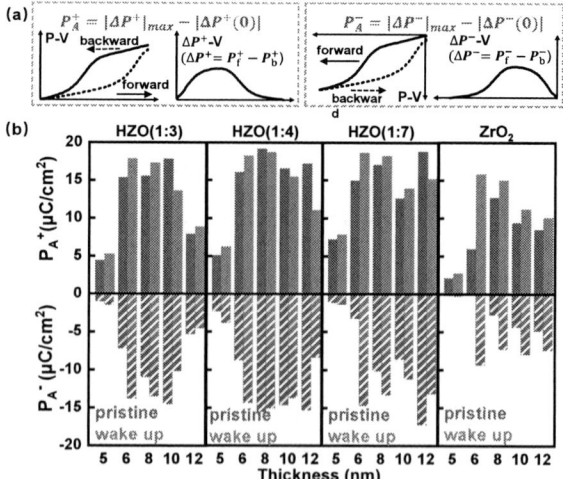

Fig. 4. (a) Definition of P_A^+ and P_A^-. (b) The distribution of P_A^+ and P_A^- values as functions of film thickness and Hf:Zr ratios.

peaks. Particularly in pure ZrO_2, films of various thicknesses exhibit dielectric (DE) behavior. When the electric field reaches 4 MV/cm, the AFE properties of all films become pronounced. For HZO(1:3) films with thickness exceeding 6 nm, the *C-V* curves likely show a superposition of AFE and FE peaks. HZO(1:7) and pure ZrO_2 films show more DE properties when the thickness is below 8 nm. Within the broad thickness range of 6 to 10 nm, HZO(1:4) films clearly exhibit two pairs of symmetrical butterfly-like peaks.

Fig. 3(a) depicts the polarization-voltage (*P-E*) curves of all films before and after cycling under the electric field of 4 MV/cm. The waveform used for measurement is a bipolar triangular wave with a frequency of 25 kHz, while a square wave with a frequency of 250 kHz is used for cycling. According to the results of the *P-E* loops (pristine state), all devices can be roughly classified into three types as shown in Fig. 3(b): (i) FE with partial AFE, (ii) totally AFE, and (iii) AFE with partial DE. It is evident that only the 8 nm and 10 nm HZO(1:4) films exhibit stable AFE properties. In contrast, thinner films or those with higher Zr content require varying degrees of wake-up operations to display good AFE properties. On the other hand, thicker films or those with lower Zr content tend to exhibit partial FE characteristics.

To quantitatively evaluate the polarization of AFE dipoles in different AFE HZO films, two additional parameters, namely positive net polarization (P_A^+) and negative net polarization (P_A^-), were defined (Fig. 4(a)). Fig. 4(b) shows the P_A^+ and P_A^- values of all films. It can be observed that, compared to other films, HZO(1:4) films with

a thickness range of 6-10 nm have relatively higher P_A^+ and P_A^- values, with almost no change before and after cycling.

979-8-3503-6184-1/24 $31.00 © 2024 IEEE

Specifically, the 8 nm HZO(1:4) film has the highest P_A^+ and P_A^- values, at 19 and 15 µC/cm², respectively.

B. Origin of Anti-ferroelectricity in HZO Films

To understand the mechanism behind the different AFE behaviors, we analyzed the lattice structures of 6 nm and 10 nm HZO(1:4) films (Fig. 5(a)). It was found that the 6 nm HZO(1:4) film shows the *o-I* phase lattice structure, whereas the 10 nm HZO(1:4) film exhibits the *o-III* phase lattice structure. Fig. 5(b) provides a comparison of the *t-*, *o-I*, and *o-III* phases in HZO films. These results indicates that the *o-I* phase is the direct source of the AFE properties in HZO films.

Based on the above findings, we summarize two pathways for the AFE properties of HZO films (Fig. 5(c)). The first pathway involves HZO films with the *o-I* phase as the dominant crystalline phase, which exhibit good AFE properties without extra wake-up process and feature large P_A^+/P_A^- values and low polarization switching fields. The second pathway involves HZO films predominantly in the *t-*phase, which require large electric field wake-up operations to exhibit good AFE behavior. Notably, for films that are too thick or have low Zr content, the *o-I* phase is not very stable and can transform into the *o-III* phase during cycling, leading to degradation of AFE properties.

C. Asymmetric properties induced by electrode stress

The asymmetric loop is observed in the *C-V* and *P-E* curves, with the polarization capability of the AFE dipoles being notably better in the positive loops compared to the negative loops. This phenomenon may be closely related to the stress from the electrode. Owing to the lower thermal expansion coefficient of W, the top electrode can induce significant tensile stress on the HZO layer during the cooling stage of annealing [6].

Fig. 6(a) shows the displacement of oxygen atoms in the sub-lattices (SLs) and the corresponding polarization states under different voltage biases. The stability of the *o-I* and *o-III* is primarily determined by the combined effects of bond stretching energy (U_{bond}) and dipole-dipole interaction energy ($U_{P-P,int}$), as shown in Fig. 6(b). When an electric field is applied, the displacement of oxygen atoms polarizes the film, increasing the $U_{P-P,int}$. This effect causes dipole repulsion, increasing the spacing between the two SLs and stabilizing

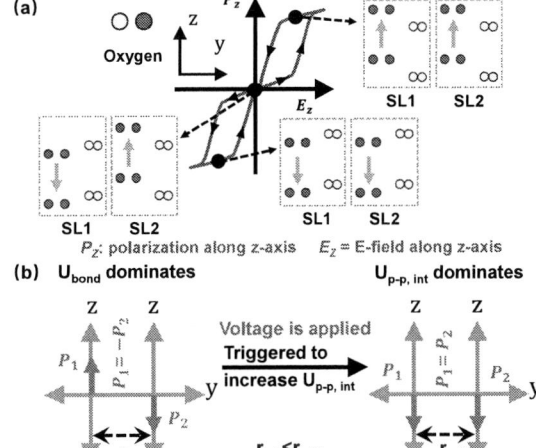

Fig. 6. (a) The internal oxygen lattice configuration and the corresponding P_z versus E_z characteristics in the AFE HZO. (b) Diagram of the transition from *o-I* phase to *o-III* phase.

the *o-III* phase under high electric fields [7]. Considering the larger tensile stress near the top electrode, the lattice spacing near the top electrode is greater than near the bottom electrod. Thus, oxygen atoms can more easily move upward and overcome U_{bond}, explaining why the P_A^+ values are higher than the overall P_A^- values for all films.

IV. CONCLUSION

In this study, it was found that HZO films with an Hf/Zr atomic ratio of 1:4 and a thickness between 6 and 10 nm exhibit optimal AFE characteristics. Additionally, our results indicate that the AFE properties of HZO films mainly originate from the *o-I* phase, and the different AFE behaviors is closely related to phase transitions among the *t-*, *o-I*, and *o-III* phases. Furthermore, it was determined that the asymmetry in the *C-V* and *P-E* curves is caused by significant tensile stress from the top electrode, making the films more easily polarized under positive voltage.

REFERENCES

[1] T. S. Böscke, J. Müller, D. Bräuhaus, U. Schröder, and U. Böttger, "Ferroelectricity in hafnium oxide thin films," *Appl. Phys. Lett.*, vol. 99, no. 10, p. 102903, Sep. 2011.

[2] J. Müller *et al.*, "Ferroelectric hafnium oxide: A CMOS-compatible and highly scalable approach to future ferroelectric memories," *in 2013 IEEE International Electron Devices Meeting (IEDM)*, Washington, DC, USA, Dec. 2013, p. 10.8.1-10.8.4.

[3] M. Pesic *et al.*, "Built-In Bias Generation in Anti-Ferroelectric Stacks: Methods and Device Applications," *IEEE J. Electron Devices Soc.*, vol. 6, pp. 1019–1025, 2018.

[4] M. H. Park *et al.*, "Surface and grain boundary energy as the key enabler of ferroelectricity in nanoscale hafnia-zirconia: a comparison of model and experiment," *Nanoscale*, vol. 9, no. 28, pp. 9973–9986, 2017.

[5] Zhejiang Liryder Technologies Co., LTD, "Zhejiang Liryder Technologies Co., LTD," 2024. [Online]. Available: https://www.liryder.com/en/.

[6] P. Jiang *et al.*, "Stress Effects of Interconnecting Metals on Back-End-of-Line Compatible Hf$_{0.5}$Zr$_{0.5}$O$_2$ Ferroelectric Capacitors," *IEEE Electron Device Lett.*, vol. 44, no. 4, pp. 602–605, Apr. 2023.

[7] A. K. Saha, B. Grisafe, S. Datta, and S. K. Gupta, "Microscopic Crystal Phase Inspired Modeling of Zr Concentration Effects in Hf$_{1-x}$Zr$_x$O$_2$ Thin Films," *in 2019 Symposium on VLSI Technology*, Kyoto, Japan, Jun. 2019, pp. T226–T227.

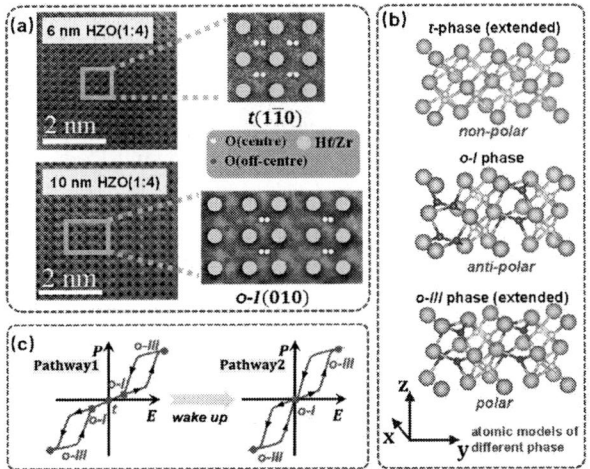

Fig. 5. (a) STEM-ABF images of 6 nm and 10 nm HZO(1:4) films. (b) Different phase structures of HZO. (c) Two pathways of AFE behavior in HZO films.

The maximum storage capacity of open-loop written RRAM is around 4 bits

Yongxiang Li [1], Shiqing Wang [1], Zhong Sun*[1,2]

[1] School of Integrated Circuits, Institute for Artificial Intelligence, Peking University
[2] Beijing Advanced Innovation Center for Integrated Circuits

* Email: yongxiang.li@stu.pku.edu.cn, zhong.sun@pku.edu.cn

Abstract—There have been a plethora of research on multi-level memory devices, where the resistive random-access memory (RRAM) is a prominent example. Although it is easy to write an RRAM device into multiple (even quasi-continuous) states, it suffers from the inherent variations that should limit the storage capacity, especially in the open-loop writing scenario. There have been many experimental results in this regard, however, it lacks a comprehensive analysis of the valid multi-bit storage capability, especially in theoretical terms. The absence of such an insight usually results in misleading conclusions that either exaggerate or underestimate the storage capacity of RRAM devices. Here, by the concept of information theory, we present a model for evaluating the storage capacity of open-loop written RRAM. Based on the experimental results in the literature and the test results of our own devices, we have carefully examined the effects of number of pre-defined levels, conductance variation, and conductance range, on the storage capacity. The analysis leads to a conclusion that the maximum capacity of RRAM devices is around 4 bits.

Keywords—multi-level memory, information theory, storage capacity, RRAM, variation

I. INTRODUCTION

With the rapid development of technologies such as internet of things, artificial intelligence, and advanced wireless communications, the amount of data generated have been increasing exponentially [1]. It has caused a strong demand for high-density information storage, which, in turn, flourishes the study on emerging memory technologies, such as RRAM, and phase-change memory (PCM), *etc.* [2]. They combine the nonvolatility property and other high-performance indicators [3]. In particular, the underlying physical mechanisms allow these devices to be programmed in an analog manner, through continuous modulation of the internal state variables [4]. Such a capability is also highly desired in emerging computing paradigms, such as in-memory computing and analog matrix computing [5, 6].

The multi-level RRAM device may be programmed by using various schemes, such as those based on the modulation of compliance current, magnitude or width of gate or electrode voltage, or number of voltage pulses. Ideally, the device conductance may be quasi-continuously modulated, resulting in several tens to hundreds of discrete levels [7, 8]. However, in the open-loop writing with no repeated corrections, such results are prone to be overestimated, due to the inherent device variations. On the other hand, the test results of large-volume devices often give relatively conservative estimations (~3 bits), by using programming strategies such as incremental step pulse and incremental gate voltage [9-13]. Such results are purely empirical, lack of sufficient theoretical guidance for continuous optimization.

The storage capacity of RRAM devices may indeed be enhanced by using the strict while tedious closed-loop verify scheme [7, 14], where each target level with a pre-defined conductance window can be achieved through tens of set or reset operations, resulting in well separated states with no overlap. However, multiple cycles of verification cause high costs of programing latency and energy dissipation, and repeated switching transitions are also harmful to the endurance performance of RRAM devices.

In order to explore the achievable maximum storage capacity of RRAM devices with open-loop writing (which in turn may be used as an optimization guideline), in this work we have studied this issue with the concept of information theory. A comprehensive analysis has been conducted to reveal the impact of number of levels, conductance variation and conductance range on the storage capacity. The results suggest that, under the consideration of common conditions of RRAM devices, particularly the obstacle set by the conductance variation, the maximum storage capacity of open-loop written RRAM is around 4 bits.

II. STORAGE CAPACITY MODEL

A. Definition of Storage Capacity

Given that the RRAM device conductance can be determined by external parameters, we consider the open-loop writing of RRAM devices as an analog coding problem. As shown in Fig. 1, the external electrical input is encoded as the conductance output stored in the RRAM device through an operation with inherent noise that finally present an output distribution [15]. Each level features a specific mean conductance value G and a standard deviation σ_G. This probabilistic correspondence between input and output can be described by the concept of information theory, obtaining the storage capacity of the open-loop writing of RRAM is,

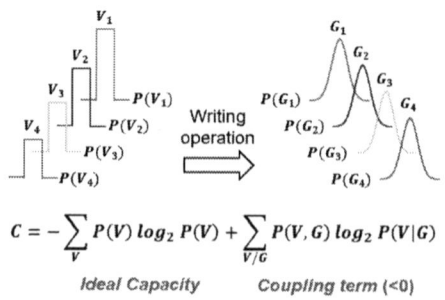

$$C = -\sum_V P(V)\,log_2\,P(V) + \sum_{V/G} P(V,G)\,log_2\,P(V|G)$$

Ideal Capacity   Coupling term (<0)

Fig. 1. Storage capacity model and calculation equation.

$$C = -\sum_V P(V)\log_2 P(V) + \sum_{V,G} P(V,G)\log_2 P(V|G), \quad (1)$$

where $P(V)$ is the probability of taking value of V among all input voltages, $P(G)$ is the probability of G among all output conductances, $P(V|G)$ is the conditional probability of the input voltage V on the output conductance G, and $P(V,G) = P(G)P(V|G)$. Eq. 1 consists of two terms, where the first term indicate the ideal situation with no noise distortion. The second term represents the noise coupling relationship between the voltage input and the conductance output, which is always less than 0. Intuitively, it is due to the overlaps between neighboring levels that causes the loss of effective bits of information. It should be stressed that although the input is considered as voltage magnitude in this model, it could also be voltage width, pulse number, or compliance current [14, 16].

B. Survey of Experimental Results and RRAM Model

To evaluate the storage capacity of open-loop written RRAM devices, we have conducted a survey of the conductance ranges and variations of multi-level RRAM disclosed in the literature, which are based on frequently-used metal oxides as the resistive switching layer, such as HfO₂, TiO₂, and Al₂O₃ [9-13, 16-17]. Fig. 2a shows that although the relationship between σ_G and G is noisy, there is a trend that σ_G is independent on G. The conductance range $G_{\min}\sim G_{\max}$ integrating the results of the RRAM devices is [1~250] μS, and the conductance variation σ_G fluctuates in the range of [3~20] μS.

Based on the observation of constant σ_G, we have established a simple but reasonable multi-level RRAM model, as shown in Fig. 2b. Different conductance ranges and σ_G's will be considered, resulting in different degrees of inter-level overlaps. All conductance levels are evenly distributed in a given conductance range, following normal distributions with the same σ_G for each level. The mean values of every two neighboring levels are separated by a conductance difference Δ_G. For some conductance levels at the edge of range, their distributions may overflow beyond the available range, the distribution probabilities are therefore truncated and normalized. To alleviate this issue, we have also reserved a certain amount of space at the edge, that is, the highest and the lowest levels are shifted from $G_{\min}$ and $G_{\max}$ to the left and right by $\Delta_G/2$, respectively.

C. Storage Capacity Analysis

Based on the multi-level RRAM model, the equivalent

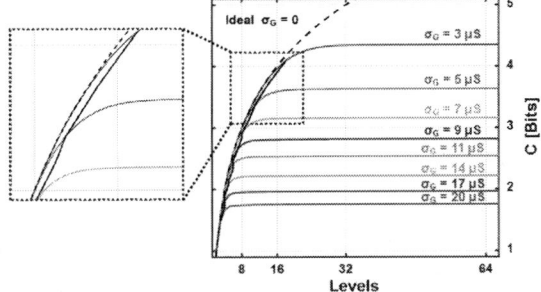

Fig. 3. Growth curve of storage capacity against the pre-defined number of discrete conductance levels. The magnified inset highlights the transitions to saturation capacities.

storage capacity C of RRAM device is calculated by using Eq. 1. First, the full range of [1~250] μS is considered, and different numbers of discrete levels are assumed. For each given σ_G, the evolvement of C increasing with the presumed L is calculated, as shown in Fig. 3. In all cases, the curve increases rapidly when L is small, where the first term in Eq. 1 dominates, and the intervals are sufficiently large to prevent the inter-level overlaps. When L exceeds a certain number (L_T), the curve quickly saturates at the final storage capacity $C_{\max}$. Because of the growing overlap between conductance levels, the storage capacity is suppressed by the second term in Eq. 1.

We provide a quantitative analysis of $C_{\max}$ and L_T. $C_{\max}$ is obtained by calculating with a sufficiently large L, and L_T is defined as then number when the capacity $C = 0.95C_{\max}$, which are marked by the black line in the inset of Fig. 3. Before the transition at L_T, the capacity curve under different σ_G conditions are all close to the ideal situation (black dotted line), suggesting the rationality of the definition of L_T for identifying the transition place. After L_T, the curve increases quickly from $0.95C_{\max}$ to $C_{\max}$, within only a few extra levels. The values of $C_{\max}$ and L_T are plotted in Fig. 4a. the maximum storage capacity is 4.3 bits under the most modest assumption of conductance variations. It also evidences that $C_{\max}$ decreases as σ_G increases, which shows a nonlinear behavior, namely the decrease of $C_{\max}$ is particularly sensitive when σ_G is relatively small. This nonlinear behavior is even stronger for L_T, suggesting that when σ_G is relatively large, it is not helpful to increase the pre-defined number of levels for storage. We have also calculated the parameter Δ_G/σ_G upon the usage of L_T, which are almost constant around 4 (Fig. 4b). It means that the coupling between conductance levels starts to occur approximately when the conductance variation $2\sigma_G$ occupy the inter-level space.

In addition to the full range of [1~250] μS, we have conducted calculations for different conductance ranges (Fig.

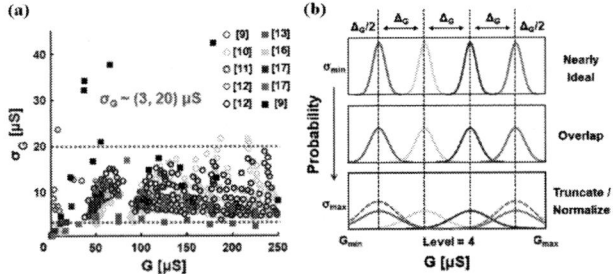

Fig. 2. (a) The σ_G-G correspondence data in Refs. [9-13, 16-17], which show roughly a constant behavior. (b) Multi-level RRAM model with identical conductance variation for each level.

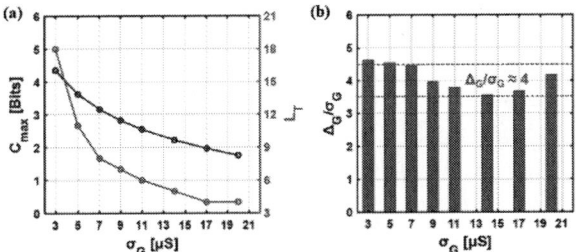

Fig. 4. (a) Dependences of $C_{\max}$ and L_T on σ_G, for the conductance range of [1~250] μS. (b) Relationship between ratio Δ_G/σ_G and σ_G.

979-8-3503-6184-1/24 $31.00 © 2024 IEEE   502

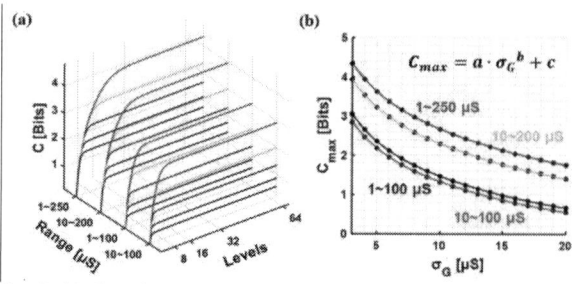

Fig. 5. (a) Growth curve of storage capacity for different conductance ranges. In each case, the considered σ_G's from top to bottom are [3, 6, 10, 15, 20] μS. (b) Dependences of C_{max} on σ_G. For all kinds of conductance range, the relationship between C_{max} and σ_G can be fittedby a power function distribution, which suitable for different conductance range.

5a). They all show similar behaviors, demonstrating the universality of the above analysis. For each conductance range, the C_{max} values under different conductance variations σ_G's are calculated, resulting in the curves shown in Fig. 5b. As the latter three ranges are narrower, the C_{max} values become smaller, all less than 4 bits. Noticeably, for all the four situations, the nonlinear correspondence between C_{max} and σ_G can be perfectly fit by a power function (Fig. 6b), but with different fitting parameters, which again supports the analysis of storage capacity of RRAM devices by the information theory.

III. EXPERIMENT RESULTS

To support the multi-level RRAM behaviors and the related storage capacity analysis, we have fabricated HfO$_2$-based RRAM devices and carried out extensive measurements of the multi-level characteristics of multiple devices [18]. The multi-level characteristics was studied by externally limiting the compliance current (Icc) during the current-voltage (I-V) sweeps (Fig. 6a). For each device, 19 Icc's in the range of [20~200] μA were set for test, resulting in 19 sets of data (30,000 in total). For each Icc, the corresponding mean value of conductance and the standard deviation are calculated. The data are summarized in Fig. 6b, which is plotted in the same manner as Fig. 2a. The resulting conductance is in the range [10-200] μS, and the variation of conductance states is in the range of [3~20] μS, which are consistent with the results in the literature, thus supporting the applicability of the above analysis.

Given the correspondence between σ_G and G is

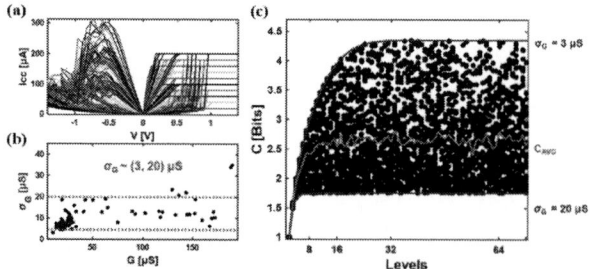

Fig. 6. (a) I-V characteristics of RRAM devices, where the compliance current I$_{CC}$ is used to control the device conductance. The inset illustrates the device structure. (b) σ_G of each conductance level in experiment. (c) Comparison of storage capacity curves, under the consideration of uniform distribution (red line) and normal distribution (blue line) for RRAM conductance levels.

considerably random (in both the reported conductance standard deviation and the experimental ones), we performed simulations with randomly-assumed σ_G for each conductance level. 100 simulations were performed, and the distribution of storage capacity and their average values are shown in Fig. 6c. These data points are all within the envelope of $\sigma_G = 3$ and 20 μS, which corresponds to the best or worst situation of all conductance levels. The average results tell that the real storage capacity of RRAM considering distribution of variations would be less than 3 bits per device. By observing the density of data points, it is concluded that the capacity is sensitive to the deterioration of σ_G, which can be explained by the exponential decrease of storage capacity in Fig. 5b.

IV. CONCLUSION

In this work, we have established a framework for evaluating the storage capacity of multi-level RRAM devices, by using the concept of information theory. Based on the reported conductance ranges and variations in the literature and those measured in our own RRAM devices, the evaluation results show that the maximum storage capacity of open-loop written RRAM devices is around 4 bits. It provides an effective estimation and an optimization guideline for maximizing the storage capacity of RRAM. We expect that such a framework should also be applicable to other nonvolatile resistive memory devices, to provide a theoretical insight into the capacity limit of storage capacity, thus supporting the delivery of storage-class memory and novel computing paradigms.

ACKNOWLEDGMENT

This work was supported in part by the National Key R&D Program of China under Grant 2020YFB2206001, in part by NSFC under Grant 92064004 and Grant 61927901, and in part by the 111 Project under Grant B18001.

REFERENCES

[1] K. Shafique et al., IEEE Access, vol. 8, 2020.

[2] Z. Wang et al., Nat. Rev. Mater., vol. 5, no. 3, Jan. 2020.

[3] D. Ielmini and H.-S. P. Wong, Nat. Electron, vol. 1, no. 6, Jun. 2018.

[4] D. Ielmini, Semicond. Sci. Technol., vol. 31, no. 6, May 2016.

[5] Z. Sun et al., Nat. Electron., vol. 6, no. 11, Nov. 2023.

[6] Z. Sun et al., Proc. Natl. Acad. Sci. U. S. A., vol. 116, no. 10, Mar. 2019.

[7] M. Rao et al., Nature, vol. 615, no. 7954, Mar. 2023.

[8] C. Li et al., Nat. Electron., vol. 1, no. 1, Dec. 2017.

[9] S. Balatti et al., in 2015 IEEE International Reliability Physics Symposium (IRPS), IEEE, Apr. 2015.

[10] V. Milo et al., APL Mater., vol. 7, no. 8, Aug. 2019.

[11] T. Zanotti et al., IEEE Trans. Electron Devices, vol. 67, no. 11, Nov. 2020.

[12] E. Perez et al., IEEE Trans. Electron Devices, vol. 68, no. 6, Jun. 2021.

[13] V. Milo et al., in 2021 IEEE International Reliability Physics Symposium (IRPS), IEEE, Mar. 2021.

[14] P. Yao et al., Nature, vol. 577, no. 7792, Jan. 2020.

[15] R. V. Zarcone et al., Sci. Rep., vol. 10, no. 1, Apr. 2020.

[16] S. Kim et al., Appl. Phys. Lett., vol. 117, Nov. 2020.

[17] P. Bousoulas et al., IEEE Electron Device Lett., vol. 37, no. 7, Jul. 2016.

[18] S. Wang et al., Sci Adv, vol. 9, no. 50, Dec. 2023.

High-Density and High-Reliability RRAM for Memory and Computing Applications

Yimao Cai*, Xiahong Zhou, Zongwei Wang, Lin Bao, Ling Liang, Cuimei Wang, Ru Huang

School of Integrated Circuits, Peking University, Beijing, China
Beijing Advanced Innovation Center for Integrated Circuits, Beijing, China

* Email: caiyimao@pku.edu.cn

Abstract—**For pursuing the implementation of high-density storage and high energy efficiency in-memory computing (IMC), we have designed many strategies to enhance the density and reliability of resistive random access memory (RRAM). In this paper, we present our high-density and high-reliability RRAM, and its different applications, especially in the fields of memory and IMC. Their opportunities for various applications are also highlighted.**

Keywords—RRAM, high density, high reliability, memory, in-memory computing

I. INTRODUCTION

The rapid developments of artificial intelligence (AI), big data and Internet of things (IoT) industries have triggered huge demands of high storage capacity and datacentric computing[1, 2]. However, the scaling down of conventional complementary metal oxide semiconductor (CMOS) technology node is now facing the limit of Moore's law, thus slowing down the pace of progress in high-density memory and high-speed computing. Moreover, the conventional von Neumann computing architecture has to separate memory and computing units, which leads to data transmission and further hardware overhead and power consumption issues [3]. As a consequence, novel memory and computing paradigms are imperative to be investigated, emphasizing those with high-density and high-reliability features.

Thus far, a variety of emerging memory technologies have been studied for memory and computing applications [4, 5]. In these techniques, RRAM has presented its superiority with high density, integrated memory with computing functions, low power consumption and so forth [6]. Therefore, it is regarded as a potential candidate for high-density memory and high-performance IMC. In recent years , studies on high-density and high-reliability RRAM for memory and computing have been widely conducted [7]. Based on our research, we present and summarize our systematic works of RRAM in enhancing integration density and reliability, high-performance IMC and its further application in machine vision calibration, highlighting the opportunities and challenges.

II. HIGH-DENSITY AND HIGH-RELIABILITY RRAM FOR MEMORY AND COMPUTING

A. High-density and high-reliability RRAM for memory

With the technology node scaling down, increasing integration density have encountered some challenges, such as the mismatch between switching voltage and Vdd in typical

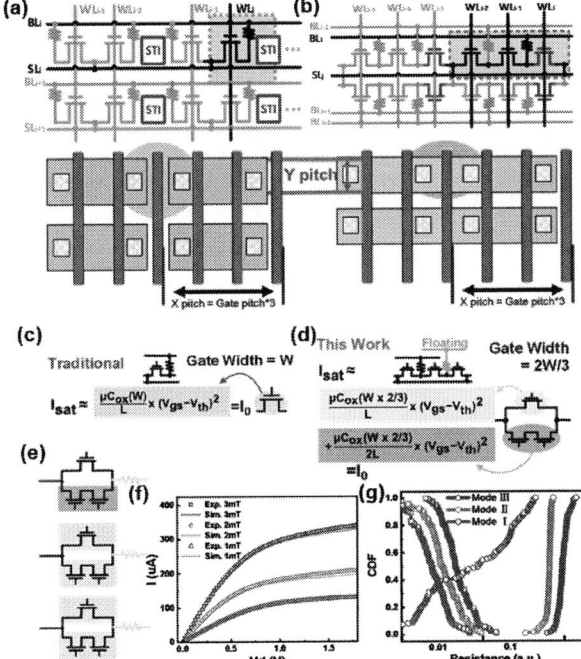

Fig. 1. (a) The schematic and layout of the conventional 2T2R cell and (b) the proposed STI-less 3T2R cell. Calculation of (c) equivalent gate width and (d) drive current of traditional and proposed STI-less 3T2R structures. Schematic of cell configurations , (f) measured and simulated drive current, and (g) typical DC switching behaviors (100 cycles) of the STI-less 3T2R under different operation modes [8].

1T1R structures. Since STI can be removed and replaced by active transistors, the STI-less 3T2R cell (gate width: 120nm) are realized to reduce the cell size [8]. **Figure 1a** shows the layout of the designed cell with evidently reduced width of transistor compared with conventional 1T1R structure (**Figure 1b**), thus enhancing ~24% integration density. Intriguingly, the drive current is almost equivalent to a conventional 1T1R cell (gate width: 200nm) (**Figure 1c & d**). Furthermore, such a design can also implement dynamic gate control and three kinds of operation modes with adjustable drive currents (**Figure 1e & f**). It renders the designed structure more diverse application scenarios. At the same time, the memory window is also enhanced to a great extent by adjusting the operation modes (**Figure 1g**).

On the other hand, since the traditional etching may result in sidewall damage, oxygen vacancies are thus liable to spillage or exchange. And it can even cause reliability

979-8-3503-6184-1/24 $31.00 © 2024 IEEE

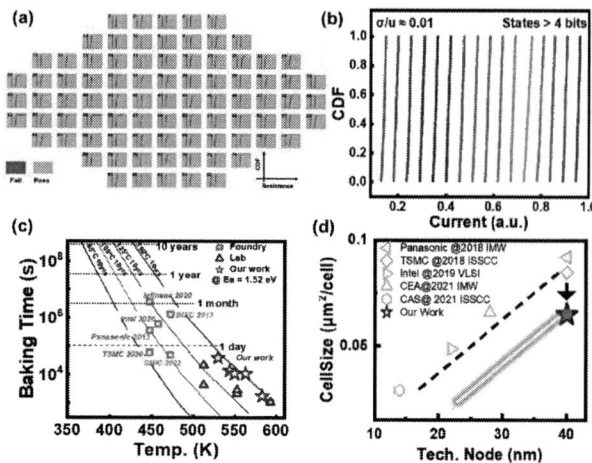

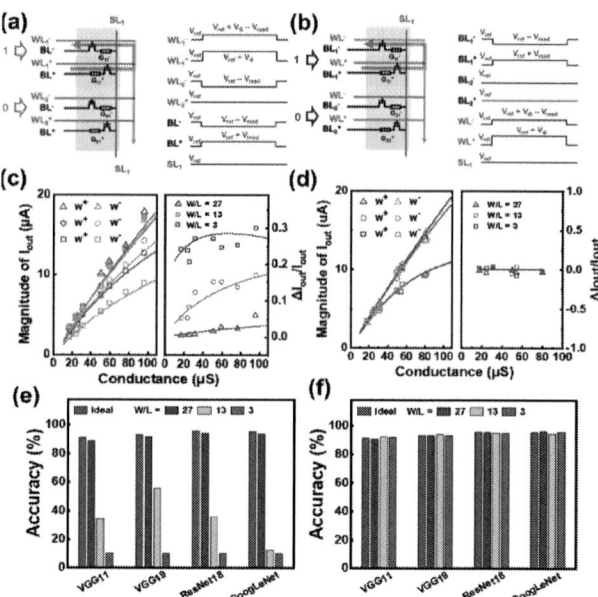

Fig. 2. (a) CDF of HRS and LRS from 85 PCM dies across the entire wafer. (b) The multilevel characteristics of SPS RRAM cells measured in 4K test chip. (c) The extrapolation of retention. (d) Benchmark and scalability of the STI-less DG cell size with the state-of-the-art foundry-based RRAM. Adopted from Ref.[8].

degradation. Given its hazard, self-passivation sidewall (SPS) forming during etching treatment is utilized to protect RRAM cell. It is worth noting that the as-produced passivation layer forms at room temperature rather than high-temperature annealing. As a result, this SPS layer improves its reliability a lot on both wafer and chip level. For example, process control monitor (PCM) data exhibits that all dies complete forming and cycling for 50 times, indicating the uniformity of cumulative distribution across the 12-inch wafer (**Figure 2a**). Moreover, the chip-level verification results also demonstrated the multilevel characteristic (**Figure 2b**) and high retention with 10 years at 150°C (**Figure 2c**). Finally, the chip with integration density of 15.43 Mb/mm² is implemented (**Figure 2d**), which is the record density thus far.

B. High-density and high-reliability RRAM for IMC

Other than memory application, high-density and high-reliability for IMC is also studied. Two typical array architectures, column subtraction 2T2R (CS-2T2R) and row subtraction 2T2R (RS-2T2R), are conventionally employed to store signed weights (W^+ and W^-) [9]. In CS-2T2R cell, signed weights are stored at different source lines, while RS-2T2R stores them at different bit lines. In fact, RS-2T2R structure has several merits over CS-2T2R, such as without peripheral subtractors and with smaller output current. These advantages contribute to the hardware overhead, power consumption reduction and even array density enhancement.

Nonetheless, the conventional RS-2T2R structure have a fatal flow for IMC originating from the intrinsic structure, especially the asymmetrical weight sensing issue owing to the electrical asymmetry. Even the recognition accuracy is related to the ratio of gate width and length (W/L) (**Figure 3c**), where the asymmetrical sensing can be alleviated by increasing W/L in some degree. However, it still can give rise to severe deviation of output current (I_{out}).

To crack the conundrum, isolated symmetrical 2T2R (IS-2T2R) structure with two complementary 1T1R is designed (**Figure 3a & b**). Compared with the two mentioned structures, IS-2T2R with different operation modes can not only effectively eliminate the asymmetrical weight sensing, but also significantly improve the integration density of array (42% improvements). Therefore, the structure is predicted to be

Fig. 3. (a & b) Schematic of IS-2T2R in two different schemes. (c) Output current for different conductance states (left) in RS-2T2R and the difference between W^+ and W^- (right) when W/L=27, 13 and 3. (d) IS-2T2R structure output current for different stored conductance states. Lines represents the simulated data and dots represents the experimental data. (e & f) Cifar 10 recognition accuracy for different networks under various gate width W in (e) conventional 2T2R and (f) IS-2T2R structures [9].

practical to utilize for IMC. In fact, the obvious deviation of I_{out} is not observed (**Figure 3d**), manifesting that the performance enhancement of this structure. Furthermore, unlike the recognition accuracy drops in conventional 2T2R structures (**Figure 3e**), the high recognition accuracy of network simulation results gives further confirmation of the validity of the innovative IS-2T2R structure (**Figure 3f**). Moreover, the resistance map of array also shows the minority of RRAMs in high resistance state. Apart from those, the source line current of proposed IS-2T2R is the lowest among the mentioned three structures, implying its energy efficiency improvements for IMC.

III. MODELING AND EVALUATION OF NON-IDEALITIES AND THERMAL ISSUES

Many factors affect the reliability or accuracy of IMC by using high-density RRAM, involving non-idealities, thermal cross-talk, and so forth. Therefore, it is of significance to delve into their physical origins and thus benefiting to build model for reliability improvements.

A. Non-idealities of device

There are many detrimental effects on RRAM devices, which degrades its accuracy and reliability, finally obstructing its applications. Noise-triggered time-dependent variability (TDV) is one of the adverse factors for RRAM used in neuromorphic circuits, owing to the consequence of resistance dynamic variation [10, 11]. It is found that the resistance changes more obviously as the resistance increases within a certain range. Therefore, it is easy to cause accuracy loss under low-power operation and further affects the synaptic weights.

Owing to the negative effects of TDV, the inherent mechanism and model need to be studied. As shown in **Figure 4a**, the physical origin of TDV is diverse, such as the vacancy's and ion's generation, recombination, and migration.

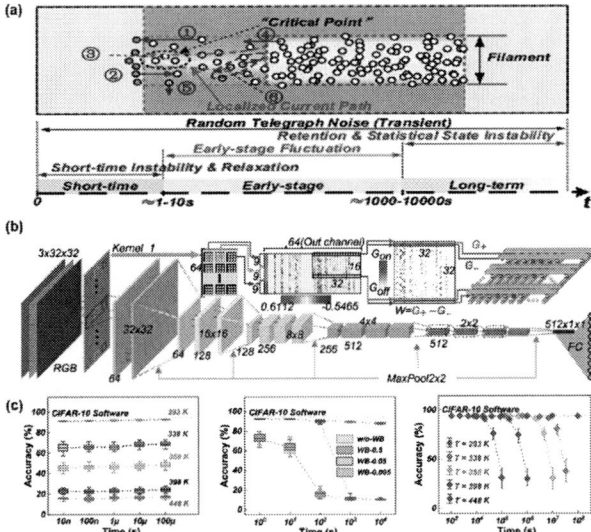

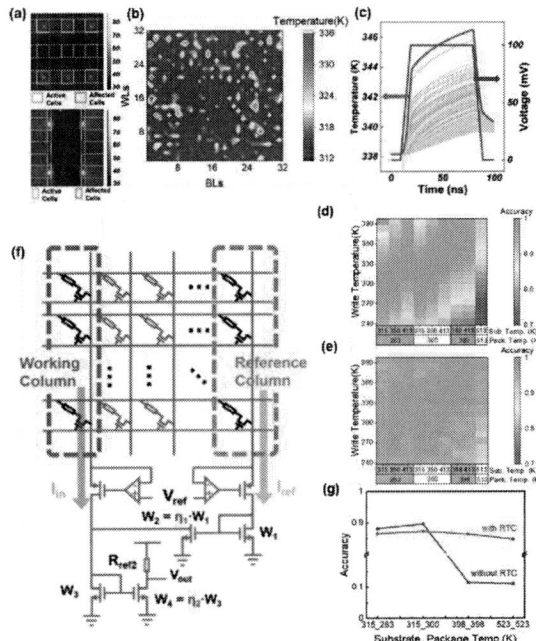

Fig. 4. (a) Schematic physical origins of the device non-idealities and the time scale difference of the device non-idealities. (b) Schematics of CNN network topology and corresponding scheme of mapping weights to RRAM arrays. (c) Impact of RTN (left), ERF (middle) and (right) RD on recognition accuracy under different temperatures [10, 11].

Fig. 5. (a) Calculated temperature profile when programming 2D and 3D RRAM array. (b) Temperature profile of an RRAM array with programmed weight from a pretrained NN. (c) Transient temperature evolution versus time in each cell during inference. Accuracy of DeepSets recognizing Modelnet10/MLP recognizing MNIST (d) without temperature effect correction scheme and (e) with RTC scheme. (f) RTC circuit for current compensation. (g) Recognition accuracy of CIFAR10 using VGG11 [11, 12].

As a consequence, such diversity indeed contributes to the variety of device non-idealities [11]. Three typical types of TDV are focused here and crudely classified by time scale difference, containing random telegraph noise (RTN), retention degradation (RD) and early-stage fluctuation (ERF) (**Figure 4a**). In early times, we have investigated and realized the simulation of producing TDV behavior by monitoring continuous current and analyzing the intrinsic parameters, which is similar to the measured ones.

Apart from this, circuit level analysis in neuromorphic system is conducted, indicating that the pattern recognition accuracy can be degraded by TDV under high resistance state to achieve low power operation. Even increasing the synapse numbers and response time, the recognition accuracy drop is still not obviously improved, demonstrating the challenge of suppressing TDV effect. Nevertheless, the ambient temperature has an influence on TDV suppression. Increasing temperature can gradually improve the recognition accuracy. Evaluations are realized on our simulation platform, which is based on MLP and CNN for recognizing MNIST and CIFAR-10 (**Figure 4b**). In addition, the accuracy difference between two training algorithms is depicted in **Figure 4c,** suggesting the accuracies of different training algorithms have different sensitivities to TDV. Importantly, we also use the simulation platform to evaluate the recognition of MNIST and CIFAR-10, which is based on multilayer perceptron (MLP) and convolution neural networks (CNN). With the simulation platform, the TDV impacts are assessed.

B. Thermal cross-talk of array

Another key issue is thermal cross-talk that directly have an impact on operation reliability [11]. Finite element analysis is conducted in order to explore the array-level temperature simulation. It is noted that there are severe thermal issues in either 2D or 3D architectures of RRAM during programming (**Figure 5a & c**), the simulations are conducted under the consideration of translated algorithm weights stored in the array. In fact, the non-ideal issues originate from the accelerated ion/vacancy diffusions ascribed to the comprehensive effects of both concentration gradient and heat accumulation.

We further investigate the thermal issues of a RRAM IMC array based on commercial 40nm CMOS process, aiming to realize a precise electro-thermal modeling framework [12]. The temperature profile of a practical RRAM array (**Figure 5b**) indicating the self-heating effect during inference. Meanwhile, the DeepSets and MLP recognition accuracy under different package and substrate temperatures shows the significant degradation at high temperature (exceeds ~398 K), implying the vulnerability of RRAM-based IMC under extreme working circumstance. Recognition accuracy affected by the temperature-aware programming is also shown in **Figure 5d & e**, indicating that the recognition accuracy without temperature effect correction scheme is affected by the programming temperature (**Figure 5d**).

To enhancing reliability, we develop a resistance-temperature compensation (RTC) protocol to mitigate thermal-induced variations in resistance in RRAM-based IMC systems. This RTC circuit include an LRS reference resistor and an additional HRS reference column, as shown in **Figure 5f**. With our proposed RTC scheme, the recognition accuracy is improved a lot in most cases (**Figure 5e**) and the enhanced performance of IMC systems under extreme temperature condition is witnessed (**Figure 5g**).

IV. MACHINE VISION CALIBRATION APPLICATION BASED ON HIGH-DENSITY AND HIGH-RELIABILITY RRAM

Based on the density and reliability enhancement of RRAM, we also investigate the further applications. Machine vision calibration is widely used in various industry field. However, lens distortion correction as an important step of

979-8-3503-6184-1/24 $31.00 © 2024 IEEE

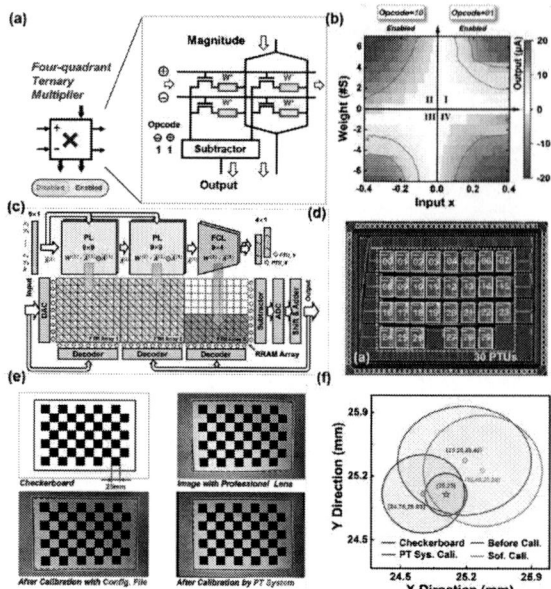

Fig. 6. (a) The schematic circuit and function table of the RRAM-based FTM. (b) The distribution of FTM output current with different storage data W and input signal x. (c) The parallel PT model and the corresponding PT MAC unit. (d) The image of test chip with fifteen PT MAC unit pairs. (e) The comparison between the images w/ and w/o LDC. (f) The LDC system shows better performance in both average distance deviation (distance between dots and asterisk) and distance dispersion (elliptical area). Adopted from Ref. [13].

machine vision for serving high-quality optical images is facing some challenges by using conventional CMOS ternary multiplication operators, such as energy and hardware overhead. Anchoring this issue, a hybrid-domain ternary multiplication accelerated computing strategy based on 40nm RRAM is proposed.[13]

Figure 6a shows the four-quadrant ternary multiplier (FTM) with 4T4R structure, realizing parallel polynomial transformation and multiplication of two signed operands (**Figure 6b**). And the FTM is further integrated into high-density array, thus implementing high parallelism. **Figure 6c** shows the polynomial transformation MAC unit with three network layers used for high order parallel polynomial transformation. And the test chip based on 40nm RRAM is shown in **Figure 6d**, exhibiting 15 pairs of polynomial transformation MAC unit. After correcting by the accelerator, the image with lens distortion correction is evidently different from that without the correction process (**Figure 6e**). Furthermore, the proposed correction system shows software-comparable performance (**Figure 6f**). More importantly, this chip achieves the throughput and energy efficiency of 15M pixels/s and 3.81G pixels/s, respectively.

V. SUMMURY

High-density and high-reliability RRAM technology holds tremendous promise across various application fields, particularly in memory and computing. Its efficient data storage capabilities coupled with lower power consumption mark it as a significant advancement. By potentially reducing hardware overhead and enhancing energy efficiency, RRAM emerges as an exciting prospect for the future technology.

ACKNOWLEDGMENT

This work was supported by National Natural Science Foundation of China under Grant 62025401, 62341407, 61927901, in part by Beijing Natural Science Foundation under Grant L223004, Beijing Nova Program under Grant 20220484113, in part by "111" Project under grant B18001, and in part by Xiaomi Foundation.

REFERENCES

[1] Y. H. Huang, Y. C. Hsieh, Y. C. Lin, Y. D. Chih, E. Wang, J. Chang, Y. C. King, and C. J. Lin, "High Density Embedded 3D Stackable Via RRAM in Advanced MCU Applications." 2023 IEEE Symposium on VLSI Technology and Circuits (VLSI Technology and Circuits), 2023, pp. 1-2.

[2] Z. Wang, Y. Cai, "Memory Technology: Development, Fundamentals, and Future Trends." in Advanced Memory Technology: Functional Materials and Devices ed. Y. Zhou, Royal Society of Chemistry, 2023, vol. 1, pp. 1-36.

[3] P. Jiang, H. Jiang, Y. Yang, L. Tai, W. Wei, T. Gong, Y. Wang, P. Xu, S. Lv, B. Wang, J. Gao, J. Li, J. Luo, J. Yang, Q. Luo, M. Liu, "A 256 Kbit Hf$_{0.5}$Zr$_{0.5}$O$_2$-based FeRAM Chip with Scaled Film Thickness (sub-8nm), Low Thermal Budget (350°C), 100% Initial Chip Yield, Low Power Consumption (0.7 pJ/bit at 2V write voltage), and Prominent Endurance (>10^{12})."2023 IEEE IEDM, pp. 1-4

[4] S.-T. Wei, B. Gao, D. Wu, J.-S. Tang, H. Qian, and H.-Q. Wu, "Trends and challenges in the circuit and macro of RRAM-based computing-in-memory systems," Chip, vol. 1, no. 1, pp. 100004, 2022.

[5] Y. Ling, Z. Wang, L. Wu, Y. Cai and R. Huang, "An RRAM-Based Hierarchical Computing-in-Memory Architecture With Synchronous Parallelism for 3D Point Cloud Recognition." IEEE Transactions on Circuits and Systems II: Express Briefs. Doi: 10.1109/TCSII.2024.3396815.

[6] T. Xie, S. Yu, and S. Li, "A High-Parallelism RRAM-Based Compute-In-Memory Macro With Intrinsic Impedance Boosting and In-ADC Computing," IEEE Journal on Exploratory Solid-State Computational Devices and Circuits, vol. 9, no. 1, pp. 38-46, 2023.

[7] Q. Zheng, X. Li, Z. Wang, G. Sun, Y. Cai, R. Huang, Y. Chen, H. Li, "MobiLattice: A Depth-wise DCNN Accelerator with Hybrid Digital/Analog Nonvolatile Processing-In-Memory Block." 2020 IEEE/ACM International Conference On Computer Aided Design (ICCAD), pp. 1-9

[8] Q. Wang, Y. Yang, Z. Wang, S. Bao, J. Sun, L. Shan, L. Bao, Y. Gao, H. Zhang, Y. Ling, W. Zhang, Y. Wang, Y. Cai, and R. Huang, "A Logic-Process Compatible RRAM with 15.43 Mb/mm^2 Density and 10years@150°C retention using STI-less Dynamic-Gate and Self-Passivation Sidewall." 2023 IEEE IEDM, pp. 1-4.

[9] Y. Ling, Z. Wang, Y. Yang, L. Bao, S. Bao, Q. Wang, Y. Cai, and R. Huang, "An Isolated Symmetrical 2T2R Cell Enabling High Precision and High Density for RRAM-based In-Memory Computing," SCIENCE CHINA Information Sciences, vol. 67, no. 5, pp. 152402, 2023.

[10] J. Kang, Z. Yu, L. Wu, Y. Fang, Z. Wang, Y. Cai, Z. Ji, J. Zhang, R. Wang, Y. Yang, and R. Huang, "Time-dependent variability in RRAM-based analog neuromorphic system for pattern recognition." 2017 IEEE IEDM, pp. 6.4.1-6.4.4.

[11] Y. Cai, Z. Wang, Z. Yu, Y. Ling, Q. Chen, Y. Yang, S. Bao, L. Wu, L. Bao, R. Wang, and R. Huang, "Technology-Array-Algorithm Co-Optimization of RRAM for Storage and Neuromorphic Computing: Device Non-idealities and Thermal Cross-talk." 2020 IEEE IEDM, pp. 13.4.1-13.4.4.

[12] Y. Ling, Z. Wang, Z. Yu, S. Bao, Y. Yang, L. Bao, Y. Sun, Y. Cai, and R. Huang, "Temperature-Dependent Accuracy Analysis and Resistance Temperature Correction in RRAM-Based In-Memory Computing," IEEE Transactions on Electron Devices, vol. 71, no. 1, pp. 294-300, 2024.

[13] L. Bao, Z. Wang, Q. Wang, Y. Yang, Y. Gao, L. Shan, J. Sun, Y. Yang, Y. Ling, H. Zhang, C. Wang, H. Xiao, L. Ye, A. Guo, L. Shen, W. Gu, G. Feng, C. Li, S. Chen, Y. Zhao, S. Huang, Y. Cai, and R. Huang, "Hybrid-Domain In-Memory Polynomial Acceleration based on 40nm RRAM Multi-Core Chip for Machine Vision Calibration." 2023 IEEE IEDM, pp. 1-4

Impact of Different MAC Schemes on Computing In Memory based on 1T1R Array

Ruiqing Xie [1], Gaoqi Yang [1], Zongwei Wang*[1,2], Linbo Shan [1], Jinshan Li [1], Chaoyi Ban [1], Lin Bao [1], Ling Liang [1], Cuimei Wang [1], Yimao Cai*[1,2] and Ru Huang [1,2]

[1] School of Integrated Circuits, Peking University, Beijing, China.
[2] Beijing Advanced Innovation Center for Integrated Circuits, Peking University, Beijing, China.

* Email: {wangzongwei, caiyimao}@pku.edu.cn

Abstract—Conventional computing architecture faces challenges due to the Von Neumann bottleneck and memory wall, which fails to meet the growing computation and storage demands of artificial intelligence (AI) applications caused by latency and energy. Computing in Memory (CIM) architecture integrates computation within memory, which can address such problems. Among novel memory-based CIM, Resistive random-access memory (RRAM)-based CIM could consume significantly less power and area when processing equivalent data volumes, widely used to accelerate multiply-accumulate (MAC) operations in neural networks. Currently, there exist two schemes of RRAM arrays for accelerating matrix computations: vertical scheme (word line parallel to source line) and horizontal scheme (word line parallel to bit line). However, a comprehensive evaluation of the strengths and weaknesses of the two schemes in matrix operations is lacking. To fill this gap, we conduct detailed tests on these two schemes using RRAM chip, evaluating the accuracy and power consumption of MAC computations. Our results show that the vertical scheme achieves a 1.98X improvement in computational accuracy while offering up to an 8.4X power decrease.

Keywords—CIM, RRAM, MAC schemes, sneak-paths

I. INTRODUCTION

Artificial Intelligence (AI) has brought significant impacts to all aspects of human society. Multiply-accumulate (MAC), as a fundamental operation in neural networks, is indispensable for AI applications such as robotic, automatic driving and intelligent manufacturing. The MAC operation based on emerging devices such as RRAM, Phase Change Random Access Memory (PCRAM), Magnetoresistive Random Access Memory (MRAM), and Ferroelectric Random Access Memory (FERAM) are gaining increasing attention because they allow computations to be performed within memory units and results to be stored in situ due to their non-volatility, which can address the memory wall issue and improve energy efficiency [1-4]. Among these technologies, RRAM with superb scalability, high speed, and compatibility with logic processes, is an excellent device for implementing CIM [5-7]. When signals are applied to the rows of the array, the total column currents, accumulated via Ohm's law and Kirchhoff's law, represent the arithmetic results of the matrix-vector multiplication operation. Compared to traditional digital arithmetic, one-step operation enables matrix-vector multiplication (**Fig. 1**), enhancing computational efficiency.

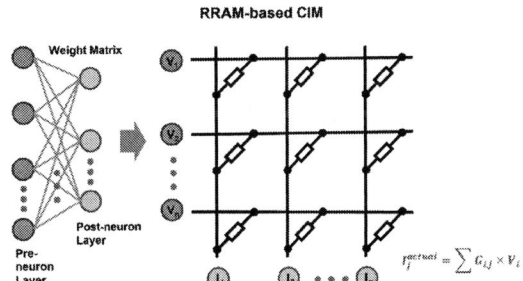

Figure 1. Structure of RRAM-based CIM and the strategy of mapping Neural Networks on RRAM crossbar, conductance of RRAM cells representing weight matrix.

In recent years, many studies have utilized RRAM arrays to accelerate MAC operation. With inputs from the bit line (BL), these studies mainly involve two array schemes: Vertical Scheme, the word line (WL) parallel to a source line (SL) (**Fig. 2(a)**); Horizon Scheme, WL parallel to BL (**Fig. 2(b)**) [8-10]. However, comprehensive evaluations of these two operation schemes are relatively lacking in the literature. In this study, we conduct linearity tests of MAC operations in a 1T1R array on a 1Mb RRAM chip, comparing the computational accuracy of the two operation schemes. Furthermore, we analyze the sneak-paths and effects in the context of the array structure and use simulation to comprehensively evaluate the impact of leakage on computational accuracy and power consumption. Our results indicate that the leakage significantly affects computational accuracy and leads to higher power consumption. Compared to horizontal operation, the vertical scheme achieves a 1.98X improvement in computational accuracy while offering up to 8.4X power reduction.

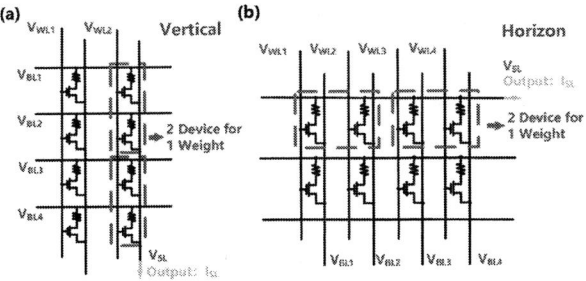

Figure 2. Two schemes for MAC operation (a) Vertical scheme (b) Horizontal scheme.

II. RRAM ARRAY AND CELL CHARACTERISTICS

Fig. 3a displays a transmission electron microscopy (TEM) cross-section image of the one-transistor-one-resistor (1T1R) cell in RRAM array fabricated with commercial 40nm CMOS technology [11]. **Fig. 3b** illustrates the 100 cycles DC I-V curves of TaOx-based RRAM, which exhibits a stable memory window of more than five. During the set operation, an abrupt current increase appeared at V_{Set} around 1.9 V, indicating a resistance switch from high resistance state (HRS) to low resistance state (LRS). Subsequently, during the reset operation, the device switches back to HRS at V_{Reset} of approximately -2.0 V.

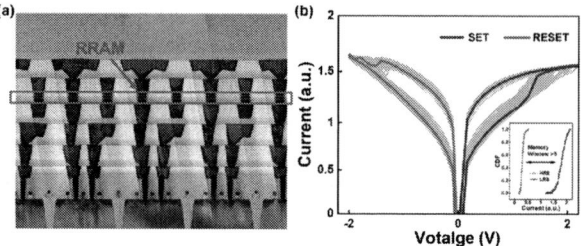

Figure 3. **(a)** Cross-sectional TEM image of the RRAM array in 1Mb Chip. **(b)** Typical IV characteristics of RRAM cell. The insert illustrates that the memory window of our TaOx-based RRAM is more than 5.

The multi-level property of RRAM enables it to store a greater number of discrete weight values. This capability provides a finer granularity of numerical representation, thereby enhancing computational accuracy. Our TaOx-based RRAM has 16 distinct conductance states with excellent uniformity ($\sigma/\mu < 0.1$) (**Fig. 4a**). Each conductance state is tested with more than 50 devices. Furthermore, frequent operation of the array requires devices with stable electrical characteristics. One TaOx cell can store at least 4-bit information (16 states), and its read endurance is evaluated to be more than 1k cycles (**Fig. 4b**). During 1000 read cycles, the device's 16 conductance states remain distinguishable, demonstrating that the fabricated RRAM array can support high-precision network computations.

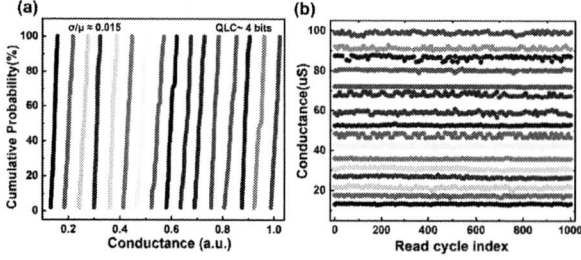

Figure 4. **(a)** The multilevel characteristics of RRAM cells measured in 1Mb Chip. Each state consists of more than 50 devices. **(b)** Read cycle endurance test of the device of 16 states for 1K switching cycles.

III. RESULTS AND DISCUSSION

A. MAC Calculation Accuracy Test

The differential conductance pairs are utilized to represent positive and negative weights [12]. In the vertical and horizontal schemes, each weight is encoded as different conductance between two adjacent RRAM cells in the same column or row. Compared to storing each weight in a single cell, this differential conductance method also avoids the impact of high-resistive state leakage on the computation results.

To evaluate the performance of the two schemes, calculation accuracy tests are conducted with varying numbers of parallel lines on a 1Mb chip. The differential conductance is set from -35 µS to 35 µS in 5 µS intervals to represent 4-bit weights ranging from -7 to 7. These weights are sequentially and cyclically written in a row or column, and the linearity of MAC computation is tested with both 32 and 64 lines in parallel for both vertical and horizontal schemes.

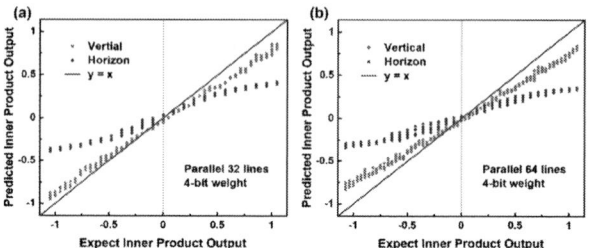

Figure 5. The MAC outputs versus ideal outputs of the (a) parallel 32 lines and (b) parallel 64 lines in 1Mb Chip.

Fig. 5 illustrates the measured MAC outputs for varying numbers of parallel lines compared to the ideal outputs under 4-bit weight conditions. The results show that the outputs of the vertical scheme are closer to the ideal values for the same computations compared to the horizontal scheme. As the computation results increase, the calculation accuracy of the horizontal scheme drops rapidly. Conversely, the vertical scheme demonstrates only a marginal decline in calculation accuracy, exhibiting better linearity. As the number of parallel lines increases, the linearity of both schemes deteriorates, and the disparity between the two becomes more pronounced.

Thus, vertical scheme could offer better accuracy when dealing with massive MAC operations on the 1T1R array.

B. Discussion

Due to the voltage drop across parasitic resistance and the non-ideal effects of transistors, the MAC output is lower than the ideal value, which degrades the computation accuracy [13]. In the horizontal scheme, the WL controls the switching of an entire column of transistors. When a selected cell is being processed, cells in the same column are also opened, and the pulse voltage on the BL is applied to these cells, causing leakage. As shown in the sneak-paths in **Fig. 6**, the leakage is mainly influenced by the resistance of the RRAM cells in the same column; lower resistance results in higher leakage. Furthermore, larger array sizes introduce more sneak-paths, increasing the total leakage. A portion of this leakage flows into the SL of the computation row, thereby impacting calculation accuracy and increasing the operational power consumption.

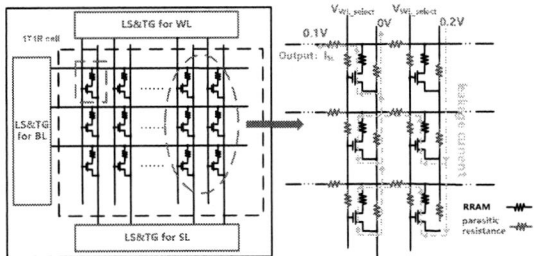

Figure 6. The arrays' Sneak-paths in horizontal scheme.

C. Accuracy and Power Consumption Simulation

To evaluate the impact of leakage on MAC operation accuracy and power consumption, simulations are carried out on two schemes in 8×8, 16×16, 32×32, and 64×64 1T1R RRAM arrays. The simulation is conducted using Cadence Virtuoso, a widely-used tool for simulating the magnitudes of node voltages and currents during operation. The array is mainly composed of transistors and resistors. We utilize fixed resistors ranging from 7.5 µS to 42.5 µS to represent the eight distinct resistive states of RRAM cells, and differential resistive pairs are employed to represent positive and negative weights. Considering the line resistance, non-idealities in the transistors, and the pull-up resistance on the SL, we compare the leakage magnitude, computational accuracy, and power consumption between the two schemes.

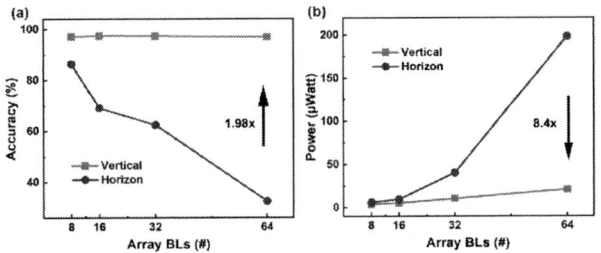

Figure 7. The comparison of accuracy and power between Vertical and Horizon schemes for different array sizes.

Fig. 7 shows the computational accuracy and power consumption of the vertical and horizontal schemes for varying array sizes. For 64 BLs, the vertical scheme achieves 1.98X computing accuracy (**Fig. 7(a)**). Additionally, the vertical scheme's power consumption is 8.4X lower than that of the horizontal scheme (**Fig. 7(b)**). The results suggest that the vertical scheme delivers superior MAC operation performance in neural networks, owing to its lower sneak-paths.

It should be noted that, due to the limited scale of the simulated arrays, the impact of parasitic line resistance is relatively low, allowing the vertical scheme to achieve over 95% computational accuracy in simulation. However, in practical array implementations, parasitic line resistance has a more significant influence. This causes increased voltage drops, leading to a decrease in computational accuracy as the computation current rises, resulting in nonlinearity.

IV. SUMMARY

In this work, we conduct a comprehensive evaluation of two schemes for RRAM array MAC operations through computational accuracy tests and simulations. The horizontal scheme introduces substantial leakage, and as the size of the array involved in the computation increases, the sneak-paths multiply, leading to a rapid decline in computational accuracy and a substantial increase in power consumption. In contrast, the vertical scheme minimizes leakage by disabling transistors outside the computing units, thereby enhancing computational accuracy while avoiding additional power consumption. Our results show that the vertical scheme achieves a 1.98X computational accuracy improvement and an 8.4X decrease in power consumption, which promises to further advance the application prospects of emerging devices in compute-in-memory paradigms.

ACKNOWLEDGMENTS

R. Xie and G. Yang contributed equally to this paper. This work was supported by the National Natural Science Foundation of China under Grant 62341407, 62025401, 92164205, 61927901, in part by the Beijing Municipal Science and Technology Program under Grant Z231100007423019, Beijing Nova Program under Grant 20220484113, "111" Project under grant B18001, and Xiaomi Foundation.

REFERENCES

[1] S. Yu, "Neuro-inspired computing with emerging nonvolatile memorys," Proceedings of the IEEE, vol. 106, no. 2, pp. 260-285, 2018.

[2] A. Sebastian, M. Le Gallo, R. Khaddam-Aljameh, et al., "Memory devices and applications for in-memory computing," Nature Nanotechnology, vol. 15, no. 7, pp. 529-544, 2020.

[3] D. Ielmini and H-S. P. Wong, "In-memory computing with resistive switching devices," Nature Electronics, vol. 1, no. 6, pp. 333-343, 2018.

[4] Q. Wei, et al., "Emerging Memory-Based Chip Development for Neuromorphic Computing: Status, Challenges, and Perspectives," IEEE Electron Devices Magazine, vol. 1, no. 2, pp. 33-49, 2023.

[5] Long Y, Na T, Mukhopadhyay S. ReRAM-based processing-in-memory architecture for recurrent neural network acceleration[J]. IEEE Transactions on Very Large Scale Integration (VLSI) Systems, 2018, 26(12): 2781-2794.

[6] Bao L, Wang Z, Wang Q, et al. Hybrid-Domain In-Memory Polynomial Acceleration based on 40nm RRAM Multi-Core Chip for Machine Vision Calibration[C]//2023 International Electron Devices Meeting (IEDM). IEEE, 2023: 1-4.

[7] L. Han et al., "A Convolution Neural Network Accelerator Design with Weight Map and Pipeline Optimization," in Proceedings of the 2023 60th ACM/IEEE Design Automation Conference (DAC), IEEE, 2023.

[8] P. Yao, H. Wu, B. Gao, et al., "Fully hardware-implemented memristor convolutional neural network," Nature, vol. 577, no. 7792, pp. 641-646, 2020.

[9] W. Wan, R. Kubendran, C. Schaefer, et al., "A compute-in-memory chip based on resistive random-access memory," Nature, vol. 608, no. 7923, pp. 504-512, 2022.

[10] Y. Gao, Z. Wang, Z. Yu, L. Bao, Y. Cai, R. Huang, "Compensation of Conductance Mismatch with Redundant Bit-lines for RRAM-based Voltage Sensing Mode Computing-in-Memory," in Proceedings of the 2024 8th IEEE Electron Devices Technology & Manufacturing Conference (EDTM), IEEE, 2024, pp. 1-3.

[11] Y. Ling, Z. Wang, Z. Yu, et al., "Temperature-dependent accuracy analysis and resistance temperature correction in RRAM-based in-memory computing," IEEE Transactions on Electron Devices, vol. 71, no. 1, pp. 294-300, 2023.

[12] X. Zhang, Z. Wang, W. Song, et al., "Experimental demonstration of conversion-based SNNs with 1T1R Mott neurons for neuromorphic inference," in Proceedings of the 2019 IEEE International Electron Devices Meeting (IEDM), IEEE, 2019, pp. 6.7.1-6.7.4.

[13] Y. Ling, Z. Wang, Y. Yang, et al., "An isolated symmetrical 2T2R cell enabling high precision and high density for RRAM-based in-memory computing," Science China Information Sciences, vol. 67, no. 5, article 152402, 2024.

Investigation of Gate Injection Charges Behavior on FeFETs with TiN/Al$_2$O$_3$/Hf$_{0.5}$Zr$_{0.5}$O$_2$/SiON/Si Structure by analyzing ISPP/ISPE

Jia Yang [1,2,3], Runhao Han [1,2,3], Saifei Dai [1,2,3], Tao Hu [1,2,3], Xianzhou Shao [1,2,3], Kanyi Li[1,2,3], Wenbo Fan[1,2,3], Xiaoqing Sun [1,2,3], Junshuai Chai [1,2,3], Hao Xu [1,2,3], Kai Han [4], Xiaolei Wang [1,2,3], Wenwu Wang [1,2,3] and Tianchun Ye [1,2,3]

[1] Key Laboratory of Fabrication Technologies for Integrated Circuits, Chinese Academy of Sciences, Beijing 100029.
[2] Institute of Microelectronics, Chinese Academy of Sciences, Beijing 100029, China.
[3] the School of Integrated Circuits, University of Chinese Academy of Sciences, Beijing 100049, China.
[4] School of Physics and Electronic Information, Weifang University, Shandong, 261061, China.

* Email: yangjia@ime.ac.cn, wangxiaolei@ime.ac.cn

Abstract—In this work, we study the behavior of the gate injection charges during the programming/erasing (PGM/ERS) operation in Hf$_{0.5}$Zr$_{0.5}$O$_2$-based ferroelectric field-effect transistors (FeFETs) with TiN/Al$_2$O$_3$/Hf$_{0.5}$Zr$_{0.2}$/SiON/Si (MAFIS) structure. The maximum memory window (MW$_{max}$) increases with an increasing Al$_2$O$_3$ top dielectric layer thickness. This is caused by the increase of the equivalent oxide thickness (EOT), and not the increased gate injection charge. Besides, the ISPP/ISPE slope of the MAFIS increases as the thickness of the structure increases, which can reduce the write voltage and reduce the power consumption. In addition, the holes injection from the metal gate after PGM operation is increased after FE layer wake-up. However, electron injection cannot increase as the holes during ERS operation. The key to improving the retention of the MW$_{max}$ is to make the electrons injected from the gate retained at ERS operation.

Keywords—FeFET, memory window, interlayer, injection charges, endurance.

I. INTRODUCTION

The ability to achieve additional z-scaling within the same cell height stores data by ferroelectric polarization compared to 3D NAND based on Charge Trap Flash (CTF) [1]. In addition, HfO$_2$-based ferroelectric silicon channel field-effect transistors (HfO$_2$ Si-FeFETs) have the advantages of high speed, low write power, and CMOS compatibility. So the HfO$_2$ Si-FeFETs show great potential in 3D NAND devices [2]. However, the memory window (MW) of HfO$_2$ Si-FeFETs must be significantly enhanced for the multi-bit operation [1]. Recently, many studies have found that the metal/top interlayer (T.IL)/ferroelectric/bottom interlayer (B.IL)/Si (MIFIS) structure can combine the effects of gate injection charges and ferroelectric polarization, resulting in a larger MW [3]. Yoon et al. from SK Hynix used a top dielectric layer in the ferroelectric (FE) layer and realized an MW of 10.54 V [4]. Lim et al. from Samsung Electronics used a top dielectric layer and realized an MW of 5.5 V [5]. Das et al. realized an MW of 7.3 V by inserting a 3 nm Al$_2$O$_3$ between the FE layer and the metal [6]. Our group adopted Al$_2$O$_3$ as the top layer and achieved an MW of 4.19 V [7, 8].

Although the MIFIS structure has significantly improved the MW, the reliability of this MIFIS device is poor. This is one of the important issues for the application of HfO$_2$ Si-FeFETs in 3D NAND [5, 7]. The degradation of MW is most likely due to the tunneling of the charge injected from the gate,

which cannot be effectively retained between the top dielectric interlayer and the FE layer. However, there is a lack of understanding of the physical mechanisms behind this problem. Therefore, it is necessary to investigate the behavior of the gate injection charge during the PGM/ERS (P/E) process to make the MW$_{max}$ to be held stable.

In this work, we fabricate an Al$_2$O$_3$ top dielectric interface layer with different thicknesses (Dev.A, Dev.B, Dev.C is 1.8nm, 3.6nm, 5.5nm) on the TiN/Al$_2$O$_3$/Hf$_{0.5}$Zr$_{0.5}$O$_2$/SiON/Si (MAFIS) structure. We investigate the effect of Al$_2$O$_3$ thickness on the performance of FeFETs. Analyzing the evolution of MW with the thickness of Al$_2$O$_3$ top dielectric layer increasing. That results in the MW$_{max}$ increases due to the increased EOT of the device. At the same time, it also increases the slope of ISPP/ISPE. By analyzing the threshold voltage (V$_{th}$) variation during the P/E process, we also found that gate injection electrons are more prone to tunneling during ERS operation. Therefore, the key to improving the MW$_{max}$ retention is to make the electrons injected from the gate be retained during the erase operation. This work provides some references for further optimization of the gate stack structure in the future.

II. DEVICE STRUCTURE AND FABRICATION

Fig. 1(a) and (b) show the schematic and fabrication process of the MAFIS structure gate stacks of HfO$_2$ Si-FeFETs. The devices were fabricated using the gate-last process on an 8-inch p-type silicon wafer (8~12 Ω·cm). The SiON interfacial layer was obtained by subjecting the SiO$_x$ layer (grown by O$_3$ oxidation) to decoupled plasma nitridation (DPN) followed by post-nitridation anneal (PNA). Then, the Hf$_{0.5}$Zr$_{0.5}$O$_2$ and different thicknesses interlayer Al$_2$O$_3$ were grown by atomic layer deposition (ALD) at 300 ℃. After that, TiN and W were grown. Finally, the devices were annealed at 400 ℃ in an N$_2$ atmosphere for 60 s to form the orthorhombic phase. Fig. 1(c) shows High Resolution Transmission Electron Microscopy (HRTEM) images for the device with a 2nm Al$_2$O$_3$ top dielectric interface layer. The gate width/length (L/W) of these devices in this work is 5/150 um. The electrical characteristics were measured by Keysight B1500A. The V$_{th}$ is defined by the constant current method.

979-8-3503-6184-1/24 $31.00 © 2024 IEEE

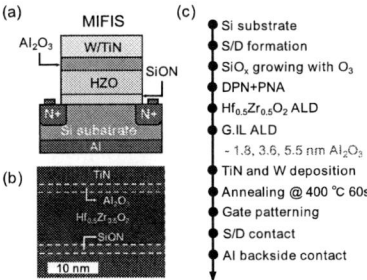

Fig. 1. (a) The structure schematic of the MAFIS structure. (b) The fabrication process flows of devices with different thicknesses (1.8nm, 3.6nm, 5.5nm). (c) HRTEM image of MAFIS device with 2nm Al_2O_3 top dielectric layer.

III. RESULT AND DISCUSSION

Table. 1 shows the performance of MAFIS devices with different thicknesses of Al_2O_3 top dielectric layer. The memory performances are shown, respectively, such as the operating voltage (V_{op}), the maximum MW, and the MW degradations all more than 50% after 3k P/E operation.

TABLE I. MEMORY PERFORMANCE

Device	Memory performance		
	Dev.A	Dev.B	Dev.C
T.IL (Al_2O_3)	1.8 nm	3.6 nm	5.5 nm
V_{PGM}/V_{ERS}	+9/-6V	+11/-7.5	+12.5/-9
MW_{max}	2.34V	3.51V	4.67V
MW_{max} after 3k cycle	1.16V	1.61V	1.91V
MW loss	50%	54%	59%

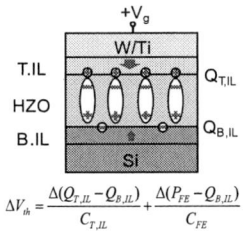

$$\Delta V_{th} = \frac{\Delta(Q_{T,IL} - Q_{B,IL})}{C_{T,IL}} + \frac{\Delta(P_{FE} - Q_{B,IL})}{C_{FE}}$$

Fig. 2. Schematic of MAFIS gate stack and charge distribution after program operation

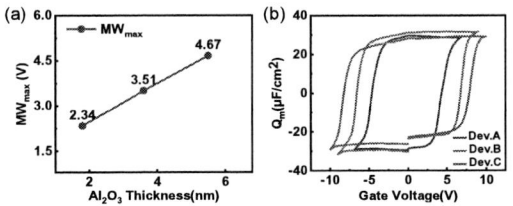

Fig. 3. (a) MW_{max} change v.s the T.IL thickness change in MAFIS structure. (b) Q_m-V curves of Dev.A, B, C.

Fig. 2 shows the charge distribution of the top and bottom dielectric layers in the MAFIS structure with an applied voltage. When program voltage (V_{PGM}) is applied at the gate, the ferroelectric layer is polarized downward. The holes are injected from the gate and electrons are injected from the silicon channel. On the contrary, when applied erase voltage

(V_{ERS}), the polarization of the FE layer is reversed. The electrons and holes are injected from the gate side and silicon channel side, respectively. There is a positive feedback effect between the charge injected from the gate and the ferroelectric polarization. The mechanism helps to achieve a larger MW, and the analytic model of MW is also given as [5]:

$$\Delta V_{th} = \frac{\Delta(Q_{T,IL} - Q_{B,IL})}{C_{T,IL}} + \frac{\Delta(P_{FE} - Q_{B,IL})}{C_{FE}} \tag{1}$$

where $C_{T,IL}$ and C_{FE} are the top interlayer capacitance and FE layer capacitance, respectively. $Q_{T,IL}$ is charges injected from the metal gate. $Q_{B,IL}$ is the charges injected from the silicon channel. P_{FE} is ferroelectric polarization.

Dev.A, Dev.B, and Dev.C. have the similar structure. The B.IL and FE layers are in a similar situation (similar polarization of the FE layer and effects of the B.IL) when the device achieves a MW_{max}. As shown in Fig. 3(a), the MW_{max} of the devices in the MAFIS structure increases linearly with the Al_2O_3 layer increasing. Fig. 3(b) shows the charge density vs. voltage (Q_m - V) curves of Dev.A, Dev.B, and Dev.C. Notably, although the thickness of T.IL is increases, there is no significant change in Q_m. This means that Q_m is not the significant cause for the change of MW. According to the MW calculation formula (1), the change of MW with Al_2O_3 thickness is likely to be due to the increase of equivalent oxide thickness (EOT). That decreases the $C_{T,IL}$ and makes MW increase.

Next, we investigate the behavior of the charge injected from the gate ($Q_{T,IL}$) during cycles. In the traditional structure, e.g., the metal/$Hf_{0.5}Zr_{0.2}$/bottom interlayer/Si (MFIS), it is difficult to distinguish between P and $Q_{B,IL}$. The situation is further complicated with the introduction of the T.IL and the addition of $Q_{T,IL}$. The incremental step pulse programming/the incremental step pulse erase (ISPP/ISPE) can help to qualitatively understand different charge variations. This is because the ferroelectric contributes a small window from the previous work, so the significant change in MW during ISPP and ISPE should be due to the charge injected from the gate. The comparisons of g (ISPP /ISPE) behaviors of Dev.A, Dev.B, and Dev.C are shown in Fig. 4(a-b). Interestingly, the $V_{th,PGM}$ decreases with the increase of voltage, while $V_{th,ERS}$ increases first and then decrease with the increase of voltage. This means that during the programming operation, the holes injected from the gate are increasing with V_{PGM} increasing. While during the erasing operation, there is a maximum amount injection electrons, and the injection electrons will gradually decrease due to tunneling after reaching the V_{ERS} of the maximum $V_{th,ERS}$.

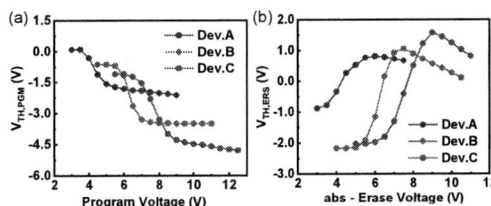

Fig. 4. (a) Comparison of ISPP behaviors of Dev. A, B, C. (b) Comparison of ISPP behaviors of Dev. A, B, C.

In 3D NAND, the MW is required to be at least 6V for multi-bit operation. The ISPP/ISPE slope reflects the gate voltage required to change the multi-bit storage state. Fig. 5. shows the ISPP slope and ISPE slope of Dev.A, Dev.B, and

Dev.C. Both the ISPP slope and ISPE slope show an upward trend with the Al_2O_3 top dielectric interface layer increasing. Therefore, in the MAFIS structure, increasing the thickness means less gate voltage required to change the storage state of the FeFETs, and also less voltage power consumption.

$$ISPP\ slope = \frac{V_{TH\_PGM,max} - V_{TH\_PGM,min}}{\Delta V_{PGM}} \quad (2)$$

$$ISPE\ slope = \frac{V_{TH\_ERS,max} - V_{TH\_ERS,min}}{\Delta V_{ERS}} \quad (3)$$

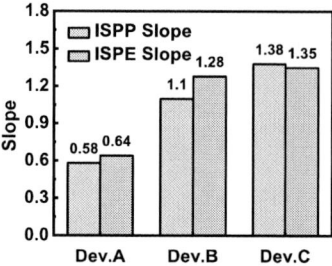

Fig. 5.   (a) The comparisons of ISPP and ISPE slope of Dev.A, B, C.

Fig. 6. (a-f) shows the change of ISPP and ISPE of the devices with different Al_2O_3 thicknesses from the initial stage to the 3k cycle, which can reflect the behavior of the gate injection charges during the endurance test. We can find that $V_{th,PGM}$ will reach the lowest after 100 cycles, while $V_{th,ERS}$ always has the maximum value in the initial state. Based on previous research work, in the case of holes that are injected from the gate side in program operation, the injection efficiency is lower than that of electrons due to the high energy band offset [8].

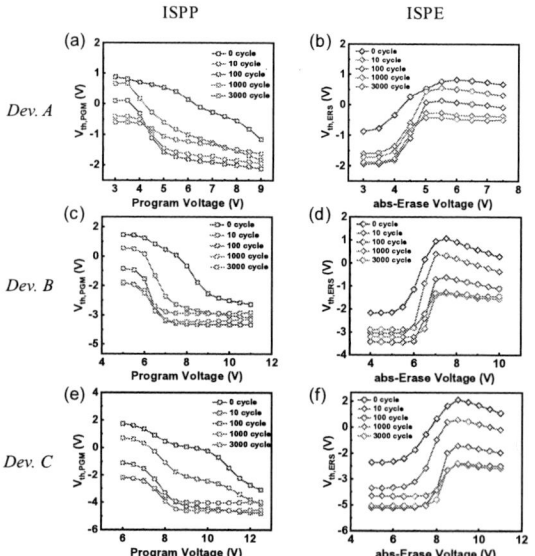

Fig. 6.   (a) Change of ISPP and ISPE of Dev. A, B, C. from the initial stage to 3k cycle.

The FE layer will be waken-up after P/E cycles, the polarization of FE layer will result the band bending of the B.IL. This promotes hole inject from the gate. Compared with hole injection during PGM operation, electron injection from the gate is easier at the ERS state, so the FE layer polarization does not have much effect on the electron injection during P/E

cycles. The ISPE of the three devices shifts downward all with cycle increasing, which is likely due to the electrons are tunnel. That is not conducive to the maintenance of large MW.

IV.   CONCLUSION

In the $TiN/Al_2O_3/Hf_{0.5}Zr_{0.5}O_2/SiON/Si$ structure, the MW_{max} increases with increasing Al_2O_3 top dielectric layer thickness. That phenomenon is caused by the increase of the EOT of gate stack, not by the increase of gate injection charge. Besides, the ISPP/ISPE slope of the FeFETs increases as the Al_2O_3 thickness of the structure increases, which can reduce the gate voltage required for the device to change the storage state and result in less voltage power consumption. In addition, the holes injected from gate will increase after a P/E process of FE layer wake-up in PGM operation. While there is no process of electron injection with the FE layer awakening during ERS operation. The electrons are easier to inject and tunnel from the gate in ERS operation compare to holes in PGM operation. Therefore, the key to improving the MW_{max} retention performance during P/E operation is to decrease the drift of $V_{th,ERS}$, making the gate injected electrons can be retained stably.

ACKNOWLEDGMENT

This work was supported in part by the National Natural Science Foundation of China under Grant Nos. 92264104 and 52350195, in part by National Key Research and Development Program of China under Grant No. 2022YFB4400300, and in part by the R&D Program of Beijing Municipal Education Commission under Grant No.KZ202210009014, in part by the Young Elite Scientists Sponsorship Program under Grant No. BYESS2023033, and Supported by the Postdoctoral Fellowship Program of CPSF under Grant No. GZC20232925.

REFERENCES

[1] L. Breuil, R. Izmailov, M. Popovici, J. Stiers, A. Arreghini, S. Ramesh et al., "Gate Side Injection Operating Mode for 3D NAND Flash Memories," in 2024 IEEE International Memory Workshop (IMW), 2024, pp. 1-4.

[2] K. Florent, S. Lavizzari, L. Di Piazza, M. Popovici, E. Vecchio, G. Potoms et al., "First demonstration of vertically stacked ferroelectric Al doped HfO_2 devices for NAND applications," in 2017 IEEE Symposium on VLSI Technology and Circuits (VLSI Technology and Circuits), 2017, pp. T158-T159.

[3] J.-G. Lee, J. Kim, D. I. Suh, I. Kim, G. D. Han, S. W. Ryu et al., "Memory Window Expansion for Ferroelectric FET based Multilevel NVM: Hybrid Solution with Combination of Polarization and Injected Charges," in 2022 IEEE International Memory Workshop (IMW), 2022, pp. 1-4.

[4] S. Yoon, S.-I. Hong, D. Kim, G. Choi, Y. M. Kim, K. Min et al., "QLC programmable 3D ferroelectric NAND Flash memory by memory window expansion using cell stack engineering," in 2023 IEEE Symposium on VLSI Technology and Circuits (VLSI Technology and Circuits), 2023, pp. 1-2.

[5] S. Lim, T. Kim, I. Myeong, S. Park, S. Noh, S. M. Lee et al., "Comprehensive Design Guidelines of Gate Stack for QLC and Highly Reliable Ferroelectric VNAND," in 2023 International Electron Devices Meeting (IEDM), 2023, pp. 1-4.

[6] D. Das, H. Park, Z. Wang, C. Zhang, P. V. Ravindran, C. Park et al., "Experimental demonstration and modeling of a ferroelectric gate stack with a tunnel dielectric insert for NAND applications," in 2023 International Electron Devices Meeting (IEDM), 2023, pp. 1-4.

[7] T. Hu, X. Sun, M. Bai, X. Jia, S. Dai, T. Li et al., "Enlargement of Memory Window of Si Channel FeFET by Inserting Al_2O_3 Interlayer on Ferroelectric $Hf_{0.5}Zr_{0.5}O_2$," IEEE Electron Device Lett., vol. 45, no. 5, pp. 825-828, 2024.

[8] I. Myeong, H. Kim, S. Kim, S. Lim, K. Kim, W. Kim et al., "Strategies for a wide Memory Window of Ferroelectric FET for multilevel Ferroelectric VNAND operation," IEEE Electron Device Letters, 2024.

Electric-Thermal Characteristics of Bottom P-i-N isolated Nanosheet Gate-All-Around FETs

Chunlei Wu[1,2,3*], Hanzhi Gu[1], Jian Ma[1], Boqian Shen[1], Fei Zhao[1], Yueyuan Yu[1], Yiming Xia[1], Qingqing Sun[1,2,3], David Wei Zhang[1,2,3]

[1] School of Microelectronics, Fudan University, Shanghai 200433, China
[2] Shanghai Integrated Manufacturing Innovation Center Co., Ltd, Shanghai 201203, China
[3] Jiashan Fudan Institute, Jiaxing 314100, China

* Email: wuchunlei@fudan.edu.cn

Abstract—The electrical and thermal characteristics of a novel bottom P-i-N isolation GAA NSFET has been studied. Through introducing a reverse biased p-i-n sub-TFET channel, the proposed NSFET with bottom PIN isolation scheme can reduce off leakage current to the same level as NSFET with full BDI scheme. Moreover, owing to the all silicon-based isolation instead of buried dielectric isolation in BDI scheme, the proposed bottom PIN NSFETs exhibits superior thermal properties at the same time.

Keywords—Gate-All-Around (GAA), Stacked nanosheet, Parasitic channel, Reverse biased P-i-N, Self-heating effect (SHE)

I. INTRODUCTION

Gate-All-Around (GAA) nanosheet FET has been widely recognized as the mainstream logic device structure for beyond 3 nm node applications, owing to its superior gate electrostatics[1]. To meet the requirement of low power applications, however, sub-fin leakage in GAA nanosheet FET is one of the most important problems to be addressed. Different from sub-fin leakage in FinFET, GAA nanosheet has a tri-gated sub-fin with much wider "fin width", the gated sub-fin region acts as a parasitic channel with poor gate control and accordingly undesired leakage problems.

Multiple approaches have been discussed to address the parasitic channel leakage issue, including punch through stoppers (PTS) [2], partial/full bottom dielectric isolation (BDI) [3] and narrowed sub-fin technique [4]. For aggressively scaled GAA nanosheet FETs, however, PTS suffers from increased process difficulty and high sensitivity to process variations. BDI scheme, which blocks the parasitic channel leakage path with dielectrics, can achieve much lower leakage current. While introducing dielectric layers underneath the source /drain (S/D) regions also brings about severe self-heating effects (SHE) [5]. The high device temperature leads to degraded carrier mobility and reduced drain current. Temperature rise further introduces reliability issues, such as hot carrier injection (HCI) or bias temperature instability (BTI). In this case, new approaches for parasitic channel isolation are still in ample necessity for low power applications of GAA NSFETs.

In this work, we propose a novel Bottom PIN isolation scheme for parasitic channel leakage control and self-heating effects suppression in GAA NSFETs. The TCAD simulation results show that GAA NSFETs with bottom PIN isolation configuration can achieve suppressed parasitic channel

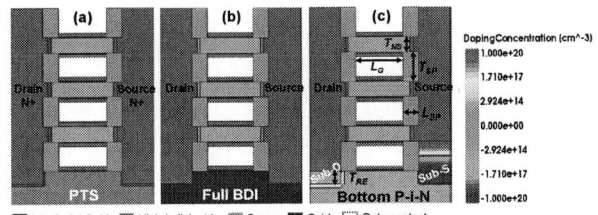

Fig. 1 Schematics of GAA FETs with different parasitic channel isolation schemes, (a) PTS, (b) Full BDI, and (c) Bottom P-i-N. (n-type FETs)

TABLE I.   STRUCTURE PAREMETER USED IN SIMULATION

Parameters	Values [unit]
Channel length, L_G	18 (nm)
Spacer length, L_{SP}	6 (nm)
Nanosheet width, W_{NS}	20 (nm)
Nanosheet thickness, T_{NS}	6 (nm)
Nanosheet space, T_{SP}	12 (nm)
Interfacial oxide thickness, T_{IL}	0.5 (nm)
High-k dielectric thickness, T_{OX}	1.5 (nm)
S/D recess depth, T_{RE}	5 (nm)
Sub-gate height, H_{SUB}	5 (nm)
S/D doping, N_{SD}	1×10^{20} (cm^{-3})
Channel doping, N_{CH}	5×10^{14} (cm^{-3})
Sub-source doping, N_{SUBS}	1×10^{20} (cm^{-3})
Sub-drain doping, N_{SUBD}	1×10^{18} (cm^{-3})
PTS doping, N_{PT}	2×10^{18} (cm^{-3})

leakage and improved thermal properties at the same time. The results suggest great potential of the proposed bottom PIN isolation scheme in the prospective low power logic applications.

II. DEVICE STRUCTURE AND SIMULATION MEHODOLOGY

The structure of the studied GAA NSFET with bottom PIN isolation scheme is schematically shown in Fig. 1. Compared with NSFETs with PTS and full BDI schemes, NSFETs with bottom PIN has asymmetrically doped sub-source and sub-drain regions underneath the source and drain regions respectively. In NSFET with bottom PIN, the p⁺ sub-source of $N_{SUBS} = 1 \times 10^{20}$ cm⁻³, and the n⁻ sub-drain of $N_{SUBD} = 1 \times 10^{18}$ cm⁻³ is used to suppress the ambipolar behavior in the BTBT

979-8-3503-6184-1/24 $31.00 © 2024 IEEE

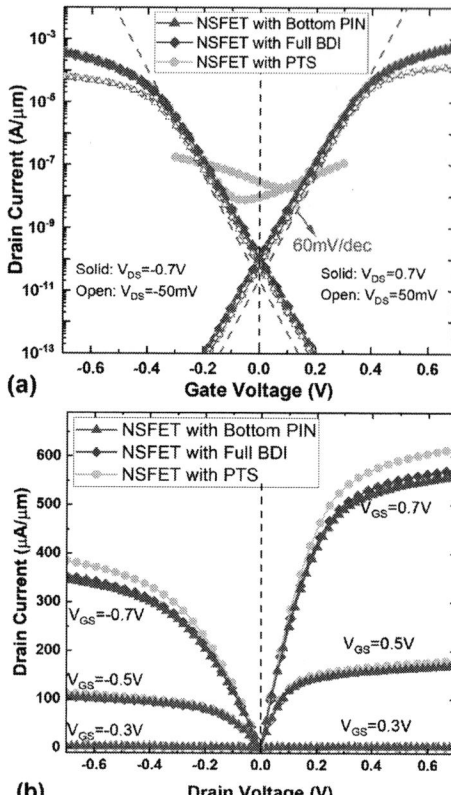

Fig. 2 (a) Transfer characteristics of NSFETs with bottom PIN in comparison with NSFETs with full BDI and NSFETs with PTS. The gate work function is set as 4.49 eV for n-type devices and 4.81eV for p-type devices to fix the OFF-current of NSFET with full BDI (I_{off} = 100 pA/μm at V_{GS} = 0 V). (b) Output characteristics of NSFETs with bottom PIN in comparison with NSFETs with PTS, and NSFETs with full BDI.

sub-channel. The doping concentrations of the n^+ doped source/drain, p^- doped channel is N_{SD} = 1 × 10^{20} cm^{-3}, N_{CH} = 5 × 10^{14} cm^{-3}. In NSFET with PTS, the PTS doping concentration is N_{PT} = 2 × 10^{18} cm^{-3}. Doping decay length of L_{decay} =1 nm is assumed if not otherwise stated. The structure dimensions including gate length (L_G) of 18 nm, NS width (W_{NS}) of 20 nm, NS thickness (T_{NS}) of 6 nm, interfacial oxide thickness (T_{IL}) of 0.5 nm, high-k dielectric thickness (T_{OX}) of 1.5 nm is used, following 3 nm node ground rules [6], as shown in Table I. The S/D recess depth (T_{RE}) and sub-gate height (H_{SUB}) of 5 nm are used if not otherwise stated.

Device simulations are carried out using Synopsys TCAD Sentaurus tools [7]. Self-consistent calculation is achieved through transport equation combined with Poisson equation and density-gradient quantum correction models. Low field ballistic mobility, auto-orientation inversion and accumulation layer mobility (IALMob) as well as high field saturation velocity are included in Mobility models for DD simulation. Dynamic nonlocal path BTBT, Shockley-Read-Hall and Auger recombination are included in Recombination models for BTBT simulation. The mobility models used have been calibrated with the latest NSFETs experimental data [8], while the BTBT model has been calibrated with the tri-gated TFET experimental data [9]. The electron-thermal coupled simulation is performed using Thermodynamic model accelerate the thermal behavior simulation without losing

TABLE II. THERMAL PAREMETER USED IN SIMULATION

Thermal Conductivity	Values [unit]
Channel Region, κ_{ch}	7.5 (W/K·m)
Source/Drain Region, $\kappa_{S/D}$	25 (W/K·m)
Substrate (Bulk Silicon), κ_{sub}	148 (W/K·m)
PTS doped Region, κ_{pts}	40 (W/K·m)
Oxide (SiO2), κ_{SiO2}	1.4 (W/K·m)
Metal Gate (W), κ_W	175 (W/K·m)
Interconnect (W), κ_W	175 (W/K·m)
Thermal Contact Resistance	**Values [unit]**
Si/SiO2 interface, $\rho_{th, Si/SiO2}$	2×10^{-4} (cm^2·K/W) [10]
Back End of Line, $\rho_{th, BEOL}$	4.8×10^{-4} (cm^2·K/W) [10]
Substrate, $\rho_{th, Sub}$	5.4×10^{-2} (cm^2·K/W) [10]
Thermal Boundary condition	300 K

accuracy [10]. Different thermal conductivity and thermal contact resistance for different area was considered as shown in Table II. The thermal boundary conditions settings are set to ambient temperature (300 K).

III. RESULTS AND DISCUSSION

A. Parasitic Channel Leakage Control

The transfer characteristics of NSFET with bottom PIN are compared with that of conventional NSFET, NSFET with PTS, and NSFET with full BDI in Fig. 2 (a). It can be seen that, by introducing the sub-TFET channel, bottom PIN NSFET can effectively reduce the leakage current to the same level as full BDI NSFET, achieving perfectly controlled parasitic channel leakage and a near-60 mV/dec SS. The excellent parasitic channel leakage block capability of bottom PIN scheme, as shown in Fig.2 a), can be attributed to the small off-state leakage dictated by the reverse-biased PIN diode, which is about 4 decades lower compared to a typical Si MOS channel. The results suggest that the NSFETs with bottom PIN isolation can achieve parasitic channel leakage control as effective as full BDI NSFETs.

Fig. 2 (b) shows the output curves of NSFET with bottom PIN isolation, on-current of 558 μA/μm (n-type NSFET) and 347 μA/μm (p-type NSFET) has been obtained at V_{GS} = V_{DS} = 0.7 V. As can be noticed in the figure, the saturation drain current of three different NSFETs is close. The saturation current of BDI NSFETs and bottom PIN NSFETs is slightly lower than that of PTS NSFETs, owing to the absence of parasitic channel current. While the saturation current of bottom PIN NSFETs is slightly lower than that of BDI NSFETs, owing to small difference of source/drain series resistance.

B. Thermal Properties Improvement

In gate-all-round devices such as NSFETs, it is difficult for heat generated in the channel to flow out to the substrate as effective as in planar FETs or FinFETs. Therefore, self-heating effects become one of the greatest challenges in advanced GAA device optimizations. In BDI NSFETs, the introduction of bottom dielectric layer blocks the heat dissipation path from S/D region to the substrate, resulting in

979-8-3503-6184-1/24 $31.00 © 2024 IEEE

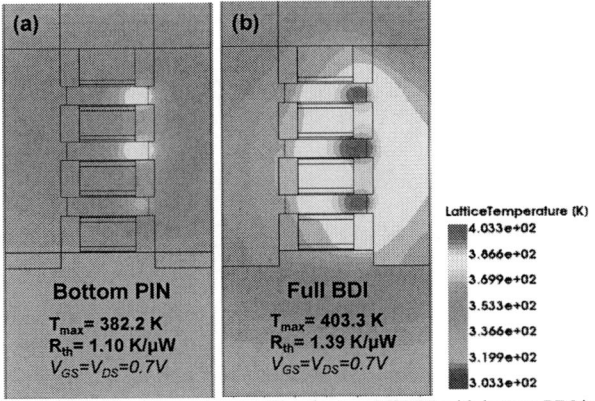

Fig. 3 (a) Lattice temperature profile of n-type NSFET with bottom PIN in comparison with, (b) NSFET with full BDI at supply voltage $V_{DD} = 0.7$ V.

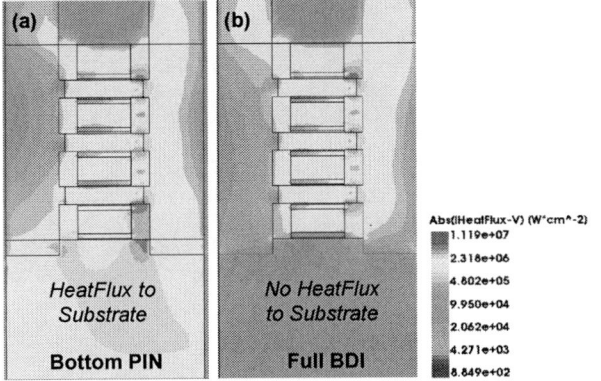

Fig. 4 Heat flux profile of n-type NSFET with bottom PIN in comparison with, (b) NSFET with full BDI at supply voltage $V_{DD} = 0.7$ V.

aggravated self-heating effects. Compared to BDI NSFETs, the main advantage of the proposed bottom PIN NSFETs is the all silicon-based isolation with much higher thermal conductivity, thus improved thermal properties can be achieved.

Fig.3 and Fig.4 compares the lattice temperature and heat flux profiles of NSFETs with bottom PIN isolation and NSFETs with full BDI respectively. As can be seen in Fig. 3 (a) and (b), the peak lattice temperature (T_{max}) locates at the drain extension region of the NS channel, and the middle NS channel shows the highest lattice temperature, since it has the longest dissipation path to the heat sink at both backend-of-line (BEOL) and substrate. A much lower T_{max} of 382.2 K has been achieved in the bottom PIN NSFETs, compared to T_{max} of 403.3 K in full BDI NSFETs. Reduced equivalent thermal resistance R_{th} of 1.1K/µW has been obtained as well. Note that the equivalent thermal resistance R_{th} here is defined as (T_{max}-T_{amb})/($I_{ON} \cdot V_{DD}$), in which T_{amb} is the ambient temperature (300 K), and $I_{ON} \cdot V_{DD}$ is the power consumption.

The significant improved thermal behavior can be attributed to the all-silicon substrate that plays a important role in heat dissipation. As clearly shown in Fig.4 (a) and (b), NSFETs with bottom PIN and full BDI shows different heat fluxes. In full BDI NSFETs, the heat dissipation path to the substrate is eliminated and the only heat sink remained is the BEOL region. In bottom PIN NSFETs, however, the self-generated heat in NS channel can be well dissipated through both the BEOL and substrate. The results suggest that the bottom PIN isolation can help to realize lower device temperature and smaller channel thermal resistance.

IV. CONCLUSION

The electrical and thermal characteristics of a novel bottom P-i-N isolation GAA NSFET has been studied. By introducing a reverse biased p-i-n configuration, bottom PIN NSFETs can achieve not only full BDI-comparable parasitic channel leakage but also improved self-heating effects. The conclusions would provide insights regarding the exploitation of novel GAA NSFETs design for ultra-low power operation.

ACKNOWLEDGMENT

This work was supported in part by NSFC under Grant 62304048, in part by the Science and Technology Commission of Shanghai Municipality under Grant 21TS1400800.

REFERENCES

[1] G. Bae et al., "3nm GAA Technology featuring Multi-Bridge-Channel FET for Low Power and High Performance Applications," 2018 IEEE International Electron Devices Meeting (IEDM), San Francisco, CA, USA, 2018, pp. 28.7.1-28.7.4.V.

[2] Jegadheesan, K. Sivasankaran and A. Konar, "Optimized Substrate for Improved Performance of Stacked Nanosheet Field-Effect Transistor," in IEEE Transactions on Electron Devices, vol. 67, no. 10, pp. 4079-4084, Oct. 2020.

[3] J. Zhang et al., "Full Bottom Dielectric Isolation to Enable Stacked Nanosheet Transistor for Low Power and High Performance Applications," 2019 IEEE International Electron Devices Meeting (IEDM), San Francisco, CA, USA, 2019, pp. 11.6.1-11.6.4.

[4] J. Gu et al., "Narrow Sub-Fin Technique for Suppressing Parasitic-Channel Effect in Stacked Nanosheet Transistors," in IEEE Journal of the Electron Devices Society, vol. 10, pp. 35-39, 2022.

[5] M. Saleh, et al., "Impact of Bottom Dielectric Isolation of Si-Stacked Nanosheet Transistor on Stress and Self-Heating at 3-nm Node and Beyond," IEEE Trans. Electron Devices, vol. 70, no. 11, pp. 5535-5542, Nov. 2023, doi: 10.1109/TED.2023.3318554.

[6] International Roadmap for Devices and Systems (IRDS) [Online]. Available: https://irds.ieee.org/editions/2022.

[7] TCAD Sentaurus User Guide Version O-2018.06, June 2018, Synopsys, Inc., MountainView CA, USA, 2018..

[8] N. Loubet et al., "Stacked nanosheet gate-all-around transistor to enable scaling beyond FinFET," in Proc. Symp. VLSI Technol., Jun. 2017, pp. T230–T231.

[9] L. Knoll et al., "Inverters With Strained Si Nanowire Complementary Tunnel Field-Effect Transistors," in IEEE Electron Device Letters, vol. 34, no. 6, pp. 813-815, June 2013.

[10] C. Yoo, et al., "Analysis of Self-Heating Effects in Multi-Nanosheet FET Considering Bottom Isolation and Package Options," in IEEE Transactions on Electron Devices, vol. 69, no. 3, pp. 1524-1531, Mar. 2022.

Surface Potential-Based Compact Model for ITO Thin-film Transistors with Ultra-thin Channel

Wenting Xu [1], Xinxin Shen [1], Zuoxu Yu [1], Tingrui Huang [1], Yuzhen Zhang [1], Weifeng Sun [1], and Wangran Wu*[1]

[1] National ASIC System Engineering Research Center, Southeast University, Nanjing 210096, China

* Email: 220216001@seu.edu.cn, wrwu@seu.edu.cn

Abstract—In this paper, we propose a novel analytical surface potential-based compact model for Indium Tin Oxide (ITO) thin-film transistors (TFT) with ultra-thin channel. We develop an equivalent capacitance model that accounts for the thickness of the active layer to derive the surface potential, which reveals the coupling of both sides of the nano-meter ITO layer. By applying Schroder series modification, the analytical solution of surface potential is derived for the ITO TFTs with varying channel thicknesses. Subsequently, the carriers' mobility model is formulated in ITO films, grounded on the transport mechanisms of variable range hopping, trap limited conduction, and percolation conduction. Finally, leveraging the surface potential and mobility models, we derive a drain current model for ITO TFTs using the Sah equation. The resulting model is then integrated using Verilog-A language. The proposed model is applied to the simulation of an ITO TFT-based inverter, which agrees well with the experimental results.

Keywords—indium tin oxide, thin film transistor, compact model, surface potential, ultra-thin-channel

I. INTRODUCTION

ITO film possesses the characteristics of a wide band gap, exceptional transparency, and a high electron density surpassing 10^{20} cm^{-3} [1-7]. The researches [8-12] point out that TFTs utilizing nano-meter thick ITO films as semiconductor channels exhibit outstanding current performance, with higher mobility than the extensively studied a-IGZO. The electrical properties of ITO TFTs are strongly correlated with the channel thickness. The study of circuits based on ITO devices urgently requires an accurate compact model to describe the devices' behavior. Compared to the V_{th}-based model, the surface potential-based model possesses inherent continuity and could accurately simulate higher-order effects. However, the reported compact models for TFTs do not contain the parameter of channel thickness. Thus, the existing model cannot precisely simulate the electrical properties of ITO TFTs, which are extremely sensitive to the channel thickness.

In this study, we establish a compact model for ITO-TFTs with ultra-thin channel grounded in surface potential. The impact of channel thickness is taken into account in the surface potential model. The carriers' mobility model is accomplished by considering the transport mechanisms of variable range hopping (VRH), trap limited conduction (TLC), and percolation conduction (PERC). The compact model is then derived by combining the surface potential and

The work is supported in part by the National Natural Science Foundation of China (62274033, 62074034), Natural Science Foundation of Jiangsu Province (BK20221453)，and Fundamental Research Funds for the Central Universities (Corresponding author: Wangran Wu, wrwu@seu.edu.cn).

mobility models. After integrating the model by Verilog-A language, it is implemented in the simulation of ITO TFT-based inverter and compared with the experimental results.

II. MODELING

A. Surface Potential Solving

The bottom-gate staggered ITO TFTs with different channel thicknesses are examined (Fig. 1(a)). The device's fabrication process can be found in [13], and Fig. 1(b) shows the optical view of ITO TFT. The flow diagram of the surface potential-based compact model is given in Fig. 1(c). To establish a surface potential model for ITO TFT that incorporates the influence of channel thickness, we commence with the Poisson equation, as in (1). By employing the gradual channel approximation (GCA), and Gaussian boundary conditions, we derive (2). Where Φ_T is the thermal potential

$$\frac{\partial D}{\partial z}(y,z) = -q n_i e^{\frac{(\varphi(y,z)-\phi_{im}(y))}{\phi_T}} \tag{1}$$

$$\begin{cases} \left(k_1 q_1 + \alpha \coth\left(\frac{\alpha}{2}\right)\right)(k_1 q_1 + k_2 q_2) = k_1^2 q_1^2 - \alpha^2 \\ \left(k_2 q_2 + \alpha \coth\left(\frac{\alpha}{2}\right)\right)(k_1 q_1 + k_2 q_2) = k_2^2 q_2^2 - \alpha^2 \\ \left(k_1 q_1 + \alpha \coth\left(\frac{\alpha}{2}\right)\right)\left(k_2 q_2 + \alpha \coth\left(\frac{\alpha}{2}\right)\right) = \frac{\alpha^2}{\sinh(\alpha/2)^2} \end{cases} \tag{2}$$

$$F(x_1) = \left[k_1 q_1 + \alpha \cdot \coth\left(\frac{\alpha}{2} + t_{ITO} \cdot \delta\right)\right](k_1 q_1 + k_2 q_2) - A_0 e^{x_1 - x_n} - B_0 e^{t(x_1 - x_n)} \tag{3}$$

$$A_0 = 2q t_{ITO} N_t \Gamma(1 - T/T_0)\Gamma(1 + T/T_0)/(C_{ITO}\phi_{T0})$$

$$B_0 = 2q t_{ITO} N_t v_0 \tau_0 / t C_{ITO}\phi_{T0}$$

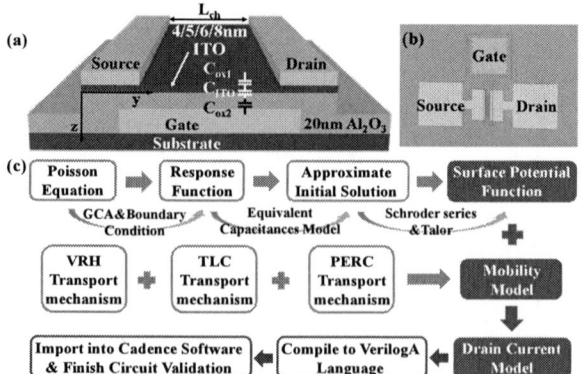

Fig. 1. (a) Bottom-gate interleaved structure diagram and equivalent capacitances diagram of ITO TFT; (b) The optical view of ITO TFT; (c) A flow diagram of surface potential-based compact model for ITO TFT.

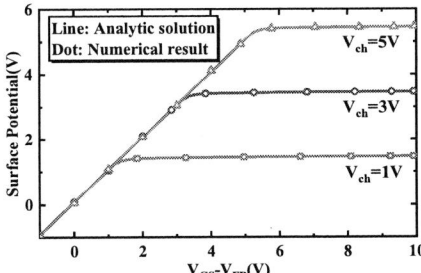

Fig. 2. Comparison of analytical and numerical solutions of surface potential under different channel voltages.

at room temperature, Φ_{T0} is the thermal potential at characteristic temperature, t_{ITO} represents the thickness of the ITO active layer, $k=C_{OX}/C_{ITO}$, and $\alpha=Q/(C_{ITO}\Phi_{T0})$.

After algebraic manipulations, and utilizing the Schroder series to incorporate the free electrons in the extended state. An equation was established to be solved with only $x1$ as the variable in (3).

$$
\varphi_s = \phi_{T0}\left\{ 2x_1 - \frac{F(x_1)}{F'(x_1) - \frac{F(x_1)}{F'(x_1)}\frac{F''(x_1)}{2}} - \frac{F(x_{11})}{F'(x_{11}) - \frac{F(x_{11})}{F'(x_{11})}\frac{F''(x_{11})}{2}} \right.
$$
$$
\left. -2\ln\left(k_1q_1 + \alpha \cdot \coth\left(\frac{\alpha}{2} + t_{ITO}\cdot\delta\right)\right) + 2\ln\left(\alpha \cdot csch\left(\frac{\alpha}{2} + t_{ITO}\cdot\delta\right)\right) \right\} \quad (4)
$$

Finally, based on the calculations above, and by applying a second-order Taylor expansion to $x1$ with two consecutive corrections, we can obtain more precise analytical solutions for the surface potentials φ_s in (4). Figure 2 shows the fit results, validating the consistency between the theoretical formula and the analytical model.

B. Mobility Model

At lower bias and temperature, the VRH transport domains. The mobility is given in (5).

$$
\mu_{VRH} = \mu_{R\_VRH}A^* \exp\left(\left(\frac{T_{CT}}{T}\right)^{\frac{1}{4}}\right)(V_g - V_T)^{\gamma} \quad (5)
$$

Where the defect states have an exponential distribution, μ_{R_VRH} represents the reference mobility for the VRH transport, A^* is a coefficient of VRH transport, T_{CT} signifies the characteristic temperature of the VRH transport, and γ is an exponential parameter linked to the density of tail states.

At lower bias voltages, TLC emerges as the dominant carrier transport mechanism across a wide temperature range. The mobility corresponding to the TLC transport mechanism is provided in (6).

$$
\mu_{TLC} = \mu_{BM}^* B^*\left(V_g - V_{FB}\right)^{2\left(\frac{T_0}{T}-1\right)} \quad (6)
$$

Where B^* is a coefficient of TLC transport, μ_{BM}^* represents the band mobility.

Under higher gate voltage bias, the Fermi level enters the conduction band, and PERC becomes the dominant carrier transport mechanism. In this model, the mobility expression associated with PERC is given in (7). Where, C^* a coefficient of PERC transport, V_{G_TRAP} is the gate voltage corresponding

$$
\mu_{PERC} = \mu_{BM}^* C^*\left(V_g - V_{G\_TRAP}\right)^{4\left[\frac{(D_B-W_B)}{D_B}\right]} \quad (7)
$$

to the transition from trap confinement to percolation, D_B and W_B represent the gap and width of the barrier, respectively. Finally, carrier transport in ITO is achieved through the combination of traps in local states and free electrons in the extended state.

C. Current Model

In this model, the Sah equation under the asymptotic channel approximation is utilized to establish the drain current model in (8) and (9).

$$
I_{ds} = -\mu_{eff}\frac{W}{L}\int_0^{V_{ds}}Q_i dV_{ch} = \mu_{eff}\frac{W}{L}\int_0^{V_{ds}}\int_{\varphi_{s2}}^{\varphi_{s1}}\frac{qn}{E}d\varphi dV_{ch}
$$
$$
= \mu_{eff}\frac{W}{L}\int_0^{V_{ds}}\int_{\varphi_{s2}}^{\varphi_{s1}}\left(\frac{1}{2\varepsilon_{ITO}E}C_{ITO}^2\phi_{T0}^2\frac{\partial\alpha^2}{\partial V_{ch}} - \varepsilon_{ITO}\frac{\partial E}{\partial V}\right)d\varphi dV_{ch} \quad (8)
$$

$$
E = \pm\frac{1}{\varepsilon_{ITO}}\sqrt{\frac{QC_{ITO}\phi_{T0} + A_0 e^{\frac{\varphi-V_{ch}}{\phi_{T0}}} + B_0 e^{\frac{\varphi-V_{ch}}{\phi_T}}}{\left(C_{ITO}\phi_{T0}\right)^2}} \quad (9)
$$

Where W represents the channel width, L stands for the channel length, ε_{ITO} is the dielectric constant of ITO, and Q_i denotes the total cumulative layer charge density. The carrier concentration is indicated by n in (8). E represents the electric field intensity, derived by considering the coupled charge in (9). By substituting the analytical expressions for the channel's front and back surface potentials and algebraic operations, we arrive at the final drain current expression for the ITO TFT in (10). This expression comprises three distinct components: the total channel surface current (I_s), the interface coupling current (I_c), and the subthreshold current (I_{sub}), as detailed in (10).

$$
I_{ds} = \mu_{eff}\frac{W}{L}(I_s + I_c + I_{sub})
$$
$$
= \mu_{eff}\frac{W}{L}\left\{ C_{oxt}\left[\left(\varphi_{s1d}-\varphi_{s1s}\right)\left(V_{g1}-V_{FB1}+2\phi_T\right) - \frac{1}{2}\left(\varphi_{s1d}^2-\varphi_{s1s}^2\right)\right] + \frac{C_{ITO}}{2\phi_{T0}^2}\left(\alpha_s^2-\alpha_d^2\right) \right.
$$
$$
\left. +2qN_t t_{ITO}\Gamma\left(1-\frac{T}{T_0}\right)\Gamma\left(1+\frac{T}{T_0}\right)\left(\phi_{T0}-\phi_T\right)\left(1-e^{\frac{V_{ds}}{\phi_{T0}}}\right)e^{\frac{\varphi_{s1s}+\varphi_{s2s}}{2\phi_{T0}}} \right\} \quad (10)
$$

III. MODEL VALIDATION

We utilized the Keysight B1500A and Keysight E4980A to conduct tests on ITO TFTs. Additionally, simulation verification of the inverter circuit is also performed and compared with the experimental results to ensure its functionality.

A. Device Validation

The mobility model was fitted and compared to the

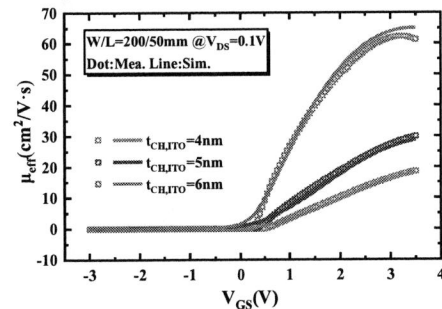

Fig. 3. The experimental and simulated carrier mobility of ITO TFTs with 4 nm, 5 nm, and 6 nm channel thicknesses.

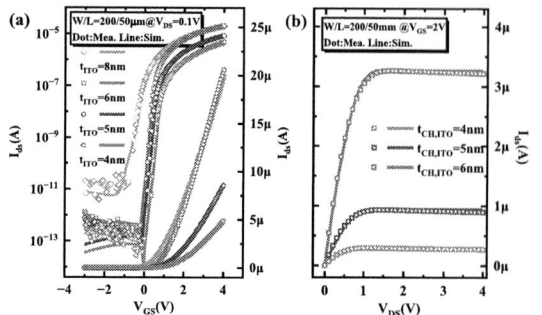

Fig. 4. The experimental and simulated (a) transfer and (b) output characteristics of ITO TFTs with various channel thicknesses.

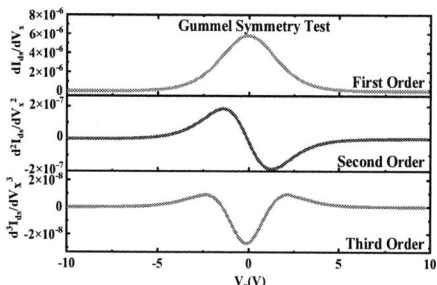

Fig. 5. Gummel symmetry test of the proposed compact model.

experimental results for ITO TFTs with different channel thicknesses in Fig. 3. It is observed that the mobility decreases in the device with thinner channel thickness. The mobility results of the compact model align well with the experimental results.

Combining the carrier density and mobility models, the transfer and output characteristics of ITO TFTs with various channel thicknesses were also fitted (Fig. 4(a) and (b)). The drain current of the compact model aligns well with the experimental measurements, conforming to the validation of the proposed compact model. The V_{th} decreases and I_{ds} increases with the increasing channel thicknesses.

The Gummel Symmetry Test (GST) is illustrated in Fig. 5, the current of the proposed model is relatively smooth and continuous at the zero points of the first- and third-order derivatives. The second-order derivative curve also passes through the origin as an odd function, and the curves are generally smooth. This demonstrates the proposed model possesses excellent symmetry and continuity characteristics.

B. Circuit Simulation Verification

After completing the modeling process, we encoded the

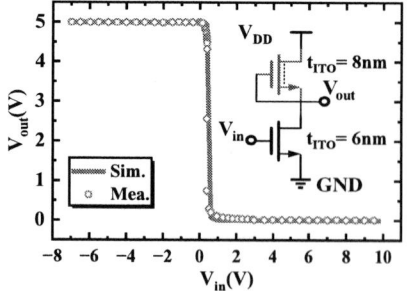

Fig. 6. Voltage transfer curve of ITO TFT-based inverter with depletion load.

model using the Verilog-A language. Then, we imported it into the Cadence platform for the inverter circuit with depletion load to verify the effectiveness of the model. The upper transistor is a depletion-mode device with a channel thickness of 8 nm, while the lower transistor is an enhancement-mode device with a channel thickness of 6 nm, as shown in the inset of Fig. 6. The circuit simulation and experimental results align well, indicating the accuracy the proposed compact model.

IV. CONCLUSION

The surface potential-based compact model for ITO TFTs with ultra-thin channel is developed in this paper. The effect of channel thickness on the electrical properties was involved in the model by considering the coupling of top and bottom potential of thin ITO layer. The analytical solution of surface potential is derived using Schroder series modification. The carrier mobility model contains all transport mechanisms, including VRH, TLC, and PERC, encompassing a wide range of temperatures and bias conditions. Finally, the compact model was derived and integrated using the Verilog-A language. The proposed model passed GST. The simulation and experimental results are consistent well in ITO TFTs with various channel thicknesses and ITO TFT-based inverter. Thus, the proposed model could advance the development and application of ITO TFTs in displays and other circuits.

REFERENCES

[1] K. Nomura, A. Takagi, T. Kamiya, H. Ohta, M. Hirano, H. Hosono, "Amorphous oxide semiconductors for high-performance flexible thin-film transistors," Japanese Journal of Applied Physics, vol. 45, pp. 4303, May 2006.

[2] T. Hirao, M. Furuta, H. Furuta, T. Matsuda, T. Hiramatsu, H. Hokari, et al., "Novel top-gate zinc oxide thin-film transistors (ZnO TFTs) for AMLCDs," Journal of the Society for Information Display, vol. 15, pp. 17-22, Jan 2007.

[3] I. M. Choi, M. J. Kim, N. On, A. Song, K. B. Chung, H. Jeong, et al., "Achieving high mobility and excellent stability in amorphous In–Ga–Zn–Sn–O thin-film transistors," IEEE Transactions on Electron Devices, vol. 67, pp. 1014-1020, Feb 2020.

[4] K. Myny, "The development of flexible integrated circuits based on thin-film transistors," Nature Electronics, vol. 1, pp. 30-39, Jan 2018.

[5] K. Kandpal, N. Gupta, "Perspective of zinc oxide based thin film transistors: a comprehensive review," Microelectronics International, vol. 35, pp. 52-63, Jan 2018.

[6] J. W. Borchert, U. Zschieschang, F. Letzkus, M. Giorgio, R. T. Weitz, M. Caironi, et al., "Flexible low-voltage high-frequency organic thin-film transistors," Science advances, vol. 6, pp. eaaz5156, May 2020.

[7] F. Chen, M. Zhang, Y. Wan, X. Xu, M. Wong, H. S. Kwok, "Advances in mobility enhancement of ITZO thin-film transistors: a review," Journal of Semiconductors, vol. 44, pp. 091602, Sep 2023.

[8] Q. Gao, T. Cao, J. Li, et al., "Enhancing electrical performance and stability of nanometer-thin ITO transistors via thermally oxidized alumina passivation layer," AIP Advances, vol. 13, pp. 075111, 2023.

[9] M. Si, J. Andler, X. Lyu, et al., "Indium-tin-oxide transistors with one nanometer thick channel and ferroelectric gating," ACS nano, vol. 14, pp. 11542-11547, 2020.

[10] Q. Li, J. Dong, D. Han, et al., "Effects of channel thickness on electrical performance and stability of high-performance InSnO thin-film transistors," Membranes, vol. 11, pp. 929, 2021.

[11] M. Hu, L. Xu, X. Zhang, et al., "High mobility amorphous InSnO thin film transistors via low-temperature annealing," Applied Physics Letters, vol. 122, pp. 033503, 2023.

[12] D. Xu, M. Zhang, R. Duan, et al., "High-performance nano-scale InSnO transistors," Japanese Journal of Applied Physics, vol. 63, pp. 02SP48, 2024.

[13] W. Wu, T. Huang, G. Yang, et al., "High-voltage Indium-Tin-Oxide Thin-film Transistors Possessing Drift Region Capped with Indium-Tin-Oxide Layer," IEEE Electron Device Letters, pp. 1-1, 2024.

A Continuous Full Channel Potential Model for Accurate Synthetic Electricfield Calculating in Gate-All-Around Devices

Fei Zhao[1], Chunlei Wu[1,2,3*], Yumin Xu[1], Boqian Shen[1], Jian Ma[1], Hanzhi Gu[1], Yueyuan Yu[1], Yiming Xia[1], Qingqing Sun[1,2,3], David Wei Zhang[1,2,3]

[1] School of Microelectronics, Fudan University, Shanghai 200433, China
[2] Shanghai Integrated Manufacturing Innovation Center Co., Ltd, Shanghai 201203, China
[3] Jiashan Fudan Institute, Jiaxing 314100, China

* Email: wuchunlei@fudan.edu.cn

Abstract—In this paper, a continuous full-channel potential model for Gate-All-Around devices is established and verified, which can give accurate description of channel potential profiles of the whole channel region. The model predicted surface and in-channel potential profiles of nanosheet channels with different width/thickness ratio agrees well with the TCAD simulation. Furthermore, great accuracy of the model has been verified through successful prediction of the synthetic electric field effects in scaled GAA channel. The results suggest that this full channel potential model is promising in future device modeling and circuit simulation studies of GAA devices.

Keywords—*Analytical model, Gate-All-Around (GAA), Nanosheet, surface potential, synthetic electric field effects*

I. INTRODUCTION

Gate-All-Around (GAA) FET architecture has become the most promising mainstream logic device structure in sub-3-nm integrated circuit (IC) design due to its superior gate control of the highly scaled channel. The GAA structure can offer the best solution to short-channel effects (SCEs) as well as high current drivability per layout footprint owing to 3D stacked channels [1]. Massive work has been reported involving the adoption of GAA technology in terms of both experimental studies [2] and theoretical simulation studies [3]. Meanwhile, circuit applications exploring of GAA devices has also received widely interest.

Circuit applications of GAA devices require a compact SPICE model for accurate description of devices' electrical characteristics [4]. Since logic devices' operation including the subthreshold conduction in MOSFET and the band-to-band tunneling in TFET are governed by the channel potential distribution, accurate full channel potential modeling is in ample necessity especially in extremely scaled GAA devices. Most reported potential-based GAA device modeling, however, the potential models are incomplete and can only give prediction at the channel surface [5-7]. The growing channel-scaling demand challenges the accuracy of classical potential models due to nonnegligible size effects including synthetic electric field effects and quantum effects [10].

In this paper, a continuous full channel potential model based on modified pseudo-2D Poisson equation is proposed, which is able to accurately describe the potential distribution of the whole channel region in GAA devices. Great accuracy

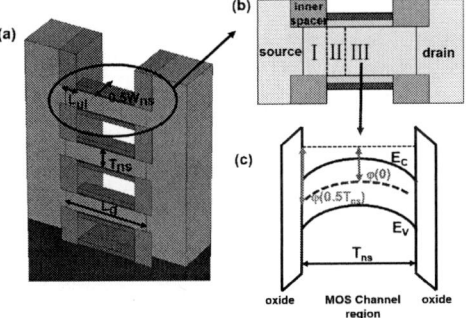

Figure 1. a) The structure of the studied GAA device; b) Cross-sectional schematic of a GAA nanosheet channel; c) A band diagram along nanosheet thickness T_{ns} in MOS Channel region, $\varphi(0.5T_{ns})$ is the NS Channel surface potential, and $\varphi(0)$ is the NS Channel center potential.

of the model has been verified through successful prediction of the synthetic electric field effects in scaled GAA channel. The derivation of the presented full channel potential model takes an n-type GAA nanosheet TFET as an example, but it can be easily extended to GAA nanosheet MOSFETs, as well as other device structures such as double-gate or tri-gated structures.

II. LIMITS OF CLASSICAL CHANNEL POTENTIAL MODEL

As a typical example, TCAD simulations of Si based three-layer stacked nanosheet GAA TFETs are performed. The cross-sectional schematic of a nanosheet GAA TFET is shown in Fig.1 a). L_g is the gate length (20 nm), T_{ns} is the channel thickness, W_{ns} is the nanosheet channel width, L_{ul} is the gate-to-source underlap length (4nm). And the highly p+ doped source N_s, lightly p-doped channel N_{ch}, and the n+ doped drain N_D are 5×10^{20} cm^{-3}, 5×10^{14} cm^{-3}, and 5×10^{18} cm^{-3} respectively. The gate work function is set as 4.61 eV if not otherwise stated. A GAA TFET can be regarded as a serial combination of a gated tunnel diode (regions I and II) and a MOS Channel (region III) (Fig.1 b)). A band diagram along nanosheet thickness T_{ns} in region III is shown in Fig.1 c), in which the parabolic-like potential profiles can be observed.

The channel potential distribution $\varphi(x, y)$ in MOS-channel is governed by 2D Poisson equation [8]. In order to reduce the 2D equation into an analytically solvable 1D equation, the

979-8-3503-6184-1/24 $31.00 © 2024 IEEE

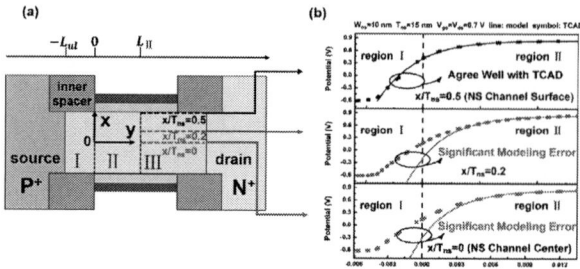

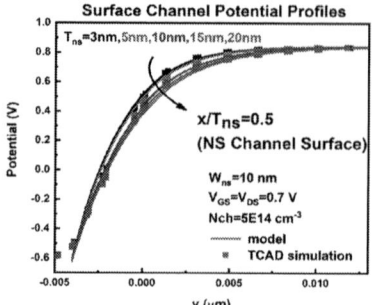

Figure 2. a) Cross-sectional schematic of a GAA TFET, the physical parameters and a coordinate system for modeling are defined; b) Potential profiles modeling results corresponding to channel surface and varied channel depths respectively.

parabolic potential approximation has been widely used in previous device modeling [7], using a second-order polynomial function to describe the potential profiles along the channel depth direction,

$$\varphi(x,y) = c_0(y) + c_1(y)x + c_2(y)x^2 \quad (1)$$

With the boundary conditions and the continuity of potential and electric field considered, the channel potential $\varphi(x, y)$ can be solved and expressed as

$$\varphi(x,y) = \varphi_s(y) + \left(\frac{T_{ns}}{4} - \frac{x^2}{T_{ns}}\right)\frac{\varepsilon_{ox}(\varphi_s(y) - V_{gs} + V_{fb})}{\varepsilon_{si}t_{ox}} \quad (2)$$

in which $\varphi_s(y)$ is the obtained potential profiles at channel surface ($x=T_{ns}/2$).

As shown in Fig.2 b), the above widely used potential model is only suitable for potential calculating at the channel surface. Significant modeling error would occur along with varied channel depth x/T_{ns}, especially in regions I and II. This is due to the arbitrary assumption of fixed parabolic coefficients for the whole channel region, fails to take different gate electrostatics in regions I and II into consideration.

III. FULL CHANNEL POTENTIAL MODEL

A. Full channel Potential modeling

Matching the boundary conditions including surface potential and electric field at edges of regions I and II, the lengths of region I and II, the surface potential profiles across regions I and II can be calculated by solving the pseudo-2D Poisson equation and expressed as

$$\varphi_{s,\text{I}}(y) = \frac{\varphi_s(0) - \Phi_{\text{I}}}{\sinh\left(\frac{L_{ul}}{\lambda_{\text{I}}}\right)}\sinh\left(\frac{L_{ul}}{\lambda_{\text{I}}} + \frac{y}{\lambda_{\text{I}}}\right)$$
$$- \frac{V_S - \Phi_{\text{I}}}{\sinh\left(\frac{L_{ul}}{\lambda_{\text{I}}}\right)}\sinh\left(\frac{y}{\lambda_{\text{I}}}\right) + \Phi_{\text{I}} \quad (3)$$

$$\varphi_{s,\text{II}}(y) = (\varphi_{ch} - \Phi_{\text{II}})\cosh\left(\frac{y - L_{\text{II}}}{\lambda_{\text{II}}}\right) + \Phi_{\text{II}} \quad (4)$$

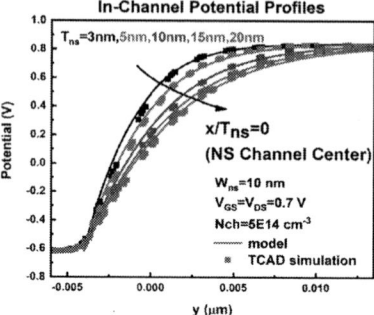

Figure 3. Model predicted (line) and TCAD simulated (symbol) results of the channel surface potential profiles with varied nanosheet thickness T_{ns}.

Figure 4. Model predicted (line) and TCAD simulated (symbol) results of the channel center potential profiles with varied nanosheet thickness T_{ns}.

$$L_{\text{II}} = \lambda_{\text{II}}\cosh^{-1}\left(\frac{\varphi_s(0) - \Phi_{\text{II}}}{\varphi_{ch} - \Phi_{\text{II}}}\right) \quad (5)$$

$$\Phi_j = V_{gs} - V_{fb} - \frac{qN_{ch}}{\varepsilon_{si}}\lambda_j^2 \quad (j = \text{I, II}) \quad (6)$$

$$\varphi_s(0) = \Phi_{\text{I}} + \left(\frac{1}{\lambda_{\text{I}}^2 \tanh^2\left(\frac{L_{ul}}{\lambda_{\text{I}}}\right)} - \frac{1}{\lambda_{\text{II}}^2}\right)^{-1} \cdot \left(\frac{(\Phi_{\text{I}} - \Phi_{\text{II}})}{\lambda_{\text{II}}^2}\right.$$
$$+ \frac{(V_S - \Phi_{\text{I}})}{\lambda_{\text{I}}^2 \tanh\left(\frac{L_{ul}}{\lambda_{\text{I}}}\right)\sinh\left(\frac{L_{ul}}{\lambda_{\text{I}}}\right)}$$
$$+ \left(\frac{\left((\Phi_{\text{I}} - \Phi_{\text{II}})\cosh(\frac{L_{ul}}{\lambda_{\text{I}}}) + (V_S - \Phi_{\text{I}})\right)^2}{\lambda_{\text{I}}^2\lambda_{\text{II}}^2 \sinh^2\left(\frac{L_{ul}}{\lambda_{\text{I}}}\right)}\right.$$
$$\left.\left. + \frac{(\varphi_{ch} - \Phi_{\text{II}})^2}{\lambda_{\text{I}}^2\lambda_{\text{II}}^2 \tanh^2\left(\frac{L_{ul}}{\lambda_{\text{I}}}\right)} - \frac{(\varphi_{ch} - \Phi_{\text{II}})^2}{\lambda_{\text{II}}^4}\right)^{1/2}\right) \quad (7)$$

where $\lambda_{\text{I}} = \sqrt{\varepsilon_{Si}T_{ns}(t_{ox}\pi/2)/(n\varepsilon_{ox})}$, $\lambda_{\text{II}} = \sqrt{\varepsilon_{Si}T_{ns}t_{ox}/(n\varepsilon_{ox})}$ are the natural length of the gate fringing region I and MOS-gated region II respectively. V_S is the source potential. φ_{ch} is the channel potential in region III, which is obtained by solving 1D Poisson equation with Fermi-Dirac statistics [9].

979-8-3503-6184-1/24 $31.00 © 2024 IEEE

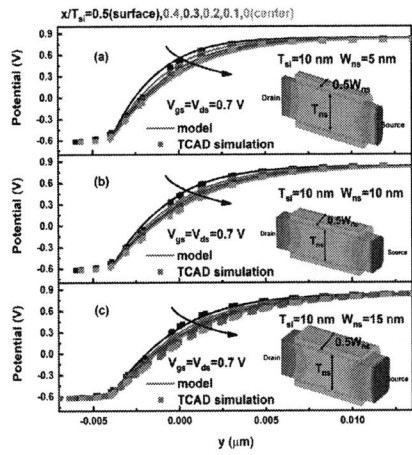

x/T$_{ns}$=0.5(surface),0.4,0.3,0.2,0.1,0(center)

Figure 5. Continuous full channel potential profiles under different W$_{ns}$/T$_{ns}$ conditions of a) T$_{ns}$=10nm, W$_{ns}$ = 5 nm; b) T$_{ns}$=10nm, W$_{ns}$ = 10 nm; c) T$_{ns}$=10nm, W$_{ns}$ = 15 nm.

In order to accurately calculate the channel potential profiles at any channel depths, this work has adopted an additional dynamic boundary condition

$$\varphi\left(x',y\right)=\varphi\left(y\right)=c_0'\left(y\right)+c_1'\left(y\right)x'+c_2'\left(y\right)x'^2 \quad (8)$$

in which a changing $\varphi(y)$ corresponds to channel depth $x = x'$ is used instead of the fixed $\varphi_s(y)$ at channel surface (x=T$_{ns}$/2). Then solving the pseudo-2D Poisson equation with modified boundary condition and a continuous full channel potential profiles with varying parabolic coefficients can be obtained

$$\varphi\left(x,y\right)=\varphi_s\left(y\right)$$
$$+\left(\frac{T_{ns}}{4}-\frac{x^2}{T_{ns}}\right)\frac{\varepsilon_{ox}\left(\varphi_{ch}-V_{gs}+V_{fb}\right)}{\varepsilon_{si}t_{ox}}\frac{\left(y+L_{ul}\right)}{\left(L_{ul}+L_{II}\right)} \quad (9)$$

Fig. 3 and Fig. 4 show the calculated channel potential profiles at nanosheet channel surface and channel center respectively, and good agreements with TCAD simulation results have been achieved for different nanosheet thickness T$_{ns}$. To further validate the accuracy of the presented potential model, Fig. 5 shows the model predicted full channel potential profiles for GAA nanosheet with different W$_{ns}$/T$_{ns}$ ratio in comparison with TCAD simulations.

B. Synthetic Electricfield modeling

In extremely scaled GAA devices with ultra-thin-channel, there's a unique synthetic electric field effect, that the electric field in GAA channel tends to increase with channel dimension downsizing [10], as shown in Fig.6. The enhanced channel electric field would then lead to different effects such as extra carrier mobility degradation in MOS-channel and boosted band-to-band tunneling rate in GAA tunnel junction. With the presented full channel potential model, the electric field distribution in the channel can be accurately calculated. As shown in Fig.7, the synthetic electric field effect has been well predicted.

IV. SUMMARY

In this paper, a continuous full channel potential modeling method for GAA devices is presented, which can accurately describe the potential distribution of the whole channel region in scaled GAA devices. The model results of surface and in-channel potential profiles of nanosheet channels under

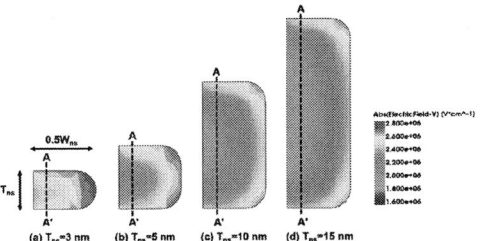

Figure 6. Distribution of electric field in GAA nanosheet channel cross-section at source junction with different channel thickness, a) T$_{ns}$=3 nm, b) T$_{ns}$=5 nm, c) T$_{ns}$=10 nm, d) T$_{ns}$=15 nm.

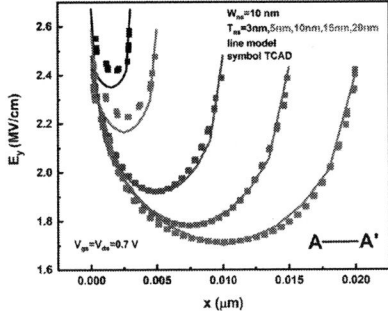

Figure 7. Modeled and simulated Synthetic electric field effects in GAA nanosheet channel, with varied nanosheet thickness T$_{ns}$ from 3 to 20 nm.

different W$_{ns}$/T$_{ns}$ ratio agrees well with the TCAD simulation, indicating its good applicability for future device modeling and circuit simulation studies of scaled GAA devices.

ACKNOWLEDGMENT

This work was supported in part by NSFC under Grant 62304048, in part by the Science and Technology Commission of Shanghai Municipality under Grant 21TS1400800.

REFERENCES

[1] N. Loubet et al., "Stacked nanosheet gate-all-around transistor to enable scaling beyond FinFET," Proc. Symp. VLSI Technol., pp. T230-T231, Jun. 2017.

[2] S. Barraud et al., "7-Levels-Stacked Nanosheet GAA Transistors for High Performance Computing," Proc. Symp. VLSI Technol., pp. 1-2, 2020.

[3] Y. Xu et al., "A Novel Hybrid-Channel Gate-All-Around Nanosheet Transistor for Leakage Control and Subthreshold Slope Reduction," CSTIC, pp. 1-3, 2024.

[4] Y. S. Chauhan et al., "FinFET Modeling for IC Simulation and Design: Using the BSIM-CMG Standard," NY, USA: Academic, 2015.

[5] L. Zhang and M. Chan, "SPICE modeling of double-gate tunnelFETs including channel transports," IEEE Trans. Electron Devices, vol. 61, no. 2, pp. 300–307, Feb. 2014.

[6] B. Lu et al., "A compact model for nanowire Tunneling-FETs," IEEE Trans. Electron Devices, vol. 69, no. 1, pp. 419–426, Jan. 2022.

[7] R. -H. Yan, A. Ourmazd and K. F. Lee, "Scaling the Si MOSFET: from bulk to SOI to bulk," IEEE Trans. Electron Devices, vol. 39, no. 7, pp. 1704-1710, Jul. 1992.

[8] K. K. Young, "Short-channel effect in fully depleted SOI MOSFETs," IEEE Trans. Electron Devices, vol. 36, no. 2, pp. 399-402, Feb. 1989.

[9] C. Wu, R. Huang, Q. Huang, C. Wang, J. Wang, and Y. Wang, "An analytical surface potential model accounting for the dual-modulation effects in tunnel FETs," IEEE Trans. Electron Devices, vol. 61, no. 8, -pp. 2690–2696, Jun. 2014.

[10] Y. Xu et al., "Study of synthetic electric field effects and quantum confinement effects in extremely scaled Gate-All-Around Tunnel FET," ICSICT, pp. 1-3, 2022.

979-8-3503-6184-1/24 $31.00 © 2024 IEEE

Deep Learning and Adaptive Pattern Search Based BSIM-CMG Parameter Extraction Applicable to Process Migration

Xingyu Li, Wangyong Chen*, Linlin Cai

School of Microelectronics Science and Technology, Sun Yat-sen University, Guangzhou, 510275, China

* Email: chenwangy@mail.sysu.edu.cn

Abstract—**Efficient and accurate parameter extraction of compact models across different processes is critical but time-consuming. In this paper, we propose a deep learning and adaptive pattern search based method to accelerate the process of BSIM-CMG parameter extraction and achieve the efficient migration of model parameters between different processes. The proposed method is validated on BSIM-CMG models from ASAP 7nm and SMIC 14nm PDKs, including the electrical characteristics of Cgg, Gm and Ids within a temperature range of -40 to 125°C. The robustness tests on the extracted model and its parameters are conducted, showing the excellent fit with target data. Furthermore, the fitting error (RMSPE) of the simulated electrical characteristic data of the BSIM-CMG model obtained is less than 1%.**

Keywords—*compact model, machine learning (ML), FinFET, parameter extraction*

I. Introduction

In the Design Technology Co-optimization (DTCO) flow, the compact model of semiconductor devices plays a key role in the bridge between device manufacturing and circuit design. The parameter exaction of the compact model is also critical for the model application in the circuit design as it strongly impacts the predictive capability of the compact model and helps optimize the power, performance, area, and cost (PPAC) metrics of the circuit [1]. Currently, the IV/CV characteristics of advanced multi-gate MOSFETs down to 5nm, e.g., FinFET, have been accurately captured by industry-standard BSIM-CMG [2]. However, extracting a large number of technology-related model parameters in BSIM-CMG, and applying them to the circuit simulator still relies on device modeling engineers, where most of extraction steps are semi-automated. Therefore, the indispensable parameter extraction process is highly time-consuming, which requires the specific expertise and experience in general.

To address this issue, data-driven methods, e.g., deep learning (DL), have gained great popularity because of the huge growth in computing power and data. The deep learning algorithm has been proven effective in handling the parameter extraction task [3], [4]. However, migrating the BSIM-CMG model from one process to another requires re-performing the training process of a DL model with a complex structure, which is time-consuming and hinders the wide application of the method. Therefore, in this paper, we propose a DL and adaptive pattern search based method to implement the parameter extraction of BSIM-CMG model with a simple DL model, and it is also applicable to the process migration of BSIM-CMG model parameters with an efficient manner. The effectiveness of this proposed method is demonstrated on BSIM-CMG models from PDK ASAP 7nm and SMIC 14nm.

II. Method description

The BSIM-CMG parameter extraction method based on DL and adaptive pattern search strategy is shown in Figure 1. A simple DL model is trained with target data including Cgg-Vgs, Gm-Vgs and Ids-Vds curves generated by Monte Carlo simulation, and the DL model is used to roughly predict BSIM-CMG global parameters. The BSIM-CMG global parameters predicted by DL model are then optimized based on adaptive pattern search to obtain the final parameter extraction results. We generated 300,000 sample datasets via Monte Carlo simulations, allocating 80% of the data for training the DL model and reserving 20% for testing. Since the number of internal parameters of the proposed DL model in this work is about 11 times less than that of typical DL structure [5], the training takes about 3 hours in an Intel(R) Core(TM) i5-1035G1 at 1 GHz laptop using only the CPU. The training efficiency of the proposed DL model is significantly improved compared to the 3 hours required to complete the training in an Intel(R) Core(TM) i7-11700 at 2.5 GHz, with NVIDIA GeForce RTX 3060 Ti desktop computer using GPU-accelerated computing [5]. As a result, high time costs due to the need to retrain DL models during process migration are reduced.

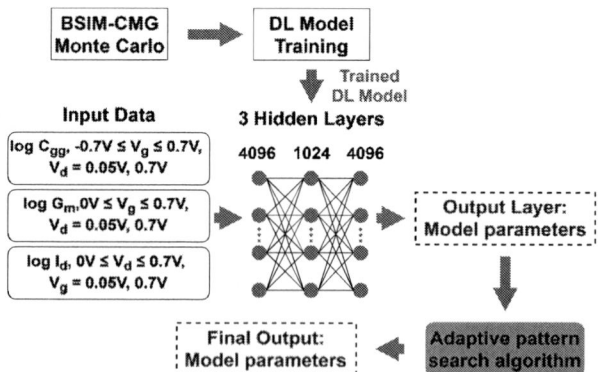

Fig.1. The workflow for the BSIM-CMG model parameter extraction.

We use Cgg-Vgs, Gm-Vgs and Ids-Vds curves at 25°C ambient temperature as input data for DL model training. The model parameters extracted by the simple DL model have been able to make the simulated electrical characteristics of the BSIM-CMG model roughly match the target electrical characteristics, but the fitting error (RMSPE) is still greater than the expected value. Then, we input the parameters extracted from the simple DL model into the parameter optimizer based on the adaptive pattern search algorithm to further optimize the parameters iteratively. Considering the high computational complexity, large memory consumption, and the curse of dimensionality faced by search algorithms when dealing with high-dimensional search problems, we design an adaptive pattern search method based on a one-dimensional adaptive pattern search and a search dimension rotation mechanism for model parameter extraction. The approach is presented in Figure 2.

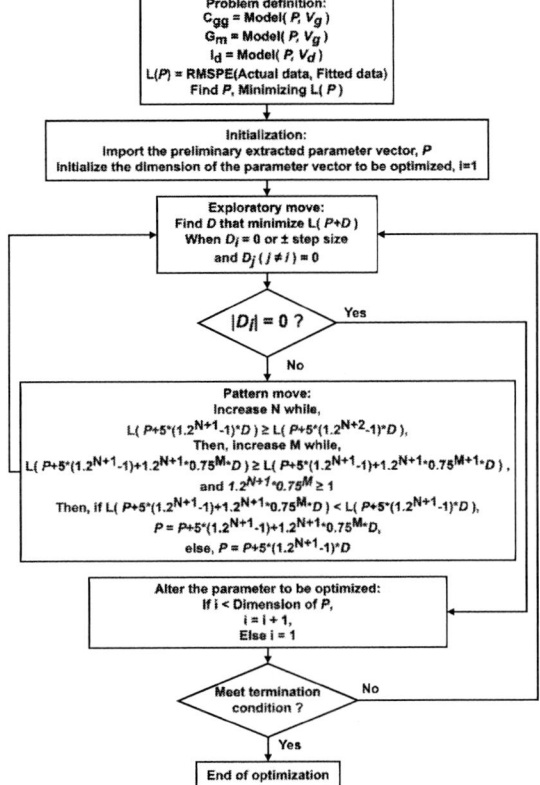

Fig.2 Flow chart of adaptive pattern search based on one-dimensional search and dimension rotation.

In addition, in order to speed up the efficiency of parameter optimization, the optimization is divided into 10 steps. The process is shown in Figure 3. In step 1, the Cgg-Vgs curve in linear model (Vds = 0.05V) fitting is performed to optimize the parameters related to EOT, gate work function (PHIG), length parameters (XL, LINT). In step 2, the Cgg-Vgs curve in saturation model (Vds = 0.7V) fitting is performed to optimize the parameters related to DIBL effect (DSUB and ETA0). In step 3, the subthreshold region parameter (CDSC), mobility parameter (U0, UA, EU and UTL), length parameters (LINT), and the parameters related to saturation velocity (VSAT, KSATIV and MEXP) and effective field (ETAMOB) are extracted by fitting Gm-Vgs curve in linear model (Vds = 0.05V). In step 4, the subthreshold region parameters (CDSC, CDSCD), and the parameters related to EOT, channel length modulation (PCLM), saturation velocity (KSATIV) and DIBL effect (DSUB and ETA0) are extracted by fitting Gm-Vgs curve in saturation model (Vds = 0.7V). In steps 5 to 6, the fitting of the Cgg-Vgs curve is repeated once. In step 7 to 8, the Ids-Vds curve fitting is performed to optimize the subthreshold region parameters (CDSC, CDSCD) and the parameters related to EOT, channel length modulation

(PCLM), saturation velocity (VSAT, KSATIV, AT) and DIBL effect (DSUB and ETA0). Finally, in steps 9 to 10, the fitting of the Gm-Vgs curve is repeated once.

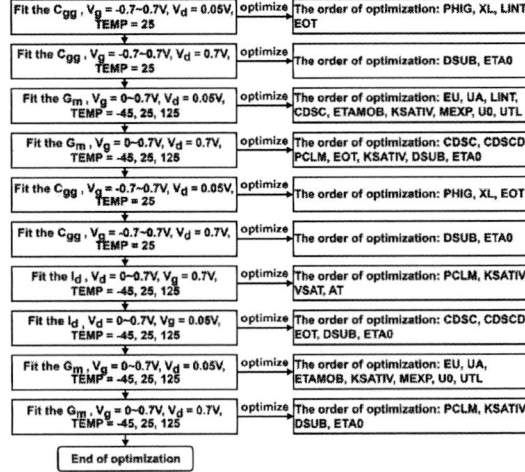

Fig.3 Flow chart of model parameter optimization.

III. RESULTS AND DISCUSSIONS

The extracted parameters make the BSIM-CMG model fit the electrical characteristic data from ASAP 7nm PDK [2] well, as shown in Figures 4-5. The fitting error (RMSPE) of Cgg, Gm and Ids curves under two bias cases (Vds = 0.05V, 0.7V) and three kinds of ambient temperatures (Temp = -40°C, 25°C and 125°C) are 0.18%, 0.45% and 0.38%, respectively.

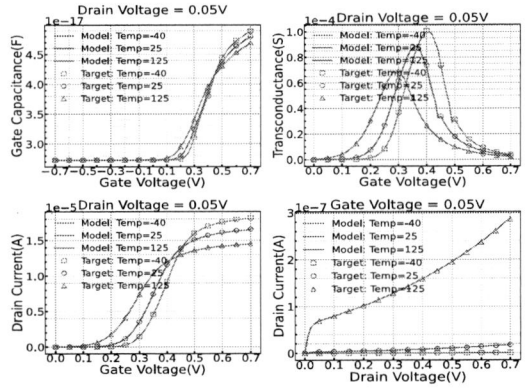

Fig.4 Cgg, Gm, Ids-Vgs (Vds = 0.05V) fitting, Ids-Vds fitting (Vgs = 0.05V) of ASAP 7nm PDK at different Temp.

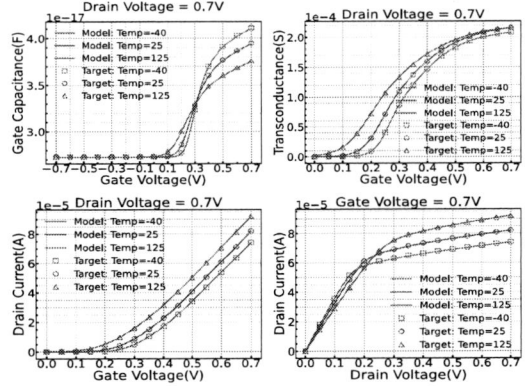

Fig.5 Cgg, Gm, Ids-Vgs (Vds = 0.7V) fitting, Ids-Vds fitting (Vgs = 0.7V) of ASAP 7nm PDK at different Temp.

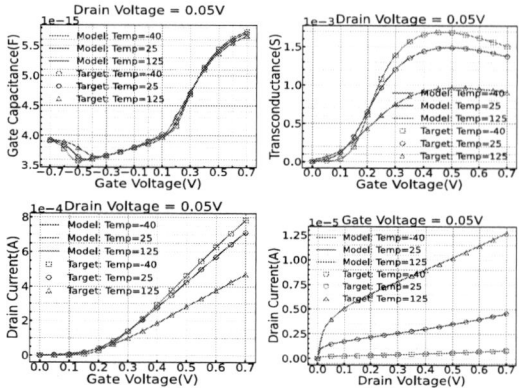

Fig.6 Cgg, Gm, Ids-Vgs (Vds = 0.05V) fitting, Ids-Vds fitting (Vgs = 0.05V) of SMIC 14nm PDK at different Temp.

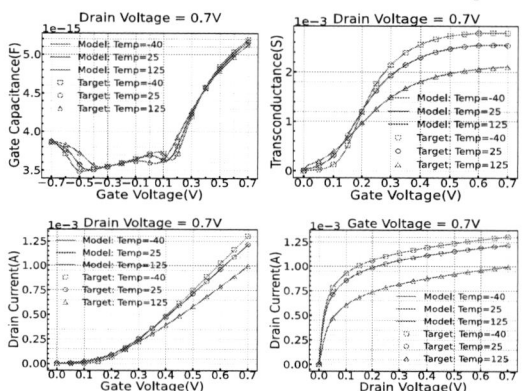

Fig.7 Cgg, Gm, Ids-Vgs (Vds = 0.7V) fitting, Ids-Vds fitting (Vgs = 0.7V) of SMIC 14nm PDK at different Temp.

Table I. Fitting error (RMSPE) before and after optimization

ASAP 7nm	Cgg-Vgs	Gm-Vgs	Ids-Vgs	Ids-Vds
Before optimization	0.25%	7.27%	5.82%	6.20%
After optimization	0.18%	0.45%	0.38%	0.45%
SMIC 14nm	Cgg-Vgs	Gm-Vgs	Ids-Vgs	Ids-Vds
Before optimization	0.12%	4.81%	3.93%	3.79%
After optimization	0.02%	0.30%	0.25%	0.22%

When the process transfers to SIMC 14nm, the model parameters extracted by the proposed method also demonstrate excellent matching, as shown in Figures 6-7. The fitting error (RMSPE) of Cgg-Vgs, Gm-Vgs and Ids-Vds curves under two bias cases (Vds = 0.05V, 0.7V) and three kinds of ambient temperatures (Temp = -40°C, 25°C and 125°C) are 0.02%, 0.30% and 0.25%, respectively. The fitting errors of the electrical characteristic data from ASAP 7nm and SMIC 14nm before and after optimization by the model parameter optimizer designed in this work are listed in Table I. DL model can predict the parameters of BSIM-CMG model very quickly. The trained simple DL model can complete the prediction of BSIM-CMG model parameters within 1s, but the fitting error of the simulated electrical characteristics of the prediction model is larger than the expected value. Therefore, the BSIM-CMG model parameters predicted by DL model need to be further optimized. The model parameter optimizer designed in this work can efficiently optimize the BSIM-CMG model parameters predicted by DL model, so that the fitting error of the simulated electrical characteristics of the extracted

BSIM-CMG model can reach the expected value. It requires approximately 600s to optimize the model parameters of ASAP 7nm PDK and approximately 3600s to optimize the model parameters of SMIC 14nm PDK.

Table II. The fitting error (RMSPE) after the fluctuation of parameter PHIG

ASAP 7nm	Cgg-Vgs	Gm-Vgs	Ids-Vgs	Ids-Vds
PHIG - 0.2eV	0.19%	2.81%	0.34%	0.45%
PHIG + 0.2eV	0.17%	0.50%	0.48%	0.47%
SMIC 14nm	Cgg-Vgs	Gm-Vgs	Ids-Vgs	Ids-Vds
PHIG - 0.2eV	0.02%	0.28%	0.20%	0.15%
PHIG + 0.2eV	0.02%	1.68%	0.46%	0.40%

The robustness of the model extracted by the method designed in this work is examined by making the PHIG values of the target model and the extracted model fluctuate ±0.2eV at the same time, and then measuring the fitting error between the simulated results and the target electrical characteristics. The robustness test results of the extracted model are listed in Table II. It can be seen that the average fitting error fluctuation between the simulated electrical characteristic data of BSIM-CMG model and the target data is less than 3% after PHIG fluctuation.

IV. CONCLUSION

In this paper, a BSIM-CMG model parameter extraction method based on the combination of deep learning and adaptive pattern search algorithms is proposed. By greatly simplifying the structure of DL model and using adaptive pattern search to optimize model parameters, it is capable of migrating efficiently model parameters between different processes with a high accuracy. The deep learning algorithm is used to extract the rough model parameters of the BSIM-CMG, and then the proposed model parameter optimizer is used to further optimize the extracted model parameters. The fitting error (RMSPE) of the simulated electrical characteristic data of the BSIM-CMG model obtained is less than 1%, which is reduced by more than 90% at most compared with the model extracted by only using the deep learning model. In addition, the extracted model parameters have good robustness.

ACKNOWLEDGMENT

This work was supported in part by the National Natural Science Foundation of China under Grants 62304263, 62204269 and in part by the Guangdong Basic and Applied Basic Research Foundation under Grants 2023A1515011418, 2024A1515010349.

REFERENCES

[1] X Li, et al. Overview of emerging semiconductor device model methodologies: From device physics to machine learning engines. Fundamental Research, 2024, 22(1).

[2] V Vashishtha et al., ASAP7 predictive design kit development and cell design technology co-optimization: Invited paper. IEEE/ACM International Conference on Computer-Aided Design, 2017, 992-998.

[3] A Ashai, et al. Deep Learning-Based Fast BSIM-CMG Parameter Extraction for General Input Dataset. IEEE Transactions on Electron Devices, 2023, 70(7): 3437-3441.

[4] F Chavez, et al. I-V Global Parameter Extraction for Industry Standard FinFET Compact Model using Deep Learning. IEEE International Symposium on Radio-Frequency Integration Technology, 2023, 20-22.

[5] F Chavez, et al. Deep learning-based I-V Global Parameter Extraction for BSIM-CMG[J]. Solid-State Electronics, 2023, 209: 108766.

979-8-3503-6184-1/24 $31.00 © 2024 IEEE

Vertical Channel Transistor (VCT) for Advanced Logic and Memory Applications

Mingmin Shi[1], Ran Bi[1], Ming Li[1,2*]

[1] School of Integrated Circuits, Peking University, China, 100871
[2] Beijing Advanced Innovation Center for Integrated Circuits, China, 100871

* Email: liming.ime@pku.edu.cn

Abstract—With the device feature size scaling down continuously, vertical channel transistor (VCT) becomes promisingly attractive for IC manufacturing thanks to its structural benefits in decoupling gate patterning and junction engineering. On one hand, the vertically aligned bitline configuration makes it naturally suit for $4F^2$ DRAM cell application. On the other hand, its offerings of variant junction engineering and tight layout design provides VCT potential for SoC applications. In this paper, we will discuss the VCT design considerations for advanced logic and DRAM applications from viewpoints of asymmetrical source/drain engineering and off-state leakage mechanism by TCAD.

Keywords—VCT, DRAM, GIDL, asymmetry engineering

I. INTRODUCTION

The technology node has been shrunk to 3 nm and below nowadays. The vertical channel transistor (VCT) provides a natural advantage in scaling down as it allows for a significant reduction in the footprint size compared to traditional devices with lateral channels[1,2]. The variant junction engineering and compact layout design offered by VCT also hold potential for logic especially SoC applications, as shown in Fig. 1(a)[3]. On the other hand, the vertical aligned bit line arrangement of VCT naturally suit for the requirements of $4F^2$ DRAM cell[4-8]. Generally, inversion-mode VCT (IM-VCT) is required for performance guarantee but suffers from source/drain asymmetry and gate-to-channel misalignment[9,10]. Compared with IM-VCT, junction-less vertical channel transistor (JL-VCT) offers greater convenience in process manufacturing and enables gate-to-channel self-alignment, which can effectively relax the constrict on the layout design but lack of current drivability. Therefore, JL-VCT is regarded more suitable for the DRAM applications, as shown in Fig. 1(b). No matter IM-VCT and JL-VCT, the fully depleted channel and source/drain asymmetry will challenge the optimal design for considering performance-power-area tradeoff. Actually, the source/drain asymmetry will not only affect on-state performance but also cause more complicated off-state leakage mechanisms, which strongly affect the static power consumption in logic and retention characteristics of DRAM. Besides, the floating body effect (FBE) in JL-VCT also needs to be studied for better performance in memory applications[11,12].

In this work, based on the computer-aided-design (TCAD) tools of Synopsys Sentaurus, the VCT model is constructed and calibrated to the experimental results. The impact of source/drain asymmetry on IM-VCT is investigated firstly in terms of gate-to-channel alignment and ion implantation diffusion distance into channel. Furthermore, the off-state leakage mechanism and FBE of JL-VCT is studied and an optimization strategy is proposed for JL-VCT.

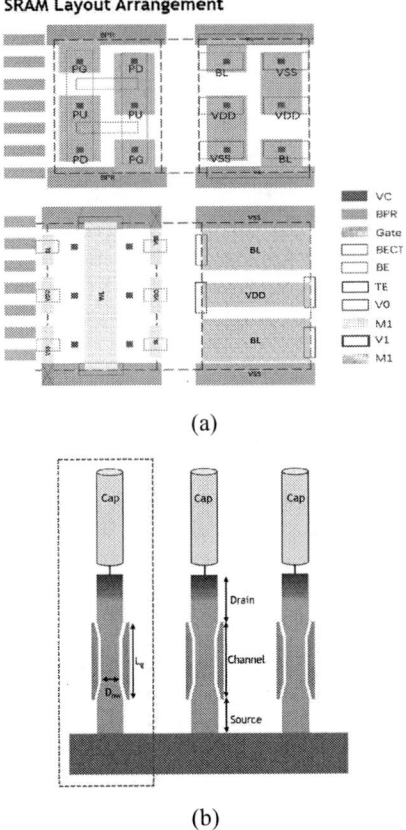

Fig. 1 (a) SRAM layout arrangement of IM-VCT (b) DRAM cell with JL-VCT

II. TCAD SIMULATION METHODOLOGY

The simulated VCT is assumed to be consisted of a vertical pillar with metal gate all-around. According to the experiment studies, VCT channel usually exhibits a dumbbell shape. The doping profile in VCT will be changed with the application goals such as logic or DRAM. For example, in $4F^2$ DRAM, the JL-VCT is doped with uniform profile in vertical pillar but heavily doped on the top to act as the node contact region (NC). As illustrated in Fig.2 (a), a typical JL-VCT can be determined by the critical structure parameters including L_{ch}, D_{nw}, G_{ox}, L_{taper}. It has to be noted that in JL-VCT, the overlap length (L_{ov}) is redefined as the distance between gate edge to the position when NC doping drops to the level of channel doping. As L_{ov} is negative, it means the NC doping far away from channel. The default key parameters and doping concentrations in channel, source and drain are listed in table1. The doping profile along channel direction is shown in Fig.2 (b).

979-8-3503-6184-1/24 $31.00 © 2024 IEEE

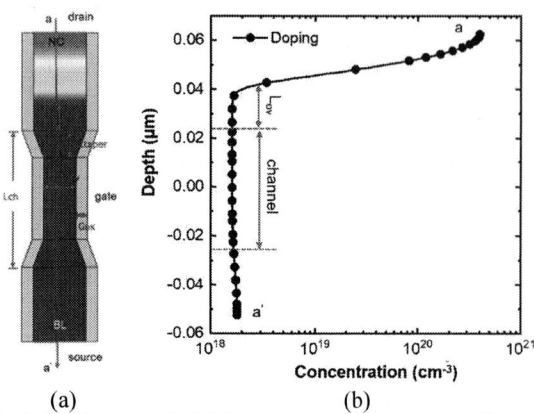

(a)   (b)

Fig. 2 (a) schematic JL-VCT structure and (b) doping profile along channel direction from NC to source.

In the device simulation, to correctly consider the transport in VCT, quantum potential correction is applied to the transport equation with thin layer mobility model and non-local band-to-band tunneling (BTBT) [13]. By calibration of physical model, the TCAD simulated I_d-V_g curve is compared to the experimental result in Fig.3. The fitting error of simulated curve to experimental data is less than 5%.

Table 1 Default simulation parameters

Parameters	Default Value	Doping Profile	Default Value
L_{ch} (nm)	45	N_{ch} (cm^{-3})	1.6e18
G_{ox} (nm)	5	N_{source} (cm^{-3})	1.8e18
D_{nw} (nm)	15	N_{ch} (cm^{-3})	4e20
L_{ov} (nm)	20		

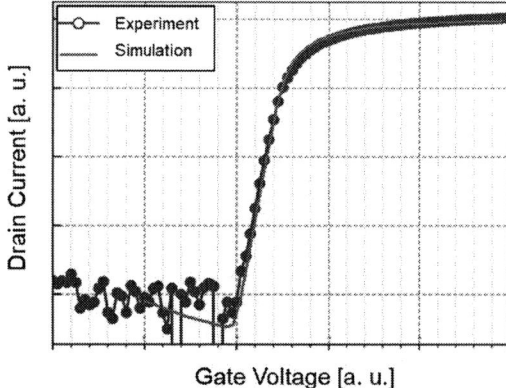

Fig. 3 Calibrated JL-VCT I_d-V_g curve compared to experimental result.

III. RESULTS AND DISCUSSIONS

A. Asymmetry analysis of IM-VCT for logic applications

As we reported before[10], the natural asymmetry of source/drain configuration of VCT will strongly affect the stress distribution in channel and thus the performance of VCT CMOS devices. It is found that the uniaxial stress in VCT is opposite to that in lateral channel devices, i.e. tensile for SiGe stressor but compressive for Si source/drain.

Beside that stress asymmetry, the structural asymmetry due to gate-to-channel misalignment and dopant diffusion at source and drain is also important in VCT design for logic especially SoC applications.

As shown in Fig.4, the gate-to-channel misalignment (L_d) can be divided into three cases: overlap to drain ($L_d>0$), just alignment ($L_d=0$), overlap to source ($L_d<0$).

With the calibrated physical model, I_{on} and I_{off} characteristics are simulated with L_d from -15 nm to 15 nm. It is surprisingly that the peak I_{on} is reached not at the just alignment position but shifting to source side. It should be noted that the source/drain and channel doping profiles are kept unchanged during L_d variation. Different from I_{on} asymmetry, I_{off} seems to reach the minimum as the gate is just aligned to channel.

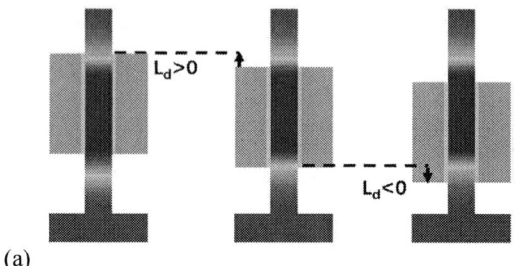

(a)

(b)   (c)

Fig. 4 (a) Gate alignment deviation diagram. Effect of gate alignment position change on (b) on-state current and (c) off-state current.

The performance peak can be explained as the result of maximum effective mobility in channel achieved at the designated gate-to-source overlap. As gate moves away from drain, the high field induced by gate-to-drain voltage drop is reduced so as to improve the mobility at the drain side. As a result, the average mobility in channel will be enhanced. As gate moves continuously toward source, the channel mobility increases continuously but the underlap between gate and drain will cause a large parasitic series resistance. As a result, I_{on} achieves the peak value at a designated position where the mobility gain and parasitic series resistance reach a balance.

For I_{off} trend, no matter gate-to-drain overlap or gate-to-source overlap, band-to-band tunneling will be enhanced so as to make I_{off} increase.

On the other hand, the ion implantation and annealing will cause different dopant diffusion from top side and bottom side so as to result in asymmetrical doping profile. Here we just consider the asymmetry induced by the top dopant diffusion. As shown Fig.5, the underlap and overlap to gate is formed with different dopant diffusion length L_{diff}. The simulated results show that I_{on} increases from underlap to overlap. If the pillar top is used as drain, the I_{on} is much higher than that as the bottom as drain. It indicates that the low-resistance source will help to improve performance. Comparative to I_{on}, I_{off} also increases with L_{diff}, i.e. from underlap to overlap. It can be

979-8-3503-6184-1/24 $31.00 © 2024 IEEE

easily understood as the result of effective channel length shortening.

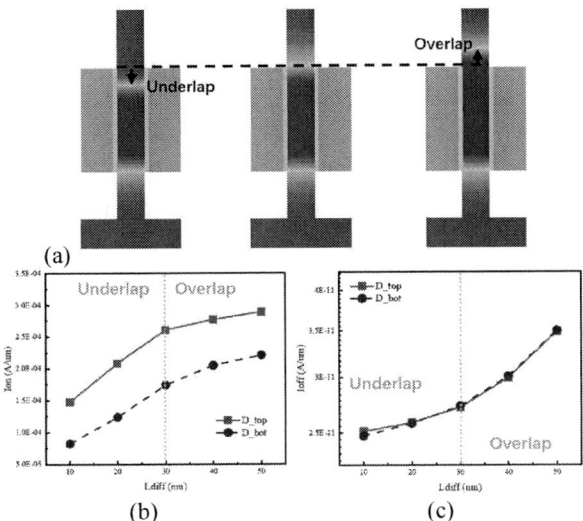

(a)

(b)   (c)

Fig. 5(a) Schematic diagram of changes in ion implantation diffusion depth. Effect of ion implantation diffusion depth on (b) on-state current and (c) off-state current.

B. Off-state mechanism analysis of JL-VCT

In order to understand in-depth the off-state leakage mechanisms in VCT especially JL-VCT for DRAM applications, the carrier generation rate distribution along channel direction under different reverse biases are extracted by simulation, as shown in Fig.6. The generation rate profiles are extracted at the channel surface and center in terms of electron generation by BTBT, hole generation by BTBT and Shockley–Read–Hall (SRH) generation. Due to non-local nature, the electron and hole generation rates due to BTBT are separate. As reported in previous work [11], the BTBT at channel surface can be regarded as transverse BTBT (T-BTBT) and that at channel center as longitudinal (L-BTBT).

As V_g=0 V, both the T-BTBT and L-BTBT are not triggered and the off-state leakage is dominated by SRH generation. As reverse V_g increases, the L-BTBT generation at channel center starts to take effect and become comparable to SRH. As V_g reversely increases, T-BTBT increases fast and becomes comparable but still less than L-BTBT. The total BTBT generation will overpass the SRH and become dominant.

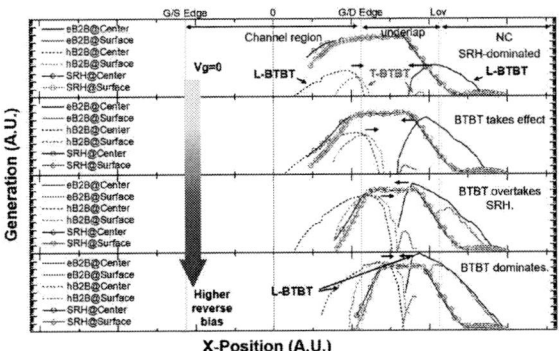

Fig. 6 The distribution of BTBT and SRH generation.

C. Optimization of JL-VCT

To optimize the JL-VCT structure in terms of performance and off-state leakage current, the dependance of I_{off}, I_{GIDL} with key device parameters and process parameters are simulated. A sensitivity value S_X is defined as below:

$$S_X = \frac{2(X_{max} - X_{min})}{X_{max} + X_{min}} \quad (1)$$

where X means the investigated electrical characteristics such as I_{on}, I_{off} and so on, X_{max} and X_{min} the upper and lower level of X. The sensitivity of V_{th}, SS, I_{off}, I_{min}, I_{GIDL} and I_{on} to device parameters (N_{ch}, D_{nw}, L_{ov}, L_g, L_{drain}, L_{source}, G_{ox}) and process parameters (N_{drain}, N_{source}) of JL-VCT is listed in table 2.

Table 2 Sensitivity of JL-VCT to key device parameters

Sx(X=	Vth	SS	Ioff	Imin	GIDL	Ion
Nch	37.9%	19.4%	130.4%	139.3%	132.9%	45.8%
Dnw	30.3%	12.8%	34.8%	137.9%	74.9%	107.7%
Lov	4.1%	4.0%	195.5%	188.7%	199.5%	9.8%
Lg	24.2%	18.9%	41.7%	44.1%	33.8%	8.0%
Ldrain	0.2%	0.2%	6.3%	4.2%	67.2%	6.9%
Lsource	0.8%	5.2%	3.1%	2.1%	64.6%	92.3%
Gox	9.0%	8.0%	29.5%	16.9%	53.7%	7.1%
Ndrain	0.0%	0.0%	0.8%	1.3%	31.0%	3.3%
Nsource	0.6%	5.9%	0.0%	0.8%	29.9%	55.0%

It can be found that N_{ch} and D_{nw} impact on off-state and on-state characteristics simultaneously. Dimension size and doping concentration at source side mainly affect the on-state performance but slightly on off-state current. In contrast, drain side mainly affect on off-state current not the on-state performance. It provides a strategy for optimizing JL-VCT in terms of on/off tradeoff.

Firstly, the optimization was carried to achieve a relatively high I_{on} and low I_{off} with cross matrix of N_{ch} and D_{nw}. Then, L_{source} and N_{source} are optimized in term of I_{on}. L_{drain} and N_{drain} are optimized in term of I_{off}. Thanks to the decoupling of I_{on} and I_{off} sensitivity due to source/drain asymmetry, JL-VCT can be optimized to meet the spec of I_{on} and I_{off}. Fig.7 shows an optimization case for JL-VCT, where the gray area represents the allowed region for N_{ch} and D_{nw} to meet the V_{th}, I_{on}, I_{GIDL}, I_{min} simultaneously.

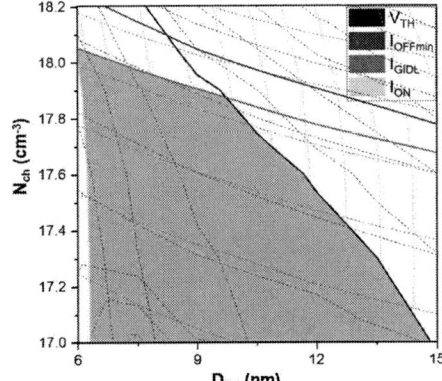

Fig.7 An optimization case for JL-VCT

D. FBE of JL-VCT

Usually, FBE is regarded ignorable in JL-VCT since the body is not "floated". However, as JL-VCT is biased from on-

979-8-3503-6184-1/24 $31.00 © 2024 IEEE   528

state to off-state, the channel should be depleted so as to produce an electrically floated body again. To investigate FBE of JL-VCT, the gate voltage was swept from positive to negative with different speed. As shown in Fig. 8(a), as the gate voltage sweeps faster, the JL-VCT shows increased off-state current during the shutdown process. It's the direct evidence for the FBE. Compared to bulk device as shown in Fig. 8(b), FBE of JL-VCT is much stronger.

The strong FBE of JL-VCT can be explained as the result of transient carrier generation and accumulation during transient sweeping. As device is turned off, the junction-less channel is depleted and the electron-hole pairs are generated through BTBT and SRH. The generated electrons and holes will be swept into drain and channel, respectively. For quick sweeping, the holes in channel will not be annihilated immediately but accumulated at source side to lower the source potential barrier to induce large diffusion current. It should be noted that hole in JL-VCT is minor carrier but major carrier in IM-VCT channel. In JL-VCT channel, since the major carrier has been full depleted so that the recombination will be difficult to happen. For IM-VCT, however, both the minor and major carriers are not depleted so that the recombination is easier. It can be deduced that FBE of JL-VCT is possibly stronger than IM-VCT. Actually, in the simulated results not shown here, as the sweeping time is reduced from 1 ms to 1 μs, the I_{min} gain due FBE is about 973 for JL-VCT but 827 for IM-VCT.

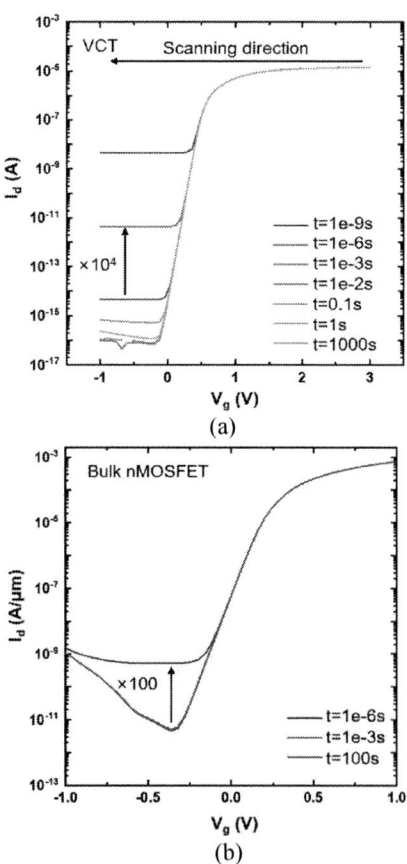

Fig. 8 Effect of gate voltage sweep speed on off-state leakage of (a) JL-VCT and (b) bulk MOSFET.

IV. CONCLUSION

This paper investigates the key challenges associated with VCT design for advanced logic and memory applications. The source/drain asymmetry is studied in terms of gate alignment and dopant diffusion for IM-VCT. It's found that the slight gate alignment to source will help to increase drivability but not good for off-state leakage. The off-state leakage mechanism in JL-VCT is also investigated for DRAM application. The results show that L-BTBT and SRH dominate the off-state current in JL-VCT. With the decoupling sensitivity of device characteristic to source and drain, the JL-VCT is optimized to meet the spec of DRAM cell transistor. At last, the FBE of JL-VCT is discussed.

ACKNOWLEDGMENT

This work was supported by SAMT project (SAMT-BD-KT-22030103) and NSFC Project (61927901).

REFERENCES

[1] J. Ryckaert et al., "Enabling Sub-5nm CMOS Technology Scaling Thinner and Taller!," 2019 IEEE International Electron Devices Meeting (IEDM), San Francisco, CA, USA, 2019, pp. 29.4.1-29.4.4.

[2] C. Lee, W. Choi, M. Kwak, S. Kim and H. Hwang, "Excellent Synapse Characteristics of 50 nm Vertical Transistor with WOx channel for High Density Neuromorphic system," 2021 Symposium on VLSI Technology, Kyoto, Japan, 2021, pp. 1-2.

[3] L Capodieci, J P Cain, T Huynh-Bao et al., "Toward the 5nm technology: layout optimization and performance benchmark for logic/SRAMs using lateral and vertical GAA FETs", Spie Advanced Lithography, pp. 978102, 2016.

[4] A. Yoo et al., "High-Performance Gate-all-around Junctionless Vertical-Channel Transistors with the Ultra-low Sub-threshold Swing for Next-generation 4F2 DRAM," 2023 International Electron Devices Meeting (IEDM), San Francisco, CA, USA, 2023, pp. 1-4.

[5] H. Chung et al., "Novel 4F2 DRAM cell with Vertical Pillar Transistor(VPT)," 2011 Proceedings of the European Solid-State Device Research Conference (ESSDERC), Helsinki, Finland, 2011, pp. 211-214.

[6] C. Chen et al., "First Demonstration of Stacked 2T0C-DRAM Bit-Cell Constructed by Two-Layers of Vertical Channel-All-Around IGZO FETs Realizing 4F2 Area Cost," 2023 International Electron Devices Meeting (IEDM), San Francisco, CA, USA, 2023, pp. 1-4.

[7] M. H. R. Ansari, N. Navlakha, J. -T. Lin and A. Kranti, "High Retention With n -Oxide- p Junctionless Architecture for 1T DRAM," in IEEE Transactions on Electron Devices, vol. 65, no. 7, 2018, pp. 2797-2803.

[8] Li M, Chen G, Huang R, "High Performance GAA SNWT with a Triangular Cross Section: Simulation and Experiments," Applied Sciences, 2018, pp.1553.

[9] S Reboh, R Coquand, S Barraud et al., "Strain stress and mechanical relaxation in fin-patterned Si/SiGe multilayers for sub-7 nm nanosheet gate-all-around device technology", Applied Physics Letters, vol. 112, 2018, pp. 051901.

[10] R. Bi, B. Zhang, J. Wang, et al., "TCAD Study on Strain Engineering in Vertical Channel Gate-all-around Transistor," 2023 IEEE 15th International Conference on ASIC (ASICON), Nanjing, China, 2023, pp. 1-4.

[11] J. Fan, M. Li, X. Xu, et al., "Insight Into Gate-Induced Drain Leakage in Silicon Nanowire Transistors," Transactions on Electron Devices, 2015, pp 213-219.

[12] Y. Cho et al., "Suppression of the Floating-Body Effect of Vertical-Cell DRAM With the Buried Body Engineering Method," in IEEE Transactions on Electron Devices, vol. 65, no. 8, 2018, pp. 3237-3242.

[13] Sentaurus™ Device User Guide, pp.529, Version T-2022.03, March 2022

High Precision I-V Characteristics SPICE Model for Silicon Carbide MOSFET

Jinhong Shi[1], Yongxi Li[1], Jincheng Shi[1], Ruguan Li[2], Haimeng Huang[1]*, and Hongqiang Yang[1]*

[1]*Glasgow College, University of Electronic Science and Technology of China, Chengdu 610054, China*
[2]*GRG Metrology & Test Group Co., Ltd., Guangzhou 510656, China*
* Email: hqyang@uestc.edu.cn;hmhuang@uestc.edu.cn

Abstract—An enhanced SPICE model for Silicon Carbide (SiC) MOSFETs designed to match these devices' current-voltage characteristics precisely is proposed in this study. Compared to existing commercial SiC SPICE models, this model achieves higher fitting accuracy and thoroughly accounts for the non-ideal effects in small-scale devices. Specifically, the model focuses on the influence of channel length modulation to accurately depict the first quadrant $I_{DS} - V_{DS}$ characteristic curves. Ultimately, the model demonstrates significant improvements over existing models in terms of R^2 value, RMSE value, and MAPE value, establishing its superior fitting capabilities.

I. INTRODUCTION

Silicon Carbide (SiC) MOSFETs have become indispensable in numerous high-end applications due to their high breakdown voltage, short switching times, and high power density [1-4]. Accurate modeling of these devices, particularly the current-voltage characteristics, is crucial for circuit simulation and design [4]. Numerous circuit simulation researchers have proposed various methods to improve the JFET model from SPICE 2 [1]. For instance, study [1] optimized the model's expression for current dependence on drain-source voltage, $I_D[V_{DS}]$, significantly enhancing the model's usefulness as the output current approaches saturation. Another research optimized the current's dependence on gate-source voltage, $I_D[V_{GS}]$, effectively describing the relationship between the output current and gate voltage [3]. However, these optimized models do not fit well with unique phenomena like short-channel effects at high drain-source voltages (V_{DS}) [4]. These inaccuracies can significantly impact the precision of simulation results in circuit design.

Therefore, to enhance the accuracy of simulation results, research [4] has built upon existing SPICE models by incorporating the influence of gate-source voltage V_{GS} on fitting parameters. However, it still has not fully considered the impact of gate-source voltage on channel length modulation effects in short-channel devices. Consequently, this study focuses on optimizing the channel length modulation term $(1 + \lambda V_{DS})$ in the original SPICE model and simultaneously revises the functional types of $I_{DS}[V_{DS}]$ and $I_{DS}[V_{GS}]$ to achieve higher fitting accuracy.

II. BEHAVIORAL SPICE MODEL DESCRIPTION

A. The Model of VDMOS

As illustrated in Fig. 1(a), the VDMOS structure resembles the conventional MOSFET architecture and can be divided

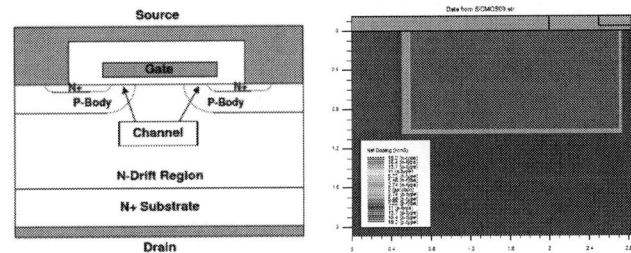

Fig. 1. The (a) structure of VDMOS and its (b) device Simulation result in Silvaco TCAD.

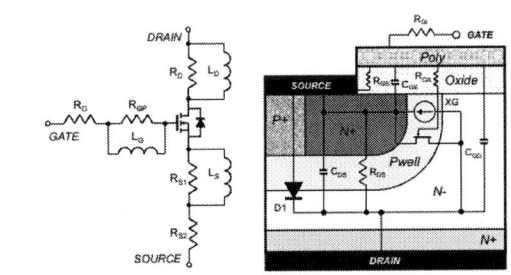

Fig. 2. The (a) parasitic components (b) structure and an equivalent circuit of a packaged SiC power D-MOSFET [4].

into four distinct regions. Based on this structure, the device configuration of the Silicon Carbide MOSFET H1M120F060 was simulated using the process simulation software Silvaco TCAD, with the results depicted in Fig. 1(b). Additionally, the generic SPICE model for this type of device, as shown in Fig. 2, comprises three main parts: (i.) thermal grid based on the package, (ii.) parasitic parameters of wire bonding and die soldering, and (iii.) bare die characteristics model [4].

B. Optimization Results and Principles

Based on the form used in the general circuit analysis program SPICE 2, the final drain current expression optimized by research [1], [3], and [4] is shown in Equation (1).

$$I_{DS} = \beta(1 + \tanh(\psi))(1 + \lambda V_{DS})\tanh(\alpha V_{DS}) \quad (1)$$

979-8-3503-6184-1/24 $31.00 © 2024 IEEE

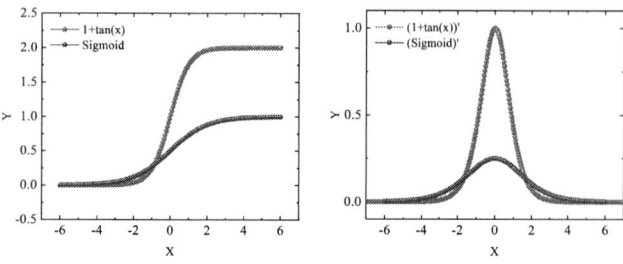

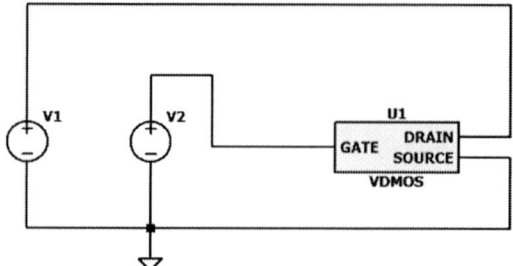

Fig. 3. Comparison of (a) original curve and (b) first derivative curve.

Fig. 4. Test circuit in LTspice.

The final modification results proposed in this study are shown in Equation (2).

$$I_{DS} = \beta \frac{3(1 + \lambda V_{DS} + x V_{GS})^{\kappa}}{1 + e^{(-\psi \cdot V_{GS})}} erf(\tanh(\alpha V_{DS})(1 - k V_{DS})) \quad (2)$$

In this equation, the optimization results of fitting parameters ψ, α, β, and λ are expressed in Equation (3) to (6).

$$\psi = \sum_{i=1}^{5} P_i (V_{GS} - V_{pk0})^i \quad (3)$$

$$\alpha = \sum_{i=1}^{4} \alpha_i \cdot (V_{GS})^{4-i} \quad (4)$$

$$\beta = \sum_{m=1}^{4} \beta_m \cdot (V_{GS})^{4-m} \quad (5)$$

$$\lambda = \sum_{k=1}^{8} \lambda_k \cdot (V_{GS})^{8-k} \quad (6)$$

The model denoted as $I_{dA}[V_{GS}]$ is optimized in a similar principle to research [3] in this study. We propose the substitution of the original hyperbolic tangent function $(1 + \tanh(\psi))$, with an optimized Sigmoid function. As shown in Fig. 3 both the hyperbolic tangent and Sigmoid functions are commonly used as activation functions in Deep Neural Networks (DNNs), possessing similar intrinsic curve saturation characteristics and a "bell-shaped" structure in their first-order derivatives [3, 5, 7].

Also, the model denoted as $I_{dB}[V_{DS}]$ is optimized based on study [1], which incorporates the model with the hyperbolic tangent function $\tanh(\alpha V_{DS})$ to enhance the smooth saturation characteristics below the pinch-off point. To reduce numerical instability arising from the transitions of tanh function, the error function is employed to $\tanh(\alpha V_{DS})$ [6, 7]. This approach effectively smooths the curve and ensures accurate simulation and analysis in semiconductor device characterization.

For the channel length modulation effect, also considered as output conductance, is ideally determined solely by V_{DS}. However, we introduce the impact of gate-source voltage V_{GS} the same as study [8] on output conductance into the model for the following reasons:

(i) Beyond the point of output saturation, V_{GS} significantly affects the thickness of the depletion zone, which may thin the thickness of the drift region, and then reduce its conductivity.

(ii) V_{GS} can also significantly affect the substrate current. The large electric field increases the energy of carriers near the drain region and makes them more likely to enter the substrate, thus increasing the substrate current and altering the output conductance.

At the same time, an exponential term κ has also been incorporated in this study, indicating device characteristics and manufacturing process parameters. Additionally, to suppress the upward bending of the fitted output characteristic curve relative to the actual values near saturation, an attenuation factor $(1 - k V_{DS})$ has been added to the $I_{dB}[V_{DS}]$ component. These optimized equations facilitate more accurate and easier fitting of characteristics.

III. SIMULATION METHODS AND RESULTS

A. Simulation Methods

The simulation results from Silvaco TCAD on the current-voltage characteristics of devices in the first quadrant were considered as real data. In contrast, the commercial SiC MOSFET and updated model in this study were considered test data. Optimization effects from each component were independently compared by fitting them into the SPICE simulator software LTspice. Then, predictive performance metrics including R^2 value, RMSE value, and MAPE value were extracted using MATLAB.

B. Simulation Results

This study introduces a more comprehensive approach incorporating physical characteristics to match the $I_D - V_{DS}$ characteristics of SiC MOSFETs. Validation of optimization effects for each component is conducted using test circuits in LTspice, as illustrated in Fig. 4. As an example of the optimization of $I_{dB}[V_{DS}]$, fitting results comparing the original model (as per Equation (1)) and the test model (as per Equation (7)) are shown in Fig. 5.

$$I_{DS} = \beta \frac{3}{1 + e^{(-\psi V_{GS})}} (1 + \lambda V_{DS}) \tanh(\alpha V_{DS}) \quad (7)$$

Following the integration of optimized components, curve fitting using least square method in MATLAB yields the results and is shown in Fig. 6, which are compared with the original model. Fig. 7 presents the comparative results of the coefficient of determination R^2. It is evident that the model proposed in

979-8-3503-6184-1/24 $31.00 © 2024 IEEE

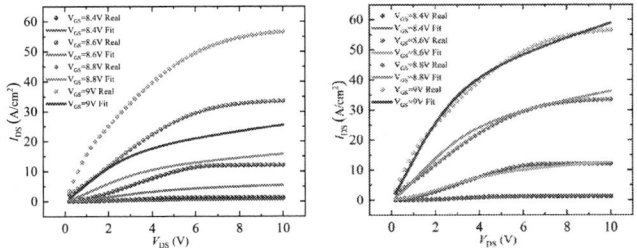

Fig. 5. Comparison of (a) original model and (b) test model.

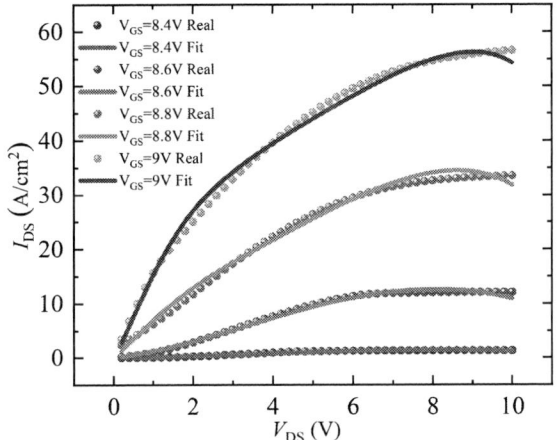

Fig. 6. Comparison of latest model.

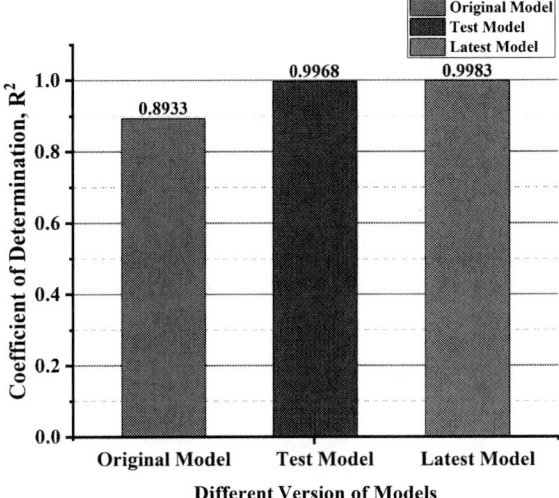

Fig. 7. Comparison of R^2 values.

characteristics in 1st quadrant. By optimizing the model function combinations and focusing on the influential factors of channel length modulation effects, the proposed model significantly outperforms existing commercial models across multiple predictive performance metrics. This advancement provides circuit designers and simulation engineers with a more precise and reliable tool for circuit design and analysis.

this study exhibits significant improvement compared to both the original model and the validation model.

$$\text{RMSE} = \sqrt{\frac{1}{n}\sum_{i=1}^{n}\left(A_{i,\text{model}} - A_{i,\text{meas.}}\right)^2} \quad (8)$$

$$\text{MAPE} = \left(\frac{100}{n}\right)\sum_{i=1}^{n}\left|\frac{A_{i,\text{meas.}} - A_{i,\text{model}}}{A_{i,\text{model}}}\right| \quad (9)$$

In addition to the coefficient of determination R^2, significant reductions are observed in RMSE values (computed using Equation (8)) and MAPE values (computed using Equation (9)), as detailed in Table I.

TABLE I
RMSE AND MAPE OF DIFFERENT VERSIONS

Predictive Performance Metrics	RMSE	MAPE
Original Model	5.84	0.17
Test Model	1.00	0.083
Latest Model	0.72	0.063

IV. CONCLUSION

This study successfully introduces an enhanced SPICE model for Silicon Carbide (SiC) MOSFETs, demonstrating excellent performance in accurately matching current-voltage

REFERENCES

[1] W. R. Curtice, "A MESFET Model for Use in the Design of GaAs Integrated Circuits," in IEEE Transactions on Microwave Theory and Techniques, vol. 28, no. 5, pp. 448-456, May 1980, doi: 10.1109/TMTT.1980.1130099.

[2] M. Mudholkar, S. Ahmed, M. N. Ericson, S. S. Frank, C. L. Britton and H. A. Mantooth, "Datasheet Driven Silicon Carbide Power MOSFET Model," in IEEE Transactions on Power Electronics, vol. 29, no. 5, pp. 2220-2228, May 2014, doi: 10.1109/TPEL.2013.2295774.

[3] I. Angelov, H. Zirath and N. Rosman, "A new empirical nonlinear model for HEMT and MESFET devices," in IEEE Transactions on Microwave Theory and Techniques, vol. 40, no. 12, pp. 2258-2266, Dec. 1992, doi: 10.1109/22.179888.

[4] F. -J. Hsu et al., "High accuracy large-signal SPICE model for silicon carbide MOSFET," 2018 IEEE 30th International Symposium on Power Semiconductor Devices and ICs (ISPSD), Chicago, IL, USA, 2018, pp. 403-406. doi: 10.1109/ISPSD.2018.8393688.

[5] B. Wang, Z. Duan, Z. Shen, Y. Zhao, L. Gao and C. Wang, "A Reconfigurable High-Precision and Energy-Efficient Circuit Design of Sigmoid, Tanh and Softmax Activation Functions," 2023 IEEE International Conference on Integrated Circuits, Technologies and Applications (ICTA), Hefei, China, 2023, pp. 118-119, doi: 10.1109/ICTA60488.2023.10364285.

[6] Rajanand and P. Singh, "ErfReLU: Adaptive Activation Function for Deep Neural Network," Pattern Analysis and Applications, vol. 27, p. 68, 2024. [Online]. Available: https://doi.org/10.1007/s10044-024-01277-w

[7] Bagul, Y. J. and Chesneau, C., "Sigmoid functions for the smooth approximation to the absolute value function," Moroccan Journal of Pure and Applied Analysis (MJPAA), vol. 7, no. 1, pp. 12-19, 2021. doi: 10.2478/mjpaa-2021-0002.

[8] M. S. Islam, M. M. Zaman, "A seven-parameter nonlinear I–V characteristics model for sub-lm range GaAs MESFETs," Solid-State Electronics, vol.48, pp. 1111–1117, 2004. doi:10.1016/j.sse.2004.01.007.

An Analytical Model for Characterizing Density of States of Oxide Transistors

Siyuan Hu[1], Chuanlin Sun[2], Junchen Dong[2,*], Zhensong Li[2], Kai Zhao[2], Dedong Han[1,*], Xing Zhang[1,3]

[1]School of Integrated Circuits, Beijing Advanced Innovation Center for Integrated Circuits, Peking University, Beijing 100871, China
[2]School of Information & Communication Engineering, Beijing Information Science & Technology University, Beijing 100101, China
[3]Peking University Shenzhen Graduate School, Shenzhen 518055, China

* Email: jcdong@bistu.edu.cn, handedong@pku.edu.cn

Abstract—Herein, an analytical model for analyzing subgap density of states (DOS) of the ITO transistors is proposed. The potential distribution within the active layer of oxide transistors is fully described by this model. Based on capacitance-voltage characteristics, an extraction method of density of states (DOS) in the ITO active layer is proposed. The TCAD simulation results based on DOS parameters show a good consistency with the measured transfer curves, which achieves a goodness of fit (R^2) of over 0.97. This work propels the practical application of oxide transistors.

Keywords—density of states, extraction method, analytical model, TCAD simulation, thin film transistors

I. INTRODUCTION

Due to high performance, preferable uniformity, low cost, and low temperature process, the oxide transistors draw much attention in the field of integrated circuits (IC) [1, 2]. So far, a variety of novel IC applications based on the oxide transistors have been developed, such as capacitor-less DRAM, flexible microprocessors, and RFID circuits [3-5]. To further apply the oxide transistors, it is necessary to have a deep understanding of transport characteristics of them [6, 7]. Normally, transport characteristics are closely related with subgap density of states (DOS) of the oxide semiconductor active layer [8]. Therefore, an effective DOS extraction method is of great importance for analyzation of transport mechanisms of the oxide transistors [9, 10]. Previous works reported a feasible extraction method based on the capacitance-voltage (*C-V*) characteristics of the oxide transistors, of which subgap DOS was considered to be uniformly ionized [11]. For the practical oxide transistors, however, DOS varies with potential distribution in the active layer normally, and there is a lack of research into this mechanism.

Herein, we fabricate InSnO (ITO) transistors and propose an analytical model to fully describe potential distribution in the ITO active layer. Combining with the *C-V* characteristics, a DOS extraction method is established based on the potential distribution model. The simulation results using the extracted DOS parameters show a good consistency with the measured transfer curves.

II. ANALYTICAL MODEL

A. Potential distribution in active layer

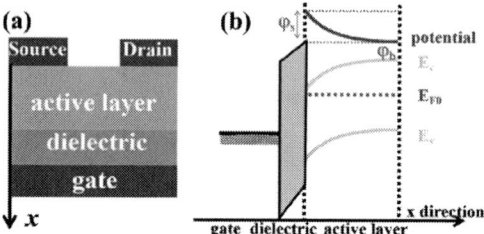

Fig. 1. (a) Structure of oxide transistors with bottom gate. (b) Energy band and potential diagram along the vertical direction of oxide transistors ($V_G > 0$).

Fig. 1 (a) presents device structure of the oxide transistors. Fig. 1(b) shows energy band and electric potential diagrams of the ITO transistors. φ_s and φ_b are surface potential and boundary potential, respectively. Electric potential varies from φ_s to φ_b is dependent on Poisson's equation, which is given by:

$$\frac{d^2\varphi}{dx^2} = -\frac{\rho(x)}{\varepsilon_s} = \frac{q}{\varepsilon_s}[n_{\text{free}}(x) + n_{\text{acceptor}}(x) - n_{\text{donor}}(x)] \tag{1}$$

where ε_s is permittivity of the active layer. n_{free}, n_{acceptor} and n_{donor} are concentration of free electrons, ionized acceptor states, and donor states, respectively. Normally, n_{acceptor} and n_{donor} are much larger than n_{free} in most part of the active layer. Additionally, n_{acceptor} and n_{donor} dominates under positive and negative gate bias, respectively. Therefore, it is reasonable to consider only one of the three terms (n_{acceptor} or n_{donor}) on the right-hand side of (1).

Density of acceptor states and donor states are assumed to be exponential distribution, which can be modeled as:

$$g_A(E) = N_{TA} \times \exp(\frac{E - E_C}{W_{TA}}) + N_{DA} \times \exp(\frac{E - E_C}{W_{DA}}) \tag{2}$$

$$g_D(E) = N_{TD} \times \exp(\frac{E - E_C}{W_{TD}}) + N_{DD} \times \exp(\frac{E - E_C}{W_{DD}}) \tag{3}$$

physical significance of each variable are shown in TABLE I.

For the oxide transistors, subthreshold and on-state regions of the transfer curves are primarily influenced by the acceptor states, while off-state region is mainly influenced by the donor states. Take n_{acceptor} as an example, it can be obtained from:

$$n_{\text{acceptor}} = \int_{E_V}^{E_{F0} + q\varphi} g_A(E)dE = \int_{E_V}^{E_{F0}' + q\varphi} N_{TA/DA} \exp(\frac{E_{\text{trap}}}{W_{TA/DA}})dE_{\text{trap}}$$

$$\approx N_A W_A \exp(\frac{E_{F0}' + q\varphi}{W_A}) \tag{4}$$

979-8-3503-6184-1/24 $31.00 © 2024 IEEE   533

TABLE I.   MEANINGS OF EACH VARIABLE

Variable	Meaning
N_{TA}/N_{DA}	Density of tail/deep acceptor states
N_{TA}/N_{DA}	Distribution energy of tail/deep acceptor states
N_{TD}/N_{DD}	Density of tail/deep donor states
W_{TD}/W_{DD}	Distribution energy of tail/deep donor states
E_c	Conduction band energy

Here, $E-E_c$ is replaced by E_{trap}, which means E_c is considered as an energy reference. Thus, E_{F0} and E_V, representing Fermi energy level and valence band energy respectively, should be replaced by $E_{F0}{}'$ and $E_V{}'$. For convenience, E_{F0} and E_V are still used in the following content. N_A and W_A are dominant parameter couple between N_{TA}, W_{TA} and N_{DA}, W_{DA}.

Using the following formula:

$$\frac{d^2\varphi}{dx^2} = \frac{1}{2}\frac{d}{d\varphi}\left(\frac{d\varphi}{dx}\right)^2 \tag{5}$$

of which the specific relationship between φ and x is given by (6), and it can be solved in an analytical form, and given by (7). Then TCAD simulation is employed to validate the rationality of hypothesis and analytical model as shown in Fig. 2. The arctan function exhibits a strong liner relationship with x, indicating that potential distribution along vertical direction follows (7). Combined with gradual channel approximation, which indicates potential varies linearly along the channel direction. The potential distribution in the active layer of oxide transistors can be fully depicted.

$$\frac{d\varphi}{dx} = \sqrt{\frac{2}{\varepsilon_s}\int_{\varphi_b}^{\varphi}\rho(\varphi)d\varphi} = \sqrt{\frac{2q}{\varepsilon_s}\int_{\varphi_b}^{\varphi}N_A W_A\exp(\frac{E_{F0}+q\varphi}{W_A})d\varphi}$$
$$= \sqrt{\frac{2W_A^2}{\varepsilon_s}N_A\exp(\frac{E_{F0}+q\varphi_b}{W_A})(\exp(\frac{q\varphi-q\varphi_b}{W_A})-1)} = \sqrt{\frac{2W_A^2}{\varepsilon_s}N_{Ab}(\exp(\frac{q\varphi-q\varphi_b}{W_A})-1)} \tag{6}$$

$$\sqrt{\frac{2\varepsilon_s}{qN_{Ab}}}\arctan(\sqrt{\exp(\frac{q\varphi-q\varphi_b}{W_A})-1}) = x \tag{7}$$

B. Extraction of subgap density of states

An intermediate parameter θ is introduced to simplify calculations. It is defined as:

$$\theta = \sqrt{\frac{qN_{Ab}}{2\varepsilon_s}}t_s \tag{8}$$

where t_s is the thickness of active layer. θ ranges from 0 to $\pi/2$. It should be noticed that N_{Ab} is exactly the DOS at the potential of φ_b. Thus φ_s can be obtained from (7) and (8), expressed as follows:

$$\varphi_s = \varphi_b - 2W_A\ln(\cos\theta) \tag{9}$$

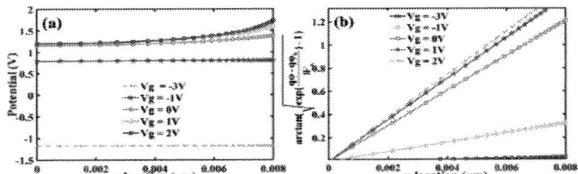

Fig. 2.   (a) Potential distribution along vertical direction obtained from TCAD simulation. (b) Arctan function versus x.

$$\frac{d\varphi_s}{d\varphi_b} = 1 + \theta\tan\theta \tag{10}$$

Based on Gaussian's law and (6), the total charge per unit (Q_s) area is given by:

$$Q_s = \varepsilon_s\frac{d\varphi}{dx}\Big|_{\varphi=\varphi_s} = \sqrt{2\varepsilon_s\int_{\varphi_b}^{\varphi_s}\rho(\varphi)d\varphi} = \frac{2W_A\varepsilon_s}{t_s}\theta\tan\theta \tag{11}$$

The gate-to-source capacitance (C_g) is a series connection of the active layer capacitance (C_s) and gate dielectric capacitance (C_{ox}), which can be expressed as:

$$C_g = \frac{1}{\frac{1}{C_s}+\frac{1}{C_{ox}}} = \frac{1}{\frac{1}{\frac{dQ_s}{d\varphi_s}}+\frac{1}{C_{ox}}} \tag{12}$$

C_s can be obtained from (8), (10) and (11):

$$C_s = \frac{dQ_s}{d\varphi_s} = \frac{dQ_s}{d\theta}\frac{d\theta}{d\varphi_b}\frac{d\varphi_b}{d\varphi_s} = \frac{\varepsilon_s}{t_s}\frac{\theta\tan\theta+\theta^2\sec^2\theta}{1+\theta\tan\theta} \tag{13}$$

With the C-V characteristics measured from the devices, θ varying with gate bias is obtained by (12) and (13). Thus, N_{Ab}, that is $g(E_{F0}+q\varphi_b)$, can be calculated by (8).

To fully depict the energy distribution of DOS, the flat band Fermi level (E_{F0}) is extracted by transfer characteristics of the devices with a small drain voltage. The surface potential φ_s is obtained:

$$\varphi_s = \int_{V_{fb}}^{V_g}1-\frac{C_g}{C_{ox}}dV_g \tag{14}$$

where V_{fb} is the flat band voltage. V_g is the gate bias. The drain current can be expressed as:

$$I = q\mu\frac{W}{L}V_d\int_0^{t_s}n_{free}(x)dx = q\mu\frac{W}{L}V\int_{\varphi_b}^{\varphi_s}\frac{N_c\exp(\frac{E_{F0}+q\varphi}{kT})}{\frac{d\varphi}{dx}}d\varphi \tag{15}$$

where μ is the field effect mobility. W and L is the channel width and length, respectively. V_d is the drain voltage. N_c is the effective density of the conduction band states. Taking the derivative of both sides of the (15) with respect to φ_s yields:

$$\frac{d\left(\frac{I}{q\mu W/L}\right)}{d\varphi_s} = \frac{N_c\exp(\frac{E_{F0}+q\varphi_s}{kT})}{\frac{d\varphi}{dx}\Big|_{\varphi=\varphi_s}} = \frac{N_c\exp(\frac{E_{F0}+q\varphi_s}{kT})}{\frac{Q_s}{\varepsilon_s}} \tag{16}$$

Substituting (10), (11) into (16) yields:

$$N_c\exp(\frac{E_{F0}+q\varphi_s}{kT}) = \frac{d\left(\frac{I}{q\mu W/L}\right)}{d\theta}\frac{d\theta}{d\varphi_b}\frac{d\varphi_b}{d\varphi_s}\cdot\frac{Q_s}{\varepsilon_s}$$
$$= \frac{d\left(\frac{I}{q\mu W/L}\right)}{d\theta}\frac{\theta^2\tan\theta}{t_s(1+\theta\tan\theta)} \tag{17}$$

With θ calculated by (13), right-hand side of (17) can be obtained. Thus, $N_c\exp(E_{F0}+q\varphi_s/kT)$ versus φ_s can be plotted. Using linear fitting method and a known value for N_c, E_{F0} can be extracted, as shown in Fig. 3(a). Thus, the energy distribution of DOS can be acquired. Fig. 3 (b) shows the φ_s and φ_b versus V_g obtained from the extraction process.

979-8-3503-6184-1/24 $31.00 © 2024 IEEE

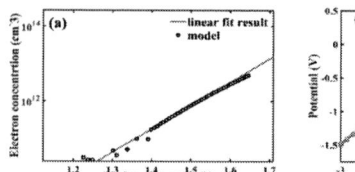 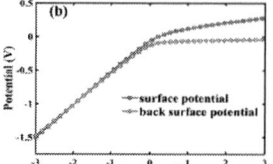

Fig. 3. (a) The $N_c \exp(E_{F0} + q\varphi_s/kT)$ (electron concentration) versus φ_s and the linear fit result. (b) φ_s and φ_b versus V_g.

III. RESULTS AND DISCUSSION

A. Device Fabrication

The fabrication process of the oxide transistor is described as follows. A 100-nm Al film is deposited as the gate electrode by sputtering process. Then the gate dielectric of 20-nm Al_2O_3 is deposited by atomic layer deposition (ALD) process. An active layer of 6-nm ITO is deposited by sputtering. Finally, a 100-nm Al film is deposited by sputtering as source/drain (S/D) electrodes.

B. Extraction results

The C-V and transfer characteristics of the ITO transistor with W/L = 100 μm/100 μm are measured by probe stations (Keithley 4200-SCS semiconductor characterization system, Agilent B1500A semiconductor device analyzer). The C-V characteristics are measured by applying AC voltage signal with a relative low frequency to approximate the quasi-static capacitance. Using the method proposed earlier, the energy distribution of DOS is obtained from the measured data. By fitting the DOS data using (2), the exponential distribution parameters can be extracted, which shows in Fig. 4. The extracted values are $N_{TA} = 9.885 \times 10^{19}$ cm^{-3} eV^{-1}, $W_{TA} = 0.1165$ eV, $N_{DA} = 2.567 \times 10^{17}$ cm^{-3} eV^{-1}, and $W_{DA} = 0.27$ eV.

The extracted DOS parameters are imported into TCAD while the other device parameters in simulation are the same as the fabricated devices. The transfer curve and C-V curve are generated by TCAD and compared with the measured data as shown in Fig. 5. Both of two simulation curves exhibit good agreement with measured data. By calculating goodness of fit (R^2), we obtained $R^2 = 0.975$ from transfer characteristics and $R^2 = 0.979$ from C-V characteristics. Thus, the accuracy of proposed extraction method is validated.

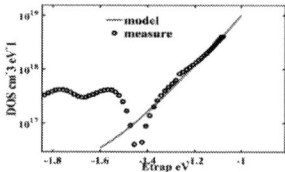

Fig. 4. The energy distribution of DOS and fit results.

IV. CONCLUSION

In this study, we proposed an analytical model deduced from Poisson's equation. The electric potential distribution in the active layer is expressed analytically, and an extraction method of subgap DOS is proposed. We utilize this extraction method for obtaining DOS parameters of the ITO transistors. Furthermore, TCAD simulation is performed with the DOS parameters. A good consistency between simulation results and measured transfer and C-V curves is obtained, showing R^2 of 0.975 and 0.979, respectively. Our method is efficient and

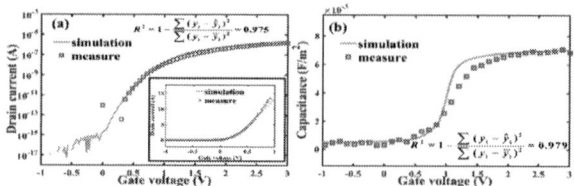

Fig. 5. Experiment data (symbols) and simulated results (lines) of ITO transistors. (a) Transfer characteristics. The inset shows the results in linear coordinates. (b) C-V characteristics at the frequency of 3 kHz.

practical for extracting the DOS parameters of oxide transistors.

ACKNOWLEDGMENT

This work was supported by the R&D Program of Beijing Municipal Education Commission (KM20231123201), the Beijing Municipal Natural Science Foundation (4232067), Shenzhen Science and Technology Innovation Committee under Grant KQTD 2020082011310-5004

REFERENCES

[1] G. W. Yang, J. Park, S. Choi, C. Kim, D. M. Kim, S. Choi, J. Bae, I. H. Cho, and D. H. Kim, "Total Subgap Range Density of States-Based Analysis of the Effect of Oxygen Flow Rate on the Bias Stress Instabilities in a-IGZO TFTs," IEEE Transactions on Electron Devices, vol. 69, pp. 166-173, 2022.

[2] S. Yan, Z. Cong, N. Lu, J. Yue, and Q. Luo, "Recent progress in InGaZnO FETs for high-density 2T0C DRAM applications," Science China-information Sciences, vol. 66, 2023.

[3] D. Saito, J. Doevenspeck, S. Cosemans, H. Oh, M. Perumkunnil, I. A. Papistas, A. Belmonte, N. Rassoul, R. Delhougne, G. Kar, P. Debacker, A. Mallik, D. Verkest, and M. H. Na, "IGZO-Based Compute Cell for Analog In-Memory Computing—DTCO Analysis to Enable Ultralow-Power AI at Edge," IEEE Transactions on Electron Devices, vol. 67, pp. 4616-4620, 2020.

[4] S. R. Bhalerao, D. Lupo and P. R. Berger, "Flexible Thin Film Transistor (TFT) and Circuits for Internet of Things (IoT) based on Solution Processed Indium Gallium Zinc Oxide (IGZO)," IEEE International Flexible Electronics Technology Conference, pp. 0023-0025, 2021.

[5] B. Tiwari, P. G. Bahubalindruni, A. Santa, J. Martins, P. Mittal, J. Goes, R. Martins, E. Fortunato, and P. Barquinha, "Oxide TFT Rectifiers on Flexible Substrates Operating at NFC Frequency Range," IEEE Journal of the Electron Devices Society, vol. 7, pp. 329-334, 2019.

[6] P. Sihapitak, J. P. Bermundo, E. Bestelink, R. A. Sporea, and Y. Uraoka, "Optimizing a-IGZO Source-Gated Transistor Current by Structure Alteration via TCAD Simulation and Experiment," IEEE Transactions on Electron Devices, vol. 71, pp. 2431-2437, 2024.

[7] C. Hsu, J. Li, P. Huang, W. Jhang, and M. Joodaki, "Study of Electrical Characteristics for Dual-Gate TFTs With Asymmetric Defect Distributions and Gate Work Functions," IEEE Transactions on Electron Devices, vol. 70, pp. 1-4, 2023.

[8] C. Hsu, J. Li, P. Huang, W. Jhang, and M. Joodaki, "Study of Electrical Characteristics for Dual-Gate TFTs With Asymmetric Defect Distributions and Gate Work Functions," IEEE Transactions on Electron Devices, vol. 70, pp. 1-4, 2023.

[9] C. Chen, K. Abe, H. Kumomi, and J. Kanicki, "Density of States of a-InGaZnO From Temperature-Dependent Field-Effect Studies," IEEE Transactions on Electron Devices, vol. 56, pp. 1177-1183, 2009.

[10] H. Bae, H. Choi, S. Jun, C. Jo, Y. H. Kim, J. S. Hwang, J. Ahn, S. Oh, J. Bae, S. Choi, D. H. Kim, and D. M. Kim, "Single-Scan Monochromatic Photonic Capacitance-Voltage Technique for Extraction of Subgap DOS Over the Bandgap in Amorphous Semiconductor TFTs," IEEE Electron Device Letters, vol. 34, pp. 1524-1526, 2013.

[11] S. Jun, C. Jo, H. Bae, H. Choi, D. H. Kim, and D. M. Kim, "Unified Subthreshold Coupling Factor Technique for Surface Potential and Subgap Density-of-States in Amorphous Thin Film Transistors," IEEE Electron Device Letters, vol. 34, pp. 641-643, 2013.

979-8-3503-6184-1/24 $31.00 © 2024 IEEE

A Physics-Informed Neural Network Model for Body Potential Distribution in MOSFETs down to 50 K

Honglin Wu [1], Fangxing Zhang [1], Xinyue Zhang [1], Baokang Peng [1], Wu Dai [1], Lining Zhang [1]*,
Runsheng Wang [2], Ru Huang [2]

[1] School of Electronic and Computer Engineering, Peking University, China
[2] School of Integrated Circuits, Peking University, China

* Email: eelnzhang@pku.edu.cn

Abstract—A potential model using the physics-informed neural network (PINN) is developed for the channel potential profile in the temperature range of 50K to 300K. By parameterizing the PINN, the voltage and spatial dependence can be incorporated into the model from accumulation to the inversion region. The results show that the PINN model achieves high accuracy for calculations of the gate-controlled charge density. Compared to iterative solutions, the method proposed in this paper solves the device's potential more quickly. Further, the model considers the incomplete ionization effects under low temperatures, beyond the capability of the full-depletion approximation. The PINN-based potential model could serve as a basis for cryogenic device modeling.

Keywords—PINN, MOSFET, body potential distribution

I. INTRODUCTION

Metal-oxide-semiconductor field-effect transistors (MOSFETs) have been utilized in aerospace applications and the CMOS hybrid circuits of emerging quantum computing (QC) hardware systems. The integration of CMOS circuits with QC systems requires operation at cryogenic temperatures to maintain the functionality of quantum bits in silicon and enhance computational efficiency [1]-[2].

Currently, solving the Poisson's equation at low temperatures typically uses the Newton iterative method [3], which is highly dependent on initial guess values for convergence speed and computationally intensive due to complex low-temperature physics. Physics-Informed Neural Networks (PINNs) have been widely reported as an effective method for solving partial differential equations [4]-[5]. PINNs can solve equations without data [6], and parameterized PINNs provide a faster and more scalable alternative to traditional iterative methods.

In this work, we propose a PINN model to model the body potential MOSFETs at different temperatures, effectively overcoming the limitations of the traditional Newton iterative method for low-temperature Poisson's equation solutions. The trained PINN model accurately captures the potential distribution along the vertical channel direction, facilitating precise computation of the mobile charge within the silicon body. This method significantly enhances computational efficiency while maintaining the accuracy of the physical model. And it holds significant implications for accurately modeling MOSFET currents under low-temperature conditions.

II. MODELING THE POTENTIAL PROFILE WITH PINN

A. Physical Problem Setting

To solve the body potential distribution in MOSFET, we consider the one-dimensional Poisson's equation as Eq. (1):

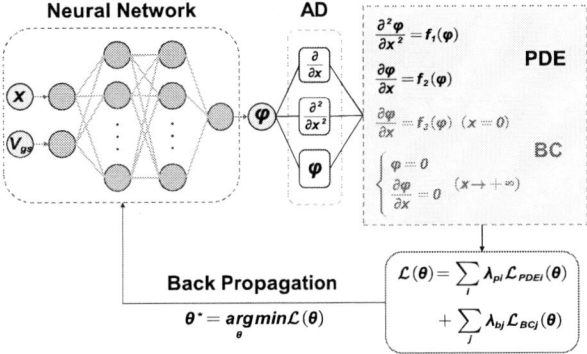

Fig. 1. An physics-informed neural network architecture.

$$PDE1: \frac{\partial^2 \varphi(x)}{\partial x^2} = -\frac{q}{\varepsilon_{Si}} \left\{ \begin{array}{l} p_{p0}\left[\exp\left(-\frac{\varphi(x)}{kT/q}\right)-1\right] \\ -n_{p0}\left[\exp\left(\frac{\varphi(x)}{kT/q}\right)-1\right] \end{array} \right\} \quad (1)$$

where $\varphi(x)$ is the potential perpendicular to the channel direction, ε_{Si} is the silicon material's dielectric constant, q is the quantity of electric charge, as well as p_{p0} and n_{p0} are the equilibrium electron and hole concentrations respectively. By integrating both sides of Eq. (1) with respect to x, we obtain an expression for the electric field within the body, as shown in Eq. (2). We consider incomplete ionization when calculating p_{p0} and n_{p0}, making the model suitable for low temperatures.

$$PDE2: \frac{\partial \varphi(x)}{\partial x} = -sgn(V_{gs}-V_{fb})$$
$$\times \left\{ 2\frac{kT}{\varepsilon_{Si}} \left\{ \begin{array}{l} p_{p0}\left[exp\left(-\frac{\varphi(x)}{kT/q}\right)+\frac{\varphi(x)}{kT/q}-1\right] \\ +n_{p0}\left[exp\left(\frac{\varphi(x)}{kT/q}\right)-\frac{\varphi(x)}{kT/q}-1\right] \end{array} \right\} \right\}^{\frac{1}{2}} \quad (2)$$

The boundary conditions, specified in Eq. (3) - (5), represent the gate control equation and the conditions that both the potential and electric field approach zero at infinity:

$$BC1: \left.\frac{\partial \varphi(x)}{\partial x}\right|_{x=0} = -\frac{\varepsilon_{ox}}{\varepsilon_{Si}} \frac{V_{gs}-V_{fb}-\varphi(x=0)}{T_{ox}} \quad (3)$$

$$BC2: \varphi(x \to +\infty) = 0 \quad (4)$$

$$BC3: \left.\frac{\partial \varphi(x)}{\partial x}\right|_{x \to +\infty} = 0 \quad (5)$$

where ε_{ox} is the oxide's dielectric constant, T_{ox} is the thickness of the oxide layer.

B. PINN Framework

The model based on PINN we propose in this work is shown in Fig. 1. This model consists of a fully connected

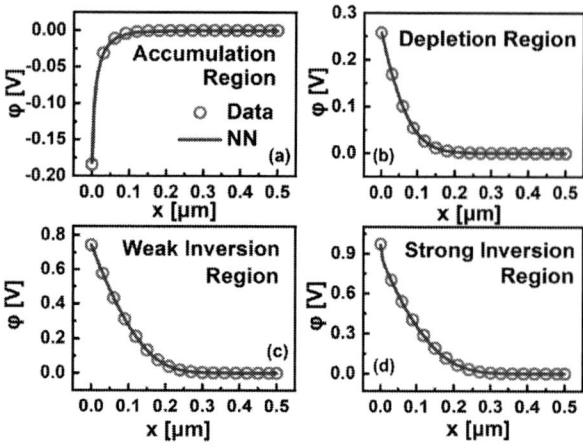

Fig. 2. The outputs of the NN model show good consistency with the data at 300K. In all operational regions, the RMSE of the four NN models is less than 0.004. It should be noted that the V_{gs} values corresponding to (a) - (d) are -1.0V, -0.3V, 0.2V, 1.0V, respectively.

neural network. For any trained network, the input is the position coordinate x along the vertical channel direction of the NMOS. Through forward propagation, the network outputs the corresponding potential value $\varphi(x)$. The predicted $\varphi(x)$ satisfies the two PDE equations (Eq. (1) and Eq. (2)), and multiple boundary conditions. Eq. (6) presents the loss function, where L_{PDE1} represents the loss of Poisson's equation, and L_{PDE2} represents the loss of the electric field equation.

$$Loss = \lambda_{p1} L_{PDE1} + \lambda_{p2} L_{PDE2} + \lambda_{b1} L_{BC1} \\ + \lambda_{b2} L_{BC2} + \lambda_{b3} L_{BC3} \tag{6}$$

$$L_{PDE1} = MAE\left(\frac{\partial^2 \varphi_{NN(x)}}{\partial x^2} - f_{PDE1}\left(\varphi_{NN(x)}\right)\right)$$

$$L_{PDE2} = MAE\left(\frac{\partial \varphi_{NN(x)}}{\partial x} - f_{PDE2}\left(\varphi_{NN(x)}\right)\right)$$

$$L_{BC1} = MAE\left(\frac{\partial \varphi_{NN(x)}}{\partial x} - f_{BC1}\left(\varphi_{NN(x)}\right)\Big|_{x=0}\right)$$

$$L_{BC2} = MAE\left(\varphi_{NN(x)}\big|_{x \to x_{max}}\right)$$

$$L_{BC3} = MAE\left(\frac{\partial \varphi_{NN(x)}}{\partial x}\Big|_{x \to x_{max}}\right)$$

The network is implemented within the PyTorch framework. During the training process, the tanh activation function is employed, and the Adam optimizer is utilized for optimization.

III. RESULTS AND DISCUSSIONS

In this section, we discuss the ability of PINN model on body potential distribution and inversion charge model in different temperatures and discuss the impact of network parameter size and the choice of activation functions on model accuracy.

All the data for comparison in this work were obtained through the numerical solution method. This method involves first using the electric field equation and the gate control equation to perform Newton iteration, thereby obtaining the surface potential V_{sc} at $x=0$. Subsequently, the body potential distribution is determined iteratively based on the electric field equation. In Fig. 2, the potential solutions obtained by the

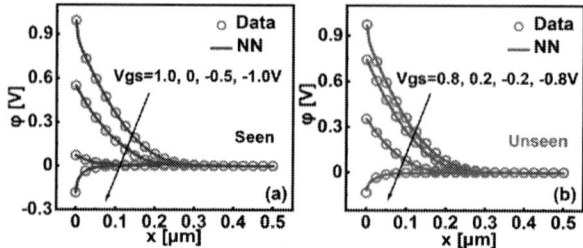

Fig. 3: Agreement between the PINN Model and the data at 300K. (a) Results of the PINN model trained on various V_{gs} values, the RMSE of training is 0.0086. (b) Results of the PINN model tested on various interpolated V_{gs} values, the RMSE of testing is 0.0060.

PINN model at 300K are compared with those from numerical solutions. It can be seen from Fig. 2(a) to Fig. 2(d) that the PINN model, without using any data, effectively solves the body potential distribution of NMOS operating in different regions with high accuracy, including accumulation region, depletion region, weak inversion region, and strong inversion region.

Considering the dependence of body potential distribution on the gate voltage, we add V_{gs} as an additional input dimension to the network. During training, the V_{gs} values are distributed across the four operational regions of the MOSFET. The surface potentials obtained from iterative calculations for different V_{gs} are used as label data for training. After training, the PINN model can accurately determine the potential at any position coordinate x along the vertical channel direction for any V_{gs}. Compared to iterative solutions, the inference speed of the PINN model is faster.

As illustrated in Fig. 3, the PINN model demonstrates excellent consistency with the data, both in training under multiple V_{gs} and in testing with interpolated V_{gs}. The model achieves high accuracy across the accumulation, depletion, and inversion regions.

When the temperature decreases, the intrinsic carrier concentration n_i in MOSFETs reduces, particularly at lower temperatures where the value of n_{p0} becomes extremely small. During training, it was observed that PDE equations incorporating the n_{p0} term led to convergence difficulties. Therefore, in this work, we initially train the model using two PDE equations that neglect the n_{p0} term. After a specified number of training epochs, the model completes the initial training phase. Subsequently, the n_{p0} term is reintroduced into the training process. This approach effectively addresses the convergence issue during training. We train the PINN model at 50K, and as shown in Fig. 4, it accurately models the body potential distribution across four different operational regions at low temperatures which is significant for charge modeling at low temperatures.

The physics-trained PINN model effectively captures the underlying physical principles. As shown in Fig. 5(a), the model accurately solves for the surface potential across different V_{gs} values. The surface potentials calculated by the PINN model are in good agreement with the data, both at room temperature and at temperatures as low as 50K, with the RMSE error of 0.00182. This method can quickly solve the body potential, providing accurate initial guess values for the surface potential at low temperatures, thereby significantly reducing the time required for physical modeling.

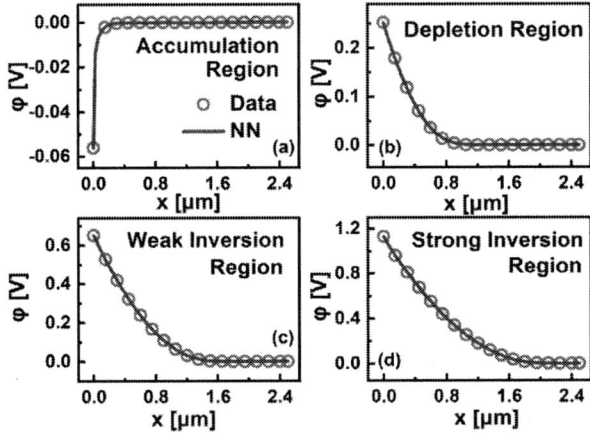

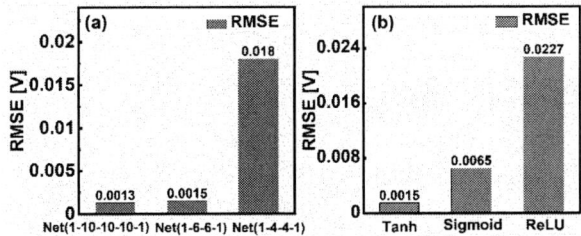

Fig. 6. RMSE of PINN models with different sizes and activation functions.

Fig. 4. The outputs of the NN model show good consistency with the data at 50K. In all operational regions, the RMSE of the four NN models is less than 0.006. It should be noted that the V_{gs} values corresponding to (a) - (d) are -1.0V, -0.4V, 0V, 0.5V, respectively.

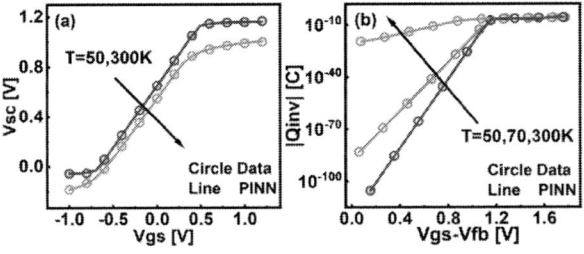

Fig. 5. (a) Comparison between V_{sc} calculated by the PINN model and data from 50K to 300K. (b) Comparison between Q_{inv} calculated from the body potential obtained by the PINN model and data from 50K to 300K.

We discuss the calculation of the inversion region charge with the method that obtains Q_{inv} by integrating the body potential along the vertical channel direction [7], as shown in Eq. (8).

$$Q_{inv} = \int_0^{+\infty} q n_{p0} \exp\left(\frac{\varphi(x)}{kT/q}\right) dx \qquad (8)$$

Fig. 5(b) compares the inversion region charge calculated by the PINN model with the results obtained using traditional iterative methods. At different temperatures, the PINN model accurately models the inversion region charge, demonstrating its effectiveness for low-temperature device modeling. Our PINN model considers the incomplete ionization effects under low temperatures, so we can accurately model the body potential distribution, which enables precise calculation of Q_{inv} and facilitates subsequent current modeling. Further, the PINN-based potential model could serve as a basis for cryogenic device modeling.

The root mean square error (RMSE) of the potential model is a function of network size. Fig. 6(a) shows the RMSE for different network sizes, while Fig. 6(b) shows the RMSE for models with different activation functions. Generally, the model accuracy is higher with more network parameters.

For solving the body potential at a given V_{gs}, the iterative method takes 6.53 seconds. In contrast, the trained PINN model takes only 0.0051 seconds, demonstrating its effectiveness in accelerating the modeling process for low-temperature devices.

IV. SUMMARY

In this work, a physics-informed neural network potential model is proposed to model the body potential distribution at temperatures ranging from 50K to 300K. The network takes the vertical channel direction coordinate x and the gate voltage V_{gs} as inputs, and outputs the potential at the corresponding position for the given gate voltage. This method has been discussed and compared with the Newton iterative method. The results demonstrate that the PINN model can accurately model the potential and has a faster inference speed than iterative solutions. The PINN model considers incomplete ionization, resulting in a more accurate Q_{inv} obtained through body potential integration compared to traditional physical approximations. Based on the above advantages, the PINN model is expected to facilitate rapid and accurate current modeling at low temperatures.

ACKNOWLEDGMENT

This work is supported in part by the National Key Research and Development Program of China (2023YFB4402204), and in part by the Guangdong Basic and Applied Basic Research Foundation (2024B1515020064).

REFERENCES

[1] B. Patra et al., "Cryo-CMOS circuits and systems for quantum computing applications," in IEEE Journal of Solid-State Circuits, vol. 53, no. 1, pp. 309-321, Jan. 2018.

[2] E. Charbon et al., "Cryo-CMOS for quantum computing," 2016 IEEE International Electron Devices Meeting (IEDM), San Francisco, CA, USA, 2016, pp. 13.5.1-13.5.4.

[3] Chao Luo, Zhen Li, Teng-Teng Lu, Jun Xu, Guo-Ping Guo, "MOSFET characterization and modeling at cryogenic temperatures", Cryogenics, Volume 98, 2019, pp. 12-17.

[4] M. Raissi, P. Perdikaris, G.E. Karniadakis, "Physics-informed neural networks: A deep learning framework for solving forward and inverse problems involving nonlinear partial differential equations", Journal of Computational Physics, vol. 378, 2019, pp. 686-707.

[5] Long, Zichao, Yiping Lu, and Bin Dong. "PDE-Net 2.0: Learning PDEs from data with a numeric-symbolic hybrid deep network." Journal of Computational Physics 399 (2019): 108925.

[6] P. Gaire and S. Bhardwaj, "Data-free, physics-informed solution of poisson equation through artificial feed-forward neural network," 2023 International Applied Computational Electromagnetics Society Symposium, Monterey/Seaside, CA, USA, 2023, pp. 1-2.

[7] H.C. Pao and C.T. Sah, "Effects of diffusion current on characteristics of metal oxide (insulator)-semiconductor transistors," Solid-State Electronics, vol. 9, no. 10, pp. 927–937, 1966.

979-8-3503-6184-1/24 $31.00 © 2024 IEEE

Modeling the Transient Characteristics with Trap Behaviors in LTPS-TFTs

Haolin Li [1], Zheng Zhou[1,2], Xiaoyan Liu*[1,2]

[1] School of Integrated Circuits, Peking University, Beijing 100871, China
[2] Beijing Advanced Innovation Center for Integrated Circuits, Beijing 100871, China.

* Email: liuxiaoyan@pku.edu.cn

Abstract—**Transient drain current (Ids) overshoot is generally observed during the switching process of low-temperature poly-silicon thin-film transistors (LTPS-TFTs), followed by a gradual recovery to the steady-state afterwards. To model this behavior and to predict its impact at the circuit level, a surface-potential based LTPS-TFT compact model is established first, considering the trap occupation in poly-silicon (poly-Si) channel and the gate oxide. Then a time-dependent trapping and detrapping model is brought forward to describe the Ids transient characteristics of LTPS-TFTs. The model can be used in circuit simulation to optimize the circuits in display.**

Keywords—LTPS-TFT, compact model, trapping, detrapping

I. INTRODUCTION

TFTs are the basic components of integrated circuits (ICs) on panel, where LTPS-TFTs stand out for its high mobility and reliability, feasibility of complementary metal-oxide-semiconductor (CMOS) logic and low parasitic effect. Nowadays LTPS-TFTs have been widely utilized in flat panel display (FPD) technologies, especially active-matrix organic light emitting diodes (AMOLEDs)[1]. Acting as the driving device of OLEDs, the drain current instability of LTPS-TFTs will disturb the illumination of the AMOLED pixel[2]. Therefore, it is important to figure out and model the mechanism of TFT transient characteristics. Previous transient trap models are lack of trap behaviors and fail to fully reproduce the Ids characteristics, while TCAD simulation is time-consuming and the parameters cannot be flexibly adjusted[3-6]. In this work, a surface-potential based TFT model with trapping and detrapping behavior is established to describe the quasistatic and transient characteristics of LTPS-TFTs. Then a dual-type trap model with the details of trap distribution is developed to model the transient Ids characteristics. Our model provides an efficient and accurate solution for the device and circuit-level LTPS-TFTs simulation and can be used to optimize the circuits in display.

II. MECHANISM AND MODELING

A. The Mechanism of Transient Current Overshoot

Due to the existence of grain boundary in poly-Si structure, considerable traps can be detected in the channel of LTPS-TFTs, which is one of the possible origins of threshold voltage (Vth) shift and transient Ids characteristics. For instance, when an LTPS -TFT is switched from off to on-state at the constant drain-to-source voltage (Vds), its Ids will first overshoot the anticipated current and undergo a gradual recovery afterwards[4]. On the contrary, when the TFT turns off, its drain current will start with a relatively lower current after the switching process. In AMOLEDs, both of the behaviors lead to the image sticking phenomena, which will gradually vanish in tens of seconds.

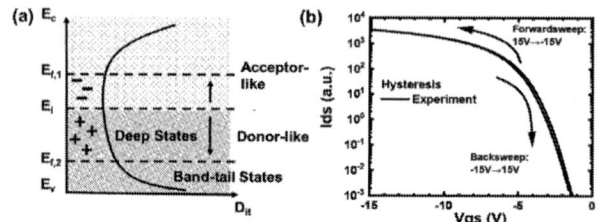

Fig. 1. (a) Diagram of the trap occupation process of poly-Si;
(b) TFT Ids-Vgs Hysteresis effect.

The diagram of charge trapping and detrapping process is shown in Fig. 1. The acceptor-like traps are mainly distributed in the upper band whereas the donor-like traps are mainly distributed in the lower band. For the n-type poly-Si, when the Fermi level (Ef) moves towards the valence band and the inversion layer forms, more donor-like traps lie above Ef and holes are trapped (whereas electrons are detrapped). The ionized positively-charged traps will lead to a negative shift of Vth and larger gate-source voltage (Vgs) to reach the same surface potential (ϕ_s).Therefore, a p-type LTPS-TFT which undergoes a large negative bias tend to have a larger |Vth| and lower Ids than those without such high stress voltage. When a smaller Vgs is continually set to the TFT's gate after the large bias, the Vth will gradually recover to the level near its quasistatic state. The trapping and detrapping behavior not only lead to the transient Ids overshoot, but also the common hysteresis effect of LTPS-TFTs, which is shown in Fig.1 (b).

Apart from the electric property, both of the acceptor and donor-like traps in poly-Si can be further classified by their position of energy levels. For donor-like traps which have a stronger impact on the carriers of p-type TFTs, the band tail states locate near the valence band whereas the deep states lie in the middle of the bandgap. During the process that a p-type TFT is switched from accumulation to the inversion region, it is mainly the deep states that are first occupied and cause the negative shift of Vth, accompanied by the degradation of subthreshold swing (SS) . At larger bias it is the band tail states that dominate and affect the on-state current (Ion). In poly-Si, density of states (DOS) distribution function of the two types of traps can be approximated as:

$$g_{cd}\left(E\right) = g_{cd0}\exp\left(-\frac{E_v - E}{E_{sigd}}\right) \tag{1}$$

B. Surface-Potential Based Model of LTPS-TFTs

In order to model the transient Ids characteristics of an LTPS-TFT, a compact model is required to obtain its DC operation points first. Compared to Vth or charge based models, a surface-potential based compact model contains more physical effects and is capable for all the working areas

979-8-3503-6184-1/24 $31.00 © 2024 IEEE

without any smooth function, which is adequate for the introduction of the effect of traps. Considering that the band tail and deep states determine SS and Ion respectively, two exponential distribution functions with different characteristic energy are involved for the practical device modeling.

Unlike a conventional MOS structure with the body contact generally grounded, the back side of TFT is floating with a thin active layer. Therefore, the surface potential backside (ϕ_b) along x-axis perpendicular to the channel cannot be approximated to 0 and extra terms considering the effect of ϕ_b are involved in the conventional Poisson Equation[3]. To iteratively solve the derivative continuity equation, an approximation connecting ϕ_s and ϕ_b is required. The entire calculation process is shown in Table I.

Table I. The process of calculating surface-potential based Ids

List of Equations for Surface-Potential Based TFT Model	
$-\varepsilon_0\varepsilon_{Si}\dfrac{\partial^2\psi(x)}{\partial x^2}=\rho(x)=p-n+N_D^+-N_A^-+N_{TD,empty}^+-N_{TA,full}^-$	Poisson Equation
$N_{TD,empty}^+=\int(1-f(E))\cdot g(E)dE$ $\approx g_{cd0}E_{sigd}\left(\dfrac{E_i}{E_{sigd}}\pi\right)\Big/\sin\left(\dfrac{E_i}{E_{sigd}}\pi\right)\cdot\exp\left(-\dfrac{q\psi(x)}{E_{sigd}}\right)=N_{TD0}\exp\left(-\dfrac{q\psi(x)}{E_{sigd}}\right)$	
$\Delta\phi_m=\dfrac{qN_{sub}t_{Si}^2}{2\varepsilon_0\varepsilon_{Si}}+\dfrac{E_{sigd}}{E_{sigd}/E_{sigd}}\cdot\dfrac{N_{TD0}}{N_{sub}}\cdot\exp\left(\dfrac{\phi_s}{E_{sigd}}\right),\ \ \phi_b=V_t\ln\left(1+\exp\left(\dfrac{\phi_s-\Delta\phi_b}{V_t}\right)\right)$	
$C_{ox}(V_{gs}-V_{fb}-\phi_s)=\sqrt{2q\varepsilon_0\varepsilon_{Si}N_{sub}V_t}\left\{\exp\left(-\dfrac{\psi}{V_t}\right)+\left(\dfrac{\psi}{V_t}\right)+\dfrac{n_i^2}{N_{sub}^2}\exp\left(\dfrac{\psi}{V_t}\right)\right.$ $\left.+\sum_i\left(\dfrac{E_{sigd}}{qV_t}\dfrac{N_{TD0}}{N_{sub}}\exp\left(\dfrac{\psi}{V_t}\right)\right)\right\}^{\frac{1}{2}}\Big|_{\phi_b}^{\phi_s}$	Continuity Equation
$\beta\phi_{sL}=\beta\phi_{s0}+\beta Vds+\ln(Q_{sL}/Q_{s0})$	Gradual channel approximation
$I=\mu C_{ox}\dfrac{W}{L}\left[(V_{gs}-V_{fb})(\phi_{sL}-\phi_{s0})-\dfrac{1}{2}(\phi_{sL}^2-\phi_{s0}^2)-\dfrac{2}{3}\gamma(\phi_{sL}^{3/2}-\phi_{s0}^{3/2})\right]$ $+\mu C_{ox}\dfrac{W}{L}V_t\left[(\phi_{sL}-\phi_{s0})+\gamma(\phi_{sL}^{1/2}-\phi_{s0}^{1/2})\right]$	Charge sheet approximation

C. Transient Ids Model of LTPS-TFTs

Once the Poisson equation is solved, further modeling for transient Ids characteristics is available. Note that the Ids curve versus time after switching generally shows a quasi-exponential (not fully exponential) form, its corresponding differential equation of the behavior in long-term can be first written as

$$\Delta q(t)=\dfrac{\Delta t}{TAU}\cdot(Q_{eq}-q(t))\cdot TRAPA \qquad (2)$$

where TRAPA and TAU are both fitting parameters. The dynamic trap density q(t) is iterated at each timestep and initialized with the quasistatic ϕ_s when the calculation start. Then an extra term independent of surface-potential is involved in the continuity equation. A similar ϕ_s/ϕ_b calculation to the DC simulation is executed considering the extra time-dependent disturbance and then the transient Ids is achieved by our model by iterative solution.

Different from the previous transient models which are lack of trap distribution information, in our model the transient behavior of deep and tail states are calculated respectively based on ϕ_s and with different time constant TAU, which is also valid for the actual devices. Both quasistatic and transient models are converted into Verilog-A version, which can be calculated by HSPICE and saves the cost of calculation time compared to the self-consistent TCAD simulation. In addition, although some TCAD provides PMIs including extended non-radiative multi-phonon (ENMP) model which is capable of modeling Ids instability, it is time-consuming to adjust plenty

of parameters whereas other built-in models are not open for user-defined parameters, such as negative bias temperature instability (NBTI) model.

III. Results and Discussion

A. Quasistatic Simulation

The comparison of calculated transfer characteristics between our surface-potential based model and Sentaurus TCAD using the same mobility model are shown in Fig. 2. Ids-Vgs curve of LTPS-TFTs with different trap DOS and Vds are simulated. It is demonstrated that our compact model is in good agreement with the TCAD simulation results that start from the subthreshold region, which proves the reliability of our 1-D approximations and initial conditions for Newton iteration.

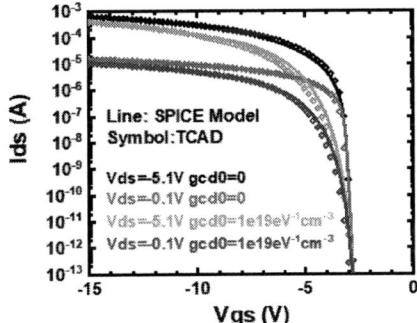

Fig. 2. Comparison of transfer characteristics between the proposed compact model and TCAD simulation.

To confirm the validity of our model with practical devices, we further establish a dual-type trap distribution model based on the experimental data, including both band tail states and deep states. As is shown in Fig.3, our model is well fitted with the measured Ids-Vgs only when two types of traps are both taken into consideration, where $g_{cd, tail} = 4\times10^{20}$ eV^{-1}cm^{-3}, $E_{sigd,tail} = 0.05$eV and $g_{cd,deep} = 2\times10^{17}$ eV^{-1}cm^{-3}, $E_{sigd,deep}= 1$eV. The fitting parameters share the same orders of magnitude with the actual devices[7].

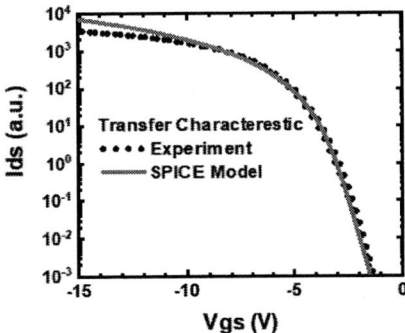

Fig. 3. Measured data fitted by the proposed compact model.

B. Transient Simulation

To further verify the feasibility of our transient model, a time-dependent AC sweep simulation is firstly carried out to reproduce the hysteresis effect. Fig.4 shows the calculated hysteresis of the LTPS-TFT whose transfer curve has been shown in Fig. 3. The calculated hysteresis shows the same orientation with the experiment result which demonstrates a negative Vth shift (~ -0.15V) when the p-type TFT undergoes a forward sweep (15V to -15V).

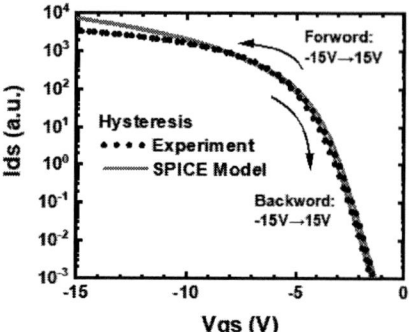

Fig. 4. Calculated hysteresis by the proposed model.

Then the calculation results of transient Ids overshoot are demonstrated in Fig. 5 and 6. Vds remains a constant value during the calculation process and various step signals with different pulse width or amplitude are applied to the gate electrode. Since LTPS-TFTs are generally used as a current driving device in AMOLEDs, the gate bias during the stress and recover period is represented by the Ids it drives. Fig. 5 shows the transient Ids after 10s stress under different gate voltage. The values of the previous 10s Ids are normalized from 0.01 to 1000. With the increase of the Ids or (Vgs) gap between the stress and recovery period, the reverse Ids overshoot is more pronounced, which is attributed to the larger ΔVth. The fitting curve shows an error rate less than 10%, indicating that our model is valid for TFT transient Ids modeling.

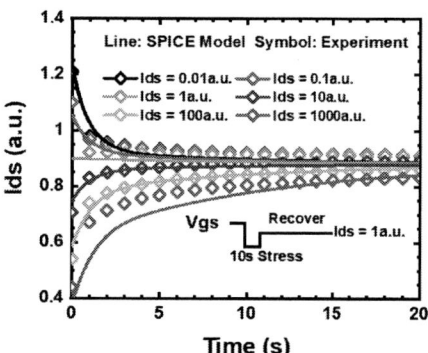

Fig. 5 Transient Ids after 10s stress whose voltage ranging from 0.01 to 1000a.u..

Fig. 6 shows the transient Ids of TFT after driving 100a.u. Ids for different bias time. It is demonstrated that the Ids overshoot vanishes with less stress time, while the transient Ids curve shows a similar trends to that of switching from lower Ids instead. The mechanism behind the phenomenon is that the stress time is too short to shift the |Vth| larger than that of equivalent Ids = 1a.u., so the Ids will continue to fall after switching from a larger Vgs.

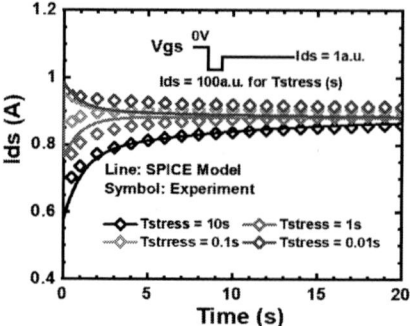

Fig. 6. Transient Ids after 100a.u. stress whose duration ranging from 0.01s to 10s.

Note that the transient Ids and trap behavior are highly correlated with the initial state, especially for the circumstance of slight disturbance before switching, such as stress with small bias or short time which is shown in Fig. 5 and 6. It is rather difficult to keep the initial condition identical during realistic experiment, which will lead to inevitable errors and even different trends. In addition, the initial condition is hard to extract from the measured data. In our model the initial condition can be controlled by setting initial Vgs, which is benefit for modeling from experimental data compared to existing ENMP and compact models.

IV. SUMMARY

Based on the surface-potential based TFT model, a novel dual-type transient model is established in this paper to model the transient characteristics of LTPS-TFTs. By setting the time constant for tail and deep states respectively, the model shows a good agreement with the measured data with the error rate less than 10%. Compared to TCAD ENMP models, our model is low time-cost and the initial condition of our trap model can be modulated flexibly. By integrating our model to a SPICE model, further circuit-level simulation is available.

REFERENCES

[1] A. Yan et al., "Thin‐Film Transistors for Integrated Circuits: Fundamentals and Recent Progress," Advanced Functional Materials, vol. 34, no. 3, 2023

[2] H. Kim, J. Park, T. Khim, S. Bak, J. Song, and B. Choi, "Threshold voltage instability and polyimide charging effects of LTPS TFTs for flexible displays," Scientific Reports, vol. 11, no. 1, 2021.

[3] H. Tsuji et al., "A New Surface Potential Based Poly-Si TFT Model for Circuit Simulation," in 2006 International Electron Devices Meeting, 11-13 Dec. 2006 2006, pp. 1-4.

[4] Y. Oodate, Y. Tanimoto, H. Tanoue, H. Kikuchihara, H. J. Mattausch, and M. Miura-Mattausch, "Compact Modeling of the Transient Carrier Trap/Detrap Characteristics in Polysilicon TFTs," IEEE Transactions on Electron Devices, vol. 62, no. 3, pp. 862-868, 2015.

[5] G. Kawachi, "P‐193: Late‐News‐Poster: A Novel Charge Based TFT Compact Model Applicable to Image‐Retention Simulation of AMOLEDs," SID Symposium Digest of Technical Papers, vol. 51, no. 1, pp. 1390-1393, 2020.

[6] "Sentaurus Device User Manual," CA, USA: Synopsys, Synopsys, 2022.

[7] J. Lee and B. Choi, "Effects of Channel Type and Doping on Hysteresis in Low-Temperature Poly-Si Thin-Film Transistors," IEEE Transactions on Electron Devices, vol. 65, no. 3, pp. 986-994, 2018.

A Novel β-Ga$_2$O$_3$-Based Enhancement-Mode Transistor Combining Heterojunction Gate and Fin shaped Gate

Yu Shao[1], Yunlong He[*1], Xiaoli Lu[*1], Songyao Wang[1], Xuefeng Zheng[1], Xiaohua Ma[1], Yue Hao[1]

[1] National Engineering Research Center of Wide Band-gap Semiconductor, School of Microelectronics, Xidian University, Xian, China

* Email: ylhe@xidian.edu.cn, xllu@xidian.edu.cn

Abstract—In this work, we propose a β-gallium oxide (β-Ga$_2$O$_3$)-based enhancement-mode transistor combining metal-oxide-β-Ga$_2$O$_3$/NiO$_X$ heterojunction gate and fin shaped gate (MOHJ-FinFET). Using the Sentaurus TCAD software, we have verified that the structure improved mutual constraints between breakdown voltage (BV) and on-resistance (R$_{on}$), allowing the device to have a combination of high BV and low R$_{on}$. The four devices with different structures were simulated to investigate the physical mechanisms that reduce the R$_{on}$ and increase the BV of MOHJ-FinFET. The threshold voltage (V$_{TH}$) of MOHJ-FinFET is 1.55 V with a Baliga figure of merit (BFOM=BV^2/R$_{on}$) up to 5.51 GW/cm^2, this study provides important guidance for the design of β-Ga$_2$O$_3$-based enhancement-mode transistors with high BFOM in the future.

Keywords—β-gallium oxide, Baliga figure of merit, Enhancement-mode

I. INTRODUCTION

β-gallium oxide (β-Ga$_2$O$_3$) material has an ultra-wide bandgap (4.6V-4.9V), which gives it a very high theoretical breakdown field strength of 8MV/cm. The Baliga figure of merit (BFOM= BV^2/R$_{on}$) used to evaluate the switching characteristics of the power device is 3444, which is 3444 times that of Silicon (Si), 4 times that of gallium nitride (GaN) and 10 times that of silicon carbide (SiC) [1]. Recently, power diodes based on β-Ga$_2$O$_3$ have achieved BFOM of 13.2GW/cm^2, exceeding the theoretical limits of SiC and GaN materials and demonstrating the great potential of β-Ga$_2$O$_3$ in the application of next-generation power electronics [2]. At present, due to the lack of effective P-type doping method of β-Ga$_2$O$_3$, conventional transistors often exhibit depletion-mode, while in practical circuits it is necessary to achieve enhancement-mode transistors to reduce static power consumption, simplify circuit topology and for safety considerations. The current enhancement-mode of β-Ga$_2$O$_3$ based devices are mainly fabricated by recessed gate and heterojunction gate. However, all these methods essentially require a reduction in the number of channel carriers, which inevitably leads to a significant increase in on-resistance and makes it difficult to reach the theoretical limit of material performance. In this work, a metal-oxide-β-Ga$_2$O$_3$/NiO$_X$ heterojunction gate with fin shaped gate field-effect transistor (MOHJ-FinFET) is proposed, which effectively increases the output current of the device while maintaining enhancement-mode performance. The breakdown voltage (BV) of the device is also improved. The four devices with different structures were simulated by Sentaurus TCAD simulation software to investigate the physical mechanisms that reduce the on-resistance (R$_{on}$) and increase the BV.

II. DEVICE DESIGN AND ANALYSIS OF RESULTS

In the channel of the actual device, Si as a doped impurity, the energy level is not shallow enough to be fully ionized, and the actual number of ionized impurities affects the electron mobility due to Coulomb scattering. To make the simulation result more close to the actual situation, the PhuMob mobility model with incomplete ionization model is used, and the parameters of the model are exactly the same as in the reference [3]. In addition, Fermi-Dirac statistics, high field saturation of mobility, SRH, Impaction model are used in simulation of the forward and reverse properties. The other simulation parameters of β-Ga$_2$O$_3$ and NiO$_X$ in the simulation refer to the literature [1] [4]. The model is used to build a device with exactly the same structure as in [5]. The simulated transfer characteristics are compared with the actual reported results as shown in Fig.1, which verifies the correctness of the model.

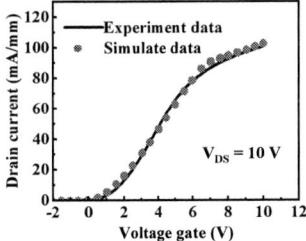

Fig.1. Comparison of simulation results with actual reported results [5].

In order to verify the excellent performance of MOHJ-FinFET, we have designed planar devices, metal-oxide-β-Ga$_2$O$_3$/NiO$_X$ heterojunction gate with recessed gate (MOHJ-FET) and metal-β-Ga$_2$O$_3$/NiO$_X$ heterojunction gate with recessed gate (MHJ-FET). As well as MOHJ-FinFET and metal -β-Ga$_2$O$_3$/NiO$_X$ heterojunction gate with fin shaped gate (MHJ-FinFET). Fabrication of devices can start from isolation, the doping concentration of Ga$_2$O$_3$ channel is 1×10^{17} cm^{-3} followed by photolithography to transfer the fin pattern to the epitaxial wafer, and then inductively coupled plasma (ICP) etching for 300 nm to the unintentional doping (UID) layer, the devices adopt a slanted fin structure to effectively increase the breakdown voltage of the devices [6], with an average width (w$_{fin}$) of 100 nm and a length of 3 μm. For the recessed devices, the etching depth is 250 nm and retained 50 nm thickness channel (t$_{ch}$) (Since the depletion effect of Fin shaped gate comes mainly from both sides, setting w$_{fin}$ = 2·t$_{ch}$ makes the depletion similar for all devices), and the length of recess is also 3 μm. The ohmic contacts are formed by growing the ohmic metal Ti/Au in the source-drain region and annealing. For MHJ-FinFET and MHJ-FET, the gate is formed by sputtering NiO$_X$ with doping concentration of $1 \times$

979-8-3503-6184-1/24 $31.00 © 2024 IEEE

10^{17} cm^{-3} and depositing Ni/Au, while for MOHJ-FinFET and MOHJ-FET, the devices can be fabricated by depositing SiO$_2$ through ALD after sputtering of NiO$_X$, and then finally depositing the gate metal. The length of gate (L$_g$), NiO$_X$ and SiO$_2$ is 1.5 µm, the thickness of SiO$_2$ and NiO$_x$ is 50 nm, the length from gate to drain (L$_{gd}$) is 5 µm, and the length from gate to source (L$_{gs}$) is 2.5 µm, as shown in Fig. 2.

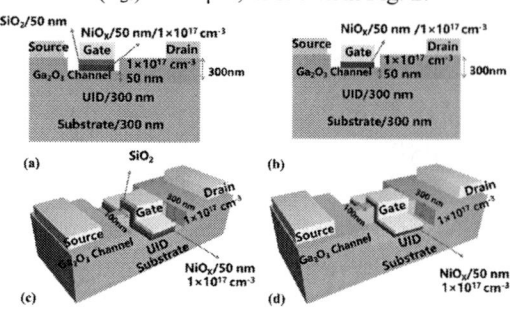

Fig. 2. Structural schematic of (a) MOHJ-FET (b) MHJ-FET (c) MOHJ-FinFET (d) MHJ-FinFET structure.

The transfer characteristics of the four devices are shown in Fig. 3(a), and the threshold voltage (V$_{TH}$) is defined as the gate voltage when the drain current reaches 1 mA/mm. The corresponding V$_{TH}$ of MOHJ-FET, MHJ-FET, MOHJ-FET and MHJ-FinFET are 0.28 V, 3.08 V, 1.55 V and 3.58 V, respectively. For the planar devices MOHJ-FET and MHJ-FET, the enhancement-mode can be attributed to the depletion effect of β-Ga$_2$O$_3$/NiO$_x$ pn heterojunction, the carriers in the channel, which are only 50 nm after etching, are completely depleted. For fin shaped devices MOHJ-FinFET and MHJ-FinFET, the channel thickness is 300 nm, while the w$_{fin}$ 100 nm, and the devices also realize enhancement-mode under the combined effect of heterojunction depletion on both sidewalls. The V$_{TH}$ of the fin shaped devices is larger than that of the planar devices, which is related to the depletion effect of the Fin gate in the three-dimensional direction, and the mode of depletion effect of the two structures of the MHJ-FET and the MHJ-FinFET, for example, is shown in Fig. 3(b). At the same applied gate voltage, the fin shaped gate has a top-down depletion effect in addition to the side-to-inside depletion effect, whereas the conventional structure has only a top-down depletion effect. Therefore, due to the stronger depletion effect of the fin shaped structure, the fin shaped structure depletes a larger proportion of the area at the same gate voltage, which corresponds to a higher gate voltage required to reach the V$_{TH}$ point (when the depletion area is exactly equal to the channel area), and thus has a higher V$_{TH}$.

For heterojunction structure, the depletion effect can be described as eq. (1). Where N$_d$ and N$_a$ are the doping concentrations in β-Ga$_2$O$_3$ and NiO$_x$ respectively, V$_{bi}$ is the built-in potential difference of the PN junction, and ε$_n$ and ε$_p$ are the dielectric constants of β-Ga$_2$O$_3$ and NiO$_x$ respectively. If the width of depletion region W$_n$ is exactly equal to the thickness of the channel, it can be assumed that the V$_{PN}$ at this point is exactly equal to the V$_{TH}$ of the device. As for the device with oxide layer, the metal-oxide-semiconductor structure also forms a barrier, and its energy band structure is shown in Fig.3(c). Here, Φ$_m$ is the workfunction of the metal, χ is the affinity of electron, Φ$_m'$ is the modified workfunction of the metal, χ' is the modified affinity, Φ$_{s0}$ is the surface potential, Φ$_{fp}$ is the barrier height between the intrinsic Fermi level E$_{Fi}$ and the Fermi level E$_F$. The potential barrier through the oxide layer V$_{oxo}$ satisfies the eq. (2).

$$W_n = \left[\frac{2\varepsilon_n \varepsilon_p N_a (V_{bi} - V_{PN})}{eN_d(\varepsilon_n N_d + \varepsilon_p N_a)} \right]^{\frac{1}{2}} \quad (1)$$

$$e\phi_m' + eV_{oX0} = e\chi' + \frac{E_g}{2} - e\phi_{s0} + e\phi_{fp} \quad (2)$$

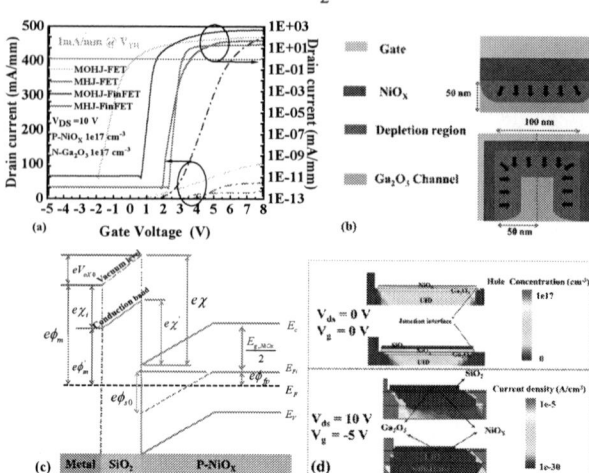

Fig.3(a) Transfer characteristic curves (where the solid line indicates logarithmic coordinates and the dashed line indicates linear coordinates); (b) Depletion effects of fin shaped gate and planar gate; (c) Energy band diagram of Gate-oxide-NiO$_X$ structure; (d) Hole concentration distributions at 0 bias and current distributions in the off-state of MOHJ-FET and MHJ-FET.

It can be seen that for a simple heterojunction structure: V$_{TH}$=V$_{PN}$. While for a heterojunction gate with an oxide layer, part of the voltage applied to the gate will drop to the oxide layer, satisfying the expression: V$_{TH}$=V$_{PN}$+V$_{oxo}$, so the V$_{TH}$ of devices with oxide layer should be higher than that of devices without oxide layer. However, the simulation results show that the devices with oxide layer have lower V$_{TH}$, which can be attributed to the gate-oxide-semiconductor structure forming a depletion region in the NiO$_x$, which makes the effective hole concentration of NiO$_x$ less than 1×10^{17} cm^{-3} as shown in Fig.3(d). Therefore, the V$_{TH}$ of the device with oxide layer is the result of the combination of the above two conditions, showing a weaker depletion effect than that of the simple heterojunction structure. Therefore, in Fig. 3(d), under the same reverse bias condition, the current path of the MHJ-FET is completely switched off, whereas the MOHJ-FinFET still has a very small amount of leakage current path in the substrate, leading to a relatively high off-state current.

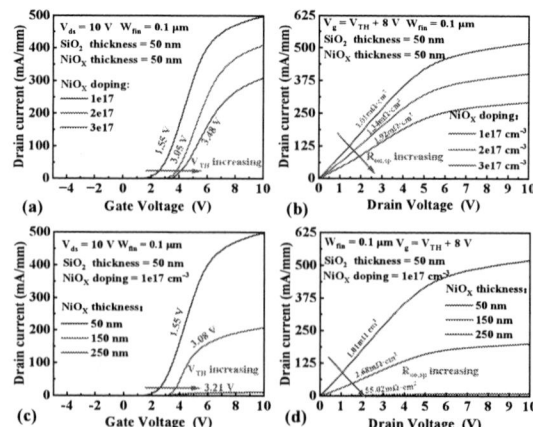

Fig. 4 Transfer curves (a) and output curves (b) of MOHJ-FinFET with different NiO$_x$ concentrations; Transfer curves (c) and output curves (d) of MOHJ-FinFET with different NiO$_x$ thicknesses.

979-8-3503-6184-1/24 $31.00 © 2024 IEEE

Depletion of holes in NiO_x can be attenuated by increasing the NiO_x thickness or NiO_x concentration and thus increases the V_{TH} of the device, but this results in an increase in R_{on}, as shown in Fig.4, which requires further research.

In addition, as can be seen in Fig. 5(a), the device without oxide layer forms current under the positive gate voltage, which is similar to the forward conduction process of a PN junction diode. As a result, a large gate current is generated at the gate, leading to gate breakdown, which complicates circuit design by limiting the gate voltage to a certain range [7]. While the SiO_2 oxide layer due to the large bandgap, so that a large conduction band offset is formed between NiO_x and SiO_2, electrons in the NiO_x are difficult to cross the barrier to the gate to form a leakage current. The maximum output currents of the four devices, MOHJ-FET, MHJ-FET, MOHJ-FinFET and MHJ-FinET, are 192 mA/mm, 35 mA/mm, 520 mA/mm and 70 mA/mm, respectively, as shown in Fig.5(b). Among them, the MHJ-FET and MHJ-FinFET have leakage currents due to the conduction of the PN junction caused by a large positive bias on the gate, and thus I_d has a negative value when V_{ds} is small [8]. The R_{on} of these four devices is 22.71 $\Omega \cdot mm$, 265.6 $\Omega \cdot mm$, 11.22 $\Omega \cdot mm$ and 47.8 $\Omega \cdot mm$, respectively, corresponding to special on-resistance $R_{ON,SP}$ = $R_{on} \times (L_{gs}+L_{gd}+L_g)$ of 2.04 $m\Omega \cdot cm^2$, 23.1 $m\Omega \cdot cm^2$, 1.01 $m\Omega \cdot cm^2$ and 4.3 $m\Omega \cdot cm^2$. The oxide layer significantly decreases gate leakage current, resulting in a device with desirable output characteristics. The combination of fin shaped gate and heterojunction gate with respect to the planar structure achieves an increase in current while maintaining a positive V_{TH}.

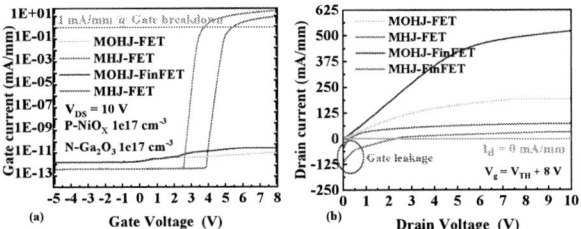

Fig. 5. (a) Ig-V_g curves in logarithmic coordinates for four devices; (b) Output curves for four devices.

In addition, the breakdown voltages of the four devices, MOHJ-FET, MHJ-FET, MOHJ-FinFET and MHJ-FinFET, are 2130 V, 1825 V, 2358 V and 2231 V, respectively, as shown in Fig.6(a). The fin shaped devices exhibit higher breakdown voltages compared to the planar devices, and the SiO_2 layer further enhances the breakdown voltage of the devices. The electric field crowding effect of the fin shaped structure is at the edge of the gate near the drain and the peak field strength occurs in the centre of the gate, so the peak field strength for the fin shaped device can be extracted as shown in the Fig.6(b), whereas for the planar device it can be extracted under the NiO_x. From Fig.6(b) it can be seen that the combination of metal-oxide-β-Ga_2O_3/NiO_x heterojunction gate and fin shaped gate significantly reduces the peak field strength of the device and has better breakdown characteristics. The corresponding BFOM of the four devices, MOHJ-FET, MHJ-FET, MOHJ-FinFET and MHJ-FinET, are 2.22GW/cm^2, 0.14 GW/cm^2, 5.51 GW/cm^2 and 1.15 GW/cm^2, respectively. The proposed MOHJ-FinFET structure improves the mutual constraints between R_{on} and BV of the device, and is able to combine ultra-high BV with low R_{on}, and the very high BFOM is closer to the theory of β-Ga_2O_3 material itself.

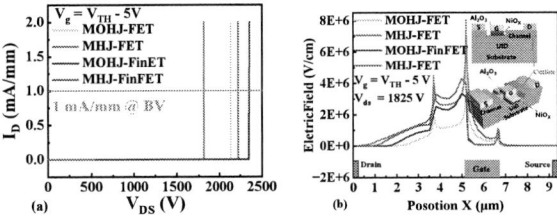

Fig. 6. (a)Breakdown characteristic curves of the four devices; (b)Electric field distribution curves of the four devices.

III. CONCLUSION

In conclusion, we propose a new β-Ga_2O_3-based field effect enhancement-mode transistor combining metal-oxide-β-Ga_2O_3/NiO_x heterojunction gate and fin shaped gate, and the TCAD simulation results show that the combination of the fin shaped structure and the heterojunction structure improves the output characteristics of the device while maintaining the enhancement-mode, and the gate oxide layer effectively suppresses the gate breakdown under forward bias. The physical mechanisms that reduce the on-resistance and increase the breakdown voltage is also analyzed. This study provides important guidance for the design of β-Ga_2O_3-based enhancement-mode devices with high BFOM in the future.

ACKNOWLEDGMENT

This work was supported by the fundamental research funds for the central universities of China under Grant No. ZDRC2002, the National Innovation Center of Radiation Application under Grant No. KFZC2022020401, and the Fundamental Research Funds for the Central Universities No. YJSJ24020.

REFERENCES

[1] Higashiwaki M, Kuramata A, Murakami H, et al. State-of-the-art technologies of gallium oxide power devices[J]. Journal of Physics D: Applied Physics, 2017, 50(33): 333002.

[2] Zhang J, Dong P, Dang K, et al. Ultra-wide bandgap semiconductor Ga_2O_3 power diodes[J]. Nature communications, 2022, 13(1): 3900.

[3] Wong H Y. TCAD Simulation Models, Parameters, and Methodologies for β-Ga_2O_3 Power Devices[J]. ECS Journal of Solid State Science and Technology, 2023, 12(5): 055002.

[4] He J, Liao F, Zhu K, et al. Design of a 10 kV and 16.5 GW cm^{-2} NiO/β-Ga_2O_3 heterojunction diode on a complete wafer with a positive beveled-mesa[J]. ECS Journal of Solid State Science and Technology, 2023, 12(1): 015001.

[5] Wang X, Lu X, He Y, et al. An E-mode β-Ga_2O_3 metal-heterojunction composite field effect transistor with a record high P-FOM of 0.73 GW/cm^2[C]//2023 35th International Symposium on Power Semiconductor Devices and ICs (ISPSD). IEEE, 2023: 390-393.

[6] Liu H, Li J, Lv Y, et al. Improved electrical performance of lateral β-Ga_2O_3 MOSFETs utilizing slanted fin channel structure[J]. Applied Physics Letters, 2022, 121(20).

[7] Wang C, Yan Q, Su C, et al. Demonstration of the β-Ga_2O_3 MOS-JFETs With Suppressed Gate Leakage Current and Large Gate Swing[J]. IEEE Electron Device Letters, 2023, 44(3): 380-383.

[8] W. Lei et al., "Proposal and Simulation of Ga_2O_3 MOSFET With PN Heterojunction Structure for High-Performance E-Mode Operation," in IEEE Transactions on Electron Devices, vol. 69, no. 7, pp. 3617-3622, July 2022, doi: 10.1109/TED.2022.3172919.

Electrical Characteristics and Thermal Reliability Investigation of TreeFET, FishboneFET, CombLikeFET and NSFET

Mingyu Ma [1], Wenbin Wang[1], Haokun Li[1], Shujun Gao[1], Hailong You[1], Cong Li *[1]

[1] Xidian University. Taibai No2 Road Xi'an Shann'Xi China

Email: m593247311@gmail.com, licong@xidian.edu.cn

Abstract—**This work compares the electrical characteristics and thermal reliability of various TreeLikeFET(TreeFET, FishboneFET, CombLikeFET) and Nano-SheetFET(NSFET) devices at advanced technology node. In terms of DC characteristics, although various TreeLikeFET achieve higher I_{ON}, NSFET often have smaller I_{OFF} and DIBL. In terms of Thermal Reliability, the equivalent thermal resistance of various TreeLikeFET is significantly reduced compared to NSFET due to the presence of heat dissipation channels. In addition, the different geometric structures (whether there is an NS-Bulk channel and the shape of the overlap region) result in significant differences in the electrical characteristics and self-heating performance of TreeLikeFET devices.**

Keywords—TreeLikeFET, NSFET, DC characteristics, Thermal Reliability

I. INTRODUCTION

As the semiconductor process node develops below 5nm, continuing to scale down the size of FinFET structure faces increasingly challenges, such as increasingly complex process flows and higher process variation sensitivity. However, unlike FinFET, the main channel surface of NSFET is {100}, which exhibits a significant difference in electron mobility and hole mobility. Consequently, this leads to severe current mismatch between NFET and PFET[1]. To solve this issue, new device structures that combine channels of different surface orientation have been proposed. These devices that combine channels of different surface orientation include TreeFET[2], FishboneFET[3], CombLikeFET[4], etc., which are collectively referred to as TreeLikeFET[3].

In addition to improving the current matching between NFET and PFET, the existence of vertical channels in TreeLikeFET greatly alleviates the geometric confinement and thus mitigates the self-heating effect[5]. Except for TreeFET, the self-heating effects of FishboneFET and CombLikeFET, which have NS-Bulk channel, are more valuable for research and analysis. However, there are few papers on the self-heating effect and thermal reliability of TreelikeFET devices currently.

For TreeLikeFET devices, there is an area where the horizontal and vertical channels overlap, and the gate has weaker control over this area. Due to the different geometric characteristics of the overlapping area of different TreeLikeFET structures, there is a significant variation in carrier distribution, which leads to a significant impact on the output current of the device (on current and leakage current) [2,3,4].

II. DEVICE STRUCTURE AND SIMULATION CONDITION

Fig. 1.(a) shows various structures compared in this work under the 3nm node, including TreeFET, FishboneFET, CombLikeFET, 3-NSFET and 4-NSFET. And Fig. 1.(b) shows the channel profile and doping of the devices. As can been seen from the images, in this work, TreeFET consists of three NS and two Inter-bridge (IB) channel that connect adjacent NS located in the middle of NS. Compared with TreeFET, FishboneFET has an additional NS-Bulk Channel in the middle of NS. And CombLikeFET has a vertical channel from the top NS to the Bulk substrate on one side of the NS. The main parameters related to size and doping in this work are based on the 2020 International Roadmap for Devices and Systems (IRDSs)[6], and the specific parameters are shown in Table I.

Simulation is performed in Sentaurus TCAD. Quantization confinement is considered by adopting density gradient quantization model. Simulation uses the inversion and accumulation layer mobility model to simulate the mobility degradation caused by impurity, phonon, and surface roughness scattering. SRH recombination, Auger recombination, Philips unified mobility model are also adopted. Besides, the channel crystal orientation is precisely set and the anisotropy of heat propagation is considered. In order to compare the self-heating effects of NSFET and TreeLikeFET, the thermal conductivity of different regions are set according to literature reference[5], as shown in Table II. To ensure the accuracy of TCAD simulation, the physical model and material parameters were calibrated using 3-NSFET experimental data, as shown in Fig. 2.

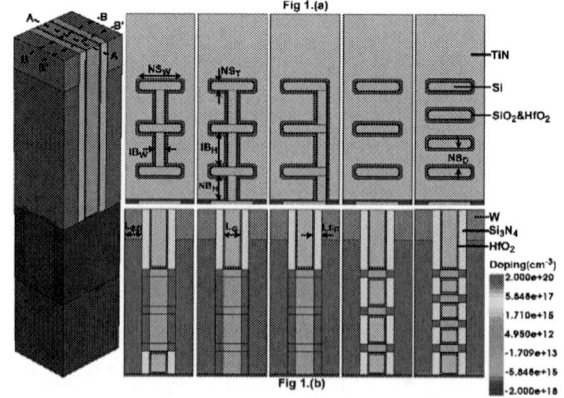

Fig. 1. (a)Structure of TreeFET, FishboneFET, CombLikeFET, 3-NSFET and 4-NSFET channel(Cutline A-A). (b)Doping of TreeFET, FishboneFET, CombLikeFET, 3-NSFET and 4- NSFET(Cutline B-B for TreeFET, FishboneFET, 3-NSFET, 4- NSFET and B'-B` for CombLikeFET).

TABLE I.   DEVICE PHYSICS PARAMETER

Parameter	Symbol	Value
Gate Length	L_g	12 nm
Spacer Length	L_{sp}	5 nm
S/D Length	L_{sd}	12 nm
Nano-Sheet Width	NS_W	25 nm
Nano-Sheet Thickness	NS_T	5 nm
Inter-Bridge Width	IB_W	5 nm
Inter-Bridge Height	IB_H	20 nm
NS-Bulk Channel Height	IB_{Bot}	15 nm
NS Distance(For NSFET)	NS_D	11.7 nm
Equivalent Oxide Thickness	E_{OT}	0.7 nm
Channel Doping	N_{Ch}	1e16 cm^{-3}
S/D Doping	N_{SD}	2e20 cm^{-3}
Substrate Doping	N_{Sub}	2e18 cm^{-3}
Work Function	W_F	4.50 eV

TABLE II.   THERMAL CONDUCTIVITY OF VARIOUS REGIONS

Region	Material	Thermal Conductivity $(W \cdot K^{-1} \cdot m^{-1})$
Channel	Si	8.3
Source/Drain	Si	16
Substrate	Si	148
Oxide	SiO$_2$	1.4
Oxide	HfO$_2$	2.3
Spacer	Si$_3$N$_4$	18.5
Gate Metal	TiN	19.2
Source/Drain Metal	W	170

III. COMPARISON BETWEEN VARIOUS STRUCTURES

For a direct and accurate comparison of the performance between NSFET and TreeLikeFET, in this section, the distance between the NSs in 4-NSFET has also been finely adjusted to ensure the same device height.

A. Electrical characteristics

Fig. 3 shows the I_D-V_G characteristics curves of various structures at V_{DS}=0.7V and V_{DS}=0.05V. The detailed information of on-current(I_{ON}), leakage current(I_{OFF}) and Drain Induced Barrier Lowering(DIBL) for these curves is shown in Table III. Here, the I_{ON} is obtained at V_{GS}=0.7V and V_{DS}=0.7V, the I_{OFF} is obtained at V_{GS}=0V and V_{DS}=0.7V. And DIBL is calculated using the following formula:

$$DIBL = (V_{tlin} - V_{tsat}) / (V_{Dlin} - V_{Dsat}) \qquad (1)$$

Here, V_{tsat} is the threshold voltage calculated at saturation apply voltage, V_{Dsat}=0.7V and V_{tlin} is the threshold voltage calculated at linear apply voltage, V_{Dlin}=0.05V.

As can be seen from Fig. 3 and Table III, the TreeLikeFET exhibits higher I_{ON} compared to NSFET. Compared with 4-NSFET, TreeFET, FishboneFET and CombLikeFET have 2.4%, 4.3% and 12.9% increase in I_{ON}. However, due to the better gate control ability, 3-NSFET and 4-NSFET exhibit significantly lower I_{OFF} and DIBL compared to TreeLikeFET. Therefore, 3-NSFET and 4-NSFET have higher I_{ON}/I_{OFF} ratio.

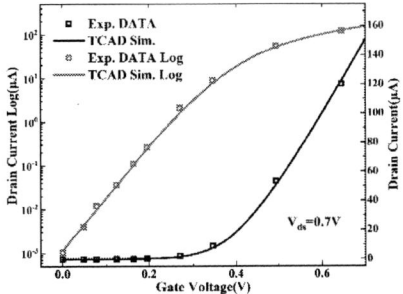

Fig. 2.   TCAD calibration results (line) and the experimental data (symbols) for the 3Stack-NSFET.

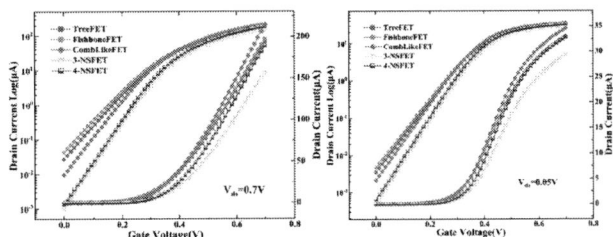

Fig. 3. I_D -V_G curves of different device structures at (a)V_{DS} = 0.7V and (b)V_{DS} = 0.05V.

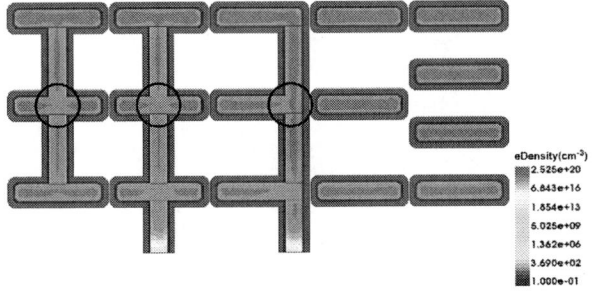

Fig. 4.   eDensity of different device structures at V_{DS} = 0.7V.

TABLE III.   DC CHARACTERISTICS OF DIFFERENT STRUCTURES

Structure	Ion(μA)	Ioff(μA)	Ion/Ioff	DIBL(mV)
TreeFET	193.5	0.0280	6.92×10^3	146.7
FishboneFET	197.0	0.0435	4.52×10^3	163.1
CombLikeFET	213.2	0.00970	2.20×10^4	108.2
3-NSFET	156.4	0.00101	1.54×10^5	67.7
4-NSFET	188.9	0.00135	1.40×10^5	69.4

It should be noticed that CombLikeFET has the largest I_{ON}, the smallest I_{OFF} and the smallest DIBL among various TreeLikeFET devices, which is caused by the different structure at the overlap of horizontal and vertical channels, as shown in Fig. 4. Because the channel overlap of CombLikeFET is a T-shaped structure, it has stronger channel control ability than TreeFET and FishboneFET, so it has higher current density in the channel overlap region and thus obtains higher I_{ON}. It can be expected that the stronger channel control ability of CombLikeFET can also reduce I_{OFF} and DIBL. In addition, the different distribution of horizontal and vertical resistance may also be the reason for the significant change of electrical characteristics of CombLikeFET compared with TreeFET and FishboneFET.

979-8-3503-6184-1/24 $31.00 © 2024 IEEE

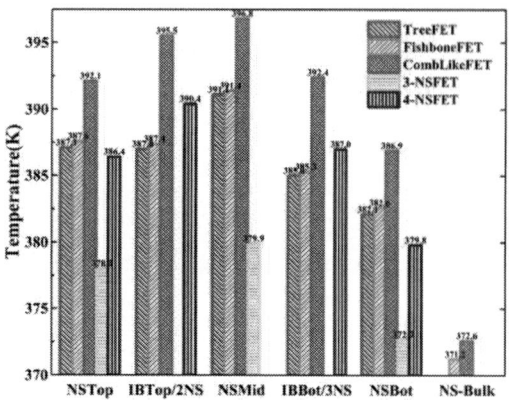

Fig. 5. Temperatures in various regions of different devices.

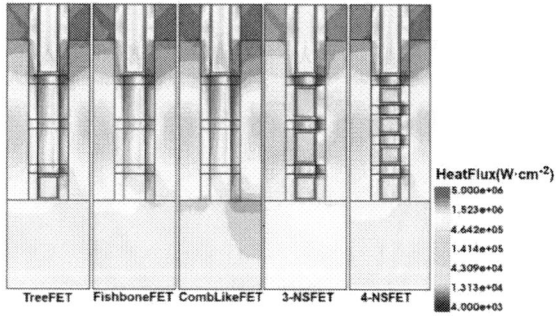

Fig. 6. HeatFlux of different device structures at $V_{DS} = 0.7$V

TABLE IV. EQUIVALENT THERMAL RESISTANCE OF DIFFERENT STRUCTURES

Structure	Equivalent Thermal Resistance $(K \cdot \mu A^{-1} \cdot V^{-1})$
TreeFET	2.888
FishboneFET	2.838
CombLikeFET	2.660
3-NSFET	3.470
4-NSFET	2.952

B. Thermal reliability

When $V_{DS} = V_{GS} = 0.7$V, the temperature of different regions of the channel and equivalent thermal resistance are shown in Fig. 5 and Table IV. The equivalent thermal resistance R_{th} is calculated by the following formula:

$$R_{th} = T_{MAX} / P = T_{MAX} / (I_{on} \cdot V_{DS}) \qquad (2)$$

As can be seen from Fig. 5 and Table IV, 3-NSFET has the lowest lattice temperature, but it's extremely low I_{ON} results in the highest equivalent thermal resistance. Due to the same effective channel length, TreeFET and 4-NSFET exhibit similar I_{ON} and lattice temperature. However, because the vertical channel in TreeFET can efficiently transfer heat, TreeFET has a more uniform temperature distribution, resulting in a lower equivalent thermal resistance. For FishboneFET and CombLikeFET, although higher I_{ON} leads to higher lattice temperature, due to the existence of heat dissipation channel from channel to substrate, their equivalent thermal resistance is significantly reduced, compared with

NSFET, by 3.9% and 9.9%. It should be noted that the equivalent thermal resistance of CombLikeFET is significantly lower than that of TreeFET and FishboneFET. The cause of this phenomenon may be related to differences in device structure. Considering that the temperature at the NS Channel edge is lower when the device is operating, the vertical channels (IB Channel and NS-Bulk Channel) located at the NS edge in the Comb-LikeFET will be more conducive to heat dissipation, thereby reducing thermal resistance. The heat transfer scenarios of different device structures are illustrated in Fig. 6.

IV. CONCLUSION

This article compares the performance of various TreeLikeFET and NSFET devices at advanced technology nodes, and explores the impact of several parameters on device performance. In terms of electrical characteristics, compared to 4-NSFET, TreeFET, FishboneFET, and CombLikeFET exhibit I_{ON} enhancements of 2.4%, 4.3%, and 12.9%, respectively. When compared to 3-NSFET, the enhancement reaches over 23.7%. However, for 3-NSFET and 4-NSFET, their stronger gate control capability results in lower I_{OFF} and DIBL. In terms of self-heating effects, the low I_{ON} of 3-NSFET results in an equivalent thermal resistance significantly higher than that of other structures. However, for TreeLikeFET, due to the efficient heat transfer capability of the vertical channel, TreeFET, FishboneFET, and CombLikeFET exhibit thermal resistance reductions of 2.2%, 3.9%, and 9.9% compared to 4-NSFET.

ACKNOWLEDGMENT *(Heading 5)*

Project Supported by the National Key Research and Development Program of China: Research on Industrial Analog Chip Designs and Process Compatibility and Standards for Reliability Technologies (grant number: 2022YFF0605800), and partly supported by the Science and Technology Development Program of Shaanxi (Grant No.2023-YBGY-273).

REFERENCES

[1] Fischetti, M.V., Ren, Z., Solomon, P.M., Yang, M., Rim, K., 2003. Six-band k·p calculation of the hole mobility in silicon inversion layers: Dependence on surface orientation, strain, and silicon thickness. Journal of Applied Physics 94, 1079－1095.

[2] Ye, H.Y., Liu, C.W., 2020. On-current enhancement in treefet by combining vertically stacked nanosheets and interbridges. IEEE Electron Device Letters 41, 1292–1295.I. S. Jacobs and C. P. Bean, "Fine particles, thin films and exchange anisotropy," in Magnetism, vol. III, G. T. Rado and H. Suhl, Eds. New York: Academic, 1963, pp. 271–350.

[3] Cao, L., Zhang, Q., Yao, J., Li, J., Liu, Y., Luo, Y., Kong, Z., Zhou, N., Gao, J., Lu, Y., et al., 2023. Investigation of fabricated cmos fishbonefets and treefets with strained sige nano-fins on bulk-si substrate. IEEE Electron Device Letters 44, 1396–1399.R. Nicole, "Title of paper with only first word capitalized," J. Name Stand. Abbrev., in press.

[4] Li, X., Zhu, H., Gan, W., Huang, W., Wu, Z., 2022. A three-dimensional simulation study of the novel comb-like-channel field-effect transistors for the 5-nm technology node and beyond. IEEE Transactions on Electron Devices 69, 4786–4790.

[5] Tsen, C.J., Chung, C.C., Liu, C.W., 2022. Self-heating mitigation of treefets by interbridges. IEEE Transactions on Electron Devices 69, 4123–4128.

[6] (2020). International roadmap for devices and systems more moore white paper. https://irds.ieee.org/editions/2020.

979-8-3503-6184-1/24 $31.00 © 2024 IEEE

A Fast-Response Current Source with High-Impedance for Zero-Crossing-Based Circuits

Ruoyu Li [1], Xianglong Wang [1], Jianqiang Xu [1], Yintang Yang*[1]

[1] Faculty of Integrated Circuit, Xidian University, Xi'an 710071, China

* Email: liruoyu@stu.xidian.edu.cn, ytyang@xidian.edu.cn

Abstract—In this paper, a high-impedance current source is developed to rapidly establish the output current. Under the framework of low bandwidth and large load capacitor, the output current can be quickly established by adding an auxiliary loop structure. Compared with the conventional current source circuit, the developed circuit reduces the design complexity and avoids the nonlinearity in zero-crossing-based pipelined Analog-to-Digital Converters (ADCs). Based on the TSMC 65 nm CMOS technology, the simulation results show that the established time is 974.2 ps with the output current of 608.8 μA. Therefore, the faster zero crossing detection can be achieved by the developed circuit, which can be applied in the high-speed and high-resolution pipelined ADCs.

Keywords—analog-to-digital converter, current source, rapidly establish, zero crossing detection

I. INTRODUCTION

With the development of CMOS process feature size, intrinsic device impedance and supply voltage, it becomes more and more difficult to design high-performance operational amplifiers based on the conventional switched capacitor circuits. To overcome this difficulty, Fiorenza et al. [1] proposed comparator-based switched-capacitor (CBSC), using a comparator instead of an operational amplifier in conventional switched capacitor circuits, which can be applied in pipelined ADCs to improve its power efficiency. In addition, Brooks et al. [2] extended the CBSC to zero-crossing-based circuits (ZCBC), which uses a dynamic zero-crossing detector (ZCD) to further improve power efficiency.

At transfer phase, the current source charges the load capacitor. When ZCD detects the virtual ground condition reached, the current source turns off. The common turn-off mechanism of current source is divided into gate-switching structure and drain-switching structure [1,3]. In the gate-switching structure, the large parasitic capacitance of the gate limits speed, especially in the case of high-speed clock switching. The nonlinear on-resistance of the switch, charge injection and clock feedthrough effects introduce nonlinearities in drain-switching structure. Meanwhile, the limited output impedance of the current source and the delay of the ZCD also introduce the nonlinear overshoot voltage into the residual voltage, affecting the resolution. In order to reduce the influence of limited output impedance of current source, Wilson current mirror (CM) [4], improved Wilson CM [5] and cascode CM structures [6] are often used in ZCBC. Due to larger load capacitance and shorter transfer time, the output impedance of those circuits for high-resolution zero-crossing-based Pipelined ADCs is too small. Therefore, it is valuable to improve the output impedance and turn-off mechanism of current source.

In this paper, a current source circuit providing high output impedance with new turn-off mechanism is proposed. By adding an auxiliary loop, the voltage of each node of the current source can be quickly established to improve the building speed. Section II shows the proposed current source circuit structure and the simulation results are presented in Section III. Section IV presents the conclusion.

II. CURRENT SOURCE CIRCUIT MODEL

The schematic of ZCBC is shown in Fig. 1. During the sampling phase, φ_S is high and the input voltage is sampled onto capacitor C_1 and C_2. During the transfer phase, a short pulse φ_R is used to pull output voltage to the ground firstly. Then φ_A is high and the controlled current source charges C_L, generating a voltage ramp on the output voltage VOUT. This causes V_X to ramp with it via the capacitor divider. When ZCD detects the virtual ground condition reached, φ_1 becomes low and the current source is disconnected with C_L.

The schematic of the developed current source is shown in Fig. 2. An auxiliary loop consisting of M4, M6, M7 and M2 is added. Its small signal circuit schematic of the current source is shown in Fig. 3. The KCL equation is

$$[(V_1 g_{m4} r_{o4} + V_1)g_{m6} r_{o6} + V1]g_{m5} = \frac{V_X + V_1}{r_{o5}} = \frac{-V_1}{r_{o2}} \quad (1)$$

$$\frac{-V_1}{r_{o2}} = I_X \quad (2)$$

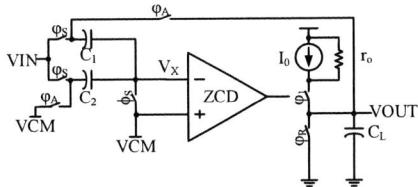

Fig. 1. Schematic of the ZCBC

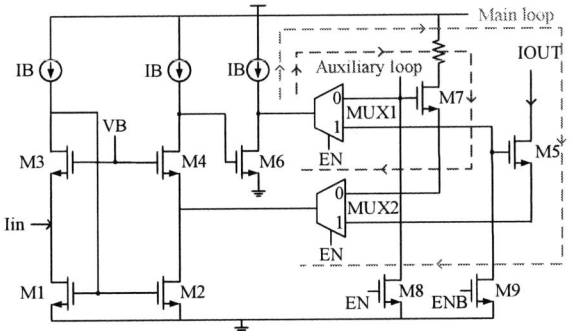

Fig. 2. Schematic of the developed fast-response high-impedance current source circuit.

979-8-3503-6184-1/24 $31.00 © 2024 IEEE

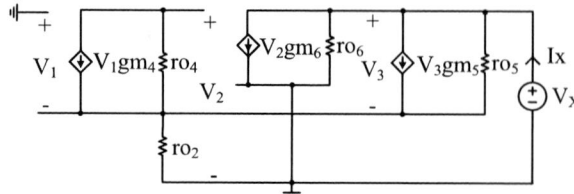

Fig. 3. Small signal circuit schematic of high-impedance current source circuit.

Thus, the output impedance expression is

$$R_{out} = \frac{V_X}{I_X} \approx g_{m4}r_{o4}g_{m6}r_{o6}g_{m5}r_{o5}r_{o2} \tag{3}$$

The high-impedance current source can provide high output impedance increased by the intrinsic device gain of the cascode transistor and inverting transistor based on the cascode structure.

The working principle of the improved current source is as follows: At sampling phase, the current source is disconnected with load capacitor and the enable signal EN is low. The drain of M6 is connected to the gate of M7, and the source of M4 is connected to the source of M7. In this way, the auxiliary loop is in working state, and M5 is turned off by the pull-down transistor. When EN goes high, the current source is connected to ZCD at transfer phase. The drain of M6 is connected to the gate of M5 from the gate of M7 through the MUX1, while the source of M4 is connected to the source of M5 from the source of M7 through the MUX2. The current source works in main loop state instated of the auxiliary loop state. At the same time, and M7 is turned off by M8.

By setting the voltage of each node in the main loop as same as that in the auxiliary loop, when switching between the two loops, the output current can be established fast. The pipelined ADCs based on zero crossing detection technology can achieve higher speed and accuracy.

III. SIMULATION RESULT

In this research, the simulation model of the developed fast-response and high-impedance current source circuit is established based on the TSMC 65 nm CMOS technology and Virtuoso toolbox.

The power supply voltage is 1.2 V, and the changes of the output voltage and output current are shown in Fig. 4. Obviously, when the output voltage changes by 0.8 V, the change of output current is only 2.07 μA. Thus, the output impedance of the developed current source is larger than that of the conventional structure. In addition, the transient establishment times of output current with conventional gate switching structure and developed current source structure are shown in Fig. 5. It can be seen that the developed current source establishes 608.8 μA output current within 1 ns, while the conventional structure needs 255 ns. Therefore, the developed current source is much faster than the conventional structure.

Moreover, the performances of different current sources are presented in Table I. Due to the auxiliary loop, the power consumption of this developed current source is slightly higher than that of the conventional structure, but much less than that of the current source based on gate driver [7]. And the establishment time is improved significantly.

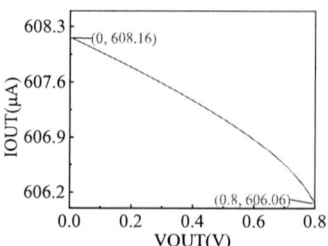

Fig. 4. Output current versus output voltage

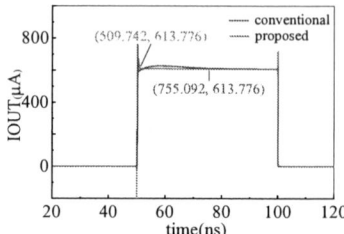

Fig. 5. Output current transient establishment waveform

TABLE I. PERFORMANCE OF DIFFERENT CURRENT SOURCE STRUCTURES

	Conventional	*[7]*	*This work*
Technology	65 nm CMOS	180 nm BCD	65 nm CMOS
Power supply (V)	1.2	5	1.2
Output current (μA)	608.8	5600	608.8
Establishment time (nA)	255	40.5	0.9742
Power consumption (μA)	477.8	4120	761.5

IV. CONCLUSION

This paper presents a fast-response and high-impedance current source, which can be applied in ZCBC. The auxiliary loop can hold the current source loop opening, so the output current can be established quickly. Meanwhile, the developed current source circuit can avoid the conventional turn-off mechanism of current source, reducing the non-ideal factor introduced by the switch. The output current of the developed current source only causes 2.09 μA change when the output voltage changes by 0.8 V. Compared with the conventional current source, the current source can build 608.8 μA current in 974.2 ps. Therefore, this design can realize the faster zero crossing detection, and it can be applied in high-speed and high-resolution zero-crossing-based pipelined ADCs

REFERENCES

[1] J. K. Fiorenza, T. Sepke, P. Holloway, C. G. Sodini, and H.-S. Lee, "Comparator-Based Switched-Capacitor Circuits for Scaled CMOS Technologies," IEEE Journal of Solid-State Circuits, vol. 41, no. 12, pp. 2658-2668, 2006.G.

[2] L. Brooks and H.-S. Lee, "A Zero-Crossing-Based 8-bit 200 MS/s Pipelinedd ADC," IEEE Journal of Solid-State Circuits, vol. 42, no. 12, pp. 2677-2687, 2007.

[3] S. Lee, A. P. Chandrakasan, and H.-S. Lee, "A 12 b 5-to-50 MS/s 0.5-to-1 V Voltage Scalable Zero-Crossing Based Pipelinedd ADC," IEEE Journal of Solid-State Circuits, vol. 47, no. 7, pp. 1603-1614, 2012.

[4] S. Lee, A. P. Chandrakasan, and H.-S. Lee, "A 12 b 5-to-50 MS/s 0.5-to-1 V Voltage Scalable Zero-Crossing Based Pipelinedd ADC," IEEE Journal of Solid-State Circuits, vol. 47, no. 7, pp. 1603-1614, 2012.

[5] B. A. Minch, "Low-Voltage Wilson Current Mirrors in CMOS," in 2007 IEEE International Symposium on Circuits and Systems (ISCAS), 2007, pp. 2220-2223

979-8-3503-6184-1/24 $31.00 © 2024 IEEE

[6] M. Gupta, B. Aggarwal, and A. K. Gupta, "A very high performance self-biased cascode current mirror for CMOS technology," Analog Integrated Circuits and Signal Processing, vol. 75, no. 1, pp. 67-74, 2013/04/01 2013.

[7] T. Zekorn, E. Wehr, K. Vohl, R. Wunderlich, and S. Heinen, "A Fast-Response Reference Current Source for High- and Low-Side High-Voltage Current Mirrors for Gate-Shaping Digital Gate Drivers," in 2023 18th Conference on Ph.D Research in Microelectronics and Electronics (PRIME), 2023, pp. 189-192.

A PPG Analog Front-End With PVT-Insensitive High-Pass Frequency

Zhaofeng Huang, Zepeng Huang, Hengchang Bi, Xing Wu, Liangjian Lyu*

In Situ Devices Center, School of Integrated Circuits, East China Normal University, Shanghai, 200241, China

*Email: ljlv@cee.ecnu.edu.cn

Abstract—This paper presents an analog front-end (AFE) circuit for photoplethysmography measurement that features a PVT-insensitive high-pass frequency. The DC photocurrent is cancelled out by a DC servo loop which consists of a very-large time constant integrator with a PVT-insensitive pseudo resistor and a resistive feedback current source. Designed in 65 nm CMOS, the proposed AFE consumes 20 µW from a 1.2 V power supply, and is capable of eliminating the DC current ranging from 100 nA to 10 µA while stabilizing the high-pass pole at around 0.34 Hz with a maximum variation of 11.7%.

Keywords—*photoplethysmography, analog front-end, PVT-insensitive, DC servo loop*

I. INTRODUCTION

Photoplethysmography (PPG) is a widely adopted medical method that uses light-emitting diodes (LEDs) and photodetectors (PDs) to measure light absorption in tissues such as bones and blood vessels. PPG captures the pulsation state of vessels, thereby measuring vital signs including heart rate (HR), blood pressure, and oxygen saturation (SpO_2) [1].

As depicted in Fig. 1(a), a typical PPG signal waveform comprises of DC and AC parts. The DC photocurrent generated by the reflected photoelectric signals of venous blood and other tissues accounts for over 90% of the overall PPG signal. While the AC photocurrent, generated by the pulsating components in the arteries, is much smaller in amplitude. Besides, the PPG signal generally occupies a frequency range of 0.5 Hz to 10 Hz [2], which implies a bandpass filter is necessary in the readout system.

Fig. 1(b) illustrates the PPG signal readout system which consists of a transmitter and a receiver. The light generated from the LED in the transmitter traverses or get reflected by human tissues, and is then captured by the photodiode (PD) in the receiver. The receiver further processes photocurrent signal by the subsequent analog front-end (AFE), ADC, and digital processor. The AFE consists of a trans-impedance amplifier (TIA) connected to the PD to convert the current signal into a voltage signal. An anti-aliasing filter then effectively filter out noise except the frequency range of the PPG signal, and drives the ADC. Finally, the digitized recorded signals are further processed.

This widely adopted architecture for the PPG signal readout system confronts two main challenges. Firstly, low power consumption is crucial for wearable systems. A duty-cycled LED driver is beneficial for significantly lowering the transmitter power. Methods based on compressive sensing have been proposed to further reduce the system power to reach the microwatt level [3]. Another challenge is the demand for a high dynamic range (HDR) in the readout circuit to simultaneously handle small AC signals along with large DC signals. A common approach is to employ high-resolution ADCs at the cost of higher power consumption. Therefore, an

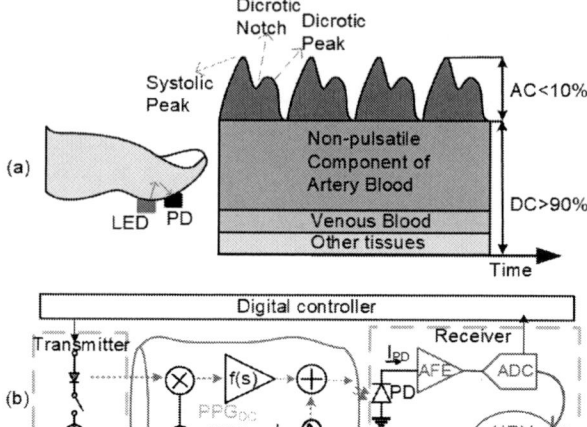

Fig. 1. (a) A typical PPG waveform detecting by an PD and an LED. (b) The block diagram of PPG signal monitoring system.

alternative method is the utilization of analog or digital DC servo loops (DSLs) to remove the DC component [4]-[5]. The digital DSL exhibits a precise transfer function and a fast-responding transient response, but tends to have poor noise performance due to its quantization noise [4]. The analog counterparts generally have better noise performance, but their bandwidth usually vary greatly with process, supply voltage, and temperature (PVT) variations [5]. Additionally, in implementations with transistor-based current feedback loops, the noise and bandwidth are affected by the DC current change in the PPG signal [5].

This paper presents a low-power HDR PPG readout circuit. Compared to conventional DSLs implementations, the proposed design exhibits improved robustness in high-pass frequency and noise performance across variations in PVT and DC input photocurrent amplitude. The rest of this paper is organized as follows: Section II provides an overview of the circuit structure, Section III presents the simulation results, and Section IV concludes the paper.

II. THE PROPOSED PPG AFE

Fig. 2 illustrates the overall architecture of the PPG signal processing system and its frequency response. The system incorporates a photodiode as the detector, a TIA with DSL to convert AC photocurrent to voltage and suppress DC parts, a low-pass filter to reject out-band noise, and a post-amplifier to provide extra gain and drives the following ADC.

A. The Comparison of MOS-DSL and R-DSL

In this paper, the RC-TIA architecture is adopted to convert the photo-current to voltage as shown in Fig. 3, where its gain

979-8-3503-6184-1/24 $31.00 © 2024 IEEE

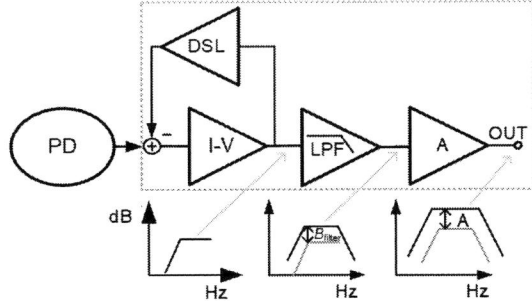

Fig. 2. The system block diagram of the PPG AFE.

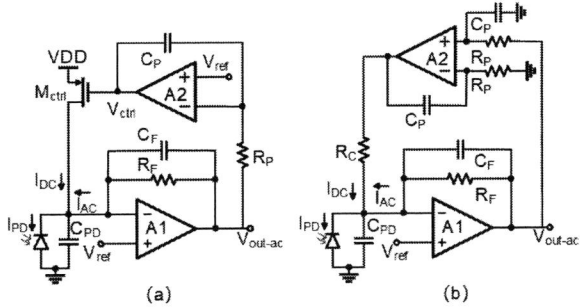

Fig. 3. Two types of the DSL structure. (a) MOS-DSL. (b) R-DSL.

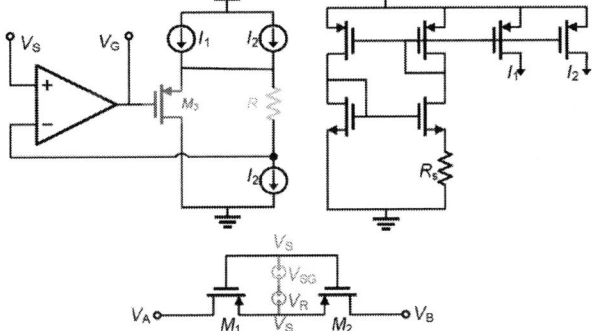

is set by the feedback resistor R_F. However, a significant DC component may cause output saturation. To address this, a MOS-DSL was proposed to eliminate the DC component as shown in Fig. 3(a) [5]. This solution incorporates a DSL into the TIA, consisting of an integrator and a transistor M_{ctrl}. The integrator added to the feedback loop introduces a low-pass cutoff frequency lower than 0.5 Hz. The transfer function of the entire system is described as follows:

$$Z_{TIA}(s) = -\frac{1 + sC_PR_PA_2}{1 + sC_PR_PA_2 + R_Fg_{m,ctrl}A_2} \cdot \frac{R_F}{1 + sR_FC_F}, \quad (1)$$

where C_P and R_P is the capacitances and resistor of the integrator, C_F is the compensation capacitor of TIA, $g_{m,ctrl}$ is the transconductance of M_{ctrl}, which varies with its DC current. (1) shows that the entire TIA with MOS-DSL is a high-pass system with one pole and one zero as shown in (2) and (3). The high-pass pole of the circuit changes greatly with the transconductance of the M_{ctrl}, and thereby greatly affected by the DC current of the TIA input.

$$\omega_z(s) = \frac{1}{C_PR_PA_2} \quad (2)$$

$$\omega_p(s) = \frac{1 + R_Fg_{m,ctrl}A_2}{C_PR_PA_2} \quad (3)$$

Neglecting the noise contribution from the integrator, the overall power spectral density (PSD) of the input-referred noise is expressed as follows:

$$\overline{I_n^2} = \overline{I_{PD}^2} + \frac{4kT}{R_F} + \overline{e_{n,A1}^2}\left[\frac{1}{R_F^2} + 4\pi f(C_F + C_{PD})\right] + \overline{I_{n,M_{ctrl}}^2} \quad (4)$$

where C_{PD} is capacitances of the PD, $\overline{e_{n,A1}^2}$ is the input referred noise of A1. $\overline{I_{n,M_{ctrl}}^2}$ is the noise of the current canceling transistor M_{ctrl} given by

$$\overline{I_{n,M_{ctrl}}^2} = \frac{8kTg_{m,ctrl}}{3} + \frac{2K_F}{C_{OX}WL} \cdot \frac{1}{f} \cdot g_{m,ctrl}^2, \quad (5)$$

where k is the Boltzmann constant, T is the temperature, $g_{m,ctrl}$ is the transconductance of the M_{ctrl}, K_F is the flicker noise coefficient, and C_{OX} is the oxide capacitance. According to (4) and (5), the input referred noise of the TIA also changes greatly with the gm of the MOS.

In this system, any changes in the DC input current will influence the transconductance of the transistor M_{ctrl}, thus

leading to the changes in both high-pass pole and noise. To address these issues, this paper adopts a resistive DSL (R-DSL). As depicted in Fig. 3(b), the feedback loop utilizes the resistor R_C to convert the non-inverting integrator output voltage to the DC cancellation current. Its transfer function is as follows:

$$Z_{TIA}(s) = -\frac{1 + sC_CR_PA_2}{1 + sC_CR_PA_2 + \dfrac{R_FA_2}{R_C}} \cdot \frac{R_F}{1 + sR_FC_F}. \quad (6)$$

The entire TIA is also a high-pass system with one pole and one zero as shown in (7) and (8).

$$w_z = \frac{1}{C_CR_PA_2} \quad (7)$$

$$w_p = \frac{1 + \dfrac{R_FA_2}{R_C}}{C_CR_PA_2} \quad (8)$$

The overall PSD of the input-referred noise of the TIA is expressed as follows:

$$\overline{I_n^2} = \overline{I_{PD}^2} + \frac{4kT}{R_F} + \overline{e_{n,A1}^2}\left[\frac{1}{R_F^2} + 4\pi f(C_f + C_{PD})\right] + \frac{4kT}{R_c}. \quad (9)$$

Comparing the two types of DSLs, it is apparent that the proposed R-DSL has a more robust high-pass poles and superior noise performance, which are independent of the DC input current.

B. PVT-Insensitive Very-Large Time Constant Integrator

Achieving a sub-hertz high-pass frequency requires a large time constant integrator in the DSL, which is usually implemented with TΩ-level pseudo-resistors and pF-level capacitors. However, conventional pseudo-resistors with large PVT variations are unable to provide robust operations. The

979-8-3503-6184-1/24 $31.00 © 2024 IEEE

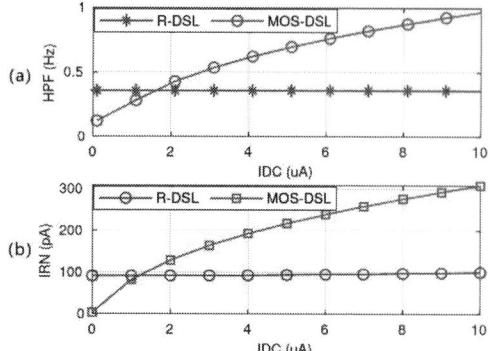

Fig. 5. Simulated (a) high-pass frequency and (b) input referred noise of the proposed AFE at different DC currents from 100nA to 10μA.

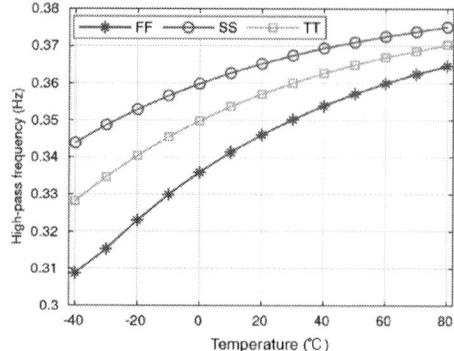

Fig. 6. Simulated high-pass frequency of the proposed DSL at different process corners and temperatures and DC current is 5 μA.

integrator of this design adopts the PVT-insensitive tunable pseudo-resistance as depicted in Fig. 4 [6]. The constant-gm bias operating at the weak inversion region generates a proportional-to-absolute-temperature (PTAT) current defined by

$$I = \frac{nV_T ln(K)}{R}, \qquad (10)$$

where K is the ratio of the current mirror.

The bias voltage generation circuit of the pseudo-resistor is composed of a replica transistor, three adjustable current sources, a resistor, and an amplifier, forming a feedback loop to ensure V_G-V_S equals $V_{GS,M3} - V_R$. Therefore, the resistance of the pseudo-resistor is derived as:

$$R_{PR} = \frac{U_T}{I_{FT}} \exp\left(\frac{RI_{CT}}{\eta U_T}\right) \propto R_s \exp\left(\frac{R}{R_s}\right). \qquad (11)$$

The resistance depends only on R_s and the exponential term defined by the resistor ratio. The integrator then implemented together with PVT-insensitive on-chip capacitors.

III. SIMULATION RESULTS

The proposed AFE is implemented in a 65 nm CMOS process, and the simulation results indicate a total power consumption of 20 μW from a single voltage supply of 1.2 V.

To validate the influence of different DC inputs on the transfer response and noise performance of the proposed AFE. Fig. 5 (a) and (b) respectively compare the frequency response and input referred noise of MOS-DSL and the proposed R-DSL with different input photocurrents ranging from 100 nA to 10 μA. The simulation results indicate that the R-DSL exhibits better characteristics under varying DC currents, maintaining a stable high-pass pole within the required frequency range and consistent noise performance. Fig. 6 shows the high-pass frequency of the proposed DSL at different process corners and temperatures with a fixed DC photocurrent of 5 μA. The PVT corner simulation shows that the high-pass pole of the entire system varies slightly between 300 and 380 mHz over a temperature range of -40 to 80 °C.

IV. CONCLUSION

A PPG analog front-end with PVT-insensitive high-pass frequency and noise performance is introduced in this paper. The robustness operation is achieved by a resistive DC servo loop and a very-large time constant integrator with PVT-insensitive tunable pseudo-resistors. The simulation results show that the high-pass pole is stabilized at 0.34 Hz with a

TABLE I. COMPARISON WITH STATE-OF-THE-ART WORKS

	This Work	[3]	[4]	[5]
Process (nm)	65	130	180	180
Power Supply (V)	1.2	1.5	1.8	1.8
High-pass freq. (Hz)	0.3-0.38	N/A	N/A	0.1-0.5
Power (μW)	20	83	216	180
Input noise (pA)	90	N/A	480	260
DC rejection (μA)	10	12	20	20

maximum variation of 11.7% and a input noise of around 90 pA, with DC input current ranging from 100 nA to 10 μA.

ACKNOWLEDGEMENT

This work was supported by the National Natural Science Foundation of China (No. 62204085). This work was also supported by the Strategic Priority Research Program of the Chinese Academy of Sciences under Grant No. XDB44000000-11 and the Lingang Laboratory under Grant No. LG-QS-202202-12.

REFERENCES

[1] Y. Khan, D. Han, J. Ting, M. Ahmed, R. Nagisetty and A. C. Arias, "Organic Multi-Channel Optoelectronic Sensors for Wearable Health Monitoring," IEEE Access, vol. 7, pp. 128114-128124, 2019.

[2] Q. Lin, W. Sijbers, C. Avidikou, C. Van Hoof, F. Tavernier, and N. Van Helleputte, "Photoplethysmography (PPG) Sensor Circuit Design Techniques," in Proc. IEEE Custom Int. Circuits Conf. (CICC), Newport Beach, CA, USA, 2022, pp. 01-08.

[3] P. V. Rajesh et al., "22.4 A 172 μW compressive sampling photoplethysmographic readout with embedded direct heart-rate and variability extraction from compressively sampled data," in IEEE Int. Solid-State Circuits Conf. (ISSCC) Dig. Tech. Papers, Jan. 2016, pp. 386–387.

[4] E. S. Winokur, T. O'Dwyer, and C. G. Sodini, "A low-power, dual-wavelength photoplethysmogram (PPG) SoC with static and time varying interferer removal," IEEE Trans. Biomed. Circuits Syst., vol. 9, no. 4, pp. 581–589, Aug. 2015.

[5] R. K. Pandey and P. C. -P. Chao, "A Dual-Channel PPG Readout System With Motion-Tolerant Adaptability for OLED-OPD Sensors," in IEEE Trans. Biomed. Circuits Syst, vol. 16, no. 1, pp. 36-51, Feb. 2022,

[6] R. Gan, L. Lyu, G. Mu and C. -J. R. Shi, "A Neural Recording Analog Front-End with Exponentially Tunable Pseudo Resistors and On-Chip Digital Frequency Calibration Loop Achieving 3.4% Deviation of High-Pass Cutoff Frequency in 5-to-500 Hz Range," in Proc. IEEE Custom Int. Circuits Conf. (CICC), Apr. 2022, pp. 1–2.

979-8-3503-6184-1/24 $31.00 © 2024 IEEE

A Fully integrated FVF based low-noise voltage buffer for ADC reference

Ikhwan Kim, Yajie Qin*

School of Information Science and Technology, Fudan University, Shanghai 200433, China
* Email: 20110720122@fudan.edu.cn, yajieqin@fudan.edu.cn

Abstract—This paper presents a Flipped Voltage Follower (FVF) based voltage buffer designed to provide a low-noise, high-precision reference voltage for ADC applications. The FVF based voltage buffer features load tracking zero compensation to reduce power consumption and eliminate the need for large off-chip capacitor. To achieve low frequency noise, chopper stabilization is employed in the error amplifier (EA) to mitigate flicker noise and thermal noise. The proposed design is using 180nm CMOS process, with a target output voltage of 2.5 V from a 5 V supply voltage. When using the chopper, the low-frequency noise (0.1 Hz – 10 Hz) is $0.40\mu V_{P-P}$, and the wideband noise (10 Hz – 1 kHz) is $1.10\mu V_{rms}$. The power consumption is $56\mu A$ across an I_{Load} range of 0 mA to 10 mA and a load capacitance range of 0 pF to 20 pF.

Keywords—low noise ADC reference, FVF based voltage reference, capacitor-free

I. INTRODUCTION

The demand for high precision ADCs is rapidly increasing due to the swift development of sectors such as energy, robotics, electric vehicles, healthcare, and IoT. This surge in demand underscores the critical role of high-precision, low-noise voltage reference in ensuring the accuracy and reliability of the system. As advanced electric devices continue to miniaturize, there is a growing need for the development of chips that not only deliver high performance but also meet the requirements for a smaller footprint.

Fig. 1 illustrates the block diagram of conventional types of voltage reference for ADC. In type (a), a voltage buffer is added to the Bandgap Reference (BGR) to enhance output driving capability. Type (b) employs an LDO regulator to design the ADC voltage reference. Both types can achieve high driving capability to handle dynamic load current, and they can be designed to have very low noise and offset. However, techniques such as auto-zeroing may be required to address the offset of the error amplifier, and an off-chip large capacitor is typically used to mitigate wideband thermal noise.

In this paper, we propose a FVF based voltage buffer for ADC reference that can eliminate $1/f$ noise and the offset of the EA, which critically affect ADC performance, as well as remove the need for an off-chip capacitor.

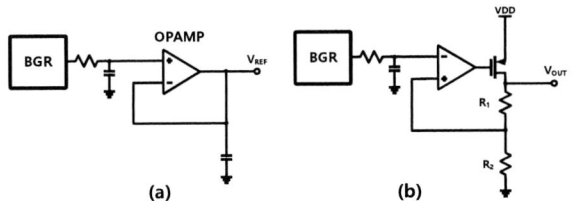

Fig. 1. Block diagram of a conventional ADC references (a) BGR with voltage buffer (b) LDO regulator based voltage buffer

This work was supported by the National Natural Science Foundation of China(82227803).

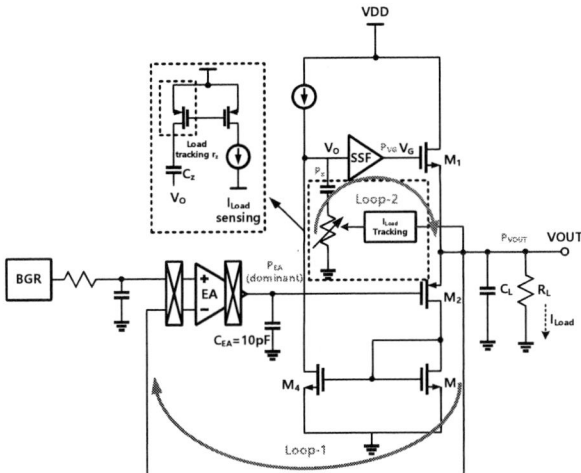

Fig. 2. Architecture of the proposed FVF based voltage buffer

II. SYSTEM ARCHITECTURE

Fig. 2 shows the architecture of the proposed FVF based voltage buffer for ADC reference. Unlike conventional LDO regulators, the FVF structure does not require feedback resistors, making it much more suitable for low-noise systems. The proposed architecture achieves high driving capability while applying a chopper stabilize technique to attain low $1/f$ noise and offset. Additionally, the power consumption required to ensure the stability of the feedback loop is minimized through the I_{Load} tracking circuit.

This architecture contains two feedback loops: Loop-1, which is passed back from V_{OUT} to V_{OUT} via EA, and Loop-2 which is passed back from V_{OUT} to V_{OUT} via Super Source Follower (SSF). Loop-1 has two poles, located at V_{OUT} and the output of EA. Due to the sufficiently large C_z in the I_{Load} sensing circuit and the Miller effect [1], the dominant pole is P_{VG}. The presence of the I_{Load} tracking zero cancels out P_z, thereby ensuring sufficient phase margin (PM) for Loop-2, independent of I_{Load} and C_L.

Given that the ratio of I_{Load} to I_{Load} sensing is N:1, and the size of the load tracking PMOS is W/L, r_z is expressed as follows:

$$r_z = \sqrt{\frac{kN}{2I_{load}\mu C_{xop}(W/L)}} \tag{1}$$

A. N-type Flipped Voltage Follower

An FVF-based design for use as a reference buffer is advantageous for achieving fast transient response and low output impedance [2]. However, to design for low noise in an ADC reference application, it requires a large off-chip capacitor and high power consumption [3].

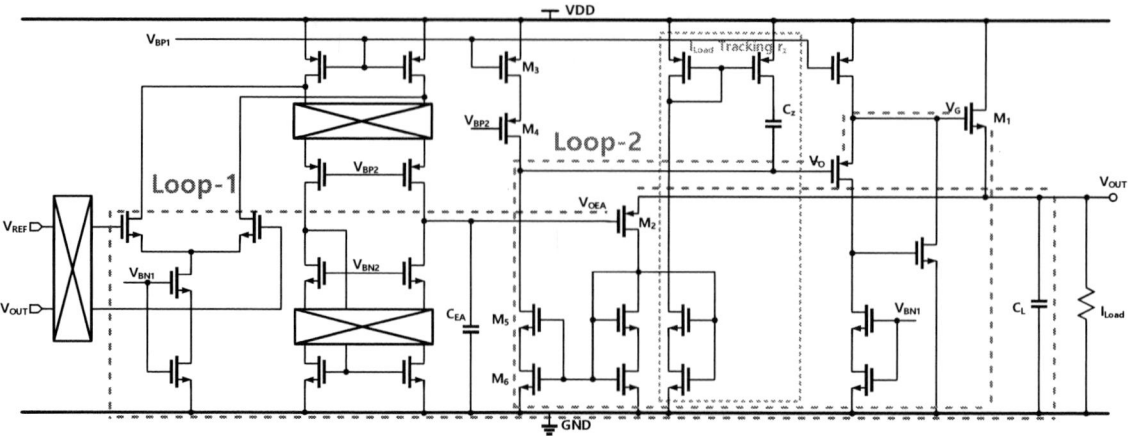

Fig. 3. Detailed schematic of the propsed FVF based voltage buffer

Although an output capacitorless FVF structure has been proposed to minimize chip size, considerations regarding low noise and offset, which significantly affect high-precision ADC reference designs, have not been addressed [4].

Fig. 3 shows the transistor level schematic of the proposed FVF based voltage buffer. The circuit demonstrates Loop-1 passing through EA and Loop-2 passing through SSF and I_{Load} tracking circuit.

Compared to the conventional P-type FVF structure, the N-type FVF structure adopted in this design is expected to have a relatively faster transient response due to the presence of a fast feedback loop (Loop-2). Simultaneously, sufficient phase margin for the system stability of the feedback loop can be secured with low power consumption through the SSF. In this design, the power consumption of the SSF is 28μA. The loop gain of the FVF is as follows:

$$A_{FVF} \approx 3g_{m2}(g_{m4}r_{o3}r_{o4}//g_{m5}r_{o5}r_{o6}) \qquad (2)$$

For $A_{FVF} \gg 1$, the loop gain of Loop-1 is equal to the gain of the EA.

B. Feedback loop stability

Fig. 4 shows the position changes of poles and zeros transistor level schematic of the proposed FVF based voltage buffer. The circuit demonstrates Loop-1 passing through EA and Loop-2 passing through SSF and I_{Load} tracking circuits.

To prevent oscillation at V_{OUT} and obtain a high-precision output, it is essential to ensure sufficient phase margin in both Loop-1 and Loop-2. In this design, for C_L range of 0 pF to 20 pF and I_{Load} range of 0 mA to 10 mA, the PM of Loop-1 ranged from 72.7 to 72.8, while the PM of Loop-2 ranged from 59.2 to 63.5.

The positions of poles in Loop-1 are given by (3), and the positions of the poles and zero in Loop-2 are given by (4):

$$\omega_{P_{EA}} = \frac{1}{r_{EA}C_{EA}}, \quad \omega_{P_{VOUT}} = \frac{1}{r_{VOUT}C_L} \qquad (3)$$

$$\omega_{P_{VG}} = \frac{1}{r_{SSF}C_{VG}}, \quad \omega_{P_z} = \frac{1}{C_z((r_{M1}||r_{M2})+r_z)}, \qquad (4)$$

$$\omega_{Z_z} = \frac{1}{C_z r_z}$$

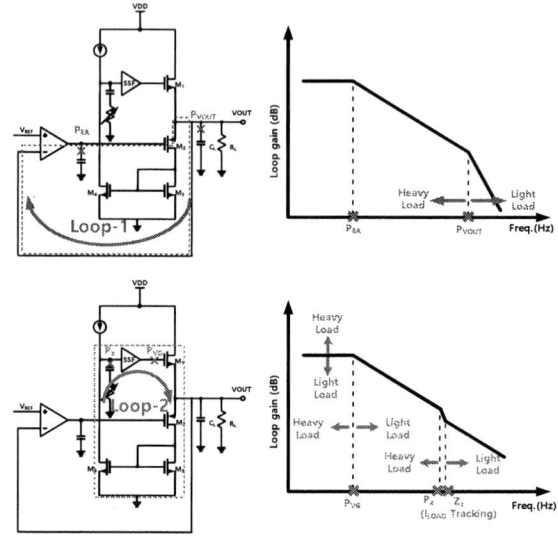

Fig. 4. Poles and zeros in Loop-1 and Loop-2

III. SIMULATION RESULTS

This circuit is designed and simulated using a 180 nm CMOS process. Fig. 5 shows the transient response of the proposed voltage buffer. Under the worst-case scenario ($C_L = 0$ pF, $I_{Load} = 10$ mA), the maximum V_{PP} of V_{OUT} is 115.87 mV in case (a) and 88.47 mV in case (b). The static error due to the offset of the voltage buffer in this design is 75μV.

For a 2.5 V reference, the 1-bit accuracy of a 12-bit ADC is ±0.61mV. Therefore, defining the settling time as the time required to reach 1-bit accuracy of 12-bit ADC, it is 2.45μs in case (a) and 1.81μs in case (b).

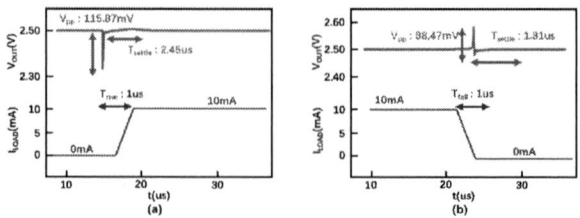

Fig. 5. Transient response of proposed buffer (a) I_{Load} changes from 0 mA to 10 mA (b) I_{Load} changes from 10 mA to 0 mA

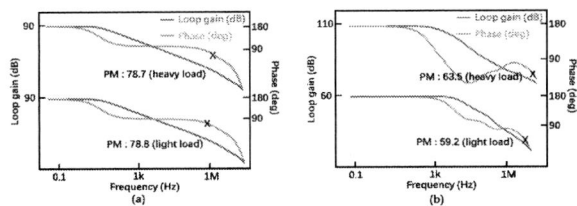

Fig. 6. Bode plot of feedback loop (a) Bode plot of Loop-1 (b) Bode plot of Loop-2

Fig. 6 shows the change in PM for Loop-1 and Loop-2 under light load (0 mA) and heavy load (10 mA) conditions. For Loop-1, the dominant pole P_{EA} is located at 16.5 Hz and is unaffected by changes in I_{Load}. Since P_{VOUT} is always positioned outside the unit gain frequency (UGF) (>1.4 MHz) within the range of I_{Load} variation (0 mA to 10 mA), the PM remains nearly constant despite changes in I_{Load}. For Loop-2, while P_{VG} varies with I_{Load}, P_z and Z_z cancel each other around 100 kHz under light load and around 85 kHz under heavy load, maintaining the PM of Loop-2 close to 60 in both cases.

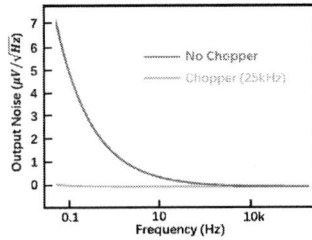

Fig. 7. Output noise simulation

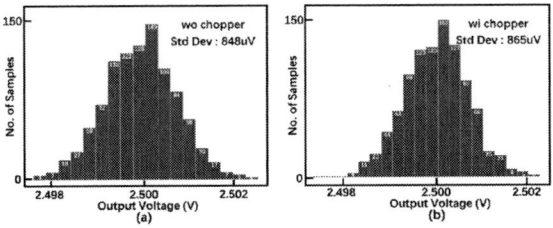

Fig. 8. Monte carlo simulation

Fig. 7 show the output noise of the FVF voltage buffer. For ADC references, two frequency bands of noise are of particular importance: low-frequency noise (0.1 Hz to 10 Hz) and wideband noise (10 Hz to 1 kHz). Generally, wideband noise can be reduced with an off-chip capacitor, while low-frequency noise is influenced by circuit design optimization. In this design, under the conditions of $C_L = 0$ pF and $I_{Load} = 10$ mA, the output noise from 0.1 Hz − 10 Hz is 12.03μ$V_{P\text{-}P}$ without using a chopper, and the noise from 10 Hz − 1 kHz is 3.76μV_{rms}. With a chopper, the output noise from 0.1 Hz − 10 Hz is 0.40μ$V_{P\text{-}P}$, and the noise from 10 Hz − 1kHz is 1.10μV_{rms}.

Fig. 8 shows the results of a Monte Carlo simulation with 1000 samples for the FVF buffer. The initial accuracy of the ADC reference is also a critical indicator for implementing high-precision references. Without using a chopper, the 3-

sigma value is 2.544 mV, resulting in σ/μ of ±0.102% at a 2.5 V output. With a chopper, the 3-sigma value is 2.595 mV, resulting in σ/μ of ±0.104% at a 2.5 V output.

TABLE I.   PERFORMANCE COMPARISON

	[5]	[6]	**This work**
Supply Voltage (V)	4.0 ~ 50	1.2	**5**
Output voltage (V)	2.5, 3, 5, 10	0.6	**2.5**
Power	1.2 mA	69 μA	**56 μA**
σ/μ (%)	±0.04	±0.21	**±0.10**
Noise (0.1 -10 Hz)	10 μ$V_{P\text{-}P}$	9.9 μ$V_{P\text{-}P}$	**0.40 μ$V_{P\text{-}P}$**
Max. Load (mA)	10	N/A	**10**
Load Cap.	0.1 μF	N/A	**0 pF – 20 pF**

Table I shows the comparison between the proposed design and state-of-the-art works. Compared to previous designs, the proposed design achieves the lowest low-frequency noise with significantly lower power consumption and high driving capability without using an external capacitor.

IV. CONCLUSION

This paper proposes a capacitor-free N-type FVF based voltage buffer suitable for high-precision ADC references. By leveraging a chopper-stabilized error amplifier and a load-tracking zero compensation technique, the design achieves minimal low-frequency noise and low power consumption at the same time. The design ensures sufficient feedback loop stability without needing an external large capacitor, making it ideal for minimizing chip area. Implemented using a 180nm CMOS process, the FVF based voltage buffer consumes 56 μA of power and achieves a low-frequency (0.1 Hz − 10 Hz) noise level of 0.40μ$V_{P\text{-}P}$, with σ/μ of ±0.10%.

REFERENCES

[1] G. A. Rincon-Mora, "Active capacitor multiplier in Miller-compensated circuits," IEEE J. Solid-State Circuits, vol. 35, no. 1, pp. 26–32, Jan. 2000.

[2] Tsz Yin Man, Ka Nang Leung, Chi Yat Leung, P. K. T. Mok, and Mansun Chan, "Development of Single-Transistor-Control LDO Based on Flipped Voltage Follower for SoC," IEEE Trans. Circuits Syst. Regul. Pap., vol. 55, no. 5, pp. 1392–1401, Jun. 2008.

[3] C. Li, C.-H. Chan, Y. Zhu, and R. P. Martins, "Analysis of Reference Error in High-Speed SAR ADCs With Capacitive DAC," IEEE Trans. Circuits Syst. Regul. Pap., vol. 66, no. 1, pp. 82–93, Jan. 2019.

[4] K. C. Koay, S. S. Chong, and P. K. Chan, "A FVF based output capacitorless LDO regulator with wide load capacitance range," in 2013 IEEE International Symposium on Circuits and Systems (ISCAS2013), Beijing: IEEE, May 2013, pp. 1488–1491.

[5] Gongmosemi, "GM7400, " datasheet, 2023 [Rev.0] [Online]; http://www.gongmosemi.com/upload/2024/06/20240605172139.pdf (accessed July 12, 2024).

[6] X. Liao, X. Liu, Y. Wang, and L. Liu, "A High-Precision Current-Mode Bandgap Reference With Low-Frequency Noise/Offset Elimination," IEEE Trans. Circuits Syst. II Express Briefs, vol. 70, no. 11, pp. 3993–3997, Nov. 2023.

979-8-3503-6184-1/24 $31.00 © 2024 IEEE

A Resistor-Free Grounded High-Frequency Memristor Emulator

Xinying Su, Bingjun Xiong, Junjie Yu and Jingjing Liu*

School of Electronics and Communication Engineering, Sun Yat-Sen University, Shenzhen, China

* suxy55@mail2.sysu.edu.cn, liujj77@mail.sysu.edu.cn

Abstract—The memristor emulator circuit is an economical and effective tool that can simulate various circuit characteristics of the corresponding solid-state memristor. This paper proposes a resistor-free high frequency grounded incremental/decremental memristor emulators (MRE). It consists of a voltage-current conversion (VCC) module, an OTA and a grounded capacitor. The theoretical analysis of the proposed memristor emulator is conducted. The memristor emulator is designed using a standard 180nm CMOS technology. The simulation results show that the proposed memristor emulator demonstrate pinched hysteresis loops for the voltage-current curves with a power supply voltage of $\pm$ 0.9V and prove its memristor characteristics. The proposed design exhibits nonlinearity over a wide input voltage range and performs well at frequencies of up to 50MHz.

Keywords—*operational transconductance amplifier (OTA), memristor emulator (MRE), pinched hysteresis loop (PHL)*

I. INTRODUCTION

The memristor is a nonlinear resistor with charge memory function, whose resistance value can change according to the history of the input current or voltage, enabling it to memorize the charge or magnetic flux by changing its resistance. Utilizing the above properties, memristors can be applied in various application fields, including high-speed memory arrays such as RRAM [1], analog and digital circuits [2], neuromorphic circuits [3], etc.

Due to the high manufacturing cost and structural complexity, physical memristors have not yet been commercialized. Therefore, simple and accurate memristor simulation circuits are needed to mimic the characteristics of real memristors. Currently, most memristor simulators consist of active modules (CFTA [4], VDTA[5], VDCC[6], etc.) and passive components such as capacitors. However, most existing memristor simulators have a maximum operating frequency of up to 1MHz and relatively complex circuit structures [4], [6].

This paper proposes a resistsor-free high frequency memristor emulator with a voltage-current conversion (VCC) module. The VCC is connected with an operational transconductance amplifier (OTA) active module and a grounded passive capacitor to form a complete high-performance memristor emulator (MRE) circuit based on a standard 180nm CMOS technology. The simulation results verify the pinched hysteresis loop of the proposed memristor emulator circuit. The structure of this paper is as follows: Section II gives the design and mathematical derivation principles of the proposed MRE circuit. Section III contains the simulation results and discussion, and finally Section IV concludes the paper.

II. THE PROPOSED MEMRISTOR EMULATOR

A. The VCC and OTA Modules

The circuit structure of the VCC module is shown in Fig. 1(a). The transconductance of the VCC module G_{m1} can be derived from (1)-(3).

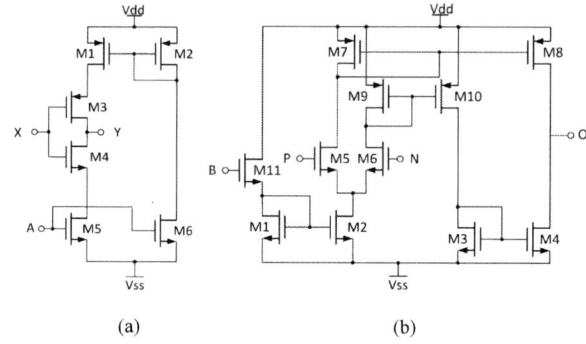

Fig. 1. The circuit structures of (a) VCC, (b) OTA.

$$G_{m1} = \frac{g_{M3} + g_{M4}}{2} \tag{1}$$

$$g_{M3/4} = \sqrt{2 I_D K_{M3/4}} \tag{2}$$

$$K_{M3/4} = \mu C_{ox} \left(\frac{W}{L}\right)_{M3/4} \tag{3}$$

where g_{M3} and g_{M4} are the transconductance of M3 and M4 respectively, I_D is drain current, μ is the mobility of carriers, C_{OX} is the gate oxide capacitance per unit area and W/L is the aspect ratio of transistors. Transistors M5 and M6 operate in the saturation region, and according to (1)-(3), and the drain current formula in the saturation region, (4) can be derived.

$$G_{m1} = k_1 (V_A - V_{ss} - V_{th}) \tag{4}$$

where V_{th} is the threshold voltage of the transistor, and $k_1 = \left(\sqrt{K_{M3}} + \sqrt{K_{M4}}\right)\sqrt{K_{M5,6}}/2\sqrt{2}$. The relationship between the input voltage V_X and the output current I_Y is given by

$$I_Y = G_{m1} V_X \tag{5}$$

This work is supported by National Natural Science Foundation of China with project number 62174181.

The schematic circuit of the OTA module is shown in Fig. 1(b), and the output current of the OTA can be obtained from (6).

$$I_O = G_{m2}(V_P - V_N) \qquad (6)$$

where the transconductance of the OTA is $G_{m2} = \dfrac{k}{\sqrt{2}}(V_B - V_{ss} - V_{th})$. When the input voltage V_{in} is connected to the P terminal, the N terminal is grounded, and vice versa. So the output current of the OTA can be written as follows:

$$I_O = \pm G_{m2}V_{in} \qquad (7)$$

B. The Proposed Memristor Emulator Circuit

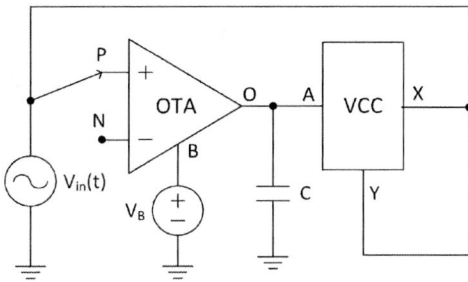

Fig. 2. Structure of the proposed memristor emulator.

The structure of the memristor emulator is shown in Fig. 2. A capacitor C is connected between the OTA and VCC module, leading to the following equation:

$$V_C = V_A = V_O = \pm \frac{1}{C}\int G_{m2}V_{in}dt = \pm \frac{G_{m2}A_m \cos \omega t}{\omega C} \qquad (8)$$

where input voltage V_{in} is a sinusoidal signal $V_{in} = A_m \sin \omega t$. Substituting (8) into (4), we obtain (9).

$$G_{m1} = k_1\left(\frac{\pm G_{m2}A_m \cos \omega t}{\omega C} - V_{ss} - V_{th}\right) \qquad (9)$$

The X port is connected to the Y port and the input voltage V_{in}, and the Y port outputs current I_Y. Therefore, $I_Y = I_X = I_{in}$, $V_X = V_{in}$. The missing element memristor relates the charge and flux, as $dq = Wd\Phi$, which defines memductance W for flux-controlled memristor.

$$W(\Phi_{in}) = \frac{I_{in}}{V_{in}} = \frac{I_Y}{V_X} = G_{m1} \qquad (10)$$

Substituting (9) into (10), we obtain

$$W(\Phi_{in}) = k_1(-V_{ss} - V_{th}) \pm \frac{k_1 G_{m2}A_m \cos \omega t}{\omega C} \qquad (11)$$

III. SIMULATION RESULTS AND DISCUSSION

To verify the performance of the proposed grounded memristor emulator, simulations were carried out based on a standard 180nm CMOS process. In the simulation DC supply voltages $V_{dd} = -V_{ss} = 0.9\text{V}$, bias voltage $V_B = 0.5\text{V}$, and input voltage $V_{in} = 0.25 \sin \omega t$. The transistor sizes in the memristor emulator circuit are shown in Table I.Fig. 3 shows the transient response with an input voltage frequency of 2 MHz and a capacitance of 50pF.

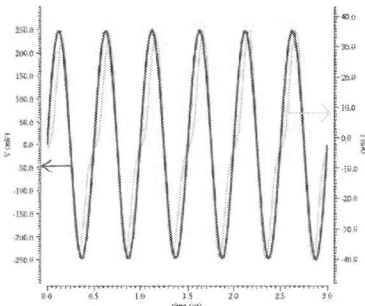

Fig. 3. Transient analysis at 2MHz with 50pF capacitor.

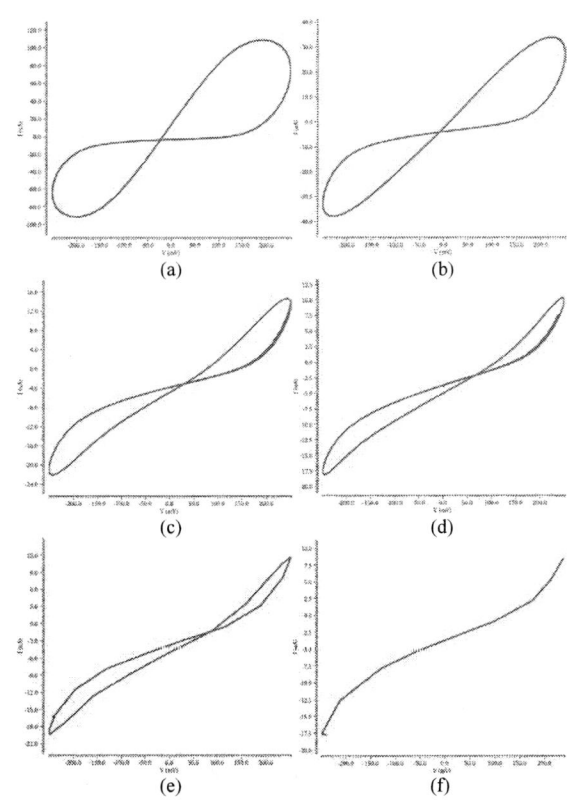

Fig. 4. Simulation results of the pinched hysteresis loops for the proposed memristor emulator circuit with different frequencies, (a) 900KHz, (b) 2MHz, (c) 4MHz, (d) 6MHz, (e) 8MHz, (f) 10MHz.

By plotting the input current versus the input voltage, the pinched hysteresis loop (PHL) characteristic of a memristor was obtained. To determine the effect of the input voltage frequency on the memristor, a 50pF capacitor was connected, and only the input signal frequency was varied, resulting in pinched hysteresis loops at different frequencies. As shown in Fig.4 , as the frequency increases, the area of the PHL gradually decreases, and when the frequency reaches 10MHz, the memristor behaves like a linear resistor.

TABLE I.   ASPECT RATIO OF ALL MOS TRANSISTORS

	Transistors	W/L
OTA	M1-M6	1.44μm/180nm
	M7-M10	4.5μm/180nm
	M11	1.8μm/180nm
VCC	M1、M2	6.5μm/180nm
	M3	3.25μm/180nm
	M4	4μm/180nm
	M5、M6	2μm/180nm

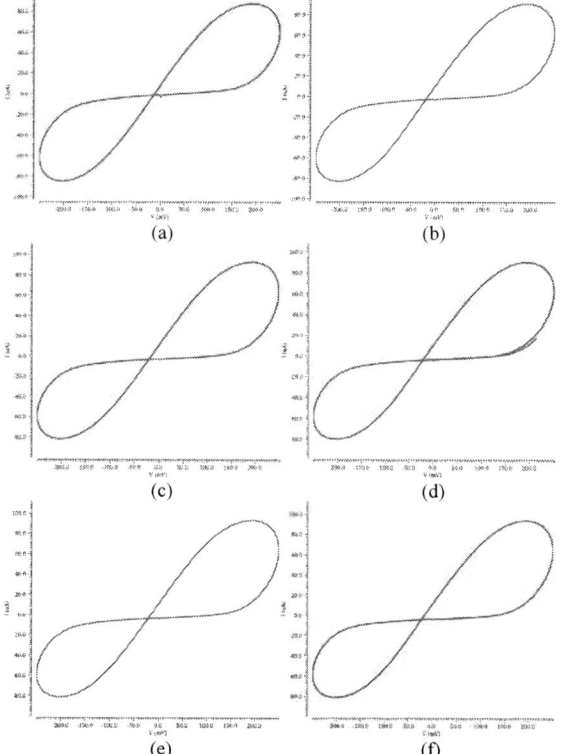

Fig. 5. Current-voltage curve of the proposed memristor emulator, at (a) 1MHz, 50pF, (b) 2MHz, 25pF, (c) 5MHz, 10pF, (d) 7.14MHz, 7.14pF, (e) 10MHz, 5pF, (f) 50MHz, 1pF.

Fig. 5 shows the simulation results of the memristor emulator when the capacitance frequency product is constant. It can be seen that the capacitance and frequency are inversely proportional, and when their product is a constant, the PHL remains essentially the same.

In Fig. 6, the maximum value of PHL at 25MHz with 2pF capacitor for the two memristor in parallel is approximately 300μA, while for the single memristor it is around 90μA. Similarly, PHL at 5MHz with 10pF capacitor for the parallel memristors is approximately 530μm, while for the single memristor it is around 90μm. This clearly indicates that the parallel memristors has a larger PHL area.

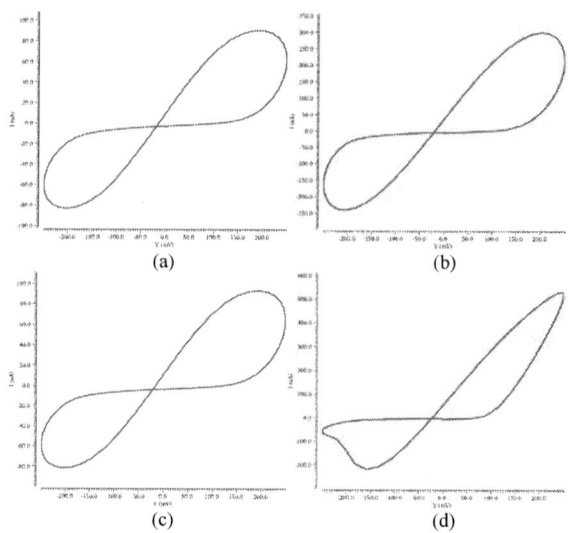

Fig. 6. Current-voltage curves for (a) single memristor emulator at 25MHz with 2pF capacitor, (b) two memristor emulators in parallel at 25MHz with 2pF capacitor, (c) single memristor emulator at 5MHz with 10pF capacitor, (d) two memristor emulators in parallel at 5MHz with 10pF capacitor.

IV. CONCLUSION

This paper proposes a resistor-free grounded incrementing/decrementing memristor emulator based on OTA and VCC modules. Through simulations results, the mathematical analysis of the proposed memristor model is validated. Various parameter analysis, such as large changes in frequency and capacitance, are conducted to demonstrate the circuit's robustness. The proposed memristor's advantages include: i) electronic tunability, ii) resistor-free structure, iii) incrementing/decrementing modes, and iv) an operating range of up to 50 MHz. These observations confirm that the proposed MRE design is easy to implement and can be a good circuit model for the memristors.

REFERENCES

[1] R. M. Claure, T. Tan Nguyen and B. Cambou, "Low Complexity Memristor-based RRAM Design for IoT Applications," 2023 7th International Conference on I-SMAC (IoT in Social, Mobile, Analytics and Cloud) (I-SMAC), Kirtipur, Nepal, 2023, pp. 42-45.

[2] S. Kirilov and V. Mladenov, "Application of New Metal-Oxide Memristor Models in Digital and Analog Electronic Circuits," 2023 19th International Conference on Synthesis, Modeling, Analysis and Simulation Methods and Applications to Circuit Design (SMACD), Funchal, Portugal, 2023, pp. 1-4.

[3] P. Nune and S. Mandal, "TaOx based memristor model and its emulator design for future neuromorphic computing," 2021 13th International Conference on Electronics, Computers and Artificial Intelligence (ECAI), Pitesti, Romania, 2021, pp. 1-5.

[4] S. S. Prasad, P. Kumar and R. K. Ranjan, "Resistorless memristor emulator using CFTA and its experimental verification", IEEE Access, vol. 9, pp. 64065-64075, 2021.

[5] J. Vista and A. Ranjan, "Flux controlled floating memristor employing VDTA: Incremental or decremental operation", IEEE Trans. Comput.-Aided Design Integr. Circuits Syst., vol. 40, no. 2, pp. 364-372, Feb. 2021.

[6] K. Bhardwaj and M. Srivastava, "Floating memristor and inverse memristor emulation igurations with electronic/resistance controllability", IET Circuits Devices Syst., vol. 14, pp. 1065-1076, Oct. 2020.

979-8-3503-6184-1/24 $31.00 © 2024 IEEE

An Ultra-Low-Leakage Current Sensing Interface for Wide Temperature Range

Jinsheng Tang, Chun Zhao, Lin He*

College of Integrated Circuit Science and Engineering,
Nanjing University of Posts and Telecommunications(NJUPT),
9 Wenyuan Road, Nanjing 210023, Jiangsu, P.R. China

* Email: helin@njupt.edu.cn

Abstract—**A current sensing interface circuit based on a 180nm deep n-well CMOS technology is presented in this paper. It achieves ultra-low leakage across a wide temperature range. This interface circuit is essentially a resistive transimpedance amplifier with a pseudo-resistor acting as the feedback element, which comprises two series nMOS transistors in a deep well. The two nMOS transistors operate in sub-threshold region to realize a high resistive value. The leakage current from the inner node of the pseudo-resistor is minimized by using an auxiliary feedback loop. The leakage current associated with the ESD diodes is minimized using another auxiliary feedback loop. The prototype amplifier with the proposed leakage current suppression technique reduced the leakage current from 3.5pA to 54.9fA, when compared to a conventional transimpedance current amplifier of similar dimensions.**

Keywords—transimpedance amplifier, current sensing interface, ultra-low-leakage current

I. INTRODUCTION

Detecting extremely weak currents is crucial for industrial applications, such as tracking trace amounts of toxins or pathogens[1] and detecting nuclear radiation[2]. These tasks necessitate circuits capable of precise real-time detection, even in extreme temperatures or other adverse environmental conditions. Unfortunately, high temperatures can cause severe leakage currents, making conventional amplification circuits unsuitable for sensitive applications.

Current sensing interfaces are divided into two categories: resistive transimpedance amplifiers (RTIAs) and capacitive transimpedance amplifiers (CTIAs)[3]. CTIAs often operate in discrete time domains. In continuous time detection applications, RTIAs are preferred. RTIAs convert weak current into detectable voltage through a huge-value feedback resistor, which is usually implemented with a MOS transistor operating in subthreshold region[4].

As temperature rises, the leakage current associated with the current sensing interface rises as well, which may limit the RTIA's current detection capability. To address this issue, a novel sensing interface circuit is presented that offers precise, low-leakage current sensing across a wide temperature range. The paper is organized as follows. Section II analyzes the leakage mechanism of the current sensing interface. Details of the circuit implementation are presented in Section III. Section IV gives the simulation result, and the conclusion is drawn in Section V.

II. LEAKAGE MECHANISM

A. Pseudo-resistor

When detecting extremely weak currents, the direct current leakage of pseudo-resistors is a significant consideration, primarily arising from the parasitic diodes of MOS devices. Fig. 1 shows a conventional symmetric pseudo-resistor with parasitic diodes. An independent N-well and a P-type substrate form a reverse-biased PN junction, referred to as D1. P+/N-well junction diodes(D2-D5) are also formed between the source/drain of the MOS transistors and the N-wells. D1 inevitably generates reverse junction leakage currents (illustrated by the red path in Fig. 1), which can make up a significant part of the detected current (represented as the blue path or yellow path in Fig. 1(b)). At room temperature, the DC leakage usually stays at the picoampere level. Furthermore, leakage current rises exponentially as temperature rises. In contrast, the leakage from D2 to D5 diodes is typically minimal because they operate in parallel with the pseudo-resistor conduction channel and generally exhibit much lower leakage compared to D1.

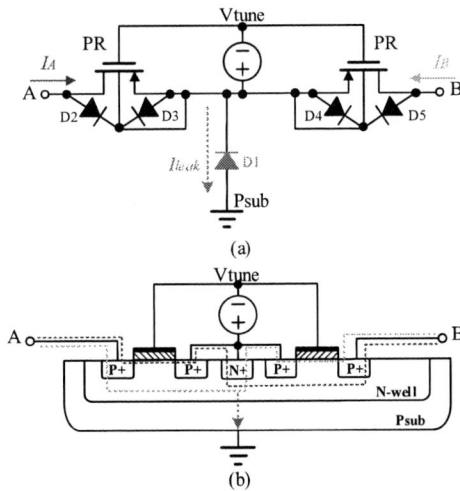

Fig. 1. (a) Schematic diagram of the floating pseudo-resistor with parasitic diodes; (b) illustrates the pseudo-resistor's current flow path; blue and yellow indicate the current path, while the red path indicates the current leakage path.

B. ESD Protection Circuits

Electrostatic Discharge (ESD) events usually occur when electrostatic charges move between materials, causing chip failure. To protect an analog signal input, on-chip ESD protection circuits are used as shown in Fig. 2[5]. This protection circuit includes resistors, two stages of diodes, and power-rail clamps. During a positive ESD event, the voltage at the I/O pin rises quickly, making the first diode (Dp1) forward-biased. This redirects most of the current from the core to the VDD. The rest of the current flows through the resistor (R1) and the second diode (Dp2), adding a voltage

979-8-3503-6184-1/24 $31.00 © 2024 IEEE

drop to protect the gate oxide. The fast change in voltage also triggers the NMOS clamp, which creates a low-impedance path to the ground, protecting the core circuits.

ESD protection also contributes to leakage. When the core circuit operates, the input voltage lies between the power rails, causing the ESD diodes to be reverse-biased. The reverse-biased diodes generate a reverse saturation current. The reverse saturation current mismatches cause a leakage current, which exacerbates significantly at high temperatures.

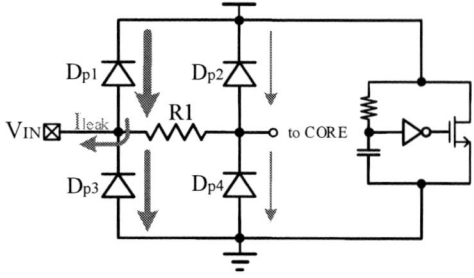

Fig. 2. Schematic diagram of ESD protection with leakage path

III. CIRCUIT IMPLEMENTATION

The leakage current may pollute the input current. In this design, auxiliary feedback loops are introduced to the pseudo-resistors and the ESD protection circuits to clamp the voltages across those leaky diodes to zero, theoretically eliminating the reverse saturation leakage and its fluctuations.

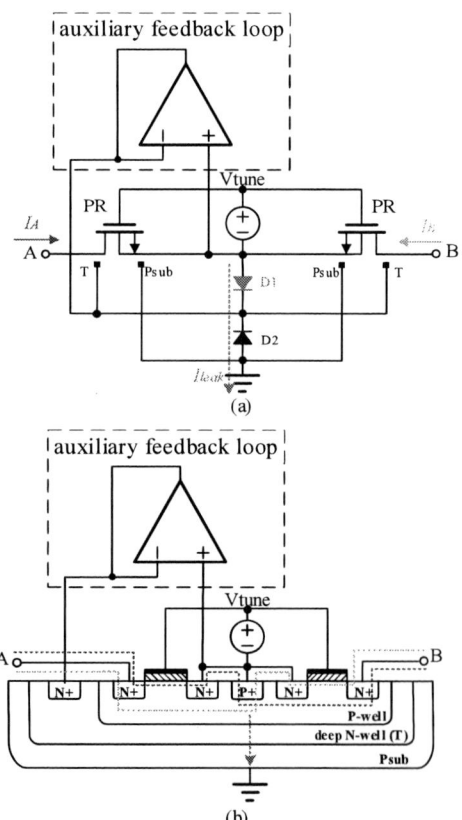

Fig. 3. (a) Schematic diagram of the auxiliary feedback loop pseudo-resistor with parasitic diodes; (b) illustrates the pseudo-resistor's current flow path; blue and yellow indicate the current path, while the red path indicates the current leakage path.

Leakage suppressing feedback loop associated with the pseudo-resistor is illustrated in Fig.3(a). Different from its PMOS counterpart (Fig.1(a)), this design utilizes nMOS symmetric pseudo-resistors based on a deep N-well process. The leaky diode D1 in Fig.1(a) has one terminal fixed to the ground, but the leaky diode D1 in Fig.3(a) has two adjustable terminals which make the leakage suppressing feedback loop possible.

Fig.4(a) shows the implementation of the leakage suppressing feedback loop associated with the ESD. When the R-TIA is operating, the auxiliary feedback loop clamps the voltage across $Dn_{1,2}$ and $Dp_{1,2}$ to zero, ensuring that the input current flows into the RTIA without loss. Specifically, Dn_1 and Dn_2 are deep N-well N-type diodes, which prevent input leakage current from the Nwell/Psub junction diodes, while the other diodes use P-type diodes. This selection maximizes the effectiveness of the auxiliary feedback loop, resulting in low leakage current. Additionally, Fig.4(b) shows the current path during a positive ESD event at the Iin pin.

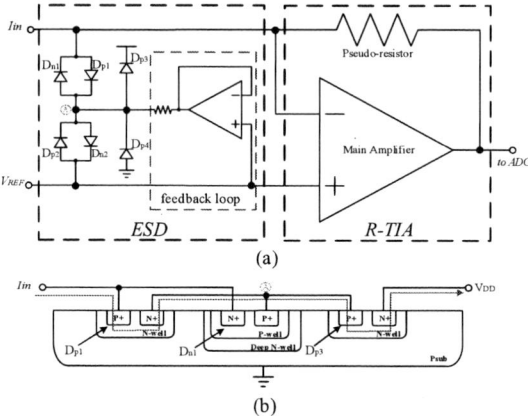

Fig. 4. (a) shows the circuit interface with a leakage suppressing feedback loop associated with ESD; (b) blue illustrates the current flow when a positive ESD event occurs at the Iin port.

The main amplifier utilizes a folded cascode architecture with a pMOS differential input pair to enhance gain and reduce 1/f noise. It features a complementary common-source output stage in a Class AB topology for rail-to-rail output swing, thereby increasing the input dynamic range. The auxiliary operational amplifier implemented in this work is a traditional two-stage Miller operational amplifier.

IV. SIMULATION RESULTS

The proposed weak current interface based on R-TIA is designed in a 180nm deep n-well CMOS process, whose layout is shown in Fig. 5, occupying a core area of 0.2mm×0.13mm.

Fig. 5. Layout of the proposed weak current interface.

979-8-3503-6184-1/24 $31.00 © 2024 IEEE

The input capacitance is 5pF, and the main amplifier has a gain-bandwidth product of 15M and a compensation capacitor of 100fF. At 85°C, the pseudo-resistor can be tuned by adjusting the external bias. As shown in Fig. 6, simulations predict the frequency response for transimpedance values of 100M, 1G, 10G, and 100G, achieving corner frequencies of approximately 16.4kHz, 1.6kHz, 162.4Hz, and 13.9Hz, respectively.

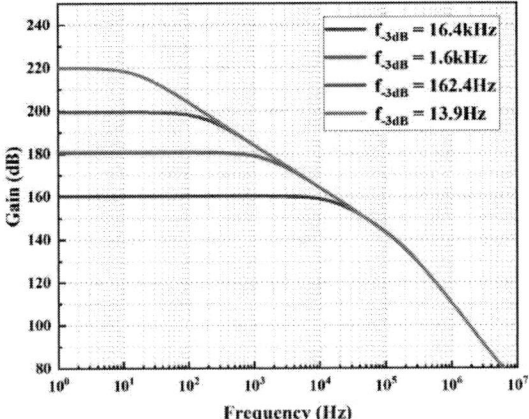

Fig. 6. Simulated frequency response of proposed TIA

Fig. 7 depicts the input-referred current noise of the TIA at 85 °C for transimpedance values of 100M, 1G, 10G, and 100G. The simulations were conducted under 1pA input current conditions. At the TIA gain of 100 GΩ, the simulated integrated input-referred RMS current noise is 9.97 fA$_{RMS}$ within the corner frequency range.

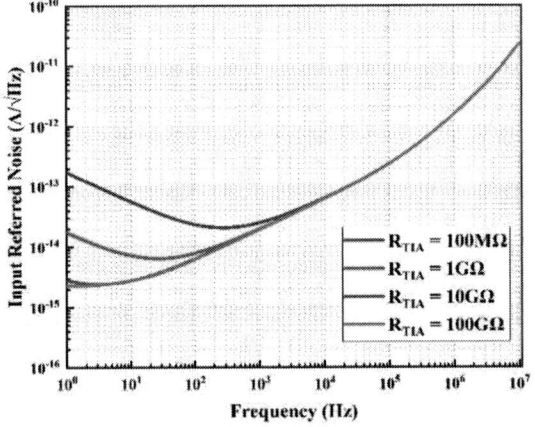

Fig. 7. Simulated PSD of the input-referred current noise of proposed TIA

The proposed circuit's current leakage levels were compared against those of conventional circuits under different temperature scenarios. The simulation results are shown in Fig. 8. At the entire commercial temperature range, the proposed circuit exhibits significantly lower leakage currents than conventional circuits. Additionally, the proposed circuit shows stable leakage performance under temperature variations.

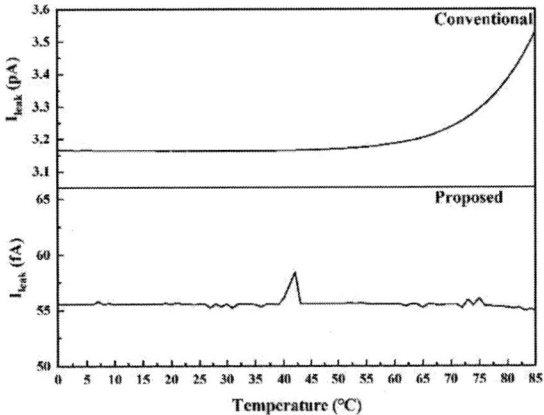

Fig. 8. Simulation comparing the leakage levels of the conventional circuit and the proposed circuit.

V. Conclusion

This paper presents a current sensing interface circuit based on 180nm deep n-well CMOS technology, which achieves ultra-low leakage. The circuit uses a pseudo-resistor as the feedback component, allowing adjustable gain to detect different current levels. Auxiliary feedback loops are used to minimize leakage current, making it unaffected by external temperature changes. This design enables precise detection of extremely weak currents over a wide temperature range.

References

[1] Lambrou T P, Anastasiou C C, Panayiotou C G, et al. A low-cost sensor network for real-time monitoring and contamination detection in drinking water distribution systems[J]. IEEE Sensors Journal, 2014, 14(8): 2765-2772.

[2] Okazaki Y, Tanaka T, Saito N, et al. Subfemtoampere resolved ionization current measurements using a high-resistance transimpedance amplifier[J]. IEEE Transactions on Instrumentation and Measurement, 2022, 71: 1-8.

[3] Awan M A, Wang B, Quadir N A, et al. Review and analysis of CMOS current readout circuits for biosensing applications[C]//2021 IEEE International Symposium on Circuits and Systems (ISCAS). IEEE, 2021: 1-5.

[4] Guglielmi E, Toso F, Zanetto F, et al. High-value tunable pseudo-resistors design[J]. IEEE Journal of Solid-State Circuits, 2020, 55(8): 2094-2105.

[5] Ker M D. Whole-chip ESD protection design with efficient VDD-to-VSS ESD clamp circuits for submicron CMOS VLSI[J]. IEEE Transactions on Electron Devices, 1999, 46(1): 173-183.

A Global Threshold Voltage Finder Technology for the Readout Circuit of Event-based Vision Sensor

Yanwen Su[1], Hao Li[2*], Dongsheng Liu [1], Ang Hu[1], Kaiyue Li[1]

1 School of Integrated Circuit, Huazhong University of Science and Technology, Wuhan 430074, China
2 Hubei Optics Valley Laboratory, Wuhan 430074, China

* Email: hli@hust.edu.cn

Abstract

The event-based vision sensor (EVS) is a new image sensor that detects the change in light intensity instead of the value of light. The mechanism of EVS has many advantages such as higher dynamic range, higher detection speed, and lower data redundancy. However, traditional EVSs use pre-defined contrast threshold voltage to detect events, which is risky because of the influence of process voltage temperature (PVT) fluctuation and array mismatch. To solve this problem, this paper presents a global threshold voltage finder technology for the EVS readout circuit. The binary search technology is used to acquire the flip voltage of comparators in the EVS pixel array. The simulation results show that the voltage error between the threshold finder result and the average of the pixel array is 3.83 mV. Compared with a single reference pixel finder, the voltage error reduces by 61% to the standard deviation of pixel arrays.

1. Introduction

With the development of CMOS image sensors (CIS), the resolution of CIS becomes larger to get more information in the field of view. Although the CIS devices have been used widely, there are some problems with traditional CIS such as motion blur and overexposure. To overcome this problem, the event-based vision sensor (EVS) has been proposed[1]. Compared with traditional CIS image sensors by frame, the EVS detects the change in input light intensity[2]. By measuring the trend of light change, the EVS will generate two kinds of event information which are called ON event and OFF event. On event represents the light increasing and OFF event represents the light decreasing. If the input light intensity is changed rapidly, the event information will be generated fast too. On the contrary, the EVS will not generate any event information when the input light intensity is constant. Because of the unique image characteristic, EVS avoids the time of current integration so that it can detect the contour of high-speed moving objects. Therefore, the EVS has great advances in machine vision, autonomous driving, and remote monitoring regions.

Comparing the amplified voltage with two pre-defined threshold voltages, the EVS generates different polarity events. However, the same as CIS, the EVS also has the problem of pixel arrays mismatch. The processing mismatch of pixels will cause these pixels to have different event thresholds. Furthermore, in the highly sensitive detection condition, the threshold mismatch will make the pixels generate error events. To solve the problem, this paper proposes a global threshold voltage finder technology, which determines the threshold voltage by measuring the flip voltage of the comparators in different pixels. Although the measured flip voltages are different under the influence of the mismatch, the average value can be calculated by equal interval sampling.

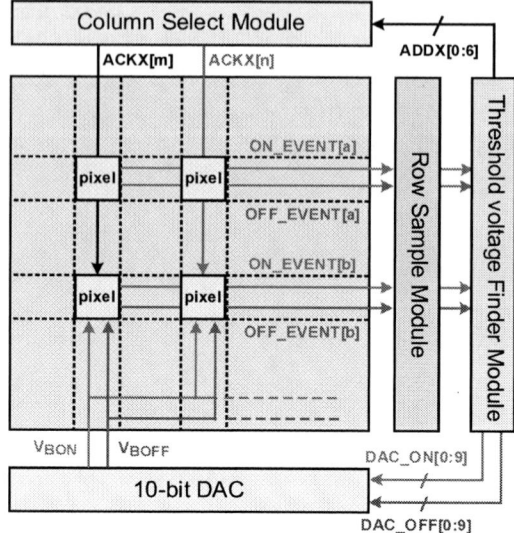

Fig.1 The architecture of the global threshold voltage finder module.

The structure of this paper is as follows: Section 2 introduces the architecture of the global threshold voltage finder technology. Section 3 analyses the principle of the technology and the simulation results are given in Section 4. Section 5 gives the conclusion for the paper.

2. The Architecture of Global Threshold Voltage Finder Technology

The architecture of the global threshold voltage finder technology is shown in Fig.1. The EVS system contains a column select module, row sample module, pixel arrays, and voltage bias modules such as DAC. In this paper, before the EVS pixels start to work, a threshold voltage finder module is used to generate the input voltage of comparators. Firstly, a column address ADDX is given to the column select module. The event information in the responding column pixels transfers to the row sample module when all the pixels are in the reset state. The pixel in the m column and a row is regarded as the reference pixel. According to the reference pixel event information, the DAC control words DAC_ON[0:9] and DAC_OFF[0:9] are adjusted from high bit to low bit. After one reference pixel threshold voltage finder process is complete, the other pixel in the m column and b row will be regarded as a new reference pixel. Based on the new reference pixel, the DAC control words will be adjusted in order again. By repeating the threshold voltage finder process such times, several threshold voltages of different pixels can be saved. The final result is the average value of these voltages.

979-8-3503-6184-1/24 $31.00 © 2024 IEEE 563

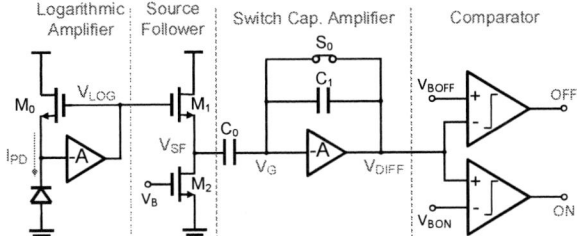

Fig.2 The pixel circuit of the EVS readout circuit.

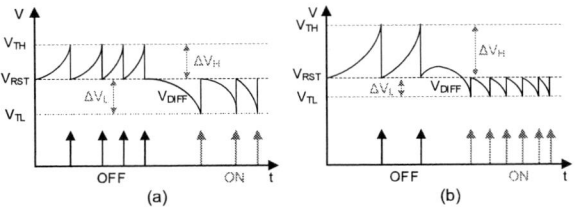

Fig.3 (a)The event trigged waveform in normal threshold voltage conditions. (b)The event triggered waveform when the threshold voltage fluctuated.

The pixel circuit of EVS is shown in Fig.2. There are four parts in the pixels, which are called logarithmic amplifier, source follower, switch capacitor amplifier, and comparator. The first stage of the EVS pixel is a trans-impedance logarithmic amplifier. The input photocurrent is transferred to V_{LOG} as the logarithmic relationship. The second stage is source follower which is used to isolate the noise from the posterior circuit. The signal voltage V_{SF} is magnified to V_{DIFF} by the switch capacitor amplifier. The comparators detect the V_{DIFF} with the threshold voltages and generate corresponding pole events.

Formula (1) shows the voltage express of V_{DIFF}. In the formula, n is the subthreshold slope factor of M_0 and U_T is the thermal voltage. C_0 and C_1 are the proportional capacitors in the switch capacitor amplifier. I_{PD} is the input photocurrent. The formula shows that the V_{DIFF} is not decided by the value of value I_{PD}, but deicide by the variation of I_{PD}.

$$\Delta V_{DIFF} = nU_T \frac{C_0}{C_1} \ln\left(\frac{I_{PD} + \Delta I_{PD}}{I_{PD}}\right) \quad (1)$$

When the V_{DIFF} is over V_{TH} or below V_{TL}, the OFF or ON event will be generated as Fig.3(a). The ΔV_H and ΔV_L decide the variation of light current that can trigger an event. Unfortunately, the threshold voltage would not remain constant because of the mismatch. Fig.3(b) shows that ΔV_H becomes larger and ΔV_L becomes smaller. Correspondingly, the number of OFF events decreases and ON events increases. Fig.4 shows the circuit of the proposed DAC. The DAC is composed of the resistance array and the current-steering module. The higher 4-bit words control the resistance array to generate V_{RA}. The current-steering module is controlled by the lower 6-bit words, which provide the current through the feedback resistance R_f. The DAC output voltage V_{DAC} is

$$V_{DAC} = V_{RA} + I_{cs}R_f \quad (2)$$

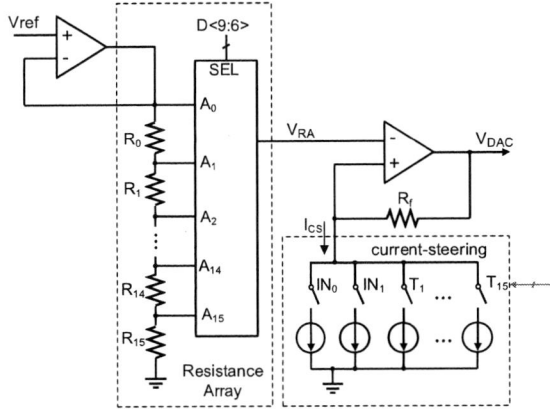

Fig.4 The circuit of the proposed 10-bit DAC.

3. The Principle of Global Threshold Voltage Finder Technology

Although threshold voltage mismatches exist in the pixel arrays, the threshold voltage values meet normal distribution [3]. Therefore, it is reasonable to set the input voltage of comparators as the average value of threshold voltage in the pixel arrays. The global threshold voltage finder technology is the way to obtain the average value.

The pixel in the m column and a row is set as a reference pixel. The threshold voltage finder module uses binary search technology to find the flip voltage of the reference comparator. Fig.5(a) shows the waveform of the comparator that the output voltage ON flips to a low level at the V_{FLIP}. When the output voltage ON is low level after the first voltage V_0 is set, the second voltage V_1 will become half of V_0. Similarly, the ON becomes high level due to the lower V_1, so the V_2 increases half voltage than V_1. Fig.5(b) shows the transient waveform of the threshold voltage finder process. The DAC output voltage approaches V_{FLIP} by each time cycle. By using binary search technology, the finder process can get the flip voltage in ten clock cycles with 10-bit DAC. The next step is to find the flip voltage of the other pixels. For example, choose the pixel in the n column and a row for the next reference pixel, and repeat the voltage finder process. And so on, by searching the voltage in different positions, the average of the value is close to the value of pixel arrays.

The global threshold voltage finder technology has a great advantage for event detection. Finding the flip voltage in pixel arrays can reduce the influence of different processes, voltages, and temperatures (PVT). Because of the threshold voltage mismatch in the pixel array, sampling only one voltage is unreliable. The global threshold voltage finder technology uses the average value of different position pixels to approximate the average value of the pixel array. Although the voltage mismatch cannot be eliminated, setting the input voltage of comparators as the average value of normal distribution can minimize the influence of the mismatch.

Traditional EVS chips use pre-defined threshold voltage, which would be influenced by different PVT [4]. The V_{TH} and V_{TL} are likely to deviate from the pre-defined value, such as the condition in Fig.3(b). A part of EVS designs uses DAC to generate threshold voltage by outside control words [5], so the voltage can be adjusted as needed. However, manually

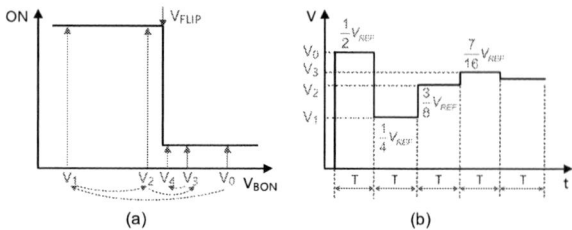

(a)   (b)

Fig.5 (a)The threshold voltage finder process with binary search technology. (b)The transient waveform of threshold voltage finder process.

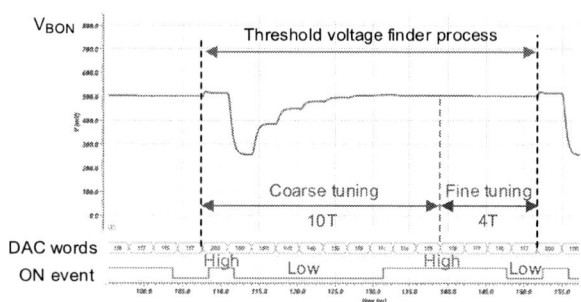

Fig.6 The simulation waveform of the threshold finder process

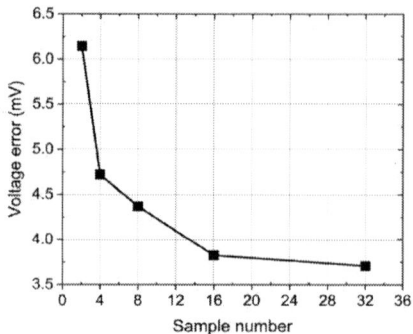

Fig.7 The voltage error of different sample numbers.

5. Conclusion

In this paper, the global threshold voltage finder technology is proposed. Compared with the pre-defined threshold voltage, the found voltage increases the influence of PVT fluctuation and pixel mismatch. By using 16 sample pixels to find the threshold voltage, the voltage error between the found voltage and the average of arrays reduces to 3.83mV. The δ of the threshold voltage is 9.52mV, and the voltage error of 16 pixels global threshold finder reduces by 61%.

Acknowledgments

This work is supported by the National Key Research and Development Program of China (No. 2021YFA0715502, No. 2023YFB4502100), the National Natural Science Foundation of China (No.62374064, No. 62104076, No. 62134002), the National Key Analog Integrated Circuit Laboratory Project of China. (No. JCKY2021210C004)

adjusting threshold voltage without the information of pixel arrays is inaccurate and inefficient.

4. Simulation Result

Fig.6 shows the simulation waveform of the threshold finder process. According to the voltage of the ON event output, the DAC output V_{BON} adjusts correspondingly. The threshold finder process takes fourteen clock cycles to decide the threshold voltage for one pixel. The coarse-tuning process contains ten clock cycles, which are used to adjust the DAC control words from high to low bit with binary search technology. The fine-tuning process contains the last four clock cycles, which are only used to adjust the lowest bit.

To analyze the proposed global threshold voltage finder technology, an 8×8 pixel array with the mismatch model is given to test. By Monte Carlo analysis, the δ of the threshold voltage is 9.52mV. The difference between the result of the threshold voltage finder and the average value of pixel arrays in different sampling numbers is shown in Fig.7. With the increase of sample number, the finder voltage error reduces to a lower value. Although increasing the sample numbers can improve the accuracy of the result, the cost of hardware and the time of the finder process also need to be considered. The voltage error of the 16 sample numbers is 3.83mV and the value of the 32 sample numbers is 3.71mV. By using more sample pixels, the accuracy increases insignificantly(0.12mV). Therefore, it is rational to use 16 pixels to sample.

The simulation result shows that threshold voltage finder technology can search the average value of the pixel array with a lower 0.39δ voltage error. The voltage error of 16 pixels global threshold finder reduces by 61% compared to only one reference pixel.

References

[1] P. Lichtsteiner, C. Posch and T. Delbruck, "A 128 × 128 120 dB 15 μs Latency Asynchronous Temporal Contrast Vision Sensor," in IEEE Journal of Solid-State Circuits, vol. 43, no. 2, pp. 566-576, Feb. 2008

[2] Z. Ding et al., "A Multimode Neuromorphic Vision Sensor With Improved Brightness Measurement Performance by Pulse Coding Method," in IEEE Internet of Things Journal, vol. 11, no. 4, pp. 6266-6277, 15 Feb.15, 2024

[3] Y. Hu, S. -C. Liu and T. Delbruck, "v2e: From Video Frames to Realistic DVS Events," 2021 IEEE/CVF Conference on Computer Vision and Pattern Recognition Workshops (CVPRW), Nashville, TN, USA, 2021, pp. 1312-1321

[4] Li H, Li G, Shi L. Super-resolution of spatiotemporal event-stream image captured by the asynchronous temporal contrast vision sensor[J]. Neurocomputing, 2018.

[5] M. Yang, S. -C. Liu and T. Delbruck, "A Dynamic Vision Sensor With 1% Temporal Contrast Sensitivity and In-Pixel Asynchronous Delta Modulator for Event Encoding," in IEEE Journal of Solid-State Circuits, vol. 50, no. 9, pp. 2149-2160, Sept. 2015

A Residue Amplifier With 72.27 dB Loop-Gain and 4.64 GHz Closed-Loop Bandwidth consuming 6.4 mW for 12-Bit 1-Gsps Pipelined ADC

Jiangbo Wei[1], Jin Liu[1], Wei Tian[1], Chao Wang[1]*, Maliang Liu[2]

[1]Xi'an Microelectronics Technology Institute, Xi'an, China, [2]Xidian University, Xi'an, China
*Email: 12207935@qq.com

Abstract

A two-stage Miller-free compensation op-amp using multi-power domain for 1-GS/s 12-bit pipelined ADC is proposed in this paper. The first stage based on complementary push-pull structure and the second stage of telescopic cascode with gain-boosting eliminate the Miller compensation and fulfill the 72.27dB loop-gain (LG) and 4.64GHz closed-loop bandwidth (CLBW) with a phase margin (PM) of 67.6° while consuming 6.4mW at 0.9/1.8V. The MDAC (Multiplying Digital-to-Analog Converter) based on this presented op-amp occupies an active area of 0.0342 mm^2 in 28 nm process. Considering the load capacitor C_L=250fF.

1. Introduction

Op-amp is the most versatile building block of RF-Analog microsystem, which is mainly used in instrumentation amplifier systems, DC-DC converters, ADCs, active filters, high-speed digital-to-analog interfaces, audio power amplifiers and so on [1]. Cutting down the power supply voltage and working current of the chip is particularly urgent in the chip design, but this will curb the dynamic range and noise performance of the op-amp, making the low-voltage and low-power system mandate higher requirements for the op-amp [2]. Pipelined ADC has the unparalleled superiority in achieving fine resolution and high speed, compared with other ADCs. As the most critical and power-hungry module of pipelined ADC, the input and output swing, LG, CLBW and settling time of the residue amplifier (RA) have a drastic impact on the overall performance of ADC. Especially in a multi-stage pipelined architecture, if the design of the RA does not meet the requirements, the performance of each stage of ADC will be deviated. The realization of high gain as well as high bandwidth op-amp with considerable power consumption has captured academic community's attention. This work presents a two-stage amplifier without Miller-compensation in multi-power domain for a 12-bit 1-GS/s pipelined ADC, which achieves a 4.64 GHz CLBW and 72.27 dB LG with 6.4mW power consumption.

2. Analysis of Closed-loop RA in Pipelined ADC

Fig. 1 shows the diagram of proposed closed-loop RA-based MDAC, and describes the different working processes of the MDAC controlled by Φ_1, Φ_{1e} and Φ_2 in detail. We achieve the output stage signal-to-noise ratio requirement by using a high-supply to obtain a large output swing without Miller-compensation, shown in Fig. 1(d). Under the full swing input, the THD of RA is obtained by FFT analysis of the sampled and amplified data, as shown in the Fig. 2.

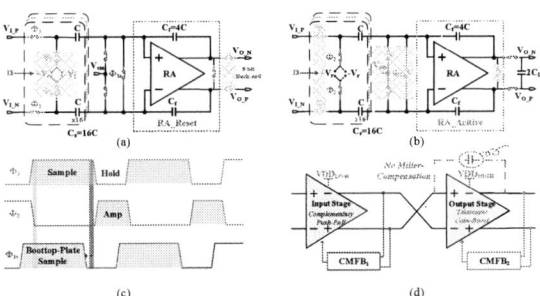

Fig. 1 Schematic of the MDAC: (a) Sample phase, (b) Hold Phase, (c) Timing diagram, (d) RA block diagram

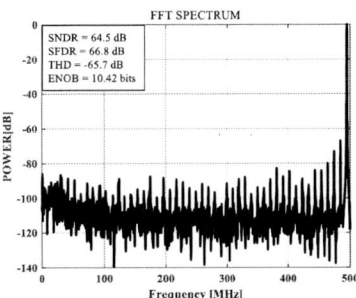

Fig. 2 THD of the proposed RA

3. Circuits details of the proposed op-amp

Fig. 3 shows the detailed schematic of the proposed op-amp. The input stage is a low-supply 0.9V pre-amplifier based on complementary push-pull structure that provides wide bandwidth for fast settling. On top of that, approximately twice the transconductance value of conventional differential pairs can be achieved. The second stage employs a telescopic gain-boost cascode structure with high-supply 1.8V to achieve high dc gain, large output swing and high linearity. Besides it, the inputs and outputs of the two telescopic auxiliary amplifiers are respectively connected with the sources

979-8-3503-6184-1/24 $31.00 © 2024 IEEE

and drains of the cascode transistors. Thus, a high gain, wide bandwidth Miller-free compensation two stage op-amp is realized. Both stages employ a conventional switched capacitor common mode feedback structure to maintain good stability of the RA [3].

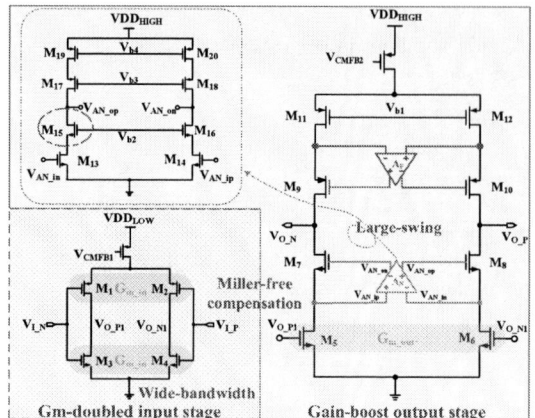

Fig. 3 The proposed op-amp with multi-power domain

The whole op-amp with multi-power domain is implemented with thin-oxide MOSFET transistors for high performance, but breakdown and reliability issues during power-on cannot be ignored. During the multi-power domain power-on process, the voltage fluctuations at folded nodes should be carefully considered, especially for M_{15} where V_{ds15} fluctuates the most. We present a specific power-on sequence to ensure that the circuit does not suffer from the above issues. It can be seen from Fig. 4 that the method of starting the high supply after a delay can improve the reliability of the op-amp across 27 PVT corners.

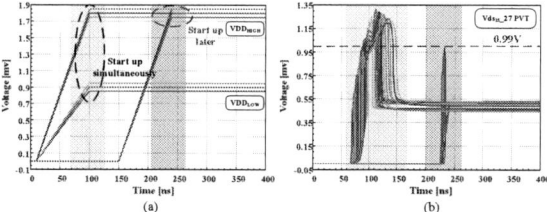

Fig. 4 V_{ds15} fluctuation of M_{15} under different power-on sequences across PVT variations

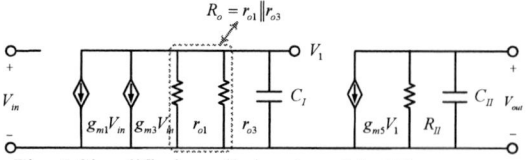

Fig. 5 Simplified small signal model of the op-amp

From Fig. 5, the transfer function of the whole amplifier can be described as:

$$H(s) = -(g_{m1} + g_{m3}) \cdot R_o \left\| 1/sC_I \cdot g_{m5} \cdot R_{II} \right\| 1/sC_{II}$$

$$= -\frac{(g_{m1} + g_{m3}) \cdot g_{m5} \cdot R_o R_{II}}{R_o R_{II} s^2 C_I C_{II} + s(R_o C_I + R_{II} C_{II}) + 1} \quad (1)$$

where C_I consists of $C_{gd1,2}$, $C_{gd3,4}$, and $C_{gs5,6}$. The loading C_L=250fF is the main component of C_{II} and $C_I \ll C_{II}$.

4. Post-layout Simulation Results

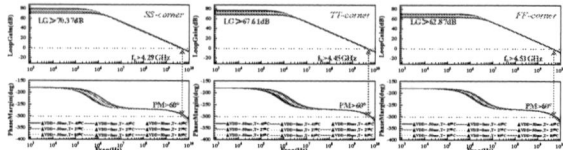

Fig. 6 Loop Frequency of the RA across PVT variations

To verify the frequency characteristics of the op-amp more comprehensively, Fig. 6. shows the loop frequency response of RA under different PVT variations. The post-layout simulation results illustrate that, in all cases, the op-amp can still achieve a LG greater than 62.87dB and a CLBW greater than 4.29GHz, while meeting a PM higher than 60°.

To pick up a good circuit yield, it is necessary to use Monte Carlo analysis to obtain the probability of the process corners and local device mismatch change. The PM distribution obtained by 100 times of Monte Carlo analysis is shown in Fig. 7. Obviously, benefited from our special op-amp structure, the stability of the RA without Miller compensation is still well guaranteed. The average PM is 65.89°, and standard deviation is 2.16°.

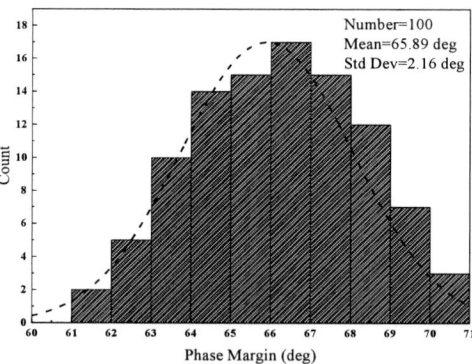

Fig. 7 Monte Carlo analysis of the RA's PM distribution

The nonlinearity of op-amp, depending on the gain and swing range, produces harmonic frequency component, which drastically deteriorates the linearity of ADC. By scanning the DC parameters of the input differential V_{in}, the corresponding output differential swing V_{out} is obtained, and then the slope of V_{out} is calculated to obtain the output Gain. Through coordinate transformation, the relationship between gain and output swing is finally obtained, as shown in Fig. 8(c). Thanks

979-8-3503-6184-1/24 $31.00 © 2024 IEEE   567

to the high-supply of second stage, the op-amp OLG can remain stable in the swing of 0.5V at the single end across PVT variations.

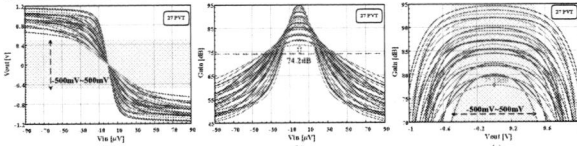

Fig. 8 The OLG vs output swing across PVT variations

Settling time is an important indicator of op-amp which is significant for GS/s ADCs. It includes both the time required for the output to slew as well as the time to settle within the specified error band [4]. The first part is determined by the op-amp big signal performance which is called SR. And the other is up to the zero-pole position which is reflection of the small signal performance.

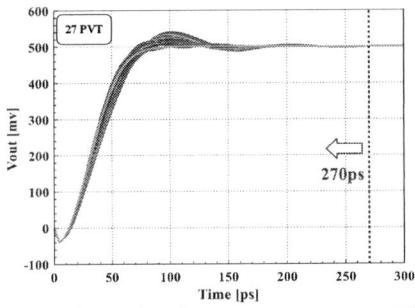

Fig. 9 The large signal step response across PVTs

As a result, the short settling time mainly relies on the rise in the SR and the proper set of the zero-pole position. The main solution to shorting the slew time is to increase the tail current. There is a trade-off between the settling time and power consumption. Because the load capacitance, which is exactly the sampling capacitance of the second stage, is around a small value of 250fF. With a quiescent current of 1.5mA, the SR can achieve 6GV/s. When it comes to the first stage, the output resistance is small and the quiescent current does not have to be too high for a wide bandwidth. With a good control of the PM, no matter the differential signal loop or the common signal loop, this circuit gets rid of the harm from overshoot and oscillator ring. Simulation results, as shown in Fig. 9, prove that the closed-loop amplifier can stabilize the signal within 270ps under different PVT conditions. In addition to the settling time to meet the requirements of less than 280ps, the error of the op-amp after the settling and stability is less than $1/2^{10}$. The settling error ε_{set} can be expressed more intuitively by the following formula:

$$\varepsilon_{set} = 20\lg[\frac{\left|V_{set} - V_{out@270ps}\right|}{V_{set}}] \quad (2)$$

where V_{set}=500mv, $V_{out@270ps}$ refers to the Vout value after 270ps stabilization. Fig. 10. illustrates that at all PVT corners, the settling error can be less than -61dB, ensuring that the performance of the back-end ADC is not affected. Fig. 11 shows the microphotograph of the MDAC based on the proposed op-amp.

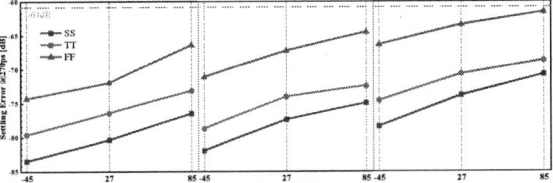

Fig. 10 The large signal step response across PVTs

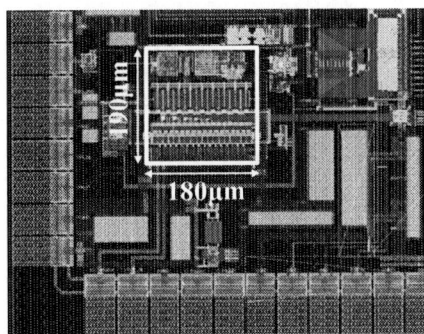

Fig. 11 Layout of the proposed MDAC

5. Summary

With the gradual popularity of portable electronic products, low-power, high-performance op-amp has become a research hotspot in the field of analog circuits. A two-stage no Miller-compensated dual-supply op-amp with complementary push-pull amplifier is proposed for high-speed and high-precision pipelined ADC which achieves a LG of 72.27dB and a closed-loop bandwidth of 4.64GHz, while meeting a PM higher than 60° simulated by Monte Carlo analysis, within a rather lower budget power of 6.4mW.

References

[1] H. Ju and M. Lee, IEEE Transactions on Very Large Scale Integration Systems, 28, p. 1770-1781 (2020).

[2] S. Liu, Z. Zhu, J. Wang, L. Liu and Y. Yang, 2017 IEEE Transactions on Circuits and Systems I: Regular Papers, 66, p.20-30 (2019).

[3] O. Choksi and L. R. Carley, IEEE Transactions on Circuits and Systems II: Analog and Digital Signal Processing, 50, p.906-917 (2003).

[4] M. H. Naderi, S. Prakash and J. Silva-Martinez, IEEE Transactions on Circuits and Systems I: Regular Papers, 65, p.3769-3779 (2018).

A 180 mV–1.6 V Thermoelectric Energy Harvesting Converter with Low-Voltage Cold Start and Less than 1 µW Power Loss

Chunlin Wang [1], Anzhi Yan [1], Tianyu Guo [1], Peng Wan [1], Houfang Liu [1], Yi Yang [1], Tianling Ren*[1]

[1] School of Integrated Circuits and Beijing National Research Center forInformation Science and Technology (BNRist), Tsinghua University

* Email: wangcl21@outlook.com, RenTL@tsinghua.edu.cn

Abstract—The extraction of thermal gradient energy from ambient sources to power wireless sensors has emerged as a competitive solution. However, its micro-watt-level input power and milli-volt-level input voltage present obstacles to energy utilization and circuit start. In this work, we have designed a low-voltage cold-start DC-DC thermoelectric energy harvesting converter. By constructing ring oscillators utilizing stacked inverters, charge pumps with a dynamic gate bias technique, and a clock amplification circuit, the converter successfully boosts the voltage from 180 mV to 1.6 V within 3.2 ms while maintaining a power loss of only 999.7 nW. Without off-chip components, the proposed topology improves circuit integration and power density. It is valuable for the battery-less operation of wearable sensors powered by thermoelectric generators.

Keywords—*thermoelectric energy harvesting, cold start, stacked inverter, dynamic gate bias technique, clock amplification*

I. INTRODUCTION

Wireless sensors, acting as an important bridge connecting humans and the environment, play a crucial role in the Internet of Things (IoT) [1]. However, powering these sensors remains a major challenge. Compared with other energy sources, thermoelectric generators (TEGs) can directly convert heat originated from ubiquitous temperature differences into electric energy without interference from environmental factors. Their structures offer advantages such as small volume, light-weight, and vibration-free operation, which have attracted increasing attention as a renewable energy source. These advantages significantly enhance the comfort of sensors, making the use of TEGs highly promising [2].

In wearable and implantable IoT applications, human body heat is a better energy source for TEGs to continuously operate unaffected by environmental fluctuations. However, TEGs generate a low output voltage, which is insufficient to directly drive circuits [3]. Therefore, a suitable voltage converter is needed as a link between TEGs and loads. In the past few years, many researchers have done a lot of work in this area. Baek-Min Lim et al. have designed a thermoelectric self-starting boost converter based on a Colpitts oscillator, but this system requires two off-chip inductors, which means low integration [4]. A battery-less thermoelectric energy harvesting power management integrated circuit designed by Cui Peng et al. has a minimum operating voltage of 10 mV. However, the power consumption of it is relatively high [5].

In this work, a low-voltage, cold-start DC-DC thermoelectric energy harvesting converter is designed. The converter can boost voltage from 180 mV to 1.6 V with power loss less than 1 µW. It can achieve a cold start without the help of other external energy sources or off-chip components. The

rest of the paper is organized as follows: Section II presents the system design of the proposed converter. Section III shows the circuit implementation and section IV focuses on the demonstration of simulation results. Finally, the conclusion is given in Section IV.

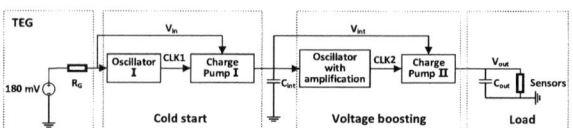

Fig. 1. The low-voltage cold-start DC-DC thermoelectric energy harvesting converter circuit structure, which uses a TEG as an input.

II. DESIGN OF THE PROPOSED CONVERTER

Fig. 1 illustrates the overall block diagram of the proposed DC-DC converter. The input voltage from the TEG is much lower than the threshold voltage of standard MOSFETs. As a result, it cannot directly drive the circuit to start. To address this challenge without requiring additional assistance, the cold start is a good solution. In contrast to alternative designs relying on off-chip components or auxiliary power supplies, this cold start block enables higher integration. It is mainly composed of two key parts: a ring oscillator composed of stacked inverters and a three-stage dual-branch charge pump with a dynamic gate bias technique (DGB). Both of them utilize Low-Voltage-Threshold (LVT) MOSFETs. The cold start output not only supplies a voltage rail but also provides the necessary power for the subsequent block.

The latter block primarily consists of a ring oscillator composed of double-stacked inverters, a clock amplification circuit, and a two-stage charge pump having the same structure as Charge Pump I. Inverters with more layers stacked increase the inherent latency per stage. This reduces the oscillation frequency of the clock, thereby minimizing the dynamic loss of the system. In this block, we design a clock amplification circuit that can provide almost twice the clock oscillation of the input voltage, exceeding the threshold voltage of standard MOSFETs. A high-amplitude clock can better conduct MOSFETs and reduce the conduction loss of the charge pump connected behind it.

III. CIRCUIT IMPLEMENTATION

A. Ring Oscillator Composed of Stacked Inverters

Oscillator I is a ring oscillator composed of 13 stacked inverters, each of which is a delay cell, and its detailed structure is shown in Fig. 2(a). The input signal is sent to the gate of each MOSFET, and the source of M_{p1} and N_{n1} is

979-8-3503-6184-1/24 $31.00 © 2024 IEEE

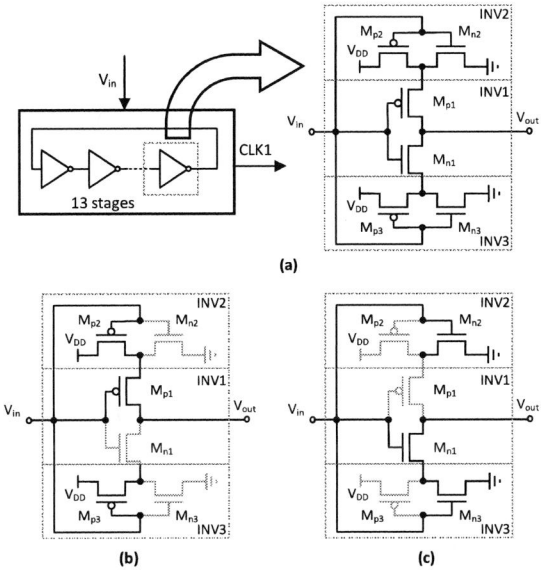

Fig. 2. Oscillator I detailed circuit design structure: (a) ring oscillator composed of stacked inverter delay cells, (b) low V_{in}, (c) high V_{in}.

connected to the output of the other two inverters. Among them, the gate widths of M_{p2} and M_{n3} are slightly larger compared with the other MOSFETs to stabilize the DC operating point of the delay cell. The signal input to INV1 is amplified by INV2 and INV3. As a result, the gain of this structure are greater than those obtained by a single inverter alone [6]. Fig. 2(b) and Fig. 2(c) show the working situation when V_{in} is low and V_{in} is high in sequence.

Due to the low input voltage, LVT MOSFETs are used here. Compared with standard ones, they have stronger driving ability in this condition. In addition, the size of the transistors in INV1(M_{p1} and M_{n1}) is designed to be slightly larger than the smallest one. In this way, it can increase the parasitic capacitance, prolong the charging and discharging time, and thus reduce the oscillation frequency. This design offers two advantages. Firstly, it can relatively relax the time requirements for the rising and falling edges. Secondly, it can reduce the number of times parasitic capacitors are charged and discharged per unit time, thereby reducing the dynamic power consumption of this ring oscillator.

B. Charge Pump with DGB

In traditional circuit design, Dickson charge pump are used to achieve voltage boosting. However, the main limitation of it is the high voltage loss and it is unsuitable for low-voltage circuit designs. To solve the problem, we have implemented a dual-branch charge pump, utilizing DGB to optimize performance. This design decreases both forward conduction loss and reverse leakage current loss, which are the two main factors affecting boosting ability.

The structure of Charge Pump I is shown in Fig. 3. It is a cascade structure characterized by a PMOS-NMOS dual switch design (M_{px2} and M_{nx2}, x = 1 or 2 in Fig. 3(b)). The application of DGB has enhanced the effectiveness and the two branches have been isolated to prevent reverse current crosstalk. The basic principle is to apply different voltages to the gate of the PMOS (M_{px1}, x = 1 or 2) used for charge transfer at different times in one clock cycle.

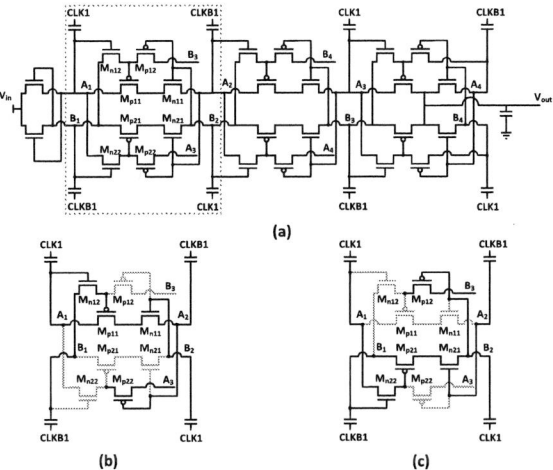

Fig. 3. Charge Pump I with DGB to improve charge transfer ability and reduce energy loss: (a) structure of Charge Pump I, (b) high CLK and low CLKB input, (c) low CLK and high CLKB input.

The conduction resistance of MOSFETs is inversely proportional to the overdrive voltage. In the forward conduction time, adjusting the gate voltage of PMOS (M_{px1}, x = 1 or 2) to increase V_{gs} reduces the MOSFET's conduction resistance, thereby enhancing the current driving ability of the charge pump. Conversely, in the reverse conduction time, decreasing V_{gs} leads to an increase in the conduction resistance, correspondingly attenuating the reverse current flow of the branch. M_{px2} and M_{nx2} (x = 1 or 2) share a synchronized off time, which prevents current from reversely flowing through this path [7].

C. Voltage Boosting Block

This block mainly consists of a ring oscillator composed of double-stacked inverters, a clock amplification circuit, and a two-stage charge pump with DGB. The working principle of this oscillator is similar to Oscillator I. However, compared with the previous one, this design stacks more layers to further enhance the pull-up and pull-down capabilities.

In order to improve the voltage-boosting ability of the charge pump, a clock amplification circuit is designed. The schematic of it is shown in Fig. 4. It utilizes the charge storage capacity of capacitors. In the initial state, the voltage of C_c is zero. When CLK is high and CLKB is low shown in Fig. 5(a), the pull-up PMOS M_{Hp1} is turned on, and the capacitor is

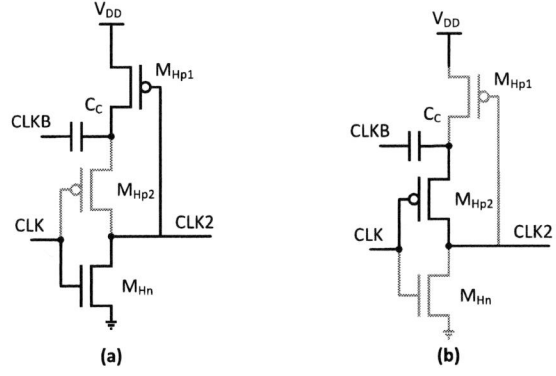

Fig. 4. Clock amplification circuit designed to increase the amplitude of the clock: (a) high CLK and low CLKB, (b) low CLK and high CLKB.

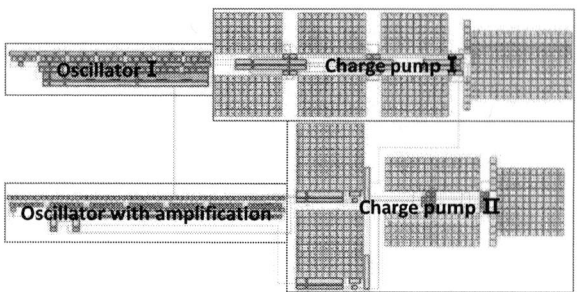

Fig. 5. Layout of the proposed thermoelectric energy harvesting converter designed in a 0.25 μm process.

charged to VDD by the power supply. When the input clock CLK is low shown in Fig. 5(b), M_{Hp1} is turned off, and M_{Hp2} is turned on. At this time, with CLKB being high, CLK2 gets charged to 2VDD, thereby accomplishing an amplitude amplification function.

Charge Pump II in the voltage boosting block is a two-stage structure that uses the same topology as Charge Pump I. The voltage of the clock exceeds the threshold voltage, making MOSFETs within it have better charge transfer capability.

IV. SIMULATION RESULTS

The proposed thermoelectric energy harvesting converter is designed in a 0.25 μm process, and its layout is shown in Fig. 5. The threshold voltage for LVT MOSFETs is approximately 200 mV, while for standard MOSFETs, it is around 550 mV. After the operation of the clock amplification circuit, the voltage amplitude of the clock has been increased from 427 mV to 730 mV. The simulation voltage waveforms of CLK and CLK2 in Fig. 4 are shown in Fig. 6(a). The schematic-level and post-layout transient simulation waveforms of the output voltage are presented in Fig. 6(b). Due to the influence of parasitic parameters, the time for the output voltage of the proposed converter to reach steady state is extended from 3.2 ms to 4.7 ms given a 50 MΩ load. This converter can realize a cold start at all process corners and achieve a cold start at as low as 100 mV. Under an input voltage of 180 mV, the converter reliably outputs a voltage of 1.6 V with a low ripple voltage of 14.9 mV, indicating its stability and ability to operate effectively even under low current conditions. The power loss of the entire circuit is merely 999.7 nW. The performance of this proposed converter is compared with the state-of-the-art works in Table I.

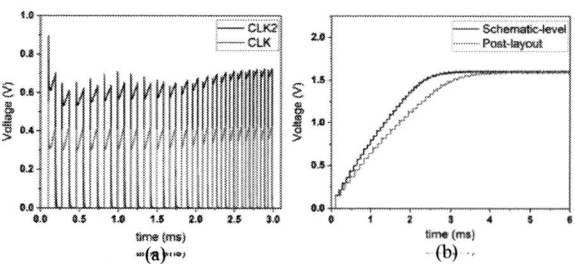

Fig. 6. Simulation waveforms of the converter: (a) Comparison of transient simulation voltage waveforms between CLK and CLK2; (b) schematic-level and post-layout transient simulation waveforms of the output voltage.

TABLE I. COMPARISON TABLE

	This work	[4]	[5]	[8]	[9]
Energy source	TEG	TEG	TEG	TEG+ PEG	TEG
Process (nm)	250	65	180	180	180
Min V_{TEG} for a cold start (mV)	100	40	180	10 mV +PEG	170
Output voltage (V)	1.6	1.1	1.7	1.2	1
Power loss (μW)	0.9997	636.8	24.6	37.5	--[a]
Inductor (μH)	no	3.3*2 [b]	4.7	220	no
Auxiliary start unit	no	no	no	yes	no

[a]. -- means not mention.

[b]. *2 means there are two off-chip inductors.

V. CONCLUSION

A low-voltage cold-start DC-DC converter for thermoelectric energy harvesting applications is presented in this paper. With an input of 180 mV, the converter can output a voltage of 1.6 V, and the time for the output to reach a stable state is only 3.2 ms. The power loss is merely 999.7 nW. These simulation results verify the feasibility of this approach. This small current requirement enables the converter to perform excellently in various applications. It can be integrated with multiple sensors to achieve more convenient detection of human physiological signals.

REFERENCES

[1] S. Fan, R. Wei, L. Zhao, X. Yang, L. Geng and P. X. . -L. Feng, "An Ultralow Quiescent Current Power Management System With Maximum Power Point Tracking (MPPT) for Batteryless Wireless Sensor Applications," in IEEE Transactions on Power Electronics, vol. 33, no. 9, pp. 7326-7337, Sept. 2018.

[2] Y. Zhang et al., "A Batteryless 19 μW MICS/ISM-Band Energy Harvesting Body Sensor Node SoC for ExG Applications," in IEEE Journal of Solid-State Circuits, vol. 48, no. 1, pp. 199-213, Jan. 2013.

[3] Q. Wan, Y. -K. Teh, Y. Gao and P. K. T. Mok, "Analysis and Design of a Thermoelectric Energy Harvesting System With Reconfigurable Array of Thermoelectric Generators for IoT Applications," in IEEE Transactions on Circuits and Systems I: Regular Papers, vol. 64, no. 9, pp. 2346-2358, Sept. 2017.

[4] B. -M. Lim, J. -I. Seo and S. -G. Lee, "A Colpitts Oscillator-Based Self-Starting Boost Converter for Thermoelectric Energy Harvesting With 40-mV Startup Voltage and 75% Maximum Efficiency," in IEEE Journal of Solid-State Circuits, vol. 53, no. 11, pp. 3293-3302, Nov. 2018.

[5] P. Cui, B. Wei, Z. Liang, X. Wei and W. Xu, "A Power Management Unit for Battery-Less TEG Energy Harvesting With Low Voltage Self-Startup," 2021 6th International Conference on Integrated Circuits and Microsystems (ICICM), Oct. 2021, pp. 160-165.

[6] S. Bose and M. L. Johnston, "A Stacked-Inverter Ring Oscillator for 50 mV Fully-Integrated Cold-Start of Energy Harvesters," 2018 IEEE International Symposium on Circuits and Systems (ISCAS), May 2018, pp. 1-5.

[7] J. K. Yong et al., "A 0.1-V VIN Subthreshold 3-Stage Dual-Branch Charge Pump With 43.4% Peak Power Conversion Efficiency Using Advanced Dynamic Gate-Bias," in IEEE Transactions on Circuits and Systems II: Express Briefs, vol. 69, no. 9, pp. 3929-3933, Sept. 2022.

[8] R. Wang, Y. Liang and S. Du, "A 10-mV-Startup-Voltage Thermoelectric Energy Harvesting System With a Piezoelectric Starter," 2022 IEEE International Symposium on Circuits and Systems (ISCAS), May 2022, pp. 1482-1486.

[9] H. O. Tabrizi, H. M. P. C. Jayaweera and A. Muhtaroğlu, "Fully Integrated Autonomous Interface With Maximum Power Point Tracking for Energy Harvesting TEGs With High Power Capacity," in IEEE Transactions on Power Electronics, vol. 35, no. 5, pp. 4905-4914, May 2020.

979-8-3503-6184-1/24 $31.00 © 2024 IEEE

A Signal Conditioning ASIC With High Precision and Low Noise for MEMS Accelerometers

Quan Sun [1], Rui Liu [2], Zhe Zheng [2], Lei Dong [1], Ji-jiang Wang*[1]

[1] Xi 'an Aerosemi Technology. Co., Ltd. Xi'an, 710075, China
[2] Beijing Smart-chip Microelectronics Technology Co., Ltd. Beijing, 100192, China

* Email: wjj@aerosemi.com

Abstract—This paper proposes a high-precision signal conditioning Application-Specific Integrated Circuit (ASIC) for capacitive microelectromechanical systems (MEMS) accelerometer sensors. The capacitor-voltage(C-V) converter adopts a time-multiplexed method, thus indirectly realizing full-differential signal processing and suppressing the effects of common-mode signals and parasitic capacitance. An adjustable array of compensation capacitors is also incorporated in the C-V converter to effectively suppress the offset voltage and nonlinearity of the circuit with the correction of the sensing element. The main signal path of the system is based on a fully differential design, and to effectively suppress the offset voltage and low-frequency noise in the system, an auto-zeroing technique is used with correlated double sampling. The ASIC is fabricated in a 0.18 μm CMOS process. The results show that the overall acceleration system has a range of ±2g, a supply voltage of 3.3V, a total system power consumption of 4.95mW, a nonlinearity error of 0.075%, a noise floor of 0.76μg/√Hz, and a bandwidth of 10kHz. The application-specific integrated circuit for signal conditioning in this thesis realizes the performance of high precision, low offset, and low noise in an open-loop topology. It can be applied to many types of MEMS accelerometers.

Keywords—MEMS accelerometer, High precision, Signal Conditioning ASIC, Fully differential.

I. INTRODUCTION

MEMS accelerometers, as sensors for detecting the acceleration of targets, have a wide range of applications in vibration detection and motion recognition. The signal conditioning ASIC is supposed to pick it up and convert it as a suitable range voltage signal after C-V conversion, amplification, noise reduction, and filtering. A highly customized signal conditioning ASIC is needed to match the MEMS accelerometer [1]. The working mode of the accelerometer signal conditioning ASIC can be categorized into two modes: open-loop and closed-loop.

Ippei Akita et al. and Longjie Zhong et al. adopted the open-loop structure [3-4]. The open-loop structure is relatively simple and can achieve low-power circuits. Still, the linearity is generally poor, which has a more significant impact on the system's accuracy.

Xiangyu Li et al. and Yuki Furubayashi et al. adopted the closed-loop structure [5-6]. The capacitive accelerometer of the closed-loop system has the characteristics of high sensitivity, high precision, and high linearity. Still, the acceleration measurement range is small, and the complex structure leads to high power consumption. Therefore, this paper does not design a closed-loop accelerometer signal conditioning ASIC.

The signal conditioning ASIC designed in this paper has a main signal path, including a capacitor-voltage converter, a switched capacitor programmable gain amplifier (PGA), and a fully differential Sallen Key low-pass filter (LPF). The ASIC adopts a low-power open-loop structure and various measures to improve linearity and overall performance. The C-V converter adopts a time-division multiplexing method to achieve fully differential signal processing, suppressing the influence of common-mode signals and parasitic capacitance. We also add an adjustable compensation capacitor array to the C-V converter, which can effectively suppress the offset voltage and nonlinearity of the circuit while correcting the sensing elements. Based on the fully differential design, the main channel of the system uses auto-zero technology to effectively suppress offset voltage and low-frequency noise in the system and perform correlated double sampling.

This paper is organized as follows. Section II describes the system structure and C-V converter. Section III presents the simulation results. Section IV concludes this paper.

II. CAPACITIVE MEMS ACCELEROMETER SENSORS

A. System architecture

This paper designs a high-precision MEMS accelerometer signal conditioning ASIC, as shown in Fig. 1 of the system architecture. The ASIC includes a fully differential C-V conversion circuit, a fully differential programmable gain amplifier, and a fully differential programmable low-pass filter.

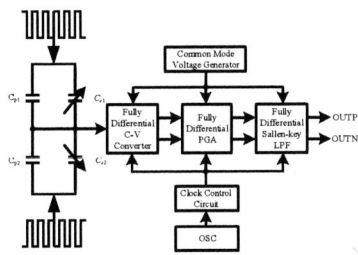

Fig. 1. System architecture

The C-V converter converts the differential change of two capacitors into a detectable voltage. This paper proposes a time-division multiplexing fully differential C-V converter with a half-bridge sensing structure and two parallel single-ended charge integrators. To reduce the offset and nonlinearity of the sensor, we add a capacitive compensation array to the C-V converter. The differential output of the capacitive voltage converter is connected to a programmable gain

---

Supported by The Scientific Research Programs for High Level Talents of Beijing Smart-chip Microelectronics Technology Co., Ltd.

979-8-3503-6184-1/24 $31.00 © 2024 IEEE

amplifier, which amplifies the signal to a reasonable range, thus accommodating various accelerometers and enabling multi-range measurements.

We have designed a time-division multiplexing fully differential switched capacitor programmable gain amplifier with continuous output to provide continuous signals for subsequent low-pass filters. The differential output of the programmable gain amplifier is sent to the low-pass filter, which filters out the high-frequency ripple of the input signal to obtain a low-frequency signal with a continuous output that meets the requirements. We have also designed a fully differential Sallen Key LPF with a fully differential amplifier. The clock circuit provides a clock signal for the entire signal conditioning ASIC.

B. Capacitor-voltage converter

The common-mode rejection and anti-mismatch design of C-V converters play an important role in improving the accuracy of the entire accelerometer signal conditioning ASIC. Fig. 2 shows the structure of the C-V converter. VCM is the common-mode voltage. Among them, capacitors C_3, C_4, C_7, and C_8 are auto-zero capacitors, which can eliminate the offset of op amps OP1 and OP2 and the low-frequency noise of the circuit. C_1, C_2, C_9, and C_{10} are integral feedback capacitors. The overall correlated double sampling structure is adopted, and C_{S1} and C_{S2} are sampled twice, output by OUTP and OUTN, respectively. We subtract the two signals to eliminate low-frequency mechanical noise and the noise generated by sampling switches s4 and s5. The function of OP3 is to protect the circuit under strong impact. R_1, C_{11}, R_2, and C_{12} form a filtering circuit to filter the output once and input it into OP3. If the C-V converter receives a solid impact, the ASIC will reset when the output voltage exceeds the threshold voltage set by the operational amplifier.

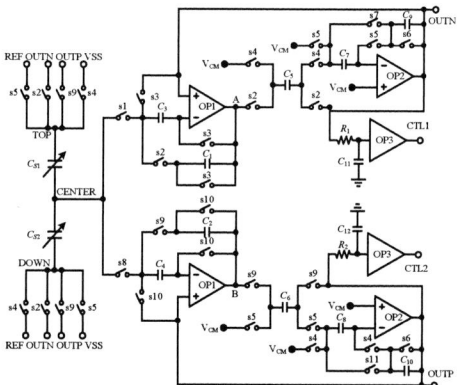

Fig. 2. Structure of C-V converter

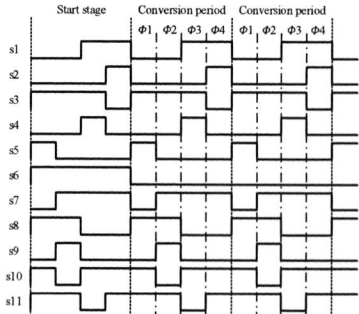

Fig. 3. Switch signals of C-V converter

The C-V converter adopts a switched-capacitor circuit. We control the operation states of the upper and lower parts of the circuit by managing the conduction and disconnection of the switches. The system's clock frequency is set at 250 kHz. Fig. 3 depicts the switch signals of the C-V converter. Two symmetrical single-ended C-V converters operate alternately. The two converters operate during s8 conduction and s1 conduction in one period, respectively.

At the start stage, when s6 is turned on, the charge at both ends of the capacitors C_9 and C_{10} is cleared to zero. When s4 or s5 is turned on, it will form a unit gain auto-zero with the operational amplifier OP2, C_7, and C_8. Except for the start stage, s6 is in the disconnected state in other periods.

In stages $\Phi 1$ and $\Phi 2$, the lower half of the single-ended C-V converter works. In stages $\Phi 3$ and $\Phi 4$, the upper half of the single-ended C-V converter works.

In the NT period, when the C-V converter changes from working state $\Phi 1$ to $\Phi 2$, the charges on C_{S1} and C_{S2} transfer to C_2. At this time, due to the change in V_A voltage, charge is also stored on C_6. In stages $\Phi 3$ and $\Phi 4$, C_6 is not connected to the circuit until the $\Phi 1$ stage of the (N+1)T period, when the charge of C_6 transfers to capacitor C_{10}. Similarly, in the NT period, when the C-V converter changes from working state $\Phi 3$ to $\Phi 4$, the charges on C_{S1} and C_{S2} transfer to the upper part of C_1. Since $C_1 = C_2$, $C_5 = C_6$, $C_9 = C_{10}$, and output $V_{OUT}[n] = V_{OUTP}[n] - V_{OUTN}[n]$, we can obtain

$$-\frac{C_6}{C_2}[V_{OUT}[n] \cdot (C_{S1} + C_{S2}) - (V_{REF} - V_{SS}) \cdot (C_{S1} - C_{S2})] = (V_{OUT}[n+1] - V_{OUT}[n])C_{10}$$
(1)

Let $q = 1 - \dfrac{C_6 \cdot (C_{S1} + C_{S2})}{C_2 \cdot C_{10}}$, $b = (V_{REF} - V_{SS}) \cdot \dfrac{C_6 \cdot (C_{S1} - C_{S2})}{C_2 \cdot C_{10}}$, and because $V_{OUT}[0] = 0$, and by design, we can make 0<q<1, so we can obtain

$$
\begin{aligned}
V_{OUT}[n] &= b \cdot \frac{1 - q^{n-1}}{1 - q} \approx b \cdot \frac{1}{1 - q} \\
&= (V_{REF} - V_{SS}) \cdot \frac{C_6 \cdot (C_{S1} - C_{S2})}{C_2 \cdot C_{10}} \cdot \frac{1}{1 - [1 - \frac{C_6 \cdot (C_{S1} + C_{S2})}{C_2 \cdot C_{10}}]} \\
&= \frac{C_{S1} - C_{S2}}{C_{S1} + C_{S2}} \cdot (V_{REF} - V_{SS})
\end{aligned}
$$
(2)

According to (2), we can deduce that the output of the designed fully differential time-multiplexed C-V converter depends only on the sizes of the MEMS mechanical capacitors C_{S1} and C_{S2}, as well as the voltages VREF and VSS, and is independent of the sizes of other capacitors within the C-V converter.

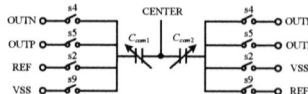

Fig. 4. Capacitor compensation for C-V converter

Because of the manufacturing variations in the MEMS mechanical capacitors, which are the sensitive structures of the sensor, there can be changes in the DC component of the signal. Due to the difficulty in avoiding manufacturing errors,

979-8-3503-6184-1/24 $31.00 © 2024 IEEE   573

we have designed a capacitive compensation array, as shown in Fig. 4.

After adding compensation capacitors C_{com1} and C_{com2}, we can obtain

$$V_{OUT}[n] = \frac{(C_{S1} - C_{com2}) - (C_{S2} - C_{com1})}{(C_{S1} - C_{com2}) + (C_{S2} - C_{com1})} \cdot (V_{REF} - V_{SS}) \quad (3)$$

Therefore, by adjusting the sizes of the compensation capacitors C_{com1} and C_{com2}, we eliminated the manufacturing variations present in the sensor-sensitive structure, reducing both offset and circuit noise. As a result, the resolution and linearity of the proposed front-end circuit improve obviously due to the low common-mode disturbance and high match with the sensing element.

III. SIMULATION RESULTS

We have designed the high-precision MEMS accelerometer signal conditioning ASIC using 0.18μm CMOS technology. We conduct simulations to verify the improvement in the ASIC's linearity and the reduction of noise.

We tested the linearity of the accelerometer through a tumble test, using a meter header with a configured sensitivity of 1.5V/g. The acceleration range generated by the vibration table is ±2g.

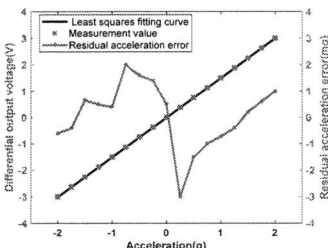

Fig. 5. Differential output voltage and corresponding least squares fitting curve with residual acceleration error

From Fig. 5, we can obtain the experimental differential output voltage and the corresponding least-squares fitting curve, as well as the residual acceleration error. It can be seen from the figure that the maximum error of the residual acceleration is -3mg, and the ASIC measurement range is ±2g, with a total of 4g. 3mg/4g=0.075%. Therefore, the DC nonlinearity of the accelerometer signal conditioning ASIC is 0.075%. In addition, the circuit design of the C-V converter, fully differential programmable gain amplifier, and Sallen Key low-pass filter solve the problems of mismatch, mechanical thermal noise, and superimposed high-frequency noise, thereby further improving the linearity of the circuit and significantly reducing the nonlinearity of the front-end circuit.

The measured noise power spectral density is shown in Fig. 6, from which we can determine that the accelerometer's base noise is 0.76μg/√Hz. Owing to the accelerometer's sensitivity of 1.5V/g, the base noise equates to 1.14μV/√Hz. Through conducting noise measurements, we observe that the auto-zero and correlated double sampling techniques significantly eliminate 1/f noise.

Table I concludes the specific parameters about the accelerometer system of this research. Moreover, the performance parameters of the accelerometer systems reported in other recent related studies are listed for comparison. The ASIC designed in this paper achieves significantly lower DC nonlinearity than other circuit structures. Compared with the smallest DC nonlinearity in comparative papers (0.15%), the DC nonlinearity of the ASIC designed in this paper dropped by nearly 50%. Compared to other open-loop structures, while achieving minimal noise, nonlinearity is also tiny. Compared with closed-loop structures, the range is more extensive, and the structure is more straightforward.

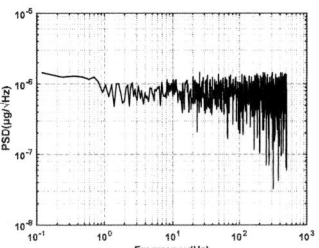

Fig. 6. Noise spectrogram of accelerometer

TABLE I. COMPARISON TABLE

	[2]	[3]	[4]	[5]	This work
Operation Mode	open loop	open loop	closed loop	closed loop	open loop
Range	±5g	±4g	±1g	±0.55g	±2g
Noise (μg/√Hz)	290	112	0.48	0.022	0.76
Nonlinear	1%	1.5%	0.15%	-	0.075%
Bandwidth (Hz)	50	12.5k	300	400	10k
Voltage (V)	1	1.8	5	1.4/1.8/12	3.3

IV. CONCLUSION

This paper designs a new type of high-precision fully differential capacitive accelerometer signal conditioning ASIC. By introducing capacitance compensation arrays to correct sensor errors, we improve the system's accuracy and reduce offset and nonlinearity. We have entirely designed the main channel of the system with a fully differential approach to effectively reduce common-mode interference. To effectively suppress the offset voltage and low-frequency noise in the system, we utilize auto-zero technology and subject it to correlated double sampling. The final simulation results indicate that this structure enables the ASIC to have high accuracy and low noise.

REFERENCES

[1] Smart Sensors and MEMS: Intelligent Sensing Devices and Microsystems for Industrial Applications[M]. Woodhead Publishing, 2018.

[2] Akita I, Okazawa T, Kurui Y, et al. A feedforward noise reduction technique in capacitive MEMS accelerometer analog front-end for ultra-low-power IoT applications[J]. IEEE Journal of Solid-State Circuits, 2019, 55(6): 1599-1609.

[3] Zhong L, Liu S, Xu D. Correlated double amplifying readout technique for low-noise power-efficient MEMS capacitive accelerometer[J]. IEEE Transactions on Instrumentation and Measurement, 2022, 71: 1-11.

[4] Li X, Hu J, Liu X. A high-performance digital interface circuit for a high-Q micro-electromechanical system accelerometer[J]. Micromachines, 2018, 9(12): 675.

[5] Furubayashi Y, Oshima T, Yamawaki T, et al. A 22-ng/√Hz 17-mW Capacitive MEMS Accelerometer With Electrically Separated Mass Structure and Digital Noise-Reduction Techniques[J]. IEEE Journal of Solid-State Circuits, 2020, 55(9): 2539-2552.

Design of a high-precision self-calibration readout circuit for CMOS microbolometer

Qianhao Zhang [1], Jie Liu [1], Sheng Xu[1], Yiming Liao*[2], Feng Yan[1], and Xiaoli Ji[1]

[1] *School of Electronic Science and Engineering, Nanjing University, Jiangsu 210046, China*
[2] *School of Electronic and Optical Engineering, Nanjing University of Science and Technology, Jiangsu 210094, China*

* Corresponding Author's Email: liaoyiming@njust.edu.cn

Abstract—Microbolometer has a wide range of applications in military detection, fire warning, security monitoring and biomedical detection. However, because the response current signal of polycrystalline silicon microbolometer is very weak and difficult to detect, the traditional readout circuit structure is difficult to meet the needs of signal reading. In this paper, a CTIA infrared readout circuit with self-calibration function is proposed to solve the fixed mode noise of microbolometer. The circuit adopts 14-bit hybrid scaling DAC to achieve real-time and adaptive control, and improves calibration accuracy. The test results show that for pixel deviations of less than 0.5%, the self-calibration circuit can eliminate 87.5% fixed-mode noise.

Keywords—*CMOS infrared detector; Fixed mode noise; Self-calibration readout circuit*

I. INTRODUCTION

Microbolometer is the most widely used uncooled infrared detector. The microbolometer based on CMOS process has the advantages of small size, low power consumption, wide spectral response, low cost and integrated design with readout circuit, so it has a good application prospect [1]. Since the output signal of CMOS microbolometer is very weak, the high performance and low noise of the readout circuit directly affect the sensitivity and dynamic range of the imaging system. In order to reduce the noise, Chen et al. [2] based on the background suppression circuit proposed by the French company ULIS, proposed a blind element compensation method to add blind pixels to the CTIA circuit, but due to the process deviation in the manufacturing process of CMOS microbolometer, the resistance of different pixels is not exactly equal, resulting in fixed mode noise (FPN). For FPN, one method is to use CDS, but the CDS circuit uses amplifiers and capacitors to amplify the suppression function of FPN, resulting in the injection of additional noise [3-4]. Another method to reduce FPN is pixel calibration technology, Xu et al. [5] used a self-calibration circuit to automatically suppress pixel FPN, but the 6-bit DAC cannot guarantee the calibration accuracy. Kim et al. [6] used high-precision DAC for calibration, but it still needed external control and was not practical. Although these methods have certain calibration effects, they still have great limitations.

Aiming at the fixed mode noise caused by pixel non-uniformity, an infrared readout circuit with pixel noise self-calibration function is designed in this paper. The 14-bit hybrid zoom DAC is introduced in this circuit, which improves the calibration accuracy and conversion rate while saving the electric road surface. At the same time, digital memory is introduced at the same time. When the system is working, it only needs one calibration and can provide stable compensation current for a long time. It has real-time performance and self-adaptability.

II. CMOS MICROBOLOMETER NOISE ANALYSIS

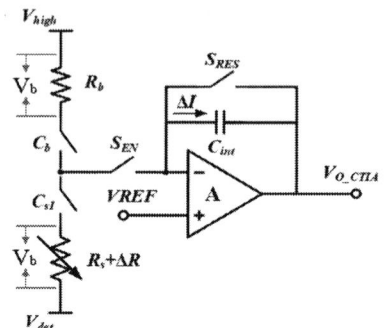

(a)

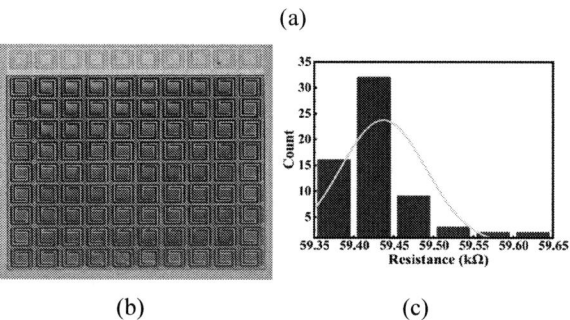

(b)   (c)

Fig. 1.   (a) single pixel readout circuit.   (b) 8×8 pixel array micrograph. (c) Resistance distribution diagram of 8×8 pixel array.

Fig.1(a) shows the structure of a single-pixel readout circuit for a CMOS microbolometer. In the absence of infrared radiation, the resistance of the blind pixel R_b and the probe pixel R_s is equal, and the current flowing into the CTIA is zero.

However, due to the process deviation, the pixel resistance deviation is ΔR. Taking the prepared pixel array as an example, the resistance of the 8×8 polysilicon microbolometer array was characterized. The micrograph of the 8×8 pixel array was shown in Fig.1(b), and the test results were shown in Fig.1(c). The pixel resistance value was distributed in the range of 59.35 kΩ-59.65 kΩ, and the pixel deviation was in the range of 0-0.5%.

Due to the variation of the resistance pixels, even in the absence of infrared radiation, there is still noise current ΔI flowing into the CTIA circuit, noise current ΔI can be expressed by the following formula:

$$\Delta I = \frac{V_{high} - VREF}{R_b} - \frac{VREF - V_{det}}{R_s + \Delta R} \qquad (1)$$

979-8-3503-6184-1/24 $31.00 © 2024 IEEE

According to formula (1), under the condition of no infrared radiation, the same bias V_b is set at both ends of the blind pixel and the detecting pixel respectively, that is, keep $(V_{high}-VREF)$ and $(VREF-V_{det})$ equal to V_b, and the integrated current flowing into CTIA is the noise current caused by pixel deviation ΔR:

$$\Delta I = V_b \times \frac{\Delta R}{R(R+\Delta R)} \qquad (2)$$

At this time, the noise voltage V_n caused by the noise current is:

$$V_n = \frac{\Delta I \cdot t}{C_{\text{int}}} \qquad (3)$$

III. SELF-CALIBRATION READOUT CIRCUIT DESIGN

A. Overall architecture of self-calibration circuit

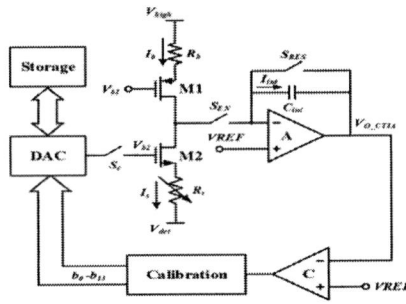

Fig. 2. Overall Architecture of Self-Calibration Readout Circuit.

The structure of self-calibration readout circuit is shown in Fig.2. The main structure includes CTIA readout circuit, offset tube M1 and M2, sampling and holding circuit, comparator C, digital calibration circuit and DAC circuit. The current flowing through the blind pixel R_b is I_b, the current flowing through the probe pixel R_s is I_s, and the resistance deviation causes the current I_s and I_b to be unequal, resulting in noise current (I_b-I_s).

The expression for currents I_b and I_s are as follows:

$$I_b = \frac{V_{high} - V_{gs1} - V_{b1}}{R_b} \qquad (4)$$

$$I_s = \frac{V_{b2} - V_{gs2} - V_{det}}{R_s} \qquad (5)$$

After the noise current is integrated and amplified by the CTIA circuit, it is converted into voltage V_{O_CTIA} at the output end. The output voltage V_{O_CTIA} is compared with the reference voltage $VREF$. The comparison result is transmitted to the digital calibration circuit. The digital calibration circuit controls and adjusts the digital signal b0~b13 according to the output result of the comparator C and feeds back to the DAC. The DAC changes the gate voltage V_{b2} of the transistor M2 according to the feedback digital signal, so as to change the voltage at both ends of the detection pixel R_s, and achieve the purpose of suppressing the detection pixel branch noise

current. After the calibration, the DAC switching signal is stored in the Storage, which is convenient for the array to call directly when it is used next time, eliminating the steps of secondary calibration.

B. DAC Design

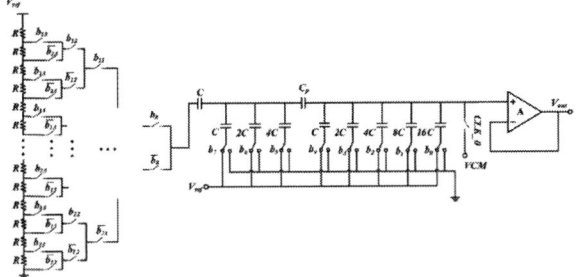

Fig. 3. 3-segment combined DAC with different scaling.

In order to improve the calibration accuracy, a 14-bit DAC module circuit with different scaling types is proposed in this paper. The DAC circuit structure is shown in Fig. 3. The main structure of the whole 14-bit DAC is composed of two different types of DAC, which can be divided into high level segment, middle level segment and low level segment. The high section adopts the segmented capacitor DAC structure. The low section DAC adopts the form of resistance voltage division. The whole DAC structure not only guarantees the resolution and conversion speed, but also solves the problem of too much area caused by too many unit devices.

C. Design of Digital Control circuit

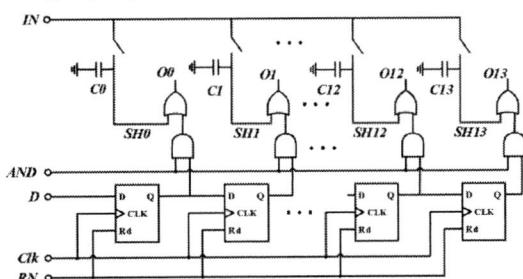

Fig. 4. Digital calibration circuit structure.

According to the comparison result between the CTIA output V_{O_CTIA} and the reference voltage VREF, the DAC is controlled to adjust the bias voltage V_{b2}. The circuit structure is shown in Fig. 4. The whole calibration process adopts a bit-by-bit calibration working mode, from high to low O0 to O13 output ports are connected to the input switches b0 to b13 of the DAC. In the calibration process, the highest bit b0 is first set to "1" and then adjusted according to the output result of the comparator. After the first calibration is complete, set the second O1 to "1" and calibrate again until the last one is finished.

D. Simulation Result

The self-calibration circuit in Fig. 2 was simulated. The power supply voltage was 5.0 V, the reference voltage $VREF$ was set to 2.5 V, the integration time was 34 μs, the blind pixel R_b and the probe pixel R_s were set to 60.0015kΩ and 60.3011 kΩ respectively, and the pixel deviation was 0.5%. When the pixel bias is 1.5 V, the introduced fixed mode noise current is 353.9 nA.

979-8-3503-6184-1/24 $31.00 © 2024 IEEE   576

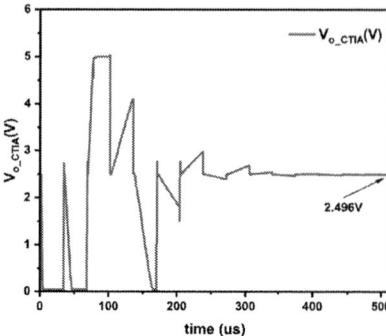

Fig. 5. Calibration curve of CTIA output voltage V_{O_CTIA}.

Fig.5 is the working curve of the self-calibration circuit. It can be seen from the figure that under the control of the calibration circuit, the output signal of CTIA constantly feedbacks the comparison result between the output signal and the reference voltage *VREF*, and controls the DAC to adjust the compensation current of the detection branch. After the calibration is completed, the output voltage is stable at 2.496 V, and the noise current is 1.1 nA, and the noise contrast is suppressed by 99.69% before calibration.

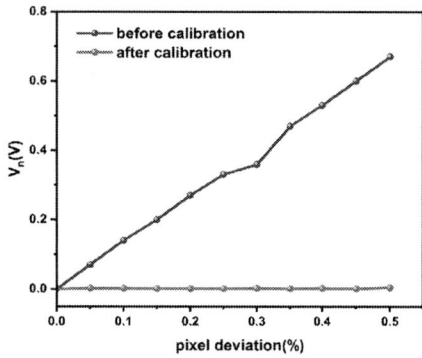

Fig. 6. CTIA noise voltage under different pixel deviations.

Fig.6 shows the comparison of noise voltage before and after calibration when the pixel deviation is less than 0.5%. It can be seen that the calibrated noise voltage is suppressed below 4 mA, the maximum is 4 mA, the minimum is 1mA. The self-calibration circuit can effectively suppress the fixed mode noise current introduced by pixel deviation, solve the problem of non-real-time and manual calibration of traditional noise reduction circuit structure, and improve the detector performance.

IV. CHIP TEST RESULTS

Due to the limitation of the preparation process and circuit structure, it is not possible to test the exact value of the pixel deviation of the chip, but the overall pixel deviation range is within 0.5%. A chip is selected from the prepared chips for testing. Set the VREF to 2.5 V, the pixel bias to 1.5 V, and observe the CTIA output voltage V_{O_CTIA} with an oscilloscope. The results show that the output voltage of the CTIA circuit before calibration is 2.58 V, the noise voltage is 80 mV, and the output voltage after calibration is 2.49 V, the noise voltage is 10 mV, and the noise suppression rate reaches

87.5%, which meets the detection demand. However, there is still a gap between the theoretical results and the measured results, which may be due to the deviation of the process, resulting in the noise caused by the imbalance of the CTIA circuit.

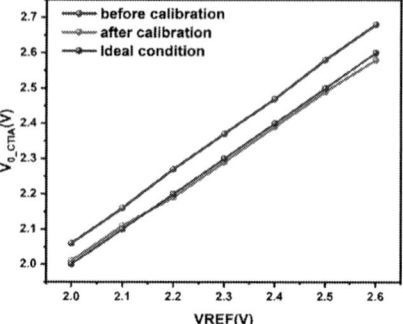

Fig. 7. CTIA output voltage V_{O_CTIA} under different VREFs.

The fixed pixel bias is 1.5 V, and the calibration of CTIA output voltage V_{O_CTIA} with *VREF* between 2.0 V-2.6 V is tested. Fig.7 shows the voltage output diagram of CTIA output voltage V_{O_CTIA} before and after calibration under different VREFs. It can be seen that when the VREF is 2.0 V-2.6 V, the noise suppression rate is above 75%, and the average noise suppression rate is 83.4%. The results show that the self-calibration circuit presented in this paper has excellent noise reduction effect in the range of operating voltage.

V. SUMMARY

This paper presents a readout circuit with the function of pixel noise self-calibration, which can effectively suppress the fixed mode noise caused by pixel deviation. For 0.5% pixel deviation, the simulation results show that the self-calibration circuit can eliminate 99.69% of the fixed mode noise. For pixel deviations of less than 0.5%, the experimental results show that the noise reduction effect of the chip can reach 87.5%.

REFERENCES

[1] Ambrosio R, Moreno M, Mireles Jr J, et al. An overview of uncooled infrared sensors technology based on amorphous silicon and silicon germanium alloys[J]. physica status solidic, 2010, 7(3-4): 1180-3.

[2] Xiqu Chen, Qiang Lv. A versatile CMOS readout integrated circuit for microbolometric infrared focal plane arrays[J]. Optik, 2013, 124(20): 4639-4641.

[3] Im S, Park S G. Thermal noise analysis of switched-capacitor integrators with correlated double sampling[J]. International Journal of Circuit Theory & Applications, 2016, 44(12) : 2101-2113.

[4] Altun O, Tasdemir F. Low-noise readout circuit for SWIR focal plane arrays[C]. SPIE Defense + Security, 2017: 1017707.

[5] S. Xu, Y. Z. Guo, X. S. Kong, H. Y. Zhu, et. al. Self-Calibration Readout Circuits for CMOS Microbolometers (in Chinese), International Conference on Solid-State & Integrated Circuit Technology2022, Nangjing, pp. 1-3.

[6] Kim G, Lim S, Kim Y, et al. High-uniformity post-CMOS uncooled microbolometer focal plane array integrated with active matrix circuit[C]. Transducers & Eurosensors Xxvii: the , International Conference on Solid-State Sensors, Actuators and Microsystems. IEEE, 2013: 2361-2

A Smooth Two-Stage Soft Start Method for Current Mode Boost Converter

Yue Shi [1,2], Shi-dong Wang [1], Zekun Zhou [1]*, Bo Zhang [1], and Zhigang Qin [3]

[1] State key Laboratory of Electronic Thin Films and Integrated Devices, University of Electronic Science and Technology of China, Chengdu, China

[2] Chengdu University of Information Technology, Chengdu, China

[3] Department of Electronic Engineering, Saitama Institute of Technology, Saitama, Japan

* Email: zkzhou@uestc.edu.cn

Abstract—Current-mode boost converter is widely used in portable devices powered by battery. However, the difficulty lies in suppressing its overshoot voltage and inrush current during startup phase. This paper proposes a novel two-stage soft start method to greatly suppress overshoot current and achieve smooth startup under wide range of load currents. Using body selection circuit to flexibly change the direction of the freewheeling transistor's body diode. When output voltage is lower than input voltage, linearly charge output capacitor by the method like low-dropout regulator. Once the output voltage is close to input voltage, start switching to step up output to its steady value. The proposed soft start method is verified by cadence using 180 nm BCD technology. Simulation results demonstrate that the proposed method can achieve a very smooth startup and obtain a stable soft start time, which is unaffected by load variations.

Keywords—soft start, boost converter, current mode, body selection

I. INTRODUCTION

Boost converter is a basic DC-DC step-up circuit and widely used in portable devices powered by lithium battery to supply a stable voltage. Besides, current mode control is commonly adopted in boost converters due to its advantage of fast line transient response, easy to compensate and set over-current limit [1]. But how to avoid the inrush current during startup phase is an annoying problem, since it is quite different from buck converter. Fig. 1 shows the structure of boost converter and current waveforms under different cases. During steady state operation, the inductor current increases during on-time and decreases during off-time, the current is around its average value and follows volt-second balance. But when output voltage is lower than input voltage, inductor current increases in both on and off time, resulting large inrush current. Keep closing M_N when $V_{out} < V_{in}$ can reduce overshoot current, but the diode still delivers power from V_{in} to V_{out}. The inductor, capacitor, and resistor can form a resonant network which can lead V_{out} be twice of V_{in}, this will cause serious damage especially in small step up ratio applications.

There are two main soft start methods for voltage regulator. One is generating a ramp voltage that rises from zero to keep small difference between voltage loop inputs [2], and the other is to limit the maximum current during startup [3]. Either the requirement of reducing current in advance to avoid output overshoot, or longer soft start time in heavy load conditions, is the main drawback [4]. Several technologies have been used to optimize soft start performance of boost converter. [3-4] Considering the initial condition is $V_{out} = V_{in}$ and raising output voltage from V_{in} to its steady state which is problematic, Ref. [5] uses a high pass filter to add a large resistor to output capacitor dynamically. When overshoot current surges in output, this method can suppress inrush current partly but it needs a lot of off-chip components which increases printed circuit board (PCB) area and cost.

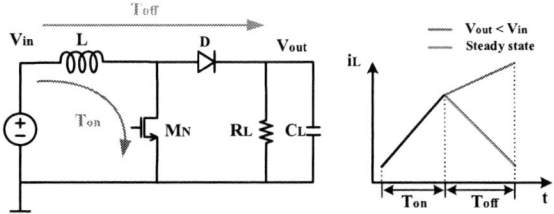

Fig. 1. Structure of boost converter and its current waveforms.

Using synchronous rectifier and body selection can avoid initial overcurrent from the root cause [6]. Ref. [7] adopts body selection and uses current mirror to charge output with constant current until output voltage is close to input voltage and then gradually increase duty cycle to step up output to its steady value. Current mirror method has the same issues with current limit method as mentioned above, and gradually increase duty cycle only suits pulse width modulation (PWM) converter.

Based on the issues mentioned above, this paper proposes a novel two-stage soft start method to greatly suppress inrush current and output overshoot while achieving same soft start time under different load conditions for current mode boost converter.

II. PROPOSED TWO-STAGE SOFT START METHOD

Fig. 2 illustrates the structure of proposed two-stage soft start strategy, where the modulation adopts valley current mode adaptive on-time (AOT) and the current sampling circuit employs SenseFET-based current sensor [1]. Using synchronous rectifier and body selection to block unexpected current path and improve efficiency. By applying a fixed slope ramp voltage to make a constant soft start time, which will not change due to load variations.

A. Stage One–Linear Precharge

When $V_{out} < V_{in}$, keeping turning off M_{NS} and using an operational amplifier (OPA) to linearly control M_{PS} is quite similar with low-dropout regulator (LDO). OPA behaves an inner loop to modulate equal value of V_C and current sense voltage. If the reference of EA raises, V_C will raise and then the gate of M_{PS} will be pulled lower to increase inductor current and keep the current sense voltage is same with V_C. Finally, output voltage is charged to a higher value to follow the increase of reference voltage. This is the negative feedback mechanism of linear charge stage.

B. Stage Two–Switching Step Up

When output voltage is close to input voltage, switching procedure is started to step V_{out} to its steady state. This stage is the same with normal operation of boost converter. At each cycle begins, M_{NS} turns on and the adaptive on-time module is activated at the same time. M_{NS} will be turned off when on-time is ended. Then M_{PS} turns on, and current sense circuit senses the current of M_{PS}. When current sense voltage is beneath V_C, turn off M_{PS} and start next cycle.

979-8-3503-6184-1/24 $31.00 © 2024 IEEE   578

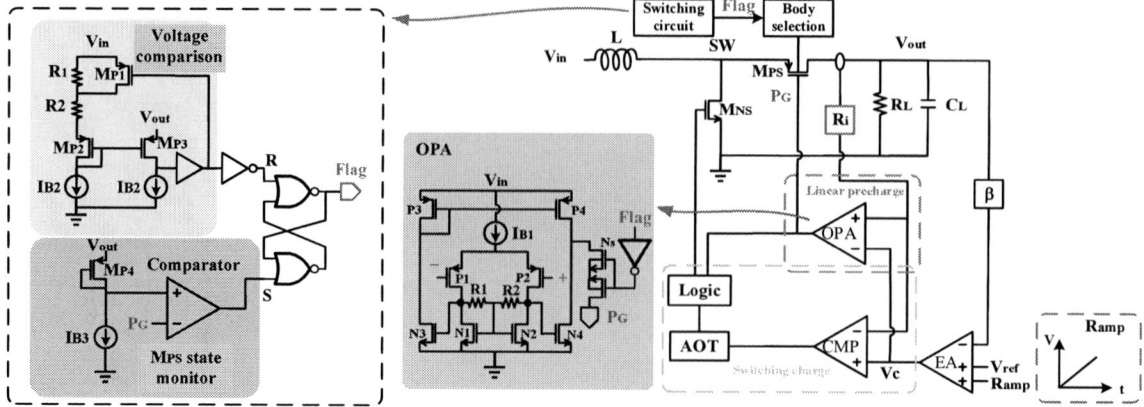

Fig. 2. The structure of proposed two-stage soft start method and its specific circuit implementation.

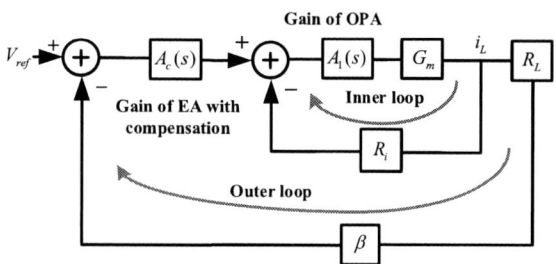

Fig. 3. Block diagram of stage one for small signal calculation.

C. Switching between Two Stages

Under dual loop modulation, output voltage depends on the reference voltage and V_C is equal to current sense voltage. By using same current sense circuit and voltage outer loop, smooth transition between two stages can be obtained, but switching point should be carefully designed.

The left of Fig. 2 shows the switching control circuit for the two stages. When Flag is low, the converter works in stage one and the body of M_{PS} connects to SW. When Flag is high, the converter works in stage two and the body of M_{PS} connects to V_{out}. Voltage comparison circuit compares the value of V_{in} and V_{out}, and uses R2 to introduce offset. When V_{out} increases to the value of V_{in} minus the offset voltage, the reset of RS latch will be low. This offset voltage is to make sure that the reset port can turn low under all load conditions and this is a rough judgment. Then suitable switching point is obtained by using another comparator to determine whether M_{PS} enters linear region or not. If it enters linear region, the difference between V_{out} and V_{in} is small, then the set of RS latch will turn high to lock the Flag to high level. According to this method, smooth transition and no overshoot current can be obtained during startup.

III. STABILITY DESIGN

Fig. 3 illustrates the small signal diagram of precharge, whose inner loop gain is,

$$T_{inner,1} = A_1(s)G_mR_i \qquad (1)$$

where G_m is the transconductance from P_G to M_{PS}'s channel current. Then the following expression can be obtained,

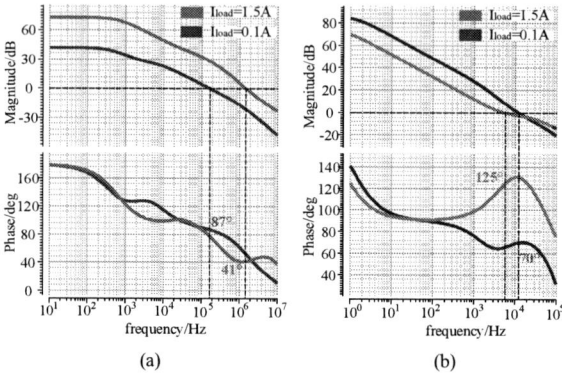

Fig. 4. Simulation results of close loop transfer function under different load conditions. (a) Inner loop. (b) Outer loop.

$$A_1(s) = \frac{g_m r_o}{1 + s r_o C_{gs}} \frac{1 + s g_{mp} L}{1 + s g_{mp} L C_{gd} C_{gs}^{-1}} \qquad (2)$$

$$G_m = \frac{g_{mp}\left(1 + s L r_{op}^{-1}\right)}{1 + s\left(g_{mp}L + L r_{op}^{-1}\right) + s^2 L C_{gs}} \qquad (3)$$

where g_m and r_o is the transconductance and output resistance of OPA, respectively; g_{mp}, r_{op}, C_{gs} and C_{gd} is transconductance, output resistance, gate-to-source and gate-to-drain capacitor of M_{PS}, respectively. Therefore,

$$T_{inner,1} = \frac{g_m r_o}{1 + s r_o C_{gs}} \frac{g_{mp} R_i \left(1 + s L r_{op}^{-1}\right)}{1 + s g_{mp} L C_{gd} C_{gs}^{-1}} \qquad (4)$$

Based on the deep negative feedback of inner loop, the loop transfer function of outer loop can be expressed as,

$$T_{outer,1} = \frac{\beta A_c(s) R_L}{(1 + s R_L C_L) R_i} \qquad (5)$$

As load current increases, the bandwidth of inner loop increases but the non-dominant pole decreases. Since the non-dominant zero is located at a relatively high frequency, the worst case of stability occurs under heavy load.

In stage two, the close loop gain of outer loop in this stage is given by [8],

979-8-3503-6184-1/24 $31.00 © 2024 IEEE

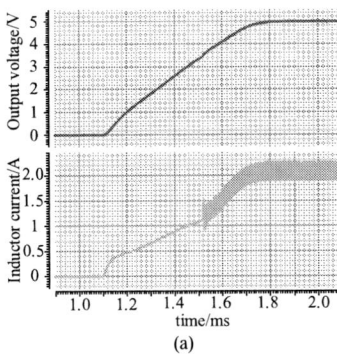

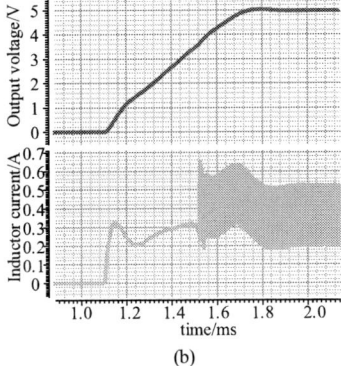

Fig. 5. Soft start simulation results. (a) RL = 3.5 Ω. (b) RL = 20 Ω.

$$T_{outer,2} = \frac{\beta A_c(s)(1-D)R_L}{1+s(Q\omega_{Ton})^{-1}+(s\omega_{Ton}^{-1})^2} \frac{1-sLR_L^{-1}(1-D)^{-2}}{R_i(2+R_LC_L)} \quad (6)$$

The outer loop in both stages can be easily compensated by using the same type-II compensator, while the inner loop of stage one should be carefully designed to achieve high bandwidth and adequate phase margin.

Fig. 4 shows the inner and outer loop frequency response of stage one under heavy and light loads. The simulation results show that the inner loop gets a far higher bandwidth than outer loop, with adequate phase margin in both loops.

TABLE I.   KEY PATAMETER OF BOOST CONVERTER

Technology	SMIC 0.18 μm 1P4M
Input voltage	2.4 ~ 4.2 V
Output voltage	5 V
Switching frequency	Quasi-fix frequency of 4 MHz
Maximum load current	1.5 A
Feedback voltage	1 V
Soft start time	600us

IV. SIMULATION RESULTS

The soft start method is verified by cadence simulation, the inductor and capacitor of boost converter is 560 nH and 20 μF, respectively. Table I sums some key parameters of boost converter and the simulation results are shown in Fig. 5. There is no inrush current and output voltage overshoot during the whole soft start period. When output is low, the inductor current is linear and no switching occurs, while output capacitor is gradually charged. When output is close to input, switching procedure is started and current ripple appears in inductor current. Although there is a certain response process during state switching and slight jitter in inductor current, the current change is too small which can be neglected. When load current changes, the converter can adaptively adjust the inductor current to achieve constant soft start slope, which is unaffected by load conditions. The soft start time is determined by the ramp voltage raises from zero to feedback voltage and is set to 600 μs in this design.

V. CONCLUSION

In this paper, two-stage soft start strategy is proposed to suppress inrush current during startup of boost converter. LDO mode is introduced to linearly charge output at initial, while` using the same outer loop to achieve smooth transition. Both theoretical analysis and simulation are used to verify the feasibility of two-stage soft start method, and simulation result shows that proposed method can achieve excellent soft start while almost no current or voltage overshoot occurs.

ACKNOWLEDGMENT

This work was supported by the National Natural Science Foundation of China under Grant 62074028, Sichuan Natural Science Foundation under Grant 23NSFSC0359, Chunhui Cooperative Research Program of the Ministry of Education of China under Grant HZKY20220583, and Central Government's Major Project for Transforming Significant Scientific and Technological Achievements in Universities and Research Institutions in Sichuan under Grant 2022ZHCG0117.

REFERENCES

[1] W. Hong and M. Lee, "A 10-MHz Current-Mode AOT Boost Converter With Dual-Ramp Modulation Scheme and Translinear Loop-Based Current Sensor for WiFi IoT Applications," in IEEE Journal of Solid-State Circuits, vol. 56, no. 8, pp. 2388-2401, Aug. 2021, doi: 10.1109/JSSC.2020.3047000.

[2] A. Song, W. Tang, H. Li and Z. Liu, "A Pulse-Skipping Soft-Start Circuit for DC-DC Converters," 2023 3rd International Conference on Electrical Engineering and Control Science (IC2ECS), Hangzhou, China, 2023, pp. 505-509.

[3] L. Huang, P. Luo, C. Wang and X. Zhou, "A High Speed On-Chip Soft-Start Technique With High Start-Up Stability for Current-Mode DC-DC Converter," in IEEE Access, vol. 7, pp. 27579-27585, 2019, doi: 10.1109/ACCESS.2019.2901529.

[4] A. Vasilica, V. Anghel, G. Pristavu and G. Brezeanu, "Suppressing start-up time variation versus load current — Adaptive soft-start in boost LED drivers," ESSCIRC Conference 2015 - 41st European Solid-State Circuits Conference (ESSCIRC), Graz, Austria, 2015, pp. 192-195, doi: 10.1109/ESSCIRC.2015.7313861.

[5] S. Fan, Z. Xue, Z. Guo, Y. Wang and L. Geng, "VRSPV Soft-Start Strategy and AICS Technique for Boost Converters to Improve the Start-Up Performance," in IEEE Transactions on Power Electronics, vol. 31, no. 5, pp. 3663-3672, May 2016.

[6] F. Luo and D. Ma, "Design of Digital Tri-mode Adaptive-Output Buck–Boost Power Converter for Power-Efficient Integrated Systems," in IEEE Transactions on Industrial Electronics, vol. 57, no. 6, pp. 2151-2160, June 2010, doi: 10.1109/TIE.2009.2034170.

[7] W. -R. Liou et al., "Monolithic Low-EMI CMOS DC–DC Boost Converter for Portable Applications," in IEEE Transactions on Very Large Scale Integration (VLSI) Systems, vol. 22, no. 2, pp. 420-424, Feb. 2014, doi: 10.1109/TVLSI.2013.2243927.

[8] Y. -C. Hsu, D. Chen, S. -F. Hsiao, H. -Y. Cheng and C. -S. Huang, "Modeling of the Control Behavior of Current-Mode Constant On-Time Boost Converters," in IEEE Transactions on Industry Applications, vol. 52, no. 6, pp. 4919-4927, Nov.-Dec. 2016, doi: 10.1109/TIA.2016.2597280.

Design of 12-bit Low-Power Single-Slope ADC with 2048 Columns for Infrared Focal Plane Array

Lixiang Han [1], Hao Li [2], Yihui Zhang [3], Liang Gao [1], Dongsheng Liu [3]*

[1] Wuhan National Laboratory for Optoelectronics, Huazhong University of Science and Technology, Wuhan 430074, China
[2] Hubei Optics Valley Laboratory, Wuhan 430074, China
[3] School of Integrated Circuits, Huazhong University of Science and Technology, Wuhan 430074, China

* Email: dsliu@hust.edu.cn

Abstract

This paper presents 2048-column 12-bit single-slope (SS) analog-to-digital converters (ADCs) with low power consumption for infrared focal plane array. The SS ADC consists of a sampling circuit, a comparator, and a logic circuit. The 4-bit coarse ramp generator employs a resistor-divider DAC and the 8-bit fine ramp generator employs a current-steering DAC for coarse and fine quantization of the SS ADC, respectively. The coarse and fine ramp generators are shared by 2048 columns of SS ADCs, which reduces power consumption and has good consistency between columns. The test results show that the DNL is -0.54/+0.51 LSB and the INL is -1.52/+2.25 LSB. The power of single-column SS ADC is 31.2μW and the FoM is 0.123pJ/conv.

1. Introduction

In recent years, the infrared focal plane array has been widely studied because of its ability to realize night imaging and thermal radiation detection. It has a wide range of applications in automobile driving, industrial testing, military, and other fields. SS ADCs have become the mainstream structure in the readout circuit, which is the core module of the infrared focal plane array due to its simple structure and low power consumption. In addition, ADCs commonly used in readout circuits also include successive approximation register (SAR) ADC [1], and cyclic ADC [2], which have large area and high power consumption respectively. Traditional SS ADC has a slower conversion speed, which goes against the trend of increasing the size of the readout circuit array. In [3], to improve the conversion speed, statistical and coded prediction methods are used to reduce the scanning range of the ramp generator. However, this increases the power consumption and design complexity of digital circuits. In [4], a two-step structure is employed to divide the quantization process into two phases, and the total time is reduced from 2^{M+N} clock cycles to 2^M+2^N clock cycles, which is more obvious for medium-high precision ADCs. Two-step structures that require two ramp generators reduce the linearity of the SS ADCs.

In this research, 2048-column 12-bit low-power SS ADCs are proposed. The quantization process is mainly divided into half-interval conversion, coarse quantization, and fine quantization by using a multi-step structure. Calibration circuits are designed for both coarse and fine ramp generators to improve linearity. The 4-bit coarse ramp generator and the 8-bit fine ramp generator adopt a resistor-divider DAC and a current-steering DAC for coarse and fine quantization, respectively. A 1-bit redundancy is also designed to reduce comparison errors caused by offset and noise. To ensure the quality of each column's clock signal, a multi-stage inverter is inserted every 512 columns to generate a new clock for the subordinate ADC, and the maximum delay of each stage clock is less than one clock cycle.

2. Architecture and principle of the proposed SS ADC

Fig. 1 shows the architecture of the proposed SS ADC. Each ADC unit consists of a comparator, a sampling circuit, and a logic circuit. The comparator utilizes a two-stage amplifier and one-stage latch structure, resulting in noise reduction and increased output turnover speed. The sampling circuit consists of switches and sampling capacitors. The logic circuit is used to detect the output voltage of the comparator and transfer the digital code at the end of the quantization, consisting of three different counters and a 4-16 decoder.

The coarse ramp generator resistor-divider DAC and the fine ramp generator current-steering DAC are connected to the 2048 columns of SS ADCs through different switches, which provide a coarse ramp signal and a fine ramp signal during the quantization, respectively.

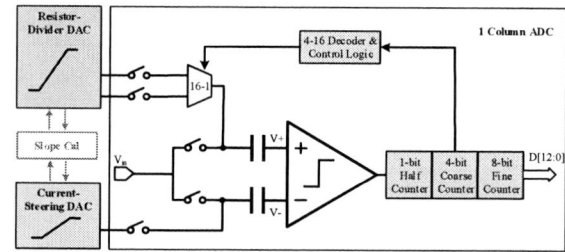

Fig. 1 Architecture diagram of the proposed SS ADC

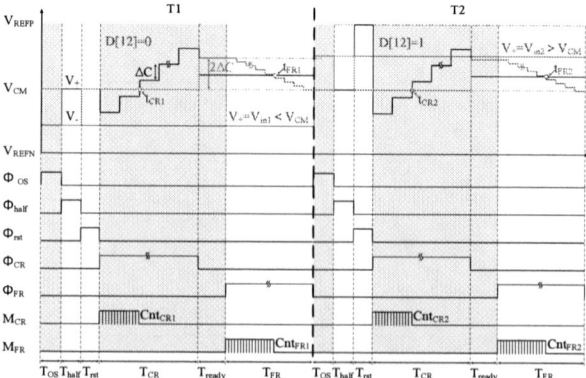

Fig. 2 Timing diagram of the proposed SS ADC

979-8-3503-6184-1/24 $31.00 © 2024 IEEE

Fig. 2 shows the timing of the proposed SS ADC. The ADC operation may be described as follows : 1) the offset voltage of the first-stage amplifier is stored during the T_{OS} phase; 2) in the half-interval conversion phase T_{half} and T_{rst}, the input signal V_{in} is compared with the common mode voltage V_{CM}. As shown in $T1$ and $T2$, if $V_{CM} > V_{in}$, $D[12]$ is stored as 0, conversely, it is stored as 1; 3) taking $T1$ as an example, with the continuous increase of the coarse ramp voltage, when $V+ > V-$, the output of the comparator flips and the coarse ramp counter saves the 4-bit coarse quantization result $D[11:8]$; 4) finally, in the fine quantization phase T_{FR}, When $V-<V+$, the fine slope counter stores the 8-bit fine quantization result $D[7:0]$.

On this basis, it is necessary to subtract the correction code of the redundant bit, and a 13-bit digital code $D[12:0]$ containing 1-bit redundancy can be obtained.

3. Design of 2048-column SS ADCs

The single-column ADC differs from the 2048-column ADCs in design mainly in the consistency of each column. The main module design and 2048-column design methods of the SS ADC are as follows.

3.1 4-bit coarse ramp generator

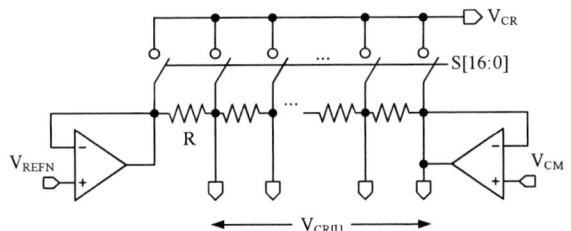

Fig. 3 Architecture diagram of the 4-bit coarse ramp generator

Fig. 3 shows the architecture of the 4-bit coarse ramp generator. The reference voltages V_{CM} and V_{REFN} are provided through the output of the operational amplifiers, increasing the driving capability of the ramp signal V_{CR} and achieving voltage division through the resistance string.

3.2 8-bit fine ramp generator

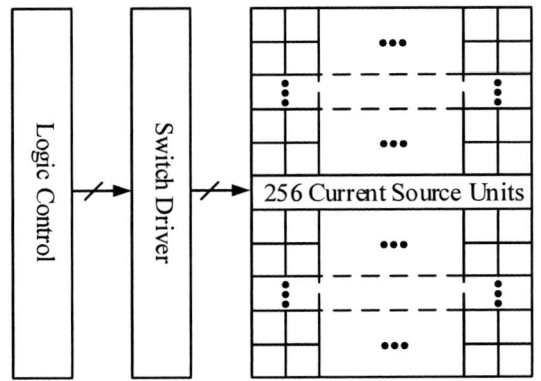

Fig. 4 Architecture diagram of the 8-bit fine ramp generator

Fig. 4 shows the architecture of the 8-bit fine ramp generator. It consists of a current source array, switch driver, and control circuits, among which the current source unit and the switch driver circuit are shown in Fig. 5. The current source unit adopts a cascode structure to improve the accuracy of the mirrored current, and the switch driver adopts a low crossover structure to ensure that the current source is always in the on-state during the switching process, which is also conducive to improving the accuracy of the mirrored current.

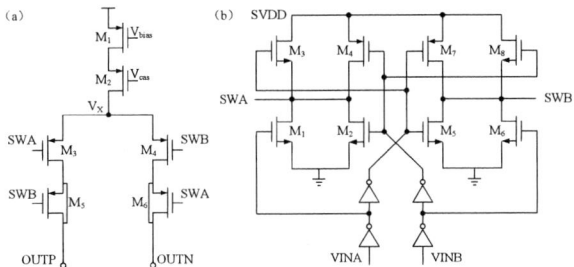

Fig. 5 Current source unit and switch driver

3.3 2048-column SS ADCs

The layout width of a single-column ADC is 10μm, which matches the common pixel size for infrared focal plane array. Therefore, the layout width of the 2048-column ADCs will reach 20.48mm, which will cause the power/ground, control signal, and clock deviation between the different columns of ADCs. The establishment of the RC delay model is shown in Fig. 6. Based on the parasitic parameter extraction results, capacitors $C<0:2047>$ and resistors $R<0:2047>$ are all the same, respectively. $C<0>$ ranges from 0.5 to 10fF in the model and the value of the resistance $R<0>$ is between 10-100mΩ due to the difference in the width of the metal wire.

Also, the following methods are employed as shown in Fig. 7: multi-stage inverters are inserted before the common control signals; the clock signal adopts a grouping structure, and multi-stage inverters are inserted between different groups.

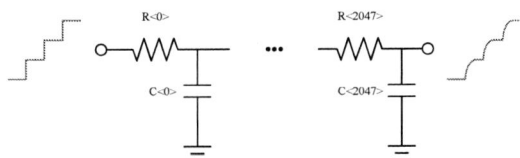

Fig. 6 RC delay model of 2048-column ADCs

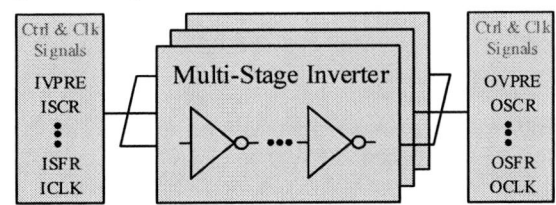

Fig. 7 Multi-stage inverters between digital and analog signals

4. Results

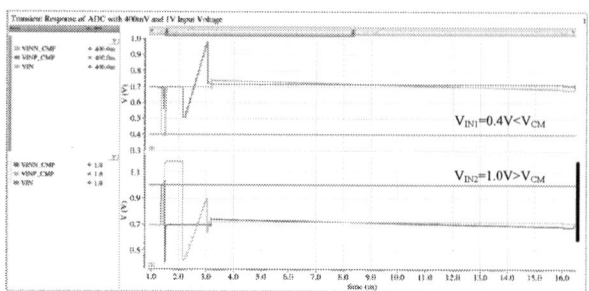

Fig. 8 Simulation results of a single-column ADC quantization process

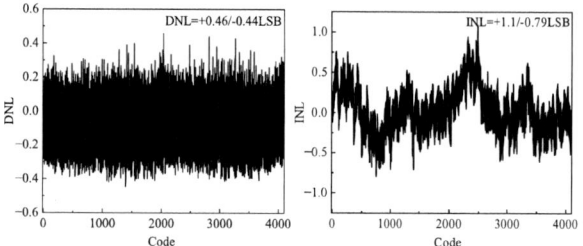

Fig. 9 DNL and INL test results of a single-column ADC

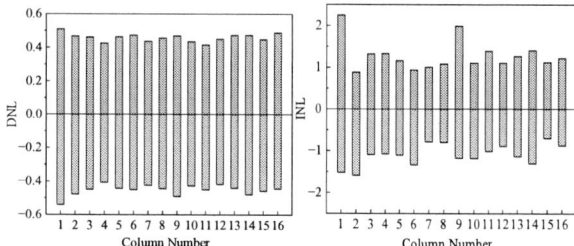

Fig. 10 DNL and INL test results of 16-column ADCs

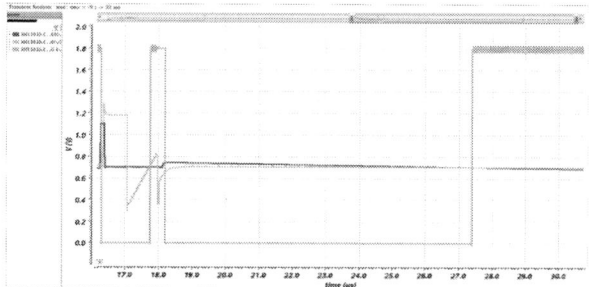

Fig. 11 Simulation results of ADC in column 2048 quantization process

Fig. 8 shows the simulation results of a single-column ADC in both $V_{in}<V_{CM}$ and $V_{in}>V_{CM}$ cases at the slow corner. Half-interval comparison, coarse quantization, and fine quantization processes are established correctly in both cases. 16-column and 2048-column ADCs were fabricated and simulated respectively. Fig. 9 and Fig. 10 show the DNL and INL test results of single-column and 16-column ADCs respectively. The worst DNL and INL appear in the first column, with a DNL of -0.54/+0.51 LSB and an INL of -1.52/+2.25 LSB, resulting in an average DNL of +0.46/-

0.45 LSB and an INL of +1.29/-1.09 LSB for the 16 columns ADCs.

The quantization process at the slow corner of the ADC in column 2048, which is furthest away from the clock and control signals and has the largest RC delay and signal load, is shown in Fig. 11. The power consumption of single-column ADC is 31.2μW at a sampling rate of 61.7kS/s, and the FoM is 0.123pJ/conv which is calculated by power consumption required for single-step conversion. The performance summary and the comparison with other research on the proposed ADC are presented in Table 1 showing advantages in power and FoM.

Table 1. Performance Summary and Comparison

	This work*	[4]	[5]
Process	180nm CMOS	130nm CMOS	130nm CMOS
Resolution	12-bit	12-bit	12-bit
Sample rate	61.7kS/s	156.25kS/s	100kS/s
DNL	-0.54/+0.51 LSB	-0.49/+1.34 LSB	-1/+0.83 LSB
INL	-1.52/+2.25 LSB	-2.47/+2.44 LSB	-3.31/+4.78 LSB
Power	31.2μW	90μW	62μW
FoM	0.123pJ/conv	0.141pJ/conv	0.151pJ/conv

*Average of 16-column ADCs

5. Summary

In this paper, 2048-column 12-bit low-power SS ADCs have been designed for infrared focal plane array. The multi-step structure with coarse and fine ramp generators and multi-stage comparator are adopted for high speed and low power consumption. The worst DNL and INL of 16-column ADCs are -0.54/+0.51 LSB and -1.52/+2.25 LSB respectively. At a sampling rate of 61.7kS/s, the power consumption of single-column ADC is 31.2μW and the FoM is 0.123pJ/conv. The design and simulation verification of 2048-column ADCs are verified by inserting multi-stage inverters between control signals and ADC circuits and generating clock signal in groups.

Acknowledgment

This work is supported by the National Key Research and Development Program of China (No. 2021YFA0715502, No. 2023YFB4502100), the National Natural Science Foundation of China (No.62374064，No. 62104076, No. 62134002), the National Key Analog Integrated Circuit Laboratory Project of China. (No. JCKY2021210C004)

References

[1] Sun-Il Hwang, Jae-Hyun Chung, Hyeon-June Kim, et al. IEEE Transactions on Electron Devices, pp. 1119–1126 (2018).

[2] Fang Tang, Bo Wang, AmineBermak, et al. IEEE Transactions on Electron Devices, pp. 162–167 (2015).

[3] Mohamed R. Elmezayen , Bingxing Wu and Suat Utku Ay. IEEE Transactions on Circuits and Systems–I: Regular Papers, pp. 4484–4493 (2020).

[4] Himchan Park, Junan Lee, Jinwoo Kim, et al. IEEE Transactions on Circuits and Systems–I: Regular Papers, pp. 2147–2155 (2015).

[5] Qihui Zhang, NingNing, Zhong Zhang, et al. IEEE Transactions on Very Large Scale Integration (VLSI) Systems, pp. 644–655 (2022

A High-Voltage Smooth Self-Starting Reference Current Source Circuit

Dongyan Zhao[1], Jie Pan[1], Chenghao Zhang[2], Yidong Yuan[1], Yi Hu[1], Hongwei Shen[1], Zekun Zhou*[2], *Member, IEEE*

[1] Beijing Smart-chip Microelectronics Technology Co., Ltd., Beijing, China
[2] State key Laboratory of Electronic Thin Films and Integrated Devices, University of Electronic Science and Technology of China, Chengdu, China

*Email: zkzhou@uestc.edu.cn

Abstract—This paper presents a high-performance self-starting reference current source capable of operating in high-voltage scenarios. By employing loop control and segmented current control, it addresses the issues of poor performance in conventional high-voltage references and the need for external startup signals. In terms of voltage endurance, the reference source utilizes high-voltage transistor isolation to maintain high-voltage tolerance, while the core reference operates using low-voltage devices, thus avoiding the performance drawbacks associated with high-voltage components. Moreover, the startup sequence involves constant current charging followed by constant voltage charging, facilitating smooth and controllable self-starting. The design is implemented using a 0.18 μm process. The circuit's maximum input voltage is 26V, with a power supply rejection ratio (PSRR) of 78.8dB, a line regulation of 0.22%/V, a temperature coefficient (TC) of 8.96 ppm/℃ for the -40 ℃ to 125°C temperature range and a power consumption of 7.5μA.

Keywords—high-voltage reference, smooth self-starting, PSRR, line regulation

I. INTRODUCTION

The reference current source is a core module in analog circuits, providing a fixed and stable bias current for operational amplifiers, ADCs, and other circuits. In applications such as switch-mode power supplies and high-voltage sensors, where the system supply voltage is high, a high-voltage reference circuit is required. However, due to the poor performance of high-voltage transistors, achieving high performance with conventional reference structures directly using high-voltage components is challenging.

Moreover, since the reference source has a zero current state during power-up, a start-up circuit is required to break the degenerate state [1]. Such circuits are commonly used for self-starting references [2-5]. However, these circuits are typically applied in low-voltage scenarios. When it comes to high-voltage power-up, due to the absence of corresponding low-voltage power rails, direct application of these circuits is not feasible, necessitating the design of new startup circuits.

The high-voltage reference current source proposed in this paper focuses on improving the performance of high-voltage references and achieving on-chip smooth self-starting. Section II describes the complete current reference, the self-start circuit, and the loop control. Finally, schematic and post-layout simulation results are presented in Section III.

II. THE PROPOSED REFERENCE STRUCTURE

The schematic of proposed current reference block diagram is illustrated in Fig. 1. The circuit consists of three main components, which are loop control, smooth self-start circuit, and low-voltage reference core.

A. Loop control cirtcuit

The proposed loop control circuit is depicted in Fig. 2. MP5, MN7, MN6, and R3 constitute the feedback network of the control loop. This loop consists of two feedback paths, which

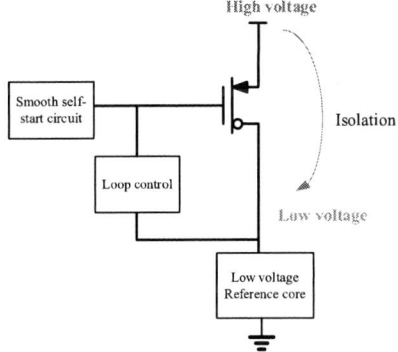

Figure 1. Current reference block diagram

are Loop1 and Loop2 shown in the left part of Fig. 2. The expression for the loop gain is as follows,

$$T = \frac{gm_{MP5} \, gm_{MN6} \, gm_{MP4} \, ro_{MP6}}{gm_{MN7}(1 + gm_{MN6} R_3)} \qquad (1)$$

where gm represents the transconductance of transistors, ro denotes the small-signal equivalent output impedance of transistors.

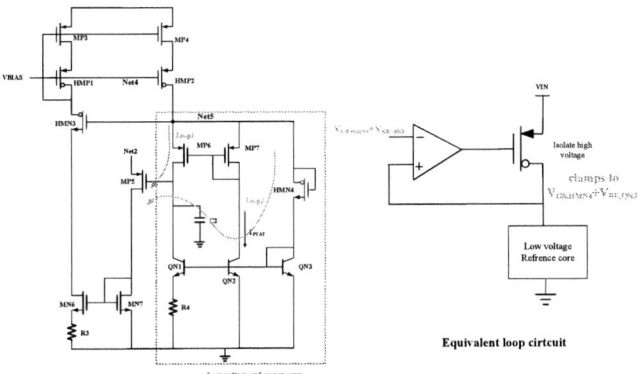

Figure 2. Loop control circuit

In the loop, only the gate of MP5 serves as a high-impedance node, requiring Type I compensation for stability assurance. It's worth noting that R3 here can adjust the current through MP5 at steady state. With appropriate design, MP5 and MP7 currents can be made identical, ensuring $V_{GS,MP5} = V_{GS,MP7}$, thus resulting in $V_{DS,MP6} = V_{DS,MP7}$, enhancing the accuracy of the current mirror between MP6 and MP7 to improve reference precision.

979-8-3503-6184-1/24 $31.00 © 2024 IEEE

Consequently, the loop control circuit can clamp the voltage at Net5, ensuring that the power supply for the reference core does not exceed the voltage endurance range. Functionally, the loop control circuit can be considered as an equivalent low dropout regulator (LDO) referenced to $V_{GS,HMN4} + V_{BE,QN3}$, providing stable low-voltage power supply for the low-voltage reference core. The equivalent control principle is shown in the right part of Fig. 2.

B. Smooth self-start circuit

The designed smooth self-start circuit is illustrated in Fig. 3. The circuit consists of MN1-MN5 and R2 for generating a constant current, labelled as I_{fix}. Upon power-up, as the power supply VIN rises, MN2 and MN3 conduct, causing NET2 to increase in voltage and thereby leading MN4 and MN5 to turn on. Given that MN2, MN4, and MN5 all operate in the saturation region and their currents are close, they can be approximated as $V_{GS,MN2} = V_{GS,MN4(MN5)}$. Consequently, I_{fix} can be expressed as,

$$I_{fix} = \frac{V_{GS,MN3}}{2R_2} \qquad (2)$$

I_{fix} passes through current mirrors MP3 and MP4 to charge Net5, causing the supply voltage of the reference to increase. When the voltage of Net5 reaches $V_{GS,HMN4} + V_{BE,QN3}$, the loop control circuit is triggered to generate a dynamic current that is proportional to the reference current I_R. This dynamic current ensures that the charging current is equal to the reference current, maintaining the Net5 voltage constant until the reference is fully established.

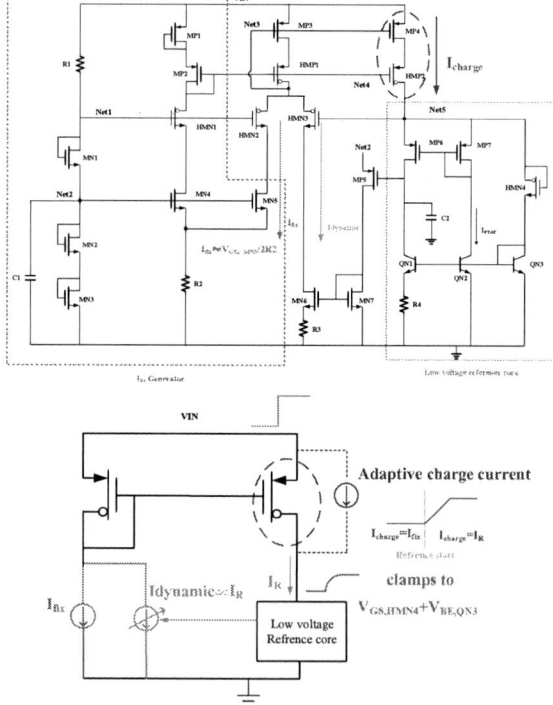

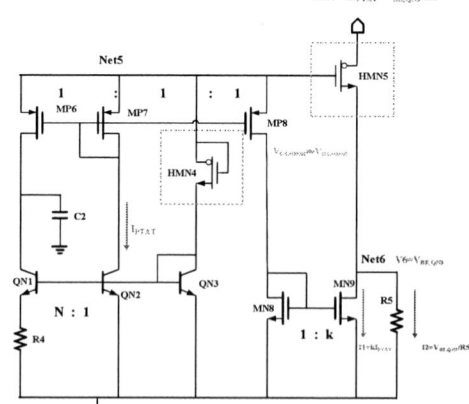

Equivalent smooth self-start cirtcuit

Figure 3. Smooth self-start circuit

In summary, the self-starting process of the circuit, depending on whether the reference is working. The power-

on process can be divided into two stages: constant current charging and constant voltage charging. At the beginning of power on, when the reference is not working, a constant current I_{fix} is used to charge the reference circuit, causing the supply voltage of the reference to continuously rise. When the supply is sufficient for the reference to operate, the supply voltage will be clamped by the loop control circuit, switching to a constant voltage charging mode, with the charging current following the reference current until the reference is fully established, thus achieving smooth self-starting.

C. Low voltage reference core

The low voltage reference core, as shown in Fig. 4, consists of MP6, MP7, QN1, QN2, and R4, forming the PTAT current generation circuit. The PTAT current can be expressed as follows:

$$I_{PTAT} = \frac{V_T lnN}{R_4} \qquad (3)$$

The PTAT current passes through the current mirror to MN9, resulting in the MN9 current I1,

$$I1 = kI_{PTAT} = \frac{kV_T lnN}{R_4} \qquad (4)$$

On the other hand, the CTAT current is generated through QN3, HMN4, HMN5, and R5. By design parameters, $V_{GS,HMN4} \approx V_{GS,HMN5}$ can be achieved, making the NET6 voltage $V_{Net6} = V_{BE,QN3}$. Therefore, the R5 current I2 can be expressed as follows,

$$I2 = \frac{V_{BE,QN3}}{R_5} \qquad (5)$$

Thus, the reference current IREF can be written as follows,

$$IREF = \frac{kV_T lnN}{R_4} + \frac{V_{BE,QN3}}{R_5} \qquad (6)$$

By adjusting the current mirror coefficient k and the resistance value of R4, R5, zero-temperature current can be achieved.

Figure 4. Low voltage reference core

III. IMPLEMENTATION

Fig. 5 shows the complete schematic diagram of the proposed reference current source in this paper, consisting of three parts: the smooth self-starting circuit, loop control circuit, and low voltage reference core. Fig. 6 displays the layout of this reference source, occupying an area of $270\ \mu m \times 185\ \mu m$.

979-8-3503-6184-1/24 $31.00 © 2024 IEEE   585

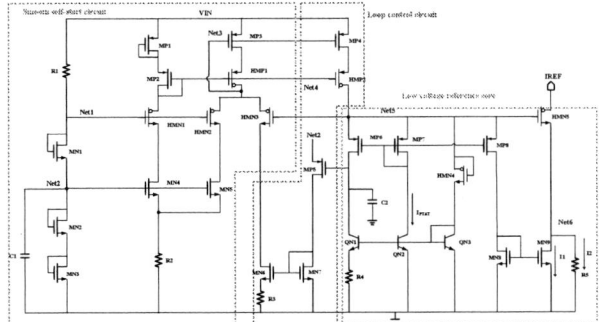

Figure 5. The complete reference schematic diagram

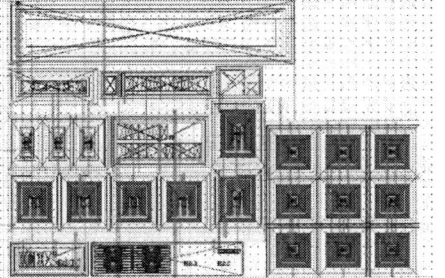

Figure 6. The proposed reference layout with 270×185 μm^2 area

The PSRR results of proposed current reference circuit are shown in Fig. 7. After the implementation of loop control circuit, the PSRR has been enhanced by 18dB, achieving improved performance. The power-on result of the circuit is shown in Fig. 8. When powered on at a rate of 20V/ μ s, the supply voltage of the reference can rise synchronously, with only an overshoot voltage of 78mV, achieving smooth self-start. That is about half of the structure without smooth self-start technique, while no oscillation occurs. The line regulation and temperature coefficient result of proposed current reference circuit are shown in Fig. 9. When Vin varies from 5 to 21V, the change in IREF is 47nA, resulting in a line regulation of 0.22%/V. With a temperature variation from -40°C to 125°C, the change in IREF is 1.6nA, and the temperature coefficient 8.96ppm/°C.

IV. SUMMARY

This paper presents a high-voltage reference current source with a loop control circuit and smooth self-start circuit. By utilizing voltage control and segmented power-on strategies, it achieves improved performance and on-chip smooth self-start capability, addressing the issues of poor performance and the need for an external start-up power source in traditional high-voltage reference sources.

ACKNOWLEDGMENT

This work was supported by the Laboratory Open Fund of Beijing Smart-chip Microelectronics Technology Co., Ltd. (NO. SGSC0000MNQT2401427).

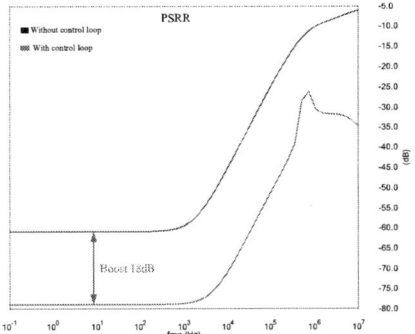

Figure 7. PSRR simulation result of the proposed reference

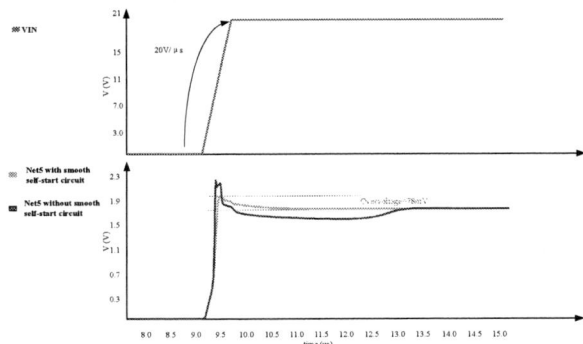

Figure 8. Self-start simulation result of the proposed reference

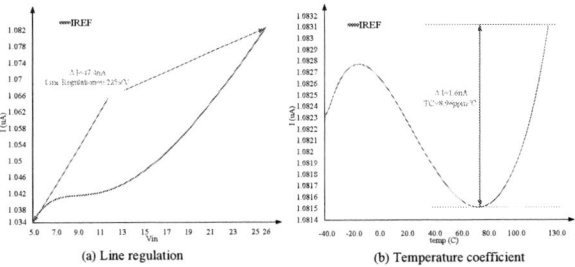

(a) Line regulation   (b) Temperature coefficient

Figure 9. Line Regulation and Temperature coefficient simulation of the proposed reference

REFERENCES

[1] O, T. V., Wisland, D. T., Lande, T. S., & Moradi, F. "Novel startup circuit with enhanced power-up characteristic for bandgap references." SOC Conference, 2008 IEEE International IEEE, 2008:123-126.

[2] Waltari, Mikko, and Halonen, Kari, "Reference Voltage Driver for Low-Voltage CMOS A/D Converters," Proceedings of ICECS 2000, Vol. 1, pp. 28-31, 2000.

[3] A. Boni, "Op-amps and startup circuits for CMOS bandgap references with near 1–V supply," IEEE Journal of Solid State Circuits, Vol. 37, NO. 10, pp. 1339-1343, Oct. 2002.

[4] Shrimali, Hitesh, and V. Liberali. "The start-up circuit for a low voltage bandgap reference." IEEE International Conference on Electronics, Circuits and Systems, 2015:92-95.

[5] Z. Zhou, Y. Wang and Y. Shi, "A Bandgap Reference Using a Novel Soft Self-start Bias Circuit," 2018 IEEE Asia Pacific Conference on Circuits and Systems (APCCAS), Chengdu, China, 2018, pp. 484-488, doi: 10.1109/APCCAS.2018.8605671.

A Temporal and Spatial Reuse Interpolation Hardware for VVC Motion Compensation

Huanxiang He[1], Shushi Chen[1], Leilei Huang[2]*, Yibo Fan[1]*

[1]State Key Laboratory of Integrated Chips and Systems, Fudan University, Shanghai 200433, China

[2]Institute of Microelectronic Circuits and Systems, East China Normal University, Shanghai 200241, China

*Email:llhuang@cee.ecnu.edu.cn, fanyibo@fudan.edu.cn

Abstract—To improve the inter-frame prediction effect, Versatile Video Coding (VVC) advances motion compensation with 1/16 pixel accuracy, which substantially elevates the computational demands. In this paper, a temporal and spatial reuse interpolation hardware for VVC motion compensation is designed and implemented. The proposed hardware treats the interpolation circuit as a whole and designs an 8×8 motion compensation circuit, which further reduces the number of adders. Compared to previous studies, this design has shown improvements in both area and performance. The hardware is synthesized using 28nm process, and the results show that its area consumption is only 25K, and it can realize 8K@40fps throughput at 500MHZ.

Keywords—VVC, FME, motion compensation, temporal and spatial reuse

I. INTRODUCTION

ITU and ISO released the Versatile Video Coding (VVC) standard[1]-[2], which was expected to provide higher compression efficiency than the High Efficiency Video Coding standard[3]-[5]. This improvement comes at the cost of increased computational complexity, as VVC utilizes 15 8-tap Fractional Interpolation (FI) filters for fractional interpolation, allowing for more detailed pixel interpolation than High Efficiency Video Coding's 3 8-tap filters[3].

This paper presents a VVC fractional interpolation hardware for motion compensation. From the perspective of pixel points in the input interpolation circuit, a universal circuit structure is designed for different motion vector (MV) precision based on a specific input pixel. Additionally, by considering the reuse of certain input pixels themselves during the computation of various output pixels, circuit reuse is implemented, which further reduces the number of adders.

In the work [6], the VVC FI hardware design is proposed with an emphasis on filter coefficients, advocating for the creation of configurable filters. This design employs eight identical filters to address the interpolation for 8×8 sub-blocks, which results in built-in redundancy in circuit logic. The work [7] introduces a VVC FI hardware design that operates all fifteen fractional interpolation filters in parallel. The design does not take advantage of filter reuse, thereby increasing the overall area needed for the chip. In the work cited [8], the approximate VVC FI hardware introduces a simplification of the filter coefficients, which consequently leads to a degradation in rate-distortion performance.

The VVC fractional interpolation hardware proposed is realized with Verilog HDL and synthesized using the SMIC 28nm technology node, and achieves a top operational frequency of 521 MHz. Under the most stringent conditions, it can process 4K@174fps. The hardware occupies a compact area of merely 25K and exhibits low power consumption, amounting to only 2.5 mW.

The structure of the rest parts is organized as follows. Section II offers an overview of the proposed VVC fractional interpolation algorithm. Section III elaborates on the proposed hardware design and presents a comparative evaluation of our design against existing approaches. Finally, the conclusions are given in Section IV.

II. METHOD

For each integer pixel, VVC interpolates 15 horizontal half pixels, 15 vertical half pixels, and 225 quarter pixels. Since last 7 filters are symmetric of the first 7 filters, coefficients of only first 8 filters are shown in Table I. As a result, only the implementation of the circuit design for the 8 filters is required.

This paper primarily concentrates on the algorithms for motion compensation, so one fractional pixel is inserted at each integer pixel location dictated by the MV. As shown in Fig. 1, all prediction units are partitioned into 8×8 sub-blocks. During the interpolation process executed at an 8×8 sub-block, considering the augmented set of 7 extended pixels, a total of 15 pixels per row, designated as $A_{(-3,0)}$ to $A_{(11,0)}$, are input into the interpolation circuit. Taking the fractional pixels $b_{(0,0)}$ and $c_{(0,0)}$ in Fig. 1 as examples, if the MV is $i_{(7/16,0)}$, the formula for calculating b(0, 0) is shown in (1). Pixels from $b_{(0,0)}$ to $b_{(7,0)}$ are calculated in 1 clock cycle, so an entire 8×8 block requires 8 clock cycles for computation. This situation arises when either the horizontal or the vertical coordinate of the MV is 0.

$$b_{(0,0)} = \begin{pmatrix} -A_{(-3,0)} + 4 \times A_{(-2,0)} \\ -11 \times A_{(-2,0)} + 45 \times A_{(-1,0)} \\ +34 \times A_{(1,0)} - 10 \times A_{(2,0)} \\ +4 \times A_{(3,0)} - A_{(4,0)} \end{pmatrix} >> 6 \quad (1)$$

TABLE I.   VVC FI FILTER COEFFICIENTS

MV	$A_{(-3,0)}$	$A_{(-2,0)}$	$A_{(-1,0)}$	$A_{(0,0)}$	$A_{(1,0)}$	$A_{(2,0)}$	$A_{(3,0)}$	$A_{(4,0)}$
1	0	1	-3	63	4	-2	1	0
2	-1	2	-5	62	8	-3	1	0
3	-1	3	-8	60	13	-4	1	0
4	-1	4	-10	58	17	-5	1	0
5	-1	4	-11	52	26	-8	3	-1
6	-1	3	-9	47	31	-10	4	-1
7	-1	4	-11	45	34	-10	4	-1
8	-1	4	-11	40	40	-11	4	-1

979-8-3503-6184-1/24 $31.00 © 2024 IEEE

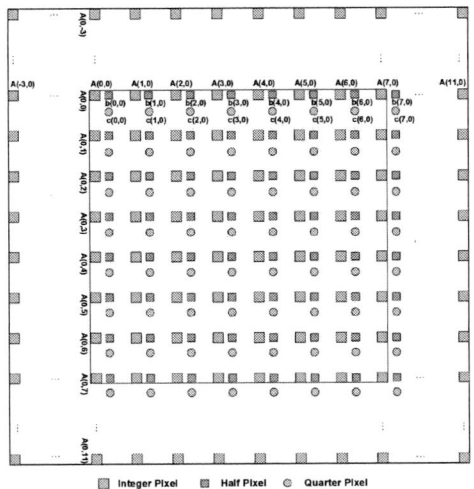

Fig. 1. Integer, half and quarter pixels.

In the other condition, if neither the horizontal nor the vertical coordinate of the MV is 0, such as MV i(7/16 , 5/16), the formula for calculating c(0, 0) is shown in (2). The values $b_{(0,-3)}$ to $b_{(0,11)}$ need to be calculated and transposed first, which costs 15 clocks. The computation is the same with (1). Then, compute values c with the values b, which costs another 8 clock cycles, so an entire block requires 23 clocks.

$$c_{(0,0)} = \begin{pmatrix} -b_{(0,-3)} + 4 \times b_{(0,-2)} \\ -11 \times b_{(0,-1)} + 52 \times b_{(0,0)} \\ +26 \times b_{(0,1)} - 8 \times b_{(0,2)} \\ +3 \times b_{(0,3)} - b_{(0,4)} \end{pmatrix} >>> 6 \quad (2)$$

From the perspective of input pixels, this paper explores the pattern between the input pixels and their associated filter coefficients under varying MV precision. The proposed hardware interpolates one row of pixels in each clock cycle.

Each row of coefficients in Table I represents the multiplication factors for a given pixel under varying MV precision. As depicted in Fig. 1, during the interpolation of the first row of fractional pixels, the pixel point $A_{(-3,0)}$ is exclusively used in the computation of $b_{(0,0)}$. Similarly, pixel point $A_{(-2,0)}$ is employed in interpolations for $b_{(0,0)}$ and $b_{(1,0)}$, corresponding to the first and second columns of Table I. By extending this logic, we can deduce the complete set of

coefficients needed for each input pixel across different MV precision. Subsequently, we design with the minimum number of adders. Moreover, for different columns, where the same input pixel contributes to multiple output pixels at some certain MV precision, we optimize spatial reuse to minimize the total number of adders, always balancing latency against resource utilization.

The circuit structure is illustrated in Fig. 2, and some detailed circuit implementations for data-paths are depicted in Fig. 3. Based on the temporal reuse design, for the coefficients corresponding to different MVs, which are represented by different rows in Table I, a circuit consisting of components labeled "datapath_1" through "datapath_8" can be devised to implement the required function. Taking the coefficients 26 and 52 as an example, they can be decomposed as (3) and (4).

$$52 = (x << 6) - (x << 4) + (x << 2) \quad (3)$$

$$26 = (x << 4) + (x << 2) + (x << 1) \quad (4)$$

Based on the Spatial reuse design, the coefficients 26 can be decomposed as (5).

$$26 = 52 >> 1 \quad (5)$$

It can be observed that this algorithm saves two adders and one shift operation. As illustrated by the red lines in Fig. 2, the structure of coefficient 52 from datapath_3 is reused in datapath_4 for coefficient 26. In datapath_2 to datapath_5, similar designs are adopted, sharing partial output results among the various datapaths. Taking pixel $A_{(0,0)}$ as an example, initially four adders would be necessary to compute it. However, when the MV precision is 7, employing horizontal circuit reuse reduces the adder count to just one. Considering the adder count for each column, a total of three adders are sufficient. Additionally, when the MV precision is 5, reuse is possible, but it does not reduce the total number of adders for the entire column and introduces unnecessary latency due to horizontal reuse, so we opt for reuse only when the MV precision is 7.

This methodology is applied consistently across other input pixels during the design phase. This sharing reduces the number of adders from 11 to 8, and also decreases the number of shift operations required.

The circuit structures for certain datapaths that utilize both temporal and spatial reuse are shown in Fig. 3.

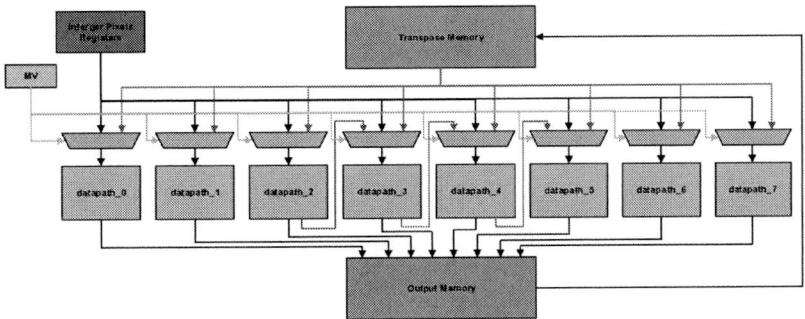

Fig. 2. VVC FI motion compensation hardware design

979-8-3503-6184-1/24 $31.00 © 2024 IEEE

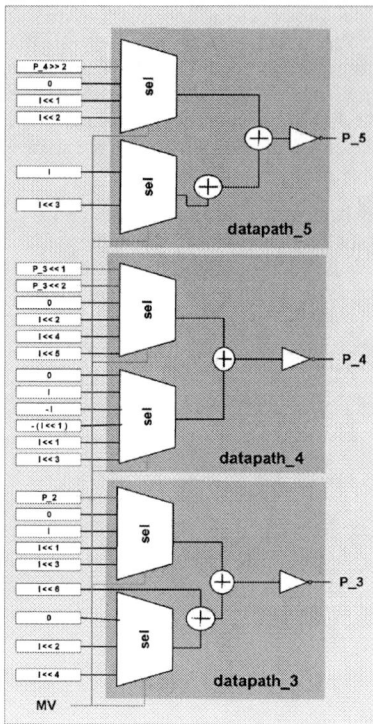

Fig. 3. Circuit structures of datapaths utilizing both temporal and spatial reuse

III. EXPERIMENTS

The synthesis of the design for the proposed VVC fractional interpolation hardware is carried out using the sc9mc_cmos28slp_base_hvt_c30_ff_nominal_min_1p10v_8 5c standard cell library. Following synthesis, the generated netlists were subjected to placement and routing procedures to ensure optimal layout for fabrication. The synthesis results, as reported by the Design Compiler, are summarized in Table II. The proposed solution achieves a maximum frequency of 521 MHz, which is higher than the work [6]-[7]. It means that the proposed solution has the potential for higher throughput and better performance. The design demonstrated the capability to process up to 174 frames per second for 3840×2160 resolution video frames. Apart from the improvements made to the algorithm, it is important to note that while the proposed fractional interpolation hardware and [6] focus solely on motion compensation, while [7] can be used for both motion estimation and compensation.

TABLE II.   HARDWARE COMPARISON

	Baseline[6]	[6]	[7]	Proposed
Technology	TSMC 90nm	TSMC 90nm	TSMC 90nm	SMIC 28nm
Slice/Gate Count	48.3K	11.7K	37.6K	25.4K
Max. Freq. (MHz)	417	357	435	521
Total power (mW)	264.35	62.88	375.3	2.53
Frames per Second	110 3840×2160	95 3840×2160	88 1920×1080	174 3840×2160

The gate count of the proposed hardware is 25.4K, which is significantly lower than the solution in work [6] and [7]. Part of the reason is due to differences in the technology libraries employed, but primarily it is a result of the temporal and spatial reuse design strategies implemented. This suggests that the proposed solution is more area-efficient.

Power consumption is a pivotal factor, especially for battery-operated devices. The synthesized design exhibited a total power consumption of 2.5327 mW, which is considered low for a hardware accelerator capable of processing high-resolution video content.

IV. CONCLUSION

In this paper, an interpolation hardware design for VVC motion compensation is proposed. Utilizing temporal and spatial reuse techniques, the proposed VVC fractional interpolation hardware is capable of processing 4K @174fps at a clock frequency of 521 MHz. Compared to the original VVC fractional interpolation hardware operating at the same frequency, our design demonstrates an increase in throughput of up to 26%.

ACKNOWLEDGMENTS

This work was supported in part by the National Key R&D Program of China (2023YFB4502802), in part by the National Natural Science Foundation of China (62031009), in part by Alibaba Innovative Research (AIR) Program, in part by Alibaba Research Fellow (ARF) Program.

REFERENCES

[1] JVET. Versatile video coding (vvc). [Online]. Available: https://jvet.hhi.fraunhofer.de/

[2] A. Browne, Y. Ye, and S. Kim, "Algorithm description for versatile video coding and test model 17 (vtm 17)," JVET-Z2002, July 2022.

[3] J. Vanne, M. Viitanen, T. D. Hamalainen and A. Hallapuro, "Comparative Rate-Distortion-Complexity Analysis of HEVC and AVC Video Codecs," IEEE Transactions on Circuits and Systems for Video Technology, vol. 22, no. 12, pp. 1885-1898, Dec. 2012.

[4] Í. Siqueira, G. Correa and M. Grellert, "Rate-Distortion and Complexity Comparison of HEVC and VVC Video Encoders," 2020 IEEE 11th Latin American Symposium on Circuits & Systems (LASCAS), San Jose, Costa Rica, 2020, pp. 1-4.

[5] C. M. Diniz, M. Shafique, S. Bampi and J. Henkel, "A Reconfigurable Hardware Architecture for Fractional Pixel Interpolation in High Efficiency Video Coding," IEEE Transactions on Computer-Aided Design of Integrated Circuits and Systems, vol. 34, no. 2, pp. 238-251, Feb. 2015.

[6] H. Azgin, A. C. Mert, E. Kalali and I. Hamzaoglu, "A Reconfigurable Fractional Interpolation Hardware for VVC Motion Compensation," 2018 21st Euromicro Conference on Digital System Design (DSD), Prague, Czech Republic, 2018, pp. 99-103.

[7] A. CanMert, E. Kalali and I. Hamzaoglu, "A Low Power Versatile Video Coding (VVC) Fractional Interpolation Hardware," 2018 Conference on Design and Architectures for Signal and Image Processing (DASIP), Porto, Portugal, 2018, pp. 43-47.

[8] H. Azgin, E. Kalali and I. Hamzaoglu, "An Approximate Versatile Video Coding Fractional Interpolation Hardware," 2020 IEEE International Conference on Consumer Electronics (ICCE), Las Vegas, NV, USA, 2020, pp. 1-4.

A Broadband Digital Beamforming Method Based on FPGA

Yiwen Tang, Guowen Jia, Zhen Zhang and Yue Zhang*

Sun Yat-Sen University, Shenzhen, China

* tangyw5@mail2.sysu.edu.cn, zhangyue8@mail.sysu.edu.cn

Abstract—This paper introduces a wideband digital beamforming (DBF) method leveraging stretch processing to mitigate the impact of the Aperture Fill Time (AFT) phenomenon in wideband radar systems. The proposed method involves mixing the received echo signals with a specific reference signal, converting them into narrowband signals with phase differences across channels, followed by phase compensation to minimize these differences. Additionally, the implementation of this DBF method on a Field-Programmable Gate Array (FPGA) is explored, utilizing MicroBlaze to calculate the parameters for the reference signal and phase compensation, which are then stored in FIFO buffers. The reference signal parameters are employed to generate the necessary reference signals for each channel using the DDS IP core. Stretch processing is realized through signal mixing, and the phase compensation parameters stored in the FIFOs are used to correct the stretch processed signals. The signals from each channel are subsequently summed to complete the DBF process. Simulation results demonstrate that the phase differences between signals from different channels were reduced from a maximum of 262.8 degrees to 0.7188 degrees after compensation. Consequently, the signal power obtained after DBF is significantly enhanced.

Keywords—digital beamforming, stretch processing, phase compensation, FPGA, MicroBlaze

I. INTRODUCTION

The utilization of wideband phased arrays is deemed superior for target detection and radar imaging applications, primarily owing to their remarkable range resolution capabilities, which offer critical technical enhancements to contemporary radar systems [1]. Nonetheless, the aperture fill time (AFT) phenomenon, an inherent characteristic of wideband arrays, presents a notable impediment to the optimization of beamforming performance [2]. Therefore, investigating methods to address the AFT phenomenon holds significant research value. Furthermore, Field-Programmable Gate Arrays (FPGAs) are extensively employed in modern radar systems [3], thus, implementing solutions to mitigate AFT on FPGAs carries considerable practical significance.

This paper presents a wideband signal DBF method based on stretch processing, aiming to remove the influence of the AFT phenomenon for wideband radar systems. The study employs stretch processing to mix the received echo signal with a specific reference signal, resulting in narrowband signals with phase differences for each channel, followed by phase compensation to reduce phase differences between channels. Furthermore, an FPGA-based implementation of this digital beamforming (DBF) method is proposed, utilizing the MicroBlaze in the FPGA [4] to compute the parameters for the reference signal and phase compensation, which are stored in the FIFO buffers. The reference signal parameters are then used to generate the required reference signals for

This work is supported by National Natural Science Foundation of China with project number 62174181. This research was funded by the National Natural Science Foundation of China under Grant U2133216 and the Science and Technology Planning Project of Key Laboratory of Advanced IntelliSense Technology, Guangdong Science and Technology Department under Grant 2023B1212060024.

each channel using the DDS IP core. Stretch processing is achieved through mixing [5], and the phase compensation parameters stored in the FIFOs are utilized to compensate the stretch processed signals, followed by the summation of signals from each channel to complete the DBF process. The structure of this paper is as follows. Section II begins by presenting a wideband DBF method based on the stretch processing, followed by the proposal of its implementation on an FPGA. Section III presents the experimental results of the FPGA implementation of the wideband DBF. Finally, Section IV concludes the paper.

II. THE PROPOSED WIDEBAND DBF METHOD AND IMPLEMENTATION

A. Wideband DBF Method Based on Stretch Processing

Fig. 1 illustrates the phased array echo signal model, with the array configured as a linear array consisting of M elements. The spacing between adjacent elements is d, and the array receives target echo signals in the far field, with an incident angle of θ.

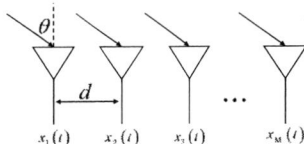

Fig. 1. The phased array echo signal model consisting of M elements.

By taking the first element as the reference element, the propagation delay of the echo signal to the m-th element relative to the reference element can be obtained:

$$\tau_m = \frac{(m-1)d\sin\theta}{c} \quad m = 1, 2, \cdots, M \quad (1)$$

The linear frequency modulation (LFM) signal is the most commonly used wideband signal, widely employed in radar and communication systems. This paper focuses on DBF tailored to the characteristics of LFM signals, converting the wideband signal to a narrowband signal by stretch processing. Subsequently, phase compensation is applied to the individual channel signals to achieve DBF of wideband signals.

Assuming the radar transmits a LFM signal with a starting frequency of f_0, spanning a bandwidth B over a time period T. Hence, we can define the chirp rate μ as $\mu = B/T$. When the radar detects a target, the delay of the echo signal received by the reference array element relative to the transmitted signal is t_0. The echo signal received by the m-th array element can be obtained:

$$s_m(t) = e^{j\left(\pi\mu(t-t_0-\tau_m)^2 + 2\pi f_0(t-t_0-\tau_m)\right)} \text{rect}\left(\frac{t-t_0-\tau_m}{T}\right) \quad (2)$$

To ensure that the stretch processing signal of each channel remains consistent, differing only in phase, the required reference signal corresponding to the m-th channel can be expressed as follows:

$$s_{\text{ref}m}(t) = e^{j\left(\pi\mu\left(t - t_{\text{ref}} - \tau_m\right)^2 + 2\pi f_0\left(t - t_{\text{ref}} - \tau_m\right)\right)} \text{rect}\left(\frac{t - t_{\text{ref}} - \tau_m}{T}\right) \quad (3)$$

By mixing echo signals of each channel with the corresponding reference signal, the resulting signal for the m-th channel can be obtained. The stretch processing signal $x_m(t)$ with frequency f_b and phase Φ_m in the m-th channel can be described as follows:

$$f_b = \mu\left(t_{\text{ref}} - t_0\right) \quad (4)$$

$$\phi_m = 2\pi f_0\left(t_{\text{ref}} - t_0\right) - \pi\mu\left(t_{\text{ref}}^2 - t_0^2\right) - 2\pi\mu\left(t_{\text{ref}} - t_0\right)\tau_m \quad (5)$$

$$x_m(t) = e^{j\left(2\pi f_b t + \phi_m\right)} \quad (6)$$

Therefore, the signals of each channel have the same frequency after stretch processing, with a phase difference of $\Delta\Phi_m = 2\pi(t_{\text{ref}} - t_0)\tau_m$ relative to the reference channel. To achieve frequency and phase alignment of the signals across all channels, the phase of the signals of each channel must be compensated by $\Delta\Phi_m$. After phase compensation, the final signal obtained for each channel is denoted as follows:

$$y_m(t) = e^{j\left(2\pi f_b t + 2\pi f_0\left(t_{\text{ref}} - t_0\right) - \pi\mu\left(t_{\text{ref}}^2 - t_0^2\right)\right)} \quad (7)$$

By accumulating the signals from the M channels, the DBF process is completed, the signal after DBF processing can be obtained:

$$y(t) = M e^{j\left(2\pi f_b t + 2\pi f_0\left(t_{\text{ref}} - t_0\right) - \pi\mu\left(t_{\text{ref}}^2 - t_0^2\right)\right)} \quad (8)$$

B. FPGA-based Wideband Digital Beamforming

Based on the theory laid out above, it is necessary to mix the echo signals with different reference signals across various channels. On the FPGA, the bandwidth and pulse width parameters of the reference LFM signal can be dynamically configured using MicroBlaze, and these parameters are subsequently stored in 8 FIFO buffers. By dynamically modifying the input parameters of the DDS IP core during each clock cycle, the LFM reference signals are generated. Fig. 2 depicts the generation of the LFM reference signals of 8 channels.

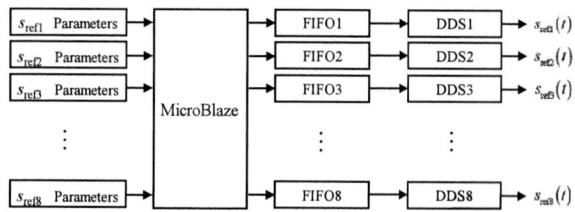

Fig. 2. Generation of the LFM reference signals of 8 channels.

Analogous to the DDS input parameters, the phase compensation parameters for each channel, denoted as $\Delta\Phi_m$, are generated by the MicroBlaze processor and subsequently stored in the corresponding FIFO buffers. These parameters are retrieved from the corresponding FIFOs and used to compensate the phase of the stretch processing signal $x_m(t)$. By aggregating the compensated signal $y_m(t)$ from 8 channels, the

DBF process is concluded, resulting in the acquisition of the final signal $y(t)$. Fig. 3 depicts the process of the 8-channel DBF process based on stretch processing.

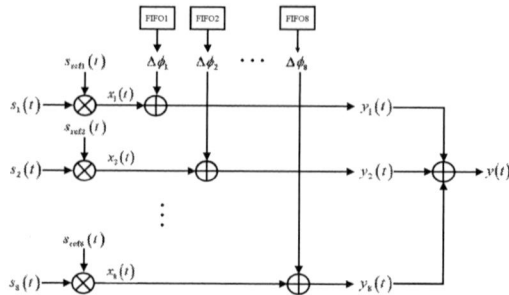

Fig. 3. 8-Channel DBF process based on stretch processing.

The MicroBlaze in FPGA is employed to calculate the parameters required for the DDS IP core, which generates the reference signals, and phase compensation parameters. These parameters are then stored in the corresponding FIFO buffers for each channel. Subsequently, the FPGA performs the stretch processing and phase compensation. Finally, the phase-compensated signals of 8 channels are accumulated, thereby completing the DBF process based on LFM signals.

III. SIMULATION RESULTS AND DISCUSSION

Deploy the broadband DBF process proposed above on FPGA. The experimental conditions are set as follows: the number of channels M=8, adjacent elements spacing d=0.015m, echo signal bandwidth B=400MHz, pulse width PW=40us, target echo angle θ=30°, target echo delay t_0= 300us, reference signal delay t_{ref} = 2us.

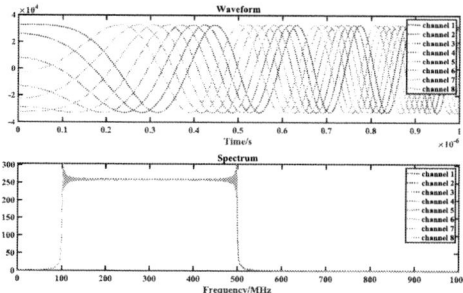

Fig. 4. The waveform and spectrum of echo signals of 8 channels.

As described in Fig. 4, due to the presence of unit spacing, there exists a certain phase difference between the echo signals of the other channels and the echo signal of the reference channel. In order to generate the LFM reference signal through DDS IP and to perform phase compensation for the stretch signal, MicroBlaze in FPGA stores the DDS input parameters and phase compensation parameters into 8 FIFO buffers.

The echo signals are mixed with the reference signals generated by DDS, resulting in stretch processed signals. Due to the inherent phase differences present in the stretch processed signals of 8 channels, it is necessary to compensate for the phase of each channel signal using the phase compensation values stored in the FIFOs. Fig. 6 illustrates the stretch processed signals of each channel prior to phase compensation, as well as the signals of each channel after phase compensation has been applied.

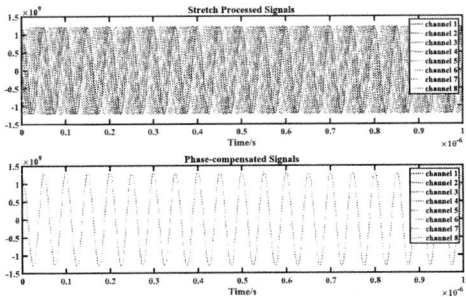

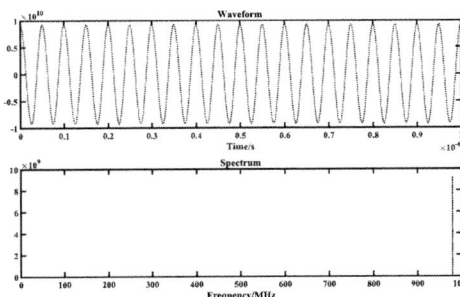

Fig. 5. The waveforms of stretch processed and phase-compensated signals.

As illustrated in Fig. 5, the phase differences between the stretch processed signals of each channel are significantly reduced after phase compensation. The phase differences of signals from 8 channels relative to channel 1 are shown in Table I, both before and after phase compensation.

TABLE I. PERFORMANCE SUMMARY AND COMPARISON

Channel	Phase difference before compensation (°)	Phase difference after compensation (°)
Channel 1	0	0
Channel 2	37.5480	0.0928
Channel 3	75.0960	0.1412
Channel 4	112.6440	0.2444
Channel 5	150.1920	0.4292
Channel 6	187.7399	0.4896
Channel 7	225.2879	0.576
Channel 8	262.8359	0.7188

Fig. 6 illustrates the signal waveform and spectrum obtained by summing the signals after phase compensation from 8 channels. It can be observed that compared to the signal of single channel, the power of this signal is significantly increased, and the frequency remains unchanged, achieving the effect of wideband Digital Beamforming.

Fig. 6. The waveform and spectrum of the signal after performing DBF.

The conventional broadband beamformer is implemented by applying delay, weighting, and summation to the outputs of each element in the array. However, when a broadband signal passes through this beamformer, the beam pattern formed by signals of different frequencies varies, with only the responses in the center direction of the main lobe being consistent. Therefore, only when the center of the main lobe aligns with the target, the beamformer output signal will not be distorted.

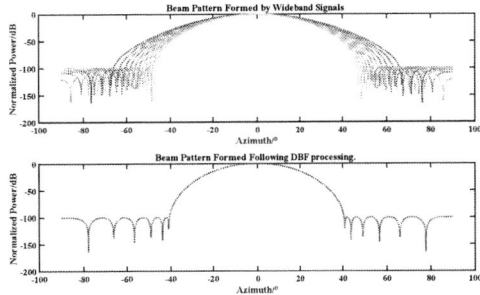

Fig. 7. The beam patterns formed before and after the application of DBF to the broadband signal.

As illustrated in Fig. 7, the beam pattern for the formation of a broadband signal exhibits a significant variation in main lobe width. Following the application of DBF to the broadband signal, it is transformed into a narrowband signal, eliminating the aforementioned issue, resulting in the beam pattern depicted in Fig. 7.

IV. CONCLUSION

This paper proposes a wideband DBF scheme based on stretch processing. When the echo signal is an LFM signal, a specific reference signal is employed to complete stretch processing. The resulting signal is then subjected to phase compensation to mitigate phase differences between channels. Subsequently, the signals from each channel are coherently combined to achieve wideband DBF. Furthermore, a FPGA-based DBF implementation is proposed. It utilizes the MicroBlaze to calculate the parameters necessary for the generation of reference signals by the DDS IP core, as well as the values required to compensate for the phase differences between the stretch processed signals of each channel and those of the reference channel. These parameters are stored in the FIFO buffers. The DDS input parameters within the FIFO are used to produce the reference signals needed for the stretch processing. Ultimately, the phase compensation values stored in the FIFO buffers are applied to compensate the phase of the stretch processed signals from each channel, significantly reducing the phase differences between channels, enabling their combination. This process realizes a wideband DBF operation based on FPGA technology.

REFERENCES

[1] Z. Liang, Q. Liu and T. Long, "A Novel Subarray Digital Modulation Technique for Wideband Phased Array Radar," in *IEEE Transactions on Instrumentation and Measurement*, vol. 69, no. 10, pp. 7365-7376, Oct. 2020.

[2] Zhang, C. and Q. Lai, "Research on phased array radar affected by aperture fill time, "*Journal of Microwave Science*, Vol. 33, No. 4, 67-69, 2017.

[3] H. Wang, T. Shan, X. Qiao and H. Cheng, "Research on the design of FPGA-based radar signal processing system," *2021 7th International Conference on Condition Monitoring of Machinery in Non-Stationary Operations (CMMNO)*, Guangzhou, China, 2021, pp. 36-39.

[4] E. H. El Mimouni and M. Karim, "A MicroBlaze-based Multiprocessor System on Chip for real-time cardiac monitoring," *2014 International Conference on Multimedia Computing and Systems (ICMCS)*, Marrakech, Morocco, 2014, pp. 331-336.

[5] H. A. Krichene, M. J. Pekala, M. D. Sharp, K. C. Lauritzen, D. G. Lucarelli and I. -J. Wang, "Compressive sensing and stretch processing," *2011 IEEE RadarCon (RADAR)*, Kansas City, MO, USA, 2011, pp. 362-367.

Conditional cycle termination RANSAC

Tong Jiang [1,2], Yujie Huang [*1,2], Liyuan Peng [1,2], Mingyu Wang [*1], Wenhong Li [1], Minge Jing [1], Xiaoyang Zeng [*1]

[1] State Key Lab of ASIC & System, Fudan University, 825 Zhangheng Rd, 201203, Shanghai, China
[2] Shanghai ExploreX Technology Co., Ltd., 188 Shengrong Rd, 200120, Shanghai, China

* Email:{mywang, huangyj19, xyzeng}@fudan.edu.cn

Abstract—In the field of computer vision, the acquired dataset usually contains a certain number of outliers and noise, which leads to errors in the estimated mathematical model. RANSAC estimates model parameters by randomly selecting a subset of the data, reducing the impact of these outliers and noise. However, for datasets with a high proportion of inliers, the traditional RANSAC has to perform a large number of cyclic modeling operations after finding the correct model. In this paper, a conditional termination verification strategy for RANSAC is presented. We boldly terminated the cycle operation early which reduced the amount of computation by 17.78% and increased the computation speed by 34.09%.

Keywords—*RANSAC, Model evaluation, Inliers, Outliers, Conditional termination*

I. INTRODUCTION

The RANSAC (Random Sample Consensus) algorithm introduced by Fishler and Bolles in 1981 [1] is a widely used robust estimator that has become a de facto standard in the field of computer vision. RANSAC and related hypothesize-and-verify methods[2~7] have been applied to many vision problems.

It is generally accepted that the dataset has outliers. In this case, compared to the least squares method[12] to fit the model based on all data points, RANSAC can effectively identify and eliminate outliers in the data points first, and then fit the optimal model that can adapt to most of the data points, which greatly improves the accuracy of the model. Fig. 1 shows the superiority of RANSAC by using data points to fit a straight line as an example. Compared to Least Squares fitted model, RANSAC fitted model has a higher accuracy (the correct data points are closer to the straight line) after removing the two outliers.

The RANSAC algorithm proceeds as follows:

1. Randomly select m points, m is the minimum number of points required to establish a model.

2. Use the m points to fit the model.

3. Calculate the distance from other points to the fitted model. If it is less than a certain threshold, the point is regarded as an internal point, and the number of internal points is counted.

4. Repeat N times to select the model with the largest number of points in the interior. N satisfies Equation 1,

$$N=log_{(1-a\char94 m)}(1-P) \qquad (1)$$

where a is the proportion of inliers in the input data, and P is the probability that the sample points are all drawn from inliers at least once in N times.

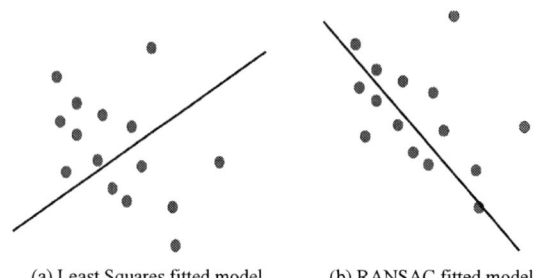

(a) Least Squares fitted model   (b) RANSAC fitted model

Fig. 1 Comparison of the least-squares fitting model with the RANSAC fitted model.

It can be observed that due to the lack of clear selection rules, some modeling operations will be performed by RANSAC even after the correct model has been found. For example, in the field of image stitching, the overlap of different images is large, so the proportion of inliers is high, and it is easy to select m inliers. At this point, the model is sufficient, even though it has fewer inliers.

J.Matas and O.Chum et al. have proposed the Progressive Sample Consensus (PROSAC) algorithm[9], which evaluates each pair of matching points through similarity, and preferentially selects the data points with high scores for modeling to reduce the number of sampling times and improve the operation speed. However, scores are manually defined and this method isn't fully scripted. They also propose the locally optimized RANSAC (LO-RANSAC)[8]. Due to the influence of noise and other factors, even if the points taken are correct, the number of inliers of the fitted model may still be low, which will indirectly lead to an increase in the number of cycles. To make the fitting model more accurate and increase the number of inliers, LO-RANSAC performs multiple calculations, and each fitting uses all the inliers points determined in the previous step as the fitting data and gradually expands the number of inliers. This method relies too much on the randomly taken data points, and if it contains the wrong point, a lot of invalid operations will be performed. Randomized RANSAC[10] is an idea for faster model evaluation. Among different evaluation methods, the $T_{d,d}$ test randomly selects d data points, if all are the inliers of the fitting model, then continues to judge whether the other data points are inliers, if there are non-inliers in d points, then re-select the points to carry out the next model fitting. This method's randomness is too high, and there is no basis for treating all randomly selected d points as correct points, which may misjudge the correct model. Sequential Probability Ratio Test(SPRT)[11] solves a global optimization problem, minimizing a single real number, the time, to decision, which introduces an overly complex computational process.

979-8-3503-6184-1/24 $31.00 © 2024 IEEE

It can be seen that the existing methods cannot reduce the amount of computation on the premise of ensuring the quality of the model. In this paper, we propose a conditional cycle termination verification strategy for the situation with a high proportion of intra-class points. It exploits the fact that due to the influence of noise and other factors, it is inaccurate to evaluate a model based solely on the number of inliers. Therefore, it is not necessary to look for the largest number of inliers and the method we propose can obtain a sharp reduction in the amount of computation and a significant increase in computing speed.

II. ALGORITHM

This chapter will prove the inaccuracy of evaluating the model by the number of inliers alone, so it is not necessary to obsess about finding the model with the most inliers. We propose a strategy of conditional cycle termination to improve the speed of the algorithm and reduce the amount of computation.

A. Deficiencies evaluated by the model

Traditional RANSAC only uses the number of inliers as the only criterion for judging the correctness of the model, but this is unreasonable.

Take a linear model as an example and set it to $y=3x+2$. Select 15 data points on it, as shown in Fig. 2(a). A piece of noise is added at random, as shown in Fig. 2(b). Applying RANSAC to the points after adding noise, setting the threshold to 1.5 and the confidence level to 0.99, the resulting linear model is $y=2.75x+4.75$, and the number of inliers is 10, while the number of inliers of the correct model $y=3x+2$ is only 6 with the same setting. It can be seen that it is inaccurate to use only the number of inliers as the evaluation criterion of the model.

B. Conditional cycle termination RANSAC

Specifically, as soon as the number of inliers is greater than the number related to the total number of data points, the loop is terminated. In this regard, we introduce a termination percentage β satisfying the following formula,

$$num\_termination = num\_total * \beta \qquad (2)$$

where *num_total* is the total number of data points, and *num_termination* is the number of inliers that terminate the loop. The improved RANSAC algorithm is shown in Fig. 3.

Applying this strategy to the example in Fig. 2, since the number of inliers of the correct model is 6, it can be assumed that the results of the number of inliers greater than 6 are acceptable, so the $\beta=6/15=0.4$ is taken. The result is $y=2.82875x+4.72625$, in this case, the format of the in-class point is 7. At this point, the proposed approach reduces the computation time by a factor of nearly 10 without increasing the error.

III. EXPERIMENTAL RESULTS

The RANSAC algorithm has a wide range of applications in the field of computer vision, and here we test its application in the field of image stitching. In the image stitching algorithm, RANSAC is applied to the data points after brute-force matching to remove the mismatched points.

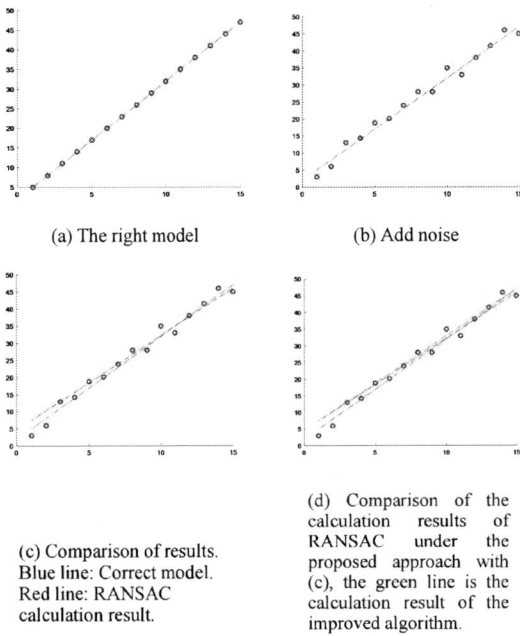

(a) The right model (b) Add noise

(c) Comparison of results. Blue line: Correct model. Red line: RANSAC calculation result.

(d) Comparison of the calculation results of RANSAC under the proposed approach with (c), the green line is the calculation result of the improved algorithm.

Fig. 2 Fitting comparison of different algorithms.

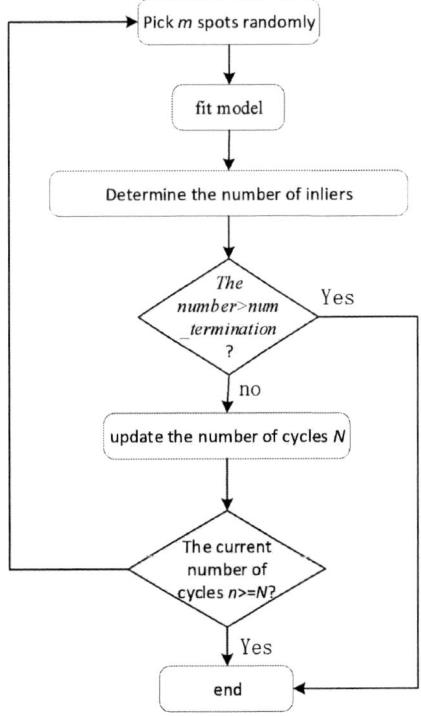

Fig. 3 Algorithm flow diagram

First, select a model as the correct model. Then select a complete image, divide it into two local images with overlapping regions, invert the selected model and use it to distort one of the local images, and then use different methods to fit the original model. The parameters are set as follows: the threshold is set to 7, the confidence level is 0.99, the maximum number of cycles is 1000, and the

979-8-3503-6184-1/24 $31.00 © 2024 IEEE

TABLE I. Comparison of results from different methods.

average	RANSAC	$T_{d,d}$ test	Ours
Time/s	0.088	0.075	0.058
cycles	45	94	37
inliers	1607	1518	1606
error	20.46%	18.76%	19.53%

termination percentage β is 0.4. The calculated model is compared with the correct model, and every parameter in the model is added by weight to obtain the sum of the errors and the sum of the correct model. Divide the two to get the percentage of error. After 20 replicates of the experiment, the average results are shown in Table 1, and

(a) Traditional RANSAC-right

(b) The proposed approach -right

(c) Traditional RANSAC-left

(d) The proposed approach -left

Fig. 4 Comparison of the results obtained by the traditional RANSAC with the proposed approach on the image.

the distribution of the screened match points on the image is shown in Fig. 4.

As can be seen from the pictures, the improved method has about the same distribution of matching points as the traditional method. In the table, the conditional cycle termination strategy effectively reduces the number of cycles and the running time. The number of cycles is reduced by 17.78% and the running time is reduced by 34.09%.

IV. CONCLUSION

The conditional cycle termination RANSAC is proposed in this paper. Based on the fact that the correct model may not have the highest number of inliers, it is not necessary to be obsessed with the pursuit of the highest number of points in the model. So we boldly terminate the loop early when the number of inliers reaches a certain percentage of the total number of points. The experimental results show that the number of cycles is reduced by 17.78% and the running time by 34.09% without introducing large errors.

REFERENCES

[1] M.A. Fischler and R.C. Bolles. Random sample consensus: A paradigm for model fitting with applications to image analysis and automated cartography. CACM, 24(6):381–395, June 1981.

[2] O. Chum, J. Matas, and J. Kittler. Locally optimized RANSAC. In Proc. DAGM. Springer-Verlag, 2003..

[3] D.R. Myatt, P.H.S. Torr, S.J. Nasuto, J.M. Bishop, and R. Craddock. Napsac: High noise, high dimensional robust estimation - it's in the bag. In BMVC02, volume 2, pages 458–467, 2002.

[4] D. Nister. Preemptive RANSAC for live structure and motion estimation. In Proc. ICCV03, volume I, pages 199–206, October 2003.

[5] B. Tordoff and D.W. Murray. Guided sampling and consensus for motion estimation. In Proc. 7th ECCV, volume 1, pages 82–96. Springer-Verlag, 2002.

[6] P. H. S. Torr and A. Zisserman. MLESAC: A new robust estimator with application to estimating image geometry. CVIU, 78:138–156, 2000.

[7] P.H.S. Torr, A. Zisserman, and S.J. Maybank. Robust detection of degenerate configurations while estimating the fundamental matrix. CVIU, 71(3):312–333, Sep. 1998.

[8] Chum, O., Matas, J., Kittler, J. (2003). Locally Optimized RANSAC. In: Michaelis, B., Krell, G. (eds) Pattern Recognition. DAGM 2003. Lecture Notes in Computer Science, vol 2781. Springer, Berlin, Heidelberg. https://doi.org/10.1007/978-3-540-45243-0_31

[9] O. Chum and J. Matas, "Matching with PROSAC - progressive sample consensus," 2005 IEEE Computer Society Conference on Computer Vision and Pattern Recognition (CVPR'05), San Diego, CA, USA, 2005, pp. 220-226 vol. 1, doi: 10.1109/CVPR.2005.221.

[10] J Matas, O Chum, Randomized RANSAC with Td,d test, Image and Vision Computing, Volume 22, Issue 10, 2004, Pages 837-842, ISSN 0262-8856

[11] J. Matas and O. Chum, "Randomized RANSAC with sequential probability ratio test," Tenth IEEE International Conference on Computer Vision (ICCV'05) Volume 1, Beijing, China, 2005, pp. 1727-1732 Vol. 2, doi: 10.1109/ICCV.2005.198.

[12] PLACKETT R L. The Discovery of the Method of Least Squares[J]. Biometrika, 1972, (59) : 2392251

979-8-3503-6184-1/24 $31.00 © 2024 IEEE

A Multi-Resolution Propagation Algorithm and Pixel Grouping Storage Strategy for PatchMatch Stereo

Kai Liu, Zhenyu Zhang, Haiwei Wang, Leilei Huang, Chunqi Shi, Long Xu, Runxi Zhang

Institute of Microelectronics Circuits and Systems
East China Normal University
Shanghai, 200241, P. R. China
Email: llhuang@cee.ecnu.edu.cn, cqshi@ee.ecnu.edu.cn, lxu@cee.ecnu.edu.cn, rxzhang@ee.ecnu.edu.cn

Abstract—As an effective means of measuring spatial distance, stereo matching has broad application prospects, especially in real-time application scenarios such as autonomous driving. Compared with the commonly used local matching method, which requires cost calculation for all disparities within the disparity range, the random search idea of the PatchMatch Stereo (PMS) can effectively reduce the amount of calculation and is more competent for matching under a large disparity range. This paper proposes a multi-resolution iterative propagation algorithm for PMS to further reduce its computational workload while improving the matching accuracy. In addition, in response to the multi-resolution image reading requirements of this algorithm, a pixel grouping storage strategy is proposed, which can reduce the bandwidth occupancy when reading low-resolution images from DDR, and can be flexibly configured according to the multi-resolution propagation strategy.

Index Terms—stereo match, PatchMatch, bandwidth

I. INTRODUCTION

Stereo matching is a method for estimating information of depth through a pair of images from binocular camera. It can be used in scenarios such as autonomous driving and remote sensing imaging [1]. Compared with other 3D distance estimating methods such as RADAR, it has obvious advantages in terms of cost, efficiency, and power consumption.

Stereo matching usually includes four steps: cost calculation, cost aggregation, disparity calculation and post-processing. Cost calculation is to calculate the initial cost value for each pixel based on the cost function and disparity. Then, the costs are aggregated according to a certain pixel range to obtain the aggregated cost. The aggregation cost is used to calculate the disparity of each pixel. The most commonly used method is Winner-Takes-All (WTA), which takes the disparity value with the lowest cost as the disparity of the current pixel. At last, a series of post-processing, such as LRC and filtering, is performed on the obtained disparity map to obtain the final disparity map.

There are mainly 3 methods for stereo matching: local method, global method and Semi-Global Matching (SGM) [3]. The global method obtains the disparity by minimizing the global energy instead of cost aggregation, while the local method usually calculate in a local window centered on the pixel to be calculated. SGM combines the global method with the local method, and it aggregates the cost by minimizing

the energy in multiple one-dimensional directions. Therefore, the performance and cost of SGM are also between these two methods. However, all of them compute the cost for all disparities within the disparity range, which has a larger hardware cost when the disparity range is large.

Differently, PatchMatch Stereo (PMS) [3] performs a random search for disparities instead of calculating the cost for all disparities, which consists of three steps: random initialization, iterative propagation and post-processing. First, a disparity value within the disparity range is randomly assigned to each pixel. Iterative propagation obtains the disparity values of other pixels, calculates and compares the aggregation cost, retaining the one with the lower cost as the new disparity of the current pixel. The new disparity may be obtained from neighboring pixels, corresponding pixels in another view, or randomly generated. Since the time of propagation is fixed, the computational complexity of PMS is independent of the parallax range. Therefore, it has great potential in large disparity realtime stereo matching. However, the computational complexity of each iteration of the PMS algorithm is still high, and reading the same image at each calculation stage will occupy more memory bandwidth.

In order to solve these problems, we proposed a multiresolution propagation strategy based on PMS. It can effectively reduce the amount of computation in the iterative propagation stage and improve computational efficiency. In addition, in view of the need to obtain images of different resolutions when calculating the cost under multiple resolutions, we proposed a pixel grouping storage strategy, which can reduce the bandwidth occupancy when obtaining multiresolution images from DDR. It is also friendly to resourceconstrained hardware implementation systems. The rest of this paper is organized as follows. Section II proposes the multiresolution propagation algorithm and pixel grouping storage strategy for it. Section III discusses the experimental results. Finally, the conclusion is concluded in Section VI.

II. PROPOSED METHODS

A. Multi-resolution propagation

The multi-resolution propagation algorithm proposed in this work is shown in Fig. 1. For the resolution of propagation, we chose two strategies: Quarter-Half-Full and Half-Full.

979-8-3503-6184-1/24 $31.00 © 2024 IEEE

Quarter-Half-Full performs random initialization and the first propagation at a resolution of $1/4 \times 1/4$, and performs Search & Propagation once at three resolutions: $1/4 \times 1/4$ (Quarter), $1/2 \times 1/2$ (Half), and 1×1 (Full). After completing a Search & Propagation at a low resolution, the obtained disparity map is enlarged to a high resolution, and this operation is repeated until the Full resolution is reached. At this time, the disparity map is post-processed and output as the final disparity map. Half-Full completes random initialization and the first propagation at a Half resolution, and then performs two Search & Propagation operations at Half and Full resolutions and post-processes to obtain the disparity results.

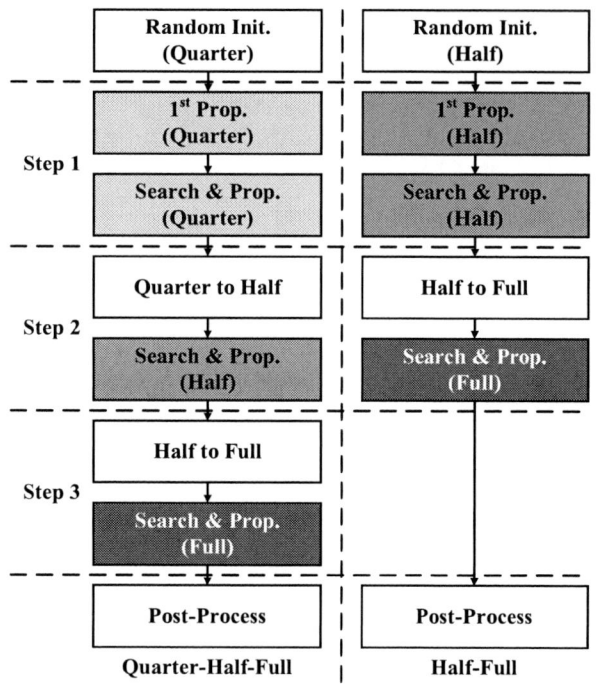

Fig. 1. Overflow of the multi-resolution propagation algorithm

In the first propagation, the candidate disparities for each pixel include 3 initial disparities and the updated disparities of the pixels on its left and above:

$$d_{cand}(i,j) = \{d_{init}(i, j-2), d_{init}(i, j-1),$$
$$d_{init}(i,j), d_{new}(i, j-1), d_{new}(i-1, j)\} \quad (1)$$

The Census cost is used for cost calculation. It performs Census transformation on the image and calculates the Hamming distance of the corresponding pixels. The aggregation cost is calculated by summing the rectangular area costs. It aggregates the costs of all candidate disparities and selects the one with the smallest cost to update the disparity.

$$d_{new}(i,j) = \underset{d \in d_{cand}}{\arg\min} Cost(i,j,d) \quad (2)$$

The Search & Propagation step combines random search and spatial propagation in PMS. The step size of the random search depends on the disparity range corresponding to the current image resolution and the number of random searches between it. Generally speaking, for a certain image resolution, the disparity range is also certain. In this case, the step size of the first random search is 1/4 of the disparity range, and each subsequent random search is 1/2 of the previous step size. When the image is downsampled, the disparity range will also decrease. For example, at half resolution, the disparity range is also reduced to half of the original. Therefore, after the first search is completed at a low resolution, the increase in resolution will offset the reduction in the search step size. Based on this, we use the step size of the first search in all searches. After each pixel completes the search, it will perform spatial propagation, that is, check whether the disparity of the pixels on its left and above is more suitable for itself:

$$d_{cand}(i,j) = \{d_{new}(i, j-1), d_{new}(i-1, j),$$
$$d_{last} + rand(-\Delta d, \Delta d)\} \quad (3)$$

where Δd is the search step size, which is set to 1/4 of the disparity value corresponding to the initial resolution.

Finally, after completing the full-resolution Search & Propagation step, the obtained disparity map is subjected to LRC and weighted median filtering to obtain the final disparity map.

B. Pixel grouping storage strategy

The multi-resolution propagation requires reading images of different resolutions. For its hardware implementation, images are often read from DDR by row. For the acquisition of multi-resolution images, the simplest method is to cache the entire image and obtain images of different resolutions. This method can ensure the instant acquisition of data, but it requires a lot of hardware resources. Another way is to input the image data row by row for each calculation step. This can save hardware resources, but at low resolution, each row of data needs to be processed to remove unnecessary pixels, which will cause bandwidth waste. The lower the required resolution, the more serious the bandwidth waste problem.

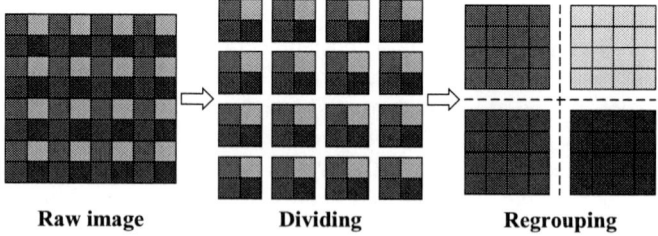

Fig. 2. The proposed pixel grouping storage strategy

In order to solve the bandwidth waste problem when reading low-resolution images, we proposed a pixel grouping storage strategy, as shown in Fig. 2. First, the image stored in DDR is divided into 2×2 pixel blocks. The 4 pixels in each 2×2 block are grouped to obtain 4 images with 1/4 resolution, which are stored separately. When there is a need to access a low-resolution image, the image with a specific index can be read directly from the DDR. When the original resolution image is needed, all four groups of images are read and reassembled to obtain the required image. In addition, this block storage strategy can also achieve image downsampling in a single direction, and can be flexibly configured according to the multi-resolution iterative propagation strategy.

III. EXPERIMENTAL RESULTS

We evaluate our proposed multi-resolution propagation on the KITTI2015 [5] stereo matching dataset. In the evaluation, we use 7×7 Census cost and perform cost aggregation in a window of 11×29, and the original disparity range is set to 128.

TABLE I
EVALUATION RESULTS

Propagation		Full	Half-Full	Quarter-Half-Full
Error rate		10.37%	7.83%	7.83%
Calculation amount (Iteration per pixel)		11	4.75	4.19
		100%	43.18%	38.09%
DDR Bandwidth (Frame)	Before grouping	4	2	2
		100%	50%	50%
	After grouping	4	1.5	1.375
		100%	37.5%	34.4%

Based on the above parameter settings, we evaluated three scenarios: Full resolution propagation (search 3 times at the same resolution), Half-Full propagation, and Quarter-Half-Full propagation. Fig. 3 shows the error rate comparison of each step.

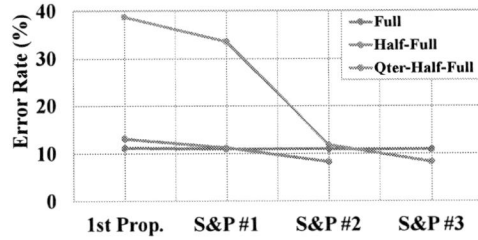

Fig. 3. Comparison of average error rate based on the KITTI2015 dataset

It can be observed that the accuracy of propagation at multi resolution is significantly better than the result of propagation only at Full resolution, decreased 2.54% error rate. For Quarter-Half-Full Propagation, its error rate is very high at quarter resolution, but drops rapidly at half resolution, which shows that compared with quarter resolution, half resolution search can more effectively reduce the error rate. The results of Half-Full Propagation also illustrate this point. In addition, Half-Full method achieves the same accuracy as Quarter-Half-Full method while reducing one search, which shows that it has higher matching efficiency. The computational complexity of the above cases is evaluated as well. The results show

that low-resolution propagation can also effectively reduce the computational complexity of the iterative propagation stage. Compared with the common Full-resolution propagation method, the computational complexity of the Half-Full method and the Quarter-Half-Full method is reduced by 56.81% and 61.91%.

The bandwidth usage of reading images from DDR is also evaluated here. The results show that the pixel grouping strategy can effectively reduce the bandwidth usage for image reading at low resolution, with the Half-Full method reducing it by 25% and the Quarter-Half-Full method reducing it by 31%.

IV. CONCLUSION

In this paper, we proposed a multi-resolution propagation strategy based on PMS and a pixel grouping storage strategy for multi-resolution. The proposed multi-resolution propagation method can significantly reduce the amount of computation while improving the matching accuracy. The pixel block storage strategy reduces the bandwidth usage when reading low-resolution images from DDR, and can be flexibly configured according to the multi-resolution iterative propagation strategy. The proposed algorithm was evaluated on the KITTI2015 dataset. The error rate of multi-resolution propagation was reduced by about 2.54% compared with full resolution results. Half-Full method shows a higher matching efficiency than Quarter-Half-Full method. The computational complexity was reduced by about 60%. The proposed pixel grouping storage strategy reduced the bandwidth usage of the Half-Full method by 25%, and the Quarter-Half-Full method by 31% respectively when reading low-resolution images from DDR.

ACKNOWLEDGMENT

This work is supported by the Fundamental Research Funds for the Central Universities, Science and Technology Commission of Shanghai Municipality (Grant No. 22DZ2229004).

REFERENCES

[1] S. Sivaraman and M. M. Trivedi, "Looking at vehicles on the road: A survey of vision-based vehicle detection, tracking, and behavior analysis," IEEE Trans. Intell. Transp. Syst., vol. 14, no. 4, pp. 1773–1795, Dec. 2013.
[2] H. Hirschmuller, "Stereo Processing by Semiglobal Matching and Mutual Information," in IEEE Transactions on Pattern Analysis and Machine Intelligence, doi: 10.1109/TPAMI.2007.1166, vol. 30, no. 2, pp. 328-341, Feb. 2008.
[3] M. Bleyer, C. Rhemann, and C. Rother, "Patchmatch stereo—Stereo matching with slanted support windows," in Proc. Brit. Mach. Vis. Conf., 2011, pp. 1–11.
[4] Z. Ma, K. He, Y. Wei, J. Sun, and E. Wu, "Constant time weighted median filtering for stereo matching and beyond," in Proc. IEEE Int. Conf. Comput. Vis., Dec. 2013, pp. 49–56.
[5] M. Menze and A. Geiger, "Object scene flow for autonomous vehicles," in Proc. IEEE Conf. Comput. Vis. Pattern Recognit., Jun. 2015, pp. 3061–3070.
[6] M. Tomasi, M. Vanegas, F. Barranco, J. Diaz, and E. Ros, "Realtime architecture for a robust multi-scale stereo engine on FPGA," IEEE Trans. Very Large Scale Integr. (VLSI) Syst., vol. 20, no. 12, pp. 2208–2219, Dec. 2012.
[7] W. Wang, J. Yan, N. Xu, Y. Wang, and F.-H. Hsu, "Real-time highquality stereo vision system in FPGA," IEEE Trans. Circuits Syst. Video Technol., vol. 25, no. 10, pp. 1696–1708, Oct. 2015.

979-8-3503-6184-1/24 $31.00 © 2024 IEEE

An XOR Arbiter PUF based on the IGZO TFT Devices

Xiang Chen, Yongliang Chen, Xiaole Cui *

School of ECE, Peking University Shenzhen Graduate School

* Email: cuixl@pkusz.edu.cn

Abstract—**Physical Unclonable Function (PUF) is of great concern in recent years as a hardware security primitive. PUF generates unique identifiers for each chip based on the random variations in physical parameters during the manufacturing process. The indium-gallium-zinc-oxide thin film transistor (IGZO TFT) exhibits obvious variations in device parameters, and it has the potential as the entropy source of PUF. In this work, the XOR Arbiter PUF based on the IGZO TFT devices is designed. The Monte Carlo simulations show that the randomness and uniqueness of the proposed TFT-based PUF scheme are better than that of the corresponding PUF scheme using MOSFET.**

Keywords—*PUF, IGZO TFT, XOR Arbiter PUF, Pseudo CMOS logic*

I. INTRODUCTION

Physical Unclonable Function (PUF) is a hardware security primitive, which is able to generate the chip-specific identifiers. The input and output of PUF are noted as challenge and response, respectively. The challenge and its corresponding response form a challenge response pair (CRP). The CRP is generated based on the random fluctuations in the physical parameters of devices. The devices providing the random variations are referred to as entropy sources. The entropy source with larger variations benefits the randomness and uniqueness of PUF. Currently, PUF has been widely used in hardware security applications, such as the key management, authentication and anti-counterfeiting. Compared with the traditional key storage techniques, PUF does not explicitly store the key in the non-volatile memory, thereby improving the confidentiality of the secret information [1].

The IGZO TFT is widely applied in the modern display driver integrated circuits (DDICs). Compared with the traditional amorphous silicon-based TFT devices, IGZO TFT has the relatively higher mobility, larger current driving ability, lower turn-off current, better manufacturing uniformity, higher transmittance, and it can be manufactured at the room temperature [2]. On the other hand, the smaller transistor usually has the larger relative parameter variations. The channel length variation of a 5 μm IGZO TFT reaches 0.7 μm, and the relative variation is about 1.4×10^{-1}, whereas the channel length variation of a MOSFET with a 40 nm process is only about 9.44×10^{-4} nm [3], and the relative variation is about 2.36×10^{-5}. The IGZO TFT has larger variation in physical parameters than MOSFET. Furthermore, due to the aging factors, such as the positive bias temperature stress and hot carrier injection, the threshold voltage fluctuation in MOSFETs is about 50 mV [4], whereas the threshold voltage fluctuation in IGZO TFTs reaches ±3 V. It implies that the IGZO TFTs have potential to be used in the design of PUFs.

In [5-7], IGZO TFTs are utilized to build the weak PUFs, such as the Cap-PUF [5], Bio-PUF [6] and resistive PUF [7]. To the best of our knowledge, the TFT devices have not been applied to the delay-based PUFs, such as the Arbiter PUF. Only the n-type IGZO TFTs are usually applied to the circuit designs. For the traditional circuit designers familiar with the CMOS circuits, it is inconvenient to design the complex logic circuits with these n-type IGZO TFTs. This may be the reason of the absence of the delay-based PUFs based on IGZO TFTs.

This work designs an XOR Arbiter PUF based on the IGZO TFT devices, for the XOR Arbiter PUF has better security performance than that of the traditional Arbiter PUF. Simulation results demonstrate that the randomness and uniqueness of the IGZO TFT-based XOR Arbiter PUF are better than those of the CMOS-based XOR Arbiter PUF. Whereas, the reliability of the TFT-based scheme is not good as that of the CMOS-based scheme.

II. THE IGZO TFT-BASED XOR ARBITER PUF

A. The TFT-based Logic Gates

The XOR Arbiter PUF consists of two Arbiter PUFs and one XOR gate, and each Arbiter PUF consists of a MUX (multiplexer) chain and an arbiter. Each stage of the MUX chain includes two multiplexers. These circuit blocks are discussed in this section. Unlike the traditional CMOS circuit, the IGZO TFT circuit only uses the n-channel devices. The basic gate circuits are required in the design of the IGZO TFT-based XOR Arbiter PUF.

It is the most common idea of the IGZO TFT based logic gates to use the diode connected IGZO TFTs as the pull-up load devices, which is similar as the NMOS logic gates based on the MOSFET devices. Taking the inverter as an example, an n-channel TFT with the diode connection is utilized as the load transistor, and another n-channel TFT is utilized as the driver transistor. Fig. 1 (a) shows that when the input signal is LOW, the driving transistor M2 turns off, and the output level is affected by the threshold voltage loss of the load transistor M1. When the input signal becomes HIGH, M1 and M2 turn on simultaneously. Therefore, the output low-level V_{OL} is determined by the conduction resistance of the driving transistor and the load transistor. Typically, V_{OL} is designed to turn off the pull-up TFT below the threshold voltage, thereby reducing the output voltage to VSS. But this design requires a large pull-down TFT, and it only has a narrow output voltage range [8]. Therefore, it is necessary to improve the design of the structure to solve the above problems.

The pseudo CMOS logic circuit is one of the method to mitigate the high steady-state power consumption problem [8], as shown in Fig. 1 (b). A pseudo CMOS logic circuit consists of two stages. The first stage is the same as the NMOS logic gate. The second stage consists of the pull-up TFT M3 and the pull-down TFT M4. The output of the first stage is connected

979-8-3503-6184-1/24 $31.00 © 2024 IEEE

to the gate of M3, and the input IN is connected to the gate of M4. When the input signal is LOW, M2 and M4 turn off, and the output level of the first stage is affected by the threshold voltage loss of M1. Usually, the power supply VB is stronger than the power supply VDD, so that the output level of the gate circuit is able to reach VDD. Therefore, compared to the NMOS logic, the pseudo CMOS logic circuits provide a larger output range. On the basis of the inverter, the similar circuits of the NAND, NOR gates are designed, as shown in Fig. 2 (a) and (b), respectively.

The pseudo CMOS logic based on the IGZO TFT is adopted in this work. The inverters, NAND gates and NOR gates are designed with the 5 μm IGZO TFT process. The simulated waveforms of these logic gates are presented in Fig. 3, where the temperature is 27 °C, the VDD, VB and VSS are 10 V, 12 V and -10 V, respectively.

The simulation results of the inverter show that, the output voltage of inverter achieves 10 V (V_{OH}), when its input voltage

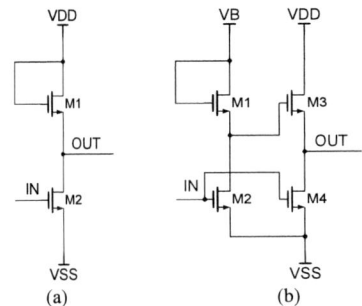

Fig. 1. Schematic of two TFT-based inverter circuits (a) NMOS logic inverter (b) pesudo CMOS logic inverter.

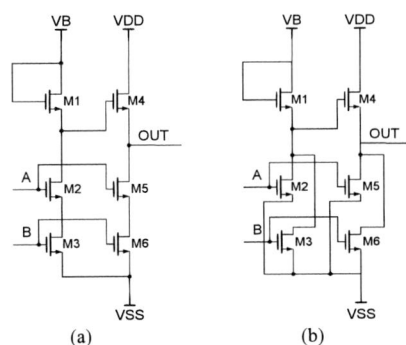

Fig. 2. The schematic of logic gates with the pesudo CMOS logic (a) NAND gate (b)NOR gate.

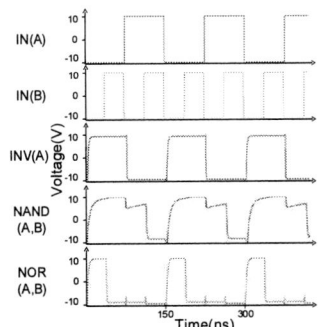

Fig. 3. The waveforms of pesudo CMOS logic based Inverter, NAND and NOR gates.

is LOW. And the output voltage of inverter achieves -9.659 V (V_{OL}) when its input voltage is HIGH. The delay of the inverter is 2.6 ns, and the average current is 210 μA within one cycle. The average power consumption of a pseudo CMOS inverter is 4200 μW. The waveform of the NAND gate indicates that, OUT becomes LOW when IN(A) and IN(B) are 10 V, and the other input combinations result in the high level of OUT. V_{OH} and V_{OL} are measured as 10 V and -8.46 V, respectively. The delay of NAND gate is 7.4 ns, and the average current is 125 μA. The power consumption of NAND gate is 2500 μW. The waveform of the NOR gate presents that, OUT becomes HIGH when IN(A) and IN(B) are LOW, and OUT are LOW with other input combinations. V_{OH} and V_{OL} are 10 V and -9.213 V, respectively. The delay of NOR gate is 3.4 ns, and an average current of 75 μA. The average power consumption of NOR gate is 1500 μW.

B. The Structure of the TFT-based XOR Arbiter PUF

As shown in Fig. 4 (a), the XOR Arbiter PUF consists of a series of MUX chain, the arbiter and the XOR gate[9]. The 2-to-1 multiplexers (MUXs) in the same stage are connected with the same input, and their selection signals are opposite. The structure of TFT-based MUX is presented in Fig. 4 (b). Fig. 4 (c) designs the XOR gate based on the pseudo CMOS logic circuit as mentioned in Section II (A). The arbiter is implemented by a DFF (D flip-flop), because the output of DFF depends on whether the '0→1' transition reaches before or after the '0→1' CLK-edge. The structure of DFF is shown in Fig. 4 (d). A 6-bit XOR Arbiter PUF is constructed based on these circuit blocks with the 5 μm IGZO TFT process.

C. Discussion

The size of TFT-based XOR Arbiter PUF can be represented as (n, k), where n is the bit-width of challenge and k is the number of XORed sub-Arbiter PUFs.

The MUX chain consists of several cascaded multiplexers. The maximum number of stage of the MUX chain determines the maximum bit-width of challenge n. The circuit fails when V_{OH} of the previous stage multiplexer is lower than the V_{IH} of its next stage multiplexer. The V_{IH} of the multiplexer is 0.97 V according to the simulation. As shown in Fig. 5, the V_{OH} of

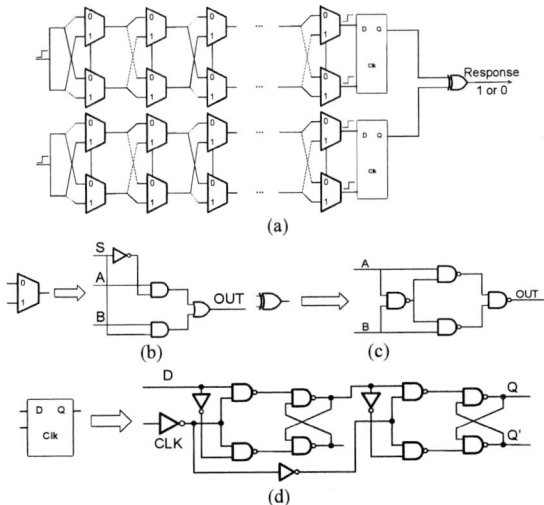

Fig. 4. The XOR Arbiter PUF (a) The architecure of the XOR Arbiter PUF (b)The MUX (c) The XOR gate (d) The DFF. All logic gates are implemented by the pesudo CMOS logic based on the IGZO TFTs.

the first stage multiplexer is 5.85 V, and the voltage loss is 4.15 V. Whereas the V_{OH} of multiplexers after the second stage is around 5.81 V, which is higher than the 0.97 V. It shows that the MUX chain has good cascading capability.

The XOR operation with k inputs can be realized by cascading $\lceil k/2 \rceil$ stages of 2-input XOR gates, where $\lceil x \rceil$ means round up of x. So the maximum number of k is determined by the maximum stage of cascaded XOR gates. The output error of the XOR gate occurs if V_{OH} of the previous stage of XOR gate is lower than V_{IH} of the next stage of XOR gate. The simulation results show that the V_{IH} of the XOR gate is 0.88 V. As shown in Fig. 6, the V_{OH} of the first stage of the XOR gate is 4.85 V. The voltage loss after the second stage of XOR gate is about 0.12 V. So the maximum number of the cascaded XOR gates is $(4.85 - 0.88)/0.12 = 33$ stages. This indicates that TFT-based PUF has good scalability.

D. The Performance of the Proposed XOR Arbiter PUF

The simulations are carried out on the 6-bit XOR Arbiter PUF based on the IGZO TFT devices. The average current, the power consumption and the delay are 41 mA, 820 mW and 325.2 ns, respectively, when 1-bit response is generated.

Randomness, uniqueness, and reliability are the three main indicators of PUF. The randomness describe the proportion of "1" in the response bits of a specific PUF circuit. The ideal value is 50%. The uniqueness refers to the ratio of different bits in the response of different PUF circuits, with an ideal value of 50%. The reliability measures the ratio of the changed

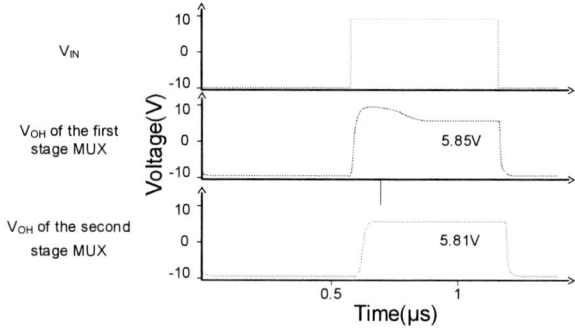

Fig. 5. The voltage waveforms of cascaded MUXes.

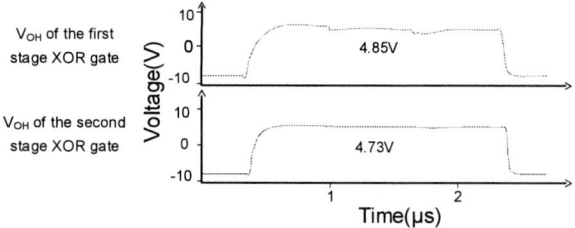

Fig. 6. The voltage waveforms of cascaded XOR gates.

TABLE I.   THE PERFORMANCE OF THE IGZO TFT-BASED XOR ARBITER PUF

PUF Type	Randomness /Standard deviation	Uniqueness /Standard deviation	Reliability
CMOS XOR Arbiter PUF	49.625%/0.103	49.967%/0.0512	95.367%
IGZO TFT XOR Arbiter PUF	49.711%/0.090	50.048%/0.0501	90.845%

response bits in different operating environments for a specific PUF circuit. The ideal value is 100%.

In the simulation, the entropy sources of CMOS-based PUF and IGZO TFT based PUF are assigned with random values, which follow the normal distributions $N(0,0.7)$ and $N(0,9.44 \times 10^{-4})$, respectively. And the threshold voltage drift of the MOSFET and TFT after aging is 50 mV and ± 3 V, respectively. The Monte Carlo analysis is applied to the randomness, uniqueness, and reliability of the XOR Arbiter PUFs, and the results are shown in Table I. Table I shows that the randomness and uniqueness of IGZO TFT based XOR Arbiter PUF are near 50.0%, and their standard deviations are smaller than those of CMOS XOR Arbiter PUF. It indicates that the IGZO TFT XOR Arbiter PUF has better randomness and uniqueness compared with of the CMOS-based counterpart. However, the reliability of the IGZO TFT XOR Arbiter PUF is not as good as that of the CMOS-based counterpart.

III. CONCLUSIONS

This work designs the IGZO TFT based XOR Arbiter PUF circuit. The simulation results show that, the IGZO TFT XOR Arbiter PUF has better randomness and uniqueness than those of the CMOS XOR Arbiter PUF, because the IGZO TFT devices have larger variations. The IGZO TFT XOR arbiter has good scalability, because the MUX chain and XOR has good cascading capability.

ACKNOWLEDGMENT

This work was partly supported by the NSFC under Grant No. 92373206, Shenzhen Science and Technology Program under Grant No. JCYJ20220818100814033, SGDX20230116093303006, KJZD20231023100201003 and KQTD20200820113105004.

REFERENCES

[1] K. Dey, M. Kule and H. Rahaman, "PUF Based Hardware Security: A Review," 2021 International Symposium on Devices, Circuits and Systems (ISDCS), pp. 1-6, 2021.

[2] Zhu, Ying, et al. "Indium–gallium–zinc–oxide thin-film transistors: Materials, devices, and applications, " Journal of Semiconductors, vol. 42, no. 3, 2021.

[3] Zhou, Hui, et al., "PSP Statistical Modeling for 40nm MOSFET," Applied Mechanics and Materials, vol. 236–237, pp. 144–148, Nov 2012.

[4] P. Magnone et al., "Impact of Hot Carriers on nMOSFET Variability in 45- and 65-nm CMOS Technologies," IEEE Transactions on Electron Devices, vol. 58, no. 8, pp. 2347-2353, Aug 2011.

[5] Lee D, Lee J, Shin M, et al., "Sol-gel processed Y2O3 embedded capacitor based physically unclonable function," Materials Science in Semiconductor Processing, vol.168, 2023.

[6] Y. Cao, K. Zhang, Z. Huang and S. Li, "A Unique and Robust Physically Unclonable Function Based on Bionic Tunable Ion Gel-Gated Synaptic Transistors," IEEE Electron Device Letters, vol. 44, no. 12, pp. 1995-1998, Dec 2023.

[7] Y. Han, S. Lee, E. K. Lee, H. Yoo, B. C. Jang, "Strengthening Multi-Factor Authentication Through Physically Unclonable Functions in PVDF-HFP-Phase-Dependent a-IGZO Thin-Film Transistors," Adv. Sci. , 11, 2024.

[8] T. -C. Huang, et al., "Pseudo-CMOS: A Design Style for Low-Cost and Robust Flexible Electronics," IEEE Transactions on Electron Devices, vol. 58, no. 1, pp. 141-150, Jan 2011.

[9] G. E. Suh and S. Devadas, "Physical unclonable functions for device authentication and secret key generation," Proc 44th ACM/IEEE Design Autom Conf, pp. 9–14, Jun 2007

Design and Implementation of Hierarchical Storage Structure for MCCSIP-RAA

Longmei Nan*[1], Yu Jin[1], Yiran Du[1], Tao Chen[1], Yanjiang Liu[1], Wei Li[1]

[1]Institute of Information Science and Technology, Zhengzhou 450001, China

* Email: lnan13@fudan.edu.cn

Abstract—**The Multi-core cryptographic specific instruction processor with reconfigurable accelerated array (MCCSIP-RAA) combines flexibility, efficiency, and resource utilization, but its high computational throughput also puts enormous pressure on storage structures. Based on the analysis of MCCSIP-RAA structure and multi-core processor storage system, combined with the flow processing characteristics of cryptographic data address continuity and data flow, a hierarchical storage structure combining flow storage mechanism and shared storage mechanism was designed. The first layer of the hierarchical storage structure is the input/output buffer of the MCCSIP-RAA, which adopts FIFO structure and supports burst transmission of batch data. The second layer of the hierarchical storage structure consists of input/output storage for various cryptographic specific processors, input/output storage for cryptographic acceleration arrays, and storage for related task control data. The FIFO structure is used. The third layer of the hierarchical storage structure includes internal storage of multi-core cryptographic processors, internal storage of acceleration arrays, and shared memory between processing cores, ensuring efficient processing of cryptographic calculations. A shared storage structure based on RAM was designed for subkey data with a large amount of data that is repeatedly and continuously used. A shared storage structure based on FIFO was designed for continuous and large-scale stream data processing between different cores. A separate /clustered storage structure was designed for each core, including a universal register heap, dedicated bit registers, subkey shadow registers, etc. The experimental results show that the hierarchical storage structure can effectively alleviate the contradiction between cryptographic processing flexibility, efficiency, and resource utilization in the target architecture MCCSIP-RAA. Adding smaller resources to cryptographic algorithm processing can achieve processing performance of 3-7 times for different cryptographic algorithms, effectively improving resource utilization.**

Keywords—MCCSIP-RAA, Flow storage, inter-core shared storage, Hierarchical storage, Separate and clustered storage

I. INTRODUCTION

The implementation of cryptographic algorithms plays an important role in information security. It is effective to design a flexible and efficient MCCSIP-RAA. So it is meaningful to design a flexible and efficient storage structure that meets the characteristics of cryptographic processing for it.

II. ANALYSIS OF MCCSIP-RAA ARCHITECTURE AND MULTI-CORE PROCESSOR STORAGE SYSTEMS

A. Analysis of MCCSIP-RAA Architecture

The storage structure designed in this paper is oriented towards to the MCCSIP-RAA, so it is necessary to analyze the MCCSIP-RAA, firstly.

This work is supported by Natural Science Foundation of Henan (No. 232300421393).

MCCSIP-RAA is proposed based on the fusion analysis of multi-core cryptographic processors and acceleration arrays. It draws on the analysis of cryptographic parallel processing structural parameters such as computing granularity, parallelism, and number of computing cores. It contains four cryptographic processors and a shared acceleration array, data shared storage and fast exchange storage, input microprogram controller and output microprogram controller. The four cryptographic processors are designed in an isomorphic manner and are the core computing components of task level parallel processing. They are named SPcore1-SPcore4, respectively. The reconfigurable acceleration array is the main computational component for nonlinear operations in stream cipher algorithms, named Array. It is shared by four instruction processors in a shared form and can also work independently. A fast data exchange area and a data transmission path need to be set up between the four instruction cryptographic processors. A fast data exchange storage is also required between the four instruction cryptographic processors and the reconfigurable acceleration array. So, it can be seen that designing efficient storage structures that meet the requirements of fast data exchange is an important module of MCCSIP-RAA.

B. Analysis of multi-core processor storage systems

The quality of a storage system has a significant impact on the throughput of multi-core processors. Designing a reasonable storage system has become an important issue in improving the processing speed of multi-core processors. Therefore, this paper analyzes typical processor storage systems [1-3]. Firstly, in a centralized shared storage structure, each processor has its own independent register file and cache, which need to be connected to the on-chip shared cache when used. Secondly, in a distributed storage architecture, each processor has its own private register file, on-chip cache, or on-chip storage, which is then connected to the on-chip network. The problem with these two storage structures is that it is difficult to support the access requirements of a large number of computing units to parallel data in multi-core processors, and the lack of targeted parallel processing features for cryptographic tasks. Thirdly, stream storage is a high-performance processor storage structure for stream processing. In this storage structure, the processor core accesses local storage in FIFO mode, and data exchange between local storage and data transfer between local storage and off chip storage are carried out through DMA.

III. DESIGN OF HIERARCHICAL STORAGE STRUCTURE FOR MCCSIP-RAA

A. The hierarchical organizational form of MCCSIP-RAA 's storage structure

Considering the poor data reuse, good address continuity,

and strong data flow characteristics in cryptographic tasks, this article adopts a combination of flow storage mechanism and shared storage mechanism to improve memory access efficiency and strengthen the connection in related cores.

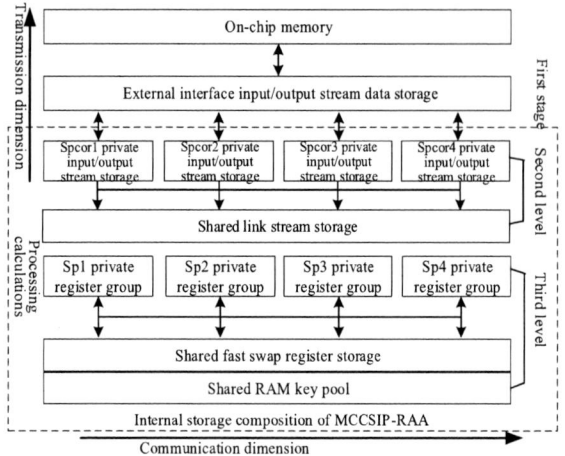

Fig. 1. Hierarchical organizational form of storage structure

As shown in Fig. 1, a three-level storage organization is adopted. This organizational form can conveniently and effectively manage and utilize storage resources at all levels. The setting of the first level storage resources improves the transmission efficiency of interface data. The setting of the second level storage resources improved the constraint of multitasking cryptographic processor data transmission speed on cryptographic computing performance for each processor. The setting of the third level storage resources ensures data exchange between cryptographic calculation tasks within each cryptographic processor. The FIFO storage resources at the first level are mainly used for the transmission of batch stream data between the upper and lower levels, the shared fast exchange storage and key pool RAM storage are mainly used for data exchange and communication between cores, and the private register heap, sub key shadow register, dedicated bit register, etc. of the third processor are mainly used to support various internal computing operations of each processor.

B. The Hierarchical Storage Structure of MCCSIP-RAA

On the basis of the hierarchical organizational form mentioned above, this section provides a detailed design of the hierarchical storage structure, as shown in Fig. 2.

The first level storage structure is the input/output buffer, implemented using FIFO structure, with capacities of 512x32 bits, labeled as PUSHFIFO and POPFIFO respectively. It is designed for transmission and reception of external data in the system. The DMA controller quickly transfers the pending data from the SoC system memory to the PUSHFIFO. After processing, the DMA controller quickly transfers the completed data from the POPFIFO to the SoC system memory. The DMA controller not only completes the timing matching between FIFO and AXI buses, but also supports burst transmission of batch data. One burst can effectively transmit 32x32 bit data, thereby accelerating the transmission efficiency of interface data in embedded systems.

The second level storage structure includes three parts.(a) The input/output storage of each processor adopts a FIFO structure design, with an input capacity of 128x37 bits (due to

the large amount of data processed by dedicated processors when processing grouping algorithms, the depth is set to 128 in order to reduce data allocation time, in the 37 bits width, 32 bits is the input data of the dedicated processor, and 5 bits is the port address of the input data of the dedicated processor), and the output cache capacity is 128x32 bits, which is sequentially identified as IN-FIFO1-IN-FIFO4, OUT-FIFO1-OUT-FIFO4. (b) The input and output storage of the acceleration array is designed using a FIFO structure, with an input capacity of 32x35 bits (the amount of input data that needs to be interacted is small, so the depth is set to 32, and 32 bits of 35 bits are the acceleration array data, and 3 bits are the acceleration array address). The stream cipher algorithm generates relatively small amounts of data each time, so the output cache capacity is set to 32x32 bits, labeled as ArrayinFIFO or ArrayoutFIFO. (c) Task control data storage, identified as linked data FIFO, Linktask FIFO, Outtask FIFO, its main function is to link data buffer with input/output task control data. The linked data buffer FIFO is used to cache the link processing of large amounts of data between different cores, requiring a large storage capacity of 128x32 bits. The storage required for input/output task control data is small, so both Linktask FIFO and Outtask FIFO are set to 32x32 bits.

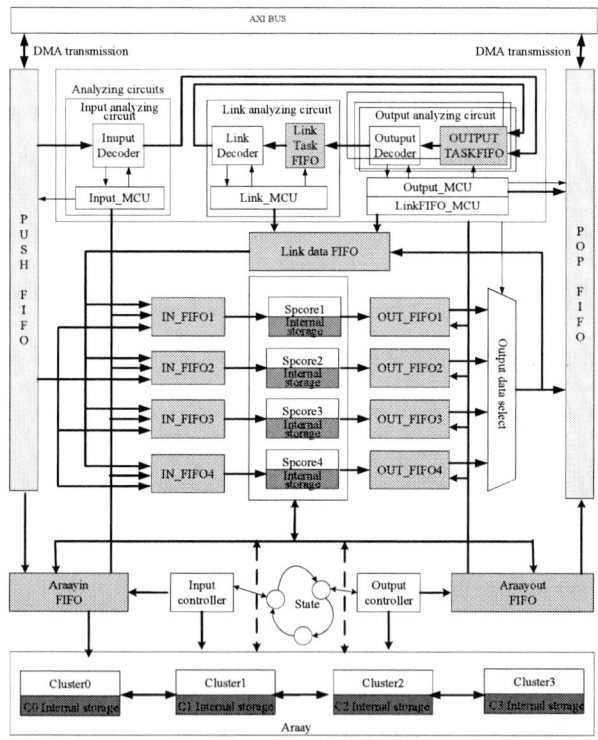

Fig. 2. The Hierarchical Storage Structure for MCCSIP-RAA

The third level storage structure includes three parts. (a) The internal memory of a multi-core cryptographic processor consists of five parts: a general register heap, a dedicated bit register, a sub key shadow register for each cryptographic processor, and a fast exchange storage and key pool shared by four cores. A separate and clustered storage structure was designed for each processing core, as shown in Fig. 3. The general-purpose register heap is used to store input, output, temporary data, with a capacity of 64x32 bits. The dedicated bit register for memory unit updates, with a capacity of 8×1 bits, the sub key shadow register is used to store wheel subkey, with a capacity of 16x32 bits. Different types of storage within each core are evenly allocate to four clusters, each specific sub storage is only belongs to corresponding specific cluster, so as to improve memory access efficiency. The fast exchange storage area provides a storage exchange area for communication between small batch data cores, with a capacity of 64x32 bits, shared by four cryptographic processors. It uses register mapping to quickly exchange data with other cores, thereby achieving the interaction of small batch data of different cores. The key pool provides storage areas for each wheel key of block cipher algorithms. After analyzing the sub key storage of each block algorithm, its storage capacity is set to 512x32 bits, which is also shared by four cores. RAM mapping is used to exchange data with other cores. (b) Internal memory of the accelerate array includes the general registers and dedicated bit registers within each cluster. Due to the fact that the acceleration array is an acceleration component for stream cipher, the data storage capacity is small. Additionally, since the maximum data processing capacity of each cluster in the acceleration array is 128 bits, considering data backup, the general register storage capacity for a single cluster is set to 32x32 bits. A dedicated bit register is designed for nonlinear Boolean function operations, and each bank in a single cluster requires at least 2 bits of register feedback. Therefore, a dedicated bit register of 8 bits is set for a single cluster. (c) The fast data exchange storage between four cryptographic processors and the acceleration array provides a storage exchange area for small batch data communication between each processor and the acceleration array (the acceleration array is designed for nonlinear operations of sequence ciphers, and its data interaction with each cryptographic processor is mainly the state sequence of shift registers, so the data volume is relatively small). Its capacity is set to 32x32 bits, which is shared between various cryptographic processors and the acceleration array, and register mapping is adopted.

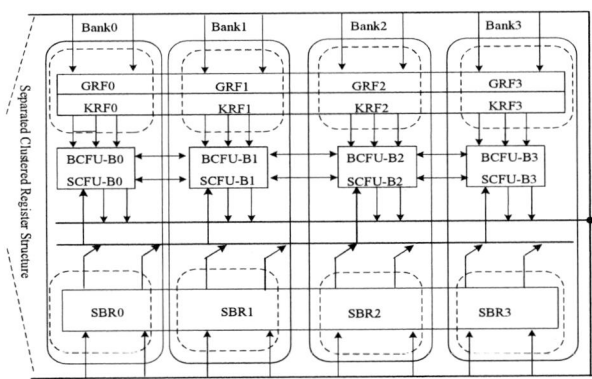

Fig. 3. Separate clustered storage structure of a cryptographic processor

IV. Algorithm Mapping and Performance Evaluation

Taking the SM4 algorithm mapping as an example, it can be mapped in a task parallel manner across four processor cores for multi task parallel execution. The mapping process mainly includes the configuration process, data calculation process, and data output process, as shown in Table 1.

The executive performance of typical algorithms in each platforms [4-6] and this work is shown in Table 2. It can be seen that with the support of the hierarchical storage structure

designed in this article, MCCSIP-RAA can efficiently and flexibly implement different cryptographic algorithms, and its processing performance is significantly better than other platforms.

TABLE I. SM4 Algorithm Mappin

Instructions
Parallel injection main program
MOVI < H>,<Ri2>,<0x2fa>   CFG <Ri2>;
DMAP.128 < P_Ins>,<SPCore1>,<SPCore2>,< SPCore3>, < SPCore4>
DMA.R < P_Ins>,< SPCore1>,< SPCore2>,< SPCore3>,< SPCore4>
Task input packet allocation main program
Ddtainput : DMAI.128 < Dataport>, < SPCore1>
DMAI.128<Dataport>, <SPCore2> DMAI.128< Dataport>,<SPCore3>
DMAI.128< Dataport>,<SPCore4>   INC <Ri4>   CMP <Ri4>#8   JUMP
Task output packet scheduling main program
Ddtaoutput: CF <SPCore1>,< Output1> // INC <Ri4> CMP <Ri4>#8
Ddtaoutput1： CHECK Fifooutfull _ SPCore1 DMAO.128 < SPCore1>
Ddtaoutput2： CHECK Fifooutfull_SPCore2 DMAO.128 < SPCore2> ...
Ddtaoutput4： Ddtaoutput1： CHECK Fifooutfull_ SPCore1
DMAO.128 < SPCore1>   JUMP Ddtaoutput
Encryption phase main program
XOR RgA0, RgB10, RgC10, RgD10 →XORRgA0, RgA0, RgA1 → SBOX8T8.
P0 RgA0, RgA0 IROL32 RgA11, RgA0, #2	IROL32 RgB11, RgA0, #10
	IROL32 RgC11, RgA0, #18| IROL32 RgD11, RgA0, #24
XOR RgA12, RgA11, RgA10, RgA0 → XOR RgB12, RgA12, RgB11, RgC11
→ XOR RgC12, RgB12, RgD11

TABLE II. Performance Comparison Analysis of MCCSIP-RAA

Design	Frequency (MHz)	Area (mm²)	Process (nm)	Algorithm name	Throughput rate (Mbps)
[4]	1000	16	65	AES	2160
				IDEA/SM4	700/670
[5]	500	1.89	65	AES	2181
				DES/ IDEA	491/563
[6]	350	18	180	AES	1587
				DES/ IDEA	460/400
This Work	300	17	65	SM4/AES	4978
				DES /IDEA	3079
				SM3/ Shink	2214/4756
				A5/W7	394
				Grain/ LILI	300
				TRIVIUM	240
				Toyocrypt/E0	1200

V. Conclusions

In summary, based on the typical multi-core storage structure, combined with the characteristics of cryptographic data address continuity and data flow, this article designs a hierarchical storage structure that combines stream storage mechanism and shared storage mechanism, which can effectively support the efficient processing of MCCSIP-RAA and provide reference for other multi-core systems.

EFERENCES

[1] Li Junwei, Research on Multi Core Cryptoprocessor Storage System and Task Mapping Mechanism [D], Master's thesis, Zhengzhou: Information Engineering University, 2014.

[2] Stream Processing: Enabling the new generation of easy to use, high Performance DSPs White Paper, Stream Proeessors, Inc, 2007.

[3] Xiao Ruijin, Research on Hierarchical Storage System for Multicore Processors [D], Shanghai: Fudan University, 2012.

[4] Fengxiao, Reconfigurable Asymmetrical Multi-core Architecture for Block Cipher[J]. Chinese Journal of Electronics, In press, 2016.

[5] Li Gongli, Research on Key Technologies of Block Cipher Stream Processor [D], Zhengzhou: Information Engineering University, 2018.

[6] Li Wei, A reconfigurable block cryptographic processor based on VLIW architecture[J]. China Communications, 2016, 13(1):91-99.

A Highly Scalable Hardware HEVC Encoder Based on FPGA

Guohao Xu[1], Chenlong He[1], Shiyan Yi[1], Leilei Huang[2*], Xiaoyang Zeng[1], and Yibo Fan[1*]

[1]State Key Laboratory of Integrated Chips and Systems, Fudan University, Shanghai, China
[2]Institute of Microelectronic Circuits and Systems, East China Normal University, Shanghai, China
* Email: llhuang@cee.ecnu.edu.cn, fanyibo@fudan.edu.cn

Abstract—The proposal of HEVC greatly reduces the bandwidth of video transmission and video storage. Oriented to different scenarios, the requirements of resources, throughput, and compression rate of hardware encoder are different. Therefore, we design a highly scalable HEVC encoder consisting of top-level and module-level and these scalable features enable the encoder to be flexibly changed to fit different devices and scenarios. Furthermore, we propose ping-pong address mapping and shadow registers to improve the overall system's throughput. Finally, we implement the two extreme profiles: best compression performance and best hardware performance. Experiments show that these two profiles achieve 1080p@60/1080p@30 fps with 128K and 405K LUTs in Xilinx ZCU102 and Xilinx Alveo U250, respectively. The BD-Rate compared to x265 achieves -9.13%/14.45% respectively. To the best of our knowledge, this paper is the first to present a scalable architecture.

Index Terms—scalable hardware encoder, FPGA, HEVC, ping-pong address mapping

I. INTRODUCTION

HEVC [1] greatly improves video compression ratio through complex tools such as quad-tree coding structure, 35 intra-prediction modes, and context adaptive binary arithmetic coding (CABAC). However, this makes the internal computational complexity of the video encoder extremely high, leading to a significant increase in the computation time of the software and making it difficult to meet the real-time requirements of video encoding.

To support the real-time requirements of video encoding, previous works [2]–[5] design application-specific integrated circuits (ASICs) to implement HEVC hardware. However, video coding for different device platforms has different requirements for hardware resources, throughput, and compression rates. For example, for surveillance devices, the encoder's resources and throughput become more sensitive, and the encoder's quality requirements are relatively loose. While the encoder's quality requirements are more strict, and the encoder's resources and throughput are looser for data center devices. Other device platforms may trade-off between these requirements. As a result, different ASICs often need to be designed for different platforms, leading to a sharp increase in Non-Recurring Engineering (NRE) cost and Time To Market (TTM) cycle.

FPGA-based encoders have short development cycles, low development costs, and risks. There are many works [6]–[10] implement HEVC based on FPGA. Works [6], [7] designed part hardware of video coding based on FPGA to speed up software coding. Although high throughput can be achieved, a large amount of computation in the CPU still makes the system dependent on CPU performance and the speed of data interaction with the FPGA. Works [8]–[10] offloaded the entire encoder to the FPGA. The encoder proposed by Pastuszak *et al.* [10] proposed 5 Profile for the number of different prediction units (PUs) to improve computational scalability. However, they implement the trade-off between throughput and BD-Rate without considering the area.

Therefore, we design the highly scalable encoder hardware architecture in which the main encoder modules are all offloaded to the FPGA, and the CPU only controls the encoder behavior. Our contribution can be summarised as follows:

- We propose the hardware architecture of the encoder and analyze the throughput bottleneck on the system.
- We present ping-pong address mapping and shadow register to optimize the throughput of the overall system.
- We design the rotating cache and scalable prediction engine to achieve the top-level and module-level scalability for the encoder.

II. SYSTEM ARCHITECTURE

The architecture and execution steps of the proposed HEVC hardware system are shown in Fig. 1. The dash lines represent the execution at the frame level. Specifically, the CPU determines the QP of the next frame based on the number of bit streams in the previous frame and also controls the GOP structure in case of a bidirectional frame (B frame). For each frame, the CPU writes original pixels to the DDR (step ①), and then the CPU configures the encoder with many scalable registers and enables the START register (step ②). Then, the encoder reads the original pixels from the DDR and the reference pixels (step ③) for P-frames or B-frames. After the completion of a frame, the bitstream (BS) and the reconstructed pixels (REC) are written out (step ④). Finally, the CPU reads the BS from DDR (step ⑤).

Inside the encoder, the main modules, denoted as white boxes, are all pipelined with a granularity of the largest coding unit (LCU). FTH is responsible for acquiring the input pixels, and RMD, IME, and FME are rough selection modules for intra-frame prediction modes and inter-frame motion vectors (MVs). RDO, the decision module, calculates the rate and distortion cost to get the final division and modes. Finally,

979-8-3503-6184-1/24 $31.00 © 2024 IEEE

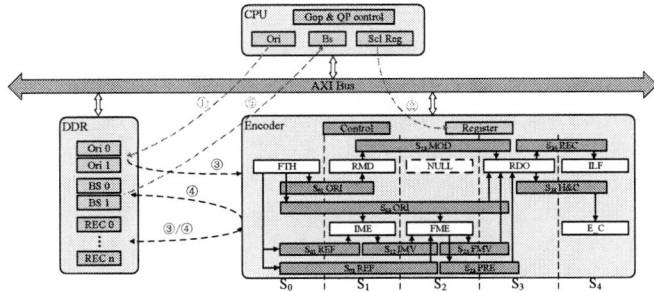

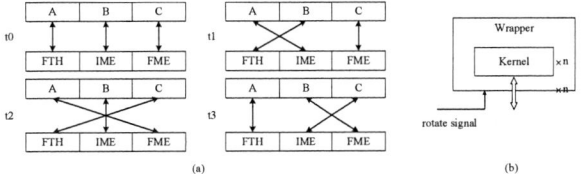

Fig. 1. System architecture and execution steps of proposed HEVC hardware encoder. ORI: Original pixel; IMV: Integer-pixels motion vectors; FMV: Fractional-pixels motion vectors; MOD: RMD-select modes; PRE: Prediction pixel; H&C: Header and quantization coefficients; REF: Reference pixel; REC: Reconstruction pixels.

Fig. 2. Schematic of Rotating cache with respect to (a) connectivity and (b) architecture.

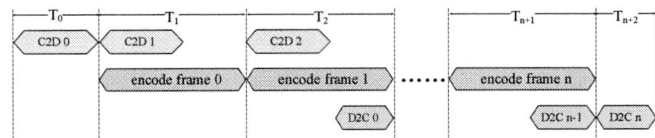

Fig. 3. Pipeline of cache to DDR, Encoder, and DDR to cache.

entropy coding is performed to generate the final bit stream. The gray part represents the rotating cache, which stores the intermediate data between the various pipeline stages.

In addition, the above architectural execution is a serial process. Steps ③ and ④ only involve the interaction between the Encoder and the DDR, which has less impact on the system throughput. Step ①, ② and ⑤ involve the interaction between CPU cache and DDR/Encoder, which has a larger impact on system throughput, so we optimize these three steps.

III. SYSTEM OPTIMIZATION

A. The Scalable in Encoder

The top-level scalability is achieved through the rotating cache, which serves as a storage mechanism that constantly rotates connection relationships between multiple-stage pipelines, as illustrated in Fig. 2(a). The rotating caches can all be implemented using the structure depicted in Fig. 2(b). The caches denoted as A, B, and C can be called the Kernel. The number of these kernels, denoted as 'n', depends on the number of stages the rotating caches span. The Wrapper portion includes the logic for rotation, involving the connections from interfaces to the Kernels. When all pipeline stages are completed, the rotating signal is set to 1. Once set to 1, the Kernel connections will switch at n intervals. Therefore, the logic of the rotating cache can be easily implemented through scripting or block generation. In this way, we can make the flexibility of the main modules' pipeline stages changeable, achieving top-level scalability.

Many modules in the encoder have similar prediction or cost computation engines. We design these engines as scalable engines. For example, there are three types of prediction engines for RMD, which support DC&Planner, mode 2~18, and mode 18~34. The balance of area, throughput, and BD-Rate can be achieved by changing the number of these three types of engines. For the Cost engines, we split the SATD cost of the large-size CU into the accumulation of the small-size SATD cost. Although this loses some BD-Rate, it guarantees the same throughput and scalability as the prediction engine.

B. Ping-pong Address Mapping

As mentioned above, moving the original pixels and the bit stream (steps ① and ⑤) has a significant impact on the throughput of the system. We propose a flow stage as shown in Fig. 3 to optimize the impact that ① and ⑤ bring to the overall timing.

C2D represents the data movement from the CPU cache to DDR, D2C is the data transfer from DDR to the CPU, and the number represents the current frame index. Specifically, at T_0, the frame 0 original pixels are moved from CPU to Ori 0 (green area in Fig 1) in DDR. At T_1, memory in the CPU, which stores the frame 1 original pixels, is mapped to DDR. To prevent data overwriting, the address mapping location is Ori 1. This process is repeated for the following moments, and the original pixel address mapping location is always ping-pong. The data transform of BS is also similar to Ori.

After the ping-pong address map, the system makes full use of DDR bandwidth when the encoder is working, greatly reducing the throughput impact of ① and ⑤ to the system.

C. Shadow Register

The effect of ② on the timing of the system is shown in Fig 4(a). Only after each configuration of the Regular Registers (RR) can the dynamically scalable items be identified, and then the encoder can work. The more dynamically scalable items there are, the greater the impact on the system's throughput. Therefore, we propose the shadow registers (SRs), a set of register arrays belonging to Encoder, which are identical to the RRs.

The gains brought by SR can be illustrated in Fig 4(b). Before frame 0 encoding, it is still necessary to configure RR. However, during the execution of frame 0 encoding, the CPU configures the SR which is prepared for frame 1 encoding. When T_1 arrives, only one cycle is needed to transfer the data in SR to the RR. Ultimately, SR removes almost all of the impact of configuration registers on system throughput at the expense of increasing the area of the register array.

979-8-3503-6184-1/24 $31.00 © 2024 IEEE

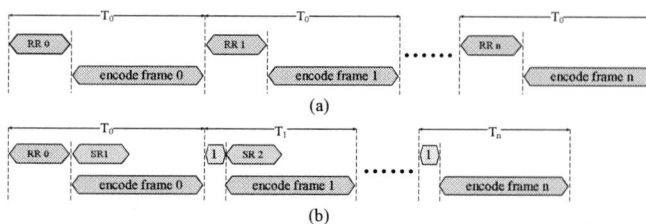

Fig. 4. Timing diagram of configure register and Encoder with respect to (a) before optimization and (b) after optimization.

TABLE I
COMPARISON OF TWO EXTREME PROFILES

	HB		CB	
	LUT	DSP	LUT	DSP
RMD	7882	0	54280	166
IME	6214	0	68142	2
FME	15758	123	74778	666
RDO	42323	647	91404	1160
DBF	5118	0	5398	0
E_C	21534	0	26234	0

TABLE II
COMPARISON WITH PREVIOUS WORKS

Encoder System	[8]	[9]	[10]	ours	
				HB	CB
Dynamic scalability	✗	✗	✓	✓	✓
Statically scalability	✗	✗	✗	✓	✓
Frequency (MHz)	-	140	100	150	120
Resolution	1080p	1080p	1080p	1080p	1080p
Throughput (frame/s)	60	30	60	60	30
LUT count	-	83548	93184	127680	405236
DSP count	-	34	481	813	1996

IV. RESULT

To illustrate the scalability of our proposed architecture, we implemented two extreme profiles for the encoder: best compression performance (BC) and best hardware performance (BH) on the Xilinx Aleno U250 and Xilinx ZCU102, respectively. BH and BC achieve synthesis at 150 MHz and 120 MHz, respectively. Table I presents the resource utilization. The compression rate and quality can be measured by BD-Rate, which is -9.13%/14.45% compared to the x265. We can arbitrarily adjust the scalable features to realize an encoder that trades between BC and BH profile performance for different scenarios and devices.

Table II shows our comparison with other work. Both CB and HB have larger hardware resources than other work. This is because the other work only supports I-frames, while HB supports P-frames, and HB even supports B-frames. It is worth noting that we are the only ones that support P-frames for FPGA, which have much higher compression rates than the encoders that support I-frames. Only our proposed encoder supports static extensibility, so our encoder can be flexibly changed to fit different devices and scenarios.

V. CONCLUSION

In this paper, we implement a highly scalable HEVC encoder based on FPGA. We propose top-level and module-level scalability for the hardware architecture. These features allow the encoder to adapt to different devices and scenarios flexibly. In addition, we optimize the throughput of the whole system through ping-pong address mapping and shadow register and finally achieve 1080p@30fps and 1080p@60fps in U250 and ZCU102 FPGAs, respectively.

ACKNOWLEDGMENT

This work was supported in part by the National Key R&D Program of China (2023YFB4502802), in part by the National Natural Science Foundation of China under Grant 62031009, in part by the "Ling Yan" Program for Tackling Key Problems in Zhejiang Province (No.2022C01098), in part by Alibaba Innovative Research (AIR) Program, in part by Alibaba Research Fellow (ARF) Program, in part by the Fudan-ZTE Joint Lab, in part by CCF-Alibaba Innovative Research Fund For Young Scholars.

REFERENCES

[1] G. J. Sullivan, J. -R. Ohm, W. -J. Han and T. Wiegand, "Overview of the High Efficiency Video Coding (HEVC) Standard," in IEEE Transactions on Circuits and Systems for Video Technology, vol. 22, no. 12, pp. 1649-1668, Dec. 2012.

[2] X. Huang, H. Jia, B. Cai, C. Zhu, J. Liu, M. Yang, D. Xie, and W. Gao, "Fast algorithms and vlsi architecture design for hevc intra-mode decision," J. Real-Time Image Process., vol. 12, pp. 285–302, 2016

[3] Y. Omori, T. Onishi, H. Iwasaki, and A. Shimizu, "A 120 fps high frame rate real-time hevc video encoder with parallel configuration scalable to 4k," in Proc. IEEE Symp. Low-Power High-Speed Chips (COOL CHIPS), 2017, pp. 1–3.

[4] Y. Zhang and C. Lu, "High-Performance Algorithm Adaptations and Hardware Architecture for HEVC Intra Encoders," in IEEE Transactions on Circuits and Systems for Video Technology, vol. 29, no. 7, pp. 2138-2145, July 2019.

[5] T. Onishi et al., "A Single-Chip 4K 60-fps 4:2:2 HEVC Video Encoder LSI Employing Efficient Motion Estimation and Mode Decision Framework With Scalability to 8K," in IEEE Transactions on Very Large Scale Integration (VLSI) Systems, vol. 26, no. 10, pp. 1930-1938, Oct. 2018.

[6] P. Sjövall, V. Viitamäki, A. Oinonen, J. Vanne, T. D. Hämäläinen and A. Kulmala, "Kvazaar 4K HEVC intra encoder on FPGA accelerated airframe server," 2017 IEEE International Workshop on Signal Processing Systems (SiPS), Lorient, France, 2017, pp. 1-6.

[7] Y. Qiu et al., "A Heterogeneous HEVC Video Encoder System Based on Two-Level CPU-FPGA Computing Architecture," 2021 IEEE 14th International Conference on ASIC (ASICON), Kunming, China, 2021.

[8] K. Miyazawa et al., "Real-time hardware implementation of HEVC video encoder for 1080p HD video," 2013 Picture Coding Symposium (PCS), San Jose, CA, USA, 2013, pp. 225-228, doi: 10.1109/PCS.2013.6737724. keywords: Video coding;Streaming media;Real-time systems;Encoding;Hardware;Bit rate;High definition video,

[9] S. Atapattu, N. Liyanage, N. Menuka, I. Perera and A. Pasqual, "Real time all intra HEVC HD encoder on FPGA," 2016 IEEE 27th International Conference on Application-specific Systems, Architectures and Processors (ASAP), London, UK, 2016, pp. 191-195.

[10] G. Pastuszak and A. Abramowski, "Algorithm and Architecture Design of the H.265/HEVC Intra Encoder," in IEEE Transactions on Circuits and Systems for Video Technology, vol. 26, no. 1, pp. 210-222, Jan. 2016.

[11] K. Suehring and X. Li, "Jvet common test conditions and software reference configurations," JVET-B1010, 2016.

A Hardware-friendly Fast Block Partition Decision Algorithm Based on Histogram of Oriented Gradient for AV1

Guohao Xu[1], Shiyan Yi[1], Zhijian Hao[1], Leilei Huang[2]*, Hao Zhang[1], Xiaoyang Zeng[1], and Yibo Fan[1]*

[1]State Key Laboratory of Integrated Chips and Systems, Fudan University, Shanghai, China
[2]Institute of Microelectronic Circuits and Systems, East China Normal University, Shanghai, China
* Email: llhuang@cee.ecnu.edu.cn, fanyibo@fudan.edu.cn

Abstract—AV1 is a recently proposed video coding standard that significantly improves the video compression ratio at the same quality as HEVC but at the cost of substantially increasing the computational complexity. To address this challenge, we propose a fast hardware-oriented algorithm based on the histogram of the oriented gradient (HOG) to make a fast block partition decision, which is the highest time complexity step in AV1. Firstly, we utilize HOG to prune the vertical or horizontal partition. Secondly, the sum of absolute differences (SAD) of the child and parent blocks HOG is used to further accelerate the partition decision. Finally, we also make hardware-oriented optimizations for HOG calculation. Experimental results show that the proposed algorithm can save up to 44.32% time with only 1.46% BD-Rate increase.

Index Terms—AV1, hardware-oriented, fast block partition decision, histogram of the oriented gradient

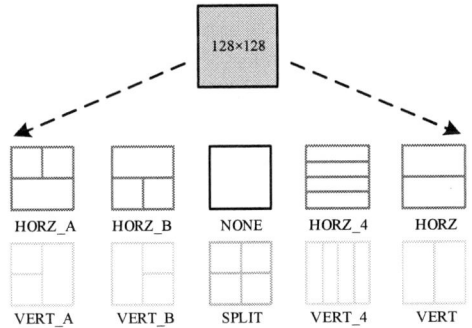

Fig. 1. The coding block partition type in AV1.

I. INTRODUCTION

AV1 [1] introduces larger coding units, more block partition types, more prediction modes, and multi-symbol arithmetic coding compared to HEVC [2], and these new tools greatly improve compression efficiency. At the same time, these new features bring huge complexity. Taking intra-coding as an example, calculating the rate-distortion cost (RDC) of a coding block requires evaluating 56 directional prediction modes and 9 non-directional prediction modes. Then, each transform size and its corresponding 16 transform kernels need to be evaluated as well. Therefore, it takes about 1000 times to calculate the RDC of a coding block.

Fig. 1 shows 10 partition types of the coding block in AV1, where the size of the coding unit in AV1 is up to 128×128. On the one hand, 10 partition types generate 29 different coding blocks, and the above computational complexity increases to 29 times. On the other hand, the square block indicated as blue in the figure, can be partitioned recursively until 8×8 block, which means 85 (1 + 4 + 16 + 64) times computational complexity. The final RDC calculation can be up to 2 million times, greatly hindering the encoder speed as well as increasing the hardware implementation complexity.

To reduce the computational complexity, many works [3]–[14] proposed fast algorithms for AV1. Among them, [3] used a CDEF-like direction detection algorithm to achieve fast direction-based prediction mode decision, which achieves 22.56% time savings at the cost of losing 1.26% BD-Rate.

[5] designed fast algorithms for the transform kernel decision, achieving 20.09% time saving by the frequency matching factors and Gaussian distributions. However, the above works do not target the most time-consuming step in AV1, so their time savings are not enough.

According to [6], [7], block partition decision is the most time-consuming step in AV1. [8]–[10] achieved extreme time savings using neural networks for fast partition decisions, but the hardware implementation is too expensive. The fast partition decision algorithms proposed in [11], [12] can only be used for scenarios where the same video is encoded multiple times. [14] saves 29.06% time by checking the partitions based mid-depth, resulting in 0.95% BD-Rate increase. The timing save is also not enough.

In order to solve the hardware-unfriendly and insufficient time-saving problems mentioned above, this paper proposes a fast block partition decision based on the histogram of oriented gradient (HOG), which avoids the use of neural networks while greatly reducing the computational complexity of the encoder process. Specifically, our contribution can be categorized into the following three points:

- For the parent block, we propose to prune the horizontal and vertical partition by utilizing the horizontal and vertical gradient in HOG.
- For child block, we utilize the sum of absolute differences (SAD) between child block HOGs and parent block HOG to measure the degree of difference after partition, further

979-8-3503-6184-1/24 $31.00 © 2024 IEEE

reducing the complexity of the partition decision.

- We make hardware-oriented optimizations for the decision of the HOG index and the computation of the child blocks HOG.

Our proposed hardware-friendly fast block partition decision algorithm is tested under the AV1 reference model [15], saving up to 44.32% time with only 1.46% BD-Rate increase.

II. Proposed Hardware-friendly Algorithm

A. Overall Algorithm Process

Fig. 2 shows the flowchart of the proposed HOG-based algorithm. Firstly, HOG is calculated based on the original pixels of the input square block (the parent block). Vertical and horizontal gradients (V_g, H_g) in HOG are used to skip the partition. When V_g is much larger than H_g, the texture of the block is horizontal, which means the block tends to horizontal partition. In this case, four vertical partitions (orange blocks) in Fig. 1 can be skipped without further cost computation. When H_g is much larger than V_g, four partitions (green blocks) are skipped.

Then, the HOGs of the child block generated by each partition type are calculated, and the SAD is calculated between these child block HOGs and parent HOGs. The higher the SAD, the more different information each of the sub-blocks contains, i.e., the corresponding partition is the best. Therefore, the partition with the largest SAD will be selected, and the 8 partitions (green and yellow blocks) in Fig. 1 will be filtered

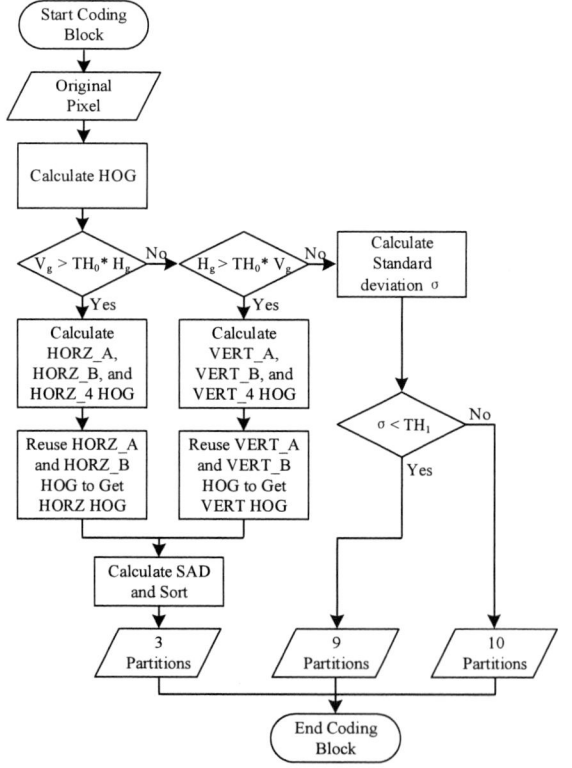

Fig. 2. Flowchart of proposed HOG-based algorithm.

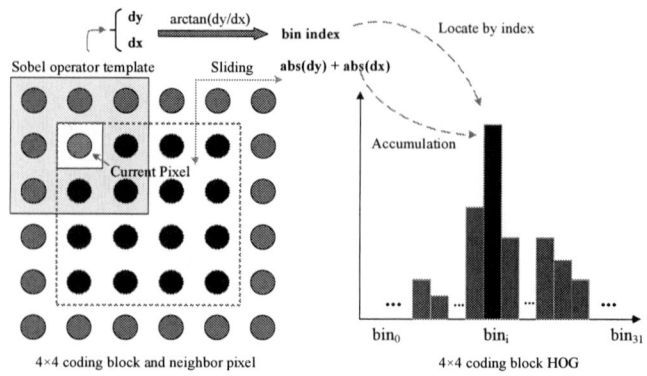

Fig. 3. The calculation process of HOG.

to 1, which greatly reduces the complexity of the partition decision.

Finally, when the difference between the horizontal and vertical gradients is not large, we compute the standard deviation of the original pixels and use the flatness to judge whether to continue the recursive partition or not. When the standard deviation is small, representing that the current coding block is very flat, then the block will not be recursively partitioned, i.e., the partition in the blue color of Fig. 1 will be pruned. The thresholds for gradient (TH_0) and variance (TH_1) are 2.9 and 9.5, respectively.

B. Hardware-oriented Optimization

Fig. 3 shows the computation process of HOG. Firstly, dy and dx are obtained by Sobel operator. Then, the bin index of the gradient is computed by *arctan*, and one of the 32 bins is localized according to this bin index. Finally, the gradient value of the localized bin is updated by the accumulation of dy and dx. The HOG of the coding block is obtained until all pixels are traversed by sliding the Sobel operator template.

In the above process, calculating *arctan(dy/dx)* involves not only the division but also the inverse tangent, which poses a great challenge to the hardware implementation. The inverse tangent function is a monotonic function, so we calculate in advance the tangent value corresponding to the 32-bin boundary value, which we call the threshold value. The boundary can be confirmed by comparing the magnitude of dy/dx with these constant threshold values. Thus, the original inverse tangent function is replaced by a table lookup operation, and division can be replaced by a constant coefficient multiplication operation, which greatly reduces the hardware implementation cost of the proposed algorithm.

In addition, 32 comparisons are required in the above process of localization. To reduce the number of comparisons, we utilize the following two properties of the inverse tangent function:

$$\arctan(-x) = -\arctan(x) \quad (1)$$

$$\arctan\left(\frac{1}{x}\right) = \frac{\pi}{2} - \arctan(x) \quad (2)$$

979-8-3503-6184-1/24 $31.00 © 2024 IEEE

According to (1), we can have only the tangent values of the boundary angles in the range $[0, \pi/2]$. According to (2), we can save only the tangent values of the boundary angles in the range $[0, \pi/4]$. The final localization is accomplished with only 8 comparators and the sign bits of dy and dx.

The child block HOG for each partition is calculated as shown in the following equation:

$$child_{SAD} = \sum_{j=1}^{n} \sum_{i=1}^{32} |bin_{child}[j][i] - bin_{parent}[i]|, \quad (3)$$

where bin_{child} and bin_{parent} represent the bin in the HOG of the parent and child blocks, respectively. i represents the index of the bin, j represents the index number of child blocks in the current block partition, and n represents the number of child blocks in the current block partition.

By observing the three partition of $HORZ_A$, $HORZ_B$ and HORZ in Fig. 1, we can find that the two child blocks of HORZ are composed of $HORZ_A$ and $HORZ_B$. Therefore, we don't need to calculate the HOG of HORZ, and we also don't need to calculate the HOG of VERT, which saves 1/4 of the computation.

III. EXPERIMENTAL RESULT

To showcase the efficiency of our proposed algorithm, it is integrated into AOM for encoding test sequences in CTC [16] with QPs of 22, 27, 32, and 37. The RD performance and time saving are measured by BD-Rate [17] and CPU time. The final result and comparison with other works are shown in Table I. The proposed algorithm saves 44.32% of the time, which is twice as much as [5] and achieves the highest saving time. In terms of compression rate, we have only 1.46% BD-Rate increase. Due to our HOG lookup table and comparator reduction, our algorithm is very hardware-friendly.

IV. CONCLUSION

In order to reduce the complexity of the encoder, this paper proposes a fast partition decision algorithm based on HOG. The algorithm uses vertical and horizontal gradients of HOG to skip nearly half of the number of partitions and further selects three partitions using the SAD of the HOGs of the parent and child blocks, which greatly accelerates the internal computation of the encoder. In addition, we also make hardware-oriented optimizations for the judgment of the HOG index and the computation of the child blocks HOG,

TABLE I
COMPARISON WITH OTHER WORKS

	2018 SPL [13]	2019 ICASSP [14]	2024 TBC [5]	Ours
Module	Partition Decision	Partition Decision	Transform Kenel	Partition Decision
BD-Rate	2.46%	0.95%	1.15%	1.46%
Time Saving	22.62%	29.06%	20.09%	44.32%
Hardware -friendly	✔	✗	✔	✔

respectively. Finally, our proposed algorithm saves 44.32% time with only 1.46% BD-Rate improvement.

ACKNOWLEDGMENT

This work was supported in part by the National Key R&D Program of China (2023YFB4502802), in part by the National Natural Science Foundation of China under Grant 62031009, in part by the "Ling Yan" Program for Tackling Key Problems in Zhejiang Province (No.2022C01098), in part by Alibaba Innovative Research (AIR) Program, in part by Alibaba Research Fellow (ARF) Program, in part by the Fudan-ZTE Joint Lab, in part by CCF-Alibaba Innovative Research Fund For Young Scholars.

REFERENCES

[1] J. Han et al., "A Technical Overview of AV1," in Proceedings of the IEEE, vol. 109, no. 9, pp. 1435-1462, Sept. 2021, doi: 10.1109/JPROC.2021.3058584.

[2] G. J. Sullivan, J. -R. Ohm, W. -J. Han and T. Wiegand, "Overview of the High Efficiency Video Coding (HEVC) Standard," in IEEE Trans. Circuits Syst. Video Technol., vol. 22, no. 12, pp. 1649-1668, Dec. 2012.

[3] M. Corrêa, D. Palomino, G. Corrêa and L. Agostini, "Direction-Based Fast Mode Decision and Hardware Design for the AV1 Intra Prediction," 2022 35th SBC/SBMicro/IEEE/ACM Symposium on Integrated Circuits and Systems Design (SBCCI), Porto Alegre, Brazil, 2022, pp. 1-6.

[4] H. Su, M. Chen, A. Bokov, D. Mukherjee, Y. Wang and Y. Chen, "Machine Learning Accelerated Transform Search For AV1," 2019 Picture Coding Symposium (PCS), Ningbo, China, 2019, pp. 1-5.

[5] Z. Hao et al., "Fast Transform Kernel Selection Based on Frequency Matching and Probability Model for AV1," in IEEE Trans. Broadcast., vol. 70, no. 2, pp. 693-707, June 2024.

[6] Í. Siqueira, G. Corrêa and M. Grellert, "Complexity and Coding Efficiency Assessment of AOMedia Video 1," 2023 IEEE 14th Latin America Symposium on Circuits and Systems (LASCAS), Quito, Ecuador, 2023, pp. 1-4.

[7] X. Zhao et al., "Study On Coding Tools Beyond AV1," 2021 IEEE International Conference on Multimedia and Expo (ICME), Shenzhen, China, 2021, pp. 1-6.

[8] Y. Li, Z. Liu, X. Ji and D. Wang, "CNN Based CU Partition Mode Decision Algorithm for HEVC Inter Coding," 2018 Proc. IEEE Int. Conf. Image Process. (ICIP), Athens, Greece, 2018, pp. 993-997.

[9] J. Xu, G. Wu, C. Zhu, Y. Huang and L. Song, "CNN-Based Fast CU Partitioning Algorithm for VVC Intra Coding," 2022 Proc. IEEE Int. Conf. Image Process. (ICIP), Bordeaux, France, 2022, pp. 2706-2710.

[10] A. Feng, K. Liu, D. Liu, L. Li and F. Wu, "Partition Map Prediction for Fast Block Partitioning in VVC Intra-Frame Coding," in IEEE Transactions on Image Processing, vol. 32, pp. 2237-2251, 2023.

[11] B. Guo, Y. Han and J. Wen, "Fast Block Structure Determination in Av1-Based Multiple Resolutions Video Encoding," 2018 IEEE International Conference on Multimedia and Expo (ICME), San Diego, CA, USA, 2018.

[12] B. Guo, X. Chen, J. Gu, Y. Han and J. Wen, "A Bayesian Approach to Block Structure Inference in AV1-Based Multi-Rate Video Encoding," 2018 Data Compression Conference, Snowbird, UT, USA, 2018, pp. 383-392.

[13] M. Tang, X. Chen, J. Gu, Y. Han, J. Wen and S. Yang, "Accelerating HEVC Encoding Using Early-Split," in IEEE Signal Processing Letters, vol. 25, no. 2, pp. 209-213, Feb. 2018.

[14] J. Gu and J. Wen, "Mid-depth Based Block Structure Determination for AV1," Proc. IEEE Int. Conf. Acoust., Speech Signal Process. (ICASSP), Brighton, UK, 2019, pp. 1617-1621.

[15] "Alliance for Open Media Video Codec reference implementation," https://github.com/mozilla/aom, accessed 2024.

[16] Frank Bossen et al., "Common test conditions and software reference configurations," JCTVC-L1100, vol. 12, 2013.

[17] Gisle Bjøntegaard, "Improvements of the bd-psnr model, vcegai11," in ITU-T Q. 6/SG16, 34th VCEG Meeting, Berlin, Germany (July 2008), 2008.

979-8-3503-6184-1/24 $31.00 © 2024 IEEE

A Low Power Narrow-Band Complex-Bandpass Filter Based on Feedforward Compensation Amplifiers for NB-IoT Applications

Xu Zhao [1], Ziqiang Wang*[2], Jie Gan[1]......

[1] Beijing Smart-chip Microelectronics Technology Co.,Ltd
[2] School of Integrated Circuits, Tsinghua University

* Email: wangziq@tsinghua.edu.cn

Abstract—**This paper presents a low power narrow-band complex-bandpass (CBP) filter based on feedforward compensation amplifiers for NB-IoT applications. It consists of two 3rd-order complex-bandpass filters centered at 120kHz with a bandwidth of 180kHz and one programmable gain amplifier (PGA) with a gain range of 0-25dB and a step of 5dB. The feedforward frequency compensation is utilized to lower down the power consumption of the embedded amplifiers. The CBP filter has been implemented in 180nm CMOS, and the simulation results show that the CBP filter has achieved 6th-order filtering characteristics, with accurate center frequency and bandwidth control by utilizing the filter tuning technique. The current consumption is only 2.5mA. The CBP filter could be used in low power NB-IoT receivers.**

Keywords—*Complex-bandpass filter, CMOS, analog baseband, low power*

I. INTRODUCTION

NB-IoT protocol is widely used to implement the wireless communication in various Internet-of-Things (IoT) applications. Due to narrow signal bandwidth, the NB-IoT receiver usually utilizes Low-IF architecture where the analog baseband is a crucial part. It should provide enough complex-bandpass filtering to reject the adjacent channel interference and provide programmable gain amplification, aiming to lower down the pressure on the ADC. However, the analog baseband should consume as little as possible power consumption to length the life time of the battery which used to power the NB-IoT equipment.

This paper presents a low power narrow-band complex-bandpass (CBP) filter based on feedforward compensation amplifiers for NB-IoT applications. It consists of two 3rd-order complex-bandpass filters centered at 120kHz with a bandwidth of 180kHz and one programmable gain amplifier (PGA) with a gain range of 0-25dB and a step of 5dB. The feedforward frequency compensation is utilized to lower down the power consumption of the embedded amplifiers. The CBP filter has been implemented in 180nm CMOS, and the simulation results show that the CBP filter has achieved 6th-order filtering characteristics, with accurate center frequency and bandwidth control by utilizing the filter tuning technique. The current consumption is only 2.5mA. The CBP filter could be used in low power NB-IoT receivers.

II. CIRCUIT IMPLEMENTATION

A. Filter Architecture

Fig.1 shows the presented filter architecture. It consists of two 3rd-order complex-bandpass filters centered at 120kHz with a bandwidth of 180kHz and one programmable gain

amplifier (PGA) with a gain range of 0-25dB and a step of 5dB. The PGA is inserted between two 3rd-order filters, which could reduce the noise contribution from the second filter since it could provide enough amplification, which help improve the whole CBP filter noise performance. It is a conventional resistor feedback amplifier with variable feedback resistance to provide the programable gain.

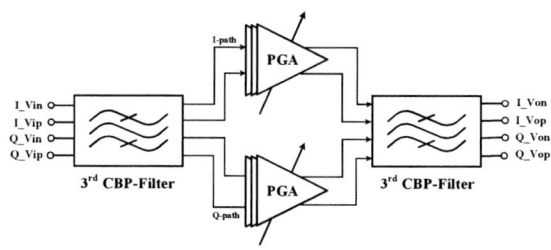

Fig. 1. The complex-bandpass filer architecture.

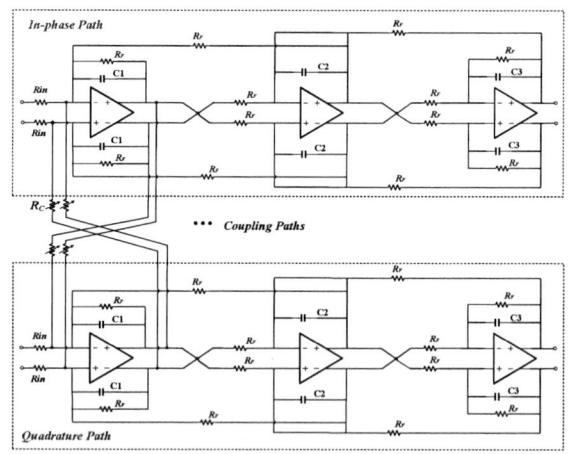

Fig. 2. The 3rd-order complex band-pass filter.

The 3rd-order CBP filter is formed by two 3rd-order low-pass leapfrog filters (I and Q), cross-coupled by coupled resistors R_C, as shown in Fig.2. By controlling the resistance of the cross-coupled resistors, the center frequency of the CBP filter is shifted to the desired positive or negative frequency, implementing the complex band-pass filtering.

$$\omega_c = 1/(R_c C_1)$$

Where C1 is the feedback capacitance of the integrator in I or Q path.

979-8-3503-6184-1/24 $31.00 © 2024 IEEE

B. Embedded Amplifier Design

The performance and power consumption of the CBP filter is dependent on the embedded amplifier. The amplifier should provide enough loop gain at the corner frequency of the filtering characteristics, to maintain the well-defined filtering characteristics and achieving good linearity, while consuming as little as possible power consumption. Usually the above requirements results in enough high GBW (unity-gain bandwidth), and most amplifier design optimizes the GBW under the power consumption limit.

However, it is more important to optimize the loop gain at the corner frequency of the filtering characteristics, since this gain actually decides the filter performance. As shown in Fig. 3, the loop gain at the corner frequency ω_c is higher with the design of wider bandwidth and lower GBW, compared with the design of narrower bandwidth and wider GBW. So in our design, we optimize the amplifier bandwidth under the power consumption limit.

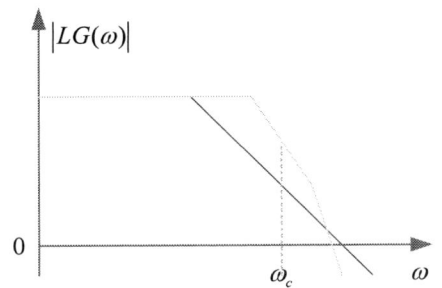

Fig. 3. The loop gain optimization comparison.

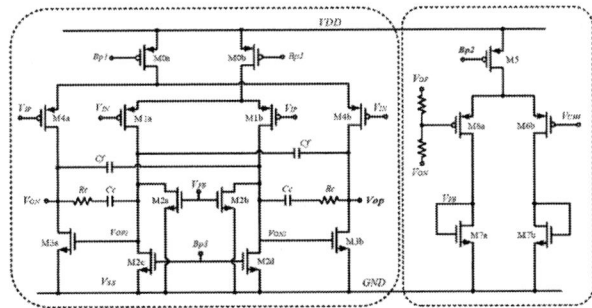

Fig. 4. The schematic of the embedded amplifier.

The schematic of the embedded amplifier is shown in Fig. 4. It is based on the conditional two stage operational transconductance amplifier (OTA), with the PMOS input differential pair to lower down the $1/f$ noise contribution. A common-mode feedback loop with the diode-load error amplifier is inserted to stabilize the output common-mode level. The feedback transistors are splitted into two transistor pairs (M2a and M2c, M2b and M2d), with M2c and M2d is biased with one bias circuit, which is helpful to lower down the oscillation risk. C_c is the miller frequency compensation capacitor to provide enough phase margin. R_c is series with C_c to remove the effect of the right-half plane zero, which would degrade the phase margin.

Besides, the feedforward compensation capacitance C_f is introduced. Its anti-pole splitting effect counter-acts the pole-splitting of the miller-compensation capacitance C_c, and puts the dominant pole to the higher frequency, which extends the bandwidth of the amplifier, as shown in Fig. 5. Another role of the feedforward compensation capacitance C_f is to help maintain the stability of the common-mode feedback loop. Usually, the stability design of the common-mode feedback loop should not affect the differential-mode loop, which leaves little design freedom. In this design, the feedforward compensation capacitance C_f acts the same miller compensation role with the C_c in the common-mode state, while it counter-acts the pole-splitting of the miller-compensation capacitance C_c in the differential-mode state, which increases the design freedom for the common-mode loop stability.

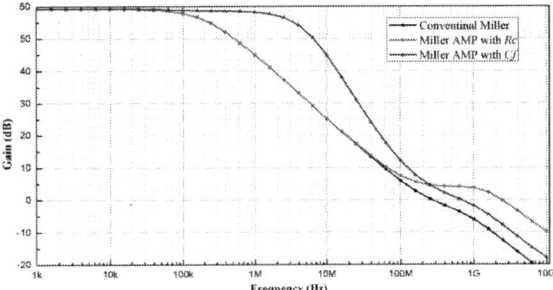

Fig. 5. The effect of the anti-pole splitting capacitance C_f.

However, different from the conditional two stage OTA, the second stage inserted the feedforward differential pair M4a/M4b/M0a, which directly forward the input signals V_{IP}/V_{IN} to the amplifier output V_{ON}/V_{OP}, which help extend the bandwidth of the amplifier bandwidth. Its operation principle is explained in Fig. 6. In Fig. 4, M1/M2/M3 form the high gain signal transfer path with two poles, which could be expressed as:

$$H_1(s) = \frac{A_0}{(1 + s/b_0)(1 + s/b_1)}$$

Where $b_0 \ll b_1$. And M4 forms the low gain high speed signal transfer with one pole at b_1:

$$H_2(s) = \frac{A_1}{(1 + s/b_1)}$$

The whole transfer function is:

$$H(s) \simeq \frac{A_0 \left(1 + \frac{A_1}{A_0}\frac{s}{b_0}\right)}{(1 + s/b_0)(1 + s/b_1)}$$

Where $A_0 \gg A_1$ is assumed. It introduces one left-half plane zero at the $b_0 * A_0/A_1$, which help improve the phase margin.

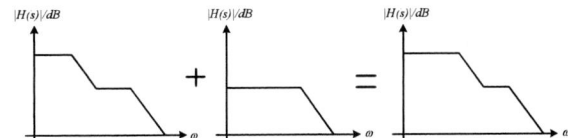

Fig. 6. The operational principle of the feedforward compensation.

Fig. 7 shows the simulation amplifier gain response, the introduced left-half plane zero obviously help improve the phase margin.

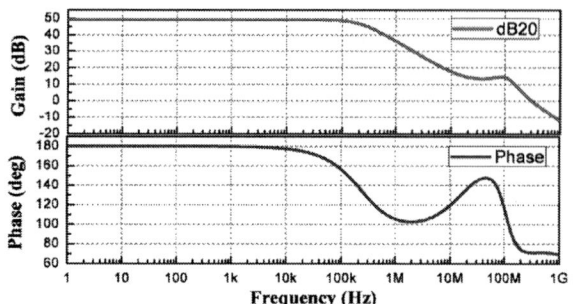

Fig. 7. The simulation amplifier gain response with the feedforward compensation.

III. SIMULATION RESULTS

The presented 6th-order CBP filter has been implemented in 180nm CMOS. Fig. 8 shows the simulated filtering response, which shows the 120dB/dec 6th-order filtering characteristics with the center frequency of 120kHz and the bandwidth of 180kHz. The center frequency and bandwidth could be tuned with the on-chip filter tuning circuit, with the accuracy error of $< 5\%$.

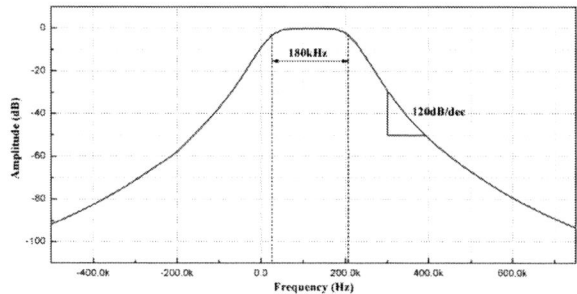

Fig. 8. The simulated filter response of the 6th-order CBP filter.

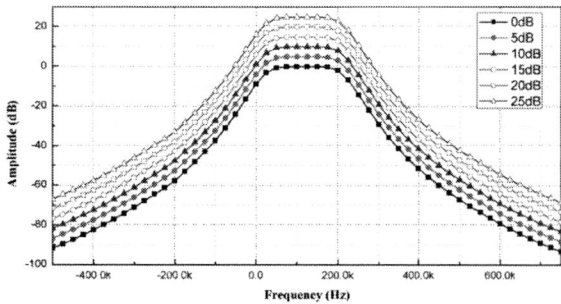

Fig. 9. The simulated filter gain response of the 6th-order CBP filter.

Fig. 9 shows the simulated filter gain response of the 6th-order CBP filter. It could provide 0-25dB programmable gain with the step of 5dB, and the gain error is less than 0.3dB. This figure also shows the filtering characteristics is maintained at different gain configurations.

Fig. 10 shows the simulated P1dB and IP3 of the presented 6th-order CBP filter at the highest gain configuration. The output P1dB is 12.26dBm and output IP3 is 35.13dBm.

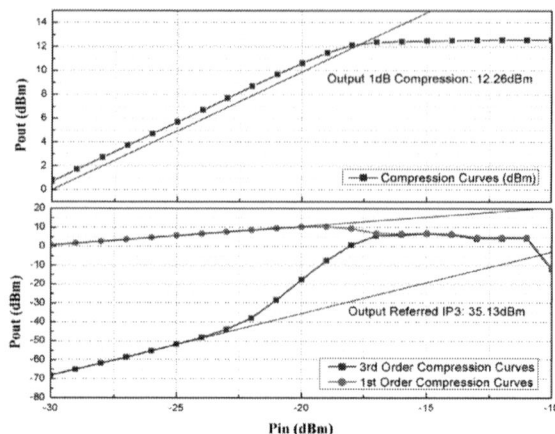

Fig. 10. The simulated P1dB and IP3 of the 6th-order CBP filter.

The simulated power consumption is only 2.5mA with the power supply of 1.8V.

IV. CONCLUSIONS

In this paper, a low power narrow-band complex-bandpass (CBP) filter based on feedforward compensation amplifiers for NB-IoT applications. It consists of two 3rd-order complex-bandpass filters centered at 120kHz with a bandwidth of 180kHz and one programmable gain amplifier (PGA) with a gain range of 0-25dB and a step of 5dB. The feedforward frequency compensation is utilized to lower down the power consumption of the embedded amplifiers. The CBP filter has been implemented in 180nm CMOS, and the simulation results show that the CBP filter has achieved 6th-order filtering characteristics, with accurate center frequency and bandwidth control by utilizing the filter tuning technique. The current consumption is only 2.5mA. The CBP filter could be used in low power NB-IoT receivers.

ACKNOWLEDGEMENTS

This paper is supported by The Scientific Research Programs for High Level Talents of Beijing Smart-chip Microelectronics Technology Co., Ltd.

REFERENCES

[1] H. R. Kooshkaki, P. P. Mercier, "A 0.75 mW Receiver Front-End for NB-IoT," *IEEE Radio Frequency Integrated Circuits Symposium (RFIC)*, pp. 173-176, CA, USA, 2023.

[2] Z. Song, X. Liu, X. Zhao, Q. Liu, Z. Jin, and B. Chi, "A Low-Power NB-IoT Transceiver With Digital-Polar Transmitter in 180-nm CMOS," *IEEE Trans. Circuits and Systems-I: Regular Papers*, vol. 64, no. 9, pp. 2569-2581, 2017.

[3] Y. Pu, W. Li, M. Li, C. Wang, F. Chen, Q. Li, and H. Xu, "A Tri-Mode Reconfigurable Receiver for GNSS/NB-IoT/BLE With 68-dB HR3 and 60-dB IMRR in 28-nm CMOS," in *IEEE Transactions on Very Large Scale Integration (VLSI) Systems*, vol. 31, no. 8, pp. 1140-1152, Aug. 2023.

[4] Y. Xu, B. Chi, *et al.* Power-scalable, complex bandpass/low-pass filter with I/Q imbalance calibration for a multimode GNSS receiver. *IEEE Transactions on Circuits and Systems II: Express Briefs*, vol. 59, no. 1, pp. 30-34, 2012.

[5] A. Vasilopoulos, G. Vitzilaios, G. Theodoratos, et al., "A Low-Power Wideband Reconfigurable Integrated Active-RC Filter With 73 dB SFDR," IEEE Journal of Solid-State Circuits, vol. 41, pp. 1997-2008, 2006.

A High-Performance MTJ-LUT Circuit Using 4T1M Architecture

Yu Pan [1], Yuejun Zhang*[1], Shuaicheng Guo [1], Yuanxin Tian [1], Bo Hong [1], Rui Fang [1], Liang Wen*[2], Yong Ding*[3]

[1] Faculty of Electrical Engineering and Computer Science, Ningbo University, Ningbo, 315211, China
[2] Department of Electronic Technology, China Coast Guard Academy, Ningbo, 315211, China
[3] College of Integrated Circuits, Zhejiang University, Hangzhou, 310058, China

* Email: zhangyuejun@nbu.edu.cn, lwen13@fudan.edu.cn, Dingyong09@zju.edu.cn

Abstract—**As fundamental components of programmable logic circuits, Lookup Table (LUT) circuits enable the implementation of arbitrary combinational logic. The volatility, standby power dissipation, and propagation delay of LUT circuits critically affect the performance of programmable logic circuits. In order to address the problems of circuit volatility, propagation delay and standby power dissipation, a high-performance Magnetic Tunnel Junction LUT (MTJ-LUT) circuit design scheme is proposed based on the 4T1M structure. According to the characteristics of MTJ principle, and comparing with the classical MTJ-LUT memory cell, MTJ-LUT memory module based on 4T1M structure is proposed; Based on the LUT circuit principle of 4T1M structure, a circuit architecture with current mirror bias circuitry is proposed to adopt the compliant reference resistor, and the steady-current-powered reading mode is used to achieve the optimal reading operation and reduce the delay; The circuit function is implemented in SMIC 40nm process using SPICE simulation tool. The experimental results show that the propagation delay and power dissipation of the MTJ-LUT circuit proposed are reduced by 72% and 15%, respectively, compared with the conventional MTJ-LUT circuit.**

Keywords—4T1M, MTJ-LUT, low delay, circuit design

I. INTRODUCTION

Based on the development of integrated circuits, Field-Programmable Gate Arrays (FPGAs) are widely used due to their high speed and flexibility, and lookup table (LUT) circuits, which are capable of implementing combinational logic functions, constitute a critical component of FPGAs. However, Static Random Access Memory (SRAM) based LUT circuits are not suitable for applications relying on limited power supply due to their volatility, high standby power dissipation etc. In addition, non-volatile Magnetic Tunnel Junction (MTJ) based LUT circuits offer lower power dissipation and better applicability [1-2].MTJ not only contributes to the implementation of efficient LUT circuits, but also improves the non-volatility and energy efficiency of the circuits. In recent years, different design methods and techniques have been proposed in related studies to enhance the performance of MTJ-based LUT circuits. Masanori N proposed a single-ended non-volatile LUT circuit based on a Three-Terminal Magnetic Tunnel Junction (3T-MTJ) structure [3] .This design simplifies the circuit structure by separating the read and write circuits. Ramtin Z. proposed a writing scheme based on Transmission Gate (TG) circuits, which features a symmetrical switch structure, improving the performance in terms of switching energy and device count [4] . Daisuke S. proposed a Split Magnetic Tunnel Junction (SMTJ) lookup table circuit, which separates the selection tree structure for combinational logic and the read circuit, enhancing the circuit read speed[5]. The existing designs provide valuable references for the research of MTJ-LUT circuits, but still suffer from propagation delay and power dissipation problems. Therefore, proposes an MTJ-LUT circuit improvement scheme based on 4T1M structure. According to the characteristics of MTJ principle, and comparing with the classical MTJ-LUT memory cell, we propose the MTJ-LUT memory cell with 4T1M structure, and optimize the read circuit module and the reference array cells to propose the high-performance MTJ-LUT circuit structure based on 4T1M structure.

II. DESIGN OF MTJ-LUT MEMORY CELL

The central component of the MTJ-LUT is the magnetic tunnel junction (MTJ), which consists of three layers of material: the reference layer, the free layer and barrier, as shown in Fig. 1(a). Fig. 1(b) describes the MTJ resistive hysteresis curve when the voltage $V_{(P,N)}>0.9V$ between MTJ terminals, the MTJ resistance is changed from high resistance to low resistance; when the voltage $V_{(P,N)}<-0.6V$ between MTJ terminals, the MTJ resistance is changed from low resistance to high resistance.

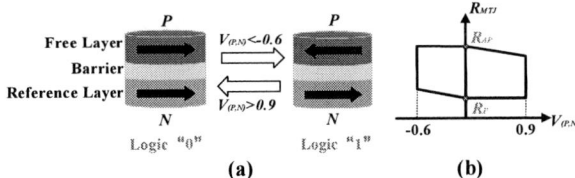

Fig. 1. The structural function of MTJ. (a) The basic operating principle of MTJ, (b) MTJ resistive hysteresis curve.

The memory cell is the basic cell for the MTJ-LUT to store information. Different MTJ-LUT memory cells are composed of different numbers of MTJs and transistors and are named by the number of MTJs and transistors, Fig. 2(a) is the architecture of a classical 1T1M memory cell(T stands for transistor and M stands for MTJ),When writing different storage information, the voltage is applied in different directions on both terminals of MTJ, positive phase voltage is written as '0', and negative phase voltage is written as '1'. The threshold voltage of transistor is expressed by (1) as:

$$V_{th}=V_{th0}+\gamma(\sqrt{|V_{SB}+2\phi_F|}-\sqrt{|2\phi_F|}) \qquad (1)$$

The threshold voltage V_{th0} is the threshold voltage when $V_{SB}=0$, γ is the body effect coefficient, V_{SB} is the source-to-substrate voltage, and ϕ_F is the Fermi potential. In writing '0' operation, BL is connected to high level, SL is grounded, the V_{SB} of transistor N1 is 0, $V_{th}=V_{th0}$. In writing '1' operation, SL is connected to high level, BL is grounded, the V_{SB} of transistor

979-8-3503-6184-1/24 $31.00 © 2024 IEEE

N1 and the threshold voltage V_{th} are both raised. The resistance of N1 increases and is significantly higher than the resistance during the writing '0' operation, resulting in a decrease in the voltage at both terminals of MTJ, which makes it difficult to achieve the voltage difference required for the write '1' operation, resulting in a write failure and affecting the reading reliability.

To address the problems faced by 1T1M memory cell, we propose 4T1M memory cell. As shown in Fig. 2(b), the 4T1M memory cell consists of two transmission gates and MTJ. Signal WL controls the writing of signals *BL* and *SL*, and *BL* is not directly connected to MTJ to ensure that the reading circuit is not affected by the writing process. During the write operation, the two transistors of the transmission gate are turned on at the same time, and they are connected in parallel with each other, so that the impedance of the transmission gate is lower than the resistance of either transistor, This solution solves the problem of transistor resistance increase due to the Body Effect, and improves the correction and stability of the written information.

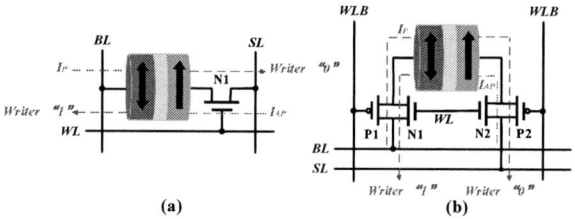

Fig. 2.   The operating schematic of the memory cell. (a) 1T1M memory cell, (b) 4T1M memory cell.

III. STRUCTURE OF 4T1M MTJ-LUT

In this section, we present an overview of the top-level architecture of the 4T1M MTJ-LUT. Then, we introduce the structure of the MTJ cell and the architecture of the read circuits.

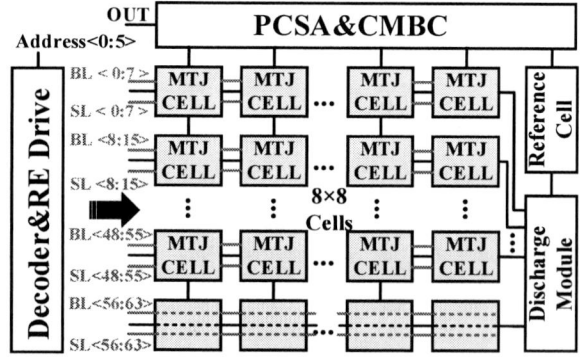

Fig. 3.   The top architecture of 4T1M MTJ-LUT.

A. The Top Architecture of 4T1M MTJ-LUT

Fig. 3 depicts the top-level architecture of the proposed 4T1M MTJ-LUT, which comprises a MTJ cell array, a decoder (with the RE drive unit), reference unit, Pre-Charge Sense Amplifier (PCSA), Current Mirror Bias Circuit (CMBC), and Discharge Module. The MTJ cell array consists of 64 MTJ cells, each configured in a high or low resistance state according to the *BL* signal and *SL* signal. Then, the address signal controls the decoder to enable the RE drive

unit to activate the MTJ cell, the voltages at both terminals of the MYJ cell and the Reference unit are output by CMBC, and then amplified and output by the PCSA. To save the area of the MTJ-LUT, the entire MTJ cell array shares a read amplifier and a reference unit.

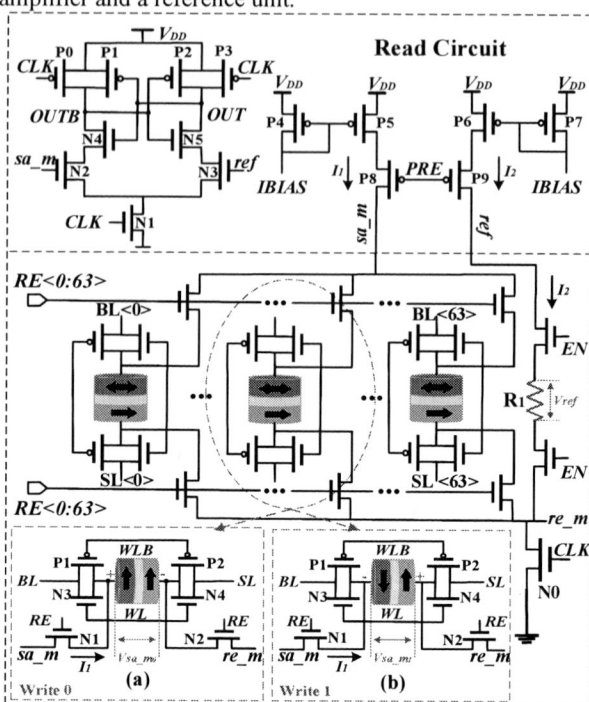

Fig. 4.   4T1M MTJ-LUT Circuit. (a) The operating principle of writing '0' to MTJ cell, (b) The operating principle of writing '1' to MTJ cell .

B. MTJ cell

Fig. 4(a) (b) is the circuit architecture of the MTJ cell for storing the logic '0' and logic '1' states, respectively. P1, P2, N1, N2, and the MTJ device consist of the 4T1M memory cell, As shown in Fig. 4(a), When the signal *WL* is set to high level, the input voltage of the signals *BL* and *SL* is transferred to the two terminals of MTJ; when the signal *BL* is set to high level and *SL* is set to low level, a positive-phase voltage is formed in the two terminals of MTJ, and the logic '0' is stored in the 4T1M memory cell. Meanwhile, when the reading driving signal *RE* is set to a high level, the current I_l through the port *sa_m* flows into the MTJ and is discharged to the ground through the port *re_m*, enabling the MTJ to generate the voltage V_{sa_m0}. On the contrary, Fig. 4(b) shows an inverted voltage applied to the MTJ terminals, which generates a voltage V_{sa_m1} in the high resistance state, and a logic '1' is stored in the 4T1M memory cell.

C. Design of the 4T1M MTJ-LUT Read Circuit

As shown in Fig. 4, the reading circuit consists of CMBC and PCSA. The current mirror bias circuit consists of two symmetrical current mirrors to provide stable currents to the MTJ cell and the reference unit, respectively. When *RE* controls the conduction of N3 and N4, the read enable signal PRE is connected to a low level, and the currents I_l and I_2 pass through *sa_m* and *ref* to flow to the MTJ unit and the reference unit, respectively, enabling the reference resistor R1 in the reference unit to generate a voltage V_{ref}, The reference resistor R1 resistance value is between the high and low resistance values of the MTJ device as shown in (2):

979-8-3503-6184-1/24 $31.00 © 2024 IEEE   615

$$R_{ref} \cong \frac{1}{2}(R_{AP} + R_P) \qquad (2)$$

Due to the asymmetric architecture of the MTJ cell array and reference cell circuits, the correct V_{sa_m1} and V_{ref} is not generated at both terminals of the MTJ and R1 without a stable bias current supply. Therefore, a current mirror bias technique is added to the reading circuit to ensure the correction of the V_{sa_m} and V_{ref}, and to improve the stability of the circuit. There is a pre-charging phase and a discharging phase in the PCSA, and when the clock signal CLK is set low, P0 and P3 pre-charge OUT and $OUTB$ to VDD, N1 is turned off, and the circuit fails to conduct. When CLK is set high, the charging process ends, and V_{sa_m} and V_{ref} turn on N2 and N3. If V_{sa_m} > V_{ref}, the N2 conduction ability is greater than that of N3, and the voltage discharge rate at $OUTB$ is greater than that of OUT, and finally, after the inverter positive feedback latching architecture, the OUT is rapidly charged to a high level, and outputs '1'. Conversely, OUT rapidly discharges to a low level and outputs '0'.

IV. EXPERIMENTAL RESULTS AND ANALYSIS OF MTJ-LUT STRUCTURE

The design is implemented in SMIC 40nm with the SPICE circuit simulation tool. The design is completed using the MTJ SPICE model at 1.1V rated voltage. Fig. 5 shows the output function of input signal $A5\text{-}A0 = '011111'$ with $A5\text{-}A0 = '100000'$ MTJ-LUT circuit. As shown in Fig. 5 (a), when BL is set high and SL is set low, the MTJ<31> cell is writing logic '0'. When CLK is high and PRE is low, the read circuit is turned on, and the output signal OUT outputs a logic '0' at less than 20 ps. As shown in Fig. 5 (b), the MTJ<32> cell is writing logic '1', the WL signal is set low after completion of the write operation, and when the PRE signal is low, OUT outputs logic '1' at less than 30 ps. Fig. 6 shows the Monte Carlo simulation results of MTJ-LUT under 800 process deviations, showing the delay results of outputting logic '0' and logic '1' respectively. It is shown that the propagation delay of our proposed MTJ-LUT is low under process deviations, and the small standard deviation indicates the good stability of the performance of the proposed.

Table I illustrates the results of this design in comparison with the related literature. Compared with the propagation delay of MTJ-LUT in other literature, we proposed differential read MTJ-LUT architecture based on current mirror bias circuit with 72% reduction in propagation delay. The output power consumption of MTJ-LUT based on 4T1M architecture is reduced by 15% with improved circuit delays compared to the literature [5] differential type MTJ-LUT circuit.

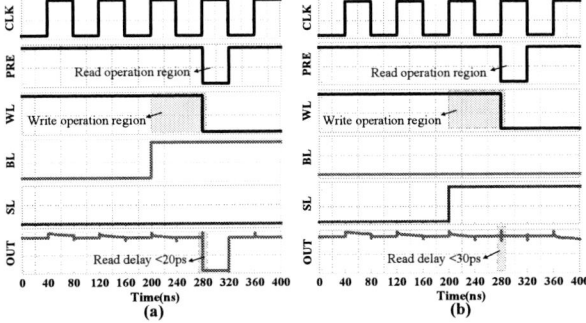

Fig. 5. Transient response of MTJ-LUT implementing operation. (a) $A5 - A0 = $ "011111", (b) $A5 - A0 = $ "100000".

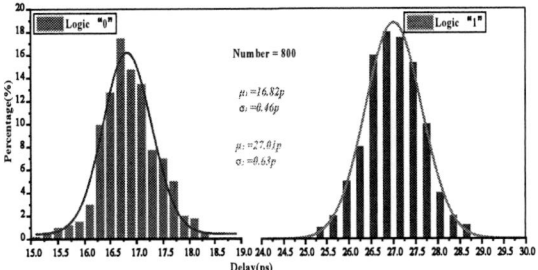

Fig. 6. Monte Carlo simulation results of MTJ-LUT propagation delay.

TABLE I. RESULTS OF COMPARISONS WITH OTHER WORKS

Reference	[3]	[4]	[5]	This work
Sensing type	Single-ended	Differential	Single-ended	Differential
Active power [μW]	6.26	9.05	14.5	7.7
Delay [ps]	195	122	97.4	27

V. CONCLUSION

This paper proposes a low delay MTJ-LUT circuit design based on 4T1M structure. The 4TIM memory cell is used to enhance the writing capability, optimize the read circuit and reference array, reduce the propagation delay, and improve the performance of the MTJ-LUT circuit. The simulation is performed in SMIC 40nm process using Cadence EDA tool, and the experimental results show that the propagation delay and power dissipation of MTJ-LUT circuit based on 4T1M structure are reduced by 72% and 15%, respectively, compared with the traditional MTJ-LUT circuit.

ACKNOWLEDGMENT

This work is supported by the National Natural Science Foundation of China (62474100, 62174121, 62134002), the Science and Technology Innovation 2025 Major Project of Ningbo (2022Z203), Ningbo University and Ningbo Yongxin Microelectronics Technology Co., LTD. Digital Integrated Circuit Design Joint Laboratory (XQ2022000005), Ningbo University Graduate Education Practice Base (YJD202305). the Ningbo University Student Research and Innovation Program (2024SRIP1311). Yangtze River Delta Science and Technology Innovation Community Joint Research Project (2022CSJGG1100). Central Guided Local Science and Technology Development Funding Program (2023ZY1069).

REFERENCES

[1] R. Zand and R. F. DeMara, "MRAM-Enhanced low power reconfigurable fabric with multi-level variation tolerance," IEEE Transactions on Circuits and Systems I: Regular Papers, vol. 66, no. 12, pp. 4662-4672, August 2019.

[2] G. Kolhe, et al, "Securing hardware via dynamic obfuscation utilizing reconfigurable interconnect and logic blocks," 2021 58th ACM/IEEE Design Automation Conference (DAC), San Francisco, CA, USA, pp. 229-234, December 2021.

[3] M. Natsui, D.Suzuki, Y. Lin and T Hanyu, "A 71%-Area-Reduced six-input nonvolatile lookup-table circuit using a three-terminal magnetic-tunnel-junction-based single-ended structure," Japanese Journal of Applied Physics. vol.52. March 2013.

[4] R. F. DeMara, A. Roohi, R. Zand and S. D. Pyle, "Heterogeneous technology configurable fabrics for field-programmable co-design of CMOS and spin-based devices," IEEE International Conference on Rebooting Computing (ICRC), Washington, pp. 1-4, November 2017.

[5] D.Suzuki and T. Hanyu, "Design of a highly reliable, high-speed MTJ-Based lookup table circuit using fractured logic-in-memory structure," Japanese Journal of Applied Physics. vol.58.April 2019.

Optimizing Communication Efficiency of GNN Inference in Distributed Systems

Wenqian Zhou [1], Qiaosha Zou*[2]

[1] the Institute of Brain-inspired Circuits and Systems, Fudan University, Shanghai 200437, China
[2] Zhejiang Lab, Hangzhou, China

* Email: qiaoshazou@zhejianglab.com

Abstract—**With the widespread adoption of Graph Neural Network (GNN) models, the demand for GNN inference acceleration has increased due to their slow runtime, especially when handling large-scale graphs. While distributed systems have shown promising results in computation acceleration, their potential for optimizing communication remains largely unexplored. In this paper, we propose novel feature pairing and feature compression methods for efficient and reduced data communication. Moreover, we co-design a multi-core system with Network-on-Chip (NoC) for evaluation. Experimental results show that our techniques can achieve up to 54.1% reduction in communication overhead, and 1.65x speedup in runtime.**

Keywords—GNN, distributed deep learning, communication, interconnection network, algorithm-architecture co-design.

I. INTRODUCTION

Graph Neural Networks (GNNs) have recently demonstrated promising results in various real-world applications, including recommendation systems and classification tasks. As the scale of graphs and models in these applications continue to grow, the inference time also increases, which is unacceptable for time-sensitive tasks. Therefore, accelerating the inference process becomes increasingly vital. Distributed GNNs offer a solution by leveraging multiple computing machines working in parallel. Due to the complex dependencies among graph data, it is impractical to partition the data into irrelevant subsets, necessitating the transmission of attributes between neighbouring vertices across machines. Therefore, the inference phase in a distributed system can be abstracted into three key stages: aggregation, combination, and communication phase.

Although significant efforts have been made to accelerate aggregation and combination [1], in distributed systems, it is inter-machine communication that constitutes more than half of the total runtime [2]. Moreover, the communication overhead increases with the number of machines, making it the bottleneck in distributed GNNs. Therefore, more attention should be paid to optimizing communication. Several techniques have been proposed for optimizing communication. CNS [2] proposed a code-based method to reduce redundant communication. However, the Parallel Server architecture lacks in scalability. Other works focus on compression-based techniques. Graph partition combined with sparsification, and polarizing [3] can create subgraphs with fewer boundary connections, leading to less communication overhead. However, these partition methods are time-consuming and impractical for time-sensitive tasks. And pruning feature data [4] might decrease accuracy.

This study was supported by funds from the National Science and Technology Innovation 2030 Major Projects of China (STI2030-Major Projects-2022ZD0206500).

In this paper, we propose a communication optimization method that introduces lossless feature compression and pairing techniques to GNN inference in a decentralized system to alleviate the communication bottleneck and enhance performance. Our method introduces little computation overhead. And our simulation results based on a cycle-accurate simulator on distributed systems demonstrate that the proposed techniques can improve the performance of distributed GNN inference with real-world datasets. Our contributions can be summarized as follows.

1) Based on the profiling results from real-world datasets, we propose a lossless layer-wise feature compression method using an optimized compressed sparse column (CSC) format, significantly reducing communication overhead.

2) By exploiting the natural community properties of graphs, we introduce a feature pairing method with multicast packages to mitigate redundant communication and enhancing communication efficiency.

3) We implement a cycle-accurate distributed simulation system for GNN inference to evaluate the proposed techniques on real-world datasets. The experimental results demonstrate up to 54.1% reductions in communication overhead and 1.65x speedup in runtime on a torus topology.

II. METHOD

A. GNN execution pattern and graph partition

$$X^{(l+1)}=\sigma\big(AX^{(l)}W^{(l)}\big) \qquad (1)$$

Equation (1) shows the layer-wise forward propagation of a multi-layer GCN used in the classification tasks. A is the graph adjacency matrix, $X^{(l)}$ is the input feature matrix for layer l, and $W^{(l)}$ is the weight matrix for layer l. The non-linear activation function is denoted by $\sigma(\cdot)$. Multiplying A and $X^{(l)}$ aggregates information from 1-hop neighboring nodes. The product $AX^{(l)}W^{(l)}$ perform combination operation, which yields the output of this layer, and also the input feature matrix for the next layer $X^{(l+1)}$.

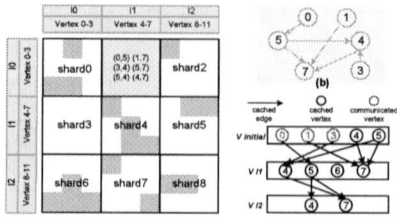

Fig. 1. Toy example of an adjacency matrix (a) 3×3 sharding partition. (b) Sub-graph of S_1. (c) GNN inference in $core_2$.

To deploy the model in a distributed system using data parallel computation, data partitioning is necessary. We adopt a static sharding partitioning method, as illustrated in Fig. 1 (a). The vertices V of the graph G=(V,E) are sorted in ascending order based on vertiex IDs and divided into P disjoint segments, referred to as intervals (I), each containing consecutive indices. Both the source and destination vertices are partitioned into these intervals, dividing all edges E into P×P shards (S). Thus each shard contains a subset of graph. In GNN tasks, data samples (i.e., vertices) have complex dependencies. Thus subsets of nodes and their k-hop neighbors' IDs and features must be all allocated to each worker in order to achieve similarly data parallelism as DNN. However, this approach introduces significant memory overhead and computation redundancy because the k-hop neighbors of different vertices often overlap substantially.

Therefore, our system is designed to communicate remote feature data between cores as needed and caches only neighbors IDs at local. We assign edges in S_i along with attributes in the destination I to $core_i$, while source vertex features are retrieved from other cores through communication. For example $core_1$ stores the subgraph shown in Fig. 1 (b), and the attributes of vertices 4, 5, 6 and 7. It needs to retrieve the attributes of source vertices 0,1, and 3 from S_0, as shown in Fig. 1(c), before proceeding with computation. For diagonal shards, all necessary data for computation is already available, thereby reducing communication overhead. This method hybrids DepCommu and DepCache [5], balancing redundant computation and communication overhead.

B. Feature Pairing

Real-world graphs naturally exhibit community properties, where vertices tend to form clusters or communities with dense internal connections. Inspired by this observation, we find that sharded subgraphs have common neighbors, meaning that different shards often need to retrieve the same source vertces. We quantify this property through experimental profiling. As shown in Fig. 3, the percentage of shared neighbors based on our sharding partition ranges from 19% to 58%, 6% to 22%, and 20% to 23% for the Cora, Citeseer, and Pubmed datasets, respectively.

To address this, we propose a feature pairing technique. Once a core receives requests from all other cores, it detects duplicate elements and sends features of these vertices in a multicast package. We also modified the packet format to support this technique. For example in a 3×3 NoC network, as shown in Fig. 2, we use the multi_flag, to indicate a multicast package. Additionally, We added a pairing routing module in router, where the multicast communication table is stored. When a packet with multi_flag set to 1 is routed from source node to the first destination, this module changes the destination bits in headflit to the second destination according to the address of current router, and sets the multi_flag to 0. Meanwhile, a copy of the packet is transmitted to the local port of current node, and will be transmitted to the related IP after depacketized. This technique optimizes the communication process by reducing redundant data transmission and enhancing overall efficiency.

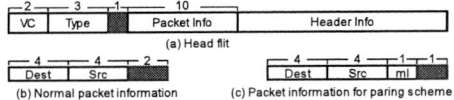

Fig. 2. Flit formatting in a 3×3 network.

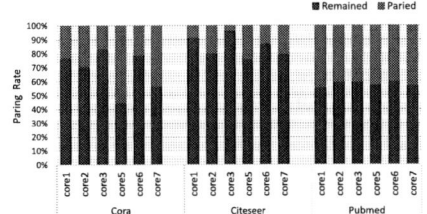

Fig. 3. Pairing rate of different datasets.

TABLE I. CHARACTERISTICS OF 3 WIDLY-USED GCN DATASETS

Characteristics		Datasets		
		Cora	Citeseer	Pubmed
Sparsity	A	99.82%	99.89%	99.97%
	W	0%	0%	0%
	X^1	98.73%	99.15%	90.00%
	X^2	22.00%	10.90%	22.40%
Dimensions	Node	2708	3327	19717
	Feature	1433	3703	500

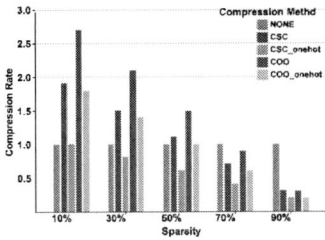

Fig. 4. Compression rate with different compression method.

C. Feature Compression

Compression-based methods are widely used in distributed deep learning systems to mitigate communication overheads. We first experimentally profiled the datasets and found that adjacency matrices exhibit high sparsity as expected. Additionally, feature matrices show varying degrees of sparsity depending on the dataset and layer. As shown in Characteristics of 3 widly-used GCN datasets, the initial feature matrix X^1 is very sparse, with 90.00%-99.15% sparsity, while X^2 becomes much denser, with 10.90%-22.40%. Furthermore, Fig. indicates that the commonly used compression methods, such as CSC, are effective in reducing data movement only at very high sparsity levels (over 60%).

We also observed that in classification tasks, X^1 is encoded in a one-hot manner. This inspired us to introduce CSC-onehot compression for the first layer. Typical CSC compression uses three arrays to represent non-zero values: an indices array for column coordinates, a data array for the values of non-zero elements, and an indptr array to show the offset of the first non-zero element in the indices array. For one-hot encoded matrices, we can omit the data array. The results in Fig. 4 show that CSC-onehot achieves a higher compression rate, thereby reducing communication overhead.

III. EVALUATION

A. System Modeling and Configuration

In this paper, we implemented a distributed system simulator for GNN. The overall framework is shown in Fig. 6. BookSim2 [6] is a cycle-accurate NoC simulator based on C++. We extended it to handle packets with specific sizes at specific cycles, and use it for interconnection modeling. We

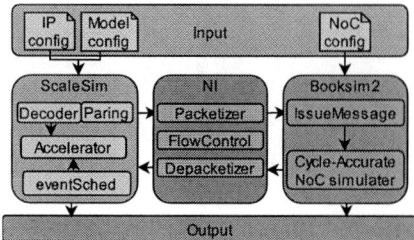

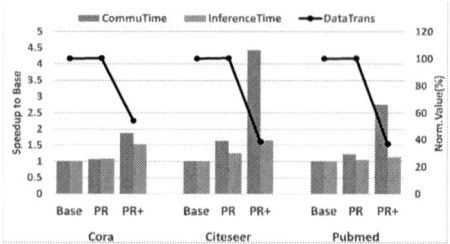

Fig. 5. Overall simulation system framework

TABLE II.   System Configurations

	Parameter	Configuration
PE	MAC array	256×256
	Dataflow	Output Stationary
	Precision	32 bit
Accelerator	Number of PEs	16
	Clock Frequency	1 GHz
Network	Number of Accelerators	9
	Topology	Torus
	Flow Control	Virtual Cut-Through
	Router Clock Frequency	1 GHz
	Number of VCs	4
	VC Buffer Depth	318 flits
	Packet Size	8192 Bytes
	Per Flit Latency	21 ns
	Bandwidth	25 GBps

configure a TPU-like accelerator with 16 processing elements (PEs), where each PE has a systolic array simulated using ScaleSim [7]. To support our feature compression and pairing methods, we added a CSC decoder module and a pairing module. We also designed an event scheduler module to arrange communication and computation events. Furthermore, we imple-mented a Network Interface (NI) in python to facilitate interaction between the accelerator and NoC, which includes packetizer, depacketizer and credit-based flow control moccludes. Detailed system configurations are provided in System Configurations. Each Accelerator is assumed to have double buffering and sufficient memory bandwidth to achieve the highest compute throughput. The network buffer size is configured to cover the credit round-trip loop. The network bandwidth is set to be consistent with modern training systems, and the link is adjusted to match this target bandwidth [8].

B. Results

$$T_{total} = T_{prep} + I_{commu} + T_{comm} + T_{proc} \qquad (2)$$

Focusing on communication improvement, inference time of a node can by represented by (2), where I_{commu} represents the idle time during the communication phase. We run GCN models for paper classification tasks on the implemented system, and quantify the improvement by measuring the amount of transmission data and runtime according to (2). As shown in Fig. 6, our proposed method achieves 1.88x, 4.43x and 2.76x communication speedup, 1.52x, 1.65x and 1.13x inference speedup, along with 54.1%, 38.5% and 37.7% reductions in transmission amounts on Cora, Citeseer and Pubmed dataset, respectively. Although the feature paring technique introduces a communication overhead of 0.11%-0.39%, it helps reduce runtime by improving efficiency. Moreover, the results in Fig. 7 demonstrate that our techniques achieve a 66.8% reduction in I_{commu} on Citeseer, significantly improved communication efficiency. And the reduction for Cora and Pubmed are 23.4% and 53.8%.

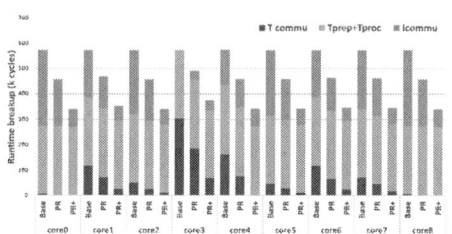

Fig. 6. Comparation of different communication schemes on three datasets. PR and PR+ stand for using pairing and combined techniques. Norm.Value denotes the amount of transmitted data normalized to base scheme.

Fig. 7. Inference runtime breakdown on Citeseer.

IV. CONCLUSION

In this paper, we identify the inefficiencies in the GNN inference process in distributed systems and opportunity of algorithm-architecture co-design. To mitigate communication overhead and imbalance, we introduce feature pairing and feature compression techniques to leverage the data redundancy among compute cores inherent in unstructured graph data. We implemented a simulation system and ran paper classification tasks with real-world datasets. Experimental results show that our proposed techniques can achieve 54.1%, 38.5% and 37.7% reduction in communication overhead, and a 1.52x, 1.65x, 1.13x speedup in runtime.

REFERENCES

[1] M. Yan et al., "HyGCN: A GCN accelerator with hybrid architecture," 2020 IEEE International Symposium on High Performance Computer Architecture (HPCA), 2020, pp. 15-29.

[2] Y. Wang et al., "Accelerating distributed GNN training by codes," in IEEE Transactions on Parallel and Distributed Systems, vol. 34, no. 9, pp. 2598-2614, Sept. 2023.

[3] H. You, T. Geng, Y. Zhang, A. Li and Y. Lin, "GCoD: Graph convolutional network acceleration via dedicated algorithm and accelerator co-design," 2022 IEEE International Symposium on High-Performance Computer Architecture (HPCA), 2022, pp. 460-474.

[4] W. Zhang, J. Sun and G. Sun, "Accelerating GNN inference by soft channel pruning," 2022 IEEE 13th International Symposium on Parallel Architectures, Algorithms and Programming (PAAP), 2022, pp. 1-6.

[5] Q. Wang et al., "NeutronStar: Distributed GNN training with hybrid dependency management," In Proceedings of the 2022 International Conference on Management of Data (SIGMOD '22), 2022.

[6] N. Jiang et al., "A detailed and flexible cycle-accurate network-on-chip simulator," in International Symposium on Performance Analysis of Systems and Software (ISPASS), 2013, pp. 86–96.

[7] A. Samajdar et al., "A Systematic Methodology for Characterizing Scalability of DNN Accelerators using SCALE-Sim," 2020 IEEE International Symposium on Performance Analysis of Systems and Software (ISPASS), Boston, MA, USA, 2020, pp. 58-68.

[8] S. Laskar et al., "Enhancing collective communication in MCM accelerators for deep learning training," 2024 IEEE International Symposium on High-Performance Computer Architecture (HPCA), 2024, pp. 1-16.

SST: Simplified Space-Time Transformer based on Time-assisted Spatial MSA for 3D Human Pose Estimation

Sheng Lu [1], Qiyun Dong [1], Zhenyin Zhang [1], Gengsheng Chen [1,2], Yinna Zhu *[1], Wei Xu *[1]

[1] School of Microelectronics, Fudan University, Shanghai, China
[2] Jiashan Fudan Institute, Jiaxing, Zhejiang Province 314100, China

Abstract—Depth ambiguity in 2D human joint estimation is a persistent issue for 2D-3D human pose estimation networks. To cope with this challenge, temporal dimensions are adopted in existing models, however, none of them are able to fully utilize the information embedded in the input data. In this paper, we present a Time-assisted Spatial (TaS) MSA & Simplified Space-Time Transformer (SST) to better capture the spatial-temporal relationships. First, we design a new Time-assisted Spatial (TaS) MSA to comprehensively model spatial-temporal relationships. Secondly, we combine TaS MSA and Temporal MSA in parallel to enhance modeling capability and to build Simplified Space-Time Transformer (SST) model. Thirdly, we find an optimal pipeline of SST through contrasting the impact of parallel blocks and intermediate feature dimensions on the model's performance. Experimental results show that our model achieves the highest accuracy on Human3.6M dataset, with 0.4mm gain against current methods and 9% improvement in difficult positions.

Keywords—3D Human Pose Estimation, Transformer, Spatial-temporal relationship

I. INTRODUCTION

3D human pose estimation is pervasively utilized in human-machine interaction, autonomous driving, auxiliary medical care, etc. Traditional monocular 3D human pose estimation methods normally utilize convolutional and fully connected layers to predict 3D human joints. To better utilize 2D human pose estimation and to improve the accuracy, a typical process of 3D human pose estimation generally has two stages. First, obtaining 2D human joint positions with a 2D pose estimation. Second, mapping them to the corresponding 3D joint positions, e.g. SimpleBaseline3D [1] and VideoPose3D [2]. With the introduction of PoseFormer [3], Transformer [4] becomes a promising foundational architecture with ascendant performance [1, 5, 6, 7]. However, the persistent challenges of location uncertainty and depth ambiguity remain unsolved due to the absence of depth information.

To mitigate the impact of depth ambiguity to the accuracy, contemporary research efforts opt for Temporal Multi-head Self Attention (MSA) to model temporal relationships of human joint motions from consecutive video images [8, 9, 10, 11]. However, these methods use temporal MSA and spatial MSA in a cascaded manner with each level focusing merely on either the temporal or the spatial relationships, which largely limits their ability to fully utilize the spatial-temporal feature information that transmitted from the upper levels, leading to a lessened effect on feature learning. Meanwhile, due to the strict constraints of resource-limited edge computing devices on both the

computingcomplexity and the training speed, it is also impracticable for them to comprehensively build spatial-temporal relationship between every human joint throughout an entire video.

To address the above issues, in this paper, we propose a new Time-assisted Spatial (TaS) MSA using Limited KQV calculations, to pursuit a better utilization of the spatial-temporal information from 2D human joint motions while at the same time reduce the model complexity as well. Our contributions are summarized as follows:

1) We bring in a new self-attention block called TaS MSA to better explore spatial-temporal relationships during the motions of human joints. We bulid a new Simplified Space-Time Transformer (SST) model by combining TaS MSA with Temporal MSA and Merging Cell. SST significantly mitigates depth ambiguity in 3D human pose estimation and improves the data utilization.

2) We design a pipeline work of SST, by comparing different network depths, intermediate feature dimensions and parallelism. We tackle the over-fitting issue bysetting an ideal training epoch.

3) Experimental results show that the SST model achieves a state-of-the-art performance against the existing peer works with a 0.4mm decrease in average MPJPE and a roughly 9% improvement in performance.

II. RELATED WORKS

A. 3D Human Pose Estimation

According to estimation processes, existing CNN-based 3D human pose estimation methods falls into two categories, single-stage and double-stage.

A single-stage method takes 2D human pose images as its inputs to estimate 3D positions of human joints directly by using one end-to-end network. Based on Hourglass structure [12] of 2D HPE, Pavlakos et al. use 3D heat maps in their C2F-Vol [13] to represent the 3D positions of human joints and gradually increases the resolution in depth to reduce the storage consumption. Due to the large span in representation between 2D images and 3D human joints, it is difficult for a single end-to-end model to perceptively discover multiple types of information. Therefore single-

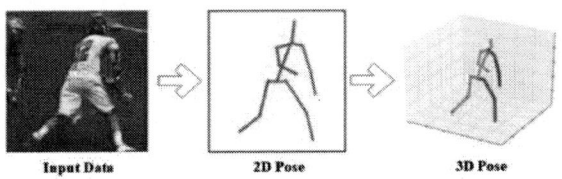

Fig. 1. An example of double-stage approach

---

* Correspondence to: Fudan Univ., 825 Zhangheng Rd., Shanghai, China.
  E-mail: zhuyn22@m.fudan.edu.cn (Yinna Zhu)
  Wei_xu@fudan.edu.cn (Wei Xu)

stage methods usually require longer inference time and more training epochs, while using more sophisticated flow designs.

For a double-stage method, e.g. SimpleBaseline3D [1], its first stage is used to estimate 2D joints according to the input images, and its second stage estimates the positions of 3D joints according to the estimated 2D joints. The advantages of the double-stage method is the ability to fully utilize the superior performance of the fully trained 2D human pose estimation network to achieve accurate 3D pose estimation. However, using the 2D human pose estimation leads to the loss of depth information in input images. Such kind of loss could be particularly severe when the input of the initial stage only contains one 2D image.

B. Vision Transformers

Traditional 3D pose estimation primarily uses CNN as their backbone. In recent works, Transformer [4] becomes a more promising backbone architecture due to its distinguished performance. In 2021, Zhen et al. first introduce Transformer to human pose estimation in PoseFormer [3], which predicts the 3D human joints from the center frame of the video. However, their model does not fully explore the inherent temporal information in the images. In 2022, Zhang et al., in their MixSTE [9], place their focus on the spatial-temporal relationships to produce feature codes with greater expression ability. Meanwhile, MixSTE expands the output from discrete images to a consecutive video. But its cascading structure of time and space encoders still makes it inefficient to reveal the inherent spatial-temporal linkages. MotionBert [11], raised by Zhu et al. in 2022, uses a two-flow spatial-temporal converter on encoding human joints, however is still limited in modeling the global spatial-temporal relationships clearly by discrete modeling of temporal and spatial relationships. Most of the earlier structures could only separately construct temporal or spatial relations because of the non-negligible computing cost. As a result, they are unable to accurately estimate the position of 3D human joints which are in relation to their previous position and the current position of their adjacent joints.

III. METHOD

A. Architecture

SST follows the paradigm of double-stage 3D human pose estimation methods. The first stage of SST uses whatever 2D human pose estimation network to estimate 2D joints. The 2D joints are used as the input data $X_{in} \in \mathbb{R}^{F \times J \times C_{in}}$ of the second stage. Fig. 2 illustrates the main architecture of the second stage, where F represents the total number of thevideo frames; J represents the number of joints, C_{in} represents the dimension of input joint features, and C_f represents the dimension of intermediate features. First, X_{in} is mapped to intermediate feature $X_f \in \mathbb{R}^{F \times J \times C_f}$ through a full connection layer. Then the learnable spatial embedding $E_S \in \mathbb{R}^{1 \times J \times C_f}$ and temporal embedding $E_T \in \mathbb{R}^{F \times 1 \times C_f}$ are added to X_f through a broadcast mechanism to produce the output $X_{mid} \in \mathbb{R}^{F \times J \times C_f}$. The embedded spatial-temporal relationships are learnt through the subsequent N layers of SST. To take the advantages of residual connections, the input and output shape of each SST layer are kept as $F \times J \times C_f$. Finally, the output 3D data $X_{out} \in \mathbb{R}^{F \times J \times C_{out}}$ is obtained using a full connection layer.

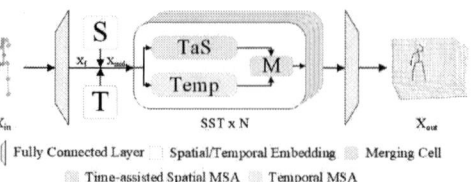

Fig. 2. Main Architecture of SST

|| Fully Connected Layer   Spatial/Temporal Embedding   Merging Cell

Time-assisted Spatial MSA   Temporal MSA

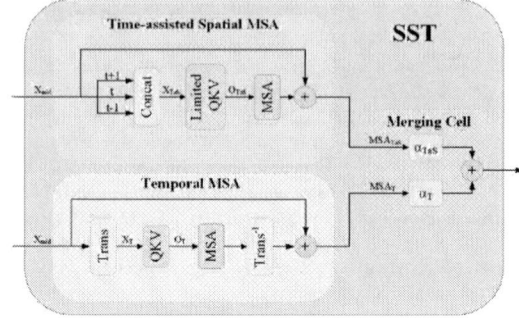

Fig. 3. Key Components of SST

B. Simplified Space-Time Transformer

As shown in Fig. 3, SST consists of 3 modules: Temporal MSA, TaS MSA and a Merging cell. The Temporal MSA module and the TaS MSA module works in parallel.

1) Time-assisted Spatial MSA:

To mitigate depth ambiguity, TaS MSA introduces temporal dimension information to traditional Spatial MSA. Considering the complexity of MSA calculation, TaS MSA only takes the information from the two nearest frames into account. Therefore, the transformed input $X_{TaS} \in \mathbb{R}^{F \times J \times 3C_f}$ is defined as:

$$X_{TaS}^t = \text{Concat}(X_{mid}^{t-1}, X_{mid}^t, X_{mid}^{t+1}) \quad (1)$$

where data image at time t is denoted by superscript. In order to reduce the computational complexity of self-attention of long sequence, we suggest Limited KQV calculation module. For Q_{TaS}^i, K_{TaS}^i, V_{TaS}^i in i-th head, the calculation is described as:

$$Q_{TaS}^i = X_{TaS}W_{TaS}^{(Q,i)}, K_{TaS}^i = X_{TaS}W_{TaS}^{(K,i)}, V_{TaS}^i = X_{TaS}W_{TaS}^{(V,i)} \quad (2)$$

where $W_{TaS}^{(Q,i)}$, $W_{TaS}^{(K,i)}$, $W_{TaS}^{(V,i)} \in \mathbb{R}^{3C_f \times C_f}$, $i \in [1, h]$, h represents the number of heads in MSA. The extra computing consumption brought by time dimension is therefore limited by maintaining the shape of Q_{TaS}^i, K_{TaS}^i and V_{TaS}^i same as traditional Spatial MSA. The output of Limited KQV of i-th head, i.e., O_{TaS}^i, and MSA still follows the traditional definition:

$$O_{TaS}^i = \text{Softmax}\left(\frac{Q_{TaS}^i(K_{TaS}^i)^T}{\sqrt{d_m}}\right)V_{TaS}^i \quad (3)$$

$$MSA_{TaS} = \text{Concat}(O_{TaS}^1, \dots, O_{TaS}^h)W_{TaS}^O + X_{mid} \quad (4)$$

where $W_{TaS}^O \in \mathbb{R}^{d_m \times d_m}$. A residual connection is constructed between the MSA output and the module input, and the addition result $MSA_{TaS} \in \mathbb{R}^{F \times J \times C_f}$ is passed to the Merging Cell.

TABLE I. QUANTITATIVE COMPARISON WITH THE SOTA METHOD ON HUMAN 3.6M DATASETS, USING REAL 2D DATA (GT-2D).

MPJPE	Dir.	Disc.	Eat	Greet	Phone	Photo	Pose	Pur.	Sit	SitD.	Smoke	Wait	WalkD	Walk	WalkT.	Avg
PoseFormer[3]	30.0	33.6	29.9	31.0	30.2	33.3	34.8	31.4	37.8	38.6	31.7	31.5	29.0	23.3	23.1	31.3
MHFormer[17]	27.7	32.1	29.1	28.9	30.0	33.9	33.0	31.2	37.0	39.3	30.0	31.0	29.4	22.2	23.0	30.5
MixSTE[9]	21.6	22.0	20.4	21.0	20.8	24.3	24.7	21.9	26.9	24.9	21.2	21.5	20.8	14.7	15.6	21.6
P-STMO[10]	28.5	30.1	28.6	27.9	29.8	33.2	31.3	27.8	36.0	37.4	29.7	29.5	28.1	21.0	21.0	29.3
MotionBert[11]	**15.9**	_17.3_	_16.9_	**14.6**	_16.8_	_18.6_	_18.6_	**18.4**	_22.0_	_21.8_	_17.3_	_16.9_	_16.1_	_10.5_	_11.4_	_16.9_
Ours	_16.3_	**17.2**	**16.8**	_15.3_	**15.7**	**17.8**	**18.5**	_19.2_	**21.8**	**19.9**	**17.0**	**15.8**	**15.6**	**10.2**	**10.7**	**16.5**

Bold: optimal, <u>underline</u>: suboptimal.

Depth ambiguity can be effectively alleviated by adding time dimension, because the relationships during the motions of human joints in consecutive video images can be comprehensively modeled using TaS MSA. Meanwhile, by utilizing the Limited KQV calculation method, training becomes more controllable and consumes less time, and deployment requires less hardware resources.

2) Temporal MSA:

Temporal MSA compensates for TaS MSA's shortcoming and improves the model's performance by focusing on individual human joints across consecutive video images. Thus, we regroup the data in X_{mid}, organizing the data of the same joint in different video frames together as $X_T \in \mathbb{R}^{J \times F \times C_f}$. The transformation from X_{mid} to X_T is described as

$$X_T = \text{Trans}(X_{mid}) \qquad (5)$$

MSA is also carried out for the X_T:

$$Q_T^i = X_T W_T^{(Q, i)}, K_T^i = X_T W_T^{(K, i)}, V_{TaS}^i = X_T W_T^{(V, i)} \qquad (6)$$

$$O_T^i = \text{Softmax}\left(\frac{Q_T^i (K_T^i)^T}{\sqrt{d_m}}\right) V_T^i \qquad (7)$$

$$MSA_T = \text{Trans}^{-1}(\text{Concat}(O_T^1, \dots, O_T^h) W_T^O) + X_{mid} \qquad (8)$$

where $W_T^{(Q, i)}, W_T^{(K, i)}, W_T^{(V, i)} \in \mathbb{R}^{C_f \times C_f}$. Same to TaS MSA, a residual connection is constructed between the output and input of the Temporal MSA. The result is passed to the Merging Cell.

3) Merging Cell

A couple of weight factors are utilized to combine the outputs of two MSA. The factors are learnable during training for a better fit. The merging cell can be described as follows:

$$\alpha = \text{Softmax}(\text{Split}(\text{Concat}(MSA_{TaS}, MSA_T) W_\alpha)) \qquad (9)$$

where $W_\alpha \in \mathbb{R}^{2C_f \times 2C_f}$ is the learnable parameter, and the Split() function regroups data of $F \times J \times 2C_f$ into $2 \times F \times J \times C_f$. Based on α, the iteration between input and output of SST of depth p and p+1 can be described as:

$$X_{mid}^{p+1} = \alpha^p[0] \odot MSA_{TaS}^p + \alpha^p[1] \odot MSA_T^p \qquad (10)$$

where the symbol '$\odot$' denotes the Hadamard product.

IV. EXPERIMENTS

A. Implementation Details

We train our SST model on Human3.6M dataset using 4 Nvidia Tesla V100 GPUs, with learning rate (lr) of 0.0002,

batch size of 16, number of attention heads (h) of 8, and intermediate feature dimension (C_f) of 512. We use subjects 1, 5, 6, 7, 8 for training and subjects 9, 11 for testing. We train SST for 120 epochs using Adam optimizer. Human 3.6M is split up into 15 daily actions. Actions of directions, discussion, greet, etc. are relatively distinct because these actions are usually standing, with less cover between the limbs and the torso. Actions of photo, pose, sitting down, etc. are more difficult due to estimate because of the extent of obscuration that can occur during these actions. In photo, hands often cover the head. During sitting down, legs may cover the torso.

B. Comparison with State-of-the-Art Methods

The model's performance is measured in millimeters using Mean Per Joint Position Error (MPJPE).

$$MPJPE = \frac{1}{N} \sum_{i=1}^{N} ||P_i - P_i^*||_2 \qquad (11)$$

The i_{th} anticipated 3D human joint position for a human pose representation with N joints is P_i, the i_{th} ground-truth (GT) value is P_i^*.

Table 1 shows the test results. GT 2D human joints are used as standard input of the second stage, so that the comparison among the performance of networks will not be influenced by different 2D human pose estimation in the first stage.

The best MPJPE (mm) of the prediction result of our SST method is 16.5 on the Human3.6M datasets, which is 0.4 more accurate than the best of MotionBert model (fine-tuned). SST achieves the SOTA (state-of-the-art) performance in accuracy, and outperforms the prior models in challenging poses like Photo and Sitting Down, reducing the MPJPE by 0.8 (4%) and 1.9 (9%) respectively. The experiment results demonstrate that our SST is quite robust in the face of extreme self-occlusion and depth ambiguity.

C. Ablations and Discussion

Fig. 4 displays the epoch-changing curve of MPJPE (as well as modifications of several Transformer modules) with regard to the epoch. Sea represents a network using the Seaformer [18] structure. Space_Cf512 represents a network using a parallel structure of the Spacial MSA and the Temporal MSA (hereinafter referred to as ST structure) with 512 intermediate feature dimensions. Space_Cf256 represents a network using ST structure with 256 intermediate feature dimensions. Space_ll3 combines 3 ST structure parallelly with intermediate feature dimension of 64, 128, 256 respectively. The black dots are the optimal MPJPE results for each model during training.

Fig. 4 shows that the ideal outcome occurs between epochs 40~60. The model exhibits a clear overfitting tendency in later epochs, and the MPJPE of the anticipated outcomes evaluated on Human 3.6M grows overall. In order

979-8-3503-6184-1/24 $31.00 © 2024 IEEE

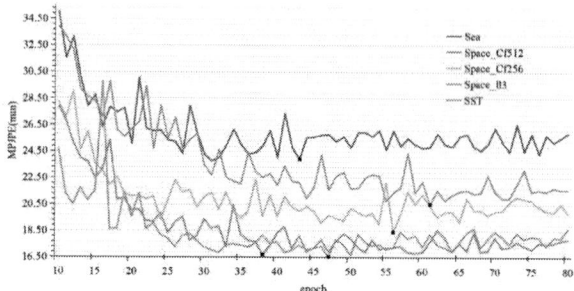

Fig. 4. MPJPE(mm) of different models in different training epochs

to avoid wasting computing resources, it is appropriate to set the training epoch to 70.

Based on the comparison shown in Fig. 4, it can also be concluded that SST uses the optimal network structure. Comparison between SST and Space_Cf512 shows that removing TaS MSA will cause a degradation of performance. Furthermore, comparing against Space_ll3, Space_Cf256, and Space_Cf512 reveals that the use of independent pipeline with higher C_f prominently improves the SST's inference capability. Accordingly, a method commonly used in general computer vision tasks, paralleling with lower C_f modules to get better model performance in order to reduce computation cost, is probably not suitable for this type of network structure.

V. CONCLUSION

In this paper we propose a novel TaS MSA and an SST model for 3D human pose estimation. This model realizes a better utilization of spatial-temporal information in 2D human joint motions. Besides, SST avoid excessive computing consumption by Limited KQV calculation. On the Human 3.6M dataset, this model achieves better performance than the current SOTA. The average MPJPE fell to 16.5 mm, which is 0.4 mm below the present SOTA technique, and the MPJPE in challenging poses, like Sitting Down, is reduced by 1.9 mm (9%). Additionally, the over-fitting issue that arose during model training is examined, and a better way of choosing the training epoch is suggested.

REFERENCES

[1] Julieta Martinez, Rayat Hossain, Javier Romero and James J. Little; "A simple yet effective baseline for 3dhuman pose estimation," in Proceedings of the IEEE International Conference on Computer Vision (ICCV), 2017, pp. 2640–2649.

[2] Dario Pavllo, Christoph Feichtenhofer, David Grangier and Michael Auli, "3D Human pose estimation in video with temporal convolutions and semi-supervised training," in Proceedings of the IEEE/CVF Conference on Computer Vision and Pattern Recognition (CVPR), 2019, pp. 7753–7762.

[3] Ce Zheng, Sijie Zhu, Matias Mendieta, Taojiannan Yang, Chen Chen and Zhengming Ding, "3D Human Pose Estimation with Spatial and Temporal Transformers," in Proceedings of the IEEE/CVF International Conference on Computer Vision (ICCV), 2021, pp. 11656–11665.

[4] Ashish Vaswani, Noam Shazeer, Niki Parmar, Jakob Uszkoreit, Llion Jones, Aidan N. Gomez, Łukasz Kaiser and Illia Polosukhin,

"Attention is All you Need," Neural Information Processing Systems, 2017, pp.5998–6008.

[5] Tianlang Chen, Chen Fang, Xiaohui Shen, Yiheng Zhu, Zhili Chen and Jiebo Luo, "Anatomy-Aware 3D Human Pose Estimation With Bone-Based Pose Decomposition," in IEEE Transactions on Circuits and Systems for Video Technology, 2022, vol. 32, no. 1, pp. 198–209.

[6] Kehong Gong, Jianfeng Zhang and Jiashi Feng, "PoseAug: A Differentiable Pose Augmentation Framework for 3D Human Pose Estimation," in Proceedings of the IEEE/CVF Conference on Computer Vision and Pattern Recognition (CVPR), 2021, pp. 8575–8584.

[7] Dario Pavllo, Christoph Feichtenhofer, David Grangier and Michael Auli, "3D Human Pose Estimation in Video with Temporal Convolutions and Semi-Supervised Training." in Proceedings of the IEEE/CVF Conference on Computer Vision and Pattern Recognition (CVPR), 2019, pp. 7753–7762.

[8] Wenkang Shan, Zhenhua Liu, Xinfeng Zhang, Zhao Wang, Kai Han, Shanshe Wang, Siwei Ma and Wen Gao, "Diffusion-Based 3D Human Pose Estimation with Multi-Hypothesis Aggrgation," in Proceedings of the IEEE/CVF International Conference on Computer Vision (ICCV), 2023, pp. 14761–14771.

[9] Jinlu Zhang, Zhigang Tu, Jianyu Yang, Yujin Chen and Junsong Yuan, "MixSTE: Seq2seq Mixed Spatio-Temporal Encoder for 3D Human Pose Estimation in Video," in Proceedings of the IEEE/CVF Conference on Computer Vision and Pattern Recognition (CVPR), 2022, pp. 13232–13242

[10] Wenkang Shan, Zhenhua Liu, Xinfeng Zhang, Shanshe Wang, Siwei Ma and Wen Gao , "P-STMO: Pre-Trained Spatial Temporal Many-to-One Model for 3D Human Pose Estimation," in In: Avidan, S., Brostow, G., Cissé, M., Farinella, G.M., Hassner, T. (eds) Computer Vision – ECCV 2022, vol 13665, pp. 461–478.

[11] Wentao Zhu, Xiaoxuan Ma, Zhaoyang Liu, Libin Liu, Wayne Wu, and Yizhou Wang. "Motionbert: Unified pretraining for human motion analysis," arXiv preprint arXiv:2210.06551, 2022.

[12] Alejandro Newell, Kaiyu Yang and Jia Deng, "Stacked Hourglass Networks for Human Pose Estimation," In: Leibe, B., Matas, J., Sebe, N., Welling, M. (eds) Computer Vision - ECCV 2016, vol 9912, pp. 483–499

[13] Georgios Pavlakos, Xiaowei Zhou, Konstantinos G. Derpanis and Kostas Daniilidis, "Coarse-to-fine volumetric prediction for single-image 3D human pose," in Proceedings of the IEEE Conference on Computer Vision and Pattern Recognition (CVPR), 2017, pp. 7025–7034.

[14] Zitian Wang, Xuecheng Nie, Xiaochao Qu, Yunpeng Chen and Si Liu, "Distribution-Aware Single-Stage Models for Multi-Person 3D Pose Estimation," in Proceedings of the IEEE/CVF Conference on Computer Vision and Pattern Recognition (CVPR), 2022, pp. 13096–13105.

[15] Alexey Dosovitskiy, Lucas Beyer, Alexander Kolesnikov, Dirk Weissenborn, Xiaohua Zhai, Thomas Unterthiner, Mostafa Dehghani, Matthias Minderer, Georg Heigold, Sylvain Gelly, Jakob Uszkoreit and Neil Houlsby, "An image is worth 16x16 words: Transformers for image recognition at scale." arXiv preprint arXiv:2010.11929, 2020.

[16] Catalin Ionescu, Dragos Papava, Vlad Olaru and Cristian Sminchisescu, "Human3.6M: Large Scale Datasets and Predictive Methods for 3D Human Sensing in Natural Environments," in IEEE Transactions on Pattern Analysis and Machine Intelligence, 2014, vol. 36, no. 7, pp. 1325–1339.

[17] Wenhao Li, Hong Liu, Hao Tang, Pichao Wang, and Luc Van Gool. "Mhformer: Multi-hypothesis transformer for 3d human pose estimation," In Proceedings of the IEEE/CVF Conference on Computer Vision and Pattern Recognition (CVPR), 2022, pp. 13147–13156.

[18] Qiang Wan, Zilong Huang, Jiachen Lu, Gang Yu and Li Zhang, "Seaformer: Squeeze-enhanced axial transformer for mobile semantic segmentation." arXiv preprint arXiv:2301.13156, 2023.

979-8-3503-6184-1/24 $31.00 © 2024 IEEE

SALTS: An Efficient and Flexible Self-Attention Accelerator with Long Token Support on FPGA

Kaiqi Chen[*1], Xinhua Shi[1], Jun Han[*1]

[1] State Key Laboratory of Integrated Chips and Systems, Fudan University

* Email: kqchen23@m.fudan.edu.cn, junhan@fudan.edu.cn

Abstract—**Transformers are trending in multiple machine-learning areas. However, as the state-of-the-art Transformer models call for a longer input token series, the performance degradation in self-attention modules becomes indispensable. In this paper, we propose SALTS, an FPGA-based hardware accelerator tailored for self-attention modules with long input tokens. We developed a highly flexible and reconfigurable hardware template for agile accelerator generation and deployment, featuring integer-only inference and long token support. Our best accelerator deployed on the Xilinx ZCU102 FPGA board reached 385.5 GOPS inference throughput and 39.08 GOPS/W energy efficiency. The energy efficiency outperforms CPU, GPU, and similar work by 81.42×, 5.99×, and 6.40×, respectively.**

Keywords—*Transformer, neural network accelerator, agile design*

I. INTRODUCTION

Recent years have witnessed an ever-rising popularity of using Transformer-based models in various computer vision and natural language processing tasks. The amazing performance of Transformers originates from its self-attention [1] mechanism, which enables Transformers to capture the global context information. However, the self-attention module comes with heavy computation workloads, for its complexity grows quadratically with the number of input tokens. This is gradually transforming into a problem as the required number of input tokens for state-of-the-art large language model (LLM) applications is getting considerably larger (e.g., a maximum of 8,192 tokens for GPT-4), making the self-attention module a performance bottleneck in Transformer inference. On top of that, a large number of tokens is detrimental to the system throughput. The softmax operation requires a complete row of the previous product matrix before it continues its computation. It is impractical to expand the size of the matrix multiplication unit to token length considering the limited hardware resources. Therefore, stalling the pipeline for multiple cycles to wait for a complete row of the product matrix is inevitable, which results in throughput degradation.

Meanwhile, the rapid iteration of algorithms brings about various Transformer architectures, which pose challenges for agile hardware design and deployment. To alleviate these problems, plenty of research has been conducted on designing hardware templates and using Field Programmable Gate Arrays (FPGAs) to quickly deploy hardware accelerators tailored for specific Transformers or self-attention modules [2-3]. However, previous works overlooked the stalling problem introduced by softmax and failed to achieve ideal throughput when dealing with extra-long input token series.

In this paper, we propose SALTS: an FPGA-based **S**elf-attention **A**ccelerator with **L**ong **T**oken **S**upport. The main contributions can be summarized as follows:

- **Long token support.** We inherited the design of [4] to avoid pipeline stalling introduced by the softmax operation, thus enabling our design to deal with long input token series without performance degradation.

- **Reconfigurable efficient template.** We implemented a flexible hardware template to accelerate self-attention modules with different sizes. This template can generate appropriate accelerator hardware designs for different Transformer models, achieving maximum throughput and fully exploiting FPGA onboard resources by simply adjusting design parameters.

- **Integer-only inference.** The hardware template only contains integral operations, with neglectable precision loss compared with float point inference, providing an extra performance boost.

Our system achieved 385.50 GOPS inference throughput and 39.08 GOPS/W energy efficiency on Xilinx ZCU102 at 100MHz.

II. HARDWARE DESIGN

A. Overview

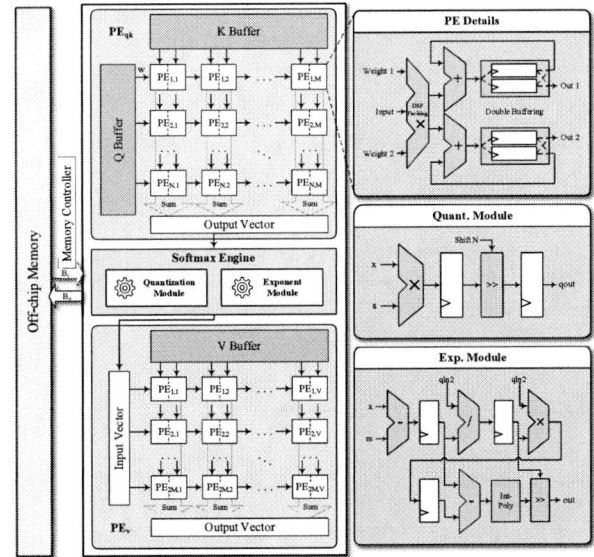

Fig. 1. The overall hardware architecture of SALTS

Fig. 1 depicts the overall architecture of our design, including two homogeneous processing element (PE) arrays, an improved softmax engine, and memory controllers. The system is designed as a template with abundant reconfigurable parameters, to accommodate self-attention modules with different sizes.

979-8-3503-6184-1/24 $31.00 © 2024 IEEE

B. Softmax Engine with Long Token Support

As in (1) and (2), the original softmax operation requires traversing the input vector three times: element-wise exponent, reduced summation, and element-wise division. Q_k stands for the k-th row of matrix Q. This process is expensive for hardware resources and detrimental to throughput.

$$x_i = Q_k[:]K^T[:,i] \quad (1)$$

$$softmax_k(x_i) = e^{x_i} / \Sigma_i e^{x_i} \quad (2)$$

We incorporated the algorithm optimization proposed by [4], in which softmax can be completed along with self-attention in a single traverse on the input vector, as presented in (3) - (5), where m and d are both initialized as 0.

$$m_i = max(m_{i-1} - x_i) \quad (3)$$

$$d_i = d_{i-1}e^{m_{i-1}-m_i} + e^{x_i-m_i} \quad (4)$$

$$o_i = (o_{i-1}d_{i-1}e^{m_{i-1}-m_i} + e^{x_i-m_i} V[i,:]) / d_i \quad (5)$$

This method enables a fully pipelined hardware design. i) PE_{qk} generates a segment in a row of the output matrix; ii) the segment performs calculations along the data path; iii) PE_v accumulates its output to the partial sum. The self-attention operation for a row finishes once the last segment is accumulated to the partial sum. In summary, a long series of input tokens can be divided into several segments and processed individually, the segment length can be adjusted according to hardware resource limitations, thus saving hardware resources and increasing throughput.

C. Integer-only Inference

Thanks to the mature quantization technique, Transformer model parameters can be converted from 32-bit floating point (FP-32) to 8-bit integers (INT-8) or even lower bit-width without significant precision loss. We assume the self-attention module was already quantized to INT-8 or lower precision with symmetric quantization, as in (6):

$$q = \lfloor x \times s \rceil \quad (6)$$

where x is the original floating-point value, s is a floating-point scaling factor, and q is the quantized integer. For hardware simplicity, the floating-point operation is eliminated by bit shifting as shown in (7) and (8), in which s' is the new scaling factor calculated offline, and C is an adequately large positive integer constant. All other floating-point divisions or multiplications encountered in our design utilize this method.

$$s' = \lfloor s \times 2^C \rceil \quad (7)$$

$$q = (x \times s') \gg C \quad (8)$$

Another challenge is to represent the exponent function with integer operations. We followed the practice of I-BERT [5], in which an integer polynomial and a series of extra integer operations fit the exponent function. The "Exp. Module" in Fig. 1 shows the detailed implementation.

D. PE Design

Our system contains two homogeneous PE arrays, performing vector-matrix multiplication. The two PE arrays are implemented as weight-stationary systolic arrays. The operands are directly loaded from the according buffer, and the input vectors stream in and pass along the systolic arrays to perform vector-matrix multiplication with each column of the weights. Finally, an adder tree accumulates the results, and output through ping-pong buffering to support pipelining.

As mentioned in Section II.C, the model parameters are quantized to INT-8 or lower precision integers. Yet, the Digital Signal Processors (DSPs) on FPGA boards are often designed for larger bit-width operations (e.g., an 18×27 multiplier for Xilinx devices). To fully utilize the DSP resources on FPGA, we adopted the DSP packing technique to combine two INT-8×INT-8 multiplications into one INT-8×INT-16 multiplication that can be handled by one DSP operation. Specifically, in alignment with our weight-stationary PE design, we merged two weight columns to form the INT-16 multiplicands and the inputs as the original INT-8 multipliers (likewise for PE_v).

E. Hardware Template

Our design is highly flexible, the reconfigurable design parameters are listed in Table I. These parameters are categorized into three classes: PE Size (PS), Memory Controller (MC), and Timing Constraints (TC). PS determines the size of the two systolic arrays and quantization bit width; MC parameters are configured based on the communication bandwidth with off-chip memory, and they decide how fast data can be transferred from/to off-chip memory; TC focuses on meeting the timing requirements of large combinational logic in the hardware design, such as the number of pipeline stages in adder trees and multipliers.

The template design provides finer-grained control over resource utilization and accelerator performance, enabling agile accelerator development tailored for different models.

TABLE I.   HARDWARE TEMPLATE DESIGN PARAMETERS

Category	Notation	Description
PE Size	N	Number of rows in PE_{qk}.
	M	Number of columns in PE_{qk}.
	V	Number of columns in PE_v.
	w	Quantization bit width.
Memory Controller	B_i	Number of Bytes loaded from off-chip memory at once.
	B_o	Number of Bytes stored to off-chip memory at once.
Timing Constraints	$RegLst^a$	Number of pipeline stages in adder trees/multipliers and so on.

a. A set of parameters, details omitted for simplicity.

III. EXPERIMENTS

A. Experimental Setup

In this work, we described our hardware template with Chisel and then generated it to Verilog to support our highly reconfigurable design. The FPGA board we chose was Xilinx ZCU102, we used Vivado 2019.2 to synthesize and implement our design. For Transformer models and datasets, we used the BERT-base [6] and the ViT [7] as our example models, with THUCNews [8] and CIFAR-10 as the datasets, respectively.

B. Integer Inference Precision Analysis

To evaluate the precision of our integer-only inference, we conducted two experiments comparing the conventional FP-32 inference results and our integer-only inference results. We evaluated text and image classification tasks on BERT-base and ViT, respectively. Table II shows that our integer-only inference method introduced neglectable precision loss compared with the float point inference.

979-8-3503-6184-1/24 $31.00 ©2024 IEEE

TABLE II. PRECISION COMPARISON WITH FLOAT-POINT INFERENCE

Model	FP Precision	Int Precision	Precision Loss
BERT	89.16%	88.65%	0.51%
ViT	98.94%	98.29%	0.65%

C. Comparison between Different Design Parameters

We explored the relationship between resource utilization and the corresponding accelerator throughput with different design parameters. Each configuration was denoted with a three-element tuple, representing N, M, and V, respectively. Bit-width was set to 8, MC and TC parameters were set to reasonable values and stayed unchanged throughout the experiments.

Fig. 2 shows the comparison results. Resource utilization is positively correlated to hardware throughput when utilization is relatively low. However, accelerators with close resource utilization can also vary in performance. Take (16, 32, 16) and (32, 32, 8) as an example, (32, 32, 8) has a large PE_{qk} and small PE_v, resulting in unreasonable pipelining as shown in Fig. 3. In contrast, (16, 32, 16) is a much more reasonable configuration with balanced PE array sizes, the operations of softmax and PE_v are covered by PE_{qk}, thus reaching a higher throughput.

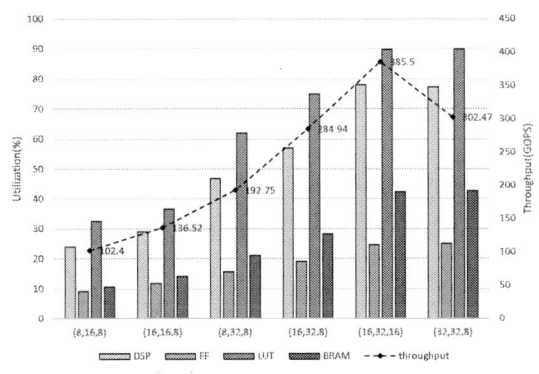

Fig. 2. Comparison between different design parameters

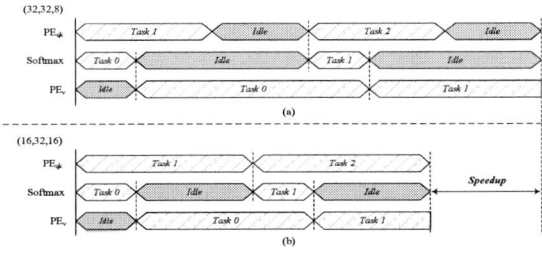

Fig. 3. Pipelining comparison between two design configurations

D. Performance on Long Token Series

We evaluated our hardware (design parameters are set to (16, 32, 16)) performance under different input token lengths. Table III shows that the throughput of our design is not affected by the number of input tokens across a span of a thousand tokens, owing to the softmax design and appropriate pipelining.

TABLE III. PERFORMANCE WITH DIFFERENT TOKEN LENGTHS

Token Length	192	512	1024
Throughput (GOPS)	385.50	385.50	385.50

E. Comparison with other Hardware

We selected the design parameters with the highest throughput, i.e. (16, 32, 16), and compared its performance with Intel Core i7-12700 CPU, NVIDIA GeForce RTX 3090, and similar work on self-attention hardware accelerator design proposed by [2]. We set the working frequency of our design at 100MHz. Throughput was calculated as the ratio of the number of operations to the execution time. We used HWINFO to measure CPU power, the Nvidia-smi tool to measure GPU power, and the Xilinx Power Estimator to measure the power of our design. Our work reached 10.74×, 0.59×, and 2.12× throughput, and 81.42×, 5.99×, and 6.40× energy efficiency compared to CPU, GPU, and FPGA, respectively.

TABLE IV. COMPARISON WITH OTHER HARDWARE

Hardware	Throughput (GOPS)	Energy Efficiency (GOPS/W)
CPU	33.00	0.48
GPU	651.51	6.52
FPGA [2]	182.22	6.11
Ours	385.50	**39.08**

IV. CONCLUSION

In this work, we proposed SALTS, a highly flexible and reconfigurable hardware template tailored for self-attention modules with long token support, featuring integer-only inference and can adapt to different Transformer models. Our design fully utilized FPGA onboard resources and reached higher throughput with proper pipelining technique. SALTS reached 385.50 GOPS throughput and 39.08 GOPS/W energy efficiency on Xilinx ZCU102 FPGA board at 100MHz. Its energy efficiency outperforms CPU, GPU, and similar work by 81.42×, 5.99×, and 6.40×, respectively.

ACKNOWLEDGMENT

This work was supported by the National Natural Science Foundation of China under Grant 61934002 and 62234008.

REFERENCES

[1] Vaswani A, Shazeer N, Parmar N, et al. Attention is all you need[J]. Advances in neural information processing systems, 2017, 30.

[2] Liu X, Jiang J, Xu J, et al. Evaluating a New Attention Framework Based on Matrix Blocking for Attention Models on FPGAs[C]. 2022 IEEE 34th International Conference on Tools with Artificial Intelligence (ICTAI). IEEE, 2022: 607-615.

[3] Gong Y, Xu Z, He Z, et al. N3h-core: Neuron-designed neural network accelerator via fpga-based heterogeneous computing cores[C]. Proceedings of the 2022 ACM/SIGDA International Symposium on Field-Programmable Gate Arrays. 2022: 112-122.

[4] Dao T, Fu D, Ermon S, et al. Flashattention: Fast and memory-efficient exact attention with io-awareness[J]. Advances in Neural Information Processing Systems, 2022, 35: 16344-16359.

[5] Kim S, Gholami A, Yao Z, et al. I-bert: Integer-only bert quantization[C]. International conference on machine learning. PMLR, 2021: 5506-5518.

[6] Devlin J, Chang M W, Lee K, et al. Bert: Pre-training of deep bidirectional transformers for language understanding[J]. arXiv preprint arXiv:1810.04805, 2018.

[7] Dosovitskiy A, Beyer L, Kolesnikov A, et al. An image is worth 16x16 words: Transformers for image recognition at scale[J]. arXiv preprint arXiv:2010.11929, 2020.

[8] Maosong Sun, Jingyang Li, Zhipeng Guo, et al. THUCTC: An Efficient Chinese Text Classifier. 2016.

RISC-V Neural Network Instruction Design and Simulation with Cache Scheduling via ROCC Interface

Siyao Dai[1], Zikang Zhou[1], Jun Han*[1]

[1] State Key Laboratory of Integrated Chips and Systems, Fudan University, Shanghai 200433, China

Email: junhan@fudan.edu.cn

Abstract—In today's era of information technology, Convolutional Neural Networks (CNNs) have found wide applications in fields such as object detection and object recognition. However, due to the massive computational requirements of CNNs, hardware acceleration and harnessing the parallelism of convolutional calculations have become research hotspots in contemporary society. Compared to FPGA and ASIC, the combination of General-Purpose Processors (Central Processing Unit, CPU) and CNN accelerators (coprocessors) offers both versatility and flexibility. Therefore, we propose a RISC-V-based neural network accelerator design. We customized the DSA interfaces to communicate with the CPU and cache. Besides, we adopted a bottom-up method to find the best mapping scheme and explored mapping schemes under different cache sizes. Our experiments show that the proposed system can improve performance by over 90% compared to performing convolution operations with the O3CPU.

Keywords—RISC-V, accelerator, CNN, mapping

I. INTRODUCTION

With the rapid development of AI technology, the development and deployment of AI applications on Internet of Things (IoT) node devices has become a prevailing trend. The Artificial Intelligence of Things (AIoT) combines the technical advantages of AI and IoT, aiming to create more efficient IoT operations and improve human-computer interaction. To reduce computing latency and network congestion, researchers have transferred a large number of computing tasks from cloud servers to edge devices. How to improve the AI computing performance while meeting the requirements of device power consumption, area constraints, and cost is an important research topic.

Due to the simplicity of the RISC-V architecture instruction set and its fast processing speed, it can provide high performance at relatively low power consumption, making it suitable for edge devices. In such edge devices, due to the limited on-chip buffer capacity, data needs to be partitioned into multiple tiles to be loaded into the buffer. For a given CNN layer, the accelerator can employ different methods to schedule memory access operations, such as selecting different tile sizes. We call the process of data partitioning and distribution as mapping in this paper. Mapping schemes have an implicit impact on performance and energy efficiency. To fully reduce memory access latency and energy consumption, we need to find an efficient scheduling strategy.

Convolutional layers and matrix multiplication account for most computations in convolutional neural networks. For example, in the Inception network, convolution operations constitute more than 98% of the computations [1]. So we need

to explore the scheduling optimization of convolutional operators. Previous mapping schemes mainly focus on maximizing PE utilization [2]. However, DRAM costs show a strong correlation with the overall accelerator-level energy and cycle results, whereas the PE utilization has a relatively weaker correlation [3]. Therefore, chasing after the PE utilization sometimes leads to a less efficient mapping scheme.

To address the above problem, this paper proposes a neural network accelerator based on RISC-V extended instruction set and explores memory access scheduling schemes under different cache sizes. The accelerator we designed supports the compilation and execution of common networks such as GoogLeNet and ResNet. The following highlights the key contributions of our work:

- We designed a CNN accelerator based on RISC-V and customized two DSA interfaces to communicate with the CPU and cache.

- We explored mapping schemes under different cache sizes. We generated a scheduling table and filled it from bottom to top, selecting the solutions that meet the constraints and have the minimum cost.

The rest of this paper is organized as follows. Section II describes the details of our accelerator. Section III shows the mapping method. Section IV shows the experimental results. Section V makes the summary.

II. PROPOSED ARCHITECTURE

A. Instruction design proposal

In this paper, we proposed some customized instructions based on the RISC-V Specifications. We designed custom instructions for six types of operators: convolution, relu, maxpool, adaptivepool, linear, and softmax. For the convolution instruction, the im2col operation is eliminated at the instruction level, leaving space for accelerators that can directly compute convolutions or include other optimization operations. We use fixed-length instructions for RISC-V instruction set extension, encoding with the Custom-1 instruction [4]. We set func7 as the hint field, which indicates whether the registers corresponding to rs1, rs2, and rd are CPU registers or DSA registers. For operators with more information, such as convolution, multiple instructions are used to convey information.

For operators represented by multiple instructions, a field should be added to the instructions to differentiate these instructions. Since the custom DSA instructions do not require register write-back, the rd field in the instructions is also used to transmit input information. The following figure illustrates the custom convolution instruction.

979-8-3503-6184-1/24 $31.00 © 2024 IEEE

31	28 27	25 24	20 19	15 14	12 11	7 6	0
hint_conv	hint_matrix	param2	param1	hint	param3	opcode	

Fig. 1. Custom convolution instruction.

B. Co-Processor Architecture

We implemented a DSA module on the Gem5 platform. The CPU and DSA form a heterogeneous system. The DSA includes memory access logic, decoding logic, computation logic, and buffers.

The communication interface between the DSA accelerator and the CPU is designed based on the ROCC (Rocket Custom Coprocessor) interface of the Rocket Core. The ROCC interface is tightly integrated with the processor, enabling direct access to the processor's registers and memory space without the need for complex intermediary layers.

The DSA module in this paper implements two ports: the dsa module port, which connects to the CPU, and the dsa memory port, which connects to the memory system. In Gem5, different components use packets to interact with each other [5]. The interface connected to the memory system is connected to the L2 Cache.

TABLE I.   CUSTOM DSA PORTS AND THEIR FUNCTIONAL FUNCTIONS

Port name	Port type	Function	Functionality
dsa module port	Response Port	recvTiming Req	Receive the request sent by the CPU
		sendTiming Resp	Send response to cpu
		recvResp Retry	If sending the response fails, resend it.
dsa memory port	Request Port	recvTiming Resp	Receive the response sent by the memory system.
		sendTiming Req	Send a request to the memory system to read/write memory.
		recvReq Retry	If sending the request fails, resend it.

Fig. 2 illustrates the workflow of DSA. In this paper, when an operator corresponds to multiple instructions, the instruction that needs to access memory to write back results returns a Complete signal after the DSA module finishes the computation and writes the result to the output address. Other instructions return a Complete signal after decoding.

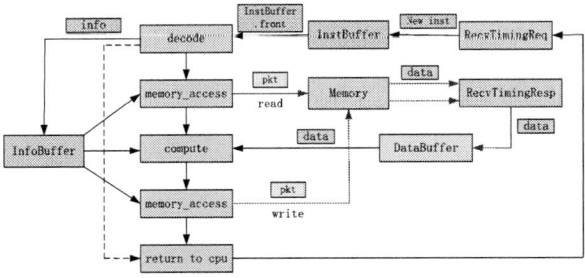

Fig. 2. The workflow of DSA module.

III. MEMORY ACCESS SCHEDULING SCHEME

In this paper, we try to find a dataflow and mapping combination that gives the minimum energy or cycle cost [3].

In CNN, a convolutional layer includes three types of data: input, weight and output. A convolutional layer includes seven nested for-loops. So the data in a layer consists of seven dimensional parameters including output channel(K), batch size(B), output width(P), output height(Q), input channel(C), kernel width(R), and kernel height(S). Input height and width can be calculated using the parameters mentioned above.

In order to minimize the energy and latency costs, we need to find an efficient mapping scheme to reduce data movement. Fig. 3 shows the memory hierarchy of our accelerator. The accelerator includes three levels of memory: off-chip memory(DRAM), L2 cache and PE array.

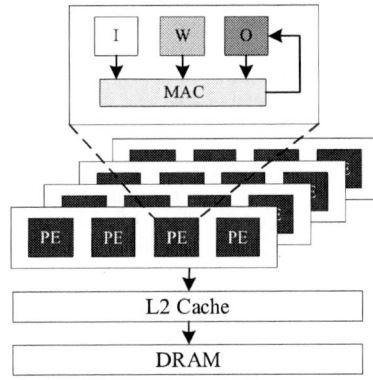

Fig. 3. The accelerator has three levels of memory hierarchy.

In order to find the best mapping scheme, we made a scheduling table based on the hierarchical level of accelerator components and layer parameters. The data in the table represents the size of data that needs to be stored in the corresponding buffer. The data in the table must satisfy both hardware constraints and computational parameter constraints. The size of input tile, kernel tile, and output tile can be calculated according to the table and can not exceed the buffer size of the level. For a certain dimension, the multiplication of the data sizes in each buffer must equal the parameter value in that dimension.

TABLE II. shows a simplified example of the scheduling table with a 2*2 output tile and a 3*3 kernel tile. Other parameters are set to 1. In this case, the product of all numbers in the second column must be equal to P = 2. Using the values in the second row, we can calculate that the sizes of the input tile, weight tile, and output tile in the L2 CACHE are 3, 3, and 1, respectively.

TABLE II.   A SIMPLIFIED EXAMPLE OF THE SCHEDULING TABLE

Parameters	P	Q	R	S
L2 cache	1	1	3	1
DRAM	2	2	1	3

We filled in the table mentioned above from the bottom row to minimize the costs of moving data from DRAM to the L2 cache. We generated all possible combinations of mapped numbers for the bottom row. Then we computed the size of data tiles and checked the hardware constraints. We excluded all cases that did not meet the hardware constraints. Then we evaluated the cost of the remaining cases and selected the one with the minimum cost. For the next target level, we repeated the steps mentioned above. To avoid a large number of prime numbers in the mapping table, at each optimization step, we selected another scheduling candidate that offered the largest factorization degree. When processing the top of the scheduling table, we selected the scheduling strategy with the fewest memory accesses. In our experiment, the input and output data are arranged in memory in the NCHW format. To improve the cache hit rate, we place P in the innermost loop. Similarly, since the weights are stored in KCSR format, we place R inside the loop of S.

IV. Experimental Results and Discussion

We used HHB to compile ONNX files, resulting in network weights and binary files. To pass custom DSA instructions, we added inline assembly code in the C++ functions. Besides, we performed preprocessing on the images before feeding them into the network, including cropping, resizing, and normalization. We used the Gem5 platform for simulation, employing the O3CPU model.

Since convolutional layers account for the majority of computation time in CNNs, we focus our discussion on convolutional layers due to the limited length of this article. We selected the first convolutional layer of AlexNet (called conv1) and ResNet18 (called conv2) for testing. We tested the computation time with the accelerator disabled, during which the O3 CPU performed the convolution operations. In our simulation model, the parallelism on both the input channels and output channels is set to 32. We chose 128KB and 256KB cache sizes for testing. The loop order found by our mapping scheme is shown in Fig. 4.

```
for (q1=0; q1<Q1; q1++)        // different output tiles
  for (p1=0; p1<P1; p1++)
    for (k=0; k<K; k++)        // output channel
      for (s=0; s<S; s++)      // kernel height
        for (r=0; r<R; r++)    // kernel width
          for (q=0; q<Q; q++)  // output height
            for (p=0; p<P; p++) // output width
```

Fig. 4.   The loop order found by our mapping scheme.

As shown in the table below, different cache sizes correspond to different sizes of data tiles. The larger the cache, the larger the data tile it can hold, reducing the number of times data needs to be transferred from DRAM to cache and thereby decreasing computation time.

TABLE III.   Mapping results for different cache sizes.

Cache size/KB	Layer	Output size	Output tile
128KB	conv1	64*55*55	32*11*5
	conv2	64*112*112	32*16*16
256KB	conv1	64*55*55	32*11*11
	conv2	64*112*112	32*28*16

We tested the runtime of the convolution layer in each case. Fig. 5 shows the computation time of convolutional layers under different cases. When computations are performed by the CPU, the convolutional layers take a longer time. Enabling the accelerator can reduce computation time by more than 90%. Compared to the situation where the accelerator is enabled but mapping is not utilized, employing the mapping strategy can improve performance by 15%.

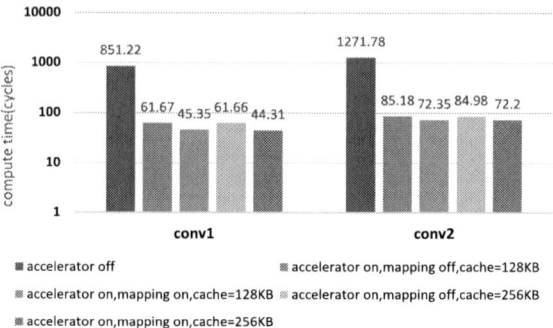

Fig. 5.   The computation time of convolutional layers under different cases.

V. Summary

In this paper, we propose a RISC-V based neural network accelerator that supports operators such as convolution, pooling, and fully connected layers. Besides, we optimize the data flow using a mapping strategy. For convolution operators, the accelerator we designed can reduce computation time by over 90%, demonstrating excellent acceleration performance.

VI. Acknowledgement

This work was supported by the National Natural Science Foundation of China under Grant 61934002 and 62234008.

References

[1] E. Talpes et al., "Compute Solution for Tesla's Full Self-Driving Computer," in IEEE Micro, vol. 40, no. 2, pp. 25-35, 1 March-April 2020, doi: 10.1109/MM.2020.2975764.

[2] S. Dave, Y. Kim, S. Avancha, K. Lee, and A. Shrivastava, "dMazeRunner: Executing perfectly nested loops on dataflow accelerators," ACM Trans.Embedded Comput. Syst., vol. 18, no. 5s, pp. 1–27, Oct. 2019.

[3] C. Park, B. Kim, S. Ryu and W. J. Song, "NeuroSpector: Systematic Optimization of Dataflow Scheduling in DNN Accelerators," in IEEE Transactions on Parallel and Distributed Systems, vol. 34, no. 8, pp. 2279-2294, Aug. 2023, doi: 10.1109/TPDS.2023.3283491.

[4] ANDREW WATERMAN S I, Krste Asanović. The RISC-V Instruction Set Manual Volume I: Unprivileged ISA[EB/OL]. [2019-12-13].https://riscv.org/wp-content/uploads/2019/12/riscv-spec-20191213.pdf.

[5] LOWE-POWER J. Creating SimObjects in the memory system[EB/OL].[2024-05-14]. https://www.gem5.org/documentation/learning_gem5/part2/memoryobject/.

Impact of external magnetic interference on the performance of MRAM-based neuromorphic computing

Yingtong He [1], Suihuan An [1], Yu Chen [1], Xue Zhou [1], Xihui Yuan [1], Weidong Zhang [2], Zheng Chai *[1], and Tai Min [1]

[1] State Key Laboratory for Mechanical Behavior of Materials, and School of Materials Science and Engineering, Xi'an Jiaotong University, Xi'an, China, [2] School of Engineering, Liverpool John Moores University, UK

* Email: zheng.chai@xjtu.edu.cn

Abstract—The Magneto-resistive random access memories (MRAMs) have the potential to be used in neuromorphic computing based on the computing in memory (CiM) architecture. However, as a spintronic device, the resistance state will be interfered by the external magnetic field, thus causing inference degradation. Error correction code or magnetic shielding could increase its immunity, but at additional cost. Neural networks have intrinsic error tolerance, which could help relax the standard of device-level immunity, but that's seldom investigated. In this work, we comprehensively conducted an evaluation of the impact of external magnetic interference on the performance of MRAM-based neuromorphic computing, and found that the error tolerance could help protect a multilayer perceptron (MLP) neural network from showing degraded accuracy. Moreover, such tolerance differs between weights in different layers and between bits with different significance. Based on those observations, we further proposed an optimization method for immunity against field interference with reduced cost. This work helps improve the usefulness of MRAM-based neural networks in practical environments.

Keywords—*spintronics, MRAM, magnetic field, neural network, error tolerance*

I. INTRODUCTION

As artificial intelligence (AI) and neuromorphic computing continue to revolutionize people's lives, the development of robust and reliable memory technologies becomes increasingly crucial. Novel memory technologies, such as resistive random access memory (RRAM), phase change memory (PCM), and magneto-resistive random access memory (MRAM), are drawing widespread attention due to their unique structure and outstanding performance. MRAM, in particular, receives increasing interest due to its compelling features, including "unlimited" endurance, fast operating speed, excellent stability and compatibility with CMOS process[1][2]. For example, in 2022, Samsung demonstrated the MRAM based in-memory computing, which achieved an accuracy of 98% in classification of hand-written digits and a 93.4% accuracy in face detection[3]. The advancements in MRAM-based neuromorphic computing exemplify the ongoing efforts to create highly capable and reliable AI systems, ultimately enhancing the reliability and trustworthiness of these transformative technologies.

However, as a spintronic memory technology, MRAM is inherently susceptible to external magnetic field interference. The resistance state of MRAM, which represents the stored data, can be inadvertently altered by the interaction with

external magnetic fields. It may cause reverse magnetization in an uncontrolled manner, potentially leading to readout errors. Methods such as error correction code (ECC) and shielding structures have been proposed to mitigate the impact of magnetic interference for the memory application of MRAMs so as to maintain the reliability and integrity of the overall system[4][5], at the cost of increased complexity and overhead. On the other hand, fault tolerance (FT) is an important property of neural networks that ensures their reliability even when significant portions of a network are lost or faulty. Such FT could potentially allow neural network to tolerate the faulty switching of one or a few MRAMs, thus easing the requirement for magnetic field immunity against external interference. In addition to ECC and shielding, leveraging the inherent FT of neural networks could be an alternative way to reduce the cost and improve energy efficiency of MRAM-based neuromorphic chips. Nevertheless, research in this field is still in its early stages, especially on the evaluation of chip's FT against external magnetic field.

In this paper, we employ the NeuroSim framework to simulate and evaluate the performance of an MRAM-based multilayer perceptron (MLP) neural network. We investigate the impact of external magnetic disturbance on different weight layers and on different weight bits. Furthermore, we propose optimization strategies for the design of MRAM in magnetic interference environments, aiming to provide support for practical implementation of MRAM-based neuromorphic computing at minimum cost. This work provides a new method to enhance the application of MRAM-based neurocomputing chips in practical environments.

II. DEVICE AND SIMULATION

A. Device

In this work, MRAM devices function as the weights in the MLP. Specifically, the MRAM belongs to the spin transfer torque MRAM (STT-MRAM) type, which is the mainstream MRAM technology based on the STT effect and offers improved scalability compared to conventional MRAM technologies[6]. The magnetic reversal can be affected by electric field or external magnetic field. When the intensity of the external magnetic field exceeds a specific threshold, the magnetic moment of the free layer in the magnetic tunnel junction (MTJ) will switch, causing a change in the state of the STT-MRAM. The STT effect enables a highly efficient and fast writing scheme[7], making STT-MRAM suitable for hardware neural network systems where higher speed and lower energy consumption are always desirable. Naik demonstrated the magnetic immunity of 22nm FD-SOI 40Mb eMRAM, whose tolerable magnetic field for the stand-by state is around 2,200 Oe, corresponding to a predicted bit error rate (BER) of 1E+4 ppm at 25℃[8]. Additionally, when the magnetic field intensity increases to 2,500 Oe, BER is 1E+5 ppm

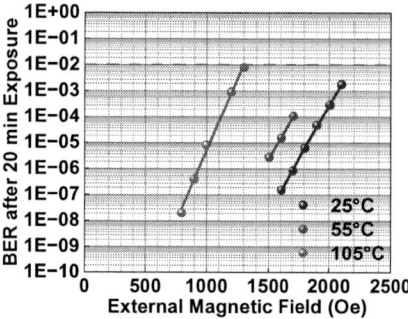

Fig.1 BER of stand-by magnetic immunity under varied field at 25,55 and 105 ℃ for 20 min exposure. The predicted magnetic field of BER=1E+4ppm at 25℃ is 2,200 Oe[8].

approximately, predictively.

B. NeuroSim

NeuroSim[9][10][11] is a comprehensive C++ based neural network simulation software platform which supports a wide range of mainstream and emerging memory technologies, including digital non-volatile memories such as MRAM. The platform also accommodates various machine learning neural networks architectures. This paper uses the MLP+NeuroSim simulator to evaluate the classification performance on the MNIST handwritten datasets with three-layer MLP networks. The MNIST dataset consists of 70,000 samples, with a ratio of training-to-testing standing at 6:1. In this work, the MLP is represented by a neural network where each weight is constituted by 4 bits, and each bit is represented by an MTJ device, as illustrated in Fig.2.

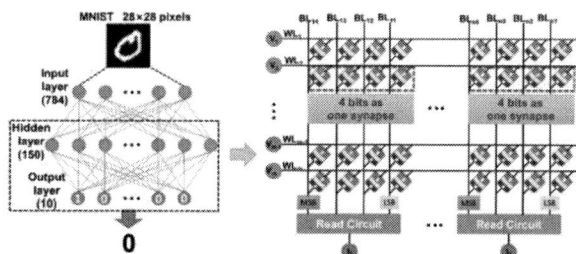

Fig.2 MLP hardware-scheme for recognizing MNIST handwritten numbers.

III. MODELLING AND SIMULATION

A. Device-level magnetic field immunity

For the commercialization of MRAM memory chips, the magnetic field immunity is always a critical consideration, since external magnetic field can affect the BER of MRAM devices, and potentially leading to read-out failure. The magnetic field immunity refers to the ability of a single MRAM memory cell or device to withstand external magnetic fields and maintain its resistance state unchanged. It is usually characterized and reported in terms of the intensity of magnetic field and duration that can be tolerated before a bit error occurs.

B. The simulation method to evaluate the impact of external magnetic field interference on MLP

We evaluate the impact of magnetic field interference (P→AP or AP→P) on MLP's accuracy, via applying the interference at all weights in the layer, in the following procedure:

(1) The MLP is trained using 60000 28×28 pixels handwritten digits, and then tested using 10000 handwritten digits, for the accuracy without interference.

(2) The direction of interference (P→AP or AP→P) will determine if the "0" bits are switched to "1" or the other way around.

(3) During interference, the probability of a weight bit being switched is determined by the strength of the interference.

(4) The interfered MLP is tested again for the accuracy after interference, which will be compared with the accuracy without interference obtained in (1).

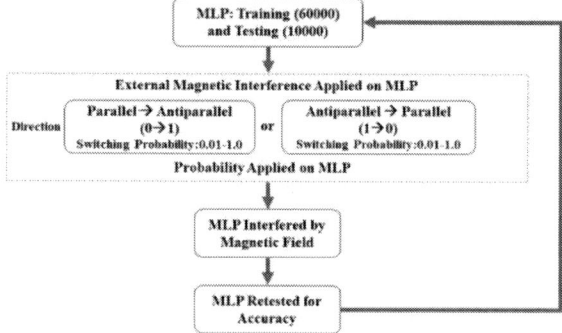

Fig.3 Overview of the proposed simulation scheme.

C. Impact of external magnetic field interference on the weights in MLP

First of all, we investigate the impact of external magnetic field interference on the weights in MLP. Assuming all the AP-states will be probabilistically switched to the P-state (or vice versa), the probability is determined by the external magnetic field. Specifically, when a 2,200 Oe magnetic field is applied to the MLP at room temperature, the interference-induced switching probability is approximately 0.01. P_{on} represents the probability of switching the MTJ state from AP-state to P-state, while P_{off} denotes the probability of switching from P-state to AP-state. As illustrated in Fig.4(a) and (b), when changing P_{on}, P_{off} remains 0 (and vice versa), the post-training distribution ratio between the AP-state and the P-state can be observed. When both P_{on} and P_{off} are set to 0, the proportion of the P-state is 41.95% while the AP-state is 59.05%. As the magnetic field intensity increases, and thus the probability of switching (either P_{on} or P_{off}) increases, the proportion of P-states or AP-states will correspondingly increase, depending on the direction of the magnetic field. Fig.4(c) demonstrates that as the switching probability increases, both interference directions exhibit similar trends: the accuracy of the MLP decreases. It is worth noting that an accuracy of approximately 90% could still be maintained if 1% of the P-state or AP-state weight bits are altered to the other state. This suggests that even if the MRAMs are suffering a ~2,200 Oe magnetic field with high BER of 10^4

ppm, the actual accuracy degradation of the neural network will only be less than 1%. It strongly indicates that the correction/shielding methods could be relaxed for the MRAM-based neural networks implementations, compared to the requirements for MRAM memory chips. However, as the magnetic field intensity increases to around 2500Oe, i.e. the switching probability is approximately 0.1, the accuracy of the MLP drastically drops to less than 50%.

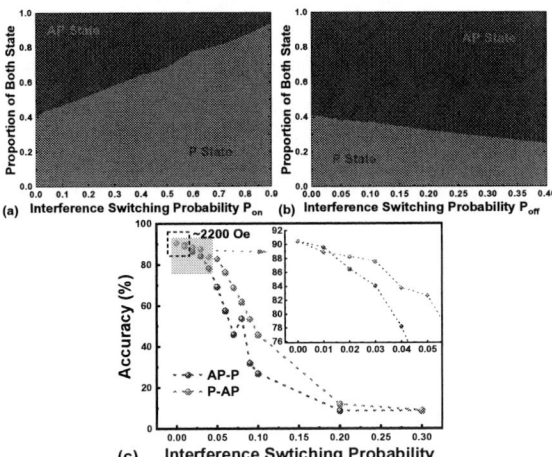

Fig.4 (a)Proportion of both P-state and AP-state when changing P_{on}, P_{off} remains 0; (b) Proportion of both P-state and AP-state when changing P_{off}, P_{on} remains 0; (c)Accuracy of switching probability of all bits on MLP, with a magnified small-probability region as an inset.

D. Impact of external magnetic field interference on weights in different layers in the MLP

We evaluate the impact of magnetic field interference (P→AP or AP→P) on MLP's accuracy, via applying the interference at weights in different layers, i.e. the weights between the input layer and the hidden layer (IH) and the weights between the hidden layer and the output layer (HO).

As Fig. 5 shows, as the interference-induced switching probability of MRAMs increases, the neural network's accuracy will drop, in both interference directions and on both weight layers. While the interference direction does not have a significant impact, the specific layer in which the weights are affected by magnetic field shows a major difference in the neural network's accuracy after interference. At lower switching probability, for example 20%, the neural network's accuracy will sharply decline to less than 20% if the IH weights are interfered, while the accuracy remains above 85% when the interference affects the HO weights, in both interference directions. It indicates that the HO weights exhibit a higher degree of immunity to magnetic field exceeding 2,400 Oe.

This could be explained by that the impact of altering the weights in each layer varies significantly due to the hierarchical nature of feature extraction. The first layer, i.e. the IH layer, directly connected to the input features, is primarily responsible for learning the most fundamental patterns from the raw data. These initial patterns form the basis for all subsequent processing. Therefore, modifying the weights in this layer disrupts the foundational feature extraction process, leading to a substantial degradation in the model's performance. In contrast, the second layer, i.e. the

HO layer, receives its input from the first layer, working on already extracted features rather than the raw data. This layer's role is to refine and abstract the features provided by the first layer, contributing to the model's ability to discern more complex patterns. Changes in the weights of the second layer, while still impactful, generally result in a less severe drop in accuracy.

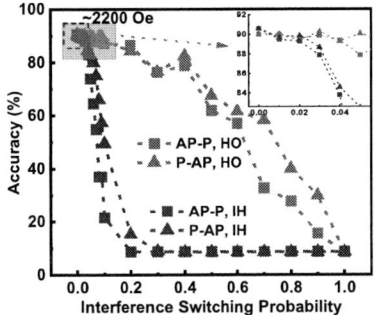

Fig.5 Accuracy of switching probability of bits between layers (IH and HO) on MLP, with a magnified small-probability region as an inset.

E. Impact of external magnetic field interference on bits with different significance in a weight

We evaluate the impact of magnetic field interference on MLP's accuracy, via applying the interference at different bits in the weight, from the least significant bit (LSB, 1st bit) to the most significant bit (MSB, 4th bit).

As Fig.6a shows, under an AP→P magnetic field, the accuracy gradually drops as more AP state bits are interfered to P states under a stronger magnetic field or longer exposure time. Among the 4 bits comprising a weight, the neural network exhibits the weakest immunity to external interference on its MSBs: if 20% AP-state bits are switched to P-state, the accuracy sharply plummets to around 20%. This can be attributed to the fact that the MSB has the highest weight and any disturbance to it would result in a relatively large change in the overall weight value. Therefore, the MSB part in the neural network chip should be carefully protected.

However, the neural network becomes less vulnerable to external magnetic field interference as the bits with lower significance are affected: it only shows around 60% accuracy even if all (100%) AP-state bits in the LSBs are converted to the P-state. The similar trend can be observed in the P→AP case, where an 80% accuracy is maintained if all P-state bits in the LSBs are interfered to the AP-state.

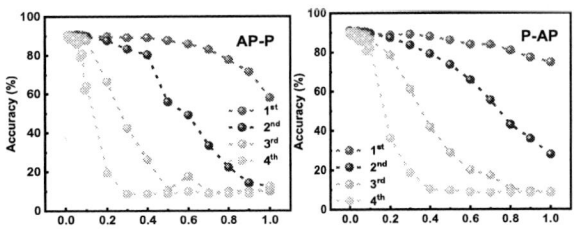

Fig.6 Accuracy of switching probability of bits at different locations on MLP. (a)AP state switching to P state. (b) P state switching to AP state.

Therefore, we can conclude that external magnetic field interference on bits of different significance within a weight

979-8-3503-6184-1/24 $31.00 © 2024 IEEE

has varying impact on the neural network's accuracy. This can be explained by the fact that bits with lower significance, though altered more frequently during training, contribute less to the overall value of a weight.

F. Optimization Strategies

Based on the previous discussion, the inherent FT of MLP can alleviate the requirements for magnetic resistance of MRAM. According to the results of Fig.4(c) and Fig.5, the accuracy of MLP maintains around 90% when a 2200Oe magnetic field is applied to all bits or the HO weights. Furthermore, Fig.6 illustrates that the LSB can withstand higher intensity magnetic fields compared to other bits. Their immunity is sufficient to resist the magnetic field strengths encountered in daily work environments, for instance, the magnetic field of the magnetic closures for purses is around 100 Oe, and of wireless chargers is 50~300 Oe. Considering the inherent FT of neural networks, these findings suggest that the ECC or shielding requirements of HO weights and LSBs might be, to a certain extent, relaxed. This could enable the use of MRAM devices with less magnetic shielding, leading to potential cost saving and reduced design complexity of MRAM-based hardware neural networks.

IV. Conclusion

In this work, we construct a three layer MLP to recognize the MNIST dataset to evaluate the impact of external magnetic interference on the performance of MRAM-based neuromorphic computing, and found that the error tolerance could help protect a multilayer perceptron (MLP) neural network from showing degraded accuracy. Moreover, such tolerance differs between weights in different layers and between bits with different significance. Based on those observations, we further proposed an optimization method for immunity against field interference with reduced cost. This work helps improve the usefulness of MRAM-based neural networks in practical environments.

Acknowledgment

This work is supported by the National Key R&D Program of China (Grant No. 2022YFB4400200), the National Natural Science Foundation of China (No. 62104188 and 12327806), and in part by the major key project of Peng Cheng Laboratory under grant PCL2023AS1-2.

References

[1] W. J. Gallagher, et al., "22nm STT-MRAM for reflow and automotive uses with high yield, reliability, and magnetic immunity and with performance and shielding options," 2019 IEEE International Electron Devices Meeting (IEDM), 2019, pp. 2.7.1-2.7.4.

[2] C. H. Chen, et al., "Reliability and magnetic immunity of reflow-capable embedded STT-MRAM in 16nm FinFET CMOS process," 2021 Symposium on VLSI Technology, 2021, pp. 1-2.

[3] S. Jung, et al. "A crossbar array of magnetoresistive memory devices for in-memory computing," Nature, 2022, 601(7892), pp. 211–216.

[4] V. B. Naik, et al., "Extended MTJ TDDB model, and improved STT-MRAM reliability with reduced circuit and process variabilities," 2022 IEEE International Reliability Physics Symposium (IRPS), 2022, pp. 6B.3-1-6B.3-6.

[5] Y. D. Chih, et al., "13.3 A 22nm 32Mb embedded STT-MRAM with 10ns read speed, 1M cycle write endurance, 10 years retention at 150°C and high immunity to magnetic field interference," 2020 IEEE International Solid-State Circuits Conference (ISSCC), 2020, pp. 222-224.

[6] J. S. Meena, S. M. Sze, U. Chand, T. Tseng, "Overview of emerging nonvolatile memory technologies," Nanoscale Res Lett, 2014, 9, pp. 1-13.

[7] D. Apalkov, B. Dieny and J. M. Slaughter, "Magnetoresistive random access memory," in Proceedings of the IEEE, 2016, vol. 104, no. 10, pp. 1796-1830.

[8] V. B. Naik et al., "Manufacturable 22nm FD-SOI embedded MRAM technology for industrial-grade MCU and IOT applications," 2019 IEEE International Electron Devices Meeting (IEDM), 2019, pp. 2.3.1-2.3.4.

[9] P. Chen, X. Peng and S. Yu, "NeuroSim+: An integrated device-to-algorithm framework for benchmarking synaptic devices and array architectures," 2017 IEEE International Electron Devices Meeting (IEDM), 2017, pp. 6.1.1-6.1.4.

[10] A. Lu, X. Peng, W. Li, H.Jiang, S. Yu, "NeuroSim simulator for compute-in-memory hardware accelerator: Validation and benchmark," Frontiers in artificial intelligence, 2021, 4: 659060.

[11] Y. Luo, X. Peng, S. Yu. "MLP+ NeuroSimV3. 0: Improving on-chip learning performance with device to algorithm optimizations," Proceedings of the international conference on neuromorphic systems. 2019, pp. 1-7.

979-8-3503-6184-1/24 $31.00 © 2024 IEEE

A Hardware Accelerator for Image Super-Resolution with Algorithm Lightweighting and Custom Fusion Engine

Menghan Li*[1], Sheng Lu, Jun Han*[1]

[1] State Key Laboratory of Integrated Chips and Systems, Fudan University, Shanghai 200433, China

* Email: limh21@m.fudan.edu.cn, junhan@fudan.edu.cn

Abstract—Super-resolution (SR) is a crucial component of end-side image processing tasks in constrained sensor environments. However, the existing convolutional neural networks (CNNs) used for SR have significant computational and parameter requirements, necessitating the use of specifically optimized acceleration hardware for the deployment of SR tasks. Accordingly, in this paper, we initially adopt effective lightweight strategies and mixed-precision quantization to obtain the hardware-friendly Light-FSRCNN, which reduces storage consumption by 73.4% in comparison to the original network. Furthermore, we devise a mixed-precision computation engine with a reduced area overhead, which is 15.9% more compact than traditional engines. The hardware processor constructed with this engine exhibits an energy efficiency of 1750.9 GOPS/W under TSMC 12nm synthesis, outperforming CPU, GPU, and analogous SR accelerators.

Keywords—Deep learning, image super-resolution, mixed precision quantization, neural network accelerator

I. INTRODUCTION

Super-resolution (SR) is a form of image processing that enhances image resolution through the use of techniques such as interpolation or deep learning. As a front-end image processing workflow, SR is not constrained by end-side hardware conditions, which makes it particularly valuable in scenarios where sensor performance is insufficient.

In recent years, CNN-based deep learning methods such as FSRCNN[1] have become important SR task implementation algorithms. However, since the final layer output of SR neural network has a higher resolution than the input image, it will have larger intermediate feature maps to maintain the correlation information between pixels than neural network algorithms applied in image classification and other fields. This leads to a larger number of parameters and intermediate data, and higher bit width requirement. This indicates that a substantial on-chip memory capacity is necessary to maintain the reconstruction accuracy, which is challenging to achieve for the acceleration hardware on the end side.

Accordingly, we implement lightweighting algorithm and specialized hardware. In order to reduce the amount of data, this paper implements operator replacement and regularization pruning on the FSRCNN network. Furthermore, in order to reduce the bit width of the data, we perform mixed-precision quantization of the network with minimal accuracy loss prior to hardware deployment. Additionally, we design a custom fusion engine with a smaller area to support mixed-precision convolution operations. In light of these advancements, we develop an accelerator with high data reuse for the deployment of end-to-end image SR tasks.

The contributions of this paper are summarized as follows:

- We propose a hardware-friendly SR network, Light-FSRCNN, which achieves comparable accuracy to commonly used SR networks while reducing storage consumption by 73.4%.

- We design a mixed-precision convolution engine that reduces area by 15.9% and develop a high-efficiency hardware architecture to deploy the optimized algorithm.

- We implement the proposed SR processor under TSMC 12nm technology, achieving 32.1 FPS for x4 scale Full HD image SR tasks. The energy efficiency reaches 1750.9 GOPS/W, outperforming CPU, GPU, and SR accelerators in similar hardware comparisons.

II. THE PROPOSED LIGHTWEIGHT SR NETWORK

In this section, we present our approach to lightweight the original SR network.

A. Operator Replacement and Regularization Prunning

Operator replacement (OR) is performed on the original FSRCNN network in accordance with the requirements of hardware deployment. Specifically, to circumvent the generation of checkerboard artifact resulting from transposed convolution during upsampling, we substitute transposed convolution with pixelshuffle, a compatible alternative to convolution operators. Additionally, we utilize structured pruning to eliminate superfluous channels from the high-computational convolutional layers of the original network. Furthermore, to avert network overfitting and augment network sparsity, we deploy L2 regularization pruning (RP), thereby enhancing the algorithm's accuracy.

B. Mixed-Precision Quantizatioin

In SR tasks, the network output is composed of high-resolution image pixels. However, computations involving high single bit-width data are resource-intensive in terms of hardware computation and storage. Our experiments demonstrate that the perceptual capabilities of the SR network's layers vary. Furthermore, the application of low-bit quantization to certain less critical layers does not significantly impact SR accuracy. In order to trade off computational precision and hardware efficiency, we implement mixed-precision quantization with 8-bit and 16-bit precision prior to deployment. By meticulously controlling the quantization of each layer, we are able to achieve a 58.04% reduction in memory consumption while maintaining computational accuracy. In conclusion, the outcomes of the ablation study on network lightweighting techniques are illustrated in Table 1.

979-8-3503-6184-1/24 $31.00 © 2024 IEEE

TABLE I. ABLATION STUDY FOR LIGHT-FSRCNN

Model	Metrics			
	Params	*Mem (kB)*	*MAdd (G)*	*PSNR*
Baseline	12809	51.24	3.32	30.55
Baseline+OR	8116	32.46	2.1	30.44
Baseline+OR+RP	8116	32.46	2.1	30.47
Light-FSRCNN (Mixed-Precision)	8116	13.62	2.1	30.47

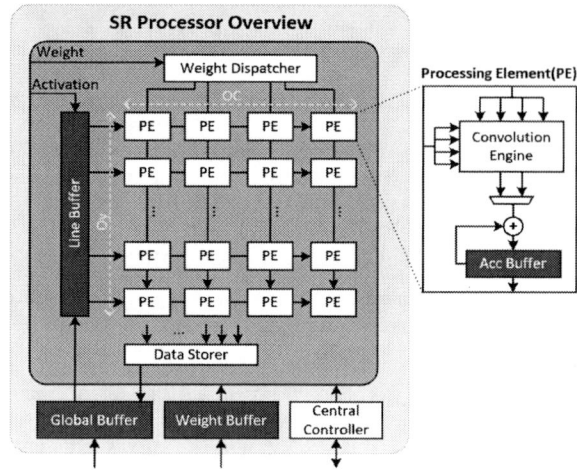

Fig. 1. SR accelerator architecture overview

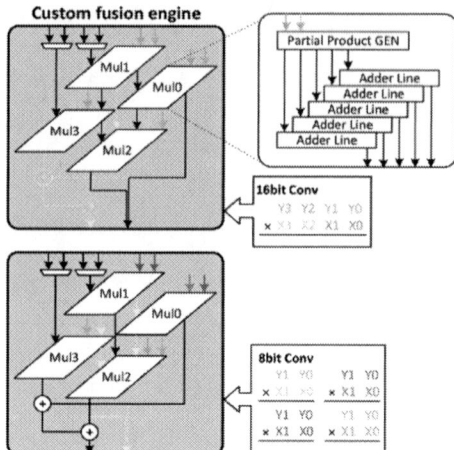

Fig. 2. Microarchitecture of custom fusion engine

III. HARDWARE ARCHITECTURE

This section will introduce the hardware architecture designed for the deployment of the SR network. A custom fusion engine is developed for mixed-precision operations, with the objective of accelerating convolution operations within the network. This approach represents a key strategy for achieving lightweight SR deployment. In addition, a high-utilization accelerator architecture has been implemented.

A. Architecture Overview

The architecture overview of the proposed SR accelerator is illustrated in Fig. 1. The accelerator incorporates a multitude of mixed-precision processing elements (PE), a unified global buffer, line buffers for caching activations, and a central controller.

The SR accelerator is capable of performing either 8-bit or 16-bit convolution operations. Each PE is capable of simultaneously mapping either 8-bit operations for four input channels (IC) or 16-bit operations for a single input channel. Moreover, convolution operations are unfolded along the array's row and column directions, corresponding to the vertical feature map (Oy) and output channel (OC) dimensions, respectively, in order to achieve parallel computation.

B. Mixed-Precision Convolution Engine Microarchitecture

The implementation of mixed-precision computation primarily entails two methodologies: the initial approach is fission, which leverages a portion of the hardware resources of a high-precision multiplier to facilitate low-precision multiplication operations. This method has a minimal impact on the hardware, but it does result in a reduction of half the hardware computing power in low-precision mode. The second method is fusion, which employs the output bit stitching of low-precision multipliers to achieve high-

precision multiplication. This method does not result in a loss of computing power during precision switching. Nevertheless, in order to achieve fusion for low-precision computation results, it is essential to accumulate the corresponding bits of the results. Furthermore, pre- and post-processing of input and output data is necessary. This redundant hardware occupies a significant area (approximately 36% in original Synopsys DesignWare implementation). In light of these considerations, our computation engine employs fusion in its calculation process. To reduce area occupation, a more refined microarchitecture design is required, based on the particularities of convolution operations.

A reconfigurable custom fusion engine for mixed-precision convolution operations is designed, with the circuit structure of each multiplier customized. Fig. 2 illustrates the microarchitecture and reconfigurable scheme in both precision modes. In the high-precision mode, the result of the high-precision multiplication does not require the complete output of four low-precision multiplication results before accumulation. In lieu of this, the output of the preceding multiplier can be employed as the input for the subsequent multiplier, thereby facilitating the early summation of partial products from the low-precision multipliers. This approach obviates the necessity for the hardware logic that will otherwise be required for the fusion of the low-precision multipliers' results. Moreover, the internal circuitry of the multipliers is custom-designed and highly flexible, allowing for the offline accumulation of the partial products' sign bits. In practice, the coding scheme of the low-precision multiplication units can be controlled via compensation signals, thereby enabling high-precision signed number two's complement operations. This obviates the necessity for intricate pre- and post-processing logic for disparate coding schemes.

Furthermore, in low-precision mode, the accumulation of the products of feature maps and weights from different input channels within the same convolution sliding window is required. Therefore, the feature maps and corresponding weights of four input channels are assigned to the same custom fusion engine for accumulation. This is achieved through the implementation of a dedicated hardware dataflow and corresponding instruction scheme, which will be introduced subsequently. By reusing the accumulation and control logic from the high-precision mode, the custom fusion

979-8-3503-6184-1/24 $31.00 © 2024 IEEE

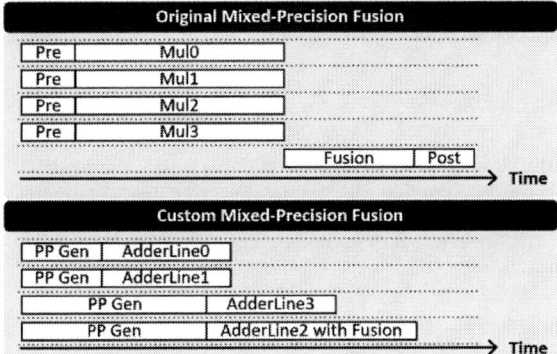

Fig. 3. Timing comparison of original and custom fusion process

TABLE II.   SR Performance Comparison

Metric	FSRCNN			Light-FSRCNN(ours)		
	×2	×3	×4	×2	×3	×4
PSNR	36.94	33.06	30.55	36.93	33.12	30.47

TABLE III.   Area Comparison

	Original Fusion			Custom Fusion (ours)		
	Mul	Fusion	Total	PP GEN	AdderLine	Total
Area (um²)	57	130	359	24	36	302

TABLE IV.   Energy Efficiency Comparison

	12th i9-12900h	Titan RTX	TCAS-I 20`[3]	Jestcs 20`[4]	Ours
Technology	-	-	65nm	65nm	12nm
Supply Voltage	-	-	1.2V	1.1V	0.8V
Frequency (MHz)	-	-	200	200	200
SR speed (FPS)	15.4	215.4	-	88.3	32.1
Power (W)	80	280	0.206	0.211	0.0385
Energy Efficiency (GOPS/W)	0.639	2.554	471.04	1126.4	1750.9

engine is able to minimize the addition of hardware. The accumulation stage of partial sums is capable of directly completing the accumulation of results from the Mul1 and Mul2 multipliers, thereby improving energy efficiency.

By optimizing the microarchitecture of the fusion engine, our hardware occupies less area than DesignWare and has a shorter critical path. Fig. 3 illustrates the timing comparison of internal modules between our custom fusion engine and the original engine. Table 3 indicates that the custom fusion engine reduces the area by 15.9%.

C. High-utilization Dataflow

The accelerator array is capable of fully realizing data reuse in the spatial dimension. Weight data from disparate OC dimensions are broadcast to disparate columns of the PE array, whereas input activation data from distinct Oy dimensions are broadcast to disparate rows of the PE array. This results in a reduction in the number of data loads. In high-precision computation mode, the role of each compute engine is to

perform the convolution calculation for a single IC. In low-precision computation mode, each compute engine is capable of executing calculations for four ICs. The design employs layer-fusion[2], whereby all compute engines share an on-chip global buffer for the storage of intermediate feature map data. This results in a reduction of off-chip interactions and a lowering of system bandwidth pressure. Furthermore, the accelerator is equipped with a line buffer for the caching of input feature maps. This buffer is capable of storing multiple rows of input feature maps corresponding to the convolution slide window, thereby enabling data reuse in the temporal dimension.

IV. Experiment

A. Light-FSRCNN performance

The Light-FSRCNN model is implemented using PyTorch, trained using the same method as FSRCNN, and tested on the Set5 dataset. As evidenced in Table 2, our network exhibits enhanced SR representation capabilities. Furthermore, as illustrated in Table 1, for the x4 scale Full HD SR task, the memory requirements of Light-FSRCNN are reduced by 73.4%, while its peak signal-to-noise ratio (PSNR) is comparable to that of the larger FSRCNN.

B. Hardware Implementation

The proposed SR accelerator is implemented using Chisel and subsequently synthesized using a 12nm TSMC process with a clock frequency of 200MHz. In terms of area, as illustrated in Table 3, our custom fusion engine demonstrates a reduction in area of 15.9% in comparison to the original engine utilizing Synopsys DesignWare.

In terms of performance, as illustrated in Table 4, the accelerator achieves 32.1 FPS for the x4 scale Full HD SR task deployment. Additionally, it achieves an energy efficiency of 1750.9 GOPS/W, which is 2740 times that of the Intel 12th i9-12900 CPU and 686 times that of the Nvidia Titan RTX GPU. Furthermore, it achieves a higher energy efficiency compared to other similar accelerators[3][4].

V. Conclusion

The paper presents a high-efficiency SR accelerator based on a hardware-friendly Light-FSRCNN. By implementing a custom fusion engine and high-utilization dataflow, the accelerator achieves 32.1 FPS for x4 scale full HD SR with an energy efficiency of 1750.9 GOPS/W.

Acknowledgment

This work was supported by the National Natural Science Foundation of China under Grant 61934002 and 62234008.

References

[1]   Dong, Chao, Chen Change Loy, and Xiaoou Tang. "Accelerating the super-resolution convolutional neural network." Computer Vision–ECCV 2016: 14th European Conference, 2016, pp.391–407.

[2]   Alwani, Manoj, et al. "Fused-layer CNN accelerators." 2016 49th Annual IEEE/ACM International Symposium on Microarchitecture (MICRO). IEEE, 2016.

[3]   Im, Dongseok, et al. "DT-CNN: An energy-efficient dilated and transposed convolutional neural network processor for region of interest based image segmentation." IEEE Transactions on Circuits and Systems I: Regular Papers 67.10 (2020): 3471-3483.

[4]   Lee, Juhyoung, Jinsu Lee, and Hoi-Jun Yoo. "SRNPU: An energy-efficient CNN-based super-resolution processor with tile-based selective super-resolution in mobile devices." IEEE Journal on Emerging and Selected Topics in Circuits and Systems 10.3 (2020)

Hardware Implementation of High Speed Fault-Tolerant Parallel Accelerator

Wenzhe Ma [1], Wenzhe Ma*[1]

State Key Laboratory of Integrated Chips and Systems, Fudan University
School of Microelectronics, Fudan University, Shanghai 201203, CHINA

* Email: 22212020022@m.fudan.edu.cn, 22212020022@m.fudan.edu.cn

Abstract—Process advancements and single-event effects in terrestrial environments have heightened the importance of fault-tolerant design for FPGA-based digital circuits. This paper presents a hardware implementation of a fault-tolerant parallel matrix multiplication accelerator. Through speed-optimized design of fault correction circuits and the addition of custom combinational logic to the controller module, a high-speed hardware system capable of tolerating dual faults is realized. Compared to similar works from JJ Davis in 2014, results demonstrate a 185% increase in maximum operating frequency under dual-fault conditions, while incurring only a 4.88% increase in area overhead.

Keywords—*Fault Tolerance, FPGA, Accelerator, High Speed*

I. INTRODUCTION

With the advent of the intelligent era, artificial intelligence is emerging as the decisive force propelling humanity into the intelligent age. As a method for realizing artificial intelligence, machine learning's workload comprises various neural networks, ranging from convolutional and fully connected to Transformer and recommendation models. Matrix-matrix multiplication lies at the core of these neural network computations [1][2].

Leveraging its flexible hardware programmability, FPGA has found widespread applications in high-performance computing, image processing, and various other domains with the advancement of microelectronic technologies[3]. Advanced commercial FPGAs have entered the 7nm era, and while process advancements bring about increased transistor density and switching speed, they also lead to increased process variations, device degradation, and susceptibility to faults[4]. During configuration, FPGAs rely on SRAM programmable points to determine the circuit functionality of the design inputs, and these SRAM programmable points are susceptible to degradation and single-event effects in terrestrial environments, causing stored logic values to flip[5]. Consequently, there is a surging demand for fault-tolerant design in FPGA-based digital circuit designs.

J.J. Davis et al. previously conducted area and performance evaluations for a fault-tolerant matrix multiplication Parallel accelerator[6] , but this design only addressed single-fault tolerance for Multiply-Accumulate units(MACs), indicating that further research is required to extend the fault tolerance to realistic scenario of double-fault MACs. J.J. Davis et al. also employed Dynamic Partial Reconfiguration(DPR) technology to reconfigure fault-tolerant hardware for low overhead double-fault MAC tolerance[7] , but this introduced additional software overhead and reduced operating speed. Consequently, the maximum achievable operating frequency for this dual-fault MAC fault-tolerant approach, integrating both hardware and software methodologies, was limited to only 53.101MHz.

This paper achieves double-fault MAC tolerance and performance enhancement through an optimized design incorporating Circular Shifter circuit and Datapath Controller compared to hardware and software collaboration method. Circular Shifter circuit is designed to operate with a signle clock cycle, while Datapath Controller incorporate customized combinational logic to enhance fault tolerance of both adjacent and non-adjacent MAC with high speed. These hardware methods ensure that the area penalty remains within a reasonable range.

The main contributions of this work are:

- Application of a single-cycle Circular Shifter designed for accelerating the fault-tolerant accelerator system.

- Implementation of dual-fault hardware fualt tolerance through customized combinatorial logic within Datapath Controller.

- Quantitative analysis of the performance and area overhead metrics for a hardware apporach to dual-fualt tolerant accelerator implementation.

II. ABFT PRINCIPLE OF MATRIX MULTIPLICATION

A. Algorithm Based Fault Tolerance

Algorithm-Based Fault Tolerance (ABFT) can be applied to many algebraic operations, including matrix operations[8] and Fourier transforms[9]. ABFT is an online fault tolerance technique tailored for specific algorithms. Compared to Built-In Self-Test (BIST) techniques, it avoids offline overhead. Unlike Roving STARs techniques[10] that detect FPGA chip faults online through roving but suffer from path delay effects forcing clock slowdown, ABFT circumvents this issue. Compared to Triple Modular Redundancy (TMR)[11], ABFT significantly reduces area overhead. ABFT combines the advantages of high fault coverage and low detection latency while maintaining a relatively low area cost.

B. ABFT in Matrix Multiplication Operator

For the application of ABFT in fault-tolerant matrix multiplication, when the original inputs are N-order square matrices A and B, matrix A undergoes column encoding to yield the column-encoded matrix Ac, and matrix B undergoes row encoding to yield the row-encoded matrix Br. The multiplication of Ac and Br results in the fully encoded matrix Cf. Under correct matrix multiplication conditions, each element of the (N+1)th row of Cf should sum to the elements of the corresponding column in the first N rows. If an element in the (N+1)th row does not meet this criterion, an error has occurred in the computation of the corresponding column [6].

979-8-3503-6184-1/24 $31.00 © 2024 IEEE

III. IMPLEMENTATION

A. System Overview

To ensure the fairness of subsequent comparisons in this design, the entire fault-tolerant accelerator system is deployed on the Zynq XC7Z020 chip, which is a programmable SoC consisting of a Processing System (PS) part suitable for ARM core development and a Programmable Logic (PL) part suitable for FPGA development. The overall block diagram of the system is shown in Figure 1, where the fault-tolerant matrix multiplication parallel accelerator is deployed on the PL side, with speed optimization applied to the circular shifter circuit and the datapath controller optimized to adapt to dual-fault MAC fault tolerance. To facilitate the data transfer from off-chip DDR3 to the Fault Tolerant Accelerator, a DMA (Direct Memory Access) circuit is deployed within the system to achieve high-speed data transfer.

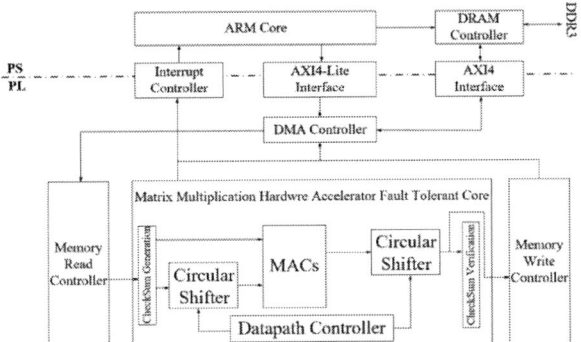

Fig. 1. System block diagram with Optimized Circular and Datapath Controller.

B. Hardware Overview

Figure 2 illustrates the system hardware architecture, divided into two parts: the data fault-tolerant computation core and the interaction between the fault-tolerant core and BRAMs. The matrix multiplication hardware accelerator MACs module is designed with parallelization, where the

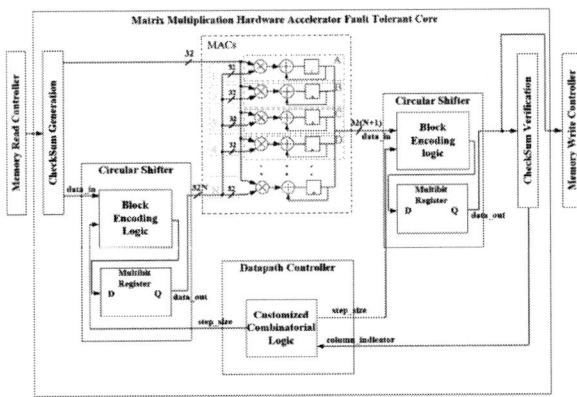

Fig. 2. System hardware block diagram.

Latin letters in Figure 2 represent individual multiply-accumulate units, and the Arabic numerals indicate the indices of the corresponding elements in the input matrix rows.

The fault tolerance aspect consists of fault detection and fault correction. During fault detection, the checksum generation module is responsible for row and column encoding of the input initial matrix data (in this design, a 32-by-32 matrix with 32-bit element data width). The checksum verification module performs detection on the output matrix results. As described in Section II B, when the corresponding column of the fully-coded matrix Cf violates the required condition, the corresponding MAC is determined to be faulty, and the fault information is fed back to the Datapath Controller module.

For fault correction, the Circular Shifter module performs shift operations on the input and output data of the MACs module based on the step size and shift direction signals sent by the Datapath Controller module, correcting the faulty multiplication results. To achieve speed enhancement, the Circular Shifter module employs a Block Encoding logic for the input data, completing the shift operation within a single clock cycle.

The data interaction between the fault-tolerant core and BRAMs is primarily facilitated by the BRAM Controller module for timing coordination.

C. Fault Avoidance

Once the CheckSum Verification module detects a faulty MAC, the hardware system enters a fault avoidance phase under the control of the Datapath Controller module. Figure 3 illustrates the fault-tolerant steps when two MACs fail in a matrix of order N=4. Latin letters represent the various multiply-accumulate units, and Arabic numerals indicate the row elements of the matrix.

Fault avoidance is divided into two steps. The first step involves shifting the input matrix row data downward through the Circular Shifter module on the input side of the MACs module, while simultaneously shifting the output matrix row data upward through the Circular Shifter module on the output side of the MACs module. This avoids the erroneous calculation output from the faulty MACs. In the second step, no upward or downward shifting of input and output data is performed. Instead, the output from the normal MACs is used to fill the gaps left by the faulty MACs in the first step.

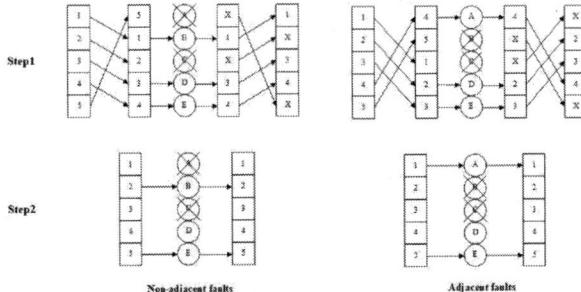

Fig. 3. Resource reallocation steps for N=4 accelerator with double faulty MAC.

To simulate real-world scenarios of adjacent and non-adjacent dual MAC faults, this design incorporates custom combinational logic circuits into the Datapath Controller module. This allows it to handle both adjacent and non-adjacent dual MAC faults simultaneously. The fault tolerance for adjacent and non-adjacent dual MAC faults is managed by different shift step signals output from the Datapath Controller module.

979-8-3503-6184-1/24 $31.00 © 2024 IEEE

IV. EXPERIMENTS AND RESULTS

A. Experiment Condition

To validate the high performance of this design, a matrix of order 32 with a data width of 32 bits was selected. Concurrently, NOT gates were added to the least significant bit outputs of two MAC units to simulate the occurrence of dual faults in order to obtain the correct functional simulation waveform of fault tolerance. The hardware system was deployed on the PL side of the Zynq xc7z020 chip, and its maximum operating frequency and resource utilization were obtained through implementation with the Vivado STA tool. In this experiment, the maximum clock frequency was defined as the point at which the setup time slack on the critical path was less than 0.01 ns. The functional correctness is ascertained through FPGA board-level validation.

B. Results and Evaluation

The outcomes of the FPGA prototype verification for this design are depicted in Figures 4 and 5. Figure 4 illustrates the waveforms of the data bus observed in real-time using an Integrated Logic Analyzer (ILA), while Figure 5 presents the physical board photograph. Upon in-line debugging with the ILA, the counter count_A_row correctly initiated the second fault-tolerant computation process after reaching the value of 67, thereby confirming the successful functional validation at the board level.

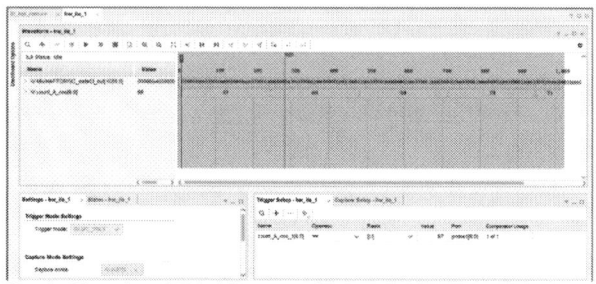

Fig. 4. Upper board ILA debugs waveform.

Fig. 5. Board level experiment diagram.

The resource utilization and maximum operating frequency of the fault-tolerant accelerator in this design are presented in TABLE I. The DSP resource usage aligns with J.J. Davis's work, while the BRAM resource consumption is lower. However, the LUT resource usage is approximately four times higher, which can be attributed to the increased utilization of LUTs as LUTRAMs during circuit implementation. The arithmetic mean of the utilization rates for the four types of FPGA resources yields a total resource occupation of approximately 24.48%. Compared to J.J.

Davis's work, this represents an increase of 4.88% in overall on-chip resource utilization, which is a consequence of the area penalty incurred by the high-speed-prioritized circular shifter circuit.

The fault-tolerant accelerator achieves a maximum operating frequency of 151.515 MHz. This represents a substantial performance enhancement of 185% compared to J.J. Davis's work, aligning well with the objective of high-performance design.

TABLE I.   RESOURCE USAGE AND FMAX FOR N=32

Design	Registers	LUTs	BRAMs	DSPs	Total resources	fmax (MHz)
JJ Davis[7]	3675 (3.45%)	4363 (8.2%)	61 (21.8%)	99 (45%)	19.6%	53.101
Our work	11952 (11.23)	16668 (31.33%)	14.5 (10.36%)	99 (45%)	24.48%	151.515

This significant improvement in operating frequency demonstrates the efficacy of the design optimizations implemented in this study, particularly in addressing the challenges of fault tolerance while maintaining high speed.

V. CONCLUSION

In this paper, we present a hardware design for a fault-tolerant parallel matrix multiplication accelerator based on the ABFT concept, utilizing 32 MAC units to perform 32x32 matrix multiplication while tolerating dual MAC faults. By optimizing the circular shifter circuit and the data path controller, we achieved significant performance improvements with a reasonable area penalty, effectively eliminating software overhead. This work provides a valuable reference for the development of high-performance fault-tolerant matrix multiplication accelerators.

REFERENCES

[1] J. Welser, J. W. Pitera, C. Goldberg, "Future computing hardware for AI," *IEEE International Electron Devices Meeting*, pp. 1.3.1-6, 2018.

[2] S. Kim, C. Hooper, T. Wattanawong, "Full stack optimization of transformer inference: a survey," unpublished.

[3] S. P. Park, D. Lee, K. Roy, "Soft-error-resilient FPGAs using built-in 2-D Hamming product code," *IEEE transactions on very large scale integration (VLSI) systems*, pp. 248-256, 2011.

[4] J. J. Davis, P. Y. K. Cheung, "Reduced-precision Algorithm-based Fault Tolerance for FPGA-implemented Accelerators," *International Symposium on Applied Reconfigurable Computing*, pp. 361-368, 2016.

[5] S. Rihui, "Soft Error Fault Simulation Study for FPGAs," *Harbin Institute of Technology*, 2019.

[6] J. J. Davis, P. Y. K. Cheung, "Datapath fault tolerance for parallel accelerators," *International Conference on Field-Programmable Technology*, pp. 366-369, 2013.

[7] J. J. Davis, P. Y. K. Cheung, "Achieving low-overhead fault tolerance for parallel accelerators with dynamic partial reconfiguration," *24th International Conference on Field Programmable Logic and Applications*, pp. 1-6, 2014.

[8] K. H. Huang, J. A. Abraham, "Algorithm-based fault tolerance for matrix operations," *IEEE transactions on computers*, pp. 518-528, 1984.

[9] S. J. Wang, N. K. Jha, "Algorithm-based fault tolerance for FFT networks," *IEEE Transactions on Computers*, pp. 849-854, 1994.

[10] M. Abramovici, C. Strond, C. Hamilton, "Using roving STARs for on-line testing and diagnosis of FPGAs in fault-tolerant applications," *International Test Conference 1999. Proceedings*, pp. 973-982, 1999.

[11] S. D'Angelo, C. Metra, S. Pastore, "Fault-tolerant voting mechanism and recovery scheme for TMR FPGA-based systems," *Proceedings 1998 IEEE International Symposium on Defect and Fault Tolerance in VLSI Systems*, pp.233-240, 1998.

979-8-3503-6184-1/24 $31.00 © 2024 IEEE

Composite Filter-based Bicubic Interpolation Method and FPGA Implementation

Li Zhang[1], Jingjing Liu[1]*, Yujie Zhu[1], Jianhua Zhang[1]*

[1] the Shanghai Key Laboratory of Chips and Systems for Intelligent Connected Vehicle, School of Microelectronics, Shanghai University, Shanghai 200444, China

* Email: jjliu@shu.edu.cn

Abstract—This paper proposes a bicubic interpolation method based on Clamp-Laplace composite filters, to solve the problems of blurring and distortion in interpolated images generated by traditional interpolation methods. In addition, the proposed method is implemented in hardware based on ZCU104 FPGA platform. The qualitative and quantitative experimental results show that the proposed method improves PSNR by 1.76% and SSIM by 1.02% compared to the traditional bicubic interpolation method. The proposed processor has been implemented on FPGA platform, with a processing speed of 144 Mpixels/s, an energy efficiency of 157.3 Mpixels/J, a decrease of 5.23% in PSNR and 1.06% in SSIM compare with software implementation.

Keywords—Composite filters, Bicubic interpolation method, Image scaling, FPGA

I. INTRODUCTION

With the continuous development of micro-display devices such as smartphones, tablets, and wearable devices, higher requirements have been imposed on image display, and image scaling has gradually become a research hotspot[1]. Image scaling is a technique that adjusts the size of an image from one dimension to another and can be broadly categorized into two types: interpolation-based methods and deep learning-based methods.

In interpolation based image scaling methods, the nearest neighbor method only uses the nearest pixel values, which is the easiest to implement. However, the images generated by the nearest neighbor method are prone to jagged edges and detail distortion, resulting in poor quality. The bilinear interpolation method calculates the value of the target pixel by weighted averaging the four nearest pixels around it. Although bilinear interpolation methods are more accurate than nearest neighbor interpolation methods, compared to more advanced interpolation methods, bilinear interpolation methods may still have a certain degree of blur or distortion in processing edges and details. For more complex image structures or large-scale scaling, they may not be able to fully capture subtle changes. Bicubic interpolation is an advanced interpolation method that considers the weighted average of the 16 nearest pixels around each pixel, which can better smooth the image and preserve details. Compared with other methods, bicubic interpolation is considered the best solution to achieve scaling between interpolation quality, computational complexity, or hardware resource consumption. However, when display devices require better visual quality and higher pixel density, the interpolated images generated by the standard bicubic interpolation method cannot satisfied higher standards.

In recent years, there are many scaling methods based on deep learning have been proposed by researchers with the flourishing development of deep learning, which can utilize the contextual information of images for scaling, rather than relying solely on local pixel interpolation, thereby providing more accurate and natural scaling effects and effectively improving image scaling quality. Sun et al. [2] proposed a hardware efficient method utilized residual recursive neural network based on FPGA for video super resolution, which supports up-scaling from full-high-definition to 4K ultra-high-definition at $2\times$. Although this method utilized normalization and fixed point quantization to reduce memory consumption, and utilized group convolution to reduce complexity of network, it still required a large amount of hardware resources and memory consumption. Deep learning based methods can learn the features and structures of images and generate high-quality scaling results. However, deep learning based methods are generally more complex and require large storage space and computing resources for inference, which may pose challenges in inference speed in real-time applications or resource limited environments.

This paper proposes a bicubic interpolation method combined with Clamp-Laplace composite filters which named CLIM, and implements the method in hardware based on ZCU104 FPGA platform. The process of CLIM is shown in Figure 1.

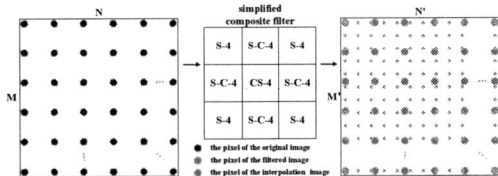

Figure 1. The process of CLIM

II. PROPOSED METHOD

A. The proposed CLIM

The bicubic interpolation method uses the 16 nearest pixels around the interpolation point for calculation. The relationship between the interpolation point and the nearest pixel is shown in Figure 2, and the calculation of interpolation coefficients is as follows,

$$h(s) = \begin{cases} 1.5|s|^3 - 2.5|s|^2 + 1, & 0 \le |s| \le 1 \\ -0.5|s|^3 + 2.5|s|^2 - 4|s| + 2, & 1 \le |s| \le 2 \\ 0, & 2 \le |s| \end{cases} \quad (1)$$

where s represents the distance between the interpolation point and the reference point, and $h(s)$ represents the interpolation kernel function.

The pixel value of interpolation point is calculated as follows,

979-8-3503-6184-1/24 $31.00 © 2024 IEEE

$$f(x,y) = YFX^T \qquad (2)$$

$$Y = [h(1+\Delta y) \quad h(\Delta y) \quad h(1-\Delta y) \quad h(2-\Delta y)] \qquad (3)$$

$$F = \begin{bmatrix} f(i-1,j-1) & f(i-1,j) & f(i-1,j+1) & f(i-1,j+2) \\ f(i,j-1) & f(i,j) & f(i,j+1) & f(i,j+2) \\ f(i+1,j-1) & f(i+1,j) & f(i+1,j+1) & f(i+1,j+2) \\ f(i+2,j-1) & f(i+2,j) & f(i+2,j+1) & f(i+2,j+2) \end{bmatrix}$$
$$(4)$$

$$X = [h(1+\Delta x) \quad h(\Delta x) \quad h(1-\Delta x) \quad h(2-\Delta x)] \qquad (5)$$

where $f(x,y)$ represents the pixel value of interpolation point, Δx and Δy represent horizontal and vertical distance between the insertion point and its nearest pixel on the upper left.

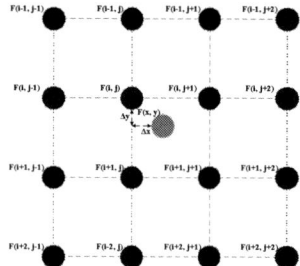

Figure 2. The relationship between the interpolation and the nearest point.

However, when the original image scale is large, the interpolation images obtained by commonly used bicubic interpolation methods still have blurring effects. To address the above issues, this article introduces a Laplacian high-pass filter, which enhances the high-frequency components of the image to highlight the edge information[3]. At the same time, to suppress transition sharpening, Clamp low-pass filter is introduced to retain the low-frequency components of the image. In order to reduce the hardware cost caused by the two filters introduced, a 5×5 combination filter is utilized to equivalent 3×3 high-pass filter and 3×3 low-pass filter, and the combination filter is simplified to further reduce hardware resource consumption and improve interpolation speed. The detailed calculation is as follows,

$$A'(i,j) = A(i,j) \times \frac{\begin{bmatrix} 1 & 1 & 1 \\ 1 & C & 1 \\ 1 & 1 & 1 \end{bmatrix}}{C+8} \times \frac{\begin{bmatrix} 0 & -1 & 0 \\ -1 & S & -1 \\ 0 & -1 & 0 \end{bmatrix}}{S-4}$$

$$= A(i,j) \times \frac{\begin{bmatrix} 0 & -1 & -1 & -1 & 0 \\ -1 & S-2 & S-C-2 & S-2 & -1 \\ -1 & S-C-2 & CS-4 & S-C-2 & -1 \\ -1 & S-2 & S-C-2 & S-2 & -1 \\ 0 & -1 & -1 & -1 & 0 \end{bmatrix}}{(C+8)(S-4)} \qquad (6)$$

where $A'(i,j)$ represents the filtered pixel matrix, $A(i,j)$ represents the original pixel matrix, C represents the low-pass parameter, and S represents the high pass parameter.

The simplified composite filter is as follow,

$$Kernel_{CSF} = \begin{bmatrix} S-4 & S-C-4 & S-4 \\ S-C-4 & CS-4 & S-C-4 \\ S-4 & S-C-4 & S-4 \end{bmatrix} \qquad (7)$$

$$gain_{CSF} = (C+8)(S-4)-4$$

where $Kernel_{CSF}$ represents the kernel of simplified composite filter, $gain_{CSF}$ represents the gain of simplified composite filter.

B. FPGA Implementation

After weighing the hardware resource consumption, method implementation complexity, and method implementation effectiveness, this paper utilizes ZCU104 FPGA platform to implement the proposed method in hardware. The overall framework of the proposed hardware implementation is shown in Figure 3.

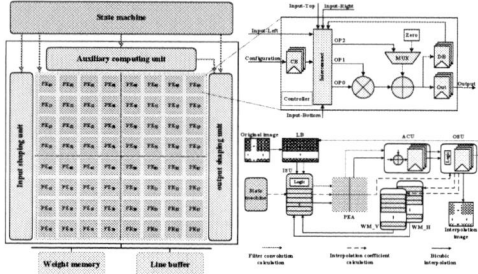

Figure 3. The overall framework of the proposed hardware implementation

Considering that bicubic interpolation requires the calculation of the 16 nearest neighboring pixels for interpolation, this paper proposes a 4×4 PE array. In order to enhance the parallelism of the interpolation method and improve the interpolation speed, the four sub 4×4 PE arrays are combined into an overall 8×8 PE array. In addition, the proposed hardware implementation framework also includes two shaping units, two memory units, an auxiliary computing unit (ACU), and a state machine (SM). Specifically, the input integer unit (ISU) is utilized to read and consolidate the input data, ensuring that the correct input data is fed into the PE array, and the output integer unit (OSU) is utilized to reorganize and transfer the output data to LB for caching. Additionally, the line buffer (LB) is used to cache the integrated input data and filtered data, and the weight memory (WB) unit is used to store interpolation coefficients.

The overall computation process can be divided into three stages, including composite filter convolution calculation, interpolation coefficient calculation, and bicubic interpolation.

In the stage of composite filter convolution calculation, six rows of original data are read through the ISU each time. Four 3×3 PE arrays are utilized simultaneously to compute the filtered data for four rows. Subsequently, the filtered data is transmitted to the ACU for summation and normalization by the gain factor. After integration through the OSU, the filtered data is transmitted to LB for caching.

In the interpolation coefficient calculation stage, the 16 nearest pixel points corresponding to the interpolation point

979-8-3503-6184-1/24 $31.00 © 2024 IEEE

are determined by SM, and then recombined by ISU and input into the PE array. Four PE sub arrays are utilized to calculate the coefficients of the intervals (-2, -1), (-1, 0), (0, 1) and (1, 2), which is shown in (1), and the calculated interpolation coefficients are stored in WM through OSU.

In the bicubic interpolation stage, the interpolation coefficients stored in WM and the filtered data stored in LB are read by the ISU. Then two multiplication operations are performed through the PE array to obtain the results of each sub array. The ACU is utilized for summation, and the interpolation result is finally output through the OSU.

III. Experimental Results and Discussion

A. Qualitative analysis

To demonstrate the interpolation effect of the proposed interpolation method, experiments are conducted on images of different resolutions. First, the test image is downsampled. Then, different interpolation methods are employed to interpolate the image, including BL[4], BC[5], FSRCNN[6] and IFIA[1]. The qualitative results of different methods are shown in Figure 4. The peak signal-to-noise ratio (PSNR) and structural similarity index metrics (SSIM) are measured between the interpolated image and the original image for quantitative analysis. TABLE I. shows the quantitative indicators for interpolation using different methods, the optimal value of every indicator is represented in bold, and the suboptimal value is indicated by an underlined.

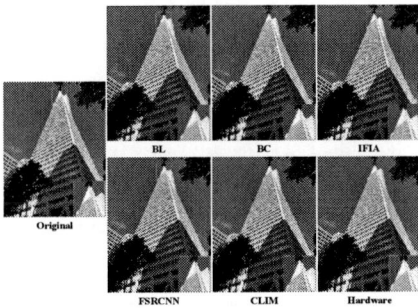

Figure 4. The qualitative results of different methods

TABLE I. AVERAGE VALUES OF TWO QUANTITATIVE INDEXES USING DIFFERENTS INTERPOLATION AGORITHMS

Model	PSNR	SSIM
BL[4], 2001	24.15	0.8837
BC[5], 1981	24.98	0.9031
FSRCNN[6], 2016	**29.06**	**0.9535**
IFIA[1], 2024	24.17	0.9067
CLIM	25.42	0.9123
Hardware	24.09	0.9026

The experimental results indicate that the proposed method in this paper outperforms bilinear interpolation (BL) and traditional bicubic interpolation (BC) methods in terms of PSNR and SSIM scores, and is only surpassed by the deep learning-based interpolation method (FSRCNN). Although deep learning-based interpolation method achieves the best PSNR and SSIM scores, it requires significant hardware resources and are challenging to implement. In addition, the average PSNR achieved by hardware implementation is reduced by 5.23%, and SSIM is reduced by 1.06%. The proposed method has better performance indicators compared to other interpolation based methods. Although there are still shortcomings compared to deep learning based methods, the proposed method utilizes fewer hardware resources.

B. Implementation results

The method proposed in this article is implemented at a frequency of 250MHz on ZCU104 and compared with other interpolation methods in terms of hardware implementation, as shown in TABLE II.

TABLE II. COMPARISON OF HARDWARE IMPLEMENTATIONS

Model	BL[4]	BC[5]	IFIA[1]	SRCNN[6]	CLIM
FPGA devices or CMOS tech	TSMC 0.13μm	TSMC 0.13μm	HLMC 55nm	Xilinx XCKU040	Xilinx ZCU104
FPGA resources or Equivalent gate count	**9.28K**	30.6K	98K	LUTs 151K Regs 121K	LUTs 18.4K Regs 5.6K
Line buffer	**4**	6	10	46	10
Memory size (Bytes)	**7680**	11520	62.4K	194K	63.1K
Fmax (MHz)	280	279	333	150	250
Throughput (Mpixels/s)	280	–	194.5	**600**	144
Energy efficiency (Mpixels/J)	–	29166	23661	105.5	157.3

IV. Conclusion

This paper proposes a bicubic interpolation method combined with Clamp-Laplace composite filters and the experimental results show that the proposed method improves PSNR by 1.76% and SSIM by 1.02% compared to the traditional bicubic interpolation method. The proposed method is implemented on ZCU104 FPGA platform, with a reduction of 5.23% in PSNR and 1.06% in SSIM compare with software implementation, a processing speed of 144 Mpixels/s, and energy efficiency of 157.3 Mpixels/J.

Acknowledgment

This work was supported in part by the National Natural Science Foundation of China under Grant 62204044, and in part by the State Key Laboratory of Integrated Chips and Systems under Grant SKLICS-K202302, and in part by the Special funds for promoting high-quality industrial development in Shanghai under Grant JJ-ZDHYLY-01-23-0004.

References

[1] A. Guo, E. Lin, J. Zhang, and J. Liu, "An energy-efficient image filtering interpolation algorithm using domain-specific dynamic reconfigurable array processor," Integration, vol. 96, p. 102167, 2024.

[2] K. Sun, M. Koch, Z. Wang, S. Jovanovic, H. Rabah, and S. Simon, "An fpga-based residual recurrent neural network for real-time video super-resolution," IEEE Transactions on Circuits and Systems for Video Technology, vol. 32, no. 4, pp. 1739–1750, 2021.

[3] P. Zhi-Yong, H. Z. Tan, and C. Di-Hu, "An improved low-cost adaptive bicubic interpolation arithmetic and vlsi implementation," Acta Automatica Sinica, vol. 39, no. 4, pp. 407–417, 2013.

[4] W. Y. V. Leung, P. J. Bones, and R. G. Lane, "Statistical interpolation of sampled images," Optical Engineering, vol. 40, pp. 547–553, 2001.

[5] R. Keys, "Cubic convolution interpolation for digital image processing," IEEE transactions on acoustics, speech, and signal processing, vol. 29, no. 6, pp. 1153–1160, 1981.

[6] C. Dong, C. C. Loy, and X. Tang, "Accelerating the super-resolution convolutional neural network," in Computer Vision–ECCV 2016: 14th European Conference, Amsterdam, The Netherlands, October 11-14, 2016, Proceedings, Part II 14. Springer, 2016, pp. 391–407.

979-8-3503-6184-1/24 $31.00 © 2024 IEEE

MTJ based Temperature Tracking Read/Write Assist for High Speed SRAM Bitcell

Yongliang Zhou*†, Chengxing Dai, Jingxue Zhong, Yingxue Sun, Xin Li, Chunyu Peng

*School of integrated circuits, Anhui University, HeFei, China
†Anhui Anxin Electronic Technology Co., Ltd
* Email: zhouyongliang@ahu.edu.cn

Abstract—**In this paper, we present the implementation of a temperature-adaptive assist circuit (TAA) aimed at stabilizing the read and write times of high-speed SRAM cells over a wide temperature range. The assist circuitry is composed of Magnetic Tunnel Junction (MTJ) and 22nm Fully Depleted Silicon-On-Insulator (FDSOI) technology (VDD=0.8). Over a temperature range of 230K to 370K, the variation in write access time is significantly reduced by 87%, from 0.547 ps to 0.072 ps, while the variation in read access time is reduced by 62.5%, from 0.8 ps to 0.3 ps. The maximum read and write access time are reduced by 0.988ps and 2.21ps, respectively.**

Keywords—*High Speed SRAM, Temperature, MTJ, Write/Read assist.*

I. INTRODUCTION

Embedded SRAM is a key component that usually determines the overall performance of the processor, and read and write latency is a key parameter that determines the performance of SRAM. The mismatch in High-Speed SRAM cell caused by temperature is a problem that cannot be ignored. As shown in Fig 1, an increase in temperature may lead to fail to write or read correctly in a too high temperature. The lower cell-VDD (LCV) , using negative BL (NBL) technology [1] and the WL underdrive scheme (WLUD) are widely used for read/write [2]. But few of them involve temperature considerations. The threshold voltage and mobility are influenced by the temperature, and the expression of the MOSFET on-current is shown below

$$I_D = \frac{1}{2}\mu C_{OX}\frac{W}{L}(V_{GS} - V_{TH})^2 \quad (1)$$

Here, I_D is the on-current in the saturation region, μ is the low field mobility, V_{GS} is the voltage difference between the gate and the source, C_{OX} is front gate oxide capacitance, W and L are the width and length of the MOSFET, respectively. According to (1), the on-current of the MOSFET is affected by both the threshold voltage (V_{TH}) and the μ. But the influence of the mobility is greater [3]. Then the write/read speed of SRAM cell will decrease, and leading to a large change in the write/read speed at a wide temperature, as shown

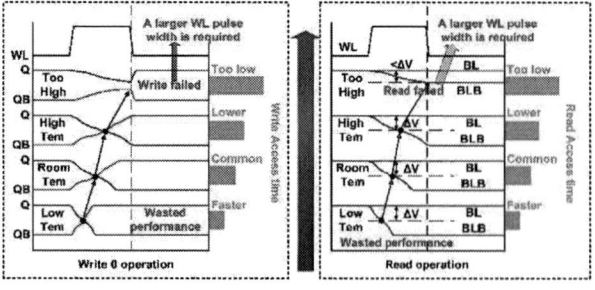

Fig. 1. Variation in write/read time due to temperature

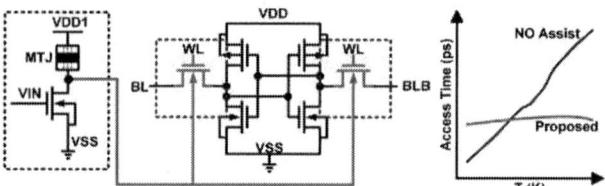

Fig. 2. proposed assist circuit

in Fig 1, too fast write/read speed will cause performance waste, and too slow write/read speed may lead to the final write/read failure.

To alleviate the aforementioned issues and increase the read/write speed, this paper proposes a temperature auto-tracking assist circuit shown in Fig 2.

II. CHANGES IN FDSOI V_{TH} AND MTJ RESISTANCE

A. FDSOI MOSFET

Fig 4 (a) shows the relationship between the V_{TH} of small size (L=20nm) and large size (L=800nm) N-MOSFET with the back gate bias voltage (V_B) under different operating conditions, while Fig 4 (b) shows the relationship between the threshold voltage of N-MOSFET with temperature at different sizes, where V_{lin} and V_{sat} are the threshold voltage in the linear region and the saturation region, respectively. The threshold voltage decreases as the temperature rises. Generally speaking, the relationship between V_{TH} and temperature can be expressed by the following expression [4]

$$V_{TH} = V_{FB} + \frac{kT}{q}\cdot\ln\left(\frac{Q_{ith}}{q\cdot n_i\cdot t_{si}}\right) + \frac{Q_{ith} + q\cdot N_a\cdot t_{si}}{C_{ox}} \quad (2)$$

Where n_i is the intrinsic carrier concentration. Among them the n_i is the main factor affecting the temperature characteristics of V_{TH}. And n_i increases as the temperature rises, so the threshold voltage decreases as the temperature rises. And due to the occurrence of the short channel effect, the value of the V_{TH} (L) will be smaller at a small size. The relationship between the threshold voltage and the back gate bias voltage (V_B) can be obtained as [5]

$$V_{TH2} - V_{TH1} =$$
$$U_T \cdot \left(\frac{1}{n_{ff}} - \frac{1}{n_{fb}}\right) \cdot \left[-\frac{A\cdot\log\left(e^{B\theta} + \frac{A+C-1}{C-1}\right)}{B(C-1)(A+C-1)} + \frac{\theta\cdot(A+C)}{A+C-1}\right]_{\theta_1}^{\theta_2^{-1}} \quad (3)$$

The formula (3) can obtain the subsequent change result by only giving an initial value.

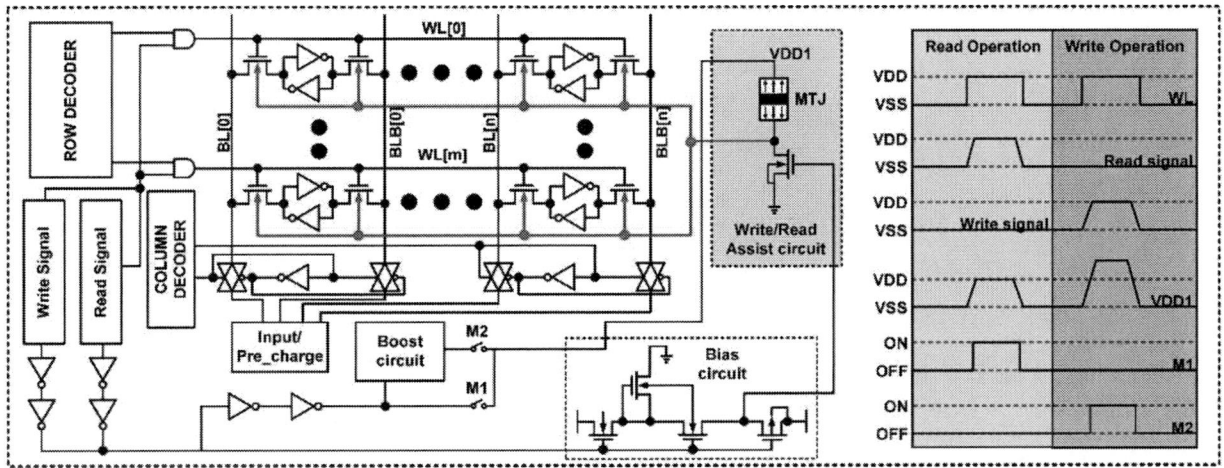

Fig. 3. TAA implementation in the read/write operation

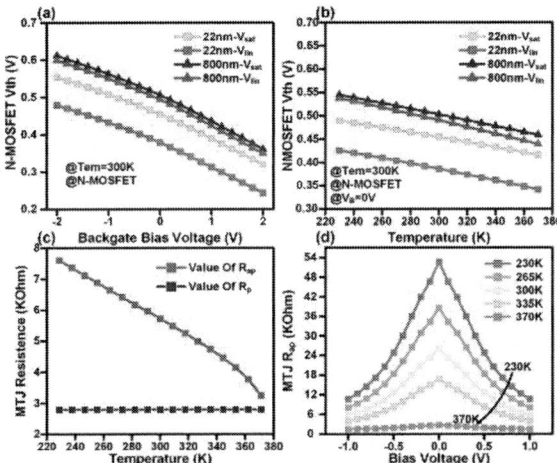

Fig. 4. Threshold voltage variations and MTJ resistance changes under different conditions

B. MTJ

Fig 4 (c) and (d) show the variation of the resistance of MTJ at different temperatures and voltages, where R_{ap} and R_p correspond to the high (antiparallel states) and low (parallel states) resistance states of MTJ, respectively. As can be seen from Fig 4 (c), the R_{ap} is greatly affected by temperature, while the resistance of MTJ in the low resistance state is basically not affected by temperature. Fig 4 (d) shows the resistance of MTJ increases as the absolute value of the applied voltage decreases. The temperature and voltage dependence of the resistance in parallel R_p and antiparallel R_{ap} state according to [3] and [6] can be expressed as

$$R_{ap}(T,V) = R_{ap}(0,0) \cdot \left\{ \begin{array}{l} \left[1 + Q\frac{1}{\xi}\frac{2S}{E_m}kTln\left(\frac{kT}{E_c}\right)\right]^{-1} \\ \cdot \left[1 + \left(\frac{V}{V_h}\right)^2\right] \end{array} \right\} \quad (4)$$

$$R_p(T,V) = R_p(0,0) \cdot \left\{ \begin{array}{l} \left[1 + Q\xi\frac{2S}{E_m}kTln\left(\frac{kT}{E_c}\right)\right]^{-1} \\ \cdot \left[1 + \left(\frac{V}{V_h}\right)^2\right] \end{array} \right\} \quad (5)$$

Where V_h is a voltage parameter for which TMR becomes half the value of TMR$_0$. The value of V_h is defined by tunneling magnetoresistance (TMR), which is defined as the ratio between the high resistance in the AP state (R_{ap}) and the low resistance in the P state (R_p) of the MTJ, it can be expressed as

$$TMR = TMR_0 \cdot \left[1 + \left(\frac{V}{V_h}\right)^2\right]^{-1} \quad (6)$$

The TMR here does not take into account the effect of temperature, but only the change in TMR caused by the bias voltage, where TMR$_0$ is at the zero bias. Based on the magnitude of the parameters mentioned [3] and combined with the resistance expression of the MTJ above, we can know that R_{ap} is more affected by temperature and that as the temperature increases, R_{ap} shows a decreasing trend. The assist circuit of this design is expected to change the bias voltage of the MTJ by about 0.3 V and the temperature to change by 140 K. Based on these parameters, it can be calculated that the change in resistance caused by the bias voltage in the circuit proposed in this paper is negligible. All subsequent designs only need to take into account the effects of temperature.

III. TEMPERATURE-ADAPTIVE WRITE ASSIST CIRCUITRY

The application of the write-assist circuit proposed in this paper in the read/write operation is shown in Fig 3, where the assist circuit consists of MTJ and FDSOI N-MOSFET. According to [3] and [7], the resistance of the MOSFET increases with the increase of temperature, while the R_{ap}

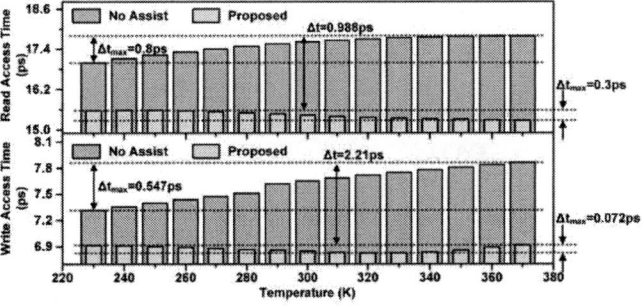

Fig. 5. The comparison of write access time before and after improvement (TT proccess corner)

979-8-3503-6184-1/24 $31.00 © 2024 IEEE   644

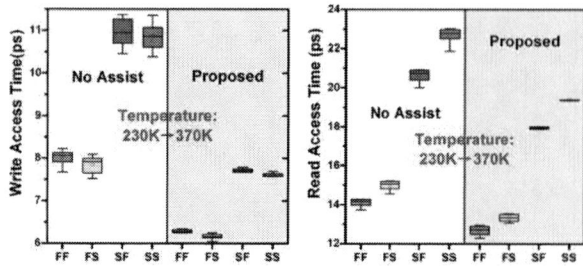

Fig. 6. The relationship between read/write access time of 6T SRAM and temperature across different process corners

decreases with the increase of temperature, so the output voltage of the assist circuit increases with the increase of temperature. Then the V_{TH} decreases with increasing temperature and V_B. So the V_{TH} change ratio with temperature can be increased by applying the back gate bias voltage to resist the effect of the mobility change, thereby reducing the influence of temperature on SRAM write/read access time.

As shown in Fig 5, the maximum variation in read and write access time is reduced to 0.3ps and 0.072ps, respectively. The maximum read and write access time are reduced by 0.988ps and 2.21ps, respectively. Proposed assist circuit also exhibits excellent stability and highspeed characteristics at different process corner as shown in Fig. 6, especially in SS and SF. The 5000 Monte Carlo (MC) distribution of the read and write access time are shown in Fig 7. The minimum WSNM and minimum WM in Fig 8 increased from 0.3006V and 0.234V to 0.3262V and 0.3152V, respectively.

IV. SUMMARY

As presented in Table 1, this paper proposes the temperature-adaptive assist circuit (TAA) from the perspectives of temperature management and access time optimization, which can greatly reduce the change of write/read speed caused by temperature change. Under TT process corner, the minimum WSNM and minimum WM increase from 0.3006V and 0.234V to 0.3262V and 0.3152V, respectively. The maximum read and write access time are reduced by 0.988ps and 2.21ps, respectively. The maximum change in write access time decreases from 0.547 ps to 0.072 ps by 13%, the maximum change in read access time decreases from 0.8 ps to 0.3 ps by 37.5%.

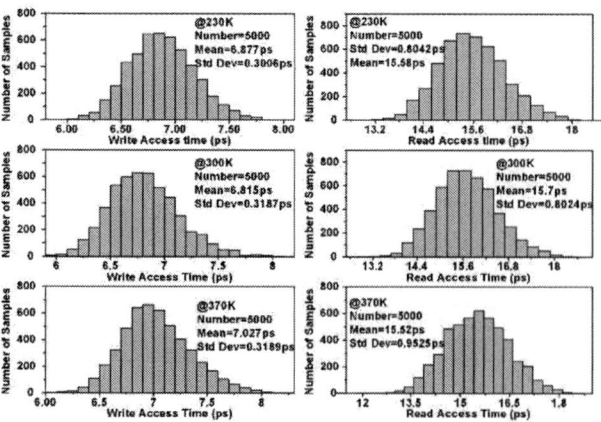

Fig. 7. The 5000 Monte-Carlo distribution of proposed assist circuit for read and write access time

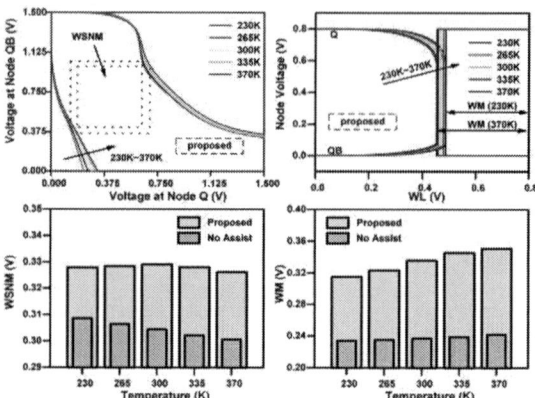

Fig. 8. WSNM and WM 6T SRAM cell (TT proccess corner)

TABLE I. SRAM READ/WRITE ASSIST CIRCUITS

Scheme	Research on SRAM read/write assist circuits		
	Focus	Factors	Methods
[1]	Write	V_{MIN}	LCV; NBL
[2]	Read;Write	V_{MIN}; Write-ability; Variation tolerance	WLUD; TVC
[8]	Read;Write	Voltage; Temperature	VATA; TATA
Proposed	Read;Write	Temperature; Access time	TAA

ACKNOWLEDGMENT

This work is supported by Anhui Provincial Natural Science Foundation 2308085QF214, Natural Science Foundation of the Higher Education Institutions of Anhui Province under Grant 2023AH040011 and National Natural Science Foundation of China under Grant 62274001.

REFERENCES

[1] Y. -H. Chen et al., "A 16 nm 128 Mb SRAM in High-κ Metal-Gate FinFET Technology With Write-Assist Circuitry for Low-VMIN Applications," in IEEE Journal of Solid-State Circuits, vol. 50, no. 1, 2015, pp. 170-177.

[2] V. P. -H. Hu, M. -L. Fan, P. Su and C. -T. Chuang, "Analysis of GeOI FinFET 6T SRAM Cells With Variation-Tolerant WLUD Read-Assist and TVC Write-Assist," in IEEE Transactions on Electron Devices, vol. 62, no. 6, 2015, pp. 1710-1715.

[3] Drewello V , Schmalhorst J , Thomas A ,et al. "Evidence for strong magnon contribution to the TMR temperature dependence in MgO based tunnel junctions," in Physical Review B, vol. 77, no. 1, 2008, pp. 014440.

[4] M. Shin et al., "Low temperature characterization of 14nm FDSOI CMOS devices," 2014 11th International Workshop on Low Temperature Electronics (WOLTE), Grenoble, France, 2014, pp.29-32.

[5] H. -C. Han, Z. Zhao, S. Lehmann, E. Charbon and C. Enz, "Novel Approach to FDSOI Threshold Voltage Model Validated at Cryogenic Temperatures," in IEEE Access, vol. 11, 2023, pp. 56951-56957.

[6] H. Lim, S. Lee and H. Shin, "Advanced Circuit-Level Model for Temperature-Sensitive Read/Write Operation of a Magnetic Tunnel Junction," in IEEE Transactions on Electron Devices, vol. 62, no. 2, 2015, pp. 666-672.

[7] J. Liu, H. Ohsato, B. Da and Y. Koide, "Investigation of Ohmic Contact Resistance, Surface Resistance, and Channel Resistance for Hydrogen-Terminated Diamond MOSFETs," in IEEE Transactions on Electron Devices, vol. 69, no. 3, 2022, pp. 1181-1185.

[8] I. Lee et al., "24.3 A Voltage and Temperature Tracking SRAM Assist Supporting 740mV Dual-Rail Offset for Low-Power and High-Performance Applications in 7nm EUV FinFET Technology," 2019 IEEE International Solid-State Circuits Conference - (ISSCC), San Francisco, CA, USA, 2019, pp. 392-394

A 12V to 1V Tri-state DSD Hybrid Converter by Self-Balanced Dual Flying Capacitors with 0.3mV Output Ripple and 90.09% Peak Efficiency

Yixing Wang[1], Qianhui Liu[1], Yuhua Chen[1], Yizhe Yang[1], Yimeng Zhang[1,2]*, Yuming Zhang[1,2]

[1]School of Microelectronics, Xidian University
[2]Shaanxi Key Lab of Integrated Circuits and Systems, Xidian University
Xian 710000, China
* Email: zhangyimeng@xidian.edu.cn

Abstract—**This paper presents a tri-state Double Step-Down (DSD) hybrid converter that enables direct 12 V to 1 V DC-DC power transfer with high voltage conversion ratio and high power density. The converter in this paper combines and improves the advantages of the 3-level and DSD, enabling the use of low-voltage power devices. The dual-loop output reduces the output voltage ripple while expanding the minimum on-time by a factor of six compared to conventional buck converter, facilitating high switching frequency operation. A mirroring device is used to avoid control delays during the master-slave phase. The converter is designed based on a 0.18μm BCD process and can convert 12.0 V to 1.0 V at 1MHz. The converter achieves 90.09% peak efficiency for 12 V to 1 V conversion. It also has a 0.3 mV output voltage ripple at a 1 A current load.**

I. INTRODUCTION

The growth of the automotive and data centre industries has led to a rise in demand for high-performance power converters. Among these, 12 V power systems are particularly prevalent. In such systems, there is a pressing necessity for single-stage buck DC-DC converters to directly supply current to the point of load, in order to achieve high efficiency and high power density. Conventional buck converters have short on-time at high voltage Conversion Ratios (CR), limiting high frequency operation. Double Step-Down (DSD) converters can have a longer on-time [1-3], however, this converter needs a switch with a voltage of V_{IN}. New research has led to the development of mixed-mode DC-DC converters, such as the tri-state DSD [4], which combines a 3-level structure with a DSD structure that reduces the switch in the DSD with a voltage stress of V_{IN} to $V_{IN}/2$ [5]. However, this converter has more states and requires seven power switches and additional flying capacitor balancing circuits. It also has a voltage CR of only D/2, but there is still room for improvement.

This paper proposes a new tri-state DSD hybrid converter to solve the above problems. The converter combines a 3-level buck [6] and a modified DSD. The tri-state DSD hybrid converter has the following advantages: it can operate in the MHz range; it has no additional voltage balancing circuits, reducing complexity; and the converter voltage CR reaches D/3; and the low-side power switches have a voltage stress of $V_{IN}/3$, while the high-side switches in the traditional DSD structure has a voltage stress of $2\,V_{IN}/3$, which reduces the switching loss of the converter.

II. TOPOLOGY AND WORKING PRINCIPLE

A. Power stage structure

To make the high-voltage conversion ratio circuit suitable for high-frequency operation and achieve high power density, we combine the 3-level buck converter with the DSD structure and improve the converter. For the 3-level structure, shown in Fig. 1. A flying capacitor is added to the stacked buck converter to reduce the voltage stress to $V_{IN}/2$ for each power switch. Two power switches in the discharge path mean more conduction losses. The DSD converter is a modified two-phase buck using a four-switches dual inductor and a flying capacitor, reducing voltage stress on most power switches to $V_{IN}/2$. Concurrently, an internal feedback mechanism enables the flying capacitor to maintain a state of equilibrium without the necessity for supplementary balancing circuits. For a specific F_{SW}, the DSD converter exhibits a T_{ON} that is twice that of a conventional buck. The two-phase parallel structure improves current delivery and reduces voltage ripple. However, the DSD structure has a high voltage stress on one of the high-side power switches, which is a challenge for high power density design. The voltage CR of the two converters are both D/2, which can be improved.

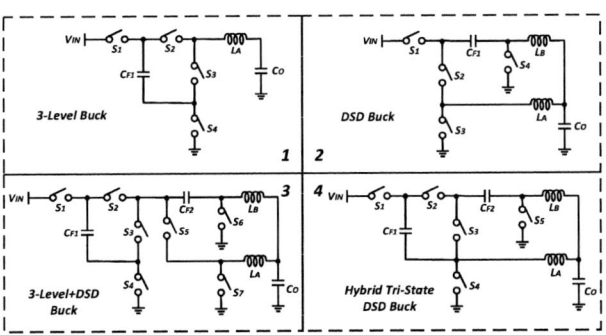

Fig. 1. The derivation of the hybrid buck converter proposed.

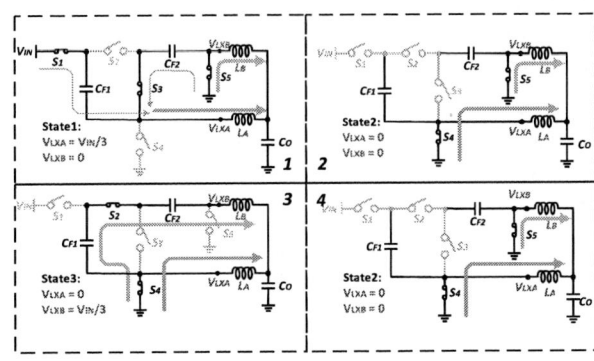

Fig. 2. Tri-state DSD hybrid converter principle of operation.

979-8-3503-6184-1/24 $31.00 © 2024 IEEE

This study combines the advantages of the two converters by using a 3-level structure to reduce voltage stress and a DSD structure to reduce output ripple. It proposes a tri-state Double Step-Down (DSD) hybrid converter, shown in Fig. 1. It has five power switches, two flying capacitors, and two inductors. The voltages of the flying capacitors are $2 V_{IN}/3$ and $V_{IN}/3$, which makes the voltage stress of the power switches $2 V_{IN}/3$ and $V_{IN}/3$, respectively. This effectively solves the problem that the voltage stress of the power switches on the high side in the DSD structure is V_{IN}.

B. Operating states

This tri-state DSD hybrid converter has three states. See Fig. 2. In state 1, power switches S_1, S_3, and S_5 conduct, C_{F1} is charged, C_{F2} is discharged, inductor L_A is charged, and L_B is discharged. In state 2, S_4 and S_5 conduct, while the two flying capacitor voltages remain unchanged. Both L_A and L_B are discharged. In state 3, S_2 and S_4 conduct, with C_{F2} and L_B being charged by the flying capacitor C_{F1}, while L_A discharges. In a single cycle, the operating sequence of each state is state 1, state 2, state 3, state 2. The corresponding operating waveforms are illustrated in Fig. 3. In the steady state, the phase difference between P_A and P_B is 180° and the duty cycle is less than 50%.

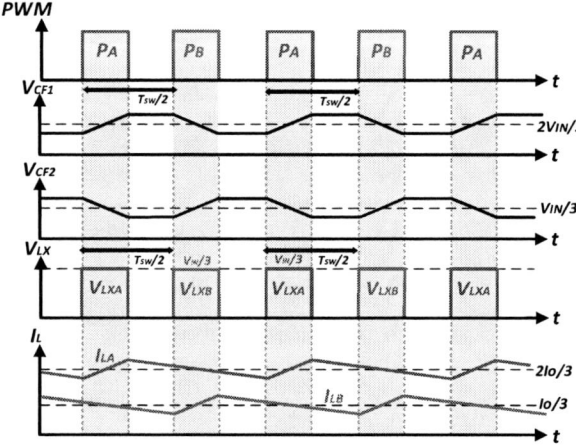

Fig. 3. Tri-state DSD hybrid converter operating waveforms.

III. SYSTEM AND CRITICAL CIRCUIT DESIGN

A. System Design

The system block diagram of the tri-state DSD hybrid converter is shown in Fig. 4. It consists mainly of a power stage, a bootstrap driver module, a level shifter, a dead time controller and a master phase mirror. To meet the high frequency and high power density requirements, all power switches are implemented in NMOS and driven by floating rail bootstrap gate driver modules. For a 12 V input, M_{N1}, M_{N4} and M_{N5} have a voltage stress of 4 V, allowing smaller and faster low voltage devices to be used. To reduce the complexity of the controller, PWM peak current mode control is used and the controller contains two feedback loops: Phase A and Phase B. Phase A is the master phase and Phase B is the slave phase. V_{OUT} and V_{REF} are compared with the sampled inductor current after the Type-II EA output to produce the master phase duty cycle PA, which is then mirrored by the master phase to produce the slave phase duty cycle PB.

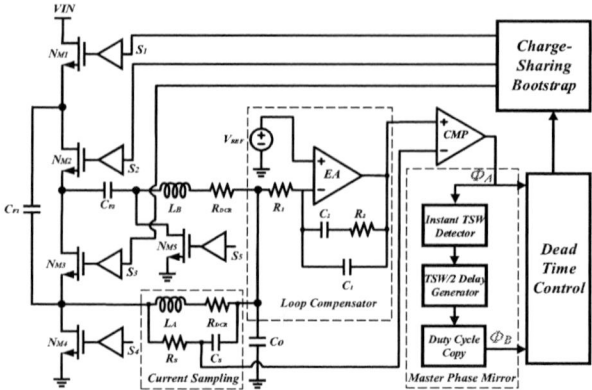

Fig. 4. System block diagram of hybrid tri-state buck converter.

B. Adaptive delay-compensated master phase mirrorsr

An adaptive delay-compensated master phase mirroring device is capable of generating a duty cycle for a slave phase based on the duty cycle information of the master phase. Additionally, the device incorporates an adaptive delay-compensated $T/2$ delay generator and a duty cycle replicator. Fig. 5 shows the schematic of the $T/2$ delay generator, which is divided into two parts: cycle detection and $T/2$ delay generation. When the rising edge of the main phase P_A arrives, $\Phi 1$ is activated and the current I_D charges C_{S1}, after which it passes through the peak voltage detection module to obtain V_{S2}, which contains the cycle. When M_{N1} is off, $\Phi 2$ works to charge C_{S2}. $\Phi 1$ and $\Phi 2$ operate alternately to maintain a constant voltage across C_{T1}. The delay compensation current I_{OS} is used to compensate for the loop delay. After a delay of $T/2$, the comparator outputs a pulse signal. The duty cycle replicator is divided into two parts, duty cycle time capture and duty cycle replication, and operates similarly to the duty cycle detector. When the $T/2$ delay generates a high potential flag signal, the duty cycle replicator module is enabled, generating the P_B master phase mirror module with the same on-time as the P_A.

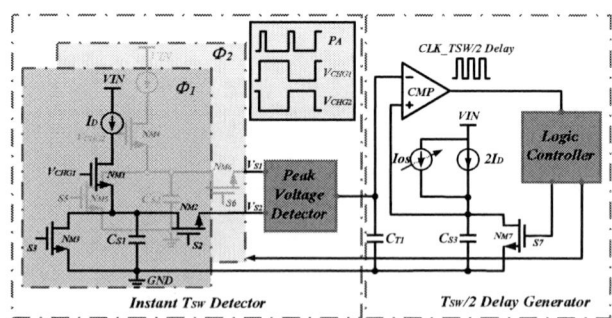

Fig. 5. T/2 delay generator with adaptive delay compensation.

IV. SIMULATION RESULTS ANALYSIS

The tri-state DSD hybrid converter is implemented in a 0.18 μm BCD process with DC-DC voltage conversion from 12 V to 1.0 V and a test load current of 0.2 A-3.0 A. The parameters are as follows: V_{IN}=12 V, V_O=1 V, F_{SW}=1 MHz, L_1 = L_2= 4.7 μH, C_{F1}= C_{F2}= 2.2 μF, Co = 22 μF. The voltage stress of N_{M2} and N_{M3} is 8 V and the ratings are chosen to be 12 V to withstand the voltage stress during pre-charging. N_{M1},

N_{M4} & N_{M5} have a voltage stress of 4 V and the ratings are chosen to be 5 V for the low voltage devices.

Fig. 6 depicts the steady-state waveforms of the proposed hybrid converter under a load of 1 A. The average values of the two-phase currents I_{LA} and I_{LB} are 2 I_o/3 and I_o/3, respectively, equating to 666.6 mA and 333.3 mA, with a ripple of 165 mA. The average value of the output current I_o is 1 A. The ripple of the output voltage is 99 mV, while the ripple of the output current is 0.3 mV. The P_A is duty-cycled to the master phase after half a cycle, and a mirror copy of the P_A is made to obtain the P_B. Fig. 7 shows the efficiency of the circuit at 1 MHz switching frequency with load current Io varying from 0.1 A to 3 A, where the peak efficiency is 90.09%.

Finally, Table 1 shows the performance comparison with the state-of-the-art converters, and the proposed converter has higher voltage conversion ratio and power density to meet the high switching frequency application requirements.

TABLE I. PERFORMANCE COMPARISON WITH PREVIOUSLY PUBLISHED WORKSS

	JSSC 2020 [1]	JSSC 2023 [2]	JSSC 2022 [3]	JSSC 2021 [4]	**This Work**
Topology	DSD	CCC	DSD	Tri-State DSD	Hybrid Tri-State DSD
Input Voltage	48 V	12 V	12 V	12 V	12 V
Output Voltage	1 V	0.9-1.8 V	1 V	1 V	1 V
Switching Frequency	2 MHz	2 MHz	1 MHz	1 MHz	1 MHz
Flying Capacitor	1 uF	2×2.2 uF	4.7 uF	2×1 uF	2×2.2 uF
Load Current	1.5 A	1 A	4 A	3 A	1 A
CR	D/2	D/2	D/2	D/2	**D/3**
Max Voltage Stress	V_{IN}	V_{IN}/2	V_{IN}	V_{IN}/2	**2 V_{IN}/3**
Peak Efficiency	56.8%	86.8%	88.3%	88.3%	**90.09%**

V. CONCLUSION

This paper describes a hybrid tri-state DSD converter that enables direct high voltage conversion ratio power conversion. The design proposed in this study combines the advantages of a 3-level and a DSD converter, thereby greatly mitigating some serious challenges, including high-voltage stress, an extremely short duty cycle time, and the need for additional flying capacitor balancing circuits. In order to avoid the control delay of master-slave phase, the master-phase mirrors with adaptive delay compensation are designed. The experimental results provide compelling evidence that the design is effective.

ACKNOWLEDGMENT

This work was supported by the National Natural Science Foundation of China (Grant No. 62234010).

REFERENCES

[1] D. Yan, X. Ke and D. B. Ma, "Direct 48-/1-V GaN-Based DC–DC Power Converter With Double Step-Down Architecture and Master–Slave AO2T Control," in IEEE Journal of Solid-State Circuits, vol. 55, no. 4, pp. 988-998.

[2] T. Hu, M. Huang, R. P. Martins and Y. Lu, "A 12-to-1 Flying Capacitor Cross-Connected Buck Converter With Inserted D > 0.5 Control for Fast Transient Response," in IEEE Journal of Solid-State Circuits, vol. 58, no. 11, pp. 3207-3218.

[3] Z. Liu, J. Yuan, F. Wu and L. Cheng, "A 12V/24V-to-1V PWM-Controlled DSD Converter With Delay-Insensitive and Dual-Phase Charging Techniques for Fast Transient Responses," in IEEE Journal of Solid-State Circuits, vol. 57, no. 12, pp. 3853-3864.

[4] K. Wei, Y. Ramadass and D. B. Ma, "Direct 12V/24V-to-1V Tri-State Double Step-Down Power Converter With Online VCF Rebalancing and In-Situ Precharge Rate Regulation," in IEEE Journal of Solid-State Circuits, vol. 56, no. 8, pp. 2416-2426.

[5] P. S. Shenoy et al., "A 5 MHz, 12 V, 10 A, monolithically integrated two-phase series capacitor buck converter," 2016 IEEE Applied Power Electronics Conference and Exposition (APEC), Long Beach, CA, USA, 2016, pp. 66-72.

[6] W. Jung et al., "Dual-Path Three-Level Buck Converter With Loop-Free Autocalibration for Flying Capacitor Self-Balancing," in IEEE Transactions on Power Electronics, vol. 36, no. 1, pp. 51-55, Jan. 2021.

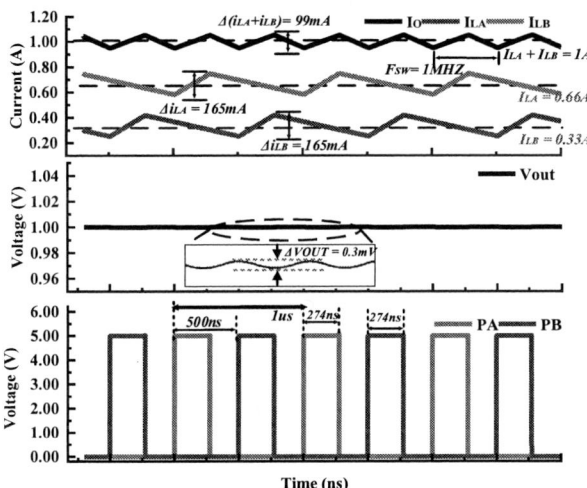

Fig. 6. System simulation results.

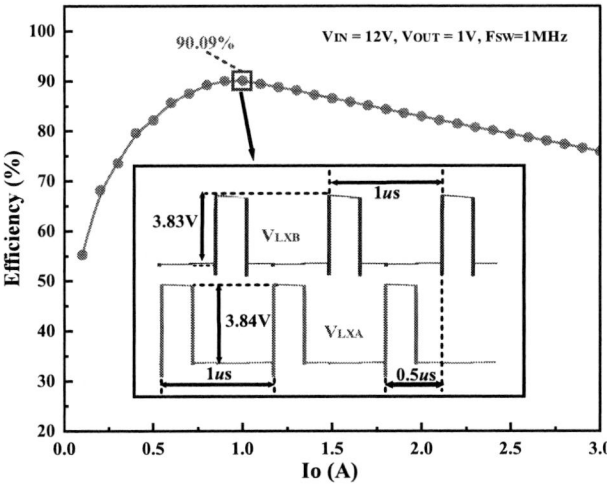

Fig. 7. Measured efficiency versus load current.

System-level Evaluation of AOS Gain Cell eDRAMs for Low-power Normally-off Computing

Long Chen [1,2], Yecheng Yang [2], Wei Li [2], and Shao Hao Wang [2*]

[1] School of Advanced Manufacturing, Fuzhou University, Fujian Jinjiang 362251, China
[2] FZU-Jinjiang Joint Institute of Microelectronics, Fuzhou University, Fujian Jinjiang 362251, China
* Email: shwang@fzu.edu.cn

Abstract—**Recent gain cell (GC) embedded dynamic random access memory (eDRAM) using short channel length amorphous oxide semiconductor (AOS) field-effect transistors (FETs) has demonstrated extremely low leakage current and promising three-dimensional integrating capacity. This makes AOS-GC eDRAM a promising technology for normally-off computing applications. However, the lack of comprehensive models and evaluation approaches prevents the exploration and optimization of its design space. We presented a system-level simulation approach, *GCSim*, to perform power analysis for low-power scenarios. The results indicate that AOS-GC eDRAM can provide higher density, lower static power, and dynamic energy compared to static random access memory (SRAM) and silicon-based GC (Si-GC) eDRAM. For normally-off computing applications, such as the intermittent wake-up deep neural network (DNN), the quasi-nonvolatile AOS-GC eDRAM can achieve up to 96.7% energy savings compared to SRAM when the wake-up interval exceeds one second.**

Keywords—*amorphous oxide semiconductor (AOS), eDRAM, low-power, gain cell (GC), normally-off computing*

I. INTRODUCTION

With the rise of the Internet of Things (IoT), there is a growing demand for low-power system-on-chip solutions to support IoT sensor nodes and edge artificial intelligence (AI) nodes. To meet the reducing power consumption limitations, various techniques such as the Normally-off computing have been proposed to enhance the energy efficiency by activating the system only during computing [1]. Such scenarios also require lower power memory modules. However, conventional static random access memory (SRAM) and silicon-based gain cell (Si-GC) embedded dynamic random access memory (eDRAM) suffer from significant static power issues caused by leakage current and refresh power. Emerging nonvolatile memories have lower static power but higher dynamic energy during computing [2].

Amorphous oxide semiconductor (AOS) GC eDRAM based on AOS field-effect transistors (AOSFET) offers low leakage power and low dynamic energy, making it a promising quasi-nonvolatile memory for Normally-off computing systems. The AOSFETs can be categorized by structure into planar and vertical types. The back-gate (BG) structure is a typical planar type, which offers high on-current, but it has a large footprint [3]. The channel-all-around (CAA) structure in the vertical type offers higher memory density but lower on-current [4]. The structures of the BG and CAA types are shown in Fig. 1(a) and 1(b), respectively.

For different structures, AOSFETs have to be optimized at the array levels for various applications. A system-level evaluation approach is thus required to rapidly explore this design space, which can quickly assess among multiple design options without the need for long and complex full design iterations. NVSim tools can effectively models the circuit-level behaviors of caches and main memories composed of

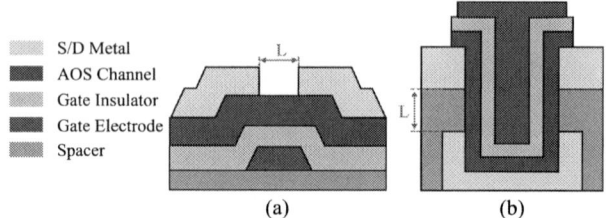

Fig. 1. The structures of AOSFETs. (a) Planar back-gate type. (b) Vertical channel-all-around type.

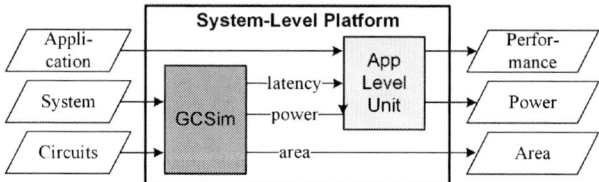

Fig. 2. Diagram of the proposed system-level evaluation platform.

SRAM and non-volatile memories [5]. NVMExplore extends NVSim by enhancing the end-to-end evaluation process and providing application-level evaluation [6]. GEMTOO is the first circuit-level simulator to support Si-GC eDRAM, can simulate the timing and area of memory arrays [7]. However, GEMTOO still lacks a power analysis model and cannot simulate the quasi-nonvolatile nature of AOS-GC eDRAM.

In this paper, we introduce GCSim, a circuit-level model based on NVSim, to provide power analysis for GC arrays, as well as to simulate the behaviors of dual-port operation, word line boosting, and read bit line saturation. By integrating GCSim with NVMExplore, a system-level simulation can be performed for AOS-GC eDRAM in terms of power, area, and latency. Moreover, the power efficiency and latency of these memory arrays will also be evaluated in intermittent wake-up deep neural network (DNN) inference applications.

II. SYSTEM-LEVEL SIMULATION PLATFORM

Fig. 2 shows the block diagram of the proposed system-level evaluation platform, which consists of the GCSim module and the application-level unit (ALU). There are three categories of input parameters: (1) Application-level parameters include application-specific workload information. (2) Systems-level parameters include memory array capacity and subarray organization configurations. (3) Circuit-level parameters include the process nodes, the specifications of the memory cell, and specific circuit design parameters.

The proposed GCSim module consists of the power model, the area model, and the latency model of GC eDRAM. It can thus calculate the area, latency, and power of the target memory array based on the input parameters and passes the calculated latency and power consumption to ALU module. The ALU will further evaluates the array's performance and power consumption in specific application scenarios.

979-8-3503-6184-1/24 $31.00 © 2024 IEEE   649

The GC eDRAM cell and read/write circuits are shown in Fig. 3(a). The cell can be a two transistors no capacitor (2T0C) or other GC structures. During the write operation, the write word line (WWL) is biased to a boosted voltage, allowing data (V_{Write}) from the write bit line (RBL) to be written to the storage node (SN) through the write transistor (WT). For the read operation, the voltage at the SN is converted into read current (I_{RT}) by the read transistor (RT), and the sense amplifier (SA) reads the data by comparing the swing of the read bit line (RBL) with the reference voltage (V_{REF}). In the hold state, all control signals are biased to fixed levels. The refresh operation consists of read and write-back steps, refreshing all subarrays in parallel. The retention time of AOS-GC is a few kilo-seconds and 8 orders of magnitude longer than that of Si-GC. Due to its quasi-nonvolatile property, AOS-GC eDRAM can be considered always available and does not require frequent refresh operations [7].

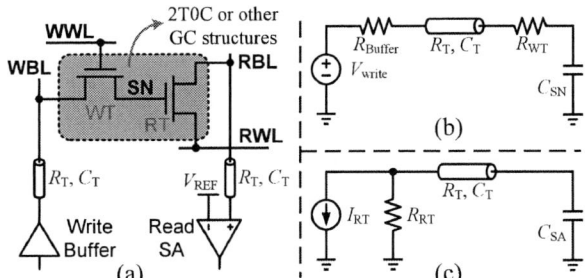

Fig. 3. (a) Gain cell and the basic read/write circuits. The equivalent circuits of the write path (b) and read path (c).

TABLE I. OMPARSION FOR SIMULATED AREA, LATENCY, AND ENERGIES OF THE GCSIM

Parameters	2T1C AOS-GC [1]			2T0C Si-GC [8]		
	Mea.	Sim.	Diff.	Mea.	Sim.	Diff.
Macro area (mm²)	1.75	1.30	−25.7%	0.17	0.16	−4.8%
Access latency (ns)	1.00	1.05	5.1%	1.65	1.62	−1.8%
Read energy (pJ)	56.0	54.4	−2.9%	/	11.9	/
Write energy (pJ)	64.0	62.0	−3.1%	/	10.3	/
Refresh power (μW)	/	52.0		217	195	−10.4%

A. Power Model of 2T0C GC eDRAM

In the read operation, the dynamic energy originates from the RBL swing, the RT conduction, as shown below

$$E_r = (C_{RBL}V_{RBL}\Delta V_{RBL} + V_{DS}I_{RT}t_{SA})N_c + E_{other} \quad (1)$$

where, C_{RBL} is the parasitic capacitance on RBL, N_c is the number of columns, V_{RBL} is the precharge voltage on RBL at the beginning of the read operation, ΔV_{RBL} is the voltage swing on RBL during the read operation, V_{DS} is the drain-source voltage of the RT, I_{RT} is the on-state current of the RT, and E_{other} is the dynamic energy consumption from the read-related peripheral circuits and interconnection network.

In the write operation, the dynamic energy mainly comes from the swing of WBL, as shown below [5]

$$E_w = (C_{WBL} + C_{SN})V_{WBL}^2 N_c + E_{other} \quad (2)$$

where, C_{WBL} is the parasitic capacitance on the WBL, C_{SN} is the capacitance of the SN, and V_{WBL} is the voltage of the WBL during the write operation.

A refresh operation consists of a read operation and a write operation. Therefore, its dynamic energy can be written as

$$E_{ref} = (E_r + E_w - 2E_{route})N_r \quad (3)$$

where, E_{route} is the interconnection network energy, and N_r is the number of rows in the array. Once the arrays are refreshed, the average static power can be written as

$$P_{static} = E_{ref}/t_{DRT} + P_{leakage} \quad (4)$$

where t_{DRT} is the data retention time and $P_{leakage}$ is the leakage power of the peripheral circuits.

B. GCSim Area Model

Different from single-layer Si-GC memory arrays, vertical AOS-GC memory allows for three-dimensional (3D) stacking above the layer of CMOS processes in peripheral circuits [4]. GCSim provides configuration interfaces for both of single layer and 3D stacking subarray area models.

C. GCSim Latency Model

The equivalent circuits for the read and write paths of the GC memory array are shown in Figs. 3(b) and 3(c), respectively. The read and write latency can be written as

$$t_w = t_{WWL} - \tau_w ln(1 - \alpha) + t_{route} \quad (5)$$

$$t_r = t_{RWL} - \tau_r ln(1 - \beta) + t_{SA} + t_{route} \quad (6)$$

where, τ_w and τ_r are the durations of writing and reading, respectively. t_{WWL} is the WWL latency and t_{RWL} is the RWL latency. α and β are swing ratios of WBL and RBL, respectively, which vary from 0 to 1. t_{route} and t_{SA} are the interconnection network latency between subarrays and the SA read latency, respectively [7].

III. VERIFICATION OF GCSIM AND EVALUATION

In this section, we will use the GCSim model to evaluate the performance of Si-GC arrays and 2T0C AOS-GC arrays.

A. Validation of GCSim

According to the reported AOS-GC and Si-GC array macros [1], [8]. The capacity and channel length of the AOS-GC array are 1 Mb and $F = 60$ nm, respectively. The cell area and retention time of 2T1C AOS-GC are 234 F^2 and 6000 s, respectively. The capacity of the 2T0C Si-GC array is 192 kb, and the CMOS process is $F = 65$ nm. Its cell area and the retention time are 113 F^2 and 110 μs, respectively. We used GCSim to evaluate the area, latency, and energies of these arrays. Table I compares the simulated results with the measured results. Except for a 25.7% difference in the AOS-GC area, GCSim can accurately estimate the latency, read/write, and refresh energies of these GC arrays with an accuracy of less than 10.4%. The area of AOS-GC is underestimated due to the lack of information on the overhead of actual peripheral circuits.

B. Evaluation of Different Memory Arraies

GCSim was used to evaluate the performance of Si-GC, BG AOS-GC, and CAA AOS-GC arrays. The corresponding parameters are shown in Table II. A 65 nm SRAM with default configuration is used for benchmarking. Here, both the SRAM and Si-GC arrays use a planar design for the area model, while the AOS-GC arrays use a two-layer 3D stacking

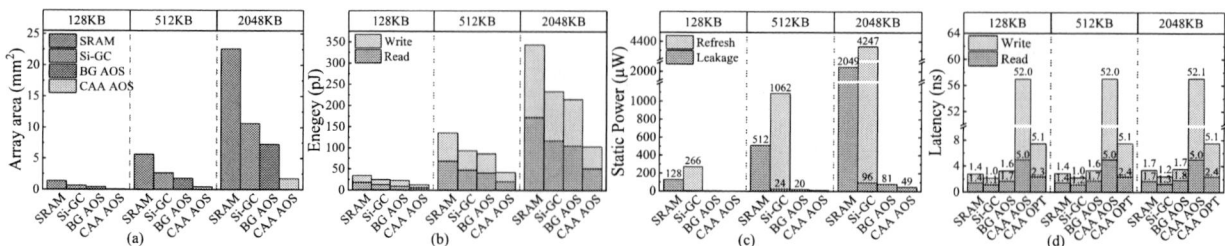

Fig. 4. Comparison of (a) the areas, (b) dynamic energies, (c) static powers, and (d) access latencies of SRAM and GC memories at different capacities.

design [4]. To emphasize the differences among these GC arrays, the subarray size is fixed at 128 rows and 256 columns with a word width of 64 bits. Fig. 4(a) compares the array areas of SRAM and these GC memories at different capacities. The CAA array has the smallest area, averaging 92.2% smaller than SRAM and 83.5% smaller than Si-GC. Fig. 4(b) compares the dynamic energies of these arrays. The CAA array has the smallest dynamic energy, averaging 66.7% smaller than SRAM and 52.4% smaller than Si-GC. This is due to the relatively low routing energy in a smaller array area. Fig. 4(c) compares the static power of these arrays. Since the AOS-GC almost no need for the refresh operation, the refresh power is significantly reduced. The static power of the CAA array is 97.6% and 98.9% smaller than that of SRAM and Si-GC, respectively. Fig. 4(d) compares the access latencies of these arrays. The BG array is on average 50% slower than Si-GC, while the CAA array is 24.5 times slower. This significant increase in latency is due to the relatively low on-current density of CAA AOS. We expect that if the WT and RT on-current densities of the CAA AOS can be increased to 5 and 4 A/m through process optimization, the read and write latencies could be reduced by 52.8% and 90.2%, respectively, approaching the performance of BG AOS arrays.

C. Evalution of an Intermittent Wake-up Scenario

In intermittent wake-up normally-off DNN applications, AOS-GC array can store the DNN weight data with extremely low static power consumption. By using the estimated area, latency, and energies by GCSim, Fig. 5 shows the total energy consumption for one day using the ResNet26 DNN model, with each array having a capacity of 2 MB. The results indicate that when the wake-up interval becomes longer than one second (10^5 times per day), the total energy consumption of either SRAM or Si-GC arrays no longer reduces. This is due to their relatively high static powers. In this scenario, the quasi-nonvolatile AOS-GC array can still gain energy efficiency as the wake-up interval increases. When compared to SRAM, the BG and CAA AOS-GC arrays can achieve up to 96.0% and 97.6% savings in total energy consumption, respectively. As shown in Fig. 5, when the wake-up interval becomes less than one second, the read and write energies become dominate, and the total energy increases when the interval reduces. In this case, the activation rate of the CAA AOS-GC array increases to 80%. Once the CAA AOSFET can provide higher on current as shown in Table II, the activation rate of the array can be reduced to 20%, allowing it to operate at shorter wake-up intervals.

IV. CONCLUTION

We have proposed a circuit-level model, GCSim, which can effectively estimate the area, latency, and power consumption of various GC memory arrays. The simulation

TABLE II. GC EDRAM CELL PARAMETERS

Parameter	Si-GC [8]	BG AOS [3]	CAA AOS [4]
Process (nm)	65 (CMOS)	70 (AOS)	70 (AOS)
Cell area (F^2)	113	110	4
Retention time (s)	1.1×10^{-4}	1.0×10^4	75
WT on-current (A/m)	117	15	0.42 (5*)
WT off-current (A/m)	3.9×10^{-6}	1.0×10^{-14}	1.8×10^{-11}
RT on-current (A/m)	230	29	1.5 (4*)
RT off-current (A/m)	3.0×10^{-5}	0.7	2.0×10^{-3}

* Target large on-current CAA AOS FETs for normally-off computing.

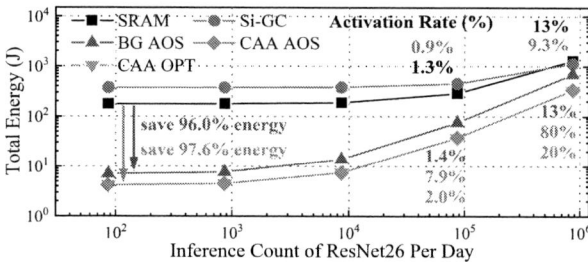

Fig. 5. The total energy consumption of different 2MB memory arrays as a functions of the intermittent wake-up times of the ResNet26 per day. The coresposning activation rates of these arrays are also shown.

results of GCSim indicate that AOS-GC arrays can provide higher memory density, lower static power, and dynamic power compared to SRAM and Si-GC arrays, but with larger read and write latencies. We also conducted a system-level evaluation using GCSim and ALU for intermittent wake-up DNN applications. The results demonstrate that only the quasi-nonvolatile AOS-GC arrays can achieve consistent energy savings across wake-up intervals, with the optimized CAA AOS-GC arrays providing both low power consumption and high performance.

ACKNOWLEDGMENT

This work was supported by the National Natural Science Foundation of China (No. 62474044).

REFERENCES

[1] T. Ishizu et al. Symp. VLSI Circuits, pp. C162–C163, 2017.
[2] T. Nakada et al., Normally-Off Computing. Springer Japan, 2017.
[3] Q. Hu et al., IEDM Tech. Dig., pp. 26.6.1-26.6.4, 2022.
[4] C. Chen et al., IEDM Tech. Dig., pp. 1-4, 2023.
[5] X. Dong et al., IEEE Trans. Comput. Aided Des. Integr. Circuits Syst., 31, pp. 994-1007, 2012.
[6] L. Pentecost et al., Proc. Int. Symp. HPCA, pp. 938-956, 2022.
[7] A. Bonetti et al., IEEE Trans. VLSI Syst., 28, pp. 646-659, 2020.
[8] K. C. Chun et al., IEEE J. Solid-State Circuits, 47, pp. 547-559, 2012.

A High Sigma Monte Carlo Analysis Solution Via Machine Learning for SRAM Margin Signoff

Amy Rao*

Tower 1, EBA Center, 387 Huimin Road Lot 10 Yangpu District of Shanghai

* Email: jierao@synopsys.com mhni@synopsys.com

Abstract

Memory lies at the heart of every electronic application, and demand is growing all the time. Today SRAM plays a crucial role for AI, particularly for high-performance applications in the cloud. In SRAM development Monte Carlo simulation is necessary and it helps designers to achieve the desired high sigma characterization and ensure design robustness. However, with contemporary devices, the cost in time and resources for brute-force Monte Carlo is prohibitive. Synopsys PrimeSim™ suite provides advanced Monte Carlo analysis solution to help designers achieve their goal on more memory, faster memory, and more reliable memory with shrinking time to market (TTM). This paper introduces PrimeSim™ XA developing a new monte carlo technology aims at reducing memory footprint and turnaround time while maintaining accuracy. PrimeSim™ AVA integrated with PrimeSim™ simulators to provide Machine-Learning-based Monte Carlo analysis by using highly accurate surrogate models of the design which can be built and trained to predict high sigma circuit behavior, thus greatly reducing the run time.

I. Introduction

The SRAM market is growing rapidly, and it is the backbone of AI and Machine Learning architectures for neuromorphic computing. The market is expected to reach $20 billion by 2025. As AI design increasingly demands more internal memory access, but as each new node came online the SRAM capacity grew and the cell sizes shrink more slowly than processes. That brings significant challenges to designer and manufacturing process, like the increasing effect of leakage currents and the slowdown scaling of supply voltage becomes a limit for power reduction in new nodes, and the layout parasitic and electronic migration becomes more pronounced in small nodes. This causes circuit parameters like maximum clock frequency and leakage power to be tuned iteratively until it is converging to an optimal point, and circuit size with large RC info increases as well. Consequently, the SRAM simulation in signoff cycle is more resource consuming than before and results in a big TTM risk.

In AI driven EDA flow, process node migration in analog design is not as easy as in digital design. Normally, lots of manual and iterative effort are required to tune analog design to meet specifications. However, analog migration automatization is becoming not just important but necessary. On 2023 Synopsys announced its analog design migration flow was enabled across TSMC's advanced process technologies, including N4P, N3E, and N2, and this was for all TSMC advanced FinFET technologies and delivered a performance advantage in SPICE, FastSPICE, and mixed-signal simulations [1]. The circuit simulation is a key component of analog migration flow. Hence, the EDA simulator is forced to be more efficient and capacity expansion.

Blocks	Counts (Repetition)	Sigma (99% yield)	No of MC Simulations
Bitcells	64 M	6.3	6.4 B*
Local I/O	30 K	5.0	3.0 M
Row Decoders	125 K	5.2	12.5 M
Local Control	2 K	4.4	0.2 M
Global I/O	4 K	4.6	0.4 M
Global Control	1K	4.3	0.1 M

* All values considering uncertainty of 10%

Figure 1, Monte Carlo simulation required for Sram *[3]*

In SRAM development, to ensure consistent and reliable quality the process variation must be considered. Process variation is the naturally occurring variation in the attributes of transistors, like length, width, oxide thickness, when integrated circuits are fabricated. SRAM is susceptible to process variation due to its minimized circuit size, especially for advanced nodes. For example, Intel observed 30% variation in chip frequency and 20X variation in chip leakage in 1,000 sample chips fabricated in 180nm technology *[2]*. While global variation is captured by analyzing the design at different PVT corners, local process variation cannot be handled effectively with just the traditional corner-based static timing analysis. To perform SRAM design margin analysis, modeling process variation by monte carlo simulation is necessary. This statistical method can help designers to achieve the desired high sigma characterization and ensure design robustness.

Typically, for a SoC memory the bit counts range from 300M to 500M. Let's assume 500MB in one chip, and SoC is built-up with HDSP compiler and an average of 512 bits are connected to 1 Sense Amplifier. Total numbers of SA on chip would be 1Mb. There would be a total of (500M/512) ~ 1M array units, each array unit containing 512 bits connected to 1 SA. The probability of finding 1 read-faulty array unit from 1M units will correspond to 4.8 Sigma. As a result, SRAM Monte Carlo simulation is extremely slow, and millions or billions of simulations are needed to capture a failure event. That brings significant challenges to simulators due to the resource cost.

II. High-Capacity Variation Analysis

Synopsys PrimeSim™ XA, it provides transistor level simulation to deliver superior performance while meeting stringent accuracy for SRAM timing, power and margin analysis, leading Fast SPICE for SRAM verification. To solve the resource consuming issue in brute-force Monte Carlo, PrimeSim™ XA developed an advanced technology called High-Capacity Variation Analysis. It uses an alternative selective variation injection approach and significantly reduces the simulation memory and turnaround time. By circuit profiling, the active devices are identified by checking the currents in MOSFET terminals with certain threshold value. These active devices are varied under circuit definition by simulator, as the way in brute-force monte carlo. For the inactive devices which are not likely to result in any circuit failure, there is no variation, and thereby reduce the required resources of simulations. The advantages of HCVA can be observed in Synopsys SG SRAM cases, Micron flash cases, etc.

Meanwhile, based on the monte carlo measurement distribution result, PrimeSim™ XA integrated with PrimeSim™ DR, a design robustness platform, to offer a high sigma margin analysis. It answers which area is weak or sensitive. In SRAM design or any design with high frequency and high accuracy requirements, to guarantee a stable read or write operation, a sufficient timing margin must be preserved. Hense, the top level's targ-trig delay measurement results are isolated from monte carlo distribution and collecting the outliers for the margin analyzer to focus on. It uses a Machine Learning methodology to check how far the worst targ-trig delay meas' sample distribution (non-gaussian) from nominal, and the safe region is defined as 1-stddev+mean – nominal < 5ps (absolute delay) or 5% (delay percent) vary of nominal. The 5 is an experimental value or it can refer to the margin sign-off spec if any. It identifies the sample index exhibiting high variation on key measures, and profiles the device characteristics on the specific sample to explore the devices which are impacting the key measures. That allows SRAM designers to assess the weakness of their designs.

With selective variation monte carlo approach, PrimeSim™ XA can achieve higher monte carlo coverage with fewer sample, and significantly expands the circuit capacity. The efficient and economics large sample simulations are allowed for SRAM components, SRAM instances, and full analog IP.

III. High Sigma Monte Carlo Analysis

Robust, accurate, and fast Monte Carlo simulation of small circuits like standard cells, SRAM bit cells, sense amplifiers, etc., in the $3.5 - 6.5\sigma$ range which requires 20K – 50B samples with standard Monte Carlo [4]. Besides selective variation injection, there is an area where Machine Learning can optimize.

Synopsys PrimeSim™ Advanced Variation Analysis, integrated with simulators like PrimeSim™ XA to perform high sigma monte carlo analysis. As Figure 1 shows, the highly accurate surrogate model of critical SRAM components such as bit cells, sense amplifier, read write assist circuitry, self-timed clocks, IO multiplexers and buffers, etc., is built and trained to predict high sigma circuit behavior, and the samples required for the desired sigma which are likely to be in the tail region and capture the non-gaussian behavior are selected by the model, and with these samples the simulator is invoked to do regular simulation. Then AVA collects the distributions and analysis the failure probability. It provides excellent accuracy across 4 to 6 sigma with 100 to 1000 times faster throughput, compared to brute-force monte carlo.

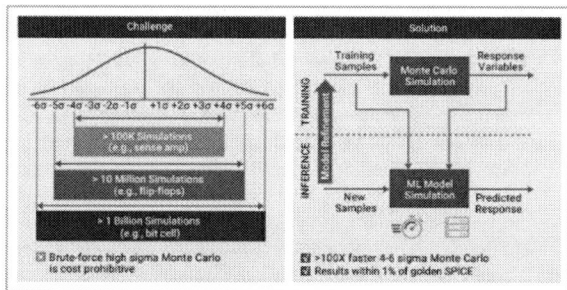

Figure 1, A high sigma monte carlo analysis solution via Machine Learning

For example, a SRAM bit cell in 5nm has 6 MOSFET, 65 resistors and 152 capacitors. To achieve 5.3 sigma, if you use brute-force monte Carlo simulation with Simple Random Sampling method, it takes almost 16 days and 100 million samples simulation. If use AVA high sigma monte carlo analysis, it only takes a few minutes, and the Quantile accuracy is better than 1%.

Case (Tech)	# MOS	# R	# C	Comment
INV (16)	2	116	184	Standard Cell
INV (7)	6	151	259	Standard Cell
SRAM (7)	6	0	0	Bit Cell
OR (16)	8	195	604	Standard Cell
Nand (16)	11	219	435	Standard Cell
Majority gate (16)	13	252	629	Standard Cell
Latch (16)	14	309	715	Standard Cell (S)
AOI (7)	22	481	1,109	Standard Cell
Ring Osc (7)	26	0	0	Ring Osc
MUX (16)	76	1,136	2,288	Standard Cell
AOI (16)	80	1,064	2,600	Standard Cell
OAI (16)	84	1,090	2,995	Standard Cell
Clock gate (16)	122	1,879	3,578	Standard Cell

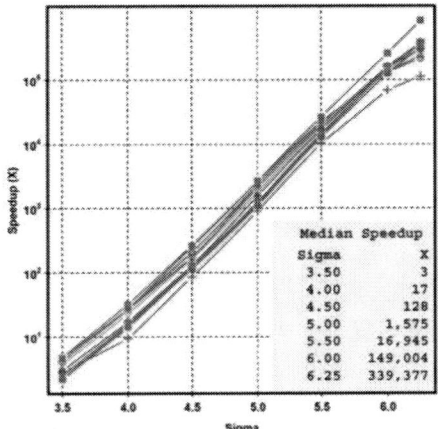

Figure 2, AVA high sigma analysis speedup vs brute-force monte carlo *[4]*

For larger circuit like SRAM slice, memory cut or analog IP, PrimeSim™ AVA with XA can perform the high sigma analysis in an accurate and efficient manner.

signal circuit designs, which is crucial for ensuring robust, high-quality products. For SRAM components the turnaround time is extremely fast, and the Quantile accuracy is better than 1%. For larger analog IP, the accuracy is comparable to brute-force monte carlo.

IV. Summary

Synopsys PrimeSim™ AVA&XA high sigma analysis solution provides an accurate and efficient approach to perform assessing the reliability and manufacturing yield of SRAM and complex, large-scale analog and mixed-

V. Acknowledgments

Thanks to Amelia, Horace, Antony, Sumon, Karthik, Kishore, Manju, and the entire Synopsys PrimeSim™ XA and AVA team. Your excellent work has been truly inspiring to me! Thanks to Synopsys Shanghai as well. It is a great workplace!

References

[1] Synopsys and TSMC Advance Analog Design Migration with Reference Flow Across Advanced TSMC Processes, Kelli Wheeler, 2023

[2] S. Borkar, T. Karnik, S. Narendra, J. Tschanz, A. Keshavarzi and V. De, "Parameter Variations and Impact on Circuits and Microarchitecture", Proc. DAC, 2003, pp. 338-342.

[3] Memory Yield Estimation Flow using Techniques in PrimeSim HSPICE AVA and XA, Ashish Kumar, STMicroelectronics, Greater Noida Shubham Varshney, Zia Semiconductors, Greater Noida Rakesh Shenoy, Synopsys, Noida, 2021

[4] HSPICE Advanced Variability Analysis, Kishore Singhal, March 30, 2020

Enhanced Multi-bit Computation using CIM SRAM Technology

Ruiyong Zhao [1], Yibo Hu [1], Zhipeng Ren [1], Yizhe Yin [1], Jing Chen*[1]

[1,2,3,4] Shanghai Institute of Microsystem and Information Technology, Chinese Academy of Sciences, Shanghai 200031, China;

* Email: zry@mail.sim.ac.cn (R.Z.), jchen@mail.sim.ac.cn

Abstract—**Tackling the challenges of substantial data loading latency and excessive computational circuit area overhead prevalent in contemporary multi-bit Compute-In-Memory (CIM) SRAM macro designs, this paper introduces a strategy based on "complementary cell computation architectures coupled with complementary data inputs." This approach establishes a high-bandwidth, high-compute-performance, and expansible multi-bit CIM SRAM macro, particularly apt for the highly parallel matrix multiplication operations encountered in neural network algorithms. The proposed CIM SRAM macro unit executes efficient binary multiplications up to 8 bits width without reliance on additional circuits such as ADC or incurring data loading delays, thereby reinforcing computational precision and system bandwidth. Simulation outcomes illustrate that, when operated at a frequency of 1GHz, the HSBC SRAM achieves an area efficiency uplift to 0.556MB/mm ^{2} and 186TOPS/mm^2 by expanding both input and weight data widths to 8 bits.**

Keywords—*multi-bit Computation, CIM, SRAM*

I. INTRODUCTION

SRAM-based in-memory computing exploits SRAM's innate properties to execute computations directly at storage sites, curtailing data movement between storage and processing units. This significantly reduces energy consumption and accelerates computation. However, fundamental SRAM constraints typically confine each cell to storing a single bit, making SRAM-based in-memory computing apt for binary neural networks or models with binary weights due to their reduced storage needs and computational complexity. While efficient for these simplified models, they may face limitations in precision for complex deep learning applications. To enhance performance in advanced machine learning tasks, research focuses on expanding bit capacity for increased inference accuracy, but this amplifies design challenges as higher bit widths necessitate more intricate circuits, leading to larger chip areas and potential latency issues, as depicted in Fig.1.

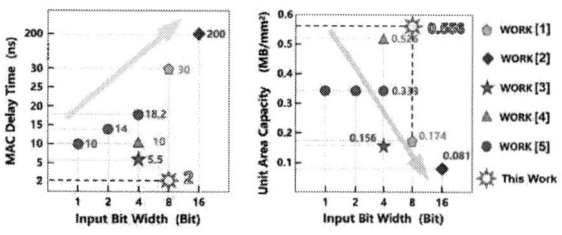

Fig. 1. Latency&Circuit Scale Trend in Multi-bit CIM SRAM .

Addressing these trade-offs, We propose the HSBC

This research was funded by the Science and Technology Commission of Shanghai Municipality under Grant No. 21TS1401100, and by the Zhangjiang National Laboratory under Grant No. Z-W-22-007.

SRAM macro in Fig.2, featuring a computing array, dynamic differential current sensing amplifiers (DISA), control, decoder, driver, and buffer circuits. Engineered with a 28nm process, this architecture boasts high bandwidth and computing power. Simulated at 1GHz, it broadens input and weight bit-width to 8 bits, enhancing area efficiency to 0.556 MB/mm^2 and delivering a compute throughput of 186 TOPS/mm^2 .

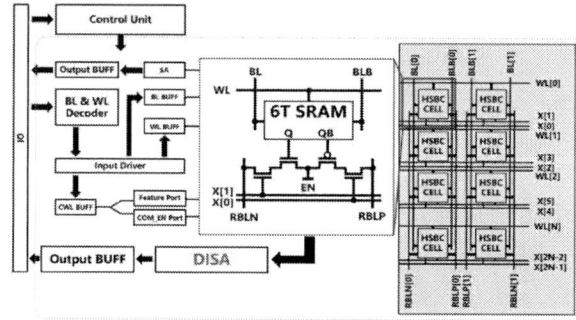

Fig. 2. The Structure of HSBC SRAM.

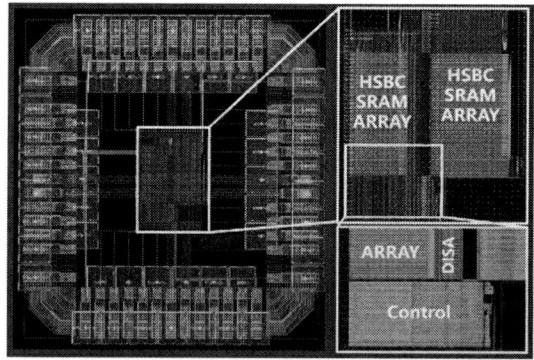

Fig. 3. The Layout of HSBC SRAM.

II. PROPOSED CIM ARCHITECTURE AND COMPUTING MODE

A. Proposed Bit-Cell

Conventional 6T SRAM cells, using six transistors, store one bit of data. As shown in Fig.4, the HSBC SRAM cell adds four computing transistors to generate dual cell currents, IN and IP, for computational tasks. These reflect varied data charge states during computations and are precisely read via separate lines—I_N for RBLN and I_P for RBLP. This dual-read mechanism enables in-cell computations without compromising SRAM's read/write speed.

979-8-3503-6184-1/24 $31.00 © 2024 IEEE

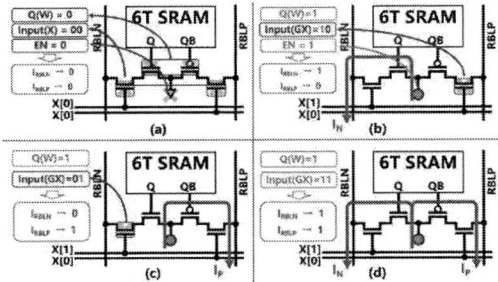

Fig. 4. The computational mode of the HSBC SRAM cell and results of current.

With a fixed weight (W=1), HSBC SRAM generates four unique current scenarios based on input pair X[1:0], executing binary multiplications. Results are derived from RBLN and RBLP currents, following analog computation principles where X[1:0] serves as binary inputs. Table presents the HSBC SRAM current truth table, mapping input-weight combinations to ideal Output[1:0] currents, symbolizing the product of X[1:0] and W.

TABLE I.   TRUTH TABLE OF COMPUTATION FOR HSBC SRAM CELL IN COMPUTATION MODE

	EN	X[1]	X[0]	Q	QB	RBLN	RBLP
①	0	-	-	-	-	0	0
②	1	0	0	-	-	0	0
③	1	0	1	1	0	0	1
④	1	0	1	0	1	0	0
⑤	1	1	0	1	0	1	0
⑥	1	1	0	0	1	0	0
⑦	1	1	1	1	0	1	1
⑧	1	1	1	0	1	0	0

The HSBC SRAM SRAM unit innovates by performing 2-bit input with 1-bit weight multiplication in a single cell, doubling computational power. In terms of area occupancy, compared to traditional 6T SRAM cells (designed in accordance with standard logical design rules), the HSBC SRAM cell incurs an area overhead ratio of 1.9:1.

B. DISA

As illustrated in Fig.5, the DISA circuit comprises three main components: a precharge circuit (PCHG), a cross-sharing sensitive amplifier (DHSA), and a dynamic current logic output amplifier (DLSA). Unlike the precharge voltage typically employed in conventional SRAM, the precharge circuit in the HSBC SRAM's computing region utilizes half of the supply voltage to precharge the computation small signal (Sense BL) bitline, aimed at facilitating the DHSA to swiftly convert minute current signals into voltage outputs. Prior to the processing of computation current signals by DISA, PCHG initializes the computation small signal bitline to 0.6V. Upon entry of the computation current signal from the sum bitline RBLP1 into DHSA, it transforms the bitline current signals into a pair of complementary logic levels, which are then conveyed through Sense BL to DLSA for the calculation of output current signals across different columns, yielding the computation results C and R. Operated under clock control, DLSA disconnects the computation path and activates the reset path during the high phase of the clock to

discharge residual charges post-computation; conversely, during the clock's low phase, it opens the computation path, disables the reset path, and accomplishes the logical operation on the complementary levels outputted by DHSA.

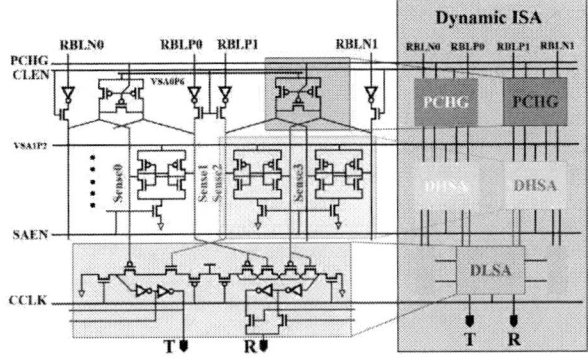

Fig. 5. DISA Circuit in HSBC SRAM.

C. Multi-channel Computation in Convolutional Operation

In deep learning and computer vision applications, convolution kernels often feature multiple channels to extract various aspects of input features. Taking a three-channel convolution as an example, the weight datasets W2, W1, and W0 correspond to three distinct kernel channels, embodying the weight information across different feature dimensions of the input data. As depicted in Figure 5-17, when the weight data width is set to 2-bit binary, the combined total of 6 bits for W2-W0 aligns with a single byte. For a 4-bit binary width, the 12 bits required by W2-W0 equate to two bytes.

The HSBC SRAM array's storage layout compactly accommodates all these values within the space of one or two bytes, respectively. Fig.6 presents the designed for parallel computation and efficiency, with each row dedicating a single byte to store the weights of all three channels at a given position. Consequently, as input data X traverses the entire input domain in a fixed format, every row of the array conducts bit-wise multiplication of the stored weights with the corresponding input data. This arrangement leverages the parallel computation capability of the HSBC SRAM array to rapidly execute dense multiply-accumulate operations on input X across the three channel weights W2-W0, significantly accelerating convolutional computations.

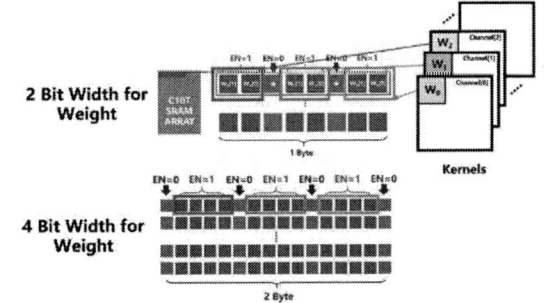

Fig. 6. Weight Storage Strategy in the HSBC SRAM Array.

D. Parallel Computation of Matrix Multiplication

As illustrated in Fig.7, during matrix multiplication operations, the HSBC SRAM array partitions Matrix A into vector pairs [A00, A01] and [A10, A11], leveraging the HSBC SRAM architecture as a storage domain to preserve these weight information components. Concurrently, Matrix B is correspondingly decomposed into vector pairs [B00, B10] and [B01, B11], which are fed into the computation as input data sequences through port X. The upper portion of the diagram exhibits the encoded sub-vectors of Matrix A within the HSBC SRAM (represented in blue) alongside the input sub-vectors of Matrix B (depicted in red), while the lower section aligns with another set of sub-vectors for both Matrices A and B. Within a single clock cycle, the array facilitates element-wise multiplication of corresponding sub-vectors housed in the HSBC SRAM, culminating in the summation of these products at the array's terminus. For instance, the multiply-accumulate (MAC) operation between [A00, A01] and [B00, B10] yields C10, with analogous treatments applied to other combinations to ensure the comprehensive output of all elements in the resultant product matrix C within that solitary period. This highlights the efficacy and computational prowess of the HSBC SRAM framework in accelerating matrix multiplication tasks.

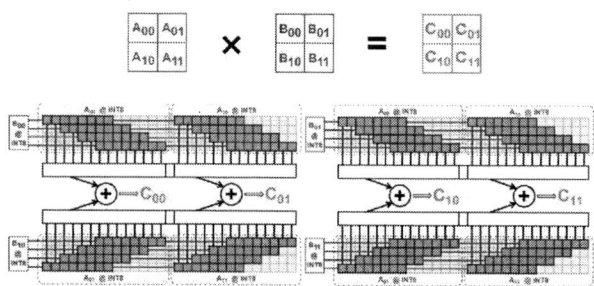

Fig. 7. Matrix multiplication mode in the HSBC SRAM array.

III. SIMULATION AND DISCUSSION

To validate the functionality of the HSBC SRAM array architecture in both convolutional and matrix multiplication computing modes, a 32x16 cell array was realized with additional peripheral circuits. Fig.8 depicts the functional waveform of the HSBC SRAM macro under convolutional computing mode. In Phase ①, DISA operation initiates with precharging; prior to processing computation current signals, PCHG precharge the computation small signal bitline to 0.6V. Following precharge, DISA receives computation result current signals generated by the HSBC SRAM array. The computation phase, central to DISA's function, individually addresses current signal combinations {0,0}, {0,1}, {1,0}, and {1,1} provided by the HSBC SRAM array, with Steps ③, ⑤, ⑥, and ⑦ correlating to each signal's processing sequence. DISA employs its internal circuits and logic units to analyze, compare, and integrate input current signals, yielding the final computation outcomes. Upon completing a computation cycle, DISA enters a reset phase, discharging charges and resetting node potentials to ensure all sections return to the predefined state set during precharge. Detailed analysis of the simulation waveform reveals a clear concurrence between DISA's handling of the computation currents produced by the HSBC SRAM array and theoretical expectations. The waveform encapsulate the state transitions of the resultant bit R and carry indicator bit T, affirming the accurate and efficient execution of required computations by HSBC SRAM when employing multi-bit CIM schemes.

Furthermore, this simulation outcome verifies the HSBC SRAM's capability to reliably transform computation results into signal forms suitable for subsequent circuitry processing through the DISA circuit.

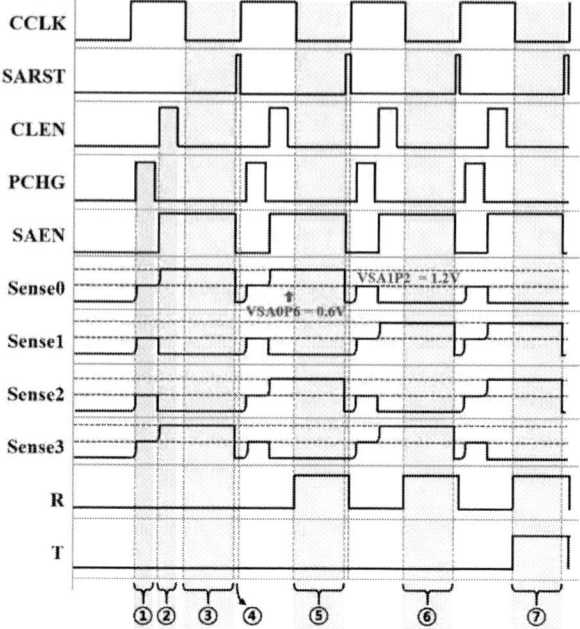

Fig. 8. Function waveform in matrix multiplication mode.

IV. SUMMARY

This paper introduces a high-speed, high-bandwidth, and high-compute-power HSBC SRAM. Operating at 1 GHz, the HSBC SRAM extends the bit-width of both input and weight data to 8 bits, achieving an area efficiency of 0.556 MB/mm² and computility density of 186 TOPS/mm².

REFERENCES

[1] MOCHIDA R, KOUNO K, HAYATA Y. A 4M Synapses integrated Analog ReRAM based 66.5 TOPS/W Neural-Network Processor with Cell Current Controlled Writing and Flexible Network Architecture; proceedings of the 2018 IEEE Symposium on VLSI Technology, F 18-22 June 2018, 2018 [C].

[2] WANG B, XUE C, FENG Z. A 28nm Horizontal-Weight-Shift and Verticalfeature-Shift-Based Separate-WL 6T-SRAM Computation-in-Memory Unit- Macro for Edge Depthwise Neural-Networks; proceedings of the 2023 IEEE International Solid-State Circuits Conference (ISSCC), F 19-23 Feb. 2023, 2023[C].

[3] LIU S, LI P, ZHANG J. 16.2 A 28nm 53.8TOPS/W 8b Sparse Transformer Accelerator with In-Memory Butterfly Zero Skipper for Unstructured-Pruned NN and CIM-Based Local-Attention-Reusable Engine; proceedings of the 2023 IEEE International Solid-State Circuits Conference (ISSCC), F 19-23 Feb. 2023, 2023 [C].

[4] CHEN P, WU M, ZHAO W, et al. 7.8 A 22nm Delta-Sigma Computing-In- Memory ($\Delta\Sigma$ CIM) SRAM Macro with Near-Zero-Mean Outputs and LSB-First ADCs Achieving 21.38TOPS/W for 8b-MAC Edge AI Processing; proceedings of the 2023 IEEE International Solid-State Circuits Conference (ISSCC), F 19-23 Feb. 2023, 2023 [C].

[5] CHIU Y C, KHWA W S, LI C Y, et al. A 22nm 8Mb STT-MRAM Near-Memory- Computing Macro with 8b-Precision and 46.4-160.1TOPS/W for Edge-AI Devices; proceedings of the 2023 IEEE International Solid-State Circuits Conference (ISSCC), F 19-23 Feb. 2023, 2023 [C].

979-8-3503-6184-1/24 $31.00 © 2024 IEEE

A Compute-in-Memory Macro Based on Complementary 2T2C FeRAM Cell for BNNs

Jinyu Li [1], Mingzhang Xie [2], Shisheng Xiong*[1]

[1] School of Information Science and Technology, Fudan University, Shanghai 200438, China
[2] China Resources Microelectronics Co., Ltd., Wuxi 214061, China

* Email: 22210720163@m.fudan.edu.cn, sxiong@fudan.edu.cn

Abstract—The growing demand for efficient AI algorithms in edge devices necessitates the exploration of novel memory architectures. Ferroelectric RAM (FeRAM) emerges as a promising candidate for Compute-In-Memory (CIM) applications owing to its advantageous properties. This study proposes a design utilizing complementary 2T2C FeRAM cells specifically tailored for XNOR computation in binary neural networks. The design leverages simultaneous activation of multiple wordlines for efficient bit-counting of XNOR results, significantly accelerating computations. To optimize readout performance, the effect of bitline capacitance is investigated. Simulations demonstrate that a 3×3 binary convolution and bit-counting consume an average of 3.7 pJ and take 32.9 ns across various process corners, while for a single XNOR and bit-counting operation, the energy consumption is 71.9 fJ. Compared to traditional single XNOR and write-back operations, its computation speed is increased by nearly 2x, significantly enhancing computing efficiency.

Keywords—*FeRAM, complementary cells, XNOR, Compute-In-Memory*

I. INTRODUCTION

With the rapid advancement of artificial intelligence, edge devices are increasingly tasked with executing complex AI algorithms [1]. However, conventional Von Neumann computing architectures encounter significant challenges, including data transfer delays and excessive energy consumption, collectively referred to as the "memory wall" and "power wall" [2]. Compute-in-Memory (CIM) has emerged as a groundbreaking approach [3], integrating storage and computation onto a single chip, thereby significantly enhancing computational efficiency.

Recently, hafnium zirconium oxide (HZO)-based Ferroelectric Random Access Memory (FeRAM) has gained prominence [4]. FeRAM boasts a relatively simple manufacturing process, compatibility with CMOS logic, non-volatility, low power consumption, high durability [5], and reliability, making it an attractive candidate for CIM applications [6]. By exploiting the charge-sharing capability of FeCAP units, FeRAM can perform XNOR logic operations, effectively replacing numerous multiplications and accumulations typically required in neural networks. This makes it compatible with Binary Neural Networks (BNNs) [7], which use only two values to represent weights and activations, saving memory and computing resources.

Utilizing FeRAM as the storage medium to realize XNOR logic operations, and deploying BNNs on this basis, significantly reduces algorithm power consumption and improves operation speed and reliability [8]. This approach offers distinctive advantages and vast application potential in edge computing scenarios. Currently, there is relatively limited research on utilizing FeRAM for CIM, and existing studies have not yet reached a mature stage. Additionally, significant modifications to traditional memory structures pose implementation challenges, with some designs suffering from limited parallelism.

To address these challenges, this study introduces a CIM macro for BNNs utilizing 2T2C FeRAM, emphasizing a complementary unit-based XNOR calculation method, optimizing parallelism, and analyzing bitline capacitance's impact on readout accuracy, all achieved without the need for modifications to traditional FeRAM cells.

II. PROPOSED CIM MACRO

A. Complementary 2T2C FeRAM Cell

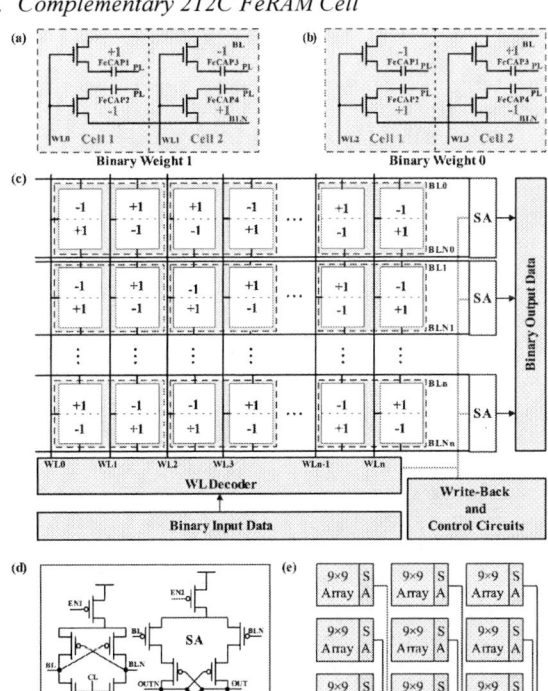

Fig. 1. Complementary cells and compute macro. (a) Complementary cell with weight 1. (b) Complementary cell with weight 0. (c) Compute array. (d) Sense amplifier. (e) Overall architecture.

TABLE I. XNOR COMPUTATION TRUTH TABLE

Input			Weight			XNOR-Output		
Value	*WL0*	*WL1*	*Value*	*Cell 1*	*Cell 2*	*Value*	*BL*	*BLN*
1	1	0	1	+1	-1	1	High	Low
1	1	0	0	-1	+1	0	Low	High
0	0	1	1	+1	-1	0	Low	High
0	0	1	0	-1	+1	1	High	Low

979-8-3503-6184-1/24 $31.00 © 2024 IEEE

To implement XNOR computation, this paper proposes a complementary compute unit based on 2T2C FeRAM. In the design, as illustrated in Fig. 1, adjacent pairs of 2T2C FeRAM cells, controlled by two wordlines (WLs), under the same BL or BLN, are utilized as a complementary compute unit without requiring significant modifications to the traditional 2T2C FeRAM array. This circuit can operate in two modes: a standard read/write mode for data access, and an XNOR computation mode for CIM. In the XNOR computation mode, the binary weights of the BNN reside within complementary compute units, exploiting the inherent complementarity of the data. As depicted in Fig. 1(a), if Cell1 stores "1", FeCAP1 and FeCAP2 are polarized positively(+1) and negatively(-1), respectively, while Cell2 is polarized oppositely. In this configuration, Cell1 and Cell2 jointly represent "1". During computation, combinations of whether two WLs are activated represent "0" or "1". For instance, if WL0 is activated and WL1 is deactivated, it represents input as "1"; otherwise, it represents "0". This enables the external input data to control the reading of weight data, generating the XNOR computation result, which is output in the form of charge quantities on BL and BLN. The SA then amplifies the voltage difference between the BL and BLN, converting the charge quantities into a digital output, which represents the XNOR computation result. The truth table illustrating the XNOR operation is shown in TABLE I.

B. Parallel Optimization of the CIM Circuit

By utilizing the principle of charge conservation, multiple complementary WLs can be activated simultaneously, causing the charges on the same BL or BLN to accumulate. After a read operation, the SA compares the resulting voltage difference between BL and BLN to determine the binary output (0 or 1). This process parallelizes XNOR computations, performs bit-counting, and binarizes the results efficiently. To enhance accuracy, a multi-bit ADC can also be employed instead of a rail-to-rail SA for reading, offering greater precision in the output.

To match convolution operations, we designed the circuit for a 3×3 convolution kernel. Each processing element (PE) consists of a 9×9 array of 2T2C complementary cells, totaling 9×18 traditional 2T2C cells per PE, as shown in Fig. 1(c). And 9 or more PEs form a group connected to a set of register circuits, as shown in Fig. 1(e). The 3×3 convolution kernel is

sequentially mapped into a single row of complementary cells. This design requires that an odd number of weights be stored in each row to ensure that a voltage difference exists between the BL and BLN, allowing for proper amplification.

Activating multiple WLs simultaneously can disrupt normal memory write-back, resulting in data corruption. To overcome this challenge, we propose a complementary cell reverse copy method illustrated in Fig. 2. The entire write-back process is divided into two steps: the first step involves activating the WL of the corresponding complementary cell and reading out the data through an amplifier; the second step involves inverting the read data through an inverter and then writing it back to the cells where the data was destroyed. The control of WL0 and WL1 is opposite in two steps, and their control signals are generated by EN and Input together, as shown by WL and WLN in Fig. 3.

III. THE IMPACT OF BITLINE CAPACITANCE

The WL capacitance, governed by Q=CV, directly impacts voltage readout, as demonstrated in Fig. 4. Experimental P-V curves of ferroelectric capacitors under varying cycling counts are presented in Fig. 5(a), informing our Verilog-A model development and circuit simulations.

Within the 9×9 array, we investigated the impact of varying bitline capacitance on the voltage difference between BL and BLN during a 3×3 convolution operation, which encompasses 9 concurrent XNOR computations followed by bit-counting and binarization. The results are illustrated in Fig. 5(b). Here, "4/5" signifies a scenario where there are four "1"s and five "0"s in the result, and similarly for other instances. The voltage difference for results of "4/5" and "5/4" is similar in the circuit, differing only in polarity, hence this figure represents the voltage difference for all possible computation results.

Optimal bitline capacitance is found to be within 340fF to 440fF. The maximum voltage difference (around 473mV) occurs when all convolution results are either all "1"s or all "0"s. Conversely, the minimum difference (around 56mV) appears with four or five "1"s. Both scenarios ensure sufficient amplification by the sense amplifier. Taking into account parasitic capacitance, a bitline capacitance of 350fF is chosen for this design.

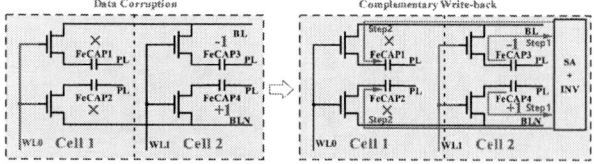

Fig. 2. Complementary write-back process

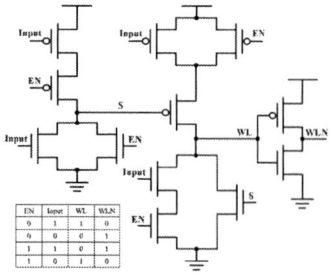

Fig. 3. Write-back control circuit

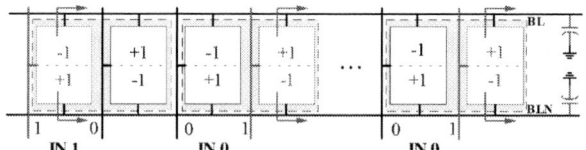

Fig. 4. Influence of bitline capacitance on readout.

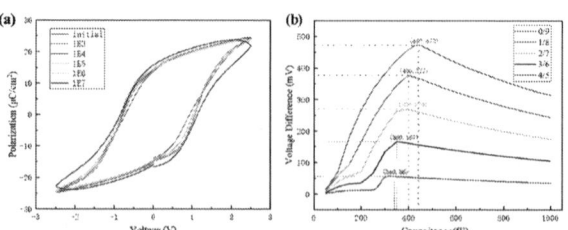

Fig. 5. (a) Actual P-V curves under different cycling counts. (b) Influence of different bitline capacitances on the voltage difference between BL and BLN.

979-8-3503-6184-1/24 $31.00 © 2024 IEEE

IV. Evaluation

This work presents a circuit-level simulation using the 25°C, includes a 9×9 FeRAM array, sense amplifier, and readout paths. The experimental simulation results, as depicted in Fig. 6, validate the CIM functionality in BNNs by demonstrating stepped bitline voltage differences during XNOR operations and varying bit-count outcomes. Furthermore, energy consumption and delay under tt, ss, ff corners were evaluated for XNOR and bit-counting, as shown in Fig. 7. During convolution and bit-counting operations across the entire array, SA, and data transmission path, the average energy consumption and delay under tt, ss, and ff process corners are 3.7 pJ and 32.9 ns, respectively, while for a single XNOR and bit-counting operation, the energy consumption is 71.9 fJ. Considering the write-back operation, excluding peripheral decoding and control circuits, the total energy consumption is about 111.7 pJ. By simultaneously activating multiple pairs of WLs for XNOR computations, the operation speed is approximately doubled compared to the traditional approach of sequentially activating a single pair of WLs per computation.

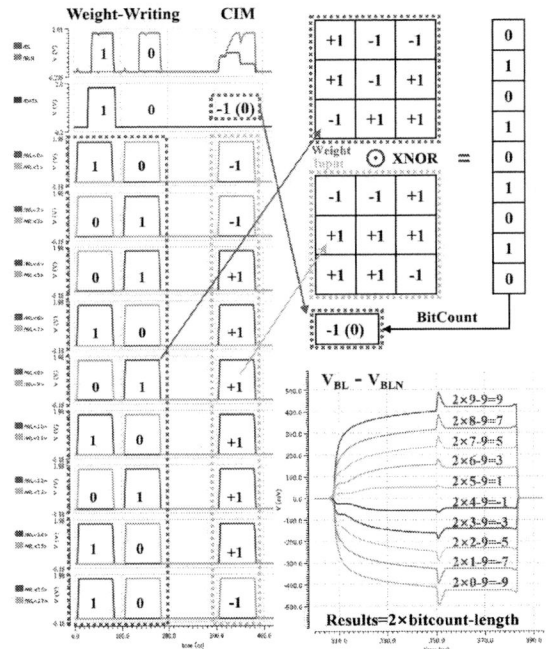

Fig. 6.   Testing of CIM functionality.

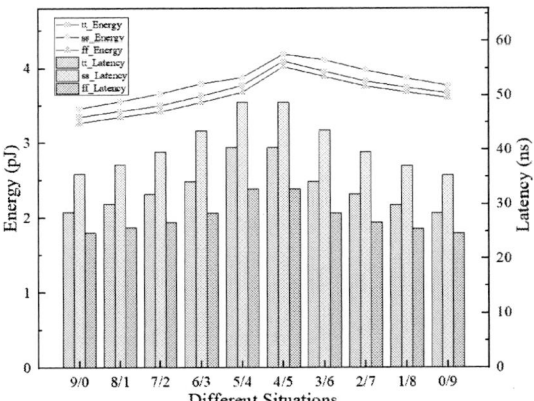

Fig. 7.   Energy consumption and delay under different scenarios.

TABLE II.   COMPARISON WITH OTHER STUDIES

	This paper	FeRAM[8]	RRAM[9]	SRAM[10]
Nonvolatile	Yes	Yes	Yes	No
Cell	2T2C FRAM	1T2C FRAM	1T1R RRAM	10T1C SRAM
Parallelism	High	Medium	High	High
Operations	XNOR Bit-Count	XNOR	XNOR Bit-Count	XNOR Bit-Count
Endurance	High	High	Medium	High

A comparative analysis of this study with related works, as summarized in TABLE II, highlights the distinct advantages of the proposed CIM macro, including its non-volatility, high parallelism, and superior endurance characteristics.

V. Conclusion

This work introduces a novel in-memory computing (CIM) design optimized for binary neural networks, utilizing the Complementary 2T2C FeRAM cell and leveraging simultaneous activation of multiple wordlines for efficient bit-counting on XNOR results. Additionally, it incorporates a dedicated write-back scheme to address the cell's destructive read operation. Simulations demonstrate high efficiency: a 3×3 binary convolution with bit-counting consumes only 3.7 pJ on average and takes 32.9 ns, while a single XNOR and bit-counting operation requires just 71.9 fJ. Furthermore, This method delivers nearly 2x faster computations than traditional methods that activate and write back to a single pair of WLs per operation.

Acknowledgment

The processes and materials used in this research were supported by China Resources Microelectronics Co., Ltd.

References

[1] R. Singh and S. S. Gill, "Edge AI: A survey," Internet of Things and Cyber-Physical Systems, vol. 3, pp. 71–92, Jan. 2023.

[2] M. M. Waldrop, "The chips are down for Moore's law," Nature News, vol. 530, no. 7589, p. 144, Feb. 2016.

[3] S. Yu, H. Jiang, S. Huang, X. Peng, and A. Lu, "Compute-in-Memory Chips for Deep Learning: Recent Trends and Prospects," IEEE Circuits and Systems Magazine, vol. 21, no. 3, pp. 31–56, 2021.

[4] S. Deng et al., "Overview of Ferroelectric Memory Devices and Reliability Aware Design Optimization," in Proceedings of the 2021 on Great Lakes Symposium on VLSI, in GLSVLSI '21. New York, NY, USA: Association for Computing Machinery, Jun. 2021, pp. 473–478.

[5] J. Yang et al., "A 9Mb HZO-Based Embedded FeRAM with 1012-Cycle Endurance and 5/7ns Read/Write using ECC-Assisted Data Refresh and Offset-Canceled Sense Amplifier," in 2023 IEEE International Solid- State Circuits Conference (ISSCC), Feb. 2023, pp. 1–3.

[6] Z. Sun, S. Kvatinsky, X. Si, A. Mehonic, Y. Cai, and R. Huang, "A full spectrum of computing-in-memory technologies," Nat Electron, vol. 6, no. 11, Art. no. 11, Nov. 2023.

[7] M. Rastegari, V. Ordonez, J. Redmon, and A. Farhadi, "XNOR-Net: ImageNet Classification Using Binary Convolutional Neural Networks," in Computer Vision – ECCV 2016, B. Leibe, J. Matas, N. Sebe, and M. Welling, Eds., Cham: Springer International Publishing, 2016, pp. 525–542.

[8] Q. Wang et al., "A 1T2C FeCAP-Based In-Situ Bitwise X(N)OR Logic Operation with Two-Step Write-Back Circuit for Accelerating Compute-In-Memory," Micromachines, vol. 12, no. 4, Art. no. 4, Apr. 2021.

[9] X. Sun, S. Yin, X. Peng, R. Liu, J. Seo, and S. Yu, "XNOR-RRAM: A scalable and parallel resistive synaptic architecture for binary neural networks," in 2018 Design, Automation & Test in Europe Conference & Exhibition (DATE), Mar. 2018, pp. 1423–1428.

[10] D. Kushwaha et al., "An Energy-Efficient High CSNR XNOR and Accumulation Scheme for BNN," IEEE Transactions on Circuits and Systems II: Express Briefs, vol. 69, no. 4, pp. 2311–2315, Apr. 2022.

979-8-3503-6184-1/24 $31.00 © 2024 IEEE

A Novel High Speed Low Power Differential Circuit-Based FRAM Read Scheme

Qiuyu Tao[1#], Jiabao Ye[2#], Xuecheng Cui[2], Nan Jiang[1], Jiangtao Cao[3], Xibo Chen[3], Zhangbin Yang[4], Daixiao Peng[4], Xi Cai[4], Xueguang Lian[4], Yong Ding[2], Jiuren Zhou[3], Bing Chen*[2,3], Genquan Han[3]

[1]College of Information Science & Electronic Engineering, Zhejiang University, Hangzhou, China

[2]School of Integrated Circuits, Zhejiang University, Hangzhou, China

[3] Hangzhou Institute of Technology, Xidian University, Hangzhou, China

[4] China Three Gorges Construction Engineering Corporation, Beijing, China

*Email: chenbing@xidian.edu.cn

# These authors contributed to this work equally.

Abstract—**In recent years, ferroelectric random access memory (FRAM) has received huge attention for its excellent low power consumption, tiny cell size and inherent non-volatile nature. However, the high readout delay and power consumption of traditional FRAM read circuits have significantly impacted their readout performance. This paper proposes a novel FRAM read scheme based on differential circuit, which exploits the differences in the rate of bit line (BL) voltage changes across different data states of FRAM for data reading. Compared to conventional approaches, our novel read circuit improves readout speed and significantly reduces power consumption, making it a feasible low-power reading solution for FRAM arrays.**

Keywords—FRAM, differential circuit, read scheme

I. INTRODUCTION

With the explosive growth in the amount of data stored worldwide, there is an increasing demand for high performance non-volatile memory (NVM) technologies [1-2]. Traditional NVMs like FLASH and Dynamic random access memory (DRAM) demonstrate limitations in cell area, power consumption, and speed, prompting the academic community to explore new NVMs to overcome these challenges [3]. In recent years, FRAM has garnered widespread attention due to its low power consumption, strong scalability, and small cell area, making it a strong candidate for the next generation of high-performance memory technologies [4-5]. The discovery of scalable ferroelectric materials, doped hafnium oxide, has accelerated the switching speed of FRAM cells, enabling the application of large-scale FRAM arrays [6]. The read circuit plays a crucial role in enhancing the reliability of FRAM arrays. Therefore, the development of a high-performance and reliable FRAM read scheme is essential, which will be beneficial for the future application of hafnium-based ultra-large-scale FRAM arrays.

In a traditional FRAM read scheme, as shown in Fig .1, a reference cell for FRAM is required to assist with reading [7]. The polarization state of this reference cell is fixed, typically set to the positively polarized state to reduce power consumption. During the read operation, BL and bit line reference (BLR) are first pre-charged to 0 V, and then disconnected from the 0 V connection, allowing BL and BLR to float. A read pulse V_{read} is then applied to Plate Line (PL) and Plate Line Reference (PLR). If the polarization state in the memory cell is positive, indicating the stored information is "0", the polarization state remains unchanged, and the current

on BL is very small, producing a low voltage V_{BL} after passing through R_{sin1}. Since the reference cell is also in a positive polarization state, the voltage V_{BLR} on BLR is also very low and very close to V_{BL}. These two voltages, after being processed by a sense amplifier (SA), output a low voltage, indicating the read information is "0". Conversely, if the polarization state in the memory cell is negative, indicating the stored information is "1", then after applying the V_{read} pulse on PL, the polarization state of the memory cell flips, generating a larger polarization current on BL. This creates a higher voltage on the floating BL through R_{sin1}. At this point, the difference between V_{BL} and V_{BLR} is significant, and after processing by the SA, a high voltage is output, indicating the read information is "1", thus correctly completing the read operation of the FRAM memory cell.

Obviously, this traditional read scheme has some issues. Firstly, the circuit requires a reference unit to assist in reading, and as the number of readings increases, the reference unit will suffer from fatigue degradation, affecting the accuracy of the readout results. Additionally, the circuit requires two large sensing resistors to bring the voltage on BL into a range detectable by the SA, which increases the power consumption of the read circuit. Furthermore, due to the offset voltage of the SA and the influence of parasitic resistances and capacitances on BL during the readout process, the voltage difference between BL and BLR needs to reach a certain level to enable the SA to output correct results, resulting in relatively high readout delay and slower readout speed for this read scheme.

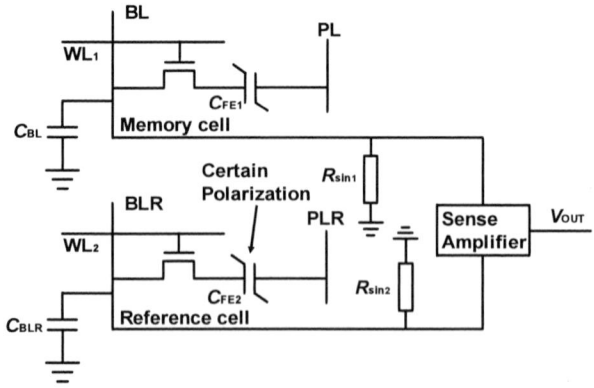

Fig .1 Schematic of a conventional FRAM read scheme.

Therefore, in this paper, we proposed a novel FRAM read scheme based on differential circuit technology. During the reading process, the polarization-reversed ferroelectric cells cause a rapid rise in the corresponding BL voltage, while the polarization-unreversed ferroelectric cells have a slower BL voltage rise rate. The differential circuit module can differentiate the voltage on the BL, detecting different rates of voltage change and outputting corresponding reading results. Our proposed novel FRAM read scheme eliminates the need for a reference unit required in traditional circuits, thus saving power. Additionally, it does not require the BL voltage to rise to a stable detection window for reading; instead, it can trigger and recognize information at the onset of a sudden voltage rise, reducing readout delay and speeding up the FRAM read circuit.

II. CIRCUIT IMPLEMENTATION

The schematic of the novel FRAM read scheme based on the differential circuit is shown in Fig .2. The proposed differential circuit consists of an operational amplifier, two resistors and two capacitors.

After the read operation, the voltage signal on BL serves as the input signal V_{IN} to the differential readout circuit, which outputs the signal V_{OUT} after processing. First, an s-domain analysis of this differential circuit based on the ideal operational amplifier is conducted. The relationship between V_{IN} and V_{OUT} can be expressed by (1).

$$\frac{V_{IN}}{R_1 + \dfrac{1}{sC_1}} = \frac{0 - V_{OUT}}{R_2 // \dfrac{1}{sC_2}} \tag{1}$$

By simplifying equation 1, (2) can be obtained.

$$\frac{V_{OUT}}{V_{IN}} = -R_2 C_1 \frac{s}{(sC_1 R_1 + 1)(sC_2 R_2 + 1)} \tag{2}$$

In the case where the frequency is not very high, the term $(sC_1 R_1 + 1)(sC_2 R_2 + 1)$ in (2) is much larger than the term s. Therefore, (2) can be approximated by (3). By taking the inverse Laplace transform, (3) can be transformed into (4).

$$\frac{V_{OUT}}{V_{IN}} = -sR_2 C_1 \tag{3}$$

$$V_{OUT} = -R_2 C_1 \frac{dV_{IN}}{dt} \tag{4}$$

From (4), it can be seen that the output of our proposed novel FRAM readout scheme based on the differential circuit is a function of the rate of change of the input signal, enabling it to perform the function of differentiation on the input signal. In order for the proposed new read scheme to function properly, a five-transistor OTA operational amplifier with high gain, high gain-bandwidth product (GBW), and sufficient phase margin is designed, as shown in Fig. 2b. This circuit uses PMOS input pairs as the input terminals. To meet the negative voltage characteristics of the differential circuit output and the readout characteristics of the FRAM unit, the lowest potential of the operational amplifier is set to $-V_{DD}$.

The simulated Bode plot of the operational amplifier is shown in Fig .3. The simulation indicates that the amplifier has a gain of 39.3 dB, providing a significant gain for small signals. In the designed circuit, the dominant pole is located at the point where the low-frequency gain decreases by 3 dB, which is about 648 kHz. Therefore, the GBW of this amplifier

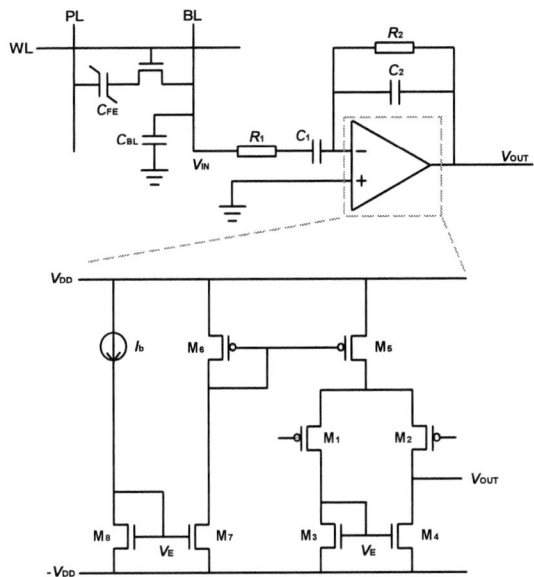

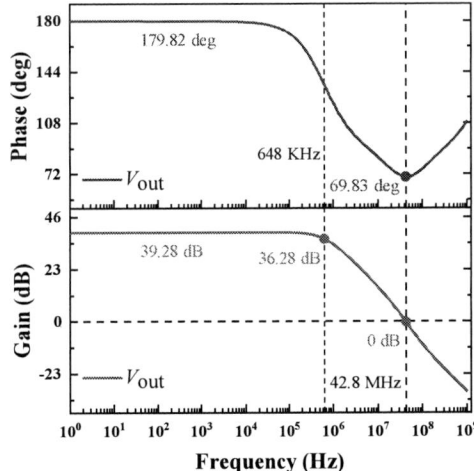

Fig .2 Schematic of the proposed novel FRAM read scheme based on differential circuit.

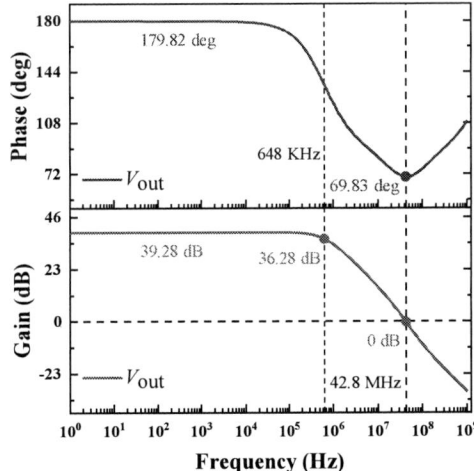

Fig .3 The simulated Bode plot of the five-transistor OTA operational amplifier, including phase margin and gain of the circuit at different frequencies.

is approximately 59.7 MHz, which can meet the speed requirements for the output signal of the FRAM units. Since the designed amplifier is a single-stage amplifier and the location of the secondary pole is relatively distant, with a phase margin of approximately 69.8°, the closed-loop application is stable without oscillations, and the circuit can operate normally.

III. RESULTS AND DISCUSSION

To validate the feasibility of the proposed novel FRAM read scheme, we conducted simulations using EDA tools. In the circuit simulation, we set the circuit parameters to $R_1 = 1K\Omega$, $R_2 = 300K\Omega$, $C_1 = 15pF$, and $C_2 = 25fF$ to ensure the proper functioning of the novel circuit. The simulation results are shown in Fig. 4. Fig. 4 (a) illustrates the writing process of the FRAM cell. Fig. 4 (b) demonstrates the readout process of the novel read scheme based on the differential circuit.

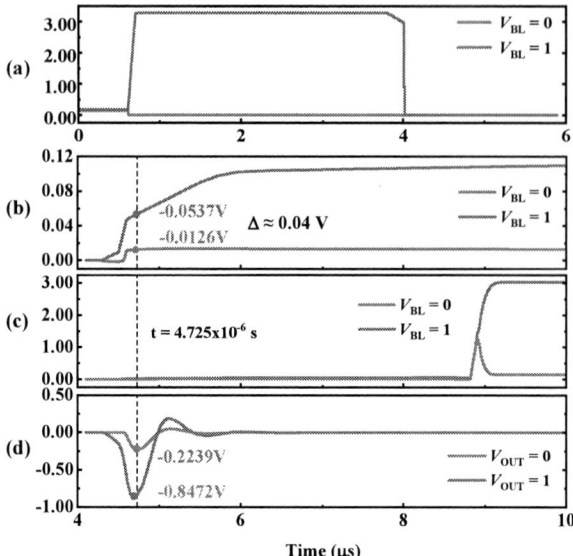

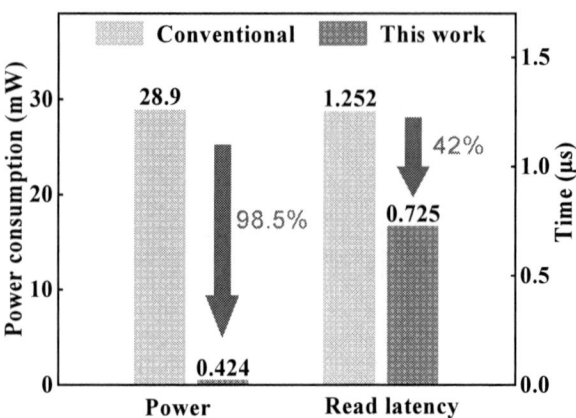

Fig .5 Comparison of power consumption and read latency of the conventional read scheme and the proposed novel read scheme.

Fig .4 Simulation results: (a) writing process of the FRAM cell, (b) readout process of the proposed novel read scheme, (c) readout process of the conventional read scheme, and (d) output results of the novel read scheme. $V_{BL} = 0$ represents the FRAM in the positive polarization state (Stored Information "0"), and $V_{BL} = 1$ represents the FRAM in the negative polarization state (Stored Information "1"). $V_{OUT} = 0$ corresponds to the output result of $V_{BL} = 0$; $V_{OUT} = 1$ corresponds to the output result of $V_{BL} = 1$.

Fig. 4 (c) demonstrates the readout process of the conventional read scheme. Fig. 4 (d) demonstrates the output results of the novel read scheme. The voltage changes on BL vary according to the information stored in the FRAM cell. During the readout process of the novel scheme, for FRAM cells in the negative polarization state storing "1", the polarization state flips during reading, resulting in a larger slope of voltage change on BL, as indicated by $V_{BL} = 1$. Consequently, the amplitude of $V_{OUT} = 1$ after passing through the differential read circuit is larger. Conversely, for FRAM cells in the positive polarization state storing "0", where the polarization state remains unchanged during reading, the slope of voltage change on BL is smaller, as indicated by $V_{BL} = 0$. Accordingly, the amplitude of $V_{OUT} = 0$ is smaller. From the simulation results, it can be observed that the proposed novel FRAM read scheme can accurately read the state of FRAM cells and achieves significantly faster readout speed compared to the conventional read scheme.

Important circuit technical indicators of readout speed and power consumption are compared between the proposed novel FRAM read scheme and the traditional FRAM read scheme, as shown in Fig. 5. In the traditional read scheme, the voltage difference between BL and BLR needs to reach a certain level for the SA to output correct results. However, our novel read scheme can recognize the information at the early stage of the voltage rise on BL. Therefore, our proposed novel read scheme exhibits significant improvements in readout delay compared to the traditional read scheme, reducing readout delay by approximately 42%. Additionally, since our novel read scheme does not require the assistance of traditional reference units, it has a significant advantage in power consumption compared to the traditional read scheme, reducing power consumption by approximately 98.5%, enabling the novel read scheme to operate in ultra-low power conditions.

IV. SUMMARY

In this work, a novel FRAM read scheme based on a differential circuit is proposed. The differential circuit is utilized to detect the differences in the rate of BL voltage changes of FRAM cells in different states for information retrieval. Compared to the traditional FRAM read scheme, the novel readout scheme has been optimized for readout speed, reducing readout delay by approximately 42%, and achieves a significant reduction in power consumption by approximately 98.5%. This work is poised to provide a low-latency, ultra-low power read scheme solution for FRAM arrays.

ACKNOWLEDGMENT

This work was supported by the National Key R&D Program of China (Grant No. 2022ZD0119002), the National Natural Science Foundation of China (Grant No. 92064001 & 62174146), the Major Program of Zhejiang Natural Science Foundation (Grant No. LZ23F040001), the Zhejiang Province Key R & D programs (Grant No. 2024C01010), the China Three Gorges Construction Engineering Corporation (Grant No. JGD0323003), the China Three Gorges corporation (Grant No.202203269), the Yangtze River Delta Science and Technology Innovation Community Joint Research Project (Grant No. 2022CSJGG1100), and the Central Guided Local Science and Technology Development Funding Program (Grant No. 2023ZY1069).

REFERENCES

[1] H.-S. P. Wong and S. Salahuddin, "Memory leads the way to better computing," Nature Nanotech, vol. 10, no. 3, pp. 191–194, Mar. 2015.

[2] S. Yu and P.-Y. Chen, "Emerging Memory Technologies: Recent Trends and Prospects," IEEE Solid-State Circuits Mag., vol. 8, no. 2, pp. 43–56, 2016.

[3] S. Yu, "Neuro-Inspired Computing With Emerging Nonvolatile Memorys," Proc. IEEE, vol. 106, no. 2, pp. 260–285, Feb. 2018.

[4] H. Ishiwara, "Ferroelectric random access memories," J. Nanosci. Nanotechnol., vol. 12, pp. 7619-7627, October 2012.

[5] T. Mikolajick, U. Schroeder and S. Slesazeck, "The Past, the Present, and the Future of Ferroelectric Memories," IEEE Transactions on Electron Devices, vol. 67, no. 4, pp. 1434-1443, April 2020.

[6] X. Lyu and M. Si, "Ferroelectric and Anti-Ferroelectric Hafnium Zirconium Oxide: Scaling Limit, Switching Speed and Record High Polarization Density," 2019 Symposium on VLSI Technology, Kyoto, Japan, pp. T44-T45, June 2019.

[7] Mitra. S, "A method of characterizing sense amplifier imbalance issues on a 2T2C FRAM memory," Integrated Ferroelectrics, 27(1–4), pp. 325–336, Aug 2006.

A 13-bit,1 MS/s Cyclic ADC, for high-speed CMOS Image sensor

Qi Lv, Rensheng Shen, Yu Cheng, Guoqiang Zhong, Yang Qu, Yuchun Chang *

School of Integrated Circuits, Dalian University of Technology, Dalian 116024, China

* Email: cyc@dlut.edu.cn

Abstract—In this paper, a high-speed and low-power column-level Cyclic ADC design scheme is proposed for CMOS image sensors (CIS). The proposed 13-bit Cyclic ADC is made up of a four-loop implementation of the 3.5-bit single slope ADC (SS-ADC) as a sub-ADC, which only contains analog circuits of the amplifier and comparator. In addition, the correlated-double-sampling (CDS) circuit is used to pre-amplify the voltage in cycle when idle, thus improving the speed of the circuit. Using the 0.18μm standard CMOS process, the results indicate that a 13-bit ADC with a sampling rate of 1MS/s has been implemented and has a signal-to-noise and distortion ratio (SNDR) of 70.24 dB and a spurious free dynamic range (SFDR) of 74.10dBc. The power dissipation is 120μW with a 1.8V supply.

Keywords—*Analog-to-digital converter (ADC), cyclic ADC, single-slope (SS) quantizer, CMOS image sensor, high-speed conversion.*

I. INTRODUCTION

With the increasing demand for image sensors in various applications, there is a growing need for high-speed and high-resolution CMOS image sensors (CISs), consequently requiring higher bandwidth readout circuits. However, this increased demand also brings about greater costs such as elevated power consumption, larger area requirements, and complex structures. The design of the analog-to-digital converters (ADCs) plays a crucial role in determining the overall performance of the readout circuit. The single-slope ADCs (SS-ADCs) is widely used due to its simplicity and ease of design, while its shared ramp structure aligns well with column-parallel ADC designs. However, its conversion speed is limited by the speed of clock, like Flash ADCs which require $2N-1$ comparison period to achieve N-bit accuracy. On the other hand, successive approximation register (SAR) ADCs offer significant improvements in speed as they only require N comparison period for N-bit accuracy, nevertheless, their digital-to-analog converter (DAC) often necessitates a large capacitor array for implementation despite some designs managing to reduce total capacitor count through redundant capacitor configurations. Furthermore, SAR ADCs accuracy heavily relies on layout and process considerations even though certain studies have made substantial progress in addressing these limitations through separate capacitor designs that impose more complex process requirements [1]. Cyclic ADCs are derived from Pipeline ADCs and strike a balance between speed, power consumption, and area requirements while offering comparable speeds to SAR ADCs but with smaller area without excessive process demands. The method of redundant signed digital (RSD) algorithm can correct the error caused by the offset of the comparator [2]. But as shown in Equation (1), the band-gain-width (GBW) of the amplifier (AMP) in the multiplying digital-to-analog

converter (MDAC) circuit needs to be satisfied (1) to ensure that the ADC achieves conversion accuracy, witch 'n' is the accuracy of the loop, 't' is the conversion time, and 'f' is the loop feedback coefficient. It means that higher loop amplification requires greater GBW, which not only limits the speed, requires higher power consumption, but also increases the difficulty of AMP design.

$$GBW > \frac{(n+1)\ln 2}{2\pi f \cdot t} \qquad (1)$$

In this design, the idle time multiplexing of the correlated-double-sampling (CDS) circuit for a double pre-amplification circuit, which reduces the requirements about the GBW of the AMP in the MDAC circuit and enhances ADC working speed without adding extra power consumption. The sub-ADC employs a single-slope structure, which effectively saves power consumption and area through voltage of the ramp ($VRAMP$) sharing. A 13-bit ADC with a sampling rate of 1MS/s has been implemented with a power consumption of 120μW.

Section II presents the operating principle of the proposed ADC and its operation sequences with the timing diagram. The simulation results of the proposed ADC are analyzed and compared with prior works in Section III. Finally, conclusions are given in Section IV.

II. ARCHITECTURE OF THE PROPOSED CYCLIC ADC

The block diagram of the proposed ADC is shown in Fig.1.

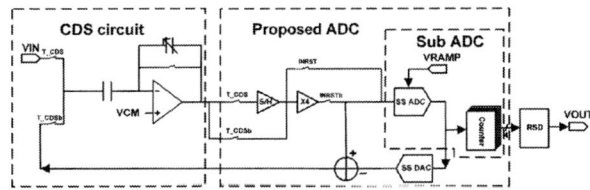

Fig.1 The block diagram of the proposed ADC.

Compared with the traditional readout circuit, the CDS circuit is given additional work in this design. By controlling the switch T_CDS and T_CDSb, the CDS circuit changes to be a double pre-amplification circuit to pre-amplify the output voltage of last cycle for next period of quantization. The sub-ADC is implemented through a 3.5-bit SS-ADC, comparing the $VRAMP$ with the voltage values after each cycle calculation including the amplifier circuit and sub-DAC .In addition to the first cycle to complete 4-bit quantization, the remaining three cycles achieve 3-bit quantization and 1-bit redundant bit for RSD correction, and finally output 13-bit quantization result. The schematic of the proposed ADC is shown in Fig.2.

979-8-3503-6184-1/24 $31.00 © 2024 IEEE

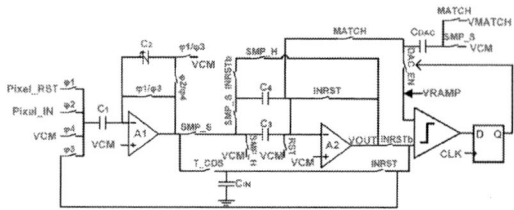

Fig.2. The schematic of the proposed ADC.

The amplifier (A1) and its associated capacitors with switches constitute the CDS and the preventive amplification capacitor circuit. By adjusting the capacitance of the variable capacitor (C_2), the magnification of the circuit is changed. The main amplifier (A2) together with the capacitors C_3, C_4 and their switches constitute the 4-times (×4) multiplier circuit, where capacitor C_3 and C_4 have capacitance values of $3 C_u$ and C_u, respectively. The 'Double-Tail' structure of comparer we used, have a good balance between power consumption and accuracy, but the performance of the comparator is greatly affected by the input voltage. So, a D flip-flop (DFF) is connected in series to ensure the stability of the comparator output signal [3]. The timing diagram of a complete data conversion is shown in Fig.3 and Fig.4 (a) and (b) show the sequential operation phases of the CDS circuit.

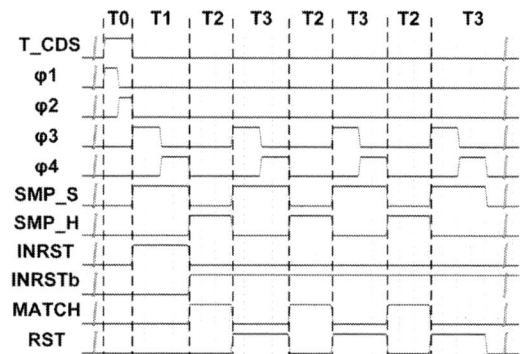

Fig.3 The timing diagram of a complete data conversion.

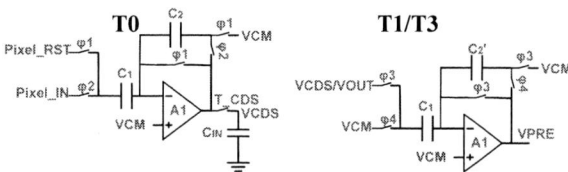

Fig.4 (a) CDS phase.   Fig.4 (b) two-times multiplier phase.

When the circuit working at the CDS operation phase (T0) in Fig.4 (a). By controlling the switch φ1 and φ2, capacitance (C_{in}) collects the analog voltage from the pixel through the CDS circuit and S/H circuit. When the ADC is working at comparing phase (T1/T3), the CDS function turns to be the pre-amplification as shown in Fig.4 (b). Changing the capacitance C_2 as $C_2' = C_2/2$, the circuit amplifies the $VCDS$ or $VOUT$ from last cycle by a factor of two to $VPRE$. And at this point, after pre-amplified, the voltage of $VPRE$ can be expressed as Equation (2), and 'i' is the number of circuit cycles, from 0 to 3.

$$VPRE^{i+1} = 2VOUT^i - VCM \qquad (2)$$

Fig.5 (a) to (c) show the sequential operation phases of the proposed ADC circuit. In the 1st operation phase of the proposed ADC (T1) as shown in Fig.5 (a).

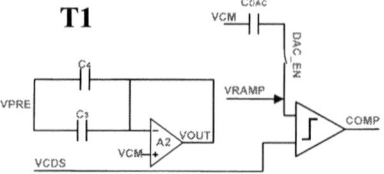

Fig.5 (a) The 1st operation phase.

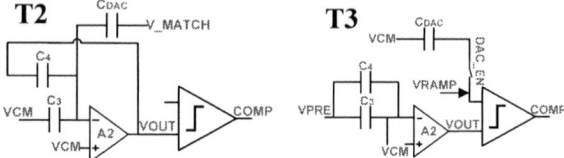

Fig.5 (b) The 2nd operation phase.   Fig.5 (c) The 3rd operation phase.

The sub-ADC performs the first cycle of the quantized $VCDS$ to compare with the $VRAMP$ which is generated by the ramp generator and produce a 4-bit quantization result. At the same time, capacitance C_3 and C_4 sample the $VPRE$ which the CDS circuit has already multiplied $VCDS$ in two and store it. At the same time, capacitance C_{DAC} samples the $VRAMP$ voltage according to the signal DAC_EN sent by DFF. The capacitance of the C_{DAC} have capacitance values of $8C_u$ to achieve a 3-bit, or 8-times (×8), amplification. The total amount of charge stored in the circuit at this time Q_1 can be expressed as Equation (3).

$$Q_1 = (2VOUT^i - 2VCM) \times 4C_u - (VDAC - VCM) \times 8C_u \quad (3)$$

In the 2nd operation phase of the proposed ADC (T2) as shown in Fig.5 (b). In this sampling phase(T1), the charge sampled by the capacitor C_3, C_4 and C_{DAC} is transferred to the C_4, so as the voltage has completed a multiplicative (X8) transition at the $VOUT$. When the top plate of C_{DAC} has triggered from VCM to $VMATCH$ which is equal to $VCM - VREF/16$, so that $VOUT$ and $VRAMP$ are compared on the same reference while voltage subtraction is realized. The total charge Q_2 at this point can be expressed as Equation (4).

$$Q_2 = (VOUT^{i+1} - VCM) \times C_u - (VMATCH - VCM) \times 8C_u \quad (4)$$

After the conversion, the total charge stays the same, so Q_1 equals to Q_2. The output voltage $VOUT$ can be expressed as Equation (5).

$$VOUT^{i+1} = 8(VOUT^i - VDAC - VREF/16) + VCM \qquad (5)$$

The voltage conversion is completed, a 3-bit A/D conversion result and an extra 1-bit redundant for error correction are obtained. And then, this cycle's $VOUT$ will be sampled by the pre-amplification circuit for next cycle.

The 1-bit redundant bit provides an additional margin of $VREF/16$ for the offset brought by the comparator. Fig. 6 shows the adjusted residue voltage in the range between $-VREF/2$ and $+VREF/2$.

In the 3rd operation phase of the proposed ADC (T3) as shown in Fig.3 (d). Like the 1st (T1), capacitors C_3 and C_4 sample the $VOUT$ after being pre-amplified, while the sampling capacitor of sub-DAC also samples the $VRAMP$ according to the output signal of the comparator. After that, the circuit enters the cycle of T3 to T2, and then T2 to T3 again.

979-8-3503-6184-1/24 $31.00 © 2024 IEEE   665

Until four cycles are completed, a 13-bit quantization result is obtained.

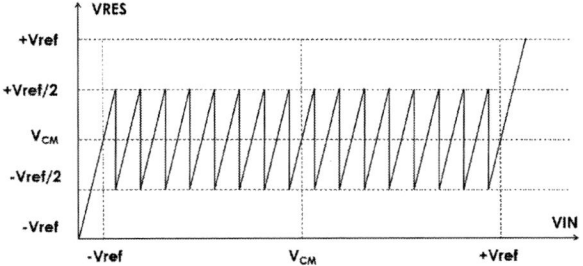

Fig.6 Residue voltage of the proposed ADC.

III. RESULTS AND ANALYSIS

Fig.7 shows the process simulation diagram of one A/D conversion, Through the calculation of the MDAC circuit, *VOUT* changes in each cycle according to the (5), then generate signal COMP by compare whit *VRAMP*. Counter quantizes the digital code by the width of COMP, after four cycles, complete the 13-bit quantization finally.

The FFT simulation results are shown in Fig.8, the ENOB=10.38bit, SFDR=74.10dBc, and SNDR =70.24dB. Compared with the advanced ADC, the results are shown in Table.1.

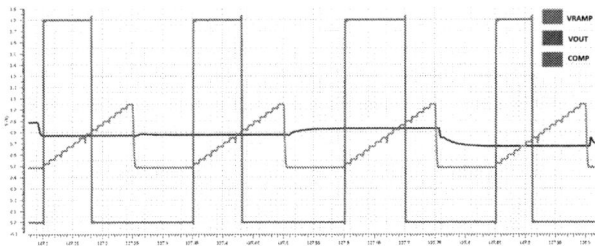

Fig.7 The process simulation diagram of one A/D conversion.

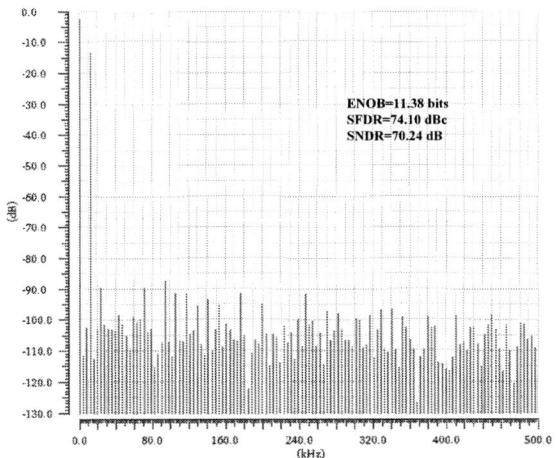

Fig.8 The FFT simulation results.

The comparator completes the overall 13-bit quantization after 4 cycles, which saves power consumption compared with [4][5], who use more than eight cycles to quantize. Using SS-ADC as sub-ADC to completes the 3.5-bit quantization, compared with [6], only one comparator is needed so that the area is also reduced. By reusing the CDS circuit, the proposed ADC has a higher speed than [7].

TABLE I. PERFORMANCE COMPARISON

Parameters	[5]	[6]	[7]	This work
Technology[nm]	180	180	180	180
Samp.rate[MS/s]	1	1.67	0.78	1
Supply[V]	1.8/3.3	1.8/3.3	/	1.8
Power consumption per channel [μW]	800	256	87	120
FoM[pJ/conv-step]	0.19	0.037	0.0299	0.045

IV. CONCLUSION

A 13-bit, 1 MS/s Cyclic ADC is designed in a standard 180nm CMOS technology. A 3.5-bit SS-ADC is used as sub-ADC. By reusing the CDS circuit, which improves the speed of the circuit and reduces the power consumption. The total power consumed by the ADC is 120μW, resulting in the SFDR=74.10dBc, SNDR=70.24dB and FoM of 0.045pJ/conversion-step.

ACKNOWLEDGMENT

This work is supported by National Key R&D Program of China (2023YFB4503003) and Special projects for industrial foundation reconstruction and high-quality development of manufacturing industry in 2022(No. TC220A04A-49).

REFERENCES

[1] B. Ginetti and P. G. A. Jespers, "A 1.5 Ms/s 8-Bit Pipelined RSD A/D Converter," ESSCIRC '90: Sixteenth European Solid-State Circuits Conference, Grenoble, France, 1990, pp. 137-140.

[2] D. Schinkel, E. Mensink, E. Klumperink, E. van Tuijl and B. Nauta, "A Double-Tail Latch-Type Voltage Sense Amplifier with 18ps Setup+Hold Time," 2007 IEEE International Solid-State Circuits Conference. Digest of Technical Papers, San Francisco, CA, USA, 2007, pp. 314-605, doi: 10.1109/ISSCC.2007.373420.

[3] M. Li et al., "A 6.94-fJ/Conversion-Step 12-bit 100-MS/s Asynchronous SAR ADC Exploiting Split-CDAC in 65-nm CMOS," in IEEE Access, vol. 9, pp. 77545-77554, 2021, doi: 10.1109/ACCESS.2021.3079406.

[4] A. Kaur and M. Sarkar, "An input folding high speed cyclic ADC for column-parallel readout in CMOS image sensors," 2022 IEEE International Symposium on Circuits and Systems (ISCAS), Austin, TX, USA, 2022, pp. 910-914, doi: 10.1109/ISCAS48785.2022.9937720.

[5] A. Kaur, D. Mishra and M. Sarkar, "A 12-bit, 2.5-bit/cycle, 1 MS/s two-stage cyclic ADC, for high-speed CMOS Image sensors," 2018 IEEE International Symposium on Circuits and Systems (ISCAS), Florence, Italy, 2018, pp. 1-5, doi: 10.1109/ISCAS.2018.8350899.

[6] J. -Y. Jeong, J. Shim, S. -K. Hong and O. -K. Kwon, "A High-Speed and Energy-Efficient Multi-Bit Cyclic ADC Using Single-Slope Quantizer for CMOS Image Sensors," in IEEE Transactions on Circuits and Systems II: Express Briefs, vol. 68, no. 7, pp. 2322-2326, July 2021, doi: 10.1109/TCSII.2021.3062139.

[7] A. Kaur, D. Mishra and M. Sarkar, "A 12-bit, 2.5-bit/Phase Column-Parallel Cyclic ADC," in IEEE Transactions on Very Large Scale Integration (VLSI) Systems, vol. 27, no. 1, pp. 248-252, Jan. 2019, doi: 10.1109/TVLSI.2018.287134.

An Area-Efficient 16-bit Four-channel R-2R DAC Based on Switching On-resistance Adaptive Calibration Technique

Kejun Wu [1], Yuchen Liu [1], Yuhan Hu [1], Yu He [1], Zhen Yu [1], Ning Ning*[1]

[1] University of Electronic Science and Technology of China

* Email: ning_ning@uestc.edu.cn

Abstract—**In this paper, an area-efficient 16-bit four-channel R-2R DAC based on switching on-resistance adaptive calibration technique is proposed. Through the switching on-resistance adaptive calibration technique, all MOS transistor switches in the R-2R DAC resistance network are of the same size, which will reduce the area of the switch array. Since the MOS transistor switches have the same size, the equivalent resistance seen by each node of R-2R resistance network is the same. This will eliminate the nonlinear effects caused by the switching on-resistance and improve the effective resolution of the DAC. The on-resistance of the switch that is easily affected by PVT change is equivalent to a resistance that is not easily affected, which improves the PVT change suppression ability of the DAC. The 16-bit four-channel DAC designed with the switching on-resistance adaptive calibration technique is verified based on the 180nm CMOS process. The post-simulation results show that the ENOB of the single-channel DAC under TT corner is 15.35 bit, and the SFDR is 101.5 dBc@1MHz. The single-channel DAC area is only 0.0265 mm^2 and the power consumption is only 0.021mW at 1.8V power supply.**

Keywords—*R-2R DAC, 16-bit, area-efficient, switching on-resistance adaptive calibration technique*

I. INTRODUCTION

In the traditional R-2R DAC design, as the DAC resolution increases, it faces a series of problems such as extremely large area and difficulty in improving the effective resolution. Firstly, the area of resistor array and switch array increases dramatically as the resolution of the R-2R DAC increase. Secondly, the impact of on-resistance on DAC performance increases with the increase in conversion resolution. Finally, PVT changes become more crucial for high precision R-2R DAC performance, directly affecting conversion resolution.

In order to solve the above problems, existing high precision R-2R DACs have adopted a variety of technical means to improve some performance indicators[1-3]. For example, digital calibration technique is used to improve the performance of DAC under mismatch to improve DAC resolution. H. Fan uses the ordering technique and Y. Li uses the ordered element matching (OEM) to improve DAC resolution. Esmaili, A. use the foreground calibration technique to circumvent both current source and resistor mismatches at the output node. But they require additional digital circuit design, which increase design complexity, chip power consumption and chip area.

This paper uses the switching on-resistance adaptive calibration technique to achieve the same size of all MOS switches, which will reduce the area of the switch array. Since the MOS transistor switches have the same size, the equivalent resistance seen by each node of R-2R resistance network is the

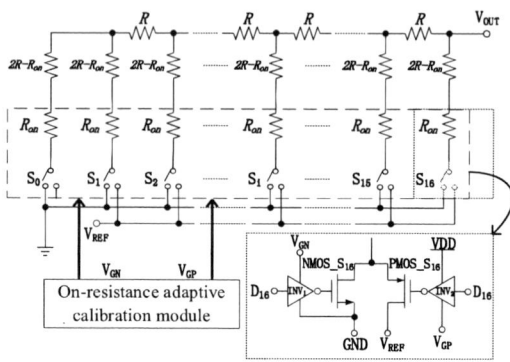

Fig. 1. The overall structure of single-channel 16-bit R-2R DAC.

same. This will eliminate the nonlinear effects caused by the switching on-resistance and improve the effective resolution of the DAC. The on-resistance of the switch that is easily affected by PVT change is equivalent to a resistance that is not easily affected, which improves the PVT change suppression ability of the DAC.

II. 16-BIT R-2R DAC ARCHITECTURE

The overall structure of the small-area 16-bit R-2R DAC based on the switching on-resistance adaptive calibration technique proposed in this paper is shown in Fig. 1. It includes R-2R resistor network, switch array and on-resistance adaptive calibration module. The resistance value of 2R in the R-2R resistor network is ($2R$-R_{on}), where R_{on} is the equivalent resistance when the MOS switch is turned on. The switches in the switch array have the same geometric size and the same on-resistance value. Therefore, the resistance value seen to the left of each node in the R-2R resistor network is R, which is not affected by the on-resistance of the switch. Compared with the traditional R-2R DAC where the MOS switch area needs to be doubled with the increasing one bit of DAC resolution, all switches in our architecture have the same area, which can significantly reduce the area of the switch array in the DAC.

As shown in Fig. 1, the on-resistance adaptive calibration technique realizes that the NMOS switch transistor or PMOS switch transistor in the on-state has the same on-resistance, and its value is R_{on}. All switches have the same on-resistance when they are in the on state. This makes the R-2R resistor network well matched, achieving high precision DAC characteristics. The above calibration technique not only makes the on-resistance of the MOS switch independent of its own characteristics, but also has adaptive stability with PVT changes. This optimizes DAC performance and improves DAC resolution while reducing the impact of PVT changes on DAC performance.

979-8-3503-6184-1/24 $31.00 © 2024 IEEE

III. CIRCUITS IMPLEMENTATION

A. Design of switch

As shown in Fig. 1, the structures of switches S_0 to S_{16} are the same geometric size. The switch structure is explained by taking the MSB switch S_{16}. The switch consists of two inverters, one NMOS transistor and one PMOS transistor.

The two inverters are used to transmit the calibration voltages V_{GN} and V_{GP}. Taking the NMOS transistor switch as an example, its on-resistance value in the linear region can be expressed as

$$Ron = \frac{1}{\mu Cox(W/L)(V_{GS} - V_{TH})} \qquad (1)$$

As shown in (1), when the influence of PVT changes is not considered, NMOS switches of the same geometric size have the same on-resistance. At the same time, by changing the V_{GS} voltage of the NMOS switch and the PMOS switch respectively, they can have the same on-resistance. Furthermore, assuming that the switching on-resistance R_{on} can be made equivalent to an actual resistor, this actual resistor can better match the resistor in the resistor network. Then the switching on-resistance R_{on} and the 2R resistor in the resistor network (its resistance value is $2R-R_{on}$) can be connected in series to be equivalent to a resistor with a resistance value of $2R$. In this way, the new resistor network matching will no longer be affected by the switching on-resistance value and PVT changes.

B. Design of on-resistance adaptive calibration technique

The principle of on-resistance adaptive calibration technique is to generate the gate voltage of the switch MOS transistor. When the gate voltage turns on the switch transistor which works in the linear region, its on-resistance is equal to R_{on}. The gate voltage generation method of the above switch transistor is shown in Fig. 2.

As shown in Fig. 2, V_{GP} and V_{GN} are the gate voltages of the PMOS switch transistor and the NMOS switch transistor, respectively. The on-resistance adaptive calibration circuit has three resistors of the same type with different resistance values, namely R_{on}, R1, and R2. The design uses the rpposab resistor (P-type polycrystalline silicon resistor with salicide block). Among them, R_{on} has a smaller resistance value, which is equal to the MOS transistor on-resistance in the DAC. Compare with resistor R_{on}, resistor R1 and R2 are design with larger resistance values to reducing current. In this design, by adjusting the resistance values of R1 and R2, operational amplifier A1 and A2 are the same. This multiplexing allows A1 and A2 to share the same biases circuit which can reduce design complexity and power consumption.

The function of operational amplifier A1 and A2 is used to clamp the voltage. A1 clamps the voltage between points A and B, and A2 clamps the voltage between points C and D. When points A and B have the same voltage, the voltage drop and current on PMOS and R_{on} are the same, which can make the on-resistance of PMOS equal to the resistance value of R_{on}. Similarly, the on-resistance of NMOS is equal to the resistance value of R_{on}. And A1 and A2 form a negative feedback structure, which can stabilize the output voltages V_{GP} and V_{GN}.

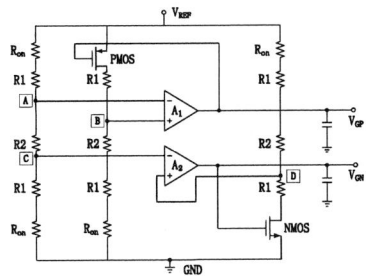

Fig. 2. The structure of switching on-resistance adaptive calibration circuit.

There is a capacitor at both output points V_{GN} and V_{GP}, which is used to further stabilize the output point voltage. The switch conduction state will change continuously as the input code of the DAC changes. This will cause the load of the nodes V_{GN} and V_{GP} to change continuously. Its essence is the change of load capacitance. By adding capacitors to the above nodes, the impact of this change on the calibration voltage can be reduced. This structure uses MOS transistor as capacitors, which can further reduce the area.

C. PVT stability analysis

Compared with the traditional R-2R resistor network, the on-resistance adaptive calibration circuit not only makes the R-2R resistor network better matched, but also allows the voltages V_{GP} and V_{GN} to change with PVT changes. This makes the MOS on-resistance have adaptive stability.

For the change of process, in the traditional R-2R DAC, it can be seen from (1) that the change of process corner causes the change of parameters such as threshold voltage V_{TH}. This change has a particularly significant impact on the on-resistance of MOS transistors. It will further deteriorate the performance of DAC. In this structure, the on-resistance R_{on} of the MOS switch transistor is equivalent to the actual resistance. The actual resistance value is much less affected by the process corner than the MOS switch transistor, which can achieve the purpose of optimizing the DAC performance.

For the change of temperature, in the R-2R resistor network in this paper, the first-order temperature coefficients of the resistors can cancel each other, which makes the DAC insensitive to temperature changes.

For the change of power supply voltage, since the R-2R resistor network and the on-resistance adaptive calibration circuit are both powered by the same reference voltage V_{REF}, the influence of power supply voltage fluctuation on the DAC is reduced.

IV. LAYOUT AND POST SIMULATION RESULTS

As shown in Fig. 3, the layout design of the 4-channel 16-bit R-2R DAC with switching on-resistance adaptive calibration technique is completed. The overall layout design adopts a central symmetric structure. The resistor array of each channel DAC uses co-centroid matching to improve the matching of the R-2R network. To reduce area and power consumption, the four channel DACs share one switching on-resistance adaptive calibration circuit. The area of the 4-channel DAC is 0.53*0.2mm^2. The power consumption of the 4-channel DAC is 0.0828mW when power supply is 1.8V.

979-8-3503-6184-1/24 $31.00 © 2024 IEEE

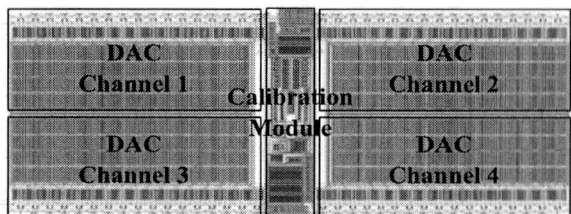

Fig. 3. The layout design of the 4-channel 16-bit R-2R DAC.

The post-simulation results of SFDR as shown in Fig. 4, when the input frequency is 1 MHz under TT corner, the SFDR of the single-channel DAC is 101.5 dBc that the ENOB is 15.34 bits. The performance of ENOB and SFDR with process corner changes is shown in Fig. 5. As shown in Fig. 5(a), without calibration, ENOB performance varies from 13.83bit to 15.34bit at different process corners. The variation range is greater than 1.51bit. With calibration, ENOB performance varies from 15.26bit to 15.352bit. The variation range is less than 0.1bit. As shown in Fig. 5(b), the SFDR performance without calibration varies from 85.64dBc to 101.56dBc at different process corners. The variation range is greater than 15.9dBc. With calibration, the SFDR performance varies from 100.5dBc to 101.57dBc. The variation range is less than 1.1dBc.

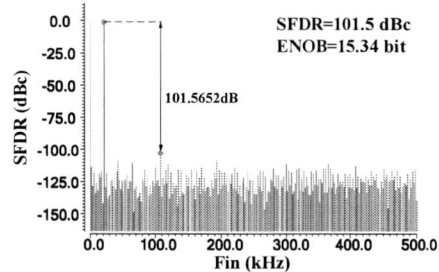

Fig. 4. Dynamic performance at TT corner.

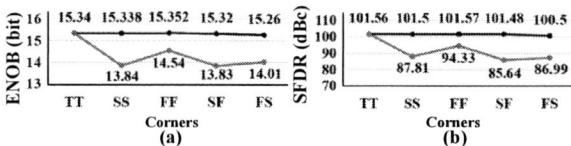

Fig. 5. Dynamic performance with different process corners.

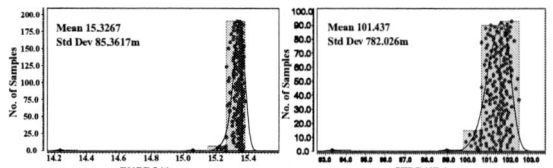

Fig. 6. The dynamic performance with 200 Monte Carlo post-simulation.

The performance of ENOB and SFDR with 200 Monte Carlo simulations were performed on the single-channel DAC as shown in Fig. 6. The simulation results show that the mean of ENOB is 15.32bit with proposed calibration technique. The standard deviation of ENOB is 0.085 bit. The mean of SFDR is 101.4dBc that the standard deviation is 0.782 dB. It can be seen that the calibration technique proposed in this paper is beneficial to reduce the impact of process mismatch on the performance of high resolution DAC. The static performance of the single-channel DAC is shown in Fig. 7. The DNL is less than ±0.5 LSB and the INL is less than ±1 LSB. The comparison of the DAC parameters with other publications is shown in Table I.

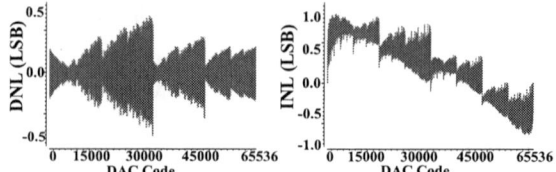

Fig. 7. The static performance of the proposed DAC.

V. CONCLUSION

This paper proposes an area-efficient 16-bit four-channel R-2R DAC based on the switching on-resistance adaptive calibration technique. Through the switching on-resistance adaptive calibration technique, all MOS transistor switches in the R-2R DAC resistor network have the same size. From post-simulation verification, the calibration technique proposed in this paper can not only significantly reduce the area of DAC, but also significantly improve the dynamic performance and PVT change suppression ability. The 16-bit four-channel DAC designed is verified based on the 180nm process. The variation range of ENOB and SFDR is reduced from 1.51bit to 0.1bit and from 15.9dBc to 1.1dBc, respectively. The area of single-channel DAC is only 0.0265mm^2.

TABLE I.   THE COMPARISON OF THE DAC PARAMETERS WITH OTHER PUBLICATIONS

	[4]	[5]	This work
Technique	65nm	500nm	**180nm**
Architecture	R-2R DAC	R-2R DAC	**R-2R DAC**
Sample rate (MS/s)	4	0.1	**1**
Resolution (bit)	16	12	**16**
SFDR/SNR (dBc/dB)	65/NA	NA/NA	**101.5/94.1**
ENOB (bit)	N/A	N/A	**15.34**
DNL/INL (LSB)	-0.04/-0.26	0.23/0.67	**±0.5/±1**
Power consumption (mW)	1.2mW @1.2V	525mW @15V	**0.021mW @1.8V**
Area (mm^2)	N/A	20	**0.0265**

REFERENCES

[1] H. Fan, J. Li and F. Maloberti, "Order Statistics and Optimal Selection of Unit Elements in DACs to Enhance the Static Linearity," in IEEE Transactions on Circuits and Systems I: Regular Papers, vol. 67, no. 7, pp. 2193-2203, July 2020.

[2] Y. Li and D. Chen, "Low-Cost, High-Precision DAC Design Based on Ordered Element Matching," in IEEE Transactions on Circuits and Systems I: Regular Papers, vol. 66, no. 2, pp. 502-512, Feb. 2019.

[3] Esmaili, A., Babazadeh, H., "A Foreground Self-calibration Technique for High-Resolution Switched-Current R-2R Digital-to-Analog Converters," in Circuits Syst Signal Process 39, 2307–2327 (2020).

[4] A. A. Noorwali, S. M. Qasim, A. S. Doost and A. Huynh, "A 16-bit 4 MSPS DAC for lock-in amplifier in 65nm CMOS," 2016 IEEE 13th International Conference on Networking, Sensing, and Control (ICNSC), Mexico City, Mexico, 2016, pp. 1-5.

[5] H. Fan et al., "A 4-Channel 12-Bit High-Voltage Radiation-Hardened Digital-to-Analog Converter for Low Orbit Satellite Applications," in IEEE Transactions on Circuits and Systems I: Regular Papers, vol. 65, no. 11, pp. 3698-3706, Nov. 2018.

A Background Calibration Method of Bandwidth Mismatch for Time-Interleaved ADCs Based on Neural Network

Tianqi Yang, Longsheng Wang, Xin Zhao, Shubin Liu, Dengquan Li*, and Zhangming Zhu

Key Laboratory of Analog Integrated Circuits and Systems (Ministry of Education), School of Integrated Circuits,
Xidian University, China

* Email: dqli@xidian.edu.cn

Abstract—This paper proposes a background calibration algorithm based on neural network (NN) for bandwidth mismatch in TI-ADCs. The calibration method uses a three-layer neural network to achieve lower training resource consumption and higher calibration accuracy by processing the input and target output data of the TI-ADC. The optimal structure of the neural network is determined by simulation. A 12-bit, 3.2GS/s TI-ADC model is used to validate our proposed method. Simulation results show that SNDR and SFDR are improved from 49.75dB and 53.37dB to 69.51dB and 88.75dB after calibration, respectively. The calibration method provides good calibration performance for different types of signals, including single-tone and multi-tone signals, and it is applicable over a wide frequency range. Meanwhile, it has a high effective bandwidth that covers the overall Nyquist frequency range.

Keywords—Time-interleaved ADC, bandwidth mismatch, neural network, mismatch calibration.

I. INTRODUCTION

In the era of information, ADC is being utilized in more and more fields. Time-interleaved ADC (TI-ADC) achieves higher sampling frequencies by using multiple sub-ADCs to sample alternately. Although the sub-ADCs are theoretically identical, in reality, the parameters of the multiple sub-ADCs are not perfectly the same, leading to inter-channel mismatch like offset mismatch, gain mismatch, timing mismatch, bandwidth mismatch etc.

Bandwidth mismatch is one of the unescapable mechanisms that reduce linearity in TI-ADC. Calibrating bandwidth mismatch poses a unique design challenge for its nonlinear relation between input frequency and phase [1]. For the bandwidth mismatch, some methods have been proposed to calibrate it in the past few years [1], [4], [6]. However, most of the methods focus on the circuit details like sampling resistances and capacitors to suppress the bandwidth mismatch, and inevitably extra circuit components will be introduced, leading to an increase in circuit area and cost [1]. Besides, although the introduction of extra circuit components can suppress the bandwidth mismatch, it also has side-effect like narrowing the sub-ADCs' bandwidth and so on [1]. In [2], the different types of distortion in TI-ADC are divided into two parts, each part have one specific neural network for calibration, but the bandwidth mismatch is not included in the calibrated distortion in [2].

In this paper, we present a neural network-based calibration method for bandwidth mismatch in TI-ADC. The advantages of the method are that it can relax algorithms requirements, applicable to multiple types of TI-ADC, capable of calibrating various signals, and usable over a wide

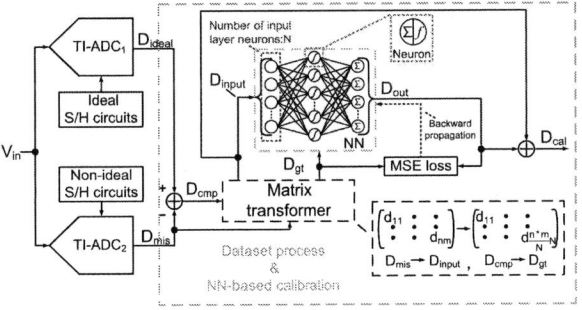

Fig. 1 The structure of proposed method.

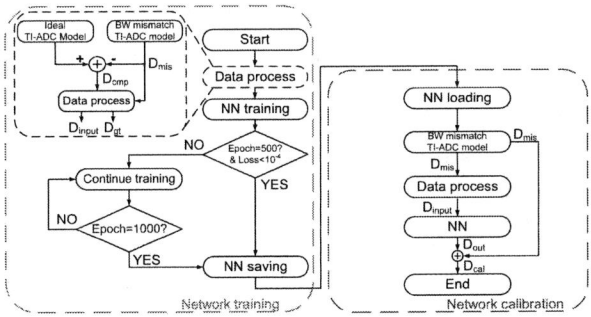

Fig. 2 The flowchart of calibration scheme.

frequency range. The rest of this paper is organized as follows. The overall calibration scheme is detailed in Section II. Section III shows the simulation results, and Section IV concludes the paper.

II. PROPOSED NEURAL NETWORK-BASED CALIBRATION

The overall calibration scheme of this paper is shown in Fig. 1. The analog signal V_{in} is transformed into the ideal signal (D_{ideal}) and the signal with bandwidth mismatch (D_{mis}) for network training, then the well-trained network can be used for mismatch calibration. The flowchart of the calibration scheme is illustrated in Fig. 2. The basic idea is to train and use a neural network to automatically reduce the difference between the signal with bandwidth mismatch and the ideal signal. The details of the calibration scheme are as follows.

A. Ground Truth for Network

A suitable network ground truth (Ideal ADC output) is an important factor for an effective calibration. There are two approaches to obtain the ground truth dataset. One needs a relatively simple algorithm but has low accuracy and poor

---

This work was supported in part by the National Natural Science Foundation of China (62274129, 62021004), in part by the Scientific Research Plan Projects of Shaanxi Education Department (20JY020), and in part by the Fundamental Research Funds for the Central Universities (QTZX23075).

adaptability, another requires additional calculations but has higher accuracy and is adaptable to various data types. For the first approach, it uses data without mismatch as the ground truth for the neural network which can directly obtain the calibrated result. However, in the training process, the network loss can be dominated by the components of the original signal, which may impose limitations on the signal type in terms of signal amplitude. The second approach uses the compensation values of data with bandwidth mismatch as the ground truth. By eliminating the components of the original signal, this approach does not impose restrictions on signal types. Additionally, the training speed of the network can be greatly improved without the components of the original signal. Therefore, the second approach described above is used in this paper. The D_{cmp} represent the compensation values, as shown in Fig. 1.

B. Dataset Process

The process of training a neural network is a process of constantly feeding a large amount of data to the network. In order to pursue the speed and accuracy of training, it is necessary for the dataset format to be the most compatible with the network structure, which requires that the dataset must be processed before training.

Matrix transformer completes the dataset processing task. It is capable of changing the format of the ADC output data based on the number of points N in the input layer, transforming D_{cmp} and D_{mis} into D_{gt} and D_{input} that are compatible with the network structure, respectively. This is a prerequisite for smooth network training, and in the following text it's also the prerequisite for quickly testing the optimal structure of the neural network.

C. Neural Network for Calibration

Neural network is the main body of the calibration scheme. Its structure and parameters can greatly affect the calibration accuracy. An optimal neural network structure can effectively suppress bandwidth mismatch with minimal computation resources [5]. The structure of a neural network must first determine the number of input and output neurons, the number of hidden layers and the number of neurons in each hidden layer. Fig. 3 (a) shows that the training performance of a single hidden layer and double hidden layers is almost identical. A significant improvement in SNDR and SFDR appears when updating the number of input neurons from 4 to 8. Further increasing the number of input neurons beyond 8 does not bring significant parameter improvement anymore. Fig. 3 (b) shows that the SNDR increases and then remains mostly constant with the increase in the number of hidden layer neurons under three different input neurons. Fig. 3 (c) shows the SFDR changes. Considering the trade-off between calibration accuracy, network training speed and computation resources, it is decided to use 8 input (output) neurons and 1 hidden layer and set the number of hidden layer neurons to 8.

Note that in this work, the number of output neurons is equal to the number of input neurons, this is because the TI-ADC output in this work is transformed to decimal form already, therefore, the network completes decimal to decimal conversion, not binary to decimal, so there is no need for the network to complete many-to-one function. This may sacrifice some computation resources, but it can greatly improve the speed of network training and calibration.

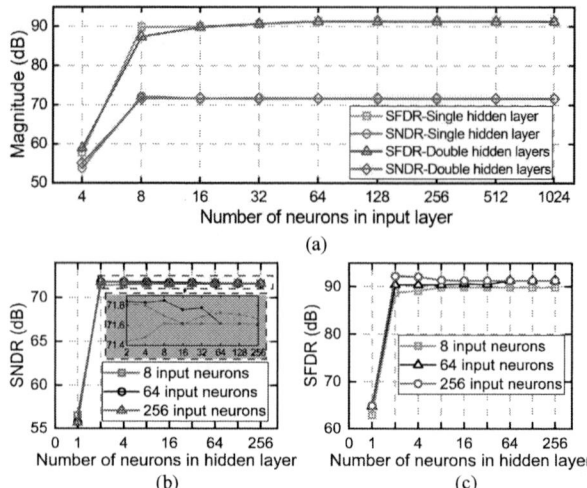

(a)

(b)   (c)

Fig. 3 Relationship between (a) SNDR, SFDR and the input neurons and hidden layers (b) SNDR and hidden layer neurons (c) SFDR and hidden layer neurons.

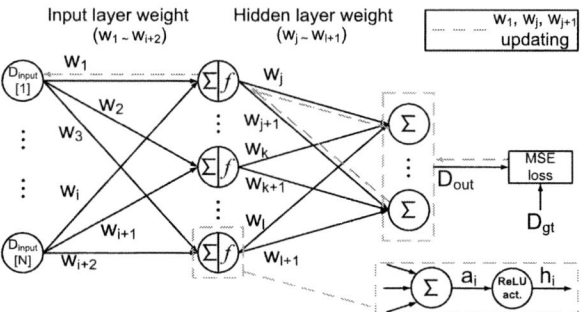

Fig. 4 Forward and backward propagation of the proposed network.

The hidden layer uses the Rectified Linear Unit (ReLU) activation function to add non-linearity to enhance fitting effects. The ReLU activation function can be represented by $ReLU(x)=max(0, x)$. The action of the ReLU function in a neural network can be represented by (1)

$$Output_{ReLU}[i] = ReLU[(\sum_{j=1}^{N} w_{ij} D_{input}[j]) + b_{ij}] \qquad (1)$$

where $D_{input}[j]$ represents the j-th input of the neural network, $Output_{ReLU}[i]$ is the output of the i-th neuron in the hidden layer, w_{ij} represents the coefficients corresponding to the j-th input for the i-th neuron in the hidden layer, and b_{ij} denotes the bias corresponding to the j-th input for the i-th neuron in the hidden layer. The detail of the neurons in hidden layers in this work is shown in Fig. 4.

It is important to choose an appropriate loss function for it plays a key role in reflecting the performance of neural networks as well as in backward propagation. The loss function utilized in this network is the Mean Squared Error (MSE), represented by (2)

$$MSE(D_{out}, D_{gt}) = \frac{\sum_{i=1}^{n} (D_{out} - D_{gt})^2}{n} \qquad (2)$$

A small MSE value indicates that the neural network's output is close to the ground truth.

The loss function plays a more important role in being able to calculate the gradients of the loss function with respect to each parameter through the chain rule of derivatives, and updating the parameters based on these gradients [2]. This process is known as backward propagation in neural networks as shown in Fig. 4. Taking the parameter input layer weight w in network training as an example, the backward propagation equation can be represented by (3)

$$w_{new} = w_{old} - l \cdot \nabla_w MSE(D_{out}, D_{gt}) \qquad (3)$$

where l represents the learning rate of the network. (3) indicates that updating w_{old} can be achieved by computing the gradient of the loss function with respect to w. The term $\nabla_w MSE(f(x), y)$ in (3) can be represented by (4)

$$\nabla_w MSE(D_{out}, D_{gt}) = \frac{\partial MSE}{\partial w_i} = \frac{\partial MSE}{\partial D_{out}} \cdot \frac{\partial D_{out}}{\partial a_i} \cdot \frac{\partial a_i}{\partial h_i} \cdot \frac{\partial h_i}{\partial w_i} \qquad (4)$$

where h_i refers to the value of D_{input} after undergoing a linear transformation, and a_i represents the value of h_i after passing through the activation function. The process of updating the weights w mentioned above can be represented by the blue dashed lines in Fig. 4. For other parameters in the neural network, such as bias b, updating them can be achieved using the same method. The neural network optimizer is chosen to be Adam, with an initial learning rate of 0.001. The minimum number of epochs used for network training is 500.

III. SIMULATION AND RESULT ANALYSIS

To validate the calibration effect of the calibration scheme, an 8-channel 12-bit 3.2GS/s TI-ADC with the bandwidth mismatch is constructed. The bandwidth mismatch follows a Gaussian distribution $B_m \sim N(0, 0.003^2)$, where B_m refers to the bandwidth mismatch of the m-th channel.

Fig. 5 (a) illustrates the simulation result of single-tone signal. Due to bandwidth mismatch, respectively, the SNDR of TI-ADC is only 57.32dB and SFDR is only 63.75dB. The SNDR and SFDR of TI-ADC are improved to 68.03dB and 86.83dB after calibration, respectively. The output spectrum before and after calibration for a two-tone input signal is shown in Fig. 5(b). After calibration, all spurs are suppressed by at least 11.99dB.

Fig. 6 illustrates the SNDR and SFDR variation with input frequency. It can be seen that the SNDR and SFDR after calibration can remain around 70dB and 90dB over all Nyquist frequency, respectively The SNDR improvement is relatively small at lower frequencies for the bandwidth mismatch issue of the TI-ADC is weak at low frequencies.

The comparison with other works on bandwidth mismatch calibration or on neural network-based ADC calibration schemes is shown in Table. I.

IV. CONCLUSION

This paper presents a neural network-based calibration scheme for the bandwidth mismatch in TI-ADC. The proposed method utilizes a three-layer neural network that optimizes the input and output data format, bringing reduced computation resources and higher calibration accuracy. The simulation results show the SNDR and SFDR of a 12-bit, 3.2GS/s TI-ADC from 49.75dB and 53.37dB to 69.51dB and 88.75dB, respectively. Furthermore, for the two-tone signal, it can at least suppress the spurs by 11.99dB, and its dynamic performance shows that after calibration, both SNDR and SFDR have improvement at all tested frequencies and can remain at a good level, around 70dB and 90dB, respectively. Therefore, it is suitable for signals of various types over a wide frequency range as well.

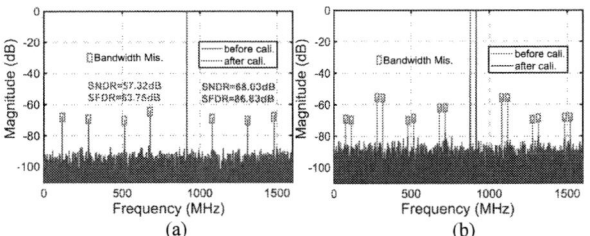

Fig. 5 Simulation output spectrum before and after calibration (a) single-tone signal (b) multi-tone signal.

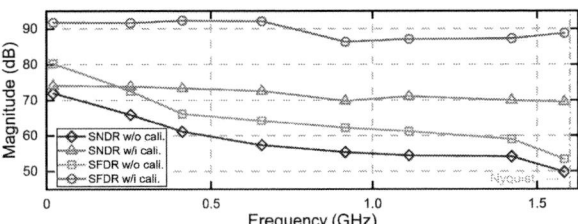

Fig. 6 SNDR and SFDR versus input frequency before and after calibration.

TABLE I.   COMPARISON WITH OTHER WORKS

		This Work	TCAS-I [1]†	TCAS-I [2]*	TVLSI [4]†
Resolution [bit]		**12**	10	12	12
Fs [GS/s]		**3.2**	12.8	5	3.6
SNDR @Nyq. [dB]	w/o cali.	49.75	39.9	48.5	N/A
	w/i cali.	**69.51**	59.6	58.6	N/A
SFDR @Nyq. [dB]	w/o cali.	53.37	50	59.6	64.16
	w/i cali.	**88.75**	72.8	70.7	78.62
Cali. Method		**NN**	Foreground signal injection	NN in chip	Derivative-based

*Measurement results. †Work on bandwidth mismatch calibration.

REFERENCES

[1] Y. Park, J. Kim and C. Kim, "A scalable bandwidth mismatch calibration technique for time-interleaved ADCs," *IEEE Trans. Circuits Syst. I, Reg Papers.*, vol. 63, no. 11, pp. 1889-1897, Nov. 2016.

[2] D. Zhai et al., "High-speed and time-interleaved ADCs using additive-neural-network-based calibration for nonlinear amplitude and phase distortion," *IEEE Trans. Circuits Syst. I, Reg Papers.*, vol. 69, no. 12, pp. 4944-4957, Dec. 2022.

[3] T. Zhang, Y. Cao, S. Zhang, C. Chen, F. Ye, and J. Ren, "Machine learning based prior-knowledge-free calibration for split pipelined-SAR ADCs with open-loop amplifiers achieving 93.7-dB SFDR," in *Proc. IEEE 45th Eur. Solid State Circuits Conf. (ESSCIRC)*, Sep. 2019, pp. 189-192.

[4] Y. A. Tavares, K. -Y. Lee and M. Lee, "All-digital bandwidth mismatch calibration of TI-ADCs based on optimally induced minimization," *IEEE Trans. Very Large Scale Integr. (VLSI) Syst.*, vol. 28, no. 5, pp. 1175-1184, May 2020.

[5] Z. Luo, S. Du, Z. Zhang, F. Lv, Q. Hong and M. Lai, "Artificial neural network based on memristive circuit for high-speed equalization," *IEEE Trans. Circuits Syst. I, Reg Papers.*, vol. 71, no. 4, pp. 1745-1756, April 2024.

[6] P. Satarzadeh, B. Levy, and P. Hurst, "Adaptive semiblind calibration of bandwidth mismatch for two-channel time-interleaved ADCs," *IEEE Trans. Circuits Syst. I*, vol. 56, no. 9, pp. 2075–2088, Sep. 2009.

A Second-Order Dual-Charge-Pump Passive Noise-Shaping SAR ADC for Medical Implant Devices

Kangkang Sun, Xuanxin Ke, Haoning Sun, Yuchen Wang, Feng Yan, Jingjing Liu*

School of Electronics and Communication Engineering, Sun Yat-Sen University, Shenzhen, China

* Email: sunkk3@mail2.sysu.edu.cn, liujj77@mail.sysu.edu.cn

Abstract—A 14.14bit 2MS/s Second-Order Passive Noise-Shaping Successive Approximation Register Analog-to-Digital Converter (NS SAR ADC) is proposed in this paper. The structure utilizes techniques such as charge pumps and multi-input comparator with a gain to compensate for the signal loss during the noise shaping process. It incorporates two zero points at 0.8 in the noise transfer function (NTF), which enhances the noise shaping capability. This ADC is designed using a standard 180nm CMOS process. The simulation results show that the ADC consumes 56.8 μW, achieving a signal-to-noise-and-distortion ratio (SNDR) of 86.75 dB and a spurious-free dynamic range (SFDR) of 97.16 dB with an oversampling ratio (OSR) of 8 at 2MS/s, resulting in an Schreier figure of merit (FoMs) of 180.18dB.

Keywords—*SAR ADC, Noise-Shaping, charge pump, multi-input comparator.*

I. INTRODUCTION

Recently, Electrocardiogram (ECG) devices with low power consumption, high resolution, and portability have gradually been reported. However, the power consumption of the front-end module, the ADC module, and the DSP module greatly limits the application of the portable ECG devices for the long-term heart monitoring [1]. As shown in Fig. 1, the ADC is a fundamental and critical component between the sensor and the analog front-end circuit in biomedical applications. Its power consumption and accuracy greatly determine the application scenarios and scope of the entire system.

Due to the inherent low power consumption advantage of SAR ADCs, their precision is difficult to achieve beyond 10 bits, and even if it does exceed 10 bits, the energy efficiency becomes very poor [2]. To improve accuracy, researchers have proposed a noise-shaping SAR ADC structure based on sigma-delta modulation. Therefore, noise-shaping SAR ADCs inherently possess the advantages of low power consumption and high precision. In the NS-SAR ADC structure, an active integrator filter based on the amplifier structure effectively constructs a clear noise transfer function, bringing more significant low-frequency suppression effects for the design of NTF, thereby enhancing the performance of noise shaping and the precision of the ADC. However, the amplifier circuit increases the complexity of the entire system design and additional power consumption. At the same time, the amplifier is more sensitive to process variations, and the changes of the gain could reduce the stability of the circuit.

To solve the mentioned issues, this paper proposes a second-order passive noise-shaping structure utilizing a charge pump, achieving 14.12 bits at a 2MS/s sampling rate. The rest of the paper is structured as follows: Section II analyzes the challenges faced by the passive noise shaping. Section III describes the proposed second-order dual-charge-

This work is supported by National Natural Science Foundation of China with project number 62174181.

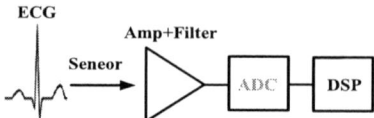

Fig. 1 The ECG signal acquisition and processing system.

pump passive Cascade of Integrators with Feed-Forward (CIFF) NS-SAR ADC. Section IV gives the simulation results, and Section V concludes the paper.

II. ANALYSIS OF CHALLENGES FACED BY PASSIVE NOISE SHAPING

In the passive noise shaping circuits, the sharing and transfer of charges inevitably result in signal loss, causing the non-ideal behavior of the passive integrators and affecting the noise shaping effect, which ultimately impacts the accuracy of the ADC. As shown in Fig. 2, Guo presents a classic design of a first-order passive loop filter [3], which employs a multi-input comparator with a gain to compensate for the signal loss during the transfer of residual voltage between capacitors. Assuming the comparator provides a relative gain of g, $C_1 = C_3 = C$, $C_2 = a/(1-a)C$, where C is the total capacitance of the DAC and a is capacitance ratio. The expression for the integrated voltage V_{int} after charge sharing between the sampling capacitor and the integration capacitor in the z-domain for the residual voltage V_{RES} can be derived as follows:

$$V_{int}(z) = \frac{(1-a)a}{1-(1-a)z^{-1}}V_{RES}(z). \quad (1)$$

So the noise transfer function NTF is

$$NTF = \frac{1-(1-a)z^{-1}}{1-(1-a)(1-ga)z^{-1}}. \quad (2)$$

The NTF reveals a zero at $z=1-a$ and a pole at $z=(1-a)(1-ga)$.

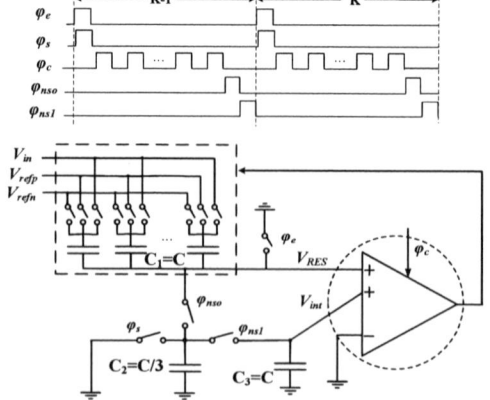

Fig. 2 First-Order Passive NS-SAR ADC Structure.

With the gain g set to $g=1/a$, and given $a=1/4$, the first-order NTF can be simplified to be $(1-0.75z^{-1})$. The presence of a zero at 0.75 leads to a -12 dB suppression at direct current (DC).

To improve noise shaping, the second-order passive NS-SAR ADC incorporates an additional set of integrator capacitors C_{int2} and upgrades the comparator to a six-input design [4]. This configuration creates two zeros at 0.75 in the NTF, enabling second-order noise shaping with -24 dB suppression at DC. However, it necessitates precise aspect ratios of 1:4:16 for the comparator's input pairs. The large input pairs increase power consumption and introduce parasitic capacitance, potentially degrading the ADC accuracy and comparator speed, representing a key limitation in passive noise shaping.

III. THE PROPOSED SECOND-ORDER DUAL-CHARGE-PUMP PASSIVE NS-SAR ADC

This paper proposes a second-order passive noise-shaping structure using the charge pump principle, which achieves high quantization precision by creating two zeros at 0.8 with a comparator gain ratio of 1:2.5:6.25.

A. Modulator Structure

Fig. 3 illustrates the proposed second-order dual-charge-pump passive CIFF NS-SAR ADC structure introduced. For simplicity of description, the single-ended configuration is shown here, whereas the actual circuit is a fully differential 10-bit SAR ADC. Compared to the architecture proposed in [4], this work has two improvements. First, a charge pump with 2× gain is integrated at both the sampling capacitor C_{RES} and the second-order integrator capacitor C_{INT2}. The relative gain for the passive integrators is supplied by the two charge pumps in conjunction with the multi-input comparator. The charge pump at the sampling capacitor provides a 2× gain to compensate for the signal loss, and the second charge pump on C_{INT2} offers a 2× gain to compensate for the integration loss. As depicted in Fig. 4, this results in a reduced aspect ratio of 1:2.5:6.25 for the input pairs of the strong ARM latch comparator, thereby eliminating the large size of the input pairs. Second, to enhance the noise shaping effect, the capacitance values are set to $C_{RES} = C_{INT1} = C_{INT2} = 1/4C$, with C being the total capacitance of the CDAC array. The proposed second-order NS-SAR ADC architecture adds only an additional $0.75C$ compared to the fundamental SAR ADC without NS, and this modest increase in capacitance

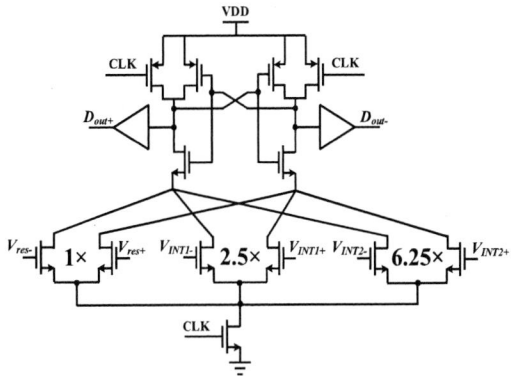

Fig. 4 The comparator with the aspect ratio of 1:2.5:6.25 for the three input pairs.

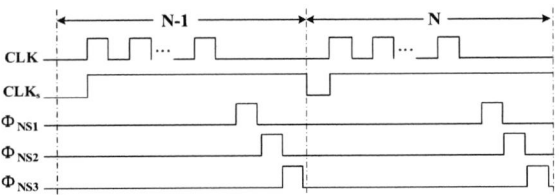

Fig. 5 The timing diagram of the proposed second-order NS-SAR ADC.

contributes to power consumption reduction and less chip footprint compared to [4].

Fig. 5 illustrates the operational timing of this structure. During the $(N\text{-}1)$th quantization cycle, the ADC first performs a standard 10-bit SAR ADC conversion process. After the DAC completes the final voltage flip, Φ_{NS1} is activated, allowing C_{RES} to sample the residual voltage. Φ_{NS1} then turns off, and the charge pump initiates the boosting operation. Φ_{NS2} and Φ_{NS3} are sequentially activated, enabling the sampling capacitor C_{RES} to perform the integration with C_{INT1} and C_{INT2}. In the Nth quantization cycle, the residual integration voltage from the $(N\text{-}1)$th cycle is engaged in the entire SAR ADC quantization process.

B. Transfer Function Analysis

Assuming $2C_{RES1} = 2C_{RES2} = C_{INT1} = 2C_{INT2a} = 2C_{INT2b} = a/(1-a)C$, with C being the DAC's total capacitance, and given the first-order integration path's relative gain is g_1 and the second-order's is g_2, the corresponding signal flow graph for this structure is depicted in Fig. 6. The NTF's general expression can be derived as follows:

$$\text{NTF} = \frac{[1-(1-a)z^{-1}]^2}{1+(1-a)(4g_2a^2+2g_1a-2)z^{-1}+(1-a)^2(1-2g_1a)z^{-2}} \quad (3)$$

With $g_1=2.5$, $g_2=6.25$, and $a=0.2$, these values are substituted into (3) and it can be simplified to be (4).

$$\text{NTF} = (1-0.8z^{-1})^2 \quad (4)$$

According to (4), the NTF of this design has two zeros located at 0.8, leading to a more significant reduction in low-frequency noise. The position of these zeros is entirely determined by the ratio of the capacitors, making this NTF minimally affected by PVT (Process, Voltage, Temperature) variations and highly robust. Compared to the first-order NS-SAR ADC architecture in [3], this structure's NTF does not have poles, thus offering greater system stability. Compared to [4], this design has larger zeros, indicating better noise

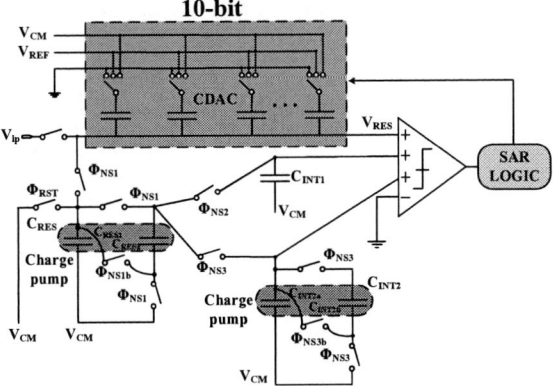

Fig. 3 The proposed second-order dual-charge-pump passive NS-SAR ADC.

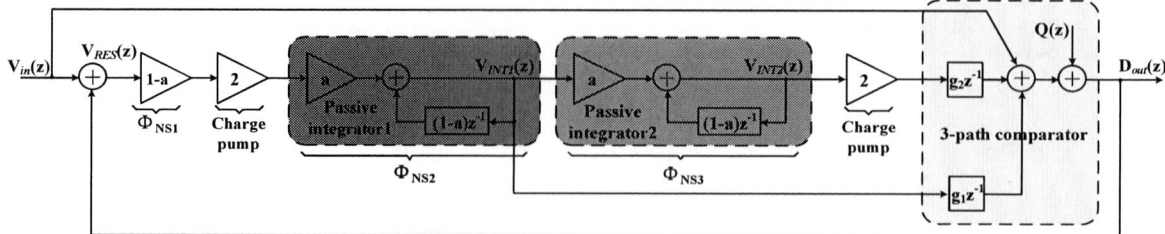

Fig. 6 The signal flow diagram of the proposed second-order NS-SAR ADC.

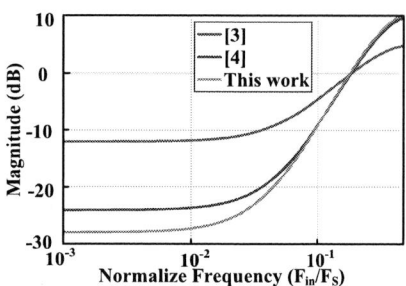

Fig. 7 The NTF frequency responses of different architectures.

shaping. Fig. 7 compares the NTF frequency responses, showing the proposed structure's NTF provides a more pronounced high-pass filtering effect, with approximate -28 dB suppression at DC.

IV. SIMULATION RESULTS OF THE PROPOSED ADC

This paper presents a noise-shaping SAR ADC architecture designed in a 180nm process. The ADC is simulated with a 1.2V power supply, a sampling frequency of 2MHz, and an OSR of 8, Fig. 8 displays the ADC's output spectrum for a 93.75kHz sine wave input with -0.06dBFS. The output spectrum clearly demonstrates the second-order noise shaping, pushing low-frequency quantization noise to the high-frequency range. Within the target bandwidth (125kHz), the ADC achieves an SNDR of 86.75dB, an effective bit resolution (ENOB) of 14.12 bits, and an SFDR of 97.16dB. The total power consumption of the proposed ADC is 56.8μW. According to the FoM definition, the Walden FoM for the NS-SAR ADC is calculated to be 3.19 fJ/Conv.-step, and the corresponding Schreier FoM is 180.18dB. Table I lists and compares the simulation results with previous works, indicating that the proposed second-order dual-charge-pump passive NS-SAR ADC not only achieves higher precision but also exhibits superior energy efficiency, which is particularly advantageous for bio-medical electronic applications.

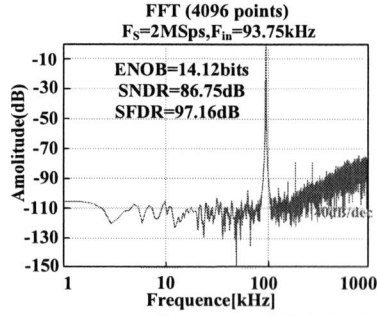

Fig. 8 The output spectrum of the proposed NS-SAR ADC.

TABLE I
PERFORMANCE SUMMARY AND COMPARISON.

Specifications	[4]	[5]	[6][a]	**This work[a]**
Process	40nm	65nm	180nm	**180nm**
Supply (V)	1.1	1	0.8	**1.2**
ADC Type	NS-SAR	SAR	Time	**NS-SAR**
Fs (MHz/s)	8.4	0.5	0.004	**2**
BW(kHz)	262	-	-	**125**
ENOB (bit)	12.7	-	10.9	**14.12**
SNDR (dB)	78.4	71.8	67.4	**86.75**
SFDR (dB)	-	101.2	-	**97.16**
Power (μW)	143	9.14	0.06812	**56.8**
FoM$_W$ (fJ/Conv.-step)	-	5.7	8.91	**3.19**
FoM$_S$ (dB)	171	176.2	-	**180.18**

[a.] Simulation Results

V. CONCLUSION

This paper presents a 14.12-bit 2MS/s NS SAR ADC in 180nm CMOS technology. The proposed structure utilizes a charge pump and a multi-input comparator with a gain to compensate for signal loss during the noise shaping process, thereby enhancing the noise shaping capability and improving the quantization accuracy. The simulation result shows that the proposed ADC achieves a Schreier FoM of 180.18dB, which is highly suitable for medical electronic applications.

REFERENCES

[1] W. Yang, H. Jiang, Y. Yin and Z. Wang, "A 4-μW Analog Front End Achieving 2.4 NEF for Long-Term ECG Monitoring," in IEEE Transactions on Biomedical Circuits and Systems, vol. 15, no. 4, pp. 655-665, Aug. 2021.

[2] W. -E. Lee and T. -H. Lin, "A 0.6-V 12-bit Set-and-Down SAR ADC With a DAC-Based Bypass Window Switching Method," in IEEE Transactions on Circuits and Systems II: Express Briefs, vol. 70, no. 9, pp. 3223-3227, Sept. 2023, doi: 10.1109/TCSII.2023.3266605.

[3] W. Guo and N. Sun, "A 12b-ENOB 61μW noise-shaping SAR ADC with a passive integrator," ESSCIRC Conference 2016: 42nd European Solid-State Circuits Conference, Lausanne, Switzerland, 2016, pp. 405-408,

[4] H. Zhuang et al., "A Second-Order Noise-Shaping SAR ADC With Passive Integrator and Tri-Level Voting," in IEEE Journal of Solid-State Circuits, vol. 54, no. 6, pp. 1636-1647, June 2019.

[5] M. Venkatesh, G. M. Salgado, K. G. McCarthy and I. O'Connell, "A 500 kS/s 71.8 dB 5.7 fJ/Conv-step Switch Supply Based Comparator SAR ADC for Biomedical Portable Devices," 2023 IEEE Biomedical Circuits and Systems Conference (BioCAS), Toronto, ON, Canada, 2023, pp. 1-5.

[6] A. Karimlou and M. Yavari, "A Low-Power Delta-Modulation-Based ADC for Wearable Electrocardiogram Sensors," in IEEE Transactions on Circuits and Systems II: Express Briefs, vol. 69, no. 9, pp. 3670-3674, Sept. 2022.

A 114.4-dB DR, 26-kHz BW Discrete-Time Incremental Zoom ADC

Yuanhong Ding [1], Longjiang Jia [1], Jian Mei [2], Lei Deng [2], Rui Yin*[1,3]

[1] Institute of Microelectronics, State Key Laboratory of Integrated Chip and Systems, Fudan University, Shanghai, China
[2] National Integrated Circuit Innovation Center, Shanghai, China
[3] Jiashan Fudan Institute, Jiaxing, China

* Email: yhding23@m.fudan.edu.cn, yinrui@fudan.edu.cn

Abstract—**This article details a discrete-time zoom analog-to-digital converter (ADC) designed for analog-front-end (AFE) applications. The ADC integrates a 5-bit SAR ADC with a fine two-order incremental sigma-delta modulator (ISDM). This combination allows for dynamic adjustment of the ISDM references to achieve a high signal-to-noise-and-distortion ratio (SNDR). To enhance the linearity of the fine ADC, which is constrained by component matching, a differential element mismatch (DEM) technique is employed. The prototype, built using 0.18-μm technology, covers a footprint of 0.25 mm². It delivers a 104.4-dB SNDR and a 114.4-dB dynamic range (DR) across a 26-kHz bandwidth, while drawing 5.61 mW from a 1.8V supply. The device boasts a Schreier figure-of-merit (FoM) of 181.1 dB and an SNDR FoM of 171.1 dB.**

Keywords—*A/D conversion, SAR ADC, discrete-time delta-sigma ADC, incremental ADC, zoom ADC*

I. INTRODUCTION

Instrumentation applications, such as those involving analog-front-end readout circuits for bridge transducers and sensors, demand analog-to-digital converters (ADCs) with high absolute accuracy, excellent linearity, and substantial resolution, all while maintaining a compact form factor[1]. Meeting these criteria can often lead to ADCs that suffer from low energy efficiency or high power consumption. Moreover, in some scenarios, ADCs are required to monitor multiple signals at the same time, and the residual voltage from the previous cycle can have an effect on the signal in the next cycle, thus reducing accuracy [1].

The Incremental Zoom ADC satisfies all of these requirements through the integration of a SAR ADC with low power consumption, which determines a coarse reference for the fine ISDM, with a significantly reduced loop filter swing for energy-efficient designs and decreased quantization noise, with a high resolution ISDM. The outputs of the two converters are then simply summed for a total digital output. The integrator and the digital registers in the ISDM have a reset switch that removes the residual voltage that has accumulated on the capacitor from the previous cycle, thus separating the input signals of the two cycles.

This paper introduces a 26-kHz BW Zoom ADC with a DR of 114.4 dB, a SNDR of 104.4 dB. The structure of the paper is as follows: Section 2 details the overall system and circuit architecture, Section 3 discusses the simulation outcomes, Section 4 offers a summary and conclusions.

II. PROPOSED ZOOM ADC WITH MULTIPLEXER

A. Zoom ADC

Fig. 1 illustrates the fundamental block architecture of the Zoom ADC, incorporating an N-bit SAR ADC for coarse quantization and an ISDM for fine quantization. Unlike other two-step quantization ADCs, the coarse quantizer in the Zoom ADC is only used to provide a reference voltage for the fine quantizer, instead of feeding the quantization error it generates into the fine quantizer to quantize it again. This avoids errors arising from the phase reduction process.

The coarse SAR ADC and the fine ISDM both working at a sampling frequency f_s, deliver an output code m, where the former produces an N-bit result. This output is sent to the zoom logic for addition or subtraction operations to, depending on the output code of the ISDM, generate the reference voltage of the ISDM

$$V_{REF+} = (m+M+1) \times V_{lsb.coarse} \tag{1}$$

$$V_{REF-} = (m-M) \times V_{lsb.coarse} \tag{2}$$

In an N-bit DAC, the quantization step is denoted as $V_{LSB.Coarse}$, with M acting as the over-ranging factor. This factor is essential for addressing SAR ADC imperfections and quantization noise, ensuring the modulator remains within a stable operational range.

In Zoom ADCs, the primary determinants of accuracy can be classified as follows, the number of coarse-quantized ADC bits, the number of ISDM orders, the oversampling rate, the KT/C noise, mismatch of capacitors and the non-ideality of the first-stage integrator.

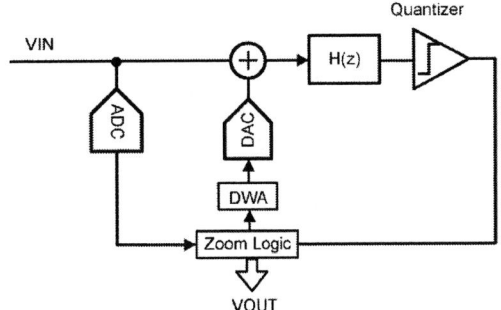

Fig. 1 Fundamental block diagram of a zoom ADC incorporating an SAR ADC, a ISDM with 1-bit quantizer, a DAC, and a DWA algorithm.

B. Coarse SAR ADC and Sigma-Delta Modulator

As in previous work, the target SQNR for the zoom ADC is set to around 115 dB to achieve a dynamic range close to 110 dB. The choice of a second-order feedforward modulator is a favorable trade-off between SQNR and design difficulty, while ensuring low power consumption. Fig. 3 illustrates the relationship between SQNR and OSR across various coarse ADCs resolutions (N). A 5-bit SAR ADC was selected as the coarse quantizer, which guarantees sufficient SQNR and

979-8-3503-6184-1/24 $31.00 © 2024 IEEE

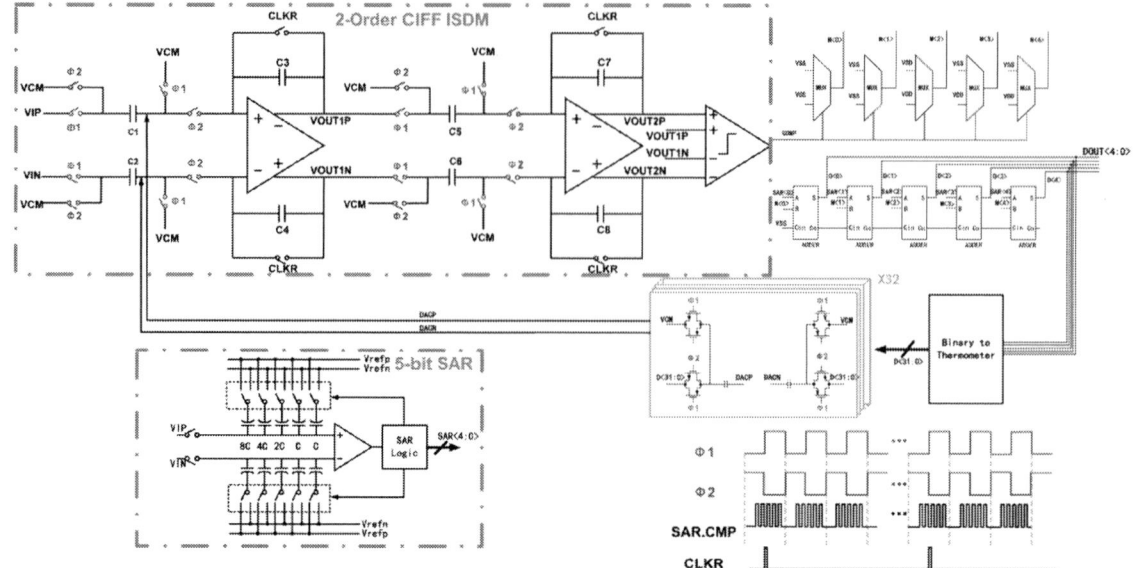

Fig. 2 Schematic representation of the circuit proposed for a DT zoom ADC, accompanied by its respective timing chart.

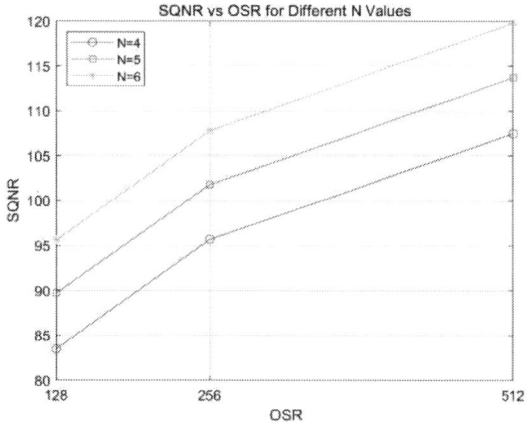

Fig. 3 SQNR as a function of OSR for various coarse resolution (N) when using an single-bit quantizer.

reduces the design difficulty of the coarse quantizer to some extent. The OSR is 512 times(f_s=26.624MHz), which ensures that SQNR of about 115 dB is achieved over the bandwidth (BW=26kHz)we are interested in.

The ISDM features two switched-capacitor integrators. The input sampling capacitor $C_{1,2}$ which is designed to address KT/C noise and is composed of 32 unit capacitors (C_0=128 fF). The capacitors in the second stage is significantly smaller, and its size is limited by matching requirements. To maintain the linearity of the operational transconductance amplifiers (OTAs), the integration capacitors in both stages are selected to keep the output swing within the amplifiers' linear range. For this, $C_{1,2}$=$C_{3,4}$=32×C_0=4.096pF, $C_{5,6}$=$C_{7,8}$=2×C_0=256fF.

A fully differential sampling network is used to capture the input. To ensure high linearity, the sampling switches are bootstrapped, while the remaining switches utilize transmission gate configurations to streamline circuit design.

The input signal is being sampled simultaneously with the output of OTA$_1$ during phase-ϕ_1. In phase-ϕ_2, both stages of the integrator start integrating, at the same time, and the DAC feeds back the reference voltage in the Zoom logic to the input

of the first stage integrator at the rising edge of phase 2. This results in a reduction of the input swing of the integrator, so we use a differential telescopic cascode as OTA. This is a simple unipolar system and can easily achieve high gain, thus minimizing the non-ideal of NTF due to the limited gain of the amplifiers. In phase-ϕ_{CLKR}, reset switches of the integrator and registers in zoom ADC are closed, completely erasing the information stored in the registers and the capacitor from the previous cycle.

The SAR ADC is controlled by the clock signal ϕ_2. At each rising edge of this clock, the ADC begins by sampling the input and moves into the conversion phase. It generates a 5-bit output m, before reverting to tracking mode for the next cycle. Meanwhile, the zoom logic combines this output with the quantizer's result to precompute the 5-bit output. At the beginning of ϕ_2, this precomputed result is encoded in a thermometer format, passed through a DWA block, and then sent to the capacitive DAC.

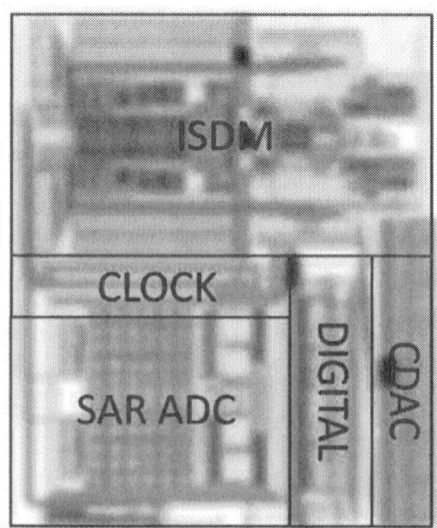

Fig. 4 Layout of the proposed ADC.

979-8-3503-6184-1/24 $31.00 © 2024 IEEE

III. SIMULATION RESULTS

The ADC is constructed with 180 nm technology and has been simulated in Cadence Virtuoso, with consideration given to factors such as transient noise, component mismatch and parasitic capacitance. The layout of the entire ADC, illustrated in Fig. 4, occupies a core area of approximately 0.25 mm^2.

Fig. 5 presents the output spectrum for both the ISDM and zoom ADC models, with the SNDR of 105.6dB for the Zoom ADC and 93.4dB for ISDM when the modulators are both SDMs with a 2nd order CIFF structure. This indicates that the zoom architecture achieves a higher accuracy compared to ISDM.

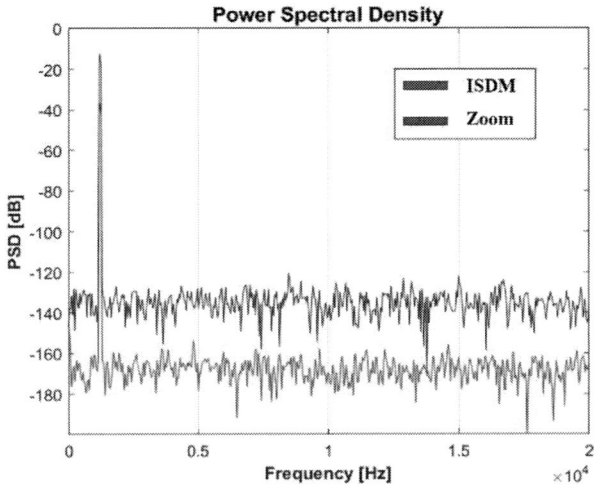

Fig. 5 Output spectrum of ISDM and Zoom ADC model.

Fig. 6 illustrates that with a sample frequency of 26.624 MHz, a 1.8 V supply and -3.5 dBFS singal input, the proposed zoom ADC can achieve 104.4 dB SNDR and 110.9dB SNR at 26-kHz bandwidth and consumes 5.6mW (the analog, digital part consume 75.9% and 24.1% of the total power respectively).This results in a Schreier figure-of-merit (FoM) of 181.1 dB and an SNDR FoM of 171.1dB.

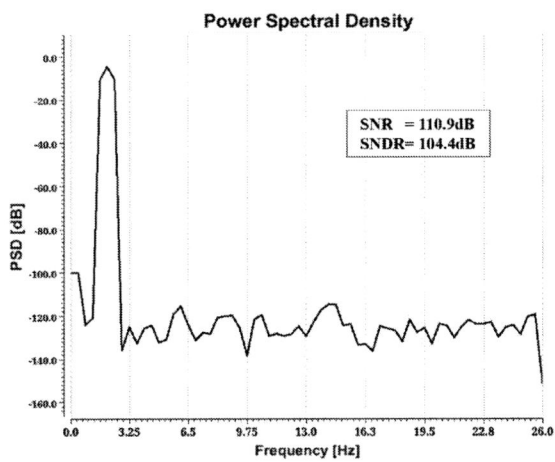

Fig. 6 Output spectrum for a -3.5dBFS input signal at 9.75kHz.

TABLE I. COMPARISON WITH STATE-OF-THE-ART WORKS

WORK	This work	JSSC 2016[3]	ISSCC 2022[4]	TCASII 2023[5]	TCASI 2023[6]
Architecture	DT Zoom	CT ΔΣM	Cnt.+ CT IΔΣM	ISDM	DT Zoom
Tech(nm)	180	65	28	180	180
Supply(V)	1.8	1	1.8/1.1	3.3	1.8
Power(uW)	5609	800	590	820	1300
f_S(MHz)	26.624	6.4	500	6.25	10
BW(kHz)	26	25	25	8.78	20
DR(dB)	114.4	103	102.2	91	104.4
SNDR(dB)	104.4	95.2	100.1	80	102.8
FoM_{SNDR}*	171.1	170.1	176.2	154.7	174.6
FoM_S**	181.1	177.9	177.3	165.7	176.2

* $FoM_{SNDR} = SNDR + 10\log_{10}(BW/Power)$

** $FoM_S = DR + 10\log_{10}(BW/Power)$

IV. CONCLUSION

A discrete time Zoom ADC has been presented for use in situations where it is required for processing sensor data. This design features a 5-bit SAR ADC, complemented by a second-order loop filter and a single-bit quantizer. This configuration results in a system primarily constrained by thermal noise, achieving a dynamic range (DR) of 114.4 dB. The loop filter's minimal internal voltage swing facilitates the use of telescope cascode OTAs, which offer both high gain and high bandwidth. Additionally, the utilisation of fully differential sampling enhances the CMRR. These advancements culminate in an ADC with a Schreier FoM of 171.1 dB and a SNDR FoM of 181.1 dB.

REFERENCES

[1] E. Eland, S. Karmakar, B. Gönen, R. van Veldhoven and K. A. A. Makinwa, "A 440-μW, 109.8-dB DR, 106.5-dB SNDR Discrete-Time Zoom ADC With a 20-kHz BW," in IEEE Journal of Solid-State Circuits, vol. 56, no. 4, pp. 1207-1215, April 2021.

[2] S. Lee, W. Jo, S. Song and Y. Chae, "A 300-μW Audio ΔΣ odulator With 100.5-dB DR Using Dynamic Bias Inverter," in IEEE Transactions on Circuits and Systems I: Regular Papers, vol. 63, no. 11, pp. 1866-1875, Nov. 2016.

[3] Y. H. Leow, H. Tang, Z. C. Sun and L. Siek, "A 1 V 103 dB 3rd-Order Audio Continuous-Time ΔΣ ADC With Enhanced Noise Shaping in 65 nm CMOS," in IEEE Journal of Solid-State Circuits, vol. 51, no. 11, pp. 2625-2638, Nov. 2016.

[4] L. Jie, M. Zhan, X. Tang and N. Sun, "A 0.014mm2 10kHz-BW Zoom-Incremental-Counting ADC Achieving 103dB SNDR and 100dB Full-Scale CMRR," 2022 IEEE International Solid-State Circuits Conference (ISSCC), San Francisco, CA, USA, 2022.

[5] Y. Liu et al., "A Fully Dynamic 4-Channel 13b Simultaneous Sampling Incremental ΔΣ ADC Using Cascoded Floating-Inverter-Amplifier and Code Division Multiplexing Technique," in IEEE Transactions on Circuits and Systems II: Express Briefs, vol. 70, no. 10, pp. 3737-3741, Oct. 2023.

[6] Y. Liang et al., "A Reconfigurable 12-to-18-Bit Dynamic Zoom ADC With Pole-Optimized Technique," in IEEE Transactions on Circuits and Systems I: Regular Papers, vol. 70, no. 5, pp. 1940-1948, May 2023.

A Multi-Phase Clock Self-Calibrating Circuit

Zhihuai Li*, Li Jiang, Xueming Wei, Zilu Cai, Jiami Tang

Guangxi Key Laboratory of Wireless Wideband Communication and Signal Processing, Guilin University of Electronic Technology, Guilin 541004, China

*Email: aickid@163.com

Abstract-This paper introduces a self-calibrating circuit for multi-phase signals that enables autonomous phase offset calibration in a multi-phase delay-locked loop (MDLL). This paper refines multi-phase signals mismatches using an adjacent phase spacing error extraction technique and a high-precision phase detection circuit. The calibration algorithm addresses the static phase mismatch in the MDLL and mitigates voltage controlled delay line (VCDL) mismatches. The adjacent phase spacing difference extraction technique, by eliminating the need for an actual phase spacing quantization, the adjacent phase spacing difference extraction technique reduces the digital time converter's (DTC) required dynamic range. Utilizing a 40nm CMOS process with a core layout area of 0.02856 mm², the design simulation results demonstrate a reduction in maximum phase error from 5.96° pre-calibration to 0.718° post-calibration within an input frequency range of 1~4GHz.

Keywords-Phase Calibration, Phase Detection, MDLL, DTC

I. INTRODUCTION

Time-interleaved analog-to-digital converters (TI-ADC), due to their high speed and high precision, have become a focal point in high-speed ADC research. The performance of a single-pass ADC dictates the TI-ADC's power consumption and speed. Interleaved sampling of the TI-ADC's single channel, controlled by the multiphase clock, can lead to a reduction in the effective number of bits of the entire TI-ADC due to sampling clock offset. Therefore, precise control of the phase accuracy of multiphase signals is a critical technology in TI-ADC design.

Multi-phase signals, primarily generated by MDLL, can experience mismatches in the voltage-controlled delay cells (VCDC) due to process fluctuations that occur during integrated circuit manufacturing. This ultimately leads to uneven phase differences among the output multi-phase signals.

Literature [4] utilizes a cascaded current-split charge pump (CP) technique to mitigate the static phase error of MDLL, effectively suppressing the 16-phase delay error. Nonetheless, it does not address the delay discrepancies arising from mismatches among delay cells. In contrast, Literature [2] adopts a 4-bit digital control output buffer to calibrate mismatches in DLL multiphase outputs. The calibration process employs a digital phase detector, responsible for measuring the phase difference's magnitude between consecutive phases and subsequently adjusting the digital control settings accordingly. Nevertheless, the incorporation of a D flip-flop (DFF) within the digital phase detector framework imposes limitations on the precision of phase calibration.

In this paper, we propose a self-calibration technique to address the output phase signal mismatch in MDLL. Through the use of an adjacent phase spacing error extraction method combined with a calibration algorithm, we regulate the

mismatch among DTC-calibrated multiphase signals, leading to enhanced phase uniformity.

II. MULTI-PHASE DELAY-LOCKED LOOP

The fundamental components of an MDLL comprise a phase discriminator (PD), a charge pump, a loop filter (LF), and a voltage-controlled delay line (VCDL), as illustrated in the simplified diagram presented in Fig. 1.

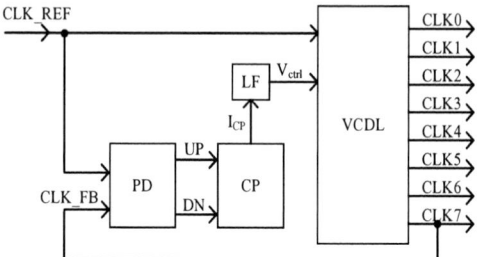

Fig. 1. Simple delay-locked loop structure

The block diagram of the multi-phase signal self-calibration circuit is depicted in Fig. 2. The multi-phase mismatch calibration circuit primarily comprises a digital-to-time converter (DTC), a phase selector, a phase magnitude detector, a frequency divider, and a digital control circuit.

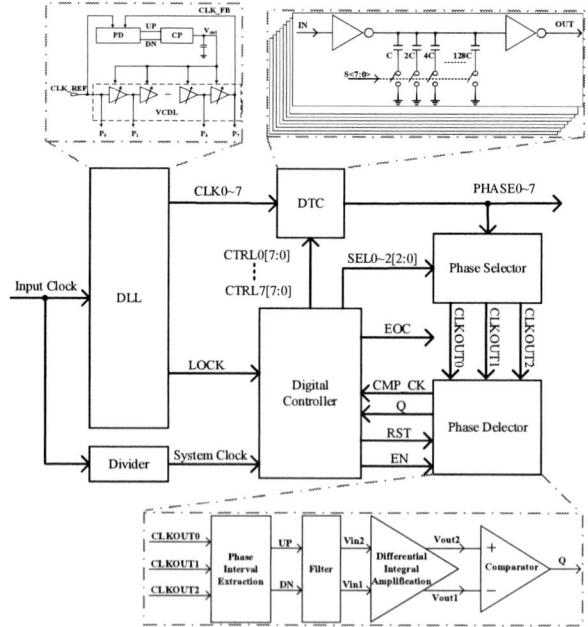

Fig. 2. Block diagram of multi-phase signal self-calibration circuit system

A. Phase Mismatch

The two types of phase mismatch in MDLL are static phase error and multiphase mismatch. Primary causes of static phase error include a locked-in phase discrepancy, which can be attributed to mismatches in the phase discriminator and charge pump circuitry.

This work was supported by the National Nature Science Foundation under Grant No. 62164003, Guangxi Natural Science Foundation under Grant No. 2024GXNSFAA010487, and the Foundation of Guangxi Key Laboratory of Wireless Wideband Communication and Signal Processing under Grant No.GXKL06230106 and No.GXKL06200131.

979-8-3503-6184-1/24 $31.00 © 2024 IEEE

In this paper, we conduct a Monte Carlo simulation of the voltage-controlled delay cell (VCDC), as illustrated in Fig. 3. Under the conditions of a 40nm CMOS process and an input frequency of 1.5GHz, the simulation-spanning 1000 data points-reveals a phase mismatch of 24ps. The static phase error intrinsic to the MDLL manifests as uneven phases in the multiphase signal output. Consequently, once the multiphase signals have been calibrated, no further calibration operations are necessary to address static phase errors.

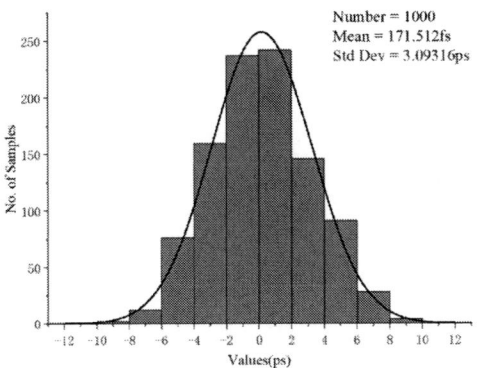

Fig. 3. Plot of Monte Carlo simulation results for delay cell

B. Calibration Algorithm

The calibration algorithm primarily utilizes relative phase error data, sampled from each individual phase signal, to ascertain the requisite calibration magnitude. Specifically, Fig. 4 illustrates the calibration procedure applied to a 4-phase signal.

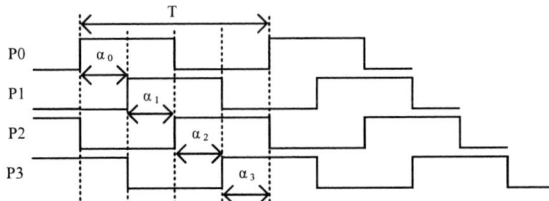

Fig. 4. Schematic diagram of four-phase DLL output signals

The four-phase signals, P0, P1, P2, and P3, respectively, exhibit phase spacing magnitudes of α_0, α_1, α_2, and α_3, whose sum equals the magnitude of the period T. Specifically, the phase detection circuit can sample the difference between the average of α_0 and α_1 and α_0 itself, a value that possesses the property of transmissibility. Consequently, in measuring phase errors, it suffices to sample pairs (α_0, α_1), (α_1, α_2), (α_2, α_3), and (α_3, α_0). These four sets of phase error samples are adequate to elucidate the disparity in magnitude between any two phase spacings. This information permits the calculation of the deviation of each phase spacing from the mean phase spacing. Thereafter, the Digital-to-Time Converter (DTC) can be manipulated to adjust the corresponding error amounts accordingly.

C. Phase Detection

Traditional phase detection circuits utilize DFF to determine phase relationships. Yet, they face limitations due to the inherent hold time constraints imposed by DFF, which restrict the precision of phase error sampling. This challenge impedes achieving the high accuracy demanded for calibrating high-frequency phase signals.

Consequently, this paper presents a high-precision phase detection circuit, as depicted in Fig. 5. This circuit fundamentally comprises a phase discriminator, a filter, an integrator, and a comparator. The underlying principle of phase detection involves translating the disparity in magnitude between two successive phase intervals into a disparity represented by two direct current (DC) voltage levels. Following this, the integrator serves to amplify the disparity in these DC voltage levels. Ultimately, the comparator assesses the relative magnitudes of the integrated voltage outputs, allowing for the determination of the comparative magnitudes of the initial phase spacings.

The phase discriminator PD2 extracts the phase difference between CLKOUT0 and CLKOUT1, converting it into the pulse width signal UP. Similarly, PD1 extracts the phase difference between CLKOUT1 and CLKOUT2 and transforms it into the pulse width signal DN. These pulse width signals, UP and DN, are subsequently converted by the filter into DC voltage signals Vin2 and Vin1, respectively. The integrator processes the voltage difference between Vin2 and Vin1, generating output voltages Vout1 and Vout2. The ratio of the magnitudes of integral outputs Vout1 and Vout2 mirrors the ratio of the pulse widths of UP and DN. The comparator then transforms the integrator's analog output into a digital signal. Through manipulation of the phase selector, the phase detection circuit enables sampling of the discrepancy between the mean of any two contiguous phase spacings and an individual phase spacing within the eight-phase signal. This capability facilitates the digital circuit's calibration of each phase signal according to the computed calibration magnitude.

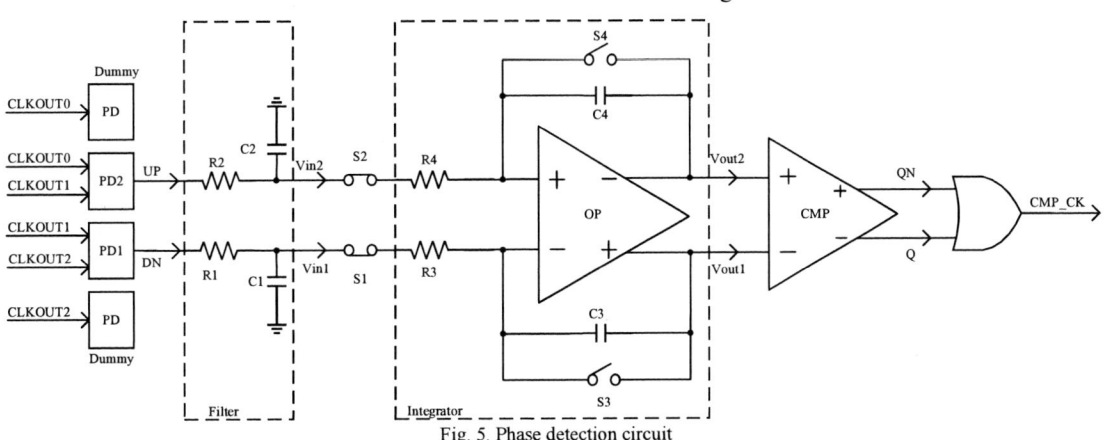

Fig. 5. Phase detection circuit

979-8-3503-6184-1/24 $31.00 © 2024 IEEE

III. SIMULATION RESULTS

This paper presents a 40nm CMOS design (Fig. 6), occupying 0.272x0.105 mm², operating at 1.2V, with MDLL outputs ranging from 1GHz to 4GHz. It compensates MDLL phase errors, depending on VCDC delay mismatches, using an ideal signal. Calibration is performed across process, temperature, and frequency variables, validated through extensive simulations.

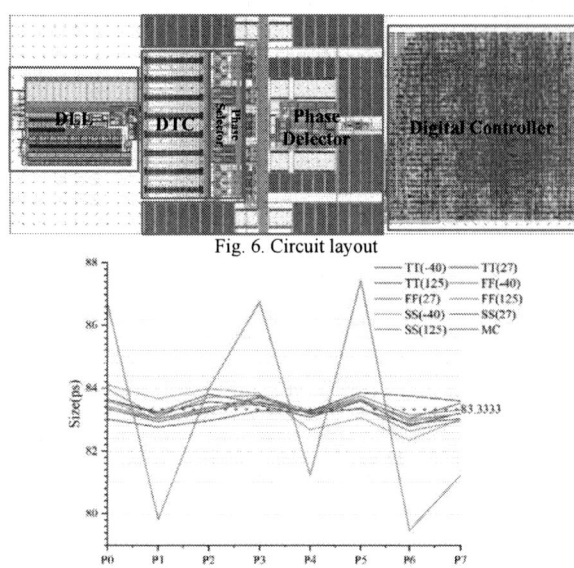

Fig. 6. Circuit layout

Fig. 7. Comparison of phase difference before and after phase calibration

Figure 7 compares the sizes of eight phase intervals pre- and post-calibration of the MDLL 1.5GHz multiphase output,

across process corners and temperatures. The x-axis shows intervals P0-P7; the y-axis, their widths. 'MC' indicates initial mismatches, other values denote calibrated widths. 'TT(27)' represents a 'TT' process corner at 27°C. Remarkably, calibration slashes the maximum phase error from 2.218° to 0.532°, validating the proposed method's efficacy in boosting phase precision.

Figure 8 exhibits the simulation outcomes, illustrating the maximum phase error reductions achieved following calibration, across various input reference frequencies of the DLL. These simulations were conducted under standardized conditions comprising a supply voltage of 1.2 volts, a 'TT' process corner, and an ambient temperature of 27°C.

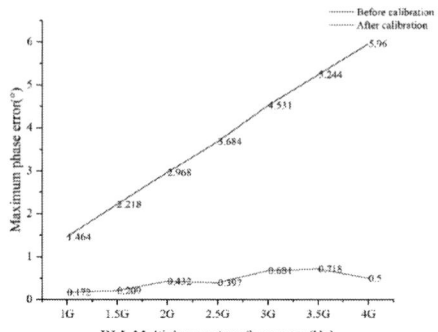

Fig. 8. Calibration effect curve of multi-phase calibration at different frequency points

Table 1 provides a performance comparison of this study against other literature. The phase calibration accuracy of this research surpasses that reported in references [2-3] and [5-6], and it operates over a broader frequency range compared to the studies referenced in [2], [5], and [6].

Table 1. Comparison of the performance of this paper with other literature

Parameters	[5] JSSC'21	[4] TMTT'23	[3] SSC-L'23	[2] JSSC'06	[6] ISSCC'03	This work
CMOS process	28nm	40nm	40nm	0.18um	0.35um	40nm
Voltage	1V	1.1V	1V	1.8V	3.3V	1.2V
Design Methods	Analog	Analog	Digital	Digital	Analog	Digital and Analog
Phase error(°)	2.82	1.55	1.68	1.26	6.1	0.532@1.5GHz
Frequency range(GHz)	1.3~4	2~7.4	7~10	0.7~2	0.13~0.165	1~4
Area(mm²)	0.0056	0.0168	0.023	1.03	2.25	0.02856
RMSJ(ps)	1.820	0.496	0.837	2.450	3.411	0.796@1.3GHz
Power(mW)	6.5	18.3	7.2	81	18	7.51

IV. CONCLUSION

In this paper, we employ a hybrid digital-analog design methodology to self-calibrate delay mismatches within the VCDC. We utilize a technique that extracts adjacent phase spacing differences, coupled with a rapid calibration algorithm, to achieve precise calibration of multiphase signals. Simulation results under the 40nm CMOS process model demonstrate that the calibration algorithm proposed herein achieves high calibration accuracy, exhibiting a phase calibration error of no more than 0.532° at 1.5GHz and an RMS jitter of 0.796ps at 1.333GHz. Following calibration, the maximum phase error is reduced from 5.96° to 0.718° across the frequency range of 1~4GHz.

REFERENCES

[1] Park, Ch , O. Kim , and B. S. Kim . "A 1.8-GHz self-calibrated phase-locked loop with precise I/Q matching." IEEE Asia Pacific Conference on Asics IEEE, 2001.

[2] Chang, H. H. , Chang, J. Y. , Kuo, C.Y. , and Liu, S. I. . "A 0.7-2-GHz self-calibrated multiphase delay-locked loop." IEEE Journal of Solid-State Circuits 41.5(2006):1051-1061.

[3] Park, G. J. , Lee, D. J. , Han, J. , and Bae, W. . "A High-Frequency and Low-Jitter DLL With Quadrature Error and Duty-Cycle Corrections Based on Asynchronous Sampling." IEEE Solid-State Circuits Letters 6 2023: 41–44.

[4] Yang, J. , Pan, Q. , Yin J. , and Mak, P. I. . "A 2.0-to-7.4-GHz 16-Phase Delay-Locked Loop With a Sub-0.6-ps Phase-Delay Error in 40-nm CMOS." IEEE Transactions on Microwave Theory and Techniques PP.99(2023):1-9.

[5] Park, H. , Sim, J. , Choi, Y. , Choi, J. , and Kim C. . "A 1.3–4-GHz quadrature-phase digital DLL using sequential delay control and reconfigurable delay line." IEEE Journal of Solid-State Circuits 56.6(2021):1886-1896.

[6] Chang, H. H. , C. H. Sun , and S. I. Liu . "A low-jitter and precise multiphase delay-locked loop using shifted averaging VCDL." IEEE International Solid-state Circuits Conference IEEE, 2003.

[7] Liu, H. , Lv, S. , Huang, X. , Rhee, W. , and Wang, Z. . "A fractional-NBB-DPLL with auto-tuned DTC and FIR filter for noise and spur reduction." 2017 IEEE International Symposium on Radio-Frequency Integration Technology (RFIT) IEEE, 2017.

A High-Resolution Low-Power Extended-Range Incremental ΣΔ ADC For Battery Management System

Long Zhang [1], Quan Sun [1], Rui Liu [2], Zhe Zheng [2], Jingjing Zhang [2], Haitao Liu*[1]

[1] Xi'an Aerosemi Technology Company Ltd. Xi'an 710076, China

[2] Beijing Smart-Chip Microelectronics Technology Company Ltd. Beijing 102200, China

* Email: zhangl@aerosemi.com, lht@aerosemi.com

Abstract—**In this paper, a high-resolution analog-to-digital converter designed for battery management system is proposed. This ADC employs an extended-range incremental sigma-delta (ΣΔ) architecture. The resolution of first-order incremental ΣΔ modulator is significantly improved by means of extended counting which uses a successive approximation ADC to encode the residual error of the modulator. By this means, it is possible to achieve enhanced resolution and reduced power consumption. The ADC is designed with a 0.5μm BCD process under a power supply voltage of 5V and sampling clock frequency of 4.1MHz. the simulation results show that maximum absolute conversion error of battery voltage, ranging from 0V to 5V, is less than 0.5mV at a conversion rate of 1 kSample/s, with a power consumption of 300uA.**

Keywords—*incremental ΣΔ ADC, extended-range, successive approximation ADC, battery management system*

I. INTRODUCTION

The battery management system (BMS) is the core component of electric vehicles. Its basic function is to convert information such as battery voltage and temperature, which plays an extremely important role in battery safety and system operation [1].

In order to ensure the accuracy of the battery pack's state of charge (SOC) calculation and battery protection, a high-resolution analog-to-digital converter (ADC) is required. Successive approximation register (SAR) ADC and Sigma-delta (ΣΔ) ADC are commonly used in high-resolution signal acquisition systems. SAR ADC is characterized by fast conversion speeds and low power consumption, but it lacks robustness, is easily interfered by noise, and requires complex calibration methods [2]. ΣΔ ADC is based on oversampling and noise-shaping technology, achieves higher precision and has fewer requirements on the front-end anti-aliasing filter circuit. but it cannot provide a sample-to-sample mapping between individual input and output samples [3]. Incremen-tal ΣΔ ADC has the same general topology as ΣΔ ADC, but unlike the traditional structure, the integrator resets after each conversion, allowing for a one-to-one mapping between input and output, which meets the application requirements of BMS.

This paper presents an extended-range incremental ΣΔ ADC that uses a SAR ADC to encode the residual error of first-order incremental ΣΔ ADC. Simulation results show that the maximum conversion error is less than 0.5mV, with a power consumption of only 300μA.

II. EXTENDED-RANGE INCREMENTAL ΣΔ ADC ARCHITECTURE

The process of battery voltage measurement in BMS shown in Fig. 1. High voltage multiplexer samples the battery voltage signal, which is then quantized by a 16-bit ADC. Finally, calibration and transmission are handled by a digital control circuit. The resolution and linearity of the ADC determine the accuracy of battery voltage measurement.

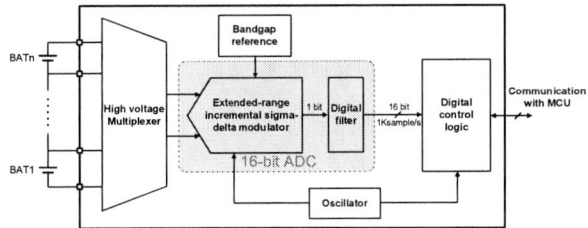

Fig. 1. Architecture of battery voltage measurement chip for the BMS

A. Structure of Incremental ΣΔ ADC with Extended-Range

In order to reduce system power consumption and improve stability, a first-order incremental ΣΔ ADC is used in this prototype, but to achieve n-bit resolution, up to 2^n clock cycles are required. So this prototype uses an extended-counting method [4]. the architecture is shown in Fig. 2. The ADC workflow includes two stages. In the first stage, the ADC operates as an incremental ΣΔ modulator, and the output bitstream is captured by a digital filter, providing the most significant bits (MSBs). In the second stage, the residual error of the integrator output at the end of first stage is used as the input of the residual ADC to obtain the least significant bits (LSBs). This second stage, called the "extended conversion" [5], is implemented as a successive approximation algorithm in this paper.

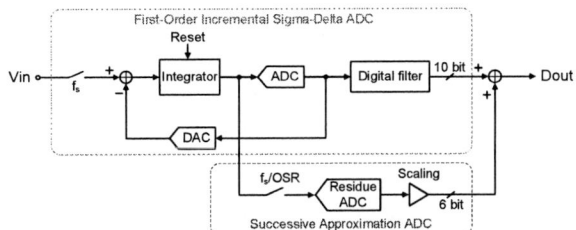

Fig. 2. First-order extended-range incremental ΣΔ ADC architecture

B. First-Order incremental ΣΔ modulator

Shown in Fig. 3 is the block diagram of first-order incremental modulator with single-bit quantizer. The coeffici-ent b is the feedback signal gain ranging from 0 to 1.

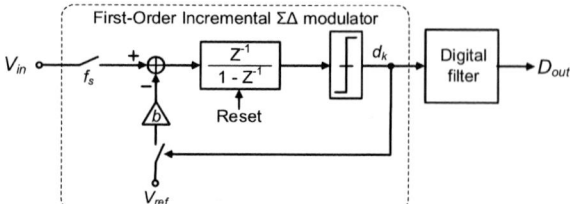

Fig. 3. First-order incremental ΣΔ modulator architecture

The integrator output sequence can be derived as:

$$V[0] = 0$$
$$V[1] = b(V_{in}[0] - d_0 V_{ref})$$
$$V[2] = b(V_{in}[1] - d_1 V_{ref}) + V[1]$$
$$= b(V_{in}[0] + V_{in}[1] - d_0 V_{ref} - d_1 V_{ref}) \quad (1)$$
$$\vdots$$
$$V[n] = b(\sum_{k=0}^{n-1}(V_{in}[k] - d_k V_{ref}))$$

where V_{ref} is the DAC feedback voltage, d_k is the output of modulator in kth cycle, n is the clock cycle number. Because the input is the battery voltage, it can be regarded as a constant DC signal during the conversion, resulting in

$$b\sum_{k=0}^{n-1} V_{in}[k] = b(n-1)V_{in} \quad (2)$$

With reasonable scaling parameters, $V[n]$ is bounded in the range of $\pm V_{ref}$, resulting in the following relationship from Eq. (1) and Eq. (2):

$$\left| V_{in} - \frac{\sum_{k=0}^{n-1} d_k V_{ref}}{n-1} \right| < \frac{V_{ref}}{b(n-1)} \quad (3)$$

Usually n is large enough, so input signal can be represented as:

$$V_{in} = \frac{\sum_{k=0}^{n-1} d_k V_{ref}}{n-1} \quad (4)$$

The minimum quantization step size of first-order incremental ΣΔ ADC can be expressed as:

$$V_{LSB} = \frac{V_{ref}}{b(n-1)} \quad (5)$$

From Eq. (4) and Eq. (5), it can be concluded that the resolution of first-order incremental ΣΔ ADC is positively correlated with the number of conversion cycles n. In this prototype , the incremental ΣΔ ADC is 10-bit and the conversion cycle n is 1024.

C. Residue ADC design

As can be seen from Eq. (3) that there is still an error less than V_{LSB} between the quantized output and the ideal voltage value, which is the residual error of ΣΔ modulator. In this paper, a resistive SAR is used as the residual ADC.

After the m-bit SAR ADC conversion, the final minimum quantiza-tion step size $V_{LSB(NEW)}$ can be expressed as:

$$V_{LSB(NEW)} = \frac{V_{ref}}{b(n-1)2^m} \quad (6)$$

It can be seen from Eq. (6) that the quantization step is significantly reduced. In this prototy, the effective number of bits (ENOB) of the SAR ADC is 6-bit, and a 2um width poly resistor sheet is selected to ensure the conversion result is not affected by mismatch [6].

D. Circuit implementation

The proposed extended-range incremental ΣΔ ADC implemented with a differential switched-capacitor circuit is shown in Fig. 4. During incremental ΣΔ ADC conversion, the feedback voltage VH and VL are equal to V_{ref} and GND respectively. When trigger to SAR ADC conversion, the input sample switch is open and VL remains connected to GND. After each clock cycle, the binary-weighted resistor string is selected in turn, and VH gradually changes from V_{ref} to 1/64 V_{ref}, quantizing the residual error on the integration capacitor in a successive approximation manner.

The integrator ouput voltage and comparator data output diagram is shown in Fig. 5. In the first stage, incremental ΣΔ ADC works for up to 1024 clock cycles. In the second stage, SAR ADC digitizes the residual error within 6 clock cycles. The comparator output code is processed and combined by digital filter can obtain a 16-bit conversion result.

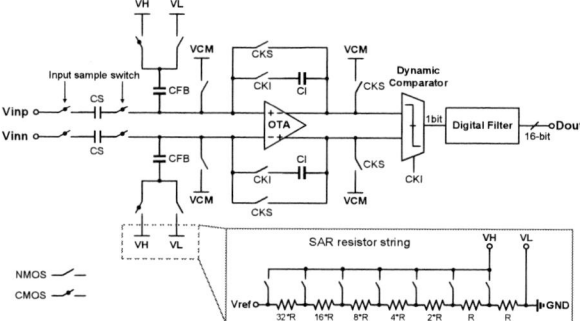

Fig. 4. implementation of extended-range incremental ΣΔ ADC

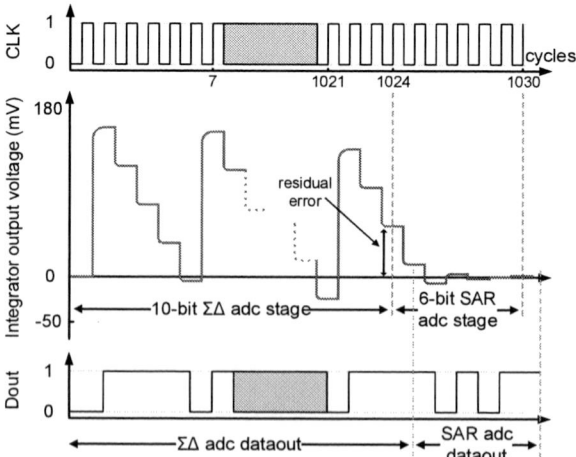

Fig. 5. Integrator output voltage and comparator dataout

The operational amplifier (OTA) is implemented using a gain-boosted single-stage folded-cascode topoloty shown in Fig. 6. Compared with telescopic cascode OTA, this structure obtains wider input range, higher voltage gain, higher pole frequency and greater output impedance, these advantages come at the cost of higher power dissipation, which is still very worthwhile. The OTA adopts a traditional switched capacitor common-mode feedback structure to generate Vcmfb voltage, which is not given in Fig. 6.

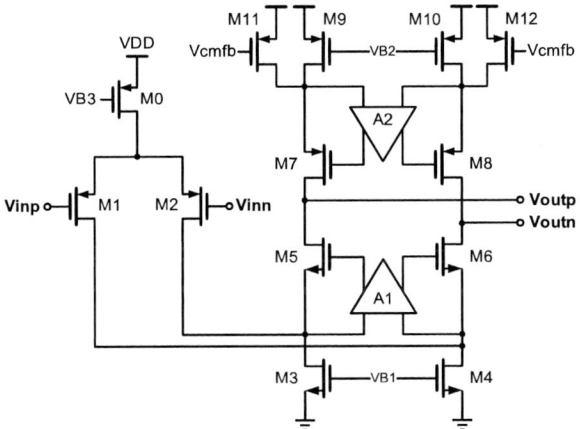

Fig. 6. The single-stage folded-cascode OTA in integrator

The dynamic comparator used in modulator adopts the structure of pre-amplification and latching is shown in Fig. 7. the pre-amplification can reduce the influence of kickback noise on the input and also reduce the equivalent offset voltage. The latching comparator can maintain a high speed with very low power consumption. CLK1 and CLK2 are two-phase non-overlapping clock signals.

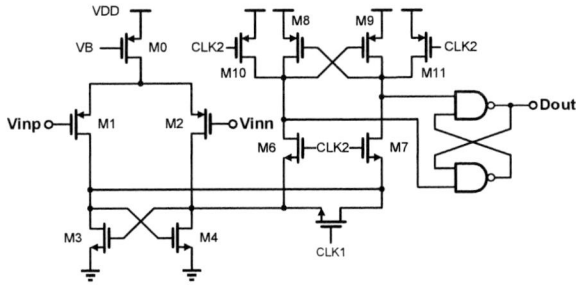

Fig. 7. The dynamic comparator in modulator

III. SIMULATION RESULTS

The proposed first-order incremental ΣΔ ADC with extended-range is designed with a 0.5 μm BCD process. The prototype operates from a 5.0V supply and 4.1MHz sample clock with total power dissipation of 300μA.

Fig. 8 shows the total conversion error at room temperature for battery voltage ranging from 0V to 5V, the maximum absolute error is less than 0.5mV. The time takes for the adc to perform a battery conversion is only 1 ms.

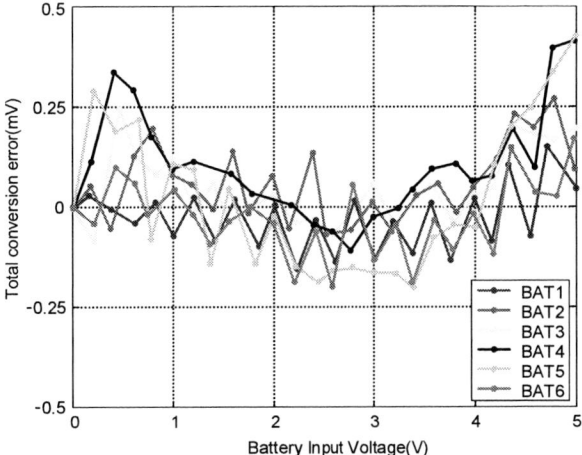

Fig. 8. Total conversion error of battery voltage

IV. CONCLUSION

This paper introduces an extended-range incremental ΣΔ ADC for battery management system. A first-order incremental ΣΔ ADC is employed to convert the input voltage firstly, and then a SAR ADC is used to digitize the residue charge at the output of integrator. This architecture takes advantage of oversampling and extended-counting, significantly increasing resolution and reduces conversion time. The ADC is designed with a 0.5μm BCD process, simulation results under a power supply of 5V and a sampling clock frequency of 4.1MHz show that the maximum absolute conversion error of input voltage ranging from 0V to 5V is less than 0.5mV. The ADC only needs 1ms to finish one-time conversion, with a power consumption of 300μA. This design is suitable for high-precision battery voltage measurement applications, meeting the demands of modern electric vehicles.

ACKNOWLEDGMENT

This paper is supported by The Scientific Research Programs for High Level Talents of Beijing Smart-chip Microelectron-ics Technology Co., Ltd.

REFERENCES

[1] C. Zhang, K. Li, S. Mcloone and Z. Yang, "Battery modelling methods for electric vehicles - A review," in *Proc. 2014 European Control Conf. (ECC)*, Strasbourg, France, 2014, pp. 2673-2678.

[2] H. Zhang et al., "A 1.25-MHz-BW, 83-dB SNDR Pipelined Noise-Shaping SAR ADC With MASH 2-2 Structure and kT/C Noise Cancellation," *IEEE Trans. Circuits Syst. II, Exp. Briefs,* vol. 70, no. 10, pp. 3872-3876, Oct. 2023.

[3] X. Wang, H. Zhang, G. Yang, C. Li and Y. Hao, "A 12-bit incremental ΣΔ ADC for battery management system in electric vehicles," in *Proc. 2014 IEEE Int. Conf. Solid-State Integr. Circuit Technol. (ICSICT)*, Guilin, China, 2014, pp. 1-3.

[4] J. De Maeyer, P. Rombouts and L. Weyten, "A double-sampling extended-counting ADC," *IEEE J. Solid-State Circuits,* vol. 39, no. 3, pp. 411-418, March 2004.

[5] A. Agah, K. Vleugels, P. B. Griffin, M. Ronaghi, J. D. Plummer and B. A. Wooley, "A High-Resolution Low-Power Incremental ΣΔ ADC With Extended Range for Biosensor Arrays," *IEEE J. Solid-State Circuits,* vol. 45, no. 6, pp. 1099-1110, June 2010.

[6] Y. Wang, J. Sun, M. Yu and F. Lai, "Low power 12bit 50KS/s R-C SAR ADC implemented based on mismatch analysis," in *Proc. 2011 Academic Int. Symp. Optoelectron. Microelectron. Technol.*, Harbin, China, 2011, pp. 340-343.

An Infrared AFE Chip and System with Non-Invasive Blood Glucose Detection Output

Bin Li [1], Jiyuan Guo [1], Chengzhen Xie [1], Jian Mei [2], Lei Deng [2], Rui Yin*[1,3]

[1] School of Microelectronics, Fudan University, Shanghai, China
[2] National Integrated Circuit Innovation Center, Shanghai, China
[3] Jiashan Fudan Institute, Jiaxing, China

* Email: b_li20@fudan.edu.cn, yinrui@fudan.edu.cn

Abstract—**China, with the highest diabetic patient count globally as of 2021 stats, presents substantial market potential and broad prospects for non-invasive blood glucose detection (NIBGD) in healthcare. Infrared absorption spectroscopy (IAS) stands out among NIBGD methods for its advantages. Leveraging CMOS for its high integration, reliability, low power, and compactness, this study targets sensor exploration. The paper presents a long-wave IAS-based glucose detection scheme using multispectral imaging, with an AFE chip and system achieved via circuit design and Verilog-A modeling. Simulation results show the superior performance of PGA in gain deviation, and accurately discerning 50 mg/dL and 138 mg/dL glucose levels.**

Keywords—*non-invasive blood glucose detection (NIBGD), infrared multispectral imaging, analog front-end (AFE), Verilog-A modelling*

I. INTRODUCTION

In its 10th edition of the IDF Diabetes Atlas, the International Diabetes Federation (IDF) predicted that the global prevalence of diabetes would reach 537 million by 2021. Of these individuals, China would have the highest number of cases, at 141 million (12.8% of the global total) [1]. This serves to illustrate the significant global public health implications of diabetes.

Monitoring blood glucose levels is of paramount importance for the effective management of diabetes, the delivery of clinical treatment, and the improvement of quality of life in those affected. Current methods for testing blood glucose levels are invasive, accurate, but painful, risky, and costly. Consequently, non-invasive techniques have gained attention due to their comfort, safety, affordability, and potential applications. Infrared absorption spectroscopy (IAS) plays a pivotal role in non-invasive blood glucose detection (NIBGD), enabling the identification and measurement of molecule concentration via light interaction [2].

The development of microelectronics has led to the emergence of CMOS-based sensor chips, which have been identified as offering a number of advantages, including integration, reliability, low power consumption, and miniaturization. The limited availability of CMOS-based IAS sensors indicates a clear need for innovation in this field. A novel CMOS spectral detection AFE chip and system has been developed in this paper using multispectral imaging. The CMOS-based scheme provides glucose concentration readings with high accuracy and miniaturization, which are essential for portable and wearable medical devices.

This paper presents the design and implementation of a 4 × 4 array-scale multispectral imaging AFE chip and system.

Section 2 outlines the system's architecture and key circuit design and Section 3 presents simulation results validating the AFE chip and system performance. The research is summarized in Section 4.

II. OVERALL ARCHITECTURE AND CIRCUIT IMPLEMENTATION

The architecture of the AFE chip and system is shown in Fig. 1, which consists of a 4×4 photodetector array, pixel readout circuit, PGA, ADC, DSP and TCU (Timing Control Unit). The final blood glucose monitoring data is processed by the DSP module and output via IIC interface protocol.

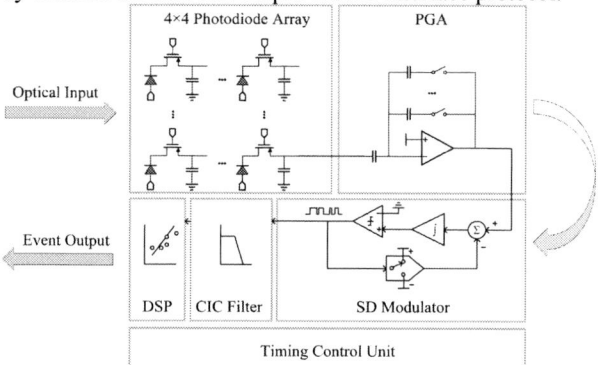

Fig. 1. Architecture of the AFE chip and system.

A. Verilog-A Modelling

Verilog-AMS HDL is used for the design of analog and mixed-signal circuits and helps in the design and verification of mixed-signal systems [3]. In this paper, an infrared detector and a Sigma-Delta modulator are modelled using Verilog-A. A mercury-cadmium-telluride (MCT) photodiode is chosen for the IR detector, taking into account the dark current and photocurrent non-linearity at -50mV bias. The paper uses an Incremental Sigma-Delta ADC for high precision Analog-to-Digital conversion, shifting the quantization noise to higher frequency. The ADC implementation consists of a modulator (modelled in Verilog-A) and a digital filter (implemented in Verilog).

B. Photocurrent Integration

The pixel readout circuit is connected to the IR detector and transforms the photocurrent into a voltage signal with spectral information via several switch-controlled capacitors. The circuit can operate in ITR (Integrate Then Read) and IWR (Integrate While Read) modes and supports both snapshot and rolling exposure methods; in this work, rolling exposure is selected to operate in ITR mode. The pixel readout circuit used in this paper is shown in Fig. 2.

979-8-3503-6184-1/24 $31.00 © 2024 IEEE

When the pixel readout circuit is in the ITR mode, the equivalent capacitance of the parallel transistor capacitors C1 and C2 serves as the pixel readout circuit integration capacitance, M3 is always on and M4 is always off. In the reset stage, M1 and M2 are closed and M6 is open to reset the photodiode parasitic capacitance, C1 and C2 capacitance; in the integral sampling stage, M1 is closed, M2 and M6 are open and the photo response current is integrated over the equivalent capacitance of C1 and C2 in parallel; and in the output stage, M1 and M2 are open and M6 is closed and the photocurrent integral voltage outputs.

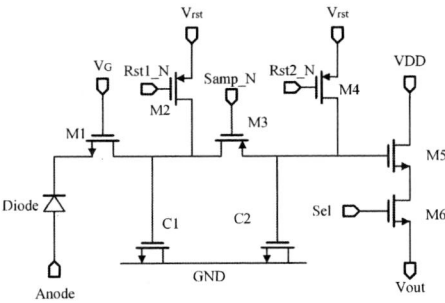

Fig. 2. Pixel integration read circuit.

C. Programmable Gain Amplification

The programmable gain amplifier (PGA), adjusts the input signal amplitude for ADC compatibility, improving conversion accuracy and accommodating a wide range of input signal strengths for the AFE chip and system. Fig. 3 shows the PGA circuit, with VREF as the amplifier reset voltage. The PGA operates in RST (reset) and INT (integration) stages: at RST stage, S1 and S3 close while S2 opens, setting V1 input and Vout1 = VREF output; during INT stage, S1 and S3 open while S2 closes, taking V2 input and producing Vout2 = VREF - C1 / C2(V2 - V1) output. The gain of the PGA is set by the C1 / C2 ratio, which is adjustable by the capacitance of C2. This design corrects for distortion from the pixel readout and the operational amplifier, improving signal accuracy and simplifying ADC design.

In the context of the PGA circuit, C1 represents a constant capacitor with a capacitance value of 800 fF, while C2 is a constant capacitor with a capacitance value of 50 fF connected in parallel with four capacitors and the capacitors are 100 fF, 200 fF, and 400 fF, respectively, as illustrated in Fig. 3. This configuration allows the PGA to have as many as 16 amplification slots for detecting a larger dynamic range of glucose concentrations.

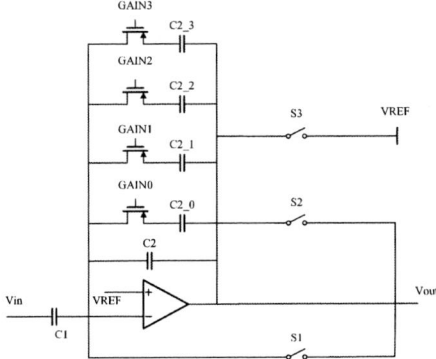

Fig. 3. The PGA circuit.

Faced with the challenge of achieving the required gain with a single-stage operational amplifier for the PGA, the operational amplifier within the PGA is configured as a two-stage architecture, the schematic of which is shown in Fig. 4.

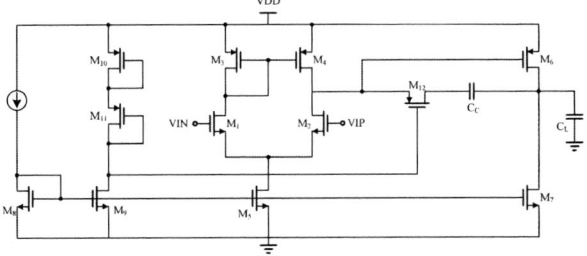

Fig. 4. The schematic of the two-stage OTA in PGA.

D. Digital Filter and DSP Module

Sigma-Delta ADC samples at a rate above the Nyquist frequency, spreading the quantization noise over a wider bandwidth and minimizing the noise power per Hertz. The oversampling induced high frequency noise is effectively attenuated by a subsequent digital filter, resulting in a highly accurate digital signal. The design uses a multi-stage CIC filter for digital low-pass filtering, characterized by an integrator and comb filter without multiplication, which significantly reduces power consumption. To increase the versatility of the filter, the design exploits the parameter configurability of Verilog.

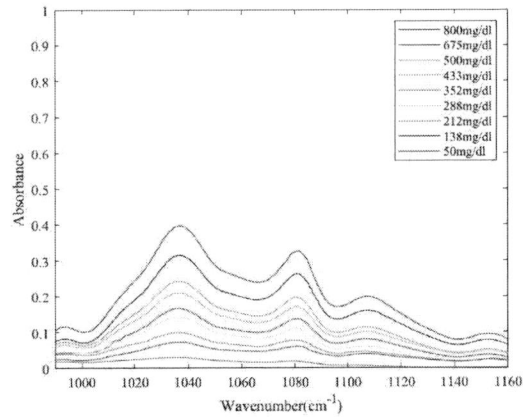

Fig. 5. Absorbance spectra of different glucose concentrations.

In the field of scientific investigation, specific research projects have demonstrated that the absorption of glucose solutions displays heightened sensitivity to variations in concentration within the specific infrared spectral region extending from 950 to 1200 cm^{-1} [4]. A transmission spectrum of a glucose solution, characterized by a concentration gradient spanning from 0 to 807 mg/dL, has been meticulously selected for the DSP design, as illustrated in Fig. 5. The spectral analysis revealed the presence of distinct waveform absorption peaks, which were found to be localized at four discrete wavenumbers. The identified wavenumbers were 1035 cm^{-1}, 1080 cm^{-1}, 1105 cm^{-1}, and 1150 cm^{-1}. These wavenumbers, which are indicative of the glucose solution's inherent molecular vibrations, were subsequently employed in a sophisticated multivariate linear regression analysis. The rationale behind the selection of these specific wavenumbers was to anchor the regression model with reference points that closely approximate the normative blood glucose levels observed in clinical settings, specifically 50 mg/dL and 138 mg/dL, as shown in Fig. 6.

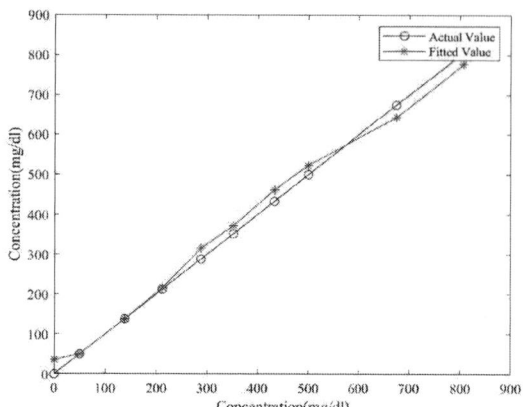

Fig. 6. Fitted concentration and actual concentration.

The efficacy of this regression model was quantified through the determination of the coefficient of determination (R^2), which was calculated to be an impressive 0.9909. This value underscores the model's robust predictive capabilities. In the translation of this analytical model to a practical hardware implementation, the regression coefficients were meticulously quantified to a 16-bit integer format, yielding a coefficient of determination R^2 of 0.9908. This minor reduction in R^2 is attributed to the inherent limitations of digital representation in hardware systems.

For the implementation of the model in the Verilog programming language, which is commonly utilized in the design of digital circuits, the coefficients were converted to a 16-bit integer format. This conversion resulted in a reduction in the coefficient of determination R^2 from 0.9908 to 0.9907.

III. SIMULATION RESULTS

The simulation results of the PGA are shown in TABLE I, and it can be seen that the 16 amplification gears can be amplified normally, and the deviations are less than 0.1%, with high amplification performance and accuracy.

TABLE I.   THE LOSSY VOLTAGE AND DEVIATION OF PGA

Gain	Input Voltage (V)	Lossy Voltage (μV)	Deviation (%)
1	1	240	0.024
16/15	0.9	280	0.029
8/7	0.7	244	0.031
16/13	1.3	452	0.028
4/3	0.9	300	0.025

Gain	Input Voltage (V)	Lossy Voltage (μV)	Deviation (%)
16/11	1.1	367	0.023
8/5	1	417	0.026
16/9	0.9	375	0.023
2	1	763	0.038
16/7	0.7	618	0.039
8/3	0.6	677	0.042
16/5	0.5	681	0.043
4	0.5	719	0.036
16/3	0.3	519	0.032
8	0.2	794	0.050
16	0.1	1225	0.077

The infrared spectral information in the vicinity of the human blood glucose concentration range is detected by the AFE chip and system designed in this paper. The simulation results demonstrate that the detection deviation is negligible. The glucose long-wave infrared spectral information of 50 mg/dL and 138 mg/dL was inputted and following digital signal processing, the concentrations of 49 mg/dL and 139 mg/dL were obtained, respectively.

IV. CONCLUSION

This paper presents a AFE chip and system for non-invasive blood glucose detection, designed based on multispectral imaging technology. The system employs a 4×4 array-scale HgCdTe infrared detector, a pixel readout circuit utilizing roll-off exposures in ITR mode, and data processing techniques such as multivariate linear fitting. The signal is amplified by a PGA that can achieve 16 amplification steps, and the AD conversion is completed. The accurate detection of glucose concentrations of 50 mg/dL and 138 mg/dL was successfully achieved.

REFERENCES

[1] H. Song et.al, "IDF Diabetes Atlas: Global, regional and country-level diabetes prevalence estimates for 2021 and projections for 2045," Diabetes research and clinical practice, vol. 183, pp. 109-119, January 2022.

[2] S.K. Chamoli, S.C. Singh, and C. Guo, "Design of extremely sensitive refractive index sensors in infrared for blood glucose detection," IEEE Sensors Journal, vol. 20, pp. 4628-4634, January 2020.

[3] F. Pêcheux, C. Lallement and A. Vacnoux, "VHDL-AMS and Verilog-AMS as alternative hardware description languages for efficient modeling of multidiscipline systems," IEEE transactions on Computer-Aided design of integrated Circuits and Systems, vol. 24, pp. 204-225, January 2005.

[4] I.L. Jernelv, K. Strøm, D.R. Hjelme and A. Aksnes, "Infrared spectroscopy with a fiber-coupled quantum cascade laser for attenuated total reflection measurements towards biomedical applications," Sensors, vol. 19, pp. 5130, November 2019.

A 12-BIT 8GS/S TIME-INTERLEAVED PIPELINE-SAR ADC WITH CALIBRATION

Jie Pu[1], Jinda Yang[1], Yuanjun Cen[1], Jianwen Li[1], Rong Han[1], Xing Zhu[1] and Lei Chen[1]
[1] Chengdu Sino Microelectronics Technology Co., Ltd.,
Chengdu 610041, CHINA
Email: pj1112@163.com

ABSTRACT

A low power 12bit 8GS/s time-interleaved ADC is described in this paper，which is fabricated in a 28nm mixed-signal CMOS process. The sampling rate of 8GS/s is achieved by interleaving eight sub-ADC operating at 1GHz. The sub-ADC is realized with two stage pipeline-SAR structure for saving power. An input buffer is implemented with improved flipped source follower structure to obtain better linearity and bandwidth performance. Foreground and background calibration is further employed to correct the non-linearity of sub-ADC and eliminate interleaving spurs. Measure results show that the ADC achieves 8GS/s sampling rate with an input bandwidth of 6GHz, as well as 68.6dB SFDR and 54.9dB SNDR with 3.9GHz，-1dBFS signal input. Total power consumption is 1.8W from 1.0V/2.5V supply.

INTRODUCTION

In the field of communication applications, high-speed high-precision ADCs are widely used, however single channel ADC speed is limited for a given resolution with acceptable power. Thus time-interleaving (TI) structure is widely used [1], which is an important means of improving the sampling rate by alternating the conversions of multiple single-channel sub-ADCs with acceptable power. Typically, the more interleaving channels, the higher the correction complexity, and the fewer channels, the higher the sub-ADC power consumption, affecting the overall power efficiency, which needs to be balanced. In this paper, an eight-channel time interleaving structure is selected, and the sub-ADC uses a two-stage pipeline-SAR structure to save power while having a relatively low correction complexity. The full-scale input signal range is 1.2Vpp. An improved flipped source follower structure is used as input buffer, which maintains high linearity and low output impedance, while having a large signal range. Any mismatch in the sub-ADC and between the channels will introduce nonlinear distortion [1,2,3], causing spurious tones to cause a decline in SFDR performance. This paper uses a combination of foreground and background calibration methods to correct the spurious tones caused by the mismatch in the sub-ADC and between the channels for further improving SFDR.

TI ADC ARCHITECTURE

The ADC structure designed in this paper is shown in Fig.1, which is implemented with an 8-channel time-interleaved pipeline-SAR structure. An improved flipped source follower (SF) structure is adopted as input buffer to improve input bandwidth, linearity, and output impedance, and to drive the 8 sub-ADC loads and routing networks. The sub-ADC is implemented with a two-stage pipeline-SAR structure operating at 1GHz.

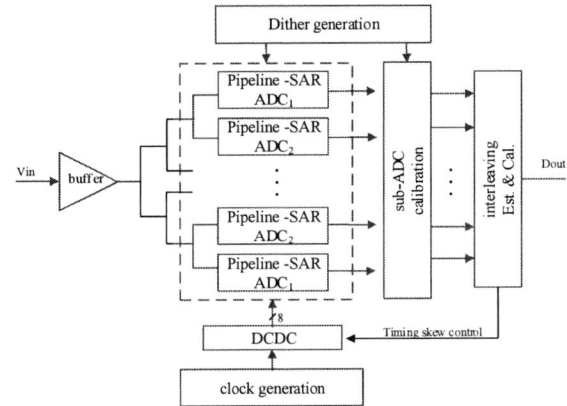

Fig.1: Architecture of proposed TI ADC

A digital correction of capacitance mismatch and inter-gain error in sub-ADC is employed to improve the sub-ADC performance. A background estimation method is used to eliminate the time-interleaving mismatch such as offset, gain, and clock phase errors, and to improve the overall SFDR. The offset and gain error are corrected in the digital domain, and a digital controlled delay cell (DCDC) is added to the main clock circuit for clock phase error correction as shown in Fig.1.

INPUT BUFFER

The input buffer [4] performance directly affects the overall linearity and bandwidth of the ADC, and for a multi-channel time-interleaved ADC, the input buffer bandwidth requirement is higher than that of a single-channel ADC because the output signal routing from the buffer to the sub-ADC is longer and the parasitic capacitance is larger.

To achieve this goal, this paper proposes an improved flipped SF buffer structure, as shown in Fig.2, which has high linearity and low output impedance, and a larger output signal range. As shown in the figure, the MOS M_{P1} and M_{P2} work in source follower mode, and the input signal is coupled to the gate of M_{P1} through capacitor c_{1p},

---

979-8-3503-6184-1/24 $31.00 © 2024 IEEE

that reducing the V_{ds} variation of input MOS M_{P2} which make small variation of the output impedance(r_o), thus improve its linearity. The drain signal of M_{P1} is connected to the gate of current source M_{P3} through capacitor C_{2p}, forming a negative feedback loop to compensate for the leakage current of output load C_L and improve linearity. Note that there is a parasitic capacitance at the gate of input MOS M_{P2}, that is varying with the input signal which makes nonlinearity. So we add a MOS capacitor (C_{var}) at the gate of M_{P2} with properly designed size biased at an appropriate point (V_{bvar}) to reverse cancel the parasitic nonlinear capacitance of M_{P2}, the linearity can be further improved. Using the linearity optimization technologies mentioned above, the input buffer linearity can reach 75dB.

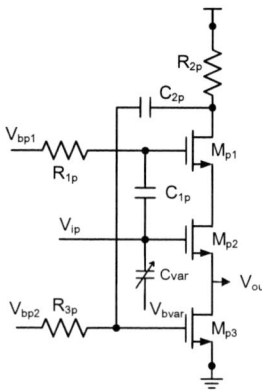

Fig.2: The proposed CMOS input buffer

PIPELINE-SAR SUB-ADC

The architecture of the proposed 12-bit 1GSPS pipeline-SAR sub-ADC is shown in Fig.3(a), which shows a pipelined architecture with 6-bit first stage and 8-bit second stage. Both stages use 2b/cycle SAR ADC structure [5]. The sub-ADC work timing margin is shown in Fig.3 (b). The sampling and holding function are built independently in SAR ADC and MDAC of the first stage.

The MDAC and SAR ADC bandwidth of the first stage may have slightly RC mismatched, and the SAR ADC sample clock will be trimmed to compensate for residual mismatches to minimize correction range usage and increasing linearity. The SAR ADC comparator offsets are also background calibrated to minimize correction range usage.

The inter-stage residual amplifier uses an open-loop dynamic amplifier structure with low power consumption. A dither signal is injected into the MDAC of the first-stage for inter-stage dynamic amplifier gain error correction with a background estimation method in digital domain. Since the first-stage quantizes 6bits, the correction range is small, a reverse dither signal is injected into the DAC of the second-stage SAR ADC to reduce the dither injection occupying correction range while also keeping the dynamic amplifier gain error information.

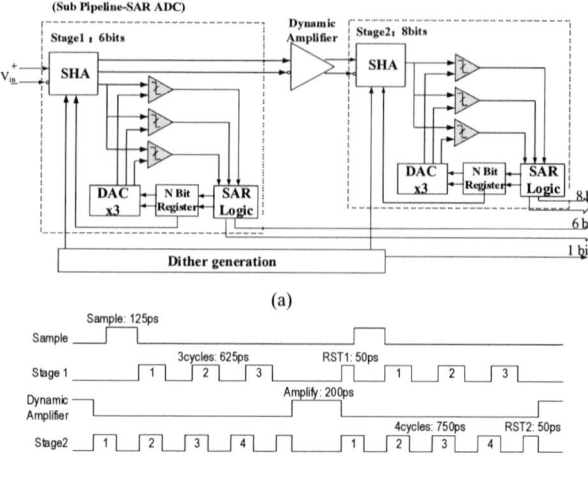

Fig.3: Architecture of proposed Pipeline-SAR sub-ADC

CALIBRATION TECHNIQUE

To improve the dynamic performance of SFDR, various nonlinear correction techniques are used.

A foreground estimation and correction method is used to eliminate the capacitance mismatch error of the first stage SAR ADC. The capacitance error estimation procedure is realized in the test mode with the input connect to common voltage, by controlling the capacitor of MDAC inversion in the first stage, and used the second stage quantize the residue signal, and then we can use the quantizing code to calculate the relative error of capacitor weight. In normal work mode, the measured capacitor weight error can be corrected in the digital domain.

The dynamic amplifier inter-gain error between the first and second stages is estimated and eliminated by dither injection with the background method, where the dither injection capacitor also has an error, which can be compensated together with the capacitor error calibration procedure as described above. The inter-gain error parameter estimation expression is as follows.

$$\Delta G^{i+1} = \Delta G^i + \mu \cdot (D(Vin) \times (1 + \Delta G^i) - D(Vdither)) \cdot D(Vdither)$$

Where ΔG^i is the ith iteration of inter-gain error, $D(Vin)$ and $D(Vdither)$ are the ADC output code and residue dither code.

A background correction technique is used to eliminate the time interleaving offset, gain, and clock phase error. The offset and gain error are obtained by statistically averaging the amplitude and power of each sub-ADC channel, and compensated in the digital domain. The clock phase error is obtained by calculating the cross-correlation function between adjacent sub-ADC channels, and eliminated by adjusting DCDC and iteratively converging.

Both the estimation and correction aspects of foreground and background calibration describe above are implemented on-chip.

DESIGN VERIFICATION

The prototype ADC is implemented in 28nm mixed-signal CMOS process with an area of 21.6mm² which include analog part, digital part and Serdes interface with layout shown in Fig.4. The analog part is powered by 1.0/2.5V supply while digital circuit operates under 1V supply. Operating at 8GS/s, the analog part consumes 950mW, and the digital part consume 850mW. The measured spectrum before and after calibration with -1.08dBFS, 3.905GHz input signal at a sample rate of 8GS/s are depicted in Fig.5, while the red cycles represent interleaving spur. The proposed ADC can achieve an SFDR of 68.6 dB and an SNDR of 54.9 dB at 3.905GHz input after the on-chip calibration. We can see that the calibration technique developed in this paper can get an improvement of 15dB for SFDR and 6dB for SNDR. The detailed performance summary of the ADC is described in Table I with comparison to other high-speed ADCs.

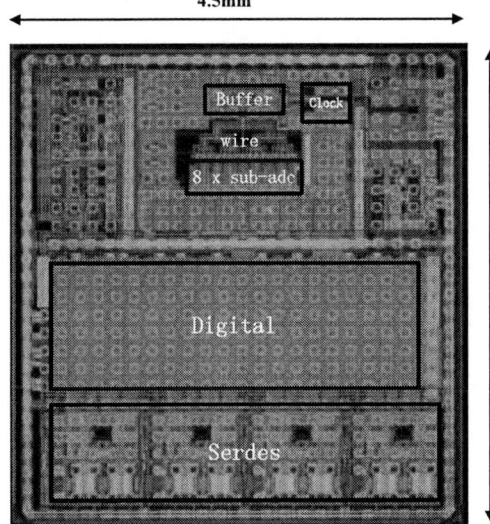

4.5mm

4.8mm

Fig.4: Layout of TI ADC in 28nm process

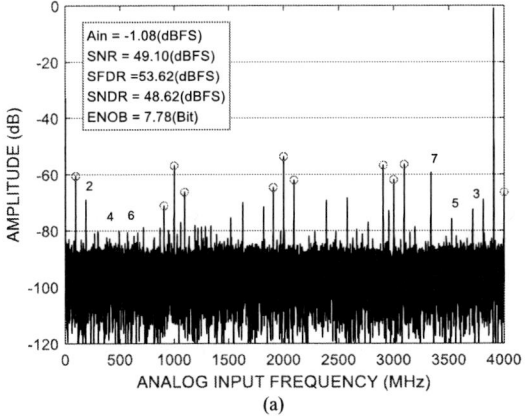

(a)

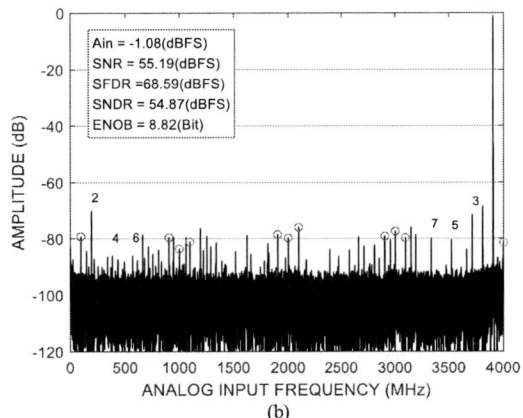

(b)

Fig.5: measured results (a) before calibration (only partial trim) (b) after calibration

TABLE 1
PERFORMANCE SUMMARY AND COMPARISON

	[1]	[2]	[3]	This work
Architecture	TI pipeline	TI SAR	TI pipeline	TI pipe-SAR
Calibration	foreground /background	foreground	foreground /background	foreground /background
Technique	28nm	28nm FDSOI	65nm	28nm
Resolution	12 bit	10 bit	-	12 bit
bandwidth	7.4GHz	4GHz	4GHz	6GHz
Sample rate	10 GS/s	8 GS/s	4 GS/s	8 GS/s
NSD	-157 dBFS/Hz	-154 dBFS/Hz	-154 dBFS/Hz	-155 dBFS/Hz
Supply	-	1.9/1.1/0.9 V	1.8/1.0 V	1.0/2.5 V
Power	2.9 W	0.3 W(1)	2.2 W	1.8W(2)
Input Freq.	4GHz	3.82GHz	1.842GHz	3.905GHz
Input Ampl.	-1 dBFS	-4.5 dBFS	-4 dBFS	-1 dBFS
SNDR	55 dBFS	49 dBFS	55.5 dBFS	54.9 dBFS
SFDR	64 dBFS	60.3 dBFS	64 dBFS	68.6 dBFS
FoMS@HF	147.4dB	150.2 dB	145.1 dB	148.4 dB

$FoMS = SNDRdB + 10log(fsnyq/2P)$, fsnyq is Nyquist frequency, P is power.
(1) only ADC core
(2) include analog part and digital part, without Serdes

REFERENCES

[1] S. Devarajan et al., "A 12b 10GS/s Interleaved Pipeline ADC in 28nm CMOS Technology", ISSCC, Dig. Tech. Papers, pp. 288-289, Feb. 2017.

[2] J. P. Keane et al. "An 8GS/s Time-Interleaved SAR ADC with Unresolved Decision Detection Achieving -58dBFS Noise and 4GHz Bandwidth in 28nm CMOS", ISSCC, pp. 284-285, Feb. 2017.

[3] M. Straayer et al., "A 4 GS/s time-interleaved RF ADC in 65 nm CMOS with 4 GHz input bandwidth," ISSCC, pp. 464–465, Feb. 2016.

[4] J. Ramirez-Angulo et al., "The flipped source followers: A useful cell for low-voltage low-power circuit design," IEEE Transactions on Circuits and Systems I, Vol. 52, pp. 1276-1291, July. 2005.

[5] Zhiheng Cao et al., "A 32mW 1.25GSs 6b 2bstep SAR ADC in 0.13μm CMOS," ISSCC, pp. 542-543, Feb. 2008.

An Ultra-High Frame Rate ROIC for Hyperspectral Detection

Angyang Li [1], Ningning Li [2], Jian Mei [1], Lei Deng [1], Rui Yin [*2,3]

[1] National Integrated Circuit Innovation Center, Shanghai, China
[2] Jiashan Fudan Institute, Jiaxing, China
[3] Institute of Microelectronics, State Key Laboratory of Integrated Chip and Systems, Fudan University, Shanghai, China

* Email: liangyang@shnicic.com, yinrui@fudan.edu.cn

Abstract—In hyperspectral detection, it is typically sufficient to extract the spectral information of the target substance's characteristic bands to complete the analysis of the target component. With the increasing demands for the number of spectra and spectral accuracy in spectral testing, the resolution of spectral cameras has multiplied, which improves the performance of spectral detection but also leads to a reduction in the camera's frame rate. To address this issue, this paper designs an ultra-high frame rate readout integrated circuit (ROIC) with an array size of 640×512 and a multiple regions of interest (ROI) function. This circuit can read out data only from the regions of interest, effectively reducing the readout time. Additionally, the two-step analog-to-digital converter (ADC) processing structure significantly improves the signal conversion rate. Simulation results indicate that the proposed ROIC can successfully read multiple characteristic bands of kaolinite and achieve a frame rate of 2691Hz, making it highly suitable for hyperspectral detection tasks.

Keywords—hyperspectral detection, readout integrated circuit, multi-region of interest, ultra-high frame rate

I. INTRODUCTION

Hyperspectral detection technology can acquire both image information and continuous spectral information of a target. By examining the spectral information of the characteristic bands of the target and combining it with machine learning algorithms, it is possible to detect the ingredients of the studied object [1]. With the development of hyperspectral technology, there has been a growing demand for multi-band and high spectral accuracy [2], leading to a significant increase in the array size of ROIC. However, the increase in array size substantially raises the image data volume of the circuit, consequently reducing the frame rate. To address this issue, the traditional solution has been to apply windowing to the readout integrated circuit (ROIC), also known as region of interest (ROI) readout [3]. Nevertheless, when dealing with targets whose characteristic bands have a wide span, the frame rate problem remains inadequately resolved.

Based on the aforementioned situation, this paper designs an ultra-high frame rate ROIC with multiple ROI readout functionality for hyperspectral detection. The circuit has an array size of 640×512 and processes the sampled voltage using a two-step analog-to-digital converter (ADC), effectively enhancing the signal conversion rate. Additionally, the application of multiple ROI readout technology enables the circuit to skip uninterested regions during readout, significantly reducing the image data readout time. Simulation results indicate that the proposed ROIC can successfully read

multiple characteristic bands of kaolinite and achieve a frame rate of 2691Hz, making it suitable for hyperspectral detection.

II. DESIGN OF AN ULTRA-HIGH FRAME RATE ROIC

A. The ROIC Architecture

The spectral camera receives optical signals through its optical system and converts them into image data with spectral information using the ROIC [4]. As shown in Fig. 1, the overall architecture of the ROIC in this paper includes pixel circuits, column-level processing circuits, and logic control circuits.

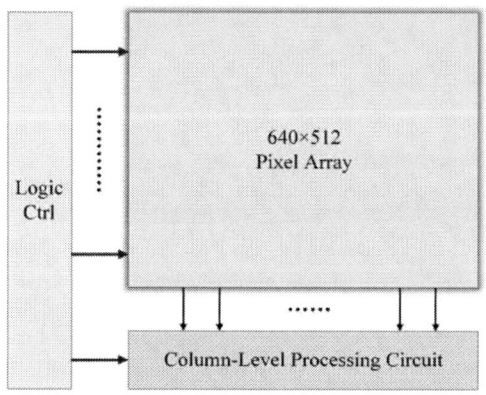

Fig. 1.   Architecture diagram of ROIC.

The pixel circuit has an array size of 640×512. In spectral testing, each row of pixels corresponds to a spectral channel responsible for converting photocurrents. The overall spectral range covered is from 900 to 3000 nm. The logic control circuit handles analog control tasks, including integration, sampling, and readout operations. The column-level processing circuit is responsible for transmitting and processing integrated voltages. As shown in Fig. 2, it illustrates the overall structure of the column-level link.

The column-level link consists of three parts: a column-level buffer, a correlated double sampling (CDS) circuit, and an ADC. Firstly, the column-level buffer isolates the parasitic capacitance of the column-level bus and the CDS sampling capacitor, accelerating signal establishment. The correlated double sampling circuit performs the sampling and holding of the integrated voltage, and eliminates the fixed noise of the column-level circuit. Lastly, the ADC module utilizes a two-step architecture, combining 7-bit coarse quantization with 9-bit fine quantization to convert analog voltages into digital signals. This approach ensures accuracy while accelerating the signal conversion rate [5].

979-8-3503-6184-1/24 $31.00 © 2024 IEEE

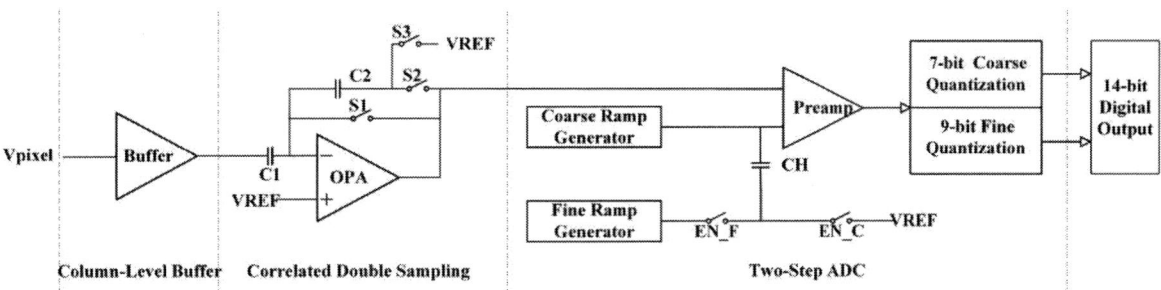

Fig. 2. Diagram of the column-level circuit.

B. Frame Rate Analysis of ROIC

The ROIC in this paper has an array size of 640×512 and operates at integration while read(IWR) mode. In this mode, the integration is performed at the same time as the data readout, so that the time required for the ROIC to run one frame can be approximated as the readout time for all image data. Therefore, the frame rate of the ROIC can be expressed as:

$$F=\frac{1}{a \cdot T_{row}} \quad (1)$$

Where a represents the total number of rows to be read out, T_{row} denotes the time required to read out data from each row of pixels, primarily constrained by the operating speed of the column-level link. To meet power consumption requirements, t the row read time is limited to 6.5 us through the column-level circuit structure mentioned above. Therefore, the improvement in frame rate of the ROIC in this paper can only be achieved by reducing the number of rows.

Below is an analysis of the frame rate of the ROIC using kaolinite mineral detection as an example. When performing spectral analysis of target components using a spectral camera, it is typically necessary to collect spectral information at peak and trough wavelengths. This section simulates the acquisition of kaolinite spectral information by extracting data from three bands of the ROIC: 1385-1420 nm, 1805-1890 nm, and 2145-2270 nm. Subsequently, through theoretical calculations, the frame rate of the ROIC will be analyzed in the following three usage scenarios:1. Full readout of all bands. 2. Single-window readout. 3. Multi-window readout. These analyses aim to assess the frame rate performance of the ROIC under different operational scenarios for detecting kaolinite minerals.

TABLE I. FRAME RATE CALCULATION RESULTS UNDER DIFFERENT USAGE SCENARIOS

Parameter	Scenario 1	Scenario 2	Scenario 3
Row Number	512	178	52
Frame(Hz)	300	864	2985

As shown in TABLE I, during kaolinite detection, the frame rate of the ROIC using multi-ROI technology can reach 2985Hz. This represents an 895% increase compared to Scenario 1 (full readout), and a 245.5% increase compared to traditional single-window technology, which prove it highly suitable for hyperspectral detection applications.

C. Design of Multi-ROI Technology

The implementation of multi-window technology in this paper mainly consists of two parts: 1. generation of pre-control words; 2. logical processing of row selection signals based on the pre-control words.

The pre-control word (EN_ROW) is used to define the working rows, generated using a shift register. As shown in Fig. 3, CLK_EN_ROW serves as the clock for the shift register, providing 512 clock edges. SEQ_EN_ROW acts as the data input for the shift register, undergoing logical transformations controlled by external window information and an internal counter. For instance, if the configured row windows are (0,9) and (502,511), SEQ_EN_ROW will be raised when the counter reaches 0 and 502, and lowered when it reaches 10 and 512.

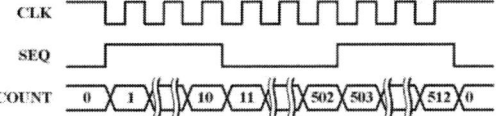

Fig. 3. Timing diagram for pre-control word generation.

Below is an explanation of the logic for generating row selection signals. As shown in Fig. 2, the control subunit of the multi-window scheme in this paper consists of three parts. The first part is the first row determination, used to discern whether the current row is the first row and to switch the input data of the corresponding control word unit accordingly. The SEL_NEG[N-1] signal is the output signal from the previous row, used as the input data for the control word unit of the current row. The second part involves the logic for generating row selection signals, using D flip-flops triggered by rising and falling edges to latch input data, and performing AND logic operations to obtain the row selection signal SEL[N]. The final part is a skip logic. It determines the input data passed to the next row's control unit based on the EN_ROW signal. If EN_ROW is high, indicating the current row is active, the output of the inverted D flip-flop triggered by the falling edge is passed to the next row. Otherwise, SEL[N-1] is outputted.

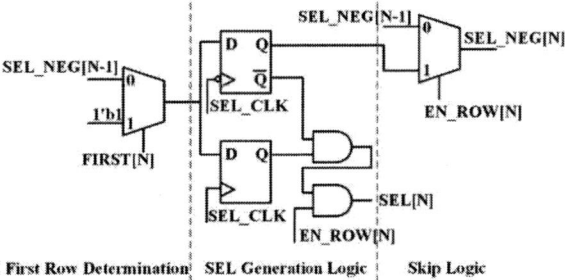

Fig. 4. Control subunit architecture.

Fig. 5 illustrates the timing diagram for row selection signals using the first two rows as an example. The first row is designated as the initial row with active windows, while the second row is inactive. As seen, due to EN_ROW[2] being low, SEL[2] remains low, indicating that the second row is not selected for readout. Following the skip logic, SEL_NEG[2] passed to the third row is consistent with SEL_NEG[1].

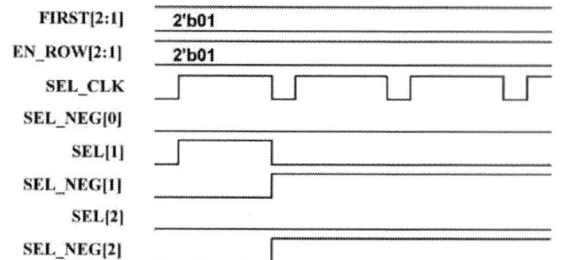

Fig. 5.   Timing diagram of row selection signals generation.

III. SIMULATION RESULT

This paper validated the multi-ROI function of the ROIC through simulated kaolinite spectral detection. The photocurrent received by the infrared spectral sensor is influenced by factors such as light source intensity, spectral transmittance, and quantum efficiency. For a more intuitive analysis of simulation results, it is assumed that the photocurrent received by the ROIC is only related to the spectral transmittance of the infrared spectrum and the intensity of the light source. With 100% transmittance, the photocurrent generated by the light source in the sensor is 1nA. Finally, the photocurrent for each spectral segment is determined by combining the spectral transmittance of kaolinite. The ROIC is configured with a spectral resolution of 5nm, covering 512 spectral channels, with a total wavelength range of 900 to 3000nm. Each row of the ROIC inputs the photocurrent corresponding to its respective spectral segment.

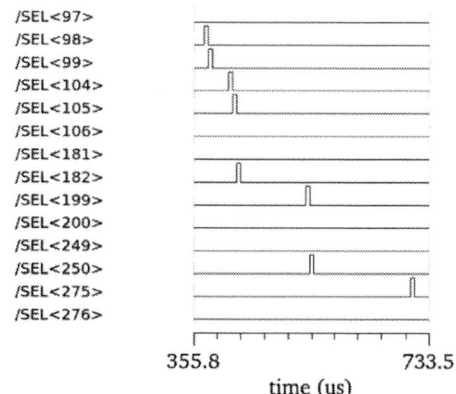

Fig. 6.   Multi-ROI simulation results.

As shown in Fig. 6, this simulation presents the multi-ROI results tailored for the characteristic bands of kaolinite. The window coordinates are (98,105), (182,199), and (250,275). Due to the large number of windows, only the simulation results for selected row signals are displayed. It can be observed that row 97 corresponding to the row select signal SEL[97] remains low throughout, indicating that spectral data for this row is not being read out. SEL[98], however, is pulled high for a period of time to read out data. At this point, the

ROIC operates at a frame rate of 2691Hz, which is highly suitable for high-speed capture scenarios.

The comparison between the original spectrum of kaolinite and the readout spectrum is shown in Fig. 7. Simulation results indicate that the ROIC in this study can effectively capture the spectral information of kaolinite's characteristic bands. Comparing with the original spectrum, the error is only 0.1%, meeting the performance requirements of the ROIC for spectral detection.

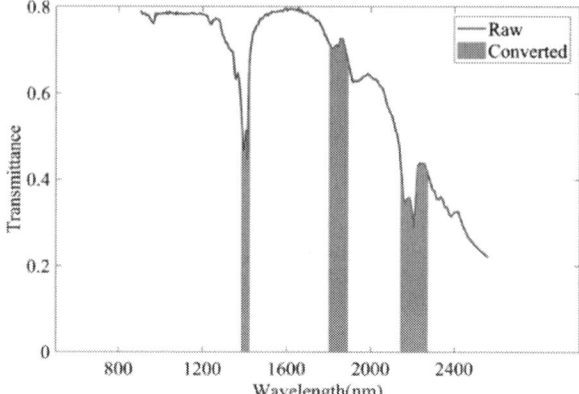

Fig. 7.   Kaolinite spectrum comparison chart.

IV. CONCLUSION

To address the issue of frame rate reduction due to increased resolution in hyperspectral cameras, this paper designs an ultra-high frame rate ROIC. The circuit has an array size of 640×512 and employs a two-step ADC architecture to effectively enhance the conversion speed of the ROIC. The multi-ROI technology based on control subunits allows the ROIC to directly skip uninterested regions during data readout, significantly reducing the image data readout time. Simulation results validate that the ROIC achieves a frame rate of 2691Hz when reading the spectral information of kaolinite's characteristic bands, making it highly suitable for hyperspectral testing scenarios.

REFERENCES

[1] O. Nevalainen, T. Hakala, J. Suomalainen, R. Mäkipää, M. Peltoniemi and A. Krooks, "Fast and nondestructive method for leaf level chlorophyll estimation using hyperspectral LiDAR," Agricultural and Forest Meteorology, vol. 198–199, pp. 250-258, August 2014.

[2] N. M. Nasrabadi, "Hyperspectral Target Detection : An Overview of Current and Future Challenges," in IEEE Signal Processing Magazine, vol. 31, no. 1, pp. 34-44, January 2014.

[3] O. Schrey, J. Huppertz, G. Filimonovic, A. Bussmann, W. Brockherde and B. J. Hosticka, "A 1 K/spl times/1 K high dynamic range CMOS image sensor with on-chip programmable region-of-interest readout," in IEEE Journal of Solid-State Circuits, vol. 37, no. 7, pp. 911-915, July 2002.

[4] X. Shi, D. Liu, Z. Chen, G. Chen, S. Huang and W. Lu, "A Low-Power Single-Slope based 14-bit Column-Level ADC for 384×288 Uncooled Infrared Imager," 2019 IEEE 13th International Conference on ASIC (ASICON), Chongqing, China, pp. 1-4, November 2019.

[5] Q. Zhang, N. Ning, J. Li, Q. Yu, K. Wu and Z. Zhang, "A 12-Bit Column-Parallel Two-Step Single-Slope ADC With a Foreground Calibration for CMOS Image Sensors," in IEEE Access, vol. 8, pp. 172467-172480, 2020.

979-8-3503-6184-1/24 $31.00 © 2024 IEEE

A BackgroundDigital Calibration Method for DTCs Used in Digital PLL Employing Dual-Path DTC

Renxuan Li [1], Xiaoyu Shan [1], Li Wang [1], Dongsheng Liu [1], Ang Hu*[1]

[1] School of Integrated Circuit, Huazhong University of Science and Technology, Wuhan, 430074, China

* Email: ang_hu@hust.edu.cn

Abstract—This paper presents a background digital calibration method for the nonlinear digital-to-time converters (DTCs) used in Fractional-N digital phase-locked loops (DPLL) employing dual-path DTC. The method utilizes polynomial predistortion and piecewise linear predistortion to calibrate the first-order and high-order nonlinearity of DTCs, respectively. It addresses the limitations of traditional calibration methods, which are not applicable to the DPLL using dual-path DTC, and the challenge of achieving convergence of high-order coefficients in the polynomial calibration method. Additionally, it demonstrates faster convergence rate compared to using only piecewise linear predistortion. The simulation results show the fractional spur at 26MHz frequency offset was reduced by 49dB, and the loop phase noise at 7.44MHz frequency offset was reduced by 50dB, when the calibration was turned on.

Keywords—*DPLL, DTC, digital calibration, fractional spur, phase noise*

I. INTRODUCTION

Thanks to the advancement of advanced CMOS technologies, the use of Fractional-N DPLL to replace analog phase-locked loops has become a prevailing trend in recent years. Typically, Fractional-N DPLL utilizes a sigma-delta modulator (SDM) to achieve the fractional average division ratio and a DTC to compensate for the instantaneous phase error [1]. However, the DTC following the currently mainstream technical route usually exhibits insurmountable nonlinear characteristics, leading to difficulties in accurately compensating for the instantaneous phase error and resulting in the generation of fractional spurs.

Traditional methods to reduce nonlinearity in DTCs involves digital calibration, typically achieved by predistorting the DTC control words. Ideally, the predistortion value corresponding to each possible value of the DTC control words should be recorded [2]. However, this method requires significant resource consumption and is therefore impractical. A compromise solution is to approximate the predistortion function by a piecewise linear function, which can effectively reduce the resource overhead. In [3], a sub-range DTC is employed, assuming that the fine-tuning range of the DTC is approximately linear. Calibration of the coarse-tuning range of a DTC can be achieved by incorporating the integral nonlinearity (INL) of the coarse-tuning range of the DTC into the fine-tuning control word.

However, the digital calibration methods mentioned above are not directly applicable to the Fractional-N DPLL employing dual-path DTC. This is because it is not possible to directly determine which DTC's nonlinearity causes the phase difference between the TDC's input signals. In [4], a third-order polynomial function is used as the predistortion function of the DTC to be compatible with the dual-path DTC architecture. However, in some special cases, the convergence of the coefficients of the quadratic and cubic terms of the predistortion function will be significantly slower.

This paper presents a background digital calibration method for the nonlinear DTCs. Compared to the existing methods, this method is compatible with the DPLL employing dual-path DTC and demonstrates a faster convergence rate.

II. THE PROPOSED CALIBRATION METHOD

As mentioned above, the predistortion functions used in DTC nonlinear calibration mainly consist of piecewise linear functions and polynomial functions. The first order term coefficient of the polynomial predistortion function converges rapidly and the piecewise nonlinear predistortion function can be well compatible with DTCs with complex nonlinearity. After comprehensively considering factors such as resource expenditure and the convergence rate, this paper selects a combination of the polynomial predistortion function and the piecewise linear predistortion function. The former is utilized to eliminate the first-order nonlinearity, while the latter is employed to eliminate the second-order and higher nonlinearity.

A. Overall architecture of the DPLL

Before specifically describing the calibration method for DTCs, it is necessary to first introduce the DPLL architecture used in this paper.

Fig. 1 shows the overall architecture of the DPLL. DST-NLC is the DTC nonlinear calibration module, D_{PDSD} and D_{PDSR} are the original DTC control words, while D_{DCWD} and D_{DCWR} are the actual DTC control words output by DST-NLC. The delay difference between DTC_D and DTC_R corresponds to the instantaneous phase error that needs to be compensated with respect to the instantaneous frequency division ratio D_{MMD} (the average value of D_{MMD} is equal to the target frequency division ratio D_{FRAC}).

This DPLL is a modification of the one mentioned in [4]. PDS-SDM has the capability to perform probability density shaping for D_{PDSD} and D_{PDSR}. Prior to the completion of the calibration of DTC_D and DTC_R, PDS-SDM can suppress the fractional spurs of the loops, although this results in a certain degree of deterioration of the phase noise.

B. Implementation of digital calibration

Fig. 2 shows the detailed structure of DST-NLC. Considering the nonlinearity of DTC_D and DTC_R, it can be inferred that DTC_D and DTC_R exhibit similar nonlinearity due to the same fabrication process and structure. Therefore, this paper assumes that the nonlinear functions of DTC_D and DTC_R are both $f(x)$. It implies that the delay time of DTC_D, denoted as t_D, is equal to $Cons1 \cdot f(D_{DCWD})$, where Cons1 is a configurable constant representing the minimum time step of the DTCs. Ideally, $f(x)$ should satisfy $f(x) = x$.

979-8-3503-6184-1/24 $31.00 © 2024 IEEE

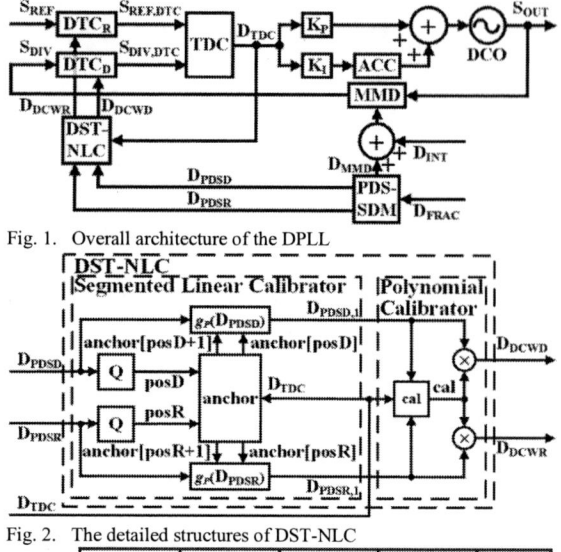

Fig. 1. Overall architecture of the DPLL

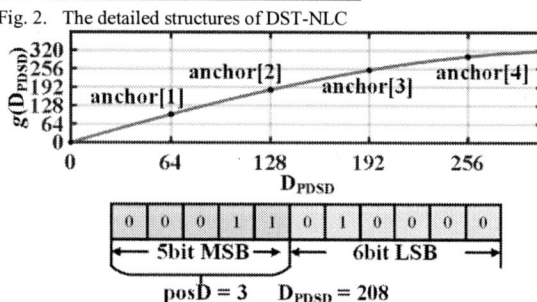

Fig. 2. The detailed structures of DST-NLC

Fig. 3. The relationship between posD, D_{PDSD} and anchor[k]

The function of DST-NLC is to predistort D_{PDSD} and D_{PDSR}. DST-NLC can be divided into two sub-modules, as shown in Fig. 2. The output of the module Segmented Linear Calibrator (SLC) is the input of the module Polynomial Calibrator (PC). We use the D_{PDSD} path as an example to explain how to compute D_{DCWD} and D_{DCWR}. Obviously, the range of D_{PDSD} and D_{PDSR} is [0, 2047]. The predistortion function of PC, denoted as $h(x)$, is a polynomial function. Only the first-order term of $h(x)$, denoted as cal, is retained, so that $h(x)$ is equivalent to horizontally scaling the DTC output curve. $h(D_{PDSD,1})$ can be represented as follows:

$$D_{DCWD} = h(D_{PDSD,1}) = cal \cdot D_{PDSD,1} \qquad (1)$$

Where $D_{PDSD,1}$ is the output of SLC, and D_{DCWD} is the input of DTC_D. It can minimize the difference between the scaled DTC output curve and the ideal DTC output curve. The inverse function of the function corresponding to the scaled curve is defined as $g(x)$, so that $g(x) = f^{-1}(x) / cal$, where $f^{-1}(x)$ is the inverse function of $f(x)$. The predistortion function of SLC is a piecewise linear function $g_P(x)$ which is a piecewise linear fitting of $g(x)$. Therefore, $g_P(x)$ can be described by various endpoints and connection lines, and $g_P(D_{PDSD})$ can be represented as follows:

$$D_{PDSD,1} = g_P(D_{PDSD}) = anchor[posD]$$
$$+ (D_{PDSD} - 64 \cdot posD) \qquad (2)$$
$$\cdot (anchor[posD+1] - anchor[posD])$$

Where anchor[k]s (k is an integer that falls in the range of [0, 32]) is a sequence, with the values shown in Fig. 3. When k<32, there is anchor[k] = $g(64 \cdot k)$, and when k = 32, there is anchor[k] = $g(2047)$. Besides, as shown in Fig. 3, posD is the 5bit MSB of D_{PDSD}. Therefore, the value of posD falls in the range of [0, 31]. Obviously, the point (D_{PDSD}, $g_P(D_{PDSD})$) lies in the (posD + 1)th segment of the curve of $g_P(D_{PDSD})$, and the points (64 · posD, anchor[posD]) and (64 · (posD + 1), anchor[posD + 1]) are the two endpoints of this segment.

This paper uses an iterative method to compute anchor[k]s and cal, referring to the algorithm in [4]. For anchor[k]s, it is first determined whether posD is equal to posR (posR is the 5bit MSB of D_{PDSR}). If they are equal, anchor[posD+1] can be updated as follows:

$$anchor[posD+1] = anchor[posD+1] \cdot z^{-1}$$
$$+ (D_{PDSD} \cdot z^{-1} - D_{PDSR} \cdot z^{-1}) \cdot (-D_{TDC}) \cdot Cons2 \qquad (3)$$

Where anchor[posD + 1] · z^{-1} is the value of anchor[posD + 1] in the previous phase detection period (the initial values of anchor[k]s satisfy anchor[k] = $64 \cdot k$). The output value of TDC, denoted as D_{TDC}, is positive when $S_{DIV,DTC}$ lags behind $S_{REF,DTC}$. Cons2 is a configurable constant controlling the iteration step of anchor[k]s. For cal, there is:

$$cal = cal \cdot z^{-1}$$
$$+ (D_{PDSD,1} \cdot z^{-1} - D_{PDSR,1} \cdot z^{-1}) \cdot (-D_{TDC}) \cdot Cons3 \qquad (4)$$

Where cal · z^{-1} is the value of cal in the previous phase detection period (the initial value of cal satisfies cal = 1). D_{TDC} is as mentioned above, and Cons3 is a configurable controlling the iteration step of cal.

III. SYSTEM SIMULATION AND RESULT ANALYSIS

The validity of the calibration method mentioned above is verified through simulating a DPLL Verilog model with the DST-NLC module. The DST-NLC module in the DPLL is synthesizable. The reference frequency of the DPLL is 26MHz, and the output frequency is 2.44GHz. The noise characteristics of the DCO used in the DPLL are set according to [5], where the noise values were −50dBc@1KHz, −130dBc@1MHz and −150dBc@10MHz. Additionally, the delay time of the nonlinear DTCs can be represented as follows:

$$t_D = Cons1 \cdot 1.4 \cdot \frac{D_{DCWD}}{2 \cdot 2047}$$
$$+ 0.4 \cdot (\frac{D_{DCWD}}{2 \cdot 2047})^2 + 0.2 \cdot (\frac{D_{DCWD}}{2 \cdot 2047})^3 \qquad (5)$$

Fig. 4 shows the stabilization processes of the parameters anchor[k]s and cal. The red line in Fig. 4 represents the results of transient simulations on anchor[k]s when using the method presented by this paper with independent first-order nonlinear calibration module, PC, and the blue line in Fig. 4 represents the results of transient simulations on anchor[k]s when using the traditional method without PC.

979-8-3503-6184-1/24 $31.00 © 2024 IEEE

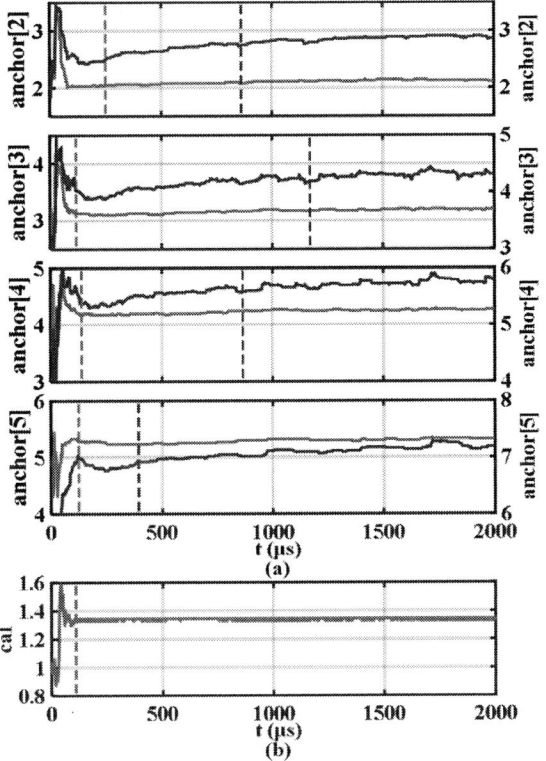

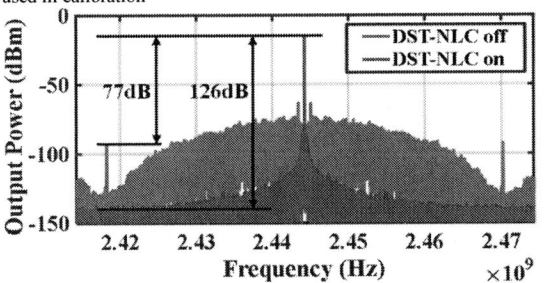

Fig. 4. The stabilization process of the parameters anchor[k]s (a) and cal (b) used in calibration

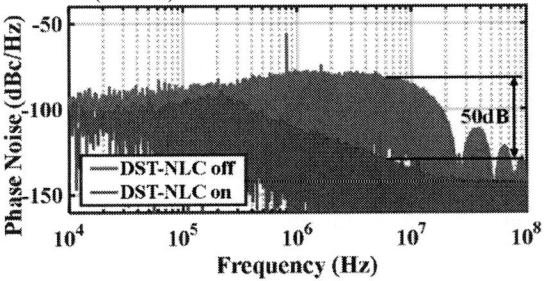

Fig. 5. The output spectrum comparison with calibration disabled (red line) and enabled (blue line)

Fig. 6. The phase noise spectrum comparison with calibration disable (red line) and enable (blue line)

Defining that an anchor[k] is convergent when the range of fluctuations is less than ±3%, it is clear that anchor[2], anchor[3], anchor[4], and anchor[5] converge within 250μs, 120μs, 150μs, and 120μs, respectively, when using the new method. Compared to the convergence times of the anchor[k]s (850μs, 1180μs, 880μs, and 400μs, respectively) when using the traditional method, the convergence times of the

anchor[k]s when using the new method are significantly shorter. Besides, another parameter, cal, which is only used in the new method, also converges within 100μs. Therefore, we believe that the convergence rate of the new method is faster.

Fig. 5 shows the output spectrum comparison with calibration disabled and enabled. The red line in Fig. 5 represents the output spectrum with calibration disabled. It is evident that a −77dBc spur at the 26MHz frequency offset still exists. The blue line in Fig. 5 represents the output spectrum with calibration enabled, and it can be observed that the spur at the 26MHz frequency offset is reduced by 49dB to −126dBc. Additionally, Fig. 5 also demonstrates that the phase noise is significantly suppressed when calibration is enabled.

Fig. 6 shows the phase noise spectrum comparison with calibration disabled and enabled. It is evident that the phase noise is suppressed by up to 50dB at a frequency offset of 7.44MHz.

IV. CONCLUSION

This paper introduces a background digital calibration method for the nonlinear DTCs used in Fractional-N DPLL employing dual-path DTC. The validity of the calibration method mentioned above is verified through simulating a DPLL Verilog model. This method demonstrates several advantages, including faster convergence rate, and compatibility with the DPLL employing dual-path DTC. Simulation results shows that this method can reduce the fractional spur at 26MHz frequency offset by 49dB, the phase noise at 7.44MHz frequency offset by 50dB, and the key parameters in the calibration system can be converged within 250μs, which is faster than the traditional method. Therefore, we believe that this method enhances the performance of the DPLL.

ACKNOWLEDGMENT

This work is supported by the National Key Research and Development Program of China (No. 2021YFA0715502, No. 2023YFB4502100), the National Natural Science Foundation of China (No.62374064, No. 62104076, No. 62134002), the National Key Analog Integrated Circuit Laboratory Project of China. (No. JCKY2021210C004), the Innovation Project of JinYinHu Laboratory. (Grant No. 2024JYH011401)

REFERENCES

[1] D. Tasca, M. Zanuso, G. Marzin, S. Levantino, C. Samori and A. L. Lacaita, "A 2.9–4.0-GHz Fractional-N Digital PLL With Bang-Bang Phase Detector and 560-${\rm fs}_{\rm rms}$ Integrated Jitter at 4.5-mW Power," in IEEE Journal of Solid-State Circuits, vol. 46, no. 12, pp. 2745-2758, Dec. 2011.

[2] S. Levantino, G. Marzin and C. Samori, "An Adaptive Pre-Distortion Technique to Mitigate the DTC Nonlinearity in Digital PLLs," in IEEE Journal of Solid-State Circuits, vol. 49, no. 8, pp. 1762-1772, Aug. 2014.

[3] B. Liu et al., "A 1.2ps-jitter fully-synthesizable fully-calibrated fractional-N injection-locked PLL using true arbitrary nonlinearity calibration technique," 2018 IEEE Custom Integrated Circuits Conference (CICC), San Diego, CA, USA, 2018, pp. 1-4.

[4] C. Hwang, H. Park, Y. Lee, T. Seong and J. Choi, "A Low-Jitter and Low-Fractional-Spur Ring-DCO-Based Fractional-N Digital PLL Using a DTC's Second-/Third-Order Nonlinearity Cancellation and a Probability-Density-Shaping ΔΣM," in IEEE Journal of Solid-State Circuits, vol. 57, no. 9, pp. 2841-2855, Sept. 2022.

[5] Hu, Yizhe. Flicker noise upconversion and reduction mechanisms in RF/millimeter-wave oscillators for 5G communications. Diss. 2019.

A High-Precision Sigma-Delta ADC for Battery Management System

Hao Xue[1], Liji Wu*[2,3], Jing Hu*[1], Zhiwei Li[2,3], Xiangmin Zhang[2,3]

[1] Electronic Engineering College, Heilongjiang University, Harbin, China
[2] School of Integrated Circuit, Tsinghua University, Beijing, China
[3] Beijing National Research Center for Information Science and Technology, Beijing, China

*lijiwu@mail.tsinghua.edu.cn, *hjlyh@126.com.

Abstract— A high-precision Sigma-Delta ADC is presented for battery management system (BMS) chips in new energy vehicles. It is designed with 180nm BCD (Bipolar CMOS DMOS) technology. Sigma-Delta ADC consists of a modulator and a digital filter. The modulator adopts a three-order cascaded integrator feedforward (CIFF) structure, where a 4-bit Flash ADC, working as a quantizer, can reduce quantization noise, improve dynamic performance and stability, and propose an active summation method to overcome the coefficient attenuation problem. The simulation results show that the ADC signal-to-noise distort ratio (SNDR) of 122.6 dB and a signal bandwidth (BW) of 4 KHz at a sampling frequency of 1.024 MHz, with a power consumption of 3.87 mW at a power supply voltage of 5 V, the design is feasible for the BMS chips.

Keywords—*Sigma-Delta ADC, Multi-bit quantization, BMS.*

I. INTRODUCTION

The demand for BMS chips is increasing in new energy vehicles [1]. The high-precision characteristics of Sigma-Delta ADC are feasible for BMS chips, where precision, signal bandwidth, and power consumption are the main challenges for Sigma-Delta ADC [2]. This paper proposes a high-precision Sigma-Delta ADC for BMS chips. Fig. 1 shows the structure of the BMS chips, where the Sigma-Delta ADC detects the voltage of each battery cell and transmits the results to the cascade communication circuit.

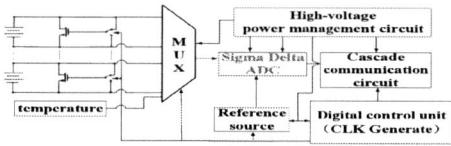

Fig. 1. The structure of the BMS chip

The voltage range of a single battery cell in new energy vehicles is between 1.0V and 4.2V depending on their chemicals. The high-voltage power management circuit provides a power supply voltage of 5V for the ADC, the reference source circuit can provide a reference voltage of 2.5V, and regarding the bandwidth requirements and power consumption, 1.024 MHz of clock frequency for the ADC is chosen in this design.

II. STRUCTURE OF SIGMA-DELTA ADC

Sigma-Delta ADC consists of modulator and digital filter. The modulator designed in this paper is shown in Fig. 2. It consists of switched-capacitor integrator, clock circuit, signal summation circuit, 4-bit Flash quantizer, and data weighted average (DWA).

Fig. 3. The structure of digital filter.

The structure of the digital filter is shown in Fig. 3. It consists of a cascaded integrator comb (CIC) filter, a compensated cascaded integrator comb (CCIC) filter, and a half band (HB) filter.

III. CIRCUITS DESIGN

A. Switched-capacitor integrator

In the design of switched-capacitor integrators, the sampling and integration switches are controlled by clocks CLK1 and CLK2, respectively. CLK1 and CLK2 are non-overlapping clocks, while CLK1D and CLK2D represent their delays. V_{CM} is the common mode voltage, C_s is the sampling capacitor, and C_f is the integrating one.

The working process of an integrator within one cycle can be divided into two stages, the sampling phase and integration phase, respectively.

In the sampling phase, the right pole of C_s is connected to V_{CM}, the input signal charges C_s. Due to the negative feedback effect of the operational amplifier, the voltage at both ends of C_f remains invariable. In the Integral phase, the left pole plate of C_s is connected to V_{CM}, and the integral switch is connected to C_s and C_f. Due to the virtual ground characteristic of the negative feedback operational amplifier, the charge of the right pole plate of C_s is transferred to the left pole plate of C_f, causing the output voltage to change by a quantity proportional to C_s/C_f.

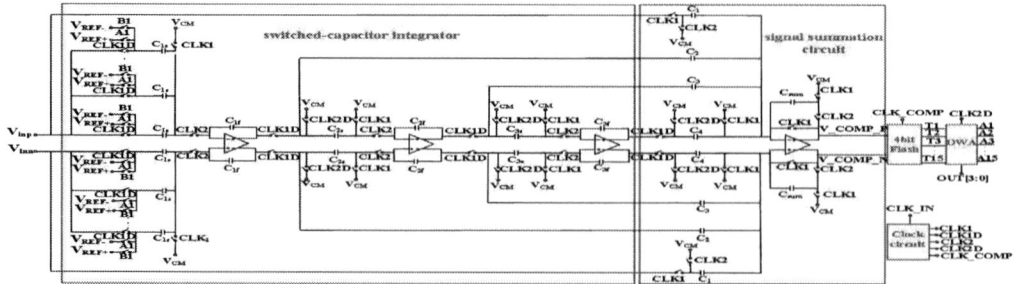

Fig. 2. The structure of modulator.

979-8-3503-6184-1/24 $31.00 © 2024 IEEE

B. Signal summation circuit

In the design of signal summation circuit, when CLK1 is at high level, the charges stored in the capacitors in the summation circuit are cleared to ensure that they do not affect the next summation operation. When CLK2 is at a high level, capacitors C_1-C_4, C_{sum}, and operational amplifiers form an inverse summation circuit to perform reverse summation operations on each input signal, this process is called active summation. The output V_{sum} of the active summation circuit is (1)

$$V_{sum} = \frac{\sum_f C_f V_f}{C_{sum}} \tag{1}$$

The advantage of an active summation circuit is that it can accurately complete the sum operation of multiple input signals without any additional attenuation or sensitivity to parasitic capacitance.

C. Clock circuit

The clock circuit is used to control the working timing of the modulator, and its structure is shown in Fig. 4.

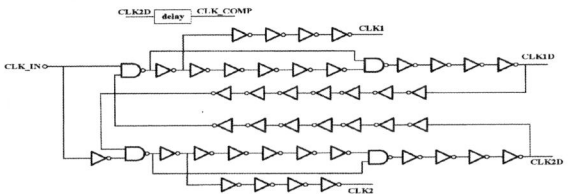

Fig. 4. The structure of clock circuit.

The main functions of clock circuits with different phase clocks are:

- Provide a non overlapping clock to ensure the normal operation of each switch capacitor integrator in the modulator.

- To eliminate the clock feedthrough and channel injection effects of the sampling switch, a clock with a falling edge delay is provided on the basis of the original two-phase non overlapping clock.

- Provide a reset clock for the dynamic comparator.

D. 4-bit Flash Quantizer

The structure of a 4-bit Flash quantizer is shown in Fig. 5, where a fully differential Flash ADC is selected as the quantizer. Due to the fact that N-bit flash ADC requires 2^n-1 comparators, 4-bit ADC requires 15 identical comparators.

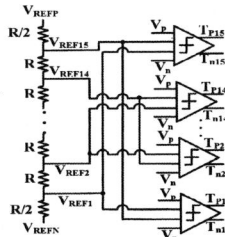

Fig. 5. The structure of 4-bit Flash quantization circuit .

The external reference voltage is generated by the resistance voltage divider circuit to generate reference voltages for each comparator, which are sent to 15 comparators for comparison with the results V_p and V_n

generated by the signal summation circuit. Finally, a 15-bit thermometer code is generated, which is shifted by the DWA module and used as the input signal for the feedback DAC.

E. Data weighted average

Due to the inherent elements mismatch of multi feedback DACs, it will significantly affect the system performance, so some additional dynamic element matching technology is required [3]. The DWA circuit structure is shown in Fig. 6. The core idea of DWA is to periodically select feedback units for the DAC, thereby averaging the nonlinear effects caused by mismatch and having a shaping effect on the mismatch of the DAC.

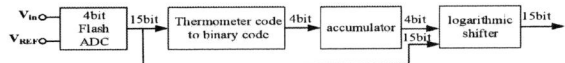

Fig. 6. The structure of DWA circuit

The DWA first converts the thermometer code into a binary code, which serves as the control signal of the accumulator and is added to the binary code of the previous clock cycle retained by the accumulator to update the pointer. The function of an accumulator in a DWA circuit is as a pointer to control the shift operation of a logarithmic shifter. The input of the logarithmic shifter is the thermometer code output by the Flash ADC during this clock cycle. Under the instruction of the accumulator, the shifted thermometer code is output as the input control signal for the feedback DAC.

F. digital filter

The CIC filter structure is shown in Fig. 7, which can complete 32 times down-sampling, but may experience pass-band attenuation.

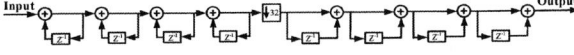

Fig. 7. The structure of CIC filter.

The function of CCIC filter is to solve the passband attenuation problem of CIC filter. It needs to complete 2 down-sampling, and its structure is shown in Fig. 8, where K represents the filter coefficients.

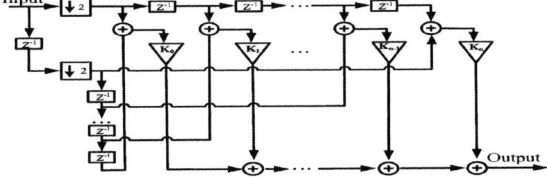

Fig. 8. The structure of CCIC filter.

The final stage uses an HB filter, Its structure is shown in Fig. 9, which can halve the order of the filter and complete 2 times down-sampling. Finally, the three filters designed can be cascaded to achieve filtering and 128 times down-sampling functionality.

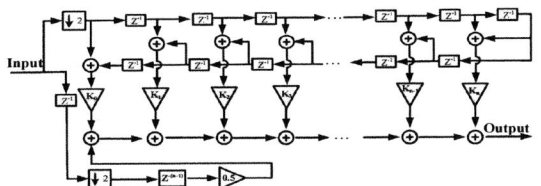

Fig. 9. The structure of HB filter.

IV. VERIFICATION

Fig. 10 shows the output quantization waveform of the modulator, indicating that the modulator has achieved multi bit quantization function, and the input signal amplitude range of the modulator is 0.6 V to 4.4 V, with an input frequency of 1156.25 Hz. Fig. 11 shows the simulation results of the proposed modulator without 4-bit Flash ADC as a quantizer. Fig. 12 shows the simulation results with 4-bit Flash ADC as a quantizer, indicating a significant improvement in performance of the latter. Fig. 13 shows the simulation results of Sigma-Delta ADC, and the proposed Sigma-Delta ADC achieves an ENOB of 20.07 bits, and SNDR of 122.6 dB, and a power consumption of 3.87 mW.

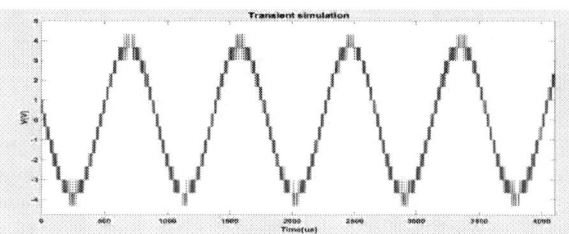

Fig. 10. The output quantization waveform of the modulator

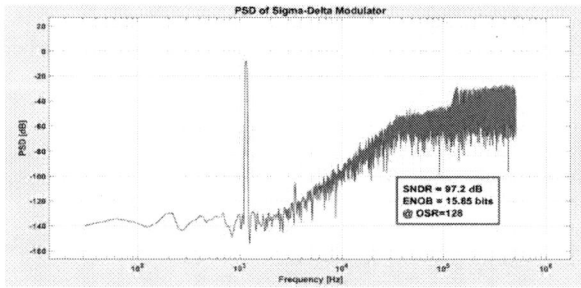

Fig. 11. Simulation results of modulator (without 4-bit Flash ADC) .

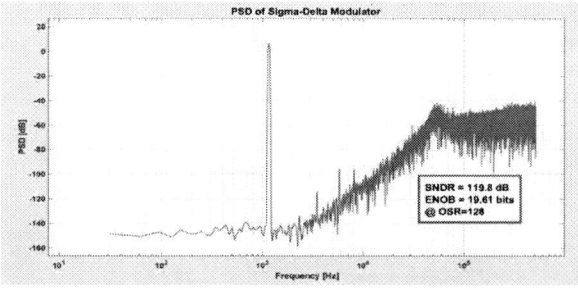

Fig. 12. Simulation results of modulator (with 4-bit Flash ADC) .

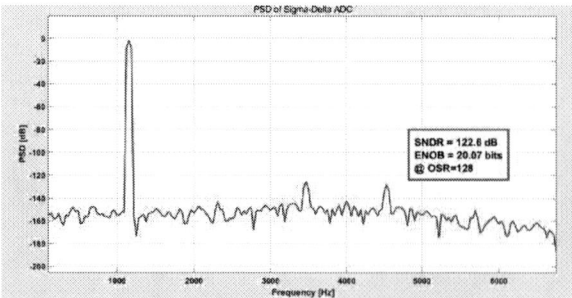

Fig. 13. Simulation results of Sigma-Delta ADC.

The simulation results with different process corners and temperature conditions are shown in Table I.

TABLE I. SIMULATION RESULTS WITH DIFFERENT CORNERS CONDITIONS

Simulation conditions	SNDR[dB]	ENOB[bit]
tt_27°C	122.6	20.07
ss_125°C	116.0	18.98
ff_-40°C	125.8	20.61

Comparisons with some published literature are listed in Table II. The calculation of FOM$_S$ (Schreier figure-of-merit) is (2), which is formula used to evaluate the performance of ADC. with higher values indicating better performance.

$$FOM_S = SNDR + 10\lg(\frac{BW}{Power}) \qquad (2)$$

Compared with other published ADCs, this ADC shows good performance in terms of FOM$_S$.

TABLE II. COMPARISON THE PERFORMANCE

Reference	[2]	[4]	[5]	[6]	Proposed work
Supply(V)	3.3	1.8	1.8	3.3	5
BW(KHz)	4	16	20	1	4
SNDR(dB)	90.3	93.0	90.4	109.3	122.6
ENOB(bit)	14.71	15.16	14.73	17.90	20.07
Power(mW)	5.0	8.28	12.99	0.48	3.87
FOMs(dB)	149.3	165.0	152.3	172.5	182.7

V. CONCLUSION

This paper introduces a Sigma-Delta ADC for BMS chips. A 4-bit Flash ADC was proposed as a quantizer, significantly reducing quantization noise and increasing system stability. A method for active summation is proposed to address the issue of coefficient attenuation. The ADC can operate within the temperature range of -40°C to 125°C, and the detection voltage range includes the working range of each battery voltage. Therefore, this Sigma-Delta ADC is feasible in the BMS chips for the new energy vehicles.

REFERENCES

[1] G. Shi, C. Yang, L. Sun, T. Tao, X. Ren and J. Chen, "Interface Design for Analog Front-End Chips in Battery Management Systems," 2023 9th International Conference on Computer and Communications (ICCC), Chengdu, China, 2023, pp. 2673-2677

[2] T. L. Cao, Y. Han , S. F. Zhang, X. X. Han and Y. Y. Chen, "A 0.18 μm high resolution bandpass delta-sigma modulator for acceleration transducer applications," 2016 13th IEEE International Conference on Solid-State and Integrated Circuit Technology (ICSICT), Hangzhou, 2016, pp. 897-899.

[3] O. Ismail, P. Kaesser, J. G. Kauffman and M. Ortmanns, "DAC Element Mismatch Shaping Algorithms in Incremental Delta-Sigma ADCs," 2024 IEEE International Symposium on Circuits and Systems (ISCAS), Singapore, Singapore, 2024, pp. 1-5.

[4] Y. Chen, Z. Wang, Y. Zhuang and H. Tang, "Analysis and Design of Sigma-Delta ADCs for Automotive Control Systems," 2021 IEEE 3rd International Conference on Circuits and Systems (ICCS), Chengdu, China, 2021, pp. 235-241.

[5] L. C. Gunnam, G. -M. Sung and L. -W. Weng, "2+1- order Switched-current MASH Delta-Sigma ADC with the digital cancellation circuit," 2017 International Conference on Applied System Innovation (ICASI), Sapporo, Japan, 2017, pp. 1801-1804.

[6] B. Mou, W. Zhu, Z. Wang, B. Zhang, H. Feng and Y. Liang, "A High-Resolution Low-Power Delta-Sigma Modulator ADC for Biosensors," 2023 8th International Conference on Intelligent Computing and Signal Processing (ICSP), Xi'an, China, 2023, pp. 850-853.

Multi-Sampling Mode CDAC Design for a 12-bit 200MS/s Pipelined-SAR ADC

Tianyu Zhang, Fan Ye and Shunli Ma*

State-key Laboratory of ASIC and System, Fudan University, Shanghai 200433, China
*Email: fanye@fudan.edu.cn, shunlima@fudan.edu.cn

Abstract

This paper proposes a multi-sampling mode capacitor DAC (CDAC) for a 12-bit 200MS/s pipelined-SAR ADC, addressing the issue of overfitting in neural network-based calibrations. By implementing normal, offset, and proportional sampling modes, the design ensures the linearity of the ADC's transfer function. The proposed CDAC utilizes a bottom-plate sampling method and a high-linearity bootstrap circuit. Simulation results demonstrate that the proposed ADC achieves an SFDR of around 75dB and an ENOB of approximately 10.8bits across all sampling modes, validating the effectiveness of the design in enhancing ADC performance.

1. Introduction

With the development of wireless communication, there is an increasing demand for high-speed and high-resolution analog-to-digital converters (ADCs), among which pipelined-SAR ADCs are a better choice for this indicator. However, considering the sensitivity of parameters such as gain and linearity of open-loop amplifiers to PVT conditions, adaptive calibration is necessary.

Previously reported deterministic digital calibrations are often based on split ADCs, which mostly use the least mean square (LMS) method [1]. However, the nonlinear calibration of multistage ADCs is a nonlinear least squares problem, making the LMS algorithm suboptimal [2]. Inspired by artificial intelligence, neural network-based calibrations have gradually become alternatives. Due to the strong fitting capability of the nonlinear filters constructed by neural network-based calibrations, there is a potential for overfitting.

This paper proposes a multi-sampling mode CDAC for a 12bit 200MS/s pipelined-SAR ADC, which can effectively avoid the issue of overfitting. Section II introduces the conditions for determining the linearity of ADC to explain why designing CDAC with multiple sampling modes is necessary, Section III introduces the implementation method of the multi-sampling mode CDAC, Section IV presents the simulation results of this design, and Section V concludes the paper.

2. Conditions for ADC Linearity

Fig.1 shows the block diagram of a traditional split pipelined-SAR ADC, which divides the conventional pipelined-SAR ADC into two identical channels, A and B. Both channels sample the same input signal simultaneously and perform the conversion independently. Non-ideality in channels A and B cause their transfer functions to become nonlinear functions, linear functions with different slopes, or functions with a fixed difference. An adaptive filter based on the LMS methods is commonly used to calibrate these non-idealities, adjusting the weights to minimize the mean square value of the output difference e between the two channels.

However, there may be multiple local optima for e, with only one set of weights corresponding to a linear ADC system. Therefore, it is necessary to have the two channels convert the input x and the dithered input $x + a$. Pseudo-random noise sequences PN_A and PN_B are used to obtain different residual modes. Assuming the transfer functions of channels A and B are f_A and f_B, respectively, if the following conditions are met:

$$f_A(x) - f_B(x) = 0, f_A(x + a) - f_B(x) = b \quad (1)$$

it indicates that the transfer function of the calibrated ADC is linear. Since a linear function's ability to fit a nonlinear curve is limited, mis-convergence is almost impossible.

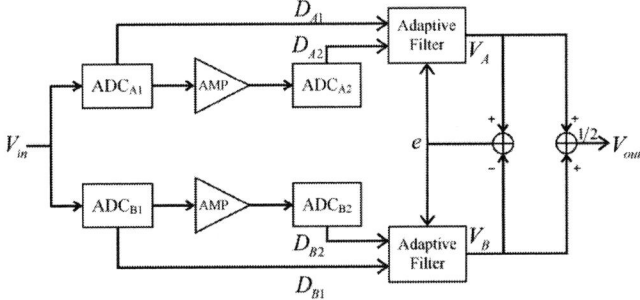

Figure 1. Block diagram of the traditional split pipelined-SAR ADC

However, if a neural network calibration algorithm is used, relying solely on Eq.(1) to determine whether the ADC transfer function is linear may lead to overfitting. A simple example is a step function with width Δ and height c, which satisfies Eq.(1) but is not a linear function. Therefore, when using neural network algorithms for the digital calibration of Split-ADCs, the criteria for determining that the transfer function has been corrected to linearity must be more stringent. We introduce a new criterion involving the input signal scaled proportionally, as follows:

979-8-3503-6184-1/24 $31.00 © 2024 IEEE

$$f_A(x) - f_B(x) = 0, f_A(kx) = kf_B(x), k \neq 0 \quad (2)$$

If both Eq.(1) and Eq.(2) are satisfied, it can be conclusively stated that the calibrated ADC system has a linear transfer function.

The specific argument is as follows. First, consider the form of the transfer function $f(x)$ that only satisfies Eq.(1). We construct the function $g(x) = f(x)-bx/a$. It can be verified that $g(x)$ is a periodic function with period a. Therefore, $f(x)$ is a superposition of a periodic function and a linear function, i.e.,

$$f(x) = g(x) + bx/a, g(x) = g(x+a) \quad (3)$$

Only when $g(x)$ is a constant function, $f(x)$ is a linear function.

Next, consider the form of the transfer function $f(x)$ that only satisfies Eq.(2). Considering only the case where $x>0$, we construct the function $h(x)$, which is a periodic function with a period of 1. Let

$$f(x) = xh(\log_k x), h(x) = h(x+1) \quad (4)$$

It can be verified that $f(kx) = kf(x)$, thus satisfying the condition. Similarly, only when $h(x)$ is a constant function is $f(x)$ a linear function.

Finally, consider the case where both Eq.(1) and Eq.(2) are satisfied simultaneously, meaning $f(x)$ can be expressed in the forms of both Eq.(3) and Eq.(4). By substituting kx for x in Eq.(3), we get $g(kx) = kg(x)$. Extending this, we obtain $g(k^n x) = k^n g(x)$, where n is a positive integer.

Since $g(x)$ is a continuous periodic function, it must be bounded. Let M be its upper bound. Assuming $k > 1$, we have $0 \leq |g(k^n x)| \leq M$. Therefore, $0 \leq |g(x)| \leq M/k^n$, and taking the limit as n approaches $+\infty$, according to the squeeze theorem, we must have $g(x) = 0$. Hence, $f(x)$ must be a linear function in this case.

In summary, using a neural network-based digital calibration algorithm requires the ADC to implement at least three different sampling modes, namely:

1. Normal Sampling Mode: where the input signal is x and the output signal is $f(x)$.

2. Offset Sampling Mode: where the input signal is $x + a$ and the output signal is $f(x + a)$ - b.

3. Proportional Sampling Mode: where the input signal is scaled, such as kx, to satisfy the condition $f(kx) = kf(x)$.

These modes ensure that the transfer function of the calibrated ADC is linear by fulfilling both the necessary and sufficient conditions for linearity.

3. Implementation of Multiple Sampling Modes

A. Implementation of Capacitor Array

The implementation of the first-stage capacitor array is shown in Fig.2(a). It uses bottom-plate sampling to ensure the accuracy of the comparator's operation. Additionally, a Vcm-Based capacitor flipping method based on split capacitors is used to reduce the power consumption of capacitor flipping [3].

As shown in Fig.2, to achieve proportional sampling, we exclude the LSB capacitor from the sampling process. Its bottom plate is connected to V_P, V_N, and V_{CM} during the sampling phase. This structure implements the sampling mode $kV_{in}+ \Delta$. If kV_{in} is considered a completely new input signal V_x, then the three input modes can be represented as V_x, $V_x+\Delta$, and $k'V_x$, respectively.

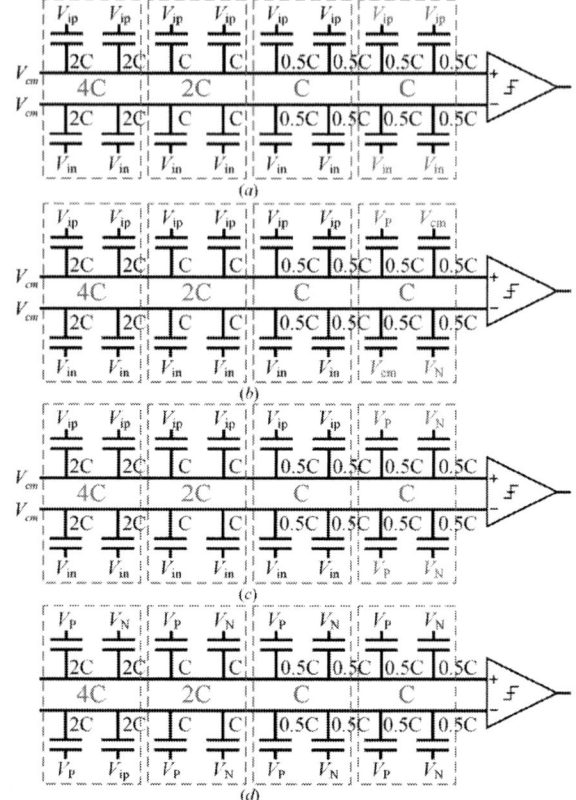

Figure 2. Capacitor array design with three sampling modes. (a) Normal mode; (b) Offset mode; (c) Proportional mode; (d) ADC flipping.

When the flipping signal D=1, the voltage changes in the normal mode, offset mode, and proportional mode are as follows:

$$\begin{cases} V_{N+} - V_{N-} = \dfrac{8}{7} \cdot \dfrac{7}{8}(V_{ip} - V_{in}) \\ V_{O+} - V_{O-} = \dfrac{7}{8}(V_{ip} - V_{in}) + \dfrac{1}{16}(V_P - V_N) \\ V_{P+} - V_{P-} = \dfrac{7}{8}(V_{ip} - V_{in}) \end{cases} \quad (5)$$

where $V_+ + V_- = 2V_{CM}$, the subscripts N, O, and P are abbreviations for the three modes.

This implementation of the first-stage capacitor array does not require the introduction of new capacitors, nor does it need additional switches on the top plate of the capacitor array, thereby reducing area overhead and minimizing the introduction of nonlinear parasitic capacitances in the signal path. Additionally, the capacitor array flipping and successive approximation logic remain unchanged across all sampling modes, simplifying the circuit design.

However, this approach transfers control to the sampling switches, which places demands on the bootstrap circuit module controlling the sampling switches for matching and linearity.

B. *Implementation of Bootstrap Circuit*

For our multi-sampling mode capacitor array, since the participation of the LSB capacitor in sampling is determined by the pseudo-random control signal PN, PN needs to control the MOS switch corresponding to the LSB capacitor. This overlaps with the clock CK. To ensure matching as much as possible, we propose a selectable gate voltage bootstrap circuit for sampling mode as shown in Fig.3.

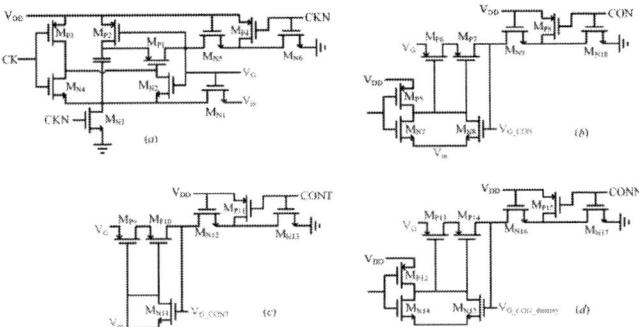

Figure 3. (a) Traditional gate voltage bootstrap; (b) controllable gate voltage bootstrap circuit; (c) Voltage matching circuit for high-side sampling capacitor; (d) dummy matching circuit.

This circuit consists of four parts. Fig.3(a) is a traditional gate voltage bootstrap circuit [4], which generates a bootstrap voltage V_G that is higher than the input voltage by V_{DD} during sampling. In Fig.3(b), M_{P6} and M_{P7} are conduction switches of the same size, used to connect the bootstrapped gate voltage V_G to the gate of the actual sampling switch. Their conduction is controlled by a pseudorandom number PN. M_{N9}, M_{N10}, and M_{P8} are three pull-down transistors, and the pull-down control signal CON can quickly pull the gate voltage of the NMOS sampling switch to GND at the end of sampling.

If the V_G signal is directly applied to the gate of the NMOS sampling switch, the gate voltage signal will arrive earlier compared to the gate voltage output by V_{G-CON}. Without the voltage loss caused by M_{P6} and M_{P7} and other MOS transistors introducing circuit noise, the gate signals of the two different NMOS sampling switches under the same signal will be inconsistent, introducing nonlinearity. Therefore, Fig.3(c) is used to provide the bootstrap gate voltage for the high-bit capacitor that is constantly involved in sampling. M_{P9} and M_{P10} are always on, and the sizes of the corresponding MOS transistors in Fig.3(b) and Fig.3(c) are proportional, with this ratio equal to the ratio of the capacitance driven by the two modules.

Fig.3(d) has the same structure and MOS transistor sizes as Fig.3(b), but its control signals PNN and CONN are exactly opposite to PN and CON in Fig.3(b). Therefore, when Fig.3(b) is operating normally, Fig.3(d) is not; when the output voltage of Fig.3(b) is GND, Fig.3(d) is engaged. By dynamic matching,

the load on the output voltage V_G of Fig.3(a) does not change, further improving the linearity of V_G.

4. Simulation Results

The post-simulation results of the proposed ADC are shown in Table 1, where the first-stage SAR ADC utilizes the previously proposed CDAC. It can be seen that the circuit's performance is maintained across all four sampling modes. Fig.4 shows the layout of the first-stage SAR ADC and the gate voltage bootstrap circuit.

Table 1. Simulation results of dynamic parameters of the proposed ADC

Sampling Mode	f_s= 100MS/s, f_{in}= 2.4MHz		
	SNR	SFDR	ENOB
Normal Sampling	67.30	75.30	10.89
Proportional Sampling	65.65	72.96	10.61
Positive Offset Sampling	67.14	73.65	10.86
Negative Offset Sampling	66.12	76.94	10.69

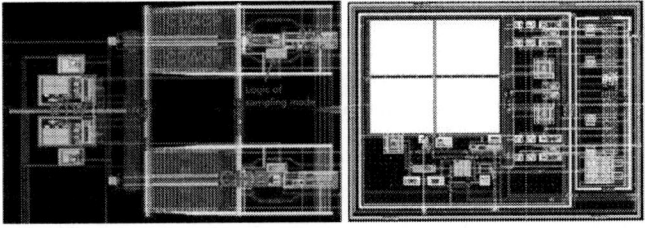

Figure 4. Layout of first-stage SAR ADC (left) and gate voltage bootstrap circuit (right).

5. Conclusion

This paper demonstrates that using a neural network-based digital calibration algorithm requires the ADC to implement at least three different sampling modes. For a 12-bit 200MS/s pipelined-SAR ADC designed in 28nm CMOS, various sampling modes for the CDAC and corresponding high-linearity gate voltage bootstrap switches were implemented. Measurement results show that the proposed ADC achieves an SFDR of around 75dB and an ENOB of around 10.8bits.

References

[1] J. McNeill, M. C. W. Coln and B. J. Larivee, ""Split ADC" architecture for deterministic digital background calibration of a 16-bit 1-MS/s ADC," in IEEE Journal of Solid-State Circuits, vol. 40, no. 12, pp. 2437-2445.

[2] T. Zhang, Y. Cao, S. Zhang, C. Chen, F. Ye and J. Ren, "Machine Learning Based Prior-Knowledge-Free Calibration for Split Pipelined-SAR ADCs with Open-Loop Amplifiers Achieving 93.7-dB SFDR," ESSCIRC 2019 - IEEE 45th European Solid State Circuits Conference (ESSCIRC), Cracow, Poland, 2019, pp. 189-192.0

[3] Y. Zhu et al., "A 10-bit 100-MS/s Reference-Free SAR ADC in 90 nm CMOS," in IEEE Journal of Solid-State Circuits , vol. 45, no. 6, pp. 1111-1121, June 2010.

[4] B. Razavi, "The Bootstrapped Switch [A Circuit for All Seasons]," in IEEE Solid-State Circuits Magazine, vol. 7, no. 3, pp. 12-15, Summer 2015.

DSP-PUF: A Software PUF Based on Digital Signal Processor for IoT Security

Tengfei Yuan [1], Pengjun Wang*[2], Yuejun Zhang *[1], Mingze Ren, Shuang Hu

[1] Faculty of Electrical Engineering and Computer Science, Ningbo University, Zhejiang, 315211, China.
[2] Electrical and Electronic Engineering, Wenzhou University, Wenzhou, 325035, China.

* Email: wangpengjun@wzu.edu.cn, zhangyuejun@nbu.edu.cn

Abstract—**Physical unclonable function (PUF) utilizes the tiny deviations in the chip manufacturing to generate unique identity keys. Traditional PUFs are generally implemented on dedicated circuit structures, which may not be effective in resource-constrained Internet environments. To solve the above problems, this paper proposes a software PUF based on Digital Signal Processor (DSP-PUF). The DSP-PUF uses the timing violation of the internal registers of the chip to generate random response. Primarily, the mechanism of generating abnormal information between timing paths and registers in DSP chip is analyzed. Secondly, the overclocking clock is applied to make the DSP work abnormally, and the unique and unclonable random response is generated according to the dependence of the response on the path delay. Eventually, DSP-28335 is selected as the DSP-PUF carrier, and the DSP-PUF response extraction is completed by computer, oscilloscope, DSP and other equipment. The test results indicate that the 12000 bit responses generated by DSP-PUF has passed the NIST randomness test, and the worst case reliability is 88.01% in the temperature range of 20 ℃-80 ℃. Moreover, the proposed DSP-PUF does not use additional circuit structure, and does not need to change the existing circuit structure. It is suitable for privacy information encryption in resource constrained Internet of things (IoT).**

Keywords—*PUF, timing violation, DSP, privacy information, dedicated circuit.*

I. INTRODUCTION

With the wide application of Internet of things technology, the security problems it faces are also gradually highlighted, including the security of Internet of things devices, privacy leakage, etc. In addition, due to the limitation of computing power, storage capacity and other resources in the Internet of things (IOT), it is particularly important to provide secure and reliable solutions for privacy data. The physical unclonable function (PUF) [1] extracts the unique identity key from the chip manufacturing deviation. Ideally, PUF security stems from the non replicability of the internal structure of the chip. PUF produces the same output response under a given specific challenge, and this challenge response mechanism cannot be predicted.

According to whether additional hardware design is required, PUF is divided into software PUF (SPUF) and hardware PUF. Hardware PUF usually needs additional circuit to generate response, and typical hardware PUF are Ring Oscillator PUF (RO-PUF) and Arbiter-PUF(APUF) [2]. The SPUF does not need to change its circuit structure or add additional circuit resources. Static random access memory PUF (SRAM PUF) [3] and Magnetic random access memory PUF (MRAM PUF) [4] are typical SPUFs. In the resource constrained Internet of things, hard PUF is often difficult to play a role in it because it requires additional hardware resources. In addition, the randomness of APUF and RO-PUF stem from

the deviation of two identical timing paths, which makes APUF and RO-PUF no longer safe in the face of modeling attacks. SRAM PUF and MRAM PUF are weak PUF, when they are applied in a large number of data scenarios, the limited challenge-response pairs (CRPs) space will limit the performance of SRAM PUF and MRAM PUF.

To solve these problems, this paper proposes a Digital Signal Processor PUF (DSP-PUF). The entropy source of the proposed DSP-PUF comes from the random response generated by the timing violation of the internal registers in the chip. Its advantage is that it does not need to change the original circuit structure and does not increase the additional hardware resource overhead. DSP-PUF has two working modes: generating PUF response under overclocking clock signal and processing input data under normal clock signal. The common advantage of these two modes is that they can use PUF response to confuse the privacy data in the Internet of things to ensure the security of these data in the transmission. The experimental results show that the randomness of DSP-PUF reaches 49.11%, and the worst reliability of DSP-PUF is 88.01% in the temperature range of 20-80 ℃.

II. THE PROCESSED DSP-PUF

In this section, we first discuss the DSP-PUF design; Then the generation mechanism of chip timing violation abnormal response is described. Finally, according to the critical path of DSP, the appropriate overclocking clock signal is selected to simulate the DSP-PUF.

A. DSP-PUF design process

DSP has higher performance than general microprocessor in processing digital signals. In the field of Internet, the application scope of DSP covers multimedia communication, industrial automation control, automobile safety and automatic driving technology, etc. The proposed DSP-PUF uses DSP as the PUF carrier, computer as the DSP control center, and oscilloscope to display the DSP-PUF output response. The DSP-PUF design scheme is shown in Fig. 1.

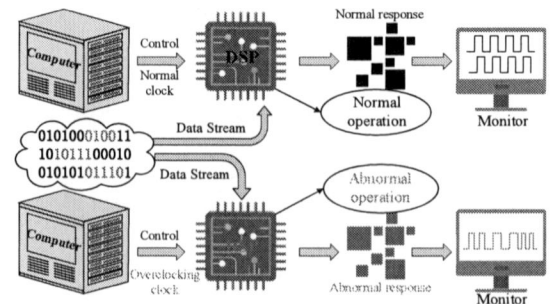

Fig. 1. DSP-PUF design scheme.

The computer controls the running state of DSP, and makes DSP in two working modes by changing the input clock signal of DSP. When the applied clock is the normal clock, DSP works normally to process the input data stream; When the applied clock is an overclocking clock, the DSP is in abnormal working mode, and the output is a random abnormal response.

B. Abnormal information generation mechanism

The core component of SPUF is how to generate abnormal information. DSP-PUF uses register timing violation and data path delay deviation as PUF entropy source. DSP data path deviation is mainly affected by manufacturing fluctuations, and this deviation will become more significant with the increase of the number of sequential paths.

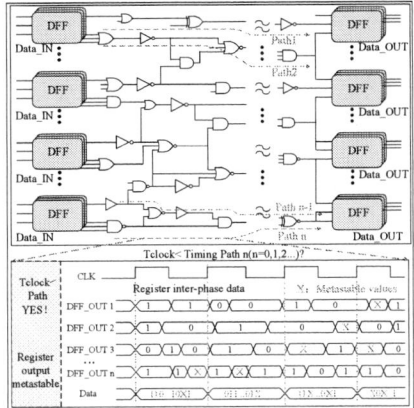

Fig. 2. Register timing violation analysis.

To ensure the normal operation of the circuit, it is necessary to analyze the timing of the circuit to meet the requirements of the set-up and hold-up time between the internal registers of the chip, so as to output the normal response. The set-up and hold-up time between registers must meet (1) and (2):

$$T_{clk} > T_{co} + T_{logic} + T_{setup} + T_{skew} + T_{jitter} \tag{1}$$

$$T_{co} + T_{logic} - T_{skew} - T_{jitter} > T_{hold} \tag{2}$$

Where T_{clk}, T_{jitter}, T_{co}, T_{logic}, T_{skew}, T_{setup}, and T_{hold} are clock signal period, clock jitter, the transmission time between two registers, combined logic delay, two clock delays, set-up time and hold-up time respectively.

The timing violation between registers is shown in Fig. 2. As the clock frequency increases, the corresponding clock cycle is shortened. When the clock cycle is lower than the time required for the internal register path of the chip, the set-up and hold-up time requirements between registers are not met, and the data sampling deviation between registers will lead to the random metastable value at the output of registers, and the random metastable value will be transmitted between registers. These metastable values are transmitted in the register chain, which will cause more register sampling exceptions.

C. Delay analysis and clock frequency selection

DSP-PUF needs to apply overclocking clock to generate abnormal information, and the applied overclocking clock signal cannot cause functional failure to DSP. When overclocking clock is applied, DSP will generate PUF

response, and when normal clock signal is applied, DSP will operate normally. The internal register group of DSP is shown in Fig. 3. A large number of combinational logic units and register arrays are integrated in DSP. Timing violation can trigger DSP abnormal response, which is closely related to data path change and register arrangement. In addition, the data path deviation is affected by the chip manufacturing, and the complex timing path and register set provide an ideal entropy source for the design of soft PUF.

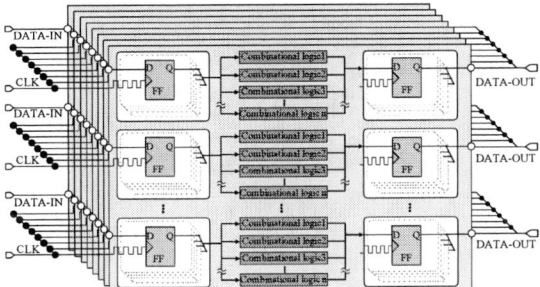

Fig. 3. DSP internal register.

The working frequency of the DSP used is 150MHz. In this experiment, the overclocking clock is 300MHz. The reason why the clock frequency is 300MHz is given below. The experimental results show that the clock frequency is slightly higher than 150MHz, and the output response effect of DSP-PUF is not ideal, showing poor randomness; When the clock frequency is far greater than 150MHz, too high clock frequency will cause the DSP to fail to work or even damage the DSP. In addition, the DSP clock is provided by the phase-locked loop, and only one frequency doubling is needed to convert the clock from 150MHz to 300MHz, which can quickly realize the switching of DSP working mode, and is very suitable for application in complex Internet of things environment

III. EXPERIMENTAL RESULTS

In this section, we describe the test platform of DSP-PUF in detail and comprehensively evaluate its performance. The evaluation includes the randomness, reliability and safety of DSP-PUF. In addition, we also compare the proposed DSP-PUF with the current advanced PUF technology.

A. Test platform

The proposed DSP-PUF technology uses DSP-28335 as the carrier, and the processor is manufactured based on 90nm CMOS. As shown in Fig. 4. The test platform is utilized to evaluate the performance of the proposed DSP-PUF. On this platform, the computer is connected with DSP through serial port to control the input clock frequency of DSP and switch the processor mode. At the same time, the DSP-PUF response signal is generated through the DSP port and transmitted to the oscilloscope by the probe.

Fig. 4. Test platform.

979-8-3503-6184-1/24 $31.00 © 2024 IEEE

B. Randomness

The randomness of PUF is usually determined by the distribution of PUF response logic 0 and logic 1. Ideally, the probability of PUF response "0" and "1" is 50%. Fig. 5. shows the two-dimensional gray scale diagram of 2400 bit DSP-PUF response. In 2400 bit response, the probability of logic 0 is 49.11%, and the probability of logic 1 is 50.89%. From Fig 5, it can be seen that the 2400 bit DSP-PUF response presents a random distribution.

Fig. 5. DSP-PUF response grayscale image.

C. Reliability

PUF reliability requires that PUF still maintain the same incentive ability given the same challenge in different external environments. Generally, the bit error rate is used to evaluate the reliability of PUF. As shown in Fig. 6, the reliability of the proposed DSP-PUF is evaluated at different temperatures. Within the range of 20℃-80℃, the worst reliability of the proposed DSP-PUF is 88.01%, and the reliability of the DSP-PUF is close to the ideal 1.

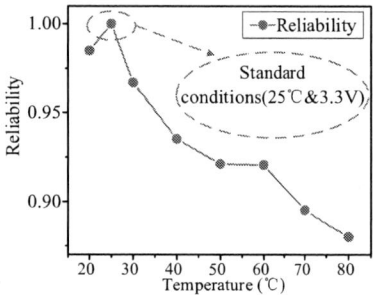

Fig. 6. DSP-PUF reliability.

D. Security

Security requirements the PUF response generated by the designed PUF circuit is difficult to predict the DSP-PUF challenge response mechanism even if the attacker knows part of the CPRs. In the advanced PUF work, NIST tests are used to evaluate the PUF response security. The 12000 bit response to DSP-PUF is divided into 10 groups as NIST input. The test results are shown in Table I. When the p value is greater than 0.01, the test response can be considered to have high security. The results show that the proposed DSP-PUF performs well in security.

TABLE I.   NIST TEST

Test	Stream length	P-value	Pass?
Frequency	1200	0.5415	YES
Block Frequency	1200	0.5567	YES
Rank	1200	0.3833	YES
FFT	1200	0.2001	YES
Serial	1200	0.5687	YES
Non-overlapping	1200	0.2172	YES
Cumulative Sums	1200	0.4218	YES
Longest Runs	1200	0.2316	YES

E. Comparison with advanced work

In Table II, the comparison between the proposed DSP-PUF and the advanced work is listed. It can be seen from Table II that the proposed DSP-PUF has good randomness and unpredictability, and does not need to change the circuit.

TABLE II.   COMPARISON WITH ADVANCED WORK

Type	[5] Processor	[6] Processor	[7] Scan	This Paper Processor
Randomness	0.304	-	0.4457	0.491
Reliability (%)	96.7	0.74	0.79	88.01
Temp Range (℃)	20~100	-	10~70	20~80
Vdd Range (V)	-	5	1.1~1.3	3.3V
NIST	NO	NO	YES	Yes
Additional Cost	YES	NO	YES	NO

IV. CONCLUSION

In this paper, we propose DSP-PUF. The proposed DSP-PUF uses the random metastable value generated by DSP in the case of timing path violation as the entropy source. Compared with the traditional hardware PUF, the proposed DSP-PUF does not need additional hardware resource overhead, nor does it need to make any changes to the existing circuit structure. It is more suitable for privacy information protection in the resource constrained Internet of things environment.

ACKNOWLEDGMENT

This work is supported by the National Natural Science Foundation of China (62174121, 62234008, 62134002), the Science and Technology Innovation 2025 Major Project of Ningbo (2022Z203), Ningbo University and Ningbo Yongxin Microelectronics Technology Co., LTD. Digital Integrated Circuit Design Joint Laboratory (XQ2022000005), Ningbo University Graduate Education Practice Base (YJD202305), the Technology Innovation Project of Ningbo University under Grant (2024SRIP1302).

REFERENCES

[1] L. Ni and J. Zhang, "S2RAM PUF: An ultra-low power subthreshold SRAM PUF with zero bit error rate," ProceeDSngs of the 61th ACM/IEEE Design Automation Conference, 2024: 23-27.

[2] X. Li, P. Wang, G. Li, L. Ni and Y. Zhang, "Design of interface circuits and lightweight PUF for TMR sensors," IEEE Sensors Journal, 2023, 23(11): 11754-11761.

[3] Z. Zhou, P. Wang and G. Li, "Bagua Protocol: A Whole-Process configurable protocol for IoT sensing devices security based on strong PUF," IEEE Internet of Things Journal, 2024, 11(1): 805-819.

[4] L. Ni, P. Wang, Y. Zhang, G. Li and J. Zhang, "PI PUF: A Processor-intrinsic PUF for IoT," Computers and Electrical Engineering, 2023, 105: 108540.

[5] A. Aysu and P. Schaumont, "PASC: Physically authenticated stable-clocked SoC platform on low-cost FPGAs," 2013 International Conference on Reconfigurable Computing and FPGAs, Cancun, Mexico, 2013: 1-6.

[6] I. Tsiokanos, J. Miskelly and C. Gu, "DTA-PUF: Dynamic timing-aware physical unclonable function for resource-constrained devices," ACM Emerg Technol Comput System, 2021, 17(3): 1-24.

[7] Y. Zheng, F. Zhang and S. Bhunia, "DScanPUF: A Delay-Based physical unclonable function built into scan chain," IEEE Transactions on Very Large Scale Integration (VLSI) Systems, 2016, 24(3): 1059-1070.

A Q/V Band 49.6-54.5GHz,3.53dB NF,45dB Gain,2.09° Phase Error,2-Way Phased-Array Receiver for Satellite Application

Congrui Li, Qi Zhao, Ruolan Chen, Shulan Chen, Yan Wang, Lei Zhang*

School of Integrated Circuits, Tsinghua University, Beijing 100084, China

* Email: zhang.lei@tsinghua.edu.cn

Abstract—**A Q/V-band 2-way phased-array receiver with high resolution, low amplitude and phase error is presented. The receiver includes optimized low-noise amplifier (LNA), driving amplifier (DA), phase shifter (PS) and attenuator (ATT), as well as the power divider, with a compact layout. The optimization of phase shifter and attenuator manages to minimize the phase and amplitude errors. Implemented in a 65nm CMOS process, the maximum small gain at 52GHz of the two-way phased array receiver is 45dB, it also achieves a 3dB bandwidth of 4.9GHz, a minimal noise figure of 3.53dB, the phase error and amplitude error at 52GHz are 2.09° and 1.13dB, respectively.**

Keywords—mm-wave, 2-way phased-array receiver, CMOS, phase shifter, attenuator.

I. INTRODUCTION

Recently, as the demand of satellite communication increases, the utilization of electromagnetic waves in the Q/V band is becoming increasingly popular due to the wide bandwidth, large capacity, and low cost. However, the wavelength of the Q/V band wave is short, signals from a single channel transceiver are easily absorbed and blocked during transmission, which causes high transmission loss, limited transmission distance. To alleviate the issues, phased-array technology has been introduced since it can achieve beamforming, control gain phase through electrical signals, eliminate interference signals, spatial filtering, etc. This paper presents a 2-way phased-array receiver operating from 49.6GHz to 54.5 GHz, achieving a gain of 45dB, a minimal noise figure (NF) of 3.53dB, with a 2.09° phase error and a 1.13dB amplitude error, respectively.

II. PHASE ARRAY SYSTEM DESIGN AND ANALYSIS

A. Architecture of the Proposed Phased-Array Receiver

The architecture of the proposed millimeter wave phased array receiver is shown in Fig. 1., which consists a low-noise amplifier (LNA), two driver amplifiers (DA), a power divider and a power combiner, two phase shifters (PS) and two attenuators (ATT).

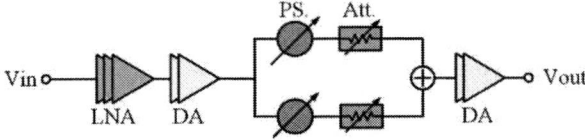

Fig. 1. The architecture of the proposed phased-array receiver.

B. Analysis of the Phased-Array Receiver

The main spec of a phased-array receiver includes gain, bandwidth, NF, and amplitude/phase RMS errors in phase shifting and attenuation.

The gain and bandwidth of the phased-array receiver are determined by the gain and bandwidth of LNA and DA. According to the formula of NF of a cascaded system, the NF of the receiver is mainly determined by the NF of the LNA. The amplitude error and phase error mainly depend on the amplitude and phase errors of the phase shifter and attenuator. The Wilkinson power divider is adopted as power divider and power combine network, and the isolation of the network has to be less than -20dB in the frequency band of interest for satellite applications.

III. DESIGN OF BUILDING BLOCKS

A. Low Noise Amplifier

The schematic of the proposed LNA design is illustrated in Fig. 2. The LNA adopts three-stage common-source pseudo-differential transistors to achieve a high gain, which is also able to suppress the noise of the other stages. To minimize the insertion loss, the matching network is carefully optimized. For the input matching network, a 1:2 balun is carefully designed to achieve both a good power gain and noise figure. The matching impedance of the first stage trades between the optimal noise matching impedance and the optimal power matching impedance. The matching center frequencies of the middle stages are interlaced to extend the bandwidth of the LNA. The bias voltage is chosen as 0.6V to ensure the LNA and the receiver with enough gain and relatively low power consumption.

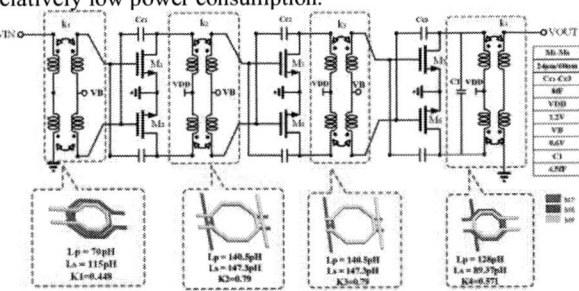

Fig. 2. The architecture of the proposed LNA.

The simulation results show that the three stage LNA has a gain of 28.21dB at 52GHz, with its 3dB bandwidth from 48.2GHz to 56.4GHz, while the NF at 52GHz is 3.42dB. The total DC power consumption of the LNA is 47.86 mW.

B. Driving Amplifier

The DA is to increase gain of the receiver while suppressing the noises of the following stages. The schematic of DA is shown in Fig. 3. The two stage DA has a gain of

979-8-3503-6184-1/24 $31.00 © 2024 IEEE

22.88dB at 52GHz. The 3dB bandwidth covers from 48GHz to 56GHz.The LNA and the two DAs can provide 72dB gain.

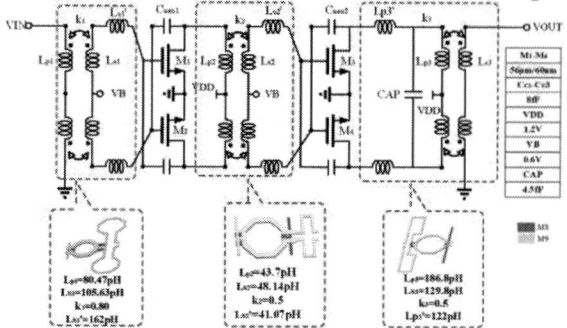

Fig 3. The architecture of the proposed DA.

C. The Power Divider

The structure of the power divider is a multistage one-to-two Wilkinson power divider, the layout is shown in Fig. 4(a). The main part of the power combiner and divider is identical. The isolation is optimized by adjusting the resistance between the two output ports.

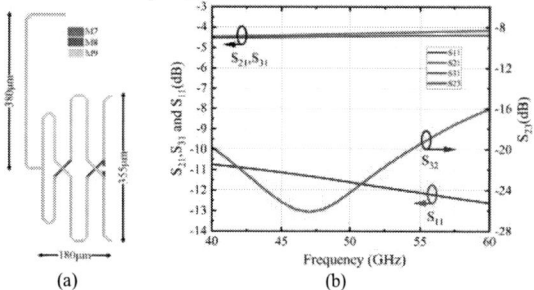

(a)   (b)

Fig. 4. (a) The 3D model, (b)The S-parameter of the network.

The simulation results in Fig. 4. (b) indicate that the insertion loss is 1.28dB at 52GHz, the S_{11} is less than -10dB in the entire frequency range, the S_{32} is less than -20dB from 40GHz to 55GHz.

D. Phase Shifter

The phase resolution of the phase shifter in the phased array receiver determines the scanning resolution of the phased array. The architecture is a vector interpolation phase shifter, as shown in the Fig. 5. The phase shifter combines power divider, the matching balun at the input/output stage, two variable gain amplifiers (VGA) with redundant bits, and a quadrature hybrid. By properly controlling the gain of the VGA of the I-path and Q-path, the amplitude and phase errors of the phase shifter can be minimized.

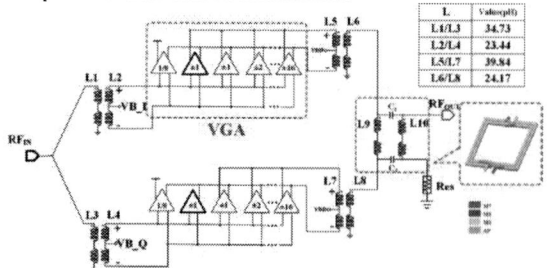

Fig. 5. The structure of the phase shifter.

The schematic of the basic cell of the VGA is shown in the Fig. 6. The size of MOS in the weight 1 cell are as follows: M1-M4:4μM/1μm.M5-M6:2μm/1μm.M7:1μm/1μm. For the cell weight of 2, there are two cells in parallel, etc.

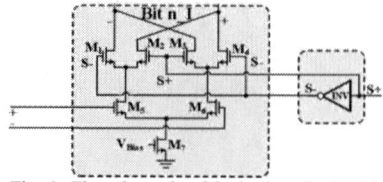

Fig. 6. The schematic of the basic cell of VGA

The simulation result indicates a phase RMS error of 0.90° and an amplitude RMS error of 1.13dB at 52GHz, and the gain of the phase shifter is -13.7dB at 52GHz.

E. Attenuator

The proposed attenuator can offer a 31.5dB range for attenuation, whose schematic is shown in Fig.7. (a). The structure of 0.5,1dB,2dB and 4dB attenuation cells are bridge T-shaped attenuation units. The 8dB and 16dB attenuation units are π-type attenuation units, The specific sizes of the transistors in the attenuator are shown in Fig. 7. (b).

Since the resistance or capacitance of the attenuator would fluctuate with the process, voltage and temperature, the accuracy of the attenuator could be exacerbated. In order to alleviate this deviation, 12 phase correction cells are introduced to the 8dB and 16dB attenuation cells, respectively. From node 1 to 6, we parallel two congruent cells. Each phase correction cell essentially a switch capacitor cell. The value of the capacitor Cs is 73.25fF, whose structure is MOM capacitor using the vertical parallel plate from M1 to M7, and the area of the capacitor is 60μm^2.Each cell can offer about 5° phase redundancy.

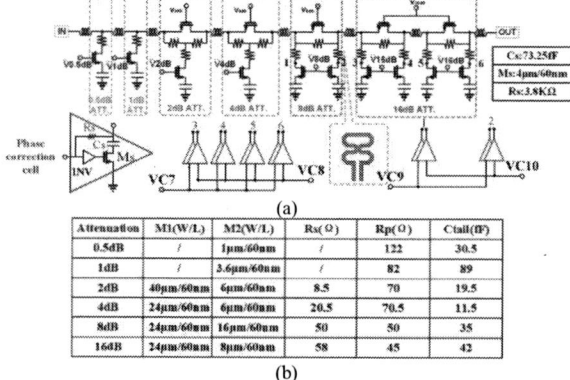

(a)

Attenuation	M1(W/L)	M2(W/L)	Rs(Ω)	Rp(Ω)	Ctail(fF)
0.5dB	/	1μm/60nm	/	122	30.5
1dB	/	3.6μm/60nm	/	82	89
2dB	40μm/60nm	6μm/60nm	8.5	76	19.5
4dB	24μm/60nm	6μm/60nm	20.5	76.5	11.5
8dB	24μm/60nm	16μm/60nm	50	50	35
16dB	24μm/60nm	8μm/60nm	58	45	42

(b)

Fig. 7. (a) The architecture of the attenuator, (b) the sizes of transistors and resistance of the attenuator.

The simulation result in Fig. 8. (a) and (b) shows the gain and the phase of the 64 states of the attenuator. Fig. 8 (c) shows that the phase error of the attenuator is 0.75° and the amplitude error is 0.11dB. Fig. 8. (d) shows about 20° of redundancy can be provided by the 12 phase correction cells to overcome PVT variations. The insertion loss of the reference state is -10.12dB.

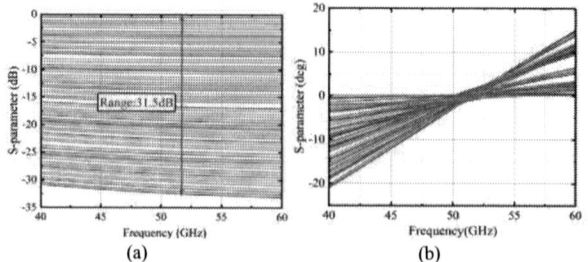

(a)   (b)

979-8-3503-6184-1/24 $31.00 © 2024 IEEE

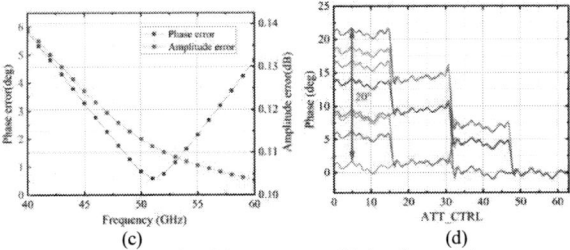

(c)   (d)

Fig. 8. (a) The amplitude of the attenuator, (b) the phase of the attenuator (c) the amplitude error and phase error of the attenuator, (d) The effect of introducing the phase correction cell to overcome PVT variations.

IV. RESULTS

The post simulation results of the 2-way phased-array receiver are shown in the Fig. 9(a)-(d). The max gain at 52GHz of the 2-way phased-array receiver is 45dB, the 3dB-bandwidth covers from 49.6GHz to 54.5GHz, the NF at 52GHz can reach 3.53dB. The phase RMS error of phase control is 1.02° and the gain RMS error of phase control is 1.13dB, the phase RMS error of gain control at 52GHz is 2.09° and the amplitude RMS error of gain control at 52GHz is 0.2dB of the attenuator. The range of NF variation with temperature from -45℃ to 125℃ is 2dB, which shows the robustness of the receiver.

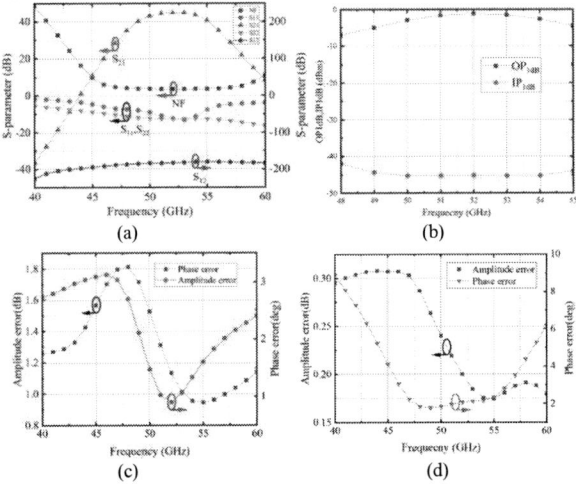

Fig. 9. (a) the S-parameter of the 2-way phased array, (b) the P_{sat} and OP_{1dB} of the 2-way phased array, (c)The amplitude/phase errors of phase shifting, (d)The amplitude/phase errors of attenuation.

The layout of the 2-way phased-array receiver is shown in Fig. 10. The whole layout area of the receiver is 3.672mm² and the total DC power consumption is 208.7mW.

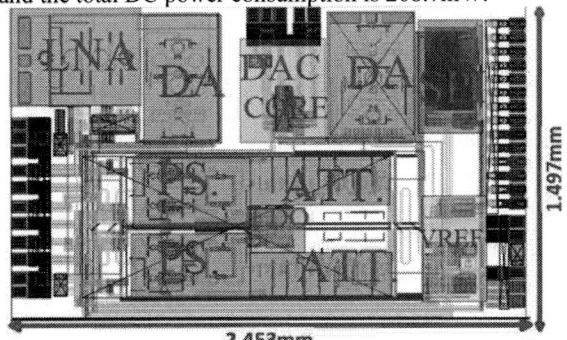

Fig. 10. The layout of the proposed 2-way phased-array ,the length of the layout is 2.453mm,the width of the layout is 1.497mm.

The comparison table is shown in Table I.

Table I. Comparison table

Ref.	This work	[3]	[4]	[5]	[6]
Tech.	65nm CMOS	45nm RFSOI	40nm CMOS	TowerJazzS BC18S3	45nm CMOS
Elements	2-way RX	4-way TRX	144 TRX	2×128-way TRX	4-way RX
Frequency (GHz)	52	60	60	62	55
Supply (V)	1.2	---	0.9	2	1.1
Gain (dB)	45	22*	38*	24.5*	26.2
BW (GHz)	4.9	8	9	7*	>20
Amplitude Error (dB)	1.13	---	1.5	1	0.9*
Phase Error (deg)	2.09	---	--	6.5	4.8*
Gain Range (dB)	33	12*	30	20	---
Phase Range (deg)	360	>360	360	360	360
NF (dB)	3.53	10	6.5	8	5.5
OP_{1dB} (dBm)	-1.12	---	---	-3.5*	-1.8*

*Estimate from the figure.

V. SUMMARY

This paper presents a compact 2-way phased array in a 65nm CMOS process. The 2-way phased array exhibited a 3-dB bandwidth of 4.9 GHz from 49.6 GHz to 54.5GHz with a 45dB gain. At 52GHz, the whole receiver achieves a NF of 3.53dB, a RMS phase error of 2.09°, and a RMS amplitude error of 1.13dB. The receiver also has a robustness to PVT variations due to the proposed compensation technique.

ACKNOWLEDGEMENT

This work is supported by the National Key R&D Program of China 2021YFB2900403.

REFERENCES

[1] M. Boers et al., "A 16TX/16RX 60 GHz 802.11ad Chipset With Single Coaxial Interface and Polarization Diversity," *IEEE J. Solid-State Circuits*, vol. 49, no. 12, pp. 3031-3045, Dec. 2014.
[2] A. A. Alhamed, O. Kazan and G. M. Rebeiz, "A Multi-Standard 15-57 GHz 4-Channel Receive Beamformer with 4.8 dB Midband NF for 5G Applications," in Proc. *IEEE/MTT-S Int. Microw. Symp. (IMS)*, Los Angeles, CA, USA, 2020, pp. 1011-1014.
[3] A. Dascurcu, S. Ahasan, A. Binaie, K. J. Lu, A. Natarajan and H. Krishnaswamy, "A 60GHz Phased Array Transceiver Chipset in 45nm RF SOI Featuring Channel Aggregation Using HRM-Based Frequency Interleaving," in *Proc. IEEE Radio Symp. Integr. Circuits (RFIC)*, Denver, CO, USA, 2022, pp. 67-70,
[4] T. Sowlati *et al.*, "A 60-GHz 144-Element Phased-Array Transceiver for Backhaul Application," *IEEE J. Solid-State Circuits*, vol. 53, no. 12, pp. 3640-3659, Dec. 2018.
[5] U. Kodak, B. Rupakula, S. Zihir and G. M. Rebeiz, "A 62 GHz Tx/Rx 2x128-Element Dual-Polarized Dual-Beam Wafer-Scale Phased-Array Transceiver with Minimal Reticle-to-Reticle Stitching," in *Proc. IEEE Radio Symp. Integr. Circuits (RFIC)*, Boston, MA, USA, 2019, pp. 335-338.
[6] S. Kundu and J. Paramesh, "A Compact, Supply-Voltage Scalable 45–66 GHz Baseband-Combining CMOS Phased-Array Receiver," *IEEE J. Solid-State Circuits*, vol. 50, no. 2, pp. 527-542, Feb. 2015.
[7] B. -H. Ku and S. Hong, "6-bit CMOS Digital Attenuators With Low Phase Variations for X-Band Phased-Array Systems," in *IEEE Trans. Microw. Theory Techn.*, vol. 58, no. 7, pp. 1651-1663, July 2010.

979-8-3503-6184-1/24 $31.00 © 2024 IEEE

A Fractional-N SPLL Using Space-time Averaging and Phase Interpolator for Quantization Noise Reduction

Shengxiang Liu[1], Ke Sun[1], Chengyu Yang[1], Dongsheng Liu[1], Ang Hu[1*]

1 School of Integrated Circuit, Huazhong University of Science and Technology, Wuhan 430074, China

* Email: ang_hu@hust.edu.cn

Abstract

The techniques of Space-Time Averaging (STA) and Phase Interpolator (PI)-based phase compensation are both widely used methods for Fractional-N phase locked loop (PLL) quantization noise reduction. The STA method has good PVT robustness but low fractional division resolution, while high-frequency PI circuits have high precision but poor PVT robustness. To address these issues, this paper proposes a quantization noise reduction technique that combines STA and PI methods, leveraging the high robustness of STA and the high precision of PI. Based on SMIC 28nm CMOS technology, a 0.9V low phase noise 5 to 8GHz Sampling PLL is designed. Pre-simulation results show that in fractional-N mode, the proposed PLL achieves 646fs rms jitter and −59 dBc fractional spur, validating the effectiveness of the proposed technique.

1. Introduction

A traditional Fractional-N PLL mainly includes a Fractional-N divider, a Phase-Frequency Detector (PFD), a Charge Pump (CP), a Loop Filter (LF), and a Voltage-Controlled Oscillator (VCO). Figure 1 shows the structure of the basic Fractional-N PLL, where the Delta-Sigma Modulator (DSM) performs time averaging of the division ratio, forming the Fractional-N divider in conjunction with the Multi-Modulus Divider (MMDIV).

The DSM can perform time averaging of the input division ratio, ensuring that the DSM output's average value is consistent with the input value. However, since the MMDIV can only achieve integer division, the output of the DSM must be an integer. For example, if the input to a first-order DSM is 3.25, the output will be a periodic sequence like [3, 3, 3, 4, 3, 3, 3, 4, ...]. Figure 2 shows the time-domain waveform of the output signal of the basic traditional fractional-N phase-locked loop when the input division ratio is 3.25. It can be observed that the divider output clock aligns with the reference clock every 13 VCO cycles. However, within these 13 cycles, there will be a certain deviation between the divider output clock and the reference clock. This deviation is converted into the control voltage of the VCO through the PFD, CP, and LF, ultimately manifesting as fluctuations in the output frequency of the PLL, which results in phase noise [1].

To address the cause of quantization noise, this paper proposes a quantization noise reduction method that combines STA and PI. By reducing the division step of the divider, this approach reduces quantization noise at its source. The structure of this paper is as follows:Section 2 introduces the architecture of the fractional-N SPLL combining STA and PI.Section 3 presents the design of the key circuit modules.Section 4 provides the simulation results.Section 5 concludes the paper.

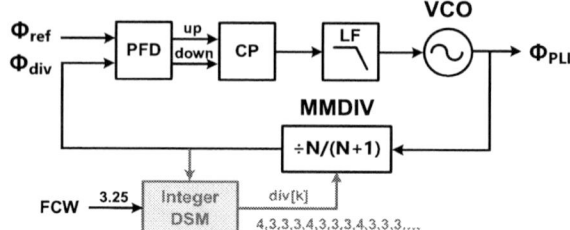

Fig. 1　Structure of the basic Fractional-N PLL

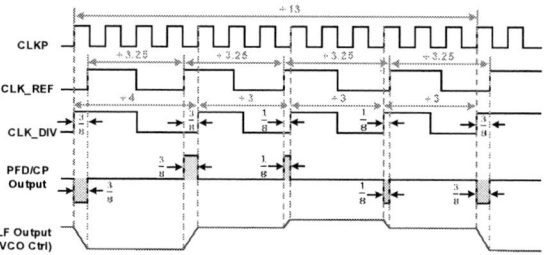

Fig. 2　Time-Domain waveform of the basic Fractional-N PLL with a Division Ratio of 3.25

2. The architecture of the Fractional-N SPLL combining STA and PI

The proposed Fractional-N SPLL architecture combining STA and PI is shown in Figure 3. The input frequency control word (FCW) is converted into the M-bit frequency divider array $\overrightarrow{DivN_t[k]}$ through the DSM and dynamic element matching (DEM) module. And $\overrightarrow{DivN_t[k]}$ is then input into the vector divider to achieve space averaging of the division ratio. Additionally, a high-precision PI circuit is used to perform phase interpolation on the output of the frequency divider array. The final output phase is the arithmetic mean of the phases from the vector divider.

Since the PI only requires two-phase clocks as input, the Vector Divider in the STA method can be replaced by a single divider. The Division ratio array $\overrightarrow{DivN_t[k]}$ is processed through the Phase Select (PS) module, which provides the PI control word. After above improvements, the block diagram of proposed STA+PI technique is shown in Figure 4. The proposed STA+PI technique works as follows:Firstly FCW is divided into integer and fractional parts, represented as $d_{inte}[k]$ and $d_{frac}[k]$ respectively.The fractional part $d_{frac}[k]$ is processed through the DSM and DEM modules, converting it into a M-bit integer value vector $\overrightarrow{N_{DEM}[k]}$. And then it converted by the PS module into the Delay Control Word (DCW).Finally, $\overrightarrow{N_{DEM}[k]}$ and $d_{inte}[k]$ are added to

979-8-3503-6184-1/24 $31.00 © 2024 IEEE

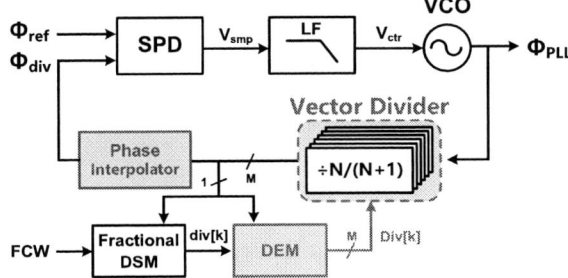

Fig.3 Block Diagram of the STA+PI SPLL Architecture

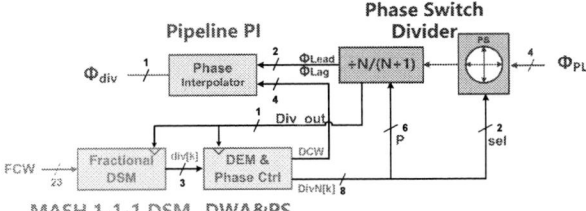

Fig.4 Block Diagram of the proposed STA+PI technique

obtain the vector division ratio $\overrightarrow{DivN_i\,[\,k\,]}$. Mathematically, we have:

$$DivN_i[k] = d_{inte}[k] + N_{DEM,i}[k] \qquad (1)$$

$\overrightarrow{DivN_i\,[\,k\,]}$ and the four quadrature clocks $\overrightarrow{\varphi_{PLL}}$ are sent to the phase-switching divider, producing three output clocks: Div_out, φ_{lead} and φ_{lag}. Div_out serves as the system clock for the divider's digital module, while φ_{lead} and φ_{lag} along with the DCW, are sent to the PI circuit, ultimately outputting a low phase noise clock signal φ_{div}.

The time-domain waveform in Figure 5 demonstrates how the PI circuit interpolates the output clocks ($CLK\_\#1$ to $CLK\_\#4$) from the four channels, resulting in a final clock (CLK_PI) with a phase ideally matching the phase of the reference clock. Consequently, the VCO control voltage, obtained after the SPD and LF, remains stable without fluctuations, indicating that the divider does not introduce additional phase noise to the SPLL.

Moreover, we can also derive the effectiveness of quantization noise reduction in STA+PI from the frequency domain. By modeling the quantization error of the DSM as e_q, which is high-pass shaped by its noise transfer function (NTF) $NTF_{DSM}(z)$, we can understand its impact on the system. Since the quantization step of the DSM in the proposed STA+PI SPLL is $\frac{1}{M}$, the magnitude of e_q is $\frac{1}{M}$ of that in a traditional Fractional-N SPLL. Assuming e_q follows uniform distribution and $NTF_{DSM}(z) = (1 - z^{-1})^L$, the power spectral density (PSD) of the quantization noise at the output of the STA+PI PLL can be derived as follows:

$$S_{QN}(f) = \frac{1}{M^2}\frac{\pi^2 T_{ref}}{3}\left|\frac{G_{close}(f)}{N+\alpha}\right|^2 |1 - z^{-1}|^{2L-2} \qquad (2)$$

This indicates that comparing to a traditional Fractional-N SPLL, the M channels STA+PI Fractional-N SPLL reduces quantization noise by $20\log_{10} M$ dB at all frequencies.

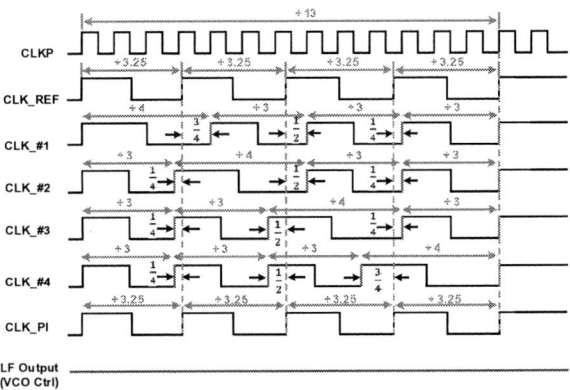

Fig.5 Time-Domain Waveform of the STA+PI SPLL

The proposed STA+PI architecture replaces the PFD and CP array in the STA PLL with a high-precision PI circuit, achieving higher frequency division accuracy compared to the STA PLL [1] and better mismatch robustness compared to PI-based PLLs [2]. Unlike traditional PI-based PLLs, the PI circuit in this architecture is working under the low frequency clock divided by the MMDIV, significantly reducing the power consumption of the PI circuit. The main advantage of this architecture ensures low quantization noise while achieving good robustness, and at the same time has a moderate area and power consumption.

3. The design of the key modules

The key modules of the proposed STA+PI technique include the phase-switching divider, MASH 1-1-1 DSM, pipelined PI, and Data Weighted Averaging & Phase Select (DWA&PS) module.

The structure of the phase-switching divider [3] is shown in Figure 6. We uses forward phase-switching logic to achieve switching between 1/4 phase steps. This allows for the minimum division step of 1/4 to be obtained with the addition of simple digital circuits.

The structure of the MASH 1-1-1 DSM used in this paper is shown in Figure 7. In addition, the Linear Feedback Shift Register (LSFR) dither is added to the inputs of the second and third stages, which greatly suppresses the periodicity of the DSM output and reduces the peak value of fractional spurs[4].

Figure 8 shows the block diagram of the DWA&PS module. The input of the module is the instantaneous division ratio $div[k]$ obtained from the DSM. The module outputs two signals, $DivN[k]$ and DCW. $DivN[k]$ serves as the instantaneous division ratio input to the phase-switching divider, while DCW is used as the phase shift control word input to the PI circuit. This work employs a simple first-order DEM, also known as DWA technology, to achieve space averaging of the crossover ratio.

Figure 9 shows the structure of the 4-bit pipelined PI circuit [5] used in this paper. The idea of the pipeline structure is that the PI circuit only needs to output one phase that is used at any sampling time. The number of PI units used in the N-bit PI circuit is only N, which greatly reduces the area and power consumption of the PI circuit.

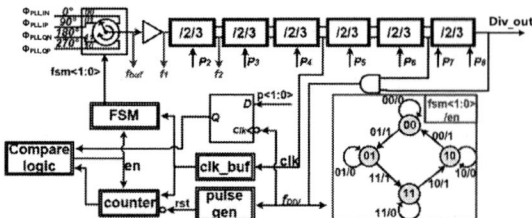

Fig. 6 Structure of the Phase-Switching Divider

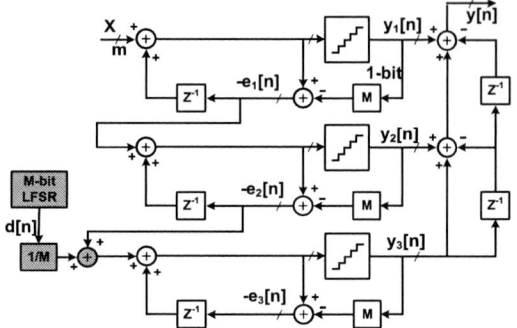

Fig. 7 Structure of the MASH 1-1-1 DSM

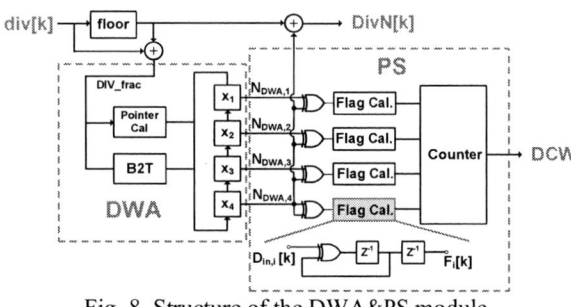

Fig. 8 Structure of the DWA&PS module

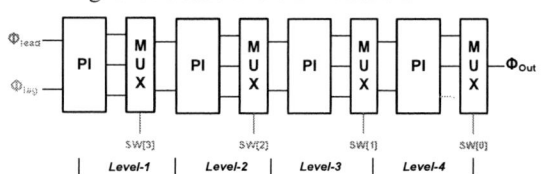

Fig.9 The structure of the 4-bit pipelined PI circuit

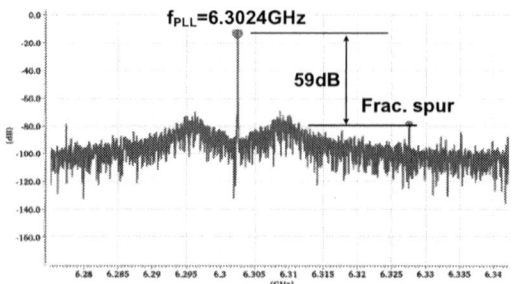

Fig.10 Output Signal Spectrum of the STA+PI SPLL

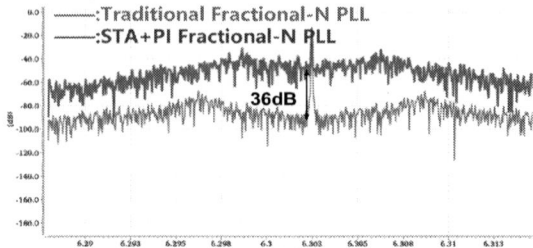

Fig.11 Comparison of Output Spectrum Between STA+PI SPLL and Traditional SPLL

5. Conclusion

This paper proposes a quantization noise reduction technique that combines STA and PI. By using SMIC 28nm technology, the SPLL achieved a fractional spur magnitude of -59 dBc and an integrated jitter of 646 fs from 1 kHz to 50 MHz, validating the effectiveness of the STA+PI technique.

Acknowledgement

This work is supported by the National Key Research and Development Program of China (No. 2021YFA0715502), the National Natural Science Foundation of China (No.62374064, No. 62104076, 62134002), the National Key Analog Integrated Circuit Laboratory Project of China (No. JCKY2021210C004), the Introduced Innovative R&D Team of Dongguan (No.201760712600139)

References

[1] Y. Zhang et al., "A Fractional- N PLL With Space–Time Averaging for Quantization Noise Reduction," in IEEE Journal of Solid-State Circuits, vol. 55, no. 3, pp. 602-614, March 2020

[2] R. K. Nandwana et al., "A Calibration-Free Fractional-N Ring PLL Using Hybrid Phase/Current-Mode Phase Interpolation Method," in IEEE Journal of Solid-State Circuits, vol. 50, no. 4, pp. 882-895, April 2015

[3] A. Hu, D. Liu, K. Zhang, L. Liu and X. Zou, "A 0.045- to 2.5-GHz Frequency Synthesizer With TDC-Based AFC and Phase Switching Multi-Modulus Divider," in IEEE Transactions on Circuits and Systems I: Regular Papers, vol. 67, no. 12, pp. 4470-4483, Dec. 2020

[4] V. R. Gonzalez-Diaz, M. A. Garcia-Andrade, G. E. Flores-Verdad and F. Maloberti, "Efficient Dithering in MASH Sigma-Delta Modulators for Fractional Frequency Synthesizers," in IEEE Transactions on Circuits and Systems I: Regular Papers, vol. 57, no. 9, pp. 2394-2403, Sept. 2010

[5] A. Tharayil Narayanan et al., "A Fractional-N Sub-Sampling PLL using a Pipelined Phase-Interpolator With an FoM of -250 dB," in IEEE Journal of Solid-State Circuits, vol. 51, no. 7, pp. 1630-1640, July 2016

4. Simulation Result

The input reference frequency of the designed STA+PI SPLL is 195 MHz, and the output frequency range from 5 to 8 GHz. Figure 10 shows the output signal spectrum of the STA+PI SPLL. The maximum fractional spur is located at 25 MHz from the center frequency, with a magnitude of -59 dBc.

Figure 11 shows the comparison of the spectrum between traditional fractional-N SPLL and the SPLL using the 16-channel STA+PI method, at an output frequency of 6.3024 GHz and a division ratio of 32.32. It can be seen that the noise floor of the SPLL using the 16-channel STA+PI method is reduced by approximately 36 dB compared to the traditional fractional-N SPLL. The simulation results are similar to the result $20\log_{10} 64\, dB = 36dB$ derived from the formula , validating the effectiveness of the STA+PI technology. And the simulated results indicate 0.86ps integrated rms jitter from 1kHz to 50MHz at 6.3024GHz.

A 18V, 600mA Load Current, 22MHz High-Voltage Power Amplifier with Over-Temperature Protection and Bidirection Enable Logic

Yuan Ren , Xin'an Wang*

Shenzhen Graduate School, Peking University, Shenzhen, China

* Email: anxinwang@pku.edu.cn

Abstract—This paper presents a rail-to-rail high-voltage power operational amplifier for industrial driving applications. The quiescent consumption of it is 4mA and 0.28mA when it is externally shutdown. The chip is designed with high Gain-Bandwidth Product and high Slew Rate which are 22MHz and 61V/us, respectively. Besides, many special circuit blocks have been mentioned and some application problems have been discussed in this paper. The design chip is based on 180-nm CMOS process, its supply voltage ranging from 4.5V to 18V and has good performance from -40°C to 125°C. Test results show that it has excellent over-temperature protecting and continuous current outputting capability.

Keywords—high voltage, rail-to-rail amplifier, protection circuits, output current limitation design, over-temperature protection

I. INTRODUCTION

Signal chain chip products are widely used in many industrial fields, and amplifiers, as the cornerstone of analog circuits, have more complex application situations and needs[1]. The power op-amps have both strong output driving capability and signal processing function, which can freely output current to meet the application requirements.

The power op-amp in this paper is designed using a high-voltage BCD process. In addition, it will have serious heat generation while working, so it puts strict requirements on the package thermal resistance and heat dissipation ability[2].The design difficulties of this chip are mainly in the following three respects:

1) Chip power consumption and stability.
Parasitic capacitance severely affect loop phase margin. Requiring a compromise between power consumption and chip area budgets.

2) Chip high voltage operating state.
The gate-substrate breakdown voltage of those devices is still around 5V, so special attention needs to be paid to protect these internal circuit nodes from device breakdown[3].

3) Chip over-temperature function and logic read/write.
The low-power standby control port has an input/output function. The chip also enters a low-power shutdown state in the over-temperature condition, which requires logic protection and leakage protection[4], [5].

II. THE PROPOSED POWER OP-AMP

A. Overview architecture
The proposed op-amp consists of Bandgap, LDO, Core Amplifier, Gain Boost, Slew Rate Enhancement, Input Protection, Over-temperature Protection, Output Current Limitation, as well as functional modules such as Enable logic and Soft-start. Fig.1 illustrates the overall block diagram.

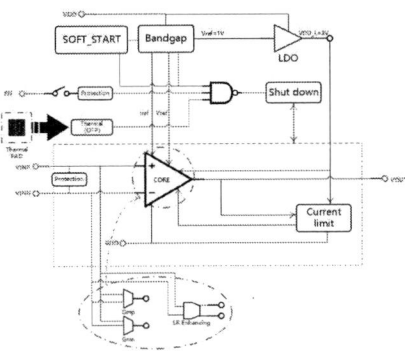

Fig. 1. The overview architecture of the proposed op-amp

Bandgap block provides the reference voltage and reference current for the whole chip. The reference voltage is processed and outputted through LDO, and the low voltage potential obtained is used as the power supply source for the low voltage reference module and the low voltage logic circuits. The signal flows into the core amplifier and passes through the input protection module to reach the input stage. The output stage has a CLASS AB structure and is protected by POR signal and output limitation signal. The Over-temperature Protection block contains a separate bandgap that ensures the on-time, stable, and accurate temperature detection.

B. Bandgap block design
Bandgap block should be protected strictly from possible high-voltage breakdown. Three protection structures are shown in Fig. 2. To simplify the image, the protection modules in the circuit are indicated by arrows and pointed out the direction. If it is no longer mentioned, the protection of subsequent circuits is no longer plotted. Fig. 3 shows the specific structure of Bandgap. High-voltage MOS devices in the figure drain extended to indicate.

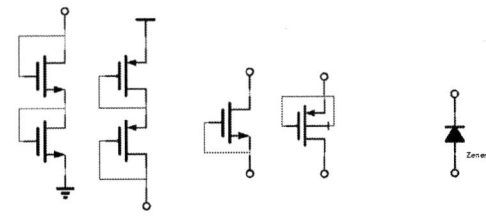

Fig. 2 High voltage protection modules

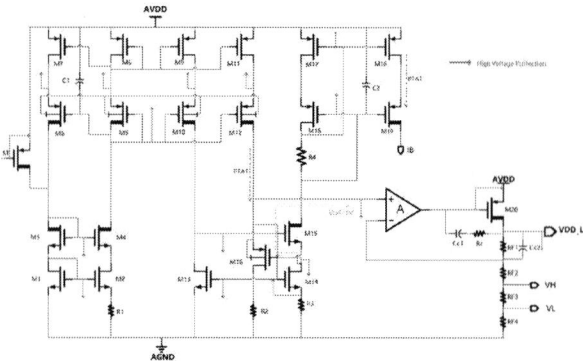

Fig. 3 Bandgap block

The ST signal is given by the startup circuit. M1 and M2 operate in the subthreshold region with triode characteristics. M3 and M4 further reduce the influence of the channel-length modulation effect, and the voltage difference of V_{gs} generates PTAT current across the resistor R1, as shown in equation (1):

$$I_{ref} = \ln N * \xi V_T / R1 \qquad (1)$$

N represents the multiplicity of M2 over M1. The LDO adopts high-voltage devices as the input ports, inputting 1V zero-temperature coefficient voltage, outputting 3V low-voltage potential, as well as the logic hysteresis interval reference potentials of 1.2V and 0.5V for the Enable Logic. The LDO is powered individually, with internal high-voltage structures. The Bode figure and the simulation waveform of the power-up process is shown in Fig. 4.

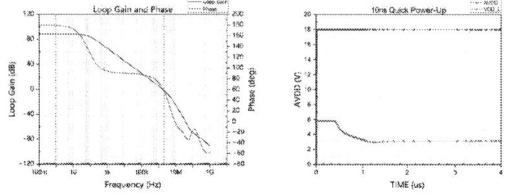

Fig. 4 Bode and Power-up waveform

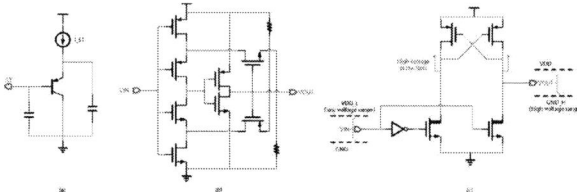

Fig. 5 Logic modules

C. Function block design

The design of functional blocks mainly includes Current Distribution, Soft-start, Schmitt Hysteresis Comparator, Level Shifter, I/O Logic and so on, as shown in Fig. 5. Each module uses the local current mirror and the low voltage potential of the LDO for power supply to output reference current.

D. Rail-to-rai OPA block design

The input common-mode voltage range is designed from VSS-0.5V to VDD+0.5V. The final V_{os} voltage for the input stage is calculated to be 1.2mV.

The intermediate stage circuit needs to keep the current stable over the full input common-mode voltage range. In this

work, the feedback loop is used, one of its OTA is shown in Fig. 6(a). M25 and M26 are used to further enhance the OTA gain, so that this loop always has a good clamping capability under the variation of PVT. In addition, the intermediate stage circuit has Gain Boost modules, one of which is structured as shown in Fig. 6(b).

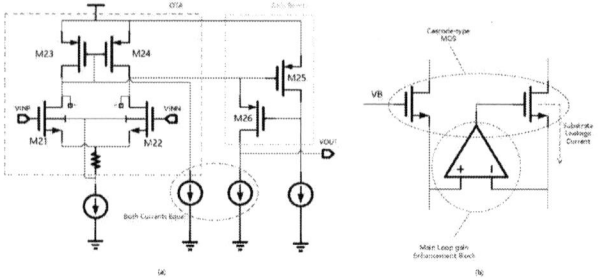

Fig. 6 The OTA and Gain Boost loop of the intermediate stage

The output stage is CLASS AB structure. The simulated output short-circuit current is shown in Fig. 7, and the instantaneous output current of the chip can reach 800mA in load response.

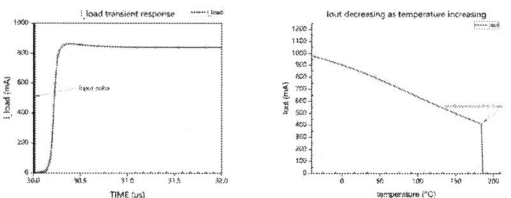

Fig. 7 The output current characteristics

E. Protection and logic block design

The Over-temperature Protection block consists of separate reference, temperature sensor. The EN port has a discharge path and current limiting resistor. The maximum output current of a microcontroller port is usually 5mA, EN port's sinking current capability must stronger than this number, so the chip's external logic unit can be driven down. The *STANDBY* signal is derived from the core amplifier to prevent logic contention during power-up process.

III. VERIFICATION RESULTS

Fig. 8 shows the overall layout of the chip, including blocks such as ESD, Bandgap, and Core.

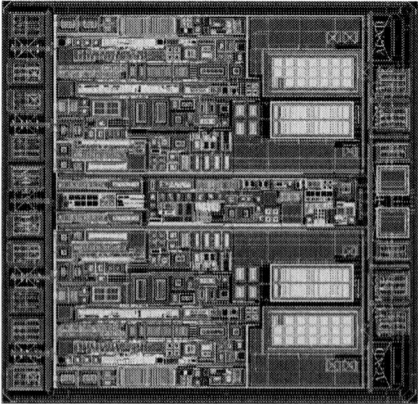

Fig. 8 Layout of the OPA

979-8-3503-6184-1/24 $31.00 © 2024 IEEE

TABLE. 1 PERFORMANCE PARAMETER

Parameter	Condition	Test (TYP)	Unit
Offset voltage	VS=18V	0.33	mV
Quiescent current	VS=18V, RL=10KΩ, CL=80pF, VCM=VS/2	4.05	mA
Gain-Bandwidth Product	VS=18V, RL=10KΩ, CL=80pF, VCM=VS/2	22	MHz
Phase Margin	VS=18V, RL=10KΩ, CL=80pF, VCM=VS/2	67	deg
Slew Rate	VS=18V, RL=10KΩ, CL=80pF, VCM=VS/2	61	V/us
CMRR	VS=18V, VCM=(V-) - 0.2V to (V+) + 0.2V	118	dB
	VS=18V, VCM=(V-) - 0.5V to (V+) + 0.5V	96	dB
PSRR	VS=4.5V to 18V, VCM=VS/2	113	dB
AOL	VS=18V, RL=10KΩ, VO=(V-) + 0.1V to (V+) - 0.1V	152	dB
Output current	VS=18V	576	mA
Noise	VS=18V	11.3	uV$_{pp}$
Thermal shutdown/recovery	VS=18V	175/150	°C
Shutdown current	VS=18V, EN=GND	0.28	mA

The package is ESOP16 with a grounded heat sink pad. The EN pin logic level is given with reference to the test environment VSS. Fig. 9 and Fig. 10 give the sustained output waveforms of the chip while over-temperature. The performance data of the chip are in Table. 1. The test data are the average values of 5 randomly selected samples.

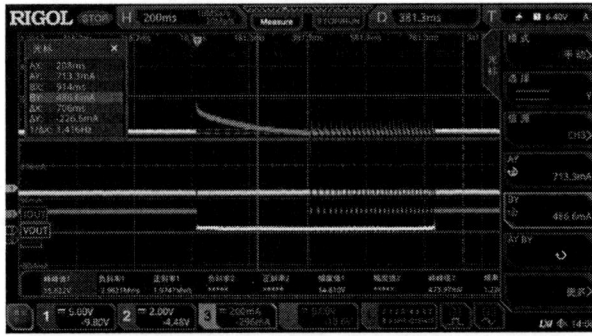

Fig. 9 Over-temperature output source current waveforms

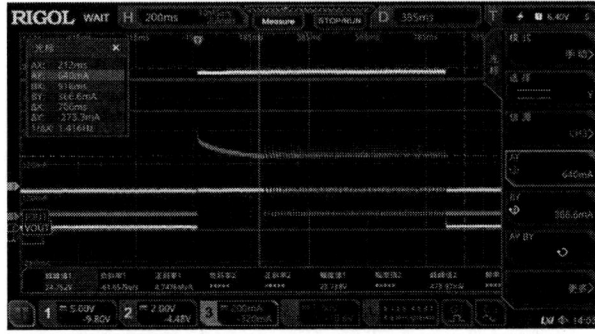

Fig. 10 Over-temperature output sink current waveforms

IV. CONCLUSION

The high-voltage power operational amplifier described in this paper is designed using 180nm high-voltage BCD process, operates with 18V supply voltage, has overall power consumption of 4mA and a shutdown power consumption of 0.28mA, has low offset voltage and low noise characteristics. It operates at temperatures ranging from -40°C to 125°C, and has an over-temperature protection hysteresis interval of 150°C to 175°C. The performance of the chip has been verified through simulation design, layout drawing, post-simulation design and chip test. The signal processing capability of 22MHz bandwidth and the outputting capacity of 600mA enable the chip to be applied in several power driving and signal sensing situations and controlled by logic units such as microcontrollers.

ACKNOWLEDGMENT

This work is supported by Science and Technology Planning Project of Shenzhen Municipality, Project No.KQTD20200820113105004.

REFERENCES

[1] X. Wang, X. Zhu and S. Diao, "A Unity-Gain Buffer with 1.6GHz Bandwidth and 1900V/μs Slew Rate based on an OPA Adopting Cascaded Class-AB Structure," 2022 15th International Congress on Image and Signal Processing, BioMedical Engineering and Informatics (CISP-BMEI), Beijing, China, 2022, pp. 1-4, doi: 10.1109/CISP-BMEI56279.2022.9979885.

[2] D. Bianchi, F. Quaglia, A. Mazzanti and F. Svelto, "A 90Vpp 720MHz GBW linear power amplifier for ultrasound imaging transmitters in BCD6-SOI," 2012 IEEE International Solid-State Circuits Conference, San Francisco, CA, USA, 2012, pp. 370-372, doi: 10.1109/ISSCC.2012.6177054.

[3] Q. Wu and J. Wu, "High Precision Voltage Sampling Circuit for Battery Management System," 2024 IEEE 7th Advanced Information Technology, Electronic and Automation Control Conference (IAEAC), Chongqing, China, 2024, pp. 713-717, doi: 10.1109/IAEAC59436.2024.10503982.

[4] M. H. Hedayati, J. Wang, H. C. P. Dymond, D. Liu and B. H. Stark, "Overtemperature Protection Circuit for GaN Devices Using a di/dt Sensor," in IEEE Transactions on Power Electronics, vol. 36, no. 7, pp. 7417-7428, July 2021, doi: 10.1109/TPEL.2020.3041594.

[5] S. Pashmineh, S. Bramburger and D. Killat, "Design of a high-voltage rail-to-rail error amplifier based on standard CMOS used in an LDO," 2016 IEEE Canadian Conference on Electrical and Computer Engineering (CCECE), Vancouver, BC, Canada, 2016, pp. 1-5, doi: 10.1109/CCECE.2016.7726693

A 47μW Wake-Up Receiver With -77dBm Sensitivity Using a Mixer-First Architecture

Weitao He [1], Yaxin Zeng [1], Bin Jia [2], Hao Min [1], Hao Xu [1], Na Yan[*1]

[1] State Key Laboratory of Integrated Chips and Systems, Fudan University, Shanghai 200433, China
[2] EPIC MEMS Corporation, Xiamen, 361026, China

* Email: yanna@fudan.edu.cn

Abstract—This paper presents a wake-up receiver for IoT(Internet-of-Things) applications operating at 500kbps and basing on OOK (on-off keying) modulation. The system uses a mixer-first structure combined with a high-Q off-chip matching network including a BAW filter. A single balanced active mixer is utilized to amplify and down-convert the signal, enhancing the power efficiency of the IF amplifier and envelope detector. The circuit is implemented in 40nm CMOS technology and operates in the 2.4GHz IoT frequency band. The simulation results using extracted layout shows that the receiver core consumes only 47μW and achieves a sensitivity of -77dBm at 500kbps(10^{-3} bit error rate) with a 1.1V supply. The chip area is only $0.032mm^2$.

Keywords—wake-up receiver, Internet of Things, ultra-low power wireless, BAW resonator, mixer-first, amplifier, envelope detector

I. INTRODUCTION

The rapid advancement of the Internet of Things (IoT) has significantly increased the deployment of Wireless Sensor Networks (WSNs), essential for applications ranging from home automation to industrial monitoring. A critical challenge in WSNs is extending the limited battery life of each sensor node, which is often deployed in large numbers and hard-to-reach locations. Since each node's RF receiver consumes energy when it is turned on, the most effective way to reduce power consumption is to turn off the receiver and wake it up only when needed. A dedicated wake-up receiver can be used to continuously monitor the channel and activate the main receiver after detecting the wake-up signal, thereby greatly extend the battery life of each node [1].

Direct envelope detection receiver can be implemented with RF amplification and an envelope detector, which consumes tens or hundreds of nanowatts at the expense of low sensitivity [2, 3]. In order to further balance sensitivity and power consumption, super-regenerative receiver [4], uncertain-IF(Intermediate Frequency) receiver [5] and low-IF receiver [6] have been proposed. Super-regenerative receiver is based on injection locking and can provide large gain through a high-Q oscillator, but this architecture is susceptible to large signal blocking. The uncertain-IF receiver reduces power consumption by simplifying clock generation circuit, but requires larger IF bandwidth, resulting in higher power consumption of the IF amplifier compared to the low-IF receiver. In this paper, an ultra-low power wake-up receiver with a low IF structure is implemented.

The structure of this paper is organized as follows. Section II introduces the detailed circuit design of the wake-up receiver. Section III presents the post-simulation results of the proposed receiver circuit in 40nm CMOS technology. Section IV presents the conclusion.

II. CIRCUIT DESIGN

As shown in Fig.1, the ultra-low power receiver consists of an active mixer, an intermediate frequency (IF) amplifier and an envelope detector. Since the active mixer used in this design has poor selectivity, a matching network with an off-chip BAW filter are required. Channel 7 in the WiFi 802.11 protocol is selected, thus the matching network is tuned to a center frequency of 2.442GHz. The signal is sent to the computer for further digital processing. As shown in Fig.2, the mixer receives the high-frequency signal and moves it to the intermediate frequency of 2M-6MHz. The amplifier amplifies the signal on the intermediate frequency to ensure that the envelope detector can demodulate a recognizable envelope signal.

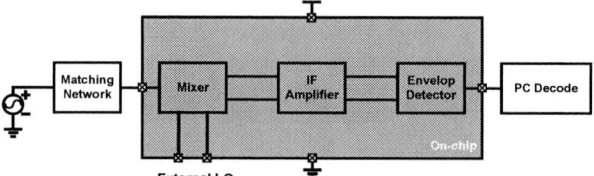

Fig. 1. Structure of the proposed wake-up receiver.

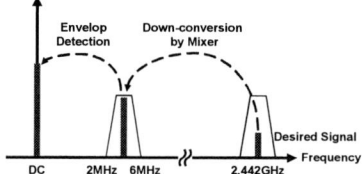

Fig. 2. Frequency translation of the wake-up receiver

A. Matching Network and Active Mixer

In order to suppress more out-of-band interferences, an off-chip BAW filter is introduced, and off-chip inductors are used to match the input as shown in Fig.3 (a). While improving the selectivity of the input, it also provides a passive voltage gain for the input signal. The BAW filter model of EPIC EP7EW2 is used here.

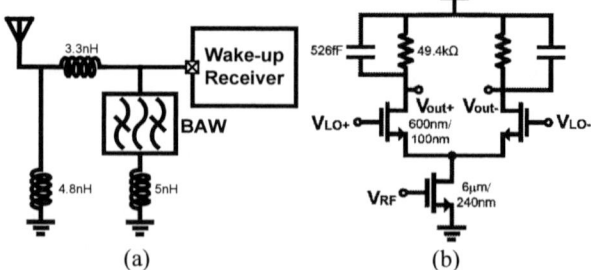

Fig. 3. (a) Matching network of the wake-up receiver, (b) Circuit structure of the single balanced mixer.

979-8-3503-6184-1/24 $31.00 © 2024 IEEE

In order to obtain high voltage gain and linearity, a single balanced mixer as shown in the Fig.3 (b) is selected. The input stage transistor is sized of 6μm/240nm and operates in the subthreshold region. The switching stage transistors are sized of 600nm/100nm to reduce the gate capacitance. In order to suppress out-of-band signals, a 526fF capacitor and a 49.4kΩ resistor are added to the load stage of the mixer to form a low-pass filter. At the same time, the high-frequency harmonics can be filtered out by the multi-stage amplifier and the envelope detector.

B. IF Amplifier

As shown in Fig.4, the IF amplifier is based on three-stage fully differential operation amplifiers to provide a gain of more than 50dB. In order to reduce power consumption, the fully differential operation amplifier with a resistive load is utilized to avoid the extra power consumption caused by common-mode feedback and bias circuits. The first and third stage operation amplifiers has a tail current source, and the intermediate coupling capacitor introduces a DC zero point in the transfer function to achieve a bandpass amplification function. The second stage operation amplifier with only one tail current source to provide a stable gain. The three-stage amplifiers together form a bandpass filter with a bandwidth of 2-6MHz.

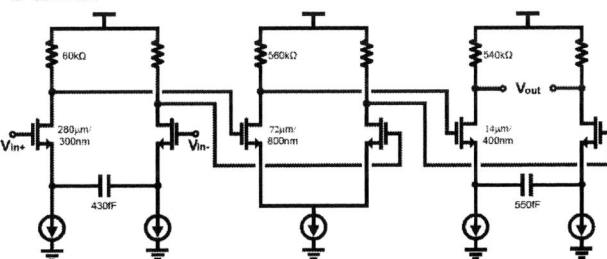

Fig. 4.　Circuit structure of the IF amplifier.

C. Envelope Detector

As the last stage of the system, the envelope detector uses the nonlinearity of MOS to demodulate the signal and extract the envelope of the useful signal. Considering the need for the detection efficiency, a common-source envelope detector is selected. As shown in Fig.5, the envelope detector is made of a pair 2μm/200nm transistors and a 300kΩ resistor.

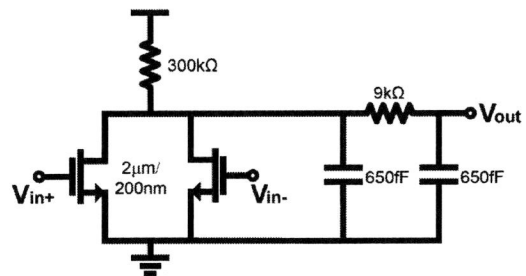

Fig. 5.　Circuit structure of the envelope detector.

Assume that the input signal is a small signal and can be expressed as:

$$v_i(t) = V_{cm}(1 + m_a \cos\Omega t)\cos\omega_c t \qquad (1)$$

Where V_{cm} represents the common mode voltage, Ω is the modulation frequency, m_a is the modulation coefficient, and ω_c is the carrier frequency. The transistors of the envelope detector works in the subthreshold region, we can get:

$$i_D = a_0 + a_1 v_i + a_2 v_i^2 + a_3 v_i^3 + \cdots \qquad (2)$$

Substituting formula (1) into (2), we can get the envelope signal we need:

$$v_{AV\Omega} = a_2 m_a V_{cm}^2 R\cos\Omega t \qquad (3)$$

R in the formula represents the load resistance. In addition to the envelope signal, nonlinearity will also generate high-frequency signals of $\cos 2\Omega t$. In order to filter out high-order harmonics and local oscillator penetration signals, a low-pass filter consisting of two 650fF capacitors and a 9kΩ resistor is set at the output of the envelope detector.

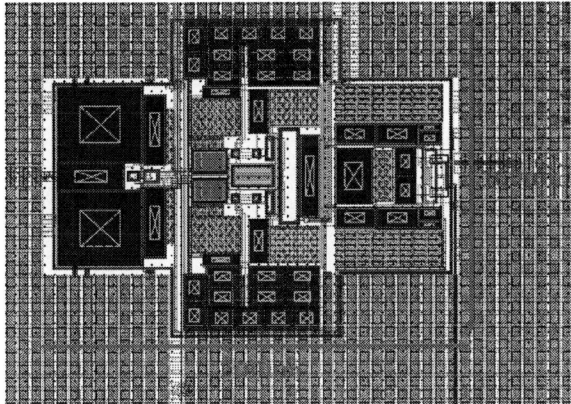

Fig. 6.　Layout of the wake-up receiver

III. SIMULATION RESULTS

The proposed wake-up receiver is designed in 40nm CMOS technology. The layout is shown in Fig.6. The ports reserved for the differential clock signals are located at the top of the layout. The clock signals are connected to the gates of the mixers through symmetrical routing.

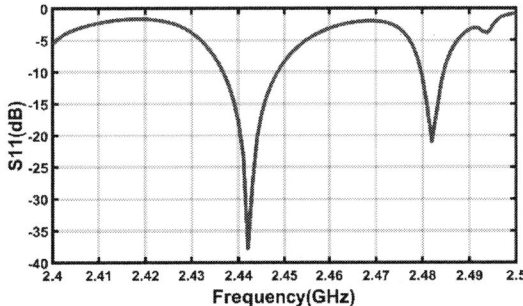

Fig. 7.　Input return loss of the wake-up receiver.

Simulation result of input return loss is shown in Fig.7. Through the matching network composed of off-chip inductors and BAW filter, the 200fF pad capacitance and 800pH bonding inductance are considered at the same time. The simulation shows that S11 is less than -10dB in the frequency range of 2.436GHz-2.448GHz. The input return loss at the center frequency of 2.442 GHz is -38dB.

In order to consider the actual load conditions, a 1pF capacitor is connected at the output of the wake-up receiver. The transient simulation result of -77dBm input of the wake-up receiver is shown in Fig. 8. The final output of the envelope

979-8-3503-6184-1/24 $31.00 © 2024 IEEE　　　716

detector has an amplitude of 340mV and the SNR of the output is 15.5dB. The comparison table is shown in table1.

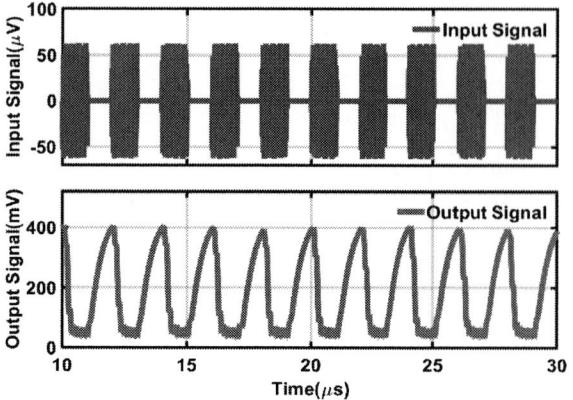

Fig. 8. Transient simulation result of the wake-up receiver

TABLE I. COMPARISON WITH THE STATE-OF-THE-ART

		This Work	[7]	[8]
CMOS Tech.(nm)		40	14	180
Architecture		Mixer-First	Mixer-First	Mixer-First
Supply(V)		1.1	0.95	1
Modulation		OOK	OOK	FSK
Carrier Frequency(GHz)		2.4	2.4	0.9
Data Rate(kbps)		500	62.5	/
Sensitivity(dBm)@10^{-3}		-77	-72	-87
Rx Gain(dB)		68	37	41.5
Power (uW)	Total	47	95*	220
	Mixer	21.5	/	/
	IF-Amplifier	22.2	/	/
	Envelope Detector	3.3	/	/
Required External Components		BAW	32kHz Real Time Clock	Transform

*With Local Oscillator

IV. CONCLUSION

A 1.1V wake-up receiver implemented in 40nm CMOS is proposed in this paper. The wake-up receiver uses an off-chip BAW filter and inductors as a matching network. The receiver consists of a single balanced mixer, an IF amplifier, and an envelope detector. The receiver occupies an area of 0.032mm². From the post-layout simulation results, the receiver core consumes only 47µW and achieves a sensitivity of -77dBm at 500kbps and OOK modulation, which is suitable for IoT applications.

REFERENCES

[1] J. M. Rabaey, J. Ammer, T. Karalar, Suetfei Li, B. Otis, M. Sheets, T. Tuan, "PicoRadios for wireless sensor networks: the next challenge in ultra-low power design," *2002 IEEE International Solid-State Circuits Conference. Digest of Technical Papers (Cat. No.02CH37315)*, San Francisco, CA, USA, 2002, pp. 200-201 vol.1.

[2] X. Huang, A. Ba, P. Harpe, G. Dolmans, H. de Groot and J. R. Long, "A 915 MHz, Ultra-Low Power 2-Tone Transceiver With Enhanced Interference Resilience," in IEEE Journal of Solid-State Circuits, vol. 47, no. 12, pp. 3197-3207, Dec. 2012.

[3] P. -H. P. Wang et al., "A Near-Zero-Power Wake-Up Receiver Achieving −69-dBm Sensitivity," in IEEE Journal of Solid-State Circuits, vol. 53, no. 6, pp. 1640-1652, June 2018.

[4] V. Dabbagh Rezaei and K. Entesari, "A Fully On-Chip 80-pJ/b OOK Super-Regenerative Receiver With Sensitivity-Data Rate Tradeoff Capability," in IEEE Journal of Solid-State Circuits, vol. 53, no. 5, pp. 1443-1456, May 2018.

[5] N. M. Pletcher, S. Gambini and J. Rabaey, "A 52 µ W Wake-Up Receiver With − 72 dBm Sensitivity Using an Uncertain-IF Architecture," in IEEE Journal of Solid-State Circuits, vol. 44, no. 1, pp. 269-280, Jan. 2009.

[6] K. -M. Kim *et al.*, "A −124-dBm Sensitivity Interference-Resilient Direct-Conversion Duty-Cycled Wake-Up Receiver Achieving 0.114 mW at 1.966-s Wake-Up Latency," in *IEEE Journal of Solid-State Circuits*, vol. 58, no. 6, pp. 1667-1680, June 2023.

[7] E. Alpman *et al.*, "802.11g/n Compliant Fully Integrated Wake-Up Receiver With −72-dBm Sensitivity in 14-nm FinFET CMOS," in *IEEE Journal of Solid-State Circuits*, vol. 53, no. 5, pp. 1411-1422, May 2018.

[8] O. Elsayed, M. Abouzied, V. Vaidya, K. Ravichandran and E. Sánchez-Sinencio, "An Ultralow-Power RF Wireless Receiver With RF Blocker Energy Recycling for IoT Applications," in *IEEE Transactions on Microwave Theory and Techniques*, vol. 66, no. 11, pp. 4927-4942, Nov. 2018.

979-8-3503-6184-1/24 $31.00 © 2024 IEEE

A Ka-Band CMOS Broadband Power Amplifier with 35.3% PAE for SATCOM Applications

Zhiqing Liu*, Yu Chu, Yuting Sun

Southwest China Institute of Electronic Technology, Chengdu 610036, China

* Email: jxlzq2023@163.com

Abstract—**This paper presents a Ka-band power amplifier (PA) for satellite communication (SATCOM) applications in 65nm CMOS process. In this design, active amplification stages adopt an optimized power cell topology with the design method of source-drain swap and the neutralization technique, leading to a low loss and a high efficiency. For the passive matching network, a transformer-based magnetically coupled resonator (MCR) is utilized. By constructing the transmission response (S21) with two peaks in operation frequency band, the circuit realizes a large bandwidth and an excellent in-band flatness. According to the measurement results, the PA achieves a gain of 23dB, a peak power added efficiency (PAE) of 35.3%, a gain flatness of less than 0.5dB within the application frequency band. The fractional bandwidth of the PA is 30% from 24GHz to 32.5GHz.**

Keywords—*CMOS PA, SATCOM, source-drain swap, MCR, broadband*

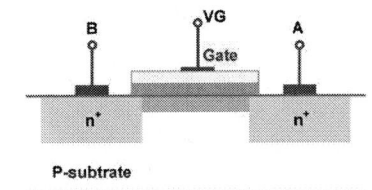

Fig. 1 The cross-section view of nFETs.

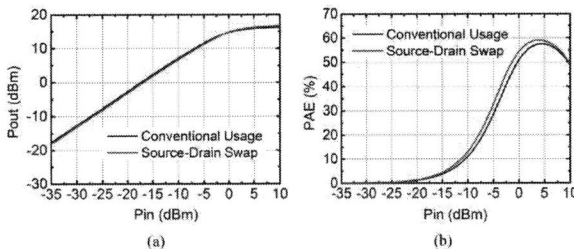

Fig. 2 Performance comparison of nmos_rf between the conventional usage and the source-drain swap: (a) Pout, (b) PAE.

I. INTRODUCTION

In recent years, large-scale millimeter wave (mm-Wave) phased array systems using silicon-based solutions have provided low-cost, high-speed, and reliable electronically controlled scanning antenna (ESA) terminals for low earth orbit high-throughput broadband SATCOM applications [1]. For the transmit (TX) link, using the Ka band as the operating frequency of its ground terminal provides more abundant available frequency resources than the C band and Ku band, resulting in faster transmission rates, and the terminal size can be made smaller. However, due to the long communication distance (e.g., ~500km for the low earth orbit satellites, ~36000km for the geostationary orbit satellites), a larger number of array elements are needed in satellite ground ESA terminals to ensure communication quality.

As one of the most critical blocks in its phased array system, PA urgently needs to improve the output power and efficiency to reduce system size and power consumption, while increasing communication distance. In addition, to fully utilize the spectrum resources of the Ka band, more stringent requirements have been put forward for the PA bandwidth. However, the standard CMOS process has disadvantages such as low breakdown voltage and large parasitic effects, making it difficult to achieve broadband high-power output. At the same time, due to the lower resistivity of the silicon-based substrate (≈10Ω·cm), the passive device loss is large, resulting in a decrease in the overall efficiency. Therefore, how to achieve broadband and high-efficiency CMOS PA simultaneously in mm-Wave still faces significant challenges.

Regarding the above issues, this paper analyzes the impact of parasitics and layout placement on efficiency. By optimizing the topology of power cell and adopting the neutralization technique to obtain a low loss, a high gain and an improved circuit stability. Moreover, a broadband matching design of the transformer-based MCR is employed

for quickly achieving an optimal balance between bandwidth and efficiency, while providing a compact and symmetrical layout design.

II. CIRCUIT ANALYSIS AND DESIGN

A. Optimization of Power Cell Topology

For typical CMOS process, nFETs in the PDK are divided into two types: core-type nch and RF-type nmos_rf. The nch only contains the active area (source, drain, gate) and cannot be used directly. It has a high degree of design freedom, but heavily relies on accurate modeling. In contrast, the nmos_rf provides a definite interconnect relationship and parasitic model, and has passed verification by the process manufacturer, so it is widely used in mm-Wave design. However, for PA designs that usually need multiple transistors in parallel to ensure output power, directly using nmos_rf inevitably introduces long interconnects around the transistor, which are not scaled with process, degenerate the transistor and thus restrict the power gain and maximum achievable output power. Thus, it is critical to minimize the parasitic losses during layout optimization.

Fig. 1 shows a cross-section view of nFETs. Notice that there is no difference in the physical composition of the source and drain in the active area. Even though the process manufacturer has already defined it, in practical applications, the working state of the transistor is more determined by the voltage between the electrodes: if the level of point A is higher than that of point B, point A can be considered as the drain, then point B is the source, and vice versa [2]. As a result, swapping the source and drain of a transistor does not affect its function. Fig. 2 shows the performance comparison.

979-8-3503-6184-1/24 $31.00 © 2024 IEEE

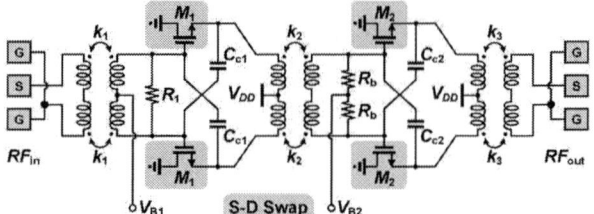

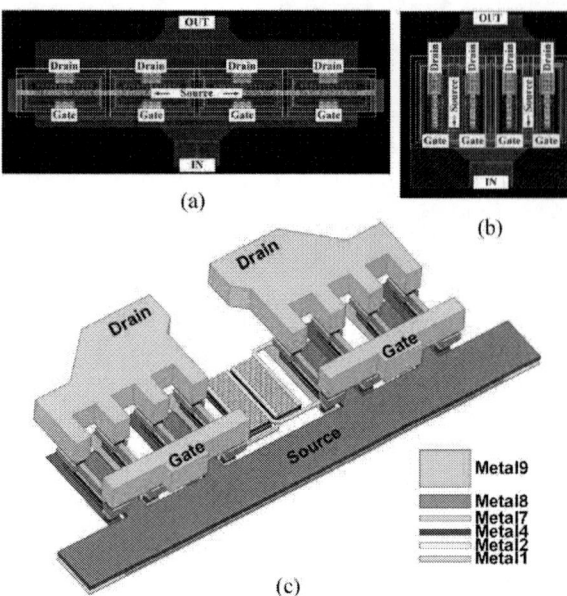

Fig. 4 Schematic of the proposed PA.

Fig. 3 The power cell layout of (a) the conventional method, (b) the source-drain swap. (c) The 3D layout topology of output stage.

It can be seen that after source-drain swap, the Pout of the nmos_rf remains consistent, and the maximum PAE has a slight improvement. This difference mainly lies in the layout of nmos_rf itself. It indicates that the pre-given parasitics of the nmos_rf in the drain network have relatively lower impact than that of source network. Therefore, the method of source-drain swap can serve as a starting point for improving efficiency.

The key advantage of the source-drain swap is that the obtained layout of the power cell is more compact, with minimized parasitics caused by external interconnects around transistors. Fig. 3 shows the corresponding layout of the two methods. For the conventional one as shown in Fig. 3(a), four nmos_rfs are arranged in a line, and the gate/drain node of each nmos_rf needs to be connected together before linking to the front/rear matching circuit though a simple one-divided-four/four-in-one network. Obviously, compared to the method of source-drain swap as shown in Fig. 3(b), the power division/synthesis network in this method has longer interconnects. Meanwhile, the outermost transistor of the power cell in this method has a longest input-to-output path, leading to a phase difference with the central transistor. At mm-Wave bands, the out-phase loss is also one of unfavorable factors affecting efficiency. Although using a network based on binary-tree structure can avoid this issue, it introduces longer passive routing and occupies a larger chip area, leading to more transmission losses. Fig. 3(b) shows a optimized layout by using the source-drain swap. Since the needed output power of each channel is only about 9dBm for SATCOM applications, the number of parallel transistors inside the power cell is relatively small. When higher power output is required, this method can be easily duplicated to construct a large transistor with minimized interconnect parasitics. In addition, due to the vertical arrangement of each transistor in the power cell after source-drain swap, the distance between the gate and source (original PDK definition) is farther, and the parasitic capacitance C_{gs} generated by upper metal interconnections is naturally smaller than that C_{gd} in conventional method. Thus, the reduction of coupling leakage could also improves the efficiency of the power cell.

Besides to the layout optimization that minimizes the device and interconnects parasitics, the neutralization technique is also used in the design to improve the transistor gain and stability [3]. Thanks to the use of source-drain swap, the lateral length of the power cell is reduced, and the arrangement of neutralization capacitors between the differential pair can evenly accommodate each transistor, as shown in Fig. 3(c). In this design, the MOM neutralization capacitors adopt the vertical parallel plate (VPP) structure [4]. The structure purely makes use of lateral capacitance component and no vertical component to obtain higher precision, which is the key to achieve the consistency between simulation and measurement. Considering the substrate coupling effect, the capacitors are implemented by metal 4 to metal 7. The value of a single capacitor and the coupling characteristics between capacitors are modeled and simulated using 3D EM simulators.

B. Design of Broadband Matching Network

Fig. 4 shows the schematic of the prosposed Ka-band PA, which has two class-AB common-source (CS) amplification stages. The transistor sizes of the first and second stage is 80μm/60nm and 256μm/60nm, respectively. The value of neutralization capacitance is designed to be 15*f*F for the first stage and 40*f*F for the second stage, with optimal reverse isolation and unconditional stability as the priority factors for value selection. Three customized MCRs are used for the input, interstage, and output matching networks. Among them, the input and and interstage matching networks usually have higher *Q* than the output matching network due to the smaller transistor size and the higher impedance looking into the gate of the output transistor. Similar to an *RLC* resonator, there is a trade-off between the bandwidth and *Q* in the MCR. Therefore, the design focus of broadband and flatness is on the input and the inter-stage MCRs, while the output MCR mainly optimizes its efficiency as much as possible.

In this design, two methods are used to extend bandwidth of the input and inter-stage MCRs. The first one is to directly reduce the *Q* by simply placing a small resistor in parallel at the higher impedance side of the MCR. Here, this method is only used in the first stage with a resistance of 400 ohm, considering that it can also provide an improved input return loss without affecting the implementation size of on-chip inductors. The second one is to use a weakly-coupled transformer [5]. A smaller coupling factor *k* can realize large impedance transformation ratio with moderate inductance ratio, which helps to realize actual inductors. However, the value of *k* also affect the in-band flatness and gain ripple. For practical engineering applications, we would not like that *k* to be too small, since a moderate value can push the two

979-8-3503-6184-1/24 $31.00 © 2024 IEEE

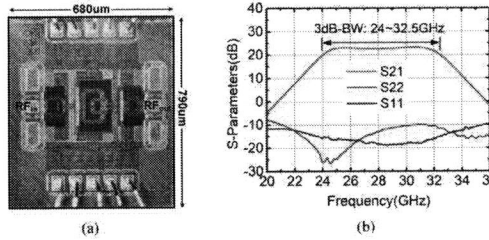

(a)　　　　　　　　(b)

Fig. 5 (a) Die microphotograph and (b) measured S-parameters.

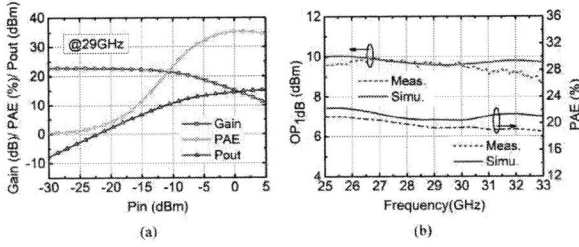

(a)　　　　　　　　(b)

Fig. 6 Measured large signal performance (a) vs. input power at 29GHz, and (b) vs. frequency.

amplitude peaks of Z21 in the MCR away from each other by some distance [6], so that the gain (Z21) at the center frequency can present a valley amplitude.

Both the input and inter-stage MCRs use a slightly asymmetric configuration [6] to ensure that the overall gain curve (S21) under simulation exhibits a slight positive slope versus frequency (the gain at the high-frequency peak is slightly higher than that at the low-frequency peak). It can avoid the gain rise of mm-Wave design at low operating frequencies that usually occurs after manufacturing.

III. MEASUREMENT RESULTS

Fig. 5(a) shows a chip microphotograph of the proposed Ka-band PA. Fabricated in 65-nm CMOS process, it occupies a silicon area of 790×680μm², including all pads and ESD. The circuit has been measured by Chip-on-Board (COB). The dc, tuning control pads are wire-bonded to a PCB, while the RF pads are accessed through the GSG probes. Under a 1.2V power supply, the PA consumes a static dc power of 48mW.

Fig. 5(b) shows the measured S-parameters. The curve of small signal gain (S21) presents two peaks of 23dB at 25.5GHz and 30.5GHz, with a 3-dB bandwidth of 8.5GHz from 24GHz to 32.5GHz. Meanwhile, it indicates an excellent gain flatness of less than 0.5dB within the frequency range between the two peaks. The return loss (S11/S22) is better than -10dB over the corresponding band. The features of S-parameters including bandwidth and flatness are beneficial for preventing the impact of frequency shift on the cascading of the required SATCOM TX link (27.5~30GHz). Fig. 6(a) shows the large signal performance at the center frequency 29GHz. The PA achieves an OP$_{1dB}$ of 9.7dBm with a PAE of 19.3% and a saturated P$_{sat}$ of 14.8dBm with a peak PAE of 35.3%. Fig. 6(b) shows the OP$_{1dB}$ and PAE versus frequency. It can be seen that the obtained measurements coincide well with the simulations. The output capability within the entire frequency band meets the design requirements.

Table I summarizes the performance of this design and compares it with other state-of-art works. The proposed PA achieves a better efficiency and in-band gain flatness due to

the optimized topology of power cell and broadband matching network.

IV. CONCLUSION

This paper presents a Ka-band two-stage class AB PA in 65nm CMOS process, with an optimized power cell topology and a broadband matching network based on the MCR. The measured results show that the PA achieves a gain of 23dB, a peak PAE of 35.3%, an excellent gain flatness of less than 0.5dB and a wide operation frequency from from 24GHz to 32.5GHz, which can well satisfy the requirements of the low-cost high-efficiency SATCOM applications. At the same time, the entire design considerations and optimization procedures of this circuit also have a good practicality in engineering .

TABLE I. SUMMARY AND COMPARISON OF PAS

Ref.	This work	[7]	[8]	[9]
Tech.	65nm CMOS	40nm CMOS	65nm CMOS	65nm SOI CMOS
Freq. (GHz)	29	27	30	26
Supply (V)	1.2	1.1	1	2
Gain (dB)	23	22.4	18	19
S11/S22 (dB)	≤-10dB	≤-7dB	≤-4dB	≤-4.6dB
P$_{1dB}$ (dBm)	9.7	13.7	12.67	20.8
P$_{sat}$ (dBm)	14.8	15.1	13.77	21.3
PAE$_{max}$ (%)	35.3	33.7	29.8	26.15
Fra. BW (%)	30	15.6	7.6	24.8
Area (um²)	790*680	250*900ª	245*723ª	460*1300

ª Without pads

REFERENCES

[1] Y. Wang, D. You, X. Fu, et al., "A Ka-band SATCOM transceiver in 65-nm CMOS with high-linearity TX and dual-channel wide-dynamic-range RX for terrestrial terminal," IEEE J. Solid-State Circuits, vol. 57, no. 2, pp. 356-370, Feb. 2022.

[2] B. Razavi, Design of Analog CMOS Integrated Circuits, 2nd ed., New York: McGraw Hill, 2017.

[3] W. L. Chan and J. R. Long, "A 58–65 GHz neutralized CMOS power amplifier with PAE above 10% at 1-V supply," IEEE J. Solid-State Circuits, vol. 45, no. 3, pp. 554-564, Jun. 2010.

[4] R. Aparicio and A. Hajimiri, "Capacity limits and matching properties of integrated capacitors," IEEE J. Solid-State Circuits, vol. 37, no. 3, pp. 384-393, Mar. 2002.

[5] X. Cao, T. Wu, and S. Ma, "A 26-39.5 GHz two-path voltage-combined power amplifier with bandwidth broadening technique in 22nm FD-SOI," in Proc. IEEE International Conference on Solid-State & Integrated Circuit Technology (ICSICT), 2022, pp. 1-3.

[6] H. Jia, C. C. Prawoto, B. Chi, Z. Wang and P. Yue, "A full kaband power amplifier with 32.9% PAE and 15.3-dBm power in 65-nm CMOS," IEEE Transactions on Circuits and Systems-I: Regular Papers, vol. 65, no. 9, pp. 2657-2668, Sep. 2018.

[7] S. Shakib, M. Elkholy, J. Dunworth, et al., "A wideband 28 GHz power amplifier supporting 8×100 MHz carrier aggregation for 5G in 40 nm CMOS," in Proc. IEEE Int. Solid-State Circuits Conf. (ISSCC) Dig. Tech. Papers, 2017, pp. 44-45.

[8] C. Yu, J. Feng, and D. zhao, "A Ka-band 65-nm CMOS neutralized medium power amplifier for 5G phased-array applications," in Proc. IEEE MTT-S International Wireless Symposium (IWS), 2018, pp. 1-3.

[9] Z. Chen, X. Wang, X. Ma, et al., " A 26-GHz linear power amplifier with 20.8-dBm OP1dB sSupporting 256-QAM wideband 5G NR OFDM for 5G base station equipment" in Proc. IEEE Radio Frequency Integrated Circuits Symposium (RFIC), 2023, pp. 193-196.

RF Front-End Chip Design for Ku-Band with 130nm CMOS Technology

Huiquan Xie [1], Ziyu Wang, Tianrui Wang, Yifei Chen, Maliang Liu [*1], Yintang Yang

[1] School of Microelectronics, Xidian University, Xi'an 710071, China

* Email: hqxie@stu.xidian.edu.cn, mlliu@xidian.edu.cn

Abstract—**Ku-band front-end chip with integrated switch on 130nm CMOS technology is implemented for satellite communication at 14-18 GHz bands. The single-pole-double-throw (SPDT) switch with 46dB isolation reduces front-end performance only by 1dB at 18GHz. The receiver (RX) path has 2.73dB noise figure at 16.5GHz and 17.6±0.5dB gain in band. Saturation output power (Psat) of transmitter (TX) with stacked power amplifier (PA) is 14.3dBm with 21.0±0.7dB gain and 16.5% peak power added efficiency (PAE). The PA and low noise amplifier (LNA) dissipate 206mW and 43mW from 1.2V supplies, and occupying 1.62mm×1.04mm in total.**

Keywords—**LNA, PA, mmWave, CMOS**

I. INTRODUCTION

Satellite communication refers to the communication method where satellites are used as relay stations to transmit information between two locations on Earth. Satellite communication services can cover a wide area and are used to provide services such as telephone communication, data transmission, and broadcasting. They are also applied in fields such as military, aviation, and transportation. The Ku band is a commonly used frequency band in satellite communication[1]. Due to the significant loss in satellite communication, phased array antennas are used to enhance signal gain and directivity, requiring multiple amplitude and phase control units for beamforming[2].

This paper presents the design of an RF front-end chip for the Ku band using 130nm CMOS process. The chip consists of a PA, a SPDT switch, and a LNA. The position of the design in a phased array antenna system are shown in Fig. 1[3], It serves as a bidirectional amplifier between the antenna and the phase control unit.

II. CIRCUIT DESIGN

The colored area in Fig. 1 represents the scope of our study, the input of PA and the output of LNA share a common port, allowing the transmit and receive channels to share a beamforming module, thus reducing system design costs.

A. Single-Pole Double-Throw Switch Design

This section describes the design of the SPDT RF switch. This switch offers advantages such as high breakdown voltage and low leakage current. The schematic diagram of the switch is shown in Fig. 2. It consists of four NMOS stacks and four inductors. The control voltages V_G and V_B are connected to the gate and body of the NMOS transistor, respectively, through a

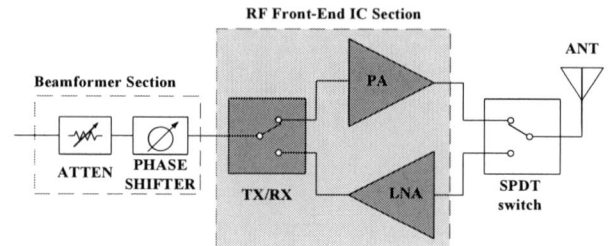

Fig. 1. RF front end of a phase array system

large resistor to control the on/off state of the switch. The resistor ensures that the variation in NMOS V_{GS} is minimal compared to the DC bias, enabling low insertion loss in the ON state and high isolation in the OFF state, even under high-frequency and large-signal input conditions. The stacking of two NMOS transistors improves the voltage withstand capability of the switch, and the resistor between the source and drain evenly distributes the voltage between the input and output during the OFF state, enhancing the linearity of the switch. The series inductors in the switch help to compensate for the gate-source capacitance of the NMOS transistor and improve impedance matching. When selecting between the upper and lower frequency channels, the NMOS stack in series is turned on, and the one in parallel is turned off, providing low insertion loss. Conversely, the NMOS stack in parallel is turned on, and the one in series is turned off, providing high isolation.

B. Low-Noise Amplifier Design

The LNA consists of three stages, each serving a specific function. The first stage utilizes gate-series inductors and source-degeneration inductors to achieve input power matching and noise matching. The second stage is designed for broadband amplification, while the third stage is responsible for output matching to 50 ohms. The schematic of LNA is shown in Fig. 3.

The width of the first-stage NMOS transistor is determined using (1)[4]. For long-channel devices, with Q_{SP} approximately 4.5, $L = 100nm$, and $R_s = 50\Omega$, we can estimate the width of the NMOS transistor, i.e. W_{opt}. Once the width is determined, C_{gs1} is also determined. Then, by setting the real part of the input impedance in (2) to 50Ω, we can find the value of L_{s1}. Next, we set the imaginary part of the resonance frequency to 0 to determine the value of L_g. Afterward, L_{d1} and C_2 are adjusted to ensure that the peak gain of the first stage falls at 14GHz, while achieving a relatively stable response within the band when combined with the second and third stages.

979-8-3503-6184-1/24 $31.00 © 2024 IEEE

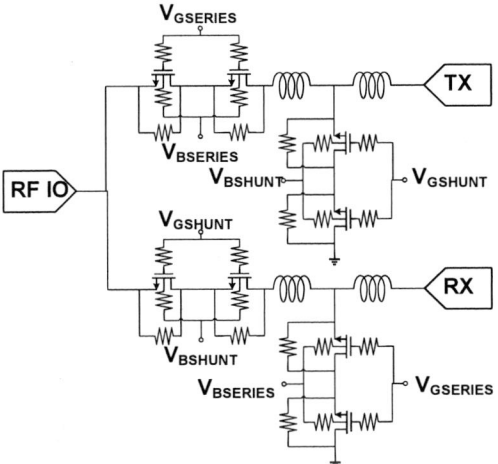

Fig. 2. Detailed schematic of SPDT

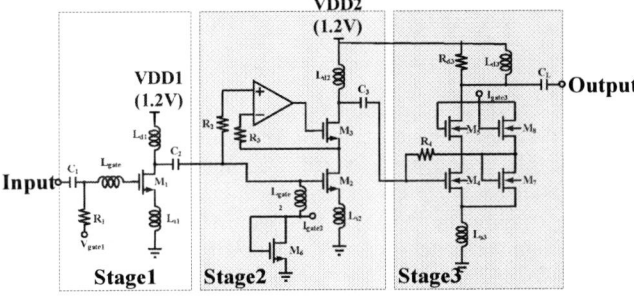

Fig. 3. Detailed schematic of LNA

$$W_{\text{opt}P} = \frac{3}{2} \frac{1}{\omega L C_{\text{ox}} R_s Q_{sP}} \approx \frac{1}{3\omega L C_{\text{ox}} R_s} \qquad (1)$$

$$Z_{\text{in}} = s\left(L_{gate} + L_{s1}\right) + \frac{1}{sC_{gs1}} + \frac{g_{m1}}{C_{gs1}} L_{s1} \qquad (2)$$

In the first stage, external level biasing is employed, allowing the measurement of the VDD1 current to obtain the DC operating point of the M1 transistor. The second and third stages utilize current biasing, generated by an internal self-biased current mirror, reducing the number of required power supply pads and improving PVT stability.

To ensure stable operation and reduced noise, the first and the second stage incorporates a body-floating device. The threshold voltage of this device is influenced by the drain voltage. To address this, an operational amplifier is used to provide bias to the source of M3, while also equalizing the drain voltages of M2 and M6. This configuration ensures a more stable mirrored current. The third stage utilizes high linearity body-contact transistors and employs a common-source/common-gate current mirror biasing scheme for improved performance.

C. Power Amplifier Design

The PA structure shown in Fig. 4 uses a differential structure and is divided into a driver amplifier and a power amplifier to meet the gain requirements. The presence of the gate-drain capacitance C_{gd} will reduce the gain, worsen the

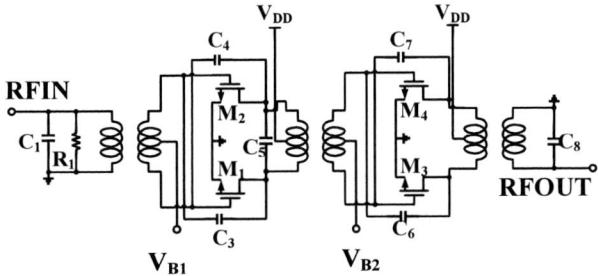

Fig. 4. Detailed schematic of PA

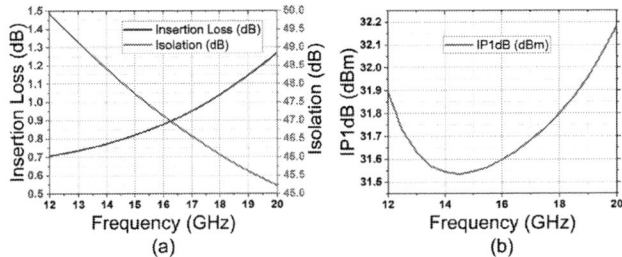

Fig. 5. (a) Insertion loss and isolation of SPDT. (b)IP1dB versus Frequency

reverse isolation, and even cause stability issues. Therefore, a cross-coupled capacitor with the same capacitance as C_{gd} is used to counteract the effect of parasitic capacitance. The three coupling inductors are implemented using a stacked structure to obtain the maximum coupling coefficient and improve the gain. VDD is 1.2V, and the bias VB1 and VB2 are generated by a current mirror to improve PVT stability.

III. POST SIMULATION RESULTS

The following post-simulation results are obtained based on full-wave electromagnetic (EM) analysis of the entire layout. For the RF switch, it is separately extracted and simulated to verify its linearity, insertion loss, and isolation to meet the design requirements. Perform post-simulation on the receiving and transmitting paths separately, the results are as follows.

A. SPDT Simulation Result

The simulation results of the RF switch are shown in Fig. 5(a) and Fig. 5(b). In the frequency range of 14~18GHz, the insertion loss is less than 1.04dB, the isolation in the OFF state is above 46dB, and the 1dB compression point is 31.4dBm, which is higher than the maximum output power of the PA.

B. RX path Simulation Result

Fig. 6(a) shows the simulated s-parameters of RX path. within the frequency range of 14GHz to 18GHz, the flat gain is 17.6±0.5dB, the S11 is less than -9.40dB within the band, and S22 is less than -11.62dB. The simulated noise figure. reaches its minimum value of 2.73dB around 16.5GHz, and it remains below 3.01dB within the bandwidth, as shown in Fig. 6(b). The simulated IP1dB in band is -21.9~-19.4dBm. The LNA operates at a supply voltage of 1.2V with a static current of 35.5mA, resulting in a total power consumption of 42.6mW.

C. TX path Simulation Result

The simulated s-parameters of TX path is shown in Fig. 7. In TX mode, the measured flatness of the gain within the 14GHz

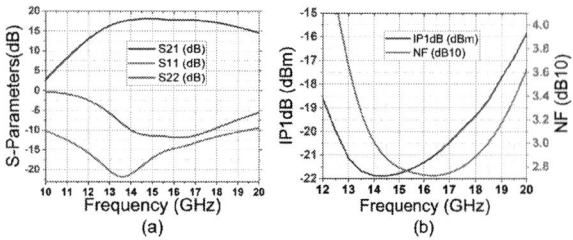

Fig. 6. (a)Simulated S-Parameters of RX path(b)Simulated NF of RX path

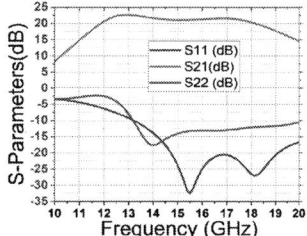

Fig. 7. Simulated S-Parameters of TX path

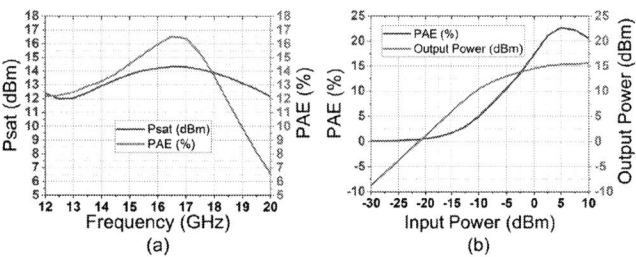

Fig. 8. (a)Simulated Psat and PAE of TX path versus frequency.(b) Simulated PAE and output power of TX path at 16GHz input frequency

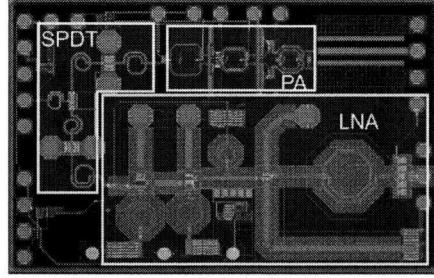

Fig. 9. Layout of RF-front end chip

TABLE 1. PERFORMANCE COMPARISON

	This work	[5]	[6]	[7]	[8]
Process	130nm CMOS	0.25um GaAs	0.25um GAN	40nm CMOS	0.13um SiGe
Frequency (GHz)	14~18	14~18	8-12	17.5~32.5	18~21
Chip Size (mm²)	1.68	3.33	4.67	N/A	1.74
RX path					
Gain(dB)	17.1~18.1	12.4-13.8	19.2-21.2	13.2	17
NF(dB10)	2.73~3.01	4.2~4.7	4.5-5	3.8-4.8	2
IP1dB(dBm)	-21.9~ -19.4	10.3	-8	-3	-16.2
Pdc(mW)	42.6	695	1000	41.6	20.7
TX path					
Gain(dB)	20.3~21.7	23.6~25	24-30	14.3	30.2
PAE(%)	13.3~16.5	22~28.3	10.5~15.4	23.2~34.7	27.6
Psat(dBm)	13.0~14.3	21.8~23.6	30.5~32.5	16.1	15.5
Pdc(mW)	206.4	670	4400	65.0	76

Fig 9. shows the layout of this chip. The performance comparison is shown in TABLE 1.This work exhibits a small footprint, low noise figure, and low power consumption.

IV. CONCLUSION

A Ku-band RF front-end chip has been developed, featuring an SPDT switch for selecting the transmit or receive channel. The chip has a compact size of 1.62mm×1.04mm. Based on simulation results, both channels achieve good input-output matching within the 14-18GHz range. When switching to the receive channel, it achieves a flat gain of 17.6±0.5dB, a noise figure of 2.73~3.01dB, and an input 1dB compression point of -20.9dBm at 16GHz. When switching to the transmit channel, it achieves a flat gain of 21.0±0.7dB, a Psat of 14.2dBm at 16GHz input, and a PAE of 16.0% at 0dBm input.

REFERENCES

[1] N. Saydam and K. Yegin, "Ku band amplifier design for satcom systems," 2016 24th Telecommunications Forum (TELFOR), Belgrade, Serbia, 2016, pp. 1-4.

[2] F. Boulos, U. Johannsen and S. Caizzone, "Customizable Phased Array Antenna based on Domino Tiles for Satcom Applications," 2022 IEEE International Symposium on Phased Array Systems & Technology (PAST), Waltham, MA, USA, 2022, pp. 01-05.

[3] P. Gkoutis, G. Konidas and G. Kalivas, "30 GHz Front-End with Adaptively Biased PA and Current Steering LNA for Phased Array Systems," 2022 IFIP/IEEE 30th International Conference on Very Large Scale Integration (VLSI-SoC), Patras, Greece, 2022, pp. 1-6.

[4] Lee T H. The design of CMOS radio-frequency integrated circuits[M]. Cambridge university press, 2003.

[5] Y. Li and S. Mou, "A Ku-Band Self-Biased Bidirectional Amplifier in 0.25 μm PHEMT Technology," 2020 IEEE 6th International Conference on Computer and Communications (ICCC), Chengdu, China, 2020, pp. 1166-1170.

[6] D. Kim, D. -H. Lee, S. Sim, L. Jeon and S. Hong, "An X-Band Switchless Bidirectional GaN MMIC Amplifier for Phased Array Systems," in IEEE Microwave and Wireless Components Letters, vol. 24, no. 12, pp. 878-880, Dec. 2014.

[7] H. Gao, J. Jin, Z. Yang, J. Zhou and X. Liu, "A 30GHz Bidirectional PA/LNA with Transformer-Based Switchable RC Matching Network," 2023 IEEE 15th International Conference on ASIC (ASICON), Nanjing, China, 2023, pp. 1-4.

[8] A. Colzani, M. Fumagalli and A. Fonte, "SiGe BiCMOS building blocks for a K/Ka-band flexible phased array system for SatCom applications," 2022 Microwave Mediterranean Symposium (MMS), Pizzo Calabro, Italy, 2022, pp. 1-5.

to 18GHz range is 21.0±0.7dB. The output return loss is less than -12dB, and the input return loss is less than -13.8dB. The power saturation output point is measured to be the 6dB output compression point, with the Psat ranging from 13.0dBm to 14.3dBm within the band. At an input of 0dBm, the power added efficiency (PAE) ranges from 13.3% to 16.5%, as shown in Fig. 8(a). Fig. 8(b) shows when input frequency is 16GHz, the Psat of TX path is about 14.2dBm and the peak PAE is 16.0%. The PA consumes a quiescent current of 171.9mA and a static power consumption of 206.4mW at a voltage of 1.2V.

Back-gate Bias Assisting VCRO Design

Chenglin Ye , Zheng Zhou*, Xiaoyan Liu

[1]School of Integrated Circuits, Peking University
Beijing, 100871, China

* Email: zhouzime@pku.edu.cn

Abstract—A novel design method of voltage-controlled ring oscillator (VCRO) based on back-gate bias using fully depleted silicon-on-insulator (FDSOI) technology is proposed. This work compares it with traditional front-gate bias controlled oscillators in terms of frequency tuning, power consumption, phase noise as well as figure of merit (FoM). The results demonstrate the novel back-gate-controlled ring oscillator can reach 26.8× frequency tuning range compared with the front-gate controlled ring oscillator. The frequency tuning linearity and phase noise are optimized as well. Through back-gate bias, we can achieve a better frequency-power-noise tradeoffs of VCRO.

Keywords—back-gate bias control, voltage-controlled ring oscillator (VCRO), fully depleted silicon-on-insulator (FDSOI)

I. INTRODUCTION

Ring oscillators are widely adopted currently as an integral part of many systems because of their wide tuning range, compact layout, and ability to generate multiple phases at the cost of phase noise compared with inductance-capacitance (LC) oscillators.[1] In order to control the RO frequency and obtain a VCRO, many techniques are applied, among which attaching a tail current source to the delay stages is the most common technique.[2] A tunable signal is added to the front gate of the tail current transistor for varying the current throughout the delay stage, thereby controlling the RO frequency. However, due to the jitter of output signals, it has been a challenging issue to achieve VCRO design of low power consumption, low phase noise, and high frequency tuning range.

FDSOI with ultrathin Si body and ultrathin buried oxide is an alternative technology to scaling down because of its advantages of low power consumption, good frequency characteristics, high immunity to process fluctuation, low process complexity, etc.[3]-[5] Moreover, its unique back-gate bias capability can assist circuits to meet both high performance as well as low power requirements in various application. In this work, a novel design method of VCRO based on back-gate bias instead of front-gate bias is proposed. In order to assess the comprehensive performance of the novel back-gate bias design of VCRO, based on the method for evaluating inverter type RO in [6], we compare frequency tuning characteristics, power consumption, as well as phase noise between front-gate bias structure and back-gate bias structure VCRO at 6 GHz oscillation frequency level. Through theoretical analysis and simulation verification, the back-gate bias controlled ring oscillators demonstrate a better frequency-power-noise tradeoffs.

II. VOLTAGE-CONTROLLED RING OSCILLATORS

Fig. 1 shows the VCRO structures. As Fig. 1(a), a ring oscillator comprises several delay stages in a loop, each of which is composed of a differential structure and has two input signals and two output signals. Usually, the delay stage

of VCRO is implemented using a classical differential amplifier and a front-gate bias voltage (FBV) signal is added to the tail transistor for tuning oscillation frequency as shown in Fig. 1(b). However, for wider frequency tuning range, new back-gate bias voltages (BBV) instead of FBV can be used to each transistor of the delay stage as shown in Fig. 1(c). When VCRO is working properly, Vout1-Vout2 will output the oscillating signal as shown in Fig. 2.

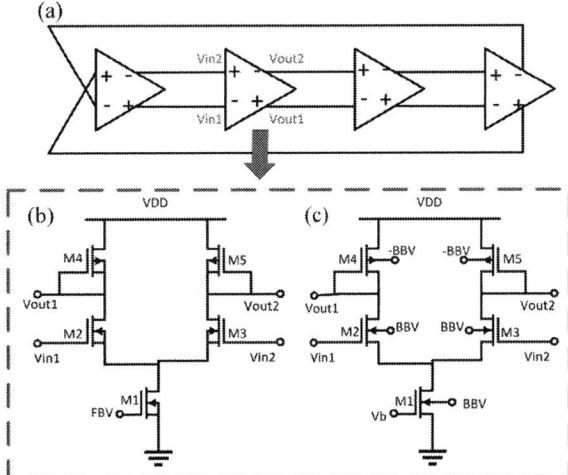

Fig. 1. (a) Schematics of a typical 4-stages differential VCRO, (b) single delay stage using front-gate bias, (c) single delay stage using back-gate bias

In this work, GlobalFoundries 22FDX® PDK is used and L of M1~M5 is fixed to 30nm. To confirm that output signals of VCRO can oscillate, the ratio of W/L of NMOS and PMOS is fixed to 3 for simplification.

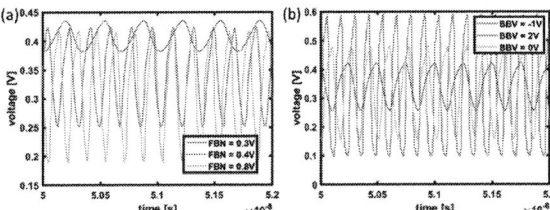

Fig. 2. Output waveform of (a) front-gate bias and (b) back-gate VCRO.

III. PERFORMANCE OF VCRO BASED ON BACK-GATE BIAS

To evaluation the performance of VCRO biased by back-gate, VCROs are simulated with various design parameters and biases conditions. The frequency of output signal is influenced by a few parameters, including number of delay stages M, VDD, W and L of transistors as well as controlling voltage (FBV & BBV). To obtain a specific frequency, M, VDD and L are first fixed. Then a specific controlling voltage is selected as a nominal voltage, and W is fixed accordingly. In this section, for a target frequency of 6 GHz, using fixed parameters of $M = 5$, $VDD = 0.8\ V$, $L =$

$30\ nm$, different nominal voltages are selected for different design sizes of W in front-gate bias and back-gate bias VCROs. Frequency, power, noise properties are evaluated using the series of designed VCROs.

A. Frequency tuning

Fig. 3(a) and Fig. 3(b) demonstrate frequency tuning characteristics of 6 GHz VCRO using front-gate bias structure with nominal voltage of $FBV = 0.6\ V$ and back-gate bias structure with nominal voltage of $BBV = 0\ V$, respectively. It can be seen that back-gate bias VCRO has better frequency tuning range and linearity. As shown in Fig. 3(c) and Fig. 3(d), the maximum frequency tuning ranges of front-gate bias and back-gate bias VCROs are 0.593 GHz and 15.91 GHz, while the minimum frequency tuning range of these two VCROs are 0.489 GHz and 2.59 GHz respectively, revealing a **5.3×~26.8×** tuning range improvement of back-gate bias structure than front-gate bias.

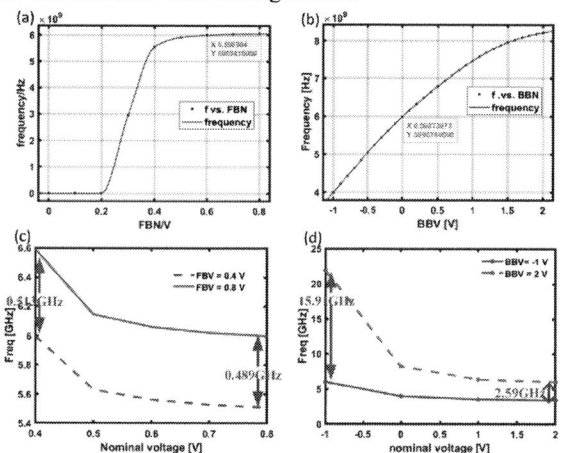

Fig. 3. Frequency tuning characteristics of 6-GHz VCRO using (a) front-gate bias and (b) back-gate bias structures; frequency tuning ranges with different nominal voltages using (c) front-gate bias and (d) back-gate bias structures.

B. Power Comsumption

Fig. 4 demonstrates power consumption distribution with different nominal voltages using front-gate bias and back-gate bias structures respectively. As shown in Fig. 4, for 6 GHz oscillation frequency, the minimum power consumption of front-gate bias structure is 0.291 mW when the nominal voltage is 0.4 V, and the minimum power consumption of back-gate bias structure is 0.071 mW when the nominal voltage is 2 V. The latter is only **24.4%** of the former.

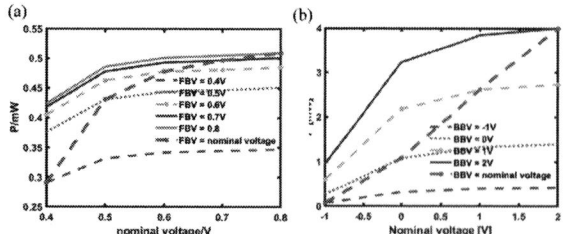

Fig. 4. Power consumption distribution with different nominal voltages using (a) front-gate bias and (b) back-gate structures.

The oscillation frequency of a M-stage VCRO can be obtained as follows:

$$f_0 = \frac{g_{m1}}{2Mln2C} = \frac{\sqrt{\beta I}}{2Mln2C} \propto I^{\frac{1}{2}} W^{-\frac{1}{2}} \qquad (1)$$

where C denotes the total capacitance seen at the output of each delay stage, I denotes the tail current, and β regroups the devices' dimensions W and L, the mobility and the gate-oxide capacitance per unit of area. Then the power consumption is proportional to W at a given frequency as follows:

$$P \propto I \propto f_0^2 W \qquad (2)$$

Due to the larger regulated voltage range of the back-gate bias structure, the design scale of lower W can be provided for same oscillation frequency compare with front-gate bias structure, thus providing the possibility of lower power consumption design.

C. Phase Noise and Figure of Merit (FoM)

The phase noise spectrum is induced by both white noise and flicker noise. From [7], [8], White noise and flicker noise are related to transistor size and threshold voltage. As shown in Fig. 5(a) and Fig. 5(b), when offset frequency is 100 MHz, the minimum phase noise spectrum of front-gate and back-gate bias structure are -69.52 dBc/Hz and −87.55 dBc/Hz respectively. Back-gate bias can change the threshold voltage of the transistor, and the back-gate bias structure can provide a larger W design, both which can effectively reduce the phase noise, which provides a possibility for designing a ring oscillator with lower phase noise.

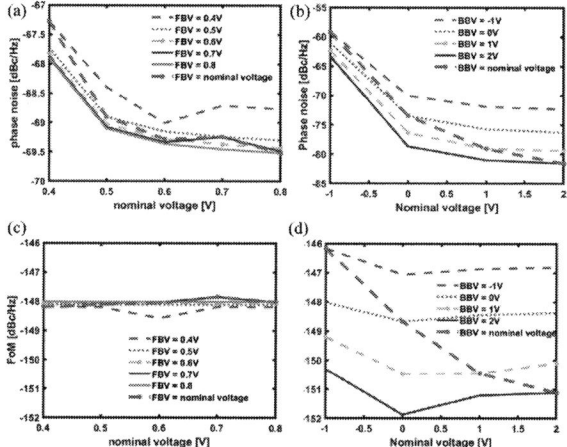

Fig. 5. Phase noise distribution with different nominal voltage using (a) front-gate bias and (b) back-gate bias structures; FoM distribution with different nominal voltages using (c) front-gate bias and (d) back-gate bias structures.

Figure of merit (FoM) focus particularly on the trade-off between power consumption and phase noise and it can be obtained as follows:

$$FoM = \mathcal{L}(\Delta f) \left(\frac{P}{1mW}\right) \left(\frac{f_0}{\Delta f}\right)^2 \qquad (3)$$

where $\mathcal{L}(\Delta f)$ denotes the phase noise spectrum, Δf denotes the offset frequency. FoM is not related to the gate width and frequency, and only the FoM induced by white noise depends on the threshold voltage change.[7] As shown in Fig.4(c) and Fig.4(d), FoM of front-gate bias structure remains constant while FoM of back-gate structure decreases as the nominal voltage increases. The main reason for this effect is that the back-gate bias changes the threshold voltage of the differential pair, thus improving the FoM induced by white noise part.

IV. VCRO DESIGN BASED ON FDSOI BACK-GATE BIAS

The oscillation frequency of the RO is determined by the number of delay stages M, devices' dimensions, controlling voltage, VDD, etc., which will affect the performance of the ring oscillator, such as power consumption, phase noise, etc. Fig. 6 shows the effect of different M and VDD on the performance of the VCRO. As shown in Fig. 6, for the same oscillation frequency, M has no effect on VCRO phase noise and power consumption, while the reduction of VDD will optimize VCRO phase noise and power consumption. But M will affect the layout of the VCRO, increasing M at the same oscillation frequency makes W smaller but the number of cascades increases, which makes the layout larger aspect ratio.

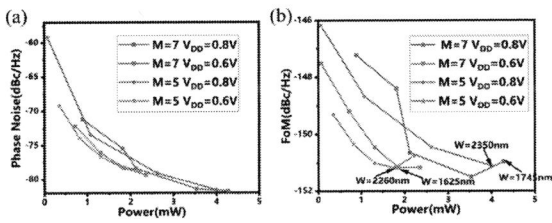

Fig. 6. (a) Phase noise-power curve and (b) FOM-power curve of back-gate RO under different VDD and M

Fig.7 summarizes the design point of VCRO using back-gate bias structure. For the target oscillation frequency, the layout, frequency tuning range, power consumption and phase noise are considered to determine M and BBV. At the same time, VDD and L are selected reasonably, and finally W is determined with the determination of the preceding parameters. Compared with the conventional front-gate bias structure VCRO, the back-gate bias structure VCRO can provide designers with a wider choice of BBV to meet the tradeoffs of frequency, power consumption and phase noise under different application requirements.

VCRO design point

Determining the oscillation frequency f_o, device gate length L → f_o, L

Considering layout → Determining the number of stages M:
for layout aspect ratio ↗, M ↗
for layout aspect ratio ↘, M ↘ → M

Determining the operating voltage: Reduce VDD as much as possible for better VCRO phase noise and power performance → VDD

Considering frequency tuning range → Determining nominal voltage BBV:
for frequency tuning range ↗, BBV ↘

Considering power consumption → Determining nominal voltage BBV:
for power consumption ↘, BBV ↘ → BBV

Considering phase noise → Determining nominal voltage BBV:
for phase noise ↘, BBV ↗

The device gate width W is determined

Fig. 7. VCRO design point: for the target oscillation frequency, the layout, frequency tuning range, power consumption and phase noise are considered to determine M and BBV, and VDD and L are selected reasonably to determine W finally.

V. CONCLUSION

In this work, extensive evaluations between VCRO based front-gate and back-gate bias structures are presented. The evaluation results demonstrate that VCROs using back-gate bias structures can achieve 26.8 × maximum frequency tuning range, 4.1 × minimum power consumption, better linearity as well as lower phase noise and FoM for the same oscillation frequency compared with the VCROs using front-gate bias structures. In addition, the effect of a range of design parameters on the VCRO is considered, including devices' dimensions W and L, operating voltage VDD, nominal voltage BBV, and number of stages M, based on which the VCRO design point is presented to guide the design of VCROs using back-gate bias structures for better frequency-power-noise tradeoffs compared with front-gate bias structures.

ACKNOWLEDGMENT

This work was supported in part by National Key Research and Development 2022YFB4401704 and the NSFC 92364104.

REFERENCES

[1] B. Razavi, "The Ring Oscillator [A Circuit for All Seasons]," *IEEE Solid-State Circuits Mag.*, vol. 11, no. 4, pp. 10–81, 2019.

[2] M. Moghavvemi and A. Attaran, "Recent Advances in Delay Cell VCOs [Application Notes]," *IEEE Microw. Mag.*, vol. 12, no. 5, pp. 110–118, Aug. 2011.

[3] W. Cao *et al.*, "The future transistors," *Nature*, vol. 620, no. 7974, pp. 501–515, Aug. 2023.

[4] K. Cheng and A. Khakifirooz, "Fully depleted SOI (FDSOI) technology," *Sci. China Inf. Sci.*, vol. 59, no. 6, p. 061402, Jun. 2016.

[5] M. Casse *et al.*, "FDSOI for cryoCMOS electronics: device characterization towards compact model," in *2022 International Electron Devices Meeting (IEDM)*, San Francisco, CA, USA: IEEE, Dec. 2022, p. 34.6.1-34.6.4.

[6] M. Schramme, L. Van Brandt, D. Flandre, and D. Bol, "Comprehensive Analytical Comparison of Ring Oscillators in FDSOI Technology: Current Starving Versus Back-Bias Control," *IEEE Trans. Circuits Syst. Regul. Pap.*, vol. 69, no. 5, pp. 1883–1895, May 2022.

[7] A. A. Abidi, "Phase Noise and Jitter in CMOS Ring Oscillators," *IEEE J. Solid-State Circuits*, vol. 41, no. 8, pp. 1803–1816, Aug. 2006.

[8] W.-J. Zhu and J.-G. Ma, "Investigating the effects of the number of stages on phase noise in CMOS ring oscillators".

A 3.2-to-7.1GHz Quad-Core Dual-Mode Oscillator Achieving 193.6 dBc/Hz Peak FoM

Xiaoyu Shan[1], Renxuan Li[1], Mengming Zhang[1], Ang Hu[1], Dongsheng Liu*[1]

[1] School of Integrated Circuits, Huazhong University of Science and Technology, Wuhan 430074, China

* Email: dsliu@hust.edu.cn

Abstract—In this paper, a low-phase-noise mode-switching oscillator using a modified quad-winding transformer and a hybrid bias switched capacitor array is proposed. The adopted transformer has the following improvements compared with the original design: 1) the core size is reduced by 45%; 2) the frequency spacing between modes 1 and 2 is increased by 44%; 3) the equivalent parallel resistance in mode 2 is increased by 27%. The above three features together result in a tinier area, wider frequency tuning range (TR), and lower phase noise. Low-flicker-noise resistive bias and compact bottom-pinned bias switched capacitors are used in the capacitor array to achieve low phase noise and small frequency tuning steps. Designed in a 40-nm CMOS process, the proposed oscillator exhibits frequency tuning from 3.2 to 7.1 GHz while consuming 21.2 to 25.4 mW power dissipation. The phase noise at 4.48 GHz oscillation frequency is -134.6 dBc/Hz at 1-MHz offset, corresponding to a figure-of-merit (FoM) of 193.6 dBc/Hz and FoM_T of 211.2 dBc/Hz. The $1/f^3$ phase noise corner is 131-417 kHz over 76.2% TR.

Keywords—*Coupled oscillator, low phase noise, mode switching, muti-core, transformer, voltage-controlled oscillator (VCO), wide tuning range.*

I. INTRODUCTION

Multiprotocol compatibility and elevated communication rates drive the oscillator towards both a wide frequency tuning range and low phase noise. Harmonically tuned tanks engage extra harmonic resonance (second and/or third) to reduce the dc and/or root-mean-square (rms) value of the impulse sensitivity function (ISF) [1], therefore improving the phase noise performance. However, the harmonic tuning method requires complex tank designs and is not easy to implement in a wide-tuning range oscillator. Drain or source damping resistors are engaged to suppress the flicker noise up-conversion, nevertheless, the start-up margin and voltage swing are impaired. The switched capacitor, switched inductor [2] and mode-switching [3] [4] [5] are three approaches to achieve a wide tuning range. The first two employ MOSFET switches to adjust the capacitance or inductance of the tank. However, the tank quality factor is degraded by the non-ideal MOS switch, which results in poor phase noise performance. Mode switching techniques are widely used to achieve a wide frequency tuning range without quality factor penalty, whereas its phase noise improvement relies on increasing the number of oscillator cores, which increases the circuit size and power consumption.

Our previous work that uses multi-magnetic-coupling and active-source-degenerating techniques to achieve low phase noise and wide tuning range is proposed in [6]. In this paper, an improved design is presented to further improve the phase noise performance and frequency tuning range while occupying a smaller area. In Section II, the proposed oscillator design is described. Section III demonstrates the post-simulation results. A conclusion is drawn in Section IV.

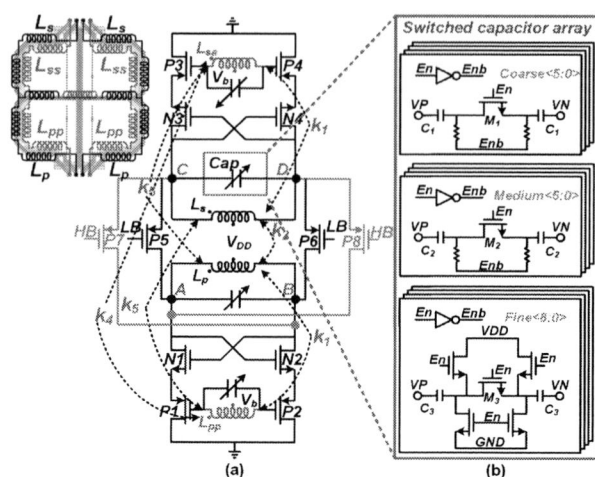

Fig. 1. Schematic of the oscillator (a) and the switched capacitor array (b).

II. THE OSCILLATOR IMPLEMENTATION

A. Design of the Quad-Winding-Coupling Transformer

Fig. 1 shows the oscillator design. The proposed quad-winding transformer is also presented at the top-left of Fig. 1 (a). The design of switched capacitor array in each oscillator core is identical and shown in Fig. 1 (b). Conventionally, dual-winding single-coupling transformer is adopted in a mode-switching oscillator to expand the frequency tuning range. The oscillation mode is controlled by the switch array or the negative transconductance. The previously proposed quad-winding transformer consists of four windings (L_p, L_s, L_{pp}, L_{ss}) and magnetic couplings between them (k_1, k_2, k_4, k_5). Two primary coupling cores (L_p, L_s) can be controlled by switch array to oscillate in phase (A & C are in phase, even mode, low frequency band) or out of phase (A & C are out of phase, odd mode, high frequency band) which mimics the conventional dual-mode oscillator. The secondary coupling core (L_{pp}, L_{ss}) drives the corresponding PMOS (P1, P2, P3, P4) to oscillate synchronously with the corresponding NMOS (N1, N2, N3, N4).

Fig. 2 (a)(b) show the layout of conventional and the proposed transformer design. The core size of the proposed transformer (0.31 mm²) is reduced by 45% compared with the original design (0.56 mm²). The simulated magnitude and phase response of resonator's input impedance in modes 1 and 2 for conventional and the proposed transformer design is shown in Fig. 2 (c)-(f). A 1.5 pF capacitor is added in each tank as a capacitive part to sustain the oscillations. The resonant frequencies of modes 1 and 2 are 3.32 GHz & 4.23 GHz in the original design and 3.21 GHz & 4.51 GHz in the proposed design, respectively. The frequency spacing is

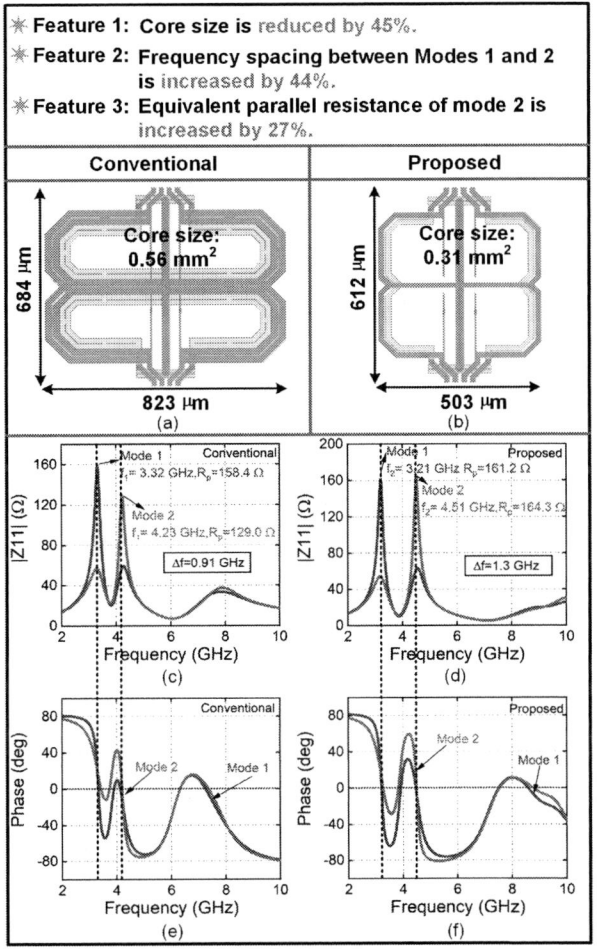

‖ **Feature 1:** Core size is reduced by 45%.

‖ **Feature 2:** Frequency spacing between Modes 1 and 2 is increased by 44%.

‖ **Feature 3:** Equivalent parallel resistance of mode 2 is increased by 27%.

Fig. 2. The layout of conventional (a) and the proposed (b) quad-winding transformer design. Simulated magnitude and phase response of resonator's input impedance in mode 1 (blue line) and mode 2 (red line) for conventional (c)(e) and the proposed (d)(f) transformer design. A 1.5 pF capacitor is added in each tank as a capacitive part to sustain the oscillations.

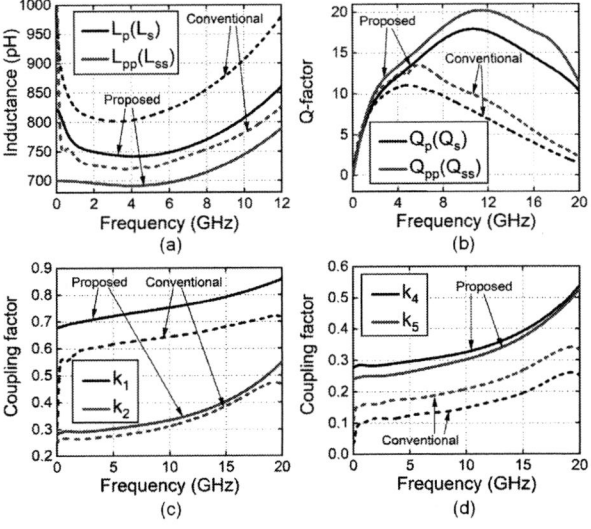

Fig. 3. Simulated (a) inductance and (b) quality factor of each winding in the transformer. (c) (d) Simulated coupling factor between windings of the multi-magnetic-coupling transformer. The conventional and proposed designs are shown in dashed and solid lines, respectively.

increased from 0.91 GHz to 1.3 GHz (increased by 44%). As a result, the frequency overlap can be significantly reduced, leading to a wider frequency tuning range. The magnitude of the resonator's input impedance in mode 2 is 129.0 Ω and 164.3 Ω in the original and proposed design, respectively. The larger equivalent parallel resistance in the proposed design results in lower phase noise.

The simulated inductance, quality factor and coupling factor between windings are presented in Fig. 3 (a)-(d). Simulated transformer parameters for the proposed and conventional designs are indicated by solid and dashed lines, respectively. There are four inspirations in the proposed oscillator topology:

(1) larger k_2, k_4, k_5 can increase the effective inductance ratio of mode 1 ($L_{eff,1}$) to mode 2 ($L_{eff,2}$);

(2) the closer k_0 is to 1 (the closer L_p or L_s is to L_{pp} or L_{ss}), the larger the effective inductance ratio of mode 1 ($L_{eff,1}$) to mode 2 ($L_{eff,2}$);

(3) larger k_1 (stronger coupling between L_p or L_s and L_{pp} or L_{ss}) results in larger effective inductance in both modes 1 and 2.

(4) larger k_2, k_4, k_5 can push the optimum k_1 to a higher value that achieves the peak of $L_{eff,1}/L_{eff,2}$.

Thus, boosting k_2, k_4, k_5 and reducing the intrinsic inductance deviation between L_p (L_s) and L_{pp} (L_{ss}) can increase the frequency spacing between modes 1 and 2. Whereas, a large k_2 leads to a reduction of the effective inductance in mode 2, resulting in a deterioration of the phase noise. On the other hand, by increasing k_1, the effective inductance is enhanced in both modes 1 and 2. Observing that larger k_2, k_4, k_5 can push the optimum k_1 to a higher value that achieves the peak of $L_{eff,1}/L_{eff,2}$, therefore a larger k_1 can compensate the negative impact of stronger k_2 on effective inductance in mode 2. Thus, by increasing k_1, k_2, k_4, k_5 and bringing L_p (L_s) closer to L_{pp} (L_{ss}), both the frequency spacing between modes 1 and 2 and $L_{eff,1}$ & $L_{eff,2}$ are enhanced. By shrinking the transformer linewidth, the coil length and area occupation are reduced, while the intrinsic quality factor is higher. As a conclusion, the optimized transformer results in smaller area occupancy, wider frequency tuning range, and lower phase noise for the proposed oscillator compared to the original design.

B. Design of the Switched Capacitor Array and Active Core

The lack of fine-tuning in the original oscillator design limits its application. The switched capacitor array in the proposed oscillator consists of 6 bits coarse, 6 bits medium and 9 bits fine capacitor unit (Fig. 1 (b)). The coarse and medium capacitor arrays are binary-weighted, while the fine capacitor array is thermometer-decoded. Low-flicker-noise resistive bias structures are used in coarse and medium arrays, while bottom-pinned bias topologies [7] are used in fine arrays to minimize area consumption. The capacitor array achieves a minimum tuning step of 18 aF. The simulated (post-layout) frequency tuning step is between 16.1 kHz (achieved at the lowest output frequency in mode 1) and 76.6 kHz (achieved at the highest output frequency in mode 2). The size of NMOS (N1-N4) and PMOS (P1-P4) are 26 μm/40 nm and 62.4 μm/40 nm, respectively, to guarantee a robust start-up condition and reducing the parasitic capacitance (the channel

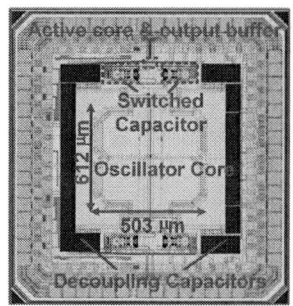

Fig. 4. Chip layout

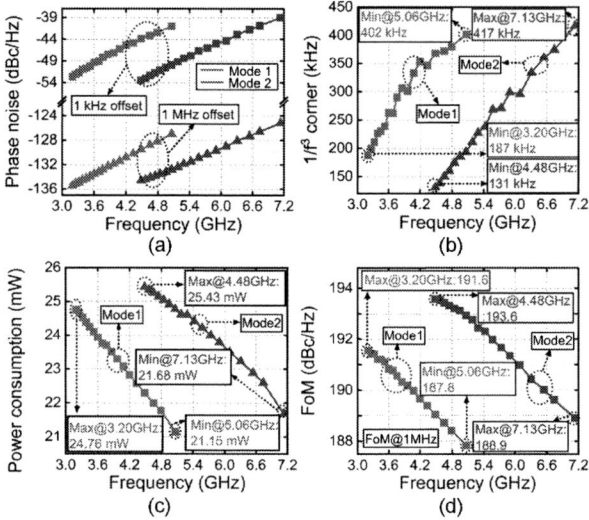

Fig. 5. Simulated (a) phase noise, (b) 1/f³ corner, (c) power consumption and (d) FoM over tuning range at the two oscillation modes.

length is reduced from 270 nm in the original design to 40 nm in the proposed design). The chip layout is shown in Fig. 4.

III. POST-LAYOUT SIMULATION RESULTS

Fig. 5 shows the phase noise, 1/f³ corner, power consumption, and FoM of the post-layout simulation. The simulations show a 76.2% continuum tuning from 3.20 GHz to 7.13 GHz. Modes 1 and 2 cover the frequency range 3.20 GHz to 5.06 GHz and 4.48 GHz to 7.13 GHz. The best FoM at 1 MHz offset is 193.6 dBc/Hz at 4.48 GHz oscillation frequency. The simulation results are summarized and compared with the state-of-the-art in Table I. The proposed oscillator achieves the best FoM$_T$ of 211.1 dBc/Hz compared to other works.

IV. CONCLUSION

In this paper, we propose a low-phase-noise wide-tuning-range oscillator using an optimized quad-winding transformer and a hybrid bias switched capacitor array. The proposed transformer achieves smaller area occupancy, larger frequency spacing between modes 1 and 2, and larger equivalent parallel resistance in mode 2. The capacitor array combines a low-flicker-noise resistive bias structure with a low-parasitic-capacitance bottom-pinned bias topology. With the optimized transformer and switched capacitor array, the oscillator achieves 76.2% tuning range, 193.6 dBc/Hz peak FoM and 16.1 kHz minimum tuning step. The best FoM$_T$ is 211.2 dBc/Hz at 1 MHz offset, which is the best performance compared to the state of the art.

ACKNOWLEDGMENT

This work is supported by the National Key Research and Development Program of China (No. 2021YFA0715502, No. 2023YFB4502100), the National Natural Science Foundation of China (No.62374064，No. 62104076, No. 62134002), the National Key Analog Integrated Circuit Laboratory Project of China. (No. JCKY2021210C004).

REFERENCES

[1] Y. Shu, H. J. Qian, X. Gao, and X. Luo, "A Low Phase Noise and High FoM Distributed-Swing-Boosting Multi-Core Oscillator Using Harmonic-Impedance-Expanding Technique," *IEEE Journal of Solid-State Circuits*, vol. 56, no. 12, pp. 3728-3740, Dec. 2021.

[2] S. L. Liu, K. H. Chen, and A. Chin, "A Dual-Resonant Mode 10/22-GHz VCO With a Novel Inductive Switching Approach," *IEEE Transactions on Microwave Theory and Techniques*, vol. 60, no. 7, pp. 2165-2177, July. 2012.

[3] M. Taghivand, K. Aggarwal, and A. S. Y. Poon, "21.5 A 3.24-to-8.45GHz low-phase-noise mode-switching oscillator," *IEEE Int. Solid-State Circuits Conf. (ISSCC) Dig. Tech. Papers.* pp. 368-369, Mar. 2014.

[4] G. Li, L. Liu, Y. Tang, and E. Afshari, "A Low-Phase-Noise Wide-Tuning-Range Oscillator Based on Resonant Mode Switching," *IEEE Journal of Solid-State Circuits*, vol. 47, no. 6, pp. 1295-1308, June. 2012.

[5] C. C. Lim, H. Ramiah, J. Yin, N. Kumar, P. I. Mak, and R. P. Martins, "A 5.1-to-7.3 mW, 2.4-to-5 GHz Class-C Mode-Switching Single-Ended-Complementary VCO Achieving >190 dBc/Hz FoM," *IEEE Transactions on Circuits and Systems II: Express Briefs*, vol. 66, no. 2, pp. 237-241, Feb. 2019.

[6] X. Shan, D. Liu, A. Hu, Z. Jin, J. Lu, A. Li, K. Li, and X. Zou, "A Low-Phase-Noise Wide-Tuning-Range Mode-Switching Oscillator Using Multi-Magnetic-Coupling and Active-Source-Degenerating Techniques," *IEEE Journal of Solid-State Circuits*, pp. 1-14. 2024.

[7] B. Hershberg, K. Raczkowski, K. Vaesen, and J. Craninckx, "A 9.1–12.7 GHz VCO in 28nm CMOS with a bottom-pinning bias technique for digital varactor stress reduction," *European Solid State Circuits Conference (ESSCIRC)*. pp. 83-86, Sept. 2014.

TABLE I
PERFORMANCE COMPARISON WITH STATE-OF-THE-ART WIDEBAND CMOS VCOs

	This work	ISSCC' 14 [3]	JSSC' 12 [4]	JSSC' 21 [1]	TCAS-II'19 [5]	JSSC' 24 [6]
Technology	**40nm CMOS**	40nm CMOS	65nm CMOS	40nm CMOS	130nm CMOS	40nm CMOS
V_{DD} (V)	**1.2**	0.8	0.6	1.1	1.2	1.5
Frequency (GHz)	**3.20 ~ 7.13**	3.24 ~ 8.45	2.52 ~ 5.52	3.09 ~ 4.04	2.4 ~ 5	2.88 ~ 5.87
Tuning Range (%)	**76.2**	89.1	74.6	26.6	70.6	68.3
Power (mW)	**21.1 ~ 25.4**	14 ~ 16.5	9.8 ~ 14.2	20.9 ~ 23.1	5.1 ~ 7.3	20.5 ~ 27.9
PN@1MHz (dBc/Hz)	**-135.4 ~ -125.2**	-129 ~ -120	-128.6 ~ -121.3	-138.9 ~ -132.2	-130 ~ -122.3	-135.8 ~ -124.4
FoM @ 1MHz (dBc/Hz)	**187.8 ~ 193.6**	187 ~ 187.1	185.1 ~ 186.2	191.1 ~ 195.1	188.9 ~ 189.2	183.4 ~ 191.5
FoM$_T$ @ 1MHz (dBc/Hz)	**205.5 ~ 211.2**	206 ~ 206.1	202.6 ~ 203.7	199.6 ~ 203.6	205.9 ~ 206.2	200.1 ~ 208.2
1/f³ corner (kHz)	**131 ~ 417**	214 ~ 439	363 ~ 676	100 ~ 130	398 ~ 501	68 ~ 234
Core size (mm²)	**0.31**	0.432	0.294	0.36	0.33	0.55

$$\text{FoM} = |L(\Delta f)| + 20 \log_{10}\left(\frac{f_0}{\Delta f}\right) - 10 \log_{10}(P_{DC}/1mW)$$

$$\text{FoM}_T = \text{FoM} + 20 \log_{10}(Tuning\ Range(TR)/10)$$

979-8-3503-6184-1/24 $31.00 © 2024 IEEE

A 20.6 to 30.5 GHz Two Stage Cascode LNA in 40nm CMOS for Phase Array Tranceiver

Lei wang [1], Kefeng Han[2], Hao Xu [1], Rui Yin [1], Na Yan*[1]

[1]Fudan University, Shanghai, China
[2]Jiashan Fudan Institute, Zhejiang, China

* Email: yanna@fudan.edu.cn

Abstract—A 20.6 to 30.5 GHz two stage cascode LNA is proposed for phased array transceivers. The design employs a two-stage inductively degenerated cascode LNA, achieving simultaneous noise and input matching through inductive degeneration. An additional inductor at the cascode stage, placed between the common-source (CS) and common-gate (CG) stages, cancels parasitic capacitance at this node, effectively reducing LNA noise and phase variation. Variable gain steps are implemented using a current steering technique, which ensures precise gain control and flexibility in signal amplification. A transformer and a Marchand balun are employed for interstage and output matching, respectively, optimizing signal integrity and performance. The proposed LNA achieves an NF < 4.1 dB from 20.6 to 30.5 GHz range, a gain adjustment range of 12 dB with 3 dB step, IP1dB > -17 dBm and IIP3 > -9 dBm. The LNA exhibits an off-state impedance of 250 Ω, ensuring adequate isolation. Power consumption is 18 mW from a 1.1V supply, with a compact active area of 0.5 mm² .

Keywords— inductively degenerated LNA, variable gain, CMOS

I. Introduction

The fifth-generation mobile network (5G) achieves high-throughput and low-latency data transmission, applied in intelligent driving, artificial intelligence, and the internet of things (IoT). Within frequency range 2 (FR2) defined by the 5G New Radio (NR) standard, the 28 GHz band is available in various countries [1]. Due to its smaller wavelength, millimeter wave has inevitably higher free space path loss (FSPL) than other frequency bands [2]. The beamforming technique has been widely utilized in millimeter-wave transceivers [3]. It enhances signal strength in a fixed direction, providing an additional gain factor of $10log_{10}(N)$ to both the transmitter power and the receiver signal to noise (SNR) ratio. LNA typically serves as the first stage of the receiver. The noise figure (NF) of the LNA plays a crucial role in the receiver's performance, as indicated by the Friis formula. Many approaches have been published to optimize the NF of LNA. Some approaches focus on noise canceling methods [4], utilizing auxiliary amplifier to cancel noise generated by the main amplifier. However, the power consumption and additional noise from the auxiliary amplifiers have become limiting factors. More advanced approaches suggest transformer coupling to alleviate noise while maintaining a compact silicon footprint. [5][6] utilize a transformer between the gate and source of a common-gate (CG) LNA to alleviate noise and enhance gain. [7] proposed a noise-canceling CG LNA employing a transformer with source and drain inductors. [8] proposed a common-source (CS) LNA with a transformer between the gate and drain. Utilizing a drain to gate feedback path, the LNA achieves wideband input matching and enables noise canceling simultaneously.

This brief proposes a two stage inductively degenerated cascode LNA. To the best of the authors' knowledge, the proposed LNA provides a more balanced performance in terms of gain, NF, and power consumption compared to other reported approaches. The rest of this brief is organized as follows: Section II describes the proposed LNA and provides a detailed analysis of the technique. Section III presents the simulation and measurement results.

II. Analysis Of Proposed LNA

Inductively degenerated common source (IDCS) LNA are widely used in LNA design for their superior noise performance and input matching. The proposed LNA demonstrated in this brief utilizes a two-stage inductively degenerated CS LNA. In the following section, we will briefly derive the noise performance and input matching of the proposed LNA.

A. Input Impedance Matching

The input impedance of a CS LNA is typically capacitive, primarily due to the gate-source capacitance C_{gs} . An inductively degenerated CS LNA uses an inductor between the source of the CS LNA and ground to achieve a real part of the input impedance, thereby facilitating the impedance matching to 50 Ω. This results in improved noise performance [8]. Fig.1 shows the first stage of the LNA. The input impedance Z_{in} of the inductively degenerated CS LNA is derived as follows:

$$Z_{in}(s) \approx s\left(L_g + L_s\right) + sC_{gs} + \frac{g_m L_s}{R_s} \qquad (1)$$

The expression $\frac{g_m L_s}{R_s}$ is designed to achieve an input matching of 50 Ω. The inductor L_g inserted between the input and gate of the inductively degenerated CS LNA is used to resonate out the capacitance C_{gs} with the assistance of L_s.

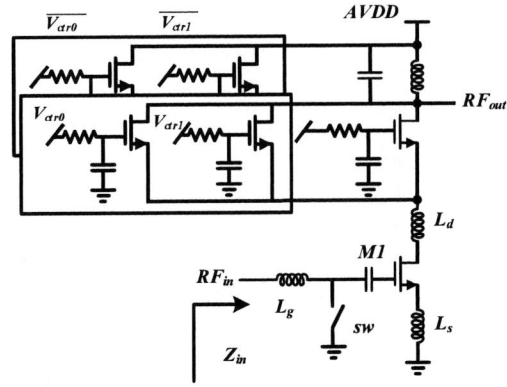

Fig.1. The first stage of proposed LNA

B. Noise Analysis

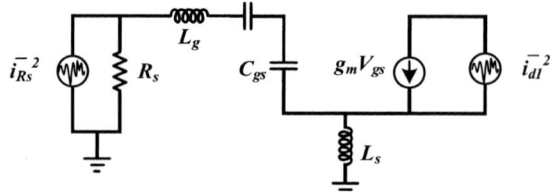

Fig. 2. Small signal diagram of the simplified IDCS LNA

For simplicity, C_{gd} and g_{mb} are omitted in the transistor model, as shown in Fig. 2. The noise contribution of the CG transistor in the cascode LNA is relatively lower due to the amplification provided by the CS transistor. For the sake of simplicity in the derivation, only the thermal channel noise of the CS MOS transistor $\overline{i_{d1}^2}$ (2) and the thermal noise of the source resistance R_s $\overline{i_{Rs}^2}$ (3) are taken into consideration.

$$\overline{i_{d1}^2} = 4kT\gamma g_m \Delta f \qquad (2)$$

$$\overline{i_{Rs}^2} = \frac{4kT}{R_s} \Delta f \qquad (3)$$

$$\overline{i_{Rs\_out}^2} = \frac{g_m^2 \, 4kTR_s}{A^2} \Delta f \qquad (4)$$

$$\overline{i_{d1\_out}^2} = \frac{|C_{gs}L_g s^2 + C_{gs}L_s s^2 + C_{gs}R_s s + 1|^2 \, g_m \, 4kT\gamma}{A^2} \Delta f \qquad (5)$$

$$A = C_{gs}L_g s^2 + C_{gs}L_s s^2 + C_{gs}R_s s + gmL_s s + 1 \qquad (6)$$

Equations (4) and (5) are derived from Kirchhoff's Current Law (KCL) and Kirchhoff's Voltage Law (KVL), providing the output noise current expressions of $\overline{i_{Rs\_out}^2}$ and $\overline{i_{d1\_out}^2}$. The NF of the typology in Fig. 2 is as follows:

$$NF = 1 + \frac{\overline{i_{d1\_out}^2}}{\overline{i_{Rs\_out}^2}} \qquad (7)$$

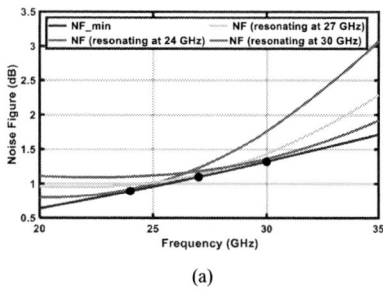

(a)

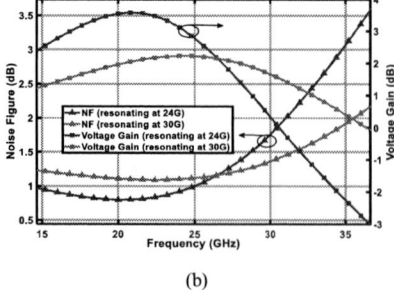

(b)

Fig. 3. (a) NF min and NF resonating at different frequencies (b) Voltage gain and NF resonate at 24 GHz and 30 GHz

$$NF_{min} = 1 + \gamma \frac{|\omega C_{gs}|^2 R_s}{g_m} \qquad (8)$$

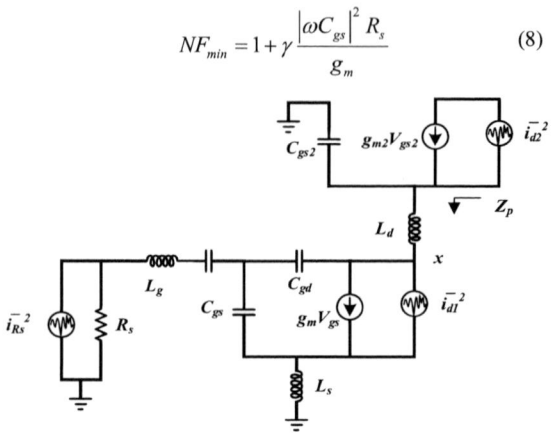

Fig. 4. Small signal diagram of proposed typology LNA

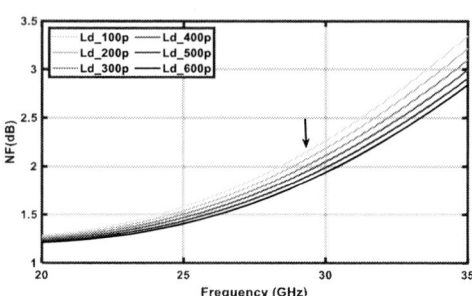

Fig. 5. Corresponding NF with different L_d

Equation (8) derives the NF_min of the circuit in Fig. 2. Considering the losses in the inductors, the series resistances R_{Ls} and R_{Lg} represent the losses in the source and gate inductor, respectively. As shown in (5) by replacing L_g with $L_g + R_{Lg}$ and L_s with $L_s + R_{Ls}$, we obtain the new expression of NF. Intuitively, NF will increase due to R_{Ls} and R_{Lg}. It is important to note that the series resistance is inversely proportional to the Q factor of the inductor. The Q values of both L_s and L_g play a crucial role in noise suppression, necessitating careful design consideration. Fig. 3 illustrates the relationship between NF and NF_min. In Fig. 3(a), the plot shows NF_min and NF (with input impedance RLC resonating at 24GHz, 27GHz, 30GHz, respectively) versus frequency. Fig. 3(b) demonstrates the voltage gain from the input of the inductively degenerated CS LNA to the gate of the MOS, along with the NF resonating at 24GHz and 30GHz. From Fig. 3, the following conclusions can be drawn:

1. The NF_min curve is obtained by tuning each frequency at its resonant frequency and is a monotonically increasing function with respect to frequency.

2. NF and NF_min intersect at their resonant points. Typically, the minimum value of NF is located just to the left of the resonant frequency. This is because the voltage gain is higher to the left of the resonance point, as shown in Fig. 3(b). Therefore, selecting the resonance point is a trade-off between input matching and NF.

Consider the noise generated by CG transistor. The subsection aims to demonstrate how an additional inductor L_d suppresses the thermal noise of the M2 channel.

$$\overline{i_{o,n,M2}^{2}} = \frac{4kT\gamma g_{m2}}{|\,1 + g_{m2}Z_p\,|^{2}} \quad (9)$$

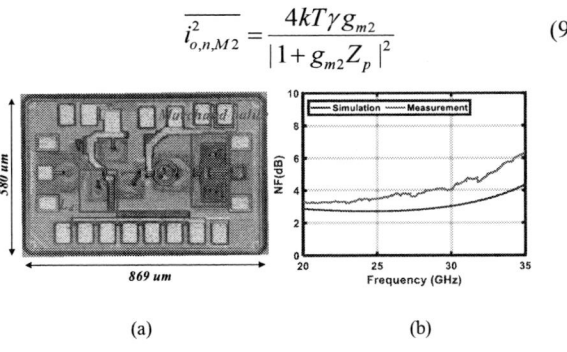

(a)   (b)

Fig. 6. (a) Die photo of proposed LNA (b) measured noise figure and simulated noise figure

Fig. 4 illustrates the proposed LNA featuring an additional inductor L_d. To facilitate an intuitive understanding of noise suppression, the gate-drain capacitance (C_{gd}) of M1 is taken into consideration. The impedance denoted by Z_p originates from the source of the CG MOS transistor. As indicated in Equation (9), the capacitance at point x resonates with the additional inductor L_d, effectively reducing the noise contribution of M2. As a proof of concept, Fig. 5 displays NF with varying values of L_d. A larger L_d is required to achieve a lower NF.

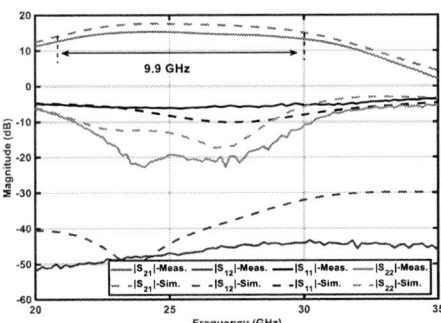

Fig. 7. Measured and simulated S parameter

III. MEASUREMENT RESULTS

The LNA was fabricated using 40-nm CMOS technology. As depicted in Fig. 6(a), the LNA occupies a chip area of 869 um × 580 um, including G-S-G pads DC pads and seal ring. The S-parameters and NF were measured using the Keysight N5247B PNA-X Network Analyzer. The measured and simulated S-parameters and NF are presented in Fig. 6(b). and Fig. 7, respectively. The proposed LNA achieves a peak gain of 15.35 dB at 24.27 GHz with a -3dB bandwidth of 9.9 GHz. Additionally, S11 is below -4.8 dB and S22 is below -6.7 dB from 20.6 GHz to 30.5 GHz, respectively. The lowest NF is 3.23 dB at 23 GHz, and the NF is below 4.1 dB from 20.6 GHz to 30.5 GHz, as depicted in Fig. 6(b). And the overall circuit consumes 18 mW from a 1.1 V supply. Table I presents the measurement results of proposed LNA and compares it with other CMOS LNA. The relatively high NF is due to a switch connected in parallel at the input, as shown in Fig. 1, which is used to switch between the transmit and receive channels.

TABLE I. COMPARSION OF STATE-OF-ART CMOS LNA

	This work	MWCL'18 [9]	TCASII'18 [10]	RFIC'07 [11]
Frequency (GHz)	20.6-30.5	26-33	33	28.5-30
Tech.	40 nm	40 nm	28 nm	90 nm
BW$_{3dB}$ (GHz)	9.9	7	4.4	2.6
Gain (dB)	15.4	27.1/18.4	18.6	20
NF (dB)	3.23-4.1	3.3-4.4	4.9	2.9-4.2
Power (mW)	18	31.4/21.5	9.7	16.25
Total area (mm^2)	0.5	0.54	0.51	0.67

ACKNOWLEDGMENT

This work was supported by the National Key Research and Development Program of China (No.2023YFB4403302).

REFERENCES

[1] J. Pang *et al.*, "A 28-GHz CMOS Phased-Array Transceiver Based on LO Phase-Shifting Architecture With Gain Invariant Phase Tuning for 5G New Radio," in *IEEE Journal of Solid-State Circuits*, vol. 54, no. 5, pp. 1228-1242, May 2019

[2] Y. Wang *et al.*, "A 39-GHz 64-Element Phased-Array Transceiver With Built-In Phase and Amplitude Calibrations for Large-Array 5G NR in 65-nm CMOS," in *IEEE Journal of Solid-State Circuits*, vol. 55, no. 5, pp. 1249-1269, May 2020

[3] M. Boers *et al.*, "A 16TX/16RX 60 GHz 802.11ad Chipset With Single Coaxial Interface and Polarization Diversity," in *IEEE Journal of Solid-State Circuits*, vol. 49, no. 12, pp. 3031-3045, Dec. 2014

[4] F. Bruccoleri, E. A. M. Klumperink and B. Nauta, "Wide-band CMOS low-noise amplifier exploiting thermal noise canceling," in *IEEE Journal of Solid-State Circuits*, vol. 39, no. 2, pp. 275-282, Feb. 2004

[5] J. Zhang and D. Zhao, "A 20-GHz Ultra-Low-Power LNA Using gm-Boosted and Current-Reuse Techniques in 65-nm CMOS for Satellite Communication Terminals," *2019 IEEE Asian Solid-State Circuits Conference (A-SSCC)*, Macau, Macao, 2019

[6] S. Guo, T. Xi, P. Gui, D. Huang, Y. Fan and M. Morgan, "A Transformer Feedback Gm-Boosting Technique for Gain Improvement and Noise Reduction in mm-Wave Cascode LNAs," in *IEEE Transactions on Microwave Theory and Techniques*, vol. 64, no. 7, pp. 2080-2090, July 2016

[7] Takao Kihara, Toshimasa Matsuoka and Kenji Taniguchi, "A 1.0 V, 2.5 mW, transformer noise-canceling UWB CMOS LNA," *2008 IEEE Radio Frequency Integrated Circuits Symposium*, Atlanta, GA, USA, 2008

[8] X. Fan, H. Zhang and E. SÁnchez-Sinencio, "A Noise Reduction and Linearity Improvement Technique for a Differential Cascode LNA," in *IEEE Journal of Solid-State Circuits*, vol. 43, no. 3, pp. 588-599, March 2008

[9] M. Elkholy, S. Shakib, J. Dunworth, V. Aparin and K. Entesari, "A Wideband Variable Gain LNA With High OIP3 for 5G Using 40-nm Bulk CMOS," in *IEEE Microwave and Wireless Components Letters*, vol. 28, no. 1, pp. 64-66, Jan. 2018

[10] M. Keshavarz Hedayati, A. Abdipour, R. Sarraf Shirazi, C. Cetintepe and R. B. Staszewski, "A 33-GHz LNA for 5G Wireless Systems in 28-nm Bulk CMOS," in *IEEE Transactions on Circuits and Systems II: Express Briefs*, vol. 65, no. 10, pp. 1460-1464, Oct. 2018

[11] E. Adabi, B. Heydari, M. Bohsali and A. M. Niknejad, "30 GHz CMOS Low Noise Amplifier," *2007 IEEE Radio Frequency Integrated Circuits (RFIC) Symposium*, Honolulu, HI, USA, 2007

979-8-3503-6184-1/24 $31.00 © 2024 IEEE

A 12-32 GHz Power Amplifier with 32-dBm Psat and 25% PAE in 0.15μm GaN

Xiangran Ni [1,2], Chunyue Bo [1,2], Tianyu Li [1,2], Qingyang Dong [1,2], Xin Jiang [1,2], Weijun Luo*[1,2]

[1] School of Integrated Circuits, University of Chinese Academy of Sciences, Beijing 100049, China
[2] Institute of Microelectronics of Chinese Academy of Sciences, Beijing 100029, China

* Email: nixiangran@ime.ac.cn , luoweijun@ime.ac.cn

Abstract—**This paper presents the design and simulation results of an ultra-wideband (UWB) power amplifier (PA) MMIC fabricated using 0.15μm GaN on SiC HEMT technology. The PA achieves a remarkable 91% relative bandwidth from 12 GHz to 32 GHz. It features a two-stage cascade structure, delivering a power-added efficiency (PAE) exceeding 24% and a saturation output power of over 31 dBm. The amplifier exhibits a gain ranging from 14.5 dB to 17.1 dB across the entire band, occupying a compact chip size of 4.5 mm². Compared to existing designs, this PA demonstrates outstanding performance, making it highly suitable for UWB communication systems.**

Keywords—power amplifier, ultra wideband, 0.15μm GaN

I. INTRODUCTION

With the widespread adoption of 5G communication technology in millimeter-wave bands, the demand for high-frequency and broadband capabilities has accelerated the development of power amplifier designs. Currently, research on millimeter-wave power amplifiers predominantly utilizes SiGe, GaAs, and CMOS processes, typically offering lower output power and focusing on narrowband applications. These technologies fail to meet the future communication systems' demands for high power, high efficiency, and ultra-wideband capabilities. Gallium nitride (GaN) devices present several advantages, such as high power density, temperature resilience, and low on-resistance, making them well-suited for high-power and high-efficiency RF circuits. Advancements in process technology have expanded GaN devices' operating frequencies from sub-6 GHz to the millimeter-wave bands, thereby bolstering research into high-power and high-efficiency millimeter-wave RF front-end chips. However, GaN devices exhibit significant losses and impedance variations in millimeter-wave bands, posing challenges in realizing ultra-wideband GaN millimeter-wave RF front-ends. Developing ultra-wideband power amplifiers for millimeter-wave frequencies is crucial to advancing and popularizing millimeter-wave communication technology.

The distributed structure offers significant advantages in achieving ultra-wideband performance, thereby enhancing the bandwidth of amplifier circuits [1]. The fundamental principle involves connecting gate-source parasitic capacitors (Cgs) and drain-source parasitic capacitors (Cds) to gate and drain inductors, forming an artificial transmission line. This configuration transforms them into distributed capacitors, facilitating wideband input-output matching and thereby extending the frequency response range of the amplifier circuit. Theoretically, the distributed amplifier circuit structure can achieve the widest bandwidth and maintain good gain flatness. However, due to its inherent complexity, traditional distributed amplifiers suffer from lower power efficiency, posing challenges in realizing high-performance power amplification. C. F. Campbell and S. Nayak had design

a 16-40GHz GaN distributed power amplifier MMICs utilizing an advanced 0.15μm GaN process, which peforms relative bandwidth of 85.7%, PAE of less than 20%, peak output powers of 36 dBm[2]. The efficiency of the power amplifier remains a concern and requires a deeper optimization.

In this work, we propose a two-stage cascaded 12-32GHz power amplifier with a relative bandwidth of 91%, a saturation output power of 31dBm, a gain of 14.5-17.1dB, and a maximum power increase efficiency of 33%. The PA design are given in Section II. Simulation results are presented in Section III. Conclusions are presented in Section IV.

II. POWER AMPLIFIER DESIGN

A. Process Technology

This UWB millimeter-wave PA is designed and simulated based on the 0.15μm depletion GaN/SiC HEMT (NP15-00) process provided by WIN. NP15-00 is a high-power 0.15μm gate GaN technology manufactured on a 100mm SiC substrate for millimeter-wave high-power applications. It exhibits a maximum transconductance peak of 390 Ms/mm, a maximum output current density of 860 mA/mm, a Vdg breakdown voltage test of >120V in MP15-00 technology, and a saturation output power density (Psat) of about 4.4 W/mm..

B. Circuit Design and Analysis

Due to the low gain at high frequency, a two-stage cascade structure is used to meet the gain requirements. The input matching of a UWB power amplifier is crucial for bandwidth limitation, whereas the output power matching is essential for optimizing efficiency and power delivery. So the first stage adopts a distributed structure to achieve good input matching, and the second stage adopts a two-way synthesis structure to achieve a larger output power. The hybrid topology combines the bandwidth advantages of distributed UWB with the high efficiency and power advantages of the reactance matching structure. This approach strategically allocates resources, focusing on the input stage of the power amplifier as the primary bandwidth-limiting factor and the output stage as critical for optimizing power and efficiency.

The first stage distributed amplifier is connected in parallel with 4-stage GaN transistors, each with a GaN transistor size of 2x60μm. Input matching tunes the impedance points to the optimal source impedance and load impedance by adjusting the transmission line Lg1-Lg4 and Ld1-Ld4 in first stage. Capacitors Cc1 and Cc2 are input and output isolation capacitors respectively. Adjusting the value of resistor Rg can obtain the appropriate stability factor, generally the stability factor needs to be controlled in the whole band greater than 1 to determine that the amplifier is in a stable state. Fig.1 shows the input matching S11 is basically less than -10dB, and most

979-8-3503-6184-1/24 $31.00 © 2024 IEEE

band is less than -15dB, which verifies that the structure has good matching performance. It can be seen from Fig. 2 that the stability factor of the distributed amplifier is greater than 1 in the whole frequency band and is in an absolutely stable state.

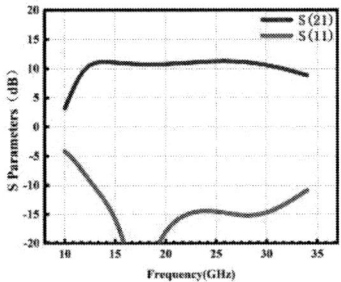

Fig. 1. S parameters of first stage.

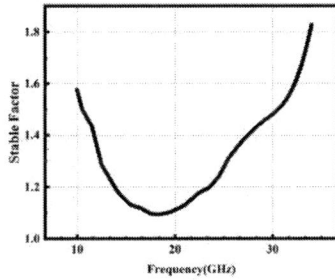

Fig. 2. Stability factor first stage.

The second stage uses two 8x60um GaN HEMTs in parallel, which can reduce the size of individual transistors, improve their frequency performance and the heat dissipation performance. Both transistors use the same bias circuit to simplify the circuit and reduces the chip area. In broadband design, the optimum impedance points of different frequency points vary greatly, the impedance trend of the output matching network with frequency needs to be matched with the optimal load impedance trend of GaN transistors. Simultaneously, it is crucial to ensure that the matching network exhibits minimal insertion loss within the operational frequency band. The microstrip line output matching structure is adopted to achieve this purpose. Fig.3 shows the PAE of the output amplifier is basically remained above 25%, and the saturation output power exceeded 31dBm.

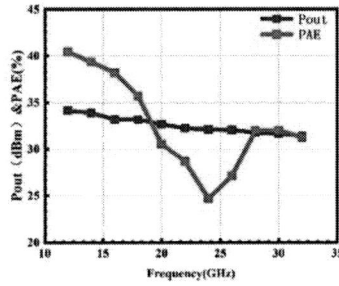

Fig. 3. Pout , PAE of second stage

In the prior design, the output impedance of the first stage and the input impedance of the second stage were both aligned with the 10Ω load impedance. Subsequently, interstage matching was optimized across the entire amplifier..

After both stages are complete, the entire circuit requires further adjustment and optimization. Fig.4 illustrates the schematic design of the ultra-wideband (UWB) power amplifier MMIC, while Fig.5 depicts the layout of the power amplifier.

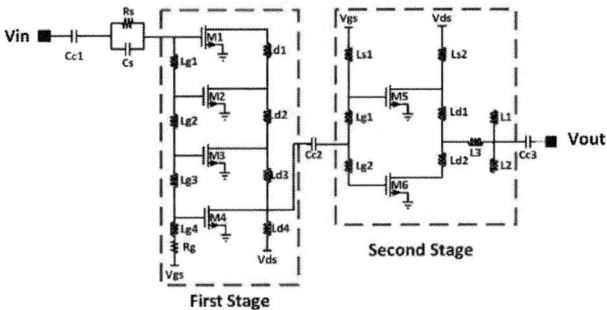

Fig. 4. Schematic of the power amplifier MMIC.

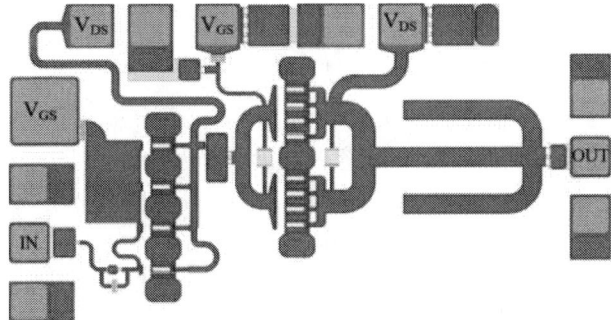

Fig. 5. Layout of the power amplifier MMIC.

III. SIMULATION RESULTS

Electromagnetic field (EM) simulations were conducted for the layout. Fig.6 presents the S-parameters of the layout, while Fig.7 displays the electromagnetic field simulation results for large signals. As depicted in Fig.6, the power amplifier demonstrates a gain ranging from 14.3 to 17.5 dB across the frequency band of 12-32 GHz, with fluctuations within the band remaining below 3.2 dB. The overall gain throughout the frequency range exceeds 14.3 dB. In comparison with the schematic simulation results, there is a slight reduction of 0.5 dB in small signal gain due to additional losses introduced by parasitic parameters in the layout. The S11 parameter is below -10 dB within the 12-32 GHz, while S22 remains below -10 dB in 13.5-17 GHz and 29-31GHz and below -6 dB from 12 to 32 GHz. Through optimization in layout design, S22 has been improved, thus enhancing the matching effectiveness compared to the schematic design. As illustrated in Fig.7, within the 12-32 GHz range, the power added efficiency (PAE) remains consistently above 24%, with saturation output power exceeding 32 dBm. However, at certain frequencies, the saturation output power reaches only 31 dBm due to additional losses caused by parasitic parameters introduced in the layout. These factors also impact the precise matching of the optimal load impedance. The simulation results affirm that the layout of the designed power amplifier satisfies the design criteria for high efficiency and ultra-wideband operation.

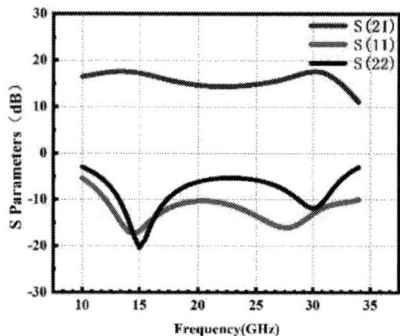

Fig. 6.   S parameters of PA.

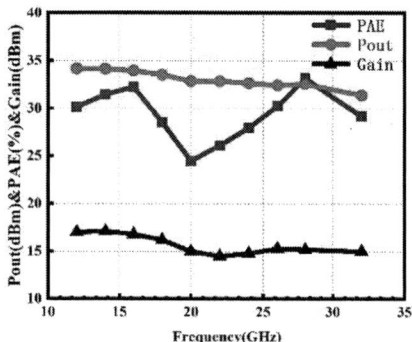

Fig. 7.   Pout,PAE,Gain of PA.

TABLE I.   SUMMARY AND COMPARISON OF GAN PAS

Ref.	Freq. [GHz]	Fra. BW [%]	Gain [dB]	Pout [dBm]	PAE [%]	Size [mm²]
[3]	17-35	69.2	0-12	22.5-23.5	30-40	1.5
[4]	33-48	37	19.6-21.6	26-28	25	2.9
[5]	16-40	85.7	25	36	10-20	6.3
[6]	17-40	80.7	19	40	13-17	45.65
This work	12-32	90.9	>15	>31	>25	1.3

IV. SUMMARY

This design describes the design and simulation performance of an ultra-wideband power amplifier (PA) MMIC operating in12-32GHz. The circuit utilizes a two-stage cascade topology. The input stage features a distributed matching structure known for its superior broadband characteristics, while the output stage employs a parallel transistor configuration to achieve efficient output matching. The EM simulation results indicate that the first stage achieves favorable input matching, while the second stage demonstrates high output power and efficiency performance. Ultimately, the design achieves the following characteristics: the relative bandwidth is 91%, the PAE is greater than 24%, and the saturated output power is basically more than 32dBm. The simulation results indicate that the power amplifier (PA) exhibits ultra-wideband characteristics and high efficiency, thereby suggesting significant potential for applications in the millimeter-wave band.

REFERENCES

[1] K. W. Kobayashi, D. Denninghoff and D. Miller, "A Novel 100 MHz–45 GHz Input-Termination-Less Distributed Amplifier Design With Low-Frequency LowNoise and High Linearity Implemented With A 6 Inch 0.15 μm GaN-SiC Wafer Process Technology," in IEEE Journal of Solid-State Circuits, vol. 51, no. 9, pp. 2017-2026, Sept. 2016.

[2] C. F. Campbell, S. Nayak, M. Y. Kao, and S. Q. Chen, "Design and Performance of 16-40GHz GaN Distributed Power Amplifier MMICs Utilizing an Advanced 0.15μm GaN Process," IEEE MTT-S International Microwaves Sysposium(IMS), 2016, San Francisco.

[3] P. -C. Huang, Z. -M. Tsai, K. -Y. Lin and H. Wang, "A 17–35 GHz Broadband, High Efficiency PHEMT Power Amplifier Using Synthesized Transformer Matching Technique," in IEEE Transactions on Microwave Theory and Techniques, vol. 60, no. 1, pp. 112-119, Jan. 2012.

[4] S. Chen and S. Nayak, "A 1/2 Watt high linearity and wide bandwidth PHEMT driver amplifier MMIC for millimeter-wave applications," in IEEE Int. Microw. Symp. Dig., Jun. 11–16, 2006, pp. 1863–1866.

[5] C. F. Campbell, S. Nayak, M. Y. Kao, and S. Q. Chen, "Design and Performance of 16-40GHz GaN Distributed Power Amplifier MMICs Utilizing an Advanced 0.15μm GaN Process," IEEE MTT-S International Microwaves Sysposium(IMS), 2016, San Francisco.

[6] P. M. Smith, et. Al., "Next-Generation Wideband GaN and MHEMT MMICs for K/Ka-Band Transmit and Receive Applications," GOMACTech 2015.

A source-driven push-push doubler with wideband 2nd harmonic feedback

Yuyang Chen [1], Ao Zhang [2], Jianjun Gao [3], Jianjun Zhou*[1]

[1] Department of Micro/Nano Electronics, Shanghai Jiao Tong University, Shanghai, China
[2] School of Microelectronics, Nantong University, Nantong, China
[3] School of Physics and Electronic Science, East China Normal University, Shanghai, China

* Email: zhoujianjun@sjtu.edu.cn

Abstract—This paper proposes a source-driven push-push frequency doubler with second harmonic feedback. The innovative second harmonic feedback technique is subjected to detailed analysis and simulation, thereby demonstrating its potential to enhance conversion gain across a broad operational bandwidth. A prototype push-push frequency doubler is implemented in a 40nm CMOS process. The broadband matching of the source-driven multiplier and the proposed band second harmonic feedback technique allows for a 3dB bandwidth of 55GHz-93GHz with 51% fractional bandwidth, as demonstrated by the simulation.

Keywords—frequency doubler, conversion gain, second harmonic feedback, source-driven

I. INTRODUCTION

The utilization of voltage-controlled oscillators and phase-locked loops for the direct production of local oscillators in millimeter-wave transceivers is associated with two principal disadvantages: namely, high power consumption and high noise. An alternative structure, which employs low-frequency sources in conjunction with a frequency multiplier chain, has been developed to overcome these issues and has become the mainstream approach in recent years [1][2].

There are three principal types of frequency doublers (FD): push-push[3], Gilbert[4], and injection lock[5]. Among these, the Gilbert type is based on the multiplication of two fundamental frequency signals. The additional intermediate nodes result in a reduction in bandwidth. Furthermore, the output swing is relatively limited, which restricts its application in the CMOS process. Injection-locked FD generates a second harmonic by injecting a fundamental signal into a subharmonic resonator. The bandwidth of the injected locking doublers is also constrained by the limited locking range of a resonator.

Push-push FD employs the secondary nonlinearity of a pair of transistors to generate a second harmonic at the drain. The simplicity of this configuration renders it particularly well-suited to high-frequency, wideband applications. However, push-push FD is subject to the detrimental effects of negative feedback, which is caused by the gate-drain parasitic capacitance. This results in a reduction in conversion gain and efficiency. The current solutions, such as the second harmonic trap at the input, are not optimal as the impact on the bandwidth is introduced by that trap.

In this paper, a source-driven push-push FD with a novel second harmonic feedback technique is proposed. By connecting the gate of a source-driven push-push FD to a series inductor, the feedback signal is inverted thereby converting the negative feedback into positive feedback. A prototype frequency multiplier is implemented on a 40nm CMOS process, and simulations show that the bandwidth is not largely influenced by the proposed second harmonic feedback technique.

II. ANALYSIS OF THE PROPOSED FD

A. Traditional push-push FD

A common concern among push-push FDs is the issue of second harmonic negative feedback. The second harmonic signal generated by the multiplier is fed back to the transistor gate through the gate-drain parasitic capacitance, whereupon the transconductance produces an out-of-phase second harmonic current. Given the differential-to-signal-ended nature of push-push FD, it is not possible to mitigate this feedback using a cross-coupled capacitor, as is common in mm-wave amplifier design. Consequently, the conversion gain and efficiency of the multiplier are adversely affected.

It is conventional practice to install a second harmonic trap, comprising either a transmission line at a higher frequency or a series resonator at a lower frequency, at the doubler input to minimize second harmonic feedback. However, the presence of a second harmonic short at the transistor gate results in an in-band notch in the output impedance of the doubler, which in turn restricts the doubler's bandwidth. The second harmonic feedback technique entails the introduction of an inductor in series with the transistor gate, which results in a 180-degree phase shift and the conversion of negative feedback into positive feedback, thereby enhancing the output power. However, the condition of positive feedback is largely determined by the series inductor, which also serves as a component of the input-matching network. The second harmonic feedback and input matching have a significant impact on each other, making the design procedure quite challenging.

B. proposed source-driven push-push FD

In this paper, a source-driven FD with second harmonic feedback is proposed. To decouple second harmonic feedback and input matching, the transistor gate is no longer connected to signal input and the transistor is driven only at its source. In comparison to the conventional gate-driven push-push FD, the proposed source-driven FD incorporates a resistive input, thereby facilitating wideband input matching. To invert the phase of the leakage signal, a series inductor is incorporated into the circuitry of each transistor gate. The transistors are biased at the ground to enhance efficiency.

As illustrated in Fig. 1(a), a single-ended input signal is fed to a transformer balun, after which the differential input signal is introduced to the transistor source. The transistor gate is connected in series with a series inductor. The load and output matching network is constituted by a series capacitor and a shunt inductor.

979-8-3503-6184-1/24 $31.00 © 2024 IEEE

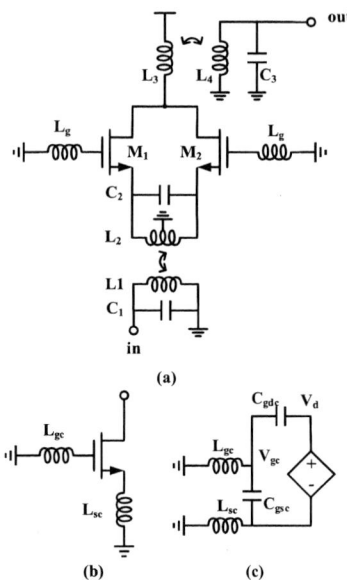

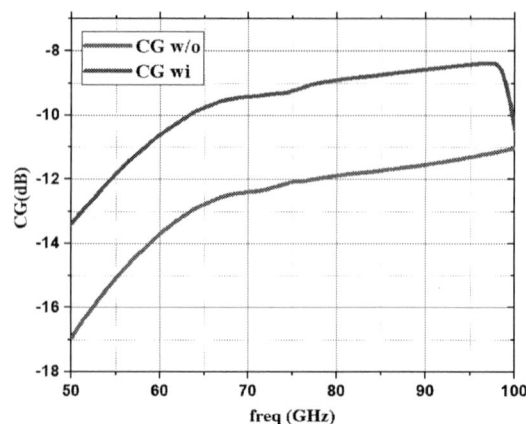

Fig.2 Simulated conversion gain of a source-driven FD with proposed 2nd harmonic feedback technique(blue line) or without proposed 2nd harmonic feedback technique(red line)

Fig.1 (a) Proposed source-driven push-push FD, (b) proposed FD in common mode and (c) small signal model of drain-gate feedback

In order to analyze the proposed second harmonic feedback technique, it is necessary to simplify the doubler in common mode, as illustrated in Fig. 1(b). In this figure, components with a suffix of "c" refer to the common mode inductor or capacitor. The small-signal mode of the common-mode doubler is illustrated in Fig. 1(c).

The transfer function from drain to V_{gs} can be derived as:

$$H(s) = \frac{s^2 L_{gc} C_{gdc}}{s^4 L_{gc} L_{sc} C_{gdc} C_{gsc} + s^2 \left(L_{gc} C_{gdc} + L_{gc} C_{gsc} + L_{sc} C_{gsc} \right) + 1} \quad (1)$$

As illustrated in Equation (1), the transfer function exhibits two zeros at DC, resulting in a 180-degree phase shift. The denominator, on the other hand, has two pairs of poles that locate slightly below $1/\sqrt{L_{gc} C_{gdc}}$ and $1/\sqrt{L_{sc} C_{gsc}}$. In the frequency range below the two pole pairs, the doubler exhibits the behavior of a common gate amplifier at the fundamental frequency and a common source amplifier at the second harmonic frequency. In both cases, the output second harmonic power is enhanced.

The strength of positive feedback is largely contingent upon the value of L_{gc}. At a relatively low-frequency band an increase in L_{gc} results in an expansion of out-of-phase swing of V_{gs} and a concomitant enhancement of output power. However, to ensure the positive feedback of the multiplier within desired frequency band, the multiplier should operate under $1/\sqrt{L_{gc} C_{gdc}}$ and $1/\sqrt{L_{sc} C_{gsc}}$. Therefore, a tradeoff exists between conversion gain and bandwidth.

As illustrated in Fig. 2, the proposed technique has the effect of increasing the conversion gain by approximately 3 dB. A notch is observed at about 100GHz, but thanks to careful sizing of both the transistor and input transformer, it's kept away from the desired frequency band of 60GHz to 77GHz and has a mild impact on in-band conversion gain.

III. IMPLEMENT OF PROPOSED FD

The proposed source-driven FD is implemented in a 40nm CMOS process. The layout is illustrated in Fig. 3 with a core area of 0.09mm^2 and a total area of 0.375mm^2 including both GSG and pad-ring. The frequency multiplier is supplied with a 1.1V supply voltage.

With an input power of 12dBm, the conversion gain and efficiency of the frequency multiplier are simulated as shown in Fig. 4. The 3dB bandwidth of the frequency multiplier is 38GHz, spanning a range of 55 GHz to 93 GHz. Accordingly, the fractional bandwidth (FBW) is calculated to be 51.3%. As illustrated in the accompanying graph, the peak drain efficiency (DE) is 15.2%.

The fundamental rejection and fourth harmonic rejection of the frequency multiplier are illustrated in Fig. 5, which has also been simulated with an input power of 12 dBm. Among the operation bandwidth, the proposed frequency doubler achieves a minimum fundamental rejection of >28dB.

The input and output return loss of the frequency multiplier are depicted in Fig 6. The matching is adequate within the operational bandwidth.

A comparison of the proposed frequency multiplier with state-of-the-art mm-wave frequency doublers is presented in Table I below. It can be observed that the proposed frequency multiplier exhibits a satisfactory bandwidth, attributable to the broadband nature of the proposed second-harmonic feedback and wideband input matching.

TABLE I.   SUMMARY AND COMPARISON OF MM-WAVE FREQUENCY DOUBLERS

Ref	This work*	[6]	[7]	[8]	[9]
Process	40nm CMOS	65nm CMOS	28nm CMOS	45nm SOI	40nm CMOS
Freq (GHz)	55-93	56-67	48-64	54-75	50-57
FBW (%)	51.3	13	28	32	12.7
CG (dB)	-5	9.1	10.2	-4.1	0
DE (%)	15.2	7.2	18.1	33	18.1
FRR (dB)	>28	>33	>24	>32	N.A.

*Simulation

979-8-3503-6184-1/24 $31.00 © 2024 IEEE

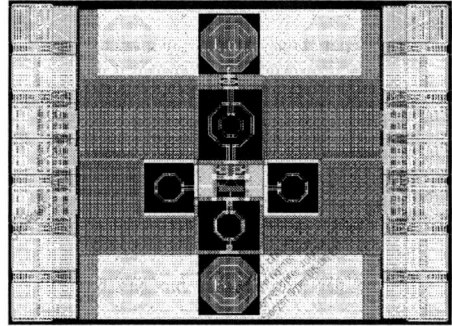

Fig. 3 Layout of the proposed FD

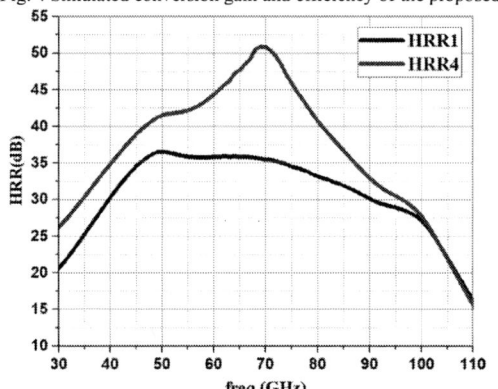

Fig. 4 Simulated conversion gain and efficiency of the proposed FD

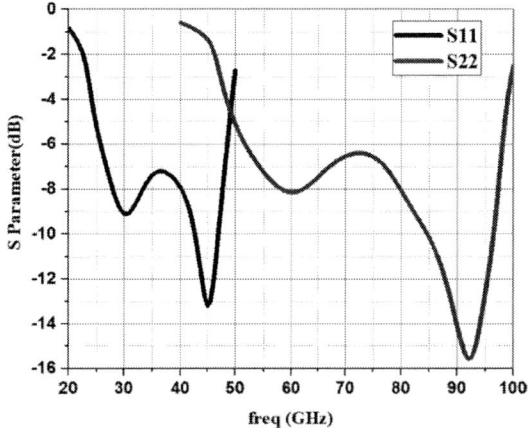

Fig. 6 Simulated return loss of the proposed FD

ACKNOWLEDGMENT

This work was supported by the Chinese National Natural Science Foundation under Grant 62122051.

REFERENCES

[1] A. Visweswaran et al., "A 28-nm-CMOS based 145-GHz FMCW radar: System, circuits, and characterization," IEEE J. Solid-State Circuits, vol. 56, no. 7, pp. 1975–1993, Jul. 2021.

[2] K. Okada et al., "A full 4-channel 6.3 Gb/s 60 GHz direct-conversion transceiver with low-power analog and digital baseband circuitry," in IEEE Int. Solid-State Circuits Conf. (ISSCC) Dig. Tech. Papers, Feb. 2012

[3] A. Aghighi, M. Essawy and A. S. Natarajan, "A Frequency Doubler With Second Harmonic Feedback for Wideband, Efficient Frequency Multiplication at Millimeter-Wave," in IEEE Transactions on Microwave Theory and Techniques, vol. 72, no. 5, pp. 2704-2715, May 2024

[4] L. Piotto, G. De Filippi, D. D. Maistro, S. Erba and A. Mazzanti, "A K-band Gilbert-Cell Frequency Doubler with Self-Adjusted 25% LO Duty-Cycle in SiGe BiCMOS Technology," ESSCIRC 2022- IEEE 48th European Solid State Circuits Conference (ESSCIRC), Milan, Italy, 2022

[5] Z. Wang, K. Ma, Z. Ma, H. Fu and J. Xu, "A 20.7–43.8-GHz Low Power Reconfigurable ×2/× 3 Frequency Multiplier for Multiple 5G-mm-Wave Bands," in IEEE Journal of Solid-State Circuits, vol. 57, no. 8, pp. 2348-2361, Aug. 2022

[6] C. So and S. Hong, "A V-Band Differential Push–Push Frequency Doubler With a Current-Reuse gm-Boosted Buffer," in IEEE Microwave and Wireless Technology Letters, vol. 33, no. 3, pp. 299-302, March 2023

[7] H. Fu, H. Wang and K. Ma, "A High Conversion Gain Frequency Doubler Using Transformer-Based Second Harmonic Feedback Technique in 28-nm CMOS," in IEEE Microwave and Wireless Components Letters, vol. 32, no. 9, pp. 1071-1074, Sept. 2022, doi: 10.1109/LMWC.2022.3168580

[8] J. Moody, "A Double Balanced Frequency Doubler Achieving 70% Drain Efficiency and 25 % Total Efficiency," 2023 IEEE Radio Frequency Integrated Circuits Symposium (RFIC), San Diego, CA, USA, 2023, pp. 157-160

[9] J. Yoo and S. Hong, "Highly Efficient Differential Frequency Doubler With Output Resistance Boosting Feedback," in IEEE Journal of Solid-State Circuits, vol. 59, no. 2, pp. 414-423, Feb. 2024

Fig. 5 Simulated harmonic rejection of the proposed FD

IV. CONCLUSION

In this paper, a source-driven push-push FD is proposed and a novel second harmonic feedback technique is applied to improve its conversion gain. Analysis and simulation demonstrate that the proposed second harmonic feedback technique can effectively enhance conversion gain without a significant impact on the bandwidth. To provide further verification, a prototype push-push FD is implemented in a 40nm CMOS process, and the simulation results show a 3dB bandwidth of 55GHz-93GHz with a fractional bandwidth of 51.3%, a peak efficiency of 15.2%, and an in-band fundamental rejection of 28dB or more.

Low Power Processor For IoT Device

Jincheng Li [1], Jiyuan Bai [1], Zelin Wang [1], GengSheng Chen[1,2] , Xiaofang Zhou*[1]

[1] School of Microelectronics, Fudan University, Shanghai, China
[2] Jiashan Fudan Institute, Jiaxing, Zhejiang Province, China

* Email:xiaofangzhou@fudan.edu.cn

Abstract—**Internet-of-Things(IoT) devices demand features such as long runtime, low heat dissipation and high reliability, as collecting and transmitting data under limited power supply. Therefore, there is a pressing demands for low power consumption and multi operation modes. Based on above requirements, we proposes ZCLP, a processor deploys clock gating and sleep mode. Prototypes are fabricated under TSMC 40nm CMOS process with an area of 0.74 mm². Measurements indicate that under 1.2V supply, our design achieves a maximum operating frequency of 30 MHz with an average 18.16 microwatts per megahertz of active mode and 13.45 microwatts per megahertz of sleep mode.**

Keywords—low power, RISC-V, processor, embedded application

I. INTRODUCTION

The embedded design has been widely applied in fields such as video surveillance, medical data collection, smart home technology, and more. Such devices frequently operate in a low-active state and are powered by batteries, demanding an ultra-low power design.

Ultra-low power strategies include multiple operating modes, clock gating, multi-voltage, and multi-VTH. Such methods have been adopted by many designs and gained promising optimization. OpenRISC1200[1] applies multi-voltage and multi-VTH. Ref[2] applies multi-voltage and multiple operating modes.

ZeroCore[3] is a processor based on the RISC-V with configurable feature, which indicates optimization potential in terms of power. We have made adjustments as follows:

1. Sleep mode is introduced. The design has active and sleep modes for different scenarios. During sleep mode, the processor stops instruction fetching, flushes the pipeline, and halts unnecessary clocks.

2. Clock gating is applied. Gated clocks are generated from the enable signal and raw clock. Ref[5][6] shows different structures of clock gating cells and we choose latch-based design, which has an advantage in area and power over flip-flop design. In addition, the structure causes no glitches in the gated clock and promote stability.

3. The number of pins is reduced. Each pin needs to connect to a PAD. The number of PADs restricts the minimum area of the chip. We use the time-division multiplexing method to simplify the pins.

Considering IoT devices give priority to power over performance, all cells except IO are using the standard cell library with high threshold voltage to minimize power consumption. We also provide a primitive version(called ZC) without sleep mode and clock gating for comparison. Through Multi Project Wafer(MPW), ZCLP and ZC, taped out using TSMC 40nm process, are tested in both operation modes.

The test results show that the chip can operate at a maximum frequency of around 30 MHz. In the active mode, the power consumption is 494 μ W, while in sleep mode, it consumes 367 μ W. The power consumption per megahertz of active mode across various frequencies is approximately 18.16 μ W/MHz. In sleep mode, the power consumption is 13.45 μ W/MHz.

II. LOW POWER DESIGN

A. Sleep Mode

The processor is waiting for the user interruption most of the execution time in reality. By introducing sleep mode, the clocks of inactive modules are shut down to decrease the dynamic power.

We set the WFI(Wait For Interrupt) instruction as the entrance of sleep mode. The clock control module controls the switch between modes. When the processor receives a WFI instruction, the clock control module drives the processor into sleep mode. After the arrival of an interrupt, the processor is awakened by the CSR (Control and Status Register) module and the clock control module.

In sleep mode, the processor does not execute instructions, halts the pipeline, and sends a stop signal to the clock control module. Except for the interrupt handling and the processor awakening, the clocks of all other circuits stop toggling. During sleep mode, any interrupt can wake up the processor. The processor can then choose to either enter the interrupt service routine or continue executing the next instruction before entering sleep mode. The register mstate in CSR module determines which method is adapted to return to active mode.

Using nop and wfi instructions as waiting scenario, while running Dhrystone program as the working scenario. Given a 1.21 V supply and a 33.3 MHz clock, the power report is shown in Table I. Though the real test may differ from the simulation as the load is not accurate, the data can indicate the significance of sleep mode. The power consumption of sleep mode decreased by 30.3% compared to NOP. The power of sleep mode is 62% of the active mode, because sleep mode only controls the core, the peripherals still work in sleep mode.

TABLE I. POWER UNDER DIFFERENT STATE[4]

Instruction or Program	Internal power / μ W	Switching power / μ W	Leakage power / μ W	Total power / μ W
WFI	153	78	2.66	234
NOP	234	99	2.75	336
Dhrystone	258	117	2.65	377

B. Clock Gating

Gated clocks are inserted automatically using Synopsys Design Compiler 2016.03(DC). For sequential logic that meets the specifications, the tool inserts clock gating cells at points corresponding to preset parameters for clock gating style. Sequential logic that supports clock gating insertion requires that when the enable is inactive, apart from reset, the output of registers must remain unchanged, as follows:

```
always @ (posedge clk or negedge rstn)
  if(~rstn) begin
          dout <= 1'b0;
  end
  else if (enable) begin
          dout <= .. .. ..;
  end
end
```

For clocks with small fan-outs, the additional power consumption introduced by clock gating cells may exceed the power that is saved through gating. Therefore, there exists an optimal threshold parameter for gating. By setting the minimum bit width in DC, the number of gated clocks and corresponding power consumption for different parameters is shown in Table II.

TABLE II.   EFFECT FOR DIFFERENT BIT WIDTH[4]

Minimum bitwidth	Num of gated clocks	Num of gated registers	Power /μW
1	206	1949	244
2	91	1882	221
3	78	1848	216
4	72	1836	221
8	70	1839	222
16	40	1695	239
32	39	1694	239
64	6	1089	332
128	2	716	394
none	0	0	536

From the table, the optimal minimum bit width of clock gating style for this design is 3.

C. Pin Simplification

ZeroCore has a total of 209 IO[3]. IO pads consume considerable power, so reducing the number of IOs can significantly lower power consumption.

For memory access pins, the address and data use 32 pins respectively. These pins are simplified through time-division multiplexing. The address is reduced to 16 bits, 64 KB in total. Data will be divided into the MSB of 16-bit and the LSB of 16-bits. Then the address and data share the same pins of 16 bits with an additional selection signal.

III. IMPLEMENTATION

A. Synthesis

The synthesis tool used is Synopsys Design Compiler 2016.03. Setting 5% of clock period as the clock uncertainty. IO pad units are selected as standard input PDIDGZ and standard output PDO04CDG.

The clock gating style is specified as follows.

```
set_clock_gating_style -sequential_cell latch
    -positive_edge_logic {integrated:CKLNQD2BWPHVT}
    -num_stages 2 -minimum_bitwidth 3 -setup 25 -hold 15
```

The sequential cell is specified as a latch. According to TABLE II, a minimum_bitwidth of 3 is optimal. The num_stages is set to 2, ensuring that manually added gated clocks can also be covered.

The area report shows that the design contains 11,131 cells, with a total area of 20,885 square micrometers. At a utilization rate of 0.5, the core is $205\mu m$ x $205\mu m$.

B. Place and Route

The back-end tool used is Synopsys IC Compiler 2016.03.

A square layout is adopted, expanding the IO to 64 in total by adding extra power and ground pads, with 16 IO pads per side. According to the documents, the size of the IO pad units is 190μm x 30μm. Calculating based on this, the core size is 480μm per side, and the chip size is 860μm per side. Due to the constraints of IO quantity and size, the actual chip area exceeds the area reported by DC, in other words, the PAD limit exists.

The design uses the 1p7m5x1z technology file, which includes 7 metal layers (M1 to M7) and 1 AP layer. The power ring is set on M6 and M7 layers, connecting to the power straps on these layers. Specifically, the power straps on M6 are vertical, with a width of 2μm and a spacing of 30μm. The power straps on M7 are horizontal, with a width of 10μm and a spacing of 30μm.

The layout strategy prioritizes power efficiency. For clock tree synthesis, the routing rules include double-row spacing.

C. Verification

The verification tool used is Mentor Calibre. After LVS and DRC checks, the final layout is shown in Figure 1, where the left is ZC, and the right is ZCLP.

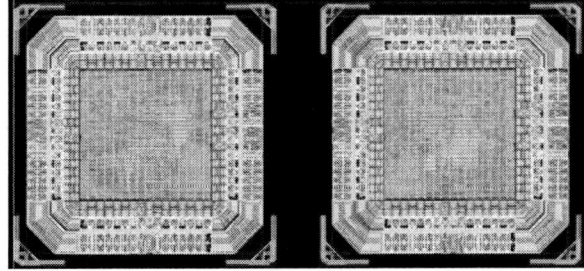

Fig. 1. Layout view

D. Overview

From post lay-out simulation to get feature of ZCLP. The data is listed in Table III

TABLE III.   FEATURE SUMMERY

Feature	Data
Process	TSMC40LP
Efficiency	0.26DMIPS/MHz

Feature	Data
Maximum frequency	33MHz
Number of logic cell	12,492
Area	0.74 mm^2

IV. TEST AND ANALYSIS

A. Test Platform

Power is measured indirectly by measuring the supply current at a given voltage. The operating voltage of the chip can be read from the digital screen on the board, and the current operating current can be obtained by serially connecting an ammeter.

The overall testing environment is shown in Figure 4.

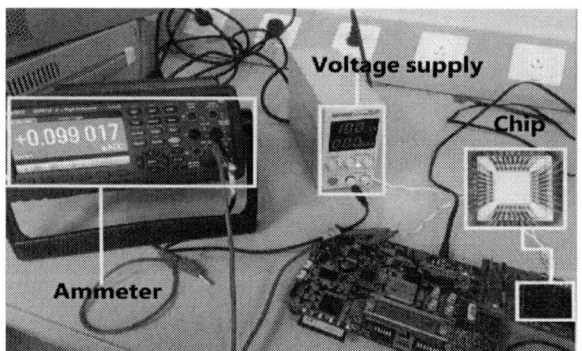

Fig. 2. Test platform

B. Measurement

1) Sleep Mode

Under 1.2 V supply voltage, as the clock frequency increases from 5MHz to 30MHz, the power is shown in Fig 5.

The power per megahertz of ZC and ZCLP has a minimum at 20 MHz, measuring 24.18 μ W/MHz and 10.8 μ W/MHz. The average power per megahertz is 26.3 μ W/MHz for ZC and 13.45 μ W/MHz for ZCLP. The data indicates that introducing sleep mode reduces the total power of the processor when inactive by 48.9%.

POWER (UW)

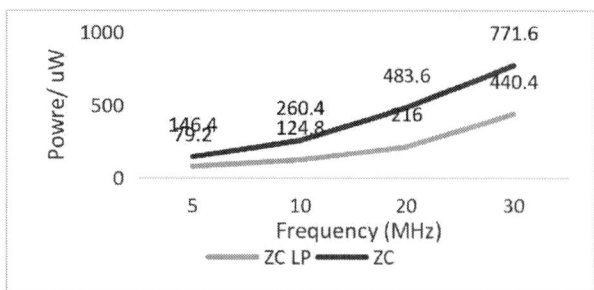

Fig. 3. Sleep power under different frequency

2) Active Mode

Under 1.2 V supply voltage, as the clock frequency increases from 5MHz to 30MHz, the power is shown in Fig 6.

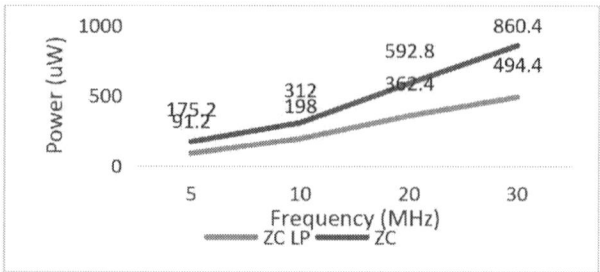

Fig. 4. Active power under different frequency

The power per megahertz of ZC and ZCLP has a minimum decrease as the frequency increases and has a minimum on 30 MHz, 28.68 μ W/MHz for ZC, and 16.68 μ W/MHz for ZCLP. The average power per megahertz is 31.14 μ W/MHz for ZC and 19.36 μ W/MHz for ZCLP. The power consumption of ZCLP is approximately 55% of ZC.

The data also shows that as the frequency increases, the power of active mode rises faster than sleep mode except on 30 MHz. The sleep mode generally consumes 40% less power than the active mode.

Compared to OpenRISC1200[1] whose total power is 12.481 mW at 200 MHz or 62.4 μ W/MHz, and the design in Ref[2] achieves a minimum energy operation of 190 pJ/cycle or 190 μ W/MHz, our design can only operate below 30MHz but consume far less power.

V. CONCLUSION:

This paper proposes a processor ZCLP for embedded devices. Prototypes of ZCLP and its primitive version ZC are fabricated under TSMC 40 nm process by the MPW project. Test the power of different modes and compare the power of the designs to get precise optimization.

The design deploys sleep modes and clock gating, pins are simplified by multiplexing. Power consumption in sleep mode has been reduced to approximately 54.2%, while active power is reduced to around 55%. Across the functional frequency, power consumption per megahertz measures 18.16μW in active mode and 13.45μW in sleep mode, which caters to the low-power and small area requirements of embedded applications.

REFERENCES

[1] V. Melikyan, T. Hakhverdyan, S. Manukyan, A. Gevorgyan and D. Babayan, "Low power OpenRISC processor with power gating, multi-VTH and multi-voltage techniques," *2016 IEEE East-West Design & Test Symposium (EWDTS)*, Yerevan, Armenia, 2016, pp. 1-4, doi: 10.1109/EWDTS.2016.7807678.

[2] C. Duran et al., "An Energy-Efficient RISC-V RV32IMAC Microcontroller for Periodical-Driven Sensing Applications," 2020 IEEE Custom Integrated Circuits Conference (CICC), Boston, MA, USA, 2020, pp. 1-4, doi: 10.1109/CICC48029.2020.9075877.

[3] Ziwei Liu. Design and Implementation of an Embedded Processor Core Based on RISC-V ISA[D].Fudan University,2021.

[4] Zelin Wang, Low Power Design and Implementation of Embedded RISC-V Processors[D].Fudan University,2020.

[5] J. Shinde and S. S. Salankar, "Clock gating — A power optimizing technique for VLSI circuits," *2011 Annual IEEE India Conference*, Hyderabad, India, 2011, pp. 1-4, doi: 10.1109/INDCON.2011.6139440.

A Heterogeneous Integration System of Analog In-Memory Computing and Field-Programmable Gate Array

Hua Chen [1,2], Yiming Qu [1], Wenhao Wu [1,2], Yi Zhao *[1,2,3]

[1] School of Integrated Circuits, East China Normal University, Shanghai 200241, China
[2] China Nanhu Academy of Electronics and Information Technology, Jiaxing 314001, China
[3] College of Information Science and Electronic Engineering, Zhejiang University, Hangzhou 310027, China

* Email: yizhao@zju.edu.cn

Abstract—With the development of deep learning models, deploying neural networks to resource-limited edge devices has become a trend. However, the high computation requirements and energy consumption of neural network models have limited their development. Analog In-Memory Computing (AIMC) features high energy efficiency, providing new opportunities for the edge deployment of neural networks. The system needs to support different types of neural network models in practical applications, and the system-level application is still immature. This paper proposes a heterogeneous system integrating analog in-memory computing and Field-Programmable Gate Array (FPGA). The system reserves multiple sensor interfaces for access and preprocessing peripherals, including digital NN accelerators and AIMC, which can effectively execute various layers in neural network models, achieving complete end-to-end neural network inference. Depending on the size and number of FC layers, the system has varying degrees of improvement in inference time and RAM occupation.

Keywords—*Analog in-memory computing, heterogeneous integration system, accelerator.*

I. INTRODUCTION

Deep learning models have been widely applied to various real-life scenarios, including language translation, computer vision, autonomous vehicles, etc. [1,2,3]. Deploying neural networks to resource-limited edge devices can significantly enhance the intelligence level of the system, while also protecting privacy by avoiding data transmission to the cloud. However, the high computational power required for neural network inference and the resulting high energy consumption hinder this trend. The main part of neural network inference computing is matrix-vector multiplication, and its execution efficiency is limited by the output bandwidth between the data and computation units. The traditional von Neumann architecture struggles to overcome the memory wall problem. In-memory computing (IMC) is an emerging technology that is expected to solve the aforementioned issues by performing computations directly where the data is stored, thereby reducing the high energy consumption of data transfer between memory and computation units in data-intensive applications like ML [4]. Previous research has shown that analog In-memory computing technology exhibits exceptional peak energy efficiency [5]. When implementing dense fully connected (FC) layers in DNNs, the AIMC architecture can achieve orders of magnitude reductions in latency and energy consumption [6]. However, when executing Convolutional Neural Networks (CNNs), it is susceptible to reliability issues caused by noise and interconnect parasitics. On the other hand,

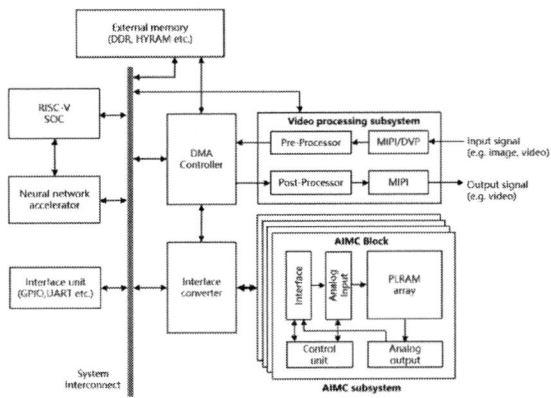

Fig. 1. Overview of Heterogeneous Integration System.

in practical applications, neural network models consist of various types of layers (e.g., fully connected, convolution, depthwise separable convolution, etc.), and the system needs to support the aforementioned operations to complete end-to-end inference [7]. Moreover, when considering edge system-level applications, the requirements may be diverse. For example, the system may need to support multiple sensor interfaces to access various sensors and preprocess sensor signals before performing neural network inference.

To address the aforementioned challenges, this paper proposes a heterogeneous system that integrates AIMC and FPGA. Firstly, it combines the advantages of traditional digital processing circuits and AIMC to effectively execute various types of layers in neural network models (fully connected, convolution, depthwise separable convolution, etc.). Secondly, the system integrates a rich variety of peripheral modules to facilitate the connection of various sensors. Moreover, since the system uses FPGA to construct circuits except AIMC, it endows the system with greater flexibility and scalability.

The structure of this paper is as follows: Section 2 provides a detailed description of the proposed heterogeneous system. Section 3 describes the AIMC subsystem used in this study. Section 4 conducts experiments and analyzes the results. Finally, Section 5 summarizes this paper.

II. SYSTEM OVERVIEW

Figure 1 illustrates the heterogeneous integrated system proposed in this paper, which includes a RISC-V SOC, a

---

This work was supported in part by the National Key Research and Development Program of China under Grant 2020AAA0109001, and the National Natural Science Foundation of China under Grant U23B2040.

979-8-3503-6184-1/24 $31.00 © 2024 IEEE

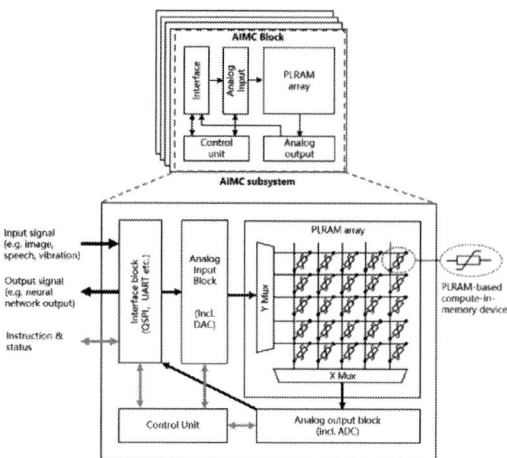

Fig. 2. Block diagram of AIMC subsystem.

neural network accelerator based on all-digital design, an image processing subsystem, an AIMC subsystem, and peripherals for efficient handling of transmission and interfaces, such as Direct Memory Access (DMA), GPIO, UART, etc. Except for the AIMC subsystem, the rest of the system's circuits are built using Efinix's Titanium series FPGA [8].

The task scheduling and parameter configuration of the system are mainly accomplished by the RISC-V SOC built on VexRiscv, with a working frequency ranging from 20MHz to 400MHz. The video processing subsystem receives external images or video inputs in the input direction, which are transmitted to the preprocessing module via the MIPI/DVP interface. The preprocessing module (Pre-Processor) can process the input signals in terms of frame rate, resolution, and format conversion according to the configuration of the RISC-V processor. In addition, it can also perform mean filtering, median filtering, and other processing as configured, before finally storing the processed data in the external memory via DMA. In the output direction, the video processing subsystem reads data from the external memory through DMA and sends it to the post-processing module (Post-Processor). The post-processing module adjusts the video data resolution and frame rate according to the configuration information of the RISC-V processor for proper display.

The neural network accelerator and AIMC subsystem are two main modules for accelerating neural network inference. The neural network accelerator is built based on Efinix's tinyML platform [9] In this system, operators such as convolution, depthwise separable convolution, ADD, and multiplication are mainly implemented. Since the fully connected layer (FC) consists of vector-matrix multiplication operations, the AIMC subsystem is used to complete it.

In addition to supporting FC, the AIMC subsystem also supports operations such as conv, depthwise conv, and relu. Although analog computing may cause some precision loss, its peak energy efficiency can reach more than 10 TOPS/W, making it an independent low-power subsystem suitable for operations with low precision requirements (such as the FC layer in neural network inference) and tasks that are not power-sensitive but have low computational requirements (such as voice keyword wake-up). Another advantage of using AIMC for neural network inference acceleration is that

weights can be directly stored within the memory computing unit, without occupying additional space in the system's SRAM and external memory.

In terms of external memory, low-power Double Data Rate (LPDDR) memory or Hyper RAM is used, which is particularly suitable for edge devices due to its lower operating voltage and power-saving modes.

III. AIMC SUBSYSTEM ARCHITECTURE

Figure 2 presents the AIMC subsystem used in this design, which consists of four AIMC blocks. Each block includes an integrated in-memory computing array composed of PLRAM [10] devices, an analog input module (including a Digital-to-Analog Converter, DAC), an analog output module (including an Analog-to-Digital Converter, ADC), a control unit, and an interface module.

The core of the AIMC subsystem is an array with a crossbar structure formed by PLRAM devices. The array consists of 4096 columns and 1024 rows, with a PLRAM device at each intersection of a row and column, used to store 7-bit neural network weights. AIMC accelerates the key operation of vector and matrix multiplication in neural networks based on Kirchhoff's laws of voltage and current. Specifically, the input voltage on the word line represents the value of the input vector, and the result of the multiplication and accumulation operation in each dimension is obtained by measuring the total current accumulated on the bit line.

The analog input module uses DAC circuits to convert the input digital signals into the analog voltage signals required by the array, while the output module uses current mirrors and ADC circuits to convert the analog signals into digital signals and then transmits these signals to the interface module. The interface module includes digital interfaces such as QSPI, SPI, UART, I2C, and I2S for instruction and data transmission. The configuration and data flow control of the aforementioned modules are completed by the control unit, which also has certain arithmetic-logical capabilities, such as addition, subtraction, shifting, etc., that can be used to implement activation operations like relu in neural networks.

Combining the integrated in-memory computing array based on PLRAM devices and logical operation functions, the AIMC subsystem supports operations such as convolution, depthwise separable convolution, fully connected, relu, DFT, etc. It has the capability to independently deploy neural network models and exhibits high energy efficiency.

IV. RESULTS AND DISCUSSIONN

The digital circuit part of the integration system proposed is constructed using FPGA, and the AIMC is integrated into the system based on previous research results. Both are systematically packaged to form the overall system, as shown in Figure 3. This section first introduces the software toolchain used for neural network deployment on the host and the integration system, followed by an analysis of the experimental results.

A. Software Toolchain

To enable user-friendly neural network model deployment on this system, customized network parser has been developed.

Specifically, the FC layer will be executed by the AMIC, while the remaining layers will be executed in the CPU and the Neural Network Accelerator. On the host processor side,

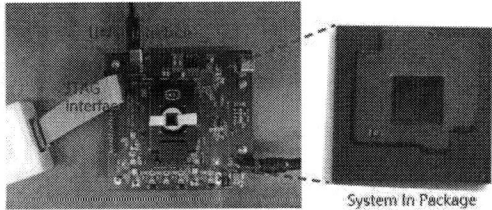
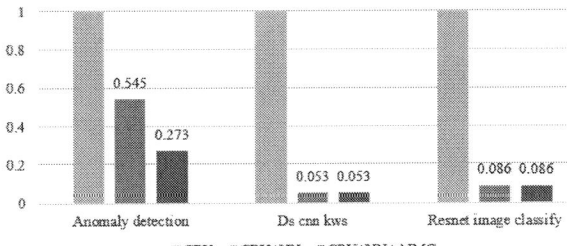

System In Package

Fig. 3. Block diagram of AIMC subsystem.

the neural network model is trained using TensorFlow or PyTorch, and then the trained model is converted into TFLite format data. The model is quantized using TensorFlow Lite.

B. Results Analysis

We selected three tasks for testing and conducted comparative experiments under the CPU (RISC-V), CPU+NN, and CPU+NN+AIMC systems, with results shown in Figures 4, 5, and Table 1. Task 1's neural network model is a fully connected (FC) network, while FC layers in Tasks 2 and 3 occupy a smaller proportion. From the experimental results, it is evident that the inference time is longest when using only the CPU. By enabling the digital NN accelerator, the execution of Tasks1, 2 and 3 was accelerated by 83%, 17.8 times, and 10.6 times, respectively. When enabling AMIC to execute Task 1, the inference speed further increased by 100%, while the acceleration for Tasks 2 and 3 was not significant. When using AMIC for inference, since the weights are stored in PLRAM, there is no need to occupy RAM to store weights during runtime, resulting in reduced RAM usage for Tasks 1, 2, and 3 by 258KB, 0.75KB, and 0.625KB, respectively. In addition, the inference accuracy for all three tasks hardly decreased.

V. CONCLUSION

This paper designed a heterogeneous system which integrates an AIMC and FPGA. Since the system has many sensor interfaces and processing peripherals, it can effectively execute various types of neural network models, achieving complete end-to-end neural network inferences. Specifically, it is recommended to use AIMC for executing the FC layer, while the remaining layers should be handled by digital NN accelerators and CPUs. Experimental results show that by integrating AIMC computation, the system's inference accuracy could be hardly affected, and depending on the size and number of FC layers, the system exhibits various degrees of improvement in the inference time and RAM usage.

REFERENCES

[1] T. Hemed, N. Lavie and R. Kaplan, "Distributed Deep Learning on Wimpy Smartphone Nodes," 2018 IEEE International Conference on the Science of Electrical Engineering in Israel (ICSEE), Eilat, Israel, 2018, pp. 1-5.

[2] Zhenxing Zhou, Yisiang Neo, King-Shan Lui, Vincent W.L. Tam, Edmund Y. Lam, Ngai Wong, "A portable hong kong sign language translation platform with deep learning and jetson nano," The 22nd International ACM SIGACCESS Conference on Computers and Accessibility, 2020, pp.1-4

[3] Miotto R, Wang F, Wang S, Jiang X, Dudley JT, "Deep learning for healthcare: review, opportunities and challenges," Brief Bioinform, 2018, 19(6), pp1236-1246.

[4] Abu Sebastian, Manuel Le Gallo, Riduan Khaddam-Aljameh, Evangelos Eleftheriou, "Memory devices and applications for in-memory computing," Nature Nanotechnology, 2020, 15, pp. 529–544.

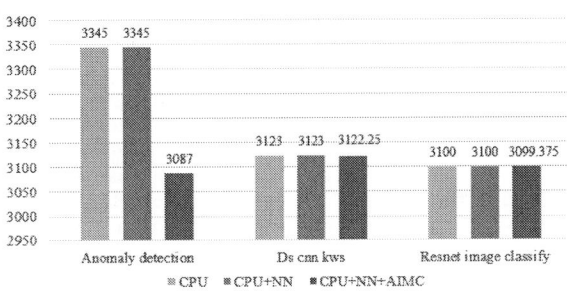

Fig. 4. Inference time on three tasks.(normalized to CPU baseline)

Fig. 5. RAM used on three tasks in KB.

TABLE I. RAM REDUCTION, AND ACCURACY DIFFERENCE UNDER DIFFERENT TASKS.

No.	Tasks	RAM reduction (KB)	Accuracy difference
1	Anomaly detection	258	-0.01%
2	Ds cnn kws	0.75	-0.02%
3	Resnet image classify	0.625	-0.02%

[5] Naveen Verma, Hongyang Jia, Hossein Valavi, Yinqi Tang, Murat Ozatay, Lung-Yen Chen, et al., "In-memory computing: Advances and prospects," IEEE Solid-State Circuits Magazine, 2019, vol. 11, no. 3, pp. 43–55.

[6] Mohammed Elbtity, Abhishek Singh, Brendan Reidy, Xiaochen Guo, Ramtin Zand, "An in-memory analog computing co-processor for energy-efficient cnn inference on mobile devices," 2021 IEEE Computer Society Annual Symposium on VLSI (ISVLSI), 2021, pp. 188–193.

[7] Sunil Shukla, Bruce Fleischer, Matthew Ziegler, Joel Silberman, Jinwook Oh, Vijavalakshmi Srinivasan, et al., "A scalable multi-TeraOPS core for AI training and inference," IEEE Solid-State Circuits Lett., 2018, vol. 1, no. 12, pp. 217–220.

[8] https://www.efinixinc.com/products-titanium.html

[9] https://www.efinixinc.com/solutions-tinyml.html

[10] Shifan Gao, Guangjun Yang, Xiang Qiu, Chun Yang, Cheng Zhang, Binhan Li, et al., "Programmable Linear RAM: A New Flash Memory-based Memristor for Artificial Synapses and Its Application to Speech Recognition System," 2019 IEEE International Electron Devices Meeting (IEDM), 2019, pp. 14.1.1-14.1.4

A 10-MHz Four-Phase Hysteretic Control DC-DC Converter with Inductor Current Self-balancing

Yushen Zhang[1], Ningning Li[1], Yibo Zhang[1], Yizhe Yang[1], Yimeng Zhang*[1,2], Yuming Zhang[1,2]

[1] School of Microelectronics,Xidian University

[2] Shaanxi Key Lab of Integrated Circuits and Systems,Xidian University

Xi'an,China

*Email:zhangyimeng@xidian.edu.cn

Abstract—In order to meet the stringent requirements of modern application CPU for fast transient response and high current density, a 10-MHz four-phase dc–dc power converter is presented. It employs a hysteretic current control mode to increase the ultra-fast transient response as much as possible, and in order to minimize severe heat distribution issues in integrated multiphase converters in CCM mode with high currents, the multi-phase current self-calibration technology is also equipped in the loop. Thus, the phase currents will be balanced efficiently and precisely, and the heat loss due to inductive mismatch will be reduced. The hysteretic current buck converter that is designed in a 0.18μm CMOS process.The hysteretic current buck converter that is designed in a 0.18μm CMOS process. Achieved a maximum error of only 30mA (6%) in each phase current at a load current Io of 5.5A and a load current step of 2.5A/1ns.

Keywords—*multi-phase converter, hysteretic current control, high current density, current self-balance*

I. INTRODUCTION

With the iteration and development of CPUs, their main frequency has reached the GHz level. Considering the greater power density, a faster operating frequency has also become a major issue in the design of CPU power management chips. To handle larger operating currents, a multiphase buck converter has been proposed. Subsequently, a hysteretic current window control technique [1] has been introduced to cope with faster load transient response. However, this technique is overly dependent on the detection accuracy of the inductor current. Compared to a single-phase converter, the current information V_{sens} ignores the ripple effect of the output voltage [2], rendering the control loop ineffective as shown in Fig. 1.

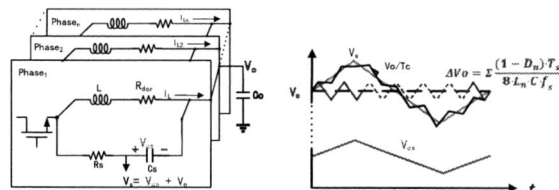

Fig. 1. Interference to the sampled currents in multiphase converters

In the multi-phase converter, due to manufacturing processes, there is a 20% deviation between the inductance L and the DCR resistance [1]. This inevitable mismatch, caused by actual process parameters, ultimately leads to an imbalance in the distribution of current I_L among the various phases of the inductors L_n, as shown in Fig. 2. This not only significantly affects efficiency but can also cause damage to the chip.

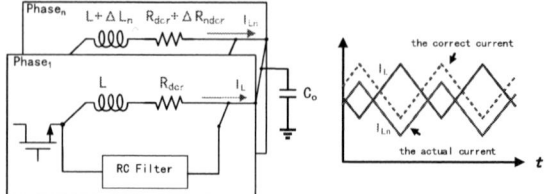

Fig. 2. Inductor mismatch causes current imbalance

In this paper, a hysteretic current self-balancing technique is proposed to solve the current imbalance problem in multiphase circuits with respect to the above problem.

II. PROPOSED DESIGN

A. Multi-phase Hysteretic Current Control Buck Converter

Fig. 3 shows the top block diagram of the proposed multiphase buck converter. It consists of four parts: inductor current sensor, current self-balance circuit, hysteretic controller, and ZVS soft-switching.

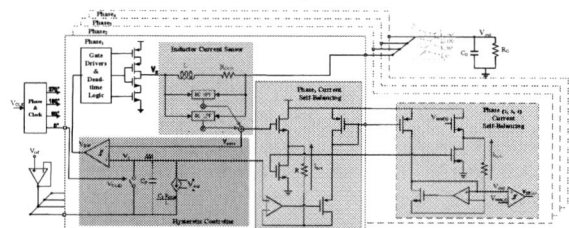

Fig. 3. Block diagram of proposed design

The inductor current sensor is used to detect the current information current V_{sens} of the power inductors, and V_{sens} is sent to the controller to compare with the sawtooth wave signal V_c to generate the hysteretic window signal V_{hys} and the switch signal V_{sw}, while the hysteretic window signal is sent to the current self-balancing circuit to adjust V_c to compensate for the mismatches occurring in the inductors of each phase, and make sure the chip can work properly.

B. Inductor Current Sensor

The sampling in the diagram is composed of DC amplification and AC amplification parts. First, in the DC part, the sampling branch for the inductor's V_{dcr} consists of R_s and C_s. R_{dc} and the M_l transistor form a common-source amplifier with a tail current source, ensuring that the voltage across the operational amplifier is V_{dcr}. The operational amplifier is used to compress the small signal on V_{dcr}, reducing the output Vo ripple interference（1）.

979-8-3503-6184-1/24 $31.00 © 2024 IEEE

$$\Delta V_o = \sum_1^n \frac{(1-D_n)\cdot T_s}{8\cdot L_n \cdot C \cdot f_s} \tag{1}$$

The resistor R_{dc} only samples the DC component of V_{dcr} and converts it into a current I_{dc}, which is then mirrored to R_{sens} by the current mirror composed of M_1-M_3. The amplification factor of the DC value V_{sens_dc} is determined by R_{sens}/R_{dc}.

The AC part consists of the branch formed by R_{ac} and C_{ac}. By changing the ratio of R_{ac}/R_s, the sampling ripple of the inductor current can be amplified and filtered for AC signals, resulting in V_{sens_ac}. Finally, these two components are superimposed on R_{sens} to obtain the V_{sens} signal. M_4 acts as a voltage buffer.

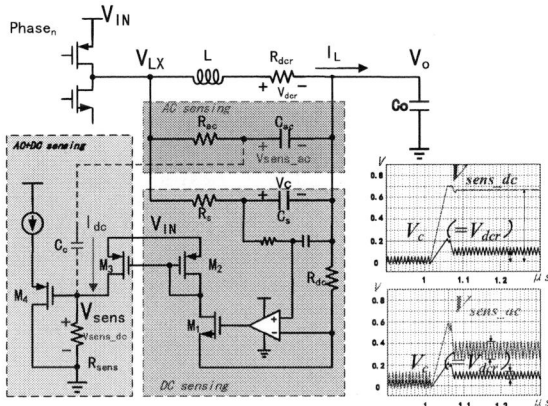

Fig. 4. Block diagram of proposed current Sensor

C. Current Self-balancing Technology

The V_{sens} signal obtained in the previous step, due to the relationship （2）in the DCR sampling,

$$\frac{R_{dcr}}{L} = R_c \times C_c \tag{2}$$

it has a falling slope usually determined by R_c and C_c in the second half of the T/N cycle. The sawtooth wave signal V_c generated in the hysteretic controller does the same matching of R_{dcr}/L, so the falling slope of the V_c signal remains essentially the same as that of V_{sens}. In the ideal equivalent model, there is no error in the currents of each phase, but as mentioned in Section 1, the inductor mismatch phenomenon is unavoidable and the hysteretic controller is unable to correct the DC error present in the currents of each phase. In order to solve this problem, a current self-calibration technique is introduced in Fig. 5, in hysteretic controllers, V_{sens} is usually compared with V_c, and the difference between the two is regarded as a hysteretic window, V_{hys}, when the value of the window is zero, and the output of the hysteretic comparator is changed, and it is considered that the power MOSFETs are turned off and the inductor is discharged and when the window opens, the inductor is charged. Although the inductor is charged and discharged with the same magnitude and period in each phase, there is no way for the controller to detect the presence of error current. However, each window voltage V_{hys} can reflect the DC information of the inductor current on that phase, which means that the comparison between the individual window voltages can be used to obtain the information of the current error existing between each phase, as shown in Fig. 5(b) the inductor current error ΔI_L in each

phase can be reflected as the difference of the average value of the window voltage in each phase. So the window V_{hys} of the main phase is converted to the window current I_{hys} by the resistor R in the circuit frame diagram of Fig. 5. For the window current I_{hysn} of the other phases is connected to the main phase window current I_{hys} through the current mirror, so that make $I_{hysn}=I_{hys}$, and $\Delta I_{hys}=0$, and the window size V_{hysn} of the other phases is changed, then V_{cn} of the rest of the phases is also changed, and the hysteretic boundaries of the V_c signals are changed, The calibrated signal V_{cn} will be compensated by the hysteretic controller to the loop, which finally makes the current error $\Delta I_L=0$ in each phase, as shown in Fig.5.(c1).

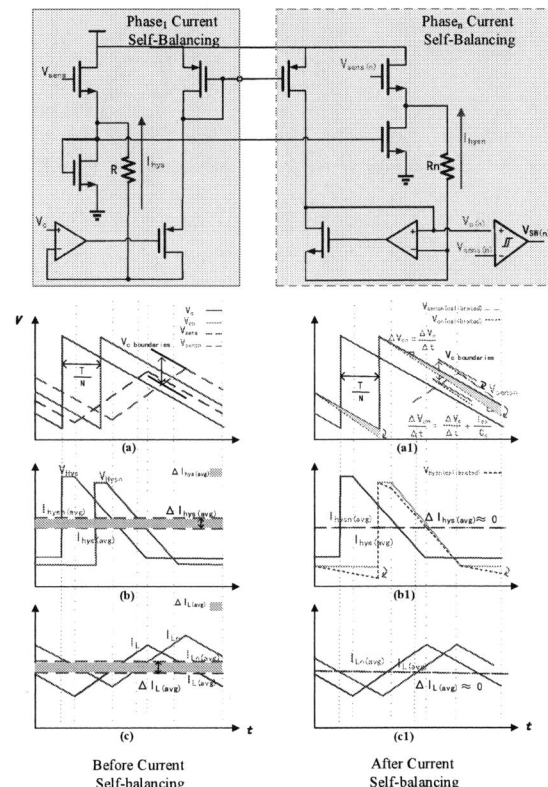

Fig. 5. Proposed current self-balancing circuit and timing

III. SIMULATION RESULTS

The proposed multiphase buck converter incorporates a modified current sensor and a current self-balancing technique and is simulated using a 0.18μm CMOS process. The converter operates with four phases at a 10 MHz switching frequency, utilizing 100 nH inductors per phase. The equivalent impedance R_{dcr} of the inductors is 30 mΩ, with an industry standard ±20% mismatch in L and DCR between phases. Simulation results, shown in Fig. 6, indicate that without the proposed circuit, the maximum current mismatch is 186 mA (31.3%) following a load step from 1A to 5.5A. With the proposed circuit enabled, the maximum current mismatch is significantly reduced to 30 mA (6.7%), demonstrating effective sampling and current equalization capabilities. These results validate the effectiveness of the proposed technique.

979-8-3503-6184-1/24 $31.00 © 2024 IEEE

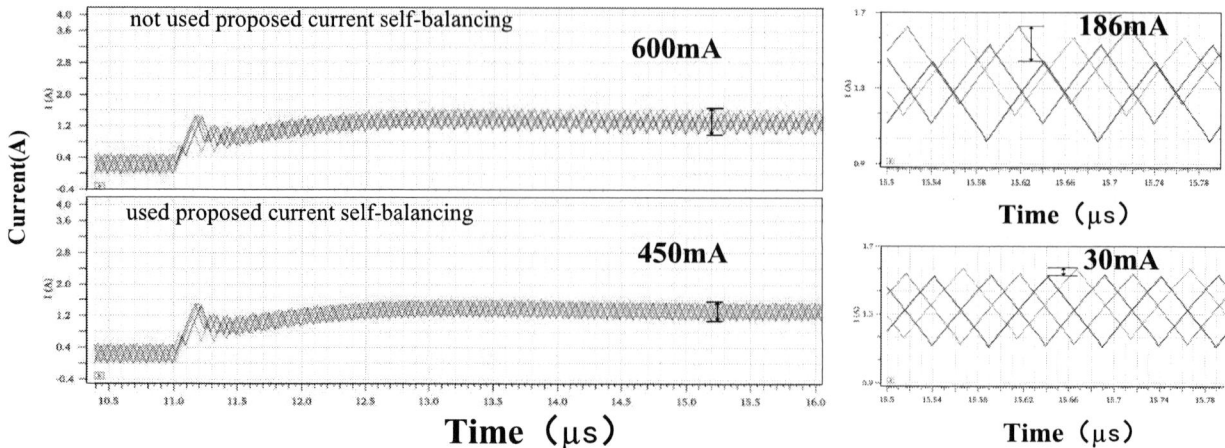

Fig. 6. Inductor current waveforms with and without the proposed circuit

In order to verify the effect of the proposed circuit on the load transient of the entire system, significant load step changes ranging from 0.1 to 2.6 A were simulated. In response to a 2.5 A/1 ns step-up in Io, as shown in Fig. 7, the converter achieves an undershoot voltage below 49 mV. Table 1 compares the performance of this work with results from related papers in recent years. This work demonstrates the most accurate current balance and smaller V_{out} droop compared to other recent articles.

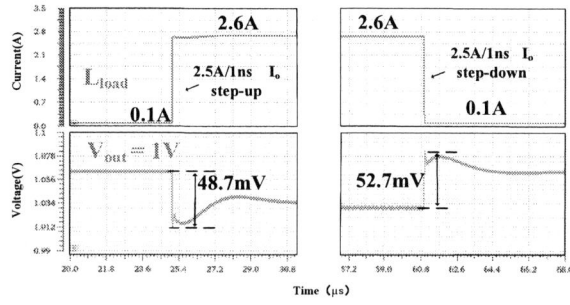

Fig. 7. The simulation waveform for load transient

TABLE I. PERFORMANCE COMPARISON TABLE

Parameter	Reference			
	[1][1]	[2][1]	[3][1]	This work[2]
Technology	350nm	180nm	180nm	180nm
Control	SAW Hysteretic	ZDS Hysteretic	DAB Hysteretic	CM Hysteretic
# of Phase	4	4	4	4
Ind. (nH)	200	78	18/60/100	100
Switching Frequency	25MHz	40MHz	25/25/10 MHz	10MHz
Ind. Matching[3]	High	High	Low	Low
Current Balance	No	No	Yes	Yes
Vout Droop	103mv @1.6V	118mV @1.2V	100mV @1.2V	48.7mV @1V

Parameter	Reference			
	[1][1]	[2][1]	[3][1]	This work[2]
@Current Step	4A/5ns	5A/5ns	2.85A/2ns	2.5A/1ns
Max.Load Current	6A	6A	4A	5.5A

[1]Measured[2]Simulated[3]the Matching Requirement of RC & L/DCR

IV. CONCLUSION

This paper introduces a 10-MHz four-phase DC-DC power converter incorporating hysteretic current control and an inductor current self-balancing technique. The converter maintains phase current balance consistently during operation. Simulation results validate effective mitigation of phase inductor current imbalances and address limitations of conventional current sampling methods. This approach represents significant advancements in current sampling and balancing methodologies, making it well-suited for high-power CPU applications.

ACKNOWLEDGMENT

This work was supported by the National Natural Science Foundation of China (Grant No. 62234010).

REFERENCES

[1] B. Lee, M. K. Song, A. Maity and D. B. Ma, "10.7 A 25MHz 4-phase SAW hysteretic DC-DC converter with 1-cycle APC achieving 190ns tsettle to 4A load transient and above 80% efficiency in 96.7% of the power range," 2017 IEEE International Solid-State Circuits Conference (ISSCC), San Francisco, CA, USA, 2017, pp. 190-191

[2] M. K. Song, J. Sankman and D. Ma, "4.2 A 6A 40MHz four-phase ZDS hysteretic DC-DC converter with 118mV droop and 230ns response time for a 5A/5ns load transient," 2014 IEEE International Solid-State Circuits Conference Digest of Technical Papers (ISSCC), San Francisco, CA, USA, 2014, pp. 80-81

[3] L. Zhao, J. Tang, K. Wei and C. Huang, "A 4-Phase DAB Current-Mode Hysteretic Controlled Buck Converter With Relaxed Inductor Requirements and Enhanced DC and Dynamic Performance," in IEEE Journal of Solid-State Circuits, vol. 59, no. 5, pp. 1556-1566, May 2024,

[4] E. A. Burton et al., "FIVR — Fully integrated voltage regulators on 4th generation Intel® Core™ SoCs," 2014 IEEE Applied Power Electronics Conference and Exposition - APEC 2014, Fort Worth, TX, USA, 2014, pp. 432-439

A High Precision -40 °C to 150 °C Bandgap Reference with Dual Temperature Compensation

Yuhan Zhang, Jianzheng Li, Xiaomeng An, Lina Wang, Yajie Qin*

School of Information Science and Technology, Fudan University, Shanghai 200433, China

* Email: u202014700@hust.edu.cn, yajieqin@fudan.edu.cn

Abstract—This paper presents a novel bandgap reference circuit with a dual temperature compensation method that obtains a high precision voltage over a wide temperature range. The proposed compensation method combines a conventional calibration method with a multiband curvature compensation technique to further suppress the effects of temperature drift caused by transistors, resistors, and other devices. The proposed bandgap is implemented in a 110 nm CMOS process with a supply voltage of 1.5 V. Post layout simulation results show an optimal temperature coefficient (TC) of 4.8 ppm/°C from -40 °C to 150 °C, with a power consumption of 14.8 μA.

Keywords—wide temperature range, dual temperature compensation (DTC), multiband curvature compensation

I. INTRODUCTION

To function reliably under harsh conditions, chips for automotive and space satellite applications must maintain stability over a wide temperature range. Bandgap reference circuits with low temperature drift performance over a wide temperature range are essential for such analog chips. Numerous compensation methods aim to minimize the errors caused by the nonlinear dependence between temperature and emitter–base voltage (V_{EB}). For example, multiband curvature compensation as used in [1], but due to the large peak to peak values of the original output curve, it takes a relatively large number of compensation branches to achieve a low temperature coefficient (TC). Some designs use resistors to extract the nonlinear term [2], but this approach cannot avoid approximation errors and the intrinsic temperature coefficient of the device. Other methods exploit the zero temperature coefficient point of a MOSFET to design the bandgap reference, but the operating temperature range may be limited. To achieve lower temperature coefficients over a wide temperature range, this design employs the core bandgap reference structure of the current mode, using resistors to extract the nonlinear term, and integrates multiband curvature compensation technique, which further improves the accuracy of the bandgap reference with lower power consumption.

II. TECHNICAL PRINCIPLE AND CIRCUIT STRUCTURE

A. Current Mode Bandgap Reference Principle

Fig. 1 shows the current mode bandgap reference, which is the core structure of this design. Two folded cascode amplifiers, A_1 and A_2, are used to set the voltages at A, B, and C to V_{EB2}. The start up circuit pulls down the voltage at V_p during initialization to avoid the degenerate operating point.

The ratio of R_1 and R_2 is $\frac{\partial |V_{BE}|}{\partial T} : \frac{\partial V_T \ln N}{\partial T}$ for offsetting first order temperature coefficients. Amplifier A_1 together with M_1, M_2, Q_1, and Q_2 generates a proportional to absolute temperature (PTAT) current, while amplifier A_2, M_3, and R_2 generate a complementary to absolute temperature (CTAT) current. These two currents are replicated and added together,

This work was supported by the National Key Research and Development Program of China (2023YFB4704000).

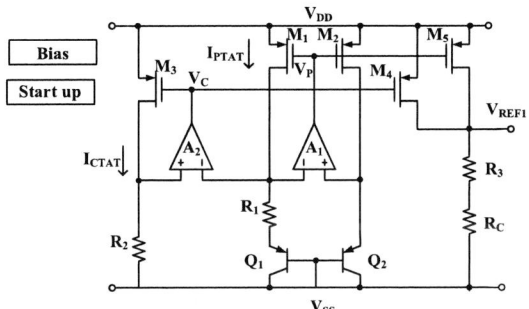

Fig. 1. Core bandgap reference structure.

generating a reference voltage with zero temperature drift through R_3 and R_C.

The expression for the output voltage is as follows:

$$V_{REF1} = \frac{R_3 + R_C}{R_2}\left(|V_{BE2}| + \frac{R_2}{R_1}V_T \ln N\right) \quad (1)$$

where N is the ratio of the areas of Q_1 and Q_2, V_T is the thermal voltage.

According to previous work [3], the relationship between V_{EB} and temperature can be expressed as follows:

$$V_{EB}(T) = V_{BG} - (V_{BG} - V_{EB}(T_r))\frac{T}{T_r} - (\eta - m)\, V_T \ln\frac{T}{T_r} \quad (2)$$

where V_{BG} is the silicon bandgap voltage at the reference temperature T_r, η is a process related constant, and m is the order of the collector current temperature dependence, *e.g.*, $m = 1$ when the BJT is biased by the PTAT current, and $m = 0$ when it is biased by a temperature independent current.

Since V_{EB} is not a purely linear function of temperature, the output voltage is highly influenced by temperature. The nonlinear error can be reduced by using resistance to extract the nonlinear term. However, the above results do not take the variation of $V_{EB}(T_r)$ into account, and the resistance will introduce its own temperature coefficient, as well as the amplifier offset and so on. Therefore, the output curve still exhibits a convex or concave shape. To achieve a more accurate output voltage, this paper introduces compensation currents with different amplitudes and temperature thresholds into the expression above, directly optimizing the curve. The combination of these methods significantly reduces the temperature coefficient.

B. Proposed Multi Compensated Bandgap Reference

This paper proposes a dual temperature compensation (DTC) method, which employs resistors to extract high order nonlinearity compensation, thereby reducing the peak to peak value of the output curve. Moreover, by incorporating multiple stages of compensation currents, local optimization is applied to the output curve, achieving optimal compensation effect. Compared with the separate introduction of resistance extraction nonlinear term and multiband curvature compensation in [1, 2], the temperature coefficient of this technique is further reduced, as shown in Fig. 2. The specific circuit implementation of each approach is described in the following sections.

Firstly, to eliminate the influence of high order terms in the V_{EB} expression, a compensation technique utilizing resistor extraction of high order nonlinearity is employed, similar to [2]. As depicted in Fig. 3, Q_3 is biased with an approximately zero temperature drift current, while Q_2 is biased with a positive temperature coefficient current. The expressions for V_{EB2} and V_{EB3} can be obtained as follows:

$$V_{BE2} = V_{BG} - (V_{BG} - V_{BE}(T_r)) \frac{T}{T_r} - (\eta - 1) \ V_T \ln \frac{T}{T_r} \quad (3)$$

$$V_{BE3} = V_{BG} - (V_{BG} - V_{BE}(T_r)) \frac{T}{T_r} - \eta V_T \ln \frac{T}{T_r} \quad (4)$$

Therefore，the voltage V_{NL} across R_5 is:

$$V_{NL} = V_T \ln \frac{T}{T_r} \quad (5)$$

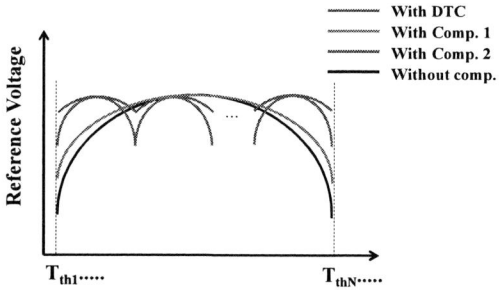

Fig. 2. Comparison of various compensation effect.

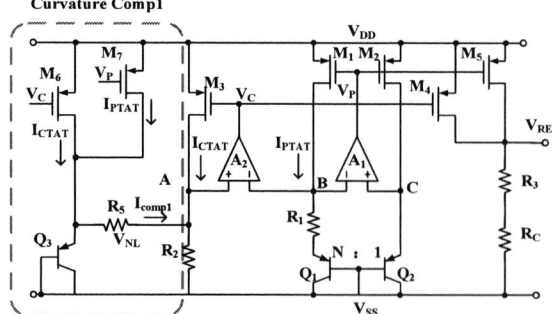

Fig. 3. Resistive extraction of nonlinear terms.

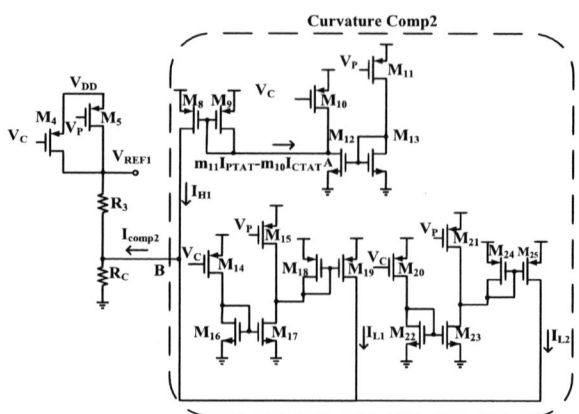

Fig. 4. Multiband curvature compensation.

V_{NL} is converted to current across R_5 and added to the output branch, thereby altering the expression for the output voltage to:

$$V_{REF1} = \frac{R_3 + R_C}{R_2} (|V_{BE2}| + \frac{R_2}{R_1} V_T \ln n + \frac{R_2}{R_5} V_T \ln \frac{T}{T_r}) \quad (6)$$

When R_5 is satisfied:

$$R_5 = \frac{R_2}{\eta - 1} \quad (7)$$

The nonlinearity with temperature of V_{BE} can be cancelled out, while the linear term can be adjusted by the ratio of R_2 and R_1.

Fig. 4 shows the specific implementation circuit of the multiband curvature compensation, which allows for a trade off between overall power consumption and output accuracy by adjusting the number of branches for adding or subtracting compensation currents. Since the resistor ratio introduces a positive temperature coefficient, the peak of the curve shifts towards higher temperatures after the first compensation method is introduced. This design adopts two low temperature compensations and one high temperature compensation to achieve better overall compensation effects. By adjusting the parallel number ratio of transistors, both the temperature threshold and the amplitude of the compensation current can be modified (for example, adjusting $m_{10} : m_{11}$ can alter the temperature threshold of the compensation current; adjusting $m_8 : m_9$ can change the amplitude of the compensation current). The final output reference voltage is determined as follows:

$$V_{REF2} = \frac{R_3 + R_C}{R_2} (|V_{BE2}| + \frac{R_2}{R_1} V_T \ln n + \frac{R_2}{R_5} V_T \ln \frac{T}{T_r})$$
$$+ R_C (I_{H1} + I_{L1} + I_{L2}) \quad (8)$$

By comparing equations (1) and (9), it becomes evident that the DTC initially offsets the nonlinear relationship between output voltage and temperature. Subsequently, it introduces the multiband curvature compensation for localized optimization, ultimately enhancing output stability.

979-8-3503-6184-1/24 $31.00 © 2024 IEEE

III. Layout Design And Simulation Results

This circuit is designed in a 110 nm CMOS process and the layout is shown in Fig. 5, with an area of 165 μm × 270 μm.

Fig. 6 shows the temperature variation of the simulated reference voltage at TT corner. The TC is 13.8 ppm/°C when only using the conventional compensation method, whereas the TC is only 4.8 ppm/°C with the DTC method. The residual TC is reduced by 65% with the proposed DTC method.

Fig. 7 shows the simulated reference voltage versus temperature across various process corners without trimming. The simulated TC varies between 4.8 ppm/°C (at TT corner) to 7.0 ppm/°C (at FS corner) over the temperature range of

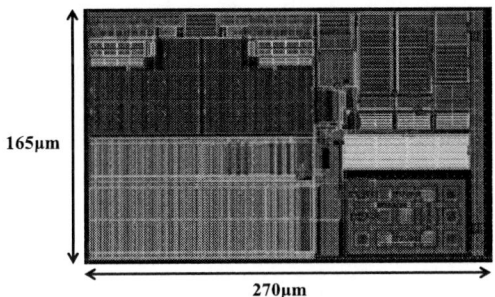

Fig. 5. Layout of the bandgap reference circuit.

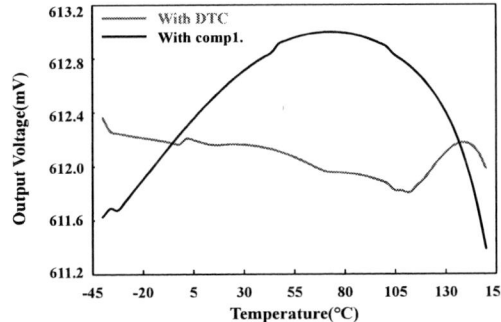

Fig. 6. V_{ref} under the TT corner.

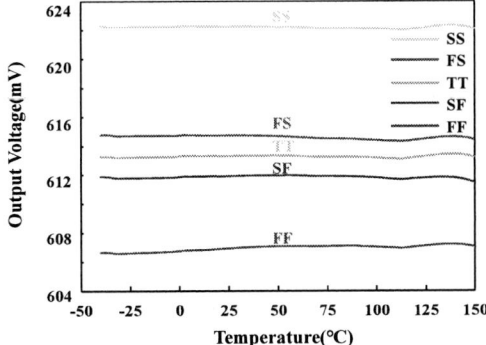

Fig. 7. V_{ref} under different corners.

TABLE I. PERFORMANCE COMPARISON

	[1]	[4]	[5]	This work
Supply Voltage (V)	1.3~1.8	1.0~1.8	1.2~2.5	1.5
Current Consumption (μA)	120	192 (pW)	0.81	14.8
Reference Voltage (mV)	1170	692	1001.1	612
Optimum TC (ppm/°C)	5.78	25	18.6	4.8
Temperature Range (°C)	-40~150	-20~100	-40~120	-40~150
Technology (μm)	0.13	0.18	0.065	0.11

-40 °C to 150 °C. An 8 bit resistive DAC is implemented on R_3 to compensate for the static error due to process variations. The trimming step is 7 mV and the static error is only 0.45% after process compensation.

The bandgap reference consumes 14.8 μA with a supply voltage of 1.5 V. Table I compares this design to state of the art works. Compared to bandgap references with similar output voltage, the proposed design achieves the lowest temperature drift over a wide temperature range, while also maintaining lower power consumption.

IV. Conclusion

This paper describes a design of a wide temperature range, low temperature drift bandgap reference circuit. A novel DTC method with resistive extraction of nonlinear terms, as well as multi segment curvature compensation is employed, greatly improving the accuracy of the bandgap reference. The bandgap reference is implemented in a 110 nm CMOS process, and post simulation results show that the design achieves an optimal temperature coefficient of 4.8 ppm/°C over the temperature range -40 °C to 150 °C, with a current consumption of only 14.8 μA. This design provides an output voltage of 612 mV and exhibits excellent performance across the wide temperature range.

References

[1] K. Chen, L. Petruzzi, R. Hulfachor, and M. Onabajo, "A 1.16-V 5.8-to-13.5-ppm/°C Curvature-Compensated CMOS Bandgap Reference Circuit With a Shared Offset-Cancellation Method for Internal Amplifiers," IEEE J. Solid-State Circuits, vol. 56, no. 1, pp. 267–276, 2021.

[2] P. Malcovati, F. Maloberti, C. Fiocchi, and M. Pruzzi, "Curvature-compensated BiCMOS bandgap with 1-V supply voltage," IEEE Journal of Solid-State Circuits, vol. 36, no. 7, pp. 1076–1081, 2001.

[3] G. C. M. Meijer, "Thermal sensors based on transistors," Sensors and Actuators, vol. 10, no. 1–2, pp. 103–125, 1986.

[4] Y. Ji, J. Lee, B. Kim, H. -J. Park and J. -Y. Sim, "A 192-pW Voltage Reference Generating Bandgap- Vth With Process and Temperature Dependence Compensation," IEEE Journal of Solid-State Circuits, vol. 54, no. 12, pp. 3281–3291, 2019.

[5] C.-W. U, M.-K. Law, R. P. Martins, and C.-S. Lam, "Sub-μW Auto-Calibration Bandgap Voltage Reference With 1σ Inaccuracy of ± 0.12% Within − 40°C to 120°C," IEEE Journal of Solid-State Circuits, vol. 59, no. 2, pp. 540–550, 2024.

979-8-3503-6184-1/24 $31.00 © 2024 IEEE

A Biphasic Neural Stimulator with Adaptive Pulse-Width Modulation Charge Balancer

Hailong Tang, Wenxian Gu, Yifan Song, Hengchang Bi, Xing Wu, Liangjian Lyu*

In Situ Devices Center, School of Integrated Circuits, East China Normal University, Shanghai 200241, China

* Email: ljlv@cee.ecnu.edu.cn

Abstract—Ensuring a balanced charge is crucial for a safe neural stimulation with biphasic current pulses. In this paper, we present a novel charge-balancing technique based on adaptive pulse width modulation. The pulse-width of anodic stimulation is modulated by the residual voltage sampled from the stimulation electrode and the injected charge sampled from the ratioed replica branch. Additionally, we employ an adaptive threshold voltage generator to mitigate the current mismatch introduced by the replica branch. The proposed stimulator is implemented in a HV 0.18-µm BCD CMOS process. Simulation results demonstrate that the proposed method achieves a charge balancing precision of ±10 mV, even under a 100% mismatch.

Keywords—neural stimulator, charge balancer, adaptive pulse width modulation, adaptive threshold voltage generator

I. INTRODUCTION

For a closed-loop Brain-Machine-Interface (BMI) system, the current-mode stimulator is widely adopted due to its superior control over charge injection and enhanced safety. Usually, a source-sink topology consisting of two independent P/N current sources is used to generate biphasic rectangular current pulses that activate neurons in nearby tissues. It is crucial to ensure that no residual charge accumulation occurs after the stimulation, as this may cause tissue damage. However, the P/N current sources cannot be precisely matched owing to process variations [1]. Therefore, additional charge balance circuits are required to eliminate residual charge within the tissue.

Fig. 1 presents several widely-adopted charge-balancing techniques. The passive charge balance involves a large DC-blocking capacitor and electrode shorting to eliminate the residual charge. However, a large off-chip DC-blocking capacitor in the order of µF is not feasible for integrated circuits and multichannel applications. Moreover, electrode shorting loosens the space limitation for the stimulator by turning on a low resistive path to discharge the residual charge [2]. Nevertheless, the required discharging time and current are not well-controlled as they are determined by the time constant of the electrode-tissue impedance and the discharge path resistance. The pulse insertion technique uses pre-defined current pulses to remove the residual charge during the inter-pulse time [2]. However, this method may cause overcompensation due to the fixed current pulses. Additionally, the compensated current pulses during the inter-pulse interval may interfere with the stimulation [3]. Inter-Pulse charge control avoids inserting compensated current pulses but requires an amplifier with a large bandwidth. Moreover, the system stability may be affected by the electrode-tissue impedance as it introduces a pole and a zero [4]. Pulse width modulation is another effective method to eliminate the residual charge. Current amplitude and phase mismatch leads to charge imbalance. Therefore, it is possible to modulate a matched anodic phase width to realize the charge balancing. Compared to the methods mentioned above,

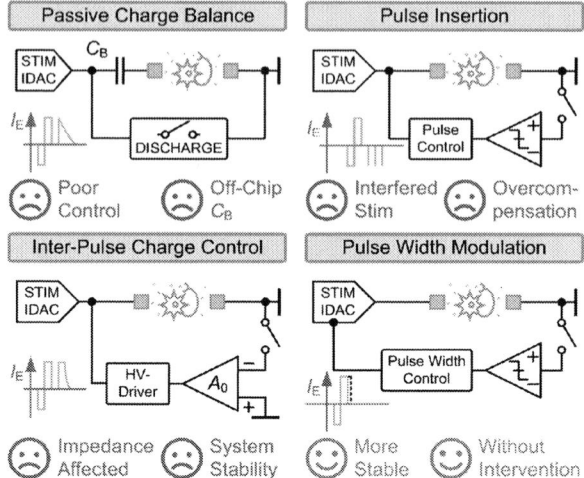

Fig. 1. Comparison of several types of charge balancing techniques.

this approach avoids interfered stimulation and is more stable [5]. Once the matched pulse width is located, no further intervention is required. A stimulus-synchronized charge balancing method is proposed for switched-capacitor stimulation based on pulse width modulation [6]. However, it ignores the mismatch between the current mirrors which may lead to imbalanced charge transformation. Details will be discussed in Section II.

In this work, we present a novel charge balancing method based on adaptive pulse width modulation (APWM). This paper is organized as follows, Section II explains the mechanism of the proposed approach and the detailed circuits. Section III shows the simulation results of the circuits. Lastly, Section IV draws the conclusion.

II. SYSTEM OVERVIEW AND CIRCUIT DESIGN

A. System Architecture

Fig. 2 (a) depicts the detailed system architecture of the proposed charge balance method. A source-sink topology is used to conduct the stimulation current I_{ANO} and I_{CAT} generated from two independent current-mode DACs. The stimulated current is mirrored with the ratio of N:M to the switched capacitors for charge balancing. C_{MEM} is used to record stimulated charge during the cathodic phase while C_{sample} stores the residual charge after each stimulation. Fig. 2 (b) shows the operation timing diagram of the switches and the voltage variations on the capacitor C_{MEM}.

B. Charge Balancing Mechanism

The whole operation of charge balancing can be divided into three phases: sampling (Φ_1), stimulating (Φ_2) and calibrating (Φ_3). In the sampling phase, S_{sample} goes high and S_{MEM} stays low. C_{sample} is connected to the electrode and

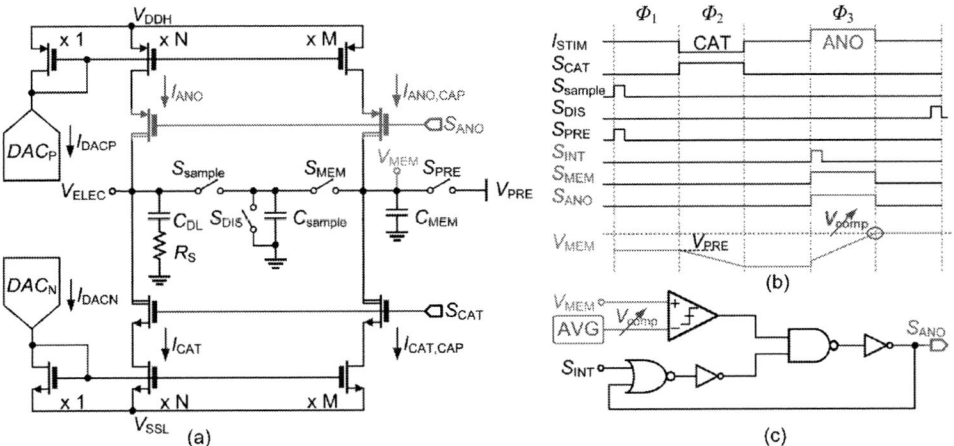

Fig. 2. (a) Schematic of the proposed charge balancing method and its (b) operation timing diagram. (c) S_{CAT} signal generator.

samples the residual voltage V_{ELEC}. Meanwhile, S_{PRE} goes high to pre-charge C_{MEM} to V_{PRE}. After a short period of sampling, S_{CAT} becomes high and starts to perform the stimulation. During this period, stimulating current I_{CAT} is mirrored with the ratio of N:M to current $I_{CAT,CAP}$ and connected to C_{MEM}. Currently, the capacitor voltage V_{MEM} begins to decrease due to $I_{CAT,CAP}$ as shown in Fig. 2 (b). Thus, we can consider that the stimulated charge is stored by C_{MEM} through current mirrors. In the case of calibrating phase, S_{INT} first starts high for a short while to enable S_{ANO} shown in Fig. 2 (c). C_{MEM} and C_{sample} are connected by S_{MEM} to share the residual charge and stimulus charge. The capacitor voltage V_{MEM} is compared with the threshold voltage (V_{comp}). Once V_{MEM} reaches V_{comp}, S_{ANO} goes low indicating that the calibrating phase is finished as shown in Fig. 2 (b). The value of V_{comp} is given by [6] as:

$$V_{comp} = \frac{C_{MEM} \cdot V_{PRE}}{C_{MEM} + C_{sample}} . \quad (1)$$

Lastly, at the end of each stimulation phase, S_{DIS} becomes high to discharge C_{sample} and prepare for the next sampling.

C. Mismatch Analysis

Assuming T_{CAT} and T_{ANO} are the pulse width of the cathodic and anodic phase, I_{CAT} and I_{ANO} are the stimulus current amplitude during the cathodic and anodic phase. $I_{CAT,CAP}$ and $I_{ANO,CAP}$ represent the absolute value of stimulus current to the capacitors C_{MEM} as depicted in Fig. 2 (a). $V_{REST,0}$ is the residual voltage of the electrode after last stimulation cycle. Usually, T_{CAT} is set by the digital controller. Thus anodic phase T_{ANO} can be calculated as (2) by observing the voltage variations on C_{MEM} (which is V_{MEM}) as shown in Fig. 2 (b).

$$T_{ANO} = \frac{V_{comp} \cdot (C_{MEM} + C_{sample}) - C_{MEM} \cdot V_{PRE}}{I_{ANO,CAP}}$$
$$+ \frac{I_{CAT,CAP} \cdot T_{CAT} - C_{sample} \cdot V_{REST,0}}{I_{ANO,CAP}} \quad (2)$$

The first term of (2) can be canceled out due to (1). So (2) can be simplified as:

$$T_{ANO} = \frac{I_{CAT,CAP}}{I_{ANO,CAP}} \cdot T_{CAT} - \frac{Q_{REST,0}}{I_{ANO,CAP}}. \quad (3)$$

Accordingly, the rest charge after this stimulation $Q_{REST,1}$ can be derived as:

$$Q_{REST,1} = I_{ANO} \cdot T_{ANO} - I_{CAT} \cdot T_{CAT} + Q_{REST,0}$$
$$= I_{ANO} \cdot T_{CAT} \cdot \frac{I_{CAT,CAP}}{I_{ANO,CAP}} - I_{CAT} \cdot T_{CAT}$$
$$- Q_{REST,0} \cdot \frac{I_{ANO}}{I_{ANO,CAP}} + Q_{REST,0} \cdot \quad (4)$$

If $\frac{I_{ANO}}{I_{ANO,CAP}} = \frac{I_{CAT}}{I_{CAT,CAP}}$, which means a perfect replication of current between I_{CAT} and $I_{CAT,CAP}$ (also I_{ANO} and $I_{ANO,CAP}$), $Q_{REST,1}$ can be further derived as:

$$Q_{REST,1} = Q_{REST,0} \cdot (1 - \frac{I_{ANO}}{I_{ANO,CAP}}) = Q_{REST,0} \cdot (1 - \frac{N}{M}). \quad (5)$$

Therefore, the residual charge after the nth stimulation period ($Q_{REST,n}$) can be summarized as:

$$Q_{REST,n} = 0 \ (N = M)$$
$$= \frac{M}{N} \cdot Q_{REST,0} \cdot \left[1 - \left(1 - \frac{N}{M} \right)^n \right] (N \neq M) . \quad (6)$$

However, a perfect current mirror is unattainable due to the fabrication imperfection, causing the variation in the current ratio N:M. Thus the first term and second term of (4) cannot be canceled out, thereby induces a large charge accumulation of Q_{REST}.

D. Adaptive Threshold Voltage Generator

Here, we proposed a novel approach to overcome the charge accumulation induced by current mismatch in the replica branch. Fig. 3 shows the proposed adaptive threshold voltage generator (AVG). Two comparator circuits compare the residual voltage of the electrode with safe window voltage ($\pm V_{safe}$) and generates the output error logic signals. Once the residual voltage exceeds the safe window the AVG automatically increases or decreases the threshold voltage of the comparator (V_{comp}) in Fig. 2 (b) by specific steps to achieve charge balancing. To ensure optimal charge balancing

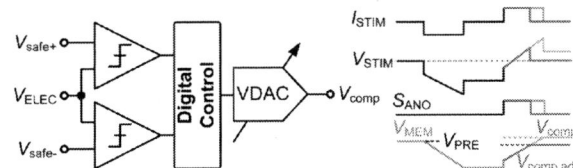

Fig. 3. The proposed adaptive threshold voltage generator.

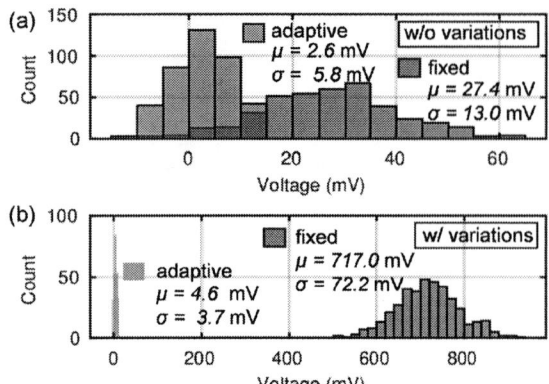

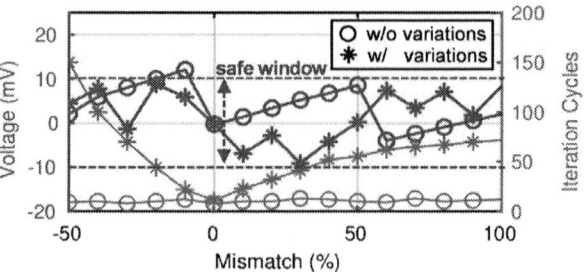

Fig. 5. The simulated residual vloatge under different current mismatches with the APWM charge balancer.

Fig. 4. The Monto Carlo simulation results of the residual voltage V_{REST}.

conditions, step size in AVG can be dynamically adjusted by digital controller and VDAC. However, it is necessary to consider that there is a trade-off between the convergence speed and charge balancing accuracy when choosing the proper step size.

III. SIMULATION RESULT

The stimulator with the proposed APWM charge balancer is implemented in a 0.18-μm HV BCD CMOS process. The electrode-tissue impedance is configured as follows: $C_{DL} = 100$ nF, $R_S = 1$ kΩ. The pulse width of cathodic phase is 50 μs. For simplicity, the ratio of N:M is set to 1:1. Consequently the stimulus amplitude of I_{CAT} and $I_{CAT,CAP}$ are both configured to 1 mA. To simulate a 10% mismatch between two P/N DACs, the anodic current I_{ANO} is set to 1.1 mA.

To validate the mismatch introduced by the difference between $\dfrac{I_{ANO}}{I_{ANO,CAP}}$ and $\dfrac{I_{CAT}}{I_{CAT,CAP}}$ discussed in Section II, the residual voltage of electrode (V_{REST}) is compared with and without variations between I_{ANO} and $I_{ANO,CAP}$ under fixed and adaptive V_{comp}. Fig. 4 shows the histogram of V_{REST} after 400 Monto Carlo simulations without current variations ($I_{ANO,CAP} = I_{ANO} = 1.1$ mA) and with current variations ($I_{ANO} = 1.1$ mA, $I_{ANO,CAP} = I_{CAT} = 1$ mA). Note that the safe window is set to ± 10 mV in the proposed AVG. It shows that the mismatch between $\dfrac{I_{ANO}}{I_{ANO,CAP}}$ and $\dfrac{I_{CAT}}{I_{CAT,CAP}}$ leads to large residual charge accumulation with a fixed comparator threshold V_{comp} in Fig. 4 (b). Meanwhile, the results also show that the proposed adaptive threshold voltage V_{comp} has better performance than the fixed V_{comp} as shown in Fig. 4 (a) (b).

Fig. 5 shows the residual voltage with different current mismatches. Note that the mismatch is calculate as:

$$Mismatch = \frac{I_{ANO} - I_{CAT}}{I_{CAT}}. \qquad (7)$$

Results show that the residual voltage can be controlled within the safe window regardless of variations between I_{ANO} and $I_{ANO,CAP}$. Obviously, The proposed APWM charge balancer presents a good robustness even under a large current mismatches. It is noteworthy that large mismatches leads to long convergence time for charge balancing.

IV. CONCLUSION

This paper presents a charge balancer with adaptive pulse width modulation. The proposal achieves a charge balancing precision within ±10 mV range under a maximum current

mismatch of 100%, when the stimulation current is 1 mA. Besides, it also achieves a good robust performance against large current mismatches. Compared with the state-of-the-art works, the proposed design is much more robust over component mismatches and avoids interfering the stimulation.

ACKNOWLEDGMENT

This work was supported by the National Natural Science Foundation of China (No. 62204085). This work was also supported by the Strategic Priority Research Program of the Chinese Academy of Sciences under Grant No. XDB44000000-11 and the Lingang Laboratory under Grant No. LG-QS-202202-12.

TABLE I.　COMPARISION WITH THE STATE-OF-THE-ART

	This Work	[2]	[3]	[4]	[5]
Process	0.18μm	PCB	0.18μm	0.35μm	0.18μm
Supply (V)	±15	–	40	38	3.3
I_{stim} Range (mA)	< 10	0.5	< 12.75	< 0.5	< 1
CB Strategy	APWM	Pulse Insert	Time Based	IPCC[a]	Pulse Adjust
CB Error (mV)	< 10	< 100	< 2	< 50	< 4

[a] Inter-Pulse Charge Control

REFERENCES

[1] Y. Wu, D. Jiang and A. Demosthenous, "A Multi-Channel Stimulator With High-Resolution Time-to-Current Conversion for Vagal-Cardiac Neuromodulation," IEEE Trans Biomed Circuits Syst, vol. 15, no. 6, pp. 1186-1195, Dec. 2021.

[2] K. Sooksood, T. Stieglitz and M. Ortmanns, "An Active Approach for Charge Balancing in Functional Electrical Stimulation," IEEE Trans Biomed Circuits Syst, vol. 4, no. 3, pp. 162-170, June 2010.

[3] H. Pu, O. Malekzadeh-Arasteh, A. R. Danesh, Z. Nenadic, A. H. Do and P. Heydari, "A CMOS Dual-Mode Brain-Computer Interface Chipset With 2-mV Precision Time-Based Charge Balancing and Stimulation-Side Artifact Suppression," IEEE J. Solid-State Circuits, vol. 57, no. 6, pp. 1824-1840, June 2022.

[4] N. Butz, U. Kalita and Y. Manoli, "Active Charge Balancer With Adaptive 3.3 V to 38 V Supply Compliance for Neural Stimulators," IEEE Trans Circuits Syst. I Regul. Pap, vol. 68, no. 10, pp. 4013-4024, Oct. 2021.

[5] F. Eshaghi, T. Moeinfard, E. Najafiaghdam and H. Kassiri, "A Neurostimulator IC With Impedance-Aware Dynamic-Precision One-Shot Charge Balancing," IEEE Solid-State Circuits Lett, vol. 4, pp. 202-205, 2021.

[6] M. Park, K. Eom, H. -S. Lee, S. -B. Ku and H. -M. Lee, "A 9-V-Tolerant Stacked-Switched-Capacitor Stimulation System With Level-Adaptive Switch Control and Rapid Stimulus-Synchronized Charge Balancing for Implantable Devices," IEEE J. Solid-State Circuits, vol. 59, no. 3, pp. 817-829, March 2024.

Assembly of Oxidized/Intrinsic 2D MXene Film for Improved Absorption Electromagnetic Shielding

Yulin Guo[1], Siteng Li[1], Jiafeng Song[1], Yilin Sun[1], Zhifang Liu*[1]; Weijia Luo*[2]

[1] School of Integrated Circuits and Electronics, Beijing Institute of Technology, Beijing, China
[2] State Key Laboratory of New Ceramics and Fine Processing, School of Materials Science and Engineering, Tsinghua University, Beijing, China

* Email: gylbit1404@163.com, bit_lst_6@163.com, sjf@bit.edu.cn; lzf@bit.edu.cn, luoweijia@tsinghua.edu.cn

Abstract—In recent years, two-dimensional transition metal carbon/nitride (MXene) has been widely used in the field of electromagnetic shielding due to its excellent solution processability and high conductivity. This article applies the intrinsic MXene material to avoid introducing other materials, and combines pure MXene with oxidized MXene to prepare composite films. The reason why the introduction of oxidized MXene can effectively improve the absorption rate of intrinsic MXene electromagnetic shielding films is due to its loose and porous structure. It is worth noting that the prepared composite film can still maintain a shielding efficiency of over 99.97% in the R-band (22-33GHz); In addition, compared to pure MXene film, the composite MXene film can increase the absorption rate of electromagnetic waves by 18.69%, achieving better green and environmentally friendly shielding characteristics.

Keywords—Composite films, Electromagnetic shielding, MXene, Loose and porous structure

I. INTRODUCTION

With the development and popularization of smart devices, various electronic products have been widely used in daily life. Electronic products not only bring convenience to people, but also bring numerous problems, among which electromagnetic interference (EMI) has become an issue which cannot be ignored. EMI has adverse effects on human health, information security and precision equipment[1]. Therefore, the development of lightweight, high-performance and green EMI shielding materials has received extensive attention.

In 2011, Yuri *et al.* synthesized two-dimensional transition metal carbons/nitrides (MXene), with a general chemical formula of $M_{n+1}X_nT_x$, where M represents 3-6 transition metals, X represents carbon and/or nitrogen, and T is the terminating group (usually -OH, -F, or =O)[2]. The most widely studied MXene, $Ti_3C_2T_x$, which is suitable for achieving high EMI surface reflection due to its high density of Fermi level states and high conductivity, making it an ideal material for achieving high EMI shielding performance[3]. Besides, MXene has good mechanical durability and excellent fabricability. Regardless of whether it is an organic solvent system or an aqueous solution system, MXene flakes have good solution dispersibility and are easy to manufacture EMI shielding films through spraying, greatly simplifying the process complexity[4]. In addition, the two-dimensional layered structure, variable surface chemical groups, and adjustable interlayer spacing of MXene make its conductivity adjustable, providing convenience for us to optimize the conductivity of EMI shielding materials[5-6].

Due to the high conductivity of MXene materials, their shielding of electromagnetic waves is mainly based on the reflection principle, which can easily lead to the dissipation and secondary pollution of electromagnetic waves. Therefore, improving their absorption rate while ensuring the shielding efficiency of electromagnetic waves has important environmental significance. To solve this problem, the combination of MXene and other materials has become an effective strategy[7]. However, this increases the complexity of the system and the cost of process.

Herein, we have done a groundbreaking work to optimize the EMI shielding performance of MXene by using intrinsic MXene materials without introducing external materials. We constructed a sandwich structure composite EMI shielding film by using oxidized $Ti_3C_2T_x$ and pure $Ti_3C_2T_x$. We found the oxidized MXene layer is completely wrapped by the pure MXene film and the composite of loose oxidized MXene layers and pure MXene layers would produce a large number of microstructural scale voids and gaps. As the total number of layers in the composite film increases, stronger multiple reflections and scattering will be caused inside it, and the EMI shielding performance of the film is significantly improved. As a result, the introduction of the oxidized MXene layer can increase the absorption rate of pure MXene film for electromagnetic waves by 18.69%.

Compared with conventional MXene-based EMI shielding studies, we proposed a strategy intrinsic MXene materials for oxidation, treatment, and preparation, instead of compositing with other functional materials, reducing the complexity of the system and facilitating optimization and analysis; Additionally, the manufacturing process is simple and cost-effective providing broad prospects for future applications.

II. EXPERIMENT METHODS

A. Preparation of the solution and the composite films for EMI sheilding

In this study, the spray coating method was mainly used to prepare MXene films and pure/oxidized MXene composite films. Firstly, pure MXene ($Ti_3C_2T_x$) solution was prepared via a liquid exfoliation approach. $Ti_3C_2T_x$ nanoflakes were obtained by liquid chemical exfoliating MAX phase precursor (powder) with a mixed solution of LiF/HCl. Then, 6 parts of 2mL MXene solution (10 mg ml^{-1}) were extracted from above. These 6 parts of MXene solution were oxidized by adding 300, 400, 500, 600, 700 and 800 μL H_2O_2 solution (35%). Subsequently, the MXene solution containing H_2O_2 was subjected to 15 minutes of ultrasonic treatment to ensure its uniformity and promote complete oxidation reaction. Polyethylene terephthalate (PET) was used as the substrate for electromagnetic shielding films. The O_2 plasma cleaned PET substrate was placed on a heating plate, and then MXene and oxidized MXene were evenly sprayed layer by layer on the PET substrate using a spray gun. The prepared composite film

979-8-3503-6184-1/24 $31.00 © 2024 IEEE

structure and finished product are shown in the following Figure 1(a),(b), respectively.

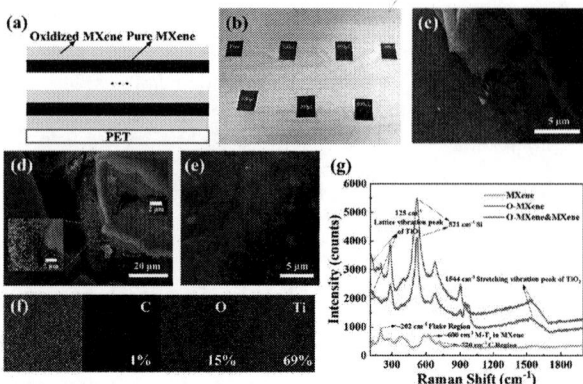

Fig.1. (a) Schematic diagram of the structure of pure/oxidized MXene composite films; (b) Thin films prepared by spraying pure MXene solution and MXene solution with different amounts of oxidized H_2O_2 added. This includes pure MXene films and composite films prepared by pure MXene and oxidized MXene solutions which added 300, 400, 500, 600, 700 and 800 μL of H_2O_2, respectively; (c) Scanning electron microscopy images of pure MXene thin film which show clear layered structures; (d) Scanning electron microscopy images of cracks in pure/oxidized MXene composite films. It shows clear composite structures of oxidized MXene layers and pure MXene layers; (e) Scanning electron microscopy images of the surface of pure/oxidized MXene composite films; (f) In situ mapping of the surface of pure/oxidized MXene composite films. Mapping reveals the main elements of composite films: C, O, Ti; (g) Raman spectrum of pure MXene film, oxidized MXene film, and pure/oxidized MXene composite film

B. Characterization of the Composite Films

The prepared pure MXene film and the pure/oxidized MXene composite film were characterized by scanning electron microscopy (SEM), and the surface of the pure/oxidized MXene composite film was analyzed by elemental maps in situ. As shown in Figure 1(c), the pure MXene film exhibits a typical layered structure, and the clear layered stacking can be seen in the image. In Figure 1(d), the composite structure of pure/oxidized MXene is shown. Through the cracks, it can be seen that oxidized MXene (mainly TiO_2) particles are uniformly distributed in the lower layer, and a uniformly stacked layered structure, namely pure MXene, can be seen in the upper layer, proving the successful preparation of the composite film of pure/oxidized MXene. Figure 1(e) and (f) show the mapping results of the composite film, which show that the oxygen content in the composite film reaches 16% and is evenly distributed on the surface of the film, further confirming the success of the preparation.

Raman spectroscopy analysis was performed on the prepared thin film using a Raman spectrometer, and the results are shown in Figure 1(g). By analyzing the spectrum, it can be concluded that the Raman spectrum of pure MXene film has a typical characteristic peak at 202 cm^{-1}, which is consistent with the Flake Region peak in the MXene standard Raman spectrum. Meanwhile, there is a significant peak in the spectral line around 600 cm^{-1}, which caused by the M-T_x bond in MXene. In addition, there is a peak at 720 cm^{-1} that is consistent with the standard spectrum, mainly caused by the carbon structure in MXene[8].

Further observation of the Raman spectra of pure/oxidized MXene composite film and comparison with the Raman spectra of completely oxidized MXene(i.e. TiO_2) films, reveals that both films have a peak at 125 cm^{-1}, which caused

by the lattice vibration of TiO_2. There is also a peak at 1544 cm^{-1}, which is caused by the stretching vibration of O-Ti-O bonds or Ti-O bonds in TiO_2[9-10]. The above results fully demonstrate the successful introduction of oxidized MXene, i.e. TiO_2, into the pure/oxidized MXene composite film. The Raman spectra of pure/oxidized MXene composite films still retain the characteristic peaks of pure MXene, such as 202 cm^{-1} peak and 600 cm^{-1} peak, compared to the Raman spectra of completely oxidized MXene films, indirectly indicating that this study has successfully achieved the composite of MXene layer and oxidized MXene layer.

III. EXPERIMENTAL RESULTS

Figure 2(a) shows the total electromagnetic shielding effectiveness (SE) lines of pure MXene film and different composite films in the R-band. The total SE of all films reaches 35dB or even higher, indicating that the shielding efficiency of all films for electromagnetic waves can reach over 99.97%. The above results indicate that the MXene layer in the composite film is the main source of high SE, which can ensure sufficient reflection of electromagnetic waves. In addition, it can be seen from the figure that the total SE of the film prepared with MXene added 500 μL H_2O_2 is relatively low, which is due to the introduction of oxidized layer leading to changes in the structure of the composite film and a decrease in conductivity. In summary, the introduction of an oxidized MXene layer will not decrease the shielding effectiveness of the composite film, indicating the full feasibility of this study.

In electromagnetic shielding research, the total shielding effectiveness (SE) is usually divided into transmission shielding effectiveness (SET) and reflection shielding effectiveness (SER). From Figure 2(b), it can be seen that the shielding film prepared by the MXene added 500 μL H_2O_2 has the best SET, which is much higher than its SER, indicating that the film has good absorption characteristics for electromagnetic waves.

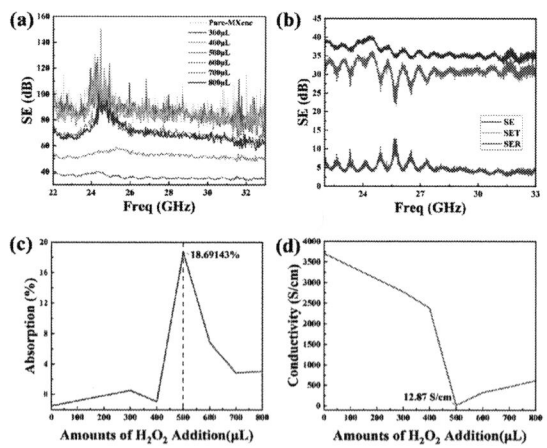

Fig.2. (a) Electromagnetic shielding performance lines of different composite films measured by a vector network analyzer (VNA) in the R-band (22-33GHz); (b) The reflection shielding efficiency (SER), transmission shielding efficiency (SET), and total shielding efficiency (SE) images of the composite film prepared by MXene added 500 μL H_2O_2; (c) Images of electromagnetic wave absorption rates of different composite films; (d) Conductivities of different composite films

Furthermore, we calculated the electromagnetic wave absorption rates of different films, and the results obtained are shown in Figure 2(c). It can be seen that the shielding film

979-8-3503-6184-1/24 $31.00 © 2024 IEEE

prepared by MXene added 500 μL H_2O_2 has the highest electromagnetic wave absorption rate, reaching 18.69% (absorption limit value of 50%[11]), indicating that the thin film exhibits excellent electromagnetic wave absorption performance. The negative absorption rate is mainly caused by errors caused by measurement noise. From Figure 2(d), it can be seen that the conductivity of the film exhibits a completely opposite change curve to the electromagnetic absorption rate of the films, proving that conductivity is one of the influencing factors on the electromagnetic absorption characteristics of the film. The lower the conductivity, the higher the electromagnetic absorption rate of the film, but it is not the only factor. The microstructure of the film also plays a crucial role in improving the electromagnetic absorption rate.

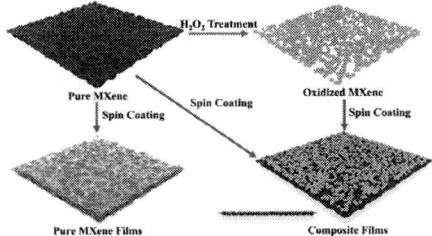

Fig.3. Schematic diagram of the assembly process of composite films prepared by pure/oxidized MXene

Further explain the basic principle of achieving high electromagnetic absorption through MXene and oxidized MXene composite films through Figure 3. As shown in the figure, pure MXene can achieve sufficient oxidation by adding H_2O_2, thereby generating spherical oxidized MXene (i.e. TiO_2). When the spraying amount is large enough, the oxidized MXene itself can also form a thin film. This study aims to construct a composite film by combining pure MXene film with oxidized MXene layer. When pure MXene and oxidized MXene form a composite film, the pure MXene layer still has a complete structure, which can provide sufficient reflection of electromagnetic waves. Furthermore, it can be seen that pure/oxidized MXene form a composite structure filled with pores. When electromagnetic waves are incident into the film, strong multiple reflections and scattering occur inside the film, thereby improving the absorption performance of electromagnetic waves. Through the above methods, a composite thin film with high electromagnetic shielding performance and high electromagnetic wave absorption rate has been achieved.

When using MXene added 500 μL H_2O_2 to prepare shielding films, the solution introduces more pores into the composite film while ensuring the integrity of the pure MXene film, thus achieving a higher electromagnetic absorption rate; When there is less H_2O_2 added, MXene fails to fully oxidize and there are still a large number of layered structures present in the solution. After spraying, the pores introduced by the composite film are less, and the electromagnetic shielding is still mainly led by reflecting, resulting in a lower absorption rate. When too much H_2O_2 is added, MXene is completely oxidized, and there is a large amount of oxidized MXene in the solution system. After spraying, a large amount of oxidized MXene accumulates in the formed film, and the pore structure is reduced. Although its electromagnetic absorption

rate is improved compared to pure MXene film, it shows a significant decrease compared to the shielding film prepared by MXene added 500 μL H_2O_2 to.

IV. CONCLUSION

In this study, we successfully prepared a class of pure/oxidized MXene composite films using PET as the substrate. The prepared composite film can achieve an electromagnetic shielding efficiency no less than 35dB in the R-band, with a shielding efficiency of up to 99.97%. Furthermore, we have achieved the preparation of high electromagnetic absorption thin films by introducing an oxidized MXene layer, with a maximum absorption rate of 18.69%. The preparation of thin films only uses intrinsic MXene without introducing other materials, greatly reducing the process complexity. This thin film not only achieves high electromagnetic shielding and absorption rate, but also demonstrates a green and sustainable environmental concept, and has the characteristic of flexibility, demonstrating broad application prospects in future electromagnetic shielding and other fields.

ACKNOWLEDGMENT

Yulin Guo, Siteng Li and Jiafeng Song contributed equally to this work. This work was funded by State Key Laboratory of New Ceramic and Fine Processing Tsinghua University (Grant No. KF202307), the National Natural Science Foundation of China under Grant No. 62304020 and No. 52202370，China Postdoctoral Science Foundation (Grant No. 2023T160359)，National Key R&D Program of China (Grant No. 2023YFB3811300).

REFERENCES

[1] S. Li, Y. Sun, Z. Liu, Y. Ding, and Z. Chen, "Flexible MXene/Ag nanowires composite film for high-performance electromagnetic shielding," in 2023 IEEE 7th International Symposium on Electromagnetic Compatibility (ISEMC), 2023, pp. 1-3: IEEE.

[2] M. Naguib et al., "Two-dimensional transition metal carbides," ACS Nano, vol. 6, no. 2, pp. 1322-1331, 2012.

[3] J. L. Hart et al., "Control of MXenes' electronic properties through termination and intercalation," Nat. Commun., vol. 10, no. 1, p. 522, 2019.

[4] T. S. Mathis et al., "Modified MAX phase synthesis for environmentally stable and highly conductive Ti_3C_2 MXene," ACS Nano, vol. 15, no. 4, pp. 6420-6429, 2021.

[5] D. Kim, T. Y. Ko, H. Kim, G. H. Lee, S. Cho, and C. M. Koo, "Nonpolar organic dispersion of 2D $Ti_3C_2T_x$ MXene flakes via simultaneous interfacial chemical grafting and phase transfer method," ACS Nano, vol. 13, no. 12, pp. 13818-13828, 2019.

[6] M. Han et al., "Solution-processed $Ti_3C_2T_x$ MXene antennas for radio-frequency communication," Adv. Mater., vol. 33, no. 1, p. 2003225, 2021.

[7] J. Liu et al., "Hydrophobic, flexible, and lightweight MXene foams for high-performance electromagnetic-interference shielding," Adv. Mater., vol. 29, no. 38, p. 1702367, 2017.

[8] A. Sarycheva and Y. Gogotsi, "Raman spectroscopy analysis of the structure and surface chemistry of $Ti_3C_2T_x$ MXene," Chem. Mater., vol. 32, no. 8, pp. 3480-3488, 2020.

[9] W. Ma, Z. Lu, and M. Zhang, "Investigation of structural transformations in nanophase titanium dioxide by Raman spectroscopy," Appl. Phys. A, vol. 66, pp. 621-627, 1998.

[10] W. Zhang, Y. He, M. Zhang, Z. Yin, and Q. Chen, "Raman scattering study on anatase TiO_2 nanocrystals," J. Phys. D, vol. 33, no. 8, p. 912, 2000.

[11] T. Zhao et al., "Ultrathin MXene assemblies approach the intrinsic absorption limit in the 0.5-10 THz band," Nat. Photonics, vol. 17, no. 7, pp. 622-628, 2023.

Improved Channel Width and Morphology of Epi Silicon FinFET via Low Thermal Budgets Fin Thinning Technology

Peng Wang1,2,3, Yupeng Lu1,2,3, Guanqiao Sang1,2, Renjie Jiang1,2,3, Lei Cao1,2,3, QingKun Li1,2,3, Lianlian Li1,2,3, hang zhang1,2,3, zhonrui wang1,2,3, meihe zhang1,2,3, Qingzhu Zhang1,2,3*, Junfeng Li 1,2,3* and Huaxiang Yin1,2,3

1 Key Laboratory of Fabrication Technologies for Integrated Circuits, Chinese Academy of Sciences, Beijing 100029;

2Institute of Microelectronics, Chinese Academy of Sciences, Beijing 100029, China;

3School of Integrated Circuits, University of Chinese Academy of Sciences, Beijing 100049, China.

*Corresponding Author's Email: lijunfeng@ime.ac.cn, zhangqingzhu@ime.ac.cn

Abstract— In this paper, Atomic level low-temperature ozone (LTO) treatment were utilized to successfully thin the width of fin and improve the morphology. Meanwhile, the drain current of and subthreshold swing of Epitaxial Si FinFET were compared under three scheme. The result show that after 8 cycles of LTO and HF rinsing, the width of fin reduce to 25.16nm, the drain current and subthreshold swing of FinFET device decreased by 51.99% (from 1.46×10^{10} A/um to 7.01×10^{11} A/um) and 6.84% (from 72.12 mV/dec to 67.19 mV/dec), respectively. In addition , after 12 cycles of LTO and HF rinsing, the width of fin reduce to 20.16nm, the drain current and subthreshold swing of FinFET device decreased by 71.09% (from 1.46×10^{10} A/um to 4.22×10^{11} A/um) and 6.79% ((from 72.12 mV/dec to 67.22 mV/dec), respectively. The result provide a guidance for channel morphology optimization and low-temperature integration of FinFET devices at advanced technology nodes.

Keywords—FinFET, fin-width effect, LTO, drain current, subthreshold swing

Introduction

As the demand for integrated circuits (ICs) with critical dimensions continues to grow, finfet-field-effect transistors (FinFETs) remain the most influential in the IC market due to their mature process and powerful driving capability. The dimensions of fin width (W_{fin}) influence the device's performance. In advanced FinFET technology, fins are currently made by a self-aligned multipatterned process, which brings about wiggling, larger surface roughness. The organic molecular chains of the photoresist cannot be completely broken during the lithography process, which will also result the high fin linear roughness. The surface roughness plays a crucial role in drain current, subthreshold swing and d mobility degradation of FinFET. As a result, narrow-channel FinFET devices formed by one-time lithography and etching tend to have poor leakage and subthreshold characteristics.[1-3]

In this work, we investigated the effects of atomic level low temperature ozone (LTO) treatment on the width and morphology of fins, In addition, We investigate the leakage current and subthreshold characteristics of FinFET devices. The results show that LTO are effective in Optimization morphology of Fin, and can significantly reduce FinFET leakage and subthreshold swing.[4-6] This provides a guidance for realizing FinFET devices with high gate control and high mobility in low-temperature processes.

Experiment

The fabrication flow of the epitaxial Si FinFET device is illustrated in Fig. 2. Firstly, a p-type Si (100) wafer with a resistivity of 8-12 Ω-cm was utilized as the starting substrate. Then CMOS based ground plane (GP) doping process, and spike anneal of 1050 °C were carried out to reduce parasitic channel. In the following step, a 70 nm Si film was epitaxial grown in a reduced pressure chemical vapor deposition (RPCVD) chamber. Subsequently, the fins with various widths were formed utilizing the spacer image transfer (SIT) and the reactive ion etching. The reference of the fin was defined with an average width of 35 nm, The Fin channel was then thinned to 25 nm and 20 nm, respectively, using low-temperature ozone stripping. Then shallow trench isolation (STI) filling and planarization, followed by low-temperature rapid thermal annealing (RTA) at 850 °C for 30 s. The heights of the fins were controlled by using a 1:100 diluted HF solution. The α-Si dummy gate was formed with a steep sidewall over 87 °. Subsequently, lightly doped drain (LDD) and highly doped drain (HDD) were implemented after the spacer 1 and spacer 2 formation. After removing the dummy gate, a high quality interfacial layer was formed using O3 oxidation. CMOS HKMG processes were formed through atomic layer deposition (ALD). Subsequently, SD metal-contact and backend-of-line (BEOL) processes for device formation were performed in subsequent steps. [7-9]

The top views of the device's structures were observed using S-4800 scanning electron microscopes (SEM). The electrical characterization was performed using an Agilent 4156 semiconductor parameter analyzer.

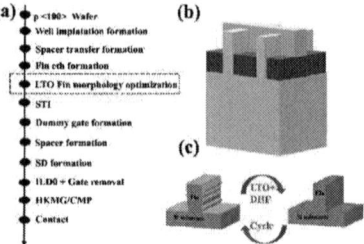

979-8-3503-6184-1/24 $31.00 © 2024 IEEE

Fig. 1. (a) Fin preparation process flow. (b) structure schematic of FinFET. (c) Schematic diagram of fin width and morphology.

Results and discussion

A. Optimization of channel morphology and width

Fig. 2(a) shows the SEM of fin structure formed by sidewall transfer and reactive ion etching(RIE) with an average width of 35 nm. The surface of the fin trench is the result of an imperfect fracture of the photoresist, which has resulted in a higher level of roughness. Fig. 2(b) and (c) show the SEM of Fin structure after 8 and 12

time LTO treatments, respectively, with average widths of 25 nm and 20 nm, respectively. After LTO thinning, Fin channel surface roughness is significantly reduced.

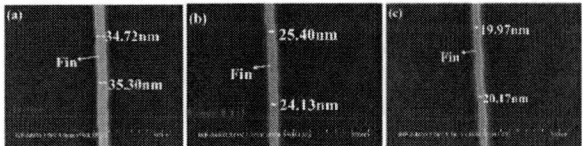

Fig. 2. SEM image of fin width. (a) initial fin of LTO. (b) after 8 time LTO treatments. (c) after 12 time LTO treatments.

Fig. 3(a) demonstrates the variation curve of Fin thickness with time for low-temperature ozone oxidation, and Fig. 3(b) demonstrates the low-temperature ozone oxidation mechanism. Since there is a self-limitation of low-temperature ozone oxidation, it is difficult to continue oxidizing when the thickness of a single oxidation reaches 6.3 nm, therefore, the LTO can accurately control the Fin channel width, and Fig. 3(c) demonstrates the variation curve of Fin width with the number of times of LTO treatment.

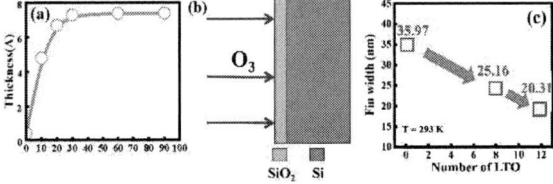

Fig. 3. (a) Ozone oxidation self-limiting curve. (b) Ozone Oxidation Thinning Fin. (c) the relationship between fin width and the number of low temperature atomic level ozone treatment.

Fig. 4 (a) and (b) show the fin width of 50 randomly selected points after multiple LTOs treatments. And the LWR were calculated based on the following equation:

$$LWR = \sqrt{1/N \sum_{i=1}^{N} \left[w_i(t) - \overline{w(t)} \right]^2}$$

Where $w_i(t)$ is the width of FinFET channel at the line$i$ on the specific fin structure, and $w(t)$ is the average value in the width of all the lines on the specific fin structure.

The result show that after 8 cycles and 12 cycles of LTO treatment, the LWR of fin decreased by 43.24% (from 2.66 nm to 1.51 nm) and 61.3% (from 2.66 nm to 1.03 nm), respectively,

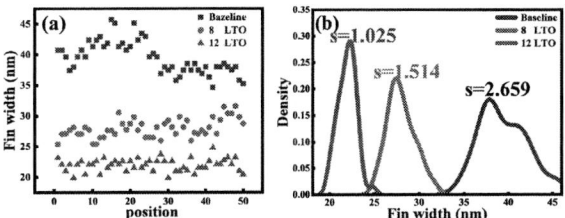

Fig. 4. (a) Fin width of 50 random points. (b). Changes in Surface line width roughness after LTO treatment.

b. Electrical Characteristics Analysis

Fig. 4 shows the device transfer characteristics of the n-type FinFET with Lg = 40 nm and Vd=0.9V under different W_{fin} (35nm,25nm,20nm).

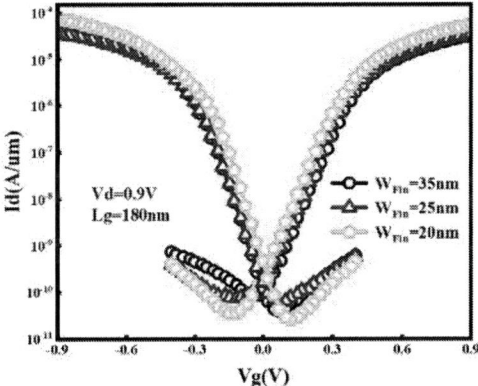

Fig. 4. Transfer characteristic curves of epitaxial Si FinFET devices with different channel widths.

Figure 5(a) (b) illustrates the N/PFET ion/Ioff mapping plots for different fin widths. For a NMOS, the Ioff is 1.46×10^{10} A/um for W_{Fin}=35 nm, and 7.01×10^{11} A/um and 4.22×10^{11} A/um for W_{Fin}=25 nm and 20 nm, respectively. The Ioff of the NMOS has been reduced by 51.99% and 71.09%, respectively. For PMOS, the Ioff is 2.36×10^{10} A/um for W_{Fin}=35nm, and 1.57×10^{10} A/um and 1.49×10^{10} A/um for W_{Fin}=25nm and 20nm, respectively. This represents a reduction of 33.47% and 36.86% for PMOS. The drain current decreases with a decreasing fin width.

Figure 5(c) (d) show the subthreshold swing (SS) of N/PFETs with differentfin widths. For NMOS, the SS is 72.12 mV/dec for W_{Fin}=35 nm and 67.19 mV/dec. The SS is 67.22 mV/dec for W_{Fin}=25 nm and 20 nm, respectively. For NMOS with W_{Fin}=25 nm and 20 nm, the SS is reduced by 6.84% and 6.79% for W_{Fin}=25 nm and 20 nm, respectively. For PMOS, the SS is observed to be 71.82 mV/dec for W_{Fin}=35 nm, and 68.88 mV/dec and 67.19 mV/dec for W_{Fin}=25 nm and 20 nm, respectively. The subthreshold swing of PMOS is observed to be reduced by 4.09% and 6.45% for W_{Fin}=25 nm and 20 nm, respectively. This phenomenon can be attributed to the fact that the LTO treatment process effectively removes the impurity ions and dangling bonds that are generated during the reactive ionization process. This reduction in the interfacial defect density of the device is a direct result of the LTO treatment process.

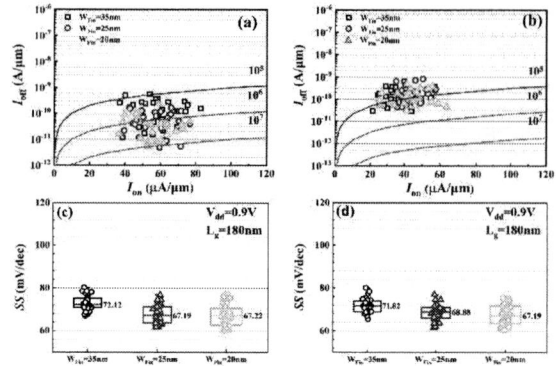

Fig. 5. FinFET device Ion/Ioff statistics and SS comparison.

Conclusions

In this paper, LTO methods were significantly effective in Optimization of Fin channel width and morphology. In addition, after LTO treatment, the FinFET device leakage and subthreshold swing are significantly reduced. The results provide a further understanding of the optimization of fin structures and can serve the development of nano FinFET and GAAFET technology.

[1] Sun, Lei , et al. "Application of frequency domain line edge roughness characterization methodology in lithography." Proceedings of SPIE - The International Society for Optical Engineering 9424(2015).K. Elissa, "Title of paper if known," unpublished.

[2] Boutchacha, Touati , and Gérard Ghibaudo. "Improved Modeling of Low-Frequency Noise in MOSFETs—Focus on Surface Roughness Effect and Saturation Region." IEEE Transactions on Electron Devices 58.9(2011):3156-3161.Y. Yorozu, M. Hirano, K. Oka, and Y. Tagawa, "Electron spectroscopy studies on magneto-optical media and plastic substrate interface," IEEE Transl. J. Magn. Japan, vol. 2, pp. 740–741, August 1987 [Digests 9th Annual Conf. Magnetics Japan, p. 301, 1982].

[3] Kiselev, A. , et al. "Morphological characterization of soot aerosol particles during LACIS Experiment in November (LExNo)." Journal of Geophysical Research: Atmospheres (2010).

[4] 刘昊炎, 李永亮, and 王文武. "Narrowed Si_(0.7)Ge_(0.3)channel FinFET with subthreshold swing of64 mV/Dec using cyclic self-limited oxidation and removal process." 中国物理 B: 英文版 32.7(2023):500-503.

[5] He, X. , et al. "Impact of aggressive fin width scaling on FinFET device characteristics." 2017 IEEE International Electron Devices Meeting (IEDM) IEEE, 2017.

[6] Guo, Xinjie , et al. "A novel approach to simulate Fin-width Line Edge Roughness effect of FinFET performance." 2010 IEEE International Conference of Electron Devices and Solid-State Circuits (EDSSC) 0.

[7] Zhang, Qingzhu , et al. "Optimization of Structure and Electrical Characteristics for Four-Layer Vertically-Stacked Horizontal Gate-All-Around Si Nanosheets Devices." Nanomaterials 3(2021).

[8] Choi, Yang Kyu , et al. "FinFET process refinements for improved mobility and gate work function engineering." International Electron Devices Meeting IEEE, 2002.

[9] Zhang, Zhaohao , et al. "FinFET with Improved Subthreshold Swing and Drain Current using 3 nm Ferroelectric Hf0.5Zr0.5O2." IEEE Electron Device Letters (2019):1-1.

Deep Investigation into Variability of Complementary Dopant Segregated Tunneling FET Based on Foundry Platform

Rundong Jia[1], Jianfeng Hang[1], Kaifeng Wang[1], Yongqin Wu[2], Hongyan Han[2], Ye Ren[2], Weihai Bu[2], Runsheng Wang[1,3], Qianqian Huang[1,3*] and Ru Huang[1,3*]

[1] School of Integrated Circuits, Peking University, Beijing 100871, China
[2] Seimiconductor Technology Innovation Center (Beijing), Beijing 100176, China
[3] Beijing Advanced Innovation Center for Integrated Circuits, Beijing 100871, China
*Email: hqq@pku.edu.cn, ruhuang@pku.edu.cn

Abstract—In this work, the variability of complementary novel dopant-segregated tunneling FET (C-DS-TFET) based on CMOS baseline platform are systematically and experimentally investigated. Different from conventional MOSFET, with the channel length decreasing, average values of V_{th} and I_{ON} show negligible dependence, while I_{OFF} of nDS-TFET and pDS-TFET show different trend due to different area dependence of gate leakage current and reverse biased p-i-n current. Additionally, the variation of V_{th}, I_{ON} and I_{OFF} show negligible dependence on channel length due to the channel-length-independent tunneling area. Moreover, with channel width scaling down, average values of V_{th} show non-monotonic dependence, and the inverse narrow width effect begins to make difference when channel width is smaller than 120nm. Variation of these three device metrics increase as channel width decreases due to reduced tunneling area. Furthermore, current variation also increases as voltage decreases, indicating the importance of variation-aware design of circuits for low voltage applications.

Keywords—*variability, tunneling FET, dopant segregation, gate leakage current*

I. INTRODUCTION

Tunneling FET (TFET) has become the most promising candidates for post-Moore low power application [1-3]. Benefiting from the band-to-band tunneling (BTBT) mechanism, sub-60mV/decade subthreshold slope can be achieved in TFET, thus enabling potential of ultra-low supply voltage. Due to its p-i-n structure, low leakage current can be obtained simultaneously in Si TFET devices [2-3]. High manufacturability and compatibility with CMOS process also make Si TFET a strong competitor for complementary devices among all kinds of TFETs based on different materials, such as III-V and 2D materials [4-5]. However, on-state current (I_{ON}) of conventional Si TFET is relatively low due to low BTBT probability which hinders its practical application. In order to address this issue, novel complementary dopant-segregated TFETs (C-DS-TFETs) with superior characteristics have been demonstrated on CMOS baseline platform successfully [6-7]. By leveraging most of process steps in CMOS production and some optimization flow especially for TFET, fabricated DS-TFETs can achieve 3 decades I_{ON} enhancement than conventional TFET, and the highest I_{ON}/I_{OFF} ratio of 7 decades among all industry-manufactures[6]. Nevertheless, apart from device performance, performance variation is also vital for high-volume production and circuit application. In our previous work, variability of conventional complementary TFETs (C-TFETs) have been analyzed and experimentally investigated in standard 12-inch Si CMOS platform using 0.13μm process flow [8]. In this work, based on 55nm CMOS foundry baseline platform, variability of novel DS-TFET are experimentally investigated. The impacts of device geometric parameters and supply voltage on the variation of device performance metrics are investigated.

II. DEVICE AND EXPERIMENT

The schematic device structures of n-type and p-type DS-TFET are shown in Fig. 1. Sidewall structure are optimized to be asymmetric for device performance enhancement in TFET, where the spacer is thick at drain side but thin at source side. Self-aligned dopant segregation can be realized by utilizing sidewall. NiSi pocket layer aligned with gate edge at source side can be formed to obtain relatively larger electric field for a higher BTBT probability than conventional TFET, and self-aligned underlap can also be simultaneously realized at drain side with thick sidewall to suppress ambipolar current. C-DS-TFETs are fabricated by utilizing the integration process flow developed on baseline CMOS foundry platform in our previous work [6].

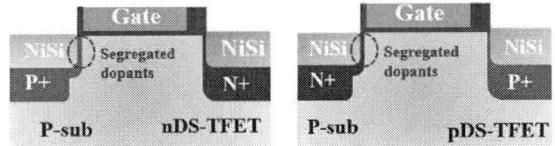

Fig. 1 Device structure of C-DS-TFETs

III. RESULTS AND DISCUSSION

A. Device Performance and its Variation

Transfer characteristics of more than 60 transistors across the whole 300mm wafer are measured and shown in Fig. 2. I_{ON} is extracted at $|V_{GT}|=2.0V$, and off current (I_{OFF}) is extracted at V_{BTBT}, which is the turn-on voltage of tunneling current. The threshold voltage (V_{th}) is obtained by using constant current method. Due to the more abrupt tunnel junction originated from the lower diffusion capability of As than BF$_2$, larger I_{ON} can be obtained in pDS-TFET than nDS-TFET. Fig 4 shows the histogram distribution of extracted V_{th} of C-DS-TFET, and it can be fitted well using Gaussian distribution.

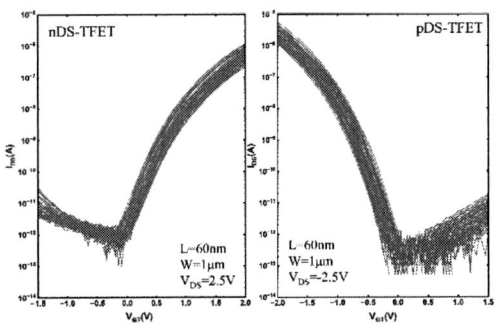

Fig. 2 Measured transfer curves of nDS-TFET and pDS-TFET fabricated on 55nm CMOS baseline platform

979-8-3503-6184-1/24 $31.00 © 2024 IEEE

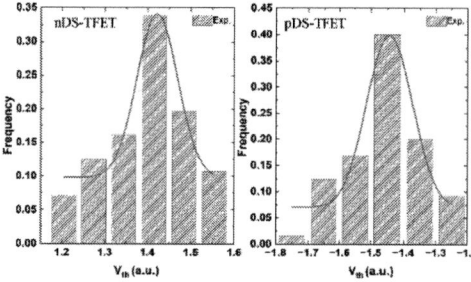

Fig. 3 Histogram of V_{th} of nDS-TFET and pDS-TFET

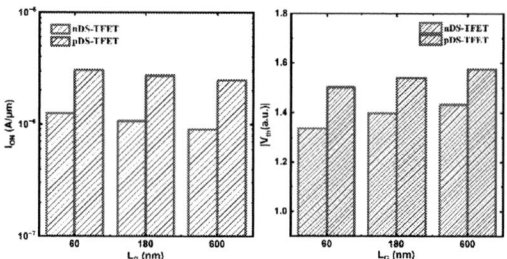

Fig. 4 Average values of I_{ON} and V_{th} dependence on L_G

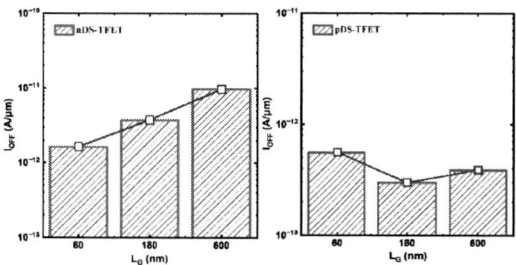

Fig. 5 Average values of measured I_{OFF} dependence on L_G

B. Impact of Channel Length

Average values of measured I_{ON} and V_{th} for both nDS-TFET and pDS-TFET show weak dependence on channel length (L_G), as shown in Fig. 4. The tunneling current of DS-TFET is mainly controlled by the electric field at source junction, and the limited turn-on area and large junction resistance at the source tunnel junction result in negligible dependence on L_G for Si TFET.

For average values of measured I_{OFF}, nDS-TFET and pDS-TFET show different dependence on L_G. As shown in Fig. 5, with channel length decrease, I_{OFF} of nDS-TFET decreases, while I_{OFF} of pDS-TFET shows no obvious trend. By further investigating the transfer curves and gate leakage current (I_G), dominating I_{OFF} mechanisms are found to be different in nDS-TFET and pDS-TFET as shown in Fig. 6. In nDS-TFET, I_{OFF} is mainly controlled by gate leakage current which is expected to be area-dependent, however, in pDS-TFET, I_{OFF} is dominated by reverse biased p-i-n current which is expected to be area-independent.

In nDS-TFET, I_G normalized to channel width and channel area are separately shown in Fig. 7. Surface current density $I_{G,surface}$ is almost the same across different channel length, while line current density $I_{G,line}$ decreases with reduced channel length due to $I_{G,line} = I_{G,surface} \times L_G$.

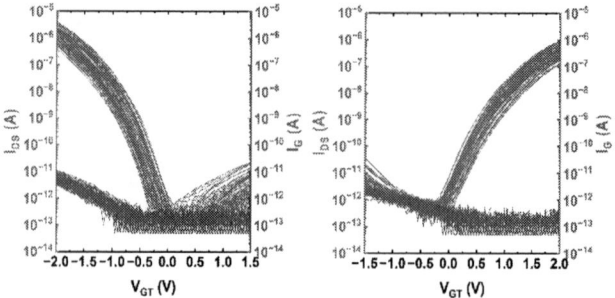

Fig. 6 Measured transfer characteristics and corresponding gate current of nDS-TFET and pDS-TFET

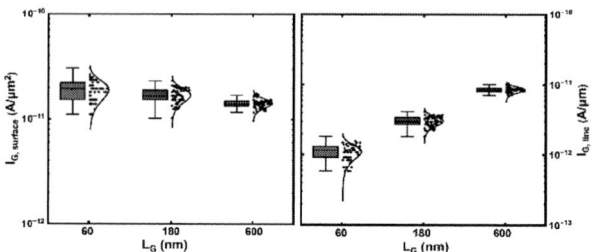

Fig. 7 Distribution of surface current density $I_{G,surface}$ and line current density $I_{G,line}$ in nDS-TFET

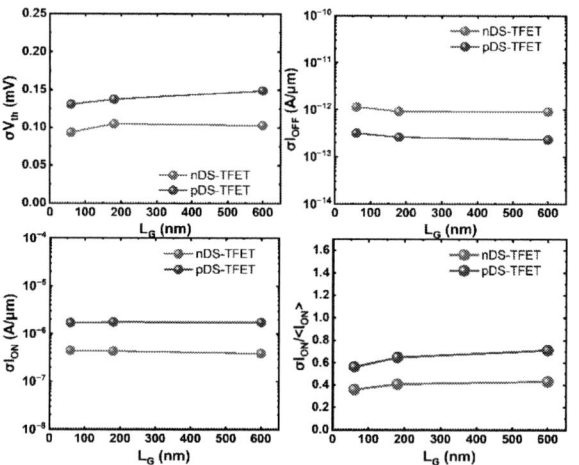

Fig. 8 Measured σV_{th}, σI_{OFF}, σI_{ON} and calculated $\sigma I_{ON}/\langle I_{ON}\rangle$ dependence on L_G

Fig. 8 shows variations of the main device metrics in C-DS-TFETs, and sigma values of measured V_{th}, I_{ON} and I_{OFF} also show negligible dependence on L_G, which are caused by limited tunneling area. Tunneling area is directly related to the regions where tunneling electrons and holes generate at source junction, and the critical factor affecting TFET variation is the BTBT generation area (A_{BTBT}, $A_{BTBT} = L_{BTBT} \times W$). L_{BTBT} is channel-length-independent, thus resulting in little sensitivity of variation behavior to L_G.

σV_{th}, σI_{ON} and $\sigma I_{ON}/\langle I_{ON}\rangle$ of pDS-TFET is larger than nDS-TFET, and this may be caused by different variability of segregation coefficient of As and BF_2 during silicide process. For σI_{OFF}, it is relatively low in pDS-TFET than in nDS-TFET, and this is induced by different variability of gate leakage current and reverse biased p-i-n current.

979-8-3503-6184-1/24 $31.00 © 2024 IEEE

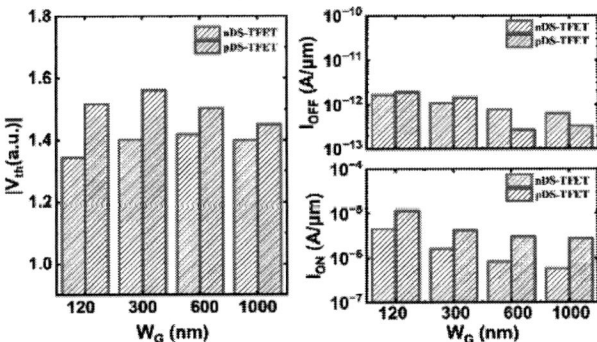

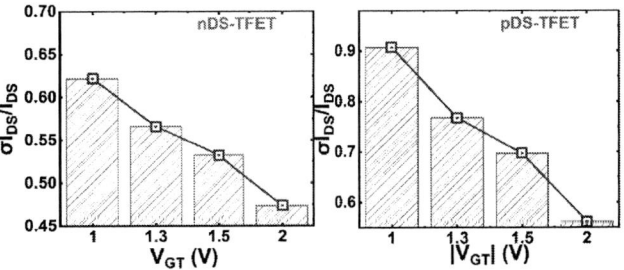

Fig. 11 Measured $\sigma I_{DS}/\langle I_{DS}\rangle$ dependence of V_{GT}

Fig. 9 Average values of measured V_{th}, I_{OFF} and I_{ON} dependence on W_G

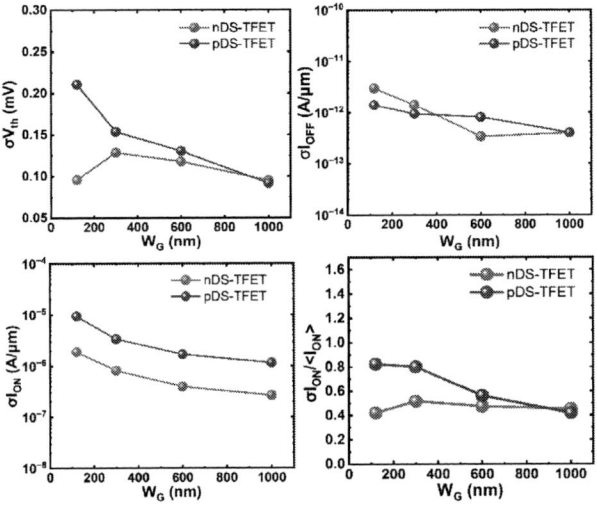

Fig. 10 Measured σV_{th}, σI_{OFF} and σI_{ON} dependence on W_G

C. Impact of Channel Width and gate voltage

Average values of measured V_{th} for both nDS-TFET and pDS-TFET show non-monotonic dependence on channel width (W_G), as shown in Fig. 9. When W_G is within 300nm~1μm range, $\langle V_{th}\rangle$ shows weak dependence, but when W_G reduces from 300nm to 120nm, $\langle V_{th}\rangle$ reduces. This could be caused by inverse narrow width effect (INWE)[9]. Both $\langle I_{OFF}\rangle$ and $\langle I_{ON}\rangle$ increase as W_G reduces.

Fig. 10 shows variations of these three metrics in C-DS-TFETs with different W_G. As the W_G scales down, σV_{th}, σI_{OFF}, σI_{ON} and $\sigma I_{ON}/\langle I_{ON}\rangle$ increase, which is mainly induced by reduced A_{BTBT}.

Fig. 11 shows the dependence of $\sigma I_{DS}/\langle I_{DS}\rangle$ on different gate voltage. As the gate voltage increases, both σI_{DS} and $\langle I_{DS}\rangle$ increase, but the calculated $\sigma I_{DS}/\langle I_{DS}\rangle$ reduces due to the increment of $\langle I_{DS}\rangle$ is larger than that of σI_{DS}. This means the circuit needs to be carefully designed under low supply voltage.

IV. CONCLUSIONS

In this work, variability of C-DS-TFETs based on 55nm CMOS foundry baseline platform are experimentally investigated. Average values of V_{th}, I_{ON}, and sigma values of V_{th}, I_{ON}, I_{OFF} show negligible dependence on channel length, while average value of I_{OFF} in nDS-TFET and pDS-TFET show different trend due to different dominating leakage current. Average values of V_{th} show non-monotonic dependence on channel width due to inverse narrow width effect, and variation of these three device metrics increase as channel width decreases. Current variation increases as voltage decreasing, which indicating the indispensability of variation-aware design in low-voltage application.

ACKNOWLEDGMENT

This work was supported by NSFC (61927901, 62374009), 111 Project (B18001).

REFERENCES

[1] A. Seabaugh, Q. Zhang, "Low-voltage tunnel transistors for beyond CMOS logic", Proceedings of the IEEE, 2020, Vol. 98, 2095-2110

[2] Q. Huang, R. Huang, Z. Zhan, Y. Qiu, W. Jiang, C. Wu, et al, "A novel Si tunnel FET with 36mV/dec subthreshold slope based on junction depleted-modulation through striped gate configuration", IEDM, 2012, pp. 8.5.1-8.5.4

[3] Q. Huang, R. Huang, C. Wu, H. Zhu, C. Chen, J. Wang, et al, "Comprehensive performance re-assessment of TFETs with a novel design by gate and source engineering from device/circuit perspective", IEDM, 2014, pp.13.3.1-13.3.4

[4] G. Dewey, B. Chu-Kung, J. Boardman, J. M. Fastenau, J. Kavalieros, R. Kotlyar, et al, "Fabrication, characterization, and physics of III-V heterojunction tunneling field effect transistors (H-TFET) for steep sub-threshold swing", IEDM, 2011, pp. 33.6.1-33.6.4

[5] D. Sarkar, X. Xie, W. Liu, W. Cao, J. Kang, Y. Gong, et al, "A subthermionic tunnel field-effect transistor with an atomically thin channel", Nature, 2015, vol. 526, pp. 91 -95

[6] K. Wang, Q. Huang, Y. Wu, Y. Ren, R. Wei, Z. Wang, et al, "A novel energy-efficient salicide-enhanced tunnel device technology based on 300 mm foundry platform towards AIoT applications", IEEE ESSDERC, 2022, pp. 360-363

[7] Y. Li, Q. Huang, M. Yang, T. Li, Z. Wang, W. Bu, et al, "A novel self-aligned dopant-segregated schottky tunnel-FET with asymmetry sidewall based on standard CMOS technology", IEEE ICSICT, 2020, pp. 1-3

[8] Q. Huang, R. Jia, C. Chen, H. Zhu, L. Guo, J. Wang, et al, "First foundry platform of complementary tunnel-FETs in CMOS baseline technology for ultralow-power IoT applications: manufacturability, variability and technology roadmap", IEDM, 2015, pp. 22.2.1-22.2.4

[9] W. Lau, K. See, C. Eng, W. Awl, K. Jo, K. Tee, et al, "Anomalous narrow width effect in NMOS and PMOS surface channel transistors using shallow trench isolation", IEEE ESSDERC, 2005, pp. 773-776

Investigation of Common-Gate and Split-Gate Structures Based on CFET Standard Cells

Peishun Tang, Rongzheng Ding, Xiaona Zhu*, Shaofeng Yu*

School of Microelectronics, Fudan University, Shanghai 200433, China

* Email: pstang22@m.fudan.edu.cn, {xiaona_zhu, shaofeng_yu}@fudan.edu.cn

Abstract—This study focuses on two basic types of stacked Complementary-FET (CFET) devices: Common-Gate (CG) and Split-Gate (SG) structures. CG structures stand out due to their simplified fabrication process, while SG structures allow for independent control of each gate, thus enhancing the flexibility of circuit design. This paper aims to provide an in-depth exploration of the fabrication processes, performance characteristics, and circuit-level analyses of these two structures, revealing their importance and potential in modern integrated circuit design. Through comparative studies, we aim to better understand and optimize the application effectiveness of CFET stacked devices in various scenarios, offering theoretical support and practical guidance for future circuit design and technology optimization.

Keywords—Complementary-FET（CFET）, performance-power-area(PPA)

I. INTRODUCTION

With the challenges posed by Moore's Law, the rise of three-dimensional integrated circuits, and increasing demands for low-power and high-performance electronic devices, Complementary FET (CFET) as a stacked device model has garnered widespread interest in both academic and industrial sectors [1]. Its unique vertical complementary structure opens new possibilities for integrated circuit design and manufacturing. The inherent structure of CFET may help reduce parasitic effects, gradually enhancing performance and power efficiency [2]. Combined with adaptive design and innovations like back-side power supply, CFET is poised to potentially succeed Gate-All-Around (GAA) devices as the next generation of advanced processes within the next decade.

According to their gate types, CFET units can be categorized into Common Gate (CG) and Split Gate (SG) structures [3]. In comparison, each type has its pros and cons. CG structure feature simpler fabrication processes and relatively lower costs, but may encounter issues in circuit applications where gate signals cannot pair, necessitating structures like Dummy Gate (DG) and Diffusion Break (DB), which introduce commonality issues. Conversely, SG structure, although more challenging in fabrication, can optimize circuit structure based on CFET units for optimal area efficiency and process integration.Currently, CFET technology is still in the exploration phase, and there is no definitive conclusion on which device structure holds greater development potential. As CFET technology advances, it holds promise for addressing these issues and providing clearer directions for its process optimization in the future.

Therefore, this study involves the construction of two structures. Initially, we analyze the most suitable transistor structures for chip design at the device level. Then, we conduct circuit-level simulations to compare the two structures from four aspects: Performance, Power and Area (PPA) .

II. PROCESS EMULATION

A. CFET Process Flow

Fig. 1 displays the key steps of the CFET structure, with the basic process framework derived from our research group's laboratory [4-5]. First, isolation between channels is achieved through the stacking of multiple layers of Si and SiGe. Next, the positions for the Nanosheets, referred to as Fins, are etched. To ensure separation between power and signal lines, buried power rails (BPR) and backside power delivery techniques are employed [6]. The Inner Spacer in step six effectively reduces parasitic capacitance on both sides of the device structure, enhancing device characteristics. Finally, the source, drain, and gate of the upper and lower devices are fabricated sequentially. The source and drain are selectively grown at the Inner Spacer using epitaxial techniques, and the gate is formed within the dummy gate created in the earlier processes.

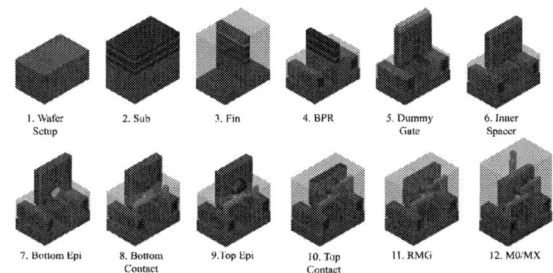

Fig. 1. Key Steps in the CFET Fabrication Process.

Fig. 2 presents the CFET structural diagram and the cross-sectional view of the gate. In practical fabrication processes, the gate metal layer requires multiple Atomic Layer Deposition (ALD) depositions of various metals to achieve the desired work function. For simplicity, we substitute this with a single adjustable work function metal layer. Additionally, to highlight the different metal work functions for PMOS and NMOS, distinct colors are used in the illustration.

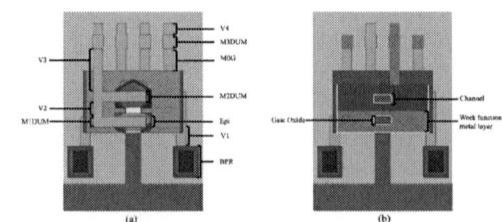

Fig. 2. (a) CFET structure; (b) Cross section of gate.

B. CG and SG Structures

To distinguish the process complexity of SG structures compared to CG structures, we modified the CG process flow to develop the SG structure [3]. In the SG process, after depositing the work function metal layer for the lower device, the gate oxide layer for the upper device is created using ALD technology. This achieves the isolation of the gates between the upper and lower devices. Finally, due to the isolation of the devices, additional process steps are required to extract the gate signals of the lower devices, thereby increasing the manufacturing cost of the SG structure.

In the SG structure, since the isolation between devices is achieved using the oxide layer of the upper device, a significant drawback is the formation of a large capacitance between the upper and lower gates, which adversely affects device performance.

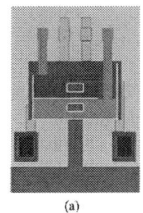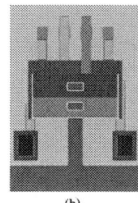

Fig. 3. (a) The cross-sectional views of the gate structures for both CG; (b) SG (right) structures.

III. RESULTS AND PERFORMANCE

Fig. 4 (left) illustrates several possible scenarios for the CG structure, including Common Real Gate, Common Dummy Gate, P Dummy Gate with N Real Gate, and P Real Gate with N Dummy Gate configurations. In these cases, source-drain shorting is used to deactivate devices, but this increases wiring complexity. In contrast, an SG structure can be easily converted into any of these CG configurations.

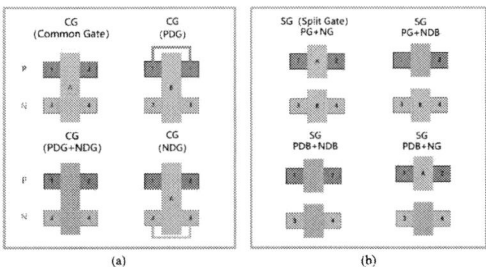

Fig. 4. (a) CG structure; (b) SG structure.

To analyze the differences between CG and SG structures in circuit applications, it is necessary for the NMOS and PMOS gates to connect to different electrical signals. Therefore, we selected four small-scale circuits: an inverter circuit, a transmission gate circuit, a tri-state gate circuit, and a MUX21 circuit, as shown in Fig. 5. Prior to simulation, to ensure the accuracy of the simulation data, the compact models of the transistors used in the circuit simulations were calibrated through numerical simulations.

In the inverter circuit, both single CG and SG structures can achieve the desired circuit functionality. In the TG and TBUF circuits, due to the gates connecting to different electrical signals, using the CG structure would require adding two dummy devices, P Dummy Gate and N Dummy

Gate, to match the gate signals. In the MUX21 circuit, besides the differing gate signals, the source and drain signals also cannot be paired, necessitating the addition of a Common Dummy Gate structure. Figure 6 illustrates the CFET configurations for the four types of circuits.

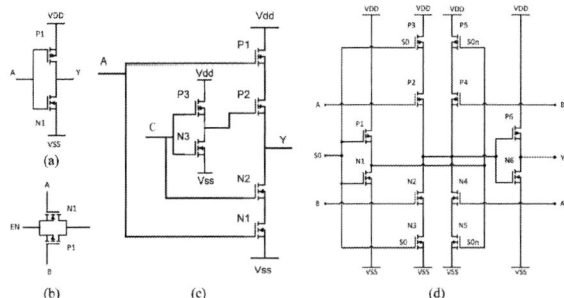

Fig. 5. Schematic: (a) INV; (b) TG; (c) TBUF; (d) MUX21.

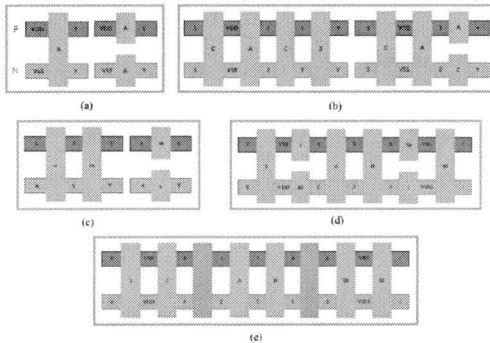

Fig. 6. CFET configurations : (a) INV made of CG (left) and INV made of SG (right); (b) TBUF made of CG (left) and TBUF made of SG (right); (c) TG made of CG (left) and TG made of SG (right); (d) MUX21 made of SG; (e) MUX21 made of CG.

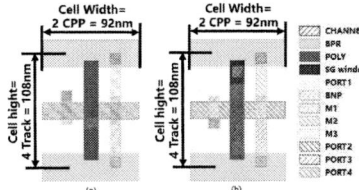

Fig. 7. (a) layout of INV_CG; (b) layout of INV_SG.

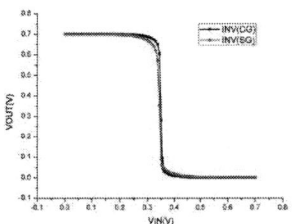

Fig. 8. Voltage transfer characteristics of INV_CG and INV_SG.

First, we constructed inverter circuits using a single CG structure or a single SG structure, respectively, with a supply voltage of 0.7V, and simulated their voltage transfer characteristics. As shown in the fig. 8, both structures switch around half of VDD, indicating that both can function as inverters. Additionally, through prior adjustments to the gate

979-8-3503-6184-1/24 $31.00 © 2024 IEEE

metal work functions, good symmetry between NMOS and PMOS was achieved.

A. Area

To ensure that the track height meets the routing requirements of the four circuits and to facilitate area comparison, we uniformly adopted a 4T design standard. As shown in Fig. 7, the inverter circuit uses only a pair of CFET devices, resulting in the same area for both structures. Fig. 9, 10, and 11 present the layouts of TG, TBUF, and MUX21 circuits constructed using CG and SG structures, respectively. It can be observed that the layouts using SG structures are more area-efficient.

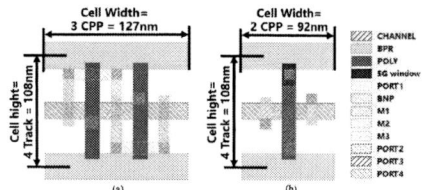

Fig. 9. (a) layout of TG_CG; (b) layout of TG_SG.

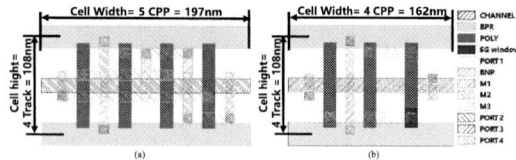

Fig. 10. (a) layout of TBUF_CG; (b) layout of TBUF_SG.

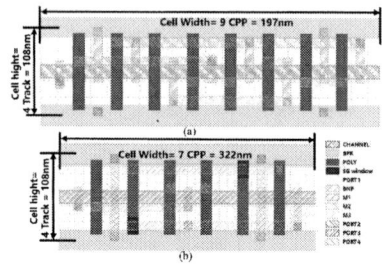

Fig. 11. (a) layout of MUX21_CG; (b) layout of MUX21_SG.

B. Delay and Power

- Applying the same input signals to identical circuits, first verify the correct functionality of the logic gates. Then, measure the transmission delay from each input to the output, averaging multiple data sets. Here, the transmission delay is defined as the difference in time when both the input and output signals reach half of VDD.

- Due to the continuously varying circuit voltage, we only calculate the dynamic power consumption, as shown in Equation (1). Here, α represents the frequency of capacitor charging and discharging, C is the equivalent capacitance, V is the supply voltage, and f is the clock frequency at which the circuit operates.

$$P_{dynamic} = \alpha C V^2 f \qquad (1)$$

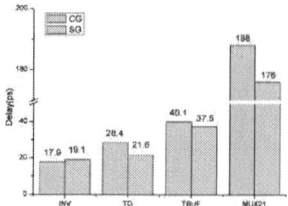

Fig. 12. delay of each component.

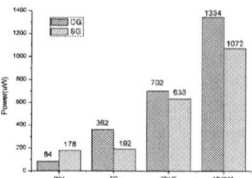

Fig. 13. Power analyses of each component.

In the circuit layout, the SG structure demonstrates flexibility, allowing for a reduction in the number of transistors used, thereby improving circuit performance. From these four circuits, as shown in Figures 12 and 13, the inverter circuit constructed with the CG structure has an advantage. However, the other three circuits, constructed with the SG structure, demonstrate superior performance.

IV. CONCLUSION

This paper investigates two structures of CFET and compares them at both the device and circuit levels. For individual CG and SG structures, the CG structure exhibits less pronounced parasitic effects compared to the SG structure, resulting in superior voltage transfer performance, delay, and power consumption for INV_CG. For the TG, TBUF, and MUX21 circuits, the use of the SG structure reduces the number of transistors, giving TG_SG, TBUF_SG, and MUX21_SG an advantage in terms of area, delay, and power consumption.

REFERENCES

[1] J. Ryckaert et al., "The Complementary FET (CFET) for CMOS scaling beyond N3," 2018 IEEE Symposium on VLSI Technology, Honolulu, HI, USA, 2018, pp. 141-142, doi: 10.1109/VLSIT.2018.8510618.

[2] R. Ding et al., "A Novel Zigzag SRAM Bitcell Design in the Complementary FET Framework," in IEEE Transactions on Electron Devices, vol. 70, no. 9, pp. 4622-4627, Sept. 2023, doi: 10.1109/TED.2023.3289476.

[3] L. Liebmann, J. Smith, D. Chanemougame and P. Gutwin, "CFET Design Options, Challenges, and Opportunities for 3D Integration," 2021 IEEE International Electron Devices Meeting (IEDM), San Francisco, CA, USA, 2021, pp. 3.1.1-3.1.4, doi: 10.1109/IEDM19574.2021.9720577.

[4] R. Ding, Y. Li, Y. Liu, Q. Wu, X. Zhu and S. Yu, "HD SRAM bitcell size shrink beyond 7nm node by CFET without EUV," 2021 International Workshop on Advanced Patterning Solutions (IWAPS), Foshan, China, 2021, pp. 1-4, doi: 10.1109/IWAPS54037.2021.9671236.

[5] X. Zhu, R. Ding, Y. Li, Q. Wu and S. Yu, "CFET 6T HD SRAM Designs with 3nm Design Rule," 2022 China Semiconductor Technology International Conference (CSTIC), Shanghai, China, 2022, pp. 1-4, doi: 10.1109/CSTIC55103.2022.9856851.

[6] B. Vincent, J. Boemmels, J. Ryckaert and J. Ervin, "A Benchmark Study of Complementary-Field Effect Transistor (CFET) Process Integration Options Done by Virtual Fabrication," in IEEE Journal of the Electron Devices Society, vol. 8, pp. 668-673, 2020, doi: 10.1109/JEDS.2020.2990718.

979-8-3503-6184-1/24 $31.00 © 2024 IEEE

Exploration of the effect of silver impurity on the minority carrier lifetime of semiconductor

Xin Tian [*1,2], Peizhi Zhao [2], Yudong Li [2], Jun Xu [1], Tianling Ren [1]

[1] School of Integrated Circuits, Tsinghua University. Haidian District, 100084, Beijing
[2] Jiangsu Xinhua Semiconductor Technology Co., Ltd. Xu Zhou, 221000, China

[*] Email: tianxin@sinvar.com

Abstract—This paper elucidates the effect of silver impurities on the minority carrier lifetime within semiconductor materials. The minority carrier lifetimes in polysilicon samples, incorporating varying concentrations of silver impurities, were meticulously measured and analyzed. Findings reveal a robust and discernible correlation between the concentration of silver impurities and minority carrier lifetime; notably, an incremental increase in silver impurity concentration precipitates a marked reduction in the minority carrier lifetime. These insights hold substantial relevance for guiding the performance modulation of electronic-grade polysilicon and enhancing the design optimization of associated devices.

Keywords—Minority carrier lifetime, electronic grade polysilicon, silver impurity

I. INTRODUCTION

In the fabrication of electronic-grade polysilicon, the minority carrier lifetime serves as a critical indicator of the material's quality and performance within semiconductor devices. Impurities in polysilicon, particularly metallic contaminants, exert a profound influence on this lifetime. The incursion of metal impurities into the semiconductor lattice engenders novel energy states, thereby establishing recombination centers for charge carriers. For instance, metallic elements such as copper, iron, and silver instigate the formation of deep-level energy complexes within silicon, facilitating the recombination of electrons and holes, and consequently diminishing the minority carrier lifetime. Additionally, these impurities can perturb the intrinsic energy band structure of semiconductors, potentially introducing extraneous gap states that alter the energy landscape across the conduction and valence bands. Such alterations may hinder the mobility of minority carriers, elevate the likelihood of collisions with impurities or defects, and thereby abbreviate the minority carrier lifetime. Within the polysilicon manufacturing pipeline, the prevalence of silver impurities merits particular attention due to their introduction via raw materials and during equipment processing stages. This study aims to delineate the relationship between silver impurities and the minority carrier lifetime in electronic-grade polysilicon, thereby informing strategies for product quality enhancement and process optimization.EXPERIMENT.

II. EFFECT OF DIFFERENT IMPURITIES ON THE CONCENTRATION OF ZONE MELTED MONOCRYSTALLINE SILICON RODS.

The zone-melting process of polysilicon is utilized for the purification and removal of impurities from silicon. This method is predicated on the differential solubility of impurities in the solid and liquid phases, as illustrated in Fig. 1.

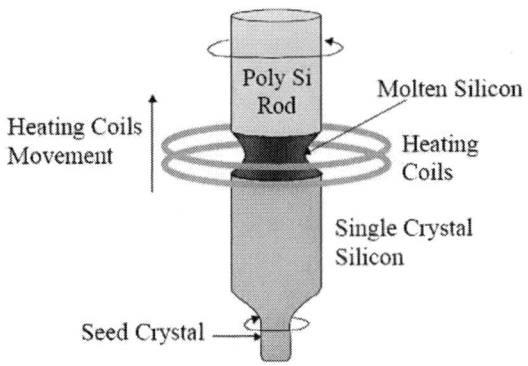

Fig.1 Schematic diagram of the melt recrystallization process of polysilicon by zone melting method.

In a two-phase system characterized by low impurity concentrations, the relationship between the concentrations in the solid and liquid phases at the solid-liquid interface is governed by the equilibrium distribution coefficient, known as the segregation coefficient, K0. This coefficient is given by the equation:

$$K0=Cs/CL$$

where CL is the concentration of the solute on the liquid side at the liquid-solid interface, and Cs represents the solute concentration on the solid side of the interface. Notably, K0 is invariant with respect to temperature and concentration, dependent solely on the properties of the solute and the solvent.

The implications of K0 values are as follows:

When K0<1 and Cs<CL, the concentration of impurities in the initially solidified section is lower than in the later solidified sections.

When K0=1, Cs=CL, indicating an equal impurity concentration throughout the solidified sections.

When K0>1, Cs>CL, the impurity concentration in the initially solidified section exceeds that in subsequent sections.

Table 1 presents the segregation coefficients for common elements. According to the segregation principle of impurities in silicon, metal elements such as Fe, Zn, Cu, Ag, and P predominantly accumulate at the tail of the monocrystalline.

silicon rod, while the concentration of oxygen is primarily located at the rod's head.

Element	B	P	C	O	Fe	Cu	Zn	Ag
Equilibrium distribution coefficient	0.8	0.35	0.07	1.27	1.0×10^{-7}	4.0×10^{-4}	1.0×10^{-5}	1.0×10^{-6}

Table 1. Element segregation coefficient.

III. EXPERIMENTAL SECTION

Experiments in this paper, the minority lifetime test required for the zone-fused monocrystalline silicon rods and polycrystalline silicon were provided by Jiangsu Xinhua Semiconductor Technology Co., Ltd (Xuzhou, China). The monocrystalline silicon rods with a diameter of about 12 mm. Zone melting monocrystalline silicon rod cutting using the SYJ-150 Precision Cutting Machine of Shenyang Kejing Instrument Co., Ltd (Shenyang, China). The minority lifetime was tested for 3 times, and the results were averaged. The minority lifetime tester is the direct current photoconductivity minority lifetime tester of HF-100DCA model (Napson, Japan). Polysilicon matrix impurity analysis using Agilent's inductively coupled plasma mass spectrometer model number ICP-MS 8900.

IV. RESULTS AND DISCUSSION

In the conducted experiment, data from twelve sample groups were analyzed, revealing that the concentration of polysilicon matrix metal impurities—including Fe, Cu, Zn, and three additional metals—was less than 5 ng/kg.

Employing the control variable method facilitated a comparative analysis of the minority carrier lifetime in polysilicon as influenced by variations in the Ag (silver) content. As illustrated in Figure 2, there was a discernible decline in the minority carrier lifetime of monocrystalline silicon rods with an increasing concentration of Ag in the polysilicon matrix. Specifically, a dataset where the Ag element was undetectable was chosen as a reference point. In this instance, despite the presence of other metal impurities, the minority carrier lifetime for the monocrystalline silicon rod was recorded at 4579 µs. In stark contrast, for samples where the Ag content approximated 27 ng/kg, the minority carrier lifetime dramatically reduced to less than 1000 µs. Further analysis of the remaining 10 groups, excluding the one with no detectable Ag, indicated a clear logarithmic relationship between the Ag concentration and the minority carrier lifetime. This relationship can be mathematically expressed by the functional equation $y=-1391\ln(x)+5352$. It is important to note that this equation is specifically applicable for low total metal samples (<10 ng/kg) analyzed using an inductively coupled plasma mass spectrometer, with Ag concentrations not exceeding 50 ng/kg. This functional relationship serves as a crucial reference for understanding the impact of Ag on the carrier dynamics in silicon-based materials.

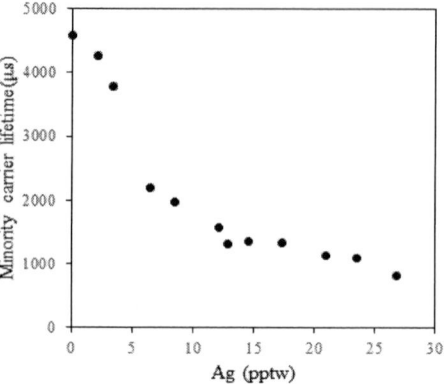

Fig 2. Variation of minority carrier lifetime with silver impurity content.

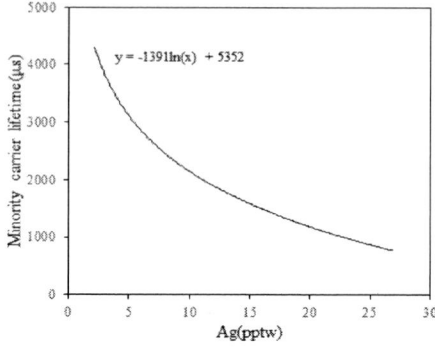

Fig 3. Schematic diagram of the function relationship of minority carrier lifetime with silver impurity content.

V. CONCLUSIONS

This study investigated the correlation between the minority carrier lifetime of zone-melted monocrystalline silicon rods and the silver content in the polysilicon matrix. Through the analysis of twelve data sets, a relationship was established between the low-concentration Ag content and the minority carrier lifetime of monocrystalline silicon rods. This relationship enables a bidirectional determination of the minority carrier lifetime of monocrystalline silicon rods based on the Ag content in the polysilicon matrix, and vice versa, providing valuable insights for optimizing semiconductor material properties.

REFERENCES

[1] Yu Zhanliang, Liu Yike, Tang Yaqin, et al. Distribution and morphology of metal impurities in metallurgical silicon[J]. Journal of Xinyu Univerty, 2014, 19(3):4.

[2] Tian Xin, Wu Feng, Yu Yue. The Production Technology of Electronic Grade Polysilicon[J]. Shandong Chemical Industy, 2017, 46(18):3.

[3] YU Lijun, Duan Jinsheng. Research and Analysis of Influence Factor of Lifetime of Polysilicon[J], Equipment for Electronic Products Manufacturing, 2012, 41(6):5.

Fabrication and Electrical Characterization of Mo/Hf$_x$Zr$_{1-x}$O$_2$/Mo ferroelectric capacitors

Chunsheng Jiang[1,2], Wencai Du[1,2], Qin Xie[1,2]*

[1] Guangxi Key Laboratory of Brain-inspired Computing and Intelligent Chips, Guangxi Normal University
[2] Key Laboratory of Integrated Circuits and Microsystems (Guangxi Normal University), Education Department of
Guangxi Zhuang Autonomous Region, Guilin, China
*Email: xieqin2816@163.com

Abstract

Since their discovery in 2011, Hafnia-based ferroelectrics have shown significant promise for both logic and memory applications. In this study, the hafnium zirconium oxide (Hf$_x$Zr$_{1-x}$O$_2$) capacitors with molybdenum (Mo) electrodes (Mo/HZO/Mo) were fabricated using magnetron sputtering and atomic layer deposition (ALD). Polarization-electric field (*P-E*) loop measurements revealed their excellent ferroelectric properties, which were further confirmed by the Grazing Incidence X-ray Diffraction (GIXRD) analysis. Notably, the Hf$_{0.5}$Zr$_{0.5}$O$_2$ samples annealed at 600 ℃ exhibited high remnant polarization (2P_r > 40 μC/cm²), a minimal wake-up effect, and stable ferroelectricity, enduring over 10⁶ cycles in fatigue tests.

1. Introduction

The Mo/HZO/Mo capacitor has recently attracted significant attention from the academic community due to its exceptional ferroelectric performance, high scalability, remarkable endurance, low deposition temperature, and compatibility with mainstream CMOS process[1-2]. The fundamental reason for its excellent ferroelectric performance is that Mo material has a smaller coefficient of thermal expansion compared to other electrode materials such as TiN and nickel, resulting in higher tensile stress in the HZO film. This strong tensile stress promotes the formation of the ferroelectric orthorhombic phase (o-phase)[3]. In this paper, we first fabricate and characterize Mo/HZO/Mo capacitors. We then explore in detail the effects of annealing temperature and the composition ratio of Hf and Zr on their ferroelectric performance. Finally, we conduct fatigue tests to assess the reliability of the Mo/HZO/Mo capacitors.

2. Experiment Fabrication and Characterization

The device is manufactured on an n-type silicon wafer oriented in the (100) crystal direction. The detailed fabrication process for the Mo-HZO-Mo capacitors, illustrated in Fig. 1, can be described as follows.

• Step 1: Use magnetron sputtering equipment to deposit a 50 nm layer of molybdenum onto the silicon substrate, serving as the bottom electrode.

• Step 2: Use an atomic layer deposition (ALD) system to deposit a 10 nm Hf$_x$Zr$_{1-x}$O$_2$ film at a temperature of 200°C.

• Step 3: Lithography and development processes are performed to define the electrode pattern, resulting in electrodes that are circular with a diameter of 50 um.

• Step 4: Depositing a 150 nm layer of Mo as the top electrode using magnetron sputtering.

• Step 5: The capacitors are prepared with sandwich capacitor structures of Mo/HZO/Mo, using ultrasonic cleaning and stripping processes.

• Step 6: The capacitor reaches the desired temperature within 60 seconds in a nitrogen atmosphere for high-temperature rapid thermal annealing (RTA), followed by a 30-second annealing period.

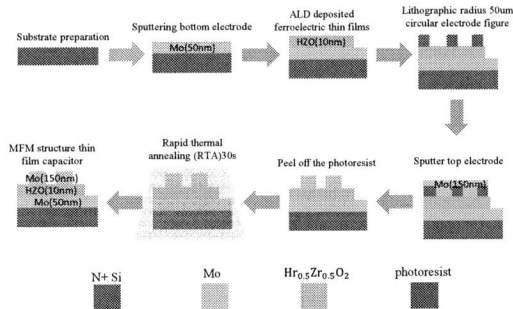

Figure 1. Fabrication process of Mo/HZO/Mo capacitors

The surface morphology of Hf$_{0.5}$Zr$_{0.5}$O$_2$ thin films was examined using an atomic force microscope (AFM). Structural analysis of Hf$_{0.5}$Zr$_{0.5}$O$_2$ capacitors annealed at 400°C, 500°C, and 600°C was conducted using high-power low-angle grazing incidence X-ray diffraction (GIXRD). The electrical properties of Mo/HZO/Mo capacitors were analyzed using a ferroelectric analyzer manufactured by Beijing Huatsu company. Fig. 2(a) shows a two-dimensional (2D) AFM image of the Hf$_{0.5}$Zr$_{0.5}$O$_2$ thin film, while Fig. 2(b) presents a three-dimensional (3D) AFM image of the same film. The mean surface roughness (Ra) and root mean square roughness (RMS) of the Hf$_{0.5}$Zr$_{0.5}$O$_2$ film are 0.85 nm and 1.086 nm, respectively. This suggests that the thin Hf$_{0.5}$Zr$_{0.5}$O$_2$ films we produced have large grain sizes and demonstrate excellent atomic-scale uniformity.

979-8-3503-6184-1/24 $31.00 © 2024 IEEE

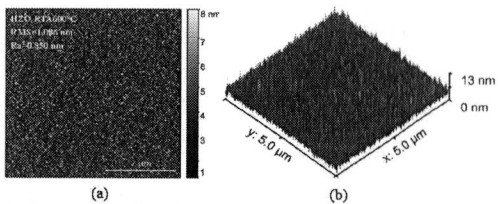

(a)   (b)

Figure 2 The atomic surface topography diagrams of $Hf_{0.5}Zr_{0.5}O_2$ films: (a) a two-dimensional representation and (b) a three-dimensional representation.

Fig. 3 presents GIXRD patterns of the polycrystalline $Hf_{0.5}Zr_{0.5}O_2$ films annealed at 400°C, 500°C, and 600°C. The measurements were taken at a slight incidence angle of 0.5°, with 2θ values ranging from 28° to 34°. Upon analysis, peaks at 2θ values of 28.5° and 31.5° correspond to the M (-111) and (111) phases, respectively, while the peak around 30.5° stands for a mixture of O (111) and T (011) phases, making it challenging to distinguish between O and T phases [4-6]. The M phase exhibits the widest peak width for the sample annealed at 400°C and the narrowest for the sample annealed at 600°C. Consequently, the proportion of the orthorhombic phase (O phase) is largest in the sample annealed at 600°C, which contributes significantly to the high remnant polarization of the HZO thin film [7].

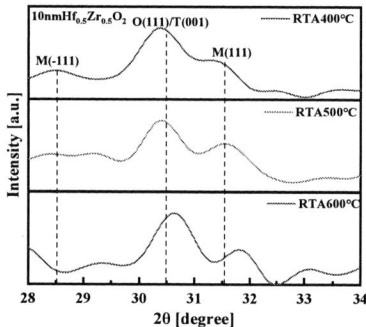

Figure 3. GIXRD patterns of 10 nm $Hf_{0.5}Zr_{0.5}O_2$ films after annealing at 400°C, 500°C and 600°C.

3. Result and discussion

In this section, we will study the influence of various process parameters (e.g. annealing temperature and the composition ratio of elements) on the ferroelectric performance and reliability of Mo/HZO/Mo capacitors.

3.1 Impact of annealing temperature on electrical characteristics of HZO capacitors

Figs. 4 (a) and (b) present the polarization-electric field (P-E) curves and the transient current density vs. electric field (J_s-E) curves of 10 nm $Mo/Hf_{0.5}Zr_{0.5}O_2/Mo$ capacitors. The sample annealed at 300℃ shows no ferroelectric properties due to the lack of the orthorhombic phase. However, capacitors annealed at temperatures ranging from 400℃ to 600℃ exhibit clear ferroelectric hysteresis. The remnant polarization values

$(2P_r)$ are 30.96 μC/cm², 34.48 μC/cm², and 40.90 μC/cm², respectively, while the coercive electric field (E_c) values are 1.53 MV/cm, 1.41 MV/cm, and 1.37 MV/cm, respectively. These P-E measurements were conducted using a bipolar triangular wave pulse with a frequency of 1 kHz and an electric field of 3 MV/cm. In summary, as the annealing temperature increases, the remnant polarization improves and the coercive electric field decreases. This is due to the increasing proportion of the orthorhombic phase with higher annealing temperatures.

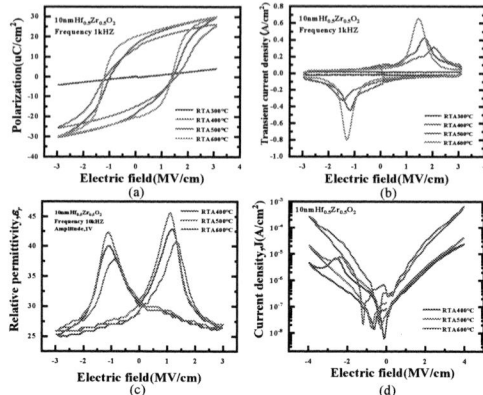

Figure 4. The influence of annealing temperature on electrical characteristics of $Hf_{0.5}Zr_{0.5}O_2$ films. (a) Polarization vs. electric field (P-E) curves; (b) Transient current density vs. electric field (J_s-E) curves; (c) The relative dielectric constant (ε_r) vs. electric field (E) curves; (d) The static leakage current vs. electric field (E) curves.

Figure 4(c) displays the relative dielectric constant (ε_r) versus electric field (E) curves for $Hf_{0.5}Zr_{0.5}O_2$ films. These measurements were obtained by applying a small signal with a frequency of 10 kHz and an amplitude of 1 V across the capacitors. The ε_r-E curves present a "butterfly" shape with double peaks, resulting from polarization reversal as the electric field increases. Notably, the capacitor annealed at 500℃ shows the highest relative dielectric constant, whereas the one annealed at 600℃ has the lowest value. This is because the T-phase, which appears more prominently at 500℃, has a higher dielectric constant compared to the ferroelectric O-phase[2], which is more prevalent in the capacitor annealed at 600℃. Consequently, the increased proportion of the O-phase at 600℃ results in the lowest relative dielectric constant for that capacitor. Fig. 4(d) depicts the leakage current of $Mo/Hf_{0.5}Zr_{0.5}O_2/Mo$ capacitors at various annealing temperatures. The data indicates that higher annealing temperatures result in increased leakage current density. This increase is attributed to the larger grain size observed in samples annealed at 600℃, which leads to more defects or leakage paths in the thin $Hf_{0.5}Zr_{0.5}O_2$ film[8].

3.2 Polarization characteristics of different composition ratio of Hf to Zr

To optimize the process conditions, this section examines the effect of the Hf to Zr composition ratio on the polarization characteristics of $Hf_xZr_{1-x}O_2$ films. HZO films with the ratio of Hf:Zr ranging from 3:7 to 7:3 and a thickness of 10 nm were prepared. The P-E and J_s-E curves were measured using a triangular pulse with a frequency of 1 kHz and an amplitude of 3V, as shown in Fig. 5(a) and Fig. 5(b). As depicted in Fig. 5(a), the P-E curves for $Hf_{0.3}Zr_{0.7}O_2$ and $Hf_{0.4}Zr_{0.6}O_2$ films exhibit obvious anti-ferroelectric behavior, with their J_s-E curves showing double peaks. When the Hf: Zr ratio is 0.5:0.5, the anti-ferroelectric property disappears, and the J_s-E curve presents a single peak, indicating good ferroelectric performance for the $Hf_{0.5}Zr_{0.5}O_2$ films. However, as the Hf content continues to increase, the remnant polarization value from the P-E curves gradually decreases, as does the peak value of transient current density. Thus, it is concluded that the optimal composition ratio of Hf to Zr should be 0.5:0.5.

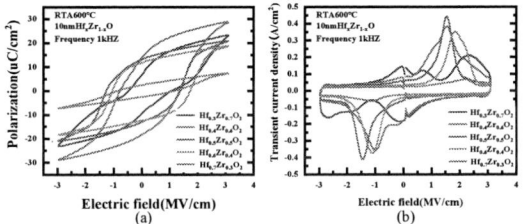

Figure 5. (a) Measured P-E curves of Mo/$Hf_xZr_{1-x}O_2$/Mo capacitors with different composition ratio of Hf to Zr; (b) Transient J_s-E curves of the same Mo/$Hf_xZr_{1-x}O_2$/Mo capacitors with different composition ratio of Hf to Zr.

3.3 Effect of annealing temperature on the switching endurance of $Hf_{0.5}Zr_{0.5}O_2$ films

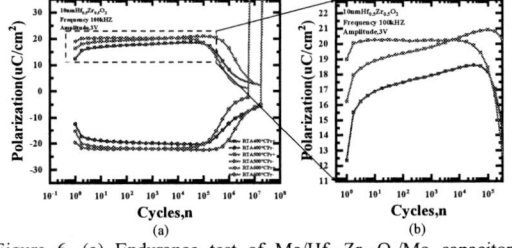

Figure 6. (a) Endurance test of Mo/$Hf_{0.5}Zr_{0.5}O_2$/Mo capacitors for different annealing temperatures. (b) Enlarged region derived from the selected area marked in (a).

As illustrated in Figs. 6(a) and 6(b), the endurance of thin $Hf_{0.5}Zr_{0.5}O_2$ films annealed at 400℃, 500℃, and 600℃ were measured using square wave pulse cycles with an amplitude of 3 V and a frequency of 100 kHz. All samples demonstrated switching endurance for over 10^6 cycles. The cyclic measurements reveal that samples annealed at 400℃ and 500℃ exhibit a clear wake-up effect, with noticeable fatigue appearing around 1×10^5 cycles. In contrast, annealing at 600°C significantly reduces the

wake-up effect across a wide range of stress cycles. However, this process leads to earlier fatigue and dielectric breakdown.

4. Summary

In this paper, capacitors with a Mo/HZO/Mo structure were fabricated using ALD and magnetron sputtering equipment. The effects of various process parameters on polarization characteristics and reliability of $Hf_xZr_{1-x}O_2$ films were further investigated. The results demonstrate that $Hf_{0.5}Zr_{0.5}O_2$ film exhibits excellent both ferroelectricity and endurance. Our work can contribute to optimizing the performance of both logic and memory devices based on HZO film.

Acknowledgments

This work is supported in part by the National Science Foundation of Guangxi province (No. 2023GXNSFBA026021), Guangxi Science and Technology Base and Talent Special Project (No. AD22035213), Guilin Innovation Platform and Talent Special Plan (No. 20220125-1), Outstanding Youth Fund of Guangxi Normal University (No. 2022TD005), and the National Science Foundation of China (No. 61904164).

References

[1] Y. Lee, Y. Goh, J. Hwang, et al. IEEE Trans. Electron Devices, p. 523(2021).

[2] P. R. S. Reddy, et al. Phys. B: Condens. Matter., p. 416024 (2024).

[3] F. Huang, B. Saini, L. Wan, et al. ACS nano, 2024.

[4] M. H. Park, Y. H. Lee, H. J. Kim, et al. Adv. Mater., p. 1811(2015).

[5] Y. Wei, P. Nukala, M. Salverda, et al. Nat. Mater., p. 1095(2018).

[6] J. Muller, T. S. Boscke, U. Schroder, et al. Nano Lett., p. 4318(2012).

[7] L. Xu, T. Nishimura, S. Shibayama, et al. J. Appl. Phys., p. 124104(2017).

[8] H. H. Hsu, et al. Coat., p. 733(2020).

Effect of Cascade Current Density and Plating Time on TSV Filling Effect in DC Power Supply

Weifeng Chen [12], Lijuan Peng [1], Xiaohui Wang*[1], Fangzhou Wang [1], Guojian Ding [1], Qi Feng [1], Ping Yu [1], Peng Zuo [1], Feng Liu [1], Jiang Ma*[2], Yang Wang [1], Haiqiang Jia*[1], Hong Chen [1]

[1] Songshan Lake Materials Laboratory, Dongguan, 523808, China
[2] Shenzhen University, Shenzhen, 518055, China

* Email: 815512388@qq.com, wangxiaohui@sslab.org.cn, majiang@szu.edu.cn, jiahaiqiang@sslab.org.cn

Abstract—This study investigates the influence of different current densities and plating times on the filling effect of TSVs (Through-Silicon Vias) of different sizes. A titrator and recirculating cooling water system were employed to ensure the variation in additive concentration and maintain the electrolyte temperature during TSV copper plating. Through multiple rounds of experiments, the effects of process parameters such as current density and plating time were systematically studied, exploring their impacts on the filling effect of TSVs. The results indicate that excessive current density and duty cycle (plating time: idle time) can lead to defects in the copper pillars. In the step staircase DC plating, as the current density increases and then decreases, the TSV copper plating effect is further optimized, with a gradual reduction in the duty cycle strengthening the optimization effect of TSV copper plating. The findings provide important experimental support for optimizing TSV copper plating technology, reducing TSV copper plating costs, and further investigating the growth mechanism of TSV copper plating, thereby contributing to the improvement of the reliability and filling effect of TSV copper plating.

Keywords—TSV, Direct current electroplating, Filling effect, Flawless copper plating component

I. INTRODUCTION

As electronic products advance, manufacturers and researchers are developing advanced packaging technologies. Three-dimensional silicon integration, particularly through Through-Silicon Via (TSV) technology, promises significant benefits like volume reduction, increased bandwidth, and reduced power consumption. In the manufacturing of three-dimensional integrated circuits (3D-ICs), electroplating copper through TSVs is key[1-2]. While research often focuses on additives and metals for filling, understanding basic parameters like current density and plating time remains crucial. Although Pulse Plating Rectifier (PPR) electroplating can achieve higher filling rates, direct current (DC) electroplating has stable output voltage and lower equipment costs, making it suitable for applications with high requirements for voltage stability[3-4]. Therefore, it is very necessary to understand the mechanism of how parameters such as current density, plating time, and idle time affect the copper plating situation of TSVs[5].

This study investigates the influence of different current densities and plating times on the filling effect of TSVs (Through-Silicon Vias) of different sizes. A titrator and recirculating cooling water system were employed to ensure the variation in additive concentration and maintain the electrolyte temperature during TSV copper plating. Through multiple rounds of experiments, the effects of process parameters such as current density and plating time were

systematically studied, exploring their impacts on the filling effect of TSVs. The results indicate that excessive current density and duty cycle (plating time: idle time) can lead to defects in the copper pillars. In the six-step staircase DC plating, as the current density increases and then decreases, the TSV copper plating effect is further optimized, with a gradual reduction in the duty cycle strengthening the optimization effect of TSV copper plating. The findings provide important experimental support for optimizing TSV copper plating technology, reducing TSV copper plating costs, and further investigating the growth mechanism of TSV copper plating, thereby contributing to the improvement of the reliability and filling effect of TSV copper plating.

II. EXPERIMENTAL

This study presents a comprehensive experimental procedure for the fabrication of TSV, meticulously incorporating the following process steps along with associated data:

A. Pre-treatment Process Stage

- Silicon wafers were first cleaned and treated with HMDS vapor to enhance surface adhesion, followed by spin-coating a 10μm thick layer of 4620 photoresist with optimized parameters for uniform coverage. The wafers were then aligned with a 50μm photomask and exposed under controlled parameters for precise pattern transfer, followed by development, baking, and spinning to remove unexposed photoresist and enhance adhesion. The patterned photoresist was etched using a dry etching process with a deep-to-width aspect ratio of 5:1, with precise control over etch parameters to achieve the desired profile and uniformity. An asher treatment at 250°C for 10 minutes was removed residual photoresist and contaminants. Furnace annealing was conducted to form high-aspect-ratio TSVs with precise dimensional control and smooth sidewalls. Titanium/copper layers were then deposited using magnetron sputtering with optimized parameters for complete coverage. The TSVs were filled with conductive metal via electroplating, with parameters tuned for uniform filling and robust electrical interconnectivity. Processed wafers were diced into individual chips and underwent wet processes, including sample preparation, thinning, and chemical polishing, followed by meticulous microscopic inspection to evaluate the morphology, integrity, and dimensional accuracy of the TSV structures.

979-8-3503-6184-1/24 $31.00 © 2024 IEEE

III. RESULTS AND DISCUSSION

In direct current deposition, the current density and plating time were key factors influencing the copper filling characteristics and microstructure. Through varying the plating time and idle time at different current densities, we have identified optimal plating process parameters for achieving improved filling effects and systematically studied their impact on filling characteristics. To give a specific description of the deposition rate distribution, three locations on the surface of the electrodeposited Cu were marked, as shown in Fig. 2. A new variables V was introduced:

$$V = \frac{L}{t} \tag{1}$$

where V_a, V_b and V_c were the average deposition rates which were contributed by the TSV sidewall only at locations a, b and c, respectively; their direction was marked in Fig. 2.

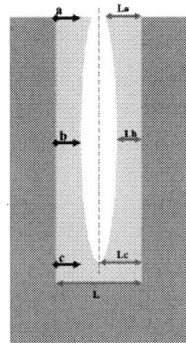

Fig. 2. Three locations (a, b and c) on the surface of the electrodeposited Cu and the direction of the increased thickness which was contributed by the TSV sidewall. Locations u, m and l represented the upper, middle and lower parts of the electrodeposited Cu on the TSV sidewall, respectively

Thus the ratio of the average deposition rates μ can be represented by the ratio of increased thickness of electrodeposited Cu in a certain period of electroplating time. Two new variables μ_x and μ_y were introduced:

$$\mu_x = \frac{V_b}{V_a} \tag{2}$$

$$\mu_y = \frac{V_c}{V_b} \tag{3}$$

Table 1 lists the dynamic μ_x and μ_y values of Cu filled TSVs in different duty cycle respectively. It was found that, When the plating time was greater than the idle time, μ_x was greater than μ_y indicating bottom-up filling, such as Fig. 1. Compared with Fig. 1, μ_x changes from smaller than μ_y to larger than μ_y as the duty cycle decreases. Therefore, in Fig. 3, the steps of plating were increased and the duty cycle was further reduced.

TABLE I.

No.	Influence of six different duty cycle ratios on μ_x and μ_y in TSV copper plating		
	Duty cycle	μ_x	μ_y
1	1:1 4:5 3:5	0.38	0.38

No.	Influence of six different duty cycle ratios on μ_x and μ_y in TSV copper plating		
	Duty cycle	μ_x	μ_y
2	5:3 5:3 5:4 1:1	0.81	0.77
3	1:1 5:3 1:1	0.94	0.55
4	1:1 1:2 1:3 1:3	0.46	0.93
5	1:1 1:2 2:5 4:11	1.20	1.0
6	1:2 5:13 1:6 1:6 1:9 1:16	1.0	0.91

[a.] Influence of six different duty cycle ratios on μ_x and μ_y in TSV copper plating

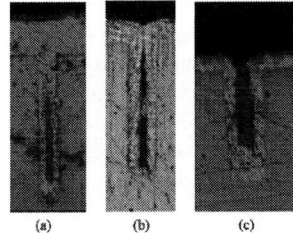

Fig. 1. TSV in No.1, (a) 1:10 TSV copper plated in caliber 10μm, (b) 1:5 TSV copper plated in caliber 25μm (c) 1:3 TSV copper plated in caliber 50μm.

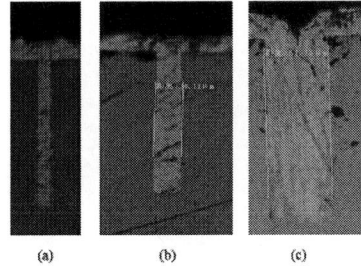

Fig. 3. TSV in No.6, (a) 1:10 TSV copper plated in caliber 10μm, (b) 1:5 TSV copper plated in caliber 25μm (c) 1:3 TSV copper plated in caliber 50μm.

At this time, the μ_x and μ_y values of TSV were close to 1, which was close to the conformal growth mode. Thus, when the accelerant dominates the electrodeposition, the deposition rate at the top of the through-hole was high, and the μx and μy values were slightly less than 1, which was not conducive to bottom-up filling. Because the growth pattern of electrodeposited Cu in this case was similar to that of the sub-conformal filling model, it was clear that the aspect ratio of the unfilled portion of TSV increases as electrodeposition proceeds. Consequently, a low duty cycle proves disadvantageous for bottom-up growth or void-free filling of TSVs with large dimensions and low aspect ratios but favors bottom-up filling of TSVs with high aspect ratios and small sizes. Thus, it finds application in the middle and late stages of electrodeposition to achieve void-free filling of various TSVs. Therefore, it can be applied in the middle and late stages of electrodeposition for void-free filling of various TSVS.

979-8-3503-6184-1/24 $31.00 © 2024 IEEE

IV. CONCLUSIONS

A method combining multi-step electroplating and cyclic electroplating was proposed to fill TSV with a deep-to-width aspect ratio of 3:1, 5:1 and 10:1. The influence of current density and plating ratio on TSV filling was investigated. At low current densities, the deposition rate at the bottom of the vias was low. For larger-sized TSV, this condition was unfavorable for bottom-up growth or void-free filling, whereas for higher aspect ratio TSV, it favored the formation of bottom-up filling patterns, which could be applied for void-free filling of various TSV at the early stage of electroplating. At higher current densities, accelerators accumulated at the bottom corners of the unfilled areas, increasing the local deposition rate and favoring bottom-up growth. However, due to mass transport limitations of copper ions, bottom-up filling could only be achieved for large-sized, low aspect ratio TSV, especially for high aspect ratio, small-sized TSV. Considering the combined effects of current density and TSV size, multi-step current density filling of TSV was adopted, resulting in basic filling without voids. Furthermore, the influence of plating ratio on TSV filling was discussed. Experimental results indicated that an appropriate plating ratio promoted the filling effect of TSV electroplating, thereby further enhancing the filling effect of TSV. Electroplating was conducted at a current density of 0.36 ASD for 20s:40s, followed by electroplating at 0.51 ASD for 25s:65s and 15s:90s, then at 0.40 ASD for 15s:90s and 10s:90s. Finally, electroplating was conducted at 0.3 ASD for 5s:80s to fill the TSV, achieving basic filling without voids.

ACKNOWLEDGMENT

This work was supported in part by the Guangdong Basic and Applied Basic Research Foundation, China, under Grant No. 2020A1515110567.

REFERENCES

[1] P. Garrou, C. Bower, P. Ramm, Handbook of 3D Integration, Wiley-VCH, Weinheim, Germany, 2008.

[2] P. Ramm, A. Klumpp, J. Weber, 3D Integration technologies for MEMS/IC Systems, IEEE Bipolar/BiCMOS Circuits and Technology Meeting (BCTM) 2009; 88: 138–141.

[3] S. Pozder, Progress of 3D Integration Technologies and 3D Interconnects, IEEE International Interconnect Technology Conference 2007; 76: 213–215.

[4] J.U. Knickerbocker, P.S. Andry, B. Dang, R.R. Horton, M.J. Interrante, C.S. Patel, et al., Three-dimensional silicon integration, IBM Journal of Research and Development 2008; 52:553–569.

[5] M. Koyanagi, T. Nakamura, Y. Yamada, H. Kikuchi, T. Fukushima, T. Tanaka, H. Kurino, Three-dimensional integration technology based on wafer bonding with vertical buried interconnections, IEEE Transactions on Electron Devices 2006; 53:2799–2808.

High-Performance Carbon Nanotube Optoelectronic Transistors for Memory Applications

Shuang Liu[1,2,3], Heyi Huang[1,2,3]*, Yanqing Li[1,2,3], Yadong Zhang[1,2,3], Feixiong Wang[1,2,3], Yupeng Lu[1,2,3],

Renjie Jiang[1,2,3], Jiali Huo[1,2,3], Huaxiang Yin[1,2,3]*

[1]Key Laboratory of Fabrication Technologies for Integrated Circuits, Chinese Academy of Sciences, Beijing 100029.
[2] Institute of Microelectronics, Chinese Academy of Sciences, Beijing 100029, China.
[3] School of Integrated Circuits, University of Chinese Academy of Sciences, Beijing 100049, China.

* Corresponding Author's Email: huangheyi@ime.ac.cn, yinhuaxiang@ime.ac.cn

Abstract—Semiconducting single-walled carbon nanotubes (SWCNTs), distinguished by their direct bandgap and one-dimensional (1D) architecture, are exceptionally well-suited for crafting nanoscale optoelectronic devices. In our study, we successfully fabricated high-performance carbon nanotube field-effect transistors (CNFETs) that exhibited a subthreshold swing as low as 129 mV/dec and an ON/OFF current ratio exceeding 10^5. These devices boasted an on-state current surpassing 100 µA/µm, coupled with an ON/OFF current ratio over 10^4 at a drain voltage of -1 V. Moreover, we demonstrated the capability of these devices for optical sensing and their capacity to achieve multi-bit memory, underscoring their potential in the burgeoning domain of in-memory computing. We further exhibited the proficiency of CNT optoelectronic transistors in executing multiple memory write and erase. These capabilities lay a robust foundation for the future multifunctional integration of these devices in advanced integrated circuit systems.

Index Terms—carbon nanotube field-effect transistors (CNFETs), optoelectronic devices, multi-bit memory, multiple memory write and erase

I. INTRODUCTION

Carbon nanotubes possess a suite of exceptional material properties, including high carrier mobility, as well as remarkable mechanical, thermal, electronic, and optical characteristics[1, 2]. Owing to these characteristics, CNTs have emerged as highly valuable materials in the nanoelectronics field. Moreover , they have attracted increasing attention because of their ability to improve the performance and extend the functionality of existing silicon technologies. In 2016, Suzuki et al. presented a study on image sensors based on carbon nanotube array devices[3]. Additionally, Jiang at el. has reported the development of a broadband photodetector utilizing N_2H_4-doped SWCNTs[4]. Furthermore, CNFETs have demonstrated potential applications in solar cells[5] and photoelectrochemical devices[6]. However, researchers exploring their memory capabilities have remained relatively scarce, presenting an opportunity for further investigation in this domain.

In the paper, we successfully fabricated high-performance CNFETs with SS as low as 129 mV/dec and an ON/OFF current ratio exceeding 10^5. Furthermore, we demonstrated the devices' capabilities in optical sensing and their potential for multi-bit memory, highlighting their promise for in-memory computing applications. Moreover, the CNT optoelectronic transistors exhibited remarkable proficiency in performing memory write and erase. These capabilities paved the way for enhanced integrated circuit performance and functionality.

II. EXPERIMENT

In our study, the overall structure of the CNFET is depicted in **Figure 1a** and it was fabricated through the following sequence of steps: Initially, a 200 nm SiO_2 layer was deposited on a P-type silicon substrate by plasma-enhanced chemical vapor deposition (PECVD). Subsequently, 20 nm TiN and 75 nm W layers were applied by physical vapor deposition (PVD), followed by lithography and reactive ion etching (RIE) to pattern the back-gate electrode. An atomic layer deposition (ALD) process at 300°C was then employed to deposit a 15-nm HfO_2 high-k dielectric layer for the back gate. A networked carbon nanotube film prepared using a solution deposition method, served as the device channel. The fabrication process entailed:(i) In an ultra-clean environment, a highly concentrated CNT dispersion was diluted with xylene at a 1:10 ratio to form a solution conducive to the precise deposition of CNT thin films. (ii) Pre-cleaned substrates were immersed in the dilute solution for 48 hours at ambient temperature under static conditions. (iii) After the immersion, the substrates were subjected to a series of 5-minute washes in acetone and anhydrous ethanol to ensure complete cleaning. (iv) The substrates were then dried using high-purity nitrogen to finalize the fabrication of the CNT films. In the next step, a process of annealing at 300 °C for 15-minutes under a nitrogen atmosphere was applied to enhance the properties of the CNT films. This was followed by an electron beam evaporation (EBE) to deposit a 2 nm Ti layer and a 230 nm Pd layer, utilizing the lift-off technique. Photolithography defined the channel area, and dry etching was applied to the CNTs to finalize the channel area, thereby completing the fabrication of the CNFETs.

979-8-3503-6184-1/24 $31.00 © 2024 IEEE

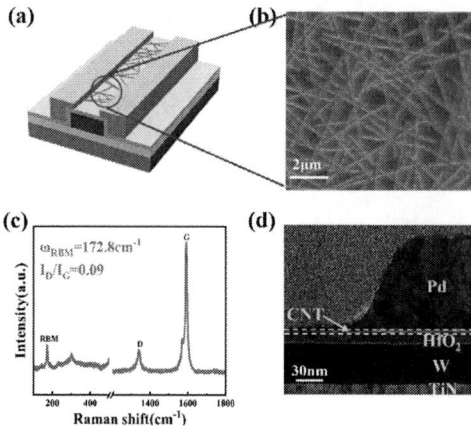

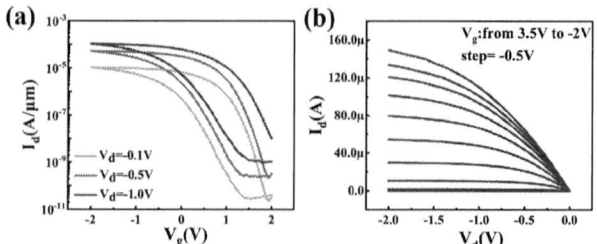

Fig 2 Electrical characteristics of the typical CNFET. (a) The transfer characteristic curve of the typical CNFET. (b) The output characteristic curve of the typical CNFET. The output characteristic curve illustrates V_d sweeping from 0V to -2V, while V_g is stepped from 3.5V to -2V in intervals of -0.5V.

Fig 1 Structural characterization of carbon nanotube thin-film field-effect transistors. (a) Schematic structure of the CNFETs. (b) The SEM image of CNT film topography at the device channel. (The scale bar is 2 μm.) (c) The RAMAN characterization of the CNT film corresponds to a radial breathing mode position of 172.8 cm^{-1} and D/G band ratio of 0.09 for CNT film. (d) TEM schematic of CNFET's source/drain electrode and channel contact position.

Figure 1b presents a Scanning Electron Microscope (SEM) image of the CNT film in the channel region, which exhibits minimal polymer residue and displays uniform CNT diameters and extended lengths. These characteristics are indicative of a high-quality CNT film, thus laying a robust foundation for the fabrication of subsequent high-performance devices. Further characterization of the CNT film is performed using Raman spectroscopy, identifying a radial breathing mode (RBM) band at 172.8 cm^{-1} and the D/G band ratio of 0.09, as shown in **Figure 1c**. Utilizing the empirical formula for CNT diameter calculation based on the RBM band:

$$\omega_{RBM} = \frac{234}{d} + 10 \qquad (1)$$

The CNT diameter is calculated to be 1.4 nm. And the CNT is a semiconducting type. The low D/G band ratio observed in the Raman spectrum suggests a minimal presence of defects and amorphous carbon within the CNTs, which is a promise of high-quality CNT material. Subsequently, we utilized Transmission Electron Microscopy (TEM) to meticulously examine the source/drain electrode and channel contact positions of the device, as depicted in **Figure 1d**. Additionally, Energy- Dispersive X-ray Spectroscopy (EDS) was conducted for elemental analysis. The findings confirm that our device exhibits favorable source-drain contact integrity, which is imperative for its superior electrical performance.

III. RESULTS AND DISCUSSION

We conducted comprehensive electrical measurements on all devices under ambient conditions to discern the electronic properties of the CNFETs. **Figure 2** illustrates the electrical characteristics of a typical device. Specifically, **Figure 2a** delineates the dependence of the drain-source current on the gate voltage, ranging from 2 V to -2 V, at various drain-source voltages of -0.1 V, -0.5 V, and -1.0 V, respectively. Using the transfer characteristic curve at a drain voltage of -0.5 V as a reference, the device has an on-state current of 52.8 μA/μm and an ON/OFF current ratio of more than 10^5. In addition, the device has a subthreshold swing as

low as 129 mV/dec, a threshold voltage of 1.01 V, and a hysteresis voltage of about 1 V. The on-state current is more than 100 μA/μm and the ON/OFF current ratio exceeds 10^4 at a drain voltage of -1.0 V. The output characteristic curve shown in **Figure 2b** indicates that our device has excellent source-drain contact. These exceptional electrical properties suggest that our devices are exceptionally well-suited for a spectrum of high-performance applications and are indeed promising contenders for future multifunctional technologies.

Subsequently, we conducted an exhaustive investigation of the CNFETs, to assess their viability for practical applications. **Figure 3** illustrates the modulation of I_{ds} by adjusting the pulse width under a constant light intensity of 14 mW/cm² at a 365 nm wavelength. The results indicate that an increase in light pulse width corresponds to a higher change in I_{ds} (ΔI_{ds}). Furthermore, the device exhibits an increase in current during light stimulation, which is indicative of its sensing capabilities. Remarkably, the current sustains an elevated level post-stimulation, suggesting a memory effect within the device. This persistent photocurrent is attributed to the creation of electron-hole pairs within the CNT layer upon light illumination. The holes are efficiently transferred, contributing to the enhanced photocurrent, while electrons are captured at the channel-gate dielectric interface[7]. This electron-trapping mechanism is indicative of the device's capacity for information retention, underscoring its potential for applications in data memorization.

We further investigated the suitability of these devices for memory applications and the ability to enable multi-bit memory. **Figure 4** depicts the response of CNT optoelectronic transistors to a sequence of light pulses and electrical pulses. The devices demonstrated the capacity for dynamic modulation of current states. A systematic series of measurements was meticulously executed, utilizing 15 light pulses at a wavelength of 365 nm, each with an intensity of 14 mW/cm², to effectively modulate the current of the device. The light pulse durations were adjusted from 0.5s to 1.85s, with a uniform interval of 6.2s between pulses, as detailed in **Figure 4a**. In addition, a series of electrical pulses were applied for suppression purposes, with each pulse lasting 1.4s and separated by intervals of 5.8s. The applied voltages spanned from -0.4 V to -2.3 V, as illustrated in **Figure 4b**. Our findings demonstrate that the progressive enhancement of both optical and electrical stimulus signals notably improves the uniformity of the current step width and enhances the linearity of current modulation [8, 9]. This discovery lays a robust groundwork for the advancement of in-memory computing capabilities.

979-8-3503-6184-1/24 $31.00 © 2024 IEEE

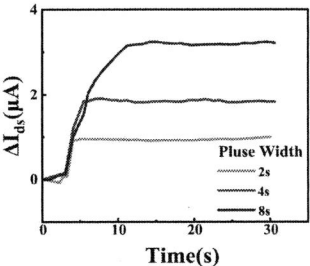

Fig 3 Optoelectronic characteristics of the CNT optoelectronic transistors. ΔI_{ds} triggered by different optical pulse widths.

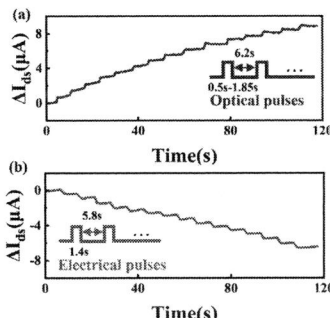

Fig 4 Memory characteristics of the CNT optoelectronic transistors. (a)Accumulation of optical potentiation in progressive multi-bit memory, with different optical durations and an interval of 6.2s. (b)Accumulation of electrical depression in progressive multi-bit memory, with electrical pulse of 1.4s and an interval of 5.8s.

Subsequently, we elucidated the "memory write and erase" capability of the CNT optoelectronic transistors, employing the numeral "6" as a representative case study. **Figure 5a** depicts the specific process during training, demonstrating the changes in I_{ds} with light pulse and electrical stimulation. Initially, the pixel transistors were subjected to a 365 nm light pulse with an intensity of 14 mW/cm², a duration of 4.1s, and a gate voltage of 2V. This stimulation led to a change in the drain current of approximately 2.4 μA, manifesting the numeral "6" clearly at **stage t_1**. Post-stimulation, the numeral remained clear for 8s at **stage t_2**, demonstrating the device's effective memory retention capability. Subsequently, the application of a negative gate voltage of -2V with a pulse width of 0.2s effectively erased the numeral "6," as observed at **stage t_3**. This erasing process is attributed to the de-trapping of electrons, where the negative gate voltage attracts holes to the channel-gate dielectric interface, neutralizing the trapped electrons and facilitating the erasing mechanism. Finally, the reapplication of a light pulse under identical conditions to the initial stimulus (14 mW/cm², 4.1s, $V_g = 2V$) enabled the transistor array to distinctly display the numeral "6" once more at **stage t_4**. This capability for rewriting underscores the device's potential for multiple cycles of memory writing and erasing. These observations collectively indicate that the CNT optoelectronic transistors exhibit the essential functions of memory writing, erasing, and rewriting, highlighting their suitability for applications in advanced memory systems.

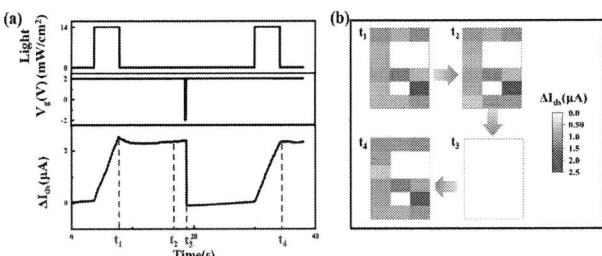

Fig 5 Memory function emulation in CNT optoelectronic transistors. a) Schematic diagram of the light and gate control signal sequences and the corresponding ΔI_{ds} outputs for the 6×3 array device simulating the "learning and forgetting" behavior. (b) Evolution of source-drain current during the learning and forgetting process. The number ("6") was selected to illustrate the intensity of memorization.

IV. CONCLUSION

In this study, we succeeded in developing high-performance carbon nanotube field effect transistors (CNFETs) with on-state currents exceeding 100 μA/μm and a high ON/OFF current ratio. Moreover, we demonstrated the optical sensing capabilities of these devices, as well as their potential for multi-bit memory, highlighting their promise for in-memory computing applications. Furthermore, the carbon nanotube optoelectronic transistors are capable of performing multiple memory write and erase operations. These attributes establish a robust foundation, rendering these devices highly suitable candidates for multifunctional integration in advanced technological systems.

V. ACKNOWLEDGMENT

This work was financially supported by the National Key R & D plan "nano frontier" key special project (grant no.2021YFA1200502).

REFERENCES

[1] G. Hills et al., "Understanding Energy Efficiency Benefits of Carbon Nanotube Field-Effect Transistors for Digital VLSI," IEEE Transactions on Nanotechnology, vol. 17, no. 6, pp. 1259-1269, 2018.

[2] P. Avouris, Chen, Z. & Perebeinos, V., "Carbon-based electronics.," Nature Nanotechnology, vol. 2, 605–615, 2007.

[3] D. Suzuki, S. Oda, and Y. Kawano, "A flexible and wearable terahertz scanner," Nature Photonics, vol. 10, no. 12, pp. 809-813, 2016.

[4] H. Zhou et al., "Polarimetric Vis-NIR photodetector based on self-aligned single-walled carbon nanotubes," Carbon, vol. 143, pp. 844-850, 2019.

[5] H. Zhu, J. Wei, K. Wang, and D. Wu, "Applications of carbon materials in photovoltaic solar cells," Solar Energy Materials and Solar Cells, vol. 93, no. 9, pp. 1461-1470, 2009.

[6] M. Grätzel, "Photoelectrochemical cells.," Nature, vol. 414, 338–344, 2001.

[7] N. Li et al., "Gate-tunable large-scale flexible monolayer MoS2 devices for photodetectors and optoelectronic synapses," Nano Research, vol. 15, no. 6, pp. 5418-5424, 2022.

[8] C. Liu et al., "Two-dimensional materials for next-generation computing technologies," Nature Nanotechnology, vol. 15, no. 7, pp. 545-557, 2020.

[9] Z. Zhang, S. Wang, C. Liu, R. Xie, W. Hu, and P. Zhou, "All-in-one two-dimensional retinomorphic hardware device for motion detection and recognition," Nature Nanotechnology, vol. 17, no. 1, pp. 27-32, 2021.

Investigation of the channel width dependence of IGZO TFT by experiment and TCAD simulation

Yanyu Yang [1,2,3], Yupeng Lu [1,2,3], Shuang Liu[1,2,3], Renjie Jiang[1,2,3], Jie Luo[1,2,3], Yunjiao Bao[1,2,3], Peng Wang[1,2,3], Gaobo Xu*[1,2], Huaxiang Yin[1,2,3]

[1.] Key Laboratory of Fabrication Technologies for Integrated Circuits, Chinese Academy of Sciences, Beijing 100029.
[2.] Institute of Microelectronics, Chinese Academy of Sciences, Beijing 100029, China.

[3.] School of Integrated Circuits, University of Chinese Academy of Sciences, Beijing 100049, China

*Corresponding Author's Email: xugaobo@ime.ac.cn

Abstract—**In this study a inverted stagger IGZO TFTs with 350nm/ 250nm/ 100nm/ 50nm channel width was realized by dry etching during an active layer pattern. A severe degradation was observed with threshold voltage shift positively, SS enlarge, and mobility degradation when channel width was down to 50nm. Assisted by TCAD simulation and previous report, we confirm this degradation is attributed to insufficient reacted methyl ion and by-product absorbed on channel sidewall from the dry etching process. The methyl ion acts as negative interface trap and induces depletion region in the channel margin. This degradation effect is particularly severe when the channel width is below 100 nm.**

Keywords—*IGZO TFT, channel width, dry etch, etch damage*

I. INTRODUCTION

A three-dimensional integrated circuit (3D IC) is a promising technology in the semiconductor industry due to the multifunctional integration of different layers such as power memory, logic, and sense. The amorphous indium gallium zinc oxide (a-IGZO) in recent years has been considered as a competitive candidate material for 3D-IC due to its higher mobility (10~30 cm²/Vs) and its ultra-low thermal budget with back-end-of-line (BEOL)-compatible process compatibility [1-6]. Many studies are focused on IGZO devices with mobility enhancement or reliability improved by plasma treatment or post-annealing [7, 8]. Besides, high density with high performance and low cost are coordinated. So far devices with channel width larger than 500nm have been widely investigated. Still, in practical applications device width of less than 500nm is mainstream and easy to achieve high-density structure. It was reported that channel dimensions have a significant effect on threshold voltage and on-current. Using the energy band theory, the adsorption of oxygen or water vapor on the IGZO top surface leads to large band bending with low effective carrier concentration in the channel and a higher effective channel resistance [9]. However, when the gate width is reduced by the same order of magnitude as the thickness, the effect of the sidewalls on the channel performance is just as important as the effect of the top surface.

In this study, an inverted stagger IGZO TFTs with 350nm/ 250nm/ 100nm/ 50nm channel width was realized respectively by dry etching during an active layer pattern. The device demonstrates high performance including a high Ion/Ioff over 10^7 and positive threshold voltage and SS down to 400mV/dec when the channel width is larger than 100nm. The device illustrates severe degradation with threshold voltage shift positively and SS enlarge when channel width is below 100nm. This degradation may be attributed to insufficient reacted methyl ion absorbed on the channel sidewall which acts as a negative interface trap and induces depletion region by positive oxygen vacancy, and SS enlarges when the channel width is below 100nm. This degradation may be attributed to insufficient reacted methyl ion absorbed on the channel sidewall which act as a negative interface trap and induces depletion region by positive oxygen vacancy.

II. EXPERIMENT

In this paper a conventional inverted stagger TFT was fabricated with key process flow shown in Figure 1. (a) and schematic image shown in Figure 1. (b).

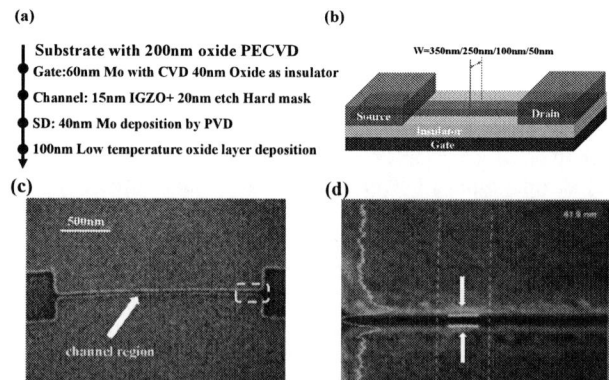

Figure 1. (a) Fabrication process flow (b) Schematic image of IGZO TFT with shrinking channel width down to 50nm (c) Top view SEM image of the channel pattern after lithography (d) Top view SEM image showing an α-IGZO TFT with a smallest width with 42nm.

First, the 60nm molybdenum metal layer was deposited by physical vapor deposition (PVD) process as back gate metal and followed by photolithography process and dry etching with Fluorine-based gases to pattern gate layer. Next a 40nm oxide layer was served as an insulator layer using the plasma-enhanced CVD (PECVD) method at a temperature of 400°C. Then a 15nm IGZO active layer (AL) was sputtered by magnetron RF sputtering approach at room temperature (RT), using IGZO target with atomic ratio of In Ga: Zn = 1:1:1 with Ar: O$_2$ at gas flow for 30:3. Before patterning active channel layer by electron beam lithography (EBL), a 20nm molybdenum was utilized as a hard mask layer to ensure the integrity of the pattern transfer process. Due to the isotropy etch rate of wet etching, it is hard to guarantee active layer width under 200nm. Hence, we adopted a dry etching process to pattern the IGZO active layer with a gas flow of CH$_4$ and Ar for 80sccm:20sccm at room temperature. Accompanied by the dry etch process, the 50nm/100nm/250nm/350nm channel width was obtained. Figure 1.(c) and (d) illustrate the SEM top view of the IGZO channel layer after electron beam lithography including a channel width of 42nm. After SD contact was deposited by PVD for 40nm molybdenum, dry etching was adopted to etch molybdenum for 60nm, and both the SD layer and hard mask layer was etched. Finally, a 100nm oxide passivation layer was grown by chemistry vapor deposition below 300℃。

All electrical measurements of the transform current–voltage were performed using an 4156C semiconductor parameter analyzer.

III. RESULT AND DISSICUSION

The threshold voltage is defined as gate voltage at a fixed normalization drain current equal to 1E-9A. And on-current is confirmed to the drain current under V$_{GS}$-V$_{TH}$=10V and V$_{DS}$=10V. Also, SS and μ_{FE} are extracted by the following Equation (1) (2) to evaluate device performance.

$$SS = \frac{d(\log(I_{DS}))}{dV_{GS}} \quad (1)$$

$$\mu = \frac{G_m}{V_{DS} * C_{ox} * \frac{W}{L}} \quad (2)$$

Figure 2. (a) shows transfer characteristics for IGZO TFTs with a fixed channel length of 1 um and various channel widths ranging from 50 nm to 350 nm at a fixed drain bias of 10 V. As a result of normalization width/length ratio less than 0.25, the device with raised value for I$_{off}$ more than 100pA shown on/off ratio over 10^6. It shows good operations performance even with width scaling to 100nm. Threshold voltages for all devices exhibit positive values, implying good performance for practical application. The output curves with 100nm channel width are summarized in Figure 2 (b) in which V$_{GS}$ varied from 0 to 16 V with a step of 2 V. Under small drain voltage, the device exhibits a rectification characteristic, demonstrating the source/drain contact is ohmic contact. As drain voltage and gate voltage both raised over 12V, drain current present up to 21uA.

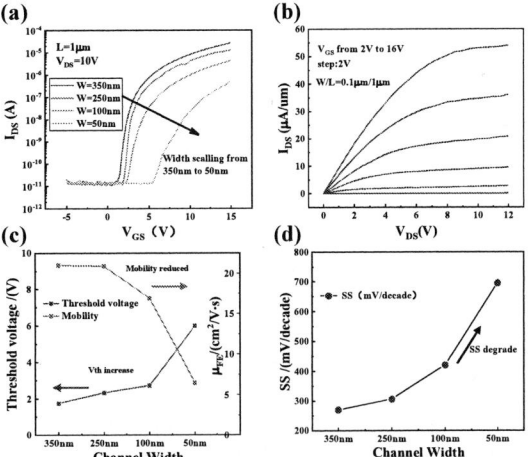

Figure 2. (a) Transfer characteristics on V$_{DS}$=10V with channel width of 350nm/250nm/100nm/50nm and channel length of 1um (b) output characteristics with width/length=0.1um/1um (c) (d) VTH and SS mobility variation with channel width decrease.

Figure 2 (c) shows the extracted threshold voltage field-effect mobility and (d) SS with channel width scaling from 350nm to 50nm. It can be observed that when width shrinks from 350nm to 100nm, device performance degenerates slightly with V$_{TH}$ gradually increasing from 2V to 2.7V and SS exhibiting the same evolving trends that gradually increase from 271mV/dec to 421mV/dec and mobility is consistently maintained at 18cm^2/V·s. But when width scaling to 50nm, it demonstrates a large degradation about V$_{TH}$ increase to 6V and SS abruptly increase from 421mV/dec to 700mV/dec which demonstrates the SS and threshold voltage as a function of channel width. With the structure and components of each layer being stable during fabrication process, the interface between each layer is highly influenced by the process or environment. The SS degradation is likely due to the additional interface trap state. In turn, the additional interface trap could be calculated with Equation (3)

$$D_{it} = \frac{C_{OX}}{q} \frac{\Delta SS(\log_{10} e)}{k_B T / q} \quad (3)$$

Here, e is the base of the natural logarithm, T is the temperature, and q is the electronic charge. As SS increases, the interface trap with a width of 50nm is estimated to be 2.5 × 10^{12}cm^{-2}. There are multiple interface charge states that influence IGZO device performance such as oxygen vacancy, hydrogen ion, and various gas ions involved in etching process. Previous reports have focused on the passivation layer on top of the channel which induces electron concentration increase by generating extra oxygen vacancy [10]. Their effects such as V$_{TH}$ shift and on-current increase are determined by charge state, Fermi level, and local structure environment. In our fabrication process, CH$_4$ is used as a gas ingredient to pattern active channel layer by ICP (inductively coupled plasma etch), the etching mechanism for CH4-based etchant is as follows

$$CH_4 \Rightarrow CH_3^- + H^+ (plasma\ active)$$

$$3CH_3^- + In^{3+}/Ga^{3+} \Rightarrow In(CH_3)_3/Ga(CH_3)_3$$

$$2CH_3^- + Zn^{2+} \Rightarrow Zn(CH_3)_2$$

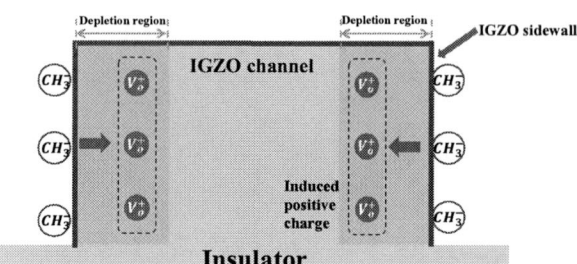

Figure 3. Schematic image of cross-section IGZO TFT with depletion region generated by negative methyl ion on channel sidewall

Previous study reports that by-products would be generated during the reaction at room temperature which is induced by the high-boiling point of CH_4 dry etchant by-product [11]. After completing the etching process, some reactive methyl molecules are adsorbed on IGZO sidewalls due to insufficient reaction or by-product generated as shown in Figure 3, and these negatively charged ions (CH_3^-) caused the device to generate a deep depletion region with positive charge in channel margin. Since the insufficient reacted charge acts in the depletion region only ten nanometers away from the line sidewalls boundary, the threshold voltage and channel resistance would increase as the channel width is decreased below a critical size.

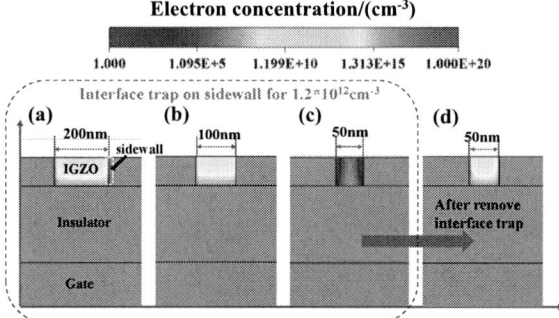

Figure 4. Cross-section of electron concentration distribution for a device with interface trap and channel width of (a)200nm (b)100nm(c)50nm and for a device without interface trap and channel width of 50nm.

To better explain the reason for the degradation of the reduced channel width, we also explored a TCAD simulation on IGZO TFT with or without a negative interface trap of $2.5*10^{12}cm^{-3}$. The density of state used in simulated transfer characteristics could have a good fit with the experiment result. Figure 4 depicts the electron concentration distribution in a cross-section of (a)200nm/(b)100nm/(c)50nm channel width separate with negative sidewall interface trap state when $V_G= 4V$. The channel width of 200 nm or 100nm observed a uniform distribution of electron concentration with approximately $10^{13}cm^{-3}$, but a severe decrease in electron concentration happened when the width was scaled to 50nm. However, the device returns to normal when removing the sidewall interface trap as shown in Figure 4. (d). Hence, our TCAD simulation results show that the sidewall trap of negative methyl ions plays an important role in electron

concentration decrease when the channel width down to 50nm.

IV. CONCLUSION

Based on the dry etching fabrication process used in the IGZO pattern, an inverted stagger IGZO TFTs with 350nm/ 250nm/ 100nm/ 50nm channel width was realized. The good operation performance with Ion/Ioff over 106 and positive threshold voltage was observed with channel width larger than 100nm, but an abrupt degradation existed when channel width scaling from 100nm to 50nm with threshold voltage shift positively from 3V to 6V and SS enlarge from 421mV/dec to 700mV/dec. Assisted by TCAD simulation and previous report, we indicate this degradation may be attribute to insufficient reacted methyl ion absorbed on channel sidewall from dry etching process which act as negative interface trap and induced depletion region with positive oxygen vacancy. And this degradation effect is particularly severe when the channel width below 100 nm.

ACKNOWLEDGMENT

This work was supported by BJSAMT Project (SAMT- ZK-KT-22030102). And this work is supported by the National Key Research and Development Program (No. 2022YFE0124200).

V. REFERENCE

[1] K. Nomura *et al.*, "Room-temperature fabrication of transparent flexible thin-film transistors using amorphous oxide semiconductors," (in English), *Nature,* Article vol. 432, no. 7016, pp. 488-492, Nov 2004.

[2] J. M. Zhou *et al.*, "Energy-Efficient Artificial Synapses Based on Flexible IGZO Electric-Double-Layer Transistors," (in English), *Ieee Electron Device Letters,* Article vol. 36, no. 2, pp. 198-200, Feb 2015.

[3] A. Daus *et al.*, "Flexible CMOS electronics based on p-type Ge2Sb2Te5 and n-type InGaZnO4 semiconductors," in *63rd IEEE Annual International Electron Devices Meeting (IEDM)*, San Francisco, CA, Dec 02-06 2017, NEW YORK: Ieee, in IEEE International Electron Devices Meeting, 2017,

[4] R. An *et al.*, "A Hybrid Computing-In-Memory Architecture by Monolithic 3D Integration of BEOL CNT/IGZO-based CFET Logic and Analog RRAM," presented at the 2022 International Electron Devices Meeting (IEDM), San Francisco, CA, USA, 2022

[5] A. Belmonte *et al.*, "Tailoring IGZO-TFT architecture for capacitorless DRAM, demonstrating 103s retention, > 1011 cycles endurance and Lg scalability down to 14nm," in *IEEE International Electron Devices Meeting (IEDM)*, San Francisco, CA, Dec 11-16 2021, NEW YORK: Ieee, in IEEE International Electron Devices Meeting, 2021,

[6] Y. X. Sui *et al.*, "Room-Temperature Ozone Sensing Capability of IGZO-Decorated Amorphous Ga2O3 Films," (in English), *ACS Appl. Mater. Interfaces,* Article vol. 12, no. 7, pp. 8929-8934, Feb 2020.

[7] G. Yan *et al.*, "First Demonstration of True 4-bit Memory with Record High Multibit Retention >103s and Read Window >105 by Hydrogen Self-Adaptive-Doping for IGZO DRAM Arrays," presented at the 2023 International Electron Devices Meeting (IEDM), San Francisco, CA, USA, 2023

[8] Y. T. Chien *et al.*, "Performance Enhancement of InGaZnO Top-Gate Thin Film Transistor With Low-Temperature High-Pressure Fluorine Treatment," (in English), *Ieee Electron Device Letters,* Article vol. 42, no. 11, pp. 1611-1614, Nov 2021.

[9] Y. W. Heo *et al.*, "Effects of channel dimensions on performance of a-InGaZnO4 thin-film transistors," (in English), *J. Vac. Sci. Technol. B,* Article vol. 29, no. 2, p. 7, Mar 2011, Art no. 021203.

[10] P. P. Zhang *et al.*, "Physical Insights Into the Mobility Enhancement in Amorphous InGaZnO Thin-Film Transistor by SiO2 Passivation Layer," (in English), *Ieee Transactions on Electron Devices,* Article vol. 67, no. 6, pp. 2352-2358, Jun 2020.

[11] S. Kundu *et al.*, "High-Density Patterning of InGaZnO by CH4: a Comparative Study of RIE and Pulsed Plasma ALE," (in English), *ACS Appl. Mater. Interfaces,* Article vol. 14, no. 29, pp. 34029-34039, Jul 2022.

979-8-3503-6184-1/24 $31.00 © 2024 IEEE

A Test and Evaluation Platform for Quantitative Analysis of High-Reliability Designs

Yifeng Huang [1], Wenqing Wan [2], Chang Wu*[1]

[1] School of Microelectronics, Fudan University, Shanghai, China
[2] Fudan Microelectronics Group, Shanghai, China

* Email: 20300750032@fudan.edu.cn, wanwenqing@fmsh.com.cn, wuchang@fudan.edu.cn

Abstract — Charged particles in space can induce Single Event Effects (SEE), with Single Event Upset (SEU) being the most common. SEU can lead to data errors and system failures, highlighting the need for high-reliable circuits design. This paper presents a comprehensive error injection testing platform and quantitative analysis model for error rate, error detection rate, and area overhead of various self-checking logic designs under SEU. We propose a novel and efficient error injection and testing approach based on netlist editing for statistical analysis on the effectiveness of various hardening schemes of logic cell designs. This quantitative analysis enables advanced low cost SEU mitigation designs.

Keywords—Single Event Upset (SEU), SEU Hardening, FPGA, High-Reliability Design

I. INTRODUCTION

When charged particles in space interact with integrated circuits on spacecraft, they may induce single event effects (SEEs), such as single event upset (SEU) and single event latch-up (SEL) [1]. These effects can lead to data errors, system reboots, or even permanent damage, resulting in the failure of integrated circuit systems [2]. To address these challenges, it is crucial to develop hardened integrated circuits that maintain high performance and reliability in radiation environments.

There are various logic design technologies are proposed for SEU mitigation, like TMR (Triple Modular Redundancy) [3], DMR (Dual Modular Redundancy) [4] and ECC (Error Correction Code) [5]. Logic design based SEU mitigation can have much lower cost than fabrication and layout design methods. Quantitative analysis on the SEU mitigative effect and logic error detection capability as well the area cost is very important for low cost and high reliable designs.

In this paper, we present an efficient simulation based logic error detection platform with quantitative analysis model for SEU mitigation FPGA designs. This platform enables detailed analysis on the error rate and error detection rate as well as area cost under SEU for various error mitigation designs. This analysis enables further study on low cost and high reliable radiation resistant designs, which is of high importance for star-link and in-vehicle electronics.

II. REVIEW OF ERROR INJECTION SCHEMES

To perform quantitative analysis of SEUs on FPGA, one important technology is error injection for a given netlist. The current mainstream SEU injection schemes can be classified into three types: error injection based on SEM IP, error injection based on error generation circuits, and error injection based on Force command simulation errors.

A common method of error injection uses Xilinx's SEM (Soft Error Mitigation) IP core to directly inject errors into the corresponding LUT by addressing, and monitors the test results via serial port communication with the host computer. The main drawback is its low testing convenience and efficiency. Additionally, the error injection might disrupt the configuration registers used by SEM IP itself, disrupting the fault injection process [6].

Another approach directly performs error injection and testing using error generation circuits. LFSR and Built-in Self-Test (BIST) technology greatly improve the test efficiency. However, editing the wire value differs from the mechanism of actual SEU, which may result in significant discrepancies in quantitative analysis results, thereby failing to accurately reflect the actual situation [7].

A third scheme involves using the Force command in Xsim to modify the value of a wire to simulate SEU, followed by pre-synthesis simulation. It has a short workflow that allows quick and simple evaluation of hardening performance, but also fails to accurately replicate the mechanism of SEU effects.

The problem with those methods is that they do not allow accurate control of the location of error injection, thus, the coverage of error injection is low. In the next chapter, we will propose an error injection technology with direct netlist modification so that we can insert SEU error in any specified logic cell, including flip-flops and LUTs. Thus, we can guarantee our error injection covers the entire design based on a given SEU rate. Then, we can perform simulation to report the logic error rate and error detection rate statistically.

III. QUANTITATIVE SEU EFFECT EVALUATION PLATFORM

For a given netlist, our SEU effect evaluation platform can perform error injection at any specific logic cell and configuration bit. Third party simulation is performed to report the logic error rate and error detection rate for various SEU mitigation designs.

A. Accurate Error injection and testing based on Netlist editing and Simulation

Our design is based on AMD Vivado tool [8]. We import a given netlist into Vivado and use the following tcl command to get all the LUTs and their logic configurations.

    set lut_list [get_cells -hierarchical -filter { PRIMITIVE_TYPE =~ LUT* }]

The logic configuration of LUTs can be got as follows.

    set init_value [get_property INIT $lut]

It is important to note that init_value is now a hexadecimal number, which need to be changed into binary to complete the randomly flipping. Perform simulation on the circuit netlist and record the results as golden data.

979-8-3503-6184-1/24 $31.00 © 2024 IEEE

Flip a random configuration bit of the selected LUT to simulate SEU occurrence. The process of updating the initial value of LUT can be performed by following command:

To inject a SEU error on a particular configuration bit of a LUT can be done as follows.

set_property INIT $flipped_init $lut

After an SEU injection, we can perform simulation on the revised netlist, and compare results with the original netlist to see of the SEU will cause logic output error or not.

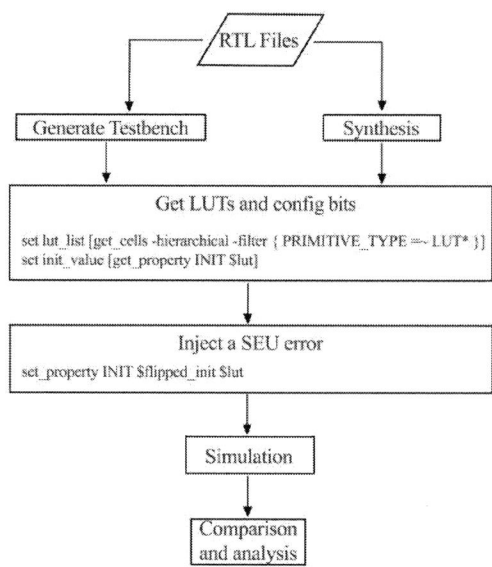

Fig. 1. Error injection test platform process

The advantage of our SEU injection over SEM IP is that we can control the exact of locations of error injection. For a given error injection ratio, our platform and generate a set of error injections to cover all logic cells and all logic configuration bits evenly and report meaningful statistic results for quantitative analysis of different SEU mitigation designs.

B. Four-metrics guided Quantitative Analysis and Evaluation Model

In order to evaluate the stabilization and error-checking ability of a hardened logic unit in terms of SEU resistance and the resource overhead of the unit, we introduce four metrics to evaluate it.

1) Error rate

The error rate reflects the probability of actual errors occurring in a circuit due to SEU. The error rate is obtained by dividing the number of actual computation result errors by the total number of single-bit flip-flop injections. Since not every single bit flip-flop triggers an error in the final result, the error rate characterizes the passive reliability of an individual logic cell in the face of SEUs.

2) Reporting accuracy

The reporting accuracy is the ratio of correctly reported errors to the total number of reported errors. In SEU-resistant systems, if a refresh scheme is used instead of an error correction scheme (DMR and ECC in this study can only detect errors but not correct them), every reported error

requires the system to pause and refresh through reconfiguration. Hence, the reporting accuracy significantly impacts overall system efficiency, as unnecessary refreshes due to false error reports can lead to performance loss.

3) Detection rates

The error detection rate represents the proportion of circuit errors that are successfully detected and reported by the reinforced logic. Since the error indication signals are also realized by LUTs, they may also be affected by SEU errors. As a result, we propose to compute the error-checking rates for any SEU checking logic design. An error-checking rate below 100% means the design may not be able to identify some of the logic errors.

4) Areas

Any logic checking and correction design will incur logic area overhead to the design. Measure the area overhead is very important for low cost and high reliable designs. Synthesis can be performed for any SEU mitigation designs to report their area overhead.

C. Guide highly reliable design through testing and evaluation of various logic units

Based on the four metrics mentioned above, the automated design tool can automatically call logic units with different reinforcement schemes in different modules according to different needs, in order to achieve the optimal cost performance of the overall design. For example, for modules requiring extremely high reliability, the logic unit with low error rate and the highest possible error checking rate should be selected to form the module, and the series-parallel realization should be dynamically adjusted according to the parameter performance of different bit widths in order to meet the demand for the highest reliability; while for low-power design, the logic unit with advantageous area parameters can be selected to form the design.

Automated design tools according to the parameters of the automatic realization of the selection can maximize the pursuit of the overall optimal cost-effective, and according to the actual needs of the rapid call of different logic units to form a design, you can accelerate the design process and the pursuit of high-reliability and low-cost balance.

IV. EXPERIMENTAL RESULTS

Using the test approach mentioned above, this part conducts error injection experiments on 1, 8, 16, 64, and 128-bit adders/bitwise AND/bitwise OR/bitwise XOR hardened with DMR and ECC separately. The experimental process is as follows:

1) Automatically generate a Testbench for each unit case, and produce a total of 20,000 test sets by randomly setting each bit of the test vector to 0 or 1.

2) Perform multiple rounds of error injection and simulation testing for each logic unit, with the number of rounds depending on the number of LUTs in the netlist and the bit width of each LUT.

The number of test vectors within a single testbench's is set after repeated testing. Fewer rounds will result in low coverage of the input situation, which will cause the results to exhibit poor statistical characteristics, while too large a number of test vectors will result in slower and less efficient

979-8-3503-6184-1/24 $31.00 © 2024 IEEE   781

testing. Therefore, the number of groups of test vectors within each testbench was set to 20,000 in this experiment.

It is important to note that LUTs with larger bit widths have more configuration bits, resulting in a higher probability of SEU occurrence. Hence more rounds of test are required to simulate and evaluate the effect. For a LUTn, the number of rounds should be $2^{(n-2)}$.

In this study, the DMR-reinforced unit includes an additional check-bit module that simultaneously computes the input data and compares its result with that of the original logic unit. This comparison generates a one-bit validity signal, where a value of 1 indicates that the computation is correct, otherwise signifies a possible fault. The ECC scheme, on the other hand, performs reduction XOR operations separately on the outputs of two identical logic operation modules, and then further XORs the two results to generate a single check bit. If the ECC result is 0, the output is considered reliable, or the output is considered unreliable.

A portion of the test results are shown in the table below.

TABLE I.   A PORTION OF TEST RESULTS

Unit	Error Rate		Reporting Accuracy		Detection Rate		Area	
	DMR	ECC	DMR	ECC	DMR	ECC	DMR	ECC
add_64	3.44%	3.07%	23.88%	16.68%	100.00%	66.80%	150	154
and_16	3.40%	3.84%	18.69%	28.93%	100.00%	100.00%	27	29
or_16	3.39%	3.85%	22.42%	31.07%	100.00%	100.00%	27	29
xor_64	3.29%	2.38%	26.70%	18.88%	100.00%	100.00%	107	103

Observing the experimental results, it can be found that for add_64, it is almost undesirable to use this ECC reinforcement scheme because its detection rate cannot reach 100%. and_16 and or_16 ECC reinforcement schemes have higher reporting accuracy, but they also require a larger area, and the error rate is slightly higher than that of the DMR scheme. The DMR scheme of xor_64 shows better performance in terms of reporting accuracy, but the ECC scheme consumes less resources and has a lower error rate.

We can increase the function of the automated design tool to call the experimental results database, and according to the actual design requirements, based on the performance of the four metrics, we can realize the automatic or manual designation of the reinforcement scheme of the logic unit in a module, so as to achieve the optimal cost-effectiveness of the overall design.

V. CONCLUSIONS

As the growth of star-link technology and reliable in-vehicle electronics, the design of low cost and high reliable designs with SEU mitigation capability is very important. In this paper, we present a efficient simulation-based error injection and performance evaluation platform quantitative analysis for various SEU mitigation designs, like DMR, TMR and ECC. Such quantitative analysis enables low cost and high reliable synthesis technology for SEU mitigation designs to be used in star-link and in-vehicle electronics.

REFERENCES

[1] Guantai Zhao. "The impact of single event upsets on relay protection devices." North China Electric Power University (Beijing), 2024. DOI:10.27140/d.cnki.ghbbu.2023.002031.

[2] Changhe Wang. "Impact of single event effects on the reliability of satellite space operations." Semiconductor Intelligence, no. 01, 1998, pp. 1-8.

[3] Pratt, Brian, et al. "Improving FPGA design robustness with partial TMR." 2006 IEEE International Reliability Physics Symposium Proceedings. IEEE, 2006.

[4] G. L. Nazar, L. P. Santos and L. Carro, "Accelerated FPGA repair through shifted scrubbing," 2013 23rd International Conference on Field programmable Logic and Applications, Porto, Portugal, 2013, pp. 1-6, doi: 10.1109/FPL.2013.6645533.

[5] Bolchini, Cristiana, Fabio Salice, and Donatella Sciuto. "Designing self-checking FPGAs through error detection codes." 17th IEEE International Symposium on Defect and Fault Tolerance in VLSI Systems, 2002. DFT 2002. Proceedings. IEEE, 2002.

[6] Mengru Wang, Shan Zhou, Panpan Xue, Lu Kong, Jinbo Wang. "Single event fault injection of SRAM-based FPGA using SEM IP and partial reconfiguration." Microelectronics and Computers, vol. 38, no. 08, 2021, pp. 8-12. DOI:10.19304/j.cnki.issn1000-7180.2021.08.002.

[7] Xiaohui Lin, Weikun Xie, Guodong Song. "Research on LUT testing methods based on new BIST." Modern Electronic Technology, vol. 47, no. 04, 2024, pp. 23-27. DOI:10.16652/j.issn.1004-373x.2024.04.005.

[8] Feist, Tom. "Vivado design suite." White Paper 5.30 (2012): 24.

Hot-Carrier Injection Characterization of n-LDMOS Transistors and Stress Tests in a Buck Converter Configuration

Chun Yee Chu*, Wai Tung Ng

Department of Electrical and Computer Engineering, University of Toronto, Toronto, Canada
* Email: cy.chu@mail.utoronto.ca

Abstract— In this work, on-resistance degradation caused by hot-carrier injection (HCI) in 16 V n-type LDMOS devices is studied and characterized. The HCI susceptibility of these devices provides guidelines for circuit engineers to avoid operating devices under unsafe conditions. Resistance increases over stress time in a logarithmic relationship, while an exponential relationship between the operational lifetime and the inverse of drain-to-source voltage (V_{ds}) is observed. Midrange gate-to-source voltage (V_{gs}) levels above threshold voltage (V_{th}), such as 1.9 and 3 V, cause the most damage to devices. Stressing is also performed on the n-LDMOS devices in a buck converter configuration. After 400 hours of stress, the maximum efficiency decreased by 0.3%, suggesting traditional failure criteria may be too conservative.

Keywords—LDMOS, HCI, degradation, buck converter

I. INTRODUCTION

Lateral Double Diffused Metal Oxide Semiconductor (LDMOS) Field Effect Transistors, with device structure shown in Fig. 1, are widely used in switched-mode power supplies. Specific on-resistance ($R_{on,sp}$) is a critical metric in LDMOS devices because it affects conduction loss and the power conversion efficiency of the converters.

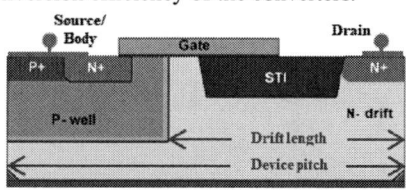

Fig. 1. n-LDMOS device structure with shallow trench isolation (STI) in the drift region.

For n-type LDMOS devices, hot-carrier injection (HCI) is the dominant aging effect. HCI can degrade V_{th}, and linear or saturation transconductance [1]. The degradation is caused by the generation of acceptor-type interface traps (N_{it}) [1]. In a MOSFET, the lateral electric field accelerates carriers toward the drain. Hot carriers with high enough energy can break silicon-hydrogen bonds at the Si/SiO2 interface to form N_{it}. These N_{it} are located near the drain terminal and can reduce local carrier density and mobility. Hot carriers can also generate secondary carriers near the drain by impact ionization, and these secondary carriers can also be injected into the oxide, degrading device characteristics [2].

Conventionally, power device engineers aim to optimize the trade-off between $R_{on,sp}$ and breakdown voltage (BV). However, the long-term performance of LDMOS devices should not be neglected. In this work, reliability will be characterized by constructing safe operating area (SOA) with HCI taken into consideration. Stressing will also be performed in a buck converter configuration to observe degradation in switching operations.

II. RELIABILITY OF LDMOS TRANSISTORS

To study the HCI reliability of LDMOS devices, two experiments are conducted. The first experiment involves stressing LDMOS devices under DC conditions, with stress continuously applied to the devices. In the second experiment, LDMOS devices are stressed in a buck converter configuration through switching operations.

A. DC Stress Test

DC stress tests are conducted on standalone LDMOS devices. The goal is to determine if there is any observable trend in the HCI degradation of these devices and to characterize the degradation with HCI. In the stress test, two types of devices under test (DUTs) are selected. The specifications of the devices are presented in TABLE I. The 16 V Modified device has a shorter drift region than the 16 V Standard device, resulting in a smaller $R_{on,sp}$, a smaller BV and a shorter device pitch.

TABLE I. SPECIFICATIONS OF DUTs

Parameters	16V Standard LDMOS	16V Modified LDMOS
V_{th}	1.60 V	1.60 V
$R_{on,sp}$	1.00 AU	0.78 AU
BV	32 V	26 V
Width	10 μm	10 μm
Pitch	1.00 AU	0.90 AU

AU = arbitrary unit

To conduct stress tests, various combinations of $V_{gs,stress}$ and $V_{ds,stress}$ are required. A PCB, with schematic diagram shown in Fig. 2 (a), is designed to supply different $V_{gs,stress}$ to separate packages. As a result, more devices can be stressed simultaneously. Another PCB is designed to be used with the Keysight B2902B source/measure unit for measuring IV characteristics of devices.

Before starting the stress test, it is essential to perform device characterization for several reasons. First, we need to obtain the initial $R_{on,sp}$ and V_{th} values of the fresh devices to compare with those of the degraded devices in the stress test. Second, we need to confirm that the $R_{on,sp}$ and V_{th} values match the values from the wafer acceptance test (WAT) to ensure the devices are in good condition. To obtain the $R_{on,sp}$ and V_{th} values, I_{ds}-V_{gs} curves of the devices are measured at $V_{ds} = 0.1$ V. V_{th} is extracted graphically and $R_{on,sp}$ is calculated with I_{ds} at $V_{gs} = 5$ V. To conduct stress tests with a high accuracy, devices with more than 5% difference in either $R_{on,sp}$ or V_{th} compared to WAT values are filtered out.

The chosen $V_{gs,stress}$ values for the stress test are 0 V, 1.3 V, 1.6 V, 1.9 V, 3 V, and 4 V. Since 1.6 V is the V_{th} of the devices, more values are chosen around this threshold to monitor degradation when the devices are switching. As we anticipate a higher $V_{ds,stress}$ will accelerate degradation, we start with a $V_{ds,stress}$ of 24 V. $V_{ds,stress}$ is then reduced by 2 V at a time until sufficient data is obtained.

979-8-3503-6184-1/24 $31.00 © 2024 IEEE

Fig. 2 (b) shows the setup of DC stress test. Devices are stressed until they meet the common failure criteria of 10% degradation in $R_{on,sp}$ [3], [4], then they will be replaced by "fresh" devices which will be stressed under different stressing conditions. If the devices do not reach the failure criteria after 6 weeks of stressing, they are still replaced due to time constraints.

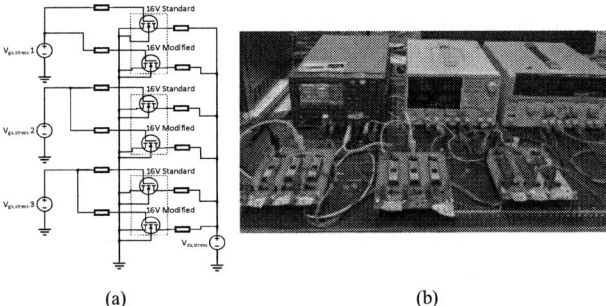

Fig. 2. (a) Schematic diagram of a stressing PCB. (b) DC stress test setup with three stressing PCBs.

Stress periods are in exponential scale, with each subsequent period being approximately twice as long as the previous one. At the end of each stress period, stressing is paused, devices are left idle for at least 15 minutes for devices to cool down before measuring the I_{ds}-V_{gs} characteristics to obtain $R_{on,sp}$ and V_{th} values. All stressing and measurements in the experiment are conducted at room temperature.

B. Buck Converter Stress Test

Fig. 3 shows the setup for stress testing with monolithic synchronous buck converters. A DC power supply provides the input voltage of the converter (V_{in}) and the supply voltage to the IC (V_{cc}), a digital multimeter measures the output voltage (V_{out}), and an electronic load in constant current mode provides the load when measuring efficiency. Detailed specifications are shown in TABLE II. The buck converter being tested is designed with the 16 V Standard device with a larger device area to withstand high current.

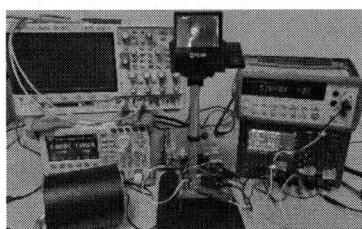

Fig. 3. Buck converter stress test setup.

TABLE II. SPECIFICATIONS OF BUCK CONVERTERS

Parameter	Value	Parameters	Value
V_{in} (stress)	25.75 V	f_{sw}	1 MHz
V_{in} (efficiency measurement)	6 V	Capacitor	22 μF
V_{in} (R_{on} measurement)	5 V	Inductor	2.2 μH
V_{cc}	5 V	Duty cycle	50%

Before stressing, R_{on} measurements are performed on the highside (HS) and lowside (LS) devices separately. When measuring R_{on}, the electronic load is configured to draw a current of 0.2 A. The efficiency of the converter is also measured before stressing, with V_{in} = 6 V and output current (I_{out}) at various levels. During measurements, a fan is used to avoid overheating. The input current (I_{in}) can be recorded

from the power supply while V_{out} can be read from the multimeter. Efficiency versus output power (P_{out}) curves are then plotted. Stressing is performed with V_{in} = 25.75 V with no load. Between stress periods, R_{on} and efficiency are measured.

III. RESULTS AND DISCUSSION

A. DC Stress Test

Fig. 4 presents the percentage change in $R_{on,sp}$ when $V_{ds,stress}$ is approximately 20 V for various $V_{gs,stress}$ levels. Chart (a) and (b) correspond to the 16 V Standard and Modified devices respectively. Stress time in hours is displayed on the x-axis using a logarithmic scale. Each data point in the graphs represents a measurement done after a certain stress period. The red dotted line represents a 10% change in $R_{on,sp}$, marking the threshold for device failure.

Logarithmic trend lines fit well with the degradation relationship between $R_{on,sp}$ and stress time, and we can generalize it with the following expression:

$$\Delta R_{on,sp} = A + B \ln(t) \tag{1}$$

where $\Delta R_{on,sp}$ is the percentage change in $R_{on,sp}$, t is the stress time, A and B are constants depending on the stressing conditions and device type.

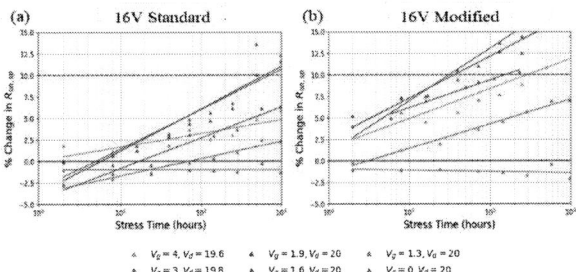

Fig. 4. $R_{on,sp}$ degradation against stress time when $V_{ds,stress}$ is approximately 20 V: (a) 16 V Standard device (b) 16 V Modified device.

The degradation of $R_{on,sp}$ under other $V_{ds,stress}$ levels are recorded but not displayed. To summarize, degradation is generally most rapid when $V_{gs,stress}$ is 1.9 V and 3 V, followed by 1.6 V and 4 V. This matches with findings in previous literatures that higher $V_{gs,stress}$ does not always result in faster degradation [5]. Degradation is even slower when $V_{gs,stress}$ is 1.3 V, which is below V_{th}. When $V_{gs,stress}$ is 0 V, no significant increase in $R_{on,sp}$ is observed. Given the same $V_{gs,stress}$ level, higher $V_{ds,stress}$ levels result in faster degradation. Under the same stress condition, the 16 V Modified device degrades faster than the 16 V Standard device.

Similar to $R_{on,sp}$, V_{th} of the devices also increases over stress time in a logarithmic manner. As ΔV_{th} can affect channel resistance, the effect of ΔV_{th} is accounted for in $\Delta R_{on,sp}$, which impacts conduction losses and the efficiency of converters.

In Fig. 4, the points where the red dotted lines intersect with the plots indicate the lifetime of devices under different stress conditions. Lifetime versus $1/V_{ds}$ plots are shown in Fig. 5. An exponential relationship between operational lifetime and $1/V_{ds}$ is observed, which can be expressed as:

$$Lifetime = A \cdot e^{B/V_{ds}} \tag{2}$$

979-8-3503-6184-1/24 $31.00 © 2024 IEEE

where *Lifetime* is the time required for a 10% increase in $R_{on,sp}$. A and B are constants that depend on the stress condition and the device type. Plots for $V_{gs} = 0$ V is not shown because the $R_{on,sp}$ change is too insignificant. The red and blue dotted lines in the figure represent a DC lifetime of 0.2 and 10 years respectively.

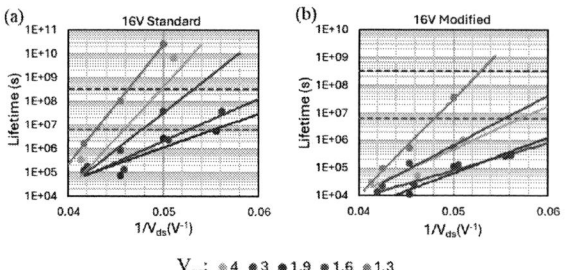

V_{gs}: • 4 • 3 • 1.9 • 1.6 • 1.3

Fig. 5. DC lifetime prediction plots. (a) 16 V Standard device (b) 16 V Modified device.

Fig. 6 shows both the 0.2-year and 10-year HCI reliability for both devices under DC operations. If the device is operated within the 0.2-year HCI reliability window, it can operate continuously without HCI reliability concerns for at least 0.2 years. For both devices, because when $V_{gs} = 0$ V, the change in $R_{on,sp}$ is too insignificant, we can assume the reliability concern is not limited by HCI but by the *BV*. Because the stress test was not performed at $V_{gs} = 5$ V, the HCI reliability was extrapolated.

It might be too stringent to specify a 10-year DC lifetime in buck converters because V_{ds} and V_{gs} are only high simultaneously for a short period of time during switching. Since many in the industry consider that a 0.2-year lifetime under DC stress is roughly equivalent to a 10-year lifetime in an alternating current application [6], the 0.2-year HCI reliability window may be more relevant if we want to predict how the LDMOS devices would degrade in buck converters.

Assuming the input voltage of the buck converter is 16 V, for the 16 V Standard device, there is safety margin between the operating voltage and the 0.2-year window so it can operate without HCI reliability concern. However, for the 16 V Modified device, the switching locus of the device will likely lie in unsafe operating regions. This provides warnings that the design of the device may need to be adjusted, such as extending the drift region, to improve HCI reliability. For circuit design engineers, this information can guide them in operating the device to ensure sufficient longevity.

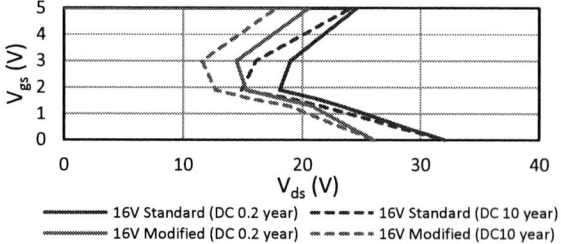

Fig. 6. Reliable HCI operating range for 16 V Standard and Modified devices.

B. Buck Converter Stress Test

Fig. 7 (a) shows that at high P_{out}, there is a drop in maximum efficiency after stressing. However, this drop is not significant, with only about 0.3% of change after 409 hours of stress. Fig. 7 (b) shows the change in R_{on} of the HS and LS

devices, along with 3 trials of DC stress results of the same device under $V_{ds,stress} = 25.75$ V and $V_{gs,stress} = 5$ V for comparison. The HS and LS devices show 3.5% and 2.5% of increase in R_{on} respectively after 409 hours of stress. HS has a different switching locus from LS device [6], and is reasonable to have more severe HCI degradation. The DC to buck converter stressing ratio is observed to be bigger than 50:1, meaning the 0.2 to 10-year lifetime ratio might be too conservative, but more research should be done with various stress conditions before drawing a conclusion.

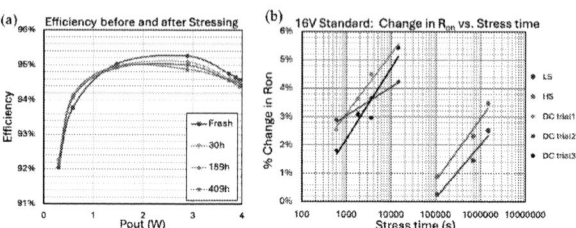

Fig. 7. (a) Efficiency of buck converter before and after stressing. (b) On-resistance change over stress time for highside and lowside devices in buck converter compared with DC stressing.

R_{on} change is more significant than change in efficiency because R_{on} only affects conduction loss, but there are other sources of power loss, such as switching loss, that can affect efficiency. For low voltage rating LDMOS devices, R_{on} is relatively small and the degradation in R_{on} has less pronounced effects compared to high voltage rating devices. The failure criteria of 10% change in R_{on} may be too stringent as the impact on buck converter performance is not severe.

IV. CONCLUSIONS

From the DC stress test, an approach of characterizing degradation with HCI is introduced. The HCI stress shows that the device with a shorter drift region and device pitch is more susceptible to HCI degradation. Midrange V_{gs} above V_{th} are shown to cause the most damage to the devices. $\Delta R_{on,sp}$ increases over stress time in a logarithmic relationship, while an exponential relationship between lifetime and $1/V_{ds}$ is observed. Stressing on buck converters does not show significant drop in efficiency, showing the traditional failure criteria may be too conservative for low R_{on} devices. R_{on} change in a buck converter is significantly slower than in a DC stress condition, but the conversion ratio depends on multiple factors and requires further research to establish.

REFERENCES

[1] Chenming Hu *et al.*, "Hot-electron-induced MOSFET degradation—Model, monitor, and improvement," *IEEE Trans. Electron Devices*, vol. 32, no. 2, pp. 375–385, Feb. 1985.

[2] J. Keane *et al.*, "An All-In-One Silicon Odometer for Separately Monitoring HCI, BTI, and TDDB," *IEEE J. Solid-State Circuits*, vol. 45, no. 4, pp. 817–829, Apr. 2010.

[3] P. L. Hower and S. Pendharkar, "Short and long-term safe operating area considerations in LDMOS transistors," in *2005 IEEE International Reliability Physics Symposium, 2005. Proceedings. 43rd Annual.*, Apr. 2005, pp. 545–550.

[4] N. Soin *et al.*, "Measurement and characterization of hot carrier safe operating area (HCI-SOA) in 24V n-type lateral DMOS transistors," in *2012 10th IEEE International Conference on Semiconductor Electronics (ICSE)*, Sep. 2012, pp. 659–663.

[5] Z. Zheng *et al.*, "Mechanism Analysis and Improved Model for HCI in 200V STI-based Triple RESURF LDMOS With n-p-n Layer," in *2023 35th International Symposium on Power Semiconductor Devices and ICs (ISPSD)*, May 2023, pp. 258–261.

[6] A. Kalnitsky, "Structure and method for a switched circuit device," U.S. Patent 9209683B2, Dec. 08, 2015.

Semimetal Alloy Contact with Low Resistivity and Enhanced Thermal Budget for MoS₂ FETs

Kwok-Ho WONG* and Mansun CHAN

Department of Electronic and Computer Engineering, The Hong Kong University of Science and Technology,
Clear water bay, Kowloon, Hong Kong
*Email: khwongby@connect.ust.hk

Abstract—**The issue of thermal instability in Bismuth (Bi) contacts hinders the reliability of high-performance molybdenum disulfide (MoS₂) transistors in practical applications. Here we present a new method using a semimetal alloy to improve the thermal stability and contact resistance of MoS₂ transistors. Through this approach, the devices with TiBi alloy contacts withstand the backend-of-the-line (BEOL) processes, maintaining a high on-state current after 1 hour of 400 °C annealing. Bi-alloy contacts hold promise for improving MoS₂-based electronic devices, contributing to future developments in semiconductor technology.**

Keywords—MoS₂, transition metal dichalcogenides, two-dimensional materials, semimetal-semiconductor contact, semimetal, field-effect transistor, contact resistivity, thermal stability

I. INTRODUCTION

Two-dimensional (2D) materials, such as graphene and transition metal dichalcogenides (TMDs), have attracted considerable attention for advanced electronic devices due to their high electron mobility, direct bandgap, mechanical flexibility and atomic-scale thickness [1]–[3]. These distinctive characteristics position them as potential candidates for a variety of electronic applications, including high-performance field-effect transistors, photodetectors, and flexible electronics [4]-[7]. Despite their impressive capabilities, their implementation has practical challenges, particularly when it comes to establishing effective contacts between semiconductors and metal electrodes in 2D devices. One major issue is Fermi-level pinning, which significantly impedes charge carrier injection and increases contact resistance. This underscores the importance of developing innovative contact engineering strategies to fully harness the potential of 2D materials in electronic applications [8]-[11].

Recently, researchers have observed that metals with low melting points and high vapor pressures, including Sn [12], Bi [13] and Sb [14], exhibit remarkably low contact resistance when interacting with molybdenum sulfide (MoS₂). Among them, the Bi contact to monolayer MoS₂ achieves the lowest contact resistance (R_C = 123 Ω·μm), which is attributed to the aligned energy level with the electron affinity of MoS₂ and the semimetallic nature [13]. However, low-melting-point metals are not suitable for practical use in silicon-based BEOL processes due to their thermal instability. Moreover, Sb exhibits a higher melting point of 630.6°C but shows inferior on-state current under identical external voltage bias conditions [14]. Therefore, improving the thermal stability of the Bi contact remains one of the main issues confronting high-performance MoS₂ FETs in practical applications.

To tackle the problem of low thermal stability in Bi-contacted devices, one possible approach is to use a semimetal alloy as the contact. With this method, a higher melting point would be achieved while preserving the semimetal properties. As a result, MoS₂-based devices with semimetal alloy contact are expected to achieve both ohmic contact behavior and improved thermal stability.

II. ENHANCED THERMAL STABILITY BY TiBi ALLOY

We initially fabricated Bi devices for reference, followed by laminated TiBi devices to evaluate the thermal stability enhancements offered by the TiBi alloy. The fabrication procedures and the physical structure of back-gated MoS₂ transistors are depicted in Fig. 1. A 13 nm Al₂O₃ is used as a high-k gate dielectric. After mechanical exfoliation, the samples are cleaned using acetone and IPA/DI water, and then baked on the hotplate. Photolithography is then applied to define the top S/D contacts. Results for the 3 μm channel length will be presented. For Bi devices, 20nm Bi is deposited on MoS₂ at a slow deposition rate of 0.5 A/s and at a high vacuum pressure of 3x10⁻⁷ Torr. In the case of laminated TiBi devices, the contact layer is formed by evaporating 2nm Bi first, followed by 2nm Ti in 5 cycles. Finally, 30nm Au is evaporated and the liftoff process is performed afterwards. Vacuum thermal annealing is performed at 400 °C for 1 hour.

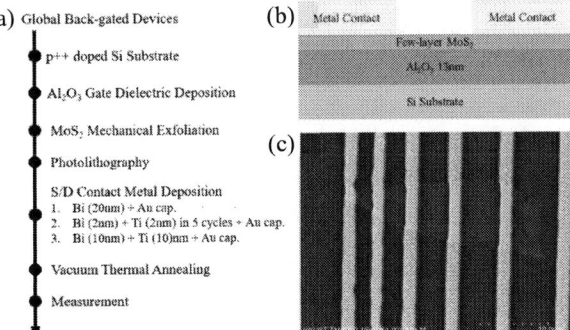

Fig. 1. (a) and (b) Fabrication process and physical structure of the MoS₂ FET test structures for contact workfunction engineering. (c) The SEM image of the MoS₂ FET after annealing is also given.

The thermal instability of Bi devices is examined in Fig. 2 by comparing the transfer (I_D-V_{GS}) and output (I_D-V_{DS}) characteristics of the Bi devices before and after 400 °C annealing. MoS₂ FETs with Bi contact (L_{CH} = 3 μm) obtain an I_{ON} of ~18 μA at V_{DS} = 1 V, with an I_{ON}/I_{OFF} ratio of ~10⁴. Linear I_D-V_{DS} characteristics at low bias are indicative of ohmic contact behavior in Bi-contacted devices, indicating that low Schottky barrier height is formed, which is consistent with the reported result [13]. After undergoing annealing at 400 ℃, the Bi device does not fail completely, but it experiences severe degradation due to its low melting point. Despite this, the device continues to function, likely because the Au layer acts as a partial contact. Additionally, the Schottky behavior observed in Fig. 2b may be attributed to the Au-MoS₂ contact.

979-8-3503-6184-1/24 $31.00 © 2024 IEEE

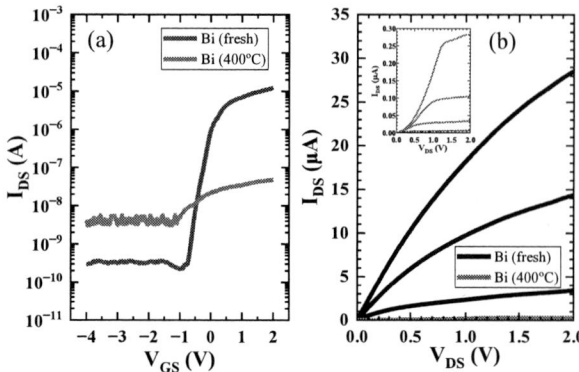

Fig. 2. (a) and (b) Transfer and output characteristics of the Bi-contacted MoS$_2$ FET before and after annealing at 400°C

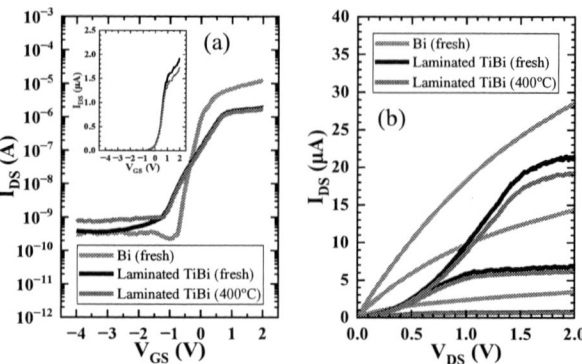

Fig. 3. (a) and (b) Transfer and output characteristics of the fresh Bi- and laminated TiBi devices before and after annealing (V$_{GS}$ = 0V, 1V and 2V)

Ti is intentionally selected to form an alloy for improving the thermal stability of the Bi contact due to its high melting point and matched energy level. Additionally, stacking thin layers facilitates the formation of the desired semimetal alloy. The electrical characteristics of fresh Bi devices, and laminated TiBi devices before and after 400°C annealing are compared in Fig. 3. In contrast to the Bi devices, the laminated TiBi devices show inferior on-state current, but they excellently maintain 90% of their performance after annealing at 400°C. The enhanced thermal resilience observed in contacts made of TiBi alloy, as opposed to those made of pure Bi, is attributed to the alloying effect. During the high-temperature annealing, the alloying process creates a more robust crystalline structure, thereby offering a higher melting point and resistance to deformation. Collectively, these attributes render the TiBi alloy contacts exceptionally durable, ensuring consistent performance of the devices they are part of, even when subjected to conditions of elevated thermal budget. This robustness makes them advantageous for applications where thermal endurance is paramount for device reliability. Despite this, laminated TiBi devices experience a non-linear behavior at low bias. Therefore, it is necessary to adjust the alloy formation process such that the thermal stability and device performance are balanced.

III. CONTACT ENHANCEMENT WITH DUAL LAYER TIBI CONTACT

A non-uniform semimetal alloy is suggested to reduce the contact resistance. In this way, Bi contacts with MoS$_2$ and TiBi alloy forms on top of the Bi layer, maintaining semimetal-TMD contacts and enhanced thermal resilience from the Ti element. Such alloy is developed by evaporating 10nm Bi followed by 10nm Ti. Different from laminated TiBi devices, dual-layer TiBi devices are named since they only involve two layers of metal deposition. Fig. 4 presents the electrical characteristics of dual-layer TiBi MoS$_2$ devices before and after 400°C annealing. Before annealing, dual-layer TiBi devices achieve an I$_{ON}$ of ~15 μA at V$_{DS}$ = 1 V, with an I$_{ON}$/I$_{OFF}$ ratio of ~10^5. After annealing, dual-layer TiBi devices still obtain 75% of their performance. When comparing the laminated TiBi devices in Fig. 3, linear behavior at low bias is observed in dual-layer TiBi devices, indicating that the contact resistance is improved by modulating the process of contact alloy deposition. The major reason for performance improvement in dual-layer TiBi devices is explained by the absence of Ti. Considering the laminated TiBi layer, it can be regarded as a layer, where the thin Bi layer incorporates Ti atoms as impurities. Originally,

the contact metal with MoS$_2$ is Bi, which is a semimetal with no density of states (DOS) at the Fermi level. When the Bi layer is thin, Ti deposition introduces DOS to the contact interface due to its metallic nature. Additionally, a Ti-Bi mix occurs during the evaporation process, thus the contact metal with MoS$_2$ becomes a Ti-Bi alloy rather than pure Bi. Consequently, metal-induced gap states (MIGS) arise from the Ti atoms and the interface between the Ti-Bi alloy and MoS$_2$ exhibits Fermi-level pinning, leading to a degradation in device performance. To mitigate the deterioration induced by Ti, a thicker Bi contact layer helps decrease the proportion of Ti atoms relative to Bi in the contact interface, resulting in reduced MIGS generation. Therefore, dual-layer TiBi devices perform similarly to Bi devices in their initial state.

To further evaluate the formation of TiBi alloy, Fig. 5 presents the TEM images and corresponding EDS mapping of a metal-MoS$_2$ contact region for both laminated TiBi devices and dual-layer TiBi devices after annealing. It is noted that good adhesion between TiBi alloy and FL-MoS$_2$ is maintained after the thermal processing without any noticeable damage in both devices. For laminated TiBi, there is a mixing layer of Bi and Ti at the metal-MoS$_2$ interface, indicating that the effective work function is dependent on both Bi and Ti, which is consistent with the aforementioned results. For dual-layer TiBi devices, only a small number of Ti atoms are found to be deposited on the MoS$_2$ surface, indicating that the contact interface is mainly between Bi and MoS$_2$. In addition, the Bi content being lower than the preset thickness is attributed to the resublimation of Bi during the Ti layer deposition process.

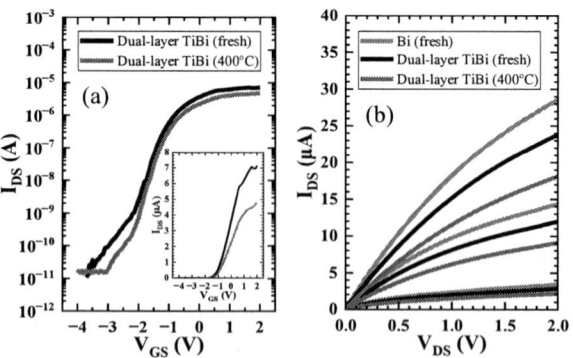

Fig. 4. (a) and (b) Transfer and output characteristics of the dual-layer TiBi-contacted MoS$_2$ FET before and after annealing (V$_{GS}$ = 0V, 1V and 2V)

979-8-3503-6184-1/24 $31.00 © 2024 IEEE

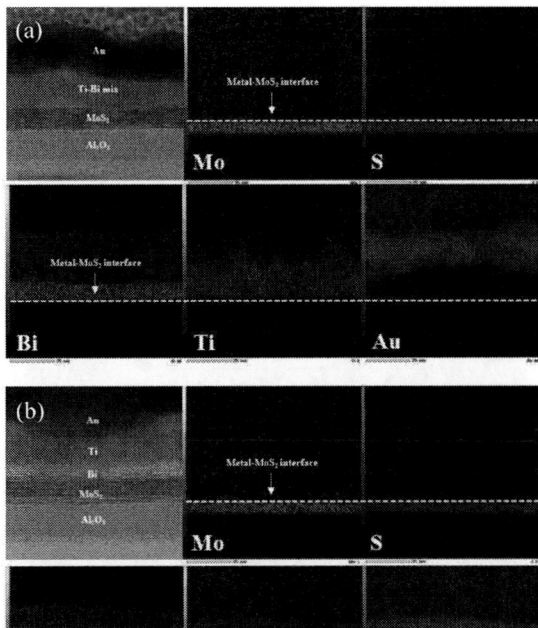

Fig. 5. High-resolution cross-sectional TEM images of a) laminated TiBi/Au and b) dual-layer TiBi/Au contact on FL-MoS$_2$ after 400°C thermal annealing, and EDS mapping of Mo, S, Ti, Bi, and Au elemental distribution.

IV. CONCLUSION

In conclusion, the implementation of semimetal alloy contacts in MoS$_2$ FETs demonstrates significant improvement in both electrical performance and thermal stability. The device with TiBi alloy not only achieves a high on-state current but also retains 75% of its performance after annealing at 400°C, exceeding the thermal stability of pure Bi contacts. This enhanced thermal resilience is attributed to the robust crystalline structure formed during the alloying process. Overall, this study highlights the potential of Bi alloy contacts for enhancing MoS$_2$-based electronic devices, paving the way for future advancements in semiconductor technology.

ACKNOWLEDGEMENT

This work was supported by the General Research Fund (GRF) 16201223 from the Research Grants Council (RGC) of Hong Kong. We acknowledge the Nanosystem Fabrication Facility (CWB) (NFF (CWB)) of the HKUST for the device/system fabrication.

REFERENCES

[1] M. Chhowalla, D. Jena, & H. Zhang, "Two-dimensional semiconductors for transistors". Nat. Rev. Mater. 1, 16052 (2016), doi: 10.1038/natrevmats.2016.52.

[2] A. Allain, J. Kang, K. Banerjee & A. Kis, "Electrical contacts to two-dimensional semiconductors". Nat. Mater. 14, 1195–1205 (2015). doi: 10.1038/nmat4452.

[3] D. Akinwande et al., "Graphene and two-dimensional materials for silicon technology". Nature 573, 507–518 (2019). doi: 10.1038/s41586-019-1573-9.

[4] B. Radisavljevic, A. Radenovic, J. Brivio et al., "Single-layer MoS2 transistors". Nature Nanotech 6, 147–150 (2011). doi: 10.1038/nnano.2010.279.

[5] L. Britnell et al., "Strong light-matter interactions in heterostructures of atomically thin films". Science 340, 1311–1314 (2013). doi: 10.1126/science.1235547.

[6] C. Xie, C. Mak, X. Tao and F. Yan, "Photodetectors based on two-dimensional layered materials beyond graphene". Adv. Funct. Mater. 27, 1603886 (2017). doi: 10.1002/adfm.201603886.

[7] A. Nathan et al., "Flexible Electronics: The Next Ubiquitous Platform," in Proceedings of the IEEE, vol. 100, no. Special Centennial Issue, pp. 1486-1517 (2012). doi: 10.1109/JPROC.2012.2190168.

[8] S.G. Louie & M. L. Cohen, "Electronic structure of a metal-semiconductor interface". Phys. Rev. B 13, 2461–2469 (1976). doi: 10.1103/PhysRevB.13.2461.

[9] T. Nishimura, K. Kita & A. Toriumi, "Evidence for strong Fermi-level pinning due to metal-induced gap states at metal/germanium interface". Appl. Phys. Lett. 91, 123123 (2007). doi: 10.1063/1.2789701.

[10] K. Sotthewes et al., "Universal Fermi-level pinning in transition-metal dichalcogenides". J. Phys. Chem. C 123, 5411–5420 (2019). doi: 10.1021/acs.jpcc.8b10971.

[11] R. T. Tung, "The physics and chemistry of the Schottky barrier height". Appl. Phys. Rev. 1, 011304 (2014). doi: 10.1063/1.4858400.

[12] A. S. Chou et al., "High On-State Current in Chemical Vapor Deposited Monolayer MoS2 nFETs with Sn Ohmic Contacts". Ieee Electron Device Letters, 42(2), 272-275 (2021). doi: 10.1109/LED.2020.3048371.

[13] P. C. Shen PC et al., "Ultralow contact resistance between semimetal and monolayer semiconductors". Nature. 593(7858), 211-217 (2021). doi: 10.1038/s41586-021-03472-9.

[14] A. -S. Chou et al., "Antimony Semimetal Contact with Enhanced Thermal Stability for High Performance 2D Electronics," 2021 IEEE International Electron Devices Meeting (IEDM), San Francisco, CA, USA, 2021, pp. 7.2.1-7.2.4, doi: 10.1109/IEDM19574.2021.9720608.

979-8-3503-6184-1/24 $31.00 © 2024 IEEE

Automated Verification of Functional Interface Connections in Circuit Schematics

Keli Long [1], Xingyu Gao [2], Lei Li [2], Jinxiang Wang [1], Fangfa Fu*[1], Liangquan Qiao [1], Jinghan Zhou [1]

[1] Department of Microelectronics Center, Harbin Institute of Technology, Harbin 150000 China
[2] 58th Research Institute of China Electronics Technology Group Corporation, Wuxi 214000 China

* Email: fff1984292@hit.edu.cn

Abstract—As transistor scaling approaches its physical limits, performance improvement of modern computing systems is driven by high integration and functional specialization. However, the resulting rapid growth of functional interface connections within computing systems brings challenges to verification efficiency. Current EDA tools are often inefficient at verifying these connections, requiring manual intervention to extract connection information and define verification rules. In this paper, we propose an automated verification method. Our approach designs a general schematic data structure to represent component connection information, describes processes for the automatic extraction of interface connections, and proposes a flexible rule description method to enhance extensibility. Experimental results show that our method reduces verification time from minutes to microseconds, up to 10 milliseconds, and improves the coverage of automatic schematic verification.

Keywords—EDA, Circuit Schematic, Verification

I. INTRODUCTION

As transistor size scaling approaches physical limits, the performance growth rate of modern computing systems dominated by transistor size scaling gradually slows down [1]. Furthermore, single-chip integration is increasingly unable to meet the demands of modern computing, due to the limitations of single-wafer size, yield rates, and the dark silicon problem. To address these challenges, heterogeneous integrated systems have emerged as a new direction for performance growth [2]. This approach involves two key aspects: integrating a large number of functional components into a single package to enhance overall performance, energy efficiency, and reliability; and decomposing different functional modules of SoCs into multiple chiplets to improve production yield, manufacturing efficiency, reduce design and verification costs, and increase design flexibility. Both aspects significantly increase the number of functional interfaces within heterogeneous integrated systems, necessitating more advanced verification methods.

Existing EDA tools, such as Altium Designer, Siemens EDA, Cadence's OrCAD, and KiCAD [3], primarily focus on electrical rule checks (ERC) and design rule checks (DRC) for low-level port connections. When performing high-level functional interface connection rule verification (FRV), these tools often require manual extraction of complex functional interface connection information and the creation of connection rules that are deeply coupled to specific devices and less reusable. This process is labor-intensive and prone to errors, leading to inefficiencies in the design verification process. Table I presents several mainstream EDA tools that include connectivity checking functions along with the types of rule checks they support and their different implementation methods.

TABLE I. MAINSTREAM EDA TOOLS AND THEIR SUPPORTED RULE CHECKS

Check Type	Tool Type			
	Cadence OrCAD	Altium Designer	Mentor Graphics PADS	KiCAD
DRC	Auto	Auto	Auto	Auto
ERC	Auto	Auto	Auto	Auto
FRV	Semi-Auto	Semi-Auto	Semi-Auto	Semi-Auto

The organization of this paper is as follows: Section II presents the framework for interface-oriented schematic inspection, detailing the main stages of the verification process. Section III describes the specific methods of the key functions within the framework. This includes the process of capturing interface connection data, defining connection rules, and the techniques for rule execution. Section IV provides experimental validation of the proposed methods, demonstrating their effectiveness and efficiency. Section V concludes the paper.

II. FRAMEWORK FOR INTERFACE-ORIENTED SCHEMATIC INSPECTION

This section presents the framework for interface-oriented schematic inspection, detailing the main stages involved in the verification process. The proposed framework consists of three primary stages: capturing functional interface connection information, matching the captured information against relevant connection rules, and reviewing the corresponding connection information based on these rules, as shown in Figure 1.

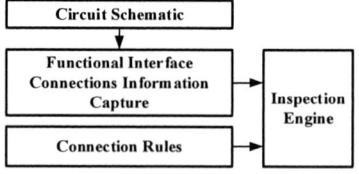

Figure 1. Framework for Interface-Oriented Schematic Inspection

The first stage involves accurately capturing the functional interface connection information from the circuit schematics. This step ensures that all relevant data regarding the interfaces and their connections are systematically collected and documented. The process uses advanced techniques, including machine learning, to automate the extraction of connection information, thereby reducing the need for manual intervention and minimizing errors.

In the second stage, the captured connection information is systematically matched against a predefined set of connection rules. These rules are designed to ensure that the connections meet specified criteria and standards. By using a specialized language for rule description, the process allows for flexible and reusable rule definitions. The language incorporates Python keywords, regular expressions, and custom operators to create robust and adaptable connection rules.

The final stage involves a thorough review of the connection information. A dedicated rule execution engine processes the defined rules, verifying the captured interface connections against the predefined criteria. This step ensures that all connections comply with the established rules and highlights any discrepancies for further investigation. By automating this verification process, the framework enhances the overall reliability and accuracy of schematic reviews.

III. FUNCTIONAL INTERFACE CONNECTION INFORMATION CAPTURE AND CONNECTION RULES DESCRIPTION

A. Circuit Schematic Presentation

When examining the connection rules of a circuit system, connection information is typically represented using circuit schematics. These schematics are the fundamental blueprints of an electronic system's electrical design, depicting system composition and interconnections. To facilitate the expression of schematic information, this paper proposes a public schematic representation data structure. The data structure, illustrated in Figure 2, includes three parts: device information, port information, and connection information, represented by the `DeviceInfo`, `PortInfo`, and `ConnInfo` classes respectively. The `DeviceInfo` class describes each device, including its name, description, and type. The `PortInfo` class describes each port, including its name, number, function, associated device, and other electrical properties. The `ConnInfo` class describes each connection, including its name, type, and associated ports.

```
// Schematic Structure
class Schematic:
    List<DeviceInfo> devices
    List<PortInfo> ports
    List<ConnectionInfo> connections
```

Figure 2. Schematic data structure expressed in pseudocode

B. Functional Interface Connection Information Capture

In circuit schematics, the functions of low-level ports are not always singular, multi-functional ports often exist. Therefore, automatically extracting high-level functional interface connection information from these low-level port connections in schematics presents a significant challenge. In the paper, the method for capturing interface connection information is shown in Figure 2. Initially, interface pairs with strong connection relationships are screened out. The formula for calculating the connection strength coefficient is as follows:

$$\delta = E/V_{conn} \tag{1}$$

Where E represents the expected number of connections and V_{conn} represents the actual number of connections. When $\delta \geq 0.5$, the interface pairs are considered strongly connected.

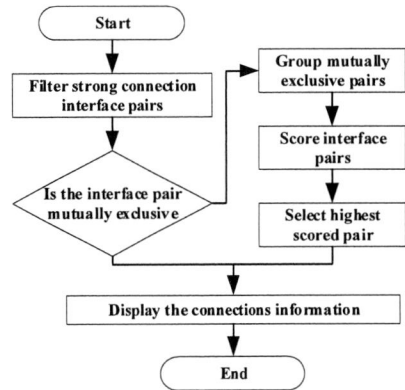

Figure 2. Functional Interface Connection Information Capture Process

Further, interface pairs that are strongly connected but not mutually exclusive are selected, while mutually exclusive pairs are grouped. Typically, there are multiple groups of mutually exclusive interface pairs. In these relationships, the same underlying port is shared between pairs, meaning one port has two functions, which is not allowed in actual circuits.

Finally, the probability score P of each interface pair is calculated, and the interface pair with the highest probability score in each group is selected. The probability score reflects the likelihood of an actual connection relationship in each group. Generally, P is positively correlated with both the actual number of connections V_{Conn} and the connection strength coefficient δ. Empirically, the formula for calculating P is given by:

$$P = \delta^2 \times (V_{conn}/100 + 1) \tag{2}$$

Here, δ is squared to increase its correlation with P, and V_{conn} is scaled and then incremented by 1 to reduce its correlation with P.

C. Connection Rules Description

To ensure extensibility, this paper describes rules using Python keywords, regular expressions, and custom operators and operands, some of which are listed in Table II. For example, in Xilinx7 series FPGAs connecting to DDR3 interfaces, the DQS signals in DDR3 must connect to the DQS signals in the FPGA, expressed as:

Filt(s_inf='DDR3', s_func='DDR3.DATA.DQS', d_inf='BANK', get_func='*')=={'dqs'}.

TABLE II. CUSTOM OPERATORS AND OPERANDS

Custom symbol	Description
Filt()	An operator for filtering, contains four operands: source interface name, source port function, destination interface name, and destination port function.
Count()	An operator for counting.
Pin.AttriName	`AttriName` indicates a port's attribute name. For example, `Pin.Vol` indicates the voltage attribute, and `Pin.Direction` indicates the input and output directions of the current interface.
Inf.AttriName	`AttriName` indicates an interface attribute name. For example, `Inf.Type` indicates the type attribute of the current interface.

D. Inspection Engine

The inspection engine mainly includes two functions: rule matching and rule execution. Rule matching refers to the process of matching rules with connections of specific functional interfaces, as each connection rule is valid for one or a few types of connections. Rule execution involves parsing and executing a rule to check the matching interface connection.

Python is used to describe the meaning or function of key characters in the rule. Therefore, rules can be parsed using the function `eval(rule_expr, envs)` built into the Python language. The parameter `rule_expr` indicates the specific regular expression, and `envs` indicates the semantic environment of the regular expression. The `eval()` function converts the specified string `rule_expr` into executable Python code based on `envs` enhanced semantic environment and Python keywords. By running this Python code, the rule is executed, and the validity of the corresponding connection relationship is checked.

IV. EVALUATION

This section details the experimental verification of the proposed interface-oriented circuit schematic verification method. The verification process involved using a schematic file containing known types of errors as a test case. The schematic was subjected to fully automatic rules verification using both the proposed method and OrCAD 16.6, a schematic verification tool by Cadence, without any human intervention. In addition, the experiment was conducted on a personal computer with Intel's core i7-7700HQ CPU.

The experimental setup included a series of test cases designed to evaluate the accuracy and efficiency of the proposed method. These test cases encompassed various error types, including electrical rule check (ERC) errors and functional rule verification (FRV) errors. The known errors in the schematic file provided a controlled environment to assess the effectiveness of the rule checks.

The verification process was conducted as follows:

1. Schematic Analysis: The schematic file was analyzed using the proposed method to automatically capture interface connection information and apply the predefined rules.

2. Rule Checking: The same schematic file was analyzed using OrCAD 16.6 to perform rule checks based on its built-in functionalities.

3. Error Reporting: For both tools, the ability to detect and report the known errors in the schematic was evaluated. If the corresponding errors were reported, the test was considered to pass; otherwise, it was considered to fail.

The verification results are presented in Table III. The table compares the performance of OrCAD 16.6 and the proposed method across different types of fully automated rule checks, including ERC and FRV. The results indicate that the proposed method outperforms OrCAD 16.6 in automatically detecting functional interface connection errors.

In particular, OrCAD's semi-automatic verification for each functional interface requires manual intervention and takes at least a few minutes, while our automatic verification method takes milliseconds, up to 10 milliseconds.

TABLE III. COMPARISON OF ORCAD 16.6 AND THE PROPOSED METHOD IN SCHEMATIC VALIDATION WITHOUT MANUAL INTERVENTION

Check Type	Check Item		Num. of test cases	Tool Type	
				OrCAD V16.6	*Ours*
				Test Pass Rate	*Test Pass Rate*
ERC	Single Node Net		68	100%	100%
	Driverless Net		68	100%	100%
	Drive Confict Net		68	100%	100%
	Net Name Repetition		68	100%	100%
	Consistency between the VCC/GND port name and the net name		68	0	100%
	Unconnected Ports		68	100%	100%
	Refs Missing Ports		68	100%	100%
	Power Supply and Ground Connection		68	100%	100%
FRV	Interface Missing Signal		68	0	100%
	Port polarity matches the differential signals		68	0	100%
	Function interface Special connection rule	DDR	18	0	100%
		DDR2	18	0	100%
		DDR3	18	0	100%
		SPI	18	0	100%
		PCIe	18	0	100%
		SRIO	18	0	100%
		JTAG	18	0	100%
		SGMII	18	0	100%
		EMIF	18	0	100%
		1553B	18	0	100%
		Xilinx 7 series FPGA memory interface	38	0	100%
		NAND Flash Interface	18	0	100%
		NOR Flash Interface	18	0	100%
		I2C	18	0	100%
		UART	18	0	100%

V. SUMMARY

In this paper, we propose a method for automatically verifying functional interface connections. Our primary contributions include designing a general schematic data structure to represent component connections, utilizing empirical formulas and processes for automated interface connection extraction, and developing a rule description language incorporating Python keywords, regular expressions, and custom operators for reusable and adaptable connection rules. Experimental results show that our method improves verification coverage and reduces verification time from minutes to seconds. In conclusion, our method offers a significant advancement in design verification by providing an automated approach to verifying functional interface connections.

REFERENCES

[1] R. Munoz, 'Furthering Moore's Law Integration Benefits in the Chiplet Era', IEEE Des. Test, vol. 41, pp. 81–90, August 2023.

[2] P. Wesling, 'The Heterogeneous Integration Roadmap: Enabling Technology for Systems of the Future', IEEE Pan Pacific Microelectronics Symposium, pp. 1–4, Feb. 2020.

[3] A. A. Gautam and V. Laxmi, 'Gate Drive for Power Electronic Converters : An Insight into KiCAD's PCB design !', Digests 2nd Global Conference for Advancement in Technology, pp. 1–6, October 202

Co-Optimization Design Method of Temperature Variation and Circuit Aging in Digital Circuits

Songxuan He, Wangyong Chen*, Ling Xiong, Linlin Cai*

School of Microelectronics Science and Technology, Sun Yat-sen University, Guangzhou, 510275, China

*Email: chenwangy@mail.sysu.edu.cn, caillin3@mail.sysu.edu.cn

Abstract—With the advancement of technology, there is a growing demand for digital designs with high robustness against temperature variation and circuit aging, especially in the field of autonomous driving. In this paper, we propose a co-optimization design method to enhance the immunity to temperature variation and aging of circuits utilizing the existing EDA tools. The seamless integration with existing design processes enables the proposed method to be easily adopted and utilized. The key features of the co-design strategy include utilizing a penalty standard cell library with the awareness of zero temperature delay points for improving temperature insensitivity and creating a subset of the standard cell library to enhance circuit reliability. The proposed method is validated using multiple benchmark circuits, showing that the temperature variation can be reduced to 45.6%, and the delay degradation caused by circuit aging can be reduced to 76.1%.

Keywords—circuit reliability, transistor aging, logic synthesis, temperature sensitivity

I. INTRODUCTION

As technology advances, the demand for high computing performance in digital integrated circuits is increasing. This is accompanied by higher power consumption density and thermal effects, especially with the emergence of 3D devices[1]. Higher on-chip temperatures accelerate circuit aging, reducing reliability and performance[2-3]. Different applications have varying reliability requirements[4]. Temperature variations and circuit aging pose critical challenges for advanced devices, threatening the reliability and stability of electronic systems. It is crucial to ensure the robustness of integrated circuits across a wide temperature range throughout their expected lifetime.

The current approach to addressing temperature variation and circuit aging in digital circuits relies on timing guardband[5], which sacrifices performance without fundamentally improving stability. In our work, we aim to integrate an active co-optimization design method for temperature variation and circuit aging into the existing digital design flow. This involves modifying cell types and delay tables to reduce temperature variation and circuit aging in circuits designed by the logic synthesis tool.

II. TEMPERATURE-IMMUNE DESIGN USING ZTD

Researchers have proposed a method called zero temperature delay (ZTD)-based design to minimize the impact of temperature on circuit delay[6]. It aims to find the optimal power supply voltage to reduce temperature sensitivity on the critical path. Operating the circuit with this method at a lower voltage minimizes the adverse effects of temperature variations but may increase delay and impact performance. Therefore, this paper chooses to work at the nominal voltage of the cells, and the zero temperature delay point is only used as a penalty factor[7]. When multiple temperatures are considered, the zero-temperature delay will be an area. The cost function[8] helps determine the zero-temperature delay point. This is shown in Figure 1.

$$C(V_{DD}) = \frac{1}{N^2} \sum_{i=1}^{N} |\tau_{D,i}(V_{DD})| \sum_{i=1}^{N} \frac{1}{|\tau_{D,i}(V_{DD})|} \quad (1)$$

$$V_{ZTD} = \min \{C(V_{DD})\} \quad (2)$$

where N is the number of measured temperatures, $\tau_{D,i}$ is the delay at the corresponding voltage, and V_{ZTD} is the zero temperature delay point.

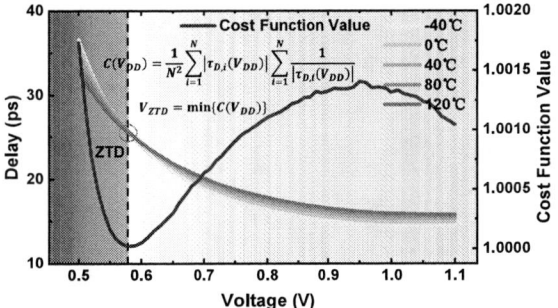

Fig. 1. Delay-Voltage curves at different temperatures and cost function

In order to explain the function of the zero temperature delay point as a penalty factor, we first need to introduce the standard cell library. This library contains many cells, each with various properties such as maximum capacitance, function, timing, and power consumption. We are particularly interested in the timing of the cells in this article. The delay of a cell is influenced by factors such as temperature, voltage, process, inherent device delay, input transition time, and output load. The standard cell library includes a two-dimensional delay lookup table that can be used to find the corresponding delays for specific input transition times and output load combinations. In simpler terms, different combinations of input transition times and output loads correspond to specific delay values.

Replacing the delay value is the same as altering the value corresponding to the input transition time and output load. Therefore, as long as the relationship between the delay value and the replacement value is established, it is equivalent to replacing the delay library in the standard cell library with another library, while retaining the lookup table relationship between the input transition time and output load and the replacement value. In this article, we need to find the corresponding temperature fluctuation (i.e. the slope of the fitting curve between delay and temperature) for each delay value, replace it, and establish a temperature fluctuation (TF) library. Find the corresponding zero temperature delay point for each delay value, calculate the penalty value through a penalty function, replace it, and establish a penalty cell library. Calculate the aging degree of each delay value and then statistically analyze the proportion of delays with an aging degree below a certain threshold within the delay library to help filter cell types.

III. CO-DESIGN METHOD DESCRIPTION

The general framework of the whole process is shown in Figure 2. In the preparation stage, the threshold voltage offset of each MOSFET of the corresponding cell is obtained through circuit aging. The transistor model is then modified and re-characterized to obtain the aging cell library. Subsequently, we compare delay degradation in the cell library before and after aging to identify and characterize the subset of the standard cell library. Additionally, the zero temperature delay point for each slew-load combination and the temperature variation at the operating voltage are derived from the standard cell library collection. After that, the penalty cell library subset is created based on the standard cell library subset, and the temperature fluctuation library is established based on the standard cell library. Finally, the penalty cell library subset is used to perform logic synthesis on the circuit to obtain the corresponding circuit gate-level netlist, the standard cell library is used to perform timing analysis on the synthesized circuit netlist, and the temperature fluctuation library is used to perform temperature variation analysis on the circuit. Comprehensively compare the delay, power consumption, area, delay increased by circuit aging, and temperature variation of the circuit obtained by the standard cell library and the circuit obtained by applying the method in this paper.

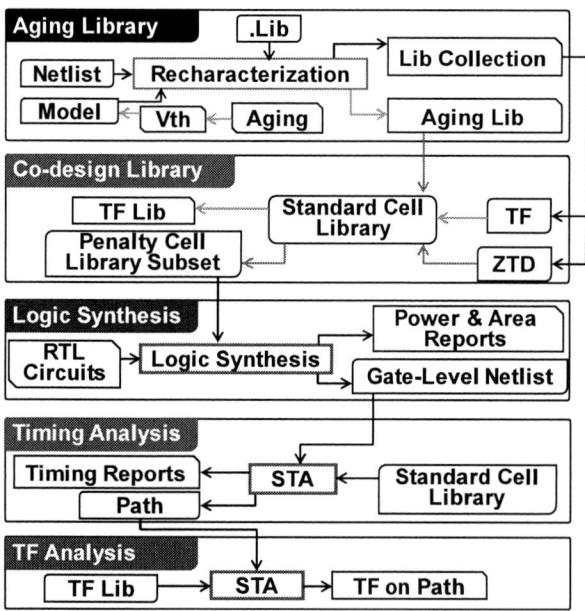

Fig. 2. Overall flow chart: establishment of penalty cell library subset and multiple evaluations

Due to aging effects such as bias temperature instability (BTI) in digital circuits, the threshold voltage will shift as the usage time increases. This paper proposes a method to suppress the aging of digital circuits by using a standard cell library subset consisting of cells with less delay degradation. The aging considered in this paper is the threshold voltage shift under long-term operation [9] and does not consider the transient errors caused by aging [10-11]. Two criteria need to be met for filtering: First, the increase in the aging ratio of the

chosen cell must be lower than the set threshold. Second, the subset must encompass all cell types, meaning that filtering is based on the aging ratio of the delay of the same type of cells with different driving capabilities. After the filtering is completed, the obtained cells are re-characterized as a standard cell library subset for the subsequent construction of the penalty cell library subset.

To incorporate the zero-temperature delay point into the traditional digital design flow, the zero-temperature delay information can be ideally included in the standard cell library as an additional quality factor. However, this would require a significant amount of work, so the zero temperature delay information is used as a delay penalty factor in the original standard cell timing library. In this paper, the method used is to increase the delay of the cell. This means that the delay is increased as a penalty for cells whose zero temperature delay point is far from the operating voltage under a specific slew-load combination. To meet timing constraints, this will guide the logic synthesis tool to use cells that are less affected by temperature and readjust the driving strength of the cells.

The penalty functions used in this paper are linear function and square function:

$$linear(x, V_{ZTC}) = x \times (1 + |V_{DD} - V_{ZTC}| \times weight) \quad (3)$$

$$square(x, V_{ZTC}) = x \times (1 + (V_{DD} - V_{ZTC})^2 \times weight) \quad (4)$$

where x represents the delay value in the 2D lookup table, V_{ZTC} is the zero temperature delay point corresponding to the slew-load combination, V_{DD} is the circuit operating voltage.

IV. RESULTS AND DISCUSSION

It is worth noting that in the synthesis, we need first to use the standard cell library to perform logic synthesis under the timing constraint of 0ps to obtain the minimum delay of the corresponding RTL circuit. Multiply it by 1.6 as the subsequent timing constraint, so that the synthesis tool can optimize the power consumption and area, and strive to align with the timing constraints in the actual design process. Under the same timing constraints, logic synthesis is performed using a standard cell library and different penalty cell library subsets respectively. We can get the gate-level circuit netlist obtained by the traditional EDA workflow and the method in this work.

In the comparison provided in Table 1, the penalty functions presented are those with smaller increases in delay due to circuit aging and larger reductions in average temperature variation along paths. In general, the circuit synthesized by the method in this paper has lower delay, lower delay added by circuit aging, and lower average temperature variation. However, there are varying degrees of increase in area and power consumption. This is because the method proposed in this article not only considers timing, power consumption, and area in the logic synthesis process but also needs to consider temperature variation and circuit aging. Therefore, power consumption and area may be large. We can also see from the table that the circuit area and power consumption obtained by applying the penalty cell library subset increase. The increase is larger for smaller circuits. This approach incurs higher area and power consumption costs. However, in fields where area and power constraints are less critical, chip stability and reliability are paramount, this approach is highly valuable.

979-8-3503-6184-1/24 $31.00 © 2024 IEEE

TABLE I.

MULTIPLE EVALUATIONS OF CIRCUITS SYNTHESIZED FROM STANDARD CELL LIBRARY AND PENALTY CELL LIBRARY SUBSET

Circuits	Function	Area(μm²)	Power(mW)	Delay(ps)	ΔDelay(%)	Average TF(fs/K)
c499	origin	18.69	0.06	311.67	7.187	80.76
	$1+\Delta^2\times100$	25.46	0.12	239.67	5.470	70.81
c2670	origin	23.68	0.05	351.13	7.775	91.01
	$1+|\Delta|\times10$	31.38	0.08	207.79	6.073	47.00
c7552	origin	71.19	0.19	353.73	7.712	81.47
	$1+\Delta^2\times100$	106.05	0.28	253.12	6.914	61.37
voter	origin	504.77	0.49	1063.61	7.423	257.54
	$1+|\Delta|\times100$	509.00	0.61	833.15	6.916	211.04
memctrl	origin	1205.68	0.78	725.94	8.878	290.25
	$1+|\Delta|\times30$	1508.77	1.08	607.21	7.093	143.94

where Δ equals $V_{DD} - V_{ZTC}$, ΔDelay is $(\tau_{aged} - \tau_{fresh})/\tau_{fresh}$.

The critical path of a circuit may become less critical after aging[12]. Here, we are considering the critical path of the circuit after aging, rather than the aging of the critical path of the circuit. As shown in Figure 3, the delay of the circuit using the penalty cell library subset after aging is lower than that of the circuit using the standard cell library. Figure 4 shows that the method can reduce the temperature variation of the entire circuit. The maximum value decreased from 338.01fs/k to 170.75 fs/k, and the average value decreased from 290.25fs/K to 143.94 fs/K.

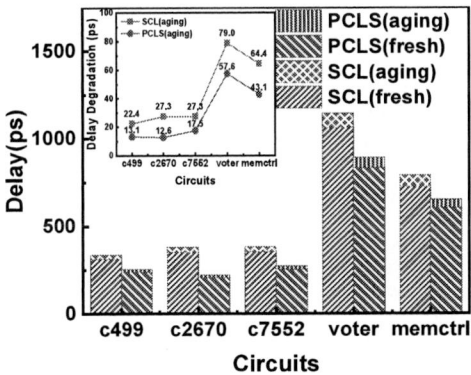

Fig. 3. Comparison of the delay before aging and the increased delay after aging of the circuit synthesized using the standard cell library (SCL) and the penalty cell library subset (PCLS)

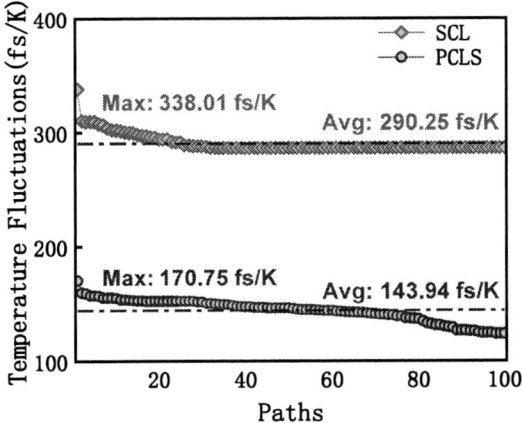

Fig. 4. Top 100 paths with the largest temperature variation in memctrl circuits synthesized using the standard cell library (SCL) and the penalty cell library subset (PCLS)

V. CONCLUSION

The method proposed in this paper integrates both the zero temperature delay point information and the aging delay information of the cells into the existing delay library of the standard cell library. Through the analysis of the results, it can be known that using the zero temperature delay point to create a penalty cell library can prevent the increase of delay and reduce temperature variation (can be reduced to 45.6%). Using the delay degradation to establish a subset of standard cell libraries can reduce the delay degradation caused by circuit aging (can be reduced to 76.1%).

ACKNOWLEDGMENT

This work was supported in part by the National Natural Science Foundation of China under Grants 62304263, 62204269 and in part by the Guangdong Basic and Applied Basic Research Foundation under Grants 2023A1515011418, 2024A1515010349.

REFERENCES

[1] X. Huang et al., "Sub 50-nm FinFET: PMOS," in International Electron Devices Meeting 1999. Technical Digest (Cat. No.99CH36318), Dec. 1999, pp. 67–70. doi: 10.1109/IEDM.1999.823848.

[2] C. Prasad et al., "Self-heat reliability considerations on Intel's 22nm Tri-Gate technology," in 2013 IEEE International Reliability Physics Symposium (IRPS), Apr. 2013, p. 5D.1.1-5D.1.5. doi: 10.1109/IRPS.2013.6532036.

[3] G. Naima and S. B. Rahi, "Low Power Circuit and System Design Hierarchy and Thermal Reliability of Tunnel Field Effect Transistor," Silicon, vol. 14, no. 7, pp. 3233–3243, May 2022, doi: 10.1007/s12633-021-01088-2.

[4] G. Previati, G. Mastinu, and M. Gobbi, "Thermal Management of Electrified Vehicles—A Review," Energies, vol. 15, no. 4, 2022, doi: 10.3390/en15041326.

[5] K. Jeong, A. B. Kahng, and K. Samadi, "Quantified Impacts of Guardband Reduction on Design Process Outcomes," in 9th International Symposium on Quality Electronic Design (isqed 2008), Mar. 2008, pp. 790–797. doi: 10.1109/ISQED.2008.4479839.

[6] S. Salamin, V. M. Van Santen, M. Rapp, J. Henkel, and H. Amrouch, "Minimizing Excess Timing Guard Banding Under Transistor Self-Heating Through Biasing at Zero-Temperature Coefficient," IEEE Access, vol. 9, pp. 30687–30697, 2021, doi: 10.1109/ACCESS.2021.3057900.

[7] F. Klemme and H. Amrouch, "Transistor Self-Heating-Aware Synthesis for Reliable Digital Circuit Designs," IEEE Transactions on Circuits and Systems I: Regular Papers, vol. 70, no. 12, pp. 5366–5379, Dec. 2023, doi: 10.1109/TCSI.2023.3315293.

[8] M. Runge, S. Linnhoff, and F. Gerfers, "A Temperature and Process Corner Insensitive Design Method for Digital Circuits in 40nm CMOS," in 2018 IEEE 61st International Midwest Symposium on Circuits and Systems (MWSCAS), Aug. 2018, pp. 779–782. doi: 10.1109/MWSCAS.2018.8623863.

[9] H. Amrouch, V. M. van Santen, T. Ebi, V. Wenzel, and J. Henkel, "Towards interdependencies of aging mechanisms," in 2014 IEEE/ACM International Conference on Computer-Aided Design (ICCAD), Nov. 2014, pp. 478–485. doi: 10.1109/ICCAD.2014.7001394.

[10] V. M. van Santen, H. Amrouch, N. Parihar, S. Mahapatra, and J. Henkel, "Aging-aware voltage scaling," in 2016 Design, Automation & Test in Europe Conference & Exhibition (DATE), Mar. 2016, pp. 576–581. Accessed: Sep. 25, 2023. [Online]. Available: https://ieeexplore.ieee.org/document/7459378

[11] V. M. van Santen, H. Amrouch, J. Martin-Martinez, M. Nafria, and J. Henkel, "Designing guardbands for instantaneous aging effects," in 2016 53nd ACM/EDAC/IEEE Design Automation Conference (DAC), Jun. 2016, pp. 1–6. doi: 10.1145/2897937.2898006.

[12] H. Amrouch, B. Khaleghi, A. Gerstlauer, and J. Henkel, "Reliability-aware design to suppress aging," in 2016 53nd ACM/EDAC/IEEE Design Automation Conference (DAC), Jun. 2016, pp. 1–6. doi: 10.1145/2897937.2898082.

Boolean Matrix Factorization Algorithm based on Error Shaping Technique and its Application on Approximate Logic Synthesis

Runhua Yang [1], Rensheng Shen*[1]

School of Integrated circuits, Dalian University of Technology, Dalian 116024, China

* 2023485575@qq.com, shjiank@dlut.edu.cn

Abstract—**Boolean matrix factorization method can be used to create an approximate circuit from a given circuit. The accuracy loss of approximate circuit originates from the factorization error of each subcircuit. Hence, reducing factorization error of Boolean matrix has significant value in the approximate logic synthesis. In the case of the truth table of the logic circuit, the number of rows is much larger than the number of columns that can be utilized to further optimize the factorization method. In this brief, the error shaping technique is proposed to concentrate the factorization errors in several specific columns, which can be easily cleared by the proposed column error clear scheme. Compared with the typical factorization methods, for the truth table of an n-input, m-output logic circuit, the accuracy of proposed factorization method can be improved significantly. Finally, the proposed Boolean matrix factorization algorithm is integrated into the approximate logic synthesis tool and compared with BLASYS. The synthesis results demonstrate state-of-the-art performance.**

Keywords—*Boolean Matrix Factorization, Approximate Logic Synthesis, Approximate Computing, Error Shaping*

I. INTRODUCTION

Approximate computing has attracted a lot of attention from industry and academia. By introducing some errors, many handcrafted arithmetic circuits have been optimized so that the hardware performance is improved significantly. To reduce the design efforts, several approximate logic synthesis (ALS) techniques for arbitrary logic circuits are discussed in[1]. BLASYS [2]~[4], an ALS tool based on Boolean Matrix Factorization (BMF), has been proved to be very effective in exploring the design space, scale to large circuits.

BLASYS partitions a large circuit into several small circuits and then enumerates the truth table of each subcircuit that is denoted as Boolean matrix M. Moreover, the Boolean matrix M can be approximately factorized into two Boolean matrices S and B so that the difference between M and SB is minimized. The logic resources of the approximate circuit are dominated by the compressor circuit whose truth table is given by matrix S. The outputs of the compressor circuit can be connected by the decompressor circuit according to matrix B, which occupies a small amount of logic resources. As the size of matrix S is far less than matrix M, the logic resources are reduced. Obviously, the accuracy loss of the approximate circuit originates from the BMF of each small subcircuit. Hence, it is highly desired to reduce the factorization error of BMF [2][3].

BMF is an NP-hard problem solved with many heuristic algorithms [5]~[8]. The ASSO algorithm uses binary bases from a column-wise correlation matrix in a heuristic manner to perform factorization [5]. Then, an improved BMF

algorithm comparing to ASSO called Nassua optimizes the initialization of the matrix factorization by locating dense seeds hidden in the matrix [6]. Recently, an efficient BMF algorithm, called MEBF [7], permutates the columns and rows in each iteration such that the permutated matrix is an upper triangular like matrix approximately. In the case of bioinformatics, a new BMF algorithm via expectation maximization (BEM) is proposed in [8].

However, in the case of ALS application, there exist several different characteristics. **(1)** The size of Boolean matrix M of an n-input, m-output circuit is $2^n \times m$. Hence, the number of rows is far larger than the number of columns. Some methods, such as MEBF [7], are not good at performing factorization in this case. **(2)** The addition in BMF is generally realized by the **OR** operator. For ALS, the decompressor circuit (matrix B) is just used to connect the outputs of the compressor circuit that can be realized by many different logic gates such as **OR/XOR**. Due to these two reasons, a novel BMF algorithm is proposed to achieve small factorization error for the Boolean matrix of circuits.

The rest of this brief is organized as follows. The ESBMF algorithms is discussed in Section II. The performance of ESBMF in ALS tool are shown in Section III. The conclusion is summarized in Section IV.

II. ERROR SHAPING TECHNIQUE BASED BMF ALGORITHM

A. Overview of Proposed Boolean Matrix Factorization

The factorization process of Boolean matrix can be depicted in Fig. 1, where the factorization degree is 3. The first column (red) of S and the first row of B produce a red matrix. Hence, a preliminary estimation of M denoted as $S(:,1) \times B(1,:)$ is obtained. Then, the second column (blue) of S and the second row of B produce a blue matrix and update the estimation of M written as $S(:,1) \times B(1,:) + S(:,2) \times B(2,:)$. Repeat this process until a satisfactory solution is found. It is easy to conclude that the smaller factorization degree results in more logic resources reduction and more accuracy loss.

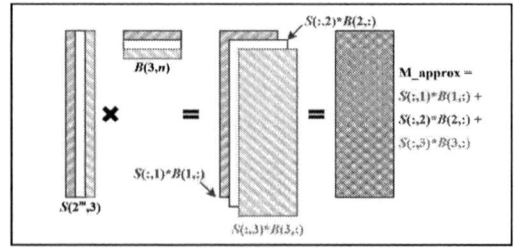

Fig. 1. Overview of Boolean Matrix Factorization Process.

In the k-th iteration, the challenge is to find a pair of $S(:, k)$ and $B(k,:)$ so that the factorization error is reduced. Three schemes are discussed in the following subsections.

B. Matrix Error Minimize (MEM) Scheme

In the k-th iteration, if $B(k, :)$ is given, an optimum $S(:, k)$ is selected in Algorithm-1 so that the factorization error written as $\|M-S(:,1:k)\times B(1:k,:)\|$ is minimized. For MEM scheme, the addition operators are realized by **OR** gates.

The estimation of M in the $(k-1)$-th iteration is calculated in Line 1. Since $M(:, j)$ and $B(k, j)$ are given, an optimal column vector $S\_opt(:, j)$ is selected by making the formula in Line 3 to be true. In the end, each element $S(i, k)$ is determined by $S\_opt(i, :)$ shown in Line 5~10.

Algorithm 1: $S(:, k) = \text{MEM}(S(:,1:k-1), B(1:k-1,:), M, B(k,:))$
Input: $S(:,1:k-1)$, $B(1:k-1,:)$, M, $B(k,:)$
Output: $S(:,k)$
1    $M\_approx = S(:,1:k-1)\times B(1:k-1,:)$
2   for $j =1$: size(M, 2)
3      $M(:,j) \leftarrow \text{OR}(S\_opt(:,j)\times B(k,j), M\_approx(:,j))$
4   end
5   for $i=1$: size(M, 1)
6   if (number of zeros in $S\_opt(i,:)$ <= number of ones in $S\_opt(i,:)$)
7        $S(i,k) = 1$
8   else
9        $S(i,k) = 0$
10   end

C. Column Error Clear (CEC) Scheme

Algorithm-2 shows that the estimation of M in the $(k-1)$-th iteration is calculated in Line 1. The factorization error in the t-th column is calculated in Line 2 that can be cleared with the **XOR** operator. That is, $S(:, 1:k-1)\times B(1:k-1, t)+S(:, k)$ is equal to $M(:, t)$, where the addition operator is realized by the XOR gate. In the end, once $S(:, k)$ is obtained, $B(k, :)$ is calculated in Line 3~8.

Algorithm 2: $[S(:,k), B(k,:)] = \text{CEC}(S(:,1:k-1), B(1:k-1,:), M, t)$
Input: $S(:,1:k-1)$, $B(1:k-1,:)$, M, t
Output: $S(:,k)$, $B(k,:)$
1    $M\_approx = S(:,1:k-1)\times B(1:k-1,:)$
2    $S(:,k) = \text{XOR}(M\_approx(:,t), M(:,t))$
3   for $j=1$: size(M, 2)
4   if ($j = t$)
5        $B(k, j) = 1$
6   else
7        $M(:,j) \leftarrow \text{OR}(S(:, k)\times B(k, j), M\_approx(:,j))$
8   end

D. BMF based on Error Shaping Technique

Algorithm-3 shows that the proposed ESBMF algorithm is divided into two steps. **Step-1:** the MEM and CEC schemes are used to find an optimal $B(k, :)$ that provides the maximum factorization error reduction in the k-th iteration. However, this method is easy to be trapped in a local optimum point. **Step-2:** utilize the MEM scheme to find an optimal $B(k, :)$ that concentrates the most errors in several columns. These errors can be cleared by the CEC scheme in the following iterations.

Step-1: A feasible solution for $B(k, :)$ is selected from the unique rows in M_target and the related optimal $S(:, k)$ can be decided by MEM scheme. The corresponding error reduction is obtained in Line 9. Repeat this process, the best $B(k, :)$ from $B\_feasible$ can be found so that the maximum error reduction of MEM scheme is obtained. In addition, the errors produced by $S(:,1:k-1)\times B(1:k-1,:)$ in different columns is calculated in Line 12. The maximum column error is found in Line 14 that is the minimum error reduction of CEC scheme. Therefore, as shown in Line 15~19, if the error reduction of MEM scheme is less than CEC scheme, we

should use CEC scheme to clear the t-th column error. Otherwise, MEM scheme is utilized. In summary, the maximum error reduction caused by MEM or CEC schemes can be obtained in each iteration. In the end, the minimum error of Step-1 is stored in Line 22.

Step-2: A new criterion is utilized to evaluate the best $B(k, :)$ from $B\_feasible$ such that the factorization error can be further reduced. In the k-th iteration, if the most factorization errors can be concentrated in the $(f-k)$ columns, these errors can be cleared by the CEC scheme in the $\{k+1,\cdots f\}$-th iterations. As a result, the errors in different columns of different feasible $B(k, :)$ are calculated in Line 27. The number (r) of columns that cannot be cleared by CEC is calculated in Line 28 and the residual error (the sum of the smallest r column errors) is calculated in Line 29. If the residual error is smaller than the minimum error in Step-1, the optimal $S(:, k)$ and $B(k, :)$ should be re-selected and the CEC scheme should be employed in the following iterations.

Algorithm 3: $[S(:, 1: f), B(1: f, :)] = \text{ESBMF}(M, f)$
Input: M, f (factorization degree)
Output: S, B
1   STEP-1
2    $M\_approx = 0$
3   for $k =1$: f
4      $M\_target = \text{XOR}(M\_approx, M)$
5      $B\_feasible = \text{unique}(M\_target, \text{'rows'})$
6   for $i =1$: size($B\_feasible$, 1)
7        $B(k, :) = B\_feasible(i, :)$
8        $S(:, k) = \text{MEM}(S(:,1:k-1), B(1:k-1,:), M, B(k,:))$
9        $\Delta\text{Error}(i) = \|M-S(:,1:k-1)\times B(1:k-1,:)\| - \|M-S(:,1:k)\times B(1:k,:)\|$
10   end
11   for $j =1$: size(M, 2)
12       $\text{ErrorColumn\_1}(j) = \|M(:, j)-S(:,1:k-1)\times B(1:k-1, j)\|$
13   end
14     $t = \text{find}(\text{ErrorColumn\_1} = \max(\text{ErrorColumn\_1}))$
15   if ($\text{ErrorColumn\_1}(t) > \max(\Delta\text{Error})$)
16       $[S(:,k), B(k,:)] = \text{CEC}(S(:,1:k-1), B(1:k-1,:), M, t)$
17   else
18   Select the best $\{S(:, k), B(k, :)\}$ according to $\max(\Delta\text{Error})$
19   end
20   updates M_approx
21   end
22    $\text{LocalMinError} = \|M-S(:,1:f)\times B(1:f, :)\|$
23   STEP-2
24   In the k-th iteration, the errors concentrated in the $(f-k)$ columns can be
25   easily cleared by the CEC scheme in the $\{k+1,\cdots f\}$-th iterations.
26   for $i =1$: size($B\_feasible$, 1)
27      $\text{ErrorColumn\_2}(i, j) = \|M(:, j)-S(:,1:k)\times B(1:k, j)\|$
28      $r = \text{size}(M, 2) - (f-k)$
29      $\text{ResError}(i) = \text{sum}(\text{sort}(\text{ErrorColumn\_2}(i, :)), r)$
30   end
31   if (min(ResError) < LocalMinError)
32   Select the best $\{S(:, k), B(k, :)\}$ according to min(ResError)
33   Then, start **CEC** scheme in the following iterations
34   else
35   Use the results in STEP-1
36   end

III. RESULTS DISCUSSION

A. Comparison of Different BMF Algorithms

The proposed ESBMF has been compared to the improved ASSO based on mixed **OR/XOR** [2], MEBF [7], and BEM [8]. Generate 1000 random Boolean matrices with three different dense levels (the probability of Logic one) to imitate the truth table of the 10-input 10-output logic circuit. The Hamming distance is used to measure the accuracy levels of various BMF algorithms written as

$$\text{Error} = \frac{\sum \text{xor}(M, SB)}{2^{10}\times 10} \quad (1)$$

The maximum error is shown in Tab. I. Simulation results demonstrate that the proposed ESBMF can achieve the minimum factorization error.

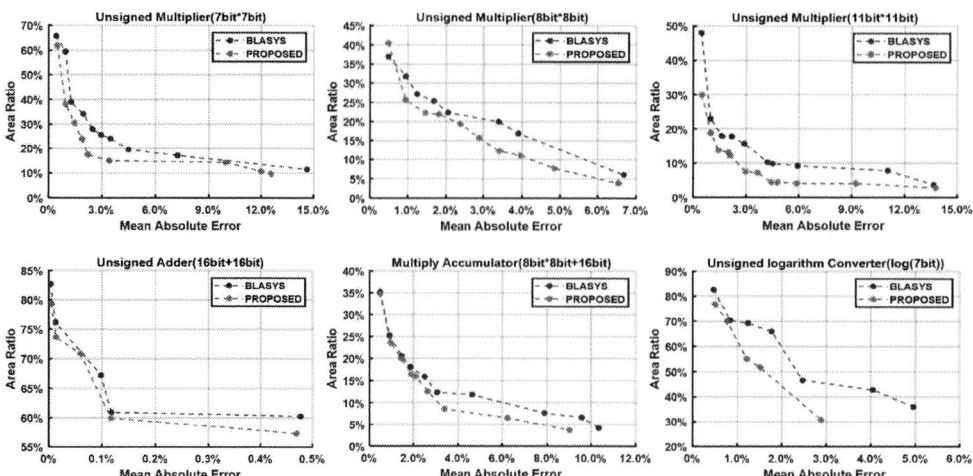

Fig. 2. Comparison of BLASYS [4] and proposed ESBMF algorithms.

TABLE I. MAXIMUM FACTORIZATION ERRORS OF DIFFERENT BMF ALGORITHMS

1000× random M (1024×10) P(Logic 1) = 0.2					
	f=1	f=3	f=5	f=7	f=9
ASSO+[2]	18.96%	14.51%	10.21%	6.07%	2.01%
MEBF [7]	19.10%	14.89%	10.64%	6.47%	2.01%
BEM [8]	20.09%	18.48%	16.20%	12.85%	9.60%
ESBMF	18.96%	14.51%	10.21%	6.07%	2.01%
1000× random M (1024×10) P(Logic 1) = 0.5					
	f=1	f=3	f=5	f=7	f=9
ASSO+[2]	38.68%	34.43%	27.58%	14.06%	4.66%
MEBF [7]	46.44%	35.55%	25.91%	20.76%	15.54%
BEM [8]	38.68%	30.16%	25.84%	20.64%	15.57%
ESBMF	38.32%	29.25%	20.47%	12.22%	4.82%
1000× random M (1024×10) P(Logic 1) = 0.8					
	f=1	f=3	f=5	f=7	f=9
ASSO+[2]	21.01%	21.01%	20.87%	19.20%	8.88%
MEBF [7]	66.88%	44.73%	26.43%	16.85%	13.49%
BEM [8]	21.01%	17.74%	11.57%	7.31%	3.75%
ESBMF	21.01%	16.37%	11.95%	7.54%	3.79%

B. Compare of Different ALS Algorithms

The proposed ESBMF algorithm had been integrated into the ALS tool (BLASYS) [4]. The exploration method for the hyper-parameters in ALS tool remains the same [2]. Fig. 2 reports the ALS results of six arithmetic circuits (Three multipliers and an adder from [2], a multiply accumulator from [1], and a logarithmic converter from [9]). The x-axis is the mean absolute error (MAE) written as (2) of the approximate circuits. The y-axis is the area utilization of the approximate circuits with different calculation errors.

$$\text{MAE} = \frac{\sum_{all\ inputs} \frac{\left| O_{accurate} - O_{approximate} \right|}{\text{Max output value}}}{\text{Total number of inputs}} \quad (2)$$

Fig. 2 demonstrates the proposed ESBMF can reduce the accuracy loss of approximate circuits under the similar area reduction as the factorization error of truth table is relatively low. In addition, the comparison of BLASYS and other ALS methods are well discussed in [2]. Therefore, the proposed ESBMF algorithm can be used in ALS tool and reach state of the art performance in many commonly used circuits.

IV. CONCLUSION

In the case of the Boolean matrix of the practical circuit, the number of columns is much less than the number of rows.

The proposed error shaping scheme concentrates the most errors in several columns that can be effectively cleared by the column error clear scheme with the aid of the **XOR** gate. As a result, compared with state of the art, the proposed method is very effective in performing the factorization of truth table. In addition, the performance of the approximate logic synthesis tool (BLASYS) also can be promoted by replacing ASSO+ with the proposed factorization method (ESBMF).

ACKNOWLEDGMENT

This work is supported by National Key R&D Program of China (2023YFB4503003) and Special projects for industrial foundation reconstruction and high-quality development of manufacturing industry in 2022(No. TC220A04A-49).

REFERENCES

[1] I. Scarabottolo, G. Ansaloni, G. A. Constantinides, L. Pozzi and S. Reda, "Approximate Logic Synthesis: A Survey," in Proceedings of the IEEE, vol. 108, no. 12, pp. 2195-2213, Dec. 2020.

[2] J. Ma, S. Hashemi and S. Reda, "Approximate Logic Synthesis Using Boolean Matrix Factorization," in IEEE Transactions on Computer-Aided Design of Integrated Circuits and Systems, vol. 41, no. 1, pp. 15-28, Jan. 2022.

[3] Soheil Hashemi, "Approximate Computing Techniques for AccuracyEnergy Trade-offs", in PHD Dissertation of Brown University, pp. 42-58. 2015.

[4] BLASYS: Approximate Logic Synthesis Using Boolean Matrix Factorization, https://github.com/scale-lab/blasys.

[5] P. Miettinen, T. Mielikäinen, A. Gionis, G. Das and H. Mannila, "The Discrete Basis Problem," in IEEE Transactions on Knowledge and Data Engineering, vol. 20, no. 10, pp. 1348-1362, Oct. 2008.

[6] Sanjar Karaev, Pauli Miettinen, and Jilles Vreeken, "Getting to Know the Unknown Unknowns: Destructive-Noise Resistant Boolean Matrix Factorization," Proceedings of the 2015 SIAM International Conference on Data Mining (SDM). 2015, 325-333.

[7] Wan, Changlin, Wennan Chang, Tong Zhao, Mengya Li, Sha Cao and Chi Zhang. "Fast and Efficient Boolean Matrix Factorization by Geometric Segmentation." pp. 6086-6093, in The Thirty-Fourth AAAI Conference on Artificial Intelligence (AAAI-20).

[8] Lifan Liang, Kunju Zhu, Songjian Lu, BEM: Mining Coregulation Patterns in Transcriptomics via Boolean Matrix Factorization, Bioinformatics, Volume 36, Issue 13, July 2020, Pages 4030–4037.

[9] B. Xiong, Y. Li, S. Li, S. Fan and Y. Chang, "Half-Precision Logarithmic Arithmetic Unit Based on the Fused Logarithmic and Antilogarithmic Converter," in IEEE Transactions on Very Large Scale Integration (VLSI) Systems, vol. 30, no. 2, pp. 243-247, Feb. 2022.

Automatically Device Sizing of Analog Circuit through Sequential Model-Based Optimization with Circuit Recognition

Shun-Qi DAI *[1], Xiao WANG [1], Yuan LEI [1], Bei-Ping YAN [1]

[1] AECS Department, Hong Kong Applied Science and Technology Research Institute (ASTRI), Hong Kong, P.R. China

* Email: shunqidai@astri.org

Abstract— **This paper presents a new approach to improve the efficiency of automatically device sizing in analog circuits using a sequential model-based optimization technique combined with circuit recognition. The proposed method effectively reduces the design space dimension automatically and achieves convergence within a significantly reduced number of iterations. To validate the effectiveness of this approach, two commonly used CMOS amplifier circuits have been successfully optimized within fewer simulations.**

Keywords— *Analog design automation, Device sizing, Circuit recognition*

I. INTRODUCTION

Analog circuits play a crucial role in Mixed-signal Integrated Circuits. However, unlike their digital counterparts, the design flow of analog circuits still requires significant manual effort, which becomes a bottleneck for achieving efficient system chip design. In current analog design flow, a considerable amount of time is spent on determining the dimensions of transistors [1-2]. To reduce the analog design cycle time, it is crucial to develop methods that can automatically determine the appropriate size of the device.

The traditional device sizing method can be classified into two main categories: equation-based approach and simulation-based approach. The equation-based approach tries to use simple equations and regression models to express the performance of the circuit [3-4]. Despite their avoidance of circuit simulations, these methods encounter challenges in accurately representing the performance of real circuits and efficiently generating equations for each new topology. Simulation-based methods approach circuit performance as black-box functions that are evaluated through circuit simulations. Global optimization algorithms are employed to achieve the optimal design point [5-6]. However, despite their independence from circuit topology, these methods often require a significant amount of time to achieve convergence and need designers to determine the dimension of the design space.

To achieve more efficient automatically device sizing in analog circuit, we present a new sequential model-based optimization method combined with circuit recognition, which mimics human designer to establish the relationship between the transistors to reduce the dimension of design space.

II. SEQUENTIAL MODEL-BASED OPTIMIZATION WITH CIRCUIT RECOGNITION

A. Problem Definition of Device Sizing

When designing an analog circuit, designers typically start by selecting the circuit topology and then determine the appropriate values for the design variables based on the device model, aiming to achieve the design specifications. However, in this paper, we assume that circuit topology has been chosen, and our focus is on automatically identifying the values of design variables. Thus, device sizing problem can be formulated as a multiple objective black-box optimization problem with constraints:

$$\begin{aligned} \text{minimize} \quad & f_i(x) \\ \text{subject to} \quad & g_i(x) \leq 0 \\ & \forall i \in 1,2,3 \cdots N, \end{aligned} \tag{1}$$

where $x \in R^d$, R^d and d represents design space and the number of design variables. $f_i(x)$ and $g_i(x)$ denotes the ith performance metric and constrain. A multiple objective problem can be converted into a single-objective problem by combining multiple objectives into a single cost function. This is commonly achieved by employing an aggregate function, as follows:

$$L(x) = \sum_{k=1}^{N} w_k \times -ReLU(f_k(x) - G_k) \tag{2}$$

where w_k is the weight factor of the kth objective function, G_k denotes the kth specification of performance metrics and $L(x)$ represents the new cost function. Therefore, the analog device sizing problem can be formulated as a single objective black-box optimization problem as follows:

$$\text{minimize} \quad L(x) \tag{3}$$

B. Sequential Mode-Based Optimization Framework with Circuit Recognition

In analog design, designers typically rely on their previous knowledge to establish the relationships between transistor dimensions to reduce design variables. To mimic human designer, we proposed a sequential mode-based optimization framework with circuit recognition as described in Fig. 1(a). It is composed of three parts: circuit recognition module, optimization module and SPICE simulator. At beginning, the original netlist containing circuit topology and design variables is passed on to the circuit recognition module. The circuit recognition module is responsible for identifying commonly used circuit building blocks in analog circuits, such as the differential stage, differential pair, current mirror array, and current mirror. An extensible circuit pattern library is also established, where new circuit patterns and their corresponding device relationships can be added by analog designers themselves, as shown in Fig. 1(b). Based on the recognition result, the relationship of the transistors can be achieve and a new annotated netlist with fewer design variables is generated. Subsequently, the annotated netlist, along with input definitions including circuit specifications and the bounds of design variables, is transmitted to the sequential model-based optimization (SMBO) module for an

979-8-3503-6184-1/24 $31.00 © 2024 IEEE

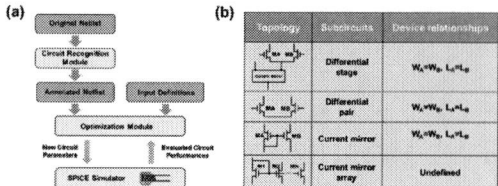

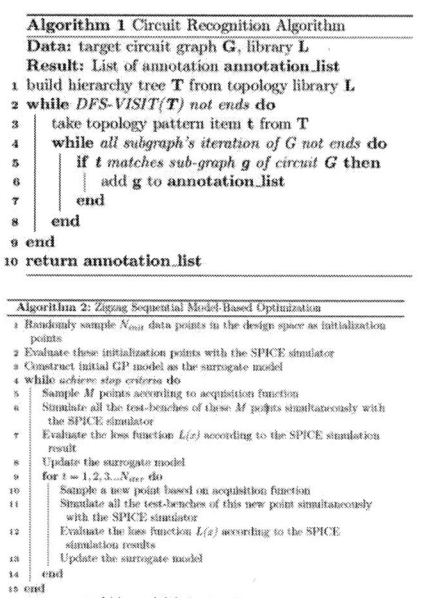

Figure 1(a) The proposed workflow of sequential model-based optimization equipped with circuit hierarchy recognition module. (b) Circuit Pattern Library with selected circuit pattern and their corresponding device relationships.

Algorithm 1 Circuit Recognition Algorithm

Data: target circuit graph **G**, library **L**
Result: List of annotation annotation_list
1 build hierarchy tree **T** from topology library **L**
2 **while** *DFS-VISIT(T) not ends* **do**
3   take topology pattern item t from **T**
4   **while** *all subgraph's iteration of G not ends* **do**
5   **if** *t matches sub-graph g of circuit G* **then**
6   add g to **annotation_list**
7   **end**
8   **end**
9 **end**
10 **return annotation_list**

Algorithm 2: Zigzag Sequential Model-Based Optimization

1 Randomly sample N_{init} data points in the design space as initialization points
2 Evaluate these initialization points with the SPICE simulator
3 Construct initial GP model as the surrogate model
4 **while** *achieve step criteria* **do**
5   Sample M points according to acquisition function
6   Simulate all the test-benches of these M points simultaneously with the SPICE simulator
7   Evaluate the loss function $L(x)$ according to the SPICE simulation result
8   Update the surrogate model
9   **for** $t = 1, 2, 3, ... N_{iter}$ **do**
10   Sample a new point based on acquisition function
11   Simulate all the test-benches of this new point simultaneously with the SPICE simulator
12   Evaluate the loss function $L(x)$ according to the SPICE simulation results
13   Update the surrogate model
14   **end**
15 **end**
16 **return** $argmin\ L(x)$ recorded during iterations

---

iterative optimization procedure. The optimization module is tasked with exploring the design space by assigning values of new points to the design variables, while the SPICE simulator is responsible for evaluating these assigned values. The details about algorithm in the circuit recognition module and optimization module is explained as follows:

C. Circuit Recognition Algorithm

The circuit recognition method is described in Algorithm 1. In the first step, a hierarchy tree T is constructed according to the circuit pattern library. Then, T is traversed using a depth-first search method (DFS-VIST) to extract the pattern t [7]. Following this, a subgraph search is performed to check whether t exists within the entire schematic graph G. The search process is done by isomorphic graph check. If g and t are isomorphic pair, g will be annotated according to t and appended to the resulting annotation list. Finally, subcircuits with circuit pattern annotations and transistor locations are returned. This list of annotated subcircuits serves as the basis for generating a new netlist with fewer design variables.

D. Zigzag Sequential Mode-Based Optimization

The proposed zigzag sequential model-based optimization, implemented in the optimization module, is described in Algorithm 2. It begins by randomly sampling N_{init} initial data points within design space. These generated data points are then evaluated using a SPICE simulator to construct the initial Gaussian Process (GP) surrogate model.

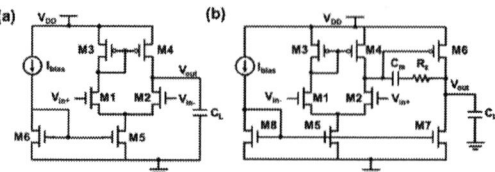

Figure 2. Schematic of 5-Transistor Amplifier and Two-stage Operational Amplifier.

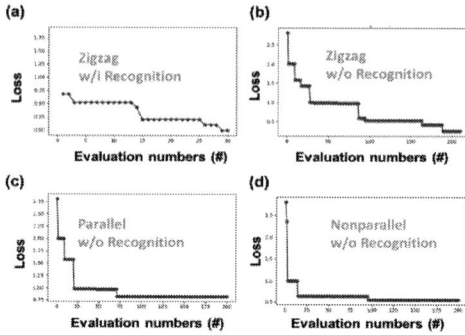

Figure 3. Optimization convergence of 5-T Amplifier through different sequential model-based optimizations with and without circuit recognition module.

After that, it follows a nested iteration loop. During each iteration in the outer loop, it consists of a parallel stage and a sequential stage, namely the inner iteration loop. In the parallel stage, the multi-point Expected Improvement (qEI) is employed as acquisition function to select M data points simultaneously [8]. Next, these M points are passed to SPICE simulator for cost function evaluation and training the GP model. To minimize the overall simulation time, all testbenches of these M points are simulated concurrently. Then, it comes the sequential iteration loop. At each iteration, a data point is selected each time based on Expected Improvement (EI) and passed to SPICE simulator for updating the GP model [9]. This loop will end until the maximal iterations number N_{iter} reached.

III. EXPERIMENTAL RESULTS

In this section, we demonstrate the efficiency of the proposed approach utilizing two CMOS amplifier circuits which designed on a 0.18-μm CMOS foundry process with 3.3 V supply voltage and 2.0 pF load capacitance. The circuit simulations were conducted utilizing NgSpice simulator.

A. Single-Stage 5-Transistor Amplifier

We firstly evaluated our method using a 5-transistor amplifier as described in Fig. 2(a). This design has 13 design variables, including the lengths and widths of transistors and bias current I_{bias}. The cost function $L(x)$ are listed as follows:

$$\text{minimize} \quad L(x) = -GAIN - GBW - PM - SR \quad (4)$$

In this test circuit, we run sequential model-based optimization with and the without circuit recognition module in zigzag, parallel and nonparallel mode. The initialization sample number N_{init} and maximal evaluation N_{max} is set to 10 and 200, respectively. The convergence behavior for the zigzag, parallel and nonparallel runs with and without circuit recognition module are given in Fig. 3. Compared with parallel and nonparallel mode, the zigzag mode has the fastest

979-8-3503-6184-1/24 $31.00 © 2024 IEEE

TABLE II. Recognition result of 5-T Amplifier.

Topology Pattern	Recognized Transistor Group	Transistor Relationship
Differential Stage NMOS	M1, M2, M6, M5	W1=W2, L1=L2, W5=W6, L5=L6
Differential Pair NMOS	M1, M2	W1=W2, L1=L2,
Current Mirror NMOS	M6, M5	W5=W6, L5=L6
Current Mirror PMOS	M3, M4	W3=W4, L3=L4

TABLE III. Optimization result of 5-T Amplifier.

Specification	Target	Case A	Case B	Case C	Case D
Method		Zigzag w/i Reconginition	Zigzag w/o Reconginition	Parallel w/o Reconginition	Nonparallel w/o Reconginition
Technology	180 nm	180 nm	180 nm	180 nm	180 nm
Supply voltage [V]	3.3 V	3.3 V	3.3 V	3.3 V	3.3 V
Load capacitance [pF]	2.0	2.0	2.0	2.0	2.0
DC gain [dB]	50.0	50.8	48.3	35.1	51.7
Slew rate [V/us]	10.0	19.2	21.0	127.3	58.7
GBW [MHz]	30.0	32.6	27.6	165.0	29.3
Phase margin [deg]	60	78.7	83.8	84.3	82.3
Number of unknowns		7	13	13	13
Number of evaluations		30	200	200	200
Pass score		100.0	50.0	75.0	75.0

convergence rate. However, without the help of circuit recognition module, all of these modes fail to achieve pass score of 100, which defines as the ratio of the number of achieved metrics to the total number of the metrics, within 200 evaluations. This is primarily attributed to the presence of numerous design variables, resulting in a vast and challenging design space to explore. To reduce design space, circuit recognition module is utilized before optimization. The circuit recognition result of 5-T amplifier is illustrated in TABLE II. With the help of circuit recognition, the number of design variables is reduced from 13 to 7, leading to a significant improvement, as all the design specifications can now be achieved within a mere 30 evaluations.

B. Two-Stage Operational Amplifier

The second test circuit of the proposed method is a two-stage operational amplifier as shown in Fig. 2(b). It has 19 design variables, including lengths and widths of transistors, bias current I_{bias}, Miller capacitor C_m and zero compensation resistor R_z. The cost function $L(x)$ are listed as follows:

$$\text{minimize} \quad L(x) = -GAIN - GBW - PM - SR \quad (5)$$

In the case of two-stage operational amplifier, we compare the performance of zigzag sequential model-based optimization with and without the circuit recognition module. The N_{init}, N_{max} are set as 10 and 500, respectively. Fig. 4 illustrates the convergence behavior of the zigzag sequential model-based optimization with and the without circuit recognition. The circuit recognition result is illustrated in TABLE IV. The inclusion of the circuit recognition module reduces design variables from 19 to 13, leading to a faster convergence. Consequently, the optimization process is able to satisfy all the design specifications within 350 evaluations.

IV. CONCLUSION

In this paper, we introduce a new approach for analog circuit sizing called zigzag sequential model-based optimization with circuit recognition. We apply this method to the design of two commonly used analog circuits to demonstrate its effectiveness. By utilizing the proposed method, we successfully optimize a 5-T amplifier and a two-stage operational amplifier, achieving all the design specifications within 30 and 350 evaluations, respectively. These results indicate the potential of zigzag sequential

TABLE IV. Recognition result of Two-stage Operational Amplifier

Topology Pattern	Recognized Transistor Group	Transistor Relationship
Differential Stage NMOS	M1, M2, M8, M5	W1=W2, L1=L2, W5=W8, L5=L8
Differential Pair NMOS	M1, M2	W1=W2, L1=L2,
Current Mirror Array NMOS	M8, M5, M7	
Current Mirror NMOS	M8, M5,	W5=W8, L5=L8
Current Mirror PMOS	M3, M4	W3=W4, L3=L4

TABLE V. Optimization result of Two-stage Operational Amplifier

Specification	Target	Case A	Case B
Method		Zigzag w/i Reconginition	Zigzag w/o Reconginition
Technology	180 nm	180 nm	180 nm
Supply voltage [V]	3.3 V	3.3 V	3.3 V
Load capacitance [pF]	2.0	2.0	2.0
Cm [pF]	0.1-3.0	2.91	0.96
Rz [kohm]	0.1-100	13.9	59.1
Ibias [uA]	5-50	50.0	17.0
DC gain [dB]	60.0	66.3	7.9
Slew rate [V/us]	10.0	68.4	52.7
GBW [MHz]	30.0	49.3	25.5
Phase margin [deg]	60	68.4	120.6
Number of evaluations		350	>500
Number of unknowns		13	19
Pass score		100	50

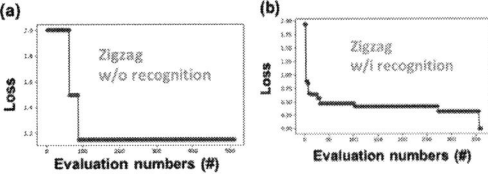

Figure 4. The convergence behavior of Two-stage Operational Amplifier with and the without circuit recognition.

model-based optimization with circuit recognition as a promising method to achieve device sizing automation for analog circuits in the future.

REFERENCES

[1] A. Girardi, T. De-Oliveira and S. Ghissoni, "A comprehensive review on automation-based sizing techniques for analog IC design," in *Journal of Integrated Circuits and Systems*, vol. 17(3), pp. 1-14, 2022.

[2] R. Martins, N. Lourenço and N. Horta., Analog Integrated Circuit Design Automation. Cham, Switzerland: Springer, 2017.

[3] Y. Wang, M. Orshansky, C. Caramanis, "Enabling efficient analog synthesis by coupling sparse regression and polynomial optimization," in *IEEE Proceedings of the 51st Annual Design Automation Conference*, pp. 1-6, 2014.

[4] A. Sayed, A. Mohieldin, M. Mahroos, "A fast and accurate geometric programming technique for analog circuits sizing," in *31st International Conference on Microelectronics (ICM) IEEE*, pp. 316-319, 2019.

[5] B. Manuel, J. Guilherme, and N. Horta, "Analog circuits optimization based on evolutionary computation techniques," in *Integration*, vol. 43(1), pp. 136-155, 2011.

[6] S. Mallick, R. Kar, D. Mandal, and S. P. Ghoshal, "Optimal sizing of CMOS analog circuits using gravitational search algorithm with particle swarm optimization," in *International Journal of Machine Learning and Cybernetics*, vol. 8, pp. 309-331, 2017.

[7] F. E. Sandnes and O. Sinnen, "Stochastic DFS for multiprocessor scheduling of cyclic taskgraphs," in *International Conference on Parallel and Distributed Computing: Applications and Technologies (LCPDC) 2004*, pp. 354-362, 2004.

[8] C. Chevalier and D. Ginsbourger, "Fast computation of the multi-points expected improvement with applications in batch selection," in *International Conference on Learning and Intelligent Optimization (LION) 2013*, pp. 59-69, 2013.

[9] R. Astudillo and P. Frazier, "Bayesian optimization of composite functions," in *International Conference on Machine Learning (PMLR) 2019*, pp. 354-363, 2019.

Vanadium Oxide-Based Artificial Synapses for Construction of Artificial Neural System

Zhuoling Zhou [1], Libin Liang [1], Hongzhi Chen [2], Changjiu Teng*[1], Shilong Zhao*[1], Wenjun Chen*[1]

[1] School of Electronic Information Engineering, Foshan University, Foshan, 528000, China
[2] School of Mechatronic Engineering and Automation, Foshan University, Foshan, 528000, China

Email: first_author@email.address_zhuoling_zhou@163.com
corresponding_author@email.address_tengchangjiu@sina.com, shilongzhao@fosu.edu.cn, chenwj46@fosu.edu.cn

Abstract-To construct an artificial neural system, three types of devices are required: volatile synapses, non-volatile synapses, and volatile neurons. For process simplification, finding one appropriate material that can achieve all three functionalities is significant , but such a material has yet to be reported. VO_2 as a special memristive material, has several unique properties for neuromorphic devices, including Mott insulator transition, high temperature coefficient of resistance and controllability of grain boundaries. Therefore, researchers usually focus on its volatile resistive switch used as an artificial neuron. In this study, we fabricated an volatile artificial synapse with a Ag/VO_x/FTO sandwich-like architecture. These devices exhibit continuous I-V potentiation and depression with overall enhancement exceeding 200% and depression around 500% Additionally, they emulate synaptic plasticity such as Short-Term Depression (STD) and Short-Term Potentiation (STP), complementary with VO_x neurons and paving the way for the development of an all-VO_x artificial neural system.

Keywords—Artificial synapse, VO_x materials, Resistive switching, Synaptic plasticity

I. INTRODUCTION

An artificial neural system requires three types of devices: non-volatile synapses, volatile synapses, and volatile neurons[1]. For process simplification, discovering one kind of material to realize the three functions is urgently required. Memristors as the most likely candidate for realization of artificial neural units simulation, were proposed by Professor Leon Chua proposed the theory of the memristor in 1971[2] and realized by HP Labs using TiO_x in 2008[3]. In biological brain, neurons and synapses coordinate to store and transmit information. Synapses, linking neurons involved in learning and memory, enhance information integration and transmission within neural networks[4, 5]. Presynaptic neurons respond to stimuli, prompting postsynaptic neurons to release neurotransmitters, which bind to receptors and trigger synaptic enhancement or suppression responses (Figure 1a).Vanadium, as an adjacent element to titanium in the periodic table, exhibits multiple valence states and possesses characteristics such as insulator-metal transition (IMT)[6], controllable amorphous or crystalline growth (controllable grain boundaries)[7], and high temperature coefficient of resistance (TCR)[8]. These properties make it a potential candidate for both first-order artificial neuron devices and higher-order memristor devices. Wang et al. utilized laser direct writing technology to rapidly and single-step synthesize and pattern vanadium disulfide (V_5S_8) at room temperature, successfully converting V_5S_8 to VO_2 within seconds, and applied this method to construct heterostructure devices based on Mott memristors[8]. Brain et al. achieved rapid response and information storage by laser-induced local phase transition of VO_2, demonstrating various neuromorphic functions such as binary threshold, memory effect, and spike encoding[9]. Deng et al. demonstrated the potential of this material system in neural network hardware by achieving monolithic integration of VO_2 volatile neurons and non-volatile synapses on the same substrate through selective regional hydrogen doping technology[10]. Here, we report volatile synaptic devices based on VO_x(Figure 1b). The Ag/VO_x/FTO devices fabricated using magnetron sputtering can simulate synaptic inhibition and potentiation, as well as synaptic plasticity like STD and STP. These devices are expected to complement previously reported neuron devices(Figure 1c), forming an all-VO_x artificial neural system and contributing to memristor integration and process compatibility.

VO_x, as a Mott-type metal oxide device, primarily relies on its IMT for its memristive effects. At approximately 68° C (341K), VO_x transitions from a monoclinic insulator phase to a rutile metallic phase, significantly reducing its resistance[10]. The excellent conductivity of Ag electrodes and the temperature resistance of FTO provide stable electrical contact.

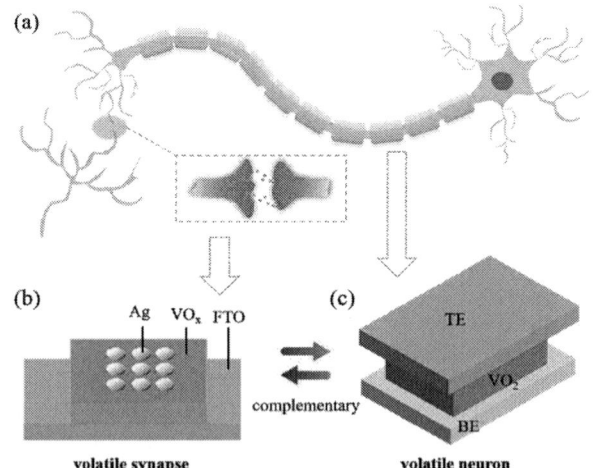

Fig. 1.Schematic diagram of synaptic structure and Ag/VO_x/FTO memristor device structure.a) Schematic diagram of synapse-neuron interactions.b) Schematic diagram of the Ag/VO_x/FTO device structure.c)Volatile VO_2-Based Neuron Schematic Diagram.

II. EXPERIMENT

We employed a JCP350 magnetron sputtering system from Beijing Technol Science Co., Ltd. to deposit centimeter-scale VO_x thin films onto FTO substrates serving as the bottom electrode. The process operated at a power of 100 watts with a gas ratio of 19.5:0.5 for argon to oxygen,

979-8-3503-6184-1/24 $31.00 © 2024 IEEE

under a working pressure of 0.50 Pa, at room temperature (25°C), and a sputtering duration of 10 minutes. Subsequently, conductive silver paste was applied as the top electrode.The device's electrical properties, including the I-V characteristic curves and pulse measurements, were analyzed using a semiconductor analyzer, model Primariu FS-Pro. The negative terminal was connected to the FTO, and the positive terminal was connected to the Ag electrode. To prevent overheating of the device, the current was limited to 0.1A. Pulse measurements were conducted in the I-V-T format, where pulse width and intervals were adjusted by setting the excitation time and intervals.

III. RESULTS AND DISCUSSION

The negative bias sweep I-V characteristics of the device are shown in Figure 2a, with a set voltage of -0.5V for 15 consecutive cycles. As the number of sweeps increases, the current of the device gradually increases, indicating an increase in conductance, simulating the behavior of synaptic potentiation[11].Subsequently, a positive bias sweep was performed on the device as shown in Figure 2b, with a set voltage of +0.5V, also for 15 consecutive cycles. The current decreases with the number of sweeps increases, simulating the behavior of synaptic depression[11]. The conductance under potentiation (-0.4V) and depression (+0.4V) bias sweep states was plotted into a scatter plot, clearly showing the trend of conductance changes. Subsequently, continuous sweeping was performed under ± 0.8V bias (0V → -0.8V → 0V → 0.8V → 0V) as shown in Figure 2d. The device transitions from HRS to LRS under negative bias, and similarly from HRS to LRS under positive bias, exhibiting a distinct unipolar memory window with an on/off ratio of approximately 16.

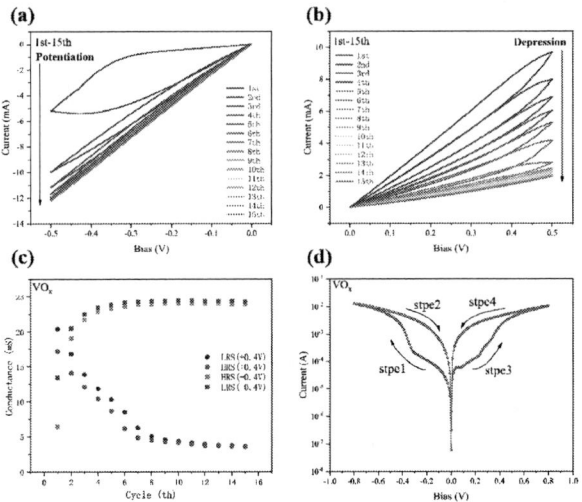

Fig. 2. I-V characteristics of the Ag/VO$_x$/FTO device. a, b) The I-V characteristics under 15 cycles of sweeping. c) Conductance of the device under ±0.4V bias. d) The I-V characteristics under ±0.8V scanning.

Next, we investigated the synaptic plasticity of the device under pulsed signals (Figure 3). We applied a series of 10 positive pulses with an amplitude of +0.5V, with both the pulse interval and pulse width set to 100 microseconds. It was observed that the current gradually decreased with the number of pulses applied. Similarly, we applied a series of 10 negative pulses with an amplitude of -0.7V, using the same parameters. In this case, the current gradually increased

with the number of pulses applied (Figure 3a). The average conductance corresponding to each pulse was calculated to observe its variation (Figure 3c), showing a decrease and an increase, respectively. This result simulates the STD and STP behaviors of biological synaptic plasticity[12]. We altered the pulse width, increasing it to 500 microseconds while keeping other parameters unchanged. An increase in current was observed under both positive and negative pulses, with an amplitude increase of approximately 24.7% (Figure 3b). The synaptic weight change was calculated for the first and second pulses of both positive and negative pulses. For the 100-microsecond pulses, the weight change rates for positive and negative pulses were -59.4% and 133%, respectively. For the 500-microsecond pulses, the weight change rates for positive and negative pulses were -63.9% and 48.6% (Figure 3d), indicating that the device possesses the characteristic of adjusting nonlinear conductance and its weight by modifying the pulse width.

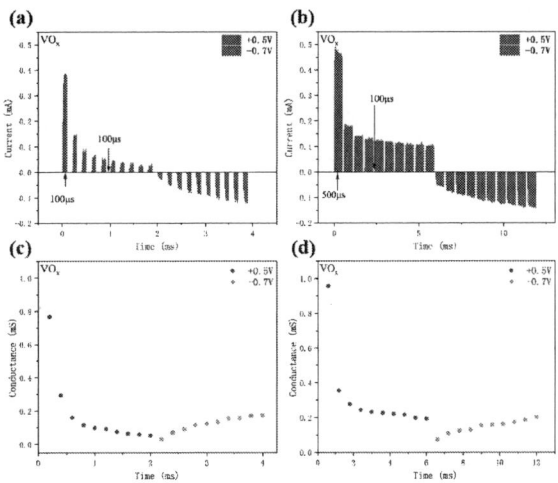

Fig. 3. Synaptic behavior of the Ag/VO$_x$/FTO memristor under pulsed voltage stimulation. a, b) Pulse characteristics of the device under different pulse widths. c, d) Scatter plots of nonlinear conductance of the device corresponding to different pulse widths.

The resistive switching mechanism of our fabricated Ag/VO$_x$/FTO devices is attributed to the combined effects of the IMT transition and redox reactions.The device remains in a monoclinic phase without applied bias, corresponding to the HRS (Figure 4a). Upon applying a bias voltage, the local temperature increases, causing an IMT phase transition in the resistive layer. VO$_x$ transforms into the rutile phase. Concurrently, silver loses electrons to form Ag^+ and is gradually reduced at the cathode back to metallic Ag atoms. Oxygen vacancies (V$_o$) also migrate towards the cathode, forming a conductive pathway that lowers the resistance, reaching the LRS (Figure 4b). When the bias direction reverses, the generated heat decreases accordingly. The rutile phase gradually reverts to the monoclinic phase, Ag is re-ionized to Ag^+, oxygen vacancies revert to oxygen atoms, and ultimately, the conductive pathway disappears, returning to HRS (Figure 4c, d).

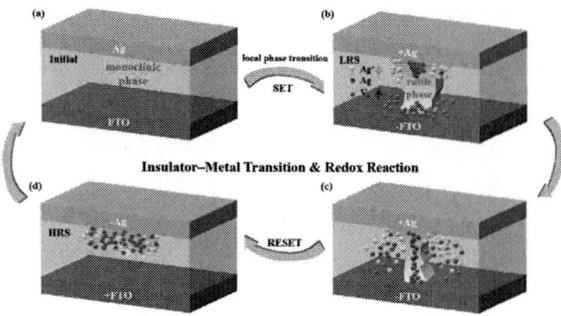

Fig. 4. Ag/VO$_x$/FTO Schematic diagram of the insulator-metal transition principle & redox reaction of the device.

IV. SUMMARY

In summary, we fabricated an artificial synaptic memristor with an Ag/VO$_x$/FTO structure using a magnetron sputtering process. The device exhibited inhibition and potentiation behavior under forward and reverse voltage sweeps, respectively, demonstrating significant bidirectional conductance changes. Under pulsed conditions, it successfully emulated biological synaptic behaviors such as STD and STP. This study holds promise for advancements in memristor integration and process compatibility and is expected to complement previously reported neuron devices, collectively contributing to the construction of a complete all-VO$_x$ artificial neural system.

ACKNOWLEDGMENT

The authors acknowledge the supports by the Guangdong Basic and Applied Basic Research Foundation (Nos. 2021A1515110980, 2022A1515140158 and 2023A1515110759,), the National Natural Science Foundation of China (Nos. 12304212), Foshan University School of Physics and Optoelectronics, Laboratories 116-120 and School of Electronic Information Engineering, Laboratories 212-214.

REFERENCES

[1] He, K., Wang, C., He, Y., Su, J., and Chen, X.: 'Artificial Neuron Devices', Chem Rev, 2023, 123, (23), pp. 13796-13865

[2] Chua, L.: 'Memristor-The missing circuit element', IEEE Transactions on Circuit Theory, 1971, 18, (5), pp. 507-519

[3] Strukov, D.B., Snider, G.S., Stewart, D.R., and Williams, R.S.: 'The missing memristor found (vol 453, pg 80, 2008)', Nature, 2009, 459, (7250)

[4] Zhou, Z.Y., Yan, X.B., Zhao, J.H., Lu, C., Ren, D.L., Lu, N.D., Wang, J.J., Zhang, L., Li, X.Y., Wang, H., and Zhao, M.L.: 'Synapse behavior characterization and physics mechanism of a

TiN/SiOx/p-Si tunneling memristor device', Journal of Materials Chemistry C, 2019, 7, (6), pp. 1561-1567

[5] Yan, X., Li, X., Zhou, Z., Zhao, J., Wang, H., Wang, J., Zhang, L., Ren, D., Zhang, X., Chen, J., Lu, C., Zhou, P., and Liu, Q.: 'Flexible Transparent Organic Artificial Synapse Based on the Tungsten/Egg Albumen/Indium Tin Oxide/Polyethylene Terephthalate Memristor', ACS Appl Mater Interfaces, 2019, 11, (20), pp. 18654-18661

[6] Yuan, R., Duan, Q., Tiw, P.J., Li, G., Xiao, Z., Jing, Z., Yang, K., Liu, C., Ge, C., Huang, R., and Yang, Y.: 'A calibratable sensory neuron based on epitaxial VO$_2$ for spike-based neuromorphic multisensory system', Nature Communications, 2022, 13, (1)

[7] Li, Z., Zhang, Z., and Zhou, X.: 'Chemical Modulation of Metal－Insulator Transition toward Multifunctional Applications in Vanadium Dioxide Nanostructures', Small, 2023, 19, (44)

[8] Wang, B., Peng, R., Wang, X., Yang, Y., Wang, E., Xin, Z., Sun, Y., Li, C., Wu, Y., Wei, J., Sun, J., and Liu, K.: 'Ultrafast, Kinetically Limited, Ambient Synthesis of Vanadium Dioxides through Laser Direct Writing on Ultrathin Chalcogenide Matrix', ACS Nano, 2021, 15, (6), pp. 10502-10513

[9] Blankenship, B.W., Li, R., Guo, R., Zhao, N., Shin, J., Yang, R., Ko, S.H., Wu, J., Rho, Y., and Grigoropoulos, C.: 'Photothermally Activated Artificial Neuromorphic Synapses', Nano Lett, 2023, 23, (19), pp. 9020-9025

[10] Deng, S.B., Yu, H.M., Park, T.J., Islam, A., Manna, S., Pofelski, A., Wang, Q., Zhu, Y.M., Sankaranarayanan, S., Sengupta, A., and Ramanathan, S.: 'Selective area doping for Mott neuromorphic electronics', Science Advances, 2023, 9, (11)

[11] Tang, L., Teng, C., Xu, R., Zhang, Z., Khan, U., Zhang, R., Luo, Y., Nong, H., Liu, B., and Cheng, H.M.: 'Controlled Growth of Wafer-Scale Transition Metal Dichalcogenides with a Vertical Composition Gradient for Artificial Synapses with High Linearity', ACS Nano, 2022, 16, (8), pp. 12318-12327

[12] Huang, J., Yang, S., Tang, X., Yang, L., Chen, W., Chen, Z., Li, X., Zeng, Z., Tang, Z., and Gui, X.: 'Flexible, Transparent, and Wafer-Scale Artificial Synapse Array Based on TiO(x) /Ti(3) C(2) T(x) Film for Neuromorphic Computing', Adv Mater, 2023, 35, (33), pp. e2303737

High performance FeFET with α-IGZO Channel Enabled by Atomic-Layer-Deposited HfO₂ Interfacial Layer

Yinchi Liu[1,2], Hao Zhang[1], Xinlong Zhou[1], Dmitriy Anatolyevich Golosov[3], Chenjie Gu[4], Hongliang Lu[1], Shijin Ding[1], and Wenjun Liu*[1,2,5]

[1]School of Microelectronics, Fudan University, Shanghai 200433, China; [2]Research Institute of Fudan University in Ningbo, Ningbo 315300, China; [3]Belarusian State University of Informatics and Radioelectronics, Minsk 20013, Republic of Belarus; [4]Department of Microelectronics, Ningbo University, Ningbo 315200, China; [5]Zhangjiang Fudan International Innovation Center, Shanghai China *Email: wjliu@fudan.edu.cn

Abstract—The ferroelectric filed-effect transistor (FeFET) with amorphous Indium-Gallium-Zinc-Oxide (α-IGZO) channel and atomic-layer-deposited 2 nm HfO₂ interfacial layer (IL) was designed and fabricated for optimizing both the memory window (MW) and reliability. Compared to the FeFET without IL, an improved reliable operation voltage and enhanced MW of 1.69 V at the reliable operating bias was achieved. Additionally, a ~1000x improvement of endurance was demonstrated after exerting pulses around 10^7 cycles without hard breakdown, while maintaining retention > 10 years. This work proposes an effective strategy to enhance the MW and reliability for future non-volatile memory applications.

Keywords—atomic layer deposited, interfacial layer, *Hf₀.₅Zr₀.₅O₂, memory window, reliability.*

I. INTRODUCTION

Ferroelectric field-effect transistors (FeFETs) based on metal oxide channel materials, such as amorphous Indium-Gallium-Zinc-Oxide (α-IGZO), Indium Oxide (In₂O₃), W-doped In₂O₃, have been considered as promising candidates for prospective applications in monolithic 3D integration toward large scale memory [1]-[3]. This is motivated in part by the low power consumption, high writing speed, complementary metal-oxide-semiconductor (CMOS) technology compatibility as well as outstanding scalability of HfO₂-based FeFET. Unfortunately, the application of FeFET is still restricted by its narrow memory window (MW) and poor endurance, which are not only associated with ferroelectric layer itself but also linked to the interface between the channel and gate insulator [4].

Significant efforts and progresses have been made to optimize memory properties of FeFETs, such as interfacial layer (IL) thickness reduction via oxygen scavenging, IL-free gate stack [5]-[6]. However, a large gate voltage drop across the low-*k* IL, an inferior MW in IL-free gate structure are unexpectedly obtained. Comparatively, combining the metal sacrificial layer with high-*k* IL in oxide FeFET could be an effective way for the device property enhancement. Nevertheless, this approach has not been extensively studied and its mechanism behind is still under debate.

In this work, α-IGZO FeFETs with incorporating 2 nm HfO₂ IL were constructed. The fabricated FeFET with HfO₂ IL reflects a competitive MW of 1.69 V and excellent reliability with continuous loading among 10^7 cycles, and an extrapolated retention of > 10 years was achieved.

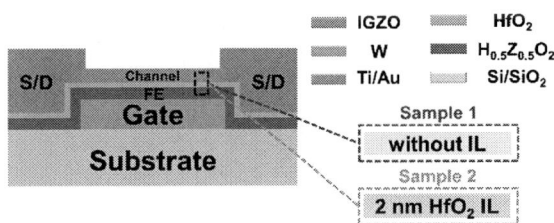

Fig. 1 The device structure of the fabricated FeFET without and with 2 nm HfO₂ interfacial layer.

II. EXPERIMENT

To understand the influence of interfacial layer on the electric characteristics and reliability of FeFET, the FeFETs without and with 2 nm HfO₂ interfacial layer were fabricated and the device structure were shown in Fig. 1. First, a 50 nm tungsten (W) thin film was deposited onto the SiO₂/p-Si substrate using physical vapor deposition (PVD), followed by photolithography and SF₆/CHF₃ inductively coupled plasma (ICP) dry etching for gate isolation. Next, 12 nm HZO was prepared by atomic layer deposition (ALD) at 280 °C with Hf[N(CH₃)₂]₄, Zr[N(CH₃)₂]₄ and O₂ as Hf, Zr and oxygen precursors. Then, the W was deposited onto the thin film as capping layer since it can effectively induce HZO crystallization. After that, the thin films went through rapid thermal annealing (RTA) at 500 °C for 30 s in N₂ atmosphere for the crystallization and then the W capping layer was removed by wet etching. Afterward, 2 nm HfO₂ was then grown by ALD at 250 °C as IL and a control device without HfO₂ IL was also prepared. Subsequently, 8 nm α-IGZO was constructed by RF magnetron sputtering at room temperature and the active channel layer was patterned by photolithography and wet etching (diluted HCl). Finally, photolithography and electron beam evaporation (EBE) were adopted to pattern and form source and drain contacts of Ti/Au (10/50 nm).

III. PERFORMANCE

A. Ferroelectric characteristics

Fig. 2(a) displays the *P-V* characteristics at reliable voltage ranges of metal-ferroelectric-insulator-metal (MFIM)-structure ferroelectric capacitors without and with 2 nm HfO₂ IL. Compared with the control device, larger spontaneous

979-8-3503-6184-1/24 $31.00 © 2024 IEEE

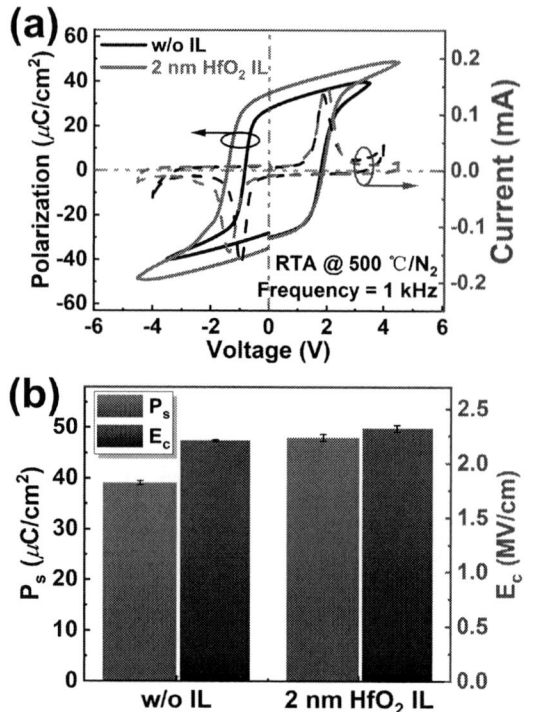

Fig. 2. (a) Polarization and current loops of ferroelectric capacitor based on the MFIM structure without and with 2 nm HfO_2 IL under a reliable operating bias. (b) The P_s and V_c of MFM and MFIM structure.

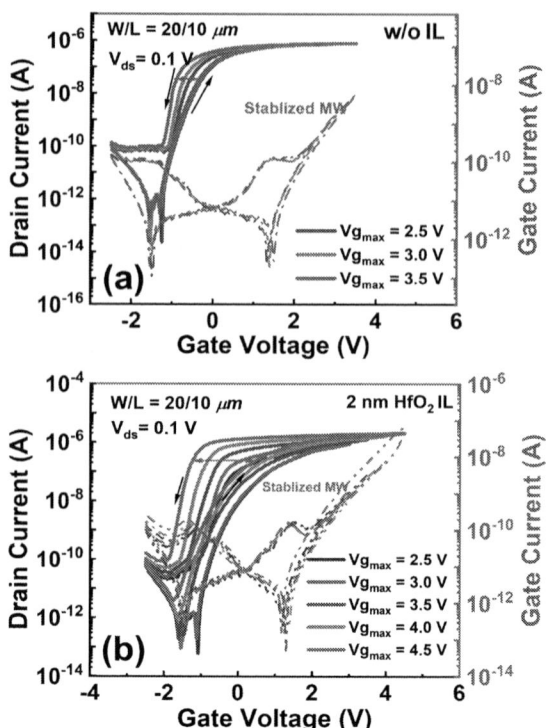

Fig. 3 The I_d-V_g and I_g-V_g curves of FeFETs (a) without and (b) with HfO_2 IL based on the DC measurement.

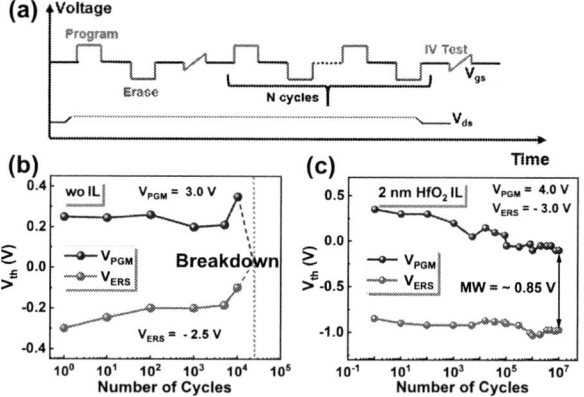

Fig. 4. (a) Pulse sequence for endurance testing used. Evolution of the V_{th} with program and erase cycling for the device (b) without and (c) with HfO_2 IL.

polarization density (P_s) and doubled coercive voltage ($2V_c$) were observed in the capacitor with HfO_2 IL as shown in Fig. 3(b). It is illustrated that, although the HfO_2 IL could induce undesired voltage distribution, it substantially increases the reliable voltage range and enhances the P_s under higher operating bias. Note that, large of P_s and V_c could improve memory properties of FeFET.

B. Memory windows

Fig. 3(a) and (b) display the transfer characteristics and gate current obtained with DC measurement for the FeFETs without and with HfO_2 IL under the drain voltage (V_d) of 0.1 V and different gate voltage (V_g), respectively. The channel width and length of the FeFETs are 20 and 10 μm, respectively. Both the stable FE type counterclockwise hysteresis and negative differential resistance-type behavior were obtained in the two FeFETs under different V_g. Compared to the FeFET without IL, the FeFET with HfO_2 IL shows an enhanced MWs up to 1.69 V under the V_g of 4.5 V. The improvement of MW is primarily attributed to the optimization of the electric field distribution within the gate-stack and effectively suppress the electron trapping through the utilization of HfO_2 IL. In contrast, the FeFET without HfO_2 IL experiences a reduction in on-state current and early dielectric breakdown of below 4.0 V (not shown) due to a decline in interface quality and an increase in trap density caused by etching process damage to the interface.

C. Reliability and retention characteristics

To future explore the influence of HfO_2 IL on the memory properties of α-IGZO FeFET, the endurance characteristics of

FeFETs without and with HfO_2 IL were measured and analyzed. The pulse sequence for endurance testing used are shown in Fig. 4(a). The pulse width of program and erase processes are 10 μs and 50 ms. Fig. 4(b) and Fig. 4(c) display the cycling characteristics of FeFET by loading fatigue cycles. Note that, considering the device has asymmetrical breakdown voltage, different program and erase pulse magnitudes were employed while ensuring complete flipping of the ferroelectric domains. Compared to the control device, the FeFET with HfO_2 IL shows more excellent reliability, lighter degradation of MW, and even breakdown until ~10^7

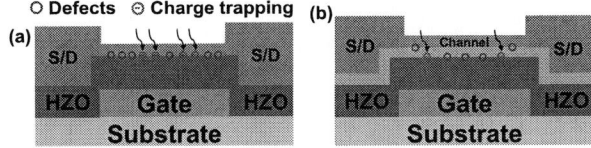

Fig. 5. The summarized schematic of the α-IGZO FeFET (a) without and (b) with HfO$_2$ IL.

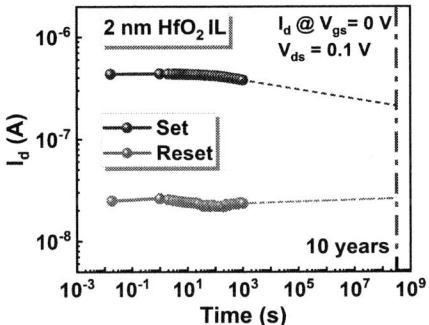

Fig. 6. The summarized schematic of the α-IGZO FeFET (a) without and (b) with HfO$_2$ IL.

cycles. The enhanced reliability is believed to be associated with the lower original defects formed during the process and the charge-injection induced defects during program and erase operation as shown in Fig. 5 (a) and (b) [7]. Finally, the memory properties of FeFET with HfO$_2$ IL were compared with reported oxide FeFETs of similar W_{ch} and L_{ch}, as listed in Table I [8]-[11]. Our fabricated devices in this work exhibit a more considerable MW and superior memory performance in terms of endurance and retention.

Fig. 6 presents the retention characterization of the FeFET with 2 nm HfO$_2$ IL. It is demonstrated that the FeFET with 2 nm HfO$_2$ IL expresses a stable drain current window of ~ 100 after program and erase, and retention of > 10 years is obtained by linear extrapolation.

IV. CONCLUSION

In summary, we have designed and fabricated an enhanced MW and reliability of α-IGZO FeFET with atomic-layer-deposited HfO$_2$ IL. deposited HfO$_2$ IL. By integrating 2 nm HfO$_2$ IL between α-IGZO and HZO dielectric, a larger MW of ~1.1 V is demonstrated under the UFIV method compared to the FeFET without IL, which is competitive to the existing memory devices. Moreover, a superior MW is still maintained after ~10^7 cycles and extrapolated retention of > 10 years is obtained. These findings indicate an alternative method to boost the MW and reliability for FeFETs by interfacial engineering.

TABLE I. BENCHMARK OF THE KEY PERFORMANCES OF THE FEFET

	Key Structure and Performance				
	This Work	*Ref [8]*	*Ref [9]*	*Ref [10]*	*Ref [11]*
Gate Structure	HZO + HfO$_2$	Double gate	HZO + Al$_2$O$_3$	HZO + Al$_2$O$_3$	HZO + Al$_2$O$_3$
Channel material	IGZO	IGZO	ZnO	IZO	IGZO
Channel length	10 μm	10 μm	10 μm	40 μm	10 μm
P/E Condition	+4 V/ -3 V	+5 V/ -3 V	±6 V	±3 V	±8 V
Memory Window	~1.69 V	1.1 V	~1.7 V	~0.8 V	~0.85 V
Endurance	~10^7	*NA*	NA	NA	NA
Retention	10 yrs	10^8 s	10^4 s	NA	10 yrs

REFERENCES

[1] C. -K. Chen, Z. Fang, S. Hooda, M. Lal, U. Chand, Z. Xu, et al, "First demonstration of ultra-low dit top-gated ferroelectric oxide-semiconductor memtransistor with record performance by channel defect self-compensation effect for BEOL-compatible non-volatile logic switch," *in IEDM Tech. Dig.*, Dec. 2022, pp. 114-117.

[2] Z. Lin, M. Si and P. D. Ye, "ultra-fast operation of BEOL-compatible atomic-layer-deposited In$_2$O$_3$ Fe-FETs: cchieving memory performance enhancement with memory window of 2.5 V and high endurance > 10^9 cycles without V$_T$ drift penalty," *in Proc. Symp. VLSI Technol*, Jun. 2022, pp. 391-392.

[3] S. Dutta, H. Ye, W. Chakraborty, Y.-C. Luo, M. San Jose, B. Grisafe, et al, "Monolithic 3D integration of high endurance multi-bit ferroelectric FET for accelerating compute-in-memory," *in IEDM Tech. Dig.*, Dec. 2020, pp. 801-804.

[4] U. Schroeder, M. H. Park, T. Mikolajick, and C. S. Hwang, "The fundamentals and applications of ferroelectric HfO$_2$," *Nature Rev. Mater.*, vol. 7, no. 8, pp. 653-669, Mar. 2022.

[5] B. H. Kim, S. -H. Kuk, S. K. Kim, J. P. Kim, Y. -J. Suh, J. Jeong, et al, "Effect of scandium insertion into the gate-stack of ferroelectric field-effect transistors," *IEEE Trans. Electron Devices*, vol. 70, no. 4, pp. 1996-2000, Apr. 2023.

[6] F. Mo, Y. Tagawa, C. Jin, M. Ahn, T. Saraya, T. Hiramoto, et al, "Low-voltage operating ferroelectric FET with ultrathin IGZO channel for high-density memory application," *IEEE Journal of the Electron Devices Society*, vol. 8, pp. 717-723, Jul. 2020.

[7] Y. Zhou, Z. Liang, W. Luo, M. Yu, R. Zhu, X. Lv, J. Li, Q. Huang, F. Liu, K. Tang, R. Huang, "Ferroelectric and Interlayer Co-optimization with In-depth Analysis for High Endurance FeFET," *in IEDM Tech. Dig.*, Dec. 2022, pp. 118-121.

[8] F. Mo, Y. Tagawa, C. Jin, M. Ahn, T. Saraya, T. Hiramoto, et al, "Low-Voltage Operating Ferroelectric FET with Ultrathin IGZO Channel for High-Density Memory Application," *IEEE Journal of the Electron Devices Society*, vol. 8, pp. 717-723, Jul. 2020.

[9] M. M. Hasan, C. W. Ahn, T. H. kim, and J. Jang, "Solution processed high performance ferroelectric Hf$_{0.5}$Zr$_{0.5}$O$_2$ thin film transistor on glass substrate," *Appl. Phys. Lett.*, vol. 118, no. 15, Apr. 2021, Art. no. 152901.

[10] Y. Li, R. Liang, J. Wang, Y. Zhang, H. Tian, H. Liu, et al, "A Ferroelectric Thin Film Transistor Based on Annealing-Free HfZrO Film," *in IEEE Journal of the Electron Devices Society*, vol. 5, no. 5, pp. 378-383, Sept. 2017.

[11] D. Lehninger, M. Ellinger, T. Ali, S. Li, K. Mertens, M. Lederer, et al, "A Fully Integrated Ferroelectric Thin-Film-Transistor – Influence of Device Scaling on Threshold Voltage Compensation in Displays," *Adv. Electron. Mater.*, vol. 7, no. 6, Jun. 2021, Art. no. 2100082.

A Simulation Study on Cell Scaling Impacts in 3D Charge-trapping (CT) Flash Memory

Wanyu Li[1], Haitao Dong[1], Qianwen Wang[2], Yang Feng[1], Xuepeng Zhan[1], Jixuan Wu[1,*] and Jiezhi Chen[1,*]

[1] School of Information Science and Engineering, Shandong University, Qingdao, China
[2] School of Information Science & Technology, Qingdao University of Science & Technology, Qingdao, China

* Email: jixuanwu@sdu.edu.cn, chen.jiezhi@sdu.edu.cn

Abstract—To provide feasible scaling strategies of 3D NAND flash memory with high operation speed and robust endurance, a systematical study has been conducted on the design of flash cell unit, with a main focus on the impacts of size: the thickness of tunneling layer (T_{TNL}) and charge trapping layer (T_{ct}). It is observed that, as scaling T_{TNL} and T_{ct}, we can obtain a higher programming efficiency, which is a key for low-power and high-speed memory, while the reliabilities like read disturbs and retention degrade, which should be well optimized to meet the needs of read intensive high-capacity memories. By the simulation data, several strategies are discussed to build the fully-flash architecture as an alternative solution for high-efficient data processing.

Keywords—Flash memory, Charge trapping, Reliability

I. INTRODUCTION

3D NAND flash memory is the main stream of non-volatile memory (NVM), providing a cost-effective solution to large capacity memory and has been widely used in mobile, data center, etc. Recently, many impressive progresses of 3D NAND flash memory have been reported, such as 321 stacked storage layers [1], as high as 3.6 GT/s I/O speed in 28.5Gb/mm² 1Tb 3D NAND flash chip [2], and impressive 7-bit/cell bit density [3]. It is believed that 3D NAND will continue to dominate NVM markets in near future. However, due to the limited cost benefit comparing to HDD, the hybrid system of SSD and HDD will continue in a long time; besides, the low endurance (10^3~10^5 cycles) and slow Program/Erase (PE) speed largely limit its applications in storage-class memory (SCM). To tackle the gap between DRAM and Flash memory, there are some solutions by optimizing the designs of circuits and planes, like Z-NAND from Samsung and XL-Flash from Kioxia. [4]. Also, compute express link (CXL) turns to be the mainstream in storage and datacenter, driving an essential evolution to enhance the big data processing efficiency. However, it is still limited work to think about the tradeoff of performances and reliabilities in various kinds of applications in compute and storage. In our previous work, by adopting the method of hot carrier injection, it was shown that PE speed of floating-gate (FG) flash can be enhanced to sub-100ns and the endurance can be improved to over 10^8 cycles [5-6]. Though this, for charge-trapping (CT) flash, charges are stored in the trapping centers in a common CT layer wherein the lateral charge migration is a serious concern, the optimization methods of operation modes are difficult to achieve a large enhancement to PE speed.

In this work, a systematical study has been done on flash memory for design-technology co-optimizations (DTCO). Based on Silicon-oxide-nitride-oxide-silicon (SONOS) cell, we analyzed the effects of tunneling layer (T_{ox}) and charge trapping layer (T_{ct}). Besides, aiming at usage as SCM, some other strategies are also discussed.

II. CELL STRUCTURE AND SIMULATION METHOD

Figure 1 shows the simulated structure of 3D NAND flash memory. In this study, SONOS is simulated and the key parameters are listed in Table 1. The layers from the outside to the inside are control gate (CG), blocking layer (BLK), charge trapping layer (CTL), tunneling oxide layer (TNL), thin poly-Si channel, core-oxide, respectively. The materials and thicknesses of each layer are also denoted. To include the lateral charge migration (LCM) effects that is an important factor that degrades reliabilities in 3D NAND [7]. three connected cells are constructed with a common silicon nitride (SiN) CTL, wherein the middle cell is the target cell in this study and the other are two neighbor cells, all of which have 40nm cell height (L_g), 40nm space region (L_s), and 8nm BLK thickness.

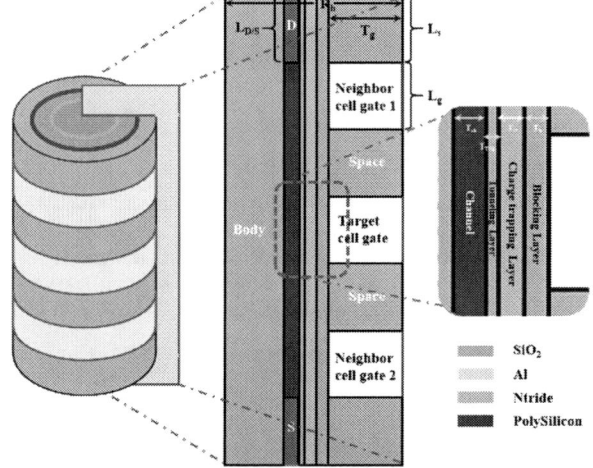

Fig. 1. The cross-sectional view of the NAND string with one target cell and two neighbor cells, the cell is constructed by the standard SONOS structure.

Table I List of simulation parameters

Device parameters	Value
Gate length(L_g)	40nm
Space length	40nm
Gate thickness	50nm
Blocking thickness	8nm
Charge trapping thickness(T_{ct})	7~9nm
Tunneling thickness(T_{TNL})	3~5nm
Channel thickness	10nm

979-8-3503-6184-1/24 $31.00 © 2024 IEEE

Moreover, at the interface of CTL/TNL, defects are set with the densities of 3E19cm^{-3} for donor/acceptor impurity with energy levels of 1.08eV and 2eV that are related to trap centers to store electrons [7-8]. In the simulations, models of Fermi-statistical, Shockley-Read-Hall (SRH), mobility, and Auger recombination are adopted. To investigate the charge diffusions, the models that related to longitudinal diffusion are also added in our simulations: trap-assisted tunneling (TAT) model, direct tunneling model, and FN tunneling model.

III. RESULTS AND DISCUSSIONS

We apply the programming voltage (V_{pgm}) to the target CG and the pass voltage (V_{pass}) to neighbor cells according to the row stripe pattern [9-11]. Incremental step pulse program (ISPP) scheme is adopted for the programming and the retention properties are then studied in our simulations. In Fig.2(a), we set 1ms pulse width (t_{pulse}) in ISPP sequence to the target cell with 4nm-thick TNL (T_{TNL}) and 8nm-thick CTL (T_{ct}). As V_{pgm} increases, more charges will be stored in the CTL, which can be observed from the right-shift threshold-voltage (V_{th}) of the target cell. By adopting the incremental pulse amplitudes in ISPP, the right-shift V_{th} could be compensated by the higher V_{pgm} to ensure the electric field for effective charge injection. Also, in Fig.2(b), we use constant V_{pgm} of 18V but various T_{pulse} to study its I_d-V_g characteristics. It is known that a poorer retention can be observed in high-V_{th} cells because the cell with higher charged electrons is easier to loss its electrons. Thereby, considering that the same ISPP can cause different V_{th} shifts in the cells with various T_{TNL} and T_{ct}, it is necessary to keep the same program levels when we study the impacts of T_{TNL} and T_{ct} in retention.

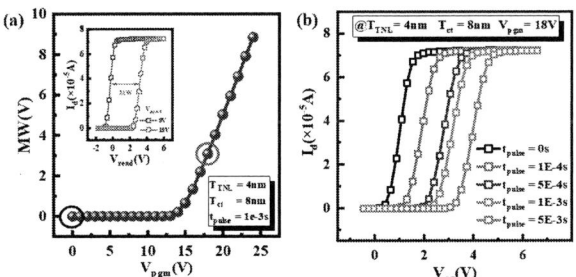

Fig.2. (a) The memory window increases as increasing V_{pgm}, and (b) a larger t_{pulse} can result in a larger V_{th} shift.

In Fig.3, we compare the programming efficiencies in the cells with various T_{ct}, ranging from 7nm to 9nm, while keep the same T_{TNL} of 4nm and t_{pulse} of 1ms. As shown in the figure, as scaling T_{ct}, ISPP slopes are almost identical, indicating that the programming efficiency is stable. However, for the same memory window (MW), we can shrink the pulse number with ~1V lower V_{pgm}, which means less power consumption. As for the retention properties, by keeping the same V_{th} shifts in cells with various T_{ct}, it is observed that scaling T_{ct} could accelerate the charge loss. On the one side, this could be explained by taking the band bending into account; on the other side, more trapped electrons could distribute closer to the space region, which is much easier for lateral migration. Besides, it could be more serious in real cells as scaling T_{ct} because there could be more electron charges locate at the interface between CTL and TNL/BKL. In our previous work, it is known that oxygen-rich region besides the interface can generate more shallow traps under the electrical stress, and shallow traps can contribute to worse charge migration as well as the vertical charge loss.

Next, the impacts of T_{TNL} are studied. As shown in Fig.4, after programming to the same V_{th} of 0.5V in the cells with a fixed T_{ct} of 7nm, it is found that the charge loss degrades obviously in the cell with a thinner T_{TNL}, which could be explained by enhanced vertical charge loss. On the one side, scaling T_{TNL} is a simple way to obtain a better programming speed; on the other side, it is a serious challenge no matter for the read-intensive hot data due to stronger read disturb (RD) and retention-required cold data due to a faster charge loss. To process data reading, V_{cc} is applied to the bit line to ensure that the string conducts, while V_{pass} is applied to word lines (WLs) over unselected cells. Since there is a difference between the cell V_{th} and V_{pass}, on the one side, a small amount of charge migrates laterally between the cell region and the space region; on the other side, vertical charge injection happens and V_{th} of the cell shifts. As shown in Fig.5, we can observe obvious RD degradation in the cell with a thinner T_{ct} or T_{TNL}. Specially, the cells with a lower V_{th} will subject to stronger RD due to the larger electric field at the same V_{pass} applied to un-selected cells. In Fig.6, from the trapped charge perspective, the reason for RD-induced V_{th} shifts can be clearly understood.

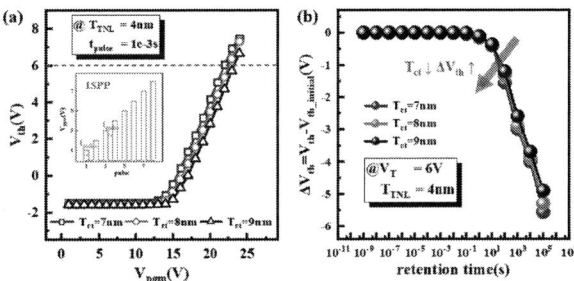

Fig.3. Simulated V_{th} shifts of (a) ISPP program and (b) retention properties in flash cells with various T_{ct} ranging from 7nm to 9nm.

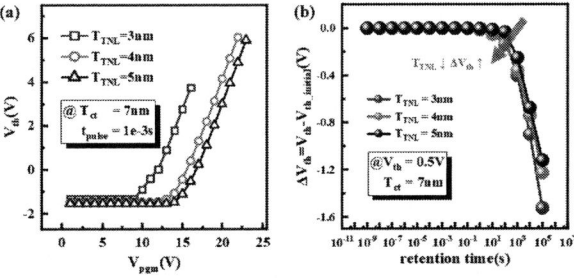

Fig.4. Simulated V_{th} shifts of (a) ISPP program and (b) retention properties in flash cells with various T_{TNL} ranging from 3nm to 5nm.

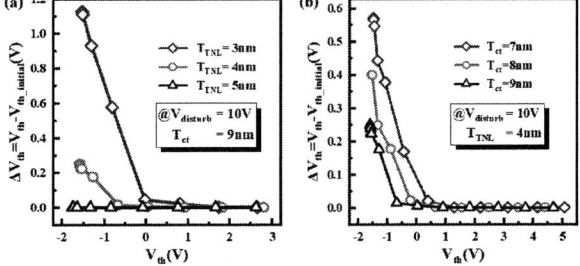

Fig.5 Simulated read disturbs in cells with (a) a fixed T_{ct} of 9nm but various T_{TNL}, (b) a fixed T_{TNL} of 4nm but various T_{ct}.

979-8-3503-6184-1/24 $31.00 © 2024 IEEE   808

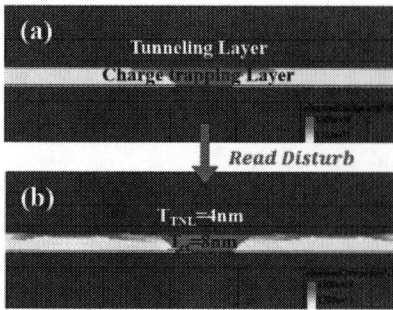

Fig.6. Simulated trapped electron distributions (a) before and (b) after read disturb, which causes the observed V_{th} shifts in the target cell from additional charge injection and re-distribution.

On the basis of our simulation data, to fill the gap between DRAM and large-capacity flash memory, we can design the fast-speed and high-endurance flash dell, while sacrificing the capacity. Nevertheless, we can have serval strategies to do optimizations: Stage I, we can adopt CHEI and HHI methods in NOR array to achieve $10^8 \sim 10^{10}$ high endurance and sub-50ns operation speed; Stage II, we can scaling T_{TNL} and T_{ct} with multiple planes in NAND array to enhance the speed to sub-100ns as the buffer to the next stage; Stage III, CXL combined XL-Flash or Z-NAND can be utilized as random R/W access memory expansion; Stage IV, CXL combined large-capacity 3D NAND flash for read-intensive high bandwidth memory. It deserves to be noted that, by taking the energy efficiency and design flexibility into account, using the same memory but various design could be a more feasible way to assure the reliability in actual applications. Thereby, building fully-flash architectures for flash-native solutions around compute and storage could be the near future.

IV. CONCLUSIONS

A systematical study has been done by focusing on the DTCO strategies of flash cell size to meet the requirements of SCM. We analyzed the impacts of the thickness of TNL and CTL and show the tradeoff between the performance (PE speed and efficiency) and the reliability (disturb and data retention), which are mainly related to the vertical charge loss and lateral charge diffusion. It is believed that flash memory cell also can be well optimized to provide simple and feasible approaches to realize flash-native architectures.

ACKNOWLEDGMENT

This work was supported by China Key Research and Development Program under Grant (Nos. 2023YFB4402500, 2023YFB4402400), National Natural Science Foundation of China (Nos. 62034006, 92264201, U23B2040), and Natural Science Foundation of Shandong Province (ZR2023QF054, ZR2023LZH007, TSQN202306059).

REFERENCES

[1] B. Kim et al., "28.2 A High-Performance 1Tb 3b/Cell 3D-NAND Flash with a 194MB/s Write Throughput on over 300 Layers," 2023 IEEE International Solid-State Circuits Conference (ISSCC), San Francisco, CA, USA, 2023, pp. 27-29;

[2] W. Jung et al., "13.3 A 280-Layer 1Tb 4b/cell 3D-NAND Flash Memory with a 28.5Gb/mm2 Areal Density and a 3.2GB/s High-Speed IO Rate," IEEE International Solid-State Circuits Conference (ISSCC), San Francisco, CA, USA, 2024, pp. 236-237;

[3] H. Tanaka et al., "Toward 7 Bits per Cell: Synergistic Improvement of 3D Flash Memory by Combination of Single-crystal Channel and Cryogenic Operation," 2022 IEEE International Memory Workshop (IMW), Dresden, Germany, 2022, pp. 1-4;

[4] T. Shiozawa, et al., "Emerging Usage and Evaluation of Low Latency FLASH," 2020 IEEE International Memory Workshop (IMW), Dresden, Germany, 2020, pp. 1-4;

[5] Y. Feng, et al., "Design-Technology Co-Optimizations (DTCO) for General-Purpose Computing In-Memory Based on 55nm NOR Flash Technology," 2021 IEEE International Electron Devices Meeting (IEDM), San Francisco, CA, USA, 2021, pp. 12.1.1-12.1.4;

[6] Y. Feng, et al., "A Novel Array Programming Scheme for Large Matrix Processing in Flash-Based Computing-in-Memory (CIM) With Ultrahigh Bit Density," in IEEE Transactions on Electron Devices, vol. 70, no. 2, pp. 461-467, Feb. 2023;

[7] J. Wu, et al., "Comprehensive investigations on charge diffusion physics in SiN-based 3D NAND flash memory through systematical Ab initio calculations," 2017 IEEE International Electron Devices Meeting (IEDM), San Francisco, CA, USA, 2017, pp. 4.5.1-4.5.4;

[8] H.T. Lue, et al., " Charge-Trapping Memories: From the Fundamental Device Physics to 3D Memory Architectures (3D NAND, 3D NOR, 3D DRAM) and Computing in Memory (CIM)," 2023 IEEE International Electron Devices Meeting (IEDM), San Francisco, CA, USA, 2023.

[9] K. Mizoguchi, et al. "Lateral charge migration suppression of 3D-NAND flash by Vth nearing for near data computing." 2017 IEEE International Electron Devices Meeting (IEDM). San Francisco, CA, USA, 2017, pp. 19.2. 1-19.2. 4;

[10] J. Park, et al. "Extraction of Nitride Trap Profile in 3-D NAND Flash Memory Using Intercell Program Pattern," in IEEE Access, vol. 9, pp. 118794-118800, 2021.

[11] B. Choi et al. "Comprehensive evaluation of early retention (fast charge loss within a few seconds) characteristics in tube-type 3-D nand flash memory," in Proc. IEEE Symp. VLSI Technol., 2016, pp. 1-2.

Comprehensive Charaterizations on Read Disturbs in QLC Charge-Trap (CT) 3D NAND Flash

Shaoqi Yang[1], Xiaohuan Zhao[1], Peng Guo[2], Qianwen Wang[3, *], Guangkuo Yang[1], Xinyi Guo[1], Pengpeng Sang[1], Jixuan Wu[1], Xuepeng Zhan[1], and Jiezhi Chen[1, *]

[1]School of Information Science and Engineering, Shandong University, Qingdao, P. R. China; [2]Shandong Sinochip Semiconductors Co., Ltd, Jinan, P. R. China; [3]School of Information Science & Technology, Qingdao University of Science & Technology, Qingdao, P. R. China.

*Email: qw.wang@qust.edu.cn, chen.jiezhi@sdu.edu.cn

Abstract—To address the concern of serious read disturb (RD) in 3D NAND flash with lateral charge migration (LCM), Quad-level-cell (QLC) Charge-trap (CT) type 3D NAND flash memory is characterized systematically on a NAND chip tester, including the properties of block read disturb (BRD) and single- page read disturb (SPRD). More importantly, the correlations between BRD and SPRD are investigated via simple linear equivalence, segmented linear equivalence, and interpolation equivalence. It is found that the method of the interpolated equivalence could be the best choice due to the complex physics in 3D NAND. It is observed that the ratio of SPRD to BRD decreases at the initial stage with a low fail-bit count (FBC), while it converges to a constant after PE cycling with a high FBC. The experiment data and the analysis approach in this work can be referred to design NAND-based storage systems with robust reliabilities.

Keywords—QLC, 3D NAND flash, Read Disturb, Reliability

I. INTRODUCTION

Large-capacity 3D NAND flash memory is the dominant non-volatile memory (NVM) in stand-alone storages and it has been widely used in various electronic devices. However, along with higher bit-density from SLC (1-bit/cell), MLC (2-bit/cell) to TLC (3-bit/cell) and QLC (4-bit/cell), the reliability degradation turns to be worse due to the limited read margins. For read-intensive applications, read disturb (RD) is a critical concern in 3D NAND due to complex mechanisms, including later-charge-migration (LCM), vertical charge re-distribution, electric-field-induced charge injection, and unstable trapped charge emission, etc. [1-4] Thereby, it is strongly required to perform systematical characterizations on RD in high bit-density 3D NAND and develop optimization strategies.

For read-intensive hot data, it is stored in a specific region. Depending on the data size and block size, there are two types of RD, single page read disturb (SPRD) and block read disturb (BRD). SPRD is a special scenario that a single page is read multiple times and this can cause read disturbance in the entire block. According to the controller's internal RD algorithm, the garbage collection mechanism will be activated when the counter for the number of read exceeds a certain threshold. However, a drawback of this processing algorithm is that there exist a significant storage overhead for the counters in the drive. [5] So far, the studies on RD in NAND flash are mostly focused on BRD in MLC and TLC flash memory, while the studies in QLC 3D NAND flash memory are still quite limited.

In this work, a comprehensive characterizations on BRD and SPRD, as well as their correlations, are studied in QLC charge-trap (CT) 3D NAND flash memory. Several methods are used to estimate the equivalence ratios that can be used for system design in solid-state drivers (SSD) [6].

II. CHARACTERIZATION OF SINGLE PAGE READ DISTURB

The NAND chip tested in this work is 112-layer CT-type 3D QLC NAND flash memory. The specific operation setup is as follows: First, bad block screening is performed to exclude bad block interference; second, to get the effect of RD in memory cells with various degradation, some of the target blocks are subjected to different levels of program/erase (P/E) cycling with generated random numbers, a maximum cycle specification of 3K. Then, at various cycling conditions, BRD is characterized by setting the maximum number of read cycles to 3K. As to SPRD, it is performed on fixed pages with a maximum number of 4500K. All operations are done at room temperature (RT). Here, at each node where the data is read, we compare the dumped read data with the original data to analyze fail-bit-count (FBC).

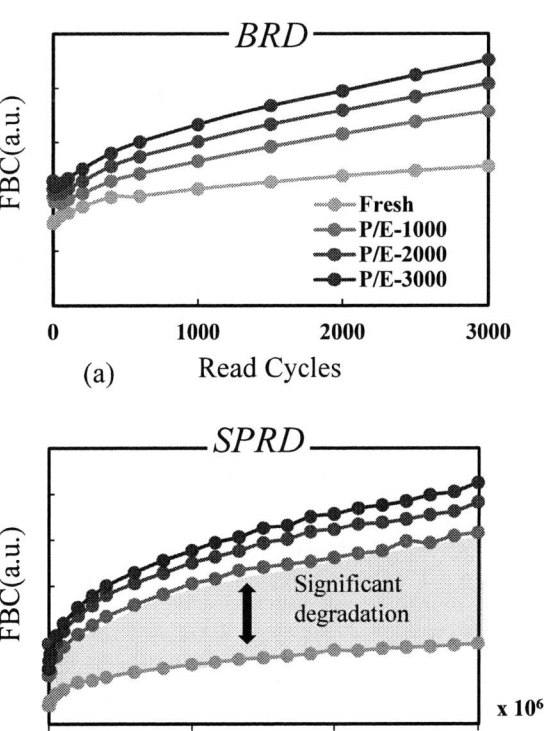

Fig1. Characterization of (a) BRD and (b) SPRD with P/E cycling.

979-8-3503-6184-1/24 $31.00 © 2024 IEEE

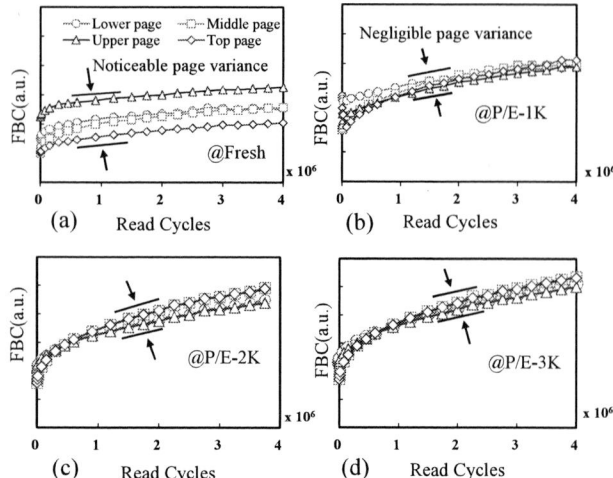

Fig2. Differences between pages in SPRD with incremental P/E cycling: (a) intial stage; (b) P/E-1K; (c) P/E-2K; (d) P/E-3K.

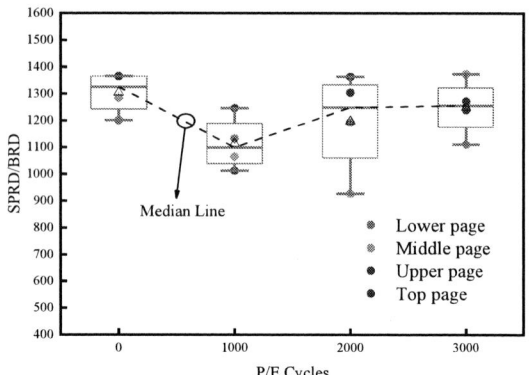

Fig3. Extracted equivalent ratios from the linear fit to the measured data.

The results are shown in Fig. 1, where the disturbances are more severe along with P/E cycling, wherein it is found that SPRD is more sensitive to PE cycling. Fig. 2 shows the results of SPRD at each specific page: lower page, middle page, upper page, and top page. It is observed that the differences between the pages are not significant.

III. CORRELATIONS OF SPRD AND BRD

To estimate the equivalence ratio of BRD and SPRD, three different methods are studied in this work. First, we used a relatively simple linear fit to study above measurement data. Then the ratio of the growth rates of the FBC of BRD and SPRD with the number of readings can be taken as the equivalence ratio. The results are summarized in Fig. 3. With the increase of the degree of cycling between the pages, the equivalent ratios do not have a clear pattern of change. Still, they are roughly around 1200, that is to say, using this equivalent method, 1200 times SPRD is roughly equal to one-time BRD. In addition, we varied the maximum number of BRD readings to 1500, 2500, and 3000. It is found that the equivalent ratios are not stable, as shown in Fig. 4. In detail, the ratio decreases as the maximum number of BRD increases in each category of page. It is considered that this comes from the inaccuracy or over fitting by the simplified method, which can only give a rough approximation.

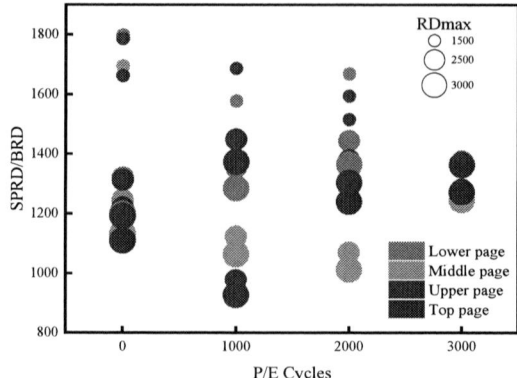

Fig4. Effect of the read cycles of BRD on equivalent ratios on various pages.

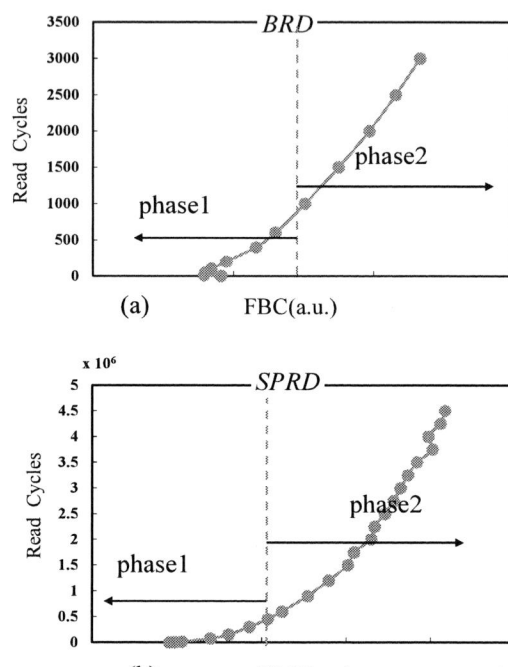

Fig5. Segmented fitting of (a) BRD and (b) SPRD.

Then, another fitting method is proposed. We selected the data after a certain degree of P/E cycling and performed segmental fitting of BRD and SPRD (lower page) respectively (Fig. 5), resulting in Eq. 1 and Eq. 2 ($FBC_1 < FBC_2$). Then, we can calculate the equivalent ratios by Eq. 3.

$$BRD = \begin{cases} a_{11}x + b_{11}, & 0 \leq x < FBC_1 \\ a_{12}x + b_{12}, & FBC_1 \leq x \end{cases} \quad (1)$$

$$SPRD = \begin{cases} a_{21}x + b_{21}, & 0 \leq x < FBC_2 \\ a_{22}x + b_{22}, & FBC_2 \leq x \end{cases} \quad (2)$$

$$SPRD/_{BRD} = \begin{cases} \dfrac{a_{21}}{a_{11}}, & 0 \leq x < FBC_1 \\ \dfrac{a_{21}}{a_{12}}, & FBC_1 \leq x < FBC_2 \\ \dfrac{a_{22}}{a_{12}}, & FBC_2 \leq x \end{cases} \quad (3)$$

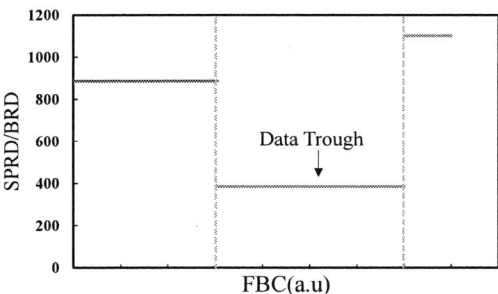

Fig6. Equivalent results for segmented fitting.

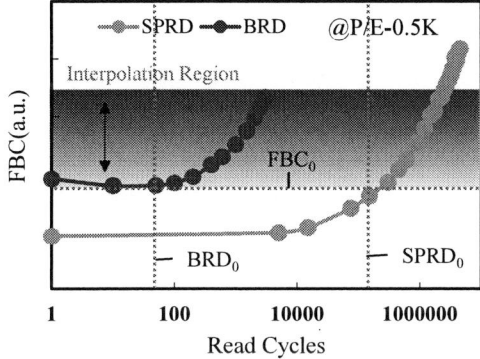

Fig7. Interpolation between BRD and SPRD (lower page) at the same FBC.

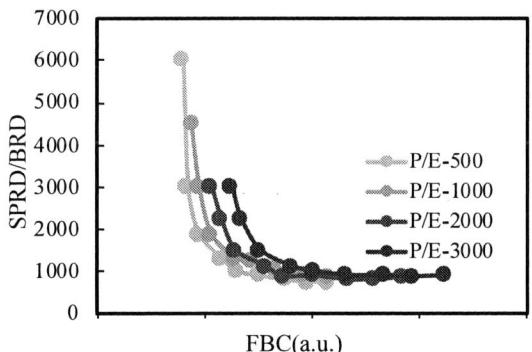

Fig8. Extracted equivalent ratios from the interpolation data in Fig.7.

The results are shown in Fig. 6, staircase-like equivalent ratios can be observed from low FBC to high FBC. It is worth noting that, in comparison to the results of simple linear fitting, these equivalence ratios are all below 1200 (First stage ~900, middle stage ~400, and final stage ~1100). Although the segmented fitting to the basic data is more accurate, it is not easy to be utilized in actual usage since we need to define the critical values of FBC_0, FBC_1 and FBC_2, which depends on the tested chips of various technology nodes.

To find a more suitable method to estimate the correlation between SPRD and BRD as well as its dependence on FBC, direct calculation of the equivalent ratios is performed by interpolation, as shown in Fig. 7. In this way, the estimated ratio of SPRD and BRD are summarized in Fig. 8. Using the interpolation method, the ratio of SPRD to BRD tends to be a constant (~800) at high FBC. It is interesting to notice that

this constant is quite lower than the calculated value from the ratio (~1680) of read times between BRD (15 levels at QLC * 448 WL = 6720 read times in each time of block reading) and SPRD (4 read times for the lower page read at QLC). This could be understood by taking complex effects into account, including LCM and vertical charge re-distribution in read cycling. As reported in our previous work [7], there exist a strong correlation in data retention (DR) and RD during the long-time test. For SPRD characterization, the test time is much longer than BRD, which means that DR in SPRD test has a larger contribution that can accelerate FBC degradation. More importantly, DR degrades with P/E cycling. Thereby, on the one side, our result indicates that SPRD could be much worse than our expectation and should be well considered; on the other side, we can use the proposed method to estimate the approximate ratio of SPRD and BRD and utilize it to simplify the system design.

IV. CONCLUSIONS

In this paper, different methods are utilized to study the correlations between SPRD and BRD in QLC (4-bit/cell) CT 3D NAND flash memory with 112 storage layers. Although the linear fitting method is simple, the accuracy is not good and it could largely depend on technology nodes. By adopting the interpolated equivalence method, it is shown that the ratio of SPRD to BRD decreases with P/E cycling (high FBC) and finally converges to a constant value. This method can be easily utilized for actual reliability test and guide the design of NAND-based storage systems.

ACKNOWLEDGMENT

This work was supported in part by the National Natural Science Foundation of China under Grants U23B2040 and 62034006; in part by the Natural Science Foundation of Shandong Province under Grants ZR2023LZH007, ZR2023QF054, TSQN202306059.

REFERENCES

[1] F. Wang, et al., "Charge Loss Induced by Defects of Transition Layer in Charge-Trap 3D NAND Flash Memory," in IEEE Access, vol. 9, pp. 47391-47398, 2021.

[2] H. -H. Wang et al., "A New Read-Disturb Failure Mechanism Caused by Boosting Hot-Carrier Injection Effect in MLC NAND Flash Memory," 2009 IEEE International Memory Workshop, Monterey, CA, USA, 2009, pp. 1-2.

[3] Y. Zhang, et al., "A Novel Read Scheme for Read Disturbance Suppression in 3D NAND Flash Memory," in IEEE Electron Device Letters, vol. 38, no. 12, pp. 1669-1672, Dec. 2017.

[4] C. Zambelli, et al., "Uniform and concentrated read disturb effects in mid-1X TLC NAND flash memories for enterprise solid state drives," 2017 IEEE International Reliability Physics Symposium (IRPS), Monterey, CA, USA, 2017, pp. PM-5.1-PM-5.4.

[5] Micron Technology, Inc. NAND Flash Design and Use Considerations Introduction. [2006]

[6] Y. Deguchi, et al., "System-level error correction by read-disturb error model of 1Xnm TLC NAND Flash memory for read-intensive enterprise solid-state drives (SSDs)," 2016 IEEE International Reliability Physics Symposium (IRPS), Pasadena, CA, USA, 2016, pp. MY-6-1-MY-6-4.

[7] Y. Kong et al., "Retention Correlated Read Disturb Errors in 3-D Charge Trap NAND Flash Memory: Observations, Analysis, and Solutions," in IEEE Transactions on Computer-Aided Design of Integrated Circuits and Systems, vol. 39, no. 11, pp. 4042-4051, Nov. 2020.

Aspect Ratio Dependent Optimization and Comparison of Specific ON-Resistance of SJ and H*k* MOSFETs with Extremely High Permittivity

Chenxing Wang[1], Zhentao Xiao[1], Zonghao Zhang[1], Zhenghao Jin[1], Zhiwan Liu[1], Zonglin Li[1], Zhemin An[1],
Yunteng Jiang[1], Ruguan Li[2*], Haimeng Huang[1], and Hongqiang Yang[1*]

[1]*Glasgow College, University of Electronic Science and Technology of China, Chengdu 610054, China*
[2]*GRG Metrology & Test Group Co., Ltd., Guangzhou 510656, China*
* Email: hqyang@uestc.edu.cn;lirg@grgtest.com

Abstract—Considering the fabrication difficulty of the high-permittivity (H*k*) MOSFETS, this study proposes two aspect ratios, AR_S for the *n*-pillars and AR_I for the H*k*-pillars, to optimize the specific on-resistance ($R_{on,sp}$) under breakdown voltage of 1000 V. Numerical calculations by MATLAB and simulations by MEDICI demonstrate and validate that with sufficiently large permittivity ratios (K_r) and optimal AR_I, H*k* MOSFETs can achieve lower $R_{on,sp}$ compared to conventional superjunction (SJ) MOSFETs. The accurate fitting function can be used to providing design instructions efficiently to make trade-offs between fabrication difficulty and required $R_{on,sp}$ and BV.

Index Terms—Aspect ratio (AR), breakdown voltage (BV), high permittivity (H*k*), optimization methodology, specific ON-resistance ($R_{on,sp}$), superjunction (SJ).

I. INTRODUCTION

Superjunction (SJ) has been widely applied due to its ability to maintain low specific on-resistance ($R_{on,sp}$) at high breakdown voltages (BV) [1], [2]. Excellent charge compensation can be achieved through the alternating *n*- and *p*-pillars in the SJ devices. However, the identical dopant implantation for *n*- and *p*-pillars is difficult to realize [3]. To address this, a novel device called the high-permittivity superjunction (H*k*) replaces the conventional *p*-pillar with a high permittivity insulator material, which can also effectively absorb the electric field line (EFL) laterally from the *n*-pillars. This will enable higher doping implantation in the *n*-pillar, ultimately leading to a reduction in $R_{on,sp}$. Research on extremely high-permittivity materials such as BaTiO₃ and SrTiO₃ [4] has gained attention for their exceptional properties used in electronic devices.

In this study, the impact of two types of aspect ratios, namely AR_I aand AR_S, respectively, are proposed to represent the fabrication parameters of H*k*- and *n*-pillars in H*k* MOS-FETs. Optimization of $R_{on,sp}$ based on AR_I and AR_S will be analyzed, comparing with the conventional optimization of SJ devices. Meanwhile, the research on the variation in permittivity, especially in extremely high permittivity, will be included in the work to explore their impact on optimization.

II. OPTIMIZATION METHODOLOGY AND FITTING

Fig. 1(a) and (b) illustrate the cross-sectional schematics of the half-cell structure for the SJ and H*k* MOSFETs, respectively, including the coordinate system for the drift regions. In

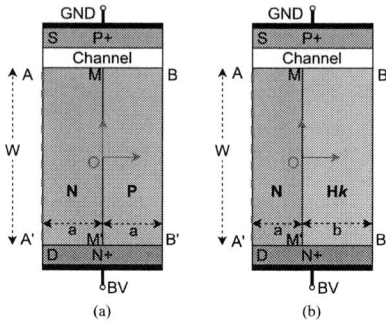

Fig. 1. Cross-sectional schematics of the half-cell structure and the coordinate system for the drift region design of (a) SJ MOSFET and (b) H*k* MOSFET.

the SJ MOSFET, the *n*- and *p*-pillars both have a width and doping concentration of *a* and *N*, with a drift region depth *W* and an aspect ratio (AR) of *W*/2*b*. In the H*k* MOSFET, the *n*-pillar width, depth, and doping concentration are *a*, *W*, and *N*, respectively, while the H*k* region has a width *b*. Therefore, the aspect ratio of the *n*-pillar (AR_S) and the H*k* region (AR_I) are defined as *W*/2*a* and *W*/2*b*, respectively.

Based on the formula of specific on-resistance, the equation for SJ and H*k* MOSFETs can be expressed as

$$R_{on,sp} = \begin{cases} \frac{W}{q\mu_n N} \cdot \frac{2a}{a_{eff}}, & \text{for SJ} \\ \frac{W}{q\mu_n N} \cdot \left(1 + \frac{b}{a}\right) = \frac{W}{q\mu_n N} \cdot \left(1 + \frac{AR_I}{AR_S}\right), & \text{for H}k. \end{cases} \quad (1)$$

The mobility of electrons (μ_n) is determined by the Caughey-Thomas model, i.e., $\mu_n = 55.2 + 1373.8/[1 + (N/1.02 \times 10^{17})^{0.73}]$ (cm² · V⁻¹ · s⁻¹) [5]. The a_{eff} in SJ is the effective width of the *n*- or *p*-pillar considering the JFET effect.

A. Optimization Methodology of $R_{on,sp}(AR, K_r)$

According to [3], to obtain the optimized $R_{on,sp}$ of SJ and H*k* MOSFETs, two constraints are considered to solve the *N* and *W* in Eq. (1).

1) Critical Depletion & Zero Electric Filed at Point A': As verified in [2], AA' shown in Fig. 1 is the easiest breakdown path in the SJ-like power MOSFETs. The EFL along AA 'can

979-8-3503-6184-1/24 $31.00 © 2024 IEEE

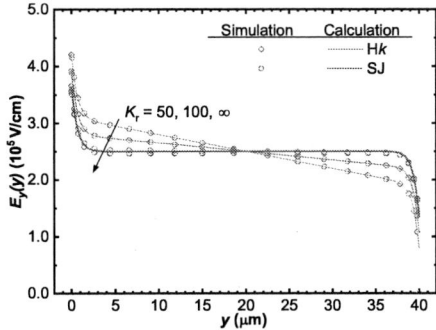

Fig. 2. Calculated and simulated results of the electric field's profile of E_y along the line AA' in Fig. 1 for SJ and Hk (K_r = 50, 100, ∞), where BV = 1000 V, W = 40 μm, $a = b = 1$ μm, and $N = 1 \times 10^{16}$ cm^{-3} are used.

be obtained by solving the Poisson equation with boundary conditions [1], [6], which can be formulated as

$$
E_{Sy}(y) = \begin{cases} \dfrac{\text{BV}}{W} - \dfrac{4qNW}{\varepsilon_S \text{AR}} \displaystyle\sum_{n=\text{odd}} \dfrac{(-1)^n}{(n\pi)^2} \dfrac{\sinh\frac{n\pi\text{AR}y}{W}}{\cosh\frac{n\pi\text{AR}}{2}}, & \text{for SJ} \\[4ex] \dfrac{\text{BV}}{W} + \dfrac{qNy}{\varepsilon_S} + \dfrac{4qNW}{\varepsilon_S} \times \\[2ex] \displaystyle\sum_{n=\text{odd}} \dfrac{(-1)^n}{(n\pi)^2} \dfrac{\text{sech}\frac{n\pi}{2\text{AR}_S}\sin\frac{n\pi y}{W}}{1+\frac{1}{K_r}\tanh\frac{n\pi}{2\text{AR}_S}\coth\frac{n\pi}{2\text{AR}_I}}, & \text{for H}k. \end{cases} \tag{2}
$$

$K_r = \varepsilon_I/\varepsilon_S$ is the permittivity ratio between the insulator and the semiconductor region of an Hk MOSFET. When considering extremely high permittivity materials, i.e., $K_r \to \infty$, the expression of electric field for Hk can be simplified as

$$
E_{Sy}(y) = \frac{\text{BV}}{W} + \frac{qNy}{\varepsilon_S} + \frac{4qNW}{\varepsilon_S} \sum_{n=\text{odd}} \frac{(-1)^n}{(n\pi)^2} \text{sech}\frac{n\pi}{2\text{AR}_S}\sin\frac{n\pi y}{W}. \tag{3}
$$

From Eq. (3), it can be observed that the distribution of electric is unrelated to AR$_I$. Fig. 2 demonstrates an example of the y-direction component of the electric field intensity along the line AA'. The values of E_y in this figure are computed using MATLAB with the given values of BV, K_r, W, a, b and N, which are validated by MEDICI.

The physical meaning of Eq. (2) is explained in detail in previous literature [1], [6], which indicates that the drift region can be fully utilized when the n-pillar is critically depleted, corresponding to zero electric field at A' in Fig. 1.

2) Critical Breakdown & Avalanche Breakdown Condition: This condition is considered to optimize the $R_{\text{on,sp}}$ of the drift region [3], [7]. Therefore, to meet the condition of avalanche breakdown, the impact ionization integral I_n should satisfy

$$
I_n = \int \alpha_n \exp\left[\int_0^y (\alpha_p - \alpha_n)dy'\right] dy = 1, \tag{4}
$$

where y and y' are the distances along AA' in Fig. 1. α_n and α_p are impact ionization rates in the Chynoweth model [8].

The transcendental equations built by two constraints are modeled and solved using MATLAB. Based on the optimization methodology proposed in [3], the optimized $R_{\text{on,sp(opt)}}$, $W_{(\text{opt})}$ and $N_{(\text{opt})}$ can be calculated regarding BV, K_r, AR$_S$ and AR$_I$.

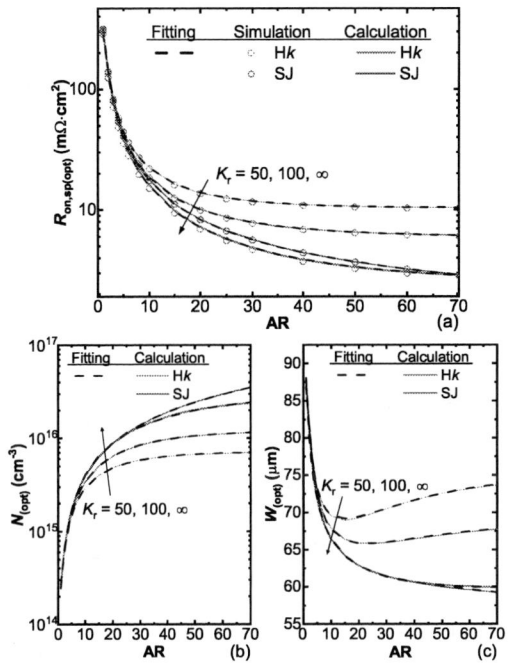

Fig. 3. Dependence of AR on (a) $R_{\text{on,sp (opt)}}$, (b) $N_{(\text{opt})}$ and (c) $W_{(\text{opt})}$ for the SJ and Hk (K_r = 50, 100, ∞) MOSFETs.

B. Empirical Function Fitting

An empirical function for $R_{\text{on,sp(opt)}}$ with respect to AR$_S$, AR$_I$ is fitted to explore further because these two parameters can indicate the process difficulty of n-pillar and Hk-pillar. The parameters are taken logarithm and polynomial surface expression is fitted as

$$
R_{\text{on,sp(opt)}} = \exp\left(\sum_{0\le i+j\le 4} p_{ij}(\ln \text{AR}_S)^i(\ln \text{AR}_I)^j\right). \tag{5}
$$

The values of coefficients in this fitting formula are presented in TABLE I. This method can simplify calculations and extend to other designing parameters, instructing further applications.

TABLE I
THE VALUES OF COEFFICIENTS IN THE FITTING FORMULA

p_{00}	p_{10}	p_{01}	p_{20}	p_{11}	p_{02}	p_{30}
5.733	-0.571	-0.435	-0.041	-0.311	0.059	0.027

p_{21}	p_{12}	p_{03}	p_{31}	p_{22}	p_{13}	p_{04}
0.048	0.024	0.029	0.002	-0.021	0.014	-0.006

III. RESULTS & DISCUSSION

A. Numerical Calculation Results and Comparison between SJ and Hk MOSFETs with Identical AR

Fig. 3 shows the relationship of (a) $R_{\text{on,sp(opt)}}$, (b) $N_{(\text{opt})}$ and $W_{(\text{opt})}$ with respect to AR for the SJ and Hk (K_r = 50, 100 and ∞) MOSFETs at BV = 1000 V. In this case, AR$_I$ and AR$_S$ are considered to be identical of magnitude AR. Obviously, the fitting lines are in excellent align with the calculation results. The optimized $R_{\text{on,sp(opt)}}$ values are also

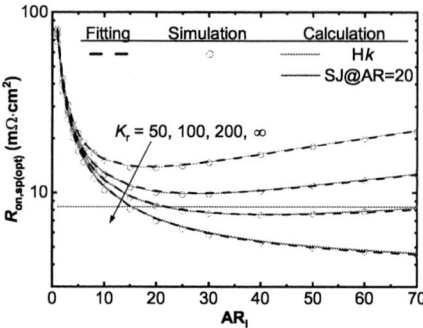

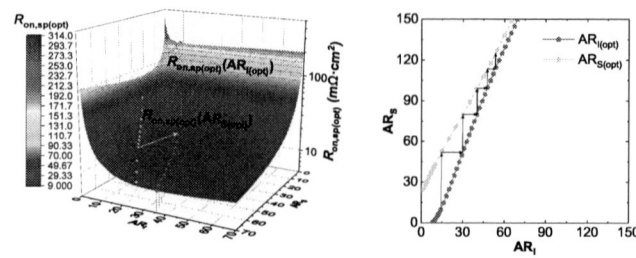

Fig. 4. Dependence of AR_I on (a) $R_{on,sp(opt)}$, (b) $N_{(opt)}$ and (c) $W_{(opt)}$ for the SJ MOSFET and Hk MOSFET (K_r=50, 100, 200, ∞).

Fig. 5. (a) $R_{on,sp}$ as a function of AR_I and AR_S for the Hk MOSFET and (b) the projection of $AR_{I(opt)}$ (red dots) and $AR_{S(opt)}$ (green dots).

verified by the MEDICI simulation. It can be seen that as AR increases, $R_{on,sp(opt)}$ decreases since the higher AR causes the thinner width of pillars in SJ and Hk MOSFETs, facilitating the absorption of EFLs. When K_r is sufficiently large and AR is small, Hk exhibits smaller $R_{on,sp(opt)}$ than that of SJ due to the smaller effective on-area of SJ. However, for larger AR, due to the weaker EFLs absorption by Hk compared to the charge compensation effect of SJ, Hk must reduce $N_{(opt)}$ and increase $W_{(opt)}$ to avoid breakdown before A in Fig. 1, as shown in Fig. 3(b) and (c). This will slightly increase the $R_{on,sp(opt)}$ of Hk.

B. Optimization for Aspect Ratio of Hk Pillar

As shown in Fig. 4, the relation between $R_{on,sp(opt)}$ and AR_I for the Hk ($K_r = 50$, 100, 200 and ∞) MOSFETs with a fixed $AR_S = 20$ is explored further. The AR of n-pillar for SJ is set equally with Hk, that is, $AR = AR_s = 20$, and used as a reference shown in Fig. 4. With a finite K_r, the $R_{on,sp(opt)}$ decreases when AR_I is small and increases slightly with the increase of AR_I. For Hk with small AR_I, the large portion of Hk-pillar causes the reduction of effective on-area portion in Eq. (1), leading to the large $R_{on,sp(opt)}$, and with large AR_I, the narrow Hk-pillar fails to absorb the EFLs adequately, leading to increasing trend in $R_{on,sp(opt)}$. This phenomenon does not exist in infinite K_r. The trend of $R_{on,sp(opt)}$ with finite K_r indicates that an optimal AR_I exists to obtain the minimal $R_{on,sp(opt)}$, and it can be further optimized to be better than that of SJ by sufficiently large K_r and suitable AR_I.

C. Optimization for Double Aspect Ratios of Hk MOSFETs

To further analyze the optimization of AR_S, Fig. 5(a) reveals the optimized $R_{on,sp(opt)}$ as a function of AR_S and AR_I. The green dots in Fig. 5 shows the optimized AR_S under fixed AR_I to obtained a minimal $R_{on,sp(opt)}$, so as red dots for optimized AR_I under fixed AR_S. The small AR_S causes the inadequate absorption of the EFLs by the Hk-pillar, while the large AR_S reduces the effective on-area portion, indicating the existence of optimal AR_S to obtain a minimal $R_{on,sp(opt)}$. The projection at the bottom of these two symbols is shown in Fig. 5(b), demonstrating the minimal $R_{on,sp(opt)}$ can be approximated by progressively iterating over AR_I and AR_S. Although the best combination of AR_S and AR_I may be limited by the fabrication

process, the design instructions can be provided using the emperical fitting function to achieve an optimal performance of 1000 V Hk MOSFET with limited AR_S and AR_I.

IV. CONCLUSION

To consider the fabrication limit of SJ-like devices, this study investigates the optimization and comparison for the $R_{on,sp}$ of SJ and Hk MOSFETs with varying high permittivity materials at different aspect ratios. By introducing two aspect ratios, AR_I and AR_S, for the Hk MOSFET, the study highlights how the dimensions of n- and Hk-pillars affect device optimization. The results reveal that Hk MOSFET with extremely high permittivity materials can achieve lower $R_{on,sp}$ compared to traditional SJ MOSFET, especially at optimal aspect ratios. An empirical function fitting approach facilitates practical optimization, offering design for Hk at different aspect ratios. The proposed double aspect ratio optimization provides new insights for comparing of imbalanced asymmetric SJ and Hk MOSFETs, and can be extended on wide bandgap Hk devices, such as the silicon carbide, gallium nitride, and gallium oxide.

REFERENCES

[1] X. Chen, P. Mawby, K. Board, and C. Salama, "Theory of a novel voltage-sustaining layer for power devices," *Microelectron. J.*, vol. 29, no. 12, pp. 1005–1011, Dec. 1998.

[2] H. Huang, K. Hu, W. Xu, S. Xu, W. Cui, W. Zhang, and W. T. Ng, "Numerical solutions for electric field lines and breakdown voltages in superjunction-like power devices," *IEEE Trans. Electron Devices*, vol. 67, no. 9, pp. 3898–3902, Sept. 2020.

[3] H. Huang, S. Xu, W. Xu, K. Hu, J. Cheng, H. Hu, and B. Yi, "Optimization and comparison of drift region specific ON-resistance for vertical power Hk MOSFETs and SJ MOSFETs with identical aspect ratio," *IEEE Trans. Electron Devices*, vol. 67, no. 6, pp. 2463–2470, Jun. 2020.

[4] H. Abdelkefi, H. Khemakhem, G. Vélu, J. C. Carru, and R. Von der Mühll, "Dielectric properties and ferroelectric phase transitions in Ba$_x$Sr$_{1-x}$TiO$_3$ solid solution," *Journal of Alloys and Compounds*, vol. 399, no. 1, pp. 1–6, 2005.

[5] D. Caughey and R. Thomas, "Carrier mobilities in silicon empirically related to doping and field," *Proc. IEEE*, vol. 55, no. 12, pp. 2192–2193, Dec. 1967.

[6] X. Chen and M. Huang, "A vertical power MOSFET with an interdigitated drift region using high-k insulator," *IEEE Trans. Electron Devices*, vol. 59, no. 9, pp. 2430–2437, Sep. 2012.

[7] H. Huang and X. Chen, "Optimization of specific on-resistance of balanced symmetric superjunction MOSFETs based on a better approximation of ionization integral," *IEEE Trans. Electron Devices*, vol. 59, no. 10, pp. 2742–2747, Oct. 2012.

[8] A. G. Chynoweth, "Ionization rates for electrons and holes in silicon," *Physical Review*, vol. 109, no. 5, p. 1537, May 1958.

Copper Ion Migration in van der Waals CuInP$_2$S$_6$ Devices with Vertical and Lateral Structures

Jie Li, Yirong Guo, Pengying Chang*

Key Laboratory of Optoelectronics Technology, Ministry of Education, Beijing University of Technology,
Beijing, 100124, China.
* Email: pychang@bjut.edu.cn

Abstract—Van der Waals CuInP$_2$S$_6$ (CIPS) exhibits interesting functional behaviors, including reversible ferroelectric polarization, copper ion (Cu+) migration and negative capacitance effect. Besides, CIPS are found to persist in both major out-of-plane polarization and minor in-plane polarization. In this work, we fabricate two-terminal CIPS devices based on Au/CIPS/Au heterostructure with vertical and lateral conduction channels respectively. The vertical device exhibits bipolar switching characteristic, while lateral one shows only unipolar switching along with rectification. Moreover, the current is significantly higher in vertical case than that in lateral case, which is explained by carrier transport and Cu+ migration along short and long channels. These experimental results provide physical insight into CIPS conductions as a promising candidate for future information storage devices.

Keywords—CuInP$_2$S$_6$, two-dimensional ferroelectric materials, carrier transport, Cu+ migration

I. INTRODUCTION

In order to adapt to the rapid development of today's information technology, device sizes need to be continuously miniaturized to meet the requirements of high integration. However, in traditional 3D devices, the performance of the device is often limited by the length and volume of the electron channel, which hinders the performance of the device at a small size. The two-dimensional materials have attracted a lot of attention due to their atomic thickness and good surface properties, showing great potential in continually shrinking device sizes. At the same time, vertical stacking of two-dimensional materials with different electronic, optical, and magnetic properties can form van der Waals (vdW) heterostructures with unique properties.

Among these 2D materials, CuInP$_2$S$_6$ (CIPS) has attracted considerable attention due to its strong intrinsic coupling of ionic activity and ferroelectric polarization [1],[2], leading to a myriad of interesting properties, including ferroelectric, ion conductivity, and pyroelectricity. The copper ion (Cu+) position and migration behavior in the sub-crystal lattice play a crucial role in controlling these properties, which offer an opportunity to design devices with field-tunable properties.

In this work, inspired by the long-range migration of Cu+ ions, we fabricate layered CIPS-based vertical and lateral structures respectively. The I-V characteristic and related conduction mechanism are strongly dependent on the device structures. This is because that the voltage-driven ion migration of ferroelectric CIPS (Cu+ ion migration energy of about 0.65 – 0.75 eV) [3],[4] will greatly affect the Schottky-like barrier and the space charge region of the metal/CIPS interface [5]. It is also found that the in-plane vdW homojunction can be achieved by the field-driven ion process without chemical doping or complex transfer process in lateral case, which is of great significance in the electronic and optical device applications.

II. EXPERIMENT AND METHOD

A. Au/CIPS/Au vertical structure

Fig. 1(a) shows the device schematic of vertically stacked Au/CIPS/Au. Photolithography is carried out on Si/SiO$_2$ substrate with reverse glue, and then the photoresist is used as a reticle to carry out RIE etching on SiO$_2$. A groove of about 52nm is generated with etching time 1 min. Then the electrode of Ti/Au 5/55nm is sputtered with magnetron sputtering, and then is peeled off to produce a bottom electrode. The CIPS sheet is mechanically stripped from the CIPS single crystal and transferred to the bottom electrode. Finally, the top electrode is sputtered alike bottom electrode.

B. Au/CIPS/Au lateral structure

Fig. 1(b) shows the lateral Au/CIPS/Au structure. Layered CIPS 2D material is transferred on top of Si/SiO$_2$ using a transfer platform. By photolithography, both left and right electrodes are patterned on the CIPS 2D material, and then the electrodes of Ti/Au 5/55nm were sputtered by magnetron sputtering, and then peeled off.

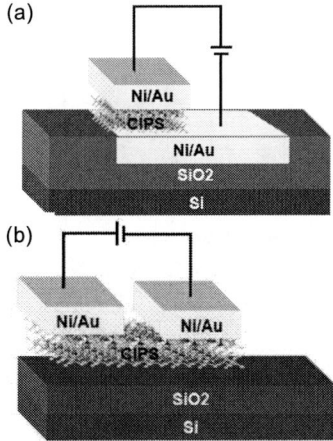

Fig. 1 Schematics of fabricated CIPS device. (a) Vertical device, with top electrode biased and bottom electrode grounded. (b) Lateral device, with left electrode biased and right electrode grounded.

979-8-3503-6184-1/24 $31.00 © 2024 IEEE

C. Electrical measurements

Electrical measurements of fabricated CIPS devices are carried out by a semiconductor parameter analyzer (B1500). During an impulse test with a two-terminal device, an impulse voltage is applied to one electrode and the other electrode is grounded, as illustrated in Fig. 1(a) and (b).

III. RESULTS AND DISCUSSION

A. I-V characteristics of vertical and lateral CIPS devices

As known from [3],[4], CIPS has not only out-of-plane ferroelectric polarization, but also in-plane ferroelectric polarization. Fig. 2(a) and (b) show the I-V characteristics of vertical and lateral CIPS devices. It is found that the I−V curves of vertically structured device shows switchable bidirectional diode behavior accompanied with resistive switching (RS). Conversely, the lateral structure presents an obvious asymmetrical relationship, showing the self-rectification characteristics of the device. Therefore, it is reasonable to suspect that the I-V characteristic are strongly dependent on the device structures, due to different internal conductive mechanisms.

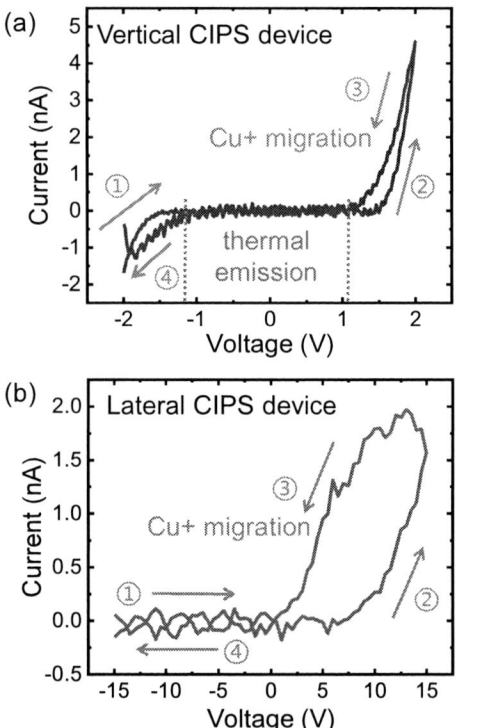

Fig. 2 I-V curves of Au/CIPS/Au with vertical (a) and lateral (b) structure.

B. Conductive mechanism of vertical device

As shown in Fig. 2(a), the conduction mechanism of vertical CIPS device is analyzed from the aspects of electron transport and Cu+ migration.

Firstly, from the analysis of the energy band, in the process of 1, as the negative bias voltage gradually becomes close to zero, the effective barrier height gradually increases, and the electron tunneling is more difficult, so the current gradually decreases. In the process 2, the polarity of the

applied voltage changes. As the positive voltage gradually increases, the effective barrier height gradually decreases, and the current gradually increases. Similarly, it can be seen that the current in process 3 and 4 is gradually decreasing and increasing, respectively.

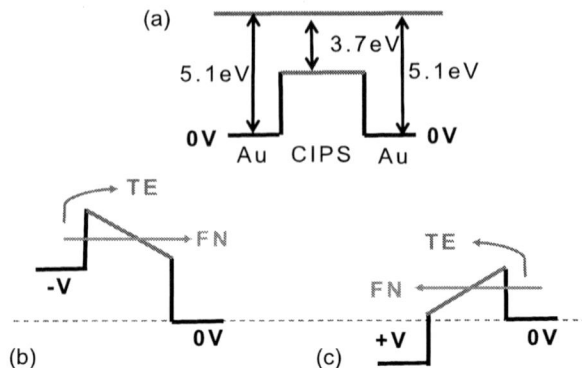

Fig. 3 Energy band diagrams of the electronic transport of vertical CIPS device. (a) At equilibrium (0V), (b) at negative bias, and (c) at positive bias.

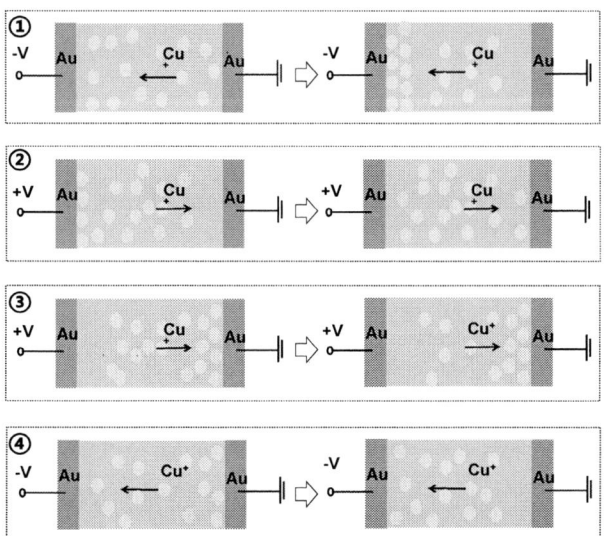

Fig. 4 Schematics of Cu+ migration process in vertical CIPS device.

Next, from the perspective of copper ion motion, in the process of 1, due to the large negative voltage at the beginning, the electric field inside Cu+ produces a large force on the Cu +, which makes Cu+ produce a small amount of accumulation on the left side. In the process 2, with the change of the polarity of the voltage applied to the left gold electrode, the copper ions began to migrate to the right end, but did not cause the accumulation of Cu+ on the right side. In the process 3, the voltage applied by the left gold electrode is still a positive electrode, and the copper ions still migrate to the right end. As the positive voltage of the left Au electrode gradually decreases, the moving Cu+ continues but becomes slower, and still accumulates on the right side. It is observed that the threshold voltage for Cu+ migration at forward bias (process 2) is relatively large, and as Cu+ still moves violently, so the current is still in a large range under the high voltage. On the other hand, the threshold voltage for

979-8-3503-6184-1/24 $31.00 © 2024 IEEE   817

Cu+ migration at reverse bias (process 3) is relatively small. As the movement of Cu+ becomes slow, the current shows a downward trend; In the 4 process, the left Au electrode is connected to the negative voltage, and Cu+ begins to move to the left. Although the negative voltage of the left Au electrode gradually increases, it is difficult for Cu+ to accumulate at the left, because in previous process the polarity of the voltage makes the Cu+ located much right. Therefore, the current under negative bias is smaller than that of the positive bias. Anyway, it is not difficult to find that the vertical structure of Au/CIPS/Au exhibits the threshold RS characteristic [8].

C. Conductive mechanism of lateral device

For the lateral structure, due to the limitation of the structure of the device itself, the length of the channel between the two metal electrodes is too long (about 5um), and no tunneling effect is found, so the conductive mechanism of this structure is mainly Cu+ migration.

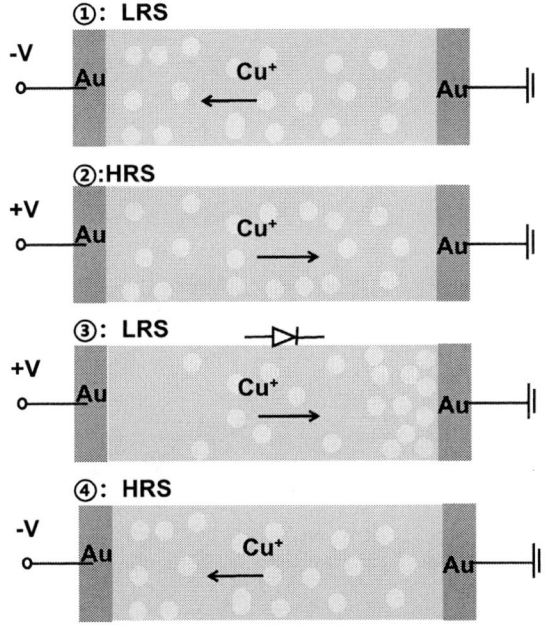

Fig. 5 Schematics of Cu+ migration process in lateral CIPS device.

When the voltage is swept from -2V → 0V (process 1), the Cu+ ions can gradually migrate to the left. Then the voltage becomes positive, from 0V → 2V (process), the internal Cu+ changes the direction of movement and begins to move to the right, and aggregates near the cathode surface. This results in abundant Cu+ ions near the cathode and lack near the anode, forming a p-n homology junction in CIPS, and the device switches from HRS to LRS (process 2→3).

When further increasing the opposite electric field, the device is converted to HRS (process 3→4), and a reversed p-n homology junction is established. Therefore, more mobile copper ions will be driven to the cathode and may be reduced to copper atoms. As a result, an asymmetrical interface barrier is formed, which leads to the whole behavior.

IV. CONCLUSION

CIPS is a special ferroelectric that couples ferroelectricity and ionic conductance, in which the migration of Cu+ ions in the lattice triggers a series of novel characteristics. The conductive mechanism of the Au/CIPS/Au vertical structure includes Cu+ migration and electronic thermal emission mechanism, while the lateral structure device is mainly dominated by Cu+ migration, and no tunneling mechanism is observed. Comparing the vertical and lateral devices, it can be found that the vertical structure exhibits hysteresis under both positive and negative voltages, while the lateral structure device shows hysteresis only at positive voltages. In addition, it can be found that the turn-on voltage required in the vertical structure is much smaller than that in the horizontal structure, while current level is higher in vertical case than in lateral case.

ACKNOWLEDGMENT

This work was supported in part by the Beijing Natural Science Foundation under Grant 4232061, and in part by the National Natural Science Foundation of China under Grant 61804003, and in part by the BPHR under Grant 202203024.

REFERENCES

[1] A.Simon,J. Ravez, V. Maisonneuve, C. Payen, V. B. Cajipe.Paraelectric-ferroelectric transition in the lamellar thiophosphate CuInP$_2$S$_6$.Chem.Mater. 1994, 6, 1575.

[2] V.Maisonneuve,V.B.Cajipe,A.Simon,R.VonDerMuhll,J.Ravez.Ferri electric ordering in lamellar CuInP$_2$S$_6$.Phys. Rev. B 1997, 56, 10860.

[3] Maisonneuve,V.;Reau,J.M.;Dong,M.;Cajipe,V.B.;Payen,C.;Ravez,J.I o-nic conductivity in ferroic CuInP$_2$S$_6$ and CuCrP$_2$S$_6$.Ferroelectrics 1997, 196, 257-260.

[4] Balke, N.; Neumayer, S.M.; Brehm, J.A.; Susner, M.A.; Rodriguez, B.J.; Jesse, S.;Kalinin, S.V.;Pantelides,S.T.; McGuire, M.A.;Maksymovych, P. Locally Controlled Cu-Ion Transport in Layered Ferroelectric CuInP$_2$S$_6$. ACS Appl. Mater. Interfaces 2018, 10,27188-27194.

[5] Fan, Z.; Fan, H.; Lu, Z.; Li, P.; Huang, Z.; Tian, G.; Yang, L.; Yao, J.; Chen,C.; Chen, D.Ferroelectric Diodes with Charge Injection and Trapping. Phys. Rev. Appl. 2017, 7, 014020.

[6] Maisonneuve, V.; Evain, M.; Payen, C.;Cajipe, V.B.;Molini é, P. Room-temperature crystal structure of the layered phase Cu$_I$In$_{III}$P$_2$S$_6$. J. Alloys Compd. 1995, 218, 157-164.

[7] Niu, L.; Liu, F.; Zeng, Q.; Zhu, X.; Wang, Y.; Yu, P.; Shi, J.; Lin, J.; Zhou, J.; Fu, Q.Controlled synthesis and room-temperature pyroelectricity of CuInP$_2$S$_6$ ultrathin flakes. Nano Energy 2019, 58, 596-603.

[8] Munjal, S.; Khare, N. Advances in resistive switching based memory devices. J. Phys. D: Appl. Phys. 2019, 52 (43), 433002.

Simulation Study of the Impact of Split Gate on SiC DTMOS Short Circuit Withstand Capability

Zixun Chen [1], Jinping Zhang*[1,2], Yang Liu [2], Xudong Ma [2] and Bo Zhang [1]

[1] National Key Laboratory of Electronic Thin Films and Integrated Devices, University of Electronic Science and Technology of China, Chengdu, 610054, P. R. China.
[2] Weihai Singa Electronics CO.LTD, Weihai, 264209, P. R. China.

* Email: jinpingzhang@uestc.edu.cn

Abstract—Simulation study of the impact of split gate (SG) on silicon carbide double trench MOSFET short circuit (SC) withstand capability is conducted in this paper. A proper electro-thermal model is built to study SC transient and evaluate SC withstand capability of the devices. The simulation results show that after introducing SG, the saturation current decreases by about 70% in certain condition. As a result, it effectively mitigates the self-heating during SC transient and extends the SC time from 1.5μs to 6μs. Furthermore, both SG oxide layer thickness and current spreading layer doping concentration of the SG device determine the trade-off relationship between SC withstand capability and conduction loss.

Keywords—*silicon carbide, double trench MOSFET, split gate, current spreading layer, short circuit, electro-thermal model*

I. INTRODUCTION

SiC MOSFETs are promising to replace Si IGBTs in high efficiency and compact applications [1]. Compared with SiC planar gate MOSFETs, SiC trench gate MOSFETs have higher channel density and lower on-state resistance (R_{on}). Reference [2] proposes SiC double trench MOSFET (DTMOS) which achieves ultra low conduction loss and alleviates the electric field in gate oxide layer. However, the large reverse capacitance of DTMOS slows down switching speed and leads to higher switching loss [3]. Moreover, DTMOS has inferior short circuit (SC) withstand capability owing to its higher channel density and higher saturation current. For SiC MOSFETs in the system, they should endure the severe electro-thermal stress during SC transient before the detection circuit reaction and safely turned off. As a result, the SiC MOSFETs with enhanced SC withstand capability are beneficial to the overall system design. Reducing the saturation current is an effective way to reduce the self-heating during SC transient and it consequently enhances the SC withstand capability. In this paper, simulation study of the impact of split gate (SG) on SiC DTMOS SC withstand capability is conducted. The corresponding electro-thermal model is built to study the SC transient and evaluate the SC withstand capability of the devices. The simulation results show that the SG device has considerable SC withstand capability enhancement compared with the conventional DTMOS.

II. DEVICE STRUCTURES AND TCAD MODELING

The DTMOS with SG (SG-DT for short) is shown in Fig. 1(a). As comparison, the conventional DTMOS (C-DT for short) is shown in Fig. 1(b). For fair comparison, the area, cell width (W_{cell}), N- drift region thickness (H_{N-}) and doping concentration of them are same. They are 1cm², 2.5μm, 12μm and 8×10^{15}cm⁻³ respectively. The P+ shielding region (P+S), which extends downward from the source trench and

exceeds the SG trench of SG-DT or gate trench of C-DT, effectively decreases the oxide layer electric field. The size of P+S are illustrated in Fig. 1 and table I. The SG trench is just under the gate trench and their width are same. The SG is directly connected to the source electrode. To optimize the R_{on}, the current spreading layer (CSL) which doping concentration (N_{CSL}) is 2.5×10^{16}cm⁻³ is introduced in SG-DT. Other structural parameters and their values are summarized in Fig. 1 and table I. Silvaco TCAD simulation tools are used in this study. Fermi-Dirac carrier statics, incomplete ionization, concentration dependent mobility, Conwell-Weisskopf and surface mobility are selected in the simulation.

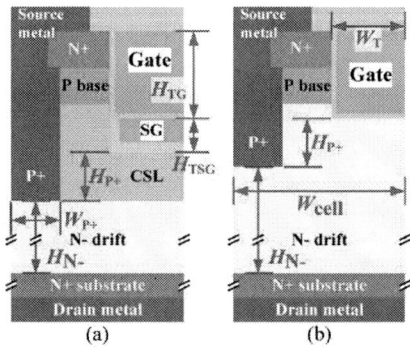

Fig. 1. Device structures, (a) SG-DT and (b) C-DT.

TABLE I. MAIN DEVICE STRUCTURE PARAMETERS AND THEIR VALUES.

Device Structural Parameter	Value	
	SG-DT	C-DT
P+S width, W_{P+}/μm	1.0	
P+S exceeding part height, H_{P+}/μm	1.0	
Trench width, W_T/μm	1.0	
Gate trench depth, H_{TG}/μm	0.8	
Gate oxide layer thickness, T_{OXG}/nm	50	
SG trench depth, H_{TSG}/μm	0.6	-
SG oxide layer thickness, T_{OXSG}/nm	100	-

To study the distribution and variation of electrical and thermal parameters during SC transient, thermal model based on the actual package structure is built and shown in Fig. 2(a). To speed up the simulation process, the electrical characteristics are considered only in the device part, as shown in Fig. 2(b). That is to say, there is no current flow from SiC substrate to copper (Cu) plate in the SC simulation process. This simplification has no impact on the thermal characteristics of them. The thermal conductivity and heat capacity of each part are assigned referring to [4-7]. The

979-8-3503-6184-1/24 $31.00 © 2024 IEEE

thermal boundary condition is set to ambient temperature 300K.

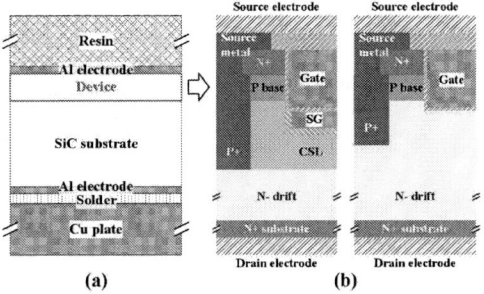

Fig. 2. (a) Thermal model based on the actual package structure and (b) corresponding device part.

III. RESULTS AND DISCUSSION

The drain current (I_D) as a function of drain-source voltage (V_{DS}) under different gate-source voltage (V_{GS}) of SG-DT and C-DT are shown in Fig. 3. It can be seen that SG-DT has much lower saturation current under high V_{DS} compared with C-DT. When V_{DS}=600V, V_{GS}=12, 15 and 18V, the saturation current (I_{SAT}) of SG-DT reaching 1.85×10^3, 2.08×10^3 and 2.24×10^3A, respectively, while the I_{SAT} of C-DT reaching 4.86×10^3, 7.21×10^3 and 9.37×10^3A, respectively. In the same condition, the I_{SAT} of SG-DT are 62%, 71% and 76% off compared with C-DT. Moreover, when V_{DS}=600V, V_{GS} increases from 12V to 15V and 18V, the I_{SAT} of SG-DT increases by 12% and 21%, respectively, while the I_{SAT} of C-DT increases by 48% and 72%, respectively. Therefore, compared with C-DT, the saturation current of SG-DT is less dependent on V_{GS} and keeps at a relatively lower value.

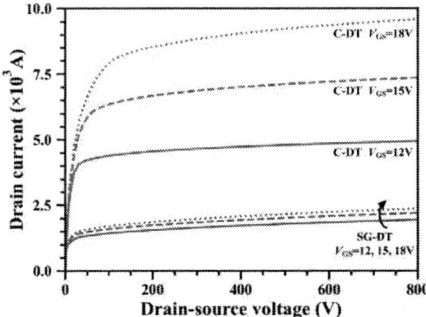

Fig. 3. The I_D as a function of V_{DS} when V_{GS}=12, 15 and 18V of SG-DT and C-DT.

The SC transient is simulated based on the thermal model built in section II. The I_D and V_{DS} waveforms of SG-DT and C-DT during SC transient are shown in Fig. 4. The SC time (t_{SC}) of SG-DT and C-DT are 6µs and 1.5µs, respectively. The peak SC current of SG-DT and C-DT are 3.10×10^3A and 1.16×10^4A, respectively. With the advantage of 73% lower peak SC current, the self-heating during SC transient of SG-DT can be mitigated and the SC withstand time (SCWT) of SG-DT can be extended. The V_{DS} spikes of C-DT is larger for its larger *di/dt* during the switching transient and larger current reduction during the SC transient. The maximum temperature in SiC region ($T_{SiC,max}$) and temperature of point A (T_A) in source metal, which locates at 1µm above the N+ source region surface and the total thickness there is 4µm, are shown in Fig. 5. Although the t_{SC} of SG-DT is 3 times

longer than C-DT, the $T_{SiC,max}$ of SG-DT is 1430K, which is 220K lower than C-DT. The T_A after SC transient of both devices reach peak value 1100K, higher than the melting point of aluminum (Al, 933K). So both devices face the risk that the metal interconnection will fail after SC. When reducing the t_{SC} of SG-DT to 4µs, the corresponding waveforms and temperature are also shown in Fig. 4 and 5, respectively. The $T_{SiC,max}$ and T_A of SG-DT in this condition are 1200K and 920K, respectively. Therefore the metal interconnection of SG-DT can withstand the self-heating during SC transient. It can be concluded that decreased saturation current of SG-DT helps mitigate the self-heating during SC transient and enhance the SC withstand capability.

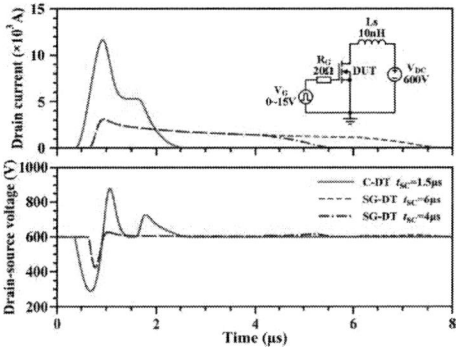

Fig. 4. The I_D and V_{DS} waveforms of SG-DT (dashed lines for t_{SC}=6µs and dash-dot lines for t_{SC}=4µs) and C-DT during SC transient. The corresponding simulation circuit is shown in the inset.

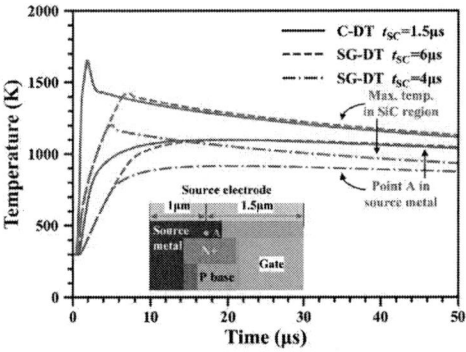

Fig. 5. The maximum temperature in SiC region (red lines) and the temperature of point A in source metal (green lines) of SG-DT (dashed lines for t_{SC}=6µs and dash-dot lines for t_{SC}=4µs) and C-DT (solid lines).

To carry out comprehensive inspections of the SC withstand capability of two devices, the distribution and variation of electrical and thermal parameters during SC transient are extracted. For SG-DT, the SC current (I_{SC}) and $T_{SiC,max}$ reaches their peak value at time point t_1=0.95µs and t_2=7.06µs, respectively. For C-DT, the I_{SC} and $T_{SiC,max}$ reaches their peak value at time point t_3=0.93µs and t_4=1.90µs, respectively. The current density and Joule heat power distribution of two devices, when each I_{SC} reaching its peak value, are shown in Fig. 6, respectively. It can be seen that unlike the C-DT with accumulation layer surrounding the gate trench, the I_{SC} of SG-DT flows along the side wall of gate trench, then it moves around the SG trench and finally it spreads into the N- drift layer. It can also be seen that the Joule heat power distribution is consistent with the current density distribution. The self-heating during SC transient in C-DT is more severe due to much larger I_{SC}. The current density and temperature distribution of two devices, when

979-8-3503-6184-1/24 $31.00 © 2024 IEEE   820

each $T_{SiC,max}$ reaching its peak value, are shown in Fig. 7, respectively. It can be seen that their current density distributions seem unchanged but their values are reduced, i.e. the parasitic bipolar junction transistors in both devices are not activated in such harsh eletro-thermal conditions. Furthermore, the heat accumulates in the upper part of device and conducts to both sides. The source metal layer will determine the actual SCWT of both devices because the source metal is closer to heat source and the melting point of Al is lower than SiC.

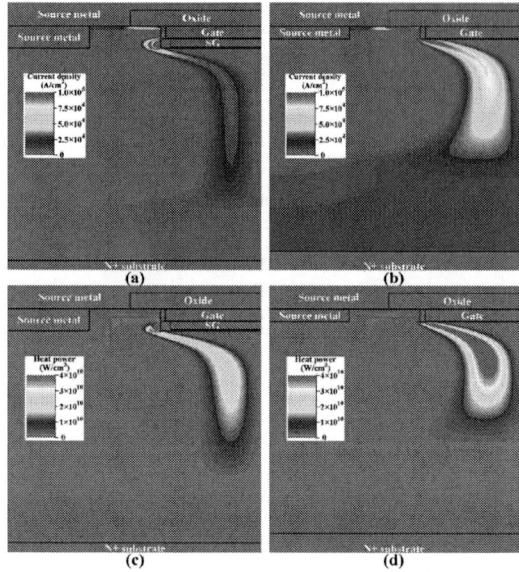

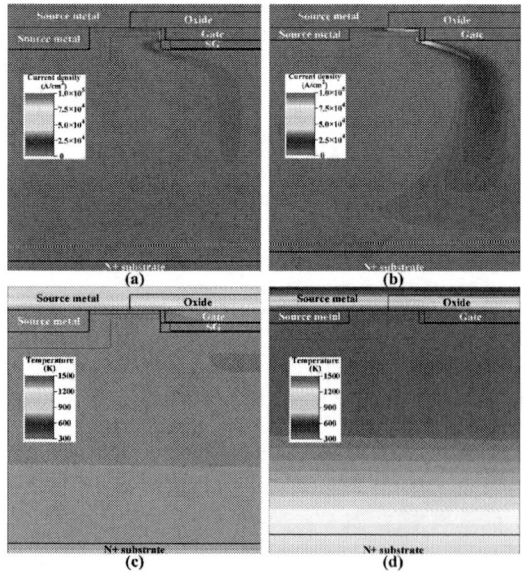

Fig. 6. The current density distribution, (a) SG-DT at t_1, (b) C-DT at t_3, and the Joule heat power distribution, (c) SG-DT at t_1, (d) C-DT at t_3.

Fig. 7. The current density distribution, (a) SG-DT at t_2, (b) C-DT at t_4, and the temperature distribution, (c) SG-DT at t_2, (d) C-DT at t_4.

Both the N_{CSL} and T_{OXSG} of SG-DT have effects on R_{on}, I_{SAT} and their trade-off relationship, as shown in Fig. 8. Each curve represents one particular N_{CSL}, and each point from upper to lower at one curve corresponds to T_{OXSG} varying from 0.3μm to 0.05μm. It can be seen that SG-DT with higher N_{CSL} can achieve a better trade-off between I_{SAT} and

R_{on}. But it is difficult to further decrease the I_{SAT} with higher N_{CSL} for the limitation of T_{OXSG}. Furthermore, thinner SG oxide layer can decrease the I_{SAT}, while thicker SG oxide layer can decrease the R_{on}. Lower I_{SAT} helps reduce self-heating and enhance SC withstand capability, while lower R_{on} helps reduce conduction loss. In conclusion, SG-DT with well selected N_{CSL} and T_{OXSG} values can meet the requirement of SC withstand capability and conduction loss.

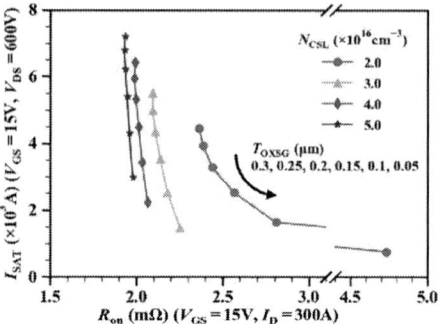

Fig. 8. The trade-off between I_{SAT} and R_{on} of SG-DT under different N_{CSL} and T_{OXSG}.

IV. SUMMARY

In this paper, simulation study of the impact of SG on SiC DTMOS SC withstand capability is conducted. The electro-thermal model is built to study the SC transient and evaluate the SC withstand capability of SG-DT and C-DT. The simulation results show that the SG helps SG-DT decrease I_{SAT} by about 70% in certain condition. It is essential for mitigating the self-heating during SC transient. As a result, SG-DT achieves enhanced SC withstand capability compared with C-DT. Furthermore, both T_{OXSG} and N_{CSL} of SG-DT determine the trade-off relationship between SC withstand capability and conduction loss.

ACKNOWLEDGMENT

This work was supported in part by the Taishan Leading Talent of Innovation in Shandong Province, China.

REFERENCES

[1] L. Zhang, X. Yuan, X. Wu, C. Shi, J. Zhang and Y. Zhang, "Performance evaluation of high-power SiC MOSFET modules in comparison to Si IGBT modules," IEEE Transactions on Power Electronics, vol. 34, no. 2, pp. 1181-1196, Feb. 2019.

[2] T. Nakamura et al., "High performance SiC trench devices with ultra-low ron," in Proc. 2011 IEDM, Washington, DC, USA, 2011, pp. 26.5.1-26.5.3.

[3] Z. Chen, J. Zhang and B. Zhang, "Simulation Study of a Novel 4H-SiC Split Gate Double Trench MOSFET with Side Wall Gate," in Proc. IEEE 16th ICSICT, Nangjing, China, 2022, pp. 1-3.

[4] D. Kim, A. J. Morgan, N. Yun, W. Sung, A. Agarwal and R. Kaplar, "Non-isothermal simulations to optimize SiC MOSFETs for enhanced short-circuit ruggedness," in Proc. 2020 IEEE IRPS, Dallas, TX, USA, 2020, pp. 1-6.

[5] Tsibizov, I. Kovačević-Badstübner, B. Kakarla and U. Grossner, "Accurate temperature estimation of SiC power MOSFETs under extreme operating conditions," IEEE Transactions on Power Electronics, vol. 35, no. 2, pp. 1855-1865, Feb. 2020.

[6] Z. Bai et al., "Investigation on single pulse avalanche failure of 1200-V SiC MOSFETs via optimized thermoelectric simulation," IEEE Transactions on Electron Devices, vol. 68, no. 3, pp. 1168-1175, March 2021.

[7] K. Yao, H. Yano, H. Tadano and N. Iwamuro, "Investigations of SiC MOSFET short-circuit failure mechanisms using electrical, thermal, and mechanical stress analyses," IEEE Transactions on Electron Devices, vol. 67, no. 10, pp. 4328-4334, Oct. 2020.

979-8-3503-6184-1/24 $31.00 © 2024 IEEE

Improved Hall Mobility Measurement Distinguishing Interface Capturing Effect in 4H-SiC Inversion Channel

Xiangrui Fan, Hao Fu, Xinyu Zhang, Zilong Wu, Jiameng Sun, Jiaxing Wei*, Siyang Liu, Weifeng Sun

National ASIC System Engineering Research Center, School of Integrated Circuits, Southeast University,
Nanjing, 210096, China

*Email: jiaxingwei@seu.edu.cn, liusy2017@seu.edu.cn

Abstract—The low inversion channel electrons mobility has been one of the most serious factors constraining the development of 4H-SiC power devices. However, the mobility of the mobile electrons in the inversion channel is unable to be characterized by the conventional filed-effect mobility (μ_{FE}) measurement method due to the capturing effect of the interface traps at the 4H-SiC/SiO₂ interface. In this paper, an improved hall mobility (μ_{Hall}) measurement method is proposed to distinguish the influence of the interface traps on the inversion electron concentration and extract the mobility of the mobile electrons in the inversion channel. The established hall measurement platform allows a minimum μ_{Hall} measurement down to 10 cm²/(v·s) and a measurement error of less than 0.1 cm²/(v·s). Two types of SiC LDMOSFETs are designed and fabricated to experimentally extract the μ_{Hall} and the μ_{FE}. Dependences of the μ_{Hall} on the gate-source voltage, drain-source current, and magnetic are investigated. The μ_{Hall} and the μ_{FE} of the inversion electrons are measured to be 92.3 cm²/(v·s) and 20.6 cm²/(v·s), respectively. The density of the captured electrons and the mobile electrons are successfully extracted to be 109.28 nC/cm² and 31.80 nC/cm² respectively. This work provides helpful guidelines for further improving the inversion channel electrons mobility issues in 4H-SiC.

Key words: 4H-SiC Inversion Channel Mobility, Hall Mobility, Field-effect Mobility, Interface Traps

I. INTRODUCTION

The hexagonal polytype of silicon carbide metal-oxidation-semiconductor field effect transistor (4H-SiC MOSFET) has been widely applied in the high efficiency, high power, and high temperature electronic applications[1]. However, the specific resistance of the 4H-SiC MOSFET is still serious limited by the inversion channel electrons mobility[2]. To best of our knowledge, the inversion channel electrons mobility of most commercial 4H-SiC MOSFET products is relatively low as in the range of 15 cm²/(v·s) - 25 cm²/(v·s)[2]. However, the ideal electron mobility of the 4H-SiC material is as high as 900 cm²/(v·s). Therefore, it is significant to improve the electrons mobility, and the first of which is to accurately evaluate the inversion channel electrons mobility.

The conventional method is to measure the field effect mobility (μ_{FE}). When measuring the μ_{FE}, all electrons induced beneath the oxide are assumed to be mobile. However, 75%-90% of electrons in the inversion layer are captured by the interface traps[3]. As a result, accurate electron mobility cannot be measured through the field effect.

In this paper, an improved method is proposed to distinguish the influence of the interface traps and extract the mobile electrons density in the inversion channel by comparing the hall mobility measurement and the field effect mobility measurement. The mobile electrons mobility in the inversion channel is obtained by measuring the hall mobility (μ_{Hall}). The drawbacks of the conventional mobility measurement

method are discussed in section II. An improved μ_{Hall} measurement method is proposed in section III to measure the mobile electrons mobility on a self-designed platform. Section IV provides analysis of the test results and discussion about the influence of the nonideal factors on the experimental results. The density of the mobile electrons and the captured electrons are successfully extracted.

II. MOBILITY MEASUREMENT METHOD COMPARISON

In order to show the influence of the inversion channel electrons mobility on the channel resistance ($R_{Channel}$), as shown in the Fig.1(a), a 4H-SiC LDMOSFET is structured in the Silvaco TCAD simulation, where the oxide layer thickness is 35 nm, the concentration of the P-well is 1×10^{18} cm⁻³, the length of the channel and the overlay of gate-to-source is 4.8 μm and 0.2 μm, respectively. When the inversion channel mobility is increased from 20 cm²/(v·s) to 60 cm²/(v·s), the $R_{Channel}$ of the LDMOSFET is reduced by 59.18% from 513.03 mΩ to 209.39 mΩ, as shown in Fig.1(b). Thus, it is significant to improve the electrons mobility and measure the inversion channel electrons mobility accurately.

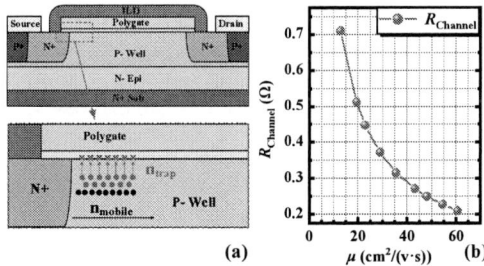

Fig.1 (a) A 4H-SiC LDMOSFET is constructed in silvaco TCAD simulation. The interface traps are located at 4H-SiC/SiO₂ interface, resulting in the decrease of the mobile electrons in the inversion channel. (b) The $R_{Channel}$ of the LDMOSFET is simulated. The R_{on} decreases with the increase of the inversion channel electrons mobility.

Fig.2(a) shows a LDMOSFET for the conventional field effect mobility measurement, which is fabricated with a n-type channel length of 740 μm and width of 160 μm. The oxide layer thickness is 40 nm and the concentration of P- well is 1×10^{18} cm⁻³. The μ_{FE} is expressed by (1) by measuring the transfer characteristics, and the total inversion electrons density (Q_{total}) is extracted as follows[1][7].

$$\mu_{FE} = \frac{L}{c_{OX}WV_{DS}} \times \frac{dI_{DS}}{dV_{GS}} \qquad (1)$$

$$Q_{total} = \frac{c_{OX}(V_{GS}-V_{TH})}{t_{inv}} \qquad (2)$$

L and W are the length and the width of the channel, C_{ox} is the oxide capacitance and t_{inv} is the thickness of the channel inversion layer. When measuring the μ_{FE}, all electrons induced in the inversion channel are assumed to be mobile. However, 75%-90% of the electrons in the inversion layer are fixed because of the interface traps[3]. Moreover, the C_{ox} and the t_{inv} cannot be calculated precisely due to some inevitable nonideal factors. Therefore, the μ_{FE} calculated by (1) and the mobile electrons density calculated by (2) are not accurate. The measured μ_{FE} is presented in Fig.2(b). At the temperature of 300 K, the μ_{FE} is measured only to be 20.51 cm²/(v·s), and is increased to 22.12 cm²/(v·s) when the temperature is increased to 423 K. The positive temperature coefficient of the μ_{FE} is believed to result from the substantial trapped electrons in the inversion channel[5]. When the temperature is increased, the trapped electrons are released into the conduction band and become mobile. Thus, the conductivity of the channel is enhanced and the μ_{FE} is increased. In order to measure the inversion channel electrons mobility accurately, it is necessary measure the mobility without the influence of the interface traps.

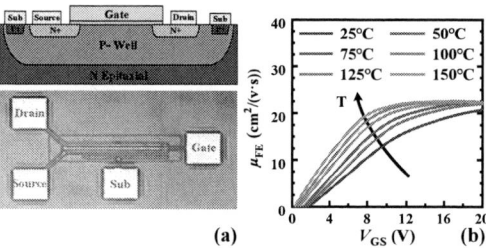

Fig.2 (a) A 4-ports LDMOSFET with a channel length of 740 μm and width of 160 μm is fabricated for field-effect mobility measurement. (b) Field-effect mobility increases with the increase of the temperature from 25 ℃ to 175 ℃.

Fig.3(a) illustrates the improved method for hall mobility measurement, which distinguish the interface capturing effect. When the inversion channel electrons are induced in the 4H-SiC LDMOSFET by the positive gate voltage applied (V_{GS}), the mobile electrons in the N+ source region will be drifted to the drain electrode under the positive drain voltage (V_{DS}). When a magnetic field (B) is applied perpendicular to the electron current flow plane, the mobile electrons is deflected by the Lorentz force (f_L) generated by B, eventually gathering at the lower side (port 3 and 4) of the channel. Then a hall voltage (V_{Hall}) is generated by the accumulated electrons between port 2 and 4. With the electrons accumulating, the V_{Hall} is enlarged until a balance between the Coulomb force generated by V_{Hall} and the f_L is achieved. Then, the mobile electrons will no longer deflect and a stable V_{Hall} is measured. The μ_{Hall} is expressed by (3), where W is the channel width, $V_{Channel}$ and L_{12} are the voltage and distance between port 1 and 2, respectively [6].

$$\mu_{Hall} = \frac{V_{Hall}L_{12}}{V_{Channel}BW} \quad (3)$$

A tiny V_{Hall} of only few nanovolts has been reported by previous researches, which is challenging to be measured[6]. Therefore, various geometrical parameters are detailly considered when designing a test graph to enlarge the V_{Hall}. However, although a wider channel will generate a larger V_{Hall}, it causes serious fluctuations in the V_{Hall} due to the misalignment between the upper and the lower ports because of the lithographic deviations. As a result, a good compromise between the magnitude and the stability of V_{Hall} is obtained when a LDMOSFET with a channel width of 500 μm and length of 100 μm is designed and fabricated, as shown in Fig.3(b), which has 8 ports including 4 hall probes. The oxide

layer thickness is 40 nm and the concentration of P-well is 1×10^{18} cm⁻³. Even under a small I_{DS} of 0.1 μA, the V_{Hall} is estimated to be over 0.1 mV with the fluctuation less than few microvolts, which is easy to be measured accurately.

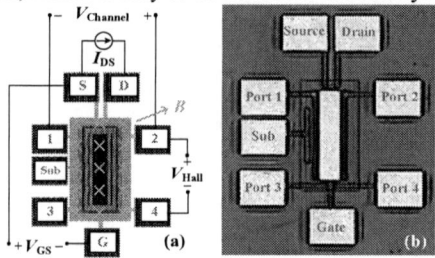

Fig.3 (a)The layout of the improved hall mobility test graph with 8-ports and the illustration of the hall mobility measurement. (b) A LDMOSFET with an inversion channel width of 500 μm and length of 100 μm is fabricated for hall mobility measurement.

As shown in Fig.4, a hall measurement platform is constructed. The maximum B up to 1 Tesla is provided by a magnetic source to induce a large V_{Hall} in the inversion channel. The minimum V_{Hall} down to 1 μV is measured and the I_{DS} is provided by a microvolts source meter simultaneously. The improved platform allows a minimum μ_{Hall} measurement down to 10 cm²/(v·s) and a measurement error of less than 0.1 cm²/(v·s).

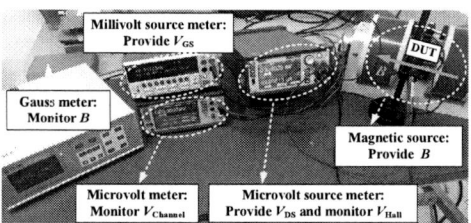

Fig.4 An improved test platform is constructed for hall mobility measurement, including a gauss meter, a magnetic source, a microvolts source meter, a millivolts source meter and a microvolts meter.

III. RESULTS AND DISCUSSION

In order to comprehensively investigate on the hall mobility of the inversion channel, the μ_{Hall} is measured through the self-constructed platform under different V_{GS}, I_{DS} and B. As shown in Fig.5, it is obviously observed that the measured μ_{Hall} increases as the V_{GS} or B is increased, while decreases with the increase of the V_{DS} or I_{DS}. The μ_{Hall} is measured to be only 13.9 cm²/(v·s) under the conditions of V_{GS} = 15 V, I_{DS} = 5 μA, and B = 377 mT. However, the μ_{Hall} is measured to be 154.4 cm²/(v·s) under the conditions of V_{GS} = 20 V, I_{DS} = 0.1 μA, and B = 573 mT. The dependence of the measured μ_{Hall} on the V_{GS} and I_{DS} has also been reported by previous studies[8].

At a low V_{GS} of 10 V or less, the electrons mobility is low because of the heavy coulomb scattering caused by the interface charges, which is the primary factor contributing to the low mobility in the inversion channel[3]. When V_{GS} is increased, the electrons captured by the interface traps are released and the coulomb scattering is reduced. As a result, the μ_{Hall} increases with the increase of V_{GS}.

When the I_{DS} is increased, the LDMOSFET operates in the pinched-off region. The V_{DS} is redistributed nonuniformly in the inversion channel because of the depletion layer near the drain electrode, as electric field in the depletion layer is larger than which in the inversion channel. Thus, the $V_{Channel}$ is overestimated and the μ_{Hall} is underestimated.

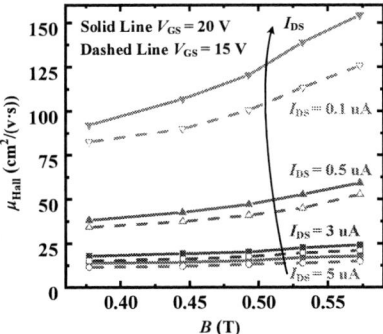

Fig.5 The μ_{Hall} is measured under different V_{GS}, I_{DS} and B. The solid lines and the dashed lines are measured at the $V_{\text{GS}} = 20$ V and $V_{\text{GS}} = 15$ V, respectively. Lines from the bottom to the top are measured at the I_{DS} of 5 μA, 3 μA, 0.5 μA and 0.1 μA, respectively. The μ_{Hall} increases when V_{GS} or B is increased, or I_{DS} is decreased.

However, few studies reported the dependence of the μ_{Hall} on the B. It is attributed to the reverse motion of the mobile electrons from the drain electrode to the source electrode, as shown in Fig.6(a). When the B is small as 300 mT, the electrons are moved into the inversion channel from the source electrode and drifted towards the lower side of the channel (port 3 and 4) under the combined work of the f_{L} and the Coulomb force (f_{C}) generated by V_{DS}. As shown in Fig.6(b), when the B is increased, a balance is achieved between f_{L} and f_{C}. Then, electrons are drifted in the opposite direction towards the source electrode, but still arrive at the lower side of the channel finally. However, when B is further increased to 800 mT, as presented in Fig.6(c), the electrons especially that far from the lower side of the channel are moved reversely towards the source electrode and finally drifted out of the channel. Then more electrons are added into the channel to compensate for the loss of the current, which causes the actual mobile electrons in the inversion channel underestimated. As a result, the measured μ_{Hall} increases with the increase of the B. Thus, the accurate μ_{Hall} is measured to be 92.3 cm^2/(v·s) under the conditions of $V_{\text{GS}} = 20$ V, $V_{\text{DS}} = 40$ mV, and $B = 377$ mT, which is 4.5 times higher than the μ_{FE} measured under the same conditions. From the analysis mentioned above, it is recommended to measure the μ_{Hall} under a small B and I_{DS}, and a large V_{GS}.

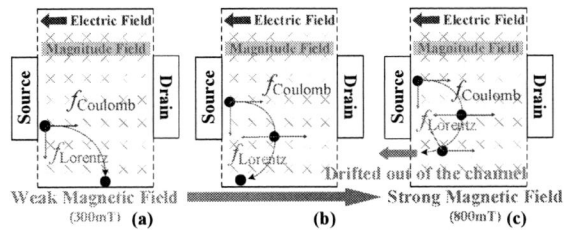

Fig.6 Dependence of the hall mobility on different B from 300 mT to 800 mT. Mobile electrons are moved reversely and drifted out of the channel when a large B is applied, resulting in the overestimation of the μ_{Hall}.

To find out the reasons for the difference between the μ_{Hall} and μ_{FE}, the mobile electrons density (Q_{mobile}) in the inversion channel is extracted to be 31.80 nC/cm^2 from $Q_{\text{mobile}} = \sigma / \mu_{\text{Hall}}$ [3], where σ is the inversion channel conductivity. The Q_{total} is extracted to be 141.08 nC/cm^2 from (2). The density of electrons captured by the interface traps (Q_{it}) is extracted to be 109.28 nC/cm^2. It is observed from Fig.7 that 77.46% of electrons in the inversion channel are captured by the interface traps, which leads to the low μ_{FE} and its positive temperature coefficient, as is illustrated above.

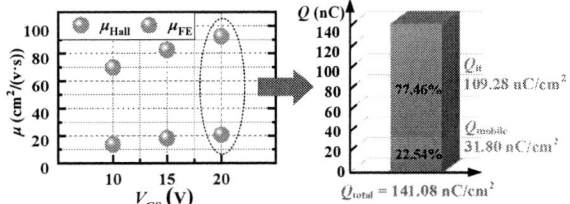

Fig.7 Comparison of μ_{Hall} and μ_{FE} under different V_{GS}. Q_{total}, Q_{mobile} and Q_{it} are extracted to be 141.08 nC/cm^2, 31.80 nC/cm^2 and 109.28 nC/cm^2 respectively.

IV. CONCLUSION

In this paper, an improved hall mobility measurement method is proposed to distinguish the influence of the interface traps at 4H-SiC/SiO$_2$ interface. Two LDMOSFETs with channel width of 500 μm and length of 100 μm, and width of 160 μm and length of 740 μm are fabricated to measure the μ_{Hall} and the μ_{FE} of the inversion channel. Based on the self-constructed measurement platform, the μ_{Hall} and μ_{FE} of the inversion electrons are measured to be 92.3 cm^2/(v·s) and 20.6 cm^2/(v·s), respectively. The Q_{mobile} is extracted to be 31.80 nC/cm^2. However, the Q_{it} is extracted to be 109.28 nC/cm^2 by subtracting the Q_{mobile} from the Q_{total}, which indicates that 77.46% of the electrons in the inversion channel induced by the MOS capacitance are captured by the interface traps and only 22.54% of electrons are mobile. The μ_{Hall} and the Q_{it} reflects on the quality of the 4H-SiC/SiO$_2$ interface and provides helpful guidelines for further improving the inversion channel electrons mobility issues in 4H-SiC power devices.

ACKNOWLEDGMENTS

This work was supported in part by the National Natural Science Foundation of China under Grant 62174029, in part by the Fund for Transformation of Scientific and Technological Achievements of Jiangsu Province under Grant BA2023001, in part by the Research and Development Plan of Jiangsu Province under Grant BE2022073 and Grant BE2022048-3, in part by the Distinguished Young Scientists Foundation of Jiangsu Province under Grant BK20230025, in part by the Fundamental Research Funds for the Central Universities under Grant 2242024RCB0028, in part by the Important Special Project of Nanjing City under Grant 2021-11004, and in part by the Fund for Transformation of Scientific and Technological Achievements of Wuxi City under Grant C20231021.

REFERENCES

[1] Baliga B J, Silicon carbide power devices. World scientific, 2006.

[2] Dhar S, Haney S, Cheng L, et al, "Inversion layer carrier concentration and mobility in 4H–SiC metal-oxide-semiconductor field-effect transistors," Journal of Applied Physics, vol.108, No.5, 2010.

[3] Kimoto, Tsunenobu, and James A. Cooper, Fundamentals of silicon carbide technology: growth, characterization, devices and applications, 2014.

[4] Saks, N. S., and A. K. Agarwal, "Hall mobility and free electron density at the SiC/SiO2 interface in 4H–SiC," Applied Physics Letters, vol.77.20, pp.3281-3283, 2000.

[5] Ouisse, T, "Electron transport at the SiC/SiO2 interface," physica status solidi (a), vol.162.1, pp.339-368, 1997.

[6] Saks, N. S, "Hall effect studies of electron mobility and trapping at the SiC/SiO 2 interface," Silicon carbide: Recent major advances, pp.387-410, 2004.

[7] E. Arnold and D. Alok, "Effect of interface states on electron transport in 4H-SiC inversion layers," IEEE Transactions on Electron Devices, vol.48, no.9, pp.1870-1877, Sept. 2001.

[8] Uhnevionak V, Burenkov A, Strenger C, et al, "Comprehensive study of the electron scattering mechanisms in 4H-SiC MOSFETs," IEEE Transactions on Electron Devices, vol.62, No.8, pp.2562-2570, 2015.

979-8-3503-6184-1/24 $31.00 © 2024 IEEE

A Superior SiC Lateral MOSFET with Patterned P-bury Layer Made on N-type Wafers

Xuke Yan[1], Junji Cheng*[1], Xiaojun Fu*[2], Bo Yi[1], Haimeng Huang[1] and Hongqiang Yang[1]

[1] State Key Laboratory of Electronic Thin Films and Integrated Devices, University of Electronic Science and Technology of China, Chengdu 611731, China
[2] The 24th Research Institute of China Electronics Technology Group Corporation, Chongqing 400060, China

* Email: chengjunji2005@126.com and xjfu2000@163.com

Abstract—**A new structure of silicon-carbide (SiC) lateral metal-oxide-semiconductor field-effect-transistor (MOSFET) is proposed. It features a patterned P-bury layer realized by ion-implantation on an N-type wafer. Compared to the conventional SiC lateral MOSFET with P-bury layer made by epitaxy process, since the process of ion-implantation provides the P-bury layer with more application space for the technology of optimum variation lateral doping (OPTVLD), the performance of the proposed device is significantly improved. According to the simulation results, the proposed device exhibits a breakdown voltage as high as 1023 V and a specific on-resistance as low as 1.81 mΩ·cm², resulting in a 43% higher Baliga's figure of merit than that of the conventional one.**

Keywords—SiC Lateral MOSFET, OPTVLD, BV, $R_{on,sp}$.

I. INTRODUCTION

Silicon carbide (SiC) lateral metal-oxide-semiconductor field-effect-transistor (MOSFET) combines the material advantages of SiC and the integrative benefits of lateral devices, presenting expansive application space for the monolithic integration in power and RF circuits [1−2].

The primary task in making a SiC lateral MOSFET is to construct a grounded P-type substrate located below its N-drift region. One of the current methods is to epitaxially grow the N-drift on an epitaxial P-type layer on top of a heavily doped P-type wafer [1−2]. Although this method contributes to the reduced-surface-field (RESURF) effect, the fabrication of P-type wafers encounters serious issues such as aluminum source depletion [3], difficulty in controlling single crystallinity and doping concentration [4], which leads to the P-type substrate with unsatisfactory quality.

To eliminate the dependency on P-type wafers, researchers have proposed epitaxially growing a grounded P-bury layer on a high voltage N-type wafer to serve as the P-type substrate [5]. This method uses N-type wafers which are feasible, but like the P-type wafers, the epitaxial P-bury layer also exhibits significant quality issues. More importantly, the epitaxial P-bury layer cannot be patterned, as well as that it is difficult to use some technologies such as the variation lateral doping (VLD) to further improve the device performance. Up to now, there is no report to achieve a SiC lateral MOSFET that neither relies on P-type wafers nor employs P-type epitaxy, which seriously restricts its development.

In this paper, a new structure of SiC lateral MOSFET that breaks the mentioned technical bottleneck is reported. Its main innovation lies in achieving a patterned P-bury layer on an N-type wafer through the ion implantation process. Additionally, by using the optimum VLD (OPTVLD) technology to optimize the P-bury layer, a significantly improved device performance that being superior than the prior art is obtained.

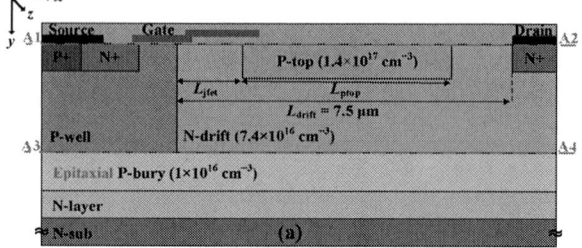

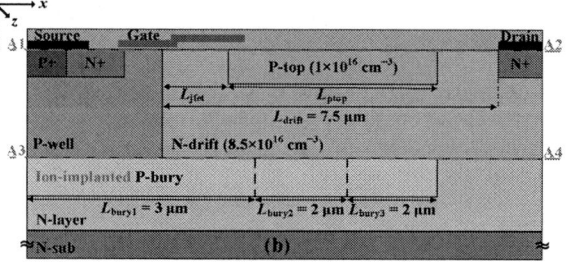

Fig. 1. Structures of SiC lateral MOSFETs: (a) the conventional device with epitaxial P-bury and (b) the proposed device with implanted OPTVLD P-bury.

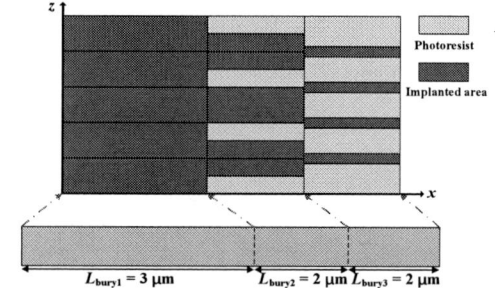

Fig. 2. Layout view of the OPTVLD mask for the P-bury layer.

II. STRUCTURES AND MECHANISM

Fig. 1(a) and Fig. 1(b) illustrates the conventional SiC lateral MOSFET with an epitaxial P-bury layer and the proposed one with a P-bury layer made by ion-implantation, respectively. The process for the latter starts with a N-type epitaxial layer on a N-type substrate wafer. Then, a P-type ion-implantation is performed while using the mask as shown in Fig. 2, which aims to form the VLD P-bury by performing single implantation. Subsequently, the N-drift is epitaxially grown and the P-top and P-well regions are formed through ion-implantation. At last, after the fabrication of active regions and electrodes, the proposed device is completed.

It is obvious that the epitaxial P-bury in the conventional device cannot be patterned [1−2]. In comparison, if the implanted P-bury layer in the proposed device is designed

979-8-3503-6184-1/24 $31.00 © 2024 IEEE   825

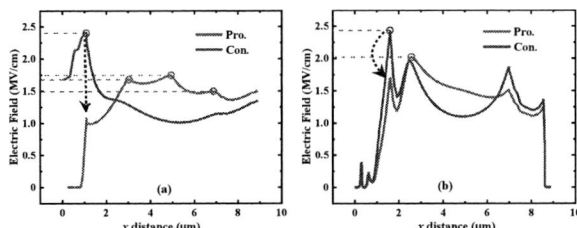

Fig. 3. Critical electric field distribution along (a) A3-A4 and (b) A1-A2.

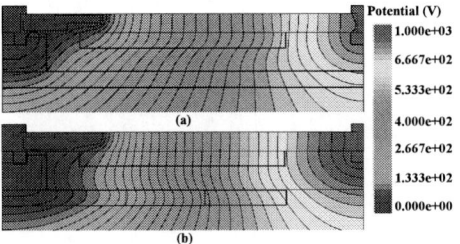

Fig. 4. Potential contours in the (a) conventional and (b) proposed devices.

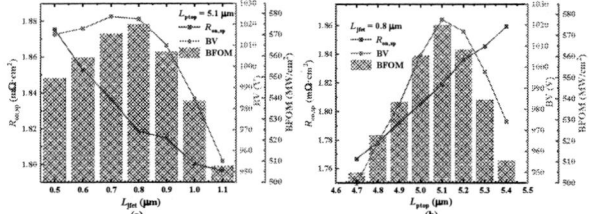

Fig. 5. Dependence of $R_{on,sp}$, BV and BFOM on (a) L_{jfet} and (b) L_{ptop}.

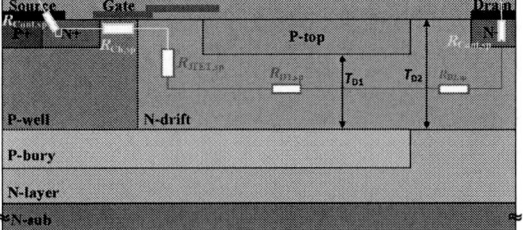

Fig. 6. The resistance distribution in the proposed device.

according to the OPTVLD theory [6], the surface electric field distribution could be optimized and the relationship between the breakdown voltage (BV) and specific on-resistance ($R_{on,sp}$) could be accordingly improved.

When the devices are working in the off-state, the P-bury layers are depleted. The VLD design in the proposed device introduces three additional electric field peaks on the underside of N-drift, as indicated by the red circles in Fig. 3(a). This alleviates the electric field concentration at the junction under the P-well, as compared in Fig. 3(a). Moreover, the high electric field at the gate end is also alleviated, as shown in Fig. 3(b). Fig. 4 compares the potential contours in the case of critical breakdown. It can be observed that because the VLD design makes the lateral distribution of the potential contours along the P-bury layer more uniform, the surface electric field distribution is significantly improved.

III. RESULTS AND DISCUSSION

Sentaurus that can simulate the processes such as ion-implantation is used for structure establishment. The mobility models are calibrated based on experimental data to achieve a more realistic channel mobility. The optimization of the conventional device is routine. But to get optimized electric field distribution in the proposed device, L_{jfet}, L_{ptop}, the doping concentration of P-top (N_{ptop}) and the parameters of VLD P-bury need to be designed. Fig. 5(a) and Fig. 5(b) show the dependence of $R_{on,sp}$ and BV on L_{jfet} and L_{ptop}, respectively. The total $R_{on,sp}$ can be expressed as:

$$R_{on,sp} = R_{Drift,sp} + R_{Ch,sp} + R_{Cont,sp} + R_{Other,sp} \quad (1)$$

where $R_{Cont,sp}$ is the specific ohmic contact resistance and $R_{Ch,sp}$ is the channel resistance. Besides, the total specific drift region resistance $R_{Drift,sp}$ consists the junction-field-effect-transistor (JFET) region resistance ($R_{JFET,sp}$), and the resistances of the areas below ($R_{D1,sp}$) and outside ($R_{D2,sp}$) of the P-top RESURF region (as shown in Fig. 6), which is given by [2].

$$R_{Drift,sp} = R_{JFET,sp} + R_{D1,sp} + R_{D2,sp} \quad (2)$$

$$R_{D1,sp} = (L_{ptop} / q\mu_N N_{drift} T_{D1}) \times \text{cell-pitch} \quad (3)$$

$$R_{D2,sp} = (L_{drift} - L_{ptop} - L_{jfet}) / (q\mu_N N_{drift} T_{D2}) \times \text{cell-pitch} \quad (4)$$

where T_{D1} is the thickness of the drift region excluding P-top. T_{D2} is the drift region thickness including P-top. N_{drift} is the doping concentration of N-drift. Symbols of q and μ_N are the electronic charge and electron mobility, respectively. Cell-pitch denotes the cellular distance. The variations in $R_{on,sp}$ with L_{ptop} and L_{jfet} can be well explained by (1) to (4). $R_{D2,sp}$ decreases with the increase in L_{jfet}, leading to a reduction in $R_{on,sp}$. As L_{ptop} increases, because T_{D1} is less than T_{D2}, the increase in $R_{D1,sp}$ is greater than the decrease in $R_{D2,sp}$, resulting in an increase in $R_{on,sp}$. The values of $L_{ptop} = 5.1$ μm and $L_{jfet} = 0.8$ μm contributes to the highest Baliga's figure of merit (BFOM, defined as BV^2/$R_{on,sp}$).

In the design of the VLD P-bury, because the three-sections VLD approximation is proved to be approximate the ideal effect of OPTVLD [6], L_{bury1}, L_{bury2} and L_{bury3} are set as 3 μm, 2 μm and 2 μm, respectively. The three-sections VLD is implemented in one step by using the layout indicated in Fig. 2. In this case, the doping concentrations of N_{pbury1}, N_{pbury2} and N_{pbury3} are proportionally configured [7]. As the potential of P-bury increases from left to right, N_{pbury3} has the smallest value. The dependence of $R_{on,sp}$ and BV on N_{pbury3} is shown in Fig. 7, wherein the maximum BV of 1023 V is achieved at $N_{pbury3} = 7 \times 10^{16}$ cm^{-3}. It should be explained that because the P-top and P-bury jointly deplete N-drift, as N_{pbury3} increases, a transition from under-compensation for N-drift to over-compensation for N-drift occurs. Consequently, N_{drift} in the proposed device can be increased while maintaining BV.

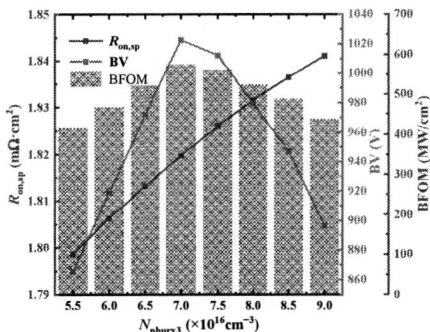

Fig. 7. Dependence of $R_{on,sp}$, BV and BFOM on N_{pbury3}.

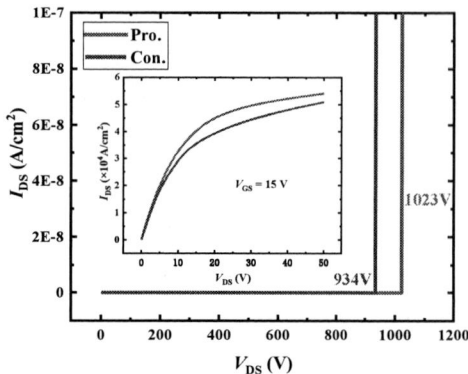

Fig. 8. Comparison of output and breakdown I_{DS}-V_{DS} characteristics.

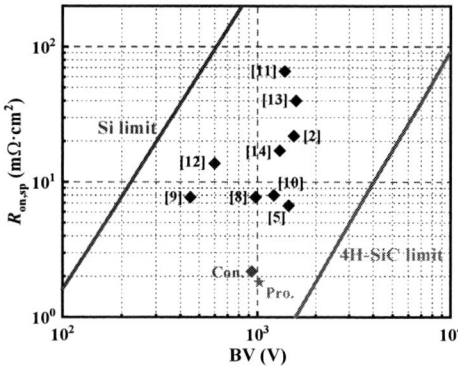

Fig. 9. Comparison of performance between this work and the prior arts.

Based on the above analyses, the optimal parameters for the proposed device are L_{ptop} = 5.1 μm, L_{jfet} = 0.8 μm and N_{pbury3} = 7×10^{16} cm^{-3}. Fig. 8 compares the output and breakdown characteristics. The proposed device obtains the BV of 1023 V, $R_{on,sp}$ of 1.81 mΩ·cm^2 and BFOM of 575 MW/cm^2, which is 9.5% higher, 16.6% lower and 43% higher than that of the conventional one, respectively. Fig. 9 compares the trade-off relationship between BV and $R_{on,sp}$. It can be found that the proposed SiC lateral MOSFET obtains better performance than the conventional device as well as the reported prior art.

IV. CONCLUSION

A new SiC lateral MOSFET with an ion-implanted P-bury layer made on the N-type wafer is proposed and investigated. It not only provides a feasible and promising approach to make a SiC lateral MOSFET that neither relies on P-type wafers nor employs P-type epitaxy process, but also provides the device with significantly improved performance. More specifically, the variation lateral doping design of the P-bury introduces multiple surface electric field peaks and makes the surface electric field distribution more uniform. Besides, because the N-drift is further compensated by the enhanced VLD doping, its doping concentration could be set higher. Therefore, a BV of 1023 V and a $R_{on,sp}$ of 1.81 mΩ·cm^2 are

finally achieved for the proposed device, which are 9.5% higher and 16.6% lower than those of the conventional one, respectively. The BFOM of the proposed device reaches 575 MW/cm^2, which is 43% higher than that of the conventional one, indicating a great improvement on device performance. It is sincerely hoped that this work can bring new promotion and help to the development of SiC lateral MOSFET.

REFERENCES

[1] H. Yu, J. Wang, L. Liu, and K. Sheng, "A Novel SiC LDMOS with Electric Field Optimization by Step Doping Technology," 2020 17th China International Forum on Solid State Lighting & 2020 International Forum on Wide Bandgap Semiconductors China (SSLChina: IFWS), pp. 23-26, Nov 2020.

[2] L. Liu, J. Wang, Z. Zhu, H. Xu, Q. Guo, N. Ren, and K. Sheng, "1540 V 21.8mΩ·cm² 4H-SiC lateral MOSFETs with DOUBLE RESURFs for power integration applications," Solid-State Electronics, vol. 211, pp. 108829, Jan 2024.

[3] N. Schulze, J. Gajowski, K. Semmelroth, M. Laube, and G. Pensl, "Growth of Highly Aluminum-Doped p-type 6H-SiC Single Crystals by the Modified Lely Method," Materials Science Forum, vol. 353-356, pp. 45-48, Jan 2001.

[4] G. Zhong, X. Xie, X. Yang, Y. Peng, X. Chen, X. Hu, and X. Xu, "Growth of High Quality and Low Resistivity Φ100mm p-type 4H-SiC Single Crystal," 2020 17th China International Forum on Solid State Lighting & 2020 International Forum on Wide Bandgap Semiconductors China (SSLChina: IFWS), pp. 15-18, Nov 2020.

[5] Y. Gu, C. Pan, X. Wang, J. Ma, S. Liu, L. Zhang, and W. Sun, "A 1400V SiC LDMOS with P-tops and P-buffer for Ultra-low Specific Resistance," Proc. 2022 16th International Conference on Solid-State and Integrated Circuit Technology (ICSICT), pp. 1-3, Oct 2022.

[6] X. B. Chen, B. Zhang, and Z. J. Li, "Theory of optimum design of reverse-biased p-n junction using resistive field plate and variation lateral doping," Solid State Electron, vol. 35, no. 9, pp. 1365–1370, Sep. 1992.

[7] M. Kong, Y. Duan, J. Gao, R. Yan, B. Zhang, and H. Yang, "A novel optimum variation lateral doping SiC lateral double-diffused metal oxide semiconductor with improved performance," Semiconductor Science and Technology, vol. 37, no. 10, pp. 105022, Sep 2022.

[8] M. Bao, Y. Wang, C. Yu, and F. Cao, "A SiC LDMOS with Electric Field Modulation by a Step Compound Drift Region," Superlattices and Microstructures, vol. 119, pp. 94-102, July 2018.

[9] N. Yun and W. Sung, "Design and Fabrication Approaches of 400–600 V 4H-SiC Lateral MOSFETs for Emerging Power ICs Application," IEEE Transactions on Electron Devices, vol. 67, no. 11, pp. 5005-5011, Nov. 2020.

[10] L. Zhang, J. Ma, Y. Cui, W. Cui, S. Yuan, J. Zhu, and W. Sun, "Simulation Study of A 1200V 4H-SiC Lateral MOSFET With Reduced Saturation Current," IEEE Electron Device Letters, vol. 42, no. 7, pp. 1037-1040, July 2021.

[11] M. Noborio, J. Suda, and T. Kimoto, "4H-SiC Lateral Double RESURF MOSFETs With Low on Resistance," IEEE Transactions on Electron Devices, vol. 54, no. 5, pp. 1216-1223, May 2007.

[12] N. Yun, J. Lynch, and W. Sung, "Demonstration and analysis of a 600 V, 10 A, 4H-SiC lateral single RESURF MOSFET for power ICs applications," Applied Physics Letters, vol. 114, no. 19, pp. 192104, May 2019.

[13] M. Noborio, J. Suda, and T. Kimoto, "1580-V–40-mΩ·cm² Double-RESURF MOSFETs on 4H-SiC (0001)," IEEE Electron Device Letters, vol. 30, no. 8, pp. 831-833, Aug. 2009.

[14] J. Weisse, H. Mitlehner, L. Frey, and T. Erlbacher, "Design of a 4H-SiC RESURF n-LDMOS Transistor for High Voltage Integrated Circuits," Materials Science Forum, vol. 963, pp. 629-632, July 2019.

High Performance Termination Design and Fabrication For SiC MOSFET Device

Lei Huang, Junhou Cao, Chenlu Wang, Hao Fu, Jiaxing Wei*, Siyang Liu*, Weifeng Sun

National ASIC System Engineering Research Center, School of Electronic Science and Engineering, Southeast University, Nanjing, 210096, China

* Email: jiaxingwei@seu.edu.cn, liusy2017@seu.edu.cn

Abstract—In this paper, a high performance termination is designed and fabricated in a SiC MOSFET. The P-Rings are formed in the junction termination extension (JTE) region. By forming P-Rings in the junction termination extension (JTE) region, a termination structure called RA-JTE with high performance can be obtained. Some parameters of the termination are simulated, and the optimal structure is decided by analyzing the simulation result. In this study, the RA-JTE structure is applied to SiC MOSFETs for the first time, and the transition region between the main junction and the JTE region is optimized by expanding the P-Well region of the last cell. Devices with the proposed termination structure have been successfully fabricated and then the HTRB reliability test is carried out. The experimental results show that the SiC MOSFET with RA-JTE termination structure has excellent reliability.

Keywords—termination, SiC, breakdown voltage(BV), electric field, junction-termination-extension(JTE), ring-assited-junction-termination-extension (RA-JTE), JTE dose sensitivity, HTRB, reliability

I. INTRODUCTION

Silicon carbide (SiC) has a larger band gap and a better thermal conductivity compared with silicon. Therefore, SiC MOSFETs are widely used in high-voltage working scenarios. However, because the electric field is crowded near the main junction edge, the breakdown voltage (BV) may drop significantly. To fully exploit the superior voltage supporting capability offered by SiC MOSFET, a highly efficient edge termination structure is needed. Among the traditional termination structures, floating field rings (FFR)[1,2]and junction termination extensions (JTE) [3,4]are most widely adopted. The P-ring injection of the FFR structure and the P-Well injection of the cell share a same step injection process, which avoid the additional injection to simplify the process. However, the performance of the FFR termination structure is significantly influenced by the fluctuation of the width and the spacing of P-rings. And the presence of a large number of floating rings will lead to an increase in the device size. As a comparison, the JTE structure can reduce the area of the termination, but it requires an additional injection process and is very sensitive to the injection dose. In order to obtain the termination structure with both stability and small size, ring-assisted JTE (RA-JTE) by injecting the P-rings into the JTE reign have been reported in SiC PiN diodes[5].

In this paper, the sensitivity of the RA-JTE and the traditional JTE to the dose injection is compared by simulations. Based on Silvaco TCAD simulation, the influence of key parameters are investigated. Moreover, the RA-JTE structure is applied to a 1700V SiC MOSFET for the first time, and the HTRB test is carried out to verify the reliability of the proposed devices.

II. DEVICE STRUCTURE

The cross-sectional structure of the RA-JTE is shown in Fig.1(a). A 11.5μm thick drift layer with a doping concentration of 9.5×10^{15} cm^{-3} grown on a 4H-SiC substrate is adopted. The JTE region is connected to the cell region through an extended P-Well region, acting as a main junction. The partial overlap between the cell region and the JTE region reduce the breakdown risk in the transition region. In order to obtain a better continuity with the cellular region, P-rings and P-wells are injected in the same step injection process. The device structure is also monitored by SEM. The cross-section diagram of a RA-JTE structure with a JTE region width of 50μm and three P-Rings is shown in Fig1.(b).

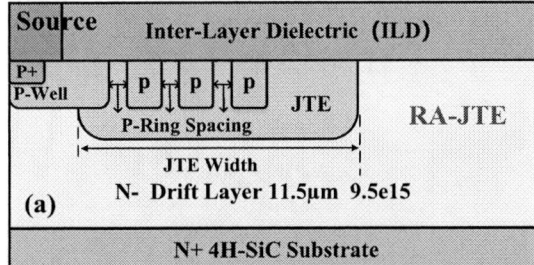

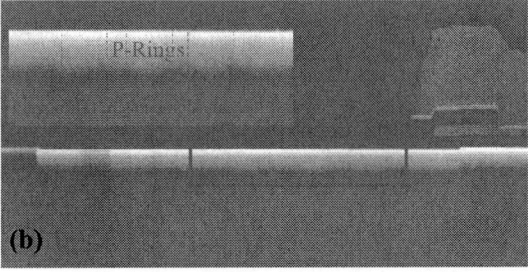

Fig.1 Schematic termination structures of **(a)** RA-JTE and **(b)** the cross -section of the RA-JTE termination with a JTE width of 50um

III. SIMULATION RESULTS AND DISCUSS

Devices with JTE and RA-JTE termination structures are designed and compared by simulations. Several key parameters which affect the performance of the RA-JTE are also discussed. To investigate the influence of the key parameters, including the JTE width, the P-rings spacing and P-Rings number of RA-JTE. The P-rings spacing is calculated as $S=S_0+\alpha(N-1)$, where N is the number of the P-ring, the S_0 is the spacing between the main junction and the first ring, the α is the multiplication coefficient and the S is the spacing between the N P-ring and the (N-1) P-ring.

979-8-3503-6184-1/24 $31.00 © 2024 IEEE

The presence of P-rings in the JTE region makes the electric field more evenly distributed, as shown in Fig.2(a) and Fig.2(b). Because the electric field distribution is more uniform, the RA-JTE termination structure has a better voltage resistance under low dose injection conditions in the JTE region. Fig.2(c) shows the specific electric field distribution values along A to A'. The peak electric field at the main junction is alleviated by the three peak electric fields introduced by the P-rings. The peak electric field of the main junction is then decreased from 2.7 (MV/cm) to 2.5(MV/cm). Similarly, the leakage current paths of the two structures are analyzed by simulation in Fig. 3. Obviously, the leakage current at the main junction of the RA-JTE structure is inhibited by introduced P-rings.

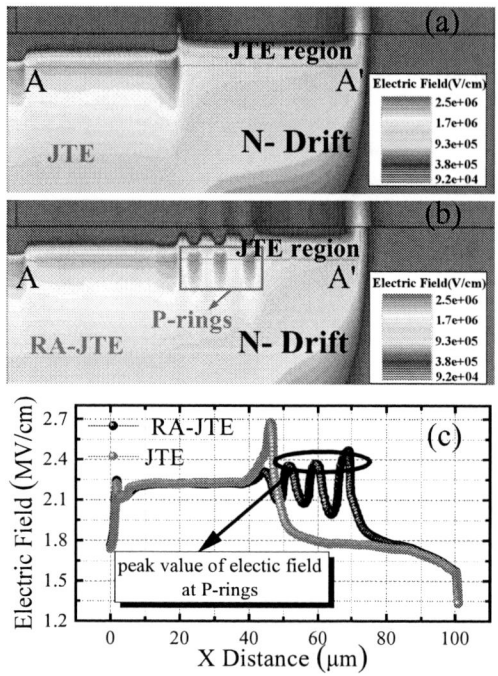

Fig.2 Simulated electric field distribution of JTE(a) and RA-JTE (b) and the electric field value from A-A' of JTE and RA-JTE under 1500V

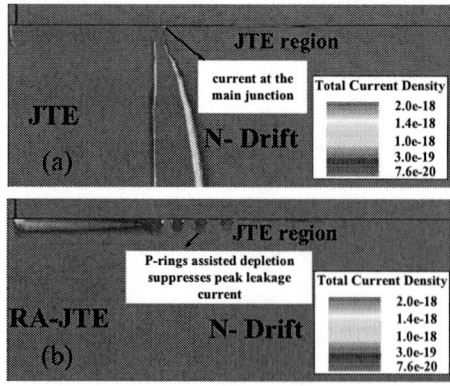

Fig.3 Simulated Leakage Current of JTE(a) and RA-JTE (b) under 1500V

As shown in Fig.4, the BV of the device with the JTE termination structure decreases sharply as the JTE region

injection dose decreases, while the BV of the RA-JTE termination structure is more uniform. The RA-JTE structure maintains BV stability over a wider injection dose range. Therefore RA-JTE structure has a wider design window. However, at high doses, the effect of P-ring injection becomes very weak. Moreover, due to the presence of P-rings, the BV of the RA-JTE will descend under the condition of high injection dose and high JTE region width.

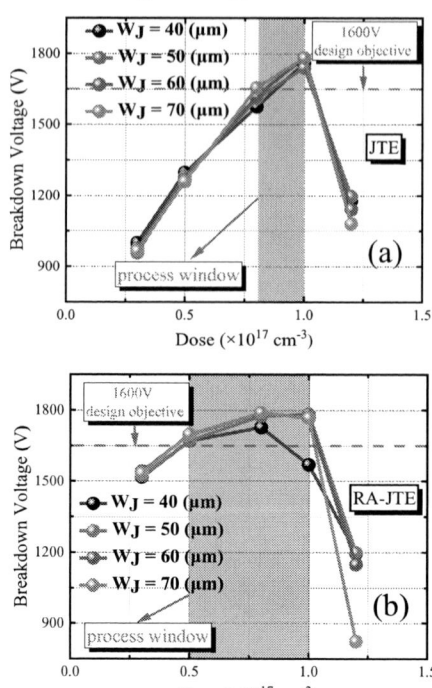

Fig.4:Simulated breakdown voltage of (a) JTE and (b) RA-JTE versus JTE dose (W_J is the width of JTE region)

The variation trend of BV with the JTE width is shown in Figure 5(a). The BV of the device increases with the increase of the termination width until it reaches 50μm and becomes stable. The reason why the BV remains stable is the influence of the JTE region width varying on the electric field distribution becomes weak probably. In Fig.5(b), the more P-rings injected into the JTE region, the higher the BV is. This is because more P-rings introduce more electric field peaks in to the termination. Similarly, when the number of P-rings increases to a certain number, the effect of numbers of P-rings on the BV becomes weaker. Therefore, in this device preparation, considering the process strip, only three P-Rings are injected into the JTE region.

In this study, the effects of the S_0 (the spacing between the first P-Ring and main junction) and α(multiplication factor) on breakdown voltage were also studied. As shown in Fig.5(c) and Fig.5(d), the fixed α is equal to 1, and the breakdown voltage is almost constant at different spacing between the first P-Ring and main junction. Similarly, when S_0 is fixed, the breakdown voltage remains almost constant with the change of α. It indicates that the RA-JTE structure has a high tolerance for P-Rings injection process.

979-8-3503-6184-1/24 $31.00 © 2024 IEEE   829

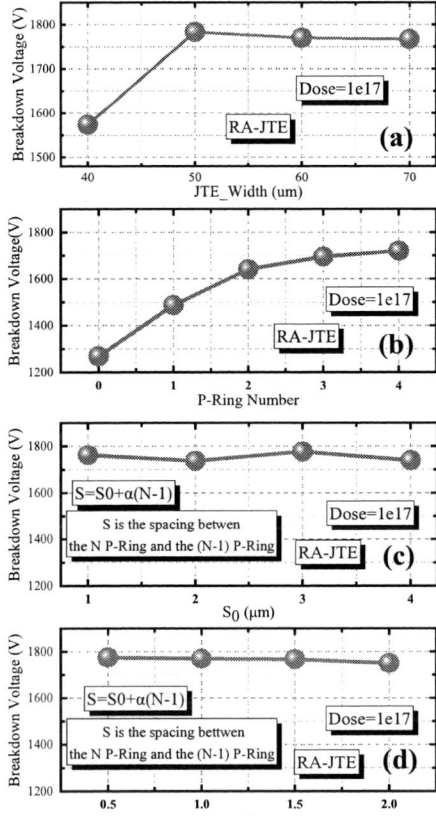

Fig.5:Simulated breakdown voltage of RA-JTE varies with (a) JTE width, (b) P-Rings number, (c) S_0(the spacing between the main junction and the first ring) and (d) α (multiplication coefficient)

IV. EXPERIMENTAL RESULTS

Based on the above analysis, terminations with different JTE widths, spacing of P-Rings and number of P-Rings are fabricated. In this study, HTRB reliability experiments are carried out on devices with RA-JTE termination structure to investigate whether the RA-JTE structure can maintain high reliability.

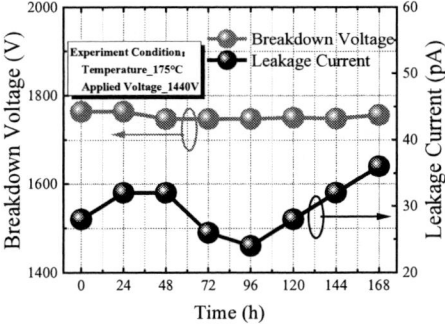

Fig.6 Breakdown voltage and current of different termination with increasing time of HTRB stress

The HTRB reliability experiments set at 175°C and V_{ds}=1440V are carried out. Fig.6 shows that the RA-JTE have almost unchanged BV and leakage current after 168h stress. It indicates that the RA-JTE has a strong robustness in the environment of high temperature and high voltage.

V. CONCLUSION

The RA-JTE termination structure is proposed and applied to SiC MOSFETs, and SiC MOSFETs using this termination structure with BV over 1700V are successfully fabricated. The TCAD simulations show that the RA-JTE structure can maintain a high BV over a larger JTE injection dose range compared to the traditional JTE structure. The influence of some key parameters on BV is also discussed. In addition, the HTRB tests are carried out, and the RA-JTE shows a high reliability. Simulation and experimental results demonstrate that the RA-JTE maintains high reliability while obtaining both high BV and small size simultaneously.

ACKNOWLEDGMENT

This work was supported in part by the National Natural Science Foundation of China under Grant 62174029, in part by the Fund for Transformation of Scientific and Technological Achievements of Jiangsu Province under Grant BA2023001, in part by the Research and Development Plan of Jiangsu Province under Grant BE2022073 and Grant BE2022048-3, in part by the Fundamental Research Funds for the Central Universities under Grant 2242024RCB0028, in part by the Important Special Project of Nanjing City under Grant 2021-11004, and in part by the Fund for Transformation of Scientific and Technological Achievements of Wuxi City under Grant C20231021.

REFERENCES

[1] S-H. Ryu, A. K. Agarwal, "Multiple Floating Guard Ring Edge Termination For Silicon Carbide Devices," U.S. Patent 7 026 650, Apr. 11, 2006 .

[2] H. Onose, S. Oikawa, T. Yatsuo, and Y. Kobayashi, "Over 2000 VFLR termination technologies for SiC high voltage devices," in Proc. ISPSD, Toulouse, France, pp. 245-248, May, 2000.

[3] R. Perez, D. Tournier, et al., "Planar Edge Termination Design andTechnology Considerations for 1.7-kV 4H-SiC PiN Diodes," IEEE Trans. Electron Devices, vol. 52, no. 10, pp. 2309-2316, Oct. 2005.

[4] T. Hiyoshi, et al, "Simulation and Experimental Study on the JunctionTermination Structure for High-Voltage 4H-SiC PiN Diodes," IEEE Trans. Electron Devices, vol. 55, no. 8, pp. 1841–1846, Aug. 2008.

[5] N. Yun and W. Sung, "Detailed Analysis on Determining Effective Dose for Various JTE-Based Edge Terminations Utilized on 4H-SiC Power Devices," in IEEE Transactions on Electron Devices, vol. 69, no. 7, pp. 3826-3832, July 2022.

[6] C. -N. Zhou, Y. Wang, R. -F. Yue, G. Dai and J. -T. Li, "Improved etched multistep JTE for UHV SiC power devices," 2017 International Conference on Electron Devices and Solid-State Circuits (EDSSC), Hsinchu, Taiwan, 2017, pp. 1-2.

[7] C. -N. Zhou, Y. Wang, R. -F. Yue, G. Dai and J. -T. Li, "Implantation-free 2-step junction termination extension with 2-space modulated buffer trench regions for UHV 4H-SÍC GTO thyristors," 2016 IEEE International Conference on Electron Devices and Solid-State Circuits (EDSSC), Hong Kong, China, 2016, pp. 414-417.

[8] X. Huang, B. J. Baliga, A. Q. Huang, A. Suvorov, et. al, "SiC Symmetric Blocking Terminations Using Orthogonal Positive Bevel Termination and Junction Termination Extension," in Proc. ISPSD, Kanazawa, Japan, pp. 179-182, May, 2013.

Analysis of The Separation Degree For P-pillar in SiC Super-Junction Structure Through "Multiple Epitaxy-Ion Implantation" Route

Hao-Bo Kang [1], Hao Yuan* [1], Feng-Yu Du [1], Yu Zhou [1], ke-Yu Liu [1], Xiao-Yan Tang [1], Chao Han [1,2], Qing-Wen Song [1,2], Yu-Ming Zhang [1,2]

[1] The Key Laboratory of Wide Band-Gap Semiconductor Materials and Devices, Xidian University, Xi'an 710071, China
[2] The Xidian-Wuhu Research Institute, wuhu 241000, China.

* Email: 21111110530@stu.xidian.edu.cn, haoyuan@xidian.edu.cn

Abstract—A new process method is analyzed to address the issues of long experimental cycles and high wafer surface stress in current "multiple epitaxy-ion implantation" process route for silicon carbide (SiC) super-junction. Due to the fact that the maximum depth of conventional Aluminum ion implantation is generally ~0.8μm (700 keV) and lower diffusion coefficient in SiC, it is advisable to increase the longitudinal spacing of the p-pillar in super-junction to reduce the number of epitaxial growth cycles. Therefore, the number of epitaxial growth cycles from 9 to 12, for a 10 μm super-junction, are selected to analyze the degree of separation for p-pillar formed by ion implantation in this paper. Simulation results show that the breakdown voltage, charge balance, specific on-resistance and anti-charge bias ability of super-junction with four different degrees of separation (0.3, 0.2, 0.1, 0μm) are not clearly changed. Furthermore, the forward recovery characteristics does not significantly degrade among 0.2, 0.1, 0μm three separation degrees. Therefore, it can be seen that this approach has high feasibility which effectively shortens the experimental period (reduced by 16.6%) of preparing super-junction and reduces the risk of crystal cracking due to stress problems. That provides valuable reference for design and production of the SiC super-junction.

Keywords—Silicon Carbide, Super-Junction, Schottly Barrier Diode, Forward Recovery

I. INTRODUCTION

With the rapid development of the power system, the withstand voltage level of semiconductor power switching devices continues to increase. Therefore, for silicon carbide power devices, as the withstand voltage level increases, the device resistance will also increase, which will lead to a significant increase in power consumption in the forward conduction state. In order to minimize the on-resistance in the high-voltage field, one way is to use bipolar devices and utilize their own conductivity modulation effect to decrease resistance, represented by IGBT, PIN, and thyristor. However, due to bring in the minority carrier injection, which will increase the switching time and reduce the switching speed of the device. At the same time, the extra pn-junction will increase the forward conduction voltage drop, which restricts the use of bipolar devices. Compared with traditional structure devices, introducing a super-junction structure in the unipolar devices can significantly trade-off the relationship between breakdown voltage and specific on-resistance [1]. At the same withstand voltage level, its specific on-resistance is obviously reduced compared to the traditional devices, which can successfully break the one-dimensional theoretical limit of silicon carbide unipolar device.

At present, there are three main technical routes for preparing silicon carbide super-junction device. The "trench etching-sidewall ion implantation" route, represented by professor Kuang Sheng Zhejiang University team, has realized a series of achievements [2-4]. Although it avoided the problems of long experiment period and increased wafer surface stress caused by multiple epitaxies, the demand for high voltage level is limited by the depth of silicon carbide etching and the quality of trench dielectric refilling. Similarly, the "trench etching-epitaxial refilling" route has also achieved some results [5], but it is also limited by factors such as etching depth, p-type epitaxial refilling quality, and grinding process. As is well known, for super-junction, the narrower cell pitch will bring higher BFOM [6-7]. While the before two routes inevitably have the problem is that along with the etching depth increasing, the opening becomes wider and wider, which seriously restricts the further reduction of super-junction cell pitch and prevents the achievement of high aspect ratio.

In response to the above issues, we have adopted the "multiple epitaxy-ion implantation" route to prepare super-junction, which has the biggest advantage of obtaining a higher aspect ratio and more effectively breaking the one dimensional theoretical limit. AISI Japan team has prepared 3300V SJ UMOS, including 16 times epitaxial of semi super-junction and 28 times epitaxial of full super-junction [8], but this will greatly prolong the experiment period and pose a risk of fragmentation during the long epitaxial process. In this paper, we propose a new structure that appropriately increases the longitudinal p-pillar spacing in the super-junction, effectively shortening the number of epitaxial growth cycles and reducing the experimental period while ensuring the static and dynamic performance of the device does not degrade.

II. DEVICE STRUCTURE

Based on most ion implantation process conditions, the ion implantation machine can provide a maximum implantation energy of 700keV, resulting in an Al implantation junction depth of ~0.8μm after compensation with N-type epitaxy. If the epitaxial thickness is exactly 0.8μm, 12 epitaxial growth cycles are required to achieve a 10 μm length of super-junction. To reduce the number of epitaxial growth cycles, the thickness of each epitaxial growth must be more than 0.8μm, resulting in slight interruptions at the bottom of the p-pillar after each ion implantation, as shown in *h* (p-pillar separation degree) in Figure 1.

979-8-3503-6184-1/24 $31.00 © 2024 IEEE

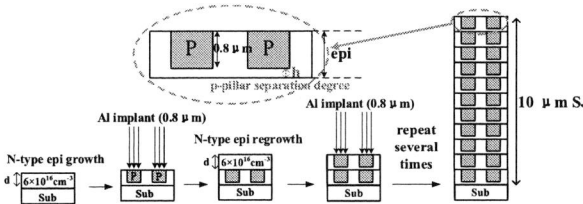

Fig. 1. Schematic diagram of prepared super-junction device using the "multiple epitaxy-ion implantation" process route

TABLE I. DEVICE TYPE

Device type	A	B	C	D
Epitaxial growth cycles	9	10	11	12
Single epitaxial growth thickness (μm)	1.1	1	0.9	0.8
Separation degree of p-pillar (μm)	0.3	0.2	0.1	0

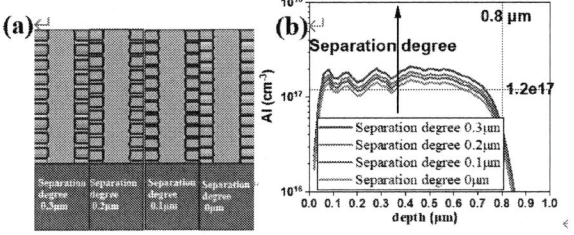

Fig. 2. (a) Schematic diagram of device structure under different separation degree of p-pillar. (b) Concentration distribution of p-pillar under different separation degree of p-pillar (one cycle).

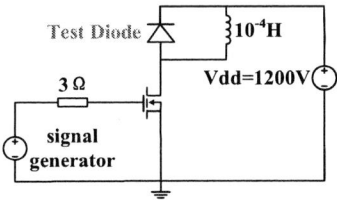

Fig.3. Double pulse test circuit diagram

The device A, B, C, D with four p-pillar separation degree of 0.3, 0.2, 0.1, 0μm are shown in Table I and Figure 2 (a), they were prepared under the premise of a 10 μm length of super-junction and a 0.8μm depth of p-pillar at each epitaxial growth. For the 10μm super-junction structure 0.3, 0.2, 0.1, and 0μm p-pillar separation degrees were selected in sequence to achieve it and explore separation degree of p-pillar on the forward recovery characteristics and anti-charge bias characteristic of the super-junction SBD. As the separation degree increases, the concentration of p-pillar is increased after each epitaxy growth as shown in Figure 2 (b). The optimal separation degree of p-pillar suitable for the 10 μm super-junction was selected based on the breakdown voltage, specific on-resistance, forward recovery (test circuit is shown in Figure 3) and anti-charge bias characteristic.

III. SIMULATION RESULTS AND DISCUSS

Figure 4(a) shows the 1D electric field distribution along the n-pillar centerline at the breakdown time of four types devices. It can be observed that their breakdown voltages are consistent maintaining at 1600V, indicating that the charge balance is not influenced. Then the forward conductivity characteristic of four devices was shown in Figure 4(b) and their I-V curves are basically overlapped, the specific on-

resistance ($R_{on,sp}$) reaching at 0.4mΩ*cm². Finally, Figure 4(c) shows the anti-charge bias characteristic, the four curves roughly overlap and have a consistent distribution. As a result, the increase of p-pillar separation degree from 0 to 0.3μm has no essential effect on breakdown voltage, charge balance, specific on-resistance, and the anti-charge bias ability.

The forward recovery curves of four types devices are shown in Figure 5. When a constant forward current (24A) is applied to the test device in reverse bias state, the forward recovery capacity of device A is the worst, with a large voltage drop peak and a stable voltage drop value up to 2.01V. In contrast, the capacity of forward recovery for B, C, D device are relatively acceptable. There are no obvious voltage drop peaks and no significant deviation from the standard value of 1.6V after stabilization, which basically meets the design requirements.

The forward recovery capability is determined by the contraction of depletion layer in drift region, and the most important factor affecting the contraction of depletion layer is the p-pillar separation degree. If the h (shown in Figure 1) is too long during the reverse bias state, which will cause the p-pillar potential far away from anode to be too high as shown

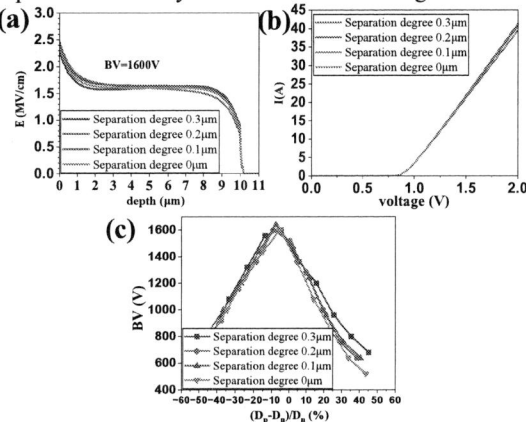

Fig.4. (a) 1D electric field distribution along the n-pillar centerline at the breakdown time of four types devices. (b) Positive I-V curves of four types devices. (c) Anti-charge bias ability of A, B, C, D device

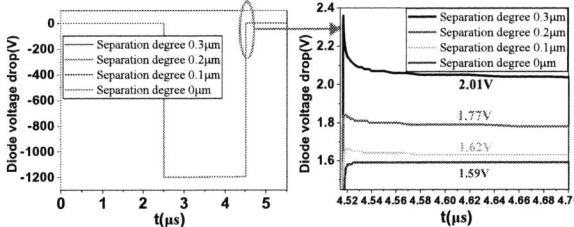

Fig.5. Forward recovery I-V curve of A, B, C, D device.

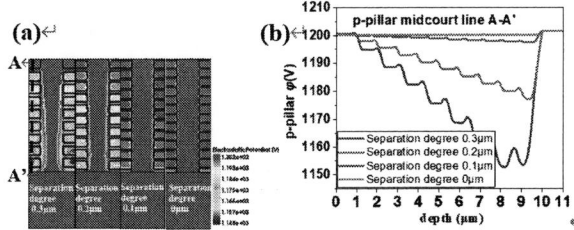

Fig.6 (a) Schematic diagram of depletion layer and 2D dimensional distribution of electric potential and (b) p-pillar potential 1D dimensional distribution after forward recovery of A, B, C, D devices.

in Figure 6. During the process of device transitioning from reverse bias to forward bias, a higher potential difference can prevent the anode holes from injecting up to the bottom of p-pillar in time, which will cause the depletion layer can not contract well. Next, the forward current path in n-pillar will be severely compressed and narrower than before, which can cause an increase of on-resistance. As a result of this, under the premise of a constant applied current of 24A, the device voltage drop shows an overshoot peak during the forward recovery process and cannot recover to the standard voltage drop (1.6V) after stabilization.

The reason for the worst forward recovery capacity of device A (in Figure 5) is due to its maximum p-pillar separation degree (h=0.3μm), which results in the least depletion layer contraction and the highest on-resistance. When the applied current is the same, the voltage drop (2.01V) is higher than other devices. Meanwhile the B, C, D device have negligible p-pillar separation degree (h=0.2, 0.1, 0μm), resulting in a more complete contraction of p-pillar depletion layer during the forward recovery process.

IV. Summary

Overall, a kind of SiC super-junction preparation method was analyzed and demonstrated which could shorten the experimental period and reduce the risk of fragmentation. As to 10μm SiC super-junction, the required number of epitaxy cycles was reduced from 12 to 10 (reduced by 16.6%) through "multiple epitaxy-ion implantation" process route. According to simulation results, the breakdown voltage, charge balance, specific on-resistance, anti-charge bias, and forward recovery ability of super-junction with 0.2, 0.1, 0μm three different degrees of separation are not apparent changed, which proves the feasibility of this scheme. Meanwhile, this method also provides valuable reference for further optimization of the preparation process to super-junction.

Acknowledgment

This work was supported in part by the National Key R&D Program of China under Grant 2021YFB3601800 and the Major Projects of Shanxi Province (Grant no. 202101030201001)，in part by the Natural Science Basic Research Program of Shaanxi under Grants 2024JC-YBMS-474 and 2024JC-YBQN-0647.

References

[1] S. Harada et al., "First Demonstration of Dynamic Characteristics for SiC Superjunction MOSFET Realized using Multi-epitaxial Growth Method," 2018 IEEE International Electron Devices Meeting (IEDM), San Francisco, CA, USA, 2018, pp. 8.2.1-8.2.4

[2] X. Zhong, B. Wang, J. Wang and K. Sheng, "Experimental Demonstration and Analysis of a 1.35-kV 0.92-m $\Omega\cdot$cm2 SiC Superjunction Schottky Diode," in IEEE Transactions on Electron Devices, vol. 65, no. 4, pp. 1458-1465, April 2018

[3] B. Wang, H. Wang, C. Wang, N. Ren, Q. Guo and K. Sheng, "Design and Fabrication of 1.92 kV 4H-SiC Super-Junction SBD With Wide-Trench Termination," in IEEE Transactions on Electron Devices, vol. 68, no. 11, pp. 5674-5681, Nov. 2021

[4] H. Wang, C. Wang, B. Wang, N. Ren and K. Sheng, "4H-SiC Super-Junction JFET: Design and Experimental Demonstration," in IEEE Electron Device Letters, vol. 41, no. 3, pp. 445-448, March 2020

[5] R. Kosugi et al., "Breaking the Theoretical Limit of 6.5 kV-Class 4H-SiC Super-Junction (SJ) MOSFETs by Trench-Filling Epitaxial Growth," 2019 31st International Symposium on Power Semiconductor Devices and ICs (ISPSD), Shanghai, China, 2019, pp. 39-42

[6] R. Ghandi, A. Bolotnikov, S. Kennerly, C. Hitchcock, P. -m. Tang and T. P. Chow, "4.5kV SiC Charge-Balanced MOSFETs with Ultra-Low On-Resistance," 2020 32nd International Symposium on Power Semiconductor Devices and ICs (ISPSD), Vienna, Austria, 2020, pp. 126-129

[7] C. Wang et al., "Performance Limit and Design Guideline of 4H-SiC Superjunction Devices Considering Anisotropy of Impact Ionization," in IEEE Electron Device Letters, vol. 43, no. 12, pp. 2025-2028, Dec. 2022

[8] M. Baba, T. Tawara, T. Morimoto, S. Harada, M. Takei and H. Kimura, "Ultra-Low Specific on-Resistance Achieved in 3.3 kV-Class SiC Superjunction MOSFET," 2021 33rd International Symposium on Power Semiconductor Devices and ICs (ISPSD), Nagoya, Japan, 2021, pp. 83-86.

Numerical Analysis of the CIBL Effect on Short-Circuit Characteristics of DG-CSTBTs with Reduced Mesa Width

Zhengyu Lang [1], Jinping Zhang*[1 2 3], Shiwei Zheng [1], Shuyang Huang [1], Haonan Deng [1], and Bo Zhang [1]

[1] National Key Laboratory of Electronic Thin Films and Integrated Devices, University of Electronic Science and Technology of China, Chengdu, 610054, P. R. China.
[2] Nanjing SilverMicro Electronics, Nanjing, 211200, P.R. China
[3] Institute of Electronic and Information Engineering of UESTC in Guangdong, Dongguan, 523808, P.R. China

* Email: jinpingzhang@uestc.edu.cn

Abstract—As the mesa width of Trench Insulated Gate Bipolar Transistor (TIGBT) decreases, the Collector Bias Induced Barrier Lowering (CIBL) effect becomes increasingly severe, negatively impacting the device performance. In this paper, short-circuit characteristics of Carrier Stored Trench Bipolar Transistor with Dummy Gate (DG-CSTBT) with various mesa widths were studied by the Sentaurus TCAD. The simulation results indicate that reducing the mesa width leads to increasing saturation current. It is caused by the carrier injection enhancement effect because of the narrower mesa. In addition, the current unsaturation phenomena occur in the narrow mesa DG-CSTBT, which is attributed to the CIBL effect. The study shows, with the same V_{ce}, DG-CSTBT with narrower mesa exhibits wider depletion regions in Pbody, which can be interpreted as a reduction in channel length, the CIBL effect is induced. The short-circuit characteristics are degraded. Moreover, with appropriate P_b and T_{ox}, the CIBL effect in DG-CSTBT is suppressed and the short-circuit failure time is increased without increasing V_{th}.

Keywords—Carrier Stored Trench Bipolar Transistor (CSTBT), dummy gate, mesa width, CIBL effect.

I. Introduction

Insulated Gate Bipolar Transistors (IGBTs) demonstrate excellent overall performance and have been widely utilized in various power electronic systems. In the on-state, the IGBT drift region undergoes conductivity modulation due to the injection of numerous holes and electrons, resulting in a low on-state voltage (V_{ceon}). Structures such as Carrier Stored Trench Bipolar Transistor (CSTBT) [1], Floating-P IGBT (FP-IGBT) [2], Injection Enhanced Gate Transistor (IEGT) [3], and Partially-Narrow-Mesa (PNM) IGBT [4] have been proposed to enhance carrier injection effect and reduce the V_{ceon}. Currently, narrowing the mesa width is a commonly employed method to meet the industry's demand for IGBT with lower V_{ceon} [5]. However, reducing the mesa width increases the saturation current due to the enhanced channel density in the IGBT. This results in the degradation of the short-circuit capability, which is a critical characteristic of IGBT. To address this issue, a dummy gate structure has been proposed to reduce the saturation current and thereby improve the short-circuit capability [6]. However, when the mesa width is sufficiently small, the reduction of mesa width will trigger the Collector Bias Induced Barrier Lowering (CIBL) effect, leading to a non-saturated current [7]. In this paper, to demonstrate the mechanism of CIBL with extremely narrow mesa width (W_m) and its impact on short-circuit characteristics, a CSTBT with Dummy Gate (DG-CSTBT) is analyzed with Sentaurus TCAD.

II. Device Structures

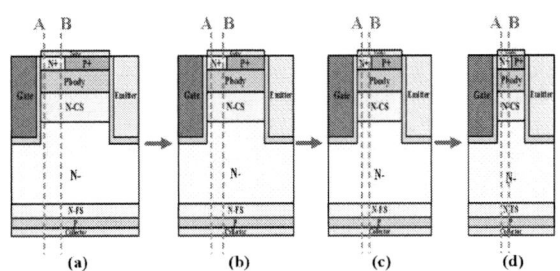

Fig. 1. Device structures, mesa width (a) 0.6μm, (b) 0.4μm. (c) 0.3μm, (d) 0.2μm

TABLE I. Main Device Structure Parameters and Their Values.

Device Structural Parameter	Value			
	a	*b*	*c*	*d*
cell width /μm	1.6	1.4	1.3	1.2
mesa width /μm	0.6	0.4	0.3	0.2
Trench width /μm	0.5			
Trench depth /μm	5			
N+ depth /μm	0.5			
Pbody depth /μm	1.5			

Figure 1 (a) to (d) illustrate cross section views of the DG-CSTBT with mesa widths of 0.6μm, 0.4μm, 0.3μm, and 0.2μm, respectively. The DG-CSTBT features an N-type doped CS layer between the P-body region and the N-region. One of its trenches is connected to the emitter instead of the gate as the dummy gate. In the on-state, the CS layer suppresses the extraction of holes on the emitter side, thereby increasing the conductivity modulation. Consequently, the V_{ceon} is reduced. In addition, thanks to the dummy gate structure, the short-circuit capability has been optimized. The switching loss has been reduced by decreasing the Miller capacitance. In the simulation, the trench width of the DG-CSTBT is kept constant at 0.5μm, and the oxide layer thickness is maintained at 100nm. The doping concentration distribution is identical. The major parameters of the devices are summarized in Table I.

III. Results and Discussion

The output curves of DG-CSTBT with various mesa widths are shown in Figure 2. A DG-CSTBT cell with a W_m of 0.6 μm and a cell longitudinal length (L_c) of 1 μm was used as a baseline. The output characteristics of different mesa widths are compared under the same cell area (S) and under the same L_c, respectively. With the same S, a smaller W_m corresponds to an increased saturation current owing to the increase of channel density. With the W_m decreased, a gradual transition toward unsaturation is observed in the collector current. As the V_{ce} transits from 0V to the rated voltage of 600V, in the case of W_m=0.6μm, 0.4μm, 0.3μm, and 0.2μm, the increases of non-saturated current are 8.9%, 9.8%, 15.2%, and 30.1% respectively. At the same L_c, with V_{ce}=20V, the currents for W_m =0.6μm and 0.4μm are nearly identical. The currents for W_m =0.6μm and 0.4μm are 1.5×10^{-4}A and 1.52×10^{-4}A, respectively. However, the currents for W_m=0.3μm and 0.2μm are 1.6×10^{-4}A and 2.1×10^{-4}A with V_{ce}=20V, respectively. With the same L_c, DG-CSTBT has identical channel lengths, implying that their saturation currents should be nearly identical. However, this trend does not apply to the case of W_m=0.2μm and W_m=0.3μm due to the stronger conductivity modulation in DG-CSTBTs of the narrower mesa. For W_m values of 0.6μm, 0.4μm, 0.3μm, and 0.2 μm, the non-saturated current increases are 8.9%, 9.6%, 14.2%, and 29.1%, respectively. The current unsaturation is caused by the CIBL effect.

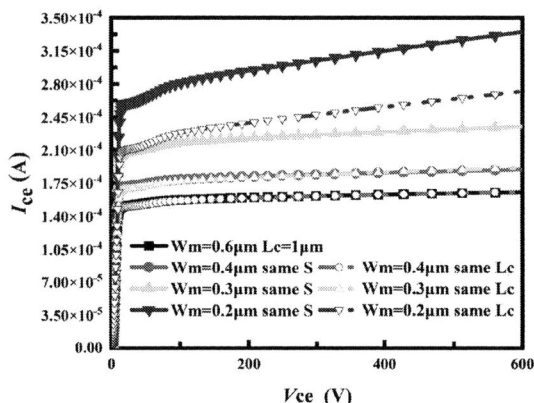

Fig. 2. Output characteristic of DG-CSTBT with various mesa width

The on-state carrier distribution of DG-CSTBT with the same L_c along line A is shown in Figure 3. In the on-state, electrical conductivity modulation occurs within the drift region, improving the on-state resistance of DG-CSTBT. As the mesa width decreases, the conductivity modulation effect is enhanced. Therefore, the saturation current of narrow mesa DG-CSTBT increases.

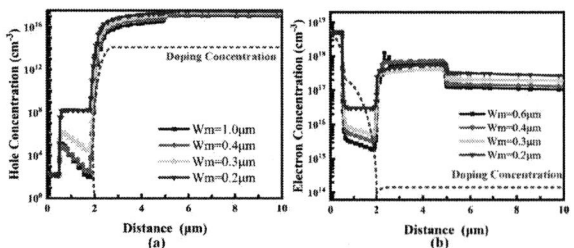

Fig. 3. Carrier distribution in on-state along line A with the same L_c (a) Hole Concentration (b) Electron Concentration

Figure 4 shows a potential distribution along line B with the same L_c for four DG-CSTBTs under various V_{ce}. When the V_{ce} is increased from 2V to 8V, the barriers at the Pbody/N+ junction of the DG-CSTBTs with 0.6μm and 0.4μm mesa width remain unchanged, indicating that the CIBL effect has not manifested. The barriers at the Pbody/N+ junction with 0.3μm and 0.2μm are reduced significantly. In the on-state, a

depletion region is established at the Pbody/NCS junction, which withstands the forward voltage. For a given V_{ce}, the quantity of space charge within the Pbody depletion region remains constant. As the mesa width decreases, a wider depletion region is required to provide the same amount of space charge. It can be interpreted as a reduction in channel length. When the N+/Pbody junction is sufficiently close to the depletion region, the barrier between N+ and Pbody decreases, resulting in the onset of the CIBL effect.

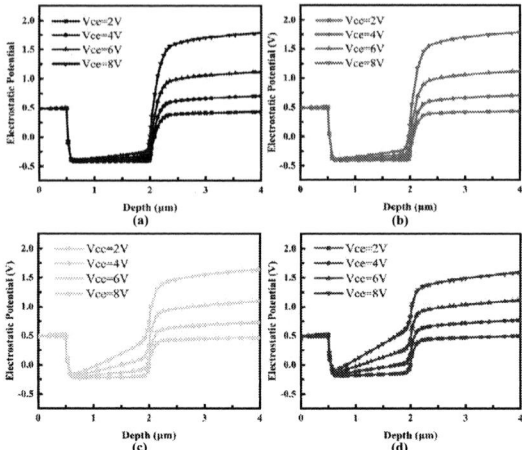

Fig. 4. The electrostatic potential of DG-CSTBT with the same L_c along line B with mesa width (a) 0.6μm (b) 0.4μm (c) 0.3μm (d) 0.2μm

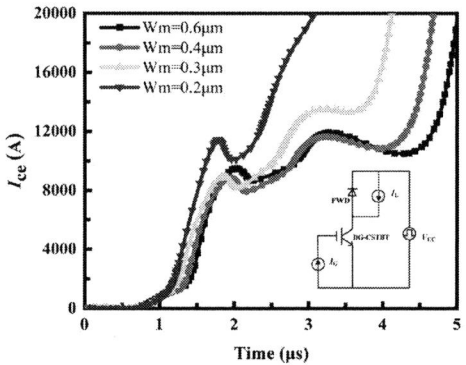

Fig. 5. The short-circuit characteristic curves of DG-CSTBT under different mesa widths and the same L_c=1×10⁸μm.

Figure 5 shows the short-circuit characteristic of DG-CSTBTs with different mesa widths and the same L_c of 1×10^8μm. In the circuit, V_G is 15V, V_{CC} is 600V, R_G is 10Ω and L_s is 10nH. It can be seen that the short-circuit failure times of the DG-CSTBTs are measured at 4.4μs for a mesa width of 0.6μm, 3.8μs for 0.4μm, 3.1μs for 0.3μm, and 2.1μs for 0.2μm. With the reduction of mesa width, it shows a sharp decline in the short-circuit capability of DG-CSTBT. This is due to the fact that narrow mesa DG-CSTBT exhibits a higher current caused by CIBL during short-circuit conditions.

Figure 6 shows the output characteristics of the DG-CSTBT with 0.2μm mesa width under different P_b conditions. It can be observed that as P_b rises from 1×10^{17}cm⁻³, 1.5×10^{17}cm⁻³, 2.5×10^{17}cm⁻³, 3.5×10^{17}cm⁻³, to 5×10^{17}cm⁻³, the increments in saturation current are, 463.2%, 30.1%, 13.1%, 8.6%, 6.1%, respectively. It can be observed that increasing the P_b can effectively weaken the CIBL effect in narrow mesa DG-CSTBT with a narrower depletion region under the same V_{ce}. However, the higher P_b resulted in increasing V_{th}. With an increase in P_b from 1×10^{17}cm^{-3} to 5×10^{17}, the V_{th} is 3.53V, 4.89V, 6.50V, 7.99V, and 9.97V, respectively. Figure 7 compares the electrostatic potential under various V_{ce} with different P_b. As the P_b increases, the barrier of N+/Pbody is raised. The CIBL effect of DG-CSTBT with narrow mesa diminishes.

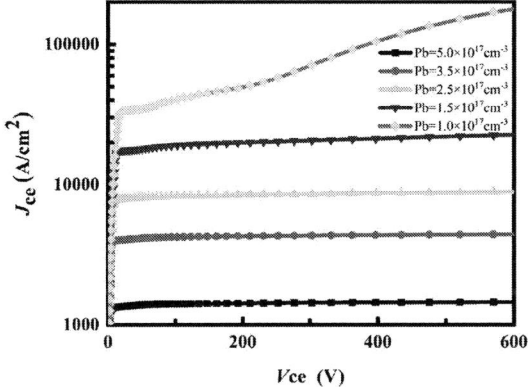

Fig. 6.   The output characteristic curves of DG-CSTBT with a mesa width of 0.2μm under different P_b.

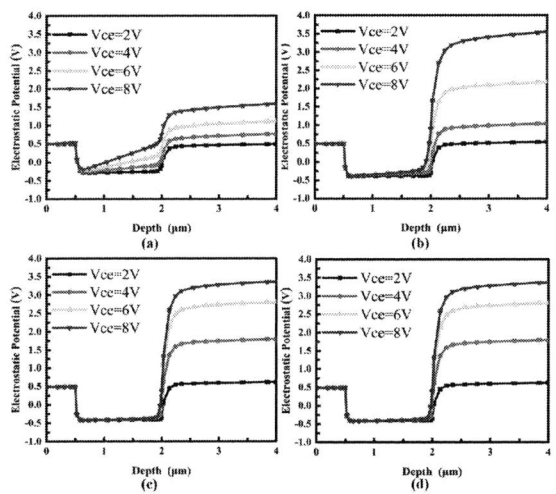

Fig. 7.   The electrostatic potential of DG-CSTBR with 0.2μm mesa width when V_{ce} rise with various P_b (a) 1×10^{17} cm^{-3} (b) 2.5×10^{17} cm^{-3} (c) 3.5×10^{17} cm^{-3} (d) 5×10^{17} cm^{-3}

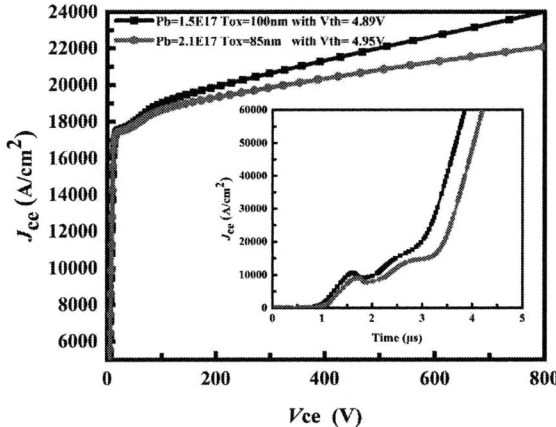

Fig. 8.   DG-CSTBT with 0.2μm mesa width output and the short-circuit curves of (a) P_b=1.5$\times10^{17}$cm^{-3}, T_{ox}=100nm (b) P_b=2.1$\times10^{17}$cm^{-3}, T_{ox}=85nm

Reducing the gate oxide thickness (T_{ox}) will alleviate the increase in V_{th} caused by the rising P_b. Figure 8 compares the output and the short-circuit curves of P_b=1.5$\times10^{17}$cm^{-3}, T_{ox}=100nm and P_b=3.5$\times10^{17}$cm^{-3}, T_{ox}=85nm with similar V_{th}. The result shows that the CIBL effect will be suppressed with V_{th} unchanged by controlling the Tox and Pb. The short-circuit characteristics of CSTBT with narrow mesa have been optimized.

IV. SUMMARY

The DG-CSTBT with various mesa widths are analyzed by TCAD simulation. Results show that the narrow mesa DG-CSTBT exhibits CIBL effect. The narrow mesa enhances carrier injection, leading to a higher saturation current of DG-CSTBT. As V_{ce} increases, DG-CSTBT with the narrower mesa width shows a wider depletion region. When the N+/P-body junction is in close proximity to the depletion region, the barrier height between the N+ and P-body decreases, initiating the CIBL effect. Additionally, because of narrower depletion region under the same V_{ce}, the CIBL effect of DG-CSTBT with higher Pb is weakened. However, the higher P_b results in an increase in the V_{th}. With appropriate P_b and T_{ox}, the CIBL effect in DG-CSTBT is suppressed and the short-circuit failure time is increased without increasing V_{th}.

ACKNOWLEDGMENT

This work was supported in part by the Nanjing Science and Technology Program, China under Grant No.202209009 and Guangdong Basic and Applied Basic Research Foundation, China under Grant No.2023A1515010068.

REFERENCES

[1] H. Takahashi, H. Haruguchi, H. Hagino and T. Yamada, "Carrier stored trench-gate bipolar transistor (CSTBT)-a novel power device for high voltage application," 8th International Symposium on Power Semiconductor Devices and ICs. ISPSD '96. Proceedings, Maui, HI, USA, 1996, pp. 349-352, doi: 10.1109/ISPSD.1996.509513.

[2] K. Oyama et al., "Novel 600-V trench high-conductivity IGBT (Trench HiGT) with short-circuit capability," Proceedings of the 13th International Symposium on Power Semiconductor Devices & ICs. IPSD '01 (IEEE Cat. No.01CH37216), Osaka, Japan, 2001, pp. 417-420, doi: 10.1109/ISPSD.2001.934642.

[3] M. Kitagawa, I. Omura, S. Hasegawa, T. Inoue and A. Nakagawa, "A 4500 V injection enhanced insulated gate bipolar transistor (IEGT) operating in a mode similar to a thyristor," Proceedings of IEEE International Electron Devices Meeting, Washington, DC, USA, 1993, pp. 679-682, doi: 10.1109/IEDM.1993.347221.

[4] M. Sumitomo, J. Asai, H. Sakane, K. Arakawa, Y. Higuchi and M. Matsui, "Low loss IGBT with Partially Narrow Mesa Structure (PNM-IGBT)," 2012 24th International Symposium on Power Semiconductor Devices and ICs, Bruges, Belgium, 2012, pp. 17-20, doi: 10.1109/ISPSD.2012.6229012.

[5] M. Tanaka and I. Omura, "Scaling rule for very shallow trench IGBT toward CMOS process compatibility," 2012 24th International Symposium on Power Semiconductor Devices and ICs, Bruges, Belgium, 2012, pp. 177-180, doi: 10.1109/ISPSD.2012.6229052.

[6] F. Wolter, W. Roesner, M. Cotorogea, T. Geinzer, M. Seider-Schmidt and K. -H. Wang, "Multi-dimensional trade-off considerations of the 750V micro pattern trench IGBT for electric drive train applications," 2015 IEEE 27th International Symposium on Power Semiconductor Devices & IC's (ISPSD), Hong Kong, China, 2015, pp. 105-108, doi: 10.1109/ISPSD.2015.7123400.

[7] K. Eikyu et al., "On the scaling limit of the Si-IGBTs with very narrow mesa structure," 2016 28th International Symposium on Power Semiconductor Devices and ICs (ISPSD), Prague, Czech Republic, 2016, pp. 211-214, doi: 10.1109/ISPSD.2016.7520815.

979-8-3503-6184-1/24 $31.00 © 2024 IEEE

Innovations in GaN HEMT Design: Achieving Superior Power Output and Thermal Management

Shiming Li, [1,] * Bowen Yang, [1] Mei Wu, [1] Ling Yang, [1] Bin Hou, [1] Meng Zhang, [1] Xiaohua Ma, [1] and Yue Hao [1]

[1] School of Microelectronics, Xidian University, Xi'an 710071, China
E-mail: smli@stu.xidian.edu.cn

Abstract—**This study introduces a novel doping design of the GaN buffer layer using Fe/C co-doping on a SiC substrate, aimed at optimizing both thermal and electrical performance in high-power applications. The unique approach involves reducing the average doping concentration, which enhances the thermal conductivity of the GaN buffer layer, thus facilitating more efficient heat dissipation. We compare our novel device structure with conventional Fe/C co-doped GaN HEMTs, demonstrating that our design achieves an output power of 18.5 W (42.67 dBm) at 3.6 GHz with a Power-Added Efficiency (PAE) of 61%. Comparative analysis of the thermal and electrical characteristics underscores the benefits of our approach, with our devices showing lower peak temperatures and improved performance under high-power conditions. his study not only presents a breakthrough in GaN HEMT design but also significantly contributes to the development of more robust and efficient devices for military, satellite, and communication base station applications.**

Keywords—**GaN HEMTs, Fe/C co-doped GaN buffer, self-heating, radio frequency, output power**

I. INTRODUCTION (*HEADING 1*)

Gallium Nitride (GaN) high electron mobility transistors (HEMTs) have rapidly evolved over the past two decades, driven by their extensive bandwidth, high breakdown field, and exceptional electron mobility [1], [2]. GaN HEMTs, which are based on heterojunctions formed between GaN and various nitrides (e.g., AlN, AlGaN, InAlN, InAlGaN), continue to advance in the RF domain [3]. Notable milestones include Wu et al.'s achievement of 30.6 W/mm output power density (P_{out}) at 8 GHz in 2004 using a field plate technique [4], Hao et al.'s attainment of 73% Power-Added Efficiency (PAE) and 13 W/mm P_{out} at 4 GHz in 2011 [5], and Moon et al.'s success in surpassing 70% PAE at 30 GHz in 2021 using a graded-channel technique [6]. More recently, Yang et al. (2023) updated X-band (P_{out} =33.1 W/mm) and Ka-band (P_{out} =14.4 W/mm) results [7], and Li et al. (2024) observed a linear increase in X-band P_{out} through passivation [8]. These advancements underscore the rapid progress of GaN HEMTs and their pivotal role in enhancing military radars, satellites, and communication base stations [9], [10].

Despite these achievements, high power densities in GaN HEMTs lead to significant heat production. As Miccoli reports, peak temperatures can exceed 200 °C at just 4 W/mm of dissipation power (P_{diss}) in GaN-on-Si HEMTs [11]. Thus,

optimizing heat dissipation while maintaining high power output remains a critical challenge. Key factors include the thermal conductivity of the buffer layer, the thermal resistance at the interface between the epitaxial layer and the substrate, and the substrate's thermal conductivity [12].

In response, we propose a novel doping design for the GaN buffer layer on a SiC substrate, aimed at balancing electrical and thermal performance. The SiC substrate, known for its high thermal conductivity, facilitates rapid heat transfer from the device to the package shell. Moreover, we've refined the Fe/C co-doped GaN buffer layer to reduce average doping concentration and enhance thermal conductivity, enabling quicker heat propagation from the trench. Our proposed low average concentration Fe/C co-doped GaN HEMT achieves an output power of 18.5 W (42.67 dBm) at 3.6 GHz with a PAE exceeding 60%.

II. DEVICE STRUCTURE AND FABRICATION

The device structures under study are depicted in Fig. 1(a) and (b). Both structures share a common growth design,

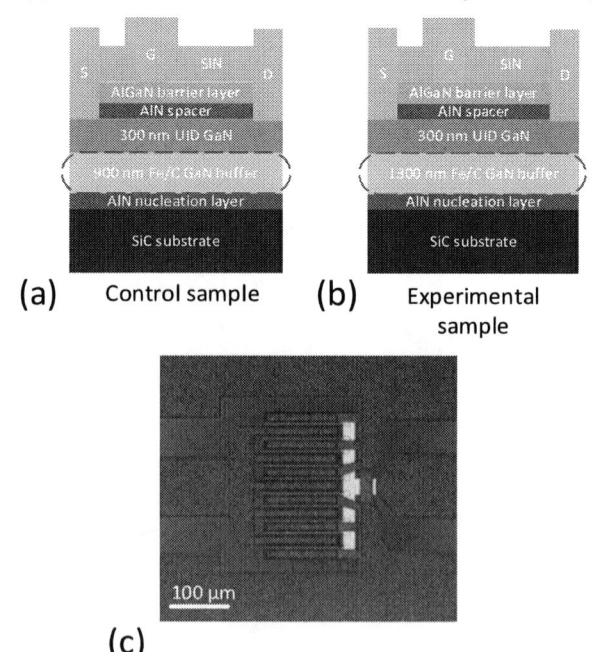

Figure 1. Schematic structure of the (a) control sample and (b) experimental sample. (c) Optical microscope of the device with air bridge finished.

979-8-3503-6184-1/24 $31.00 © 2024 IEEE

differing only in the GaN buffer layer. The epitaxial stack comprises a 350 μm 4H-SiC substrate, a 120 nm AlN nucleation layer, a GaN buffer layer, a 400 nm unintentionally doped GaN channel layer, a 1 nm AlN insertion layer, and a 20 nm Al$_{25}$Ga$_{75}$N barrier layer, sequentially from bottom to top.

As shown in fig. 2, device fabrication begins with ohmic lithography, followed by deposition of Ti/Al/Ni/Au stacked metals. Rapid thermal annealing at 850°C for 60 s in a nitrogen atmosphere forms the source and drain electrodes. The active region is protected by photoresist, and electrical isolation is achieved by bombarding the wafer with high-energy nitrogen ions. Surface pre-treatment reduces device surface states before Si$_3$N$_4$ passivation layer deposition. Stepper lithography defines the gate shape between source and drain, and after etching, Ni/Au stacked metal forms a Schottky gate. The metal interconnection pattern is photolithographed around the device, followed by the deposition of Ti/Au stacked metal. The last step of the process is the fabrication of the air bridge. Ni/Au was sputtered as a seed metal, followed by electroplating to achieve the connection between the source and the interconnect pad. The finished device is observed under an optical microscope as shown in Fig. 1 (c). The gate length (L_G) of the device is 0.5 μm, the gate width (W_G) is 10×125 μm (1.25 mm), and the source-drain spacing (L_{SD}) is 5.5 μm.

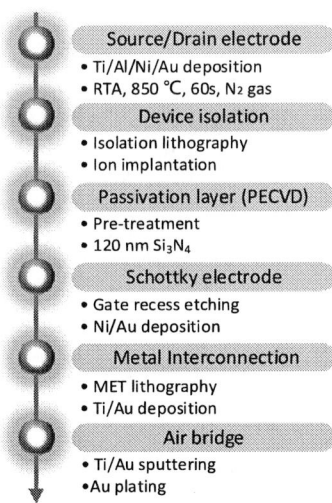

Figure 2. The process of fabricating devices in our lab. For a device with multiple finger gates, an air bridge is an essential step.

III. RESULTS AND DISCUSSION

A. Epitaxial design

The control samples used the Fe/C co-doped structure that we proposed in the past [13], on the basis of which we proposed a new Fe/C co-doped structure in this work. The revised doping strategy serves dual purposes: it inherits the advantages of Fe/C co-doping to maintain excellent electrical properties, while also reducing the average doping concentration in the buffer layer to enhance the material quality of the GaN buffer layer .

Fe and C in the epitaxial structure of the two samples were characterized using secondary ion mass spectrometry (SIMS), as shown in Fig. 3. The GaN buffer layer in the control sample

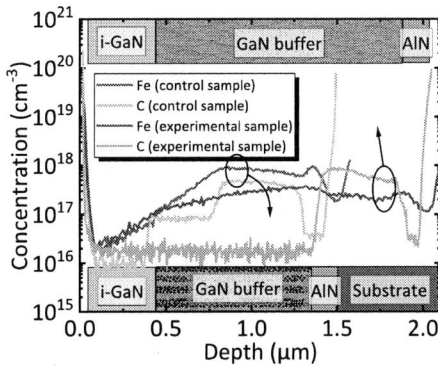

Fig. 3.   SIMS profiles of Fe and C for both samples.

comprised two sections: 500 nm of GaN doped with Fe and C, followed by 400 nm of weaker C doped GaN buffer. Similarly, the experimental sample's doping design also included two sections: 500 nm GaN buffer layer with Fe and C doped, and an additional 800 nm of Fe-only doped (lower concentration than the reference sample) GaN buffer layer. This configuration increased the total buffer thickness by 400 nm, positioning the C impurities further from the 2DEG in the channel, thereby mitigating their adverse effects. Additionally, Fig. 3 reveals that the concentration of Fe impurities at the heterogeneous interface was reduced to below 2×10^{16} cm^{-3} for both samples, which minimizing Fe tailing effect.

The experimental samples featured a thicker GaN buffer layer, thus lowering the average concentration of Fe and C impurities. Area integration of the buffer impurities concentration in the two samples, followed by division by their total thicknesses, revealed that the average doping concentration in the control sample was three times higher than that in the experimental sample. This reduction in impurity concentration led to improved lattice quality, enhancing phonon propagation throughout the crystal [14]. Thermophysical properties of each material layer in the two samples were assessed using the Transducer-Less Thermoreflectance Technique previously proposed [14], with results presented in Table I. As anticipated, reducing the average doping concentration in the GaN buffer layer enhanced its quality, yielding a significant increase in thermal conductivity by 30 W/mK. This improvement is crucial for rapid heat dissipation in high-power RF GaN HEMTs.

TABLE I.

THERMAL CONDUCTIVITY OF EACH MATERIAL LAYER CHARACTERIZED BY THERMOREFLECTANCE TECHNIQUE.

Materials	K (W/m · K)	
	Control sample	*Experimental sample*
GaN buffer	61	91
UID-GaN	157	160
SiC substrate	320	315

B. DC characterization

The basic electrical properties of the two samples were assessed using a 1500 A semiconductor analyzer. Multiple data sets were collected for each sample to ensure statistical reliability and mitigate the influence of outlier observations. The

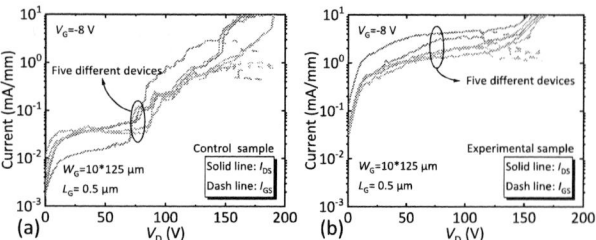

Fig. 4   Three-terminal breakdown characterization of the (a) control sample and (b) experimental sample with 1.25 mm W_G and 5.5 μm L_{SD}. The BV was extracted under leakage current criteria of 10 mA/mm.

electrical characteristics such as saturation current, peak transconductance, and current collapse ratio showed no significant differences between the two samples (data not shown). To evaluate the three-terminal breakdown characteristics, the current was limited to 10 mA/mm, and results for multiple devices from both samples are depicted in Fig. 4 (a) and (b). The average off-state leakage current of the control sample was significantly lower, by an order of magnitude, compared to the experimental sample—a difference likely attributable to the higher average doping concentration in the control sample which enhances electrical insulation [15]. Additionally, the average breakdown voltage of the control sample exceeded that of the experimental sample by 20 V, further evidencing superior electrical insulation in the GaN buffer layer of the control sample. Nevertheless, the BV of the experimental samples all exceeded 150 V and were in good agreement. Both sample types initially reached the drain current limit, particularly as the gate width was consistent at 1.25 mm. As the voltage increased, the self-heating effect exacerbated, leading to thermal breakdown in the devices. DC results indicated that the proposed doping design inherits the advantages of conventional Fe/C co-doping.

Furthermore, the output characteristics of the devices were tested with bias settings of V_D from 0 V to 20 V (V_G from -4 V to 0 V in 1 V increments). Due to the current limit of the testing equipment, which was set at 1 A, V_G testing was capped at 0 V. The results, displayed in Figs. 5 (a) and (b), revealed that as the V_D increased, the drain current initially surged and then began to decline upon reaching saturation. This decline in drain current with increased V_D is attributed to inadequate heat dissipation, which in turn degrades the electrical performance of the device, manifesting as self-heating effects. At V_D =20 V, the degradation ratio of current was 20.2 % for the control sample and 16.5 % for the experimental sample, indicating that the heat dissipation issues were more pronounced in the control sample, resulting in more significant device performance degradation. Although the

saturation currents of the two samples are comparable, the GaN buffer layer in the experimental sample exhibits higher thermal conductivity, thereby enhancing heat dissipation and effectively mitigating performance degradation due to self-heating. This observation underscores the potential of the proposed Fe/C co-doped design in addressing the thermal challenges encountered by GaN HEMTs in high-power RF applications.

C. Thermal Simulation

Steady-state heat simulations for both samples were conducted using COMSOL Multiphysics simulation software. The thermal conductivity parameters for each material layer, along with the thermal boundary resistance (TBR) between the epitaxial structure and the substrate, were calibrated based on the results from Thermoreflectance tests. The three-dimensional thermal simulation model was constructed based on the actual dimensions of the prepared device, with a size of 400 μm × 400 μm. In the simulation, the device was mounted on a copper block, with a boundary condition of 300 K applied to the block's bottom surface. All other surfaces of the model were treated as thermally insulated.

Different P_{diss} of 1 W, 2 W, 5 W, 10 W, and 15 W were applied to the simulation model. The surface temperature was extracted at the center of the active region, perpendicular to the gate width. The variations in surface temperature with increasing P_{diss} for both samples are depicted in Fig. 6. It is evident that the temperature difference between the two samples widens as the P_{diss} increases, reaching a peak difference of 32.3°C at 15 W. This substantial temperature disparity validates the current degradation observed in the output characteristics, primarily attributed to the escalating self-heating effect with increased operating voltage. The experimental sample's lower peak temperature, under identical P_{diss} conditions, is due to its GaN buffer's high thermal conductivity. This finding confirms that the structure proposed in this work ensures highly reliable operation of GaN HEMTs.

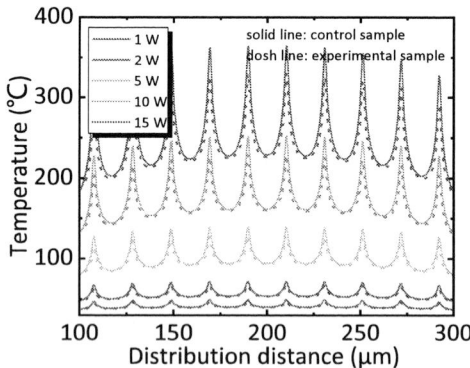

Fig. 6.   Temperature distribution profiles of two samples along the device surface, perpendicular to the gate width direction. The dimensions of the simulation models are consistent with those actually fabricated in this work.

D. RF Characterization

The large-signal characteristics of the two samples were evaluated at a frequency of 3.6 GHz using a pulsed mode with a 10% duty cycle. To facilitate Class AB operation, the quiescent current was set at 45 mA, approximately 3% of the saturation current. The power results for the control sample and

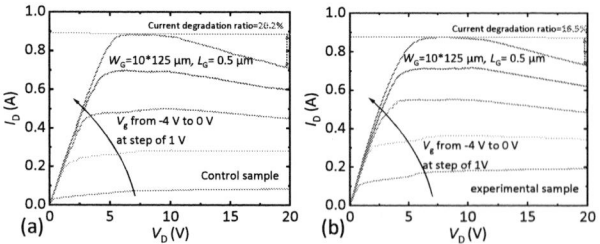

Fig. 5   Output characterization of the (a) control sample and (b) experimental sample at V_D from 0 V to 20 V. The device has 1.25 mm W_G and 5.5 μm L_{SD}

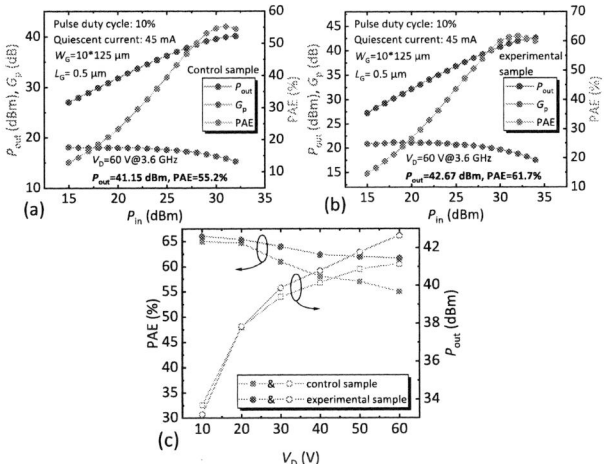

Fig. 7. Large-signal characteristics of the (a) control and (b) experimental samples with 1.25 mm W_G and 5.5 μm L_{SD} at $V_D = 60V$. The RF output power and PAE of both samples varied in response to changes in V_D, V_D from 10 V to 60 V in 10 V increments.

experimental sample at $V_D = 60$ V are displayed in Fig. 7 (a) and (b), respectively. The control sample achieved an output power of 41.15 dBm (13 W) and a Power-Added Efficiency (PAE) of 55%, whereas the experimental sample reached an output power of up to 42.67 dBm (18.5 W) with a PAE of 61%. As depicted in Fig. 7 (c), the 3.6 GHz RF results of the two samples, plotted as a function of V_D, shows similar output power and PAE at low voltage ($V_D \leq 20$ V) where device heating is minimal. However, as the drain voltage increases ($V_D > 20$ V), the self-heating effect intensifies, and the disparity in their heat dissipation capabilities becomes more pronounced, leading to noticeable differences in RF performance.

These findings substantiate the beneficial impact of the proposed low average concentration Fe/C co-doped GaN buffer layer on the electrical and thermal characteristics of the devices, enhancing the viability of GaN HEMTs for high-power applications.

IV. CONCLUSION

This study introduces a novel Fe/C co-doping method for GaN HEMTs to improve thermal and electrical performance in high-power RF applications. Our modified GaN buffer layer with reduced doping concentration significantly enhances thermal conductivity and heat dissipation, reducing self-heating effects under high operational loads. At the same time, the change in doping did not deteriorate the electrical characteristics of the device. In large-signal tests at 3.6 GHz, the experimental sample outperformed, achieving an output power of 18.5 W and a PAE of 61%, compared to 13 W and 55% PAE in the control sample. Thermal simulations corroborated these results, showing that better thermal management correlates with lower peak temperatures in the device. This breakthrough enhances device reliability and efficiency, offering considerable advantages for advanced high-power RF systems.

ACKNOWLEDGMENT

This work was supported in part by the National Natural Science Foundation of China under Grant 62234009, 62188102, 62090014, 62104178, 62104179, and 62104184.

REFERENCES

[1] U. K. Mishra, L. Shen, T. E. Kazior, and Y.-F. Wu, "GaN-Based RF Power Devices and Amplifiers," *Proc. IEEE*, vol. 96, no. 2, pp. 287–305, Feb. 2008, doi: 10.1109/JPROC.2007.911060.

[2] M. Meneghini *et al.*, "GaN-based power devices: Physics, reliability, and perspectives," *J. Appl. Phys.*, vol. 130, no. 18, p. 181101, Nov. 2021, doi: 10.1063/5.0061354.

[3] Y. Hao, X. Ma, M. Mi, and L.-A. Yang, "Research on GaN-Based RF Devices: High-Frequency Gate Structure Design, Submicrometer-Length Gate Fabrication, Suppressed SCE, Low Parasitic Resistance, Minimized Current Collapse, and Lower Gate Leakage," *IEEE Microw. Mag.*, vol. 22, no. 4, pp. 34–48, Apr. 2021, doi: 10.1109/MMM.2020.3047746.

[4] Y.-F. Wu *et al.*, "30-W/mm GaN HEMTs by field plate optimization," *IEEE Electron Device Lett.*, vol. 25, no. 3, pp. 117–119, Mar. 2004, doi: 10.1109/LED.2003.822667.

[5] Y. Hao *et al.*, "High-Performance Microwave Gate-Recessed AlGaN/AlN/GaN MOS-HEMT With 73% Power-Added Efficiency," *IEEE Electron Device Lett.*, vol. 32, no. 5, pp. 626–628, May 2011, doi: 10.1109/LED.2011.2118736.

[6] J.-S. Moon *et al.*, "Power Scaling of Graded-Channel GaN HEMTs With Mini-Field-Plate T-gate and 156 GHz fT," *IEEE Electron Device Lett.*, vol. 42, no. 6, pp. 796–799, Jun. 2021, doi: 10.1109/LED.2021.3075926.

[7] L. Yang *et al.*, "Record Power Performance of 33.1 W/mm with 62.9% PAE at X-band and 14.4 W/mm at Ka-band from AlGaN/GaN/AlN:Fe Heterostucture," in *2023 International Electron Devices Meeting (IEDM)*, Dec. 2023, pp. 1–4. doi: 10.1109/IEDM45741.2023.10413780.

[8] S. Li *et al.*, "Enhanced Performance of GaN HEMTs in X-band Applications Using SixN/Si3N4 Bilayer Passivation Technique," *Phys. Status Solidi A*, p. 2400047, May 2024, doi: 10.1002/pssa.202400047.

[9] S. Nakajima, "GaN HEMTs for 5G Base Station Applications," in *2018 IEEE International Electron Devices Meeting (IEDM)*, Dec. 2018, p. 14.2.1-14.2.4. doi: 10.1109/IEDM.2018.8614588.

[10] J. He, W.-C. Cheng, Q. Wang, K. Cheng, H. Yu, and Y. Chai, "Recent Advances in GaN-Based Power HEMT Devices," *Adv. Electron. Mater.*, vol. 7, no. 4, p. 2001045, 2021, doi: 10.1002/aelm.202001045.

[11] C. Miccoli, L. Gervasi, V. Cerantonio, J. Pomeroy, M. Kuball, and F. Iucolano, "Peak channel temperature determination for an AlGaN/GaN HEMT with Raman Thermography and MTTF extraction for long term reliability," in *2022 IEEE 9th Workshop on Wide Bandgap Power Devices & Applications (WiPDA)*, Nov. 2022, pp. 35–39. doi: 10.1109/WiPDA56483.2022.9955286.

[12] D. Francis and M. Kuball, "GaN-on-diamond materials and device technology: A review," in *Thermal Management of Gallium Nitride Electronics*, Elsevier, 2022, pp. 295–331. doi: 10.1016/B978-0-12-821084-0.00006-8.

[13] L. Yang *et al.*, "The DC Performance and RF Characteristics of GaN-Based HEMTs Improvement Using Graded AlGaN Back Barrier and Fe/C Co-Doped Buffer," *IEEE Trans. Electron Devices*, vol. 69, no. 8, pp. 4170–4174, Aug. 2022, doi: 10.1109/TED.2022.3179675.

[14] C. Yuan *et al.*, "Transducer-Less Thermoreflectance Technique for Measuring Thermal Properties of the Buried Buffer Layer and Interface in GaN-based HEMTs," *ACS Appl. Electron. Mater.*, vol. 4, no. 12, pp. 5984–5995, Dec. 2022, doi: 10.1021/acsaelm.2c01163.

[15] S. Tanabe, N. Watanabe, M. Uchida, and H. Matsuzaki, "Effects of surface morphology and C concentration in C-doped GaN buffer on breakdown voltage of AlGaN/GaN HEMTs on free-standing GaN substrate: Breakdown voltage of AlGaN/GaN HEMTs on free-standing GaN substrate," *Phys. Status Solidi A*, vol. 213, no. 5, pp. 1236–1240, May 2016, doi: 10.1002/pssa.201532781.

979-8-3503-6184-1/24 $31.00 © 2024 IEEE

An Enhanced RC-IGBT Incorporating Superjunction and Discontinuous Field Stop Layers for Improved Efficiency

Yiming Jia[1], Jieyu Long[1], Zhiwei Jing[1], Haimeng Huang[1*], and Hongqiang Yang[1*]

[1]*Glasgow College, University of Electronic Science and Technology of China, Chengdu 610054, China*
* Email: hqyang@uestc.edu.cn; hmhuang@uestc.edu.cn

Abstract—This paper introduces a 1200V-rated DFS-SJ RC-IGBT, integrating a discontinuous field-stop (DFS) layer and a super junction (SJ) structure. Extensive simulations using Silvaco TCAD reveal that the DFS-SJ RC-IGBT eliminates the snapback phenomenon at an LC of 60 micrometers, enhancing performance without affecting breakdown voltage. The novel design demonstrates reduced turn-off losses compared to conventional RC-IGBTs. The enhanced plasma depletion during turn-off, facilitated by the SJ structure, results in faster electron extraction and minimized tail current. These improvements highlight the DFS-SJ RC-IGBT as a viable solution for advanced power semiconductor applications, addressing key limitations of traditional RC-IGBTs.

Keywords—Discontinuous field-stop layer, super junction, reverse conducting, insulated gate bipolar transistor (IGBT), snapback phenomenon

I. INTRODUCTION

The Reverse-Conducting Insulated Gate Bipolar Transistor (RC-IGBT) integrates the functionalities of an IGBT and a Freewheeling Diode (FWD) into a single structure [1], [2], enabled by advances in thin wafer processing technologies allowing distinct doping profiles on the anode side [3]. Compared to the conventional IGBT-FWD combination, RC-IGBTs offer lower manufacturing costs, mitigated parasitic effects, reduced temperature fluctuations, higher power density, lower switching losses, and smaller chip size [4]–[6]. However, during the unipolar to bipolar transition, intense switching induces an undesirable snapback phenomenon in the forward conduction state, impeding parallel chip operation and compromising reliability [7]. Traditional RC-IGBTs mitigate snapback through a low-impedance path via the N+ field stop layer and a wide P+ collector region. Yet, an excessive P+/N+ ratio can cause uneven current distribution, leading to current crowding, localized hotspots, and potential thermal failure [8]. To address these issues, innovative structures have been proposed: super junctions to reduce drift region resistance [9], floating p-layers [10], dual-gate structures [6], and modified buffer layers [11] to suppress the snapback phenomenon.

A 1200V-rated RC-IGBT with Discontinuous Field Stop (DFS) and Super junction (SJ) structures is presented. Extensive Silvaco TCAD simulations reveal that the DFS and SJ effectively eliminate snapback by reducing the half-cell pitch to $60\mu m$ without compromising the breakdown voltage (BV). The DFS-SJ RC-IGBT also exhibits lower turn-off losses and

higher short-circuit withstand capability at the same forward voltage drop, offering significant advantages over conventional RC-IGBTs.

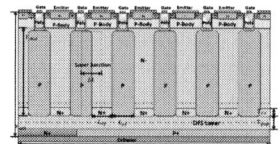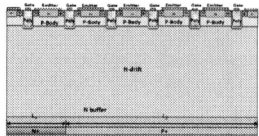

Fig. 1. (a) DFS-SJ RC-IGBT structure schematic diagram (Left) (b) Conventional RC-IGBT structure schematic diagram (Right).

II. OPERATION AND MECHANISM

Figure. 1 presents the cross-sectional views of the DFS-SJ RC-IGBT and the conventional RC-IGBT. The schematic cross-section of the DFS-SJ RC-IGBT is shown in Figure. 1(a). Six MOS cells are uniformly distributed along the x-direction on the emitter side. The drift region features a superjunction structure with alternating N and P columns, having a width of ΔL and a column depth of Y_{mid}. The total device depth is Y_{cell}. In the collector region, the original N^+ field stop layer is replaced by discontinuous N islands, uniformly spaced along the x-direction and located a few micrometers above the P^+ collector region. These N islands are interspersed with P columns, with the width of each N island, L_{nf}, matching the N columns in the drift region, and the spacing, l_{pf}, matching the P columns. Additionally, L_C represents half the collector cell length, L_N and L_P are the widths of the back surface half-cell N^+ and P^+, respectively, T_{bf} is the buffer layer thickness, and T_{gap} is the distance between the collector and buffer layer.

Figures. 2 depicts the electron current distribution during forward conduction in unipolar mode. The P columns act as electron barriers, forcing electrons to flow laterally through the low-resistance N-buffer layer after passing through the superjunction. Subsequently, electrons traverse the high-resistance N-drift region between the buffer and the collector, finally being collected by the N^+ collector. As shown in Figure. 2(a), when the voltage drop between the P^+ collector and the DFS exceeds the built-in potential ($V_{pn} \approx 0.7V$), the PN junction is triggered, and holes are injected into the drift region. V_{SB} is the transition voltage from unipolar to bipolar mode, and V_H is the voltage at the end of this transition. The difference ΔV_{SB} indicates the severity of the snapback effect, with $\Delta V_{SB} = 0$ signifying no snapback. The compact model of V_{SB} is detailed in [11]

$$V_{SB} = \left(1 + \frac{R_{drift} + R_{ch}}{R_{CS}}\right)V_{pn}. \tag{1}$$

Here, R_{drift} and R_{ch} are the resistances of the drift region and the channel, respectively. R_{CS} is the equivalent resistance of the N-drift region, which combines the discontinuous field stop layer and the field stop region with the collector in unipolar mode. It is mainly contributed by the high-resistance N-drift region. Compared to the highly doped field stop region in conventional structures, the structure with a low-doped drift region has a higher R_{CS}. In traditional IGBT structures, increasing R_{drift} is limited by breakdown voltage ratings, as the doping and thickness of the drift region determine the device's breakdown. The use of SJ technology adds an additional degree of freedom. The balance of charges between the alternating p/n columns allows the device's breakdown voltage to no longer be determined by the low doping concentration of the drift region. Thus, N_{drift} doping can now be significantly higher, reducing the resistance of this region under unipolar conduction. In this case, the drift region comprises $R_{n\text{-}drift}$ and $R_{p\text{-}pillar}$, and the total drift resistance depends on Y_{mid}. With the same Y_{mid}, the drift resistance using an SJ structure is much lower than that of a conventional structure. Therefore, reducing R_{drift} and increasing R_{CS} effectively lowers V_{SB}, as V_{pn} occupies a larger proportion of V_{SB} in unipolar mode, making it easier for the PN junction to conduct.

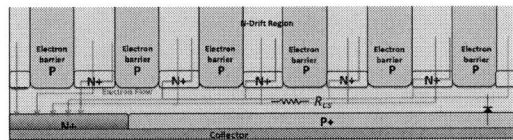

Fig. 2. Operation mechanism of the DFS-SJ RC-IGBT in the unipolar mode.

III. RESULTS AND DISSCUSSION

Table. 1 lists the parameters used for simulating both the DFS-SJ RC-IGBT and the conventional RC-IGBT.

TABLE I
DEVICE PARAMETER SPECIFICATION

Parameters	DFS-SJ	Conv.
MOS cell dimension, L_M	$10\mu m$	$10\mu m$
Gate oxide thickness, t_{ox}	$100nm$	$100nm$
Gate trench depth, d_G	$4\mu m$	$4\mu m$
Drift region doping, N_d	$2.42 \times 10^{14}\,cm^{-3}$	$2.42 \times 10^{14}\,cm^{-3}$
Pillar region doping, P_d	$2.42 \times 10^{14}\,cm^{-3}$	$2.42 \times 10^{14}\,cm^{-3}$
Superjunction width, ΔL	$5\mu m$	—
Buffer region doping, N_b	$1 \times 10^{17}\,cm^{-3}$	$1 \times 10^{17}\,cm^{-3}$
N-collector length, Y_N	$7.5\mu m$	$7.5\mu m$
Half collector cell, L_C	$60\mu m$	$60\mu m$
Wafer thickness, Y_{Cell}	$130\mu m$	$130\mu m$
Carrier life time , τ	$1\mu s$	$1\mu s$
buffer layer thickness, T_{bf}	$5\mu s$	—

Figure. 3 illustrates the forward characteristics of the DFS-SJ RC-IGBT. The DFS-SJ structure eliminates the snapback

effect at an LC of 60 micrometers. In contrast, the conventional RC-IGBT exhibits significant snapback under the same conditions.

The effect of varying values of Y_{mid} on snapback voltage was studied through numerical simulations. Figure. 3 shows the on-state characteristics of the TFS RC IGBT and the semi-SJ IGBT for different Y_{mid} values. As expected, the thicker the pillars, the lower the snapback voltage, as the overall unipolar resistance of the drift region is decreased. In fact, when the SJ structure length reaches 120 μm or higher, the snapback voltage completely disappears even when operating at very low temperatures. Figure. 4 illustrates the trend between snapback voltage and Y_{mid} values. As anticipated, the relationship between these two variables is linear, as the length of the pillars is proportional to the device's resistance.

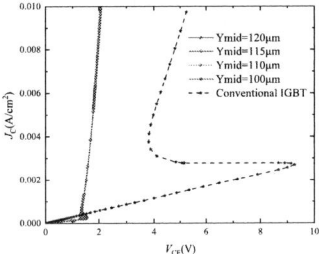

Fig. 3. Forward conduction characteristics of the Conventional RC-IGBT and designed DFS-SJ RC-IGBT for different Y_{mid} values.

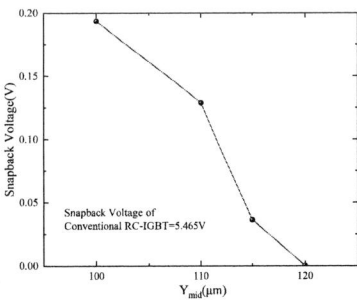

Fig. 4. Snapback Voltage of the Conventional RC-IGBT and designed DFS-SJ RC-IGBT for different Y_{mid} values.

Figure. 5 illustrates the blocking characteristics of the DFS-SJ RC-IGBT. In Figure. 5(a), the collector current is nearly zero at low voltages, but rises sharply near 1200V, indicating breakdown. Figure. 5(b) shows the blocking mechanism. In the drift region [12], the electric field for alternating P and N pillars is a combination of E_p and $E_q(x,y)$, as described in

$$E(x,y) = E_p + E_q(x,y). \tag{2}$$

E_p is the potential field from the applied voltage, a constant with a one-dimensional y component in the SJ region, as modeled in [13]. $E_q(x,y)$ is the charge field from ionized charges in the N and P regions, with two-dimensional x and y components. Multiple space charge regions form between the P pillars, adjacent N pillars, N buffer layer, and gap N drift region, with the field terminating in the alternating buffer layers. A parasitic PNP transistor, formed by P pillars, gap N drift region, and P+ collector, must avoid base punch-through to maintain high breakdown voltage.

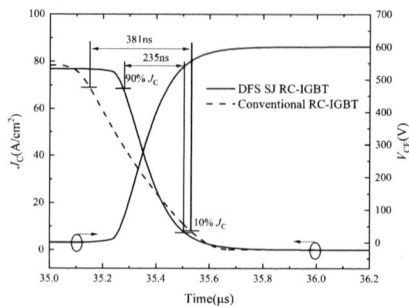

Fig. 6. Turn-off characteristics of the designed DFS-SJ RC-IGBT.

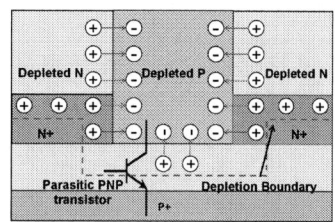

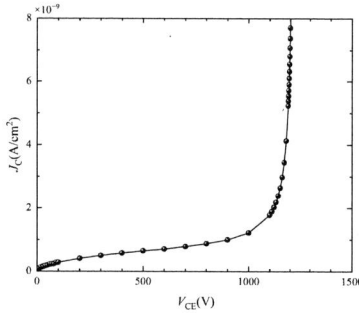

Fig. 5. (a) Mechanism of Depletion Region Formation (b) Breakdown voltage of the designed DFS-SJ RC-IGBT.

Figure. 6 shows the inductive turn-off curves of DFS-SJ RC-IGBT and conventional RC-IGBT at 300K. The gate voltage switches from 600V to 0V at 78A forward current. Compared with the conventional structure, the turn-off time of DFS-SJ structure is shortened by 146 μs. This is due to the fact that the DFS-SJ structure with smaller L_P and P columns in the SJ junction aids to speed up the plasma depletion during IGBT turn-off, the excess electrons in the drift region are extracted more quickly through a shortened path, thus shortening the turn-off time.

IV. Conclusion

In this study, a novel 1200V-rated DFS-SJ RC-IGBT, integrating a discontinuous field-stop (DFS) layer and a super junction (SJ) structure, was proposed. Extensive simulations using Silvaco TCAD revealed that the DFS-SJ RC-IGBT significantly outperforms conventional RC-IGBTs. By reducing the half-cell pitch to 60μm, the snapback phenomenon was effectively eliminated without affecting the breakdown voltage. The introduction of the SJ structure and DFS layer reduced the ratio of unipolar resistance in the drift region, thereby

decreasing the snapback voltage. When the SJ structure length reached 120μm or higher, the snapback voltage completely disappeared even at low temperatures. Furthermore, the DFS-SJ RC-IGBT exhibited excellent performance in terms of lower turn-off losses, demonstrating faster switching speeds and lower tail currents. The results showed a linear relationship between the snapback voltage and Y_{mid} value. Overall, the DFS-SJ RC-IGBT provides better on-state performance, minimal snapback voltage, and lower switching losses, offering a promising solution for power semiconductor applications.

References

[1] E. Griebl, L. Lorenz, and M. Purschel, "Lightmos a new power semiconductor concept dedicated for lamp ballast application," in *Proc. IEEE Ind. Appl. Conf.*, Salt Lake City, UT, USA, Oct. 2003, pp. 768–772.

[2] Takahashi, Yamamoto, Aono, and Minato, "1200 v reverse conducting igbt," in *Proc. 16th Int. Symp. Power Semiconductor Devices*, 2004, pp. 133–136.

[3] T. Minato and H. Takahashi, "New power-element technology," *Mitsubishi Electric ADVANCE*, vol. 105, pp. 24–27, March 2004.

[4] E. Findlay and F. Udrea, "Reverse-conducting insulated gate bipolar transistor: A review of current technologies," *IEEE Trans. Electron Devices*, vol. 66, no. 1, pp. 219–231, Jan. 2019.

[5] E. Napoli, P. Spirito, A. G. M. Strollo, F. Frisina, L. Fragapane, and D. Fagone, "Design of igbt with integral freewheeling diode," *IEEE Electron Device Letters*, vol. 23, no. 9, pp. 532–534, 2002.

[6] L. Zhu and X. Chen, "An investigation of a novel snapback-free reverse-conducting igbt and with dual gates," *IEEE Trans. Electron Devices*, vol. 59, no. 11, pp. 3048–3053, Nov. 2012.

[7] L. Storasta, A. Kopta, and M. Rahimo, "A comparison of charge dynamics in the reverse-conducting rc igbt and bi-mode insulated gate transistor bigt," in *Proc. Int. Symp. Power Semiconductor Devices ICs (ISPSD)*, Jun. 2010, pp. 391–394.

[8] W. Chen, Z. Li, M. Ren, J. Zhang, B. Zhang, Y. Liu, Q. Hua, K. Mao, and Z. Li, "A high reliable reverse-conducting igbt with a floating p-plug," in *2013 25th International Symposium on Power Semiconductor Devices IC's (ISPSD)*, 2013, pp. 265–268.

[9] M. Antoniou, F. Udrea, F. Bauer, and I. Nistor, "A new way to alleviate the rc igbt snapback phenomenon: The super junction solution," in *Proc. ISPSD*, Jun. 2010, pp. 153–156.

[10] W. Chen, Z. Li, M. Ren, J. Zhang, B. Zhang, Y. Liu, Q. Hua, K. Mao, and Z. Li, "A snapback suppressed reverse-conducting igbt with a floating p-region in trench collector," *IEEE Electron Device Letters*, vol. 33, no. 3, pp. 417–419, Mar. 2012.

[11] G. Deng, X. R. Luo, J. Wei, K. Zhou, L. Huang, T. Sun, Q. Liu, and B. Zhang, "A snapback-free reverse conducting insulated-gate bipolar transistor with discontinuous field-stop layer," *IEEE transactions on electron devices*, vol. 65, no. 5, pp. 1856–1861, 2018.

[12] B. Zhang, W. Zhang, M. Qiao, Z. Zhan, and Z. Li, "Concept and design of super junction devices," *J. Semicond.*, vol. 39, no. 2, 2018.

979-8-3503-6184-1/24 $31.00 © 2024 IEEE

Simulation Study on 1200V CS-SemiSJ-IGBT for Reduced Switching Loss and Fast Switching

Luping Li [1,2], Zehong Li [1,2,3,*], Peng Chen[1,4,5,*], Yuzhou Wu[6], Qiansheng Rao[1,2,5], Ming Li[1,2,5],
Haifeng Qin[5], Li Wan[5], Yang Yang[5], Wei Li[5], Min Ren[1,2,3,*]

[1] *State Key Laboratory of Electronic Thin Films and Integrated Devices, University of Electronic Science and Technology of China (UESTC), Chengdu 611731, China*
[2] *Chongqing Institute of Microelectronics Industry Technology, UESTC, Chongqing 401331, China*
[3] *Shenzhen Institute for Advanced Study, UESTC, Shenzhen 518110, China*
[4] *School of Aeronautics and Astronautics, UESTC, Chengdu 611731, China*
[5] *China Resources Microelectronics (Chongqing) Ltd., Chongqing 401331, China*
[6] *Shanghai Super Semiconductor Technology Company Ltd., Shanghai 201203, China*
* Email: lizh@uestc.edu.cn, chenp@uestc.edu.cn, renmin@uestc.edu.cn

Abstract—**Characteristics of 1200V CS-SemiSJ-IGBT and its pillar thickness influence on key performance are investigated though TCAD simulation in this work. Firstly, 1200V CS-SemiSJ-IGBT with 45 μm thick SJ-pillar is illustrated to present performance characteristics and advantage of CS-SemiSJ-IGBT. Secondly, 1200V CS-SemiSJ-IGBTs with different SJ-pillar thickness are investigated to reveal its optimal SJ-pillar thickness. As a conclusion, the 1200V CS-SemiSJ-IGBT with 45 μm thick SJ-pillar can provide comparative trade-off as FullSJ-IGBT with the same voltage level.**

Index Terms—**Carrier-Storage Insulate Gate Bipolar Transistor (CS-IGBT), Semi-Super-Junction (SemiSJ), Low Switching Loss**

I. INTRODUCTION

The SemiSJ-IGBT was early proposed to optimize the trade-off of IGBT [1], [2], but it has not been applied in commercial products. One of the main reason is that the performance improvement level and intrinsic mechanism of SemiSJBT is still unclear. Since 650 V/750 V rated FullSJ-IGBT has been tried to be commercialized [3], [4], more efforts have been done on 1200 V rated SemiSJ-IGBT [5]. In this work, 1200V SemiSJBT with N-type Carrier-Storage (N-CS) layer is investigated to theoretically verify prospect of SemiSJBT with current popular SJ pillar thickness.

II. 1200V CS-SEMISJ-IGBT WITH 45 μM SJ-PILLAR

A. Device Structure

Cell structure of the CS-N-drift/SemiSJ/FullSJ-IGBTs (CS-NDBT/SemiSJBT/FullSJBT) and their structural parameters are presented in Fig.1 and Table.I. The sole structural difference among the three IGBTs locates at their drift region, where N-type drift region in CS-NDBT and CS-SemiSJBT both are 5e13 cm^{-3} doped, and SJ-pillar are 3e15 cm^{-3} doped. Thickness of the N-drift region in CS-NDBT is 110 μm,

This work was supported in part by the Natural Science Foundation of Chongqing, China, under Grant 2022NSCQ-MSX3071; and in part by the Natural Science Foundation of Guangdong Province under Grant 2023A1515012652.

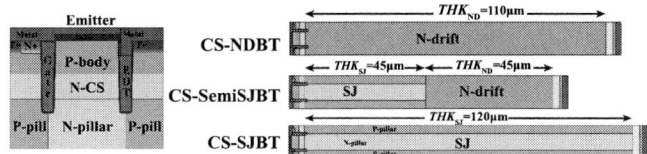

Fig. 1: Device structure of the simulated IGBTs

TABLE I: Structural parameters of simulated 1200V IGBTs

Parameter	CS-NDBT	CS-SemiSJBT	CS-FullSJBT
SJ Pitch		12 μm	
Gate Pitch		6 μm	
P-body Thickness/Doping		1.5 μm/1.5e17 cm^{-3}	
Trench Dept/Widthh		5.5 μm/1 μm	
Gate Oxide Thickness		1000 Å	
N-CS Thickness/Doping		2 μm/3e15 cm^{-3}	
FS Layer Thickness/Doping		2.5 μm/1e16 cm^{-3}	
P-sub Thickness/Doping		1 μm/1e17 cm^{-3}	
N-drift/SJ-pillar Doping		5e13 cm^{-3}/3e15 cm^{-3}	
N-drift Thickness	110 μm	45 μm	0 μm
SJ-pillar Thickness	0 μm	45 μm	120 μm

length of SJ-pillar in CS-FullSJBT is 120 μm, and thickness of N-drift/SJ-pillar in CS-SemiSJBT both are 45 μm, whose *BV* all are about 1350 V as illustrated in Fig.2a.

B. Output Characteristics and On-state Loss Reduction

Simulated key electric parameters of the IGBTs in Fig.1 are presented in Fig.2. As the output curves in Fig.2b shows, their output J_C-V_C are roughly coincident and their maximal current density ($J_{C,max}$) are approximate, which means they have comparative short-circuit resistance. But on-state loss (V_{on}) at J_C=200 A/cm^2 meets that $V_{on,FullSJ} > V_{on,ND} > V_{on,SemiSJ}$, thus V_{on} of CS-SemiSJBT is optimal.

C. C-V Characteristics

Simulated *C-V* curves of three IGBTs are plotted in Fig.2c. In NDBT, C_{out} and C_{res} of CS-NDBT starts to drop at $V_{CE} \approx 3$ V since longitudinal (*y*) depletion zone early isolated

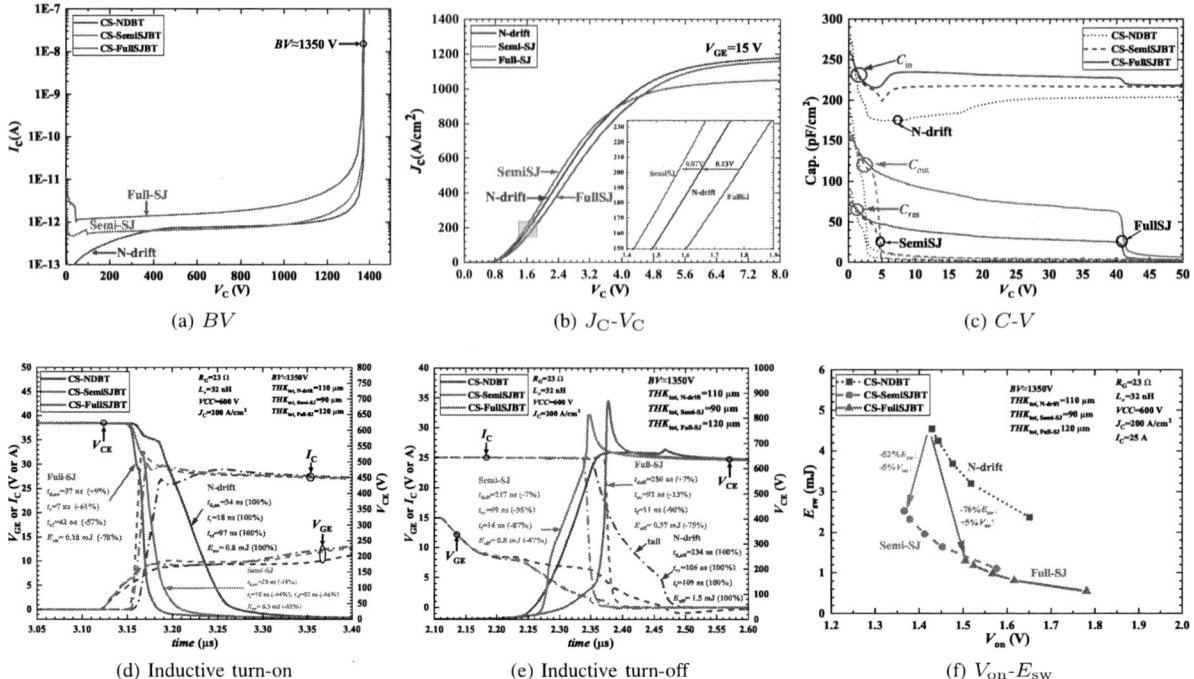

Fig. 2: Simulation results of the 1200V CS-ND/SemiSJ/FullSJ-IGBTs

Collector and Gate at small V_{CE}, where only longitudinal electric field (E_y) works and no horizontal electric field (E_x) exists. While in FullSJBT, C_{out} and C_{res} of CS-NDBT starts to drop until $V_{CE} \approx 40$ V, where horizontal (x) depletion zone has been built and longitudinal (y) depletion zone starts to establish. The sudden C_{out} and C_{res} drop at large V_{CE} is unwanted since it cause more serious Electromagnetic Interference (EMI) to application systems. In SemiSJBT, the V_{CE} where C_{out} and C_{res} drop is approximate to that of NDBT and the problem of EMI can be much reduced from FullSJBT.

D. Inductive Switching and its Loss Reduction

Inductive switching of the IGBTs are simulated under V_{CC} = 600 V and $J_C = 200$ A/cm^2. During the turn-on process in Fig.2d, $t_{d,on}$ of three IGBTs are approximate, while current raise time t_r of FullSJBT and SemiSJBT are -61% and -44% smaller than that of NDBT, and voltage fall time (t_{vf}) of FullSJBT and SemiSJBT are -56% and -47% smaller than that of NDBT. Consequently, turn-on speed of SemiSJBT is equivalent to FullSJBT and it is faster than NDBT, thus E_{on} of SemiSJBT is reduced by -62% from NDBT.

During the turn-off process in Fig.2e, main difference of three IGBTs happens in both voltage raise (t_{vr}) and current fall (t_f) periods. t_{vr} of FullSJBT and NDBT are approximate, while V_{CE} of FullSJBT at each time point is smaller than NDBT, which provides the precondition for E_{on} reduction of FullSJBT from NDBT during t_{vr}. Differently, V_{CE} of SemiSJBT at each time point of t_{vr} is not less than NDBT, rather than its t_{vr} is -35% less than NDBT. During t_f periods, the tail current does not appear in both FullSJBT and SemiSJBT, consequently t_f of SemiSJBT and FullSJBT are reduced by -87% and -90% from NDBT, and E_{off} are reduced by -47% and -75%, respectively.

E. V_{on}-E_{sw} Trade-off Improvement

V_{on}-E_{sw} trade-off of three IGBTs at different P-sub concentration are presented in Fig.2f. As can be seen, V_{on} and E_{sw} of SemiSJBT are reduced by -5% and -52% from NDBT, respectively. While V_{on} of FullSJBT is increased by +5% from NDBT and its E_{sw} is reduced by -76% from NDBT. Since current output ability of Carrier-storage SJ-pillar/N-drift with identical thickness are roughly equivalent [6], the V_{on} reduction of SemiSJBT and increase of FullSJBT from NDBT can be attributed to their thinner/thicker drift region. As a result, although E_{sw} reduction rate of FullSJBT is higher than SemiSJBT, their V_{on}-E_{sw} trade-off are finally approximately equivalent due to the increased V_{on} in FullSJBT.

III. 1200V CS-SEMISJ-IGBT WITH DIFFERENT SJ-PILLAR THICKNESS

A. BV vs. Thickness

The reason for increased thickness of SJ-pillar in FullSJBT is investigated though BV-$Thickness$ relationship of NDBT and FullSJBT in Fig.3a. As is shown, at small thickness period, total thickness (THK_{tot}) of FullSJBT takes is less than NDBT at the same BV, but this trend is ended at $THK_{tot} = 90$ μm where BV of FullSJBT and NDBT are equivalent. After that, THK_{tot} of FullSJBT takes is higher than NDBT with the same BV, which is the reason for increased thickness in 1200 V FullSJBT than NDBT.

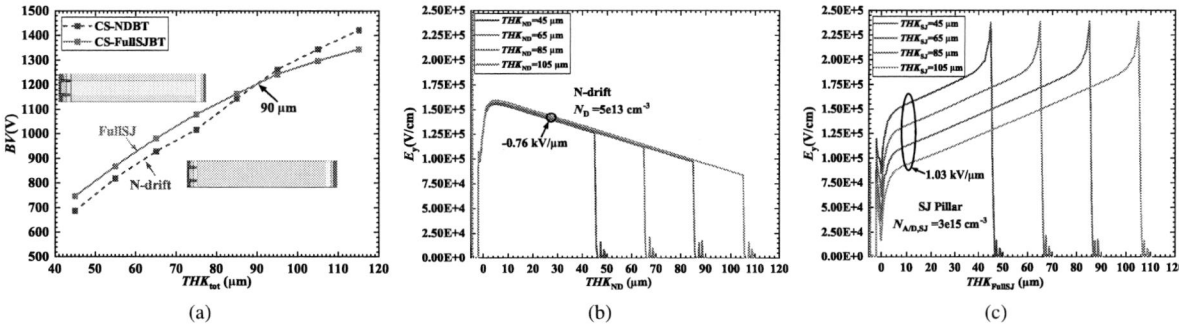

Fig. 3: Simulated BV and E_y at different N-drift/SJ-pillar thickness

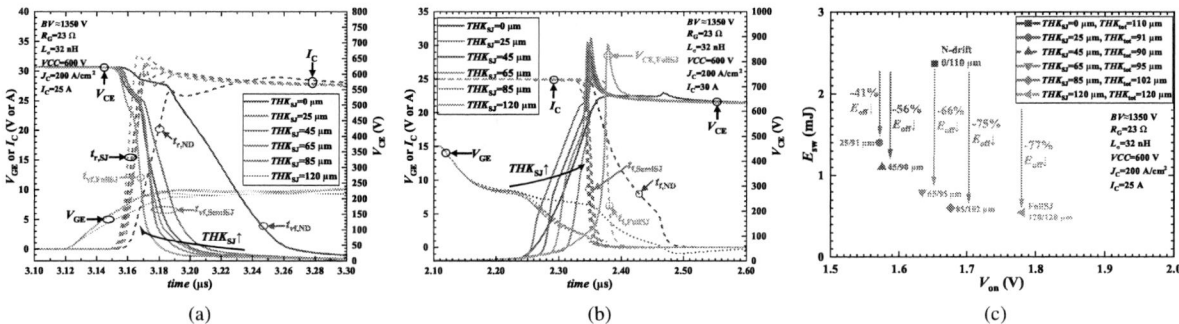

Fig. 4: Simulation results of CS-SemiSJBT at different SJ-pillar thickness ratio

Longitudinal electric field (E_y) of FullSJBT and NDBT at different N-drift/SJ-pillar thickness are presented in Fig.3b and Fig.3c to reveal the intrinsic reason for increased thickness of SJ-pillar at high BV period. $|dE_y/dy|$ of the 5e13 cm^{-3} doped NDBT and 3e15 cm^{-3} doped FullSJBT are 0.76 kV/μm and 1.03 kV/μm, respectively. Thus reduction rate of E_y of FullSJBT is larger than NDBT, leads to effect of thickness works on E_y/BV of FullSJBT is weaker than NDBT, thus FullSJBT takes higher thickness to support the same BV.

B. Inductive Switching and V_{on}-E_{sw} Trade-off

Simulated inductive switching and V_{on}-E_{sw} trade-off 1200 V SemiSJBT with different SJ-pillar/N-drift thickness ration ($R_{SJ} = THK_{SJ}/THK_{ND}$) are also presented in Fig.4 to observe the influence of SJ-pillar thickness ration (R_{SJ}) on device performance. With the increase of R_{SJ}, effect of SJ-pillar is gradually increased, and the turn-on/off characteristics are also gradually tends to FullSJBT, and the optimal V_{on}-E_{sw} trade-off appears at THK_{SJ} =45/65/85 μm as shown in Fig.4c.

IV. CONCLUSION

With the introduction of N-CS layer and SemiSJ-pillar, V_{on}-E_{sw} trade-off of SemiSJBT is optimized from NDBT and comparative to FullSJBT. In addition, CS-SemiSJBT with different SJ-pillar thickness reveals that SemiSJBT with 45 μm thick SJ-pillar is one of the preferable solution to obtain the best V_{on}-E_{sw} among 1200 V SemiSJ/FullSJ-IGBTs. This work

indicates the worth of CS-SemiSJBT to be further developed with current mainstream SJ-pillar thickness like 45 μm.

REFERENCES

[1] Marina Antoniou, Florin Udrea, Friedhelm Bauer, and Iulian Nistor, "The Semi-Superjunction IGBT," *IEEE Electron Device Letters*, vol. 31, no. 6, pp. 591–593, Jun. 2010.

[2] Yangjie Ou, Shan Lu, Wie Coa, and Dong Liu, "A Bottom Semi-Superjunction IGBT With Improving Relationship Between P/N doping and Vce,sat," in *2022 IEEE 17th Conference on Industrial Electronics and Applications (ICIEA)*, Dec. 2022, pp. 841–844.

[3] Yuzhou Wu, Zehong Li, Jia Pan, Chong Chen, Jiuying Yu, Min Ren, and Bo Zhang, "650 V Super-Junction Insulated Gate Bipolar Transistor Based on 45 Mm Ultrathin Wafer Technology," *IEEE Electron Device Letters*, vol. 43, no. 4, pp. 592–595, Apr. 2022.

[4] Arthur Su, Jianing Guan, Rongzhen Qin, Tanya Trajkovic, and Florin Udrea, "Experimental Investigation of a Novel 750V SJ-RETIGBTs (Superjunction Recessed Emitter Trench IGBTs) for the Automotive Application," in *2024 36th International Symposium on Power Semiconductor Devices and ICs (ISPSD)*. Bremen, Germany: IEEE, 2024-06-02/0006.

[5] Tomohiro Tamaki, Atsufumi Inoue, and Masayuki Furuhashi, "Dynamic Charge Imbalance in Superjunction IGBTs: Design, Simulation, and Experimental Validation," in *2024 36th International Symposium on Power Semiconductor Devices and ICs (ISPSD)*, 2024.

[6] Luping Li, Zehong Li, Yuzhou Wu, Peng Chen, Qiansheng Rao, Yuanzhen Yang, Qiang Yuan, Rong Zhou, and Min Ren, "Investigation on the carrier-storage super-junction IGBT: Characteristics, mechanism, and advantages," *Microelectronics Journal*, vol. 142, p. 105993, Dec. 2023.

A Dual-Gate Trigger Thyristor for Reducing the Probability of False Triggering

Pengcheng Xing[1], Qingbo Wan[1], Jie Huang[1], Ruize Sun[1], Chao Liu[1], and Wanjun Chen[1]*

[1]State Key Laboratory of Electronic Thin Films and Integrated Devices, University of Electronic Science and Technology of China (UESTC), Chengdu 610054, China

* Email: wjchen@uestc.edu.cn

Abstract—To reduce the probability of false triggering, a novel Dual-Gate Trigger Thyristor (DGTT) is proposed. This device is characterized by two equivalent series-connected gates and is triggered to turn on only when both gates are activated simultaneously. Theoretical analysis based on probability theory shows that, under the same conditions of erroneous drive signals, the DGTT shows a lower probability of false triggering compared to the Insulated Gate Trigger Thyristor. Subsequent simulation experiments have also verified this, and show good agreement with the theoretical analysis.

Keywords— DGTT, thyristor, false triggering, dual gate

I. INTRODUCTION

Pulse power systems, characterized by long-term energy storage and short-term energy release, have found extensive applications in medical, environmental, and industrial fields [1]. Currently, pulse power systems are trending towards high power, miniaturization, and full solid-state development. Novel solid-state pulse power devices have become a research hotspot and core component of some power systems due to their excellent pulse performance and simple drive circuits [2]-[5].

As pulse power system volume decrease, the density of switching elements increases, and operating voltage rises, the Electro-Magnetic Interference (EMI) during system operation also increases [6]. Increased EMI more easily interferes with signal transmission within the system, sometimes causing control circuits to output erroneous control signals, leading to false triggering of pulse power switches. For pulse power systems characterized by high power, false triggering of pulse power switches often results in catastrophic consequences, easily causing the pulse power switches to burn out or the systems to undergo destructive failure. To reduce the probability of false triggering in pulse power systems, filter circuits have been integrated into the systems, achieving good results [6]. However, this inevitably increases the complexity of the system, offsetting the advantage of the simple drive circuits of novel solid-state pulse power devices and limiting further reduction in system size.

In this work, we propose a Dual-Gate Trigger Thyristor (DGTT). This device features two equivalent series-connected gates, and it is only triggered on when both gates are simultaneously activated. Both theoretical analysis and numerical simulations have verified that this triggering mode endows the device with good resistance to erroneous drive signals, significantly reducing the probability of false triggering. This is conducive to reducing reliance on filter circuits and further miniaturizing pulse power systems.

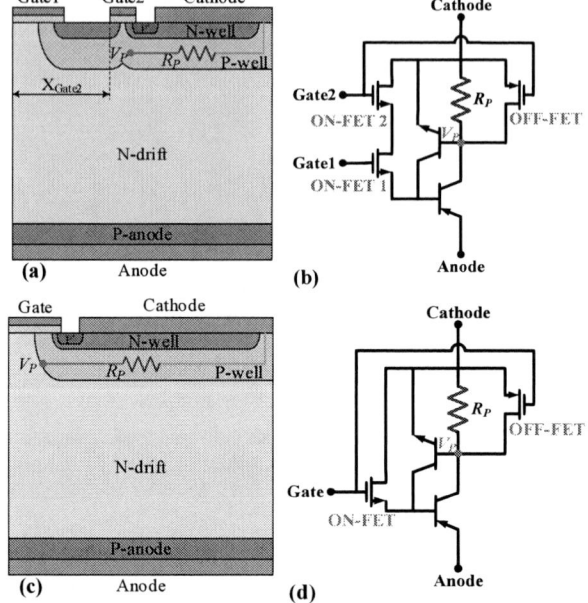

Fig. 1. DGTT (a) structure and (b) equivalent circuit. IGTT (c) structure and (d) equivalent circuit

II. STRUCTURE AND MECHANISM

Compared to the Insulated Gate Trigger Thyristor (IGTT) structure[1][2], the DGTT adds Gate2, with the P$^+$ region adjusted to be located between Gate2 and the Cathode. Two equivalent series-connected gates ensure that when only one gate is activated, no electrons can be injected into the N-drift region, keeping the device in the off-state. When both gates are activated simultaneously, the electron can be injected from the Cathode through ON-FET2 and ON-FET1 into the N-drift region. This provides base current to the parasitic PNP transistor formed by the P-anode/N-drift/P-well structure, turning on the transistor, as shown in segment A-B of Fig. 2. As the current further increases, the voltage drop across the equivalent resistance R_P in the P-well region gradually rises until it exceeds the turn-on voltage of the P-well/N-well junction. This activates the parasitic NPN transistor formed by the N-well/P-well/N-drift structure, with a large current injected from the N-well, as shown in segment B-C of Fig. 2. At this point, the NPN and PNP transistors begin to form regenerative feedback, and the DGTT gradually transitions into thyristor mode. As the current further increases, a large number of carriers are injected into the N-drift region, resulting in strong conductivity modulation within the device. Shown in segment C-D, the device operates in thyristor mode, capable of handling pulse current.

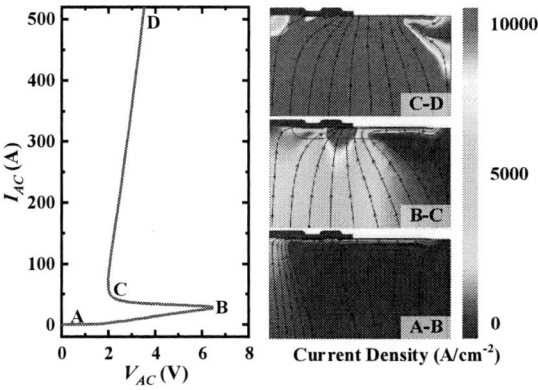

Fig. 2. The output characteristic of the DGTT and the current distribution of the device in each segment of the output curve. Segment A-B corresponds to Insulated Gate Bipolar Transistor (IGBT) mode operation, segment B-C illustrates the transition from IGBT mode to thyristor mode, and segment C-D indicates thyristor mode operation.

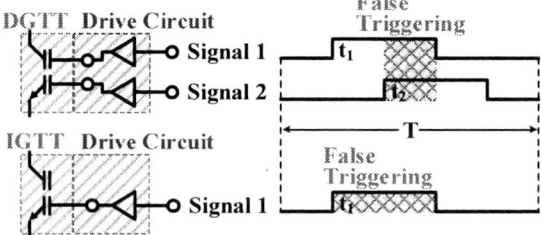

Fig. 3. Schematic of the drive circuits and erroneous signal for DGTT and IGTT in the pulse power system.

As shown in Fig. 3, since the DGTT has two gates, it correspondingly requires two sets of independent drive circuits to achieve a low probability of false triggering. It can be assumed that under EMI, the probabilities that Signal1 and Signal2, which should be low-level signals, turn into erroneous high-level signals with durations of t_1 and t_2 within the operating time T are P_1 and P_2 respectively. These signals are interfered with and become high-level randomly and independently. For ease of analysis, we first define several events during the device operation.

- Event A: During the operating time T, Signal1 is randomly interfered with, resulting in an erroneous high-level signal lasting t_1.

- Event B: During the operating time T, Signal2 is randomly interfered with, resulting in an erroneous high-level signal lasting t_2.

- Event C: During the operating time T, Signal1 and Signal2 are randomly interfered with, resulting in erroneous high-level signals lasting t_1 and t_2 respectively.

- Event D: The erroneous high-level signals from Signal1 and Signal2 overlap during the operating time T.

- Event E: During the operating time T, Signal1 and Signal2 are randomly interfered with, resulting in erroneous high-level signals lasting t_1 and t_2 respectively, and these erroneous high-level signals overlap during the operating time T.

Thus, it can be concluded that

$$P(A) = P_1 \qquad P(B) = P_2 \tag{1}$$

$$P(C) = P(A \cap B) \tag{2}$$

Since the two drive circuits of the DGTT are independent, events A and B are also independent. Therefore, we can derive the following

$$P(C) = P(A)P(B) = P_1 P_2 \tag{3}$$

For erroneous high-level signals that occur randomly within the operating time T, if the sum of the durations of the two erroneous high-level signals is greater than T, then the two signals must overlap. Therefore, we have

$$P(D) = \begin{cases} \dfrac{t_1 + t_2}{T} & 0 < t_1 + t_2 \le T \\ 1 & T < t_1 + t_2 \end{cases} \tag{4}$$

Because the occurrence of erroneous signals and the timing of these signals are independent, the probability of the DGTT being falsely triggered is

$$P_{DGTT,FT} = P(E) = P(C \cap D) = P(C)P(D) \tag{5}$$

$$P_{DGTT,FT} = \begin{cases} P_1 P_2 \dfrac{t_1 + t_2}{T} & 0 < t_1 + t_2 \le T \\ P_1 P_2 & T < t_1 + t_2 \end{cases} \tag{6}$$

Similarly, the probability of the IGTT being falsely triggered is

$$P_{IGTT,FT} = P(A) = P_1 \tag{7}$$

For Signal1 and Signal2 in the same pulse power system, it can be further assumed that they are subject to similar EMI conditions. The probability that each signal independently becomes an erroneous high-level signal with a pulse width of t due to random interference within the operating time T is P. Thus, equations (6) and (7) can be simplified as follows

$$P_{DGTT,FT} = \begin{cases} P^2 \dfrac{2t}{T} & 0 < t \le \dfrac{T}{2} \\ P^2 & \dfrac{T}{2} < t \end{cases} \tag{8}$$

$$P_{IGTT,FT} = P \tag{9}$$

From the formulas, it is evident that $P_{DGTT,FT} \le P_{IGTT,FT}$, indicating that the DGTT has better resistance to false triggering compared to the IGTT. Additionally, the shorter the pulse width t of the interference signal, the lower the probability of the DGTT being falsely triggered.

III. SIMULATION RESULTS AND DISCUSSION

We designed a 1200V DGTT based on the IGTT process and conducted numerical simulations. The structure parameters are shown in TABLE I., and the simulation circuits and results are shown in Fig. 4. As illustrated, the DGTT, as analyzed previously, is triggered to initiate pulse discharge only when both gates are simultaneously being activated. Additionally, the peak pulse discharge current (I_{peak}) and the current rising slew rate (di/dt) show no significant difference compared to the IGTT. These results indicate that the DGTT has good resistance to false triggering and good pulse discharge capability.

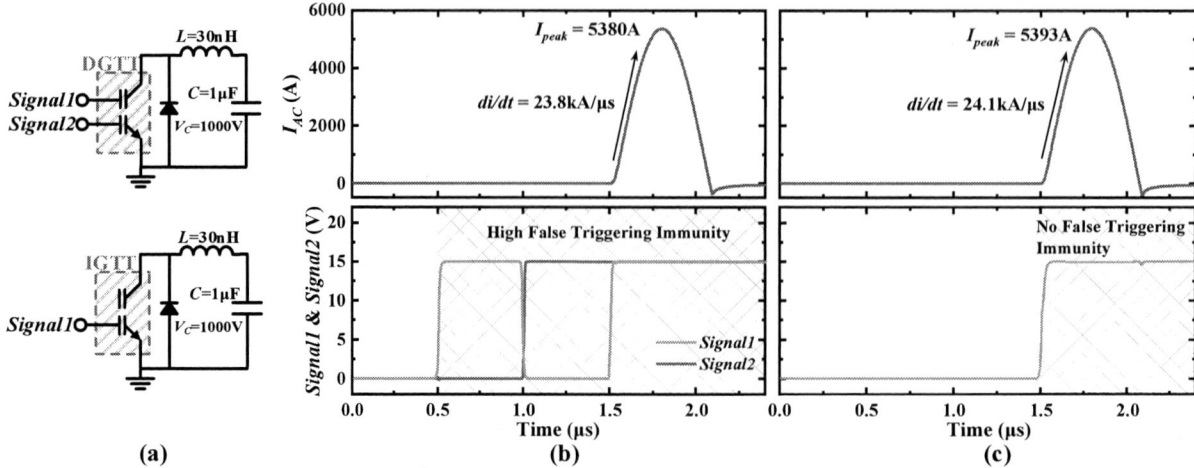

Fig. 4 (a) Test circuits for the devices. (b) Gate voltage and pulse current for DGTT. (c) Gate voltage and pulse current for IGTT.

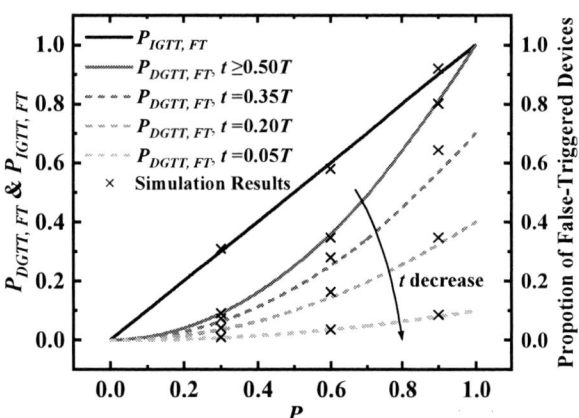

Fig. 5. The analytical results of false trigger probabilities for DGTT and IGTT under different probabilities of erroneous signal generation, as well as the simulation-derived proportion of falsely triggered device.

TABLE I. STRUCTURE PARAMETERS

Structure Parameters	DGTT	IGTT
Cell pitch (μm)	50	50
Gate1 width (μm)	10	10
Gate2 width (μm)	4	-
Xgate2 (μm)	15	-
Gate oxide thickness (nm)	100	100
Wafer thickness (μm)	200	200
N-drift doping (cm^{-3})	4×10^{13}	4×10^{13}
Pwell doping (cm^{-3})	3×10^{17}	3×10^{17}
Nwell doping (cm^{-3})	1×10^{20}	1×10^{20}
P+ doping (cm^{-3})	2×10^{20}	2×10^{20}
Pcollector doping (cm^{-3})	1×10^{19}	1×10^{19}

Next, utilizing numerical simulation tools, we generated random erroneous high-level signals, Signal1 and Signal2, with a pulse width of t within the time T as the driving signals for the DGTT and IGTT. We conducted 50 sets of simulations for each pulse width and obtained the proportion of false-triggered devices, as shown in Fig. 5. The simulation results of the proportion of false-triggered devices align well with the analytical results from equation (8) and (9), verifying the accuracy of the theoretical analysis and confirming that the DGTT has a lower probability of false triggering compared to the IGTT in response to erroneous driving signals.

IV. CONCLUSION

A DGTT with high resistance to erroneous signals is proposed. The device features two gates and is triggered on only when both gates are activated. Theoretical analysis and simulation results have shown that the proposed DGTT significantly reduces the probability of false triggering under short-term erroneous driving signal compared to the IGTT. Furthermore, the DGTT is compatible with IGTT technology and exhibits pulse discharge capability comparable to IGTT, indicating promising potential.

ACKNOWLEDGEMENT

This work was supported in part by the National Natural Science Foundation of China under Grant U21A20499 and 62334003, the University-Industry Collaborative Education Program (231004866165126).

REFERENCES

[1] W. Chen et al., "Design and Characterization of High di/dt CS-MCT for Pulse Power Applications," *IEEE Transactions on Electron Devices*, vol. 64, no. 10, pp. 4206–4212, Oct. 2017.

[2] W. Chen et al., "High Peak Current MOS Gate-Triggered Thyristor With Fast Turn-On Characteristics for Solid-State Closing Switch Applications," *IEEE Electron Device Letters*, vol. 37, no. 2, pp. 205–208, Feb. 2016.

[3] X. Xu et al., "A Novel Thyristor-Based Bidirectional SSCB With Controllable Current Breaking Capability," *IEEE Transactions on Power Electronics*, vol. 37, no. 4, pp. 4526–4534, Apr. 2022.

[4] X. Xu, W. Chen, H. Tao, Q. Zhou, Z. Li, and B. Zhang, "Design and Experimental Verification of an Efficient SSCB Based on CS-MCT," *IEEE Transactions on Power Electronics*, vol. 35, no. 11, pp. 11682–11693, Nov. 2020.

[5] C. Liu et al., "High Voltage Insulated Gate Trigger Thyristor With High-Efficiency Injection for Fast Turn-on and High Current Pulse," *IEEE Electron Device Letters*, vol. 40, no. 12, pp. 1965–1968, Dec. 2019.

[6] L. Zi, H. Liu, S. Jiang, and J. Rao, "A novel drive circuit with overcurrent protection for solid state pulse generators," *IEEE Transactions on Dielectrics and Electrical Insulation*, vol. 26, no. 2, pp. 361–366, Apr. 2019.

979-8-3503-6184-1/24 $31.00 © 2024 IEEE

Edge-Dependence of Threshold Voltage in MoS₂ Nanoribbon-Based 2D FETs

Zhirong Peng* and Mansun Chan

Department of Electronic and Computer Engineering, The Hong Kong University of Science and Technology, Clear Water Bay, Hong Kong SAR, China

* Email: zhirong.peng@connect.ust.hk

Abstract—As devices scale down to the nanoscale, their behavior deviates significantly from bulk materials, but the influence of edge effects on device performance has rarely been investigated, necessitating a deeper understanding. This work investigates the edge-dependent threshold voltage (V_T) in molybdenum disulfide (MoS₂) nanoribbon-based two-dimensional field-effect transistors (2D FETs). The influence of orientation, termination, and width of MoS₂ nanoribbons on the V_T are studied. These three variables are found to affect V_T to varying extents, providing important insights into the design of MoS₂ nanoribbon-based 2D FETs, and highlighting the importance of edge characteristics on device performance.

Keywords— MoS₂ nanoribbon, edge-dependence, threshold voltage

I. INTRODUCTION

Molybdenum disulfide (MoS₂) is recognized as a two-dimensional (2D) material that can be used in a next-generation nanoscale electronic device. Researchers have successfully demonstrated that monolayer MoS₂ field-effect transistors (FETs) have superior performance, including high on/off current ratio of over 10^8, nearly ideal subthreshold swing, and excellent mobility of ~200 cm² V⁻¹ s⁻¹ [1], [2]. However, as devices are scaled down to the nanoscale, their behavior differs significantly from that of bulk materials. Simulation studies show that armchair MoS₂ nanoribbons (AC MoS₂-NRs) show semiconducting properties with a bandgap of ~0.56 eV, and the magnitude of mobility in the AC MoS₂-NR is comparable to its sheet counterpart [3]. Whereas zigzag MoS₂ nanoribbons (ZZ MoS₂-NRs) are magnetic materials with zero bandgap, which exists in ZZ MoS₂-NRs with widths ranging from 1 nm to 10 nm [4].

As the channel material of advanced FETs is scaled down to a few nanometers in width, the atomic details of the material at this dimension become critical. While numerous simulation studies have analyzed the effects of orientation, width, and edge passivation of MoS₂ nanoribbons on their electrical properties [4], [5], there has been insufficient focus on the relationship between various edges of the MoS₂ nanoribbons and the performance of 2D FETs fabricated from them.

To evaluate the edge effects on the performance of 2D FETs, this work examines how the edge characteristics influence the threshold voltage (V_T) of the FETs, which is a crucial parameter that determines the on/off characteristics of a FET. The edge effects on the V_T of device will be investigated in three aspects, including orientations, terminations and widths. Through this investigation, we aim to provide design guidelines for performance variations in MoS₂ FETs resulting from the use of extremely scaled-down nanoribbons with specific edges.

II. COMPUTATINAL DETAILS OF THRESHOLD VOLTAGE

The change in orientation may affect the way carriers are injected into the channel and thus the turn-on condition of the device. According to the arrangement of atoms along the z-direction, i.e., current flow direction, the orientation of MoS₂ nanoribbons can be categorized into two categories, AC MoS₂-NRs and ZZ MoS₂-NRs. For the sake of completeness of consideration, eight MoS₂ nanoribbons constructions are covered in this work as shown in Fig. 1. Depending on whether the S atoms in the top and bottom edges are symmetric or not, the AC MoS₂-NRs have two constructions shown in (a-b), called AC MoS₂-NR-1 (AC-1) and AC MoS₂-NR-2 (AC-2). The S atoms used to determine symmetry are labelled with small triangles. In addition to considering the S atoms symmetry, the ZZ MoS₂-NRs are distinguished by the atomic compositions of the two edges, i.e., S atoms and Mo atoms (c-d), all Mo atoms (e-f), and all S atoms (g-h). They are denoted as ZZ MoS₂-NR-1 (ZZ-1) to ZZ MoS₂-NR-6 (ZZ-6) in order.

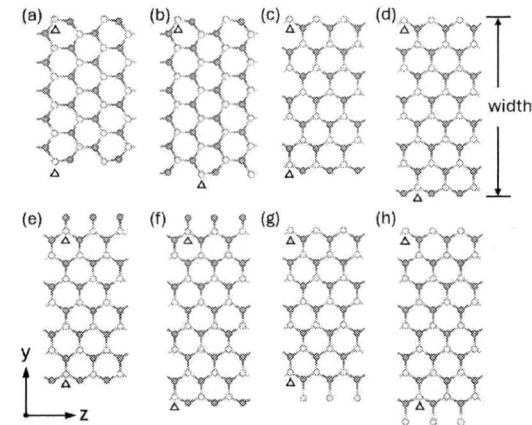

Fig. 1. The atomic structures of (a-b) AC MoS₂ nanoribbons, and (c-h) ZZ MoS₂ nanoribbons. The z-direction is a periodic direction.

In the traditional FETs, the V_T integrates various factors, including the flat band voltage (V_{FB}) for equilibrium condition, the voltage responsible for surface band bending, and the voltage drop across the dielectric layer (V_{OX}) that influences carrier concentration within the channel. However, in the nanoribbon-based device, the thickness of the active layer is extremely thin, so the band bending phenomenon is negligible. Therefore, the V_T can be defined as the gate voltage that make the inversion carrier density in the channel reach a threshold value n_{th}, and the analyses on V_T can be realized by focusing on the change trend for V_{FB} and V_{OX}.

The V_{FB} is the difference between the work function of gate metal (ψ_M) and channel material (ψ_S). It is not determined until the gate metal and channel material are

979-8-3503-6184-1/24 $31.00 © 2024 IEEE

selected. On the other hand, the V_{OX} is determined by the number of carriers and the capacitance of dielectric layer. Thus, the V_T can be compared in a straightforward way:

$$V_T = \psi_M/q - \psi_S/q + \frac{q(n_{th}-n_i)}{\varepsilon_{ox}/t_{ox}} \qquad (1)$$

where q is the electronic charge, n_i is the intrinsic carrier density of the channel, ε_{ox} is the permittivity of dielectric, and t_{ox} is the thickness of dielectric. In order to clearly demonstrate the dependence of V_T on the edge, we use Au as the gate metal, which is one of the widely used gate materials. The ψ_M of Au <111> is 5.31 eV. The dielectric layer is selected as SiO$_2$, and the n_{th} is

$$n_{th} = 1 \times 10^{12} \text{ cm}^{-2} \qquad (2)$$

The n_i of a semiconductor can be calculated based on the density of states (DOS) by:

$$n_i{}^2 = \int_{E_{CB}} DOS(E)f_e(E)\, dE \cdot \int_{E_{VB}} DOS(E)f_h(E)\, dE \qquad (3)$$

where E_{CB} and E_{VB} are energy in conduction band and valance band respectively. The $f_{e/h}(E)$ is the Fermi-Dirac function that define the probably of a state being filled by an electron/hole.

III. ORIENTATION-DEPENDENCE OF THRESHOLD VOLTAGE

To investigate the orientation-dependence of V_T, we take AC-1 and ZZ-1 as representatives of the two orientations, because they don't have dangling bonds and have similar width (~1.3 nm). The influence of termination and width will be discussed in the following two sections. Both ψ_S and DOS can be obtained based on the density functional theory (DFT). The DOS of the two types of nanoribbons are shown in Fig. 2. The computational parameters used in this work are same with [6], and the calculated results are in high agreement with it as well. It can be seen from Fig. 2 (a) that the AC MoS$_2$-NRs have bandgap, and the DOS near the Fermi level is lightly higher in AC-2 than in AC-1. The DOS of ZZ MoS$_2$-NRs shown in Fig. 2 (b) are much higher than those of the AC MoS$_2$-NRs, as well as many states available around Fermi level, allowing them to exhibit metallicity.

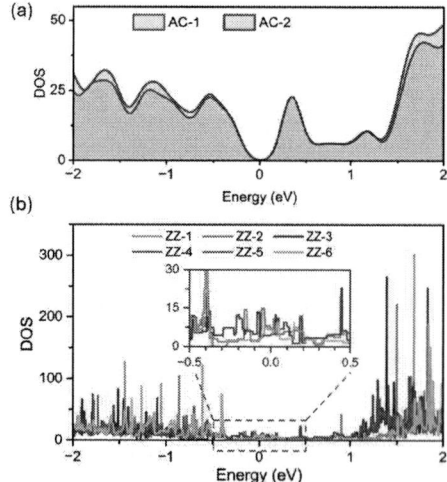

Fig. 2. The DOS of (a) AC MoS$_2$-NRs, and (b) ZZ MoS$_2$-NRs, inset with a enlarged DOS image near the Fermi level.

Based on the obtained DOS and (3), the calculated n_i is in the range of 1×10^8 cm^{-2} to 1×10^9 cm^{-2} for AC MoS$_2$-NRs. Since ZZ MoS$_2$-NRs have no bandgap and exhibit metallic

properties, their n_i should be high among 2D material family, which can be considered is in the range of 5×10^{12} cm^{-2} to 5×10^{13} cm^{-2}. The parameters that are related when considering the change of V_T with different orientations are listed in the Table 1. Here, the t_{ox} is set as 5 nm. Such orientation-dependent V_T shown in Table 1 illustrates the quite different threshold condition of the devices with AC MoS$_2$-NR and ZZ MoS$_2$-NR as channel material.

TABLE I. THE WORK FUNCTION ψ_S, INTRINSIC CARRIER CONCENTRATION n_i, AND THRESHLOD VOLTAGE V_T OF AC-1 AND ZZ-1 BASED DEVICE.

Orientation	ψ_S (eV)	n_i (cm^{-2})	V_T (V)
Armchair (AC-1)	~5.26	10^8 to 10^9	~ 0.28
Zigzag (ZZ-1)	~5.03	5×10^{12} to 5×10^{13}	~ −2.97 to −0.65

Since the t_{ox} has a significant effect on the V_T, the trends for V_T at various t_{ox} are shown in Fig. 3. Here, assuming the carrier concentration of ZZ MoS$_2$-NR is 1×10^{13}. It can be seen that when the dielectric layer is very thin, the difference in V_T values between two devices are relatively small. As the dielectric layer thickens, the V_T changes much faster in ZZ-1 based device than that in AC-1 based device. When the t_{ox} increases from 1 nm to 9 nm, the V_T for AC-1 based device increases from 0.10 V to 0.47 V (increases by 3.7 times), while the V_T for ZZ-1 based device changes from -0.14 V to -3.47 V (the absolute value of V_T increases by 24.34 times).

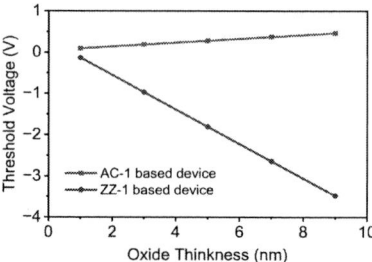

Fig. 3. Trend for V_T of AC-1 and ZZ-1 based device at various oxide thicknesses.

IV. TERMINATION-DEPENDENCE OF THRESHOLD VOLTAGE

Changing the termination atoms at MoS$_2$ nanoribbons' edges may significantly affect the electronic properties of the channel materials. Having understood that 2D FETs with armchair and zigzag terminated MoS$_2$ nanoribbon have different threshold conditions, it is necessary to further explore whether the same oriented material with different termination leads to different V_T.

Two different terminations may occur in the AC MoS$_2$-NR, which are AC-1 and AC-2. The lightly higher DOS of AC-2 indicates a bit higher n_i in AC-2, but the difference is too small to be reflected in the value of $n_{th} - n_i$. Moreover, AC-1 and AC-2 have similar ψ_S. Thus, the AC MoS$_2$-NR with different terminations do not significantly affect the V_T, which may be attributed to the fact that the armchair orientation determines its edges are all composed of Mo-S bonds.

On the contrary, different terminations in the ZZ MoS$_2$-NRs will have a more noticeable effect on the V_T, because there are more possibilities for the location of the edge termination. As explained in section II, the ZZ MoS$_2$-NRs are divided into three groups according to the terminated elements, e.g., ZZ-1 & ZZ-2, ZZ-3 & ZZ-4, ZZ-5 & ZZ-6. The two configurations within each group have almost the same electrical properties [6]. Therefore, we take one of each group, e.g., ZZ-1, ZZ-3, and ZZ-5 as a representative to analyze how the V_T of change with terminations.

Although the nanoribbons with different terminations are all metallic, they have different electrical properties. The DOS in Fig. 2 (b) show that the n_i is relatively high for ZZ-3 and comparatively small for ZZ-5. Therefore, we can take a gradient of n_i for three representatives, as shown in Table 2. In addition, the ψ_S of ZZ-3 and ZZ-5 are markedly different from that of ZZ-1 due to the dangling bonds at their edges. Moreover, the opposite electronegativity of the Mo and S atoms make their ψ_S one larger and one smaller than that of ZZ-1, which are ~4.79 eV and ~5.25 eV, respectively.

The calculated V_T of the three at $t_{ox} = 5$ nm are listed in Table 2. The ZZ-3 based device has the largest absolute value of V_T, mainly because its n_i is the largest among the three, which means a greater voltage is required to fully deplete it. Both ψ_S and n_i of the termination-varying ZZ MoS$_2$-NR are different, which would lead to various magnitudes of device-to-device variation at different t_{ox}, which can be seen in Fig. 4. When the t_{ox} is as thin as 1 nm, the V_T of all three devices is around -0.13 V. But as the oxide layer is as thick as 9 nm, the V_T of the ZZ-3 based device is 1.53 times that of the ZZ-1 based device, and 3.31 times that of the ZZ-5 based device.

TABLE II.   THE WORK FUNCTION ψ_S, INTRINSIC CARRIER CONCENTRATION n_i, AND THRESHLOD VOLTAGE V_T OF ZZ-1, ZZ-3, AND ZZ-5 BASED DEVICE.

Termination	ψ_S (eV)	n_i (cm^{-2})	V_T (V)
ZZ-1	~5.03	1×10^{13}	~ −1.81
ZZ-3	~4.79	5×10^{13}	~ −2.73
ZZ-5	~5.25	5×10^{12}	~ −0.87

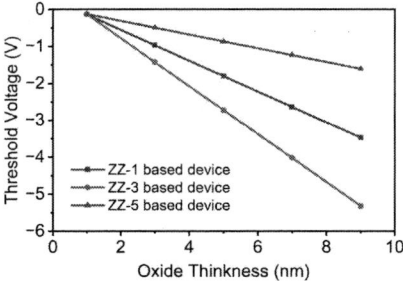

Fig. 4. Trend for V_T of different terminated ZZ MoS$_2$-NR based device at various oxide thinkness.

V. WIDTH-DEPENDENCE OF THRESHOLD VOLTAGE

With the decrease of nanoribbon width, the edge effects may become more prominent. A study in [7] shows that the band gap of AC MoS$_2$ NR oscillates at widths less than 3 nm. According to the trend of band gap variation embodied in [7], four widths that can represent the changing characteristics in the range of 1 nm to 2 nm are selected for analysing the width-dependence of V_T. The fluctuation of the calculated ψ_S with the width of AC MoS$_2$ NRs is shown in Fig. 5 (a), which shows an inverse correlation with the band gap. When the width of AC MoS$_2$ NR is ~1.59 nm, its band gap (~0.57 eV) is 9.62% larger than that of the 1.26 nm one, and its ψ_S becomes smaller by 0.03 eV.

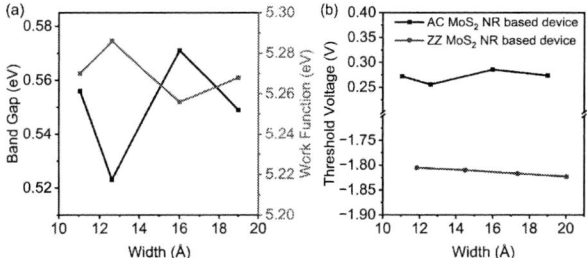

Fig. 5. (a) Trend for band gap and work function of AC MoS$_2$ NR at different width. (b) Trend for V_T at different width.

Regarding the width dependence of V_T in ZZ MoS$_2$ NR based device, we use the structures without dangling bonds. The trends of V_T with width in both AC and ZZ MoS$_2$ NR based device are shown in Fig. 5b. Here, the t_{ox} is set as 5 nm. The changing trend of V_T in AC MoS$_2$ NR based device is consistent with the trend for band gap, and the value of V_T differs by at most 0.03 V. Whereas the absolute value of V_T in ZZ MoS$_2$ NR based device is slightly larger. This is consistent with the result shown in [7] that the band structure of ZZ MoS$_2$ NR does not vary much from 1.2 nm to 2.6 nm in width. reflecting its stability with respect to the width variation.

VI. CONCLUSION

In this work, the influences of V_T in MoS$_2$ nanoribbon-based 2D FETs by various orientation, termination, and width are investigated. The AC MoS$_2$ NR based device and ZZ MoS$_2$ NR based device exhibit significant V_T difference due to their completely different electrical properties. The effects of the terminations are reflected in the work function and the intrinsic carrier concentration of the ZZ MoS$_2$ NR. While the width mainly changes the work function of channel, which in turn affects V_T. These findings highlight the need to consider edge effects when designing and optimizing nanoscale electronic devices.

ACKNOWLEDGMENT

This work was supported by the General Research Fund (GRF) 16203222 from the Research Grants Council (RGC) of Hong Kong.

REFERENCES

[1] B. Radisavljevic, and A. Kis, "Mobility engineering and a metal–insulator transition in monolayer MoS$_2$," *Nat. Mater.*, vol. 12, no. 9, pp. 815-820, 2013.

[2] B. Radisavljevic, A. Radenovic, J. Brivio, V. Giacometti, and A. Kis, "Single-layer MoS$_2$ transistors," *Nat. Nanotechnol.*, vol. 6, no. 3, pp. 147-150, 2011.

[3] Y. Cai, G. Zhang, and Y. M. Zhang, "Polarity-reversed robust carrier mobility in monolayer MoS$_2$ nanoribbons," *J. Am. Chem. Soc.*, vol. 136, no. 17, pp. 6269-6275, 2014.

[4] Y. Li, Z. Zhou, S. Zhang, and Z. Chen, "MoS$_2$ nanoribbons: high stability and unusual electronic and magnetic properties," *J. Am. Chem. Soc.*, vol. 130, no. 49, pp. 16739-16744, 2008.

[5] M. Sagynbaeva, P. Panigrahi, L. Yunguo, M. Ramzanet, and R. Ahuja, "Tweaking the magnetism of MoS$_2$ nanoribbon with hydrogen and carbon passivation," *Nanotechnology*, vol. 25, no. 16, pp. 165703, 2014.

[6] H. Pan, and Y. W. Zhang, "Edge-dependent structural, electronic and magnetic properties of MoS$_2$ nanoribbons," *J. Mater. Chem.*, vol. 22, no. 15, pp. 7280-7290, 2012.

[7] Y. Li, Z. Zhou, S. Zhang, and Z. Chen, "MoS$_2$ nanoribbons: high stability and unusual electronic and magnetic properties," *J. Am. Chem. Soc.*, vol. 130, no. 49, pp. 16739-16744, 2008.

ultra fast diode avalanche shaper with floating junction

Zhen Yang [1,2], Yu Zhou* [1,2], Xiao-Yan Tang [1], Chao Han [2], Qing-Wen Song [1,2], Yu-Ming Zhang [1,2]

[1] Xidian University School of Microelectronics, Xi'an 710071, Shaanxi, China
[2] Xidian-Wuhu Research Institute, Wuhu 241000, Anhui, China

* Email: yzwuhu@stu.xidian.edu.cn, zhouyu01@xidian.edu.cn

Abstract—This paper introduces the structure of a silicon carbide (SiC) floating junction Diode Avalanche Shaper (FJ-DAS). Unlike traditional diode avalanche sharper (DAS) structures, FJ-DAS incorporates a floating junction to modulate the internal electric field. This design overcomes the limitations imposed by low doping in the base region for quasi-uniform triggering, thereby enhancing device performance. Additionally, the introduction of the floating junction mitigates electric field concentration effects at the junction, providing theoretical support for designing N-base region DAS terminals. Through computer-aided design (TCAD) simulations, we demonstrate that FJ-DAS outperforms conventional DAS structures under identical device parameters. Specifically, FJ-DAS exhibits a turn-on time of 9 ps and a switching speed of 256.7 kV/ns, compared to the traditional DAS with a turn-on time of 20 ps and a switching speed of 85.5 kV/ns. FJ-DAS achieves approximately 55% faster turn-on time than its conventional counterpart.

Keywords— 4H-SiC, Pulse Power, Diode Avalanche Shaper, floating junction

I. INTRODUCTION

DAS (Diode Avalanche Shaper), as a type of pulsed power device, finds extensive applications in pulsed power systems. Based on the Delayed Avalanche Breakdown (DAB) effect, DAS can rapidly turn on within picoseconds, allowing it to generate pulses with steep leading edges. The DAB effect is the fastest non-optical method for producing abundant electron-hole plasma in semiconductors[1].

Recent research[2] indicates that the quasi-uniform triggering mechanism at low doping levels results in faster conduction speed compared to the TRAPATT (Trapped Plasma Avalanche Triggered Transit) triggering mechanism. In silicon-based DAS, reducing the doping concentration in the drift region leads to a trapezoidal electric field distribution inside the device when biased. This extends the range of delayed avalanche breakdown, accelerating the device's conduction speed.

In pulsed power applications, Si-based devices are commonly used for DAS. However, 4H-SiC offers several advantages over Si. With a wider bandgap, higher critical breakdown field, and superior thermal conductivity, 4H-SiC allows for thinner epitaxial layers, more uniform internal electric fields, and higher plasma wave densities at the same breakdown voltage[3-4]. This enhances device switching speed while reducing dynamic on-state resistance and switching losses. However, compared to silicon-based DAS, SiC DAS achieves quasi-uniform triggering with a lower base doping concentration. Simulation studies have also verified this quasi-uniform triggering mode in SiC DAS, where the base region doping concentration needs to be reduced to $1 \times 10^{15} \text{cm}^{-3}$.[5]

However, achieving quasi-uniform triggering in SiC DAS presents challenges in terminal design. In their pioneering work, reference[1] experimentally fabricated $P^+/P^-/N^+$ type DAS devices. However, at low doping concentrations, there was a concentration of electric field near the junction, making it challenging to prepare faster N^- type base regions in practical processes.

To overcome this limitation, we introduced a floating junction to modulate the internal electric field of SiC DAS, allowing uniform triggering independent of base region doping concentration. The structure, as shown in Figure 1, extended the range of delayed avalanche breakdown due to the electric field distribution created by the floating junction.Using TCAD (Through Computer-Aided Design) software, we compared the performance and internal avalanche mechanisms of traditional SiC DAS and floating junction DAS under identical device parameters. Simulation results demonstrated that the introduction of the floating junction altered the conventional TRAPATT triggering mechanism, enhancing DAS performance.

II. SIMULATION AND ANALYSIS

A. SIC FJ- DAS Model

In Figure 1(a), the conventional SiC DAS structure exhibits a trapezoidal electric field distribution during delayed avalanche breakdown, where avalanche occurs within the region of electric field intensity greater than E_b. The FJ-DAS (Floating Junction Diode Avalanche Shaper) structure, as shown in Figure 1(b), features a p-type buried layer within the n-type drift region. When reverse bias is applied to FJ-DAS, the depletion layer extends from the anode electrode to the II-drift region. After depletion, it extends from the bottom of the buried p-layer to the I-drift region. Comparing the electric field distributions, conventional DAS has a single trapezoidal distribution, while FJ-DAS divides the electric field into two trapezoids. In their work[6], Jingyu Li et al. mention that the internal electric field strength during static breakdown depends on the doping concentration of the floating junction. According to the principle of uniform triggering, an appropriate floating junction doping concentration should be chosen to ensure equal electric field strength (E_m) in the upper and lower drift regions, allowing simultaneous avalanche breakdown.

B. Advantages of FJ-DAS Over Conventional DAS

To investigate the impact of introducing a floating junction on DAS performance, this study focuses on a $P^+/N^-/N^+$ type ideal 4H-SiC DAS device with a breakdown voltage of 900 V. The designed structure includes a P^+ region with a thickness of 2 μm, a drift (base) region with a thickness of 6 μm, and a substrate thickness of 5 μm, as shown in Figure 1(b). Additionally, an FJ-DAS structure with the same

979-8-3503-6184-1/24 $31.00 © 2024 IEEE

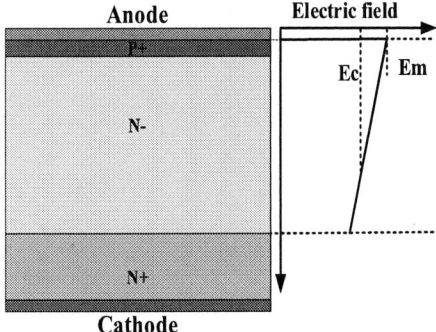

(a)

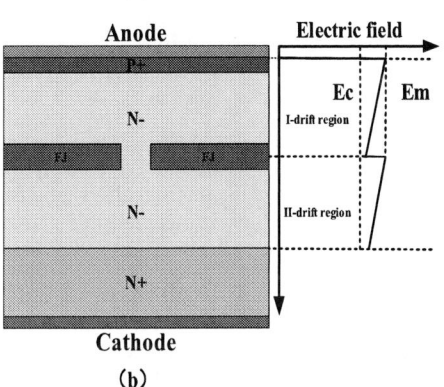

(b)

Fig. 1. illustrates the schematic diagrams of conventional DAS and FJ-DAS. The addition of the floating junction modifies the electric field distribution, extending the range of avalanche breakdown. (a) Conventional DAS. (b) FJ-DAS.

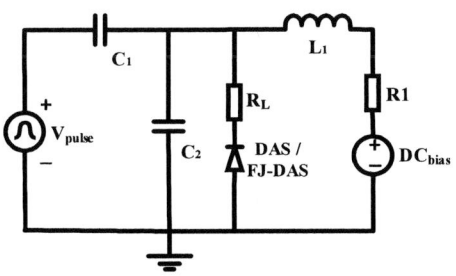

Fig. 2. shows the schematic diagram of the simulation circuit.

parameters is considered, where a floating junction is added at the midpoint of the drift region, as depicted in Figure 1(b).

The doping concentrations in the P^+ and N^+ regions remain constant at $1\times10^{19}cm^{-3}$ and 5×10^{18} cm^{-3}, respectively, while the base region doping concentration is chosen to be relatively high to meet subsequent process design requirements. By selecting an appropriate floating junction doping concentration, the electric field (E_m) in the upper and lower drift regions can be equal. The subsequent sections will explore the simulation study of these two structures.

DAS and FJ-DAS both utilize the simulation circuit shown in Figure 2. An AC voltage source (V_{ac}) provides a triangular waveform signal, resulting in an instantaneous voltage change rate (A) for the device and R_L. The DC voltage source (V_{bias}) supplies a reverse DC bias (V_{pulse}) to the device. These sources are isolated by capacitors C_1, inductor L_1, and resistor R_1, with values of 0.1 nF, 1 mH, and 1

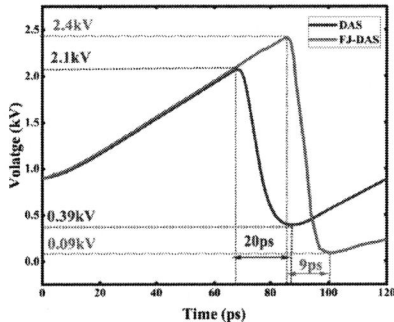

Fig. 3. shows the switching characteristics of conventional DAS and FJ-DAS devices. The turn-on time for conventional DAS is 20 ps, while FJ-DAS achieves a turn-on time of 9 ps.

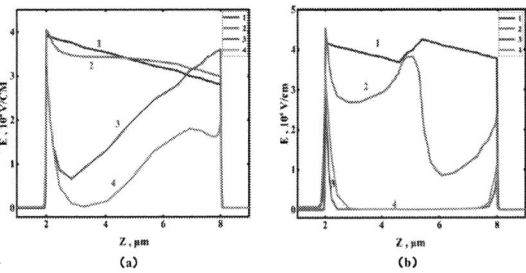

Fig. 4. shows the electric field distribution E(Z) during delayed avalanche breakdown for conventional DAS and FJ-DAS. The brown, red, blue, and green curves represent time steps of 3 ps. (a) Conventional DAS. (b) FJ-DAS.

kΩ, respectively. Capacitor C2 and load resistance RL have values of 3 pF and 50 Ω. In the simulation, the devicevoltage is initially raised to V_{bias} using quasi-static simulation, followed by mixed-circuit simulation to determine the time-dependent voltage across the device.

In Figure 3, when the base doping concentration of FJ-DAS is $1\times10^{16}cm^{-3}$ and the floating junction doping concentration is $4.8\times10^{16}cm^{-3}$, with a width of 0.8 μm and a spacing of 3 μm, the turn-on time of conventional DAS is 20 ps, with a turn-on rate of 85.5 kV/ns. In contrast, the turn-on time of FJ-DAS is 9 ps, with a turn-on rate of 256.7 kV/ns. It is evident that FJ-DAS outperforms the ideal DAS structure with the same base doping concentration.

As shown in Figure 4(a)(b), during conduction, the internal electric field of conventional DAS remains close to the TRAPATT triggering mechanism. Plasma accumulates in the high-field region and extends toward the low-field region.Then the high-field front moves along the depletion region, causing localized field collapse. In FJ-DAS, the internal electric field behaves differently, resembling quasi-uniform triggering. When reaching the maximum delayed avalanche electric field (E_m), FJ-DAS exhibits a longitudinal field distribution with two inverted trapezoids. Avalanche ionization occurs in regions where the electric field exceeds the critical field (E_c). Both the I-drift region and II-drift region in FJ-DAS accumulate plasma due to their high fields surpassing E_c simultaneously. As shown in Figure 5, in the FJ-DAS, the longitudinal electric field near the middle floating junction is closer to the ideal DAS field. Compared to the transverse field outside the opening, the field strength in the floating junction region is higher. As shown in Figure 6(a), this introduces a transverse peak electric field within the floating junction area. When the transverse peak electric field

979-8-3503-6184-1/24 $31.00 © 2024 IEEE

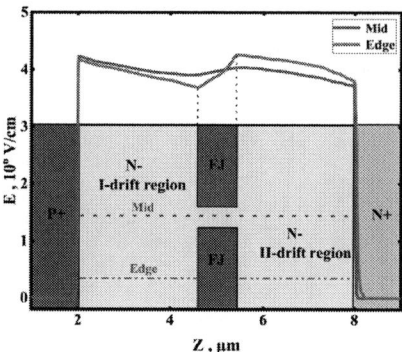

Fig. 5. Schematic diagram illustrating the electric field selection positions for FJ-DAS. The distribution of electric field E(Z) between the central and edge positions.

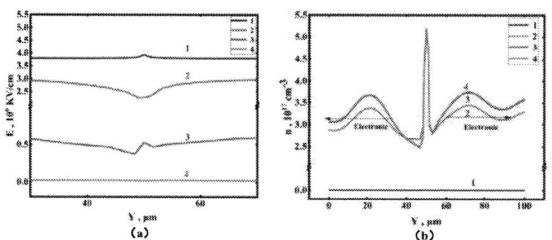

Fig. 6. (a) Evolution of the lateral electric field distribution E(Y) over time. (b) Variation of the lateral electron distribution n(Y) with time. The brown, red, blue, and green curves represent time increments of 3 ps.

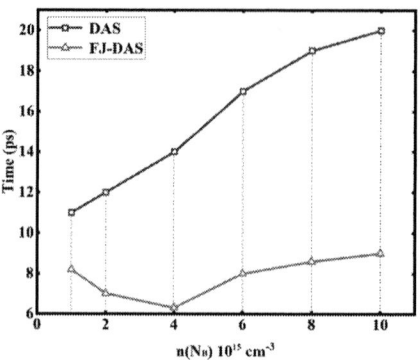

Fig. 7. Comparison of turn-on time between traditional DAS and FJ-DAS at base region doping concentrations ranging from 1×10^{15} cm^{-3} to 1×10^{16} cm^{-3}.

reaches a value greater than the threshold electric field (Ec), it triggers a delayed avalanche breakdown. The accumulated high-density plasma extends to both sides of the device, providing additional plasma for the longitudinal avalanche breakdown. Following the generation of high-density plasma due to the longitudinal high electric field, a reverse electrostatic field forms, shielding the local high electric field.Consequently, the electric fields of I-drift region and II-drift region regions collapse synchronously. Subsequently, the plasma extends toward the lower electric field region at the center, quickly filling the entire base region. This collapse of the overall electric field to zero results in the device switching to a low-resistance state, completing the turn-on process. In this way, the FJ-DAS can be thought of as two short-base ideal DAS devices connected in series for synchronous triggering, significantly accelerating the turn-on speed. Simultaneously, the introduction of the floating junction enhances the breakdown voltage of the device. Therefore, compared to the traditional structure, the turn-on rate of FJ-DAS is improved.

In our simulations, we compared the performance of FJ-DAS with traditional DAS under lower base doping conditions. By varying the base concentration and selecting the optimal FJ doping level, we compared the turn-on time of FJ-DAS with that of the ideal structure. As shown in the Figure 7, within the base concentration range of 1×10^{15}cm^{-3} to 1×10^{16}cm^{-3}, the turn-on time of FJ-DAS is significantly shorter than that of the same doping concentration in the ideal structure. Furthermore, as the base region doping concentration increases, the advantage of FJ-DAS becomes more pronounced. This indicates that even as ideal DAS approaches more uniform triggering, FJ-DAS still maintains a significant advantage. Additionally, this provides theoretical assistance for the subsequent preparation of N-type base region terminals.

III. CONCLUSION

This study investigates the performance improvement of SiC DAS by introducing FJ structure. By incorporating the FJ in SiC DAS, we optimize the internal electric field distribution, achieving a uniform triggering mechanism that overcomes the limitations imposed by traditional DAS due to base region doping concentration.

TCAD simulations demonstrate that FJ-DAS outperforms conventional DAS in terms of turn-on time and switching speed. Specifically, FJ-DAS exhibits a turn-on time of 9 ps and a switching speed of 256.7 kV/ns, whereas traditional DAS has a turn-on time of 20 ps and a switching speed of 85.5 kV/ns. FJ-DAS achieves approximately 55% faster turn-on time than its conventional counterpart

In summary, the combination of structural optimization and material advantages positions FJ-DAS as a promising candidate for high-performance power devices in pulse power systems, providing valuable insights for future designs.

[1] Y. Zhou *et al.*, "Demonstration of Picosecond 4H-SiC Diode Avalanche Shaper With Voltage Rise Rate of 11.14 kV/ns and Peak Power Density of 62 MW/cm $\$^\wedge 2\$$," *IEEE Transactions on Power Electronics,* vol. 37, no. 4, pp. 3724-3727, 2021.

[2] A. S. Kesar *et al.*, "A fast avalanche Si diode with a 517 μm low-doped region," *Applied Physics Letters,* vol. 117, no. 1, 2020.

[3] M. S. Ivanov, P. B. Rodin, P. A. Ivanov, and I. V. Grekhov, "Parameters of silicon carbide diode avalanche shapers for the picosecond range," *Technical Physics Letters,* vol. 42, pp. 43-46, 2016.

[4] P. Rodin, P. Ivanov, and I. Grekhov, "Performance evaluation of picosecond high-voltage power switches based on propagation of superfast impact ionization fronts in SiC structures," *Journal of applied physics,* vol. 99, no. 4, 2006.

[5] D. Guo *et al.*, "Investigation on triggering mode and criterion of 4H-SiC diode avalanche shaper," *IEEE Transactions on Electron Devices,* vol. 70, no. 8, pp. 4075-4080, 2023.

[6] J. Li *et al.*, "Analytical model and optimization strategy for SiC floating junction JBS diodes," *Microelectronics Journal,* vol. 137, p. 105800, 2023.

Silicon Carbide Diode Avalanche Shaper with Multi-Point Quasi-Uniform Triggering

Lin Cheng [1,2], Yu Zhou* [1,2], Xiao-Yan Tang [1], Chao Han [2], Yu-Ming Zhang [1,2], Qing-Wen Song [1,2]

[1] Xidian University School of Microelectronics, Xi'an 710071, Shaanxi, China
[2] Xidian-Wuhu Research Institute, Wuhu 241000, Anhui, China

* Email: cll@stu.xidian.edu.cn, zhouyu01@xidian.edu.cn

Abstract—This paper proposes a novel structure for manufacturing $P^+N^-N^+$ type silicon carbide Diode Avalanche Shaper (DAS) devices. This innovative structure introduces multiple points of electric field concentration within the active region, achieving "quasi-uniform" distribution of the electric field across multiple points. This approach addresses the issue of local electric field concentration within the device, enabling "quasi-uniform" triggering during the DAS device's turn-on process. The process of multi-point "quasi-uniform" triggering significantly reduces thermal accumulation caused by local electric field concentration, thereby preventing the device from burning out due to local thermal accumulation during the turn-on process. Simulation results show that the temperature rise of the multi-trench structure device can be controlled within 10K per cycle, which is 1.4% of that of positive bevel mesa structure with arc-shaped injection.

Keywords—*Silicon Carbide, Pulse Power, Diode Avalanche Shaper, Terminal Technology*

I. INTRODUCTION

With the continuous development of pulsed power technology, it has found extensive applications in defense, military, fundamental research, and medicine. Pulse switch devices are the core of pulsed power technology. I.V. Grekhov from the Ioffe Institute first discovered the delayed avalanche breakdown (DAB) phenomenon[1]. Based on this characteristic, the diode avalanche shaper (DAS) can effectively sharpen pulses, making it an outstanding pulse power switch.

The simplified version of the DAS working circuit is shown in Figure 1(a). The DAS operates under a DC bias below the static breakdown voltage. When a voltage pulse ($A \times t$) arrives, the device reaches the delayed breakdown voltage (V_{DB}), which is higher than the static breakdown voltage (V_{SB}), and the DAS will undergoes a delayed avalanche breakdown and rapidly conducts. Its voltage (V_{DAS}) is transferred to R_L, resulting in an output pulse (V_{out}) that is steeper than the input voltage (V_{in}), as shown in Figure 1(b).

4H-SiC has a high critical electric field, high electron saturation velocity, and high thermal conductivity, it is an excellent material for fabricating DAS devices.Current research on 4H-SiC DAS mainly focuses on theoretical and simulation studies. The literature[2] indicate that 4H-SiC DAS can reduce the device turn-on time to the picosecond range.

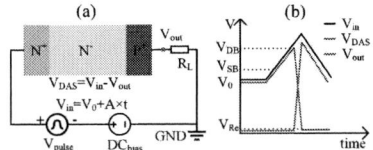

Fig. 1. (a) Schematic diagram of DAS operating circuit (b) Schematic diagram of pulse sharpening.

According to the doping type in the device base region, 4H-SiC DAS can be classified into $P^+P^-N^+$ structure DAS (p-DAS) and $P^+N^-N^+$ structure DAS (n-DAS).In SiC, the diffusion coefficient and saturation drift velocity of electrons are both greater than holes. This results in the n-DAS having a faster turn-on speed than the p-DAS, which enables the generation of sharper output pulses.

When DAS triggers delayed avalanche breakdown (DAB), the maximum electric field within the device greatly exceeds the static critical electric field. The delayed breakdown voltage (V_{DB}) is approximately 1.4 times higher than the static breakdown voltage (V_{SB})[3], and any local field concentration can lead to diode failure[4]. Therefore, experimental research on SiC DAS is currently in its initial stages.

The article [5] reports the first experimental fabrication of high-performance picosecond 4H-SiC DAS using a $P^+P^-N^+$ multilayer epitaxial structure. This study validates the feasibility and advantages of using SiC for DAS fabrication. However, there have been no successful reports on the fabrication of faster n-DAS devices.

In this paper, we using ISE-TCAD software to investigate the terminal structure design of n-DAS devices. The study analyzed the reasons why traditional flat terminal structures such as Junction Termination Extension (JTE) and Field Limiting Rings (FLRs), as well as conventional positive bevel mesa structures with arc-shaped injection, are unsuitable for n-DAS devices. Furthermore, a novel structure design suitable for n-DAS devices was proposed.

II. SIMULATION PRESENTATION

A. Simulation Parameter Setting

The coupled solutions of the current density equation, Poisson's equation, carrier continuity equation, and thermodynamic equation were used to obtain simulation results. Due to different device packaging forms, their surface thermal resistance also varies. The surface thermal resistance is set to 0.3 cm²K/W in this paper[6]. Models including high-field velocity saturation mobility, Shockley-Read-Hall (SRH) recombination, Auger recombination, incomplete ionization and were considered to ensure the reliability of the simulation results.

B. Device Structure Model

The four device structures are shown in Figure 2. The first structure uses a non-uniform field limiting rings (FLRs) structure, the second structure employs a junction termination extension (JTE) structure, the third structure features a positive bevel mesa structure with arc-shaped injection, and the fourth structure utilizes a active region with multiple trenches coupled positive bevel mesa structure with arc-shaped injection. To discuss the thermal effects of different structures on the devices, their active area is 0.001

979-8-3503-6184-1/24 $31.00 © 2024 IEEE

cm². The P⁺ region uses the same ion implantation scheme, the N⁻ region has a thickness of 10 μm with a doping concentration of 1×10^{15}cm⁻³.

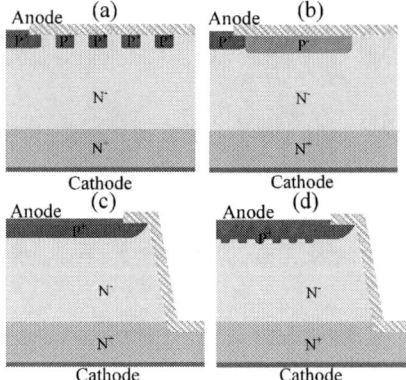

Fig. 2. (a) non-uniform field limiting rings (FLRs) structure. (b) junction termination extension (JTE) structure. (c) positive bevel mesa structure with arc-shaped injection. (d) active region with multiple trenches coupled positive bevel mesa structure with arc-shaped injection.

III. ANALYSIS OF SIMULATION RESULTS

A. Disadvantages of the traditional planar terminal structure.

The electric field distribution in conventional termination structures during static breakdown is shown in Figure 3. The peak electric field is located at the boundary between the active region and the termination area, or within the termination region outside the active area of the device. Both structures rely on the expansion of the depletion region to increase the breakdown voltage. When a trigger pulse is applied to the device, the edge of the active area experiences the earliest electric field concentration. This localized electric field concentration can cause an avalanche breakdown in this area, generating a large number of electron-hole pairs in a short time. The avalanche current's flow path is concentrated, leading to significant local thermal accumulation effects, making the device highly susceptible to burnout and failure in this region.Therefore, conventional termination structures are not suitable for silicon carbide DAS devices.

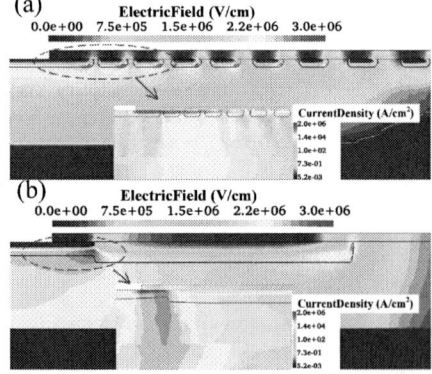

Fig. 3. (a) The electric field distribution during static breakdown and the current distribution during dynamic turn-on for non-uniform field limiting rings (FLRs) structures. (b) The electric field distribution during static breakdown and the current distribution during dynamic turn-on for junction termination extension (JTE) structures.

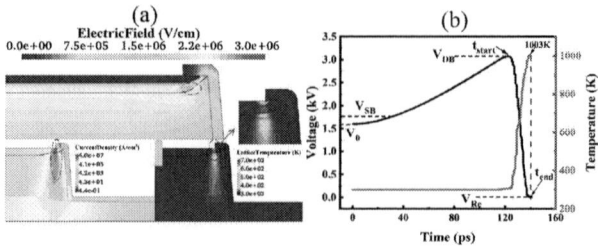

Fig. 4. (a) Electric field distribution during static breakdown and current density distribution and lattice temperature at the end of dynamic turn-on for the positive bevel mesa structure with arc-shaped injection. (b) Voltage and temperature rise curves during the dynamic process of the device.

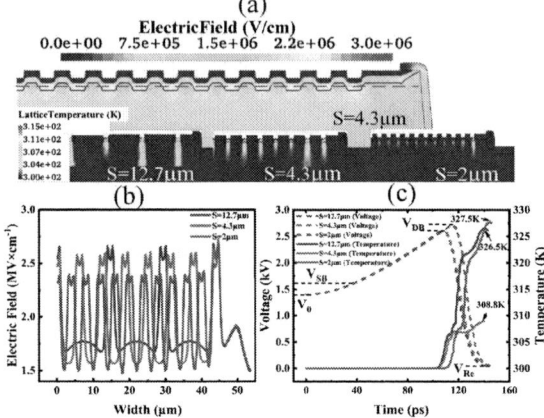

Fig. 5. (a) Electric field distribution during static breakdown and lattice temperature at the end of dynamic turn-on for the active region with multi-trench coupled positive bevel mesa structure with arc-shaped injection. (b) Tangential electric field distribution of muti-trench structures with different spacings. (c) Voltage and temperature rise curves during the dynamic process of devices with different spacings.

B. Advantages and disadvantages of the positive bevel mesa structure with arc-shaped injection.

The device with a positive bevel mesa structure with arc-shaped injection exhibits the electric field distribution during static breakdown as shown in Figure 4. This structure effectively avoids the concentration of electric fields at the surface, confining high electric field regions within the active area.

As shown in Figure 4(a), there is an unavoidable local electric field concentration at the pn junction bend of the device. The electric field at this point reaches the delayed avalanche threshold earlier than within the bulk, resulting in localized avalanche initiation. A significant avalanche current concentrates at this location, causing a rapid increase in lattice temperature.As depicted in Figure 4(b), the device temperature begins to rise gradually after the delayed avalanche start time (t_{start}). By the time the device fully opens (t_{end}), the maximum lattice temperature reaches 1003K, with a single avalanche temperature rise of 703K. Such high temperature rise can easily lead to device burnout, making this structure unsuitable for SiC DAS devices.

C. Advantages of the active region multi-trench structure.

First, multiple trenches are etched in the active region. Then, a trapezoidal silicon dioxide (SiO₂) blocking layer is applied. Using ion implantation technology, both an arc-shaped pn junction in the edge region and multiple trench pn junctions within the active region are simultaneously formed. This approach leverages the arc-shaped pn junction to mitigate the boundary electric field concentration caused by surface effects. Additionally, the trench structure introduces multiple localized electric field concentrations within the active region, suppressing electric field concentration at the inflection point of the arc-shaped pn junction.

After introducing multiple localized electric field concentrations within the active region, we optimized the trench parameters and distribution through simulation. This allowed achieving a 'quasi-uniform' distribution of electric fields at the trench during reverse voltage operation, effectively addressing the issue of premature breakdown caused by localized electric field concentration in the device.

When the device is subjected to a reverse trigger pulse, the multiple electric field concentrations within the active region will trigger avalanche breakdown almost synchronously and uniformly, achieving a "quasi-uniform triggering" effect. As shown in Figure 5(a), each trench serves as a conduction path for the avalanche current, and the multi-trench structure prevents localized temperature rise. This resolves the issue of excessive local temperature rise. As illustrated in Figure 5(c), the maximum temperature rise of the device remains below 30K at different trench spacings.

Under a given active region area, when the trench width is fixed, the spacing of trenches is a crucial parameter in a multi-trench structure. It determines the effectiveness in mitigating thermal accumulation caused by local electric field concentration. In this study, the active region length is set to 50μm. The trench width is fixed at 2 μm, and the trench depth is fixed at 1 μm.Simulations compared the results for trench spacings of 12.7μm, 4.3μm, and 2μm.

As shown in Figure 5(b), when the trench spacing is 12.7 μm, the internal electric field is very uniform; however, the trench structure cannot provide sufficient avalanche current conduction paths, leading to a significant increase in thermal accumulation. Conversely, when the trench spacing is 2 μm, the electric fields at the bottoms of the trenches couple with each other, resulting in a pronounced electric field concentration at the bottom corner of the outermost trench. At this point, the electric field distribution within the active region of the device resembles that of a single arc-shaped junction, causing localized avalanching at the outermost trench and a noticeable rise in thermal accumulation. With the trench spacing is 4.3 μm, a sufficient number of conduction paths is provided, avoiding excessive local electric field concentration. The avalanche current is evenly distributed, resulting in minimal temperature rise.

As shown in Figure 5(c), the maximum temperatures at trench spacings of 12.7 μm and 2 μm reached 326.5K and 327.5K, respectively. However, with a trench spacing of 4.3 μm, the maximum temperature of the device was only 308.8K, resulting in a net temperature rise of just 8.8K. The temperature increase is nearly uniformly distributed across the entire active region of the device, preventing burnout due to localized heat concentration and thus enhancing the device's reliability.

IV. SUMMARY

In this study, a comparative simulation was performed on three structures: non-uniform field limiting rings, junction termination extension, and positive bevel mesa structure. The simulation results indicate that the limitations of these structures in SiC DAS devices are primarily due to the introduction of localized electric field concentration, leading to significant localized heat accumulation and potential burnout during device operation. To address the issue of localized electric field concentration, a novel structure with multi-trench coupling in the active region and positive bevel mesa structure was proposed. Simulations verified the feasibility of this structure in mitigating thermal accumulation during device operation. By appropriately configuring the trench spacing, with a trench width of 2 μm and spacing of 4.3 μm, the maximum temperature rise at the end of device operation was limited to just 8.8K, demonstrating the viability of this structure for application in SiC DAS devices.

REFERENCES

[1] I. Grekhov and A. Kardo-Sysoev, "Subnanosecond current drops in delayed breakdown of silicon pn junctions," Sov. Tech. Phys. Lett, vol. 5, no. 8, pp. 395-396, 1979.

[2] M. S. Ivanov, P. B. Rodin, P. A. Ivanov, and I. V. Grekhov, "Parameters of silicon carbide diode avalanche shapers for the picosecond range," Technical Physics Letters, vol. 42, pp. 43-46, 2016.

[3] V. Brylevskiy, I. Smirnova, A. Gutkin, P. Brunkov, P. Rodin, and I. Grekhov, "Delayed avalanche breakdown of high-voltage silicon diodes: Various structures exhibit different picosecond-range switching behavior," Journal of applied physics, vol. 122, no. 18, 2017.

[4] M. Ivanov, N. Podolska, and P. Rodin, "Quasi-streamer mode of delayed avalanche breakdown initiated by technological imperfections," in Journal of Physics: Conference Series, 2017, vol. 816, no. 1: IOP Publishing, p. 012033.

[5] Y. Zhou et al., "Demonstration of Picosecond 4H-SiC Diode Avalanche Shaper With Voltage Rise Rate of 11.14 kV/ns and Peak Power Density of 62 MW/cm^2," IEEE Transactions on Power Electronics, vol. 37, no. 4, pp. 3724-3727, 2021.

[6] D. Guo, Y. Zhou, X. Tang, and Y. Zhang, "Direct comparison of silicon carbide and silicon diode avalanche shaper in multi-pulse applications," Journal of Crystal Growth, vol. 603, p. 127007, 2023.

Super Field Plate LIGBT with Improved Performance for Both Cell and Terminal Region

Weihao Lu [1], Jing Li [1], Jitong Wang [1], Chaoyang Peng [1], Chunwei Zhang *[1]

[1] School of Information Science and Engineering, University of Jinan, Jinan 250022, China

* Email: 202221100392@stu.ujn.edu.cn, ise_zhangcw@ujn.edu.cn

Abstract—**In this study, the performance of the Super Field Plate (SuFP) technology proposed in our previous work is experimentally verified based on a lateral insulated gate bipolar transistor (LIGBT). The results demonstrate that the SuFP-LIGBT with a 30μm drift region length (L_{drift}) achieved 563.6V breakdown voltage (BV), which is 38.1% larger than that of conventional device. Notably, the SuFP also effectively reduces the L_{drift} of the terminal region due to its excellent electric field optimization effect. In addition, the SuFP is proved to have better immunity to the influence of high-voltage interconnection (HVI) than conventional field plate (FP) technology because it can cover a larger drift region and provide a better shielding effect. Therefore, the SuFP technology is very promising for LIGBT applications.**

Keywords—*lateral insulated gate bipolar transistor, Super Field Plate, breakdown voltage, high-voltage interconnection*

I. INTRODUCTION

Power devices play a pivotal role in modern electronic systems and are extensively utilized in fields such as automotive electronics and wireless communications, functioning as essential components for the control and conversion of electrical energy [1-2]. High breakdown voltage (BV) and low on-state resistance ($R_{on,sp}$) are required to improve the efficiency of power management systems. However, the BV and $R_{on,sp}$ are two competitive performances for power devices [3]. Therefore, optimizing the trade-off between BV and $R_{on,sp}$ is always a hot topic for power device researchers and is widely investigated.

As is well known, field plate (FP) technology is a widely adopted technique to optimize electric fields, which is compatible with all manufacturing processes and would not increase production costs. However, the optimization effect of FP technology on electric field distribution cannot reach the ideal result [4]. In contrast, the superjunction (SJ) technology can achieve excellent electric field optimization effect by introducing p-type doing region. However, the application of SJ technology requires precise doping processes and additional manufacturing costs [5]. Thereby, the Super Field Plate (SuFP) that combined the advantages of SJ and FP technologies are theoretically proposed in our previous work [6]. However, its performance has not yet been experimentally validated. Moreover, the design of the power device includes both the cell region and the terminal region. Meanwhile, the impact of the high-voltage interconnection (HVI) also needs to be considered during the application of SuFP [7-8]. Therefore, the effect of the SuFP on the device still requires further investigation to facilitate their application.

In this study, the SuFP technology is applied on a 500V a

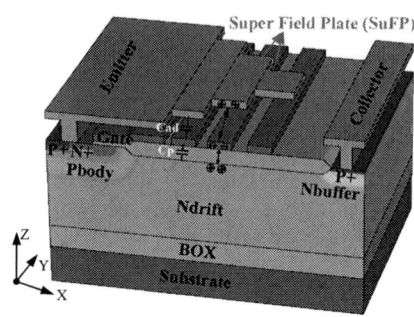

Fig. 1. The schematic diagram of the SuFP-LIGBT. (The passivation layer between adjustment plate and field plate is hidden.)

lateral insulated gate bipolar transistor (LIGBT) to evaluate its performance experimentally. The influences of SuFP on both cell and terminal regions are discussed. Moreover, the influence of SuFP technology on HVI is also investigated. The investigation results illustrate that the SuFP is a promising technology in the field of LIGBT.

II. DEVICE STRUCTURE AND EXPERIMENTS

The schematic diagram of the studied SuFP-LIGBT is shown in Fig. 1. It is evident that each SuFP is comprised of two series-connected capacitances: the capacitance between the adjustment plate and the induction plate (C_{ad}) and the capacitance between the induction plate and the N-drift (C_p). The induced charge amount (Q_{FP}) on the adjustable field plate can be expressed as follows:

$$Q_{FP} = (V_{FP} - V_{surf}) \cdot C_P = -\frac{C_{ad}}{C_P + C_{ad}} \cdot V_{surf} \cdot C_P \qquad (1)$$

Where V_{FP} is the induced potential of the adjustable field plate, and V_{surf} is the potential at the corresponding surface of the drift region.

According to (1), the induced charge on the adjustable field plate is influenced by the size of the C_{ad}. Therefore, by designing C_{ad} appropriately, the induced negative charge on the adjustable field plate can be adjusted to achieve balance with the positive space charge in the drift region and the substrate depletion charge. The horizontal arrangement of multiple SuFP enables the entire drift region to achieve charge balance with the existence of the substrate-assisted depletion effects, thereby ensuring a uniform electric field distribution throughout the device's drift region.

To validate the effect of SuFP, a series of SuFP-LIGBT were fabricated using 0.5μm Silicon-On-Insulator (SOI) technology. The key process parameters are as follows: the N-drift region length (L_{drift}) is 30.0μm with a thickness of 18.0μm, the buried oxide layer thickness is 3.5μm, the gate oxide layer thickness is 50nm, the field oxide layer thickness

---

This work was supported in part by the Natural Science Foundation of Shandong Province under Grant ZR2021MF010, Grant ZR2023ZD03, and Grant 2023KJ109, in part by the National Natural Science Foundation of China under Grant 62104080 and Grant 62174068.

979-8-3503-6184-1/24 $31.00 © 2024 IEEE

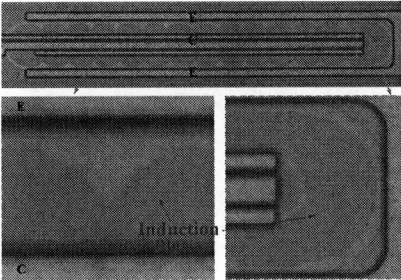

Fig. 2. The microscopic image of a fabricated SuFP-LIGBT.

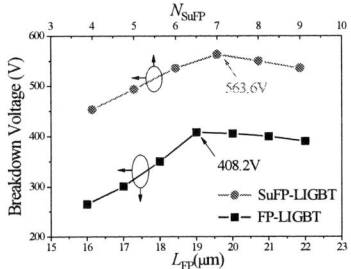

Fig. 3. The measured BV of different L_{FP} and N_{SuFP}.

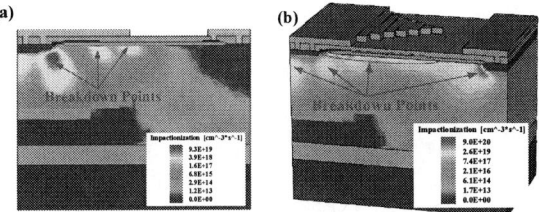

Fig. 4. The impact ionization rate distributions of (a) FP-LIGBT and (b) SuFP-LIGBT under breakdown condition.

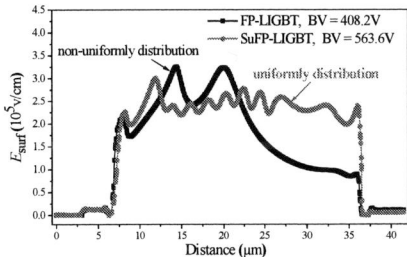

Fig. 5. The E_{surf} distributions of FP-LIGBT and SuFP-LIGBT under breakdown condition.

is 0.6μm, the passivation layer thickness is 1.0μm, and the channel length is 1.5μm. The aforementioned process parameters represent optimized values that have been determined through consideration of various factors, including voltage blocking capability, current density, thermal resistance, and process cost. Furthermore, the length and spacing of the field plates are set as the minimum values allowed by design rules (1.0μm). Fig. 2 shows the micrograph of the manufactured SuFP-LIGBT. It is noteworthy that the field plates are continuous in the width direction, including the terminal region, thus ensuring consistent device characteristics across the width. The adjustment plates consist of multiple units, each unit being 50μm wide.

III. RESULTS AND DISCUSSIONS

The BV of SuFP-LIGBT with varying numbers of SuFP (N_{SuFP}) was quantified and compared with that of FP-LIGBT, as illustrated in Fig. 3. It is noteworthy that the BV of FP-LIGBT increases rapidly with the increase of field plate length (L_{FP}) when $L_{FP} < 19$μm and decreases slowly when $L_{FP} > 19$μm. Similarly, the BV of SuFP-LIGBT increases rapidly with N_{SuFP} at $N_{SuFP} < 7$ and decreases slowly with N_{SuFP} when $N_{SuFP} > 7$. To understand these results, the impact ionization rate distributions of the FP-LIGBT and SuFP-LIGBT at their respective maximum BV were extracted, as shown in Fig. 4. Fig. 4 (a) shows that the FP-LIGBT with $L_{FP} = 19$μm undergoes breakdown at both the junction corner and the FP edge. The breakdown points of SuFP-LIGBT are further extracted in Fig. 4 (b). It is evident that the breakdown occurs simultaneously at the junction corner and several other regions on the surface of the drift region, which validates the effectiveness of SuFP as it precisely fulfills the requirements.

To understand the BV improvement of SuFP-LIGBT, the surface electric field (E_{surf}) distributions of FP-LIGBT and SuFP-LIGBT under breakdown condition are extracted, as shown in Fig. 5. It is evident that the E_{surf} of SuFP-LIGBT is considerably more uniform than that of FP-LIGBT. Consequently, the SuFP-LIGBT exhibits a 38.1% higher BV than the FP-LIGBT. And the SuFP-LIGBT retains the same

current handling capability as FP-LIGBT, as the potential difference between the field plate and drift region can be considered negligible, exerting no impact on current handling capacity. Hence, the proposed SuFP-LIGBT demonstrates outstanding voltage blocking capability in microelectronics applications.

The SuFP technology applied to the terminal structure of LIGBT is illustrated in Fig. 6, where the L_{drift} at the junction angles is increased to mitigate the effects of electric field concentration. The E_{surf} distributions of SuFP-LIGBT and terminals under breakdown condition are extracted, as shown in Fig. 7. The BV of SuFP-LIGBT terminal is 560.2V when the L_{drift} is 35μm, which is only 1% smaller than the BV of the cell region. This illustrates that the SuFP effectively eliminates the effects of electric field concentration at the junction angles. Fig. 8 compares the BV of the FP-LIGBT and SuFP-LIGBT terminals with different L_{drift}. It is noted that, for SuFP- LIGBT, the L_{drift} of the terminal region only require 5μm longer than that of the cell region. In contrast, for FP-LIGBT, the terminal region need 10μm larger L_{drift} than cell region. Consequently, the application of SuFP technology would decrease the area of terminal region as well as the cost of the device.

It is noted that the SuFP consists of two layers: the adjustment field plate and the charge induction plate. Due to the continuity of the charge induction plate in the width direction, the potential of the charge induction plate is fixed at a constant value across the width direction, which is determined by the inductive control layer in the cell region, so it can shield the drift region from the effects of external signals. In this way, the SuFP would have good shielding effect on the HVI. Fig. 9 shows the SuFP-LIGBT with HVI, which is connected to the collector and covers the top of the device. To verify the shielding effect of charge induction plate, the E_{surf} of SuFP-LIGBT and FP-LIGBT with HVI across the drift region are extracted, as shown in Fig. 10. It is evident that the electric field in the region where the SuFP is present remains constant and therefore has a relatively small effect on the BV. The results show that, with the influence of HVI, FP-LIGBT

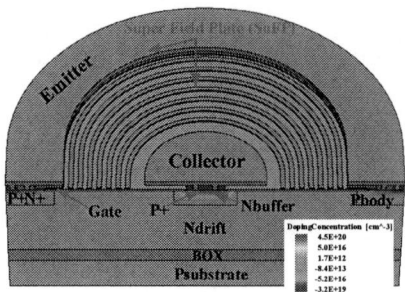

Fig. 6. The terminal schematic diagram of SuFP-LIGBT. (The passivation layer between the adjustment plate and the induction plate is hidden.)

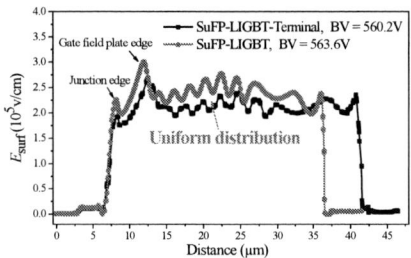

Fig. 7. The E_{surf} distributions of SuFP-LIGBT and SuFP-LIGBT terminal under breakdown condition.

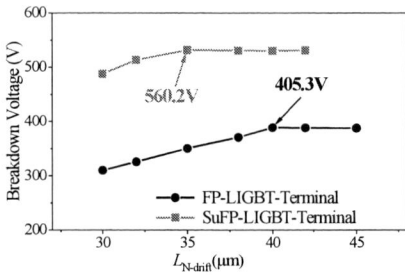

Fig. 8. The BV of FP-LIGBT and SuFP-LIGBT terminals with different L_{drift}.

shows a 20.3% reduction in BV, whereas the BV of SuFP-LIGBT is only decreased by 7.8%. Therefore, compared to FP-LIGBT, SuFP-LIGBT exhibits higher immunity to HVI.

IV. CONCLUSION

This study investigates the performance of SuFP-LIGBT experimentally. The experimental results show that we have achieved a BV of 563.6V for a LIGBT with 30μm drift region, which is improved by 38.1% compared to the conventional FP technology. Meanwhile, the SuFP technology can also optimize the electric field distribution of terminal region, as a result, the L_{drift} of terminal region is reduced from 40μm to 35μm. Moreover, our investigations prove that the SuFP exhibit improved immunity to the influence of HVI relative to conventional field plate because it can cover more drift region. Therefore, the SuFP-LIGBT exhibit high performance and high immunity to HVI, and is a promising device structure.

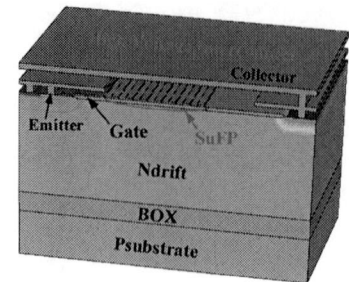

Fig. 9. The SuFP-LIGBT with HVI. (The passivation layer between the adjustment plate and the induction plate is hidden.)

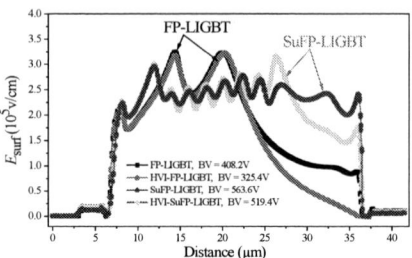

Fig. 10. The E_{surf} distributions of FP-LIGBT and SuFP-LIGBT with their HVI structures under breakdown condition.

REFERENCES

[1] J. Yang, M. Zhang, Y. Wu, M. Wang, and J. Wei, "Double-gate RESURF lateral insulated gate bipolar transistor with built-in p-channel MOSFET for active conductivity modulation control throughout drift region," IEEE Electron Device Lett., vol. 43, no. 2, pp. 272–275, Feb. 2022.

[2] J. Lim, J. Seong, J. Jeon, Y. Kim, H. Im, et al., "Design and experimental evaluation of transfer-molded 650 V super-junction MOSFET power module for industrial applications," IEEE Trans. Ind. Appl., vol. 57, no. 6, pp. 6295–6305, Nov. 2021.

[3] Y. Wang, B. Duan, H. Song, and Y. Yang, "Accumulation-mode lateral double-diffused MOSFET breaking silicon limit by eliminating dependence of specific ON-resistance on doping concentration," IEEE Trans. Electron Devices, vol. 68, no. 5, pp. 2414–2419, May 2021.

[4] S. Karmalkar, M. Shur, G. Simin, and M. Khan, "Field-plate engineering for HFETs," IEEE Trans. Electron Devices, vol. 52, no. 12, pp. 2534–2540, Dec. 2005.

[5] F. Udrea, G. Deboy, and T. Fujihira, "Superjunction power devices, history, development, and future prospects," IEEE Trans. Electron Devices, vol. 64, no. 3, pp. 713–727, Mar. 2017.

[6] C. Zhang, H. Guo, Z. Chen, W. Yue, Y. Li, et al., "Super field plate technique that can provide charge balance effect for lateral power devices without occupying drift region," IEEE Trans. Electron Devices, vol. 67, no. 5, pp. 2218–2222, May 2020.

[7] K. Yew, R. Ong, H. Yap, W. Yi, J. Phang, et al., "Insights on inter-metal reliability assessment of high voltage interconnects," 2022 IEEE International Reliability Physics Symposium (IRPS), Dallas, TX, USA, 2022, pp. P49-1-P49-4.

[8] L. Du, Y. Guo, J. Zhang, J. Yao, K. Yang and M. Li, "The new structure and analytical model of a high-voltage interconnection shielding structure with high-k dielectric pillar," IEEE Trans. Electron Devices, vol. 67, no. 4, pp. 1745-1750, April 2020.

High-temperature oxidation of 4H-SiC and gate oxide reliability dependence on oxidation temperature

Baoyan Feng [1], Xiaoyan Tang *[1], Yi bo Zhang[1], Chao Han [2], Hao Yuan[1], Qing wen Song[1]

[1] The key Laboratory of Wide Band Gap Semiconductor Materials and Devices, Xidian University, Xi'an 710071, China
2.Xidian-Wuhu Research Institute, Wuhu 241002, China

* Email:byff@stu.xidian.edu.cn, Corresponding author: xytang@mail.xidian.edu.cn

Abstract—**In this study, the MOS capacitors were fabricated under different oxidation process, Analyzed the effect of oxidation temperature on interface traps and electrical performance of SiC MOS capacitor. In addition, the long-term lifetime of SiO_2 was studied through time-dependent dielectric breakdown (TDDB) measurements. The experimental results show that the density of interface traps (Dit) is the lowest at 1400 ℃, and it can be as low as 1.1×10^{11} $cm^2\cdot eV^{-1}$ at a distance of 0.2eV from the conduction band, At the same time, the gate oxide formed under this condition has a longer lifetime, which provides a research basis for the long-term and dynamic reliability of 4H-SiC MOSFET.**

Keywords—4H-SiC, Oxidation, MOS capacitor, Dit, TDDB

I. INTRODUCTION

Compared with silicon (Si) materials, silicon carbide (SiC) has the advantages of wide bandgap width and high breakdown field strength, which has a broad application prospect in high-power and high-temperature applications [1, 2]. In addition, SiC is a compound semiconductor material that can be grown directly by thermal oxidation of silicon dioxide (SiO_2) [3]. However, The molecular structure of 4H-SiC is different from Si, leading to various types of traps near the interface and in the oxide [5], which poses a serious challenge to the reliability of the gate oxide of SiC MOSFET.

During the thermal oxidation process of 4H-SiC materials, due to the presence of carbon containing by-products and dangling bonds, a large number of interface states exist near the interface [6, 7], which reduce the channel mobility of the device and increase the specific on-resistance [8, 9]. Meanwhile, the interfacial transition consisting of silicon oxycarbide (SiOxCy) [10, 11] causes serious threshold instability problems [12]. 4H-SiC MOSFET usually operate under strong electric field conditions, and the relative dielectric constant of SiO_2 is much smaller than that of 4H-SiC, so according to Gauss' theorem, it is known that the gate oxide layer needs to withstand a larger electric field intensity, which can lead to the premature failure of the oxide layer [13]. Therefore, it is important to systematically study the thermal oxidation reaction of SiC to improve the reliability of the oxide. In this study, it was found that increasing the thermal oxidation temperature has a significant improvement effect on the reliability of 4H-SiC MOSFET.

II. EXPERIMENT

The experimental design adopts an orthogonal comparison scheme, with different oxidation temperatures as the research elements, and investigates the influence of different oxidation process on the capacitance characteristics of 4H-SiC MOS.

The substrates were 4° off-axis n-type 4H–SiC (0001) with a 13 μm epitaxial layer doped with N (about 7.8×10^{16} cm^{-3}), The substrate thickness is about 380μm, and the substrate doping concentration is $5\times10^{18}cm^{-3}$.The diameter of the circular gate electrode was 300 μm.

The thickness of the oxide was measured using a J.A. Woollam M2000D ellipsometer. The oxidation process and thickness of the experimental samples are shown in Table 1.

TABLE I.   OXIDATION PROCESS AND THICKNESS OF EXPERIMENTAL SAMPLESS

Samples Number	Oxidation conditions	Oxide Thickness	Average oxidation rate (nm/h)	NO annealing
A	1300°C/37 min	54 nm	89.38	1300 °C
B	1350°C/17min	54 nm	190.59	30
C	1400°C/7min	56 nm	480.00	minutes

The electrical parameters of the Samples in this study were characterized using CASCADE's MPS150 Triax probe stage coupled with Keysight's B1505A semiconductor parameter analysis system. The surface morphology of the oxide was studied using Agilent 5500 Atomic Force Microscopy (AFM).

III. MEASUREMENT AND DISCUSSION

A. Surface topography of SiO_2 film

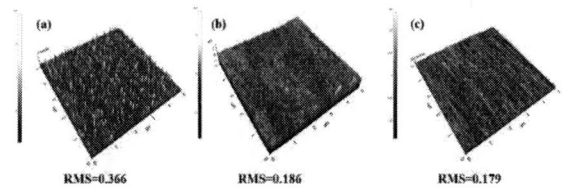

Fig. 1. surface morphology of samples at (a) 1300°C (b) 1350°C and (c) 1400°C oxidation temperatures observed by AFM.

Fig. 1 shows that there are varying degrees of protrusions on the surface of the film after oxidation reaction. As the oxidation temperature varies, the peak height, diameter, and distribution also different. The root mean square (RMS) roughness of samples A, B, and C are 0.366nm, 0.186nm, 0.179nm, respectively. It can be found that increasing the thermal oxidation temperature effectively improves the surface roughness of the sample. After analysis, it can be considered that the protrusions on the surface of SiO_2 films grown by thermal oxidation may be related to some incompletely reacted atomic lattice mismatches. At higher oxidation temperatures, the oxidation reaction rate is faster, which can promote oxidation reactions and suppress the

979-8-3503-6184-1/24 $31.00 © 2024 IEEE

formation of surface protrusions.

B. The density of interface traps

Density of interface traps (Dit) was measured within the range of 0.2–0.6 eV under conduction band (E_C) through high-low frequency capacitance–voltage (Hi-Lo C–V) measurement. The calculation formula for this method is as (1). The high-frequency curve is measured at 1MHz, while the low-frequency curve is measured at 1KHz. The peak conductivity is extracted from the C-V curve at 100KHz. The voltage range during testing is from -10V to 10V, and the scanning method is from depletion to accumulation.

$$D_{it} = \frac{C_{ox}}{q}\left(\frac{C_{LF}/C_{ox}}{1-C_{LF}/C_{ox}} - \frac{C_{HF}/C_{ox}}{1-C_{HF}/C_{ox}}\right) \quad (1)$$

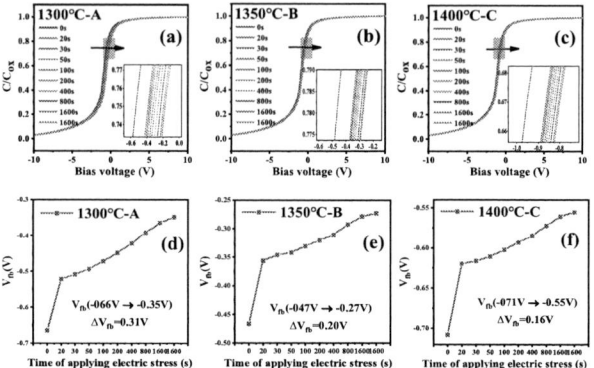

Fig. 2. (a) Hi-Lo C–V curves of three samples. (b) peak conductivity curves of three samples

Fig. 2 (a) shows the Hi-Lo C–V curves of three samples. Dit within the range of 0.2–0.6 eV under Ec was extracted from the Hi-Lo C–V curves. Dit is large when the energy level is close to 4H–SiC Ec, thus have a substantial effect on channel mobility. Through experiments, it can be found that high thermal oxidation temperature is beneficial for reducing the Dit of samples , especially at 1400℃. The Dit of the sample C can be as low as 1.1×10^{11}cm^2·eV^{-1} at the distance of 0.2eV from the E_C. Fig. 2 (b) shows the peak conductivity curves of three samples. The results of the two methods are consistent. So high-temperature thermal oxidation can effectively reduce the traps caused by structural defects.

C. The density of near interface traps

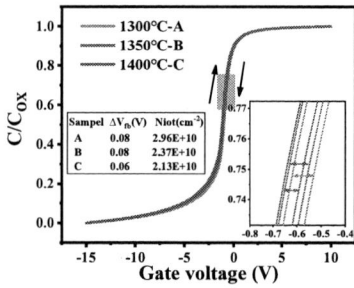

Fig. 3. Comparison of flat-band voltage(V_{fb}) drift of three samples after CV hysteresis test

When the gate voltage is scanned from the depleted state to the accumulated state, near interface oxide traps(NIOT) near the SiC/SiO$_2$ interface will capture charges. However, when the gate voltage is scanned back from the accumulated state to the depleted state, NIOT will cause a shift in the C-V curve, resulting in hysteresis phenomenon. In the experiment, the voltage scanning method is from accumulation to depletion, and then from depletion to accumulation. The C-V testing frequency is 100 KHz.

Fig. 3 shows that Hysteresis C-V curves of three samples .Using a higher oxidation temperature can effectively reduce the hysteresis phenomenon. Sample C has the lowest near-interfacial trap density compared to Samples A and B, about 2.13×10^{11}cm^2·eV^{-1}. The high-temperature thermal oxidation process can effectively reduce the density of near interface traps.

D. The density of oxide traps

Time Dependent Bias Stress (TDBS) testing was used to compare and analyze the electron trap density in the oxide layers of samples A, B, and C under different oxidation temperatures. The voltage stress applied to the three samples is +20V, and the stress time varies from(20s, 30s, 50s, 100s, 200s, 400s, 800s, 1600s, 1600s). C-V tests were conducted on the samples before and after applying voltage stress. The parameters of the C-V test program remained consistent throughout the TDBS testing process, Comparing the drift of V_{fb} before and after stressing.

Fig. 4. (a) (b) (c) Under different stress times, the C-V curves of the three samples(A, B, and C) appear different shifting. (d) (e) (f) Variation of V_{fb} drift in three samples A, B and C.

Fig. 4 (a) (b) (c) shows C-V curves of the three samples(A, B, and C) appear different shifting. It can be observed that the forward drift of V_{fb} gradually saturates, this means that the oxide layer traps are gradually filled and tend to saturate. Fig. 4 (d) (e) (f) shows under different electrical stress times, drift of V_{fb} for three samples. eventually, when the cumulative electrical stress time reaches 4800s, the $\triangle V_{fb}$ of three samples are 0.31V, 0.20V, and 0.16V, respectively. The $\triangle V_{fb}$ of sample C has the smallest drift, which means it has fewer oxide layer traps. in other words, The increase in thermal oxidation temperature reduces the density of oxide layer traps.

E. Leakage characteristics of oxide layer

The gate leakage characteristics of three samples were characterized using Current Voltage (I-V) test, During the I-V test, a voltage was applied to the gate electrodes. The voltage was increased in steps of +0.1 V till the gate oxide breakdown oxide electric field (E) used in these experiments is given by (2).

$$E = \frac{V-V_{fb}}{T_{ox}} \quad (2)$$

where V is the gate voltage, V_{fb} is the flat band voltage, V_{fb} was also determined from high-frequency C-V characteristics and T_{ox} is the thickness of the oxide layer.

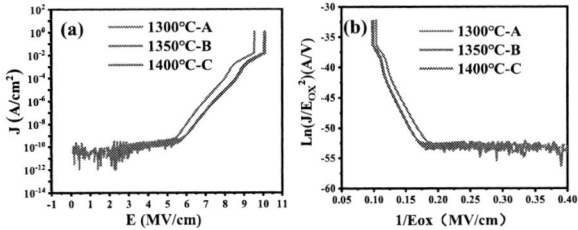

Fig. 5. (a) the current density-electric field (J-E) curves of SiC MOS capacitors under different thermal oxidation conditions. (b) Barrier height extracted from J-E curve.

The J-E relationship curve shows the breakdown of electric field and F-N tunneling barrier heights of three samples prepared at different thermal oxidation temperatures. The F-N tunneling points of samples A, B, and C are 5.3MV/cm, 5.78MV/cm, and 5.67MV/cm, respectively. It can be observed that with the increase of thermal oxidation temperature, the breakdown electric field and barrier height exhibit a folding phenomenon, showing the optimal value under 1350 ℃ thermal oxidation conditions. Beyond this temperature, the barrier height shows a downward trend.

F. Long-term reliability of the oxide layer

To investigate the effect of electric field strength on the time-dependent dielectric breakdown (TDDB) effect of three samples. a constant electric field of 9MV/cm was applied to each group of samples for TDDB. The V_{fb} of the three samples are -0.55V, -0.55V, and -0.58V, respectively. Experimental setup 10 devices per group, and the experimental temperature condition is set at 150 ℃.

The gate current and experimental time curve of the devices were recorded in the experiment, as well as the breakdown time T_{BD} of each sample, As shown in Fig.6(a).

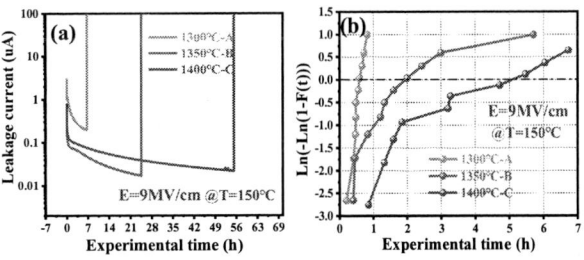

Fig. 6.(a) Leakage current and experimental time curve for SiC MOS capacitors fabricated by gate oxidation at various temperatures.(b)Weibull distribution of T_{BD} obtained from TDDB measurement for SiC MOS capacitors fabricated by gate oxidation at various temperatures.

Fig.6 (a) shows that when the gate leakage current suddenly increases to the specified current of 100uA, The sample has experienced breakdown. Fig.6(b) shows Weibull distribution of breakdown time (T_{BD}) obtained from TDDB measurement for SiC MOS capacitors fabricated by gate oxidation at various temperatures. When Ln(-ln (1-F (t))=0, the abscissa corresponding to the two parameter Weibull distribution function is the characteristic lifetime of the group. It can be seen that the T_{BD} distribution of samples A, B, and C gradually shifts to the right, indicating that the characteristic lifetime of the sample is larger, while the characteristic lifetime of sample C is the longest. This means that high-temperature thermal oxidation improves the long-term lifetime of the sample.

IV. CONCLUSIONS

In summary, we improved the quality of the SiO$_2$ film and SiC/SiO$_2$ interface and the voltage stability of SiC MOS capacitors through optimization of the thermal oxidation process. We found that the thermal oxidation temperature of 1400°C improves the quality of 4H-SiC/ SiO$_2$ interface more than that of 1300°C and 1350°C. This study not only reduced the interface state of 4H SiC/ SiO$_2$, but also improved its long-term reliability. The Dit at 0.02 eV under E_C was reduced to 1.1×10^{12}eV^{-1}·cm^{-2}. In addition, under the thermal oxidation condition of 1400 ℃, the voltage stability of MOS capacitors is improved. Therefore, 1400 °C is the best thermal oxidation process to optimize the performance of silicon carbide MOS capacitors compared to 1300 °C, 1350 °C.

V. ACKNOWLEDGMENT

This work was supported by the National Key R&D Program of China (Program No. 2023YFB3609503).

REFERENCES

[1] K. Suganuma, "Future technology trends," in Wide Bandgap Power Semiconductor Packaging: Materials, Components, and Reliability,Oxford, U.K.: Woodhead, 2018, pp. 3–53.

[2] T. Kimoto, "Material science and device physics in SiC technology for high-voltage power devices," Jpn. J. Appl. Phys., vol. 54, no. 4, Apr. 2015, Art. no. 040103, doi: 10.7567/jjap.54.040103.

[3] T. Kimoto, A. Iijima, H. Tsuchida, T. Miyazawa, T. Tawara, A. Otsuki, T. Kato,Y. Yonezawa, Reliab. Phys. Symp. (2017) pp. 2A–1.1-2A-1.7.

[4] T. Hosoi, D. Nagai, M. Sometani, Y. Katsu, H. Takeda, T. Shimura, M. Takei,H. Watanabe, Appl. Phys. Lett. 109 (2016) 040103.

[5] Afanas'ev V V, Ciobanu F, Dimitrijev S, et al. Band alignment and defect states at SiC/oxide interfaces[J]. Journal of Physics: Condensed Matter, 2004, 16(17): S1839.

[6] Afanasev V V, Bassler M, Pensl G, et al. Intrinsic SiC/SiO2 interface states[J]. physica status solidi (a), 1997, 162(1): 321-337.

[7] Chung G Y, Tin C C, Williams J R, et al. Improved inversion channel mobility for 4H-SiC MOSFETs following high temperature anneals in nitric oxide[J]. IEEE Electron Device Letters,2001, 22(4): 176-178.

[8] Williams J R, Chung G Y, Tin C C, et al. Passivation of the 4H-SiC/SiO2 interface with nitric oxide[J]. Materials Science Forum, 2002, 389: 967-972.

[9] Taillon J A, Hyuk Yang J, Ahyi C A, et al. Systematic structural and chemical characterization of the transition layer at the interface of NO-annealed 4H-SiC/SiO2 metal-oxide-semiconductor field-effect transistors[J]. Journal of Applied Physics, 2013, 113(4): 044517.

[10] Li W, Zhao J, Wang D. Structural and electronic properties of the transition layer at theSiO2/4H-SiC interface[J]. AIP Advances, 2015, 5(1): 017122.

[11] Green R, Lelis A, El M, et al. Bias-Temperature-Stress Response of Commercially-Available SiC Power MOSFETs[J]. Materials Science Forum, 2015, 821: 677-680.

[12] Chbili Z, Cheung K P, Campbell J P, et al. Time dependent dielectric breakdown in high quality SiC MOS capacitors[J]. Materials Science Forum, 2016, 858: 615-618.

Optimization for a High-voltage Recessed-gate β-Ga_2O_3 MOSFET by Gate and Drain Field Plate Technology

Bo Yi*[1,2], Yuan Qiao[1], Ming Dai[1], Fan Xu[1], JunJi Cheng[1], HaiMeng Huang[1], MouFu Kong[1], XingLi Jiang[3], HongQiang Yang*[1]

[1] University of Electronic Science and Technology of China, China
[2] Chongqing Institute of Microelectronics Industry Technology, UEST, China
3 Chengdu Semi-Future Technology Co., Ltd, China

* Email: yb@uestc.edu.cn, hqyang@uestc.edu.cn

Abstract—In this paper, a recessed-gate β-Ga_2O_3 MOSFET is investigated by TCAD. Enhanced mode is obtained by etching the channel layer to be 40 nm. High breakdown voltage and low specific on-resistance ($R_{on,sp}$) is obtained by optimizing the field plate length and SiO_2 thickness for the gate and drain field plates, as well as the doping concentration of the channel layer. Simulation results show that when the gate and drain field plates are 4 μm and 2 μm with SiO_2 thickness being 500 nm, the device obtains a BV of 4256 V and $R_{on,sp}$ = 23.3 mΩ·cm², showing a high Baliga's Figure of merit of 777.4 MW/cm².

Keywords—*enhancement-mode, β-Ga_2O_3 MOSFET, high breakdown voltage, field plate.*

I. INTRODUCTION

Currently, silicon-based power semiconductors are approaching the limits of the physical properties and are hard to meet the requirements for miniaturization and high efficiency of power semiconductors. Gallium oxide (Ga_2O_3), with its excellent material characteristics such as ultra-wide bandgap, high breakdown electric field, is expected to become the main material for the next generation of power semiconductor devices [1]-[2]. β-Ga_2O_3 exhibits high thermal stability and yields the best quality in single-crystal growth. Hence β-Ga_2O_3 is the closest to commercial application in the semiconductor industry. According to previous reports, β-Ga_2O_3 MOSFETs fabricated by the field plate technology have obtained high breakdown voltages [3]-[8], up to 8000V. β-Ga_2O_3 can be doped with many elements to form n-type semiconductors [9]-[10]. Due to the current absence of a mature method for p-type doping of gallium oxide, it hinders the development of β-Ga_2O_3-based enhancement-mode (e-mode) devices. A recessed-gate technique is usually utilized [8] to acquire enhancement-mode β-Ga_2O_3 MOSFETs. Nonetheless, it is still hard to approach to the Ga_2O_3 limit. Thus, in this paper, we discussed the influence of gate and drain field plates on the BV of the Lateral β-Ga_2O_3 MOSFETs to order to realize high BV and low $R_{on,sp}$.

II. DEVICE STRUCTURE AND MECHANISM

The schematic cross section for the investigated lateral recessed-gate β-Ga_2O_3 MOSFET is shown in Fig. 1. The device consists of a 10 μm substrate and a 400 nm unintentionally doped buffer layer with a donor doping concentration of 1.5×10^{15} cm⁻³. The channel thickness is set to 200 nm. To fully deplete the channel in the recess region, the recess depth is determined to be 160 nm leaving 40 nm of active channel thickness under the gate. A 50 nm Al_2O_3 layer and SiO_2 dielectric layer are designed to withstand high reverse bias voltages. Gate field plate and drain field plate are designed to improve the breakdown voltage of the device. Both the gate field plate and the drain field plate range from 2μm to 4μm. The source-to-gate distance (L_{GS}), gate length (L_G), and gate-to-drain distance (L_{GD}) are designed to be 3, 2, and 13.5 μm, respectively. The source region and drain region are set to 2×10^{19} cm⁻³ for Ohmic contact. The doping concentration in the channel region has a great influence on BV and $R_{on,sp}$. Therefore, it is varied from 5×10^{17} cm⁻³ to 9×10^{17} cm⁻³ to investigate the influence.

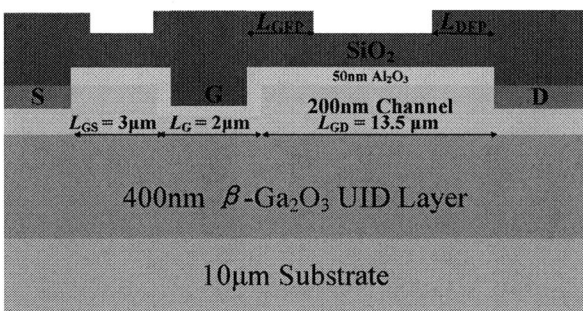

Fig. 1 The investigated device structure for the recessed-gate β-Ga_2O_3 MOSFET.

III. SIMULATION AND DISCUSSION

The characteristics of the device are investigated using TCAD simulation tools. The work function of gate electrode is set as 5.1 eV. The interface charge between the β-Ga_2O_3 channel and the insulating layer is set to -5×10^{12} cm⁻³ [8].

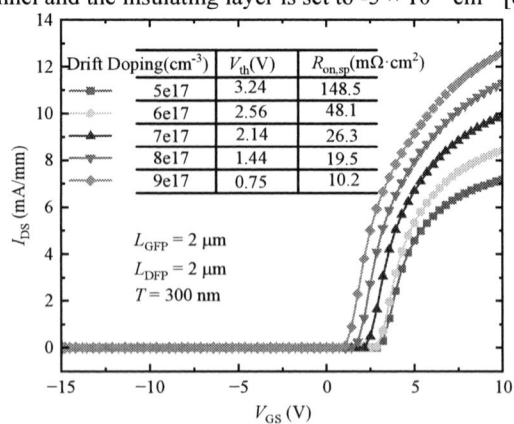

Drift Doping(cm⁻³)	V_{th}(V)	$R_{on,sp}$(mΩ·cm²)
5e17	3.24	148.5
6e17	2.56	48.1
7e17	2.14	26.3
8e17	1.44	19.5
9e17	0.75	10.2

$L_{GFP} = 2$ μm
$L_{DFP} = 2$ μm
$T = 300$ nm

Fig. 2 Transfer curves for the devices with different channel (drift) doping concentrations. V_{DS}= 1 V.

979-8-3503-6184-1/24 $31.00 © 2024 IEEE

Fig. 2 shows the transfer characteristics of the device. The gate field plate length (L_{GFP}) and drain field plate length (L_{DFP}) are set to 2μm and SiO_2 thickness (T) is set to 300 nm. Devices with doping concentrations in the drift region ranging from 5×10^{17} cm^{-3} to 9×10^{17} cm^{-3} all exhibit enhanced mode. As the doping concentration increases, threshold voltage (V_{th}) decreases. Besides, $R_{on,sp}$ decreases due to that the resistance of drain and source access decreases.

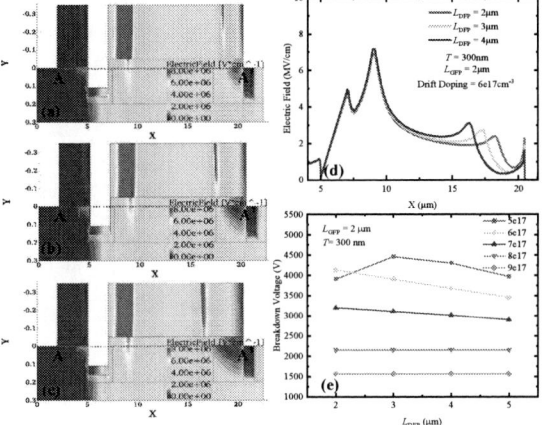

Fig. 3 (a)-(c) Two-dimensional electric field distribution at $V_{DS} = 3000$ V with drain field plate ranging from 2μm to 4μm and (d) Electric field strength of the devices at 1 nm below the insulating layer (line AA') (e) Breakdown voltages for a fixed gate field plate of 2μm under different doping concentrations.

Fig.3 (a)-(c) show the two-dimensional electric field distribution for devices with drain field plates ranging from 2 μm to 4 μm at $V_{DS} = 3000$ V for a fixed gate field plate of 2 μm. As Fig. 3(a)-(d) show that the highest electric field occurs at the edge of gate field plate. The length of drain field plate only increases the peak electric field (at the edge of gate field plate) a little, resulting in slight decrease of BV. Also, the influence of drift doping concentration is investigated and the results are summarized in Fig. 3(e). As we can see, BV increases when the drift layer is lightly doped. However, this results in significant increase in $R_{on,sp}$ as shown in Fig. 2.

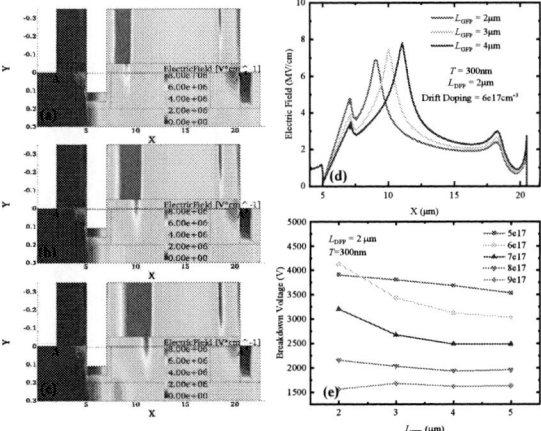

Fig. 4 (a)-(c) Two-dimensional electric field distribution at $V_{DS} = 3000$ V with gate field plates ranging from 2 μm to 4 μm and (d) Electric field strength of the devices at 1 nm below the insulating layer (line AA') (e) Breakdown voltages for a fixed drain field plate of 2 μm under different doping concentrations.

Fig.4 (a)-(c) show the two-dimensional electric field distribution for devices with gate field plates ranging from 2 μm to 4 μm at $V_{DS} = 3000$ V for a fixed drain field plate of 2 μm. As Fig. 4(a)-(d) show that the highest electric field occurs at the edge of gate field plate. And with the increase of L_{GFP}, the peak electric field increases, resulting in decrease of BV. Also, the influence of drift doping concentration is investigated and the results are summarized in Fig. 4(e). As we can see, BV decreases with the increase of L_{GFP} under different drift doping concentrations. Considering the increase of $R_{on,sp}$, in the next investigations, the drift doping is set to 7e17 cm^{-3}.

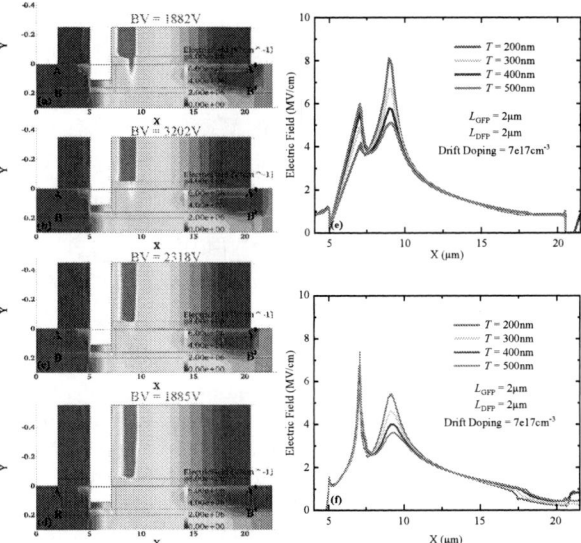

Fig. 5 (a)-(d) Two-dimensional electric field distribution at $V_{DS} = 2000$ V under different dielectric thicknesses and (e)-(f) Electric field strength of the four devices at 1 nm and 160 nm below the gate dielectric layer (line AA' and BB').

The thickness of the SiO_2 dielectric layer also has a significant effect on the BV of the device. Fig. 5 (a)-(d) illustrate the two-dimensional electric field distribution of the device with different SiO_2 layers, ranging from 200 nm to 500 nm. The simulation results show that BV increases from 1883 to 3203 when SiO_2 layer increases from 200 nm to 300 nm. While BV starts to decrease when SiO_2 layer increases from 300 to 500 nm. This is because with the increase of SiO_2, the peak electric field at the edge of gate field plate decreased, as shown in Fig. 5(e). While the peak electric field at the edge of the recessed gate increases with the increase of SiO_2, as shown in Fig. 5(f). Thus, the breakdown transfers from the edge of gate field plate to the edge of the recessed gate.

Table I. Highest BV and corresponding $R_{on,sp}$ under different SiO_2 thicknesses

parameter	N_{Drift} = 7e17 cm^{-3}			
T	200	300	400	500
BV (V)	1883	3203	3864	4256
L_{FP}	L_{GFP}= 2μm L_{DFP}= 2μm	L_{GFP}= 2μm L_{DFP}= 2μm	L_{GFP}= 3μm L_{DFP}= 2μm	L_{GFP}= 4μm L_{DFP}= 2μm
$R_{on,sp}$ (mΩ·cm^2)	20.5	26.3	20.8	23.3

To obtain the highest BV under different SiO_2 thicknesses, different lengths of field plate are simulated. The results are summarized in Table I. When the SiO_2 is 500 nm, the highest

BV of 4256 V is obtained with L_{GFP} and L_{DFP} being 4 μm and 2 μm. Fig. 6 shows a comparison of the investigated devices with recently reported lateral Ga_2O_3 MOSFETs. As it shows, by optimizing the gate and drain field plates, better trade-off between BV and $R_{on,sp}$ can be obtained.

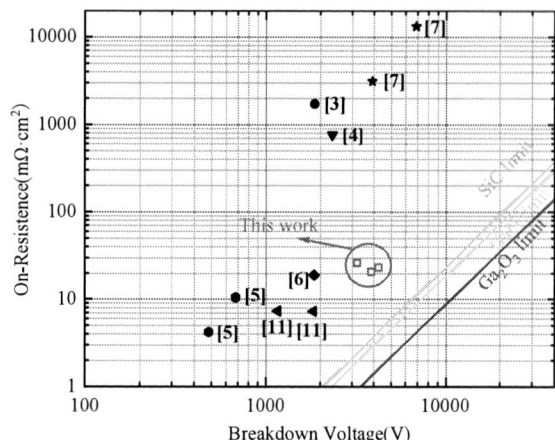

Fig. 6 $R_{on,sp}$ vs breakdown voltage plot of lateral Ga_2O_3 MOSFETs from this study and previously reported researches.

IV. CONCLUSION

A high performance enhancement-mode recessed-gate β-Ga_2O_3 MOSFET device has been obtained by optimizing the gate and drain field plates. High threshold voltage of 2.14 V is obtained using recess-gate technology. Moreover, when the SiO_2 layer for field plate is 500 nm, high breakdown voltage of 4256 V with $R_{on,sp}$ = 23.3 mΩ·cm² is obtained with gate and drain field plates being 4 μm and 2 μm. The field plate technology shows a promising method to obtain high-voltage lateral β-Ga_2O_3 MOSFET with low $R_{on,sp}$.

REFERENCES

[1] M. Higashiwaki and G. H. Jessen, "Guest editorial: The dawn of gallium oxide microelectronics," Appl. Phys. Lett., vol. 112, no. 6, 2018, Art. no. 060401. doi: 10.1063/1.5017845.

[2] S. J. Pearton, J. Yang, P. H. Cary, F. Ren, J. Kim, M. J. Tadjer, and M. A. Mastro, "A review of Ga_2O_3 materials, processing, and devices," Appl. Phys. Rev., vol. 5, Mar. 2018, Art. no. 011301. doi: 10.1063/1.5006941.

[3] K. Zeng, A. Vaidya and U. Singisetti, "1.85 kV Breakdown Voltage in Lateral Field-Plated Ga_2O_3 MOSFETs," in IEEE Electron Device Letters, vol. 39, no. 9, pp. 1385-1388, Sept. 2018, doi: 10.1109/LED.2018.2859049.

[4] J. K. Mun, K. Cho, W. Chang, H.-W. Jung, and J. Do, "Editors' Choice—2.32 kV breakdown voltage lateral β-Ga_2O_3 MOSFETs with source-connected field plate," ECS J. Solid State Sci. Technol., vol. 8, no. 7, pp. Q3079–Q3082, 2019.

[5] K. Tetzner et al., "Lateral 1.8 kV β -Ga_2O_3 MOSFET With 155 MW/cm2 Power Figure of Merit," in IEEE Electron Device Letters, vol. 40, no. 9, pp. 1503-1506, Sept. 2019, doi: 10.1109/LED.2019.2930189.

[6] Y. Lv et al., "Source-Field-Plated β -Ga_2O_3 MOSFET With Record Power Figure of Merit of 50.4 MW/cm2," in IEEE Electron Device Letters, vol. 40, no. 1, pp. 83-86, Jan. 2019, doi: 10.1109/LED.2018.2881274.

[7] S. Sharma, K. Zeng, S. Saha and U. Singisetti, "Field-Plated Lateral Ga_2O_3 MOSFETs With Polymer Passivation and 8.03 kV Breakdown Voltage," in IEEE Electron Device Letters, vol. 41, no. 6, pp. 836-839, June 2020, doi: 10.1109/LED.2020.2991146.

[8] K. D. Chabak et al., "Recessed-Gate Enhancement-Mode β -Ga_2O_3 MOSFETs," in IEEE Electron Device Letters, vol. 39, no. 1, pp. 67-70, Jan. 2018, doi: 10.1109/LED.2017.2779867.

[9] K. Zeng et al., "Ga2O3 MOSFETs Using Spin-On-Glass Source/Drain Doping Technology," in IEEE Electron Device Letters, vol. 38, no. 4, pp. 513-516, April 2017, doi: 10.1109/LED.2017.2675544.

[10] N. Moser et al., "Ge-Doped β -Ga2O3 MOSFETs," in IEEE Electron Device Letters, vol. 38, no. 6, pp. 775-778, June 2017, doi: 10.1109/LED.2017.2697359.

[11] X. Zhou et al., "Realizing High-Performance β-Ga_2O_3 MOSFET by Using Variation of Lateral Doping: A TCAD Study," in IEEE Transactions on Electron Devices, vol. 68, no. 4, pp. 1501-1506, April 2021, doi: 10.1109/TED.2021.3056326.

A Novel Voltage Sensor with Composite Trench Structure for High Voltage IGBT

Yang Yang[1,2], Ze-Hong Li[1,3*], *Senior Member, IEEE*, Li-Hang Dong[3], Wei Li[1,2], Peng-Fei Jia[2], Zhi-Yu Yang[2], Li Wan[1,2], Yi-Shang Zhao[1], Tong-Yang Wang[1], and Zi-Ming Xia[1]

[1] State Key Laboratory of Electronic Thin Films and Integrated Devices, University of Electronic Science and Technology of China, Chengdu 610054, China
[2] China Resources Microelectronics (Chongqing) Limited, Chongqing 401331, China
[3] Chongqing Institute of Microelectronics Industry Technology, University of Electronic Science and Technology of China, Chengdu 610054, China
* Email: lizh@uestc.edu.cn

Abstract

A novel voltage sensor with composite trench structure for high voltage IGBT (insulated gate bipolar transistor) is proposed and numerically investigated in this paper. Based on the capacitance matching between the parasitic capacitance of the composite trench structure and the load capacitance (C_L), the sensor voltage (V_s) can replicate a scaled-down version of the IGBT voltage (V_{ce}). By varying the internal parasitic capacitance and external load capacitance, the designability and controllability of the sensor can be realized, respectively. Simulation results show that, as the structural parameter W_{mesa} increases from 0.5 μm to 2.5 μm, V_s increases from 1.45 V to 2.08 V (when V_{ce}= 1200 V). Besides, with the increase of C_L (from 0.01 pF to 10 pF), V_s decreases from 2.0 V to 0.92 V at V_{ce}=1200 V. The results are consistent with the analysis.

1. Introduction

IGBTs (insulated gate bipolar transistors), as the core components of modern power electronic systems, have been developing towards integration and intelligence to enhance the performance and stability of the systems [1]. Voltage sensing and protection in these applications have received significant attention from the industry and academia [2-5]. Currently, voltage sensing techniques are usually realized through peripheral sensing circuits, which suffer from cost, volume, and power loss. Integrated solutions are therefore desirable.

For the on-chip voltage sensor, the sense terminal can provide a sensing signal that is a scaled version of the large anode voltage. Since the sensing signal level increases as the anode voltage increases, high levels of the signal can lead to safety issues, especially in high voltage situations [6, 7]. Therefore, designability and controllability of the range of sensor voltage are critical for integrated sensors.

In this paper, a novel voltage sensor with composite trench structure for high voltage IGBT is proposed. An analytical model is derived to understand the mechanism of the sensor. Meanwhile, the effect of internal parasitic

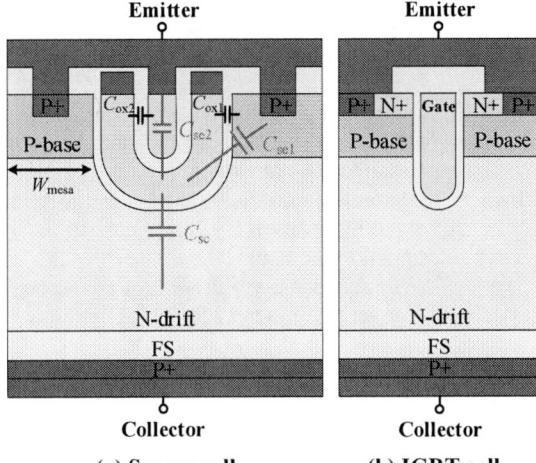

(a) Sensor cell **(b) IGBT cell**

Figure 1. Schematic cross section of (a) the voltage sensor cell and (b) the last IGBT core cell.

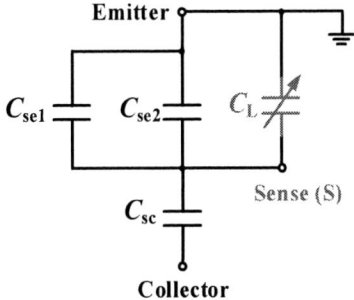

Figure 2. Equivalent circuit for the voltage sensor.

capacitance and external load capacitance (C_L) on the sensing characteristics is analyzed, which is consistent with the analytical model. The results show that the designability and controllability of the range of sensor voltage can be realized.

979-8-3503-6184-1/24 $31.00 © 2024 IEEE

2. Device Structure and Mechanism

The voltage sensor incorporated into the IGBT is presented in Figure 1. The sensor cell shares the P$^+$ collector region, the FS layer, the N-drift region, and the P-base region with the IGBT cell. The IGBT and voltage sensor is separated by P-type ring isolation. For the voltage sensor, the trench width of the sensor structure is larger than that of the IGBT. Besides, the sensor trench features additional etching of polysilicon, growth of the oxide layer and deposition of polysilicon. Thus, two oxide layers and two polysilicon layers are formed, where one polysilicon layer is in the shape of a "U" and the other polysilicon layer is in the shape of an "I". The I-shaped polysilicon is clamped to the emitter potential and the U-shaped polysilicon is connected to the sense terminal (S). The parasitic capacitors C_{se1}, C_{se2}, and C_{sc} are shown in Figure 1, and the equivalent circuit is shown in Figure 2. The load capacitance (C_L) is connected to the sense electrode, so the sensor voltage (V_s) can be obtained. The analytical model of the V_s is shown in equation (1), and the voltage sensing ratio k can be expressed in equation (2). The range of V_s and k can be designed by adjusting the value of C_{se1}, C_{se2}, C_{sc}, and C_L. The functional verification of the sensor is carried out based on a designed 1200 V IGBT structure.

$$V_s = \frac{V_{ce}}{1 + \left(C_{se1} + C_{se2} + C_L\right)/C_{sc}} \quad (1)$$

$$k = C_{sc} / \left(C_{se1} + C_{se2} + C_L + C_{sc}\right) \quad (2)$$

3. Results and Discussion

Based on TCAD software [8], the sensor and IGBT structures are designed, and the electrical characteristics are simulated. With C_L=0.01 pF, the voltage sensing characteristics is shown in Figure 3. It is worth noting that the voltage sensing curve can be divided into two stages. In the first sensing stage, the sensor voltage increases extremely fast as the IGBT voltage increases. Whereas, in the second sensing stage, the sensor voltage increases approximately in a linear trend as the IGBT voltage increases. Thus, the sensing ratio of the first stage is greater than the second stage ($k_1 > k_2$), and k_2 is approximately constant. From equation (2), if all capacitance values remain constant, then k is constant, i.e., V_s and V_{ce} are linearly related. In fact, as V_{ce} changes, the parasitic capacitance C_{sc} inside the device also changes. To analyze the two-stage phenomenon of the voltage sensing curve, the sensor structures are extracted under different V_{ce} conditions, as shown in Figure 4, where the white line is the boundary of the depletion region. When

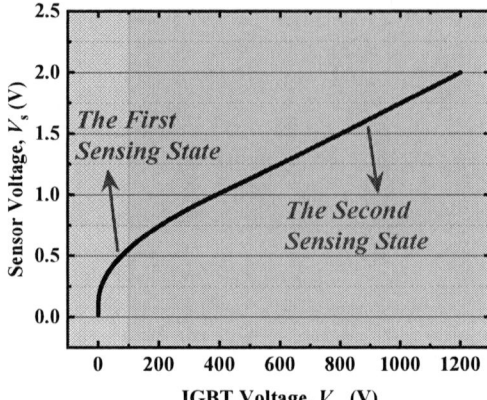

Figure 3. Voltage sensing characteristics.

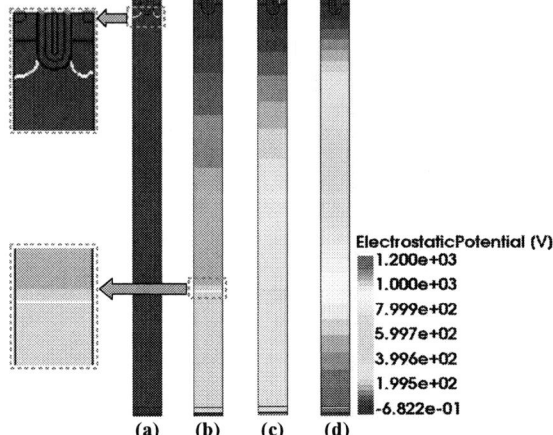

Figure 4. Electrostatic potential distribution and the boundary of the depletion region for sensor structure under the conditions of (a) V_{ce}= 2 V, (b) V_{ce}= 200 V, (c) V_{ce}= 400 V, and (d) V_{ce}= 1200 V.

V_{ce} increases, the boundary of the depletion region extends from the bottom of the P-base to the bottom of the trench, and then continues to extend toward the FS layer, leading to a decrease in C_{sc}. From equation (2), a decrease in C_{sc} leads to a decrease in k, so $k_1 > k_2$. In the second sensing state, the boundary of the depletion region is located below the trench, and the movement of the boundary has less effect on C_{sc}, so k_2 tends to be a constant.

Designability and controllability of key parameters are important for voltage sensors. Designability refers to the targeted design of the structural parameters so that the sensing characteristics meet the expected. The influence of the structural parameter W_{mesa} on the sensing characteristics is shown in Figure 5. The parameter W_{mesa}

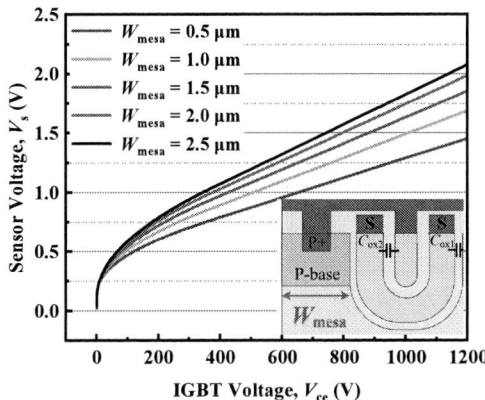

Figure 5. Influence of the structural parameter W_{mesa} on the sensing characteristics.

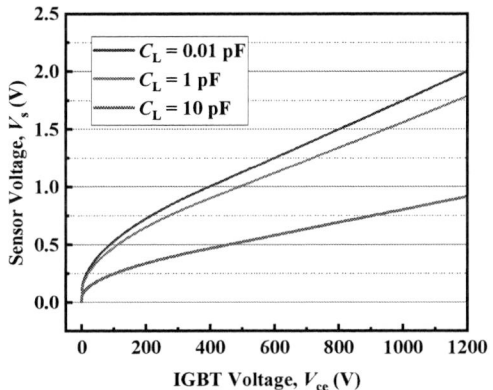

Figure 6. Influence of the load capacitance C_L on the sensing characteristics.

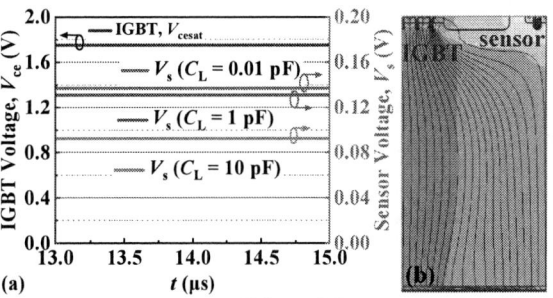

Figure 7. On-state condition of the IGBT cell and its integrated sensor. (a) V_{ce} and V_s for different C_L conditions. (b) Current lines under the conditions of V_{ce}=1.75 V and C_L= 0.01 pF.

0.092 V, respectively, as shown in Figure 7(a). The simulation results are consistent with the analysis. As can be seen from Figure 7(b), the current line flows from the IGBT cells, and the sensor structure has almost no negative effect on IGBT cells.

4. Summary

This paper proposes a novel voltage sensor with composite trench structure for high voltage IGBT. The sensing mechanism is explained with an equivalent capacitive divider model. In addition, the voltage sensing curve can be divided into two stages, which are analyzed by comparing the changes in the boundaries of the depletion region. The simulation results under different internal parasitic capacitance and external C_L show that the range of V_s is designable and controllable, which improves the flexibility of the sensor in applications.

References

[1] U. Choi et al., IEEE Transactions on Power Electronics, pp. 2517-2533, (2015).

[2] B. J. Baliga, The IGBT Device: Physics, Design and Applications of the Insulated Gate Bipolar Transistor (2015).

[3] A. Volke and M. Hornkamp, IGBT Modules: Technologies, Driver, and Application (2017).

[4] S. Ji, T. Lu, Z. Zhao, et al., IEEE Transactions on Power Electronics, pp. 4165-4174, (2015).

[5] W. Zhang et al., IEEE 31st Int. Symp. Power Semiconductor Devices and ICs, pp. 83-86 (2019).

[6] Yang Yang and Zehong Li, IEEE Transactions on Power Electronics, pp. 2491-2495 (2022).

[7] C. Caramel et al., Semicond. Sci. Technol., pp. 115014-115018, (2010).

[8] Synopsys TCAD Sentaurus, Version L-2016.03.

is half the distance between the two trenches, as shown in Figure 1. As W_{mesa} increases from 0.5 μm to 2.5 μm, the sensing ratio k_2 increases, and V_s increases from 1.45 V to 2.08 V at V_{ce}= 1200 V. Thus, changing W_{mesa} allows for designability of k_2, which enables the designability of the dynamic range of V_s.

Controllability refers to the fact that the sensor can be adjusted in the application even after it has been manufactured. To illustrate the controllability of the sensor voltage, C_L with different values is designed and simulated, as shown in Figure 6. As C_L increase from 0.01 pF to 10 pF, V_s decreases from 2.08 V to 0.92 V at V_{ce}=1200 V, which shows that the range of V_s is controllable under different C_L. Besides, the reduced percentage of the first sensing stage means that the voltage sensor can switch faster to the second sensing stage, where the sensing characteristic curve features better linearity. For the on-state of IGBT, V_{ce} = 1.75 V, and V_s features a low level. Under the conditions of C_L= 0.01 pF, 1 pF, and 10 pF, V_s = 0.137 V, 0.132 V, and

A Novel Triggered Voltage Sensing Structure for High Voltage IGBT

Yang Yang[1,2], Ze-Hong Li[2,3*], *Senior Member, IEEE*, Li-Hang Dong[3], Wei Li[1,2], Peng-Fei Jia[1], Zhi-Yu Yang[1], Li Wan[1,2], Yi-Shang Zhao[2], Lu-Ping Li[2,3], Zi-Ming Xia[2], and Tong-Yang Wang[2]

[1] China Resources Microelectronics (Chongqing) Limited, Chongqing 401331, China
[2] State Key Laboratory of Electronic Thin Films and Integrated Devices, University of Electronic Science and Technology of China, Chengdu 610054, China
[3] Chongqing Institute of Microelectronics Industry Technology, University of Electronic Science and Technology of China, Chengdu 610054, China
* Email: lizh@uestc.edu.cn

Abstract

A novel triggered voltage sensing structure for high voltage insulated gate bipolar transistor (IGBT) is proposed and simulated in this paper. The voltage sensor is characterized by a controllable starting point for the sensed voltage (V_{st}). Specifically, when the IGBT voltage (V_{ce}) is less than V_{st}, the sensor is in the blocking state; but when $V_{ce} \geq V_{st}$, the sensor is triggered to the voltage sensing state. With R_{se} connected to the sense electrode, the sensor voltage (V_s) can replicate a scaled-down version of the V_{ce}. Based on the 1200 V IGBT, the simulation results show that under the condition of $R_{se} = 100\ \Omega$, V_{st} is designed to be 300 V, and $V_s = 2.74$ V at $V_{ce} = 1200$ V. Besides, as R_{se} increases from 1 Ω to 10 kΩ, V_s (at $V_{ce} = 1200$ V) increases from 0.26 V to 3.88 V, and V_{st} decreases from 342 V to 162 V. The results are consistent with the analysis and indicate that the voltage sensor is controllable.

1. Introduction

Insulated gate bipolar transistor (IGBT) is the core device in industrial control, rail transportation and power grid systems [1]. In many scenarios, such as series voltage, online performance optimization [2, 3], and device health monitoring [4, 5], the support of IGBT voltage sensing technology is highly needed to meet the requirements of integrated circuits.

In terms of integrated voltage sensing and protection technologies, solutions can be divided into two categories, i.e. trigger protection solutions [5, 6] and sensor solutions [3, 7, 8]. For the trigger protection solution, the protection strategy takes corrective action before a catastrophic fault occurs. The protection strategy focuses on a specific voltage state rather than the detection of the entire voltage range. In work [5] and [6], the specific voltage state is set to the breakdown voltage of the avalanche diode to prevent the risk of over voltage of the IGBT. For the on-chip voltage sensor solution, the sense terminal can provide a sensing signal that is a scaled version of the large anode voltage. The focus of this type of solution is to monitor the voltage and transmit the information to the system. The system

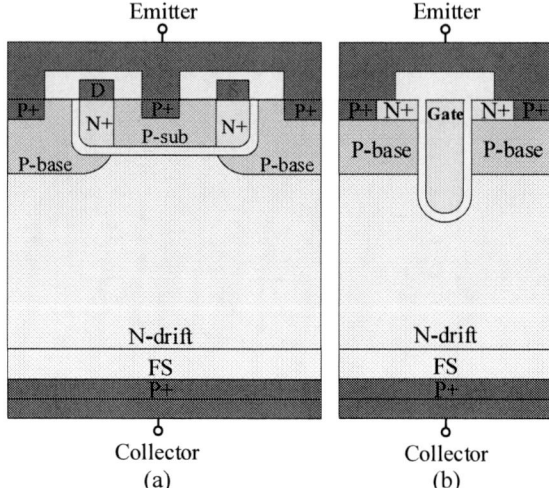

Figure 1. Schematic cross section of (a) the voltage sensor cell and (b) the last IGBT core cell.

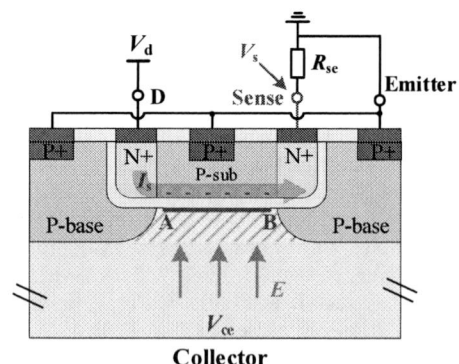

Figure 2. Mechanism of the voltage sensor.

makes decisions based on the working conditions. Since the level of the sensing signal increases as the anode voltage increases, high levels of the signal can lead to safety issues, especially in high voltage situations [7]. Thus, a novel local voltage sensing (LVS) mode has been

979-8-3503-6184-1/24 $31.00 © 2024 IEEE   871

proposed in previous work [8], which focuses on solving the contradiction between the dynamic range of the signal and the ADC resolution.

Based on the LVS mode, a novel triggered voltage sensing structure for high voltage IGBT is proposed in this paper, providing more options for the solution. The voltage sensor is characterized by two states: a blocking state and a voltage sensing state. The switching between the two states is realized by a trigger voltage, which is the starting point for the sensed voltage (V_{st}). The mechanism and characteristics of the voltage sensor are illustrated through simulations, and the results are consistent with the analysis.

2. Device Structure and Mechanism

As shown in Figure 1, the voltage sensor and IGBT regions are interconnected through the shared use of the collector, FS layer, and N-drift region. The sensor region consists of the sensor structure and the isolation, where the isolation (P-base region) is located at the periphery of the sensor structure. The sensor structure is designed with the trench oxide layer and the polysilicon layer. The polysilicon layer features N^+ and P-sub regions through the ion implantation process. Thus, the two layers form an upside-down MOSFET structure. The mechanism of the voltage sensor is shown in Figure 2. As the IGBT voltage (V_{ce}) increases, the potential of the N-drift region correspondingly increases, thereby elevating the potential at the bottom of the trench, denoted as the potential of the AB line (V_{ab}). Thus, there exists a strong inversion layer in the P-sub polysilicon near the bottom of the trench oxide. When a constant positive voltage (V_d) is applied to the drain electrode of the sensor, current (I_s) flows through the drain electrode, the N^+ region, and the inverse layer to the sense electrode. The sensing resistance (R_{se}) is connected to the sense electrode, so the sensor voltage (V_s) can be obtained.

3. Results and Discussion

Based on TCAD software Sentaurus [9], the 1200 V IGBT and its integrated sensor are designed, and the electrical characteristics are simulated. Under the condition of $R_{se} = 100\ \Omega$ and $V_d = 5V$, the voltage sensing characteristics are shown in Figure 3. The sensing curve can be divided into two states: the blocking state and the voltage sensing state. When V_{ce} is less than 300 V, $V_s = 0$ V, and this state is named the blocking state. When V_{ce} exceeds 300 V, the voltage sensor is triggered to the voltage sensing state, in which state V_s begins to follow the rise of V_{ce}, and $V_s = 2.74$ V at $V_{ce} = 1200$ V. Thus, $V_{st} = 300$ V is the starting

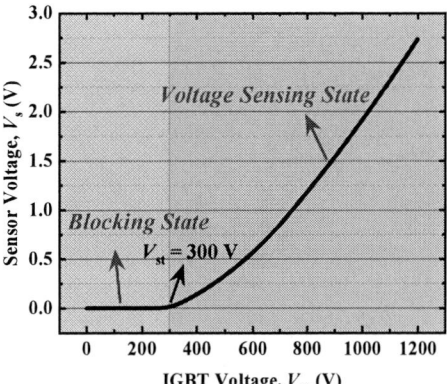

Figure 3. Voltage sensing characteristics.

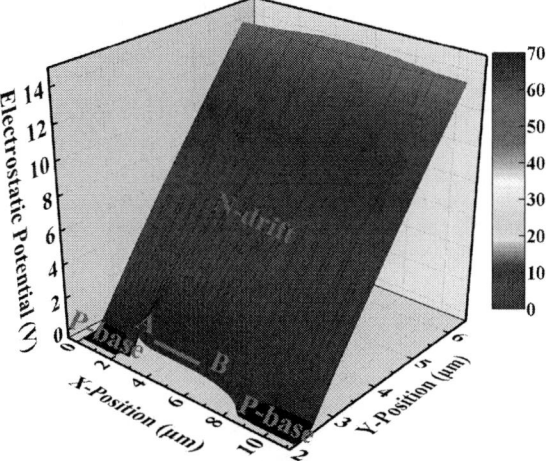

Figure 4. Electrostatic potential distribution of sensor structure under the conditions of $V_{ce} = 100$ V.

point for the sensed voltage.

To illustrate how changes in external voltage are conducted to the internal potential, and therefore affects the current within the polysilicon, the sensor structure is extracted. The electrostatic potential distribution of sensor is shown in Figure 4. As V_{ce} increases, the potential in the N-drift region rises and V_{ab} increases, but the P-base region and the P-sub region are clamped to ground potential through the connected electrodes. As a result, a potential difference ΔV is formed between V_{ab} and the P-sub region, and when ΔV cannot enable the MOSFET to form an inversion layer, the sensor is in the blocking state. When V_{ab} reaches the threshold voltage of MOSFET, the sensor enters the voltage sensing state.

The sensors are designed with controllability in mind. Controllability refers to the fact that the sensor can be adjusted in the application even after it has been

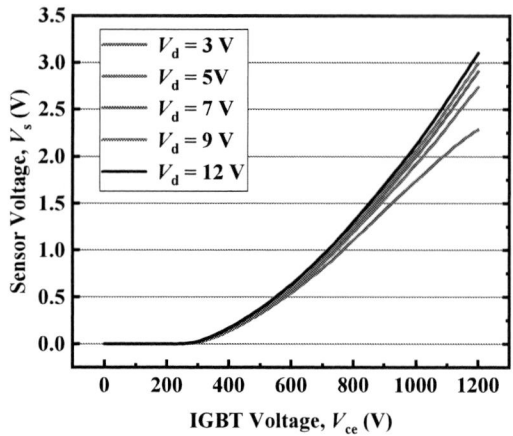

Figure 5. Influence of V_d on sensing characteristics

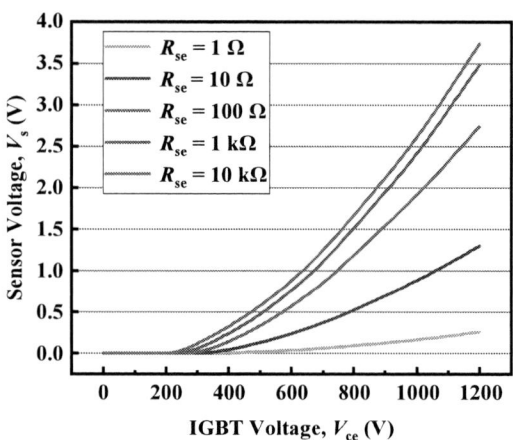

Figure 6. Influence of sensing resistance R_{se} on sensing characteristics.

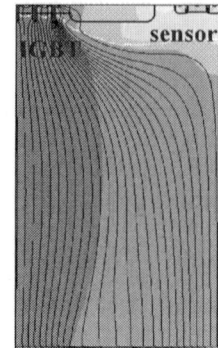

Figure 7. On-state condition of the IGBT cell and its integrated sensor at $V_d = 5$ V and $R_{se} = 100$ Ω.

IGBT is out of influence from the sensor during the on-state.

4. Summary

This paper proposes a novel triggered voltage sensing structure for high voltage IGBT. The voltage sensor features a blocking state and a voltage sensing state, which is analyzed through the conduction relationship between the external voltage and the internal potential. Under different conditions of R_{se} and V_d, the simulation results show that the range of V_{st} and V_s is controllable, which improves the flexibility of the sensor in applications.

References

[1] S. Yang et al. IEEE Transactions on Industry Applications, pp. 1441-1451, (2011).

[2] S. Ji, T. Lu, Z. Zhao, et al., IEEE Transactions on Power Electronics, pp. 4165-4174, (2015).

[3] W. Zhang et al., IEEE 31st Int. Symp. Power Semiconductor Devices and ICs, pp. 83-86 (2019).

[4] B. J. Baliga, The IGBT Device: Physics, Design and Applications of the Insulated Gate Bipolar Transistor (2015).

[5] A. P. Hsieh, F. Udrea, and W. Lin, IEEE 24th Int. Symp. Power Semiconductor Devices and ICs, Bruges, pp. 61-64, (2012).

[6] F. Alkayal et al., IEEE 18th Int. Symp. Power Semiconductor Devices ICs, pp. 1-4, (2006).

[7] C. Caramel et al., Semicond. Sci. Technol., pp. 115014-115018, (2010).

[8] Yang Yang and Zehong Li, IEEE Transactions on Power Electronics, pp. 2491-2495 (2022).

[9] Synopsys TCAD Sentaurus, Version L-2016.03.

manufactured. The influence of V_d on sensing characteristics is shown in Figure 5. For $V_d = 3$ V, when V_{ce} increases to about 1000 V, the curve exhibits a saturation trend, and the linearity of the sensing curve decreases. As V_d varies from 5 V to 12 V, the sensor voltage increases slightly, indicating that V_d can control the sensor, albeit limited. The influence of R_{se} on sensing characteristics is shown in Figure 6. Under the condition of $V_d = 5$ V, as R_{se} increases from 1 Ω to 10 kΩ, V_s (at $V_{ce} = 1200$ V) increases from 0.26 V to 3.88 V, and V_{st} decreases from 342 V to 162 V. Compared to the parameter V_d, the parameter R_{se} enables more efficient control of V_{st} and V_s.

Figure 7 shows the on-state condition of the IGBT cell and its integrated sensor. As can be seen from the distribution of the current lines (black lines), the isolation structure realizes the electrical blocking between the sensing area and the IGBT area, so that the

Investigation of Threshold Voltage Instability in p-GaN Gate HEMTs under Surge Current Stress

Xiaoming Wang[1], Yu Shi[2], David Zhou[2], Haizhao Zhi[2], Yun Xia[2], Yuxi Wan[2],
Ruize Sun[1,3], Xinghuan Chen[4], Wanjun Chen[1,3]*, Bo Zhang[1]

[1] State Key Laboratory of Electronic Thin Films and Integrated Devices, University of Electronic Science and
Technology of China, Chengdu 610054, China
[2] Shenzhen Pinghu Laboratory, Shenzhen 518111, China
[3] Institute of Electronic and Information Engineering of UESTC in Guangdong, Dongguan 523808, China
[4] China Electronic Product Reliability and Environmental Testing Research Institute, Guangzhou 510610, China

* Email: wjchen@uestc.edu.cn

Abstract—In this paper, the threshold voltage (V_{TH}) instability mechanism in p-GaN gate HEMTs under surge current stress is revealed. The significant positive shift of V_{TH} (ΔV_{TH}), which has a strong I_{peak} (i.e., the peak value of surge current) dependence, is observed after the surge current passes through the device. As the I_{peak} exceeds a critical value ($I_{critical}$), the ΔV_{TH} will saturate and may decrease. Furthermore, the $I_{critical}$ varies with the V_{GS}, initial gate leakage current ($I_{g, init}$), and substrate termination. It is suggested that electron trapping and hole trapping would take place during the stress. The electron trapping is more dependent on the current level, while the hole trapping is more dependent on the surge voltage, mainly the gate-to-channel voltage (V_{GC}). It is confirmed by mix-mode simulation and further experiments.

Keywords—*p-GaN gate HEMTs, surge current, surge voltage, V_{TH} instability, trapping effect*

I. INTRODUCTION

GaN normally-off devices are promising candidates for the next generation power electronic devices due to its excellent performance [1]. In recent years, p-GaN gate HEMTs have realized commercial applications. In practical applications, such as the synchronous buck converter, p-GaN HEMTs can conduct both forward and reverse currents [2]. Nevertheless, in reverse conduction mode, p-GaN HEMTs may need to withstand a high surge current [3]. The surge current reliability of Si or SiC devices has been widely studied, while GaN devices especially p-GaN HEMTs with zero reverse recovery charge are lacking attention [4], [5]. We observe a threshold voltage (V_{TH}) shift phenomenon in prior work [3]. Unfortunately, it has not been explained or further researched. Moreover, the impact of the initial gate leakage current ($I_{g, init}$) has not been reported.

In this work, a comprehensive study on the V_{TH} instability mechanism of p-GaN HEMTs under surge current stress is carried out. The single stress with various peak vlaue (I_{peak}) is conducted on devices with different substrate terminations (i.e., grounded and floating). Furthermore, the influence of V_{GS}, $I_{g, init}$ and stress cycle is studied. We confirm that the competing effect of the trapped hole and electron causes the V_{TH} instability.

II. DEVICE STRUCTURE AND EXPERIMENT SETUPS

The device studied here is commercially available 650-V/30-A Schottky-type p-GaN gate HEMTs [6]. The test method refers to the JEDEC Standard No.282B.01 [7]. The photograph of the workbench setup of the surge current test with p-GaN HEMT is shown in Fig. 1 (a). The schematics of

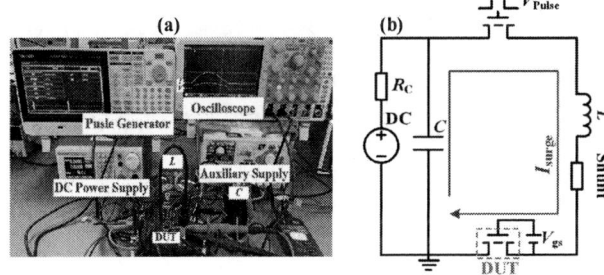

Fig. 1. Experiment setups. (a). Workbench. (b) Schematic test circuit.

the circuit are shown in Fig. 1(b). As the pulse-switching device is triggered, the surge current generated by the *LC* resonance will flow through the device from source to drain. The source pad is usually electrically connected to the substrate pad for optimal performance. Therefore, the grounded substrate refers to the substrate connected to the source in this work. The inductor of 5 mH and the capacitor of 2 mF are used to generate a surge current pulse of about 10 ms (Fig. 2(a)). To make the waveform as close to a half-sine shape as possible, the discharge circuit must have extremely low parasitic resistance. In this work, the I_{peak} is at about 4.5 ms, which is mainly limited by the on-resistance of the device under test (DUT). Nevertheless, it is very close to the actual situation. Furthermore, a diode is used to ensure the DUT is only subjected to the half-sine surge current [4].

The device will fail under the single surge current stress with an I_{peak} exceeding about 53 A, which is not shown in the picture. Therefore, the maximum I_{peak} is limited to 50 A here. The electrical properties of the device before and after stress are characterized by the Keysight B1505A curve tracer under the same conditions at room temperature. Mix-mode TCAD Sentaurus simulation is used to study the V_{TH} instability mechanism. The simulated structure is determined from the microscopic images of the decapsulated device. The simulation models are similar to the ones reported in [1].

III. RESULTS AND DISCUSSIONS

A. Threshold Voltage Shift Phenomenon

For ease of comparison, the I_{peak} is used to represent the stress level in this work. Fig. 2(a) and (b) show the typical surge current and surge voltage (i.e., V_{SD} and V_{GD} induced by surge current) waveforms of DUT at V_{GS} of 0 V. When operated in the reverse conduction mode, as the V_{GD} exceeds the V_{TH}, the 2DEG channel can be turned on and conduct

979-8-3503-6184-1/24 $31.00 © 2024 IEEE

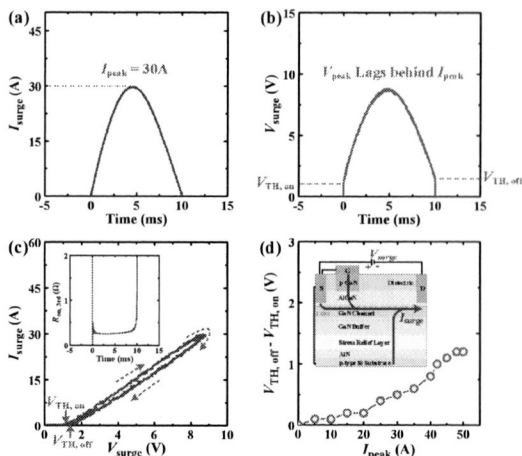

Fig. 2. (a) Surge current waveforms. (b) Surge voltage waveforms induced by the surge current at V_{GS} of 0 V. (c) I_{surge}-V_{SD} curve. (d) The evolution of the difference between the $V_{TH, off}$ and $V_{TH, on}$.

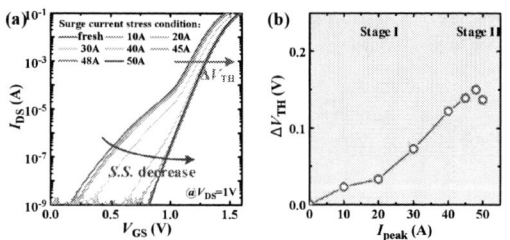

Fig. 3. Performance of the DUT after single stress with different I_{peak} at V_{GS} of 0 V. (a) Transfer characteristics. (b) ΔV_{TH}.

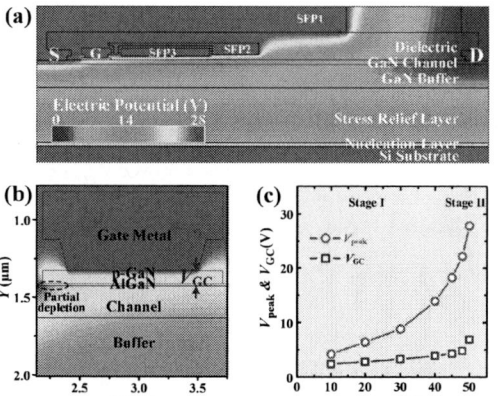

Fig. 4. (a) Simulated electric potential distribution at I_{peak} of 50 A at V_{GS} of 0V. (b) Enlarged view of (a). (c) V_{peak} and simulated V_{GC}.

surge current [5], exhibiting a channel resistance ($R_{on, 3rd}$). Notably, the $R_{on, 3rd}$ during surge current stress is several times higher than static $R_{on, 3rd}$. And a hysteresis of the I_{surge}-V_{surge} curve can be observed in Fig. 2(c). This is similar to the result in [3], which can be attributed to the accumulation of self-heating effect.

It is worth noting that the turn-off threshold ($V_{TH, off}$) is higher than the turn-on threshold ($V_{TH, on}$) under the surge current stress. This phenomenon always occurs in experiments, which shows a strong I_{peak} dependence and a critical value ($I_{critical}$) (Fig. 2(d)). Although the surge current is mainly conducted by the channel, the gate current and substrate current paths cannot be ignored. The above phenomenon may be not directly caused by thermal effects.

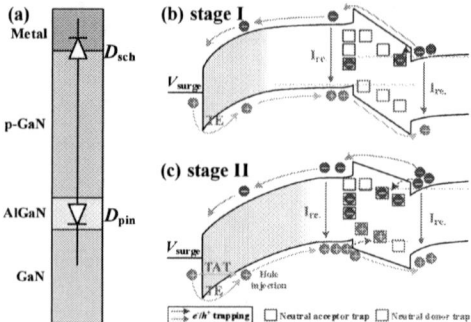

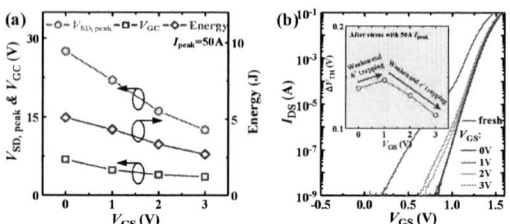

Fig. 5. (a) Equivalent circuit of the Schottky-type p-GaN gate. V_{TH} instability mechanism schematic at (b) stage I and (c) stage II.

Fig. 6. Performance of DUT under single stress with an I_{peak} of 50 A at different V_{GS}. (a) Experimental $V_{SD, peak}$, surge energy, and simulated V_{GC}. (b) Transfer characteristics and extracted ΔV_{TH}.

The results of transfer characteristics at room temperature support our judgment (Fig. 3). V_{TH} is extracted from the transfer characteristics at V_{DS} of 1 V and I_{DS} of 1 mA. After single stress, a positive shift of V_{TH} can be observed. The V_{TH} shift (ΔV_{TH}) can be divided into two stages. As the I_{peak} reaches the $I_{critical}$, in stage II, the ΔV_{TH} will saturate and even decrease with increasing I_{peak}. Furthermore, the degradation of the subthreshold swing ($S.S.$) can also be observed. The instability of V_{TH} may be related to the trapping effect.

B. Threshold Voltage Instability Mechanism

For p-GaN HEMTs, the $R_{on, 3rd}$ will increase at high I_{peak} due to the self-heating effect. It can generate an extremely high surge voltage peak value (V_{peak}) (Fig. 4). For the DUT having a bottom width of gate metal shorter than the top width of the p-GaN layer, the gate-to-channel voltage (V_{GC}) can also be very high due to the partial depletion of 2DEG under gate close to source. As the I_{peak} increases, the V_{peak} and V_{GC} increase approximately exponentially (Fig. 4(c)).

For Schottky-type p-GaN gate HEMT, the gate can be equivalent to two back-to-back diodes in a series (Fig. 5(a)). When the V_{GC} is low, the hole current can be ignored. During the surge curent stress, hot electrons may escape from the channel and be captured by the acceptor traps in the gate region (Fig. 5(b)). The net negative charge causes the positive shift of V_{TH}. As the V_{GC} increases to a certain value, numerous gate holes can be injected into the channel by trap-assisted tunneling (TAT). Holes may be captured by the donor traps on transport paths and partially offsets the electron trapping effect. The decrease of net negative charge may lead to the reduction of ΔV_{TH}.

Notably, at higher V_{GS}, the V_{GC} will be significantly reduced under the same I_{peak} (Fig. 6(a)). In this case, the ΔV_{TH} also changes (Fig. 6(b)), indicating that the root cause of the ΔV_{TH} turning phenomenon is the V_{GC}. When V_{GS} increases to 1 V, the hole trapping is suppressed due to the decrease of V_{GC}, leading to the increase in ΔV_{TH}. At higher

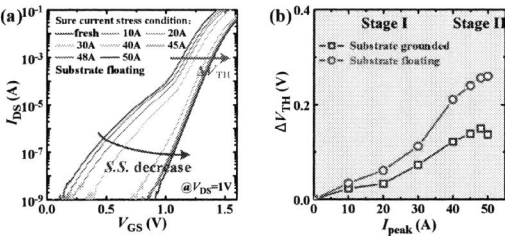

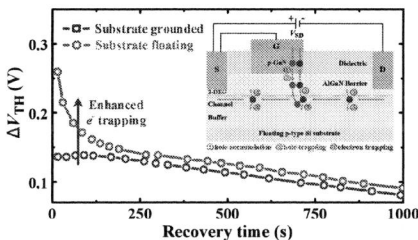

Fig. 7. (a) Transfer characteristics of the device with floating substrate after single stress with various I_{peak} at V_{GS} of 0 V. (b) Comparison of the ΔV_{TH} between different substrate terminations.

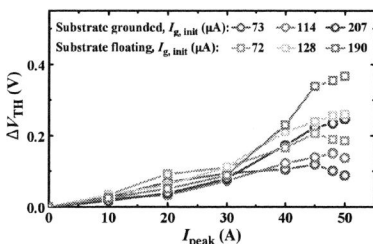

Fig. 9. The ΔV_{TH} comparison of devices with different $I_{g, init}$ after single surge current stress with different I_{peak} at V_{GS} of 0 V.

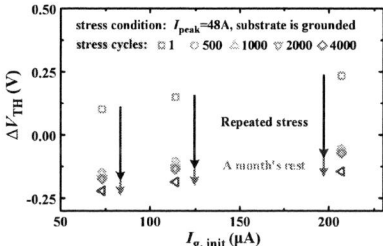

Fig. 8. Comparison of the recovery characteristics between the grounded substrate and the floating substrate after single surge current stress with an I_{peak} of 50 A at V_{GS} of 0 V.

Fig. 10. $I_{g,init}$-dependent shift of V_{TH} under repetitive surge current stress with an I_{peak} of 48 A at V_{GS} of 0 V.

V_{GS}, the electron trapping can also be weakened. This may be mainly caused by the reduction of surge energy, which will result in a decrease in junction temperature during stress. In this case, the ΔV_{TH} decreases as the V_{GS} further increases.

The above investigation is based on the device with the grounded substrate. For the devices with floating substrate, the V_{TH} shift is exacerbated after single surge current stress, especially at high I_{peak}, as shown in Fig. 7(a) and Fig. 7(b). This may be due to the elimination of the conductive path from the substrate to the drain. The enhancement of channel current results in severer electron trapping. It is confirmed by the recovery tests at room temperature (Fig. 8). When the substrate is floating, more shallow-level acceptor traps will be ionized, which can recover quickly. Moreover, the turning phenomenon of the V_{TH} recovery behavior of the device with a grounded substrate proves the simultaneous occurrence of electron and hole trapping at I_{peak} of 50 A. The V_{TH} can not recover to the initial level within 1000 s, indicating the ionization of deep-level acceptor traps in the AlGaN layer.

C. Influence of initial gate leakage current

The influence of $I_{g, init}$ (V_{GS} = 6 V) of the device under single surge current stress is further investigated. It can be observed that for the DUT with lower $I_{g, init}$, the ΔV_{TH} is higher at low I_{peak} while it is lower at high I_{peak} (Fig. 9). Obviously, substrate termination is not the cause of this phenomenon. We find that the device with lower $I_{g, init}$ typically has a larger initial V_{TH}, which may be related to the Mg-related process. In this case, the V_{SD} induced by the same I_{peak} is higher. It will cause the enhancement of electron and hole trapping. Consequently, the shift of V_{TH} in stage I is more significant, and stage II will be advanced.

Repetitive surge current stress is also carried out. As the stress cycle increases, the shift direction of V_{TH} will be reversed. Furthermore, the final degradation is irreversible. This can be attributed to the continuous enhancement of hole trapping effect and the generation of new donor traps. For the device with lower $I_{g, init}$, stronger hole trapping causes a more

negative shift of V_{TH}. On the contrary, higher $I_{g, init}$ can inhibit the trapping effect. Therefore, Ohmic gate p-GaN HEMTs may have better surge current reliability. This may be discussed in future work.

IV. Conclusion

The V_{TH} shift of p-GaN HEMTs under surge current stress with different I_{peak} is deeply investigated. We find that the competition between electron and hole trapping effect is the root cause of the degradation. The electron trapping is more related to the surge current level, while the hole trapping depends more on the surge voltage (mainly the V_{GC}). For the device with the floating substrate or lower $I_{g, init}$, the V_{TH} instability will be exacerbated. These results provide new insights for the improvement of GaN HEMTs reliability.

Acknowledgment

This work was supported in part by the National Natural Science Foundation of China under Grant 62334003 and U21A20499, and the University-Industry Collaborative Education Program (231004866165126).

References

[1] S. Yang et al., "Dynamic On-Resistance in GaN Power Devices: Mechanisms, Characterizations, and Modeling," IEEE J. Emerg. Sel. Topics Power Electron., vol. 7, no. 3, pp. 1425-1439, Sep. 2019.

[2] W.-S. Lin et al., "Toward Understanding the Failure Mechanism in p-GaN Gate HEMTs Operating in Reverse Conduction Diode Mode," IEEE Trans. Electron Devices, 2024.

[3] Y. Liu et al., "Investigation of surge current capability of GaN E-HEMTs in the third quadrant: The impact of p-GaN contact," IEEE J. Emerg. Sel. Topics Power Electron., 2019.

[4] P. Hofstetter, "Comparison of the Surge Current Ruggedness between the Body Diode of SiC MOSFETs and Si Diodes for IGBT", 2018.

[5] Z. Zhu et al., "Investigation on Surge Current Capability of 4H-SiC Trench-Gate MOSFETs in Third Quadrant Under Various V_{GS} Biases," IEEE J. Emerg. Sel. Topics Power Electron., 2021.

[6] GaN Systems. (2019). GS66508P Datasheet. [Online]. Available: http://www.gansystems.com.

[7] Silicon Rectifier Diodes, Section 4.2, document JEDEC JSD282-B Rev.01, 2002.

A Novel Ga₂O₃ High-*k* Trench MOSFET with Improved Forward and Reverse Performance

Moufu Kong[1]*, Lewei Lyu[1], Haoran Wang[1], Zhaoyu Ai[1], Xinyang Chen[1], Fanxin Meng[2], Qiang Hu[3]*

[1]State Key Laboratory of Electronic Thin Films and Integrated Devices of China, University of Electronic Science and Technology of China, Chengdu 611731, China

[2]Chengdu High-tech Development Co.Ltd, Chengdu, 610093, Sichuan, China

[3]Chengdu Semi-Future Technology Co. Ltd, Chengdu, 611730, Sichuan, China

*Corresponding author: Moufu Kong, Email: kmf@uestc.edu.cn; Qiang Hu: arlo.hu@cdp-ht.com

Abstract— In this article, a novel Ga₂O₃ power MOSFET with high permittivity (high-*k*) insulator and integrated a Schottky barrier diode (H*k*-SBD MOSFET) is proposed and compared with a conventional Ga₂O₃ MOSFET. After optimization through TCAD simulations, the proposed device achieves a breakdown voltage (*BV*) of 3320V and a specific on-resistance ($R_{on,sp}$) of 4.84 mΩ·cm², achieving a Baliga's figure of merit (BFOM) of 2.2 GW/cm². The introduction of the high-*k* insulator reduces the electric field gradient and enhances doping concentration of the N-drift region, resulting in an improved BFOM. Compared with the conventional Ga₂O₃ MOSFET, it demonstrates a 33.4% reduction in $R_{on,sp}$, a 6.6% increase in *BV*, and over 40% improvement in BFOM. Also, the integrated SBD is greatly improved the reverse conduction performance of the device.

Keywords—*Ga₂O₃ , trench MOSFET , Hk , specific on-resistance, BFOM*

I. INTRODUCTION

Compared to other wide bandgap semiconductor materials, such as SiC and GaN, *β*-Ga₂O₃ exhibits exceptional material properties, including an ultra-wide bandgap of ~ 4.9 eV and a high critical electric field (E_c) of ~8 MV/cm [1]. These characteristics yield a Baliga's Figure of Merit (BFOM = $\varepsilon\mu E_c^3$) of 3444, significantly surpassing existing wide bandgap semiconductors like SiC and GaN [2]. Moreover, *β*-Ga₂O₃ can be grown in large, uniform crystals using conventional melt growth techniques, substantially reducing production costs [2]. These advantages position *β*-Ga₂O₃ as a highly competitive candidate for the next generation of ideal power switching devices [3].

In recent years, Ga₂O₃ power devices have seen rapid advancement. For practical high-voltage applications, vertical enhancement mode (E-mode) transistors are increasingly favored. Vertical devices maximize chip area utilization for high power density chips, while thick drift layers enable high breakdown voltages. Enhancement mode devices simplify circuit designs and ensure fail-safe operation at high voltages [4],[5].

To achieve high breakdown voltage (*BV*) while minimizing specific on-resistance ($R_{on,sp}$), researchers have continuously explored and implemented various techniques, such as the Superjunction structure [6]. However, due to challenges in creating shallow acceptor states and self-trapped holes in single-crystal oxide semiconductors [7], resulting in extremely low hole mobility, there have been no reports of effective p-type Ga₂O₃ to date. Consequently, the Superjunction (SJ) structure remains impractical for Ga₂O₃ MOSFETs. In 2012, Chen et.al. [8] proposed a structure employing a high permittivity (H*k*) insulating layer, which can significantly reduce $R_{on,sp}$ by enhancing drift region concentration under the same breakdown voltage level.

Currently, research on H*k* MOSFETs based on Ga₂O₃ remains limited. This paper introduces a Ga₂O₃ High-*k* Integrated Schottky Barrier Diode (H*k*-SBD) MOSFET structure, aiming to achieve high breakdown voltage while maintaining low $R_{on,sp}$. Additionally, the proposed H*k*-SBD MOSFET incorporates an SBD, markedly enhancing reverse on-state performance. Despite a relatively intricate fabrication process, the proposed H*k*-SBD Ga₂O₃ MOSFET effectively enhances performance compared to conventional Ga₂O₃ MOSFETs. Thus, it strikes a superior balance between performance metrics and manufacturing costs.

II. DEVICE STRUCTURE AND MECHANISM

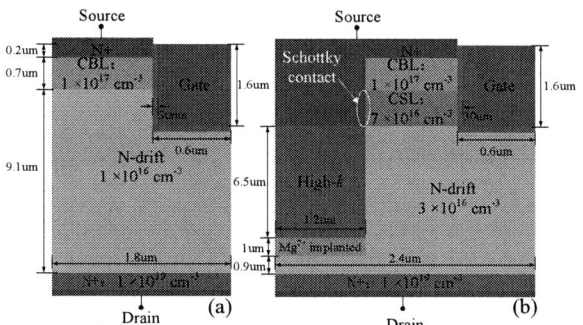

Fig.1 (a) The Conventional Ga₂O₃ MOSFET's Cell Structure Diagram, (b) The H*k*-SBD Ga₂O₃ MOSFET's Cell Structure Diagram

Fig. 1(a) and (b) depict the cell structures of the conventional Ga₂O₃ vertical MOSFET and the novel Ga₂O₃ vertical high-*k* insulator MOSFET with an integrated SBD (H*k*-SBD MOSFET), respectively. Both Ga₂O₃ MOSFETs design feature a uniform 12 μm drift region thickness. In the conventional Ga₂O₃ MOSFET (Fig. 1(a)), a current blocking layer (CBL) is implanted with high-energy, high-dose Mg²⁺ in the β- Ga₂O₃ drift layer to achieve isolation of source and drain currents, enabling enhancement-mode operation [2],[9]. In the proposed Ga₂O₃ H*k*-SBD MOSFET (Fig. 1(b)), a portion of the drift region is filled with a high-*k* insulating layer, and the doping concentration of the drift region is increased from 1×10^{16} cm⁻³ to 3×10^{16} cm⁻³. Prior to the deposition of the high-*k* dielectric layer, high-energy, high-dose Mg²⁺ implantation is performed to mitigate electric field concentration effects [9]. Additionally, a current spreading layer (CSL) suppresses the junction field-effect transistor (JFET) effect, reducing the specific on-resistance ($R_{on,sp}$) of the devices[10]. A chromium (Cr) metal Schottky contact (work function is 4.5 eV) is introduced on the trench sidewall as part of the source, forming a reverse freewheeling SBD. In the reverse conduction state, the CSL remains partially depleted, ensuring the integrated SBD operates effectively,

This work was supported in part by the Key R & D project of science and technology plan of Sichuan province (Grant No. 2023YFG0005).

979-8-3503-6184-1/24 $31.00 © 2024 IEEE

thereby enhancing the reverse freewheeling capability of the proposed Ga$_2$O$_3$ Hk-SBD MOSFET.

III. SIMULATION RESULTS AND DISCUSSION

The proposed device and the conventional device are both simulated using TCAD device design tool to verify and compare their performances. Simultaneously, the permittivity (k) of high-k insulators and the doping concentration of the drift region (N_d) can be optimized to improve the trade-off between specific on-resistance and breakdown voltage, as well as enhance reverse recovery capability. The simulation incorporates key physical models such as SRH, Avalanche (vanOverstraetendeMan), Fermi-Dirac statistics, AUGER recombination, and incomplete ionization. And the channel mobility is calibrated to ~30 cm^2/v·s [11].

A. Breakdown Characteristic

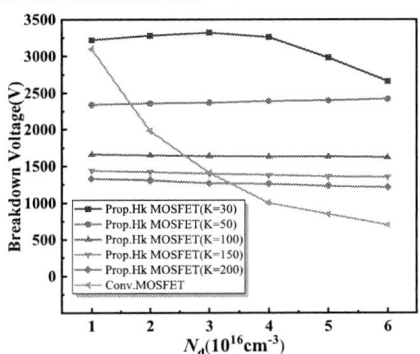

Fig.2 BVs of the conventional MOSFET and proposed Hk-SBD MOSFET with different permittivity of high-k insulators.

Fig. 2 illustrates the BV variation curves with drift region doping concentration (N_d) for both the proposed Hk-SBD MOSFET structure and the conventional structure. It is observed that as N_d increases, the breakdown voltage (BV) decreases for both the conventional and the proposed devices. However, due to the presence of the high-k (Hk) structure, most electric field lines generated by N-type doping in the drift region traverse through the Hk region, minimizing the impact of doping concentration changes on the breakdown voltage of the proposed device. The permittivity of high-k insulating materials and semiconductors is denoted by ε_I and ε_S, respectively. The Poisson equation approximated for the drift region is given by:

$$\frac{\partial E}{\partial y} = \frac{q\rho}{\varepsilon_I + \varepsilon_S} \quad (1)$$

The substantial enhancement of the effective permittivity in the drift region by the Hk insulating material minimizes the impact of doping concentration variation on the breakdown voltage of the proposed device.

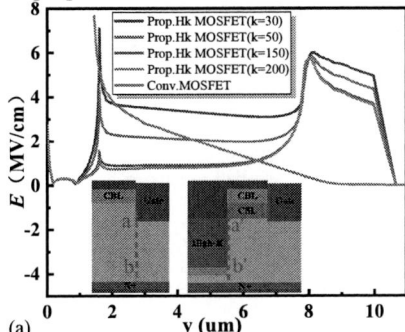

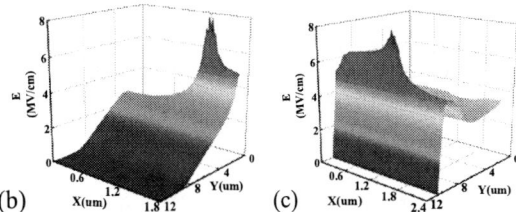

Fig. 3 (a)Electric field distributions along the dotted line ab of the conventional MOSFET and Proposed Hk-SBD MOSFET with different permittivity of high-k insulators at their own breakdown voltages. Three-dimensional electric field distributions under breakdown voltage of the (b) the conventional MOSFET, (c) the Hk-SBD MOSFET with $k = 30$

Fig. 3 illustrates the electric field distribution at their respective breakdown voltages for both device types. As depicted in Fig. 3(b) and (c), compared to the conventional MOSFET, the Hk-SBD MOSFET shows a more uniform electric field distribution in the drift region. The presence of high-k (Hk) material induces a rectangular distribution of the electric field in the drift layer, thereby achieving a higher breakdown voltage. However, when the dielectric constant (ε) is excessively high, as deduced from the formula $D=\varepsilon\underline{E}$, discontinuities in the electric displacement vector D occur at the interfaces between Hk material and semiconductor material, causing localized high electric fields within the Ga$_2$O$_3$ semiconductor material. Consequently, regions with excessively high k values may experience premature breakdown, thus reducing the overall device breakdown voltage. In Fig. 3(a), it is observed that with K=30, the device exhibits a breakdown voltage of 3320V compared to 3100V for the conventional MOSFET, marking a 6.6% increase in BV. Therefore, this study adopts a dielectric constant of 30 for further discussion.

B. Forward Characteristic

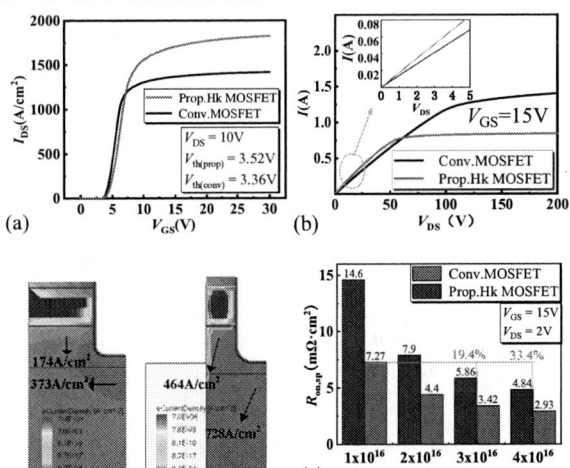

Fig.4 (a)Transfer characteristic curve and (b) output characteristic curve of conventional Ga$_2$O$_3$ MOSFET with $N_d = 1\times10^{16}$ cm^{-3} and the proposed Hk-SBD MOSFET with K = 30 and $N_d = 3\times10^{16}$ cm^{-3}. Electron current density distribution at V_{DS}=2V and V_{GS}=15 V of (c) conventional Ga$_2$O$_3$ MOSFET and (d) Hk-SBD MOSFET; (e) Variation of $R_{on,sp}$ at V_{DS} = 2V and V_{GS} = 15 V for the two devices at different N_d concentration

Fig. 4(a) shows that the threshold voltage (V_{th}) of the proposed structure is 3.52V, whereas the V_{th} of the conventional MOSFET device is 3.36V at I_{DS} = 0.1 A/cm². When comparing the performance of two different MOSFETs, it is essential to ensure their threshold voltages are similar. Fig. 4(b) compares the output characteristics and

979-8-3503-6184-1/24 $31.00 © 2024 IEEE

$R_{on,sp}$ between the conventional and proposed Hk-SBD MOSFETs, revealing a 19.4% decrease in $R_{on,sp}$. Fig. 3(c) and (d) illustrate that the current density of the proposed Hk-SBD MOSFET is significantly higher than that of the conventional MOSFET. This is due to the higher doping concentration of N_d and N_{CSL} enabled by the Hk-SBD MOSFET, leading to a corresponding increase in carrier concentration. The high-k insulating layer structure in the drift region improves the voltage endurance of the Hk-SBD MOSFET. Consequently, at the same voltage endurance level, the proposed Hk-SBD MOSFET achieves a substantial increase in electron current density and reduces its $R_{on,sp}$ by enhancing the doping concentration in the drift region. As depicted in Fig. 4(e), compared to the conventional structure, the Hk-SBD MOSFET exhibits a 19.4% reduction in specific on-resistance when $N_d = 3 \times 10^{16}$ cm^{-3}, and a reduction of 33.4% when $N_d = 4 \times 10^{16}$ cm^{-3}.

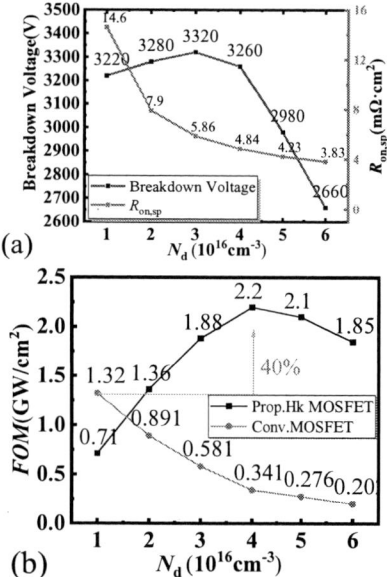

(a)

(b)

Fig. 5 (a)Variation of BV and $R_{on,sp}$ at different N_d of the proposed Hk-SBD MOSFET with K = 30 and the conventional MOSFET; (b)The variation of two devices' BFOM at different N_d concentrations.

Fig. 5 (a) illustrates that the breakdown voltage at K=30 initially increases and then decreases with doping concentration. This phenomenon arises because at lower doping concentrations of N_d, the maximum electric field strength occurs at the bottom right corner where the Hk material interfaces with the semiconductor material. As N_d increases, the maximum electric field shifts to the top right corner due to the influence of the CBL beneath the Hk material, leading to a gradual decrease in electric field in this region. Simultaneously, the electric field strength increases at the top right corner of the Hk material where there is no CBL protection, resulting in breakdown starting from this area. By balancing these electric field strengths, the maximum breakdown voltage is achieved when N_d=3×10^{16}cm^{-3}. Furthermore, the figure shows that as N_d increases, the device's $R_{on,sp}$ decreases. Fig. 5(b) presents the BFOM values of the proposed device and conventional devices across different N_d values. At a doping concentration of 4×10^{16} cm^{-3}, the device achieves a maximum BFOM of 2.2 GW/cm^2. This indicates that at this doping concentration, the device exhibits optimal breakdown characteristics and forward conduction

performance, marking a significant improvement compared to conventional MOSFETs.

C. Reverse Characteristic

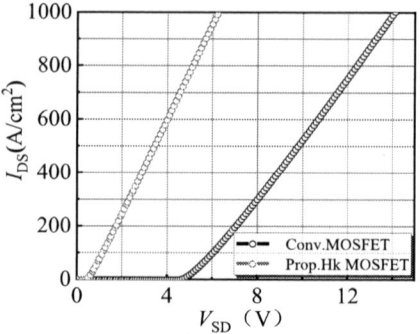

Fig.6 Reverse conduction characteristics of the two devices.

Fig. 6 compares the reverse conduction I-V characteristics of the conventional MOSFET with the Hk-SBD MOSFET. The turn-on voltage drop (knee voltage) of the integrated SBD in the proposed Ga$_2$O$_3$ MOSFET is approximately 0.77V, assuming the metal work function of 4.5 eV, whereas the knee voltage of the conventional Ga$_2$O$_3$ MOSFET is about 5.3V.

IV. CONCLUSION

This paper introduces a novel trench gate Hk-SBD Ga$_2$O$_3$ MOSFET based on TCAD numerical simulations. And the device structural parameters are optimized. According to the simulation results, we have successfully developed a device capable of sustaining a stable voltage exceeding 3300V. With the HK structure incorporated, our design achieves a high blocking voltage (up to 3320V) and a low specific on-resistance with a value of 4.84 m$\Omega \cdot$cm^2. The device exhibits an excellent figure of merit (BFOM=2.2 GW/cm^2), significantly outperforming conventional structures. Additionally, it demonstrates exceptional reverse characteristics. Therefore, this device has broad application prospects in high-power and low-loss applications.

REFERENCES

[1] M. Higashiwaki. et al. Recent progress in Ga$_2$O$_3$ power devices. *Semiconductor Sci. Technol.* 2016, vol.31, pp. 034001,

[2] Ma, Y. et al. 702.3 A·cm^{-2}/10.4 m$\Omega \cdot$cm^2 β-Ga$_2$O$_3$ U-Shape Trench Gate MOSFET With N-Ion Implantation. *IEEE Electron Device Letters.* 2023, vol.44, pp.384.

[3] M. H. Wong. et al. Field-plated Ga$_2$O$_3$ MOSFETs with a breakdownvoltage of over 750 V. *IEEE Electron Device Lett.* 2016, vol.37, pp.212.

[4] M. H. Wong, et.al., "Enhancement-Mode β-Ga$_2$O$_3$ Current Aperture Vertical MOSFETs With N-Ion-Implanted Blocker," *IEEE Electron Device Letters.*, 2020, vol.41, pp.296.

[5] X. Zhou. et al. Enhancement-mode β-Ga$_2$O$_3$ U-shaped gate trench vertical MOSFET realized by oxygen annealing. *Appl. Phys. Lett.* 2022, vol.121, pp.223501.

[6] X.B. Chen. et al. Theory of a novel voltage-sustaining layer for power devices. *Microelectronics Journal.* 1998, vol.29, pp.1005.

[7] M. Higashiwaki. et al. β-Ga$_2$O$_3$ material properties, growth technologies, and devices: A review. *AAPPS Bull*, 2022, vol.32 pp.3.

[8] X.B. Chen. et al. A Vertical Power MOSFET With an Interdigitated Drift Region Using High-k Insulator. *IEEE Transactions on Electron Devices.* 2012, vol.59, pp.2430.

[9] M. Wong. et al. Acceptor doping of β-Ga$_2$O$_3$ by Mg and N ion implantations. *Applied Physics Letters.* 2018, vol.113, pp.102103.

[10] M. Kong. et al., A novel SiC high-k superjunction power MOSFET integrated Schottky barrier diode with improved forward and reverse performance. *J. Semicond*, 2023, vol.44, pp.052801.

[11] Z. Hu, et al., Breakdown mechanism in 1 kA/cm^2 and 960V E-mode β-Ga2O3 vertical transistors. *Appl. Phys. Lett.*, 2018, vol.113, pp. 122103.

A Nonlinear Behavioral Modeling Approach for Microwave Transistors Considering Electrothermal-Aging Degradation

Lin Cheng, Hongliang Lu*, Silu Yan, Junjun Qi, Jiantao Qiao, Yuming Zhang
School of Microelectronics, Xidian University
Shaan Xi, Xi'an 710071, China
hllv@mail.xidian.edu.cn

Abstract-In this paper, a nonlinear behavioral modeling approach for microwave transistors considering electrothermal-aging degradation is explored. This approach gives an electrothermal-aging model framework based on Elman structure, in which an isothermal network with a thermal-aging correction term is introduced to account for the degradation caused by electrothermal-aging process. The validity and accuracy of this approach are verified by comparing the measured results of the devices during the degradation process with the model prediction results.

Keywords- behavioral modeling approach, Elman structure, electrothermal-aging degradation

I. INTRODUCTION

With the rapid development of microelectronics technology, the application and development of RF semiconductor devices at microwave millimeter-wave frequencies have gradually occupied an important position in high-speed communications [1]. To reduce the cost and improve the performance, some key dimensions in the process have been continuously reduced, and accordingly, the electric field strength and current density to which the device is subjected are also increasing, which exacerbates the probability and randomness of the occurrence of various failure mechanisms, and poses a serious challenge to the reliability of integrated circuits and the improvement of electronic design automation (EDA) [2].

Both the reduction of the device feature size and the increase in its frequency performance requirements have made the impact of thermal reliability on the device characteristics one of the main factors limiting the further development of the process [3]. In addition, with the great advances in the development of high-speed technology, another major issue that still remains to be completely solved is the accurate prediction and possible improvement of the lifetime of circuit designs [4]. In recent years, the studies on the electrothermal modeling of microwave devices have focused on the effect that a particular degradation phenomenon has on the device output characteristics [5-8]. The device output characteristics in the vicinity of the knee voltage are crucial for the power compression characteristics of the device, therefore, two types of methods, physically corrected normalized charge and empirically introduced additional slope factor, have been used in the literature [5] and [6], respectively, to simulate the soft-knee phenomenon occurring in the device at the operating bias, without considering the thermal effects. The severe self-heating characterization of devices under high power application conditions has always been a difficulty affecting the modeling accuracy, and many studies have improved the thermal network parameter extraction method [7-8], but the accurate characterization of thermal resistance is often accompanied by complex extraction methods making it very difficult to extract; moreover, this thermal network characterization approach is usually more accurate in smaller bias ranges and difficult to guarantee over a larger device operating range. For the aging behavior of microwave devices with nonlinear currents, researchers have mainly focused on characterizing the degradation of the Gummel properties of the devices, and then combined with TCAD simulation for explaining and describing the degradation mechanism at the physical level of the devices [9], which does not take into account the contribution of the thermal behavior during the aging process, and there are also constraints such as slow modeling speed, and not easy to be compatible with EDA software.

However, for the characterization of complex degradation behavior during electrothermal-aging process, it is necessary to fully consider the influence of degradation-inducing factors between the self-heating and aging effects. To accurately predict the degradation process, the use of the traditional equivalent circuit parameters corrected in the above way or the use of a single black-box network architecture can no longer meet the demand. In this paper, therefore, a nonlinear behavioral-level modeling approach is proposed to simultaneously consider the electrothermal-aging degradation prediction of microwave devices. This approach gives an electrothermal-aging model framework based on Elman structure, in which an isothermal network with a thermal-aging correction term is introduced. The validity and accuracy of this approach are verified by comparing the measured results of the devices during the degradation process with the model prediction results.

II. MODELING TECHNIQUES

A. Elman Neural Network Structure

Elman neural network is a typical feedback neural network. In addition to the input layer, hidden layer and output layer, on the basis of the traditional feed-forward network, the context

979-8-3503-6184-1/24 $31.00 © 2024 IEEE

layer is introduced as a one-step delay operator, so that the network structure has the ability to adapt to the time-varying characteristics of the network structure, the addition of this internal feedback structure increases the ability of the network itself to deal with the complex tasks for the strongly nonlinear prediction problems. And the prediction of the nonlinear current of devices under high-power applications in different characteristic regions is a severely nonlinear relationship with degradation-inducing factors between the self-heating and aging effects. The traditional method of a simple feedforward neural network is not applicable to this electrothermal-aging behavior. In this paper, the Elman neural network structure shown in Fig. 1 is utilized to establish the proposed nonlinear current degradation model framework.

B. Electrothermal-Aging Model Construction

Aiming at the strong nonlinearity of the interaction between the electrical aging and the thermal aging of the device, we model the isothermal degradation and construct the thermal-aging correction term for the degradation of the electrothermal characteristics of the nonlinear collector current of the device during the aging process, respectively. The analytical form of isothermal network structure is shown in Eq. (1).

$$I_{CE,ISO}^{Age} = f_{ENN}^{iso}(V_{ce}, I_b) + f_{ENN}^{\Delta iso}(V_{ce}, I_b, E_{stress}, T_{J,stress}, t_{stress}) \quad (1)$$

where E_{stress}, $T_{J,stress}$, and t_{stress} are the applied electrical stress, thermal stress, and stress time, respectively. V_{ce}, I_b are the bias voltage and input current respectively.

To consider the electrothermal degradation behavior, the thermal-aging correction term is introduced as shown in Eq. (2) to take into account the degradation due to the correlation with the self-temperature rise of the device. Among them, the power dissipation is directly related to the self-heating, therefore, we first construct the network for the degradation behavior of the power dissipation with the analytical equation shown in Eq. (3).

$$Age_{TH} = 1 - KT_{age} \cdot P_{DISS}^{Age} \quad (2)$$

$$P_{DISS}^{Age} = f_{ENN}^{p}(V_{ce}, I_b) + f_{ENN}^{\Delta p}(V_{ce}, I_b, E_{stress}, T_{J,stress}, t_{stress}) \quad (3)$$

$$KT_{age} = f_{ENN}^{k}(V_{ce}, I_b, E_{stress}, T_{J,stress}, t_{stress}) \quad (4)$$

where KT_{age} is the thermal-aging factor; P_{DISS}^{Age} is the power

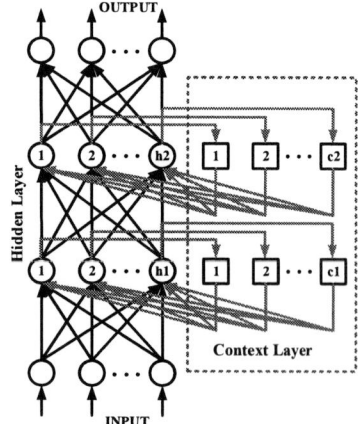

Fig. 1 Elman Neural Network.

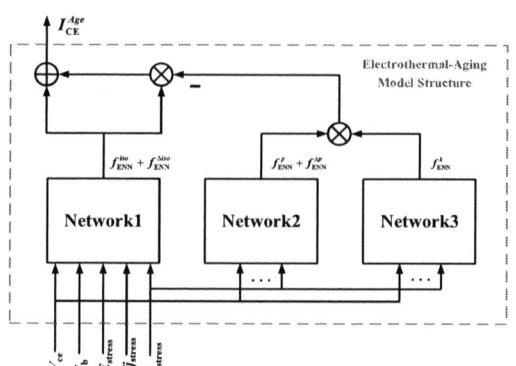

Fig. 2 Electrothermal-aging Model Framework

degradation of the device during aging.

The traditional thermal network characterization used to characterize the self-heating characteristics of RF active devices focuses on the extraction of thermal resistance, but the actual thermal resistance is not a single value but has a bias correlation, the correlation extraction of thermal resistance is often difficult to standardize and complex for the thermal characteristics of different devices. In addition, the difficulty of its extraction is increased by considering that there is also a degradation behavior of the thermal characteristics of the device during the aging process. Therefore, we introduce the thermal-aging factor KT_{age} with bias dependence and correlation with the stress state in Eq. (2), and its network resolution is shown in Eq. (4). The thermal-aging correction term constructed above is combined with the isothermal degradation term through the architecture shown in Fig.2 to achieve the degradation behavior prediction.

III. RESULTS AND DISCUSSION

To assess the validity and the accuracy of the developed modeling approach, 0.7 μm InGaAs/InP HBT devices were investigated. The combination of the PIV and DCIV measurements can be used to obtain the dataset for the nonlinear degradation of collector current induced by self-heating during the aging process. A pulse width of 500 ns is applied and the peak pulses V_{ce} were applied from 0 to 2 V in steps of 0.1 V and I_b from 100 to 1000 μA in steps of 100 μA.

In this paper, the total number of samples used for the isothermal degradation term, power dissipation degradation term and thermal-aging factor network are 410, 3690 and 2700, respectively, of which the ratio of the dataset used for training and testing is 7:3, respectively. After the network structure variables are determined, we firstly carry out training and optimization for the isothermal degradation term. The PIV characteristics of the device in the fresh state are simulated as shown in Fig. 3(a); then we simulate the degraded output characteristics after thermal decoupling, as shown in Fig. 3(b), which shows that the model can accurately describe the nonlinear degradation behavior of the PIV characteristics of the device before and after stress. The degradation process of the isothermal degradation term is mainly reflected in the intensification of the soft-knee phenomenon. Fig. 3(c)~(d)

979-8-3503-6184-1/24 $31.00 © 2024 IEEE

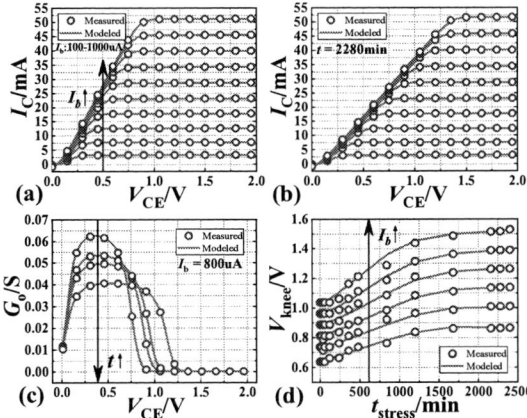

Fig. 3 Isothermal Degradation Term Modeling Results: (a)fresh PIV; (b) degraded PIV; (c)output conductance; (d)knee voltage.

gives the predicted results of the degradation process of the knee voltage, which gradually increases to saturation. The proposed model can well simulate this degradation trend and can also provide guidance for modeling device degradation in pulsed application scenarios.

The thermal-aging correction term includes the power dissipation degradation and the introduced thermal-aging factor. The power dissipation of the device is a direct influence on its internal heat accumulation and thus internal temperature rise, and the effects of increased soft-knee and self-heating degradation shown in Fig. 4(a) on it need to be considered. The thermal-aging factor with bias dependence and stress state correlation shown in Fig. 4(b) is introduced to capture the variability of the thermal effect, which is more convenient and accurate than the traditional thermal resistance extraction to characterize the thermal properties, avoiding the cumbersome physical derivation of the extraction process, and at the same time taking into account the variability caused by bias and stress. Eventually, after the above degradation terms are accurately established, the electrothermal-aging degradation process of the device is accurately predicted as shown in Fig.

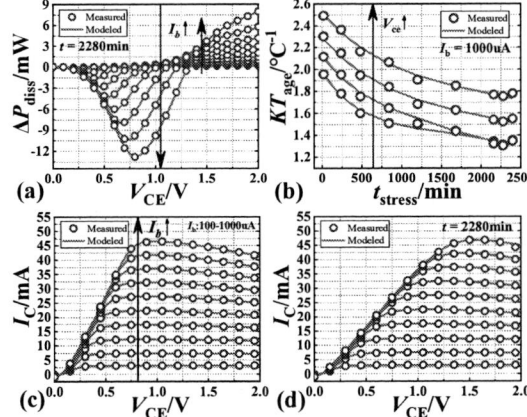

Fig. 4 Electrothermal-Aging Modeling Results: (a)P_{diss} degradation; (b) thermal-aging factor; (c)fresh DCIV; (d)degraded DCIV.

4(c)~(d) by incorporating them into the proposed structure, and the mean relative error of its prediction is kept below 2%. The proposed modeling framework has good performance advantages in solving the degradation prediction problem under multi-stress scenarios, and provides guidance for establishing efficient and accurate microwave device aging models for high-power applications.

IV. CONCLUSION

A nonlinear behavioral-level modeling method that simultaneously considers the electrothermal-aging degradation prediction of microwave devices is proposed. Through the adopted degradation modeling framework combined with the Elman structure, the isothermal degradation term and the thermal-aging correction term are constructed. An accurate description of the degradation of the devices induced by self-heating and aging is realized. Finally, the validity and accuracy of the proposed method are verified by comparing the degradation prediction results with the measurements of 0.7×15 µm^2 InGaAs/InP DHBT devices.

ACKNOWLEDGMENT

This work was supported by the National Natural Science Foundation of China under Grant 62374120.

REFERENCES

[1] R. Nericua, K. Wang, H. Zhu, R. Gómez-García and X. Zhu, "Low-Loss and Compact Millimeter-Wave Silicon-Based Filters: Overview, New Developments in Silicon-on-Insulator Technology, and Future Trends," in *IEEE Journal on Emerging and Selected Topics in Circuits and Systems*, vol. 14, no. 1, pp. 30-40, March 2024.

[2] Sanchez Daniela, Servadei Lorenzo, Kiprit Gamze Naz, Wille Robert, Ecker Wolfgang. "A Comprehensive Survey on Electronic Design Automation and Graph Neural Networks: Theory and Applications" *ACM Transactions on Design Automation of Electronic Systems*, vol. 28, no. 2, pp. 1-27, 2023.

[3] Li Geng, Wang Shang, et al., "Solder joint shape optimization and thermal-mechanical reliability improvement for microwave RF coaxial connectors," *Microelectronics Reliability*, vol.154, pp. 115345, 2024.

[4] Y. Ye, T. Chen, Z. Wang, H. Yan, B. Yu and L. Shi, "Fast and Accurate Aging-Aware Cell Timing Model via Graph Learning," in *IEEE Transactions on Circuits and Systems II: Express Briefs*, vol. 71, no. 1, pp. 156-160, Jan. 2024.

[5] V. P. Sriperumbuduri, H. Yacoub, T. K. Johansen, A. Wentzel, R. Doerner and M. Rudolph, "Modeling Base-Collector Heterojunction Barrier Effect in InP DHBTs for Improved Large Signal Performance," *2021 IEEE MTT-S International Microwave Symposium (IMS)*, Atlanta, GA, USA, 2021, pp. 355-357.

[6] Z. Hu, Q. Zhang, K. Ma, R. He and F. Feng, "An Improved Compact Large-Signal GaN HEMT Model for Switch Application," in *IEEE Transactions on Electron Devices*, vol. 69, no. 6, pp. 3061-3067, June 2022.

[7] Wang ZX, Wang H, Zhang Y, Blaabjerg F, "A multi-port thermal coupling model for multi-chip power modules suitable for circuit simulators," *Microelectronics Reliability, vol. 88-90, pp. 519-523, 2018.*

[8] L. Nyssens, M. Rack, A. Halder, J.-P. Raskin and V. Kilchytska, "On the Separate Extraction of Self-Heating and Substrate Effects in FD-SOI MOSFET," in *IEEE Electron Device Letters*, vol. 42, no. 5, pp. 665-668, May 2021.

[9] C. Mukherjee, F. Marc, et al., " A physical and versatile aging compact model for hot carrier degradation in SiGe HBTs under dynamic operating conditions," Solid-State Electronics, vol. 163, pp. 107635, 2020.

Effect of Layer Thickness on the Transport Properties of ALD-deposited ZnO/In₂O₃ Heterojunction Thin-film Transistors

Zhenwei Li [1], Tiaoyang Li*[1]

*[1]Fuzhou University-Jinjiang Joint Institute of Microelectronics and School of Physics, Information Engineering and Microelectronics, Fuzhou University, Fuzhou 350108, China

*Email: 221120012@fzu.edu.cn, tyl@fzu.edu.cn

Abstract—Oxide semiconductors have great potential for three-dimensional memory applications due to the wide bandgap, low off-state current, and low thermal budget. This work systematically investigates the effect of layer thicknesses in ZnO/In₂O₃ heterojunctions fabricated via atomic layer deposition on the performance of thin-film transistors. Initially, the effects of deposition temperature on the electrical properties of single-component In₂O₃ and ZnO thin-film transistors are explored. It was found that at a deposition temperature of 250 °C, an optimal balance between threshold voltage and mobility is achieved. Based on this, heterojunction thin-film transistors with varying thickness ratios of ZnO/In₂O₃ (total thickness of 5 nm) were designed and compared. Electrical performance analysis indicates that the heterojunction thin-film transistor with a ZnO/In₂O₃ thickness ratio of 1.6 nm/3.4 nm exhibits the best overall device performance, with a threshold voltage (V_{th}) of 0.49 V, a field-effect mobility (μ_{FE}) of up to 44.7 cm²/V·s, a subthreshold swing as low as 90 mV/dec, and an on/off current ratio as high as 10^9.

Keywords—ZnO, In₂O₃, Heterojunction, TFTs

I. INTRODUCTION

Metal oxide semiconductors have garnered extensive research attention due to their excellent electrical conductivity and optical transparency. Thin-film transistors (TFTs) that utilize metal oxide semiconductors as the conductive channel layer are widely applied across various electronic fields. For instance, TFTs based on zinc oxide (ZnO) and indium oxide (In₂O₃) have achieved significant advancements[1]–[2]. Furthermore, in-depth studies of ternary and multi-element oxides, such as indium zinc oxide (IZO), indium gallium oxide (IGO), and indium gallium zinc oxide (IGZO), have led to the proposal of numerous methods to improve the electrical characteristics and stability of metal oxide semiconductor-based TFTs. Incorporating dopants can also enhance the performance of TFTs, including indium-doped ZnO, aluminum-doped ZnO, gadolinium-doped In₂O₃, and gallium-doped In₂O₃[3]–[4]. However, the disordered structural network and inherent defects of single oxide semiconductor materials severely impact the electrical performance of oxide thin-film transistors and limit their application in next-generation electronic devices. Consequently, the method of dual-active-layer composite channels has been proposed and proven effective by many literature sources.

Among the various methods for depositing thin films of oxide semiconductors, atomic layer deposition (ALD) offers a pathway for creating low-defect, high-quality semiconductor oxide channel layers. This technique allows for precise controllability of chemical composition and physical thickness and exceptional step coverage on nanoscale trench

structures[5]. Therefore, this article utilizes ALD technology to deposit thin films of oxide semiconductors. The optimal deposition temperature for the ZnO and In₂O₃ channel layers is determined with all other conditions being consistent. Based on this, a uniform thickness of 5nm is used to fabricate heterojunction thin-film transistors with varying thicknesses of ZnO/In₂O₃. By comparing the threshold voltage(V_{th}), mobility(μ_{FE}), subthreshold swing (SS), and $I_{on/off}$ ratio, the heterojunction oxide thin-film transistor with the optimal thickness ratio is identified[6]–[8].

II. EXPERIMENTS

This study selects low-resistance silicon wafers (with a resistivity of less than 0.005 Ω·cm) as the bottom gate. They are sequentially processed through RCA-1 cleaning solution (a mixture of ammonia, hydrogen peroxide, and water) and BOE (buffered oxide etch) to remove surface impurities. Subsequently, 20 nm of HfO₂ is deposited using ALD to create the gate insulator. To ensure the quality of the insulating layer, the wafer is promptly transferred to another ALD apparatus for the active layer to be deposited to the required thickness. The extra-channel regions are patterned using electron beam lithography (EBL) and ion beam etching (IBE). Then, 15/45 nm of Ti/Au is deposited as the source and drain electrodes

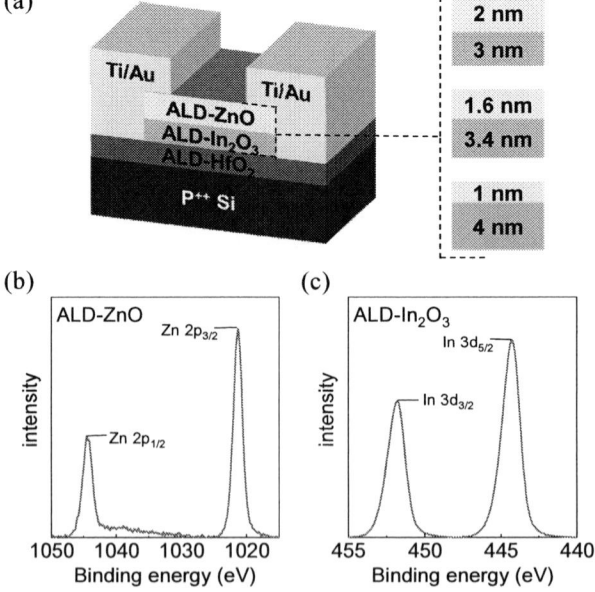

Fig. 1. (a) Schematic diagram of the complete device structure (b) (c) High-resolution XPS (X-ray Photoelectron Spectroscopy) elemental spectra for ZnO and In₂O₃

using electron beam evaporation (PVD). Finally, a post-deposition annealing process is carried out for 2 hours under a condition of 250 °C with an Ar: O_2 flow rate ratio of 80:20 sccm to complete the device fabrication. Fig.1 (a) illustrates the complete structure of the device.

Fig.1 (b)(c)presents the XPS (X-ray Photoelectron Spectroscopy) core-level spectra for ZnO and In_2O_3, respectively. The Zn $2p_{3/2}$ and $2p_{1/2}$ electron energy levels and the In $3d_{5/2}$ and $3d_{3/2}$ electron energy levels are depicted. However, the characteristic binding energy for Zn $2p_{3/2}$ is typically around 1021.6 eV, and for In $3d_{5/2}$, it is approximately 444.5 eV. The close match between the feature peaks in the figure and the typical binding energies indicates that the ALD-grown ZnO and In_2O_3 thin films exhibit a high purity level.

III. DISCUSSION

Deposition temperature plays a pivotal role in the fabrication process of thin films of oxide semiconductors, directly influencing the performance of thin-film transistor (TFT) devices. Before delving into the study of heterojunction devices, establishing the optimal deposition temperature is an essential step. This research successfully fabricated ZnO and In_2O_3 thin-film transistor devices by varying the deposition temperature from 200 °C to 300 °C. The ZnO film had a thickness of 10 nm with W/L = 40/8 μm. The In_2O_3 film had a thickness of 6 nm with W/L = 1.85/1 μm. Through device testing and data analysis, we plotted the transfer characteristic curves and 00mobility comparison graphs for these two types of thin-film transistors, as shown in Fig.2.For the ZnO thin-film transistor, experimental results indicated that the device exhibited optimal performance when the deposition temperature was set at 250 °C. At this temperature, the turn-on voltage (V_{on}) was very close to 0 V, and the field-effect mobility (μ_{FE}) of the device reached 17.8 cm²/V·s, which was particularly prominent in comparison.

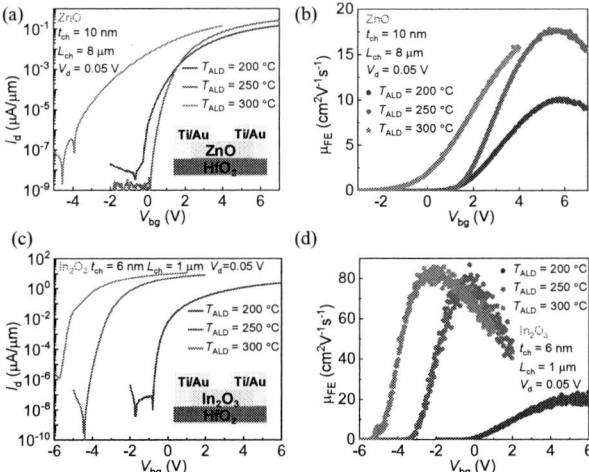

Fig. 2. (a) (b) respectively illustrate the transfer characteristic curves and mobility curves of the devices fabricated with ZnO thin films deposited at 200 °C-300 °C (W/L=40/8 μm) (c) (d) respectively illustrate the transfer characteristic curves and mobility curves of the devices fabricated with In_2O_3 thin films deposited at 200 °C-300 °C (W/L=1.85/1 μm)

For the In_2O_3 thin-film transistor, as the deposition temperature increased, a noticeable negative shift in the turn-on voltage (V_{on}) was observed, and the field-effect mobility also increased with the temperature rise. Comparative analysis also concluded that 250 °C is the optimal deposition

temperature for In_2O_3 thin films. Notably, ZnO and In_2O_3, these two different oxide semiconductor thin films, demonstrated excellent performance at the deposition temperature of 250 °C. This finding simplifies the experimental condition settings for subsequent research and provides a consistent foundation for the fabrication of heterojunction devices, facilitating further studies.

In the in-depth study of ZnO/In_2O_3 heterojunction thin-film transistors, we meticulously designed and fabricated a series of heterojunction thin-film transistors with different thickness ratios, including combinations of 2:3, 1.6:3.4, and 1:4, for a detailed performance comparison with single-component ZnO and In_2O_3 thin-film transistors. In this part of the study, a uniform 5 nm thick active layer channel was used, where the heterojunction devices were fabricated by continuously depositing two different oxide semiconductor thin films without breaking the vacuum. Taking the 2:3 ratio as an example, we first deposited 3 nm In_2O_3, followed by a 2 nm overlayer of ZnO, with the total thickness controlled at 5 nm. Through a consistent fabrication process, we completed the production of ZnO, In_2O_3, and heterojunction thin-film transistors with various ratios of ZnO/In_2O_3. After device testing, we obtained a comparative graph of the transfer characteristics, as shown in Fig.3 (a), which revealed that as the proportion of In_2O_3 thickness increased, the turn-on voltage (V_{on}) showed a significant negative shift.

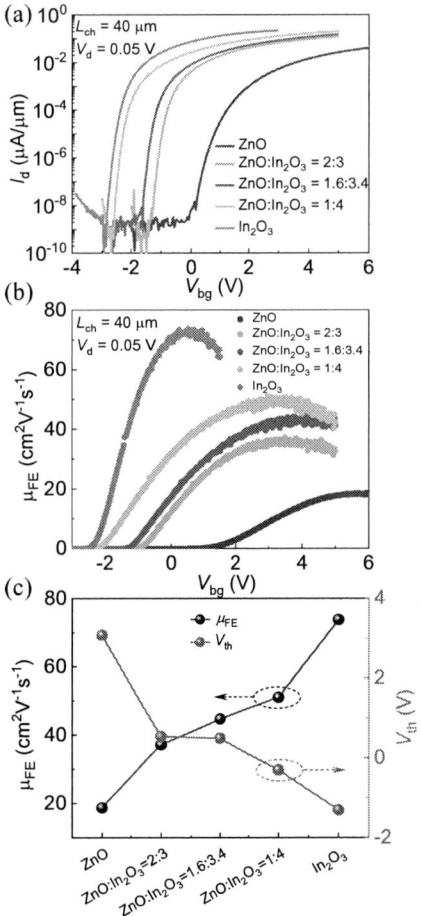

Fig. 3. (a) Comparative transfer characteristic curves of thin-film transistors with different active layers (b) Detailed comparison of field-effect mobilities (c)Plot of extracted threshold voltages and mobilities

We also analyzed the subthreshold swing and field-effect mobility. The method is as follows:

$$SS = \frac{\partial V_{gs}}{\partial (\log_{10} I_{ds})} \quad (1)$$

The drain-source current formula is given as:

$$I_{ds} = \mu_{FE} C_{ox} \frac{W}{L} \left[(V_{gs} - V_{th}) V_{ds} - \frac{1}{2} V_{ds}^2 \right] \quad (2)$$

Since the calculation of mobility uses $V_{ds} = 0.05V$, combined with the transconductance formula $g_m = \frac{\partial I_{ds}}{\partial V_{gs}}$, the mobility calculation formula can be simplified as follows:

$$\mu_{FE} = \frac{g_m L}{W C_{ox} V_{ds}} \quad (3)$$

W and L are the channel width and length, (and Cox) is the gate dielectric layer capacitance.

Data analysis shows the comparison of field-effect mobilities for various devices in Fig.3 (b), respectively. The analysis results indicate that as the thickness ratio of In_2O_3 in the heterojunction increases, the field-effect mobility correspondingly improves. By plotting the I_d-V_g curves, taking the position with the greatest slope on the linear scale of the ordinate (corresponding to the maximum transconductance), and drawing a tangent line, the intersection of this tangent with the abscissa represents the threshold voltage of each device. It can be observed that as the threshold voltage shifts negatively, the mobility rate correspondingly increases. A comprehensive comparison reveals that the performance of the ZnO/In_2O_3 heterojunction thin-film transistors is more akin to a combination of the individual performances of the two constituent single-material devices.

Table 1. Comparison of Electrical Performance for Thin-Film Transistors with Different Active Layers

Channel	V_{th} (V)	μ_{FE} ($cm^2V^{-1}s^{-1}$)	SS (mV/dec)	$I_{on/off}$
ZnO 5 nm	3.08	18.7	171	$>10^7$
ZnO:In_2O_3 2:3 nm	0.54	37.2	92	$>10^8$
ZnO:In_2O_3 1.6:3.4 nm	0.49	44.7	90	$\sim10^9$
ZnO:In_2O_3 1:4 nm	-0.3	51.0	92	$>10^8$
In_2O_3 5 nm	-1.31	73.8	88	$>10^8$

To facilitate comparison, some key electrical parameters are summarized in Table 1. Through comprehensive comparison, it is believed that the ZnO:In_2O_3=1.6:3.4 heterojunction device exhibits superior comprehensive performance. With a threshold voltage (V_{th}) of 0.49 V, greater than and closer to 0 V, it reduces power consumption. It has high mobility (44.7 cm^2/V·s), which means that electrons or holes move faster under the action of an electric field, thereby improving the switching speed of the TFT. It has a small subthreshold slope (90 mV/dec), indicating better-switching characteristics. It has a high $I_{on/off}$ current ratio (~10^9), indicating excellent switching control capability.

IV. SUMMARY

Utilizing ALD (Atomic Layer Deposition) technology for precise control of deposition conditions, high-quality ZnO and In_2O_3 thin films were successfully fabricated. Device fabrication was conducted over 200 °C to 300 °C, and comparative device testing data identified 250 °C as the optimal deposition temperature. Specifically, the ZnO thin-film transistor exhibited a turn-on voltage (V_{on}) close to 0 V, albeit with relatively low field-effect mobility of μ_{FE} = 17.8 cm^2/V·s. In contrast, the In_2O_3 thin-film transistor possessed a significantly higher mobility, approximately 80 cm^2/V·s. Building on these findings, heterojunction thin-film transistors with three different thickness ratios (2:3, 1.6:3.4, 1:4) were fabricated and compared against single-material thin-film transistor devices. The results indicated that the fabricated heterojunction devices combine the performance characteristics of both materials, demonstrating good overall performance. Detailed comparisons revealed that a thickness ratio of ZnO:In_2O_3 = 1.6:3.4 offered the best comprehensive device performance, providing valuable experimental evidence for further research and development of high-performance oxide thin-film transistors.

ACKNOWLEDGMENT

This work is funded by the National Natural Science Foundation of China (Grant No. 62204042), the Science and Technology Major Project of Fujian Province, China (Grant No. 2021HZ021027), and the Natural Science Foundation of Fujian Province, China(Grant No. 2021J05118).

REFERENCES

[1] M. Si, A. Charnas, Z. Lin, and P. D. Ye, "Enhancement-Mode Atomic-Layer-Deposited In $_2$ O $_3$ Transistors With Maximum Drain Current of 2.2 A/mm at Drain Voltage of 0.7 V by Low-Temperature Annealing and Stability in Hydrogen Environment," *IEEE Trans. Electron Devices*, vol. 68, no. 3, pp. 1075–1080, Mar. 2021.

[2] M. Wang, D. Zhan, X. Wang, Q. Hu, C. Gu, X. Li, and Y. Wu, "Performance Optimization of Atomic Layer Deposited ZnO Thin-Film Transistors by Vacuum Annealing," *IEEE Electron Device Lett.*, vol. 42, no. 5, pp. 716–719, May 2021.

[3] M. J. Kim, H. J. Park, S. Yoo, M. H. Cho, and J. K. Jeong, "Effect of Channel Thickness on Performance of Ultra-Thin Body IGZO Field-Effect Transistors," *IEEE Trans. Electron Devices*, vol. 69, no. 5, pp. 2409–2416, May 2022.

[4] T. Hong, Y. Kim, S. Choi, J. H. Lim, and J. Park, "Exploration of Chemical Composition of In–Ga–Zn–O System via PEALD Technique for Optimal Physical and Electrical Properties," *Adv. Electron. Mater.*, vol. 9, no. 4, p. 2201208, Apr. 2023.

[5] H. J. Seul, M. J. Kim, H. J. Yang, M. H. Cho, M. H. Cho, W.-B. Song, and J. K. Jeong, "Atomic Layer Deposition Process-Enabled Carrier Mobility Boosting in Field-Effect Transistors through a Nanoscale ZnO/IGO Heterojunction," *ACS Appl. Mater. Interfaces*, vol. 12, no. 30, pp. 33887–33898, Jul. 2020.

[6] P. Wen, C. Peng, Z. Chen, X. Ding, F.-H. Chen, G. Yan, L. Xu, D. Wang, X. Sun, L. Chen, J. Li, X. Li, and J. Zhang, "High mobility of IGO/IGZO double-channel thin-film transistors by atomic layer deposition," *Appl. Phys. Lett.*, vol. 124, no. 13, p. 133501, Mar. 2024.

[7] M. H. Cho, C. H. Choi, H. J. Seul, H. C. Cho, and J. K. Jeong, "Achieving a Low-Voltage, High-Mobility IGZO Transistor through an ALD-Derived Bilayer Channel and a Hafnia-Based Gate Dielectric Stack," *ACS Appl. Mater. Interfaces*, vol. 13, no. 14, pp. 16628–16640, Apr. 2021.

[8] W. S. AlGhamdi, et al. Anthopoulos, "Impact of layer thickness on the operating characteristics of In_2O_3/ZnO heterojunction thin-film transistors," *Appl. Phys. Lett.*, vol. 121, no. 23, p. 233503, Dec. 2022.

979-8-3503-6184-1/24 $31.00 © 2024 IEEE

Electrical and Thermal Analysis of CNT nTSV Applied to BS-PDN: A Modeling Study

Kai Ying[1], Baohui Xu[1], and Jie Liang[1*]

[1] School of Microelectronics, Shanghai University, Shanghai 201800, China

* Email: liangjieclair@shu.edu.cn

Abstract—**Carbon nanotube (CNT) interconnects, known for superior electrical and thermal properties, provided robust alternatives for interconnects. The proposed Backside Power Delivery Network (BS-PDN) addresses Place&Route issues, facilitating efficient power supply. In this paper, through RC-equivalent integration with Static Random Access Memory (SRAM) and BS-PDN, we analyzed the electrical properties of CNT nTSV (nanoscale Through-silicon via). Carbon-based circuits exhibit substantial power efficiency in BS-PDN, achieving over threefold power-delay product (PDP) optimization. COMSOL modeling reveals the thermal advantages of CNT nTSVs, emphasizing efficient heat dissipation.**

Keywords—CNT, Modeling, BS-PDN, nTSV

I. INTRODUCTION

The continuous advancements in the integrated circuit (IC) industry have significantly improved chip performance through size reduction, enabling more components to fit into smaller spaces. However, this miniaturization brings challenges like increased parasitic parameters, power consumption, and heat dissipation[1]. The industry is exploring solutions such as Carbon Nanotube (CNT) interconnects, which offer excellent transmission performance, thermal conductivity, and reliability[2], making them strong candidates to replace metal interconnects. Additionally, the introduction of backside power delivery networks (BS-PDN) has alleviated routing resource limitations by using the chip's backside space[3]. Despite these advancements, further research is needed to evaluate the potential of combining Carbon-based interconnects with BS-PDN to fully understand and develop Carbon-based circuits.

The diagram is shown in Fig. 1. In this paper, we use static random access memory (SRAM) cell to compare the transmission performance, power consumption, and other characteristics of metal and CNT nTSV (nanoscale Through-silicon via) in BS-PDN under different process conditions. We also deeply explored its heat dissipation, aiming to explore the characteristics and feasibility of Carbon-based circuits under BS-PDN and analyze their development trends.

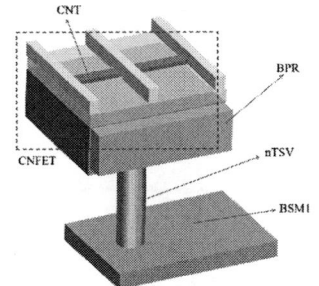

Fig. 1. Schematic of the combined application of CNFET and BS-BPR.

II. SIMULATION METHODOLOGY

As shown in Fig. 2, we establish the transmission line model to assess the basic transmission performance of CNT nTSV under different process conditions, and we also establish a 6T SRAM cell model based on PMOS for a thorough analysis of BS-PDN. The CNT TSV model data is derived from our previous work[4]. We focus on comparing the varying factors introduced by nTSV and transistors.

In this paper, we establish the SRAM cell model composed of P-type transistors rather than a regular CMOS structure. The reason is that, with the current CNT growth technology, achieving N-type end-bonded contacts for CNTs is relatively challenging compared to P-type contacts. This difficulty arises from the requirement for direct electron injection into the conduction band of CNTs at the N-type end. Metals with a low work function often tend to undergo oxidation rather than reacting with Carbon.

This study refers to the work of Lin et al.[5] to construct a 6T SRAM cell model based on P-type CNFET. Data comparisons between CNFETs and FinFETs utilize top-gate CNFETs with a CGP (contact gate pitch) of approximately 40nm, with experimental data referenced from[5, 6], and CGP of about 45nm for FinFETs (5nm process node), with data referenced from[7], Although back-gate structures can achieve CNFETs at smaller sizes (CGP =~30nm), considering the complexity of the fabrication process and the practical applicability to BS-PDN. Thus, this study still works on the top-gate structure. The structure diagram is shown in Fig. 1. Based on the work of Shulaker et al. [8], we assume that CNTs grow in a straight alignment. This article mainly uses single-walled CNTs instead of multi-walled CNTs because single-walled CNTs make it easier to achieve high-density growth and meet process requirements.

we have constructed a model of a chip module containing the BS-PDN with the BPR (Buried Power Rail) through COMSOL. A comparison is made using Ru and CNT as materials for nTSV, separately. The structure includes three frontside metal layers and two backside metal layers, and the metal layer material is Cu. The backside metal layers connect to the BPR through nTSV to provide a power supply. The nTSV is encapsulated with silicon dioxide, considered as a

TABLE I. A SUMMARY OF SIMULATION PARAMETERS FOR BS-PDN THROUGH COMSOL

Parameter	Value	Description
K_{Cu}	380 W/(m·k)	Thermal Conductivity of Copper
K_{Ru}	120 W/(m·k)	Thermal Conductivity of Ruthenium
K_{CNT}	700 W/(m·k)	Thermal Conductivity of CNT
$C_{p\,Cu}$	385 J/(kg·K)	Heat Capacity of Copper
$C_{p\,Ru}$	238 J/(kg·K)	Heat Capacity of Ruthenium
$C_{p\,CNT}$	730 J/(kg·K)	Heat Capacity of CNT

979-8-3503-6184-1/24 $31.00 © 2024 IEEE

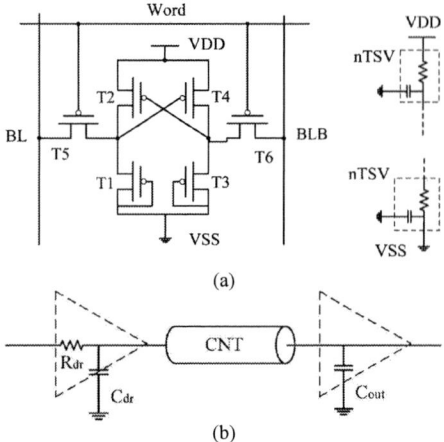

(a)

(b)

Fig. 2. (a) Combined with BS-DPN, a basic SRAM cell unit composed of 6 P-type transistors. (b) RC transmission model.

barrier layer. This structure refers to the work of Chen et al. [9], where the MOS layer is equivalent to a heat source for simulation, with a power density set at 10^8 W/m³. The first backside metal layer (BSM1) is set to connect to VSS, and the second backside metal layer (BSM2) is connected to VDD. A heat dissipation power is set on the outer surface of BSM2 to simulate natural convection boundary heat dissipation, with a power density of approximately 500 W/m². The supply voltage is set at 0.1 mV. The electrical conductivity of Ru is taken as 5×10^6 S/m. All other boundaries are maintained independently for electrical and thermal considerations. The starting temperature point for calculations is set at 293.15 K. Other parameters are summarized in Table 1[10].

III. RESULTS AND DISCUSSION

A. RC Transmission Model for Process Conditions

Using the RC transmission line model combined with Elmore delay, we calculated the delay performance of CNT nTSV under different process conditions and compared it with Cu, as shown in Fig. 3. The relevant parameters are listed in Table 2[11].

In Fig. 3, it can be observed that the electrical performance of CNTs within the common temperature range for IC (273 K - 400 K) is less affected by temperature variations compared to Copper. Furthermore, under conditions of high-density growth, the adverse effects resulting from temperature fluctuations are nearly negligible, showcasing exceptional electrothermal characteristics. This can be attributed to a paradoxical mechanism for CNTs, where elevated temperatures intensify scattering while concurrently increasing the number of conducting channels. This

distinctive trait becomes more pronounced under high-density conditions.

In Fig. 3(a), a comparative analysis is conducted for all-metallic CNTs. When CNTs exhibit full metallic properties with a density of 10^{13} tubes/cm², CNT nTSV demonstrates superior delay performance compared to Cu nTSV. Even with a contact resistance as high as 10 kΩ, there is no observable severe negative impact. Conversely, when the density of all-metallic CNTs decreases to 10^{12} tubes/cm², the transmission delay degrades by 6.52% compared to Cu nTSV, and the impact of contact resistance is more significant. High-density CNTs increase conducting channels, improving transmission capacity and mitigating contact resistance effects. However, the benefits of CNTs' long mean free path at the nanoscale are underutilized, making their electrical advantages less pronounced. In Fig. 3(b), a comparison is made for randomly chiral CNTs (approximately 1/3 metallic). Relative to Fig. 3(a), the overall delay increases. Under high-density conditions, it demonstrates a comparable transmission capability to Cu, while under low-density conditions, the overall performance is at its worst corner, with a degradation in delay of approximately 18.11% compared to Cu, indicating a substantial difference in transmission speed.

According to the state of art, conditions with a density of 10^{13} tubes/cm² can be fully realized, and contact resistance can be achieved at 10 kΩ, with theoretical potential are reaching 3.25 kΩ. Consequently, worst-case conditions entirely attainable, and best-case conditions are potentially feasible. Founded on this premise, the utilization of CNTs as nTSV proves entirely viable.

B. SRAM Cell Model for Electrical Properties

Based on the 6T SRAM in Fig. 2, we conduct in-depth comparative verification using nTSVs of different materials and various transistor combinations (FinFET & Cu nTSV, FinFET & CNT nTSV, CNFET & Cu nTSV, and CNFET & CNT nTSV). FinFET & Cu nTSV outperform CNFET in transmission delay due to CNFET's slower switching speed and lower on-current. Compared to FinFET & Cu nTSV, FinFET & CNT nTSV shows a 2.4% (read) and 5.1% (write) additional delay, increased voltage drop by 5.6 times, and no power advantage, making CNT interconnects less favorable. For CNFET & CNT nTSV, Carbon-Carbon contact reduces contact resistance. CNFET, while slower than FinFET, greatly reduces power consumption by nearly 10 times for read and 3 times for write operations, supporting low-power design trends.

Furthermore, compared to Cu nTSV, CNT nTSV shows better performance in both power consumption and transmission delay, improving by approximately 0.15% (read),

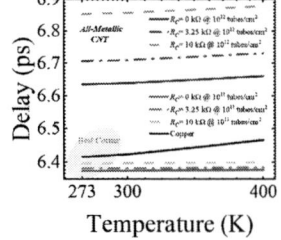

(a)

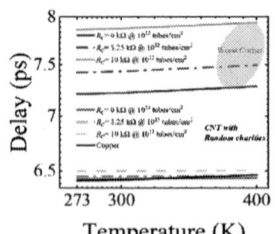

(b)

Fig. 3. (a) All-metallic CNTs and (b) random chiral CNTs under different contact resistances (0 kΩ, 3.25 kΩ, and 10 kΩ) and different densities (10^{12} tubes/cm² and 10^{13} tubes/cm²). Delay changes with the temperature and Copper is added as a reference.

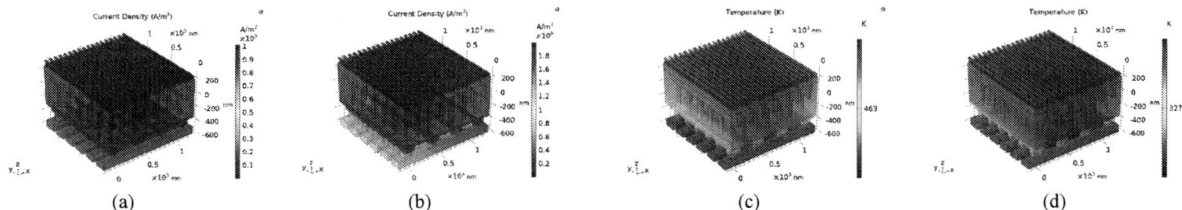

Fig. 4. A subset of models for chip modules based on BS-PDN is presented. (a) Current density and (c) temperature profiles are provided for the case where Ruthenium is used as the material for nTSV. For the scenario with CNTs as the material for TSVs, the (b) current density and (d) the temperature profile is displayed.

0.19% (write), 0.4% (read), and 1.6% (write). Carbon-Carbon contacts effectively reduce contact resistance, and Carbon nanomaterials offer higher mobility and smaller parasitic capacitance, demonstrating the multifaceted advantages of CNT nTSV in Carbon-based circuits.

C. Thermal Properties

The results of the comparative analysis in Fig. 4(a) and (c) indicate that Ru has better current transmission performance, even up to approximately ten times higher. However, both models achieved a current density of 10^8 A/m², fully meeting the requirements for common usage. In the analysis of heat dissipation shown in Fig. 4(b) and (d), the temperature of Ru nTSV can reach 426 K, while the temperature of CNT nTSV is only 327 K. The temperature difference of 99 K emphasizes the thermal advantages of CNT in the field of heat transfer. It can be observed that using CNT as nTSV makes the heat distribution relatively more uniform compared to Ru.

IV. CONCLUSION

In this paper, the study demonstrated that the combination of CNFET and CNT nTSV has a greater resistivity at the

TABLE II.   A SUMMARY OF SIMULATION PARAMETERS OF nTSV RC TRANSMISSION MODEL

Parameter	Value	Description
R_{dr}	1.522 kΩ	Driver Resistance
C_{dr}	1.12 fF	Driver Capacitance
C_{load}	4.48 fF	Load Capacitance
D_{TSV}	100 nm	TSV cross-sectional side length
D_{CNT}	2.8 nm	Diameter of CNTs
L_{TSV}	300 nm	Length of TSV
t_{ox}	10 nm	thickness of oxide layer
ε_{ox}	3.9 (SiO$_2$)	Dielectric constant of oxide layer
VDD	0.7 V	Supply Voltage

TABLE III.   A SUMMARY OF SIMULATION RESULTS

Parameter	FinFET&Cu nTSV	FinFET&CNT nTSV
Power (read) (μW)	7.733	7.735
Delay (read) (ns)	0.442	0.453
PDP (read) (μW·ns)	3.15	3.50
Power (write) (μW)	11.58	11.63
Delay (write) (ns)	0.780	0.831
PDP (write) (μW·ns)	9.03	9.67
IR-Drop	1.0 ‰	5.6‰
VDD (V)	0.5	0.5

nanoscale than metal-based interconnect, but its capacitance advantage makes it significantly improved on IR-Drop and PDP, and the power consumption can be optimized up to 3 to 10 times. The low power consumption advantage of Carbon-based circuits has also been fully verified in SRAM cells. We also used COMSOL to build a chip model including BS-PDN for thermal analysis. The results show that compared with Ru nTSV, CNT nTSV improves heat dissipation performance by about 23% with a decrease in transmission performance. In general, the low power consumption and heat dissipation advantages of Carbon-based circuits are still sacrificed at the expense of some transmission performance.

ACKNOWLEDGMENT

This work was supported in part by the National Natural Science Foundation of China under Grant 62104138.

REFERENCES

[1] Gall, D., et al., Materials for interconnects. MRS Bulletin, 2021. 46(10): p. 959-966.

[2] White, C.T. and T.N. Todorov, Carbon nanotubes as long ballistic conductors. Nature, 1998. 393(6682): p. 240-242.

[3] Chen, R., et al. Design and Optimization of SRAM Macro and Logic Using Backside Interconnects at 2nm node. in 2021 IEEE International Electron Devices Meeting (IEDM). 2021. IEEE.

[4] Xu, B., et al., A Modeling Study on Electrical and Thermal Behavior of CNT TSV for Multilayer Structure. IEEE Transactions on Electron Devices, 2023. 70(9): p. 4779-4785.

[5] Lin, Y., et al., Scaling aligned carbon nanotube transistors to a sub-10 nm node. Nature Electronics, 2023. 6: p. 1-10.

[6] Hills, G., et al., Understanding Energy Efficiency Benefits of Carbon Nanotube Field-Effect Transistors for Digital VLSI. IEEE Transactions on Nanotechnology, 2018. 17(6): p. 1259-1269.

[7] Vashishtha, V. and L.T. Clark, ASAP5: A predictive PDK for the 5 nm node. Microelectronics Journal, 2022. 126: p. 105481.

[8] Shulaker, M.M., et al. Efficient metallic carbon nanotube removal for highly-scaled technologies. in 2015 IEEE International Electron Devices Meeting (IEDM). 2015. IEEE.

[9] Chen, R., et al. Power, Performance, Area and Thermal Analysis of 2D and 3D ICs at A14 Node Designed with Back-side Power Delivery Network. in 2022 IEEE International Electron Devices Meeting (IEDM). 2022. IEEE.

[10] Kumanek, B. and D. Janas, Thermal conductivity of carbon nanotube networks: A review. Journal of materials science, 2019. 54(10): p. 7397-7427.

[11] Cheng, Z.-H., et al., Investigation of Copper–Carbon Nanotube Composites as Global VLSI Interconnects. IEEE Transactions on Nanotechnology, 2017. 16(6): p. 891-900.

979-8-3503-6184-1/24 $31.00 © 2024 IEEE

A Unified Current-Voltage Compact Model for Organic Light-Emitting Diode

Wenbin Wang[1], Mingyu Ma[1], Wangjun Yang[1], Jianghao Ma[1], Hailong You[1] and Cong Li*[1]

[1] School of Microelectronic, Xidian University, Xi'an, China

* Email: 22111213908@stu.xidian.edu.cn, licong@xidian.edu.cn

Abstract—**In this paper, we present the development of a unified OLED compact model capable of accurately describing devices with diverse structures and varying organic layer thicknesses. We apply a unification treatment to smoothly integrate the segmented models to enhance their versatility and applicability in circuit simulation. To validate the effectiveness of our proposed model, we conduct simulations and experiments on both two-layer and three-layer OLED devices using Silvaco TCAD. Additionally, a commercially available OLED device is tested to verify our model. The model fits well with both simulation data and experimental measurements. This provides great convenience for OLED-related circuit simulation, especially greatly improving the model's convergence in circuit simulation.**

Keywords—*Compact model, organic light emitting diodes(OLED), Current-voltage model, TCAD simulation*

I. INTRODUCTION

OLEDs have excellent qualities such as low power consumption and high color contrast, so they have been widely incorporated into various commercial displays. OLED devices exhibit diverse structures to handle multiple use cases. Since the first OLED with a simple monolayer structure, more layers have been used with specialized functions such as the hole injecting layer, hole transporting layer, hole blocking layer, emitting layer, and electron transporting layer. OLEDs exhibit varying current-voltage (I-V)characteristics based on their structural configurations. Factors such as the selection of organic materials and the integration of functional layers significantly influence these characteristics. With the widespread application of display circuits, the diverse IV characteristics displayed by different OLED structures pose challenges for circuit simulation[1]. Consequently, it is imperative to establish a compact model for OLEDs to simulate the characteristics of OLEDs in circuits accurately, and this model should be suitable for multiple structures of OLED devices[2]. Furthermore, considering the convergence during circuit simulation, this model needs to possess continuity, smoothness, and robustness[3].

Initially, people used the model of the p-n junction diode to replace the OLED model. However, this method is no longer applicable with the increased accuracy requirement. Some researchers have modeled OLEDs for a specific structure, but there is no general model for multiple structures. So far, the most authoritative segmented analytical model for OLEDs, derived by Zhang et al. [4] based on numerical simulations of OLED governing equations and band diagram theory, however, it lacks continuity and falls short of meeting the requirements for a compact model in circuit simulation. A good compact model needs to have such qualities as physical property, accuracy, continuity, smoothness, scalability, and robustness. Therefore, the development of a global compact model with such good qualities that can be applicable to different OLED structures is necessary[5].

In this paper, we established a unified OLED compact model applicable to various structures and organic layer thicknesses. Firstly, the model is established in different regions according to its current-voltage characteristics. Next, a unification treatment is applied to the segmented models to imbue them with smoothness, facilitating their improved application in circuit simulation. Following this, validation of the model is conducted using TCAD for both two-layer and three-layer OLED devices. Initially, validation of the electric field strength model is performed for different thicknesses, yielding satisfactory fitting results. Subsequently, simulations are conducted for different thicknesses of two-layer and three-layer OLED models to obtain current-voltage characteristic simulation data. These data are then compared with our OLED model to verify their accuracy, which is found to be satisfactory. Finally, fitting of experimental data for multi-layer OLEDs with our model shows good agreement between the model and experimental data.

II. COMPACT MODEL DEVELOPMENT

The current-voltage characteristics of OLEDs, as voltage increases, can be divided into low injection and high injection regions due to differences in current formation mechanisms. In this section, each of these regions will be modeled separately

A. Compact Modeling of Low-injection Region

As voltage increases, when $V_{pn}=V_{bi}$, a flat-band condition is reached. OLED device enters the low injection region, within which the current of the device sharply increases with voltage. However, according to research, the current within this region is still primarily diffusion current. The analytical model for the low injection region of the OLED is given by the diffusion equation as follow:

$$\frac{\partial^2 p_e}{\partial x^2} - \frac{p_e}{D_p \tau_p} = 0 \tag{1}$$

Where D_p is the hole diffusion coefficient. The hole lifetime τ_p is calculated with the Langevin recombination rate.

The analytical model of the low-injection region is obtained by solving the diffusion equation.

$$J_{LI} = J_{li0}[\frac{q(V_{pn} - V_{bi})}{nKT}] \tag{2}$$

To enable it to quickly adapt to different structures of OLED devices, flat-band voltage V_{bi} is split in the low-injection region to incorporate more process parameters into the model.

$$V_{bi} = \frac{KT}{e} \ln(\frac{NA \cdot ND}{NI^2}) = vt \cdot \ln(\frac{NA \cdot ND}{NI^2}) \quad (3)$$

The low-injection region model is as follows:

$$J_{LI} = JLI0 \cdot \exp(\frac{Vpn}{nvt})(\frac{n \cdot NI^2}{NA \cdot ND}) \quad (4)$$

Where NA and ND are the donor and acceptor concentrations, respectively, and $JLI0$ is a current density depending on the HTL and ETL density of states and their molecular level offset, which are important design parameters of multiple-layer OLEDs can be obtained from the process library. The advantage of constructing a compact model in this manner is that, while ensuring accuracy and physicality, for any OLED device, only two important parameters need to be obtained from the process library: the doping concentration and the low injection current density. Then the model can be accurately applied to circuit simulation.

B. Compact Modeling of High-injection Region

With further voltage increase, a large amount of charge accumulates at the interface between the ETL and HTL of the OLED, resulting in the formation of a large and confined electric field around it. At this point, carrier transport in the high injection region is dominated by drift, and the carrier mobility is described by the classic field-dependent Poole-Frenkel equation as follows:

$$\mu = \mu_0 \exp(\gamma\sqrt{E_x}) \quad (5)$$

The drift current formula is used to construct the current formula of the high-injection region of OLED. Considering the influence of the electric field brought by the Poole-Frenkel model on mobility, the relevant parameters of organic layer thickness are constructed in the model, so that the model has organic layer thickness dependence. So, we get the compact model in the high-injection region as follows:

$$J_{HI} = q\mu NE$$
$$= q \cdot N \cdot U0 \cdot (V_{pn} - V_{th}) \cdot \exp(\frac{\beta}{vt}\frac{\sqrt{V_{pn} - V_{th}}}{\sqrt{D_{organic}}}) \quad (6)$$

Where $\beta = \sqrt{\frac{e}{\pi\varepsilon\varepsilon_0}}$ and $U0$ represents the zero-field mobility. In this form, once the device's thickness and relevant process parameters are known, the model can be quickly applied to circuit simulation without the need for multiple parameter fittings.

III. UNIFIED MODELING

To facilitate better application and convergence of the compact model in simulations, ensuring its continuity and smoothness is imperative. Given that the luminescent operation of OLED devices is concentrated in the high injection region, the transition from the low injection region to the high injection region during the device's switching process is crucial. Hence, achieving a smooth transition from the low-injection region to high-injection region in the model is of paramount importance. Therefore, we use an innovative form to unify the segmented models.

Firstly, we present the final expression for the current density of the OLED device as Eq. (7).

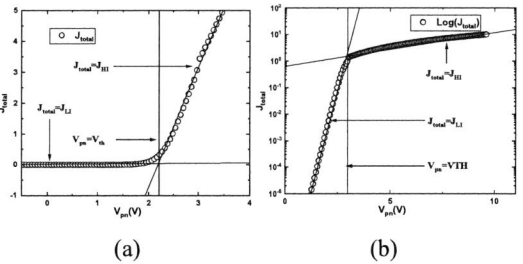

(a)   (b)

Fig. 1. In both (a) linear and (b) semilog coordinates, J_{total} transitions smoothly from J_{LI} to J_{HI} at V_{th}.

$$J_{total} = \frac{\ln(1 + \exp(G \cdot (V_{pn} - V_{th}) \cdot A))}{1 + \frac{1}{J} \cdot \exp(B \cdot G \cdot (V_{pn} - VTH) \cdot A - \frac{Vpn}{n \cdot vt})}$$
$$G = q \cdot N \cdot U0$$
$$J = JLI0 \cdot \left(\frac{n \cdot NI^2}{NA \cdot ND}\right) \quad (7)$$
$$A = \exp\left(\frac{\beta}{vt}\frac{\sqrt{(V_{pn} - V_{th})^C}}{\sqrt{D_{organic} \cdot LN}}\right)$$
$$B = 1 - \frac{2}{1 + \exp(K \cdot (V_{th} - V_{pn}))}$$
$$C = 1 + \frac{1}{1 + \exp(K \cdot (V_{pn} - V_{th} - \alpha))}$$

Where G and J are the two coefficients obtained from the segmented model. A is a coefficient derived from the Poole-Frenkel equation. B is a constructed function. The value of this function is -1 when $V_{pn} < V_{th}$, and smoothly transitions from -1 to 1 as V_{pn} increases. This function ensures the accuracy and smoothness of the transition of J_{total} from the low-injection region to the high-injection region as V_{pn} changes. Where K is a parameter representing the transition speed, with a default value of 100. C is also a function constructed to ensure that V_{pn} can take values smaller than V_{th}. Where α is a parameter constructed for smoothness with a default value of 0.01. And the parameter LN represents the number of organic layers.

As shown in Fig.1, the model maintains smooth transitions both in linear and semi-logarithmic coordinates. This not only enhances the model's accuracy but also ensures its convergence in circuit simulation applications.

IV. MODEL VALIDATION

In this section, we first calibrate the simulation data according to the experimental measured data. Then, a double-layer and a triple-layer OLED device are constructed by Silvaco TCAD, and the models are verified by adjusting the thickness of the different layers.

The device structure is shown in Fig.2. The calibration results of simulation and experimental data are shown in the Fig.3.

The double-layer OLED adopts a conventional structure comprising electrodes and two organic layers: an electron transport layer (ETL) and a hole transport layer (HTL). The cathode material is aluminum, and the ETL material is Polyfluorene (PFO). The HTL consists of polystyrene sulfonic acid or poly(3,4-ethylene dioxythiophene) (PEDOT: PSS), which is a mixture composed of two polymers. Indium

979-8-3503-6184-1/24 $31.00 © 2024 IEEE

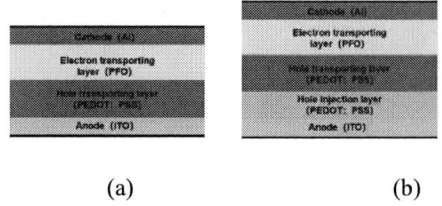

(a)   (b)

Fig. 2. (a) Two and (b) three layers OLED device structure.

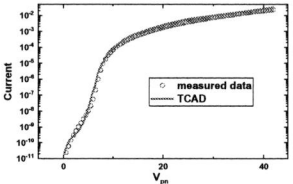

Fig. 3. TCAD simulation and experimental data calibration.

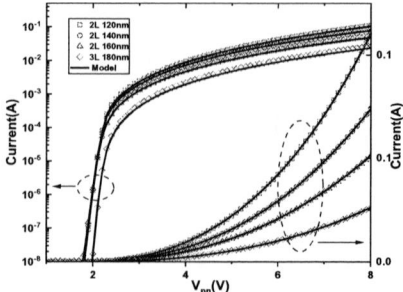

Fig. 5. The fitting of the model to the simulation results of two-layer(2L) and three-layer(3L) OLED devices with different thicknesses.

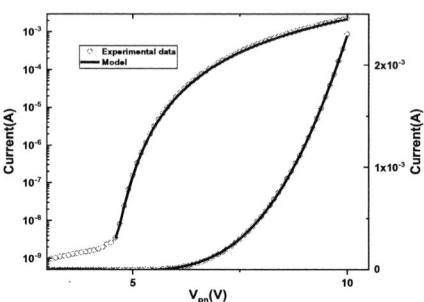

Fig. 6. The fitting results of model and experimental data.

tin oxide (ITO) is used as the anode material due to the requirement of transparency for the anode material.

Three-layer OLED is based on double-layer OLED to add a hole injection layer (HIL) between the anode and HTL layer, the material is still PEDOT: PSS, but by adjusting the proportion of the mixture to adjust its work function to the appropriate value.

According to TCAD simulation results as Fig.4, under the same thickness of the organic layer, the current in the 3-layer OLED is notably lower than that in the bilayer OLED. This difference arises because the use of a multilayer structure makes it more difficult for charge carriers to inject into the interface, resulting in a much lower charge concentration at the interface of the 3-layer OLED compared to the bilayer OLED.

Next, by varying different parameters, TCAD simulation data for bilayer and 3-layer OLEDs were obtained. Then, the model was implemented in Verilog-A for circuit simulation. Through comparison, the model exhibited good fitting with the TCAD data.

As can be seen from Fig.5, by changing the process parameters of the device, the model can be well-fitted with the simulation data of different structures and different sizes of OLED devices.

Next, we measured the current-voltage characteristics of an OLED device. We compared the experimental data with the model by varying process-related parameters such as doping concentration and thickness of organic layers in the model. We found that the model exhibited a good fit with the experimental data as shown in Fig.6.

V. CONCLUSION

A unified OLED current-voltage compact model is proposed in this paper. Based on the derivation of energy band theory, the model parameters have physical significance. The model is smooth in the whole area and can cope with the simulation of the display circuit well. The model is fitted with simulation and experimental data with good fitting accuracy.

ACKNOWLEDGMENT

Project Supported by the National Key Research and Development Program of China： Research on Industrial Analog Chip Designs and Process Compatibility and Standards for Reliability Technologies (grant number： 2022YFF0605800), and partly supported by the Science and Technology Development Program of Shaanxi (Grant No.2023-YBGY-273)

REFERENCES

[1] Güney, Arda, et al. "Experimental and modeling studies of automotive-qualified OLEDs under electrical stress." Microelectronics Reliability 111 (2020): 113704.

[2] S.-G. Lee, H.-S. Choi, C.-W. Han, S.-J. Lee, Y.-H. Tak, and B.-C.Ahn, "Numerical modeling; thickness dependence of jv characteristic for multi-layered oled device," IEICE transactions on electronics, vol. 95, no. 11, pp. 1756–1760, 2012.

[3] Cheng, Yuhua, et al. "BSIM3v3 manual." University of California, Berkeley (1995): 6-1.

[4] L. Zhang, L. Wang, W.-J. Wu, and M. Chan, "Modeling current–voltage characteristics of bilayer organic light-emitting diodes," IEEE Transactions on Electron Devices, vol. 66, no. 1, pp. 139–145, 2018.

[5] Negi, Shubham, Poornima Mittal, and Brijesh Kumar. "Analytical modeling and parameters extraction of multilayered OLED." IET Circuits, Devices & Systems 13.8 (2019): 1255-1261.

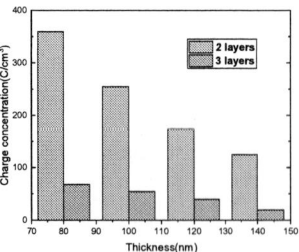

Fig. 4. Charge concentration at the interface of double-layer OLED and three-layer OLED with different thicknesses.

Threshold Voltage and Mobility Extraction of Negative Bias Temperature Instability in 22nm FD SOI MOSFETs

Yibo Hu [1], Hao Ge[1], Zhipeng Ren[1], Yizhe Yin[1], Ruiyong Zhao[1], Jing Chen*[1]

[1] State Key Laboratory of Materials for Integrated Circuits, Shanghai Institute of Microsystem and Information Technology, Chinese Academy of Sciences, 865 Changning Road, Shanghai, 200050, China

* Email: ybhu@mail.sim.ac.cn, jchen@mail.sim.ac.cn

Abstract—In this study, we report on the performance degradation of 22nm FDSOI CMOS devices due to negative bias temperature instability (NBTI). NBTI-induced oxide charge shifts the IV curves, while stress induced interface trapped charge degrades the subthreshold swing (SS) and mobility, particularly in low electric field regions as confirmed by C-V characteristics. These effects may lead to underestimation of threshold voltage changes during NBTI characterization. We propose a novel threshold voltage and mobility extraction strategy that accurately predicts output current degradation over stress time due to NBTI.

Keywords — negative bias temperature instability (NBTI), fully depleted silicon on insulator (FDSOI), parameter extraction

I. Introduction

Bias Temperature Instability (BTI) is a critical reliability concern in CMOS processes, leading to performance degradation over time. Although BTI affects both p- and n-MOSFETs, it is more pronounced in p-MOSFETs, known as Negative BTI (NBTI). Since its first study in 1966, NBTI has continually threatened the reliability of integrated circuits[1-3], especially with the adoption of high-k metal gate (HKMG) technology in sub-28nm CMOS processes [4].

Previous research has demonstrated that NBTI-induced degradation depends on gate stress voltages, defect traps, and chemical reactions [5]. However, accurate NBTI assessment remains challenging due to the absence of a comprehensive characterization procedure. NBTI stress significantly affects transconductance, output current, and leakage current, posing substantial concerns for aging measurements.

On-The-Fly (OTF) is a traditional technique for evaluating device performance degradation by holding the stress gate voltage and monitoring the output current during the stress period. While OTF can quickly measure output current, it does not account for mobility degradation, leading to incomplete results [6]. Analog circuits are particularly sensitive to threshold voltage and mobility, so underestimation can cause unexpected effects and reduce circuit lifespan. Therefore, incorporating mobility correction is crucial for accurate circuit simulation.

This work investigates the NBTI degradation of p-MOSFETs in a 22nm FDSOI process using the IV sweep method. Our findings indicate that complex degradation in the 22nm FDSOI process can lead to errors in threshold voltage extraction. We propose a new threshold voltage and mobility extraction strategy to enhance the accuracy of NBTI degradation simulation.

II. Experimental Setup

In this experiment, devices were fabricated using the HLMC 22nm HKMG FD-SOI process with dimensions of 3 μm width and 0.02 μm length. NBTI measurements involves stressing the device with a high gate voltage Vgstr (Vs=Vd=Vb=0V) and measuring degradation over time. The Id vs. Vg curve was measured at Vd = -0.05 V, Vs = 0V, and Vb = 0V. All tests were conducted at room temperature.

III. Results and Discussions

A. NBTI degradation

Figure 1(a) illustrates the drain current characteristics for p-MOSFETs as a function of gate voltage in the linear regime under NBTI stress voltage Vstr = -2.1V. The NBTI stress not only shifts the threshold voltage (Vt) but also decreases mobility. This mobility degradation results in noticeable output current degradation at low gate voltages.

Figure 1(b) shows the transconductance (Gm) curves of the device after 100, 1000, and 2000 seconds of stress. Both mobility and maximum transconductance decreases with stress time. The threshold voltage degrades from 0.32V to 0.405V after -2.1V NBTI stress for 2000 seconds, and Gm_max decreases by 10.4%, indicating that NBTI significantly deteriorates AC performance. Therefore, both threshold voltage and mobility degradations need to be accurately measured and characterized for precise NBTI analysis.

Figure 2 illustrates the effective field dependence of mobility extracted using split-CV. Before NBTI stress, mobility decreases with increasing Vg, indicating that surface roughness scattering primarily limits mobility in fresh devices. After NBTI stress, significant mobility degradation is observed at low Vg due to enhanced Coulombic scattering, while high electric field mobility remains consistent with fresh devices. This indicates that NBTI-induced interface charges in FDSOI predominantly enhance low electric field scattering, differing from BULK devices where degradation is mainly due to increased surface scattering [7].

Although the physical origin of BTI degradation is still debated, it is generally attributed to interface and oxide defects [8,9]. Figure 3 shows simulation results considering only oxide traps (Not) induced Vth shift versus measured data at 0s and 2000s. The subthreshold swing (SS) degrades by approximately 6.7mV/dec after 2000s NBTI stress. The measured drain current is lower than the simulation results, which confirms the need for a new simulation strategy.

B. Extraction of V_th, SS and Mobility

The threshold voltage, the most critical parameter in modeling MOSFETs, degrades if the interface density is presumed uniform within the bandgap, which is given by:

This work is supported in part by the Shanghai "Qi Ming Xing" Program - Sail Plan under Grant NO. 22YF1456300. In part by the Department of Science and Technology of Guangdong Province under Grant NO. 2021B0101280002.

979-8-3503-6184-1/24 $31.00 © 2024 IEEE

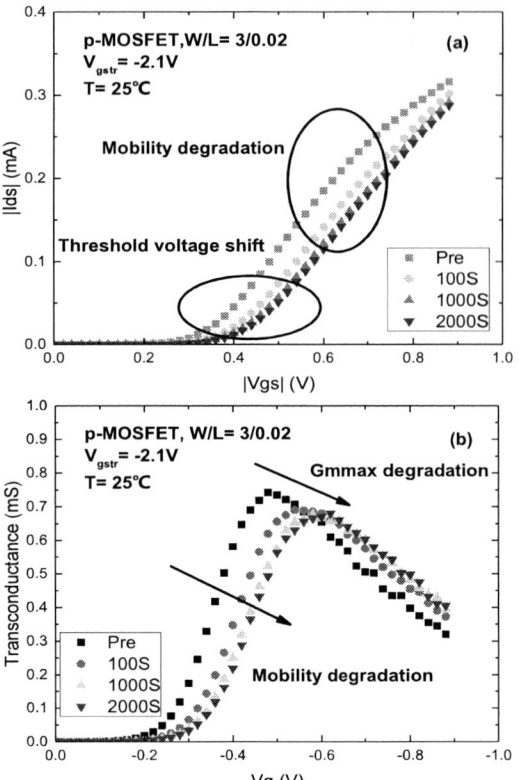

Fig. 1. Transfer characteristic curve Id-Vg (a) and transconductance curve (b) of fresh and degraded p-MOSFETs at linear regime after 0s, 100s, 1000s and 2000s NBTI stress time.

$$\Delta V_{th} = \frac{q \cdot (\Delta N_{it} + \Delta N_{ot})}{C_{ox}} \quad (1)$$

SS is relatively insensitive to trapping in oxide defects. Therefore, we could split the interface traps contribution by SS degradation. The interface trap states are extracted by: [10]

$$SS = 2.3 \cdot \frac{kT}{q} \cdot \left(1 + \frac{C_{dep}}{C_{ox}} + \frac{C_{it}}{C_{ox}}\right) \quad (2)$$

where, C_{dep}, C_{it} are the depletion layer capacitances and interface charge capacitances respectively. By introducing a ratio value $k = \Delta N_{ot}/\Delta N_{it}$, which is constant with aging, ΔV_{th} can be modeled as:

$$\Delta V_{th} = \frac{q \cdot (1 + k)\Delta N_{it}}{C_{ox}} \quad (3)$$

The stress voltage dependent ΔN_{it} degradation can be further modeled as:

$$\Delta N_{it} = A \cdot \exp(B \cdot V_{gstr}) \cdot t^n \quad (4)$$

where the A is the fitting parameter, B is the coefficient related to the stress voltage, t is the stress time and the n is the power law exponents. By linearly fitting the experimental data, the same exponent n = 0.15 is observed for all devices stressed at different gate bias. Therefore, the stress voltage does not affect the relationship indicated in *Eq.* (3), which confirms the applicability of this method in a wide voltage range.

All characterization so far relies on accurate threshold voltage extraction. Therefore, it is necessary to compare the differences among various extraction methods. Figure 4 shows threshold voltage shifts extracted by constant current [11], ELR, and OTF approximation [12] methods over stress time under identical conditions. Table I predicts device lifetimes based on a power-law distribution, with a 100 mV

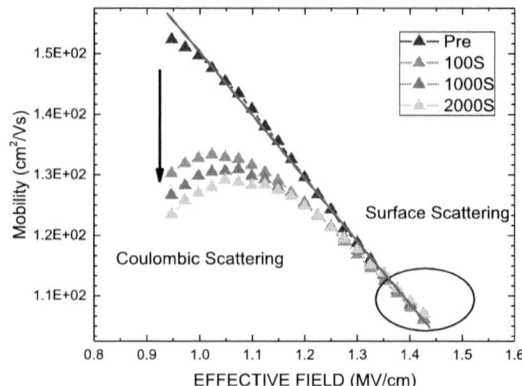

Fig. 2. Split-CV mobility as a function of stress time. After NBTI stress, the FDSOI p-MOSFET mobility is decreased at low electric field.

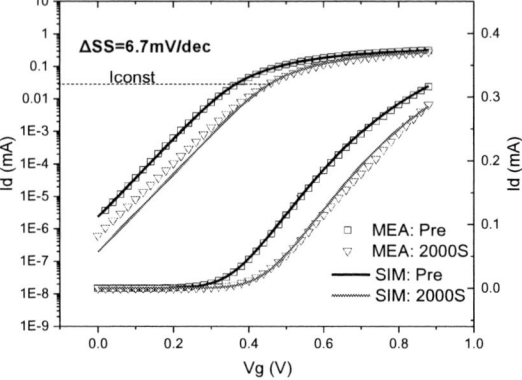

Fig. 3. The measured transfer characteristic curve at 0s and 2000s NBTI stress time. The solid line represents the simulation results only considering the threshold voltage shift.

threshold voltage offset as the failure criterion. Lifetimes under DC stress are 63,000s (OTF), 7,200s (ELR), and 11,000s (constant current). The OTF method, affected by mobility degradation, notably overestimates lifetime by more than 8 times compared to ELR. Thus, accurately characterizing subthreshold swing (SS) and mobility degradation is crucial for precise lifetime estimation and safe operation conditions in NBTI studies.

The effective mobility is another crucial parameter. The linear drain current can be modeled using a well-known empirical mobility model, represented by the following equation [13]:

$$I_{ds} = \frac{W}{L} \cdot \frac{\mu}{[1 + \theta(V_G - V_T)]} \cdot C_{ox} \cdot V_{ds} \cdot (V_G - V_T) \quad (5)$$

where W and L denote the device width and length, μ represents effective mobility, and θ is a constant related to surface scattering. However, this model's precision diminishes when mobility is influenced by Coulombic scattering in the low-field region, a primary degradation area in NBTI. Referring to the method in Ref. 26, Coulombic scattering is modeled as:

$$\mu = \mu_0/[1 + \alpha \cdot Q_{it}/(\beta + Q_{inv})] \quad (6)$$

where μ_0 is the low electric field mobility, α is the interface charge contribution constant, and β is a fitting parameter related to substrate doping concentration. Since surface scattering has minimal impact on NBTI-induced mobility

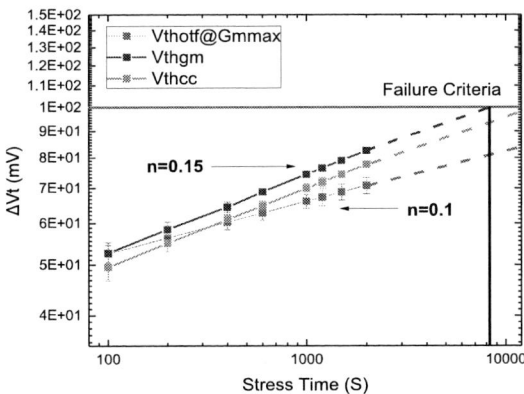

Fig. 4. BTI induced threshold voltage drifts as a function of strees time using OTF approximation (V_{thotf}), ELR (V_{thGm}) and constant current (V_{thcc}) extraction methods. The solid line represents the failure criteria of 100mV threshold voltage shift.

TABLE I
EXTRAPOLATION LIFETIME

Extraction Method	n	Lifetime @100mV (sec)
OTF@Gmmax	0.1	63000
ELR	0.15	7200
Constant Current	0.15	11000

degradation, the OTF method can be used to examine the threshold voltage shift by:

$$\Delta V_{th,hf} = |V_{gs,hf} - V_{th0}|/I_{ds,hf0} \cdot \Delta I_{ds,hf} \qquad (7)$$

where $\Delta V_{th,hf}$ is the extracted threshold voltage shift in high field after stress, $V_{gs,hf}$ is the gate bias voltage in high field, $I_{ds,hf0}$ is the drain current before stress and $\Delta I_{ds,hf}$ is the delta value of high field drain current between pre and post situation. Furthermore, the OTF measurement at a low electric field close to Gm_max determines the mobility degradation due to NBTI stress. We take the first order approximation of transconductance to calculate the mobility using:

$$\frac{\Delta \mu}{\mu} = 1 - \frac{I_{ds,lf}}{\left[I_{ds,lf0}/|V_{gs,lf} - V_{th0}| \cdot \Delta V_{th,hf} + I_{ds,lf0}\right]} \qquad (8)$$

where $I_{ds,lf0}$ and $I_{ds,lf}$ are the low field drain current before and after NBTI stress respectively.

Figure 5 illustrates mobility degradation versus gate bias conditions, showing a strong correlation with voltage bias as expected. Simulation results align closely with experimental data. Figure 6 presents NBTI stress simulation outcomes integrating threshold voltage shift, subthreshold swing, and mobility degradation modeling. Interface trap effects on threshold voltage and mobility degradation are well characterized, with simulation results exhibiting high fitting accuracy with experimental data.

IV. CONCLUSIONS

This study investigates NBTI effects in 22nm FDSOI p-MOSFETs using IV measurements. NBTI stress leads to increased subthreshold swing and significant mobility degradation under low electric fields due to enhanced Coulombic scattering. Our results indicate that both the OTF approximation and constant current method result in significant errors in threshold voltage extraction, underestimating NBTI degradation. We present a novel method for extracting threshold voltage and mobility changes

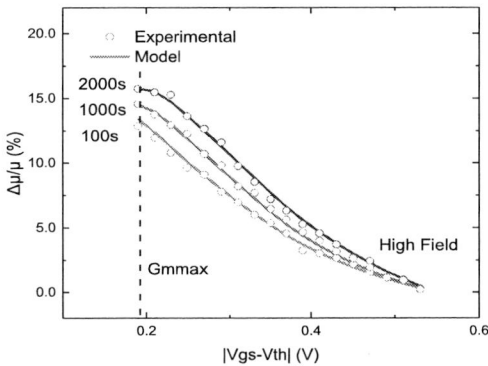

Fig. 5. Estimation of extracted mobility degradation using *Eq.* (6). The simulation results with coulombic scattering contributions are in good agreement with the measurement data.

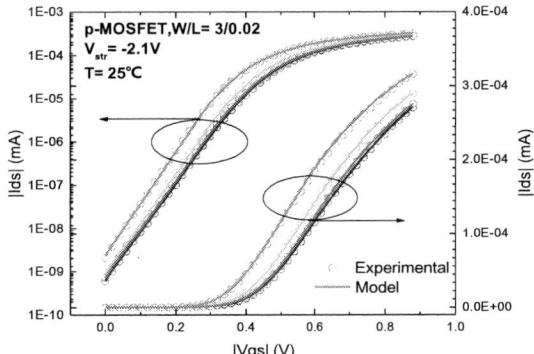

Fig. 6. Measured and simulated linear characteristics of p-MOSFETs under NBTI stress.

from NBTI degradation measurements, which shows excellent agreement with experimental data and is beneficial for device NBTI modeling and circuit aging simulations.

REFERENCES

[1] Miura, Yoshio, and Yasuo Matukura. *Japanese Journal of Applied Physics* 5.2 (1966): 180.

[2] De, Suchismita, Suchismita Tewari, and Abhijit Biswas. *Microsystem Technologies* 26 (2020): 1173-1178.

[3] Lee, Kyong Taek, et al. *2013 IEEE International Reliability Physics Symposium (IRPS)*. IEEE, 2013.

[4] Schwarzenbach, Walter, et al. *Solid-State Electronics* 117 (2016): 2-9.

[5] Liu, Q., et al. *2011 Symposium on VLSI Technology-Digest of Technical Papers*. IEEE, 2011.

[6] Maheta, Vrajesh D., et al. *IEEE transactions on electron devices* 55.7 (2008): 1630-1638.

[7] Islam, Ahmad Ehteshamul, et al. *2008 IEEE International Reliability Physics Symposium*. IEEE, 2008.

[8] Islam, Ahmad Ehteshamul, et al. *IEEE Transactions on Electron Devices* 54.9 (2007): 2143-2154.

[9] Grasser, Tibor, et al. *2009 IEEE international reliability physics symposium*. IEEE, 2009.

[10] Lun, Z., D. S. Ang, and C. H. Ling. *IEEE Electron Device Letters* 21.8 (2000): 411-413.

[11] Djezzar, Boualem, et al. *Microelectronics Reliability* 110 (2020): 113703.

[12] Kerber, Andreas, et al. *IEEE Transactions on Electron Devices* 55.11 (2008): 3175-3183.

[13] Krishnan, Anand T., et al. *IEEE international electron devices meeting 2003*. IEEE, 2003.

Modeling of Silicon Single-Photon Avalanche Diodes for Process and Design Optimization

Jing Fu [1,2], Anran Guo *[2], Hongbo Zhang [2], Guowei Li [2], Huaping Ma [2], Ruizhi Li [2], Yuwei Chen [2]

[1] National Key Laboratory of Integrated Circuits and Microsystems, Chongqing 40000
[2] Department of Solid-State Image Sensor, CETC No.44 Research Institute, Chongqing 40000

* Email: jingfu17690@163.com, guoar@cetccq.com.cn

Abstract—**Single Photon Avalanche Diode (SPAD) has been widely used in high-precision space missions due to extremely high sensitivity and dynamic range. However, the further optimization of its performance is limited by the cost of process technology. In this work, we develop and improve the Technology CAD (TCAD) simulations routine of custom SPAD, and the proposed model is calibrated under wide bias voltage range and temperature conditions. Then, the influences on the performances of various process modifications are explored. This work shows that the model can be used for the accurate design and optimization of SPAD detectors, and the proposed process modifications can effectively improve the detection efficiency and noise performance of devices.**

Keywords—*Single Photon Avalanche Diode (SPAD), TCAD simulations, Avalanche Triggering Probability, Dark Count Rate, Photon Detection Efficiency*

I. INTRODUCTION

Silicon-based optoelectronic devices have been widely used in numerous space missions with low cost, high integration, large bandwidth, high speed and high anti-interference ability[1-2]. As a solid state photo-detector, SPAD has higher sensitivity and dynamic range than traditional Photomultiplier Tube (PMT), which can realize photon detection and counting under extremely low illumination environment, with low readout noise and power consumption. Therefore, it is specially applied in more and more challenging applications such as planetary altimeters, satellite laser time transmission, and deep space detection [3-5].

SPAD usually operates in Geiger mode, the photo-generated carriers are accelerated by the extremely high electric field and multiplied by collision ionization. Therefore, the signals several orders lower than noise can be detected. However，the Dark Count Rate (DCR) is also easy to trigger due to tunneling effect under this electric field profile, which resulting in wrong counting. A compromise design is needed for the contradiction between the high Photon Detection Efficiency (PDE) and low DCR. Considering the further optimization of device performance is limited by the cost of process technology, recently, many physically based models are proposed to calculate the breakdown voltage, PDE and temporal response of SPAD [6-7]. TCAD models for CMOS SPAD pixels are also established to simulate the basic output characteristics [8-9]. But few models have been calibrated, which makes it impossible to effectively study the optimization methods of SPADs.

Our work aims to propose effective optimization methods for custom SPAD by TCAD simulations. First, the device structure and parameter models are established, and calibrated based on experimental results under wide bias voltages and temperature conditions, which covering IV characteristics, Breakdown voltage, DCR, and PDE. Then, the effects of various process options on the performances of the SPAD are analyzed based on the established models. These results show that the models presented in this paper can effectively guide the performance optimization design of the device, at the same time, the modeling method can also be used as a reference for other SPAD with different processes and structures.

II. METHOD

The SPAD is fabricated using a custom technologies, based on a reach-through structure and a diameter of 100 μm. An avalanche multiplication region is formed by the N+/P-Well junction, a guard ring formed by N-Well implants around the photosensitive area is added to avoid the breakdown, which is caused by large potential gradient at the edge of the device[10]. The parameters characterized in this paper include IV curve, Breakdown voltage, DCR, and PDE, the test methods are based on [10-11], an 850nm pulsed laser is used as the controlled light source, and the output of the SPAD is monitored with a universal counter (SR400).

TCAD is a numerical simulation tool based on semiconductor physics, which can accurately describe the internal behaviors of the device [12]. The generation of DCR mainly due to tunnel released carriers. McIntyre model and Schenk model are enabled to calculate the avalanche breakdown probability P_{tr} and tunneling rate [13], then DCR can be obtained by:

$$DCR = \int_0^{W_{dep}} P_{tr}(x) \times G_{B2B}(x)\, dx \qquad (1)$$

Where W_{dep} is the depletion region width, and the avalanche trigger probability depends on the local electric field profile within the depletion region. Since the SPAD used in this work adopts P-doped avalanche region, the avalanche probability at position (x) after a single electron enters the multiplication region is simulated by the TCAD impact ionization model.

In SPAD, PDE depends on the photon absorption probability (P_{ab}) and the avalanche breakdown probabilities (P_{tr}) within the silicon [14]. For the simulation of PDE, the incident position of the photon needs to be considered. the probability for a photon absorbed in the multiplying region is equal to one, and the delay distribution is depended on the transit time of the carriers. And for the carriers generated in the neutral regions, the probability and delay distribution are calculated by solving drift-diffusion equations [15].

979-8-3503-6184-1/24 $31.00 © 2024 IEEE

III. RESULTS

A. Pixel model

In SPAD pixel model, the precise doping and electric field profile are achieved based on layout and the PDK information, as shown in Fig.1. A reverse bias voltage （>V_{bd}） is applied to the PN junction, leading to high electric field in the depletion region. Once the electric field reach about 10^5 V/cm, the number of photo-generated carriers increase rapidly, an avalanche event follows. However, the DCR also increases significantly under the same electric field without incident photons.

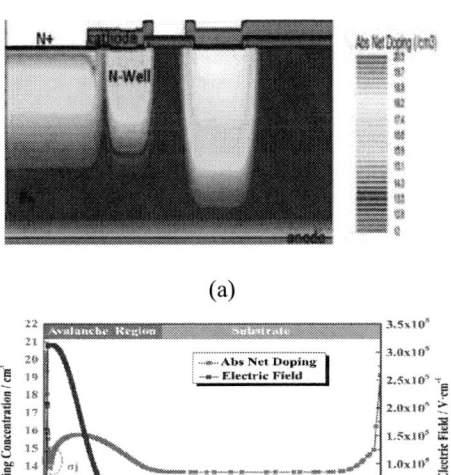

(a)

(b)

Fig.1 (a) Schematic of SPAD pixel (b) Doping concentration and electric field profile.

B. Breakdown Voltage

Based on impact ionization model and the global temperature model, the SPAD I-V characteristics and its dependence on bias voltage and temperature are simulated. As show in Fig. 2, the current increases with the bias voltage, when the applied voltage is about 110V, an extremely high electric field is formed in the multiplication region. The electron-hole pairs generated enter the multiplication region and are accelerated by the electric field, a cascade collision occurs, resulting in an avalanche current ranging from 10^{-11}A to 10^{-4}A (at T=293K). In addition, the device is affected by the temperature, and the IV curve shifts to the upper right as the temperature increases, the dark current changes from 10^{-13}A to 10^{-11}A.

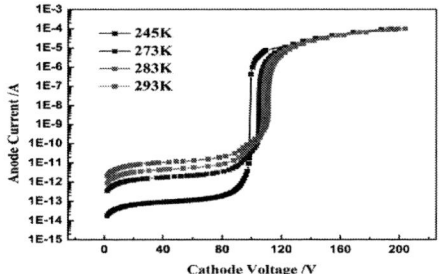

Fig. 2 Dark current varies with bias voltage under different temperatures.

Fig.3 displays the simulated Breakdown Voltage under various temperatures extracted from the IV curves. The Breakdown Voltage increases with the temperature, and the simulation and experimental results show the same growth trend. The slope d_{Vbd}/d_T represents the process of avalanche breakdown due to collision ionization, the linear fits are performed at temperatures from 245K to 293K. In test results, the $d_{Vbd}/d_T \approx 488.2$mV/K, and in simulation results, $d_{Vbd}/d_T \approx 420.8$mV/K, with an error less than 20%.

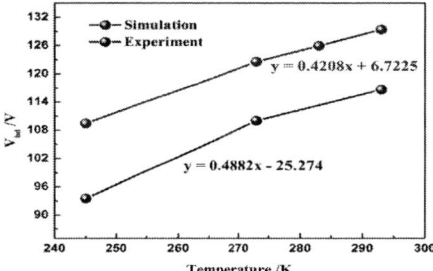

Fig.3 Comparison of experimental and simulation results on Breakdown Voltage varies with temperature.

C. Dark Count Rate

In order to simulate the DCR in SPAD, Geiger model and B_2B model are used respectively to calculate the avalanche breakdown probability (P_{tr}) and the generation rate. The generation rate is mainly distributed near the top of the PN junction, and the electron breakdown probability presents a vertical distribution due to the pull-through structure, as shown in Fig.4.

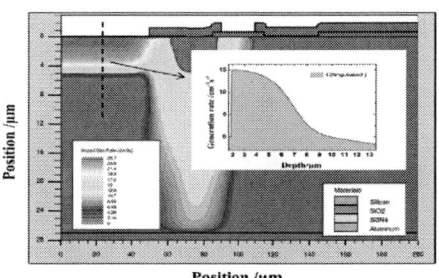

(a)

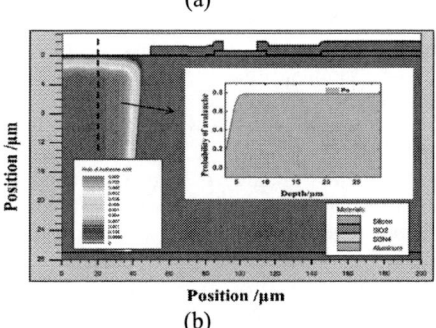

(b)

Fig.4 The simulation results of distribution for (a) generation rate and (b) breakdown probability.

Further, the integral of P_{tr} and G_{B2B} in the depletion region is calculated, and the DCR values under different bias voltages are obtained in Fig.5(a). As expected, by increasing the excess bias, the electric field across the multiplication region increases and the impact ionization coefficient of the electrons increases. Therefore, the breakdown probability of electron in the depletion region increases, resulting in higher DCR. In order to calibrate the model, the SPADs were tested under different bias voltages. Then, the measured and simulated results of DCR with different physical models are shown in Fig.5(b). It can be found that the DCR contribution is dominated by B_2B tunneling effects.

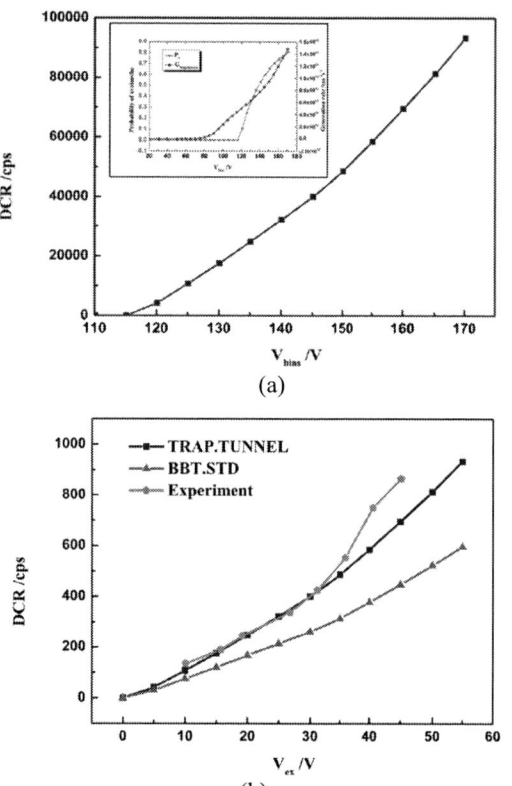

Fig.5 (a) Simulation result of DCR varies with bias voltage (insert: P_{tr} and G_{B2B} vary with bias voltage）, (b) Measured and simulated DCR as a function of excess voltage.

IV. CONCLUSION

Based on the layout and process parameters of custom SPAD, we simulate the behavior of key performance parameters under wide bias voltage range and different temperature conditions. According to the simulation results, the signal collection and carrier multiplication processes are realized, and the breakdown voltage, Dark Count Rate (DCR) models are obtained. By electrical calibration, a good agreement between model and experimental data is achieved. Therefore, the model can be reliably used for accurate design and optimization of SPAD-based detectors. This work also reveals the dominant process that contributes to DCR is the tunneling effect for SPADs operating at lower bias voltage.

ACKNOWLEDGMENT

This work are supported by the National Key Research and Development program of China (No.2022YFF0708000) and the National Natural Science Foundation of China (No.U2167208).

REFERENCES

[1] Shenjie Huang and Majid Safari. Hybrid spad/pd receiver for reliable free-space optical communication. IEEE Open Journal of the Communications Society, 1:1364–1373, 2020.

[2] Min Qian, Yi Zhang, Xiaojun Mao, Yang Gao, Xiaoyang Xuan, Min Wu, Yueping Niu, and Shangqing Gong. Flexible photoelectronic material device and investigation method for space applications. Progress in Aerospace Sciences, 139:100901, 2023.

[3] G. S. Buller and R. J. Collins. Single-photon generation and detection. Measurement Science Technology, 21(1):12002– 12028, 2009.

[4] Yuki Maruyama, Jordana Blacksberg, and Edoardo Charbon. A 1024× 8 700ps time-gated spad line sensor for laser raman spectroscopy and libs in space and rover-based planetary exploration. In 2013 IEEE International Solid-State Circuits Conference Digest of Technical Papers, pages 110–111. IEEE, 2013.

[5] Danilo Bronzi, Federica Villa, Simone Tisa, Alberto Tosi, and Franco Zappa. Spad figures of merit for photon-counting, photontiming, and imaging applications: a review. IEEE Sensors Journal, 16(1):3–12, 2015.

[6] Yue Xu, Ping Xiang, Xiaopeng Xie, and Yang Huang. A new modeling and simulation method for important statistical performance prediction of single photon avalanche diode detectors. Semiconductor Science and Technology, 31(6):065024, 2016.

[7] Angelo Gulinatti, Ivan Rech, Silvia Fumagalli, Mattia Assanelli, Massimo Ghioni, and Sergio D Cova. Modeling photon detection effciency and temporal response of single photon avalanche diodes. In Photon Counting Applications, Quantum Optics, and Quantum Information Transfer and Processing II, volume 7355, pages 161–177. SPIE, 2009.

[8] T Chaves De Albuquerque, D Issartel, R Clerc, P Pittet, R Cellier, and F Calmon. Lowering the dark count rate of spad implemented in cmos fdsoi technology. In 2019 Joint International EUROSOI Workshop and International Conference on Ultimate Integration on Silicon (EUROSOIULIS), pages 1–4. IEEE, 2019.

[9] Qian X, Jiang W, Elsharabasy A, Deen MJ. Modeling for Single-Photon Avalanche Diodes: State-of-the-Art and Research Challenges. Sensors. 2023.

[10] Anran Guo, Huaping Ma, Ruizhi Li, Yuwei Chen, Jing Fu, Guowei Li, and Renfang Lei. Fabrication and characterization of silicon spads with doping compensated avalanche region. Nuclear Instruments and Methods in Physics Research Section A: Accelerators, Spectrometers, Detectors and Associated Equipment, 1063:169274, 2024.

[11] Tsikouras Anthony, Peronio Pietro, Rech Ivan, Hirmiz Nehad, M. Deen, and Fang Qiyin. Characterization of spad array for multifocal high-content screening applications. Photonics, 3(4):56, 2016.

[12] Chinmay K Maiti. Introducing Technology Computer-Aided Design (TCAD): Fundamentals, Simulations, and Applications. Jenny Stanford Publishing, 2017.

[13] Aymeric Panglosse, Philippe Martin-Gonthier, Olivier Marcelot, Cedric Virmontois, and Pierre Magnan. Dark count rate modeling in single-photon avalanche diodes for space lidar applications. In 2019 17th IEEE International New Circuits and Systems Conference (NEWCAS), 2019.

[14] Hiwa Mahmoudi, S Saman Kohneh Poushi, Bernhard Steindl, Michael Hofbauer, and Horst Zimmermann. Optical and electrical characterization and modeling of photon detection probability in cmos single-photon avalanche diodes. IEEE Sensors Journal, 21(6):7572–7580, 2021.

[15] Angelo Gulinatti, Ivan Rech, Mattia Assanelli, Massimo Ghioni, and Sergio D. Cova. Design-oriented simulation of the photon detection effciency and temporal response of single photon avalanche diodes. In LEOS Annual Meeting, 2009.

A Novel Modeling Method for BV Characteristics of ESD Protection Devices

Ke Zhang [1,2], Yang Wang [3], Xiangliang Jin *[1,2]

[1] School of Physics and Electronics, Hunan Normal University, Changsha, China, 410081
[2] Key Laboratory of Physics and Devices in Post-Moore Era, College of Hunan Province, Changsha, China, 410081
[3] School of Integrated Circuits, Peking University, Beijing, China, 100091

* Email: Ke Zhang@zhangke20000804@163.com, Xiangliang Jin@jinxl@hunnu.edu.cn

Abstract—**Breakdown voltage (BV) is a crucial indicator used to characterize whether an electrostatic discharge (ESD) protection device can be compatible with the protected circuit. In this paper, a novel modeling method for BV characteristics of ESD protection devices is proposed, and the breakdown voltages of SCR and MSCR are accurately characterized based on 0.18μm standard CMOS process.**

Keywords—Electrostatic discharge, Breakdown voltage, Silicon-controlled rectifier, Monte Carlo

I. INTRODUCTION

Trigger voltage, holding voltage, failure current and breakdown voltage are several important performance parameters of ESD devices, which are closely related to the ESD design window of the protected circuit [1-2]. Recently, a number of researchers have employed various methods to model ESD devices with different structures. Li et al. proposed a comprehensive diode model based on several compact elements [3]; Shen et al. accurately modeled the snapback characteristic of GGNMOS by designing a voltage-controlled current source (VCCS) [4]; Cao et al. proposed a compact SCR model based on advanced BJT models and standard SPICE components [5]; Since the modeling of snapback characteristic is a significant and challenging task, most researchers have focused their interest on modeling the snapback characteristics of ESD devices. On the other hand, the BV characteristic is also a crucial indicator that used to characterize the transparency of ESD device [6]. If there is leakage drift caused by the ESD devices during the normal operation of the protected circuit, which may lead to unexpected noise signal. Therefore, it is significant and necessary to model the BV characteristics of ESD devices.

Fig. 1 illustrates the schematic diagram of the proposed novel modeling method, where the direct current (DC) analysis is first performed using Technology Computer-Aided Design (TCAD) device-level simulation, and the distributions of internal electrical parameters under different operating states could be obtained, thus facilitating the fundamental operation principle of ESD devices. By combining electric field extracted from TCAD with Monte Carlo numerical simulation, the BV characteristics of the ESD devices could be simulated. Since the electric field parameters are extracted from the snapback characteristics obtained by current injection in TCAD DC analysis, which is more in line with the actual internal electrical parameters of the ESD device. Furthermore, since the Monte Carlo numerical method predicts the probability of scattering types of particles on the breakdown surface of the device at the microscopic level, there is no convergence issue in the calculation of the

This work is supported by the National Natural Science Foundation of China (Grant No. 62174052 and Grant No.62304007).

avalanche multiplication factor M. Thus, the novel modeling method proposed in this paper has high accuracy and good convergence.

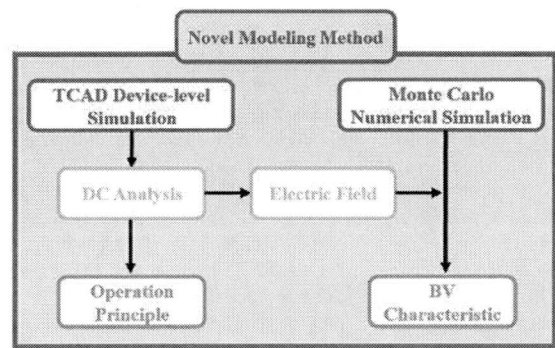

Fig. 1 The schematic diagram of the proposed novel modeling method.

II. STRUCTURE AND PRINCIPLE OF SCRs

Benefit from low parasitic capacitance, high robustness and strong discharge ability, silicon-controlled rectifier (SCR) and modified silicon-controlled rectifier (MSCR) have been widely used for on-chip ESD protection. Fig. 2(a) and Fig. 2(b) respectively show the cross sections of SCR and MSCR. In comparison to SCR, MSCR has a lower trigger voltage due to the N+ heavily doped region across the N-Well and the P-Well. However, their operating principles are basically the same. During the normal operation of the protected circuit, SCR remains off-state. Once the anode of SCR suffers from ESD stress, avalanche breakdown occurs in the reverse PN junction consisting of N-well and P-well. The voltage drops across the parasitic resistors of P-Well or N-Well generated by the avalanche current further turn on the SCR path consisting of two parasitic BJT. The parasitic NPN transistor consists of the anode N+/N-Well/P-Well/cathode N+, while the parasitic PNP transistor comprises the anode P+/N-Well/P-Well/cathode P+. When SCR is fully turn on, it primarily releases the ESD current through the SCR path formed by the parasitic NPN and PNP transistors, thereby protecting the core circuit from electrostatic damage.

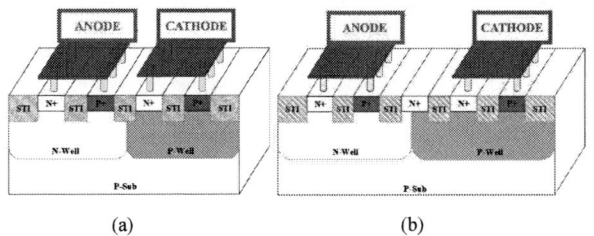

Fig. 2 The cross sections of ESD devices. (a) SCR. (b) MSCR.

III. SIMULATION AND VERIFICATION

A. TCAD Device-level Simulation

In order to extract the electric field parameters of the trigger faces and validate the fundamental operating principles of the devices, the DC analysis of the SCR and MSCR is firstly performed using Silvaco TCAD two-dimensional device simulation, and the electrical parameters distributions of the devices under different operating states are obtained. The physical models used are integrated model (cvt), Shockley-Read-Hall composite model (srh), Auger composite model (auger), band narrowing model (bgn), lattice temperature model (lat.temp), Impact ionization model (impact selb). Fig. 3(a) and Fig. 3(b) respectively show the electric field distributions when the SCR and MSCR operate at the initial turn-on state. As can be seen from the figures, when the devices are initially turned on, the electric fields of the SCR and MSCR are mainly concentrated at the PN junction composed of N-Well/P-Well and N+/P-Well, which also indicates that the conduction of the devices relies on the avalanche breakdown occurring in the reverse-biased PN junction. Fig. 4(a) and 4(b) show the current density distributions during the initial conduction of the SCR and MSCR, respectively. It's apparent that a parasitic PNP path forms within the interior of the SCR from the anode P+ to the cathode P+, while a parasitic NPN path forms within the MSCR from the anode N+ to the cathode N+. At this point, the current density inside the devices is relatively low, and the devices mainly release part of the ESD current through the parasitic BJT paths. Fig. 5(a) and Fig. 5(b) respectively show the current density distributions when the SCR and MSCR are fully turned on. It is evident from the figures that an SCR path consisting of two parasitic BJT transistors is formed inside both devices. From the magnitude and distribution range of current density, it can be seen that the devices mainly release ESD current through the SCR paths when they are fully turned on, which is consistent with the preceding analysis.

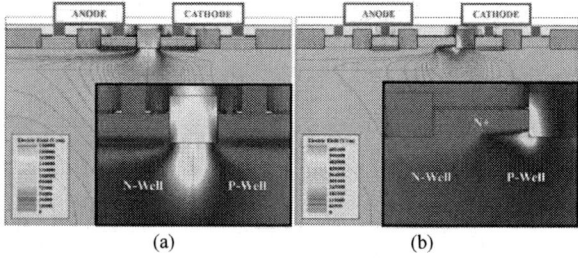

Fig. 3 The electric field distributions when devices are initially turned on. (a)SCR. (b)MSCR.

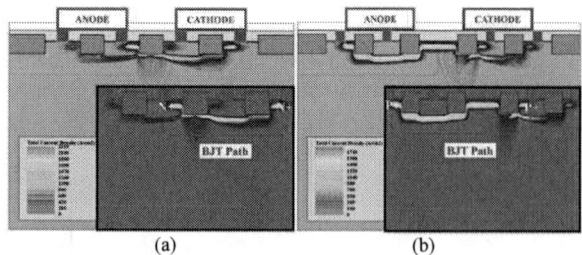

Fig. 4 The current distributions when devices are initially turned on. (a)SCR. (b)MSCR.

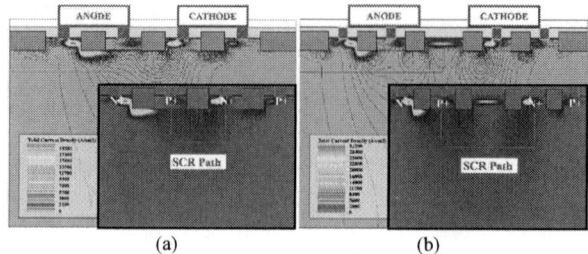

Fig. 5 The current distributions when devices are fully turned on. (a)SCR. (b)MSCR.

B. Monte Carlo Numerical Simulation

The application of Monte Carlo method in the semiconductor field is extensive and mature. Zhou et al. proposed a simple Monte Carlo model to simulate the avalanche process for p+-i-n+ diode, n+-i-p+ diode and p+n diode [7]. Plimmer et al. proposed a simple Monte Carlo model to simulate the avalanche multiplication process including the deadspace effects [8]. In order to further investigate the BV characteristics of SCR and MSCR, Monte Carlo numerical method is employed to model the avalanche breakdown process of ESD devices at the level of microscopic particle motion. The average electric field in the depletion region of the breakdown surface is a key parameter when performing Monte Carlo numerical simulation. In practice, electrical parameters such as current and voltage can be measured by testing equipment, while the electric field cannot be obtained directly by measurement, but can be calculated by TCAD.

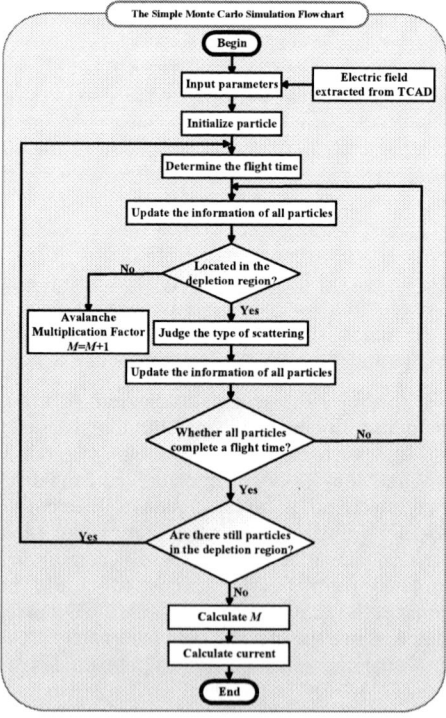

Fig. 6 The simple Monte Carlo simulation flowchart.

Fig. 6 outlines the simple Monte Carlo simulation flowchart. First, input the electric field extracted from the TCAD and the related parameters. Then, initialize particle and its flight time. After a movement of one flight time duration,

979-8-3503-6184-1/24 $31.00 © 2024 IEEE

due to particle scattering caused by the electric field, the information for all particles within the depletion region is altered, necessitating the update information of particles. Once a particle moves out of the depletion region, the avalanche multiplication factor M is considered to increase by one according to its definition. If the particle still remains within the depletion region, the scattering type is determined according to the energy of each particle. After completing the judgement of the scattering type of each particle, the information of all particles needs to be updated again. The cycle described above is performed continuously until all carriers within the depletion region have completed one cycle of a flight time. Finally, after undergoing several cycles, if all the particles in the depletion region move out of the depletion region, the simulation ends and the program exits the cycle. The output of the Monte Carlo numerical simulation provides the avalanche multiplication factor M, and the current across device can be calculated according to its defining formula. The BV characteristic curves of the SCR and MSCR simulated by the modeling method are shown in Fig. 7(a) and 7(b), respectively.

C. Experimental and Verification

In order to verify the proposed modeling method, the SCR and MSCR were fabricated based on 0.18μm standard CMOS process, and their BV characteristic curves were measured by the transmission line pulse (TLP) test platform. The comparison of measured and simulated BV characteristic curves of the SCR and MSCR are shown in Fig. 7(a) and 7(b), respectively, while the measured and simulated breakdown voltages are summarized in Table I. The experimentally measured breakdown voltage for the SCR was found to be 14.5V, while the novel modeling method yielded a value of 14.4V, resulting in a discrepancy of 0.1V. For the MSCR, the actual breakdown voltage was measured at 11V, whereas the simulated value stood at 10.8V, indicating a difference of 0.2V. Moreover, prior to breakdown, the simulated leakage current levels for both SCR and MSCR ranged between 10^{-10}A and 10^{-9}A, aligning closely with experimental findings. Comparative results underscore that modeling method proposed can precisely characterize leakage current preceding breakdown, with only minor deviation observed between simulated and experimentally determined breakdown voltages.

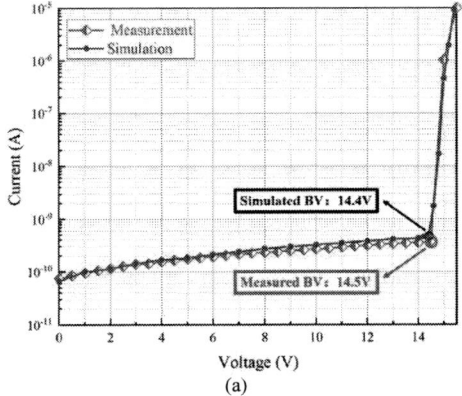

(a)

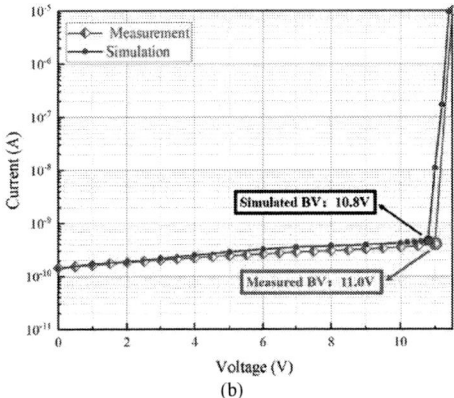

(b)

Fig. 7 The comparisons of simulated and measured BV curves. (a)SCR. (b)MSCR.

TABLE I MEASURED AND SIMULATED BV DATA

	Simulation	*Measurement*	*Error*
SCR	14.4 V	14.5 V	0.1V
MSCR	10.8 V	11.0 V	0.2V

IV. CONCLUSION

In this paper, a novel modeling method is proposed for characterizing the BV characteristics of ESD devices. By extracting the electric field parameter obtained through TCAD and combining with Monte Carlo numerical simulation, the proposed modeling method accurately simulates the BV characteristics of ESD devices. The modeling method is verified by SCR and MSCR fabricated in 0.18μm standard CMOS process, and the comparison results indicate that this modeling method offers high precision and scalability, providing valuable insights for modeling the BV characteristics of ESD devices with different structures and sizes.

REFERENCES

[1] A. Z. Wang, H. G. Feng, K. Gong, R. Y. Zhan, and J. Stine, "On-chip ESD protection design for integrated circuits: an overview for IC designers," Microelectronics Journal, vol. 32, pp. 733-747, Sep 2001.

[2] Y. F. Xi, J. A. Salcedo, Y. Z. Zhou, et al, "Design and characterization of ESD solutions with EMC robustness for automotive applications," Microelectronics Reliability, vol. 55, pp. 2236-2246, Nov 2015.

[3] H. Li, M. Miao, Y. Z. Zhou, et al, "Modeling and simulation of comprehensive diode behavior under electrostatic discharge stresses," IEEE Transactions on Device and Materials Reliability, vol. 19, pp. 90-96, Mar 2019.

[4] Z. L. Shen, Y. Z. Wang, Y. H. Li, X. Zhang, and Y. Wang, "A Scalable Model for Snapback Characteristics of Circuit-Level ESD Simulation," IEEE Transactions on Circuits and Systems II: Express Briefs, vol. 69, pp. 1547-1551, Mar 2022.

[5] J. Cao, J. Y. Xu, Y. Wang, G. Y. Lu, and X. Zhang, "A compact SCR model using advanced BJT models and standard SPICE elements," Science China-Information Sciences, vol. 59, Oct 2016.

[6] Y. Wang, Z. Y. Zhong, X. L. Jin, et al, "A Novel Gate-Controlled Dual Direction SCR With Enhanced Failure Current for On-Chip ESD Protection of Industry-Level Controller Area Network Bus," IEEE Journal of Emerging and Selected Topics in Power Electronics, vol. 10, pp. 7615-7626, Dec 2022.

[7] X. Zhou, S. Ng, and C. H. Tan, "A simple Monte Carlo model for prediction of avalanche multiplication process in Silicon," Journal of Instrumentation, vol. 7, Aug 2012.

[8] S. A. Plimmer, J. R. P. David, D. S. Ong, and K. F. Li, "A simple model for avalanche multiplication including deadspace effects," IEEE Transactions on Electron Devices, vol. 46, pp. 769-775, Apr 1999.

Analysis of The Impact of Parasitic Bipolar Amplification on Charge Sharing Based on Analytical Model

Yutao Zhang[1], Hongliang Lyu*[1], Yuming Zhang[1], Ruxue Yao[1]

[1] The State Key Discipline Laboratory of Wide Band Gap Semiconductor Technology,
School of Microelectronics, Xidian University, Xi'an 710071, China

* Email: ytzhang_xd@163.com, hllv@mail.xidian.edu.cn

Abstract—**Prior research has indicated that charge sharing mechanisms primarily involve diffusion collection and parasitic ambipolar amplification. This study introduces an analytical model for transient current at the circuit level to address single-event multiple transient effects (SEMT) resulting from charge sharing, incorporating both diffusion collection and parasitic bipolar amplification. Comparative analysis with TCAD simulation demonstrates the high accuracy of the proposed model for circuit-level SEMT assessment. Utilizing this model, the impact of parasitic bipolar amplification on charge sharing is quantitatively examined. Findings reveal that the extent of the parasitic bipolar amplification effect varies based on the linear energy transfer (LET) of the heavy ion. Specific formulas are provided to quantify the influence range of parasitic bipolar amplification across different LET values.**

Keywords—*Charge sharing, modeling method, parasitic bipolar amplification, single-event transients, single-event multi-transient effect.*

I. INTRODUCTION

Single-event effects (SEE) are significant considerations for microelectronic circuits utilized in space conditions. These effects typically arise from the impact of high-energy heavy ions on semiconductors, leading to the ionization of material atoms and the generation of numerous electron-hole pairs along the ion's trajectory. The accumulation of these charges in sensitive regions through diffusion and drift can give rise to transient currents that may disrupt the regular functioning of electronic systems. With the reduction in the size of semiconductor processes and the decrease in supply voltage, the susceptibility to SEE increases, leading to the emergence of various intricate phenomena under these circumstances [1].

The decrease in process size and the rise in circuit integration lead to the occurrence of charge sharing. When charge sharing happens, multiple sensitive areas can accumulate charges and produce transient current [2]. This, combined with the decrease in the upset threshold, can easily trigger multi-cell upset (MCU) in large-scale digital circuits and multi-bit upset (MBU) in memory circuits. This presents new obstacles for the analysis of SEE and the radiation-hardening of circuits. Numerous studies have investigated the charge sharing phenomenon, focusing on the physical mechanisms involved. Research findings suggest that charge sharing is likely caused by diffusion collection and parasitic bipolar amplification, a phenomenon observed in both Silicon-On-Insulator (SOI) and bulk devices [3]. Apart from exploring the mechanisms, analyzing the SEMT at the circuit level resulting from charge sharing is also a crucial area of study.

Various circuit-level analysis methods currently exist, all of which necessitate precise SEMT models for support. Within the realm of modeling techniques, the empirical model derived from data fitting holds significance. This approach does not depend on specific physical mechanisms and can yield precise outcomes. Nonetheless, it mandates a substantial volume of data for extracting model parameters; otherwise, model accuracy may significantly diminish. Another modeling approach is grounded in physical processes. The formulation of this model is intricately linked to physical mechanisms, enabling it to align with genuine physical trends and exhibit relatively robust scalability. However, prevailing models predominantly revolve around the diffusion collection process, with inadequate consideration given to parasitic bipolar amplification, thereby impacting model accuracy. Consequently, the quantitative analysis of parasitic bipolar amplification's influence at the circuit level poses challenges.

This study introduces a novel SEMT current model that accounts for both the parasitic bipolar amplification effect and diffusion collection. The accuracy of this model is confirmed through TCAD simulation. Utilizing this model, the study conducts a quantitative analysis of the influence of the parasitic bipolar amplification effect on charge sharing.

II. MECHANISM ANALYSIS AND MODEL DERIVATION

Previous research has typically relied on diffusion collection models to simulate the charge collection in sensitive areas. This modeling approach involves breaking down the charge collection process into two distinct subprocesses: the diffusion of charges towards the depletion region boundary and their subsequent collection through the influence of the electric field. The generated current I_{dep} can be calculated by equation (1).

$$I_{dep} = qv_n n_s(t) \qquad (1)$$

where q is the elementary charge, v_n represents the average velocity of the charge collection, which can be estimated based on the electric field within the depletion region. $n_s(t)$ represents the charge concentration diffusing towards the periphery of the sensitive area, which is expressed as equation (2):

$$n_s(t) = \iiint \frac{N_0}{(4\pi Dt)^{3/2}} \exp\left(\frac{-r^2}{4Dt}\right) dxdydl \qquad (2)$$

where N_0 is the amount of charge deposited in the unit volume, which is related to the LET of the ion, and D is the ambipolar diffusion coefficient. r is the distance between the charge trajectory and the sensitive area.

The phenomenon of parasitic bipolar amplification occurs as a result of the accumulation of holes in the well region subsequent to the collection of electrons. This accumulation is facilitated by the diffusion of holes towards the well contacts driven by a concentration gradient. Consequently, the potential within the well region rises, leading to the forward biasing of the source-well junction. Electrons from the source region are then injected into the well region and eventually gathered by the sensitive region, denoted as I_{src}. As a result, the total transient current I_{SET} comprises two constituent elements: the diffusion collection current I_{dep} and the parasitic bipolar amplification current I_{src}.

Essentially, this current I_{src} is the current generated by the forward bias of the source-well pn junction, so according to the Shockley equation (3):

$$I_{src} = I_{s0}\left[\exp(\frac{V_s}{nV_t})-1\right] \quad (3)$$

where I_{s0} represents the reverse saturation current, and V_s is the forward bias voltage across the source-body junction. The analysis indicates that the source voltage remains constant at 0V, and the forward bias voltage is induced by variations in the well potential V_{pw}. These fluctuations in the well potential are instigated by hole diffusion through the well resistance. To account for the non-uniformity in voltage distribution, an integral equation (4) is employed to describe the potential fluctuation V_{pw} of the well region:

$$V_{pw} = \int dI_w(x)dR \quad (4)$$

where R represents the resistance of the well region, which is related to the doping concentration of the well region.

I_w represents the hole diffusion current, and the corresponding diffusion current density J_w is related to the hole charge distribution in the well region, expressed by equation (5):

$$J_w = qD_p\frac{dp}{dx} \quad (5)$$

where D_p represents the hole diffusion coefficient, and p represents the charge deposition distribution caused by incident ion which is expressed by equation (6):

$$p = \frac{N_0}{4\pi Dt}\exp(-\frac{x^2}{4Dt}) \quad (6)$$

Combining expressions (4)-(6), the well region potential distribution V_{pw} can be derived. Therefore, by replacing V_s with V_{pw}, the source injection current component I_{src} caused by the parasitic bipolar amplification effect can be obtained. Finally, the current I_{SET} of the sensitive area can be obtained through equation (7):

$$I_{SET} = I_{dep} + I_{src} \quad (7)$$

III. DISCUSSION

A three-stage inverter was constructed utilizing mixed-mode simulation techniques. The design incorporates a two-dimensional layout featuring three NMOS transistors modeled using TCAD, while the PMOS transistors are based on the SPICE model sourced from the FreePDK 45nm library. The

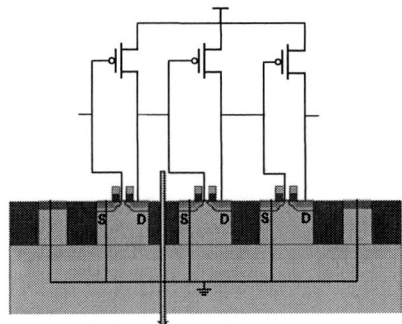

Fig. 1. The schematic of the mixed-mode simulation structure of the three-stage inverter chain

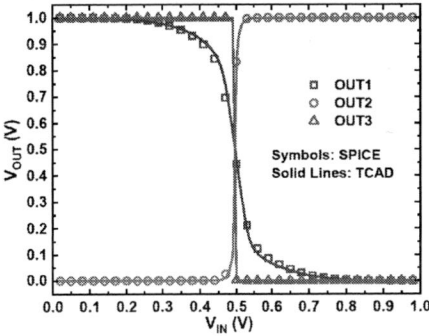

Fig. 2. Comparison of the VTC between the mixed-mode simulation and the SPICE simulation.

gate length of the NMOS is 45nm. The thickness of the gate oxide is 5nm, the doping concentration of the source and drain is 1×10^{20} cm^{-3}, and the p-well doping concentration is 1×10^{17} cm^{-3}. The simulation employs various models including Shockley-Read-Hall (SRH), Auger recombination, bandgap-narrowing, Fermi-Dirac statistics, Philips unified mobility model, and high field saturation mobility model. The mixed-mode simulation setup is illustrated in Fig. 1, with the voltage-transfer curves (VTC) depicted in Fig. 2, and the outcomes are validated through circuit-level simulation, demonstrating a strong agreement between the mixed-mode and circuit-level simulations.

The SEMT simulation was conducted to verify the accuracy of the proposed model, which is implemented in Verilog-A and based on the TCAD mixed-mode inverter chain and SPICE circuit. In the TCAD simulation, the heavy ion model parameters included a trajectory length of 2μm and an inject time of 0.02ns, with the incident location indicated in Fig. 1. The input voltage (V_{IN}) was maintained at 0V. The output voltages of three inverters were simulated by TCAD under varying LET values, represented as symbols in Fig. 3. The SPICE simulation results, derived from the proposed model, are depicted as solid lines in Fig. 3. The findings indicate that the critical LET for the first-stage inverter is approximately 0.7pC/μm as determined from the TCAD simulation. Notably, the proposed model demonstrates a high level of accuracy in predicting the critical LET, as evidenced in Fig. 3(b).

The validity of the suggested model has been demonstrated. Utilizing this model, an examination is conducted on the impact of parasitic bipolar amplification current on charge sharing collection. A configuration featuring three NMOS transistors is utilized as a case study,

979-8-3503-6184-1/24 $31.00 © 2024 IEEE

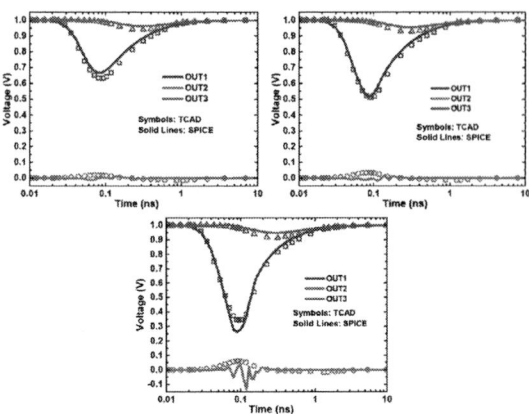

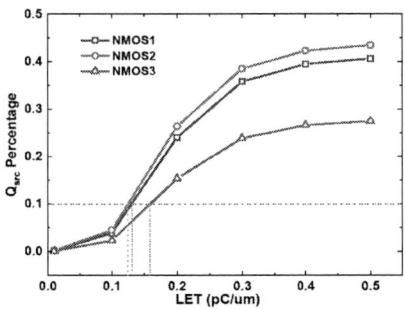

Fig. 3. Comparison of the output voltages between TCAD mixed-mode simulation and SPICE simulation under different LETs ((a)LET=0.6pC/μm, (b)LET=0.7pC/μm, (c)LET=0.8pC/μm based on the proposed model)

Fig. 4. The proportion of the charge collection by I_{src} in total charge collection in different NMOSs.

with all three NMOS transistors being OFF state (i.e. $V_G=V_S=0$, $V_D=1V$). The heavy ion characteristics remain consistent with the aforementioned simulation. The LET is specified within the range of 0.01pC/μm to 0.5pC/μm. The analysis involves determining the percentage of charge collected by I_{src} relative to the total charge collected, as illustrated in Fig. 4. As the LET value escalates, there is a corresponding increase in the proportion of charge collected by I_{src}. This trend indicates a heightened manifestation of the parasitic bipolar amplification effect across the three NMOS devices.

Additionally, there are variations in the distribution of parasitic bipolar amplification charges among the three NMOS devices. NMOS2 exhibits the highest proportion, followed by NMOS1 with a lower proportion, and NMOS3 with the smallest proportion. This discrepancy is attributed to the distance between the source and the incident location, which is identified as a critical factor. Specifically, as the distance increases at a constant LET, the potential fluctuation within the well region diminishes, resulting in a reduction of parasitic bipolar amplification.

In order to quantitatively assess the impact of distance, the distances between the source regions of three NMOSs and the injection location were determined. The critical point at which the parasitic bipolar amplified charge constitutes 10% of the total charge collection was identified, as illustrated in Fig. 4 by the purple dashed line. The corresponding LET values were identified and marked with a yellow dashed line in Fig. 4.

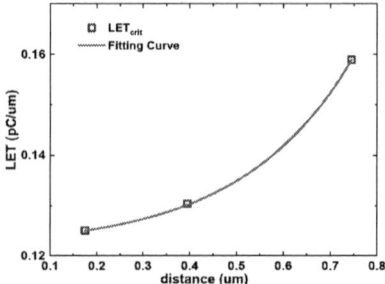

Fig. 5. The relationship between the distance and the LET when the parasitic bipolar amplification begins to take effect.

These LET values were then plotted in Fig. 5. It is evident that as the distance between the source region and the incident position increases, the LET at which the parasitic bipolar effect becomes significant also increases. The relationship between LET and distance can be determined through a fitting expression (8). Additionally, the affected range of the parasitic bipolar amplification effect resulting from particles with a specific LET can be calculated using an appropriate expression (8). If the source region falls within the calculated range, the transient current calculation for neighboring sensitive areas should account for the parasitic bipolar amplification. Otherwise, the results may deviate significantly from the actual scenario.

$$LET = 0.122 + 0.0016 \exp(\frac{dist}{0.235}) \qquad (8)$$

IV. CONCLUSION

In this study, a SEMT current model incorporating diffusion collection and parasitic bipolar amplification is introduced. The model's accuracy is validated through TCAD mixed-mode simulation. Utilizing this model, the impact of parasitic bipolar amplification on charge sharing is examined. Findings indicate that the parasitic bipolar amplification effect is dependent on the distance between sensitive areas and the incident position. Furthermore, the extent of this effect under varying LET values is quantitatively assessed, and a calculation formula is proposed.

REFERENCES

[1] D. Kobayashi, "Scaling Trends of Digital Single-Event Effects: A Survey of SEU and SET Parameters and Comparison with Transistor Performance," IEEE TRANSACTIONS ON NUCLEAR SCIENCE, pp. 1–1, 2020.

[2] W. Zhao et al., "Single-Event Double Transients in Inverter Chains Designed With Different Transistor Widths," IEEE TRANSACTIONS ON NUCLEAR SCIENCE, vol. 66, no. 7, pp. 1491–1499, Jul. 2019.

[3] D. M. Fleetwood, "Radiation Effects in a Post-Moore World," IEEE TRANSACTIONS ON NUCLEAR SCIENCE, vol. 68, no. 5, pp. 509–545, 2021.

[4] Artola, M. Gaillardin, G. Hubert, M. Raine, and P. Paillet, "Modeling Single Event Transients in Advanced Devices and ICs," IEEE TRANSACTIONS ON NUCLEAR SCIENCE, vol. 62, no. 4, pp. 1528–1539, 2015.

[5] D. A. Black, W. H. Robinson, I. Z. Wilcox, D. B. Limbrick, and J. D. Black, "Modeling of Single Event Transients With Dual Double-Exponential Current Sources: Implications for Logic Cell Characterization," IEEE TRANSACTIONS ON NUCLEAR SCIENCE, vol. 62, no. 4, pp. 1540–1549, 2015.

[6] L. Artola, G. Hubert, S. Duzellier, and F. Bezerra, "Collected Charge Analysis for a New Transient Model by TCAD Simulation in 90 nm Technology," IEEE TRANSACTIONS ON NUCLEAR SCIENCE, vol. 57, no. 4, pp. 1869–1875, Aug. 2010.

979-8-3503-6184-1/24 $31.00 © 2024 IEEE

Research on the performance degeneration of GGNMOS under total ionizing dose Radiation

Jiekai Feng [1], Ping Luo *[1,2], Chengxin Li[1], Jiaxuan Hu[1], Peng Li[3], Pengfei Liao[3]

[1] State Key Lab. of Elec. Thin Films and Inter. Dev., Univ. of Elec. Sci. and Technol. Of China, Chengdu 610054, P. R. China
[2] Chongqing Institute of Microelectronics Industry Technology, UESTC, Chongqing, 400060, P. R. China
[3]The 24th Research Institute of China Electronics Technology Group Corporation, Chongqing, 400060, China

* Email: pingl@uestc.edu.cn

Abstract—This study in combination with TCAD simulation investigates the ESD performance degeneration of gate grounded NMOS (GGNMOS) after exposure to Total ionizing dose (TID) radiation and the effect of its leakage on the internal circuits. The results show that after TID radiation, the ESD design metric V_{tr} shows a significant decrease and even the snapback behavior disappearance. And the leakage current of the GGNMOS itself can severely damage the normal functions of the internal circuits. Based on these results, an 8-type enclosed layout structure is proposed to harden the GGNMOS. This study provides a promising strategy for the failure and hardening of GGNMOS under TID radiation, which is helpful for the ESD protection of aerospace integrated circuits.

Keywords—GGNMOS, TID, ESD performance, Current leakage, Enclosed layout NMOS

I. INTRODUCTION

Electrostatic discharge (ESD) can easily breakdown integrated circuits and cause large transient currents, resulting in the failure of semiconductor devices in integrated circuits. Currently, integrated circuits need to be designed with ESD protection devices to pass the relevant ESD protection standards. Gate-Grounded NMOS (GGNMOS) is a ESD protection device used commonly by various CMOS process, and its design needs to follow the conventional ESD design window, as shown in Fig. 1. where I_{t2}, V_h, and V_{tr} are the main design metrics for ESD devices. Typically, I_{t2} represents the transient current capability of the ESD device, which needs to be as large as possible, V_h depends on the operating voltage of the protected circuit, and V_{tr} needs to be less than the breakdown voltage of the internal device[1].

Total ionizing dose (TID) effect is one of the most common irradiation damages in aerospace integrated circuits[2]. As shown in Fig. 2(a), TID induces a large number of electron-hole pairs in the STI oxide, and a portion of the positive charge accumulates on the oxide surface, which in turn induces a mirrored negative charge in the bulk Silicon[3]. These negative charges form continuous conductive channels in the NMOS, generating leakage current, and as the amount of negative charge accumulation increases gradually, the circuit function is damaged[4]. In order to improve the leakage current of NMOS devices under the TID effect, a variety of enclosed-layout NMOS devices have been investigated to replace the straight layout NMOS[5]. Fig. 2(b) shows an 8-type enclosed layout NMOS developed by our research group, demonstrated to have better TID tolerance than straight-layout NMOS, which has been used in analog DC-DC circuit design[6]. However, with the similar structure, how TID affects the performance of GGNMOS devices and the effect on the internal circuitry of the GGNMOS after irradiation by TID have been rarely reported.

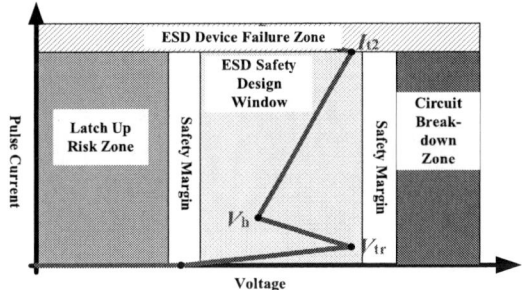

Fig. 1. ESD device design window

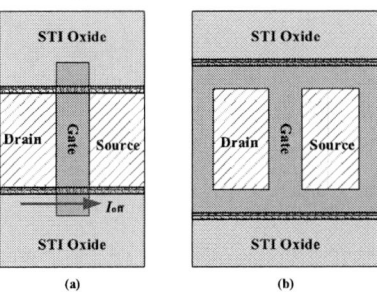

Fig. 2. (a) Conventional NMOS devices; (b) 8-type enclosed layout NMOS.

In this work, firstly, the GGNMOS TCAD simulation model is constructed. Then, the degradation of GGNMOS ESD performance under TID radiation was investigated by analyzing the current density distribution after ESD pulse injection. Furthermore, by analyzing the GGNMOS leakage current and its connection to the internal circuitry, investigating how GGNMOS affects internal circuit function under TID radiation. Finally, an 8-type enclosed layout hardening method is discussed.

II. GGNMOS STRUCTURES AND DEVICE SIMULATION METHOD

A. GGNMOS Structures Simulation

Fig. 3(a) show the layout of the multifinger GGNMOS for a 0.18um BCD process. In this study, the same device structure will be constructed in the TCAD simulator, and the number of fingers will be 2 to speed up the simulation. Fig 3(b) illustrates the 3D structure and doping concentration distribution of the GGNMOS. Fig. 3(c) shows a cross-section of the 3D structure, which shows the distribution of the oxides. Fig. 3(d) shows the electrode definition of the multifinger GGNMOS. The process parameters used in the simulator are listed in Table I.

979-8-3503-6184-1/24 $31.00 © 2024 IEEE

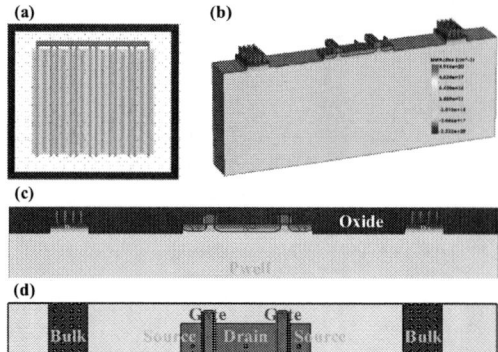

Fig. 3. (a) Multifinger GGNMOS layout; (b) 3D structure of GGNMOS; (c) Cross-section of the 3D structure; (d) Electrode definition of the GGNMOS.

Table I. Simulation process parameters

Symbol	Parameter	Values	Unit
t_{ox}	Gate oxide thickness	5	nm
h_g	PolySilicon hight	0.5	um
L_g	Channel length	0.6	um
P_s	Pwell IMP concentration	2.5e13	cm^{-3}
N_A	P+ IMP concentration	4e15	cm^{-3}
N_D	N+ IMP concentration	1e16	cm^{-3}

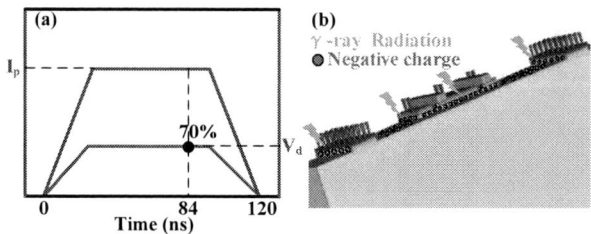

Fig. 4. Schematic diagram of (a) TLP simulation method and (b) TID injection method.

B. Device Simulation Method

Transmission line pulse (TLP) performance simulation of GGNMOS is carried out in TCAD device simulator. The simulation method is the same as the actual TLP test. As show in Fig. 4(a), a pulse current I_p is injected to the drain and the drain voltage is monitored in real time. The drain voltage is recorded when the pulse current is applied for 70% of the total time, which is considered as the drain voltage corresponding to this pulse current. The magnitude of pulse current applied increases exponentially from 2e^{-4} to 2A.

The TID effect is simulated by a physical model of injecting a fixed positive charge at the oxide and silicon substrate interfaces. The relationship between the positive charge concentration at the interface and the radiation intensity of the TID effect is referenced in Ref[7]. As shown in Fig. 4(b), when the positive charge is injected, a mirrored negative charge is induced in the silicon substrate.

III. RESULTS AND DISCUSSIONS

In this section, the impact of the TID radiation on the performance of GGNMOS is investigated by analyzing the TACD simulation results and a hardening method is proposed.

A. The ESD performance degeneration of GGNMOS under TID radiation

To investigate the effect of TID radiation on the performance of the GGNMOS, the current pulse is injected to the drain of the GGNMOS at three different TID radiation. Fig. 5 demonstrates the GGNMOS current density distribution

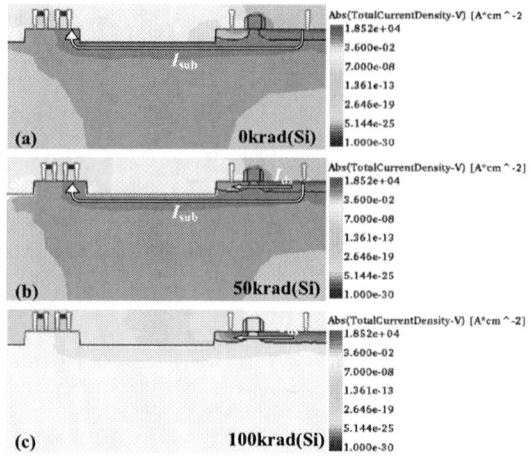

Fig. 5 TCAD-simulated current density distribution of GGNMOS under an ESD pulse at different TID radiation. (a) 0krad(Si), (b) 50krad(Si), (c) 100krad(Si).

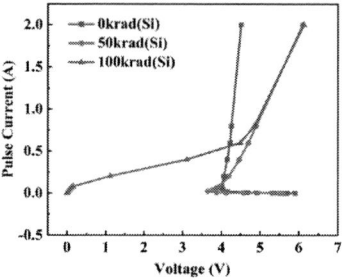

Fig. 6 TLP performance of GGNMOS obtained by TCAD simulation under different TID radiation

at the instant of pulse injection.

It can be seen from Fig. 5(a), without TID radiation, after the current pulse injection, the GGNMOS internal current is I_{sub} mainly, which flows from the drain to the substrate. Due to the existence of the substrate body resistance R_{sub}, when the product of I_{sub} and R_{sub} reaches the forward bias voltage of the emitter of the parasitic NPN transistor, the parasitic NPN transistor conduction, providing a low-resistance discharge path for the ESD current. The drain voltage at this time is the trigger voltage V_{tr} of this GGNMOS[8].

As shown in Fig. 5(b), under TID irradiation at 50krad(Si), the source-drain current I_{ds} increases significantly, which leads to the emitter forward bias voltage decrease of the parasitic NPN transistor. Since R_{sub} remains constant, V_{tr} will decreases. Especially when the leakage is larger, V_{tr} decreases more severely. For example, the area near the STI. This will result in non-uniform triggering of the GGNMOS, causing the small area to be overloaded with a large discharge current, which in turn will cause the GGNMOS to be burnt out.

As shown in Fig. 5(c), when the TID irradiation reaches 100krad(Si), the internal current is completely transformed into a source-drain current I_{ds}, the parasitic NPN transistor shows a constant open state, the snapback behavior disappears, and V_{tr} is decreased greatly. The disappearance of the snapback behavior will make the GGNMOS lose voltage clamping capability to the internal circuit, presenting diode-like ESD protection characteristics. This simulation result is similar to the experimental results reported in the Ref[9].

Figure 6 demonstrates the TLP performance curve plotted by collecting the voltage after injecting a current pulse of 2e^{-4} to 2A at the drain of the GGNMOS. It can be clearly seen that V_{tr} decreases when the TID radiation reaches 50krad(Si), and the snapback disappears when it increases to 100krad(Si), in agreement with the results of the above analyses.

B. The effect of ESD device on circuit function

Fig. 7 schematically shows the typical application of GGNOMS as an ESD protection device in circuits. The GGNMOS is connected between the IO PAD and the internal circuits. The IO PAD is connected externally to a programmable capacitor, which is charged and discharged by the M1 and M2 transistors, and the capacitor voltage V_C determines the operation mode of the circuit. Once the GGNMOS has been exposed to a TID radiation and has not been hardened, a leakage current I_{off} will be formed from PAD to ground via the GGNMOS.

Fig. 8 shows the I_d-V_g characteristics of GGNMOS with different TID radiation, and the drain voltage is fixed at 1.8V. The I_d at V_g=0V represents the leakage current due to the TID radiation. From Fig. 8, it can be seen that when the TID radiation reaches 100krad, the leakage current of the GGNMOS will be close to the μA level, which seriously affects the circuit function for μA is the normal operating current range of most circuits.

From the analysis of the performance change of GGNMOS under TID radiation, it is clear that both the ESD performance degradation and its effect on the internal circuits are due to the leakage current, and therefore the radiation hardening of GGNMOS should also be achieved by limiting the leakage current. It can be hardened using the 8-type enclosed layout studied by our research group, as shown in Fig. 9. Compared to straight gate GGNMOS, the leakage current is reduced significantly after TID radiation. And we will verify the 8-type enclosed layout GGNMOS ESD performance experimentally in the future work.

IV. SUMMARY

In summary, we constructed a 2-finger GGNMOS TCAD simulation model and set up an ESD pulse simulation method similar to the actual TLP test. Based on the analysis of TCAD simulation results, when the radiation dose is 50krad(Si), V_{tr} will decrease; increasing the TID radiation to 100krad(Si), the snapback behavior disappears, which demonstrates that the TID radiation will make the GGNMOS ESD performance degradation. By investigating how the GGNMOS is connected to the internal circuits, it is revealed that when the GGNMOS leakage current reaches the μA level under TID radiation, it will seriously impact internal circuits function, and the 8-type enclosed layout GGNMOS is proposed. This study provides the mechanism of GGNMOS degradation and hardening method under TID radiation, which is valuable for the application of ESD protection devices in aerospace integrated circuits in the future.

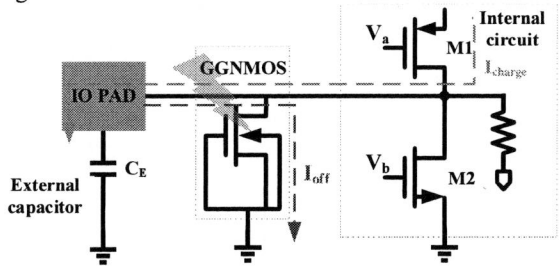

Fig. 7 Schematic diagram of the effect of ESD devices on circuit function after exposure to a TID radiation.

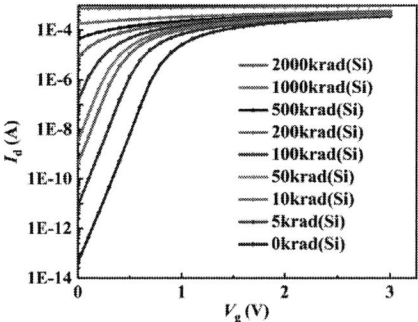

Fig. 8 I_d–V_g characteristics of GGNMOS device extracted using 3-D TCAD simulations to study impact of different TID radiation.

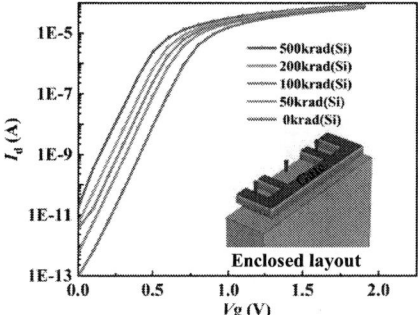

Fig. 9 I_d–V_g characteristics of 8-type enclosed layout GGNMOS device to study impact of different TID radiation.

ACKNOWLEDGMENT

This work is supported by the Chongqing Natural Science Foundation under Grant CSTB2023NSCQ-MSX0153 and project YG2404-1.

REFERENCES

[1] J. Yuxi, L. Jiao, R. Feng, C. Jialin, and Y. Dianxiong, *Journal of Semiconductors,* vol. 30, no. 8, (2009).

[2] P. E. Dodd, M. R. Shaneyfelt, J. R. Schwank, and J. A. Felix, *IEEE Transactions on Nuclear Science,* vol. 57, no. 4, pp. 1747-1763 (2010).

[3] X. Zhou, Z. a. Yuan, L. Shu, M. Qiao, Z. Lu, Y. Zhao *et al., IEEE Electron Device Letters,* vol. 40, no. 4, pp. 593-596 (2019).

[4] M. F. Bernard, L. Dusseau, J. Boch, J. R. Vaille, F. Saigne, R. D. Schrimpf *et al., IEEE Transactions on Nuclear Science,* vol. 53, no. 6, pp. 3232-3236 (2006).

[5] P. I. Vaz, T. H. Both, F. F. Vidor, R. M. Brum, and G. I. Wirth, *Journal of Electronic Testing,* vol. 34, no. 6, pp. 735-747 (2018).

[6] Y. Wu, P. Luo, and B. Zhang, *IEEE Transactions on Nuclear Science,* vol. 70, no. 8, pp. 2076-2084 (2023).

[7] H. J. Barnaby, M. L. McLain, I. S. Esqueda, and C. Xiao Jie, *IEEE Transactions on Circuits and Systems I: Regular Papers,* vol. 56, no. 8, pp. 1870-1883 (2009).

[8] R. Grisel, L. A. Coyitangiye, A. Doukkali, F. Barbier, P. Descamps, and H. Murray, *IEEE Industrial Electronics,* pp. 1817-1822 (2009).

[9] W. Liang, K. Alexandrou, M. Klebanov, C.-C. Kuo, I. Kymissis, K. B. Sundaram *et al., IPFA,* pp. 1946-1550 (2017).

The UIS Withstand Capability and Device Failure Mechanism of 650 V p-GaN Gate HEMTs

Qihang Huang[1], Luanxi Ma[1], Yanyu Nie[1], Shuting Huang[1], Jianggen Zhu[1], Yu Shi[3], Rongxin Du[3], David Zhou[3], Yuxi Wan[3], Bo Zhang[1], and Qi Zhou[1,2]*

[1] School of Integrated Circuit Science and Engineering, University of Electronic Science and Technology of China (UESTC), Chengdu, China
[2] Institute of Electronic and Information Engineering, UESTC, Dongguan, China
[3] Shenzhen Pinghu Laboratory, Shenzhen, China

* Email: zhouqi@uestc.edu.cn

Abstract—In this paper, the failure mechanism of two commercial p-GaN gate HEMTs at room and high temperature under the unclamping inductive switching (UIS) stress are studied. Both devices feature 200~300 V higher transient drain voltage withstand capability compared with their intrinsic static breakdown voltage. The average UIS failure voltage of *Dev. A & B* are ~1365 and ~1475 V, respectively. The extremely high electric-field at drain electrode induced lattice crack and associated abrupt increase in leakage current is responsible for device failure in *Dev. A*. On the other hand, two failure mechanisms were found in *Dev. B*: a) The burnt mark of thermal runaway is observed; b) the vertical microdefect is formed in the passivation dielectric induced by the high peak electric-field and hot electrons in the vicinity of the field-plate. The underlying mechanism revealed in this work provides valuable insights to understand the UIS characteristics of 650 V p-GaN gate HEMTs.

Keywords—*GaN HEMT, reliability, failure mechanism, unclamping inductive switching, electric field, hot electron*

I. INTRODUCTION

P-GaN gate HEMTs are being adopted for consumer electronics and penetrating for industrial & vehicle electronics applications where device reliability is more demanding and attracting tremendous attention. In power conversion circuit, due to the presence of inductive load, the power devices would undergo high voltage spike induced by the fast switching-off dynamics. Hence, the unclamping inductive switching (UIS) ruggedness and the corresponding device failure mechanism is of great importance. The UIS behavior of p-GaN gate HEMTs have been reported [1-7], the electron trapping in the gate and buffer layer can cause the device degradation, including the shift of V_{TH} and the increase of ON-state resistance (R_{on})[6,7]. And for the device failure, either the device failure mode is lacking [1,4,5] or only the catastrophic damage, normally the burnt out of metal electrode, is observed after device failure [2,3]. Neither the underlying device physics nor the original factor initiates the resultant device failure (e.g. electrode burn-out, catastrophic damage) is still ambiguous, which hinders further improving the UIS robustness of the device.

In this work, the UIS behavior and ruggedness of two commercial 650 V Schottky type p-GaN gate HEMTs with different inductor load at room- & high-temperature were comprehensively studied. It was found that the failure of both

devices was determined by the achieved critical peak voltage during the UIS event despite the various inductor load and temperature. Furthermore, the failure mechanism is elucidated through in-depth analysis of the failed device microstructure and associated with TCAD simulation, which gives clear insights to understand of UIS capability and device failure mechanism of 650 V-rating Schottky p-GaN gate HEMTs.

II. EXPERIMENT SETUPS AND RESULTS

A. Experimental setup

The simplified schematic of the UIS test circuit and waveforms are shown in Fig. 1. During the testing process, the load inductor L_{load} is charged by the DC power supply V_{DD} upon the turn-on of the Device Under Test (DUT). Subsequently, when the DUT is turned off, the energy stored in L_{load} is conducted through the DUT by the parasitic capacitance. Fig. 2 shows the UIS test setup and evaluation

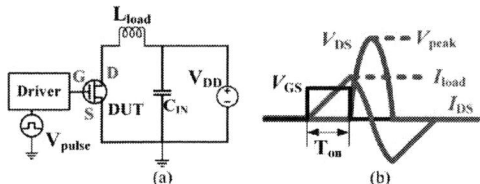

Fig. 1 (a) Schematic of UIS test circuit for GaN E-mode devices. (b) The typical waveform during the UIS test.

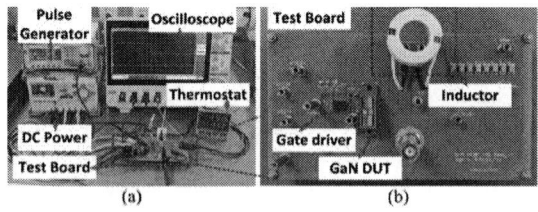

Fig. 2 (a) Photograph of the UIS test circuit setup. (b) Photograph of the UIS test board.

Table I : Typical Parameters of the DUTs

Parameter	Device A (GaN Systems)	Device B (Innoscience)
V_{DS} (V)	650	650
R_{ON} (mΩ)[a]	100	165
C_{oss} (pF)[b]	31	20

[a] V_{GS} = 6 V, I_{DS} = 3 A, T_J = 25°C.

[b] V_{DS} = 400 V, V_{GS} = 0 V, f = 100kHz.

This work was supported in part by the National Natural Science Foundation of China under Grant 62174019, in part by the Guangdong Basic and Applied Basic Research Foundation, China, under Grant 2021B1515140039, 2024A1515012139 and in part by Sichuan Science and Technology Program under Grant 2023YFG0138.

board. The driver used for GaN HEMT is ADuM4121A isolated gate driver.

The studied devices (DUTs) are 650 V commercial Schottky type p-GaN gate HEMTs from two vendors[8,9] and their typical parameters are listed in Table I.

B. UIS behavior of DUT at room temperature

The typical UIS waveforms with inductor load L_{load}=0.5 mH and device turn-on time T_{on}=10 μs are shown in Fig. 3. The peak V_{DS} is 1149 & 1228 V and the maximum load current I_{load} is 0.62 & 0.60 A for *Dev. A & B*, respectively. By increasing T_{on}, higher energy is stored in L_{load} which leads to higher I_{load} & peak V_{DS} (V_{peak}) in the DUTs during the UIS event and eventually results in device failure for T_{on}=~12.5 μs (see Fig. 4). In this case, it can be seen that the V_{DS} increased to a critical value (V_{peak}) and then exhibits a slump. Meanwhile, the leakage current I_{DS} shows a significant increase. The V_{peak} in the failed *Dev. A & B* are 1367 and 1462 V, respectively. 25 DUTs respectively for *Dev. A & Dev. B* were tested to failure under 5 different L_{load} at 25 °C. Fig. 5 (b) summarizes the failure voltage V_{peak} of the total 50 DUTs for the tested *Dev. A & Dev. B*. Comparing *Dev. A & Dev. B* and their static breakdown voltage (BV), the results exhibit the following features:

a) The DUTs' UIS failure voltage exhibit no statistically significant dependence on the L_{load};

b) *Dev. B* delivers 90~130 V higher UIS failure voltage V_{peak} than *Dev. A*;

c) Both *Dev. A & B* feature higher UIS failure voltage V_{peak} than the DUTs' intrinsic static *BV*, which is contrary to the reported commercial Cascode GaN HEMTs [10];

d) The failure voltages in both types of DUTs show a relatively small spread (i.e. ~66 V for *Dev. A* & ~28 V for *Dev. B*), which is much smaller than the commercial Cascode GaN HEMTs [10].

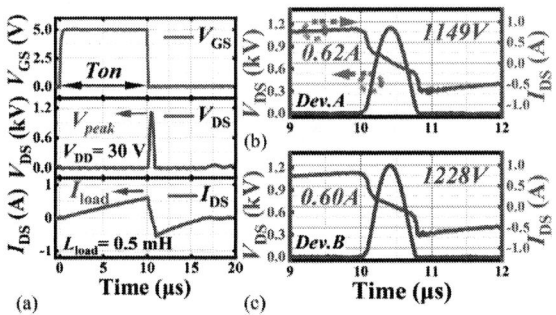

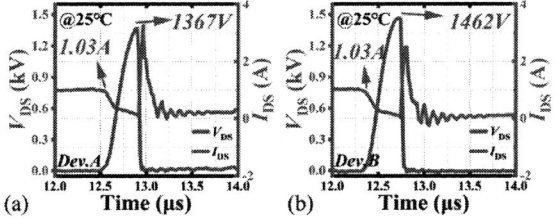

Fig. 3 The measured waveforms of V_{GS}, V_{DS}, and I_{DS} during UIS test with V_{DD}=30 V, L_{load}=0.5 mH and T_{on}=10 μs of (a) (b) Device A and (c) Device B.

Fig. 4 The measured UIS waveforms of device failed at 25°C. (a) Device A and (b) Device B.

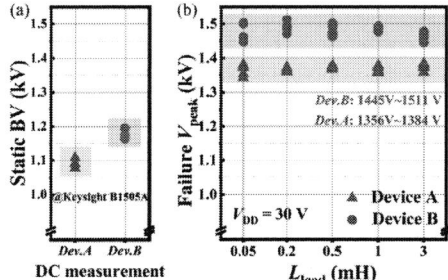

Fig. 5 Comparison of the (a) static BV and (b) UIS failure Vpeak of Device A and Device B.

C. UIS behavior of DUT at high temperature

Fig. and Fig. 7 show that the UIS waveform of the device at a temperature of 150°C. The UIS behavior of the device at high temperatures is almost indistinguishable from that at room temperature.

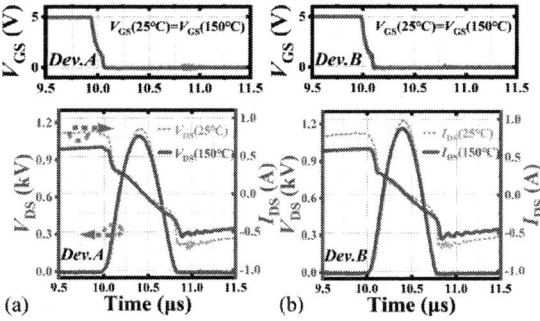

Fig. 6 The measured waveforms of UIS test at 150°C with V_{DD}=30 V, L_{load}=0.5 mH and T_{on}=10 μs. (a) Device A and (b) Device B.

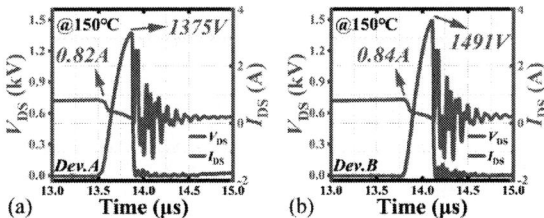

Fig. 7 The measured UIS waveforms of device failed at 150°C. (a) Device A and (b) Device B.

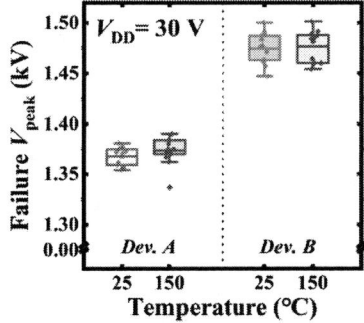

Fig. 8 Comparison of the UIS failure V_{peak} of the room and high temperature.

Following a series of failure experiments conducted under high temperature conditions (15 DUTs respectively for *Dev. A & Dev. B* were tested to failure under 5 different L_{load} at

150 °C), Fig. 8 presents a comparative analysis of the UIS failure voltage between 25 °C and 150 °C for the DUTs. The results indicate a modest enhancement in the UIS failure voltage of the DUTs when subjected to the high temperature test.

III. PHYSICAL MECHANISM OF DEVICE FAILURE AND ANALYSIS

One *Dev. A* failed at 1.37 kV & *Dev. B* failed at 1.5 kV were decapsulated. A burnt mark in *Dev. A* is observed (Fig. 9(a)). By using focused ion beam (FIB) and scanning electron microscopy (SEM), a V-shaped damage located at the drain electrode can be seen (Fig. 9(c)), beneath which a vertical lattice damage penetrating downward through the GaN epi-layer to Si substrate is observed. In addition, several void burnt marks in GaN epi-layer along with the vertical crack trace are also formed. Such a V-shaped damage could be resulted from the high electric-field (E-field) peak at the drain electrode induced by the extremely high V_{peak} (i.e. 1.37 kV) as shown in Fig. 9(e). It can be seen that the E-field strength at such high V_{peak} is as high as 12 MV/cm, which is much higher than the critical breakdown E-field of GaN and is capable of inducing the release of elastic potential energy in AlGaN/GaN hetero-structure and forms the vertical crack. More importantly, the crack forms a leaky path, the leakage current substantially increases along with the initial crack. Together with the high V_{DS} during UIS event, the proliferate transient surge energy causes the localized burnout of drain electrode metal and GaN material in the vicinity of crack trace as seen in Fig. 9(d).

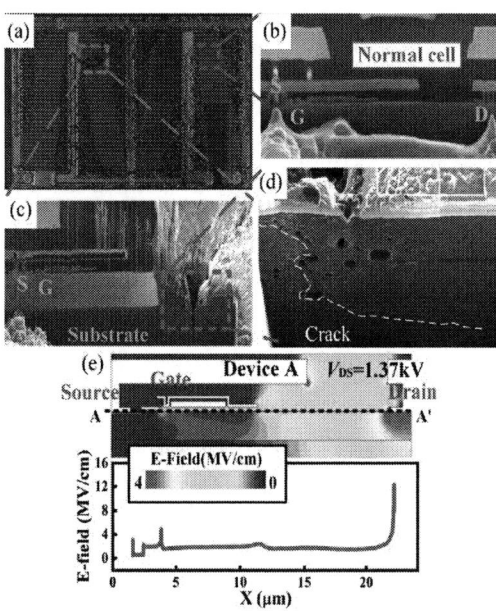

Fig. 9 (a) The photo of the failure-point position. (b) The intact device cell. (c) Cross sectional SEM image of the device. (d) The vertical crack underneath the burned drain. (e) TCAD simulated E-field distribution

By contrast, the failure mechanism of *Dev. B* is more complicated. Two burnt marks respectively locate at interdigital cell and close to the drain bonding pad are observed (Fig. 10(a)). At *Spot I*, the catastrophic thermal runaway presents at the drain electrode (Fig. 10(c)). At *Spot II*, a vertical crack at the drain-side edge of the 2nd field-plate and penetrating through the passivation dielectric as well as GaN epi-layer is observed. Such a critical spot is the location

where a significantly high peak E-field presents as seen in Fig. 10(f), which may trigger hot electron generation and the subsequent hot electron induced microdefects both in dielectric and GaN epi-layer. The higher temperature tends to mitigate the hot electron effect that leads to moderately higher UIS failure voltage V_{peak} in the DUTs tested at 150 °C (Fig. 8).

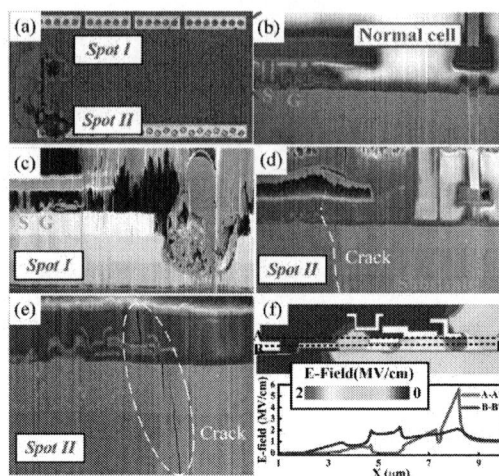

Fig. 10 (a) The photo of the failure-point position. (b) The intact device cell. (c) Cross sectional SEM image of the Spot I. (d) Cross sectional SEM image of the Spot II. (e) TCAD simulated E-field distribution.

IV. CONCLUSION

In this work, the failure mechanism of two commercial 650 V p-GaN gate HEMTs at room and high temperature under the unclamping inductive switching (UIS) stress are studied. The UIS failure voltages (~1365 and ~1475 V for *Dev. A* & *Dev. B*, respectively) are observed to be 200~300 V higher the static breakdown voltage. Further examination of the failed cells by using FIB and SEM, a crack damage be resulted from the high electric-field peak at the drain electrode located can be seen. The extremely high electric-field at drain electrode induced lattice crack and associated abrupt increase in leakage current is responsible for device failure in *Dev. A*. In addition, the failure mechanism of *Dev. B* is more complicated. The vertical microdefect is formed in the passivation dielectric induced by the high peak electric-field and hot electrons in the vicinity of the field-plate. And the diminished hot electron effect at high temperatures leads to a marginal increase in the UIS failure voltage.

REFERENCES

[1] C. Zhang *et al.*, *ISPSD.*, 66(2), pp. 125-128.(2022)

[2] S. Liu *et al.*, *IEEE Transactions on Power Electronics*, vol. 35, no. 11, pp. 11328-11331. (2020)

[3] S. Li *et al.*, *IEEE Transactions on Industrial Electronics*, vol. 69, no. 5, pp. 5041-5049.(2022)

[4] R. Sun *et al.*, *IEEE Transactions on Power Electronics*, vol. 37, no. 6, pp. 6711-6719.(2022)

[5] Q. Bao, S. Yang and K. Sheng, *ISPSD.*, pp. 337-340.(2020)

[6] Li S *et al.*, *IEEE Journal of Emerging and Selected Topics in Power Electronics*, 9(2), pp. 2227-2234. (2020)

[7] Kozak J P *et al.*, *IEEE Transactions on Power Electronics*, 38(1), pp. 435-446. (2022)

[8] "GS66504B", https://gansystems.com.

[9] "INN650DA260A", https://www.innoscience.com.

[10] Q. Song, R. Zhang, J. P. Kozak, J. Liu, Q. Li and Y. Zhang, *IEEE Transactions on Power Electronics*, 37(4), pp. 4148-4160.(2022)

Time Dependent Dielectric Breakdown in n-MOSFETs Fabricated by Low-Temperature and Low-Pressure Mild Oxidation After Plasma Solidification

Qiao Teng[1,#], Yanning Chen[2,#], Fang Liu[2], Bo Wu[2], Yongfeng Deng[2], Dawei Gao[1,*]

[1]College of Integrated Circuits, Zhejiang University, Hangzhou, China, 311200
[2]Beijing Smart-chip Microelectronics Technology Co., Ltd, Beijing, China, 100000
*Email: dawei_gao@zju.edu.cn
[#]Two authors have equal contribution to this work

Abstract—In this work, we investigate the effect of SiON manufacturing processes on TDDB behaviors for n-MOSFETs. The devices with the ultrathin SiON film fabricated by the novel low-temperature and low-pressure mild oxidation after the plasma solidification (LLMOPS) method have a lower defect generation rate and longer breakdown time than the traditional process. The kinetic Monte Carlo simulation framework is used to model TDDB behavior accurately, and it is proved that the improvement of TDDB is due to the reduction of interface states. Thus, LLMOPS technology provides a guiding principle for improving the TDDB characteristics of MOSFETs in mass production.

Keywords—MOSFETs, SiON, TDDB, Manufacturing process

I. INTRODUCTION

With the increasing demand for automotive electronics, complementary metal oxide semiconductor (CMOS) technology is widely used. Time dependent dielectric breakdown (TDDB) is still considered one of the important reliability challenges [1], especially for automotive chips. Silicon oxynitride (SiON) film has a higher dielectric constant than conventional silicon dioxide (SiO_2), making it an important material for ultra-thin gate oxide [2]. The fabrication of the high-reliability ultra-thin SiON film is crucial, as it significantly impacts the reliability of devices. However, the incorporation of high nitrogen concentrations into the SiO_2 during the manufacturing process leads to the formation of abundant interface traps, which significantly increase leakage current and accelerate dielectric breakdown. In this work, we propose a novel low-temperature and low-pressure mild oxidation after the plasma solidification (LLMOPS) technique for forming the high-quality SiON film. This method significantly suppresses the generation of interface traps and improves TDDB behaviors, providing an explicit guideline for fabricating the high-reliability SiON film.

II. DEVICES AND EXPERIMENTAL DETAILS

The n-MOSFETs for this work were fabricated by 55 nm CMOS technology with a 2 nm SiON film. The devices have a width (W) of 18 μm and a channel length (L) of 18 μm. Specifically, in process A, the SiON film was fabricated using traditional techniques, in which post-nitridation anneal (PNA) was used to repair damage after nitrogen was incorporated. Process B and Process C, utilizing the LLMOPS technique, divide the conventional PNA into two stages: plasma solidification first, followed by mild oxidation. Process B has a higher solidification temperature and lower oxidation temperature than Process C, resulting in lower interface traps. TDDB behaviors were performed on at least 10 samples for each electric field condition. The devices under test were stressed by DC stress to acquire breakdown time T_{BD} and the gate current I_G at 125 °C. The failure criterion is a sudden increase of the stress current exceeding ten times the initial value.

III. RESULTS AND DISCUSSION

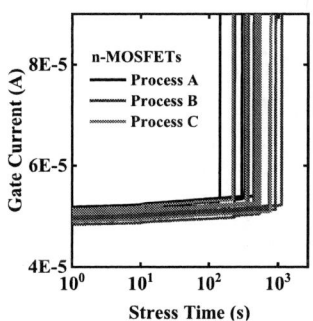

Fig. 1. Typical stress current traces measured under three process conditions

Fig. 1 shows the stress current traces of n-MOSFETs in three processes. The time to breakdown of process B is longer than that of process A and process C, indicating that different defect generation rates in various processes. To understand the trap generation rate, the median T_{BD} corresponds to the stress induced pre-breakdown leakage current (SILC) curve at three different process is shown in Fig. 2(a). The trap generation rate (n) is calculated by (1)

$$SILC \propto t^n \qquad (1)$$

The normalized slope of SILC between different processes is extracted in Fig. 2(b) for better comparison. Steeper SILC slopes are found at process A, indicating that the trap generation rate increases with higher interface traps. The SILC generation behaviors also contribute to providing further insights into the changes in trap density. The normalized increase current (J_n) is defined by (2)

$$J_n = \frac{J - J_0}{J_0} \qquad (2)$$

where J_n reflects the current change caused by the traps generated during the stress time and J_0 is the initial gate current. The higher J_n is observed in process A, as shown in

979-8-3503-6184-1/24 $31.00 © 2024 IEEE

Fig. 2(c). indicating that there are more traps in the SiON film fabricated by the traditional process. Fig. 2(c) also shows the relationship between the normalized J_n measured and the normalized injected electron (Q_{inj}). It is observed that the linear correlation for the J_n to $\log(Q_{inj})$ and the probability of SILC generation (G_{SILC}) can be estimated from the linear slope according to the following expressions (3) and (4) [3].

$$Q_{inj} = \frac{\int I(t)dt}{Area} \qquad (3)$$

$$G_{SILC} = \frac{J_n}{Q_{inj}} \qquad (4)$$

As shown in Fig. 2(d), it is found that the lower G_{SILC} is observed for the SiON film fabricated by process B, indicating a better tolerance of voltage stress compared with process A and process C. This phenomenon is attributed to the fact that the reduced interface traps improves the SILC behavior.

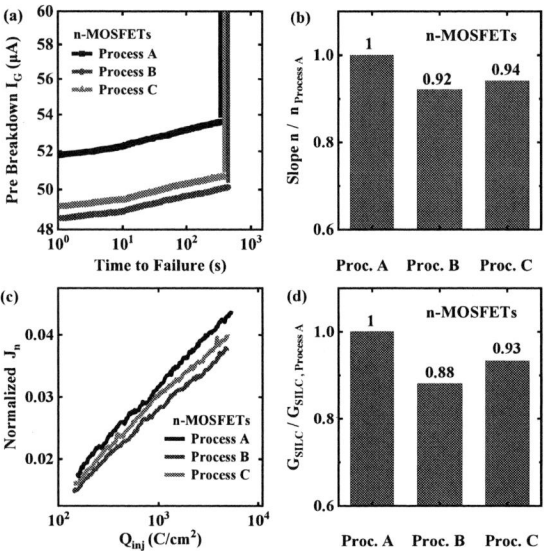

Fig. 2. (a) Pre-breakdown gate current versus stress time. (b) Comparison of the slope of pre-breakdown gate current under three processes. (c) Normalized gate current increase (J_n) as a function of normalized injected electron Q_{inj} with different processes. (d) Comparison of the SILC generation probability (G_{SILC}) under different processes.

The T_{BD} obtained from TDDB measurements are analyzed with Weibull distribution based on (5) [4]

$$\ln(-\ln(1-F)) = \beta\ln(t) - \beta\ln(t_{63.2\%}) \qquad (5)$$

where β is the slope of the Weibull distribution and $t_{63.2\%}$ is the characteristic lifetime representing the time of 63.2% failure. Fig. 3(a) represents the Weibull plots in process A, showing a good T_{BD} distribution. Process B and C have the higher T_{BD}, as shown in Fig. 3(b) and (c). The longer T_{BD} can be attributed to the need for more traps or higher trap generation rates to trigger a breakdown [5], which can be inferred from the increase in β shown in Fig. 3(d). For all the MOSFETs in three processes, the $t_{63.2\%}$ is extracted and plotted versus the E_{OX} in Fig. 4. The $t_{63.2\%}$ in the log scale is proportional to oxide electric fields with a slope of γ. The slope parameter γ is commonly referred to as the electric field acceleration factor, describing the rate of change of the

breakdown time with the electric field. The $t_{63.2\%}$ in process B is longer on each E_{OX} and the slightly improved γ increases the margin of reliability, converting to longer TDDB lifetime when projected to typical operation voltage. Lifetime predictions under 1.1 times the operating voltage at 125 °C are made based on the E-model, as shown in Fig. 5(a). In the Weibull distribution, longer T_{BD} can be observed in process B, indicating a further improvement in oxide quality. To predict the lifetime of the product (W/L=1/1 mm), the estimated lifetime of test structures (W/L=18/18 μm) is often extrapolated to product lifetimes by using the Poisson area scaling by (6) [6].

$$\tau = t_{63.2\%} * (-\ln(1-0.001))^{\frac{1}{\beta}} * (\frac{A}{A_0})^{\frac{-1}{\beta}} \qquad (6)$$

where A is the product area and A_0 is the tested structure area. Fig. 5(b) compares the normalized lifetimes of the three processes, with the process B and C improving by 1.48 times and 1.26 times relative to the process A , respectively. These results suggest that the LLMOPS technique offers a longer TDDB lifetime, which may be attributed to the reduction of interface traps suppressing the generation of trap-assisted tunneling current.

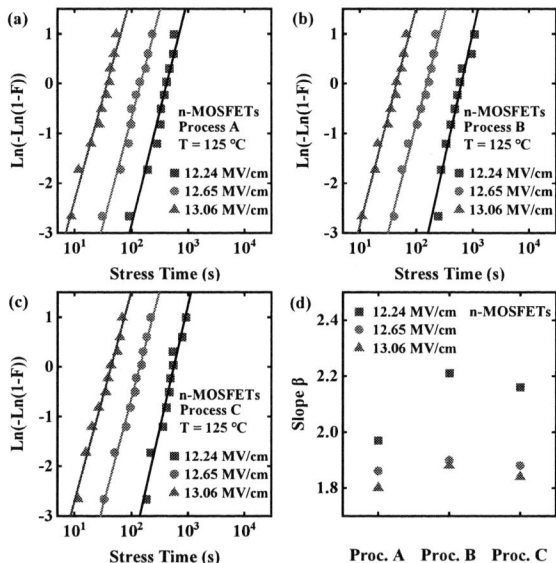

Fig. 3. Weibull distributions of T_{BD} measured at 125°C with different gate voltage stress for (a) process A, (b) process B, (c) process C. (d) Slope β of Weibull distributions in three processes.

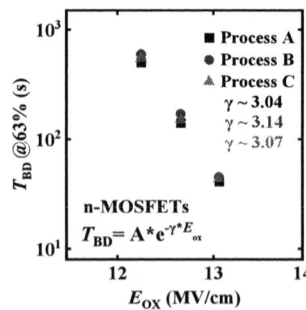

Fig. 4. Gate oxide electric field dependence of T_{BD}@63.2% described by the E-model for the three processes.

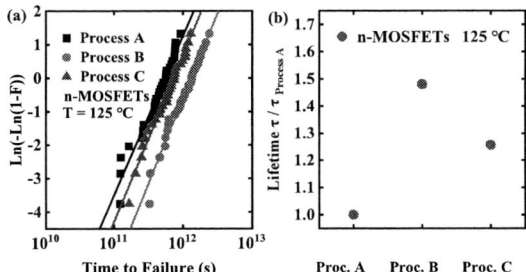

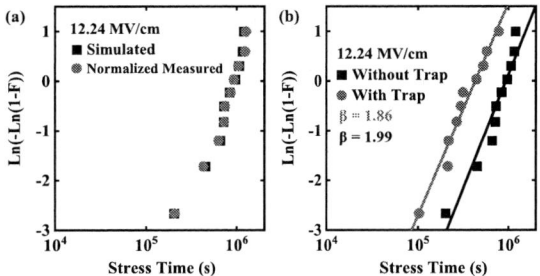

Fig. 5. (a) Weibull distributions of breakdown time extrapolated to 1.1 times the operating voltage at 125°C using the E-model for three different processes.(b) Comparison of the lifetime of three different processes based on product area.

Fig. 7. (a) Comparison of normalized measured (red) and simulated (black) TDDB distributions at a specific electric field. (b) TDDB distributions in a specific electric field with or without interface traps.

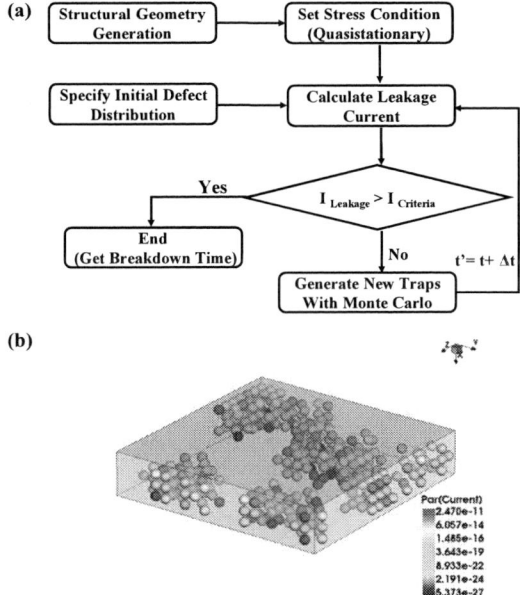

Fig. 6. (a) The KMC-TDDB simulation flowchart. (b) Defect generation in the oxide during TDDB. (Polysilicon and Silicon are hidden)

In order to investigate the interface traps on TDDB behavior, the kinetic Monte Carlo TDDB (KMC-TDDB) simulation framework is adopted to model the TDDB. As shown in Fig. 6(a), the TDDB framework predicts the breakdown time by calculating leakage current. The trap-assisted leakage generation mechanism was executed into the device simulation using the thermochemical model, which describes the defect generation rate G using (7) [7]

$$G(x,y,z) = v * e^{\left(\frac{E_A - p_0 \frac{2+k}{3} * F(x,y,z)}{K_B T(x,y,z)}\right)} \tag{1}$$

where E_A is the activation energy, p_0 is the molecular dipole moment, k is the dielectric constant, F is the local electric field, and T is the local temperature. To simplify the simulation calculation, the area scaling formula is used to calculate the Weibull distribution at 10*10 nm. Fig. 6(b) shows the generation of traps during the TDDB in a three-dimensional structure. The simulation results are consistent with the normalized measured results, as shown in Fig. 7(a), indicating that the simulation model is accurate.

Fig.7(b) shows the Weibull distribution with or without interface traps. It can be observed that the presence of traps shortens the T_{BD} and decreases the β value. This phenomenon suggests that inhibiting the formation of interface traps by the LLMOPS method significantly improves TDDB behavior.

IV. CONCLUSIONS

The influence of SiON manufacturing engineering on TDDB characteristics has been studied in this work. It is found that adopting the LLMOPS technique obtains a longer TDDB lifetime, possibly because reduced interface traps inhibits the generation of trap-assisted tunneling current. Furthermore, higher solidification temperature and lower oxidation temperature, resulting in lower interface traps, are more conducive to improving TDDB. A 3D-based kinetic Monte Carlo model is used to demonstrate that the improvement of TDDB resulted from the reduction of the interface states. Thus, LLMOPS technology is a promising method to improve the TDDB behaviors for MOSFETs.

ACKNOWLEDGMENT

The authors acknowledge support from the National Key R&D Program of China (2022YFF0605800), Zhejiang Jianbing Program (Grant No. 2022C01063).

REFERENCES

[1] X. Yu, C. Yan, Y. Ding, Y. Qu, and Y. Zhao, "GHz AC to DC TDDB Modeling with Defect Accumulation Efficiency Model," *IEEE Int. Rel. Phys. Symp.*, 2023, pp. 1-6.

[2] Z. Ke et al., "Pre-O2 treatment for LNA gate oxide leakage improvement," *IEEE Int. Rel. Phys. Symp.*, 2022, pp. P39-1-P39-4.

[3] S. Yuan, Z. Chen, J. Li, M. Tian, and R. Zhang, "Impact of Electrical Stress on Defect Generation in Thin GeO₂/Ge Gate Stacks Fabricated by Thermal Oxidation," *IEEE Trans. Electron. Devices*, vol. 67, no. 6, pp. 2516-2521, Jun. 2020.

[4] T. Liu et al., "Time-Dependent Dielectric Breakdown of Commercial 1.2 kV 4H-SiC Power MOSFETs," *IEEE J. Electron Devices Soc.*, vol. 9, pp. 633-639, Jun. 2021.

[5] R. O'Connor, G. Hughes, T. Kauerauf, and L.-A. Ragnarsson, "Time dependent dielectric breakdown and stress induced leakage current characteristics of 8Å EOT HfO2 N-MOSFETS," *IEEE Int. Rel. Phys. Symp.*, 2010, pp. 799-803.

[6] E. Y. Wu, and R. P. Vollertsen, "On the weibull shape factor of intrinsic breakdown of dielectric films and its accurate experimental determination-part I: theory, methodology, experimental techniques," *IEEE Trans. Electron. Device*, vol. 49, no. 12, pp. 2131-2140, Dec. 2002.

[7] A. Padovani, and L. Larcher, "Time-dependent dielectric breakdown statistics in SiO2 and HfO2 dielectrics: Insights from a multi-scale modeling approach," *IEEE Int. Rel. Phys. Symp.*, 2018, pp. 3A.2-1-3A.2-7.

979-8-3503-6184-1/24 $31.00 © 2024 IEEE

Simulation of BTI for GAA MOSFETs with Enhanced Parameters Extraction

Yongjia Wang [1], Yijiao Wang [2], Shuhan Wang [1], Jinghan Xu [1], Xiaoyan Liu [*1]

[1] School of Integrated Circuits, Peking University, Beijing 100871, China
[2] School of Integrated Circuit Science and Engineering, Beihang University, Beijing 100191, China

* Email: liuxiaoyan@pku.edu.cn

Abstract—**In this work, we present an enhanced parameters extraction method for simulating BTI degradation in GAA MOSFETs. Utilizing a comprehensive simulation framework based on 3D device simulations, we accurately model BTI degradation at the 5nm technology node. The proposed method involves generating defect weights maps and employing a trained deep learning network to extract the parameters of the two-state NMP model. The parameters extraction process completes within a minute and achieves a mean relative error of 6.29% compared to experimental data.**

Keywords—BTI Degradation, Parameters Extraction, Deep Learning Network, Defect Weights Map

I. INTRODUCTION

With the development of CMOS technology, non-planar MOSFETs such as FinFETs and Gate-All-Around (GAA) MOSFETs are widely used in today's advanced technology nodes. Bias Temperature Instability (BTI) continues to be a significant reliability concern in advanced semiconductor devices, especially as its effects become more pronounced with the reduction in Si channel dimensions[1]. Consequently, there is an urgent need for efficient and accurate modeling of BTI degradation characteristics.

Under the influence of BTI, defects in the gate dielectric layer trap charges, leading to shifts in device parameters such as threshold voltage. It is widely accepted that defects in the gate dielectric layer and interface states at the Si/SiO$_2$ interface play a major role in device degradation. Based on this mechanism, the two-state Nonradiative Multi-Phonon (NMP) model, which models defect capture and emission behavior, has achieved significant success in predicting BTI in planar devices[2]. However, despite the reduction in parameters compared to the original four-state NMP model, the parameters extraction process remains complex and challenging.

This work proposes a simulation framework for BTI characteristics in non-planar devices, extending the two-state NMP model to GAA MOSFETs. The framework includes a univariate sensitivity analysis of model parameters and employs deep learning methods for parameters extraction.

II. SIMULATION METHODS

A. BTI simulation method

The simulation framework for BTI in non-planar devices is depicted in Fig. 1. The process initiates with the input of device structure, defect band parameters, and stress conditions. Subsequently, a calibrated device simulation generates a defect weights map, and the NMP model is utilized to calculate the transition rates of defect states and the occupancy probability of defects, at various positions over different stress time steps. This involves a Markov process to determine the dynamic behavior of defects. By referencing the defect weights map, the cumulative impact of all defects on the

device's threshold voltage shift over stress time is computed, resulting in the ΔVth-time curve. This simulation process is demonstrated using GAA MOSFETs as shown in Fig. 1.

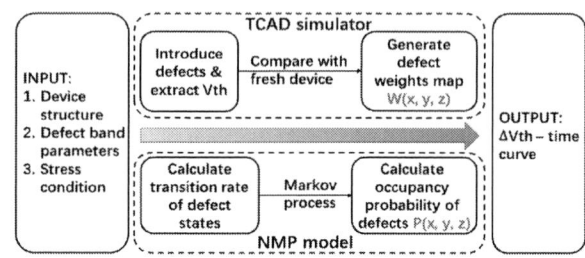

Fig. 1. Simulation framework for BTI.

As the BTI simulation process, a detailed device simulation of the GAA MOSFETs without trapped charges were simulated using a commercial device simulator[3]. Fig. 2 shows the 3D structure and cross-section of the device, and structure parameters are specified in TABLE I. . Given the sub-nanometer scale of the device dimensions, the simulations incorporated the drift-diffusion model, thermodynamic model, and quantum correction model to ensure accuracy. The Lombardi model was employed to account for the impact of impurity scattering and carrier scattering on carrier mobility. Due to the thin-channel characteristics of the device, the thin-layer model was introduced. Additionally, to accurately simulate quantum effects related to carrier concentration and its gradient, Fermi-Dirac statistics and the quantum potential model were utilized. The band narrowing model and Shockley-Read-Hall (SRH) recombination model were also applied to comprehensively consider the device's physical properties. Fig. 3 plots the calibration results of the device's transfer characteristics, with experimental data from Ref. [4].

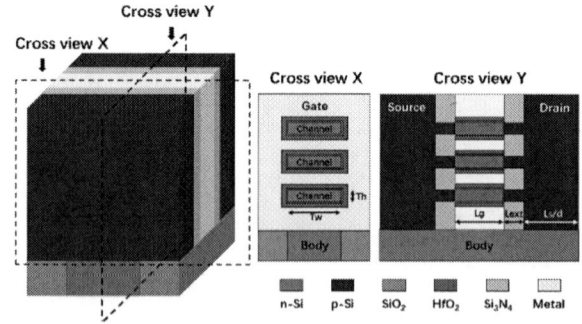

Fig. 2. From left to right: the 3D structure of the 5 nm node GAA MOSFET, cross view X, and cross view Y with channel details.

979-8-3503-6184-1/24 $31.00 © 2024 IEEE

TABLE I. DEVICE PARAMETERS

Parameters	Value	Parameters	Value
Gate length, Lg	12 nm	Channel width, Tw	25 nm
Extension length, Lext	5 nm	Channel doping concentration, Nc	1×10^{16} cm^{-3}
S/D length, Ls/d	12 nm	S/D doping concentration, Ns/d	1×10^{20} cm^{-3}
Channel height, Th	5 nm	Equivalent oxide thickness, Eot	0.9 nm

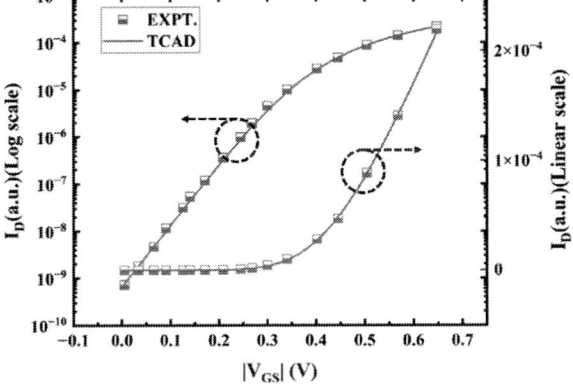

Fig. 3. Calibrated Id - Vg curve of the GAA MOSFETs with data from [4].

In generating the defect weights map, the cross-section at the center of the device channel was considered, and fixed charge defects were introduced in the gate oxide and interface at intervals of 0.1nm. After introducing the defects, the threshold voltage of the device was extracted and compared with that of a fresh device to calculate the shift in threshold voltage caused by spatial charge defects at the corresponding positions. By statistically analyzing the threshold voltage shifts caused by defects at different positions, the weights of the defects at corresponding locations were determined, generating the defect weight map as shown in Fig. 4.

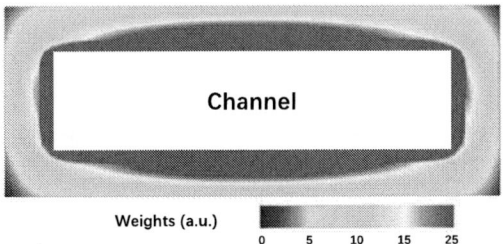

Fig. 4. Defect weights map (different colors in the figure represent the varying weights).

B. NMP model parameters extraction

The two-state NMP model requires seven parameters to describe each defect band, including defect energy level and its standard deviation, defect relaxation energy and its standard deviation, curvature ratio, defect concentration, and the length of the defect spatial distribution interval. Experimental data indicate the presence of both shallow and deep defect bands in the SiO$_2$ and HfO$_2$ layers[2]. Therefore, to fully describe the defect bands in GAA MOSFETs, excluding the defect spatial distribution interval length parameter, which can be approximated as the oxide layer thickness, 24 parameters are needed. Clearly, the relative

importance of these parameters varies. For instance, in the BTI of PMOS devices, the impact of shallow energy level defects is significantly weaker than that of deep energy level defects. Extracting all parameters using a deep learning model is not cost-effective. Hence, parameters were prioritized based on their relative importance, and only the most critical ones were selected for extraction. Through univariate sensitivity analysis, each parameter was adjusted by 5% from the default model parameters, and the changes in the model output were observed. The resulting parameter importance radar chart, as shown in Fig. 5, guided the selection of the parameters.

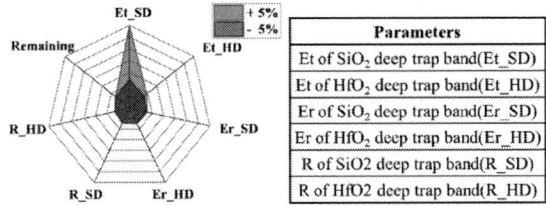

Parameters
Et of SiO$_2$ deep trap band(Et_SD)
Et of HfO$_2$ deep trap band(Et_HD)
Er of SiO$_2$ deep trap band(Er_SD)
Er of HfO$_2$ deep trap band(Er_HD)
R of SiO2 deep trap band(R_SD)
R of HfO2 deep trap band(R_HD)

Fig. 5. Relative importance of parameters (left) and selected parameters (right).

The parameters extraction process is illustrated in Fig. 6. This process employs a deep learning model implemented using the Scikit-learn Multilayer Perceptron package[5]. To balance prediction accuracy and computational efficiency, the model comprises five hidden layers, each containing 100 neurons. The input to the model consists of BTI data under various gate stress conditions, spanning a time range from 0.0001 seconds to 1000 seconds. The output of the model is the parameters of the two-state NMP model. The hidden layers utilize the hyperbolic tangent function as the activation function, and the model is optimized using the stochastic gradient descent solver. The batch size, learning rate, and regularization parameters are set to their default values. Baseline parameters for the two-state NMP model are derived from planar devices[2], and a training dataset is generated through 12,000 Monte Carlo simulations. The Monte Carlo sampling method is selected due to its superior coverage of nonlinear datasets compared to pre-assigned value sampling. During these simulations, the model parameters listed in Fig. 5 are varied randomly with a uniform distribution. The variation range for these parameters is empirically determined: Et_SD and Et_HD are varied by ±0.25, while other parameters are varied by ±20%. All other model parameters are maintained at their default values. In the model training phase, 15% of the Monte Carlo simulation data is reserved as test data. Training is halted when the loss function for the test data shows no improvement for ten consecutive cycles, thereby preventing overfitting.

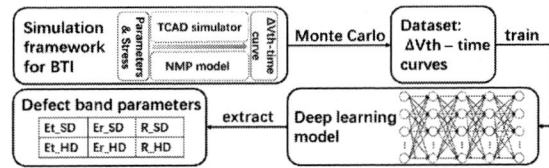

Fig. 6. Diagram of parameters extraction process.

III. RESULTS AND DISCUSSION

Initially, a set of arbitrary parameters is given, and the BTI curve is calculated using the two-state NMP model. The curve data is then input into the trained model to generate a set of parameters. The comparison between the actual parameters and the model-extracted parameters, as shown in TABLE II. , demonstrates the deep learning network's capability to learn the two-state NMP model. The relative error between the actual parameters and the model-extracted parameters is below 5%, and the relative error decreases as the importance of the parameters increases, indicating that the deep learning model has effectively captured the underlying physics of the two-state NMP model, which is crucial for reliable parameters extraction in practical applications.

TABLE II.   MODEL-EXTRACTED VS ACTUAL PARAMETERS

Parameters	Actual	Model-extracted	Relative Error
Et_SD	-5.700	-5.743	0.75%
Et_HD	-5.300	-5.275	0.47%
Er_SD	9.200	9.140	0.65%
Er_HD	10.200	9.941	2.5%
R_SD	1.600	1.661	3.8%
R_HD	1.400	1.358	3.0%

Using the trained model, BTI characteristic curve data extracted from the literature is inputted, and the deep learning network infers the defect band parameters. These extracted parameters are then input into the two-state NMP model to generate curves, which are compared with experimental data, as illustrated in Fig. 7. The average relative error between the model-generated data and the experimental data is 6.29%. The insufficient accuracy may be attributed to factors in the BTI degradation process that are not accounted for in the existing model. This issue also arises when using other predictive models[6]. For instance, when the gate voltage is -1.9V and the stress time is very short ($\sim 10^{-4}$s), the predicted threshold voltage shift is significantly less than the experimental data. In other cases, the simulation data aligns well with the experimental data.

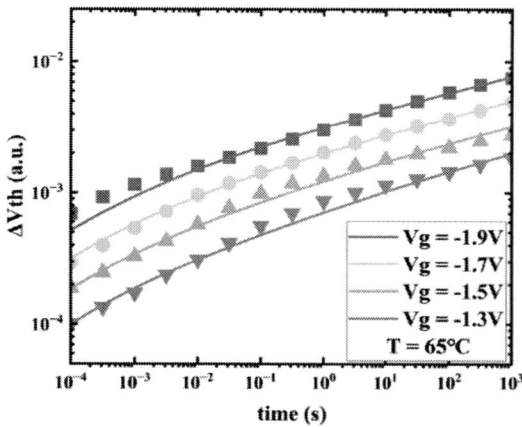

Fig. 7. Measured and modeled ΔVth-time curve during stress(data from [6]).

Furthermore, using the extracted parameters, the recovery curve after stress removal is simulated. The comparison with experimental data shows acceptable agreement, further validating the reasonableness of the parameters extraction. The recovery curve simulation demonstrates the framework's ability to accurately predict the behavior of the system post-stress removal, which is critical for understanding long-term device reliability.

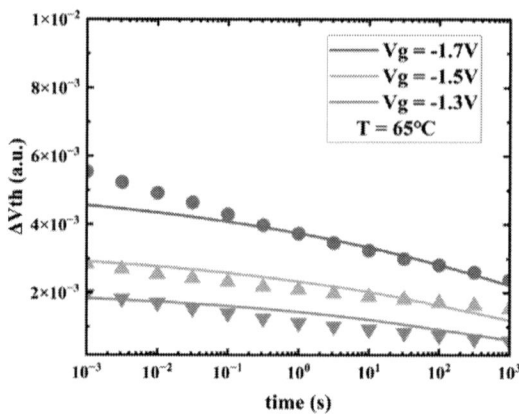

Fig. 8. Measured and modeled ΔVth - time curve after stress(data from[6]).

In this work, all model training and prediction processes were conducted on an Apple M2 CPU. The training time of the deep learning model is approximately 1 minute, while the prediction time is within 1 second. This demonstrates the superiority of the parameters extraction method and the rationality of the simulation architecture, as sufficiently accurate parameters can be obtained in a short period.

IV.   SUMMARY

This work presents an enhanced parameters extraction method for simulating BTI effects in GAA MOSFET devices, targeting the 5nm technology node. Utilizing a comprehensive 3D TCAD simulation framework, the method generates defect weights maps and employs a deep learning network to extract parameters of the two-state NMP model. The results demonstrate that the extracted parameters enable precise simulation of BTI-induced threshold voltage shifts, achieving a mean relative error of 6.29% compared to experimental data. This method significantly improves the efficiency of BTI modeling in GAA MOSFETs and facilitates further scaling of MOSFETs. Future work will include the incorporation of defect generation and recombination mechanisms, which is important to the PBTI for the NMOSFET with high K dielectric and the recovery stage. Additionally, efforts will focus on refining the deep learning model to further reduce the error margin and exploring its application to other reliability phenomena in nanoscale devices.

ACKNOWLEDGMENT

This work was supported in part by NSFC 92364104 T2293700 and T2293702 and the National Key Research and Development 2022YFB4401704.

REFERENCES

[1] Wang, M. et al., IEEE International Reliability Physics Symposium (IRPS), pp. 1-6 (2019).
[2] Rzepa et al., Microelectronics Reliability, 85, pp. 49-65 (2018).
[3] Sentaurus™ Device User Guide.
[4] N. Loubet, Symposium on VLSI Technology, pp. 230-231 (2017).
[5] F. Pedregosa et al., J. Mach.Learn. Res., vol. 12, no. 10, pp. 2825–2830 (2017).
[6] Choudhury et al., IEEE International Reliability Physics Symposium (IRPS), pp. 1-8 (2021).

Foundamentals of Low-Resistive Indium-Violet Phosphorene Top Contact: an ab-initio NEGF Study

Huaipeng Wang [1], Sicheng Liu [1], Shuaihong Li [1], Zhifang Liu [2], Yilin Sun [2], Jianlong Xu [3], Dan Xie*[1]

[1] School of Integrated Circuits, Tsinghua University, Haidian District, Beijing, 100084, P. R. China
[2] School of Integrated Circuits and Electronics, Beijing Institute of Technology, No 5 Zhongguancun South Street, Haidian District, Beijing, 100081, P. R. China
[3] Institute of Functional Nano & Soft Materials (FUNSOM), Jiangsu Key Laboratory for carbon-Based Functional Materials & Devices, Soochow University, 199 Ren'ai Road, Suzhou, Jiangsu, 215123, P. R. China

* Email: whp20@mails.tsinghua.edu.cn, xiedan@tsinghua.edu.cn

Abstract—**Two-dimensional semiconductor-metal contacts with induced mid-gap states, resulting in Fermi pinning, have been impeding the development of next-generation scaled devices. Many efforts have been devoted to find optimized evaporation methods to make ultra-clean vdW contacts. However, the fundamental mechanisms underlying the origin of induced mid-gap states and how the mid-gap states influence contact properties are still urgent to be explored. Herein, ab-initio NEGF calculations are conducted to discover the origin of mid-gap states arising in In/Au-VP vdW/bonding contacts. It is demonstrated that the bonding interactions of In-VP bonding contacts result in contact resistance as low as 0.600 kΩ·μm, which is as five orders of magnitudes as lower than case of Au-VP vdW contact. This work will pave the way for the development of next-generation devices based on 2D materials.**

Keywords—contact resistance, violet phosphorene, DFT, NEGF

I. INTRODUCTION

Semiconducting two-dimensional (2D) materials such as transition-metal dichalcogenides (TMDs) (e.g. MoS_2, WSe_2, $MoTe_2$), III-VI chalcogenides (e.g. InSe, GaSe), 2D phosphorus allotropes (e.g. black phosphorene, blue phosphorene, violet phosphorene), have been extensively studied for the mechanical, electronic, thermal, and optoelectronic properties. However, the application of 2D materials for next-generation scaled devices have been limited by the semiconductor-metal contacts which hinder the exploration of fundamental physics and the development of device applications.

For 3D metal-semiconductor contacts, ideal contacts could be realized through reducing Schottky barrier heights (SBHs) by means of controlling the work functions of metals, according to Schottky-Mott rules [1]. However, Fermi-level pinning induced by the mid-gap states arising from the metal-semiconductor interfaces impedes the Schottky-Mott rules. Many efforts have been devoted to building van der Waals (vdW) contacts in order to vanish the induced mid-gap states. Previous investigations showed that semimetal or indium contacts facilitate the controlling of mid-gap states to reduce contact resistances in n-type MoS2 and p-type WSe2 effectively [2-4]. However, the fundamental mechanisms underlying the origin of induced mid-gap states and how mid-gap states influence the contact properties are still urgent to be explored. Herein, different contact types between metals and violet phosphorene are investigated to explore the underlying physical mechanisms which influence the contact resistances.

II. METAL-SEMICONDUCTOR CONTACTS OF VP

Previous investigations have demonstrated that ultra-clean vdW contacts could be realized by direct evaporation of thin indium and thick gold electrodes on top to form alloy contacts [4, 5]. Pure gold electrode fabricated by evaporation onto few-layer 2D TMDs was also demonstrated to conduce to low-resistive contacts [6]. Herein, metal-semiconductor contacts between Au/In and VP are investigated through ab-initio non-equilibrium Green function (NEGF) methods.

Firstly, Au-VP and In-VP contact models are built as two-electrode top contact models, where metal electrodes extend along z-axis direction and terminate at the VP, as shown in Fig. 1(a-b). The distances of metal electrode separated from VP layers were calculated as the sum of vdW radii of metal atoms and phosphorus atom to retain vdW interactions between metal electrodes and VP layers. Bilayer VPs are put in the middle of supercell with periodic boundary conditions (PBCs) in the x-y plane. In addition to models with clean vdW interactions, models with bonding interactions are obtained through geometry optimization processes. Geometry optimization is conducted by means of density functional theory (DFT) calculations with SIESTA [7]. GGA-PBEsol parameterization was used as an exchange correlation functional. Dispersion corrections were used to include vdW interactions [8]. Geometries of metal layers closest to VP layers are relaxed under the maximum force tolerance of 0.01 eV/Å. Brillouin zone was sampled using Monkhorst-Pack method with a 3×3×1 k-mesh.

979-8-3503-6184-1/24 $31.00 © 2024 IEEE

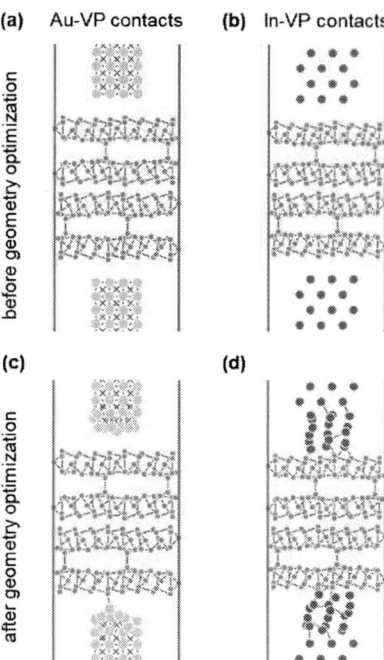

(a) Au-VP contacts   **(b)** In-VP contacts

(c)   **(d)**

before geometry optimization

after geometry optimization

Fig. 1. Schematic of different metal-VP contacts. (a) Au-VP and (b) In-VP contact structures were set up without geometry optimization, where distances between metal and VP were set according to atomic vdW radii. (c) Au-VP and (d) In-VP contact structures were set up after geometry optimization.

Metal-VP contact models after geometry optimization are shown in Fig. 1(c-d). Au atoms close to VP layers are repelled away with few Au-P bond forming while In atoms close to VP layers are attracted to VP layers and form In-P bonds. In order to probe into the electronic structures in the metal-VP interface region, density of states (DOS) and projected density of states (PDOS) are calculated in the energy range around Fermi levels, as shown in Fig. 2. The broad density of states within VP band gaps as shown in Fig. 2(a-d) originate from metal electrodes. As for Au-VP vdW contact, a mid-gap state peak (E=-1.0 eV, near VBM) within VP's band gap arises from the DOS peak of Au at the same position, attributed to the overlap between Au-6s and P-3p orbitals under vdW interactions. Such mid-gap state isolated from conduction bands and valence bands would result in Fermi-level pinning. As for In-VP vdW contact, the work function matching between In and VP lowers the CBM of VP down to near Fermi-level, and the vdW interactions result in the mid-gap state (E=-0.37 eV) in the middle of band gap. As for Au-VP bonding contact, the isolated mid-gap state (E=-1.0 eV) is weakened, however, additional mid-gap states (E=0.26 eV, 0.36 eV) exist near CBM and accrete with the edge of conduction bands, which is in favor of forming low-resistive contacts. As for In-VP bonding contact, strong In-VP bonding interactions result in the broad density of states within primitive VP band gap.

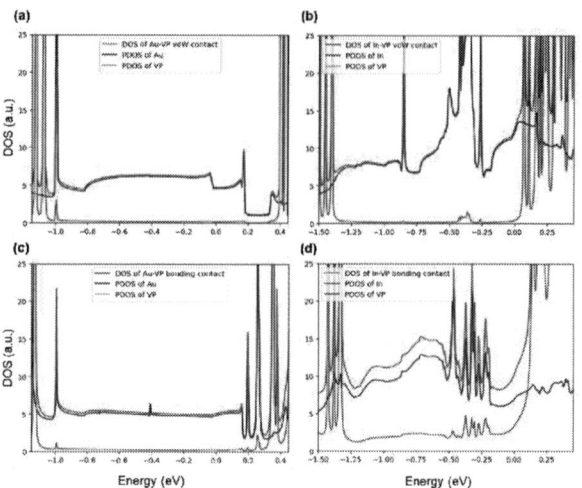

Fig. 2. DOS/PDOS of different contact types. (a) Au-VP vdW contact; (b) In-VP vdW contact; (c) Au-VP bonding contact; (d) In-VP bonding contact. Energy reference is set to Fermi level.

III. NEGF INVESTIGATION OF METAL-VP CONTACTS

A. I-V Characteristics

Electronic transport through the metal-VP contacts along z-axis direction is addressed using NEGF formalism, as implemented in the TranSIESTA and TBtrans codes [9]. The systems are separated into following regions: bottom buffer region, bottom electrode region, scattering region, top electrode region, and top buffer region. Metal atoms within a primitive cell are included in scattering region to address the metal-VP contacts. Metal atoms in buffer region are used to retain the bulk properties of electrode near vacuum layer, which are excluded from NEGF calculations. In electrode calculations, 1×1×100 k-mesh is used to ensure the precise electronic structure in the following NEGF calculations.

The transmission function is calculated as:

$$T(E) = Tr[G(E)\Gamma_{bottom}(E)G^{\dagger}(E)\Gamma_{top}(E)] \quad (1)$$

Where $G(E) / G^{\dagger}(E)$ is the retarded/advanced Green's function, and $\Gamma_{bottom/top}(E)$ is the broadening matrices:

$$G(E) = [(E+i\eta)S - H - \Sigma_{top}(E) - \Sigma_{bottom}(E)]^{-1} \quad (2)$$

$$\Gamma_{bottom/top}(E) = i[\Sigma_{bottom/top}(E) - \Sigma^{\dagger}_{bottom/top}(E)] \quad (3)$$

Here H, S, and $\Sigma_{bottom/top}$ are the Hamiltonian, the overlap matrix, and the electrode self-energies, respectively.

The electrical current is calculated from the transmission coefficients $T(E)$ according to:

$$I_{bottom \to top} = \frac{G_0}{2|e|} \int dE \int_{BZ} dk T(E)[f_{top}(E) - f_{bottom}(E)] \quad (4)$$

Where $f_{top}(E)$ and $f_{bottom}(E)$ correspond to the Fermi distribution in the bottom and top electrodes, G_0 is the quantum of conductance for the spin-degenerate case, and BZ denotes the integral over the Brillouin zone.

979-8-3503-6184-1/24 $31.00 © 2024 IEEE   917

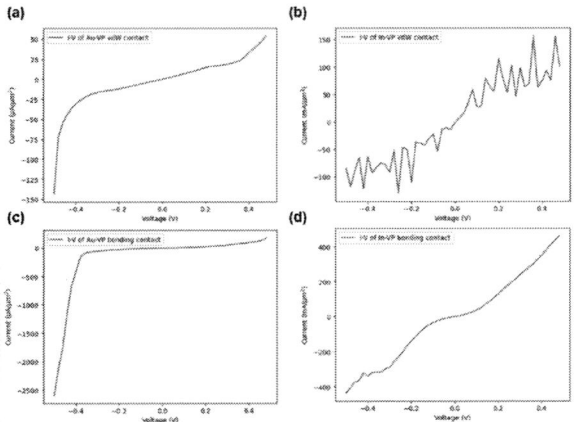

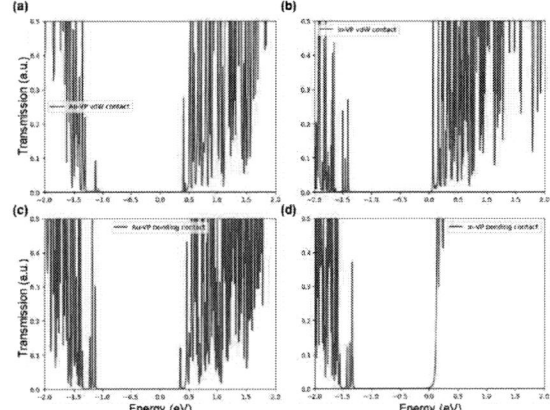

Fig. 3. Current-voltage characteristics through metal-VP vdW/bonding contact systems. (a) Au-VP vdW contact; (b) In-VP vdW contact; (c) Au-VP bonding contact; (d) In-VP bonding contact.

Current-voltage characteristics of metal-VP contacts are calculated as shown in Fig. 3. Contact resistances are retrieved by the following equations:

$$V = I(2R_{contact} + R_{VP}) \tag{5}$$

$$V = 2V_{contact}^{Hartree} + V_{VP}^{Hartree} \tag{6}$$

Where $R_{contact}$ is the contact resistance, R_{VP} is the channel resistance of VP layers, $V_{contact}^{Hartree} / V_{VP}^{Hartree}$ is the Hartree potential drop through contact/channel region. Contact resistances are calculated as shown in TABLE I.

TABLE I. CONTACT RESISTANCES OF DIFFERENT CONTACT TYPES

Contact Resistance	vdW contact	Bonding contact
Au-VP	6.68 MΩ·μm	1.60 MΩ·μm
In-VP	2.00 kΩ·μm	0.600 kΩ·μm

B. Fundamentals of low-resistive In-VP contact

As for Au-VP contacts, dominant vdW interactions result in Schottky contact in both vdW contact and bonding contact. Nevertheless, the mid-gap states near CBM arising from Au-VP bonding interactions still result in the reducing contact resistance. As for In-VP contacts, appropriate work function matching and overlap of In-5p and P-3p orbitals result in the Ohmic contacts. In-VP bonding interactions further reduce the resistances of the In-VP contacts.

In order to further explore the underlying fundamentals of metal-VP contacts, transmission functions of different contact types are calculated as shown in Fig. 4. As for Au-VP contacts, the mid-gap states near the edge of CBM result in the increasement of transmission coefficients around CBM edge, which contribute to reducing contact resistance between Au electrodes and VP layers. As for In-VP contacts, bonding interactions increase the transmission coefficients near Fermi level significantly, which result in the ideal contact between In electrode and VP layers.

Fig. 4. Transmission functions through metal-VP contacts. (a) Au-VP vdW contact; (b) In-VP vdW contact; (c) Au-VP bonding contact; (d) In-VP bonding contact.

IV. CONCLUSION

In this work, metal-VP contacts were investigated systematically using ab-initio NEGF methods. It was demonstrated that bonding interactions facilitate the formation of low-resistive contacts. In-VP bonding contact resulted in contact resistance as low as 0.600 kΩ·μm. The underlying mechanisms of low-resistive contacts were explored through DOS and transmission calculations. This discovery will enable the realization of the next generation of electronics of 2D materials.

ACKNOWLEDGMENT

The authors are grateful for financial support from the National Natural Science Foundation of China (Nos. 52072204 and 62104017). The authors also acknowledge the Beijing Super Cloud Computing Center (BSCC) for providing HPC resources that have contributed to the research results reported within this paper. URL: http://www.blsc.cn/.

REFERENCES

[1] W. Schottky, "Halbleitertheorie der sperrschicht," Naturwissenschaften, vol. 26, no. 52, pp. 843-843, 1938.

[2] G. Kwon et al., "Interaction-and defect-free van der Waals contacts between metals and two-dimensional semiconductors," Nature Electronics, vol. 5, no. 4, pp. 241-247, 2022.

[3] P.-C. Shen et al., "Ultralow contact resistance between semimetal and monolayer semiconductors," Nature, vol. 593, no. 7858, pp. 211-217, 2021.

[4] Y. Wang et al., "Van der Waals contacts between three-dimensional metals and two-dimensional semiconductors," Nature, vol. 568, no. 7750, pp. 70-74, 2019.

[5] Y. Wang and M. Chhowalla, "Making clean electrical contacts on 2D transition metal dichalcogenides," Nature Reviews Physics, vol. 4, no. 2, pp. 101-112, 2022.

[6] Y. Wang et al., "P-type electrical contacts for 2D transition-metal dichalcogenides," Nature, vol. 610, no. 7930, pp. 61-66, 2022.

[7] J. M. Soler et al., "The SIESTA method for ab initio order-N materials simulation," Journal of Physics: Condensed Matter, vol. 14, no. 11, p. 2745, 2002.

[8] S. Grimme, "Semiempirical GGA‐type density functional constructed with a long‐range dispersion correction," Journal of computational chemistry, vol. 27, no. 15, pp. 1787-1799, 2006.

[9] N. Papior, N. Lorente, T. Frederiksen, A. García, and M. Brandbyge, "Improvements on non-equilibrium and transport Green function techniques: The next-generation transiesta," Computer Physics Communications, vol. 212, pp. 8-24, 2017.

Gold Thermocompression Wafer Bonding for Quartz MEMS Applications

Ting Yang*[1], Dongxiang Han [2], Jun Xu [1], Tian-Ling Ren [1]

[1] The School of Integrated Circuit, Tsinghua University, Beijing 100084, China
[2] The State Key Laboratory of Mechanics and Control of Mechanical Structures, Nanjing University of Aeronautics and Astronautics, Nanjing 210016, China

* Email: andrewfine@semi.ac.cn

Abstract—Although the performance of quartz MEMS devices has been well proven, mass production and application are not achieved due to high packaging cost. In this paper, gold thermocompression wafer bonding is first proposed to assist the assembly of quartz MEMS devices. Ultrasonic inspection shows a void-free bonding interface while tensile test indicates the bonding strength is 21.5 ± 2.1 MPa, which confirms the feasibility of this wafer bonding process. In the end, the bonding process has been introduced to fabricate a quartz MEMS accelerometer with three layers and expected to be used for the fabrication of other MEMS devices based on various brittle materials.

Keywords—gold thermocompression, wafer bonding, quartz MEMS, MEMS accelerometer

I. INTRODUCTION

Quartz crystal resonators and their components have been widely used since 1920s in timing devices and high-performance sensors thanks to the advantages of the quartz crystal such as high quality factor and excellent frequency stability in wide range of temperature [1]. In the 1970s, J.S Staudte [2] proposed to adopt wafer-level photolithography and etch techniques to fabricate quartz resonators, which finally resulted in the concept of quartz MEMS (QMEMS) technology. During the last five decades, QMEMS technology has dramatically reduced size and fabrication cost of quartz resonators and made it possible to realize complex quartz MEMS devices especially high-performance sensors such as vibrating beam accelerometers, Coriolis vibrating gyroscopes [3], resonant pressure sensors [4] etc..

However, high-cost packaging processes always limit mass production of complex quartz MEMS devices since multi-layer assembly and a vacuum or controlled atmosphere for operation are usually required for these devices. In current packaging technology for quartz MEMS devices, each wafer of different layers has to be first diced into separate dies and then different dies are assembled together using adhesive, glass frit, braze etc. into a metal or ceramic base such as TO-series [5], which obviously indicates high manufacturing cost.

Wafer-level packaging already adopted in silicon MEMS devices also provides new ideas for quartz MEMS devices, which means multi-layer assembly and capping processes are performed at wafer level before dicing process. Wafer-level packaging is usually realized by kinds of wafer bonding techniques including fusion bonding, anodic bonding, metal thermocompression bonding etc.. Quartz/Silicon surface-activated fusion bonding was presented by C. Han [6] to realize quartz MEMS accelerometers. But different materials will lead to mismatch in thermal expansion coefficients (Silicon: ~2.6ppm [7], Quartz: ~13.6ppm [8]), which is harmful for the performance of sensors in the wide temperature range.

Gold thermocompression wafer bonding, which uses a gold film as the interlayer is a promising technique for packaging quartz MEMS devices since there are three advantages. First, quartz and gold provide similar thermal behaviors (Quartz: ~13.6ppm, Gold: ~14.2ppm [9]). Second, the bonding strength and hermeticity which are important for packaging have already been proven in other studies [10]. Third, gold film does not generally need to be prepared before bonding since gold film has been deposited on most of quartz MEMS devices as wet etch mask and excitation electrode [11]. The principle, experiments and results will be presented in the following sections.

In this paper, gold thermocompression wafer bonding is first introduced to fabricate quartz MEMS devices. Details about the bonding process, results and applications will be presented in the following sections.

II. EXPERIMENTS

A. Wafer Preparation

As shown in Fig.1, two 300μm thick, 2.5-inch Z-cut crystal quartz wafers were used in the bonding experiment. There were convexities on one side of wafer A, which were fabricated by conventional photolithography and quartz wet etching techniques. First, the chromium-gold film (20 nm Cr and 200 nm Au) was deposited on one side of the wafer. Since lower surface roughness is beneficial to bonding [12], magnetron sputtering and evaporation, as two commonly used deposition techniques were compared. Second, the film was patterned by photolithography and then used as the mask of the following quartz etching process in a HF-NH_4F solution to form the convexities. The etching depth was more than 5μm in order to keep the bonding pressure concentrate on the convexities. The surface morphology of wafer A was shown in Fig. 2. The chromium-gold mask on the upper surface of the convexities would also serve as the bonding interlayer in the next bonding process.

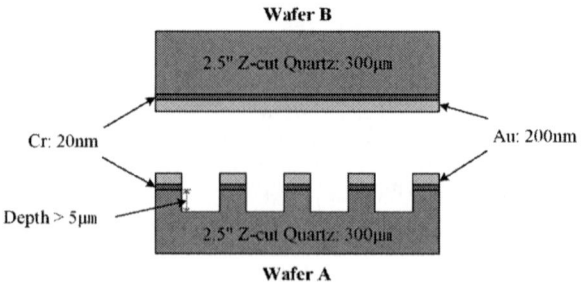

Fig. 1. Schematic cross-section of two wafers for the bonding experiment

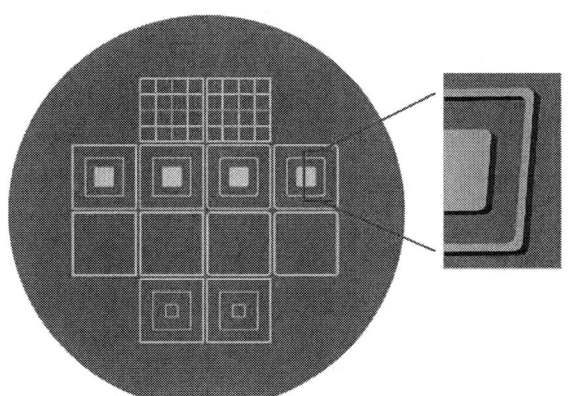

Fig. 2. The surface morphology of wafer A

B. Bonding Process

The major steps of gold thermocompression wafer bonding process were summarized in Fig. 3. Prior to wafer bonding, two wafers were immersed in a solution of H_2SO_4 : H_2O_2 = 3 : 1 at 140°C for 10 minutes in order to remove particles, organic and inorganic contaminants. A plasma treatment using Ar gas was followed in PVA TePla IoN40 plasma system to further clean the surfaces of both wafers and increase the bonding strength. The typical parameters were shown in Table I. The bonding was carried out under vacuum (lower than 10^{-3} mbar) in an EVG 510 wafer bonder. The optimized temperature and pressure profile for the wafer bonding process was presented in Fig. 4.

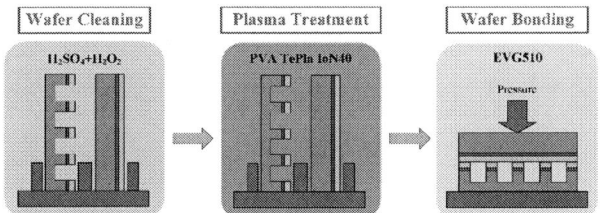

Fig. 3. The major steps of gold thermocompression wafer bonding process

TABLE I.   TYPICAL PARAMETERS OF THE PLASMA TREATMENT

Parameter	Unit	Value
Ar flow	sccm	100
RF Power	W	100
Process Time	s	50

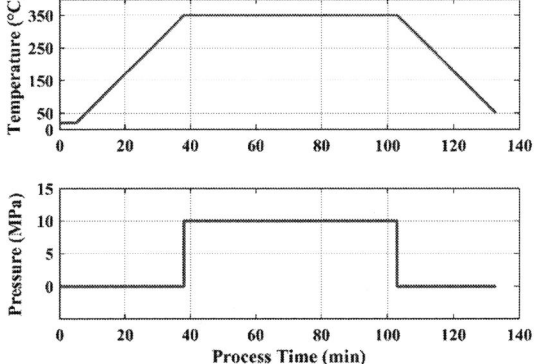

Fig. 4. The optimized temperature and pressure profile during the bonding

III.   RESULTS AND DISCUSSION

The surface profiles for different deposition techniques were both obtained from AFM (Atomic Force Microscopy) in Fig. 5, which indicated the film deposited by magnetron sputtering had lower root-mean-square roughness (R_q = 1.89nm) compared with one by evaporation (R_q = 3.21nm). So magnetron sputtering was used in the bonding experiment.

The quality of the wafer bonding in this paper was investigated by SAM (Scanning Acoustic Microscopy) and tensile test.

Fig. 6 showed the SAM image of the wafer pair. Black in the image represented well bonded area (no signal reflection) while the shallow gray area indicated lower quality bonded area (reflected signal from bonded interface). So it could be observed that the bonding was void-free with the exception of non-patterned areas.

To obtain the mechanical bonding strength, four 3mm×3mm solid-square samples in the bonded wafer pair were diced. Each sample was placed into Shan Du SLR500 tensile/compressive test tool as shown in Fig. 7 which applied a tensile force $F(t)$ until failure. The bonding strength σ was calculated from the peak tensile force F_{peak} and the bonding area A_{bond}.

$$\sigma = \frac{F_{peak}}{A_{bond}} \qquad (1)$$

The minimum attach strength σ_{std} which a 3mm×3mm die (A_{bond}=0.09cm^2) withstood could be calculated according to MIL-STD-883L Method 2027.2 in (2). As shown in Fig.8, the bonding strength σ was 21.5±2.1MPa, which was much higher than the strength criterion σ_{std} ~3.53MPa.

$$\sigma_{std} = \frac{(3.32 \times \log(A_{bond} / 6.45) + 13.3) \times 4.45}{A_{bond}} \qquad (2)$$

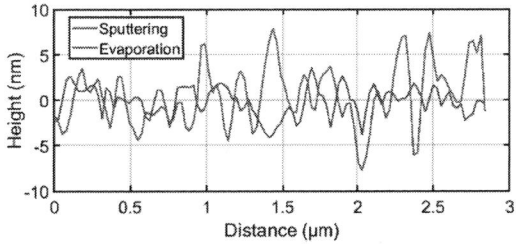

Fig. 5. Surface profiles of gold films by different deposition techniques

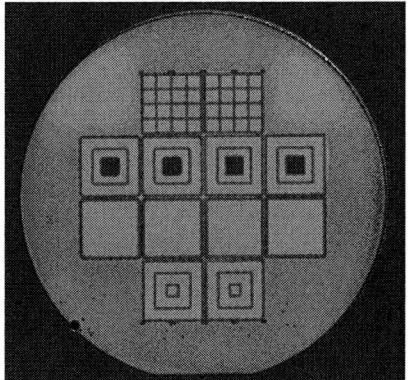

Fig. 6. The SAM image of the bonded wafer pair

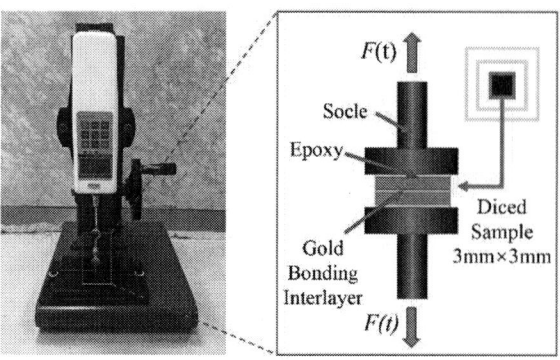

Fig. 7. The tensile test to determine the bonding strength

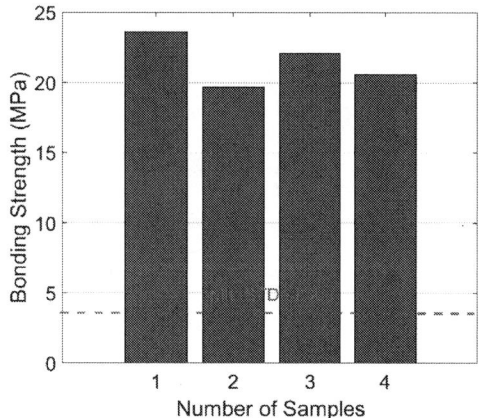

Fig. 8. Results of bonding strength test

Fig. 9 demonstrated the design of the quartz resonant MEMS accelerometer which consisted of three 2.5-inch Z-cut quartz wafers. The middle wafer with the thickness of 300μm containing a movable structure was bonded between two damping wafers (500μm). In this paper, the triple-stack assembly of this quartz accelerometer was realized by gold thermocompression wafer bonding in order to obtain high-strength interconnection and accurate damping gap. Compared with the process in Fig. 3, a triple-stack bond alignment step using EVG620NT aligner should be added between the plasma treatment and the wafer bonding. The photo of the tripe-stack bonded wafer was presented in Fig.10.

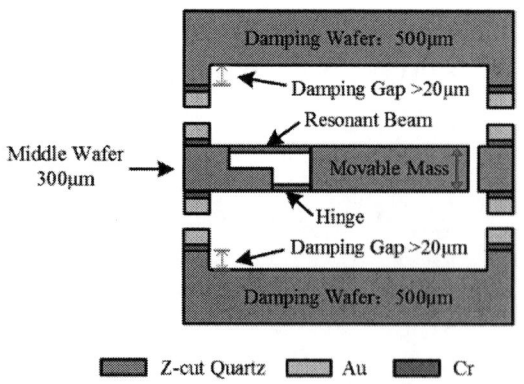

Fig. 9. Schematic cross-section of triple-stack quartz accelerometer

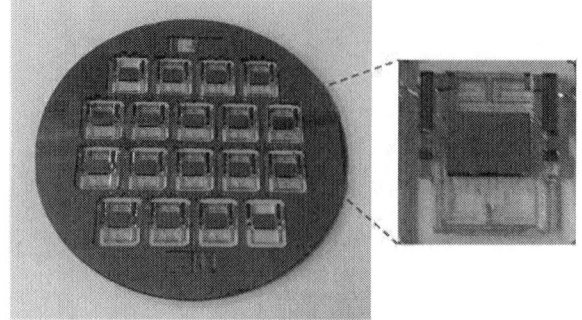

Fig. 10. The photo of triple-stack quartz accelerometer wafer

IV. CONCLUSION

Gold thermocompression wafer bonding was confirmed for its usability in the quartz-quartz interconnection. Intermediate gold layer deposited by magnetron sputtering showed excellent surface roughness. The bonding interface was void free with great bonding strength 21.5 ± 2.1 MPa, which satisfied the strength criterion of MIL-STD-883L. The process in this paper was successfully applied on the triple-stack assembly of the quartz resonant MEMS accelerometer and can be proposed in wafer-level interconnections between various brittle materials.

Future work has been focused on the vacuum wafer-level packaging based on gold thermocompression wafer bonding which will make further improvement in cost reduction and miniaturization of quartz MEMS devices.

REFERENCES

[1] J. S. Danel and G. Delapierre, "Quartz: a material for microdevices," J. Micromech. Microeng., vol. 1, no. 4, pp. 187–198, Dec. 1991.

[2] J. S. Staudte, "Microresonator of tuning fork configuration," U.S. Patent 3 683 213, Mar. 9, 1971.

[3] S. Zotov et al., "Compact in-run navigation grade IMU based on quartz MEMS," in 2020 IEEE/ION PLANS, Portland, OR, USA, Apr. 2020, pp. 728–733.

[4] W. Zhang et al., "A masterpiece of superior crystals: quartz resonant pressure sensor—a review," IEEE Sens. J., vol. 24, no. 7, pp. 9278–9298, Apr. 2024.

[5] G. R. Newell, K. S. Lewallen, S. D. Orlosky, B. D. Egley, "Accelerometer and method of manufacture," U.S. Patent 5 755 978, Dec. 5, 1996.

[6] C. Han, Y. Zhao, and C. Li, "A novel resonant accelerometer based on quartz on silicon (QoS)," in 2019 IEEE INERTIAL, Naples, FL, USA, Apr. 2019, pp. 1–4.[1]

[7] A. Masolin, P. O. Bouchard, R. Martini, and M. Bernacki, "Thermo-mechanical and fracture properties in single-crystal silicon," J. Mater. Sci., vol. 48, no. 3, pp. 979–988, Feb. 2013.

[8] J. A. Kosinski, J. G. Gualtieri, and A. Ballato, "Thermoelastic coefficients of alpha quartz," IEEE Trans. Ultrason., Ferroelect., Freq. Contr., vol. 39, no. 4, pp. 502–507, Jul. 1992.

[9] D. P. Arnold, M. Saumer, and Y. Yoon, "Additive processes for metals," in MEMS Materials and Processes Handbook, Boston, MA, USA: Springer, 2011. pp. 137-191.

[10] N. Malik, H. Tofteberg, E. Poppe, T. G. Finstad, and K. Schjolberg-Henriksen, "Hermeticity and reliability of Au-Au thermocompression bonds, realized at low temperature," ECS Trans., vol. 64, no. 5, pp. 167–176, Aug. 2014.

[11] O. Le Traon et al., "The fairy world of quartz vibrating MEMS," in 2012 European Frequency and Time Forum, Gothenburg, Sweden, Apr. 2012, pp. 214–220.

[12] B. Rebhan and K. Hingerl, "Physical mechanisms of copper-copper wafer bonding," Journal of Applied Physics, vol. 118, no. 13, p. 135301, Oct. 2015.

An Adaptive Threshold Analog Front-End Circuit for Direct ToF LiDAR

Jianping Guo [1], Xiaoyang Zeng [1], Wenhong Li [1], Mingyu Wang*[1]

[1] State-key Laboratory of ASIC and system, Fudan University, Shanghai 201203, CHINA

* Email: {20112020056, mywang}@fudan.edu.cn

Abstract—The analog front-end (AFE) circuit of the light detection and ranging (LiDAR) receivers need to provide time-of-flight (ToF) when the laser echo arrives. Accurate ToF is crucial for the ranging accuracy of LiDAR receivers. However, targets with different reflectivity will give different timing results even at the same distance due to the light intensity. This paper proposes an AFE circuit for LiDAR receiver. This AFE circuit can adaptively adjust the threshold based on photocurrent to reduce walk error under different light intensities. Proposed AFE is implemented in 55nm CMOS process. Simulation results show that within the photocurrent range of 3 to 20 *uA*, the timing error is less than 0.8 *ns*.

Keywords—analog front-end, LiDAR, walk error, DToF, linear-mode avalanche photo diode .

I. INTRODUCTION

Light Detection and Ranging (LiDAR) system is widely used in fields such as remote sensing, autonomous driving, and military applications. The Analog Front-End (AFE) circuit of the LiDAR receiver is used to detect laser echoes and convert optical signals into electrical signals. For LiDAR systems that use the Direct Time-of-Flight (DToF) method for ranging, AFE needs to mark the timing moment when the laser echo arrives. This timing information is used to obtain the Time-of-Flight (ToF) of the light to calculate the distance of the target object.

Currently, for LiDAR based on linear-mode Avalanche Photo Diode (APD) as the detector, the AFE mostly uses a comparator with constant threshold to mark timing moment (known as leading-edge detection method [1]). A typical AFE circuit is shown in Fig. 1, it consists of a transimpedance amplifier (TIA) and a comparator. TIA converts the photocurrent to a voltage V_o , and the comparator flips when V_o exceeds the comparator's threshold, AFE marks this moment as the arrival time of the laser echo. However, due to the varying strengths of the echo light intensity, the linear-mode APD generates photocurrents of different amplitudes, resulting in varying responses from the TIA. As shown in Fig. 2(a), even if the laser echoes arrive at the same time, the comparator will flip at different moments, Δt in Fig. 2(a) is called timing walk error [2].

In this paper, an adaptive threshold adjustment method is proposed, which can automatically adjust the threshold voltage of the comparator according to the magnitude of the photocurrent, achieving the effect shown in Fig. 2(b) and reducing the walk error under different light intensities.

II. PROPOSED AFE CIRCUIT

A. AFE Overview

Fig. 3(a) shows the proposed AFE circuit, it consists of a TIA (which uses a capacitor and resistor in series as a feedback path), a photocurrent copy circuit, a threshold

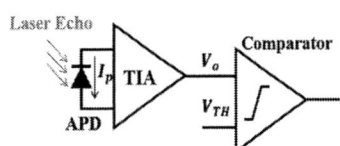

Figure 1. A typical AFE circuit for APD-based LiDAR.

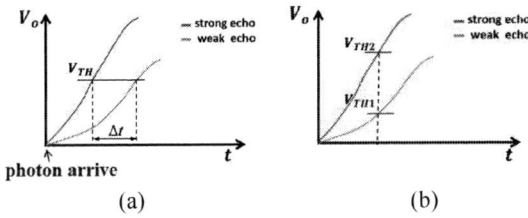

Figure 2. (a) Walk error in conventional leading-edge detection method. (b) Proposed adaptive threshold detection method.

generation circuit (THGC) and a comparator.

When the APD receives the laser echo, the photocurrent flows through the feedback path, producing a voltage on R_F, and integrating on C_F. Due to negative feedback, the voltage at the inverting input terminal of Amp_a is approximately equal to that at the non-inverting input terminal, and the same applies to Amp_b. Therefore, the voltages at the two terminals of R_{F_copy} are approximately equal to R_F. By designing $R_{F\_copy} = R_F$, I_{copy} will then track the magnitude of I_p. The bias current I_{bias} is necessary, and this will be explained later. Hence, the output current of the photocurrent copy circuit can be written as:

$$I_{TH} = I_p + I_{bias} . \qquad (1)$$

I_{TH} flows into the THGC (this work uses a transimpedance amplifier to implement it) to produce an output voltage V_p, then V_p can be expressed as:

$$V_p = V_H - I_{TH} \cdot R_{TH}$$
$$= V_H - (I_p + I_{bias}) \cdot R_{TH} \qquad (2)$$

The inverting input of the comparator is directly connected to the bottom plate of C_F, so the differential voltage input to the comparator is :

$$V_{dif} = V_p - V_L$$
$$= V_H - V_L - (I_p + I_{bias}) \cdot R_{TH} \qquad (3)$$

979-8-3503-6184-1/24 $31.00 © 2024 IEEE

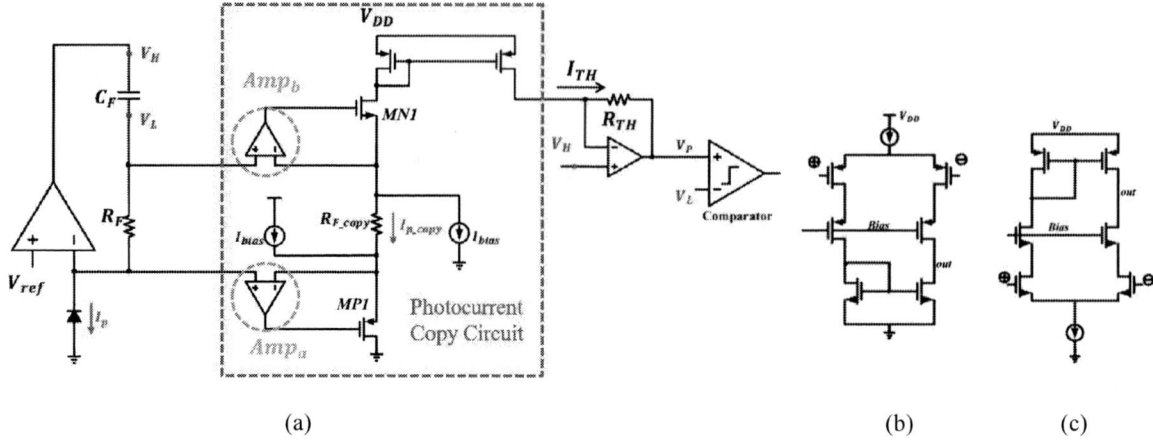

(a)   (b)   (c)

Figure 3. (a) Proposed adaptive threshold AFE Circuit. (b) Schematic of Amp_a and (c) Amp_b.

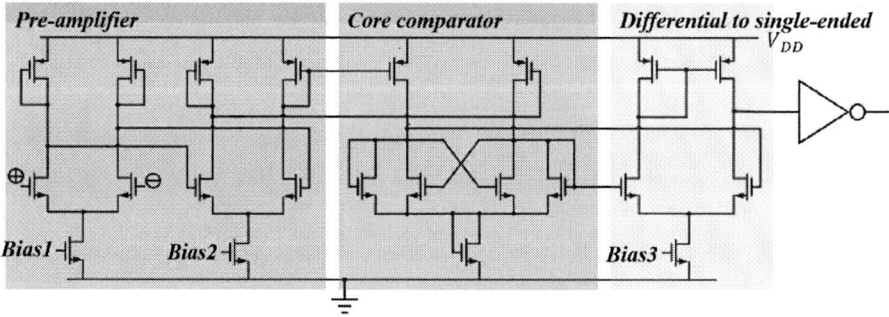

Figure 4. Schematic of comparator.

In Eq. (3), $V_H - V_L$ represents the voltage difference across C_F, which can be expressed as $\frac{I_p \cdot t}{C_F}$, so Eq. (3) can be written as:

$$V_{dif}(t) = \frac{I_p \cdot t}{C_F} - (I_p + I_{bias}) \cdot R_{TH} \qquad (4)$$

The first term of Eq.(4) is the voltage converted from the photocurrent by TIA, and the second term is the threshold generated by the THGC. When V_{dif} crosses zero, the comparator flips and the laser echo arrival moment is marked. Setting $V_{dif} = 0$, the expression for t can be obtained as :

$$t = (1 + \frac{I_{bias}}{I_p}) \cdot R_{TH} \cdot C_F \qquad (5)$$

When I_p is much larger than I_{bias}, the marked moment t is approximated to the constant $R_{TH}C_F$.

From Eq.(4), if $I_p = 0$, then when $I_p = 0$, $V_{dif} = 0$. At this point, the AFE is easily affected by noise, leading to incorrect flips. That's why I_b is necessary.

B. Implementation of Amp_a and Amp_b

The schematic of Amp_a and Amp_b is shown in Fig. 3(b) and Fig. 3(c). Taking Amp_a as an example, because the gate voltage of MP1 needs to be lower than the source, Amp_a uses PMOS as the differential input pair. Additionally, to increase the output swing towards GND, Amp_a modified the telescopic operational amplifier by removing a pair of biasing transistors. The design of Amp_b is similar to Amp_a.

C. Implementation of Comparator

The comparator consists of three main parts. The first part consists of two differential amplifiers in cascade, each using diode connected PMOS transistor as the active load. The use of multi-stage preamplifiers enables sufficient gain while maintaining a large bandwidth for the comparator [3]. The second part is core amplifier, which uses cross-coupled transistors as loads, and the positive feedback mechanism significantly increases the amplifier's gain while reducing offset voltage. The third part is a buffer amplifier that converts the differential input to a single-ended output. Finally, an inverter is cascaded for shaping the output signal of the comparator.

979-8-3503-6184-1/24 $31.00 © 2024 IEEE   923

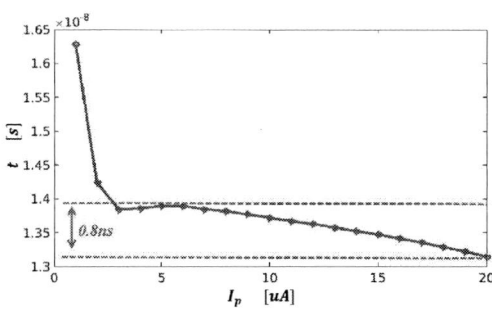

Figure 5. Timing moments of the AFE circuit reaching threshold under different photocurrent. intensities.

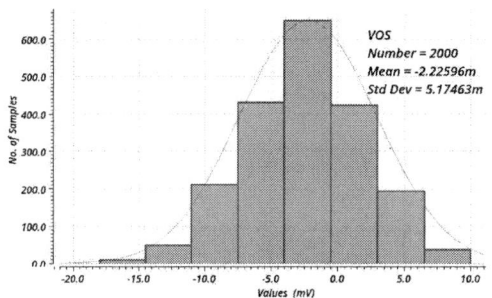

Figure 6. Monte-Carlo simulation of comparator offset voltage. (27-degrees temperature)

III. SIMULATION RESULTS

The proposed AFE circuit is implemented and simulated in 55 nm CMOS process. In this paper, I_{bias} is designed to be 500 nA, $R_{TH} \approx 10\ k\Omega$, $C_F \approx 218\ fF$. In the simulation setup, a 1 pF capacitor is loaded at the inverting input of the TIA to simulate the parasitic capacitance of the APD, and 1 pF is a common parasitic value.

Different magnitudes of photocurrent I_p are fed into the AFE at the same time, with I_p ranging from 1 to 20 uA. The moment at which the comparator's input differential voltage reaches 0 is shown in Fig. 5. From the simulation results, within the range of photocurrent from 3 to 20 uA, the relative timing error is less than 0.8 ns.

Fig.6 illustrates the offset voltage of the comparator, based on 2000 Monte Carlo simulations at 27 degrees Celsius.

IV. CONCLUSION

This paper proposes an AFE circuit for application in LiDAR receivers to address walk error caused by varying laser echo intensities. The proposed AFE circuit can automatically adjust the trigger threshold based on the intensity of light, ensuring consistent triggering time across different light intensities. Within the range of photocurrent from 3 μA to 20 μA, the walk error is less than 0.8 ns.

In addition, it should be noted that in future designs, the bias current I_{bias} and bias resistor R_{TH} need to be reasonably designed according to factors such as the offset voltage of the comparator, aiming to maintain sensitivity while reducing false triggering rates.

REFERENCES

[1] P. Palojarvi, T. Ruotsalainen and J. Kostamovaara, "A 250-MHz BiCMOS receiver channel with leading edge timing discriminator for a pulsed time-of-flight laser rangefinder," in IEEE Journal of Solid-State Circuits, vol. 40, no. 6, pp. 1341-1349, June 2005, doi: 10.1109/JSSC.2005.848022.

[2] X. Wang, R. Ma, D. Li, H. Zheng, M. Liu and Z. Zhu, "A Low Walk Error Analog Front-End Circuit With Intensity Compensation for Direct ToF LiDAR," in IEEE Transactions on Circuits and Systems I: Regular Papers, vol. 67, no. 12, pp. 4309-4321, Dec. 2020, doi: 10.1109/TCSI.2020.3022714.

[3] H. Zheng et al., "A Linear-Array Receiver AFE Circuit Embedded 8-to-1 Multiplexer for Direct ToF Imaging LiDAR Applications," in IEEE Transactions on Circuits and Systems I: Regular Papers, vol. 69, no. 12, pp. 5050-5058, Dec. 2022, doi: 10.1109/TCSI.2022.3204639.

Design of Ultra-Broadband Metamaterial Absorber from Infrared to Terahertz

Xiangze Liu [1], Wenbin Zhou [1], Tiantian Shi [1], Yiming Liao *[2], Feng Yan[1] and Xiaoli Ji [1]

[1] *School of Electronic Science and Engineering, Nanjing University, Jiangsu 210046, China*
[2] *School of Electronic and Optical Engineering, Nanjing University of Science and Technology, Jiangsu 210094, China*

* Corresponding Author's Email: liaoyiming@njust.edu.cn

Abstract—Microbolometers can operate in the infrared to terahertz regions and have important applications in high-definition imaging, temperature measurement, and ranging. However, these detectors face technical bottlenecks in the transition region from far-infrared to terahertz, such as weak spectral absorption capability, narrow bandwidth, and manufacturing difficulties. This paper designs a metamaterial absorber composed of Ti-Si-SiO$_2$-Al, which is compatible with standard CMOS processes. The absorber couples different modes of surface plasmon resonance with the intrinsic absorption of the lossy material SiO$_2$. It achieves an average absorption of 79% in the 8-52 μm range and exhibits good polarization and angle insensitivity as well as spectral broadening capabilities. This provides technical support for achieving high-detection-efficiency infrared-to-terahertz microbolometers.

Keywords—*metamaterial absorber, infrared, terahertz, standard CMOS process*

I. INTRODUCTION

Detectors in the infrared and terahertz bands play a crucial role in military reconnaissance, environmental monitoring, non-destructive testing, and medical imaging [1-2]. Among them, photothermal detectors, represented by microbolometers, operate by absorbing spectral energy and converting it into electrical signals, thus playing a significant role in the infrared and terahertz domains [3].

High-performance microbolometers require well-designed energy absorption structures to enhance the interaction between light and matter. In the terahertz band, patch antennas are commonly used for terahertz spectral absorption, with resonant frequencies typically ranging from 0.1 to 5 THz. However, at higher frequencies, the inherent losses in antennas reduce reception efficiency, and their discontinuous absorption characteristics limit their broadband applications. In the infrared band, absorbers are often designed to increase the detector's absorption rate. Commonly used metamaterial absorbers are designed with the necessary dielectric constants and magnetic permeability, but these absorbers are mostly limited to the infrared domain. To enable microbolometers to operate simultaneously in the infrared and terahertz domains, Zhang et al. [4] developed a broadband photodetector using B-phase vanadium oxide, which exhibits a certain response in the 0.405 μm to 880 μm band, but its response is primarily concentrated in the 0.405-1.5 μm range, and its average detection rate across the full band needs improvement. Yang et al. [5] used reduced graphene oxide (RGO) films to construct a suspended photodetector, which can operate from the ultraviolet to the terahertz region. The RGO films were prepared by layer-by-layer deposition of the suspension followed by thermal treatment, but this process is costly and not conducive to integration.

Therefore, it is necessary to develop absorbers that can stably absorb across the entire target spectrum and are easy to integrate into detectors. This paper designs a Ti-Si-SiO$_2$-Al metamaterial absorber using materials compatible with CMOS processes. We study the impact of structural parameters on the absorption performance in the infrared-to-terahertz frequency range and reveal the principle of surface plasmon resonance-enhanced absorption in the metamaterial. This provides insights for further broadening the frequency range and achieving high absorption rates.

II. STRUCTURE AND SIMULATION

The structure of the microbolometer integrated with the metamaterial absorber, as shown in Fig. 1(a), consists of a temperature sensor and two supporting arms. The sensor, measuring 60 μm × 60 μm, includes a top-layer absorber and a serpentine polysilicon thermistor. The supporting arms provide structural support for the sensor. When incident light irradiates the absorber, it absorbs the radiation energy, causing a change in the device temperature. This, in turn, alters the resistance of the polysilicon thermistor, generating an electrical signal that can be detected in the circuit, ultimately achieving the purpose of detection.

Figs. 1(b) and (c) show the top view and side view of the Ti-Si-SiO$_2$-Al absorber unit cell, respectively. The top layer of the absorber is a periodically arranged array of metal Ti, with its dimensional parameters listed in Table 1. The dielectric layer consists of the lossless material Si and the lossy material SiO$_2$. The fabrication of the detector can be completed by depositing Si and Ti and performing simple etching in the post-CMOS process. This four-layer absorber is compatible with polysilicon microbolometers integrated using the CMOS process.

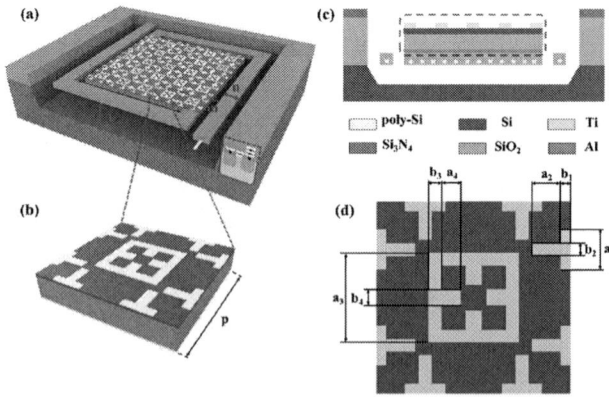

Fig. 1. (a) 3D schematic of the microbolometer integrated with the metamaterial absorber (m=4 μm, n=7 μm, l=75 μm). (b) Schematic of the absorber (thicknesses: top layer Ti-70 nm, bottom layer Al-130 nm, Si-0.22 μm, SiO$_2$-1.97 μm). (c) Cross-sectional schematic of the detector. (d) Structural dimensions of the absorber unit cell.

TABLE 1. Dimensions of the designed Absorber

Parameters	a_1	a_2	a_3	a_4	p
Value (µm)	2.5	1.8	5.6	1.2	12
Parameters	b_1	b_2	b_3	b_4	
Value (µm)	0.6	0.8	0.8	1	

The absorption spectrum of the absorber was simulated using the Finite-Difference Time-Domain (FDTD) method. Periodic boundary conditions were applied in the X and Y directions, while Perfectly Matched Layer (PML) conditions were applied in the Z direction. The refractive indices of Si, SiO_2, and Ti were obtained from the Palik's handbook, and the refractive index of Al was obtained from the CRC handbook. The grid accuracy was set to 8, and the simulation time was set to 8000 fs. The metamaterial was illuminated with a TM-polarized (electric field along the x-direction) plane wave.

III. RESULTS AND DISCUSSIONS

Fig. 2(a) shows the absorption performance of the absorber composed of Ti-Si-SiO_2-Al in the 6-55 µm band. Compared to Ti-SiO_2-Al, the introduction of the lossless dielectric Si significantly broadens the bandwidth, with the absorption rate greater than 70% extending by approximately 10 µm. Fig. 2(b) compares the absorption spectra when Ti, Au, Al, and W are used as the top metal. It can be seen that Au, Al, and W exhibit multiple narrow peak absorption characteristics in the 8-50 µm band. The high-loss characteristic of Ti provides good broadband absorption.

Figs. 2(c), (d), and (e) study the effect of different thicknesses of Ti, Si, and SiO_2 on the absorption spectrum. Increasing the thickness of the top metal reduces the structure's equivalent ohmic resistance and dynamic inductance, thereby decreasing the total impedance of the absorber and lowering the resonant frequency. As the thickness of Si increases, the long-wavelength absorption peak is somewhat enhanced and slightly red-shifted. However, thicker Si affects the coupling between different resonant modes and the intrinsic absorption of SiO_2, thereby reducing the average absorption rate. The thickness of SiO_2 primarily determines the total thickness of the dielectric layer. Within a certain range, increasing the dielectric thickness enhances the effective cavity length, thereby increasing the average absorption. Additionally, the increased phase difference causes a red shift in the absorption band. Consequently, overly thick dielectric layers adversely affect the absorption at lower wavelengths. The thickness of the dielectric layer needs to be constrained by standard CMOS processes, and a thicker absorber would increase the load on the supporting arms, significantly reducing the reliability of the device.

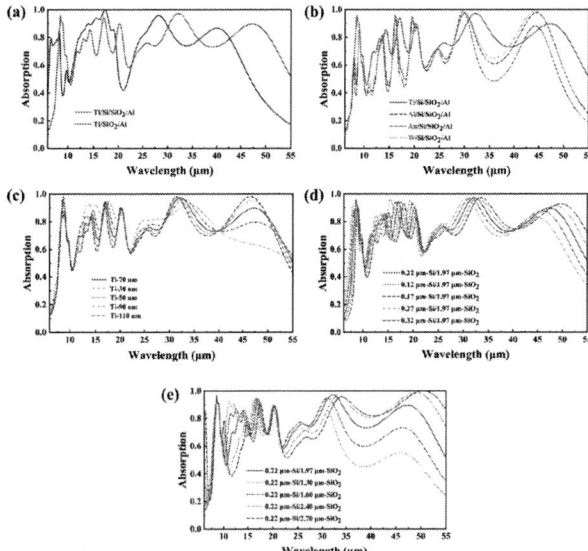

Fig. 2. (a) Comparison of absorption spectra between single-layer and bilayer dielectrics. (b) Absorption spectra for different top metals. (c) Absorption spectra for different Ti thicknesses. (d) Absorption spectra for different Si thicknesses. (e) Absorption spectra for different SiO_2 thicknesses.

Based on simulation optimization, we determined the appropriate thickness for each layer in the Ti-Si-SiO_2-Al metamaterial absorber. According to calculations, this absorber can achieve an average absorption rate of 79% in the wide wavelength range of 8-52 µm, as shown in Fig.3(a).

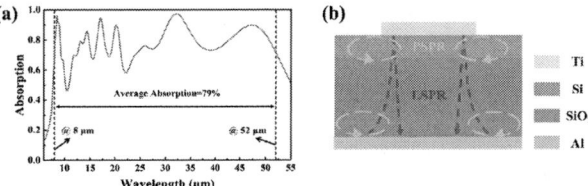

Fig. 3. (a) Average absorption rate of the absorber in the target wavelength range. (b) Resonant modes within the absorber unit cell.

Fig. 3(b) shows two resonant modes that enhance broadband absorption—propagating surface plasmon (PSP) resonance and localized surface plasmon (LSP) resonance. In the bilayer dielectric system, Si has a large dielectric constant in the infrared-to-terahertz band, which allows for the excitation of low-frequency absorption peaks dominated by LSP resonances with only a thin layer [6]. For the lossy material SiO_2, the imaginary part of its refractive index controls the intrinsic absorption in the mid-infrared region, and its lower dielectric constant makes it easier to excite high-frequency PSP resonances. The coupling between various absorption mechanisms can promote broadband absorption.

To further explain the physical mechanism of broadband absorption achieved by plasmonic resonance, we studied the electromagnetic field distribution at four typical absorption peaks: 17.07 µm, 20.24 µm, 32.18 µm, and 47.27 µm. As shown in Fig. 4, the electric field is strongly concentrated at the edges of the Ti metal, with different resonators within the structure exhibiting corresponding resonant effects at different wavelengths. At wavelengths of 17.07 µm and 20.24 µm, PSP resonances can be observed at the edges of the metal structure and within the dielectric. Before inducing LSP, the

PSP resonances in the x and y directions, along with the intrinsic absorption of SiO₂, achieve broadband absorption. As the wavelength increases, the LSP excited by the nanostructure gradually strengthens and dominates, with the magnetic field being strongly localized within the dielectric layer between the nanostructure and the bottom metal. At this stage, the combined effects of LSP and PSP enhance absorption.

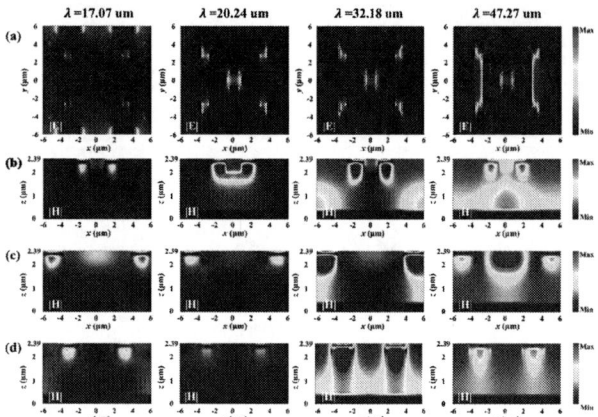

Fig. 4. (a) Electric field distribution in the XY plane at z=2.3 μm. (b) Magnetic field distribution in the XZ plane at y=0. (c) Magnetic field distribution in the XZ plane at y=3 μm. (d) Magnetic field distribution in the YZ plane at x=5.3 μm.

The symmetrical structure of the absorber provides it with good polarization insensitivity. As shown in Fig. 5, the absorption performance of the absorber is essentially identical under TM and TE polarization modes at normal incidence. When the incident light is angled, the absorber maintains high performance within an oblique incidence angle range of 40°.

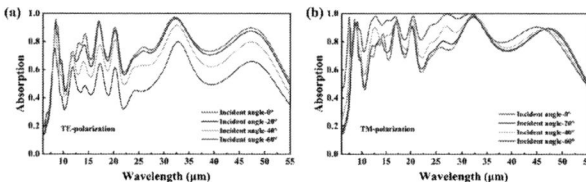

Fig. 5. (a) Absorption spectra at different oblique incidence angles under TE polarization mode. (b) Absorption spectra at different oblique incidence angles under TM polarization mode.

A comparison was made between device A, integrated with the metamaterial absorber, and device B, without the metamaterial, in Ansys Workbench under thermal simulation conditions with incident light radiation in the 8-52 μm range. The temperature rise (ΔT) over time is shown in Fig. 6. The maximum temperature rise for devices A and B was 361 mK and 154 mK, respectively, an increase of 134.4%.

The simulation environment and the initial temperature of the devices were set to 300 K, and a uniform heat flux was applied to the sensor surface. The solid line in the figure represents the fitting result using equation (1):

$$\Delta T(t) = \frac{\eta P}{G_{eff}}(1 - e^{-\frac{t}{\tau}}) = \Delta T_s(1 - e^{-\frac{t}{\tau}}) \qquad (1)$$

where η represents the average absorption rate, P represents the heat flux power, G_{eff} represents the effective thermal conductivity, ΔT_s represents the maximum temperature rise at steady state, and τ represents the thermal time constant. The inset in Fig. 6 shows the temperature distribution of device A and the concentration of the highest temperature inside the sensor. According to equation (1), the G_{eff} and τ values for device A are 2.19×10^{-7} W/K and 73.78 ms, respectively, while those for device B are 1.69×10^{-7} W/K and 76.71 ms, respectively. The results indicate that the introduction of the metamaterial absorber leads to an increase in thermal conductivity, while the simulated thermal time constant of device A is slightly lower than that of device B. These findings demonstrate that the designed absorber provides good thermal performance for CMOS microbolometers.

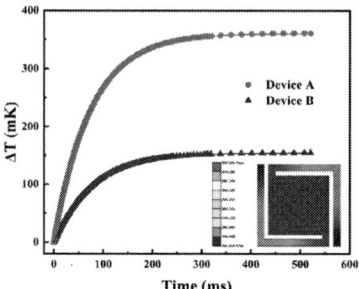

Fig. 6. Temperature variation over time for device A with the integrated metamaterial absorber and device B without the metamaterial absorber. The inset shows the temperature distribution of device A at 500 ms.

IV. CONCLUSION

This paper designs a four-layer metamaterial absorber composed of Ti-Si-SiO₂-Al. The bilayer dielectric system, consisting of the lossless material Si and the lossy material SiO₂, enhances the LSP resonance at long wavelengths. By coupling LSP, PSP, and the intrinsic absorption of the lossy dielectric, the absorber achieves an average absorption of 79% in the wide wavelength range of 8-52 μm. However, there is still room for optimization in the top metal structure, and further research is needed to extend the range to lower frequencies while maintaining a certain periodic size. The designed absorber structure is fully compatible with standard CMOS processes, paving the way for subsequent broadband detection in the infrared-to-terahertz range.

REFERENCES

[1] Zhang, Y.P.; Tang, L.B.; Liu, Y.F.; Seng, T.K.; Wu, G.; Hu, W.D.; Han, F.Z. The research progress and application of novel terahertz detectors. J. Infrared Millim. Waves 2020, 39, 191–210.

[2] Xie, L.; Yao, Y.; Ying, Y. The application of terahertz spectroscopy to protein detection: A review. Appl. Spectrosc. Rev. 2013, 49, 448–461.

[3] Zhou, W.B.; Lan, J.; Guo, Y.Z.; Liu, J.; Liu, X.Z.; Wang, K.; Yan, F.; Liao, Y.M. and Ji, X.L. High-performance microbolometers with metal-insulator-metal plasmonic absorbers in CMOS technology. Opt. Express 2024, 32, 22362-22376.

[4] Zhang, Y.; Wang, X.; Zhou, Y.; Lai, H.; Liu, P.; Chen, H.; Wang, X.; Xie, W. Highly sensitive and ultra-broadband VO₂(B) photodetector dominated by bolometric effect. Nano Lett. 2022, 22, 485–493.

[5] Yang, H.; Cao, Y.; He, J.; Zhang, Y.; Jin, B.; Sun, J.-L.; Wang, Y.; Zhao, Z. Highly conductive free-standing reduced graphene oxide thin films for fast photoelectric devices. Carbon 2017, 115, 561–570.

[6] Zhou, Y.; Qin, Z.; Liang, Z. et al. Ultra-broadband metamaterial absorbers from long to very long infrared regime. Light Sci 2021, Appl 10, 138.

979-8-3503-6184-1/24 $31.00 © 2024 IEEE

Large modulation bandwidth Si-based avalanche photodiode for visible light communications

Jiabin Wu [1], Yidi Hu [2], Chiang Zhu [2], Zhichong Wang [1], Xiaona Zhu [*2], Chao Shen[*1]

[1] School of Information Science and Technology, Fudan University, Shanghai, China
[2] School of Microelectronics, Fudan University, Shanghai, China

* Email: 23210720268@m.fudan.edu.cn, xiaona_zhu@fudan.edu.cn, chaoshen@fudan.edu.cn

Abstract—**Visible light communication (VLC) has become a rapid growing technology for 6G wireless communication. Si-based avalanche photodiode (APD) is one of the promising receivers for VLC. In this paper, we present a simulation study for high-speed Si-based APD. A modulation bandwidth of 7.2 GHz has been realized with a vertical dimension of 11.5 µm * 2 µm. We also discussed the effect of doping concentration and the dimensions of APDs on modulation bandwidth and responsivity of device. When the doping density of substrate increase to 5×10^{16} cm^{-3}, and deduce the thickness of substrate to 2 µm, we get the highest 3dB modulation bandwidth but a decline of responsivity.**

Keywords—*Avalanche Photodiode, Photodetector, Visible Light Communication*

I. INTRODUCTION

Visible Light Communication (VLC) is an emerging wireless communication technology that transmits data by loading it into visible light signals with wavelengths ranging from 375 to 780 nm. Due to the distinctive advantages of visible light, VLC communication system has a wide range of application scenarios, such as indoor, underwater[6], intra-ice, satellite communication, medical treatment, and Internet of Things (IoT)[1]. VLC is characterized by high security, high reliability, high transmission rate, and abundant spectrum resources, research on visible light communication has aroused extensive interest worldwide in recent years[2].

The VLC system, like other communication systems, is generally composed of the transmitter, the receiver and the channel. Similarly, the transmission rate of the whole communication system is determined by these three together. In VLC, the transmitting end usually uses LEDs or laser diodes(LD)[3], and the receiving end is a photodetector, according to different device structures, the photodetectors can be categorized into PINs, APDs, and so on. Among them, APD is characterized by high rate and high gain, which is a well choice for visible light communication receiving terminal. The APDs discussed in this paper are silicon-substrate based devices, and Si-APD fabrication materials are readily available, inexpensive, and have mature fabrication processes. In the future photonic integration direction, Si-APDs can be integrated with existing CMOS technology and silicon-based optoelectronics to realize multifunctional integration by taking advantage of integrated circuits due to the high compatibility of silicon materials.

II. DESIGN CONSIDERATIONS

APD operates under reverse bias, when a photon is injected, the photogenerated carriers increase rapidly in speed under the action of high electric field in the space charge region, and the high-speed moving carriers collide with the crystal lattice and ionize more carriers, thus triggering the avalanche effect. Large amounts of electrons and holes flow to the cathode and anode, respectively, resulting in a photocurrent. Compared to PIN, in APD devices, the carrier density is greater due to the avalanche effect, resulting in larger photocurrents and higher responsivity. And because of the high electric field of the PN junction, the carrier energy is higher and the drift speed is faster, so the response speed is faster. The main factors affecting the response speed of APD include:

- the equivalent terminal capacitance and load resistance formed by the internal structure of APDs.

- Carrier crossing time (drift time) in the APD depletion layer (drift region width).

- Diffusion time due to diffusion movement of carriers outside the depletion layer region.

Equivalent terminal capacitance is related to the material's dielectric constant, etc. , while carriers outside the depletion layer tend to diffuse slower due to the lower electric field strength, so consideration should be given to reducing the generation of diffusion currents or shortening the diffusion distance of diffusing carriers in the APD design. What most affects the response speed of APD devices is the drift of carriers in the depletion layer, so the width of the depletion layer and the electric field strength need to be controlled properly. Too large a width of the depletion layer may result in a long carrier drift time, and too small a width may result in a low carrier drift speed due to a short carrier acceleration time.

III. RESULTS AND DISCUSSIONS

The APD structure profiles discussed in this paper are shown in Fig. 1. It shows the basic APD structure, which contains a constant P-doped Si-substrate, and highly doped N+ and P+ regions. The peak doping densities of P+/N+ region both are about 2×10^{19} cm^{-3}, with Guass doping distribution in order to simulate ion implantation, and two P+ regions to the both sides of the APD. The APD in the Fig. 1 also has some STI(Shallow Trench Isolation) rings added to the basic structure. Important characterization for APDs applied to communication includes modulation bandwidth,

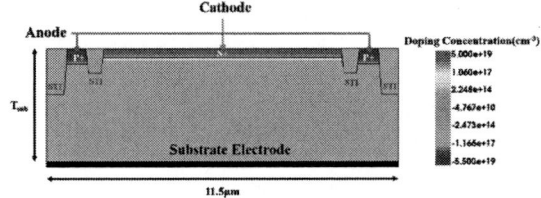

Fig. 1. The cross-sectional structure of the Si-based APD showing the doping concentration. T_{sub} represents the thickness of APD.

979-8-3503-6184-1/24 $31.00 © 2024 IEEE

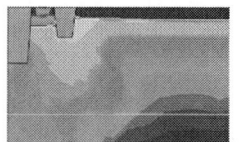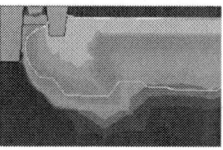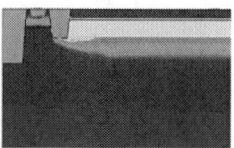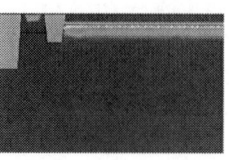

Fig.2. The velocity distribution of electrons in APDs with different doping density of substrate. The thin white lines refer to the edge of depletion regions. Their doping density of substrate are 5×10^{14} cm^{-3}, 5×10^{15} cm^{-3}, 5×10^{16} cm^{-3}, 5×10^{17} cm^{-3}, from left to right, respectively.

responsivity, etc. The main parameters and structures that need to be adjusted in this APD structure include the presence or absence of the STI ring, the doping concentration of the substrate, and the thickness of the device. The simulation results in this work were obtained from Sentaurus TCAD.

The presence of STI rings can effectively prevent early edge breakdown[4], reduce the dark current and improve the linearity of the device. The thickness of the substrate affects the drift distance of carriers between the cathode and the substrate electrodes, so the bandwidth can be increased by appropriately reducing the thickness of the substrate, but too small a thickness of the substrate will lead to insufficient absorption of photons in the device, resulting in a significant decrease in responsivity. As shown in the Fig. 3, when T_{sub} increases, the modulation bandwidth of APD drops rapidly. But responsivity improved when T_{sub} increases, shows a trade-off between bandwidth and responsivity versus the thickness of substrate. In this simulation, the doping density of substrate is 5×10^{16} cm^{-3}.

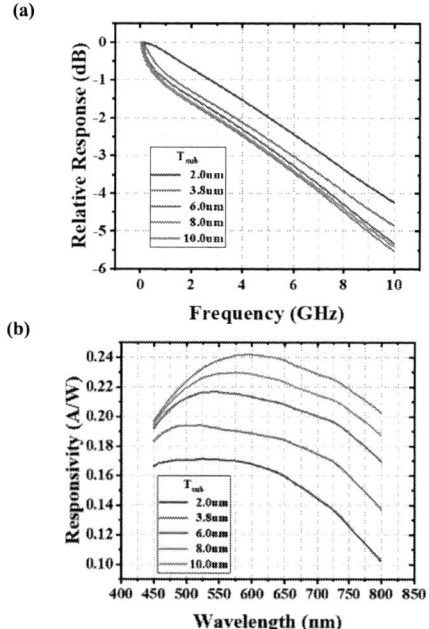

Fig.3. (a) Frequency response of APD in different thickness of substrate. (b) Responsivity of APD for multiful wavelength light with different T_{sub}.

The Fig. 4 shows the I-V curves of APDs with different doping density of substrate. We can conclude that the breakdown voltage(V_{BR}) decline when the doping of substrate increase. APDs for VLC need to operate under but near V_{BR}, in a linear mode, to reduce signal distortion in VLC. So APDs with highly doped substrate can operate in lower voltage to meet the requirement of lower power consumption.

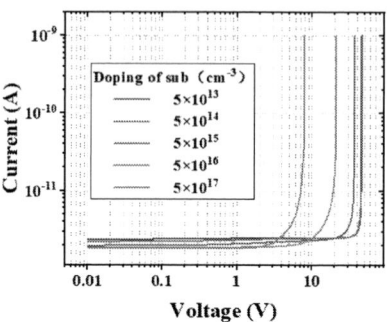

Fig.4. I-V curves of APDs with different doping density of concerntration.

The frequency response of APDs at different substrate doping concentrations and the responsivity curves for different incident light wavelengths are shown in Fig. 5. (a), (b). In the frequency response and responsivity simulation, the reverse bias applied to the device depend on the V_{BR} of APDs and T_{sub} is 3.8 µm, 3D width(z-direction) is 11.5µm. The incident light wavelength is 450 nm, light intensity is 10 mW/cm^2 for the frequency response simulation. The incident light intensity is 0.001 mW/cm^2 for the responsivity simulation. From the simulation results, it can be seen that when the substrate doping density is low, such as 5×10^{13}

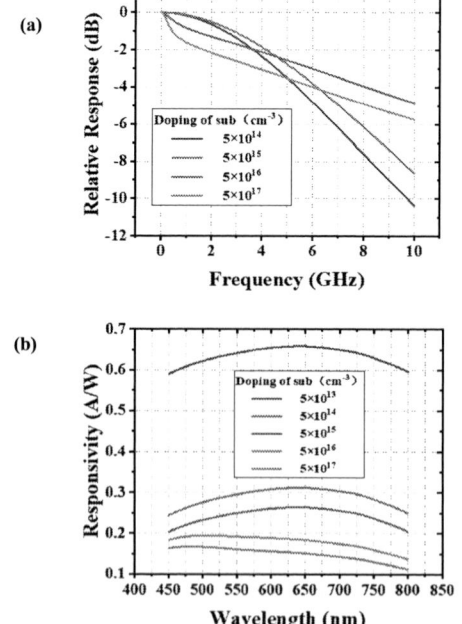

Fig. 5. (a) Frequency response of APD with different doping density of substrate. (b) Responsivity of APD for multiful wavelength light with different doping density of substrate.

cm^{-3}, 5×10^{14} cm^{-3}. the frequency response is basically the same, and the 3dB bandwidth is about 4.49 GHz. The curve of 5×10^{13} cm^{-3} is not shown in Fig. 5. (a), because it is almost overlapped by curve of 5×10^{14} cm^{-3}. When the substrate doping concentration is continued to increase, the APD modulation bandwidth is increased, when it reaches 5×10^{16} cm^{-3}, the 3dB bandwidth increase to 5.9 GHz. But the bandwidth of the device decreases when the doping concentration is too high, such as reaching 5×10^{17} cm^{-3}.

This phenomenon can be explained from the respect of the velocity of electron in APD. The electron velocity in different region of APD when the substrate at different doping concentrations is shown in Fig. 2, and the area marked by the thin white line is the depletion region. In this region carriers accelerate and collide with the lattice, triggering avalanche effect. In this APD, electrons comes from anode and substrate toward cathode. Fig. 2 shows that the average electron velocity increase when depletion area become narrower, so the electron from anode can get cathode from the edge of STI faster. But when the depletion area is too small, because of STIs, electrons from P+/anode cannot reach N+/cathode rapidly, result in the decline of response speed and bandwidth.

Similarly, the responsiveness of the device can be predicted by these results. When the doping concentration is increased, the depletion region becomes narrower, the avalanche region of the device also becomes narrower, the number of lattice collisions decreases, and the photocurrent consequently decreases. Therefore, when increasing the doping concentration of the substrate, the responsivity of the device decreases as shown in the Fig. 5. (b).

For fabrication of this APD, we can use the substrate with positive doping. And then, form the P+/N+ regions by ion implantation. As for STI circle, we can use photolithography first, and form STI circle then or cover the region with a layer of insulating material, it can also make the effect of preventing early edge breakdown.

IV. CONCLUSIONS

In the article, we design a high-speed Si-based APD with 3dB modulation bandwidth exceeding 7 GHz. And for the largest modulation bandwidth, the doping density of substrate is 5×10^{16} cm^{-3}, the thickness of substrate is 2 µm. The relationships of frequency response and responsivity versus the doping density or the thickness of substrate were discussed. The distribution of depletion regions plays a critical role in determining the modulation characteristics of the APD. A wide depletion region may lead to limited bandwidth but high responsivity when the thickness of substrate is fixed, where there is a trade-off. For the faster response speed, we should increase the doping density of substrate properly according to depletion region and shorten the thickness of sub, but it will result in lower responsivity. The paper presents new insights into the device design for constructing high-speed Si-based APDs operating in visible color regime. In practical design of APDs for VLC, we need to compromise between frequency response and responsivity.

ACKNOWLEDGEMENTS

This research is partially funded by the Natural Science Foundation of China Project, grant number 61925104, 62031011; The Key Research and Development Program of Jiangsu Province (BE2021008-5).

REFERENCES

[1] N. Chi, H. Haas, M. Kavehrad, T. D. C. Little and X. -L. Huang, "Visible light communications: demand factors, benefits and opportunities [Guest Editorial]," in IEEE Wireless Communications, vol. 22, no. 2, pp. 5-7, April 2015.

[2] L. E. M. Matheus, A. B. Vieira, L. F. M. Vieira, M. A. M. Vieira and O. Gnawali, "Visible Light Communication: Concepts, Applications and Challenges," in IEEE Communications Surveys & Tutorials, vol. 21, no. 4, pp. 3204-3237, Fourthquarter 2019.

[3] Junfei Wang, Junhui Hu, Chaowen Guan, Yuqi Hou, Zengyi Xu, Leihao Sun, et al, "High-speed GaN-based laser diode with modulation bandwidth exceeding 5 GHz for 20 Gbps visible light communication," Photon. Res. 12, 1186-1193 (2024).

[4] J. Gao and H. Wu, "A high-speed, high-sensitivity silicon avalanche photodiode in 130-nm CMOS," 2015 IEEE Photonics Conference (IPC), Reston, VA, USA, 2015.

[5] S. Ray, M. M. Hella, M. Mottaleb Hossain, P. Zarkesh-Ha and M. M. Hayat, "Speed optimized large area avalanche photodetector in standard CMOS technology for visible light communication," SENSORS, 2014 IEEE, Valencia, Spain, 2014.

[6] C. Wang, H. -Y. Yu and Y. -J. Zhu, "A Long Distance Underwater Visible Light Communication System With Single Photon Avalanche Diode," in IEEE Photonics Journal, vol. 8, no. 5, pp. 1-11, Oct. 2016.

An artificial neuromuscular synapse based on a ferroelectric Pb(Zr$_{1-x}$Ti$_x$)O$_3$/SiC floating gate transistor

Yu Liu[1*], Lin Lin[2], Xiang Wang[1], Chengyan Zhong[1], Junxiong Guo[3*], Wen Huang[2], Yufeng Guo[1*]

[1] College of Integrated Circuit Science and Engineering and National and Local Joint Engineering Laboratory of RF Integration and Micro-Assembly Technology, Nanjing University of Posts and Telecommunications, Nanjing 210023, P. R. China
[2] State Key Laboratory of Electronic Thin Films and Integrated Devices, School of Integrated Circuit Science and Engineering, University of Electronic Science and Technology of China, Chengdu 610054, P. R. China
[3] Institute of Advanced Study, School of Electronic Information and Electrical Engineering, Chengdu University, Chengdu 610106, P. R. China

* Email: liu_yu_24@126.com, guojunxiong@cdu.edu.cn, yfguo@njupt.edu.cn

Abstract— **Neuromuscular junctions (NMJs) are specialized synapses that enable the efficient transduction of neural signals into muscle contractions in biological systems. Desirable artificial NMJs aim to replicate biological synaptic plasticity and high-power actuation for advanced robotics and prosthetics, but face challenges in achieving sufficient flexibility, strength, and self-contained operation. Here, we report a novel artificial NMJ based on a silicon carbide (SiC) ferroelectric field-effect transistor. By integrating a ferroelectric floating gate dielectric with a wide-bandgap SiC transistor, we achieve a wide range of non-volatile conductance modulation, enabling the emulation of long-term synaptic plasticities. We demonstrate the PZT-SiC transistor's higher threshold voltage and lower operating currents compared to traditional SiC transistors, resulting in reduced power consumption. The device exhibits excellent synaptic plasticity, with the ability to modulate synaptic weights through successive test cycles and upon pulsatile stimulation. We showcase the potential of the PZT-SiC synaptic transistor through its application in convolutional image processing, where it successfully extracts horizontal and vertical features using Sobel kernels. Our work presents a promising hardware platform for efficient and scalable artificial NMJs, paving the way for advanced biohybrid systems and neuromorphic robots.**

Keywords—*Neuromuscular synapse, Floating gate, SiC, High power device*

I. INTRODUCTION

The somatosensory system plays a crucial role in enabling biological organisms to perceive and interact with their environment. This system consists of sensory receptors that detect various stimuli such as pressure, temperature, and vibration, afferent nerves that transmit sensory signals to the central nervous system (CNS), and efferent nerves that convey motor commands from the CNS to muscles [1]. The efferent nerves connect to muscle fibers via specialized synapses called neuromuscular junctions (NMJs), which convert action potentials into muscle contractions, thereby generating force and motion [2,3]. NMJs are highly efficient biological interfaces that can drive muscles with remarkable speed, precision, and adaptability. They can modulate the strength of muscle contractions based on the frequency and pattern of incoming action potentials, demonstrating synaptic plasticity. Moreover, NMJs can directly deliver the high power required for muscle actuation without external amplification. These capabilities have inspired researchers to develop artificial

NMJs for applications in bioinspired robotics, prosthetics, and neuromorphic engineering [4, 5].

Recent work has explored the use of ferroelectric gate dielectrics to enhance the synaptic plasticity of transistor-based artificial synapse[6-8]. Ferroelectric materials primarily exploit the differing orientations of ferroelectric domains to achieve various conductance states, concentrating on the creation of low-power synaptic devices. In parallel, there has been an increasing interest in replicating the intricate functions of biological NMJs, inspiring ongoing developments in the realm of artificial synaptic technologies. For example, Park et al. demonstrated a CuInP$_2$S$_6$ (CIPS)/GaN ferroelectric HEMT that exhibited tunable synaptic weight, spike-timing-dependent plasticity, and direct actuation of microelectromechanical system (MEMS) actuators [9]. The polarization of the CIPS ferroelectric layer coupled to the GaN channel, enabling a wide range of conductance modulation, while the high current drive of the GaN HEMT allowed for muscle-like actuation without external amplifiers. However, the relatively low bandgap and breakdown field of GaN limit the operating voltage and power handling of such devices, making them less suitable for driving high-voltage actuators or withstanding extreme environments.

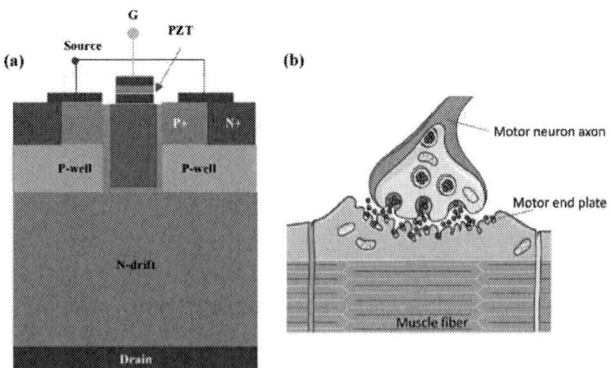

Fig. 1. (a) the structural schematic of a synaptic transistor. (b) The schematic diagram of a biological neuromuscular synapse.

To address these limitations, we propose an artificial NMJ based on a silicon carbide (SiC) ferroelectric floating gate field-effect transistor. SiC is a wide-bandgap semiconductor with a higher critical electric field and thermal conductivity than GaN, making it an ideal material for high-voltage, high-

979-8-3503-6184-1/24 $31.00 © 2024 IEEE

power, and high-temperature applications. This novel approach has the potential to overcome the limitations of previous artificial NMJs and extend their applicability to more demanding scenarios. In this work, we will present the design, fabrication, and characterization of the SiC ferroelectric field-effect transistor-based artificial NMJ, demonstrating its enhanced synaptic plasticity, high-voltage actuation capabilities, and robustness in extreme environments.

II. RESULTS AND SIGNIFICANCE

Fig. 2a and 2b present the I_{ds}-V_{ds} characteristics of both SiC and PZT-SiC transistors over multiple test cycles. While the SiC transistors maintain consistent curves across different tests, the PZT-SiC transistors exhibit a notable shift in current with each successive cycle. This shift is a clear indication of synaptic weight modulation, a crucial aspect of synaptic plasticity in neuromorphic systems. The ability of PZT-SiC transistors to dynamically adjust their synaptic weights in response to varying input stimuli is a testament to their potential in emulating the adaptive behavior of biological synapses.

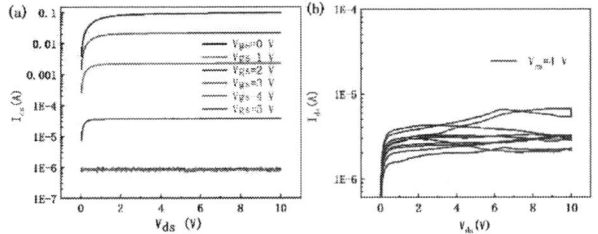

Fig. 2. The Ids-Vds characteristics of (a) SiC and (b) PZT-SiC transistors over successive test cycles.

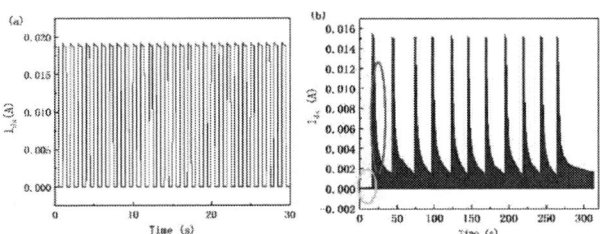

Fig. 3. The current modulation in (a) SiC and (b) PZT-SiC transistors after pulsatile stimulation

The current modulation properties of SiC and PZT-SiC transistors are further explored in Fig. 3a and 3b, which illustrate their response to pulsatile stimulation. The SiC transistors demonstrate a stable current that remains unaffected by the pulse sequences, indicating their lack of synaptic plasticity. In stark contrast, the PZT-SiC transistors exhibit a remarkable increase in current with each successive pulse, as highlighted by the green circle in Fig. 4b and expanded in Fig. 5a, suggesting their ability to undergo synaptic potentiation. This potentiation behavior is a fundamental characteristic of biological synapses, and its successful emulation by PZT-SiC transistors highlights their potential in neuromorphic computing. Conversely, the red circle in Fig. 4b, expanded in Fig. 5b, reveals that after a short-circuit discharge in the PZT film, the current gradually decreases upon subsequent pulses, mimicking synaptic depression. This depression behavior is another crucial aspect of synaptic plasticity, and its demonstration in PZT-SiC

transistors further emphasizes their versatility in emulating complex synaptic dynamics.

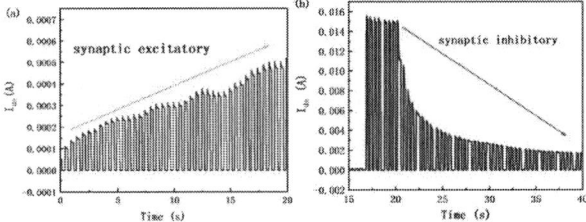

Fig. 4: (a) Synaptic excitation and (b) synaptic inhibitory of PZT-SiC synaptic transistors. A short-circuit discharge in the PZT film can easily adjust the characteristics of synapses.

To demonstrate the practical utility of PZT-SiC transistors, Fig. 5 presents the results of convolutional image processing using these devices. The pristine image shown in Fig. 5a serves as the input, while Fig. 5b and Fig. 5c showcase the processed outputs. Specifically, Fig. 5b illustrates the effective extraction of horizontal features using the Sobel-x kernel, while Fig. 5c highlights the vertical features captured by the Sobel-y kernel. These results underscore the capability of PZT-SiC transistors in feature delineation, a critical aspect of image processing applications. By successfully extracting relevant features from the input image, PZT-SiC transistors demonstrate their potential in tasks such as edge detection, object recognition, and image segmentation.

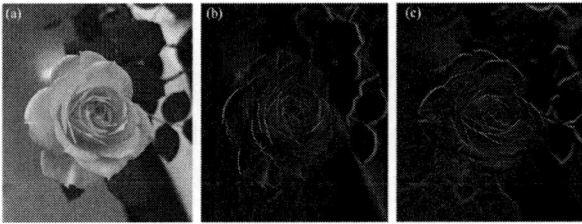

Fig. 5. (a) Original image, (b) Processed with Sobel-x kernel highlighting horizontal features, (c) Processed with Sobel-y kernel accentuating vertical features.

III. SUMMARY

In conclusion, the I_{ds}-V_{ds} characteristic curves over successive test cycles, current modulation under pulsatile stimulation, and convolutional image processing results collectively highlight the remarkable potential of PZT-SiC transistors in neuromorphic computing. Their energy efficiency, synaptic plasticity, dual functionality of potentiation and depression, and feature extraction capabilities make them a promising candidate for emulating the complex behavior of biological synapses. As research in this field continues to advance, PZT-SiC transistors are poised to play a significant role in the development of highly efficient, adaptable, and powerful neuromorphic systems that can revolutionize the way we process and interpret information.

Acknowledgments

This work was financially supported by the Natural Science Research Start-up Foundation of Recruiting Talents of Nanjing University of Posts and Telecommunications (Grant No.NY223161), and in part by the Jiangsu Provincial

979-8-3503-6184-1/24 $31.00 © 2024 IEEE

Key Research and Development Program under Grant BE2022126, , and in part by the National Natural Science Foundation of China (grant nos. 62371095), the Key R&D Program of Sichuan Province (grant nos. 2022ZHCG0041), National Key Research and Development Program of China (No. 2022YFB3206100) .

REFERENCES

[1] D. Farina, I. Vujaklija, M. Sartori, et al., "Man/machine interface based on the discharge timings of spinal motor neurons after targeted muscle reinnervation," Nature Biomedical Engineering, vol. 1, no. 2, p. 0025, 2017.

[2] M. L. Harlow, D. Ress, A. Stoschek, et al., "The architecture of active zone material at the frog's neuromuscular junction," Nature, vol. 409, no. 6819, pp. 479-484, 2001.

[3] R. A. Jones, C. Harrison, S. L. Eaton, et al., "Cellular and molecular anatomy of the human neuromuscular junction," Cell Reports, vol. 21, no. 9, pp. 2348-2356, 2017.

[4] Y. Leng, X. Li, F. Zheng, et al., "Advances in In Vitro Models of Neuromuscular Junction: Focusing on Organ‐on‐a‐Chip, Organoids, and Biohybrid Robotics," Advanced Materials, vol. 35, no. 41, p. 2211059, 2023.

[5] M. Karbalaei Akbari and S. Zhuiykov, "A bioinspired optoelectronically engineered artificial neurorobotics device with sensorimotor functionalities," Nature Communications, vol. 10, no. 1, p. 3873, 2019.

[6] S. Oh, T. Kim, M. Kwak, et al., "HfZrOx-based ferroelectric synapse device with 32 levels of conductance states for neuromorphic applications," IEEE Electron Device Letters, vol. 38, no. 6, pp. 732-735, 2017.

[7] M.-K. Kim and J.-S. Lee, "Ferroelectric analog synaptic transistors," Nano Letters, vol. 19, no. 3, pp. 2044-2050, 2019.

[8] S. Boyn, J. Grollier, G. Lecerf, et al., "Learning through ferroelectric domain dynamics in solid-state synapses," Nature Communications, vol. 8, no. 1, p. 14736, 2017.

[9] M. Park, J. Y. Yang, M. J. Yeom, et al., "An artificial neuromuscular junction for enhanced reflexes and oculomotor dynamics based on a ferroelectric $CuInP_2S_6$/GaN HEMT," Science Advances, vol. 9, no. 38, p. eadh9889, 2023.

Broadband Photodetectors Based on Graphene/Perovskite Hybrid Structure with Ferroelectric Gating

Zhongyang Liu[1], Shuangqi Dong[1], Mingjie Li[1], Huaipeng Wang[2], Dan Xie[2], Yilin Sun[*1]

[1] School of Integrated Circuits and Electronics, Beijing Institute of Technology, Beijing 100081, China
[2] School of Integrated Circuits, Beijing National Research Center for Information Science and Technology (BNRist), Tsinghua University, Beijing 100084, China

* Email: liuzy23bit@bit.edu.cn, sunyl@bit.edu.cn

Abstract—**Graphene-based optoelectronic devices have attracted much attention due to its wavelength independent light absorption and high carrier mobility. However, such graphene-based photodetectors usually exhibit lower photo responsivity due to weak light absorption. Here, we fabricated a graphene/perovskite hybrid structure, where graphene works as a channel layer in PZT-gated FETs and perovskite works as a light absorber. Due to the combination of strong polarization electric field induced by PZT and high mobility of graphene, a high photo responsivity above 10^4 AW^{-1} for visible light has been obtained in such graphene/perovskite hybrid structure. Moreover, a near-linear photocurrent with the increasing light power was also achieved, which offers a potential way to achieve high performance graphene-based optoelectronics.**

Keywords—***Graphene, Perovskite, PZT, Ferroelectric Gating, Broadband Photodetection***

I. INTRODUCTION

Graphene has drawn an enormous amount of attention for optoelectronic applications due to its unique optical and electrical properties such as broad absorption bandwidth, excellent carrier transport properties and good flexibility [1-3]. However, low absorption coefficient of graphene film greatly limits photo responsivity of graphene-based optoelectronic applications [3]. Recently, several efforts have been devoted to the graphene-based hybrid structure to achieve high photo responsivity. For example, Yang et al. reported a graphene/germanium hybrid structure for broadband photodetection [4] and Lai et al. fabricated a graphene/2D Bi_2O_2Se hybrid structure for multifunctional optoelectronics [5]. In such hybrid structures, another light absorber is used to achieve the conversion from light to charge carriers and graphene works as the channel layer to transport such photo-generated carriers. Benefitting from the high carrier mobility of graphene, a high responsivity of 10^7 AW^{-1} was widely obtained in graphene/PbS quantum dots hybrid structure for infrared photodetection [6-7].

Among various emerging light harvesting materials, methylammonium lead triiodide ($CH_3NH_3PbX_3$, X=halogen) perovskites stand out because of their suitable direct bandgap, large absorption coefficients, ease of solution-based processing and outstanding charge transport properties [8-9]. Youngbin Lee et al. have reported the fabrication of high-performance perovskite-graphene hybrid photodetectors, which exhibit a broad spectral photo responsivity between 800 and 400 nm about 180 AW^{-1} [10]. Such high photo responsivity is attributed to the enhancement of light absorption and photo-gating mechanism of hybrid structure.

In this work, bottom-gated FETs by lead zirconate-titanate (PZT) gating were fabricated consisting of a monolayer graphene channel and spin-coated with a layer of perovskite film. The ferroelectric gating provided a strong polarization field effect on the graphene channel, which was demonstrated to boost the carrier mobility of 2D materials, resulting in a high responsivity [11-12]. The optical and electrical properties of such graphene/perovskite hybrid photodetectors were investigated under the polarization of PZT for the different wavelengths between 470 nm and 740 nm. The maximum photoresponsivity has been achieved to be 3.25×10^4 A/W under the illumination with the wavelength of 660 nm, demonstrating a great potential for high-performance optoelectronic devices.

II. EXPERIMENTAL

Monolayer graphene was grown on Cu foil by CVD method and transferred onto the $PZT/Pt/Ti/SiO_2/Si$ substrates by a PMMA assistant transfer method. Here, a 260-nm thick $Pb(Zr_{0.4}Ti_{0.6})O_3$ (PZT) layer was prepared by a Sol-Gel process followed by a rapid annealing process. Then, standard lithography was used to pattern source/drain contacts, followed by electron-beam evaporation of 20/50nm Cr/Au electrodes and lift-off process. Another standard lithography process was used to form the graphene channel, which was then spin-coated by perovskite precursor solution at 4000 rpm for 60 s and heated at 90 °C for 30s. The preparation for perovskite precursor was reported in our previous work [13]. The schematic of as-prepared graphene/perovskite hybrid FETs by PZT bottom gating is shown in Fig. 1(a).

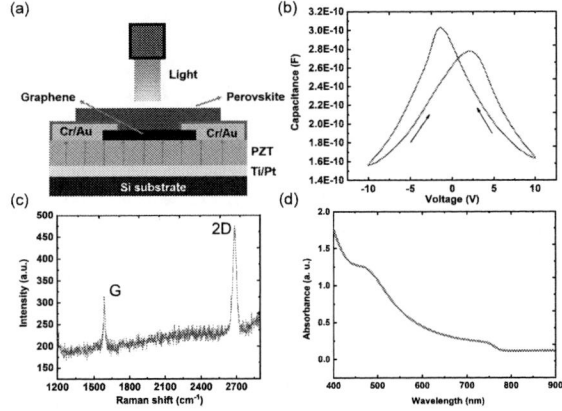

Fig. 1. (a) The schematic of graphene/perovskite hybrid FETs by PZT bottom gating. The arrows here represent the polarization direction of PZT films under a positive gate voltage; (b) Capacitance-Voltage (*C-V*) characteristics of a capacitor of Au/PZT/Pt heterostructure; (c) Raman spectrum of monolayer graphene; (d) Absorption spectra of $CH_3NH_3PbI_3$ films.

Fig. 1(b) displays the typical butterfly peak hysteresis loop in C-V curves of PZT-based capacitors, indicating a pronounced ferroelectric polarization switching characteristic. Besides, Raman spectrum of the used graphene films was shown in Fig. 1(c), which referred to two typical featured peak named by G and 2D peak. The ratio between G and 2D peal was approximately equal to 2, confirming the monolayer of graphene [14]. Finally, the absorption spectrum of as-fabricated perovskite film was measured, which showed a broadband light absorption in the visible range as shown in Fig. 1(d).

III. RESULTS AND DISCUSSIONS

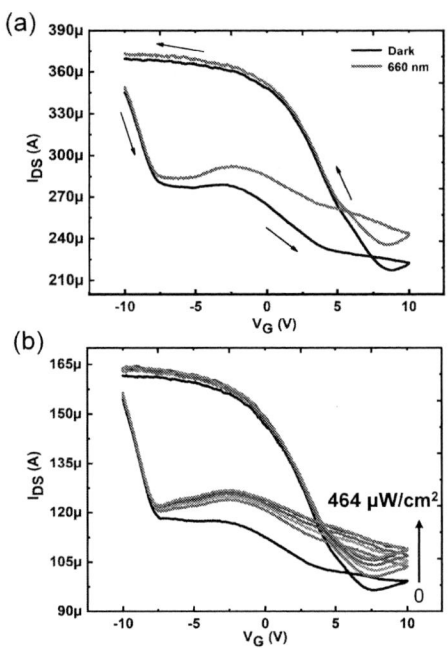

Fig. 2. (a) The transfer curves of graphene/perovskite hybrid FETs by PZT bottom gating in the dark and under illumination. (b) The transfer curves of as-fabricated PZT-gated FETs under the different power intensity.

From Fig. 2(a), the transfer curves of as-fabricated graphene/perovskite field effect transistors (FETs) with PZT bottom-gating exhibited a typical hysteresis loop. In dark, it could be inferred that graphene was slightly p-doped by the adsorbates such as water and oxygen molecules bounded at the the graphene/SiO_2 interface, which made it behave similar to an unipolar p-type semiconductor when V_G sweeping from -10 V to 10 V [15]. Another significant feature of the obtained hysteretic loop was the counterclockwise hysteresis direction, which was opposite to the hysteresis direction of ferroelectric polarization under the double-sweeping voltage. This phenomenon has been widely demonstrated in PZT bottom-gated FETs, which could be attributed to the charge trapping/de-trapping process at the interface between PZT gate dielectrics and 2D material channel layers [16-17]. In fact, for p-type semiconductor, the PZT ferroelectric polarization dominated hysteresis loop in transfer curves should be clockwise, and for n-type semiconductor. However, the interfacial charges would be trapped and de-trapped driven by the polarization field effect, which prevented the carriers in the 2D channel from directly responding to the ferroelectric switching, resulting in opposite hysteresis directions. However, considering that

charge trapping/de-trapping process is also modulated by ferroelectric gating, it can still be seen that ferroelectric polarization dominates the transport characteristics of charge carriers.

Here, an obvious photosensitive behavior can be seen when the device is exposed to a monochromatic light with wavelength of 660 nm. The result demonstrates the photo response of graphene/perovskite to monochromatic light. Moreover, the different photo response for different gate sweeping directions indicates the effect of polarization on the optical properties of our device, resulting in a photosensitive hysteresis window. Fig. 2(b) showed the transport curves of our device under the same illumination with different irradiance power, where an increased channel currents (I_{DS}) was achieved with the increasing power intensity. Interestingly, the photo response exhibited a clear dependence on gate voltage (V_G), which controlled the ferroelectric polarization states of PZT films. It could be inferred that the photoresponsivity could be effectively modulated by the ferroelectric gating.

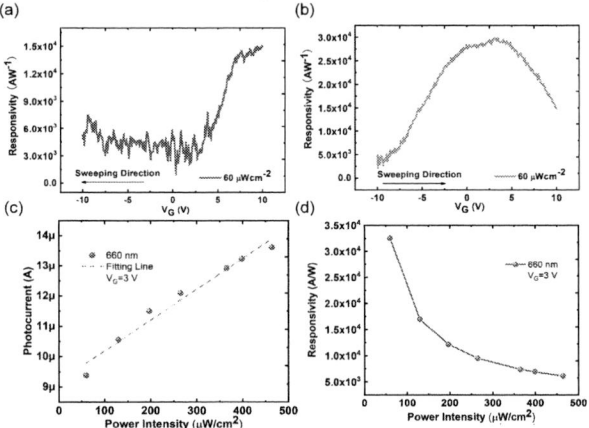

Fig. 3. The calculated responsivity under negative sweeping direction (a) and positive sweeping direction (b). (c) The calculated photocurrent was plotted as a function of power intensity, which can be fitted by the formula: y=kx+b; (d) The corresponding photoresponsivity was plotted as a function of power intensity.

To further evaluate the effect of VG on the photo responsivity of graphene/perovskite hybrid structure-based FETs with PZT gating, the responsivity, calculated by the formula: R=Photocurrent/(Power intensity × Active area), was plotted a function of V_G in both sweep directions as shown in Fig. 3(a) and 3(b). It can be seen in both sweeping directions, the responsivity showed an obvious dependence on V_G, which can be attributed to the changing ferroelectric polarization field effect at the interface. Moreover, the maximum value of responsivity was achieved at V_G of 3 V. Here, we extracted the photocurrent, namely the difference between I_{DS} under the illumination and in dark, as a function of power intensity, which exhibited a near-linear relationship. Furthermore, a maximum responsivity of 3.25 A/W was also obtained at V_G of 3 V under the illumination with the power intensity of ~60 $\mu W/cm^2$ and showed a decreasing trend with the increasing power intensity, which was similar to that of the most reported photodetectors based on 2D materials[18-20]. It is widely recognized that the increase in photocurrent is weaker than that in light power intensity due to the limited external quantum efficiency.

979-8-3503-6184-1/24 $31.00 © 2024 IEEE

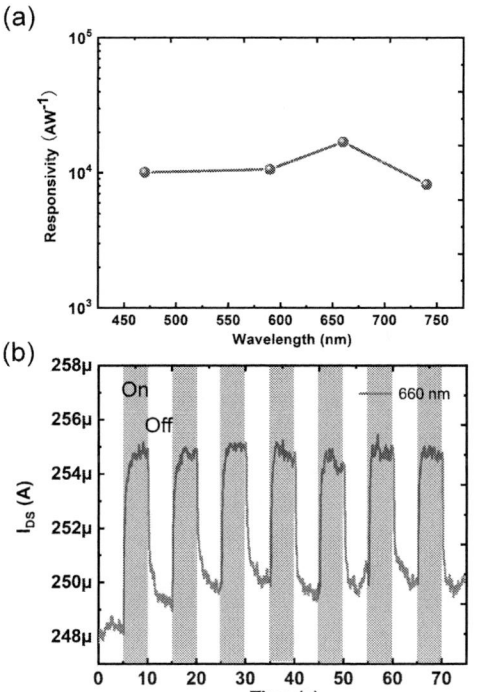

Fig. 4. (a) The responsivity under the illumination with different wavelengths; (b) The current vs time (I_{DS}-t) curves under the periodic illumination.

Finally, to demonstrate the broadband photodetection of as-fabricated device, we calculated the responsivity under the illumination with different wavelengths ranging from 470 nm to 740 nm, almost covering the entire wavelength range of visible light. From Fig. 4(a), the responsivity was obtained to be about 10^4 A/W for the visible light, which could be attributed to the excellent light absorption of perovskite in the visible range and high carrier mobility of monolayer graphene. Besides, the current-time (I_{DS}-t) was also measured by applying periodic illumination with light switching on and off for 5 s/ 5 s, respectively. From Fig. 4(b), it indicated that our device has a good repeatability, which demonstrated its great potential in high-performance photodetection.

IV. CONCLUSIONS

In summary, graphene/perovskite hybrid structure-enabled FETs by PZT bottom gating were fabricated for broadband photodetection. Such hybrid structure exhibits the high photo responsivity for the visible range from 470 nm to 740 nm, where the maximum responsivity was calculated to be 3.25×10^4 AW^{-1}. This excellent performance can be attributed to the joint effect from the unique transport properties of graphene channel, the broadband light absorption of perovskite and the ferroelectric polarization of PZT films. The results also demonstrate such hybrid photodetectors have a great potential in future graphene-based broadband photodetectors.

ACKNOWLEDGMENT

This work was supported by the National Natural Science Foundation (NSF) of China under grant 62104017 and the Beijing Institute of Technology Research and Innovation Promoting Project (Grant No.2023YCXY032).

REFERENCES

[1] F. Bonaccorso, Z. Sun, T. Hasan, et al. "Graphene photonics and optoelectronics," Nat. Photon., vol.4, pp. 611-622, Augest 2010.

[2] N. Liu, H. Tian, GSchwartz, et al., "Large-area, transparent, and flexible infrared photodetector fabricated using PN junctions formed by N-doping chemical vapor deposition grown graphene," Nano Lett., vol.14, pp. 3702-3708, June 2014.

[3] H. Guan, J. Hong, X. Wang, et al., "Broadband, High-Sensitivity Graphene Photodetector Based on Ferroelectric Polarization of Lithium Niobate", Adv. Opt. Mater., vol. 9, pp. 2100245, June 2021.

[4] F. Yang, H. Cong, K. Yu, et al. "Ultrathin Broadband Germanium–Graphene Hybrid Photodetector with High Performance", ACS Appl. Mater. Inter., vol. 9, pp. 13422–13429, March 2017.

[5] C.-M. Yang, T.-C. Chen, D. Verma. et al., "Bidirectional All-Optical Synapses Based on a 2D Bi$_2$O$_2$Se/Graphene Hybrid Structure for Multifunctional Optoelectronics", Adv. Func. Mater., vol. 30, pp. 2001598, May 2020.

[6] G. Konstantatos, M. Badioli, L.Gaudreau, et al., "Hybrid graphene-quantum dot phototransistors with ultrahigh gain," Nat. Nanotech., vol. 7, pp. 363-368, May 2012.

[7] Z. Sun, Z. Liu, J. Li, et al., "Infrared photodetectors based on CVD-grown graphene and PbS quantum dots with ultrahigh responsivity," Adv. Mater., vol. 24, pp. 5878-5883, Auguest 2012.

[8] G. Giorgi, J. -I. Fujisawa, H. Segawa, et al., "Small photocarrier effective masses featuring ambipolar transport in methylammonium lead iodide perovskite: a density functional analysis," J. Phys. Chem. Lett., vol. 4, pp. 4213-4216, November 2013.

[9] Y. Li, Y. Zhang, T. Li, et al., "Ultrabroadband, Ultraviolet to Terahertz, and High Sensitivity CH3NH3PbI3 Perovskite Photodetectors", Nano Lett., vol. 20, pp. 5646–5654, July 2020.

[10] Y. Lee, J. Kwon, E. Hwang, et al., "High-performance perovskite-graphene hybrid photodetector," Adv. Mater., vol. 27, pp. 41-46, October 2014.

[11] L. Liu, L. Wu, A. wang, et al. "Ferroelectric-Gated inSe Photodetectors with High On/Off Ratios and Photoresponsivity", Nano Lett., vol. 20, pp. 6666-6673, August 2020.

[12] H. Jiao, X. Wang, S. Wu, et al. "Ferroelectric field effect transistors for electronics andoptoelectronics," Appl. Phys. Rev., vol. 10, pp. 011310, March 2023.

[13] S. Chen, C. Teng, M. Zhang, et al., "A flexible UV-Vis-NIR photodetector based on perovskite/conjugated polymer composite," Adv. Mater., vol. 28, pp. 5969-5974, May 2016.

[14] Y. Sun, D. Xie, J. Xu, et al., "Tunable transport characteristics of double-gated graphene field-effect transistors using P(VDF-TrFE) ferroelectric gating," Carbon, vol. 96, pp. 695-700, January 2016.

[15] T. Feng, D. Xie, D. Wang, et al., "Electron-doping of graphene-based devices by hydrazine", J. Appl. Phys., vol. 116, pp. 224511, December 2014.

[16] Y. Sun, D. Xie, X. Zhang, et al., "Temperature-dependent transport and hysteretic behaviors induced by interfacial states in MoS2 field-effect transistors with lead-zirconate-titanate ferroelectric gating," Nanotechnology, vol.28, pp.045204, December 2016.

[17] W. Teng, S.-Y. Bao, Y.-Q. Hu, et al., "Ferroelectric Controlled Interfacial Effect on the Electronic Properties of PZT Gated IGZO Channel Thin-Film ransistors", ACS Appl. Electron. Mater., vol. 6, pp. 1063–1070, January 2024.

[18] C. Yin, C. Gong, J. Chu, et al., "Ultrabroadband Photodetectors up to 10.6 µm Based on 2D Fe$_3$O$_4$ Nanosheets", Adv. Mater., vol. 32, pp. 2002237, May 2020.

[19] Z. Liu, M. Li, Z. Liu, et al., "Heterostructured MXene/Si Photodiodes WithSub-1-nm h-BN Blocking Layers forSuppressing Dark Current", IEEE Electron. Dev. Lett., vol. 44, pp.476-479, March 2023.

[20] Z. Liu, M. Li, Y. Sun, et al., "Integrating surface and interface engineering to improve optoelectronic performance and environmental stability of MXene-based heterojunction towards broadband photodetection", Nano Res., vol. 16, pp. 10148–10155, March 2023.

Interconnection design of Chiplet technology

Ning Chen[1], Lei Shen[1], Chang Wu*[1]

[1] School of Microelectronics, Fudan University

Email: ningchen23@m.fudan.edu.cn, wuchang@fudan.edu.cn

Abstract—In the post Moore's law era, chiplet becomes an important technology for the continual progress of integrated circuits. Due the higher cost and larger delay of inter-die connections, interconnection design and the interface protocol are crucial technologies in chiplet design. In this paper, we discuss various interconnection designs and interface protocols for homogeneous and heterogeneous structures.

Keywords—Chiplet, Die to Die, Interconnection design

I. INTRODUCTION

As fabrication technology gradually approaches physical limit, chiplet becomes an important technology to improve circuit performance and yield. It enables large designs with smaller modules for lower costs, faster design time and higher performance [1-5].

A. More flexible process options

The processes used in digital, analog, RF, and memory chips are different. Chiplet technology allows integration of different dies with different processes for heterogeneous computing and higher performance [4]. For example, Zen 2 CPU of AMD integrates 12nm I/O module with 7nm logic operation modules together to form advanced processor [6].

B. Higher yield and lower cost

The yield of the chip is directly related to die size. Another benefit of chiplet technology is better yield and lower manufacturing cost with smaller dies. Researches show that in 14nm process, when die size is greater than 500mm², the manufacturing cost will be higher than that of chiplet design; while in the 5nm process, when die size is greater than 200mm², the manufacturing cost will be higher than that of chiplet design [7].

C. Faster design time and lower design costs

Chiplet technology can be regarded as a reusable IP. In traditional SOC design, the choice of process is determined by the functional modules with performance limit. However, in many cases, only part of the modules need advanced process, while others can be designed with lower technology for cost down. Chiplet technology enables the integration of multiple dies with different process to achieve a balance between design performance and cost. Furthermore, reuse of existing dies can reduce the design time and cost as well.

D. Higher performance

Chiplet achieves performance improvement by stacking dies more. For example, Apple's latest M1 Ultra uses Chiplet technology to integrate two M1 dies to achieve twice the computing ability [8].

In Chiplet technology, advanced packaging technology is the basis, which gradually evolves from 2D plane to 2.5 and 3D packaging. The advanced packaging process helps to improve the interconnection density, shorten the interconnection distance, and improve the integration and performance of the chip. Although the performance of Chiplet is directly affected by the development of advanced packaging and other factors, from the perspective of Chiplet design, the most important point is interconnection design.

For different chip functions and module divisions, the requirement of interconnection is different, such as homogeneous chip and heterogeneous chip.

II. INTERCONNECTION OF HOMOGENEOUS AND HETEROGENEOUS CHIPLET

According to the structure of divided modules, Chiplet can be classified into homogeneous chip and heterogeneous chip.

A. Interconnection of homogeneous chip

Homogeneous chip means that each module has the same structure, and this module can work normally alone with all the functions. The only purpose used Chiplet is improving performance.

The interconnection architecture of homogeneous chips is relatively simple, usually using ring or mesh topology. The most important point to be concerned in the design is latency.

For example, Apple's M1 Ultra chip is directly connected by two M1 Max cores, and the transmission speed between the two M1 Max chips reaches 2.5TB/s [8]. Xilinx uses homogeneous Chiplet in its FPGA product XCVU9P, which consists of three XCVU3P cores directly connected. The main advantage of the homogeneous Chiplet is that the linear increase of the performance is obtained through simple interconnection of a plurality of homogeneous modules, and the yield of the chip is improved while the manufacturing cost is reduced.

B. Interconnection of heterogeneous chip

Heterogeneous chip combines various modules with independent functions into an integrated chip. For example, it packages computing core, I/O module, memory block, and control module into an integrated chip using Chiplet technology

The impact of interconnection design on heterogeneous chip is more important, and its complex interconnection usually uses special interconnection modules to meet the requirements of complex communication path routing, network traffic bandwidth balancing and system flexible expansion [9].

Unlike Xilinx, Intel's Agilex FPGA combines AIB and EMIB technologies to establish an interconnection topology for each functional modules [10].

Heterogeneous integrated chips need more complex interconnection design to meet requirements of high bandwidth, low power consumption and low delay. Therefore, the research of D2D (die to die) interconnection design is mainly aimed at heterogeneous chips. The interconnection design in heterogeneous structures is discussed with 3 aspects in the following sections.

979-8-3503-6184-1/24 $31.00 © 2024 IEEE

III. INTERCONNECTION DESIGN OF CHIPLET TECHNOLOGY

In heterogeneous structures, interconnection performance affected by the differences in interface density, power consumption efficiency and transmission rate due to different packaging methods, different functional modules, different organic substrates, silicon substrates or other integrated materials.

Interconnection design can be analyzed from 3 aspects: high-speed interface circuit, interface protocol and interconnection network. The interconnection types can be divided into serial differential interconnection and parallel single-ended interconnection according to the type of interface signal, also it can be divided into private protocol and standard protocol according to the type of interface protocol[11-12], and it can be divided into regular network, irregular network and reconfigurable network according to the type of interconnection network topology.

A. High-speed interface circuit

High-speed interface circuits includes the following 3 types: 1) Wireline parallel communication interface oriented to 2.5D/3D integration process; 2) wireless interconnection communication interface based on inductive coupling; 3) high-bandwidth optoelectronic interconnection interface [2].

Wireless interconnection communication interface used less, because its area, power and performance is worse than wired interface. Optoelectronic interconnection interface is unable to mass production, which has many problems still to be solved in industrialization [2].

Wireline parallel communication interface gets more choices at present. Parallel single-ended signal interface transmits through a large number of parallel channels at the same time, which has obvious advantages over serial differential signal interface in bandwidth density, delay, transmission energy efficiency and design simplicity, and is more suitable for Chiplet application scenarios.

TABLE I.   CHIPLET INTERCONNECT TECHNOLOGY FEATURES

Type	protocol	Max speed	Format	Energy efficiency
Parallel single-ended	UCIe/AIB/BoW	32Gbps	NRZ	<0.5 pJ/bit
Serial differential	XSR/USR	56~112Gbps	PAM4/NRZ	>1 pJ/bit

In order to reduce the system power consumption, the equalization strategy in the signal link is mainly based on the FFE + CTLE analog hybrid architecture, which reduces the CDR circuit and places the forward clock architecture of the PLL at TX.

The signal density of 2.5D and 3D packages has been greatly improved, single-wire SerDes can support Tbps transmission bandwidth through a large number of parallel channels, and high-speed interface circuits need to focus on transmission delay and power efficiency.

B. Interface protocol

The interface protocol of Chiplet can be divided into two aspects: the physical layer interface protocol and the complete protocol stack. The physical layer mainly focuses on the pin definition, electrical characteristics, bump map and other basic characteristics, which can ensure the point-to-point transmission of data bit stream; while the protocol stack makes more detailed provisions on the basis of the physical layer, such as routing mode, data structure, reliable transmission mechanism, consistency, flow control, and establishes the end-to-end reliable data transmission [2].

Since TSMC developed CoWoS packaging technology in 2014 [3], Chiplet becomes the research focus of major leading companies. In 2016, NVIDIA proposed Nvlink technology [13] and AMD proposed Infinity Fabric technology[14], both based on SerDes; In order to meet the interconnection requirements, OIF proposed the short channel SerDes electrical physical layer specification for package interconnection [15], such as CEI-56 G-USR-NRZ; In order to solve the problems of large latency and high energy efficiency of SerDes, , more bit width parallel interconnection interface between dies be used with the progress of packaging technology becomes possible, so there are a large number of parallel interconnection interfaces, such as BOW, AIB, OpenHBI, LIPICON, Ultralink, HBM, UCIe, etc. [16-21]. As a relatively widely used standard protocol, UCIe defines the physical layer and link layer of interconnection between dies, and supports PCIe and CXL high-speed interconnection standards.

TABLE II.   INTERFACE PROTOCOL STANDARD OF CHIPLET

protocol	Initiator	Bandwidth density (Tb/mm)	Transmission rate (Gbps)	Transmission distance (mm)
NV link[13]	Nvidia	--	100	--
Infinity Fabric[14]	AMD	0.44	--	>20
AIB2.0[16]	Intel	1.24	≥2	<10
Ultralink	Cadence	0.5	40	--
BoW base[17]	ODSA	1.9	5	10
OpenHBI3.0[18]	ODSA	1.15	6.4	3
LIPINCON	TSMC	1.07	8	0.5
USR[15]	OIF	--	56	--
UCIe(std)	Intel	0.224	32	25
UCIe(adv)	Intel	1.317	32	2

C. Interconnection network

Chiplet technology forms a network (Network on Interposer) on the substrate to realize the interconnection between dies. The quality of interconnection network will directly affect the performance and power consumption of data communication, including interconnection topology, routing algorithm and fault-tolerant mechanism [2].

1) Interconnection topology:

The key factors of interconnection topology includes path routing, network traffic bandwidth balancing, and flexible expansion of system [9]. Zen2 [6] and Zen3 [22] architecture of AMD used a special interconnection structure inside the IOD to connect the CCD, DDR, PCIe and other interfaces; and research shows a special NOC structure die used to between the other dies [23-24]; In addition, the irregular topology optimizes the network links or structure according to the corresponding traffic characteristics [25-27]. On the basis of irregular topology, reconfigurable topology further improves network performance for various application traffic by dynamically reconfiguring and allocating link bandwidth according to the traffic characteristics real time [28-31].

2) Routing algorithm

The routing algorithm determines the path length and reliability of data transmission in interconnection network[2]. Router is generally used in NOC related networks. Good routing algorithm not only balances network performance, reliability and power consumption, but also takes into account universality, testability and scalability.

3) Fault tolerance

Advanced process, advanced packaging and Chiplet technology are not mature with long-term development. Considering the complexity of the interconnection and the manufacturing failure rate, a fault-tolerant mechanism can be established through redundant design to enhance the reliability and yield of the chip.

IV. DISCUSSION

The interconnection design of the Chiplet is guided by the application requirements. It depends on the data transmission characteristics of different application and environments. In general computing field, such as CPU, the data transmission has the characteristics of high randomness, multiple data structures and high cache consistency, delay optimization becomes the first priority; In parallel field, such as GPGPU, data transmission has the characteristics of large single transmission, and supporting preload data, bandwidth becomes the higher priority. Therefore, it is the development direction of future interconnection design to optimize performance for application fields and use appropriate interface networks and standardized protocols.

REFERENCES

[1] Wu H, Kang J, Chi M, et al. The technology trend of IC manufacture during Post Moore's era[C]. China Semiconductor Technology International Conference (CSTIC), Shanghai, China, 2017:1-7.

[2] White Paper on Integrated Chip and Chiplet Technology[EB/OL]. https://www.gitlink.org.cn/zone/iChips/source/12.

[3] Chen W T , Lin C C , Tsai C H ,et al. Design and Analysis of Logic-HBM2E Power Delivery System on CoWoS® Platform with Deep Trench Capacitor[C]. IEEE 70th Electronic Components and Technology Conference (ECTC). Orlando, FL, USA, 2020:380-385.

[4] Farjadrad R, Vinnakota B. A Bunch of Wires (BoW) Interface for Inter-Chiplet Communication [C]. IEEE Symposium on High-Performance Interconnects (HOTI).IEEE, 2019: 27-273.

[5] Li L Q, Liu X Y, Pang J. Key Technologies and Challenges of Chiplet[J].ZTE technology,2022,28(05):57-62.

[6] Suggs D, Subramony M, Bouvier D. The AMD "Zen 2" Processor[J]. IEEE Micro, 2020, 40(2):45-52.

[7] Feng Y, Ma K. Chiplet Actuary: A Quantitative Cost Model and Multi-Chiplet Architecture Exploration[J]. 2022.

[8] Apple unveils M1 Ultra, the world's most powerful chip for a personal computer [EB/OL]. https://www.apple.com/newsroom/2022/03/apple-unveils-m1-ultra-the-worlds-most-powerful-chip-for-a-personal-computer.

[9] Chen L, Huang L T. Chiplet functional partitioning and interconnection review[J]. Integrated Circuits and Embedded Systems,2024,24(02):41-49.

[10] Chromczak J , Wheeler M , Chiasson C ,et al. Architectural Enhancements in Intel® Agilex™ FPGAs[C]. FPGA 20' ACM/SIGDA International Symposium on Field-Programmable Gate Arrays, 2020:140-149.

[11] Sharma D D , Coughlin T , Coughlin T .Universal Chiplet Interconnect Express: An Open Industry Standard for Memory and Storage Applications[J]. Computer, 2024, 57(1):75-81.

[12] T/CESA 1248-2023, Technical requirements for chiplet interface bus[S].

[13] Whitepaper: NVIDIA® NVLink ™ High-Speed Interconnect: Application Performance[EN/OL]. http://info.nvidianews.com/rs/nvidia/images/NVIDIA NVLink High-Speed Interconnect Application Performance Brief.pdf

[14] Naffziger S , Lepak K , Paraschou M ,et al. AMD Chiplet Architecture for High-Performance Server and Desktop Products[J]. IEEE International Solid-State Circuits Conference - (ISSCC), San Francisco, CA, USA, 2020:44-45

[15] Common Electrical I/O (CEI) -Electrical and Jitter Interoperability agreements for 6G+ bps, 11G+ bps,25G+ bps, 56G+ bps and 112G+ bps I/O [EB/OL]. https://www.oiforum.com/wp-content/uploads/OIF-CEI-05.2.pdf

[16] Liu C , Botimer J , Zhang Z .A 256Gb/s/mm-shoreline AIB-Compatible 16nm FinFET CMOS Chiplet for 2.5D Integration with Stratix 10 FPGA on EMIB and Tiling on Silicon Interposer[C]. IEEE Custom Integrated Circuits Conference (CICC). Austin, TX, USA, 2021:1-2.

[17] Open compute project, OpenHBI Specification Version 1.0[EN/OL]. https://www.opencompute.org/documents/odsa-openhbi-v1-0-spec-rc-final-1-pdf.

[18] Lin M S , Goel S K , Fu C M ,et al. A 7nm 4GHz Arm®-core-based CoWoS® Chiplet Design for High Performance Computing [J].IEEE Journal of Solid-State Circuits, 2020, 55(4):956-966.

[19] Universal Chiplet Interconnect Express (UCIe) Specification Revision 1.0, Februrary 24, 2022.

[20] N. Beck, S. White, M. Paraschou and S. Naffziger, 'Zeppelin': An SoC for multichip architectures[C]. IEEE International Solid-State Circuits Conference - (ISSCC), San Francisco, CA, USA, 2018:40-42.

[21] G. Mounce, J. Lyke, S. Horan, W. Powell, R. Doyle and R. Some, "Chiplet based approach for heterogeneous processing and packaging architectures," 2016 IEEE Aerospace Conference, Big Sky, MT, USA, 2016:1-12.

[22] T. Burd et al. Zen3: The AMD 2nd-Generation 7nm x86-64 Microprocessor Core[C]. IEEE International Solid-State Circuits Conference (ISSCC), San Francisco, CA, USA, 2022:1-3.

[23] J. Lan, V. P. Nambiar, R. Sabapathy, M. D. Rotaru and A. T. Do, Chiplet-based Architecture Design for Multi-Core Neuromorphic Processor[C]. 23rd Electronics Packaging Technology Conference (EPTC), Singapore, Singapore, 2021:410-412.

[24] S. Zhu, M. Miao, Z. Zhang and X. Duan, Research on A Chiplet-based DSA (Domain-Specific Architectures) Scalable Convolutional Acceleration Architecture[C]. 23rd International Conference on Electronic Packaging Technology (ICEPT), Dalian, China, 2022:1-6.

[25] Bharadwaj, Srikant, et al. Kite: A family of heterogeneous interposer topologies enabled via accurate interconnect modeling[C]. 57th ACM/IEEE Design Automation Conference (DAC), 2020: 1-6.

[26] Kadomoto, Junichiro, Hidetsugu Irie, and Shuichi Sakai. Design of shape-changeable Chiplet-based computers using an inductively coupled wireless bus interface[C]. IEEE 38th International Conference on Computer Design (ICCD), 2020: 589-596.

[27] Kadomoto, Junichiro, et al. An Inductively Coupled Wireless Bus for Chiplet-Based Systems[C]. 25th Asia and South Pacific Design Automation Conference (ASP-DAC), 2020: 9-10.

[28] Zheng, Hao, Ke Wang, and Ahmed Louri. A versatile and flexible Chiplet-based system design for heterogeneous manycore architectures[C]. 57th ACM/IEEE Design Automation Conference (DAC), 2020: 1-6.

[29] Chen, Chia-Hsin Owen, et al. SMART: A single-cycle reconfigurable NoC for SoC applications[C] Design, Automation & Test in Europe Conference & Exhibition (DATE), 2013: 338-343.

[30] Parikh, Ritesh, Reetuparna Das, and Valeria Bertacco. Power-aware nocs through routing and topology reconfiguration[C]. Proceedings of the 51st Annual Design Automation Conference. 2014: 1-6.

[31] Wang, Mengdi, et al. Network-on-interposer design for agile neural-network processor chip customization[C]. 58th ACM/IEEE Design Automation Conference (DAC), 2021: 49-54.

Effects and Modeling Study on FDSOI MOSFETs at Cryogenic Temperature

Zhipeng Ren [1], Yibo Hu [1], Yizhe Yin [1], Ruiyong Zhao [1], Jing Chen*[1]

[1]National Key Laboratory of Integrated Circuits, Shanghai Institute of Microsystem and Information Technology, Chinese Academy of Science, Shanghai, China, 200050

* Email: renzp@mail.sim.ac.cn, jchen@mail.sim.ac.cn

Abstract—**In this work, we systematically investigate the impact of various performance metrics of 22nm FDSOI technology devices when operated at cryogenic temperatures. It establishes a cryogenic-temperature TCAD simulation model, delves into the back-gate effect and self-heating effect under cryogenic conditions, develops a parameter extraction flow applicable across the full temperature range and elucidates the temperature dependency of critical parameters. The BSIM model is revised with a fitting accuracy below 5%. Additionally, focusing on the cryogenic-temperature RF characteristics of the devices, substrate effects and gate parasitic network characteristics are characterized using an RC network, leading to the establishment of a comprehensive sub-circuit topology.**

Keywords—*Cryogenic-temperature, Fully-depleted Silicon on Insulator(FD-SOI), Radio frequency (RF) characteristics, Device model, Technology Computer Aided Design (TCAD) Simulation.*

I. INTRODUCTION

In the post-Moore era, further miniaturization of silicon-based devices encounters escalating costs and technical challenges, accompanied by complex physical phenomena and severe heat dissipation issues[1]. Cryogenic temperatures offer a transformative leap for integrated circuits, enhancing transistor characteristics towards ideal switching behavior and reducing interconnect resistance and signal delay across systems, thereby providing a new approach to overcome computational power bottlenecks. Compared to conventional bulk silicon processes, Fully Depleted Silicon-on-Insulator (FDSOI) technology not only ensures performance with lower power consumption but also possesses unique back-gate control capabilities, making it a pivotal means to extend Moore's Law[2]. Therefore, the combination of cryogenic temperatures and FDSOI devices presents even greater advantages.

This work systematically analyzes the impact of cryogenic temperatures on FDSOI device performance based on 22nm technology nodes. It establishes a cryogenic-temperature TCAD simulation model, conducts in-depth research on back-gate effects and self-heating at cryogenic temperatures, devises a full temperature range parameter extraction process and the temperature dependence of key parameters, modifies the BSIM model with fitting accuracy less than 5%, and for the RF characteristics at cryogenic temperatures, characterizes substrate effects and gate parasitic network behavior through an RC network, constructing a complete sub-circuit topology.

This work is supported in part by the Shanghai "Qi Ming Xing" Program - Sail Plan under Grant NO. 22YF1456300. In part by the Zhangjiang National Laboratory under Grant NO. Z-W-22-007.

II. DEVICE PERFORMANCE AT CRYOGENIC TEMPERATURE

A. Back-Gate Effect of FDSOI

FDSOI lacks accumulation capacitance and traditional bulk silicon extraction methods are unsuitable due to the presence of the Buried Oxide (BOX) layer. Accurate device physical parameters were obtained through the development of back-gate equivalent circuit models and parameter extraction, as illustrated in Fig.1 and Table 1. As back-gate voltages (Vbs) changes, the interval of each adjacent two curves is uniform at 300 K but non-uniform at 77 K, as shown in the black circle in Fig3.a and Fig3.b, indicating the existence of non-linear behaviors in back-gate effect at cryogenic temperatures. The possible reasons are analyzed and verified by TCAD simulation in Fig.3 and Fig.4, suggesting that normal-well devices are more susceptible to the formation of depletion regions between the buried oxide layer and the well. This phenomenon disrupts the linearity of the back-gate effect.

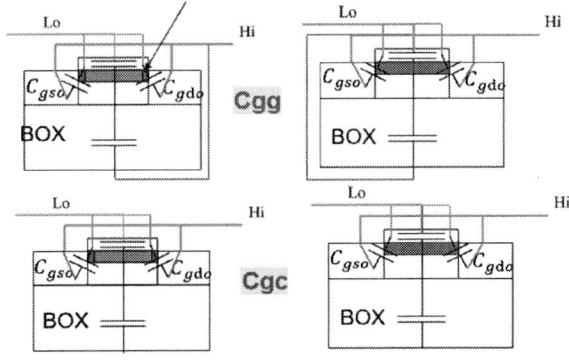

Fig.1. Equivalent circuit diagram of the capacitance model.

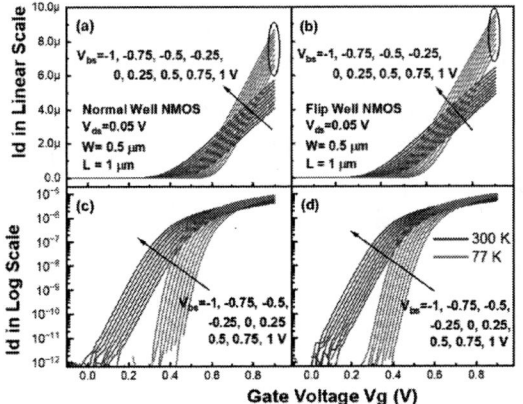

Fig.2. The Id-Vg curves of normal and flip-well N-MOSFETs at different back-gate voltages (Vbs) in linear region.

TABLE I.   COMPARISON BETWEEN DEVICE FEATURE PARAMETER EXTRACTION VALUE AND PROCESS TARGET

Process Parameters	Extracted value	Process Target
Tox	1.36nm	1.32nm
Tbox	19.8nm	20nm

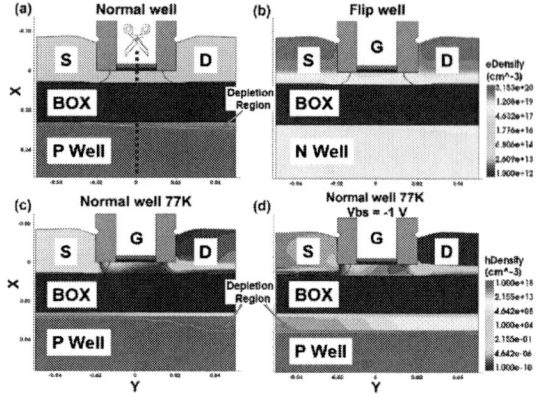

Fig. 3. The TCAD simulation of FDSOI N-MOSFETs.

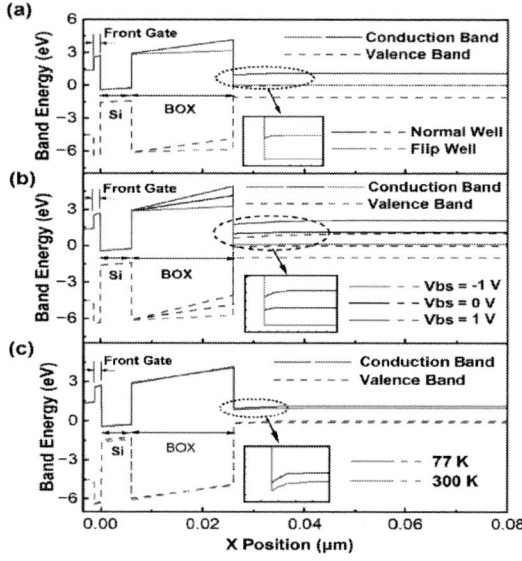

Fig.4. The calculated conduction and valence band energy distribution of N-MOSFETs along the longitudinal cross section.

B. Self-Heating Effect of FDSOI

Compared to bulk silicon, FDSOI's unique BOX layer structure exhibits lower thermal conductivity, which can lead to increased device self-heating at cryogenic temperatures. Output conductance (gds) variations extracted through RF testing are used to assess the degree of self-heating[3].We have conducted small-signal equivalent circuit modeling on the test data to ascertain the frequency range influenced by self-heating, determining it to be below 160 MHz, with excellent fitting results as depicted in Fig.5. Fig.6 and Table 2 illustrate that the self-heating effect in FDSOI increases as the temperature decreases, with the output conductance variation caused by self-heating at 77 K being 18.39% higher than that at room temperature (300K) .

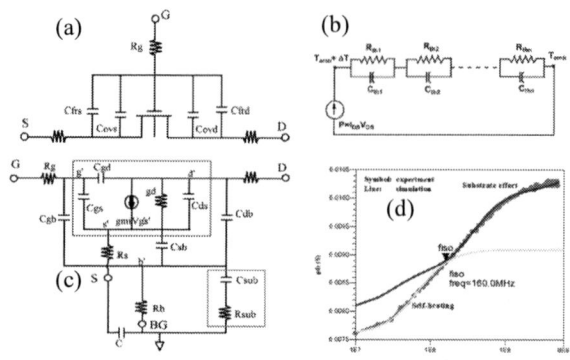

Fig.5. Small signal equivalent circuit diagram of FDSOI MOSFET including gate(a)substrate(c)and SHE network(b), Frequency versus gds using different network simulations(d).

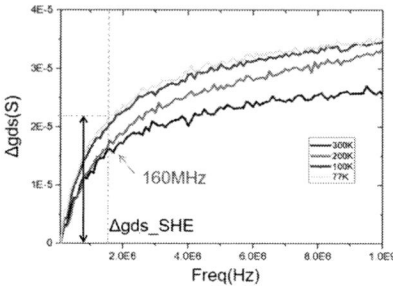

Fig.6. Frequency versus gds of FDSOI nMOSFET with Lg=32nm at different temperatures

TABLE II.   THE VALUE OF ΔGDS_SHE AT DIFFERENT TEMPERATURES

Temperature	300K	200K	100K	77K
Δgds_SHE	5.15E-05	4.80E-05	4.41E-05	4.40E-05
Δgds_SHE %	/	5.22%	14.96%	18.39%

C. RF Performance of FDSOI

Fig.7 illustrates that when the temperature decreases from 300K to 77K, the cutoff frequency (fT) experiences an increase of approximately 36.8%, while the maximum oscillation frequency (fmax) sees an enhancement of around 33.2%. Fig.8 demonstrates that although the performance of the device biased at Vdd=0.5V is slightly lower compared to when biased at Vdd=0.9V (with a decrease of within 10%), this lower bias condition reduces power consumption by roughly 50%. At cryogenic temperatures and with a reduced supply voltage, the device is capable of achieving performance akin to that observed at room temperature but with only half the power demand.

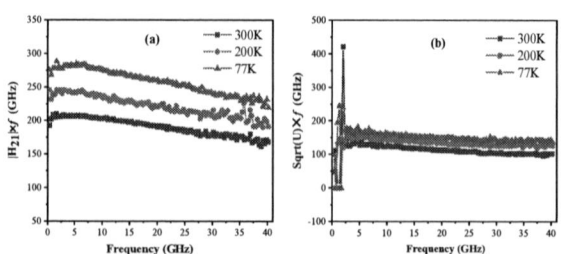

Fig.7. Frequency versus fT(a)and fmax(b) of FDSOI nMOSFET at different temperature

979-8-3503-6184-1/24 $31.00 © 2024 IEEE

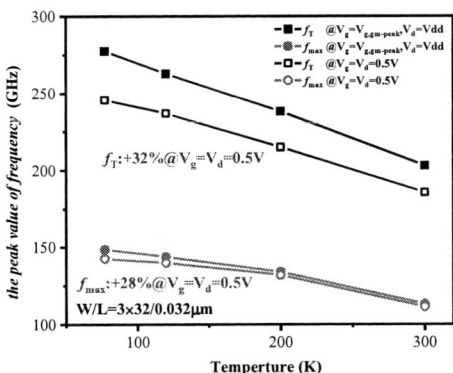

Fig.8. Temperature versus normalized gm of FDSOI nMOSFETs with different Lg.

III. FDSOI DEVICE MODEL AT CRYOGENIC TEMPERATURE

This work combines physical and mathematical modeling approaches to key device parameters, striving to retain as much of the original BSIM framework as possible while establishing a device model valid across the full temperature range from 300K to 4K. Fig.5 showcases the DC model performance across this entire temperature span, with fitting accuracy better than 5%. Building upon the cryogenic-temperature DC model, a small-signal equivalent circuit model for cryogenic conditions is developed, achieving a similarly high precision of less than 5% for RF modeling.

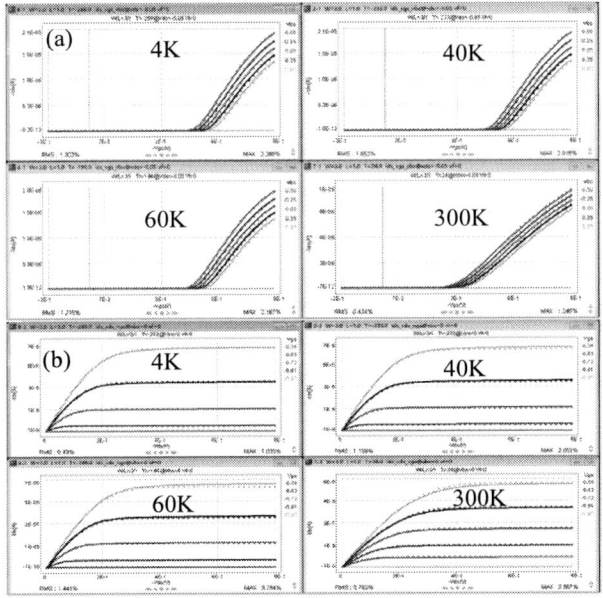

Fig.9. The Id-Vg (a) and Id-Vd (b) curves fitting results at different temperature.

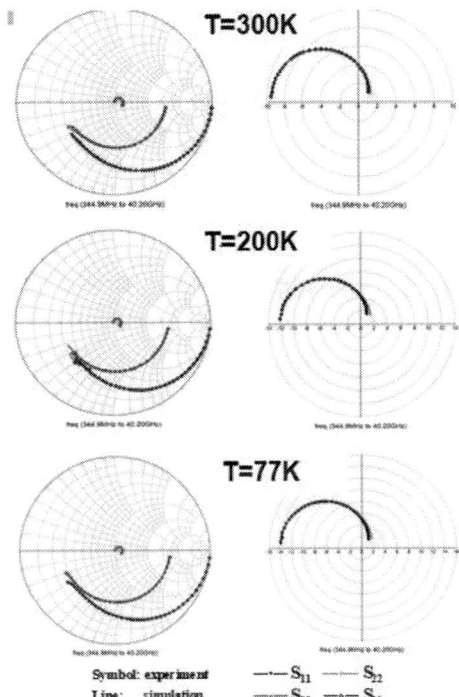

Fig.10. Measured and simulated S parameter smith charts at different temperature.

IV. SUMMARY

This study systematically investigates the effects of cryogenic temperatures on the performance of 22nm FDSOI devices, It establishes cryogenic-temperature CV equivalent circuit models and TCAD simulation models, revealing that cryogenic temperatures facilitate depletion layer formation in the well, intensifying the nonlinearity of back-gate modulation. The investigation of self-heating under cryogenic conditions shows a 18.39% increase in output conductance variation. The RF performance of FDSOI devices, including fT and fmax, improves by over 30% as temperature drops, allowing for similar performance at room temperature with just half the power consumption at 77K. A full temperature range parameter extraction process and modified BSIM model are developed, and a comprehensive sub-circuit topology capturing substrate and gate parasitic effects is established for cryogenic-temperature RF characteristics.

REFERENCES

[1] H. -L. Chiang, J. -J Wu, C. -H Chou, Y. -C Chen, L Liu, J. F. Wang, et al., "Design Technology Co-Optimization for Cold CMOS Benefits in Advanced Technologies," IEEE International Electron Devices Meeting (IEDM), pp. 13.2.1-13.2.4, 2021.

[2] S. Bonen, U. Alakusu, Y. Duan, M. J. Gong, M. S. Dadash, L. Lucci, et al. "Cryogenic characterization of 22-nm FDSOI CMOS technology for quantum computing ICs." IEEE Electron Device Letters, vol. 40, no. 1, pp. 127-130, 2018.

[3] S. Makovejev, S. Olsen, and J. Raskin. RF extraction of self-heating effects in FinFETs[J]. IEEE Trans. Electron Devices, vol. 58, no. 10, pp. 3335-3341, Oct. 2011.

AUTHOR INDEX

Ahammed, Shihab...316
Ai, Zhaoyu...877
An, Suihuan...630
An, Xiaomeng..748
An, Zhemin..813
Bai, Jiyuan...59, 739
Bai, Mingkai..484
Bai, Song...423
Ban, Chaoyi..508
Bao, Lin..504, 508
Bao, Yunjiao..777
Bi, Hengchang..551, 751
Bi, Ran..526
Bi, Shunyang...216
Bo, Chunyue..733
Brown, James..437
Bu, Weihai..364, 760
Cai, Chenxiang..273
Cai, Linlin...65, 523, 792
Cai, Rundong..83
Cai, Xi...466, 661
Cai, Yimao..504, 508
Cai, Zilu..679
Cao, Jiangtao..661
Cao, Junhou..828
Cao, Kun...338
Cao, Lei..757
Cao, Lixuan...241
Cen, Yuanjun...688
Chai, Junshuai.................460, 463, 484, 511
Chai, Z..494
Chai, Zheng...630
Chan, Mansun....................316, 325, 786, 850
Chang, Haixin..301
Chang, Hao..183
Chang, Pengying..816
Chang, Yuchun...664
Chen, Bing...661
Chen, Bofan...282
Chen, Fan..232
Chen, Gengsheng........................59, 620, 739
Chen, Gui...361
Chen, Haifeng..65
Chen, Hong..771
Chen, Hongquan...430, 451
Chen, Hongzhi..801
Chen, Hua..742
Chen, Jiezhi..............................457, 807, 810

Chen, Jing.................................655, 892, 940
Chen, Kaiqi..624
Chen, Ke..68
Chen, Kuangli..398, 454
Chen, Lei...688
Chen, Lin...368
Chen, Liying...104
Chen, Long...649
Chen, Ning...937
Chen, Peng...844
Chen, Quan...202
Chen, Rong...338
Chen, Ruolan..706
Chen, Shiyou..342
Chen, Shulan..706
Chen, Shushi..587
Chen, Sichao..219
Chen, Song..10
Chen, Tao...602
Chen, Wangyong..........................65, 523, 792
Chen, Wanjun..................395, 411, 847, 874
Chen, Weifeng..771
Chen, Wenjun...801
Chen, Wenwei...186
Chen, Xiang..599
Chen, Xibo..661
Chen, Xinghuan...874
Chen, Xingyu..466
Chen, Xinyang..877
Chen, Xu..345
Chen, Xurui..126
Chen, Yanning..910
Chen, Yifei...238, 721
Chen, Yong...122
Chen, Yongliang..599
Chen, Yu..630
Chen, Yuhua...646
Chen, Yujie..28
Chen, Yuwei...895
Chen, Yuyang...736
Chen, Zheng-Han..143
Chen, Zhikang..22
Chen, Zhong-Jian..143
Chen, Zirui..31
Chen, Zixiang...59
Chen, Zixun..819
Cheng, Chuantong..104
Cheng, Hsing-Mao..307

Cheng, Junji	825, 865
Cheng, Kai	122
Cheng, Lin	856, 880
Cheng, Nannan	258
Cheng, Yu	664
Cheng, Yuhua	447
Chenqi	31
Chu, Chun Yee	783
Chu, Junhao	178
Chu, Yu	718
Chuang, Yuan-Yu	41
Coutinho, Jose G. F.	195
Cui, Miao	408
Cui, Wentao	372
Cui, Xianghe	7
Cui, Xiaole	599
Cui, Xuecheng	661
Dai, Chengxing	107, 643
Dai, F.	433
Dai, Ming	865
Dai, Saifei	463, 484, 511
Dai, Shun-Qi	798
Dai, Siyao	627
Dai, Wu	536
Dai, Yuan	219
Deng, Chenkai	388
Deng, Gaoqiang	379
Deng, Haonan	834
Deng, Hongfei	417
Deng, Jie	235
Deng, Lei	132, 676, 685, 691
Deng, Shiyu	395
Deng, Xiaochuan	401
Deng, Yongfeng	910
Deval, Yann	267
Di, Boyuan	301
Ding, Guojian	771
Ding, Qi	392
Ding, Rongzheng	763
Ding, Runye	28
Ding, Shijin	804
Ding, Tongshu	186
Ding, X.	433
Ding, Yajing	484
Ding, Yingtao	304
Ding, Yong	466, 614, 661
Ding, Yuanhong	132, 676
Dong, Chenge	44
Dong, Feng	68
Dong, Haitao	807
Dong, Jianing	289
Dong, Junchen	264, 533

Dong, Lei	572
Dong, Li-Hang	868, 871
Dong, Qingyang	733
Dong, Qiyun	620
Dong, Shijiao	232
Dong, Shuangqi	934
Dong, Xiaoping	420
Dong, Yaoqi	345
Dong, Zuoyuan	178
Du, Feng-Yu	414, 831
Du, Gang	264
Du, Peiyuan	472
Du, Rongxin	907
Du, Wencai	768
Du, Yiran	602
Duan, Enchuan	398
Fan, Difei	388
Fan, Wenbo	511
Fan, Xiangrui	822
Fan, Yibo	77, 587, 605, 608
Fang, Jian	126
Fang, Rui	614
Fanyi, Meng	228
Feng, Baoyan	862
Feng, Chao	454
Feng, Cunfeng	289
Feng, Jiekai	904
Feng, Lichen	83, 156
Feng, Qi	771
Feng, Yang	457, 807
Fu, Chengwen	420
Fu, Fangfa	199, 789
Fu, Hao	382, 822, 828
Fu, Jing	895
Fu, Peiyuan	89
Fu, Xiaojun	825
Fukuda, Kenjiro	285
Gai, Weixin	110
Gan, Jie	611
Gao, Dan	183, 441
Gao, Dawei	910
Gao, Jianjun	736
Gao, Liang	581
Gao, Rui	437
Gao, Ruiyi	252
Gao, Shujun	545
Gao, Xingyu	199, 789
Gao, Xuchen	219
Ge, Hao	892
Ge, Ziang	208
Golosov, Dmitriy Anatolyevich	804
Gong, Sen	417

Gong, Yi	25
Gong, Yong	478
Gong, Yuehong	289
Gou, Fei	86
Gu, Chenjie	804
Gu, Hanzhi	335, 514, 520
Gu, Shengfei	74
Gu, Wenxian	751
Gu, Yong	423
Guan, Chaowen	299
Gui, Qingzhong	296
Guo, Anran	895
Guo, Huimeng	44
Guo, Jiajie	62
Guo, Jianping	922
Guo, Jiyuan	685
Guo, Junxiong	931
Guo, Mingqiang	152
Guo, Pei	379, 385
Guo, Peng	810
Guo, Qingyu	13
Guo, Shuaicheng	614
Guo, Tianyu	569
Guo, Xinyi	457, 810
Guo, Yirong	816
Guo, Yiting	35
Guo, Yuekang	175
Guo, Yufeng	931
Guo, Yulin	754
Guo, Yutuo	392
Guo, Yuzheng	296
Guo, Zhijian	469
Guo, Zhiwang	171
Guo, Zizheng	212
Han, Chao	414, 831, 853, 856, 862
Han, Dedong	533
Han, Dongxiang	919
Han, Genquan	472, 661
Han, Hongyan	760
Han, Jun	1, 16, 50, 624, 627, 634
Han, Kai	484, 511
Han, Kefeng	730
Han, Lixiang	581
Han, Rong	688
Han, Runhao	484, 511
Han, Wei-Hua	319
Hang, Jianfeng	760
Hao, Linyao	385
Hao, Sang	98
Hao, Yue	472, 542, 837
Hao, Zhijian	608
Hatayama, Kazumi	192

He, Chenlong	605
He, Huanxiang	587
He, Kunqin	392
He, Lin	560
He, Nailong	430
He, Songxuan	792
He, Weitao	715
He, Wen	289
He, Xinyu	186
He, Yanping	345
He, Yingtong	630
He, Yu	667
He, Yuhui	31
He, Yunlong	542
Hengzhou, Yuan	98
Heyuan, Liu	98
Hiramoto, Toshiro	332
Hong, Bo	614
Hongshi, Yu	228
Hou, Bin	837
Hou, Xiangyu	423
Hu, Ang	74, 563, 694, 709, 727
Hu, Haolin	454
Hu, Jiahao	401
Hu, Jianguo	202
Hu, Jianzhi	304
Hu, Jiaxuan	904
Hu, Jing	697
Hu, Jingyi	7
Hu, Qiang	877
Hu, Shuang	703
Hu, Siyuan	533
Hu, Tao	484, 511
Hu, Weiming	129
Hu, Yi	584
Hu, Yibo	655, 892, 940
Hu, Yidi	928
Hu, Yuhan	667
Hu, Yushen	329
Hu, Z.	494
Hua, Younan	228
Huang, Guang-Wei	307
Huang, Haimeng	530, 813, 825, 841, 865
Huang, Heyi	774
Huang, Jialu	178
Huang, Jian	487
Huang, Jie	847
Huang, Jun	392
Huang, Junlin	183, 441
Huang, Lei	828
Huang, Leilei	77, 587, 596, 605, 608
Huang, Menglin	342

Huang, Mingmin	420
Huang, Qianqian	760
Huang, Qihang	907
Huang, Ru	212, 355, 364, 426, 504, 508, 536, 760
Huang, Runhua	423
Huang, Shuting	398, 454, 907
Huang, Shuyang	834
Huang, Tianze	74
Huang, Tingrui	517
Huang, Wen	931
Huang, Wende	420
Huang, Yifeng	780
Huang, Yujie	4, 593
Huang, Zepeng	551
Huang, Zhaofeng	551
Huo, Jiali	774
Huo, Qiuyue	392
Ichikawa, Tamotsu	192
Ihda, Ayoub Ait	267
IImori, Daisuke	192
Imura, K.	433
Ishida, Takashi	192
Ji, Xiaoli	575, 925
Ji, Zhigang	351, 437
Jia, Bin	715
Jia, Guowen	590
Jia, Haiqiang	771
Jia, Longjiang	132, 676
Jia, Peng-Fei	868, 871
Jia, Rundong	760
Jia, Yiming	841
Jia, Yirui	430, 451
Jiang, Chen	270
Jiang, Chunsheng	768
Jiang, Dongyang	152
Jiang, Jianfei	68
Jiang, Li	679
Jiang, Nan	661
Jiang, Renjie	757, 774, 777
Jiang, Tong	593
Jiang, Xin	733
Jiang, Xingli	865
Jiang, Yunteng	813
Jiang, Zijin	408
Jianzhuang, Lu	98
Jin, Chengji	469, 472
Jin, Jing	175
Jin, Xiangliang	898
Jin, Xin	276
Jin, Yu	602
Jin, Yuhan	126
Jin, Zhenghao	813

Jing, Minge	4, 593
Jing, Zhiwei	841
Jingtian, Liu	98
Kang, Hao-Bo	831
Kang, Jin	364
Kang, Jinfeng	475
Kang, Wang	47
Kang, Yi	10
Katayama, Shogo	192
Katoh, Kentaroh	135, 192
Ke, Xiaoyu	460
Ke, Xuanxin	673
Kim, Ikhwan	554
Kobayashi, Haruo	135, 192
Kobayashi, Masaharu	332, 469
Kobayashi, Yutaro	135
Kong, Moufu	417, 865, 877
Kumar, Anuj	490
Kuo, Chia-Yo	313
Kuwana, Anna	192
Lai, Jinmei	186, 189
Lam, Sang	408
Lan, Haokun	238
Lang, Zhengyu	834
Lapuyade, Hervé	267
Lei, Tengteng	329
Lei, Wenyu	301
Lei, Yuan	798
Li, Angyang	691
Li, Bangtian	104
Li, Biao	101
Li, Bin	685
Li, Bo	113
Li, Cheng	117
Li, Chengxin	904
Li, Cong	545, 889
Li, Congrui	706
Li, Dejian	276
Li, Dengquan	149, 670
Li, Dongya	472
Li, Gang	19, 71
Li, Guowei	895
Li, Hao	563, 581
Li, Haokun	545
Li, Haolin	539
Li, Hong	101
Li, Hui	19
Li, Jianguo	498
Li, Jianwen	688
Li, Jianzheng	129, 748
Li, Jie	816
Li, Jincheng	739

Li, Jing 859
Li, Jinshan 508
Li, Jinyu 658
Li, Junfeng 757
Li, Kai 74
Li, Kaiyue 563
Li, Kanyi 511
Li, Kunyue 68
Li, Lei 199, 789
Li, Lianlian 757
Li, Luping 844
Li, Lu-Ping 871
Li, Meng 13
Li, Menghan 634
Li, Ming 364, 526, 844
Li, Mingjie 304, 934
Li, Ningning 95, 691, 745
Li, Peng 904
Li, Qingkun 757
Li, Renxiong 392
Li, Renxuan 694, 727
Li, Rophina 372
Li, Ruguan 530, 813
Li, Ruizhi 895
Li, Ruoyu 548
Li, Shangze 475
Li, Sheng 376
Li, Shiming 837
Li, Shuaihong 916
Li, Shurui 417
Li, Siteng 754
Li, Tianyu 733
Li, Tiaoyang 348, 883
Li, Wang 338
Li, Wanyu 807
Li, Wei 146, 232, 279, 602, 649, 844, 868, 871
Li, Weizhi 65
Li, Wenhong 4, 593, 922
Li, Xiangnan 385
Li, Xiaomin 228
Li, Xin 107, 643
Li, Xingyu 523
Li, Xuan 401
Li, Xueke 104
Li, Yang 481
Li, Yanqing 774
Li, Yanzhong 77
Li, Yichen 258
Li, Yongjia 258
Li, Yongxi 530
Li, Yongxiang 501
Li, Yudong 766

Li, Yujia 44
Li, Yu-Tao 255
Li, Yutao 469
Li, Zehong 844
Li, Ze-Hong 868, 871
Li, Zhaoji 451
Li, Zhenmin 7
Li, Zhensong 533
Li, Zhenwei 883
Li, Zhihuai 679
Li, Zhiqiang 222
Li, Zhiqun 282
Li, Zhiwei 697
Li, Zonglin 813
Lian, Xueguang 466, 661
Liang, Jie 886
Liang, Jingyuan 372
Liang, Libin 801
Liang, Ling 504, 508
Liao, Min 460
Liao, Pengfei 904
Liao, Xufeng 89, 276
Liao, Yiming 575, 925
Lin, Guangyao 62
Lin, Jinpeng 101
Lin, Jyi-Tsong 41, 307, 313, 322
Lin, Lin 931
Lin, Yibo 212, 364
Ling, Zipeng 1
Liou, Juin J. 228
Liu, Ao 423
Liu, Chang 345
Liu, Chao 395, 847
Liu, Chen 252
Liu, Dong 289
Liu, Donglin 222
Liu, Dongsheng 74, 563, 581, 694, 709, 727
Liu, Fang 910
Liu, Fei 92
Liu, Feng 771
Liu, Haitao 682
Liu, Houfang 569
Liu, Huajie 126
Liu, Huan 472
Liu, Jie 575
Liu, Jin 566
Liu, Jinbiao 345
Liu, Jing 457
Liu, Jingjing 56, 168, 261, 557, 640, 673
Liu, Junzhan 47
Liu, Kai 596
Liu, Ke-Yu 414, 831

Liu, Lianxi	89, 276
Liu, Linghao	159
Liu, Maliang	162, 238, 566, 721
Liu, Mingshan	345
Liu, Puyang	117
Liu, Qianhui	646
Liu, Renkuan	385
Liu, Rui	572, 682
Liu, Shengxiang	709
Liu, Shuang	774, 777
Liu, Shubin	670
Liu, Sicheng	916
Liu, Siyang	258, 376, 382, 423, 822, 828
Liu, Wei	86, 411, 427, 498
Liu, Weibing	345
Liu, Wenjuan	292
Liu, Wenjun	804
Liu, Xiangze	925
Liu, Xianping	487
Liu, Xiaotao	38
Liu, Xiaoyan	475, 539, 724, 913
Liu, Xindi	379
Liu, Yan	472
Liu, Yan-Cong	414
Liu, Yang	819
Liu, Yanheng	348
Liu, Yanjiang	602
Liu, Yao	28
Liu, Yaxin	392
Liu, Yinchi	804
Liu, Yu	931
Liu, Yuchen	667
Liu, Yuyang	28
Liu, Zhe	345
Liu, Zhifang	754, 916
Liu, Zhiqing	718
Liu, Zhiwan	813
Liu, Zhongyang	934
Liu, Zihao	332
Liu, Ziyu	368
Long, Jieyu	841
Long, Keli	199, 789
Lu, Bin	38
Lu, Cimang	478
Lu, Haoran	38, 364
Lu, Hongliang	444, 804, 880
Lu, Jiahao	74
Lu, Jinlong	379
Lu, Sheng	620, 634
Lu, Weihao	859
Lu, Wen-Gao	143
Lu, Xiaoli	542

Lu, Yupeng	757, 774, 777
Luk, Wai-Shing	219
Luk, Wayne	195
Luo, Huanlin	189
Luo, Jie	777
Luo, Jun	345
Luo, Min	289
Luo, Peng	162
Luo, Ping	904
Luo, Weijia	754
Luo, Weijun	733
Luo, Xiaorong	379, 385
Luo, Xixi	411
Lv, Hankun	466
Lv, Qi	664
Lyu, Haiyuan	358
Lyu, Hongliang	901
Lyu, Lewei	877
Lyu, Liangjian	551, 751
Ma, Bingbing	146
Ma, Huaping	895
Ma, Jian	335, 514, 520
Ma, Jiang	771
Ma, Jianghao	889
Ma, Jie	258, 423
Ma, Kaixue	228
Ma, Luanxi	907
Ma, Mingyu	545, 889
Ma, Shunli	700
Ma, Wenzhe	637
Ma, Xiaohua	542, 837
Ma, Xingyu	232
Ma, Xudong	819
Ma, Xuejiao	71
Ma, Yanfeng	376
Ma, Yao	420
Ma, Yuanxiao	469
Mahalingam, Nagarajan	228
Mahapatra, Souvik	490
Mak, Pui-In	122
Mao, Yiqing	205, 219
Mao, Zhifeng	86
Martins, Rui P.	122, 152
Mei, Jian	104, 132, 676, 685, 691
Mei, Xianqi	74
Meng, Fanxin	877
Mi, Jinyao	47
Miao, Xiangshui	31
Min, Hao	715
Min, Tai	630
Mizutani, Tomoko	332
Mo, Fei	469

Mo, Wenji 56, 168, 261
Mori, Takahiro 332
Nakatani, Takayuki 192
Nan, Longmei 602
Ng, Wai Tung 372, 783
Ni, Xiangran 733
Nie, Yanyu 907
Ning, Ning 392, 667
Niu, G. 433
Niu, Ruiting 101
Niu, Yan 289
Ogawa, Tomohiko 135
Ogihara, Gaku 192
Oka, Hiroshi 332
Okamoto, Toshiyuki 192
Ostling, Mikael 404
Ouyang, Lingyun 202
Pan, Chuanqi 376
Pan, Jie 584
Pan, Yu 614
Peng, Baokang 536
Peng, Chaoyang 859
Peng, Chunyu 107, 643
Peng, Daixiao 466, 661
Peng, Lijuan 771
Peng, Liyuan 4, 593
Peng, Lulu 392
Peng, Pan 392
Peng, Sirui 466
Peng, Xiaosong 385
Peng, Xizhu 139
Peng, Yungen 487
Peng, Zhirong 850
Pu, Jie 688
Qi, Gengzhen 235, 273
Qi, Junjun 880
Qi, Zhao 430, 451
Qiang, Hua 38
Qiao, Jiantao 880
Qiao, Liangquan 199, 789
Qiao, Ming 392, 430, 451
Qiao, Shushan 222
Qiao, Yong 68
Qiao, Yuan 865
Qin, Haifeng 844
Qin, Haiyan 47
Qin, Yajie 129, 554, 748
Qin, Zhigang 578
Qiu, Hao 113
Qiu, Xiang 478
Qu, Xin-Ping 361
Qu, Yang 664

Qu, Yiming 427, 498, 742
Qu, Yuwei 205
Que, Zhiqiang 195
Rao, Amy 652
Rao, Qiansheng 844
Ren, Junyan 159, 165
Ren, Min 844
Ren, Mingze 22, 703
Ren, Pengpeng 351
Ren, Tian-Ling 255, 919
Ren, Tianling 569, 766
Ren, Tingrui 44
Ren, Ye 760
Ren, Yuan 712
Ren, Zhipeng 655, 892, 940
Rivet, François 267
Robertson, John 296
Rochette, Stephane 267
Sang, Guanqiao 757
Sang, Pengpeng 810
Saraya, Takuya 332
Sato, Keno 192
Sawan, Mohamad 248
Sha, Yanliang 202
Shan, Hongwei 83, 156
Shan, Linbo 508
Shan, Xiaoyu 694, 727
Shang, Zi-Meng 319
Shao, Sicong 338
Shao, Xianzhou 460, 463, 484, 511
Shao, Yu 542
Shao, Yun-Hao 361
Shen, Boqian 335, 514, 520
Shen, Chao 299, 928
Shen, Hao-Yuan 255
Shen, Hongwei 584
Shen, Lei 937
Shen, Rensheng 664, 795
Shen, Xiaohan 270
Shen, Xinxin 517
Sheng, Bin 86
Shi, Chunqi 596
Shi, Jincheng 530
Shi, Jinhong 530
Shi, Mingmin 526
Shi, Shucheng 289
Shi, Tiantian 925
Shi, Xinhua 624
Shi, Yi 113
Shi, Yu 411, 874, 907
Shi, Yue 578
Shuai, Li 228

Sin, Sai-Weng 152
Someya, Takao 285
Song, Jiafeng 754
Song, Qing-Wen 414, 831, 853, 856, 862
Song, Ruiyang 244
Song, Wei 379, 385
Song, Xujin 475
Song, Yifan 751
Song, Yukun 7
Song, Zhaoxu 382
Su, Xinying 557
Su, Yanwen 563
Sun, Chengliang 292
Sun, Chuanlin 533
Sun, Dijiang 475
Sun, Haoning 56, 168, 261, 673
Sun, Jialei 68
Sun, Jiameng 822
Sun, Junyi 113
Sun, Kangkang 168, 673
Sun, Ke 709
Sun, Leihao 299
Sun, Qingqing 335, 368, 514, 520
Sun, Quan 572, 682
Sun, Ruize 395, 411, 847, 874
Sun, Weifeng 258, 376, 382, 423, 517, 822, 828
Sun, Wendi 10
Sun, Xiaofeng 258
Sun, Xiaoqing 460, 463, 484, 511
Sun, Xinqi 395
Sun, Yilin 304, 754, 916, 934
Sun, Yingxue 107, 643
Sun, Yongsheng 183, 441
Sun, Yuan 376
Sun, Yuting 718
Sun, Zhong 501
Sun, Zixuan 355, 426
Taguti, Henrique Iha 267
Tai, Wei-Heng 322
Takagi, Misaki 192
Takeuchi, Kiyoshi 332
Tan, Jialei 379
Tang, Hailong 751
Tang, He 139
Tang, Jiami 679
Tang, Jing 216
Tang, Jinsheng 560
Tang, Juan 392
Tang, Peishun 763
Tang, Xiao-Yan 414, 831, 853, 856
Tang, Xiaoyan 862
Tang, Yiwen 590

Tao, Nick 388
Tao, Qiuyu 661
Tao, Yongjin 10
Taylor, Stephen 408
Teng, Changjiu 801
Teng, Qiao 910
Thangarasu, Bharatha Kumar 228
Tian, Fengbin 463
Tian, Fengshuo 16
Tian, Wei 566
Tian, Xin 766
Tian, Yu 382
Tian, Yuanxin 614
Tiwari, Ravi 490
Tok, Kean Hong 437
Tsuchiya, Akira 310
Tu, Jiangtao 25
Tu, Ruei-Cheng 313
Wan, Li 844, 868, 871
Wan, Peng 569
Wan, Qingbo 847
Wan, Wenqing 780
Wan, Yuxi 411, 454, 874, 907
Wang, Anqing 289
Wang, Bo 228
Wang, Bohan 62
Wang, Bo-Wei 319
Wang, Chao 566
Wang, Chenlu 828
Wang, Chenxing 813
Wang, Chenxu 289
Wang, Chunlin 569
Wang, Cuimei 504, 508
Wang, Dejin 258
Wang, Denggui 376
Wang, Fangzhou 771
Wang, Feixiong 774
Wang, Haiwei 596
Wang, Haoran 877
Wang, Hongbo 355
Wang, Huaipeng 916, 934
Wang, Huaishan 392
Wang, Jiabin 89
Wang, Jian 186, 189
Wang, Ji-Jiang 572
Wang, Jinxiang 199, 789
Wang, Jitong 859
Wang, Junfei 299
Wang, Junhao 368
Wang, Kaifeng 760
Wang, Kaixuan 16
Wang, Kang 342

Wang, Langyuan ... 92
Wang, Lei ... 730
Wang, Li ... 694
Wang, Liang ... 44
Wang, Lina ... 129, 748
Wang, Lingli ... 205, 219, 225
Wang, Long ... 454
Wang, Longsheng ... 149, 670
Wang, Luda ... 244
Wang, Meng ... 289
Wang, Mengxuan ... 80
Wang, Mingbo ... 74
Wang, Mingyu ... 4, 593, 922
Wang, Peiran ... 388
Wang, Peng ... 757, 777
Wang, Pengjun ... 19, 22, 71, 703
Wang, Qianwen ... 807, 810
Wang, Qing ... 388
Wang, Ruibo ... 252
Wang, Runsheng ... 13, 178, 212, 351, 355, 364, 426, 536, 760
Wang, Shao Hao ... 649
Wang, Shi-Dong ... 578
Wang, Shiqing ... 501
Wang, Shuhan ... 913
Wang, Songyao ... 542
Wang, Teng ... 101
Wang, Tianrui ... 238, 721
Wang, Tong-Yang ... 868, 871
Wang, W. ... 433
Wang, Wenbin ... 545, 889
Wang, Wenwu ... 460, 463, 484, 511
Wang, Xiang ... 931
Wang, Xianglong ... 548
Wang, Xiao ... 798
Wang, Xiaohui ... 771
Wang, Xiaolei ... 460, 463, 484, 511
Wang, Xiaoli ... 289
Wang, Xiaoming ... 395, 411, 874
Wang, Xiaoping ... 411
Wang, Xiaowei ... 282
Wang, Xin'An ... 712
Wang, Xuebin ... 469
Wang, Yan ... 411, 706
Wang, Yang ... 771, 898
Wang, Yeliang ... 469
Wang, Yijiao ... 913
Wang, Yixing ... 646
Wang, Yixue ... 4
Wang, Yongjia ... 913
Wang, Yu ... 392
Wang, Yuan ... 13
Wang, Yuchen ... 56, 168, 261, 673

Wang, Yuhan ... 487
Wang, Zelin ... 739
Wang, Zhao ... 408
Wang, Zhengzhuo ... 202
Wang, Zhichong ... 299, 928
Wang, Zhonrui ... 757
Wang, Zijun ... 101
Wang, Ziqiang ... 611
Wang, Zirui ... 355
Wang, Ziyao ... 68
Wang, Ziyu ... 238, 721
Wang, Zongwei ... 504, 508
Wansi, Ge ... 228
Wei, Hailong ... 117
Wei, Jiangbo ... 566
Wei, Jianglin ... 192
Wei, Jiangling ... 135
Wei, Jiaxing ... 382, 822, 828
Wei, Jie ... 379, 385
Wei, Xueming ... 679
Wei, Yuxi ... 385
Wei, Zhichao ... 189
Wen, Hongyang ... 423
Wen, Liang ... 25, 35, 614
Wen, Xiaokun ... 301
Weng, Zeping ... 498
Wong, Kwok-Ho ... 786
Wong, Man ... 329
Worley, Eugene ... 447
Wu, Bo ... 910
Wu, Chang ... 80, 780, 937
Wu, Cheng ... 74
Wu, Chunlei ... 335, 514, 520
Wu, Haibo ... 401
Wu, Haiqin ... 276
Wu, Heng ... 364
Wu, Honglin ... 536
Wu, Jiabin ... 928
Wu, Jixuan ... 457, 469, 807, 810
Wu, Kejun ... 667
Wu, Liji ... 697
Wu, Lin ... 156
Wu, Mei ... 837
Wu, Ningran ... 244
Wu, Wangran ... 517
Wu, Wenhao ... 742
Wu, Xing ... 178, 551, 751
Wu, Xudong ... 19
Wu, Yi-Wen ... 255
Wu, Yongqin ... 760
Wu, Yue ... 279
Wu, Yuping ... 222

Wu, Yuzhou...844
Wu, Zilong...822
Xia, Donghao..35
Xia, Shiyu...351
Xia, Yiming.............................335, 514, 520
Xia, Yinshui...481
Xia, Yun...411, 874
Xia, Zhiying...282
Xia, Zi-Ming......................................868, 871
Xiang, Jinjuan...460
Xiang, Yuguo...165
Xiao, Shanlin...62
Xiao, Yu..351
Xiao, Zhentao...813
Xiaowen, Chen...98
Xie, Chengzhen..685
Xie, Dan...916, 934
Xie, Huiquan....................................238, 721
Xie, Jing..86
Xie, Mingzhang..658
Xie, Pei-Zhang..41
Xie, Pujin...28
Xie, Qin..768
Xie, Ruiqing..508
Xing, Linlin...38
Xing, Pengcheng...................................395, 847
Xiong, Bingjun..557
Xiong, Ling...792
Xiong, Shisheng.......................................658
Xiong, Sixing...285
Xiong, Wenjuan..345
Xu, Baohui..886
Xu, Fan...865
Xu, Gaobo...777
Xu, Guohao.......................................605, 608
Xu, Hanxi...31
Xu, Hao...............460, 463, 484, 511, 715, 730
Xu, Hongtao......................146, 232, 279
Xu, Jeffrey...345
Xu, Jianlong..916
Xu, Jianqiang...548
Xu, Jing..345
Xu, Jinghan...913
Xu, Jun..86, 766, 919
Xu, Long..596
Xu, Mengfan...35
Xu, Ruize...19
Xu, Sheng...575
Xu, Tao...401
Xu, Wei...620
Xu, Wenting...517
Xu, Wenwei...86

Xu, Xinyue..301
Xu, Yumin..335, 520
Xue, Hao..697
Xue, Xiaoyong....................................171, 466
Xv, Yawen...423
Yamamoto, Shuhei......................................192
Yan, Anzhi..569
Yan, Bei-Ping...798
Yan, Chu..427
Yan, Chuangao...162
Yan, Feng..575, 673, 925
Yan, Jin..338
Yan, Na...92, 715, 730
Yan, Shuang...264
Yan, Silu...880
Yan, Wenchao.....................................427, 498
Yan, Xuke...825
Yang, Bowen...837
Yang, Chaowei...122
Yang, Chengyu...709
Yang, Gaoqi...508
Yang, Guangkuo..810
Yang, Guanhua...469
Yang, Guo...98
Yang, Hongqiang...............417, 530, 813, 825, 841, 865
Yang, Jia..484, 511
Yang, Jie...248
Yang, Jinda...688
Yang, Ling..837
Yang, Ning.......................................398, 454
Yang, Runhua..795
Yang, Shanqiang.......................................289
Yang, Shaoqi..810
Yang, Sheng...351
Yang, Sijing..225
Yang, Tianqi..670
Yang, Ting..919
Yang, Wangjun...889
Yang, Wenhao...95
Yang, Wu..53
Yang, Yang.......................................844, 868, 871
Yang, Yanyu...777
Yang, Yecheng...649
Yang, Yi..569
Yang, Yintang....................................238, 548, 721
Yang, Yizhe.......................................95, 646, 745
Yang, Yurui...423
Yang, Yuxiao..395
Yang, Zhangbin...................................466, 661
Yang, Zhaohui...481
Yang, Zhen..853
Yang, Zhi-Yu.....................................868, 871

Yang, Zhuoyuan	1
Yang, Zixin	189
Yao, Rui Ray	408
Yao, Ruxue	444, 901
Yao, Siyuan	117
Ye, Chenglin	724
Ye, Dongxian	149
Ye, Fan	159, 165, 700
Ye, Jiabao	661
Ye, Jinhong	1
Ye, Tianchun	484, 511
Yeo, Kiat Seng	228
Yi, Bo	417, 825, 865
Yi, Shiyan	605, 608
Yin, Huaxiang	358, 757, 774, 777
Yin, Rui	132, 676, 685, 691, 730
Yin, Wenbo	205, 219
Yin, Yizhe	655, 892, 940
Yin, Yun	241
Ying, Kai	886
You, Hailong	216, 545, 889
You, Lu	117
Yu, Fei	472
Yu, Hongyu	388
Yu, Junjie	557
Yu, Ping	771
Yu, Qiang	420
Yu, Shaofeng	763
Yu, Wei-Wei	447
Yu, Xiao	472
Yu, Xinglong	50
Yu, Xinwei	427
Yu, Yueru	59
Yu, Yueyuan	335, 514, 520
Yu, Zhen	667
Yu, Zhiyi	28, 62, 487
Yu, Zhuoqing	441
Yu, Zuoxu	517
Yuan, Hao	414, 831, 862
Yuan, Minghong	392
Yuan, Shuai	264
Yuan, Tengfei	22, 703
Yuan, Xihui	630
Yuan, Yidong	584
Yuqing, Liu	228
Zeng, Wei	411, 454
Zeng, Xiaoyang	4, 171, 593, 605, 608, 922
Zeng, Xing	401
Zeng, Yaxin	715
Zhai, Ziye	292
Zhan, Xuepeng	457, 807, 810
Zhang, Ao	736

Zhang, Bo	379, 392, 395, 398, 401, 430, 451, 454, 578, 819, 834, 874, 907
Zhang, Boyang	110
Zhang, Chenghao	584
Zhang, Chenyang	351
Zhang, Chunwei	859
Zhang, David Wei	335, 514, 520
Zhang, Duoli	7
Zhang, Fangxing	536
Zhang, H.	433
Zhang, Hang	757
Zhang, Hanlu	92
Zhang, Hao	608, 804
Zhang, Haoyu	65
Zhang, He	47
Zhang, Hongbo	895
Zhang, Huihong	162
Zhang, Jiaming	74
Zhang, Jian Fu	437
Zhang, Jian	494
Zhang, Jianhua	640
Zhang, Jiayu	68
Zhang, Jide	225
Zhang, Jiming	388
Zhang, Jingjing	682
Zhang, Jingtao	1
Zhang, Jinping	819, 834
Zhang, Jun	53
Zhang, Junyu	457
Zhang, Ke	898
Zhang, Lei	706
Zhang, Li	640
Zhang, Liang	289
Zhang, Lining	351, 426, 536
Zhang, Long	258, 376, 423, 682
Zhang, Lu-Lu	255
Zhang, Meihe	757
Zhang, Meng	837
Zhang, Mengming	727
Zhang, Qianhao	575
Zhang, Qingzhu	757
Zhang, Ruikai	289
Zhang, Runxi	596
Zhang, Sen	258, 430
Zhang, Shengdong	325
Zhang, Shengnan	50
Zhang, Shutong	22
Zhang, Shuyu	92
Zhang, Tianyu	700
Zhang, W.	494
Zhang, Wei David	368
Zhang, Wei	139

Zhang, Weidong 437, 630
Zhang, Wenfeng 301
Zhang, Xiangmin 697
Zhang, Xing 533
Zhang, Xinrui 487
Zhang, Xinyu 822
Zhang, Xinyue 536
Zhang, Xuelian 222
Zhang, Xuewei 296
Zhang, Xusheng 113
Zhang, Ya-Cong 143
Zhang, Yadong 774
Zhang, Yanlong 44
Zhang, Yi Bo 862
Zhang, Yibo 95, 745
Zhang, Yihui 581
Zhang, Yimeng 95, 646, 745
Zhang, Yue 590
Zhang, Yuejun 22, 25, 35, 614, 703
Zhang, Yuhan 748
Zhang, Yu-Ming 414, 831, 853, 856
Zhang, Yuming 95, 252, 444, 646, 745, 880, 901
Zhang, Yurun 7
Zhang, Yushen 95, 745
Zhang, Yutao 444, 901
Zhang, Yuzhen 517
Zhang, Zhaohao 358
Zhang, Zhen 590
Zhang, Zhenyin 620
Zhang, Zhenyu 596
Zhang, Zhili 117, 430
Zhang, Zonghao 813
Zhao, Chun 560
Zhao, Dongyan 584
Zhao, Fei 335, 514, 520
Zhao, Hankun 304
Zhao, Kai 264, 385, 533
Zhao, Liangxiao 71
Zhao, Lu-Yu 255
Zhao, Mengyao 376
Zhao, Peizhi 766
Zhao, Qi 706
Zhao, Ruiyong 655, 892, 940
Zhao, Shilong 801
Zhao, Shulin 152
Zhao, Xiaohuan 810
Zhao, Xin 670
Zhao, Xu 611
Zhao, Yi 427, 478, 498, 742
Zhao, Yifan 50
Zhao, Yi-Shang 868, 871
Zhao, Yuanfu 44

Zhao, Yudi 264
Zhao, Yujie 192
Zhao, Zi-Ming 414
Zheng, Guiqiang 258
Zheng, Haoping 264
Zheng, Shiwei 834
Zheng, Xuefeng 542
Zheng, Yifei 376
Zheng, Zhe 572, 682
Zhenghao, Lu 228
Zhi, Haizhao 411, 874
Zhong, Chengyan 931
Zhong, Guoqiang 664
Zhong, Jingxue 107, 643
Zhong, Kun 358
Zhong, Linfeng 13
Zhong, Qingyin 258
Zhong, Tao 175
Zhong, Tianyuan 110
Zhong, Zheng-Hong 322
Zhou, David 454, 874, 907
Zhou, Dayan 165
Zhou, Hao 225
Zhou, Jianjun 175, 376, 736
Zhou, Jinggui 398
Zhou, Jinghan 199, 789
Zhou, Jingming 178
Zhou, Jiuren 661
Zhou, Mohan 279
Zhou, Pengwei 395
Zhou, Pingqiang 208
Zhou, Qi 398, 454, 907
Zhou, Qiang 171
Zhou, Ruibin 487
Zhou, Wenbin 925
Zhou, Wenqian 617
Zhou, Xiahong 504
Zhou, Xiaofang 59, 739
Zhou, Xinlong 804
Zhou, Xue 630
Zhou, Yang 289
Zhou, Yongliang 107, 643
Zhou, Yu 414, 831, 853, 856
Zhou, Zecheng 149
Zhou, Zekun 578, 584
Zhou, Zheng 539, 724
Zhou, Zhuoling 801
Zhou, Zikang 627
Zhou, Ziwei 411
Zhou, Ziyu 71
Zhu, Chiang 928
Zhu, Jianggen 398, 454, 907

Zhu, Saike...478
Zhu, Xiaona 763, 928
Zhu, Xing...688
Zhu, Yexin 149
Zhu, Yinna 620
Zhu, Yujie 640
Zhu, Zhangming 83, 149, 156, 670
Zhu, Zhiyuan 368
Zhuang, Quanrong 113
Zhuo, Tianshu 1
Zou, Qiaosha...617
Zou, Rongxin 25
Zuo, Peng..771

IEEE
445 Hoes Lane
Piscataway, NJ 08854-4141

ISBN 979-8-3503-6184-1